Our Unifying Approach to

Our approach to studying the functions of algebra allows students to make connections between graphs of functions, their associated equations and inequalities, and related applications. To demonstrate this four-part process with quadratic functions (Chapter 3), consider the following illustrations.

1 Examine the nature of the graph.

ILLUSTRATION: Graph $f(x) = 2x^2 - 4x - 6$.

Solution Because the function is quadratic, its graph is a parabola. By completing the square, it can be written in the form

$$f(x) = 2(x - 1)^2 - 8.$$

Compared with the graph of $y = x^2$, its graph is shifted horizontally 1 unit to the right, stretched by a factor of 2, and shifted vertically 8 units down. Its vertex has coordinates $(1, -8)$, and the axis of symmetry has equation $x = 1$. The domain is $(-\infty, \infty)$, and the range is $[-8, \infty)$.

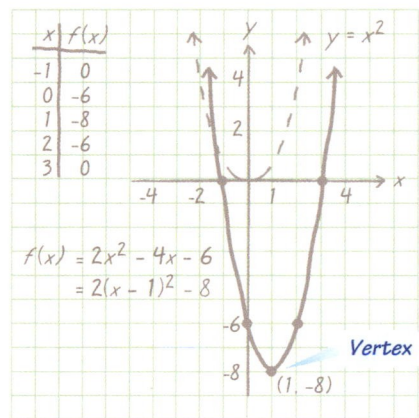

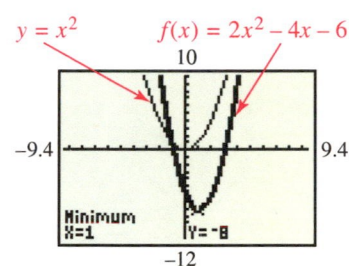

2 Solve a typical equation analytically and graphically.

ILLUSTRATION: Solve the equation $2x^2 - 4x - 6 = 0$.

Analytic Solution

$$2x^2 - 4x - 6 = 0$$
$$x^2 - 2x - 3 = 0 \quad \text{Divide by 2.}$$
$$(x + 1)(x - 3) = 0 \quad \text{Factor.}$$
$$x + 1 = 0 \quad \text{or} \quad x - 3 = 0 \quad \text{Zero-product property}$$
$$x = -1 \quad \text{or} \quad x = 3 \quad \text{Solve each equation.}$$

Check by substituting the solutions -1 and 3 for x in the original equation. The solution set is $\{-1, 3\}$.

Graphing Calculator Solution

Using the x-intercept method, we find that the zeros of f are the solutions of the equation.

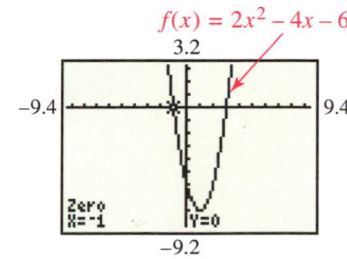

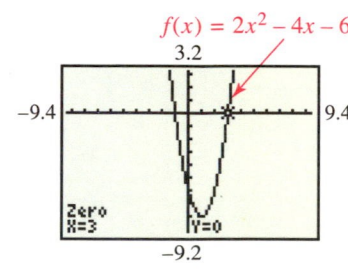

ILLUSTRATION: Solve the inequality $2x^2 - 4x - 6 \leq 0$.

Solution Divide a number line into intervals determined by the zeros of $f(x) = 2x^2 - 4x - 6$, (found in Illustration 2), which are -1 and 3. Choose a test value from each interval to identify values for which $f(x) \leq 0$.

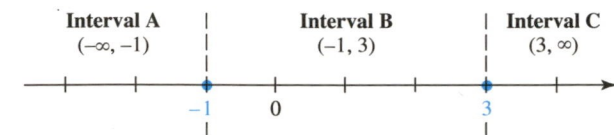

Interval	Test Value x	Is $f(x) = 2x^2 - 4x - 6 \leq 0$ True or False?	
A: $(-\infty, -1)$	-2	$f(-2) = 2(-2)^2 - 4(-2) - 6 \leq 0$ $10 \leq 0$	? False
B: $(-1, 3)$	0	$f(0) = 2(0)^2 - 4(0) - 6 \leq 0$ $-6 \leq 0$	? True
C: $(3, \infty)$	4	$f(4) = 2(4)^2 - 4(4) - 6 \leq 0$ $10 \leq 0$	? False

From the table, the polynomial $2x^2 - 4x - 6$ is negative or *zero* on the interval $[-1, 3]$. The calculator graph in Illustration 2 supports this solution, since the graph lies *on* or *below* the x-axis on this interval.

ILLUSTRATION: If an object is projected directly upward from the ground with an initial velocity of 64 feet per second, then (neglecting air resistance) the height of the object x seconds after it is projected is modeled by

$$s(x) = -16x^2 + 64x,$$

where $s(x)$ is in feet. After how many seconds does it reach a height of 28 feet?

Analytic Solution

We must solve the equation $s(x) = 28$.

$$s(x) = -16x^2 + 64x$$
$$28 = -16x^2 + 64x \quad \text{Let } s(x) = 28.$$
$$16x^2 - 64x + 28 = 0 \quad \text{Standard form}$$
$$4x^2 - 16x + 7 = 0 \quad \text{Divide by 4.}$$
$$(2x - 1)(2x - 7) = 0 \quad \text{Factor.}$$
$$x = 0.5 \quad \text{or} \quad x = 3.5 \quad \text{Zero-product property}$$

The object reaches a height of 28 feet twice, at 0.5 second (on its way up) and at 3.5 seconds (on its way down).

Graphing Calculator Solution

Using the intersection-of-graphs method, we see that the graphs of $y = -16x^2 + 64x$ and $y = 28$ intersect at points whose coordinates are $(0.5, 28)$ and $(3.5, 28)$, confirming our analytic answer.

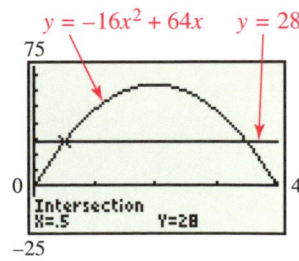

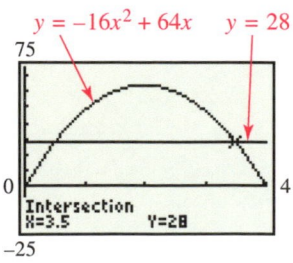

A Graphical Approach to
Algebra and Trigonometry

SIXTH EDITION

John Hornsby
University of New Orleans

Margaret L. Lial
American River College

Gary Rockswold
Minnesota State University, Mankato

with the assistance of
Jessica Rockswold

Boston   Columbus   Indianapolis   New York   San Francisco   Upper Saddle River
Amsterdam   Cape Town   Dubai   London   Madrid   Milan   Munich   Paris   Montréal   Toronto
Delhi   Mexico City   São Paulo   Sydney   Hong Kong   Seoul   Singapore   Taipei   Tokyo

Editor-in-Chief: *Anne Kelly*
Acquisitions Editor: *Kathryn O'Connor*
Project Manager: *Christine O'Brien*
Editorial Assistant: *Judith Garber*
Senior Managing Editor: *Karen Wernholm*
Senior Production Project Manager: *Kathleen A. Manley*
Digital Assets Manager: *Marianne Groth*
Media Producer: *Stephanie Green*
Software Development: *Kristina Evans, MathXL; John Flanagan, TestGen*
Senior Marketing Manager: *Peggy Sue Lucas*
Marketing Assistant: *Justine Goulart*
Senior Author Support/Technology Specialist: *Joe Vetere*
Image Manager: *Rachel Youdelman*
Procurement Specialist: *Debbie Rossi*
Associate Director of Design: *Andrea Nix*
Senior Designer: *Beth Paquin*
Text Design, Production Coordination, Composition, and Illustrations: *Cenveo® Publisher Services*
Cover Design: *Beth Paquin*
Cover Image: *David Redfern/Hulton Archive/Getty Images*

Library of Congress Cataloging-in-Publication Data
Hornsby, John, 1949-author.
  A graphical approach to algebra and trigonometry.—Sixth edition/John Hornsby, University of New Orleans, Margaret L. Lial, American River College, Gary K. Rockswold, Minnesota State University, Mankato.
  pages cm
  ISBN 0-321-92733-8 (student edition)
  1. Algebra—Textbooks. 2. Trigonometry—Textbooks. I. Lial, Margaret L., author. II. Rockswold, Gary K., author. III. Title.
  QA152.3.H68 2015
  512'.13—dc23   2013028140

6 16

www.pearsonhighered.com

ISBN-10: 0-321-92733-8
ISBN-13: 978-0-321-92733-0

On March 16, 2012, the mathematics education community lost one of its most influential members with the passing of our beloved mentor, colleague, and friend, Marge Lial. On that day, Marge lost her long battle with ALS. Throughout her illness, she showed the remarkable strength and courage that characterized her entire life.

Margaret L. Lial

We would like to share a few comments from among the many messages we received from friends, colleagues, and others whose lives were touched by our beloved Marge:

"What a lady"   "A great friend"
"A remarkable person"   "Sorely missed but so fondly remembered"
"Gracious to everyone"   "Even though our crossed path was narrow, she made an impact
"One of a kind"   and I will never forget her."
"Truly someone special"   "There is talent and there is Greatness. Marge was truly Great."
"A loss in the mathematical world" "Her true impact is almost more than we can imagine."

In the world of college mathematics publishing, Marge Lial was a rock star. People flocked to her, and she had a way of making everyone feel like they truly mattered. And to Marge, they did. She and Chuck Miller began writing for Scott Foresman in 1970. Not long ago she told us that she could no longer continue because "just getting from point A to point B" had become too challenging. That's our Marge—she even gave a geometric interpretation to her illness.

It has truly been an honor and a privilege to work with Marge Lial these past two decades. While we no longer have her wit, charm, and loving presence to guide us, so much of who we are as mathematics educators has been shaped by her influence. We will continue doing our part to make sure that the work that she and Chuck began represents excellence in mathematics education. We remember daily the little ways she impacted us, including her special expressions, "Margisms," as we like to call them. She often ended emails with one of them—the single word "Onward."

We conclude with a poem penned by another of her coauthors, Callie Daniels.

Your courage inspires me
Your strength…impressive
Your wit humors me
Your vision…progressive

Your determination motivates me
Your accomplishments pave my way
Your vision sketches images for me
Your influence will forever stay.

Thank you, dearest Marge.
Knowing you and working with you has been a divine gift.

Onward.

John Hornsby
Gary Rockswold

To Hazel, for her positive spirit and kindness,
and for always making me feel welcome.
G.R.

Foreword

The first edition of *A Graphical Approach to Algebra and Trigonometry* was published in 1996. Our experience was that the usual order in which the standard topics were covered did not foster students' understanding of the interrelationships among graphs, equations, and inequalities. The table of contents for typical college algebra texts did not allow for maximum effectiveness in implementing our philosophy because graphs were not covered early enough in the course. Thus, we reorganized the standard topics with early introduction to the graphs of functions, followed by solutions of equations, inequalities, and

applications. While the material is reorganized, *we still cover all traditional topics and skills.* The underlying theme was, and still is, to illustrate how the graph of a typical function can be used to support the solutions of equations and associated inequalities involving the function.

Using linear functions in Chapter 1 to introduce the approach that follows in later chapters, we apply a four-step process of analysis.

1. We examine the nature of the graph of the function, using both hand-drawn and calculator-generated versions. Domain and range are established, and any further characteristics are discussed.

2. We solve equations analytically, using the standard methods. Then we support our solutions graphically using the **intersection-of-graphs method** and the ***x*-intercept method** (pages 53–54).

3. We solve the associated inequalities analytically, again using standard methodology, supporting their solutions graphically as well.

4. We apply analytic and graphical methods to modeling and traditional applications involving the class of function under consideration.

After this procedure has been initially established for linear functions, we apply it to absolute value, quadratic, higher-degree polynomial, rational, root, exponential, logarithmic, and trigonometric functions in later chapters. The chapter on systems of equations ties in the concept of solving systems with the aforementioned intersection-of-graphs method of solving equations.

This presentation provides a sound pedagogical basis. Because today's students rely on visual learning more than ever, the use of graphs promotes student understanding in a manner that might not occur if only analytic approaches were used. It allows the student the opportunity to see how the graph of a function is related to equations and inequalities involving that function. The student is presented with the same approach over and over, and comes to realize that the *type* of function f defined by $y = f(x)$ under consideration does not matter when providing graphical support. For example, using the *x*-intercept method, the student sees that *x*-intercepts of the graph of $y = f(x)$ correspond to real solutions of the equation $f(x) = 0$, *x*-values of points above the *x*-axis correspond to solutions of $f(x) > 0$, and *x*-values of points below the *x*-axis correspond to solutions of $f(x) < 0$.

The final result, in conjunction with the entire package of learning tools provided by Pearson, is a course that covers the standard topics of algebra and trigonometry. It is developed in such a way that graphs are seen as pictures that can be used to interpret analytic results. We hope that you will enjoy teaching this course, and that your students will come away with an appreciation of the impact and importance of graphs in the study of algebra and trigonometry.

John Hornsby
Gary Rockswold

Contents

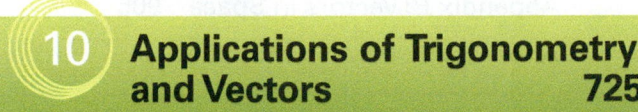

Preface

Although *A Graphical Approach to Algebra and Trigonometry* has evolved significantly from earlier editions, it retains the strengths of those editions and provides new and relevant opportunities for students and instructors alike. We realize that today's classroom experience is evolving and that technology-based teaching and learning aids have become essential to address the ever-changing needs of instructors and students. As a result, we've worked to provide support for all classroom types—traditional, hybrid, and online. In the sixth edition, text and online materials are more tightly integrated than ever before. This enhances flexibility and ease of use for instructors and increases success for students. See pages xviii–xix for descriptions of these materials.

This text incorporates an open design, helpful features, careful explanations of topics, and a comprehensive package of supplements and study aids. We continue to offer an *Annotated Instructor's Edition,* in which answers to both even- and odd-numbered exercises are provided either beside the exercises (if space permits) or in the back of the text for the instructor.

A Graphical Approach to Algebra and Trigonometry was one of the first texts to reorganize the typical college algebra table of contents to maximize the use of graphs to support solutions of equations and inequalities. It maintains its unique table of contents and functions-based approach (as outlined in the Foreword and in front of the text) and includes additional components to build skills, address critical thinking, solve applications, and apply technology to support traditional analytic solutions.

This text is part of a series that also includes the following titles:

- *A Graphical Approach to College Algebra, Sixth Edition,* by Hornsby, Lial, and Rockswold
- *A Graphical Approach to Precalculus with Limits: A Unit Circle Approach, Sixth Edition*, by Hornsby, Lial, and Rockswold

The book is written to accommodate students who have access to graphing calculators. We have chosen to use screens from the TI-84 Plus Silver Edition. However, we do not include specific keystroke instructions because of the wide variety of models available. Students should refer to the guides provided with their calculators for specific information.

New to This Edition

There are many places in the text where we have polished individual presentations and added examples, exercises, and applications based on reviewer feedback. Some of the changes you may notice include the following.

- At the request of many reviewers, we now define increasing and decreasing functions over *open* intervals, and define intercepts to be *points*, or *ordered pairs.*
- We have added more titles on graphs, captions, pointers (bubbles), color, and side comments to increase clarity and understanding for students.
- To better reflect the content covered in the exercise sets, the chapter tests have been revised.

- In several chapters, new examples and exercises have been added to better prepare students for the analytic skills necessary to be successful in calculus.
- Graphing calculator screens have been updated to the TI-84 Plus (Silver Edition) with MATHPRINT.
- Throughout the text, data have been updated to increase student interest in mathematics. Some new application topics include half-life of a Twitter link, iPads, social networks, accuracy of professional golfers, and smartphone demographics.
- Exercise sets have been revised so that odd and even exercises are paired appropriately.
- **Chapter 1** has increased emphasis on evaluating function notation, interpreting slope as a rate of change, and evaluating average rate of change using graphs.
- **Chapter 2** now has clearer explanations of how to transform graphs and also how to write transformations in terms of function notation. Additional exercises covering the domain and range of shifted functions have been included.
- **Chapter 3** includes more examples and exercises that cover curve fitting by hand, solving quadratic equations by completing the square, and solving polynomial equations and inequalities.
- **Chapter 4** includes an increased discussion of limit notation near asymptotes, circles, horizontal parabolas, rational equations and inequalities, and rational expressions with fractional exponents.
- **Chapter 5** has additional examples and exercises related to graphing inverse functions by hand, solving exponential equations with negative exponents, simplifying logarithmic expressions, and solving logarithmic equations.
- **Chapter 6** now covers matrices and linear systems. It has updated consumer spending applications, a 4-step process for solving linear systems, additional examples and exercises covering systems with no solution, and a new example to better explain the technique of finding partial fraction decompositions.
- **Chapter 7** now covers conic sections and nonlinear systems of equations and inequalities. Additional examples and exercises have been added.
- **Chapter 8** includes clearer discussions, updated figures, and more exercises related to writing angles as fractions of a revolution, determining trigonometric equations given a graph, finding transformations and phase shifts, and graphing the six trigonometric functions. It also includes additional explanations on entering trigonometric functions, their inverses, and their reciprocals into a calculator.
- **Chapter 9** now has increased clarity on just-in-time strategies for verifying identities and how graphs can be used to help identify identities. Additional examples and exercises have been added to find trigonometric function values of angles and to solve trigonometric equations, including finding all real solutions and determining whether trigonometric equations have no solution. A new application involving music has also been added.
- **Chapter 10** has new examples of how to use the law of sines and law of cosines to solve triangles, and also to solve navigation problems. Hints and comments have been added to increase understanding of vectors. It also includes more exercises involving converting complex numbers to trigonometric form and graphing parametric equations.
- **Chapter 11** has additional examples and exercises to better explain writing series in summation notation, evaluating recursive sequences, and summing series.

Features

We are pleased to offer the following enhanced features.

Chapter Openers   Chapter openers provide a chapter outline and a brief discussion related to the chapter content.

Enhanced Examples   We have replaced and included new examples in this edition, and have polished solutions and incorporated more side comments and pointers.

Hand-Drawn Graphs   We have incorporated many graphs featuring a "hand-drawn" style that simulates how a student might actually sketch a graph on grid paper.

Dual-Solution Format   Selected examples continue to provide side-by-side analytic and graphing calculator solutions, to connect traditional analytic methods for solving problems with graphical methods of solution or support.

Pointers   Comments with pointers (bubbles) provide students with on-the-spot explanations, reminders, and warnings about common pitfalls.

Highlighted Section and Figure References   Within text we use boldface type when referring to numbered sections and exercises (e.g., **Section 2.1, Exercises 15–20**), and also corresponding font when referring to numbered figures (e.g., **FIGURE 1**). We thank Gerald M. Kiser of Woodbury (New Jersey) High School for this latter suggestion.

Figures and Photos   Today's students are more visually oriented than ever. As a result, we have made a concerted effort to provide more figures, diagrams, tables, and graphs, including the "hand-drawn" style of graphs, whenever possible. We also include photos accompanying applications in examples and exercises.

Function Capsules   These special boxes offer a comprehensive, visual introduction to each class of function and serve as an excellent resource for reference and review. Each capsule includes traditional and calculator graphs and a calculator table of values, as well as the domain, range, and other specific information about the function. Abbreviated versions of function capsules are provided on the inside back cover of the text.

What Went Wrong?   This popular feature anticipates typical errors that students make when using graphing technology and provides an avenue for instructors to highlight and discuss such errors. Answers are included on the same page as the "What Went Wrong?" boxes.

Cautions and Notes   These warn students of common errors and emphasize important ideas throughout the exposition.

Looking Ahead to Calculus   These margin notes provide glimpses of how the algebraic topics currently being studied are used in calculus.

Algebra Reviews   This new feature, occurring in the margin of the text, provides "just in time" review by referring students to where they can receive additional help with important topics from algebra.

Technology Notes   Also appearing in the margin, these notes provide tips to students on how to use graphing calculators more effectively.

For Discussion   These activities appear within the exposition or in the margins and offer material on important concepts for instructors and students to investigate or discuss in class.

Exercise Sets   We have taken special care to respond to the suggestions of users and reviewers and have added hundreds of new exercises to this edition on the basis of their feedback. The text continues to provide students with ample opportunities to practice, apply, connect, and extend concepts and skills. We have included writing exercises 📄 as well as multiple-choice, matching, true/false, and completion problems. Exercises marked *Concept Check* focus on mathematical thinking and conceptual understanding, while those marked *Checking Analytic Skills* specifically are intended for students to solve *without the use of a calculator*.

Relating Concepts   These groups of exercises appear in selected exercise sets. They tie together topics and highlight relationships among various concepts and skills. All answers to these problems appear in the answer section at the back of the student book.

Reviewing Basic Concepts   These sets of exercises appear every two or three sections and allow students to review and check their understanding of the material in preceding sections. All answers to these problems are included in the answer section.

Chapter Review Material   One of the most popular features of the text, each end-of-chapter Summary features a section-by-section list of Key Terms and Symbols, in addition to Key Concepts. A comprehensive set of Chapter Review Exercises and a Chapter Test are also included.

Acknowledgments

Previous editions of this text were published after thousands of hours of work, not only by the authors, but also by reviewers, instructors, students, answer checkers, and editors. To these individuals and to all those who have worked in some way on this text over the years, we are most grateful for your contributions. We could not have done it without you.

We especially wish to thank the following individuals who provided valuable input into this and previous editions of the text.

Judy Ahrens, *Pellissippi State Technical College*
Randall Allbritton, *Daytona Beach Community College*
Jamie Ashby, *Texarkana College*
Scott E. Barnett, *Henry Ford Community College*
Gloria Bass, *Mercer University*
Pat Bassett, *Palm Beach Atlantic University*
Matthew Benander, *Pima Community College, Northwest Campus*
Daniel Biles, *Western Kentucky University*
Norma Biscula, *University of Maine, Augusta*
Linda Buchanan, *Howard College*
Jennifer Kumi Burkett, *Triton College*
Sylvia Calcano, *Lake City Community College*
Faye Childress, *Central Piedmont Community College*
Mark Crawford, *Waubonsee Community College*
Bettyann Daley, *University of Delaware*

Jacqueline Donofrio, *Monroe Community College*
Patricia Dueck, *Scottsdale Community College*
Mickle Duggan, *East Central University*
Douglas Dunbar, *Northwest Florida State College*
Nancy Eschen, *Florida Community College at Jacksonville*
Donna Fatheree, *University of Louisiana at Lafayette*
Linda Fosnaugh, *Midwestern State University*
William Frederick, *Indiana Purdue University, Fort Wayne*
Henry Graves, *Trident Technical College*
Kim Gregor, *Delaware Technical Community College*
Peter Hocking, *Brunswick Community College*
Sandee House, *Georgia Perimeter College*
W. H. Howland, *University of St. Thomas (Houston)*
Tuesday J. Johnson, *New Mexico State University*

Cheryl Kane, *University of Nebraska*

Mike Keller, *St. John's River Community College*

M. R. Khadivi, *Jackson State University*

Rosemary Kradel, *Lehigh Carbon Community College*

Pam Krompak, *Owens Community College*

Rachel Lamp, *North Iowa Area Community College*

Mary A. LaRussa, *New Mexico Tech*

Sharon Hawkins MacKendrick, *New Mexico State University at Grants*

Nancy Matthews, *University of Oklahoma*

Mary Merchant, *Cedar Valley College*

Dr. Christian R. Miller, *Glendale Community College*

Peggy Miller, *University of Nebraska at Kearney*

Phillip Miller, *Indiana University Southeast*

Stacey McNiel, *Lake City Community College*

Richard Montgomery, *The University of Connecticut*

Michael Nasab, *Long Beach City College*

Jon Odell, *Richland Community College*

Karen Pender, *Chaffey College*

Zikica Perovic, *Normandale Community College*

Mary Anne Petruska, *Pensacola State College*

Susan Pfeifer, *Butler County Community College, Andover*

John Putnam, *University of Northern Colorado*

Charles Roberts, *Mercer University*

Donna Saye, *Georgia Southern University*

Alicia Schlintz, *Meredith College*

Linda K. Schmidt, *Greenville Technical College*

Mike Shirazi, *Germanna Community College*

Jed Soifer, *Atlantic Cape Community College*

Betty Swift, *Cerritos College*

Jennifer Walsh, *Daytona Beach Community College*

Robert Woods, *Broome Community College*

Fred Worth, *Henderson State University*

Kevin Yokoyama, *College of the Redwoods*

Over the years we have come to rely on an extensive team of experienced professionals at Pearson: Greg Tobin, Anne Kelly, Katie O'Connor, Christine O'Brien, Kathy Manley, Judith Garber, Joe Vetere, Peggy Lucas, Justine Goulart, and Diahanne Lucas. Thank you to everyone.

In this edition we welcome the assistance of Jessica Rockswold, who provided excellent support throughout all phases of writing and production. Terry Krieger and Paul Lorczak deserve special recognition for their work with the answers and accuracy checking. Thanks are also due Kathy Diamond for her valuable help as project manager. Finally, we thank David Atwood, Leslie Cobar, Twin Prime Editorial, and Mark Rockswold for checking answers and page proofs and Lucie Haskins for assembling the index.

As an author team, we are committed to providing the best possible text to help instructors teach effectively and have students succeed. As we continue to work toward this goal, we would welcome any comments or suggestions you might have via e-mail to *math@pearson.com*.

John Hornsby
Gary Rockswold

Resources for Success

MyMathLab®

MyMathLab from Pearson is the world's leading online resource in mathematics, integrating interactive homework, assessment, and media in a flexible, easy-to-use format. It provides **engaging experiences** that personalize, stimulate, and measure learning for each student. And it comes from an **experienced partner** with educational expertise and an eye on the future.

To learn more about how MyMathLab combines proven learning applications with powerful assessment, visit **www.mymathlab.com** or contact your Pearson representative.

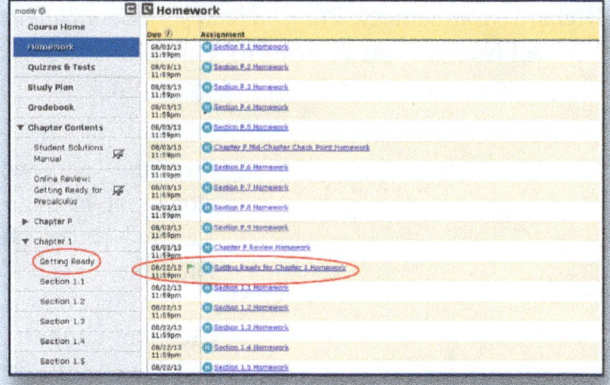

Getting Ready

Students refresh prerequisite topics through assignable skill review quizzes and personalized homework integrated in MyMathLab.

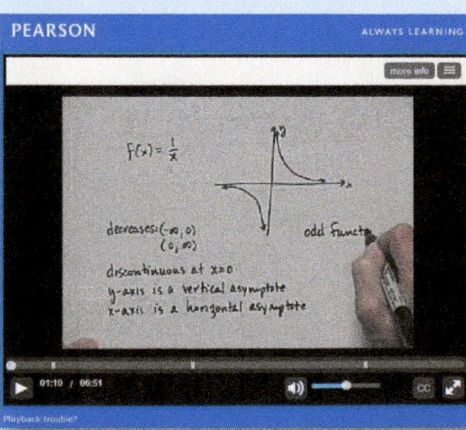

Adaptive Study Plan

The Study Plan makes studying more efficient and effective for every student. Performance and activity are assessed continually in real time. The data and analytics are used to provide personalized content, reinforcing concepts that target each student's strengths and weaknesses.

Skills for Success Module

Integrated within MyMathLab, this module helps students succeed in collegiate courses and prepare for future professions.

Ongoing Review

Reviewing Basic Concepts exercises in the text are now assignable in MyMathLab and require students to recall previously learned content and skills. These exercises help students maintain essential skills throughout the course, thereby enabling them to retain information in preparation for future math courses.

Video Assessment

Video assessment is tied to the video lecture for each section of the book to check students' understanding of important math concepts. Instructors can assign these questions as a prerequisite to homework assignments.

Enhanced Graphing Functionality

New functionality within the graphing utility allows graphing of 3-point quadratic functions, 4-point cubic functions, and transformations in exercises.

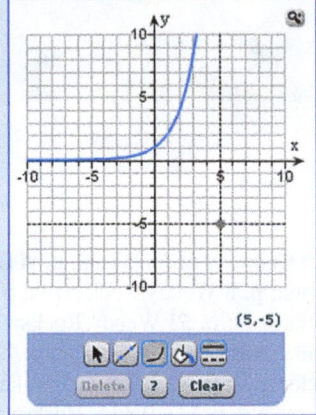

Instructor Resources

Additional resources can be downloaded from www.pearson/highered.com, or hardcopy resources can be ordered from your sales representative.

TestGen®

TestGen (www.pearsoned.com/testgen) enables instructors to build, edit, print, and administer tests using a computerized bank of questions developed to cover all the objectives of the text.

PowerPoint® Lecture Slides

Classroom presentation slides are geared specifically to sequence the text. They are available in MyMathLab.

Annotated Instructor's Edition

This edition provides answers beside the text for most exercises, and in an answer section at the back of the book for all others.

Ready to Go Courses

Now it is even easier to get started with MyMathLab. The Ready to Go MyMathLab course option includes author-chosen preassigned homework, integrated review, and more.

Instructor's Solutions Manual

This manual provides complete solutions to all text exercises.

Student Resources

Additional resources are available to help student success.

Lecture Videos

Example and content videos provide comprehensive coverage of each section and topic in the text in an engaging format that stresses student interaction. They include optional subtitles in English and Spanish. All videos are assignable within MyMathLab.

Student's Solutions Manual

This manual provides detailed solutions to odd-numbered Section and Chapter Review Exercises, as well as to all Relating Concepts, Reviewing Basic Concepts, and Chapter Test Problems.

Photo Credits

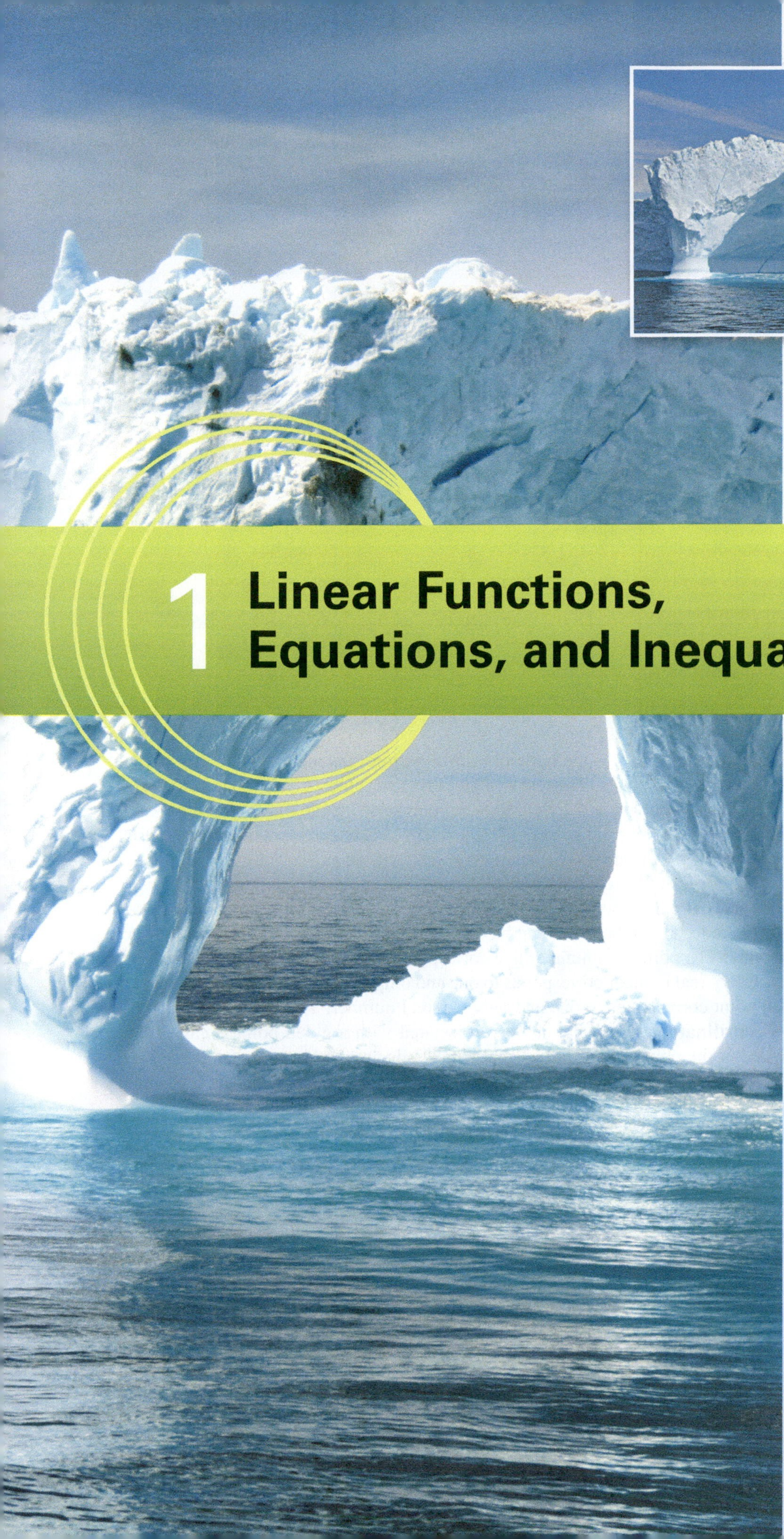

Understanding the future of ice caps in the Arctic and Antarctic regions requires the ability to describe climate change with functions and equations.

1 Linear Functions, Equations, and Inequalities

1.1 Real Numbers and the Rectangular Coordinate System

Sets of Real Numbers • The Rectangular Coordinate System • Viewing Windows • Approximations of Real Numbers • Distance and Midpoint Formulas

Sets of Real Numbers

Several important sets of numbers are used in mathematics. Some of these sets are listed in the following table.

Sets of Numbers

Set	Description	Examples
Natural Numbers	$\{1, 2, 3, 4, \dots\}$	$1, 45, 127, 10^3$
Whole Numbers	$\{0, 1, 2, 3, 4, \dots\}$	$0, 86, 345, 2^3$
Integers	$\{\dots, -2, -1, 0, 1, 2, \dots\}$	$0, -5, -10^2, 99$
Rational Numbers	$\left\{\frac{p}{q} \mid p \text{ and } q \text{ are integers}, q \neq 0\right\}$	$0, -\frac{5}{6}, -2, \frac{22}{7}, 0.5$
Irrational Numbers	$\{x \mid x \text{ is not rational}\}$	$\sqrt{2}, \pi, -\sqrt[3]{7}$
Real Numbers	$\{x \mid x \text{ is a decimal number}\}$	$-\sqrt{6}, \pi, \frac{2}{3}, \sqrt{45}, 0.41$

Whole numbers include the **natural numbers; integers** include the whole numbers and the natural numbers. The result of dividing two integers (with a nonzero divisor) is a **rational number,** or *fraction.* Rational numbers include the natural numbers, whole numbers, and integers. For example, the integer -3 is a rational number because it can be written as $\frac{-3}{1}$. Every rational number can be written as a repeating or terminating decimal. For example, $0.\overline{6} = 0.66666\dots$ represents the rational number $\frac{2}{3}$.

Numbers that can be written as decimal numbers are **real numbers.** Real numbers include rational numbers and can be shown pictorially—that is, **graphed**—on a **number line.** The point on a number line corresponding to 0 is called the **origin.** See **FIGURE 1.** Every real number corresponds to one and only one point on the number line, and each point corresponds to one and only one real number. This correspondence is called a **coordinate system.** The number associated with a given point is called the **coordinate** of the point. The set of all real numbers is graphed in **FIGURE 2.**

Some real numbers cannot be represented by quotients of integers or by repeating or terminating decimals. These numbers are called **irrational numbers.** Examples of irrational numbers include $\sqrt{3}, \sqrt{5}, \sqrt[3]{10}$, and $\sqrt[5]{20}$, but not $\sqrt{1}, \sqrt{4}, \sqrt{9}, \dots,$ which equal $1, 2, 3, \dots,$ and hence are rational numbers. If a is a natural number but $\sqrt{a}$ is not a natural number, then $\sqrt{a}$ is an irrational number. Another irrational number is π, which is approximately equal to 3.14159. In **FIGURE 3** the irrational and rational numbers in the set $\left\{-\frac{2}{3}, 0, \sqrt{2}, \sqrt{5}, \pi, 4\right\}$ are located on a number line. Note that $\sqrt{2}$ is approximately equal to 1.41, so it is located between 1 and 2, slightly closer to 1.

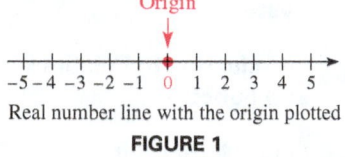

Real number line with the origin plotted

FIGURE 1

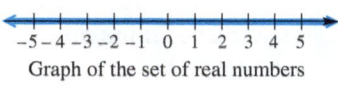

Graph of the set of real numbers

FIGURE 2

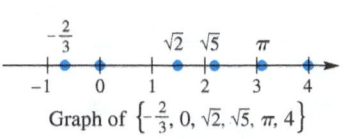

Graph of $\left\{-\frac{2}{3}, 0, \sqrt{2}, \sqrt{5}, \pi, 4\right\}$

FIGURE 3

The Rectangular Coordinate System

If we place two number lines at right angles, intersecting at their origins, we obtain a two-dimensional **rectangular coordinate system.** This rectangular coordinate system is also called the Cartesian coordinate system, which was named after

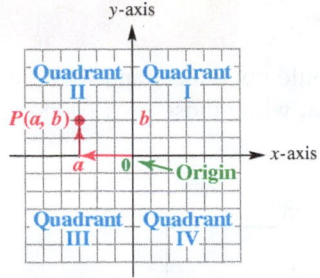

Rectangular coordinate system

FIGURE 4

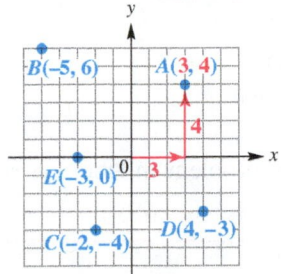

Plotting points in the *xy*-plane

FIGURE 5

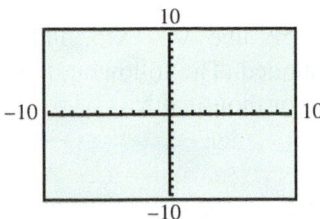

Standard viewing window

FIGURE 6

TECHNOLOGY NOTE

You should consult your owner's guide to see how to set the viewing window on your screen. Remember that different settings will result in different views of graphs.

René Descartes (1596–1650). The number lines intersect at the *origin* of the system, designated 0. The horizontal number line is called the **x-axis,** and the vertical number line is called the **y-axis.** On the *x*-axis, positive numbers are located to the right of the origin, with negative numbers to the left. On the *y*-axis, positive numbers are located above the origin, with negative numbers below.

The plane into which the coordinate system is introduced is the **coordinate plane,** or **xy-plane.** The *x*-axis and *y*-axis divide the plane into four regions, or **quadrants,** as shown in **FIGURE 4.** *The points on the x-axis or y-axis belong to no quadrant.*

Each point *P* in the *xy*-plane corresponds to a unique ordered pair (a, b) of real numbers. We call *a* the **x-coordinate** and *b* the **y-coordinate** of point *P*. The point *P* corresponding to the ordered pair (a, b) is often written as $P(a, b)$, as in **FIGURE 4,** and referred to as "the point (a, b)." **FIGURE 5** illustrates how to plot the point $A(3, 4)$. Additional points are labeled *B–E*. The coordinates of the origin are $(0, 0)$.

Viewing Windows

The rectangular (Cartesian) coordinate system extends indefinitely in all directions. We can show only a portion of such a system in a text figure. Similar limitations occur with the viewing "window" on a calculator screen. **FIGURE 6** shows a calculator screen that has been set to have a minimum *x*-value of -10, a maximum *x*-value of 10, a minimum *y*-value of -10, and a maximum *y*-value of 10. The tick marks on the axes have been set to be 1 unit apart. Thus, there are 10 tick marks on the positive *x*-axis. This window is called the **standard viewing window.**

To convey information about a viewing window, we use the following abbreviations.

Xmin:	minimum value of *x*	**Ymin:**	minimum value of *y*
Xmax:	maximum value of *x*	**Ymax:**	maximum value of *y*
Xscl:	scale (distance between tick marks) on the *x*-axis	**Yscl:**	scale (distance between tick marks) on the *y*-axis

To further condense this information, we use the following symbolism, which gives viewing information for the window in **FIGURE 6.**

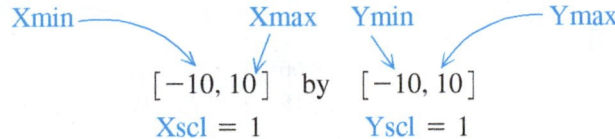

$$[-10, 10] \quad \text{by} \quad [-10, 10]$$
$$\text{Xscl} = 1 \qquad \text{Yscl} = 1$$

FIGURE 7 shows several other viewing windows. Notice that **FIGURES 7(b)** and **7(c)** look exactly alike, and unless we are told what the settings are, we have no way of distinguishing between them. In **FIGURE 7(b)** Xscl = 2.5, while in **FIGURE 7(c)** Xscl = 25. The same is true for Yscl in both.

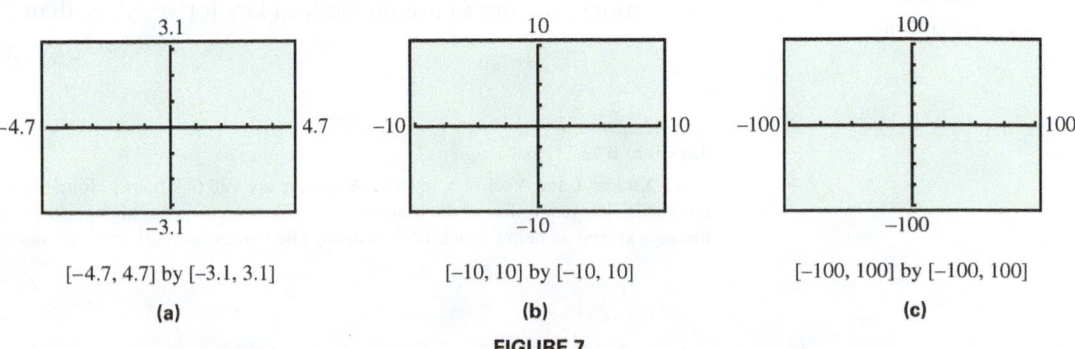

$[-4.7, 4.7]$ by $[-3.1, 3.1]$	$[-10, 10]$ by $[-10, 10]$	$[-100, 100]$ by $[-100, 100]$
(a)	**(b)**	**(c)**

FIGURE 7

WHAT WENT WRONG?

A student learning how to use a graphing calculator could not understand why the axes on the graph were so "thick," as seen in **FIGURE A,** while those on a friend's calculator were not, as seen in **FIGURE B.**

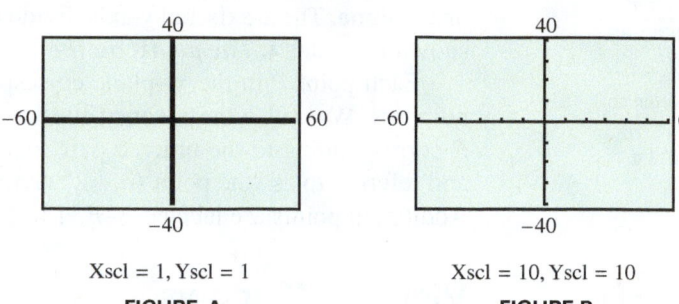

Xscl = 1, Yscl = 1

FIGURE A

Xscl = 10, Yscl = 10

FIGURE B

What Went Wrong?   How can the student correct the problem in **FIGURE A** so that the axes look like those in **FIGURE B?**

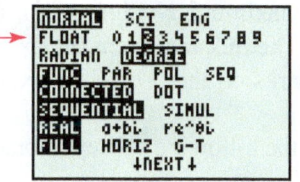

TI-84 Plus (Silver Edition)

FIGURE 8

Approximations of Real Numbers

Although calculators have the capability to express numbers like $\sqrt{2}$, $\sqrt[3]{5}$, and π to many decimal places, we often ask that answers be rounded. The following table reviews rounding numbers to the nearest tenth, hundredth, or thousandth.

Rounding Numbers

Number	Nearest Tenth	Nearest Hundredth	Nearest Thousandth
1.3782	1.4	1.38	1.378
201.6666	201.7	201.67	201.667
0.0819	0.1	0.08	0.082

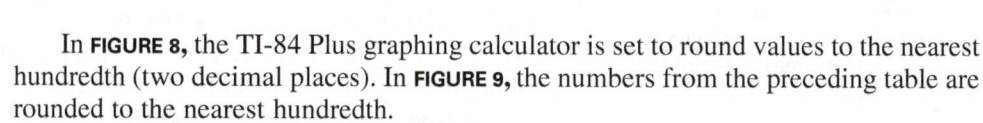

FIGURE 9

In **FIGURE 8,** the TI-84 Plus graphing calculator is set to round values to the nearest hundredth (two decimal places). In **FIGURE 9,** the numbers from the preceding table are rounded to the nearest hundredth.

The symbol $\approx$ indicates that two expressions are *approximately equal*. For example, $\pi \approx 3.14$, but $\pi \neq 3.14$, since $\pi = 3.141592654\ldots$. When using π in calculations, be sure to use the built-in key for π rather than 3.14. See **FIGURE 10.**

π

3.141592654

FIGURE 10

Answer to What Went Wrong?

Since Xscl = 1 and Yscl = 1 in **FIGURE A,** there are 120 tick marks along the *x*-axis and 80 tick marks along the *y*-axis. The resolution of the graphing calculator screen is not high enough to show all these tick marks, so the axes appear as heavy black lines instead. The values for Xscl and Yscl need to be larger, as in **FIGURE B.**

EXAMPLE 1 **Finding Roots on a Calculator**

Approximate each root to the nearest thousandth. (*Note:* You can use the fact that $\sqrt[n]{a} = a^{1/n}$ to find roots.)

(a) $\sqrt{23}$ **(b)** $\sqrt[3]{87}$ **(c)** $\sqrt[4]{12}$

Solution

(a) The screen in **FIGURE 11(a)** shows an approximation for $\sqrt{23}$. To the nearest thousandth, it is 4.796. The approximation is displayed twice, once for $\sqrt{23}$ and once for $23^{1/2}$.

(b) To the nearest thousandth, $\sqrt[3]{87} \approx 4.431$. See **FIGURE 11(b).**

(c) **FIGURE 11(c)** indicates $\sqrt[4]{12} \approx 1.861$ in three different ways.

In all the screens, note the inclusion of parentheses.

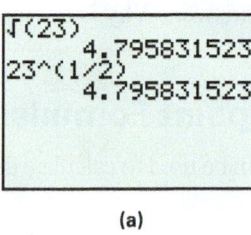

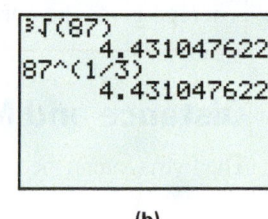

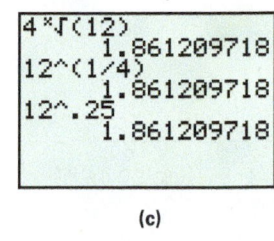

(a)   (b)   (c)

FIGURE 11

EXAMPLE 2 **Approximating Expressions with a Calculator**

Approximate each expression to the nearest hundredth.

(a) $\dfrac{3.8 - 1.4}{5.4 + 3.5}$ **(b)** $3\pi^4 - 9^2$ **(c)** $\sqrt{(4 - 1)^2 + (-3 - 2)^2}$

Solution

(a) See **FIGURE 12(a).** To the nearest hundredth,

$$\frac{3.8 - 1.4}{5.4 + 3.5} \approx 0.27.$$

(b) Many calculators also have a special key to calculate the square of a number. To the nearest hundredth, $3\pi^4 - 9^2 \approx 211.23$. See **FIGURE 12(b).**

(c) From **FIGURE 12(c)**, $\sqrt{(4 - 1)^2 + (-3 - 2)^2} \approx 5.83$.

Do not confuse the negation and subtraction symbols.

Insert parentheses around both the numerator and the denominator.

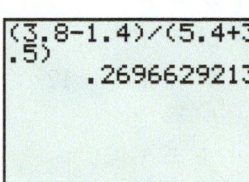

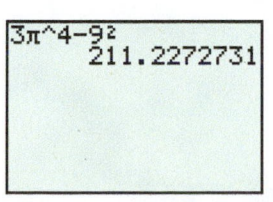

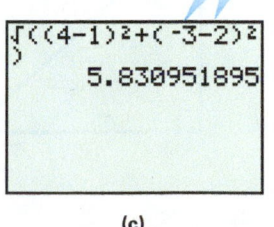

(a)   (b)   (c)

FIGURE 12

WHAT WENT WRONG?

Two students were asked to compute the expression $(2 + 9) - (8 + 13)$ on a TI-84 Plus calculator. One student obtained the answer -10, as seen in **FIGURE A,** while the other obtained -231, as seen in **FIGURE B.**

```
(2+9)-(8+13)
              -10
```

```
(2+9)·(8+13)
              -231
```

FIGURE A   **FIGURE B**

What Went Wrong?   Compute the expression by hand to determine which screen gives the correct answer. Why is the answer on the other screen incorrect?

Distance and Midpoint Formulas

The Pythagorean theorem can be used to calculate the lengths of the sides of a right triangle.

Pythagorean Theorem

In a right triangle, the sum of the squares of the lengths of the legs is equal to the square of the length of the hypotenuse.

$$a^2 + b^2 = c^2$$

Leg a, Hypotenuse c, Leg b

> **NOTE**   *The converse of the Pythagorean theorem is also true.* That is, if *a, b,* and *c* are lengths of the sides of a triangle and $a^2 + b^2 = c^2$, then the triangle is a right triangle with hypotenuse *c*. For example, if a triangle has sides with lengths 3, 4, and 5, then it is a right triangle with hypotenuse of length 5 because $3^2 + 4^2 = 5^2$.

EXAMPLE 3   **Using the Pythagorean Theorem**

Using the right triangle shown in the margin, find the length of the unknown side *b*.

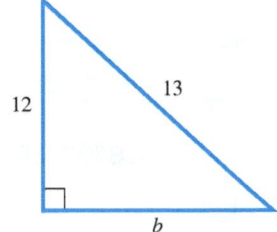

Solution   Let $a = 12$ and $c = 13$ in the Pythagorean theorem.

$a^2 + b^2 = c^2$	Pythagorean theorem
$12^2 + b^2 = 13^2$	Substitute.
$b^2 = 13^2 - 12^2$	Subtract 12^2.
$b^2 = 25$	Simplify.
$b = 5$	Take positive square root.

Answer to What Went Wrong?

The correct answer is -10, as shown in **FIGURE A.** **FIGURE B** gives an incorrect answer because the negation symbol is used, rather than the subtraction symbol. The calculator computed $2 + 9 = 11$ and then *multiplied* by the negative of $8 + 13$ (that is, -21), to obtain the incorrect answer, -231.

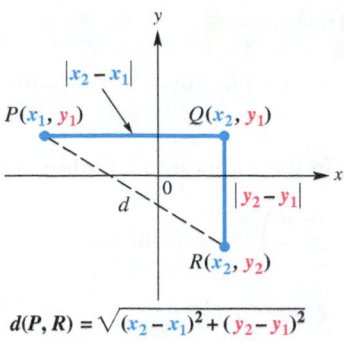

$$d(P, R) = \sqrt{(x_2 - x_1)^2 + (y_2 - y_1)^2}$$

FIGURE 13

To derive a formula to find the distance between two points in the xy-plane, let $P(x_1, y_1)$ and $R(x_2, y_2)$ be any two distinct points in the plane, as shown in **FIGURE 13.** Complete a right triangle by locating point Q with coordinates (x_2, y_1). The Pythagorean theorem gives the distance between P and R as

$$d(P, R) = \sqrt{(x_2 - x_1)^2 + (y_2 - y_1)^2}.$$

NOTE   Absolute value bars are not necessary in this formula, since for all real numbers a and b, $|a - b|^2 = (a - b)^2$.

Distance Formula

Suppose that $P(x_1, y_1)$ and $R(x_2, y_2)$ are two points in a coordinate plane. Then the distance between P and R, written $d(P, R)$, is given by the **distance formula.**

$$d(P, R) = \sqrt{(x_2 - x_1)^2 + (y_2 - y_1)^2}$$

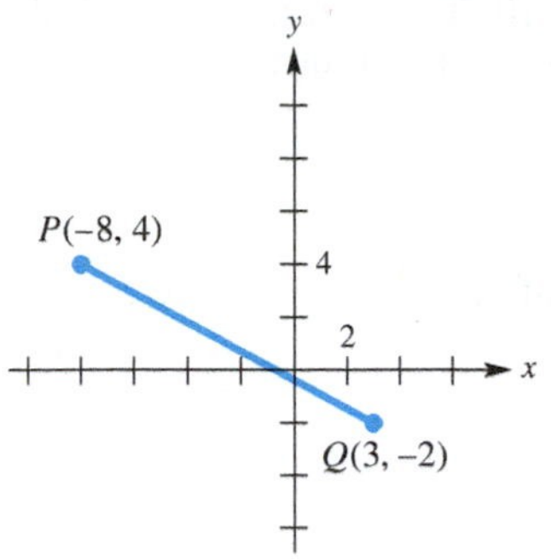

EXAMPLE 4   **Using the Distance Formula**

Use the distance formula to find $d(P, Q)$ in **FIGURE 14.**

Solution

$$d(P, Q) = \sqrt{(x_2 - x_1)^2 + (y_2 - y_1)^2} \qquad \text{Distance formula}$$

$$= \sqrt{[3 - (-8)]^2 + (-2 - 4)^2} \qquad \begin{array}{l} x_1 = -8, \ y_1 = 4, \\ x_2 = 3, \ y_2 = -2 \end{array}$$

$$= \sqrt{11^2 + (-6)^2}$$

$$= \sqrt{121 + 36} \qquad \text{Apply exponents.}$$

$$= \sqrt{157} \qquad \text{Leave in radical form.}$$

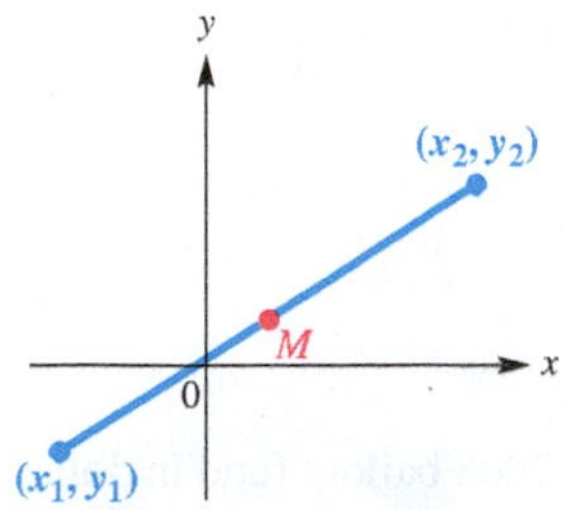

To subtract a negative number, add the opposite. That is, $3 - (-8) = 3 + 8.$

P(−8, 4)

Q(3, −2)

FIGURE 14

The *midpoint M* of a line segment is the point on the segment that lies the same distance from both endpoints. See **FIGURE 15.** *The coordinates of the midpoint are found by calculating the average of the x-coordinates and the average of the y-coordinates of the endpoints of the segment.*

(x_2, y_2)

M

(x_1, y_1)

Point M is the midpoint of the segment joining (x_1, y_1) and (x_2, y_2).

FIGURE 15

Midpoint Formula

The **midpoint** M of the line segment with endpoints (x_1, y_1) and (x_2, y_2) has the following coordinates.

$$M = \left(\frac{x_1 + x_2}{2}, \ \frac{y_1 + y_2}{2} \right)$$

EXAMPLE 5   **Using the Midpoint Formula**

Find the coordinates of the midpoint M of the segment with endpoints $(8, -4)$ and $(-9, 6)$.

Solution   Let $(x_1, y_1) = (8, -4)$ and $(x_2, y_2) = (-9, 6)$ in the midpoint formula.

$$M = \left(\frac{x_1 + x_2}{2}, \frac{y_1 + y_2}{2} \right) = \left(\frac{8 + (-9)}{2}, \frac{-4 + 6}{2} \right) \qquad \text{Substitute.}$$

$$= \left(-\frac{1}{2}, 1 \right) \qquad \text{Simplify.} \quad \bullet$$

EXAMPLE 6   **Estimating iPad Sales**

Four quarters after the launch of the iPad, about 19.5 million were sold. After 10 quarters, about 99 million iPads were sold. Use the midpoint formula to estimate how many iPads were sold 7 quarters after launch. Compare your estimate with the actual value of 50 million. (*Source:* Business Insider.)

Solution   Quarter 7 lies midway between quarters 4 and 10. Therefore, we can find the midpoint of the line segment joining the points $(4, 19.5)$ and $(10, 99)$.

$$\left(\frac{4 + 10}{2}, \frac{19.5 + 99}{2} \right) = (7, 59.25)$$

The midpoint formula estimates the number of iPads sold after 7 quarters to be 59.25 million. This is 9.25 million higher than the actual value. $\quad \bullet$

1.1 Exercises

*For each set, list all elements that belong to the (**a**) natural numbers, (**b**) whole numbers, (**c**) integers, (**d**) rational numbers, (**e**) irrational numbers, and (**f**) real numbers.*

1. $\left\{ -6, -\frac{12}{4}, -\frac{5}{8}, -\sqrt{3}, 0, 0.31, 0.\overline{3}, 2\pi, 10, \sqrt{17} \right\}$

2. $\left\{ -8, -\frac{14}{7}, -0.245, 0, \frac{6}{2}, 8, \sqrt{81}, \sqrt{12} \right\}$

3. $\left\{ -\sqrt{100}, -\frac{13}{6}, -1, 5.23, 9.\overline{14}, 3.14, \frac{22}{7} \right\}$

4. $\{ -\sqrt{49}, -0.405, -0.\overline{3}, 0.1, 3, 18, 6\pi, 56 \}$

Classify each number as one or more of the following: natural number, integer, rational number, *or* real number.

5. 16,351,000,000,000 (The federal debt in dollars in January 2013)

6. 700,000,000,000 (The federal 2008 bailout fund in dollars)

7. -25 (The percent change in the number of Yahoo searches from 2011 to 2012)

8. -3 (The annual percent change in the area of tropical rain forests)

9. $\frac{7}{3}$ (The fractional increase in online sales on Thanksgiving Day from 2006 to 2011)

10. -3.5 (The amount in billions of dollars that the Motion Picture Association of America estimates is lost annually due to piracy)

11. $5\sqrt{2}$ (The length of the diagonal of a square measuring 5 units on each side)

12. π (The ratio of the circumference of a circle to its diameter)

Concept Check   *For each measured quantity, state the set of numbers that is most appropriate to describe it. Choose from the* natural numbers, integers, *and* rational numbers.

13. Populations of cities

14. Distances to nearby cities on road signs

15. Shoe sizes

16. Prices paid (in dollars and cents) for gasoline tank fill-ups

17. Daily low winter temperatures in U.S. cities

18. Golf scores relative to par

Graph each set of numbers on a number line.

19. $\{-4, -3, -2, -1, 0, 1\}$ **20.** $\{-6, -5, -4, -3, -2\}$

21. $\left\{-0.5, 0.75, \frac{5}{3}, 3.5\right\}$ **22.** $\left\{-0.6, \frac{9}{8}, 2.5, \frac{13}{4}\right\}$

23. Explain the distinction between a rational number and an irrational number.

24. *Concept Check*   Using her calculator, a student found the decimal 1.414213562 when she evaluated $\sqrt{2}$. Is this decimal the exact value of $\sqrt{2}$ or just an approximation of $\sqrt{2}$? Should she write $\sqrt{2} = 1.414213562$ or $\sqrt{2} \approx 1.414213562$?

Locate each point on a rectangular coordinate system. Identify the quadrant, if any, in which each point lies.

25. $(2, 3)$ **26.** $(-1, 2)$ **27.** $(-3, -2)$ **28.** $(1, -4)$ **29.** $(0, 5)$

30. $(-2, -4)$ **31.** $(-2, 4)$ **32.** $(3, 0)$ **33.** $(-2, 0)$ **34.** $(3, -3)$

Name the possible quadrants in which the point (x, y) can lie if the given condition is true.

35. $xy > 0$ **36.** $xy < 0$ **37.** $\dfrac{x}{y} < 0$ **38.** $\dfrac{x}{y} > 0$

39. *Concept Check*   If the *x*-coordinate of a point is 0, the point must lie on which axis?

40. *Concept Check*   If the *y*-coordinate of a point is 0, the point must lie on which axis?

Give the values of Xmin, Xmax, Ymin, *and* Ymax *for each screen, given the values for* Xscl *and* Yscl. *Use the notation described in this section.*

41.
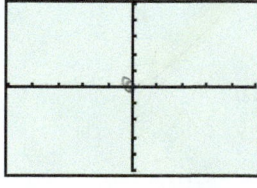
Xscl = 1, Yscl = 5

42.
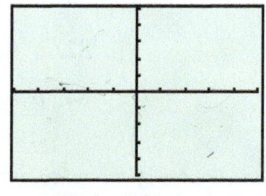
Xscl = 5, Yscl = 1

43.
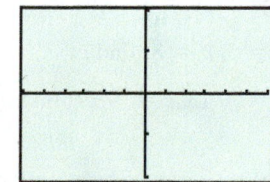
Xscl = 10, Yscl = 50

44.
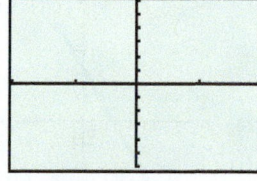
Xscl = 50, Yscl = 10

45.
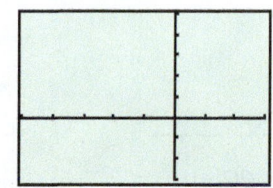
Xscl = 100, Yscl = 100

46.
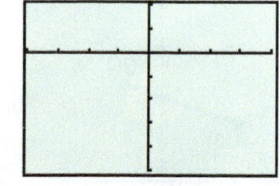
Xscl = 75, Yscl = 75

Set the viewing window of your calculator to the given specifications. Make a sketch of your window.

47. $[-10, 10]$ by $[-10, 10]$
Xscl = 1   Yscl = 1

48. $[-40, 40]$ by $[-30, 30]$
Xscl = 5   Yscl = 5

49. $[-5, 10]$ by $[-5, 10]$
Xscl = 3   Yscl = 3

50. $[-3.5, 3.5]$ by $[-4, 10]$
Xscl = 1   Yscl = 1

51. $[-100, 100]$ by $[-50, 50]$
Xscl = 20   Yscl = 25

52. $[-4.7, 4.7]$ by $[-3.1, 3.1]$
Xscl = 1   Yscl = 1

53. Set your viewing window to $[-10, 10]$ by $[-10, 10]$, and then set Xscl to 0 and Yscl to 0. What do you notice? Make a conjecture as to how to set the screen with no tick marks on the axes.

54. Set your viewing window to $[-50, 50]$ by $[-50, 50]$, Xscl to 1, and Yscl to 1. Describe the appearance of the axes compared with those seen in the standard window. Why do you think they appear this way? How can you change your scale settings so that this "problem" is alleviated?

Find a decimal approximation of each root or power. Round answers to the nearest thousandth.

55. $\sqrt{58}$

56. $\sqrt{97}$

57. $\sqrt[3]{33}$

58. $\sqrt[3]{91}$

59. $\sqrt[4]{86}$

60. $\sqrt[4]{123}$

61. $19^{1/2}$

62. $29^{1/3}$

63. $46^{1.5}$

64. $23^{2.75}$

Approximate each expression to the nearest hundredth.

65. $\dfrac{5.6 - 3.1}{8.9 + 1.3}$

66. $\dfrac{34 + 25}{23}$

67. $\sqrt{\pi^3 + 1}$

68. $\sqrt[3]{2.1 - 6^2}$

69. $3(5.9)^2 - 2(5.9) + 6$

70. $2\pi^3 - 5\pi^2 - 3$

71. $\sqrt{(4 - 6)^2 + (7 + 1)^2}$

72. $\sqrt{[-1 - (-3)]^2 + (-5 - 3)^2}$

73. $\dfrac{\sqrt{\pi - 1}}{\sqrt{1 + \pi}}$

74. $\sqrt[3]{4.5 \times 10^5 + 3.7 \times 10^2}$

75. $\dfrac{2}{1 - \sqrt[3]{5}}$

76. $1 - \dfrac{4.5}{3 - \sqrt{2}}$

Find the length of the unknown side of the right triangle. In each case, a and b represent the lengths of the legs and c represents the length of the hypotenuse.

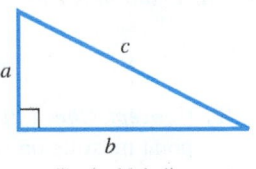

Typical labeling

77. $a = 8, b = 15$; find c

78. $a = 7, b = 24$; find c

79. $a = 13, c = 85$; find b

80. $a = 14, c = 50$; find b

81. $a = 5, b = 8$; find c

82. $a = 9, b = 10$; find c

83. $b = \sqrt{13}, c = \sqrt{29}$; find a

84. $b = \sqrt{7}, c = \sqrt{11}$; find a

*Find **(a)** the distance between P and Q and **(b)** the coordinates of the midpoint M of the segment joining P and Q.*

85.

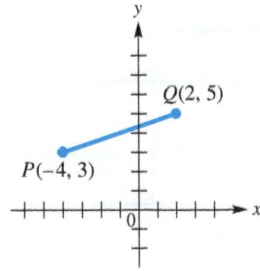

86.

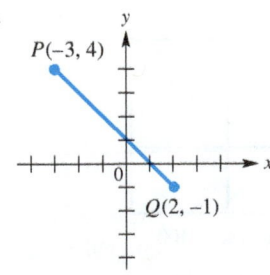

87.

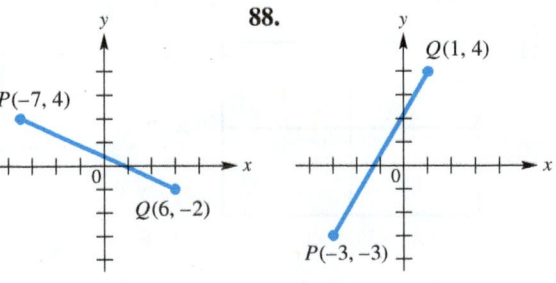

88.

89. $P(5, 7), Q(2, 11)$

90. $P(-2, 5), Q(4, -3)$

91. $P(-8, -2), Q(-3, -5)$

92. $P(-6, -10), Q(6, 5)$

93. $P(9.2, 3.4), Q(6.2, 7.4)$

94. $P(8.9, 1.6), Q(3.9, 13.6)$

95. $P(13x, -23x), Q(6x, x), \quad x > 0$

96. $P(12y, -3y), Q(20y, 12y), \quad y > 0$

Suppose that P is an endpoint of a segment PQ and M is the midpoint of PQ. Find the coordinates of endpoint Q.

97. $P(7, -4), M(8, 5)$

98. $P(13, 5), M(-2, -4)$

99. $P(5.64, 8.21), M(-4.04, 1.60)$

100. $P(-10.32, 8.55), M(1.55, -2.75)$

Solve each problem.

101. *(Modeling) Google Ad Revenue* From 2007 to 2011, worldwide Google advertising revenue (in billions of dollars) rose in an approximately linear fashion. The graph depicts this growth with a line segment. Use the midpoint formula to approximate this revenue for 2009.

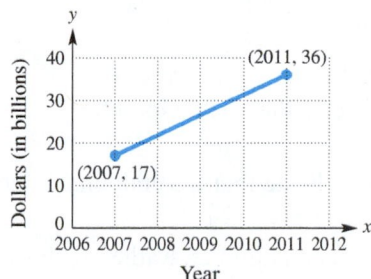

102. *(Modeling) Two-Year College Enrollment* Estimated and projected enrollments in two-year colleges for 2012, 2014, and 2016 are shown in the table. Use the midpoint formula to estimate the enrollments to the nearest thousand for 2013 and 2015.

Year	Enrollment (in thousands)
2012	7601
2014	7689
2016	7952

Source: Statistical Abstract of the United States.

103. *(Modeling) Poverty-Level Income Cutoffs* The table lists poverty-level income cutoffs for a family of four for selected years. Use the midpoint formula to estimate the poverty-level cutoffs (rounded to the nearest dollar) in 2005 and 2009.

Year	Income (in dollars)
2003	18,810
2007	21,203
2011	22,350

Source: U.S. Census Bureau.

104. *Geometry* Triangles can be classified by their sides.
(a) An **isosceles triangle** has at least two sides of equal length. Determine whether the triangle with vertices $(0, 0)$, $(3, 4)$, and $(7, 1)$ is isosceles.

(b) An **equilateral triangle** has all sides of equal length. Determine whether the triangle with vertices $(-1, -1)$, $(2, 3)$, and $(-4, 3)$ is equilateral.
(c) Determine whether a triangle having vertices $(-1, 0)$, $(1, 0)$ and $(0, \sqrt{3})$ is isosceles, equilateral, or neither.
(d) Determine whether a triangle having vertices $(-3, 3)$, $(-2, 5)$ and $(-1, 3)$ is isosceles, equilateral, or neither.

105. *Distance between Cars* At 9:00 A.M., Car A is traveling north at 50 mph and is located 50 miles south of Car B. Car B is traveling west at 20 mph.
(a) Let $(0, 0)$ be the initial coordinates of Car B in the xy-plane, where units are in miles. Plot the locations of each car at 9:00 A.M. and at 11:00 A.M.
(b) Find the distance between the cars at 11:00 A.M.

106. *Distance between Ships* Two ships leave the same harbor at the same time. The first ship heads north at 20 mph and the second ship heads west at 15 mph.
(a) Draw a sketch depicting their positions after t hours.
(b) Write an expression that gives the distance between the ships after t hours.

107. One of the most popular proofs of the Pythagorean theorem uses the figure shown here. Determine the area of the figure in two ways. First, find the area of the large square, using the formula for the area of a square. Then, find its area as the sum of the areas of the smaller square and the four right triangles. Set these expressions equal to each other and simplify to obtain

$$a^2 + b^2 = c^2.$$

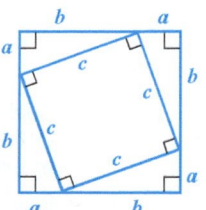

108. Prove that the midpoint M of the line segment joining endpoints $P(x_1, y_1)$ and $Q(x_2, y_2)$ has coordinates

$$\left(\frac{x_1 + x_2}{2}, \frac{y_1 + y_2}{2} \right)$$

by showing that the distance between P and M is equal to the distance between M and Q and that the sum of these distances is equal to the distance between P and Q.

1.2 Introduction to Relations and Functions

Set-Builder Notation and Interval Notation • Relations, Domain, and Range • Functions • Tables and Graphing Calculators
• Function Notation

Set-Builder Notation and Interval Notation

Suppose that we wish to symbolize the set of real numbers greater than -2. The following are two ways to do this.

1. In **set-builder notation**, we write $\{x \mid x > -2\}$, which is read "The set of all real numbers x such that x is greater than -2."

2. In **interval notation**, we write $(-2, \infty)$.

   • In this case, a left parenthesis "(" indicates that the endpoint, -2, *is not* included.

   • A square bracket, either [or], would indicate that the endpoint *is* included. For example $\{x \mid x \le -2\}$ is written as $(-\infty, -2]$.

   • The **infinity symbol** ∞ does not represent a number. Rather, it shows that the interval is unbounded and includes all numbers greater than -2. *A parenthesis is always used next to the infinity symbol.*

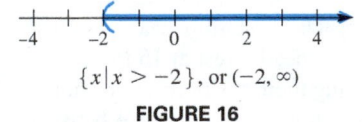

$\{x \mid x > -2\}$, or $(-2, \infty)$

FIGURE 16

A number line graph of $\{x \mid x > -2\}$ is shown in **FIGURE 16.** Note that an interval on the number line is a set of real numbers.

The following chart summarizes set-builder notation, interval notation, and graphs of intervals of real numbers. It is assumed that $a < b$.

Type of Interval	Set-Builder Notation	Interval Notation	Graph
Open interval	$\{x \mid x > a\}$	(a, ∞)	
	$\{x \mid a < x < b\}$	(a, b)	
	$\{x \mid x < b\}$	$(-\infty, b)$	
Other intervals	$\{x \mid x \ge a\}$	$[a, \infty)$	
	$\{x \mid a < x \le b\}$	$(a, b]$	
	$\{x \mid a \le x < b\}$	$[a, b)$	
	$\{x \mid x \le b\}$	$(-\infty, b]$	
Closed interval	$\{x \mid a \le x \le b\}$	$[a, b]$	
Disjoint interval	$\{x \mid x < a \text{ or } x > b\}$	$(-\infty, a) \cup (b, \infty)$	
All real numbers	$\{x \mid x \text{ is a real number}\}$	$(-\infty, \infty)$	

CAUTION   Interval notation for the open interval (*a*, *b*) looks exactly like the notation for the ordered pair (*a*, *b*). When the need arises, we will distinguish between them by referring to "the interval (*a*, *b*)" or "the point (*a*, *b*)."

EXAMPLE 1   Writing Interval Notation

Write each set of real numbers in interval notation.

(a) $\{x \mid -3 < x \le 5\}$ **(b)** $\{x \mid x \le 3\}$ **(c)** $\{x \mid 3 < x\}$

Solution

(a) Use (for $<$ and] for $\le$. Thus we write $(-3, 5]$.

(b) This interval is unbounded, so we write $(-\infty, 3]$.

(c) The inequality $3 < x$ indicates that *x* is greater than 3 (also written $x > 3$). We write the interval as $(3, \infty)$.   ●

Relations, Domain, and Range

The table shows the atmospheric carbon dioxide (CO_2) concentration in parts per million (ppm) for selected years.

Atmospheric Carbon Dioxide Concentrations

Year	1958	1975	1990	2012
CO_2 (ppm)	315	335	355	391

Source: Mauna Loa Observatory.

Since each year in the table is paired with a CO_2 concentration, we can depict this information as a set of ordered pairs in the form (year, concentration). The first component is the year, and the second component is the concentration in parts per million.

$$\{(1958, 315), (1975, 335), (1990, 355), (2012, 391)\}$$

Such a set of ordered pairs is called a *relation*.

Relation

A **relation** is a set of ordered pairs.

If we denote the ordered pairs of a relation by (*x*, *y*), then the set of all *x*-values is called the **domain** of the relation and the set of all *y*-values is called the **range** of the relation. For the relation represented by the table,

$$\text{Domain} = \{1958, 1975, 1990, 2012\}$$

and $$\text{Range} = \{315, 335, 355, 391\}.$$

Here are three other examples of relations:

$$F = \{(1, 2), (-2, 5), (3, -1)\},$$

$$G = \{(-2, 1), (-1, 0), (0, 1), (1, 2), (2, 2)\}, \text{ and}$$

$$H = \{(-4, 1), (-2, 1), (-2, 0)\}.$$

For the relations *F*, *G*, and *H*,   *All x-values*   *All y-values*

Domain of $F = \{1, -2, 3\}$,   Range of $F = \{2, 5, -1\}$;

Domain of $G = \{-2, -1, 0, 1, 2\}$,   Range of $G = \{1, 0, 2\}$;

Domain of $H = \{-4, -2\}$,   Range of $H = \{1, 0\}$.

Since a relation is a set of ordered pairs, it may be represented graphically in the rectangular coordinate system. See **FIGURE 17**.

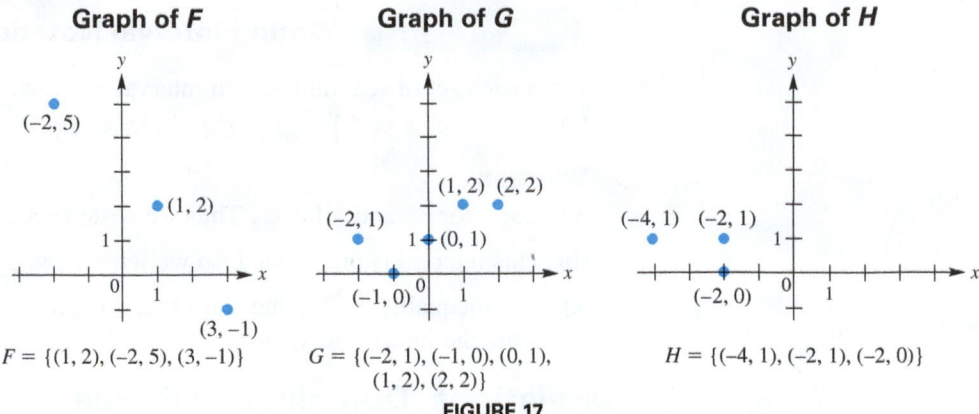

Graph of *F*   Graph of *G*   Graph of *H*

$F = \{(1, 2), (-2, 5), (3, -1)\}$ $G = \{(-2, 1), (-1, 0), (0, 1),$ $H = \{(-4, 1), (-2, 1), (-2, 0)\}$
$(1, 2), (2, 2)\}$

FIGURE 17

Some relations contain infinitely many ordered pairs. For example, let *M* represent a relation consisting of all ordered pairs having the form $(x, 2x)$, where *x* is a real number. Since there are infinitely many values for *x*, there are infinitely many ordered pairs in *M*. Five such ordered pairs are plotted in **FIGURE 18(a)** and suggest that the graph of *M* is a line. The graph of *M* includes all points (x, y), such that $y = 2x$, as shown in **FIGURE 18(b)**.

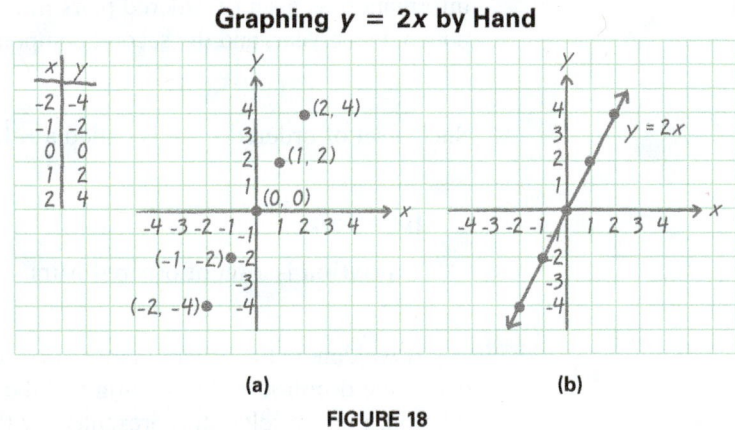

Graphing *y* = 2*x* by Hand

(a)   (b)

FIGURE 18

Mapping of *F*

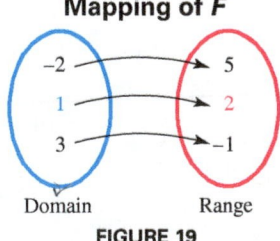

Domain   Range

FIGURE 19

A relation can be illustrated with a "mapping" diagram, as shown in **FIGURE 19** for relation *F*. The arrow from 1 to 2 indicates that the ordered pair $(1, 2)$ belongs to *F*.

In summary, a relation can be represented by any of the following.

1. A **graph,** as illustrated in **FIGURE 17** and **FIGURE 18**
2. A **table of *xy*-values,** as shown in **FIGURE 18(a)**
3. An **equation,** such as $y = 2x$ in **FIGURE 18(b)**
4. A **mapping** or **diagram,** as illustrated in **FIGURE 19**

NOTE   A graph of a relation can consist of distinct points, a line, or a curve.

Determining Domains and Ranges from Graphs

Give the domain and range of each relation from its graph.

(a)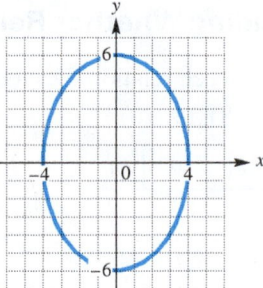

(b)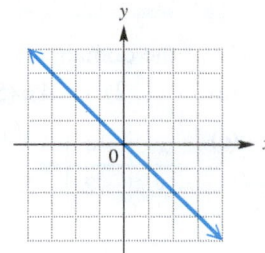

(c)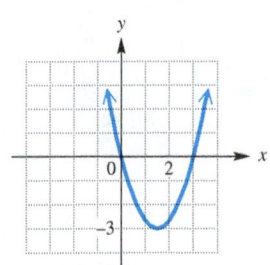

Solution

(a) See **FIGURE 20.** The x-values of the points on the graph include all numbers between -4 and 4 inclusive. The y-values include all numbers between -6 and 6 inclusive. Using interval notation, the domain is $[-4, 4]$, and the range is $[-6, 6]$.

(b) The arrowheads indicate that the line extends indefinitely left and right, as well as upward and downward. Therefore, both the domain and range are the set of all real numbers, written $(-\infty, \infty)$.

(c) The arrowheads indicate that the graph extends indefinitely left and right, as well as upward. The domain is $(-\infty, \infty)$. There is a least y-value, -3, so the range includes all numbers greater than or equal to -3, written $[-3, \infty)$.

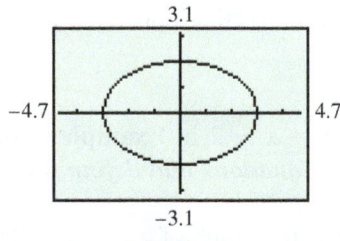

FIGURE 20

Finding Domain and Range from a Calculator Graph

FIGURE 21 shows a graph on a screen with viewing window $[-4.7, 4.7]$ by $[-3.1, 3.1]$, Xscl $= 1$, Yscl $= 1$. Give the domain and range of this relation.

Solution Since the scales on both axes are 1, the graph *appears* to have minimum x-value -3, maximum x-value 3, minimum y-value -2, and maximum y-value 2. This leads us to conclude that the domain is $[-3, 3]$ and the range is $[-2, 2]$.

FIGURE 21

Functions

Suppose that a sales tax rate is 6%. Then a purchase of $200 results in a sales tax of $0.06 \times \$200 = \$12$. This calculation can be summarized by the ordered pair $(200, 12)$. The ordered pair $(50, 3)$ indicates that a purchase of $50 results in a sales tax of $3. For each purchase of x dollars, there is exactly one sales tax amount of y dollars.

Calculating sales tax y on a purchase of x dollars results in a set of ordered pairs (x, y), where $y = 0.06x$. This set of ordered pairs represents a special type of relation called a *function.* ***In a function, each x-value must correspond to exactly one y-value.***

Function

A **function** is a relation in which each element in the domain corresponds to exactly one element in the range.*

TECHNOLOGY NOTE

In **FIGURE 21,** we see a calculator graph that is formed by a rather jagged curve. These representations are sometimes called *jaggies* and are typically found on low-resolution graphers, such as graphing calculators. In general, most curves in this book are smooth, and jaggies are just a part of the limitations of technology.

*An alternative definition of *function* based on the idea of correspondence is given later in the section.

Function

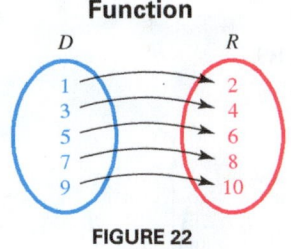

FIGURE 22

Not a Function

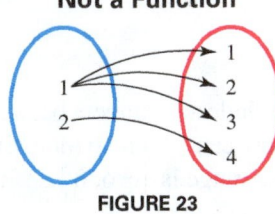

FIGURE 23

Function

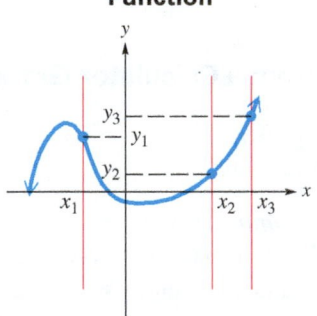

A vertical line intersects the graph
at most once.

(a)

Not a Function

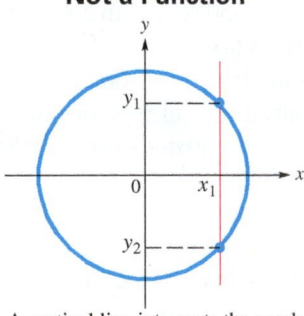

A vertical line intersects the graph
more than once.

(b)

FIGURE 24

If x represents any element in the domain, x is called the **independent variable.** If y represents any element in the range, y is called the **dependent variable,** because *the value of y depends on the value of x.* For example, sales tax depends on purchase price.

EXAMPLE 4 **Deciding Whether Relations Are Functions**

Give the domain and range of each relation. Decide whether each relation is a function.

(a) $\{(1, 2), (3, 4), (5, 6), (7, 8), (9, 10)\}$

(b) $\{(1, 1), (1, 2), (1, 3), (2, 4)\}$

(c)

x	-4	-3	-2	-1	0	1
y	2	2	2	2	2	2

(d) $\{(x, y) \mid y = x - 2\}$

Solution

(a) The domain D is $\{1, 3, 5, 7, 9\}$, and the range R is $\{2, 4, 6, 8, 10\}$. Since each element in the domain corresponds to exactly one element in the range, this set *is* a function. The correspondence is shown as a diagram in **FIGURE 22.**

(b) The domain D here is $\{1, 2\}$, and the range R is $\{1, 2, 3, 4\}$. As shown in the correspondence in **FIGURE 23,** one element in the domain, 1, has been assigned three different elements from the range, so this relation *is not* a function.

(c) In the table of ordered pairs, the domain is $\{-4, -3, -2, -1, 0, 1\}$ and the range is $\{2\}$. Although every element in the domain corresponds to the *same* element in the range, this *is* a function because each element in the domain has exactly one range element assigned to it.

(d) In $\{(x, y) \mid y = x - 2\}$, y is always found by subtracting 2 from x. Each x corresponds to just one value of y, so this relation *is* a function. Any number can be used for x, and each x will give a number that is 2 less for y. Thus, both the domain and range are the set of real numbers: $(-\infty, \infty)$. ●

Functions are often defined by equations, such as $y = x - 2$ in **Example 4(d),** where each (valid) x-value determines a unique y-value. *Equations that define functions are usually solved for the dependent variable y.*

There is a quick way to tell whether a given graph is the graph of a function. In the graph in **FIGURE 24(a),** each value of x corresponds to only one value of y, so this *is* the graph of a function. By contrast, the graph in **FIGURE 24(b)** *is not* the graph of a function. The vertical line through x_1 intersects the graph at two points, showing that there are two values of y that correspond to this x-value. This concept is known as the *vertical line test* for a function.

Vertical Line Test

If every vertical line intersects a graph in no more than one point, then the graph is the graph of a function.

EXAMPLE 5 **Using the Vertical Line Test**

(a) Is the graph in **FIGURE 25** the graph of a function?

(b) The graph in **FIGURE 26** extends right indefinitely and upward and downward indefinitely. Does it appear to be the graph of a function?

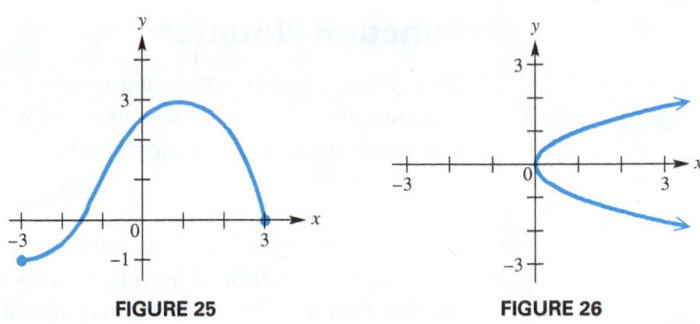

FIGURE 25   **FIGURE 26**

Solution

(a) Any vertical line will intersect the graph *at most once*. Therefore, the graph in **FIGURE 25** satisfies the vertical line test and is the graph of a function.

(b) Any vertical line that lies to the right of the origin intersects the graph *twice*. Therefore, the graph in **FIGURE 26** does not represent a function. ●

While the concept of a function is crucial to the study of mathematics, its definition may vary from text to text. We now give an alternative definition.

Function (Alternative Definition)

A function is a correspondence in which each element x from a set called the domain is paired with one and only one element y from a set called the range.

Tables and Graphing Calculators

A convenient way to display ordered pairs in a function is by using a table. An equation such as $y = 9x - 5$ describes a function. If we choose x-values to be $0, 1, 2, \ldots, 6$, then the corresponding y-values are as follows.

Use parentheses around substituted values to avoid errors.

$$y = 9(0) - 5 = -5 \qquad y = 9(1) - 5 = 4$$
$$y = 9(2) - 5 = 13 \qquad y = 9(3) - 5 = 22$$
$$y = 9(4) - 5 = 31 \qquad y = 9(5) - 5 = 40$$
$$y = 9(6) - 5 = 49$$

These ordered pairs, $(0, -5), (1, 4), \ldots, (6, 49)$, can be organized in a table, as shown below. A graphing calculator can efficiently generate this table, as shown in **FIGURE 27.**

x	y
0	−5
1	4
2	13
3	22
4	31
5	40
6	49

FIGURE 27

The same table created by a graphing calculator

TECHNOLOGY NOTE

```
TABLE SETUP
 TblStart=0
 ΔTbl=1
Indpnt: Auto Ask
Depend: Auto Ask
```

This screen indicates that the table will start with 0 and have an *increment* of 1. Both variables appear automatically. TblStart represents the initial value of the independent variable (0 in the table in **FIGURE 27**) and ΔTbl represents the difference, or increment, between successive values of the independent variable (1 in **FIGURE 27**).

Function Notation

To say that y is a function of x means that for each value of x from the domain of the function, there is exactly one value of y. To emphasize that y *is a function of x*, or that y *depends on x*, it is common to write

$$y = f(x), \qquad \text{\textit{f}(x) equals y.}$$

with $f(x)$ read **"f of x."** This notation is called **function notation.**

Function notation is used frequently when functions are defined by equations. For the function $y = 9x - 5$, we may **name** this function f and write $f(x) = 9x - 5$. If $x = 2$, then we find y, or $f(2)$, by replacing x with 2.

$$f(2) = 9 \cdot 2 - 5 = 13 \qquad \text{Substitute 2 for } x.$$

The statement "if $x = 2$, then $y = 13$" is interpreted as the ordered pair $(2, 13)$ and is abbreviated with function notation as

$$f(2) = 13. \qquad \text{"f of 2 equals 13."}$$

Also, $f(0) = 9 \cdot 0 - 5 = -5$, and $f(-3) = 9(-3) - 5 = -32$.

These ideas can be explained as follows.

Name of the function   Defining expression

$$y = f(x) = \overbrace{9x - 5}$$

Name of the dependent variable   Name of the independent variable

When f is evaluated at x, its value is $f(x)$ (or equivalently y) because $y = f(x)$.

> **CAUTION** *The symbol f(x) does not indicate "f times x,"* but represents the *y*-value for the indicated *x*-value. For example, $f(2)$ is the *y*-value that corresponds to the *x*-value 2.

We can use names other than f, such as g and h, to represent functions.

→ **Looking Ahead to Calculus**
Functions are evaluated frequently in calculus. They can be evaluated in a variety of ways, as illustrated in **Example 6**.

EXAMPLE 6   **Using Function Notation**

For each function, find $f(3)$.

(a) $f(x) = 3x - 7$

(b) The function f defined by the table

x	1	2	3	4
$f(x)$	-15	-12	-9	-6

(c) The function f depicted in **FIGURE 28**

(d) The function f graphed in **FIGURE 29**

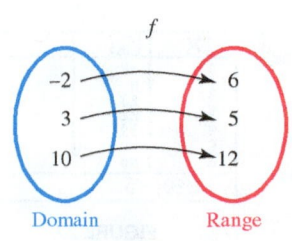

FIGURE 28

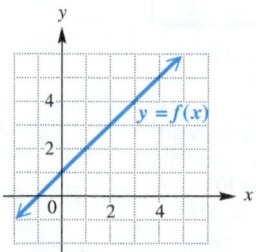

FIGURE 29

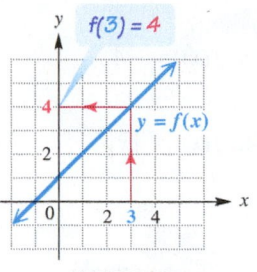

FIGURE 30

Solution

(a) Replace x with 3 to get $f(3) = 3(3) - 7 = 2$.

(b) From the table, when $x = 3$, $y = -9$. Thus, $f(3) = -9$.

(c) In **FIGURE 28**, 3 in the domain is paired with 5 in the range, so $f(3) = 5$.

(d) To evaluate $f(3)$, begin by finding 3 on the x-axis. See **FIGURE 30.** Then move upward until the graph of f is reached. Moving horizontally to the y-axis gives 4 for the corresponding y-value. Thus, $f(3) = 4$. Note that the point $(3, 4)$ lies on the graph of f.

> **NOTE**   If $f(a) = b$, then the point (a, b) lies on the graph of f. Conversely, if the point (a, b) lies on the graph of f, then $f(a) = b$. Thus, each point on the graph of f can be written in the form $(a, f(a))$.

EXAMPLE 7   **Evaluating Function Notation**

Let $f(x) = 5 - 2x$. Evaluate each expression.

(a) $f(4)$ **(b)** $f(a)$ **(c)** $f(a + h)$ **(d)** $f(5x)$

Solution   Substitute the expression inside the parentheses for x into $f(x) = 5 - 2x$ and then simplify.

(a) $f(4) = 5 - 2(4) = 5 - 8 = -3$ Substitute 4 for x.

Distributive property

(b) $f(a) = 5 - 2(a) = 5 - 2a$ Substitute a for x.

(c) $f(a + h) = 5 - 2(a + h) = 5 - 2a - 2h$ Substitute $a + h$ for x.

(d) $f(5x) = 5 - 2(5x) = 5 - 10x$ Substitute $5x$ for x.

1.2 Exercises

Using interval notation, write each set. Then graph it on a number line.

1. $\{x \mid -1 < x < 4\}$ **2.** $\{x \mid x \geq -3\}$ **3.** $\{x \mid x < 0\}$

4. $\{x \mid 8 > x > 3\}$ **5.** $\{x \mid 1 \leq x < 2\}$ **6.** $\{x \mid -5 < x \leq -4\}$

Using the variable x, write each interval using set-builder notation.

7. $(-4, 3)$ **8.** $[2, 7)$ **9.** $(-\infty, -1]$ **10.** $(3, \infty)$

11.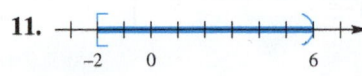

12.

13.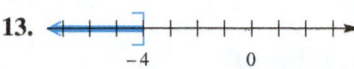

14.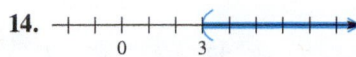

15. Explain how to determine whether a parenthesis or a square bracket is used when graphing an inequality on a number line.

16. *Concept Check*   The three-part inequality $a < x < b$ means "a is less than x, *and x is less than* b." Which one of the following inequalities is not satisfied by some real number x?

A. $-3 < x < 5$ **B.** $0 < x < 4$

C. $-3 < x < -2$ **D.** $-7 < x < -10$

Checking Analytic Skills   *Sketch the graph of f by hand.* **Do not use a calculator.**

17. $f(x) = x - 3$ **18.** $f(x) = 1 - 2x$ **19.** $f(x) = 3$ **20.** $f(x) = -4$

21. $f(x) = \dfrac{1}{2}x$ **22.** $f(x) = -\dfrac{2}{3}x$ **23.** $f(x) = x^2$ **24.** $f(x) = |x|$

Determine the domain D and range R of each relation, and tell whether the relation is a function.
Assume that a calculator graph extends indefinitely and a table includes only the points shown.

25. $\{(5, 1), (3, 2), (4, 9), (7, 6)\}$ **26.** $\{(8, 0), (5, 4), (9, 3), (3, 8)\}$ **27.** $\{(1, 6), (2, 6), (3, 6)\}$

28. $\{(-10, 5), (-20, 5), (-30, 5)\}$ **29.** $\{(4, 1), (3, -5), (-2, 3), (3, 7)\}$ **30.** $\{(0, 5), (1, 3), (0, -4)\}$

31.

x	11	12	13	14
y	-6	-6	-7	-6

32.

x	1	1	1	1
y	12	13	14	15

33.

x	0	1	2	3	4
y	$\sqrt{2}$	$\sqrt{3}$	$\sqrt{5}$	$\sqrt{6}$	$\sqrt{7}$

34.

x	1	$\frac{1}{2}$	$\frac{1}{4}$	$\frac{1}{8}$	$\frac{1}{16}$
y	0	-1	-2	-3	-4

35.
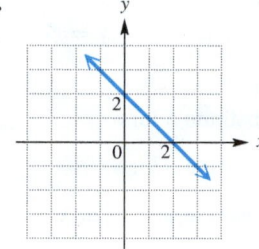

36.
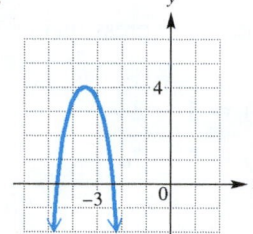

37.
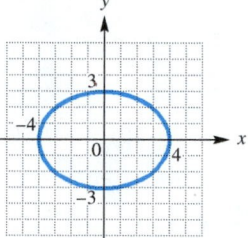

38.
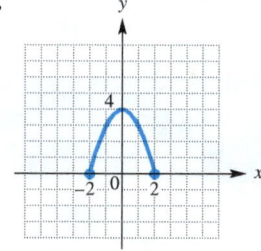

39.
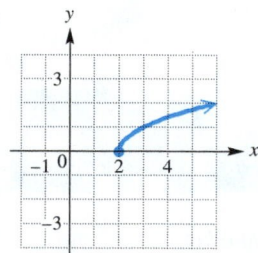

40.
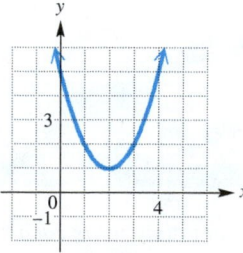

41.
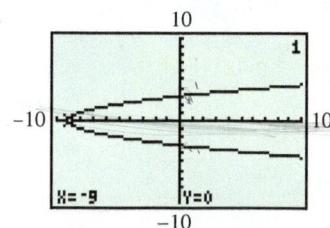

42.
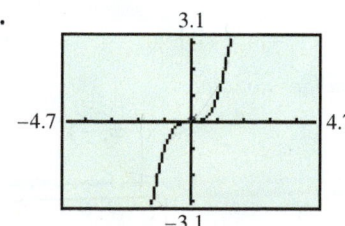

43.
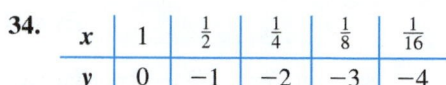

44.
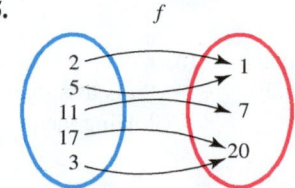

45.
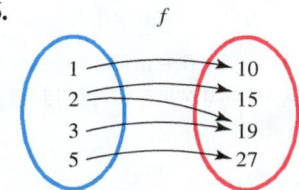

46.

Find each function value, if defined.

47. $f(-2)$ if $f(x) = Y_1$ as defined in **Exercise 43**

48. $f(5)$ if $f(x) = Y_1$ as defined in **Exercise 44**

49. $f(11)$ for the function f in **Exercise 45**

50. $f(5)$ for the function f in **Exercise 45**

51. $f(1)$ for the function in **Exercise 45**

52. $f(10)$ for the function in **Exercise 45**

Find $f(x)$ at the indicated value of x.

53. $f(x) = 3x - 4$, $x = -2$

54. $f(x) = 5x + 6$, $x = -5$

55. $f(x) = 2x^2 - x + 3$, $x = 1$

56. $f(x) = 3x^2 + 2x - 5$, $x = 2$

57. $f(x) = -x^2 + x + 2$, $x = 4$

58. $f(x) = -x^2 - x - 6$, $x = 3$

59. $f(x) = 5$, $x = 9$

60. $f(x) = -4$, $x = 12$

61. $f(x) = \sqrt{x^3 + 12}$, $x = -2$

62. $f(x) = \sqrt[3]{x^2 - x + 6}$, $x = 2$

63. $f(x) = |5 - 2x|$, $x = 8$

64. $f(x) = |6 - \frac{1}{2}x|$, $x = 20$

Find $f(a)$, $f(b + 1)$, and $f(3x)$ for the given $f(x)$.

65. $f(x) = 5x$

66. $f(x) = x - 5$

67. $f(x) = 2x - 5$

68. $f(x) = x^2$

69. $f(x) = 1 - x^2$

70. $f(x) = |x| + 4$

Concept Check   *Work each problem.*

71. If $f(-2) = 3$, identify a point on the graph of f.

72. If $f(3) = -9.7$, identify a point on the graph of f.

73. If the point $(7, 8)$ lies on the graph of f, then $f(\underline{\quad}) = \underline{\quad}$.

74. If the point $(-3, 2)$ lies on the graph of f, then $f(\underline{\quad}) = \underline{\quad}$.

Use the graph of $y = f(x)$ to find each function value: **(a)** $f(-2)$, **(b)** $f(0)$, **(c)** $f(1)$, *and* **(d)** $f(4)$.

75.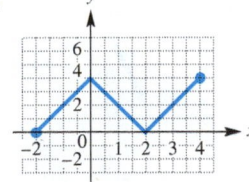

76.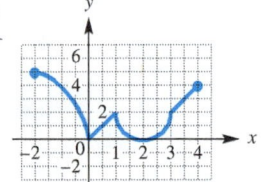

77.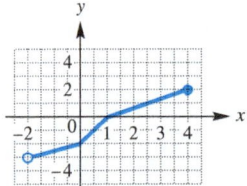

78.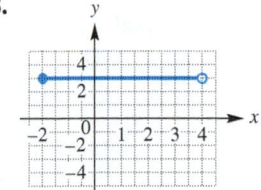

79. Explain each term in your own words.
(a) Relation
(b) Function
(c) Domain of a function
(d) Range of a function
(e) Independent variable
(f) Dependent variable

80. *Radio Stations* The function f gives the number y in thousands of radio stations on the air during year x.

$$f = \{(1975, 7.7), (1990, 10.8),$$
$$(2000, 12.8), (2005, 13.5), (2012, 15.1)\}$$

(*Source:* M. Street Corporation.)
(a) Use a mapping diagram (see **FIGURE 19**) to represent f.
(b) Evaluate $f(2000)$ and explain what it means.
(c) Identify the domain and range of f.

81. The following table lists the U.S. mobile advertising revenue in 2011 in millions of dollars for various companies.

Company	Revenue
Google	155
Apple	95
Jumptab	61
Microsoft	39

Source: Business Insider.

(a) Using ordered pairs, write a function A that gives the revenue for each company in millions of dollars. Interpret the first ordered pair.
(b) Repeat part (a) using a diagram.
(c) Identify the domain and range of A.

82. The following table lists Square's daily transactions y in millions of dollars, x months after March 2011. (Square can be used to accept credit cards on your iPhone.)

x	y
0	1.0
2	2.0
7	5.5
12	11.0

(a) Using ordered pairs, write a function T that gives the daily transactions in millions of dollars during each month. Interpret the first ordered pair.

(b) Repeat part (a) using a diagram.

(c) Identify the domain and range of T.

SECTIONS 1.1–1.2 — Reviewing Basic Concepts

1. Plot the points $(-3, 1)$, $(-2, -1)$, $(2, -3)$, $(1, 1)$, and $(0, 2)$. Label each point.

2. Find the length and midpoint M of the line segment that connect the points $P(-4, 5)$ and $Q(6, -2)$.

3. Use a calculator to approximate $\dfrac{\sqrt{5 + \pi}}{\sqrt[3]{3} + 1}$ to the nearest thousandth.

4. Find the exact distance between the points $(12, -3)$ and $(-4, 27)$.

5. The hypotenuse of a right triangle measures 61 inches, and one of its legs measures 11 inches. Find the length of the other leg.

6. Use interval notation to write the sets $\{x \mid -2 < x \le 5\}$ and $\{x \mid x \ge 4\}$.

7. Determine whether the relation shown in the graph is a function. What are its domain and range?

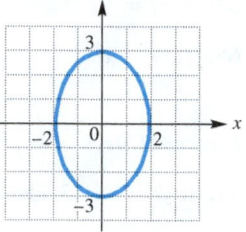

8. Graph $f(x) = \frac{1}{2}x - 1$ by hand.

9. Find $f(-5)$ and $f(a + 4)$ if $f(x) = 3 - 4x$.

10. Use the graph of $y = f(x)$ to find $f(2)$ and $f(-1)$.

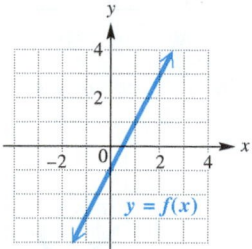

1.3 Linear Functions

Basic Concepts about Linear Functions • Slope of a Line and Average Rate of Change • Slope–Intercept Form of the Equation of a Line

Basic Concepts about Linear Functions

Suppose that by noon 2 inches of rain had fallen. Rain continued to fall at the rate of $\frac{1}{2}$ inch per hour in the afternoon. Then the total rainfall x hours past noon is given by

$$f(x) = \frac{1}{2}x + 2.$$

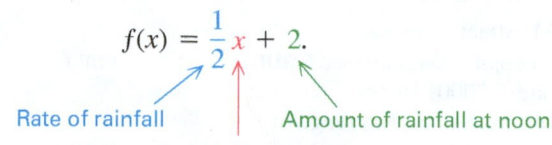

Rate of rainfall

Hours of rainfall past noon

Amount of rainfall at noon

At 3:00 P.M., the total rainfall equaled

$$f(3) = \frac{1}{2}(3) + 2 = 3.5 \text{ inches.}$$

This function f is called a *linear function*.

Linear Function

A function f defined by $\boldsymbol{f(x) = ax + b}$, where a and b are real numbers, is called a **linear function.**

NOTE For the linear function $f(x) = ax + b$, the value of a represents a **constant rate of change** and the value of b represents the **initial amount** or **initial value** of $f(x)$ when $x = 0$. In the preceding example, $a = \frac{1}{2}$ because the rain is falling at a *constant rate* of $\frac{1}{2}$ inch per hour, and $b = 2$ because the initial amount of rain that had fallen by noon ($x = 0$) was 2 inches. If something changes at a constant rate, then it can be **modeled,** or described, by a *linear* function.

The graph of a linear function is a line. Because $y = f(x)$, the graph of $f(x) = ax + b$ can be found by graphing $y = ax + b$. For example, the graph of $f(x) = 3x + 6$ is the line determined by $y = 3x + 6$. An equation such as $y = 3x + 6$ is called a **linear equation in two variables. A solution** is an ordered pair (x, y) that makes the equation true. For example, $(-2, 0)$ is a solution of $y = 3x + 6$, because $0 = 3(-2) + 6$. Verify that $(-1, 3)$, $(0, 6)$, and $(1, 9)$ are also solutions of $y = 3x + 6$ and, therefore, lie on the graph of $f(x) = 3x + 6$.

Graphing linear equations by hand involves plotting points whose coordinates are solutions of the equation and then connecting them with a straight line. **FIGURE 31(a)** shows the ordered pairs just mentioned for the linear equation $y = 3x + 6$, accompanied by a **table of values.** Notice that the points appear to lie in a straight line and that is indeed the case. Since we may substitute *any* real number for x, we connect these points with a line to obtain the graph of $f(x) = 3x + 6$, as shown in **FIGURE 31(b).**

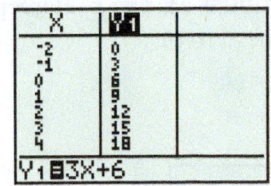
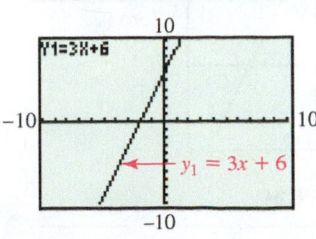

Graphing a Linear Function by Hand

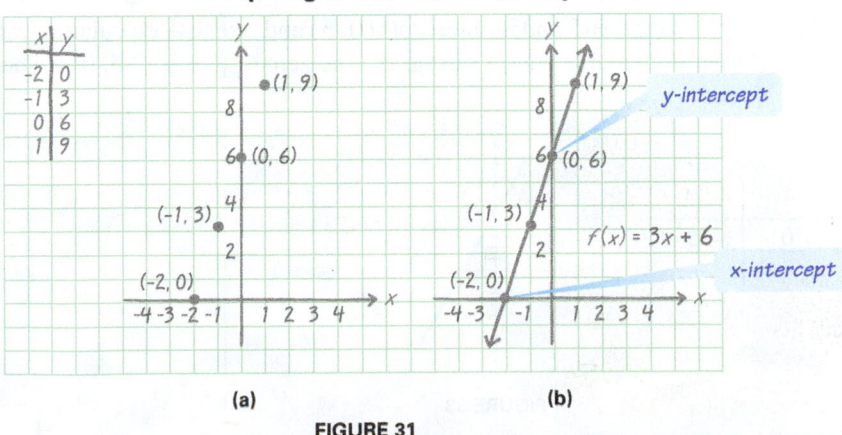

       (a)   (b)

FIGURE 31

From geometry, we know that two distinct points determine a line. ***Therefore, if we know the coordinates of two points, we can graph the line.*** For the equation $y = 3x + 6$, suppose we let $x = 0$ and solve for y. Then suppose we let $y = 0$ and solve for x, using techniques from elementary algebra.

$y = 3x + 6$			$y = 3x + 6$	
$y = 3(0) + 6$	Let $x = 0$.		$0 = 3x + 6$	Let $y = 0$.
$y = 0 + 6$	Multiply.		$-6 = 3x$	Subtract 6.
$y = 6$	Add.		$x = -2$	Divide by 3 and rewrite.

The points $(0, 6)$ and $(-2, 0)$ lie on the graph of $y = 3x + 6$ and are sufficient for obtaining the graph in **FIGURE 31(b)** on the previous page. (Sometimes it is advisable to plot a third point as a check.) The points $(0, 6)$ and $(-2, 0)$ are called the **y-** and **x-intercepts** of the line respectively.

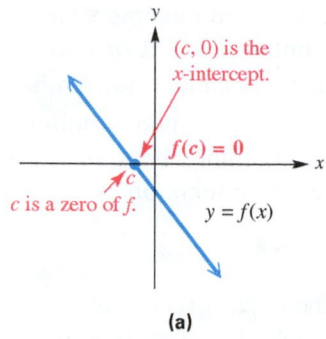

(a)

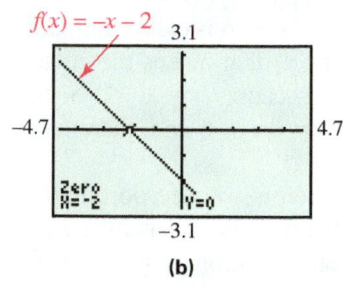

(b)

FIGURE 32

Locating x- and y-Intercepts

To find the *x*-intercept of the graph of $y = ax + b$, let $y = 0$ and solve for x (assuming that $a \neq 0$). To find the *y*-intercept, let $x = 0$ and solve for y.

We will often be interested in determining *x*-values for which a function is equal to zero.

Zero of a Function and x-Intercepts

Let f be a function. Then any number c for which $f(c) = 0$ is called a **zero** of the function f. The point $(c, 0)$ is an *x*-intercept of the graph of f. See **FIGURE 32(a).**

NOTE A calculator can be directed to find a zero of a function, such as $f(x) = -x - 2$. See **FIGURE 32(b).**

EXAMPLE 1 **Graphing a Line**

Graph the function $f(x) = -2x + 5$. What is the zero of f?

Analytic Solution

The graph of $f(x) = -2x + 5$ and its intercepts $(0, 5)$ and $(2.5, 0)$ are shown in the table and in **FIGURE 33.** The zero of f is 2.5.

x	y
0	5
2.5	0

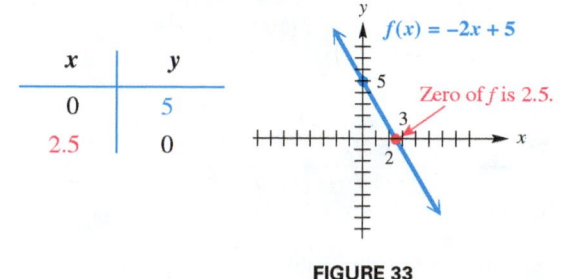

FIGURE 33

Graphing Calculator Solution

A calculator graph is shown in **FIGURE 34.** The *x*-intercept of $(2.5, 0)$ is shown, so 2.5 is the zero of f.

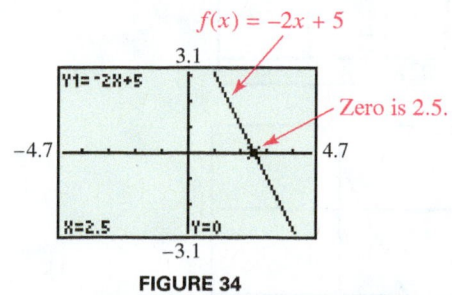

FIGURE 34

NOTE   If a linear function $f(x) = ax + b$ has a zero c, then the graph of $y = f(x)$ has x-intercept $(c, 0)$. Also, because $f(0) = a(0) + b = b$, the y-intercept is $(0, b)$.

EXAMPLE 2   **Finding a Formula for a Function**

A 100-gallon tank full of water is being drained at a rate of 5 gallons per minute.

(a) Write a formula for a linear function f that models the number of gallons of water in the tank after x minutes.

(b) How much water is in the tank after 4 minutes?

(c) Use the x- and y-intercepts to graph f. Interpret each intercept.

Solution

(a) The amount of water in the tank is *decreasing* at a constant rate of 5 gallons per minute, so the *constant rate of change* is -5. The initial amount of water is equal to 100 gallons.

$$f(x) = \text{(constant rate of change)}\,x + \text{(initial amount)}$$
$$f(x) = -5x + 100$$

(b) After 4 minutes, the tank held $f(4) = -5(4) + 100 = 80$ gallons.

(c) The graph of $y = -5x + 100$ has y-intercept $(0, 100)$ because $y = 100$ when $x = 0$. To find the x-intercept, let $y = 0$ and solve the equation $0 = -5x + 100$, obtaining $x = 20$. The x-intercept is $(20, 0)$. The graph of f is shown in **FIGURE 35.** The number 20 in the x-intercept $(20, 0)$ corresponds to the time in minutes that it takes to empty the tank, and the number 100 in the y-intercept $(0, 100)$ corresponds to the number of gallons of water in the tank initially.

Water in a Tank

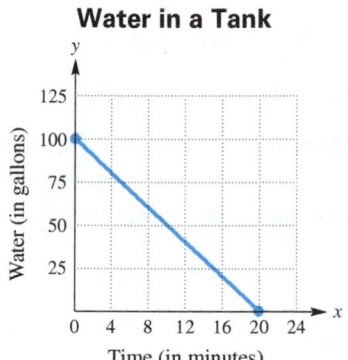

FIGURE 35

Suppose that for a linear function $f(x) = ax + b$, we have $a = 0$. Then the function becomes $f(x) = b$, where b is some real number. Because $a = 0$, function $f(x) = b$ has a zero rate of change. Thus, its graph is a horizontal line with y-intercept $(0, b)$.

EXAMPLE 3   **Sketching the Graph of $f(x) = b$**

Graph the function $f(x) = -3$.

Solution   Since y always equals -3, the y-intercept is $(0, -3)$. Since the value of y can never be 0, the graph has no x-intercept and is horizontal, as shown in **FIGURE 36.**

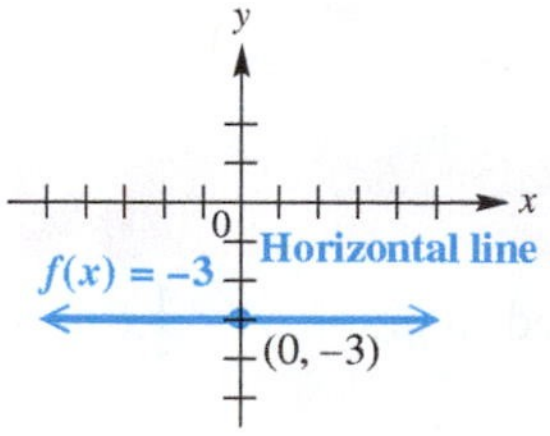

FIGURE 36

> **Constant Function**
>
> A function $f(x) = b$, where b is a real number, is called a **constant function.** Its graph is a horizontal line with y-intercept $(0, b)$. For $b \neq 0$, it has no x-intercept. (Every constant function is also linear.)

Unless otherwise specified, the domain of a linear function is the set of all real numbers. The range of a nonconstant linear function is also the set of all real numbers. The range of a constant function $f(x) = b$ is $\{b\}$.

The choice of viewing window may give drastically different views of a calculator graph, as shown in **FIGURE 37** on the next page for the graph of $f(x) = 3x + 6$.

Graphing the Same Line in Different Windows

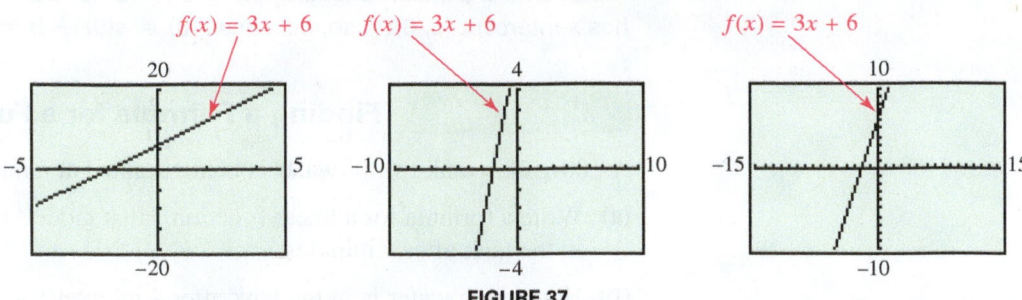

FIGURE 37

We usually want a window that shows the most important features of a particular graph. Such a graph is called a **comprehensive graph**.* The choice of window for a comprehensive graph is not unique—there are many acceptable ones. *For a line, a comprehensive graph shows all intercepts of the line.*

> **EXAMPLE 4**   **Finding a Comprehensive Graph of a Line**

Find a comprehensive graph of $g(x) = -0.75x + 12.5$.

Solution   The window $[-10, 10]$ by $[-10, 10]$ shown in **FIGURE 38(a)** does not show either intercept, so it will not work. For example, we can alter the window size to $[-10, 20]$ by $[-10, 20]$ to obtain a comprehensive graph. See **FIGURE 38(b).**

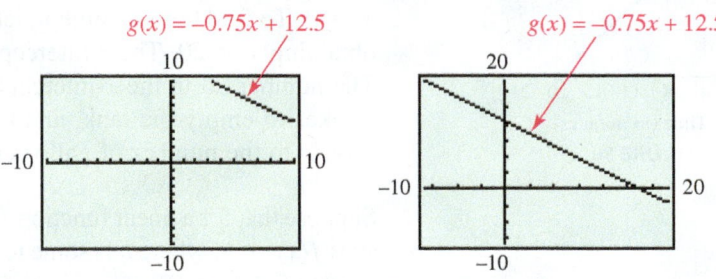

This *is not* a comprehensive graph of the line, since the intercepts are not visible.

(a)

This *is* a comprehensive graph of the line, since both intercepts are visible.

(b)

FIGURE 38

WHAT WENT WRONG?

A student learning to use a graphing calculator attempted to graph $y = \frac{1}{2}x + 15$. However, she obtained the blank screen shown here.

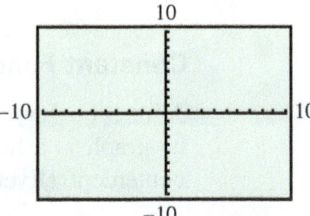

What Went Wrong?   How can she obtain a comprehensive graph of this function?

---

*The term *comprehensive graph* was coined by Shoko Aogaichi Brant and Edward A. Zeidman in the text *Intermediate Algebra: A Functional Approach* (HarperCollins College Publishers, 1996), with the assistance of Professor Brant's daughter Jennifer. The authors thank them for permission to use the terminology in this text.

Answer to What Went Wrong?
The y-intercept is $(0, 15)$ and the x-intercept is $(-30, 0)$, so the window size must be increased to show the intercepts. For example, a window of $[-40, 40]$ by $[-30, 30]$ would work.

→ **Looking Ahead to Calculus**
The concept of the slope of a line is used extensively in calculus and is generalized to include the slope of a curve at a point.

Slope of a Line and Average Rate of Change

In 2002, the average annual cost for tuition and fees at private four-year colleges was $18,780. By 2012, this cost had increased to $29,056. The line (segment) graphed in **FIGURE 39** is actually somewhat misleading, since it shows a constant rate of change and indicates that the increase in cost was the same from year to year.

We can use the graph to determine the *average* yearly increase in cost. Over the 10-year span, the cost increased $10,276. Therefore, the average yearly increase was

$$\frac{\$29,056 - \$18,780}{2012 - 2002} = \frac{\$10,276}{10} \approx \$1028.$$

This quotient is an illustration of the *slope* of the line joining $(2002, 18,780)$ and $(2012, 29,056)$. The concept of slope applies to any nonvertical line.

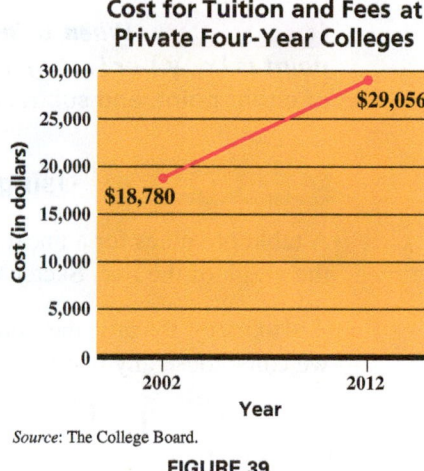

Cost for Tuition and Fees at Private Four-Year Colleges

Source: The College Board.

FIGURE 39

A table of values and the graph of the line $y = 3x + 1$ are shown in **FIGURE 40.** For each unit increase in x, the y-value increases by 3. The slope of the line is 3.

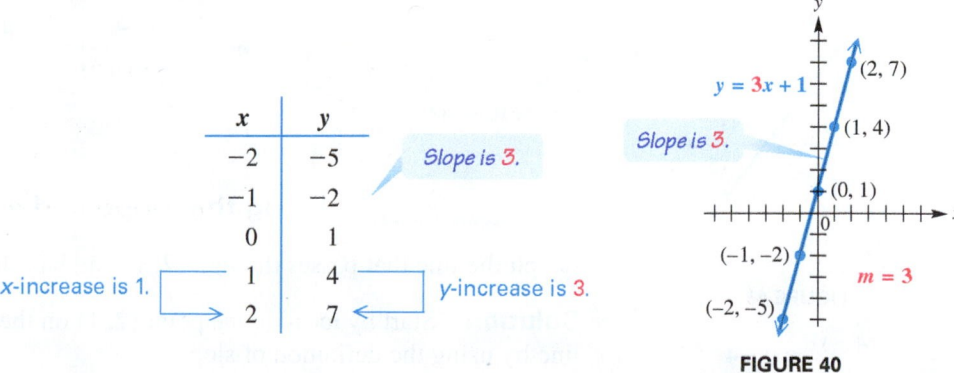

FIGURE 40

Finding Slope

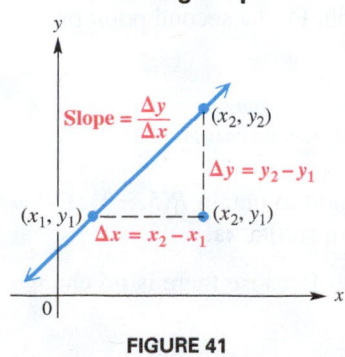

FIGURE 41

Geometrically, slope is a numerical measure of the steepness of a line and may be interpreted as the ratio of *rise* to *run*. To calculate slope, start with the line through the two distinct points (x_1, y_1) and (x_2, y_2), where $x_1 \neq x_2$. See **FIGURE 41.** The difference $x_2 - x_1$ is called the **change in x**, denoted Δx (read **"delta x"**), where Δ is the Greek letter **delta**. In the same way, the **change in y** is denoted $\Delta y = y_2 - y_1$. The *slope* of a nonvertical line is defined as the quotient of the change in y and the change in x.

> ## Slope
>
> The **slope** m of the line passing through the points (x_1, y_1) and (x_2, y_2) is
>
> $$m = \frac{\Delta y}{\Delta x} = \frac{y_2 - y_1}{x_2 - x_1}, \quad \text{where } \Delta x = x_2 - x_1 \neq 0.$$

NOTE The preceding slope definition can be used to find an equation for a nonvertical line passing through the points with coordinates (x_1, y_1) and (x_2, y_2). If $P(x, y)$ is any point on this nonvertical line, then its equation can be written $m = \dfrac{y - y_1}{x - x_1}$. See **Section 1.4**.

CAUTION *When using the slope formula, it makes no difference which point is (x_1, y_1) or (x_2, y_2); however, be consistent.* Start with the x- and y-values of either point, and subtract the corresponding values of the *other* point.

EXAMPLE 5 **Using the Slope Formula**

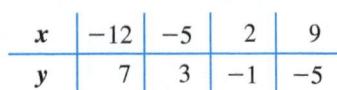

A table of values for a linear function is shown in the margin. Determine the slope of the graph of the line. Sketch the graph.

x	-12	-5	2	9
y	7	3	-1	-5

Solution Because the slope of a line is the same regardless of the two points chosen, we can choose any two points from the table. If we let

$$(2, -1) = (x_1, y_1) \quad \text{and} \quad (-5, 3) = (x_2, y_2), \quad \text{then}$$

$$m = \frac{y_2 - y_1}{x_2 - x_1} = \frac{3 - (-1)}{-5 - 2} = -\frac{4}{7}.$$

> Start with the x- and y-values of the <u>same</u> point.

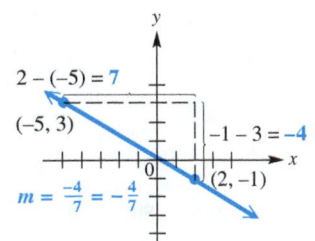

By contrast, if we let

$$(-5, 3) = (x_1, y_1) \quad \text{and} \quad (2, -1) = (x_2, y_2), \quad \text{then}$$

$$m = \frac{-1 - 3}{2 - (-5)} = -\frac{4}{7}.$$

> The slope is $-\frac{4}{7}$ no matter which point is considered first.

FIGURE 42

See **FIGURE 42**. 🟢

EXAMPLE 6 **Using the Slope and a Point to Graph a Line**

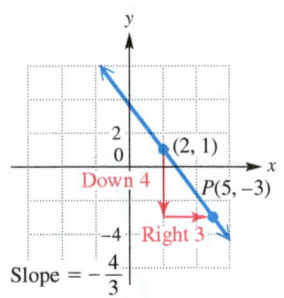

FIGURE 43

Graph the line that passes through $(2, 1)$ and has slope $-\frac{4}{3}$.

Solution Start by locating the point $(2, 1)$ on the graph. Find a second point on the line by using the definition of slope.

$$\text{slope} = \frac{\text{change in } y}{\text{change in } x} = \frac{-4}{3}$$

> Move down 4 units for every 3-unit increase in x.

Move *down* 4 units from $(2, 1)$ and then 3 units to the right to obtain $P(5, -3)$. Draw a line through this second point P and $(2, 1)$, as shown in **FIGURE 43**. 🟢

The graph of a constant function is a horizontal line. Because there is no change in y, *the slope of a horizontal line is* **0.** See **FIGURE 44.**

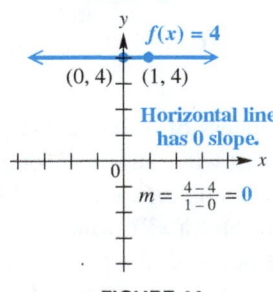

FIGURE 44

> **Geometric Orientation Based on Slope**
>
> For a line with slope m,
>
> 1. If $m > 0$ (i.e., slope is positive), the line *rises* from left to right.
> 2. If $m < 0$ (i.e., slope is negative), the line *falls* from left to right.
> 3. If $m = 0$ (i.e., slope is 0), the line is *horizontal*.

Not a Function

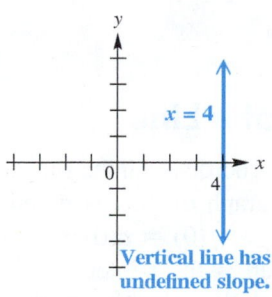

Vertical line has undefined slope.

FIGURE 45

In the slope formula, we have the condition $\Delta x = x_2 - x_1 \neq 0$. This means that $x_2 \neq x_1$. If we graph a line with two points having *equal* x-values, we get a vertical line. See **FIGURE 45**. Notice that this *is not* the graph of a function, since 4 appears as the first number in more than one ordered pair. If we were to apply the slope formula, the denominator would be 0. As a result, *the slope of a vertical line is undefined.*

> **Vertical Line**
>
> A vertical line with x-intercept $(k, 0)$ has an equation of the form $x = k$. Its slope is undefined.

We know that the slope of a line is the ratio of the vertical change in y to the horizontal change in x. ***Thus, slope gives the average rate of change in y per unit of change in x,*** where the value of y depends on the value of x. If f is a linear function defined on $[a, b]$, then we have the following.

$$\text{Average rate of change on } [a, b] = \frac{f(b) - f(a)}{b - a}$$

The next example illustrates this idea. We assume a linear relationship between x and y.

EXAMPLE 7   **Interpreting Slope as Average Rate of Change**

In 2006, online sales on Thanksgiving Day were $205 million, and in 2011 these sales reached $480 million. (*Source:* www.comScore.com)

(a) Find the average rate of change in sales in millions.

(b) Graph these sales as a line segment and interpret the average rate of change.

Solution

(a) To use the slope formula, we need two ordered pairs. Here, if $x = 2006$, then $y = 205$, and if $x = 2011$, then $y = 480$, which gives the pairs $(2006, 205)$ and $(2011, 480)$. (Note that y is in millions.)

$$\text{Average rate of change} = \frac{480 - 205}{2011 - 2006} = \frac{275}{5} = 55 \quad \textcolor{blue}{\text{Millions per year}}$$

(b) Online Thanksgiving Day sales are modeled by the line segment connecting the points $(2006, 205)$ and $(2011, 480)$ in **FIGURE 46** on the next page. The average rate of change is equal to the slope of the line segment connecting these points and indicates that sales on Thanksgiving Day *increased,* on average, by $55 million per year from 2006 to 2011. Note that this increase may not have been exactly $55 million each year, but *on average,* the rate of change was $55 million per year.

Online Thanksgiving Day Sales

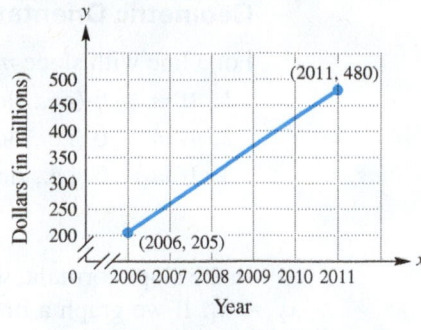

FIGURE 46

Slope–Intercept Form of the Equation of a Line

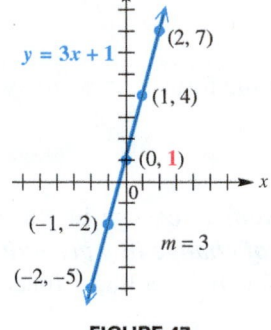

FIGURE 47

In **FIGURE 47**, the slope m of the line $y = 3x + 1$ is 3 and the y-coordinate of the y-intercept is 1. In general, if $f(x) = ax + b$, then the slope of the graph of $f(x)$ is a and the y-coordinate of the y-intercept is b. To verify this fact, notice that $f(0) = a(0) + b = b$. Thus, the graph of f passes through the point $(0, b)$, which is the y-intercept. Since $f(1) = a(1) + b = a + b$, the graph of f also passes through the point $(1, a + b)$. See **FIGURE 48.** The slope of the line that passes through the points $(0, b)$ and $(1, a + b)$ is

$$m = \frac{a + b - b}{1 - 0} = a.$$

Because the slope of the graph of $f(x) = ax + b$ is a, it is often convenient to use m rather than a in the general form of the equation. Therefore, we can write either

$$f(x) = mx + b \quad \text{or} \quad y = mx + b$$

to indicate a linear function. The slope is m, and the y-intercept is $(0, b)$. This equation for $f(x)$ is generally called the *slope–intercept form* of the equation of a line.

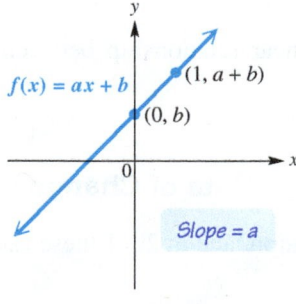

FIGURE 48

> **Slope–Intercept Form**
>
> The **slope–intercept form** of the equation of a line is $y = mx + b$, where m is the slope and $(0, b)$ is the y-intercept.

EXAMPLE 8 **Matching Graphs with Equations**

FIGURE 49 shows the graphs of four lines. Their equations are

$$y = 2x + 3, \quad y = -2x + 3, \quad y = 2x - 3, \quad \text{and} \quad y = -2x - 3,$$

but not necessarily in that order. Match each equation with its graph.

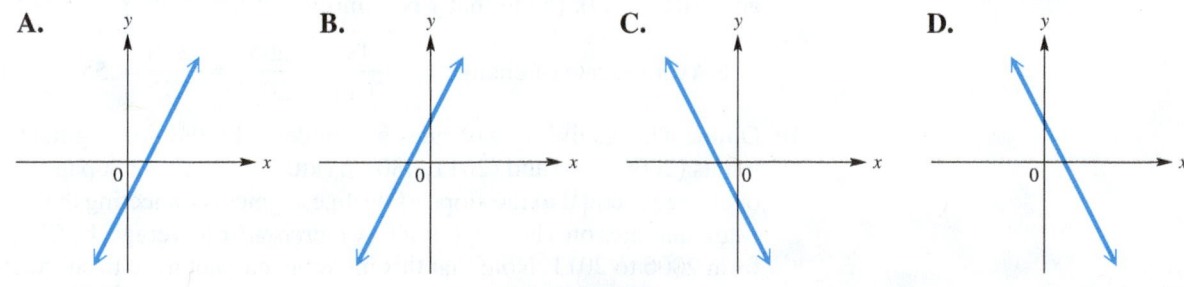

FIGURE 49

→ **Looking Ahead to Calculus**
In calculus, rates of change of nonlinear functions are studied extensively by using a type of function called the **derivative**.

Solution   The sign of m determines whether the graph rises or falls from left to right. Also, if $b > 0$, the y-intercept is *above* the x-axis, and if $b < 0$, the y-intercept is *below* the x-axis.

$y = 2x + 3$ is shown in B, since the graph rises from left to right and the y-intercept is positive.

$y = -2x + 3$ is shown in D, since the graph falls from left to right and the y-intercept is positive.

$y = 2x - 3$ is shown in A, since the graph rises from left to right and the y-intercept is negative.

$y = -2x - 3$ is shown in C, since the graph falls from left to right and the y-intercept is negative.

EXAMPLE 9   **Interpreting Slope–Intercept Form**

In 2006 about 60 billion text messages were sent in the United States. During the next 6 years, text messaging increased, on average, by 407 billion messages per year.

(a) Find values for m and b so that $y = mx + b$ models the number of text messages sent (in billions), x years after 2006.

(b) Estimate the number of text messages sent in 2012.

Solution

(a) In 2006, 60 billion text messages were sent, where $x = 0$ corresponds to 2006. Thus, the graph of $y = mx + b$ must pass through y-intercept $(0, 60)$, and so $b = 60$. The rate of change in the number of text messages was 407 billion per year, on average, so $m = 407$. Thus, the following equation models this situation.

$$y = 407x + 60 \qquad m = 407 \text{ and } b = 60$$

(b) The year 2012 is 6 years after 2006, so let $x = 6$ in $y = 407x + 60$.

$$y = 407(6) + 60 = 2502 \qquad y = 407x + 60 \text{ and } x = 6$$

Thus, about 2500 billion (2.5 trillion) text messages were sent in 2012.

EXAMPLE 10   **Finding an Equation from a Graph**

Use the graph of the linear function f in **FIGURE 50** to complete the following.

(a) Find the slope, y-intercept, and x-intercept.

(b) Write an equation defining f.

(c) Find any zeros of f.

FIGURE 50

Solution

(a) The line falls 1 unit each time the x-value increases by 3 units. Therefore, the slope is $\frac{-1}{3} = -\frac{1}{3}$. The graph intersects the y-axis at the point $(0, -1)$ and intersects the x-axis at the point $(-3, 0)$. Therefore, the y-intercept is $(0, -1)$ and the x-intercept is $(-3, 0)$.

(b) Because the slope is $-\frac{1}{3}$ and the y-intercept is $(0, -1)$, it follows that an equation defining f is

$$f(x) = -\frac{1}{3}x - 1. \qquad m = -\frac{1}{3} \text{ and } b = -1$$

(c) Zeros correspond to the x-coordinates of the x-intercepts, so the only zero of the function f is -3.

1.3 Exercises

Checking Analytic Skills   *Graph each linear function. Give the* **(a)** *x-intercept,* **(b)** *y-intercept,* **(c)** *domain,* **(d)** *range, and* **(e)** *slope of the line.* **Do not use a calculator.**

1. $f(x) = x - 4$

2. $f(x) = -x + 4$

3. $f(x) = 3x - 6$

4. $f(x) = \dfrac{2}{3}x - 2$

5. $f(x) = -\dfrac{2}{5}x + 2$

6. $f(x) = \dfrac{4}{3}x - 3$

7. $f(x) = 3x$

8. $f(x) = -0.5x$

Work each problem related to linear functions.
(a) *Evaluate* $f(-2)$ *and* $f(4)$. **(b)** *Graph f. How can the graph of f be used to determine the zero of f?*
(c) *Find the zero of f.*

9. $f(x) = x + 2$

10. $f(x) = -3x + 2$

11. $f(x) = 2 - \dfrac{1}{2}x$

12. $f(x) = \dfrac{1}{4}x + \dfrac{1}{2}$

13. $f(x) = \dfrac{1}{3}x$

14. $f(x) = -3x$

15. $f(x) = 0.4x + 0.15$

16. $f(x) = x + 0.5$

17. $f(x) = \dfrac{2 - x}{4}$

18. $f(x) = \dfrac{3 - 3x}{6}$

19. $f(x) = \dfrac{x + 5}{15}$

20. $f(x) = \dfrac{x - 4}{2}$

21. *Concept Check*   Based on your graphs of the functions in **Exercises 7** and **8,** what conclusion can you make about one particular point that *must* lie on the graph of the line $y = ax$ (where $b = 0$)?

22. *Concept Check*   Using the concept of slope and your answer in **Exercise 21,** give the equation of the line whose graph is shown to the right.

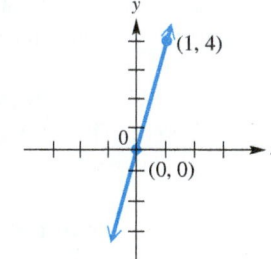

Checking Analytic Skills   *Graph each line. Also, give the* **(a)** *x-intercept (if any),* **(b)** *y-intercept (if any),* **(c)** *domain,* **(d)** *range, and* **(e)** *slope of the line (if defined).* **Do not use a calculator.**

23. $f(x) = -3$

24. $f(x) = 5$

25. $x = -1.5$

26. $f(x) = \dfrac{5}{4}$

27. $x = 2$

28. $x = -3$

29. *Concept Check*   What special name is given to the functions found in **Exercises 23, 24,** and **26?**

Give the equation of the line illustrated.

30.

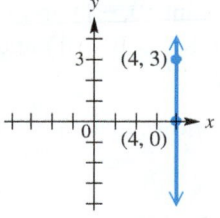

31.

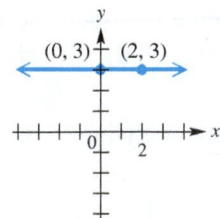

32.

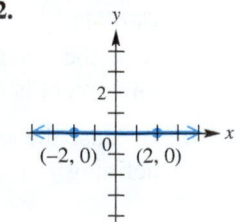

33.

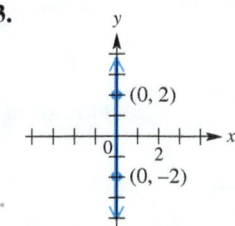

34. *Concept Check*   Answer each question.

(a) What is the equation of the *x*-axis?

(b) What is the equation of the *y*-axis?

Graph each linear function on a graphing calculator, using the two different windows given. State which window gives a comprehensive graph.

35. $f(x) = 4x + 20$
   Window A: $[-10, 10]$ by $[-10, 10]$
   Window B: $[-10, 10]$ by $[-5, 25]$

36. $f(x) = -5x + 30$
   Window A: $[-10, 10]$ by $[-10, 40]$
   Window B: $[-5, 5]$ by $[-5, 40]$

37. $f(x) = 3x + 10$
   Window A: $[-3, 3]$ by $[-5, 5]$
   Window B: $[-5, 5]$ by $[-10, 14]$

38. $f(x) = -6$
   Window A: $[-5, 5]$ by $[-5, 5]$
   Window B: $[-10, 10]$ by $[-10, 10]$

Find the slope (if defined) of the line that passes through the given points.

39. $(-2, 1)$ and $(3, 6)$

40. $(-2, 3)$ and $(-1, 2)$

41. $(8, 4)$ and $(-1, -3)$

42. $(-4, -3)$ and $(5, 0)$

43. $(-11, 3)$ and $(-11, 5)$

44. $(-8, 2)$ and $(-8, 1)$

45. $\left(\frac{2}{3}, 9\right)$ and $\left(\frac{1}{2}, 9\right)$

46. $(0.12, 0.36)$ and $(0.18, 0.36)$

47. $\left(\frac{1}{2}, -\frac{2}{3}\right)$ and $\left(-\frac{3}{4}, \frac{1}{6}\right)$

48. Given an equation having x and y as variables, explain how to determine the x- and y-intercepts.

Concept Check   Find and interpret the average rate of change illustrated in each graph.

49.
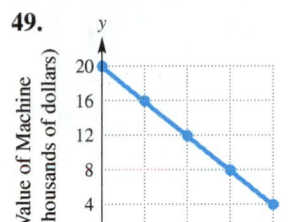

50.
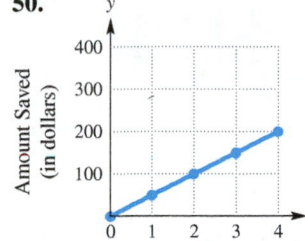

51.
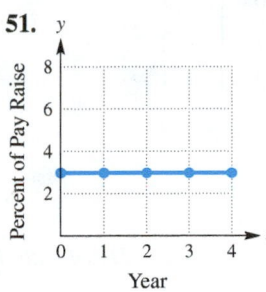

52.
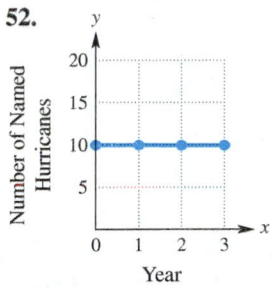

Concept Check   Match each equation with the graph that it most closely resembles.

53. $y = 3x + 6$

54. $y = -3x + 6$

55. $y = -3x - 6$

56. $y = 3x - 6$

57. $y = 3x$

58. $y = -3x$

59. $y = 3$

60. $y = -3$

A.
B.
C.
D.
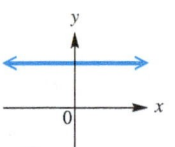
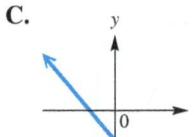
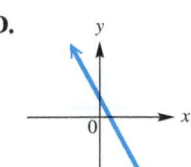

E.
F.
G.
H.
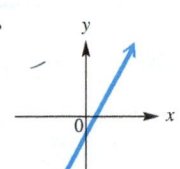
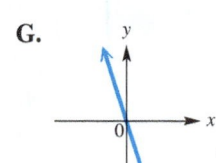
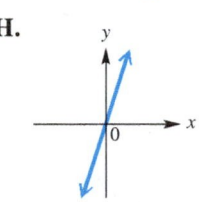

The graph of a linear function f is shown. **(a)** *Identify the slope, y-intercept, and x-intercept.* **(b)** *Write a formula for f.* **(c)** *Estimate the zero of f.*

61.
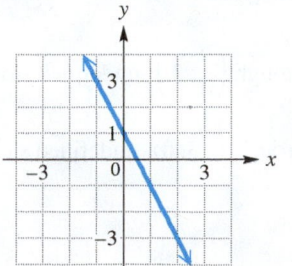

62.
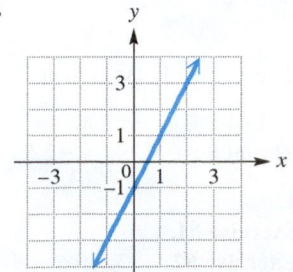

63.
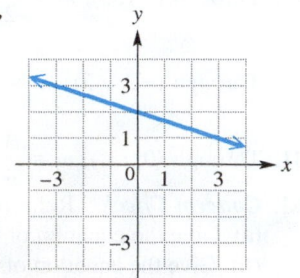

64.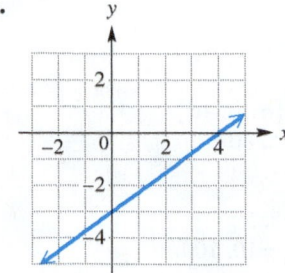

65.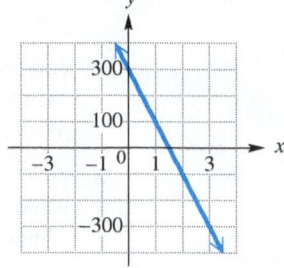

66.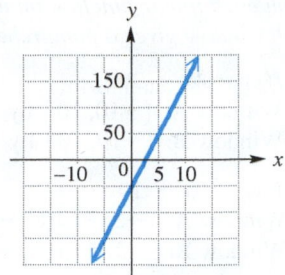

Concept Check *A linear function f has the ordered pairs listed in the table. Find the slope m of the graph of f, use the table to find the y-intercept of the line, and give an equation that defines f.*

67.

x	$f(x)$
-3	-10
-2	-6
-1	-2
0	2
1	6

68.

x	$f(x)$
-2	-11
-1	-8
0	-5
1	-2
2	1

69.

x	$f(x)$
-0.4	-2.54
-0.2	-2.82
0	-3.1
0.2	-3.38
0.4	-3.66

70.

x	$f(x)$
-100	-4
-50	-4
0	-4
50	-4
100	-4

Concept Check *Match each equation in Exercises 71–74 with the line in choices A–D that would most closely resemble its graph, where k > 0.*

71. $y = k$

72. $y = -k$

73. $x = k$

74. $x = -k$

A.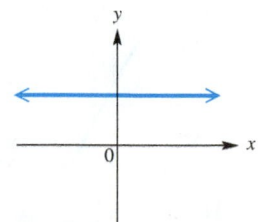

B.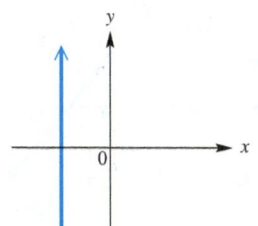

C.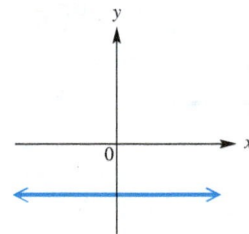

D.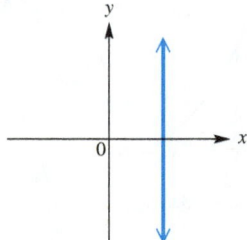

Sketch by hand the graph of the line passing through the given point and having the given slope. Label two points on the line.

75. Through $(-1, 3)$, $m = \dfrac{3}{2}$

76. Through $(-2, 8)$, $m = -1$

77. Through $(3, -4)$, $m = -\dfrac{1}{3}$

78. Through $(-2, -3)$, $m = -\dfrac{3}{4}$

79. Through $(-1, 4)$, $m = 0$

80. Through $\left(\dfrac{9}{4}, 2\right)$, undefined slope

81. Through $(0, -4)$, $m = \dfrac{3}{4}$

82. Through $(0, 5)$, $m = -2.5$

83. Through $(-3, 0)$, undefined slope

84. *Concept Check* Refer to **Exercises 81** and **82.**
   (a) Give the equation of the line described in **Exercise 81.**
   (b) Give the equation of the line described in **Exercise 82.**

(Modeling)   Solve each problem.

85. *Water Flow*   The graph gives the number of gallons of water in a small swimming pool after x hours.

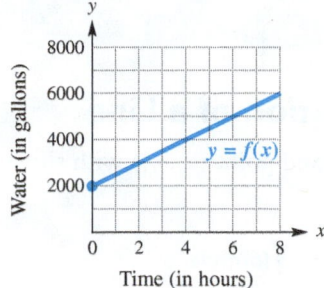

Time (in hours)

(a) Write a formula for $f(x)$.

(b) Interpret both the slope and the y-intercept.

(c) Use the graph to estimate how much water was in the pool after 7 hours. Verify your answer by evaluating $f(x)$.

86. *Fuel Consumption*   The table shows the distance y traveled in miles by a car burning x gallons of gasoline.

x (gallons)	5	10	12	16
y (miles)	115	230	276	368

(a) Find values for a and b so that $f(x) = ax + b$ models the data exactly. That is, find values for a and b so that the graph of f passes through the data points in the table.

(b) Interpret the slope of the graph of f.

(c) How many miles could be driven using 20 gallons of gasoline?

87. *Rainfall*   By noon, 3 inches of rain had fallen during a storm. Rain continued to fall at a rate of $\frac{1}{4}$ inch per hour.

(a) Find a formula for a linear function f that models the amount of rainfall x hours past noon.

(b) Find the total amount of rainfall by 2:30 P.M.

88. *U.S. HIV/AIDS Infections*   In 2010, there were approximately 1.2 million people in the United States living with HIV/AIDS. At that time the infection rate was 50,000 people per year.

(a) Find values for m and b so that $y = mx + b$ models the total number of people y in millions who were living with HIV/AIDS x years after 2010.

(b) Find y for the year 2014. Interpret your result.

89. *Distance to Lightning*   When a bolt of lightning strikes in the distance, there is often a delay between seeing the lightning and hearing the thunder. The function $f(x) = \frac{x}{5}$ computes the approximate distance in miles between an observer and a bolt of lightning when the delay is x seconds.

(a) Find $f(15)$ and interpret the result.

(b) Graph $y = f(x)$. Let the domain of f be $[0, 20]$.

90. *Air Temperature*   When the relative humidity is 100% air cools 5.8°F for every 1-mile increase in altitude. If the temperature is 80°F on the ground, then $f(x) = 80 - 5.8x$ calculates the air temperature x miles above the ground. Find $f(3)$ and interpret the result. (*Source:* Battan, L., *Weather in Your Life,* W.H. Freeman.)

91. *Sales Tax*   If the sales tax rate is 7.5%, write a function f that calculates the sales tax on a purchase of x dollars. What is the sales tax on a purchase of $86?

92. *Income and Education*   Function f gives the average 2010 individual income (in dollars) by educational attainment for people 25 years old and over. This function is defined by $f(N) = 21{,}484$, $f(H) = 31{,}286$, $f(B) = 57{,}181$, and $f(M) = 70{,}181$, where N denotes no high school diploma, H a high school diploma, B a bachelor's degree, and M a master's degree. (*Source:* U.S. Bureau of Labor Statistics.)

(a) Write f as a set of ordered pairs.

(b) Give the domain and range of f.

(c) Discuss the relationship between education and income.

93. *Tuition and Fees*   If college tuition costs $192 per credit and fees are fixed at $275, write a formula for a function f that calculates the tuition and fees for taking x credits. What is the total cost of taking 11 credits?

94. *Converting Units of Measure*   Write a formula for a function f that converts x gallons to quarts. How many quarts are there in 19 gallons?

95. *Climate Change*   During the past 50 years, the average rate of change in temperature in Antarctica has been 0.9°F per decade.

(a) Write a function W that calculates the increase in temperature after x years during this time period.

(b) Evaluate $W(15)$ and interpret the result.

96. *Climate Change*   Refer to **Exercise 95.** Use function W to estimate the time it would take to have an increase in temperature of 4.5°F at Antarctica.

1.4 Equations of Lines and Linear Models

Point–Slope Form of the Equation of a Line • Standard Form of the Equation of a Line • Parallel and Perpendicular Lines • Linear Models and Regression

Point–Slope Form of the Equation of a Line

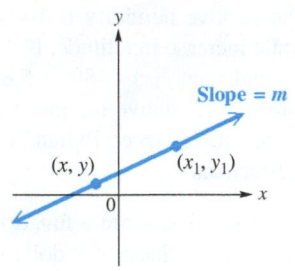

FIGURE 51

FIGURE 51 shows a line passing through the fixed point (x_1, y_1) with slope m. Let (x, y) be any other point on the line.

$$m = \frac{y - y_1}{x - x_1} \qquad \text{Slope formula}$$

$$m(x - x_1) = y - y_1 \qquad \text{Multiply each side by } x - x_1.$$

$$y - y_1 = m(x - x_1) \qquad \text{Rewrite.}$$

This result is called the **point–slope form** of the equation of a line.

→ **Looking Ahead to Calculus**

In calculus, it is often necessary to find the equation of a line given its slope and a point on the line. The point–slope form is a valuable tool in these situations.

> **Point–Slope Form**
>
> The line with slope m passing through the point (x_1, y_1) has equation
> $$y - y_1 = m(x - x_1).$$

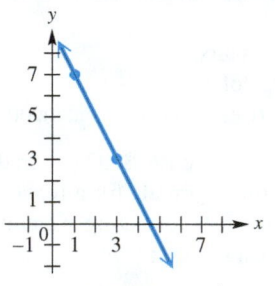

FIGURE 52

x	y
2	3
3	5
4	7
5	9
6	11
7	13
8	15

EXAMPLE 1 Using Point–Slope Form

Find the slope–intercept form of the line passing through the two points shown in **FIGURE 52**.

Solution The points shown on the line are $(1, 7)$ and $(3, 3)$. Find the slope of the line.

$$m = \frac{7 - 3}{1 - 3} = \frac{4}{-2} = -2 \qquad m = \frac{y_2 - y_1}{x_2 - x_1}$$

Now, use *either* point—say, $(1, 7)$—with $m = -2$ in the point–slope form.

$$y - y_1 = m(x - x_1) \qquad \text{Point–slope form}$$

$$y - 7 = -2(x - 1) \qquad y_1 = 7, m = -2, x_1 = 1$$

Simplify point-slope form to obtain slope-intercept form.

$$y - 7 = -2x + 2 \qquad \text{Distributive property}$$

$$y = -2x + 9 \qquad \text{Add 7 to obtain slope–intercept form.}$$

Note that we would obtain the same slope–intercept form if we used the point $(3, 3)$. Slope–intercept form is *unique*. ●

EXAMPLE 2 Using Point–Slope Form

The table in the margin lists some points found on the line $y = mx + b$. Find the slope–intercept form of the equation of the line.

Solution Choose *any* two points—such as $(2, 3)$ and $(6, 11)$—to find the slope of the line.

$$m = \frac{11 - 3}{6 - 2} = \frac{8}{4} = 2 \qquad m = \frac{y_2 - y_1}{x_2 - x_1}$$

Now, use any point on the line—say, $(2, 3)$—with $m = 2$ in the point–slope form.

$$y - y_1 = m(x - x_1) \quad \text{Point–slope form}$$

$$y - 3 = 2(x - 2) \quad y_1 = 3, \, m = 2, \, x_1 = 2$$

$$y - 3 = 2x - 4 \quad \text{Distributive property}$$

$$y = 2x - 1 \quad \text{Add 3 to obtain slope–intercept form.}$$

NOTE   Using the results of **Example 2**, we can easily write the formula for a linear function f, whose graph passes through the points in the given table, as $f(x) = 2x - 1$.

Standard Form of the Equation of a Line

Another form of the equation of a line, *standard form*, can be used to represent *any* line, including vertical lines.

> **Standard Form**
>
> A linear equation written in the form
>
> $$Ax + By = C,$$
>
> where A, B, and C are real numbers (A and B not both 0), is said to be in **standard form.**

NOTE   When writing the standard form of a line, we will usually give A, B, and C as integers with greatest common factor 1 and $A \geq 0$. If $A = 0$, we will choose $B > 0$. For example, below we use algebra to put the equation $y = -\frac{1}{2}x + 3$ in standard form.

$$y = -\frac{1}{2}x + 3 \quad \text{Slope–intercept form}$$

$$2y = -x + 6 \quad \text{Multiply by 2.}$$

$Ax + By = C$

$$x + 2y = 6 \quad \text{Add } x \text{ to obtain standard form.}$$

One advantage of standard form is that it allows the quick calculation of both intercepts. For example, given $3x + 2y = 6$, we can find the x-intercept by letting $y = 0$ and the y-intercept by letting $x = 0$.

Find x-intercept: $3x + 2(0) = 6$	Find y-intercept: $3(0) + 2y = 6$
$3x = 6$	$2y = 6$
$x = 2$	$y = 3$

Thus the x-intercept is $(2, 0)$ and the y-intercept is $(0, 3)$. This information is useful when sketching the graph of the line by hand, because the points $(2, 0)$ and $(0, 3)$ lie on the line, as demonstrated in the next example.

　Graphing an Equation in Standard Form

Graph $3x + 2y = 6$.

Analytic Solution

As just calculated, the points $(2, 0)$ and $(0, 3)$ are the x- and y-intercepts, respectively. Plot these two points and connect them with a straight line. See **FIGURE 53.** Sometimes it is advisable to plot a third point, such as $\left(1, \frac{3}{2}\right)$, to check our work.

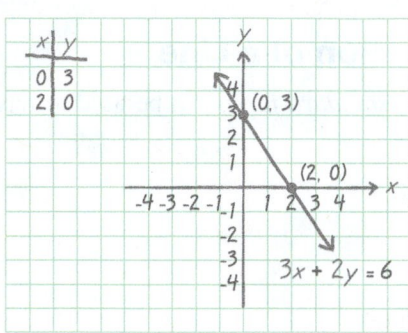

FIGURE 53

Graphing Calculator Solution

Solve the equation for y so that it can be entered into a calculator.

$$3x + 2y = 6 \qquad \text{Given equation}$$
$$2y = -3x + 6 \qquad \text{Subtract } 3x.$$
$$y = -1.5x + 3 \qquad \text{Divide by 2.}$$

The desired graph is shown in **FIGURE 54.**

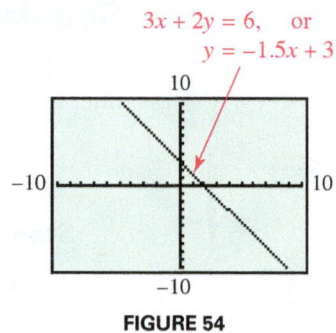

FIGURE 54

NOTE　Because of the usefulness of the slope–intercept form when graphing with a graphing calculator, we emphasize that form in much of our work.

In the standard viewing window of your calculator, graph the following four lines.

$$y_1 = 2x - 6,$$
$$y_2 = 2x - 2,$$
$$y_3 = 2x,$$
$$y_4 = 2x + 4$$

What is the slope of each line? What geometric term seems to describe the set of lines?

Parallel and Perpendicular Lines

Two lines in a plane are *parallel* if they do not intersect. Although the exercise in the "For Discussion" box in the margin does not actually prove the result that follows, it provides visual support.

> ### Parallel Lines
>
> Two distinct nonvertical lines are **parallel** if and only if they have the same slope.

　Using the Slope Relationship for Parallel Lines

Find the equation of the line that passes through the point $(3, 5)$ and is parallel to the line with equation $2x + 5y = 4$. Graph both lines in the standard viewing window.

Solution　First we need to find the slope of the given equation by writing it in slope–intercept form. (That is, solve for y.)

$$2x + 5y = 4 \qquad \text{Given equation}$$
$$y = -\frac{2}{5}x + \frac{4}{5} \qquad \text{Subtract } 2x \text{ and divide by 5.}$$

The slope is $-\frac{2}{5}$. Since the lines are parallel, $-\frac{2}{5}$ is also the slope of the line whose equation we must find. Substitute the slope and given point into the point–slope form.

$$y - y_1 = m(x - x_1) \qquad \text{Point–slope form}$$

$$y - 5 = -\frac{2}{5}(x - 3) \qquad y_1 = 5,\ m = -\frac{2}{5},\ x_1 = 3$$

$$5(y - 5) = -2(x - 3) \qquad \text{Multiply by 5 to clear fractions.}$$

$$5y - 25 = -2x + 6 \qquad \text{Distributive property}$$

$$5y = -2x + 31 \qquad \text{Add 25.}$$

$$y = -\frac{2}{5}x + \frac{31}{5} \qquad \text{Divide by 5 to obtain the slope–intercept form of the desired line.}$$

Multiply -2 by each term inside the parentheses.

An alternative method for finding this equation involves using the slope $-\frac{2}{5}$ and the point $(3, 5)$ in the slope–intercept form to find b.

$$y = mx + b \qquad \text{Slope–intercept form}$$

$$5 = -\frac{2}{5}(3) + b \qquad y = 5,\ m = -\frac{2}{5},\ x = 3$$

$$5 = -\frac{6}{5} + b \qquad \text{Multiply.}$$

$$b = \frac{31}{5} \qquad \text{Add } \frac{6}{5} \text{ and rewrite.}$$

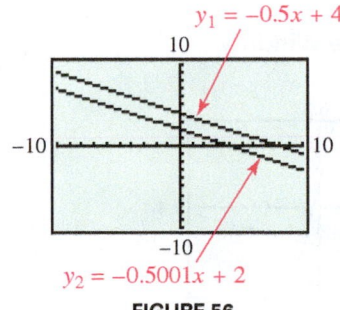

$y_2 = -\frac{2}{5}x + \frac{31}{5}$

$y_1 = -\frac{2}{5}x + \frac{4}{5}$

FIGURE 55

Therefore, the equation is $y = -\frac{2}{5}x + \frac{31}{5}$, which agrees with our earlier result. **FIGURE 55** provides support, as the lines *appear* to be parallel.

When using a graphing calculator, be aware that visual support (as seen in **Example 4**) does not necessarily prove the result. For example, **FIGURE 56** shows the graphs of

$$y_1 = -0.5x + 4 \quad \text{and} \quad y_2 = -0.5001x + 2.$$

Although they *appear* to be parallel by visual inspection, they are *not* parallel, because the slope of y_1 is -0.5 and the slope of y_2 is -0.5001.

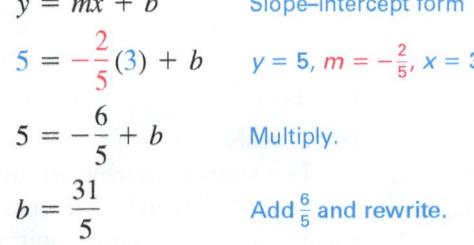

$y_1 = -0.5x + 4$

$y_2 = -0.5001x + 2$

FIGURE 56

CAUTION: RELYING TOO HEAVILY ON TECHNOLOGY Analysis of a calculator graph is often not sufficient to draw correct conclusions. While this technology is powerful and useful, we cannot rely on it alone in our work. We must understand the basic concepts of algebra as well.

FOR DISCUSSION

Using a "square" viewing window, such as $[-9.4, 9.4]$ by $[-6.2, 6.2]$ on the TI-84 Plus, graph each pair of lines. Graph each group separately.

I	II	III	IV
$y_1 = 4x + 1$	$y_1 = -\frac{2}{3}x + 3$	$y_1 = 6x - 3$	$y_1 = \frac{13}{7}x - 3$
$y_2 = -\frac{1}{4}x + 3$	$y_2 = \frac{3}{2}x - 4$	$y_2 = -\frac{1}{6}x + 4$	$y_2 = -\frac{7}{13}x + 4$

What geometric term applies to each pair of lines? What is the product of the slopes for each pair of lines?

As in the earlier "For Discussion" box, we have not proved the result that follows. Instead, we have provided visual support for it. The proof for this statement is outlined in **Exercise 59.**

Perpendicular Lines

Two lines, neither of which is vertical, are **perpendicular** if and only if their slopes have product -1.

NOTE If one line is vertical, then its slope is undefined and the statement does *not* apply.

For example, if the slope of a line is $-\frac{3}{4}$, then the slope of any line perpendicular to it is $\frac{4}{3}$, because $-\frac{3}{4}\left(\frac{4}{3}\right) = -1$. We often refer to numbers such as $-\frac{3}{4}$ and $\frac{4}{3}$ as **negative reciprocals**.

In a **square viewing window,** circles appear to be circular, squares appear to be square, and perpendicular lines appear to be perpendicular. On many calculators, a square viewing window requires that the distance along the y-axis be about two-thirds the distance along the x-axis. Examples of square viewing windows on the TI-84 Plus calculator are

$$[-4.7, 4.7] \text{ by } [-3.1, 3.1] \quad \text{and} \quad [-9.4, 9.4] \text{ by } [-6.2, 6.2].$$

FIGURE 57 illustrates the importance of square viewing windows.

TECHNOLOGY NOTE

Many calculators can set a square viewing window automatically. Check your owner's guide, or look under the ZOOM menu.

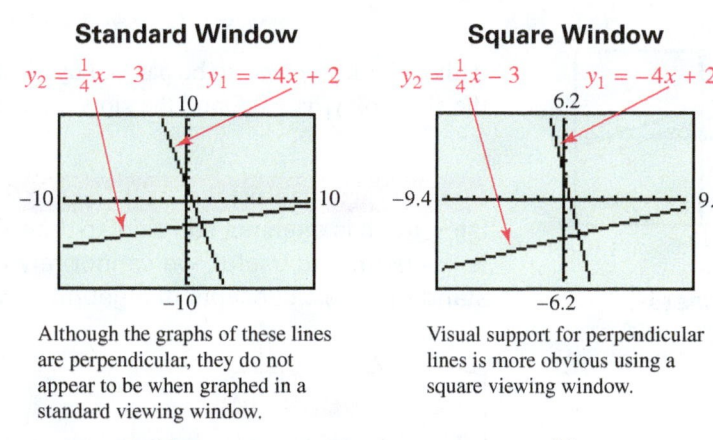

Although the graphs of these lines are perpendicular, they do not appear to be when graphed in a standard viewing window.	Visual support for perpendicular lines is more obvious using a square viewing window.
(a)	**(b)**

FIGURE 57

EXAMPLE 5 **Using the Slope Relationship for Perpendicular Lines**

Find the equation of the line that passes through the point $(3, 5)$ and is perpendicular to the line with equation $2x + 5y = 4$. Graph both lines.

Solution In **Example 4,** we found that the slope of the given line is $-\frac{2}{5}$, so the slope of any line perpendicular to it is $\frac{5}{2}$ because $\left(-\frac{2}{5}\right)\left(\frac{5}{2}\right) = -1$. We can use either

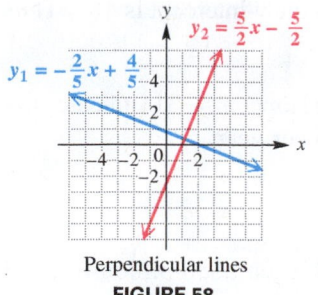

Perpendicular lines
FIGURE 58

method shown in **Example 4** to find the equation of the line. The second method gives the following result.

$$y = mx + b \qquad \text{Slope–intercept form}$$

$$5 = \frac{5}{2}(3) + b \qquad y = 5, \; m = \frac{5}{2}, \; x = 3$$

$$5 = \frac{15}{2} + b \qquad \text{Multiply.}$$

$$b = -\frac{5}{2} \qquad \text{Subtract } \tfrac{15}{2} \text{ and rewrite.}$$

Thus, the equation is $y = \frac{5}{2}x - \frac{5}{2}$. The graphs of both equations are shown in **FIGURE 58.**

Linear Models and Regression

When data points are plotted in the xy-plane, the resulting graph is sometimes called a **scatter diagram.** Scatter diagrams are often helpful for analyzing trends in data.

x (year)	y (cost)
2007	432
2008	467
2009	500
2010	525
2011	558
2012	591

Source: U.S. Center for Medicare and Medicaid Services.

EXAMPLE 6   **Modeling Medicare Costs with a Linear Function**

Estimates for Medicare costs (in billions of dollars) are shown in the table in the margin.

(a) Make a scatter diagram of the data. Let $x = 0$ correspond to 2007, $x = 1$ to 2008, and so on. What type of function might model the data?

(b) Find a linear function f that models the data. Graph f and the data in the same viewing window. Interpret the slope m.

(c) Use $f(x)$ to estimate Medicare costs in 2015.

Solution

(a) Since $x = 0$ corresponds to 2007, $x = 1$ corresponds to 2008, and so on, the data points can be expressed as the ordered pairs

$$(0, 432), \quad (1, 467), \quad (2, 500), \quad (3, 525), \quad (4, 558), \quad \text{and} \quad (5, 591).$$

Scatter diagrams are shown in **FIGURE 59.** The data appear to be approximately linear, so a linear function might be appropriate.

TECHNOLOGY NOTE

To make a scatter diagram with a graphing calculator, you may need to use the *list* feature by entering the x-values in list L_1 and the y-values in list L_2. See **FIGURE 60.**

FIGURE 60

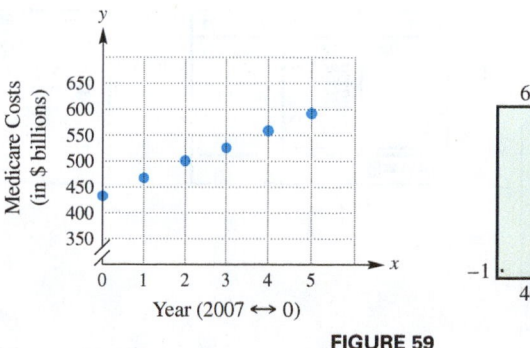

FIGURE 59

(b) We start by choosing two data points that the line should pass through. For example, use $(0, 432)$ and $(3, 525)$ to find the slope of the line.

Start with the x- and y-values of the same point.

$$m = \frac{525 - 432}{3 - 0} = 31$$

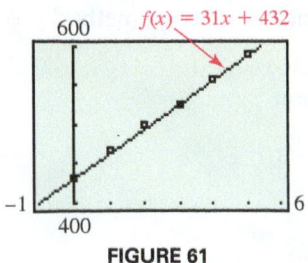

FIGURE 61

The point $(0, 432)$ indicates that the y-coordinate b of the y-intercept is 432. Thus,

$$f(x) = 31x + 432.$$

A graph of f and the data are shown in **FIGURE 61.** The slope $m = 31$ indicates that Medicare costs increased, on average, by \$31 billion per year.

(c) The value $x = 8$ corresponds to the year 2015.

$$f(8) = 31(8) + 432 = 680$$

Thus, this model projects that Medicare costs could reach \$680 billion in 2015.

> **NOTE** The formula $f(x) = 31x + 432$ found in **Example 6** is not unique. If two different points are chosen, a different formula for $f(x)$ may result. However, all such formulas for $f(x)$ should be in approximate agreement.

TECHNOLOGY NOTE

To find the equation of a least-squares regression line, refer to your owner's guide.

The method for finding the model in **Example 6** used algebraic concepts that do not always give a unique line. Graphing calculators are capable of finding the line of "best fit," called the **least-squares regression line,** by using a technique taught in statistics courses known as **least-squares regression.**

EXAMPLE 7 **Finding the Least-Squares Regression Line**

Use a graphing calculator to find the least-squares regression line that models the Medicare costs presented in **Example 6.** Graph the data and the line in the same viewing window.

Solution **FIGURE 62** shows how a TI-84 Plus graphing calculator finds the regression line for the data in **Example 6.** In **FIGURE 62(a)**, the years are entered into list L_1, and Medicare costs are entered into list L_2. In **FIGURE 62(b)**, the formula for the regression line is calculated to be

$$y \approx 31.23x + 434.10.$$

Notice that this equation is not *exactly* the same as the one found for $f(x)$ in **Example 6.** In **FIGURE 62(c)**, both the data and the regression line are graphed.

TECHNOLOGY NOTE

By choosing the DiagnosticOn option on the TI-84 Plus, the correlation coefficient r and its square r^2 are displayed. See **FIGURE 63**, and compare with **FIGURE 62(b)**.

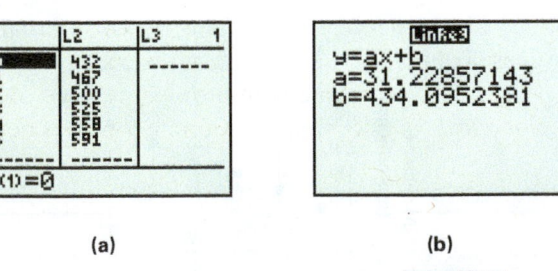

(a)   (b)   (c)

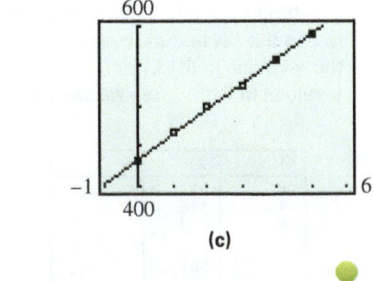

FIGURE 62

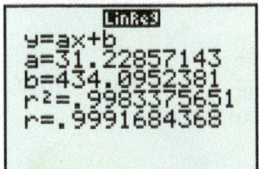

FIGURE 63

Once an equation for the least-squares regression line has been found, it is reasonable to ask, "Just how good is this line for predictive purposes?" If the line *fits* the observed data points, then future pairs of points might be expected to do so also.

One common measure of the strength of the linear relationship in a data set is called the **correlation coefficient,** denoted r, where $-1 \leq r \leq 1$. Interpretations for various values of r are given in the following table.

Correlation Coefficient r (where −1 ≤ r ≤ 1)

Value of *r*	Comments	Sample Scatter Diagram
$r = 1$	**There is an exact linear fit.** The line passes through all data points and has a positive slope.	
$r = -1$	**There is an exact linear fit.** The line passes through all data points and has a negative slope.	
$0 < r < 1$	**There is a positive correlation.** As the *x*-values increase, so do the *y*-values. The fit is not exact.	
$-1 < r < 0$	**There is a negative correlation.** As the *x*-values increase, the *y*-values decrease. The fit is not exact.	
$r = 0$	**There is no correlation.** The data have no tendency toward being linear. A regression line predicts poorly.	

There is a difference between correlation and causation. For example, when geese begin to fly north, summer is coming and the weather becomes warmer. Geese flying north correlate with warmer weather. However, geese flying north clearly do not *cause* warmer weather. Correlation does not always indicate causation.

EXAMPLE 8   **Predicting Smartphone Usage**

The table represents the percentages of persons using smartphones according to different age groups and incomes of less than $50,000 and more than $100,000.

Percentages of Persons with Smartphones

Age Group	<$50K	>$100K
25–34	60%	81%
35–44	47%	76%
45–54	31%	61%
55–64	22%	51%

Age Group	< $50K	> $100K
25–34	60%	81%
35–44	47%	76%
45–54	31%	61%
55–64	22%	51%

(table repeated)

(a) Graph the data by using the < $50K data for *x*-values and the corresponding > $100K data for *y*-values. Predict whether the correlation coefficient will be positive or negative. Does making more money at any age correlate to an increased likelihood of using a smartphone?

(b) Use a calculator to find the linear function *f* based on least-squares regression that models the data. Graph $y = f(x)$ and the data in the same viewing window.

(c) For people 65 or over, who make less than $50,000, 16% use a smartphone. Assuming that this age group follows a trend similar to that of the four age groups listed in the table, use your linear function to estimate the percentage of those 65 or over with an income of more than $100,000 who use a smartphone. Compare your result with the actual value of 42%.

Solution

(a) A scatter diagram of the data is shown in **FIGURE 64.** Because increasing *x*-values correspond to increasing *y*-values, the correlation coefficient *r* will be positive. Within every age group, those making > $100K have a higher percentage of smartphone use than those making < $50K.

(b) Because $y = f(x)$, the formula for a linear function *f* that models the data is given by

$$f(x) \approx 0.8021x + 35.1657,$$

where coefficients have been rounded to four decimal places. See **FIGURE 65.** Graphs of *f* and the data are shown in **FIGURE 66.**

(c) Because 16% of people 65 or over making less than $50,000 use a smartphone, we can use $f(x)$ to predict this percentage *y* for people 65 or over who make more than $100,000 by evaluating $f(x)$ when $x = 16$.

$$y = f(16) = 0.8021(16) + 35.1657 \approx 48\%$$

This value is more than the actual value of 42%.

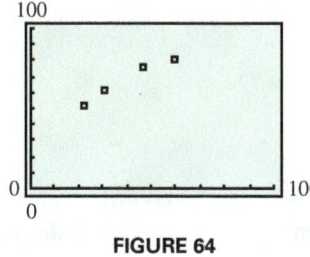

FIGURE 64

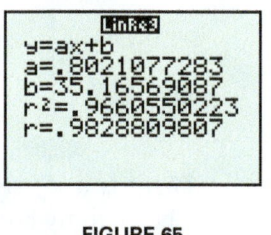

FIGURE 65

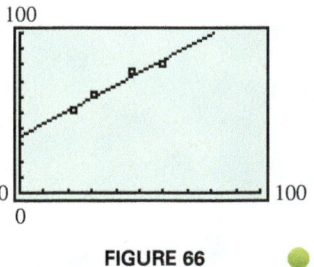

FIGURE 66

1.4 Exercises

Checking Analytic Skills   Write the slope–intercept form of the line that passes through the given point with slope m. **Do not use a calculator.**

1. Through $(1, 3)$, $m = -2$

2. Through $(2, 4)$, $m = -1$

3. Through $(-5, 4)$, $m = 1.5$

4. Through $(-4, 3)$, $m = 0.75$

5. Through $(-8, 1)$, $m = -0.5$

6. Through $(-5, 9)$, $m = -0.75$

7. Through $\left(\frac{1}{2}, -4\right)$, $m = 2$

8. Through $\left(5, -\frac{1}{3}\right)$, $m = 3$

9. Through $\left(\frac{1}{4}, \frac{2}{3}\right)$, $m = \frac{1}{2}$

10. *Concept Check*   Why is it not possible to write a slope–intercept form of the equation of the line through the points $(12, 6)$ and $(12, -2)$?

Concept Check   *Find the slope–intercept form of the equation of the line shown in each graph.*

11.   **12.**   **13.**   **14.**

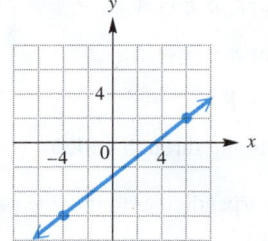   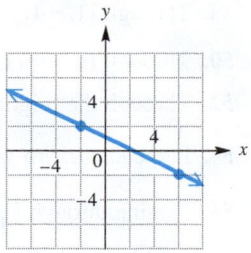   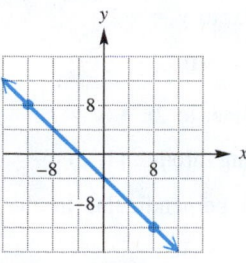   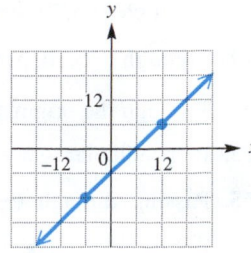

Checking Analytic Skills   *Find the slope–intercept form of the equation of the line satisfying the given conditions.* **Do not use a calculator.**

15. Through $(4, 8)$ and $(0, 4)$

16. Through $(3, 6)$ and $(0, 10)$

17. Through $(3, -8)$ and $(5, -3)$

18. Through $(-5, 4)$ and $(-3, 2)$

19. Through $(2, 3.5)$ and $(6, -2.5)$

20. Through $(-1, 6.25)$ and $(2, -4.25)$

21. Through $(0, 5)$ and $(10, 0)$

22. Through $(0, -8)$ and $(4, 0)$

23.

x	y
-7	-44
-6	-36
-5	-28
-4	-20

24.

x	y
-2.4	5.2
1.3	-24.4
1.75	-28
2.98	-37.84

25.

x	y
2	-5
3	-8
4	-11
5	-14

26.

x	y
-1.1	1.5
-1.0	2.0
-0.9	2.5
-0.8	3.0

Graph each line by hand. Give the x- and y-intercepts.

27. $x - y = 4$

28. $x + y = 4$

29. $3x - y = 6$

30. $2x - 3y = 6$

31. $2x + 5y = 10$

32. $4x - 3y = 9$

A line having an equation of the form $y = kx$, where k is a real number, $k \neq 0$, will always pass through the origin. To graph such an equation by hand, we can determine a second point and then join the origin and that second point with a straight line. Use this method to graph each line.

33. $y = 3x$ **34.** $y = -2x$ **35.** $y = -0.75x$ **36.** $y = 1.5x$

Write each equation in the form $y = mx + b$. (A suggested window for a comprehensive graph of the equation is given.)

37. $5x + 3y = 15$
  $[-10, 10]$ by $[-10, 10]$

38. $6x + 5y = 9$
  $[-10, 10]$ by $[-10, 10]$

39. $-2x + 7y = 4$
  $[-5, 5]$ by $[-5, 5]$

40. $-0.23x - 0.46y = 0.82$
  $[-5, 5]$ by $[-5, 5]$

41. $1.2x + 1.6y = 5.0$
  $[-6, 6]$ by $[-4, 4]$

42. $2y - 5x = 0$
  $[-10, 10]$ by $[-10, 10]$

Find the equation of the line satisfying the given conditions, giving it in slope–intercept form if possible.

43. Through $(-1, 4)$, parallel to $x + 3y = 5$

44. Through $(3, -2)$, parallel to $2x - y = 5$

45. Through $(1, 6)$, perpendicular to $3x + 5y = 1$

46. Through $(-2, 0)$, perpendicular to $8x - 3y = 7$

47. Through $(-5, 7)$, perpendicular to $y = -2$

48. Through $(1, -4)$, perpendicular to $x = 4$

49. Through $(-5, 8)$, parallel to $y = -0.2x + 6$

50. Through $(-4, -7)$, parallel to $x + y = 5$

51. Through the origin, perpendicular to $2x + y = 6$

52. Through the origin, parallel to $y = -3.5x + 7.4$

53. Perpendicular to $x = 3$, passing through $(1, 2)$

54. Perpendicular to $y = -1$, passing through $(-4, 5)$

55. Passing through $(-2, 4)$ and perpendicular to the line passing through $\left(-5, \frac{1}{2}\right)$ and $\left(-3, \frac{2}{3}\right)$

56. Passing through $\left(\frac{3}{4}, \frac{1}{4}\right)$ and perpendicular to the line passing through $(-3, -5)$ and $(-4, 0)$

57. Find the equation of the line that is the perpendicular bisector of the line segment connecting $(-4, 2)$ and $(2, 10)$.

58. Find the equation of the line that is the perpendicular bisector of the line segment connecting $(-3, 5)$ and $(4, 9)$

59. Refer to the given figure and complete parts (a)–(h) to prove that if two lines are perpendicular and neither line is parallel to an axis, then the lines have slopes whose product is -1.

   (a) In triangle OPQ, angle POQ is a right angle if and only if

   $$[d(O, P)]^2 + [d(O, Q)]^2 = [d(P, Q)]^2.$$

   What theorem from geometry is this?

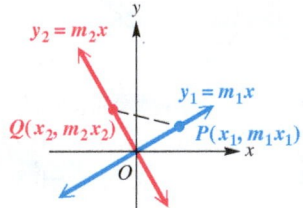

   (b) Find an expression for the distance $d(O, P)$.
   (c) Find an expression for the distance $d(O, Q)$.
   (d) Find an expression for the distance $d(P, Q)$.
   (e) Use your results from parts (b)–(d) and substitute into the equation in part (a). Simplify to show that the resulting equation is $-2m_1m_2x_1x_2 - 2x_1x_2 = 0$.
   (f) Factor $-2x_1x_2$ from the final form of the equation in part (e).
   (g) Use the zero-product property from intermediate algebra to solve the equation in part (f) to show that $m_1m_2 = -1$.
   (h) State your conclusion on the basis of parts (a)–(g).

60. *Concept Check* The following tables show ordered pairs for linear functions defined in Y_1 and Y_2. Determine whether the lines are parallel, perpendicular, or neither.

(a)

X	Y₁	Y₂
0	-3	4
1	1	3.75
2	5	3.5
3	9	3.25
4	13	3
5	17	2.75
6	21	2.5

X=0

(b)

X	Y₁	Y₂
0	-3	5
1	2	5.2
2	7	5.4
3	12	5.6
4	17	5.8
5	22	6
6	27	6.2

X=0

(c)

X	Y₁	Y₂
0	-3	12
1	2	17
2	7	22
3	12	27
4	17	32
5	22	37
6	27	42

X=0

(d)

X	Y₁	Y₂
0	2	-2
1	-2	2
2	-6	6
3	-10	10
4	-14	14
5	-18	18
6	-22	22

X=0

(Modeling) Solve each problem.

61. *Distance* A person is riding a bicycle along a straight highway. The graph to the right shows the rider's distance y in miles from an interstate highway after x hours.
   (a) Find the slope–intercept form of the line.
   (b) How fast is the biker traveling?
   (c) How far was the biker from the interstate highway initially?
   (d) How far was the biker from the interstate highway after 1 hour and 15 minutes?

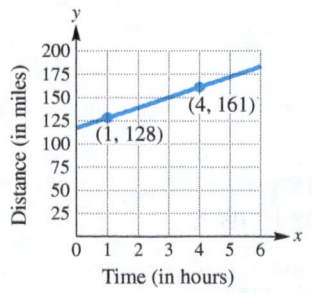

62. *Water in a Tank*   The graph shows the number of gallons of water y in a 100-gallon tank after x minutes have elapsed.

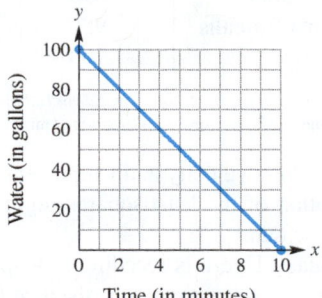

(a) Is water entering or leaving the tank? How much water is in the tank after 3 minutes?

(b) Find the x- and y-intercepts. Interpret their meanings.

(c) Find the slope–intercept form of the equation of the line. Interpret the slope.

(d) Estimate the x-coordinate of the point $(x, 50)$ that lies on the line.

63. *Online Betting*   Worldwide gambling revenue from online betting was \$18 billion in 2007 and \$24 billion in 2010. (*Source*: Christiansen Capital Advisors.)

(a) Find an equation of a line $y = mx + b$ that models this information, where y is in billions of dollars and x is the year.

(b) Use this equation to estimate online betting revenue in 2013.

64. *Average Wages*   The average hourly wage (adjusted to 1982 dollars) was \$7.66 in 1990 and \$8.27 in 2009. (*Source:* U.S. Census Bureau.)

(a) Find a point-slope form of a line that passes through the points (1990, 7.66) and (2009, 8.27).

(b) Interpret the slope.

(c) Use the equation from part (a) to approximate the hourly wage in 2005. Compare it with the actual value of \$8.18.

65. *Temperature Scales*   The table shows equivalent temperatures in degrees Celsius and degrees Fahrenheit.

°F	−40	32	59	95	212
°C	−40	0	15	35	100

(a) Plot the data by having the x-axis correspond to Fahrenheit temperature and the y-axis to Celsius temperature. What type of relation exists between the data?

(b) Find a function C that uses the Fahrenheit temperature x to calculate the corresponding Celsius temperature. Interpret the slope.

(c) What is a temperature of 83°F in degrees Celsius?

66. *Population*   Asian-American populations (in millions) are shown in the table.

Year	2003	2005	2007	2009
Population (in millions)	11.8	12.6	13.3	14.0

Source: U.S. Census Bureau.

(a) Use the points (2003, 11.8) and (2009, 14.0) to find the point–slope form of a line that models the data. Let $(x_1, y_1) = (2003, 11.8)$.

(b) Use this equation to estimate the Asian-American population in 2013 to the nearest tenth of a million.

67. *Google Ad Revenue*   The table lists the worldwide advertising revenue of Google (in billions of dollars).

Year	2005	2007	2009	2011
Revenue (\$ billions)	6	17	23	37

(a) Find the point–slope form of the line that passes through the points (2005, 6) and (2011, 37). Let (x_1, y_1) be (2005, 6).

(b) Interpret the slope of the line.

(c) Use this equation to estimate the revenue in 2007 and 2009. Compare these estimates with the actual values shown in the table.

68. *Newspaper Ad Revenue*   The table lists U.S. print newspaper advertising revenue (in billions of dollars).

Year	2006	2008	2010	2012
Revenue (\$ billions)	48	35	22	10

(a) Find the point–slope form of the line that passes though (2006, 48) and (2010, 22). Let (x_1, y_1) be (2006, 48).

(b) Find the point–slope form of the line that passes though (2008, 35) and (2012, 10). Let (x_1, y_1) be (2008, 35).

(c) Interpret the slope of the line from part (b).

(d) Use equations from parts (a) and (b) to predict the revenue for 2009.

69. *Tuition and Fees*   The table lists the average tuition and fees (in constant 2010 dollars) at private colleges and universities for selected years.

Year	1980	1990	2000	2010
Tuition and Fees (in 2010 dollars)	13,686	20,894	26,456	31,395

Source: National Center for Education Statistics.

(a) Find the equation of the least-squares regression line that models the data.

(*continued*)

(b) Graph the data and the regression line in the same viewing window.

(c) Estimate tuition and fees in 2005, and compare it with the actual value of $29,307.

70. *Tuition and Fees* The table lists the average tuition and fees (in constant 2010 dollars) at public colleges and universities for selected years.

Year	1980	1990	2000	2005	2010
Tuition and Fees (in 2010 dollars)	5938	7699	9390	11,386	13,297

Source: National Center for Education Statistics.

(a) Find the equation of the least-squares regression line that models the data.

(b) Graph the data and the regression line in the same viewing window.

(c) Estimate tuition and fees in 2007.

(d) Use the model to predict tuition and fees in 2016.

71. *Distant Galaxies* In the late 1920s, the famous observational astronomer Edwin P. Hubble (1889–1953) determined the distances to several galaxies and the velocities at which they were receding from Earth. Four galaxies with their distances in light-years and velocities in miles per second are listed in the table at the top of the next column.

Galaxy	Distance	Velocity
Virgo	50	990
Ursa Minor	650	9,300
Corona Borealis	950	15,000
Bootes	1,700	25,000

Source: Sharov, A., and I. Novikov, *Edwin Hubble, the Discoverer of the Big Bang Universe,* Cambridge University Press.

(a) Let x represent velocity and y represent distance. Find the equation of the least-squares regression line that models the data.

(b) If the galaxy Hydra is receding at a speed of 37,000 miles per second, estimate its distance from Earth.

72. *Apple Products* The table lists the worldwide average household spending (in dollars) on Apple products for selected years.

Year	2009	2011	2013	2015
Spending ($ dollars)	62	158	265	444

(a) Use regression to find a formula $f(x) = ax + b$ so that f models the data.

(b) Interpret the slope of the graph of $y = f(x)$.

(c) Estimate the average household spending on Apple products in 2014 and compare it with the actual value of $343.

(Modeling) In Exercises 73 and 74, obtain the least-squares regression line and the correlation coefficient. Make a statement about the correlation.

73. *Gestation Period and Life Span of Animals*

Animal	Average Gestation or Incubation Period, x (in days)	Record Life Span, y (in years)
Cat	63	26
Dog	63	24
Duck	28	15
Elephant	624	71
Goat	151	17
Guinea pig	68	6
Hippopotamus	240	49
Horse	336	50
Lion	108	29
Parakeet	18	12
Pig	115	22
Rabbit	31	15
Sheep	151	16

Source: Infoplease and James Doherty, The Wildlife Conservation Society.

74. *Urban Areas in the World (Population and Area)*

Urban Area	Population, x (in millions)	Area, y (in square miles)
Tokyo–Yokohama, Japan	34.2	3287
Delhi, India	23.9	747
Mumbai, India	23.3	210
Mexico City, Mexico	22.8	787
New York, United States	22.2	4477
São Paulo, Brazil	20.8	1220
Shanghai, China	18.8	1345
Los Angeles, United States	17.9	2423
Kolkata (Calcutta), India	16.6	463
Buenos Aires, Argentina	13.1	1016

Source: World Almanac and Book of Facts.

1. Write the formula for a linear function f with slope 1.4 and y-intercept $(0, -3.1)$. Find $f(1.3)$.

2. Graph $f(x) = -2x + 1$ by hand. State the x-intercept, y-intercept, domain, range, and slope.

3. Find the slope of the line passing through $(-2, 4)$ and $(5, 6)$.

4. Give the equations of a vertical line and a horizontal line passing through $(-2, 10)$.

5. Let $f(x) = 0.5x - 1.4$. Graph f in the standard viewing window of a graphing calculator, and make a table of values for f at $x = -3, -2, -1, \ldots, 3$.

6. Give the slope–intercept form of the line shown in the figure.

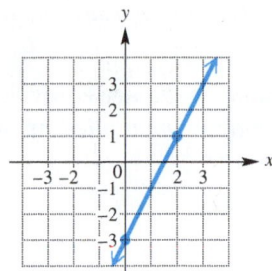

7. Find the slope–intercept form of the line passing through $(-2, 4)$ and $(5, 2)$.

8. Find the slope–intercept form of the line perpendicular to $3x - 2y = 5$ and passing through $(-1, 3)$.

9. *Average Household Size* The table lists the average number y of people per household for various years x.

x	1940	1950	1960	1970
y	3.67	3.37	3.33	3.14

x	1980	1990	2000	2010
y	2.76	2.63	2.62	2.58

Source: U.S. Census Bureau.

(a) Make a scatter diagram of the data.

(b) Decide whether the correlation coefficient is positive, negative, or zero.

(c) Find a least-squares regression line that models these data. What is the correlation coefficient?

(d) Estimate the number of people per household in 1975, and compare it with the actual value of 2.94.

1.5 Linear Equations and Inequalities

Solving Linear Equations in One Variable • Graphical Approaches to Solving Linear Equations • Identities and Contradictions • Solving Linear Inequalities in One Variable • Graphical Approaches to Solving Linear Inequalities • Three-Part Inequalities

Solving Linear Equations in One Variable

An **equation** is a statement that two expressions are equal. To *solve* an equation means to find all numbers that make the equation a true statement. Such numbers are called **solutions** or **roots** of the equation. A number that is a solution of an equation is said to *satisfy* the equation, and the solutions of an equation make up its **solution set.**

In this section, we solve *linear equations in one variable*. A linear equation in one variable has one solution.

Linear Equation in One Variable

A **linear equation in the variable** x is an equation that can be written in the form

$$ax + b = 0, \quad a \neq 0.$$

In this text, we use two distinct approaches to solving equations.

1. The *analytic approach,* in which we use paper and pencil to transform complicated equations into simpler ones.

2. The *graphical approach,* in which we often support our analytic solutions by using graphs or tables.

We begin by discussing the analytic approach.

One way to solve a given equation analytically is to rewrite it as a series of simpler **equivalent equations,** each of which has the same solution set as the given one. Equivalent equations are obtained by using the properties of equality.

Addition and Multiplication Properties of Equality

For real numbers a, b, and c, the following are true.

$a = b$ and $a + c = b + c$ **are equivalent.**
(*The same number may be added to each side of an equation without changing the solution set.*)

If $c \neq 0$, **then** $a = b$ **and** $ac = bc$ **are equivalent.**
(*Each side of an equation may be multiplied by the same nonzero number without changing the solution set.*)

Extending these two properties allows us to subtract the same number from each side of an equation and to divide each side by the same *nonzero* number.

EXAMPLE 1   Solving a Linear Equation

Solve $10 + 3(2x - 4) = 17 - (x + 5)$. Check your answer.

Solution

$$10 + 3(2x - 4) = 17 - (x + 5)$$

Distribute the minus sign to each term inside the parentheses.

$$10 + 6x - 12 = 17 - x - 5 \qquad \text{Distributive property}$$
$$-2 + 7x = 12 \qquad \text{Add } x \text{ to each side and combine like terms.}$$
$$7x = 14 \qquad \text{Add 2 to each side.}$$
$$x = 2 \qquad \text{Divide each side by 7.}$$

Check:
$$10 + 3(2x - 4) = 17 - (x + 5) \qquad \text{Original equation}$$
$$10 + 3(2 \cdot 2 - 4) = 17 - (2 + 5) \quad ? \;\text{ Let } x = 2.$$
$$10 + 3(4 - 4) = 17 - 7 \qquad ?$$
$$10 = 10 \checkmark \qquad \text{True; the answer checks.}$$

The solution set is $\{2\}$.

If an equation contains fractions, we can make an equivalent equation without fractions by multiplying each side of the equation by the least common denominator (LCD). The least common denominator is the smallest number that each denominator will divide into evenly.

EXAMPLE 2   **Solving a Linear Equation with Fractional Coefficients**

Solve $\dfrac{x + 7}{6} + \dfrac{2x - 8}{2} = -4$.

Solution   To eliminate fractions, multiply each side of the equation by the least common denominator of $\frac{x+7}{6}$ and $\frac{2x-8}{2}$, which is 6.

$$\frac{x + 7}{6} + \frac{2x - 8}{2} = -4$$

$$6\left(\frac{x + 7}{6} + \frac{2x - 8}{2}\right) = 6(-4) \qquad \text{Multiply by the LCD, 6.}$$

$$6\left(\frac{x + 7}{6}\right) + 6\left(\frac{2x - 8}{2}\right) = 6(-4) \qquad \text{Distributive property}$$

$$x + 7 + 3(2x - 8) = -24 \qquad \text{Simplify.}$$

$$x + 7 + 6x - 24 = -24 \qquad \text{Distributive property}$$

$$7x - 17 = -24 \qquad \text{Combine like terms.}$$

$$7x = -7 \qquad \text{Add 17.}$$

$$x = -1 \qquad \text{Divide by 7.}$$

An analytic check will verify that the solution set is $\{-1\}$.

We can eliminate decimal points in a linear equation by multiplying by a power of 10, as demonstrated in the next example.

EXAMPLE 3   **Solving a Linear Equation with Decimal Coefficients**

Solve $0.06x + 0.09(15 - x) = 0.07(15)$.

Solution   Since each decimal number is given in hundredths, multiply each side of the equation by 100 to clear decimal points. (To multiply a number by 100, move the decimal point two places to the right.)

$$0.06x + 0.09(15 - x) = 0.07(15)$$

$$100(0.06x + 0.09(15 - x)) = 100(0.07)(15) \qquad \text{Multiply each side by 100.}$$

$$100(0.06x) + 100(0.09)(15 - x) = 100(0.07)(15) \qquad \text{Distributive property}$$

$$6x + 9(15 - x) = 7(15) \qquad \text{Simplify.}$$

$$6x + 135 - 9x = 105 \qquad \text{Distributive property; multiply.}$$

$$-3x + 135 = 105 \qquad \text{Combine like terms.}$$

$$-3x = -30 \qquad \text{Subtract 135.}$$

$$x = 10 \qquad \text{Divide by } -3.$$

An analytic check will verify that the solution set is $\{10\}$.

The linear equations in **Examples 1–3** are called **conditional equations** because they have one solution. Later in this section we will discuss *identities* and *contradictions*, which have infinitely many solutions and no solutions, respectively.

Graphical Approaches to Solving Linear Equations

Since an equation always contains an equality symbol, we can sometimes think of a linear equation as being in the form

$$f(x) = g(x),$$

where $f(x)$ and $g(x)$ are formulas for linear functions. The equation

$$\underbrace{10 + 3(2x - 4)}_{f(x)} = \underbrace{17 - (x + 5)}_{g(x)}$$

from **Example 1** is in this form. To find the solution set graphically, we graph $y_1 = f(x)$ and $y_2 = g(x)$ and locate their point of intersection. *In general, if f and g are linear functions, then their graphs are lines that intersect at a single point, no point, or infinitely many points, as illustrated in* FIGURE 67.

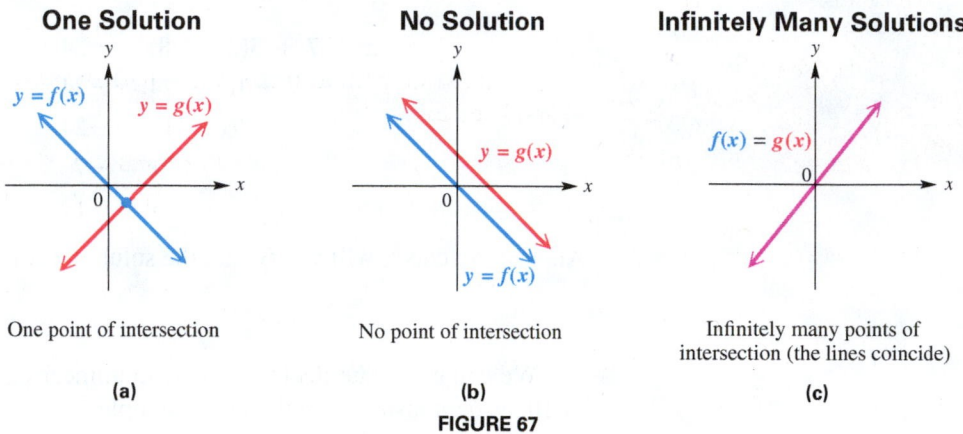

One Solution

One point of intersection

(a)

No Solution

No point of intersection

(b)

Infinitely Many Solutions

Infinitely many points of intersection (the lines coincide)

(c)

FIGURE 67

FIGURE 68(a) shows graphs of $Y_1 = 10 + 3(2X - 4)$ and $Y_2 = 17 - (X + 5)$ intersecting at the point $(2, 10)$. The x-coordinate 2 is the solution of the equation from **Example 1,** and the y-coordinate 10 is the value obtained when either Y_1 or Y_2 is evaluated at $X = 2$, as shown in the check in **Example 1.**

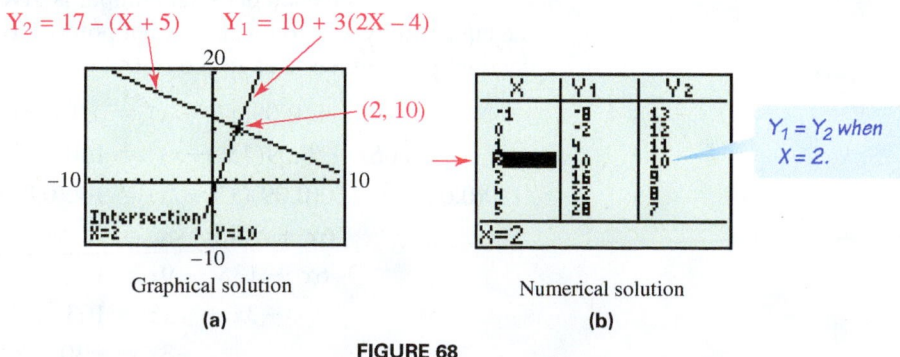

Graphical solution

(a)

Numerical solution

(b)

FIGURE 68

In FIGURE 68(b), a table of values for Y_1 and Y_2 shows that $Y_1 = Y_2 = 10$ when $X = 2$. Creating a table to find a solution of an equation gives a **numerical solution.** Locating a numerical solution involves searching a table to find an X-value for which $Y_1 = Y_2$. A table is less practical when the solution is not an integer.

The method shown in FIGURE 68(a) is called the **intersection-of-graphs method.**

TECHNOLOGY NOTE

Graphing calculators often have built-in routines to find the intersection of two graphs. For the TI-84 Plus it is located in the CALC menu.

Intersection-of-Graphs Method of Graphical Solution

To solve the equation $f(x) = g(x)$ graphically, graph

$$y_1 = f(x) \quad \text{and} \quad y_2 = g(x).$$

The x-coordinate of any point of intersection of the two graphs is a solution of the equation.

EXAMPLE 4   **Applying the Intersection-of-Graphs Method**

During the 1990s, compact discs were a new technology that replaced cassette tapes. The percent share of music sales (in dollars) that compact discs (CDs) held from 1987 to 1998 can be modeled by

$$f(x) = 5.91x + 13.7.$$

During the same period, the percent share of music sales that cassette tapes held can be modeled by

$$g(x) = -4.71x + 64.7.$$

In these formulas, $x = 0$ corresponds to 1987, $x = 1$ to 1988, and so on. Use the intersection-of-graphs method to estimate the year when sales of CDs equaled sales of cassettes. (*Source:* Recording Industry Association of America.)

Solution   We solve the linear equation $f(x) = g(x)$, that is,

$$5.91x + 13.7 = -4.71x + 64.7, \qquad \text{Let } f(x) = g(x).$$

by graphing

$$\overset{f(x)}{y_1 = 5.91x + 13.7} \text{ (CDs)} \quad \text{and} \quad \overset{g(x)}{y_2 = -4.71x + 64.7} \text{ (Tapes),}$$

as shown in **FIGURE 69(a)**. In **FIGURE 69(b)**, the two graphs intersect near the point (4.8, 42.1). Since $x = 0$ corresponds to 1987, and $1987 + 4.8 \approx 1992$, it follows that in 1992 sales of CDs and cassette tapes were approximately equal. Each shared about 42.1% of the sales that year.

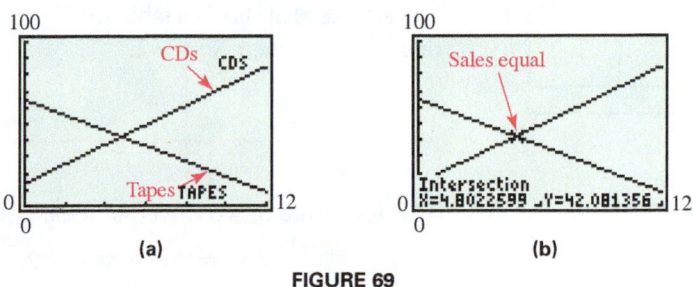

(a)   (b)

FIGURE 69

There is a second graphical method of solving equations in the form $f(x) = g(x)$.

$$f(x) = g(x)$$

$$f(x) - g(x) = 0 \qquad \text{Subtract } g(x) \text{ from each side.}$$

$$F(x) = 0 \qquad \text{Let } F(x) = f(x) - g(x).$$

We can now solve $F(x) = 0$ to obtain the solution set of the original equation. Recall from **Section 1.3** that any number that satisfies this equation is a zero of F and corresponds to the x-coordinate of an x-intercept on the graph of $y = F(x)$. The *x-intercept method* of graphical solution uses this idea.

> ### x-Intercept Method of Graphical Solution
>
> To solve the equation $f(x) = g(x)$ graphically, graph
>
> $$y = f(x) - g(x) = F(x).$$
>
> *The x-coordinate of any x-intercept of the graph of $y = F(x)$ (or zero of the function F) is a solution of the equation.*

TECHNOLOGY NOTE

Graphing calculators often have a built-in program to locate x-intercepts (roots, zeros). For the TI-84 Plus it is in the CALC menu.

The words *root, solution,* and *zero* all refer to the same basic concept. When solving equations, remember this: *The real solutions (or roots or zeros) of an equation of the form $F(x) = 0$ correspond to the x-coordinates of the x-intercepts of the graph of $y = F(x)$.*

x-Intercept Method

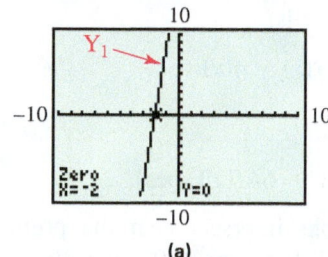

(a)

Numerical Solution

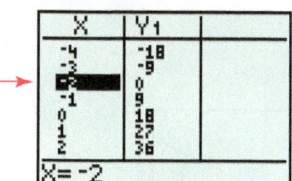

$Y_1 = 0$ when $X = -2$.

(b)

FIGURE 70

EXAMPLE 5   Using the *x*-Intercept Method

Solving analytically, the solution set of

$$6x - 4(3 - 2x) = 5(x - 4) - 10$$

is $\{-2\}$. Use the x-intercept method to solve this equation. Then use a table to obtain a numerical solution.

Solution   Begin by writing the given equation in the form $F(x) = 0$.

$$6x - 4(3 - 2x) - (5(x - 4) - 10) = 0 \qquad \text{Subtract } 5(x - 4) - 10.$$

Use parentheses.

Graph $Y_1 = 6X - 4(3 - 2X) - (5(X - 4) - 10)$, and locate any X-intercepts of Y_1. **FIGURE 70(a)** shows X-intercept $(-2, 0)$, confirming the given solution set of $\{-2\}$.

FIGURE 70(b) shows a table for $Y_1 = 6X - 4(3 - 2X) - (5(X - 4) - 10)$. Notice that $Y_1 = 0$ when $X = -2$. ●

> **FOR DISCUSSION**
>
> In **Example 5**, we solved a linear equation by letting
>
> $$f(x) = 6x - 4(3 - 2x) \quad \text{and} \quad g(x) = 5(x - 4) - 10$$
>
> and then graphing $F(x) = f(x) - g(x)$. This time, graph
>
> $$y_2 = g(x) - f(x), \quad \text{rather than} \quad y_1 = f(x) - g(x).$$
>
> Does the graph of y_2 have the same x-intercept as the graph of y_1? Try to generalize your result.

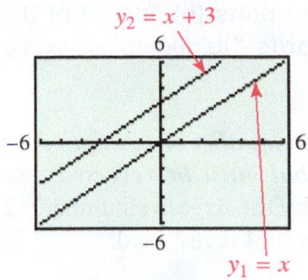

Lines are parallel.

FIGURE 71

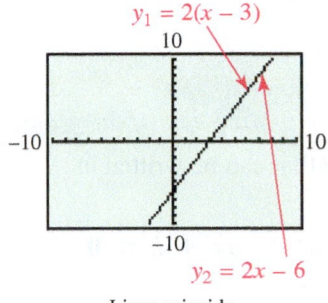

Lines coincide.

(a)

X	Y₁	Y₂
0	-6	-6
1	-4	-4
2	-2	-2
3	0	0
4	2	2
5	4	4
6	6	6

X=0

$Y_1 = Y_2$ for all X.

(b)

FIGURE 72

Identities and Contradictions

Every equation solved thus far in this section has been a conditional equation with one solution. However, because the graphs of two linear functions may not intersect or may coincide (as shown in **FIGURES 67(b)** and **67(c),** respectively), there are two other situations that may arise in solving equations.

A **contradiction** is an equation that has no solution, such as $x = x + 3$. There is no real number x equal to itself plus 3. If we try to solve this equation by subtracting x from each side, we obtain $0 = 3$, a false statement. If we try to solve $x = x + 3$ graphically by letting $y_1 = x$ and $y_2 = x + 3$, we obtain two parallel lines, as shown in **FIGURE 71**. Since the lines do not intersect, the solution set is the **empty set** or **null set,** denoted **∅.**

An **identity** is an equation that is true for all values in the domain of its variables, such as $2(x - 3) = 2x - 6$. All real numbers satisfy this equation. If we try to solve this equation, we obtain the following.

$$2(x - 3) = 2x - 6$$
$$2x - 6 = 2x - 6 \qquad \text{Distributive property}$$
$$2x = 2x \qquad \text{Add 6.}$$
$$0 = 0 \qquad \text{Subtract } 2x.$$

The equation $0 = 0$ is a true statement, indicating that the solution set is $(-\infty, \infty)$, or {all real numbers}. If we try to solve the equation graphically by letting $y_1 = 2(x - 3)$ and $y_2 = 2x - 6$, we obtain two identical lines as shown in **FIGURE 72(a).** These two lines coincide, so the solution set is {all real numbers}. The table in **FIGURE 72(b)** also suggests that $y_1 = y_2$ for all x-values.

Solving Linear Inequalities in One Variable

An equation says that two expressions are equal. An **inequality** says that one expression is greater than, greater than or equal to, less than, or less than or equal to another ($>, \geq, <, \leq$). As with equations, a value of the variable for which the inequality is true is a *solution of the inequality,* and the set of all such solutions is the *solution set of the inequality.* Two inequalities with the same solution set are **equivalent inequalities.**

Inequalities are solved using the properties of inequality.

Addition and Multiplication Properties of Inequality

For real numbers a, b, and c,

(a) $a < b$ and $a + c < b + c$ are equivalent.

(The same number may be added to each side of an inequality without changing the solution set.)

(b) If $c > 0$, then $a < b$ and $ac < bc$ are equivalent.

(Each side of an inequality may be multiplied by the same positive number without changing the solution set.)

(c) If $c < 0$, then $a < b$ and $ac > bc$ are equivalent.

(Each side of an inequality may be multiplied by the same negative number without changing the solution set, provided the direction of the inequality symbol is reversed.)

Similar properties exist for $>$, $\leq$, and $\geq$.

NOTE Because division can be defined in terms of multiplication $\left(\frac{x}{a} = x \cdot \frac{1}{a}\right)$, the word "multiplied" may be replaced by "divided" in parts (b) and (c) of the properties of inequality. Similarly, in part (a), the words "added to" may be replaced by "subtracted from."

Pay careful attention to part (c). *If each side of an inequality is multiplied by a negative number, the direction of the inequality symbol must be reversed.* For example, starting with $-3 < 5$ and multiplying each side by the *negative* number -2 gives a true result only if the direction of the inequality symbol is reversed.

$$-3 < 5 \qquad \text{\textit{Reverse the inequality symbol.}}$$
$$-3(-2) > 5(-2) \qquad \text{Multiply by } -2.$$
$$6 > -10$$

A similar situation exists when each side is divided by a negative number.

A *linear inequality in one variable* is defined as follows.

Linear Inequality in One Variable

A **linear inequality in the variable** x is an inequality that can be written in one of the following forms, where $a \neq 0$.

$$ax + b > 0, \qquad ax + b < 0, \qquad ax + b \geq 0, \qquad ax + b \leq 0$$

The solution set of a linear inequality is typically an interval of the real number line and can be expressed in interval notation.

EXAMPLE 6 **Solving a Linear Inequality**

Solve $3x - 2(2x + 4) \leq 2x + 1$.

Solution

$$3x - 2(2x + 4) \leq 2x + 1 \qquad \text{Given inequality}$$
$$3x - 4x - 8 \leq 2x + 1 \qquad \text{Distributive property}$$
$$-x - 8 \leq 2x + 1 \qquad \text{Combine like terms.}$$
$$-3x \leq 9 \qquad \text{Subtract 2x and add 8.}$$
$$x \geq -3 \qquad \text{Divide by } -3. \text{ Reverse the direction of the inequality symbol.}$$

Multiply each term in the parentheses by -2.

In interval notation, the solution set is $[-3, \infty)$.

FOR DISCUSSION

Solve each equation or inequality.

$$2(x - 3) + x = 9,$$
$$2(x - 3) + x < 9,$$
$$2(x - 3) + x > 9$$

How can the solution of the equation be used to help solve the two inequalities?

EXAMPLE 7   **Solving Linear Inequalities**

Solve each inequality.

(a) $2x - 3 < \dfrac{x + 2}{-3}$ **(b)** $-3(4x - 4) \geq 4 - (x - 1)$

Solution

(a)

$$2x - 3 < \dfrac{x + 2}{-3} \qquad \text{Given inequality}$$

$$-3(2x - 3) > -3\left(\dfrac{x + 2}{-3}\right) \qquad \text{Multiply by } -3. \text{ Reverse the inequality symbol.}$$

Multiply each term in the parentheses by -3.

$$-6x + 9 > x + 2 \qquad \text{Use the distributive property and simplify.}$$

$$7 > 7x \qquad \text{Add } 6x \text{ and subtract 2.}$$

$$1 > x \qquad \text{Divide by 7.}$$

$$x < 1 \qquad \text{Rewrite.}$$

The solution set is $(-\infty, 1)$.

(b)

$$-3(4x - 4) \geq 4 - (x - 1) \qquad \text{Given inequality}$$

Multiply each term in the parentheses by -3.

$$-12x + 12 \geq 4 - x + 1 \qquad \text{Distributive property}$$

$$-12x + 12 \geq -x + 5 \qquad \text{Simplify.}$$

$$-11x \geq -7 \qquad \text{Add } x \text{ and subtract 12.}$$

$$x \leq \dfrac{7}{11} \qquad \text{Divide by } -11. \text{ Reverse the inequality symbol.}$$

The solution set is $\left(-\infty, \dfrac{7}{11}\right]$.

Graphical Approaches to Solving Linear Inequalities

Distance from Chicago

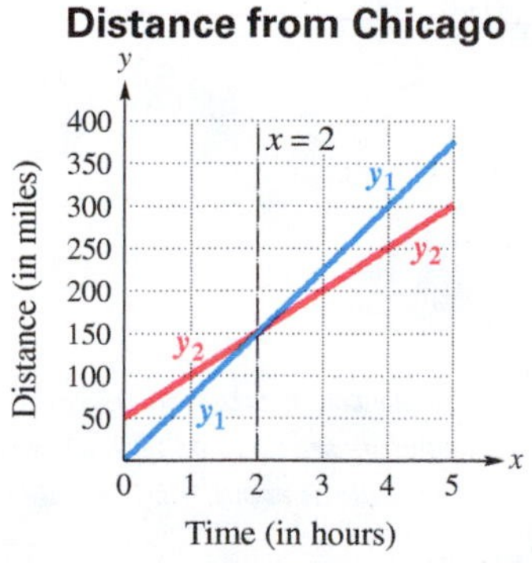

FIGURE 73

FIGURE 73 shows the distances of two cars from Chicago, Illinois, x hours after traveling in the same direction on a freeway. The distance for Car 1 is denoted y_1, and the distance for Car 2 is denoted y_2. After $x = 2$ hours, $y_1 = y_2$ and both cars are 150 miles from Chicago. To the left of the dashed vertical line $x = 2$, the graph of y_1 is below the graph of y_2, so Car 1 is closer to Chicago than Car 2. Thus,

$$y_1 < y_2 \quad \text{when} \quad x < 2.$$

To the right of the dashed vertical line $x = 2$, the graph of y_1 is above the graph of y_2, so Car 1 is farther from Chicago than Car 2. Thus,

$$y_1 > y_2 \quad \text{when} \quad x > 2.$$

This discussion leads to an extension of the intersection-of-graphs method to include inequalities.

TECHNOLOGY NOTE

When solving linear inequalities graphically, the calculator will not determine whether the endpoint of the interval is included or excluded. This must be done by considering the inequality symbol in the given inequality.

Intersection-of-Graphs Method of Solution of a Linear Inequality

Suppose that f and g are linear functions. The solution set of $f(x) > g(x)$ is the set of all real numbers x such that the graph of f is **above** the graph of g. The solution set of $f(x) < g(x)$ is the set of all real numbers x such that the graph of f is **below** the graph of g.

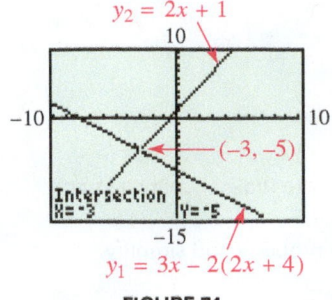

$y_2 = 2x + 1$

$(-3, -5)$

$y_1 = 3x - 2(2x + 4)$

FIGURE 74

> **NOTE**   If an inequality involves one of the symbols $\geq$ or $\leq$, the same method applies, with the solution of the corresponding equation included in the solution set.

EXAMPLE 8   **Using the Intersection-of-Graphs Method**

The solution set of the inequality $3x - 2(2x + 4) \leq 2x + 1$, solved analytically in **Example 6,** is $[-3, \infty)$. Solve this inequality graphically.

Solution   Graph

$$y_1 = 3x - 2(2x + 4) \quad \text{and} \quad y_2 = 2x + 1.$$

FIGURE 74 indicates that the point of intersection of the two lines is $(-3, -5)$. Because the graph of y_1 is *below* the graph of y_2 when x is *greater than* -3, the solution set of $y_1 \leq y_2$ is $[-3, \infty)$, agreeing with our analytic solution. (The endpoint -3 is included.)

EXAMPLE 9   **Applying the Intersection-of-Graphs Method**

When the air temperature decreases to the dew point, fog may form. This phenomenon also causes clouds to form at higher altitudes. If the ground-level temperature and dew point are T_0 and D_0, respectively, the air temperature and the dew point at an altitude of x miles can be approximated by

$$T(x) = T_0 - 19x \quad \text{and} \quad D(x) = D_0 - 5.8x.$$

If $T_0 = 75°F$ and $D_0 = 55°F$, determine the altitudes where clouds will *not* form.

Solution   Since $T_0 = 75$ and $D_0 = 55$, it follows that $T(x) = 75 - 19x$ and $D(x) = 55 - 5.8x$. Graph $y_1 = 75 - 19x$ and $y_2 = 55 - 5.8x$, as shown in **FIGURE 75.** The graphs intersect near $(1.52, 46.2)$. This means that the air temperature and dew point are both $46.2°F$ at about 1.52 miles above ground level. Clouds will not form *below* this altitude—that is, when the graph of y_1 is above the graph of y_2. The solution set is $[0, 1.52)$, where the endpoint 1.52 is approximate.

Determining Where Clouds Form

$y_1 = 75 - 19x$

$y_2 = 55 - 5.8x$

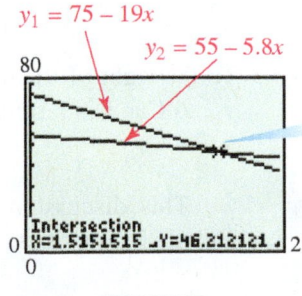

The air temperature y_1 and dew point y_2 are both 46.2°F at an altitude of about 1.52 miles.

FIGURE 75

In **Example 9,** the endpoint, 1.52 miles, is an approximation to the nearest hundredth. In solving inequalities graphically, the appropriate symbol for the endpoint, a parenthesis or a bracket, may not actually be valid because of rounding. As a result, we state an agreement to be used throughout this text.

> ### Agreement on Inclusion or Exclusion of Endpoints for Approximations
>
> When an approximation is used for an endpoint in specifying an interval, we continue to use parentheses in specifying inequalities involving $<$ or $>$ and square brackets in specifying inequalities involving $\leq$ or $\geq$.

The x-intercept method can also be used to solve inequalities. For example, to solve $f(x) > g(x)$, we can rewrite the inequality as $f(x) - g(x) > 0$ or $F(x) > 0$, where $F(x) = f(x) - g(x)$. All solutions of the given inequality correspond to the x-values where the graph of $y = F(x)$ is above the x-axis.

> ### x-Intercept Method of Solution of a Linear Inequality
>
> The solution set of $F(x) > 0$ is the set of all real numbers x such that the graph of F is **above** the x-axis. The solution set of $F(x) < 0$ is the set of all real numbers x such that the graph of F is **below** the x-axis.

FIGURE 76 illustrates this discussion and summarizes the solution sets for the appropriate inequalities.

The *x*-Intercept Method for Inequalities

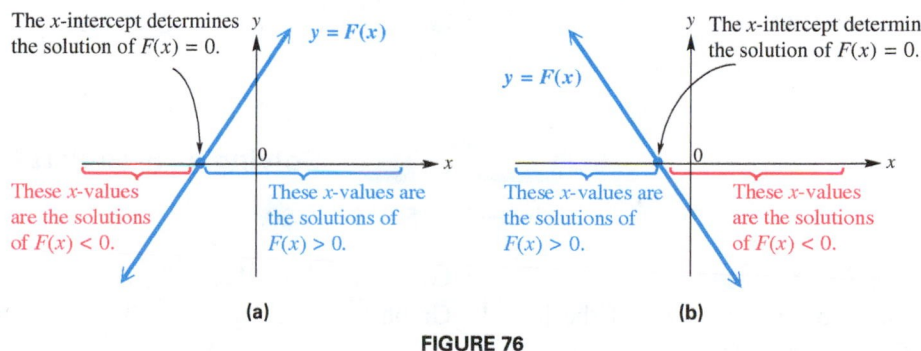

FIGURE 76

For example, in **FIGURE 76(a)** if we let $F(x) = x + 3$ then the x-intercept is $(-3, 0)$ and the solution to $F(x) = 0$ is -3. Also, $F(x) > 0$ when $x > -3$ and $F(x) < 0$ when $x < -3$.

EXAMPLE 10   **Using the *x*-Intercept Method**

Solve each inequality using the x-intercept method.

(a) $-2(3x + 1) < 4(x + 2)$ **(b)** $-2(3x + 1) \leq 4(x + 2)$

(c) $-2(3x + 1) > 4(x + 2)$ **(d)** $-2(3x + 1) \geq 4(x + 2)$

Solution

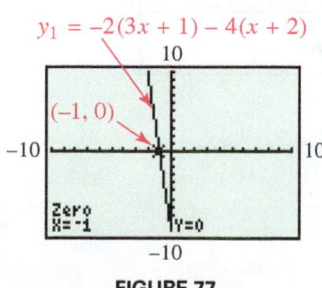

$y_1 = -2(3x + 1) - 4(x + 2)$

FIGURE 77

(a) The inequality can be written as $-2(3x + 1) - 4(x + 2) < 0$. Graph the left side of the inequality as y_1, as shown in **FIGURE 77.** The graph is below the x-axis when $x > -1$, so the solution set is $(-1, \infty)$.

(b) The solution set is $[-1, \infty)$, because the endpoint at $x = -1$ is included.

(continued)

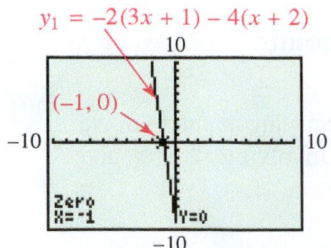

$y_1 = -2(3x + 1) - 4(x + 2)$

FIGURE 77 (repeated)

(c) The graph in **FIGURE 77** is above the x-axis when $x < -1$, so the solution set is $(-\infty, -1)$.

(d) The solution set is $(-\infty, -1\,]$, because the endpoint at $x = -1$ is included.

Notice that the same graph was used to solve all four inequalities.

Three-Part Inequalities

Sometimes a mathematical expression is limited to an interval of values. For example, suppose that $2x + 1$ must satisfy both

$$2x + 1 \geq -3 \quad and \quad 2x + 1 \leq 2.$$

These two inequalities can be combined into the **three-part inequality**

$$-3 \leq 2x + 1 \leq 2. \quad \text{Three-part inequality}$$

One place where three-part inequalities occur is manufacturing, where error tolerances are often maintained. Suppose that aluminum cans are to be constructed with radius 1.4 inches. However, because the manufacturing process is not exact, the radius r actually varies from 1.38 inches to 1.42 inches. We can express possible values for r by using the following three-part inequality.

$$1.38 \leq r \leq 1.42 \qquad \text{Possible values for } r \text{ due to errors}$$

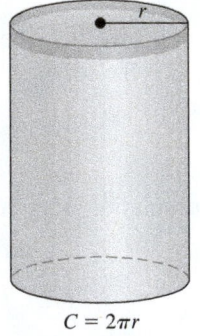

$C = 2\pi r$

FIGURE 78

See **FIGURE 78.** Then the circumference ($C = 2\pi r$) of the cylindrical can varies between $2\pi(1.38) \approx 8.67$ inches and $2\pi(1.42) \approx 8.92$ inches. This result is given (approximately) by

$$\text{Resulting range for the circumference} \qquad 8.67 \leq C \leq 8.92. \qquad C = 2\pi r$$

EXAMPLE 11 **Solving a Three-Part Inequality**

Solve $-2 < 5 + 3x < 20$.

Analytic Solution

Work with all three parts of the inequality at the same time.

$$-2 < 5 + 3x < 20$$

$$-7 < 3x < 15 \qquad \text{Subtract 5.}$$

$$-\frac{7}{3} < x < 5 \qquad \text{Divide by 3.}$$

The open interval $\left(-\frac{7}{3}, 5\right)$ is the solution set. Notice that an open interval is used because equality is *not* included in the three-part inequality.

Graphing Calculator Solution

Graph $y_1 = -2$, $y_2 = 5 + 3x$, and $y_3 = 20$, as shown in **FIGURE 79.** The x-values of the points of intersection are $-\frac{7}{3} = -2.\overline{3}$ and 5, confirming that our analytic work is correct. Notice how the slanted line, y_2, lies *between* the graphs of $y_1 = -2$ and $y_3 = 20$ for x-values between $-\frac{7}{3}$ and 5.

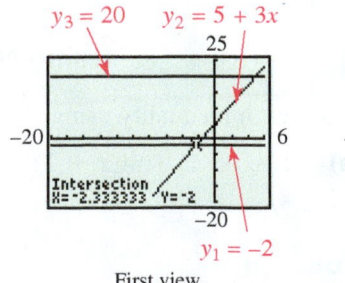

First view

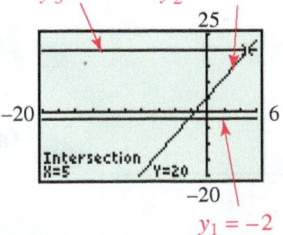

Second view

FIGURE 79

1.5 Exercises

Checking Analytic Skills *Find the zero of the function f.* **Do not use a calculator**.

1. $f(x) = -3x - 12$ **2.** $f(x) = 5x - 30$ **3.** $f(x) = 5x$

4. $f(x) = -2x$ **5.** $f(x) = 2(3x - 5) + 8(4x + 7)$ **6.** $f(x) = -4(2x - 3) + 8(2x + 1)$

7. $f(x) = 3x + 6(x - 4)$ **8.** $f(x) = -8x + 0.5(2x + 8)$ **9.** $f(x) = 1.5x + 2(x - 3) + 5.5(x + 9)$

10. *Concept Check* If c is a zero of the linear function $f(x) = mx + b$, $m \neq 0$, then the point at which the graph intersects the x-axis has coordinates (_____, _____).

In Exercises 11–13, two linear functions, y_1 and y_2, are graphed with their point of intersection indicated. Give the solution set of $y_1 = y_2$.

11.

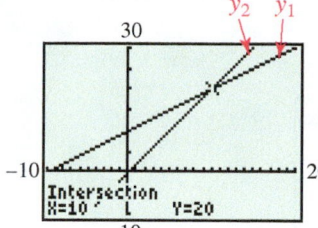

12.

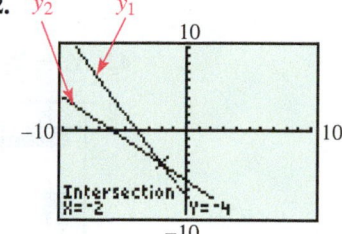

13.

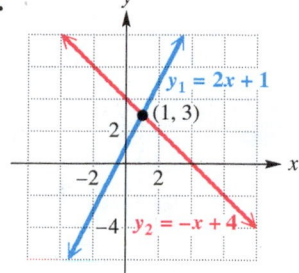

In Exercises 14–16, use the graph of $y = y_1 - y_2$ to solve the equation $y_1 = y_2$, where y_1 and y_2 represent linear functions.

14.

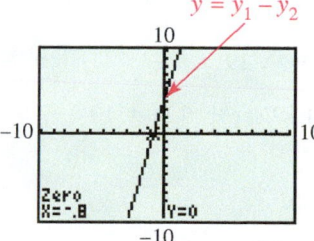

15.

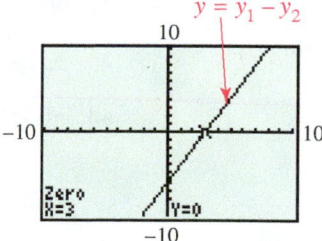

16.

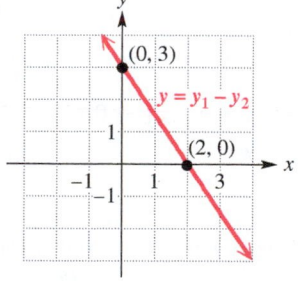

17. Interpret the y-value shown at the bottom of the screen in **Exercise 11**.

18. If you were asked to solve $2x + 3 = 4x - 12$ by the x-intercept method, why would you *not* get the correct answer by graphing $y_1 = 2x + 3 - 4x - 12$?

19. *Concept Check* If the x-intercept method leads to the graph of a horizontal line above or below the x-axis, what is the solution set of the equation? What special name is given to this kind of equation?

20. *Concept Check* If the x-intercept method leads to a horizontal line that coincides with the x-axis, what is the solution set of the equation? What special name is given to this kind of equation?

Solve each equation analytically. Check it analytically, and then support your solution graphically.

21. $2x - 5 = x + 7$ **22.** $9x - 17 = 2x + 4$

23. $0.01x + 3.1 = 2.03x - 2.96$ **24.** $0.04x + 2.1 = 0.02x + 1.92$

25. $-(x + 5) - (2 + 5x) + 8x = 3x - 5$

26. $-(8 + 3x) + 5 = 2x + 3$

27. $\dfrac{2x + 1}{3} + \dfrac{x - 1}{4} = \dfrac{13}{2}$

28. $\dfrac{x - 2}{4} + \dfrac{x + 1}{2} = 1$

29. $\dfrac{1}{2}(x - 3) = \dfrac{5}{12} + \dfrac{2}{3}(2x - 5)$

30. $\dfrac{7}{3}(2x - 1) = \dfrac{1}{5}x + \dfrac{2}{5}(4 - 3x)$

31. $0.1x - 0.05 = -0.07x$

32. $1.1x - 2.5 = 0.3(x - 2)$

33. $0.40x + 0.60(100 - x) = 0.45(100)$

34. $1.30x + 0.90(0.50 - x) = 1.00(50)$

35. $2[x - (4 + 2x) + 3] = 2x + 2$

36. $6[x - (2 - 3x) + 1] = 4x - 6$

37. $\dfrac{5}{6}x - 2x + \dfrac{1}{3} = \dfrac{1}{3}$

38. $\dfrac{3}{4} + \dfrac{1}{5}x - \dfrac{1}{2} = \dfrac{4}{5}x$

39. $5x - (8 - x) = 2[-4 - (3 + 5x - 13)]$

40. $-[x - (4x + 2)] = 2 + (2x + 7)$

Each table shows selected ordered pairs for two linear functions Y_1 and Y_2. Use the table to solve the given equation.

41. $Y_1 = Y_2$

X	Y₁	Y₂
0	4	16
1	5	14
2	6	12
3	7	10
4	8	8
5	9	6
6	10	4

X=0

42. $Y_1 - Y_2 = 0$

X	Y₁	Y₂
0	0	3
.5	1.5	3.5
1	3	4
1.5	4.5	4.5
2	6	5
2.5	7.5	5.5
3	9	6

X=0

Use the intersection-of-graphs method to approximate each solution to the nearest hundredth.

43. $4(0.23x + \sqrt{5}) = \sqrt{2}x + 1$

44. $9(-0.84x + \sqrt{17}) = \sqrt{6}x - 4$

45. $2\pi x + \sqrt[3]{4} = 0.5\pi x - \sqrt{28}$

46. $3\pi x - \sqrt[4]{3} = 0.75\pi x + \sqrt{19}$

47. $0.23(\sqrt{3} + 4x) - 0.82(\pi x + 2.3) = 5$

48. $-0.15(6 + \sqrt{2}x) + 1.4(2\pi x - 6.1) = 10$

Classify each equation as a contradiction, an identity, or a conditional equation. Give the solution set. Use a graph or table to support your answer.

49. $5x + 5 = 5(x + 3) - 3$

50. $5 - 4x = 5x - (9 + 9x)$

51. $6(2x + 1) = 4x + 8\left(x + \dfrac{3}{4}\right)$

52. $3(x + 2) - 5(x + 2) = -2x - 4$

53. $7x - 3[5x - (5 + x)] = 1 - 4x$

54. $5[1 - (3 - x)] = 3(5x + 2) - 7$

55. $0.2(5x - 4) - 0.1(6 - 3x) = 0.4$

56. $1.5(6x - 3) - 7x = 3 - (7 - x)$

57. $-4[6 - (-2 + 3x)] = 21 + 12x$

58. $-3[-5 - (-9 + 2x)] = 2(3x - 1)$

59. $\dfrac{1}{2}x - 2(x - 1) = 2 - \dfrac{3}{2}x$

60. $0.5(x - 2) + 12 = 0.5x + 11$

61. $\dfrac{x - 1}{2} = \dfrac{3x - 2}{6}$

62. $\dfrac{2x - 1}{3} = \dfrac{2x + 1}{3}$

Use the graph to the right to solve each equation or inequality.

63. $f(x) = g(x)$

64. $f(x) > g(x)$

65. $f(x) < g(x)$

66. $g(x) - f(x) \geq 0$

67. $y_1 - y_2 \geq 0$

68. $y_2 > y_1$

69. $f(x) \leq g(x)$

70. $f(x) \geq g(x)$

71. $f(x) \leq 2$

72. $g(x) \leq 2$

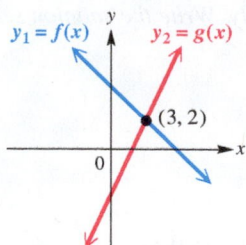

Refer to the graph of the linear function defined by $y = f(x)$ to solve each inequality. Express solution sets in interval notation.

73. **(a)** $f(x) > 0$

 (b) $f(x) < 0$

 (c) $f(x) \geq 0$

 (d) $f(x) \leq 0$

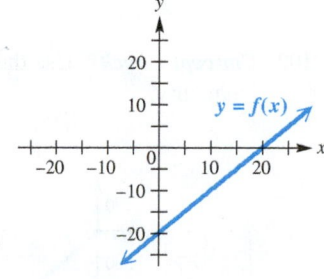

74. **(a)** $f(x) < 0$

 (b) $f(x) \leq 0$

 (c) $f(x) \geq 0$

 (d) $f(x) > 0$

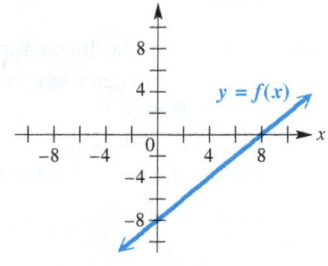

Concept Check   *In Exercises 75 and 76, f and g are linear functions.*

75. If the solution set of $f(x) \geq g(x)$ is $[4, \infty)$, what is the solution set of each equation or inequality?
 (a) $f(x) = g(x)$ **(b)** $f(x) > g(x)$ **(c)** $f(x) < g(x)$

76. If the solution set of $f(x) < g(x)$ is $(-\infty, 3)$, what is the solution set of each equation or inequality?
 (a) $f(x) = g(x)$ **(b)** $f(x) \geq g(x)$ **(c)** $f(x) \leq g(x)$

Solve each equation and inequality analytically. Use interval notation to write the solution set for each inequality.

77. **(a)** $3x - 6 = 0$
 (b) $3x - 6 > 0$
 (c) $3x - 6 < 0$

78. **(a)** $5x + 10 = 0$
 (b) $5x + 10 > 0$
 (c) $5x + 10 < 0$

79. **(a)** $1 - 2x = 0$
 (b) $1 - 2x \leq 0$
 (c) $1 - 2x \geq 0$

80. **(a)** $4 - 3x = 0$
 (b) $4 - 3x \leq 0$
 (c) $4 - 3x \geq 0$

81. **(a)** $x + 12 = 4x$
 (b) $x + 12 > 4x$
 (c) $x + 12 < 4x$

82. **(a)** $5 - 3x = x + 1$
 (b) $5 - 3x \leq x + 1$
 (c) $5 - 3x \geq x + 1$

Solve each inequality analytically, writing the solution set in interval notation. Support your answer graphically. (Hint: Once part (a) is done, the answer to part (b) follows.)

83. **(a)** $9 - (x + 1) < 0$
 (b) $9 - (x + 1) \geq 0$

84. **(a)** $6 + 3(1 - x) \geq 0$
 (b) $6 + 3(1 - x) < 0$

85. **(a)** $2x - 3 > x + 2$
 (b) $2x - 3 \leq x + 2$

86. **(a)** $5 - 3x \leq -11 + x$
 (b) $5 - 3x > -11 + x$

87. **(a)** $10x + 5 - 7x \geq 8(x + 2) + 4$
 (b) $10x + 5 - 7x < 8(x + 2) + 4$

88. **(a)** $6x + 2 + 10x > -2(2x + 4) + 10$
 (b) $6x + 2 + 10x \leq -2(2x + 4) + 10$

89. **(a)** $x + 2(-x + 4) - 3(x + 5) < -4$
 (b) $x + 2(-x + 4) - 3(x + 5) \geq -4$

90. **(a)** $-11x - (6x - 4) + 5 - 3x \leq 1$
 (b) $-11x - (6x - 4) + 5 - 3x > 1$

Solve each inequality analytically. Write the solution set in interval notation. Support your answer graphically.

91. $\frac{1}{3}x - \frac{1}{5}x \leq 2$

92. $\frac{3x}{2} + \frac{4x}{7} \geq -5$

93. $\frac{x-2}{2} - \frac{x+6}{3} > -4$

94. $\frac{2x+3}{5} - \frac{3x-1}{2} < \frac{4x+7}{2}$

95. $0.6x - 2(0.5x + 0.2) \leq 0.4 - 0.3x$

96. $-0.9x - (0.5 + 0.1x) > -0.3x - 0.5$

97. $-\frac{1}{2}x + 0.7x - 5 > 0$

98. $\frac{3}{4}x - 0.2x - 6 \leq 0$

99. $-4(3x + 2) \geq -2(6x + 1)$

100. $8(4 - 3x) \geq 6(6 - 4x)$

101. *Distance* The linear function f computes the distance y in miles between a car and the city of Omaha after x hours, where $0 \leq x \leq 6$. The graphs of f and the horizontal lines $y = 100$ and $y = 200$ are shown in the figure. Use the graphs to answer the questions that follow.

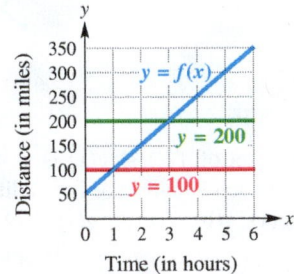

(a) Is the car moving toward or away from Omaha? Explain.
(b) At what time is the car 100 miles from Omaha? 200 miles?
(c) When is the car 100 to 200 miles (inclusive) from Omaha?
(d) When is the car's distance from Omaha greater than 100 miles?

102. *Concept Check* Use the figure to solve each equation or inequality.

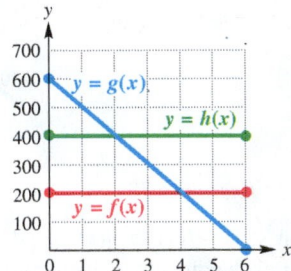

(a) $f(x) = g(x)$
(b) $g(x) = h(x)$
(c) $f(x) < g(x) < h(x)$
(d) $g(x) > h(x)$
(e) $f(x) = h(x)$
(f) $f(x) \leq h(x)$
(g) $h(x) > g(x)$
(h) $h(x) < 500$
(i) $g(x) < 200$
(j) $f(x) = 200$

Solve each three-part inequality analytically. Support your answer graphically.

103. $4 \leq 2x + 2 \leq 10$

104. $-4 \leq 2x - 1 \leq 5$

105. $-10 > 3x + 2 > -16$

106. $4 > 6x + 5 > -1$

107. $-3 \leq \frac{x-4}{-5} < 4$

108. $1 < \frac{4x-5}{-2} < 9$

109. $-\frac{1}{2} < x - 4 < \frac{1}{2}$

110. $-\frac{3}{4} < 2x - 1 < \frac{3}{4}$

111. $-4 \leq \frac{1}{2}x - 5 \leq 4$

112. $-2 < \frac{x-4}{6} < 2$

113. $\sqrt{2} \leq \frac{2x+1}{3} \leq \sqrt{5}$

114. $\pi \leq 5 - 4x < 7\pi$

(Modeling)   *Solve each problem.*

115. *Clouds and Temperature*   See **Example 9.** Suppose the ground-level temperature is 65°F and the dew point is 50°F.
  (a) Use the intersection-of-graphs method to estimate the altitudes at which clouds will not form.
  (b) Solve part (a) analytically.

116. *Amtrak Passengers*   In 2008 Amtrak had 28.7 million passengers, and in 2012 it had a record 31.2 million passengers.
  (a) Find a linear function $P(x) = ax + b$ that models the number of passengers in millions x years after 2008.
  (b) Interpret the slope of the graph of $y = P(x)$.
  (c) Use $P(x)$ to estimate the number of passengers in 2014.
  (d) Assuming trends continue, predict when Amtrak might have 35 million passengers.

117. *Error Tolerances*   Suppose that an aluminum can is manufactured so that its radius r can vary from 0.99 inches to 1.01 inches. What range of values is possible for the circumference C of the can? Express your answer by using a three-part inequality.

118. *Error Tolerances*   Suppose that a square picture frame has sides that vary between 9.9 inches and 10.1 inches. What range of values is possible for the perimeter P of the picture frame? Express your answer by using a three-part inequality.

RELATING CONCEPTS   **For individual or group discussion (Exercises 119–122)**

The solution set of a linear equation is closely related to the solution set of a linear inequality. **Work Exercises 119–122 in order** *to investigate this connection. Write answers in interval notation when appropriate.*

119. Use the x-intercept method to find the solution set of $3.7x - 11.1 = 0$. How many solutions are there? How many solutions are there to any conditional linear equation in one variable?

120. The solution from **Exercise 119** is sometimes called a **boundary number.** Find the solution sets of $3.7x - 11.1 < 0$ and $3.7x - 11.1 > 0$. Explain why the term *boundary number* is appropriate for the solution you found in **Exercise 119.**

121. Use the x-intercept method to find the solution set of the equation

$$-4x + 6 = 0.$$

Then find the solution sets of the inequalities

$$-4x + 6 < 0 \quad \text{and} \quad -4x + 6 > 0.$$

122. Generalize your results from **Exercises 119–121** by answering the questions that follow.
  (a) What is the solution set of $ax + b = 0$ if $a \neq 0$?
  (b) Suppose $a > 0$. What are the solution sets of $ax + b < 0$ and $ax + b > 0$?
  (c) Suppose $a < 0$. What are the solution sets of $ax + b < 0$ and $ax + b > 0$?

1.6   Applications of Linear Functions

Problem-Solving Strategies • Applications of Linear Equations • Break-Even Analysis • Direct Variation • Formulas

Problem-Solving Strategies

To become proficient at solving problems, we need to establish a procedure to guide our thinking. These steps may be helpful in solving application problems.

Solving Application Problems

Step 1 **Read the problem** and make sure you understand it. Assign a variable to what you are being asked to find. If necessary, write other quantities in terms of this variable.

Step 2 **Write an equation** that relates the quantities described in the problem. You may need to sketch a diagram and refer to known formulas.

Step 3 **Solve the equation** and determine the solution to the posed question.

Step 4 **Look back and check** your solution. Add units if necessary. Does it seem reasonable?

Applications of Linear Equations

EXAMPLE 1 **Determining the Dimensions of a Television Screen**

Televisions often have a $16 : 9$ **aspect ratio.** This means that the length of the rectangular screen is $\frac{16}{9}$ times its width. If the perimeter of the screen is 136 inches, find the length and the width of the screen.

Analytic Solution

Step 1 If we let x represent the width of the screen in inches, then $\frac{16}{9}x$ represents the length.

Step 2 See **FIGURE 80**.

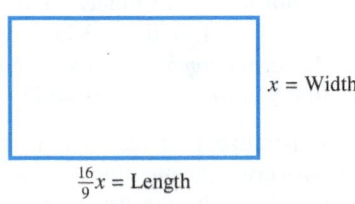

$x =$ Width

$\frac{16}{9}x =$ Length

FIGURE 80

We can use the formula for the perimeter of a rectangle to write an equation.

$$P = 2L + 2W \qquad \text{Perimeter formula}$$

$$136 = 2\left(\frac{16}{9}x\right) + 2x \qquad P = 136,\ L = \tfrac{16}{9}x,\ W = x$$

Step 3
$$136 = \frac{32}{9}x + 2x \qquad \text{Multiply.}$$

$$136 = \frac{50}{9}x \qquad \text{Combine like terms.}$$

$$x = 24.48 \qquad \text{Multiply by } \tfrac{9}{50}.\ \text{Rewrite.}$$

The width of the screen is 24.48 inches, and the length is $\frac{16}{9}(24.48) = 43.52$ inches.

Step 4 To check this answer, verify that the sum of the lengths of the four sides is 136 inches.

Graphing Calculator Solution

We graph the perimeter $P = 2L + 2W$ by letting

$$y_1 = 2\left(\frac{16}{9}x\right) + 2x$$

and

$$y_2 = 136. \quad \text{\textemdash} \quad \textit{Perimeter equals 136.}$$

As seen in **FIGURE 81**, the point of intersection of the graphs is $(24.48, 136)$. The x-coordinate supports our answer of 24.48 inches for the width.

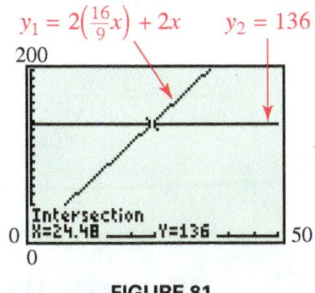

FIGURE 81

The length can be found in the same way that it was in the analytic solution.

FOR DISCUSSION

Observe **FIGURE 81,** and disregard the graph of $y_2 = 136$. What remains is the graph of a linear function that gives the perimeter y of all rectangles satisfying the 16-by-9 aspect ratio as a *function* of the rectangle's width x. Trace along the graph, back and forth, and describe what the display at the bottom represents. Why would nonpositive values of x be meaningless here? If such a television has perimeter 117 inches, what would be its approximate width?

EXAMPLE 2   **Solving a Mixture-of-Concentrations Problem**

How much pure alcohol should be added to 20 liters of 40% alcohol to increase the concentration to 50% alcohol?

Solution

Step 1   We let x represent the number of liters of pure alcohol that must be added. The information can be summarized in a "box diagram." See **FIGURE 82.**

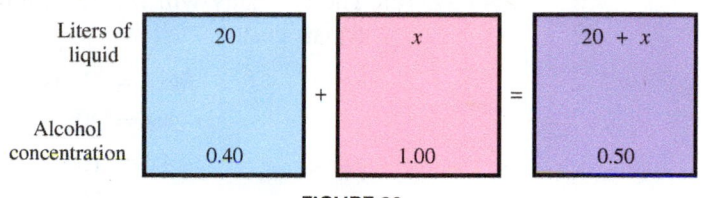

| Liters of liquid | 20 | x | 20 + x |
| Alcohol concentration | 0.40 | 1.00 | 0.50 |

+ between first two boxes, = between last two

FIGURE 82

Step 2   In each box, we have the number of liters and the alcohol concentration. We multiply the two values in each box to get the amount of pure alcohol. The amount of pure alcohol on the left side of the equation must equal the amount of pure alcohol on the right side, so the equation to solve is as follows.

$$0.40(20) \quad + \quad 1.00x \quad = \quad 0.50(20 + x)$$

Liters of pure alcohol in starting mixture   Liters of pure alcohol added   Liters of pure alcohol in final mixture

Step 3

Move the decimal points two places to the right.

$$40(20) + 100x = 50(20 + x) \qquad \text{Multiply by 100.}$$
$$800 + 100x = 1000 + 50x \qquad \text{Distributive property}$$
$$50x = 200 \qquad \text{Subtract } 50x \text{ and subtract 800.}$$
$$x = 4 \qquad \text{Divide by 50.}$$

Therefore, 4 liters of pure alcohol must be added.

Step 4   *Check* to see that $0.40(20) + 1.00(4) = 0.50(20 + 4)$ is true. ●

Break-Even Analysis

By expressing a company's cost of producing a product and the revenue the company receives from selling the product as functions, the company can determine at what point it will break even. In other words, we try to answer the question, "For what number of items sold will the revenue collected equal the cost of producing those items?"

EXAMPLE 3 **Determining the Break-Even Point**

Peripheral Visions, Inc., produces high-definition DVDs of live concerts. The company places an ad in a trade newsletter. The cost of the ad is $100. Each DVD costs $20 to produce, and the company charges $24 per DVD.

(a) Express the cost C as a function of x, the number of DVDs produced and sold.

(b) Express the revenue R as a function of x, the number of DVDs produced and sold.

(c) For what value of x does revenue equal cost?

(d) Graph $y_1 = C(x)$ and $y_2 = R(x)$ to support the answer in part (c).

(e) Use a table to support the answer in part (c).

Solution

(a) The *fixed cost* is $100, and for each DVD produced, the *variable cost* is $20. The total cost C in dollars can be expressed as a function of x, the number of DVDs produced.

$$C(x) = 20x + 100$$

(b) Since each DVD sells for $24, the revenue R in dollars is given by $R(x) = 24x$.

(c) The company will break even (earn no profit and suffer no loss) when revenue equals cost.

$R(x) = C(x)$	Revenue equals cost.
$24x = 20x + 100$	Substitute for $R(x)$ and $C(x)$.
$4x = 100$	Subtract 20.
$x = 25$	Divide by 4.

When 25 DVDs are sold, the company will break even.

(d) **FIGURE 83** confirms the solution of 25. The y-value of 600 indicates that when 25 DVDs are sold, both the cost and the revenue are $600.

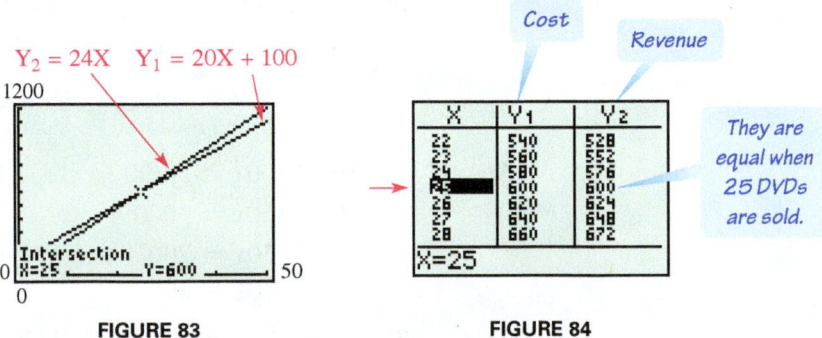

FIGURE 83   FIGURE 84

(e) The table in **FIGURE 84** shows that when the number of DVDs is 25, both function values are 600, numerically supporting our answer in part (c).

Direct Variation

A common application involving linear functions deals with quantities that *vary directly* (or are in direct proportion).

Direct Variation

A number y **varies directly** with x if there exists a nonzero number k such that

$$y = kx.$$

The number k is called the **constant of variation.**

NOTE   With direct variation, if x doubles then y doubles, if x triples then y triples, and so on.

Notice that the graph of $y = kx$ is simply a straight line with slope k, passing through the origin. See **FIGURE 85.** If we divide both sides of $y = kx$ by x, we get $\frac{y}{x} = k$, indicating that in a direct variation, the quotient (or proportion) of the two quantities is constant.

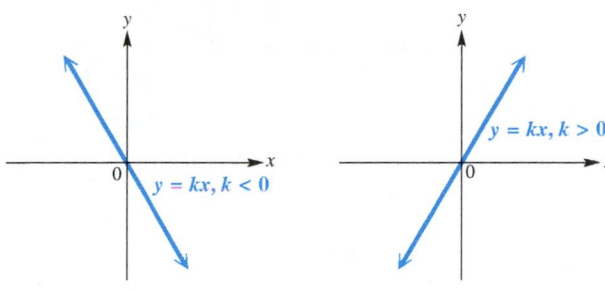

FIGURE 85

Hooke's law states that the distance y an elastic spring stretches beyond its natural length varies directly with the amount of weight x hung on the spring, as illustrated in **FIGURE 86.** This law is valid whether the spring is stretched or compressed. Thus, if a weight or force x is applied to a spring and it stretches a distance y, the equation $y = kx$ models the situation, where k is the constant of variation.

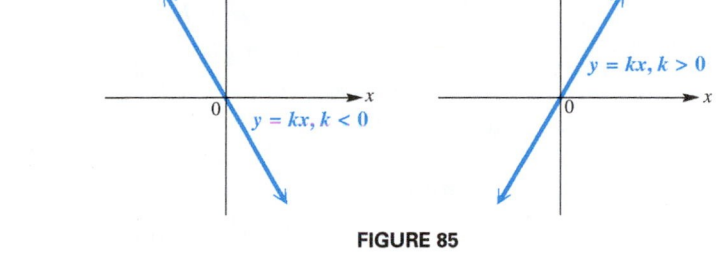

FIGURE 86

EXAMPLE 4   **Solving a Direct Variation Problem (Hooke's Law)**

If a force of 15 pounds stretches a spring 8 inches, how much will a force of 35 pounds stretch the spring? See **FIGURE 86.**

Solution   Using the fact that when $x = 15$, $y = 8$, we can find k.

$$y = kx \qquad \text{Direct variation}$$

$$8 = k \cdot 15 \qquad \text{Let } x = 15, y = 8.$$

$$k = \frac{8}{15} \qquad \text{Divide by 15 and rewrite.}$$

Therefore, the linear equation $y = \frac{8}{15}x$ describes the relationship between the force x and the distance stretched y. To answer the question stated in the problem, we let $x = 35$. The spring will stretch

$$y = \frac{8}{15}(35) = 18.\overline{6}, \quad \text{or} \quad \text{approximately 19 inches.}$$

Formulas

The formula $\mathcal{A} = \frac{1}{2}h(b_1 + b_2)$ gives the area $\mathcal{A}$ of a trapezoid in terms of its height h and its two parallel bases b_1 and b_2. See **FIGURE 87**. Suppose that we want to write the formula so that it is solved for b_1. We can use the methods of solving linear equations analytically to do this.

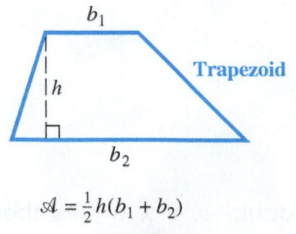

$\mathcal{A} = \frac{1}{2}h(b_1 + b_2)$

FIGURE 87

EXAMPLE 5 **Solving a Formula for a Specified Variable**

A trapezoid has area 169 square inches, height 13 inches, and base 19 inches. Find the length of the other base by first solving the formula $\mathcal{A} = \frac{1}{2}h(b_1 + b_2)$ for b_1.

Solution We treat the formula as if b_1 were the only variable and all other variables were constants.

$$\mathcal{A} = \frac{1}{2}h(b_1 + b_2) \qquad \text{Area formula}$$

Multiply both sides by 2.
$$2\mathcal{A} = h(b_1 + b_2) \qquad \text{Multiply by 2.}$$

$$\frac{2\mathcal{A}}{h} = b_1 + b_2 \qquad \text{Divide by } h.$$

$$\frac{2\mathcal{A}}{h} - b_2 = b_1 \qquad \text{Subtract } b_2.$$

$$b_1 = \frac{2\mathcal{A}}{h} - b_2 \qquad \text{Rewrite.}$$

To determine b_1, we substitute $\mathcal{A} = 169$, $h = 13$, and $b_2 = 19$.

$$b_1 = \frac{2(169)}{13} - 19 = 7 \text{ inches}$$

EXAMPLE 6 **Solving Formulas for Specified Variables**

Solve each formula for the specified variable.
(a) $P = 2W + 2L$ for L (Perimeter of a rectangle)

(b) $C = \frac{5}{9}(F - 32)$ for F (Fahrenheit to Celsius)

Solution
(a)
$$P = 2W + 2L \qquad \text{Solve for } L.$$
$$P - 2W = 2L \qquad \text{Subtract } 2W.$$
$$\frac{P - 2W}{2} = L \qquad \text{Divide by 2.}$$
$$L = \frac{P}{2} - W \qquad \text{Rewrite and simplify.}$$

(b)
$$C = \frac{5}{9}(F - 32) \qquad \text{Solve for } F.$$
$$\frac{9}{5}C = F - 32 \qquad \text{Multiply by } \frac{9}{5}.$$
$$F = \frac{9}{5}C + 32 \qquad \text{Add 32 and rewrite.}$$

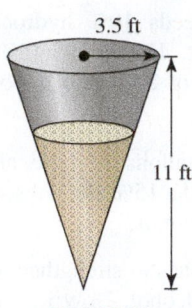

3.5 ft

11 ft

FIGURE 88

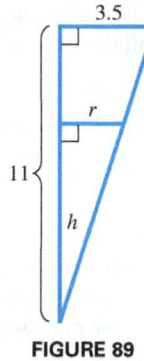

3.5

11

r

h

FIGURE 89

EXAMPLE 7   **Using Similar Triangles**

A grain bin in the shape of an inverted cone has height 11 feet and radius 3.5 feet. See **FIGURE 88.** If the grain in the bin is 7 feet high, calculate the volume of grain.

Solution   The formula for the volume V of a cone is $V = \frac{1}{3}\pi r^2 h$. The height h of the grain in the bin is 7 feet, but we are not given the radius r of the bin at that height. We can find the radius r corresponding to a height of 7 feet by using similar triangles, as illustrated in **FIGURE 89.** The sides of similar triangles are proportional.

$$\frac{r}{h} = \frac{3.5}{11} \qquad \text{Set up proportion.}$$

$$r = \frac{3.5}{11}h \qquad \text{Multiply by } h.$$

Letting $h = 7$ gives $r = \frac{3.5}{11}(7) \approx 2.227$.

The volume of the grain is

$$V = \frac{1}{3}\pi(2.227)^2(7) \qquad V = \frac{1}{3}\pi r^2 h$$

$$\approx 36.4 \text{ cubic feet.}$$

1.6 Exercises

Checking Analytic Skills   Work **Exercises 1–6** mentally. **Do not use a calculator.**

1. *Acid Solution*   If 40 L of an acid solution is 75% acid, how much pure acid is there in the mixture?

2. *Direct Variation*   If y varies directly with x, and $y = 2$ when $x = 4$, what is the value of y when $x = 12$?

3. *Acid Mixture*   Suppose two acid solutions are mixed. One is 26% acid and the other is 32% acid. Which one of the following concentrations cannot possibly be the concentration of the mixture?
A. 36%   **B.** 28%   **C.** 30%   **D.** 31%

4. *Sale Price*   Suppose that a computer that originally sold for x dollars has been discounted 30%. Which one of the following expressions does not represent the sale price of the computer?
A. $x - 0.30x$ **B.** $0.70x$ **C.** $\frac{7}{10}x$ **D.** $x - 0.30$

5. *Unknown Numbers*   Consider the following problem.

One number is three less than six times a second number. Their sum is 32. Find the numbers.

If x represents the second number, which equation is correct for solving this problem?
A. $32 - (x + 3) = 6x$ **B.** $(3 - 6x) + x = 32$
C. $32 - (3 - 6x) = x$ **D.** $(6x - 3) + x = 32$

6. *Unknown Numbers*   Consider the following problem:

The difference between six times a number and 9 is equal to five times the sum of the number and 2. Find the number.

If x represents the number, which equation is correct for solving this problem?
A. $6x - 9 = 5(x + 2)$ **B.** $9 - 6x = 5(x + 2)$
C. $6x - 9 = 5x + 2$ **D.** $9 - 6x = 5x + 2$

Solve each problem analytically, and support your solution graphically.

7. *Perimeter of a Rectangle*   The perimeter of a rectangle is 98 centimeters. The width is 19 centimeters. Find the length.

8. *Perimeter of a Storage Shed*   A carpenter must build a rectangular storage shed. She wants the length to be 3 feet greater than the width, and the perimeter must be 22 feet. Find the length and the width of the shed.

9. *Dimensions of a Label*   The length of a rectangular label is 2.5 centimeters less than twice the width. The perimeter is 40.6 centimeters. Find the width.

10. *Dimensions of a Label*   The length of a rectangular mailing label is 3 centimeters less than twice the width. The perimeter is 54 centimeters. Find its dimensions.

11. *Dimensions of a Square*   If the length of a side of a square is increased by 3 centimeters, the perimeter of the new square is 40 centimeters more than twice the length of the side of the original square. Find the length of the side of the original square.

12. *World's Largest Easel*   The painting on the world's largest easel has a perimeter of 112 feet. Its length is 96 inches more than its width. What is its length in feet? (*Source:* Roadside America.)

13. *Aspect Ratio of a Television Monitor*   The aspect ratio of older television monitors is 4 : 3. One such television has a rectangular viewing screen with perimeter 98 inches. What are the length and width of the screen? Since televisions are advertised by the diagonal measure of their screens, how would this monitor be advertised?

14. *Aspect Ratio of a Television Monitor*   Repeat **Exercise 13** for a television screen having perimeter 126 inches.

15. *Dimensions of a Puzzle Piece*   A puzzle piece in the shape of a triangle has perimeter 30 centimeters. Two sides of the triangle are each twice as long as the shortest side. Find the length of the shortest side.

16. *Perimeter of a Plot of Land*   The perimeter of a triangular plot of land is 2400 feet. The longest side is 200 feet less than twice the shortest. The middle side is 200 feet less than the longest. Find the lengths of the three sides of the triangular plot.

17. *Motion*   A car went 372 miles in 6 hours, traveling part of the time at 55 miles per hour and part of the time at 70 miles per hour. How long did the car travel at each speed?

18. *Running*   At 2:00 P.M. a runner heads north on a highway, jogging at 10 miles per hour. At 2:30 P.M. a driver heads north on the same highway to pick up the runner. If the car travels at 55 miles per hour, how long will it take the driver to catch the runner?

19. *Acid Mixture*   How many gallons of a 5% acid solution must be mixed with 5 gallons of a 10% solution to obtain a 7% solution?

20. *Acid Mixture*   Bill Charlson needs 10% hydrochloric acid for a chemistry experiment. How much 5% acid should he mix with 60 milliliters of 20% acid to obtain a 10% solution?

21. *Alcohol Mixture*   How many gallons of pure alcohol should be mixed with 20 gallons of a 15% alcohol solution to obtain a mixture that is 25% alcohol?

22. *Alcohol Mixture*   A chemist wishes to strengthen a mixture from 10% alcohol to 30% alcohol. How much pure alcohol should be added to 7 liters of the 10% mixture?

23. *Saline Solution Mixture*   How much water should be added to 8 milliliters of 6% saline solution to reduce the concentration to 4% saline?

24. *Acid Mixture*   How much water should be added to 20 liters of an 18% acid solution to reduce the concentration to 15% acid?

25. *Antifreeze Mixture*   An automobile radiator holds 16 liters of fluid. There is currently a mixture in the radiator that is 80% antifreeze and 20% water. How much of this mixture should be drained and replaced by pure antifreeze so that the resulting mixture is 90% antifreeze?

26. *Antifreeze Mixture*   An automobile radiator contains a 10-quart mixture of water and antifreeze that is 40% antifreeze. How much should the owner drain from the radiator and replace with pure antifreeze so that the liquid in the radiator will be 80% antifreeze?

Exercises 27 and 28 involve octane rating of gasoline, a measure of its antiknock qualities. In one measure of octane, a standard fuel is made with only two ingredients: heptane and isooctane. For this type of fuel, the octane rating is the percent of isooctane. An actual gasoline blend is then compared with a standard fuel. For example, a gasoline with an octane rating of 98 has the same antiknock properties as a standard fuel that is 98% isooctane.

27. *Octane Rating of Gasoline*   How many gallons of 94-octane gasoline should be mixed with 400 gallons of 99-octane gasoline to obtain a mixture that is 97-octane?

28. *Octane Rating of Gasoline*   How many gallons of 92-octane and 98-octane gasoline should be mixed together to provide 120 gallons of 96-octane gasoline?

(Modeling)   Solve each problem.

29. *Women against the Men*   For the men's Olympic 100-meter freestyle swimming event, winning times in seconds during year x can be approximated by the formula $F(x) = -\frac{5}{44}x + 276.18$, where $1948 \leq x \leq 2008$. (Assume that x is a multiple of 4 because the Olympics occur every 4 years.)
    (a) Evaluate $F(2008)$ and interpret the result.

(b) In 2008 the women's Olympic winning time for the 100-meter freestyle was about 53 seconds. Determine the years when this time would have beaten or tied the men's winning times.

30. *Women against the Men*   The men's Olympic pole vaulting winning heights in meters during year x can be approximated by $H(x) = \frac{1}{48}x - 35.83$, where $1896 \leq x \leq 2008$. (Assume that x is a multiple of 4 because the Olympics occur every 4 years.)

   (a) Evaluate $H(1920)$ and interpret the result.

   (b) In 2008 the women's Olympic winning height in the pole vault was about 5 meters. Determine the years when this height would have beaten or tied the men's winning heights.

31. *Online Holiday Shopping*   In 2011, online holiday sales were $192 billion, and in 2014, they were $249 billion. (*Source:* Digital Lifestyles.)

   (a) Find a linear function S that models these data, where x is the year.

   (b) Interpret the slope of the graph of S.

   (c) Predict when online holiday sales might reach $325 billion.

32. *Bicycle Safety*   A survey found that 76% of bicycle riders do not wear helmets. (*Source:* Opinion Research Corporation for Glaxo Wellcome, Inc.)

   (a) Write a linear function f that computes the number of people who do not wear helmets among x bicycle riders.

   (b) There are approximately 38.7 million riders of all ages who do not wear helmets. Write a linear equation whose solution gives the total number of bicycle riders. Find this number of riders.

(Modeling)   *In Exercises 33–36,*

   (a) *Express the cost C as a function of x, where x represents the number of items as described.*

   (b) *Express the revenue R as a function of x.*

   (c) *Determine analytically the value of x for which revenue equals cost.*

   (d) *Graph $y_1 = C(x)$ and $y_2 = R(x)$ on the same xy-axes and interpret the graphs.*

33. *Stuffing Envelopes*   A student stuffs envelopes for extra income during her spare time. Her initial cost to obtain the necessary information for the job was $200.00. Each envelope costs $0.02, and she gets paid $0.04 per envelope stuffed. Let x represent the number of envelopes stuffed.

34. *Copier Service*   A technician runs a copying service in his home. He paid $3500 for the copier and a lifetime service contract. Each sheet of paper costs $0.01, and he gets paid $0.05 per copy. Let x be the number of copies he makes.

35. *Delivery Service*   A truck driver operates a delivery service. His start-up costs amounted to $2300. He estimates that it costs him (in terms of gasoline, wear and tear on his truck, etc.) $3.00 per delivery. He charges $5.50 per delivery. Let x represent the number of deliveries he makes.

36. *Baking and Selling Cakes*   A baker makes cakes and sells them at county fairs. Her initial cost for the Pointe Coupee parish fair was $40.00. She figures that each cake costs $2.50 to make, and she charges $6.50 per cake. Let x represent the number of cakes sold. (Assume that there were no cakes left over.)

In Exercises 37–40, find the constant of variation k and the undetermined value in the table if y is directly proportional to x.

37.

x	3	5	6	8
y	7.5	12.5	15	?

38.

x	1.2	4.3	5.7	?
y	3.96	14.19	18.81	23.43

39. Sales tax y on a purchase of x dollars

x	$25	$55	?
y	$1.50	$3.30	$5.10

40. Cost y of buying x compact discs having the same price

x	3	4	5
y	$41.97	$55.96	?

Solve each problem.

41. *Pressure of a Liquid*   The pressure exerted by a certain liquid at a given point is directly proportional to the depth of the point beneath the surface of the liquid. If the pressure exerted at 30 feet is 13 pounds per square inch, what is the pressure exerted at 70 feet?

42. *Rate of Nerve Impulses*   The rate at which impulses are transmitted along a nerve fiber is directly proportional to the diameter of the fiber. The rate for a certain fiber is 40 meters per second when the diameter is 6 micrometers. Find the rate if the diameter is 8 micrometers.

43. *Cost of Tuition*   The cost of tuition is directly proportional to the number of credits taken. If 11 credits cost $720.50, find the cost of taking 16 credits. What is the constant of variation?

44. *Strength of a Beam* The maximum load that a horizontal beam can carry is directly proportional to its width. If a beam 1.5 inches wide can support a load of 250 pounds, find the load that a beam of the same type can support if its width is 3.5 inches. What is the constant of variation?

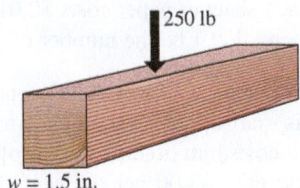

250 lb

$w = 1.5$ in.

45. *Volume of Water* A water tank in the shape of an inverted cone has height 11 feet and radius 5.0 feet, as illustrated in the accompanying figure. Find the volume of the water in the tank when the water is 6 feet deep.

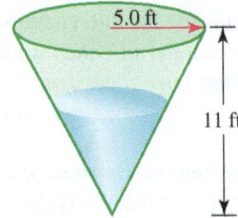

5.0 ft

11 ft

46. *Volume of Water* A water tank in the shape of an inverted cone has height 6 feet and radius 2 feet. If the water level in the tank is 3.5 feet, calculate the volume of the water.

47. *Height of a Tree* A certain tree casts a shadow 45 feet long. At the same time, the shadow cast by a vertical stick 2 feet high is 1.75 feet long. How tall is the tree? (*Hint:* Use similar triangles.)

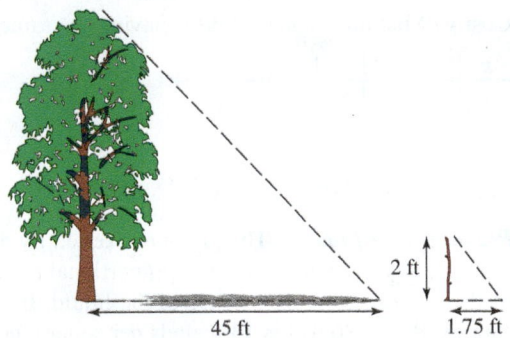

2 ft

45 ft

1.75 ft

48. *Height of a Streetlight* A person 66 inches tall is standing 15 feet from a streetlight. If the person casts a shadow 84 inches long, how tall is the streetlight?

49. *Hooke's Law* If a 3-pound weight stretches a spring 2.5 inches, how far will a 17-pound weight stretch the spring?

50. *Hooke's Law* If a 9.8-pound weight stretches a spring 0.75 inch, how much weight would be needed to stretch the spring 3.1 inches?

Biologists use direct variation to estimate the number of individuals of a species in a particular area. They first capture a sample of individuals from the area and mark each specimen with a harmless tag. Later, they return and capture another sample from the same area. They base their estimate on the theory that the proportion of tagged specimens in the new sample is the same as the proportion of tagged individuals in the entire area. Use this idea to work Exercises 51 and 52.

51. *Estimating Fish in a Lake* Biologists tagged and released 250 trout. On a later date, they found 7 tagged trout in a sample of 350. Estimate, to the nearest hundred, the total number of trout in the lake.

52. *Estimating Seal Pups in a Breeding Area* According to an actual survey in 1961, to estimate the number of seal pups in a certain breeding area in Alaska, 4963 pups were tagged in early August. In late August, a sample of 900 pups was examined and 218 of these were found to have been tagged. Use this information to estimate, to the nearest hundred, the total number of seal pups in this breeding area. (*Source:* "Estimating the Size of Wildlife Populations," Chatterjee, S., in *Statistics by Example,* obtained from data in *Transactions of the American Fisheries Society.*)

(Modeling) *In Exercises 53–56, assume that a linear relationship exists between the two quantities.*

53. *Solar Heater Production* A company produces 10 solar heaters for $7500. The cost to produce 20 heaters is $13,900.
   (a) Express the cost y as a linear function of the number of heaters, x.
   (b) Determine analytically the cost to produce 25 heaters.
   (c) Support the result of part (b) graphically.

54. *Cricket Chirping* At 68°F, a certain species of cricket chirps 112 times per minute. At 46°F, the same cricket chirps 24 times per minute.
   (a) Express the number of chirps, y, as a linear function of the Fahrenheit temperature.
   (b) If the temperature is 60°F, how many times will the cricket chirp per minute?
   (c) If you count the number of cricket chirps in one-half minute and hear 40 chirps, what is the temperature?

55. *Appraised Value of a Home*   In 2002, a house was purchased for $120,000. In 2012, it was appraised for $146,000.
 (a) If $x = 0$ represents 2002 and $x = 10$ represents 2012, express the appraised value of the house, y, as a linear function of the number of years, x, after 2002.
 (b) What was the house worth in the year 2009?
 (c) What does the slope of the line represent?

56. *Depreciation of a Photocopier*   A photocopier sold for $3000 in 2006. Its value in 2014 had depreciated to $600.
 (a) If $x = 0$ represents 2006 and $x = 8$ represents 2014, express the value of the machine, y, as a linear function of the number of years, x, after 2006.
 (b) Graph the function from part (a) in a window $[0, 10]$ by $[0, 4000]$. How would you interpret the y-intercept in terms of this particular situation?
 (c) Use your calculator to determine the value of the machine in 2010, and verify your result analytically.

57. *Climate Change*   If the global climate were to warm significantly, the Arctic ice cap would start to melt. This ice cap contains an estimated 680,000 cubic miles of water. More than 200 million people currently live on land that is less than 3 feet above sea level. In the United States several large cities have low average elevations. Two examples are Boston (14 feet) and San Diego (13 feet). In this exercise you are to estimate the rise in sea level if this cap were to melt and determine whether this event would have a significant impact on people.
 (a) The surface area of a sphere is given by the expression $4\pi r^2$, where r is its radius. Although the shape of the earth is not exactly spherical, it has an average radius of 3960 miles. Estimate the surface area of the earth.
 (b) Oceans cover approximately 71% of the total surface area of the earth. How many square miles of the earth's surface are covered by oceans?
 (c) Approximate the potential rise in sea level by dividing the total volume of the water from the ice cap by the surface area of the oceans. Convert your answer from miles to feet.
 (d) Discuss the implications of your calculation. How would cities such as Boston and San Diego be affected?
 (e) The Antarctic ice cap contains 6,300,000 cubic miles of water. Estimate how much sea level would rise if this ice cap melted. (*Source:* Department of the Interior, Geological Survey.)

58. *Speeding Fines*   Suppose that speeding fines are determined by $y = 10(x - 65) + 50$, $x > 65$, where y is the cost in dollars of the fine if a person is caught driving x miles per hour.
 (a) How much is the fine for driving 76 mph?
 (b) While balancing the checkbook, Johnny found a check that his wife Gwen had written to the Department of Motor Vehicles for a speeding fine. The check was written for $100. How fast was Gwen driving?

 (c) At what whole-number speed are tickets first given?
 (d) For what speeds is the fine greater than $200?

59. *Expansion and Contraction of Gases*   In 1787, Jacques Charles noticed that gases expand when heated and contract when cooled. A particular gas follows the model

$$y = \frac{5}{3}x + 455,$$

where x is the temperature in Celsius and y is the volume in cubic centimeters.
 (a) What is the volume when the temperature is 27°C?
 (b) What is the temperature when the volume is 605 cubic centimeters?
 (c) Determine what temperature gives a volume of 0 cubic centimeters.

60. *Sales of CRT and LCD Screens*   In the early 21st century, LCD monitors were a new technology that replaced CRT (cathode ray tube) monitors. In 2002, 75 million CRT monitors were sold and only 29 million flat LCD (liquid crystal display) monitors were sold. By 2006, the numbers were 45 million for CRT monitors and 88 million for LCD monitors. (*Source:* International Data Corporation.)
 (a) Find a linear function C that models these data for CRT monitors and another linear function L that models these data for LCD monitors. Let x be the year.
 (b) Determine the year when sales of these two types of monitors were equal.

Solve each formula for the specified variable.

61. $I = PRT$ for P (Simple interest)

62. $V = LWH$ for L (Volume of a box)

63. $P = 2L + 2W$ for W (Perimeter of a rectangle)

64. $P = a + b + c$ for c (Perimeter of a triangle)

65. $\mathcal{A} = \frac{1}{2}h(b_1 + b_2)$ for h (Area of a trapezoid)

66. $\mathcal{A} = \frac{1}{2}h(b_1 + b_2)$ for b_2 (Area of a trapezoid)

67. $S = 2LW + 2WH + 2HL$ for H (Surface area of a rectangular box)

68. $S = 2\pi rh + 2\pi r^2$ for h (Surface area of a cylinder)

69. $V = \frac{1}{3}\pi r^2 h$ for h (Volume of a cone)

70. $y = a(x - h)^2 + k$ for a (Mathematics)

71. $S = \frac{n}{2}(a_1 + a_n)$ for n (Mathematics)

72. $S = \frac{n}{2}[2a_1 + (n - 1)d]$ for a_1 (Mathematics)

73. $s = \frac{1}{2}gt^2$ for g (Distance traveled by a falling object)

74. $A = \frac{24f}{B(p + 1)}$ for p (Approximate annual interest rate)

Investment problems such as those in Exercises 75–80 can be solved by using a method similar to the one explained in **Example 2,** *along with the simple-interest formula* $I = PRT$, *where I is the interest earned, P is the initial amount of money deposited, R is the annual interest rate as a decimal, and T is the time the money is deposited in years. Solve each problem. Let* $T = 1$ *year for each exercise.*

75. *Real-Estate Financing* Cody Westmoreland wishes to sell a piece of property for $240,000. He wants the money to be paid off in two ways: a short-term note at 6% interest and a long-term note at 5%. Find the amount of each note if the total annual interest paid is $13,000.

76. *Buying and Selling Land* Bobby Aguillard bought two plots of land for a total of $120,000. When he sold the first plot, he made a profit of 15%. When he sold the second, he lost 10%. His total profit was $5500. How much did he pay for each piece of land?

77. *Retirement Planning* In planning her retirement, Mary Lynn Ellis deposits some money at 2.5% interest with twice as much deposited at 3%. Find the amount deposited at each rate if the total annual interest income is $850.

78. *Investing a Building Fund* A church building fund has invested some money in two ways: part of the money at 4% interest and four times as much at 3.5%. Find the amount invested at each rate if the total annual income from interest is $3600.

79. *Lottery Winnings* Nancy B. Kindy won $200,000 in a state lottery. She first paid income tax of 30% on the winnings. Of the rest, she invested some at 1.5% and some at 4%, earning $4350 interest per year. How much did she invest at each rate?

80. *Cookbook Royalties* Latasha Williams earned $48,000 from royalties on her cookbook. She paid a 28% income tax on these royalties. The balance was invested in two ways, some of it at 3.25% interest and some at 1.75%. The investments produced $904.80 interest the first year. Find the amount invested at each rate.

SECTIONS 1.5–1.6 **Reviewing Basic Concepts**

1. Solve $3(x - 5) + 2 = 1 - (4 + 2x)$ analytically. Support your result graphically or numerically.

2. Solve $\pi(1 - x) = 0.6(3x - 1)$ with the intersection-of-graphs method. Round your answer to the nearest thousandth.

3. Find the zero of $f(x) = \frac{1}{3}(4x - 2) + 1$ analytically. Use the x-intercept method to support your analytic result.

4. Determine whether each equation is an identity, a contradiction, or a conditional equation. Give the solution set.
(a) $4x - 5 = -2(3 - 2x) + 3$
(b) $5x - 9 = 5(-2 + x) + 1$
(c) $5x - 4 = 3(6 - x)$

5. Solve $2x + 3(x + 2) < 1 - 2x$ analytically. Express the solution set in interval notation. Support your result graphically.

6. Solve $-5 \leq 1 - 2x < 6$ analytically.

7. Use the figure at the top of the next column to solve the following.
(a) $f(x) = g(x)$ (b) $f(x) \leq g(x)$

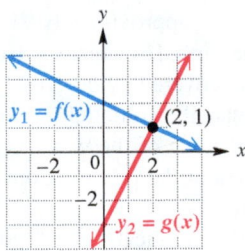

8. *Height of a Tree* A tree casts a 27-foot shadow, and a 6-foot person casts a 4-foot shadow. How tall is the tree?

9. *(Modeling) Production of DVDs* A student is planning to produce and sell music DVDs for $5.50 each. A computer with a DVD burner costs $800, and each blank DVD costs $1.50.
(a) Find a formula for a function R that calculates the revenue from selling x discs.
(b) Find a formula for a function C that calculates the cost of recording x discs. Be sure to include the fixed cost of the computer.
(c) Find the break-even point, where $R(x) = C(x)$.

10. Solve $V = \pi r^2 h$ for h.

1 Summary

1.1 Real Numbers and the Rectangular Coordinate System

natural numbers
whole numbers
integers
number line
origin
graph
coordinate system
coordinate
rational numbers
real numbers
irrational numbers
rectangular (Cartesian) coordinate system
x-axis
y-axis
coordinate (xy-) plane
quadrants
x-coordinate
y-coordinate
standard viewing window
Xmin
Xmax
Ymin
Ymax
Xscl
Yscl

Natural numbers $\{1, 2, 3, 4, 5, \dots\}$
Whole numbers $\{0, 1, 2, 3, 4, 5, \dots\}$
Integers $\{\dots, -3, -2, -1, 0, 1, 2, 3, \dots\}$
Rational numbers $\left\{\frac{p}{q} \mid p, q \text{ are integers}, q \neq 0\right\}$

Real numbers $\{x \mid x \text{ is a decimal number}\}$
(The real numbers include both the rational and the irrational numbers.)

RECTANGULAR COORDINATE SYSTEM
Points are located by using ordered pairs in the xy-plane.

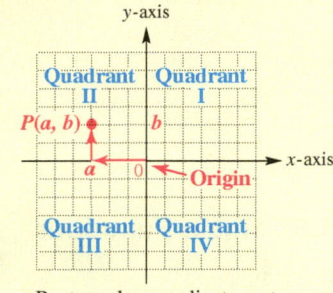

Rectangular coordinate system

ROOTS
The nth root of a real number a can be expressed as $\sqrt[n]{a}$ or $a^{1/n}$. Calculators are often used to approximate roots of real numbers.

PYTHAGOREAN THEOREM
In a right triangle, the sum of the squares of the lengths of the legs, a and b, is equal to the square of the length of the hypotenuse c. That is, $a^2 + b^2 = c^2$.

DISTANCE FORMULA
Suppose that $P(x_1, y_1)$ and $R(x_2, y_2)$ are two points in a coordinate plane. Then the distance d between P and R is given by the distance formula.
$$d(P, R) = \sqrt{(x_2 - x_1)^2 + (y_2 - y_1)^2}$$

MIDPOINT FORMULA
The midpoint M of the line segment with endpoints (x_1, y_1) and (x_2, y_2) has coordinates given by the following.
$$M = \left(\frac{x_1 + x_2}{2}, \frac{y_1 + y_2}{2}\right)$$

1.2 Introduction to Relations and Functions

set-builder notation
interval
interval notation
infinity symbol, ∞
relation
domain
range

INTERVAL NOTATION
Interval notation can be used to denote sets of real numbers.

$\{x \mid x \leq a\}$ is equivalent to $(-\infty, a]$.
$\{x \mid a \leq x < b\}$ is equivalent to $[a, b)$.
$\{x \mid x > b\}$ is equivalent to (b, ∞).

RELATION
A relation is a set of ordered pairs.

(continued)

KEY TERMS & SYMBOLS	KEY CONCEPTS

KEY TERMS & SYMBOLS

mapping
diagram
function
independent variable
dependent variable
vertical line test
function notation, $f(x)$
name of a function

KEY CONCEPTS

DOMAIN AND RANGE

If we denote the ordered pairs of a relation by (x, y), then the set of all x-values is called the domain of the relation and the set of all y-values is called the range of the relation.

FUNCTION

A function is a relation in which each element in the domain corresponds to exactly one element in the range.

VERTICAL LINE TEST

If every vertical line intersects a graph in no more than one point, then the graph is the graph of a function.

FUNCTION NOTATION

To denote that y is a function of x, we write $y = f(x)$, where f is the name of the function. The independent variable is x and the dependent variable is y.

1.3 Linear Functions

linear function
linear equation in two
  variables
solution
table of values
x-intercept
y-intercept
zero of a function
constant function
comprehensive graph
change in x, Δx
change in y, Δy
slope, m
average rate of change
slope–intercept form

LINEAR FUNCTION

A function f defined by

$$f(x) = ax + b,$$

where a and b are real numbers, is called a linear function. The graph of a linear function is a line.

INTERCEPTS

If the graph of a function f intersects the x-axis at $(a, 0)$, then $(a, 0)$ is an x-intercept.
If the graph of a function f intersects the y-axis at $(0, b)$, then $(0, b)$ is the y-intercept.

ZERO OF A FUNCTION f

Any number c for which $f(c) = 0$ is called a zero of f. The real zeros of a function f correspond to the x-coordinates of the x-intercepts of its graph.

CONSTANT FUNCTION

A function $f(x) = b$, where b is a real number, is called a constant function. A constant function is a special type of linear function whose graph is a horizontal line.

SLOPE

The slope m of the line passing through the points (x_1, y_1) and (x_2, y_2) is

$$m = \frac{\Delta y}{\Delta x} = \frac{y_2 - y_1}{x_2 - x_1}, \quad \Delta x \neq 0.$$

GEOMETRIC ORIENTATION BASED ON SLOPE

For a line with slope m, if $m > 0$ (i.e., slope is positive), the line rises from left to right. If $m < 0$ (i.e., slope is negative), it falls from left to right. If $m = 0$ (i.e., slope is zero), the line is horizontal.

VERTICAL LINE

A vertical line with x-intercept $(k, 0)$ has equation $x = k$. Its slope is undefined.

SLOPE–INTERCEPT FORM

The slope–intercept form of the equation of a line with slope m and y-intercept $(0, b)$ is

$$y = mx + b.$$

KEY TERMS & SYMBOLS	KEY CONCEPTS

1.4 Equations of Lines and Linear Models

point–slope form
standard form
parallel lines
perpendicular lines
negative reciprocals
square viewing window
scatter diagram
least-squares regression line
least-squares regression
correlation coefficient r

POINT–SLOPE FORM

The line with slope m passing through the point (x_1, y_1) has equation

$$y - y_1 = m(x - x_1).$$

STANDARD FORM

A linear equation written in the form

$$Ax + By = C,$$

where A, B, and C are real numbers (A and B not both 0), is said to be in standard form.

PARALLEL AND PERPENDICULAR LINES

Two distinct nonvertical lines are parallel if and only if they have the same slope.
Two lines, neither of which is vertical, are perpendicular if and only if their slopes have product -1.

LINEAR MODELS AND REGRESSION

If a collection of data points approximates a straight line, then we can find the equation of such a line. Choosing two data points, we apply the methods of **Sections 1.3** and **1.4** to find this equation. The equation may vary with the two points chosen.

The line of best fit can be found using the linear regression feature of a graphing calculator.

1.5 Linear Equations and Inequalities

equation
solution (root)
solution set
linear equation in one variable
equivalent equations
conditional equation
numerical solution
contradiction
empty (null) set, Ø
identity
inequality
equivalent inequalities
linear inequality in one
  variable
three-part inequality

LINEAR EQUATIONS AND INEQUALITIES

A linear equation can be written in the form $ax + b = 0$, where $a \neq 0$.
A linear inequality can be written in one of the following forms.

$$ax + b < 0, \quad ax + b > 0, \quad ax + b \leq 0, \quad \text{or} \quad ax + b \geq 0$$

ANALYTIC SOLUTIONS

To solve linear equations, use the addition and multiplication properties of equality. To solve linear inequalities, use the properties of inequality. *When multiplying or dividing each side of an inequality by a negative number, be sure to reverse the direction of the inequality symbol.*

INTERSECTION-OF-GRAPHS METHOD OF GRAPHICAL SOLUTION

To solve the equation $f(x) = g(x)$, graph

$$y_1 = f(x) \quad \text{and} \quad y_2 = g(x).$$

The x-coordinate of any point of intersection of the two graphs is a solution of the equation. The solution set of $f(x) > g(x)$ is the set of all real numbers x such that the graph of f is **above** the graph of g. The solution set of $f(x) < g(x)$ is the set of all real numbers x such that the graph of f is **below** the graph of g.

x-INTERCEPT METHOD OF GRAPHICAL SOLUTION

To solve the equation $f(x) = g(x)$, graph

$$y = f(x) - g(x) = F(x).$$

The x-coordinate of any x-intercept of the graph of $y = F(x)$ is a solution of the equation. The solution set of $F(x) > 0$ is the set of all real numbers x such that the graph of F is **above** the x-axis. The solution set of $F(x) < 0$ is the set of all real numbers x such that the graph of F is **below** the x-axis.

(continued)

1.6 Applications of Linear Functions

varies directly
constant of variation

PROBLEM-SOLVING STRATEGY

Step 1 Read the problem.

Step 2 Write an equation.

Step 3 Solve the equation.

Step 4 Look back and check.

DIRECT VARIATION

A number y varies directly with x if there exists a nonzero number k such that $y = kx$. The number k is called the constant of variation.

---

1 Review Exercises

Let A represent the point with coordinates $(-1, 16)$, *and let B represent the point with coordinates* $(5, -8)$.

1. Find the exact distance between points A and B.

2. Find the coordinates of the midpoint M of the line segment connecting points A and B.

3. Find the slope of the line segment AB.

4. Find the equation of the line passing through points A and B. Write it in $y = mx + b$ form.

Consider the graph of $3x + 4y = 144$ *in Exercises 5–8.*

5. What is the slope of the line?

6. What is the x-intercept of the line?

7. What is the y-intercept of the line?

8. Give a viewing window that will show a comprehensive graph. (There are many possible such windows.)

9. Suppose that f is a linear function such that $f(3) = 6$ and $f(-2) = 1$. Find $f(8)$. (*Hint:* Find a formula for $f(x)$.)

10. Find the equation of the line perpendicular to the graph of $y = -4x + 3$ and passing through the point $(-2, 4)$. Write it in $y = mx + b$ form.

For each line shown in Exercises 11 and 12, do the following.
  (a) *Find the slope.*
  (b) *Find the slope–intercept form of the equation.*
  (c) *Find the midpoint of the segment connecting the two points identified on the graph.*
  (d) *Find the distance between the two points identified on the graph.*

11.

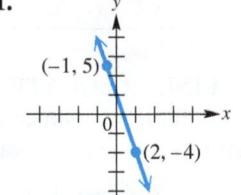

12.

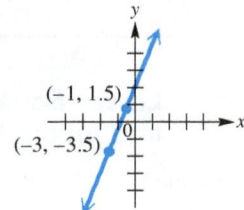

Concept Check   *Choose the letter of the graph in choices A–F that would most closely resemble the graph of* $f(x) = mx + b$, *given the conditions on m and b.*

13. $m < 0, b < 0$ **14.** $m > 0, b < 0$

15. $m < 0, b > 0$ **16.** $m > 0, b > 0$

17. $m = 0$ **18.** $b = 0$

A.

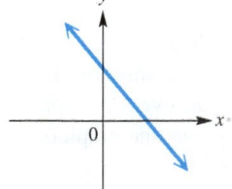

B.

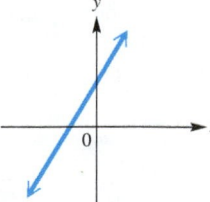

C.

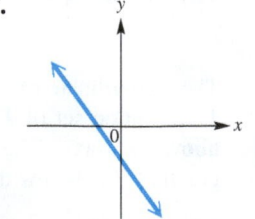

D.

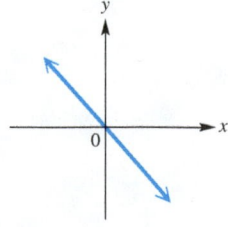

E.

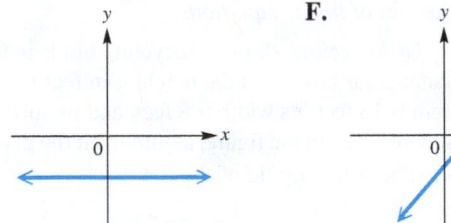

F.

19. *Cable Subscribers* In 2001, the number of basic cable TV subscribers in the United States was 66.7 million. In 2009, the number was 62.9 million. Assume a linear relationship, and find the average rate of change in the number of subscribers per year. Graph as a line segment and interpret the result. (*Source:* SNL Kagan.)

20. *Concept Check* **True or false?** The graphs of the equations $y_1 = 5.001x - 3$ and $y_2 = 5x + 6$ are shown in the screen. From this view, we may correctly conclude that these lines are parallel.

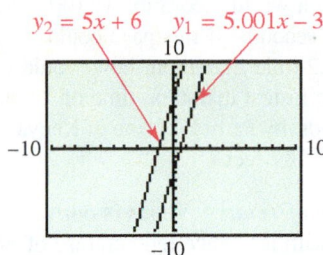

Refer to the graphs of the linear functions $y_1 = f(x)$ and $y_2 = g(x)$ in the figure. Match the solution set in Column II with the equation or inequality in Column I. Choices may be used once, more than once, or not at all.

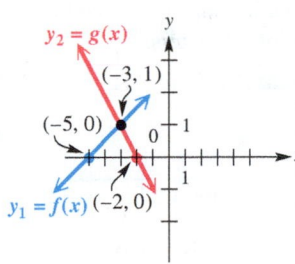

I		**II**	
21. $f(x) = g(x)$	**A.** $(-\infty, -3]$	**I.** $\{-3\}$
22. $f(x) > g(x)$	**B.** $(-\infty, -3)$	**J.** $[-3, \infty)$
23. $f(x) < g(x)$	**C.** $\{3\}$	**K.** $(-3, \infty)$
24. $g(x) \geq f(x)$	**D.** $\{2\}$	**L.** $\{1\}$
25. $y_2 - y_1 = 0$	**E.** $\{(3, 2)\}$	**M.** $(-\infty, -5)$
26. $f(x) < 0$	**F.** $\{-5\}$	**N.** $(-5, \infty)$
27. $g(x) > 0$	**G.** $\{-2\}$	**O.** $(-\infty, -2)$
28. $y_2 - y_1 < 0$	**H.** $\{0\}$	**P.** $(-2, \infty)$

Solve each equation or inequality using analytic methods. Support your answer graphically. Identify any identities or contradictions.

29. $5[3 + 2(x - 6)] = 3x + 1$

30. $\dfrac{x}{4} - \dfrac{x + 4}{3} = -2$

31. $-3x - (4x + 2) = 3$

32. $-2x + 9 + 4x = 2(x - 5) - 3$

33. $0.5x + 0.7(4 - 3x) = 0.4x$

34. $\dfrac{x}{4} - \dfrac{5x - 3}{6} = 2 - \dfrac{7x + 18}{12}$

35. $x - 8 < 1 - 2x$

36. $\dfrac{4x - 1}{3} \geq \dfrac{x}{5} - 1$

37. Solve the inequality $-6 \leq \dfrac{4 - 3x}{7} < 2$ analytically.

38. Consider the linear function

$$f(x) = 5\pi x + \left(\sqrt{3}\right)x - 6.24(x - 8.1) + \left(\sqrt[3]{9}\right)x.$$

(a) Solve the equation $f(x) = 0$ graphically. Give the solution to the nearest hundredth. Then explain how you went about solving the equation graphically.

(b) Refer to your graph, and give the solution set of $f(x) < 0$.

(c) Refer to your graph, and give the solution set of $f(x) \geq 0$.

(Modeling) Production of DVDs A company produces high-definition DVDs of live concerts. The company places an ad in a trade newsletter. The cost of the ad is $150. Each DVD costs $30 to produce, and the company charges $37.50 per DVD. Use this information to work Exercises 39–42.

39. Express the company's cost C as a function of x, the number of DVDs produced and sold.

40. Assuming that the company sells x DVDs, express the revenue as a function of x.

41. Determine analytically the value of x for which revenue equals cost.

42. The graph shows $y = C(x)$ and $y = R(x)$. Discuss how the graph illustrates when the company is losing money, when it is breaking even, and when it is making a profit.

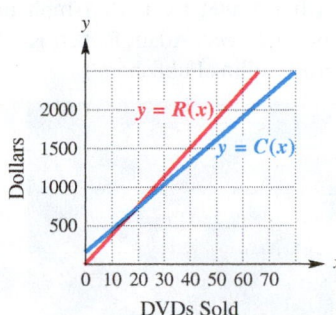

DVDs Sold

Solve the formula for approximate annual interest rate for the variable indicated.

43. $A = \dfrac{24f}{B(p+1)}$ for f **44.** $A = \dfrac{24f}{B(p+1)}$ for B

(Modeling)   Solve each problem.

45. *Temperature Levels above Earth's Surface*   On a particular day, the linear function

$$f(x) = -3.52x + 58.6$$

approximates the temperature in degrees Fahrenheit above the surface of Earth, where x is in thousands of feet and $f(x)$ gives the temperature.
(a) When the height is 5000 feet, what is the temperature? Solve analytically.
(b) When the temperature is $-15°F$, what is the height? Solve analytically.
(c) Explain how the answers in parts (a) and (b) can be supported graphically.

46. *Super Bowl Ads*   The table lists the cost in millions of dollars for a 30-second Super Bowl commercial for selected years.

Year	1994	1998	2004	2008	2012
Cost (in $ millions)	1.2	1.6	2.3	2.7	3.5

Source: MSNBC.

(a) Use regression to find a linear function f that models the data.
(b) Estimate the cost in 1990 and compare the estimate to the actual value of $0.8 million.
(c) Use f to predict the year when the cost could reach $4 million.

47. *Speed of a Batted Baseball*   Suppose a baseball is thrown at 85 mph. The ball will travel 320 feet when hit by a bat swung at 50 mph and 440 feet when hit by a bat swung at 80 mph. Let y be the number of feet traveled by the ball when hit by a bat swung at x mph. Find the equation of the line given by $y = mx + b$ that models the data. (*Note:* This function is valid for $50 \leq x \leq 90$, where the bat is 35 inches long, weighs 32 ounces, and is swung slightly upward to drive the ball at an angle of $35°$.) How much farther will a ball travel for each 1-mph increase in the speed of the bat? (*Source:* Adair, Robert K., *The Physics of Baseball*, HarperCollins Publishers.)

Solve each application of linear equations.

48. *Dimensions of a Recycling Bin*   A recycling bin is in the shape of a rectangular box. Find the height h in feet of the box if its length is 18 feet, its width is 8 feet, and its surface area is 496 square feet. (In the figure, assume that the given surface area includes the top lid of the box.)

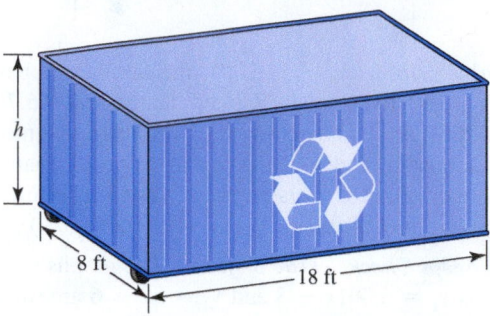

49. *Running Speeds in Track Events*   In 2009, Usain Bolt (Jamaica) set a world record in the 100-meter dash with a time of 9.58 seconds. If this pace could be maintained for an entire 26.2-mile marathon, how would that time compare with the fastest marathon time of 2 hours, 3 minutes, and 38 seconds, by Patrick Makau of Kenya in 2011? (*Hint:* 1 meter $\approx$ 3.281 feet.)

50. *Temperature of Venus*   Venus is our solar system's hottest planet, with a surface temperature of $864°F$. What is this temperature in Celsius? (*Hint:* Use $C = \dfrac{5}{9}(F - 32)$.) (*Source: Guinness Book of Records.*)

51. *Pressure in a Liquid*   The pressure on a point in a liquid is directly proportional to the distance from the surface to the point. In a certain liquid, the pressure at a depth of 4 meters is 3000 kilograms per square meter. Find the pressure at a depth of 10 meters.

52. *(Modeling) Data Analysis*   The table lists data that are exactly linear.

x	-3	-2	-1	0	1	2	3
y	6.6	5.4	4.2	3	1.8	0.6	-0.6

(a) Determine the slope–intercept form of the line that passes through these data points.
(b) Predict y when $x = -1.5$ and when $x = 3.5$.

53. *(Modeling) Math SAT Scores*   The table lists average mathematics SAT scores from past years.

Year	1996	1998	2000	2002	2004
Score	508	512	514	516	518

Source: The College Board.

(a) Use regression to find a linear function f that models these data.
(b) Use f to predict the score in 2012.
(c) The average SAT mathematics score in 2012 was actually 514. Discuss any problems with estimating data that are too far ahead of the given data.

54. *Acid Mixture*   A student needs 10% hydrochloric acid for a chemistry experiment. How much 5% acid should be mixed with 120 milliliters of 20% acid to get a 10% solution?

55. *(Modeling) DVD Production*   A company produces DVDs. The revenue from the sale of x DVDs is $R(x) = 8x$. The cost to produce x DVDs is $C(x) = 3x + 1500$. Revenue and cost are in dollars. In what interval will the company break even or make a profit?

56. *Intelligence Quotient*   A person's intelligence quotient (IQ) is found by multiplying the mental age by 100 and dividing by the chronological age.
   (a) Jack is 7 years old. His IQ is 130. Find his mental age.
   (b) If a person is 16 years old with a mental age of 20, what is the person's IQ?

57. *Predicting Weights of Athletes*   The table lists heights and weights of 10 female professional basketball players.

Female Basketball Players	
Height (in.)	**Weight (lb)**
74	169
74	181
77	199
64	139
68	135
71	161
78	187
68	150
71	181
73	194

Source: WNBA.

   (a) Find the least-squares regression line $y = mx + b$ that models this data. Let x correspond to height in inches and y correspond to weight in pounds.
   (b) Predict the weight of a female player who is 75 inches tall.

58. *Predicting Weights of Athletes*   The table lists heights and weights of 10 male professional basketball players.

Male Basketball Players	
Height (in.)	**Weight (lb)**
74	197
75	202
71	173
77	220
83	220
83	250
81	225
76	219
79	215
85	246

Source: NBA.

   (a) Find the least-squares regression line $y = mx + b$ that models this data. Let x correspond to height in inches and y correspond to weight in pounds.
   (b) Predict the weight of a male player who is 80 inches tall.

1 Test

1. Classify each number as one or more of the following: natural number, integer, rational number, or real number.
   (a) $\dfrac{4}{2}$ **(b)** π **(c)** $\sqrt{2}$ **(d)** 0.25

2. Find a decimal approximation of each expression to the nearest thousandth.
   (a) $\sqrt{5}$ **(b)** $\sqrt[3]{7}$
   (c) $3^{1/4}$ **(d)** $\dfrac{1 - 1.1^2}{2 + \pi^2}$

3. Find the length and the midpoint of the line segment connecting the points $(-2, 4)$ and $(4, -3)$.

4. Evaluate each expression if $f(x) = 4 - 7x$.
   (a) $f(-2)$ **(b)** $f(b)$
   (c) $f(a + h)$

5. Determine if $\{(3, 4), (2, -5), (1, 0), (4, -5)\}$ represents a function.

6. If possible, find the slope of the line passing through each pair of points.
   (a) $(1, 4), (-5, 1)$ **(b)** $(4, 3), (4, 9)$
   (c) $(1.2, 5), (1.7, 5)$

7. For each function in (a)–(c), determine the following.
   (i) domain   **(ii)** range
   (iii) x-intercept(s)   **(iv)** y-intercept

(a) 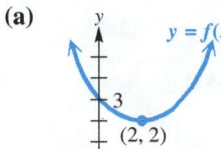   **(b)**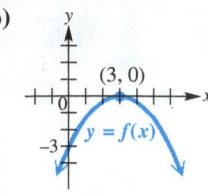

(c)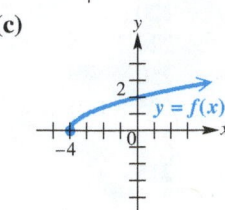

8. Use the figure to solve each equation or inequality.
  (a) $f(x) = g(x)$
  (b) $f(x) < g(x)$
  (c) $f(x) \geq g(x)$
  (d) $y_2 - y_1 = 0$

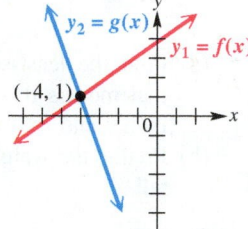

9. Use the screen to solve the equation or inequality. Here, f is a linear function defined over the domain of real numbers.
  (a) $y_1 = 0$
  (b) $y_1 < 0$
  (c) $y_1 > 0$
  (d) $y_1 \leq 0$

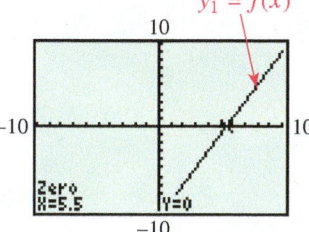

10. Consider the linear functions
$$f(x) = 3(x - 4) - 2(x - 5)$$
and $g(x) = -2(x + 1) - 3.$

  (a) Solve $f(x) = g(x)$ analytically, showing all steps. Also, check analytically.
  (b) Graph $y_1 = f(x)$ and $y_2 = g(x)$. Use your result in part (a) to find the solution set of $f(x) > g(x)$. Explain your answer.
  (c) Repeat part (b) for $f(x) < g(x)$.

11. Consider the linear function
$$f(x) = -\frac{1}{2}(8x + 4) + 3(x - 2).$$

  (a) Solve the equation $f(x) = 0$ analytically.
  (b) Solve the inequality $f(x) \leq 0$ analytically.
  (c) Graph $y = f(x)$ in an appropriate viewing window, and explain how the graph supports your answers in parts (a) and (b).

12. *(Modeling) U.S. Radio Stations*   In 2007 there were 13,837 radio stations, and in 2012 there were 15,082 radio stations.
  (a) Find a linear function $f(x) = ax + b$ that gives the number of radio stations x years after 2007.
  (b) Interpret the slope of the graph of $y = f(x)$.
  (c) Estimate the number of radio stations in 2014.

13. Find the equation of the line passing through the point $(-3, 5)$ and
  (a) parallel to the line with equation $y = -2x + 4$.
  (b) perpendicular to the line with equation $-2x + y = 0$.

14. Find the x- and y-intercepts of the line whose standard form is $3x - 4y = 6$. What is the slope of this line?

15. Give the equations of both the horizontal and vertical lines passing through the point $(-3, 7)$.

16. *(Modeling) Windchill Factor*   The table shows the windchill factor for various wind speeds when the Fahrenheit temperature is $40°$.

Wind Speed (mph)	Degrees
10	34
15	32
20	30
25	29
30	28
35	28

  (a) Find the least-squares regression line for the data. Give the correlation coefficient.
  (b) Use this line to predict the windchill factor when the wind speed is 40 mph.

17. *Motion*   A car travels a total of 275 miles in 4 hours, traveling part of the time at 60 miles per hour and part of the time at 74 miles per hour. How long did the car travel at each speed?

18. *Strength of a Beam*   The maximum load that a beam can support is directly proportional to its width. If a beam 2.25 inches wide can support a load of 510 pounds, find how large a load a similar beam that is 3.1 inches wide can support.

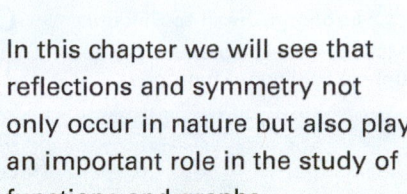

In this chapter we will see that reflections and symmetry not only occur in nature but also play an important role in the study of functions and graphs.

2 Analysis of Graphs of Functions

2.1 Graphs of Basic Functions and Relations; Symmetry

Continuity • Increasing, Decreasing, and Constant Functions • The Identity Function • The Squaring Function and Symmetry with Respect to the *y*-Axis • The Cubing Function and Symmetry with Respect to the Origin • The Square Root and Cube Root Functions • The Absolute Value Function • The Relation $x = y^2$ and Symmetry with Respect to the *x*-Axis • Even and Odd Functions

> ### → Looking Ahead to Calculus
> Many calculus theorems apply only to continuous functions.

Continuity

The graph of a linear function, a straight line, may be drawn by hand over any interval of its domain without picking the pencil up from the paper. We say that a function that can be graphed with this property is *continuous* over any interval. The graph of a continuous function has no breaks or holes in it. The formal definition of continuity requires concepts from calculus, but we can give an informal definition.

> ### Continuity (Informal Definition)
>
> A function is **continuous** over an interval of its domain if its hand-drawn graph over that interval can be sketched without lifting the pencil from the paper.

If a function is not continuous at a point, then it may have a point of *discontinuity,* as in **FIGURE 1(a),** or it may have a vertical **asymptote** (a vertical line that the graph does not intersect, as in **FIGURE 1(b)**). An asymptote is not part of a graph, but is used as an aid in drawing it. Asymptotes will be discussed further in **Chapter 4.**

Discontinuous at *x* = 2

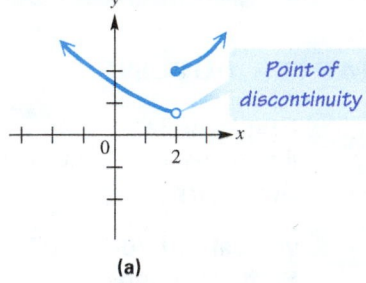

Point of discontinuity

(a)

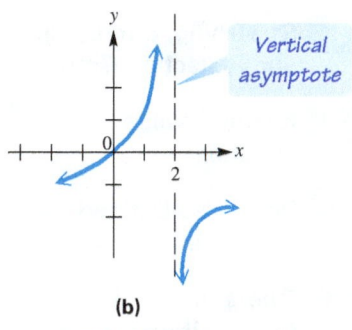

Vertical asymptote

(b)

FIGURE 1

> **EXAMPLE 1** **Determining Intervals of Continuity**

FIGURES 2, 3, and **4** show graphs of functions. Indicate the intervals of the domain over which they are continuous.

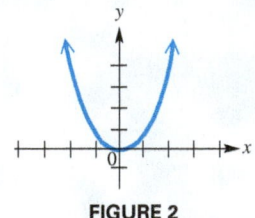

FIGURE 2

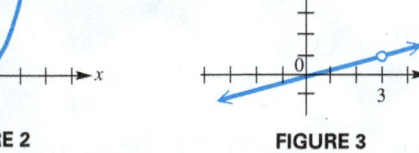

FIGURE 3

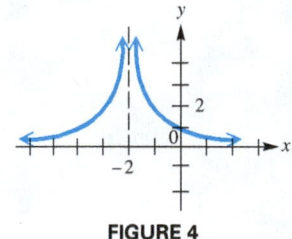

FIGURE 4

Solution The function in **FIGURE 2** is continuous over the entire domain of real numbers, $(-\infty, \infty)$.

The function in **FIGURE 3** has a point of discontinuity at $x = 3$. It is continuous over the interval $(-\infty, 3)$ and the interval $(3, \infty)$.

Finally, the function in **FIGURE 4** has a vertical asymptote at $x = -2$, as indicated by the dashed line. It is continuous over the intervals $(-\infty, -2)$ and $(-2, \infty)$. ●

Increasing, Decreasing, and Constant Functions

If a continuous function does not take on the same *y*-value throughout an *x*-interval, then its graph will either **rise** from left to right or **fall** from left to right or some combination of both. We use the words *increasing* and *decreasing* to describe this behavior.

For example, a linear function with positive slope is increasing over its entire domain, while one with negative slope is decreasing. See **FIGURE 5**.

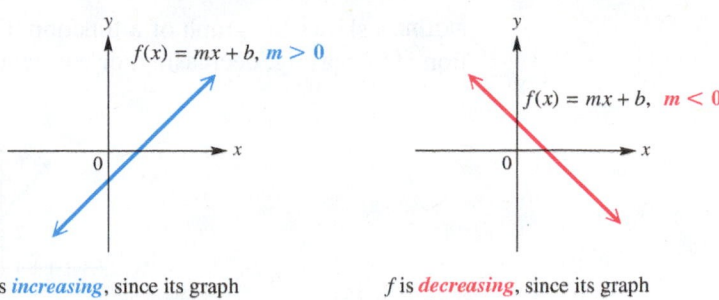

f is *increasing*, since its graph   f is *decreasing*, since its graph
rises from left to right.   falls from left to right.

FIGURE 5

Informally speaking, a function *increases* on an open interval of its domain if its graph *rises from left to right* on the interval. A function *decreases* on an open interval of its domain if its graph *falls from left to right* on the interval. A function is *constant* on an open interval of its domain if its graph is *horizontal* on the interval. While some functions are increasing, decreasing, or constant over their entire domains, many are not. Hence, we speak of increasing, decreasing, or constant over an open interval of the domain.

Criteria for Increasing, Decreasing, and Constant on an Open Interval

Suppose that a function f is defined over an open interval I.

(a) f **increases** on I if, whenever $x_1 < x_2$, $f(x_1) < f(x_2)$.

(b) f **decreases** on I if, whenever $x_1 < x_2$, $f(x_1) > f(x_2)$.

(c) f is **constant** on I if, for every x_1 and x_2, $f(x_1) = f(x_2)$.

FIGURE 6 illustrates these ideas.

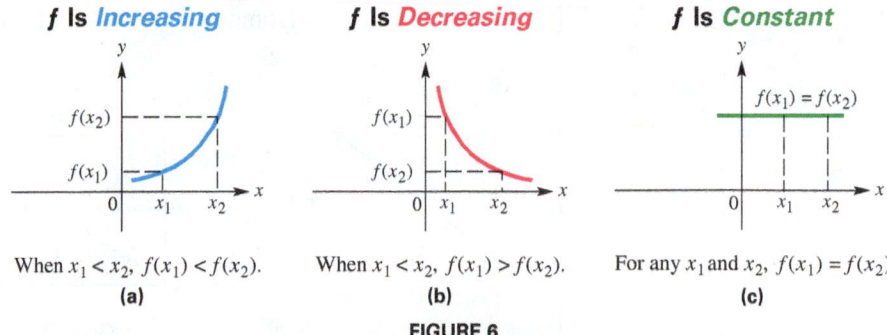

When $x_1 < x_2$, $f(x_1) < f(x_2)$.   When $x_1 < x_2$, $f(x_1) > f(x_2)$.   For any x_1 and x_2, $f(x_1) = f(x_2)$.
(a)   **(b)**   **(c)**

FIGURE 6

To decide whether a function is increasing, decreasing, or constant on an open interval, ask yourself, *"What does y do as x goes from left to right?"*

NOTE With our definition, the concepts of increasing and decreasing functions apply to intervals of the domain and not to individual points.

In this text, the definitions of increasing and decreasing functions require that the x-interval be an *open* interval. An open interval does not include endpoints. For example, with these definitions, the graph of $f(x) = x^2 + 4$, shown in **FIGURE 7**, is decreasing on $(-\infty, 0)$ and increasing on $(0, \infty)$.

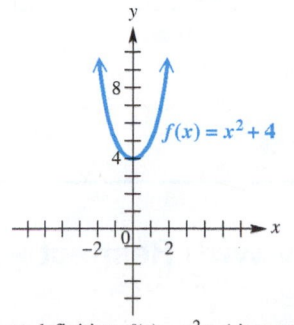

With our definition, $f(x) = x^2 + 4$ increases on $(0, \infty)$ and decreases on $(-\infty, 0)$.

FIGURE 7

Determining Intervals over Which a Function Is Increasing, Decreasing, or Constant

FIGURE 8 shows the graph of a function. Determine the intervals over which the function is increasing, decreasing, or constant.

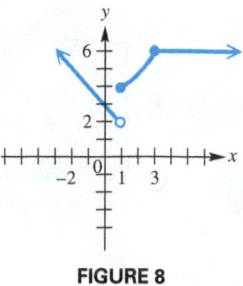

FIGURE 8

Solution   We should ask, "What is happening to the y-values as the x-values are getting larger?" Moving from left to right on the graph, we see that on the interval $(-\infty, 1)$, the y-values are *decreasing*. On the interval $(1, 3)$, the y-values are *increasing*, and on the interval $(3, \infty)$, the y-values are *constant* (and equal to 6). Therefore, the function is decreasing on $(-\infty, 1)$, increasing on $(1, 3)$, and constant on $(3, \infty)$.

CAUTION   *Remember that we are determining intervals of the domain, and thus are interested in x-values for our interval designations, not y-values.*

The Identity Function

If we let $m = 1$ and $b = 0$ in $f(x) = mx + b$ from **Section 1.3,** we have the **identity function,** or $f(x) = x$. This function pairs every real number with itself. See **FIGURE 9.**

> **Looking Ahead to Calculus**
> Determining where a function increases, decreases, or is constant is important in calculus. The **derivative** of a function provides a formula for determining the slope of a line tangent to the curve. If the slope of the tangent line is positive over an open interval, the function is increasing on that interval. Similarly, if the slope is negative, the function is decreasing. If the slope is always 0, the function is constant.

FUNCTION CAPSULE

IDENTITY FUNCTION $f(x) = x$

Domain: $(-\infty, \infty)$ Range: $(-\infty, \infty)$

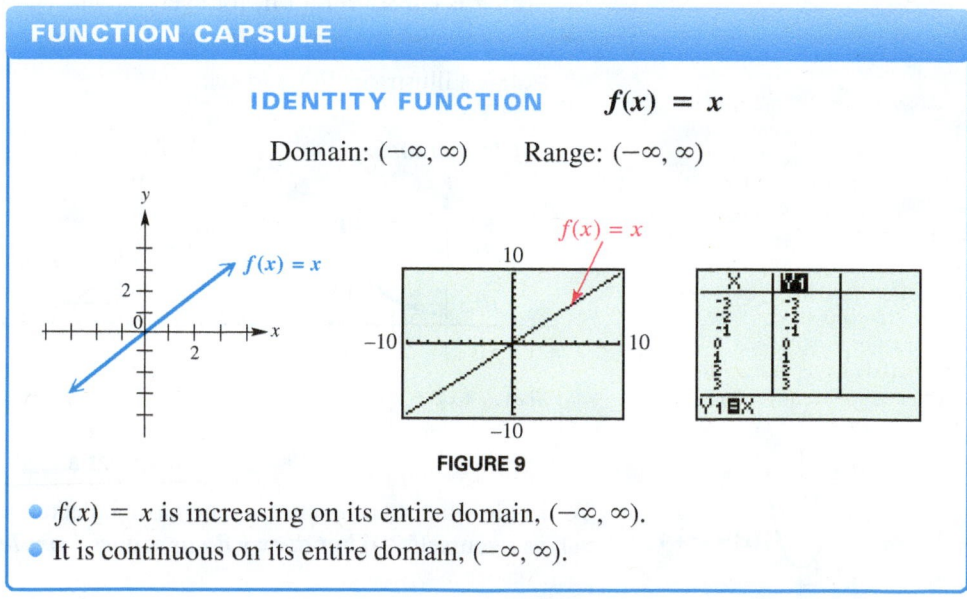

FIGURE 9

- $f(x) = x$ is increasing on its entire domain, $(-\infty, \infty)$.
- It is continuous on its entire domain, $(-\infty, \infty)$.

The Squaring Function and Symmetry with Respect to the y-Axis

The **degree** of a polynomial function in the variable x is the greatest exponent on x in the polynomial. The identity function is a degree 1 (or *linear*) function. The **squaring function,** $f(x) = x^2$, is the simplest degree 2 (or *quadratic*) function. (The word *quadratic* refers to degree 2.)

The squaring function pairs every real number with its square. Its graph is called a **parabola.** See **FIGURE 10.** The point $(0, 0)$ at which the graph changes from decreasing to increasing is called the **vertex** of the graph. (For a parabola that opens downward, the vertex is the point at which the graph changes from increasing to decreasing.)

FUNCTION CAPSULE

SQUARING FUNCTION $f(x) = x^2$

Domain: $(-\infty, \infty)$ Range: $[0, \infty)$

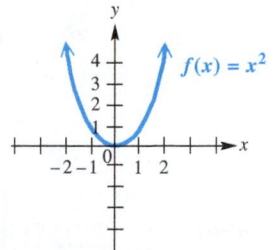 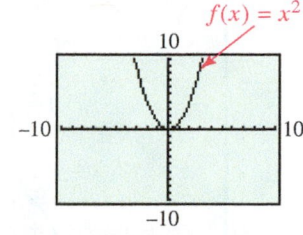 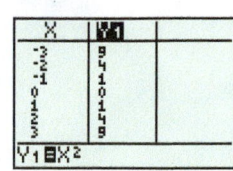

FIGURE 10

- $f(x) = x^2$ decreases on the interval $(-\infty, 0)$ and increases on the interval $(0, \infty)$.
- It is continuous on its entire domain, $(-\infty, \infty)$.

If we "fold" the graph of $f(x) = x^2$ along the y-axis, the two halves coincide exactly. In mathematics we refer to this property as symmetry, and we say that the graph of $f(x) = x^2$ is *symmetric with respect to the y-axis*.

➡ Looking Ahead to Calculus

Calculus allows us to find areas of curved regions in the plane. For example, we can find the area of the region below the graph of $y = x^2$, above the x-axis, bounded on the left by the line $x = -2$, and bounded on the right by $x = 2$. A sketch of this region is shown below. Notice that, due to the symmetry of the graph of $y = x^2$, the desired area is twice that of the area to the right of the y-axis. Thus, symmetry can be used to reduce the original problem to an easier one by simply finding the area to the right of the y-axis, and then doubling the answer.

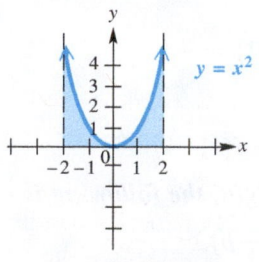

Symmetry with Respect to the y-Axis

If a function f is defined so that

$$f(-x) = f(x)$$

for all x in its domain, then the graph of f is **symmetric with respect to the y-axis.**

To illustrate that the graph of $f(x) = x^2$ is symmetric with respect to the y-axis, note that $f(-x) = f(x)$ for each x-value.

$$f(-4) = f(4) = 16 \qquad f(-3) = f(3) = 9$$
$$f(-2) = f(2) = 4 \qquad f(-1) = f(1) = 1$$

This pattern holds for any real number x, since

$$f(-x) = (-x)^2 = (-1)^2 x^2 = x^2 = f(x).$$

In general, if a graph is symmetric with respect to the y-axis, the following is true.

If (a, b) is on the graph, so is $(-a, b)$.

The Cubing Function and Symmetry with Respect to the Origin

The **cubing function, $f(x) = x^3$,** is the simplest degree 3 function. It pairs every real number with the third power, or cube, of the number. See **FIGURE 11.** The point $(0, 0)$ at which the graph of the cubing function changes from "opening downward" to "opening upward" is called an **inflection point.**

FUNCTION CAPSULE

CUBING FUNCTION $f(x) = x^3$

Domain: $(-\infty, \infty)$ Range: $(-\infty, \infty)$

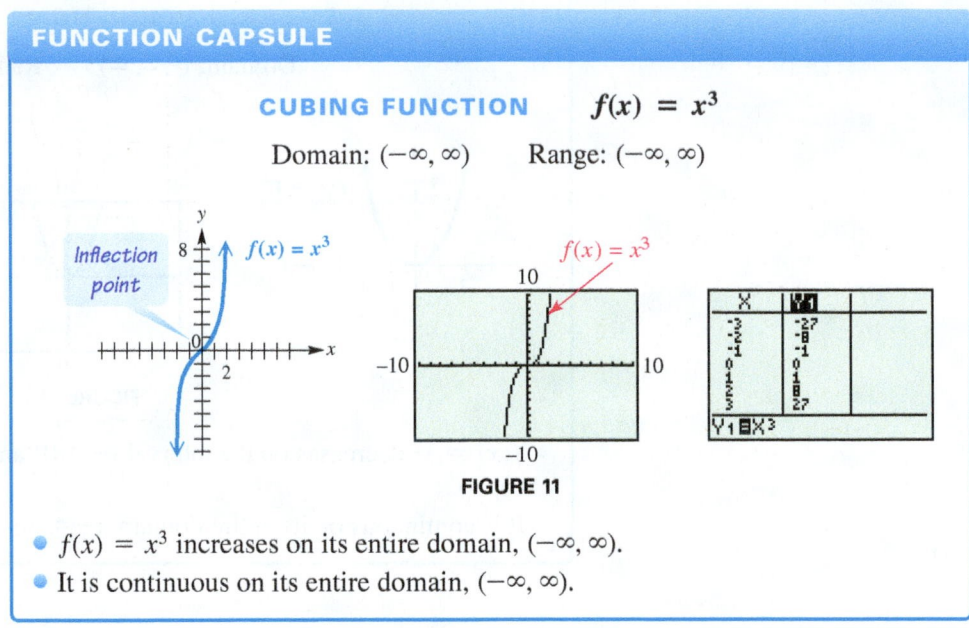

FIGURE 11

- $f(x) = x^3$ increases on its entire domain, $(-\infty, \infty)$.
- It is continuous on its entire domain, $(-\infty, \infty)$.

If we "fold" the graph of $f(x) = x^3$ along the y-axis and then along the x-axis, forming a "corner" at the origin, the two parts of the graph coincide exactly. We say that the graph of $f(x) = x^3$ is *symmetric with respect to the origin.*

FOR DISCUSSION

Suppose the graph of $f(x) = x^3$ is rotated about the origin half a turn, or 180°.

1. How does this graph compare with the original graph?
2. Explain how you can use rotation to determine whether a graph is symmetric with respect to the origin.

Symmetry with Respect to the Origin

If a function f is defined so that

$$f(-x) = -f(x)$$

for all x in its domain, then the graph of f is **symmetric with respect to the origin.**

To illustrate that the graph of $f(x) = x^3$ is symmetric with respect to the origin, note that $f(-x) = -f(x)$ for each x-value.

$$f(-2) = -f(2) = -8 \qquad f(-1) = -f(1) = -1$$

This pattern holds for any real number x, since

$$f(-x) = (-x)^3 = (-1)^3 x^3 = -x^3 = -f(x).$$

In general, if a graph is symmetric with respect to the origin, the following is true.

If (a, b) is on the graph, so is $(-a, -b)$.

EXAMPLE 3 **Determining Symmetry**

Show that

(a) $f(x) = x^4 - 3x^2 - 8$ has a graph that is symmetric with respect to the y-axis.

(b) $f(x) = x^3 - 4x$ has a graph that is symmetric with respect to the origin.

Analytic Solution

(a) We must show that $f(-x) = f(x)$ for any x.

$$
\begin{aligned}
f(-x) &= (-x)^4 - 3(-x)^2 - 8 && \text{Substitute } -x \text{ for } x. \\
&= (-1)^4 x^4 - 3(-1)^2 x^2 - 8 && (ab)^m = a^m b^m \\
&= x^4 - 3x^2 - 8 && \text{Simplify.} \\
&= f(x) && \boxed{f(-x) = f(x)}
\end{aligned}
$$

This *proves* that the graph is symmetric with respect to the y-axis.

(b) We must show that $f(-x) = -f(x)$ for any x.

$$
\begin{aligned}
f(-x) &= (-x)^3 - 4(-x) && \text{Substitute } -x \text{ for } x. \\
&= (-1)^3 x^3 + 4x && \text{Power rule} \\
&= -x^3 + 4x && * \\
&= -(x^3 - 4x) && \text{Factor out } -1. \\
&= -f(x) && \boxed{f(-x) = -f(x)}
\end{aligned}
$$

In the line denoted *, the signs of the coefficients are *opposites* of those in $f(x)$. We factored out -1 in the next line to show that the final result is $-f(x)$.

Graphing Calculator Solution

FIGURES 12 and **13** support the analytic proofs.

(a)

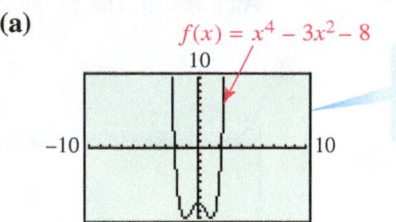

$f(x) = x^4 - 3x^2 - 8$

Symmetric with respect to the y-axis

FIGURE 12

(b)

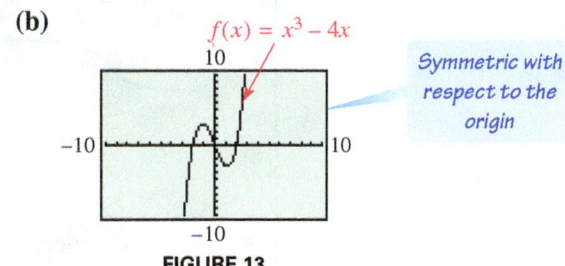

$f(x) = x^3 - 4x$

Symmetric with respect to the origin

FIGURE 13

The Square Root and Cube Root Functions

The **square root function** is $f(x) = \sqrt{x}$. See **FIGURE 14**. For the function value to be a real number, the domain must be restricted to $[0, \infty)$.

FUNCTION CAPSULE

SQUARE ROOT FUNCTION $f(x) = \sqrt{x}$

Domain: $[0, \infty)$ Range: $[0, \infty)$

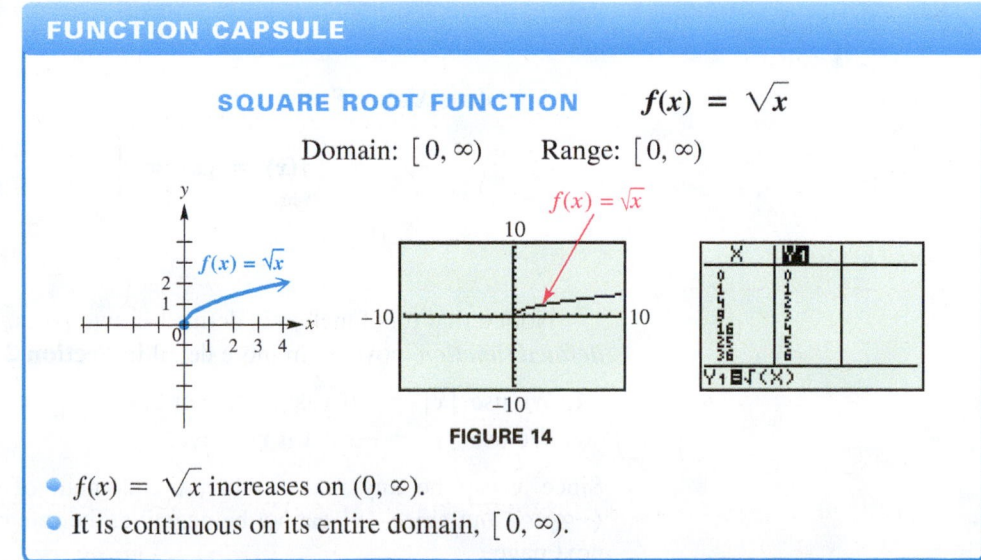

$f(x) = \sqrt{x}$

$f(x) = \sqrt{x}$

FIGURE 14

• $f(x) = \sqrt{x}$ increases on $(0, \infty)$.

• It is continuous on its entire domain, $[0, \infty)$.

TECHNOLOGY NOTE

The definition of a rational exponent allows us to enter $\sqrt{x}$ as $x^{1/2}$ and $\sqrt[3]{x}$ as $x^{1/3}$ when using a calculator.

← Algebra Review

Further discussion of rational exponents can be found in **Section R.4.**

The **cube root function, $f(x) = \sqrt[3]{x}$,** differs from the square root function in that *any* real number—positive, 0, or *negative*—has a real cube root. See **FIGURE 15.** Thus, the domain is $(-\infty, \infty)$.

$$\text{When} \quad x > 0, \quad \sqrt[3]{x} > 0.$$
$$\text{When} \quad x = 0, \quad \sqrt[3]{x} = 0.$$
$$\text{When} \quad x < 0, \quad \sqrt[3]{x} < 0.$$

As a result, the range is also $(-\infty, \infty)$.

FUNCTION CAPSULE

CUBE ROOT FUNCTION $\quad f(x) = \sqrt[3]{x}$

Domain: $(-\infty, \infty)$ Range: $(-\infty, \infty)$

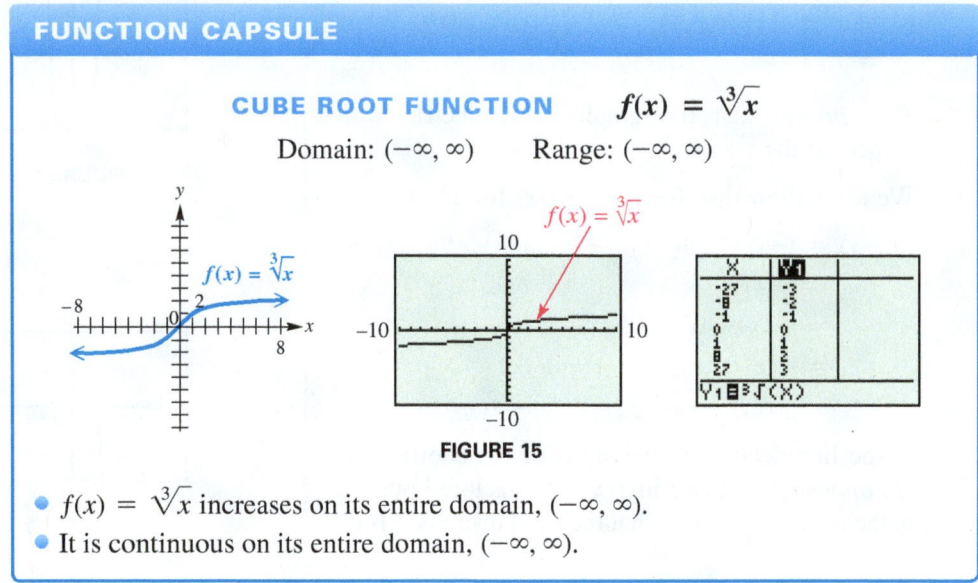

FIGURE 15

- $f(x) = \sqrt[3]{x}$ increases on its entire domain, $(-\infty, \infty)$.
- It is continuous on its entire domain, $(-\infty, \infty)$.

The Absolute Value Function

TECHNOLOGY NOTE

You should become familiar with the command on your particular calculator that allows you to graph the absolute value function.

On a number line, the absolute value of a real number x, denoted $|x|$, represents its undirected distance from the origin, 0. The **absolute value function, $f(x) = |x|$,** pairs every real number with its absolute value and is defined as follows.

Absolute Value Function

$$f(x) = |x| = \begin{cases} x & \text{if } x \geq 0 \\ -x & \text{if } x < 0 \end{cases}$$

Notice that this function is defined in two parts. It is an example of a *piecewise-defined function,* covered in more detail in **Section 2.5.**

1. We use $|x| = x$ if x is positive or 0.

2. We use $|x| = -x$ if x is negative.

Since x can be any real number, the domain of the absolute value function is $(-\infty, \infty)$, but since $|x|$ cannot be negative, the range is $[0, \infty)$. See **FIGURE 16** on the next page.

FUNCTION CAPSULE

ABSOLUTE VALUE FUNCTION $f(x) = |x|$

Domain: $(-\infty, \infty)$ Range: $[0, \infty)$

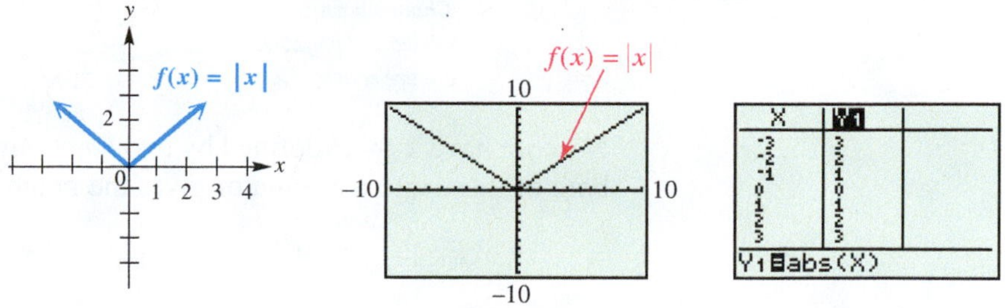

FIGURE 16

- $f(x) = |x|$ decreases on the interval $(-\infty, 0)$ and increases on the interval $(0, \infty)$.
- It is continuous on its entire domain, $(-\infty, \infty)$.

FOR DISCUSSION

Based on the functions discussed so far in this section, answer each question.

1. Which functions have graphs that are symmetric with respect to the y-axis?
2. Which functions have graphs that are symmetric with respect to the origin?
3. Which functions have graphs with neither of these symmetries?
4. Why is it not possible for the graph of a nonzero function to be symmetric with respect to the x-axis?

The Relation $x = y^2$ and Symmetry with Respect to the x-Axis

Recall that a function is a relation where every domain value is paired with one and only one range value. There are cases where we are interested in graphing relations that are *not* functions. Consider the relation defined by the equation

$$x = y^2.$$

The table of selected ordered pairs indicates that this relation has two y-values for each positive value of x, so y is *not* a function of x. If we plot these points and join them with a smooth curve, we find that the graph is a parabola opening to the right. See **FIGURE 17.**

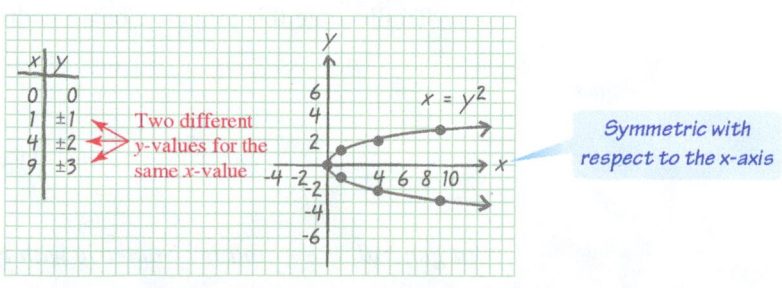

FIGURE 17

If a graphing calculator is set in function mode, it is not possible to graph $x = y^2$ directly. (However, if it is set in *parametric* mode, such a curve is possible with direct graphing.) To overcome this problem, we begin with $x = y^2$ and take the square root on each side.

$$x = y^2 \qquad \text{Given equation}$$

Choose both the positive and negative square roots of x.

$$y^2 = x \qquad \text{Rewrite so that } y \text{ is on the left.}$$

$$y = \pm\sqrt{x} \qquad \text{Take square roots.}$$

Now we have $x = y^2$ defined by two *functions:* $Y_1 = \sqrt{X}$ and $Y_2 = -\sqrt{X}$. Entering both of these into a calculator gives the graph shown in **FIGURE 18(a)**.

Graphing the Relation $x = y^2$

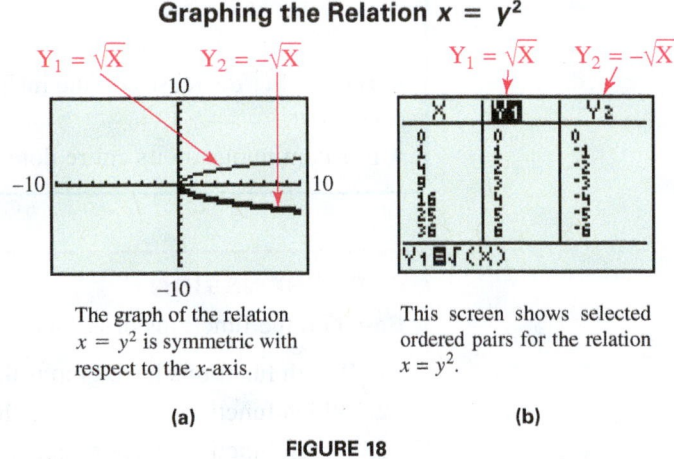

The graph of the relation $x = y^2$ is symmetric with respect to the x-axis.

(a)

This screen shows selected ordered pairs for the relation $x = y^2$.

(b)

FIGURE 18

If we "fold" the graph of $x = y^2$ along the x-axis, the two halves of the parabola coincide. We say that the graph of $x = y^2$ exhibits *symmetry with respect to the x-axis*. A table supports this symmetry, as seen in **FIGURE 18(b)**.

> **Symmetry with Respect to the x-Axis**
>
> If replacing y with $-y$ in an equation results in the same equation, then the graph is **symmetric with respect to the x-axis.**

To illustrate this property, begin with $x = y^2$ and replace y with $-y$.

$$x = y^2$$

$$x = (-y)^2 \qquad \text{Substitute } -y \text{ for } y.$$

$(ab)^m = a^m b^m$
$$x = (-1)^2 y^2 \qquad \text{Power rule}$$

$$x = y^2 \qquad \text{The result is the same equation.}$$

In general, if a graph is symmetric with respect to the x-axis, the following is true.

If (a, b) is on the graph, so is $(a, -b)$.

Even and Odd Functions

The concepts of symmetry with respect to the y-axis and symmetry with respect to the origin are closely associated with the concepts of *even* and *odd functions*.

Even and Odd Functions

A function f is called an **even function** if $f(-x) = f(x)$ for all x in the domain of f. (Its graph is symmetric with respect to the y-axis.)

A function f is called an **odd function** if $f(-x) = -f(x)$ for all x in the domain of f. (Its graph is symmetric with respect to the origin.)

As an illustration, consider the squaring and cubing functions.

$f(x) = x^2$ is an even function because

$$
\begin{aligned}
f(-x) &= (-x)^2 \\
&= (-1)^2 x^2 \quad \text{Power rule} \\
&= x^2 \\
&= f(x).
\end{aligned}
$$

$f(x) = x^3$ is an odd function because

$$
\begin{aligned}
f(-x) &= (-x)^3 \\
&= (-1)^3 x^3 \quad \text{Power rule} \\
&= -x^3 \\
&= -f(x).
\end{aligned}
$$

A function may be neither even nor odd. For example, $f(x) = \sqrt{x}$ is neither even nor odd. See **FIGURE 14** earlier in this section.

A summary of the types of symmetry just discussed follows.

Type of Symmetry	Example	Basic Fact about Points on the Graph
y-Axis symmetry	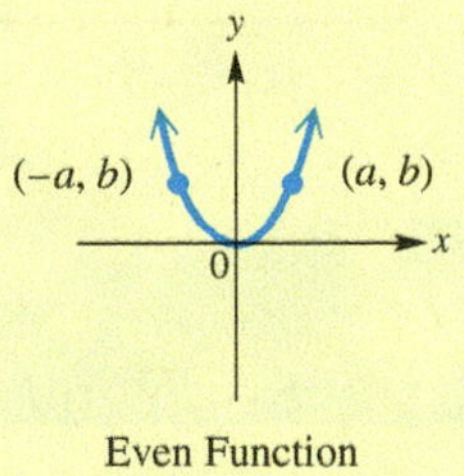 Even Function	If (a, b) is on the graph, so is $(-a, b)$.
Origin symmetry	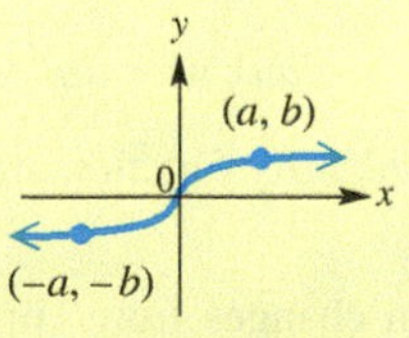 Odd Function	If (a, b) is on the graph, so is $(-a, -b)$.
x-Axis symmetry (not possible for a *nonzero function*)	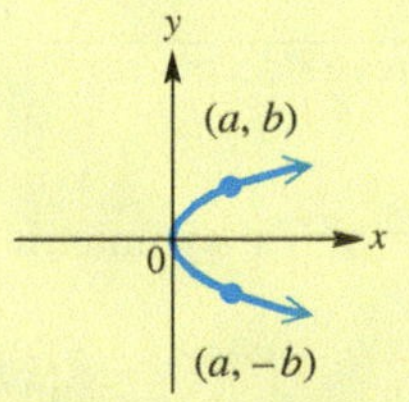 Not a Function	If (a, b) is on the graph, so is $(a, -b)$.

EXAMPLE 4 **Determining whether Functions Are Even, Odd, or Neither**

Decide whether each function is *even, odd,* or *neither.*

(a) $f(x) = 8x^4 - 3x^2$ **(b)** $f(x) = 6x^3 - 9x$ **(c)** $f(x) = 3x^2 + 5x$

Solution

(a) Replacing x with $-x$ gives

$$f(-x) = 8(-x)^4 - 3(-x)^2 = 8x^4 - 3x^2 = f(x).$$

Since $f(-x) = f(x)$ for each x in the domain of the function, f is even.

(b) $f(-x) = 6(-x)^3 - 9(-x) = -6x^3 + 9x = -(6x^3 - 9x) = -f(x)$
The function f is odd.

> *Factor out a minus sign.*

(c) $f(-x) = 3(-x)^2 + 5(-x) = 3x^2 - 5x$
Since $f(-x) \neq f(x)$ and $f(-x) \neq -f(x)$, f is neither even nor odd. ●

FOR DISCUSSION

The three functions discussed in **Example 4** are graphed in **FIGURE 19**, but not necessarily in the same order as in the example. Without actually using your calculator, identify each function.

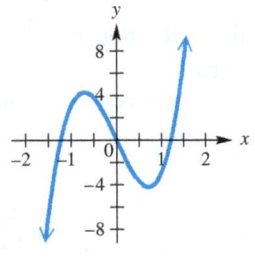 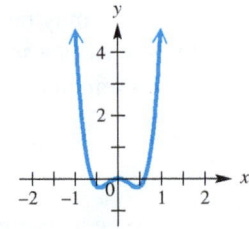 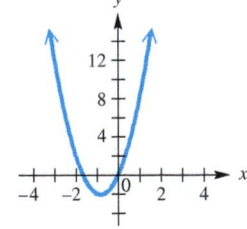

FIGURE 19

2.1 Exercises

NOTE: Write your answers using interval notation when appropriate.

Checking Analytic Skills   Fill in each blank with the correct response. **Do not use a calculator.**

1. The domain and the range of the identity function are both _____ .

2. The domain of the squaring function is _____ , and its range is _____ .

3. The graph of the cubing function changes from "opening downward" to "opening upward" at the point _____ .

4. The domain of the square root function is _____ , and its range is _____ .

5. The cube root function _____ on its entire domain.   (increases/decreases)

6. The largest open interval that the absolute value function decreases on is _____ and the largest open interval that it increases on is _____ .

7. The graph of the relation $x = y^2$ is symmetric with respect to the _____ .

8. The function $f(x) = x^4 + x^2$ is an _____ function.   (even/odd)

9. The function $f(x) = x^3 + x$ is an _____ function.   (even/odd)

10. If a function is even, its graph is symmetric with respect to the _____ . If it is odd, its graph is symmetric with respect to the _____ .

Determine the largest intervals of the domain over which each function is continuous.

11.

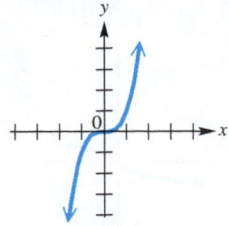

12.

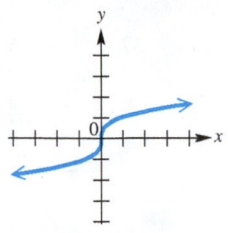

13.

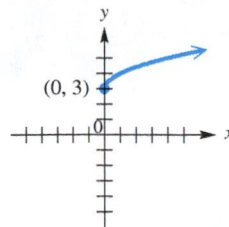

14.

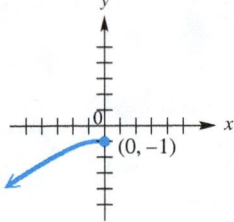

15.

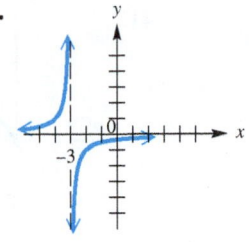

16.

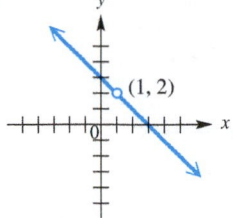

Determine the largest open intervals of the domain over which each function is **(a)** *increasing,* **(b)** *decreasing, and* **(c)** *constant. Then give the* **(d)** *domain and* **(e)** *range.*

17.

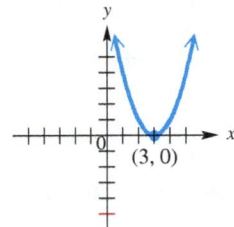

18.

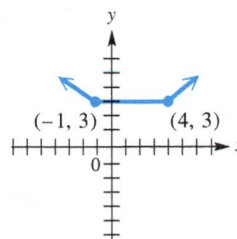

19.

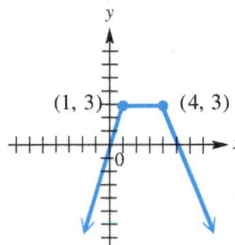

20.

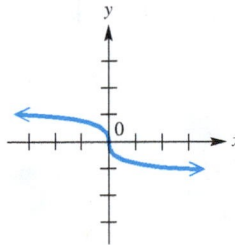

21.

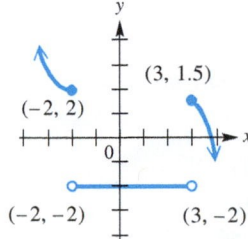

22.

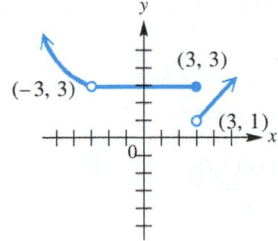

In Exercises 23–34, graph each function in the standard viewing window of your calculator, and trace from left to right along a representative portion of the specified interval. Then fill in the blank of the following sentence with either **increasing** *or* **decreasing.**

Over the interval specified, this function is _____.

23. $f(x) = x^5$; $(-\infty, \infty)$

24. $f(x) = -x^3$; $(-\infty, \infty)$

25. $f(x) = x^4$; $(-\infty, 0)$

26. $f(x) = x^4$; $(0, \infty)$

27. $f(x) = -|x|$; $(-\infty, 0)$

28. $f(x) = -|x|$; $(0, \infty)$

29. $f(x) = -\sqrt[3]{x}$; $(-\infty, \infty)$

30. $f(x) = -\sqrt{x}$; $(0, \infty)$

31. $f(x) = 1 - x^3$; $(-\infty, \infty)$

32. $f(x) = x^2 - 2x$; $(1, \infty)$

33. $f(x) = 2 - x^2$; $(-\infty, 0)$

34. $f(x) = |x + 1|$; $(-\infty, -1)$

Using visual observation, determine whether each graph is symmetric with respect to the **(a)** *x-axis,* **(b)** *y-axis, or* **(c)** *origin.*

35.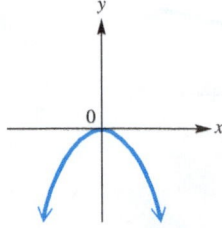

36.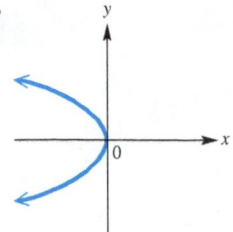

37.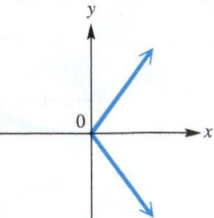

38.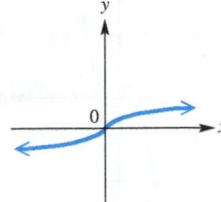

39.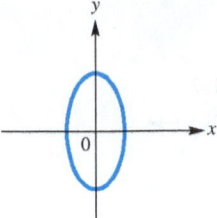

40.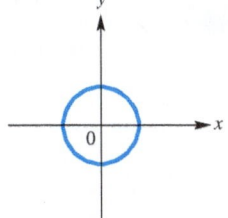

41.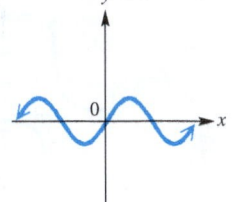

42.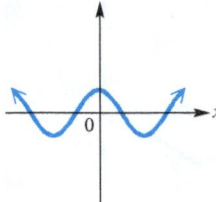

43. *Concept Check*   Complete the left half of the graph of $y = f(x)$ in the figure for each of the following conditions.
 (a) $f(-x) = f(x)$
 (b) $f(-x) = -f(x)$

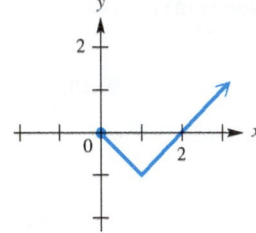

44. *Concept Check*   Complete the right half of the graph of $y = f(x)$ in the figure for each of the following conditions.
 (a) f is odd.
 (b) f is even.

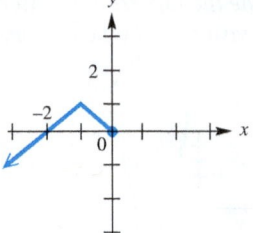

45. *Concept Check*   Complete the table, assuming that f is an even function.

x	−3	−2	−1	1	2	3
$f(x)$	21		−25		−12	

46. *Concept Check*   Complete the table, assuming that g is an odd function.

x	−5	−3	−2	0	2	3	5
$g(x)$	13		−5			−1	

Based on the ordered pairs seen in each table, make a conjecture about whether the function f is even, odd, or neither even nor odd.

47.

x	$f(x)$
−3	10
−2	5
−1	2
0	1
1	2
2	5
3	10

48.

x	$f(x)$
−3	16
−2	5
−1	1
0	−4
1	1
2	5
3	16

49.

x	$f(x)$
−3	10
−2	5
−1	2
0	0
1	−2
2	−5
3	−10

50.

x	$f(x)$
−3	−5
−2	−4
−1	−1
0	0
1	1
2	4
3	5

51.

x	$f(x)$
−3	5
−2	4
−1	3
0	2
1	1
2	0
3	−1

52.

x	$f(x)$
−3	−1
−2	0
−1	1
0	2
1	3
2	4
3	5

In Exercises 53–64, each function is either even *or* odd. *Use* $f(-x)$ *to state which situation applies.*

53. $f(x) = x^4 - 7x^2 + 6$

54. $f(x) = -2x^6 - 8x^2$

55. $f(x) = 3x^3 - x$

56. $f(x) = -x^5 + 2x^3 - 3x$

57. $f(x) = x^6 - 4x^4 + 5$

58. $f(x) = 8$

59. $f(x) = 3x^5 - x^3 + 7x$

60. $f(x) = x^3 - 4x$

61. $f(x) = |5x|$

62. $f(x) = \sqrt{x^2 + 1}$

63. $f(x) = \dfrac{1}{2x}$

64. $f(x) = 4x - \dfrac{1}{x}$

Use the analytic method of **Example 3** *to determine whether the graph of the given function is symmetric with respect to the* y-*axis, symmetric with respect to the* origin, *or neither. Use your calculator and the standard window to support your conclusion.*

65. $f(x) = -x^3 + 2x$

66. $f(x) = x^5 - 2x^3$

67. $f(x) = 0.5x^4 - 2x^2 + 1$

68. $f(x) = 0.75x^2 + |x| + 1$

69. $f(x) = x^3 - x + 3$

70. $f(x) = x^4 - 5x + 2$

71. $f(x) = x^6 - 4x^3$

72. $f(x) = x^3 - 3x$

73. $f(x) = -6$

74. $f(x) = |-x|$

75. $f(x) = \dfrac{1}{4x^3}$

76. $f(x) = \sqrt{x^2}$

2.2 Vertical and Horizontal Shifts of Graphs

Vertical Shifts • Horizontal Shifts • Combinations of Vertical and Horizontal Shifts • Effects of Shifts on Domain and Range
• Horizontal Shifts Applied to Equations for Modeling

Vertical Shifts

TECHNOLOGY NOTE

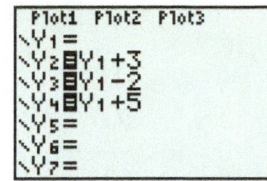

This screen shows how to minimize keystrokes on the TI-84 Plus in the activity in the "For Discussion" box. The entry for Y₁ can be altered as needed.

FOR DISCUSSION

In each group that follows, we give four related functions. Graph the functions in the first group (Group A) in the standard viewing window, and then answer the questions. Repeat the process for Group B, Group C, and Group D.

A	B	C	D
$y_1 = x^2$	$y_1 = x^3$	$y_1 = \sqrt{x}$	$y_1 = \sqrt[3]{x}$
$y_2 = x^2 + 3$	$y_2 = x^3 + 3$	$y_2 = \sqrt{x} + 3$	$y_2 = \sqrt[3]{x} + 3$
$y_3 = x^2 - 2$	$y_3 = x^3 - 2$	$y_3 = \sqrt{x} - 2$	$y_3 = \sqrt[3]{x} - 2$
$y_4 = x^2 + 5$	$y_4 = x^3 + 5$	$y_4 = \sqrt{x} + 5$	$y_4 = \sqrt[3]{x} + 5$

1. How do the graphs of y_2, y_3, and y_4 compare with the graph of y_1?

2. If $c > 0$, how do you think the graph of $y = f(x) + c$ would compare with the graph of $y = f(x)$?

3. If $c > 0$, how do you think the graph of $y = f(x) - c$ would compare with the graph of $y = f(x)$?

Choosing your own value of c, support your answers to Items 2 and 3 graphically. (Be sure that your choice is appropriate for the standard window.)

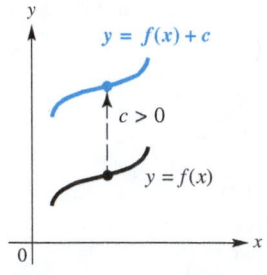

Vertical shift **upward c units**

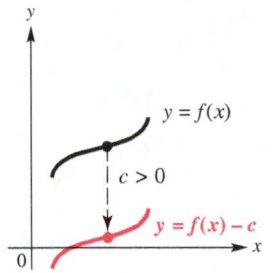

Vertical shift **downward c units**

FIGURE 20

In each group of functions in the preceding activity, we obtained a **vertical shift,** or a **vertical translation,** of the graph of the basic function with which we started. Our observations can be generalized to any function.

> **Vertical Shifting of the Graph of a Function**
>
> If $c > 0$, then the graph of $y = f(x) + c$ is obtained by shifting the graph of $y = f(x)$ *upward* a distance of c units. The graph of $y = f(x) - c$ is obtained by shifting the graph of $y = f(x)$ *downward* a distance of c units.

In **FIGURE 20**, we graphically interpret the preceding statements.

EXAMPLE 1 **Recognizing Vertical Shifts**

Give the equation of each function graphed.

(a)

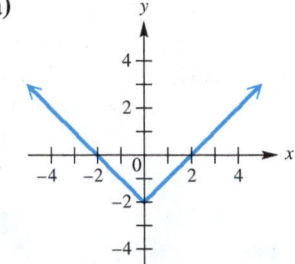

(b)

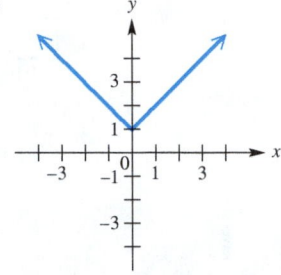

Solution   Both graphs are vertical translations of the graph of $y = |x|$. See **FIGURE 16** in **Section 2.1** to review the graph of $y = |x|$.

(a) The graph has been shifted 2 units downward, so the equation is $y = |x| - 2$.

(b) The graph has been shifted 1 unit upward, so the equation is $y = |x| + 1$.   ●

Horizontal Shifts

> **FOR DISCUSSION**
>
> This discussion parallels the one earlier in this section. Follow the same general directions.
>
> | A | B | C | D |
> |---|---|---|---|
> | $y_1 = x^2$ | $y_1 = x^3$ | $y_1 = \sqrt{x}$ | $y_1 = \sqrt[3]{x}$ |
> | $y_2 = (x - 3)^2$ | $y_2 = (x - 3)^3$ | $y_2 = \sqrt{x - 3}$ | $y_2 = \sqrt[3]{x - 3}$ |
> | $y_3 = (x - 5)^2$ | $y_3 = (x - 5)^3$ | $y_3 = \sqrt{x - 5}$ | $y_3 = \sqrt[3]{x - 5}$ |
> | $y_4 = (x + 4)^2$ | $y_4 = (x + 4)^3$ | $y_4 = \sqrt{x + 4}$ | $y_4 = \sqrt[3]{x + 4}$ |
>
> **1.** How do the graphs of y_2, y_3, and y_4 compare with the graph of y_1?
>
> **2.** If $c > 0$, how do you think the graph of $y = f(x - c)$ would compare with the graph of $y = f(x)$?
>
> **3.** If $c > 0$, how do you think the graph of $y = f(x + c)$ would compare with the graph of $y = f(x)$?
>
> Choosing your own value of c, support your answers to Items 2 and 3 graphically. (Again, be sure that your choice is appropriate for the standard window.)

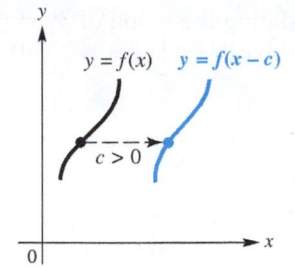

Horizontal shift to the **right** *c* **units**

In each group of functions in the preceding activity, we obtained a **horizontal shift,** or a **horizontal translation,** of the graph of the basic function with which we started. Our observations can be generalized as follows.

> ### Horizontal Shifting of the Graph of a Function
>
> If $c > 0$, the graph of $y = f(x - c)$ is obtained by shifting the graph of $y = f(x)$ to the *right* a distance of c units. The graph of $y = f(x + c)$ is obtained by shifting the graph of $y = f(x)$ to the *left* a distance of c units.

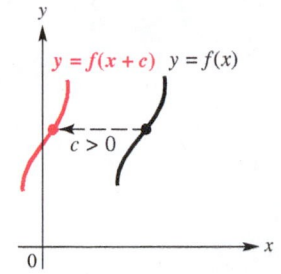

Horizontal shift to the **left** *c* **units**

FIGURE 21

In **FIGURE 21,** we graphically interpret the preceding statements.

CAUTION   *Be careful when translating graphs horizontally.* To determine the direction and magnitude of horizontal shifts, find the value of *x* that would cause the expression in parentheses to equal 0. For example, in comparison with the graph of $y = x^2$, the graph of $f(x) = (x - 5)^2$ would be shifted 5 units to the *right,* because +5 would cause $x - 5$ to equal 0. By contrast, the graph of $f(x) = (x + 5)^2$ would be shifted 5 units to the *left,* because −5 would cause $x + 5$ to equal 0.

EXAMPLE 2   **Recognizing Horizontal Shifts**

Give the equation of each function graphed.

(a) 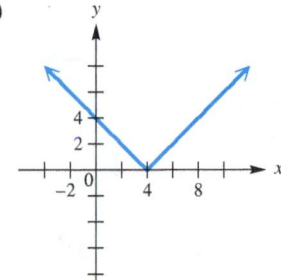   **(b)**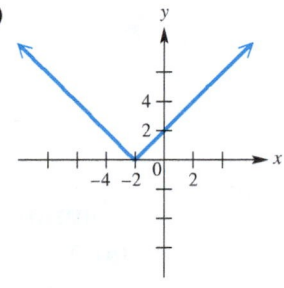

Solution   Both graphs are horizontal translations of the graph of $y = |x|$.

(a) The graph has been shifted 4 units to the right, so the equation is $y = |x - 4|$.

(b) The graph has been shifted 2 units to the left, so the equation is $y = |x + 2|$. ●

Combinations of Vertical and Horizontal Shifts

EXAMPLE 3   **Applying Both Vertical and Horizontal Shifts**

Describe how the graph of each y_2 would be obtained by translating the graph of $y_1 = |x|$. Sketch the graphs of y_1 and y_2 on the same *xy*-plane by hand.

(a) $y_2 = |x + 2| - 6$ **(b)** $y_2 = |x - 2| + 8$

Solution

(a) The function $y_2 = |x + 2| - 6$ is translated 2 units to the *left* (because of the $|x + 2|$) and 6 units *downward* (because of −6) compared with the graph of $y_1 = |x|$. See **FIGURE 22.** Notice that the point $(0, 0)$ has been translated to $(-2, -6)$.

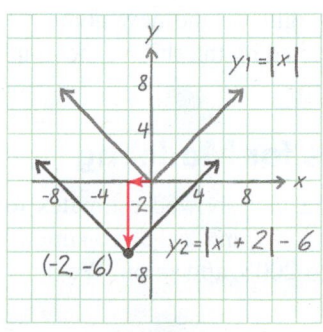

FIGURE 22

(continued)

(b) The graph of $y_2 = |x - 2| + 8$ is obtained by translating the graph of $y_1 = |x|$ 2 units to the *right* and 8 units *upward*. See **FIGURE 23**. Notice that the point $(0, 0)$ has been translated to $(2, 8)$.

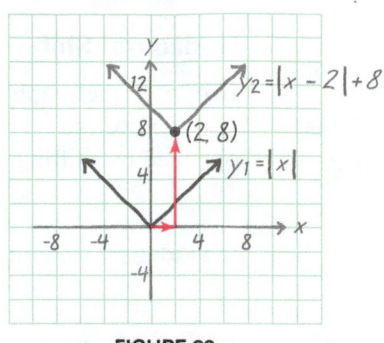

FIGURE 23

Effects of Shifts on Domain and Range

EXAMPLE 4 **Determining Domains and Ranges of Shifted Graphs**

Give the domain and range of each function.

(a)

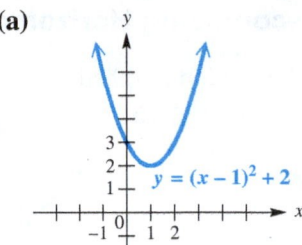

(b)

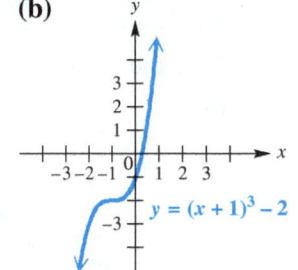

(c)

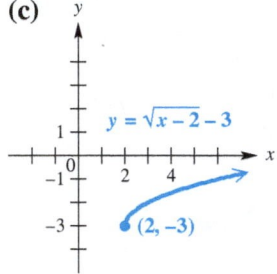

Solution

(a) To obtain the graph of $y = (x - 1)^2 + 2$, the graph of $y = x^2$ is translated 1 unit to the right and 2 units upward. The original domain $(-\infty, \infty)$ is not affected. However, the range of this function is $[2, \infty)$, because of the vertical translation. (Note that the domain and range can be also determined by direct inspection of the given graph.)

(b) The graph of $y = (x + 1)^3 - 2$ is obtained by vertical and horizontal shifts of the graph of $y = x^3$, a function that has both domain and range $(-\infty, \infty)$. Neither is affected here, so the domain and range of $y = (x + 1)^3 - 2$ are also $(-\infty, \infty)$.

(c) The function $y = \sqrt{x}$ has domain $[0, \infty)$. The graph of $y = \sqrt{x - 2} - 3$ is obtained by shifting the basic graph 2 units to the right, so the new domain is $[2, \infty)$. The original range, $[0, \infty)$, has also been affected by the shift of the graph 3 units downward, so the new range is $[-3, \infty)$.

Horizontal Shifts Applied to Equations for Modeling

In **Examples 6** and **7** of **Section 1.4**, we examined data that showed estimates for Medicare costs (in billions of dollars) between the years 2007 and 2012. To simplify our work, $x = 0$ corresponded to the year 2007, $x = 1$ to 2008, and so on. We determined that the least-squares regression line has equation

$$y \approx 31.23x + 434.10.$$

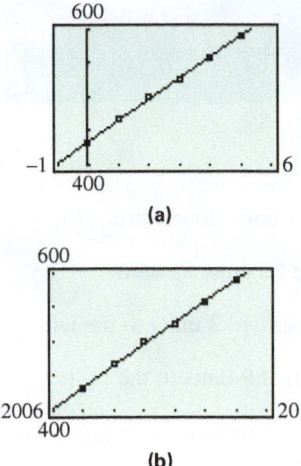

(a)

(b)

FIGURE 24

The data and graph of this line are shown in **FIGURE 24(a)**. To use the values of the years 2007 through 2012 directly in the regression equation, we must shift the graph of the line 2007 units to the right. The equation of this new line is

$$y \approx 31.23(x - 2007) + 434.10.$$

Using a graphing calculator to plot the set of points

$$\{(2007, 432), (2008, 467), (2009, 500), (2010, 525), (2011, 558), (2012, 591)\}$$

and graphing $y = 31.23(x - 2007) + 434.10$ in the viewing window $[2006, 2013]$ by $[400, 600]$ gives the graph shown in **FIGURE 24(b)**.

EXAMPLE 5 Applying a Shift to an Equation Model

The number of monthly active Facebook users F in millions from 2009 to 2013 is given by

$$F(x) = 233x + 200,$$

where x is the number of years *after* 2009.

(a) Evaluate $F(2)$, and interpret your result.

(b) Use the formula for $F(x)$ to write an equation that gives the number of monthly active Facebook users y in millions during the *actual* year x.

(c) Refer to part (b) and find y when $x = 2011$. Interpret your result.

(d) Use your equation in part (b) to determine the year when Facebook reached 1 billion monthly active users.

Solution

(a) Because $F(2) = 233(2) + 200 = 666$ and the value $x = 2$ corresponds to $2009 + 2 = 2011$, Facebook had 666 million monthly active users in 2011.

(b) Because 2009 corresponds to 0, the graph of $F(x) = 233x + 200$ should be shifted 2009 units to the right. That is, we must replace x with $(x - 2009)$ in the formula for $F(x)$.

$$y = F(x - 2009) = 233(x - 2009) + 200$$

(c) When $x = 2011$, $y = 233(2011 - 2009) + 200 = 666$, so in the year 2011 there were about 666 million monthly active Facebook users.

(d) To determine when Facebook had 1 billion (1000 million) monthly active users, we need to solve the following equation.

$$233(x - 2009) + 200 = 1000 \qquad \text{Equation to solve: } y = 1000$$

$$233(x - 2009) = 800 \qquad \text{Subtract 200 from each side.}$$

$$x - 2009 = \frac{800}{233} \qquad \text{Divide each side by 233.}$$

$$x = \frac{800}{233} + 2009 \qquad \text{Add 2009 to each side.}$$

$$x \approx 2012.4 \qquad \text{Approximate.}$$

Thus Facebook reached 1 billion monthly active users during 2012.

2.2 Exercises

Checking Analytic Skills Write the equation that results in the desired translation. **Do not use a calculator.**

1. The squaring function, shifted 3 units upward

2. The cubing function, shifted 2 units downward

3. The square root function, shifted 4 units downward

4. The cube root function, shifted 6 units upward

5. The absolute value function, shifted 4 units to the right

6. The absolute value function, shifted 3 units to the left

7. The cubing function, shifted 7 units to the left

8. The square root function, shifted 9 units to the right

9. The squaring function, shifted 2 units downward and 3 units to the right

10. The squaring function, shifted 4 units upward and 1 unit to the left

11. The square root function, shifted 3 units upward and 6 units to the left

12. The absolute value function, shifted 1 unit downward and 5 units to the right

13. The squaring function, shifted 2000 units to the right and 500 units upward

14. The squaring function, shifted 1000 units to the left and 255 units downward

15. Explain how the graph of $g(x) = f(x) + 4$ is obtained from the graph of $y = f(x)$.

16. Explain how the graph of $g(x) = f(x + 4)$ is obtained from the graph of $y = f(x)$.

Exercises 17–25 are grouped in threes: 17–19, 20–22, and 23–25. For each group, match the correct graph, A, B, or C, to the given equation.

17. $y = x^2 - 3$
A.

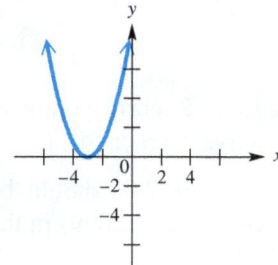

18. $y = (x - 3)^2$
B.

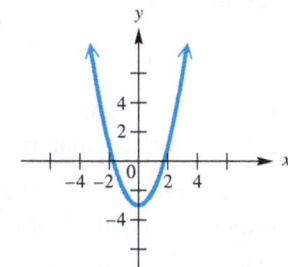

19. $y = (x + 3)^2$
C.

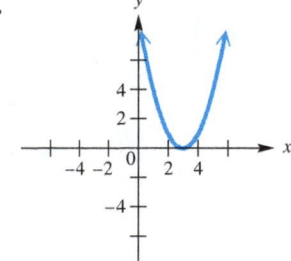

20. $y = |x| + 4$
A.

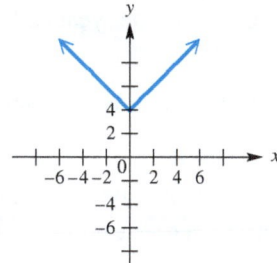

21. $y = |x + 4| - 3$
B.

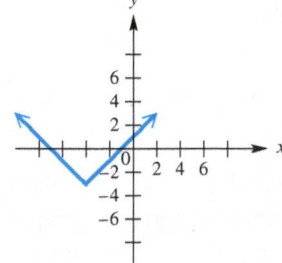

22. $y = |x - 4| - 3$
C.

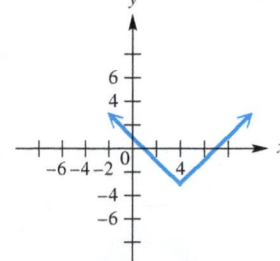

23. $y = (x - 3)^3$
A.

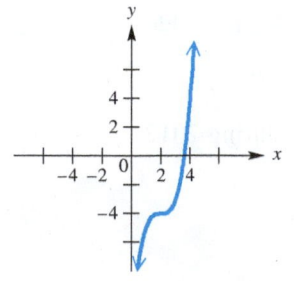

24. $y = (x - 2)^3 - 4$
B.

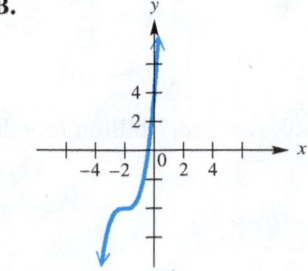

25. $y = (x + 2)^3 - 4$
C.

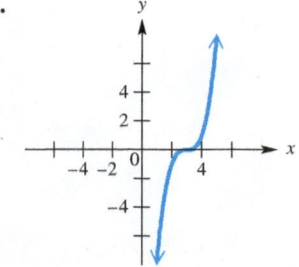

Concept Check *The function* Y_2 *is defined as* $Y_1 + k$ *for some real number k. Based on the table shown, what is the value of k?*

26.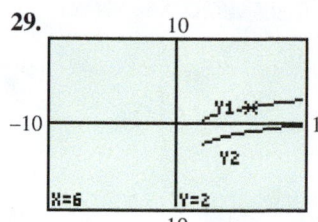

27.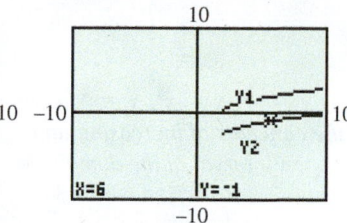

28.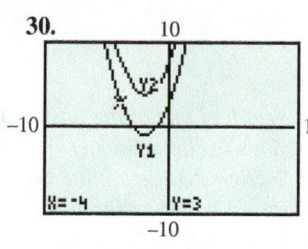

Concept Check *The function* Y_2 *is defined as* $Y_1 + k$ *for some real number k. Based on the two views of the graphs of* Y_1 *and* Y_2 *and the displays at the bottoms of the screens, what is the value of k?*

29.

(6, 2) lies on the graph of Y_1.
First view

(6, –1) lies on the graph of Y_2.
Second view

30.

(–4, 3) lies on the graph of Y_1.
First view

(–4, 8) lies on the graph of Y_2.
Second view

Use the results of the specified exercises to determine **(a)** *the domain and* **(b)** *the range of each function.*

31. $y = x^2 - 3$ (**Exercise 17**)

32. $y = (x - 3)^2$ (**Exercise 18**)

33. $y = |x + 4| - 3$ (**Exercise 21**)

34. $y = |x - 4| - 3$ (**Exercise 22**)

35. $y = (x - 3)^3$ (**Exercise 23**)

36. $y = (x - 2)^3 - 4$ (**Exercise 24**)

Without a graphing calculator, determine the domain and range of the functions.

37. $f(x) = (x - 1)^2 - 5$

38. $f(x) = (x + 8)^2 + 3$

39. $f(x) = \sqrt{x - 4}$

40. $f(x) = \sqrt{x + 1} - 10$

41. $f(x) = (x - 1)^3 + 4$

42. $f(x) = \sqrt[3]{x + 7} - 10$

Checking Analytic Skills *Use translations of one of the basic functions* $y = x^2$, $y = x^3$, $y = \sqrt{x}$, *or* $y = |x|$ *to sketch a graph of* $y = f(x)$ *by hand.* **Do not use a calculator.**

43. $y = (x - 1)^2$

44. $y = \sqrt{x + 2}$

45. $y = x^3 + 1$

46. $y = |x + 2|$

47. $y = (x - 1)^3$

48. $y = |x| - 3$

49. $y = \sqrt{x - 2} - 1$

50. $y = \sqrt{x + 3} - 4$

51. $y = (x + 2)^2 + 3$

52. $y = (x - 4)^2 - 4$

53. $y = |x + 4| - 2$

54. $y = (x + 3)^3 - 1$

Concept Check *Suppose that h and k are both positive numbers. Match each equation in Exercises 55–58 with the correct graph in choices A–D.*

55. $y = (x - h)^2 - k$

56. $y = (x + h)^2 - k$

57. $y = (x + h)^2 + k$

58. $y = (x - h)^2 + k$

A.

B.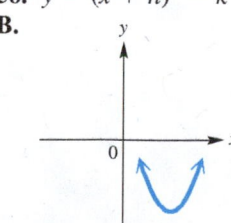

C.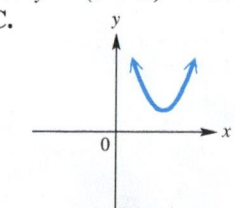

D.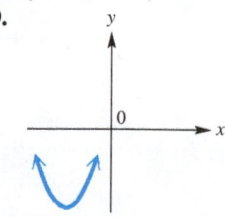

Concept Check *Given the graph shown, sketch by hand the graph of each function described, indicating how the three points labeled on the original graph have been translated.*

59. $y = f(x) + 2$ **60.** $y = f(x) - 2$ **61.** $y = f(x + 2)$ **62.** $y = f(x - 2)$

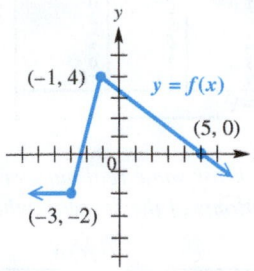

*Each graph is a translation of the graph of one of the basic functions $y = x^2$, $y = x^3$, $y = \sqrt{x}$, or $y = |x|$. Find the equation that defines each function. Then, using the concepts of increasing and decreasing functions discussed in **Section 2.1**, determine the largest open interval of the domain over which the function is **(a)** increasing and **(b)** decreasing.*

63.

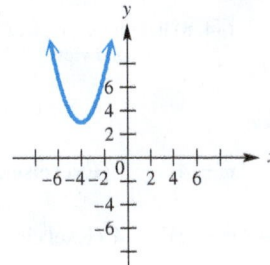

64.

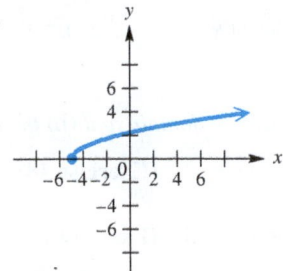

65.

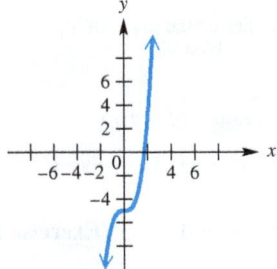

66.

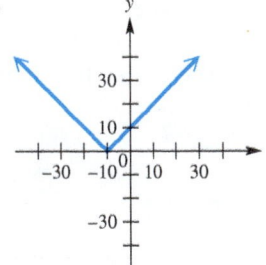

67.

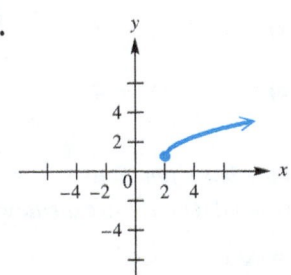

68.

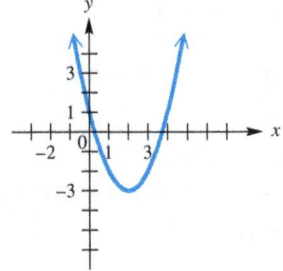

69. *Concept Check* The graph is a translation of $y = |x|$. What are the values of h and k if the equation is of the form $y = |x - h| + k$?

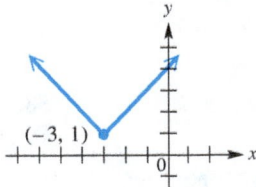

70. *Concept Check* Suppose that the graph of $y = x^2$ is translated in such a way that its domain is $(-\infty, \infty)$ and its range is $[38, \infty)$. What values of h and k can be used if the new function is of the form $y = (x - h)^2 + k$?

RELATING CONCEPTS For individual or group investigation (Exercises 71–74)

Use the x-intercept method and the given graph to solve each equation or inequality. (Hint: Extend the concepts of **Section 1.5** for graphical solution of linear equations and inequalities.)

71. (a) $f(x) = 0$
    (b) $f(x) > 0$
    (c) $f(x) < 0$

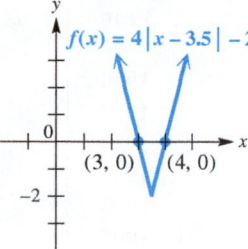

72. (a) $f(x) = 0$
    (b) $f(x) > 0$
    (c) $f(x) < 0$

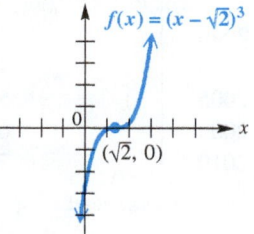

73. (a) $f(x) = 0$
    (b) $f(x) \geq 0$
    (c) $f(x) \leq 0$

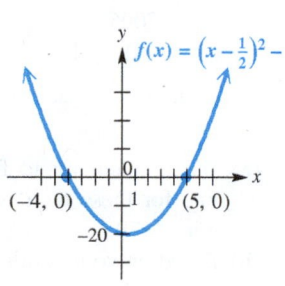

74. (a) $f(x) = 0$
    (b) $f(x) \geq 0$
    (c) $f(x) \leq 0$

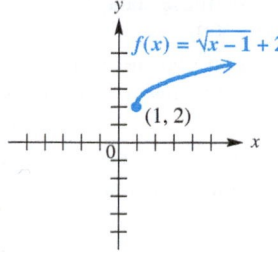

(Modeling) Solve each problem.

75. *Bankruptcies* The number of bankruptcies B filed in thousands x years *after* 2006 can be modeled by

$$y = B(x) = 66.25x + 160.$$

(a) Evaluate $B(4)$, and interpret your result.
(b) Use the formula for $B(x)$ to write an equation that gives the number of bankruptcies y filed in thousands during year x.
(c) Refer to part (b) and find y when $x = 2010$. Interpret your result.
(d) Use your equation in part (b) to determine the year when bankruptcies reached 293 thousand.

76. *U.S. Music Sales* The total music sales S from vinyl, CDs, and downloads in billions of dollars x years *after* 1999 can be modeled by

$$S(x) = -\frac{3}{7}x + 15.$$

(a) Evaluate $S(14)$, and interpret your result.
(b) Use the formula for $S(x)$ to write an equation that gives the total music sales y from vinyl, CDs, and downloads in billions of dollars during year x.
(c) Refer to part (b) and find y when $x = 2013$. Interpret your result.
(d) Use your equation in part (b) to determine the year these sales were \$12 billion.

Sales of Apple Products Average household spending on Apple products is shown in the figure for both U.S. sales and worldwide sales. Use this figure for Exercises 77–78.

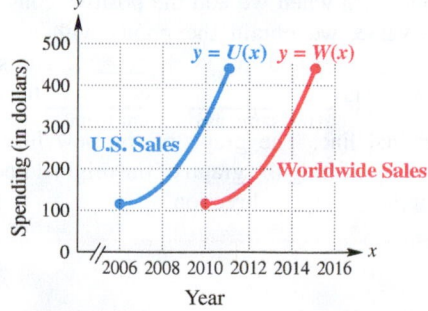

Average Household Spending on Apple Products

77. U.S. sales in dollars can be approximated during year x by

$$U(x) = 13(x - 2006)^2 + 115.$$

Evaluate $U(2011)$ and interpret your result.

78. Average worldwide household spending W on Apple products is expected to lag behind average U.S. household sales U by approximately four years. Write a formula for a function W that gives average worldwide household sales during year x. Evaluate $W(2015)$ and interpret your result.

79. *Cost of Public College Education* The table lists the average annual costs (in dollars) of tuition and fees at public four-year colleges for selected years.

Year	Tuition and Fees (in dollars)
2000	3505
2003	4632
2005	5491
2008	6532
2010	7605

Source: The College Board.

(a) Use a calculator to find the least-squares regression line for these data, where x is the number of years *after* 2000.

(b) Based on your result from part (a), write an equation that yields the same y-values when the actual year is entered.

(c) Estimate the cost of tuition and fees in 2009 to the nearest hundred dollars.

80. *Women in the Workforce* The table shows how the percent of women in the civilian workforce has changed from 1970 to 2010.

Year	Percent of Women in the Workforce
1970	43.3
1975	46.3
1980	51.5
1985	54.5
1990	57.5
1995	58.9
2000	59.9
2005	59.0
2010	58.6

Source: U.S. Bureau of Labor Statistics.

(a) Use a calculator to find the least-squares regression line for these data, where x is the number of years *after* 1970.

(b) Based on your result from part (a), write an equation that yields the same y-values when the actual year is entered.

(c) Predict the percent of population of women in the civilian workforce in 2015.

RELATING CONCEPTS For individual or group investigation (Exercises 81–88)

Recall from **Chapter 1** *that a unique line is determined by two distinct points on the line and that the values of m and b can then be determined for the general form of the linear function*

$$f(x) = mx + b.$$

Work these exercises in order.

81. Sketch by hand the line that passes through the points $(1, -2)$ and $(3, 2)$.

82. Use the slope formula to find the slope of this line.

83. Find the equation of the line, and write it in the form $y_1 = mx + b$.

84. Keeping the same two x-values as given in **Exercise 81,** add 6 to each y-value. What are the coordinates of the two new points?

85. Find the slope of the line passing through the points determined in **Exercise 84.**

86. Find the equation of this new line, and write it in the form $y_2 = mx + b$.

87. Graph both y_1 and y_2 in the standard viewing window of your calculator, and describe how the graph of y_2 can be obtained by vertically translating the graph of y_1. What is the value of the constant in this vertical translation? Where do you think it comes from?

88. Fill in the blanks with the correct responses, based on your work in **Exercises 81–87.**

If the points with coordinates (x_1, y_1) and (x_2, y_2) lie on a line, then when we add the positive constant c to each y-value, we obtain the points with coordinates $(x_1, y_1 + \underline{\quad})$ and $(x_2, y_2 + \underline{\quad})$. The slope of the new line is $\underline{\hspace{3cm}}$ the slope of
$\underline{\text{(the same as/different from)}}$
the original line. The graph of the new line can be obtained by shifting the graph of the original line $\underline{\quad}$ units in the $\underline{\quad}$ direction.

2.3 Stretching, Shrinking, and Reflecting Graphs

Vertical Stretching • Vertical Shrinking • Horizontal Stretching and Shrinking • Reflecting across an Axis • Combining Transformations of Graphs

In the previous section, we saw how adding or subtracting a constant to $f(x)$ can cause a vertical or horizontal shift to the graph of $y = f(x)$. Now we begin by seeing how multiplying by a constant alters the graph of a function.

Vertical Stretching

TECHNOLOGY NOTE

```
Plot1 Plot2 Plot3
\Y1=
\Y2=2Y1
\Y3=3Y1
\Y4=4Y1
\Y5=
\Y6=
\Y7=
```

By defining Y_1 as directed in parts A, B, and C, and defining Y_2, Y_3, and Y_4 as shown here on the TI-84 Plus, you can minimize your keystrokes. (These graphs will *not* appear unless Y_1 is defined.)

FOR DISCUSSION

In each group that follows, we give four related functions. Graph the functions in the first group (Group A), and then answer the questions. Repeat the process for Group B and Group C. Use the window specified for each group.

A	B	C
$[-5, 5]$ by $[-5, 20]$	$[-5, 15]$ by $[-5, 10]$	$[-20, 20]$ by $[-10, 10]$
$y_1 = x^2$	$y_1 = \sqrt{x}$	$y_1 = \sqrt[3]{x}$
$y_2 = 2x^2$	$y_2 = 2\sqrt{x}$	$y_2 = 2\sqrt[3]{x}$
$y_3 = 3x^2$	$y_3 = 3\sqrt{x}$	$y_3 = 3\sqrt[3]{x}$
$y_4 = 4x^2$	$y_4 = 4\sqrt{x}$	$y_4 = 4\sqrt[3]{x}$

1. How do the graphs of y_2, y_3, and y_4 compare with the graph of y_1?
2. If we choose $c > 4$, how do you think the graph of $y_5 = c \cdot y_1$ would compare with the graph of y_4? Provide support by choosing such a value of c.

In each group of functions in the preceding activity, we obtained a **vertical stretch** of the graph of the basic function with which we started. These observations can be generalized to any function.

Vertical Stretch

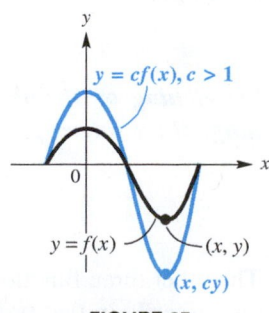

FIGURE 25

Vertical Stretching of the Graph of a Function

If a point (x, y) lies on the graph of $y = f(x)$, then the point (x, cy) lies on the graph of $y = cf(x)$. If $c > 1$, then the graph of $y = cf(x)$ is a *vertical stretching* of the graph of $y = f(x)$ by applying a factor of c.

In **FIGURE 25**, we graphically interpret the preceding statement. *Notice that the graphs have the same x-intercepts.*

Graphing $y = a|x|$, $a > 1$

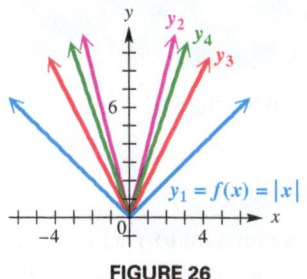

FIGURE 26

You can use a screen such as this to minimize your keystrokes in parts A, B, and C. Again, Y_1 must be defined in order to obtain the other graphs.

Vertical Shrink

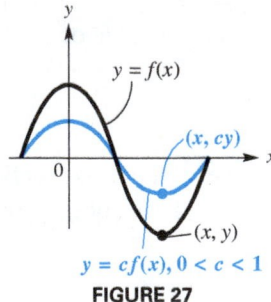

$y = cf(x), 0 < c < 1$

FIGURE 27

Graphing
$y = ax^3, 0 < a < 1$

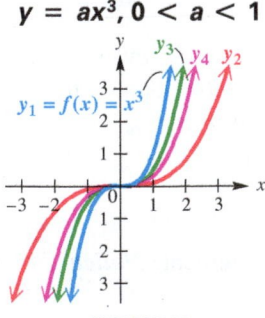

FIGURE 28

> **EXAMPLE 1** **Recognizing Vertical Stretches**

In **FIGURE 26,** the graph labeled y_1 is the function $f(x) = |x|$. The other three functions, y_2, y_3, and y_4, correspond to $2|x|$, $3|x|$, and $4|x|$, but not necessarily in that order. Determine the correct equation for each graph.

Solution The given values of c are 2, 3, and 4. The vertical heights of the points with the same x-coordinates of the three graphs will correspond to the magnitudes of these c values. Thus, the graph just above $y_1 = |x|$ will be that of $y = 2|x|$, the "highest" graph will be that of $y = 4|x|$, and the graph of $y = 3|x|$ will lie "between" the others. Therefore, $y_2 = 4|x|$, $y_3 = 2|x|$, and $y_4 = 3|x|$. ●

Vertical Shrinking

> **FOR DISCUSSION**
>
> This discussion parallels the earlier one in this section. Follow the same general directions. (*Note:* The fractions $\frac{3}{4}$, $\frac{1}{2}$, and $\frac{1}{4}$ may be entered as their decimal equivalents when plotting the graphs.)
>
> | **A** | **B** | **C** |
> |---|---|---|
> | $[-5, 5]$ by $[-5, 20]$ | $[-5, 15]$ by $[-2, 5]$ | $[-10, 10]$ by $[-2, 10]$ |
> | $y_1 = x^2$ | $y_1 = \sqrt{x}$ | $y_1 = |x|$ |
> | $y_2 = \frac{3}{4}x^2$ | $y_2 = \frac{3}{4}\sqrt{x}$ | $y_2 = \frac{3}{4}|x|$ |
> | $y_3 = \frac{1}{2}x^2$ | $y_3 = \frac{1}{2}\sqrt{x}$ | $y_3 = \frac{1}{2}|x|$ |
> | $y_4 = \frac{1}{4}x^2$ | $y_4 = \frac{1}{4}\sqrt{x}$ | $y_4 = \frac{1}{4}|x|$ |
>
> **1.** How do the graphs of y_2, y_3, and y_4 compare with the graph of y_1?
> **2.** If we choose $0 < c < \frac{1}{4}$, how do you think the graph of $y_5 = c \cdot y_1$ would compare with the graph of y_4? Provide support by choosing such a value of c.

In this "For Discussion" activity, we began with a basic function y_1, and in each case the graph of y_1 was **vertically shrunk** (or **compressed**). These observations can also be generalized to any function.

> **Vertical Shrinking of the Graph of a Function**
>
> If a point (x, y) lies on the graph of $y = f(x)$, then the point (x, cy) lies on the graph of $y = cf(x)$. If $0 < c < 1$, then the graph of $y = cf(x)$ is a *vertical shrinking* of the graph of $y = f(x)$ by applying a factor of c.

FIGURE 27 shows a vertical shrink graphically. *Vertical stretching or shrinking does not change the x-intercepts of the graph but it can change the y-intercept.*

> **EXAMPLE 2** **Recognizing Vertical Shrinks**

In **FIGURE 28,** the graph labeled y_1 is the function $f(x) = x^3$. The other three functions, y_2, y_3, and y_4, correspond to $0.5x^3$, $0.3x^3$, and $0.1x^3$, but not necessarily in that order. Determine the correct equation for each graph.

Solution   The smaller the positive value of c, where $0 < c < 1$, the more compressed toward the x-axis the graph will be. Since we have $c = 0.5, 0.3$, and 0.1, the function rules must be as follows: $y_2 = 0.1x^3$, $y_3 = 0.5x^3$, and $y_4 = 0.3x^3$. ●

Horizontal Stretching and Shrinking

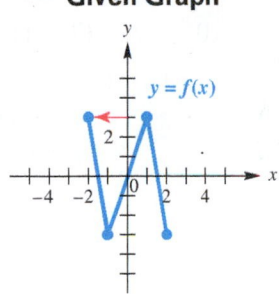

To minimize keystrokes when graphing Y_2 and Y_3, you can use a screen such as this.

Given Graph

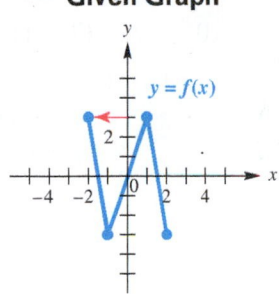

FIGURE 29

Horizontal Stretch

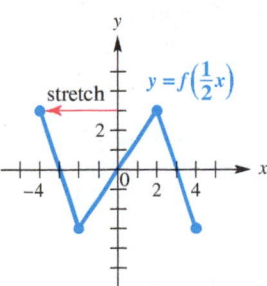

FIGURE 30

Horizontal Shrink

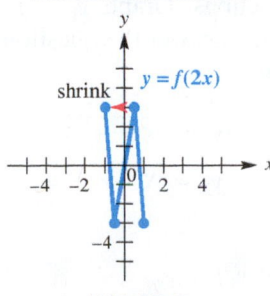

FIGURE 31

FOR DISCUSSION

This discussion parallels the two earlier ones in this section. Follow the same general directions. (*Note:* We give three related functions, $y_1 = f(x)$, $y_2 = f(2x)$, and $y_3 = f\left(\frac{1}{2}x\right)$, in each group.) Use the standard viewing window to graph the equations.

A	**B**	**C**
$y_1 = 5 - x^2$	$y_1 = x^3 - 4x$	$y_1 = \lvert x \rvert - 3$
$y_2 = 5 - (2x)^2$	$y_2 = (2x)^3 - 4(2x)$	$y_2 = \lvert 2x \rvert - 3$
$y_3 = 5 - \left(\frac{1}{2}x\right)^2$	$y_3 = \left(\frac{1}{2}x\right)^3 - 4\left(\frac{1}{2}x\right)$	$y_3 = \lvert \frac{1}{2}x \rvert - 3$

1. How do the graphs of y_2 and y_3 compare with the graph of y_1?
2. How does the graph of $y = f(cx)$ compare with the graph of $y = f(x)$ when $c > 1$ and when $0 < c < 1$? Support your answer by graphing.

In each group of functions in the preceding activity, we compared the graph of $y_1 = f(x)$ with the graphs of functions of the form $y = f(cx)$. In each case, we obtained a graph that was either a **horizontal stretch** or **shrink** of $y = f(x)$.

The line graph of $y = f(x)$ in **FIGURE 29** can be stretched or shrunk horizontally. The graph of $y = f\left(\frac{1}{2}x\right)$ in **FIGURE 30** is a horizontal stretching of the graph of $y = f(x)$, whereas the graph of $y = f(2x)$ in **FIGURE 31** is a horizontal shrinking of the graph of $y = f(x)$.

If one were to imagine the graph of $y = f(x)$ as a flexible wire, then a horizontal stretching would happen if the wire were pulled on each end, and a horizontal shrinking would happen if the wire were compressed. *Horizontal stretching or shrinking does not change the height (maximum or minimum y-values) of the graph, nor does it change the y-intercept.* These observations can be generalized to any function.

Horizontal Stretching or Shrinking of the Graph of a Function

If a point (x, y) lies on the graph of $y = f(x)$, then the point $\left(\frac{x}{c}, y\right)$ lies on the graph of $y = f(cx)$.

(a) If $0 < c < 1$, then the graph of $y = f(cx)$ is a *horizontal stretching* of the graph of $y = f(x)$.

(b) If $c > 1$, then the graph of $y = f(cx)$ is a *horizontal shrinking* of the graph of $y = f(x)$.

In **FIGURE 32** on the next page, we interpret these statements graphically. For example, if the point $(2, 0)$ is on the graph of $y = f(x)$, then the point $\left(\frac{2}{2}, 0\right) = (1, 0)$ is on the graph of $y = f(2x)$ and the point $\left(\frac{2}{\frac{1}{2}}, 0\right) = (4, 0)$ is on the graph of $y = f\left(\frac{1}{2}x\right)$. Notice in **FIGURE 32** and also in the next example that *horizontal stretching or shrinking can change the x-intercepts, but* **not** *the y-intercept.*

Horizontal Shrink   **Horizontal Stretch**

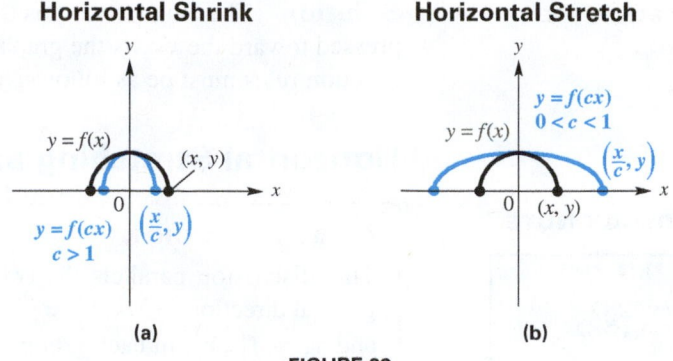

FIGURE 32

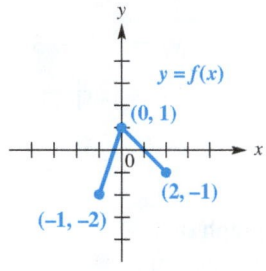

FIGURE 33

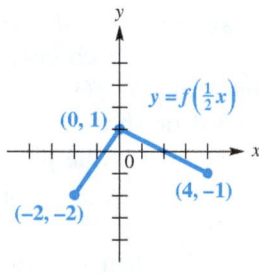

FIGURE 34

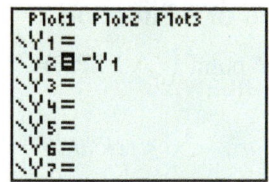

By defining Y_1 as directed and defining Y_2 as shown here, you can minimize your keystrokes.

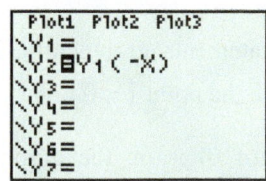

By defining Y_1 as directed and defining Y_2 as shown here (using function notation), you can minimize your keystrokes.

EXAMPLE 3 **Stretching a Graph Horizontally**

Use the graph of $y = f(x)$ in **FIGURE 33** to sketch a graph of $y = f\left(\frac{1}{2}x\right)$.

Solution Because $c = \frac{1}{2}$ and $0 < c < 1$, the graph of $y = f\left(\frac{1}{2}x\right)$ is a horizontal stretching of the graph of $y = f(x)$ and can be obtained by dividing each x-coordinate on the graph of $y = f(x)$ by $\frac{1}{2} = 0.5$. For the points $(-1, -2)$, $(0, 1)$, and $(2, -1)$, we obtain the following.

$$\left(\frac{-1}{0.5}, -2\right) = (-2, -2), \quad \left(\frac{0}{0.5}, 1\right) = (0, 1), \quad \left(\frac{2}{0.5}, -1\right) = (4, -1) \qquad \begin{array}{c}\text{Points on}\\ y = f\left(\frac{1}{2}x\right)\end{array}$$

Now sketch a line graph with these points. See **FIGURE 34.**

Reflecting across an Axis

FOR DISCUSSION

In each pair that follows, we give two related functions. Graph $y_1 = f(x)$ and $y_2 = -f(x)$ in the standard viewing window, and then answer the questions.

A	**B**	**C**	**D**
$y_1 = x^2$	$y_1 = |x|$	$y_1 = \sqrt{x}$	$y_1 = x^3$
$y_2 = -x^2$	$y_2 = -|x|$	$y_2 = -\sqrt{x}$	$y_2 = -x^3$

With respect to the x-axis, consider the following.

1. How does the graph of y_2 compare with the graph of y_1?

2. How would the graph of $y = -\sqrt[3]{x}$ compare with the graph of $y = \sqrt[3]{x}$, based on your answer to Item 1? Confirm your answer by graphing each equation.

In each pair that follows, we give two related functions. Graph $y_1 = f(x)$ and $y_2 = f(-x)$ in the standard viewing window, and then answer the questions.

E	**F**	**G**
$y_1 = \sqrt{x}$	$y_1 = \sqrt{x - 3}$	$y_1 = \sqrt[3]{x + 4}$
$y_2 = \sqrt{-x}$	$y_2 = \sqrt{-x - 3}$	$y_2 = \sqrt[3]{-x + 4}$

With respect to the y-axis, consider the following.

3. How does the graph of y_2 compare with the graph of y_1?

4. How would the graph of $y = \sqrt[3]{-x}$ compare with the graph of $y = \sqrt[3]{x}$, based on your answer to Item 3? Confirm your answer by graphing each equation.

Reflection across the *x*-Axis

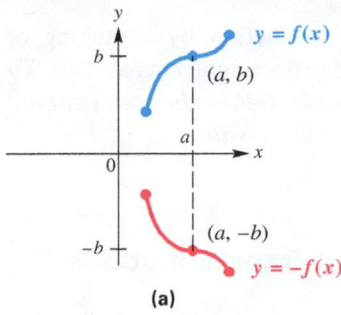

(a)

Reflection across the *y*-Axis

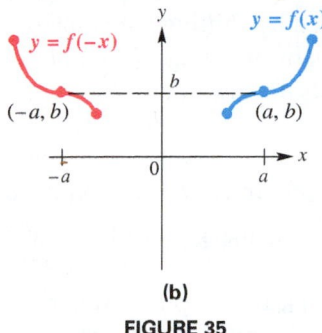

(b)

FIGURE 35

In this "For Discussion" activity, we began with a basic function y_1. In the first group of functions, we reflected y_1 across the *x*-axis. In the second group of functions, we reflected y_1 across the *y*-axis. The results can be formally summarized.

Reflecting the Graph of a Function across an Axis

For a function $y = f(x)$, the following are true.

(a) The graph of $y = -f(x)$ is a reflection of the graph of f across the *x*-axis.

(b) The graph of $y = f(-x)$ is a reflection of the graph of f across the *y*-axis.

FIGURE 35 shows how reflections affect the graph of a function in general.

EXAMPLE 4   **Applying Reflections across Axes**

FIGURE 36 shows the graph of a function $y = f(x)$.

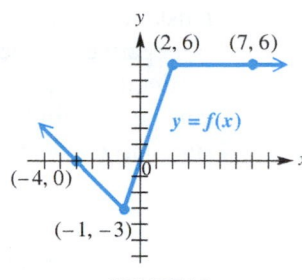

FIGURE 36

(a) Sketch the graph of $y = -f(x)$.   (b) Sketch the graph of $y = f(-x)$.

Solution

(a) We must reflect the graph across the *x*-axis. This means that if a point (a, b) lies on the graph of $y = f(x)$, then the point $(a, -b)$ must lie on the graph of $y = -f(x)$. Using the labeled points, we find the graph of $y = -f(x)$ in **FIGURE 37.**

Reflect across the *x*-Axis

For each point (a, b) on the given graph, plot the point $(a, -b)$.

FIGURE 37

Reflect across the *y*-Axis

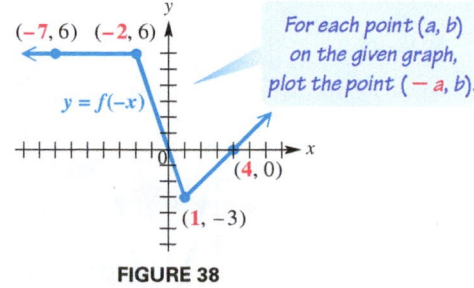

For each point (a, b) on the given graph, plot the point $(-a, b)$.

FIGURE 38

(b) We must reflect the graph across the *y*-axis, meaning that if a point (a, b) lies on the graph of $y = f(x)$, then the point $(-a, b)$ must lie on the graph of $y = f(-x)$. Thus, we obtain the graph of $y = f(-x)$ in **FIGURE 38.**

Combining Transformations of Graphs

We can create infinitely many functions from a basic function by stretching or shrinking, shifting upward, downward, left, or right, and reflecting across an axis. *To determine the order in which the transformations are made, follow the conventional order of operations as they would be applied to a particular x-value.*

EXAMPLE 5 **Describing Combinations of Transformations**

(a) Describe how the graph of $y = -3(x - 4)^2 + 5$ can be obtained by transforming the graph of $y = x^2$. Sketch its graph.

(b) Give the equation of the function that would be obtained by shifting the graph of $y = |x|$ to the left 3 units, vertically shrinking the graph by applying a factor of $\frac{2}{3}$, reflecting across the x-axis, and shifting the graph 4 units downward, in that order. Sketch its graph.

Solution

(a) In the definition of the function, $(x - 4)^2$ indicates that the graph of $y = x^2$ must be shifted 4 units to the right. Since the coefficient of $(x - 4)^2$ is -3 (a negative number with absolute value greater than 1), the graph is stretched vertically by applying a factor of 3 and then reflected across the x-axis. The constant $+5$ indicates that the graph is shifted upward 5 units. These ideas are summarized here.

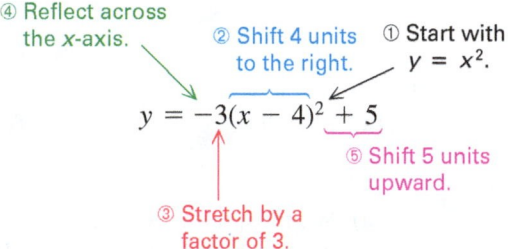

The graphs in **FIGURE 39** demonstrate Steps 2–5 to graph the transformation by hand. It is not essential to sketch all the graphs. We show them here only to illustrate the order of progression of transformation of the graph of $y = x^2$. Only the final graph is necessary.

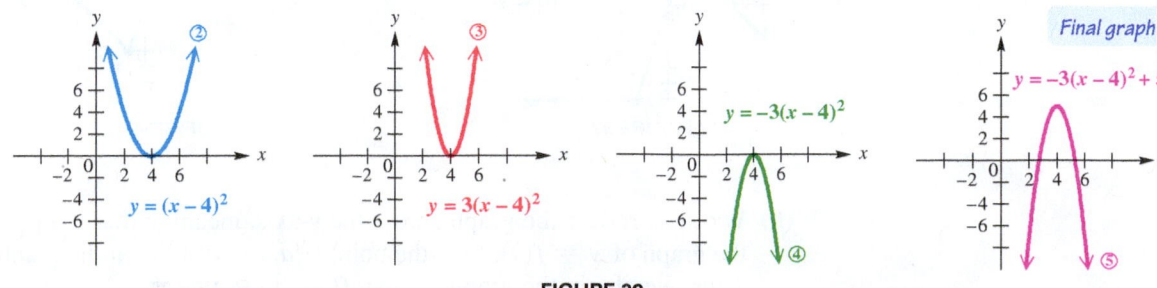

FIGURE 39

(b) Shifting the graph of $y = |x|$ to the left 3 units means that $|x|$ is transformed to $|x + 3|$. Vertically shrinking the graph by applying a factor of $\frac{2}{3}$ means multiplying by $\frac{2}{3}$, and reflecting across the x-axis changes $\frac{2}{3}$ to $-\frac{2}{3}$. Finally, shifting the graph 4 units downward means subtracting 4. Putting this all together leads to the equation

$$y = -\frac{2}{3}|x + 3| - 4.$$

The final graph is shown is **FIGURE 40**.

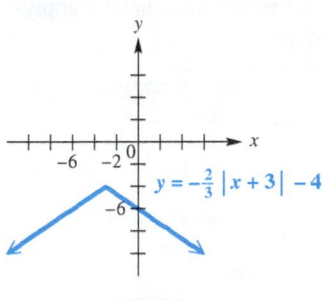

FIGURE 40

> **CAUTION**   *The order in which transformations are made can be important.* For example, changing the order of a stretch and shift can result in a different equation and graph.

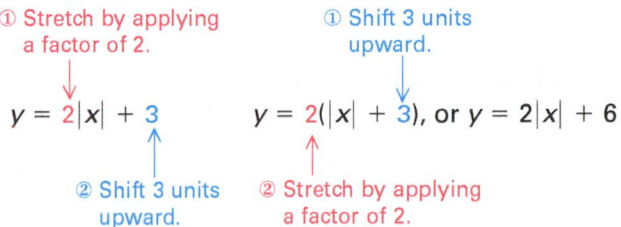

Also be careful when performing reflections and shifts. On one hand, if we reflect the graph of $y = \sqrt{x}$ across the y-axis to obtain $y = \sqrt{-x}$ and then shift it right 2 units, we obtain $y = \sqrt{-(x - 2)}$, or $y = \sqrt{-x + 2}$. On the other hand, if we shift the graph of $y = \sqrt{x}$ right 2 units to obtain $y = \sqrt{x - 2}$ and then reflect it across the y-axis, we obtain $y = \sqrt{-x - 2}$. The final equations are different and so are their graphs. (Try sketching each graph.)

> **EXAMPLE 6**   **Recognizing a Combination of Transformations**

FIGURE 41 shows the graph of $y = |x|$ in blue and another graph in red illustrating a combination of transformations. Find the equation of the transformed graph.

Solution   **FIGURE 41** shows that the lowest point on the transformed graph has coordinates $(3, -2)$, indicating that the graph has been shifted 3 units to the right and 2 units downward. Note also that a point on the right side of the transformed graph has coordinates $(4, 1)$. Thus, the slope of this ray is

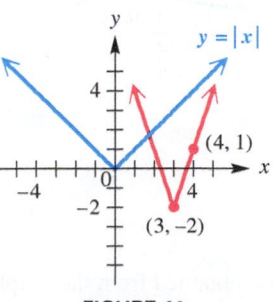

FIGURE 41

Start with the x- and y-values of the same point.

$$m = \frac{-2 - 1}{3 - 4} = \frac{-3}{-1} = 3.$$

The stretch factor is 3, and the equation of the transformed graph is

$$y = 3|x - 3| - 2.$$

2.3 Exercises

Checking Analytic Skills Write the equation that results in the desired transformation. **Do not use a calculator.**

1. The squaring function, vertically stretched by applying a factor of 2

2. The cubing function, vertically shrunk by applying a factor of $\frac{1}{2}$

3. The square root function, reflected across the y-axis

4. The cube root function, reflected across the x-axis

5. The absolute value function, vertically stretched by applying a factor of 3 and reflected across the x-axis

6. The absolute value function, vertically shrunk by applying a factor of $\frac{1}{3}$ and reflected across the y-axis

7. The cubing function, vertically shrunk by applying a factor of 0.25 and reflected across the y-axis

8. The square root function, vertically shrunk by applying a factor of 0.2 and reflected across the x-axis

Use transformations of graphs to sketch the graphs of y_1, y_2, and y_3 by hand. Check by graphing in an appropriate viewing window of your calculator.

9. $y_1 = x$, $y_2 = x + 3$, $y_3 = x - 3$

10. $y_1 = x^3$, $y_2 = x^3 + 4$, $y_3 = x^3 - 4$

11. $y_1 = |x|$, $y_2 = |x - 3|$, $y_3 = |x + 3|$

12. $y_1 = |x|$, $y_2 = |x| - 3$, $y_3 = |x| + 3$

13. $y_1 = \sqrt{x}$, $y_2 = \sqrt{x + 6}$, $y_3 = \sqrt{x - 6}$

14. $y_1 = |x|$, $y_2 = 2|x|$, $y_3 = 2.5|x|$

15. $y_1 = \sqrt[3]{x}$, $y_2 = -\sqrt[3]{x}$, $y_3 = -2\sqrt[3]{x}$

16. $y_1 = x^2$, $y_2 = (x - 2)^2 + 1$, $y_3 = -(x + 2)^2$

17. $y_1 = |x|$, $y_2 = -2|x - 1| + 1$, $y_3 = -\frac{1}{2}|x| - 4$

18. $y_1 = \sqrt{x}$, $y_2 = -\sqrt{x}$, $y_3 = \sqrt{-x}$

19. $y_1 = x^2 - 1$, $y_2 = \left(\frac{1}{2}x\right)^2 - 1$, $y_3 = (2x)^2 - 1$

20. $y_1 = 3 - |x|$, $y_2 = 3 - |3x|$, $y_3 = 3 - \left|\frac{1}{3}x\right|$

21. $y_1 = \sqrt[3]{x}$, $y_2 = \sqrt[3]{-x}$, $y_3 = \sqrt[3]{-(x - 1)}$

22. $y_1 = \sqrt[3]{x}$, $y_2 = 2 - \sqrt[3]{x}$, $y_3 = 1 + \sqrt[3]{x}$

In Exercises 23–26, the graph of $y = f(x)$ has been transformed to the graph of $y = g(x)$. No shrinking or stretching is involved. Give the equation of $y = g(x)$.

23.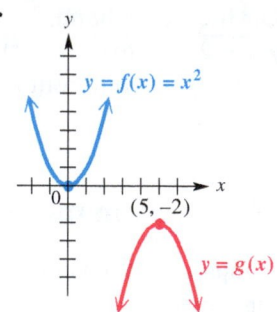

24.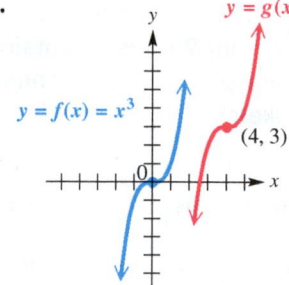

25.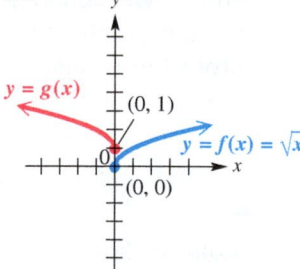

26.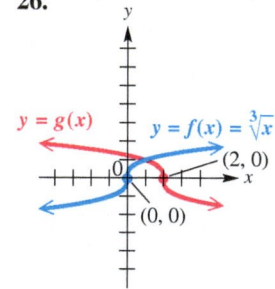

Concept Check Fill in each blank with the appropriate response. (Remember that the vertical stretch or shrink factor is positive.)

27. The graph of $y = -4x^2$ can be obtained from the graph of $y = x^2$ by vertically stretching by applying a factor of _____ and reflecting across the _____-axis.

28. The graph of $y = -6\sqrt{x}$ can be obtained from the graph of $y = \sqrt{x}$ by vertically stretching by applying a factor of _____ and reflecting across the _____-axis.

29. The graph of $y = -\frac{1}{4}|x + 2| - 3$ can be obtained from the graph of $y = |x|$ by shifting horizontally _____ units to the _____, vertically shrinking by applying a factor of _____, reflecting across the _____-axis, and shifting vertically _____ units in the _____ direction.

30. The graph of $y = -\frac{2}{5}|-x| + 6$ can be obtained from the graph of $y = |x|$ by reflecting across the _____-axis, vertically shrinking by applying a factor of _____, reflecting across the _____-axis, and shifting vertically _____ units in the _____ direction.

31. The graph of $y = 6\sqrt[3]{x} - 3$ can be obtained from the graph of $y = \sqrt[3]{x}$ by shifting horizontally _____ units to the _____ and stretching vertically by applying a factor of _____.

32. The graph of $y = 0.5\sqrt[3]{x + 2}$ can be obtained from the graph of $y = \sqrt[3]{x}$ by shifting horizontally _____ units to the _____ and shrinking vertically by applying a factor of _____.

Give the equation of each function whose graph is described.

33. The graph of $y = x^2$ is vertically shrunk by applying a factor of $\frac{1}{2}$, and the resulting graph is shifted 7 units downward.

34. The graph of $y = x^3$ is vertically stretched by applying a factor of 3. This graph is then reflected across the x-axis. Finally, the graph is shifted 8 units upward.

35. The graph of $y = \sqrt{x}$ is shifted 3 units to the right. This graph is then vertically stretched by applying a factor of 4.5. Finally, the graph is shifted 6 units downward.

36. The graph of $y = \sqrt[3]{x}$ is shifted 2 units to the left. This graph is then vertically stretched by applying a factor of 1.5. Finally, the graph is shifted 8 units upward.

Checking Analytic Skills   *Use transformations of graphs to sketch a graph of $y = f(x)$ by hand.*
Do not use a calculator.

37. $f(x) = \sqrt{x - 3} + 2$

38. $f(x) = |x + 2| - 3$

39. $f(x) = \sqrt{2x}$

40. $f(x) = \frac{1}{2}(x + 2)^2$

41. $f(x) = |2x|$

42. $f(x) = \frac{1}{2}|x|$

43. $f(x) = 1 - \sqrt{x}$

44. $f(x) = 2\sqrt{x - 2} - 1$

45. $f(x) = -\sqrt{1 - x}$

46. $f(x) = \sqrt{-x} - 1$

47. $f(x) = \sqrt{-(x + 1)}$

48. $f(x) = 2 + \sqrt{-(x - 3)}$

49. $f(x) = (x - 1)^3$

50. $f(x) = (x + 2)^3$

51. $f(x) = -x^3$

52. $f(x) = (-x)^3 + 1$

In Exercises 53–58, each figure shows the graph of $y = f(x)$. Sketch by hand the graphs of the functions in parts (a), (b), and (c), and answer the question in part (d).

53. (a) $y = -f(x)$ (b) $y = f(-x)$ (c) $y = 2f(x)$
  (d) What is $f(0)$?

54. (a) $y = -f(x)$ (b) $y = f(-x)$ (c) $y = 3f(x)$
  (d) What is $f(4)$?

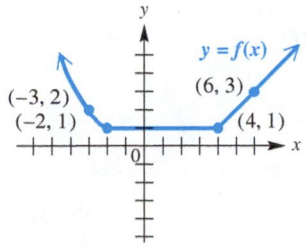

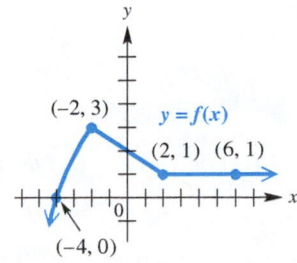

55. (a) $y = -f(x)$ **(b)** $y = f(-x)$ **(c)** $y = f(x + 1)$
   (d) What are the x-intercepts of $y = f(x - 1)$?

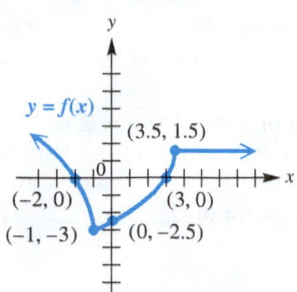

56. (a) $y = -f(x)$ **(b)** $y = f(-x)$ **(c)** $y = \frac{1}{2}f(x)$
   (d) On what largest interval of the domain is $f(x) < 0$?

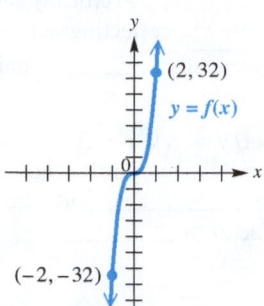

57. (a) $y = -f(x)$ **(b)** $y = f\left(\frac{1}{3}x\right)$ **(c)** $y = 0.5f(x)$
   (d) What symmetry does the graph of $y = f(x)$ exhibit?

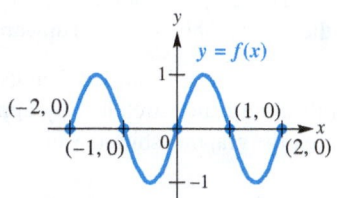

58. (a) $y = f(2x)$ **(b)** $y = f(-x)$ **(c)** $y = 3f(x)$
   (d) What symmetry does the graph of $y = f(x)$ exhibit?

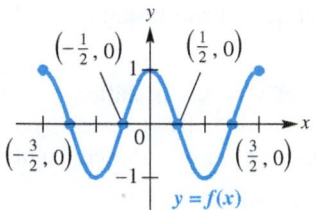

In Exercises 59–62, use the accompanying graph of $y = f(x)$ below to sketch a graph of each equation.

59. (a) $y = f(x) + 1$

  (b) $y = -f(x) - 1$

  (c) $y = 2f\left(\frac{1}{2}x\right)$

60. (a) $y = f(x) - 2$

  (b) $y = f(x - 1) + 2$

  (c) $y = 2f(x)$

61. (a) $y = f(2x) + 1$

  (b) $y = 2f\left(\frac{1}{2}x\right) + 1$

  (c) $y = \frac{1}{2}f(x - 2)$

62. (a) $y = f(2x)$

  (b) $y = f\left(\frac{1}{2}x\right) - 1$

  (c) $y = 2f(x) - 1$

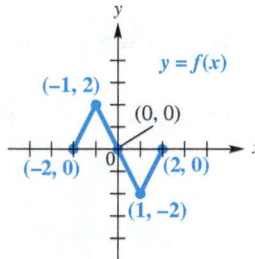

63. *Concept Check* If $(r, 0)$ is an x-intercept of the graph of $y = f(x)$, what statement can be made about an x-intercept of the graph of each function? (*Hint:* Make a sketch.)
   (a) $y = -f(x)$
   (b) $y = f(-x)$
   (c) $y = -f(-x)$

64. *Concept Check* If $(0, b)$ is the y-intercept of the graph of $y = f(x)$, what statement can be made about the y-intercept of the graph of each function? (*Hint:* Make a sketch.)
   (a) $y = -f(x)$
   (b) $y = f(-x)$
   (c) $y = 5f(x)$
   (d) $y = -3f(x)$

Let the domain of $f(x)$ be $[-1, 2]$ and the range be $[0, 3]$. Find the domain and range of the following.

65. $f(x - 2)$

66. $5f(x + 1)$

67. $-f(x)$

68. $f(x - 3) + 1$

69. $f(2x)$

70. $2f(x - 1)$

71. $3f\left(\frac{1}{4}x\right)$

72. $-2f(4x)$

73. $f(-x)$

74. $-2f(-x)$

75. $f(-3x)$

76. $\frac{1}{3}f(x - 3)$

In Exercises 77–79, each function has a graph with an endpoint (a translation of the point (0, 0).)
Enter each into your calculator in an appropriate viewing window, and using your knowledge of the
graph of $y = \sqrt{x}$, determine the domain and range of the function. (Hint: Locate the endpoint.)

77. $y = 10\sqrt{x - 20} + 5$ **78.** $y = -2\sqrt{x + 15} - 18$ **79.** $y = -0.5\sqrt{x + 10} + 5$

80. *Concept Check*   Based on your observations in **Exercise 77,** what are the domain and range of
$f(x) = a\sqrt{x - h} + k$ if $a > 0, h > 0$, and $k > 0$?

*Concept Check   The sketch shows an example of a function $y = f(x)$ that increases on the open
interval (a, b). Use this graph as a visual aid, and apply the concepts of reflection introduced in this
section to answer each question.*

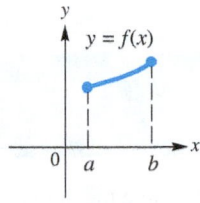

81. Does the graph of $y = -f(x)$ increase or decrease on the interval (a, b)?

82. Does the graph of $y = f(-x)$ increase or decrease on the interval $(-b, -a)$?

83. Does the graph of $y = -f(-x)$ increase or decrease on the interval $(-b, -a)$?

84. If $c > 0$, does the graph of $y = -cf(x)$ increase or decrease on the interval (a, b)?

*State the open intervals over which each function is **(a)** increasing, **(b)** decreasing, and **(c)** constant.*

85. The function graphed in **FIGURE 36**   **86.** The function graphed in **FIGURE 37**

87. The function graphed in **FIGURE 38**   **88.** $y = -\dfrac{2}{3}|x + 3| - 4$ (See **FIGURE 40.**)

*Concept Check   Exercises 89 and 90 show the graphs of $y_1 = \sqrt[3]{x}$ and $y_2 = 5\sqrt[3]{x}$. The point whose
coordinates are given at the bottom of the screen lies on the graph of y_1. Use this graph, not your calculator,
to find the coordinates of the corresponding point on the graph of y_2 for the same value of x shown.*

89.

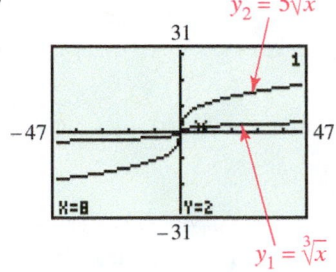

90.

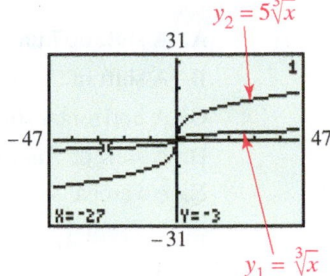

*In Exercises 91–96, the figure shows a transformation of the graph of $y = |x|$. Write the equation for
the graph. Refer to **Example 6** as needed.*

91.

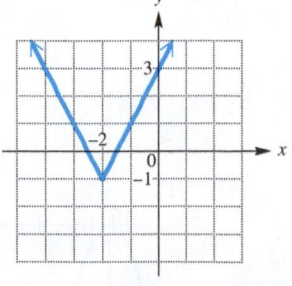

92.

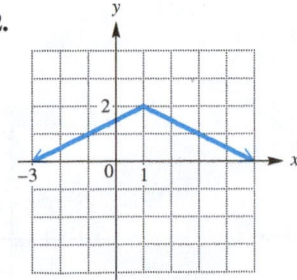

93.

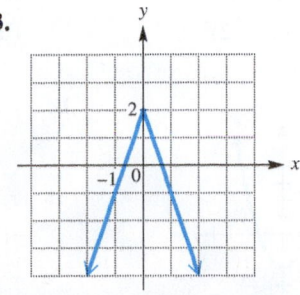

94.

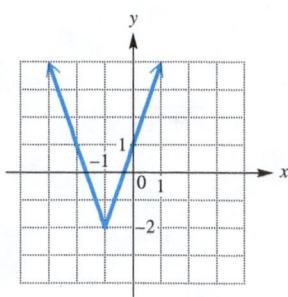

95.

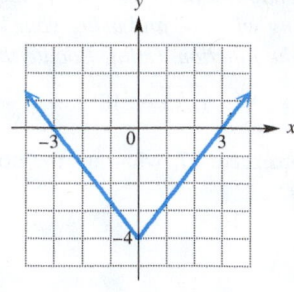

96.

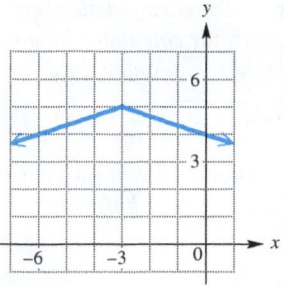

97. *Concept Check*   Suppose that the graph of $y = f(x)$ is symmetric with respect to the y-axis and is reflected across the y-axis. How will the new graph compare with the original one?

SECTIONS 2.1–2.3   Reviewing Basic Concepts

1. Give the domain and range of each function. Then determine the largest open interval of the domain over which each function f is increasing or decreasing.
   (a) $f(x) = |x| + 1$
   (b) $f(x) = (x - 2)^2$
   (c) $f(x) = -\sqrt{x}$

2. Suppose that f is defined for all real numbers and $f(3) = 6$. For each of the given assumptions, find another function value.
   (a) The graph of $y = f(x)$ is symmetric with respect to the origin.
   (b) The graph of $y = f(x)$ is symmetric with respect to the y-axis.
   (c) For all x, $f(-x) = -f(x)$.
   (d) For all x, $f(-x) = f(x)$.

3. Match each equation in Column I with a description of its graph from Column II, as it relates to the graph of $y = x^2$.

I	**II**
(a) $y = (x - 7)^2$	**A.** A shift of 7 units to the left
(b) $y = x^2 - 7$	**B.** A shift of 7 units to the right
(c) $y = 7x^2$	**C.** A horizontal stretch
(d) $y = (x + 7)^2$	**D.** A shift of 7 units downward
(e) $y = \left(\dfrac{1}{3}x\right)^2$	**E.** A vertical stretch by applying a factor of 7

4. Match each equation in parts (a)–(h) with the sketch of its graph in choices A–H below and on the next page. The graph of $y = x^2$ is shown on the right for reference.

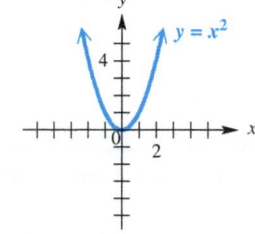

   (a) $y = x^2 + 2$
   (b) $y = x^2 - 2$
   (c) $y = (x + 2)^2$
   (d) $y = (x - 2)^2$
   (e) $y = 2x^2$
   (f) $y = -x^2$
   (g) $y = (x - 2)^2 + 1$
   (h) $y = (x + 2)^2 + 1$

A.

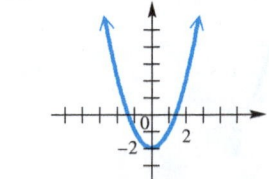

B.

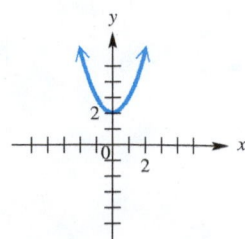

C.

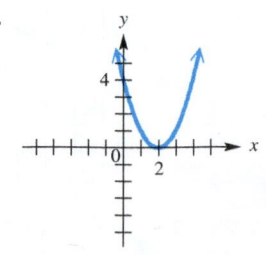

D.

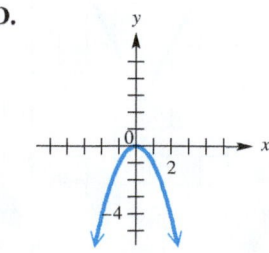

E.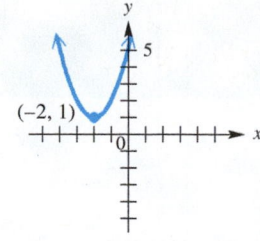

F.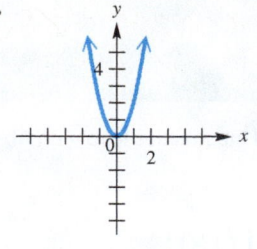

G.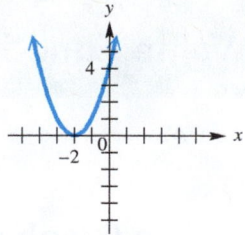

H.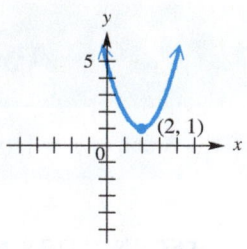

5. Sketch each graph of $y = f(x)$ by hand.
   (a) $y = |x| + 4$ (b) $y = |x + 4|$ (c) $y = |x - 4|$ (d) $y = |x + 2| - 4$ (e) $y = -|x - 2| + 4$

6. Describe the transformation of each graph of $f(x) = |x|$ or $g(x) = \sqrt{x}$, and then give the equation of the graph.

   (a)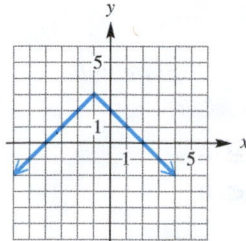

   (b)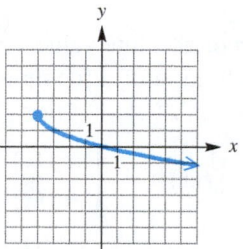

   (c)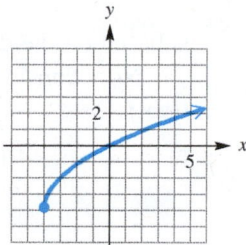

   (d)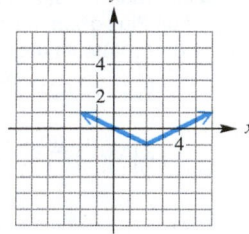

7. Consider the two functions in the figure.
   (a) Find a value of c for which $g(x) = f(x) + c$.
   (b) Find a value of c for which $g(x) = f(x + c)$.

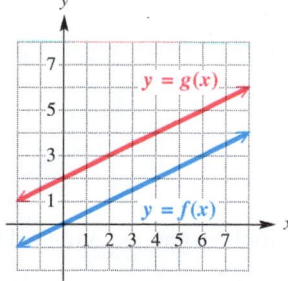

8. Suppose the equation $y = F(x)$ is changed to the equation $y = F(x + h)$. How are the graphs of these equations related? Is the graph of $y = F(x) + h$ the same as the graph of $y = F(x + h)$? If not, how do they differ?

9. Complete the table if
   (a) f is an even function and
   (b) f is an odd function.

x	$f(x)$
-3	4
-2	-6
-1	5
1	
2	
3	

10. *Google Ad Dollars*   Google's advertising revenues R in billions of dollars x years after 2004 can be modeled by $R(x) = 5x + 2$.
    (a) Evaluate $R(7)$, and interpret your result.
    (b) Use the formula for $R(x)$ to write an equation that gives the ad revenues y for Google in billions of dollars during year x.
    (c) Refer to part (b) and find y when $x = 2011$. Interpret your result.
    (d) Use your equation in part (b) to determine when Google's ad revenues first reached $27 billion.

2.4 Absolute Value Functions

The Graph of $y = |f(x)|$ • Properties of Absolute Value • Equations and Inequalities Involving Absolute Value

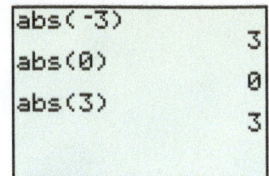

The command abs(x) is used by the TI-84 Plus graphing calculator to find absolute value.

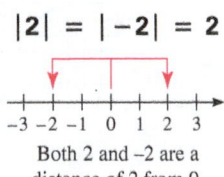

$|2| = |-2| = 2$

Both 2 and –2 are a distance of 2 from 0.

FIGURE 42

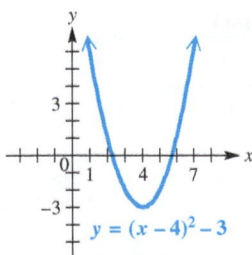

$y = (x - 4)^2 - 3$

FIGURE 43

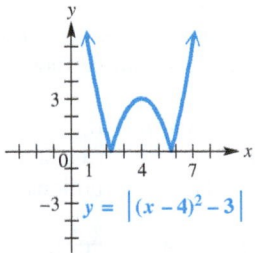

$y = |(x - 4)^2 - 3|$

FIGURE 44

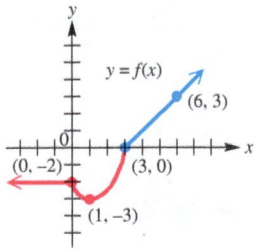

FIGURE 45

The Graph of $y = |f(x)|$

Geometrically, the absolute value of a real number is its undirected distance from 0 on the number line. See **FIGURE 42.** *As a result, the absolute value of a real number is never negative. It is always greater than or equal to 0.* Thus, the function in **Section 2.1**

$$f(x) = |x| = \begin{cases} x & \text{if } x \geq 0 \\ -x & \text{if } x < 0 \end{cases}$$

is based on the definition of absolute value.

Now consider the definition of a function that is the *absolute value of any function f.*

$$|f(x)| = \begin{cases} f(x) & \text{if } f(x) \geq 0 \\ -f(x) & \text{if } f(x) < 0 \end{cases}$$

To graph a function of the form $y = |f(x)|$, we leave the graph of f unchanged for the portion on or above the x-axis, and reflect the portion below the x-axis across the x-axis. The domain of $y = |f(x)|$ is the same as the domain of f, while the range of $y = |f(x)|$ will be a subset of $[0, \infty)$.

EXAMPLE 1 **Finding Domains and Ranges of $y = f(x)$ and $y = |f(x)|$**

FIGURE 43 shows the graph of

$$y = (x - 4)^2 - 3,$$

which is the graph of $y = x^2$ shifted 4 units to the right and 3 units downward. **FIGURE 44** shows the graph of

$$y = |(x - 4)^2 - 3|.$$

Give the domain and range of each function.

Solution All points with negative y-values in the first graph have been reflected across the x-axis in the second graph, while all points with nonnegative y-values are the same for both graphs. The domain of each function is $(-\infty, \infty)$. The range of $y = (x - 4)^2 - 3$ is $[-3, \infty)$, while the range of $y = |(x - 4)^2 - 3|$ is $[0, \infty)$. ●

EXAMPLE 2 **Graphing $y = |f(x)|$, Given the Graph of $y = f(x)$**

FIGURE 45 shows the graph of a function $y = f(x)$. Use the figure to sketch the graph of $y = |f(x)|$. Give the domain and range of each function.

Solution The graph will remain the same for points whose y-values are nonnegative, while it will be reflected across the x-axis for all other points. **FIGURE 46** on the next page shows the graph of $y = |f(x)|$. The domain of both functions is $(-\infty, \infty)$. The range of $y = f(x)$ is $[-3, \infty)$, while the range of $y = |f(x)|$ is $[0, \infty)$. ●

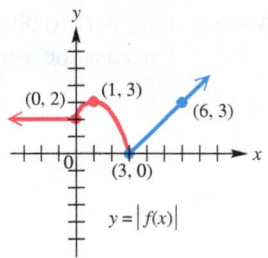

FIGURE 46

Properties of Absolute Value

> **Properties of Absolute Value**
>
> For all real numbers a and b,
>
> **1.** $|ab| = |a| \cdot |b|$
> (*The absolute value of a product is equal to the product of the absolute values.*)
>
> **2.** $\left|\dfrac{a}{b}\right| = \dfrac{|a|}{|b|}$ $(b \neq 0)$
> (*The absolute value of a quotient is equal to the quotient of the absolute values.*)
>
> **3.** $|a| = |-a|$
> (*The absolute value of a number is equal to the absolute value of its additive inverse.*)
>
> **4.** $|a| + |b| \geq |a + b|$
> **(triangle inequality)**
> (*The sum of the absolute values of two numbers is greater than or equal to the absolute value of their sum.*)

Consider the absolute value function $y = |2x + 11|$.

$$y = |2x + 11|$$

$$y = \left|2\left(x + \frac{11}{2}\right)\right| \qquad \text{Factor out 2.}$$

$$y = |2| \cdot \left|x + \frac{11}{2}\right| \qquad \text{Property 1, } |ab| = |a| \cdot |b|$$

$$y = 2\left|x + \frac{11}{2}\right| \qquad |2| = 2$$

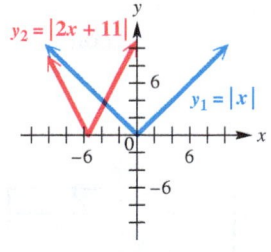

FIGURE 47

We could find the graph of $y = 2\left|x + \frac{11}{2}\right|$ by starting with the graph of $y = |x|$, shifting $\frac{11}{2}$ units to the left, and then vertically stretching by applying a factor of 2. The graphs in **FIGURE 47** support this statement.

We are often interested in absolute value functions of the form $f(x) = |ax + b|$, where the expression inside the absolute value bars is linear. *A comprehensive graph of $f(x) = |ax + b|$ will include all intercepts and the lowest point on the "V-shaped" graph.*

EXAMPLE 3   **Graphing $y = |ax + b|$ by Hand**

Graph each equation. Then comment on how the graphs are related.

(a) $y = 2x - 1$ **(b)** $y = |2x - 1|$

Solution

(a) The graph of $y = 2x - 1$ is a line with slope 2 and y-intercept $(0, -1)$, as shown in **FIGURE 48.**

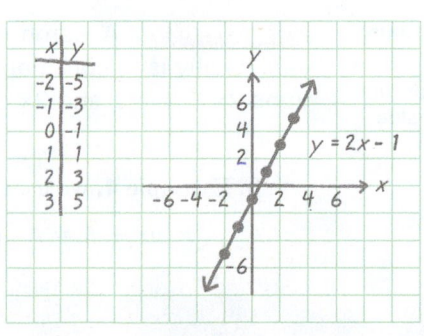

FIGURE 48

(*continued*)

(b) The graph of $y = |2x - 1|$ is V-shaped, as shown in **FIGURE 49.** One way to obtain the graph of $y = |2x - 1|$ is to reflect the graph of $y = 2x - 1$ across the x-axis whenever that graph is below that axis.

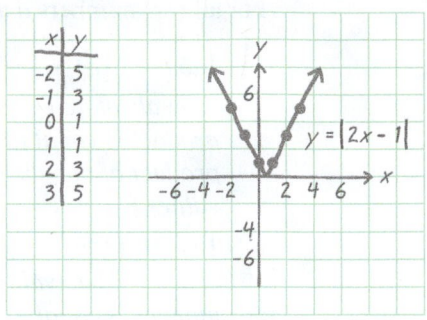

FIGURE 49

In general, the graph of $y = ax + b$ $(a \neq 0)$ is a line that crosses the x-axis, whereas the graph of $y = |ax + b|$ is V-shaped and does not cross the x-axis. Both graphs have the same x-intercept.

Equations and Inequalities Involving Absolute Value

EXAMPLE 4 **Solving an Absolute Value Equation**

Solve $|2x + 1| = 7$.

Analytic Solution

For $|2x + 1|$ to equal 7, $2x + 1$ must be 7 units from 0 on the number line. This can happen only when $2x + 1 = 7$ or $2x + 1 = -7$. Solve this compound equation as follows.

$$2x + 1 = 7 \quad \text{or} \quad 2x + 1 = -7$$
$$2x = 6 \quad \text{or} \qquad 2x = -8 \qquad \text{Subtract 1.}$$
$$x = 3 \quad \text{or} \qquad\quad x = -4 \qquad \text{Divide by 2.}$$

Check by substitution in the original equation that the solution set is $\{-4, 3\}$.

Graphing Calculator Solution

FIGURE 50 shows that the graphs of $y_1 = |2x + 1|$ and $y_2 = 7$ intersect when $x = -4$ or $x = 3$, confirming the analytic solution.

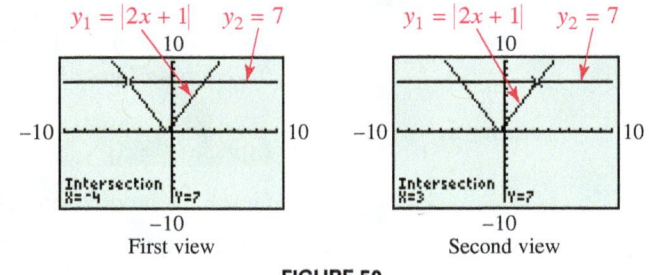

FIGURE 50

→ **Looking Ahead to Calculus**
The formal definition of limit, one of the most important concepts in calculus, uses two absolute value inequalities.

CAUTION A common error that can occur when solving $|2x + 1| = 7$ analytically is to consider only one part of the "or" statement. Both of the equations $2x + 1 = 7$ and $2x + 1 = -7$ must be considered.

In **Example 4**, we used the property that if $|ax + b| = k$, where $k > 0$, then

$$ax + b = k \quad \text{or} \quad ax + b = -k. \qquad \text{Case 1 on the next page}$$

The following summary indicates the three cases that may apply when solving absolute value equations and inequalities analytically.

Solving Absolute Value Equations and Inequalities

Let k be a positive number.

Case 1 To solve $|ax + b| = k$, solve the following compound equation.

$$ax + b = k \quad \text{or} \quad ax + b = -k$$

Case 2 To solve $|ax + b| > k$, solve the following compound inequality.

$$ax + b > k \quad \text{or} \quad ax + b < -k$$

Case 3 To solve $|ax + b| < k$, solve the following three-part inequality.

$$-k < ax + b < k$$

Inequalities involving $\leq$ or $\geq$ are solved similarly, by using the equality part of the symbol as well.

EXAMPLE 5 **Solving Absolute Value Inequalities**

Solve each inequality. **(a)** $|2x + 1| > 7$ **(b)** $|2x + 1| < 7$

Analytic Solution

(a) The inequality $|2x + 1| > 7$ can be rewritten as

$$2x + 1 > 7 \quad \text{or} \quad 2x + 1 < -7,$$

because $2x + 1$ represents a number that is *more* than 7 units from 0 on either side of the number line. Now, solve this compound inequality.

$2x + 1 > 7$ or	$2x + 1 < -7$	Case 2
$2x > 6$ or	$2x < -8$	Subtract 1.
$x > 3$ or	$x < -4$	Divide by 2.

The solution set is

$$(-\infty, -4) \cup (3, \infty).$$

(b) In $|2x + 1| < 7$, the expression $2x + 1$ must represent a number that is less than 7 units from 0 on the number line. This means that $2x + 1$ must be *between* -7 and 7. This statement is written as a three-part inequality.

$-7 < 2x + 1 < 7$	Case 3
$-8 < 2x < 6$	Subtract 1 from each part.
$-4 < x < 3$	Divide each part by 2.

The solution set is the open interval

$$(-4, 3).$$

Graphing Calculator Solution

(a) As seen in **FIGURE 51**, the graph of $y_1 = |2x + 1|$ lies *above* the graph of $y_2 = 7$ for x-values less than -4 or greater than 3, so the solution set is $(-\infty, -4) \cup (3, \infty)$.

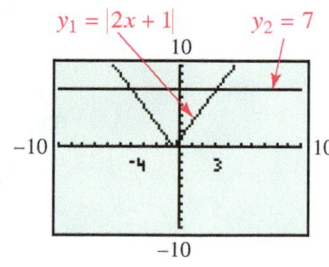

FIGURE 51

(b) **FIGURE 51** also shows that the graph of $y_1 = |2x + 1|$ lies *below* the graph of $y_2 = 7$ for x-values between -4 and 3, confirming the analytic result.

> **EXAMPLE 6** **Solving Absolute Value Equations and Inequalities**

Solve each equation or inequality.

(a) $|4 - 3x| = 2$ **(b)** $|4 - 3x| \geq 2$ **(c)** $|4 - 3x| \leq 2$

Solution

(a) $|4 - 3x| = 2$ Original equation

$4 - 3x = 2$ or $4 - 3x = -2$ Case 1

$-3x = -2$ or $-3x = -6$ Subtract 4.

$x = \dfrac{2}{3}$ or $x = 2$ Divide by -3.

The solution set is $\left\{\dfrac{2}{3}, 2\right\}$.

(b) $|4 - 3x| \geq 2$ Original inequality

$4 - 3x \geq 2$ or $4 - 3x \leq -2$ Case 2

$-3x \geq -2$ or $-3x \leq -6$ Subtract 4.

$x \leq \dfrac{2}{3}$ or $x \geq 2$ Divide by -3 and reverse the direction of the inequality symbols.

The solution set is $\left(-\infty, \dfrac{2}{3}\right] \cup [2, \infty)$.

← **Algebra Review**
Remember to always reverse the direction of the inequality symbol when multiplying or dividing by a negative number.

(c) $|4 - 3x| \leq 2$ Original inequality

$-2 \leq 4 - 3x \leq 2$ Case 3

$-6 \leq -3x \leq -2$ Subtract 4 in all expressions.

$2 \geq x \geq \dfrac{2}{3}$ Divide by -3 and reverse the direction of the inequality symbols.

$\dfrac{2}{3} \leq x \leq 2$ Rewrite.

The solution set is $\left[\dfrac{2}{3}, 2\right]$.

The next example illustrates how to solve absolute value equations and inequalities when $k < 0$.

FOR DISCUSSION

Discuss why the following are true. (Here, $k = 0$.)

1. The solution set of $|ax + b| = 0$ is $\left\{-\dfrac{b}{a}\right\}$.
2. The solution set of $|ax + b| < 0$ is $\varnothing$.
3. The solution set of $|ax + b| \geq 0$ is $(-\infty, \infty)$.

> **EXAMPLE 7** **Solving Special Cases**

Solve each equation or inequality.

(a) $|3x + 5| = -5$ **(b)** $|3x + 5| < -5$ **(c)** $|3x + 5| > -5$

Solution

(a) Because the absolute value of an expression will never be -5, this equation has no solution. The solution set is $\varnothing$.

(b) Using reasoning similar to that in part (a), the absolute value of an expression will never be less than -5. Once again, the solution set is $\varnothing$.

(c) Because the absolute value of an expression will always be greater than or equal to 0, the absolute value of an expression will always be greater than -5 (or *any* negative number). The solution set is $(-\infty, \infty)$.

If two quantities have the same absolute value, they must either be equal to each other or be negatives of each other. This fact allows us to solve absolute value equations (and related inequalities) of the form $|ax + b| = |cx + d|$.

Solving $|ax + b| = |cx + d|$

To solve the equation $|ax + b| = |cx + d|$ analytically, solve the following compound equation.

$$ax + b = cx + d \quad \text{or} \quad ax + b = -(cx + d)$$

EXAMPLE 8 **Solving an Equation Involving Two Absolute Value Expressions**

Solve $|x + 6| = |2x - 3|$.

Analytic Solution

$$\begin{array}{rcl}
x + 6 = 2x - 3 & \text{or} & x + 6 = -(2x - 3) \\
x + 9 = 2x & \text{or} & x + 6 = -2x + 3 \\
9 = x & & 3x = -3 \\
& & x = -1
\end{array}$$

Distribute negative sign.

Check: $\qquad |x + 6| = |2x - 3|$

$$\begin{array}{c|c}
|-1 + 6| = |2(-1) - 3| & |9 + 6| = |2(9) - 3| \\
5 = |-5| & 15 = |18 - 3| \\
5 = 5 \checkmark & 15 = 15 \checkmark
\end{array}$$

The solution set is $\{-1, 9\}$.

Graphing Calculator Solution

Let $y_1 = |x + 6|$ and $y_2 = |2x - 3|$. The equation $y_1 = y_2$ is equivalent to $y_1 - y_2 = 0$, so we graph

$$y_3 = |x + 6| - |2x - 3|$$

and find the x-coordinates of the x-intercepts, -1 and 9. See **FIGURE 52.**

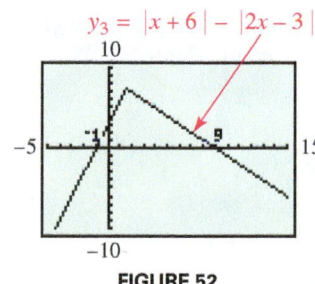

FIGURE 52

In **Example 9,** we show a method of solving two inequalities related to the absolute value equation from **Example 8.**

EXAMPLE 9 **Solving Inequalities Involving Two Absolute Value Expressions**

Solve each inequality graphically by referring to **Example 8** and **FIGURE 52.**

(a) $|x + 6| < |2x - 3|$ (b) $|x + 6| \geq |2x - 3|$

Solution

(a) The inequality $y_1 < y_2$ is equivalent to $y_1 - y_2 < 0$, or $y_3 < 0$. In **FIGURE 52,** note that the graph of y_3 is below the x-axis on the interval

$$(-\infty, -1) \cup (9, \infty).$$

(b) The inequality $y_3 \geq 0$ is satisfied on the closed interval $[-1, 9]$.

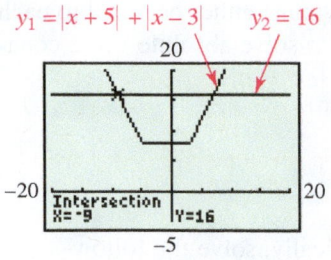

FIGURE 53

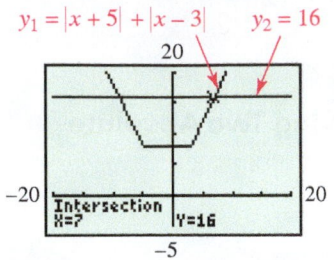

FIGURE 54

EXAMPLE 10 **Solving an Equation Involving a Sum of Absolute Values**

Solve $|x + 5| + |x - 3| = 16$ graphically by the intersection-of-graphs method. Check the solutions by substitution.

Solution Let $y_1 = |x + 5| + |x - 3|$ and $y_2 = 16$. Their graphs are shown in **FIGURES 53** and **54.** Locate the points of intersection of the graphs to find that the x-coordinates of the points are -9 and 7. To check, substitute each solution into the original equation.

Check: $|x + 5| + |x - 3| = 16$

$$|(-9) + 5| + |(-9) - 3|$$ Let $x = -9$. $\quad |7 + 5| + |7 - 3|$ Let $x = 7$.
$$= |-4| + |-12|$$ $\quad = |12| + |4|$
$$= 4 + 12$$ $\quad = 12 + 4$
$$= 16 \checkmark$$ $\quad = 16 \checkmark$

Therefore, the solution set is $\{-9, 7\}$.

2.4 Exercises

Checking Analytic Skills   *You are given graphs of functions* $y = f(x)$. *Sketch the graph of* $y = |f(x)|$ *by hand.* **Do not use a calculator.**

1.
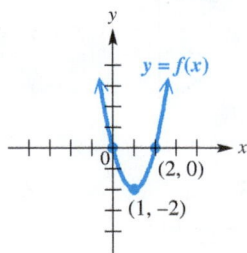

2.
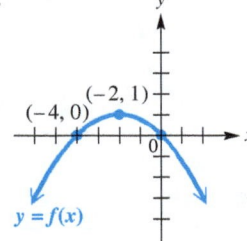

3.
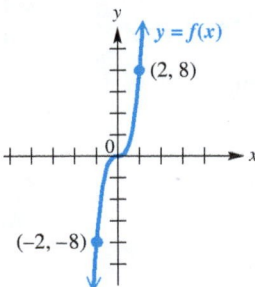

4.
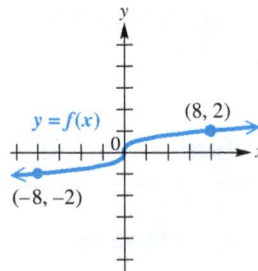

5.
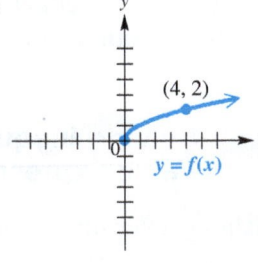

6.
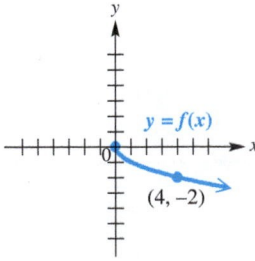

7.
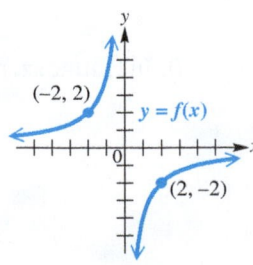

8.
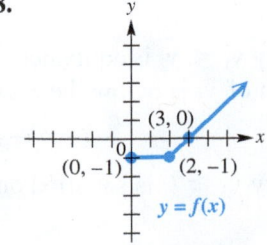

9.
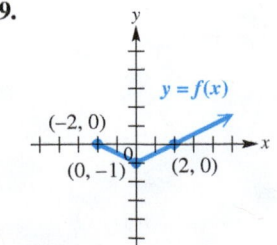

10. Explain in your own words the procedure you used to find the graphs of $y = |f(x)|$ in **Exercises 1–9.**

Concept Check   *Give a short answer to each question.*

11. If $f(a) = -5$, what is the value of $|f(a)|$?

12. How does the graph of $f(x) = x^2$ compare with the graph of $f(x) = |x^2|$?

13. What is the range of $y = |f(x)|$ if $f(x) = -x^2$?

14. If the range of $y = f(x)$ is $[-2, \infty)$, what is the range of $y = |f(x)|$?

15. If the range of $y = f(x)$ is $(-\infty, -2]$, what is the range of $y = |f(x)|$?

16. Why can't the range of $y = |f(x)|$ include -1, for any function f?

In Exercises 17–20, use graphing to determine the domain and range of $y = f(x)$ and of $y = |f(x)|$.

17. $f(x) = (x + 1)^2 - 2$
18. $f(x) = 2 - \dfrac{1}{2}x$
19. $f(x) = -1 - (x - 2)^2$
20. $f(x) = -|x + 2| - 2$

In Exercises 21–24, a line graph of $y = f(x)$ is given. Find the domain and range of $y = f(x)$ and of $y = |f(x)|$.

21.
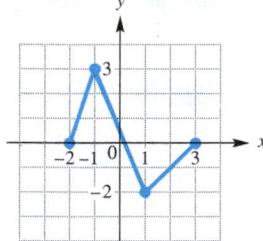

22.
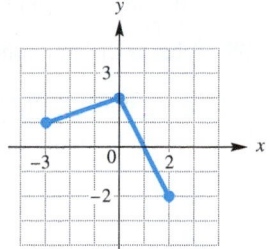

23.
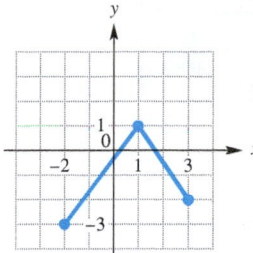

24.
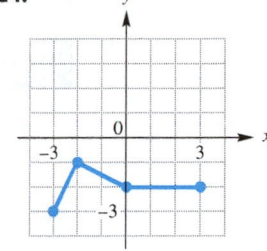

25. The graph of a function $y = f(x)$ is shown. Sketch by hand, in order, the graph of each of the functions that follow. Use the concept of reflecting introduced in **Section 2.3** and the concept of graphing $y = |f(x)|$ introduced in this section.

(a) $y = f(-x)$

(b) $y = -f(-x)$

(c) $y = |-f(-x)|$

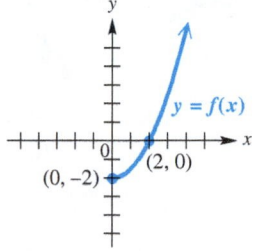

26. Repeat **Exercise 25** for the graph of $y = f(x)$ shown here.

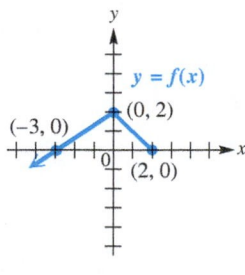

In Exercises 27 and 28, one graph is that of $y = f(x)$ and the other is that of $y = |f(x)|$. State which is which.

27. A.
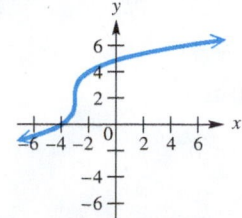

B.
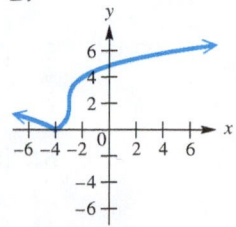

28. A.
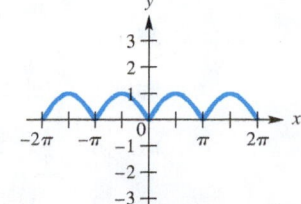

B.
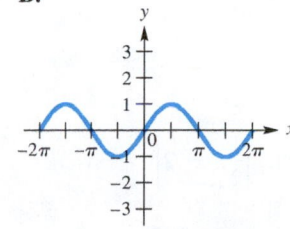

In Exercises 29–32, use the graph, along with the indicated points, to give the solution set of each of the following.

(a) $y_1 = y_2$ **(b)** $y_1 < y_2$ **(c)** $y_1 > y_2$

29.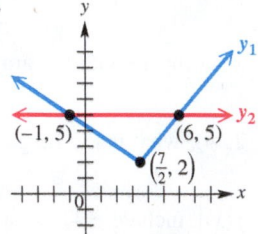

30.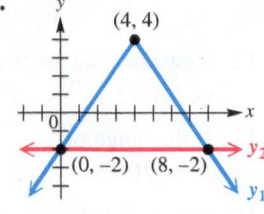

31.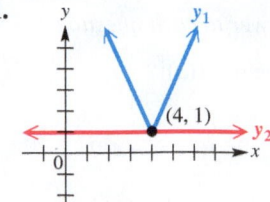

32.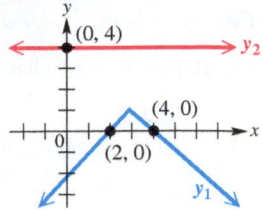

RELATING CONCEPTS **For individual or group investigation (Exercises 33–38)**

The figure shows the graphs of $f(x) = |0.5x + 6|$ *and* $g(x) = 3x - 14$. ***Work Exercises 33–38 in order,*** *without actually using your graphing calculator.*

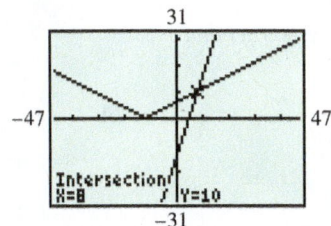

33. Which graph is that of $y = f(x)$? How do you know?

34. Which graph is that of $y = g(x)$? How do you know?

35. Solve $f(x) = g(x)$ based on the display at the bottom of the screen.

36. Solve $f(x) > g(x)$ based on the graphs and the display.

37. Solve $f(x) < g(x)$ based on the graphs and the display.

38. What is the solution set of

$$|0.5x + 6| - (3x - 14) = 0?$$

Solve each group of equations and inequalities analytically.

39. **(a)** $|x + 4| = 9$
   (b) $|x + 4| > 9$
   (c) $|x + 4| < 9$

40. **(a)** $|x - 3| = 5$
   (b) $|x - 3| > 5$
   (c) $|x - 3| < 5$

41. **(a)** $|7 - 2x| = 3$
   (b) $|7 - 2x| \geq 3$
   (c) $|7 - 2x| \leq 3$

42. **(a)** $|-9 - 3x| = 6$
   (b) $|-9 - 3x| \geq 6$
   (c) $|-9 - 3x| \leq 6$

43. **(a)** $|2x + 1| + 3 = 5$
   (b) $|2x + 1| + 3 \leq 5$
   (c) $|2x + 1| + 3 \geq 5$

44. **(a)** $|4x + 7| + 4 = 4$
   (b) $|4x + 7| + 4 > 4$
   (c) $|4x + 7| + 4 < 4$

45. **(a)** $|5 - 7x| = 0$
   (b) $|5 - 7x| \geq 0$
   (c) $|5 - 7x| \leq 0$

46. **(a)** $|\pi x + 8| = -4$
   (b) $|\pi x + 8| < -4$
   (c) $|\pi x + 8| > -4$

47. **(a)** $|\sqrt{2}x - 3.6| = -1$
   (b) $|\sqrt{2}x - 3.6| \leq -1$
   (c) $|\sqrt{2}x - 3.6| \geq -1$

Solve each equation or inequality.

48. $|2x + 4| + 2 = 10$

49. $3|4 - 3x| - 4 = 8$

50. $5|x + 3| - 2 = 18$

51. $\dfrac{1}{2}\left|-2x + \dfrac{1}{2}\right| = \dfrac{3}{4}$

52. $|3(x - 5) + 2| + 3 = 9$

53. $4.2|0.5 - x| + 1 = 3.1$

54. $|3x - 1| < 8$

55. $|15 - x| < 7$

56. $|7 - 4x| \leq 11$

57. $|2x - 3| > 1$

58. $|4 - 3x| > 1$

59. $|-3x + 8| \geq 3$

60. $\left|\dfrac{1}{2}x - 2\right| > -1$

61. $\left|6 - \dfrac{1}{3}x\right| > 0$

62. $|8x - 4| < 0$

63. $|-2x + 7| \leq -6$

64. $\left|\dfrac{3}{4}x - \dfrac{1}{2}\right| < -4$

65. $|7x - 5| \geq -5$

66. Explain how to solve an equation of the form $|ax + b| = |cx + d|$ analytically.

In Exercises 67–78, an equation of the form $|f(x)| = |g(x)|$ is given.
(a) *Solve the equation analytically and support the solution graphically.*
(b) *Solve $|f(x)| > |g(x)|$.*
(c) *Solve $|f(x)| < |g(x)|$.*

67. $|3x + 1| = |2x - 7|$

68. $|x - 4| = |7x + 12|$

69. $|-2x + 5| = |x + 3|$

70. $|-5x + 1| = |3x - 4|$

71. $\left|x - \dfrac{1}{2}\right| = \left|\dfrac{1}{2}x - 2\right|$

72. $|x + 3| = \left|\dfrac{1}{3}x + 8\right|$

73. $|4x + 1| = |4x + 6|$

74. $|6x + 9| = |6x - 3|$

75. $|0.25x + 1| = |0.75x - 3|$

76. $|0.40x + 2| = |0.60x - 5|$

77. $|3x + 10| = |-3x - 10|$

78. $|5x - 6| = |-5x + 6|$

Solve each equation graphically.

79. $|x + 1| + |x - 6| = 11$

80. $|2x + 2| + |x + 1| = 9$

81. $|x| + |x - 4| = 8$

82. $|0.5x + 2| + |0.25x + 4| = 9$

(Modeling) Average Temperature *Each inequality describes the range of average monthly temperatures T in degrees Fahrenheit at a certain location.*
(a) *Solve the inequality.*   **(b)** *Interpret the result.*

83. $|T - 50| \leq 22$, Boston, Massachusetts

84. $|T - 10| \leq 36$, Chesterfield, Canada

85. $|T - 61.5| \leq 12.5$, Buenos Aires, Argentina

86. $|T - 43.5| \leq 8.5$, Punta Arenas, Chile

(Modeling) *Solve each problem.*

87. *Weights of Babies* Dr. Cazayoux has found that, over the years, 95% of the babies he delivered weighed x pounds, where $|x - 8.0| \leq 1.5$. What range of weights corresponds to this inequality?

88. *Conversion of Methanol to Gasoline* The industrial process that is used to convert methanol to gasoline is carried out at a temperature range of 680°F to 780°F. Using F as the variable, write an absolute value inequality that corresponds to this range.

89. *Blood Pressure* Systolic blood pressure is the maximum pressure produced by each heartbeat. Both low blood pressure and high blood pressure are cause for medical concern. Therefore, health care professionals are interested in a patient's "pressure difference from normal," or P_d. If 120 is considered a normal systolic pressure, $P_d = |P - 120|$, where P is the patient's recorded systolic pressure. For example, a patient with a systolic pressure P of 113 would have a pressure difference from normal of $P_d = |P - 120| = |113 - 120| = |-7| = 7$.

(a) Calculate the P_d value for a woman whose actual systolic pressure is 116 and whose normal value should be 125.

(b) If a patient's P_d value is 17 and the normal pressure for his sex and age should be 120, what are the two possible values for his systolic blood pressure?

90. *Kite Flying* When a model kite was flown in crosswinds in tests, it attained speeds of 98 to 148 feet per second in winds of 16 to 26 feet per second. Using x as the variable in each case, write absolute value inequalities that correspond to these ranges.

Tolerances   *In quality control and other applications, we often wish to keep the difference between two quantities within some predetermined amount, or tolerance. For example, suppose $y = 2x + 1$ and we want y to be within 0.01 unit of 4. This criterion can be written as $|y - 4| < 0.01$. To find the values of x that satisfy this condition on y, we use properties of absolute value as follows.*

$$|y - 4| < 0.01$$

$|2x + 1 - 4| < 0.01$ Substitute $2x + 1$ for y.

$|2x - 3| < 0.01$ Subtract.

$-0.01 < 2x - 3 < 0.01$ Case 3: Write a three-part inequality.

$2.99 < 2x < 3.01$ Add 3 to each part.

$1.495 < x < 1.505$ Divide each part by 2.

Keeping x in the open interval (1.495, 1.505) will ensure that the difference between y and 4 is less than 0.01. In Exercises 91–94, find the open interval in which x must lie in order for the given condition to hold.

91. $y = 2x + 1$, and the difference between y and 1 is less than 0.1.

92. $y = 3x - 6$, and the difference between y and 2 is less than 0.01.

93. $y = 4x - 8$, and the difference between y and 3 is less than 0.001.

94. $y = 5x + 12$, and the difference between y and 4 is less than 0.0001.

Solve each equation or inequality graphically.

95. $|2x + 7| = 6x - 1$

96. $-|3x - 12| \geq -x - 1$

97. $|x - 4| > 0.5x - 6$

98. $2x + 8 > -|3x + 4|$

99. $|3x + 4| < -3x - 14$

100. $|x - \sqrt{13}| + \sqrt{6} \leq -x - \sqrt{10}$

2.5 Piecewise-Defined Functions

Graphing Piecewise-Defined Functions • The Greatest Integer Function • Applications of Piecewise-Defined Functions

Graphing Piecewise-Defined Functions

The absolute value function is a simple example of a function defined by different rules (formulas) over different subsets of its domain. Such a function is called a **piecewise-defined function.**

Recall that the domain of $f(x) = |x|$ is $(-\infty, \infty)$. For the interval $[0, \infty)$ of the domain, we use the rule $f(x) = x$. For the interval $(-\infty, 0)$, we use the rule $f(x) = -x$. See **FIGURE 55**. Thus, the graph of $f(x) = |x|$ is composed of two "pieces." One piece comes from the graph of $y = x$ and the other from $y = -x$.

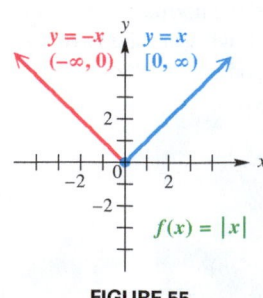

FIGURE 55

EXAMPLE 1 **Evaluating and Analyzing a Piecewise-Defined Function**

Consider the piecewise-defined function

$$f(x) = \begin{cases} 4x & \text{if } -2 \leq x \leq 2 \\ -\dfrac{1}{2}x + 9 & \text{if } 2 < x \leq 4. \end{cases}$$

Find **(a)** $f(-2)$ **(b)** $f(2)$ **(c)** $f(4)$.
(d) Sketch the graph of f. Is f continuous on its domain?

Solution

(a) Since $x = -2$ and $-2 \leq -2 \leq 2$, we use the first piece, or rule, $f(x) = 4x$. Thus,

$$f(-2) = 4(-2) = -8. \qquad \text{Let } x = -2.$$

This means that the graph of f will contain the point $(-2, -8)$.

(b) Since $x = 2$ and $-2 \leq 2 \leq 2$, we use the rule $f(x) = 4x$. Thus,

$$f(2) = 4(2) = 8. \qquad \text{Let } x = 2.$$

This means that the graph of f will contain the point $(2, 8)$.

(c) Since $x = 4$ and $2 < 4 \leq 4$, we use the rule $f(x) = -\frac{1}{2}x + 9$. Thus,

$$f(4) = -\frac{1}{2}(4) + 9 = 7. \qquad \text{Let } x = 4.$$

This means that the graph of f will contain the point $(4, 7)$.

(d) We graph the line segment defined by $y = 4x$ by plotting the solid endpoints $(-2, -8)$ and $(2, 8)$. The line segment has slope 4 and y-intercept $(0, 0)$.

Then we graph $y = -\frac{1}{2}x + 9$ for $2 < x \leq 4$, with a solid endpoint at $(4, 7)$. **FIGURE 56** shows a table of values and a hand-drawn graph of the piecewise-defined function given by $y = f(x)$. Function f is continuous on its domain $[-2, 4]$ because there are no breaks in its graph.

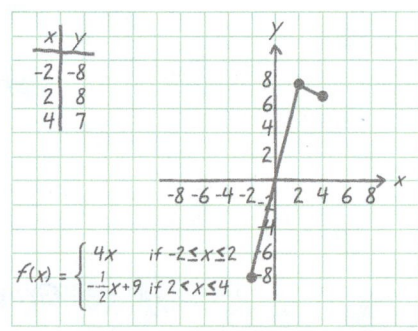

FIGURE 56

EXAMPLE 2 **Evaluating and Analyzing a Piecewise-Defined Function**

Consider the piecewise-defined function

$$f(x) = \begin{cases} x + 2 & \text{if } x \leq 0 \\ \dfrac{1}{2}x^2 & \text{if } x > 0. \end{cases}$$

(continued)

Function from Example 2

$$f(x) = \begin{cases} x + 2 & \text{if } x \le 0 \\ \dfrac{1}{2}x^2 & \text{if } x > 0 \end{cases}$$

Find **(a)** $f(-3)$ **(b)** $f(0)$ **(c)** $f(3)$.

(d) Sketch the graph of f. Is f continuous on its domain?

Solution

(a) Since $-3 \le 0$ (specifically, $-3 < 0$), we use the rule $f(x) = x + 2$. Thus,

$$f(-3) = -3 + 2 = -1. \qquad \text{Let } x = -3.$$

This means that the graph of f will contain the point $(-3, -1)$.

(b) Since $0 \le 0$ (because $0 = 0$), we again use the rule $f(x) = x + 2$.

$$f(0) = 0 + 2 = 2 \qquad \text{Let } x = 0.$$

The y-intercept of the graph is $(0, 2)$.

(c) The number 3 is greater than 0, and the second part of the rule for f indicates that we must use $f(x) = \frac{1}{2}x^2$.

$$f(3) = \frac{1}{2}(3)^2 = \frac{1}{2}(9) = 4.5 \qquad \text{Let } x = 3.$$

This means that the graph of f will contain the point $(3, 4.5)$.

(d) We graph the ray $y = x + 2$, choosing x so that $x \le 0$, with a solid endpoint at $(0, 2)$. The ray has slope 1 and y-intercept $(0, 2)$.

Then we graph $y = \frac{1}{2}x^2$ for $x > 0$. This graph will be half of a parabola with an open endpoint at $(0, 0)$. **FIGURE 57** shows a table of values and a hand-drawn graph of the piecewise-defined function given by $y = f(x)$. Function f is discontinuous at $x = 0$.

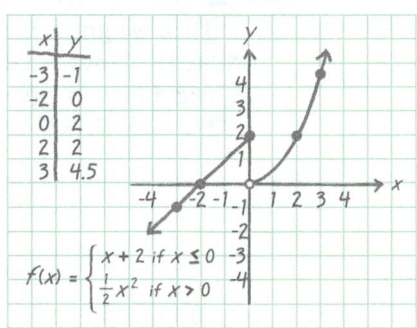

FIGURE 57

EXAMPLE 3   **Graphing a Piecewise-Defined Function**

Graph the function

$$f(x) = \begin{cases} x + 1 & \text{if } x \le 2 \\ 3 & \text{if } 2 < x \le 4 \\ -2x + 7 & \text{if } x > 4. \end{cases}$$

Give the domain and range. Is this function continuous on its domain?

Analytic Solution

To graph

$$f(x) = \begin{cases} x + 1 & \text{if } x \le 2 \\ 3 & \text{if } 2 < x \le 4 \\ -2x + 7 & \text{if } x > 4 \end{cases}$$

for x-values less than or equal to 2, we graph the portion of the line $y = x + 1$ to the left of, and including, the point $(2, 3)$. For x-values greater than 2 and less than or equal to 4, we graph the horizontal segment with equation $y = 3$. For x-values greater than 4, we graph the portion of the line $y = -2x + 7$ to the right of $(4, -1)$. See **FIGURE 58**.

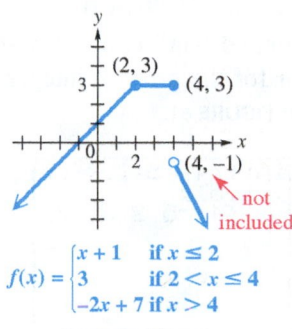

$$f(x) = \begin{cases} x + 1 & \text{if } x \le 2 \\ 3 & \text{if } 2 < x \le 4 \\ -2x + 7 & \text{if } x > 4 \end{cases}$$

FIGURE 58

The domain is $(-\infty, \infty)$, and the range is $(-\infty, 3]$. This function is not continuous on its entire domain, due to the break at $x = 4$.

Graphing Calculator Solution

To graph a piecewise-defined function with a TI-84 Plus calculator, we use the TEST and/or LOGIC features, which return a 1 for a true statement and a 0 for a false statement. **FIGURE 59(a)** shows how $f(x)$ can be entered as Y_1.† Notice that the two defining expressions are multiplied by an expression composed of an inequality involving X (which will be 1 or 0, depending on the value of X as the graphing occurs). When $X \le 2$, Y_1 is

$$(X + 1)(1) + 3(0) + (-2X + 7)(0) = X + 1.$$

When $X > 2$ and $X \le 4$, it is

$$(X + 1)(0) + 3(1) + (-2X + 7)(0) = 3.$$

And when $X > 4$, it is

$$(X + 1)(0) + 3(0) + (-2X + 7)(1) = -2X + 7.$$

The graph of Y_1 is shown in **FIGURE 59(b)**.

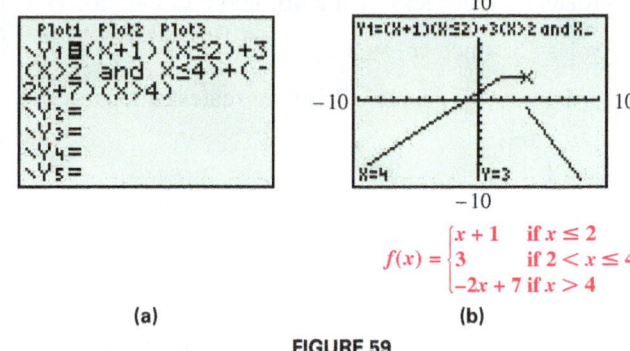

(a)   (b)

$$f(x) = \begin{cases} x + 1 & \text{if } x \le 2 \\ 3 & \text{if } 2 < x \le 4 \\ -2x + 7 & \text{if } x > 4 \end{cases}$$

FIGURE 59

EXAMPLE 4   **Finding the Formula for a Piecewise-Defined Function**

Write a formula for the piecewise-defined function f shown in **FIGURE 60**. Give the domain and range.

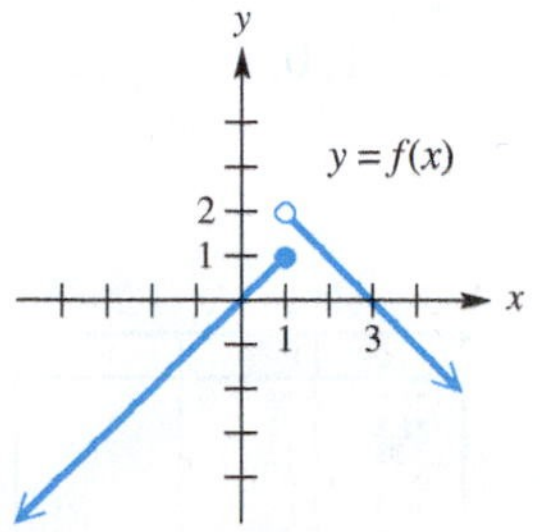

Solution   The graph of f consists of two parts. When $x \le 1$, the graph is a line with slope 1 and y-intercept $(0, 0)$. Thus, the equation of this line is $y = x$. When $x > 1$, the graph is a line with slope -1 and x-intercept $(3, 0)$. Using the point–slope form, we find that the equation of the line passing through the point $(3, 0)$ with slope -1 is

$$y - 0 = -1(x - 3), \quad \text{or} \quad y = -x + 3.$$

Thus, we can write a formula for the piecewise-defined function f as follows.

$$f(x) = \begin{cases} x & \text{if } x \le 1 \\ -x + 3 & \text{if } x > 1 \end{cases}$$

The domain is $(-\infty, \infty)$ and the range is $(-\infty, 2)$.

FIGURE 60

NOTE   Other acceptable forms exist for defining piecewise-defined functions. We give the most obvious ones.

---

†In defining the function as Y_1, the user has a choice whether to use the multiplication symbol, $*$.

The Greatest Integer Function

An important example of a piecewise-defined function is the **greatest integer function.** *The notation* $[x]$ *represents the greatest integer less than or equal to x.*

$$f(x) = [x] = \begin{cases} x & \text{if } x \text{ is an integer} \\ \text{the greatest integer less than } x & \text{if } x \text{ is not an integer} \end{cases}$$

EXAMPLE 5 Evaluating $[x]$

Evaluate $[x]$ for each value of x.

(a) 4   (b) -5 (c) 2.46   (d) π (e) $-6\frac{1}{2}$

Analytic Solution

(a) $[4] = 4$, since 4 is an integer.

(b) $[-5] = -5$, since -5 is an integer.

(c) $[2.46] = 2$, since 2.46 is not an integer and 2 is the greatest integer less than 2.46.

(d) $[\pi] = 3$, since $\pi \approx 3.14$.

(e) $[-6\frac{1}{2}] = -7$, since -7 is the greatest integer less than $-6\frac{1}{2}$.

Graphing Calculator Solution

(a)–(e) The command "int" is used by the TI-84 Plus calculator for the greatest integer function. See the list in **FIGURE 61**.

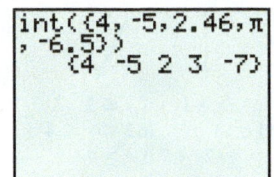

FIGURE 61

The greatest integer function (shown in **FIGURE 62**) is called a **step function.**

FUNCTION CAPSULE

GREATEST INTEGER FUNCTION $f(x) = [x]$

Domain: $(-\infty, \infty)$ Range: $\{y \mid y \text{ is an integer}\} = \{\ldots, -3, -2, -1, 0, 1, 2, 3, \ldots\}$

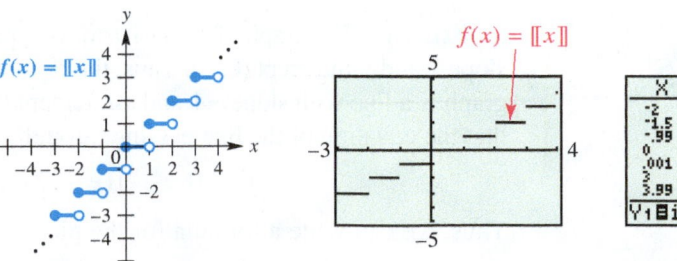

FIGURE 62

- $f(x) = [x]$ is discontinuous at integer values of its domain, $(-\infty, \infty)$.
- The x-intercepts are all points in the interval $[0, 1)$, and the y-intercept is $(0, 0)$.

Graphing a Step Function

Graph the function $f(x) = \left[\!\left[\frac{1}{2}x + 1\right]\!\right]$. Give the domain and range.

Analytic Solution

Try some values of x in the equation to see how the values of y behave. Some sample ordered pairs are given in the accompanying table. For example, if $x = -3$, then $y = \left[\!\left[\frac{1}{2}(-3) + 1\right]\!\right] = \left[\!\left[-\frac{1}{2}\right]\!\right] = -1$. These ordered pairs suggest that if x is in the interval $[0, 2)$, then $y = 1$. For x in $[2, 4)$, $y = 2$, and so on.

The graph is shown in **FIGURE 63.** The domain is $(-\infty, \infty)$, and the range is

$$\{\ldots, -2, -1, 0, 1, 2, \ldots\}.$$

x	y
-3	-1
-2	0
-1	0
0	1
1	1
2	2
3	2
4	3

The dots indicate that the graph continues indefinitely in the same pattern.

FIGURE 63

Graphing Calculator Solution

FIGURE 64(a) shows the graph of $Y_1 = \left[\!\left[\frac{1}{2}X + 1\right]\!\right]$ in the window $[-4, 4]$ by $[-4, 4]$. Scroll up and down, using the table feature to confirm that the range is $\{\ldots, -2, -1, 0, 1, 2, \ldots\}$. See **FIGURE 64(b).**

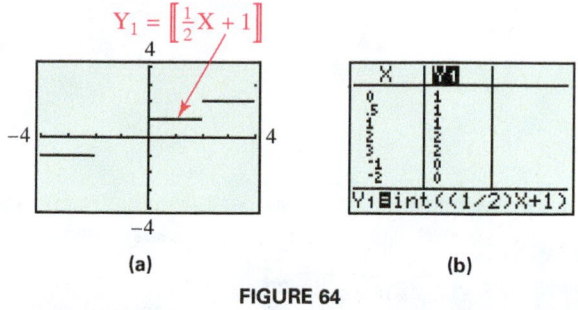

$Y_1 = \left[\!\left[\frac{1}{2}X + 1\right]\!\right]$

(a)   (b)

FIGURE 64

In **FIGURE 64(a),** the inclusion or exclusion of the endpoints for each segment is not readily determined from the calculator graph. However, the hand-drawn graph in **FIGURE 63** *does* show whether endpoints are included or excluded. ***While technology is incredibly powerful and useful, we cannot rely on it alone in our study of graphs of functions.***

When graphing piecewise-defined functions by hand, avoid the temptation to "connect" pieces that exhibit discontinuities.

Applications of Piecewise-Defined Functions

Applying the Greatest Integer Function to Parking Rates

Downtown Parking charges a $5 base fee for parking through 1 hour, and $1 for each additional hour or fraction thereof. The maximum fee for 24 hours is $15. Sketch a graph of the function that describes this pricing scheme.

Solution When the time is 0 hours, the price is $0, and so the point $(0, 0)$ lies on the graph. Then, for any amount of time during and up to the first hour, the rate is $5. Thus, some sample ordered pairs for the function in the interval $[0, 1]$ would be

$$(0, 0), (0.25, 5), (0.5, 5), (0.75, 5), \text{ and } (1, 5).$$

After the first hour, the price immediately jumps (or steps up) to $6 and remains $6 until the time equals 2 hours. It then jumps to $7 during the third hour, and so on. During the 11th hour, it will have jumped to $15 and will remain at $15 for the rest of the 24-hour period.

FIGURE 65 on the next page shows the graph of this function on the interval $[0, 24]$. The range of the function is $\{0, 5, 6, 7, \ldots, 15\}$.

(continued)

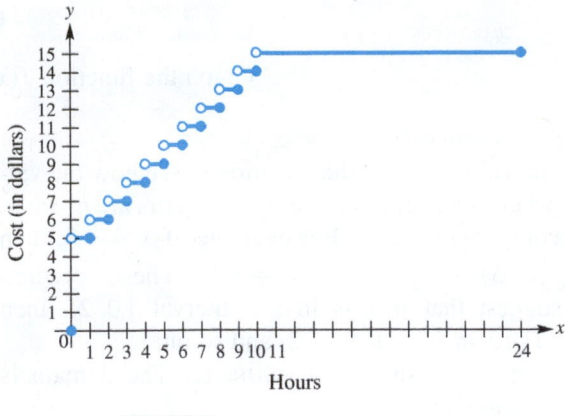

A closed endpoint is found at the extreme *right* value rather than the left, because the increase takes place *after* the hour of use has ended.

FIGURE 65

| EXAMPLE 8 | **Using a Piecewise-Defined Function to Analyze Pollution** |

Due to acid rain, the percentage of lakes in Scandinavia that had lost their population of brown trout increased dramatically between 1940 and 1975. Based on a sample of 2850 lakes, this percentage can be approximated by the piecewise-defined function

$$f(x) = \begin{cases} \dfrac{11}{20}(x - 1940) + 7 & \text{if } 1940 \leq x < 1960 \\[2mm] \dfrac{32}{15}(x - 1960) + 18 & \text{if } 1960 \leq x \leq 1975. \end{cases}$$

(*Source:* C. Mason, *Biology of Freshwater Pollution,* John Wiley and Sons.) Determine the percent of lakes that had lost brown trout by 1950 and by 1972. Interpret your results.

Solution Because $1940 \leq 1950 < 1960$, use the first formula to calculate $f(1950)$.

$$f(1950) = \frac{11}{20}(1950 - 1940) + 7$$

$$= 12.5$$

By 1950, about 12.5% of the lakes had lost their population of brown trout. In a similar manner, $f(1972)$ can be evaluated by using the second formula.

$$f(1972) = \frac{32}{15}(1972 - 1960) + 18$$

$$= 43.6$$

By 1972, about 43.6% of the lakes had lost their population of brown trout.

FOR DISCUSSION

Think of the manner in which the U.S. Postal Service determines the price to send a first-class letter. Then use the concept of the greatest integer function to explain it.

2.5 Exercises

Checking Analytic Skills   *Use the graphs to answer each item.* **Do not use a calculator.**

1. *Speed Limits*   The graph of $y = f(x)$ gives the speed limit y along a rural highway after traveling x miles.

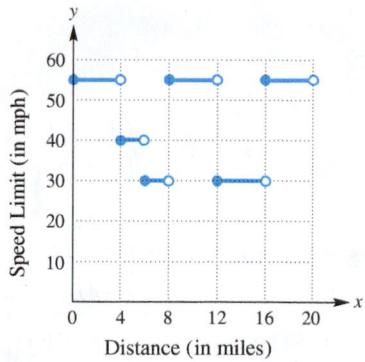

Distance (in miles)

(a) What is the speed limit at the 4-mile mark?
(b) Estimate the miles of highway with a speed limit of 30 mph.
(c) Evaluate $f(5)$, $f(13)$, and $f(19)$.
(d) At what x-values is the graph discontinuous? Interpret each discontinuity.

3. *Swimming Pool Levels*   The graph of $y = f(x)$ represents the amount of water in thousands of gallons remaining in a swimming pool after x days.

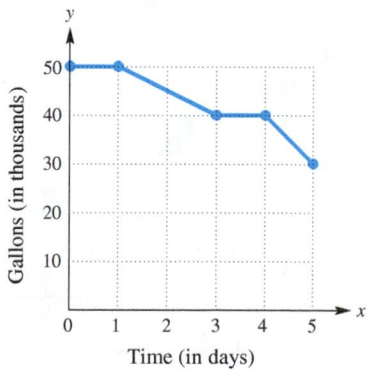

Time (in days)

(a) Estimate the initial and final amounts of water contained in the pool.
(b) When did the amount of water in the pool remain constant?
(c) Approximate $f(2)$ and $f(4)$.
(d) At what rate was water being drained from the pool when $1 \leq x \leq 3$?

2. *ATM*   The graph of $y = f(x)$ depicts the amount of money y in dollars in an automatic teller machine (ATM) after x minutes.

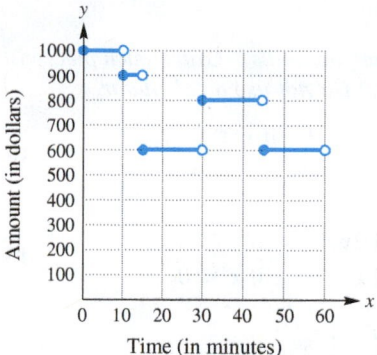

Time (in minutes)

(a) Determine the initial and final amounts of money in the ATM.
(b) Evaluate $f(10)$ and $f(50)$. Is f continuous?
(c) How many withdrawals occurred during this period?
(d) When did the largest withdrawal occur? How much was it?
(e) How much was deposited into the machine?

4. *Gasoline Usage*   The graph shows the gallons of gasoline y in the gas tank of a car after x hours.

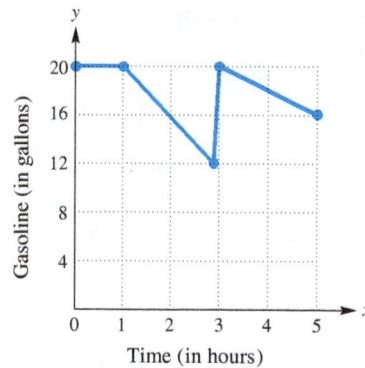

Time (in hours)

(a) Estimate how much gasoline was in the gas tank at $x = 1$.
(b) If $x = 0$ represents 3:15 P.M., at about what time was the gas tank filled?
(c) Interpret the graph.
(d) When did the car burn gasoline at the fastest rate?

Checking Analytic Skills   For each piecewise-defined function, find **(a)** $f(-5)$, **(b)** $f(-1)$, **(c)** $f(0)$, and
(d) $f(3)$. *Do not use a calculator.*

5. $f(x) = \begin{cases} 2x & \text{if } x \leq -1 \\ x - 1 & \text{if } x > -1 \end{cases}$

6. $f(x) = \begin{cases} x - 2 & \text{if } x < 3 \\ 5 - x & \text{if } x \geq 3 \end{cases}$

7. $f(x) = \begin{cases} 2 + x & \text{if } x < -4 \\ -x & \text{if } -4 \leq x \leq 2 \\ 3x & \text{if } x > 2 \end{cases}$

8. $f(x) = \begin{cases} -2x & \text{if } x < -3 \\ 3x - 1 & \text{if } -3 \leq x \leq 2 \\ -4x & \text{if } x > 2 \end{cases}$

Checking Analytic Skills   Graph each piecewise-defined function in Exercises 9–20. Is *f* continuous
on its domain? *Do not use a calculator.*

9. $f(x) = \begin{cases} x - 1 & \text{if } x \leq 3 \\ 2 & \text{if } x > 3 \end{cases}$

10. $f(x) = \begin{cases} 6 - x & \text{if } x \leq 3 \\ 3x - 6 & \text{if } x > 3 \end{cases}$

11. $f(x) = \begin{cases} 4 - x & \text{if } x < 2 \\ 1 + 2x & \text{if } x \geq 2 \end{cases}$

12. $f(x) = \begin{cases} 2x + 1 & \text{if } x \geq 0 \\ x & \text{if } x < 0 \end{cases}$

13. $f(x) = \begin{cases} 2 + x & \text{if } x < -4 \\ -x & \text{if } -4 \leq x \leq 5 \\ 3x & \text{if } x > 5 \end{cases}$

14. $f(x) = \begin{cases} -2x & \text{if } x < -3 \\ 3x - 1 & \text{if } -3 \leq x \leq 2 \\ -4x & \text{if } x > 2 \end{cases}$

15. $f(x) = \begin{cases} -\dfrac{1}{2}x^2 + 2 & \text{if } x \leq 2 \\ \dfrac{1}{2}x & \text{if } x > 2 \end{cases}$

16. $f(x) = \begin{cases} x^3 + 5 & \text{if } x \leq 0 \\ -x^2 & \text{if } x > 0 \end{cases}$

17. $f(x) = \begin{cases} 2x & \text{if } -5 \leq x < -1 \\ -2 & \text{if } -1 \leq x < 0 \\ x^2 - 2 & \text{if } 0 \leq x \leq 2 \end{cases}$

18. $f(x) = \begin{cases} 0.5x^2 & \text{if } -4 \leq x \leq -2 \\ x & \text{if } -2 < x < 2 \\ x^2 - 4 & \text{if } 2 \leq x \leq 4 \end{cases}$

19. $f(x) = \begin{cases} x^3 + 3 & \text{if } -2 \leq x \leq 0 \\ x + 3 & \text{if } 0 < x < 1 \\ 4 + x - x^2 & \text{if } 1 \leq x \leq 3 \end{cases}$

20. $f(x) = \begin{cases} -2x & \text{if } -3 \leq x < -1 \\ x^2 + 1 & \text{if } -1 \leq x \leq 2 \\ \dfrac{1}{2}x^3 + 1 & \text{if } 2 < x \leq 3 \end{cases}$

Match each piecewise-defined function with its graph in choices A–D.

21. $f(x) = \begin{cases} x^2 - 4 & \text{if } x \geq 0 \\ -x + 5 & \text{if } x < 0 \end{cases}$

22. $g(x) = \begin{cases} |x - 2| & \text{if } x \geq -1 \\ -x^2 & \text{if } x < -1 \end{cases}$

23. $h(x) = \begin{cases} 4 & \text{if } x \geq 0 \\ -4 & \text{if } x < 0 \end{cases}$

24. $k(x) = \begin{cases} \sqrt{x} & \text{if } x \geq 0 \\ -x^2 & \text{if } x < 0 \end{cases}$

A.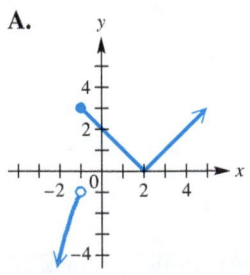

B.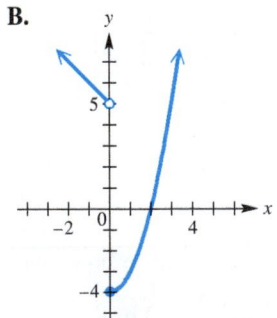

C.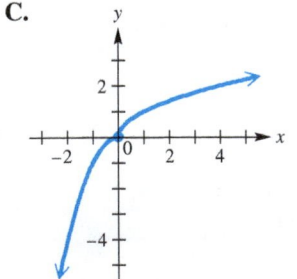

D.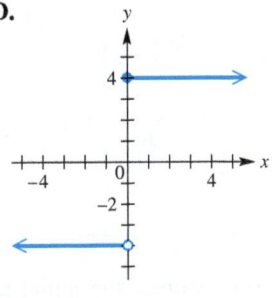

Use a graphing calculator in dot mode and the window indicated to graph each piecewise-defined function.

25. Exercise 9, window $[-4, 6]$ by $[-2, 4]$

26. Exercise 10, window $[-2, 8]$ by $[-2, 10]$

27. Exercise 11, window $[-4, 6]$ by $[-2, 8]$

28. Exercise 12, window $[-5, 4]$ by $[-3, 8]$

29. Exercise 13, window $[-12, 12]$ by $[-6, 20]$

30. Exercise 14, window $[-6, 6]$ by $[-10, 8]$

31. Exercise 15, window $[-5, 6]$ by $[-2, 4]$

32. Exercise 16, window $[-3, 4]$ by $[-3, 6]$

Write a formula for a piecewise-defined function f for each graph shown. Give the domain and range.

33.

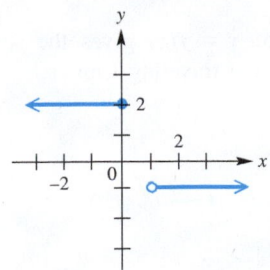

34.

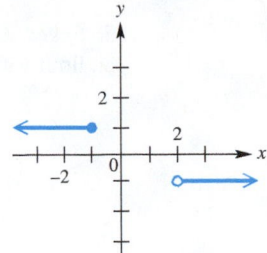

35.

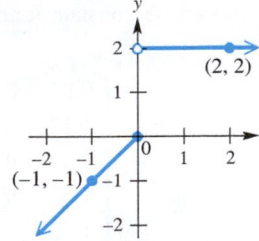

36.

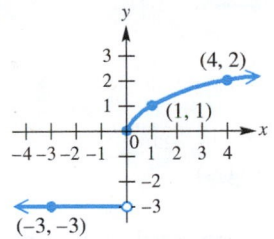

37.

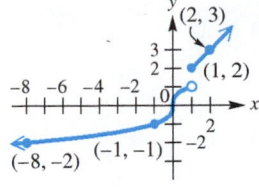

38.

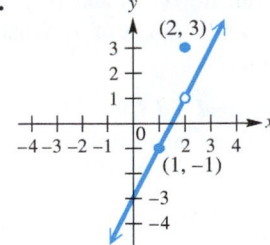

39. Why is the following not a piecewise-defined function?

$$f(x) = \begin{cases} x + 7 & \text{if } x \le 4 \\ x^2 & \text{if } x \ge 4 \end{cases}$$

40. Consider the "function" defined in **Exercise 39.** Suppose you were asked to find $f(4)$. How would you respond?

Concept Check   *Describe how the graph of the given function can be obtained from the graph of $y = [\![x]\!]$.*

41. $y = [\![x]\!] - 1.5$

42. $y = [\![-x]\!]$

43. $y = -[\![x]\!]$

44. $y = [\![x + 2]\!]$

*Use a graphing calculator in dot mode with window $[-5, 5]$ by $[-3, 3]$ to graph each equation. (Refer to your descriptions in **Exercises 41–44**.)*

45. $y = [\![x]\!] - 1.5$

46. $y = [\![-x]\!]$

47. $y = -[\![x]\!]$

48. $y = [\![x + 2]\!]$

(Modeling)   *Solve each problem.*

49. *Flow Rates*   A water tank has an inlet pipe with a flow rate of 5 gallons per minute and an outlet pipe with a flow rate of 3 gallons per minute. A pipe can be either closed or completely open. The graph shows the number of gallons of water in the tank after x minutes. Use the concept of slope to interpret each piece of this graph.

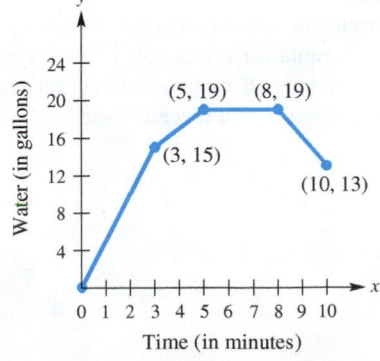

50. *Shoe Size*   Former professional basketball player Shaquille O'Neal is 7 feet, 1 inch tall and weighs 325 pounds. The table lists his shoe sizes at certain ages.

Age	20	21	22	23
Shoe Size	19	20	21	22

Source: USA Today.

(a) Write a formula that gives his shoe size y at age $x = 20, 21, 22,$ and 23.

(b) Suppose that after age 23 his shoe size did not change. Sketch a graph of a continuous, piecewise-defined function f that models his shoe size between the ages 20 and 26, inclusive.

51. *Postage Rates*   In 2013, the retail flat rate in dollars for first-class mail weighing up to 5 ounces could be computed by the piecewise constant function f, where x is the number of ounces.

$$f(x) = \begin{cases} 0.92 & \text{if } 0 < x \leq 1 \\ 1.12 & \text{if } 1 < x \leq 2 \\ 1.32 & \text{if } 2 < x \leq 3 \\ 1.52 & \text{if } 3 < x \leq 4 \\ 1.72 & \text{if } 4 < x \leq 5 \end{cases}$$

(a) Evaluate $f(1.5)$ and $f(3)$. Interpret your results.
(b) Sketch a graph of f. What is its domain? What is its range?

52. *Super Bowl Ad Cost*   The average cost of a 30-second Super Bowl ad in millions of dollars is approximated by the piecewise-defined function

$$f(x) = \begin{cases} 0.0475x - 93.3 & \text{if } 1967 \leq x \leq 1998 \\ 0.1333x - 264.7284 & \text{if } 1998 < x \leq 2013 \end{cases}$$

where x represents the year from 1967 to 2013. Find and interpret each function value.
(a) $f(1967)$ **(b)** $f(1998)$ **(c)** $f(2013)$
(d) Is f continuous on its domain?   **(e)** Graph f.

53. *Violent Crimes*   Using interval notation, the table lists the numbers of victims of violent crime per 1000 people for a recent year by age group.

(a) Sketch the graph of a piecewise-defined function that models the data, where x represents age.

(b) Discuss the impact that age has on the likelihood of being a victim of a violent crime.

Age	Crime Rate
$[12, 15)$	28
$[15, 18)$	23
$[18, 21)$	34
$[21, 25)$	27
$[25, 35)$	19
$[35, 50)$	13
$[50, 65)$	11
$[65, 90)$	2

Source: U.S Department of Justice.

54. *New Homes*   Using interval notation, the table lists the numbers in millions of houses built for various time intervals from 1950 to 2010.

Year	Houses
$[1950, 1960)$	13.6
$[1960, 1970)$	16.1
$[1970, 1980)$	11.6
$[1980, 1990)$	9.9
$[1990, 2000)$	11.0
$[2000, 2010)$	15.4

Source: U.S. Census Bureau, *American Housing Survey.*

(a) Sketch the graph of a piecewise-defined function that models the data, where x represents the year.

(b) Discuss the trends in housing starts between 1950 and 2010.

55. *Speed Limits*   The graph of $y = f(x)$ gives the speed limit y along a country road after traveling x miles.

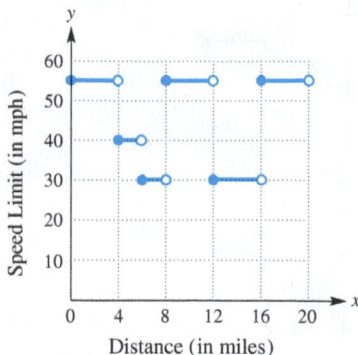

(a) What are the highest and lowest speed limits along this stretch of road?
(b) Estimate the miles of road with a speed limit of 55 mph.
(c) Evaluate $f(4)$, $f(12)$, and $f(18)$.

56. *Minimum Wage*   The table lists the federal minimum wage rates for the years 1981–2013. Sketch a graph of the data as a piecewise-defined function. (Assume that wages take effect on January 1 of the first year of the interval.)

Year(s)	Wage
1981–89	$3.35
1990	$3.80
1991–95	$4.25
1996	$4.75
1997–2006	$5.15
2007	$5.85
2008–2009	$6.55
2010–2013	$7.25

Source: U.S. Bureau of Labor Statistics.

57. *Cellular Phone Bills*   Suppose that the charges for an international cellular phone call are $0.50 for the first minute and $0.25 for each additional minute. Assume that a fraction of a minute is rounded up.
(a) Determine the cost of a phone call lasting 3.5 minutes.
(b) Find a formula for a function f that computes the cost of a telephone call x minutes long, where $0 < x \leq 5$. (*Hint:* Express f as a piecewise-defined function.)

58. *Lumber Costs*   Lumber that is used to frame walls of houses is frequently sold in lengths that are multiples of 2 feet. If the length of a board is not exactly a multiple of 2 feet, there is often no charge for the additional length. For example, if a board measures at least 8 feet, but less than 10 feet, then the consumer is charged for only 8 feet.

(a) Suppose that the cost of lumber is $0.80 every 2 feet. Find a formula for a function f that computes the cost of a board x feet long for $6 \leq x \leq 18$.

(b) Use a graphing calculator to graph f.

(c) Determine the costs of boards with lengths of 8.5 feet and 15.2 feet.

59. *Cost to Send a Package*   An express-mail company charges $25 for a package weighing up to 2 pounds. For each additional pound or fraction of a pound, there is an additional charge of $3. Let $D(x)$ represent the cost to send a package weighing x pounds. Graph $y = D(x)$ for x in the interval $(0, 6]$.

60. *Distance from Home*   Sketch a graph showing the distance a person is from home after x hours if he or she drives on a straight road at 40 mph to a park 20 miles away, remains at the park for 2 hours, and then returns home at a speed of 20 mph.

61. *Water in a Tank*   Sketch a graph that depicts the amount of water in a 100-gallon tank. The tank is initially empty and then filled at a rate of 5 gallons per minute. Immediately after it is full, a pump is used to empty the tank at 2 gallons per minute.

62. *Sub-Saharan HIV Infection Rates*   From 1990 to 2007, the number of people newly infected with HIV in Sub-Saharan Africa increased from 1.3 million to 2.7 million. From 2007 to 2012, the number fell from 2.7 million to 1.75 million.

(a) Use the data points (1990, 1.3), (2007, 2.7), and (2012, 1.75) to write equations for the two line segments describing these data in the closed intervals $[1990, 2007]$ and $[2007, 2012]$.

(b) Give a piecewise-defined function f that describes the graph.

2.6   Operations and Composition

Operations on Functions • The Difference Quotient • Composition of Functions • Applications of Operations and Composition

Operations on Functions

Just as we add, subtract, multiply, and divide real numbers, we can also perform these operations on functions.

Operations on Functions

Given two functions f and g, for all values of x for which both $f(x)$ and $g(x)$ are defined, the functions $f + g$, $f - g$, fg, and $\frac{f}{g}$ are defined as follows.

$$(f + g)(x) = f(x) + g(x) \qquad \text{Sum}$$

$$(f - g)(x) = f(x) - g(x) \qquad \text{Difference}$$

$$(fg)(x) = f(x) \cdot g(x) \qquad \text{Product}$$

$$\left(\frac{f}{g}\right)(x) = \frac{f(x)}{g(x)}, \quad g(x) \neq 0 \qquad \text{Quotient}$$

The domains of $f + g$, $f - g$, and fg include all real numbers in the intersection of the domains of f and g, while the domain of $\frac{f}{g}$ includes those real numbers in the intersection of the domains of f and g for which $g(x) \neq 0$.

EXAMPLE 1 **Using the Operations on Functions**

Let $f(x) = x^2 + 1$ and $g(x) = 3x + 5$. Perform the operations.

(a) $(f + g)(1)$ (b) $(f - g)(-3)$ (c) $(fg)(5)$ (d) $\left(\dfrac{f}{g}\right)(0)$

Analytic Solution

(a) $(f + g)(1) = f(1) + g(1)$ Definition

$\qquad = (1^2 + 1) + (3 \cdot 1 + 5)$ $f(x) = x^2 + 1;$
$\qquad\qquad\qquad\qquad\qquad\qquad g(x) = 3x + 5$

$\qquad = 2 + 8$ Simplify.

$\qquad = 10$ Add.

(b) $(f - g)(-3) = f(-3) - g(-3)$ Definition

$\qquad\qquad = 10 - (-4)$ Substitute.

$\qquad\qquad = 14$ Subtract.

(c) $(fg)(5) = f(5) \cdot g(5) = 26 \cdot 20 = 520$

(d) $\left(\dfrac{f}{g}\right)(0) = \dfrac{f(0)}{g(0)} = \dfrac{1}{5}$

Graphing Calculator Solution

FIGURE 66(a) shows f entered as Y_1 and g as Y_2.

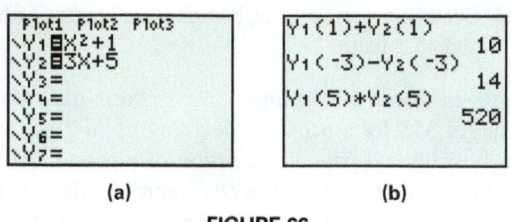

(a)   (b)

FIGURE 66

Because the calculator is capable of function notation, we can calculate parts (a)–(c) as shown in **FIGURE 66(b)**. Part (d) can be found similarly.

EXAMPLE 2 **Using the Operations on Functions**

Let $f(x) = 8x - 9$ and $g(x) = \sqrt{2x - 1}$. Perform the operations.

(a) $(f + g)(x)$ (b) $(f - g)(x)$ (c) $(fg)(x)$ (d) $\left(\dfrac{f}{g}\right)(x)$

(e) Give the domains of f, g, $f + g$, $f - g$, fg, and $\dfrac{f}{g}$.

Solution

(a) $(f + g)(x) = f(x) + g(x)$ Definition

$\qquad\qquad = (8x - 9) + \sqrt{2x - 1}$ Substitute.

(b) $(f - g)(x) = f(x) - g(x)$

$\qquad\qquad = (8x - 9) - \sqrt{2x - 1}$

(c) $(fg)(x) = f(x) \cdot g(x)$

$\qquad\qquad = (8x - 9)\sqrt{2x - 1}$

(d) $\left(\dfrac{f}{g}\right)(x) = \dfrac{f(x)}{g(x)}$

$\qquad\qquad = \dfrac{8x - 9}{\sqrt{2x - 1}}$

(e) To find the domains of the functions in parts (a)–(d), we first find the domains of f and g. The domain of f, since $f(x) = 8x - 9$, is the set of all real numbers $(-\infty, \infty)$. The domain of g, since $g(x) = \sqrt{2x - 1}$, includes only the real numbers that make $2x - 1$ nonnegative. Solve the inequality $2x - 1 \geq 0$ to find that the domain of g is $\left[\dfrac{1}{2}, \infty\right)$.

The domains of $f + g$, $f - g$, and fg are the intersection of the domains of f and g, which is

$$(-\infty, \infty) \cap \left[\frac{1}{2}, \infty\right) = \left[\frac{1}{2}, \infty\right).$$

The domain of $\frac{f}{g}$ includes those real numbers in the intersection above for which $g(x) = \sqrt{2x - 1} \neq 0$. Because $g\left(\frac{1}{2}\right) = 0$, the domain of $\frac{f}{g}$ is the open interval $\left(\frac{1}{2}, \infty\right)$.

EXAMPLE 3   **Evaluating Combinations of Functions**

If possible, use the given representations of functions f and g to evaluate

$$(f + g)(4), \quad (f - g)(-2), \quad (fg)(1), \quad \text{and} \quad \left(\frac{f}{g}\right)(0).$$

(a)

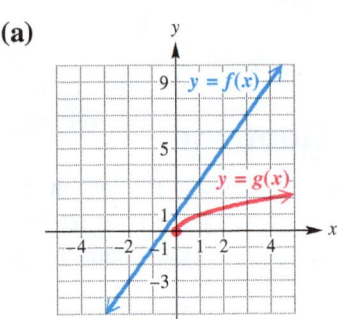

FIGURE 67

(b)

x	$f(x)$	$g(x)$
-2	-3	undefined
0	1	0
1	3	1
4	9	2

(c) $f(x) = 2x + 1$, $g(x) = \sqrt{x}$

Solution

(a) In **FIGURE 67**, $f(4) = 9$ and $g(4) = 2$.

$$\begin{aligned}
(f + g)(4) &= f(4) + g(4) \qquad &&\text{Definition} \\
&= 9 + 2 &&\text{Evaluate.} \\
&= 11 &&\text{Add.}
\end{aligned}$$

Although $f(-2) = -3$ in **FIGURE 67**, $g(-2)$ is undefined because -2 is not in the domain of g. Thus $(f - g)(-2)$ is undefined.

The domains of f and g include 1.

$$\begin{aligned}
(fg)(1) &= f(1) \cdot g(1) \qquad &&\text{Definition} \\
&= 3 \cdot 1 &&\text{Evaluate.} \\
&= 3 &&\text{Multiply.}
\end{aligned}$$

The graph of g intersects the origin, so $g(0) = 0$. Thus, $\left(\frac{f}{g}\right)(0)$ is undefined.

(b) From the table, $f(4) = 9$ and $g(4) = 2$. As in part (a),

$$(f + g)(4) = f(4) + g(4) = 9 + 2 = 11.$$

In the table $g(-2)$ is undefined, so $(f - g)(-2)$ is also undefined. Similarly,

$$(fg)(1) = f(1) \cdot g(1) = 3 \cdot 1 = 3,$$

and $\qquad\qquad \left(\frac{f}{g}\right)(0) = \dfrac{f(0)}{g(0)}$ is undefined, since $g(0) = 0$.

(continued)

(c) Use the formulas $f(x) = 2x + 1$ and $g(x) = \sqrt{x}$.

$$(f + g)(4) = f(4) + g(4) \qquad \text{Definition}$$
$$= (2 \cdot 4) + 1 + \sqrt{4} \qquad \text{Use the rules for } f \text{ and } g.$$
$$= 8 + 1 + 2 \qquad \text{Use the rules for order of operations.}$$
$$= 11 \qquad \text{Add.}$$

$$(f - g)(-2) = f(-2) - g(-2) \qquad \text{Definition}$$
$$= [2(-2) + 1] - \sqrt{-2} \qquad \text{Not a real number}$$

$$fg(1) = f(1) \cdot g(1) \qquad \text{Definition}$$
$$= (2 \cdot 1 + 1) \cdot \sqrt{1}$$
$$= 3(1)$$
$$= 3$$

$$\left(\frac{f}{g}\right)(0) = \frac{f(0)}{g(0)} \text{ is undefined, since } g(0) = 0.$$

The Difference Quotient

Suppose that the point P lies on the graph of $y = f(x)$ as in **FIGURE 68,** and suppose that h is a positive number. If we let $(x, f(x))$ denote the coordinates of P and $(x + h, f(x + h))$ denote the coordinates of Q, then the line joining P and Q has slope

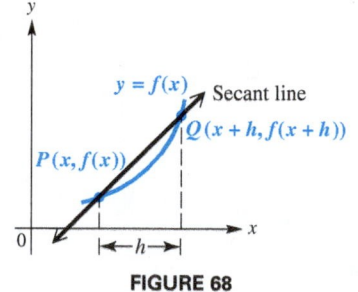

FIGURE 68

$$m = \frac{f(x + h) - f(x)}{(x + h) - x} = \frac{f(x + h) - f(x)}{h}. \qquad \text{Difference quotient}$$

This expression, called the **difference quotient,** is important in the study of calculus.

FIGURE 68 shows the graph of the line PQ (called a **secant line**). The difference quotient gives the slope of the secant line that passes through the points $P(x, f(x))$ and $Q(x + h, f(x + h))$. This slope is equal to the *average rate of change* of f from x to $x + h$.

One step in finding a difference quotient is to first evaluate $f(x + h)$. The next example demonstrates how to do this step.

EXAMPLE 4 **Finding $f(x + h)$**

Find $f(x + h)$ for each function.
(a) $f(x) = x^2$ **(b)** $f(x) = 2x - 5x^2$

Solution
(a) Substitute $(x + h)$ for x in the expression for $f(x)$.

$$f(x + h) = (x + h)^2 \qquad f(x) = x^2$$
$$= x^2 + 2xh + h^2 \qquad \text{Square the binomial.}$$

(b) Substitute $(x + h)$ for each occurrence of x in the expression for $f(x)$.

$$f(x + h) = 2(x + h) - 5(x + h)^2 \qquad f(x) = 2x - 5x^2$$
$$= 2(x + h) - 5(x^2 + 2xh + h^2) \qquad \text{Square the binomial.}$$
$$= 2x + 2h - 5x^2 - 10xh - 5h^2 \qquad \text{Distributive property}$$

⬅ Algebra Review
To review how to square a binomial, see **Section R.1.**

EXAMPLE 5   **Finding the Difference Quotient**

Let $f(x) = 2x^2 - 3x$. Find the difference quotient and simplify the expression.

Solution   To find $f(x + h)$, replace x in $f(x)$ with $x + h$ to get

$$f(x + h) = 2(x + h)^2 - 3(x + h).$$

> Distribute the minus sign to each term inside the parentheses.

$$\frac{f(x + h) - f(x)}{h} = \frac{2(x + h)^2 - 3(x + h) - (2x^2 - 3x)}{h}$$

> Remember the middle term when squaring $x + h$.

$$= \frac{2(x^2 + 2xh + h^2) - 3x - 3h - 2x^2 + 3x}{h} \qquad \text{Square } x + h; \text{ distributive property}$$

$$= \frac{2x^2 + 4xh + 2h^2 - 3x - 3h - 2x^2 + 3x}{h} \qquad \text{Distributive property}$$

$$= \frac{4xh + 2h^2 - 3h}{h} \qquad \text{Combine like terms.}$$

$$= \frac{h(4x + 2h - 3)}{h} \qquad \text{Factor out } h.$$

$$= 4x + 2h - 3 \qquad \text{Divide.} \qquad \bullet$$

Looking Ahead to Calculus
The difference quotient is essential in the definition of the **derivative of a function** in calculus. The derivative provides a formula, in function form, for finding the slope of the tangent line to the graph of the function at a given point.

The next example illustrates that the expression $f(x + h)$, which occurs in the difference quotient, is not equal to $f(x) + f(h)$.

EXAMPLE 6   **Evaluating $f(x + h)$ and $f(x) + f(h)$**

Let $f(x) = 2x - x^2$. Evaluate $f(x + h)$ and $f(x) + f(h)$. Comment on your results.

Solution   First, evaluate $f(x + h)$.

$$f(x + h) = 2(x + h) - (x + h)^2 \qquad f(x) = 2x - x^2$$
$$= 2x + 2h - (x^2 + 2xh + h^2) \qquad \text{Square binomial.}$$
$$= 2x + 2h - x^2 - 2xh - h^2 \qquad \text{Distributive property}$$

Next, evaluate $f(x) + f(h)$.

$$f(x) + f(h) = (2x - x^2) + (2h - h^2) \qquad f(x) = 2x - x^2$$
$$= 2x - x^2 + 2h - h^2 \qquad \text{Add expressions.}$$

The two expressions are *not* equivalent. The term $-2xh$ appears in the expression for $f(x + h)$, but not in that of $f(x) + f(h)$. $\bullet$

Composition of Functions

The diagram in **FIGURE 69** shows a function f that assigns, to each x in its domain, a value $f(x)$. Then another function g assigns, to each $f(x)$ in the domain of g, a value $g(f(x))$. This two-step process takes an element x and outputs an element $g(f(x))$. That is, a *sequence* of two functions, f and then g, is applied to input x.

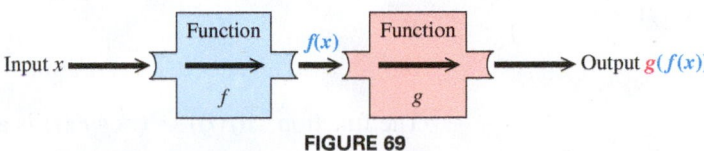

FIGURE 69

The function with y-values $g(f(x))$ is called the *composition* of function g and f, written $g \circ f$ and read "g of f."

> ### Composition of Functions
>
> If f and g are functions, then the **composite function,** or **composition,** of g and f is
>
> $$(g \circ f)(x) = g(f(x))$$
>
> for all x in the domain of f such that $f(x)$ is in the domain of g.

As a real-life example of how composite functions occur, consider the following retail situation:

A $40 pair of blue jeans is on sale for 25% off. If you purchase the jeans before noon, the retailer offers an additional 10% off. What is the final sale price of the blue jeans?

You might be tempted to say that the jeans are 35% off and calculate $40(0.35) = \$14$, giving a final sale price of $40 - \$14 = \$26$ for the jeans. **This is not correct.** To find the final sale price, we must first find the price after taking 25% off and then take an additional 10% off *that* price.

$40(0.25) = \$10$, giving a sale price of $40 - \$10 = **\$30**$. Take 25% off original price.

$30(0.10) = \$3$, giving a **final sale price** of $30 - \$3 = **\$27**$. Take additional 10% off.

This is the idea behind composition of functions.

As another example, suppose an oil well off the North Sea coast is leaking, spreading oil in a circular layer over the water's surface. See **FIGURE 70.**

FIGURE 70

At any time t, in minutes, after the beginning of the leak, the radius of the circular oil slick is $r(t) = 5t$ feet. Since $\mathcal{A}(r) = \pi r^2$ gives the area of a circle of radius r, the area can be expressed as a function of time by substituting $5t$ for r in $\mathcal{A}(r) = \pi r^2$.

$$\mathcal{A}(r) = \pi r^2$$
$$\mathcal{A}(r(t)) = \pi(5t)^2$$
$$= 25\pi t^2$$

The function $\mathcal{A}(r(t)) = (\mathcal{A} \circ r)(t)$ is a composite function of the functions $\mathcal{A}$ and r.

EXAMPLE 7   **Evaluating Composite Functions**

Let $f(x) = 2x - 1$ and $g(x) = \dfrac{4}{x - 1}$. Perform the compositions.

(a) $(f \circ g)(2)$ **(b)** $(g \circ f)(-3)$

Analytic Solution

(a) First find $g(2)$. Since $g(x) = \dfrac{4}{x - 1}$,

$$g(2) = \frac{4}{2 - 1} = \frac{4}{1} = 4.$$

Now find $(f \circ g)(2) = f(g(2))$.

$$f(g(2)) = f(4) = 2(4) - 1 = 7$$

(b) $f(-3) = 2(-3) - 1 = -7$. Use this information below.

$$(g \circ f)(-3) = g(f(-3)) = g(-7)$$

$$= \frac{4}{-7 - 1} = \frac{4}{-8} = -\frac{1}{2}$$

Graphing Calculator Solution

In **FIGURE 71,** $Y_1 = f(x)$ and $Y_2 = g(x)$.

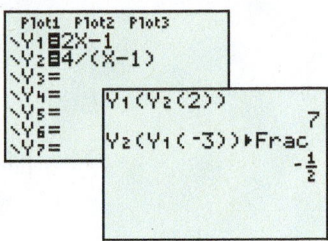

FIGURE 71

EXAMPLE 8   **Finding Composite Functions**

Let $f(x) = 4x + 1$ and $g(x) = 2x^2 + 5x$. Perform the compositions.

(a) $(g \circ f)(x)$ **(b)** $(f \circ g)(x)$

Solution

(a) $(g \circ f)(x) = g(f(x)) = g(4x + 1)$ $f(x) = 4x + 1$

$\qquad\qquad = 2(4x + 1)^2 + 5(4x + 1)$ $g(x) = 2x^2 + 5x$

$\qquad\qquad = 2(16x^2 + 8x + 1) + 20x + 5$ Square $4x + 1$; distributive property

Recall:
$(x + y)^2 = x^2 + 2xy + y^2.$ $= 32x^2 + 16x + 2 + 20x + 5$ Distributive property

$\qquad\qquad = 32x^2 + 36x + 7$ Combine like terms.

(b) $(f \circ g)(x) = f(g(x))$ Definition of composition

$\qquad\qquad = f(2x^2 + 5x)$ $g(x) = 2x^2 + 5x$

$\qquad\qquad = 4(2x^2 + 5x) + 1$ $f(x) = 4x + 1$

$\qquad\qquad = 8x^2 + 20x + 1$ Distributive property

In **Example 8,** the domain of both composite functions is $(-\infty, \infty)$. *As Example 8 shows, it is not always true that* $f \circ g = g \circ f$. However, the composite functions $f \circ g$ and $g \circ f$ are equal for a special class of functions, discussed in **Section 5.1.**

CAUTION   *In general, the composite function $f \circ g$ is not the same as the product fg.* For example, if $f(x) = 4x + 1$ and $g(x) = 2x^2 + 5x$ as defined in **Example 8,**

$$(f \circ g)(x) = 8x^2 + 20x + 1,$$

Not equal

but

$$(fg)(x) = (4x + 1)(2x^2 + 5x) = 8x^3 + 22x^2 + 5x.$$

→ **Looking Ahead to Calculus**
Finding the derivative of a function in calculus is called **differentiation.** In calculus we often differentiate composite functions, so it is important to understand composition.

EXAMPLE 9 **Finding Functions for a Composite Function**

Suppose that $h(x) = (x^2 - 5)^3 - 4(x^2 - 5) + 3$. Find functions f and g so that $(f \circ g)(x) = h(x)$.

Solution Note the repeated quantity $x^2 - 5$ in $h(x)$. If we let $g(x) = x^2 - 5$ and $f(x) = x^3 - 4x + 3$, then

$$
\begin{aligned}
(f \circ g)(x) &= f(g(x)) && \text{Definition of composition} \\
&= f(x^2 - 5) && g(x) = x^2 - 5 \\
&= (x^2 - 5)^3 - 4(x^2 - 5) + 3 && f(x) = x^3 - 4x + 3 \\
&= h(x).
\end{aligned}
$$

Other pairs of functions f and g also work. For instance, we can use

$$ f(x) = (x - 5)^3 - 4(x - 5) + 3 \quad \text{and} \quad g(x) = x^2. $$

Applications of Operations and Composition

There are many applications in business, economics, physics, and other fields that combine functions. For example, in manufacturing, the cost of producing a product usually consists of two parts: a *fixed cost* for designing the product, setting up a factory, training workers, and so on, and a *variable cost* per item for labor, materials, packaging, shipping, and so on.

A *linear cost function* has the form

$$ C(x) = mx + b, \qquad \text{Cost function} $$

where m represents the variable cost per item and b represents the fixed cost. The revenue from selling a product depends on the price per item and the number of items sold, as given by the *revenue function,*

$$ R(x) = px, \qquad \text{Revenue function} $$

where p is the price per item and $R(x)$ is the revenue from the sale of x items. The profit is described by the *profit function,* given by

$$ P(x) = R(x) - C(x). \qquad \text{Profit function} $$

FIGURE 72 shows the situation for a company that manufactures DVDs. The two lines are the graphs of the linear functions for revenue $R(x) = 168x$ and cost $C(x) = 118x + 800$, with x, $R(x)$, and $C(x)$ given in thousands. When 30,000 (that is, 30 thousand) DVDs are produced and sold, profit is

$$
\begin{aligned}
P(30) &= R(30) - C(30) \\
&= 5040 - 4340 && R(30) = 168(30);\ C(30) = 118(30) + 800 \\
&= 700. && \text{Subtract.}
\end{aligned}
$$

Thus, the profit from the sale of 30,000 DVDs is $700,000.

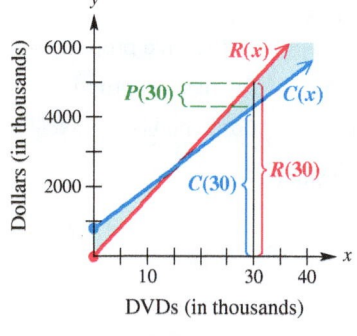

FIGURE 72

EXAMPLE 10 **Finding and Analyzing Cost, Revenue, and Profit**

Suppose that a businesswoman invests $1500 as her fixed cost in a new venture that produces and sells a device for satellite radio. Each such device costs $100 to manufacture.

(a) Write a cost function for the product if x represents the number of devices produced. Assume that the function is linear.

(b) Find the revenue function if each device in part (a) sells for $125.

(c) Give the profit function for the item in parts (a) and (b).

(d) How many items must be produced and sold before the company makes a profit?

Solution

(a) Since the cost function is linear, it will have the form $C(x) = mx + b$, with $m = 100$ and $b = 1500$.

$$C(x) = 100x + 1500$$

(b) If we let $p = 125$, then the revenue function is found as follows.

$$R(x) = px = 125x$$

(c) Using the results of parts (a) and (b), we obtain the profit function as follows.

$$
\begin{array}{ll}
P(x) = R(x) - C(x) & \text{Definition} \\
\quad = 125x - (100x + 1500) & \text{Substitute.} \\
\quad = 125x - 100x - 1500 & \text{Distributive property} \\
\quad = 25x - 1500 & \text{Combine like terms.}
\end{array}
$$

Distribute the minus sign to each term inside the parentheses.

(d) To make a profit, $P(x)$ must be positive. Set $P(x) = 25x - 1500 > 0$ and solve for x.

$$
\begin{array}{ll}
25x - 1500 > 0 & P(x) > 0 \\
25x > 1500 & \text{Add 1500.} \\
x > 60 & \text{Divide by 25.}
\end{array}
$$

Because 61 is the least integer greater than 60, at least 61 devices must be sold for the company to make a profit.

EXAMPLE 11 **Modeling the Surface Area of a Ball**

The formula for the surface area S of a sphere is $S = 4\pi r^2$, where r is the radius of the sphere. See **FIGURE 73**.

(a) Write a function $D(r)$ that gives the amount of surface area gained if the radius r of a ball is increased by 2 inches.

(b) Determine the amount of extra material needed to manufacture a ball of radius 22 inches compared with a ball of radius 20 inches.

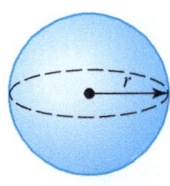

FIGURE 73

Solution

(a) This translates as a difference of functions.

$$
\begin{array}{ccc}
\text{surface} & = & \text{larger} & - & \text{smaller} \\
\text{area gained} & & \text{surface area} & & \text{surface area} \\
\downarrow & & \downarrow & & \downarrow \\
D(r) & = & 4\pi(r + 2)^2 & - & 4\pi r^2
\end{array}
$$

(b)
$$
\begin{array}{ll}
D(r) = 4\pi(r + 2)^2 - 4\pi r^2 & \text{Model from part (a)} \\
D(20) = 4\pi(20 + 2)^2 - 4\pi(20)^2 & \text{Let } r = 20. \\
\quad = 1936\pi - 1600\pi & \text{Simplify.} \\
\quad = 336\pi & \text{Subtract.} \\
\quad \approx 1056 & \text{Approximate.}
\end{array}
$$

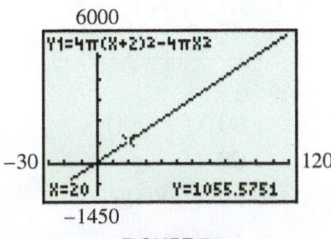

FIGURE 74

Thus, it would take about 1056 square inches of extra material. A graph of

$$Y_1 = 4\pi(X + 2)^2 - 4\pi X^2$$

is shown in **FIGURE 74**. The display at the bottom agrees with the analytic solution.

2.6 Exercises

Concept Check *Let* $f(x) = x^2$ *and* $g(x) = 2x - 5$. *Match each function in Group I with the correct expression in Group II.*

I

1. $(f + g)(x)$ **2.** $(f - g)(x)$

3. $(fg)(x)$ **4.** $\left(\dfrac{f}{g}\right)(x)$

5. $(f \circ g)(x)$ **6.** $(g \circ f)(x)$

II

A. $4x^2 - 20x + 25$ **B.** $x^2 - 2x + 5$

C. $2x^2 - 5$ **D.** $\dfrac{x^2}{2x - 5}$

E. $x^2 + 2x - 5$ **F.** $2x^3 - 5x^2$

Let $f(x) = x^2 + 3x$ *and* $g(x) = 2x - 1$. *Perform the composition or operation indicated.*

7. $(f \circ g)(3)$ **8.** $(g \circ f)(-2)$ **9.** $(f \circ g)(x)$ **10.** $(g \circ f)(x)$ **11.** $(f + g)(3)$

12. $(f + g)(-5)$ **13.** $(fg)(4)$ **14.** $(fg)(-3)$ **15.** $\left(\dfrac{f}{g}\right)(-1)$ **16.** $\left(\dfrac{f}{g}\right)(4)$

17. $(f - g)(2)$ **18.** $(f - g)(-2)$ **19.** $(g - f)(-2)$ **20.** $\left(\dfrac{f}{g}\right)\left(\dfrac{1}{2}\right)$ **21.** $\left(\dfrac{g}{f}\right)(0)$

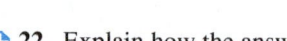 **22.** Explain how the answer to **Exercise 19** can be easily determined once the answer to **Exercise 18** is found.

Checking Analytic Skills *For each pair of functions,* **(a)** *find* $(f + g)(x)$, $(f - g)(x)$, *and* $(fg)(x)$;
(b) *give the domains of the functions in part (a);* **(c)** *find* $\dfrac{f}{g}$ *and give its domain;* **(d)** *find* $f \circ g$ *and give its domain; and* **(e)** *find* $g \circ f$ *and give its domain.* **Do not use a calculator.**

23. $f(x) = 4x - 1$, $g(x) = 6x + 3$ **24.** $f(x) = 9 - 2x$, $g(x) = -5x + 2$

25. $f(x) = |x + 3|$, $g(x) = 2x$ **26.** $f(x) = |2x - 4|$, $g(x) = x + 1$

27. $f(x) = \sqrt[3]{x + 4}$, $g(x) = x^3 + 5$ **28.** $f(x) = \sqrt[3]{6 - 3x}$, $g(x) = 2x^3 + 1$

29. $f(x) = \sqrt{x^2 + 3}$, $g(x) = x + 1$ **30.** $f(x) = \sqrt{2 + 4x^2}$, $g(x) = x$

Use the graph to evaluate each expression.

31.

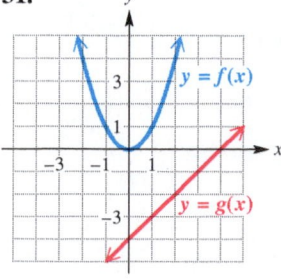

(a) $(f + g)(2)$
(b) $(f - g)(1)$
(c) $(fg)(0)$
(d) $\left(\dfrac{f}{g}\right)(1)$

32.

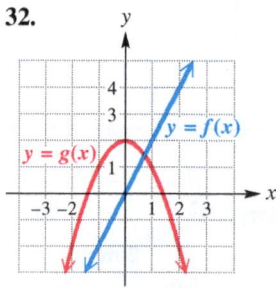

(a) $(f + g)(0)$
(b) $(f - g)(-1)$
(c) $(fg)(1)$
(d) $\left(\dfrac{f}{g}\right)(2)$

33.

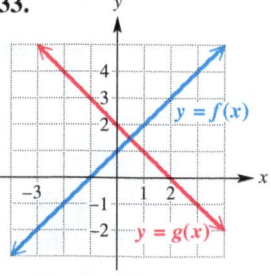

(a) $(f + g)(-1)$
(b) $(f - g)(-2)$
(c) $(fg)(0)$
(d) $\left(\dfrac{f}{g}\right)(2)$

34.

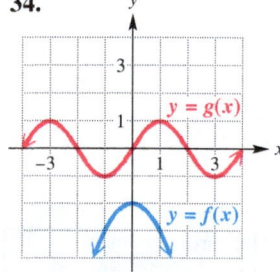

(a) $(f + g)(1)$
(b) $(f - g)(0)$
(c) $(fg)(-1)$
(d) $\left(\dfrac{f}{g}\right)(1)$

Use the table to evaluate each expression, if possible.

(a) $(f + g)(2)$ **(b)** $(f - g)(4)$ **(c)** $(fg)(-2)$ **(d)** $\left(\dfrac{f}{g}\right)(0)$

35.

x	$f(x)$	$g(x)$
−2	0	6
0	5	0
2	7	−2
4	10	5

36.

x	$f(x)$	$g(x)$
−2	−4	2
0	8	−1
2	5	4
4	0	0

37. Use the table in **Exercise 35** to complete the following table.

x	$(f + g)(x)$	$(f - g)(x)$	$(fg)(x)$	$\left(\dfrac{f}{g}\right)(x)$
−2				
0				
2				
4				

38. Use the table in **Exercise 36** to complete the following table.

x	$(f + g)(x)$	$(f - g)(x)$	$(fg)(x)$	$\left(\dfrac{f}{g}\right)(x)$
−2				
0				
2				
4				

Associate's Degrees Earned   *The graph shows the number of associate's degrees earned (in thousands) in the United States from 2000 through 2008. M(x) gives the number of degrees earned by males, F(x) gives the number earned by females, and T(x) gives the total number for both groups. Use the graph in Exercises 39–42.*

Associate's Degrees Earned

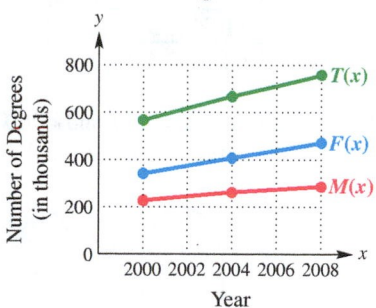

Source: U.S. National Center for Education Statistics.

39. Estimate $M(2004)$ and $F(2004)$, and use your results to estimate $T(2004)$.

40. Estimate $M(2008)$ and $F(2008)$, and use your results to estimate $T(2008)$.

41. Use the slopes of the line segments to decide in which period (2000–2004 or 2004–2008) the total number of associate's degrees earned increased more rapidly.

42. *Concept Check*   If $2000 \leq k \leq 2008$, $T(k) = r$, and $F(k) = s$, then $M(k) = $ _____.

Science and Space/Technology Spending   *The graph shows dollars (in billions) spent for general science and for space/other technologies in selected years. G(x) represents the dollars spent for general science, and S(x) represents the dollars spent for space and other technologies. T(x) represents the total expenditures for these two categories. Use the graph in Exercises 43–46.*

Science and Space Spending

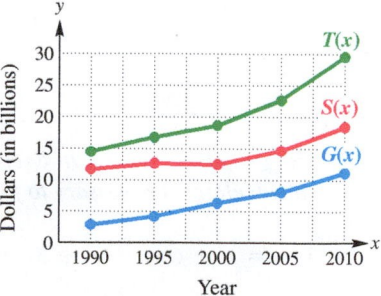

Source: U.S. Office of Management and Budget.

43. Estimate $(T - S)(2000)$. What does this function represent?

44. Estimate $(T - G)(2010)$. What does this function represent?

45. In which of the categories was spending almost static for several years? In which years did this occur?

46. In which period and which category does spending for $G(x)$ or $S(x)$ increase most?

Use the graph to evaluate each expression. (Hint: Extend the ideas of **Example 3.***)*

47.

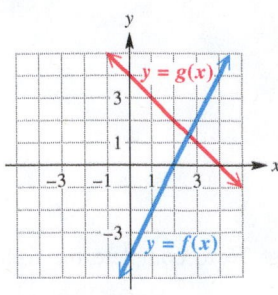

(a) $(f \circ g)(4)$

(b) $(g \circ f)(3)$

(c) $(f \circ f)(2)$

48.

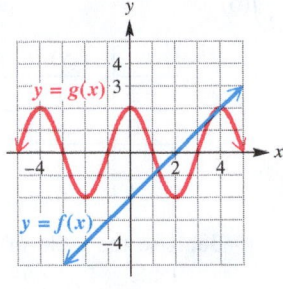

(a) $(f \circ g)(2)$

(b) $(g \circ g)(0)$

(c) $(g \circ f)(4)$

49.

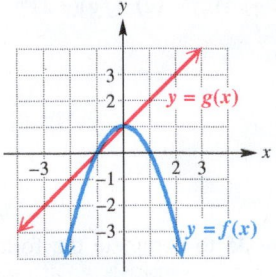

(a) $(f \circ g)(1)$

(b) $(g \circ f)(-2)$

(c) $(g \circ g)(-2)$

50.

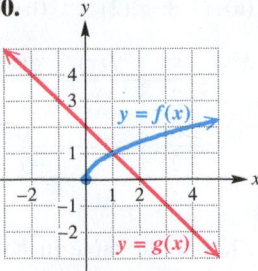

(a) $(f \circ g)(-2)$

(b) $(g \circ f)(1)$

(c) $(f \circ f)(0)$

In Exercises 51 and 52, tables for functions f and g are given. Evaluate each expression, if possible.

(a) $(g \circ f)(1)$ (b) $(f \circ g)(4)$ (c) $(f \circ f)(3)$

51.

x	$f(x)$	x	$g(x)$
1	4	1	2
2	3	2	3
3	1	3	4
4	2	4	5

52.

x	$f(x)$	x	$g(x)$
1	2	2	4
3	6	3	2
4	5	5	6
6	7	7	0

53. Use the tables for f and g in **Exercise 51** to complete the composition shown in the diagram.

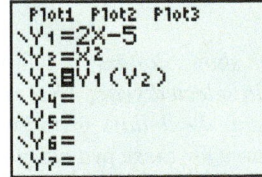

54. Use the tables for f and g in **Exercise 52** to complete the composition shown in the diagram.

The graphing calculator screen on the left shows three functions: Y_1, Y_2, *and* Y_3. *The last of these,* Y_3, *is defined as* $Y_1 \circ Y_2$, *indicated by the notation* $Y_3 = Y_1(Y_2)$. *The table on the right shows selected values of X, along with the calculated values of* Y_3. *Predict the display for* Y_3 *for the given value of X.*

55. X = −1

56. X = −2

57. X = 7

58. X = 8

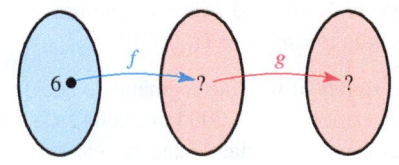

Checking Analytic Skills   In Exercises 59–67, use f(x) and g(x) to find each composition. Identify its domain.
(Use a calculator if necessary to find the domain.)   (a) $(f \circ g)(x)$ (b) $(g \circ f)(x)$ (c) $(f \circ f)(x)$

59. $f(x) = x^3$, $g(x) = x^2 + 3x - 1$

60. $f(x) = 2 - x$, $g(x) = \dfrac{1}{x^2}$

61. $f(x) = x^2$, $g(x) = \sqrt{1 - x}$

62. $f(x) = x + 2$, $g(x) = x^4 + x^2 - 3x - 4$

63. $f(x) = \dfrac{1}{x + 1}$, $g(x) = 5x$

64. $f(x) = x + 4$, $g(x) = \sqrt{4 - x^2}$

65. $f(x) = 2x + 1$, $g(x) = 4x^3 - 5x^2$

66. $f(x) = \dfrac{x - 3}{2}$, $g(x) = 2x + 3$ **67.** $f(x) = 5$, $g(x) = x$

68. *Concept Check*   If $f(x)$ defines a constant function over $(-\infty, \infty)$, how many elements are in the range of $(f \circ f)(x)$?

For certain pairs of functions f and g, $(f \circ g)(x) = x$ and $(g \circ f)(x) = x$. Show that this is true for the pairs in Exercises 69–72.

69. $f(x) = 4x + 2$, $g(x) = \dfrac{1}{4}(x - 2)$

70. $f(x) = -3x$, $g(x) = -\dfrac{1}{3}x$

71. $f(x) = \sqrt[3]{5x + 4}$, $g(x) = \dfrac{1}{5}x^3 - \dfrac{4}{5}$

72. $f(x) = \sqrt[3]{x + 1}$, $g(x) = x^3 - 1$

*Functions such as the pairs in **Exercises 69–72** are called inverse functions, because the result of composition in both directions is the identity function. (Inverse functions will be discussed in detail in **Section 5.1**.)*

73. In a square viewing window, graph $y_1 = \sqrt[3]{x - 6}$ and $y_2 = x^3 + 6$, an example of a pair of inverse functions. Now graph $y_3 = x$. Describe how the graph of y_2 can be obtained from the graph of y_1, using the graph $y_3 = x$ as a basis for your description.

74. Repeat **Exercise 73** for
$$y_1 = 5x - 3 \quad \text{and} \quad y_2 = \dfrac{1}{5}(x + 3).$$

For each function find $f(x + h)$ and $f(x) + f(h)$.

75. $f(x) = x^2 - 4$

76. $f(x) = 5x^2 + x$

77. $f(x) = 3x - x^2$

78. $f(x) = x^3$

Determine the difference quotient $\dfrac{f(x + h) - f(x)}{h}$ (where $h \neq 0$) for each function f. Simplify completely.

79. $f(x) = 4x + 3$

80. $f(x) = 5x - 6$

81. $f(x) = -6x^2 - x + 4$

82. $f(x) = \dfrac{1}{2}x^2 + 4x$

83. $f(x) = x^3$

84. $f(x) = -2x^3$

85. $f(x) = 1 - x^2$

86. $f(x) = x^2 + 2x$

87. $f(x) = 3x^2$

88. $f(x) = \sqrt{x}$
    (*Hint:* Rationalize the numerator.)

89. $f(x) = \dfrac{1}{2x}$

90. $f(x) = \dfrac{1}{x^2}$

Consider the function h as defined. Find functions f and g such that $(f \circ g)(x) = h(x)$. (There are several possible ways to do this.)

91. $h(x) = (6x - 2)^2$

92. $h(x) = (11x^2 + 12x)^2$

93. $h(x) = \sqrt{x^2 - 1}$

94. $h(x) = (2x - 3)^3$

95. $h(x) = \sqrt{6x + 12}$

96. $h(x) = \sqrt[3]{2x + 3} - 4$

*(Modeling) Cost/Revenue/Profit Analysis   For each situation, if x represents the number of items produced, **(a)** write a cost function, **(b)** find a revenue function if each item sells for the price given, **(c)** state the profit function, **(d)** determine analytically how many items must be produced before a profit is realized (assume whole numbers of items), and **(e)** support the result of part (d) graphically.*

97. The fixed cost is $500, the cost to produce an item is $10, and the selling price of the item is $35.

98. The fixed cost is $180, the cost to produce an item is $11, and the selling price of the item is $20.

99. The fixed cost is $2700, the cost to produce an item is $100, and the selling price of the item is $280.

100. The fixed cost is $1000, the cost to produce an item is $200, and the selling price of the item is $240.

(Modeling)   *Solve each application of operations and composition of functions.*

101. *Volume of a Sphere*   The formula for the volume of a sphere is $V = \frac{4}{3}\pi r^3$, where r represents the radius of the sphere.
  (a) Write a function $D(r)$ that gives the volume gained when the radius of a sphere of r inches is increased by 3 inches.
  (b) Graph $y = D(r)$ found in part (a), using x for r, in the window $[0, 10]$ by $[0, 1500]$.
  (c) Use your calculator to graphically find the amount of volume gained when a sphere with a 4-inch radius is increased to a 7-inch radius.
  (d) Verify your result in part (c) analytically.

102. *Surface Area of a Sphere*   Rework **Example 11(a)**, but consider what happens when the radius is doubled (rather than increased by 2 inches).

103. *Dimensions of a Rectangle*   Suppose that the length of a rectangle is twice its width. Let x represent the width of the rectangle.

  (a) Write a formula for the perimeter P of the rectangle in terms of x alone. Then use $P(x)$ notation to describe it as a function. What type of function is this?
  (b) Graph the function P found in part (a) in the window $[0, 10]$ by $[0, 100]$. Locate the point for which $x = 4$, and explain what x represents and what y represents.
  (c) On the graph of P, locate the point with x-value 4. Then sketch a rectangle satisfying the conditions described earlier, and evaluate its perimeter if its width is this x-value. Use the standard perimeter formula. How does the result compare with the y-value shown on your screen?
  (d) On the graph of P, find a point with an integer y-value. Interpret the x- and y-coordinates here.

104. *Perimeter of a Square*   The perimeter x of a square with side length s is given by the formula
$$x = 4s.$$

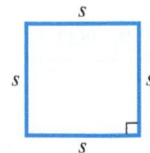

  (a) Solve for s in terms of x.
  (b) If y represents the area of this square, write y as a function of the perimeter x.
  (c) Use the composite function of part (b) to analytically find the area of a square with perimeter 6.
  (d) Support the result of part (c) graphically, and explain the result.

105. *Area of an Equilateral Triangle*   The area $\mathcal{A}$ of an equilateral triangle with sides of length x is given by
$$\mathcal{A}(x) = \frac{\sqrt{3}}{4}x^2.$$

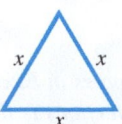

  (a) Find $\mathcal{A}(2x)$, the function representing the area of an equilateral triangle with sides of length twice the original length.
  (b) Find analytically the area of an equilateral triangle with side length 16. Use the given formula for $\mathcal{A}(x)$.
  (c) Support the result of part (b) graphically.

106. *Emission of Pollutants*   When a thermal inversion layer is over a city, pollutants cannot rise vertically, but are trapped below the layer and must disperse horizontally. Assume that a factory smokestack begins emitting a pollutant at 8 A.M. and that the pollutant disperses horizontally over a circular area. Let t represent the time in hours since the factory began emitting pollutants ($t = 0$ represents 8 A.M.) and assume that the radius of the circle of pollution is
$$r(t) = 2t \text{ miles.}$$
The area of a circle of radius r is represented by
$$\mathcal{A}(r) = \pi r^2.$$

  (a) Find $(\mathcal{A} \circ r)(t)$.
  (b) Interpret $(\mathcal{A} \circ r)(t)$.
  (c) What is the area of the circular region covered by the layer at noon?
  (d) Support your result graphically.

107. *World Population and Aggregate Age*   The following table lists the (projected) average age A for a person living during year x, and also the combined total of years T in billions lived by the current world population during year x.

x	1950	2000	2050	2100
$A(x)$	28	30	38	42
$T(x)$	80	180	360	430

Source: The U.N. Population Division.

  (a) Evaluate $A(2100)$ and $T(2100)$. Interpret your results.
  (b) Evaluate $\frac{T(2100)}{A(2100)}$. Interpret your result.
  (c) Let $P(x) = \frac{T(x)}{A(x)}$. Interpret what $P(x)$ calculates.

108. *Agriculture* The table below shows the acreage, in millions, of the total of corn and soybeans harvested annually in the United States. In the table, x represents the year and f computes the total number of acres for these two crops. The function g computes the number of acres for corn only.

x	2009	2010	2011	2012
$f(x)$	164.0	166.3	167.6	172.5
$g(x)$	86.5	88.2	92.3	96.4

Source: National Agriculture Statistics Service.

(a) Make a table for a function h that is defined by the equation $h(x) = f(x) - g(x)$.

(b) Interpret what h computes.

109. *U.S. Emissions* A common air pollutant responsible for acid rain is sulfur dioxide (SO_2). Emissions of SO_2 during year x are computed by $f(x)$ in the table. Emissions of carbon monoxide (CO) are computed by $g(x)$. Amounts are given in millions of tons.

x	1970	1980	1990	2000	2010
$f(x)$	31.2	25.9	23.1	16.3	13.0
$g(x)$	204.0	185.4	154.2	114.5	74.3

Source: EPA.

(a) Evaluate $(f + g)(2010)$.

(b) Interpret $(f + g)(x)$.

(c) Make a table for $(f + g)(x)$.

110. *Methane Emissions* The greenhouse gas methane lets sunlight into the atmosphere, but blocks heat from escaping the earth's atmosphere. Methane is a by-product of burning fossil fuels. In the table, f models the predicted methane emissions in millions of tons produced by developed countries during year x. The function g models the same emissions for developing countries.

x	1990	2000	2010	2020	2030
$f(x)$	27	28	29	30	31
$g(x)$	5	7.5	10	12.5	15

Source: Nilsson, A., *Greenhouse Earth*, John Wiley and Sons.

(a) Make a table for a function h that models the total predicted methane emissions for developed *and* developing countries.

(b) Write an equation that relates $f(x)$, $g(x)$, and $h(x)$.

111. *China's Energy Production* Estimated energy production in China is shown in the figure. The function f computes total coal production, the function g total coal *and* oil production. Energy units are in million metric tons of oil equivalent (Mtoe). Let the function h compute China's oil production. (*Source:* Priddle, R., *Coal in the Energy Supply of China*, Coal Energy Advisory Board/ International Energy Agency.)

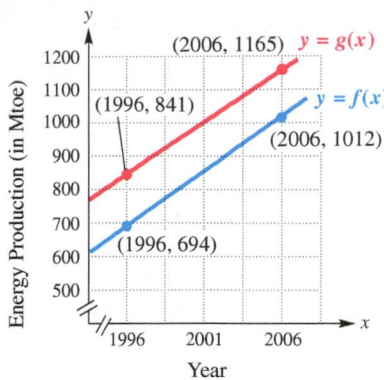

(a) Write an equation that relates $f(x)$, $g(x)$, and $h(x)$.

(b) Evaluate $h(1996)$ and $h(2006)$.

(c) Determine a formula for h. (*Hint:* Find formulas for $f(x)$ and $g(x)$ first.)

112. *U.S. AIDS* During the early years of the AIDS epidemic, cases and cumulative deaths reported for selected years x could be modeled by quadratic functions. For 1982–1994, the numbers of AIDS cases are modeled by

$$f(x) = 3200(x - 1982)^2 + 1586$$

and the numbers of deaths are modeled by

$$g(x) = 1900(x - 1982)^2 + 619.$$

Year	Cases	Deaths
1982	1,586	619
1984	10,927	5,605
1986	41,910	24,593
1988	106,304	61,911
1990	196,576	120,811
1992	329,205	196,283
1994	441,528	270,533

Source: U.S. Department of Health and Human Services.

(a) Graph $h(x) = \frac{g(x)}{f(x)}$ in the window $[1982, 1994]$ by $[0, 1]$. Interpret the graph.

(b) Compute the ratio $\frac{\text{deaths}}{\text{cases}}$ for each year. Compare the results with those from part (a).

1. Solve the equation in part (a) and the related inequalities in parts (b) and (c).

   (a) $\left|\frac{1}{2}x + 2\right| = 4$ (b) $\left|\frac{1}{2}x + 2\right| > 4$

   (c) $\left|\frac{1}{2}x + 2\right| \le 4$

2. Given the graph of $y = f(x)$ below, sketch the graph of $y = |f(x)|$.

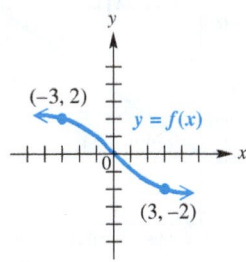

3. Solve $|2x + 4| = |1 - 3x|$.

4. For

$$f(x) = \begin{cases} 2x + 3 & \text{if } -3 \le x < 0 \\ x^2 + 4 & \text{if } x \ge 0 \end{cases},$$

   find each value.

   (a) $f(-3)$ (b) $f(0)$ (c) $f(2)$

5. Consider the following piecewise-defined function.

$$f(x) = \begin{cases} -x^2 & \text{if } x \le 0 \\ x - 4 & \text{if } x > 0 \end{cases}$$

   (a) Sketch its graph.

   (b) Use a graphing calculator in dot mode to obtain a graph in the window $[-10, 10]$ by $[-10, 10]$.

6. Given $f(x) = -3x - 4$ and $g(x) = x^2$, perform the composition or operation indicated.

   (a) $(f + g)(1)$ (b) $(f - g)(3)$

   (c) $(fg)(-2)$ (d) $\left(\dfrac{f}{g}\right)(-3)$

   (e) $(f \circ g)(x)$ (f) $(g \circ f)(x)$

7. Find functions f and g so that $h(x) = (f \circ g)(x)$, if we are given

$$h(x) = (x + 2)^4.$$

8. Find and simplify the difference quotient

$$\frac{f(x + h) - f(x)}{h}, \ h \ne 0,$$

   for $f(x) = -2x^2 + 3x - 5$.

9. *(Modeling) Musician Royalties*   A musician invests her royalties in two accounts for 1 year.

   (a) The first account pays 4% simple interest. If she invests x dollars in this account, write an expression for y_1 in terms of x, where y_1 represents the amount of interest earned.

   (b) In a second account, she invests $500 more than she invested in the first account. This second account pays 2.5% simple interest. Write an expression for y_2, the amount of interest earned.

   (c) What does $y_1 + y_2$ represent?

   (d) Graph $y_1 + y_2$ in the window $[0, 1000]$ by $[-20, 100]$. Use the graph to find the amount of interest she will receive if she invests $250 in the first account.

   (e) Support the result of part (d) analytically.

10. *Geometry*   The surface area of a cone (excluding the bottom) is given by

$$S = \pi r \sqrt{r^2 + h^2},$$

   where r is its radius and h is its height, as shown in the figure. If the height is twice the radius, write a formula for S in terms of r.

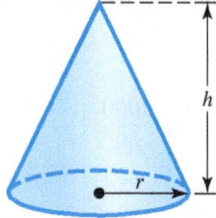

2 Summary

KEY TERMS & SYMBOLS

2.1 Graphs of Basic Functions and Relations; Symmetry

continuity
increasing function
decreasing function
constant function
identity function
degree
squaring function
parabola
vertex
symmetry
cubing function
inflection point
square root function
cube root function
absolute value function
even function
odd function

KEY CONCEPTS

CONTINUITY (Informal Definition)

A function is continuous over an interval of its domain if its hand-drawn graph over that interval can be sketched without lifting the pencil from the paper.

INCREASING, DECREASING, AND CONSTANT FUNCTIONS

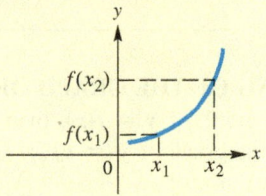

When $x_1 < x_2$, $f(x_1) < f(x_2)$.
f is **increasing**.

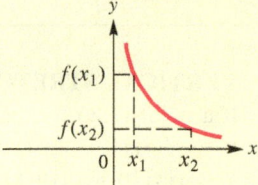

When $x_1 < x_2$, $f(x_1) > f(x_2)$.
f is **decreasing**.

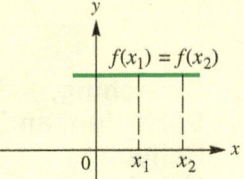

For any x_1 and x_2, $f(x_1) = f(x_2)$.
f is **constant**.

TYPES OF SYMMETRY

y-Axis Symmetry $f(-x) = f(x)$	Origin Symmetry $f(-x) = -f(x)$	x-Axis Symmetry (not possible for a nonzero function)

EVEN AND ODD FUNCTIONS

A function f is called an **even function** if $f(-x) = f(x)$ for all x in the domain of f. (Its graph is symmetric with respect to the y-axis.)

A function f is called an **odd function** if $f(-x) = -f(x)$ for all x in the domain of f. (Its graph is symmetric with respect to the origin.)

BASIC FUNCTIONS

- The **identity function,** $f(x) = x$, is increasing and continuous on its entire domain, $(-\infty, \infty)$.
- The **squaring function,** $f(x) = x^2$, decreases on the interval $(-\infty, 0)$, increases on the interval $(0, \infty)$, and is continuous on its entire domain, $(-\infty, \infty)$.
- The **cubing function,** $f(x) = x^3$, increases and is continuous on its entire domain, $(-\infty, \infty)$.
- The **square root function,** $f(x) = \sqrt{x}$, increases on $(0, \infty)$ and is continuous on its entire domain, $[0, \infty)$.
- The **cube root function,** $f(x) = \sqrt[3]{x}$, increases and is continuous on its entire domain, $(-\infty, \infty)$.
- The **absolute value function,** $f(x) = |x|$, decreases on the interval $(-\infty, 0)$, increases on the interval $(0, \infty)$, and is continuous on its entire domain, $(-\infty, \infty)$.

(continued)

KEY TERMS & SYMBOLS	KEY CONCEPTS
2.2 Vertical and Horizontal Shifts of Graphs translation	**VERTICAL SHIFTING OF THE GRAPH OF A FUNCTION** If $c > 0$, then the graph of $y = f(x) + c$ is obtained by shifting the graph of $y = f(x)$ *upward* a distance of c units. The graph of $y = f(x) - c$ is obtained by shifting the graph of $y = f(x)$ *downward* a distance of c units. **HORIZONTAL SHIFTING OF THE GRAPH OF A FUNCTION** If $c > 0$, then the graph of $y = f(x - c)$ is obtained by shifting the graph of $y = f(x)$ to the *right* a distance of c units. The graph of $y = f(x + c)$ is obtained by shifting the graph of $y = f(x)$ to the *left* a distance of c units.

2.3 Stretching, Shrinking, and Reflecting Graphs	**VERTICAL STRETCHING OF THE GRAPH OF A FUNCTION**
	If a point (x, y) lies on the graph of $y = f(x)$, then the point (x, cy) lies on the graph of $y = cf(x)$. If $c > 1$, then the graph of $y = cf(x)$ is a *vertical stretching* of the graph of $y = f(x)$ by applying a factor of c.
	
	VERTICAL SHRINKING OF THE GRAPH OF A FUNCTION
	If a point (x, y) lies on the graph of $y = f(x)$, then the point (x, cy) lies on the graph of $y = cf(x)$. If $0 < c < 1$, then the graph of $y = cf(x)$ is a *vertical shrinking* of the graph of $y = f(x)$ by applying a factor of c.
	
	HORIZONTAL STRETCHING OF THE GRAPH OF A FUNCTION
	If a point (x, y) lies on the graph of $y = f(x)$, then the point $\left(\frac{x}{c}, y\right)$ lies on the graph of $y = f(cx)$. If $0 < c < 1$, then the graph of $y = f(cx)$ is a *horizontal stretching* of the graph of $y = f(x)$.
	
	HORIZONTAL SHRINKING OF THE GRAPH OF A FUNCTION
	If a point (x, y) lies on the graph of $y = f(x)$, then the point $\left(\frac{x}{c}, y\right)$ lies on the graph of $y = f(cx)$. If $c > 1$, then the graph of $y = f(cx)$ is a *horizontal shrinking* of the graph of $y = f(x)$.
	
	REFLECTING THE GRAPH OF A FUNCTION ACROSS AN AXIS
	For a function defined by $y = f(x)$,
	  (a) the graph of $y = -f(x)$ is a reflection of the graph of f across the *x*-axis.
	  (b) the graph of $y = f(-x)$ is a reflection of the graph of f across the *y*-axis.

2.4 Absolute Value Functions	**PROPERTIES OF ABSOLUTE VALUE** For all real numbers a and b, **1.** $\|ab\| = \|a\| \cdot \|b\|$ **2.** $\left\|\dfrac{a}{b}\right\| = \dfrac{\|a\|}{\|b\|} (b \neq 0)$ **3.** $\|a\| = \|-a\|$ **4.** $\|a\| + \|b\| \geq \|a + b\|$ (triangle inequality). **GRAPH OF $y = \|f(x)\|$** The graph of $y = \|f(x)\|$ is obtained from the graph of $y = f(x)$ by leaving the graph unchanged for the portion on or above the *x*-axis and reflecting the portion of the graph below the *x*-axis across the *x*-axis.

| **KEY TERMS & SYMBOLS** | **KEY CONCEPTS** |

SOLVING ABSOLUTE VALUE EQUATIONS AND INEQUALITIES

Case 1 To solve $|ax + b| = k, k > 0$, solve the following compound equation.

$$ax + b = k \quad \text{or} \quad ax + b = -k$$

Case 2 To solve $|ax + b| > k, k > 0$, solve the following compound inequality.

$$ax + b > k \quad \text{or} \quad ax + b < -k$$

Case 3 To solve $|ax + b| < k, k > 0$, solve the following three-part inequality.

$$-k < ax + b < k$$

2.5 Piecewise-Defined Functions

piecewise-defined function
greatest integer function
step function

PIECEWISE-DEFINED FUNCTION

A piecewise-defined function is defined by different rules (formulas) over different subsets of its domain.

GREATEST INTEGER FUNCTION

$$f(x) = [\![x]\!] = \begin{cases} x & \text{if } x \text{ is an integer} \\ \text{the greatest integer less than } x & \text{if } x \text{ is not an integer} \end{cases}$$

2.6 Operations and Composition

difference quotient
composite function, $g \circ f$

OPERATIONS ON FUNCTIONS

Given two functions f and g, for all values for which both $f(x)$ and $g(x)$ are defined, the functions $f + g$, $f - g$, fg, and $\frac{f}{g}$ are defined as follows.

$$(f + g)(x) = f(x) + g(x) \qquad \text{Sum}$$
$$(f - g)(x) = f(x) - g(x) \qquad \text{Difference}$$
$$(fg)(x) = f(x) \cdot g(x) \qquad \text{Product}$$
$$\left(\frac{f}{g}\right)(x) = \frac{f(x)}{g(x)}, \quad g(x) \neq 0 \qquad \text{Quotient}$$

The domains of $f + g$, $f - g$, and fg include all real numbers in the intersection of the domains of f and g, while the domain of $\frac{f}{g}$ includes those real numbers in the intersection of the domains of f and g for which $g(x) \neq 0$.

DIFFERENCE QUOTIENT

$$\frac{f(x + h) - f(x)}{h}, \quad h \neq 0$$

The difference quotient equals the average rate of change of f from x to $x + h$, which also equals the slope of the secant line that passes through the points $(x, f(x))$ and $(x + h, f(x + h))$.

COMPOSITION OF FUNCTIONS

If f and g are functions, then the composite function, or composition, of g and f is

$$(g \circ f)(x) = g(f(x))$$

for all x in the domain of f such that $f(x)$ is in the domain of g.

2 Review Exercises

*Draw sketches of the graphs of the basic functions introduced in **Section 2.1**.*

$$f(x) = x, \qquad f(x) = x^2, \qquad f(x) = x^3,$$
$$f(x) = \sqrt{x}, \qquad f(x) = \sqrt[3]{x}, \qquad f(x) = |x|$$

Use your sketches to determine whether each statement in Exercises 1–10 is true or false. If false, tell why.

1. The range of $f(x) = x^2$ is the same as the range of $f(x) = |x|$.

2. $f(x) = x^2$ and $f(x) = |x|$ increase on the same interval.

3. $f(x) = \sqrt{x}$ and $f(x) = \sqrt[3]{x}$ have the same domain.

4. $f(x) = \sqrt[3]{x}$ decreases on its entire domain.

5. $f(x) = x$ has its domain equal to its range.

6. $f(x) = \sqrt{x}$ is continuous on the interval $(-\infty, 0)$.

7. None of the basic functions decrease on the interval $(0, \infty)$.

8. Both $f(x) = x$ and $f(x) = x^3$ have graphs that are symmetric with respect to the origin.

9. Both $f(x) = x^2$ and $f(x) = |x|$ have graphs that are symmetric with respect to the y-axis.

10. None of the graphs of the basic functions are symmetric with respect to the x-axis.

In Exercises 11–18, give the interval described.

11. Domain of $f(x) = \sqrt{x}$ **12.** Range of $f(x) = |x|$

13. Range of $f(x) = \sqrt[3]{x}$ **14.** Domain of $f(x) = x^2$

15. The largest open interval over which $f(x) = \sqrt[3]{x}$ is increasing

16. The largest open interval over which $f(x) = |x|$ is increasing

17. Domain of $x = y^2$ **18.** Range of $x = y^2$

In Exercises 19–30, graph $y = f(x)$ by hand.

19. $f(x) = (x + 3) - 1$ **20.** $f(x) = -\dfrac{1}{2}x + 1$

21. $f(x) = (x + 1)^2 - 2$ **22.** $f(x) = -2x^2 + 3$

23. $f(x) = -x^3 + 2$ **24.** $f(x) = (x - 3)^3$

25. $f(x) = \sqrt{\dfrac{1}{2}x}$ **26.** $f(x) = \sqrt{x - 2} + 1$

27. $f(x) = 2\sqrt[3]{x}$ **28.** $f(x) = \sqrt[3]{x} - 2$

29. $f(x) = |x - 2| + 1$ **30.** $f(x) = |-2x + 3|$

31. Consider the function whose graph is shown here.

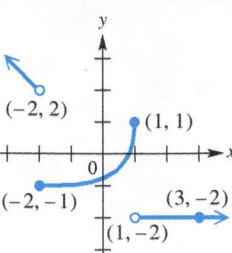

Give the largest open interval(s) over which the function

(a) is continuous.

(b) increases.

(c) decreases.

(d) is constant.

(e) What is the domain of the function?

(f) What is the range of the function?

32. The screen shows the graph of $x = y^2 - 4$. Give the two functions that must be used to graph this relation if the calculator is in function mode.

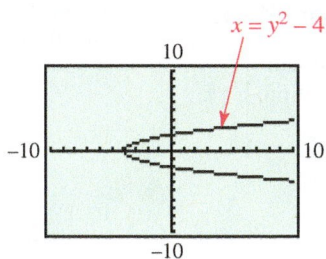

In Exercises 33–38, determine whether the given relation has x-axis symmetry, y-axis symmetry, origin symmetry, or none of these symmetries. (More than one choice is possible.) Also, if the relation is a function, determine whether it is an even function, an odd function, or neither.

33.

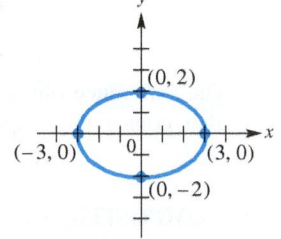

34. $F(x) = x^3 - 6$

35. $y = |x| + 4$ **36.** $f(x) = \sqrt{x - 5}$

37. $y^2 = x - 5$ **38.** $f(x) = 3x^4 + 2x^2 + 1$

Concept Check *Decide whether each statement in Exercises 39–44 is* true *or* false. *If false,* tell why.

39. The graph of a function (except for the constant function $f(x) = 0$) cannot be symmetric with respect to the x-axis.

40. The graph of an even function is symmetric with respect to the y-axis.

41. The graph of an odd function is symmetric with respect to the origin.

42. If (a, b) is on the graph of an even function, so is $(a, -b)$.

43. If (a, b) is on the graph of an odd function, so is $(-a, b)$.

44. The constant function $f(x) = 0$ is both even and odd.

45. Use the terminology of **Sections 2.2** and **2.3** to describe how the graph of $y = -3(x + 4)^2 - 8$ can be obtained from the graph of $y = x^2$.

46. Give the equation of the function whose graph is obtained by reflecting the graph of $y = \sqrt{x}$ across the y-axis, reflecting across the x-axis, shrinking vertically by applying a factor of $\frac{2}{3}$, and, finally, translating 4 units upward.

The graph of a function f is shown in the figure. Sketch the graph of each function as defined in Exercises 47–52.

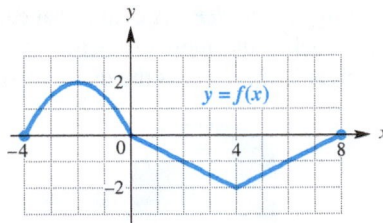

47. $y = f(x) + 3$

48. $y = f(x - 2)$

49. $y = f(x + 3) - 2$

50. $y = |f(x)|$

51. $y = f(4x)$

52. $y = f\left(\frac{1}{2}x\right)$

Let the domain of $f(x)$ be $[-3, 4]$ and the range be $[-2, 5]$. Find the domain and range of each of the following.

53. $f(x) + 4$

54. $5f(x + 10)$

55. $-f(2x)$

56. $f(x - 1) + 3$

The graph of a function $y = f(x)$ is given. Sketch the graph of $y = |f(x)|$.

57.

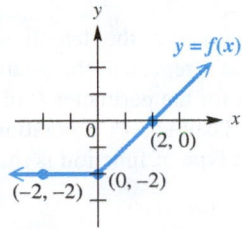

58.

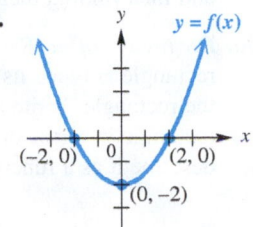

59.

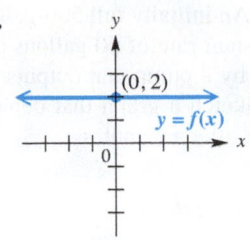

60.

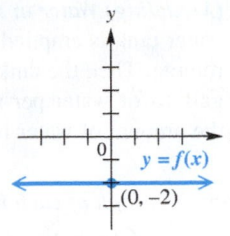

Solve each equation or inequality analytically.

61. $|4x + 3| = 12$

62. $|-2x - 6| + 4 = 1$

63. $|5x + 3| = |x + 11|$

64. $|2x + 5| = 7$

65. $|2x + 5| \le 7$

66. $|2x + 5| \ge 7$

67. $|5x - 12| > 0$

68. $|6 - x| \le -4$

69. $2|3x - 1| + 1 = 21$

70. $|2x + 1| = |-3x + 1|$

71. The graphs of $y_1 = |2x + 5|$ and $y_2 = 7$ are shown, along with the two points of intersection of the graphs. Explain how these two screens support the answers to **Exercises 64–66.**

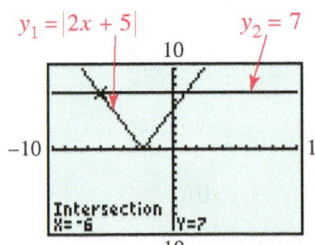

First view

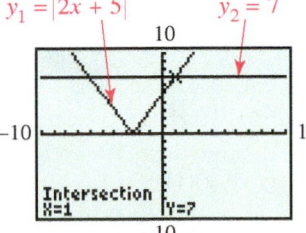

Second view

72. Solve the equation $|x + 1| + |x - 3| = 8$ graphically. Then, give an analytic check by substituting the values in the solution set directly into the left-hand side of the equation.

73. *(Modeling) Distance from Home* The graph depicts the distance y that a person driving a car on a straight road is from home after x hours. Interpret the graph. At what speeds did the car travel?

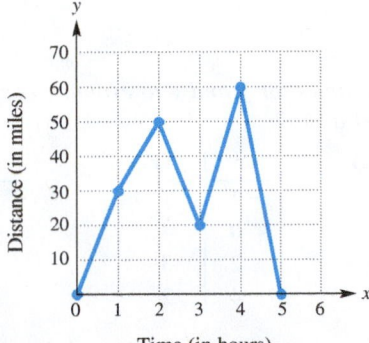

Time (in hours)

74. *(Modeling) Water in a Tank*   An initially full 500-gallon water tank is emptied at a constant rate of 50 gallons per minute. Then the tank is filled by a pump that outputs 25 gallons of water per minute. Sketch a graph that depicts the amount of water in the tank after x minutes.

Sketch the graph of each function by hand.

75. $f(x) = \begin{cases} 3x + 1 & \text{if } x < 2 \\ -x + 4 & \text{if } x \geq 2 \end{cases}$

76. $f(x) = \begin{cases} |x| & \text{if } x < 3 \\ 6 - x & \text{if } x \geq 3 \end{cases}$

77. Graph the function in **Exercise 75**, using a graphing calculator in dot mode with the window $[-10, 10]$ by $[-10, 10]$.

78. Use a graphing calculator to graph $f(x) = [\![x - 3]\!]$ in dot mode with the window $[-5, 5]$ by $[-5, 5]$.

The graphs of functions f and g are shown. Use these graphs to evaluate each expression in Exercises 79–86.

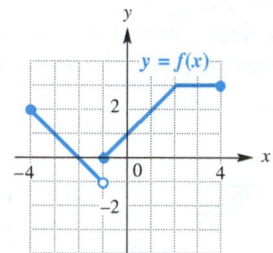
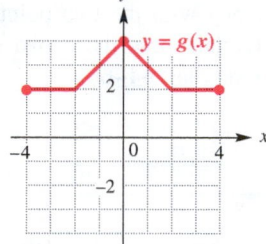

79. $(f + g)(1)$

80. $(f - g)(0)$

81. $(fg)(-1)$

82. $\left(\dfrac{f}{g}\right)(2)$

83. $(f \circ g)(2)$

84. $(g \circ f)(2)$

85. $(g \circ f)(-4)$

86. $(f \circ g)(-2)$

Use the table to evaluate each expression in Exercises 87–90, if possible.

87. $(f + g)(1)$

88. $(f - g)(3)$

89. $(fg)(-1)$

90. $\left(\dfrac{f}{g}\right)(0)$

x	$f(x)$	$g(x)$
-1	3	-2
0	5	0
1	7	1
3	9	9

Use the given tables for f and g to evaluate each expression in Exercises 91 and 92.

91. $(g \circ f)(-2)$

92. $(f \circ g)(3)$

x	$f(x)$
-2	1
0	4
2	3
4	2

x	$g(x)$
1	2
2	4
3	-2
4	0

For the given function, find and simplify

$$\frac{f(x + h) - f(x)}{h}, h \neq 0.$$

93. $f(x) = 2x + 9$

94. $f(x) = x^2 - 5x + 3$

Find functions f and g such that $(f \circ g)(x) = h(x)$.

95. $h(x) = (x^3 - 3x)^2$

96. $h(x) = \dfrac{1}{x - 5}$

(Modeling)   *Solve each problem.*

97. *Volume of a Sphere*   The formula for the volume of a sphere is $V(r) = \frac{4}{3}\pi r^3$, where r represents the radius of the sphere. Find a function D that gives the volume gained when the radius of a sphere of r inches is increased by 4 inches.

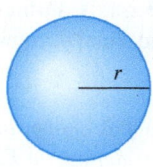

98. *Dimensions of a Cylinder*   A cylindrical can with a top and bottom makes the most efficient use of materials when its height is the same as the diameter of its top.

(a) Express the volume V of such a can as a function of the diameter d of its top.

(b) Express the surface area S of such a can as a function of the diameter d of its top. (*Hint:* The curved side is made from a rectangle whose length is the circumference of the top of the can.)

99. *Relationship of Measurement Units*   There are 36 inches in 1 yard, and there are 1760 yards in 1 mile. Express the number of inches in x miles by forming two functions and then finding their composition.

100. *Perimeter of a Rectangle*   Suppose the length of a rectangle is twice its width. Let x represent the width of the rectangle. Write a formula for the perimeter P of the rectangle in terms of x alone. Then use $P(x)$ notation to describe it as a function. What type of function is this?

2 Test

1. Match the set described in Column I with the correct interval notation from Column II. Choices in Column II may be used once, more than once, or not at all.

I	II
(a) Domain of $f(x) = \sqrt{x} + 3$	**A.** $[-3, \infty)$
(b) Range of $f(x) = \sqrt{x} - 3$	**B.** $[3, \infty)$
(c) Domain of $f(x) = x^2 - 3$	**C.** $(-\infty, \infty)$
(d) Range of $f(x) = x^2 + 3$	**D.** $[0, \infty)$
(e) Domain of $f(x) = \sqrt[3]{x} - 3$	**E.** $(-\infty, 3)$
(f) Range of $f(x) = \sqrt[3]{x} + 3$	**F.** $(-\infty, 3]$
(g) Domain of $f(x) = |x| - 3$	**G.** $(3, \infty)$
(h) Range of $f(x) = |x + 3|$	**H.** $(-\infty, 0]$
(i) Domain of $x = y^2$	
(j) Range of $x = y^2$	

2. The graph of $y = f(x)$ is shown here. Sketch the graph of each equation in parts (a)–(f). Label three points on the graph.
   (a) $y = f(x) + 2$
   (b) $y = f(x + 2)$
   (c) $y = -f(x)$
   (d) $y = f(-x)$
   (e) $y = 2f(x)$
   (f) $y = |f(x)|$

   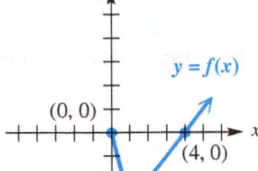

3. If the point $(-2, 4)$ lies on the graph of $y = f(x)$, determine coordinates of a point on the graph of each equation.

   (a) $y = f(2x)$ **(b)** $y = f\left(\dfrac{1}{2}x\right)$

4. Graph $y = f(x)$ by hand.
   (a) $f(x) = -(x - 2)^2 + 4$ **(b)** $y = 2\sqrt{-x}$

5. Observe the coordinates displayed at the bottom of the screen showing only the right half of the graph of $y = f(x)$. Answer the following based on your observation.

   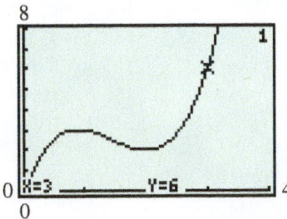

   (a) If the graph is symmetric with respect to the y-axis, what are the coordinates of another point on the graph?

   (b) If the graph is symmetric with respect to the origin, what are the coordinates of another point on the graph?

   (c) Suppose the graph is symmetric with respect to the y-axis. Sketch a typical viewing window with dimensions $[-4, 4]$ by $[0, 8]$. Then draw the graph you would expect to see in this window.

6. **(a)** Write a brief description that explains how the graph of $y = 4\sqrt[3]{x + 2} - 5$ can be obtained by transforming the graph of $y = \sqrt[3]{x}$.
   (b) Sketch by hand the graph of $y = -\frac{1}{2}|x - 3| + 2$. State the domain and range.

7. Consider the graph of the function shown here.

   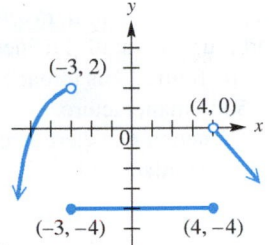

   (a) State the largest open interval over which the function is increasing.
   (b) State the largest open interval over which the function is decreasing.
   (c) State the largest open interval over which the function is constant.
   (d) State the intervals over which the function is continuous.
   (e) What is the domain of the function?
   (f) What is the range of the function?

8. Solve each equation or inequality.
   (a) $|4x + 8| = 4$
   (b) $|4x + 8| < 4$
   (c) $|4x + 8| > 4$

9. Given $f(x) = 2x^2 - 3x + 2$ and $g(x) = -2x + 1$, find the following.
   (a) $(f - g)(x)$

   (b) $\left(\dfrac{f}{g}\right)(x)$

   (c) The domain of $\dfrac{f}{g}$
   (d) $(f \circ g)(x)$
   (e) $(g \circ f)(x)$

   (f) $\dfrac{f(x + h) - f(x)}{h}$ $(h \neq 0)$

10. Consider the following piecewise-defined function.

$$f(x) = \begin{cases} -x^2 + 3 & \text{if } x \le 1 \\ \sqrt[3]{x} + 2 & \text{if } x > 1 \end{cases}$$

(a) Graph f by hand.

(b) Use a graphing calculator in dot mode to obtain a graph in the window $[-4.7, 4.7]$ by $[-5.1, 5.1]$.

(c) Give any x-values where f is not continuous.

11. *(Modeling) Long-Distance Call Charges* A certain long-distance carrier provides service between Podunk and Nowhereville. If x represents the number of minutes for the call, where $x > 0$, then the function

$$f(x) = 0.40[\![x]\!] + 0.75$$

gives the total cost of the call in dollars.

(a) Using dot mode and window $[0, 10]$ by $[0, 6]$, graph this function on a graphing calculator.

(b) Use the graph to find the cost of a call that is 5.5 minutes long.

12. *(Modeling) Cost, Revenue, and Profit Analysis* Tyler McGinnis starts up a small business manufacturing bobble-head sports figures. His initial cost is $3300. Each figure costs $4.50 to manufacture.

(a) Write a cost function C, where x represents the number of figures manufactured.

JIM THOME

(b) Find the revenue function R if each figure in part (a) sells for $10.50.

(c) Give the profit function P.

(d) How many figures must be produced and sold before Tyler earns a profit?

(e) Support the result of part (d) graphically.

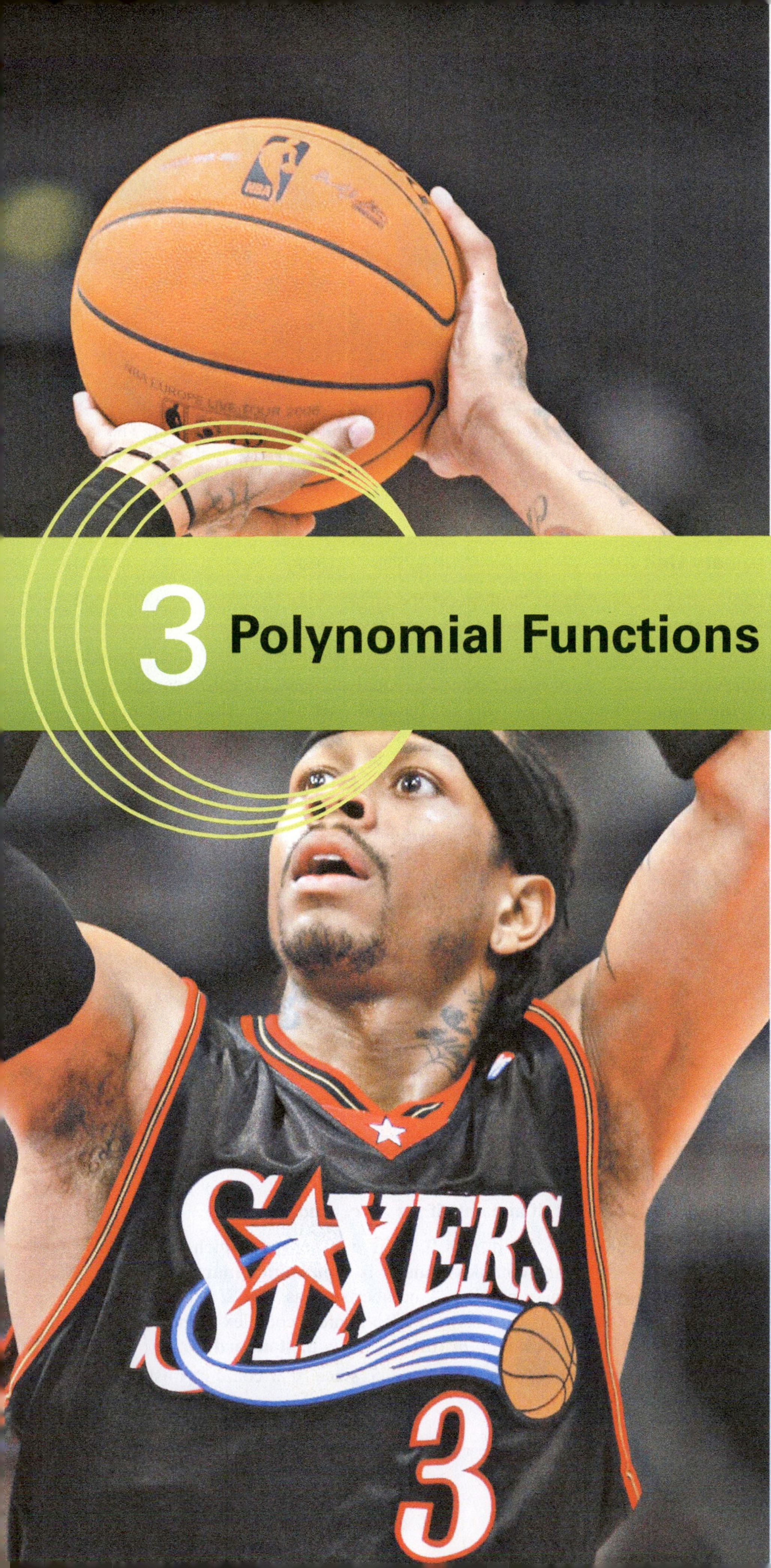

In this chapter we use *polynomial functions* to analyze applied problems, such as whether it is more effective to shoot a foul shot overhand or underhand.

3 Polynomial Functions

3.1 Complex Numbers

The Imaginary Unit *i* • Operations with Complex Numbers

No Real Solutions

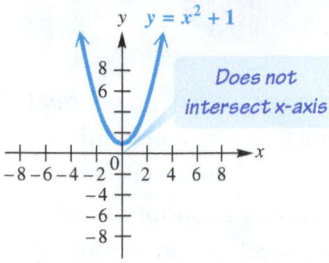

FIGURE 1

The Imaginary Unit *i*

The graph of $y = x^2 + 1$ in **FIGURE 1** does not intersect the *x*-axis, so there are no *real* solutions of the equation $x^2 + 1 = 0$. This equation is equivalent to $x^2 = -1$, and we know that no *real* number has a square of -1.

The complex number system includes the set of real numbers as a subset. The **imaginary unit *i*** is a basic unit of the complex number system and is defined as the principal square root of -1.

Imaginary Unit *i*

$$i = \sqrt{-1}, \quad \text{and therefore} \quad i^2 = -1.$$

Numbers of the form $a + bi$, where *a* and *b* are real numbers, are called **complex numbers**. In the complex number $a + bi$, *a* is the **real part** and *b* is the **imaginary part**.* Relationships among the sets of numbers are shown in **FIGURE 2.**

TECHNOLOGY NOTE

The TI-84 Plus is capable of complex number operations, as indicated by $a + bi$ here.

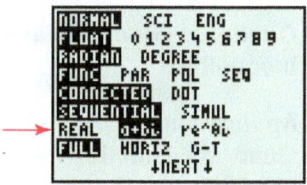

Sets of Numbers

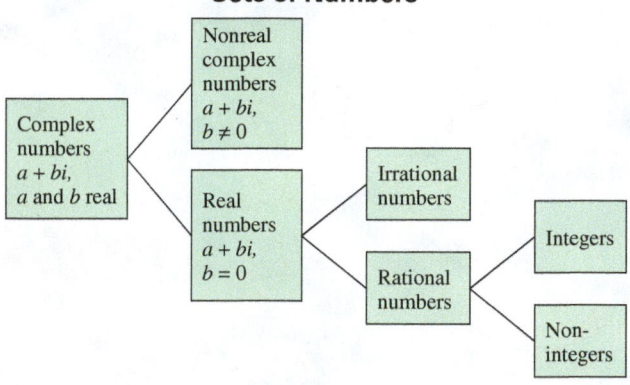

FIGURE 2

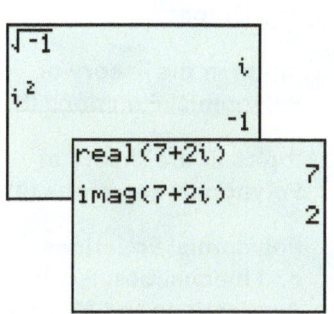

The top screen supports the definition of *i*. The bottom screen shows how the calculator returns the real and imaginary parts of $7 + 2i$.

Two complex numbers $a + bi$ and $c + di$ are equal provided that their real parts are equal and their imaginary parts are equal.

$$a + bi = c + di \quad \text{if and only if} \quad a = c \quad \text{and} \quad b = d.$$

For a complex number $a + bi$, if $b = 0$, then $a + bi = a$, which is a real number. If $a = 0$ and $b \neq 0$, the complex number is a **pure imaginary number.** For example, $3i$ is a pure imaginary number. A number such as $7 + 2i$ is a **nonreal complex number.** (Pure imaginary numbers are also nonreal complex numbers.) A complex number written in the form $a + bi$ (or $a + ib$) is in **standard form.** (The form $a + ib$ is used to write expressions such as $i\sqrt{5}$, since $\sqrt{5}i$ could be mistaken for $\sqrt{5i}$.)

---

*In some texts, the term *bi* is defined to be the imaginary part.

For a positive real number a, $\sqrt{-a}$ is defined as follows.

> **The Expression $\sqrt{-a}$**
>
> If $a > 0$, then $\qquad \sqrt{-a} = i\sqrt{a}.$

EXAMPLE 1 Writing $\sqrt{-a}$ as $i\sqrt{a}$

Write each expression as the product of i and a real number.

(a) $\sqrt{-16}$ $\qquad$ (b) $\sqrt{-75}$

Analytic Solution

(a) $\sqrt{-16} = \sqrt{-1 \cdot 16}$ $\qquad$ Factor out -1.

$\qquad = i\sqrt{16}$ $\qquad$ $\sqrt{-1} = i$

$\qquad = 4i$ $\qquad$ $\sqrt{16} = 4$

(b) $\sqrt{-75} = \sqrt{-1 \cdot 75}$ $\qquad$ Factor out -1.

$\qquad = i\sqrt{75}$ $\qquad$ $\sqrt{-1} = i$

$\qquad = i\sqrt{25 \cdot 3}$ $\qquad$ Factor.

$\qquad = 5i\sqrt{3}$ $\qquad$ $\sqrt{25} = 5$

Graphing Calculator Solution

The calculator in **FIGURE 3** is in complex number mode.

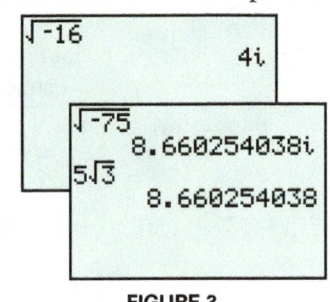

FIGURE 3

A product or quotient with a negative radicand is simplified by first rewriting $\sqrt{-a}$ as $i\sqrt{a}$ for positive a.

EXAMPLE 2 Finding Products and Quotients Involving $\sqrt{-a}$

Multiply or divide as indicated.

(a) $\sqrt{-7} \cdot \sqrt{-7}$ $\qquad$ (b) $\sqrt{-6} \cdot \sqrt{-10}$ $\qquad$ (c) $\dfrac{\sqrt{-20}}{\sqrt{-2}}$ $\qquad$ (d) $\dfrac{\sqrt{-48}}{\sqrt{24}}$

Analytic Solution

(a) $\sqrt{-7} \cdot \sqrt{-7} = i\sqrt{7} \cdot i\sqrt{7}$ $\qquad$ $\sqrt{-1} = i$

First write square roots in terms of i.

$\qquad = i^2 \cdot \left(\sqrt{7}\right)^2$ $\qquad$ $i \cdot i = i^2$

$\qquad = -1 \cdot 7$ $\qquad$ $i^2 = -1$

$\qquad = -7$ $\qquad$ Multiply.

(b) $\sqrt{-6} \cdot \sqrt{-10} = i\sqrt{6} \cdot i\sqrt{10}$ $\qquad$ $\sqrt{-1} = i$

$\qquad = i^2 \cdot \sqrt{60}$ $\qquad$ $i \cdot i = i^2$

$\qquad = -1 \cdot 2\sqrt{15}$ $\qquad$ $i^2 = -1$

$\qquad = -2\sqrt{15}$ $\qquad$ Multiply.

(c) $\dfrac{\sqrt{-20}}{\sqrt{-2}} = \dfrac{i\sqrt{20}}{i\sqrt{2}} = \sqrt{\dfrac{20}{2}} = \sqrt{10}$ $\qquad$ $\dfrac{\sqrt{a}}{\sqrt{b}} = \sqrt{\dfrac{a}{b}}$

(d) $\dfrac{\sqrt{-48}}{\sqrt{24}} = \dfrac{i\sqrt{48}}{\sqrt{24}} = i\sqrt{\dfrac{48}{24}} = i\sqrt{2}$

Graphing Calculator Solution

The screens in **FIGURE 4** show calculator verification for parts (b) and (d).

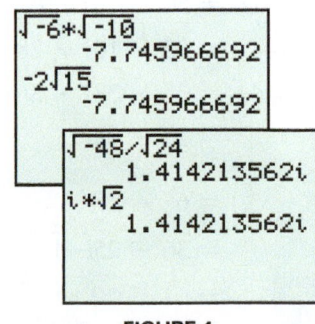

FIGURE 4

> **CAUTION** *When working with negative radicands, use the definition* $\sqrt{-a} = i\sqrt{a}$ *before using any of the other rules for radicals.* In particular, the rule $\sqrt{c} \cdot \sqrt{d} = \sqrt{cd}$ is valid only when c and d are both nonnegative. For example,
> $$\sqrt{-4} \cdot \sqrt{-9} = 2i(3i) = 6i^2 = -6, \text{ but } \sqrt{-4} \cdot \sqrt{-9} \neq \sqrt{(-4)(-9)} = \sqrt{36} = 6.$$

Operations with Complex Numbers

> **EXAMPLE 3** **Adding and Subtracting Complex Numbers**

Find each sum or difference.

(a) $(3 - 4i) + (-2 + 6i)$ **(b)** $(-9 + 7i) + (3 - 15i)$

(c) $(-4 + 3i) - (6 - 7i)$ **(d)** $(12 - 5i) - (8 - 3i)$

Analytic Solution

Add real parts. Add imaginary parts.

(a) $(3 - 4i) + (-2 + 6i) = [3 + (-2)] + (-4 + 6)i$

$= 1 + 2i$

(b) $(-9 + 7i) + (3 - 15i) = -6 - 8i$

Subtract real parts. Subtract imaginary parts.

(c) $(-4 + 3i) - (6 - 7i) = (-4 - 6) + [3 - (-7)]i$

$= -10 + 10i$

(d) $(12 - 5i) - (8 - 3i) = 4 - 2i$ ◄ $-5i - (-3i) = -2i$

Graphing Calculator Solution

FIGURE 5 shows the calculator results.

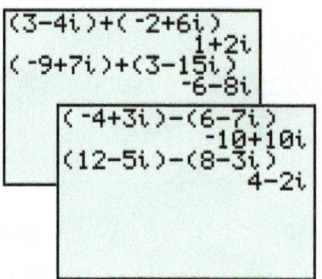

FIGURE 5

> **EXAMPLE 4** **Multiplying Complex Numbers**

Find each product.

(a) $(5 - 4i)(7 - 2i)$ **(b)** $(6 + 5i)(6 - 5i)$ **(c)** $(4 + 3i)^2$

Analytic Solution

(a) $(5 - 4i)(7 - 2i) = 5(7) + 5(-2i) - 4i(7) - 4i(-2i)$ FOIL

$= 35 - 10i - 28i + 8i^2$ Multiply.

$= 35 - 38i + 8(-1)$ $i^2 = -1$

Multiply as with binomials.

$= 27 - 38i$ Add real parts.

(b) $(6 + 5i)(6 - 5i) = 6^2 - 25i^2$ Product of the sum and difference of two terms

Remember the middle term when squaring a binomial.

$= 36 - 25(-1)$ $i^2 = -1$

$= 61$ Simplify.

(c) $(4 + 3i)^2 = 4^2 + 2(4)(3i) + (3i)^2$ Square of a binomial

$= 16 + 24i + (-9)$ $(3i)^2 = 3^2i^2 = 9(-1) = -9$

$= 7 + 24i$ Add real parts.

Graphing Calculator Solution

Calculator solutions are shown in **FIGURE 6.** The TI-84 Plus calculator does *not* need to be in complex number mode here.

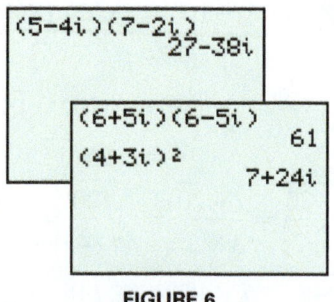

FIGURE 6

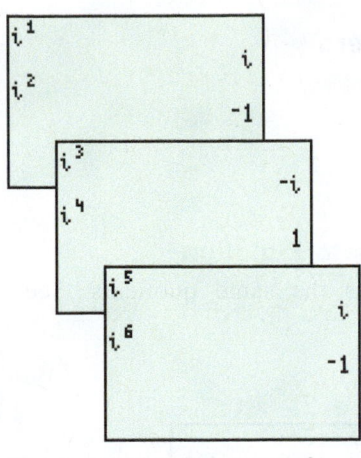

These screens show how powers of i follow a cycle.

By definition, $i^1 = i$ and $i^2 = -1$. Now, observe the following pattern.

$i^1 = i$ $i^5 = i^4 \cdot i = 1 \cdot i = i$

$i^2 = -1$ $i^6 = i^4 \cdot i^2 = 1 \cdot (-1) = -1$

$i^3 = i^2 \cdot i = -1 \cdot i = -i$ $i^7 = i^4 \cdot i^3 = 1 \cdot i^3 = -i$

$i^4 = i^3 \cdot i = -i \cdot i = -i^2 = -(-1) = 1$ $i^8 = (i^4)^2 = 1^2 = 1$

As seen for i^5, i^6, i^7, and i^8, any larger power of i may be found by writing the power as a product of two powers of i, with one exponent a multiple of 4, and then simplifying. Since

$$(i^4)^n = 1 \qquad i^4 = 1 \text{ and } (1)^n = 1$$

for all natural numbers n, we can then simplify powers of i by considering the other factor. For example,

$$i^{53} = i^{52} \cdot i^1 = (i^4)^{13} \cdot i^1 = i.$$

EXAMPLE 5   **Simplifying Powers of i**

Simplify each power of i.

(a) i^{13} **(b)** i^{56} **(c)** i^{-3}

Analytic Solution

In each case, we use the fact that $i^4 = 1$.

(a) $i^{13} = i^{12} \cdot i$

$\phantom{i^{13}} = (i^4)^3 \cdot i$ Property of exponents

$\phantom{i^{13}} = 1^3 \cdot i$ $i^4 = 1$

$\phantom{i^{13}} = i$ Simplify.

(b) $i^{56} = (i^4)^{14} = 1^{14} = 1$ $i^4 = 1$

(c) $i^{-3} = i^{-4+1}$

$\phantom{i^{-3}} = i^{-4} \cdot i$ Property of exponents

$\phantom{i^{-3}} = (i^4)^{-1} \cdot i$ Property of exponents

$\phantom{i^{-3}} = (1)^{-1} \cdot i$ $i^4 = 1$

$\phantom{i^{-3}} = i$ Simplify.

Graphing Calculator Solution

The screen in **FIGURE 7** agrees with the analytic results for parts (a) and (b). In the first case, the real part 0 is approximated as -3×10^{-13}, and in the second case, the imaginary part 0 is approximated as -4×10^{-13}.

This screen provides an example of the limitations of technology. The results may be misinterpreted if the mathematical concepts are not understood.

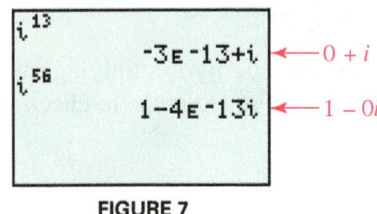

FIGURE 7

In **Example 4(b)**, $6 + 5i$ and $6 - 5i$ have equal real parts but *opposite* imaginary parts. Pairs of complex numbers satisfying these conditions are *complex conjugates*, or *conjugates*. Other examples of conjugates include: $3 - i, 3 + i$ and $-6i, 6i$. ***The product of a complex number and its conjugate is always a real number.***

Complex Conjugates

The **conjugate** of the complex number $a + bi$ is $a - bi$. Their product is the sum of the squares of their real and imaginary parts.

$$(a + bi)(a - bi) = a^2 + b^2$$

To find the quotient of two complex numbers in standard form, we multiply numerator and denominator by the conjugate of the denominator.

> **EXAMPLE 6** **Dividing Complex Numbers**

Find each quotient.

(a) $\dfrac{7 - i}{3 + i}$　　　**(b)** $\dfrac{3 + 2i}{5 - i}$　　　**(c)** $\dfrac{3}{i}$

Analytic Solution

(a) $\dfrac{7 - i}{3 + i} = \dfrac{(7 - i)(3 - i)}{(3 + i)(3 - i)}$　　　Multiply numerator and denominator by $3 - i$.

Conjugate of denominator

$= \dfrac{21 - 7i - 3i + i^2}{9 - i^2}$　　　Multiply.

$= \dfrac{21 - 10i - 1}{9 - (-1)}$　　　$i^2 = -1$

Use parentheses to avoid sign errors.

$= \dfrac{20 - 10i}{10}$

$= \dfrac{20}{10} - \dfrac{10}{10}i$, or $2 - i$　　　$\dfrac{a - bi}{c} = \dfrac{a}{c} - \dfrac{b}{c}i$

(b) $\dfrac{3 + 2i}{5 - i} = \dfrac{(3 + 2i)(5 + i)}{(5 - i)(5 + i)}$　　　Multiply numerator and denominator by $5 + i$.

$= \dfrac{15 + 3i + 10i + 2i^2}{25 - i^2}$　　　Multiply.

$= \dfrac{13 + 13i}{26}$　　　$i^2 = -1$; simplify.

$= \dfrac{13}{26} + \dfrac{13}{26}i$, or $\dfrac{1}{2} + \dfrac{1}{2}i$　　　$\dfrac{a + bi}{c} = \dfrac{a}{c} + \dfrac{b}{c}i$

(c) $\dfrac{3}{i} = \dfrac{3(-i)}{i(-i)} = \dfrac{-3i}{-i^2} = \dfrac{-3i}{-(-1)} = -3i$

In each case, check by multiplying quotient by divisor to obtain the dividend. For example, to check part (a) we multiply $(2 - i)(3 + i) = 7 - i$.

Graphing Calculator Solution

A calculator gives the same quotients. See **FIGURE 8**.

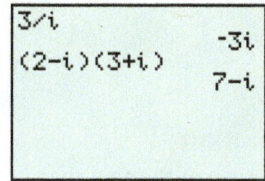

FIGURE 8

As with division of real numbers, a quotient of complex numbers can be checked by multiplying the quotient by the divisor to obtain the dividend. For example, the final entry in the bottom screen indicates a check for the result in part (a).

3.1 Exercises

Concept Check　For each complex number, **(a)** state the real part, **(b)** state the imaginary part, and **(c)** identify the number as one or more of the following: real, pure imaginary, or nonreal complex.

1. $-9i$　　　**2.** $3i$　　　**3.** π　　　**4.** $\sqrt{2}$　　　**5.** $3 + 7i$

6. $-8 + 4i$　　　**7.** $i\sqrt{7}$　　　**8.** $-i\sqrt{3}$　　　**9.** $\sqrt{-7}$　　　**10.** $\sqrt{-10}$

Checking Analytic Skills　Write each expression in standard form. **Do not use a calculator.**

11. $3i + 5i$　　　**12.** $5i - (2 - i)$　　　**13.** $(-7i)(1 + i)$　　　**14.** $\dfrac{4 + 2i}{i}$

Concept Check   *Determine whether each statement is* true *or* false. *If it is* false, *tell why.*

15. Every real number is a complex number.

16. No real number is a pure imaginary number.

17. Every pure imaginary number is a complex number.

18. A number can be both real and complex.

19. There is no real number that is a complex number.

20. A complex number might not be a pure imaginary number.

Write each number in simplest form, without a negative radicand.

21. $\sqrt{-100}$

22. $\sqrt{-169}$

23. $-\sqrt{-400}$

24. $-\sqrt{-225}$

25. $-\sqrt{-39}$

26. $-\sqrt{-95}$

27. $5 + \sqrt{-4}$

28. $-7 + \sqrt{-100}$

29. $9 - \sqrt{-50}$

30. $-11 - \sqrt{-24}$

31. $i\sqrt{-9}$

32. $i\sqrt{-16}$

Multiply or divide as indicated. Simplify each answer.

33. $\sqrt{-13} \cdot \sqrt{-13}$

34. $\sqrt{-17} \cdot \sqrt{-17}$

35. $\sqrt{-3} \cdot \sqrt{-8}$

36. $\sqrt{-5} \cdot \sqrt{-15}$

37. $\dfrac{\sqrt{-30}}{\sqrt{-10}}$

38. $\dfrac{\sqrt{-70}}{\sqrt{-7}}$

39. $\dfrac{\sqrt{-24}}{\sqrt{8}}$

40. $\dfrac{\sqrt{-54}}{\sqrt{27}}$

41. $\dfrac{\sqrt{-10}}{\sqrt{-40}}$

42. $\dfrac{\sqrt{-40}}{\sqrt{20}}$

43. $\dfrac{\sqrt{-6} \cdot \sqrt{-2}}{\sqrt{3}}$

44. $\dfrac{\sqrt{-12} \cdot \sqrt{-6}}{\sqrt{8}}$

Add or subtract as indicated. Write each sum or difference in standard form.

45. $(3 + 2i) + (4 - 3i)$

46. $(4 - i) + (2 + 5i)$

47. $(-2 + 3i) - (-4 + 3i)$

48. $(-3 + 5i) - (-4 + 5i)$

49. $(3 - 8i) + (2i + 4)$

50. $(9 - 5i) - (3i - 6)$

51. $(2 - 5i) - (3 + 4i) - (-2 + i)$

52. $(-4 - i) - (2 + 3i) + (-4 + 5i)$

53. $(-6 + 5i) + (4 - 4i) + (2 - i)$

54. $(7 + 9i) + (1 - 2i) + (-8 - 7i)$

Multiply as indicated. Write each product in standard form.

55. $(2 + i)(3 - 2i)$

56. $(-2 + 3i)(4 - 2i)$

57. $(2 + 4i)(-1 + 3i)$

58. $(1 + 3i)(2 - 5i)$

59. $(-3 + 2i)^2$

60. $(2 + i)^2$

61. $(3 + i)(-3 - i)$

62. $(-5 - i)(5 + i)$

63. $(2 + 3i)(2 - 3i)$

64. $(6 - 4i)(6 + 4i)$

65. $\left(\sqrt{6} + i\right)\left(\sqrt{6} - i\right)$

66. $\left(\sqrt{2} - 4i\right)\left(\sqrt{2} + 4i\right)$

67. $i(3 - 4i)(3 + 4i)$

68. $i(2 + 7i)(2 - 7i)$

69. $3i(2 - i)^2$

70. $-5i(4 - 3i)^2$

71. $(2 + i)(2 - i)(4 + 3i)$

72. $(3 - i)(3 + i)(2 - 6i)$

Simplify each power of i to i, 1, $-i$, or -1.

73. i^5

74. i^8

75. i^{15}

76. i^{19}

77. i^{64}

78. i^{102}

79. i^{-6}

80. i^{-15}

81. $\dfrac{1}{i^9}$

82. $\dfrac{1}{i^{12}}$

83. $\dfrac{1}{i^{-51}}$

84. $\dfrac{1}{i^{-46}}$

85. $\dfrac{-1}{-i^{12}}$

86. $\dfrac{-1}{-i^{15}}$

87. *Concept Check*   Show that $\dfrac{\sqrt{2}}{2} + \dfrac{\sqrt{2}}{2}i$ is a square root of i.

88. *Concept Check*   Show that $\dfrac{\sqrt{3}}{2} + \dfrac{1}{2}i$ is a cube root of i.

Find the conjugate of each number.

89. $5 - 3i$

90. $-3 + i$

91. $-18i$

92. $\sqrt{7}$

93. $-\sqrt{8}$

94. $8i$

Divide as indicated. Write each quotient in standard form.

95. $\dfrac{3}{-i}$

96. $\dfrac{-7}{3i}$

97. $\dfrac{-10}{i}$

98. $\dfrac{-19 - 9i}{i}$

99. $\dfrac{1 - 3i}{1 + i}$

100. $\dfrac{-12 - 5i}{3 - 2i}$

101. $\dfrac{-3 + 4i}{2 - i}$

102. $\dfrac{-6 + 8i}{1 - i}$

103. $\dfrac{4 - 3i}{4 + 3i}$

104. $\dfrac{2 - i}{2 + i}$

 105. Explain why the method of dividing complex numbers (that is, multiplying both the numerator and the denominator by the conjugate of the denominator) works. What property justifies this process?

 106. Suppose that your friend describes a method of simplifying a positive power of i. "Just divide the exponent by 4, and then look at the remainder. Then, refer to the short table of powers of i in this section. The given power of i is equal to i to the power indicated by the remainder. And if the remainder is 0, the result is $i^0 = 1$." Explain why this method works.

3.2 Quadratic Functions and Graphs

Completing the Square • Graphs of Quadratic Functions • Vertex Formula • Extreme Values • Applications and Quadratic Models

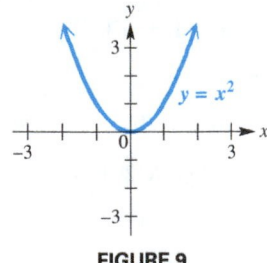

FIGURE 9

In **Chapter 2** we saw that the graph of $y = x^2$ is a parabola. See **FIGURE 9**. The function $P(x) = x^2$ is the simplest example of a *quadratic function*.

> **Quadratic Function**
>
> The function
> $$P(x) = ax^2 + bx + c,$$
> where a, b, and c are real numbers, with $a \neq 0$, is called a **quadratic function**.

Quadratic functions, as well as linear functions, are examples of **polynomial functions.**

Completing the Square

Recall from **Chapter 2** that the graph of

$$g(x) = a(x - h)^2 + k$$

has the same general shape as the graph of $f(x) = x^2$, but may be stretched, shrunk, reflected, or shifted.

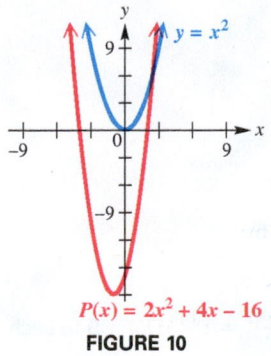

$P(x) = 2x^2 + 4x - 16$

FIGURE 10

Consider $P(x) = 2x^2 + 4x - 16$, graphed in **FIGURE 10.** It appears that the graph of P can be obtained from the graph of $y = x^2$, also shown in the figure, using a vertical stretch by applying a factor greater than 1, a shift to the left, and a shift downward. We can determine the sizes of these transformations by writing the function in the form

$$P(x) = a(x - h)^2 + k.$$

This process is called *completing the square*.

Completing the Square

To transform the quadratic function $P(x) = ax^2 + bx + c$ into the form $P(x) = a(x - h)^2 + k$, follow these steps.

Step 1 Divide each side of the equation by a so that the coefficient of x^2 is 1.

Step 2 Add $-\frac{c}{a}$ to each side.

Step 3 Add to each side the square of half the coefficient of x: $\left(\frac{b}{2a}\right)^2$.

Step 4 Factor the right side as the square of a binomial and combine terms on the left.

Step 5 Isolate the term involving $P(x)$ on the left.

Step 6 Multiply each side by a.

We can apply this procedure to $P(x) = 2x^2 + 4x - 16$.

$$P(x) = 2x^2 + 4x - 16$$

$$\frac{P(x)}{2} = x^2 + 2x - 8 \qquad \text{Divide by 2 so that the coefficient of } x^2 \text{ is 1.}$$

$$\frac{P(x)}{2} + 8 = x^2 + 2x \qquad \text{Add 8 to each side.}$$

$$\left(\frac{b}{2a}\right)^2$$

Be sure to add 1 to each side.

$$\frac{P(x)}{2} + 8 + 1 = x^2 + 2x + 1 \qquad \text{Add } \left[\tfrac{1}{2}(2)\right]^2 = 1 \text{ to each side to complete the square on the right.}$$

$$\frac{P(x)}{2} + 9 = (x + 1)^2 \qquad \text{Add on the left and factor on the right.}$$

$$\frac{P(x)}{2} = (x + 1)^2 - 9 \qquad \text{Subtract 9 from each side.}$$

$$P(x) = 2(x + 1)^2 - 18 \qquad \text{Multiply each side by } a = 2.$$

$$P(x) = 2[x - (-1)]^2 - 18 \qquad \text{Rewrite } x + 1 \text{ as } x - (-1).$$
$$\quad\quad\quad\uparrow\qquad\quad\uparrow\qquad\quad\uparrow$$
$$\quad\quad\quad a\qquad\quad h\qquad\quad k$$

From this form of the function, we see that $a = 2$, $h = -1$, and $k = -18$. Thus, the graph of $P(x) = 2x^2 + 4x - 16$ is the graph of $y = x^2$ vertically stretched by applying a factor of 2, shifted 1 unit left and 18 units downward. The vertex of the parabola has coordinates $(-1, -18)$, the domain is $(-\infty, \infty)$, and the range is $[-18, \infty)$. The function decreases on the interval $(-\infty, -1)$ and increases on the interval $(-1, \infty)$. See **FIGURE 10.**

EXAMPLE 1   **Completing the Square**

Complete the square on $P(x) = -x^2 - 6x - 8$.

Solution

$$P(x) = -x^2 - 6x - 8$$

$$-P(x) = x^2 + 6x + 8 \qquad \text{Divide by } -1.$$

$$-P(x) - 8 = x^2 + 6x \qquad \text{Subtract 8.}$$

$$-P(x) - 8 + 9 = x^2 + 6x + 9 \qquad \text{Add } \left(\frac{b}{2a}\right)^2 = \left[\frac{1}{2}(6)\right]^2 = 9 \text{ to each side.}$$

Be sure to add 9 to each side.

$$-P(x) + 1 = (x + 3)^2 \qquad \text{Add on the left and factor on the right.}$$

$$-P(x) = (x + 3)^2 - 1 \qquad \text{Subtract 1.}$$

$$P(x) = -(x + 3)^2 + 1 \qquad \text{Multiply by } -1. \qquad \bullet$$

Graphs of Quadratic Functions

The quadratic function

$$P(x) = -x^2 - 6x - 8, \quad \text{or} \quad P(x) = -(x + 3)^2 + 1$$

(from **Example 1**) can now be graphed easily because it is written in the form $P(x) = a(x - h)^2 + k$. Also, recall from **Chapter 1** that the y-intercept of the graph of an equation is the point that has x-coordinate 0. For a parabola given in the form $P(x) = ax^2 + bx + c$, the y-value of the y-intercept is $P(0) = c$. *A comprehensive graph of a quadratic function shows all intercepts and the vertex of the parabola.*

Characteristics of the Graph of $P(x) = a(x - h)^2 + k$

Consider the graph of $P(x) = a(x - h)^2 + k \ (a \neq 0)$.

(a) The graph is a parabola with vertex (h, k) and vertical line $x = h$ as its axis of symmetry.

(b) The graph opens upward if $a > 0$ and downward if $a < 0$.

(c) The graph is wider than the graph of $y = x^2$ if $0 < |a| < 1$ and narrower than the graph of $y = x^2$ if $|a| > 1$.

EXAMPLE 2   **Graphing a Quadratic Function**

Graph the function $P(x) = -(x + 3)^2 + 1$ from **Example 1.** Give the domain and range and the intervals over which the function is increasing or decreasing.

Analytic Solution

The vertex (h, k) is $(-3, 1)$, so the graph is that of $f(x) = x^2$ shifted 3 units to the left and 1 unit upward. Since

$$P(0) = -(0 + 3)^2 + 1 = -8,$$

the y-intercept is $(0, -8)$. We plot the vertex and y-intercept. To locate another point on the graph, we choose a value of x on the same side of the vertex as the y-intercept. If $x = -1$, for example, then

$$P(-1) = -(-1 + 3)^2 + 1 = -3.$$

Graphing Calculator Solution

A calculator graph of

$$P(x) = -(x - 3)^2 + 1$$

is shown in **FIGURE 12** on the next page. Since the vertex is $(-3, 1)$ and the y-intercept is $(0, -8)$, choose the viewing window $[-8, 2]$ by $[-10, 2]$, which shows a comprehensive graph of the function.

Thus the point $(-1, -3)$ lies on the graph. The line $x = -3$ through the vertex of the parabola is called the **axis of symmetry: If the graph were folded along this line, the two halves would coincide.** Using symmetry about $x = -3$, we plot the points $(-5, -3)$ and $(-6, -8)$, which also lie on the graph.

Because the coefficient of $(x + 3)^2$ is -1, the graph opens downward and has the same shape as that of $f(x) = x^2$. See **FIGURE 11.**

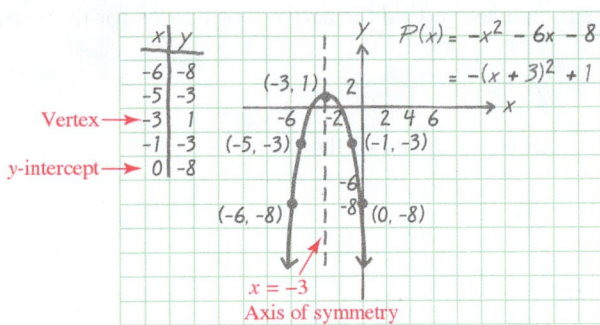

FIGURE 11

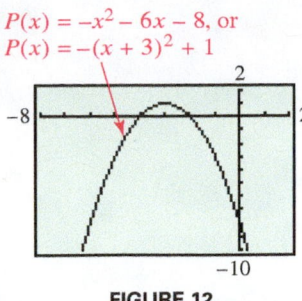

$P(x) = -x^2 - 6x - 8$, or
$P(x) = -(x + 3)^2 + 1$

FIGURE 12

The domain is $(-\infty, \infty)$. Because the graph opens downward, the vertex $(-3, 1)$ is the highest point on the graph. Thus, the range is $(-\infty, 1]$. The function increases on the interval $(-\infty, -3)$ and decreases on $(-3, \infty)$.

Vertex Formula

We can determine the coordinates of the vertex of the graph of a quadratic function by completing the square, as shown earlier. Rather than go through that procedure for each individual function, we generalize the result for the standard form of the quadratic function $P(x) = ax^2 + bx + c$, where $a \neq 0$.

$P(x) = ax^2 + bx + c$	Standard form
$y = ax^2 + bx + c$	Replace $P(x)$ with y.
$\dfrac{y}{a} = x^2 + \dfrac{b}{a}x + \dfrac{c}{a}$	Divide by a.
$\dfrac{y}{a} - \dfrac{c}{a} = x^2 + \dfrac{b}{a}x$	Add $-\dfrac{c}{a}$.
$\dfrac{y}{a} - \dfrac{c}{a} + \dfrac{b^2}{4a^2} = x^2 + \dfrac{b}{a}x + \dfrac{b^2}{4a^2}$	Add $\left[\frac{1}{2}\left(\frac{b}{a}\right)\right]^2 = \frac{b^2}{4a^2}$.
$\dfrac{y}{a} + \dfrac{b^2 - 4ac}{4a^2} = \left(x + \dfrac{b}{2a}\right)^2$	Combine terms on the left and factor on the right.
$\dfrac{y}{a} = \left(x + \dfrac{b}{2a}\right)^2 - \dfrac{b^2 - 4ac}{4a^2}$	Isolate y-term on the left.
$y = a\left(x + \dfrac{b}{2a}\right)^2 + \dfrac{4ac - b^2}{4a}$	Multiply by a.
$P(x) = a\left[x - \left(-\dfrac{b}{2a}\right)\right]^2 + \dfrac{4ac - b^2}{4a}$	Write in the form $P(x) = a(x - h)^2 + k$.

The final equation shows that the vertex (h, k) can be expressed in terms of a, b, and c. It is not necessary to memorize the expression for k, since it equals $P(h) = P\left(-\dfrac{b}{2a}\right)$.

> **Vertex Formula**
>
> The vertex of the graph of $P(x) = ax^2 + bx + c \ (a \neq 0)$ is the point
>
> $$\left(-\frac{b}{2a}, \ P\left(-\frac{b}{2a}\right)\right).$$

TECHNOLOGY NOTE

Graphing calculators are capable of finding the coordinates of the "highest point" or "lowest point" on an interval of a graph. The *approximate* coordinates for the vertex of the parabola given by $P(x) = 3x^2 - 2x - 1$ are shown in the figure below.

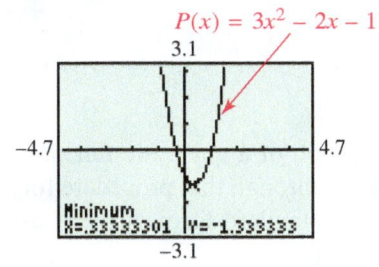

The vertex formula can be used to find the *exact* coordinates for the vertex of *any* parabola. For example, if

$$P(x) = 3x^2 - 2x - 1,$$

then $a = 3$ and $b = -2$. The x-coordinate of the vertex is

$$x = -\frac{b}{2a} = -\frac{-2}{2(3)} = \frac{1}{3},$$

and the y-coordinate of the vertex is

$$P\left(\frac{1}{3}\right) = 3\left(\frac{1}{3}\right)^2 - 2\left(\frac{1}{3}\right) - 1 = -\frac{4}{3}. \qquad P(x) = 3x^2 - 2x - 1$$

Thus the coordinates of the vertex are $\left(\frac{1}{3}, -\frac{4}{3}\right)$. The Technology Note in the margin illustrates how a calculator is able to *approximate* these coordinates. The graphing calculator supports our analytic result.

EXAMPLE 3   **Graphing a Quadratic Function**

Graph the function $P(x) = 2x^2 + 4x + 6$.

Solution   For this quadratic function, $a = 2$, $b = 4$, and $c = 6$. Find the coordinates of the vertex by using the vertex formula.

$$x = -\frac{b}{2a} = -\frac{4}{2(2)} = -1 \qquad \text{\small\textit{Use parentheses around substituted values to avoid errors.}}$$

$$y = P(-1) = 2(-1)^2 + 4(-1) + 6 = 4$$

Thus, the vertex is $(-1, 4)$. The y-intercept is $(0, 6)$ because $P(0) = 6$. The table shows five key points on the graph, as shown in **FIGURE 13**. The axis of symmetry is $x = -1$. Notice the symmetry of the y-values about the value of $x = -1$ in the table. The axis of symmetry corresponds to the x-coordinate of the vertex.

x	y
-3	12
-2	6
-1	4
0	6
1	12

Vertex → -1

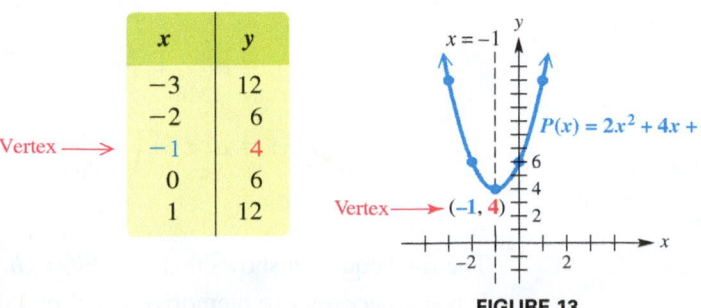

FIGURE 13

Extreme Values

The vertex of the graph of

$$P(x) = ax^2 + bx + c$$

is the lowest point on the graph of the function if $a > 0$ and the highest point if $a < 0$. Such points are called **extreme points** (also **extrema**; singular, **extremum**). Consider the quadratic function $P(x) = ax^2 + bx + c$.

(a) If $a > 0$, then the vertex (h, k) is called the **minimum point** of the graph. The **minimum value** of the function is $P(h) = k$.

(b) If $a < 0$, then the vertex (h, k) is called the **maximum point** of the graph. The **maximum value** of the function is $P(h) = k$.

FIGURE 14 illustrates these ideas.

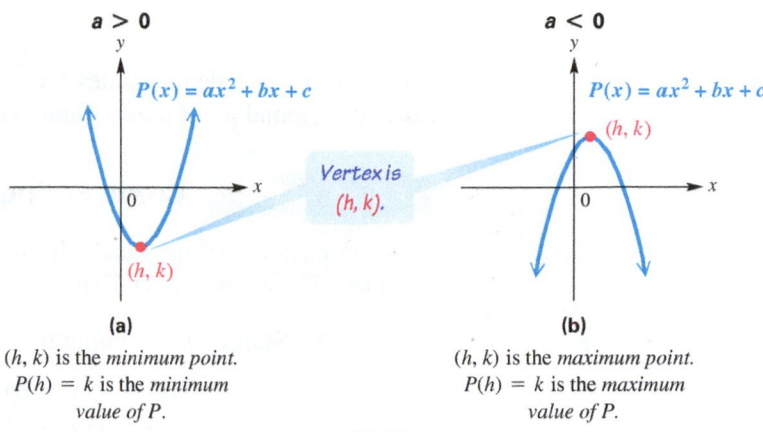

(a)
(h, k) is the *minimum point.*
$P(h) = k$ is the *minimum value of P.*

(b)
(h, k) is the *maximum point.*
$P(h) = k$ is the *maximum value of P.*

FIGURE 14

TECHNOLOGY NOTE

Graphing calculators have the capabilities of locating extrema to great accuracy. The TI-84 Plus, for example, has fMax(and fMin(capabilities, which allow the user to locate extrema without graphing. See **FIGURE 15** in **Example 4(a)**.

NOTE   For a quadratic function whose graph has two x-intercepts, the x-coordinate of the extreme point (vertex) will always be exactly halfway between those intercepts.

EXAMPLE 4   **Identifying Extreme Points and Extreme Values**

Give the coordinates of the extreme point of the graph of each function and the corresponding maximum or minimum value of the function.

(a) $P(x) = 2x^2 + 4x - 16$ **(b)** $P(x) = -x^2 - 6x - 8$

Analytic Solution

(a) In the discussion preceding **Example 1,** we found that the vertex of the graph of

$$P(x) = 2x^2 + 4x - 16,$$

or $P(x) = 2(x + 1)^2 - 18$

is $(-1, -18)$. The graph of this function opens upward, since $a = 2$ and $2 > 0$ (as seen in **FIGURE 10** earlier in this section), so the vertex $(-1, -18)$ is the minimum point and -18 is the minimum value of the function.

Graphing Calculator Solution

(a) The screens in **FIGURE 15** use the fMin(calculator function, which gives the x-value where the minimum value occurs on the graph of $P(x) = 2x^2 + 4x - 16$. The y-value, found by substitution, is the minimum value of the function.

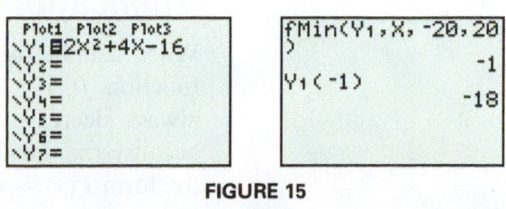

FIGURE 15

(continued)

(b) We found the vertex $(-3, 1)$ of the graph of

$$P(x) = -x^2 - 6x - 8$$

in **Example 1** and graphed the function $P(x)$ in **Example 2.** Because $a = -1$ and $-1 < 0$ in

$$P(x) = -x^2 - 6x - 8,$$
or $\qquad P(x) = -(x + 3)^2 + 1,$

the graph of the parabola opens downward (as seen in **FIGURE 11** earlier in this section), so $(-3, 1)$ is the maximum point and 1 is the maximum function value.

(b) The screen in **FIGURE 16** shows an alternative way to find the coordinates of an extreme point with a graphing calculator. The results agree with our analytic results in part (b).

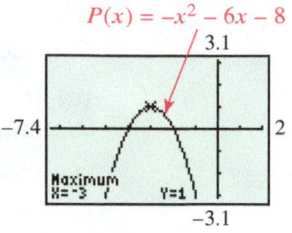

FIGURE 16

The next example demonstrates how to find the equation of a parabola, given its vertex and a second point on the parabola.

> **EXAMPLE 5** **Curve Fitting with a Parabola**

Find the equation of the quadratic function with vertex $(2, -3)$, passing through the point $(8, 69)$. See **FIGURE 17.** Express the answer in the form $P(x) = ax^2 + bx + c$.

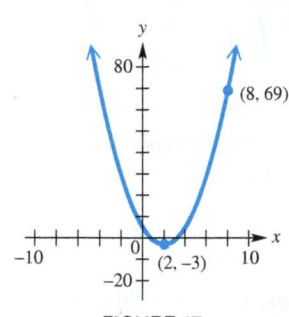

FIGURE 17

Solution   Start by substituting the vertex $(2, -3)$ into $P(x) = a(x - h)^2 + k$.

$$P(x) = a(x - h)^2 + k$$
$$P(x) = a(x - 2)^2 - 3 \qquad \text{Substitute vertex.}$$

Next substitute the given point $(8, 69)$ to determine the value of a.

$$69 = a(8 - 2)^2 - 3 \qquad \text{Substitute.}$$
$$69 = 36a - 3 \qquad \text{Simplify.}$$
$$\frac{69 + 3}{36} = a \qquad \text{Add 3 and divide by 36.}$$
$$a = 2 \qquad \text{Simplify.}$$

Thus the equation of the given parabola is

$$P(x) = 2(x - 2)^2 - 3.$$

Simplifying gives the standard form of a quadratic function.

$$P(x) = 2(x - 2)^2 - 3$$
$$P(x) = 2(x^2 - 4x + 4) - 3 \qquad \text{Square binomial.}$$
$$P(x) = 2x^2 - 8x + 8 - 3 \qquad \text{Distributive property}$$
$$P(x) = 2x^2 - 8x + 5 \qquad \text{Add.}$$

Applications and Quadratic Models

When data lie on a line or nearly lie on a line, they can often be modeled by a linear function $f(x) = ax + b$. In these situations, the data typically always increase or always decrease. In other situations, data cannot be modeled by a linear function because the data both increase and decrease over an interval. Quadratic functions of the form $P(x) = ax^2 + bx + c$ can sometimes be good models in these situations.

In the next example, we model when people are using mobile devices by using a quadratic function, where the data decrease and then increase.

EXAMPLE 6 **Modeling When People Use Mobile Devices**

The following table lists the percentage of total mobile content consumed for the hours between midnight and noon, where $x = 0$ corresponds to midnight and $x = 12$ corresponds to noon. For example, in the table, $y = 5.3$ when $x = 0$, and so 5.3% of the total mobile content consumed throughout a day is consumed at midnight. The data are plotted in a scatter diagram in **FIGURE 18.**

Time x	% of Total Consumption y
0	5.3
5	0.9
8	2.6
11	5.6

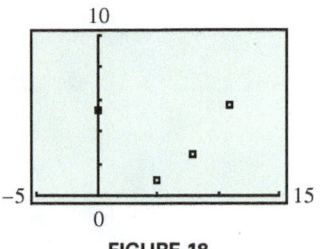

FIGURE 18

The scatter diagram suggests that a parabola that opens upward might model the data. The function

$$f(x) = 0.14x^2 - 1.5x + 5.2$$

gives an approximation of the data.

(a) Use $f(x)$ to approximate the time when mobile consumption was minimum during the period from midnight to noon.

(b) Use $f(x)$ to estimate the minimum percent of total mobile consumption.

Analytic Solution

Use the vertex formula.

(a) The x-value of the minimum point is

$$-\frac{b}{2a} = -\frac{(-1.5)}{2(0.14)} \approx 5.36 \approx 5.$$

Since $x = 5$ corresponds to 5 A.M., mobile consumption was at a minimum at 5 A.M.

(b) The minimum value is

$$f(5.36) = 0.14(5.36)^2 - 1.5(5.36) + 5.2 \approx 1.18.$$

Thus the model indicates that the minimum percent of mobile consumption was about 1.18%, slightly higher than the 5 A.M. value of 0.9% from the table.

Graphing Calculator Solution

Use a calculator to graph the function, and then use the capabilities of the calculator to find the coordinates of the minimum point. See **FIGURE 19.**

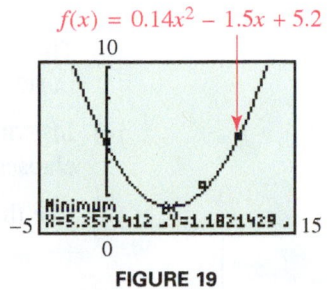

FIGURE 19

An important application of quadratic functions deals with the height of a projected object as a function of the time elapsed after it is launched.

In the formula for the height of a projected object, height is a function of time, and we use t to represent the independent variable. However, when graphing this type of function on a calculator, we use x, since graphing calculators use x for the independent variable. There is no difference between the formulas

$$s(t) = -16t^2 + v_0t + s_0$$

and

$$s(x) = -16x^2 + v_0x + s_0.$$

Height of a Projected Object

If air resistance is neglected, the height s (in feet) of an object projected directly upward from an initial height s_0 feet with initial velocity v_0 feet per second is

$$s(t) = -16t^2 + v_0t + s_0,$$

where t is the number of seconds after the object is projected.

NOTE The coefficient of t^2 (that is, -16) is a constant based on the gravitational force of Earth. This constant varies on other surfaces, such as the moon or the other planets.

EXAMPLE 7 Solving a Problem Involving Projectile Motion

A ball is thrown directly upward from an initial height of 100 feet with an initial velocity of 80 feet per second.

(a) Give the function that describes the height of the ball in terms of time t.

(b) Graph this function so that the y-intercept, the x-intercept with positive x-value, and the vertex are visible.

(c) **FIGURE 20** shows that the point (4.8, 115.36) lies on the graph of the function. What does this mean for this particular situation?

(d) After how many seconds does the projectile reach its maximum height? What is the maximum height? Solve analytically and graphically.

(e) For what interval of time is the height of the ball greater than 160 feet? Determine the answer graphically.

(f) After how many seconds will the ball hit the ground? Determine the answer graphically.

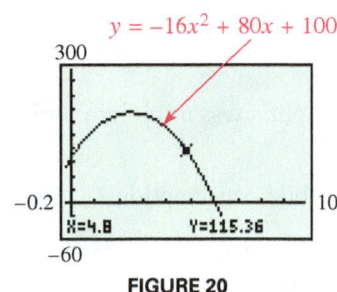

FIGURE 20

Solution

(a) Use the projectile height function with $v_0 = 80$ and $s_0 = 100$.

$$s(t) = -16t^2 + 80t + 100$$

(b) There are many suitable choices for a window, such as $[-0.2, 10]$ by $[-60, 300]$ in **FIGURE 20**, which shows the graph of $y = -16x^2 + 80x + 100$. (Here, $x = t$.)

(c) In **FIGURE 20**, when X = 4.8, Y = 115.36. Therefore, when 4.8 seconds have elapsed, the ball is at a height of 115.36 feet.

(d) Use the vertex formula to find the coordinates of the vertex of the parabola.

$$x = -\frac{b}{2a} = -\frac{80}{2(-16)} = 2.5$$
$$y = -16(2.5)^2 + 80(2.5) + 100 = 200$$

After 2.5 seconds, the ball reaches its maximum height of 200 feet.

Using the capabilities of the calculator, we see that the vertex coordinates are indeed (2.5, 200). See **FIGURE 21**.

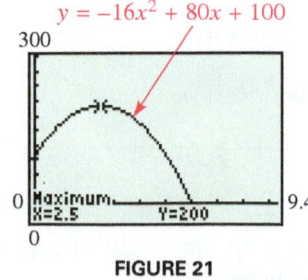

FIGURE 21

CAUTION   It is easy to misinterpret the graph in **FIGURE 21**. *This graph does not depict the path followed by the ball; it depicts height as a function of time.*

(e) With $y_1 = -16x^2 + 80x + 100$ and $y_2 = 160$ graphed, as shown in **FIGURE 22(a)** and **(b)** locate the two points of intersection. The x-coordinates of these points are approximately 0.92 and 4.08, respectively. Thus, between 0.92 and 4.08 seconds, the ball is more than 160 feet above the ground; that is, $y_1 > y_2$.

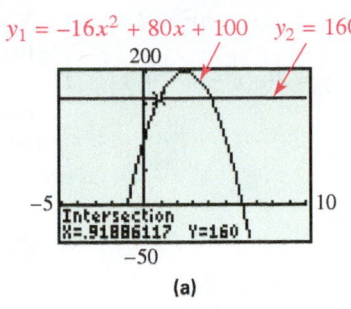
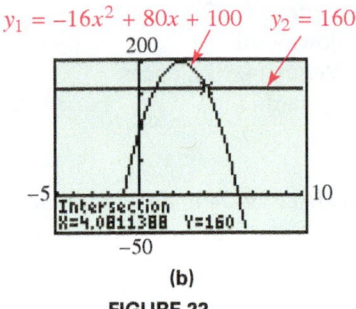
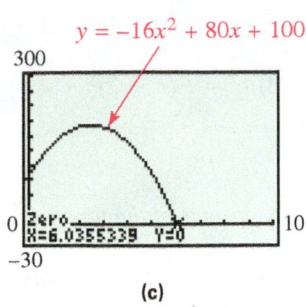

(a)   (b)   (c)

FIGURE 22

(f) When the ball hits the ground, its height will be 0 feet, so find the x-intercept with a positive x-value. As shown in **FIGURE 22(c)**, this x-value is about 6.04, which means that the ball reaches the ground about 6.04 seconds after it is thrown.

3.2 Exercises

Concept Check   *Match each function in Exercises 1–4 with its graph in A–D. Then check your answers with your calculator.*

1. $y = (x - 4)^2 - 3$ **2.** $y = -(x - 4)^2 + 3$ **3.** $y = (x + 4)^2 - 3$ **4.** $y = -(x + 4)^2 + 3$

A.
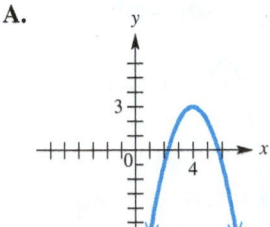

B.
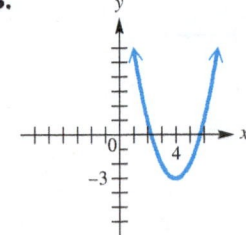

C.
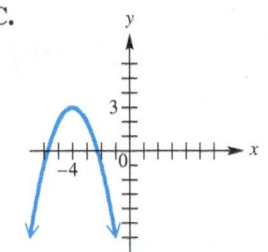

D.
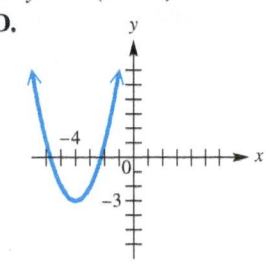

Checking Analytic Skills   *For each quadratic function defined in Exercises 5–16,* **(a)** *write the function in the form* $P(x) = a(x - h)^2 + k$, **(b)** *give the vertex of the parabola, and* **(c)** *graph the function.* ***Do not use a calculator.***

5. $P(x) = x^2 - 2x - 15$

6. $P(x) = x^2 + 2x - 15$

7. $P(x) = -x^2 - 3x + 10$

8. $P(x) = -x^2 + 3x + 10$

9. $P(x) = x^2 - 6x$

10. $P(x) = x^2 + 4x$

11. $P(x) = 2x^2 - 2x - 24$

12. $P(x) = 3x^2 + 3x - 6$

13. $P(x) = -2x^2 + 6x$

14. $P(x) = -4x^2 + 4x$

15. $P(x) = 3x^2 + 4x - 1$

16. $P(x) = 4x^2 + 3x - 1$

17. *Concept Check* Match each equation in Column I with the description of the parabola that is its graph in Column II.

I	II
(a) $y = (x - 4)^2 - 2$	**A.** Vertex $(2, -4)$, opens downward
(b) $y = (x - 2)^2 - 4$	**B.** Vertex $(2, -4)$, opens upward
(c) $y = -(x - 4)^2 - 2$	**C.** Vertex $(4, -2)$, opens downward
(d) $y = -(x - 2)^2 - 4$	**D.** Vertex $(4, -2)$, opens upward

18. *Concept Check* Match each equation in Column I with the description of the parabola that is its graph in Column II, assuming $a > 0$, $h > 0$, and $k > 0$.

I	II
(a) $y = -a(x + h)^2 + k$	**A.** Vertex in quadrant I, two x-intercepts
(b) $y = a(x - h)^2 + k$	**B.** Vertex in quadrant I, no x-intercepts
(c) $y = a(x + h)^2 + k$	**C.** Vertex in quadrant II, two x-intercepts
(d) $y = -a(x - h)^2 + k$	**D.** Vertex in quadrant II, no x-intercepts

Use the graphs of the functions in Exercises 19–24 to do each of the following:
(a) *Give the coordinates of the vertex.*
(b) *Give the domain and range.*
(c) *Give the equation of the axis of symmetry.*
(d) *Give the largest open interval over which the function is increasing.*
(e) *Give the largest open interval over which the function is decreasing.*
(f) *State whether the vertex is a maximum or minimum point, and give the corresponding maximum or minimum value of the function.*

19.

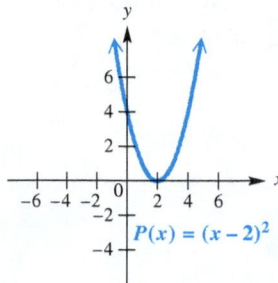

$P(x) = (x - 2)^2$

20.

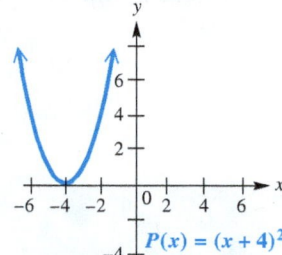

$P(x) = (x + 4)^2$

21.

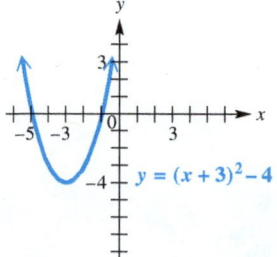

$y = (x + 3)^2 - 4$

22.

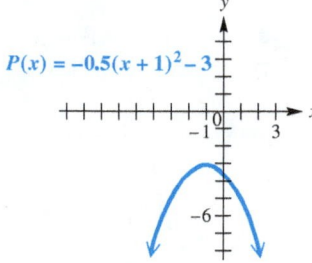

$P(x) = -0.5(x + 1)^2 - 3$

23.

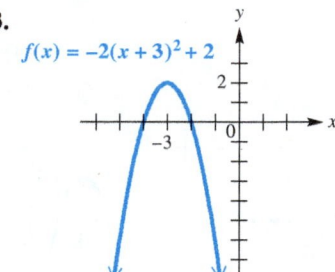

$f(x) = -2(x + 3)^2 + 2$

24.

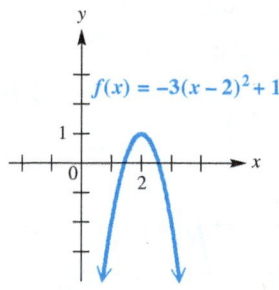

$f(x) = -3(x - 2)^2 + 1$

Checking Analytic Skills For each quadratic function defined in Exercises 25–33, **(a)** use the vertex formula to find the coordinates of the vertex and **(b)** graph the function. **Do not use a calculator.**

25. $P(x) = x^2 - 10x + 21$

26. $P(x) = x^2 - 2x + 3$

27. $y = -x^2 + 4x - 2$

28. $y = -x^2 + 2x + 1$

29. $P(x) = 2x^2 - 4x + 5$

30. $P(x) = 2x^2 - 8x + 9$

31. $P(x) = -3x^2 + 24x - 46$

32. $P(x) = -2x^2 - 6x - 5$

33. $P(x) = 2x^2 - 2x + 1$

34. *Concept Check* For $f(x) = dx^2 - \frac{1}{2}dx + k$, $d \neq 0$, find the x-coordinate of the vertex.

Graph each function in a viewing window that will allow you to use your calculator to approximate
(a) *the coordinates of the vertex and* **(b)** *the x-intercepts. Give values to the nearest hundredth.*

35. $P(x) = -0.32x^2 + \sqrt{3}\,x + 2.86$ **36.** $P(x) = -\sqrt{2}x^2 + 0.45x + 1.39$ **37.** $y = 1.34x^2 - 3x + \sqrt{5}$

38. $y = -0.55x^2 + 3.21x$ **39.** $y = -1.24x^2 + 1.68x$ **40.** $y = 2.96x^2 + 1.31$

Concept Check   *The figure shows the graph of a quadratic function* $y = f(x)$. *Use it to work
Exercises 41–44.*

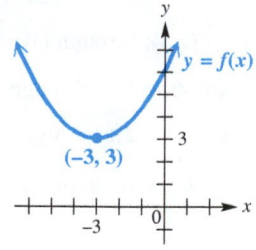

41. What is the minimum value of $f(x)$?

42. For what value of x does $f(x)$ assume its minimum value?

43. How many real solutions are there of the equation $f(x) = 1$?

44. How many real solutions are there of the equation $f(x) = 3$?

In each table, Y_1 *is defined as a quadratic function. Use the table to do the following.*
(a) *Determine the coordinates of the vertex of the graph.*
(b) *Determine whether the vertex is a minimum point or a maximum point.*
(c) *Find the minimum or maximum value of the function.*
(d) *Determine the range of the function.*

45. 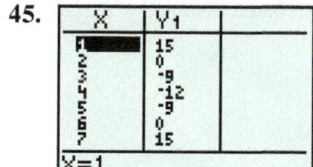   **46.** 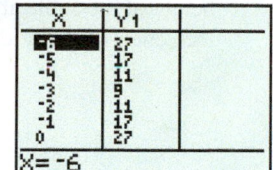   **47.** 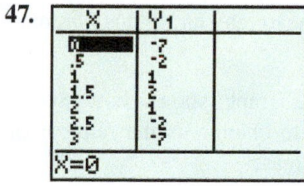   **48.**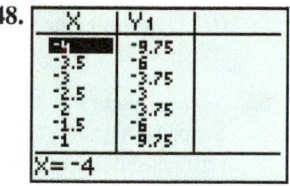

Curve Fitting   *Exercises 49–54 show scatter diagrams of sets of data. In each case, tell whether a
linear or quadratic model is appropriate for the data. For each linear model, decide whether the
slope should be positive or negative. For each quadratic model, decide whether the leading coefficient,
in this case a, should be positive or negative.*

49. Social Security assets as a function of time

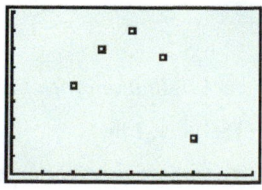

50. Growth in science centers/museums as a function of time

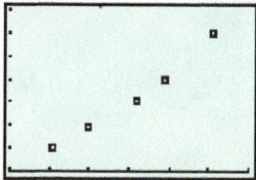

51. Value of U.S. salmon catch as a function of time

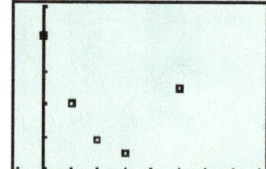

52. Height of an object thrown upward from a building as a function of time

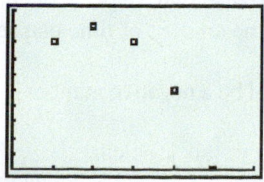

53. Number of shopping centers as a function of time

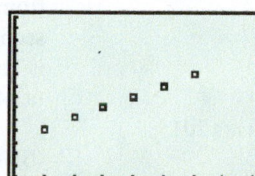

54. Newborns with AIDS as a function of time

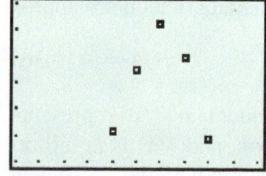

Curve Fitting *Find the equation of the quadratic function satisfying the given conditions. (Hint: Determine values of a, h, and k that satisfy $P(x) = a(x - h)^2 + k$.) Express your answer in the form $P(x) = ax^2 + bx + c$. Use your calculator to support your results.*

55. Vertex $(-1, -4)$; through $(5, 104)$

56. Vertex $(-2, -3)$; through $(0, -19)$

57. Vertex $(8, 3)$; through $(10, 5)$

58. Vertex $(-6, -12)$; through $(6, 24)$

59. Vertex $(-4, -2)$; through $(2, -26)$

60. Vertex $(5, 6)$; through $(1, -6)$

(Modeling) *Solve each problem.*

61. *Heart Rate* An athlete's heart rate R in beats per minute after x minutes is given by

$$R(x) = 2(x - 4)^2 + 90,$$

where $0 \le x \le 8$.
(a) Describe the heart rate during this period of time.
(b) Determine the minimum heart rate during this 8-minute period.

62. *Social Security Assets* The graph shows how Social Security assets are expected to change as the number of retirees receiving benefits increases.

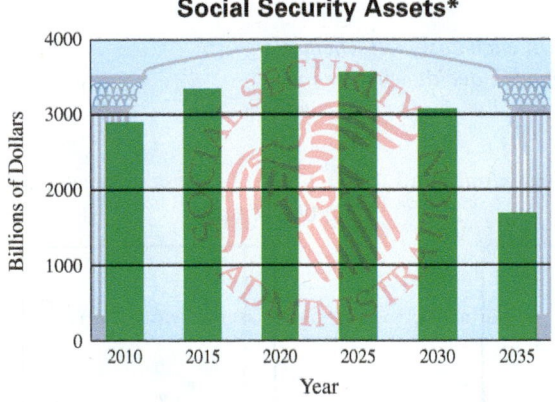

Social Security Assets*

*Projected
Source: Social Security Administration

The graph suggests that a quadratic function would be a good fit to the data, which are approximated by

$$f(x) = -10.36x^2 + 431.8x - 650.$$

In the model, $x = 10$ represents 2010, $x = 15$ represents 2015, and so on, and $f(x)$ is in billions of dollars.
(a) Explain why the coefficient of x^2 in the model is negative, based on the graph.

(b) Analytically determine the vertex of the graph.
(c) Interpret the answer to part (b) as it relates to this application.

63. *Heart Rate* The heart rate of an athlete while weight training is recorded for 4 minutes. The table lists the heart rate after x minutes.

Time (min)	0	1	2	3	4
Heart rate (bpm)	84	111	120	110	85

(a) Explain why the data are not linear.
(b) Find a quadratic function f that models the data.
(c) What is the domain of your function?

64. *Suspension Bridge* The cables that support a suspension bridge, such as the Golden Gate Bridge, can be modeled by parabolas. Suppose that a 300-foot-long suspension bridge has at its ends towers that are 120 feet tall, as shown in the figure. If the cable comes within 20 feet of the road at the center of the bridge, find a function that models the height of the cable above the road a distance of x feet from the center of the bridge.

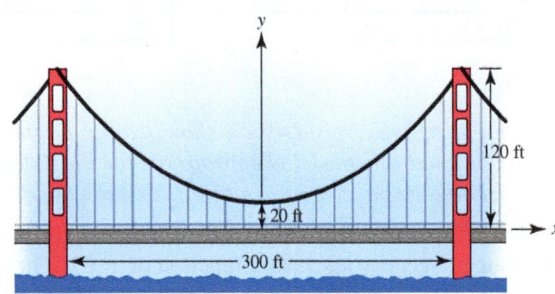

(Modeling) *The formula for the height of a projectile is*

$$s(t) = -16t^2 + v_0t + s_0,$$

where t is time in seconds, s_0 is the initial height in feet, v_0 is the initial velocity in feet per second, and s(t) is in feet. Use this formula to solve Exercises 65–68.

65. *Height of a Projected Rock* A rock is launched upward from ground level with an initial velocity of 90 feet per second. Let t represent the amount of time elapsed after it is launched.
(a) Explain why t cannot be a negative number in this situation.
(b) Explain why $s_0 = 0$ in this problem.
(c) Give the function s that describes the height of the rock as a function of t.

(d) How high will the rock be 1.5 seconds after it is launched?

(e) What is the maximum height attained by the rock? After how many seconds will this happen? Determine the answer analytically and graphically.

(f) After how many seconds will the rock hit the ground? Determine the answer graphically.

66. *Height of a Toy Rocket*   A toy rocket is launched from the top of a building 50 feet tall at an initial velocity of 200 feet per second. Let t represent the amount of time elapsed after the launch.

(a) Express the height s as a function of the time t.

(b) Determine both analytically and graphically the time at which the rocket reaches its highest point. How high will it be at that time?

(c) For what time interval will the rocket be more than 300 feet above ground level? Determine the answer graphically, and give times to the nearest tenth of a second.

(d) After how many seconds will the rocket hit the ground? Determine the answer graphically.

67. *Height of a Projected Ball*   A ball is launched upward from ground level with an initial velocity of 150 feet per second.

(a) Determine graphically whether the ball will reach a height of 355 feet. If it will, determine the time(s) when this happens. If it will not, explain why, using a graphical interpretation.

(b) Repeat part (a) for a ball launched from a height of 30 feet with an initial velocity of 250 feet per second.

68. *Height of a Projected Ball on the Moon*   An astronaut on the moon throws a baseball upward. The astronaut is 6 feet, 6 inches tall and the initial velocity of the ball is 30 feet per second. The height of the ball is approximated by the function

$$s(t) = -2.7t^2 + 30t + 6.5,$$

where t is the number of seconds after the ball was thrown.

(a) After how many seconds is the ball 12 feet above the moon's surface?

(b) How many seconds after it is thrown will the ball return to the surface?

(c) The ball will never reach a height of 100 feet. How can this be determined analytically?

Concept Check   *Sketch a graph of a quadratic function that satisfies each set of given conditions. Use symmetry to label another point on your graph.*

69. Vertex $(-2, -3)$; through $(1, 4)$

70. Vertex $(5, 6)$; through $(1, -6)$

71. Maximum value of 1 at $x = 3$; y-intercept is $(0, -4)$

72. Minimum value of -4 at $x = -3$; y-intercept is $(0, 3)$

3.3 Quadratic Equations and Inequalities

Zero-Product Property • Square Root Property and Completing the Square • Quadratic Formula and the Discriminant • Solving Quadratic Equations • Solving Quadratic Inequalities • Formulas Involving Quadratics

A *quadratic equation* is defined as follows.

> ### Quadratic Equation in One Variable
>
> An equation that can be written in the form
>
> $$ax^2 + bx + c = 0,$$
>
> where a, b, and c are real numbers, with $a \neq 0$, is a **quadratic equation in standard form.**

Zero-Product Property

Consider the following problems.

1. Find the **zeros of the quadratic function** $P(x) = 2x^2 + 4x - 16$.
2. Find the **x-coordinates of the x-intercepts of the graph** of $P(x) = 2x^2 + 4x - 16$.
3. Find the **solution set of the equation** $2x^2 + 4x - 16 = 0$.

Each of these problems is solved by finding the numbers for x that result in 0 when $2x^2 + 4x - 16$ is evaluated.

One way to solve a quadratic equation $P(x) = 0$ is to use factoring and the zero-product property.

◀ Algebra Review
To review factoring trinomials, see **Section R.2.**

Zero-Product Property

If a and b are complex numbers and $ab = 0$, then $a = 0$ or $b = 0$ or both.

CAUTION   The zero-product property applies only when $a \cdot b$ equals 0. If $a \cdot b = 1$, then it does not necessarily follow that $a = 1$ or $b = 1$ or both. For example, $a \cdot b = 1$ when $a = \frac{1}{2}$ and $b = 2$, or when $a = \frac{1}{3}$ and $b = 3$.

EXAMPLE 1   **Using the Zero-Product Property (Two Solutions)**

Solve $2x^2 + 4x - 16 = 0$.

Analytic Solution

$$2x^2 + 4x - 16 = 0 \qquad \text{Solve } P(x) = 0.$$
$$2(x^2 + 2x - 8) = 0 \qquad \text{Factor out 2.}$$
$$2(x + 4)(x - 2) = 0 \qquad \text{Factor trinomial.}$$
$$x + 4 = 0 \quad \text{or} \quad x - 2 = 0 \qquad \text{Zero-product property}$$
$$x = -4 \quad \text{or} \qquad x = 2 \qquad \text{Solve each equation.}$$

The solution set is $\{-4, 2\}$. We can check each solution by substituting it into the given equation. These results also give us the zeros of $P(x)$ and indicate that the x-intercepts of the graph of $P(x)$ are $(-4, 0)$ and $(2, 0)$.

Graphing Calculator Solution

The graph of $P(x) = 2x^2 + 4x - 16$, shown in **FIGURE 23,** has zeros -4 and 2. That is, $P(-4) = 0$ and $P(2) = 0$. Since the zeros satisfy $P(x) = 0$, they are also the solutions of the equation and correspond to the x-intercepts of the graph.

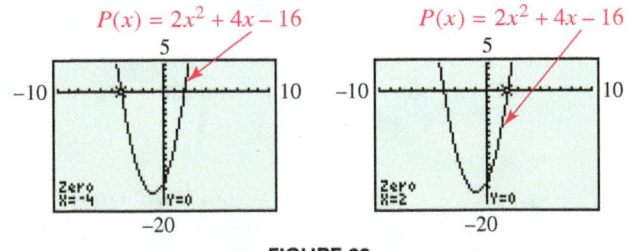

FIGURE 23

NOTE   When solving a quadratic equation analytically, the zero-product property is useful only if the quadratic expression can be factored easily, as was the case in **Example 1.**

EXAMPLE 2   **Using the Zero-Product Property (One Solution)**

Find all zeros of the quadratic function $P(x) = x^2 - 6x + 9$.

Analytic Solution

$$x^2 - 6x + 9 = 0 \quad \text{Solve } P(x) = 0.$$
$$(x - 3)^2 = 0 \quad \text{Factor.}$$
$$x - 3 = 0 \quad \text{or} \quad x - 3 = 0 \quad \text{Zero-product property}$$
$$x = 3 \quad \text{or} \quad x = 3 \quad \text{Solve each equation.}$$

There is only one *distinct* zero, 3. It is sometimes called a **double zero,** or **double solution** (**root**) of the equation. (Finding the zeros of a function defined by $P(x)$ and solving the equation $P(x) = 0$ are equivalent procedures.)

Check:
$$x^2 - 6x + 9 = 0$$
$$3^2 - 6(3) + 9 = 0 \quad ?$$
$$9 - 18 + 9 = 0 \quad ?$$
$$0 = 0 \quad \checkmark \quad \text{True}$$

Graphing Calculator Solution

The graph of P in **FIGURE 24** has only one zero and intersects the x-axis at one point $(3, 0)$, which is also the vertex of the parabola. The table shows a zero at $x = 3$, which gives a minimum value. Because of the symmetry shown in the table, 3 is the only zero.

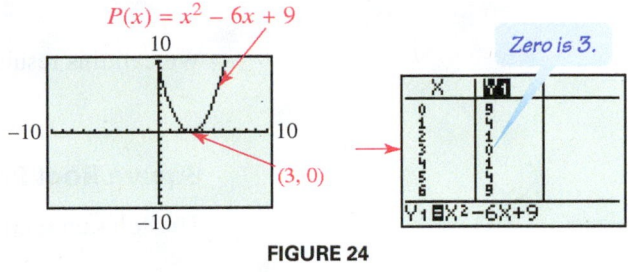

FIGURE 24

CAUTION   A purely graphical approach may not prove to be as useful with a graph like the one in **FIGURE 24**, since in some cases the vertex may be slightly above or slightly below the x-axis. *This is why we need to understand the algebraic concepts presented in this section—only when we know the mathematics can we use the technology properly.*

FIGURE 25 shows possible numbers of x-intercepts of the graph of a quadratic function that opens upward (that is, $P(x) = ax^2 + bx + c$ with $a > 0$).

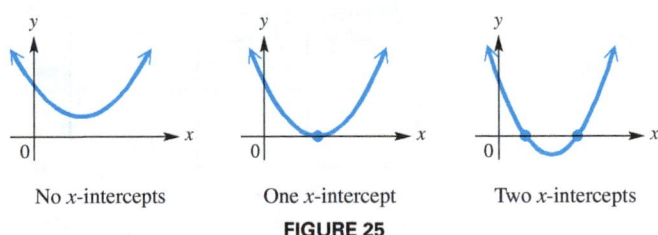

No x-intercepts   One x-intercept   Two x-intercepts

FIGURE 25

Similarly, **FIGURE 26** shows possible numbers of x-intercepts of the graph of a quadratic function that opens downward ($a < 0$). Thus, a quadratic equation can have zero, one, or two real solutions.

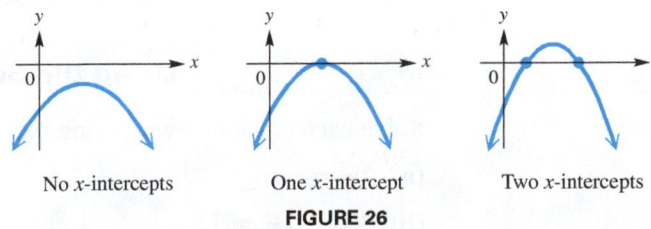

No x-intercepts   One x-intercept   Two x-intercepts

FIGURE 26

Square Root Property and Completing the Square

We can solve quadratic equations of the form $x^2 = k$, where k is a real number, by factoring, using the following sequence of equivalent equations.

$$x^2 = k$$

$$x^2 - k = 0 \qquad \text{Subtract } k.$$

$$\left(x - \sqrt{k}\right)\left(x + \sqrt{k}\right) = 0 \qquad \text{Factor.}$$

$$x - \sqrt{k} = 0 \quad \text{or} \quad x + \sqrt{k} = 0 \qquad \text{Zero-product property}$$

$$x = \sqrt{k} \quad \text{or} \quad x = -\sqrt{k} \qquad \text{Solve each equation.}$$

We call this result the **square root property** for solving quadratic equations.

Square Root Property

The solution set of $x^2 = k$ is one of the following.

(a) $\left\{ \pm\sqrt{k} \right\}$ if $k > 0$

(b) $\{0\}$ if $k = 0$

(c) $\left\{ \pm i\sqrt{|k|} \right\}$ if $k < 0$

As shown in **FIGURE 27,** the graph of $y_1 = x^2$ intersects the graph of $y_2 = k$ twice if $k > 0$, once if $k = 0$, and not at all if $k < 0$. Thus, the equation $x^2 = k$ can have zero, one, or two *real* solutions.

Solving $x^2 = k$ Graphically

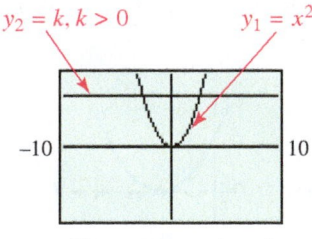

There are two points of intersection if $k > 0$.

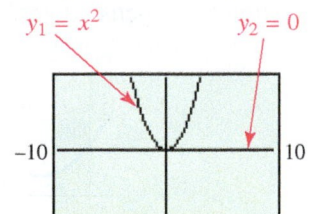

There is only one point of intersection (the origin) if $k = 0$.

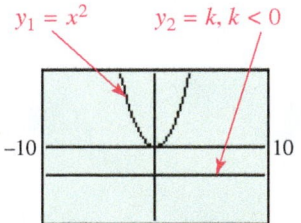

There are no points of intersection if $k < 0$.

FIGURE 27

In the next example, we use the square root property to solve some special types of quadratic equations.

EXAMPLE 3 **Using the Square Root Property**

Solve each equation by applying the square root property.

(a) $2x^2 = -10$

(b) $2(x - 1)^2 = 14$

Analytic Solution

(a) $2x^2 = -10$

$\quad x^2 = -5$ Divide by 2.

There is no real number whose square is -5, so this equation has two pure imaginary solutions. Apply the square root property.

$$x = \sqrt{-5} \quad \text{or} \quad x = -\sqrt{-5}$$

$$x = i\sqrt{5} \quad \text{or} \quad x = -i\sqrt{5}$$

The solution set is $\left\{ \pm i\sqrt{5} \right\}$. These solutions can be checked by substituting $i\sqrt{5}$ and $-i\sqrt{5}$ in the original equation.

(b) $2(x-1)^2 = 14$

$\quad (x-1)^2 = 7$ Divide by 2.

$\quad x - 1 = \pm\sqrt{7}$ Square root property

$\quad x = 1 \pm \sqrt{7}$ Add 1.

Check: $\qquad 2(x-1)^2 = 14$

$$2\left[\left(1 \pm \sqrt{7}\right) - 1\right]^2 = 14 \qquad ?$$

$$2(\pm\sqrt{7})^2 = 14 \qquad ?$$

$$2(7) = 14 \qquad ?$$

$$14 = 14 \checkmark \qquad \text{True}$$

The solution set is $\left\{1 \pm \sqrt{7}\right\}$.

Graphing Calculator Solution

(a) Notice that the graphs of $y_1 = 2x^2$ and $y_2 = -10$ in **FIGURE 28** do not intersect. Thus, there are no *real* solutions. Nonreal complex solutions cannot be found with a graph.

No Solutions

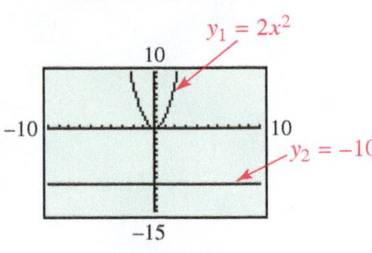

FIGURE 28

(b) Graph $y_1 = 2(x-1)^2$ and $y_2 = 14$, and locate the points of intersection. The screens in **FIGURE 29** show the *x*-coordinates of these points, which yield approximations of $1 \pm \sqrt{7}$, confirming the solution set $\left\{1 \pm \sqrt{7}\right\}$.

Two Solutions

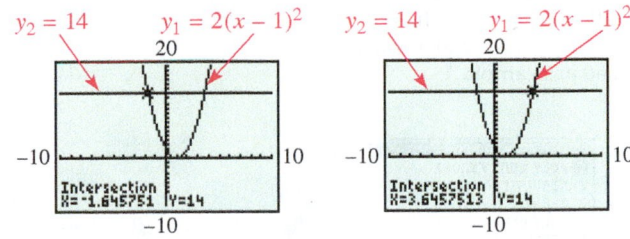

FIGURE 29

In **Section 3.2** we used a technique called *completing the square* to find the vertex of a parabola. We can also use this technique together with the square root property to solve quadratic equations, as illustrated in the next example.

> **EXAMPLE 4**   **Solving by Completing the Square**

Solve each quadratic equation by completing the square.

(a) $x^2 - 4x + 2 = 0$ **(b)** $8x^2 + 2x = 3$

Solution

(a) Because the coefficient of x^2 is 1, we begin by subtracting 2 from each side.

$$x^2 - 4x + 2 = 0 \qquad \text{Given equation}$$

$$x^2 - 4x = -2 \qquad \text{Subtract 2 from each side.}$$

$$x^2 - 4x + 4 = -2 + 4 \qquad \text{Add } \left(\tfrac{b}{2a}\right)^2 = \left(\tfrac{-4}{2(1)}\right)^2 = 4 \text{ to each side.}$$

$$(x-2)^2 = 2 \qquad \text{Factor perfect square and add.}$$

$$x - 2 = \pm\sqrt{2} \qquad \text{Square root property}$$

$$x = 2 \pm \sqrt{2} \qquad \text{Add 2 to each side.}$$

The solution set is $\{2 \pm \sqrt{2}\}$.

← **Algebra Review**

To review and factor perfect square trinomials, see **Section R.2.**

(continued)

(b) For $8x^2 + 2x = 3$, the coefficient of x^2 is 8, so we begin by dividing each side by 8.

$$8x^2 + 2x = 3 \qquad \text{Given equation}$$

$$x^2 + \frac{1}{4}x = \frac{3}{8} \qquad \text{Divide each side by 8.}$$

$$x^2 + \frac{1}{4}x + \frac{1}{64} = \frac{3}{8} + \frac{1}{64} \qquad \text{Add } \left(\tfrac{b}{2a}\right)^2 = \left(\tfrac{1/4}{2(1)}\right)^2 = \tfrac{1}{64} \text{ to each side.}$$

$$\left(x + \frac{1}{8}\right)^2 = \frac{25}{64} \qquad \text{Factor perfect square and add.}$$

$$x + \frac{1}{8} = \pm\frac{5}{8} \qquad \text{Square root property and } \sqrt{\tfrac{25}{64}} = \tfrac{5}{8}$$

$$x = -\frac{1}{8} \pm \frac{5}{8} \qquad \text{Subtract } \tfrac{1}{8} \text{ from each side.}$$

The solution set is $\left\{-\dfrac{1}{8} \pm \dfrac{5}{8}\right\}$, or $\left\{-\dfrac{3}{4}, \dfrac{1}{2}\right\}$.

TECHNOLOGY NOTE

The TI-84 Plus application PlySmlt2 can be used to solve polynomial equations. The screens below show how to solve the quadratic equation $2x^2 + 4x - 16 = 0$, which is ORDER 2 and was discussed in **Example 1**.

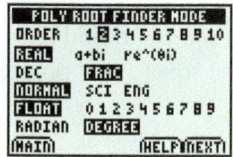

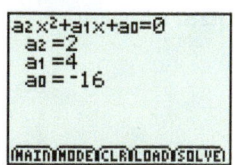

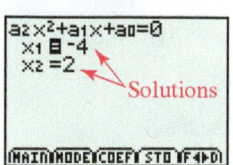

Quadratic Formula and the Discriminant

There is a formula that can be used to solve *any* quadratic equation. To find it, we complete the square on the standard form of $ax^2 + bx + c = 0$. For now, we assume $a > 0$.

$$ax^2 + bx + c = 0 \qquad \text{Standard form } (a \neq 0)$$

$$x^2 + \frac{b}{a}x + \frac{c}{a} = 0 \qquad \text{Divide by } a.$$

$$x^2 + \frac{b}{a}x = -\frac{c}{a} \qquad \text{Add } -\tfrac{c}{a}.$$

Be sure to add $\dfrac{b^2}{4a^2}$ *to each side.*

$$x^2 + \frac{b}{a}x + \frac{b^2}{4a^2} = \frac{b^2}{4a^2} - \frac{c}{a} \qquad \text{Add } \left[\tfrac{1}{2}\left(\tfrac{b}{a}\right)\right]^2 = \tfrac{b^2}{4a^2}.$$

$$\left(x + \frac{b}{2a}\right)^2 = \frac{b^2 - 4ac}{4a^2} \qquad \text{Factor on the left and simplify on the right.}$$

$$x + \frac{b}{2a} = \sqrt{\frac{b^2 - 4ac}{4a^2}} \quad \text{or} \quad x + \frac{b}{2a} = -\sqrt{\frac{b^2 - 4ac}{4a^2}} \qquad \text{Square root property}$$

$$x + \frac{b}{2a} = \frac{\sqrt{b^2 - 4ac}}{2a} \quad \text{or} \quad x + \frac{b}{2a} = \frac{-\sqrt{b^2 - 4ac}}{2a} \qquad \sqrt{4a^2} = 2a \,(a > 0)$$

$$x = \frac{-b + \sqrt{b^2 - 4ac}}{2a} \quad \text{or} \quad x = \frac{-b - \sqrt{b^2 - 4ac}}{2a} \qquad \text{Add } -\tfrac{b}{2a}.$$

These two results are also valid if $a < 0$. A compact form, called the **quadratic formula**, follows.

Quadratic Formula

The solutions of the quadratic equation $ax^2 + bx + c = 0$, where $a \neq 0$, are given by the following formula.

$$x = \frac{-b \pm \sqrt{b^2 - 4ac}}{2a}$$

CAUTION   *The fraction bar in the quadratic formula extends under the* **− b** *term in the numerator.* Be sure to evaluate the *entire* numerator before dividing by 2*a*.

The expression $b^2 - 4ac$ under the radical in the quadratic formula is called the **discriminant.** The value of the discriminant determines whether the quadratic equation has two real solutions, one real solution, or no real solutions. In the last case, there will be two nonreal complex solutions.

Meaning of the Discriminant

If *a*, *b*, and *c* are real numbers, $a \neq 0$, then the complex solutions of $ax^2 + bx + c = 0$ are described as follows, based on the value of the discriminant, $b^2 - 4ac$.

Value of $b^2 - 4ac$	Number of Solutions	Type of Solutions
Positive	Two	Real
Zero	One (a double solution)	Real
Negative	Two	Nonreal complex

Furthermore, if *a*, *b*, and *c* are *integers*, $a \neq 0$, the real solutions are *rational* if $b^2 - 4ac$ is the square of a positive integer.

NOTE   The final sentence in the preceding box suggests that a quadratic equation can be solved by factoring if $b^2 - 4ac$ is a perfect square.

Solving Quadratic Equations

EXAMPLE 5   **Using the Quadratic Formula**

Solve $x(x - 2) = 2x - 2$ with the quadratic formula.

Analytic Solution

First, write the equation in standard form.

$$x(x - 2) = 2x - 2$$
$$x^2 - 2x = 2x - 2 \qquad \text{Distributive property}$$
$$x^2 - 4x + 2 = 0 \qquad \text{Subtract 2x and add 2.}$$

Graphing Calculator Solution

In the analytic solution, the given equation becomes

$$x^2 - 4x + 2 = 0.$$

We can solve this equation graphically by using the *x*-intercept method. To do this we graph

$$y = x^2 - 4x + 2,$$

(continued)

Apply the quadratic formula to $x^2 - 4x + 2 = 0$.

$$x = \frac{-b \pm \sqrt{b^2 - 4ac}}{2a} \qquad \text{Quadratic formula}$$

$$x = \frac{-(-4) \pm \sqrt{(-4)^2 - 4(1)2}}{2(1)} \qquad a = 1, b = -4, c = 2$$

$$x = \frac{4 \pm \sqrt{8}}{2} \qquad \text{Simplify.}$$

$$x = \frac{4 \pm 2\sqrt{2}}{2} \qquad \text{Simplify the radical.}$$

$$x = \frac{2(2 \pm \sqrt{2})}{2} \qquad \text{Factor first. Then divide out the common factor.}$$

$$x = 2 \pm \sqrt{2} \qquad \text{Lowest terms}$$

The solution set is $\{2 + \sqrt{2}, 2 - \sqrt{2}\}$ and can be abbreviated as $\{2 \pm \sqrt{2}\}$.

and locate any x-intercepts, as shown in **FIGURE 30**. The x-coordinates of the x-intercepts are approximately 0.58579 and 3.4142. Note that these are approximations for the exact answers we already found, because

$$2 - \sqrt{2} \approx 0.58579 \quad \text{and} \quad 2 + \sqrt{2} \approx 3.4142.$$

Two Graphical Solutions

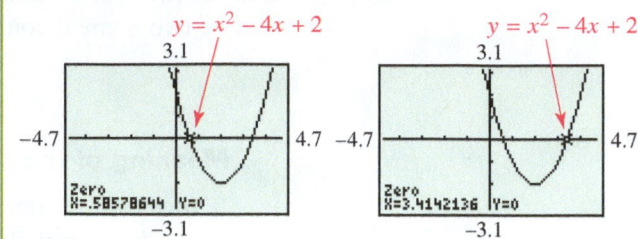

FIGURE 30

EXAMPLE 6   **Using the Quadratic Formula**

Solve $2x^2 - x + 4 = 0$ with the quadratic formula.

Solution

$$x = \frac{-b \pm \sqrt{b^2 - 4ac}}{2a} \qquad \text{Quadratic formula}$$

Use parentheses around substituted values to avoid errors.

$$x = \frac{-(-1) \pm \sqrt{(-1)^2 - 4(2)(4)}}{2(2)} \qquad a = 2, b = -1, c = 4$$

$$x = \frac{1 \pm \sqrt{1 - 32}}{4} \qquad \text{Simplify.}$$

$$x = \frac{1 \pm \sqrt{-31}}{4} \qquad \text{Work under the radical.}$$

Because the discriminant is negative, there are two nonreal complex solutions.

$$x = \frac{1 \pm i\sqrt{31}}{4} = \frac{1}{4} \pm i\frac{\sqrt{31}}{4} \qquad \text{Write in } a + bi \text{ form.}$$

The solution set is $\left\{\dfrac{1}{4} \pm i\dfrac{\sqrt{31}}{4}\right\}$.

TECHNOLOGY NOTE

The TI-84 Plus application PlySmlt2 can also be used to solve polynomial equations with nonreal complex solutions. To do this, set the calculator in $a + bi$ mode, as shown below, where the equation in **Example 6** has been solved. Note that $\frac{\sqrt{31}}{4} \approx 1.39194$ is the imaginary part.

```
 POLY ROOT FINDER MODE
ORDER   1 2 3 4 5 6 7 8 9 10
REAL   a+bi  re^(θi)
DEC     FRAC
NORMAL  SCI ENG
FLOAT   0 1 2 3 4 5 6 7 8 9
RADIAN  DEGREE
MAIN            HELP NEXT
```

```
a2x²+a1x+a0=0
  a2 =2
  a1 = -1
  a0 =4


MAIN MODE CLR LOAD SOLVE
```

```
a2x²+a1x+a0=0
  x1 =1/4+1.39194…
  x2 =1/4-1.39194…


MAIN MODE COEF STO F◄►D
```

NOTE   If we graph $y = 2x^2 - x + 4$, the graph has no x-intercepts, supporting our result in **Example 6**. There are no *real* solutions. (See **Example 3(a)** as well.)

FOR DISCUSSION

Solve $x^2 = 5x - 4$ analytically. Then, graph

$$y_1 = x^2 \quad \text{and} \quad y_2 = 5x - 4,$$

and use the intersection-of-graphs method to support your result.

1. Do you encounter a problem if you use the standard viewing window?

2. Why do you think that analytic methods of solution are essential for understanding graphical methods?

Solving Quadratic Inequalities

> ### Quadratic Inequality in One Variable
>
> A **quadratic inequality** is an inequality that can be written in the form
>
> $$ax^2 + bx + c < 0,$$
>
> where a, b, and c are real numbers with $a \neq 0$. (The symbol $<$ can be replaced with $>$, $\leq$, or $\geq$.)

We can solve a quadratic inequality graphically, using the ideas shown in the following table.

Graphical Interpretations for Quadratic Equations and Inequalities

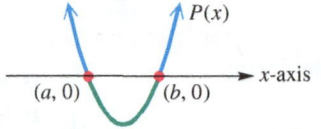

Solution Set of	Is
$P(x) = 0$	$\{a, b\}$
$P(x) < 0$	The interval (a, b)
$P(x) > 0$	$(-\infty, a) \cup (b, \infty)$

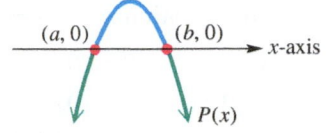

Solution Set of	Is
$P(x) = 0$	$\{a, b\}$
$P(x) < 0$	$(-\infty, a) \cup (b, \infty)$
$P(x) > 0$	The interval (a, b)

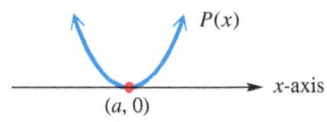

Solution Set of	Is
$P(x) = 0$	$\{a\}$
$P(x) < 0$	$\varnothing$
$P(x) > 0$	$(-\infty, a) \cup (a, \infty)$

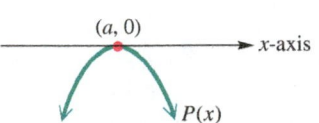

Solution Set of	Is
$P(x) = 0$	$\{a\}$
$P(x) < 0$	$(-\infty, a) \cup (a, \infty)$
$P(x) > 0$	$\varnothing$

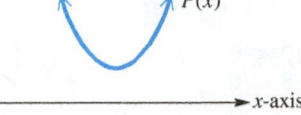

Solution Set of	Is
$P(x) = 0$	$\varnothing$
$P(x) < 0$	$\varnothing$
$P(x) > 0$	$(-\infty, \infty)$

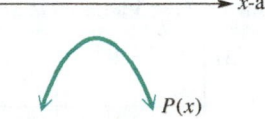

Solution Set of	Is
$P(x) = 0$	$\varnothing$
$P(x) < 0$	$(-\infty, \infty)$
$P(x) > 0$	$\varnothing$

To solve a quadratic inequality analytically, follow these steps.

Solving a Quadratic Inequality

Step 1 Solve the corresponding quadratic equation.
Step 2 Identify the intervals determined by the solutions of the equation.
Step 3 Use a test value from each interval to determine which intervals form the solution set.

Step 3 is valid because any polynomial function has a continuous graph between successive x-intercepts.

EXAMPLE 7 **Solving a Quadratic Inequality**

Solve $x^2 - x - 12 < 0$.

Analytic Solution

Step 1 Find the values of x that satisfy the corresponding quadratic equation $x^2 - x - 12 = 0$.

$x^2 - x - 12 = 0$	Standard form
$(x + 3)(x - 4) = 0$	Factor.
$x + 3 = 0$ or $x - 4 = 0$	Zero-product property
$x = -3$ or $x = 4$	Solve each equation.

Step 2 The two numbers -3 and 4 separate a number line into the three intervals shown in **FIGURE 31**. If a value in Interval A, for example, makes the polynomial $x^2 - x - 12$ negative, then all values in Interval A will make the polynomial negative.

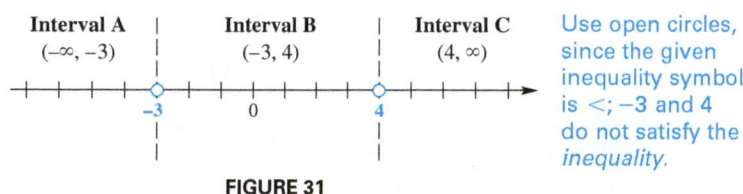

Use open circles, since the given inequality symbol is $<$; -3 and 4 do not satisfy the inequality.

FIGURE 31

Step 3 Pick a test value x in each interval to see if it satisfies $x^2 - x - 12 < 0$. If it makes the inequality true, then the entire interval belongs to the solution set.

Interval	Test Value x	Is $x^2 - x - 12 < 0$ True or False?
A: $(-\infty, -3)$	-4	$(-4)^2 - (-4) - 12 < 0$? $8 < 0$ False
B: $(-3, 4)$	0	$0^2 - 0 - 12 < 0$? $-12 < 0$ True
C: $(4, \infty)$	5	$5^2 - 5 - 12 < 0$? $8 < 0$ False

Since the values in Interval B make the inequality true, the solution set is the interval $(-3, 4)$.

Graphing Calculator Solution

The graph of

$$Y_1 = X^2 - X - 12$$

is shown in **FIGURE 32**. When the graph lies *below* the x-axis, the function values are negative. This happens in the interval $(-3, 4)$, which is the same as the analytic result.

The graph of the function is not sufficient to determine whether the endpoints should be included or excluded from the solution set. We make this decision based on the symbol $<$ in the original inequality. Thus, the endpoints are excluded.

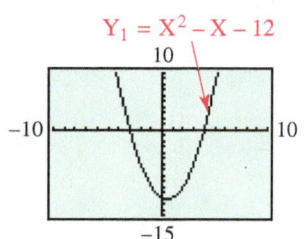

$Y_1 = X^2 - X - 12$

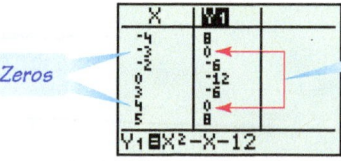

Zeros

Y_1-values are negative between the zeros.

FIGURE 32

The table in **FIGURE 32** numerically supports the analytic result. Values of X in the interval $(-3, 4)$ make $Y_1 < 0$, while values outside that interval make $Y_1 \geq 0$.

NOTE Inequalities that use the symbols $<$ and $>$ are called **strict inequalities**. The symbols $\leq$ and $\geq$ are used in **nonstrict inequalities**. The solutions of the equation

$$x^2 - x - 12 = 0$$

in **Example 7** were not included in the solution set, since the inequality was *strict*. However, in **Example 8** which follows, the solutions of the equation will be included, since the inequality there is *nonstrict*.

EXAMPLE 8   **Solving a Quadratic Inequality**

Solve $2x^2 \geq -5x + 12$.

Analytic Solution

Step 1 Find the values of x that satisfy $2x^2 = -5x + 12$.

$$2x^2 + 5x - 12 = 0 \qquad \text{Standard form}$$
$$(2x - 3)(x + 4) = 0 \qquad \text{Factor.}$$
$$2x - 3 = 0 \quad \text{or} \quad x + 4 = 0 \qquad \text{Zero-product property}$$
$$x = \frac{3}{2} \quad \text{or} \qquad x = -4 \qquad \text{Solve each equation.}$$

Step 2 The values $\frac{3}{2}$ and -4 form the intervals $(-\infty, -4)$, $\left(-4, \frac{3}{2}\right)$, and $\left(\frac{3}{2}, \infty\right)$ on the number line. See **FIGURE 33**.

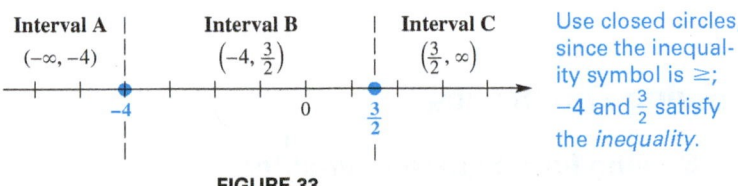

Use closed circles, since the inequality symbol is $\geq$; -4 and $\frac{3}{2}$ satisfy the *inequality*.

FIGURE 33

Step 3 Choose a test value x from each interval.

Interval	Test Value x	Is $2x^2 \geq -5x + 12$ True or False?	
A: $(-\infty, -4)$	-5	$2(-5)^2 \geq -5(-5) + 12$	?
		$50 \geq 37$	True
B: $\left(-4, \frac{3}{2}\right)$	0	$2(0)^2 \geq -5(0) + 12$	?
		$0 \geq 12$	False
C: $\left(\frac{3}{2}, \infty\right)$	2	$2(2)^2 \geq -5(2) + 12$	?
		$8 \geq 2$	True

A and **C** make this strict inequality true, so the solution set is the union of the intervals with endpoints, written $\left(-\infty, -4\right] \cup \left[\frac{3}{2}, \infty\right)$.

Graphing Calculator Solution

Let $Y_1 = 2X^2$ and $Y_2 = -5X + 12$. Use the x-intercept method here to see the intervals satisfying

$$2x^2 + 5x - 12 \geq 0.$$

Write the original inequality in the form

$$Y_1 - Y_2 = 2X^2 + 5X - 12 \geq 0.$$

The x-intercepts are the points $(-4, 0)$ and $(1.5, 0)$. Then find the domain values where the graph of $Y = Y_1 - Y_2$ lies *above or on* the x-axis. From the graph, the intervals $\left(-\infty, -4\right]$ and $\left[1.5, \infty\right)$ give these values, supporting our analytic result. See **FIGURE 34**.

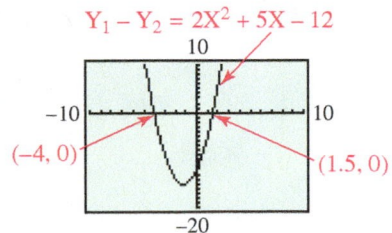

FIGURE 34

The table numerically supports the analytic result because choosing values of X less than or equal to -4 *or* greater than or equal to 1.5 makes $Y_1 \geq Y_2$ true.

FOR DISCUSSION

1. Graph $P(x) = x^2 - x - 20$ in a window that shows both x-intercepts, and without any analytic work solve each equation or inequality.

   (a) $x^2 - x - 20 = 0$ (b) $x^2 - x - 20 < 0$ (c) $x^2 - x - 20 > 0$

2. Graph $P(x) = x^2 + x + 20$ in a window that shows the vertex and the y-intercept. Then without any analytic work solve each equation or inequality for real solutions.

   (a) $x^2 + x + 20 = 0$ (b) $x^2 + x + 20 < 0$ (c) $x^2 + x + 20 > 0$

From the "For Discussion" box, we see that solving a quadratic equation leads directly to the solution sets of the corresponding inequalities.

EXAMPLE 9 Solving Quadratic Inequalities

Use the graph in **FIGURE 35** and the result of **Example 5** to solve each inequality.

(a) $x^2 - 4x + 2 \leq 0$ (b) $x^2 - 4x + 2 \geq 0$

Solution

(a) In **Example 5**, we found that the exact solutions for the equation $x^2 - 4x + 2 = 0$ are $2 - \sqrt{2}$ and $2 + \sqrt{2}$. Since the graph of $y = x^2 - 4x + 2$ lies below or intersects the x-axis (that is, $y \leq 0$) between or at these values, as shown in **FIGURE 35,** the solution set for the quadratic inequality $x^2 - 4x + 2 \leq 0$ is the closed interval $\left[2 - \sqrt{2}, 2 + \sqrt{2}\right]$.

(b) **FIGURE 35** shows that the graph of $y = x^2 - 4x + 2$ lies above or intersects the x-axis (that is, $y \geq 0$) on $\left(-\infty, 2 - \sqrt{2}\right] \cup \left[2 + \sqrt{2}, \infty\right)$, which is the solution set of $x^2 - 4x + 2 \geq 0$. ●

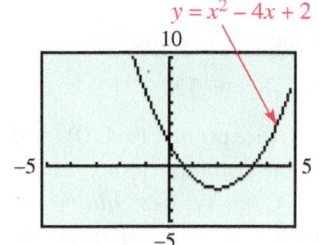

The x-values of the x-intercepts are $2 - \sqrt{2} \approx 0.5857864$ and $2 + \sqrt{2} \approx 3.4142136$.

FIGURE 35

Formulas Involving Quadratics

EXAMPLE 10 Solving for a Squared Variable

Solve each equation for the indicated variable. (Leave $\pm$ in the answers.)

(a) $\mathscr{A} = \dfrac{\pi d^2}{4}$ for d (b) $rt^2 - st = k$ for t $(r \neq 0)$

Solution

(a) $\mathscr{A} = \dfrac{\pi d^2}{4}$ *Isolate d.*

$4\mathscr{A} = \pi d^2$ Multiply by 4.

$d^2 = \dfrac{4\mathscr{A}}{\pi}$ Divide each side by π and rewrite.

Remember both the positive and negative square roots.

$d = \pm\sqrt{\dfrac{4\mathscr{A}}{\pi}}$ Square root property (Note that in applications, we are often interested only in the positive square root.)

$d = \dfrac{\pm 2\sqrt{\mathscr{A}}}{\sqrt{\pi}} \cdot \dfrac{\sqrt{\pi}}{\sqrt{\pi}}$ Rationalize the denominator.

$d = \dfrac{\pm 2\sqrt{\mathscr{A}\pi}}{\pi}$ Multiply.

(b) Because this equation has a term with t as well as one with t^2, use the quadratic formula. Subtract k from each side to get $rt^2 - st - k = 0$.

Take extra care when substituting variables in formulas.

$$t = \frac{-b \pm \sqrt{b^2 - 4ac}}{2a} \qquad \text{Quadratic formula}$$

$$t = \frac{-(-s) \pm \sqrt{(-s)^2 - 4(r)(-k)}}{2(r)} \qquad a = r,\ b = -s,\ c = -k$$

$$t = \frac{s \pm \sqrt{s^2 + 4rk}}{2r} \qquad \text{Simplify.}$$

3.3 Exercises

Checking Analytic Skills Match the equation in Column I with its solution(s) in Column II. **Do not use a calculator.**

I

1. $x^2 = 4$

3. $x^2 + 2 = 0$

5. $x^2 = -8$

7. $x - 2 = 0$

2. $x^2 = -4$

4. $x^2 - 2 = 0$

6. $x^2 = 8$

8. $x + 2 = 0$

II

A. $\pm 2i$

C. $\pm i\sqrt{2}$

E. $\pm \sqrt{2}$

G. ± 2

B. $\pm 2\sqrt{2}$

D. 2

F. -2

H. $\pm 2i\sqrt{2}$

Concept Check Answer each question.

9. Which one of the following equations is set up for direct use of the zero-product property? Solve it.
  A. $3x^2 - 17x - 6 = 0$ **B.** $(2x + 5)^2 = 7$
  C. $x^2 + x = 12$ **D.** $(3x + 1)(x - 7) = 0$

10. Which one of the following equations is set up for direct use of the square root property? Solve it.
  A. $3x^2 - 17x - 6 = 0$ **B.** $(2x + 5)^2 = 7$
  C. $x^2 + x = 12$ **D.** $(3x + 1)(x - 7) = 0$

11. Only one of the following equations does not require Step 1 of the method for completing the square. Which one is it? Solve it.
  A. $3x^2 - 17x - 6 = 0$ **B.** $(2x + 5)^2 = 7$
  C. $x^2 + x = 12$ **D.** $(3x + 1)(x - 7) = 0$

12. Only one of the following equations is set up so that the values of a, b, and c can be determined immediately. Which one is it? Solve it.
  A. $3x^2 - 17x - 6 = 0$ **B.** $(2x + 5)^2 = 7$
  C. $x^2 + x = 12$ **D.** $(3x + 1)(x - 7) = 0$

Solve each equation. For equations with real solutions, support your answers graphically.

13. $x^2 = 16$

14. $x^2 = 144$

15. $2x^2 = 90$

16. $2x^2 = 48$

17. $x^2 = -16$

18. $x^2 = -100$

19. $x^2 = -18$

20. $x^2 = -32$

21. $(3x - 1)^2 = 12$

22. $(4x + 1)^2 = 20$

23. $(5x - 3)^2 = -3$

24. $(-2x + 5)^2 = -8$

25. $x^2 = 2x + 24$

26. $x^2 = 3x + 18$

27. $3x^2 - 2x = 0$

28. $5x^2 + 3x = 0$

29. $x(14x + 1) = 3$

30. $x(12x + 11) = -2$

31. $-4 + 9x - 2x^2 = 0$

32. $-5 + 16x - 3x^2 = 0$

33. $\frac{1}{3}x^2 - \frac{1}{3}x = 24$

34. $\frac{1}{6}x^2 + \frac{1}{6}x = 5$

35. $(x + 2)(x - 1) = 7x + 5$

36. $(x + 4)(x - 1) = -5x - 4$

37. $x^2 - 2x - 4 = 0$ **38.** $x^2 + 8x + 13 = 0$ **39.** $2x^2 + 2x = -1$ **40.** $9x^2 - 12x = -8$

41. $x(x - 1) = 1$ **42.** $x(x - 3) = 2$ **43.** $x^2 - 5x = x - 7$ **44.** $11x^2 - 3x + 2 = 4x + 1$

45. $4x^2 - 12x = -11$ **46.** $x^2 = 2x - 5$ **47.** $\frac{1}{3}x^2 + \frac{1}{4}x - 3 = 0$ **48.** $\frac{2}{3}x^2 + \frac{1}{4}x = 3$

49. $(3 - x)^2 = 25$ **50.** $(2 + x)^2 = 49$ **51.** $2x^2 - 4x = 1$ **52.** $3x^2 - 6x = 4$

53. $x^2 = -1 - x$ **54.** $x^2 = -3 - 3x$ **55.** $4x^2 - 20x + 25 = 0$ **56.** $9x^2 + 12x + 4 = 0$

57. $-3x^2 + 4x + 4 = 0$ **58.** $-5x^2 + 28x + 12 = 0$

59. $(x + 5)(x - 6) = (2x - 1)(x - 4)$ **60.** $(x + 2)(3x - 4) = (x + 5)(2x - 5)$

Solve each quadratic equation by completing the square.

61. $x^2 - 2x = 2$ **62.** $x^2 + 2x = 4$ **63.** $2x^2 + 6x - 3 = 0$

64. $3x^2 - 3x - 1 = 0$ **65.** $x(x - 1) = 3$ **66.** $2x(2x - 5) = 2$

67. $2x^2 - x + 3 = 0$ **68.** $x^2 - 2x = -5$

Evaluate the discriminant, and use it to determine the number of real solutions of the equation. If the equation does have real solutions, tell whether they are rational *or* irrational. *Do not actually solve the equation.*

69. $x^2 + 8x + 16 = 0$ **70.** $8x^2 = 14x - 3$ **71.** $4x^2 = 6x + 3$

72. $2x^2 - 4x + 1 = 0$ **73.** $9x^2 + 11x + 4 = 0$ **74.** $3x^2 = 4x - 5$

Concept Check   *For each pair of numbers, find the values of a, b, and c for which the quadratic equation $ax^2 + bx + c = 0$ has the given numbers as solutions. Answers may vary. (Hint: Use the zero-product property in reverse.)*

75. $4, 5$ **76.** $-3, 2$ **77.** $1 + \sqrt{2}, 1 - \sqrt{2}$ **78.** $i, -i$

79. $2i, -2i$ **80.** $1 + \sqrt{3}, 1 - \sqrt{3}$ **81.** $2 - \sqrt{5}, 2 + \sqrt{5}$ **82.** $3i, -3i$

Concept Check   *Sketch a graph of $f(x) = ax^2 + bx + c$ that satisfies each set of conditions.*

83. $a < 0, b^2 - 4ac = 0$ **84.** $a > 0, b^2 - 4ac < 0$ **85.** $a < 0, b^2 - 4ac < 0$

86. $a < 0, b^2 - 4ac > 0$ **87.** $a > 0, b^2 - 4ac > 0$ **88.** $a > 0, b^2 - 4ac = 0$

Concept Check   *Exercises 89–102 refer to the graphs of the quadratic functions f, g, and h shown here.*

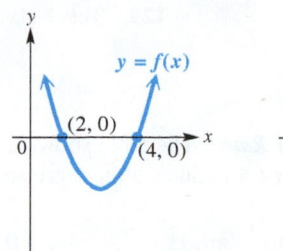 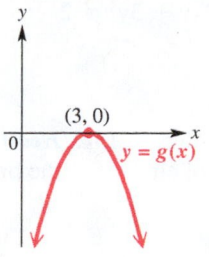 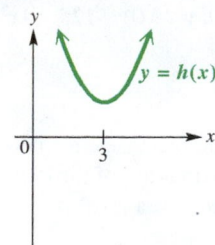

89. What is the solution set of $f(x) = 0$?

90. What is the solution set of $f(x) < 0$?

91. What is the solution set of $f(x) > 0$?

92. What is the solution set of $g(x) \geq 0$?

93. What is the solution set of $g(x) < 0$?

94. What is the solution set of $g(x) > 0$?

95. Solve $h(x) > 0$.

96. Solve $h(x) < 0$.

97. How many real solutions does $h(x) = 0$ have? How many nonreal complex solutions does it have?

98. What is the value of the discriminant of $g(x)$?

99. What is the x-coordinate of the vertex of the graph of $y = f(x)$?

100. What is the equation of the axis of symmetry of the graph of $y = g(x)$?

101. Does the graph of $y = g(x)$ have a y-intercept? If so, is the y-coordinate positive or negative?

102. Is the minimum value of h positive or negative?

Solve each inequality analytically. Support your answers graphically. Give exact values for endpoints.

103. (a) $x^2 + 4x + 3 \geq 0$
    (b) $x^2 + 4x + 3 < 0$

104. (a) $x^2 + 6x + 8 < 0$
    (b) $x^2 + 6x + 8 \geq 0$

105. (a) $2x^2 - 9x > -4$
    (b) $2x^2 - 9x \leq -4$

106. (a) $3x^2 + 13x + 10 \leq 0$
    (b) $3x^2 + 13x + 10 > 0$

107. (a) $-x^2 - x \leq 0$
    (b) $-x^2 - x > 0$

108. (a) $-x^2 + 2x \leq 0$
    (b) $-x^2 + 2x > 0$

109. (a) $x^2 - x + 1 < 0$
    (b) $x^2 - x + 1 \geq 0$

110. (a) $2x^2 - x + 3 < 0$
    (b) $2x^2 - x + 3 \geq 0$

111. (a) $2x + 1 \geq x^2$
    (b) $2x + 1 < x^2$

112. (a) $x^2 + 5x < 2$
    (b) $x^2 + 5x \geq 2$

113. (a) $x - 3x^2 > -1$
    (b) $x - 3x^2 \leq -1$

114. (a) $-2x^2 + 3x < -4$
    (b) $-2x^2 + 3x \geq -4$

Solve each formula for the indicated variable. Leave $\pm$ in answers when appropriate. Assume that no denominators are 0.

115. $s = \dfrac{1}{2}gt^2$ for t

116. $\mathcal{A} = \pi r^2$ for r

117. $a^2 + b^2 = c^2$ for a

118. $\mathcal{A} = s^2$ for s

119. $S = 4\pi r^2$ for r

120. $V = \dfrac{1}{3}\pi r^2 h$ for r

121. $V = e^3$ for e

122. $V = \dfrac{4}{3}\pi r^3$ for r

123. $F = \dfrac{kMv^4}{r}$ for v

124. $s = s_0 + gt^2 + k$ for t

125. $P = \dfrac{E^2 R}{(r + R)^2}$ for R

126. $S = 2\pi rh + 2\pi r^2$ for r

Solve each equation for x and then for y.

127. $x^2 + xy + y^2 = 0$ $(x > 0, y > 0)$ **128.** $4x^2 - 2xy + 3y^2 = 2$ **129.** $3y^2 + 4xy - 9x^2 = -1$

(Modeling) *Solve each problem.*

130. *Air Density* As the altitude increases, air becomes thinner, or less dense. An approximation of the density d of air at an altitude of x meters above sea level is

$$d(x) = (3.32 \times 10^{-9})x^2 - (1.14 \times 10^{-4})x + 1.22.$$

The output is the density of air in kilograms per cubic meter. The domain of d is $0 \le x \le 10{,}000$. (*Source:* A. Miller and J. Thompson, *Elements of Meteorology.*)
  (a) Denver is sometimes referred to as the mile-high city. Compare the density of air at sea level and in Denver. (*Hint*: 1 ft $\approx$ 0.305 m.)
  (b) Determine the altitudes where the density is greater than 1 kilogram per cubic meter.

131. *Heart Rate* Suppose that a person's heart rate, x minutes after vigorous exercise has stopped, can be modeled by

$$f(x) = \frac{4}{5}(x - 10)^2 + 80.$$

The output is in beats per minute, where the domain of f is $0 \le x \le 10$.
  (a) Evaluate $f(0)$ and $f(2)$. Interpret the result.
  (b) Estimate the times when the person's heart rate was between 100 and 120 beats per minute, inclusive.

132. *Heart Rate* The table shows a person's heart rate during the first 4 minutes after exercise has stopped.

Time (min)	0	2	4
Heart rate (bpm)	154	106	90

  (a) Find a formula $f(x) = a(x - h)^2 + k$ that models the data, where x represents time and $0 \le x \le 4$.
  (b) Evaluate $f(1)$ and interpret the result.
  (c) Estimate the times when the heart rate was from 115 to 125 beats per minute.

SECTIONS 3.1–3.3 Reviewing Basic Concepts

Perform each operation. Write answers in standard form.

1. $(5 + 6i) - (2 - 4i) - 3i$ **2.** $i(5 + i)(5 - i)$ **3.** $\dfrac{-10 - 10i}{2 + 3i}$

Give the following for the function $P(x) = 2x^2 + 8x + 5$.

4. The graph of P

5. The vertex of the graph and whether it is a maximum or minimum point

6. The equation of the axis of symmetry of the graph

7. The domain and range of P

Solve each equation or inequality analytically, and support your result graphically.

8. $9x^2 = 25$ **9.** $3x^2 - 5x = 2$ **10.** $-x^2 + x + 3 = 0$

11. $3x^2 - 5x - 2 \le 0$ **12.** $x^2 - x - 3 > 0$ **13.** $x(3x - 1) \le -4$

3.4 Applications of Quadratic Functions and Models

Applications of Quadratic Functions • A Quadratic Model

Some types of applications cannot be solved by using linear functions. Instead they require quadratic functions to be solved. In this section we solve applications and models that require us to find quadratic functions and solve quadratic equations.

Applications of Quadratic Functions

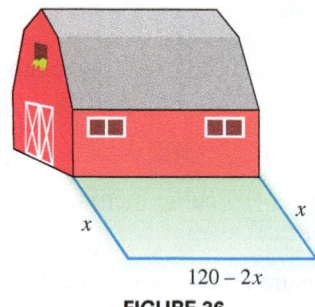

FIGURE 36

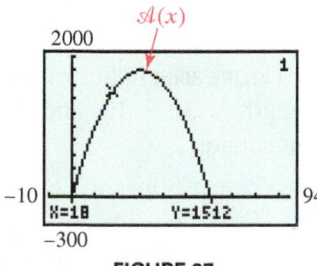

FIGURE 37

EXAMPLE 1   Finding the Area of a Rectangular Region

A farmer wishes to enclose a rectangular region. He has 120 feet of fencing and plans to use his barn as one side of the enclosure. See **FIGURE 36.** Let x represent the length of one of the parallel sides of the fencing.

(a) Determine a function A that represents the area of the region in terms of x.

(b) What are the restrictions on x?

(c) Graph the function A in a viewing window that shows both x-intercepts and the vertex of the graph.

(d) **FIGURE 37** shows the cursor at $(18, 1512)$. Interpret this information.

(e) What is the maximum area the farmer can enclose?

Solution

(a) The lengths of the sides of the region bordered by the fencing are x, x, and $120 - 2x$, as shown in **FIGURE 36.** Area = width × length, so the function is

$$A(x) = x(120 - 2x), \quad \text{or} \quad A(x) = -2x^2 + 120x.$$

(b) Since x represents length, $x > 0$. Furthermore, the side of length $120 - 2x > 0$, or $x < 60$. Putting these two restrictions together gives $0 < x < 60$ or the interval $(0, 60)$ as the domain of A in this example.

(c) See **FIGURE 37**, where $A(x) = -2x^2 + 120x$.

(d) If each parallel side of fencing measures 18 feet, the area of the enclosure is 1512 square feet. This can be written $A(18) = 1512$ and checked as follows: If the width is 18 feet, the length is $120 - 2(18) = 84$ feet, and $18 \times 84 = 1512$.

(e) The maximum value on the graph of the quadratic function occurs at the vertex. For $A(x) = -2x^2 + 120x$, use $a = -2$ and $b = 120$ in the vertex formula from **Section 3.2.**

$$x = -\frac{b}{2a} = -\frac{120}{2(-2)} = 30$$

$$A(30) = -2(30)^2 + 120(30) = 1800$$

Therefore, the farmer can enclose a maximum of 1800 square feet if the parallel sides of fencing each measure 30 feet.

Alternatively, use a calculator to locate the vertex of the parabola, and observe the x- and y-values there. The graph in **FIGURE 38(a)** confirms the analytic results. The table in **FIGURE 38(b)** provides numerical support. ●

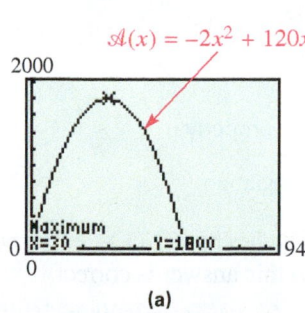

(a)

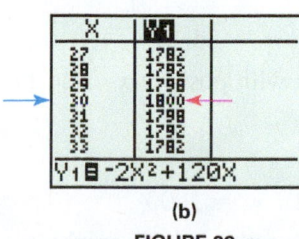

(b)

FIGURE 38

TECHNOLOGY NOTE

In the calculator graph shown in **FIGURE 38(a)** on the previous page, the display at the bottom obscures the view of the intercepts. To overcome this problem, lower the minimum values of x and y enough so that when the display appears, the x- and y-axes are visible, as in **FIGURE 37**.

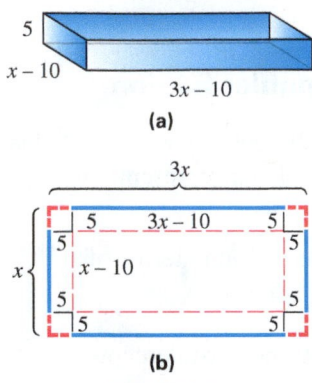

(a)

(b)

FIGURE 39

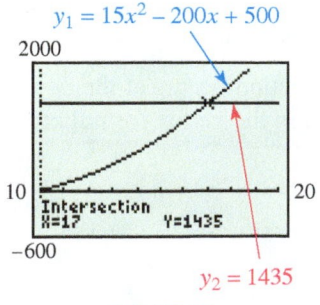

$y_1 = 15x^2 - 200x + 500$

FIGURE 40

CAUTION *Be mindful when interpreting the meanings of the coordinates of the vertex.* The first coordinate, x, gives the *domain value* for which the *function value* is a maximum or minimum. Read the problem carefully to determine whether you are asked to find the value of the independent variable x, the function value y, or both.

EXAMPLE 2 Finding the Volume of a Box

A machine produces rectangular sheets of metal satisfying the condition that the length is three times the width. Furthermore, equal-sized squares measuring 5 inches on a side can be cut from the corners so that the resulting piece of metal can be shaped into an open box by folding up the flaps. See **FIGURE 39(a)**.

(a) Determine a function V that expresses the volume of the box in terms of the width x of the original sheet of metal.

(b) What restrictions must be placed on x?

(c) If specifications call for the volume of such a box to be 1435 cubic inches, what should the dimensions of the original piece of metal be?

(d) What dimensions of the original piece of metal will ensure a volume greater than 2000, but less than 3000, cubic inches? Solve graphically.

Solution

(a) If x represents the width, then $3x$ represents the length. **FIGURE 39(b)** indicates that the width of the bottom of the box is $x - 10$, the length is $3x - 10$, and the height is 5 inches (the length of the side of each cut-out square).

$$V(x) = (3x - 10)(x - 10)(5) \qquad \text{Volume = length × width × height}$$
$$V(x) = 15x^2 - 200x + 500 \qquad \text{Multiply.}$$

(b) Since the dimensions of the box represent positive numbers, $3x - 10 > 0$ and $x - 10 > 0$, or $x > \frac{10}{3}$ and $x > 10$. Both conditions are satisfied when $x > 10$, so the domain of V is $(10, \infty)$ in this example.

(c)
$$15x^2 - 200x + 500 = 1435 \qquad \text{Let } V(x) = 1435.$$
$$15x^2 - 200x - 935 = 0 \qquad \text{Subtract 1435.}$$
$$3x^2 - 40x - 187 = 0 \qquad \text{Divide by 5.}$$
$$(3x + 11)(x - 17) = 0 \qquad \text{Factor.}$$
$$3x + 11 = 0 \quad \text{or} \quad x - 17 = 0 \qquad \text{Zero-product property}$$
$$x = -\frac{11}{3} \quad \text{or} \qquad x = 17 \qquad \text{Solve each equation.}$$

Only 17 satisfies $x > 10$. The dimensions should be 17 inches by $3(17) = 51$ inches. The product $(51 - 10) \cdot (17 - 10) \cdot 5 = 1435$, so this answer is correct.

In the same window, we can also graph
$$V(x) = (3x - 10)(x - 10)(5)$$
$$y_1 = 15x^2 - 200x + 500 \quad \text{and} \quad y_2 = 1435.$$

As shown in **FIGURE 40** the graphs intersect at the point with positive x-value 17, supporting the analytic result.

(d) Using the graphs of the functions
$$y_1 = 15x^2 - 200x + 500, \quad y_2 = 2000, \quad \text{and} \quad y_3 = 3000$$

as shown in **FIGURE 41**, we find that the points of intersection of the graphs are *approximately* (18.7, 2000) and (21.2, 3000) for $x > 10$. Therefore, the width of the rectangle should be between 18.7 and 21.2 inches, with the corresponding length three times these values (that is, between $3(18.7) \approx 56.1$ inches and $3(21.2) \approx 63.6$ inches).

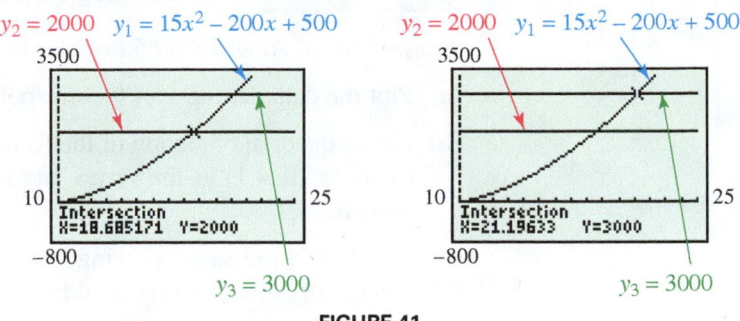

FIGURE 41

EXAMPLE 3 **Solving a Problem Requiring the Pythagorean Theorem**

A contractor finds a piece of property in the shape of a right triangle. To get some idea of its dimensions, he measures the three sides, starting with the shortest one. He finds that the longer leg is approximately 20 meters longer than twice the length of the shorter leg. The length of the hypotenuse is approximately 10 meters longer than the length of the longer leg. Find the lengths of the sides of the triangular lot.

Solution Let s be the length of the shorter leg in meters, so $2s + 20$ meters is the length of the longer leg, and $(2s + 20) + 10 = 2s + 30$ meters is the length of the hypotenuse. See **FIGURE 42.**

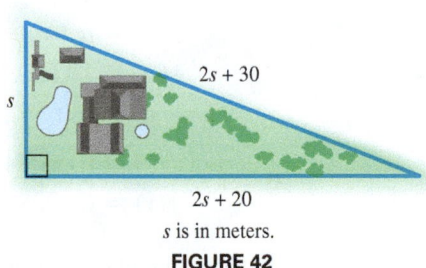

s is in meters.
FIGURE 42

The domain of s is $(0, \infty)$. Write and solve an equation.

$a^2 + b^2 = c^2$

$$s^2 + (2s + 20)^2 = (2s + 30)^2 \quad \text{Pythagorean theorem}$$
$$s^2 + 4s^2 + 80s + 400 = 4s^2 + 120s + 900$$

$(x + y)^2 = x^2 + 2xy + y^2$

$$s^2 - 40s - 500 = 0 \qquad \text{Combine like terms.}$$
$$(s - 50)(s + 10) = 0 \qquad \text{Factor.}$$
$$s - 50 = 0 \quad \text{or} \quad s + 10 = 0 \qquad \text{Zero-product property}$$
$$s = 50 \quad \text{or} \qquad s = -10 \qquad \text{Solve each equation.}$$

The solution -10 is not in the domain. The lengths of the sides of the triangular lot are 50 meters, $2(50) + 20 = 120$ meters, and $2(50) + 30 = 130$ meters.

Year, x	Percent 65 or older, y
1900	4.1
1920	4.7
1940	6.8
1960	9.3
1980	11.3
2000	12.4
2020*	16.5
2040*	20.6

*Projected
Source: U.S. Census Bureau.

A Quadratic Model

By extending the concept of regression from **Section 1.4** to polynomials, we can sometimes fit data to a quadratic function to provide a quadratic model.

EXAMPLE 4 **Modeling Population Data**

The percent of Americans 65 or older for selected years is shown in the table.

(a) Plot the data, letting $x = 0$ correspond to 1900.

(b) Find a quadratic function of the form $f(x) = a(x - h)^2 + k$ that models the data by using $(0, 4.1)$ as the vertex and a second point, such as $(140, 20.6)$, to determine a.

(c) Graph f in the same viewing window as the data. How well does f model the percent of Americans 65 or older?

(d) Use f to determine the percent of Americans 65 or older in 2010.

(e) Use the quadratic regression feature of a graphing calculator to determine the quadratic function g that provides the best fit for the data. Graph g with the data.

Solution

(a) Using the statistical feature of a graphing calculator and letting the x-list L_1 be $\{0, 20, 40, \ldots, 140\}$ and the y-list L_2 be $\{4.1, 4.7, 6.8, \ldots, 20.6\}$ gives the table shown in **FIGURE 43.** A scatter diagram is shown as well.

(b) Since $(0, 4.1)$ is the lowest point in the scatter diagram of the data, let it correspond to the vertex on the graph of a parabola that opens upward.

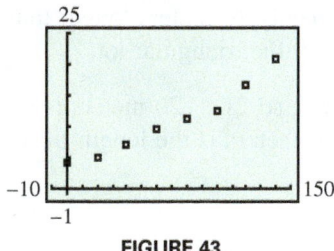

FIGURE 43

$$f(x) = a(x - h)^2 + k \qquad \text{Given form of quadratic function}$$
$$f(x) = a(x - 0)^2 + 4.1 \qquad \text{Let } h = 0 \text{ and } k = 4.1.$$

To determine a, use the given ordered pair $(140, 20.6)$.

$$20.6 = a(140)^2 + 4.1 \qquad \text{Let } x = 140 \text{ and } f(140) = 20.6.$$
$$16.5 = 19{,}600a \qquad \text{Subtract 4.1 and square 140.}$$
$$a \approx 0.0008418 \qquad \text{Rounded value of } a$$
$$\text{Let } f(x) = 0.0008418(x - 0)^2 + 4.1. \qquad \text{Substitute.}$$
$$f(x) = 0.0008418x^2 + 4.1 \qquad \text{Quadratic model}$$

(Note that choosing other second points would give different formulas for $f(x)$.)

(c) **FIGURE 44(a)** shows the graph of f plotted with the data points. There is a fairly good fit, especially for the later years.

$f(x) = 0.0008418x^2 + 4.1$

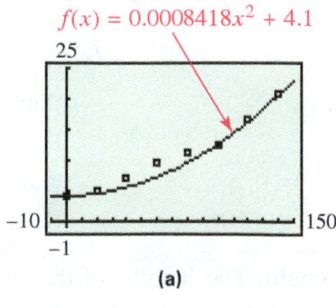

(a)

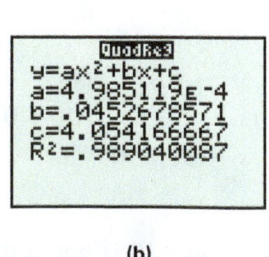

(b)

$g(x) = 0.0004985x^2 + 0.04527x + 4.054$

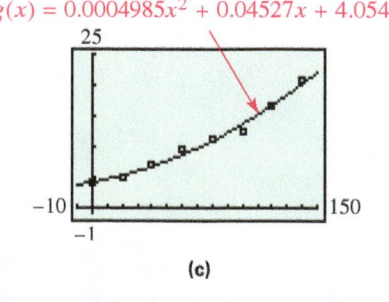

(c)

FIGURE 44

TECHNOLOGY NOTE

The TI-84 Plus computes r (the correlation coefficient) and r^2 (the coefficient of determination) for a regression that yields two estimates (a and b), such as linear regression. For a regression that yields three or more estimates (such as a, b, and c for quadratic regression, as in **FIGURE 44(b)**), the calculator computes R^2, called the coefficient of multiple determination.

(d) For the year 2010, $x = 2010 - 1900 = 110$.

$$f(110) = 0.0008418(110)^2 + 4.1 \approx 14.3$$

This model shows 14.3 percent of Americans were 65 or older in 2010.

(e) **FIGURE 44(b)** shows the calculator quadratic regression model for the data,

$$g(x) \approx 0.0004985x^2 + 0.04527x + 4.054$$

graphed in the same window as the data in **FIGURE 44(c).** Comparing this graph with the one in **FIGURE 44(a),** we see that $g(x)$ gives a closer fit to the data. This is because all data are used in determining function g, while only the first and last data points were used to determine function f. ●

3.4 Exercises

Checking Analytic Skills *Solve each problem. **Do not use a calculator.***

1. Find the maximum y-value on the graph of $y = -16x^2 + 32x + 100$.

2. Find the maximum y-value on the graph of $y = -2x^2 + 8x - 5$.

3. Find the minimum y-value on the graph of $y = 3x^2 - 24x + 50$.

4. Find the minimum y-value on the graph of $y = 5x^2 + 30x + 17$.

5. Solve $-4x^2 + 5x = 1$.

6. Solve $x^2 - 6x = 7$.

7. Solve $\frac{1}{2}x^2 + 3 = 6x$.

8. Solve $\frac{1}{4}x^2 + x = 1$.

(Modeling) *Solve each problem.*

9. *Area of a Parking Lot* For the rectangular parking area shown, which equation says that the area is 40,000 square yards?

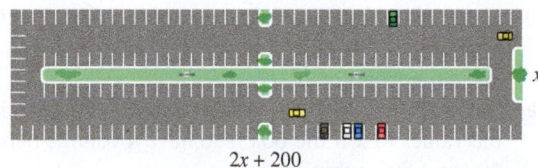

2x + 200

A. $x(2x + 200) = 40,000$

B. $2x + 2(2x + 200) = 40,000$

C. $x + (2x + 200) = 40,000$

D. None of the above

10. *Area of a Picture* The mat around the picture shown measures x inches across. Which equation says that the area of the picture itself is 600 square inches?

A. $2(34 - 2x) + 2(21 - 2x) = 600$

B. $(34 - 2x)(21 - 2x) = 600$

C. $(34 - x)(21 - x) = 600$

D. $x(34)(21) = 600$

21 in.

34 in.

x in.

x in.

11. *Sum of Two Numbers* Suppose that x represents one of two *positive* numbers whose sum is 30.
    (a) Represent the other of the two numbers in terms of x.
    (b) What are the restrictions on x?
    (c) Determine a function P that represents the product of the two numbers.
    (d) Determine analytically and support graphically the two such numbers whose product is a maximum. What is this maximum product?

12. *Sum of Two Numbers* Suppose that x represents one of two *positive* numbers whose sum is 45.
    (a) Represent the other of the two numbers in terms of x.
    (b) What are the restrictions on x?
    (c) Determine a function P that represents the product of the two numbers.
    (d) For what two such numbers is the product equal to 504? Determine analytically.
    (e) Determine analytically and support graphically the two such numbers whose product is a maximum. What is this maximum product?

13. *Maximizing Area* A farmer has 1000 feet of fence to enclose a rectangular area. What dimensions for the rectangle result in the maximum area enclosed by the fence?

14. *Maximizing Area* A homeowner has 80 feet of fence to enclose a rectangular garden. What dimensions for the garden give the maximum area?

15. *Area of a Parking Lot* American River College has plans to construct a rectangular parking lot on land bordered on one side by a highway. There are 640 feet of fencing available to fence the other three sides. Let x represent the length of each of the two parallel sides of fencing.

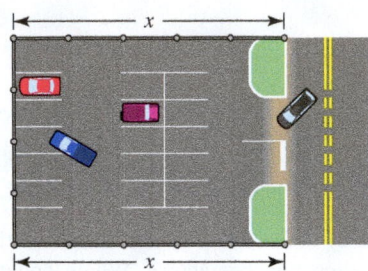

(a) Express the length of the remaining side to be fenced in terms of x.

(b) What are the restrictions on x?

(c) Determine a function $\mathcal{A}$ that represents the area of the parking lot in terms of x.

(d) Graph the function $\mathcal{A}$ from part (c) in a viewing window of $[0, 320]$ by $[0, 55{,}000]$. Determine graphically the values of x that will give an area between 30,000 and 40,000 square feet.

(e) What dimensions will give a maximum area, and what will this area be? Determine analytically and support graphically.

16. *Area of a Rectangular Region* A farmer wishes to enclose a rectangular region bordering a barn with fencing, as shown in the diagram. Suppose that x represents the length of each of the three parallel pieces of fencing. She has 600 feet of fencing available.

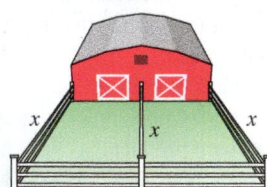

(a) What is the length of the remaining piece of fencing in terms of x?

(b) Determine a function $\mathcal{A}$ that represents the total area of the enclosed region. Give any restrictions on x.

(c) What dimensions for the total enclosed region would give an area of 22,500 square feet? Determine the answer analytically.

(d) Use a graph to find the maximum area that can be enclosed.

17. *Hitting a Baseball* A baseball is hit so that its height in feet after t seconds is

$$s(t) = -16t^2 + 44t + 4.$$

(a) How high is the baseball after 1 second?

(b) Find the maximum height of the baseball.

18. *Hitting a Golf Ball* A golf ball is hit so that its height h in feet after t seconds is

$$h(t) = -16t^2 + 60t.$$

(a) What is the initial height of the golf ball?

(b) How high is the golf ball after 1.5 seconds?

(c) Find the maximum height of the golf ball.

19. *Height of a Baseball* A baseball is dropped from a stadium seat that is 75 feet above the ground. Its height s in feet after t seconds is given by

$$s(t) = 75 - 16t^2.$$

Estimate to the nearest tenth of a second how long it takes for the baseball to strike the ground.

20. *Geometry* A cylindrical aluminum can is being constructed to have a height h of 4 inches. If the can is to have a volume of 28 cubic inches, approximate its radius r. (*Hint:* $V = \pi r^2 h$.)

21. *Volume of a Box* A piece of cardboard is twice as long as it is wide. It is to be made into a box with an open top by cutting 2-inch squares from each corner and folding up the sides. Let x represent the width of the original piece of cardboard.

(a) Represent the length of the original piece of cardboard in terms of x.

(b) What will be the dimensions of the bottom rectangular base of the box? Give the restrictions on x.

(c) Determine a function V that represents the volume of the box in terms of x.

(d) For what dimensions of the bottom of the box will the volume be 320 cubic inches? Determine analytically and support graphically.

(e) Determine graphically (to the nearest tenth of an inch) the values of x if the box is to have a volume between 400 and 500 cubic inches.

22. *Volume of a Box* A piece of sheet metal is 2.5 times as long as it is wide. It is to be made into a box with an open top by cutting 3-inch squares from each corner and folding up the sides, as shown at the top of the next page. Let x represent the width of the original piece of sheet metal.

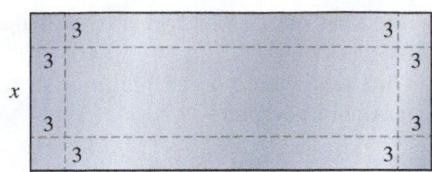

(a) Represent the length of the original piece of sheet metal in terms of x.
(b) What are the restrictions on x?
(c) Determine a function V that represents the volume of the box in terms of x.
(d) For what values of x (that is, original widths) will the volume of the box be between 600 and 800 cubic inches? Determine the answer graphically, and give values to the nearest tenth of an inch.

23. *Radius of a Can*   A can of garbanzo beans has surface area 54.19 square inches. Its height is 4.25 inches. What is the radius of the circular top? See the figure. (*Hint:* The surface area consists of the circular top and bottom and a rectangle that represents the side cut open vertically and unrolled.)

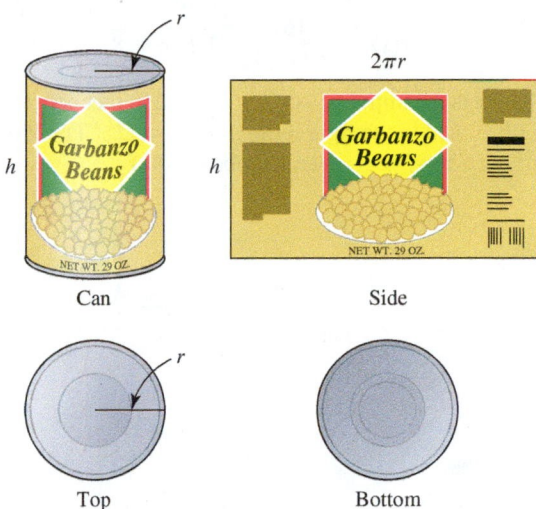

24. *Dimensions of a Cereal Box*   The volume of a 10-ounce box of cereal is 182.742 cubic inches. The width of the box is 3.1875 inches less than the length, and its depth is 2.3125 inches. Find the length and width of the box to the nearest thousandth.

25. *Radius Covered by a Circular Lawn Sprinkler*   A square lawn has area 800 square feet. A sprinkler placed at the center of the lawn sprays water in a circular pattern that just covers the lawn. What is the radius of the circle?

26. *Height of a Kite*   A kite is flying on 50 feet of string. How high is it above the ground if its height is 10 feet more than the horizontal distance from the person flying it? Assume that the string is being held at ground level.

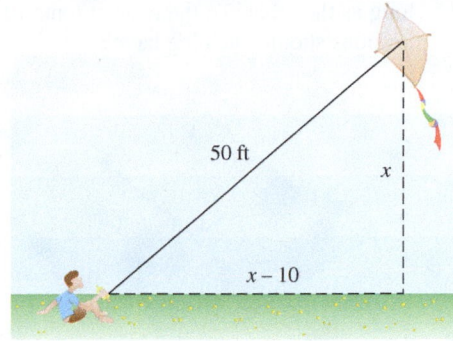

27. *Height of a Dock*   A boat with a rope attached at water level is being pulled into a dock. When the boat is 12 feet from the dock, the length of the rope is 3 feet more than twice the height of the dock above the water. Find the height of the dock.

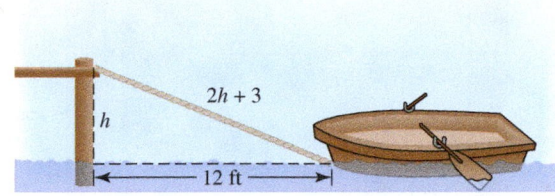

28. *Length of a Walkway*   A raised wooden walkway is being constructed through a wetland. The walkway will have the shape of a right triangle with one leg 700 yards longer than the other and the hypotenuse 100 yards longer than the longer leg. Find the total length of the walkway.

29. *Length of a Ladder*   A building is 2 feet from a 9-foot fence that surrounds the property. A worker wants to wash a window in the building 13 feet from the ground. He plans to place a ladder over the fence so that it rests against the building. He decides he should place the ladder at least 8 feet from the fence for stability. To the nearest foot, how long a ladder will he need?

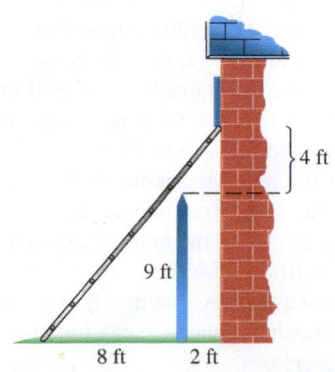

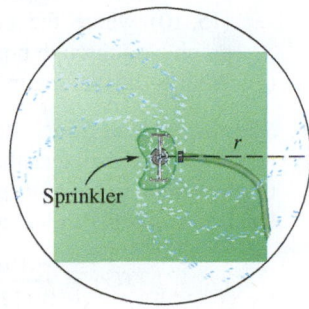

30. *Dimensions of a Solar Panel Frame* Christine has a solar panel with a width of 26 inches, as shown in the figure below. To get the proper inclination for her climate, she needs a right triangular support frame with one leg twice as long as the other. To the nearest tenth of an inch, what dimensions should each leg have?

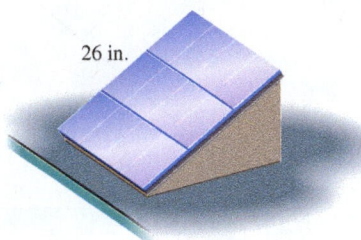

26 in.

31. *Picture Frame* A frame for a picture is 2 inches wide. The picture inside the frame is 4 inches longer than it is wide. See the figure. If the area of the picture is 320 square inches, find the outside dimensions of the picture frame.

2 in.

2 in.

32. *Maximizing Revenue* Suppose the revenue R in thousands of dollars that a company receives from producing x thousand MP3 players is $R(x) = x(40 - 2x)$.
    (a) Evaluate $R(2)$ and interpret the result.
    (b) How many MP3 players should the company produce to maximize its revenue?
    (c) What is the maximum revenue?

33. *Apartment Rental* The manager of an 80-unit apartment complex knows from experience that at a rent of $400 per month, all units will be rented. However, for each increase of $20 in rent, he can expect one unit to be vacated. Let x represent the number of $20 increases over $400.
    (a) Express, in terms of x, the number of apartments that will be rented if x increases of $20 are made. (For example, with three such increases, the number of apartments rented will be $80 - 3 = 77$.)
    (b) Express the rent per apartment if x increases of $20 are made. (For example, if he increases rent by $60 = 3 \times \$20$, the rent per apartment is given by $400 + 3(20) = \$460$.)
    (c) Determine a revenue function R in terms of x that will give the revenue generated as a function of the number of $20 increases.

(d) For what number of increases will the revenue be $37,500?
(e) What rent should he charge in order to achieve the maximum revenue?

34. *Seminar Fee* When *Respect Brings Success* charges $600 for a seminar on management techniques, it attracts 1000 people. For each decrease of $20 in the charge, an additional 100 people will attend the seminar. Let x represent the number of $20 decreases in the charge.
    (a) Determine a revenue function R that will give revenue generated as a function of the number of $20 decreases.
    (b) Find the value of x that maximizes the revenue. What should the company charge to maximize the revenue?
    (c) What is the maximum revenue the company can generate?

35. *Shooting a Foul Shot* To make a foul shot in basketball, the ball must follow a parabolic arc that depends on both the angle and velocity with which the basketball is released. If a person shoots the basketball overhand from a position 8 feet above the floor, then the path can sometimes be modeled by the quadratic function

$$f(x) = \frac{-16x^2}{0.434v^2} + 1.15x + 8,$$

where v is the initial velocity of the ball in feet per second, as illustrated in the figure. (*Source:* Rist, C., "The Physics of Foul Shots," *Discover,* October 2000.)

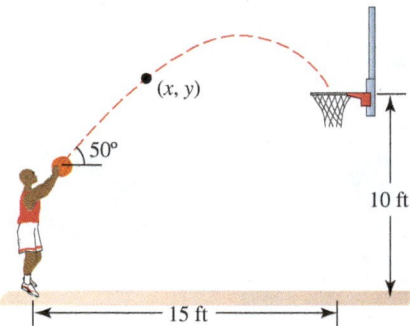

(x, y)

50°

10 ft

15 ft

(a) If the basketball hoop is 10 feet high and located 15 feet away, what initial velocity v should the basketball have?
(b) Check your answer from part (a) graphically. Plot the point $(0, 8)$ where the ball is released and the point $(15, 10)$ where the basketball hoop is. Does your graph pass through both points?
(c) What is the maximum height of the basketball?

36. *Shooting a Foul Shot* Refer to **Exercise 35.** If a person releases a basketball underhand from a position 3 feet above the floor, it often has a steeper arc than if it is released overhand, and its path can sometimes be modeled by

$$f(x) = \frac{-16x^2}{0.117v^2} + 2.75x + 3.$$

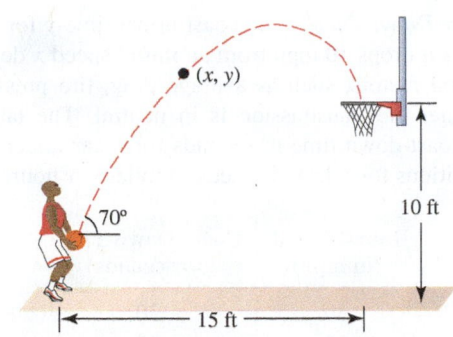

Complete parts (a), (b), and (c) of **Exercise 35.** Then compare the paths for the overhand shot and the underhand shot.

37. *Path of a Frog's Leap*   A frog leaps from a stump 3 feet high and lands 4 feet from the base of the stump. We can consider the initial position of the frog to be at (0, 3) and its landing position to be at (4, 0).

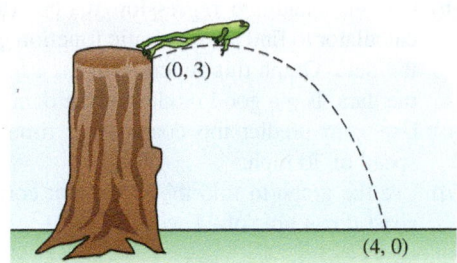

It is determined that the height *h* in feet of the frog as a function of its distance *x* from the base of the stump is given by

$$h(x) = -0.5x^2 + 1.25x + 3.$$

(a) How high was the frog when its horizontal distance *x* from the base of the stump was 2 feet?

(b) What was the horizontal distance from the base of the stump when the frog was 3.25 feet above the ground?

(c) At what horizontal distance from the base of the stump did the frog reach its highest point?

(d) What was the maximum height reached by the frog?

38. *Path of a Frog's Leap*   Refer to **Exercise 37.** Suppose that the initial position of the frog is (0, 4) and its landing position is (6, 0). The height of the frog in feet is given by

$$h(x) = -\frac{1}{3}x^2 + \frac{4}{3}x + 4.$$

(a) What was the horizontal distance *x* from the base of the stump when the frog reached maximum height?

(b) What was the maximum height?

39. *Airplane Landing Speed*   To determine the appropriate landing speed of Vangie's airplane, we might use

$$f(x) = \frac{1}{10}x^2 - 3x + 22,$$

where *x* is the initial landing speed in feet per second and *f*(*x*) is the length of the runway in feet. If the landing

speed is too fast, she may run out of runway; if the speed is too slow, the plane may stall. If the runway is 800 feet long, what is the appropriate landing speed? What is the landing speed in mph? (*Hint:* 5280 feet = 1 mile)

40. *Fatality Rate*   As a function of age group *x*, the fatality rate (per 100,000 population) for males killed in automobile accidents can be approximated by

$$f(x) = 1.8x^2 - 12x + 37.4,$$

where *x* = 0 represents ages 21–24, *x* = 1 represents ages 25–34, *x* = 2 represents ages 35–44, and so on. Find the age group at which the accident rate is a minimum, and find the minimum rate. (*Source:* National Highway Traffic Safety Administration.)

41. *Heating Costs*   The table lists the average heating bill for a natural gas consumer in Illinois during various months of the year.

Month	Bill ($)
Jan.	108
Mar.	68
May	18
July	12
Sept.	13
Nov.	54

(a) Plot the data. Let *x* = 1 correspond to January, *x* = 2 to February, and so on.

(b) Find a quadratic function $f(x) = a(x - h)^2 + k$ that models the data. Use (7, 12) as the vertex and (1, 108) as another point to determine *a*.

(c) Plot the data together with the graph of *f* in the same window. How well does *f* model the average heating bill over these months?

(d) Use the quadratic regression feature of a graphing calculator to determine the quadratic function *g* that provides the best fit for the data.

(e) Use the functions *f* and *g* to approximate the heating bill to the nearest dollar in the following months.
(i) February
(ii) June

42. *E-Book Readers*   The table lists the projected number of shipments *S* of e-book readers in millions, *x* years after 2011.

Year	S
0	23
1	15
2	11
3	8
4	7

(a) Evaluate *S*(3) and interpret the result.

(continued)

(b) Find a quadratic function f to model these data.
(c) Use f to estimate the number of shipments in 2017 to the nearest million. Do you think this is an accurate model for years past 2015?

43. *Automobile Stopping Distance* Selected values of the stopping distance y in feet of a car traveling x mph are given in the table.

Speed (in mph)	Stopping Distance (in feet)
20	46
30	87
40	140
50	240
60	282
70	371

Source: National Safety Institute Student Workbook.

(a) Plot the data.
(b) The quadratic function

$$f(x) = 0.056057x^2 + 1.06657x$$

is one model of the data. Find and interpret $f(45)$.
(c) Use a graph of the function in the same window as the data to determine how well f models the stopping distance.

44. *Coast-Down Time* The coast-down time y for a typical car as it drops 10 mph from an initial speed x depends on several factors, such as average drag, tire pressure, and whether the transmission is in neutral. The table gives the coast-down time in seconds for a car under standard conditions for selected speeds in miles per hour.

Initial Speed (in mph)	Coast-Down Time (in seconds)
30	30
35	27
40	23
45	21
50	18
55	16
60	15
65	13

Source: Scientific American

(a) Plot the data.
(b) Use the quadratic regression feature of a graphing calculator to find the quadratic function g that best fits the data. Graph this function in the same window as the data. Is g a good model for the data?
(c) Use g to predict the coast-down time at an initial speed of 70 mph.
(d) Use the graph to find the speed that corresponds to a coast-down time of 24 seconds.

3.5 Higher-Degree Polynomial Functions and Graphs

Cubic Functions • Quartic Functions • Extrema • End Behavior • *x*-Intercepts (Real Zeros) • Comprehensive Graphs
• Curve Fitting and Polynomial Models

→ Looking Ahead to Calculus

In calculus, polynomial functions are used to approximate more complicated functions, such as trigonometric, exponential, or logarithmic functions.

Linear and quadratic functions are examples of *polynomial functions*.

Polynomial Function

A **polynomial function of degree n in the variable x** is a function of the form

$$P(x) = a_n x^n + a_{n-1} x^{n-1} + \cdots + a_1 x + a_0,$$

where each a_i is a real number, $a_n \neq 0$, and n is a whole number.*

---

*While our definition requires real coefficients, the definition of a polynomial function can be extended to include nonreal complex numbers as coefficients.

The behavior of the graph of a polynomial function is due largely to the value of the coefficient a_n and the *parity* (that is, "evenness" or "oddness") of the exponent n on the term of greatest degree. For this reason, we will refer to a_n as the **leading coefficient** and to $a_n x^n$ as the **dominating term.** The term a_0 is the **constant term** of the polynomial function, and since $P(0) = a_0$, it is the y-value of the y-intercept of the graph.

As we study the graphs of polynomial functions, we use the following properties.

1. A polynomial function (unless otherwise specified) has domain $(-\infty, \infty)$.
2. The graph of a polynomial function is a smooth, continuous curve with no sharp corners.

Cubic Functions

A polynomial function of the form

$$P(x) = ax^3 + bx^2 + cx + d, \quad a \neq 0,$$

is a third-degree function, or **cubic function.** The simplest cubic function is $P(x) = x^3$. (A function capsule for $f(x) = x^3$ is shown in **Chapter 2.**) The graph of a cubic function generally resembles one of the shapes shown in **FIGURE 45.**

Quartic Functions

A polynomial function of the form

$$P(x) = ax^4 + bx^3 + cx^2 + dx + e, \quad a \neq 0,$$

is a fourth-degree function, or **quartic function.** The simplest quartic function is $P(x) = x^4$. See **FIGURE 46.** Notice that it resembles the graph of the squaring function. However, it is *not* actually a parabola.

Cubic Function Shapes

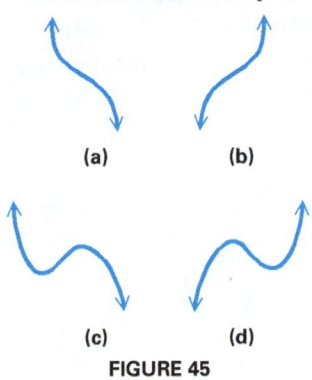

FIGURE 45

FUNCTION CAPSULE

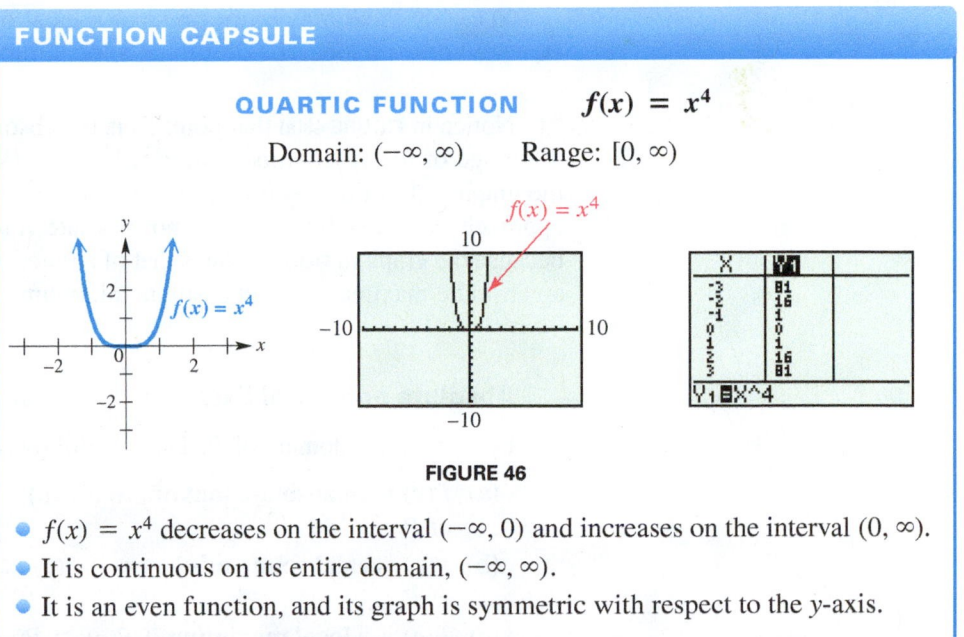

QUARTIC FUNCTION $\quad f(x) = x^4$

Domain: $(-\infty, \infty)$ Range: $[0, \infty)$

FIGURE 46

- $f(x) = x^4$ decreases on the interval $(-\infty, 0)$ and increases on the interval $(0, \infty)$.
- It is continuous on its entire domain, $(-\infty, \infty)$.
- It is an even function, and its graph is symmetric with respect to the y-axis.

Quartic Function Shapes

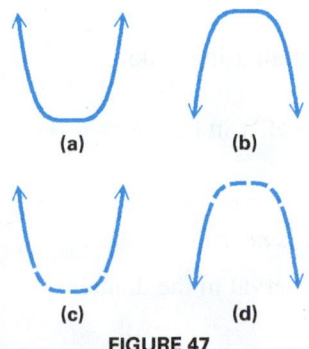

FIGURE 47

If we graph a quartic function in an appropriate window, the graph will generally resemble one of the shapes shown in **FIGURE 47.** The dashed portions in (c) and (d) indicate that there may be irregular, but smooth, behavior in those intervals.

→ **Looking Ahead to Calculus**

Suppose we need to find the x-coordinates of the two turning points of the graph of

$$f(x) = 2x^3 - 8x^2 + 9.$$

We could use the "maximum" and "minimum" capabilities of a graphing calculator and determine that, to the nearest thousandth, they are 0 and 2.667, respectively. In calculus, their exact values can be found by determining the zeros of the derivative function of $f(x)$,

$$f'(x) = 6x^2 - 16x.$$

Solving $f'(x) = 0$ would show that the two zeros are 0 and $\frac{8}{3}$, which are exact and agree with the approximations found with a graphing calculator.

Extrema

In **FIGURES 45–47** on the previous page, several graphs have **turning points** where the function changes from increasing to decreasing or vice versa. In general, the highest point at a "peak" is known as a **local maximum point,** and the lowest point at a "valley" is known as a **local minimum point.** Function values at such points are called **local maxima** (plural of *maximum*) and **local minima** (plural of *minimum*). Collectively, these values are called **extrema** (plural of *extremum*), as mentioned in **Section 3.2.**

FIGURE 48 and the accompanying chart illustrate these ideas for typical graphs.

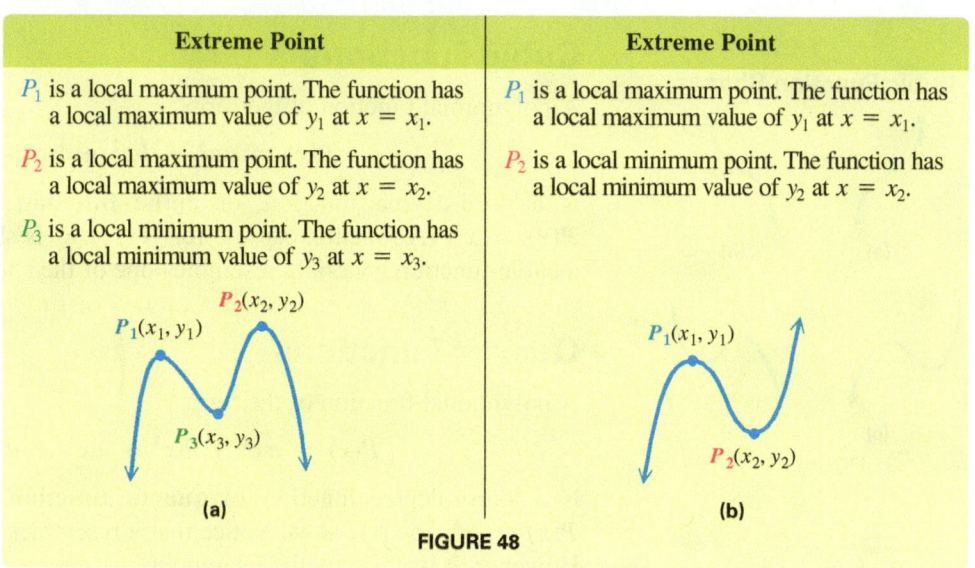

Extreme Point	Extreme Point
P_1 is a local maximum point. The function has a local maximum value of y_1 at $x = x_1$.	P_1 is a local maximum point. The function has a local maximum value of y_1 at $x = x_1$.
P_2 is a local maximum point. The function has a local maximum value of y_2 at $x = x_2$.	P_2 is a local minimum point. The function has a local minimum value of y_2 at $x = x_2$.
P_3 is a local minimum point. The function has a local minimum value of y_3 at $x = x_3$.	

FIGURE 48

NOTE A local maximum point or a local minimum point is an *ordered pair* (x, y), whereas a local maximum or a local minimum is a *y-value*.

Notice in **FIGURE 48(a)** that point P_2 is the absolute highest point on the graph, and the range of the function is $(-\infty, y_2]$. We call P_2 the **absolute maximum point** on the graph and y_2 the **absolute maximum value** of the function. Because the y-values approach $-\infty$, this function has no absolute minimum value. On the other hand, because the graph in **FIGURE 48(b)** is that of a function with range $(-\infty, \infty)$, it has neither an absolute maximum nor an absolute minimum.

Absolute and Local Extrema

Let c be in the domain of P. Then the following hold.

(a) $P(c)$ is an **absolute maximum** if $P(c) \geq P(x)$ for all x in the domain of P.

(b) $P(c)$ is an **absolute minimum** if $P(c) \leq P(x)$ for all x in the domain of P.

(c) $P(c)$ is a **local maximum** if $P(c) \geq P(x)$ when x is *near c*.

(d) $P(c)$ is a **local minimum** if $P(c) \leq P(x)$ when x is *near c*.

The expression "*near c*" means that there is an open interval in the domain of P containing c, where $P(c)$ satisfies the inequality.

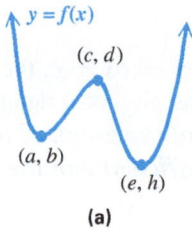

(a)

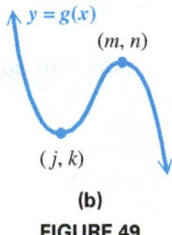

(b)

FIGURE 49

EXAMPLE 1 **Identifying Local and Absolute Extreme Points**

Consider the graphs in **FIGURE 49.**

(a) Identify and classify the local extreme points of f.

(b) Identify and classify the local extreme points of g.

(c) Describe the absolute extreme points for f and g.

Solution

(a) The points (a, b) and (e, h) are local minimum points. The point (c, d) is a local maximum point.

(b) The point (j, k) is a local minimum point and the point (m, n) is a local maximum point.

(c) The absolute minimum point of function f is (e, h), and the absolute minimum value is h, since the range of f is $[h, \infty)$. Function f has no absolute maximum value. Function g has no absolute extreme points, since its range is $(-\infty, \infty)$.

NOTE   A function may have more than one absolute maximum or minimum *point*, but only one absolute maximum or minimum *value*. (To see this, graph $f(x) = x^4 - 2x^2$. See also **FIGURE 63.**)

The graph of a polynomial function can have a maximum or a minimum point that is not apparent in a particular window. This is an example of **hidden behavior.**

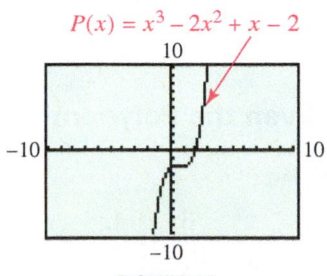

FIGURE 50

EXAMPLE 2 **Examining Hidden Behavior**

FIGURE 50 shows the graph of $P(x) = x^3 - 2x^2 + x - 2$ in the standard viewing window. Make a conjecture concerning possible hidden behavior, and verify it.

Solution   In **FIGURE 50**, because the graph levels off in the domain interval $[0, 2]$, there may be behavior that is not apparent in the given window. By changing the window to $[-2.5, 2.5]$ by $[-4.5, 0.5]$, we see that there are two extrema there. The local maximum point, as seen in **FIGURE 51**, has approximate coordinates $(0.33, -1.85)$. (There is also a local minimum point at $(1, -2)$.)

$P(x) = x^3 - 2x^2 + x - 2$

0.5

-2.5 ———————————— 2.5

Maximum
X=.33333159 Y=-1.851852

-4.5

FIGURE 51

A quadratic function has degree 2 and has only one turning point (its vertex). Extending this idea, a third-degree polynomial function has at most two turning points, a fourth-degree polynomial function has at most three turning points, and so on.

Number of Turning Points

The number of turning points of the graph of a polynomial function of degree $n \geq 1$ is at most $n - 1$.

NOTE   The above property implies that the number of local extrema is *at most* $n - 1$ for a *polynomial function* of degree $n \geq 1$. The graph may have fewer than $n - 1$ local extrema.

End Behavior

If the value of a is positive for the quadratic function $P(x) = ax^2 + bx + c$, the graph opens upward. If a is negative, the graph opens downward. The sign of a determines the **end behavior** of the graph. *In general, the end behavior of the graph of a polynomial function is determined by the sign of the leading coefficient and the parity (odd or even) of the degree.*

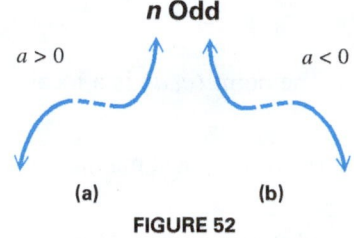

n Odd

(a)   (b)

FIGURE 52

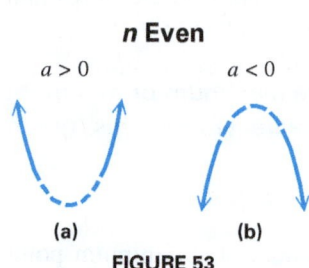

n Even

(a)   (b)

FIGURE 53

End Behavior of Graphs of Polynomial Functions

Suppose that ax^n is the dominating term of a polynomial function P of **odd degree**. (The **dominating term** is the term of greatest degree.)

1. If $a > 0$, then as $x \to \infty$, $P(x) \to \infty$, and as $x \to -\infty$, $P(x) \to -\infty$. Therefore, the end behavior of the graph is of the type shown in **FIGURE 52(a)**. We symbolize it as ↗.

2. If $a < 0$, then as $x \to \infty$, $P(x) \to -\infty$, and as $x \to -\infty$, $P(x) \to \infty$. Therefore, the end behavior of the graph is of the type shown in **FIGURE 52(b)**. We symbolize it as ↘.

Suppose that ax^n is the dominating term of a polynomial function P of **even degree**.

1. If $a > 0$, then as $|x| \to \infty$, $P(x) \to \infty$. Therefore, the end behavior of the graph is of the type shown in **FIGURE 53(a)**. We symbolize it as ⌣.

2. If $a < 0$, then as $|x| \to \infty$, $P(x) \to -\infty$. Therefore, the end behavior of the graph is of the type shown in **FIGURE 53(b)**. We symbolize it as ⌢.

EXAMPLE 3   **Determining End Behavior Given the Polynomial**

The graphs of the following functions are shown in **FIGURE 54.**

$$f(x) = x^4 - x^2 + 5x - 4, \qquad g(x) = -x^6 + x^2 - 3x - 4,$$
$$h(x) = 3x^3 - x^2 + 2x - 4, \quad \text{and} \quad k(x) = -x^7 + x - 4.$$

Based on the discussion in the preceding box, match each function with its graph.

A.
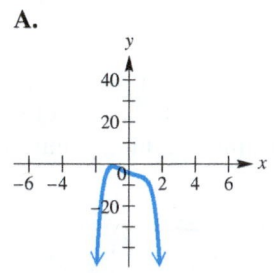

B.
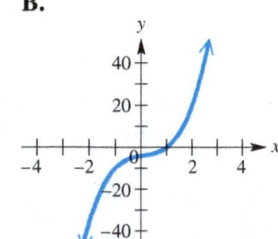

C.
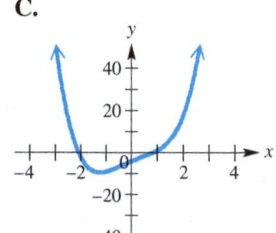

D.
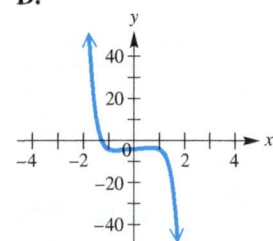

FIGURE 54

Solution   Because function f is even degree with a positive leading coefficient on its dominating term, its graph is in C.

Because function g is even degree with a negative leading coefficient on its dominating term, its graph is in A.

Because function h is odd degree with a positive leading coefficient on its dominating term, its graph is in B.

Because function k is odd degree with a negative leading coefficient on its dominating term, its graph is in D.

x-Intercepts (Real Zeros)

A nonzero linear function can have no more than one *x*-intercept, and a quadratic function can have no more than two *x*-intercepts. **FIGURE 55** shows how a cubic function may have one, two, or three *x*-intercepts. These observations suggest an important property of polynomial functions.

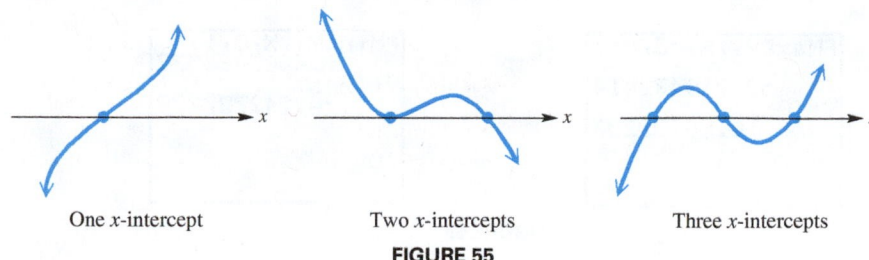

One *x*-intercept Two *x*-intercepts Three *x*-intercepts
FIGURE 55

> ### Number of *x*-Intercepts (Real Zeros) of a Polynomial Function
>
> The graph of a polynomial function of degree n will have at most n *x*-intercepts (real zeros).

$P(x) = x^3 + 5x^2 + 5x - 2$

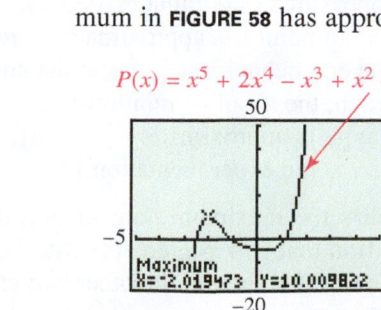

FIGURE 56

EXAMPLE 4 **Determining *x*-Intercepts Graphically**

Graphically find the *x*-intercepts of the polynomial function

$$P(x) = x^3 + 5x^2 + 5x - 2.$$

Solution The graph shown in **FIGURE 56** has three *x*-intercepts, namely $(-2, 0)$, approximately $(-3.30, 0)$ (shown in the display), and approximately $(0.30, 0)$. ●

EXAMPLE 5 **Analyzing a Polynomial Function**

Perform the following for the fifth-degree polynomial function

$$P(x) = x^5 + 2x^4 - x^3 + x^2 - x - 4.$$

(a) Determine its domain.   **(b)** Determine its range.

(c) Use its graph to find approximations of the local extreme points.

(d) Use its graph to find the approximate and/or exact *x*-intercepts.

Solution

(a) This is a polynomial function, so its domain is $(-\infty, \infty)$.

(b) This function is of odd degree. The range is $(-\infty, \infty)$.

(c) It appears that there are only two extreme points. We find that the local maximum in **FIGURE 57** has approximate coordinates $(-2.02, 10.01)$ and that the local minimum in **FIGURE 58** has approximate coordinates $(0.41, -4.24)$.

$P(x) = x^5 + 2x^4 - x^3 + x^2 - x - 4$

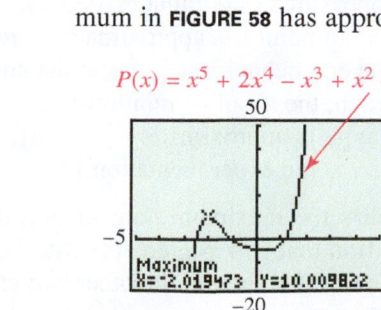

FIGURE 57

$P(x) = x^5 + 2x^4 - x^3 + x^2 - x - 4$

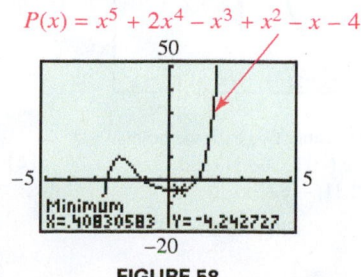

FIGURE 58

(continued)

The preceding graphical results can also be found by using a built-in utility, as shown in the two screens in **FIGURE 59.** Here Y_1 is defined by $Y_1 = P(x)$ so the x-coordinate of the local maximum point is in the interval $[-3, -1]$ while the x-coordinate of the local minimum point is in the interval $[0, 1]$.

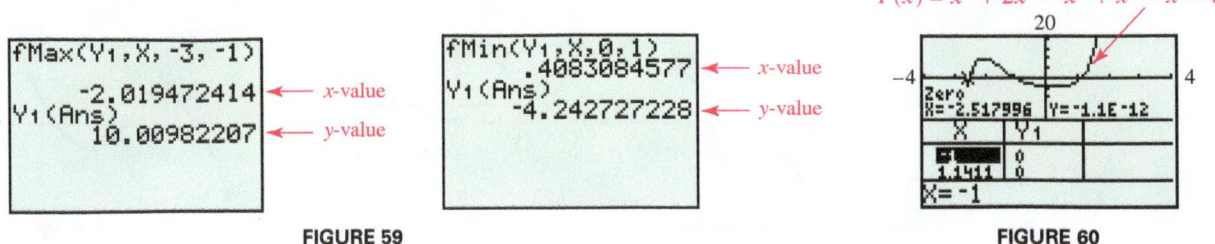

FIGURE 59   **FIGURE 60**

(d) We use calculator methods to find that the x-intercepts are $(-1, 0)$ (exact), $(1.14, 0)$ (approximate), and $(-2.52, 0)$ (approximate). See **FIGURE 60.** The first of these can be verified analytically by evaluating $P(-1)$.

$$P(-1) = (-1)^5 + 2(-1)^4 - (-1)^3 + (-1)^2 - (-1) - 4 = 0$$

This result shows again that an x-intercept of the graph of a function corresponds to a real zero of the function. This function has only three x-intercepts and thus three real zeros, which supports the fact that a polynomial function of degree n will have *at most* n x-intercepts. It may have fewer, as in this case.

In the next example, we use our knowledge of polynomials and their graphs to analyze a fourth-degree polynomial.

EXAMPLE 6   **Analyzing a Polynomial Function**

Perform the following for the fourth-degree polynomial function

$$P(x) = x^4 + 2x^3 - 15x^2 - 12x + 36.$$

(a) State the domain.

(b) Use the graph of $P(x)$ to find approximations of its local extreme points. Does it have an absolute minimum? What is the range of the function?

(c) Use its graph to find the x-intercepts.

Solution

(a) Because $P(x)$ is a polynomial function, its domain is $(-\infty, \infty)$.

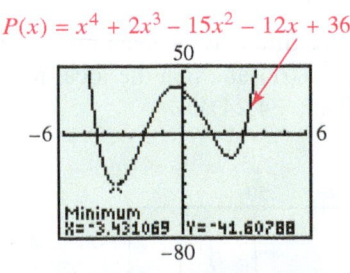

$P(x) = x^4 + 2x^3 - 15x^2 - 12x + 36$

The other two extreme points are $(-0.38, 38.31)$ and $(2.31, -18.63)$.

FIGURE 61

(b) A window of $[-6, 6]$ by $[-80, 50]$ gives the extreme points, as well as all intercepts. See **FIGURE 61.** Using a calculator, we find that the two local minimum points have approximate coordinates $(-3.43, -41.61)$ and $(2.31, -18.63)$, and the local maximum point has approximate coordinates $(-0.38, 38.31)$.

Because the end behavior is ⌣ and the point $(-3.43, -41.61)$ is the lowest point on the graph, the absolute minimum value of the function is approximately -41.61. The range is approximately $[-41.61, \infty)$. (Finding an appropriate window may require some experimentation.)

(c) This function has the maximum number of x-intercepts possible, four. Using a calculator, we find that two x-intercepts with exact values are $(-2, 0)$ and $(3, 0)$, while, to the nearest hundredth, the other two are $(-4.37, 0)$ and $(1.37, 0)$.

WHAT WENT WRONG?

A student graphed $y = 0.045x^4 - 2x^2 + 2$ in the decimal window of the TI-84 Plus calculator. Because the polynomial has even degree and positive leading coefficient, she expected to find end behavior ⌣⌣. However, this window indicates ⌢⌢ as end behavior.

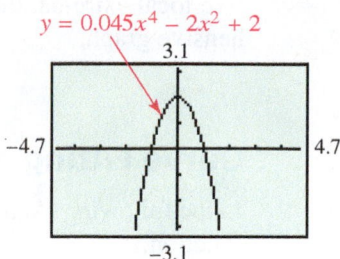

$y = 0.045x^4 - 2x^2 + 2$

What Went Wrong?   Is there a way to graph this function so that the correct end behavior is apparent?

Comprehensive Graphs

The most important features of the graph of a polynomial function are its intercepts, extrema, and end behavior. For this reason, *a comprehensive graph of a polynomial function will exhibit the following features.*

1. All x-intercepts (if any)
2. The y-intercept
3. All extreme points (if any)
4. Enough of the graph to reveal the correct end behavior

EXAMPLE 7   **Determining an Appropriate Window**

Is the graph of $P(x) = x^6 - 36x^4 + 288x^2 - 256$ in **FIGURE 62** a comprehensive graph? If not, provide one.

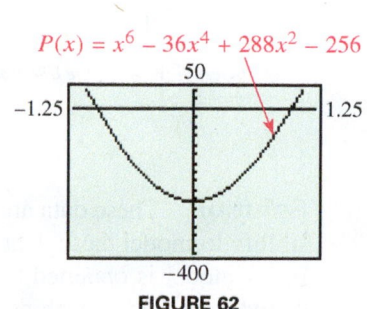

$P(x) = x^6 - 36x^4 + 288x^2 - 256$

FIGURE 62

(continued)

Answer to What Went Wrong?

The window must be enlarged. Scrolling through a table should indicate a suitable window. One example is $[-10, 10]$ by $[-25, 10]$.

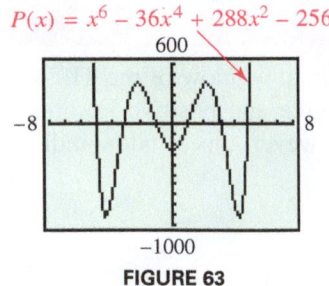

$P(x) = x^6 - 36x^4 + 288x^2 - 256$

FIGURE 63

Solution Since the function is of even degree and the dominating term has positive coefficient, the end behavior seems to be correct. The y-intercept, $(0, -256)$, is shown, as are two x-intercepts and one local minimum. This function may, however, have up to six x-intercepts, since it is of degree 6.

By experimenting with other viewing windows, we see that a window of $[-8, 8]$ by $[-1000, 600]$ shows a total of five local extrema, and four more x-intercepts that were not apparent in the earlier figure. See **FIGURE 63.** Since there can be no more than five local extrema, this second view (*not* the first view in **FIGURE 62**) gives a comprehensive graph. ●

Curve Fitting and Polynomial Models

In the following example we model temperatures at Daytona Beach with a polynomial function.

EXAMPLE 8 **Modeling Daytona Beach Temperatures**

Monthly average high temperatures at Daytona Beach are shown in the table. In this table the months have been given the standard numbering. Plot the data and determine a polynomial function to model the data. Use your function to determine the average high temperature in April and compare with the value given in the table.

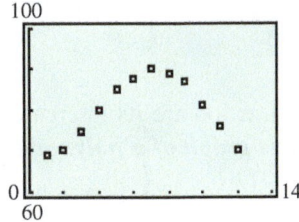

Temperature data

(a)

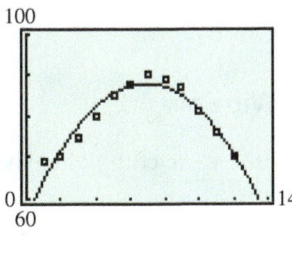

Quadratic model

(b)

Month	Temperature (°F)
1	69
2	70
3	75
4	80
5	85
6	88
7	90
8	89
9	87
10	81
11	76
12	70

Source: Williams, J., *The USA Weather Almanac.*

$y = P(x)$

Quartic model

(c)

FIGURE 64

Solution These data are plotted in **FIGURE 64(a)** and are clearly *not* linear. One possibility to model these data is to use a quadratic function, as shown in **FIGURE 64(b)**. A better model is obtained by using a fourth-degree polynomial function, as shown in **FIGURE 64(c)**. This fourth-degree polynomial can be found by using quartic regression on a calculator, and its formula is given by

$$P(x) \approx 0.0145x^4 - 0.426x^3 + 3.53x^2 - 6.23x + 72.$$

$P(4) \approx 80.0$ indicates that this model predicts the monthly average high temperature in Daytona to be about 80°F in April, which agrees with the given table of values. ●

3.5 Exercises

Concept Check   *Use the graphs shown here, which include all extrema, for Exercises 1–8.*

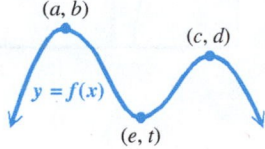
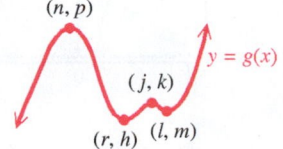

1. Use the extrema to determine the minimum degree of f.

2. Use the extrema to determine the minimum degree of g.

3. Give all local extreme points of f. Tell whether each is a maximum or minimum.

4. Give all local extreme points of g. Tell whether each is a maximum or minimum.

5. Describe all absolute extreme points of f.

6. Describe all absolute extreme points of g.

7. Give the local and absolute extreme values of f.

8. Give the local and absolute extreme values of g.

Checking Analytic Skills   *Use an end behavior diagram* ⌣, ⌢, ⤵, *or* ⤴ *to describe the end behavior of the graph of each function.* **Do not use a calculator.**

9. $P(x) = \sqrt{5}x^3 + 2x^2 - 3x + 4$

10. $P(x) = -\sqrt{7}x^3 - 4x^2 + 2x - 1$

11. $P(x) = -\pi x^5 + 3x^2 - 1$

12. $P(x) = \pi x^7 - x^5 + x - 1$

13. $P(x) = 2.74x^4 - 3x^2 + x - 2$

14. $P(x) = \sqrt{6}x^6 - x^5 + 2x - 2$

15. $P(x) = x^5 - x^4 - \pi x^6 - x + 3$

16. $P(x) = -x - 3.2x^3 + x^2 - 2.84x^4$

17. $P(x) = x^{10,000}$

18. $P(x) = -x^{104,266}$

19. $P(x) = -3x^{15,297}$

20. $P(x) = 12x^{107,499}$

Give a short written answer in Exercises 21–24.

21. The graphs of $f(x) = x^n$ for $n = 3, 5, 7, \ldots$ resemble each other. As n gets larger, what happens to the graph?

22. Repeat **Exercise 21** for $f(x) = x^n$, where $n = 2, 4, 6, \ldots$.

23. Using a window of $[-1, 1]$ by $[-1, 1]$, graph the odd-degree polynomial functions

$$y = x, \quad y = x^3, \quad \text{and} \quad y = x^5.$$

Describe the behavior of these functions relative to each other. Predict the behavior of the graph of $y = x^7$ in the same window, and then graph it to support your prediction.

24. Repeat **Exercise 23** for the even-degree polynomial functions

$$y = x^2, \quad y = x^4, \quad \text{and} \quad y = x^6.$$

Predict the behavior of the graph of $y = x^8$ in the same window, and then graph it to support your prediction.

Each function is graphed in a window that results in hidden behavior. Experiment with various windows to locate the extreme points on the graph of the function. Round values to the nearest hundredth.

25. $y = \frac{1}{3}x^3 - \frac{5}{2}x^2 + 6x - 1$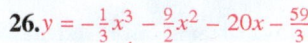

26. $y = -\frac{1}{3}x^3 - \frac{9}{2}x^2 - 20x - \frac{59}{3}$

27. $y = -x^3 - 11x^2 - 40x - 50$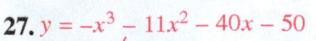

28. $y = 2x^3 - 3.3x^2 + 1.8x + 3$

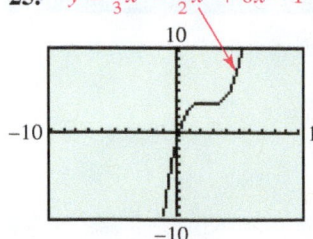
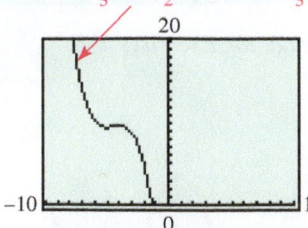
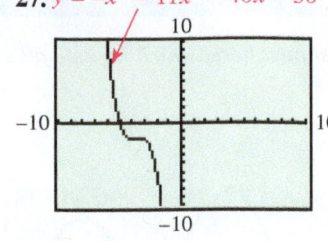
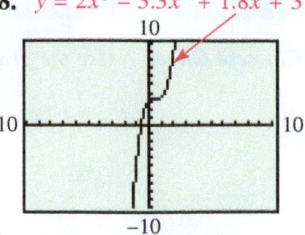

In Exercises 29–32, it is not apparent from the standard viewing window whether the graph of the quadratic function intersects the x-axis once, twice, or not at all. Experiment with various windows to find the number of x-intercepts. If there are x-intercepts, give their coordinates to the nearest hundredth.

29. $y = x^2 - 4.25x + 4.515$

30. $y = x^2 + 6.95x + 12.07$

31. $y = -x^2 + 6.5x - 10.60$

32. $y = -2x^2 + 0.2x - 0.15$

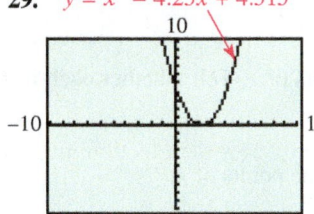
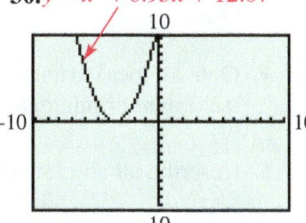
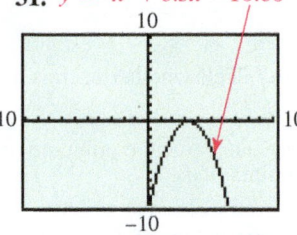
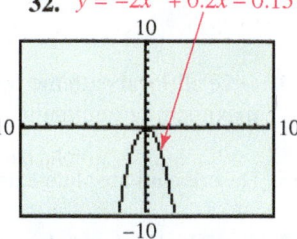

Without using a calculator, match each function in Exercises 33–36 with the correct graph in choices A–D.

33. $f(x) = 2x^3 + x^2 - x + 3$

34. $g(x) = -2x^3 - x + 3$

35. $h(x) = -2x^4 + x^3 - 2x^2 + x + 3$

36. $k(x) = 2x^4 - x^3 - 2x^2 + 3x + 3$

A.
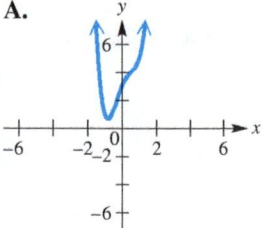

B.
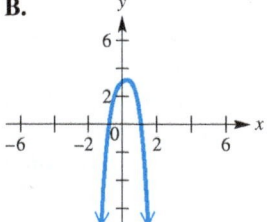

C.
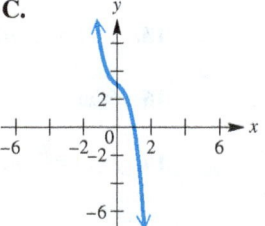

D.
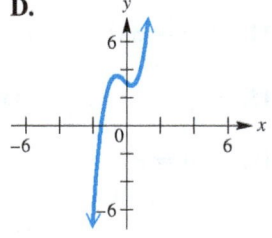

Concept Check *Without graphing, answer* true *or* false *to each statement. Then, support your answer by graphing.*

37. The function $f(x) = x^3 + 2x^2 - 4x + 3$ has four real zeros.

38. The function $f(x) = x^3 + 3x^2 + 3x + 1$ must have at least one real zero.

39. If a polynomial function of even degree has a negative leading coefficient and a positive y-value for its y-intercept, it must have at least two real zeros.

40. The function $f(x) = 3x^4 + 5$ has no real zeros.

41. The function $f(x) = -3x^4 + 5$ has two real zeros.

42. The graph of $f(x) = x^3 - 3x^2 + 3x - 1 = (x - 1)^3$ has exactly one x-intercept.

43. A fifth-degree polynomial function cannot have a single real zero.

44. An even-degree polynomial function must have at least one real zero.

Concept Check The graphs below show

$$y = x^3 - 3x^2 - 6x + 8,\qquad y = x^4 + 7x^3 - 5x^2 - 75x,$$

$$y = -x^3 + 9x^2 - 27x + 17,\quad \text{and}\quad y = -x^5 + 36x^3 - 22x^2 - 147x - 90,$$

but not necessarily in that order. Assuming that each is a comprehensive graph, answer each question in Exercises 45–54.

A. 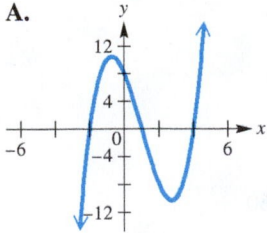   **B.** 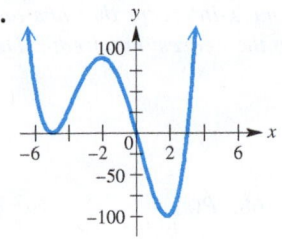   **C.** 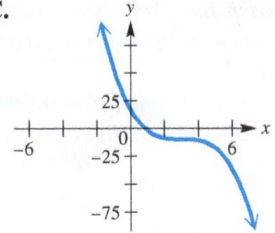   **D.**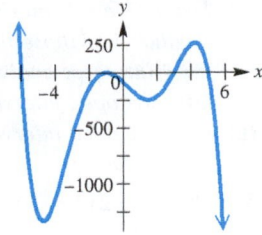

45. Which graph is that of $y = x^3 - 3x^2 - 6x + 8$?

46. Which graph is that of $y = x^4 + 7x^3 - 5x^2 - 75x$?

47. How many real zeros does the graph in C have?

48. The graph of $y = -x^3 + 9x^2 - 27x + 17$ is either C or D. Which is it?

49. Which of the graphs cannot be that of a cubic polynomial function?

50. How many positive real zeros does the function graphed in D have?

51. How many negative real zeros does the function graphed in A have?

52. Is the absolute minimum value of the function graphed in B a positive number or a negative number?

53. Which one of the graphs is that of a function whose range is *not* $(-\infty, \infty)$?

54. One of the following is an approximation for the local maximum point of the graph in A. Which one is it?
 A. $(0.73, 10.39)$ **B.** $(-0.73, 10.39)$
 C. $(-0.73, -10.39)$ **D.** $(0.73, -10.39)$

RELATING CONCEPTS   **For individual or group investigation (Exercises 55–58)**

The concepts of stretching, shrinking, translating, and reflecting graphs presented in **Sections 2.2** and **2.3** can be applied to polynomial functions of the form $P(x) = x^n$. For example, the graph of $y = -2(x + 4)^4 - 6$ can be obtained from the graph of $y = x^4$ by shifting 4 units to the left, stretching vertically by applying a factor of 2, reflecting across the x-axis, and shifting downward 6 units, so the graph should resemble the graph at the right.

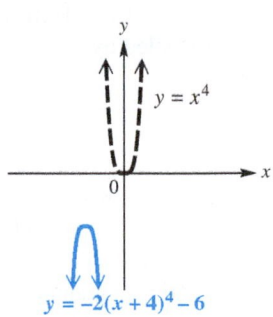

If we expand the expression $-2(x + 4)^4 - 6$ algebraically, we get

$$-2x^4 - 32x^3 - 192x^2 - 512x - 518.$$

Thus, the graph of $y = -2(x + 4)^4 - 6$ is the same as that of

$$y = -2x^4 - 32x^3 - 192x^2 - 512x - 518.$$

In Exercises 55–58, two forms of the same polynomial function are given. Sketch by hand the general shape of the graph of the function, using the concepts of **Chapter 2**, and describe the transformations. Then, support your answer by graphing it on your calculator in a suitable window.

55. $y = 2(x + 3)^4 - 7$
  $y = 2x^4 + 24x^3 + 108x^2 + 216x + 155$

56. $y = -3(x + 1)^4 + 12$
  $y = -3x^4 - 12x^3 - 18x^2 - 12x + 9$

57. $y = -3(x - 1)^3 + 12$
  $y = -3x^3 + 9x^2 - 9x + 15$

58. $y = 0.5(x - 1)^5 + 13$
  $y = 0.5x^5 - 2.5x^4 + 5x^3 - 5x^2 + 2.5x + 12.5$

For the functions in Exercises 59–66, use your graphing calculator to find a comprehensive graph and answer each of the following.

(a) *Determine the domain.*

(b) *Determine all local minimum points, and tell if any is an absolute minimum point. (Approximate coordinates to the nearest hundredth.)*

(c) *Determine all local maximum points, and tell if any is an absolute maximum point. (Approximate coordinates to the nearest hundredth.)*

(d) *Determine the range. (If an approximation is necessary, give it to the nearest hundredth.)*

(e) *Determine all intercepts. For each function, there is at least one x-intercept that has an integer x-value. For those that are not integers, give approximations to the nearest hundredth. Determine the y-intercept analytically.*

(f) *Give the open interval(s) over which the function is increasing.*

(g) *Give the open interval(s) over which the function is decreasing.*

59. $P(x) = -2x^3 - 14x^2 + 2x + 84$

60. $P(x) = -3x^3 + 6x^2 + 39x - 60$

61. $P(x) = x^5 + 4x^4 - 3x^3 - 17x^2 + 6x + 9$

62. $P(x) = -2x^5 + 7x^4 + x^3 - 20x^2 + 4x + 16$

63. $P(x) = 2x^4 + 3x^3 - 17x^2 - 6x - 72$

64. $P(x) = 3x^4 - 33x^2 + 54$

65. $P(x) = -x^6 + 24x^4 - 144x^2 + 256$

66. $P(x) = -3x^6 + 2x^5 + 9x^4 - 8x^3 + 11x^2 + 4$

Determine a window that will provide a comprehensive graph of each polynomial function. (In each case, there are many possible such windows.)

67. $P(x) = 4x^5 - x^3 + x^2 + 3x - 16$

68. $P(x) = 3x^5 - x^4 + 12x^2 - 25$

69. $P(x) = 2.9x^3 - 37x^2 + 28x - 143$

70. $P(x) = -5.9x^3 + 16x^2 - 120$

71. $P(x) = \pi x^4 - 13x^2 + 84$

72. $P(x) = 2\pi x^4 - 12x^2 + 100$

(Modeling) Solve each problem.

73. *Average High Temperatures.* The monthly average high temperatures in degrees Fahrenheit at Daytona Beach can be modeled by

$$P(x) = 0.0145x^4 - 0.426x^3 + 3.53x^2 - 6.23x + 72,$$

where $x = 1$ corresponds to January and $x = 12$ represents December.

(a) Find the average high temperature during March and July.

(b) Estimate graphically and numerically the months when the average high temperature is about 80°F.

74. *Heating Costs* In colder climates the cost for natural gas to heat homes can vary from one month to the next. The polynomial function

$$f(x) = -0.1213x^4 + 3.462x^3 - 29.22x^2 + 64.68x + 97.69$$

models the monthly cost in dollars of heating a typical home. The input x represents the month, where $x = 1$ corresponds to January and $x = 12$ to December. (*Source:* Minnegasco.)

(a) Where might the absolute extrema occur for $1 \le x \le 12$?

(b) Approximate the absolute extrema and interpret the results.

1. *Dimensions of a Garden*   An ecology center wants to set up an experimental garden using 300 meters of fencing to enclose a rectangular area of 5000 square meters.

(a) Let x meters represent the width of the garden. Why must the length be $150 - x$ meters?

(b) Write an expression $\mathcal{A}(x)$ that represents the area of the garden.

(c) What are the restrictions on x?

(d) Use the given values for area and length of fencing to find the dimensions of the garden.

2. *Research Funding*   The table gives National Science Foundation funding at the beginning of each decade since 1951. In the table, years are given as the number of years since 1900. Thus, $x = 51$ corresponds to 1951, $x = 61$ corresponds to 1961, and so on.

Year x	51	61	71	81	91	101
Dollars y (in billions)	0.1	0.2	0.6	1.1	2.3	4.7

Source: U.S. National Science Foundation.

(a) Plot the data.

(b) Find a function $f(x) = a(x - h)^2 + k$ that models the data by using $(51, 0.1)$ as the vertex and $(101, 4.7)$ as the second point.

(c) Use the quadratic regression feature of a graphing calculator to determine the quadratic function g that best fits the data.

(d) Plot f and g in the same window as the data. Discuss how well the models fit the data.

For $P(x) = 2x^3 - 9x^2 + 4x + 15$, answer the following.

3. Predict the end behavior.

4. What is the maximum number of extrema the graph of P can have? What is the maximum number of zeros?

For $P(x) = x^4 + 4x^3 - 20$, answer the following.

5. Predict the end behavior.

6. Give a comprehensive graph of function P.

7. Find all extreme points. Tell whether each one is a maximum or minimum and a local or absolute extremum.

8. Find all intercepts. Approximate values to the nearest hundredth.

3.6 Topics in the Theory of Polynomial Functions (I)

Intermediate Value Theorem • Division of Polynomials by $x - k$ and Synthetic Division • Remainder and Factor Theorems • Division of Any Two Polynomials

The topics in this section and the next complement the graphical work done in **Section 3.5** and pave the way for the work in **Section 3.8** on equations, inequalities, and applications of polynomial functions.

Intermediate Value Theorem

The intermediate value theorem applies to the zeros of every polynomial function with *real coefficients*. It uses the fact that graphs of polynomial functions are continuous curves, with no gaps or sudden jumps. The proof requires advanced methods.

> ### Intermediate Value Theorem
>
> If $P(x)$ defines a polynomial function with only real coefficients, and if, for real numbers a and b, the values $P(a)$ and $P(b)$ are opposite in sign, then there exists at least one real zero between a and b.

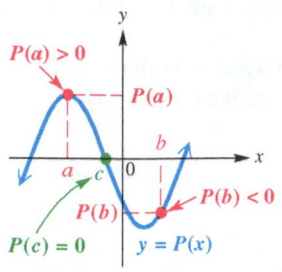

FIGURE 65

To see how the intermediate value theorem is applied, note that, in **FIGURE 65,** $P(a)$ and $P(b)$ are opposite in sign, so 0 is between $P(a)$ and $P(b)$. Then, by this theorem, there must be a number c in $[a, b]$ such that $P(c) = 0$. That is, if the graph of a polynomial $P(x)$ is above the x-axis at one point and below the x-axis at another point, then the graph of P must cross the x-axis at least once between the two points.

A simple example of the intermediate value theorem involves temperature. Suppose that at 2 A.M. the outside temperature was $-5°F$, and at 6 A.M. the outside temperature was $4°F$. Then the outside temperature must have been $0°F$ at least once between 2 A.M. and 6 A.M.

EXAMPLE 1 **Applying the Intermediate Value Theorem**

Show that the polynomial function $P(x) = x^3 - 2x^2 - x + 1$ has a real zero between 2 and 3.

Analytic Solution

Using

$$P(x) = x^3 - 2x^2 - x + 1,$$

we evaluate $P(2)$ and $P(3)$.

$P(2) = 2^3 - 2(2)^2 - 2 + 1$ Substitute.

$\quad\quad = -1$ Simplify.

$P(3) = 3^3 - 2(3)^2 - 3 + 1$ Substitute.

$\quad\quad = 7$ Simplify.

Since $P(2) = -1$ and $P(3) = 7$ differ in sign, the intermediate value theorem assures us that there is a real zero between 2 and 3.

Graphing Calculator Solution

The graph of $P(x) = x^3 - 2x^2 - x + 1$ in **FIGURE 66** shows that there is an x-intercept between $(2, 0)$ and $(3, 0)$, confirming our analytic result that there is a zero between 2 and 3. Using the table, we see that the zero lies between 2.246 and 2.247, since there is a sign change in the function values there.

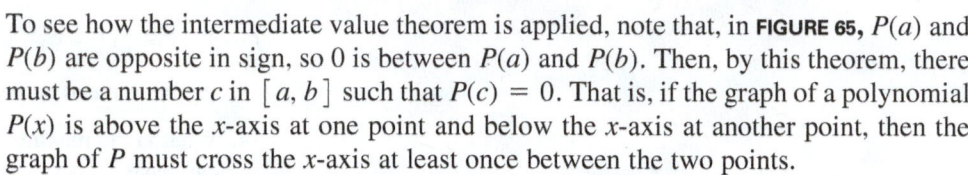
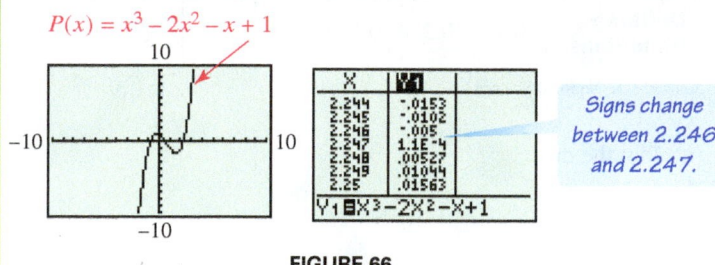

FIGURE 66

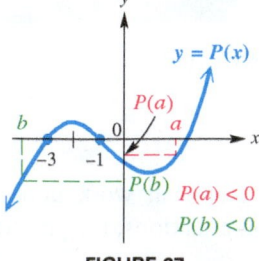

FIGURE 67

> **NOTE** In **Example 1** the intermediate value theorem does not tell us what the value of the zero is. It only tells us that there is at least one zero between $x = 2$ and $x = 3$.

> **CAUTION** *Be careful how you interpret the intermediate value theorem.* If $P(a)$ and $P(b)$ are *not* opposite in sign, it does not necessarily mean that there is no zero between a and b. For example, in **FIGURE 67**, $P(a)$ and $P(b)$ are both negative, but -3 and -1, which are between a and b, are zeros of P.

Division of Polynomials by $x - k$ and Synthetic Division

We can use long division to determine whether one whole number is a factor of another, and to determine whether one polynomial is a factor of another.

EXAMPLE 2   **Dividing a Polynomial by a Binomial**

Divide $3x^3 - 2x + 5$ by $x - 3$. Determine the quotient and the remainder.

Solution   The powers of the variable in the dividend $(3x^3 - 2x + 5)$ must be descending, which they are. Insert the term $0x^2$ to act as a placeholder.

$$\begin{array}{r} \text{Missing term} \\ \downarrow \\ x - 3\overline{)3x^3 + 0x^2 - 2x + 5} \end{array}$$

Divide as with whole numbers. Start with $\frac{3x^3}{x} = 3x^2$.

←——— Algebra Review
To review multiplying polynomials, see **Section R.1.**

To subtract, think $3x^3 - 3x^3$ *and* $0x^2 - (-9x^2).$

$$\begin{array}{r} 3x^2 \\ x - 3\overline{)3x^3 + 0x^2 - 2x + 5} \\ \underline{3x^3 - 9x^2} \\ 9x^2 \end{array}$$

←——— $\frac{3x^3}{x} = 3x^2$

←——— $3x^2(x - 3)$

←——— Subtract.

Bring down the next term.

$$\begin{array}{r} 3x^2 \\ x - 3\overline{)3x^3 + 0x^2 - 2x + 5} \\ \underline{3x^3 - 9x^2} \\ 9x^2 - 2x \end{array}$$

←——— Bring down $-2x$.

In the next step, divide: $\frac{9x^2}{x} = 9x$.

$$\begin{array}{r} 3x^2 + 9x \\ x - 3\overline{)3x^3 + 0x^2 - 2x + 5} \\ \underline{3x^3 - 9x^2} \\ 9x^2 - 2x \\ \underline{9x^2 - 27x} \\ 25x + 5 \end{array}$$

←——— $\frac{9x^2}{x} = 9x$

←——— $9x(x - 3)$

←——— Subtract and bring down 5.

Divide: $\frac{25x}{x} = 25$.

$$\begin{array}{r} 3x^2 + 9x + 25 \\ x - 3\overline{)3x^3 + 0x^2 - 2x + 5} \\ \underline{3x^3 - 9x^2} \\ 9x^2 - 2x \\ \underline{9x^2 - 27x} \\ 25x + 5 \\ \underline{25x - 75} \\ 80 \end{array}$$

←——— $\frac{25x}{x} = 25$

←——— $25(x - 3)$

←——— Subtract.

The quotient is $3x^2 + 9x + 25$ with a remainder of 80.

In **Example 2,** we divided a cubic polynomial (degree 3) by a linear polynomial (degree 1) and obtained a quadratic polynomial quotient (degree 2). Notice that $3 - 1 = 2$, so the degree of the quotient polynomial is found by subtracting the degree of the divisor from the degree of the dividend. Also, since the remainder is a nonzero constant 80, we can write it as the numerator of a fraction with denominator $x - 3$ to express the fractional part of the quotient.

$$\underset{\text{Divisor}}{\underset{\uparrow}{\underset{\text{Dividend}}{\uparrow}}}\ \frac{3x^3 - 2x + 5}{x - 3} = \underbrace{3x^2 + 9x + 25}_{\substack{\text{Quotient} \\ \text{polynomial}}} + \underbrace{\frac{80}{x - 3}}_{\substack{\text{Fractional part} \\ \text{of the quotient}}} \ \begin{matrix} \leftarrow \text{Remainder} \\ \leftarrow \text{Divisor} \end{matrix}$$

The following rules apply when dividing a polynomial by a binomial of the form $x - k$.

Division of a Polynomial $P(x)$ by $x - k$

1. If the degree n polynomial $P(x)$ (where $n \geq 1$) is divided by $x - k$, then the quotient polynomial, $Q(x)$, has degree $n - 1$.

2. The remainder R is a constant (and may be 0). The complete quotient for $\frac{P(x)}{x - k}$ may be written as

$$\frac{P(x)}{x - k} = Q(x) + \frac{R}{x - k}.$$

Long division of a polynomial by a binomial of the form $x - k$ can be condensed. Using the division performed in **Example 2,** observe the following.

$$
\begin{array}{r}
3x^2 + 9x + 25 \\
x - 3 \overline{)3x^3 + 0x^2 - 2x + 5} \\
\underline{3x^3 - 9x^2} \\
9x^2 - 2x \\
\underline{9x^2 - 27x} \\
25x + 5 \\
\underline{25x - 75} \\
80
\end{array}
\qquad
\begin{array}{r}
3 \quad\ 9 \quad 25 \\
1 - 3 \overline{)3 \quad\ 0 \quad -2 \quad 5} \\
3 \quad -9 \\
9 \quad -2 \\
9 \quad -27 \\
25 \quad\ 5 \\
25 \quad -75 \\
80
\end{array}
$$

On the right, exactly the same division is shown without the variables. All the numbers in color on the right are repetitions of the numbers directly above them, so they can be omitted, as shown below on the left. Since the coefficient of x in the divisor is always 1, it can be omitted, too.

$$
\begin{array}{r}
3 \quad\ 9 \quad 25 \\
-3 \overline{)3 \quad\ 0 \quad -2 \quad 5} \\
\underline{-9} \\
9 \quad -2 \\
\underline{-27} \\
25 \quad\ 5 \\
\underline{-75} \\
80
\end{array}
\qquad
\begin{array}{r}
3 \quad\ 9 \quad 25 \\
-3 \overline{)3 \quad\ 0 \quad -2 \quad 5} \\
\underline{-9} \\
9 \\
\underline{-27} \\
25 \\
\underline{-75} \\
80
\end{array}
$$

The numbers in color on the left are again repetitions of the numbers directly above them. They may be omitted, as shown on the right.

Now the problem can be condensed. If the 3 in the dividend is brought down to the beginning of the bottom row, the top row can be omitted, since it duplicates the bottom row.

$$
\begin{array}{r|rrrr}
-3) & 3 & 0 & -2 & 5 \\
 & & -9 & -27 & -75 \\
\hline
 & 3 & 9 & 25 & 80
\end{array}
$$

To simplify the arithmetic, we replace subtraction in the second row by addition and compensate by changing the -3 at the upper left to its additive inverse, 3.

Additive inverse →

$$
\begin{array}{r|rrrr}
3) & 3 & 0 & -2 & 5 \\
 & & 9 & 27 & 75 \\
\hline
 & 3 & 9 & 25 & 80
\end{array}
$$

← Signs changed

Quotient → $3x^2 + 9x + 25 + \dfrac{80}{x-3}$ ← Remainder

This abbreviated form of long division of polynomials is called synthetic division.

EXAMPLE 3   **Using Synthetic Division**

Use synthetic division to divide $5x^3 - 6x^2 - 28x + 8$ by $x + 2$.

Solution   Express $x + 2$ in the form $x - k$ by writing it as $x - (-2)$.

$x + 2$ leads to -2. →
$$
\begin{array}{r|rrrr}
-2) & 5 & -6 & -28 & 8
\end{array}
$$
← Coefficients of the polynomial

Bring down the 5, and multiply: $-2(5) = -10$.

$$
\begin{array}{r|rrrr}
-2) & 5 & -6 & -28 & 8 \\
 & & -10 & & \\
\hline
 & 5 & & &
\end{array}
$$

Add -6 and -10 to obtain -16. Multiply: $-2(-16) = 32$.

$$
\begin{array}{r|rrrr}
-2) & 5 & -6 & -28 & 8 \\
 & & -10 & 32 & \\
\hline
 & 5 & -16 & &
\end{array}
$$

Add -28 and 32, obtaining 4. Finally, $-2(4) = -8$.

$$
\begin{array}{r|rrrr}
-2) & 5 & -6 & -28 & 8 \\
 & & -10 & 32 & -8 \\
\hline
 & 5 & -16 & 4 &
\end{array}
$$

Add 8 and -8 to obtain 0.

$$
\begin{array}{r|rrrr}
-2) & 5 & -6 & -28 & 8 \\
 & & -10 & 32 & -8 \\
\hline
 & 5 & -16 & 4 & 0
\end{array}
$$
← Remainder

Quotient

Since the divisor $x - k$ has degree 1, the degree of the quotient will be one less than the degree of the dividend.

$$
\frac{5x^3 - 6x^2 - 28x + 8}{x + 2} = 5x^2 - 16x + 4
$$

Notice that the divisor $x + 2$ is a *factor* of $5x^3 - 6x^2 - 28x + 8$ because the remainder is 0, so $5x^3 - 6x^2 - 28x + 8 = (x + 2)(5x^2 - 16x + 4)$.   ●

Remainder and Factor Theorems

In **Example 2,** we divided $3x^3 - 2x + 5$ by $x - 3$ and obtained a remainder of 80. If we evaluate $P(x) = 3x^3 - 2x + 5$ at $x = 3$, we have the following.

$$P(3) = 3(3)^3 - 2(3) + 5$$
$$= 81 - 6 + 5$$
$$= 80$$

Notice that the remainder is equal to $P(3)$. In **Example 3,** we divided the polynomial $5x^3 - 6x^2 - 28x + 8$ by $x - (-2)$ and obtained a remainder of 0. We now evaluate $P(x) = 5x^3 - 6x^2 - 28x + 8$ at $x = -2$.

Use parentheses around substituted values to avoid errors.

$$P(-2) = 5(-2)^3 - 6(-2)^2 - 28(-2) + 8$$
$$= -40 - 24 + 56 + 8$$
$$= 0$$

The remainder is equal to $P(-2)$. These examples illustrate the *remainder theorem*.

Remainder Theorem

If a polynomial $P(x)$ is divided by $x - k$, the remainder is equal to $P(k)$.

EXAMPLE 4 **Using the Remainder Theorem**

Use the remainder theorem and synthetic division to find $P(-2)$ if

$$P(x) = -x^4 + 3x^2 - 4x - 5.$$

Analytic Solution

Use synthetic division to find the remainder when $P(x)$ is divided by $x - (-2)$.

Remember to insert 0 for the missing x^3-term.

$$-2 \overline{)\ \begin{array}{rrrrr} -1 & 0 & 3 & -4 & -5 \\ & 2 & -4 & 2 & 4 \\ \hline -1 & 2 & -1 & -2 & -1 \end{array}}$$

Remainder

Since the remainder is -1, $P(-2) = -1$ by the remainder theorem.

Graphing Calculator Solution

The graph of $P(x) = -x^4 + 3x^2 - 4x - 5$ in **FIGURE 68** indicates that the point $(-2, -1)$ lies on the graph, so $P(-2) = -1$. Alternatively, the table in **FIGURE 68** shows that $P(-2) = -1$.

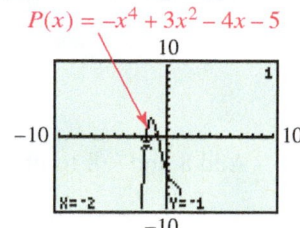
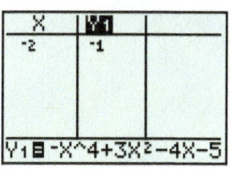

FIGURE 68

EXAMPLE 5 **Deciding whether a Number Is a Zero of a Polynomial Function**

Decide whether the given number is a zero of the function $P(x)$.

(a) 2; $P(x) = x^3 - 4x^2 + 9x - 10$

(b) -2; $P(x) = \frac{3}{2}x^3 - x^2 + \frac{3}{2}x$

Analytic Solution

(a) Proposed → $2)\overline{1 \quad -4 \quad 9 \quad -10}$ Use synthetic division.
zero

$$\begin{array}{r} \underline{2 \quad -4 \quad 10} \\ 1 \quad -2 \quad 5 \quad \quad 0 \leftarrow \text{Remainder} \end{array}$$

Since the remainder is 0, $P(2) = 0$, and 2 is a zero of the given polynomial function.

(b) Proposed → $-2)\overline{\frac{3}{2} \quad -1 \quad \frac{3}{2} \quad 0}$ Use 0 for the miss-
zero   ing constant term.

$$\begin{array}{r} \underline{-3 \quad 8 \quad -19} \\ \frac{3}{2} \quad -4 \quad \frac{19}{2} \quad -19 \leftarrow \text{Remainder} \end{array}$$

The remainder is not 0, so -2 is not a zero of P. In fact, $P(-2) = -19$. From this, we know that the point $(-2, -19)$ lies on the graph of P.

Graphing Calculator Solution

The first screen in **FIGURE 69** shows the polynomials entered as Y_1 and Y_2. The second screen shows the results of finding $Y_1(2)$ and $Y_2(-2)$ for parts (a) and (b).

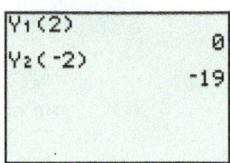

FIGURE 69

In **Example 5(a),** we showed that 2 is a zero of the polynomial function defined by $P(x) = x^3 - 4x^2 + 9x - 10$. The first three numbers in the bottom row of the synthetic division give the coefficients of the quotient polynomial. Thus,

$$\frac{P(x)}{x - 2} = x^2 - 2x + 5$$

and $P(x) = (x - 2)(x^2 - 2x + 5),$ Multiply by $x - 2$.

indicating that $x - 2$ is a *factor* of $P(x)$.

By the remainder theorem, if $P(k) = 0$, then the remainder when $P(x)$ is divided by $x - k$ is 0. This means that $x - k$ is a factor of $P(x)$. Conversely, if $x - k$ is a factor of $P(x)$, then $P(k)$ must equal 0. This is summarized in the *factor theorem*.

Factor Theorem

A polynomial $P(x)$ has a factor $x - k$ if and only if $P(k) = 0$.

EXAMPLE 6   **Using the Factor Theorem**

Determine whether the second polynomial is a factor of $P(x)$.

(a) $P(x) = 4x^3 + 24x^2 + 48x + 32;$ $x + 2$

(b) $P(x) = 2x^4 + 3x^2 - 5x + 7;$ $x - 1$

Solution

(a) $-2)\overline{4 \quad 24 \quad 48 \quad 32}$ Use synthetic division to divide
                                   $P(x)$ by $x + 2 = x - (-2)$.

$$\begin{array}{r} \underline{-8 \quad -32 \quad -32} \\ 4 \quad 16 \quad 16 \quad \quad 0 \leftarrow \text{Remainder} \end{array}$$

Since the remainder is 0, $x + 2$ is a factor of $P(x)$. A factored form (but not necessarily *completely* factored form) of the polynomial is $(x + 2)(4x^2 + 16x + 16)$.

(continued)

(b) By the factor theorem, $x - 1$ will be a factor of $P(x) = 2x^4 + 3x^2 - 5x + 7$ if $P(1) = 0$.

$$
\begin{array}{r|rrrr}
1) & 2 & 0 & 3 & -5 & 7 \\
   &   & 2 & 2 & 5 & 0 \\
\hline
   & 2 & 2 & 5 & 0 & 7
\end{array}
$$ Use synthetic division to divide $P(x)$ by $x - 1$.

$\longleftarrow P(1) = 7$

Since the remainder is 7, $P(1) = 7$, not 0, so $x - 1$ is *not* a factor of $P(x)$. ●

> **NOTE**   An easy way to determine $P(1)$ for a polynomial function P is simply to add the coefficients of $P(x)$. This method works because every power of 1 is equal to 1. For example, using
>
> $$P(x) = 2x^4 + 3x^2 - 5x + 7$$
>
> as shown in **Example 6(b)**, we have
>
> $$P(1) = 2 + 3 - 5 + 7 = 7$$
>
> confirming our result found by synthetic division.

EXAMPLE 7   Examining *x*-Intercepts, Zeros, and Solutions

Consider the polynomial function $P(x) = 2x^3 + 5x^2 - x - 6$.

(a) Show that -2, $-\frac{3}{2}$, and 1 are zeros of P, and write $P(x)$ in factored form with all factors linear.

(b) Graph P in a suitable viewing window and locate the *x*-intercepts.

(c) Solve the polynomial equation $2x^3 + 5x^2 - x - 6 = 0$.

Solution

(a)
$$
\begin{array}{r|rrrr}
-2) & 2 & 5 & -1 & -6 \\
    &   & -4 & -2 & 6 \\
\hline
    & 2 & 1 & -3 & 0
\end{array}
$$ Use synthetic division to divide $P(x)$ by $x - (-2)$.

$\longleftarrow P(-2) = 0$

Since $P(-2) = 0$, $x + 2$ is a factor; thus, $P(x) = (x + 2)(2x^2 + x - 3)$. Rather than show that $-\frac{3}{2}$ and 1 are zeros of $P(x)$, we need only show that they are zeros of $2x^2 + x - 3$ by factoring directly.

$$2x^2 + x - 3 = (2x + 3)(x - 1)$$ Factored form

The solutions of $(2x + 3)(x - 1) = 0$ are, by the zero-product property, $-\frac{3}{2}$ and 1. The completely factored form of $P(x)$ is

$$P(x) = (x + 2)(2x + 3)(x - 1).$$

(b) **FIGURE 70** shows the graph of this function. The calculator can be used to determine the *x*-coordinates of the *x*-intercepts: -2, $-\frac{3}{2}$, and 1.

(c) From part (a), the zeros of P are -2, $-\frac{3}{2}$, and 1. Because the zeros of P are the solutions of $P(x) = 0$, the solution set is $\left\{-2, -\frac{3}{2}, 1\right\}$. ●

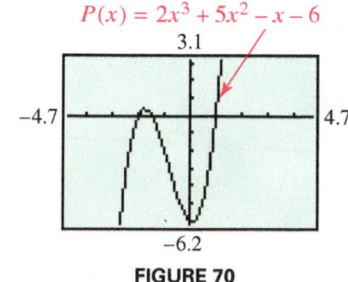

$P(x) = 2x^3 + 5x^2 - x - 6$

FIGURE 70

> **NOTE**   In **Example 7(a)**, it was easy to use factoring to find the zeros generated by the factor $2x^2 + x - 3$. It is *always* possible to use the quadratic formula at this stage of the procedure and, in fact, necessary when the quadratic factor cannot be factored further using integer coefficients.

Division of Any Two Polynomials

Thus far, we have only divided by a polynomial in the form $x - k$. However, we can use long division to divide by any polynomial.

Division Algorithm for Polynomials

Let $P(x)$ and $D(x)$ be two polynomials, with the degree of $D(x)$ greater than zero and less than the degree of $P(x)$. Then there exist unique polynomials $Q(x)$ and $R(x)$ such that

$$\frac{P(x)}{D(x)} = Q(x) + \frac{R(x)}{D(x)},$$

where either $R(x) = 0$ or the degree of $R(x)$ is less than the degree of $D(x)$.

In this algorithm $P(x)$ is the *dividend*, $D(x)$ is the *divisor*, $Q(x)$ is the *quotient*, and $R(x)$ is the *remainder*. Using this terminology, this algorithm can be written as follows.

$$\frac{\text{Dividend}}{\text{Divisor}} = \text{Quotient} + \frac{\text{Remainder}}{\text{Divisor}}$$

In the next example we demonstrate how to divide any two polynomials.

EXAMPLE 8   **Dividing Polynomials**

Divide each expression.

(a) $\dfrac{6x^2 + 5x - 10}{2x + 3}$

(b) $(5x^3 - 4x^2 + 7x - 2) \div (x^2 + 1)$

(c) $\left(x^3 + \dfrac{5}{2}x^2 + x + 2\right) \div (2x^2 + x - 1)$

Solution

(a) Begin by dividing $2x$ into $6x^2$.

$$
\begin{array}{r}
3x \phantom{{}+6x^2+5x-10} \\
2x + 3 \overline{)6x^2 + 5x - 10} \\
\underline{6x^2 + 9x} \phantom{{}- 10} \\
-4x - 10
\end{array}
$$

$\dfrac{6x^2}{2x} = 3x$

$3x(2x + 3) = 6x^2 + 9x$
Subtract: $5x - 9x = -4x$.
Bring down the -10.

In the next step, divide $2x$ into $-4x$.

$$
\begin{array}{r}
3x - 2 \phantom{{}+6x^2} \\
2x + 3 \overline{)6x^2 + 5x - 10} \\
\underline{6x^2 + 9x} \phantom{{}- 10} \\
-4x - 10 \\
\underline{-4x - 6} \\
-4
\end{array}
$$

$\dfrac{-4x}{2x} = -2$

$-2(2x + 3) = -4x - 6$
Subtract: $-10 - (-6) = -4$.

The quotient is $3x - 2$ with remainder -4. This result can also be written as

$$3x - 2 + \frac{-4}{2x + 3}.$$

Quotient $+ \dfrac{\text{Remainder}}{\text{Divisor}}$

(continued)

(b) Begin by writing $x^2 + 1$ as $x^2 + 0x + 1$.

$$\frac{5x^3}{x^2} = 5x$$

$$
\begin{array}{r}
5x - 4 \\
x^2 + 0x + 1\overline{)5x^3 - 4x^2 + 7x - 2} \\
\underline{5x^3 + 0x^2 + 5x} \\
-4x^2 + 2x - 2 \\
\underline{-4x^2 - 0x - 4} \\
2x + 2
\end{array}
$$

Insert 0x as a placeholder.

The quotient is $5x - 4$ with remainder $2x + 2$, which can be written as follows.

$$5x - 4 + \frac{2x + 2}{x^2 + 1}$$

(c) Begin by dividing $2x^2$ into x^3.

$$\frac{x^3}{2x^2} = \frac{1}{2}x$$

$$
\begin{array}{r}
\frac{1}{2}x + 1 \\
2x^2 + x - 1\overline{)x^3 + \frac{5}{2}x^2 + x + 2} \\
\underline{x^3 + \frac{1}{2}x^2 - \frac{1}{2}x} \\
2x^2 + \frac{3}{2}x + 2 \\
\underline{2x^2 + x - 1} \\
\frac{1}{2}x + 3
\end{array}
$$

The quotient is $\frac{1}{2}x + 1$ and the remainder is $\frac{1}{2}x + 3$. By the division algorithm, this result can be written as follows.

$$\frac{1}{2}x + 1 + \frac{\frac{1}{2}x + 3}{2x^2 + x - 1}$$

3.6 Exercises

Checking Analytic Skills   *Divide each expression. Apply the property* $\frac{a + b}{c} = \frac{a}{c} + \frac{b}{c}$ *if necessary.*
Do not use a calculator.

1. $\dfrac{10x^6}{5x^3}$

2. $\dfrac{6x^4}{2x^3}$

3. $\dfrac{8x^9}{3x^7}$

4. $-\dfrac{2x^5}{7x^2}$

5. $\dfrac{2x^6 + 3x^3}{2x}$

6. $\dfrac{5x^3 + x^2}{3x^2}$

7. $\dfrac{8x^3 - 5x}{2x}$

8. $\dfrac{7x^8 - 6x^3}{6x^2}$

Use the intermediate value theorem to show that each function has a real zero between the two numbers given. Then, use your calculator to approximate the zero to the nearest hundredth.

9. $P(x) = 3x^2 - 2x - 6$;   1 and 2

10. $P(x) = -x^3 - x^2 + 5x + 5$;   2 and 3

11. $P(x) = 2x^3 - 8x^2 + x + 16$;   2 and 2.5

12. $P(x) = 3x^3 + 7x^2 - 4$; $\frac{1}{2}$ and 1

13. $P(x) = 2x^4 - 4x^2 + 3x - 6$;   1.5 and 2

14. $P(x) = x^4 - 4x^3 - x + 1$;   0.3 and 1

15. $P(x) = -x^4 + 2x^3 + x + 12$;   2.7 and 2.8

16. $P(x) = -2x^4 + x^3 - x^2 + 3$;   −1 and −0.9

17. $P(x) = x^5 - 2x^3 + 1;$ -1.6 and -1.5

18. $P(x) = 2x^7 - x^4 + x - 4;$ 1.1 and 1.2

19. *Concept Check* Suppose that a polynomial function P is defined in such a way that $P(2) = -4$ and $P(2.5) = 2$. What conclusion does the intermediate value theorem allow you to make?

20. Suppose that a polynomial function P is defined in such a way that $P(3) = -4$ and $P(4) = -10$. Can we be certain that there is no zero between 3 and 4? Explain, using a graph.

Find each quotient when $P(x)$ is divided by the binomial following it.

21. $P(x) = x^3 + 2x^2 - 17x - 10;$ $x + 5$

22. $P(x) = x^4 + 4x^3 + 2x^2 + 9x + 4;$ $x + 4$

23. $P(x) = 3x^3 - 11x^2 - 20x + 3;$ $x - 5$

24. $P(x) = x^4 - 3x^3 - 5x^2 + 2x - 16;$ $x - 3$

25. $P(x) = x^4 - 3x^3 - 4x^2 + 12x;$ $x - 2$

26. $P(x) = 2x^4 + 3x^3 - 5x^2 - 18x;$ $x - 2$

27. $P(x) = x^3 + 2x^2 - 3;$ $x - 1$

28. $P(x) = x^3 - 2x^2 - 9;$ $x - 3$

29. $P(x) = -2x^3 - x - 2;$ $x + 1$

30. $P(x) = -3x^3 - x - 5;$ $x + 1$

31. $P(x) = x^5 - 1;$ $x - 1$

32. $P(x) = x^7 + 1;$ $x + 1$

Use synthetic division to find $P(k)$.

33. $k = 3;$ $P(x) = x^2 - 4x + 3$

34. $k = -2;$ $P(x) = x^2 + 5x + 6$

35. $k = -2;$ $P(x) = 5x^3 + 2x^2 - x + 5$

36. $k = 2;$ $P(x) = 2x^3 - 3x^2 - 5x + 4$

37. $k = 2;$ $P(x) = x^2 - 5x + 1$

38. $k = 3;$ $P(x) = x^2 - x + 3$

39. $k = 0.5;$ $P(x) = x^3 - x + 4$

40. $k = 1.5;$ $P(x) = x^3 + x - 3$

41. $k = \sqrt{2};$ $P(x) = x^4 - x^2 - 3$

42. $k = \sqrt{3};$ $P(x) = x^4 + 2x^2 - 10$

43. $k = \sqrt[3]{4};$ $P(x) = -x^3 + x + 4$

44. $k = \sqrt[5]{3};$ $P(x) = -x^5 + 2x + 3$

Use synthetic division to determine whether the given number is a zero of the polynomial.

45. $2;$ $P(x) = x^2 + 2x - 8$

46. $-1;$ $P(x) = x^2 + 4x - 5$

47. $4;$ $P(x) = 2x^3 - 6x^2 - 9x + 6$

48. $-4;$ $P(x) = 9x^3 + 39x^2 + 12x$

49. $-0.5;$ $P(x) = 4x^3 + 12x^2 + 7x + 1$

50. $-0.25;$ $P(x) = 8x^3 + 6x^2 - 3x - 1$

51. $-5;$ $P(x) = 8x^3 + 50x^2 + 47x + 15$

52. $-4;$ $P(x) = 6x^3 + 25x^2 + 3x - 3$

53. $\sqrt{6};$ $P(x) = -2x^6 + 5x^4 - 3x^2 + 270$

54. $\sqrt{7};$ $P(x) = -3x^6 + 7x^4 - 5x^2 + 721$

RELATING CONCEPTS   **For individual or group investigation (Exercises 55–60)**

*The close relationships among **x-intercepts of a graph of a function**, **real zeros of the function**, and **real solutions of the associated equation** should, by now, be apparent to you. Consider the graph of the polynomial function $P(x) = x^3 - 2x^2 - 11x + 12$, and **work Exercises 55–60 in order.***

55. What are the linear factors of $P(x)$?

56. What are the solutions of the equation $P(x) = 0$?

57. What are the zeros of the function P?

58. If $P(x)$ is divided by $x - 2$, what is the remainder? What is $P(2)$?

59. Give the solution set of $P(x) > 0$, using interval notation.

60. Give the solution set of $P(x) < 0$, using interval notation.

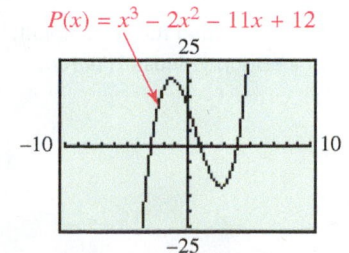

$P(x) = x^3 - 2x^2 - 11x + 12$

The x-coordinates of the x-intercepts are -3, 1, and 4.

For each polynomial, at least one zero is given. Find all others analytically.

61. $P(x) = x^3 - 2x^2 - 5x + 6;$ 3

62. $P(x) = x^3 + 2x^2 - 11x - 12;$ 3

63. $P(x) = x^3 - 2x + 1;$ 1

64. $P(x) = 2x^3 + 8x^2 - 11x - 5;$ -5

65. $P(x) = 3x^3 + 5x^2 - 3x - 2;$ -2

66. $P(x) = x^3 - 7x^2 + 13x - 3;$ 3

67. $P(x) = x^4 - 41x^2 + 180;$ -6 and 6

68. $P(x) = x^4 - 52x^2 + 147;$ -7 and 7

69. $P(x) = -x^3 + 8x^2 + 3x - 24;$ 8

70. $P(x) = -x^3 + 4x^2 + 7x - 28;$ 4

Factor $P(x)$ into linear factors given that k is a zero of P.

71. $P(x) = 2x^3 - 3x^2 - 17x + 30;$ $k = 2$

72. $P(x) = 2x^3 - 3x^2 - 5x + 6;$ $k = 1$

73. $P(x) = 6x^3 + 25x^2 + 3x - 4;$ $k = -4$

74. $P(x) = 8x^3 + 50x^2 + 47x - 15;$ $k = -5$

75. $P(x) = -6x^3 - 13x^2 + 14x - 3;$ $k = -3$

76. $P(x) = -6x^3 - 17x^2 + 63x - 10;$ $k = -5$

77. $P(x) = x^3 + 5x^2 - 3x - 15;$ $k = -5$

78. $P(x) = x^3 + 9x^2 - 7x - 63;$ $k = -9$

79. $P(x) = x^3 - 2x^2 - 7x - 4;$ $k = -1$

80. $P(x) = x^3 + x^2 - 21x - 45;$ $k = -3$

Divide.

81. $\dfrac{3x^4 - 7x^3 + 6x - 16}{3x - 7}$

82. $\dfrac{20x^4 + 6x^3 - 2x^2 + 15x - 2}{5x - 1}$

83. $\dfrac{5x^4 - 2x^2 + 6}{x^2 + 2}$

84. $\dfrac{x^3 - x^2 + 2x - 3}{x^2 + 3}$

85. $\dfrac{8x^3 + 10x^2 - 12x - 15}{2x^2 - 3}$

86. $\dfrac{3x^4 - 2x^2 - 5}{3x^2 - 5}$

87. $\dfrac{2x^4 - x^3 + 4x^2 + 8x + 7}{2x^2 + 3x + 2}$

88. $\dfrac{3x^4 + 2x^3 - x^2 + 4x - 3}{x^2 + x - 1}$

89. $\left(x^2 + \dfrac{1}{2}x - 1\right) \div (2x + 1)$

90. $(-x^2 - 1) \div (3x - 9)$

91. $(x^3 - x^2 + 1) \div (2x^2 - 1)$

92. $(-3x^3 + 2x^2 + 2x) \div (6x^2 + 2x + 1)$

3.7 Topics in the Theory of Polynomial Functions (II)

Complex Zeros and the Fundamental Theorem of Algebra • Number of Zeros • Rational Zeros Theorem • Descartes' Rule of Signs • Boundedness Theorem

Complex Zeros and the Fundamental Theorem of Algebra

In **Example 6** of **Section 3.3,** we found that the nonreal complex solutions of the equation $2x^2 - x + 4 = 0$ are $\dfrac{1}{4} + i\dfrac{\sqrt{31}}{4}$ and $\dfrac{1}{4} - i\dfrac{\sqrt{31}}{4}$. These two solutions are complex conjugates. This is not a coincidence, as given in the *conjugate zeros theorem.*

> **Conjugate Zeros Theorem**
>
> If $P(x)$ is a polynomial function having only *real* coefficients, and if $a + bi$ is a zero of $P(x)$, then the conjugate $a - bi$ is also a zero of $P(x)$.

EXAMPLE 1   **Defining a Polynomial Function Satisfying Given Conditions**

(a) Find a cubic polynomial function P in standard form with real coefficients having zeros 3 and $2 + i$.

(b) Find a polynomial function P satisfying the conditions of part (a), with the additional requirement $P(-2) = 4$. Support the result graphically.

Solution

(a) By the conjugate zeros theorem, $2 - i$ must also be a zero of the function. Since $P(x)$ will be cubic, it will have three linear factors, and by the factor theorem they must be $x - 3$, $x - (2 + i)$, and $x - (2 - i)$.

$$P(x) = (x - 3)[x - (2 + i)][x - (2 - i)] \quad \text{Factor theorem}$$

$$P(x) = (x - 3)(x - 2 - i)(x - 2 + i) \quad \text{Distributive property}$$

$$P(x) = (x - 3)(x^2 - 4x + 5) \quad \text{Multiply.}$$

$$P(x) = x^3 - 7x^2 + 17x - 15 \quad \text{Multiply.}$$

◀ **Algebra Review**
To review multiplying polynomials, see **Section R.1.**

Multiplying the polynomial by any real nonzero constant a will also yield a function satisfying the given conditions, so a more general form is

$$P(x) = a(x^3 - 7x^2 + 17x - 15).$$

(b) We must define $P(x) = a(x^3 - 7x^2 + 17x - 15)$ in such a way that $P(-2) = 4$. To find a, let $x = -2$, and set the result equal to 4. Then solve for a.

$$a[(-2)^3 - 7(-2)^2 + 17(-2) - 15] = 4 \quad \text{Substitute.}$$

$$a(-8 - 28 - 34 - 15) = 4 \quad \text{Multiply.}$$

$$-85a = 4 \quad \text{Simplify.}$$

$$a = -\frac{4}{85} \quad \text{Divide by } -85.$$

Use parentheses around substituted values to avoid errors.

$$P(x) = -\frac{4}{85}x^3 + \frac{28}{85}x^2 - \frac{4}{5}x + \frac{12}{17}$$

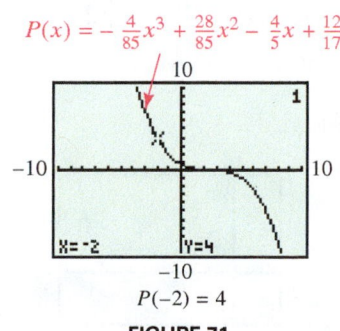

X=-2   Y=4

$P(-2) = 4$

FIGURE 71

Therefore, the desired function is

$$P(x) = -\frac{4}{85}(x^3 - 7x^2 + 17x - 15)$$

$$= -\frac{4}{85}x^3 + \frac{28}{85}x^2 - \frac{4}{5}x + \frac{12}{17}. \quad \text{Distributive property}$$

We can support this result by graphing $P(x) = -\frac{4}{85}x^3 + \frac{28}{85}x^2 - \frac{4}{5}x + \frac{12}{17}$ and showing that the point $(-2, 4)$ lies on the graph. See **FIGURE 71.** ●

Number of Zeros

The *fundamental theorem of algebra* was first proved by Carl Friedrich Gauss in his doctoral thesis in 1799, when he was 22 years old. Although many proofs of this result have been given, all involve mathematics beyond this book.

Carl Friedrich Gauss
(1777–1855)

Gauss, one of the most brilliant mathematicians of all time, also studied astronomy and physics.

Fundamental Theorem of Algebra

Every function defined by a polynomial of degree 1 or more has at least one complex zero.

NOTE Complex zeros include real zeros. A polynomial function can have only real zeros, only nonreal complex zeros, or both real zeros and nonreal complex zeros.

From the fundamental theorem, if $P(x)$ is of degree 1 or more, then there is some number k such that $P(k) = 0$. Thus, by the factor theorem, $P(x) = (x - k) \cdot Q(x)$ for some polynomial $Q(x)$. The fundamental theorem and the factor theorem can be used to factor $Q(x)$ in the same way. Assuming that $P(x)$ has degree n, repeating this process n times gives

$$P(x) = a(x - k_1)(x - k_2) \cdots (x - k_n),$$

where a is the leading coefficient of $P(x)$. Each factor leads to a zero of $P(x)$, so $P(x)$ has n zeros $k_1, k_2, k_3, \ldots, k_n$. This suggests the *number of zeros theorem*.

Number of Zeros Theorem

A function defined by a polynomial of degree n has at most n distinct (unique) complex zeros.

EXAMPLE 2 **Finding All Zeros of a Polynomial Function**

Find all complex zeros of $P(x) = x^4 - 7x^3 + 18x^2 - 22x + 12$, given that $1 - i$ is a zero.

Analytic Solution

This quartic function will have at most four complex zeros. Since $1 - i$ is a zero and the coefficients are real numbers, by the conjugate zeros theorem $1 + i$ is also a zero. The remaining zeros are found by first dividing the original polynomial by $x - (1 - i)$.

$$
\begin{array}{r|rrrrr}
1 - i) & 1 & -7 & 18 & -22 & 12 \\
 & & 1 - i & -7 + 5i & 16 - 6i & -12 \\
\hline
 & 1 & -6 - i & 11 + 5i & -6 - 6i & 0 \\
\end{array}
$$

Next, divide the quotient from the first division by $x - (1 + i)$.

$$
\begin{array}{r|rrrr}
1 + i) & 1 & -6 - i & 11 + 5i & -6 - 6i \\
 & & 1 + i & -5 - 5i & 6 + 6i \\
\hline
 & 1 & -5 & 6 & 0 \\
\end{array}
$$

Now find the zeros of the function defined by the quadratic polynomial $x^2 - 5x + 6$ by solving the equation $x^2 - 5x + 6 = 0$. By factoring, we obtain $(x - 2)(x - 3) = 0$, and so the other zeros are 2 and 3. Thus, this function has four complex zeros: $1 - i$, $1 + i$, 2, and 3.

Graphing Calculator Solution

By the conjugate zeros theorem, since $1 - i$ is a zero, $1 + i$ is also a zero. We can use a graphing calculator to find the real zeros as in **FIGURE 72**, which shows the graph of

$$P(x) = x^4 - 7x^3 + 18x^2 - 22x + 12,$$

with the real zeros identified at the bottom.

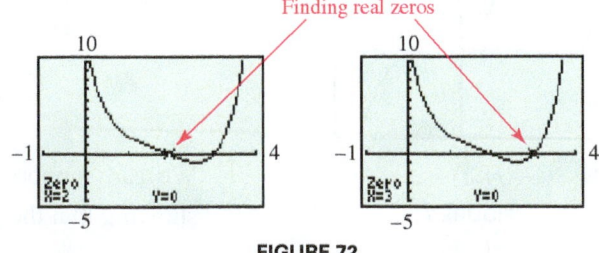

Finding real zeros

FIGURE 72

If no zero had been given for this function, one approach would be to find the real zeros using the graph, and then using synthetic division twice. Once a quadratic factor is determined, the quadratic formula can then be used.

The number of zeros theorem says that a polynomial function of degree n has *at most n* distinct zeros. In the polynomial function

$$P(x) = x^6 + x^5 - 5x^4 - x^3 + 8x^2 - 4x,$$

or

$$P(x) = x(x + 2)^2(x - 1)^3,$$

each factor leads to a zero of the function. The factor x leads to a **single zero** of 0, the factor $(x + 2)^2$ leads to a zero of -2 appearing *twice,* and the factor $(x - 1)^3$ leads to a zero of 1 appearing *three* times. The number of times a zero appears is referred to as the **multiplicity of the zero.**

EXAMPLE 3 Defining a Polynomial Function Satisfying Given Conditions

Find a polynomial function with real coefficients of least possible degree having a zero 2 of multiplicity 3, a zero 0 of multiplicity 2, and a zero i of single multiplicity.

Solution By the conjugate zeros theorem, this polynomial function must also have a zero of $-i$. This means there are seven zeros, so the least possible degree of the polynomial is 7. Therefore,

$$P(x) = x^2(x - 2)^3(x - i)(x + i) \qquad \text{Factor theorem}$$

$$P(x) = x^7 - 6x^6 + 13x^5 - 14x^4 + 12x^3 - 8x^2. \qquad \text{Multiply.}$$

This is one of infinitely many such functions. Multiplying $P(x)$ by a nonzero constant will yield another polynomial function satisfying these conditions.

The graph of this function in **FIGURE 73** shows that it has only two distinct x-intercepts, corresponding to real zeros of 0 and 2. ●

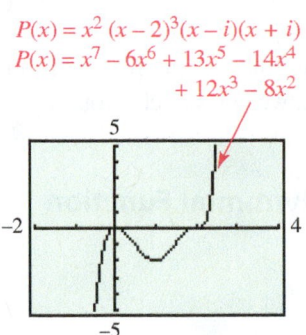

$P(x) = x^2(x - 2)^3(x - i)(x + i)$
$P(x) = x^7 - 6x^6 + 13x^5 - 14x^4 + 12x^3 - 8x^2$

The graph is tangent to the x-axis at $(0, 0)$, and crosses the x-axis at $(2, 0)$.

FIGURE 73

FOR DISCUSSION

Graph each function P in the window indicated below its formula. Then respond to the following items.

$P(x) = (x + 3)(x - 2)^2$; $P(x) = (x + 3)^2(x - 2)^3$; $P(x) = x^2(x - 1)(x + 2)^2$
$[-10, 10]$ by $[-30, 30]$ $[-4, 4]$ by $[-125, 50]$ $[-4, 4]$ by $[-5, 5]$

1. Describe the behavior of the graph at each x-intercept that corresponds to a zero of odd multiplicity.

2. Describe the behavior of the graph at each x-intercept that corresponds to a zero of even multiplicity.

The observations in the "For Discussion" box suggest that the behavior of the graph of a polynomial function near an x-intercept depends on the *parity* (odd or even) of multiplicity of the zero that leads to the x-intercept.

A zero k of a polynomial function has as multiplicity the exponent of the factor $x - k$. Determining the multiplicity of a zero aids in sketching the graph near that zero.

1. If the zero has multiplicity one, the graph crosses the x-axis at the corresponding x-intercept as seen in **FIGURE 74(a)** on the next page.

2. If the zero has even multiplicity, the graph is tangent to the x-axis at the corresponding x-intercept (that is, the graph touches but does not cross the x-axis there). See **FIGURE 74(b)** on the next page.

3. If the zero has odd multiplicity greater than one, the graph crosses the *x*-axis **and** is tangent to the *x*-axis at the corresponding *x*-intercept. This causes a change in concavity, or shape, at the *x*-intercept and the graph curves there. See **FIGURE 74(c).**

Graphs and Multiplicities

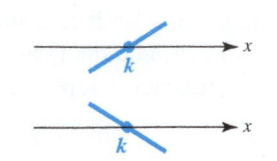

The graph crosses the *x*-axis at $(k, 0)$ if *k* is a zero of multiplicity one.

(a)

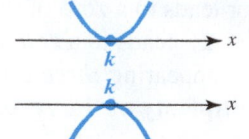

The graph is tangent to the *x*-axis at $(k, 0)$ if *k* is a zero of even multiplicity. The graph bounces, or turns, at *k*.

(b)

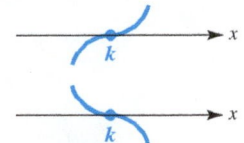

The graph crosses **and** is tangent to the *x*-axis at $(k, 0)$ if *k* is a zero of odd multiplicity greater than one. The graph curves at *k*.

(c)

FIGURE 74

By observing the dominating term, the *y*-intercept, and noting the parity of multi-plicities of zeros of a polynomial function in factored form, we can sketch a rough graph.

EXAMPLE 4 **Sketching a Graph of a Polynomial Function by Hand**

Consider the polynomial function

$$P(x) = -2x^5 - 18x^4 - 38x^3 + 42x^2 + 112x - 96,$$

or $\quad P(x) = -2(x + 4)^2(x + 3)(x - 1)^2.$ Factored form

Sketch the graph of *P* by hand. Confirm the result with a calculator.

Solution Because the dominating term is $-2x^5$, the end behavior of the graph will be ↖↘. Since $(-4, 0)$ and $(1, 0)$ are both *x*-intercepts determined by zeros of even multiplicity, the graph will be tangent to the *x*-axis at these intercepts. Because -3 is a zero of multiplicity 1, the graph will cross the *x*-axis at $(-3, 0)$. The *y*-intercept is $(0, -96)$. This information leads to the rough sketch in **FIGURE 75(a).**

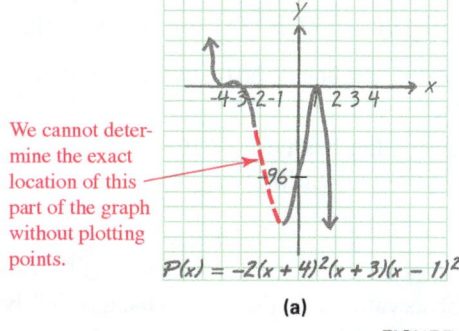

We cannot deter-mine the exact location of this part of the graph without plotting points.

$P(x) = -2(x + 4)^2(x + 3)(x - 1)^2$

(a)

$P(x) = -2(x + 4)^2(x + 3)(x - 1)^2$

(b)

FIGURE 75

The hand-drawn graph does not necessarily give a good indication of local extrema. The calculator graph shown in **FIGURE 75(b)** fills in the details.

Rational Zeros Theorem

The *rational zeros theorem* gives a method to determine all possible candidates for rational zeros of a polynomial function with integer coefficients.

Rational Zeros Theorem

Let $P(x) = a_n x^n + a_{n-1} x^{n-1} + \cdots + a_1 x + a_0$, where $a_n \neq 0$ and $a_0 \neq 0$, be a polynomial function with integer coefficients. If $\frac{p}{q}$ is a rational number written in lowest terms, and if $\frac{p}{q}$ is a zero of $P(x)$, then p is a factor of the constant term a_0, and q is a factor of the leading coefficient a_n.

Proof $P\left(\frac{p}{q}\right) = 0$, since $\frac{p}{q}$ is a zero of $P(x)$.

$$a_n\left(\frac{p}{q}\right)^n + a_{n-1}\left(\frac{p}{q}\right)^{n-1} + \cdots + a_1\left(\frac{p}{q}\right) + a_0 = 0 \qquad \text{Substitute.}$$

$$a_n\left(\frac{p^n}{q^n}\right) + a_{n-1}\left(\frac{p^{n-1}}{q^{n-1}}\right) + \cdots + a_1\left(\frac{p}{q}\right) + a_0 = 0 \qquad \text{Property of exponents}$$

$$a_n p^n + a_{n-1} p^{n-1} q + \cdots + a_1 p q^{n-1} = -a_0 q^n \qquad \text{Multiply by } q^n; \text{ add } -a_0 q^n.$$

$$p(a_n p^{n-1} + a_{n-1} p^{n-2} q + \cdots + a_1 q^{n-1}) = -a_0 q^n \qquad \text{Factor out } p.$$

Thus, $-a_0 q^n$ equals the product of the two factors, p and $(a_n p^{n-1} + \cdots + a_1 q^{n-1})$. For this reason, p must be a factor of $-a_0 q^n$. Since it was assumed that $\frac{p}{q}$ is written in lowest terms, p and q have no common factor other than 1, so p is not a factor of q^n. Thus, p must be a factor of a_0. In a similar way, it can be shown that q is a factor of a_n.

EXAMPLE 5   **Finding Rational Zeros and Factoring a Polynomial**

Find all rational zeros of $P(x) = 6x^3 - 5x^2 - 7x + 4$ and factor $P(x)$.

Solution   If $\frac{p}{q}$ is a rational zero in lowest terms, then p is a factor of the constant term 4 and q is a factor of the leading coefficient 6. Possible values for p and q are

$$p: \quad \pm 1, \quad \pm 2, \quad \pm 4$$
$$q: \quad \pm 1, \quad \pm 2, \quad \pm 3, \quad \pm 6.$$

As a result, any rational zero of $P(x)$ in the form $\frac{p}{q}$ must occur in the list

$$\pm \frac{1}{6}, \ \pm \frac{1}{3}, \ \pm \frac{1}{2}, \ \pm \frac{2}{3}, \ \pm \frac{1}{1}, \ \pm \frac{4}{3}, \ \pm \frac{2}{1}, \quad \text{or} \quad \pm \frac{4}{1}.$$

Evaluate $P(x)$ at each value in the list. See the table.

x	$P(x)$	x	$P(x)$	x	$P(x)$	x	$P(x)$
$\frac{1}{6}$	$\frac{49}{18}$	$\frac{1}{2}$	0	1	-2	2	18
$-\frac{1}{6}$	5	$-\frac{1}{2}$	$\frac{11}{2}$	-1	0	-2	-50
$\frac{1}{3}$	$\frac{4}{3}$	$\frac{2}{3}$	$-\frac{10}{9}$	$\frac{4}{3}$	0	4	280
$-\frac{1}{3}$	$\frac{50}{9}$	$-\frac{2}{3}$	$\frac{14}{3}$	$-\frac{4}{3}$	$-\frac{88}{9}$	-4	-432

(continued)

From the table on the preceding page, we see that $-1, \frac{1}{2}$, and $\frac{4}{3}$ are rational zeros of $P(x)$. Because $P(x)$ is a cubic polynomial, there are at most three distinct zeros. By the factor theorem, $(x + 1)$, $\left(x - \frac{1}{2}\right)$, and $\left(x - \frac{4}{3}\right)$ are factors of $P(x)$. The leading coefficient of $P(x)$ is 6, so we let $a = 6$.

$$P(x) = a(x + 1)\left(x - \frac{1}{2}\right)\left(x - \frac{4}{3}\right) \qquad \text{Factor theorem}$$

$$= 6(x + 1)\left(x - \frac{1}{2}\right)\left(x - \frac{4}{3}\right) \qquad \text{Let } a = 6.$$

The polynomial can be left in this form.

$$= (x + 1)(2)\left(x - \frac{1}{2}\right)(3)\left(x - \frac{4}{3}\right) \qquad \text{Factor: } 6 = 2 \cdot 3.$$

$$P(x) = (x + 1)(2x - 1)(3x - 4) \qquad \text{Multiply.} \qquad \bullet$$

EXAMPLE 6   **Using the Rational Zeros Theorem**

Perform the following for the polynomial function

$$P(x) = 6x^4 + 7x^3 - 12x^2 - 3x + 2.$$

(a) List all possible rational zeros.

(b) Use a graph to eliminate some of the possible zeros listed in part (a).

(c) Find all rational zeros and factor $P(x)$.

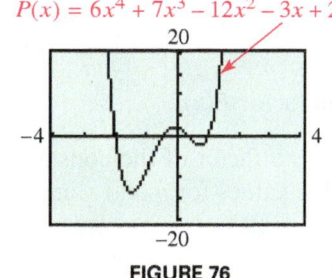

$P(x) = 6x^4 + 7x^3 - 12x^2 - 3x + 2$

FIGURE 76

Solution

(a) For a rational number $\frac{p}{q}$ to be a zero, p must be a factor of $a_0 = 2$ and q must be a factor of $a_4 = 6$. Thus, p can be ± 1 or ± 2, and q can be ± 1, ± 2, ± 3, or ± 6. The possible rational zeros, $\frac{p}{q}$, are ± 1, ± 2, $\pm \frac{1}{2}$, $\pm \frac{1}{3}$, $\pm \frac{1}{6}$, $\pm \frac{2}{3}$.

(b) From **FIGURE 76,** we see that the zeros are no less than -2 and no greater than 1, so we can eliminate 2. Furthermore, -1 is not a zero, since the graph does not intersect the x-axis at $(-1, 0)$. At this point, we have no way of knowing whether the zeros indicated on the graph are rational numbers. They may be irrational.

(c) In **Example 5,** we tested possible rational zeros by direct substitution into the formula for $P(x)$. In this example, we use synthetic division and the remainder theorem to show that 1 and -2 are zeros.

$$\begin{array}{r|rrrrr} 1) & 6 & 7 & -12 & -3 & 2 \\ & & 6 & 13 & 1 & -2 \\ \hline & 6 & 13 & 1 & -2 & 0 \end{array}$$

The 0 remainder shows that 1 is a zero. Now we use the quotient polynomial $6x^3 + 13x^2 + x - 2$ and synthetic division to find that -2 is also a zero.

$$\begin{array}{r|rrrr} -2) & 6 & 13 & 1 & -2 \\ & & -12 & -2 & 2 \\ \hline & 6 & 1 & -1 & 0 \end{array}$$

The new quotient polynomial is $6x^2 + x - 1$, which is easily factored as $(3x - 1)(2x + 1)$. Thus, the remaining two zeros are $\frac{1}{3}$ and $-\frac{1}{2}$.

Split Screen

This graph/table verifies that the four zeros of the function in **Example 6** are $-2, -\frac{1}{2}, \frac{1}{3}$, and 1.

Since the four zeros of $P(x) = 6x^4 + 7x^3 - 12x^2 - 3x + 2$ are $1, -2, \frac{1}{3}$, and $-\frac{1}{2}$, the corresponding factors are $x - 1$, $x + 2$, $x - \frac{1}{3}$, and $x + \frac{1}{2}$.

$$P(x) = a(x - 1)(x + 2)\left(x - \frac{1}{3}\right)\left(x + \frac{1}{2}\right)$$

The polynomial can be left in this form.

$$= 6(x - 1)(x + 2)\left(x - \frac{1}{3}\right)\left(x + \frac{1}{2}\right) \qquad \text{The leading coefficient of } P(x) \text{ is 6. Let } a = 6.$$

$$= (x - 1)(x + 2)(3)\left(x - \frac{1}{3}\right)(2)\left(x + \frac{1}{2}\right) \qquad \text{Factor: } 6 = 3 \cdot 2.$$

$$P(x) = (x - 1)(x + 2)(3x - 1)(2x + 1) \qquad \text{Multiply.} \qquad \bullet$$

CAUTION The rational zeros theorem gives only *possible* rational zeros; it does not tell us whether these rational numbers are *actual* zeros. Furthermore, the function must have integer coefficients. To apply the rational zeros theorem to a polynomial with fractional coefficients, multiply by the least common denominator of all the fractions. For example, any rational zeros of $P(x)$ will also be rational zeros of $Q(x)$.

$$P(x) = x^4 - \frac{1}{6}x^3 + \frac{2}{3}x^2 - \frac{1}{6}x - \frac{1}{3}$$

$$Q(x) = 6x^4 - x^3 + 4x^2 - x - 2 \qquad \text{Multiply the terms of } P(x) \text{ by 6.}$$

Descartes' Rule of Signs

Descartes' rule of signs helps to determine the number of positive and negative real zeros of a polynomial function.

Descartes' Rule of Signs

Let $P(x)$ be a polynomial function with real coefficients and a nonzero constant term, with terms in descending powers of x.

(a) The number of positive real zeros either equals the number of variations in sign occurring in the coefficients of $P(x)$ or is less than the number of variations by a positive even integer.

(b) The number of negative real zeros either equals the number of variations in sign occurring in the coefficients of $P(-x)$ or is less than the number of variations by a positive even integer.

A **variation in sign** is a change from positive to negative or negative to positive in successive terms of the polynomial when written in descending powers of the variable.

EXAMPLE 7 **Applying Descartes' Rule of Signs**

Determine the possible number of positive real zeros and negative real zeros of $P(x)$.

(a) $P(x) = x^4 - 6x^3 + 8x^2 + 2x - 1$ **(b)** $P(x) = x^5 - 3x^4 + 2x^2 + x - 1$

Solution

(a) We first observe that $P(x)$ has three variations in sign.

$$+x^4 \underset{1}{-} 6x^3 \underset{2}{+} 8x^2 + 2x \underset{3}{-} 1$$

Thus, by Descartes' rule of signs, $P(x)$ has either 3 positive real zeros or $3 - 2 = 1$ positive real zero.

(continued)

For negative zeros, consider the variations in sign for $P(-x)$.

$$P(-x) = (-x)^4 - 6(-x)^3 + 8(-x)^2 + 2(-x) - 1$$
$$= x^4 + 6x^3 + 8x^2 - 2x - 1$$

Since there is only one variation in sign, $P(x)$ has only one negative real zero.

(b) *Missing terms can be ignored,* so $P(x) = x^5 - 3x^4 + 2x^2 + x - 1$ has three variations in sign if we ignore the missing x^3-term. Thus, $P(x)$ has either three or one positive zero. Similarly $P(-x) = -x^5 - 3x^4 + 2x^2 - x - 1$ has two variations in sign, so $P(x)$ has either two or no negative zeros.

> **NOTE** When applying Descartes' rule of signs, a zero of multiplicity m is counted m times.

Boundedness Theorem

The *boundedness theorem* shows how the bottom row of a synthetic division is used to place upper and lower bounds on possible real zeros of a polynomial function.

Boundedness Theorem

Let $P(x)$ be a polynomial function of degree $n \geq 1$ with real coefficients and with a positive leading coefficient. Suppose $P(x)$ is divided synthetically by $x - c$.

(a) If $c > 0$ and all numbers in the bottom row of the synthetic division are nonnegative, then $P(x)$ has no zero greater than c.

(b) If $c < 0$ and the numbers in the bottom row of the synthetic division alternate in sign (with 0 considered positive or negative, as needed), then $P(x)$ has no zero less than c.

EXAMPLE 8 **Using the Boundedness Theorem**

Show that the real zeros of the polynomial function $P(x) = 2x^4 - 5x^3 + 3x + 1$ satisfy the following conditions.

(a) No real zero is greater than 3.   **(b)** No real zero is less than -1.

Solution

(a) Since $P(x)$ has real coefficients and the leading coefficient, 2, is positive, we can use the boundedness theorem. Divide $P(x)$ synthetically by $x - 3$.

$$c > 0 \rightarrow 3)\overline{\begin{array}{ccccc} 2 & -5 & 0 & 3 & 1 \\ & 6 & 3 & 9 & 36 \\ \hline 2 & 1 & 3 & 12 & 37 \end{array}} \quad \leftarrow \text{All are nonnegative.}$$

Thus, $P(x)$ has no real zero greater than 3.

(b) Divide $P(x) = 2x^4 - 5x^3 + 3x + 1$ synthetically by $x + 1$.

$$c < 0 \rightarrow -1)\overline{\begin{array}{ccccc} 2 & -5 & 0 & 3 & 1 \\ & -2 & 7 & -7 & 4 \\ \hline 2 & -7 & 7 & -4 & 5 \end{array}} \quad \text{Divide } P(x) \text{ by } x + 1.$$

$\leftarrow$ The numbers alternate in sign.

Thus, $P(x)$ has no zero less than -1.

3.7 Exercises

Checking Analytic Skills   *Find a cubic polynomial in standard form with real coefficients, having the given zeros. Let the leading coefficient be 1.* **Do not use a calculator.**

1. 4 and $2 + i$

2. -3 and $6 + 2i$

3. 5 and i

4. -9 and $-i$

5. 0 and $3 + i$

6. 0 and $4 - 3i$

Checking Analytic Skills   *For Exercises 7–12, find a polynomial function $P(x)$ of degree 3 with real coefficients that satisfies the given conditions.* **Do not use a calculator.**

7. Zeros of $-3, -1$, and 4; $P(2) = 5$

8. Zeros of $1, -1$, and 0; $P(2) = -3$

9. Zeros of $-2, 1$, and 0; $P(-1) = -1$

10. Zeros of 2, 5, and -3; $P(1) = -4$

11. Zeros of 4 and $1 + i$; $P(2) = 4$

12. Zeros of -7 and $2 - i$; $P(1) = 9$

For each polynomial, one or more zeros are given. Find all remaining zeros.

13. $P(x) = x^3 - x^2 - 4x - 6$;   3 is a zero.

14. $P(x) = x^3 - 5x^2 + 17x - 13$;   1 is a zero.

15. $P(x) = x^4 + 2x^3 - 10x^2 - 18x + 9$; -3 and 3 are zeros.

16. $P(x) = 2x^4 - x^3 - 27x^2 + 16x - 80$; -4 and 4 are zeros.

17. $P(x) = x^4 - x^3 + 10x^2 - 9x + 9$; $3i$ is a zero.

18. $P(x) = 2x^4 - 2x^3 + 55x^2 - 50x + 125$; $-5i$ is a zero.

For Exercises 19–30, find a polynomial function $P(x)$ having leading coefficient 1, least possible degree, real coefficients, and the given zeros.

19. 5 and -4

20. 6 and -2

21. $-3, 2$, and i

22. $1 + \sqrt{2}, 1 - \sqrt{2}$, and 3

23. $1 - \sqrt{3}, 1 + \sqrt{3}$, and 1

24. $-2 + i, -2 - i, 3$, and -3

25. $3 + 2i, -1$, and 2

26. 2 and $3i$

27. -1 and $6 - 3i$

28. $1 + 2i$ and 2 (multiplicity 2)

29. $2 + i$ and -3 (multiplicity 2)

30. 5 (multiplicity 2) and $-2i$

In Exercises 31–40, draw by hand a rough sketch of the graph of each function. (You may wish to support your answer with a calculator graph.)

31. $P(x) = 2x^3 - 5x^2 - x + 6$
$\quad = (x + 1)(2x - 3)(x - 2)$

32. $P(x) = x^3 + x^2 - 8x - 12$
$\quad = (x + 2)^2(x - 3)$

33. $P(x) = x^4 - 18x^2 + 81$
$\quad = (x - 3)^2(x + 3)^2$

34. $P(x) = x^4 - 8x^2 + 16$
$\quad = (x + 2)^2(x - 2)^2$

35. $P(x) = 2x^4 + x^3 - 6x^2 - 7x - 2$
$\quad = (2x + 1)(x - 2)(x + 1)^2$

36. $P(x) = 3x^4 - 7x^3 - 6x^2 + 12x + 8$
$\quad = (3x + 2)(x + 1)(x - 2)^2$

37. $P(x) = x^4 + 3x^3 - 3x^2 - 11x - 6$
$= (x + 3)(x + 1)^2(x - 2)$

38. $P(x) = -2x^5 + 5x^4 + 34x^3 - 30x^2 - 84x + 45$
$= (x + 3)(2x - 1)(x - 5)(3 - x^2)$

39. $P(x) = 2x^5 - 10x^4 + x^3 - 5x^2 - x + 5$
$= (x - 5)(x^2 + 1)(2x^2 - 1)$

40. $P(x) = 3x^4 - 4x^3 - 22x^2 + 15x + 18$
$= (3x + 2)(x - 3)(x^2 + x - 3)$

Concept Check Use the graphs in Exercises 41–46 to write an equation for $f(x)$ in factored form. Assume that all intercepts have integer coordinates and that $f(x)$ is either a cubic or a quartic polynomial.

41.

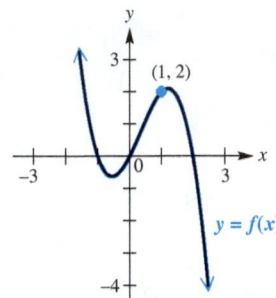

42.

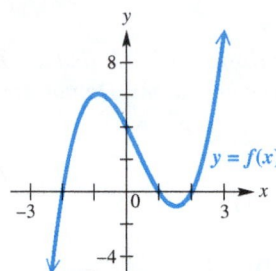

43.

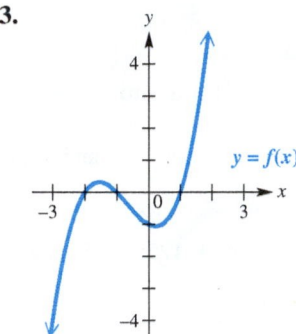

44.

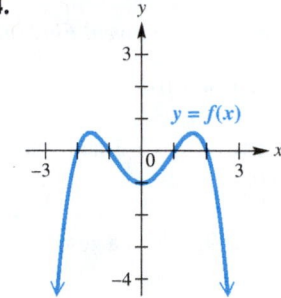

45.

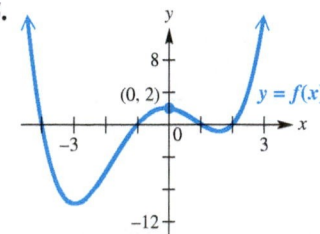

46.

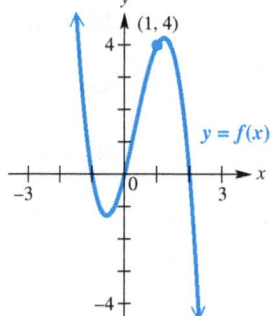

Concept Check Use the concepts of this section for Exercises 47–52.

47. Show analytically that -2 is a zero of multiplicity 2 of $P(x) = x^4 + 2x^3 - 7x^2 - 20x - 12$, and find *all* complex zeros. Then, write $P(x)$ in factored form.

48. Show analytically that -1 is a zero of multiplicity 3 of $P(x) = x^5 + 9x^4 + 33x^3 + 55x^2 + 42x + 12$, and find *all* complex zeros. Then, write $P(x)$ in factored form.

49. What are the possible numbers of real zeros (counting multiplicities) for a polynomial function with real coefficients of degree 5?

50. Explain why a polynomial function of degree 4 with real coefficients has either zero, two, or four real zeros (counting multiplicities).

51. Determine whether the description of the polynomial function $P(x)$ with real coefficients is *possible* or *not possible*.
   (a) $P(x)$ is of degree 3 and has zeros of 1, 2, and $1 + i$.
   (b) $P(x)$ is of degree 4 and has four nonreal complex zeros.
   (c) $P(x)$ is of degree 5 and -6 is a zero of multiplicity 6.
   (d) $P(x)$ has $1 + 2i$ as a zero of multiplicity 2.

52. Suppose that k, a, b, and c are real numbers, $a \neq 0$, and a polynomial function $P(x)$ may be expressed in factored form as $(x - k)(ax^2 + bx + c)$.
   (a) What is the degree of P?
   (b) What are the possible numbers of distinct *real* zeros of P?
   (c) What are the possible numbers of *nonreal complex* zeros of P?
   (d) Use the discriminant to explain how to determine the number and type of zeros of P.

*For each polynomial function, **(a)** list all possible rational zeros, **(b)** use a graph to eliminate some of the possible zeros listed in part (a), **(c)** find all rational zeros, and **(d)** factor $P(x)$.*

53. $P(x) = x^3 - 2x^2 - 13x - 10$

54. $P(x) = x^3 + 5x^2 + 2x - 8$

55. $P(x) = x^3 + 6x^2 - x - 30$

56. $P(x) = x^3 - x^2 - 10x - 8$

57. $P(x) = 6x^3 + 17x^2 - 31x - 12$

58. $P(x) = 15x^3 + 61x^2 + 2x - 8$

Use the rational zeros theorem to factor $P(x)$.

59. $P(x) = 12x^3 + 20x^2 - x - 6$

60. $P(x) = 12x^3 + 40x^2 + 41x + 12$

61. $P(x) = 24x^3 + 40x^2 - 2x - 12$

62. $P(x) = 24x^3 + 80x^2 + 82x + 24$

Find all rational zeros of each polynomial function.

63. $P(x) = x^3 + \dfrac{1}{2}x^2 - \dfrac{11}{2}x - 5$

64. $P(x) = \dfrac{10}{7}x^4 - x^3 - 7x^2 + 5x - \dfrac{5}{7}$

65. $P(x) = \dfrac{1}{6}x^4 - \dfrac{11}{12}x^3 + \dfrac{7}{6}x^2 - \dfrac{11}{12}x + 1$

66. $P(x) = x^4 - \dfrac{1}{6}x^3 + \dfrac{2}{3}x^2 - \dfrac{1}{6}x - \dfrac{1}{3}$

Use the rational zeros theorem to completely factor $P(x)$. (Hint: Not all zeros of $P(x)$ are rational.)

67. $P(x) = 6x^4 - 5x^3 - 11x^2 + 10x - 2$

68. $P(x) = 5x^4 + 8x^3 - 19x^2 - 24x + 12$

69. $P(x) = 21x^4 + 13x^3 - 103x^2 - 65x - 10$

70. $P(x) = 2x^4 + 7x^3 - 9x^2 - 49x - 35$

Use the given zero to completely factor $P(x)$ into linear factors.

71. Zero: i; $P(x) = x^5 - x^4 + 5x^3 - 5x^2 + 4x - 4$

72. Zero: $-3i$; $P(x) = x^5 + 2x^4 + 10x^3 + 20x^2 + 9x + 18$

73. Zero: $-2i$; $P(x) = x^4 + x^3 + 2x^2 + 4x - 8$

74. Zero: $5i$; $P(x) = x^4 - x^3 + 23x^2 - 25x - 50$

75. Zero: $1 + i$; $P(x) = x^4 - 2x^3 + 3x^2 - 2x + 2$

76. Zero: $2 - i$; $P(x) = x^4 - 4x^3 + 9x^2 - 16x + 20$

Use Descartes' rule of signs to determine the possible numbers of positive and negative real zeros for P(x). Then, use a graph to determine the actual numbers of positive and negative real zeros.

77. $P(x) = 2x^3 - 4x^2 + 2x + 7$

78. $P(x) = x^3 + 2x^2 + x - 10$

79. $P(x) = 5x^4 + 3x^2 + 2x - 9$

80. $P(x) = 3x^4 + 2x^3 - 8x^2 - 10x - 1$

81. $P(x) = x^5 + 3x^4 - x^3 + 2x + 3$

82. $P(x) = 2x^5 - x^4 + x^3 - x^2 + x + 5$

Use the boundedness theorem to show that the real zeros of P(x) satisfy the given conditions.

83. $P(x) = x^4 - x^3 + 3x^2 - 8x + 8$;
no real zero greater than 2

84. $P(x) = 2x^5 - x^4 + 2x^3 - 2x^2 + 4x - 4$;
no real zero greater than 1

85. $P(x) = x^4 + x^3 - x^2 + 3$;
no real zero less than −2

86. $P(x) = x^5 + 2x^3 - 2x^2 + 5x + 5$;
no real zero less than −1

87. $P(x) = 3x^4 + 2x^3 - 4x^2 + x - 1$;
no real zero greater than 1

88. $P(x) = 3x^4 + 2x^3 - 4x^2 + x - 1$;
no real zero less than −2

89. $P(x) = x^5 - 3x^3 + x + 2$;
no real zero greater than 2

90. $P(x) = x^5 - 3x^3 + x + 2$;
no real zero less than −3

Concept Check   In Exercises 91 and 92, find a cubic polynomial function having the graph shown.

91.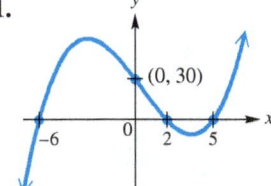

92.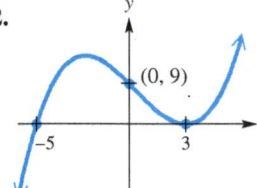

*For each polynomial function in Exercises 93–98, **do the following in order**.*
(a) *Use Descartes' rule of signs to find the possible number of positive and negative real zeros.*
(b) *Use the rational zeros theorem to determine the possible rational zeros of the function.*
(c) *Find the rational zeros, if any.*
(d) *Find all other real zeros, if any.*
(e) *Find any other nonreal complex zeros, if any.*
(f) *Find the x-intercepts of the graph, if any.*
(g) *Find the y-intercept of the graph.*
(h) *Use synthetic division to find P(4), and give the coordinates of the corresponding point on the graph.*
(i) *Determine the end behavior of the graph.*
(j) *Sketch the graph. (You may wish to support your answer with a calculator graph.)*

93. $P(x) = -2x^4 - x^3 + x + 2$

94. $P(x) = 4x^5 + 8x^4 + 9x^3 + 27x^2 + 27x$
(*Hint:* Factor out x first.)

95. $P(x) = 3x^4 - 14x^2 - 5$ (*Hint:* Factor the polynomial.)

96. $P(x) = -x^5 - x^4 + 10x^3 + 10x^2 - 9x - 9$

97. $P(x) = -3x^4 + 22x^3 - 55x^2 + 52x - 12$

98. For the polynomial functions in **Exercises 93–97** that have irrational zeros, find approximations to the nearest thousandth.

3.8 Polynomial Equations and Inequalities; Further Applications and Models

Polynomial Equations and Inequalities • Complex *n*th Roots • Applications and Polynomial Models

To solve any quadratic equation, we can use the quadratic formula. There are similar, but *very complicated*, formulas that can be used to solve third- and fourth-degree polynomial equations. (These are equations that contain x^3- and x^4-terms.) Are there formulas for fifth-degree or higher polynomial equations?

In 1824, Norwegian mathematician Niels Henrik Abel proved that it is impossible to find a formula that will yield solutions of the general *quintic* (fifth-degree) equation. A similar result holds for polynomial equations of degree greater than 5.

We use elementary methods to solve *some* higher-degree polynomial equations analytically in this section. Graphing calculators support the analytic work and enable us to find accurate approximations of real solutions of polynomial equations that cannot be solved easily, or at all, by analytic methods.

Polynomial Equations and Inequalities

EXAMPLE 1   **Solving a Polynomial Equation and Associated Inequalities**

(a) Solve $x^3 + 3x^2 - 4x - 12 = 0$ using the zero-product property.

(b) Graph $y = x^3 + 3x^2 - 4x - 12$.

(c) Use the graph from part (b) to solve the inequalities

$$x^3 + 3x^2 - 4x - 12 > 0$$
$$\text{and} \quad x^3 + 3x^2 - 4x - 12 \le 0.$$

Solution

(a)
$$x^3 + 3x^2 - 4x - 12 = 0$$
$$(x^3 + 3x^2) + (-4x - 12) = 0 \qquad \text{Group terms with common factors.}$$
$$x^2(x + 3) - 4(x + 3) = 0 \qquad \text{Factor out common factors in each group.}$$

Factor out the minus sign from each term.

$$(x + 3)(x^2 - 4) = 0 \qquad \text{Factor out } x + 3.$$
$$(x + 3)(x - 2)(x + 2) = 0 \qquad \text{Factor the difference of squares.}$$
$$x = -3 \quad \text{or} \quad x = 2 \quad \text{or} \quad x = -2 \qquad \text{Zero-product property}$$

The solution set is $\{-3, -2, 2\}$.

Algebra Review
To review factoring by grouping, see **Section R.2.**

(b) Graph $y = f(x) = x^3 + 3x^2 - 4x - 12$ and verify that the *x*-intercepts are $(-3, 0)$, $(-2, 0)$, and $(2, 0)$, supporting the analytic solution. See **FIGURE 77.**

(c) Recall that the solution set of $f(x) > 0$ consists of all numbers in the domain of f for which the graph lies *above* the *x*-axis. In **FIGURE 77,** this occurs in the intervals $(-3, -2) \cup (2, \infty)$, which is the solution set of this inequality. To solve $x^3 + 3x^2 - 4x - 12 \le 0$, locate the intervals where the graph *lies below or intersects* the *x*-axis. The figure indicates that the solution set is $(-\infty, -3] \cup [-2, 2]$. The endpoints are included here because this is a nonstrict inequality.

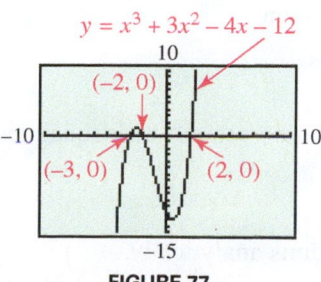
FIGURE 77

> ### Equation Quadratic in Form
>
> An equation is **quadratic in form in u** if it can be written as
>
> $$au^2 + bu + c = 0,$$
>
> where $a \neq 0$ and u is an algebraic expression.

EXAMPLE 2 **Solving an Equation Quadratic in Form and Associated Inequalities**

(a) Solve $x^4 - 6x^2 - 40 = 0$ analytically. Find all complex solutions.

(b) Graph $y = x^4 - 6x^2 - 40$, and use the graph to solve the inequalities

$$x^4 - 6x^2 - 40 \geq 0 \quad \text{and} \quad x^4 - 6x^2 - 40 < 0.$$

Give endpoints of intervals in both exact and approximate forms.

Solution

⬅ Algebra Review
To review factoring by u-substitution, see **Section R.2.**

(a)

$$x^4 - 6x^2 - 40 = 0$$
$$(x^2)^2 - 6x^2 - 40 = 0 \qquad x^4 = (x^2)^2$$
$$u^2 - 6u - 40 = 0 \qquad \text{Let } u = x^2.$$
$$(u - 10)(u + 4) = 0 \qquad \text{Factor.}$$

Remember both the positive and negative square roots.

$$u = 10 \quad \text{or} \quad u = -4 \qquad \text{Zero-product property}$$
$$x^2 = 10 \quad \text{or} \quad x^2 = -4 \qquad \text{Replace } u \text{ with } x^2.$$
$$x = \pm\sqrt{10} \quad \text{or} \quad x = \pm 2i \qquad \text{Square root property}$$

The solution set is $\left\{ -\sqrt{10},\ \sqrt{10},\ -2i,\ 2i \right\}$.

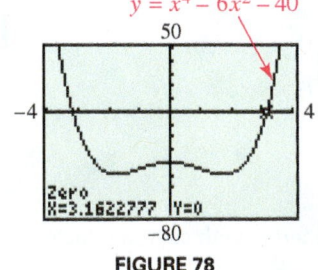

$y = x^4 - 6x^2 - 40$

FIGURE 78

(b) The graph of $y = x^4 - 6x^2 - 40$ is shown in **FIGURE 78.** The x-intercepts are approximately $(-3.16, 0)$ and $(3.16, 0)$ (x-values are approximations of $-\sqrt{10}$ and $\sqrt{10}$, respectively). ***The graph cannot support the imaginary solutions.***

Since the graph lies above or intersects the x-axis for real numbers less than or equal to $-\sqrt{10}$ and for real numbers greater than or equal to $\sqrt{10}$, the solution set of $x^4 - 6x^2 - 40 \geq 0$ includes the intervals

The graph is above or intersecting the x-axis for these x-values.

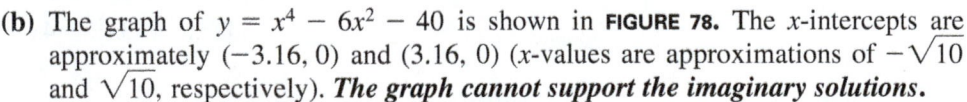

$$\left(-\infty, -\sqrt{10}\,\right] \cup \left[\,\sqrt{10}, \infty\right) \qquad \text{Exact form}$$

or

$$(-\infty, -3.16] \cup [3.16, \infty). \qquad \text{Approximate form}$$

By similar reasoning, the solution set of $x^4 - 6x^2 - 40 < 0$ is the interval

The graph is below the x-axis for these x-values.

$$\left(-\sqrt{10}, \sqrt{10}\,\right) \qquad \text{Exact form}$$

or

$$(-3.16, 3.16). \qquad \text{Approximate form}$$

The nonreal complex solutions do not affect the solution sets of the inequalities.

In the next example we solve two polynomial equations analytically.

TECHNOLOGY NOTE

The TI-84 Plus application PlySmlt2 can be used to find complex zeros of higher-degree polynomials, as illustrated below where **Example 3(a)** is solved.

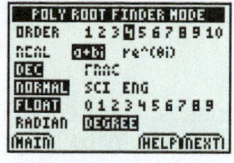

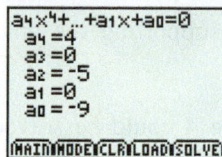

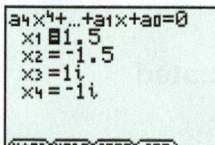

EXAMPLE 3   **Solving Polynomial Equations Analytically**

Find all solutions to each equation analytically.

(a) $4x^4 - 5x^2 - 9 = 0$ **(b)** $2x^3 + 12 = 3x^2 + 8x$

Solution

(a) The expression $4x^4 - 5x^2 - 9$ can be factored in a manner similar to the way quadratic expressions are factored. (We do not use the u-substitution here that was used in **Example 2(a)**.)

$$4x^4 - 5x^2 - 9 = 0$$

$$(4x^2 - 9)(x^2 + 1) = 0 \qquad \text{Factor.}$$

$$4x^2 - 9 = 0 \quad \text{or} \quad x^2 + 1 = 0 \qquad \text{Zero-product property}$$

$$4x^2 = 9 \quad \text{or} \quad x^2 = -1 \qquad \text{Add 9 or subtract 1.}$$

$$x^2 = \frac{9}{4} \quad \text{or} \quad x^2 = -1 \qquad \text{Divide by 4 in left equation.}$$

$$x = \pm\frac{3}{2} \quad \text{or} \quad x = \pm i \qquad \text{Square root property}$$

The solution set of the given equation is $\left\{-\frac{3}{2}, \frac{3}{2}, -i, i\right\}$.

(b) First rewrite the equation so that each term on the right side of the equation is on the left side of the equation. Then use *grouping* to factor the polynomial.

$$2x^3 + 12 = 3x^2 + 8x$$

$$2x^3 - 3x^2 - 8x + 12 = 0 \qquad \text{Subtract } 3x^2 \text{ and } 8x.$$

$$(2x^3 - 3x^2) + (-8x + 12) = 0 \qquad \text{Associative property}$$

$$x^2(2x - 3) - 4(2x - 3) = 0 \qquad \text{Factor.}$$

$$(x^2 - 4)(2x - 3) = 0 \qquad \text{Factor out } 2x - 3.$$

$$x^2 - 4 = 0 \quad \text{or} \quad 2x - 3 = 0 \qquad \text{Zero-product property}$$

$$x = \pm 2 \quad \text{or} \quad x = \frac{3}{2} \qquad \text{Solve each equation.}$$

The solution set is $\left\{-2, \frac{3}{2}, 2\right\}$.

EXAMPLE 4   **Solving a Polynomial Equation**

(a) Show that 2 is a real solution of $P(x) = x^3 + 3x^2 - 11x + 2 = 0$, and then find all solutions of this equation.

(b) Support the result of part (a) graphically.

Solution

(a)

```
2)1   3   -11    2        Use synthetic division.
        2    10   -2
   1   5    -1    0  ← P(2) = 0 by the remainder theorem.
   └─────────┘
   Coefficients of the
   quotient polynomial
```

By the factor theorem, $x - 2$ is a factor of $P(x)$.

$$P(x) = (x - 2)(x^2 + 5x - 1)$$

(continued)

To find the other zeros of P, solve $x^2 + 5x - 1 = 0$.

$$x = \frac{-b \pm \sqrt{b^2 - 4ac}}{2a} = \frac{-5 \pm \sqrt{5^2 - 4(1)(-1)}}{2(1)}$$

Quadratic formula;
$a = 1, b = 5, c = -1$

$$= \frac{-5 \pm \sqrt{29}}{2}$$

The solution set is $\left\{ \dfrac{-5 - \sqrt{29}}{2}, \dfrac{-5 + \sqrt{29}}{2}, 2 \right\}$.

(b) The graph of $P(x) = x^3 + 3x^2 - 11x + 2$ is shown in **FIGURE 79.** The x-coordinates of the x-intercepts are 2, approximately -5.19, and approximately 0.19. The latter two approximations are for $\dfrac{-5 - \sqrt{29}}{2}$ and $\dfrac{-5 + \sqrt{29}}{2}$, supporting the analytic results. ●

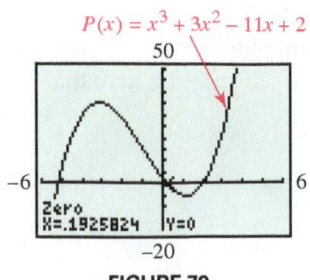

$P(x) = x^3 + 3x^2 - 11x + 2$

FIGURE 79

The associated inequalities for the equation in **Example 4** could be solved by using the solutions and graph as was done in **Examples 1** and **2**.

EXAMPLE 5 **Solving an Equation and Associated Inequalities Graphically**

Let $P(x) = 2.45x^3 - 3.14x^2 - 6.99x + 2.58$. Use a graph to solve $P(x) = 0$, $P(x) > 0$, and $P(x) < 0$. Express solutions of the equation and endpoints of the intervals for the inequalities to the nearest hundredth.

Solution The graph of P is shown in **FIGURE 80.** Using a calculator, we find that the approximate x-coordinates of the x-intercepts are -1.37, 0.33, and 2.32. Therefore, the solution set of the equation $P(x) = 0$ is $\{-1.37, 0.33, 2.32\}$. Based on the graph, the solution sets of $P(x) > 0$ and $P(x) < 0$ are, respectively,

$$(-1.37, 0.33) \cup (2.32, \infty) \quad \text{and} \quad (-\infty, -1.37) \cup (0.33, 2.32). \quad ●$$

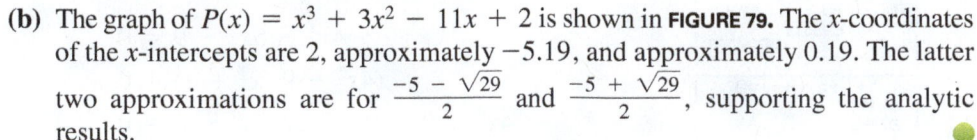

$P(x) = 2.45x^3 - 3.14x^2$
$- 6.99x + 2.58$

The other two x-intercepts are approximately (0.33, 0) and (2.32, 0).

FIGURE 80

NOTE The graphical method of solving $P(x) = 0$ in **Example 5** would not have yielded nonreal complex solutions had there been any. Only real solutions are obtained with this method.

Equations and inequalities like the ones in the next example often occur in calculus.

EXAMPLE 6 **Solving a Polynomial Equation and Inequality**

Solve the equation and inequality.

(a) $2(x^2 - 1) + 2x(2x - 1) = 0$ **(b)** $2(x^2 - 1) + 2x(2x - 1) < 0$

Solution

(a)

$2(x^2 - 1) + 2x(2x - 1) = 0$	Given equation
$(2x^2 - 2) + (4x^2 - 2x) = 0$	Multiply.
$6x^2 - 2x - 2 = 0$	Add like terms.
$3x^2 - x - 1 = 0$	Divide each side by 2.

$$x = \frac{1 \pm \sqrt{(-1)^2 - 4(3)(-1)}}{2(3)}$$

Quadratic formula;
$a = 3, b = -1, c = -1$

$$x = \frac{1 \pm \sqrt{13}}{6}$$

Simplify.

The solution set is $\left\{ \dfrac{1 \pm \sqrt{13}}{6} \right\}$.

(b) From part (a) we know that the graph of $y = 2(x^2 - 1) + 2x(2x - 1)$, or equivalently $y = 6x^2 - 2x - 2$, is a parabola that opens upward with zeros of $\frac{1 \pm \sqrt{13}}{6}$. The parabola is below the x-axis between its zeros, so the solution set of the inequality is the interval $\left(\frac{1 - \sqrt{13}}{6}, \frac{1 + \sqrt{13}}{6} \right)$. Note that it is not necessary to actually graph the parabola to solve the inequality.   ●

Complex *n*th Roots

If n is a positive integer and k is a nonzero complex number, then a solution of $x^n = k$ is called an ***n*th root of *k*.** For example, since -1 and 1 are solutions of $x^2 = 1$, they are called second, or square, roots of 1. Similarly, the pure imaginary numbers $-2i$ and $2i$ are called square roots of -4, since $(\pm 2i)^2 = -4$.

The real numbers -2 and 2 are sixth roots of 64, since $(\pm 2)^6 = 64$. However, 64 has four more complex sixth roots. While a complete discussion of the next theorem requires concepts from trigonometry, we state it and use it to solve particular problems involving nth roots.

> ### Complex *n*th Roots Theorem
>
> If n is a positive integer and k is a nonzero complex number, then the equation $x^n = k$ has *exactly* n complex roots.

EXAMPLE 7   **Finding *n*th Roots of a Number**

Find all six complex sixth roots of 64.

← Algebra Review
To review factoring the sum and difference of two cubes, see **Section R.2.**

Solution   The sixth roots must be solutions of $x^6 = 64$.

$$x^6 = 64$$
$$x^6 - 64 = 0 \qquad \text{Subtract 64.}$$
$$(x^3 - 8)(x^3 + 8) = 0 \qquad \text{Factor the difference of squares.}$$
$$(x - 2)(x^2 + 2x + 4)(x + 2)(x^2 - 2x + 4) = 0 \qquad \text{Factor the difference of cubes and the sum of cubes.}$$

Now apply the zero-product property to obtain the real sixth roots, 2 and -2. Setting the quadratic factors equal to 0 and applying the quadratic formula twice gives the remaining four complex roots, none of which are real.

$$
\begin{array}{c|c}
x^2 + 2x + 4 = 0 & x^2 - 2x + 4 = 0 \\[4pt]
x = \dfrac{-2 \pm \sqrt{2^2 - 4(1)(4)}}{2(1)} & x = \dfrac{2 \pm \sqrt{(-2)^2 - 4(1)(4)}}{2(1)} \\[10pt]
= \dfrac{-2 \pm \sqrt{-12}}{2} & = \dfrac{2 \pm \sqrt{-12}}{2} \\[10pt]
= \dfrac{-2 \pm 2i\sqrt{3}}{2} & = \dfrac{2 \pm 2i\sqrt{3}}{2} \\[10pt]
= \dfrac{2(-1 \pm i\sqrt{3})}{2} & = \dfrac{2(1 \pm i\sqrt{3})}{2} \\[10pt]
= -1 \pm i\sqrt{3} & = 1 \pm i\sqrt{3}
\end{array}
$$

Factor first. Then divide out the common factor.

Therefore, the six complex sixth roots of 64 are

$$2, \quad -2, \quad -1 + i\sqrt{3}, \quad -1 - i\sqrt{3}, \quad 1 + i\sqrt{3}, \quad \text{and} \quad 1 - i\sqrt{3}.$$ ●

Applications and Polynomial Models

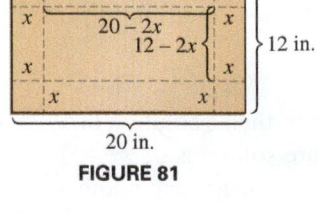

FIGURE 81

| EXAMPLE 8 | **Using a Polynomial Function to Model the Volume of a Box** |

A box with an open top is to be constructed from a rectangular 12-inch by 20-inch piece of cardboard by cutting equal-sized squares from each corner and folding up the sides. See **FIGURE 81.**

(a) If x represents the length of a side of each cut out square, determine a function V that describes the volume of the box in terms of x.

(b) Graph V in the window $[0, 6]$ by $[0, 300]$, and locate a point on the graph. Interpret the displayed values of x and y.

(c) Determine the value of x for which the volume of the box is maximized. What is this volume?

(d) For what values of x is the volume equal to 200 cubic inches? Greater than 200 cubic inches? Less than 200 cubic inches?

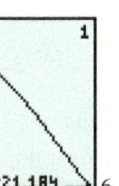

FIGURE 82

Solution

(a) As shown in **FIGURE 82,** the dimensions (in inches) of the box to be formed will be

$$\text{length} = 20 - 2x, \quad \text{width} = 12 - 2x, \quad \text{height} = x.$$

Furthermore, x must be positive, and both $20 - 2x$ and $12 - 2x$ must be positive, implying that $0 < x < 6$. The desired function is

$$V(x) = (20 - 2x)(12 - 2x)x \qquad \text{Volume} = \text{length} \times \text{width} \times \text{height}$$
$$= 4x^3 - 64x^2 + 240x.$$

(b) **FIGURE 83** shows the graph of V with the cursor at the arbitrarily chosen point $(3.6, 221.184)$. This means that when the side of each cut-out square measures 3.6 inches, the volume of the resulting box is 221.184 cubic inches.

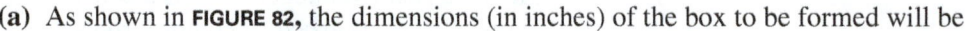

FIGURE 83

(c) Use a calculator to find the local maximum point on the graph of V. To the nearest hundredth, the coordinates of this point are $(2.43, 262.68)$. See **FIGURE 84.** Therefore, when $x \approx 2.43$ is the measure of the side of each cut-out square, the volume of the box is at its maximum, approximately 262.68 cubic inches.

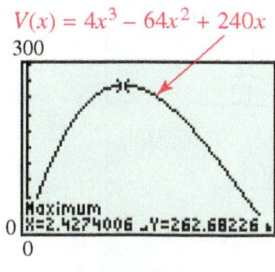

FIGURE 84

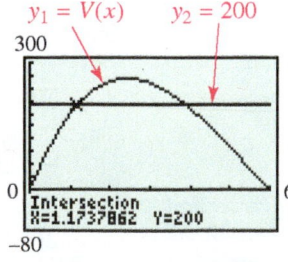

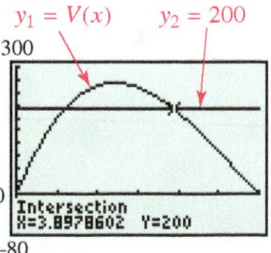

FIGURE 85

(d) The graphs of $y_1 = V(x)$ and $y_2 = 200$ are shown in **FIGURE 85.** The points of intersection of the line and the cubic curve are approximately $(1.17, 200)$ and

(3.90, 200), so the volume is equal to 200 cubic inches for $x \approx 1.17$ or 3.90, is greater than 200 cubic inches for $1.17 < x < 3.90$, and is less than 200 cubic inches for $0 < x < 1.17$ or $3.90 < x < 6$. ●

EXAMPLE 9   **Modeling with a Cubic Polynomial**

The table shows the number of transactions, in billions, by users of bank debit cards for selected years.

Year	2000	2002	2004	2007	2012
Transactions (in billions)	8.3	13.3	19.7	27.9	52.6

Source: *Statistical Abstract of the United States.*

(a) Use regression to find a cubic polynomial $P(x)$ that models the data, where x represents years after 2000. Graph P and the data together.

(b) Approximate the number of transactions in 2005 and compare it with the actual value of 22.7 billion.

(c) Estimate the year when transactions reached 68 billion.

Solution

(a) From **FIGURE 86(a)**, we see that

$$P(x) \approx 0.01698x^3 - 0.1525x^2 + 3.0877x + 8.1391.$$

A graph of P and the data are shown in **FIGURE 86(b)**. Note that $x = 0$ represents 2000, $x = 2$ represents 2002, and so on.

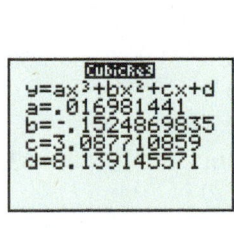

Cubic regression

(a)

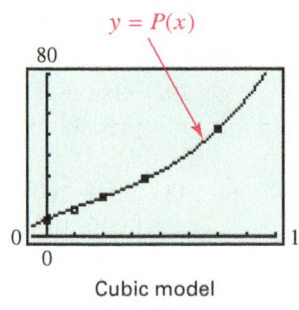

Cubic model

(b)

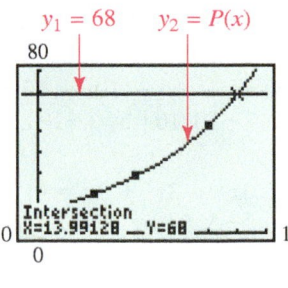

(c)

FIGURE 86

(b) The year 2005 is 5 years after 2000, so evaluate $P(5)$.

$$P(5) \approx 0.01698(5)^3 - 0.1525(5)^2 + 3.0877(5) + 8.1391 \approx 21.9$$

This model estimates 21.9 billion transactions in 2005, which is lower than the actual value of 22.7 billion transactions.

(c) We can estimate the year when there might be 68 billion transactions by determining where the graphs of $y = P(x)$ and $y = 68$ intersect. This point of intersection is near (13.99, 68) and is shown in **FIGURE 86(c)**. This model estimates about 68 billion transactions 14 years after 2000, or in 2014. ●

3.8 Exercises

Checking Analytic Skills   *Find all real solutions.* **Do not use a calculator.**

1. $x^3 - 25x = 0$

2. $x^4 - x^3 - 6x^2 = 0$

3. $x^4 - x^2 = 2x^2 + 4$

4. $x^4 + 5 = 6x^2$

5. $x^3 - 3x^2 - 18x = 0$

6. $x^4 - x^2 = 0$

7. $2x^3 = 4x^2 - 2x$

8. $x^3 = x$

9. $12x^3 = 17x^2 + 5x$

10. $3x^3 + 3x = 10x^2$

11. $2x^3 + 4 = x(x + 8)$

12. $3x^3 + 18 = x(2x + 27)$

Checking Analytic Skills   *Find all complex solutions of each equation.* **Do not use a calculator.**

13. $7x^3 + x = 0$

14. $2x^3 - 4x = 0$

15. $3x^3 + 2x^2 - 3x - 2 = 0$

16. $5x^3 - x^2 + 10x - 2 = 0$

17. $x^4 - 11x^2 + 10 = 0$

18. $x^4 + x^2 - 6 = 0$

Solve each equation analytically for all complex solutions, giving exact *forms in your solution set. Then, graph the left side of the equation as y_1 in the suggested viewing window and, using the capabilities of your calculator, support the real solutions.*

19. $4x^4 - 25x^2 + 36 = 0$;
$[-5, 5]$ by $[-5, 100]$

20. $4x^4 - 29x^2 + 25 = 0$;
$[-5, 5]$ by $[-50, 100]$

21. $x^4 - 15x^2 - 16 = 0$;
$[-5, 5]$ by $[-100, 100]$

22. $9x^4 + 35x^2 - 4 = 0$;
$[-3, 3]$ by $[-10, 100]$

23. $x^3 - x^2 - 64x + 64 = 0$;
$[-10, 10]$ by $[-300, 300]$

24. $x^3 + 6x^2 - 100x - 600 = 0$;
$[-15, 15]$ by $[-1000, 300]$

25. $-2x^3 - x^2 + 3x = 0$;
$[-4, 4]$ by $[-10, 10]$

26. $-5x^3 + 13x^2 + 6x = 0$;
$[-4, 4]$ by $[-2, 30]$

27. $x^3 + x^2 - 7x - 7 = 0$;
$[-10, 10]$ by $[-20, 20]$

28. $x^3 + 3x^2 - 19x - 57 = 0$;
$[-10, 10]$ by $[-100, 50]$

29. $-3x^3 - x^2 + 6x = 0$;
$[-4, 4]$ by $[-10, 10]$

30. $-4x^3 - x^2 + 4x = 0$;
$[-4, 4]$ by $[-10, 10]$

31. $3x^3 + 3x^2 + 3x = 0$;
$[-5, 5]$ by $[-5, 5]$

32. $2x^3 + 2x^2 + 12x = 0$;
$[-10, 10]$ by $[-20, 20]$

33. $x^4 + 17x^2 + 16 = 0$;
$[-4, 4]$ by $[-10, 40]$

34. $36x^4 + 85x^2 + 9 = 0$;
$[-4, 4]$ by $[-10, 40]$

35. $x^6 + 19x^3 - 216 = 0$;
$[-4, 4]$ by $[-350, 200]$

36. $8x^6 + 7x^3 - 1 = 0$;
$[-4, 4]$ by $[-5, 100]$

Graph each polynomial function by hand, as shown in **Section 3.7.** *Then solve each equation and inequality.*

37. $P(x) = x^3 - 3x^2 - 6x + 8$
$= (x - 4)(x - 1)(x + 2)$
   (a) $P(x) = 0$ **(b)** $P(x) < 0$ **(c)** $P(x) > 0$

38. $P(x) = x^3 + 4x^2 - 11x - 30$
$= (x - 3)(x + 2)(x + 5)$
   (a) $P(x) = 0$ **(b)** $P(x) < 0$ **(c)** $P(x) > 0$

39. $P(x) = 2x^4 - 9x^3 - 5x^2 + 57x - 45$
$= (x - 3)^2(2x + 5)(x - 1)$
   (a) $P(x) = 0$ **(b)** $P(x) < 0$ **(c)** $P(x) > 0$

40. $P(x) = 4x^4 + 27x^3 - 42x^2 - 445x - 300$
$= (x + 5)^2(4x + 3)(x - 4)$
   (a) $P(x) = 0$ **(b)** $P(x) < 0$ **(c)** $P(x) > 0$

41. $P(x) = -x^4 - 4x^3 + 3x^2 + 18x$
$\quad = -x(x - 2)(x + 3)^2$
(a) $P(x) = 0$ **(b)** $P(x) \geq 0$ **(c)** $P(x) \leq 0$

42. $P(x) = -x^4 + 2x^3 + 8x^2$
$\quad = -x^2(x - 4)(x + 2)$
(a) $P(x) = 0$ **(b)** $P(x) \geq 0$ **(c)** $P(x) \leq 0$

Solve each equation and inequality.

43. **(a)** $3(x^2 + 4) + 2x(3x - 12) = 0$
$\quad$ **(b)** $3(x^2 + 4) + 2x(3x - 12) < 0$

44. **(a)** $(x^2 + 3x - 1) + (2x + 3)(x - 5) = 0$
$\quad$ **(b)** $(x^2 + 3x - 1) + (2x + 3)(x - 5) \geq 0$

45. **(a)** $3(x + 1)^2(2x - 1)^4 + 8(x + 1)^3(2x - 1)^3 = 0$
$\quad$ **(b)** $3(x + 1)^2(2x - 1)^4 + 8(x + 1)^3(2x - 1)^3 \geq 0$

46. **(a)** $4x(x^2 + 1)(x^2 + 4)^3 + 6x(x^2 + 1)^2(x^2 + 4)^2 = 0$
$\quad$ **(b)** $4x(x^2 + 1)(x^2 + 4)^3 + 6x(x^2 + 1)^2(x^2 + 4)^2 < 0$

Solve each equation and inequality, where k is a positive constant.

47. **(a)** $3kx^2 - 7x = 0$
$\quad$ **(b)** $3kx^2 - 7x < 0$

48. **(a)** $4x^3 - kx = 0$
$\quad$ **(b)** $4x^3 - kx > 0$

Use a graphical method to find all real solutions of each equation. Express solutions to the nearest hundredth.

49. $0.86x^3 - 5.24x^2 + 3.55x + 7.84 = 0$

50. $-2.47x^3 - 6.58x^2 - 3.33x + 0.14 = 0$

51. $-\sqrt{7}x^3 + \sqrt{5}x^2 + \sqrt{17} = 0$

52. $\sqrt{10}x^3 - \sqrt{11}x - \sqrt{8} = 0$

53. $2.45x^4 - 3.22x^3 = -0.47x^2 + 6.54x + 3$

54. $\sqrt{17}x^4 - \sqrt{22}x^2 = -1$

Find all n complex solutions of each equation of the form $x^n = k$.

55. $x^2 = -1$

56. $x^2 = -4$

57. $x^3 = -1$

58. $x^3 = -8$

59. $x^3 = 27$

60. $x^3 = 64$

61. $x^4 = 16$

62. $x^4 = 81$

63. $x^3 = -64$

64. $x^3 = -27$

65. $x^2 = -18$

66. $x^2 = -52$

(Modeling) Solve each problem.

 67. *Floating Ball* The polynomial function

$$f(x) = \frac{\pi}{3}x^3 - 5\pi x^2 + \frac{500\pi d}{3}$$

can be used to find the depth that a ball 10 centimeters in diameter sinks in water. The constant d is the density

of the ball, where the density of water is 1. The smallest *positive* zero of $f(x)$ equals the depth that the ball sinks. Approximate this depth for each material and interpret the results.
(a) A wooden ball with $d = 0.8$
(b) A solid aluminum ball with $d = 2.7$
(c) A spherical water balloon with $d = 1$

68. *Floating Ball*   Refer to **Exercise 67.** Determine the depth to which a pine ball with a 10-centimeter diameter sinks in water if $d = 0.55$.

69. *Volume of a Box*   A rectangular piece of cardboard measuring 12 inches by 18 inches is to be made into a box with an open top by cutting equal-sized squares from each corner and folding up the sides. Let x represent the length of a side of each such square in inches.

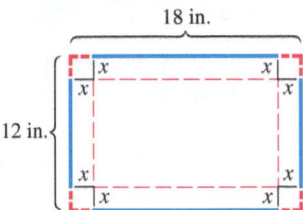

18 in.

12 in.

(a) Give the restrictions on x.
(b) Determine a function V that gives the volume of the box as a function of x.
(c) For what value of x will the volume be a maximum? What is this maximum volume?
(d) For what values of x will the volume be greater than 80 cubic inches?

70. *Construction of a Rain Gutter*   A rectangular piece of sheet metal is 20 inches wide. It is to be made into a rain gutter by turning up the edges to form parallel sides. Let x represent the length of each of the parallel sides in inches. See the figure.

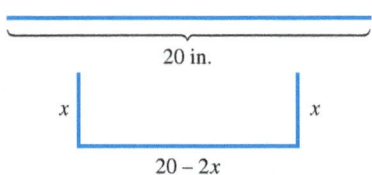

20 in.

x x

$20 - 2x$

(a) Give the restrictions on x.
(b) Determine a function $\mathcal{A}$ that gives the area of a cross section of the gutter.
(c) For what value of x will $\mathcal{A}$ be a maximum (and thus maximize the amount of water that the gutter will hold)? What is this maximum area?
(d) For what values of x to the nearest hundredth will the area of a cross section be less than 40 square inches?

71. *Buoyancy of a Spherical Object*   It has been determined that a spherical object of radius 4 inches with specific gravity 0.25 will sink in water to a depth of x inches, where x is the least positive root of the equation

$$x^3 - 12x^2 + 64 = 0.$$

To what depth will this object sink if $x < 10$?

72. *Area of a Rectangle*   Approximate the value of x in the figure that will maximize the area of rectangle $ABCD$.

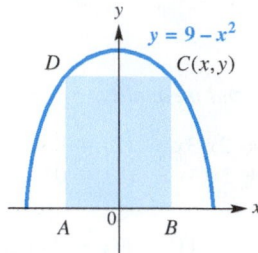

73. *Sides of a Right Triangle*   A certain right triangle has area 84 square inches. One leg of the triangle measures 1 inch less than the hypotenuse. Let x represent the length of the hypotenuse.
(a) Express the length of the leg described in terms of x.
(b) Express the length of the other leg in terms of x.
(c) Write an equation based on the information determined thus far. Square each side, and then write the equation with one side as a polynomial with integer coefficients, in descending powers, and the other side equal to 0.
(d) Solve the equation in part (c) graphically. Find the lengths of the three sides of the triangle.

74. *Butane Gas Storage*   A storage tank for butane gas is to be built in the shape of a right circular cylinder having altitude 12 feet, as shown, with a half sphere attached to each end. If x represents the radius of each half sphere, what radius should be used to cause the volume of the tank to be 144π cubic feet?

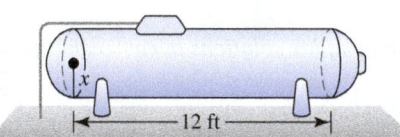

12 ft

75. *Volume of a Box*   A standard piece of notebook paper measuring 8.5 inches by 11 inches is to be made into a box with an open top by cutting equal-sized squares from each corner and folding up the sides. Let x represent the length of a side of each such square in inches.
(a) Use the table feature of your graphing calculator to find the maximum volume of the box.
(b) Use the table feature to determine to the nearest hundredth when the volume of the box will be greater than 40 cubic inches.

76. *Highway Design*   To allow enough distance for cars to pass on two-lane highways, engineers calculate minimum sight distances between curves and hills. The table at top of the next page shows the minimum sight distance y in feet for a car traveling at x mph.

x (in mph)	20	30	40	50
y (in feet)	810	1090	1480	1840

x (in mph)	60	65	70
y (in feet)	2140	2310	2490

Source: Haefner, L., *Introduction to Transportation Systems,* Holt, Rinehart and Winston.

(a) Make a scatter diagram of the data.

(b) Use the regression feature of a calculator to find the best-fitting linear function for the data. Graph the function with the data.

(c) Repeat part (b) for a cubic function.

(d) Use both functions from parts (b) and (c) to estimate the minimum sight distance for a car traveling 43 mph.

(e) Which function fits the data better?

77. *Water Pollution* Copper in high doses can be lethal to aquatic life. The table lists copper concentrations in mussels after 45 days at various distances downstream from an electroplating plant. The concentration C is measured in micrograms of copper per gram of mussel x kilometers downstream. See the table at the top of the next column.

x	5	21	37	53	59
C	20	13	9	6	5

Source: Foster, R., and J. Bates, "Use of mussels to monitor point source industrial discharges," *Environ. Sci. Technol.*

(a) Make a scatter diagram of the data.

(b) Use the regression feature of a calculator to find the best-fitting quadratic function C for the data. Graph the function with the data.

(c) Repeat part (b) for a cubic function.

(d) By comparing graphs of the functions in parts (b) and (c) with the data, decide which function best fits the given data.

(e) Concentrations above 10 are lethal to mussels. Use the cubic function to find the values of x to the nearest hundredth for which this is the case.

78. *U.S. Text Messaging* The average number of text messages sent and received each day by people of selected ages is shown in the table.

Age	20	30	40	50	60	70
Number of texts	110	42	26	14	10	5

Source: pewinternet.org

(a) Find a cubic polynomial $T(x)$ that models these data, where x is the age.

(b) Evaluate $T(35)$ and interpret the result.

(c) Determine the average age of a person who sends 85 texts per day.

SECTIONS 3.6–3.8 Reviewing Basic Concepts

For Exercises 1–3, use $P(x) = 2x^4 - 7x^3 + 29x - 30$.

1. Find $P(3)$.

2. Is 2 a zero of P?

3. One zero of P is $2 + i$. Factor $P(x)$ into linear factors.

4. Find a cubic polynomial $P(x)$ with real coefficients that has zeros of $\frac{3}{2}$ and i and for which $P(3) = 15$.

5. Find a polynomial $P(x)$ of least possible degree with real coefficients and zeros -4 (multiplicity 2) and $1 + 2i$.

6. Given $P(x) = 2x^3 + x^2 - 11x - 10$, list all zeros of P.

7. Find all solutions of $3x^4 - 12x^2 + 1 = 0$.

8. *(Modeling) Aging in America* The table lists the number N (in thousands) of Americans over 100 years old for selected years x.

x	1960	1970	1980	1990	2000	2010
N	3	5	15	37	75	129

Source: U.S. Census Bureau.

(a) Use regression to find a cubic polynomial $P(x)$ that models the data x years after 1960.

(b) Estimate N in 1994 and compare it with the actual value of 50 thousand.

(c) Estimate the year when N first reached 100 thousand.

3 Summary

| **KEY TERMS & SYMBOLS** | **KEY CONCEPTS** |

3.1 Complex Numbers

complex number system
complex number
real part
imaginary part
pure imaginary number
nonreal complex number
standard form of a complex number
complex conjugates (conjugates)

IMAGINARY UNIT i

$$i = \sqrt{-1}, \quad \text{and therefore} \quad i^2 = -1.$$

COMPLEX NUMBER

A number in the form $a + bi$, where a and b are real numbers and i is the imaginary unit, is called a complex number.

$\sqrt{-a}$

If $a > 0$, then $\sqrt{-a} = i\sqrt{a}$.

COMPLEX CONJUGATE

The complex conjugate of $a + bi$ is $a - bi$.

OPERATIONS WITH COMPLEX NUMBERS

Addition: $(a + bi) + (c + di) = (a + c) + (b + d)i$

Subtraction: $(a + bi) - (c + di) = (a - c) + (b - d)i$

Multiplication: $(a + bi)(c + di) = (ac - bd) + (ad + bc)i$

Division: To divide complex numbers, multiply both the numerator and the denominator by the conjugate of the denominator.

3.2 Quadratic Functions and Graphs

quadratic function
axis of symmetry
extreme points (extrema, extremum)
minimum point
minimum value
maximum point
maximum value

CHARACTERISTICS OF THE GRAPH OF $P(x) = a(x - h)^2 + k$

Consider the graph of $P(x) = a(x - h)^2 + k$ ($a \neq 0$).

(a) It is a parabola with vertex (h, k) and vertical line $x = h$ as axis of symmetry.

(b) It opens upward if $a > 0$ and downward if $a < 0$.

(c) It is wider than the graph of $y = x^2$ if $0 < |a| < 1$ and narrower than the graph of $y = x^2$ if $|a| > 1$.

VERTEX FORMULA

The vertex of the graph of $P(x) = ax^2 + bx + c$ ($a \neq 0$) is the point $\left(-\frac{b}{2a}, P\left(-\frac{b}{2a}\right)\right)$.

EXTREME VALUES

Consider the quadratic function $P(x) = ax^2 + bx + c$ with vertex (h, k).

(a) If $a > 0$, then the vertex (h, k) is called the **minimum point** of the graph. The **minimum value** of the function is $P(h) = k$.

(b) If $a < 0$, then the vertex (h, k) is called the **maximum point** of the graph. The **maximum value** of the function is $P(h) = k$.

HEIGHT OF A PROJECTED OBJECT

If air resistance is neglected, the height s (in feet) of an object projected directly upward from an initial height s_0 feet with initial velocity v_0 feet per second is

$$s(t) = -16t^2 + v_0 t + s_0,$$

where t is the number of seconds after the object is projected.

KEY TERMS & SYMBOLS	**KEY CONCEPTS**

3.3 Quadratic Equations and Inequalities

quadratic equation in one variable
quadratic equation in standard form
discriminant
quadratic inequality in one variable

ZERO-PRODUCT PROPERTY

If a and b are complex numbers and $ab = 0$, then $a = 0$ or $b = 0$ or both.

SQUARE ROOT PROPERTY

The solution set of $x^2 = k$ is one of the following.

(a) $\left\{ \pm\sqrt{k} \right\}$ if $k > 0$

(b) $\{0\}$ if $k = 0$

(c) $\left\{ \pm i\sqrt{|k|} \right\}$ if $k < 0$

QUADRATIC FORMULA

The solutions of the quadratic equation $ax^2 + bx + c = 0$, where $a \neq 0$, are given by the formula

$$x = \frac{-b \pm \sqrt{b^2 - 4ac}}{2a}.$$

MEANING OF THE DISCRIMINANT

If a, b, and c are real numbers, $a \neq 0$, then the complex solutions of $ax^2 + bx + c = 0$ are described as follows, based on the value of the discriminant, $b^2 - 4ac$.

Value of $b^2 - 4ac$	Number of Solutions	Type of Solutions
Positive	Two	Real
Zero	One (a double solution)	Real
Negative	Two	Nonreal complex

SOLVING A QUADRATIC INEQUALITY

Step 1 Solve the corresponding quadratic equation.

Step 2 Identify the intervals determined by the solutions of the equation.

Step 3 Use a test value from each interval to determine which intervals are in the solution set.

3.4 Applications of Quadratic Functions and Models

QUADRATIC MODEL

A quadratic function can be used to model data. A model can be determined by choosing an appropriate data point (h, k) as vertex and using another point to find the value of a in $f(x) = a(x - h)^2 + k$.

3.5 Higher-Degree Polynomial Functions and Graphs

polynomial function
leading coefficient
dominating term

EXTREMA OF POLYNOMIAL FUNCTIONS

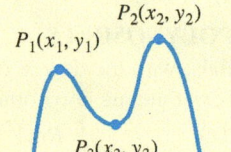

P_1 is a local maximum point.
P_2 is a local and absolute maximum point.
P_3 is a local minimum point.
There is no absolute minimum point.

(continued)

cubic function
quartic function
turning points
local maximum point
local minimum point
local maximum
local minimum
absolute maximum (mini-
   mum) point
absolute maximum (mini-
   mum) value
$x \rightarrow -\infty$
$x \rightarrow \infty$

NUMBER OF TURNING POINTS

The number of turning points of the graph of a polynomial function of degree $n \geq 1$ is at most $n - 1$.

END BEHAVIOR OF GRAPHS OF POLYNOMIAL FUNCTIONS

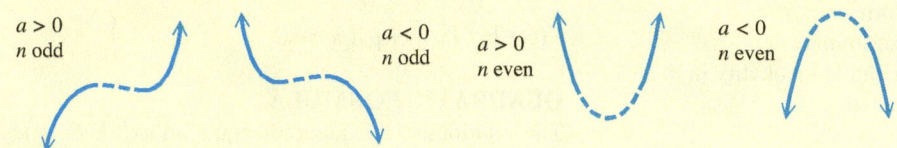

NUMBER OF x-INTERCEPTS (AND REAL ZEROS) OF A POLYNOMIAL FUNCTION

The graph of a polynomial function of degree n will have at most n x-intercepts (corresponding to its real zeros).

3.6 Topics in the Theory of Polynomial Functions (I)

synthetic division

INTERMEDIATE VALUE THEOREM

If $P(x)$ is a polynomial function with only real coefficients, and if, for real numbers a and b, the values $P(a)$ and $P(b)$ are opposite in sign, then there exists at least one real zero between a and b.

DIVISION OF A POLYNOMIAL BY $x - k$

1. If the degree n polynomial $P(x)$ (where $n \geq 1$) is divided by $x - k$, then the quotient polynomial, $Q(x)$, has degree $n - 1$.

2. The remainder R is a constant (and may be 0). The complete quotient for $\dfrac{P(x)}{x - k}$ may be written as follows.

$$\frac{P(x)}{x - k} = Q(x) + \frac{R}{x - k}$$

REMAINDER THEOREM

If a polynomial $P(x)$ is divided by $x - k$, the remainder is equal to $P(k)$.

FACTOR THEOREM

A polynomial $P(x)$ has a factor $x - k$ if and only if $P(k) = 0$.

DIVISION ALGORITHM FOR POLYNOMIALS

Let $P(x)$ and $D(x)$ be two polynomials, with the degree of $D(x)$ greater than zero and less than the degree of $P(x)$. Then there exist unique polynomials $Q(x)$ and $R(x)$ such that

$$\frac{P(x)}{D(x)} = Q(x) + \frac{R(x)}{D(x)},$$

where either $R(x) = 0$ or the degree of $R(x)$ is less than the degree of $D(x)$.

KEY TERMS & SYMBOLS	**KEY CONCEPTS**

3.7 Topics in the Theory of Polynomial Functions (II)

multiplicity of the zero
variation in sign

CONJUGATE ZEROS THEOREM

If $P(x)$ is a polynomial function having only real coefficients, and if $a + bi$ is a zero of $P(x)$, then the conjugate $a - bi$ is also a zero of $P(x)$.

FUNDAMENTAL THEOREM OF ALGEBRA

Every function defined by a polynomial of degree 1 or more has at least one complex zero.

NUMBER OF ZEROS THEOREM

1. A function defined by a polynomial of degree n has at most n distinct complex zeros.

2. A function defined by a polynomial of degree n has exactly n complex zeros if zeros of multiplicity m are counted m times.

BEHAVIOR OF THE GRAPH OF A POLYNOMIAL FUNCTION NEAR THE x-AXIS

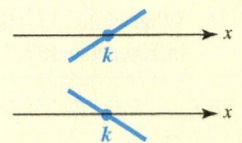

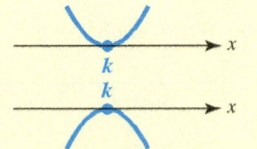

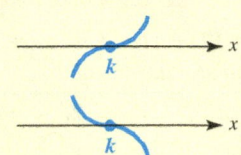

The graph crosses the x-axis at $(k, 0)$ if k is a zero of multiplicity one.

The graph is tangent to the x-axis at $(k, 0)$ if k is a zero of even multiplicity. The graph bounces, or turns, at k.

The graph crosses *and* is tangent to the x-axis at $(k, 0)$ if k is a zero of odd multiplicity greater than one. The graph curves at k.

RATIONAL ZEROS THEOREM

Let $P(x) = a_n x^n + a_{n-1} x^{n-1} + \cdots + a_1 x + a_0$, where $a_n \neq 0$ and $a_0 \neq 0$, be a polynomial function with integer coefficients. If $\frac{p}{q}$ is a rational number written in lowest terms, and if $\frac{p}{q}$ is a zero of $P(x)$, then p is a factor of the constant term a_0 and q is a factor of the leading coefficient a_n.

DESCARTES' RULE OF SIGNS

Let $P(x)$ be a polynomial function with real coefficients and a nonzero constant term, with terms in descending powers of x.

(a) The number of positive real zeros either equals the number of variations in sign occurring in the coefficients of $P(x)$ or is less than the number of variations by a positive even integer.

(b) The number of negative real zeros either equals the number of variations in sign occurring in the coefficients of $P(-x)$ or is less than the number of variations by a positive even integer.

BOUNDEDNESS THEOREM

Let $P(x)$ be a polynomial function of degree $n \geq 1$ with real coefficients and with a positive leading coefficient. Suppose $P(x)$ is divided synthetically by $x - c$.

(a) If $c > 0$ and all numbers in the bottom row of the synthetic division are nonnegative, then $P(x)$ has no zero greater than c.

(b) If $c < 0$ and the numbers in the bottom row of the synthetic division alternate in sign (with 0 considered positive or negative, as needed), then $P(x)$ has no zero less than c.

(continued)

KEY TERMS & SYMBOLS	KEY CONCEPTS
3.8 Polynomial Equations and Inequalities; Further Applications and Models equation quadratic in form	**COMPLEX nTH ROOTS THEOREM** If n is a positive integer and k is a nonzero complex number, then the equation $x^n = k$ has *exactly* n complex roots. **POLYNOMIAL MODELS** Graphing calculators can be used to find the best-fitting polynomial function models for a set of data points.

3 Review Exercises

Let $w = 17 - i$ and let $z = 1 - 3i$. Write each complex number in standard form.

1. $w + z$ **2.** $w - z$ **3.** wz

4. w^2 **5.** $\dfrac{1}{z}$ **6.** $\dfrac{w}{z}$

Consider the function $P(x) = 2x^2 - 6x - 8$ for Exercises 7–16.

7. What is the domain of P?

8. Determine analytically the coordinates of the vertex of the graph.

9. Use an end-behavior diagram to describe the end behavior of the graph of P.

10. Determine analytically the x-intercepts, if any, of the graph of P.

11. Determine analytically the y-intercept of the graph of P.

12. What is the range of P?

13. Over what largest open interval is the function increasing? Over what largest open interval is it decreasing?

14. Give the solution set of each equation or inequality.
(a) $2x^2 - 6x - 8 = 0$
(b) $2x^2 - 6x - 8 > 0$
(c) $2x^2 - 6x - 8 \leq 0$

15. The graph of P is shown here. Explain how the graph supports your solution sets in **Exercise 14.**

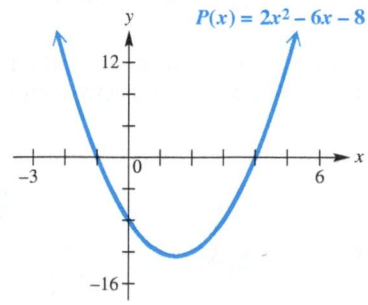

$P(x) = 2x^2 - 6x - 8$

16. What is the equation of the axis of symmetry of the graph in **Exercise 15?**

In Exercises 17–20, consider the function
$$P(x) = -2.64x^2 + 5.47x + 3.54.$$

17. Use the discriminant to explain how you can determine the number of x-intercepts the graph of P will have even before graphing it on your calculator.

18. Graph the function in the standard window of your calculator, and use the root-finding capabilities to solve the equation $P(x) = 0$. Express solutions as approximations to the nearest hundredth.

19. Use the capabilities of your calculator to find the coordinates of the vertex of the graph. Express coordinates to the nearest hundredth.

20. Verify analytically that your answer in **Exercise 19** is correct.

21. The figure below shows the graph of a quadratic function $y = f(x)$.

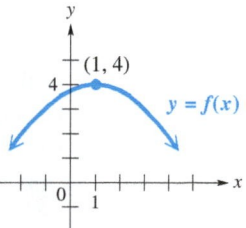

$(1, 4)$ $y = f(x)$

(a) What is the maximum value of $f(x)$?
(b) For what value of x is $f(x)$ a maximum?
(c) How many real solutions are there of the equation $f(x) = 2$?
(d) How many real solutions are there of the equation $f(x) = 6$?

(Modeling) Height of a Projectile *A projectile is fired vertically upward, and its height in feet after t seconds is given by*

$$s(t) = -16t^2 + 800t + 600.$$

Graph this function in the window $[0, 60]$ *by* $[-1000, 11{,}000]$, *and use either analytic or graphical methods to answer Exercises 22–26.*

22. From what height was the projectile fired?

23. After how many seconds will it reach its maximum height?

24. What is the maximum height it will reach?

25. Between what two times (in seconds, to the nearest tenth) will it be more than 5000 feet above the ground?

26. How long will the projectile be in the air? Give your answer to the nearest tenth of a second.

(Modeling) *Solve each problem.*

27. **Volume of a Box** A piece of cardboard is 3 times as long as it is wide. Equal-sized squares measuring 4 inches on each side are to be cut from the corners of the piece of cardboard, and the flaps will be folded up to form a box with an open top.
(a) Determine a function V that describes the volume of the box as a function of x, where x is the original width in inches.
(b) What should be the original dimensions of the piece of cardboard if the box is to have a volume of 2496 cubic inches? Solve this problem analytically.
(c) Support the answer in part (b) graphically.

28. **Concentration of Atmospheric CO_2** The International Panel on Climate Change (IPCC) stated that if current trends of burning fossil fuel and deforestation continue, then future amounts of atmospheric carbon dioxide in parts per million (ppm) will increase, as shown in the table.

Year	Carbon Dioxide
1990	353
2000	375
2075	590
2175	1090
2275	2000

Source: IPCC.

(a) Plot the data. Let $x = 0$ correspond to 1990.
(b) Find a function of the form $P(x) = a(x - h)^2 + k$ that models the data. Use $(0, 353)$ as the vertex and $(285, 2000)$ as another point to determine a.
(c) Use the regression feature of a graphing calculator to find the best-fitting quadratic function Q for the data.

29. Use the intermediate value theorem to show that the polynomial function $P(x) = -3x^3 - x^2 + 2x - 4$ has a real zero between -2 and -1.

30. Use synthetic division to find the quotient $Q(x)$ and the remainder R.
(a) $\dfrac{x^3 + x^2 - 11x - 10}{x - 3}$
(b) $\dfrac{3x^3 + 8x^2 + 5x + 10}{x + 2}$

Divide.

31. $\dfrac{6x^3 - 4x^2 + 4x + 3}{3x + 1}$

32. $\dfrac{2x^3 - 5x^2 + 1}{x^2 - 3x + 1}$

In Exercises 33–36, use synthetic division to find $P(2)$.

33. $P(x) = -x^3 + 5x^2 - 7x + 1$

34. $P(x) = 2x^3 - 3x^2 + 7x - 12$

35. $P(x) = 5x^4 - 12x^2 + 2x - 8$

36. $P(x) = x^5 + 4x^2 - 2x - 4$

37. If $P(x)$ is a polynomial function with real coefficients, and $7 + 2i$ is a zero, what other complex number must be a zero?

In Exercises 38–41, find a polynomial function with real coefficients and of least possible degree having the given zeros. Let the leading coefficient be 1.

38. $-1, 4, 7$

39. $8, 2, 3$

40. $\sqrt{3}, -\sqrt{3}, 2, 3$

41. $-2 + \sqrt{5}, -2 - \sqrt{5}, -2, 1$

42. Is -1 a zero of $P(x) = 2x^4 + x^3 - 4x^2 + 3x + 1$?

43. Is $x + 1$ a factor of $P(x) = x^3 + 2x^2 + 3x + 2$?

44. Find a polynomial function P with real coefficients of degree 4 with 3, 1, and $-1 - 3i$ as zeros and for which $P(2) = -36$.

45. Find all zeros of $P(x) = x^4 - 3x^3 - 8x^2 + 22x - 24$, given that $1 - i$ is a zero.

46. Find all zeros of $P(x) = 2x^4 - x^3 + 7x^2 - 4x - 4$, given that 1 and $2i$ are zeros.

47. Find all rational zeros of
$$P(x) = 3x^5 - 4x^4 - 26x^3 - 21x^2 - 14x + 8$$
by first listing the possible rational zeros based on the rational zeros theorem.

48. Use Descartes' rule of signs to determine the possible number of positive real zeros and negative real zeros of $P(x) = 3x^4 + x^3 - x^2 - 2x - 1$.

49. Use the boundedness theorem to show that
$$P(x) = 2x^4 + 3x^3 - 5x^2 + 8x - 10$$
has no real zero greater than 2 and no real zero less than -4.

In Exercises 50–54, consider the polynomial function

$$P(x) = x^3 - 2x^2 - 4x + 3.$$

50. Graph the function in an appropriate window to show a comprehensive graph. Based on the graph, how many real solutions does the equation $x^3 - 2x^2 - 4x + 3 = 0$ have? Use your calculator to find the real root that is an integer.

51. Use your answer in **Exercise 50** along with synthetic division to factor $x^3 - 2x^2 - 4x + 3$ so that one factor is linear and the other is quadratic.

52. Find exact values of any remaining zeros analytically.

53. Use your calculator to support your answer in **Exercise 52.**

54. Give the solution set of each inequality, using exact values.

   (a) $x^3 - 2x^2 - 4x + 3 > 0$

   (b) $x^3 - 2x^2 - 4x + 3 \le 0$

55. Use an analytic method to find all solutions of the equation $x^3 + 2x^2 + 5x = 0$. Then, without graphing, give the coordinates of all x-intercepts of the graph.

56. For the graph of $P(x) = x^4 - 5x^3 + x^2 + 21x - 18$, suppose you know that all zeros of P are integers and each zero has multiplicity 1 or 2. Give the factored form of $P(x)$.

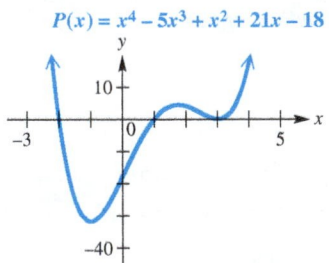

$P(x) = x^4 - 5x^3 + x^2 + 21x - 18$

Concept Check *Comprehensive graphs of polynomial functions f and g are shown here. They have only real coefficients. Answer Exercises 57–65 based on the graphs.*

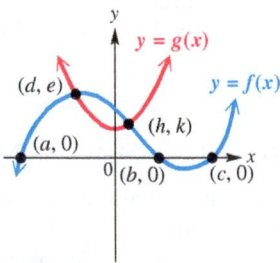

57. Is the degree of g even or odd?

58. Is the degree of f even or odd?

59. Is the leading coefficient of f positive or negative?

60. How many real solutions does $g(x) = 0$ have?

61. Express the solution set of $f(x) < 0$ in interval form.

62. What is the solution set of $f(x) > g(x)$?

63. What is the solution set of $f(x) - g(x) = 0$?

64. If $r + pi$ is a nonreal complex solution of $g(x) = 0$, what must be another nonreal complex solution?

65. Suppose that f is of degree 3. Explain why f cannot have nonreal complex zeros.

Concept Check *Answer* true *or* false *to each statement in Exercises 66–71.*

66. The function $f(x) = 3x^7 - 8x^6 + 9x^5 + 12x^4 - 18x^3 + 26x^2 - x + 500$ has eight x-intercepts.

67. The function f in **Exercise 66** may have up to six local extrema.

68. The function f in **Exercise 66** has a positive y-coordinate of its y-intercept.

69. Based on end behavior of the function f in **Exercise 66** and your answer in **Exercise 68,** the graph must have at least one negative x-coordinate for an x-intercept.

70. If a polynomial function of even degree has a positive leading coefficient and a negative y-coordinate for its y-intercept, it must have at least two real zeros.

71. Because $-\frac{1}{2} + i\frac{\sqrt{3}}{2}$ is a nonreal complex zero of

$$f(x) = x^2 + x + 1,$$

another zero must be $\frac{1}{2} + i\frac{\sqrt{3}}{2}$.

Graph the function

$$P(x) = -2x^5 + 15x^4 - 21x^3 - 32x^2 + 60x$$

in the window $[-8, 8]$ by $[-100, 200]$ to obtain a comprehensive graph. Then, use your calculator and the concepts of this chapter to answer Exercises 72–76.

72. How many local maxima does this function have?

73. One local minimum point lies on the x-axis and has an integer as its x-value. What are its coordinates?

74. The greatest x-value of an x-intercept is 5. Therefore, $x - 5$ is a factor of $P(x)$. Use synthetic division to find the quotient polynomial $Q(x)$ obtained when $P(x)$ is divided by $x - 5$.

75. What is the range of P?

76. The graph has a local minimum point with a negative x-value. Use your calculator to find its coordinates. Express them to the nearest hundredth.

77. Solve the equation

$$3x^3 + 2x^2 - 21x - 14 = 0$$

analytically for all complex solutions, giving exact values in your solution set. Then, graph the left side of the equation, and support the real solutions.

78. Consider the polynomial function

$$P(x) = -x^4 + 3x^3 + 3x^2 - 7x - 6$$
$$= (-x + 2)(x - 3)(x + 1)^2.$$

Graph the function by hand as explained in **Section 3.7,** and solve each equation or inequality.
(a) $P(x) = 0$
(b) $P(x) > 0$
(c) $P(x) < 0$

(Modeling) *Solve each problem.*

79. *Dimensions of a Cube*   After a 2-inch slice is cut off the top of a cube, the resulting solid has a volume of 32 cubic inches. Find the dimensions of the original cube.

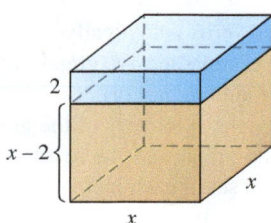

80. *Height of a Projected Object*   If air resistance is neglected, an object projected straight upward with an initial velocity of 40 meters per second from a height of 50 meters will be at a height s meters after t seconds, where

$$s(t) = -4.9t^2 + 40t + 50.$$

(a) What are the restrictions on t?
(b) Graph the function s in a suitable window.
(c) Use the graph to find the time at which the object reaches its highest point.
(d) Use the graph to find the maximum height reached by the object.
(e) Write the equation that allows you to find the time at which the object reaches the ground. Then, solve the equation, using the method of your choice. Give your answer to the nearest hundredth of a second.

81. *Military Personnel on Active Duty*   The number of military personnel on active duty in the United States during the period from 1965 through 2010 is approximated by the cubic model

$$y = -0.00002x^3 + 0.0022x^2 - 0.096x + 3.05,$$

where $x = 0$ corresponds to 1965 and y is in millions. Based on this model, about how many military personnel were on active duty in 2010? (*Source:* U.S. Department of Defense.)

82. *Medicare Beneficiary Spending*   Out-of-pocket spending projections for a typical Medicare beneficiary as a share of his or her income are given in the table.

Year	Percent of Income
1998	18.6
2000	19.3
2005	21.7
2010	24.7
2015	27.5
2020	28.3
2025	28.6

Source: Urban Institute's Analysis of Medicare Trustees' Report.

Let $x = 0$ represent 1990, so that $x = 8$ represents 1998. Use a graphing calculator to do the following.
(a) Plot the data.
(b) Find a quadratic function to model the data.
(c) Find a cubic function to model the data.
(d) Graph each function in the same viewing window as the data.
(e) Compare the two functions. Which is a better fit to the data?
(f) Would a linear model be appropriate for these data? Why or why not?

RELATING CONCEPTS

**For individual or group investigation
(Exercises 83–86)**

We can create a polynomial that represents a telephone number. For example, the 7-digit telephone number

$$123\text{--}4567$$

can be used to form the polynomial

$$\text{TEL}(x) = (x - 1)(x + 2)(x - 3)(x + 4) \cdot$$
$$(x - 5)(x + 6)(x - 7).$$

83. Make your own telephone polynomial. Give its degree and dominating term, and draw its end behavior diagram.

84. List all zeros of multiplicity 1, of multiplicity 2, and of multiplicity 3 or greater.

85. Use your graphing calculator to sketch a comprehensive graph of $\text{TEL}(x)$.

86. Identify the local extrema, domain, range, and intervals where $\text{TEL}(x)$ is increasing or decreasing.

3 Test

1. Simplify each expression. Give answers in standard form.
   (a) $(8 - 7i) - (-12 + 2i)$ (b) $\dfrac{11 + 10i}{2 + 3i}$

   (c) i^{65} (d) $2i(3 - i)^2$

   (e) $\sqrt{-36}$ (f) $\sqrt{-5} \cdot \sqrt{-20}$

2. For the function $P(x) = -2x^2 - 4x + 6$, do the following.
   (a) Find the vertex analytically.
   (b) Give a comprehensive graph, and use a calculator to support your result in part (a).
   (c) Find the zeros of P and support your result, using a graph for one zero and a table for the other.
   (d) Find the y-intercept analytically.
   (e) State the domain and range of P.
   (f) Give the largest open interval over which the function is increasing and the largest open interval over which it is decreasing.

3. Solve $6x^2 - 15x + 6 = 0$ by factoring.

4. Solve $x^2 - 4x = 2$ by completing the square.

5. (a) Solve the quadratic equation $3x^2 + 3x - 2 = 0$ analytically. Give solutions in exact form.
   (b) Graph $P(x) = 3x^2 + 3x - 2$ with a calculator. Use your results in part (a) along with the graph to give the solution set of each inequality. Express endpoints of intervals in exact form.
   (i) $P(x) < 0$

   (ii) $P(x) \geq 0$

6. *Dimensions of a Box* The width of a rectangular box is 3 times its height, and its length is 11 inches more than its height. Find the dimensions of the box if its volume is 720 cubic inches.

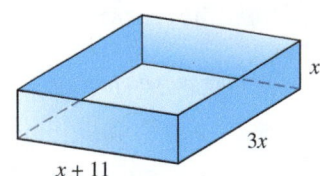

7. *(Modeling) Household Size* The table lists the average number of people age 18 or older in a U.S. household.

Year	1940	1950	1960	1970
Household size	2.53	2.31	2.12	2.05

Year	1980	1990	2000	2010
Household size	1.97	1.93	1.93	1.94

Source: U.S. Census Bureau.

   (a) Use regression to find a quadratic function that models the data. Let x represent the year.

   (b) Estimate the average number of people age 18 or older in a household during 1975. Compare your estimate to the actual value of 2.01.

8. The function
   $$f(x) = x^6 - 5x^5 + 3x^4 + x^3 + 40x^2 - 24x - 72$$
   has 3 as a zero of multiplicity 2, 2 as a single zero, and -1 as a single zero.
   (a) Find all other zeros of f.
   (b) Give an end behavior diagram. Use the information from part (a) to sketch the graph of f by hand.

9. Perform the following for the function
   $$f(x) = 4x^4 - 21x^2 - 25.$$
   (a) Find all zeros analytically.
   (b) Find a comprehensive graph of f, and support the real zeros found in part (a).
   (c) Discuss the symmetry of the graph of f.
   (d) Use the graph and the results of part (a) to find the solution set of each inequality.
   (i) $f(x) \geq 0$
   (ii) $f(x) < 0$

10. Perform the following for the function
    $$f(x) = 3x^4 + 5x^3 - 35x^2 - 55x + 22.$$
    (a) List all possible rational zeros.
    (b) Find all rational zeros.
    (c) Use the intermediate value theorem to show that there must be a zero between 3 and 4.
    (d) Use Descartes' rule of signs to determine the possible number of positive zeros and negative zeros.
    (e) Use the boundedness theorem to show that there is no zero less than -5 and no zero greater than 4.

11. (a) Use only a graphical method to find the real solutions, to the nearest thousandth if applicable, of
    $$x^5 - 4x^4 + 2x^3 - 4x^2 + 6x - 1 = 0.$$
    (b) Based on the degree and your answer in part (a), how many nonreal complex solutions does the equation have?

12. Divide.
    (a) $\dfrac{8x^3 - 4x^2}{2x}$ (b) $\dfrac{3x^3 - 5x^2 + 6}{x - 1}$
    (c) $\dfrac{2x^4 - x^3 + 4x^2 - 4x + 3}{2x - 1}$ (d) $\dfrac{x^4 - 2x + 6}{x^2 + 2}$

13. Find a cubic polynomial function $f(x)$ in standard form having real coefficients with zeros of 4 and $2i$, such that $f(1) = -15$.

14. Solve $x^3 + 3x = 0$ analytically.

In this chapter we explore phenomena that are subject to nonlinear effects, such as traffic congestion, which we will model with rational functions.

4 Rational, Power, and Root Functions

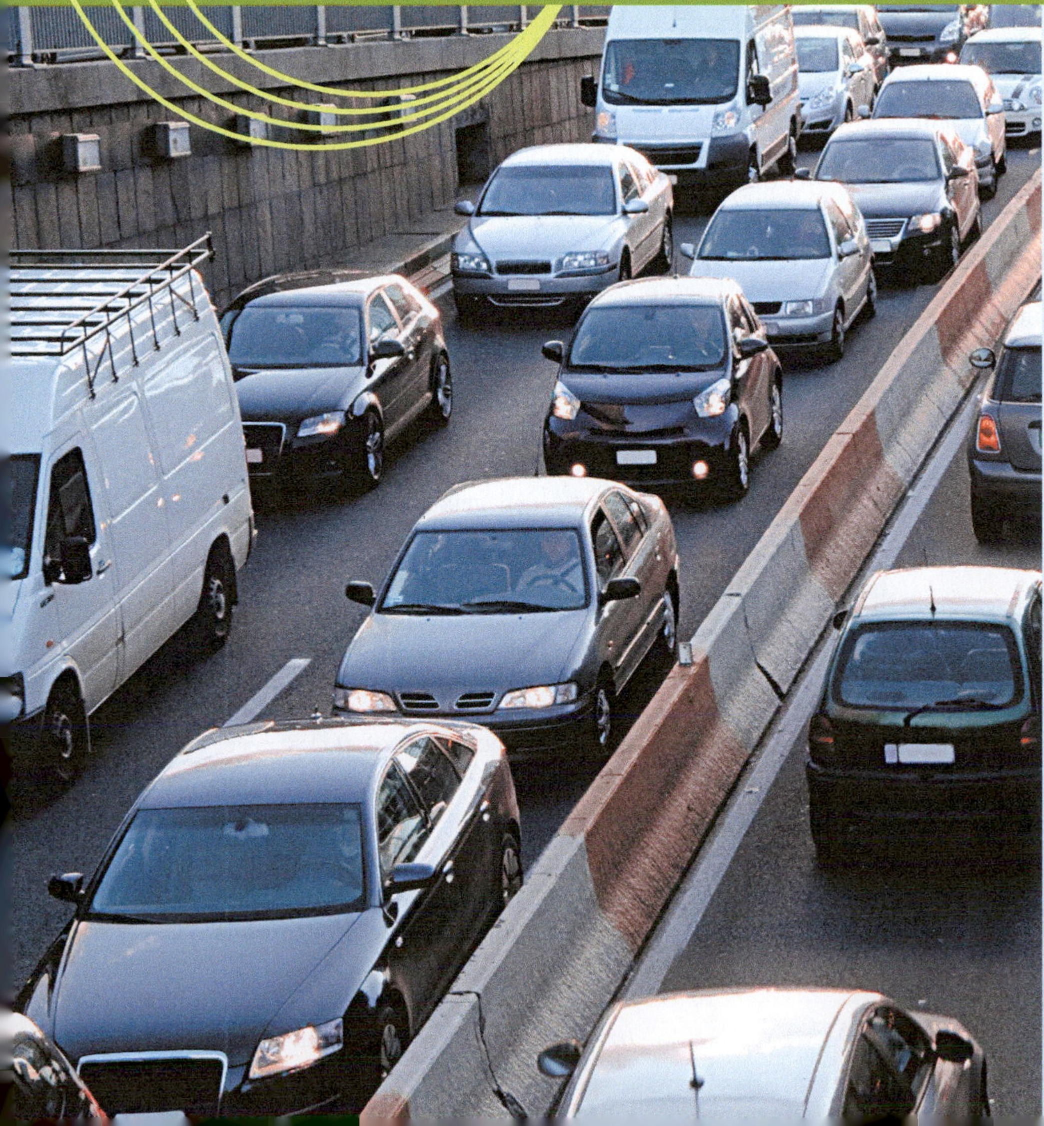

4.1   Rational Functions and Graphs (I)

The Reciprocal Function $f(x) = \dfrac{1}{x}$ • The Function $f(x) = \dfrac{1}{x^2}$

Find f (x) for some values of x close to 0.

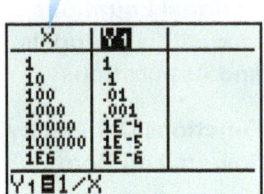

As X approaches 0 from the left, $Y_1 = \dfrac{1}{X}$ approaches $-\infty$.

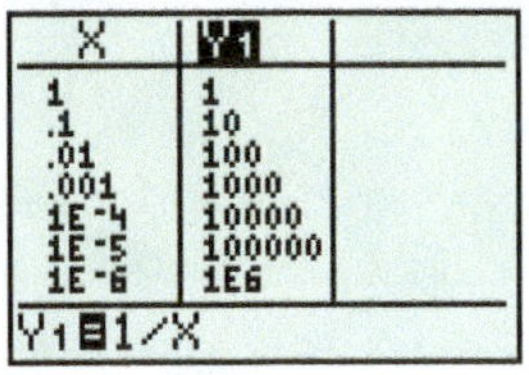

As X approaches 0 from the right, $Y_1 = \dfrac{1}{X}$ approaches ∞.

FIGURE 1

Find f (x) for increasing values of x.

As X approaches ∞, $Y_1 = \dfrac{1}{X}$ approaches 0 through positive values.

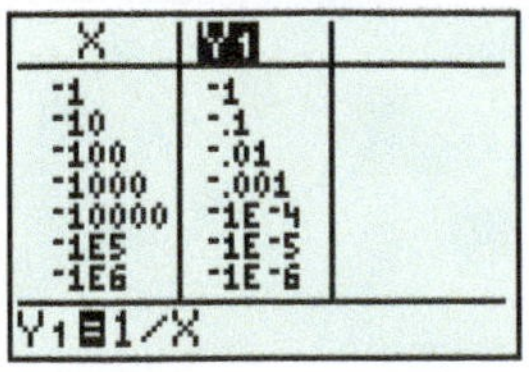

As X approaches $-\infty$, $Y_1 = \dfrac{1}{X}$ approaches 0 through negative values.

FIGURE 2

A **rational expression** is a fraction that is the quotient of two polynomials. A function defined by a rational expression is called a *rational function.*

> ### Rational Function
>
> A function f of the form
>
> $$f(x) = \frac{p(x)}{q(x)},$$
>
> where $p(x)$ and $q(x)$ are polynomials, with $q(x) \neq 0$, is called a **rational function.**

Some examples of rational functions are as follows.

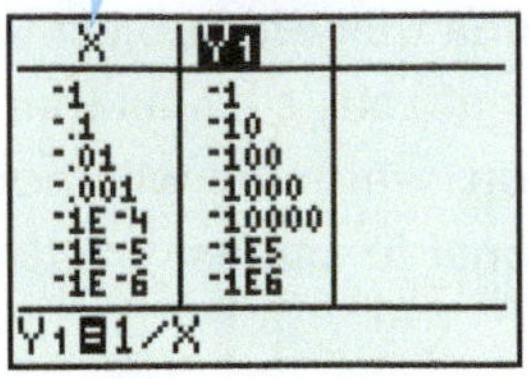

$$f(x) = \frac{1}{x}, \quad f(x) = \frac{x+1}{2x^2 + 5x - 3}, \quad \text{and} \quad f(x) = \frac{3x^2 - 3x - 6}{x^2 + 8x + 16} \qquad \text{Rational functions}$$

Because we cannot divide by zero, any values of x such that $q(x) = 0$ are excluded from the domain of a rational function. This type of function often has a *discontinuous* graph (a graph that has one or more breaks in it).

The Reciprocal Function $f(x) = \dfrac{1}{x}$

The simplest rational function with a variable denominator is the **reciprocal function.**

$$f(x) = \frac{1}{x} \qquad \text{Reciprocal function}$$

The domain of this function is the set of all real numbers except 0. The number 0 cannot be used as a value of x, but it is helpful to find values of $f(x)$ for some values of x close to 0. The tables in **FIGURE 1** suggest that $|f(x)|$ gets larger and larger as x gets closer and closer to 0.

$$|f(x)| \to \infty \quad \text{as} \quad x \to 0$$

|f (x)| approaches infinity as x approaches zero.

(The symbol $x \to 0$ means that x approaches 0, without necessarily ever being equal to 0.) Since x cannot equal 0, the graph of $f(x) = \frac{1}{x}$ will never intersect the vertical line $x = 0$ (the y-axis). This line is called a **vertical asymptote.**

On the other hand, as $|x|$ gets larger and larger, the values of $f(x) = \frac{1}{x}$ get closer and closer to 0, as shown in the tables in **FIGURE 2.** Letting $|x|$ get larger and larger without bound (written $|x| \to \infty$) causes the graph of $f(x) = \frac{1}{x}$ to move closer and closer to the horizontal line $y = 0$ (the x-axis). This line is called a **horizontal asymptote.**

CAUTION   *Asymptotes are aids in describing and sketching graphs. They are not parts of the graphs themselves.*

The graph and important features of $f(x) = \frac{1}{x}$ are shown in **FIGURE 3.**

FUNCTION CAPSULE

RECIPROCAL FUNCTION $f(x) = \frac{1}{x}$

Domain: $(-\infty, 0) \cup (0, \infty)$ Range: $(-\infty, 0) \cup (0, \infty)$

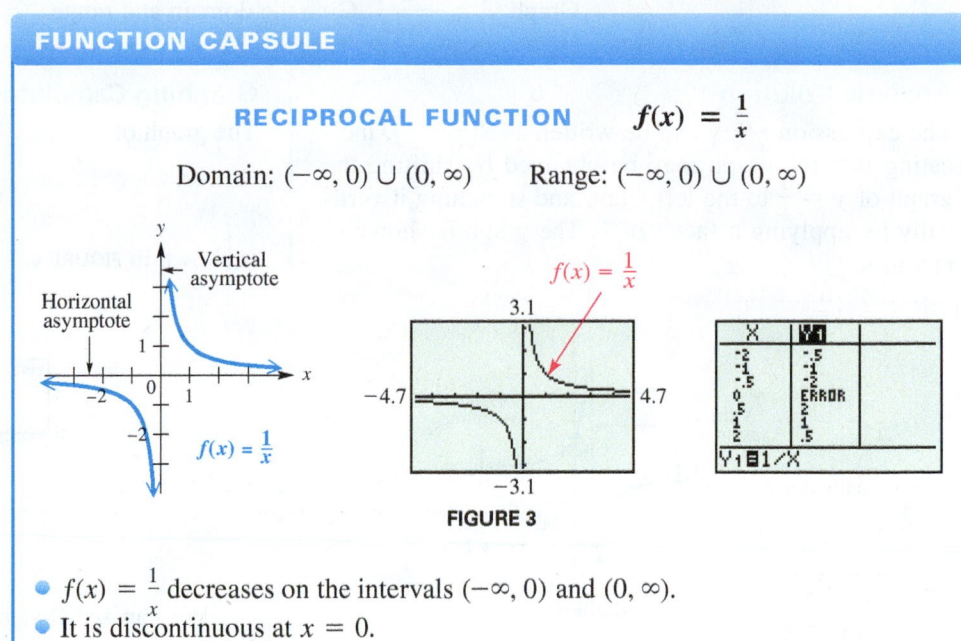

FIGURE 3

- $f(x) = \frac{1}{x}$ decreases on the intervals $(-\infty, 0)$ and $(0, \infty)$.
- It is discontinuous at $x = 0$.
- The y-axis is a vertical asymptote, and the x-axis is the horizontal asymptote.
- It is an odd function, and its graph is symmetric with respect to the origin.

EXAMPLE 1   **Graphing a Rational Function by Hand**

Graph $y = -\frac{2}{x}$. Give the domain and range.

Solution   The expression $-\frac{2}{x}$ can be written as either $2\left(\frac{1}{-x}\right)$ or $-2\left(\frac{1}{x}\right)$, indicating that the graph may be obtained by stretching the graph of $y = \frac{1}{x}$ vertically by applying a factor of 2 and reflecting it across either the y-axis or x-axis. The x- and y-axes remain the horizontal and vertical asymptotes. The domain and range are still $(-\infty, 0) \cup (0, \infty)$. See **FIGURE 4.**

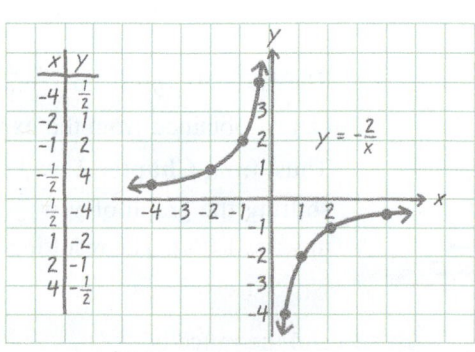

FIGURE 4

EXAMPLE 2 Graphing a Rational Function

Graph $y = \dfrac{2}{x+1}$. Give the domain and range.

Analytic Solution

The expression $\dfrac{2}{x+1}$ can be written as $2\left(\dfrac{1}{x+1}\right)$, indicating that the graph may be obtained by shifting the graph of $y = \dfrac{1}{x}$ to the left 1 unit and stretching it vertically by applying a factor of 2. The graph is shown in **FIGURE 5.**

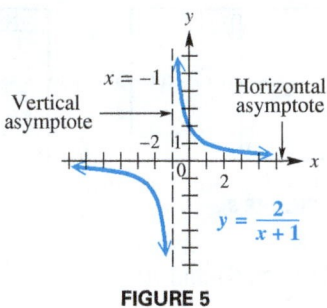

FIGURE 5

The horizontal shift affects the domain, which is now $(-\infty, -1) \cup (-1, \infty)$. The range is still $(-\infty, 0) \cup (0, \infty)$.

Graphing Calculator Solution

The graph of

$$y = \dfrac{2}{x+1}$$

is shown in **FIGURE 6.**

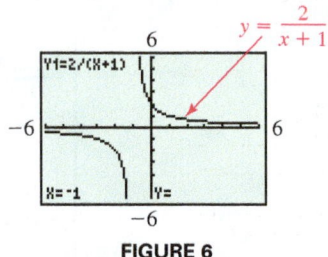

FIGURE 6

We can see that because we shifted $y = \dfrac{2}{x}$ to the *left* 1 unit, the line $x = -1$ is now a vertical asymptote. The line $y = 0$ (the x-axis) remains the horizontal asymptote.

EXAMPLE 3 Graphing a Rational Function by Hand

Use division and translations to graph $y = \dfrac{x+2}{x+1}$. Give the domain and range.

Solution First, we rewrite the given expression by dividing $x + 1$ into $x + 2$.

$$\begin{array}{r} 1 \\ x+1\overline{)x+2} \\ \underline{x+1} \\ 1 \end{array}$$

The rational expression $\dfrac{x+2}{x+1}$ is equal to $1 + \dfrac{1}{x+1}$. Note that the expression $1 + \dfrac{1}{x+1}$ can be obtained from the expression $\dfrac{1}{x}$ by replacing x with $x + 1$ and then adding 1. Thus, from **Chapter 2** we know that the graph of $y = 1 + \dfrac{1}{x+1}$ can be obtained by shifting the graph of $y = \dfrac{1}{x}$ in **FIGURE 3** on the preceding page to the left 1 unit and upward 1 unit. The resulting graph is shown in **FIGURE 7.** The domain is

$$(-\infty, -1) \cup (-1, \infty)$$

and the range is

$$(-\infty, 1) \cup (1, \infty).$$

The only vertical asymptote is $x = -1$, and the horizontal asymptote is $y = 1$.

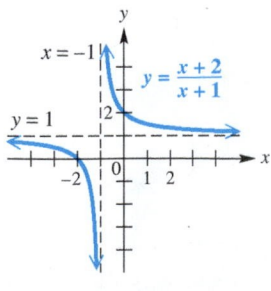

FIGURE 7

The Function $f(x) = \dfrac{1}{x^2}$

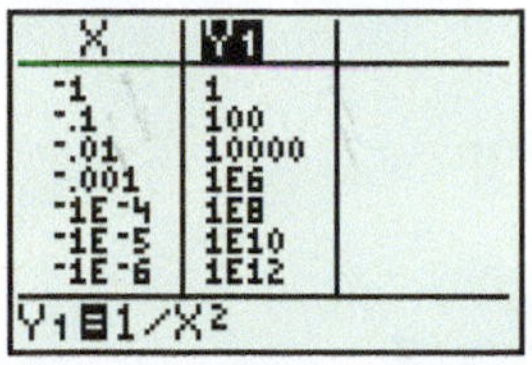

The rational function

$$f(x) = \frac{1}{x^2} \qquad \text{\color{blue}Rational function}$$

has domain $(-\infty, 0) \cup (0, \infty)$. We can use the table feature of a graphing calculator to examine values of $f(x)$ for some x-values close to 0. See **FIGURE 8**.

The tables suggest that $f(x)$ gets larger and larger as x gets closer and closer to 0. Notice that as x approaches 0 from *either* side, function values are all positive and there is symmetry with respect to the y-axis. Thus,

$$f(x) \to \infty \quad \text{as} \quad x \to 0.$$

The y-axis is a vertical asymptote.

As $|x|$ gets larger and larger, $f(x)$ approaches 0, as suggested by the tables in **FIGURE 9**. Again, function values are all positive. The x-axis is the horizontal asymptote of the graph.

The graph and important features of $f(x) = \frac{1}{x^2}$ are summarized here and shown in **FIGURE 10**.

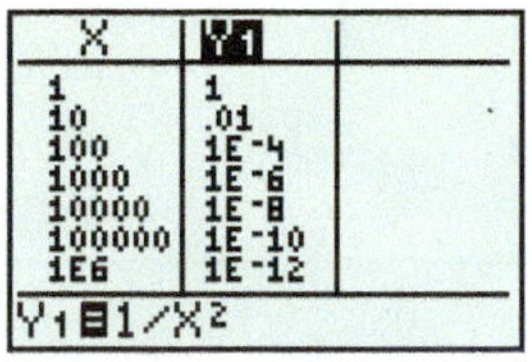

As X approaches 0 from the left, $Y_1 = \frac{1}{X^2}$ approaches ∞.

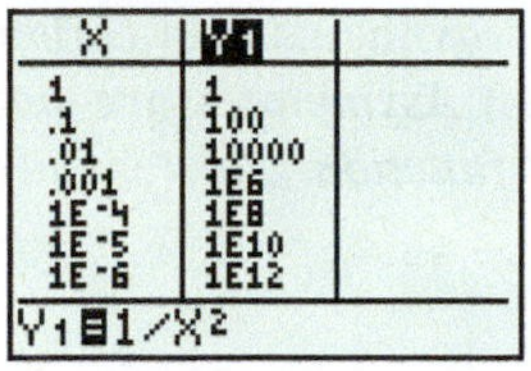

As X approaches 0 from the right, $Y_1 = \frac{1}{X^2}$ approaches ∞.

FIGURE 8

As X approaches ∞, $Y_1 = \frac{1}{X^2}$ approaches 0 through positive values.

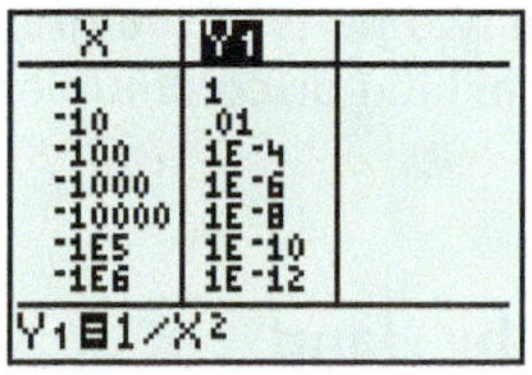

As X approaches $-\infty$, $Y_1 = \frac{1}{X^2}$ approaches 0 through positive values.

FIGURE 9

FUNCTION CAPSULE

RATIONAL FUNCTION $\qquad f(x) = \dfrac{1}{x^2}$

Domain: $(-\infty, 0) \cup (0, \infty)$ Range: $(0, \infty)$

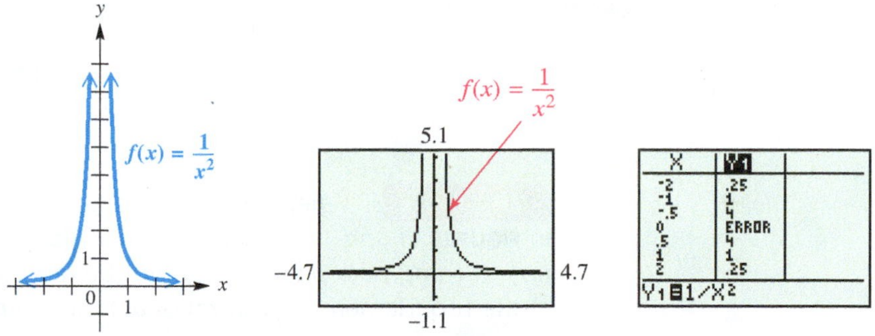

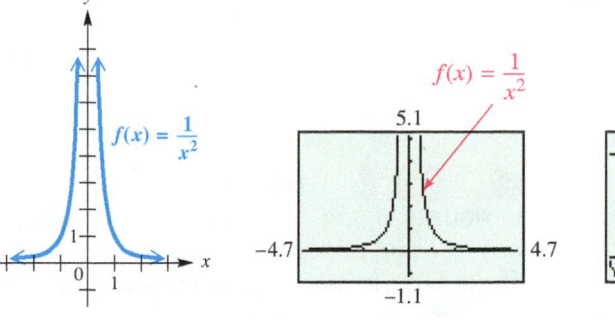

FIGURE 10

- $f(x) = \frac{1}{x^2}$ increases on the interval $(-\infty, 0)$ and decreases on the interval $(0, \infty)$.
- It is discontinuous at $x = 0$.
- The y-axis is a vertical asymptote, and the x-axis is the horizontal asymptote.
- It is an even function, and its graph is symmetric with respect to the y-axis.

FOR DISCUSSION

Graph the function $f(x) = \frac{1}{x^3}$ on a graphing calculator. Describe the behavior of the graph in each of the following situations.

  1. as $x \to \infty$ **2.** as $x \to -\infty$

  3. as $x \to 0$ from the left   **4.** as $x \to 0$ from the right

Compare this behavior with that of the graph of $y = \frac{1}{x}$. What do you notice? Repeat the exercise above for $y = \frac{1}{x^4}$, and compare to your results for $y = \frac{1}{x^2}$. Can you make a conjecture about how the exponents on x determine the results?

EXAMPLE 4 **Graphing a Rational Function**

Graph $y = \dfrac{1}{(x - 1)^2}$. Give the domain and range.

Analytic Solution

The equation $y = \dfrac{1}{(x - 1)^2}$ is equivalent to

$$y = f(x - 1), \quad \text{where} \quad f(x) = \frac{1}{x^2}.$$

This indicates that the graph of f will be shifted right 1 unit. The vertical asymptote is now $x = 1$, but the horizontal asymptote remains $y = 0$. See **FIGURE 11**. The horizontal shift affects the domain, which is now $(-\infty, 1) \cup (1, \infty)$. The range remains $(0, \infty)$.

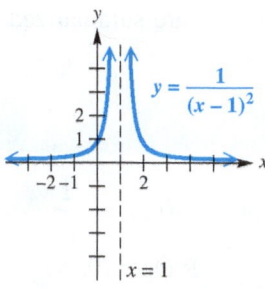

FIGURE 11

Graphing Calculator Solution

The graph of

$$y = \frac{1}{(x - 1)^2}$$

is shown in the **decimal window** of the TI-84 Plus in **FIGURE 12**. Notice that the calculator does not produce a vertical line at $x = 1$. (This *was* the case in earlier models of graphing calculators.) *Asymptotes are not part of the graph of a rational function. They simply aid us in graphing.*

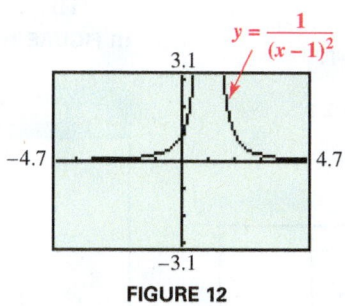

FIGURE 12

CAUTION When sketching the graph of a function by hand, such as the ones in **FIGURES 11** and **13**, be sure to show the graph *approaching the asymptote,* and not "turning back on itself." Do not be tempted to join the two parts of the graph at the top. Remember that rational function graphs often exhibit discontinuities.

EXAMPLE 5 **Graphing a Rational Function by Hand**

Graph $y = \dfrac{1}{(x + 2)^2} - 1$. Give the domain and range.

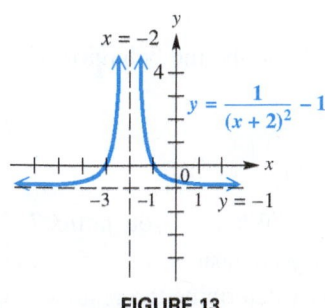

FIGURE 13

Solution The equation $y = \dfrac{1}{(x + 2)^2} - 1$ is equivalent to

$$y = f(x + 2) - 1, \quad \text{where} \quad f(x) = \frac{1}{x^2}.$$

This indicates that the graph of f in **FIGURE 10** on the preceding page will be shifted left 2 units and downward 1 unit. The horizontal shift affects the domain, which is now $(-\infty, -2) \cup (-2, \infty)$, while the vertical shift affects the range, now $(-1, \infty)$. The vertical asymptote has equation $x = -2$, and the horizontal asymptote has equation $y = -1$. The graph is shown in **FIGURE 13**.

4.1 Exercises

Checking Analytic Skills   *Provide a short answer to each question.* **Do not use a calculator.**

1. What is the domain of $f(x) = \frac{1}{x}$? What is its range?

2. What is the domain of $f(x) = \frac{1}{x^2}$? What is its range?

3. What is the largest open interval over which $f(x) = \frac{1}{x}$ increases? decreases? is constant?

4. What is the largest open interval over which $f(x) = \frac{1}{x^2}$ increases? decreases? is constant?

5. What is the equation of the vertical asymptote of the graph of $y = \frac{1}{x - 3} + 2$? of the horizontal asymptote?

6. What is the equation of the vertical asymptote of the graph of $y = \frac{1}{(x + 2)^2} - 4$? of the horizontal asymptote?

7. Is $f(x) = \frac{1}{x^2}$ an even or odd function? What symmetry does its graph exhibit?

8. Is $f(x) = \frac{1}{x}$ an even or odd function? What symmetry does its graph exhibit?

Concept Check   *Use the graphs of the rational functions in A–D below to answer each question in Exercises 9–16. Give all possible answers, as there may be more than one correct choice.*

A.
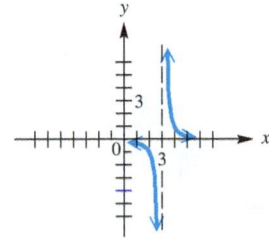

B.
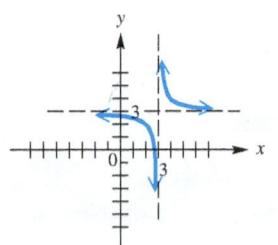

C.
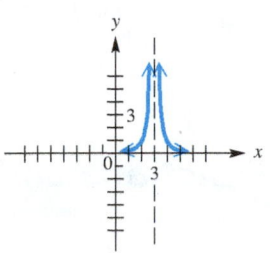

D.
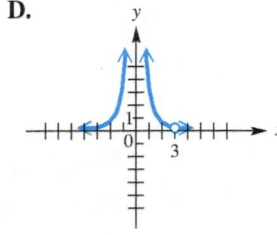

9. Which choices have domain $(-\infty, 3) \cup (3, \infty)$?

10. Which choice has range $(-\infty, 3) \cup (3, \infty)$?

11. Which choice has range $(-\infty, 0) \cup (0, \infty)$?

12. Which choices have range $(0, \infty)$?

13. If f represents the function, only one choice has a single solution of the equation $f(x) = 3$. Which one is it?

14. What is the range of the function in A?

15. Which choices have the x-axis as a horizontal asymptote?

16. Which choices are symmetric with respect to a vertical line?

Explain how the graph of f can be obtained from the graph of $y = \frac{1}{x}$ or $y = \frac{1}{x^2}$. Draw a sketch of the graph of f by hand. Then generate an accurate depiction of the graph with a graphing calculator. Finally, give the domain and range.

17. $f(x) = \frac{2}{x}$

18. $f(x) = -\frac{3}{x}$

19. $f(x) = \frac{1}{x + 2}$

20. $f(x) = \frac{1}{x - 3}$

21. $f(x) = \frac{1}{x} + 1$

22. $f(x) = \frac{1}{x} - 2$

23. $f(x) = \frac{1}{x - 1} + 1$

24. $f(x) = \frac{2}{x + 2} - 1$

25. $f(x) = \frac{1}{x^2} - 2$

26. $f(x) = \frac{1}{x^2} + 3$

27. $f(x) = -\frac{2}{x^2}$

28. $f(x) = \frac{0.5}{x^2}$

29. $f(x) = \frac{1}{(x - 3)^2}$

30. $f(x) = \frac{-2}{(x - 3)^2}$

31. $f(x) = \frac{-1}{(x + 2)^2} - 3$

32. $f(x) = \frac{-1}{(x - 4)^2} + 2$

Concept Check *The figures in A–D show the four ways that the graph of a rational function can approach the vertical line $x = 2$ as an asymptote. Identify the graph of each rational function in Exercises 33–36.*

A.

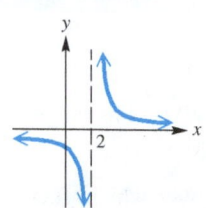

B.

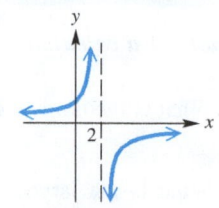

C.

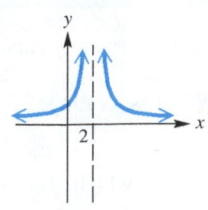

D.

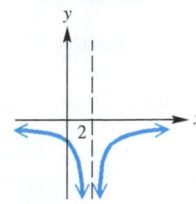

33. $f(x) = \dfrac{1}{(x-2)^2}$

34. $f(x) = \dfrac{1}{x-2}$

35. $f(x) = \dfrac{-1}{x-2}$

36. $f(x) = \dfrac{-1}{(x-2)^2}$

Concept Check *Suppose that the graph of a rational function f has vertical asymptote $x = 1$, horizontal asymptote $y = 2$, domain $(-\infty, 1) \cup (1, \infty)$, and range $(-\infty, 2) \cup (2, \infty)$. Give the vertical asymptote, horizontal asymptote, domain, and range for the graph of each shifted function.*

37. $y = f(x + 2) - 1$

38. $y = f(x - 1) + 3$

Use the method described in **Example 3** *to rewrite the rational expression, sketch the graph by hand, and generate an accurate depiction of each function with a graphing calculator.*

39. $f(x) = \dfrac{x-1}{x-2}$

40. $f(x) = \dfrac{x-2}{x-3}$

41. $f(x) = \dfrac{-2x-5}{x+3}$

42. $f(x) = \dfrac{-2x-1}{x+1}$

43. $f(x) = \dfrac{2x-5}{x-3}$

44. $f(x) = \dfrac{2x-1}{x-1}$

4.2 Rational Functions and Graphs (II)

Vertical and Horizontal Asymptotes • Graphing Techniques • Oblique Asymptotes • Graphs with Points of Discontinuity
• Graphs with No Vertical Asymptotes

Vertical and Horizontal Asymptotes

In this section we discuss more general rational functions, such as the following.

$$f(x) = \frac{x+1}{2x^2 + 5x - 3}, \quad f(x) = \frac{2x+1}{x-3}, \quad f(x) = \frac{x^2+1}{x-2}, \quad \text{and} \quad f(x) = \frac{1}{x^2+1}$$

Rational functions

We have seen how the graph of a rational function *may* approach a horizontal asymptote as $|x| \to \infty$. We also saw that the graph *may* approach a vertical asymptote as $|x| \to a$ if the function is undefined at $x = a$. We now give formal definitions for vertical and horizontal asymptotes of rational functions.

Asymptotes for Rational Functions

Let $p(x)$ and $q(x)$ be polynomials. For the rational function $f(x) = \frac{p(x)}{q(x)}$, written in *lowest terms*, and for real numbers a and b,

1. If $|f(x)| \to \infty$ as $x \to a$, then the line $x = a$ is a **vertical asymptote.**

2. If $f(x) \to b$ as $|x| \to \infty$, then the line $y = b$ is a **horizontal asymptote.**

→ **Looking Ahead to Calculus**
The rational function

$$f(x) = \frac{2}{x + 1}$$

has a vertical asymptote at $x = -1$. See **FIGURE 5** from the previous section. In calculus, the behavior of the graph of this function for values close to -1 is described by using **one-sided limits**. As x approaches -1 from the *left*, the function values decrease without bound. This is written

$$\lim_{x \to -1^-} f(x) = -\infty.$$

As x approaches -1 from the *right*, the function values increase without bound. This is written

$$\lim_{x \to -1^+} f(x) = \infty.$$

Locating asymptotes is important in graphing rational functions that are written in lowest terms.

1. ***Vertical Asymptotes: Find by determining the values of x that make the denominator, but not the numerator, equal to 0.***
2. ***Horizontal Asymptotes (and, in some cases, oblique asymptotes): Find by considering what happens to y as $|x| \to \infty$.***

These asymptotes determine the end behavior of the graph.

> **NOTE**   In **FIGURES 14–16** accompanying Examples 1–3, we show the graphs of the rational functions to illustrate asymptotes. Techniques for actually graphing such functions follow later in the section. It is not necessary to graph the function in order to find its asymptotes.

EXAMPLE 1   **Finding Asymptotes by Hand**

Find the domain and asymptotes of the graph of $f(x) = \dfrac{x + 1}{2x^2 + 5x - 3}$,

where $f(x)$ is in lowest terms.

Use this procedure in general to find vertical asymptotes.

Solution   *To find the vertical asymptotes, we set the denominator equal to 0 and solve.*

$$2x^2 + 5x - 3 = 0$$

$$(2x - 1)(x + 3) = 0 \qquad \text{Factor.}$$

$$2x - 1 = 0 \quad \text{or} \quad x + 3 = 0 \qquad \text{Zero-product property}$$

$$x = \frac{1}{2} \quad \text{or} \qquad x = -3 \qquad \text{Solve each equation.}$$

Because the numerator is not equal to 0 at these values of x, the equations of the vertical asymptotes are $x = \frac{1}{2}$ and $x = -3$. Since the domain cannot include these values, it is $(-\infty, -3) \cup \left(-3, \frac{1}{2}\right) \cup \left(\frac{1}{2}, \infty\right)$. See **FIGURE 14**.

Use this procedure in general to find horizontal asymptotes.

To find the equation of the horizontal asymptote, we divide each term in the numerator and denominator by the variable of greatest degree in the expression for $f(x)$. In this case, we divide each term by x^2.

$$f(x) = \frac{\dfrac{x}{x^2} + \dfrac{1}{x^2}}{\dfrac{2x^2}{x^2} + \dfrac{5x}{x^2} - \dfrac{3}{x^2}} = \frac{\dfrac{1}{x} + \dfrac{1}{x^2}}{2 + \dfrac{5}{x} - \dfrac{3}{x^2}}$$

As $|x| \to \infty$, the quotients $\frac{1}{x}$, $\frac{1}{x^2}$, $\frac{5}{x}$, and $\frac{3}{x^2}$ all approach 0, and the value of $f(x)$ approaches

$$\frac{0 + 0}{2 + 0 - 0} = \frac{0}{2} = 0.$$

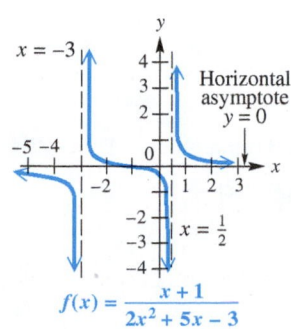

FIGURE 14

The line $y = 0$ (the x-axis) is the horizontal asymptote. See **FIGURE 14**.   ●

Finding Asymptotes by Hand

Find the domain and asymptotes of the graph of $f(x) = \dfrac{2x + 1}{x - 3}$.

Solution Set the denominator, $x - 3$, equal to 0 to find that the domain is $(-\infty, 3) \cup (3, \infty)$, and the vertical asymptote has equation $x = 3$. To find the horizontal asymptote, divide each term in the rational expression by x, since the greatest exponent on x in the expression is 1.

$$f(x) = \frac{2x + 1}{x - 3} = \frac{\dfrac{2x}{x} + \dfrac{1}{x}}{\dfrac{x}{x} - \dfrac{3}{x}} = \frac{2 + \dfrac{1}{x}}{1 - \dfrac{3}{x}}$$

As $|x| \to \infty$, both $\dfrac{1}{x}$ and $\dfrac{3}{x}$ approach 0, and $f(x)$ approaches

$$\frac{2 + 0}{1 - 0} = \frac{2}{1} = 2.$$

Thus, the line $y = 2$ is the horizontal asymptote. See **FIGURE 15**.

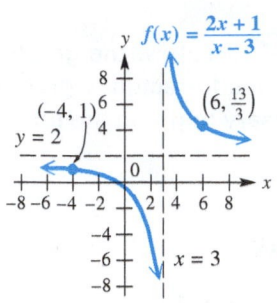

FIGURE 15

Finding Asymptotes by Hand

Find the domain and asymptotes of the graph of $f(x) = \dfrac{x^2 + 1}{x - 2}$.

Solution Setting the denominator, $x - 2$, equal to 0 shows that the domain is $(-\infty, 2) \cup (2, \infty)$, and the vertical asymptote has equation $x = 2$. If we divide by the variable of greatest degree of x as before (x^2 in this case), we see that there is no horizontal asymptote. The expression

$$f(x) = \frac{\dfrac{x^2}{x^2} + \dfrac{1}{x^2}}{\dfrac{x}{x^2} - \dfrac{2}{x^2}} = \frac{1 + \dfrac{1}{x^2}}{\dfrac{1}{x} - \dfrac{2}{x^2}} = \frac{1 + 0}{0 - 0}$$

does not approach any real number as $|x| \to \infty$, since $\dfrac{1}{0}$ is undefined. This happens whenever the degree of the numerator is *greater* than the degree of the denominator. In such cases, we divide the denominator into the numerator to write the expression in another form. (See **Section 3.6**.)

$$
\begin{array}{r}
x + 2 \\
x - 2 \overline{)\; x^2 + 0x + 1} \\
\underline{x^2 - 2x} \\
2x + 1 \\
\underline{2x - 4} \\
5
\end{array}
$$

Include 0x for the missing x-term.

Multiply $x(x - 2)$.
Subtract and bring down +1.
Multiply $2(x - 2)$.
Subtract.

(We could have used synthetic division.) We can now write the function as

$$f(x) = \frac{x^2 + 1}{x - 2} = x + 2 + \frac{5}{x - 2}.$$

As $|x| \to \infty$, $\dfrac{5}{x - 2}$ is close to 0, and the graph approaches the line $y = x + 2$. This line is an **oblique** (or **slant**) **asymptote** (slanted, neither vertical nor horizontal) for the graph of the function. See **FIGURE 16**.

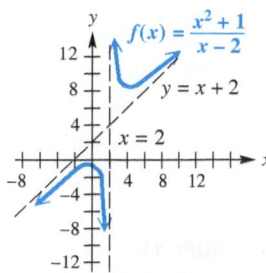

Rational function with oblique asymptote, $y = x + 2$

FIGURE 16

The procedure for determining asymptotes in **Examples 1–3** can be summarized as follows. Note that the domain of a rational function includes all real numbers except those that make the denominator equal to zero.

Determining Asymptotes

To find asymptotes of a rational function defined by a rational expression *in lowest terms*, use the following procedures.

1. **Vertical Asymptotes**
   Find any vertical asymptotes by setting the denominator equal to 0 and solving for x. If a is a zero of the denominator but not the numerator, then the line $x = a$ **is a vertical asymptote.**

2. **Other Asymptotes**
   Determine any other asymptotes. Consider three possibilities:
   (a) If the numerator has lesser degree than the denominator, then there is a **horizontal asymptote $y = 0$** (the x-axis).
   (b) If the numerator and denominator have the *same* degree and the function is of the form

   $$f(x) = \frac{a_n x^n + \cdots + a_0}{b_n x^n + \cdots + b_0}, \quad \text{where} \quad b_n \neq 0,$$

   then dividing by x^n in the numerator and denominator produces the **horizontal asymptote $y = \frac{a_n}{b_n}$.**
   (c) If the numerator is of degree exactly one greater than the denominator, then there may be an **oblique (or slant) asymptote.*** To find it, divide the numerator by the denominator and disregard any remainder. Set the (linear) polynomial portion of the quotient equal to y to find the equation of the asymptote.

NOTE

- The graph of a rational function may have more than one vertical asymptote, or it may have none at all. The graph *cannot intersect* any *vertical asymptote*.
- There can be only *one* other *nonvertical asymptote*, and the graph *may intersect* that asymptote.

This will be seen in **Examples 4** and **6**.

Graphing Techniques

A comprehensive graph of a rational function exhibits these features.

1. It shows all intercepts, both x- and y-.
2. It contains the location of all asymptotes: vertical, horizontal, and/or oblique.
3. It contains the point at which the graph intersects its nonvertical asymptote (if there is any such point).
4. Enough of the graph is visible to exhibit the correct end behavior (e.g., behavior as the graph approaches its nonvertical asymptote).

---

*When the degree of the numerator of a rational function is two or more greater than that of the denominator, there is asymptotic behavior approximating a nonlinear curve. Such functions are not covered extensively in this book.

Graphing a Rational Function

Let $f(x) = \frac{p(x)}{q(x)}$ be a function with the rational expression in lowest terms. To sketch its graph, follow these steps.

Step 1 Find the domain and all vertical asymptotes.

Step 2 Find any horizontal or oblique asymptote.

Step 3 Find the *y*-intercept, if possible, by evaluating $f(0)$.

Step 4 Find the *x*-intercepts, if any, by solving $f(x) = 0$. (These *x*-coordinates will be the zeros of the numerator $p(x)$.)

Step 5 Determine whether the graph will intersect its nonvertical asymptote $y = b$ by solving $f(x) = b$, where *b* is the *y*-value of the horizontal asymptote, or by solving $f(x) = mx + b$, where $y = mx + b$ is the equation of the oblique asymptote.

Step 6 Plot selected points as necessary. Choose an *x*-value in each interval of the domain determined by the vertical asymptotes and *x*-intercepts.

Step 7 Complete the sketch.

EXAMPLE 4 **Graphing a Rational Function with the *x*-Axis as Horizontal Asymptote**

Graph $f(x) = \dfrac{x + 1}{2x^2 + 5x - 3}$.

Solution

Step 1 From **Example 1,** the domain is $(-\infty, -3) \cup \left(-3, \frac{1}{2}\right) \cup \left(\frac{1}{2}, \infty\right)$ and the vertical asymptotes have equations $x = \frac{1}{2}$ and $x = -3$.

Step 2 Again, as shown in **Example 1,** the horizontal asymptote is the *x*-axis.

Step 3 The *y*-intercept is $\left(0, -\frac{1}{3}\right)$, since

$$f(0) = \frac{0 + 1}{2(0)^2 + 5(0) - 3} = -\frac{1}{3}.$$

Step 4 The *x*-intercept is found by solving $f(x) = 0$.

$$\frac{x + 1}{2x^2 + 5x - 3} = 0 \qquad \text{Let } f(x) = 0.$$

$$x + 1 = 0 \qquad \text{If a fraction equals 0, then its numerator must equal 0.}$$

$$x = -1 \qquad \text{The } x\text{-intercept is } (-1, 0).$$

Step 5 To determine whether the graph intersects its horizontal asymptote $y = 0$, solve

$$f(x) = 0. \leftarrow \text{ } y\text{-value of horizontal asymptote}$$

The solution to this equation was found in Step 4. The graph intersects its horizontal asymptote at $(-1, 0)$.

Step 6 The *x*-intercepts and vertical asymptotes determine the following intervals of the domain.

$$(-\infty, -3), \quad (-3, -1), \quad \left(-1, \frac{1}{2}\right), \quad \text{and} \quad \left(\frac{1}{2}, \infty\right)$$

Plot a point in each of the intervals to get an idea of how the graph behaves in each region. See **FIGURE 17** on the next page and the corresponding table.

TECHNOLOGY NOTE

This screen shows a calculator table similar to the one on the next page. By using the "Ask" feature in the Table Setup, desired test values can be evaluated.

X	Y1
-4	-.3333
-2	.2
0	-.3333
2	.2

Y₁■(X+1)/(2X²+5…

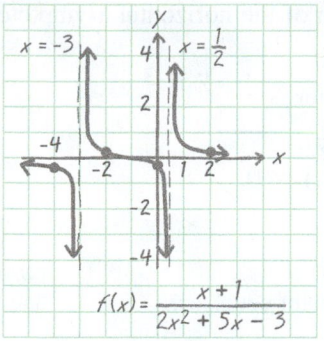

FIGURE 17

Interval	Test Value x	Value of $f(x)$	Sign of $f(x)$	Graph Above or Below x-Axis
$(-\infty, -3)$	-4	$-\frac{1}{3}$	Negative	Below
$(-3, -1)$	-2	$\frac{1}{5}$	Positive	Above
$\left(-1, \frac{1}{2}\right)$	0	$-\frac{1}{3}$	Negative	Below
$\left(\frac{1}{2}, \infty\right)$	2	$\frac{1}{5}$	Positive	Above

Step 7   Complete the sketch as shown in **FIGURE 17**.

EXAMPLE 5   **Graphing a Rational Function That Does Not Intersect Its Horizontal Asymptote**

Graph $f(x) = \dfrac{2x + 1}{x - 3}$.

Solution

Steps 1 and 2   As determined in **Example 2,** the domain is $(-\infty, 3) \cup (3, \infty)$ and the equation of the vertical asymptote is $x = 3$. The horizontal asymptote has equation $y = 2$.

Step 3 $f(0) = -\frac{1}{3}$, so the y-intercept is $\left(0, -\frac{1}{3}\right)$.

Step 4   Solve $f(x) = 0$ to find the x-intercept(s).

$$\frac{2x + 1}{x - 3} = 0 \qquad \text{Let } f(x) = 0.$$

$$2x + 1 = 0 \qquad \text{If a fraction equals 0, then its numerator must equal 0.}$$

$$x = -\frac{1}{2} \qquad \text{The } x\text{-intercept is } \left(-\frac{1}{2}, 0\right).$$

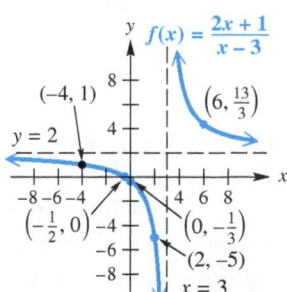

FIGURE 18

Step 5   The graph does not intersect its horizontal asymptote, since $f(x) = 2$ has no solution. (Verify this.)

Steps 6 and 7   The points $(-4, 1)$, $(2, -5)$ and $\left(6, \frac{13}{3}\right)$ are on the graph and can be used to complete the sketch, as shown in **FIGURE 18**.

EXAMPLE 6   **Graphing a Rational Function That Intersects Its Horizontal Asymptote**

Graph $f(x) = \dfrac{3x^2 - 3x - 6}{x^2 + 8x + 16}$.

Solution

Step 1   To find the domain and vertical asymptote(s), solve $x^2 + 8x + 16 = 0$.

$$x^2 + 8x + 16 = 0 \qquad \text{Set the denominator equal to 0.}$$

$$(x + 4)^2 = 0 \qquad \text{Factor as a perfect square trinomial.}$$

$$x = -4 \qquad \text{Solve.}$$

Since the numerator is not 0 when $x = -4$, the only vertical asymptote has equation $x = -4$. The domain is $(-\infty, -4) \cup (-4, \infty)$.

(continued)

→ **Looking Ahead to Calculus**

The rational function

$$f(x) = \frac{2x + 1}{x - 3}$$

in **Example 5** has horizontal asymptote $y = 2$. In calculus, the behavior of the graph of this function as $x \to -\infty$ and as $x \to \infty$ is described by using limits at infinity. As $x \to -\infty$, $f(x) \to 2$ (from below). This is written

$$\lim_{x \to -\infty} f(x) = 2.$$

As $x \to \infty$, $f(x) \to 2$ (from above). This is written

$$\lim_{x \to \infty} f(x) = 2.$$

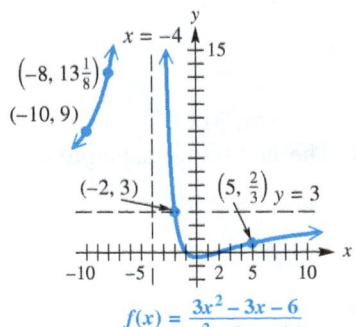

$$f(x) = \frac{3x^2 - 3x - 6}{x^2 + 8x + 16}$$

FIGURE 19

Step 2 We divide all terms by x^2 to find the equation of the horizontal asymptote.

$$y = \frac{3}{1}, \quad \begin{matrix} \leftarrow \text{Leading coefficient of numerator} \\ \leftarrow \text{Leading coefficient of denominator} \end{matrix} \quad \text{or} \quad y = 3$$

Step 3 Because $f(0) = -\frac{3}{8}$, the y-intercept is $\left(0, -\frac{3}{8}\right)$.

Step 4 To find the x-intercept(s), if any, we solve $f(x) = 0$.

$$\frac{3x^2 - 3x - 6}{x^2 + 8x + 16} = 0 \qquad \text{Let } f(x) = 0.$$

$$3x^2 - 3x - 6 = 0 \qquad \text{Set the numerator equal to 0.}$$

$$x^2 - x - 2 = 0 \qquad \text{Divide by 3.}$$

$$(x - 2)(x + 1) = 0 \qquad \text{Factor.}$$

$$x = 2 \quad \text{or} \quad x = -1 \qquad \text{Zero-product property}$$

The x-intercepts are $(-1, 0)$ and $(2, 0)$.

Step 5 We set $f(x) = 3$ and solve to locate the point where the graph intersects the horizontal asymptote.

$$\frac{3x^2 - 3x - 6}{x^2 + 8x + 16} = 3 \qquad \text{Let } f(x) = 3.$$

$$3x^2 - 3x - 6 = 3x^2 + 24x + 48 \qquad \text{Multiply by } x^2 + 8x + 16.$$

$$-3x - 6 = 24x + 48 \qquad \text{Subtract } 3x^2.$$

$$-27x = 54 \qquad \text{Subtract 24x and add 6.}$$

$$x = -2 \qquad \text{Divide by } -27.$$

The graph intersects its horizontal asymptote at $(-2, 3)$.

Steps 6 and 7 Some other points that lie on the graph are $(-10, 9)$, $\left(-8, 13\frac{1}{8}\right)$, and $\left(5, \frac{2}{3}\right)$. These can be used to complete the sketch, as shown in **FIGURE 19**. ●

Notice the behavior of the graph of the function in **FIGURE 19** near $x = -4$. As $x \to -4$ from either side, $f(x) \to \infty$. If we examine the behavior of the graph in **FIGURE 18** on the preceding page near $x = 3$, $f(x) \to -\infty$ as $x \to 3$ from the left, while $f(x) \to \infty$ as $x \to 3$ from the right. *The behavior of the graph of a rational function near a vertical asymptote $x = a$ will partially depend on the parity of the exponent on $x - a$ in the denominator (i.e., whether the exponent is odd or even).*

Behavior of Graphs of Rational Functions Near Vertical Asymptotes

Suppose that $f(x)$ is a rational expression in lowest terms. If n is the largest positive integer such that $(x - a)^n$ is a factor of the denominator of $f(x)$, the graph will behave in the manner illustrated near a.

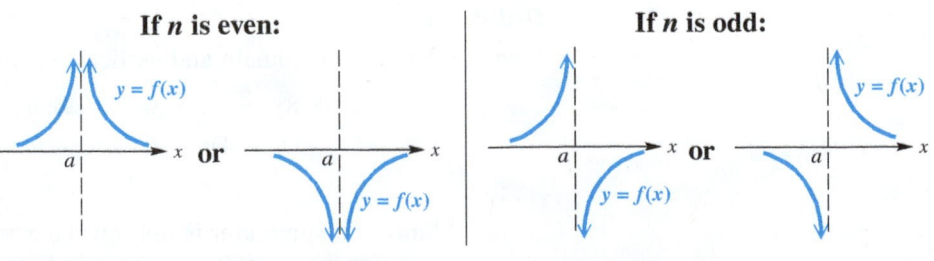

In **Section 3.7,** we observed that the behavior of the graph of a polynomial function near its zeros is dependent on the multiplicity of the zero. The same statement can be made for rational functions.

Behavior of Graphs of Rational Functions Near *x*-Intercepts

Suppose that $f(x)$ is a rational expression in lowest terms. If n is the largest positive integer such that $(x - c)^n$ is a factor of the numerator of $f(x)$, the graph will behave in the manner illustrated near c.

If *n* is even:	If *n* = 1:	If *n* is an odd integer greater than 1:
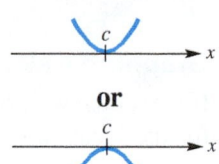	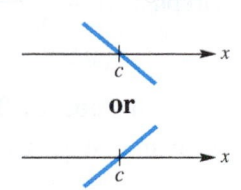	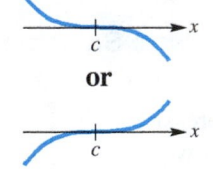
or	or	or

Oblique Asymptotes

EXAMPLE 7 **Graphing a Rational Function with an Oblique Asymptote**

Graph $f(x) = \dfrac{x^2 + 1}{x - 2}$.

Analytic Solution

As shown in **Example 3,** the vertical asymptote has equation $x = 2$, and the graph has an oblique asymptote with equation $y = x + 2$. Refer to the box on the preceding page to determine the behavior of the graph near the asymptote $x = 2$. The y-intercept is $\left(0, -\frac{1}{2}\right)$, and since the numerator $x^2 + 1$ has no real zeros, the graph has no x-intercepts. The graph does not intersect its oblique asymptote, because the equation

$$\frac{x^2 + 1}{x - 2} = x + 2$$

has no solution. (Verify this.) Using the y-intercept, asymptotes, the points $\left(4, \frac{17}{2}\right)$ and $\left(-1, -\frac{2}{3}\right)$, and the general behavior of the graph near its asymptotes leads to the graph in **FIGURE 20.**

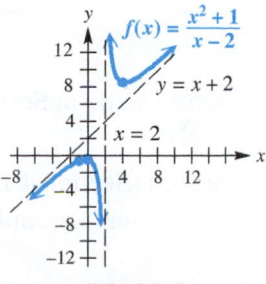

FIGURE 20

Graphing Calculator Solution

A calculator graph is shown in **FIGURE 21.** If we define

$$Y_1 = \frac{X^2 + 1}{X - 2} \quad \text{and} \quad Y_2 = X + 2$$

and observe a table of values as $|X| \to \infty$ we see that for these large values, $Y_1 \approx Y_2$. The graph of $f(x)$ approaches the oblique asymptote $y = x + 2$ as $|x| \to \infty$.

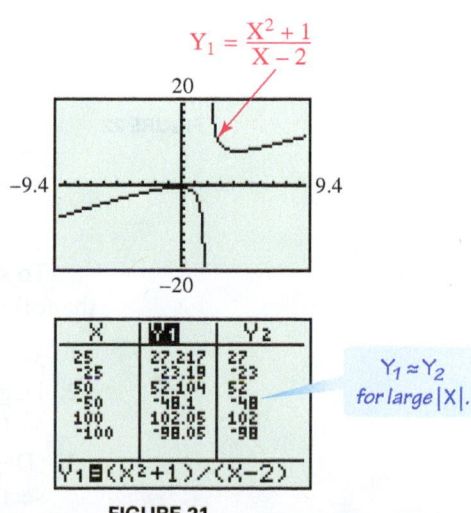

FIGURE 21

Graphs with Points of Discontinuity

A rational function must be defined by an expression in lowest terms before we can use the methods discussed thus far to hand-sketch the graph. A rational function that has a common *variable* factor in the numerator and denominator is not in lowest terms. Its graph usually has a **hole,** or **point of discontinuity.**

EXAMPLE 8 **Graphing a Rational Function Defined by an Expression That Is Not in Lowest Terms**

Graph $f(x) = \dfrac{x^2 - 4}{x - 2}$.

Analytic Solution

The domain of this function cannot include 2. The expression $\frac{x^2 - 4}{x - 2}$ is now written in lowest terms, as follows.

$$
\begin{aligned}
f(x) &= \frac{x^2 - 4}{x - 2} \\
&= \frac{(x + 2)(x - 2)}{x - 2} \qquad \text{Factor.} \\
&= x + 2 \quad (x \neq 2) \qquad \text{Lowest terms}
\end{aligned}
$$

Therefore, the graph of this function will be the same as the graph of

$$y = x + 2$$

(a straight line), with the exception of the point with x-value 2. A hole appears in the graph at $(2, 4)$. See **FIGURE 22.**

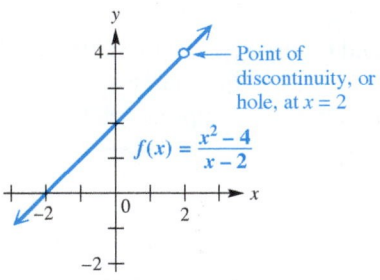

FIGURE 22

Graphing Calculator Solution

If we set the window of the TI-84 Plus so that an x-value for the location of the tracing cursor is 2, then we can see from the display that the calculator cannot determine a value for Y. We define

$$Y_1 = \frac{X^2 - 4}{X - 2}$$

and graph it in such a window, as in **FIGURE 23.** The error message in the table further supports the existence of a discontinuity at $X = 2$. (For the table, $Y_2 = X + 2$.)

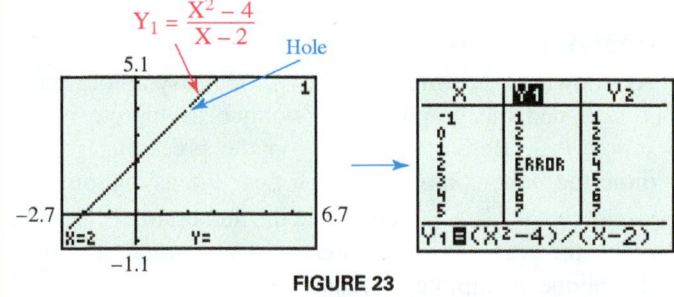

FIGURE 23

Notice the visible discontinuity at $X = 2$ in the graph. *The window was chosen so that the hole would be visible.* This requires a decimal viewing window or a window with x-values centered at $\frac{-2.7 + 6.7}{2} = 2$. *Other choices may not show this discontinuity.*

To summarize, rational functions presented so far in **Sections 4.1** and **4.2** fall into the following categories.

1. $y = \frac{1}{x}$ and $y = \frac{1}{x^2}$ and their transformations (**Section 4.1**)
2. Degree of numerator $\leq$ degree of denominator (**Examples 1, 2, 4, 5,** and **6** of this section)
3. (Degree of numerator) $- 1 =$ degree of denominator (**Examples 3** and **7** of this section), with an oblique asymptote
4. Those with common variable factors in numerator and denominator (**Example 8** of this section)

→ **Looking Ahead to Calculus**

Different types of discontinuity are discussed in calculus. The function in **Example 8,**

$$f(x) = \frac{x^2 - 4}{x - 2},$$

is said to have a **removable discontinuity** at $x = 2$, since the discontinuity can be removed by redefining f at 2. The function in **Example 7,**

$$f(x) = \frac{x^2 + 1}{x - 2},$$

has an **infinite discontinuity** at $x = 2$, as indicated by the vertical asymptote there. The greatest integer function, discussed in **Section 2.5,** has jump discontinuities, because the function values "jump" from one value to another for integer domain values.

Graphs with No Vertical Asymptotes

EXAMPLE 9 Graphing a Rational Function by Hand

Graph $f(x) = \dfrac{1}{x^2 + 1}$. Give the domain, range, and any intercepts. Discuss symmetry and any asymptotes.

Solution By plotting selected points and noting that $f(-x) = f(x)$ for all x, we see that the graph lies above the x-axis for all x. The maximum y-value is 1 when $x = 0$. The domain is $(-\infty, \infty)$ because the denominator has no real zeros, and the range is the interval $(0, 1]$. There are no x-intercepts. Because $f(0) = 1$, the y-intercept is $(0, 1)$. The graph is symmetric with respect to the y-axis. There are no vertical asymptotes, and the x-axis is the horizontal asymptote. See **FIGURE 24.**

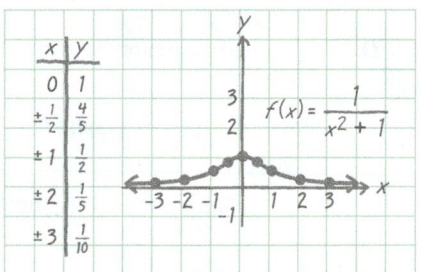

x	y
0	1
$\pm\frac{1}{2}$	$\frac{4}{5}$
± 1	$\frac{1}{2}$
± 2	$\frac{1}{5}$
± 3	$\frac{1}{10}$

$f(x) = \dfrac{1}{x^2 + 1}$

FIGURE 24

EXAMPLE 10 Analyzing and Graphing a More Complicated Rational Function

Consider the rational function

$$f(x) = \frac{2x^3 - x^2 - 2x + 1}{x^2 - x + 1}.$$

Give the x-intercepts and y-intercept. Explain why there are no vertical asymptotes. What is the equation of the oblique asymptote? Graph the function in the window $[-4.7, 4.7]$ by $[-3.1, 3.1]$. What are the domain and range?

Solution The rational expression is in lowest terms. The x-intercepts correspond to the zeros of $2x^3 - x^2 - 2x + 1$, and are $(-1, 0)$, $\left(\frac{1}{2}, 0\right)$ and $(1, 0)$. (Verify this.) Because $f(0) = 1$, the y-intercept is $(0, 1)$. There are no real zeros of the denominator $x^2 - x + 1$ because the discriminant, -3, is negative, and so there are no vertical asymptotes. When the numerator is divided by the denominator using long division (**Section 3.6**), the result is

$$2x + 1 + \frac{-3x}{x^2 - x + 1}.$$

Thus, the equation of the oblique asymptote is $y = 2x + 1$. The graph in **FIGURE 25** correctly suggests that the domain and range are both $(-\infty, \infty)$.

FOR DISCUSSION

Graph the function in **Example 10** as Y_1 and graph $Y_2 = 2X + 1$ in the same window. Then move the tracing cursor to take on large values of $|X|$. Toggle between Y_1 and Y_2. What do you notice?

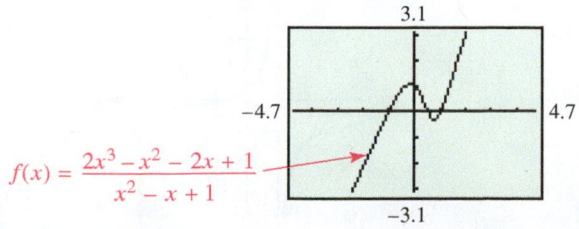

$f(x) = \dfrac{2x^3 - x^2 - 2x + 1}{x^2 - x + 1}$

FIGURE 25

4.2 Exercises

Checking Analytic Skills *Match the rational function in Column I with the appropriate description in Column II. Choices in Column II can be used only once.* **Do not use a calculator.**

I	II
1. $f(x) = \dfrac{x + 7}{x + 1}$	**A.** The x-intercept is $(-3, 0)$.
2. $f(x) = \dfrac{x + 10}{x + 2}$	**B.** The y-intercept is $(0, 5)$.
3. $f(x) = \dfrac{1}{x + 12}$	**C.** The horizontal asymptote is $y = 4$.
4. $f(x) = \dfrac{-3 + x^2}{x^2}$	**D.** The vertical asymptote is $x = -1$.
5. $f(x) = \dfrac{x^2 - 16}{x + 4}$	**E.** There is a hole in its graph at $x = -4$.
6. $f(x) = \dfrac{4x + 3}{x - 7}$	**F.** The graph has an oblique asymptote.
7. $f(x) = \dfrac{x^2 + 3x + 4}{x - 5}$	**G.** The x-axis is its horizontal asymptote.
8. $f(x) = \dfrac{x + 3}{x - 6}$	**H.** The y-axis is its vertical asymptote.

Give the equations of any vertical, horizontal, or oblique asymptotes for the graph of each rational function. State the domain of f.

9. $f(x) = \dfrac{3}{x - 5}$
10. $f(x) = \dfrac{-6}{x + 9}$
11. $f(x) = \dfrac{4 - 3x}{2x + 1}$
12. $f(x) = \dfrac{2x + 6}{x - 4}$

13. $f(x) = \dfrac{x^2 - 1}{x + 3}$
14. $f(x) = \dfrac{x^2 + 4}{x - 1}$
15. $f(x) = \dfrac{x^2 - 2x - 3}{2x^2 - x - 10}$
16. $f(x) = \dfrac{3x^2 - 6x - 24}{5x^2 - 26x + 5}$

17. *Concept Check* Which function has a graph that does not have a vertical asymptote?

A. $f(x) = \dfrac{1}{x^2 + 2}$
B. $f(x) = \dfrac{1}{x^2 - 2}$
C. $f(x) = \dfrac{3}{x^2}$
D. $f(x) = \dfrac{2x + 1}{x - 8}$

18. *Concept Check* Which function has a graph that does not have a horizontal asymptote?

A. $f(x) = \dfrac{2x - 7}{x + 3}$
B. $f(x) = \dfrac{3x}{x^2 - 9}$
C. $f(x) = \dfrac{x^2 - 9}{x + 3}$
D. $f(x) = \dfrac{x + 5}{(x + 2)(x - 3)}$

Identify any vertical, horizontal, or oblique asymptotes in the graph of $y = f(x)$. State the domain of f.

19.

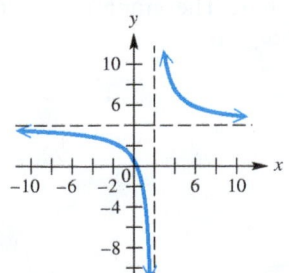

20.

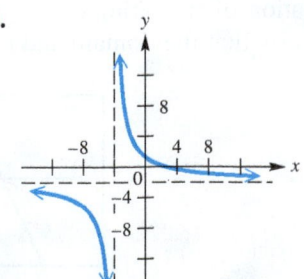

21.

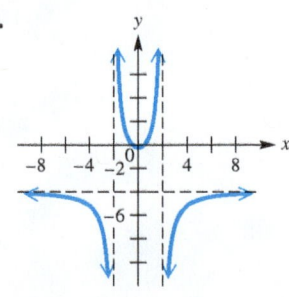

22.

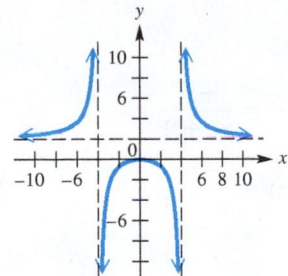

23.

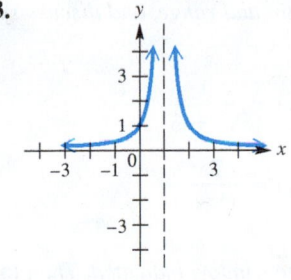

24.

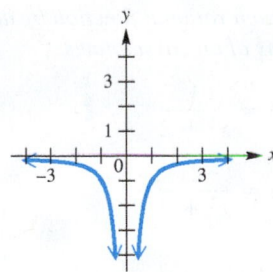

25.

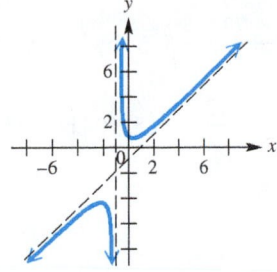

26.

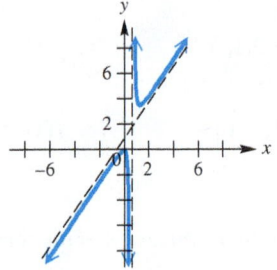

27.

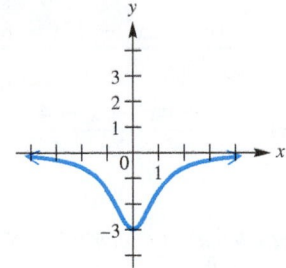

28. Explain why the graph of the rational function $f(x) = \frac{-1}{x^2 + 4}$ has no vertical asymptotes.

Checking Analytic Skills   *Sketch a graph of each rational function. Your graph should include all asymptotes.* **Do not use a calculator.**

29. $f(x) = \dfrac{x + 1}{x - 4}$

30. $f(x) = \dfrac{x - 5}{x + 3}$

31. $f(x) = \dfrac{x + 2}{x - 3}$

32. $f(x) = \dfrac{x - 3}{x + 4}$

33. $f(x) = \dfrac{4 - 2x}{8 - x}$

34. $f(x) = \dfrac{6 - 3x}{4 - x}$

35. $f(x) = \dfrac{3x}{(x + 1)(x - 2)}$

36. $f(x) = \dfrac{2x + 1}{(x + 2)(x + 4)}$

37. $f(x) = \dfrac{5x}{x^2 - 1}$

38. $f(x) = \dfrac{x}{4 - x^2}$

39. $f(x) = \dfrac{(x + 6)(x - 2)}{(x + 3)(x - 4)}$

40. $f(x) = \dfrac{(x + 3)(x - 5)}{(x + 1)(x - 4)}$

41. $f(x) = \dfrac{3x^2 + 3x - 6}{x^2 - x - 12}$

42. $f(x) = \dfrac{4x^2 + 4x - 24}{x^2 - 3x - 10}$

43. $f(x) = \dfrac{9x^2 - 1}{x^2 - 4}$

44. $f(x) = \dfrac{16x^2 - 9}{x^2 - 9}$

45. $f(x) = \dfrac{(x - 3)(x + 1)}{(x - 1)^2}$

46. $f(x) = \dfrac{x(x - 2)}{(x + 3)^2}$

47. $f(x) = \dfrac{x}{x^2 - 9}$

48. $f(x) = \dfrac{-5}{2x + 4}$

49. $f(x) = \dfrac{-4}{3x + 9}$

50. $f(x) = \dfrac{(3 - x)^2}{(1 - x)(4 + x)}$

51. $f(x) = \dfrac{(x + 4)^2}{(x - 1)(x + 5)}$

52. $f(x) = \dfrac{(x + 1)^2}{(x + 2)(x - 3)}$

53. $f(x) = \dfrac{20 + 6x - 2x^2}{8 + 6x - 2x^2}$

54. $f(x) = \dfrac{18 + 6x - 4x^2}{4 + 6x + 2x^2}$

55. $f(x) = \dfrac{x^2 + 1}{x + 3}$

56. $f(x) = \dfrac{2x^2 + 3}{x - 4}$

57. $f(x) = \dfrac{x^2 + 2x}{2x - 1}$

58. $f(x) = \dfrac{x^2 - x}{x + 2}$

59. $f(x) = \dfrac{x^2 - 9}{x + 3}$

60. $f(x) = \dfrac{x^2 - 16}{x + 4}$

61. $f(x) = \dfrac{2x^2 - 5x - 2}{x - 2}$

62. $f(x) = \dfrac{x^2 - 5}{x - 3}$

63. $f(x) = \dfrac{x^2 - 1}{x^2 - 4x + 3}$

64. $f(x) = \dfrac{x^2 - 4}{x^2 + 3x + 2}$

65. $f(x) = \dfrac{(x^2 - 9)(2 + x)}{(x^2 - 4)(3 + x)}$

66. $f(x) = \dfrac{(x^2 - 16)(3 + x)}{(x^2 - 9)(4 + x)}$

67. $f(x) = \dfrac{x^4 - 20x^2 + 64}{x^4 - 10x^2 + 9}$

68. $f(x) = \dfrac{x^4 - 5x^2 + 4}{x^4 - 24x^2 + 108}$

Graph each rational function by hand. Give the domain and range, and discuss symmetry. Give the equations of any asymptotes.

69. $f(x) = \dfrac{1}{x^2 + 2}$

70. $f(x) = \dfrac{1}{x^2 + 3}$

71. $f(x) = \dfrac{-x^2}{x^2 + 1}$

72. $f(x) = \dfrac{-2x^2}{x^2 + 2}$

73. $f(x) = \dfrac{2x^2}{x^4 + 1}$

74. $f(x) = \dfrac{-2x^2}{x^4 + 1}$

Use a calculator to graph each rational function in the window indicated. Then **(a)** *give the x- and y-intercepts,* **(b)** *explain why there are no vertical asymptotes,* **(c)** *give the equation of the oblique asymptote, and* **(d)** *give the domain and range.*

75. $f(x) = \dfrac{3x^3 + 2x^2 - 12x - 8}{x^2 + x + 4}; [-4.7, 4.7] \text{ by } [-3.1, 3.1]$

76. $f(x) = \dfrac{4x^3 + 8x^2 - 36x - 72}{2x^2 - x + 6}; [-5, 5] \text{ by } [-20, 15]$

77. $f(x) = \dfrac{x^3 + 4x^2 - x - 4}{-2x^2 - 2x - 4}; [-4.7, 4.7] \text{ by } [-3.1, 3.1]$

78. $f(x) = \dfrac{-x^3 - 7x^2 + 16x + 112}{x^2 + x + 28}; [-15, 10] \text{ by } [-5, 15]$

Concept Check   In each table, Y_1 is defined by a rational expression of the form $\dfrac{X - p}{X - q}$. Use the table to find the values of p and q.

79.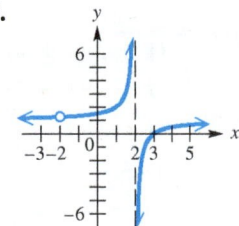

80.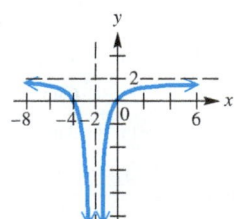

81.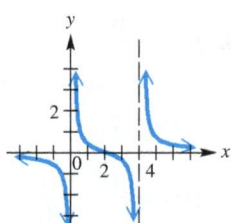

82.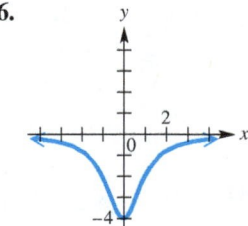

Concept Check   Find an equation for the rational function graphed. (Answers may vary.)

83.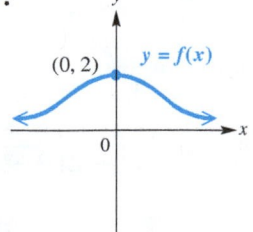

84.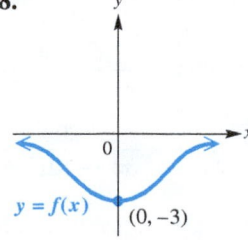

85.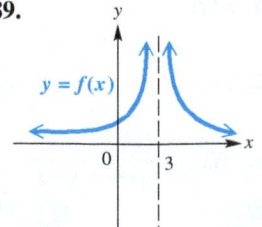

86.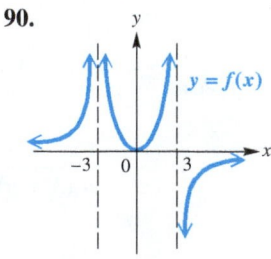

RELATING CONCEPTS   **For individual or group investigation (Exercises 87–90)**

Recall from **Section 2.3** *that if we are given the graph of $y = f(x)$, we can obtain the graph of $y = -f(x)$ by reflecting across the x-axis, and we can obtain the graph of $y = f(-x)$ by reflecting across the y-axis. In Exercises 87–90, you are given the graph of a rational function $y = f(x)$. Draw a sketch by hand of the graph of* **(a)** $y = -f(x)$ *and* **(b)** $y = f(-x)$.

87.

88.

89.

90.

Each rational function has an oblique asymptote. Determine the equation of this asymptote. Then, use a graphing calculator to graph both the function and the asymptote in the window indicated.

91. $f(x) = \dfrac{2x^2 + 3}{4 - x}$; $[-18.8, 18.8]$ by $[-50, 25]$

92. $f(x) = \dfrac{x^2 + 9}{x + 3}$; $[-9.4, 9.4]$ by $[-25, 25]$

93. $f(x) = \dfrac{x - x^2}{x + 2}$; $[-9.4, 9.4]$ by $[-15, 25]$

94. $f(x) = \dfrac{x^2 + 2x}{1 - 2x}$; $[-4.7, 4.7]$ by $[-5, 5]$

95. *Concept Check* $f(x) = \dfrac{x^5 + x^4 + x^2 + 1}{x^4 + 1}$

becomes $f(x) = x + 1 + \dfrac{x^2 - x}{x^4 + 1}$

after the numerator is divided by the denominator.
(a) What is the equation of the oblique asymptote of the graph of the function?
(b) For what x-value(s) does the graph of the function intersect its asymptote?
(c) As $x \to \infty$, does the graph of the function approach its asymptote from above or below?

96. *Concept Check*   Consider the rational function

$$f(x) = \dfrac{x^3 - 4x^2 + x + 6}{x^2 + x - 2}.$$

Divide the numerator by the denominator and use the method of **Example 3** to determine the equation of the oblique asymptote. Then, determine the coordinates of the point where the graph of f intersects its oblique asymptote. Use a calculator to support your answer.

97. Use long division of polynomials to show that for

$$f(x) = \dfrac{x^4 - 5x^2 + 4}{x^2 + x - 12},$$

if we divide the numerator by the denominator, then the quotient polynomial is $x^2 - x + 8$, and the remainder is $-20x + 100$. Graph both $f(x)$ and $g(x) = x^2 - x + 8$ in the window $[-50, 50]$ by $[0, 1000]$. Comment on the appearance of the two graphs. Explain how the graph of f approaches that of g as $|x| \to \infty$.

98. Suppose a friend tells you that the graph of

$$f(x) = \dfrac{x^2 - 25}{x + 5}$$

has a vertical asymptote with equation $x = -5$. Is this correct? If not, describe the behavior of the graph at $x = -5$.

4.3 Rational Equations, Inequalities, Models, and Applications

Solving Rational Equations and Inequalities • Models and Applications of Rational Functions • Inverse Variation • Combined and Joint Variation • Rate of Work

← Algebra Review

To review rational expressions, see **Section R.3**. To review negative exponents, see **Section R.4**.

Solving Rational Equations and Inequalities

A **rational equation** (or **rational inequality**) is an equation (or inequality) with at least one term having a variable expression in a denominator or at least one term having a variable expression raised to a negative integer power.

$$\frac{x + 2}{2x + 1} = 1, \quad \frac{x + 2}{2x + 1} \le 1, \quad \frac{x}{x - 2} + \frac{1}{x + 2} = \frac{8}{x^2 - 4}, \quad \text{and} \quad 7x^{-4} - 8x^{-2} + 1 = 0$$

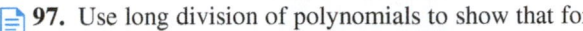
Rational equations and inequalities

CAUTION   When solving rational equations and inequalities, *remember that an expression with a variable denominator or with a negative exponent may be undefined for certain values of the variable.* We must identify these values where the associated rational function will have either a vertical asymptote or a hole.

> *To solve a rational equation, we clear fractions by multiplying each side by the least common denominator (LCD) of all rational expressions in the equation.*

EXAMPLE 1 Solving a Rational Equation

Solve $\dfrac{x + 2}{2x + 1} = 1$.

Analytic Solution

Set the denominator, $2x + 1$, equal to 0 to determine that the rational expression is undefined for $x = -\frac{1}{2}$. Thus, $-\frac{1}{2}$ cannot be a solution. Multiply each side of the equation by $2x + 1$.

$$\left(\dfrac{x + 2}{2x + 1}\right)(2x + 1) = (1)(2x + 1) \quad \text{Multiply by the LCD.}$$

$$x + 2 = 2x + 1 \quad \text{Simplify.}$$

$$1 = x \quad \text{Subtract } x \text{ and } 1.$$

Check by substituting 1 for x in the original equation.

$$\dfrac{1 + 2}{2(1) + 1} = 1 \quad ? \quad \text{Let } x = 1.$$

$$\dfrac{3}{3} = 1 \quad ? \quad \text{Simplify.}$$

$$1 = 1 \quad \checkmark \quad \text{True}$$

The solution set is $\{1\}$.

Graphing Calculator Solution

Because graphs of rational functions usually consist of several parts, using the intersection-of-graphs method can be confusing. Rewrite the equation as

$$\dfrac{x + 2}{2x + 1} - 1 = 0,$$

and graph $\qquad y = \dfrac{x + 2}{2x + 1} - 1.$

The x-intercept method shows that the zero of the function is 1 and thus the solution set is $\{1\}$. See **FIGURE 26**.

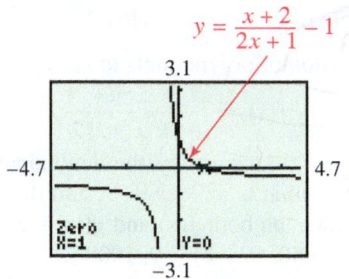

FIGURE 26

EXAMPLE 2 Solving a Rational Equation

Solve $\dfrac{x}{x - 2} + \dfrac{1}{x + 2} = \dfrac{8}{x^2 - 4}$.

Analytic Solution

Set each denominator equal to zero to determine that for this equation, $x \neq \pm 2$.

$$\dfrac{x}{x - 2} + \dfrac{1}{x + 2} = \dfrac{8}{x^2 - 4}$$

$$x(x + 2) + 1(x - 2) = 8 \quad \text{Multiply by the LCD,} \ (x - 2)(x + 2).$$

$$x^2 + 2x + x - 2 = 8 \quad \text{Distributive property}$$

$$x^2 + 3x - 10 = 0 \quad \text{Standard form}$$

$$(x + 5)(x - 2) = 0 \quad \text{Factor.}$$

$$x + 5 = 0 \quad \text{or} \quad x - 2 = 0 \quad \text{Zero-product property}$$

$$x = -5 \quad \text{or} \quad x = 2 \quad \text{2 is not in the domain.}$$

The numbers -5 and 2 are the possible solutions of the equation, but 2 is *not* in the domain of the original equation and, therefore, must be rejected. Such a value is **extraneous.** The solution set is $\{-5\}$.

Graphing Calculator Solution

Rewrite the equation as

$$\dfrac{x}{x - 2} + \dfrac{1}{x + 2} - \dfrac{8}{x^2 - 4} = 0,$$

and let y equal the left side of this equation. **FIGURE 27** shows that the zero of the function is -5, supporting the analytic solution.

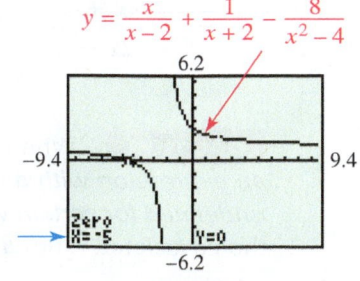

FIGURE 27

> There is no zero at $x = 2$. Graphs do not show extraneous solutions.

> **Solving a Rational Equation**
>
> **Step 1** Determine all values for which the rational equation has undefined expressions.
>
> **Step 2** To clear fractions, multiply each side of the equation by the least common denominator of all rational expressions in the equation.
>
> **Step 3** Solve the resulting equation.
>
> **Step 4** Reject any values found in Step 1.

EXAMPLE 3   **Solving a Rational Inequality by Hand**

(a) Solve $\dfrac{x + 2}{2x + 1} \le 1$.

(b) Discuss how the graph in **FIGURE 26** on the previous page supports the solution.

Solution

(a) Subtract 1 from each side and combine terms on the left side to get a single rational expression.

$$\frac{x + 2}{2x + 1} - 1 \le 0 \qquad \text{Subtract 1.}$$

$$\frac{x + 2}{2x + 1} - \frac{2x + 1}{2x + 1} \le 0 \qquad \text{The common denominator is } 2x + 1.$$

$$\frac{x + 2 - (2x + 1)}{2x + 1} \le 0 \qquad \text{Write as a single rational expression.}$$

Distribute the minus sign.

$$\frac{x + 2 - 2x - 1}{2x + 1} \le 0 \qquad \text{Distributive property}$$

$$\frac{-x + 1}{2x + 1} \le 0 \qquad \text{Combine terms in the numerator.}$$

The quotient possibly changes sign only where the x-values make the numerator or denominator 0. Set each equal to 0 and solve.

$$-x + 1 = 0 \qquad \text{or} \qquad 2x + 1 = 0$$

$$x = 1 \qquad \text{or} \qquad x = -\frac{1}{2}$$

These two x-values divide the real number line into three intervals: $\left(-\infty, -\frac{1}{2}\right)$, $\left(-\frac{1}{2}, 1\right)$, and $(1, \infty)$. See **FIGURE 28.**

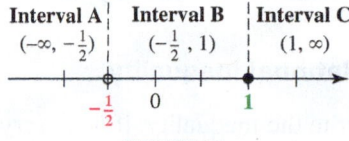

Interval A	Interval B	Interval C
$\left(-\infty, -\frac{1}{2}\right)$	$\left(-\frac{1}{2}, 1\right)$	$(1, \infty)$

FIGURE 28

Use a solid circle on 1, since the symbol $\le$ allows equality. The value $-\frac{1}{2}$ cannot be in the solution set, since it causes the denominator to equal 0. Use an open circle on $-\frac{1}{2}$.

On each open interval, the expression $\frac{-x + 1}{2x + 1}$ is either only positive or only negative. To determine the sign of the expression $\frac{-x + 1}{2x + 1}$, choose a test value from each interval and substitute it into the expression. See the table on the next page.

(continued)

Interval	Test Value x	Is $\dfrac{-x+1}{2x+1} \leq 0$ True or False?
A: $\left(-\infty, -\frac{1}{2}\right)$	-1	$\dfrac{1+1}{2(-1)+1} \leq 0$? $-2 \leq 0$ True
B: $\left(-\frac{1}{2}, 1\right)$	0	$\dfrac{0+1}{2(0)+1} \leq 0$? $1 \leq 0$ False
C: $(1, \infty)$	2	$\dfrac{-2+1}{2(2)+1} \leq 0$? $-\frac{1}{5} \leq 0$ True

We see that values of x in the intervals $\left(-\infty, -\frac{1}{2}\right)$ and $(1, \infty)$ make the quotient negative, as required. Therefore, the solution set is their union.

$$\left(-\infty, -\frac{1}{2}\right) \cup [1, \infty) \qquad \text{Solution set}$$

From our previous discussion, 1 is included in the solution set because it results in equality, whereas $-\frac{1}{2}$ is not included because it results in a denominator that is equal to 0.

(b) The original inequality is equivalent to

$$\frac{x+2}{2x+1} - 1 \leq 0.$$

From **FIGURE 26,** repeated below, we see that the graph lies *on* or *below* the x-axis for x-values to the left of the vertical asymptote $\left(x = -\frac{1}{2}\right)$ as well as for x-values greater than or equal to 1. This agrees with the solution set obtained in part (a).

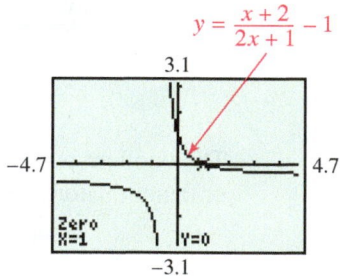

$$y = \frac{x+2}{2x+1} - 1$$

FIGURE 26 (repeated)

CAUTION *Be sure to carefully check endpoints of intervals when solving rational inequalities.*

Example 3 illustrates the general procedure for solving a rational inequality.

Solving a Rational Inequality

Step 1   Rewrite the inequality, if necessary, so that 0 is on one side and there is a single rational expression on the other side.

Step 2   Determine the values that will cause either the numerator or the denominator of the rational expression to equal 0. These values determine the intervals on the number line to consider.

Step 3   Use a test value from each interval to determine which intervals form the solution set. Be sure to check endpoints.

EXAMPLE 4   **Solving a Rational Equation and Associated Inequalities**

Consider the rational function $f(x) = 7x^{-4} - 8x^{-2} + 1$.

(a) Solve the equation $f(x) = 0$.

(b) Graph the equation $y = f(x)$. Identify the zeros of $f(x)$.

(c) Use the graph to solve the inequalities $f(x) \leq 0$ and $f(x) \geq 0$.

Solution

(a) The expression for $f(x)$ may be written equivalently as follows.

$$f(x) = \frac{7}{x^4} - \frac{8}{x^2} + 1$$

For this expression, $x \neq 0$. Solve the equation $f(x) = 0$.

$$\frac{7}{x^4} - \frac{8}{x^2} + 1 = 0 \qquad \text{Let } f(x) = 0.$$

$$7 - 8x^2 + x^4 = 0 \qquad \text{Multiply by } x^4.$$

$$x^4 - 8x^2 + 7 = 0 \qquad \text{Rewrite.}$$

$$(x^2 - 7)(x^2 - 1) = 0 \qquad \text{Factor.}$$

$$(x^2 - 7)(x + 1)(x - 1) = 0 \qquad \text{Factor again.}$$

$$x^2 - 7 = 0 \quad \text{or} \quad x + 1 = 0 \quad \text{or} \quad x - 1 = 0 \qquad \text{Zero-product property}$$

$$x = \pm\sqrt{7} \quad \text{or} \quad x = -1 \quad \text{or} \quad x = 1 \qquad \text{Solve each equation.}$$

The solution set is $\left\{ \pm\sqrt{7}, \pm 1 \right\}$.

(b) **FIGURE 29** shows that the graph of $f(x) = 7x^{-4} - 8x^{-2} + 1$ has four zeros. The display indicates that one zero is 2.6457513, an approximation for $\sqrt{7}$. The other zeros, $-\sqrt{7}, -1$, and 1, can be found similarly. The table provides support.

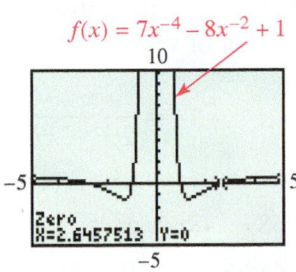
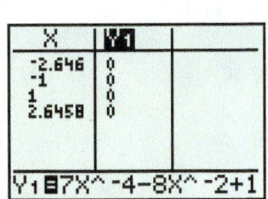

FIGURE 29

(c) The solution set of $f(x) \leq 0$ is $\left[-\sqrt{7}, -1 \right] \cup \left[1, \sqrt{7} \right]$, since these are the values for which the graph lies *on* or *below* the x-axis. Similarly, the solution set of $f(x) \geq 0$ is as follows.

$$\left(-\infty, -\sqrt{7} \right] \cup [-1, 0) \cup (0, 1] \cup \left[\sqrt{7}, \infty \right)$$

Notice that 0 is not included, because it is not in the domain of $f(x)$.

FOR DISCUSSION

In **Example 4,** by creating a table with the four solutions, as in **FIGURE 29,** we show that the function value in each case is 0. The X-values are displayed as decimal approximations, but we must enter them directly as $-\sqrt{7}$ and $\sqrt{7}$.

1. Use a table in the manner described to support the solution in **Example 1.**

2. Use a table to support the single solution in **Example 2.** What happens when you input the extraneous value 2?

Models and Applications of Rational Functions

EXAMPLE 5 **Modeling Traffic Intensity**

Vehicles arrive randomly at a parking ramp at an average rate of 2.6 per minute. The parking attendant can admit 3.2 vehicles per minute. However, since arrivals are random, lines form at various times. (*Source:* Mannering, F. and W. Kilareski, *Principles of Highway Engineering and Traffic Control,* 2d ed., John Wiley and Sons.)

(a) The **traffic intensity** x is defined as the ratio of the average arrival rate to the average admittance rate. Determine x for this parking ramp.

(b) The average number of vehicles waiting in line to enter the ramp is modeled by

$$f(x) = \frac{x^2}{2(1 - x)},$$

where $0 \leq x < 1$ is the traffic intensity. Compute $f(x)$ for this parking ramp.

(c) Graph $y = f(x)$. What happens to the number of vehicles waiting as the traffic intensity approaches 1?

(d) Solve $f(x) = 24$ and interpret the result.

Solution

(a) The average arrival rate is 2.6 vehicles per minute and the average admittance rate is 3.2 vehicles per minute.

$$x = \frac{2.6}{3.2} = 0.8125 \quad \longleftarrow \quad \text{Traffic intensity } x$$

(b) In part (a), we found that $x = 0.8125$. Find $f(0.8125)$.

$$f(0.8125) = \frac{0.8125^2}{2(1 - 0.8125)} \approx 1.76 \text{ vehicles}$$

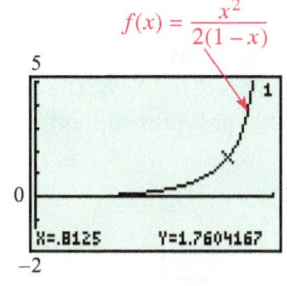

$f(x) = \frac{x^2}{2(1-x)}$

X=.8125   Y=1.7604167

FIGURE 30

(c) From the graph shown in **FIGURE 30,** we see that as x approaches 1, $y = f(x)$ gets very large. That is, the average number of waiting vehicles gets very large. A small increase in the traffic intensity can result in a dramatic increase in the average number of vehicles waiting in line.

(d)

$$24 = \frac{x^2}{2(1 - x)} \qquad \text{Let } f(x) = 24.$$

$$48(1 - x) = x^2 \qquad \text{Multiply by } 2(1 - x).$$

$$48 - 48x = x^2 \qquad \text{Distributive property}$$

$$x^2 + 48x - 48 = 0 \qquad \text{Standard form}$$

$$x = \frac{-48 \pm \sqrt{48^2 - 4(1)(-48)}}{2(1)} \qquad \text{Quadratic formula}$$

$$x \approx 0.98, -49.0 \qquad \text{Use a calculator.}$$

Reject the negative solution -49.0. When the traffic intensity x is 0.98 (near 1), the line is 24 cars, on average.

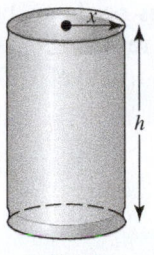

FIGURE 31

A manufacturer wants to construct cylindrical aluminum cans with volume 2000 cubic centimeters (2 liters). What radius and height of the can will minimize the amount of aluminum used? What will this amount be?

Solution   The two unknowns are the radius x and the height h of the can, as shown in **FIGURE 31.** Minimizing the amount of aluminum used requires minimizing the surface area S of the can, given by the following formula.

$$S = 2\pi xh + 2\pi x^2 \qquad \text{(Area of side)} + \text{(Area of top and bottom)}$$

The volume of the can is to be 2000 cubic centimeters, and the formula for volume is $V = \pi x^2 h$ (where x is radius and h is height).

$$V = \pi x^2 h \qquad \text{Volume of a cylinder}$$

$$2000 = \pi x^2 h \qquad \text{Let } V = 2000.$$

$$h = \frac{2000}{\pi x^2} \qquad \text{Solve for } h.$$

Now we can write the surface area S as a function of x alone.

$$S(x) = 2\pi x\left(\frac{2000}{\pi x^2}\right) + 2\pi x^2 \qquad \text{Substitute for } h.$$

$$= \frac{4000}{x} + 2\pi x^2 \qquad \text{Simplify the first term.}^*$$

$$= \frac{4000 + 2\pi x^3}{x} \qquad \text{Combine terms.}^*$$

Since x represents the radius, it must be a positive number. We graph function S and find that the local minimum point is approximately $(6.83, 878.76)$. See **FIGURE 32.** Therefore, rounded to the nearest hundredth, the radius should be 6.83 centimeters and the height should be $\frac{2000}{\pi(6.83)^2}$, or about 13.65 centimeters. These dimensions lead to a minimum amount of about 878.76 square centimeters of aluminum used.   ●

$S(x) = \dfrac{4000 + 2\pi x^3}{x}$

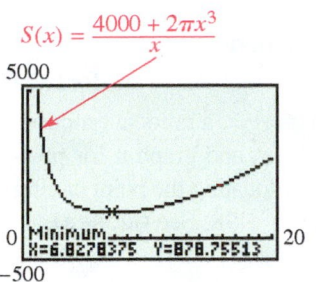

The local minimum point is approximately $(6.83, 878.76)$.

FIGURE 32

(radius, surface area)

Inverse Variation

Recall from **Section 1.6** that when two quantities x and y vary directly, an increase in one quantity results in an increase in the other. The direct variation equation is $y = kx$. When two quantities *vary inversely*, an *increase* in one quantity results in a *decrease* in the other.

Inverse Variation as the nth Power

Let x and y denote two quantities and n be a positive number. Then y is **inversely proportional to the nth power** of x, or y **varies inversely with the nth power** of x, if there exists a nonzero number k such that

$$y = \frac{k}{x^n}.$$

If $y = \frac{k}{x}$, then y is **inversely proportional** to x, or y **varies inversely** with x.

---

*These steps are not necessary to obtain the appropriate graph.

For example, it takes 4 hours to travel 100 miles at 25 miles per hour, but only 2 hours to travel 100 miles at 50 miles per hour. *Greater* speed results in *less* travel time. If s represents the average speed of a car and t is the time to travel 100 miles, then $s \cdot t = 100$, or $t = \dfrac{100}{s}$. *Doubling* the speed cuts the time in *half*. The quantities t and s vary inversely. The constant of variation here is 100.

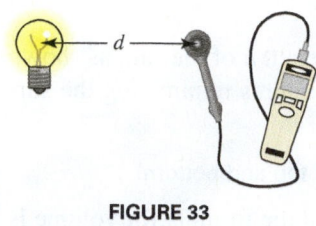

FIGURE 33

EXAMPLE 7 **Modeling the Intensity of Light**

The intensity of light I is inversely proportional to the second power of the distance d. See **FIGURE 33.** The equation

$$I = \frac{k}{d^2}$$

models this phenomenon. At a distance of 3 meters, a 100-watt bulb produces an intensity of 0.88 watt per square meter. (*Source:* Weidner, R. and R. Sells, *Elementary Classical Physics,* Volume 2, Allyn and Bacon.)

Find the constant of variation k, and then determine the intensity of the light at a distance of 2 meters.

Analytic Solution

Substitute $d = 3$ and $I = 0.88$ into the variation equation, and solve for k.

$$I = \frac{k}{d^2} \qquad \text{Variation equation}$$

$$0.88 = \frac{k}{3^2} \qquad \text{Substitute.}$$

$$k = 7.92 \qquad \text{Multiply by } 3^2, \text{ or } 9.$$

Since $k = 7.92$, the equation is

$$I = \frac{7.92}{d^2}, \quad \text{and when } d = 2, \quad I = \frac{7.92}{2^2} = 1.98.$$

The intensity at 2 meters is 1.98 watts per square meter.

Graphing Calculator Solution

From the analytic solution, $k = 7.92$. The formula $I = \frac{7.92}{d^2}$ can be written as $f(x) = \frac{7.92}{x^2}$, a rational function. Enter this function into a calculator, and graph it for positive values of x. Calculate $f(2)$ by locating the point having $x = 2$, and determine that $f(2) = 1.98$. See **FIGURE 34.**

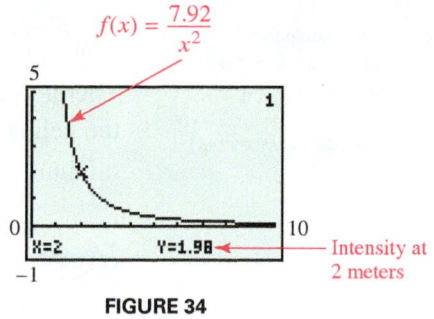

FIGURE 34

Combined and Joint Variation

One variable may depend on more than one other variable. Such variation is called **combined variation.** More specifically, when a variable depends on the *product* of two or more other variables, we refer to it as *joint variation.*

> **Joint Variation**
>
> Let m and n be real numbers. Then z **varies jointly** with the nth power of x and the mth power of y if a nonzero real number k exists such that
>
> $$z = kx^n y^m.$$

For example, the volume of a cylinder is given by $V = \pi r^2 h$. We say that V *varies jointly* with h and the square of r. The constant of variation here is π.

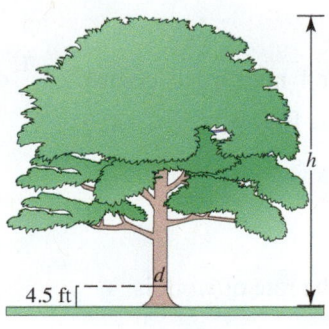

FIGURE 35

EXAMPLE 8   **Modeling the Amount of Wood in a Tree**

The volume V of wood in a tree varies jointly with the 1.12 power of its height h and the 1.98 power of its diameter d. See **FIGURE 35.** The diameter is measured 4.5 feet above the ground. (*Source:* Ryan, B., B. Joiner, and T. Ryan, *Minitab Handbook,* Duxbury Press.)

(a) Write an equation that relates V, h, and d.

(b) A tree with a 13.8-inch diameter and a 64-foot height has a volume of 25.14 cubic feet. Estimate the constant of variation k and give the variation equation.

(c) Estimate the volume of wood in a tree with $d = 11$ inches and $h = 47$ feet.

Solution

(a) Let $V = kh^{1.12}d^{1.98}$, where k is the constant of variation.

(b) $V = kh^{1.12}d^{1.98}$

$25.14 = k(64^{1.12})(13.8^{1.98})$ Substitute $d = 13.8$, $h = 64$, and $V = 25.14$.

$k = \dfrac{25.14}{(64^{1.12})(13.8^{1.98})} \approx 0.00132$ Solve for k.

Thus, $V = 0.00132h^{1.12}d^{1.98}$.   Use the approximate value of k.

(c) For the given tree, $V = 0.00132(47^{1.12})(11^{1.98}) \approx 11.4$ cubic feet.

EXAMPLE 9   **Solving a Combined Variation Problem**

The strength S of a rectangular beam varies directly with its width W and the square of its thickness T, and inversely with its length L. A beam that is 4 inches wide, 10 inches thick, and 120 inches long can support a load of 1250 pounds. Determine how much a similar beam that is 3 inches wide, 12 inches thick, and 100 inches long can support.

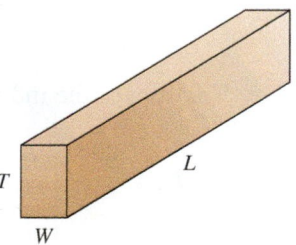

Solution   The strength S of a beam can be written as

$$S = \frac{kWT^2}{L}, \qquad (1)$$

where k is the constant of variation. To find the value of k we begin by substituting $S = 1250$, $W = 4$, $T = 10$, and $L = 120$ into equation (1) and simplifying.

$$1250 = \frac{k(4)(10)^2}{120} = \frac{400k}{120} = \frac{10k}{3}$$

Solving for k results in $k = \frac{3}{10}(1250) = 375$. Using equation (1), the strength S of a similar beam with $W = 3$, $T = 12$, and $L = 100$ is determined by the following.

$$S = \frac{375(3)(12)^2}{100} = 1620$$

The beam can support 1620 pounds.

Rate of Work

If a machine can complete a task in 4 hours, then after 1 hour it has completed $\frac{1}{4}$ of the task. Its rate is $\frac{1}{4}$ task per hour. We can generalize as follows.

Rate of Work

If 1 task can be completed in x units of time, then the rate of work is

$$\frac{1}{x} \text{ task per time unit.}$$

EXAMPLE 10   **Analyzing Work Rate**

It takes machine B one hour less to complete a task when working alone than it takes machine A working alone. If they start together, they can complete the task in 72 minutes. How long does it take each machine to complete the task when working alone?

Solution

Let x represent the number of hours it takes machine A to complete the task alone. Then, working alone, machine B takes $(x - 1)$ hours. We must have time units consistent, so convert 72 minutes to $\frac{72}{60} = \frac{6}{5}$ hour. Express the work rates per hour as follows.

$\dfrac{1}{x}$ is the rate for machine A alone.

$\dfrac{1}{x - 1}$ is the rate for machine B alone.

$\dfrac{1}{\frac{6}{5}}$, or $\dfrac{5}{6}$ is the rate for the two machines working together.

The sum of the individual rates must equal the rate together.

$$\underset{\downarrow}{\text{Rate of A}} \quad + \quad \underset{\downarrow}{\text{Rate of B}} \quad = \quad \underset{\downarrow}{\text{Rate together.}}$$

$$\frac{1}{x} \quad + \quad \frac{1}{x - 1} \quad = \quad \frac{5}{6}$$

$$6x(x - 1)\left[\frac{1}{x} + \frac{1}{x - 1}\right] = 6x(x - 1)\left(\frac{5}{6}\right) \qquad \text{Multiply by the LCD.}$$

$$6(x - 1) + 6x = 5x(x - 1) \qquad \text{Distributive property}$$

$$6x - 6 + 6x = 5x^2 - 5x \qquad \text{Distributive property}$$

$$0 = 5x^2 - 17x + 6 \qquad \text{Standard form}$$

$$0 = (5x - 2)(x - 3) \qquad \text{Factor.}$$

$$5x - 2 = 0 \quad \text{or} \quad x - 3 = 0 \qquad \text{Zero-product property}$$

$$x = \frac{2}{5} \quad \text{or} \qquad x = 3 \qquad \text{Solve.}$$

The value $\frac{2}{5}$ must be rejected, as it would cause the time of machine B to be negative. It takes machine A 3 hours to complete the task alone, and it takes machine B $3 - 1 = 2$ hours to complete the task alone.

4.3 Exercises

In Exercises 1–12, the graph of a rational function $y = f(x)$ is given. Use the graph to give the solution set of (a) $f(x) = 0$, (b) $f(x) < 0$, and (c) $f(x) > 0$. Use set notation for part (a) and interval notation for parts (b) and (c).

1.

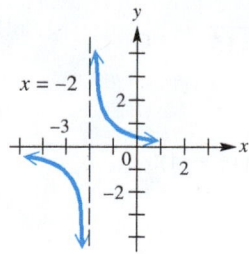

2.

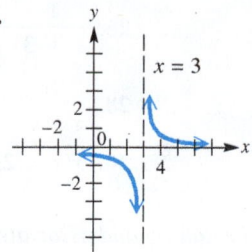

3.

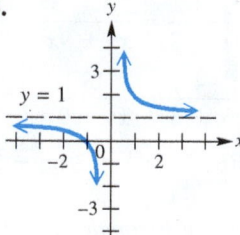

4.

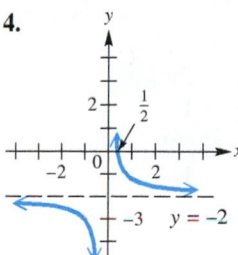

5.

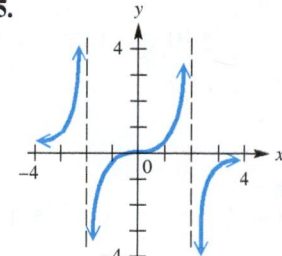

6.

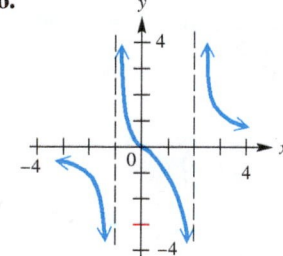

7.

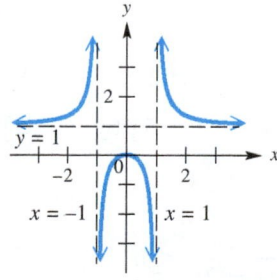

8.

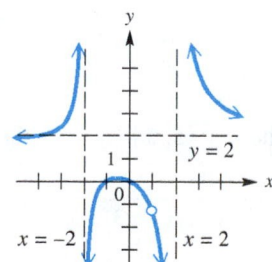

9.

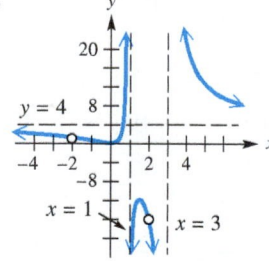

10.

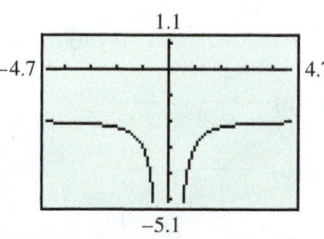

The line $x = 0$ is a vertical asymptote.

11.

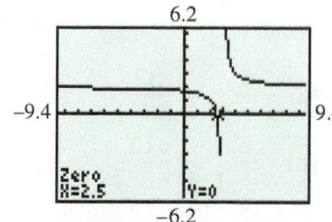

The line $x = 3$ is a vertical asymptote.

12.

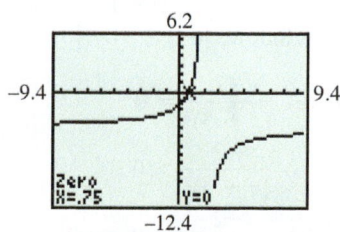

The line $x = 2$ is a vertical asymptote.

Checking Analytic Skills Find all complex solutions for each equation by hand. **Do not use a calculator.**

13. $\dfrac{2x}{x^2 - 1} = \dfrac{2}{x + 1} - \dfrac{1}{x - 1}$

14. $\dfrac{8x}{4x^2 - 1} = \dfrac{3}{2x + 1} + \dfrac{3}{2x - 1}$

15. $\dfrac{4}{x^2 - 3x} - \dfrac{1}{x^2 - 9} = 0$

16. $\dfrac{2}{x^2 - 2x} - \dfrac{3}{x^2 - x} = 0$

17. $1 - \dfrac{13}{x} + \dfrac{36}{x^2} = 0$

18. $1 - \dfrac{3}{x} - \dfrac{10}{x^2} = 0$

19. $1 + \dfrac{3}{x} = \dfrac{5}{x^2}$

20. $4 + \dfrac{7}{x} = -\dfrac{1}{x^2}$

21. $\dfrac{x}{2 - x} + \dfrac{2}{x} - 5 = 0$

22. $\dfrac{2x}{x - 3} + \dfrac{4}{x} - 6 = 0$

23. $x^{-4} - 3x^{-2} - 4 = 0$

24. $x^{-4} - 5x^{-2} - 36 = 0$

25. $\dfrac{1}{x + 2} + \dfrac{3}{x + 7} = \dfrac{5}{x^2 + 9x + 14}$

26. $\dfrac{1}{x + 3} + \dfrac{4}{x + 5} = \dfrac{2}{x^2 + 8x + 15}$

27. $\dfrac{x}{x - 3} + \dfrac{4}{x + 3} = \dfrac{18}{x^2 - 9}$

28. $\dfrac{2x}{x - 3} + \dfrac{4}{x + 3} = \dfrac{24}{9 - x^2}$

29. $9x^{-1} + 4(6x - 3)^{-1} = 2(6x - 3)^{-1}$

30. $x(x - 2)^{-1} + x(x + 2)^{-1} = 2(x^2 - 4)^{-1}$

*Use the methods of **Examples 1** and **3** to solve the rational equation and associated inequalities given in Exercises 31–42. Then, support your answer by using the x-intercept method with a calculator graph in the suggested window.*

31. (a) $\dfrac{x - 3}{x + 5} = 0$

(b) $\dfrac{x - 3}{x + 5} \le 0$

(c) $\dfrac{x - 3}{x + 5} \ge 0$

Window: $[-10, 10]$ by $[-5, 8]$

32. (a) $\dfrac{x + 1}{x - 4} = 0$

(b) $\dfrac{x + 1}{x - 4} \ge 0$

(c) $\dfrac{x + 1}{x - 4} \le 0$

Window: $[-10, 10]$ by $[-5, 10]$

33. (a) $\dfrac{x - 1}{x + 2} = 1$

(b) $\dfrac{x - 1}{x + 2} > 1$

(c) $\dfrac{x - 1}{x + 2} < 1$

Window: $[-10, 10]$ by $[-5, 10]$

34. (a) $\dfrac{x - 6}{x + 2} = -1$

(b) $\dfrac{x - 6}{x + 2} < -1$

(c) $\dfrac{x - 6}{x + 2} > -1$

Window: $[-10, 10]$ by $[-10, 10]$

35. (a) $\dfrac{1}{x - 1} = \dfrac{5}{4}$

(b) $\dfrac{1}{x - 1} < \dfrac{5}{4}$

(c) $\dfrac{1}{x - 1} > \dfrac{5}{4}$

Window: $[-5, 5]$ by $[-5, 5]$

36. (a) $\dfrac{6}{5 - 3x} = 2$

(b) $\dfrac{6}{5 - 3x} \le 2$

(c) $\dfrac{6}{5 - 3x} \ge 2$

Window: $[-5, 5]$ by $[-5, 5]$

37. (a) $\dfrac{4}{x - 2} = \dfrac{3}{x - 1}$

(b) $\dfrac{4}{x - 2} \le \dfrac{3}{x - 1}$

(c) $\dfrac{4}{x - 2} \ge \dfrac{3}{x - 1}$

Window: $[-3, 3]$ by $[-20, 20]$

38. (a) $\dfrac{4}{x + 1} = \dfrac{2}{x + 3}$

(b) $\dfrac{4}{x + 1} < \dfrac{2}{x + 3}$

(c) $\dfrac{4}{x + 1} > \dfrac{2}{x + 3}$

Window: $[-8, 5]$ by $[-10, 10]$

39. (a) $\dfrac{1}{(x - 2)^2} = 0$

(b) $\dfrac{1}{(x - 2)^2} < 0$

(c) $\dfrac{1}{(x - 2)^2} > 0$

Window: $[-5, 10]$ by $[-5, 10]$

40. (a) $\dfrac{-2}{(x + 3)^2} = 0$

(b) $\dfrac{-2}{(x + 3)^2} > 0$

(c) $\dfrac{-2}{(x + 3)^2} < 0$

Window: $[-10, 5]$ by $[-10, 5]$

41. (a) $\dfrac{5}{x + 1} = \dfrac{12}{x + 1}$

(b) $\dfrac{5}{x + 1} > \dfrac{12}{x + 1}$

(c) $\dfrac{5}{x + 1} < \dfrac{12}{x + 1}$

Window: $[-10, 10]$ by $[-10, 10]$

42. (a) $\dfrac{7}{x + 2} = \dfrac{1}{x + 2}$

(b) $\dfrac{7}{x + 2} \ge \dfrac{1}{x + 2}$

(c) $\dfrac{7}{x + 2} \le \dfrac{1}{x + 2}$

Window: $[-10, 10]$ by $[-10, 10]$

In Exercises 43–48, solve each equation and inequality.

43. (a) $\dfrac{(x - 2)(2) - (2x + 1)(1)}{(x - 2)^2} = 0$

(b) $\dfrac{(x - 2)(2) - (2x + 1)(1)}{(x - 2)^2} < 0$

44. (a) $\dfrac{(x^2 - 1)(1) - (x + 1)(2x)}{(x^2 - 1)^2} = 0$

(b) $\dfrac{(x^2 - 1)(1) - (x + 1)(2x)}{(x^2 - 1)^2} > 0$

45. (a) $\dfrac{(x^2 + 1)(2x) - (x^2 - 1)(2x)}{(x^2 + 1)^2} = 0$

(b) $\dfrac{(x^2 + 1)(2x) - (x^2 - 1)(2x)}{(x^2 + 1)^2} \geq 0$

46. (a) $\dfrac{(x^2 - 1)(3) - (3x - 1)(2x)}{(x^2 - 1)^2} = 0$

(b) $\dfrac{(x^2 - 1)(3) - (3x - 1)(2x)}{(x^2 - 1)^2} \leq 0$

47. (a) $\dfrac{(2x + 1)(2x) - (x^2 + 1)(2)}{(2x + 1)^2} = 0$

(b) $\dfrac{(2x + 1)(2x) - (x^2 + 1)(2)}{(2x + 1)^2} < 0$

48. (a) $\dfrac{(x - 1)(2x) - (x^2)(1)}{(x - 1)^2} = 0$

(b) $\dfrac{(x - 1)(2x) - (x^2)(1)}{(x - 1)^2} > 0$

In some cases, it is possible to solve a rational inequality simply by deciding what sign the numerator and the denominator must have and then using the rules for quotients of positive and negative numbers to determine the solution set. For example, consider the rational inequality

$$\frac{1}{x^2 + 1} > 0.$$

The numerator of the rational expression, 1, is positive, and the denominator, $x^2 + 1$, must always be positive because it is the sum of a nonnegative number, x^2, and a positive number, 1. Therefore, the rational expression is the quotient of two positive numbers, which is positive. Because the inequality requires that the rational expression be greater than 0, and this will always be true, the solution set is $(-\infty, \infty)$.
     Use similar reasoning to solve each inequality.

49. $\dfrac{-1}{x^2 + 2} < 0$

50. $\dfrac{5}{x^2 + 2} < 0$

51. $\dfrac{-5}{x^2 + 2} > 0$

52. $\dfrac{x^4 + 2}{-6} \leq 0$

53. $\dfrac{x^4 + 2}{-6} \geq 0$

54. $\dfrac{x^4 + x^2 + 3}{x^2 + 2} < 0$

55. $\dfrac{x^4 + x^2 + 3}{x^2 + 2} > 0$

56. $\dfrac{(x - 1)^2}{x^2 + 4} > 0$

57. $\dfrac{(x - 1)^2}{x^2 + 4} \leq 0$

58. *Concept Check*   Let $f(x) = \dfrac{x^2 - 4}{x^2 - 4}$. Solve each equation or inequality.

(a) $f(x) = 0$ **(b)** $f(x) < 0$ **(c)** $f(x) > 0$ **(d)** $f(x) = 1$

Checking Analytic Skills   *Solve each rational inequality by hand.* ***Do not use a calculator.***

59. $\dfrac{3 - 2x}{1 + x} < 0$

60. $\dfrac{3x - 3}{4 - 2x} \geq 0$

61. $\dfrac{(x + 1)(x - 2)}{(x + 3)} < 0$

62. $\dfrac{x(x - 3)}{x + 2} \geq 0$

63. $\dfrac{(x + 1)^2}{x - 2} \leq 0$

64. $\dfrac{(x - 2)^2}{2x} > 0$

65. $\dfrac{2x - 5}{x^2 - 1} \geq 0$

66. $\dfrac{5 - x}{x^2 - x - 2} < 0$

67. $\dfrac{1}{x - 3} \leq \dfrac{5}{x - 3}$

68. $\dfrac{3}{2 - x} > \dfrac{x}{2 + x}$

69. $2 - \dfrac{5}{x} + \dfrac{2}{x^2} \geq 0$

70. $\dfrac{1}{x - 1} + \dfrac{1}{x + 1} > \dfrac{3}{4}$

Solve the equation in part (a) graphically, expressing solutions to the nearest hundredth. Then, use the graph to solve the associated inequalities in parts (b) and (c), expressing endpoints to the nearest hundredth.

71. (a) $\dfrac{\sqrt{2}x + 5}{x^3 - \sqrt{3}} = 0$

(b) $\dfrac{\sqrt{2}x + 5}{x^3 - \sqrt{3}} > 0$

(c) $\dfrac{\sqrt{2}x + 5}{x^3 - \sqrt{3}} < 0$

72. (a) $\dfrac{\sqrt[3]{7}x^3 - 1}{x^2 + 2} = 0$

(b) $\dfrac{\sqrt[3]{7}x^3 - 1}{x^2 + 2} > 0$

(c) $\dfrac{\sqrt[3]{7}x^3 - 1}{x^2 + 2} < 0$

(Modeling) Solve each problem.

73. *Insect Population* Suppose that an insect population in millions is modeled by

$$f(x) = \frac{10x + 1}{x + 1},$$

where $x \geq 0$ is in months.
(a) Graph f in the window $[0, 14]$ by $[0, 14]$. Find the equation of the horizontal asymptote.
(b) Determine the initial insect population.
(c) What happens to the population after several months?
(d) Interpret the horizontal asymptote.

74. *Fish Population* Suppose that the population of a species of fish in thousands is modeled by

$$f(x) = \frac{x + 10}{0.5x^2 + 1},$$

where $x \geq 0$ is in years.
(a) Graph f in the window $[0, 12]$ by $[0, 12]$. What is the equation of the horizontal asymptote?
(b) Determine the initial population.
(c) What happens to the population after many years?
(d) Interpret the horizontal asymptote.

75. *Time Spent in Line* Suppose the average number of vehicles arriving at the main gate of an amusement park is equal to 10 per minute, while the average number of vehicles being admitted through the gate per minute is equal to x. Then the average waiting time in minutes for each vehicle at the gate is given by

$$f(x) = \frac{x - 5}{x^2 - 10x},$$

where $x > 10$. (*Source:* Mannering, F. and W. Kilareski, *Principles of Highway Engineering and Traffic Analysis*, 2d. ed., John Wiley and Sons.)
(a) Estimate the admittance rate x that results in an average wait of 15 seconds.
(b) If one attendant can serve 5 vehicles per minute, how many attendants are needed to keep the average wait to 15 seconds or less?

76. *Length of Lines* See **Example 5.** Determine the traffic intensity x when the average number of vehicles in line equals 3.

77. *Construction* Find possible dimensions for a closed box with volume 196 cubic inches, surface area 280 square inches, and length that is twice the width.

78. *Volume of a Cylindrical Can* A metal cylindrical can with an *open top* and *closed bottom* is to have volume 4 cubic feet. Approximate the dimensions that require the least amount of material. What would this amount be? (Compare this problem with **Example 6.**)

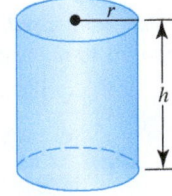

79. *Train Curves* When curves are designed for trains, sometimes the outer rail is elevated or *banked* so that a locomotive can safely negotiate the curve at a higher speed. Suppose a circular curve is being designed for a speed of 60 mph. The rational function

$$f(x) = \frac{2540}{x}$$

computes the elevation y in inches of the outer track for a curve with a radius of x feet, where $y = f(x)$. (*Source:* Haefner, L., *Introduction to Transportation Systems*, Holt, Rinehart and Winston.)

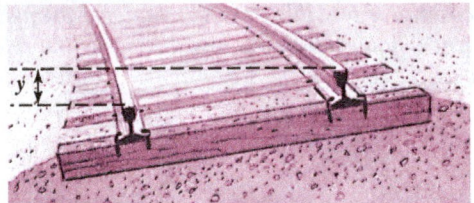

(a) Evaluate $f(400)$ and interpret its meaning.
(b) Graph f in the window $[0, 600]$ by $[0, 50]$. Discuss how the elevation of the outer rail changes with the radius x.
(c) Interpret the horizontal asymptote.
(d) What radius is associated with an elevation of 12.7 inches?

80. *Recycling* A cost–benefit function C computes the cost in millions of dollars of implementing a city recycling project when x percent of the citizens participate, where

$$C(x) = \frac{1.2x}{100 - x}.$$

(a) Graph C in the window $[0, 100]$ by $[0, 10]$. Interpret the graph as x approaches 100.
(b) If 75% participation is expected, determine the cost for the city.
(c) The city plans to spend $5 million on this recycling project. Estimate graphically the percentage of participation that they are expecting.
(d) Solve part (c) analytically.

81. *Braking Distance* The **grade** x of a hill is a measure of its steepness. For example, if a road rises 10 feet for every 100 feet of horizontal distance, then it has an uphill grade of $x = \frac{10}{100}$, or 10%. The braking (or stopping) distance D for a car traveling at 50 mph on a wet, uphill grade is given by

$$D(x) = \frac{2500}{30(0.3 + x)}.$$

(*Source:* Haefner, L., *Introduction to Transportation Systems*, Holt, Rinehart and Winston.)

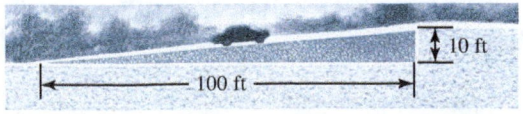

(a) Evaluate $D(0.05)$ and interpret the result.

(b) Describe what happens to braking distance as the hill becomes steeper.

(c) Estimate the grade associated with a braking distance of 220 feet.

82. *Braking Distance*　See **Exercise 81.** If a car is traveling 50 mph downhill, then its braking distance on wet pavement is given by

$$D(x) = \frac{2500}{30(0.3 + x)},$$

where $x < 0$ for a downhill grade.

(a) Evaluate $D(-0.1)$ and interpret the result.

(b) What happens to braking distance as the downhill grade becomes steeper?

(c) The graph of D has a vertical asymptote at $x = -0.3$. Give the physical significance of this asymptote.

(d) Estimate the grade associated with a braking distance of 350 feet.

Solve each problem.

83. Suppose r varies directly with the square of m and inversely with s. If $r = 12$ when $m = 6$ and $s = 4$, find r when $m = 4$ and $s = 10$.

84. Suppose p varies directly with the square of z and inversely with r. If $p = \frac{32}{5}$ when $z = 4$ and $r = 10$, find p when $z = 2$ and $r = 16$.

85. If a varies directly with m and n^2 and inversely with y^3, and $a = .9$ when $m = 4$, $n = 9$, and $y = 3$, find a if $m = 6$, $n = 2$, and $y = 5$.

86. If y varies directly with x and inversely with m^2 and r^2, and $y = \frac{5}{3}$ when $x = 1$, $m = 2$, and $r = 3$, find y if $x = 3$, $m = 1$, and $r = 8$.

87. For $k > 0$, if y varies directly with x, when x increases, y _____, and when x decreases, y _____.

88. For $k > 0$, if y varies inversely with x, when x increases, y _____, and when x decreases, y _____.

In Exercises 89–92, assume that the constant of variation is positive.

89. Let y be inversely proportional to x. If x doubles, what happens to y?

90. Let y vary inversely with the second power of x. If x doubles, what happens to y?

91. Suppose y varies directly with the third power of x. If x triples, what happens to y?

92. Suppose y is directly proportional to the second power of x. If x is halved, what happens to y?

(Modeling)　*Solve each problem.*

93. *Body Mass Index*　The federal government has developed the *body mass index* (BMI) to determine ideal weights. A person's BMI is directly proportional to his or her weight in pounds and inversely proportional to the square of his or her height in inches. (A BMI of 19 to 25 corresponds to a healthy weight.) A 6-foot-tall person weighing 177 pounds has a BMI of 24. Find the BMI (to the nearest whole number) of a person whose weight is 130 pounds and whose height is 66 inches. (*Source: Washington Post.*)

94. *Volume of a Gas*　Natural gas provides 25% of U.S. energy. The volume of a gas varies inversely with the pressure and directly with the temperature. (Temperature must be measured in degrees *Kelvin* (K), a unit of measurement used in physics.) If a certain gas occupies a volume of 1.3 liters at 300 K and a pressure of 18 newtons per square centimeter, find the volume at 340 K and a pressure of 24 newtons per square centimeter.

95. *Electrical Resistance*　The electrical resistance R of a wire varies inversely with the square of its diameter d. If a 25-foot wire with diameter 2 millimeters has resistance 0.5 ohm, find the resistance of a wire having the same length and diameter 3 millimeters.

96. *Poiseuille's Law*　According to Poiseuille's law, the resistance to flow of a blood vessel R is directly proportional to the length l and inversely proportional to the fourth power of the radius r. (*Source:* Hademenos, George J., "The Biophysics of Stroke," *American Scientist,* May–June 1997.) If $R = 25$ when $l = 12$ and $r = 0.2$, find R to the nearest hundredth as r increases to 0.3 while l is unchanged.

97. *Gravity*　The weight of an object varies inversely with the square of its distance from the center of Earth. The radius of Earth is approximately 4000 miles. If a person weighs 160 pounds on Earth's surface, what would this individual weigh 8000 miles above the surface of Earth?

98. *Hubble Telescope*　The brightness or intensity of starlight varies inversely with the square of its distance from Earth. The Hubble Telescope can see stars whose intensities are $\frac{1}{50}$ of the faintest star now seen by ground-based telescopes. Determine how much farther the Hubble Telescope can see into space than ground-based telescopes. (*Source:* National Aeronautics and Space Administration.)

99. *Volume of a Cylinder*　The volume of a right circular cylinder is jointly proportional to the square of the radius of the circular base and to the height. If the volume is 300 cubic centimeters when the height is 10.62 centimeters and the radius is 3 centimeters, approximate the volume of a cylinder with radius 4 centimeters and height 15.92 centimeters.

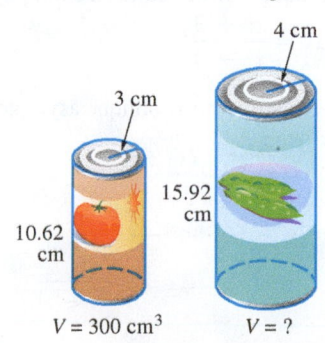

100. *Strength of a Beam* See **Example 9.** The strength S of a rectangular beam varies directly with its width W and the square of its thickness T, and inversely with its length L. A beam that is 2 inches wide, 6 inches thick, and 96 inches long can support a load of 375 pounds. Determine how much a similar beam that is 3.5 inches wide, 8 inches thick, and 128 inches long can support.

Rate of Work *Solve each problem involving rate of work.*

101. Linda and Tooney want to pick up the mess that their granddaughter, Kaylin, has made in her playroom. Tooney could do it in 15 minutes working alone. Linda, working alone, could clean it in 12 minutes. How long will it take them if they work together?

102. Johnny can groom Gary Bell's dogs in 6 hours, but it takes his business partner, "Mudcat," only 4 hours to groom the same dogs. How long will it take them to groom the dogs if they work together?

103. Mrs. Schmulen is a high school mathematics teacher. She can grade a set of chapter tests in 5 hours working alone. If her student teacher Elwyn helps her, it will take 3 hours to grade the tests. How long would it take Elwyn to grade the tests if he worked alone?

104. Tommy and Alicia are laying a tile floor. Working alone, Tommy can do the job in 20 hours. If the two of them work together, they can complete the job in 12 hours. How long would it take Alicia to lay the floor working alone?

105. If a vat of solution can be filled by an inlet pipe in 5 hours and emptied by an outlet pipe in 10 hours, how long will it take to fill an empty vat if both pipes are open?

106. A winery has a vat to hold Merlot. An inlet pipe can fill the vat in 18 hours, while an outlet pipe can empty it in 24 hours. How long will it take to fill an empty vat if both the outlet and the inlet pipes are open?

107. It takes an inlet pipe of a small swimming pool 20 minutes less to fill the pool than it takes an outlet pipe of the same pool to empty it. Through an error, starting with an empty pool, both pipes are left open, and the pool is filled after 4 hours. How long does it take the inlet pipe to fill the pool, and how long does it take the outlet pipe to empty it?

108. A sink can be filled by the hot-water tap alone in 4 minutes more than it takes the cold-water tap alone. If both taps are open, it takes 4 minutes, 48 seconds to fill an empty sink. How long does it take each tap individually to fill the sink?

SECTIONS 4.1–4.3 Reviewing Basic Concepts

1. Sketch the graph of $y = \dfrac{1}{x+2} - 3$. Then check the accuracy of your graph by using a graphing calculator.

2. What is the domain of the rational function
$$f(x) = \frac{3}{x^2 - 1}?$$

3. What is the equation of the vertical asymptote of the graph of $f(x) = \dfrac{4x+3}{x-6}$?

4. What is the equation of the horizontal asymptote of the graph of $f(x) = \dfrac{x^2+3}{x^2-4}$?

5. What is the equation of the oblique asymptote of the graph of $f(x) = \dfrac{x^2+x+5}{x+3}$?

6. Sketch the graph of $f(x) = \dfrac{3x+6}{x-4}$ by hand. Then check the accuracy of your graph by using a graphing calculator.

7. The graph of a rational function f is shown below. Give the solution set of each equation or inequality.
  (a) $f(x) = 0$ **(b)** $f(x) > 0$ **(c)** $f(x) < 0$

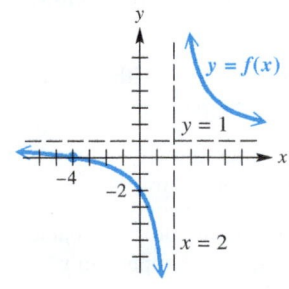

8. Find the solution set of $\dfrac{x + 4}{3x + 1} > 1$.

9. Fill in the blanks with the correct responses: $b = \dfrac{24}{h}$ is the formula for the base of a parallelogram with area 24. The base of this parallelogram varies _____ with its _____. The constant of variation is _____.

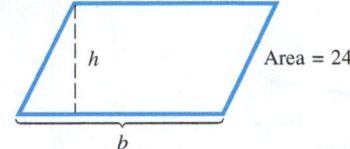

10. *Strength of a Beam*   The strength S of a rectangular beam varies directly with its width W and the square of its thickness T, and inversely with its length L. A beam that is 4 inches wide, 4 inches thick, and 50 inches long can support a load of 600 pounds. Determine how much a similar beam that is 2 inches wide, 2 inches thick, and 50 inches long can support.

4.4   Functions Defined by Powers and Roots

Power and Root Functions • Modeling Using Power Functions • Graphs of $f(x) = \sqrt[n]{ax + b}$ • Graphing Circles and Horizontal Parabolas Using Root Functions

Power and Root Functions

If a polynomial function is written as

$$P(x) = a_n x^n + a_{n-1} x^{n-1} + \cdots + a_1 x + a_0,$$

then the exponent on each power of x must be a *nonnegative* integer. Sometimes it is necessary for a function to have an exponent that can be any real number. (See **Example 4.**) These types of functions are called *power functions*. An example of a power function that is not a polynomial is $f(x) = x^{3/2}$. A *root function* is a special type of power function.

Power and Root Functions

A function f of the form

$$f(x) = x^b,$$

where b is a constant, is a **power function.**

     If $b = \dfrac{1}{n}$ for some integer $n \geq 2$, then f is a **root function.**

$$f(x) = x^{1/n}, \quad \text{or equivalently}, \quad f(x) = \sqrt[n]{x}$$

Examples of power functions include the following.

$$f(x) = x^2, \quad f(x) = x^{\pi}, \quad f(x) = x^{0.4}, \quad \text{and} \quad f(x) = \sqrt[3]{x^2}$$

Power functions

Frequently, the domain of a power function f is restricted to nonnegative numbers. Suppose the rational number $\dfrac{p}{q}$ is written in lowest terms. Then the domain of $f(x) = x^{p/q}$ is all real numbers whenever q is odd and all nonnegative real numbers whenever q is even. If b is a positive irrational number, the domain of $f(x) = x^b$ is all nonnegative real numbers.

For example, the domain of $f(x) = x^{1/3}$ $\left(f(x) = \sqrt[3]{x}\right)$ is all real numbers, whereas the domains of $g(x) = x^{1/2}$ $\left(g(x) = \sqrt{x}\right)$ and $h(x) = x^{\sqrt{2}}$ are all nonnegative numbers.

Graphs of three common power functions are shown in **FIGURES 36–38.** We first introduced the square root and cube root functions in **Section 2.1.**

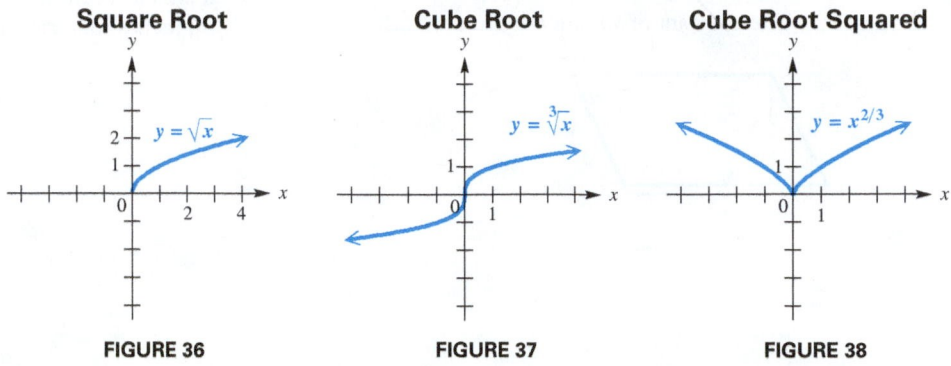

Square Root	**Cube Root**	**Cube Root Squared**
FIGURE 36	**FIGURE 37**	**FIGURE 38**

Power Functions

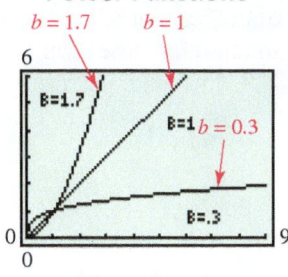

FIGURE 39

← **Algebra Review**
To review some of the properties of rational exponents and radical notation, see **Sections R.4** and **R.5.**

EXAMPLE 1 Graphing Power Functions

(a) Graph $f(x) = x^b$, where $b = 0.3$, 1, and 1.7, for $x \geq 0$.

(b) Discuss the effect that b has on the graph of f when $b > 0$ and $x \geq 1$.

Solution

(a) Calculator graphs of $y = x^{0.3}$, $y = x^1$, and $y = x^{1.7}$ are shown in **FIGURE 39.**

(b) Larger values of b cause the graph of f to increase faster for $x \geq 1$. ●

The next two examples review some of the properties of rational exponents and radical notation.

EXAMPLE 2 Applying Properties of Rational Exponents

Simplify each expression by hand.

(a) $16^{3/4}$ (b) $\left(\sqrt[3]{-64}\right)^4$ (c) $(-125)^{2/3}$

Solution

(a) $16^{3/4} = \left(\sqrt[4]{16}\right)^3 = (2)^3 = 8$ $a^{m/n} = \left(\sqrt[n]{a}\right)^m$

(b) $\left(\sqrt[3]{-64}\right)^4 = (-4)^4 = 256$

(c) $(-125)^{2/3} = \left(\sqrt[3]{-125}\right)^2 = (-5)^2 = 25$ ●

EXAMPLE 3 Writing Radicals with Rational Exponents

Use positive rational exponents to write each expression. Assume variables are positive.

(a) $\sqrt{x}$ (b) $\sqrt[3]{x^2}$ (c) $\left(\sqrt[4]{z}\right)^{-5}$ (d) $\sqrt{\sqrt[3]{y} \cdot \sqrt[4]{y}}$

Solution

(a) $\sqrt{x} = x^{1/2}$ (b) $\sqrt[3]{x^2} = (x^2)^{1/3} = x^{2/3}$

(c) $\left(\sqrt[4]{z}\right)^{-5} = (z^{1/4})^{-5} = z^{-5/4} = \dfrac{1}{z^{5/4}}$ $a^{-n} = \dfrac{1}{a^n}$

(d) $\sqrt{\sqrt[3]{y} \cdot \sqrt[4]{y}} = (y^{1/3} \cdot y^{1/4})^{1/2} = (y^{(1/3)+(1/4)})^{1/2}$ $a^m \cdot a^n = a^{m+n}$

$= (y^{7/12})^{1/2} = y^{7/24}$ $(a^m)^n = a^{mn}$ ●

Planet	x	y
Mercury	0.387	0.241
Venus	0.723	0.615
Earth	1.00	1.00
Mars	1.52	1.88
Jupiter	5.20	11.9
Saturn	9.54	29.5

Source: Ronan, C., *The Natural History of the Universe*, MacMillan.

Modeling Using Power Functions

The table in the margin lists for several planets the average distance x from the sun and the time y in Earth-years to orbit the sun. The distance x has been normalized so that Earth is 1 unit away from the sun. For example, Jupiter is 5.20 times farther from the sun than Earth is and requires 11.9 years to orbit the sun. Johannes Kepler (1571–1630) first recognized that the time it takes for a planet to orbit the sun could be modeled with a power function. In the next example we determine this function.

EXAMPLE 4   **Modeling the Period of Satellite Orbits**

Use the data in the table to complete the following.

(a) Make a scatter diagram of the data. Graphically estimate a value for b so that $f(x) = x^b$ models the data.

(b) Numerically check the accuracy of f.

(c) The average distances of Uranus, Neptune, and Pluto from the sun are 19.2, 30.1, and 39.5, respectively. Use f to estimate the periods of revolution for these satellites. Compare these answers with the actual values of 84.0, 164.8, and 248.5 years.

Solution

(a) We make a scatter diagram of the data and then graph $y = x^b$ for different values of b. From the calculator graphs of $y = x^{1.4}$, $y = x^{1.5}$, and $y = x^{1.6}$ in **FIGURES 40–42**, we see that $b \approx 1.5$.

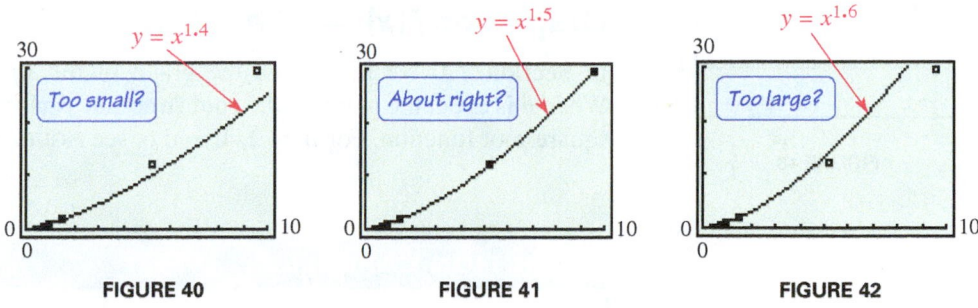

FIGURE 40   FIGURE 41   FIGURE 42

(b) The values shown in **FIGURE 43** model the data in the table remarkably well.

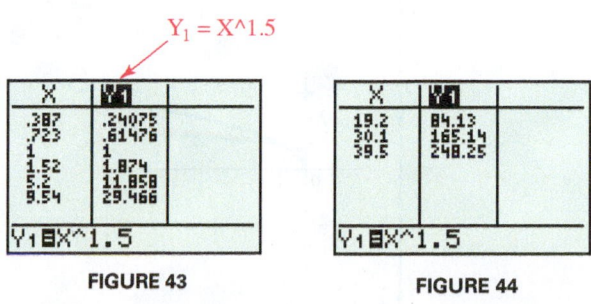

FIGURE 43   FIGURE 44

(c) To approximate the number of years for Uranus, Neptune, and Pluto to orbit the sun, we evaluate $f(x) = x^{1.5}$ at $x = 19.2$, 30.1, and 39.5, as shown in **FIGURE 44.** These values are close to the actual values.

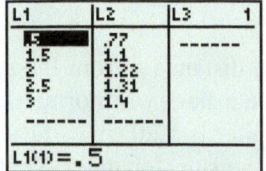

FIGURE 45

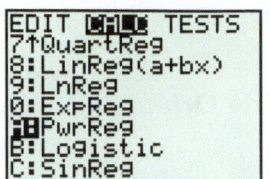

FIGURE 46

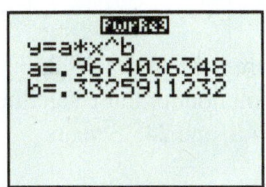

FIGURE 47

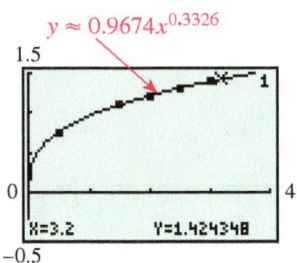

FIGURE 48

EXAMPLE 5   **Modeling the Length of a Bird's Wing**

The table lists the weight W and the wingspan L for birds of a particular species.

W (in kilograms)	0.5	1.5	2.0	2.5	3.0
L (in meters)	0.77	1.10	1.22	1.31	1.40

Source: Pennycuick, C., *Newton Rules Biology*, Oxford University Press.

(a) Use power regression to model the data, with $L = aW^b$. Graph the data and the equation.

(b) Approximate the wingspan for a bird weighing 3.2 kilograms.

Solution

(a) Since $L = aW^b$, let x be the weight W and y be the wingspan L. Enter the data from the table into a TI-84 Plus calculator, and then select power regression (PwrReg), as shown in **FIGURES 45** and **46**. Based on the results shown in **FIGURE 47**, $y \approx 0.9674x^{0.3326}$, or $L \approx 0.9674W^{0.3326}$. The data and equation are graphed in **FIGURE 48**.

(b) This model predicts the wingspan of a bird weighing 3.2 kilograms to be

$$L = 0.9674(3.2)^{0.3326} \approx 1.42 \text{ meters.}$$

See the display at the bottom of the screen in **FIGURE 48**.

Graphs of $f(x) = \sqrt[n]{ax + b}$

In **Section 2.1**, we introduced the graph of the *square* root function $f(x) = \sqrt{x}$. When n is even, the graph of the **root function** $f(x) = \sqrt[n]{x}$ resembles the graph of the square root function. For $n = 2, 4$, and 6, see **FIGURE 49**.

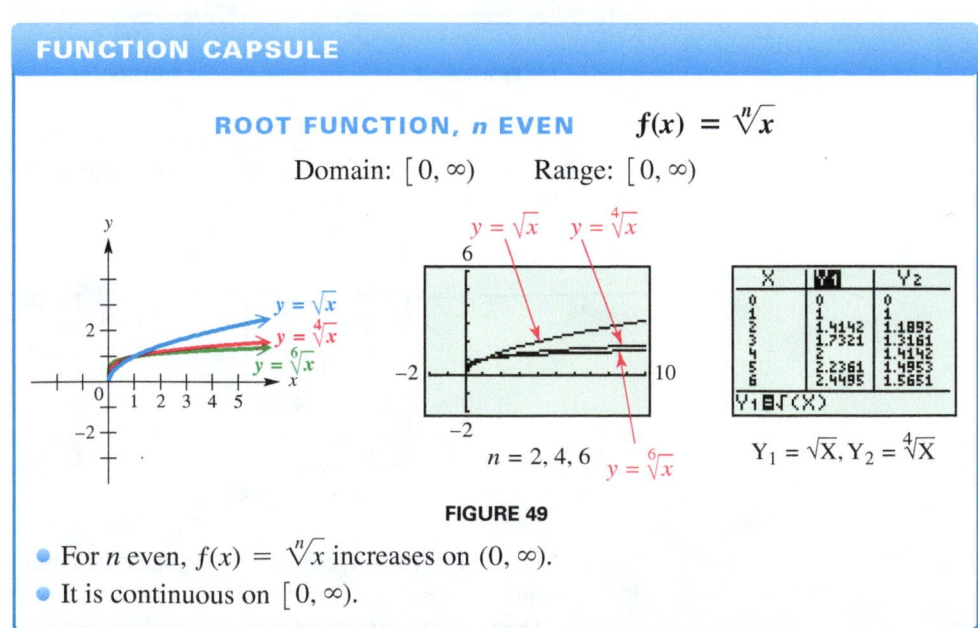

FUNCTION CAPSULE

ROOT FUNCTION, n EVEN $f(x) = \sqrt[n]{x}$

Domain: $[0, \infty)$ Range: $[0, \infty)$

$n = 2, 4, 6$

$Y_1 = \sqrt{X}, Y_2 = \sqrt[4]{X}$

FIGURE 49

- For n even, $f(x) = \sqrt[n]{x}$ increases on $(0, \infty)$.
- It is continuous on $[0, \infty)$.

Recall the graph of the *cube* root function $f(x) = \sqrt[3]{x}$ in **Section 2.1.** When n is odd, the graph of the root function $f(x) = \sqrt[n]{x}$ resembles the graph of the cube root function. For $n = 3, 5,$ and 7, see **FIGURE 50.**

FUNCTION CAPSULE

ROOT FUNCTION, n ODD $f(x) = \sqrt[n]{x}$

Domain: $(-\infty, \infty)$ Range: $(-\infty, \infty)$

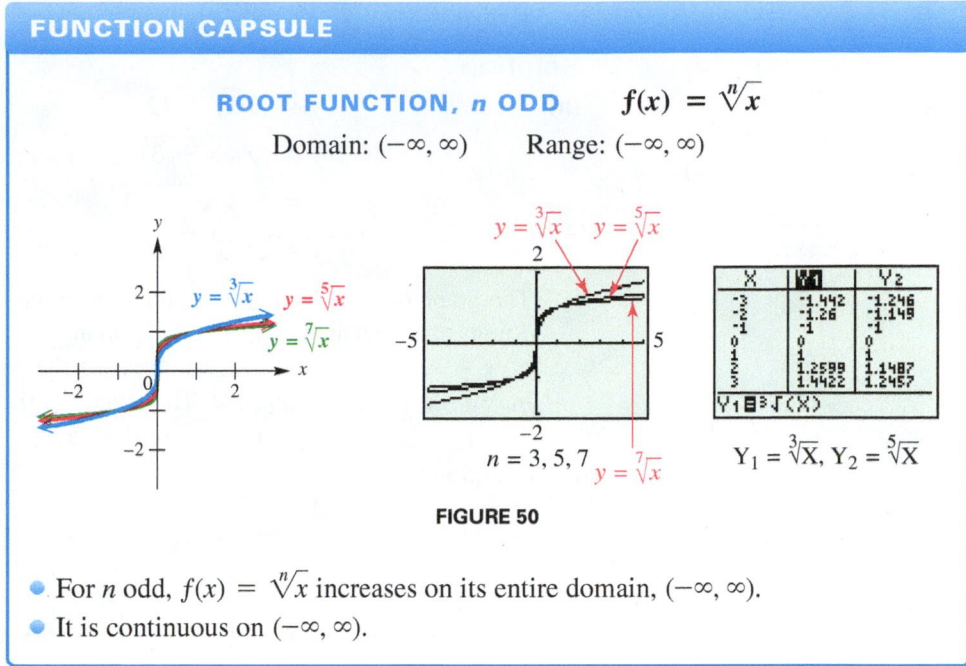

FIGURE 50

- For n odd, $f(x) = \sqrt[n]{x}$ increases on its entire domain, $(-\infty, \infty)$.
- It is continuous on $(-\infty, \infty)$.

To determine the domain of a root function of the form

$$f(x) = \sqrt[n]{ax + b},$$

we must note the parity of n (i.e., whether n is even or odd).

  1. *If n is even in $\sqrt[n]{ax + b}$, then $ax + b$ must be greater than or equal to 0.*

  2. *If n is odd in $\sqrt[n]{ax + b}$, then $ax + b$ can be any real number.*

EXAMPLE 6   **Finding Domains of Root Functions**

Find the domain of each function.
 (a) $f(x) = \sqrt[4]{4x + 12}$ **(b)** $g(x) = \sqrt[3]{-8x + 8}$

Solution

(a) For the function to be defined, $4x + 12$ must be greater than or equal to 0, since $f(x)$ is defined by an even root ($n = 4$).

$$
\begin{aligned}
4x + 12 &\geq 0 \\
4x &\geq -12 \quad &&\text{Subtract 12.} \\
x &\geq -3 \quad &&\text{Divide by 4.}
\end{aligned}
$$

The domain of f is $[-3, \infty)$.

(b) Because $n = 3$ is an odd number, the domain of g is $(-\infty, \infty)$.

EXAMPLE 7   **Transforming Graphs of Root Functions**

(a) Explain how the graph of $y = \sqrt{4x + 12}$ can be obtained from the graph of $y = \sqrt{x}$. Then graph both equations on the same coordinate axes.

(b) Repeat part (a) for the graph of $y = \sqrt[3]{-8x + 8}$, compared with the graph of $y = \sqrt[3]{x}$.

Solution

(a)
$$y = \sqrt{4x + 12} \qquad \text{Transform this equation.}$$
$$y = \sqrt{4(x + 3)} \qquad \text{Factor.}$$
$$y = \sqrt{4}\sqrt{x + 3} \qquad \sqrt{ab} = \sqrt{a} \cdot \sqrt{b}$$
$$y = 2\sqrt{x + 3} \qquad \sqrt{4} = 2$$

The graph of this function is obtained from the graph of $y = \sqrt{x}$ by shifting it left 3 units and stretching vertically by applying a factor of 2. See **FIGURE 51.** The domain of $y = \sqrt{4x + 12}$ is $[-3, \infty)$, because the root index is even, and this interval makes $4x + 12$ non-negative. The range is $[0, \infty)$.

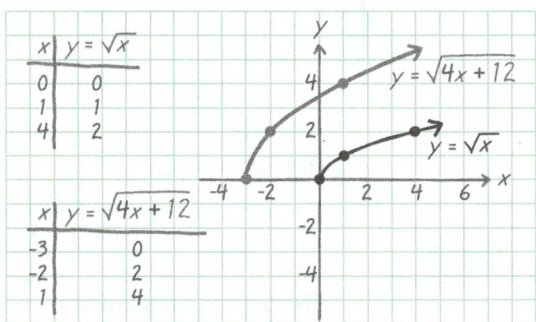

FIGURE 51

(b)
$$y = \sqrt[3]{-8x + 8} \qquad \text{Transform this equation.}$$
$$y = \sqrt[3]{-8(x - 1)} \qquad \text{Factor.}$$
Be careful with signs. $\quad y = \sqrt[3]{-8} \cdot \sqrt[3]{x - 1} \qquad \sqrt[3]{ab} = \sqrt[3]{a} \cdot \sqrt[3]{b}$
$$y = -2\sqrt[3]{x - 1} \qquad \sqrt[3]{-8} = -2$$

The graph can be obtained by shifting the graph of $y = \sqrt[3]{x}$ to the right 1 unit, stretching vertically by a factor of 2, and reflecting across the x-axis (because of the negative sign in -2). **FIGURE 52** shows the graphs and tables. The domain and range are both $(-\infty, \infty)$.

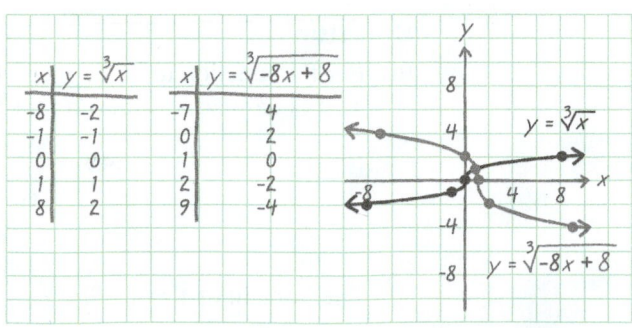

FIGURE 52

**Circle with Center (h, k)
and Radius r**

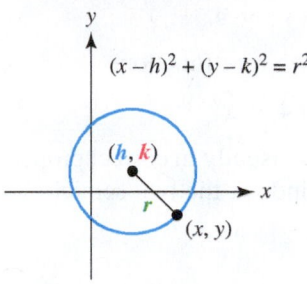

$$(x - h)^2 + (y - k)^2 = r^2$$

FIGURE 53

Graphing Circles and Horizontal Parabolas Using Root Functions

We begin by considering the circle consisting of all points (x, y) that lie a fixed distance r from a center (h, k). (See also **Section 7.1.**) The distance between (x, y) and (h, k) is given by the following equation.

Equation of a circle

$$\sqrt{(x - h)^2 + (y - k)^2} = r \qquad \text{Distance formula } (r > 0)$$

$$(x - h)^2 + (y - k)^2 = r^2 \qquad \text{Square each side.}$$

The graph of this equation is the circle shown in **FIGURE 53.**

EXAMPLE 8 **Graphing a Circle by Hand**

Sketch a graph of $(x - 1)^2 + (y + 2)^2 = 9$.

Solution The equation is given in the form $(x - h)^2 + (y - k)^2 = r^2$, so the graph is a circle with center $(h, k) = (1, -2)$ and radius $r = \sqrt{9} = 3$. This circle is shown in **FIGURE 54.**

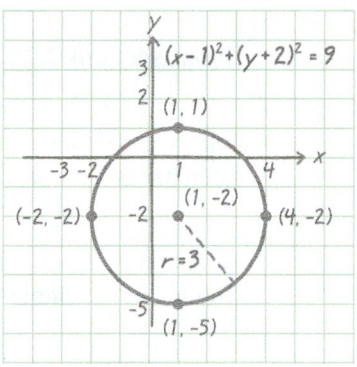

FIGURE 54

The circle with equation $x^2 + y^2 = r^2$ has center $(0, 0)$ and radius r, as shown in **FIGURE 55.** Because a circle is *not* the graph of a function, a graphing calculator in *function* mode is not appropriate for graphing this circle. However, if we imagine the circle as being the union of the graphs of two functions—one the top semicircle and the other the bottom semicircle—then we can graph both functions in the same square viewing window to obtain the circle. To accomplish this, we solve for y in $x^2 + y^2 = r^2$.

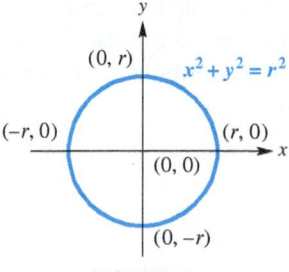

FIGURE 55

Remember both the positive and negative square roots.

$$x^2 + y^2 = r^2 \qquad \text{Equation of a circle}$$

$$y^2 = r^2 - x^2 \qquad \text{Subtract } x^2.$$

$$y = \pm\sqrt{r^2 - x^2} \qquad \text{Square root property}$$

The final result yields two equations, both of which define root functions.

$$y_1 = \sqrt{r^2 - x^2} \qquad \text{The semicircle above the } x\text{-axis}$$

and

$$y_2 = -\sqrt{r^2 - x^2} \qquad \text{The semicircle below the } x\text{-axis}$$

Since $y_2 = -y_1$, the graph of the "bottom" semicircle is simply a reflection of the graph of the "top" semicircle across the x-axis.

TECHNOLOGY NOTE

A graph of the circle

$$x^2 + y^2 = 36$$

is shown.

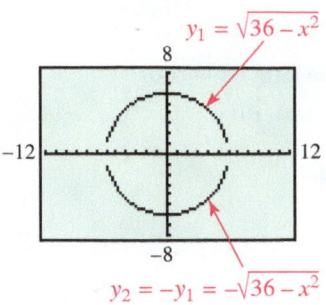

Although a square window is used, it is not a decimal window, so the two semicircles that form the graph do not completely "connect." *This is actually a complete circle. We must understand the mathematics to interpret what we see on the screen.*

EXAMPLE 9 **Graphing a Circle**

Use a calculator in function mode to graph the circle $x^2 + y^2 = 4$.

Solution This graph can be obtained by graphing the two root functions

$$y_1 = \sqrt{4 - x^2} \quad \text{and} \quad y_2 = -y_1 = -\sqrt{4 - x^2}$$

in the same window. See **FIGURE 56.** To obtain a graph that is visually in correct proportions, we use a square window. Because it is a decimal window, the two semicircles connect at the *x*-axis. (See the Technology Note.)

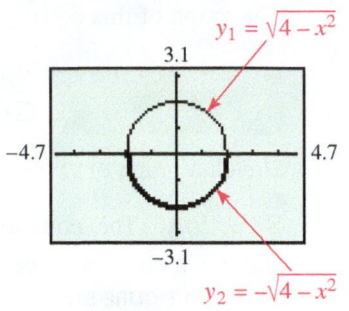

FIGURE 56

From **Sections 3.2** and **3.3,** we know that the graph of

$$y = ax^2 + bx + c \quad (a \ne 0)$$

is a parabola with a *vertical* axis of symmetry. Such an equation defines a function. If we interchange the roles of *x* and *y*, however, we obtain the equation

$$x = ay^2 + by + c,$$

whose graph is also a parabola, but with a *horizontal* axis of symmetry.

In the next example we graph a parabola with a horizontal axis by hand.

EXAMPLE 10 **Graphing a Parabola with a Horizontal Axis**

Sketch a graph of $x = y^2 - 1$. Is this parabola a graph of a function?

Solution Because there is a y^2-term and no x^2-term, this parabola has a horizontal axis. Start by making a table of values as shown and then plot these points. Connect these points with a smooth curve that opens to the right. The parabola is shown in **FIGURE 57** and is not a function because it fails the vertical line test.

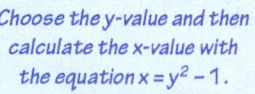

Choose the y-value and then calculate the x-value with the equation $x = y^2 - 1$.

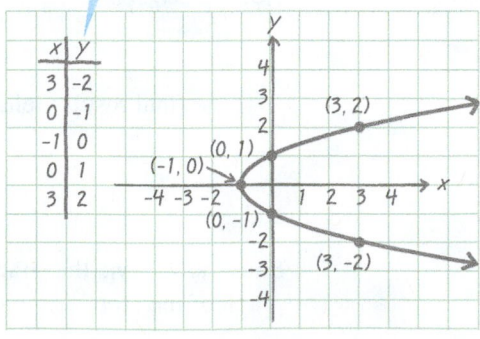

FIGURE 57

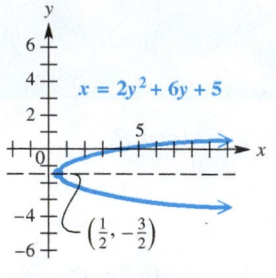

y

$x = 2y^2 + 6y + 5$

$\left(\frac{1}{2}, -\frac{3}{2}\right)$

FIGURE 58

FIGURE 58 shows a graph of $x = 2y^2 + 6y + 5$. Notice that this is also *not* the graph of a function. However, if we consider it to be the union of the graphs of two functions, one the top half-parabola and the other the bottom half-parabola, then it can be graphed by using a calculator in function mode.

EXAMPLE 11   Graphing a Horizontal Parabola

Graph $x = 2y^2 + 6y + 5$ and its axis of symmetry in the window $[-2, 8]$ by $[-4, 2]$.

Solution   We must solve for y by completing the square. (See **Section 3.2.**)

$$x = 2y^2 + 6y + 5$$

$$\frac{x}{2} = y^2 + 3y + \frac{5}{2} \qquad \text{Divide by 2.}$$

$$\frac{x}{2} - \frac{5}{2} = y^2 + 3y \qquad \text{Subtract } \tfrac{5}{2}.$$

$$\frac{x}{2} - \frac{5}{2} + \frac{9}{4} = y^2 + 3y + \frac{9}{4} \qquad \text{Add } \left[\tfrac{1}{2}(3)\right]^2 = \tfrac{9}{4}.$$

Be sure to add $\frac{9}{4}$ to each side.

$$\frac{x}{2} - \frac{1}{4} = \left(y + \frac{3}{2}\right)^2 \qquad \text{Add on the left and factor on the right.}$$

$$\left(y + \frac{3}{2}\right)^2 = \frac{x}{2} - \frac{1}{4} \qquad \text{Rewrite.}$$

$$y + \frac{3}{2} = \pm\sqrt{\frac{x}{2} - \frac{1}{4}} \qquad \text{Square root property}$$

$$y = -\frac{3}{2} \pm \sqrt{\frac{x}{2} - \frac{1}{4}} \qquad \text{Subtract } \tfrac{3}{2}.$$

Two functions are now defined. It is easier to use decimals when entering the equations into a calculator. Therefore, we define y_1 and y_2 as follows.

$$y_1 = -1.5 + \sqrt{0.5x - 0.25} \quad \text{and} \quad y_2 = -1.5 - \sqrt{0.5x - 0.25}$$

The graphs of these two functions together form the parabola with horizontal axis of symmetry $y_3 = -1.5$, shown in **FIGURE 59.**

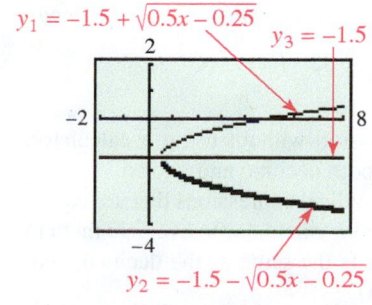

$y_1 = -1.5 + \sqrt{0.5x - 0.25}$

$y_3 = -1.5$

$y_2 = -1.5 - \sqrt{0.5x - 0.25}$

FIGURE 59

4.4 Exercises

*Note to student: At this point, you may want to review the material on radicals and rational exponents in **Chapter R.***

Checking Analytic Skills Evaluate each expression. **Do not use a calculator.**

1. $\sqrt{169}$

2. $-\sqrt[3]{64}$

3. $\sqrt[5]{-32}$

4. $\sqrt[4]{16}$

5. $81^{3/2}$

6. $27^{4/3}$

7. $125^{-2/3}$

8. $\left(\sqrt[3]{-27}\right)^2$

9. $(-1000)^{2/3}$

10. $(-125)^{-4/3}$

11. $8^{2/3}$

12. $-16^{3/2}$

13. $16^{-3/4}$

14. $25^{-3/2}$

15. $-81^{0.5}$

16. $32^{1/5}$

17. $64^{1/6}$

18. $16^{-0.25}$

19. $(-9^{3/4})^2$

20. $(4^{-1/2})^{-4}$

Use positive rational exponents to rewrite each expression. Assume variables represent positive numbers.

21. $\sqrt[3]{2x}$

22. $\sqrt{x+1}$

23. $\sqrt[3]{z^5}$

24. $\sqrt[5]{x^2}$

25. $\left(\sqrt[4]{y}\right)^{-3}$

26. $\left(\sqrt[3]{y^2}\right)^{-5}$

27. $\sqrt{x} \cdot \sqrt[3]{x}$

28. $\left(\sqrt[5]{z}\right)^{-3}$

29. $\sqrt{y} \cdot \sqrt{y}$

30. $\dfrac{\sqrt[3]{x}}{\sqrt{x}}$

Use a calculator to find each root or power. Give as many digits as your display shows.

31. $\sqrt[3]{-4}$

32. $\sqrt[5]{-3}$

33. $\sqrt[3]{-125}$

34. $\sqrt[5]{-243}$

35. $\sqrt[3]{-17}$

36. $\sqrt[5]{-8}$

37. $\sqrt[6]{\pi^2}$

38. $\sqrt[6]{\pi^{-1}}$

39. $13^{-1/3}$

40. $15^{-1/6}$

41. $32^{0.2}$

42. $81^{0.25}$

43. $\left(\dfrac{5}{6}\right)^{-1.3}$

44. $\left(\dfrac{4}{7}\right)^{-0.6}$

45. π^{-3}

46. $(2\pi)^{4/3}$

Evaluate f(x) at the given x. Approximate each result to the nearest hundredth.

47. $f(x) = x^{1.62}$, $x = 1.2$

48. $f(x) = x^{-0.71}$, $x = 3.8$

49. $f(x) = x^{3/2} - x^{1/2}$, $x = 50$

50. $f(x) = x^{5/4} - x^{-3/4}$, $x = 7$

Sketch the graph of each power function by hand, using a calculator only to evaluate y-values for your chosen x-values. On the same axes, graph $y = x^2$ for comparison. In each case, $x \geq 0$.

51. $f(x) = x^{1.05}$

52. $f(x) = x^{1.5}$

53. $f(x) = x^{2.5}$

54. $f(x) = x^{2.7}$

55. Consider the expression $16^{-3/4}$.
   (a) Simplify this expression without using a calculator. Give the answer in both decimal and $\frac{a}{b}$ form.
   (b) Write two different radical expressions that are equivalent to it, and use your calculator to evaluate them to show that the result is the same as the decimal form you found in part (a).
   (c) If your calculator has the capability to convert decimal numbers to fractions, use it to verify your results in part (a).

56. Consider the expression $5^{0.47}$.
   (a) Use the exponentiation capability of your calculator to find an approximation. Give as many digits as your calculator displays.
   (b) Use the fact that $0.47 = \frac{47}{100}$ to write the expression as a radical, and then use the root-finding capability of your calculator to find an approximation that agrees with the one found in part (a).

where x is the altitude of the plane in feet and $f(x)$ is in miles. If a plane is flying at 30,000 feet, how far can the pilot see?

65. *Trout and Pollution*   Rainbow trout are sensitive to zinc ions in the water. High concentrations are lethal. The average survival times x in minutes for trout in various concentrations of zinc ions y in milligrams per liter (mg/L) are listed in the table.

x (in minutes)	0.5	1	2	3
y (in mg/L)	4500	1960	850	525

Source: Mason, C., *Biology of Freshwater Pollution*, John Wiley and Sons.

(a) The data can be modeled by $f(x) = ax^b$, where a and b are constants. Determine a. (*Hint:* Let $f(1) = 1960$.)

(b) Estimate b.

(c) Evaluate $f(4)$ and interpret the result.

66. *Asbestos and Cancer*   Insulation workers who were exposed to asbestos and employed before 1960 experienced an increased likelihood of lung cancer. If a group of insulation workers have a cumulative total of 100,000 years of work experience, with their first date of employment x years ago, then the number of lung cancer cases occurring within the group can be modeled by

$$N(x) = 0.00437x^{3.2}.$$

(*Source:* Walker, A., *Observation and Inference: An Introduction to the Methods of Epidemiology*, Epidemiology Resources, Inc.)

(a) Calculate $N(x)$ when $x = 5$, 10, and 20. What happens to the likelihood of cancer as x increases?

(b) If x doubles, does the number of cancer cases also double?

67. *Fiddler Crab Size*   One study of the male fiddler crab showed a connection between the weight of its claws and the animal's total body weight. For a crab weighing over 0.75 gram, the weight of its claws can be estimated by

$$f(x) = 0.445x^{1.25}.$$

The input x is the weight of the crab in grams, and the output $f(x)$ is the weight of the claws in grams. Predict the weight of the claws for a crab that weighs 2 grams. (*Source:* Huxley, J., *Problems of Relative Growth*, Methuen and Co.; Brown, D. and P. Rothery, *Models in Biology: Mathematics, Statistics, and Computing*, John Wiley and Sons.)

68. *Weight and Height of Men*   The average weight for a man can sometimes be estimated by

$$f(x) = 0.117x^{1.7},$$

where x represents the man's height in inches and $f(x)$ is his weight in pounds. What is the average weight of a 68-inch-tall man?

RELATING CONCEPTS

**For individual or group investigation
(Exercises 57–60)**

Duplicate each screen on your calculator. The screens show multiple ways of finding an approximation for $\sqrt[6]{9}$. **Work Exercises 57–60 in order** *using your calculator.*

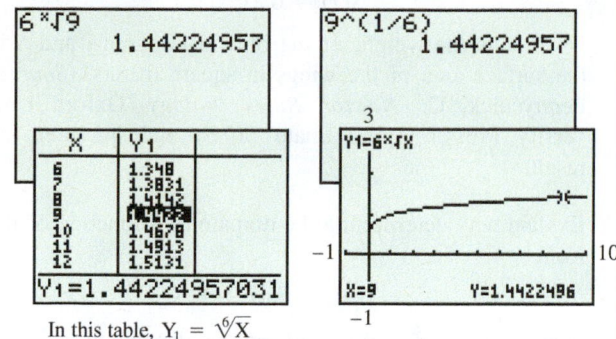

In this table, $Y_1 = \sqrt[6]{X}$.

57. Use a radical expression to approximate $\sqrt[6]{81}$ to as many decimal places as the calculator will give.

58. Use a rational exponent to repeat **Exercise 57.**

59. Use a table to repeat **Exercise 57.**

60. Use the graph of $y = \sqrt[6]{x}$ to repeat **Exercise 57.**

(Modeling)   Solve each problem.

61. *Wing Size*   Suppose that the surface area S of a bird's wings, in square feet, can be modeled by

$$S(w) = 1.27w^{2/3},$$

where w is the weight of the bird in pounds. Estimate the surface area of a bird's wings if the bird weighs 4.0 pounds.

62. *Wingspan*   Suppose that the wingspan L in feet of a bird weighing W pounds is given by

$$L = 2.43W^{0.3326}.$$

Estimate the wingspan of a bird that weighs 5.2 pounds.

63. *Planetary Orbits*   The formula

$$f(x) = x^{1.5}$$

calculates the number of years it would take for a planet to orbit the sun if its average distance from the sun is x times farther than Earth. If there were a planet located 15 times farther from the sun than Earth, how many years would it take for the planet to orbit the sun?

64. *Sight Distance*   A formula for calculating the distance one can see from an airplane to the horizon on a clear day is given by

$$f(x) = 1.22x^{0.5},$$

69. *Animal Pulse Rate and Weight*   According to one model, the rate at which an animal's heart beats varies with its weight. Smaller animals tend to have faster pulses, whereas larger animals tend to have slower pulses. The table lists average pulse rates in beats per minute (bpm) for animals with various weights in pounds (lb). Use regression (or some other method) to find values for a and b so that $f(x) = ax^b$ models these data.

Weight (in lb)	40	150	400	1000	2000
Pulse (in bpm)	140	72	44	28	20

Source: Pennycuick, C., *Newton Rules Biology,* Oxford University Press.

70. *Animal Pulse Rate and Weight*   Use the results of **Exercise 69** to calculate the pulse rates for a 60-pound dog and a 2-ton whale.

71. *Wing Size*   (See **Exercise 61.**) Heavier birds have larger wings with more surface area than do lighter birds. For some species of birds, this relationship is given by

$$S(x) = 0.2x^{2/3},$$

where x is the weight of the bird in kilograms and S is the surface area of the wings in square meters. (*Source:* Pennycuick, C., *Newton Rules Biology,* Oxford University Press.) Approximate $S(0.5)$ and interpret the result.

72. Explain why determining the domain of a function of the form

$$f(x) = \sqrt[n]{ax + b}$$

requires two different considerations, depending upon the parity of n.

Checking Analytic Skills   Determine the domain of each function. **Do not use a calculator.**

73. $f(x) = \sqrt{5 + 4x}$

74. $f(x) = \sqrt{9x + 18}$

75. $f(x) = -\sqrt[4]{6 - x}$

76. $f(x) = -\sqrt[4]{2 - 0.5x}$

77. $f(x) = \sqrt[3]{8x - 24}$

78. $f(x) = \sqrt[5]{x + 32}$

79. $f(x) = \sqrt{49 - x^2}$

80. $f(x) = \sqrt{81 - x^2}$

81. $f(x) = \sqrt{x^3 - x}$

82. Explain why the domain of $f(x) = \sqrt{x^2 + 1}$ is $(-\infty, \infty)$.

*For Exercises 83–90, the domains were determined in **Exercises 73–80**. Use a graph to **(a)** find the range, **(b)** give the largest open interval over which the function is increasing, **(c)** give the largest open interval over which the function is decreasing, and **(d)** solve the equation $f(x) = 0$ by observing the graph.*

83. $f(x) = \sqrt{5 + 4x}$

84. $f(x) = \sqrt{9x + 18}$

85. $f(x) = -\sqrt[4]{6 - x}$

86. $f(x) = -\sqrt[4]{2 - 0.5x}$

87. $f(x) = \sqrt[3]{8x - 24}$

88. $f(x) = \sqrt[5]{x + 32}$

89. $f(x) = \sqrt{49 - x^2}$

90. $f(x) = \sqrt{81 - x^2}$

Concept Check   Use transformations to explain how the graph of the given function can be obtained from the graphs of the square root function or the cube root function.

91. $y = \sqrt{9x + 27}$

92. $y = \sqrt{16x + 16}$

93. $y = \sqrt{4x + 16} + 4$

94. $y = \sqrt{32 - 4x} - 3$

95. $y = \sqrt[3]{27x + 54} - 5$

96. $y = \sqrt[3]{8x - 8}$

In Exercises 97–108, graph by hand the equation of the circle or the parabola with a horizontal axis.

97. $x^2 + y^2 = 1$

98. $x^2 + y^2 = 25$

99. $(x - 2)^2 + (y + 2)^2 = 4$

100. $(x + 1)^2 + y^2 = 9$

101. $x^2 + (y - 2)^2 = 16$

102. $(x + 3)^2 + (y - 1)^2 = 1$

103. $x = y^2 + 1$

104. $x = y^2 - 3$

105. $x = 2y^2$

106. $x = -y^2$

107. $x = -(y + 1)^2 + 2$

108. $x = (y - 2)^2 - 1$

In Exercises 109–116, describe the graph of the equation as either a circle or a parabola with a horizontal axis of symmetry. Then, determine two functions, designated by y_1 and y_2, such that their union will give the graph of the given equation. Finally, graph y_1 and y_2 in the given viewing window.

109. $x^2 + y^2 = 100$;
$[-15, 15]$ by $[-10, 10]$

110. $x^2 + y^2 = 81$;
$[-15, 15]$ by $[-10, 10]$

111. $(x - 2)^2 + y^2 = 9$;
$[-9.4, 9.4]$ by $[-6.2, 6.2]$

112. $(x + 3)^2 + y^2 = 16$;
$[-9.4, 9.4]$ by $[-6.2, 6.2]$

113. $x = y^2 + 6y + 9$;
$[-10, 10]$ by $[-10, 10]$

114. $x = y^2 - 8y + 16$;
$[-10, 10]$ by $[-10, 10]$

115. $x = 2y^2 + 8y + 1$;
$[-10, 10]$ by $[-10, 10]$

116. $x = -3y^2 - 6y + 2$;
$[-9.4, 9.4]$ by $[-6.2, 6.2]$

4.5 Equations, Inequalities, and Applications Involving Root Functions

Equations and Inequalities • An Application of Root Functions

Equations and Inequalities

To solve an equation involving roots, such as

$$\sqrt{11 - x} - x = 1,$$

in which the variable appears in a radicand, we use the following property.

← Algebra Review
To review some of the properties of radicals see **Section R.5**.

Power Property

If P and Q are algebraic expressions, then every solution of the equation $P = Q$ is among the solutions of the equation $P^n = Q^n$, for any positive integer n.

CAUTION   All solutions of the equation $P^n = Q^n$ are not necessarily solutions of the given equation $P = Q$. Therefore, we must check every solution from $P^n = Q^n$ in the *original*, or *given*, equation $P = Q$.

When the power property is used to solve equations, the new equation may have *more* solutions than the original equation. We call these **proposed solutions** of the original equation. For example, the solution set of the equation $x = -2$ is obviously $\{-2\}$. If we square each side of the equation $x = -2$, we get the new equation $x^2 = 4$, which has solution set $\{-2, 2\}$. Since the solution sets are not equal, the equations are not equivalent. *After applying the power property on equations that contain radicals or rational exponents, it is essential to check all proposed solutions in the original equation.*

To solve equations involving roots, follow these steps.

Solving an Equation Involving a Root Function

Step 1 Isolate a term involving a root on one side of the equation.

Step 2 Raise each side of the equation to a positive integer power that will eliminate the radical or rational exponent.

Step 3 Solve the resulting equation. (If a root is still present after Step 2, repeat Steps 1 and 2.)

Step 4 Check each proposed solution in the original equation.

EXAMPLE 1 **Solving an Equation Involving a Square Root**

Solve $\sqrt{11 - x} - x = 1$.

Analytic Solution

$$\sqrt{11 - x} - x = 1$$
$$\sqrt{11 - x} = 1 + x \qquad \text{Isolate the radical.}$$
$$\left(\sqrt{11 - x}\right)^2 = (1 + x)^2 \qquad \text{Square each side.}$$
$$11 - x = 1 + 2x + x^2 \qquad (x + y)^2 = x^2 + 2xy + y^2$$
$$x^2 + 3x - 10 = 0 \qquad \text{Standard form}$$
$$(x + 5)(x - 2) = 0 \qquad \text{Factor.}$$
$$x + 5 = 0 \quad \text{or} \quad x - 2 = 0 \qquad \text{Zero-product property}$$
$$x = -5 \quad \text{or} \quad x = 2 \qquad \text{Solve each equation.}$$

Check the proposed solutions, -5 and 2, in the *original* equation.

$$\sqrt{11 - x} - x = 1 \qquad \text{Original equation}$$

Let $x = -5$.

$$\sqrt{11 - (-5)} - (-5) = 1 \quad ?$$
$$\sqrt{16} + 5 = 1 \quad ?$$
$$4 + 5 = 1 \quad ?$$
$$9 = 1 \quad \text{False}$$

Let $x = 2$.

$$\sqrt{11 - 2} - 2 = 1 \quad ?$$
$$\sqrt{9} - 2 = 1 \quad ?$$
$$3 - 2 = 1 \quad ?$$
$$1 = 1 \quad ✓ \quad \text{True}$$

Squaring each side of the equation led to the *extraneous* value -5, as indicated by the false statement $9 = 1$. Therefore, the only solution of the original equation is 2. The solution set is $\{2\}$.

Graphing Calculator Solution

The equation in the second line of the analytic solution has the same solution set as the original equation, because each side of the equation has not yet been squared. If we graph

$$y_1 = \sqrt{11 - x} \quad \text{and} \quad y_2 = 1 + x$$

and observe the x-coordinate of the only point of intersection of the graphs, we see that it is 2, confirming the analytic solution. See **FIGURE 60.** Note that *extraneous* values do not occur in graphical solutions.

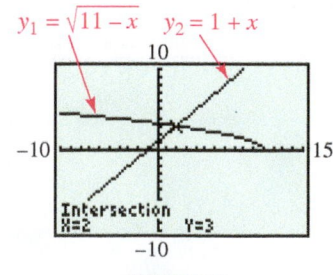

FIGURE 60

EXAMPLE 2   **Solving an Equation Involving a Rational Exponent**

Solve $(5 - 5x)^{1/2} + x = 1$.

Analytic Solution

$$(5 - 5x)^{1/2} + x = 1$$

$$(5 - 5x)^{1/2} = 1 - x \qquad \text{Isolate the term with the rational exponent.}$$

$$\left[(5 - 5x)^{1/2}\right]^2 = (1 - x)^2 \qquad \text{Square each side.}$$

$$5 - 5x = 1 - 2x + x^2 \qquad (x - y)^2 = x^2 - 2xy + y^2$$

$$x^2 + 3x - 4 = 0 \qquad \text{Standard form}$$

$$(x + 4)(x - 1) = 0 \qquad \text{Factor.}$$

$$x + 4 = 0 \quad \text{or} \quad x - 1 = 0 \qquad \text{Zero-product property}$$

$$x = -4 \quad \text{or} \qquad x = 1 \qquad \text{Solve each equation.}$$

Check the proposed solutions, -4 and 1, in the *original* equation.

$$(5 - 5x)^{1/2} + x = 1 \quad \text{Original equation}$$

Let $x = -4$.

$$[5 - 5(-4)]^{1/2} + (-4) = 1 \quad ?$$
$$25^{1/2} + (-4) = 1 \quad ?$$
$$5 + (-4) = 1 \quad ?$$
$$1 = 1 \checkmark \text{ True}$$

Let $x = 1$.

$$[5 - 5(1)]^{1/2} + (1) = 1 \quad ?$$
$$0^{1/2} + 1 = 1 \quad ?$$
$$0 + 1 = 1 \quad ?$$
$$1 = 1 \checkmark \text{ True}$$

Both proposed solutions check, so the solution set is

$$\{-4, 1\}.$$

Graphing Calculator Solution

To use the x-intercept method, we let

$$y_1 = (5 - 5x)^{1/2} + x \quad \text{and} \quad y_2 = 1.$$

We graph $y_3 = y_1 - y_2$ to produce the curve shown in **FIGURE 61**. The displays at the bottom of the screens confirm the solutions, -4 and 1.

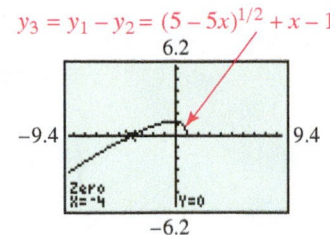

$$y_3 = y_1 - y_2 = (5 - 5x)^{1/2} + x - 1$$

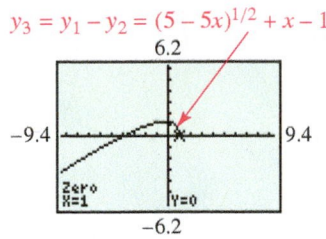

$$y_3 = y_1 - y_2 = (5 - 5x)^{1/2} + x - 1$$

FIGURE 61

EXAMPLE 3   **Solving an Equation Involving Cube Roots**

Solve $\sqrt[3]{x^2 + 3x} = \sqrt[3]{5}$.

Solution

$$\sqrt[3]{x^2 + 3x} = \sqrt[3]{5}$$

$$\left(\sqrt[3]{x^2 + 3x}\right)^3 = \left(\sqrt[3]{5}\right)^3 \qquad \text{Cube each side.}$$

$$x^2 + 3x = 5 \qquad \text{Simplify.}$$

$$x^2 + 3x - 5 = 0 \qquad \text{Standard form}$$

Since this equation cannot be solved by usual factoring methods, we use the quadratic formula, with $a = 1$, $b = 3$, and $c = -5$.

$$x = \frac{-b \pm \sqrt{b^2 - 4ac}}{2a} = \frac{-3 \pm \sqrt{3^2 - 4(1)(-5)}}{2(1)} = \frac{-3 \pm \sqrt{29}}{2}$$

NOTE When the power property is applied using an odd positive integer n, then the solution sets for $P = Q$ and $P^n = Q^n$ are typically equivalent.

While an analytic check is quite involved, it can be shown that both of these values are solutions. The solution set is

$$\left\{\frac{-3 + \sqrt{29}}{2}, \frac{-3 - \sqrt{29}}{2}\right\}.$$

WHAT WENT WRONG?

A student solved the following equation as shown.

$$x + 4 = \sqrt{x + 6}$$

$$x^2 + 8x + 16 = x + 6 \qquad \text{Square each side.}$$

$$x^2 + 7x + 10 = 0 \qquad \text{Standard form}$$

$$(x + 5)(x + 2) = 0 \qquad \text{Factor.}$$

$$x = -5 \quad \text{or} \quad x = -2 \qquad \text{Zero-product property}$$

A check shows that -5 is extraneous, and the solution set is $\{-2\}$. To support his solution, he entered $Y_1 = X + 4$ and $Y_2 = \sqrt{X + 6}$ into his calculator. Using the intersection-of-graphs method, he obtained the screen on the left below, and this assured him that he was correct.

Next, he decided to use the x-intercept method to show that -2 is a solution. He got the screen on the right below. Something was wrong.

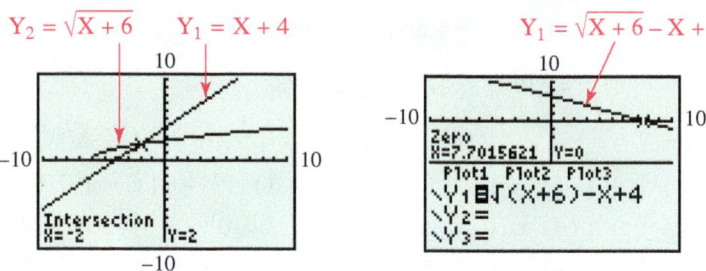

What Went Wrong? How can he correct his work so that the screen on the right shows a zero of -2?

EXAMPLE 4 **Solving an Equation Involving Roots (Squaring Twice)**

Solve $\sqrt{2x + 3} - \sqrt{x + 1} = 1$.

Solution Isolate the more complicated radical.

$$\sqrt{2x + 3} - \sqrt{x + 1} = 1$$

$$\sqrt{2x + 3} = 1 + \sqrt{x + 1} \qquad \text{Isolate } \sqrt{2x + 3}.$$

$$\left(\sqrt{2x + 3}\right)^2 = \left(1 + \sqrt{x + 1}\right)^2 \qquad \text{Square each side.}$$

$$2x + 3 = 1 + 2\sqrt{x + 1} + x + 1 \qquad \text{Simplify.}$$

Remember the middle term.

$$x + 1 = 2\sqrt{x + 1} \qquad \text{Isolate the radical term.}$$

$$(x + 1)^2 = \left(2\sqrt{x + 1}\right)^2 \qquad \text{Square each side again.}$$

$$x^2 + 2x + 1 = 4(x + 1) \qquad \text{Simplify.}$$

$(ab)^m = a^m b^m$

$$x^2 + 2x + 1 = 4x + 4 \qquad \text{Distributive property}$$

$$x^2 - 2x - 3 = 0 \qquad \text{Standard form}$$

$$(x - 3)(x + 1) = 0 \qquad \text{Factor.}$$

$$x - 3 = 0 \quad \text{or} \quad x + 1 = 0 \qquad \text{Zero-product property}$$

$$x = 3 \quad \text{or} \qquad x = -1 \qquad \text{Solve each equation.}$$

Answer to What Went Wrong?

He entered $\sqrt{(X + 6)} - X + 4$ rather than $\sqrt{(X + 6)} - (X + 4)$. He should insert parentheses around $X + 4$ in Y_1.

Check the proposed solutions, 3 and -1, in the *original* equation, as follows.

$$\sqrt{2x + 3} - \sqrt{x + 1} = 1 \qquad \text{Original equation}$$

Let $x = 3$.

$$\sqrt{2(3) + 3} - \sqrt{3 + 1} = 1 \qquad ?$$
$$\sqrt{9} - \sqrt{4} = 1 \qquad ?$$
$$1 = 1 \ \checkmark \ \text{True}$$

Let $x = -1$.

$$\sqrt{2(-1) + 3} - \sqrt{-1 + 1} = 1 \qquad ?$$
$$\sqrt{1} - \sqrt{0} = 1 \qquad ?$$
$$1 = 1 \ \checkmark \ \text{True}$$

Both solutions check, so the solution set is $\{-1, 3\}$.

> **EXAMPLE 5** **Solving Inequalities Involving Roots**

Solve each inequality.

(a) $\sqrt[3]{x^2 + 3x} \le \sqrt[3]{5}$

(b) $\sqrt{2x + 3} - \sqrt{x + 1} > 1$

Solution

(a) The associated *equation* was solved in **Example 3.** Its solutions are

$$\frac{-3 + \sqrt{29}}{2} \quad \text{and} \quad \frac{-3 - \sqrt{29}}{2}.$$

We use the *x*-intercept method to solve this inequality, which can be written $\sqrt[3]{x^2 + 3x} - \sqrt[3]{5} \le 0$. As seen in **FIGURE 62,** the graph of $y = \sqrt[3]{x^2 + 3x} - \sqrt[3]{5}$ lies on or below the *x*-axis in the interval between the two *x*-intercepts, including the endpoints. Therefore, the solution set of the given inequality is

$$\left[\frac{-3 - \sqrt{29}}{2}, \frac{-3 + \sqrt{29}}{2} \right].$$

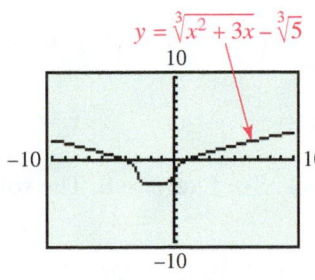

FIGURE 62

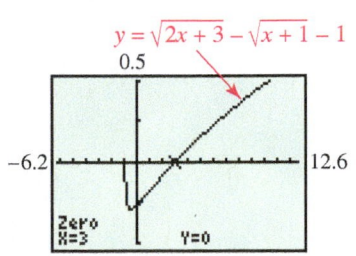

FIGURE 63

(b) This inequality is equivalent to $\sqrt{2x + 3} - \sqrt{x + 1} - 1 > 0$. The associated *equation* was solved in **Example 4;** its solutions are 3 and -1. The graph of $y = \sqrt{2x + 3} - \sqrt{x + 1} - 1$ lies above the *x*-axis for real numbers greater than 3, as we see in **FIGURE 63.** The solution set is $(3, \infty)$.

EXAMPLE 6 Solving an Equation Having Negative Exponents

Solve $15x^{-2} - 19x^{-1} + 6 = 0$.

Solution

$$15x^{-2} - 19x^{-1} + 6 = 0$$

$x^2(15x^{-2} - 19x^{-1} + 6) = x^2(0)$	Multiply each side by x^2, the LCD.
$15x^2x^{-2} - 19x^2x^{-1} + 6x^2 = 0$	Distributive property
$15 - 19x + 6x^2 = 0$	Properties of exponents
$6x^2 - 19x + 15 = 0$	Rewrite.
$(2x - 3)(3x - 5) = 0$	Factor.
$x = \dfrac{3}{2} \quad \text{or} \quad x = \dfrac{5}{3}$	Zero-product property

The solutions check. See **Exercise 7.** The solution set is $\left\{\frac{3}{2}, \frac{5}{3}\right\}$. ●

← Algebra Review
To review some of the properties of rational exponents see **Section R.4.**

EXAMPLE 7 Solving an Equation Having Fractional Exponents

Solve $2x^{2/3} + 5x^{1/3} - 3 = 0$.

Solution

$$2x^{2/3} + 5x^{1/3} - 3 = 0$$

$2(x^{1/3})^2 + 5(x^{1/3}) - 3 = 0$	Properties of exponents
$2u^2 + 5u - 3 = 0$	Let $u = x^{1/3}$ and substitute.
$(2u - 1)(u + 3) = 0$	Factor.
$u = \dfrac{1}{2} \quad \text{or} \quad u = -3$	Zero-product property
$x^{1/3} = \dfrac{1}{2} \quad \text{or} \quad x^{1/3} = -3$	Substitute $x^{1/3}$ for u.
$(x^{1/3})^3 = \left(\dfrac{1}{2}\right)^3 \quad \text{or} \quad (x^{1/3})^3 = (-3)^3$	Cube each side.
$x = \dfrac{1}{8} \quad \text{or} \quad x = -27$	Simplify.

The solutions check. See **Exercise 8.** The solution set is $\left\{\frac{1}{8}, -27\right\}$. ●

In the next example we simplify an expression and solve an equation that involves negative rational exponents.

→ Looking Ahead to Calculus
Expressions and equations like the ones in **Example 8** occur in calculus.

EXAMPLE 8 Solving an Equation with Rational Exponents

Complete the following.

(a) Write the expression $\dfrac{\frac{2}{3}(x + 1)x^{-1/3} - x^{2/3}}{(x + 1)^2}$ without any negative exponents.

(b) Use part (a) to solve the equation $\dfrac{\frac{2}{3}(x + 1)x^{-1/3} - x^{2/3}}{(x + 1)^2} = 0$.

Solution

(a) To eliminate the negative exponent, we can multiply both the numerator and denominator by the expression $x^{1/3}$. This is because $x^{1/3} \cdot x^{-1/3} = x^0 = 1$, which will eliminate the negative exponent on x in the numerator.

$$\frac{\frac{2}{3}(x+1)x^{-1/3} - x^{2/3}}{(x+1)^2} = \frac{\frac{2}{3}(x+1)x^{-1/3} - x^{2/3}}{(x+1)^2} \cdot \frac{x^{1/3}}{x^{1/3}} \qquad \text{Multiply by } \frac{x^{1/3}}{x^{1/3}} = 1.$$

$$= \frac{\frac{2}{3}(x+1)x^{-1/3} \cdot x^{1/3} - x^{2/3} \cdot x^{1/3}}{(x+1)^2 \cdot x^{1/3}} \qquad \text{Distributive property}$$

$$= \frac{\frac{2}{3}(x+1)(1) - x^1}{(x+1)^2 \cdot x^{1/3}} \qquad \begin{array}{l}\text{Properties of}\\ \text{exponents}\end{array}$$

$$= \frac{2(x+1) - 3x}{3x^{1/3}(x+1)^2} \qquad \begin{array}{l}\text{Multiply by } \frac{3}{3} \text{ to clear}\\ \text{fractions.}\end{array}$$

$$= \frac{2 - x}{3x^{1/3}(x+1)^2} \qquad \text{Combine like terms.}$$

(b) If a ratio (fraction) is equal to zero, then its numerator must equal zero.

$$\frac{2 - x}{3x^{1/3}(x+1)^2} = 0 \qquad \text{Set the expression from part (a) equal to 0.}$$

$$2 - x = 0 \qquad \text{Set numerator equal to 0.}$$

$$x = 2 \qquad \text{Solve for } x.$$

The solution set is $\{2\}$.

An Application of Root Functions

> **EXAMPLE 9**   **Solving a Cable Installation Problem Involving a Root Function**

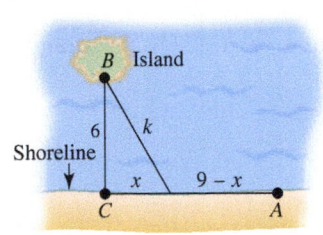

FIGURE 64

A company wishes to run a utility cable from point A on the shore (as shown in **FIGURE 64**) to an installation at point B on the island. The island is 6 miles from the shore. It costs \$400 per mile to run the cable on land and \$500 per mile under water. Assume that the cable starts at A, runs along the shoreline, and then angles and runs under water to the island. Let x represent the distance from C at which the underwater portion of the cable run begins, and let the distance between A and C be 9 miles.

(a) What are the possible values of x in this problem?

(b) Express the cost of laying the cable as a function of x.

(c) Find the total cost if 3 miles of cable are on land.

(d) Find the point at which the line should begin to angle to minimize the total cost. What is this minimum total cost? Solve graphically.

Solution

(a) The value of x must be a real number greater than or equal to 0 and less than or equal to 9, meaning that x must be in the interval $[0, 9]$.

(b) The total cost is found by adding the cost of the cable on land to the cost of the cable under water. If we let k represent the length of the cable under water, then

$$k^2 = 6^2 + x^2 \qquad \text{Use the Pythagorean theorem.}$$

$$k^2 = 36 + x^2$$

$$k = \sqrt{36 + x^2}. \qquad \text{Square root property; } k > 0$$

(continued)

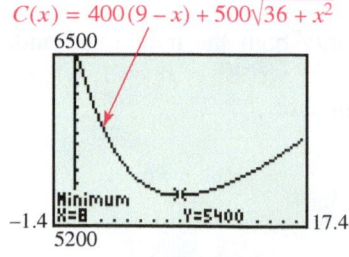

FIGURE 65

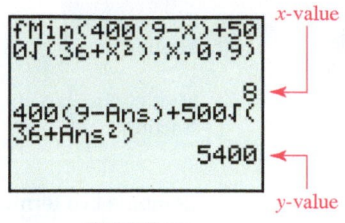

FIGURE 66

The cost of the cable on land is $400(9 - x)$ dollars, while the cost of the cable under water is $500k$, or $500\sqrt{36 + x^2}$, dollars. The total cost $C(x)$ is given by

$$C(x) = 400(9 - x) + 500\sqrt{36 + x^2}. \quad \text{Total cost in dollars}$$

(c) According to **FIGURE 64** on the preceding page, if 3 miles of cable are on land, then $3 = 9 - x$, giving $x = 6$. We must evaluate $C(6)$.

$$C(6) = 400(9 - 6) + 500\sqrt{36 + 6^2}$$
$$\approx 5442.64 \text{ dollars}$$

(d) From **FIGURE 65,** the absolute minimum value of the function on the interval $[0, 9]$ is found when $x = 8$, meaning that $9 - 8 = 1$ mile should be along land and $\sqrt{36 + 8^2} = 10$ miles should be under water. We must evaluate $C(8)$.

$$C(8) = 400(9 - 8) + 500\sqrt{36 + 8^2}$$
$$= 5400 \text{ dollars} \quad \text{Minimum total cost}$$

FIGURE 66 shows how the coordinates of the minimum point can be determined on the TI-84 Plus without actually graphing the function. ●

4.5 Exercises

Checking Analytic Skills *In Exercises 1–5, begin by drawing a rough sketch to determine the number of real solutions for the equation $y_1 = y_2$. Then solve this equation by hand. Give the solution set and any extraneous values that may occur.* **Do not use a calculator.**

1. $y_1 = \sqrt{x}$
$y_2 = 2x - 1$

2. $y_1 = \sqrt{x}$
$y_2 = x - 6$

3. $y_1 = \sqrt{x}$
$y_2 = -x + 3$

4. $y_1 = \sqrt{x}$
$y_2 = 3x$

5. $y_1 = \sqrt[3]{x}$
$y_2 = x^2$

6. Use a hand-drawn graph to explain why $\sqrt{x} = -x - 5$ has no real solutions.

7. Check that proposed solutions $\frac{3}{2}$ and $\frac{5}{3}$ from **Example 6** are solutions of $15x^{-2} - 19x^{-1} + 6 = 0$.

8. Check that proposed solutions $\frac{1}{8}$ and -27 from **Example 7** are solutions of $2x^{2/3} + 5x^{1/3} - 3 = 0$.

Checking Analytic Skills *Solve each equation by hand.* **Do not use a calculator.**

9. $x - 4 = \sqrt{3x - 8}$

10. $x - 5 = \sqrt{5x - 1}$

11. $\sqrt{x + 5} + 1 = x$

12. $\sqrt{4 - 3x} - 8 = x$

13. $\sqrt{2x + 3} - \sqrt{x + 1} = 1$

14. $\sqrt{3x + 4} - \sqrt{2x - 4} = 2$

15. $\sqrt[3]{x + 1} = -3$

16. $\sqrt[3]{x + 9} = 2$

17. $\sqrt[3]{3x^2 + 7} = \sqrt[3]{7 - 4x}$

18. $\sqrt[3]{2x^2 + 1} = \sqrt[3]{1 - x}$

19. $\sqrt[4]{x - 2} + 4 = 6$

20. $\sqrt[4]{2x + 3} = \sqrt{x + 1}$

21. $x^{2/5} = 4$

22. $x^{2/3} = 16$

23. $2x^{1/3} - 5 = 1$

24. $4x^{3/2} + 5 = 21$

25. $x^{-2} + 3x^{-1} + 2 = 0$

26. $2x^{-2} - x^{-1} = 3$

27. $5x^{-2} + 13x^{-1} = 28$

28. $3x^{-2} - 19x^{-1} + 20 = 0$

29. $x^{2/3} - x^{1/3} - 6 = 0$

30. $x^{2/3} + 9x^{1/3} + 14 = 0$

31. $x^{3/4} - x^{1/2} - x^{1/4} + 1 = 0$

32. $x^{3/4} - 2x^{1/2} - 4x^{1/4} + 8 = 0$

Use an analytic method to solve each equation in part (a). Support the solution with a graph. Then use the graph to solve the inequalities in parts (b) and (c).

33. (a) $\sqrt{3x + 7} = 2$
  (b) $\sqrt{3x + 7} > 2$
  (c) $\sqrt{3x + 7} < 2$

34. (a) $\sqrt{2x + 13} = 3$
  (b) $\sqrt{2x + 13} > 3$
  (c) $\sqrt{2x + 13} < 3$

35. (a) $\sqrt{4x + 13} = 2x - 1$
  (b) $\sqrt{4x + 13} > 2x - 1$
  (c) $\sqrt{4x + 13} < 2x - 1$

36. (a) $\sqrt{3x + 7} = 3x + 5$

(b) $\sqrt{3x + 7} > 3x + 5$

(c) $\sqrt{3x + 7} < 3x + 5$

37. (a) $\sqrt{5x + 1} + 2 = 2x$

(b) $\sqrt{5x + 1} + 2 > 2x$

(c) $\sqrt{5x + 1} + 2 < 2x$

38. (a) $\sqrt{3x + 4} + x = 8$

(b) $\sqrt{3x + 4} + x > 8$

(c) $\sqrt{3x + 4} + x < 8$

39. (a) $\sqrt{3x - 6} + 2 = \sqrt{5x - 6}$

(b) $\sqrt{3x - 6} + 2 > \sqrt{5x - 6}$

(c) $\sqrt{3x - 6} + 2 < \sqrt{5x - 6}$

40. (a) $\sqrt{2x - 4} + 2 = \sqrt{3x + 4}$

(b) $\sqrt{2x - 4} + 2 > \sqrt{3x + 4}$

(c) $\sqrt{2x - 4} + 2 < \sqrt{3x + 4}$

41. (a) $\sqrt[3]{x^2 - 2x} = \sqrt[3]{x}$

(b) $\sqrt[3]{x^2 - 2x} > \sqrt[3]{x}$

(c) $\sqrt[3]{x^2 - 2x} < \sqrt[3]{x}$

42. (a) $\sqrt[3]{4x^2 - 4x + 1} = \sqrt[3]{x}$

(b) $\sqrt[3]{4x^2 - 4x + 1} > \sqrt[3]{x}$

(c) $\sqrt[3]{4x^2 - 4x + 1} < \sqrt[3]{x}$

43. (a) $\sqrt[4]{3x + 1} = 1$

(b) $\sqrt[4]{3x + 1} > 1$

(c) $\sqrt[4]{3x + 1} < 1$

44. (a) $\sqrt[4]{x - 15} = 2$

(b) $\sqrt[4]{x - 15} > 2$

(c) $\sqrt[4]{x - 15} < 2$

45. (a) $(2x - 5)^{1/2} - 2 = (x - 2)^{1/2}$

(b) $(2x - 5)^{1/2} - 2 \geq (x - 2)^{1/2}$

(c) $(2x - 5)^{1/2} - 2 \leq (x - 2)^{1/2}$

46. (a) $(x + 5)^{1/2} - 2 = (x - 1)^{1/2}$

(b) $(x + 5)^{1/2} - 2 \geq (x - 1)^{1/2}$

(c) $(x + 5)^{1/2} - 2 \leq (x - 1)^{1/2}$

47. (a) $(x^2 + 6x)^{1/4} = 2$

(b) $(x^2 + 6x)^{1/4} > 2$

(c) $(x^2 + 6x)^{1/4} < 2$

48. (a) $(x^2 + 2x)^{1/4} = 3^{1/4}$

(b) $(x^2 + 2x)^{1/4} > 3^{1/4}$

(c) $(x^2 + 2x)^{1/4} < 3^{1/4}$

49. (a) $(2x - 1)^{2/3} = x^{1/3}$

(b) $(2x - 1)^{2/3} > x^{1/3}$

(c) $(2x - 1)^{2/3} < x^{1/3}$

50. (a) $(x - 3)^{2/5} = (4x)^{1/5}$

(b) $(x - 3)^{2/5} > (4x)^{1/5}$

(c) $(x - 3)^{2/5} < (4x)^{1/5}$

RELATING CONCEPTS   For individual or group investigation (Exercises 51–62)

Exercises 51–62 incorporate many concepts from **Chapter 3** *with the method of solving equations involving roots.* **Work them in order.** *Consider the equation*

$$\sqrt[3]{4x - 4} = \sqrt{x + 1}.$$

51. Rewrite the equation, using rational exponents.

52. What is the least common denominator of the rational exponents found in **Exercise 51?**

53. Raise each side of the equation in **Exercise 51** to the power indicated by your answer in **Exercise 52.**

54. Show that the equation in **Exercise 53** is equivalent to $x^3 - 13x^2 + 35x - 15 = 0$.

55. Graph the cubic function defined by the polynomial on the left side of the equation in **Exercise 54** in the window $[-5, 10]$ by $[-100, 100]$. How many real solutions does the equation have?

56. Use synthetic division to show that 3 is a zero of the polynomial

$$P(x) = x^3 - 13x^2 + 35x - 15.$$

57. Use the result of **Exercise 56** to factor $P(x)$ so that one factor is linear and the other is quadratic.

58. Set the quadratic factor of $P(x)$ from **Exercise 57** equal to 0, and solve the equation using the quadratic formula.

59. What are the three proposed solutions of the original equation,

$$\sqrt[3]{4x - 4} = \sqrt{x + 1}?$$

60. Let $y_1 = \sqrt[3]{4x - 4}$ and let $y_2 = \sqrt{x + 1}$. Graph $y_3 = y_1 - y_2$ in the window $[-2, 20]$ by $[-0.5, 0.5]$ to determine the number of real solutions of the original equation.

61. Use both an analytic method and your calculator to solve the original equation.

62. Write an explanation of how the solutions of the equation in **Exercise 54** relate to the solutions of the original equation. Discuss any extraneous solutions that may be involved.

*In Exercises 63–70, complete the following. See **Example 8**.*
(a) *Simplify the given expression so that it does not have negative exponents.*
(b) *Set the expression from part (a) equal to 0 and solve the resulting equation.*

63. $\dfrac{\frac{2}{3}(x-2)x^{-1/3} - x^{2/3}}{(x-2)^2}$

64. $\dfrac{\frac{2}{3}(2x-1)x^{-1/3} - 2x^{2/3}}{(2x-1)^2}$

65. $\dfrac{\frac{1}{3}(x^2+1)x^{-2/3} - 2x^{4/3}}{(x^2+1)^2}$

66. $\dfrac{\frac{2}{3}(3x+2)x^{-1/3} - 3x^{2/3}}{(3x+2)^2}$

67. $\dfrac{x^{1/4} - x^{-3/4}}{x}$

68. $\dfrac{x^{-2/3} + x^{1/3}}{x}$

69. $\dfrac{(x^2+1)^{1/2} - \frac{1}{2}x(x^2+1)^{-1/2}(2x)}{x^2+1}$

70. $\dfrac{(5x+3)^{1/2} - \frac{5}{2}x(5x+3)^{-1/2}}{5x+3}$

Solve each equation involving "nested" radicals for all real solutions analytically. Support your solutions with a graph.

71. $\sqrt{\sqrt{x}} = x$

72. $\sqrt[3]{\sqrt[3]{x}} = x$

73. $\sqrt{\sqrt{28x+8}} = \sqrt{3x+2}$

74. $\sqrt{\sqrt{2x+10}} = \sqrt{2x-2}$

75. $\sqrt[3]{\sqrt{32x}} = \sqrt[3]{x+6}$

76. $\sqrt[3]{\sqrt{x+63}} = \sqrt[3]{2x+6}$

(Modeling)   Solve each problem.

77. *Velocity of a Meteorite*   The velocity v of a meteorite approaching Earth is given by

$$v = \frac{k}{\sqrt{d}},$$

measured in kilometers per second, where d is its distance from the center of Earth and k is a constant. If $k = 350$, what is the velocity of a meteorite that is 6000 kilometers away from the center of Earth? Round to the nearest tenth.

78. *Illumination*   The illumination I in foot-candles produced by a light source is related to the distance d in feet from the light source by the equation

$$d = \sqrt{\frac{k}{I}},$$

where k is a constant. If $k = 400$, how far from the source will the illumination be 14 foot-candles? Round to the nearest hundredth of a foot.

79. *Period of a Pendulum*   The period P of a pendulum in seconds depends on its length L in feet and is given by

$$P = 2\pi\sqrt{\frac{L}{32}}.$$

If the length of a pendulum is 5 feet, what is its period? Round to the nearest tenth.

80. *Visibility from an Airplane*   A formula for calculating the distance d one can see from an airplane to the horizon on a clear day is

$$d = 1.22\sqrt{x},$$

where x is the altitude of the plane in feet and d is given in miles. How far can one see to the horizon in a plane flying at each altitude? Give answers to the nearest mile.
(a) 15,000 feet?   **(b)** 20,000 feet?

81. *Speed of a Car in an Accident*   To estimate the speed s at which a car was traveling at the time of an accident, a police officer drives a car like the one involved in the accident under conditions similar to those during which the accident took place and then skids to a stop. If the car is driven at 30 miles per hour, the speed s at the time of the accident is given by

$$s = 30\sqrt{\frac{a}{p}},$$

where a is the length of the skid marks and p is the length of the marks in the police test. Find s if $a = 900$ feet and $p = 97$ feet.

82. *Plant Species and Land Area*   A research biologist has shown that the number S of different plant species on a Galápagos Island is related to the area $\mathscr{A}$ of the island by

$$S = 28.6\sqrt[3]{\mathscr{A}}.$$

Find S for an island with each area.
(a) 100 square miles   **(b)** 1500 square miles

83. *Wire between Two Poles*   Two vertical poles of lengths 12 feet and 16 feet are situated on level ground, 20 feet apart. A piece of wire is to be strung from the top of the 12-foot pole to the top of the 16-foot pole, attached to a stake in the ground at a point P on a line formed by the vertical poles. Let x represent the distance from P to D, as shown in the diagram.

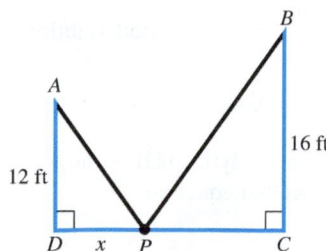

(a) Express the distance from P to C in terms of x.

(continued)

(b) What are the restrictions on the value of x?

(c) Use the Pythagorean theorem to express the lengths AP and BP in terms of x.

(d) Give a function f that expresses the total length of the wire used.

(e) Graph f in the window $[0, 20]$ by $[0, 50]$. Use your calculator to find $f(4)$, and interpret your result.

(f) Find the value of x that will minimize the amount of wire used. What is this minimum?

(g) Write a short paragraph summarizing what this problem has examined and the results you have obtained.

84. *Wire between Two Poles*   Repeat **Exercise 83** if the heights of the poles are 9 feet and 12 feet and the distance between the poles is 16 feet. Let P be x feet from the 9-foot pole. In part (e), use the window $[0, 16]$ by $[0, 50]$.

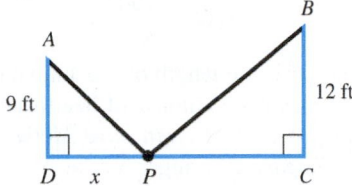

85. *Hunter Returns to His Cabin*   A hunter is at a point on a riverbank. He wants to get to his cabin, located 3 miles north and 8 miles west. He can travel 5 mph along the river but only 2 mph on this very rocky land. How far upriver, to the nearest hundredth of a mile, should he go in order to reach the cabin in a minimum amount of time? (*Hint:* distance = rate × time.)

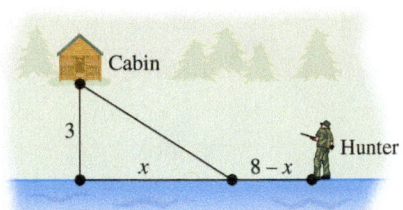

86. *Homing Pigeon Flight*   Homing pigeons avoid flying over large bodies of water, preferring to fly around them instead. (One possible explanation is the fact that extra energy is required to fly over water because air pressure drops over water in the daytime.) Assume that a pigeon released from a boat 1 mile from the shore of a lake (point B in the figure at the top of the next column) flies first to point P on the shore and then along the straight edge of the lake to reach its home at L. If L is 2 miles from point A, the point on the shore closest to the boat, and if a pigeon needs $\frac{4}{3}$ as much energy to fly over water as over land, find the

location, to the nearest hundredth of a mile, of point P that minimizes the pigeon's energy use.

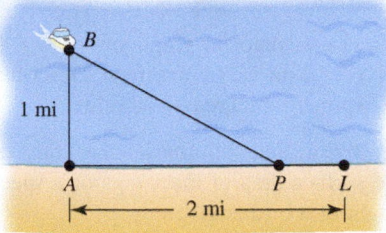

87. *Cruise Ship Travel*   At noon, the cruise ship *Celebration* is 60 miles due south of the cruise ship *Inspiration* and is sailing north at a rate of 30 mph. If the *Inspiration* is sailing west at a rate of 20 mph, find the time at which the distance d between the ships is a minimum. What is this distance to the nearest hundredth of a mile?

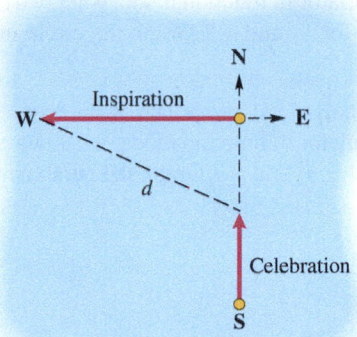

88. *Fisherman Returns to Camp*   "Mido" Simon is in his bass boat, the *Katie*, 3 miles from the nearest point on the shore. He wishes to reach his camp at Maggie Point, 6 miles farther down the shoreline. If Mido's motor is disabled, and he can row his boat at a rate of 4 mph and walk at a rate of 5 mph, find the least amount of time to the nearest hundredth of an hour that he will need to reach the camp.

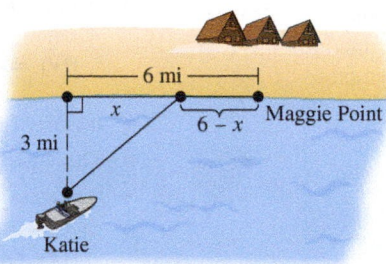

1. Use a graphing calculator to graph $y = x^{0.7}$, $y = x^{1.2}$, and $y = x^{2.4}$ for $x \geq 0$ on the same screen. Describe what happens as the exponent n increases in value.

2. *(Modeling) Wing Size* Suppose that the relationship between the surface area of the wings of a species of bird and the weight of a bird is given by

$$S(w) = 0.3w^{3/4},$$

where w is the weight of the bird in kilograms and S is the surface area of the wings in square meters. If a bird weighs 0.75 kilogram, approximate the surface area of the wings.

3. Sketch a graph of each equation by hand.
   (a) $(x - 3)^2 + (y + 1)^2 = 4$ **(b)** $x = -y^2 + 1$

4. To graph the circle with equation $x^2 + y^2 = 16$ with a graphing calculator in function mode, what two expressions must be used for y_1 and y_2? Graph the circle in a decimal window.

5. To graph the horizontal parabola $x = y^2 + 4y + 6$ with a graphing calculator in function mode, what two expressions must be used for y_1 and y_2? Graph the parabola in a square window.

6. Solve the equation $\sqrt{3x + 4} = 8 - x$ analytically.

7. Use a graph and the solution of the equation in **Exercise 6** to solve the inequality $\sqrt{3x + 4} > 8 - x$.

8. Use a graph and the solution of the equation in **Exercise 6** to solve the inequality $\sqrt{3x + 4} < 8 - x$.

9. Solve the equation $\sqrt{3x + 4} + \sqrt{5x + 6} = 2$ analytically, and support the solution with a graph.

10. *(Modeling) Animal Pulse Rates* An 18th-century study found that the pulse rate of an animal could be approximated by

$$f(x) = \frac{1607}{\sqrt[4]{x^3}}.$$

In this formula, x is the length of the animal in inches and $f(x)$ is the approximate number of heartbeats per minute. (*Source:* Lancaster, H., *Quantitative Methods in Biology and Medical Sciences,* Springer-Verlag.)
   (a) Use f to estimate the pulse rates of a 2-foot cat and a 5.5-foot person.
   (b) What length, to the nearest tenth, corresponds to a pulse rate of 400 beats per minute?

4 Summary

KEY TERMS & SYMBOLS

KEY CONCEPTS

4.1 Rational Functions and Graphs (I)

rational expression
rational function
discontinuous
reciprocal function
vertical asymptote
horizontal asymptote

RATIONAL FUNCTION
A function f of the form

$$f(x) = \frac{p(x)}{q(x)},$$

where $p(x)$ and $q(x)$ are polynomials, with $q(x) \neq 0$, is called a rational function.

$f(x) = \dfrac{1}{x}$ (the reciprocal function) decreases on the intervals $(-\infty, 0)$ and $(0, \infty)$.

- It is discontinuous at $x = 0$.
- The y-axis is a vertical asymptote, and the x-axis is the horizontal asymptote.
- It is an odd function, and its graph is symmetric with respect to the origin.

$f(x) = \dfrac{1}{x^2}$ increases on the interval $(-\infty, 0)$ and decreases on the interval $(0, \infty)$.

- It is discontinuous at $x = 0$.
- The y-axis is a vertical asymptote, and the x-axis is the horizontal asymptote.
- It is an even function, and its graph is symmetric with respect to the y-axis.

KEY TERMS & SYMBOLS	KEY CONCEPTS

4.2 Rational Functions and Graphs (II)

oblique (slant) asymptote
hole
point of discontinuity

GRAPHING A RATIONAL FUNCTION

Let $f(x) = \frac{p(x)}{q(x)}$ be a function with the rational expression in lowest terms. To sketch its graph, follow these steps.

Step 1 Find the domain and all vertical asymptotes.

Step 2 Find any horizontal or oblique asymptote.

Step 3 Find the y-intercept, if possible, by evaluating $f(0)$.

Step 4 Find the x-intercepts, if any, by solving $f(x) = 0$. (These x-values will be the zeros of the numerator $p(x)$.)

Step 5 Determine whether the graph will intersect its nonvertical asymptote by solving $f(x) = b$, where b is the y-value of the horizontal asymptote, or by solving $f(x) = mx + b$, where $y = mx + b$ is the equation of the oblique asymptote.

Step 6 Plot selected points as necessary. Choose an x-value in each interval of the domain determined by the vertical asymptotes and x-intercepts.

Step 7 Complete the sketch.

4.3 Rational Equations, Inequalities, Models, and Applications

rational equation
rational inequality
inverse variation
combined variation
joint variation

SOLVING A RATIONAL EQUATION

Step 1 Determine all values for which the rational equation has undefined expressions.

Step 2 To clear fractions, multiply each side of the equation by the LCD of all rational expressions in the equation.

Step 3 Solve the resulting equation.

Step 4 Reject any values found in Step 1.

SOLVING A RATIONAL INEQUALITY

Step 1 Rewrite the inequality, if necessary, so that 0 is on one side and there is a single fraction on the other side.

Step 2 Determine the values that will cause either the numerator or the denominator of the rational expression to equal 0. These values determine the intervals on the number line to consider.

Step 3 Use a test value from each interval to determine which intervals form the solution set. Be sure to check endpoints.

INVERSE VARIATION WITH THE nTH POWER

Let x and y denote two quantities and n be a positive number. Then y is inversely proportional to the nth power of x, or y varies inversely with the nth power of x, if there exists a nonzero number k such that

$$y = \frac{k}{x^n}.$$

If $y = \frac{k}{x}$, then y is inversely proportional to x, or y varies inversely with x.

JOINT VARIATION

Let m and n be real numbers. Then z varies jointly with the nth power of x and the mth power of y if a nonzero real number k exists such that

$$z = kx^n y^m.$$

(continued)

KEY TERMS & SYMBOLS	KEY CONCEPTS
4.4 Functions Defined by Powers and Roots power function root function	**POWER AND ROOT FUNCTIONS** A function f of the form $$f(x) = x^b,$$ where b is a constant, is a power function. If $b = \frac{1}{n}$ for some integer $n \geq 2$, then f is a root function of the form $$f(x) = x^{1/n}, \quad \text{or equivalently,} \quad f(x) = \sqrt[n]{x}.$$ ● For n even, the root function $f(x) = \sqrt[n]{x}$ increases on $(0, \infty)$ and is continuous on $[0, \infty)$. ● For n odd, the root function $f(x) = \sqrt[n]{x}$ increases on $(-\infty, \infty)$ and is continuous on $(-\infty, \infty)$.
4.5 Equations, Inequalities, and Applications Involving Root Functions proposed solution	**SOLVING AN EQUATION INVOLVING A ROOT FUNCTION** **Step 1** Isolate a term involving a root on one side of the equation. **Step 2** Raise each side of the equation to a positive integer power that will eliminate the radical or rational exponent. **Step 3** Solve the resulting equation. (If a root is still present after Step 2, repeat Steps 1 and 2.) **Step 4** Check each proposed solution in the original equation.

4 Review Exercises

For each rational function, do the following.
(a) *Explain how the graph of the function can be obtained from the graph of $y = \frac{1}{x}$ or $y = \frac{1}{x^2}$.*
(b) *Sketch the graph by hand.*
(c) *Use a graphing calculator to obtain an accurate depiction of the graph.*

1. $y = -\dfrac{1}{x} + 6$

2. $y = \dfrac{4}{x} - 3$

3. $y = -\dfrac{1}{(x-2)^2}$

4. $y = \dfrac{2}{x^2} + 1$

 5. Under what conditions will the graph of a rational function defined by an expression written in lowest terms have an oblique asymptote?

Sketch a graph of each rational function. Your graph should include all asymptotes.

6. $f(x) = \dfrac{4x - 3}{2x - 1}$

7. $f(x) = \dfrac{6x}{(x-1)(x+2)}$

8. $f(x) = \dfrac{2x}{x^2 - 1}$

9. $f(x) = \dfrac{x^2 + 4}{x + 2}$

10. $f(x) = \dfrac{x^2 - 1}{x}$

11. $f(x) = \dfrac{-2}{x^2 + 1}$

12. $f(x) = \dfrac{x^2 - 1}{x + 1}$

13. $f(x) = \dfrac{x^2 - x - 2}{x^2 + 3x + 2}$

Find an equation for each rational function graphed. (Answers may vary.)

14.

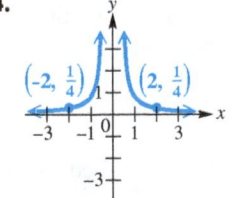

15.

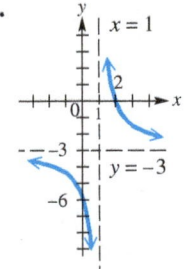

16.

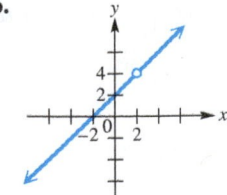

17.

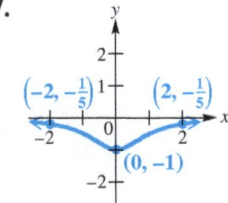

Solve the rational equation in part (a) analytically. Then, use a graph to determine the solution sets of the associated inequalities in parts (b) and (c).

18. (a) $\dfrac{5}{2x + 5} = \dfrac{3}{x + 2}$

(b) $\dfrac{5}{2x + 5} < \dfrac{3}{x + 2}$

(c) $\dfrac{5}{2x + 5} > \dfrac{3}{x + 2}$

19. (a) $\dfrac{3x - 2}{x + 1} = 0$

(b) $\dfrac{3x - 2}{x + 1} < 0$

(c) $\dfrac{3x - 2}{x + 1} > 0$

20. (a) $1 - \dfrac{5}{x} + \dfrac{6}{x^2} = 0$

(b) $1 - \dfrac{5}{x} + \dfrac{6}{x^2} \leq 0$

(c) $1 - \dfrac{5}{x} + \dfrac{6}{x^2} \geq 0$

21. (a) $\dfrac{3}{x - 2} + \dfrac{1}{x + 1} = \dfrac{1}{x^2 - x - 2}$

(b) $\dfrac{3}{x - 2} + \dfrac{1}{x + 1} \leq \dfrac{1}{x^2 - x - 2}$

(c) $\dfrac{3}{x - 2} + \dfrac{1}{x + 1} \geq \dfrac{1}{x^2 - x - 2}$

A comprehensive graph of a rational function f is shown. Use the graph to find each solution set.

22. $f(x) = 0$

23. $f(x) < 0$

24. $f(x) > 0$

25. $|f(x)| > 0$

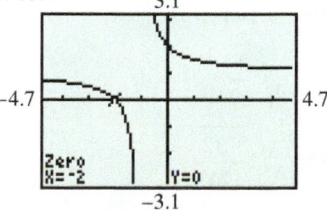

The graph has vertical asymptote $x = -1$.

(Modeling) *Solve each problem.*

26. *Environmental Pollution* In situations involving environmental pollution, a cost–benefit model expresses cost as a function of the percentage of pollutant removed from the environment. Suppose a cost–benefit model is

$$C(x) = \dfrac{6.7x}{100 - x},$$

where $C(x)$ is the cost, in thousands of dollars, of removing x percent of a certain pollutant.

(a) Use a graphing calculator to graph the function in the window $[0, 100]$ by $[0, 150]$.

(b) Find the cost to remove 95% of the pollutant.

27. *Waiting in Line* A parking lot attendant can wait on 40 cars per hour. If cars arrive randomly at a rate of x cars per hour, then the average line length to enter the ramp is given by

$$f(x) = \dfrac{x^2}{1600 - 40x}, \quad \text{where } 0 \leq x < 40.$$

(a) Solve the inequality $f(x) \leq 8$.

(b) Interpret your answer from part (a).

28. *Slippery Roads* The coefficient of friction x measures the friction between the tires of a car and the road, where $0 < x \leq 1$. A smaller value of x indicates that the road is more slippery. If a car is traveling at 60 mph, then braking distance D in feet is given by

$$D(x) = \dfrac{120}{x}.$$

(a) What happens to the braking distance as the coefficient of friction becomes smaller?

(b) Find the interval of values for the coefficient x that corresponds to a braking distance of 400 feet or more.

29. Let y be inversely proportional to x. When $x = 6$, $y = 5$. Find y when $x = 15$.

30. Let z be inversely proportional to the third power of t. When $t = 5$, $z = 0.08$. Find z when $t = 2$.

31. Suppose m varies jointly with n and the square of p, and inversely with q. If $m = 20$ when $n = 5$, $p = 6$, and $q = 18$, find m when $n = 7$, $p = 11$, and $q = 2$.

32. The formula for the height of a right circular cone with volume 100 is

$$h = \dfrac{\frac{300}{\pi}}{r^2}.$$

The height of this cone varies _____ with the _____ of the _____ of its base. The constant of variation is given by _____.

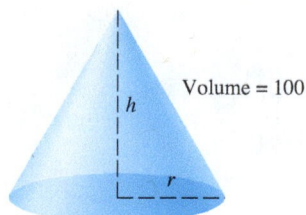

Volume = 100

33. *Illumination* The illumination produced by a light source varies inversely with the square of the distance from the source. The illumination of a light source at 5 meters is 70 candela. Approximate the illumination 12 meters from the source.

34. *Current Flow* In electric current flow, it is found that the resistance (measured in units called ohms) offered by a fixed length of wire of a given material varies inversely with the square of the diameter of the wire. If a wire 0.01 inch in diameter has a resistance of 0.4 ohm, what is the resistance of a wire of the same length and material with a diameter of 0.03 inch?

35. *Simple Interest* Simple interest varies jointly with principal and time. If $1000 left at interest for 2 years earned $110, find the amount of interest earned by $5000 for 5 years.

36. *Car Skidding* The force needed to keep a car from skidding on a curve varies inversely with the radius of the curve and jointly with the weight of the car and the square of the speed. It takes 3000 pounds of force to keep a 2000-pound car from skidding on a curve of radius 500 feet at 30 mph. What force is needed to keep the same car from skidding on a curve of radius 800 feet at 60 mph?

37. *Sports Arena Construction* The roof of a new sports arena rests on round concrete pillars. The maximum load a cylindrical column of circular cross section can hold varies directly with the fourth power of the diameter and inversely with the square of the height. The arena has 9-meter-tall columns that are 1 meter in diameter and will support a load of 8 metric tons. How many metric tons will be supported by a column 12 meters high and $\frac{2}{3}$ meter in diameter?

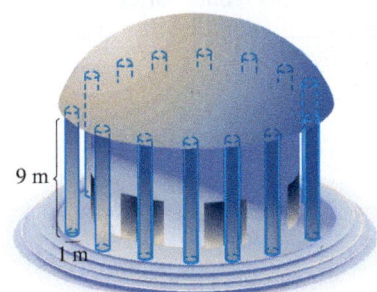

Load = 8 metric tons

38. *Sports Arena Construction* A sports arena requires a beam 16 meters long, 24 centimeters wide, and 8 centimeters high. The maximum load of a horizontal beam that is supported at both ends varies directly with the width and square of the height of the beam and inversely with the length between supports. If a beam of the same material 8 meters long, 12 centimeters wide, and 15 centimeters high can support a maximum of 400 kilograms, what is the maximum load the beam in the arena will support?

39. *Weight of an Object* The weight w of an object varies inversely with the square of the distance d between the object and the center of Earth. If a man weighs 90 kilograms on the surface of Earth, how much would he weigh 800 kilometers above the surface? (The radius of Earth is about 6400 kilometers.)

40. *Rate of Work* Louise and Keith are working on a building site cleanup. Keith can clean up all the trash in the area in 7 hours, while Louise can do the same job in 5 hours. How long will it take them if they work together?

41. *Rate of Work* Suppose that Terry and Carrie can clean their entire house in 7 hours, while Daniel, just by being around, can completely mess it up in only 2 hours. If Terry and Carrie clean the house while Daniel is at his grandma's, and then start cleaning up after Daniel the minute he gets home, how long does it take from the time Daniel gets home until the whole place resembles a disaster area?

42. *Rate of Work* Suppose it takes Jack 35 minutes to do the dishes, and together he and his partner Kevin can do them in 15 minutes. How long will it take Kevin to do the dishes alone?

Suppose that a and b are positive numbers. Sketch the general shape of the graph of each function.

43. $y = \sqrt[3]{x} + a$

44. $y = -a\sqrt{x}$

45. $y = \sqrt[3]{x} + b$

46. $y = \sqrt[3]{x} - a$

47. $y = -a\sqrt[3]{x} - b$

48. $y = \sqrt{x + a} + b$

Evaluate each expression without the use of a calculator. Then, check your results with a calculator.

49. $\sqrt{144}$

50. $\sqrt[3]{-64}$

51. $\sqrt[3]{\dfrac{1}{27}}$

52. $\sqrt[4]{81}$

53. $\sqrt[5]{-\dfrac{32}{243}}$

54. $-(-32)^{1/5}$

55. $36^{-3/2}$

56. $-1000^{2/3}$

57. $(-27)^{-4/3}$

58. $16^{3/4}$

Use your calculator to find each root or power. Give as many digits as your display shows.

59. $\sqrt[5]{84.6}$

60. $\sqrt[4]{\dfrac{1}{16}}$

61. $\left(\dfrac{1}{8}\right)^{4/3}$

62. $12^{1/3}$

Consider $f(x) = -\sqrt{2x - 4}$ for Exercises 63–65.

63. Find the domain of f analytically.

64. Use a graph to determine the range of f.

65. Give the largest open interval (if any) over which f is
(a) increasing.   **(b)** decreasing.

66. Determine the expressions for y_1 and y_2 that are required to graph the circle $x^2 + y^2 = 25$ with a graphing calculator in function mode. Then graph the circle in a square window.

Solve each equation involving radicals or rational exponents in part (a) analytically. Then solve parts (b) and (c) graphically.

67. (a) $\sqrt{5 + 2x} = x + 1$
   (b) $\sqrt{5 + 2x} > x + 1$
   (c) $\sqrt{5 + 2x} < x + 1$

68. (a) $\sqrt{2x + 1} - \sqrt{x} = 1$
   (b) $\sqrt{2x + 1} - \sqrt{x} > 1$
   (c) $\sqrt{2x + 1} - \sqrt{x} < 1$

69. (a) $\sqrt[3]{6x + 2} = \sqrt[3]{4x}$
   (b) $\sqrt[3]{6x + 2} \geq \sqrt[3]{4x}$
   (c) $\sqrt[3]{6x + 2} \leq \sqrt[3]{4x}$

70. (a) $(x - 2)^{2/3} - x^{1/3} = 0$
   (b) $(x - 2)^{2/3} - x^{1/3} \geq 0$
   (c) $(x - 2)^{2/3} - x^{1/3} \leq 0$

Solve each equation. Check your results.

71. $x^5 = 1024$ **72.** $x^{1/3} = 4$

73. $\sqrt{x - 2} = x - 4$ **74.** $x^{3/2} = 27$

75. $2x^{1/4} + 3 = 6$ **76.** $\sqrt{x - 2} = 14 - x$

77. $\sqrt[3]{2x - 3} + 1 = 4$ **78.** $x^{1/3} + 3x^{1/3} = -2$

79. $2x^{-2} - 5x^{-1} = 3$ **80.** $x^{-3} + 2x^{-2} + x^{-1} = 0$

81. $x^{2/3} - 4x^{1/3} - 5 = 0$ **82.** $x^{3/4} - 16x^{1/4} = 0$

83. $\sqrt{x + 1} + 1 = \sqrt{2x}$ **84.** $\sqrt{x - 2} = 5 - \sqrt{x + 3}$

(Modeling)   *Solve each problem.*

85. *Swing of a Pendulum*   A simple pendulum swings back and forth in regular time intervals (periods). Grandfather

clocks use pendulums to keep accurate time. The relationship between the length L of a pendulum and the time T for one complete oscillation can be expressed by the equation $L = kT^n$, where k is a constant and n is a positive integer to be determined. The following data were taken for different lengths of pendulums.

T (in seconds)	L (in feet)
1.11	1.0
1.36	1.5
1.57	2.0
1.76	2.5
1.92	3.0
2.08	3.5
2.22	4.0

(a) As L increases, what happens to T?
(b) Discuss how n and k can be found.
(c) Use the data to approximate k, and determine the best value for n.
(d) Using the values of k and n from part (c), predict T for a pendulum having a length of 5 feet.
(e) If the length L of a pendulum doubles, what happens to the period T?

86. *Volume of a Cylindrical Package*   A company plans to package its product in a cylinder that is open at one end. The cylinder is to have a volume of 27π cubic inches. What radius should the circular bottom of the cylinder have to minimize the cost of the material? (*Hint:* The volume V of a circular cylinder is $V = \pi r^2 h$, where r is the radius of the circular base and h is the height; the surface area S of a cylinder open at one end is $S = 2\pi rh + \pi r^2$.)

4   Test

For each rational function, do the following.
(a) *Sketch its graph.*
(b) *Explain how the graph is obtained from the graph of $y = \frac{1}{x}$ or $y = \frac{1}{x^2}$.*
(c) *Use a graphing calculator to obtain an accurate depiction of the graph.*

1. $f(x) = -\dfrac{1}{x}$ **2.** $f(x) = -\dfrac{1}{x^2} - 3$

3. Consider the rational function
$$f(x) = \frac{x^2 + x - 6}{x^2 - 3x - 4}.$$

Determine the answers to (a)–(f) analytically. Find
(a) the domain of f.
(b) equations of the vertical asymptotes.
(c) the equation of the horizontal asymptote.

(d) the y-intercept.
(e) x-intercepts, if any.
(f) the coordinates of the point where the graph of f intersects its horizontal asymptote.
(g) Now sketch a comprehensive graph of f.

4. Find the equation of the oblique asymptote of the graph of the rational function
$$f(x) = \frac{2x^2 + x - 3}{x - 2}.$$

Then graph the function and its asymptotes.

5. Consider the rational function
$$f(x) = \frac{x^2 - 16}{x + 4}.$$

(a) For what value of x does the graph exhibit a hole?
(b) Graph the function f and show the hole in the graph.

6. (a) Solve the following rational equation analytically.
$$\frac{3}{x-2} + \frac{21}{x^2-4} = \frac{14}{x+2}$$

(b) Use the result of part (a) and a graph to find the solution set of
$$\frac{3}{x-2} + \frac{21}{x^2-4} \geq \frac{14}{x+2}.$$

7. *(Modeling) Waiting in Line* A small section of highway that is under construction can accommodate at most 40 cars per minute. If cars arrive randomly at an average rate of x vehicles per minute, then the average wait W in minutes for a car to pass through this section of highway is approximated by
$$W(x) = \frac{1}{40-x}, \quad \text{where } 0 \leq x < 40.$$

(a) Evaluate $W(30)$, $W(39)$, and $W(39.9)$. Interpret the results.

(b) Graph W, using the window $[0, 40]$ by $[-0.5, 1]$. Identify the vertical asymptote. What happens to W as x approaches 40?

(c) Find x when the wait is 5 minutes.

8. *(Modeling) Measure of Malnutrition* The *pelidisi*, a measure of malnutrition, varies directly with the cube root of a person's weight in grams and inversely with the person's sitting height in centimeters. A person with a pelidisi below 100 is considered to be undernourished, while a pelidisi greater than 100 indicates overfeeding. A person who weighs 48,820 grams and has a sitting height of 78.7 centimeters has a pelidisi of 100. Find the pelidisi (to the nearest whole number) of a person whose weight is 54,430 grams and whose sitting height is 88.9 centimeters. Is this individual undernourished or overfed?

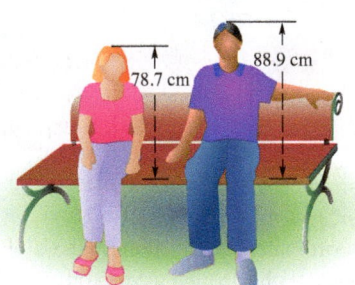

Weight: 48,820 g   Weight: 54,430 g

9. *(Modeling) Volume of a Cylindrical Can* A manufacturer wants to construct a cylindrical aluminum can with a volume of 4000 cubic centimeters. If x represents the radius of the circular top and bottom of the can, the surface area S as a function of x is given by
$$S(x) = \frac{8000 + 2\pi x^3}{x}.$$

Use a graph to find the radius that will minimize the amount of aluminum needed, and determine what this amount will be. (*Hint:* Use the window $[-5, 42]$ by $[-1000, 10,000]$.)

10. Evaluate each expression without a calculator.
(a) $\sqrt[3]{-27}$ **(b)** $25^{-3/2}$

11. Use positive rational exponents to rewrite $\left(\sqrt[3]{x}\right)^{-4}$.

12. Graph each equation.
(a) $x = y^2 + 2$ **(b)** $(x-1)^2 + (y+3)^2 = 4$

13. Graph the function $f(x) = -\sqrt{5-x}$ in the standard viewing window. Then, do the following.
(a) Determine the domain analytically.
(b) Use the graph to find the range.
(c) Give the largest open interval over which this function increases.
(d) Solve the equation $f(x) = 0$ graphically.
(e) Solve the inequality $f(x) < 0$ graphically.

14. (a) Solve the equation $\sqrt{4-x} = x + 2$ analytically, and support the solution(s) with a graph.
(b) Use the graph in part (a) to find the solution set of $\sqrt{4-x} > x + 2$.
(c) Use the graph in part (a) to find the solution set of $\sqrt{4-x} \leq x + 2$.

15. *(Modeling) Laying a Telephone Cable* A telephone company wishes to minimize the cost of laying a cable from point P to point R. Points P and Q are directly opposite each other along the banks of a straight river 300 yards wide. Point R lies on the same side of the river as point Q, but 600 yards away. If the cost per yard for the cable is \$125 per yard under the water and \$100 per yard on land, how should the company lay the cable to minimize the cost?

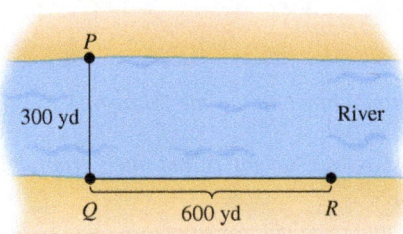

16. Suppose z varies directly with the square of x and inversely with y. If $z = 10$ when $x = 4$ and $y = 2$, find z when $x = 6$ and $y = 8$.

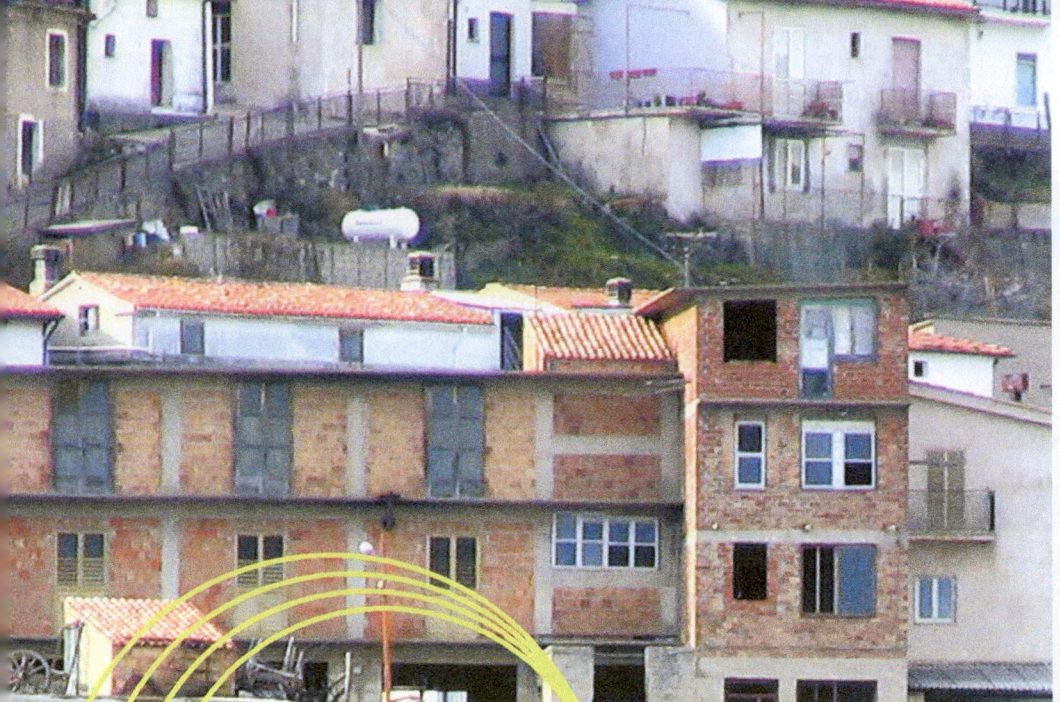

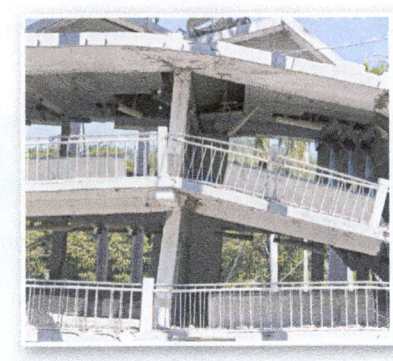

Exponential and logarithmic functions are important for our understanding of tsunamis, earthquakes, population growth, radioactivity, and financial planning.

5 Inverse, Exponential, and Logarithmic Functions

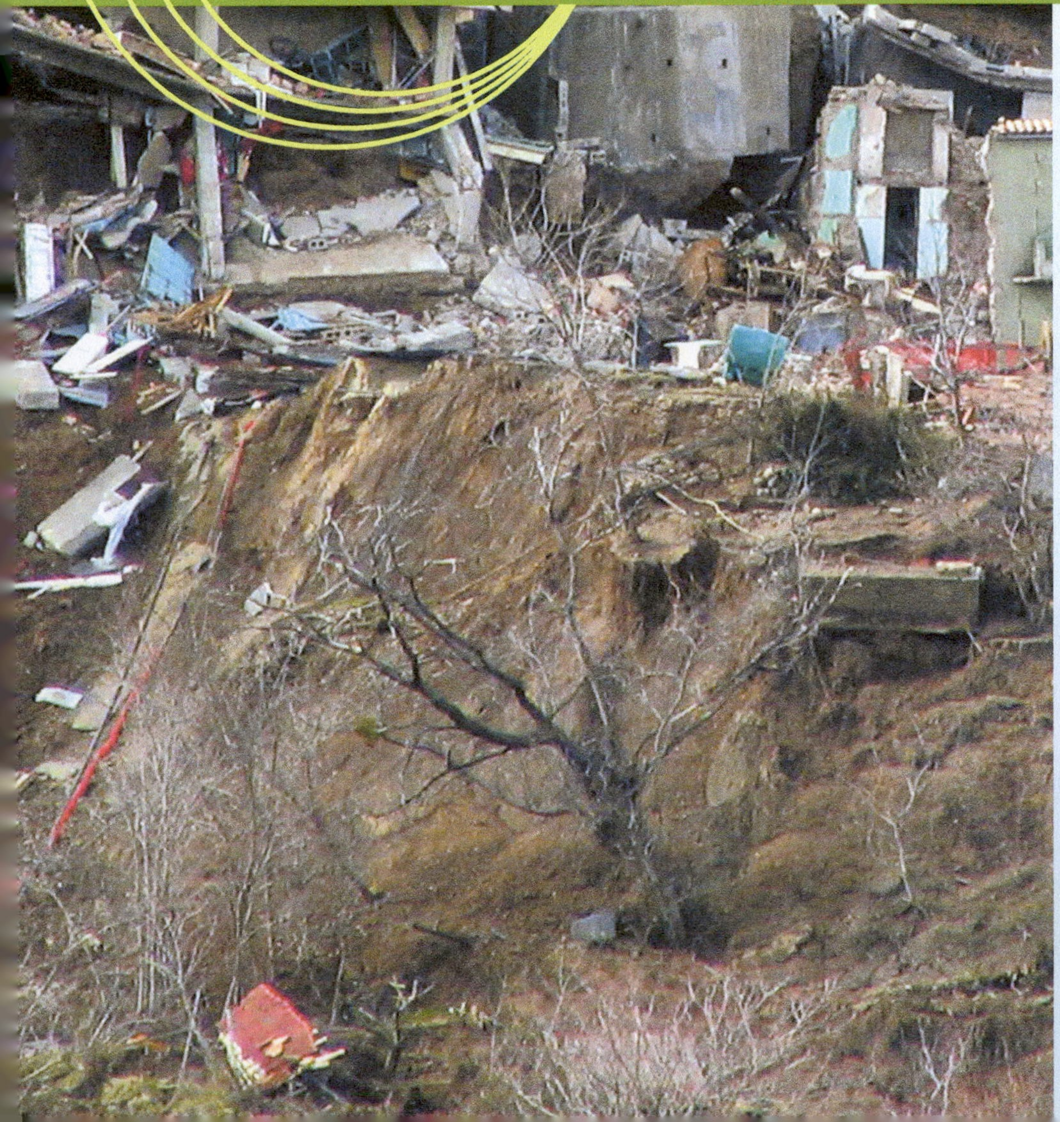

CHAPTER OUTLINE

5.1 Inverse Functions

Inverse Operations • One-to-One Functions • Inverse Functions and Their Graphs • Equations of Inverse Functions • An Application of Inverse Functions to Cryptography

Inverse Operations

Addition and subtraction are *inverse operations*, as are multiplication and division. Starting with a number x, multiplying by 8, and then dividing by 8 gives x back as the result. The functions

Multiply by 8. $\qquad f(x) = 8x \quad$ and $\quad g(x) = \dfrac{x}{8}$ *Divide by 8.*

are *inverses* of each other with respect to *function composition*. This means that if a value of x such as $x = 12$ is chosen, then

$$f(12) = 8 \cdot 12 = 96 \quad \text{and} \quad g(96) = \frac{96}{8} = 12.$$

Thus, $(g \circ f)(12) = g(f(12)) = 12$. Also, $(f \circ g)(12) = f(g(12)) = 12$ and for any value of x,

$$(f \circ g)(x) = f(g(x)) = x \quad \text{and} \quad (g \circ f)(x) = g(f(x)) = x.$$

One-to-One Functions

For the function $y = f(x) = 5x - 8$, any two different values of x produce two different values of y. However, for the function $y = g(x) = x^2$, two different values of x can lead to the *same* value of y. For example, both $x = 4$ and $x = -4$ give $y = 4^2 = (-4)^2 = 16$. A function such as $y = f(x) = 5x - 8$, for which different elements from the domain always lead to different elements from the range, is called a *one-to-one function*.

One-to-One Function

A function f is a **one-to-one function** if, for elements a and b from the domain of f,

$$a \neq b \quad \text{implies} \quad f(a) \neq f(b).$$

In part (a) of the next example we use the above statement to show that a function is *not* a one-to-one function. In part (b) we use the *contrapositive* of the above statement,

$$f(a) = f(b) \quad \text{implies} \quad a = b,$$

to show that a function is a one-to-one function. The two statements are equivalent.

EXAMPLE 1 **Deciding whether Functions Are One-to-One**

Decide whether each function is one-to-one.

(a) $f(x) = \sqrt{25 - x^2}$ **(b)** $f(x) = -4x + 12$

Solution

(a) If we choose $a = 3$ and $b = -3$, then $3 \neq -3$, but

> *$a \neq b$ leads to $f(a) = f(b)$, so f cannot be one-to-one.*

$$f(3) = \sqrt{25 - 3^2} = \sqrt{25 - 9} = \sqrt{16} = 4$$

and $f(-3) = \sqrt{25 - (-3)^2} = \sqrt{25 - 9} = 4.$

Even though $3 \neq -3$, $f(3) = f(-3) = 4$. This is *not* a one-to-one function.

(b) $f(a) = f(b)$

> *$f(a) = f(b)$ leads to $a = b$, so f is one-to-one.*

$-4a + 12 = -4b + 12$	$f(x) = -4x + 12$
$-4a = -4b$	Subtract 12.
$a = b$	Divide by -4.

By the definition, $f(x) = -4x + 12$ is one-to-one.   ●

As shown in **Example 1(a),** a way to demonstrate that a function is *not* one-to-one is to produce a pair of unequal numbers that lead to the same function value. The **horizontal line test** also tells us whether a function is one-to-one. See **FIGURE 1.**

Not One-to-One

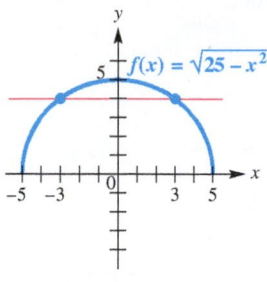

The graph of the function in **Example 1(a)** is a semicircle. The horizontal line test shows that the function is not one-to-one.

FIGURE 1

Horizontal Line Test

A function is one-to-one if every horizontal line intersects the graph of the function at most once.

> **EXAMPLE 2**   **Using the Horizontal Line Test**

Determine whether each graph is the graph of a one-to-one function.

(a) 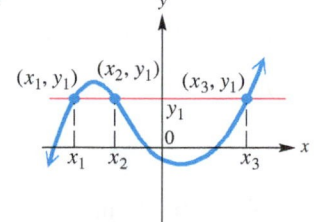   (b)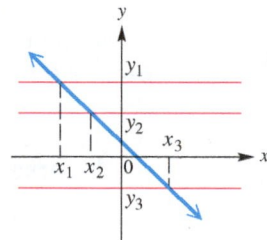

Solution

(a) Each point where the horizontal line $y = y_1$ intersects the graph has the same value of y but a different value of x. Since more than one different value of x (here, $x_1, x_2,$ and x_3) leads to the same value of y, the function is *not* one-to-one.

(b) Every horizontal line will intersect the graph at exactly one point. This function is one-to-one.   ●

> **NOTE** A function that is either always increasing or always decreasing on its domain, such as $f(x) = x^3$ or $g(x) = \sqrt{x}$, must be one-to-one.

FOR DISCUSSION

Based on your knowledge of the basic functions studied so far in this text, answer each question. In each case, assume that the function has the largest possible domain.

1. Is a nonconstant linear function always one-to-one?
2. Is an odd-degree polynomial function always one-to-one? Why or why not?
3. Is an even-degree polynomial function ever one-to-one? Why or why not?

Inverse Functions and Their Graphs

Certain pairs of one-to-one functions "undo" one another. For example, if

$$f(x) = 8x + 5 \quad \text{and} \quad g(x) = \frac{x - 5}{8},$$

then $\qquad f(10) = 8 \cdot 10 + 5 = 85 \quad \text{and} \quad g(85) = \frac{85 - 5}{8} = 10.$

That is, if we multiply 10 by 8 and add 5 to get 85, then the inverse operations are to subtract 5 from 85 and divide by 8 to get 10 back. See **FIGURE 2.**

Composition of Functions

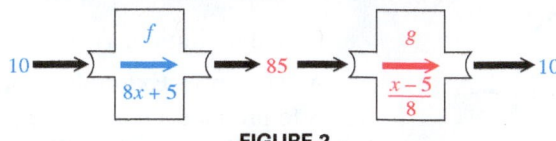

FIGURE 2

Similarly, for these same functions, check that

$$f(3) = 29 \quad \text{and} \quad g(29) = 3,$$
$$f(-5) = -35 \quad \text{and} \quad g(-35) = -5,$$
$$g(2) = -\frac{3}{8} \quad \text{and} \quad f\left(-\frac{3}{8}\right) = 2.$$

In general, if $f(a) = b$, then $g(b) = a$. In fact for *any* value of x,

$$(f \circ g)(x) = x \quad \text{and} \quad (g \circ f)(x) = x.$$

Because of this property, g is called the *inverse function* of f. **A function f has an inverse if and only if f is one-to-one.**

Inverse Function

Let f be a one-to-one function. Then g is the **inverse function** of f and f is the inverse function of g if

$$(f \circ g)(x) = x \quad \text{for every } x \text{ in the domain of } g,$$

and $\qquad (g \circ f)(x) = x \quad \text{for every } x \text{ in the domain of } f.$

EXAMPLE 3 **Deciding whether Two Functions Are Inverses**

Let $f(x) = x^3 - 1$ and $g(x) = \sqrt[3]{x + 1}$. Is g the inverse function of f?

Solution As shown in **FIGURE 3,** the horizontal line test applied to the graph indicates that f is one-to-one, so it does have an inverse. Now find $(f \circ g)(x)$ and $(g \circ f)(x)$.

One-to-One

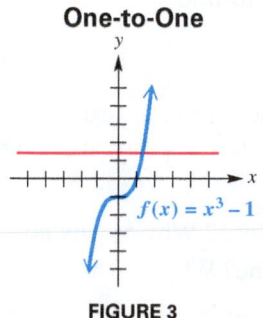

FIGURE 3

$$\begin{aligned}(f \circ g)(x) &= f(g(x)) \\ &= \left(\sqrt[3]{x + 1}\right)^3 - 1 \\ &= x + 1 - 1 \\ &= x \ \checkmark \end{aligned} \qquad \begin{aligned}(g \circ f)(x) &= g(f(x)) \\ &= \sqrt[3]{(x^3 - 1) + 1} \\ &= \sqrt[3]{x^3} \\ &= x \ \checkmark \end{aligned}$$

Since $(f \circ g)(x) = x$ and $(g \circ f)(x) = x$, function g is the inverse of function f. Also, note that f is the inverse of g.

A special notation is often used for inverse functions. If g is the inverse function of f, then g can be written as f^{-1} (read **"f-inverse"**). In **Example 3,**

$$f(x) = x^3 - 1 \quad \text{has inverse} \quad f^{-1}(x) = \sqrt[3]{x + 1}. \text{ Also,}$$

$$g(x) = \sqrt[3]{x + 1} \quad \text{has inverse} \quad g^{-1}(x) = x^3 - 1.$$

CAUTION   *Do not confuse the* -1 *in* f^{-1} *with a negative exponent.* The symbol $f^{-1}(x)$ does not represent $\frac{1}{f(x)}$. It represents the inverse function of f.

By the definition of an inverse function, the domain of f equals the range of f^{-1}, and the range of f equals the domain of f^{-1}. See **FIGURE 4.**

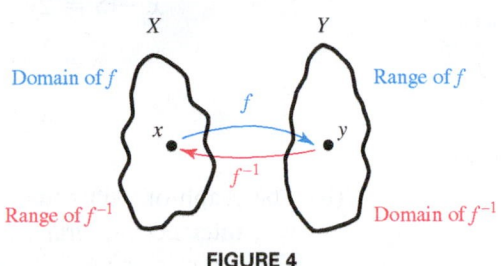

FIGURE 4

EXAMPLE 4   **Finding Inverses of One-to-One Functions**

Determine whether each function is one-to-one. If so, find its inverse.

(a) $f = \{(-2, 1), (-1, 0), (0, 1), (1, 2), (2, 2)\}$

(b) $g = \{(3, 1), (0, 2), (2, 3), (4, 0)\}$

Solution

(a) Each x-value in f corresponds to just one y-value. However, the y-value 2 corresponds to two x-values: 1 and 2. Also, the y-value 1 corresponds to both -2 and 0. Because some y-values correspond to more than one x-value, f is not one-to-one and does not have an inverse.

(b) Every x-value in g corresponds to only one y-value, and every y-value corresponds to only one x-value, so g is a one-to-one function. The inverse function is found by interchanging the x- and y-values in each ordered pair.

$$g^{-1} = \{(1, 3), (2, 0), (3, 2), (0, 4)\}$$

The domain and range of g become the range and domain, respectively, of g^{-1}.

Equations of Inverse Functions

Finding the Equation of the Inverse of $y = f(x)$

For a one-to-one function f defined by an equation $y = f(x)$, find the defining equation of the inverse as follows. (You may need to replace $f(x)$ with y first. Any restrictions on x and y should be considered.)

Step 1   Interchange x and y.

Step 2   Solve for y.

Step 3   Replace y with $f^{-1}(x)$.

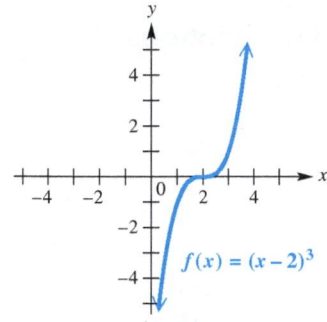

Using the method of **Example 5**, we see that the inverse of $Y_1 = (4X + 6)/5$ is $Y_2 = (5X - 6)/4$. In the tables of selected points for these functions, notice that the X- and Y-values are interchanged. This is typical for inverse functions.

EXAMPLE 5 **Finding Equations of Inverses**

Decide whether $f(x)$ is a one-to-one function. If so, find $f^{-1}(x)$.

(a) $f(x) = 2x + 5$ **(b)** $f(x) = x^2 + 2$ **(c)** $f(x) = (x - 2)^3$

Solution

(a) The graph of $f(x) = 2x + 5$ is a nonhorizontal line, so by the horizontal line test, f is a one-to-one function. Follow the steps in the preceding box.

$$f(x) = 2x + 5$$
$$y = 2x + 5 \qquad \text{Replace } f(x) \text{ with } y.$$
$$x = 2y + 5 \qquad \text{Interchange } x \text{ and } y. \text{ (Step 1)}$$
$$x - 5 = 2y \qquad \text{Subtract 5.}$$
$$y = \frac{x - 5}{2} \qquad \text{Divide by 2; rewrite.} \quad \left. \begin{array}{c} \text{Solve for } y. \\ \text{(Step 2)} \end{array} \right.$$
$$f^{-1}(x) = \frac{x - 5}{2} \qquad \text{Replace } y \text{ with } f^{-1}(x). \text{ (Step 3)}$$

(b) The graph of $f(x) = x^2 + 2$ is a parabola opening up, so some horizontal lines will intersect the graph at two points. This graph fails the horizontal line test. Thus, $f(x) = x^2 + 2$ is not one-to-one and does not have an inverse.

(c) From **FIGURE 5,** the graph of $f(x) = (x - 2)^3$ passes the horizontal line test and is one-to-one.

$$y = (x - 2)^3 \qquad \text{Replace } f(x) \text{ with } y.$$
$$x = (y - 2)^3 \qquad \text{Interchange } x \text{ and } y.$$
$$\sqrt[3]{x} = \sqrt[3]{(y - 2)^3} \qquad \text{Take the cube root of each side.}$$
$$\sqrt[3]{x} = y - 2 \qquad \text{Simplify on the right.}$$
$$\sqrt[3]{x} + 2 = y \qquad \text{Solve for } y \text{ by adding 2.}$$
$$f^{-1}(x) = \sqrt[3]{x} + 2 \qquad \text{Rewrite and replace } y \text{ with } f^{-1}(x).$$

FIGURE 5

Suppose f and f^{-1} are inverse functions and $f(a) = b$ for real numbers a and b. Then, by the definition of inverse function, $f^{-1}(b) = a$. This means that if a point (a, b) is on the graph of f, then (b, a) will belong to the graph of f^{-1}. As shown in **FIGURE 6,** the points (a, b) and (b, a) are reflections of one another across the line $y = x$.

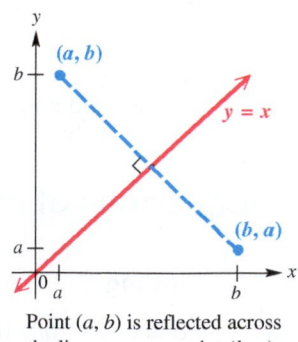

Point (a, b) is reflected across the line $y = x$ as point (b, a).

FIGURE 6

Geometric Relationship between the Graphs of f and f^{-1}

If a function f is one-to-one, then the graph of its inverse f^{-1} is a reflection of the graph of f across the line $y = x$.

FIGURE 7 illustrates this idea for each function f (in blue) and its inverse f^{-1} (in red).

Reflections across $y = x$

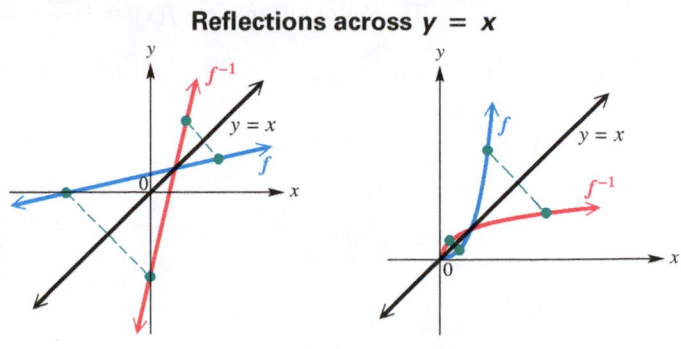

FIGURE 7

EXAMPLE 6   **Graphing an Inverse Function**

Use the graph of $y = f(x)$ to graph $y = f^{-1}(x)$.

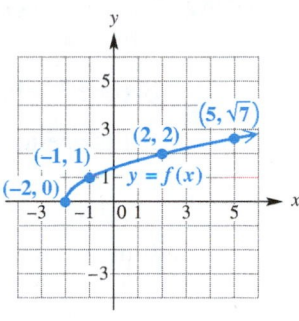

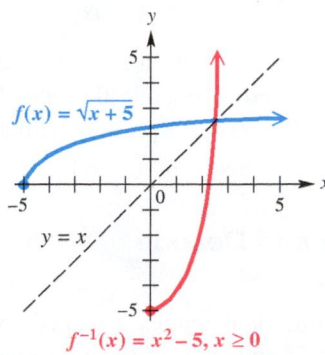

FIGURE 8

Solution   If the point (a, b) is on the graph of the function $y = f(x)$, then the point (b, a) is on the graph of $y = f^{-1}(x)$. Thus, $(-2, 0)$ is on the graph of $y = f(x)$, so $(0, -2)$ is on the graph of $y = f^{-1}(x)$. Similarly, $(1, -1)$, $(2, 2)$, and $(\sqrt{7}, 5)$, where $\sqrt{7} \approx 2.6$, are also on the graph of $y = f^{-1}(x)$. These points are plotted in **FIGURE 8** and a curve is sketched. The inverse function is a reflection across the line $y = x$. ●

EXAMPLE 7   **Finding the Inverse of a Function with a Restricted Domain**

Let $f(x) = \sqrt{x + 5}$. Find $f^{-1}(x)$.

Solution   The domain of f is restricted to the interval $[-5, \infty)$. Since f is always increasing, as shown in **FIGURE 9**, f is one-to-one and thus has an inverse function.

$$y = \sqrt{x + 5}, \quad x \geq -5 \qquad \text{Let } y = f(x).$$
$$x = \sqrt{y + 5}, \quad y \geq -5 \qquad \text{Interchange } x \text{ and } y.$$
$$x^2 = y + 5 \qquad \text{Square both sides.}$$
$$y = x^2 - 5 \qquad \text{Solve for } y.$$

We cannot simply give $f^{-1}(x)$ as $x^2 - 5$. The domain of f is $[-5, \infty)$ and its range is $[0, \infty)$. The range of f is the domain of f^{-1}, so f^{-1} must be

$$f^{-1}(x) = x^2 - 5, \quad x \geq 0.$$

The range of f^{-1}, $[-5, \infty)$, is the domain of f. Graphs of f and f^{-1} are shown in **FIGURE 9.** The line $y = x$ is included to show that the graphs of f and f^{-1} are reflections of one another—that is, mirror images with respect to this line. ●

The domain of f is $[-5, \infty)$ and the range of f is $[0, \infty)$. The domain of f^{-1} is $[0, \infty)$ and the range of f^{-1} is $[-5, \infty)$.

FIGURE 9

EXAMPLE 8 **Finding the Inverse of a Rational Function**

The rational function $f(x) = \dfrac{2x}{x-1}$ is one-to-one. Find $f^{-1}(x)$.

Solution

$$y = \frac{2x}{x-1} \qquad \text{Let } y = f(x).$$

$$x = \frac{2y}{y-1} \qquad \text{Interchange } x \text{ and } y.$$

$$x(y-1) = 2y \qquad \text{Multiply by } y-1.$$

$$xy - x = 2y \qquad \text{Distributive property}$$

Move terms with y to one side and factor to isolate y.

$$xy - 2y = x \qquad \text{Add } x \text{ and subtract } 2y.$$

$$y(x-2) = x \qquad \text{Factor out } y.$$

$$y = \frac{x}{x-2} \qquad \text{Divide by } x-2.$$

Thus $f^{-1}(x) = \dfrac{x}{x-2}$.

TECHNOLOGY NOTE

When graphing inverse functions on a calculator, a square window sometimes work best.

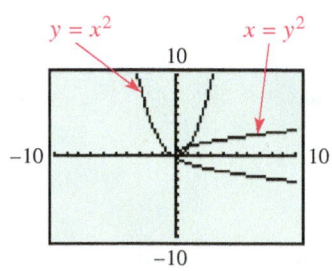

Despite the fact that $y = x^2$ is not one-to-one, the calculator will draw its "inverse," $x = y^2$.

FIGURE 10

> **Important Facts about Inverses**
>
> 1. If f is one-to-one, then f^{-1} exists.
> 2. The domain of f is equal to the range of f^{-1}, and the range of f is equal to the domain of f^{-1}.
> 3. If the point (a, b) lies on the graph of f, then (b, a) lies on the graph of f^{-1}. The graphs of f and f^{-1} are reflections of each other across the line $y = x$.

CAUTION The TI-84 Plus has the capability of "drawing the inverse" of a function. This feature does not require that the function be one-to-one. However, the resulting figure may not be the graph of a function. See **FIGURE 10.** *Again, it is necessary to understand the mathematics to interpret results correctly.*

An Application of Inverse Functions to Cryptography

A one-to-one function and its inverse can be used to make information secure. The function is used to encode a message, and its inverse is used to decode the coded message. In practice, complicated functions are used. We illustrate the process with a simple function in **Example 9.**

A	1	N	14
B	2	O	15
C	3	P	16
D	4	Q	17
E	5	R	18
F	6	S	19
G	7	T	20
H	8	U	21
I	9	V	22
J	10	W	23
K	11	X	24
L	12	Y	25
M	13	Z	26

EXAMPLE 9 **Using Functions to Encode and Decode a Message**

Use the one-to-one function $f(x) = 3x + 1$ and the numerical values assigned to each letter of the alphabet shown in the margin to encode and decode the message PLAY MY MUSIC.

Solution The message PLAY MY MUSIC would be encoded as

49	37	4	76	40	76	40	64	58	28	10

because

$$P \text{ corresponds to } 16 \quad \text{and} \quad f(16) = 3(16) + 1 = 49,$$

$$L \text{ corresponds to } 12 \quad \text{and} \quad f(12) = 3(12) + 1 = 37,$$

and so on. Using the method of earlier examples, we find that $f^{-1}(x) = \dfrac{x-1}{3}$. Using f^{-1} to decode yields

$$f^{-1}(49) = \frac{49-1}{3} = 16, \quad \text{which corresponds to } P,$$

$$f^{-1}(37) = \frac{37-1}{3} = 12, \quad \text{which corresponds to } L,$$

and so on. The table feature of the TI-84 Plus can be useful in this procedure. **FIGURE 11** shows how decoding can be accomplished for the words PLAY MY.

FIGURE 11

5.1 Exercises

Checking Analytic Skills   *Decide whether each function is one-to-one.* **Do not use a calculator.**

1. $f(x) = -3x + 5$ **2.** $f(x) = -5x + 2$ **3.** $f(x) = x^2$

4. $f(x) = -x^2$ **5.** $f(x) = \sqrt{36 - x^2}$ **6.** $f(x) = -\sqrt{100 - x^2}$

7. $f(x) = x^3$ **8.** $f(x) = \sqrt[3]{x}$ **9.** $f(x) = |2x + 1|$

10. *Concept Check*   Can a quadratic function f with domain $(-\infty, \infty)$ have an inverse function? Explain.

Decide whether each function graphed or defined is one-to-one.

11.

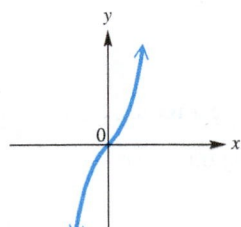

12.

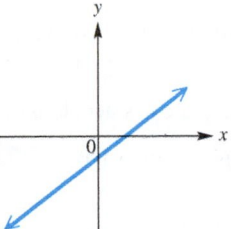

13.

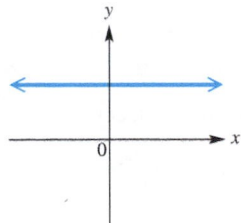

14.

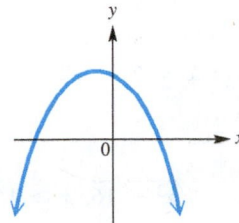

15.

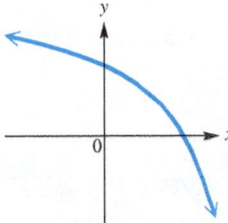

16.

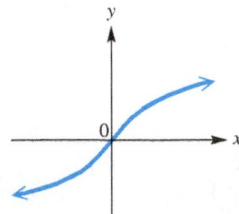

17. $y = (x - 2)^2$ **18.** $y = -(x + 3)^2 - 8$ **19.** $y = 2x^3 + 1$

20. $y = -2x^5 - 4$ **21.** $y = -\sqrt[3]{x + 5}$ **22.** $y = \dfrac{1}{x + 2}$

23. $y = \dfrac{-4}{x - 8}$ **24.** $f(x) = -7$ **25.** $f(x) = \begin{cases} 3 & \text{if } x \geq 0 \\ -x & \text{if } x < 0 \end{cases}$

26. (a) Explain why a polynomial function of even degree with domain $(-\infty, \infty)$ cannot be one-to-one.

(b) Explain why in some cases a polynomial function of odd degree with domain $(-\infty, \infty)$ is not one-to-one.

Concept Check   *Answer the following.*

27. For a function to have an inverse, it must be _____.

28. For f to be one-to-one, if $a \neq b$, then _____.

29. If f and g are inverses, then $(f \circ g)(x) =$ _____ and _____ $= x$.

30. The domain of f is equal to the _____ of f^{-1}, and the range of f is equal to the _____ of f^{-1}.

31. If the point (a, b) lies on the graph of f and f has an inverse, then the point _____ lies on the graph of f^{-1}.

32. If $f(x) = x$, then for any function g, $(f \circ g)(x) = (g \circ f)(x) =$ _____.

33. If a function f has an inverse, then the graph of f^{-1} may be obtained by reflecting the graph of f across the line with equation _____.

34. If a function f has an inverse and $f(-3) = 6$, then $f^{-1}(6) =$ _____.

35. If $f(-4) = 16$ and $f(4) = 16$, then f _____ have an inverse because _____.
(does/does not)

36. If f is a function that has an inverse and the graph of f lies completely within the second quadrant, then the graph of f^{-1} lies completely within the _____ quadrant.

Use the definition of inverse functions to show analytically that f and g are inverses.

37. $f(x) = 3x - 7$, $g(x) = \dfrac{x + 7}{3}$

38. $f(x) = 4x + 3$, $g(x) = \dfrac{x - 3}{4}$

39. $f(x) = x^3 + 4$, $g(x) = \sqrt[3]{x - 4}$

40. $f(x) = x^3 - 7$, $g(x) = \sqrt[3]{x + 7}$

41. $f(x) = -x^5$, $g(x) = -\sqrt[5]{x}$

42. $f(x) = -x^7$, $g(x) = -\sqrt[7]{x}$

Determine whether each function is one-to-one. If so, find its inverse.

43. $f = \{(10, 4), (20, 5), (30, 6), (40, 7)\}$

44. $g = \{(5, 12), (10, 22), (15, 32), (20, 42)\}$

45. $f = \{(1, 5), (2, 6), (3, 5), (4, 8)\}$

46. $g = \{(0, 10), (1, 20), (2, 10), (3, 40)\}$

47. $f = \{(0, 0^2), (1, 1^2), (2, 2^2), (3, 3^2), (4, 4^2)\}$

48. $g = \{(0, 0^4), (-1, (-1)^4), (-2, (-2)^4), (-3, (-3)^4)\}$

Concept Check   *In Exercises 49–54, an everyday activity is described. Keeping in mind that an inverse operation "undoes" what an operation does, describe the inverse activity.*

49. tying your shoelaces

50. pressing a car's accelerator

51. entering a room

52. climbing the stairs

53. wrapping a package

54. putting on a coat

For each function that is one-to-one, write an equation for the inverse function in the form $y = f^{-1}(x)$, and then graph f and f^{-1} on the same axes. Give the domain and range of f and f^{-1}. If the function is not one-to-one, say so.

55. $y = 3x - 4$

56. $y = 4x - 5$

57. $y = x^3 + 1$

58. $y = -x^3 - 2$

59. $y = x^2$

60. $y = -x^2 + 2$

61. $y = \dfrac{1}{x}$

62. $y = \dfrac{4}{x}$

63. $y = \dfrac{2}{x + 3}$

64. $y = \dfrac{3}{x - 4}$

65. $f(x) = \sqrt{6 + x}, \ x \geq -6$

66. $f(x) = -\sqrt{x^2 - 16}, \ x \geq 4$

The given function f is one-to-one. Find $f^{-1}(x)$.

67. $f(x) = \dfrac{4x}{x + 1}$

68. $f(x) = \dfrac{3x}{5 - x}$

69. $f(x) = \dfrac{1 - 2x}{3x}$

70. $f(x) = \dfrac{4 - x}{5x}$

71. $f(x) = \sqrt{x^2 - 4}, x \geq 2$

72. $f(x) = \sqrt{x - 8}, x \geq 8$

73. $f(x) = 5x^3 - 7$

74. $f(x) = 4 - 3x^3$

75. $f(x) = \dfrac{x}{4 + 3x}$

76. $f(x) = \dfrac{x}{1 - 3x}$

77. $f(x) = \dfrac{3 - x}{2x + 1}$

78. $f(x) = \dfrac{2x + 1}{x - 1}$

Concept Check   *Let $f(x) = x^3$. Evaluate each expression.*

79. $f(2)$

80. $f(0)$

81. $f(-2)$

82. $f^{-1}(8)$

83. $f^{-1}(0)$

84. $f^{-1}(-8)$

Concept Check   *The graph of a function f is shown in the figure. Use the graph to find each value.*

85. $f^{-1}(4)$

86. $f^{-1}(2)$

87. $f^{-1}(0)$

88. $f^{-1}(-2)$

89. $f^{-1}(-3)$

90. $f^{-1}(-4)$

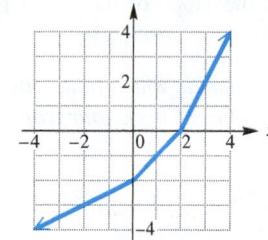

Decide whether the pair of functions f and g are inverses. In Exercises 91–93, assume axes have equal scales.

91.

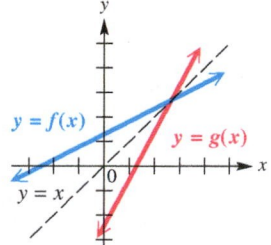

92.

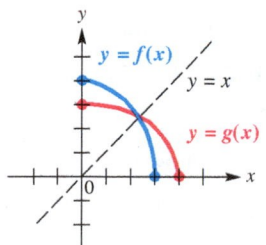

93.

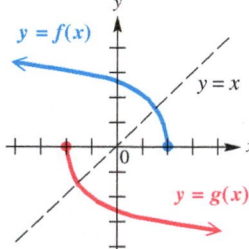

94. $f(x) = -\dfrac{3}{11}x, \quad g(x) = -\dfrac{11}{3}x$

95. $f(x) = 2x + 4, \quad g(x) = \dfrac{1}{2}x - 2$

96. $f(x) = 5x - 5, \quad g(x) = \dfrac{1}{5}x + 5$

Concept Check   *Decide whether the screens suggest that Y_1 and Y_2 are inverse functions.*

97.

X	Y₁
1	6
3	8
-4	1
9	14
3.2	8.2
0	5
-10	-5

X=1

X	Y₂
6	1
8	3
1	-4
9	9
8.2	3.2
5	0
-5	-10

X=6

98.

X	Y₁
1	7
3	9
-4	2
9	15
3.2	9.2
0	6
-10	-4

X=1

X	Y₂
7	-1
9	-3
2	4
15	-9
9.2	-3.2
6	0
-4	10

X=7

Graph the inverse of each one-to-one function.

99.

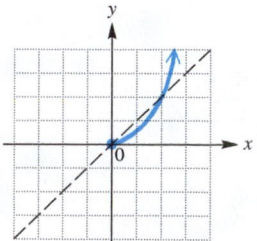

100.

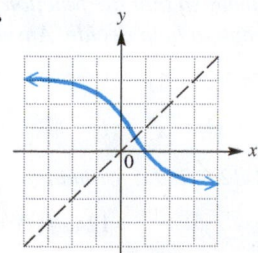

101.

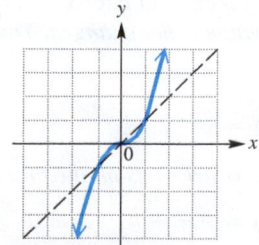

102.

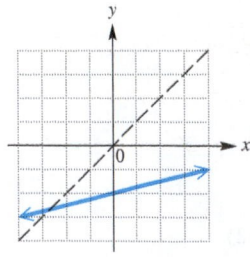

103.

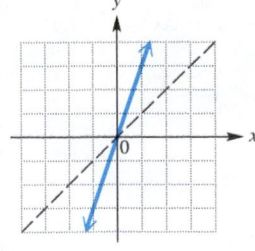

104.

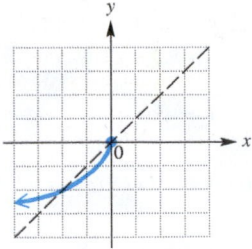

Concept Check *Answer the following.*

105. Suppose $f(x)$ is the number of cars that can be built for x dollars. What does $f^{-1}(1000)$ represent?

106. Suppose $f(r)$ is the volume (in cubic inches) of a sphere of radius r inches. What does $f^{-1}(5)$ represent?

107. If a line has nonzero slope a, what is the slope of its reflection across the line $y = x$?

108. Find $f^{-1}(f(2))$, where $f(2) = 3$.

Use a graph with the given viewing window to decide which functions are one-to-one. If a function is one-to-one, give the equation of its inverse function. Check your work by graphing the inverse function on the same coordinate axes.

109. $f(x) = 6x^3 + 11x^2 - x - 6$; $[-3, 2]$ by $[-10, 10]$

110. $f(x) = x^4 - 5x^2 + 6$; $[-3, 3]$ by $[-1, 8]$

111. $f(x) = \dfrac{x - 5}{x + 3}$; $[-9.4, 9.4]$ by $[-6.2, 6.2]$

112. $f(x) = \dfrac{-x}{x - 4}$; $[-4.7, 9.4]$ by $[-6.2, 6.2]$

*Use the alphabet coding assignment given in **Example 9** for Exercises 113–116.*

113. The function $f(x) = 4x - 1$ was used to encode the following message.

       79 71 19 3 75 83 71 19 31 83 55 79 35 75 59 55

    Find the inverse function and decode the message.

114. The function $f(x) = 3x - 3$ was used to encode the following message.

       54 12 57 54 0 24 33 57 42 9 0 72

    Find the inverse function and decode the message.

115. Encode the message NO PROBLEM, using the one-to-one function $f(x) = x^3 + 1$. Give the inverse function the decoder would need when the message is received.

116. Encode the message BEGIN OPERATIONS, using the one-to-one function $f(x) = (x + 2)^3$. Give the inverse function the decoder would need when the message is received.

While a function may not be one-to-one when defined over its "natural" domain, it may be possible to restrict the domain in such a way that it is one-to-one and the range of the function is unchanged. For example, if we restrict the domain of the function $f(x) = x^2$ (which is not one-to-one over $(-\infty, \infty)$) to $[0, \infty)$, we obtain a one-to-one function whose range is still $[0, \infty)$. See the figure to the right. Notice that we could choose to restrict the domain of $f(x) = x^2$ to $(-\infty, 0]$ and also obtain the graph of a one-to-one function, except that it would be the left half of the parabola.

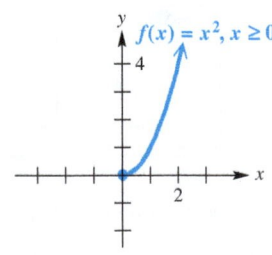

For each function in Exercises 117–122, restrict the domain so that the function is one-to-one and the range is not changed. You may wish to use a graph to help decide. Answers may vary.

117. $f(x) = -x^2 + 4$

118. $f(x) = (x - 1)^2$

119. $f(x) = |x - 6|$

120. $f(x) = x^4$

121. $f(x) = x^4 + x^2 - 6$

122. $f(x) = -\sqrt{x^2 - 16}$

Using the restrictions on the functions in Exercises 123–126, find a formula for f^{-1}.

123. $f(x) = -x^2 + 4$, $x \geq 0$

124. $f(x) = (x - 1)^2$, $x \geq 1$

125. $f(x) = |x - 6|$, $x \geq 6$

126. $f(x) = x^4$, $x \geq 0$

5.2 Exponential Functions

Real-Number Exponents • Graphs of Exponential Functions • Exponential Equations (Type 1) • Compound Interest • The Number e and Continuous Compounding • An Application of Exponential Functions

Real-Number Exponents

Recall that if $r = \frac{m}{n}$ is a rational number, then, for appropriate integer values of m and n,

$$a^{m/n} = \left(\sqrt[n]{a}\right)^m. \qquad \frac{m}{n} \text{ is in lowest terms.}$$

For example,

$$16^{3/4} = \left(\sqrt[4]{16}\right)^3 = 2^3 = 8,$$

$$27^{-1/3} = \frac{1}{27^{1/3}} = \frac{1}{\sqrt[3]{27}} = \frac{1}{3}, \quad \text{and} \quad 64^{-1/2} = \frac{1}{64^{1/2}} = \frac{1}{\sqrt{64}} = \frac{1}{8}.$$

In this section, we extend the definition of a^r to include all real (not just rational) values of the exponent r. For example, the expression $2^{\sqrt{3}}$ might be evaluated by approximating the irrational exponent $\sqrt{3}$ by the rational numbers 1.7, 1.73, 1.732, and so on. Since these decimals approach $\sqrt{3}$ more and more closely, it seems reasonable that $2^{\sqrt{3}}$ should be approximated more and more closely by the numbers $2^{1.7}$, $2^{1.73}$, $2^{1.732}$, and so on. (Recall, for example, that $2^{1.7} = 2^{17/10} = \left(\sqrt[10]{2}\right)^{17}$.) In fact, this is exactly how $2^{\sqrt{3}}$ is defined in a more advanced course.

← Algebra Review
To review properties of rational exponents see **Section R.4**.

With this interpretation of real exponents, all rules and theorems for exponents are valid for real number exponents as well as rational ones. In addition to the usual rules for exponents, we use several new properties in this chapter. For example, if $y = 2^x$, then each value of x leads to exactly one value of y. Therefore, $y = 2^x$ defines a function. Furthermore,

$$\text{if} \quad 2^x = 2^4, \quad \text{then} \quad x = 4,$$

and

$$\text{if} \quad x = 4, \quad \text{then} \quad 2^x = 2^4.$$

Also,

$$4^2 < 4^3, \quad \text{but} \quad \left(\frac{1}{4}\right)^2 > \left(\frac{1}{4}\right)^3.$$

*In general, when $a > 1$, increasing the exponent x for a^x leads to a **larger** number, but when $0 < a < 1$, increasing the exponent x for a^x leads to a **smaller** number.*

These properties are generalized below. Proofs of the properties are not given here, as they require more advanced mathematics.

FOR DISCUSSION
Do -2^4 and $(-2)^4$ evaluate to the same number? Explain how -2^n and $(-2)^n$ compare when n is a positive integer. Generalize your results for any positive base a and any positive integer n.

Additional Properties of Exponents

For any real number $a > 0$, $a \neq 1$, the following statements are true.
(a) a^x **is a unique real number for each real number** x.
(b) $a^b = a^c$ **if and only if** $b = c$.
(c) If $a > 1$ **and** $m < n$**, then** $a^m < a^n$.
(d) If $0 < a < 1$ **and** $m < n$**, then** $a^m > a^n$.

> **NOTE** Properties (a) and (b) require $a > 0$ so that a^x is always defined. For example, $(-6)^x$ is not a real number if $x = \frac{1}{2}$. This means that a^x will always be positive, since a is positive. In Property (a), $a \neq 1$ because $1^x = 1$ for every real-number value of x, and thus each value of x leads to the same real number, 1. For Property (b) to hold, a must not equal 1. For example, $1^4 = 1^5$, even though $4 \neq 5$. Also, Property (b) ensures that the function $f(x) = a^x$, $a > 0$ and $a \neq 1$, is one-to-one.

Graphs of Exponential Functions

A graphing calculator can easily find approximations of numbers raised to irrational powers. For example,

$$2^{\sqrt{6}} \approx 5.462228786, \quad \left(\frac{1}{2}\right)^{\sqrt{3}} \approx 0.3010237439, \quad \text{and} \quad 0.5^{-\sqrt{2}} \approx 2.665144143.$$

Later we find these approximations by using graphs of *exponential functions*.

Exponential Function

If $a > 0$ and $a \neq 1$, then

$$f(x) = a^x$$

is the **exponential function with base a.**

EXAMPLE 1   **Graphing an Exponential Function ($a > 1$)**

Graph $f(x) = 2^x$. Determine the domain and range of f.

Solution   Make a table of values, plot the points, and connect them with a smooth curve, as shown in **FIGURE 12.** As the graph suggests, the domain is $(-\infty, \infty)$ and the range is $(0, \infty)$.

It is difficult to graph values of a^x for $x < -2$. The calculator table in **FIGURE 13** shows that a^x approaches 0 (but remains positive) as x decreases.

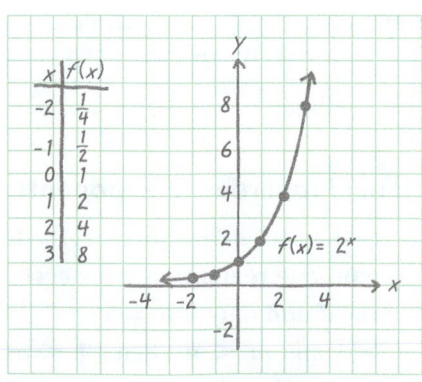

FIGURE 12

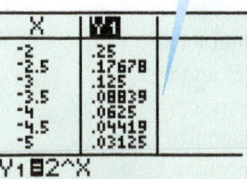

As x decreases, a^x ($a > 1$) approaches 0, but always remains positive.

FIGURE 13

Graphing an Exponential Function ($0 < a < 1$)

Graph $g(x) = \left(\frac{1}{2}\right)^x$. Determine the domain and range of g.

Solution Use the same procedure as in **Example 1.** See **FIGURES 14** and **15.** Again, the domain is $(-\infty, \infty)$ and the range is $(0, \infty)$, but here the function *decreases* on its domain. (In **Example 1,** the function *increases* on its domain.)

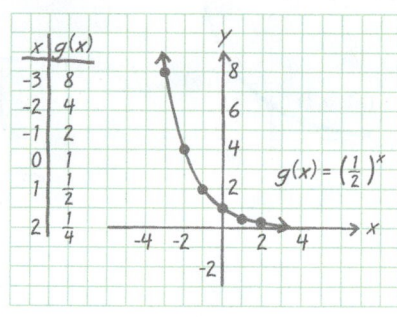

As x increases, a^x ($0 < a < 1$) approaches 0, but always remains positive.

FIGURE 14 **FIGURE 15**

TECHNOLOGY NOTE

Because of the limited resolution of the graphing calculator screen, it is difficult to interpret how the graph of the exponential function $f(x) = a^x$ behaves when the curve is close to the x-axis, as seen in **FIGURE 16**. Remember that there is no endpoint and that the curve approaches, but never touches, the x-axis. Tracing along the curve and observing the y-coordinates as $x \rightarrow -\infty$ when $a > 1$ and as $x \rightarrow \infty$ when $0 < a < 1$, or looking at a table, helps support this fact.

The behavior of the graph of an exponential function depends, in general, on the magnitude of a. As a becomes larger ($a > 1$), the graph becomes "steeper" moving to the right of the y-axis. **FIGURE 16(a)** shows the graphs of $y = a^x$ for $a = 2, 3,$ and 4. The function is *always increasing* for these values of a.

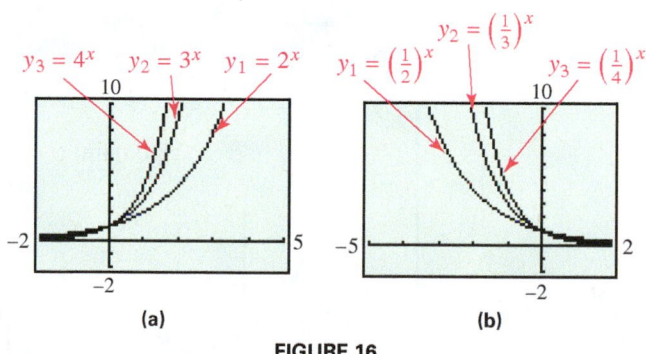

FIGURE 16

If the base a is between 0 and 1, as a gets closer to 0, the graph becomes "steeper" moving to the left of the y-axis. **FIGURE 16(b)** shows the graphs of $y = a^x$ for $a = \frac{1}{2}, \frac{1}{3},$ and $\frac{1}{4}$. The function is *always decreasing* for these values of a.

FOR DISCUSSION

1. Using a standard viewing window, we cannot observe how the graph of the exponential function $f(x) = 2^x$ behaves for values of x less than about -3. Use the window $[-10, 0]$ by $[-0.5, 0.5]$, and then trace to the left to see what happens.

2. Repeat Item 1 for the exponential function $f(x) = \left(\frac{1}{2}\right)^x$. Use the window $[0, 10]$ by $[-0.5, 0.5]$, and trace to the right.

3. Complete the following statement. For the graph of an exponential function $f(x) = a^x, a > 0, a \neq 1$, every range value appears exactly once. Thus, f is a(n) _____ function. Because of this, f has a(n) _____. (*Hint:* Recall the concepts that were studied in **Section 5.1.**)

FUNCTION CAPSULE

EXPONENTIAL FUNCTION $f(x) = a^x, \quad a > 1$

Domain: $(-\infty, \infty)$ Range: $(0, \infty)$

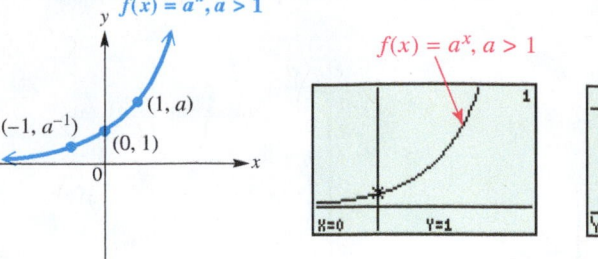

FIGURE 17

- $f(x) = a^x, a > 1$, is increasing and continuous on its entire domain, $(-\infty, \infty)$.
- The x-axis is the horizontal asymptote as $x \to -\infty$.
- The graph passes through the points $(-1, a^{-1})$, $(0, 1)$, and $(1, a)$.

FUNCTION CAPSULE

EXPONENTIAL FUNCTION $\quad f(x) = a^x, \quad 0 < a < 1$

Domain: $(-\infty, \infty)$ Range: $(0, \infty)$

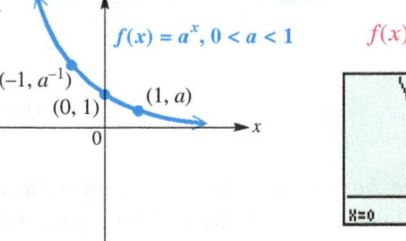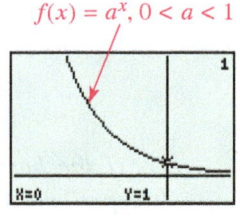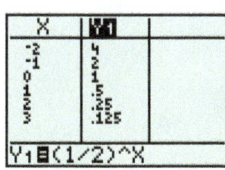

FIGURE 18

- $f(x) = a^x, 0 < a < 1$, is decreasing and continuous on its entire domain, $(-\infty, \infty)$.
- The x-axis is the horizontal asymptote as $x \to \infty$.
- The graph passes through the points $(-1, a^{-1})$, $(0, 1)$, and $(1, a)$.

Reflection across y-axis

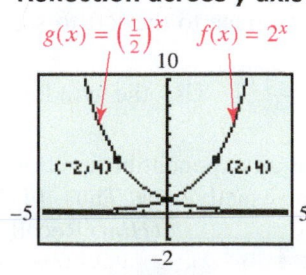

FIGURE 19

NOTE   Since $a^{-x} = (a^{-1})^x = \left(\frac{1}{a}\right)^x$, the graph of $f(x) = a^{-x}, a > 1$, is a reflection of the graph of $y = a^x$ across the y-axis. For example,

$$g(x) = 2^{-x} = \left(\frac{1}{2}\right)^x \quad \text{is a reflection of} \quad f(x) = 2^x$$

across the y-axis, as shown in **FIGURE 19.**

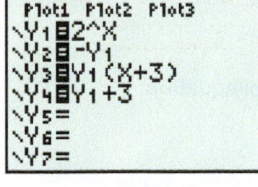

The screen shows how a graphing calculator can be directed to graph the three functions in **Example 3.** Y_1 is defined as 2^X, and Y_2, Y_3, and Y_4 are defined as reflections and/or translations of Y_1.

EXAMPLE 3   **Graphing Reflections and Translations**

Graph each function. Give the domain and range.

(a) $f(x) = -2^x$ **(b)** $f(x) = 2^{x+3}$ **(c)** $f(x) = 2^x + 3$

Solution   (In each graph, we show the graph of $y = 2^x$ for comparison.)

(a) The graph of $f(x) = -2^x$ is that of $y = 2^x$ reflected across the x-axis. The domain is $(-\infty, \infty)$, and the range is $(-\infty, 0)$. See **FIGURE 20.**

(b) The graph of $f(x) = 2^{x+3}$ is the graph of $y = 2^x$ translated 3 units to the left, as shown in **FIGURE 21.** The domain is $(-\infty, \infty)$, and the range is $(0, \infty)$.

(c) The graph of $f(x) = 2^x + 3$ is that of $y = 2^x$ translated 3 units up. See **FIGURE 22.** The domain is $(-\infty, \infty)$, and the range is $(3, \infty)$. The line $y = 3$ is a horizontal asymptote.

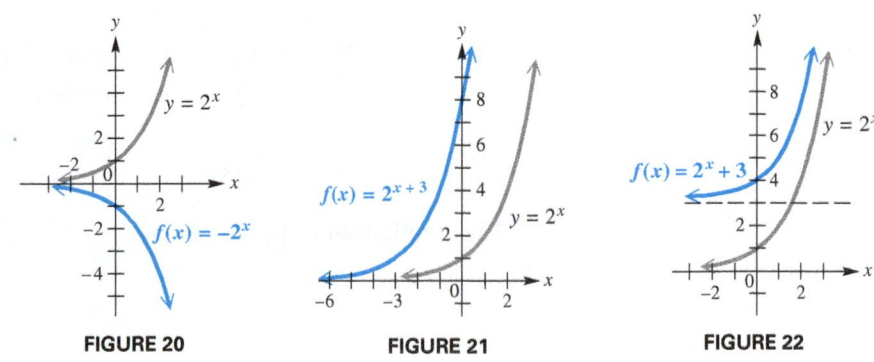

FIGURE 20   FIGURE 21   FIGURE 22

EXAMPLE 4   **Using Graphs to Evaluate Exponential Expressions**

Use a graph to evaluate each expression.

(a) $2^{\sqrt{6}}$ **(b)** $0.5^{-\sqrt{2}}$

Solution

(a) **FIGURE 23(a)** shows the graph of $Y_1 = 2^X$, with Y_1 evaluated for $X = \sqrt{6} \approx 2.4494897$. From the bottom display, we see that $Y = 2^{\sqrt{6}} \approx 5.4622288$, which confirms our earlier result when we raised numbers to irrational powers.

(b) Using the graph of $Y_1 = .5^X$, with $X = -\sqrt{2} \approx -1.414214$, we find that $Y \approx 2.6651441$, again supporting our earlier result. See **FIGURE 23(b).**

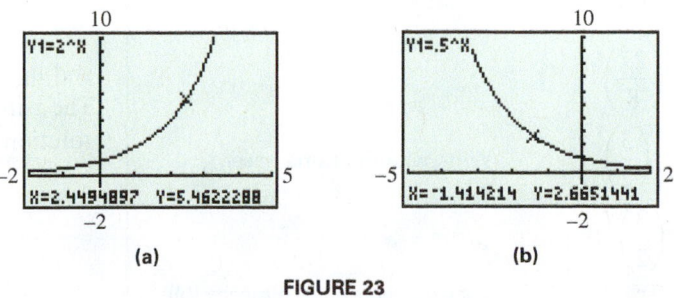

(a)   (b)

FIGURE 23

Exponential Equations (Type 1)

The equation $25^x = 125$ is different from any equation studied so far, because the variable appears in the exponent. In this book we refer to such an equation as a **Type 1 exponential equation.** (This terminology is not standard, but is rather our own method of distinguishing between two types of exponential equations.)

EXAMPLE 5 **Solving Type 1 Exponential Equations**

Solve each equation.

(a) $25^x = 125$ (b) $\left(\dfrac{1}{3}\right)^x = 81$

Solution

(a)
$$25^x = 125$$

$$(5^2)^x = 5^3$$ Write with the same base; $25 = 5^2$ and $125 = 5^3$.

$$5^{2x} = 5^3$$ $(a^m)^n = a^{mn}$

If $a^b = a^c$, then $b = c$.

$$2x = 3$$ Set exponents equal (Property (b)).

$$x = \dfrac{3}{2}$$ Divide by 2.

Check:
$$25^{3/2} = 125 \quad ? \quad \text{Let } x = \tfrac{3}{2} \text{ in the original equation.}$$

$$\left(\sqrt{25}\right)^3 = 125 \quad ? \quad a^{m/n} = \left(\sqrt[n]{a}\right)^m$$

$$5^3 = 125 \quad ? \quad \text{Simplify } \sqrt{25}.$$

$$125 = 125 \quad \checkmark \quad \text{True}$$

The solution set is $\left\{\dfrac{3}{2}\right\}$.

(b)
$$\left(\dfrac{1}{3}\right)^x = 81$$

$$(3^{-1})^x = 3^4$$ Write with the same base; $\tfrac{1}{3} = 3^{-1}$ and $81 = 3^4$.

$$3^{-x} = 3^4$$ $(a^m)^n = a^{mn}$

If $a^b = a^c$, then $b = c$.

$$-x = 4$$ Set exponents equal (Property (b)).

$$x = -4$$ Multiply by -1.

Check that the solution set is $\{-4\}$.

EXAMPLE 6 **Solving a Type 1 Exponential Equation**

Solve $1.5^{x+1} = \left(\dfrac{27}{8}\right)^x$.

Analytic Solution

Here, $1.5 = \dfrac{3}{2}$ and $\dfrac{27}{8} = \left(\dfrac{3}{2}\right)^3$, so each base can be written as a power of $\dfrac{3}{2}$.

$$1.5^{x+1} = \left(\dfrac{27}{8}\right)^x$$

$$\left(\dfrac{3}{2}\right)^{x+1} = \left[\left(\dfrac{3}{2}\right)^3\right]^x$$ Write with the same base.

$$\left(\dfrac{3}{2}\right)^{x+1} = \left(\dfrac{3}{2}\right)^{3x}$$ $(a^m)^n = a^{mn}$

$$x + 1 = 3x$$ Set exponents equal (Property (b)).

$$1 = 2x$$ Subtract x.

$$x = \dfrac{1}{2}, \quad \text{or} \quad 0.5$$ Divide by 2 and rewrite.

Check that the solution set is $\{0.5\}$.

Graphing Calculator Solution

Graph
$$y = 1.5^{x+1} - \left(\dfrac{27}{8}\right)^x,$$

and use the x-intercept method, as shown in **FIGURE 24.** The x-intercept shown is $(0.5, 0)$, confirming the analytic solution.

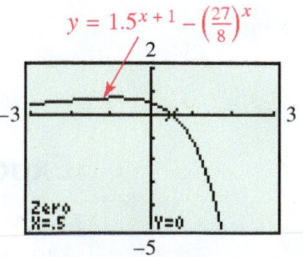

FIGURE 24

EXAMPLE 7 Using a Graph to Solve Exponential Inequalities

Use the graph in **FIGURE 24** (from **Example 6**) to solve each inequality.

(a) $1.5^{x+1} - \left(\dfrac{27}{8}\right)^x > 0$ (b) $1.5^{x+1} - \left(\dfrac{27}{8}\right)^x < 0$

Solution In **FIGURE 24,** the graph of

$$y = 1.5^{x+1} - \left(\dfrac{27}{8}\right)^x$$

lies *above* the x-axis for values of x less than 0.5 and *below* the x-axis for values of x greater than 0.5. Tracing left and right will support this observation. The solution set for part (a) is $(-\infty, 0.5)$, and the solution set for part (b) is $(0.5, \infty)$. ●

Compound Interest

The formula for **compound interest** (interest paid on both principal and interest) is an important application of exponential functions. The formula for simple interest I is $I = Prt$, where P is the principal, or initial amount of money; r is the annual rate of interest expressed as a decimal; and t is time in years that the principal earns interest. Suppose $t = 1$ year. Then, at the end of the year, the amount has grown to

$$P + Pr = P(1 + r),$$

Principal   First-year interest   First-year amount

the original principal plus the interest. If this amount is left at the same interest rate for another year, the total amount becomes

$$[P(1 + r)] + [P(1 + r)]r = [P(1 + r)](1 + r) \quad \text{Factor.}$$
$$= P(1 + r)^2.$$

First-year amount   Interest on first-year amount   Second-year amount

After the third year, this amount will grow to

$$[P(1 + r)^2] + [P(1 + r)^2]r = [P(1 + r)^2](1 + r) \quad \text{Factor.}$$
$$= P(1 + r)^3.$$

Second-year amount   Interest on second-year amount   Third-year amount

Continuing produces a general formula for interest compounded n times per year.

Compound Interest Formula

Suppose that a principal of P dollars is invested at an annual interest rate r (in decimal form), compounded n times per year. Then the amount A accumulated after t years is given by the formula

$$A = P\left(1 + \dfrac{r}{n}\right)^{nt}.$$

EXAMPLE 8 **Using the Compound Interest Formula**

If $1000 is invested at an annual rate of 4%, compounded quarterly (four times per year), how much will be in the account after 10 years if no withdrawals are made? How much interest is earned?

Analytic Solution

Use the compound interest formula with $P = 1000$, $r = 0.04$, $n = 4$, and $t = 10$.

$$A = P\left(1 + \frac{r}{n}\right)^{nt}$$ Compound interest formula

$$A = 1000\left(1 + \frac{0.04}{4}\right)^{4 \cdot 10}$$ Substitute.

$$A = 1000(1.01)^{40}$$ Simplify.

$$A \approx 1488.8637$$ Use a calculator.

To the nearest cent, there will be $1488.86 in the account after 10 years. This means that

$$\$1488.86 - \$1000 = \$488.86$$

interest was earned. (Note that there are four quarters per year, so 10 years = 40 quarters.)

Graphing Calculator Solution

The function $Y_1 = 1000\left(1 + \frac{0.04}{4}\right)^{4X}$ is graphed in **FIGURE 25** and evaluated for $X = 10$. The Y-value at the bottom of the screen is the amount in the account. (Tracing to the right on this graph gives new meaning to the phrase "watching your money grow.")

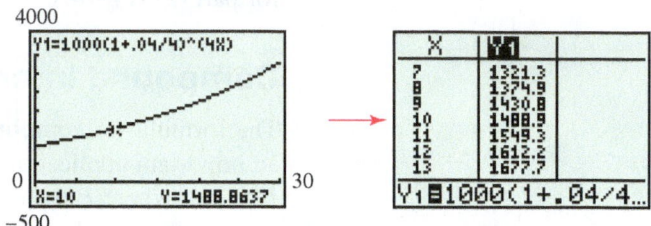

FIGURE 25

Compare the value of Y_1 when $X = 10$ in the table with the value of A in the analytic solution and the Y-value displayed at the bottom of the graph.

The Number *e* and Continuous Compounding

The more often interest is compounded within a given period, the more interest will be earned. Surprisingly, however, there is a limit on the amount of interest, no matter how often it is compounded.

Suppose that $1 is invested at 100% interest per year, compounded k times per year. Then the interest rate (in decimal form) is 1.00 and the interest rate per compounding period is $\frac{1}{k}$. According to the formula (with $P = 1$), the compound amount at the end of 1 year will be

$$A = \left(1 + \frac{1}{k}\right)^k.$$

k	$\left(1 + \frac{1}{k}\right)^k$ (rounded)
1	2
2	2.25
5	2.48832
10	2.59374
100	2.70481
1000	2.71692
10,000	2.71815
1,000,000	2.71828

A calculator gives the results in the margin for various values of k. The table suggests that as k increases, the value of $\left(1 + \frac{1}{k}\right)^k$ gets closer and closer to some fixed number. This is indeed the case. That fixed number is symbolized e, after the mathematician Euler. (Regardless of how often the interest is compounded, the $1 investment will not be worth more than $2.72 after one year.)

Value of *e*

To eleven decimal places, $e \approx 2.71828182846$.

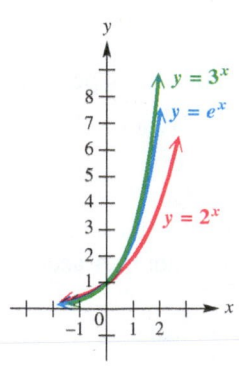

FIGURE 26

FIGURE 26 shows the functions defined by $y = 2^x$, $y = 3^x$, and $y = e^x$. Notice that because $2 < e < 3$, the graph of $y = e^x$ lies "between" the other two graphs.

As mentioned, the amount of interest earned increases with the frequency of compounding, but the value of the expression $\left(1 + \frac{1}{k}\right)^k$ approaches e as k gets larger. Consequently, the formula for compound interest approaches a limit as well, called the compound amount from **continuous compounding.** To derive the formula for continuous compounding, we begin with the compound interest formula,

$$A = P\left(1 + \frac{r}{n}\right)^{nt}.$$

Let $k = \frac{n}{r}$. Then, $n = rk$, and with these substitutions, the formula becomes

$$A = P\left(1 + \frac{r}{n}\right)^{nt} = P\left(1 + \frac{1}{k}\right)^{rkt} = P\left[\left(1 + \frac{1}{k}\right)^{k}\right]^{rt}.$$

Expression approaches e as $k \to \infty$.

If $n \to \infty$, $k \to \infty$ as well, and the expression $\left(1 + \frac{1}{k}\right)^k \to e$, as discussed earlier. This leads to the formula $A = Pe^{rt}$.

→ Looking Ahead to Calculus

In calculus, the derivative function allows us to determine the slope of a line tangent to the graph of a function. For the function $f(x) = e^x$, the derivative is the function f itself: $f'(x) = e^x$. Therefore, in calculus the exponential function with base e is much easier to work with than exponential functions having other bases.

Continuous Compounding Formula

If an amount of P dollars is deposited at a rate of interest r (in decimal form) compounded continuously for t years, then the final amount A in dollars is

$$A = Pe^{rt}.$$

EXAMPLE 9 **Solving a Continuous Compounding Problem**

Suppose \$5000 is deposited in an account paying 3% compounded continuously for 5 years. Find the total amount on deposit and the interest earned at the end of the 5 years.

Analytic Solution

Let $P = 5000$, $r = 0.03$, and $t = 5$ in the continuous compounding formula $A = Pe^{rt}$.

$A = Pe^{rt}$ Continuous compounding

$= 5000e^{0.03(5)}$ Substitute for P, r, and t.

$= 5000e^{0.15}$ Multiply.

≈ 5809.17 Approximate with a calculator.

After 5 years, the account contains \$5809.17 and so

$$\$5809.17 - \$5000.00 = \$809.17$$

in interest is earned.

Graphing Calculator Solution

As **FIGURE 27(a)** shows, the graph of $Y_1 = 5000e^{0.03X}$ indicates that when $X = 5$, $Y_1 \approx 5809.17$.

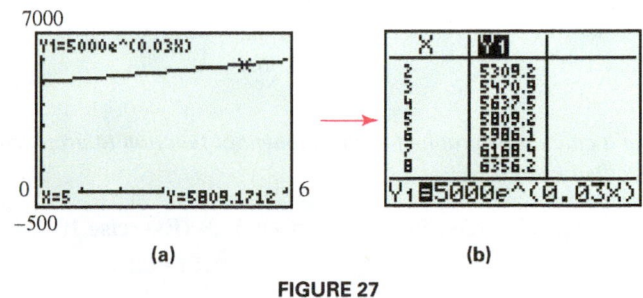

(a)   (b)

FIGURE 27

The analytic result is supported numerically by the table in **FIGURE 27(b).** When $X = 5$, the table gives $Y_1 \approx 5809.2$.

An Application of Exponential Functions

The next example shows how an exponential function can be used to model the increased risk of people dying of heart disease.

EXAMPLE 10 **Modeling Risk**

The chances of dying of heart disease increase exponentially after age 40 according to the function

$$f(x) = r(1.11)^x,$$

where r is the risk (in decimal form) at age 40 and x is the number of years after 40. (*Source:* National Center for Health Statistics.) Compare the risk at age 75 with the risk at age 40.

Solution Let $x = 75 - 40 = 35$ and write $f(x)$ in terms of r.

$$f(35) = r(1.11)^{35} \approx 38.6r \qquad \text{Use a calculator.}$$

Thus, the risk is almost 39 times as great at age 75 as at age 40.

5.2 Exercises

Checking Analytic Skills Graph each equation. **Do not use a calculator.**

1. $f(x) = 3^x$

2. $f(x) = 4^x$

3. $f(x) = \left(\dfrac{1}{3}\right)^x$

4. $f(x) = \left(\dfrac{1}{4}\right)^x$

Checking Analytic Skills Solve each equation. **Do not use a calculator.**

5. $4^x = 2$

6. $125^x = 5$

7. $\left(\dfrac{1}{2}\right)^x = 4$

8. $\left(\dfrac{2}{3}\right)^x = \dfrac{9}{4}$

Use a calculator to find an approximation for each power. Give the maximum number of decimal places that your calculator displays.

9. $2^{\sqrt{10}}$

10. $3^{\sqrt{11}}$

11. $\left(\dfrac{1}{2}\right)^{\sqrt{2}}$

12. $\left(\dfrac{1}{3}\right)^{\sqrt{6}}$

13. $4.1^{-\sqrt{3}}$

14. $6.4^{-\sqrt{3}}$

15. $\sqrt{7}^{\sqrt{7}}$

16. $\sqrt{13}^{-\sqrt{13}}$

Use a calculator graph of each exponential function to graphically support the result found in the specified exercise.

17. $y = 2^x$ (**Exercise 9**)

18. $y = 3^x$ (**Exercise 10**)

19. $y = \left(\dfrac{1}{2}\right)^x$ (**Exercise 11**)

20. $y = \left(\dfrac{1}{3}\right)^x$ (**Exercise 12**)

Graph each function by hand and support your sketch with a calculator graph. Give the domain, range, and equation of the asymptote. Determine if f is increasing or decreasing on its domain.

21. $f(x) = -1.5^x$

22. $f(x) = \left(\dfrac{2}{3}\right)^x$

23. $f(x) = e^x$

24. $f(x) = -e^x$

25. $f(x) = e^{x+1}$

26. $f(x) = e^x - 1$

27. $f(x) = 10^x$

28. $f(x) = 10^{-x}$

29. $f(x) = 4^{-x}$

30. $f(x) = 6^{-x}$

Concept Check   *In Exercises 31 and 32, refer to the graphs to the right and follow the directions in parts (a)–(f).*

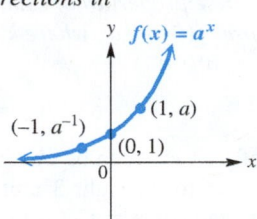

Exercise 31

31. (a) Is $a > 1$ or is $0 < a < 1$?

   (b) Give the domain and range of f, and the equation of the asymptote.

   (c) Sketch the graph of $g(x) = -a^x$.

   (d) Give the domain and range of g, and the equation of the asymptote.

   (e) Sketch the graph of $h(x) = a^{-x}$.

   (f) Give the domain and range of h, and the equation of the asymptote.

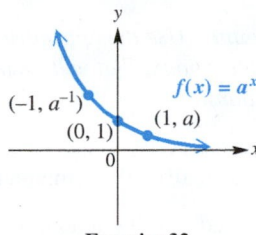

Exercise 32

32. (a) Is $a > 1$ or is $0 < a < 1$?

   (b) Give the domain and range of f, and the equation of the asymptote.

   (c) Sketch the graph of $g(x) = a^x + 2$.

   (d) Give the domain and range of g, and the equation of the asymptote.

   (e) Sketch the graph of $h(x) = a^{x+2}$.

   (f) Give the domain and range of h, and the equation of the asymptote.

*Sketch the graph of $f(x) = 2^x$. Then refer to it and use the techniques of **Chapter 2** to graph each function.*

33. $f(x) = 2^x + 1$ **34.** $f(x) = 2^x - 4$ **35.** $f(x) = 2^{x+1}$ **36.** $f(x) = 2^{x-4}$

*Sketch the graph of $f(x) = \left(\frac{1}{3}\right)^x$. Then refer to it and use the techniques of **Chapter 2** to graph each function.*

37. $f(x) = \left(\frac{1}{3}\right)^x - 2$ **38.** $f(x) = \left(\frac{1}{3}\right)^x + 4$ **39.** $f(x) = \left(\frac{1}{3}\right)^{x+2}$ **40.** $f(x) = \left(\frac{1}{3}\right)^{x-4}$

Checking Analytic Skills   *Solve each equation. **Do not use a calculator.***

41. $2^{3-x} = 8$ **42.** $5^{2x+1} = 25$ **43.** $12^{x-3} = 1$ **44.** $3^{5-x} = 1$

45. $e^{4x-1} = (e^2)^x$ **46.** $e^{3-x} = (e^3)^{-x}$ **47.** $27^{4x} = 9^{x+1}$ **48.** $32^x = 16^{1-x}$

49. $\left(\frac{1}{4}\right)^{2-x} = 2^{3x+3}$ **50.** $\left(\frac{1}{2}\right)^{3x-6} = 8^{x+1}$ **51.** $\left(\sqrt{2}\right)^{x+4} = \left(\frac{1}{4}\right)^{-x}$ **52.** $\left(\sqrt[3]{5}\right)^{-x} = \left(\frac{1}{5}\right)^{x+2}$

53. $\left(\sqrt{2}\right)^{-2x} = \left(\frac{1}{2}\right)^{2x+3}$ **54.** $\left(\sqrt[4]{3}\right)^{-x} = \left(\frac{1}{3}\right)^{x-1}$ **55.** $6^{1-x} = \left(\frac{1}{36}\right)^{2x}$ **56.** $\left(\frac{3}{5}\right)^{-x} = \left(\frac{9}{25}\right)^{1-5x}$

Solve each equation in part (a) analytically. Support your answer with a calculator graph. Then use the graph to solve the associated inequalities in parts (b) and (c).

57. (a) $2^{x+1} = 8$

   (b) $2^{x+1} > 8$

   (c) $2^{x+1} < 8$

58. (a) $3^{2-x} = 9$

   (b) $3^{2-x} > 9$

   (c) $3^{2-x} < 9$

59. (a) $27^{4x} = 9^{x+1}$

   (b) $27^{4x} > 9^{x+1}$

   (c) $27^{4x} < 9^{x+1}$

60. (a) $32^x = 16^{1-x}$

   (b) $32^x > 16^{1-x}$

   (c) $32^x < 16^{1-x}$

61. (a) $\left(\frac{1}{2}\right)^{-x} = \left(\frac{1}{4}\right)^{x+1}$

   (b) $\left(\frac{1}{2}\right)^{-x} \geq \left(\frac{1}{4}\right)^{x+1}$

   (c) $\left(\frac{1}{2}\right)^{-x} \leq \left(\frac{1}{4}\right)^{x+1}$

62. (a) $\left(\frac{2}{3}\right)^{x-1} = \left(\frac{81}{16}\right)^{x+1}$

   (b) $\left(\frac{2}{3}\right)^{x-1} \leq \left(\frac{81}{16}\right)^{x+1}$

   (c) $\left(\frac{2}{3}\right)^{x-1} \geq \left(\frac{81}{16}\right)^{x+1}$

63. *Concept Check*   If $f(x) = a^x$ and $f(3) = 27$, find each value of $f(x)$.

   (a) $f(1)$ (b) $f(-1)$

   (c) $f(2)$ (d) $f(0)$

Concept Check   *Give an exponential function in the form $f(x) = a^x$ whose graph contains the given point.*

64. $(3, 8)$ **65.** $(-3, 64)$

Concept Check *Use properties of exponents to write each function in the form $f(t) = ka^t$, where k is a constant. (Hint: Recall that $a^{x+y} = a^x \cdot a^y$.)*

66. $f(t) = 3^{2t+3}$

67. $f(t) = \left(\dfrac{1}{3}\right)^{1-2t}$

68. The graph of $y = e^{x-3}$ can be obtained by translating the graph of $y = e^x$ to the right 3 units. Find a constant C such that the graph of $y = Ce^x$ is the same as the graph of $y = e^{x-3}$. Verify your result by graphing both functions.

Compound Amount *Use the appropriate compound interest formula to find the amount that will be in each account, given the stated conditions.*

69. $20,000 invested at 3% annual interest for 4 years compounded **(a)** annually; **(b)** semiannually

70. $35,000 invested at 4.2% annual interest for 3 years compounded **(a)** annually; **(b)** quarterly

71. $27,500 invested at 3.95% annual interest for 5 years compounded **(a)** daily ($n = 365$); **(b)** continuously

72. $15,800 invested at 1.6% annual interest for 6.5 years compounded **(a)** quarterly; **(b)** continuously

Comparing Investment Plans *In Exercises 73 and 74, decide which of the two plans will provide a better yield. (Interest rates stated are annual rates.)*

73. Plan A: $40,000 invested for 3 years at 2.5%, compounded quarterly

Plan B: $40,000 invested for 3 years at 2.4%, compounded continuously

74. Plan A: $50,000 invested for 10 years at 4.75%, compounded daily ($n = 365$)

Plan B: $50,000 invested for 10 years at 4.7%, compounded continuously

Use the table capabilities of your calculator to work Exercises 75 and 76.

75. *Comparison of Two Accounts* You have the choice of investing $1000 at an annual rate of 5%, compounded either annually or monthly. Let Y_1 represent the investment compounded annually, and let Y_2 represent the investment compounded monthly. Graph both Y_1 and Y_2, and observe the slight differences in the curves. Then use a table to compare the graphs numerically. What is the difference between the returns for the investments after 1 year, 2 years, 5 years, 10 years, 20 years, 30 years, and 40 years?

76. *Comparison of Two Accounts* You have the choice of investing $1000 at an annual rate of 7.5% compounded daily or 7.75% compounded annually. Let Y_1 represent the investment at 7.5% compounded daily, and let Y_2 represent the investment at 7.75% compounded annually. Graph both Y_1 and Y_2, and observe the slight differences in the curves. Then use a table with $Y_3 = Y_1 - Y_2$ to compare

the graphs numerically. What is the difference between the returns for the investments after 1 year, 2 years, 5 years, 10 years, 20 years, 30 years, and 40 years? Why does the lower interest rate yield the greater return over time?

(Modeling) *Solve each problem.*

77. *Atmospheric Pressure* The atmospheric pressure (in millibars) at a given altitude (in meters) is shown in the table.

Altitude	Pressure	Altitude	Pressure
0	1013	6000	472
1000	899	7000	411
2000	795	8000	357
3000	-701	9000	308
4000	617	10,000	265
5000	541		

Source: Miller, A. and J. Thompson, *Elements of Meteorology,* Fourth Edition, Charles E. Merrill Publishing Company, Columbus, Ohio.

(a) Use a graphing calculator to make a scatter diagram of the data for atmospheric pressure P at altitude x.

(b) Use the concept of average rate of change to determine whether a linear or exponential function would fit the data better.

(c) The function

$$P(x) = 1013e^{-0.0001341x}$$

approximates the data. Use a graphing calculator to graph P and the data on the same coordinate axes.

(d) Use P to predict the pressures at 1500 m and 11,000 m, and compare them with the actual values of 846 millibars and 227 millibars, respectively.

78. *World Population Growth* Since 2000, world population in millions closely fits the exponential function

$$y = 6079e^{0.0126x},$$

where x is the number of years since 2000.

(a) The world population was about 6555 million in 2006. How closely does the function approximate this value?

(b) Use this model to estimate the population in 2010.

(c) Use this model to predict the population in 2025.

 (d) Explain why this model may not be accurate for 2025.

79. *Traffic Flow* At an intersection, cars arrive randomly at an average rate of 30 cars per hour. Using the function

$$f(x) = 1 - e^{-0.5x},$$

highway engineers estimate the likelihood or probability that at least one car will enter the intersection within a period of x minutes. (*Source:* Mannering, F. and W. Kilareski, *Principles of Highway Engineering and Traffic Analysis,* Second Edition, John Wiley and Sons.)

 (a) Evaluate $f(2)$ and interpret the answer.

(b) Graph f for $0 \le x \le 60$. What happens to the likelihood that at least one car enters the intersection during a 60-minute period?

80. *Growth of E. coli Bacteria* A type of bacteria that inhabits the intestines of animals is named *E. coli* (*Escherichia coli*). These bacteria are capable of rapid growth and can be dangerous to humans—especially children. In one study, *E. coli* bacteria were capable of doubling in number every 49.5 minutes. Their number after x minutes can be modeled by the function

$$N(x) = N_0 e^{0.014x}.$$

(*Source:* Stent, G. S., *Molecular Biology of Bacterial Viruses,* W. H. Freeman.) Suppose $N_0 = 500,000$ is the initial number of bacteria per milliliter.

(a) Make a conjecture about the number of bacteria per milliliter after 99 minutes. Verify your conjecture.

(b) Estimate graphically the time elapsed until there were 25 million bacteria per milliliter.

RELATING CONCEPTS **For individual or group investigation (Exercises 81–86)**

In Exercises 81–86, assume that $f(x) = a^x$, where $a > 1$. **Work these exercises in order.**

81. Is f a one-to-one function? If so, based on **Section 5.1,** what kind of related function exists for f?

82. If f has an inverse function f^{-1}, sketch f and f^{-1} on the same axes.

83. If f^{-1} exists, find an equation for $y = f^{-1}(x)$, using the method described in **Section 5.1.** (You need not solve for y.)

84. If $a = 10$, what is an equation for $y = f^{-1}(x)$? (You need not solve for y.)

85. If $a = e$, what is an equation for $y = f^{-1}(x)$? (You need not solve for y.)

86. If the point (p, q) is on the graph of f, then the point _____ is on the graph of f^{-1}.

5.3 Logarithms and Their Properties

Definition of Logarithm • Common Logarithms • Natural Logarithms • Properties of Logarithms • Change-of-Base Rule

Exponent
$\downarrow$
Exponential form: $a^y = x$
$\uparrow$
Base

Exponent
$\downarrow$
Logarithmic form: $y = \log_a x$
$\uparrow$
Base

Definition of Logarithm

In the exponential equation $2^3 = 8$, 3 is the exponent to which 2 must be raised in order to obtain 8. We call 3 the *logarithm* to the base 2 of 8, abbreviated $3 = \log_2 8$. A logarithm is an *exponent* and thus has the same properties as exponents.

> **Logarithm**
>
> For all positive numbers a, where $a \ne 1$,
>
> $$a^y = x \quad \text{is equivalent to} \quad y = \log_a x.$$
>
> *The expression $\log_a x$ represents the exponent to which the base a must be raised in order to obtain x.*

Exponential Form	Logarithmic Form
$2^3 = 8$	$\log_2 8 = 3$
$\left(\frac{1}{2}\right)^{-4} = 16$	$\log_{1/2} 16 = -4$
$10^5 = 100{,}000$	$\log_{10} 100{,}000 = 5$
$3^{-4} = \frac{1}{81}$	$\log_3\left(\frac{1}{81}\right) = -4$
$5^1 = 5$	$\log_5 5 = 1$
$\left(\frac{3}{4}\right)^0 = 1$	$\log_{3/4} 1 = 0$

In the equation $a^y = x$, y is the exponent to which a must be raised in order to obtain x. We call this exponent a **logarithm,** symbolized by the abbreviation **"log".** The expression $\log_a x$ represents the logarithm in this discussion. The number a is called the **base** of the logarithm, and x is called the **argument** of the expression. It is read **"logarithm with base a of x"** or **"logarithm of x with base a"** or **"base a logarithm of x."**

The table in the margin shows several pairs of *equivalent* statements written in both **exponential form** and **logarithmic form.** We can use these forms to solve equations.

EXAMPLE 1   Solving Logarithmic Equations

Solve each equation by first rewriting it in exponential form.

(a) $x = \log_8 4$ **(b)** $\log_x 16 = 4$ **(c)** $\log_4 x = \dfrac{3}{2}$

Solution

(a)

$$x = \log_8 4$$
$$8^x = 4 \qquad \text{Write in exponential form.}$$
$$(2^3)^x = 2^2 \qquad \text{Write as a power of the same base, 2.}$$
$$2^{3x} = 2^2 \qquad (a^m)^n = a^{mn}$$

If $a^b = a^c$, then $b = c$.

$$3x = 2 \qquad \text{Set exponents equal.}$$
$$x = \frac{2}{3} \qquad \text{Divide by 3.}$$

Since $8^{2/3} = \left(\sqrt[3]{8}\right)^2 = 2^2 = 4$, the solution set is $\left\{\frac{2}{3}\right\}$.

(b)

$$\log_x 16 = 4$$
$$x^4 = 16 \qquad \text{Write in exponential form.}$$

Remember both the positive and negative roots.

$$x = \pm\sqrt[4]{16} \qquad \text{Take fourth roots.}$$
$$x = \pm 2 \qquad \text{Simplify.}$$

Since the base x must be *positive*, reject -2. The solution set is $\{2\}$.

(c)

$$\log_4 x = \frac{3}{2}$$
$$x = 4^{3/2} \qquad \text{Write in exponential form.}$$
$$x = 8 \qquad 4^{3/2} = \left(\sqrt{4}\right)^3 = 8$$

Check that the solution set is $\{8\}$. ●

Common Logarithms

Two important bases for logarithms are 10 and e. Base 10 logarithms are called **common logarithms.** The common logarithm of x is written **log x,** where the base is understood to be 10.

Common Logarithm

For all positive numbers x, $\log x = \log_{10} x.$

Remember that the argument of a common logarithm (and any base logarithm, for that matter) must be a positive number.

FIGURE 28(a) shows a typical graphing calculator screen with several common logarithms evaluated. The first display indicates that 0.7781512504 is (approximately) the exponent to which 10 must be raised in order to obtain 6. The second says that 2 is the exponent to which 10 must be raised in order to obtain 100. This is correct, since $100 = 10^2$. The third display indicates that -4 is the exponent to which 10 must be raised in order to obtain 0.0001. Again, this is correct, since $0.0001 = 10^{-4}$.

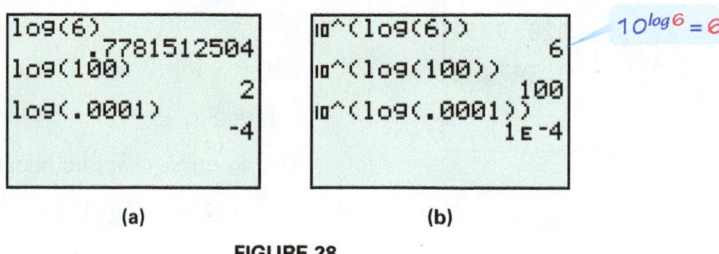

(a)   (b)

FIGURE 28

FIGURE 28(b) shows how the definition of a common logarithm can be applied. *Raising 10 to the power log x gives x as a result.*

EXAMPLE 2   **Evaluating Common Logarithms**

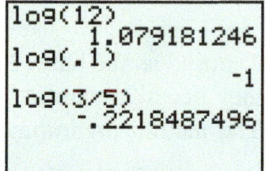

These are the results for
Example 2.

Evaluate each expression.

(a) $\log 12$ **(b)** $\log 0.1$ **(c)** $\log \dfrac{3}{5}$

Solution   A calculator is needed for (a) and (c).

(a) $\log 12 \approx 1.079181246$ **(b)** $\log 0.1 = -1$ **(c)** $\log \dfrac{3}{5} \approx -0.2218487496$ ●

FOR DISCUSSION

Use your calculator to evaluate the common logarithms of $0, -\pi, -4.3$, and -10. Discuss and generalize your results. Make sure that your calculator is set in REAL mode.

In chemistry, the **pH** of a solution is defined as

$$\text{pH} = -\log\,[\text{H}_3\text{O}^+],$$

where $[\text{H}_3\text{O}^+]$ is the hydronium ion concentration in moles per liter.* The pH value is a measure of the acidity or alkalinity of a solution. Pure water has pH 7.0. Substances with pH values greater than 7.0 are alkaline, and substances with pH values less than 7.0 are acidic. It is customary to round pH values to the nearest tenth.

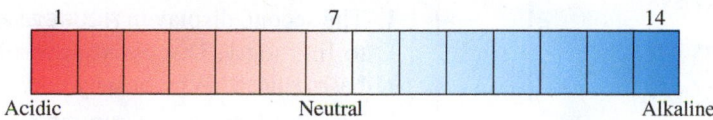

---

*A *mole* is the amount of a substance that contains the same number of molecules as the number of atoms in 12 grams of carbon 12.

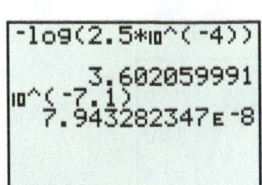

These are the calculations required in **Example 3**.

EXAMPLE 3 **Finding pH and [H₃O⁺]**

(a) Find the pH of a solution with $[H_3O^+] = 2.5 \times 10^{-4}$. Is the solution acidic or alkaline?

(b) Find the hydronium ion concentration of a solution with pH $= 7.1$.

Solution

(a) $pH = -\log[H_3O^+]$

$pH = -\log(2.5 \times 10^{-4})$ Substitute.

$pH \approx 3.6$ Use a calculator.

The solution is acidic because the pH is less than 7.0.

(b) $pH = -\log[H_3O^+]$

$7.1 = -\log[H_3O^+]$ Substitute.

$-7.1 = \log[H_3O^+]$ Multiply by -1.

$[H_3O^+] = 10^{-7.1}$ Write in exponential form.

$[H_3O^+] \approx 7.9 \times 10^{-8}$ Evaluate $10^{-7.1}$ with a calculator.

Looking Ahead to Calculus

The natural logarithmic function, $f(x) = \ln x$, and the reciprocal function, $g(x) = \frac{1}{x}$, have an important relationship in calculus. The rate of change of the natural logarithmic function is calculated by the reciprocal function.

Natural Logarithms

In many practical applications of logarithms, the number e is used as the base. Logarithms with base e are called **natural logarithms**, since they occur in natural situations from life sciences that involve growth and decay. The natural logarithm of a positive number x is written **ln x** (read "**el en x**").

> **Natural Logarithm**
>
> For all positive numbers x, $\ln x = \log_e x$.

When a calculator is used, natural logarithms of numbers are found the same way as common logarithms. The natural logarithm key is usually found in conjunction with the e^x key. **FIGURE 29** shows how the natural logarithmic function and the base e exponential function can be applied. For example, the first display in **FIGURE 29(a)** indicates that $\ln 8 \approx 2.079441542$ is the exponent to which e must be raised in order to obtain 8. This result is also supported in **FIGURE 29(b)**, where it is shown that $e^{\ln 8} = 8$.

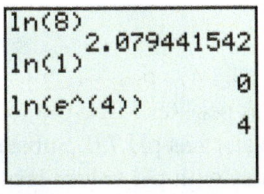

(a)

$e^{\ln x} = x, x > 0$

(b)

FIGURE 29

> **FOR DISCUSSION**
>
> Refer to **FIGURE 29(a)** to answer the following.
>
> **1.** The second display in **FIGURE 29(a)** indicates that $\ln 1 = 0$. Use your calculator to find log 1. Discuss your results. If 1 is the argument, does it matter what base is used? Why or why not?
>
> **2.** The third display in **FIGURE 29(a)** indicates that $\ln(e^4) = 4$. Use the same expression, but replace 4 with the number of letters in your last name. (If it has 4 letters, use the number 5.) What is the result? Can you make a generalization?
>
> **3.** Does your generalization apply to common logarithms as well?

The preceding "For Discussion" suggests the following inverse properties.

$$\log_a a^x = x \qquad \text{for all real numbers } x. \tag{1}$$
$$a^{\log_a x} = x \qquad \text{for all } x > 0. \tag{2}$$

This means that evaluating $\log_a x$ and a^x are inverse operations, much like adding 5 to x and then subtracting 5 are inverse operations. When applying inverse operations on x, the final value is always x.

EXAMPLE 4 **Evaluating Natural Logarithms**

Evaluate each expression.

(a) $\ln 12$ **(b)** $\ln e^{10}$ **(c)** $\ln \sqrt{e}$

Solution

(a) Use a calculator to find $\ln 12 \approx 2.48490665$.

(b) From (1) above, $\ln e^{10} = \log_e e^{10} = 10$.

(c) From (1) above, $\ln \sqrt{e} = \log_e e^{1/2} = \frac{1}{2}$.

Properties of Logarithms

> **Properties of Logarithms**
>
> For $a > 0$, $a \neq 1$, and any real number k, the following hold.
>
> **1.** $\log_a 1 = 0$ **2.** $\log_a a^k = k$ **3.** $a^{\log_a k} = k, \quad k > 0$
>
> For $x > 0$, $y > 0$, $a > 0$, $a \neq 1$, and any real number r,
>
> **Product Rule** **4.** $\log_a xy = \log_a x + \log_a y$
> (The logarithm of the product of two numbers is equal to the sum of the logarithms of the numbers.)
>
> **Quotient Rule** **5.** $\log_a \dfrac{x}{y} = \log_a x - \log_a y$
> (The logarithm of the quotient of two numbers is equal to the difference of the logarithms of the numbers.)
>
> **Power Rule** **6.** $\log_a x^r = r \log_a x$
> (The logarithm of a number raised to a power is equal to the exponent multiplied by the logarithm of the number.)

Property 1 is true because $a^0 = 1$ for any nonzero value of a. Property 2 is verified by writing the equation in exponential form, giving the identity $a^k = a^k$. Property 3 is justified by the fact that $\log_a k$ is the exponent to which a must be raised in order to obtain k. Therefore, by the definition, $a^{\log_a k}$ must equal k. The proof of Property 4, the product rule, follows.

Proof Let $\qquad\qquad m = \log_a x \qquad$ and $\qquad n = \log_a y.$

Then, $\qquad\qquad\qquad a^m = x \qquad$ and $\qquad a^n = y \qquad$ Definition of logarithm

$$a^m \cdot a^n = xy \qquad\qquad\qquad \text{Multiply.}$$
$$a^{m+n} = xy \qquad\qquad\qquad \text{Add exponents.}$$
$$\log_a xy = m + n \qquad\qquad\qquad \text{Definition of logarithm}$$
$$\log_a xy = \log_a x + \log_a y. \qquad \text{Substitute.}$$

Properties 5 and 6, the quotient and power rules, are proved in a similar way.

EXAMPLE 5 **Using the Properties of Logarithms**

Assuming that all variables represent positive real numbers, use the properties of logarithms to rewrite each expression.

(a) $\log 8x$ **(b)** $\log_9 \dfrac{15}{7}$ **(c)** $\log_5 \sqrt{8}$ **(d)** $\log 0.1$

(e) $2^{\log_2 3}$ **(f)** $\log_a \dfrac{x}{yz}$ **(g)** $\log_a \sqrt[3]{m^2}$ **(h)** $\log_b \sqrt[n]{\dfrac{x^3 y^5}{z^m}}$

Solution

(a) $\log 8x = \log 8 + \log x$ Product rule

(b) $\log_9 \dfrac{15}{7} = \log_9 15 - \log_9 7$ Quotient rule

(c) $\log_5 \sqrt{8} = \log_5 8^{1/2} = \dfrac{1}{2} \log_5 8$ Power rule

(d) $\log 0.1 = \log \dfrac{1}{10} = \log 10^{-1} = -1$ Property 2

(e) $2^{\log_2 3} = 3$ Property 3

(f) $\log_a \dfrac{x}{yz} = \log_a x - (\log_a y + \log_a z)$ Quotient and product rules

Distribute the minus sign. $= \log_a x - \log_a y - \log_a z$ Distributive property

(g) $\log_a \sqrt[3]{m^2} = \log_a m^{2/3} = \dfrac{2}{3} \log_a m$ Power rule

(h) $\log_b \sqrt[n]{\dfrac{x^3 y^5}{z^m}} = \log_b \left(\dfrac{x^3 y^5}{z^m} \right)^{1/n}$ $\sqrt[n]{a} = a^{1/n}$

$= \dfrac{1}{n} \log_b \dfrac{x^3 y^5}{z^m}$ *Use parentheses to avoid errors.*   Power rule

$= \dfrac{1}{n} \left(\log_b x^3 + \log_b y^5 - \log_b z^m \right)$ Product and quotient rules

$= \dfrac{1}{n} (3 \log_b x + 5 \log_b y - m \log_b z)$ Power rule

$= \dfrac{3}{n} \log_b x + \dfrac{5}{n} \log_b y - \dfrac{m}{n} \log_b z$ Distributive property   ●

EXAMPLE 6 **Using the Properties of Logarithms**

Use the properties of logarithms to write each expression as a single logarithm with coefficient 1. Assume that all variables represent positive real numbers.

(a) $\log_3 (x + 2) + \log_3 x - \log_3 2$ **(b)** $2 \log_a m - 3 \log_a n$

(c) $\dfrac{1}{2} \log_b m + \dfrac{3}{2} \log_b 2n - \log_b m^2 n$

Solution

(a) $\log_3 (x + 2) + \log_3 x - \log_3 2 = \log_3 \dfrac{(x + 2)x}{2}$ Product and quotient rules

(b) $2 \log_a m - 3 \log_a n = \log_a m^2 - \log_a n^3$ Power rule

$= \log_a \dfrac{m^2}{n^3}$ Quotient rule

(c) $\dfrac{1}{2}\log_b m + \dfrac{3}{2}\log_b 2n - \log_b m^2 n$

$$= \log_b m^{1/2} + \log_b (2n)^{3/2} - \log_b m^2 n \qquad \text{Power rule}$$

$$= \log_b \frac{m^{1/2}(2n)^{3/2}}{m^2 n} \qquad \text{Product and quotient rules}$$

$$= \log_b \frac{2^{3/2} n^{1/2}}{m^{3/2}} \qquad \text{Rules for exponents}$$

$$= \log_b \left(\frac{2^3 n}{m^3}\right)^{1/2} \qquad \text{Rules for exponents}$$

$$= \log_b \sqrt{\frac{8n}{m^3}} \qquad \text{Definition of } a^{1/n} \qquad \bullet$$

> **CAUTION** *There is no property of logarithms that allows us to rewrite a logarithm of a sum or difference.* That is why, in **Example 6(a)**, $\log_3(x+2)$ was *not* written as $\log_3 x + \log_3 2$. Remember, $\log_3 x + \log_3 2 = \log_3(x \cdot 2)$.
>
> The distributive property does not apply here, since $\log_3(x+2)$ is one term; "log" is a function name, *not* a factor. Also in the last two lines of **Example 6(c)** note that $\log_b a^{1/2} = \log_b \sqrt{a}$ and that $\log_b a^{1/2} \neq \sqrt{\log_b a}$.

Change-of-Base Rule

We can use a calculator to find logarithms for bases other than 10 and e.

> ### Change-of-Base Rule
>
> For any positive real numbers x, a, and b, where $a \neq 1$ and $b \neq 1$,
>
> $$\log_a x = \frac{\log_b x}{\log_b a}.$$

Proof Let

$$y = \log_a x.$$

$$a^y = x \qquad \text{Write in exponential form.}$$

$$\log_b a^y = \log_b x \qquad \text{Take the logarithm of each side.}$$

$$y \log_b a = \log_b x \qquad \text{Power rule}$$

$$y = \frac{\log_b x}{\log_b a} \qquad \text{Divide each side by } \log_b a.$$

$$\log_a x = \frac{\log_b x}{\log_b a} \qquad \text{Substitute } \log_a x \text{ for } y.$$

Any positive number other than 1 can be used for base b in the change-of-base rule, but usually the only practical bases are e and 10, since calculators typically give logarithms for only these two bases. In this case the change-of-base formula becomes

$$\log_a x = \frac{\log x}{\log a}, \quad \text{or} \quad \log_a x = \frac{\ln x}{\ln a}.$$

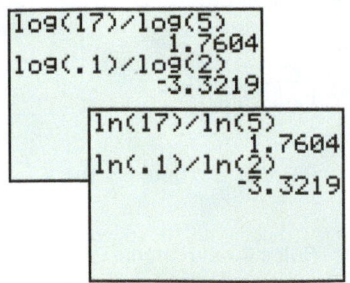

FIGURE 30

EXAMPLE 7 **Using the Change-of-Base Rule**

Use logarithms and the change-of-base rule to evaluate each expression. Round to four decimal places.

(a) $\log_5 17$ **(b)** $\log_2 0.1$

Solution With the calculator set to four decimal places, **FIGURE 30** shows how the same results are found by using either common or natural logarithms.

(a) $\log_5 17 = \dfrac{\log 17}{\log 5} = \dfrac{\ln 17}{\ln 5} \approx 1.7604$

(b) $\log_2 0.1 = \dfrac{\log 0.1}{\log 2} = \dfrac{\ln 0.1}{\ln 2} \approx -3.3219$

FOR DISCUSSION

1. Without using your calculator, determine exact values of $\log_5 5$ and $\log_5 25$. Then, use the fact that 17 is between 5 and 25 to determine between what two consecutive integers $\log_5 17$ must lie. Finally, refer to **Example 7(a)** to support your answer.

2. Without using your calculator, determine exact values of $\log_2 \left(\frac{1}{16}\right)$ and $\log_2 \left(\frac{1}{8}\right)$. Then, use the fact that 0.1 is between $\frac{1}{16}$ and $\frac{1}{8}$ to determine between what two consecutive integers $\log_2 0.1$ must lie. Finally, refer to **Example 7(b)** to support your answer.

EXAMPLE 8 **Modeling Diversity of Species**

Larger areas of land tend to have greater numbers of different species of animals than do smaller areas of land. One measure of the diversity of species in an ecological community is the **index of diversity**

$$H = -(P_1 \log_2 P_1 + P_2 \log_2 P_2 + \cdots + P_n \log_2 P_n),$$

where $P_1, P_2, \ldots, P_n$ are the proportions of a sample belonging to each of n species found in the sample. For example, if there are 15 robins, 30 sparrows, and 5 woodpeckers in a region, then there are $n = 3$ species of birds and a total of 50 birds in all. Thus $P_1 = \frac{15}{50}$, $P_2 = \frac{30}{50}$, and $P_3 = \frac{5}{50}$.

Find the index of diversity in a community with 2 species, one with 90 members and the other with 10.

Solution Since there are a total of 100 members in the community, $P_1 = \frac{90}{100} = 0.9$ and $P_2 = \frac{10}{100} = 0.1$. We use the change-of-base rule to find $\log_2 0.9$ and $\log_2 0.1$.

$$\log_2 0.9 = \left(\frac{\ln 0.9}{\ln 2}\right) \approx -0.152 \quad \text{and} \quad \log_2 0.1 = \left(\frac{\ln 0.1}{\ln 2}\right) \approx -3.32$$

$$H = -(0.9 \log_2 0.9 + 0.1 \log_2 0.1) \qquad \text{Substitute } P_1 = 0.9 \text{ and } P_2 = 0.1.$$
$$\approx -0.9(-0.152) - 0.1(-3.32) \qquad \log_2 0.9 \approx -0.152 \text{ and } \log_2 0.1 \approx -3.32$$
$$\approx 0.469 \qquad \text{Multiply and add.}$$

The interpretation of this index varies with the number of species involved. For two equally distributed species, the measure of diversity is 1. If there is little diversity, H is close to 0. Here, since $H \approx 0.5$, there is neither great nor little diversity.

5.3 Exercises

Concept Check In Exercises 1 and 2, match the logarithm in Column I with its value in Column II. Remember that $\log_a x$ is the exponent to which a must be raised in order to obtain x.

I	II		I	II
1. (a) $\log_2 16$	**A.** 0		**2.** (a) $\log_3 81$	**A.** -2
(b) $\log_3 1$	**B.** $\dfrac{1}{2}$		(b) $\log_3 \dfrac{1}{3}$	**B.** -1
(c) $\log_{10} 0.1$	**C.** 4		(c) $\log_{10} 0.01$	**C.** 0
(d) $\log_2 \sqrt{2}$	**D.** -3		(d) $\log_6 \sqrt{6}$	**D.** $\dfrac{1}{2}$
(e) $\log_e \dfrac{1}{e^2}$	**E.** -1		(e) $\log_e 1$	**E.** $\dfrac{9}{2}$
(f) $\log_{1/2} 8$	**F.** -2		(f) $\log_3 27^{3/2}$	**F.** 4

Checking Analytic Skills For each statement, write an equivalent statement in logarithmic form. **Do not use a calculator.**

3. $3^4 = 81$

4. $2^5 = 32$

5. $\left(\dfrac{1}{2}\right)^{-4} = 16$

6. $\left(\dfrac{2}{3}\right)^{-3} = \dfrac{27}{8}$

7. $10^{-4} = 0.0001$

8. $\left(\dfrac{1}{100}\right)^{-2} = 10{,}000$

9. $e^0 = 1$

10. $e^{1/3} = \sqrt[3]{e}$

Checking Analytic Skills For each statement, write an equivalent statement in exponential form. **Do not use a calculator.**

11. $\log_6 36 = 2$

12. $\log_5 5 = 1$

13. $\log_{\sqrt{3}} 81 = 8$

14. $\log_4 \dfrac{1}{64} = -3$

15. $\log_{10} 0.001 = -3$

16. $\log_3 \sqrt[3]{9} = \dfrac{2}{3}$

17. $\log \sqrt{10} = 0.5$

18. $\ln e^6 = 6$

Solve each equation. Give the exact answer.

19. $\log_5 125 = x$

20. $\log_3 81 = x$

21. $\log_x 3^{12} = 24$

22. $\log_x 5^2 = 6$

23. $\log_6 x = -3$

24. $\log_4 x = -\dfrac{1}{6}$

25. $\log_x 16 = \dfrac{4}{3}$

26. $\log_x 81 = 4$

27. $\log_2 (x + 1) = 3$

28. $\log_3 (x - 1) = 2$

29. $\log_9 \dfrac{\sqrt[4]{27}}{3} = x$

30. $\log_{1/4} \dfrac{16^2}{2^{-3}} = x$

Concept Check Simplify each expression.

31. (a) $3^{\log_3 7}$

(b) $4^{\log_4 9}$

(c) $12^{\log_{12} 4}$

(d) $a^{\log_a k}$ $(k > 0, a > 0, a \neq 1)$

32. (a) $\log_3 3^{19}$

(b) $\log_4 4^{17}$

(c) $\log_{12} 12^{1/3}$

(d) $\log_a \sqrt{a}$ $(a > 0, a \neq 1)$

33. (a) $\log_3 1$

(b) $\log_4 1$

(c) $\log_{12} 1$

(d) $\log_a 1$ $(a > 0, a \neq 1)$

34. (a) Explain in your own words the meaning of $\log_a x$. (b) In the expression $\log_a x$, why must x be nonnegative?

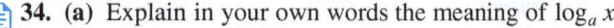

Checking Analytic Skills Evaluate each expression. **Do not use a calculator.**

35. $\log 10^{1.5}$

36. $\log 10^{4.3}$

37. $\log 10^{\sqrt{5}}$

38. $\log 10^{\sqrt{3}}$

39. $\ln e^{2/3}$

40. $\ln e^{0.5}$

41. $\ln e^{\pi}$

42. $\ln e^{\sqrt{6}}$

43. $\sqrt{7} \ln e^{\sqrt{7}}$

44. $\sqrt{2} \ln e^{\sqrt{2}}$

Use a calculator to find a decimal approximation for each common or natural logarithm.

45. log 43

46. log 1247

47. log 0.783

48. log 0.014

49. log 28³

50. log (47 × 93)

51. ln 43

52. ln 1247

53. ln 0.783

54. ln 0.014

55. ln 28³

56. ln (47 × 93)

*Refer to **Example 3**. Find the pH for each substance with the given hydronium ion $\left[\mathrm{H_3O^+}\right]$ concentration.*

57. Grapefruit, 6.3×10^{-4}

58. Limes, 1.6×10^{-2}

59. Crackers, 3.9×10^{-9}

60. Sodium hydroxide (lye), 3.2×10^{-14}

*Refer to **Example 3**. Find the hydronium ion $\left[\mathrm{H_3O^+}\right]$ concentration for each substance with the given pH.*

61. Soda pop, 2.7

62. Wine, 3.4

63. Beer, 4.8

64. Drinking water, 6.5

Use the properties of logarithms to rewrite each logarithm if possible. Assume that all variables represent positive real numbers.

65. $\log_3 \dfrac{2}{5}$

66. $\log_4 \dfrac{6}{7}$

67. $\log_2 \dfrac{6x}{y}$

68. $\log_3 \dfrac{4p}{q}$

69. $\log_5 \dfrac{5\sqrt{7}}{3m}$

70. $\log_2 \dfrac{2\sqrt{3}}{5p}$

71. $\log_4 (2x + 5y)$

72. $\log_6 (7m + 3q)$

73. $\log_k \dfrac{pq^2}{m}$

74. $\log_z \dfrac{x^5 y^3}{3}$

75. $\log_m \sqrt{\dfrac{r^3}{5z^5}}$

76. $\log_p \sqrt[3]{\dfrac{m^5}{kt^2}}$

Use the properties of logarithms to rewrite each expression as a single logarithm with coefficient 1. Assume that all variables represent positive real numbers.

77. $\log_a x + \log_a y - \log_a m$

78. $(\log_b k - \log_b m) - \log_b a$

79. $2 \log_m a - 3 \log_m b^2$

80. $\dfrac{1}{2} \log_y p^3 q^4 - \dfrac{2}{3} \log_y p^4 q^3$

81. $2 \log_a (z - 1) + \log_a (3z + 2), z > 1$

82. $\log_b (2y + 5) - \dfrac{1}{2} \log_b (y + 3)$

83. $-\dfrac{2}{3} \log_5 5m^2 + \dfrac{1}{2} \log_5 25m^2$

84. $-\dfrac{3}{4} \log_3 16p^4 - \dfrac{2}{3} \log_3 8p^3$

85. $3 \log x - 4 \log y$

86. $\dfrac{1}{2} \log x - \dfrac{1}{3} \log y - 2 \log z$

87. $\ln (a + b) + \ln a - \dfrac{1}{2} \ln 4$

88. $\dfrac{4}{3} \ln m - \dfrac{2}{3} \ln 8n - \ln m^3 n^2$

Use the change-of-base rule to find an approximation for each logarithm.

89. $\log_5 10$

90. $\log_9 12$

91. $\log_{15} 5$

92. $\log_{1/2} 3$

93. $\log_{100} 83$

94. $\log_{200} 175$

95. $\log_{2.9} 7.5$

96. $\log_{5.8} 12.7$

RELATING CONCEPTS   For individual or group investigation (Exercises 97–102)

Work Exercises 97–102 in order.

97. Use the terminology of **Chapter 2** to explain how the graph of $y = -3^x + 7$ can be obtained from the graph of $y = 3^x$.

98. Graph $y_1 = 3^x$ and $y_2 = -3^x + 7$ in the window $[-5, 5]$ by $[-10, 10]$ to support your answer in **Exercise 97**.

99. Use the capabilities of your calculator to find an approximation for the x-coordinate of the x-intercept of the graph of y_2 in **Exercise 98**.

100. Solve $0 = -3^x + 7$ for x, expressing x in terms of a base 3 logarithm.

101. Use the change-of-base rule to find an approximation for the solution of the equation in **Exercise 100**.

102. Compare your results in **Exercises 99** and **101**.

In Exercises 103–106, the equations are identities because they are true for all real numbers. Use properties of logarithms to simplify the expression on the left side of the equation so that it equals the expression on the right side, where x is any real number.

103. $\ln|x + \sqrt{x^2 + 3}| + \ln|x - \sqrt{x^2 + 3}| = \ln 3$

104. $\ln|x^2 - \sqrt{x^4 + 1}| + \ln|x^2 + \sqrt{x^4 + 1}| = 0$

105. $\frac{1}{3}\ln\left(\frac{x^2 + 1}{5}\right) - \frac{1}{3}\ln\left(\frac{x^2 + 4}{5}\right) = \ln\sqrt[3]{\frac{x^2 + 1}{x^2 + 4}}$

106. $\frac{1}{2}\ln\left(\frac{x^2}{7}\right) - \frac{1}{2}\ln\left(\frac{x^4 + x^2}{7}\right) = \ln\sqrt{\frac{1}{x^2 + 1}}$

(Modeling) *Solve each problem.*

107. *Diversity of Species*   (Refer to **Example 8.**) Suppose a sample of a small community shows two species with 50 individuals each. Find the index of diversity

$$H = -(P_1 \log_2 P_1 + P_2 \log_2 P_2 + \cdots + P_n \log_2 P_n).$$

108. *Diversity of Species*   A virgin forest in northwestern Pennsylvania has four species of large trees with the following proportions of each: hemlock, 0.521; beech, 0.324; birch, 0.081; maple, 0.074. Find the index of diversity H, using the formula in **Exercise 107.**

109. *Diversity of Species*   The number of species in a sample is approximated by

$$S(n) = a \ln\left(1 + \frac{n}{a}\right),$$

where n is the number of individuals in the sample and a is a constant that indicates the diversity of species in the community. If $a = 0.36$, find $S(n)$ for each value of n. (*Hint:* $S(n)$ must be a whole number.)

(a) 100 **(b)** 200 **(c)** 150 **(d)** 10

110. **(a)** Prove the quotient rule of logarithms.

$$\log_a \frac{x}{y} = \log_a x - \log_a y$$

(b) Prove the power rule of logarithms.

$$\log_a x^r = r \log_a x$$

SECTIONS 5.1–5.3 Reviewing Basic Concepts

1. Is the function defined by the table a one-to-one function? Explain your answer.

x	-2	2	4	6	8
y	4	4	8	12	16

2. Give the inverse of each function.

(a)

x	7	8	9	10
y	12	21	32	45

(b) $f(x) = \dfrac{x + 5}{4}$

3. Graph $f(x) = 2x + 3$ and $f^{-1}(x)$ by hand on the same axes.

4. Graph $f(x) = 3^{-x}$ by hand.

5. Graph $f(x) = 2^{x+1} - 1$.
(a) Compare the graph of f with the graph of $y = 2^x$.
(b) Give the domain and range of f.
(c) Identify any horizontal asymptotes.
(d) Determine if f is a one-to-one function.

6. Solve $4^{2x} = 8$ analytically. Use a calculator to support your answer.

7. Find the interest earned on \$600 at 4% compounded quarterly for 3 years.

8. Evaluate each logarithm without using a calculator.

(a) $\log \dfrac{1}{\sqrt{10}}$ **(b)** $2 \ln e^{1.5}$ **(c)** $\log_2 4$

9. Use the properties of logarithms to rewrite $\log \dfrac{3x^2}{5y}$.

10. Use the properties of logarithms to write

$$\ln 4 + \ln x - 3 \ln 2$$

as a single logarithm with coefficient 1.

5.4 Logarithmic Functions

Graphs of Logarithmic Functions • Finding an Inverse of an Exponential Function • A Logarithmic Model

The exponential function $f(x) = a^x$, $a > 1$, is increasing on its domain. If $0 < a < 1$, the function is decreasing on its domain. Thus, for all allowable bases a, function f is one-to-one and has an inverse. We can find the rule for f^{-1} analytically, using the three steps described in **Section 5.1.**

$f(x) = a^x$	Exponential function
$y = a^x$	Replace $f(x)$ with y.
$x = a^y$	Interchange x and y.
$y = \log_a x$	Write in logarithmic form.
$f^{-1}(x) = \log_a x$	Replace y with $f^{-1}(x)$.

The last equation indicates that the **logarithmic function** with base a is the *inverse* of the exponential function with base a. To confirm this, use properties of logarithms from **Section 5.3** to show that $(f \circ f^{-1})(x) = x$ and $(f^{-1} \circ f)(x) = x$.

$$(f \circ f^{-1})(x) = f(f^{-1}(x)) = a^{\log_a x} = x$$

$$(f^{-1} \circ f)(x) = f^{-1}(f(x)) = \log_a a^x = x$$

Thus, the functions $f(x) = a^x$ and $g(x) = \log_a x$ are inverse functions.

Logarithmic Function

If $a > 0$, $a \neq 1$, and $x > 0$, then

$$f(x) = \log_a x$$

is the **logarithmic function with base a.**

Graphing an Inverse

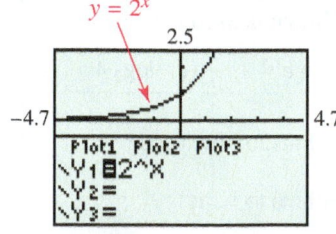

$y = 2^x$

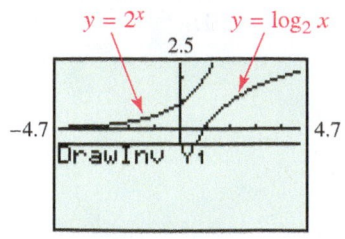

When the TI-84 Plus is directed to draw the inverse of $y = 2^x$, the calculator graphs $y = \log_2 x$.

Graphs of Logarithmic Functions

EXAMPLE 1 **Graphing Logarithmic Functions**

Graph each logarithmic function, and determine its domain and range.

(a) $F(x) = \log_2 x$ **(b)** $G(x) = \log_{1/2} x$

Solution

(a) Interchange the x- and y-coordinates in the table for $f(x) = 2^x$ from **Section 5.2,** and plot the new data to obtain the graph of $F(x) = \log_2 x$. See **FIGURE 31** on the next page. The domain of F is $(0, \infty)$, which is the range of f. Similarly, the range of F is $(-\infty, \infty)$, which is the domain of f.

(b) Use the table for $g(x) = \left(\frac{1}{2}\right)^x$ from **Section 5.2** and a similar procedure to get the graph of $G(x) = \log_{1/2} x$. See **FIGURE 32** on the next page. The domain of G is $(0, \infty)$ and the range is $(-\infty, \infty)$.

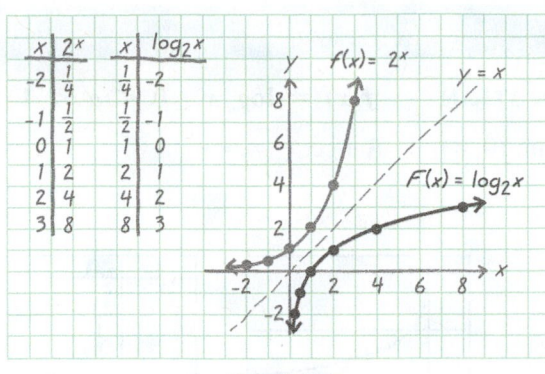

FIGURE 31

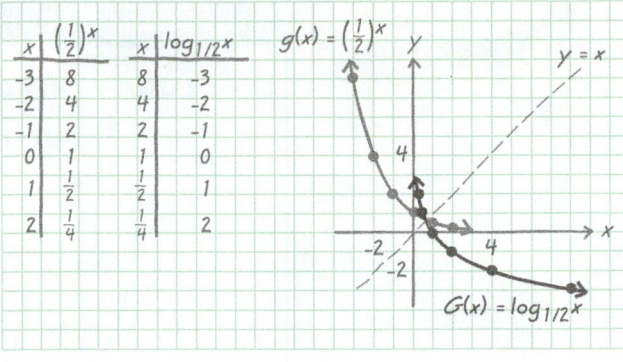

FIGURE 32

NOTE Another way to graph a logarithmic function, such as $y = \log_2 x$, is to write it in equivalent exponential form, $2^y = x$. Then substitute values for y and calculate x. *Be careful to plot the ordered pairs correctly when using this method.*

The Natural Logarithm

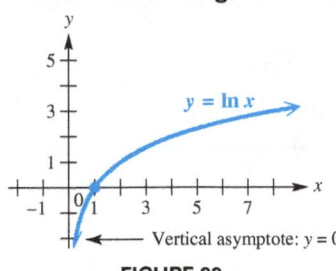

FIGURE 33

The most important logarithmic functions are $y = \ln x$ (base e) and $y = \log x$ (base 10), which are graphed in FIGURES 33 and 34.

Because e and 10 are the only logarithmic bases on most graphing calculators, we must use the change-of-base rule to graph a logarithmic function for some other base. For example, to graph $y = \log_2 x$, we graph either

$$y = \frac{\log x}{\log 2} \quad \text{or} \quad y = \frac{\ln x}{\ln 2}. \qquad \text{Change-of-base formulas}$$

We now summarize information about the graphs of logarithmic functions.

The Common Logarithm

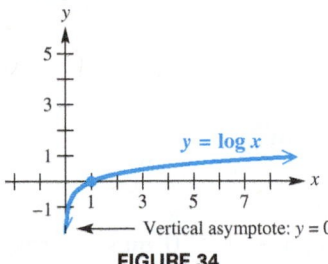

FIGURE 34

FUNCTION CAPSULE

LOGARITHMIC FUNCTION $\quad f(x) = \log_a x, \quad a > 1$

Domain: $(0, \infty)$ Range: $(-\infty, \infty)$

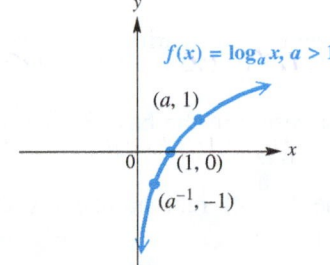

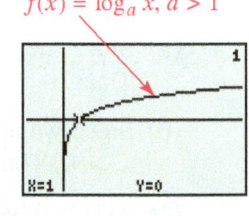

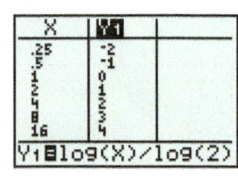

$$Y_1 = \log_2 X = \frac{\log X}{\log 2}$$

FIGURE 35

- $f(x) = \log_a x, a > 1$, is increasing and continuous on its entire domain, $(0, \infty)$.
- The y-axis is the vertical asymptote as $x \to 0$ from the right.
- The graph passes through the points $(a^{-1}, -1)$, $(1, 0)$, and $(a, 1)$.

FUNCTION CAPSULE

LOGARITHMIC FUNCTION $\quad f(x) = \log_a x, \quad 0 < a < 1$

Domain: $(0, \infty)$ $\qquad$ Range: $(-\infty, \infty)$

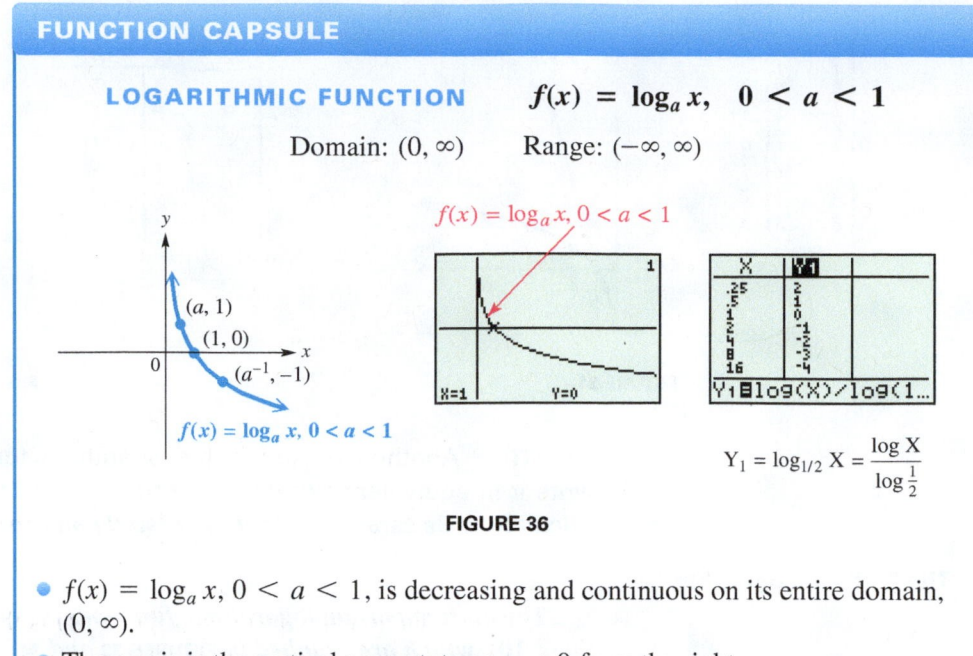

$$Y_1 = \log_{1/2} X = \frac{\log X}{\log \frac{1}{2}}$$

FIGURE 36

- $f(x) = \log_a x, 0 < a < 1$, is decreasing and continuous on its entire domain, $(0, \infty)$.
- The y-axis is the vertical asymptote as $x \rightarrow 0$ from the right.
- The graph passes through the points $(a, 1)$, $(1, 0)$, and $(a^{-1}, -1)$.

A function of the form $y = \log_a f(x)$ is defined only for values for which $f(x) > 0$. This fact is used in **Example 2,** where we discuss the domains of functions in this form.

EXAMPLE 2 **Determining Domains of Logarithmic Functions**

Find the domain of each function.

(a) $f(x) = \log_2 (x - 1)$ $\qquad$ **(b)** $f(x) = (\log_3 x) - 1$

(c) $f(x) = \log_3 |x|$ $\qquad$ **(d)** $f(x) = \ln (x^2 - 4)$

Solution

(a) Since the argument of a logarithm must be positive, $x - 1 > 0$ must be true. Thus, $x > 1$. The domain is $(1, \infty)$.

(b) Since we are interested in the base 3 logarithm of x (not $x - 1$), we must have $x > 0$. The domain is $(0, \infty)$.

(c) $|x| > 0$ is true for all real numbers x except 0. Therefore, the domain is $(-\infty, 0) \cup (0, \infty)$.

(d) We use the steps first introduced in **Section 3.3** to solve the quadratic inequality

$$x^2 - 4 > 0.$$

Step 1 $\quad$ First find the values of x that satisfy $x^2 - 4 = 0$.

$$x^2 - 4 = 0 \qquad \text{\color{blue}Corresponding quadratic equation}$$
$$(x + 2)(x - 2) = 0 \cdot \qquad \text{\color{blue}Factor.}$$
$$x + 2 = 0 \quad \text{or} \quad x - 2 = 0 \qquad \text{\color{blue}Zero-product property}$$
$$x = -2 \quad \text{or} \qquad x = 2 \qquad \text{\color{blue}Solve each equation.}$$

Step 2 The two numbers -2 and 2 divide a number line into the three intervals shown in **FIGURE 37**.

Interval A	Interval B	Interval C
$(-\infty, -2)$	$(-2, 2)$	$(2, \infty)$

Use open circles, since the inequality does not allow solutions of the equation.

FIGURE 37

Step 3 Choose a test value in each interval to see if it satisfies the original inequality $x^2 - 4 > 0$.

Interval	Test Value x	Is $x^2 - 4 > 0$ True or False?
A: $(-\infty, -2)$	-3	$(-3)^2 - 4 > 0$? $5 > 0$ True
B: $(-2, 2)$	0	$0^2 - 4 > 0$? $-4 > 0$ False
C: $(2, \infty)$	3	$3^2 - 4 > 0$? $5 > 0$ True

The domain of $f(x) = \ln(x^2 - 4)$ is the union of the values in Intervals A and C, $(-\infty, -2) \cup (2, \infty)$.

EXAMPLE 3   **Graphing Translated Logarithmic Functions**

Graph each function. Give the asymptote, x-intercept, domain, and range, and tell whether the function is increasing or decreasing on its domain.

(a) $y = \log_2(x - 1)$ **(b)** $y = (\log_3 x) - 1$

Solution

(a) Because the argument is $x - 1$, the graph of $y = \log_2(x - 1)$ is the graph of $y = \log_2 x$ shifted 1 unit to the right. The vertical asymptote also moves 1 unit to the right, so its equation is $x = 1$. The x-intercept is $(2, 0)$. The domain is $(1, \infty)$, as found in **Example 2(a)**, and the range is $(-\infty, \infty)$. The function is always increasing on its domain. See **FIGURE 38**.

This graph can also be plotted by using the equivalent exponential form.

$$y = \log_2(x - 1)$$
$$x - 1 = 2^y \qquad \text{Write in exponential form.}$$
$$x = 2^y + 1 \qquad \text{Add 1.}$$

Using this equation, we choose values for y and calculate the corresponding x-values.

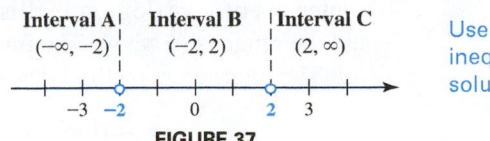

This table numerically supports the result in **Example 3(a)**. The domain of

$$y = \log_2(x - 1),$$

entered here as

$$Y_1 = \log(X - 1)/\log 2,$$

consists of real numbers greater than 1.

x	y
$\frac{5}{4}$	-2
$\frac{3}{2}$	-1
2	0
3	1
5	2

In the equation $x = 2^y + 1$, choose values for y and calculate x.

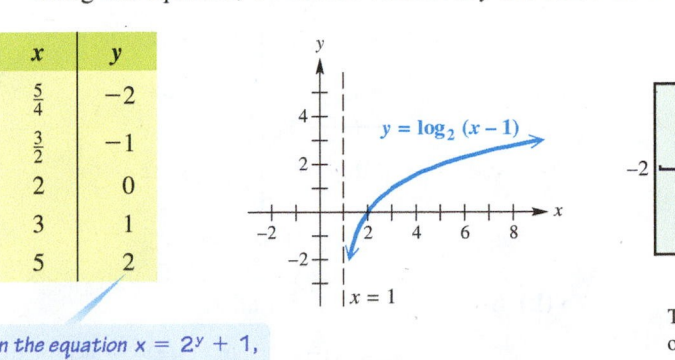

This calculator graph was obtained by using

$$y = \frac{\log(x - 1)}{\log 2}.$$

FIGURE 38

(continued)

(b) Here, 1 is subtracted from $\log_3 x$, so the graph of $y = \log_3 x$ is shifted 1 unit downward. The vertical asymptote is the y-axis, or $x = 0$, and is not affected. The x-intercept of $y = (\log_3 x) - 1$ is $(3, 0)$. The domain is $(0, \infty)$, from **Example 2(b)**, and the range is $(-\infty, \infty)$. The function is always increasing. See **FIGURE 39**.

The alternative method described in part (a) can also be used.

$$y = (\log_3 x) - 1$$
$$y + 1 = \log_3 x \qquad \text{Add 1.}$$
$$x = 3^{y+1} \qquad \text{Write in exponential form.}$$

Again, choose y-values and calculate the corresponding x-values.

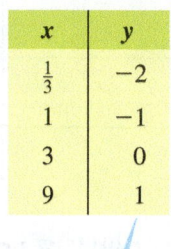

x	y
$\frac{1}{3}$	-2
1	-1
3	0
9	1

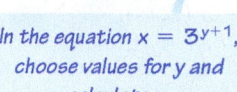

In the equation $x = 3^{y+1}$, choose values for y and calculate x.

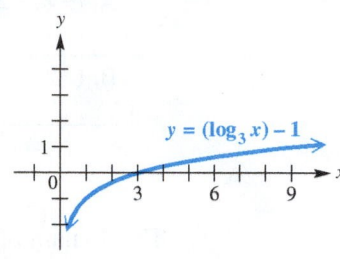

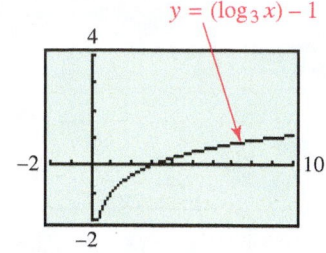

This calculator graph was obtained by using

$$y = \frac{\log x}{\log 3} - 1.$$

FIGURE 39

EXAMPLE 4 **Determining Symmetry**

Show that the graph of each function is symmetric with respect to the y-axis.

(a) $f(x) = \log_3 |x|$ **(b)** $f(x) = \ln(x^2 - 4)$

Solution

(a) Since $|-x| = |x|$ for all x,

$$f(-x) = \log_3 |-x| = \log_3 |x| = f(x).$$

Thus $f(-x) = f(x)$, and the graph is symmetric with respect to the y-axis. See **FIGURE 40**.

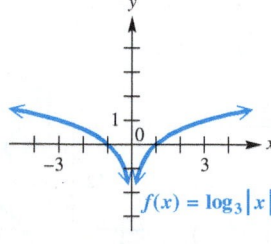

FIGURE 40

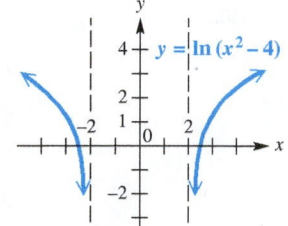

FIGURE 41

(b) Since $x^2 = (-x)^2$ for all x,

$$f(-x) = \ln\left[(-x)^2 - 4\right] = \ln(x^2 - 4) = f(x).$$

Thus, $f(-x) = f(x)$, and the graph is symmetric with respect to the y-axis. See **FIGURE 41**.

A student wanted to support the power rule for logarithms (Property 6 from **Section 5.3**): $\log_a x^r = r \log_a x$. The student defined Y_1 and Y_2, as shown in the screen on the left, and expected the two graphs to be the same. However, the graph of Y_1 was different from the graph of Y_2, as shown.

$Y_1 = \log (X)^2$

$Y_2 = 2 \log (X)$

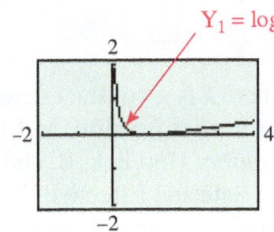
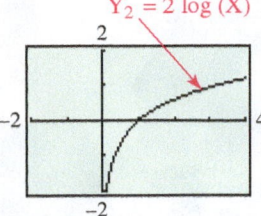

What Went Wrong? How can the student change the input for Y_1 to obtain the same graph as that of Y_2?

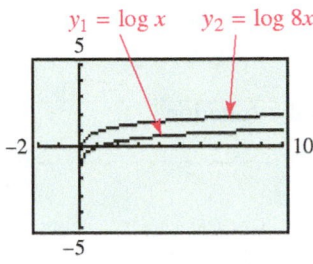

$y_1 = \log x$ $y_2 = \log 8x$

The vertical distance between y_2 and y_1 is $\log 8$.

FIGURE 42

EXAMPLE 5   Using a Property of Logarithms to Translate a Graph

Describe how to graph $f(x) = \log 8x$ by translation.

Solution By the product rule for logarithms, $\log 8x = \log 8 + \log x$, so we can obtain the graph of $y_2 = \log 8x$ by shifting the graph of $y_1 = \log x$ upward $\log 8 \approx 0.90309$ unit. **FIGURE 42** shows the graph.

Finding an Inverse of an Exponential Function

EXAMPLE 6   Finding the Inverse of an Exponential Function

Find analytically the inverse of the function $f(x) = -2^x + 3$. Then graph both functions, and discuss the relationship between the two graphs.

Solution The function is one-to-one because it is a reflection and translation of the one-to-one function $y = 2^x$. Find the equation of the inverse function.

$f^{-1}(x) = \log_2(-x + 3)$

$f(x) = -2^x + 3$

$$
\begin{array}{ll}
f(x) = -2^x + 3 & \\
y = -2^x + 3 & \text{Replace } f(x) \text{ with } y. \\
x = -2^y + 3 & \text{Interchange } x \text{ and } y. \\
2^y = -x + 3 & \text{Add } 2^y \text{ and subtract } x. \\
y = \log_2(-x + 3) & \text{Write in logarithmic form.} \\
f^{-1}(x) = \log_2(-x + 3) & \text{Replace } y \text{ with } f^{-1}(x).
\end{array}
$$

FIGURE 43

FIGURE 43 shows the graphs of both f and f^{-1}. The table compares some of the features of these inverses. Notice how the roles of x and y are reversed in f and f^{-1}.

Function	Domain	Range	x-value of the x-intercept	y-value of the y-intercept	Asymptote
$f(x) = -2^x + 3$	$(-\infty, \infty)$	$(-\infty, 3)$	$\log_2 3 \approx 1.58$	2	Horizontal: $y = 3$
$f^{-1}(x) = \log_2(-x + 3)$	$(-\infty, 3)$	$(-\infty, \infty)$	2	$\log_2 3 \approx 1.58$	Vertical: $x = 3$

Answer to What Went Wrong?

In this case, $\log (X)^2$ is being calculated as $\log (X) * \log (X)$, not $2 \log (X)$. With some calculators, the first equation should be entered as $Y_1 = (\log (X^2))(X > 0)$.

A Logarithmic Model

> **EXAMPLE 7** **Modeling Drug Concentration**

The concentration of a drug in the bloodstream decreases with time. The intervals of time in hours when the drug should be administered are given by

$$T = \frac{1}{k} \ln \frac{C_2}{C_1},$$

where k is a constant determined by the drug in use, C_2 is the concentration at which the drug is harmful, and C_1 is the concentration below which the drug is ineffective. (*Source:* Horelick, B. and S. Koont, "Applications of Calculus to Medicine: Prescribing Safe and Effective Dosage," *UMAP Module 202.*) For example, if $T = 4$, the drug should be administered every 4 hours.

Suppose that for a certain drug, $k = \frac{1}{3}$, $C_2 = 5$, and $C_1 = 2$. How often should the drug be administered?

Solution Substitute the given values in the equation of the function.

$$T = \frac{1}{k} \ln \frac{C_2}{C_1} = \frac{1}{\frac{1}{3}} \ln \frac{5}{2} \approx 2.75$$

The drug should be given about every $2\frac{3}{4}$ hours.

5.4 Exercises

Checking Analytic Skills *The graph of an exponential function f is given, with three points labeled. Sketch the graph of f^{-1} by hand, labeling three points on the graph. For f^{-1}, also state the domain, the range, whether it increases or decreases on its domain, and the equation of its vertical asymptote.*
Do not use a calculator.

1.
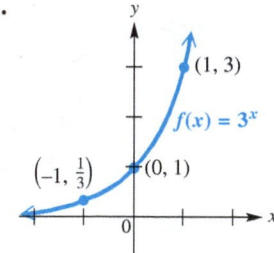

2.
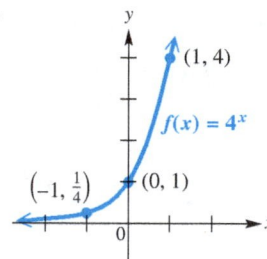

3.
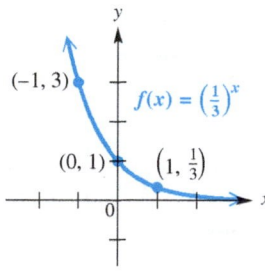

4.
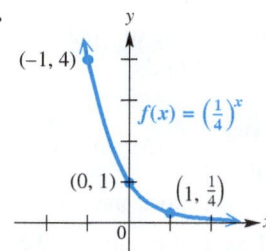

5.
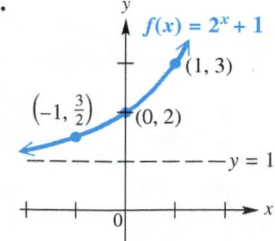

6.
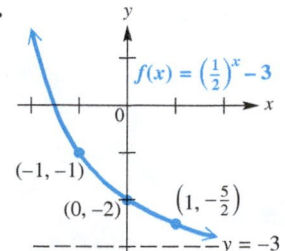

7. In **Exercises 1–6**, each function f is an exponential function. Therefore, each function f^{-1} is a(n) _____ function.

 8. Compare the characteristics of the graph of $f(x) = \log_a x$ with those of the graph of $f(x) = a^x$ in **Section 5.2.** Make a list of characteristics that reinforce the idea that these are inverse functions.

Find the domain of each logarithmic function analytically. You may wish to support your answer graphically.

9. $y = \log 2x$

10. $y = \log \dfrac{1}{2} x$

11. $y = \log (-x)$

12. $y = \log \left(-\dfrac{1}{2} x \right)$

13. $y = \ln (x^2 + 7)$

14. $y = \ln (x^4 + 8)$

15. $y = \ln (-x^2 + 4)$

16. $y = \ln (-x^2 + 16)$

17. $y = \log_4 (x^2 - 4x - 21)$

18. $y = \log_6 (2x^2 - 7x - 4)$

19. $y = \log (x^3 - x)$

20. $y = \log (x^3 - 81x)$

21. $y = \log \left(\dfrac{x + 3}{x - 4} \right)$

22. $y = \log \left(\dfrac{x + 1}{x - 5} \right)$

23. $y = \log |3x - 7|$

24. $y = \log |6x + 6|$

*Sketch the graph of $f(x) = \log_2 x$. Then refer to it and use the techniques of **Chapter 2** to graph each function.*

25. $f(x) = (\log_2 x) + 3$

26. $f(x) = \log_2 (x + 3)$

27. $f(x) = |\log_2 (x + 3)|$

*Sketch the graph of $f(x) = \log_{1/2} x$. Then refer to it and use the techniques of **Chapter 2** to graph each function.*

28. $f(x) = (\log_{1/2} x) - 2$

29. $f(x) = \log_{1/2} (x - 2)$

30. $f(x) = |\log_{1/2} (x - 2)|$

Concept Check   *In Exercises 31–38, match the correct graph in choices A–H to each equation.*

31. $y = e^x + 3$

32. $y = e^x - 3$

33. $y = e^{x+3}$

34. $y = e^{x-3}$

35. $y = \ln x + 3$

36. $y = \ln x - 3$

37. $y = \ln (x - 3)$

38. $y = \ln (x + 3)$

A.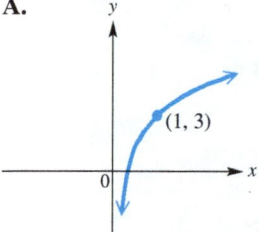

B.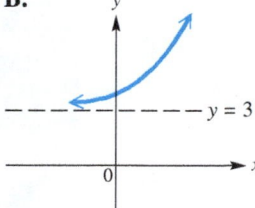

C.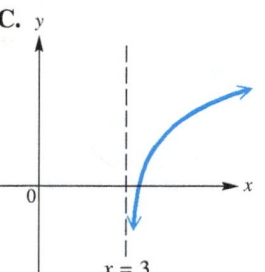

D.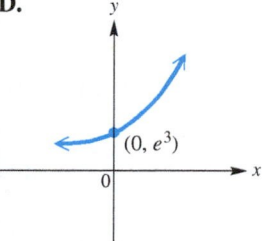

E.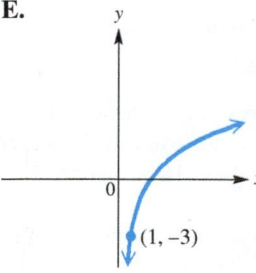

F.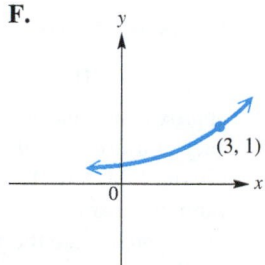

G.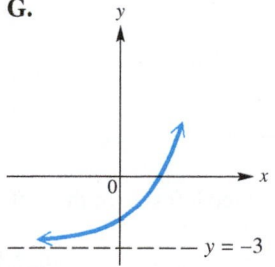

H.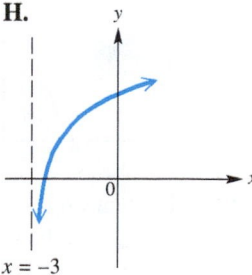

Graph each function.

39. $f(x) = \log_5 x$

40. $f(x) = \log_{10} x$

41. $f(x) = \log_{1/2} (1 - x)$

42. $f(x) = \log_{1/3} (3 - x)$

43. $f(x) = \log_3 (x - 1)$

44. $f(x) = \log_2 (x^2)$

*In Exercises 45–50, **(a)** explain how the graph of the given function can be obtained from the graph of $y = \log_2 x$, and **(b)** graph the function.*

45. $y = \log_2 (x + 4)$

46. $y = \log_2 (x - 6)$

47. $y = 3 \log_2 x + 1$

48. $y = -4 \log_2 x - 8$

49. $y = \log_2 (-x) + 1$

50. $y = -\log_2 (-x)$

51. Graph $y = \log x^2$ and $y = 2 \log x$ on separate viewing screens. It would seem, at first glance, that by applying the power rule for logarithms, these graphs should be the same. Are they? If not, why not? (*Hint:* Consider the domain in each case.)

52. Graph $f(x) = \log_3 |x|$ in the window $[-4, 4]$ by $[-4, 4]$, and compare $f(x)$ with the traditional graph in **FIGURE 40**. How might one easily misinterpret the domain of the function simply by observing the calculator graph? What is the domain of this function?

Evaluate each logarithm in three ways: (a) Use the definition of logarithm in Section 5.3 to find the exact value analytically. (b) Support the result of part (a) by using the change-of-base rule and common logarithms on your calculator. (c) Support the result of part (a) by locating the appropriate point on the graph of the function $y = \log_a x$.

53. $\log_9 27$ **54.** $\log_4\left(\dfrac{1}{8}\right)$ **55.** $\log_{16}\left(\dfrac{1}{8}\right)$ **56.** $\log_2 \sqrt{8}$

For each exponential function f, find f^{-1} analytically and graph both f and f^{-1} in the same viewing window.

57. $f(x) = 4^x - 3$ **58.** $f(x) = \left(\dfrac{1}{2}\right)^x - 5$ **59.** $f(x) = -10^x + 4$ **60.** $f(x) = -e^x + 6$

61. *Concept Check*   Suppose $f(x) = \log_a x$ and $f(3) = 2$. Determine each function value.

  (a) $f\left(\dfrac{1}{9}\right)$ **(b)** $f(27)$ **(c)** $f^{-1}(-2)$ **(d)** $f^{-1}(0)$

Use a graphing calculator to solve each equation. Give solutions to the nearest hundredth.

62. $\log_{10} x = x - 2$ **63.** $2^{-x} = \log_{10} x$ **64.** $e^x = x^2$

(Modeling)   Solve each problem.

65. *Height of the Eiffel Tower*   The right side of Paris's Eiffel Tower has a shape that can be approximated by the graph of the function

$$f(x) = -301 \ln \frac{x}{207}, \quad x > 0.$$

(Source: Banks, Robert B., *Towing Icebergs, Falling Dominoes, and Other Adventures in Applied Mathematics,* Princeton University Press.)

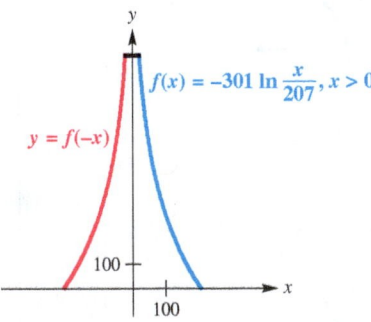

  (a) Explain why the shape of the red left side of the Eiffel Tower has the formula given by $f(-x)$.

  (b) The short black horizontal line at the top of the figure has length 15.7488 feet. Approximately how tall is the Eiffel Tower?

  (c) Approximately how far from the center of the tower is the point on the right side that is 500 feet above the ground?

66. *Age of a Whale*   The age in years of a female blue whale is approximated by

$$t = -2.57 \ln\left(\frac{87 - L}{63}\right),$$

where L is its length in feet.

  (a) How old is a female blue whale that measures 80 feet?

  (b) Estimate the length of a female blue whale that is 4 years old.

  (c) The equation that defines t has domain $24 < L < 87$. Explain why.

67. *Barometric Pressure*   The function

$$f(x) = 27 + 1.105 \log(x + 1)$$

approximates the barometric pressure in inches of mercury at a distance of x miles from the eye of a hurricane. *(Source:* Miller, A. and R. Anthes, *Meteorology,* Fifth Edition, Charles E. Merrill.)

  (a) Approximate the pressure 9 miles from the eye of the hurricane.

  (b) The ordered pair (99, 29.21) belongs to this function. What information does it convey?

68. *Sprinter's Speed and Time*   During the 100-meter dash, the elapsed time T for a sprinter to reach a speed of x meters per second is given by the following function.

$$T(x) = -1.2 \ln\left(1 - \frac{x}{11}\right)$$

  (a) How much time had elapsed when the sprinter was running 0 meters per second? Interpret your answer.

  (b) At the end of the race, the sprinter was moving at 10.998 meters per second. What was the sprinter's time for this 100-meter dash?

  (c) Find $T^{-1}(x)$ and interpret its meaning.

5.5 Exponential and Logarithmic Equations and Inequalities

Exponential Equations and Inequalities (Type 2) • Logarithmic Equations and Inequalities • Equations Involving Exponentials and Logarithms • Formulas Involving Exponentials and Logarithms

General methods for solving exponential and logarithmic equations depend on the following properties, which are based on the fact that exponential and logarithmic functions are, in general, one-to-one. Property 1 below was used in **Section 5.2** to solve Type 1 exponential equations.

> **Properties of Logarithmic and Exponential Functions**
>
> For $b > 0$ and $b \neq 1$,
> 1. $b^x = b^y$ **if and only if** $x = y$.
> 2. If $x > 0$ and $y > 0$, then $\log_b x = \log_b y$ **if and only if** $x = y$.

Exponential Equations and Inequalities (Type 2)

Unlike a Type 1 exponential equation (or inequality)—see **Section 5.2** for examples—a **Type 2 exponential equation** (or **inequality**) is one in which the exponential expressions *cannot* easily be written as powers of the same base.

$$7^x = 12 \quad \text{and} \quad 2^{3x+1} = 3^{4-x} \quad \text{Cannot be written with same base}$$

EXAMPLE 1 **Solving a Type 2 Exponential Equation**

Solve $7^x = 12$.

Analytic Solution

Property 1 cannot be used to solve this equation, so we apply Property 2. While any appropriate base b can be used to apply Property 2, we will use base e.

$$7^x = 12$$
$$\ln 7^x = \ln 12 \qquad \text{Take base } e \text{ logarithms.}$$
$$x \ln 7 = \ln 12 \qquad \text{Power rule}$$
$$x = \frac{\ln 12}{\ln 7} \qquad \text{Divide by } \ln 7.$$

The expression $\frac{\ln 12}{\ln 7}$ is the *exact* solution of $7^x = 12$. Had we used common logarithms, the solution would have the form $\frac{\log 12}{\log 7}$. In either case, we use a calculator to find a decimal approximation.

$$\frac{\ln 12}{\ln 7} = \frac{\log 12}{\log 7} \approx 1.277 \qquad \text{Nearest thousandth}$$

The exact solution set is $\left\{ \frac{\ln 12}{\ln 7} \right\}$, or $\left\{ \frac{\log 12}{\log 7} \right\}$, while $\{1.277\}$ is an approximate solution set.

Graphing Calculator Solution

Using the intersection-of-graphs method, we graph $y_1 = 7^x$ and $y_2 = 12$. The x-coordinate of the point of intersection is approximately 1.277, as illustrated in **FIGURE 44(a)**. **FIGURE 44(b)** shows that the x-coordinate of the x-intercept of $y = 7^x - 12$ is also approximately 1.277.

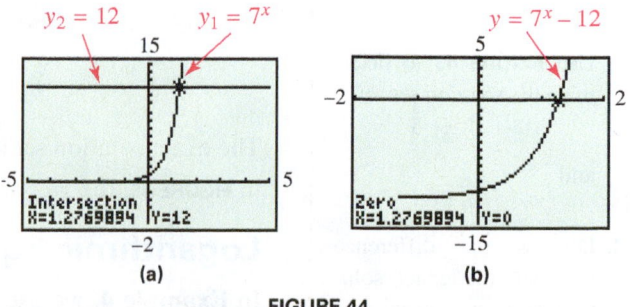

FIGURE 44

To check the analytic solution, we can also raise 7 to either of the powers $\frac{\ln 12}{\ln 7}$ or $\frac{\log 12}{\log 7}$ to obtain a result of 12. For example, $7^{\frac{\ln 12}{\ln 7}} = 12$. Verify this with a calculator.

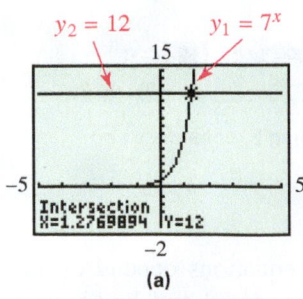

(a)

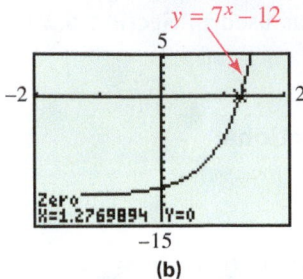

(b)

FIGURE 44 (repeated)

EXAMPLE 2 **Solving Type 2 Exponential Inequalities**

(a) Use **FIGURE 44(a)** to solve $7^x < 12$.   **(b)** Use **FIGURE 44(b)** to solve $7^x > 12$.
In each case, give both exact and approximate values.

Solution

(a) Because the graph of $y_1 = 7^x$ is below the graph of $y_2 = 12$ for all x-values less than $\frac{\ln 12}{\ln 7} \approx 1.277$, the solution set is

$$\left(-\infty, \frac{\ln 12}{\ln 7}\right), \quad \text{or} \quad (-\infty, 1.277).$$

(b) This inequality is equivalent to $7^x - 12 > 0$. Because the graph of $y = 7^x - 12$ is above the x-axis for all values of x greater than $\frac{\ln 12}{\ln 7} \approx 1.277$, the solution set is

$$\left(\frac{\ln 12}{\ln 7}, \infty\right), \quad \text{or} \quad (1.277, \infty).$$

EXAMPLE 3 **Solving a Type 2 Exponential Equation**

Solve $2^{3x+1} = 3^{4-x}$.

Solution

$$2^{3x+1} = 3^{4-x}$$

$\log 2^{3x+1} = \log 3^{4-x}$	Take common logarithms.
$(3x + 1)\log 2 = (4 - x)\log 3$	Power rule
$3x \log 2 + \log 2 = 4\log 3 - x\log 3$	Distributive property
$3x \log 2 + x\log 3 = 4\log 3 - \log 2$	Move all x-terms to one side.
$x(3\log 2 + \log 3) = 4\log 3 - \log 2$	Factor out x on the left.

Exact solution → $x = \dfrac{4\log 3 - \log 2}{3\log 2 + \log 3}$ *   Divide by $3\log 2 + \log 3$.

Approximate solution → $x \approx 1.165$ Nearest thousandth

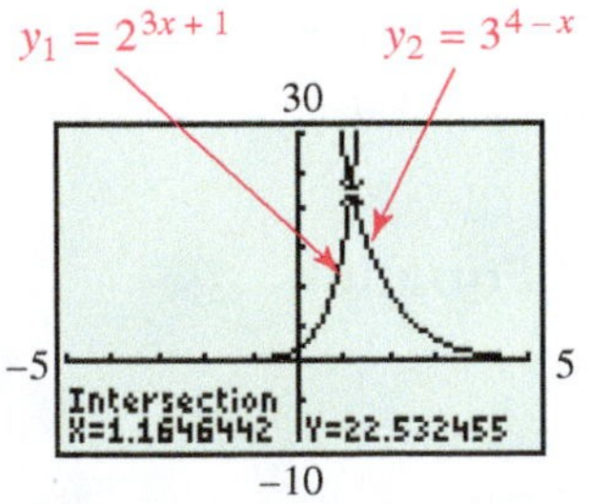

The exact solution can be written in a more compact form. Start with the equation marked * and apply the power rule.

$$x = \frac{\log 3^4 - \log 2}{\log 2^3 + \log 3}$$ Power rule

$$x = \frac{\log 81 - \log 2}{\log 8 + \log 3}$$ $3^4 = 81$ and $2^3 = 8$.

$$x = \frac{\log \frac{81}{2}}{\log 24}$$ Quotient and product rules

The exact solution set is $\left\{\frac{\log(81/2)}{\log 24}\right\}$. An approximate solution set is $\{1.165\}$. As seen in **FIGURE 45**, the x-coordinate is about 1.165, supporting the solution.

FIGURE 45

Logarithmic Equations and Inequalities

In **Example 4**, we use the quotient property of logarithms from **Section 5.3** to combine terms on the left side and then apply Property 2 (from the preceding page.) We could also solve this equation by transforming so that all logarithmic terms appear on one side. Then, by applying the properties of logarithms and rewriting in exponential form, the same solution set would result. This would also allow for solving graphically by the x-intercept method.

FOR DISCUSSION

1. Use **FIGURE 45** to determine the solution sets of
$$2^{3x+1} < 3^{4-x}$$
and
$$2^{3x+1} > 3^{4-x}.$$

2. Discuss the difference between the exact solution and an approximate solution of an exponential equation. In general, can you find an exact solution from a graph?

EXAMPLE 4   **Solving a Logarithmic Equation**

Solve $\log_3(x + 6) - \log_3(x + 2) = \log_3 x$.

Analytic Solution

The domain must satisfy $x + 6 > 0$, $x + 2 > 0$, and $x > 0$. The intersection of their solution sets gives the domain $(0, \infty)$.

$$\log_3(x + 6) - \log_3(x + 2) = \log_3 x$$

$$\log_3 \frac{x + 6}{x + 2} = \log_3 x \qquad \text{Quotient rule}$$

$$\frac{x + 6}{x + 2} = x \qquad \text{Property 2}$$

$$x + 6 = x(x + 2) \qquad \text{Multiply by } x + 2.$$

$$x + 6 = x^2 + 2x \qquad \text{Distributive property}$$

$$x^2 + x - 6 = 0 \qquad \text{Standard form}$$

$$(x + 3)(x - 2) = 0 \qquad \text{Factor.}$$

$$x = -3 \quad \text{or} \quad x = 2 \qquad \text{Zero-product property}$$

The value $x = -3$ is not in the domain of $\log_3 x$ or $\log_3(x + 2)$, so the only valid solution is 2, giving the solution set $\{2\}$.

Check: $\log_3(2 + 6) - \log_3(2 + 2) = \log_3 2$?   Let $x = 2$.

$$\log_3 \frac{8}{4} = \log_3 2 \qquad ?$$

$$\log_3 2 = \log_3 2 \qquad \checkmark \quad \text{True}$$

Graphing Calculator Solution

FIGURE 46 shows that the x-coordinate of the point of intersection of the graphs of

$$y_1 = \log_3(x + 6) - \log_3(x + 2)$$

and

$$y_2 = \log_3 x$$

is 2, which agrees with our analytic solution. Notice that the graphs do not intersect when $x = -3$, supporting the conclusion in the analytic solution that -3 is an extraneous value.

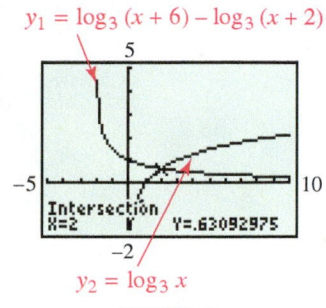

$y_1 = \log_3(x + 6) - \log_3(x + 2)$

$y_2 = \log_3 x$

FIGURE 46

EXAMPLE 5   **Solving a Logarithmic Equation**

Solve $\log(3x + 2) + \log(x - 1) = 1$.

Analytic Solution

Find the domain of each logarithm. Solve $3x + 2 > 0$ to find $x > -\frac{2}{3}$, and solve $x - 1 > 0$ to find $x > 1$. Thus the domain is $(1, \infty)$. Recall from **Section 5.3** that **$\log x$ means $\log_{10} x$.**

$$\log(3x + 2) + \log(x - 1) = 1$$

$$\log[(3x + 2)(x - 1)] = 1 \qquad \text{Product rule}$$

$$(3x + 2)(x - 1) = 10^1 \qquad \text{Write in exponential form.}$$

$$3x^2 - x - 2 = 10 \qquad \text{Multiply.}$$

$$3x^2 - x - 12 = 0 \qquad \text{Standard form}$$

The equation $3x^2 - x - 12 = 0$ cannot be solved by factoring. Use the quadratic formula.

$$x = \frac{-b \pm \sqrt{b^2 - 4ac}}{2a} = \frac{-(-1) \pm \sqrt{(-1)^2 - 4(3)(-12)}}{2(3)} = \frac{1 \pm \sqrt{145}}{6}$$

$$a = 3, b = -1, c = -12$$

Since $\frac{1 - \sqrt{145}}{6} \approx -1.840$ is less than 1, it is not in the domain and must be rejected, giving the solution set $\left\{ \frac{1 + \sqrt{145}}{6} \right\}$.

Graphing Calculator Solution

We choose to use the x-intercept method of solution. As seen in **FIGURE 47,** the x-coordinate of the x-intercept is approximately 2.174, and since $\frac{1 + \sqrt{145}}{6} \approx 2.174$, this supports the analytic result.

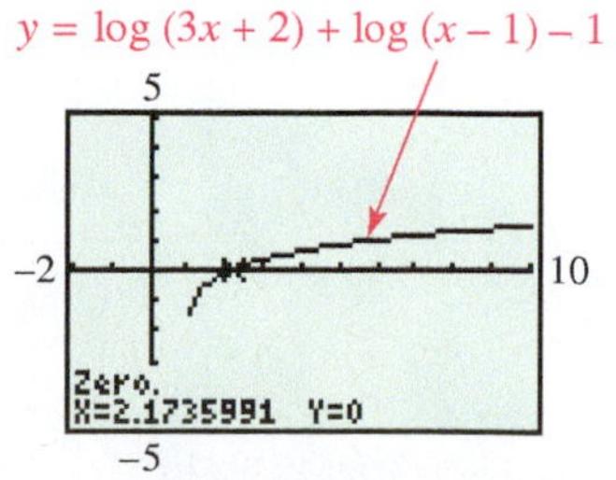

$y = \log(3x + 2) + \log(x - 1) - 1$

FIGURE 47

EXAMPLE 6 Solving an Equation and Associated Inequalities

Let $f(x) = 3\log_2(2x) - 15$. Solve the equation in (a) analytically and solve the inequalities in (b) and (c) graphically.

(a) $f(x) = 0$ **(b)** $f(x) < 0$ **(c)** $f(x) \geq 0$

Solution

(a)
$$3\log_2(2x) - 15 = 0$$
$$3\log_2(2x) = 15 \qquad \text{Add 15 to each side.}$$
$$\log_2(2x) = 5 \qquad \text{Divide each side by 3.}$$
$$2x = 2^5 \qquad \text{Write in exponential form.}$$
$$x = 16 \qquad \text{Divide by 2 and simplify.}$$

The solution set is $\{16\}$.

(b) To graph $y = f(x)$, we use the change-of-base formula and graph the equation $Y_1 = 3\log(2X)/\log(2) - 15$, as shown in **FIGURE 48**. We must have $x > 0$, due to the domain of the logarithmic function. We see that $f(x) < 0$ for $0 < x < 16$, or $(0, 16)$.

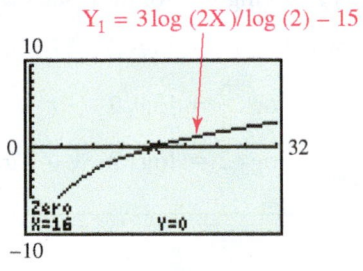

FIGURE 48

(c) From **FIGURE 48**, $f(x) \geq 0$ for $x \geq 16$. Thus, the solution set is $[16, \infty)$. ●

Equations Involving Exponentials and Logarithms

EXAMPLE 7 Solving a More Complicated Exponential Equation

Solve $e^{-2\ln x} = \dfrac{1}{16}$.

Solution $e^{-2\ln x} = \dfrac{1}{16}$ The domain is $(0, \infty)$.

$$e^{\ln x^{-2}} = \frac{1}{16} \qquad \text{Power rule}$$
$$x^{-2} = \frac{1}{16} \qquad a^{\log_a x} = x$$
$$x^{-2} = 4^{-2} \qquad \tfrac{1}{16} = \tfrac{1}{4^2} = 4^{-2};\ -4 \text{ is not a solution, since } -4 < 0 \text{ and } x > 0.$$
$$x = 4 \qquad \text{Properties of exponents}$$

An analytic check shows that 4 is indeed a solution. The solution set is $\{4\}$. ●

> **EXAMPLE 8**   **Solving a More Complicated Logarithmic Equation**

Solve $\ln e^{\ln x} - \ln (x - 3) = \ln 2$.

Solution

$\ln e^{\ln x} - \ln (x - 3) = \ln 2$	The domain is $(3, \infty)$.
$\ln x - \ln (x - 3) = \ln 2$	$e^{\ln x} = x$
$\ln \dfrac{x}{x - 3} = \ln 2$	Quotient rule
$\dfrac{x}{x - 3} = 2$	Property 2
$x = 2x - 6$	Multiply by $x - 3$.
$-x = -6$	Subtract $2x$.
$x = 6$	Multiply by -1.

Check:

$\ln e^{\ln x} - \ln (x - 3) = \ln 2$		Original equation
$\ln e^{\ln 6} - \ln (6 - 3) = \ln 2$	?	Let $x = 6$.
$\ln 6 - \ln 3 = \ln 2$	?	Simplify.
$\ln \dfrac{6}{3} = \ln 2$	?	Quotient rule
$\ln 2 = \ln 2$	✓	True

$\ln e^x = x$

The solution set is $\{6\}$.

> **EXAMPLE 9**   **Solving an Equation Quadratic in Form**

Solve $e^{2x} - 4e^x + 3 = 0$ for x.

Solution   Because $e^{2x} = (e^x)^2$, we can write and solve the given equation as follows.

$e^{2x} - 4e^x + 3 = 0$	
$(e^x)^2 - 4(e^x) + 3 = 0$	$e^{2x} = (e^x)^2$
$u^2 - 4u + 3 = 0$	Substitute u for e^x.
$(u - 1)(u - 3) = 0$	Factor the quadratic expression.
$u = 1 \quad$ or $\quad u = 3$	Zero-product property
$e^x = 1 \quad$ or $\quad e^x = 3$	Substitute e^x for u.
$x = \ln 1 \quad$ or $\quad x = \ln 3$	Write in logarithmic form.
$x = 0 \quad$ or $\quad x = \ln 3$	$\log_a 1 = 0$

Both values check, and the solution set is $\{0, \ln 3\}$.

Solving Exponential and Logarithmic Equations

An exponential or logarithmic equation can be solved by changing the equation into one of the following forms, where a and b are real numbers, $a > 0$, and $a \neq 1$.

1. $a^{f(x)} = b$
   Solve by taking a logarithm of each side. (The natural logarithm is the best choice if $a = e$.)

2. $\log_a f(x) = \log_a g(x)$
   From the given equation, $f(x) = g(x)$, which is solved analytically.

3. $\log_a f(x) = b$
   Solve by changing to exponential form, $f(x) = a^b$.

Formulas Involving Exponentials and Logarithms

EXAMPLE 10 Solving an Exponential Formula from Psychology

The strength of a habit is a function of the number of times the habit is repeated. If N is the number of repetitions and H is the strength of the habit, then, according to psychologist C. L. Hull,

$$H = 1000(1 - e^{-kN}),$$

where k is a constant. Solve this formula for k.

Solution

$$H = 1000(1 - e^{-kN})$$

$$\frac{H}{1000} = 1 - e^{-kN} \qquad \text{Divide by 1000.}$$

$$\frac{H}{1000} - 1 = -e^{-kN} \qquad \text{Subtract 1.}$$

$$e^{-kN} = 1 - \frac{H}{1000} \qquad \text{Multiply by } -1 \text{ and rewrite.}$$

$$\ln e^{-kN} = \ln\left(1 - \frac{H}{1000}\right) \qquad \text{Take the natural logarithm of each side.}$$

$$-kN = \ln\left(1 - \frac{H}{1000}\right) \qquad \ln e^x = x.$$

$$k = -\frac{1}{N}\ln\left(1 - \frac{H}{1000}\right) \qquad \text{Multiply by } -\frac{1}{N}.$$

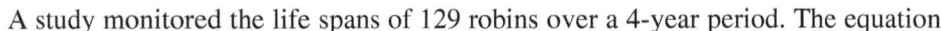

EXAMPLE 11 Modeling the Life Span of a Robin

A study monitored the life spans of 129 robins over a 4-year period. The equation

$$y = \frac{2 - \log(100 - x)}{0.42}$$

was developed to calculate the number of years y it takes for x percent of the robin population to die. (*Source:* Lack, D., *The Life of a Robin,* Collins.) How many robins had died after 6 months?

Solution Let $y = \frac{6}{12} = 0.5$ year, and solve the equation for x.

$$y = \frac{2 - \log(100 - x)}{0.42}$$

$$0.5 = \frac{2 - \log(100 - x)}{0.42} \qquad \text{Let } y = 0.5.$$

$$0.21 = 2 - \log(100 - x) \qquad \text{Multiply by 0.42.}$$

$$\log(100 - x) = 1.79 \qquad \text{Add } \log(100 - x) \text{ and subtract 0.21.}$$

$$10^{1.79} = 100 - x \qquad \text{Write in exponential form.}$$

$$x = 100 - 10^{1.79} \qquad \text{Add } x \text{ and subtract } 10^{1.79}.$$

$$x \approx 38.3 \qquad \text{Use a calculator.}$$

About 38% of the robins had died after 6 months.

5.5 Exercises

Checking Analytic Skills Solve each equation. Express all solutions in exact form. **Do not use a calculator.**

1. $3e^{2x} + 1 = 5$

2. $\dfrac{1}{2} e^x = 13$

3. $2(10^x) = 14$

4. $5(10^{3x}) - 4 = 6$

5. $\dfrac{1}{2} \log_2 x = \dfrac{3}{4}$

6. $2 \log_3 x = \dfrac{4}{5}$

7. $4 \ln 3x = 8$

8. $7 \ln 2x = 10$

Solve each exponential equation. Express the solution set so that **(a)** *solutions are in exact form and, if irrational,* **(b)** *solutions are approximated to the nearest thousandth. Support your solutions by using a calculator.*

9. $3^x = 7$

10. $5^x = 13$

11. $\left(\dfrac{1}{2}\right)^x = 5$

12. $\left(\dfrac{1}{3}\right)^x = 6$

13. $0.8^x = 4$

14. $0.6^x = 3$

15. $4^{x-1} = 3^{2x}$

16. $2^{x+3} = 5^x$

17. $6^{x+1} = 4^{2x-1}$

18. $3^{x-4} = 7^{2x+5}$

19. $2^x = -5$

20. $3^x = -4$

21. $e^{x-3} = 2^{3x}$

22. $e^{0.5x} = 3^{1-2x}$

23. $\left(\dfrac{1}{3}\right)^x = -3$

24. $\left(\dfrac{1}{9}\right)^x = -9$

25. $0.05(1.15)^x = 5$

26. $1.2(0.9)^x = 0.6$

27. $3(2)^{x-2} + 1 = 100$

28. $5(1.2)^{3x-2} + 1 = 11$

29. $2(1.05)^x + 3 = 10$

30. $3(1.4)^x - 4 = 60$

31. $5(1.015)^{x-1980} = 8$

32. $30 - 3(0.75)^{x-1} = 29$

Solve each logarithmic equation. Express all solutions in exact form. Support your solutions by using a calculator.

33. $5 \ln x = 10$

34. $3 \log x = 2$

35. $\ln (4x) = 1.5$

36. $\ln (2x) = 5$

37. $\log (2 - x) = 0.5$

38. $\ln (1 - x) = \dfrac{1}{2}$

39. $\log_6 (2x + 4) = 2$

40. $\log_5 (8 - 3x) = 3$

41. $\log_4 (x^3 + 37) = 3$

42. $\log_7 (x^3 + 65) = 0$

43. $\ln x + \ln x^2 = 3$

44. $\log x + \log x^2 = 3$

45. $2 \ln (x - 1) + 30 = 34$

46. $1 - 4 \ln (2x - 1) = -5$

47. $5 \log (x^2 - 1) + 7 = 12$

48. $8 \log (4 - x^2) - 4 = 20$

49. $3 \log_2 (3x^2 + 2) + 1 = 2$

50. $2 \log_2 (5x - 3) + 1 = 17$

51. $\log x + \log (x - 21) = 2$

52. $\log x + \log (3x - 13) = 1$

53. $\ln (4x - 2) - \ln 4 = -\ln (x - 2)$

54. $\ln (5 + 4x) - \ln (3 + x) - \ln 3 = 0$

55. $\log_5 (x + 2) + \log_5 (x - 2) = 1$

56. $\log_2 (x - 7) + \log_2 x = 3$

57. $\log_7 (4x) - \log_7 (x + 3) = \log_7 x$

58. $\log_2 (2x) + \log_2 (x + 2) = \log_2 16$

59. $\ln e^x - 2 \ln e = \ln e^4$

60. $\log_2 (\log_2 x) = 1$

61. $\log x = \sqrt{\log x}$

62. $\ln (\ln x) = 0$

63. In **Example 7,** we found that the solution of the equation $e^{-2 \ln x} = \frac{1}{16}$ is 4. Complete the following.
 (a) Solve the equation graphically.
 (b) Solve the inequality $e^{-2 \ln x} < \frac{1}{16}$, using the graph from part (a).
 (c) Solve the inequality $e^{-2 \ln x} > \frac{1}{16}$, using the graph from part (a).

64. In **Example 8,** we found that the solution of the equation $\ln e^{\ln x} - \ln (x - 3) = \ln 2$ is 6. Complete the following.
 (a) Solve the equation graphically.
 (b) Solve the inequality $\ln e^{\ln x} - \ln (x - 3) > \ln 2$, using the graph from part (a).
 (c) Solve the inequality $\ln e^{\ln x} - \ln (x - 3) < \ln 2$, using the graph from part (a).

65. A student told a friend, "You must reject any negative solution of an equation involving logarithms." Is this correct? Write an explanation of your answer.

66. Use a graph to explain why the logarithmic equation
$$\ln x - \ln (x + 1) = \ln 5$$
has no solution.

Use a graphing calculator to find the solution of each equation. Round your result to the nearest thousandth.

67. $1.5^{\log x} = e^{0.5}$

68. $1.5^{\ln x} = 10^{0.5}$

Solve each formula for the indicated variable.

69. $r = p - k \ln t$, for t

70. $p = a + \dfrac{k}{\ln x}$, for x

71. $T = T_0 + (T_1 - T_0)10^{-kt}$, for t

72. $A = \dfrac{Pi}{1 - (1 + i)^{-n}}$, for n

73. $A = T_0 + Ce^{-kt}$, for k

74. $y = \dfrac{K}{1 + ae^{-bx}}$, for b

75. $y = A + B(1 - e^{-Cx})$, for x

76. $m = 6 - 2.5 \log \left(\dfrac{M}{M_0}\right)$, for M

77. $\log A = \log B - C \log x$, for A

78. $d = 10 \log \left(\dfrac{I}{I_0}\right)$, for I

79. $A = P\left(1 + \dfrac{r}{n}\right)^{nt}$, for t

80. $D = 160 + 10 \log x$, for x

*The given equations are quadratic in form. Solve each and give exact solutions. Refer to **Example 9.***

81. $e^{2x} - 6e^x + 8 = 0$

82. $e^{2x} - 8e^x + 15 = 0$

83. $2e^{2x} + e^x = 6$

84. $3e^{2x} + 2e^x = 1$

85. $\dfrac{1}{2}e^{2x} + e^x = 1$

86. $\dfrac{1}{4}e^{2x} + 2e^x = 3$

87. $3^{2x} + 35 = 12(3^x)$

88. $5^{2x} + 3(5^x) = 28$

89. $(\log_2 x)^2 + \log_2 x = 2$

90. $(\log x)^2 - 6 \log x = 7$

91. $(\ln x)^2 + 16 = 10 \ln x$

92. $2 (\ln x)^2 + 9 \ln x = 5$

For the given f(x), solve the equation $f(x) = 0$ analytically and then use a graph of $y = f(x)$ to solve the inequalities $f(x) < 0$ and $f(x) \geq 0$.

93. $f(x) = -2e^x + 5$

94. $f(x) = -3e^x + 7$

95. $f(x) = 2(3^x) - 18$

96. $f(x) = 4^{x-2} - 2$

97. $f(x) = 3^{2x} - 9^{x+1}$

98. $f(x) = 2^{3x} - 8^{x-3}$

99. $f(x) = 8 - 4 \log_5 x$

100. $f(x) = 9 \log_3 (3x) - 18$

101. $f(x) = \ln (x + 2)$

102. $f(x) = \ln (x - 1) - \ln (x + 1)$

103. $f(x) = 7 - 5 \log x$

104. $f(x) = 3 - 2 \log_4 (x - 5)$

In general, it is not possible to find exact solutions analytically for equations that involve exponential or logarithmic functions together with polynomial, radical, and rational functions. Solve each equation using a graphical method, and express solutions to the nearest thousandth if an approximation is appropriate.

105. $x^2 = 2^x$

106. $x^2 - 4 = e^{x-4} + 4$

107. $\log x = x^2 - 8x + 14$

108. $\ln x = -\sqrt[3]{x+3}$

109. $e^x = \dfrac{1}{x+2}$

110. $3^{-x} = \sqrt{x+5}$

Use any method (analytic or graphical) to solve each equation.

111. $\log_2 \sqrt{2x^2} - 1 = 0.5$

112. $\log x^2 = (\log x)^2$

113. $\ln(\ln e^{-x}) = \ln 3$

114. $e^{x+\ln 3} = 4e^x$

(Modeling) Life Span of Robins   *Use the equation*

$$y = \frac{2 - \log(100 - x)}{0.42}$$

from **Example 11** *for Exercises 115 and 116.*

115. Estimate analytically the percentage of robins that died after 2 years.

116. Estimate the number of years elapsed for 75% of the robins to die.

(Modeling) Salinity   *The salinity of the oceans changes with latitude and depth. In the tropics, the salinity increases on the* surface of the ocean due to rapid evaporation. In the higher latitudes, there is less evaporation, and rainfall causes the salinity to be less on the surface than at lower depths. The function given by

$$f(x) = 31.5 + 1.1 \log(x + 1)$$

models salinity to depths of 1000 meters at a latitude of 57.5°N. The variable x is the depth in meters, and f(x) is in grams of salt per kilogram of seawater. (Source: Hartman, D., Global Physical Climatology, Academic Press.)

117. Approximate analytically the depth where the salinity equals 33.

118. Estimate the salinity at a depth of 500 meters.

SECTIONS 5.4–5.5   Reviewing Basic Concepts

1. Fill in the blanks: If $f(x) = 3^x$ and $g(x) = \log_3 x$, then functions f and g are _____ functions, and their graphs are _____ with respect to the line with equation _____. The domain of f is the _____ of g and vice versa.

Let $f(x) = 2 - \log_2(x - 1)$ in Exercises 2–5.

2. Graph $f(x)$.

3. Give the equation of the asymptote of the graph of $f(x)$ and any intercepts.

4. Compare the graph of $f(x)$ with the graph of $g(x) = \log_2 x$.

5. Find the inverse of $f(x)$.

Solve each equation.

6. $3^{2x-1} = 4^x$

7. $\ln 5x - \ln(x + 2) = \ln 3$

8. $10^{5 \log x} = 32$

9. $H = 1000(1 - e^{-kN})$, for N

10. *(Modeling) Caloric Intake*   The function

$$f(x) = 280 \ln(x + 1) + 1925$$

models the number of calories consumed daily by a person owning x acres of land in a developing country. (*Source:* Grigg, D., *The World Food Problem*, Blackwell Publishers.) Estimate the number of acres owned for the average intake to be 2300 calories per day.

5.6 Further Applications and Modeling with Exponential and Logarithmic Functions

Physical Science Applications • Financial and Other Applications • Modeling Data with Exponential and Logarithmic Functions

Looking Ahead to Calculus
The exponential growth and decay function formulas are studied in calculus in conjunction with the topic known as **differential equations.**

Physical Science Applications

A function of the form

$$A(t) = A_0 e^{kt}, \qquad \text{Exponential growth function}$$

where A_0 represents the initial quantity present, t represents time elapsed, $k > 0$ represents the growth constant associated with the quantity, and $A(t)$ represents the amount present at time t, is called an **exponential growth function.** (The formula $A = Pe^{rt}$ for continuous compounding of interest is an example of an exponential growth function.) This is an increasing function, because $e > 1$ and $k > 0$. For $k > 0$, a function of the form

$$A(t) = A_0 e^{-kt} \qquad \text{Exponential decay function}$$

is an **exponential decay function.** It is a decreasing function because

$$e^{-k} = (e^{-1})^k = \left(\frac{1}{e}\right)^k \quad \text{and} \quad 0 < \frac{1}{e} < 1.$$

In both cases, we usually restrict t to be nonnegative, giving domain $[0, \infty)$. **FIGURES 49** and **50** respectively show graphs of typical growth and decay functions.

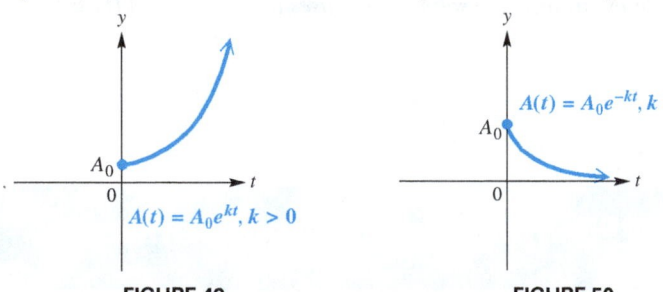

Exponential Growth Function

$A(t) = A_0 e^{kt}, k > 0$

FIGURE 49

Exponential Decay Function

$A(t) = A_0 e^{-kt}, k > 0$

FIGURE 50

If a quantity decays exponentially, the amount of time that it takes to reach one-half its original amount is called the **half-life.**

EXAMPLE 1 Finding the Age of a Fossil

Carbon 14 is a radioactive form of carbon that is found in all living plants and animals. After a plant or animal dies, the radiocarbon disintegrates. Using a technique called *carbon dating,* scientists determine the age of the remains by comparing the amount of carbon 14 present with the amount found in living plants and animals. The amount of carbon 14 present after t years is given by the exponential function $A(t) = A_0 e^{-kt}$, with $k = \frac{\ln 2}{5700}$. The half-life of carbon 14 is 5700 years.

If a fossil contains one-fifth of the carbon 14 contained in a contemporary living sample, estimate its age to the nearest thousand years.

Solution

$$A(t) = A_0 e^{-kt} \qquad \text{Exponential decay function}$$

$$\frac{1}{5}A_0 = A_0 e^{-\frac{\ln 2}{5700} t} \qquad \text{Let } A(t) = \frac{1}{5}A_0 \text{ and } k = \frac{\ln 2}{5700}.$$

$$\frac{1}{5} = e^{-\frac{\ln 2}{5700} t} \qquad \text{Divide by } A_0.$$

$$\ln \frac{1}{5} = \ln e^{-\frac{\ln 2}{5700} t} \qquad \text{Take the logarithm of each side.}$$

$$\ln \frac{1}{5} = -\frac{\ln 2}{5700} t \qquad \ln e^x = x$$

$$-\frac{5700}{\ln 2} \ln \frac{1}{5} = t \qquad \text{Multiply by } -\frac{5700}{\ln 2}.$$

$$t \approx 13,235 \qquad \text{Approximate.}$$

The age of the fossil is about 13,000 years.

EXAMPLE 2 **Finding an Exponential Decay Function, Given Half-life**

Links posted on Twitter generally experience half of their hits H within the first 2.8 hours. This pattern continues over time so that the number of hits on a link decreases by half over each 2.8-hour period. We say that Twitter links have a "half-life" of 2.8 hours. Hits on a Twitter link decay according to the function $H(t) = H_0 e^{-kt}$, where t is time in hours. Find the exact value of k and then approximate k.

Solution   The half-life tells us that when $t = 2.8$, $A(t) = \frac{1}{2}H_0$.

$$H(t) = H_0 e^{-kt}$$

$$\frac{1}{2}H_0 = H_0 e^{-k(2.8)} \qquad \text{Substitute.}$$

$$\frac{1}{2} = e^{-2.8k} \qquad \text{Divide by } H_0.$$

$$\ln \frac{1}{2} = \ln e^{-2.8k} \qquad \text{Take the logarithm of each side.}$$

$$\ln \frac{1}{2} = -2.8k \qquad \ln e^x = x$$

$$k = \frac{\ln \frac{1}{2}}{-2.8} \approx 0.24755 \qquad \text{Divide by } -2.8 \text{ and approximate.}$$

Thus, Twitter links decay according to the equation

$$H(t) = H_0 e^{-0.24755t}.$$

EXAMPLE 3 **Measuring Sound Intensity**

The loudness of sounds is measured in a unit called a **decibel.** To measure with this unit, we first assign an intensity of I_0 to a very faint sound, called the **threshold sound.** If a particular sound has intensity I, then the decibel rating of this louder sound is

$$d = 10 \log \frac{I}{I_0}.$$

The 2013 Superbowl in the Superdome reached levels of $10^{12} I_0$. (*Source:* Hollywood Reporter.) Find the decibel rating of this event, and compare it with the average rating of 80 decibels for a ringing telephone.

(continued)

Solution

$$d = 10 \log \frac{10^{12} I_0}{I_0} \qquad \text{Substitute } I = 10^{12} I_0 \text{ in the equation.}$$

$$d = 10 \log 10^{12} \qquad \text{Divide out } I_0.$$

$$d = 10(12) \qquad \log 10^{12} = 12.$$

$$d = 120 \qquad \text{Multiply.}$$

Sound levels reached 120 decibels. Since 120 decibels is equivalent to $10^{12} I_0$ and 80 decibels is equivalent to $10^{8.0} I_0$, their ratio is

$$\frac{10^{12} I_0}{10^{8.0} I_0} = 10^4 = 10{,}000.$$

The noise in the Superdome was 10,000 times the intensity of a ringing telephone. ●

Financial and Other Applications

If a quantity grows exponentially, the amount of time that it takes to become twice its original amount is called the **doubling time.** This is analogous to half-life for quantities that decay exponentially. *The initial amount present does not affect either the doubling time or the half-life.*

> **EXAMPLE 4** **Solving Compound Interest Formulas for *t***

(a) How long will it take $1000 invested at 6% interest compounded quarterly to grow to $2700?

(b) How long will it take for the money in an account that is compounded continuously at 3% interest to double?

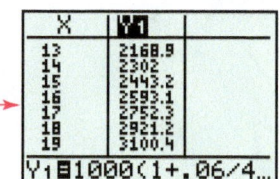

This table numerically supports the result of **Example 4(a).** Notice that $Y_1 = 2700$ for some X-value between 16 and 17. We found analytically that this value is approximately 16.678.

Solution

(a)

$$A = P\left(1 + \frac{r}{n}\right)^{nt} \qquad \text{Compound interest formula (Section 5.2)}$$

$$2700 = 1000\left(1 + \frac{0.06}{4}\right)^{4t} \qquad A = 2700, P = 1000, r = 0.06, n = 4$$

$$2700 = 1000(1.015)^{4t} \qquad \text{Simplify.}$$

$$2.7 = 1.015^{4t} \qquad \text{Divide by 1000.}$$

$$\log 2.7 = \log 1.015^{4t} \qquad \text{Take the logarithm of each side.}$$

$$\log 2.7 = 4t \log 1.015 \qquad \text{Power rule}$$

$$\frac{\log 2.7}{4 \log 1.015} = t \qquad \text{Divide by } 4 \log 1.015.$$

$$t \approx 16.678 \qquad \text{Use a calculator.}$$

It will take about $16\frac{3}{4}$ years for the initial amount to grow to $2700.

(b)

$$A = Pe^{rt} \qquad \text{Continuous compounding formula (Section 5.2)}$$

$$2P = Pe^{0.03t} \qquad A = 2P, r = 0.03$$

$$2 = e^{0.03t} \qquad \text{Divide by } P.$$

$$\ln 2 = \ln e^{0.03t} \qquad \text{Take the logarithm of each side.}$$

$$\ln 2 = 0.03t \qquad \ln e^x = x$$

$$\frac{\ln 2}{0.03} = t \qquad \text{Divide by 0.03.}$$

$$t \approx 23.1 \qquad \text{Use a calculator.}$$

The initial amount will double in about 23.1 years. ●

TECHNOLOGY NOTE

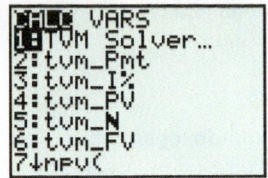

Graphing calculators such as the TI-84 Plus are capable of financial calculations. Consult your owner's guide.

A loan is **amortized** if both the principal and interest are paid by a sequence of equal periodic payments.

Amortization Payments

A loan of P dollars at interest rate i per period (as a decimal) may be amortized in n equal periodic payments of R dollars made at the end of each period, where

$$R = \frac{P}{\left[\dfrac{1 - (1 + i)^{-n}}{i}\right]}.$$

The total interest I that will be paid during the term of the loan is

$$I = nR - P.$$

EXAMPLE 5 **Using Amortization to Finance a Camper Trailer**

You purchase a camper trailer for $24,000. After a down payment of $4000, the balance will be paid off in 36 equal monthly payments at 4.5% interest per year. Find the amount of each payment. How much interest will you pay over the life of the loan?

Solution Use the amortization formula.

$$R = \frac{20{,}000}{\left[\dfrac{1 - (1 + 0.00375)^{-36}}{0.00375}\right]} \approx 594.94 \qquad \begin{aligned}&P = 24{,}000 - 4000 = 20{,}000, \\ &i = \tfrac{0.045}{12} \approx 0.00375,\ n = 36\end{aligned}$$

The monthly payment will be about $594.94. The total interest paid will be about

$$36(\$594.94) - \$20{,}000 = \$1417.84.$$

In **Example 5,** the approximate unpaid balance y after x payments is given by

$$y = R\left[\frac{1 - (1 + i)^{-(n-x)}}{i}\right].$$

For example, the unpaid balance after 12 payments is

$$y = \$594.94\left[\frac{1 - (1 + 0.00375)^{-24}}{0.00375}\right] \approx \$13{,}630.47.$$

You may be surprised that the remaining balance on a $20,000 loan after 12 payments is as large as $13,630.47. This is because most of each early payment on a loan goes toward interest.

EXAMPLE 6 **Modeling World Population**

World population in billions during year x can be modeled by $P(x) = 7(1.01)^{x-2011}$. Solve the equation $7(1.01)^{x-2011} = 8$ analytically to estimate the year when world population may reach 8 billion.

(continued)

Solution

$$7(1.01)^{x-2011} = 8 \qquad \text{Solve the given equation.}$$

$$(1.01)^{x-2011} = \frac{8}{7} \qquad \text{Divide by 7.}$$

The natural logarithm could also be used. → $$\log(1.01)^{x-2011} = \log\frac{8}{7} \qquad \text{Take the common logarithm.}$$

$$(x - 2011)\log 1.01 = \log\frac{8}{7} \qquad \text{Power rule: } \log m^r = r\log m$$

$$x - 2011 = \frac{\log\frac{8}{7}}{\log 1.01} \qquad \text{Divide by log 1.01.}$$

$$x = 2011 + \frac{\log\frac{8}{7}}{\log 1.01} \qquad \text{Add 2011.}$$

$$x \approx 2024 \qquad \text{Approximate.}$$

This model predicts that world population may reach 8 billion during 2024.

EXAMPLE 7 **Determining the Price of Solar Energy**

Solar energy currently accounts for only a small percentage of the planet's electricity, but the industry is rapidly growing as this form of energy becomes less expensive. Photovoltaic cells are needed to generate solar power, and each year since 1980 their cost has been about 90% as much as the year prior. In 1980 the price of these cells was $30 per watt produced. We can model their price in dollars with the function $f(x) = 30(0.90)^x$, where x represents years after 1980.

Predict the year when the price of photovoltaic cells might be $0.75 per watt and compare to the actual year of 2013. (*Source: The Economist.*)

Analytic Solution

We must solve the equation $30(0.90)^x = 0.75$.

$$30(0.90)^x = 0.75$$

$$(0.90)^x = 0.025 \qquad \text{Divide by 30.}$$

$$\log(0.90)^x = \log 0.025 \qquad \text{Take logarithms.}$$

$$x\log 0.90 = \log 0.025 \qquad \text{Power rule}$$

$$x = \frac{\log 0.025}{\log 0.90} \approx 35.01 \qquad \text{Divide by log 0.90.}$$

Since $x = 35$ corresponds to 35 years after 1980, our model predicts photovoltaic cells to be about $0.75 in 2015, which is two years later than the actual year of 2013.

Graphing Calculator Solution

FIGURE 51 uses the intersection-of-graphs method to confirm the result that $y = 0.75$ when $x \approx 35.01$.

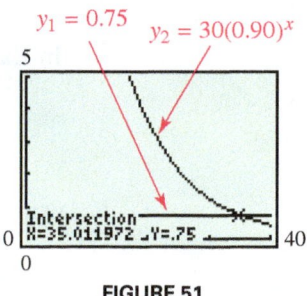

FIGURE 51

Modeling Data with Exponential and Logarithmic Functions

EXAMPLE 8 **Modeling Atmospheric CO_2 Concentrations**

Future concentrations of atmospheric carbon dioxide (CO_2) in parts per million (ppm) are shown in the table at the top of the next page. (These concentrations assume that recent trends continue.) CO_2 levels in the year 2000 were greater than they had been

Year	CO$_2$ (ppm)
2000	364
2050	467
2100	600
2150	769
2200	987

Source: Turner, R., *Environmental Economics, an Elementary Approach,* Johns Hopkins University Press.

anytime in the previous 160,000 years.

(a) Let $x = 0$ correspond to 2000 and $x = 200$ to 2200. Find values for C and a so that $f(x) = Ca^x$ models these data.

(b) Use a graphing calculator with regression capability to find an exponential function g that models all the data, and graph it with the data points. How does function g compare with function f from part (a)?

(c) Estimate CO$_2$ concentrations for the year 2025.

Solution

(a) Because $f(0) = C$ and the concentration is 364 when $x = 0$, it follows that $C = 364$.

$$f(x) = Ca^x = 364a^x$$

One possibility for determining a is to require that the graph of f pass through the point $(2200, 987)$, where $x = 200$. Thus, $f(200) = 987$.

$$364a^x = f(x)$$
$$364 \cdot a^{200} = 987 \qquad \text{\textcolor{blue}{$f(200) = 987$}}$$
$$a^{200} = \frac{987}{364} \qquad \text{\textcolor{blue}{Divide by 364.}}$$
$$(a^{200})^{1/200} = \left(\frac{987}{364}\right)^{1/200} \qquad \text{\textcolor{blue}{Raise to the $\frac{1}{200}$th power.}}$$
$$a = \left(\frac{987}{364}\right)^{1/200} \qquad \text{\textcolor{blue}{Property of exponents}}$$
$$a \approx 1.005 \qquad \text{\textcolor{blue}{Use a calculator.}}$$

Hence, $f(x) = 364(1.005)^x$. (Answers may vary slightly.)

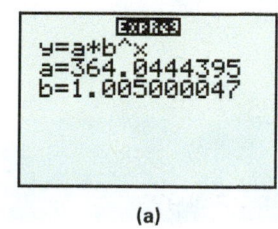

(a)

$g(x) = 364.0(1.005)^x$

(b)

FIGURE 52

(b) **FIGURE 52(a)** shows the coefficients given by the calculator for exponential regression. With these values, $g(x) \approx 364.0(1.005)^x$, which is essentially the same as $f(x)$ from part (a). The data points and the function are graphed in **FIGURE 52(b)**.

(c) Since 2025 corresponds to $x = 25$, evaluate $f(25)$.

$$f(25) = 364(1.005)^{25} \approx 412.$$

The concentration of carbon dioxide might reach 412 ppm by 2025.

Year	Social Network Users
2005	9%
2006	49%
2008	67%
2010	86%
2012	92%

EXAMPLE 9 **Modeling Social Network Use**

The table lists the proportion of Internet users aged 18–25 who reported using social networks in various years.

(a) Make a scatter diagram of the data, using $x = 1$ for 2005, $x = 2$ for 2006, and so on. Enter the y-values as percentages, *not* decimals. What type of function might model these data?

(b) Use least-squares regression to determine a and b so that $f(x) = a + b \ln x$ models the data.

(c) Graph f and the data in the same viewing window.

Solution

(a) Enter the data points $(1, 9)$, $(2, 49)$, $(4, 67)$, $(6, 86)$, and $(8, 92)$ into a calculator. A scatter diagram of the data is shown in **FIGURE 53**. A logarithmic function model may be appropriate because the data increase rapidly at first and then more slowly.

(continued)

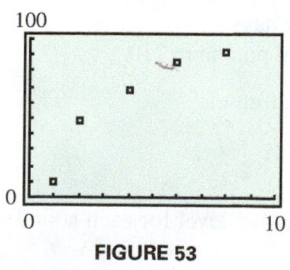

FIGURE 53

(b) In **FIGURES 54(a)** and **54(b)**, least-squares regression determines (approximately) the logarithmic function

$$f(x) = 14.06 + 39.106 \ln x.$$

(c) A graph of f and the data are shown in **FIGURE 54(c).**

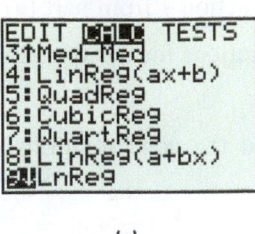

(a)

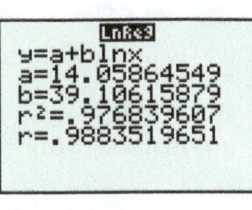

(b)

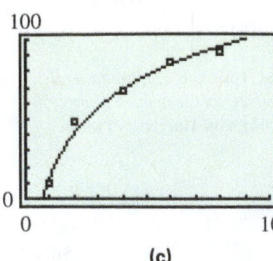

(c)

FIGURE 54

5.6 Exercises

*The information in **Example 1** allows us to use the function $A(t) = A_0 e^{-0.0001216t}$ to approximate the amount of carbon 14 remaining in a sample, where t is in years. Use this function in Exercises 1–4.*
$\left(\text{Note: } -0.0001216 \approx -\dfrac{\ln 2}{5700}.\right)$

1. *Carbon 14 Dating* Suppose an Egyptian mummy is discovered in which the amount of carbon 14 present is only about one-third the amount found in the atmosphere. About how long ago did the Egyptian die?

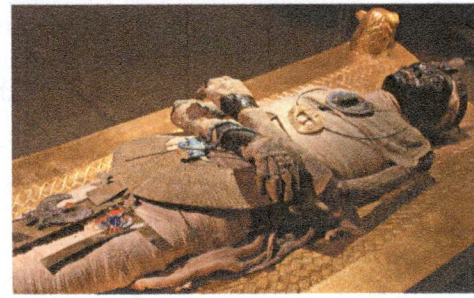

2. *Carbon 14 Dating* A sample from a refuse deposit near the Strait of Magellan had 60% of the carbon 14 of a contemporary living sample. Estimate the age of the sample.

3. *Carbon 14 Dating* A certain paint from the Lascaux caves of France contains 15% of the normal amount of carbon 14. Estimate the age of the paint.

4. *Carbon 14 Dating* Estimate the age of a specimen that contains 20% of the carbon 14 of a comparable living specimen.

5. *Radioactive Lead 210* The half-life of radioactive lead 210 is 21.7 years.
   (a) Find an exponential decay model for lead 210.
   (b) Estimate how long it will take a sample of 500 grams to decay to 400 grams.
   (c) Estimate how much of the sample of 500 grams will remain after 10 years.

6. *Radioactive Cesium 137* Radioactive cesium 137 was emitted in large amounts in the Chernobyl nuclear power station accident in Russia on April 26, 1986. The amount of cesium 137 remaining after x years in an initial sample of 100 milligrams can be described by

$$A(x) = 100e^{-0.02295x}.$$

(*Source:* Mason, C., *Biology of Freshwater Pollution,* John Wiley and Sons.)
   (a) Estimate how much is remaining after 50 years. Is the half-life of cesium 137 greater or less than 50 years?
   (b) Estimate the half-life of cesium 137.

7. *Radioactive Polonium 210* The table shows the amount y of polonium 210 remaining after t days from an initial sample of 2 milligrams.

t (days)	0	100	200	300
y (milligrams)	2	1.22	0.743	0.453

   (a) Use the table to determine whether the half-life of polonium 210 is greater or less than 200 days.
   (b) Find a formula that models the amount A of polonium 210 in the table after t days.
   (c) Estimate the half-life of polonium 210.

8. *Sound Intensity* Use the formula

$$d = 10 \log \frac{I}{I_0}$$

to estimate the average decibel level for each sound with

the given intensity I. For comparison, conversational speech has a sound level of about 60 decibels.
(a) Jackhammer: $31{,}620{,}000{,}000 I_0$
(b) iPhone 5 speakers: $10^{10} I_0$
(c) Rock singer screaming into microphone: $10^{14} I_0$

The intensity I of an earthquake, measured on the Richter scale, is given by $\log \frac{I}{I_0}$, where I_0 is the intensity of an earthquake of a certain small size. Use this information in Exercises 9 and 10.

9. *Earthquake Intensity* In November 2012 an earthquake in Guatemala measured 7.4 on the Richter scale.
   (a) Express this intensity I in terms of I_0.
   (b) In December 2012 the earthquake off the coast of California measured 6.3 on the Richter scale. Express this intensity I in terms of I_0.
   (c) How many times more intense was the earthquake in Guatemala than the earthquake in California?

10. *Earthquake Intensity* The earthquake off the coast of Northern Sumatra on Dec. 26, 2004, had a Richter scale rating of 8.9.
    (a) Express the intensity of this earthquake in terms of I_0.
    (b) Aftershocks from this quake had a Richter scale rating of 6.0. Express the intensity of these in terms of I_0.
    (c) Compare the intensities of the 8.9 earthquake to the 6.0 aftershock.

11. *Magnitude of a Star* The magnitude of a star is defined by the equation

$$M = 6 - 2.5 \log \frac{I}{I_0},$$

where I_0 is the measure of a just-visible star and I is the actual intensity of the star being measured. The dimmest stars are of magnitude 6, and the brightest are of magnitude 1. Determine the ratio of light intensities between a star of magnitude 1 and a star of magnitude 3.

Newton's law of cooling says that the rate at which an object cools is proportional to the difference C in temperature between the object and the environment around it. The temperature $f(t)$ of the object at time t in appropriate units after being introduced into an environment with a constant temperature T_0 is

$$f(t) = T_0 + Ce^{-kt},$$

where C and k are constants. Use this result in Exercises 12–14.

12. *Newton's Law of Cooling* Boiling water at 100°C is placed in a freezer at 0°C. The temperature of the water is 50°C after 24 minutes. Approximate the temperature of the water after 96 minutes.

13. *Newton's Law of Cooling* A pot of coffee with a temperature of 100°C is set down in a room with a temperature of 20°C. The coffee cools to 60°C after 1 hour.
    (a) Write an equation to model the data.
    (b) Estimate the temperature after a half hour.
    (c) About how long will it take for the coffee to cool to 50°C? Support your answer graphically.

14. *Newton's Law of Cooling* A piece of metal is heated to 300°C and then placed in a cooling liquid at 50°C. After 4 minutes, the metal has cooled to 175°C. Estimate its temperature after 12 minutes.

15. *Greenhouse Gases* Chlorofluorocarbons (CFCs) are gases that increase the greenhouse effect and damage the ozone layer. CFC 12 is one type of chlorofluorocarbon used in refrigeration, air conditioning, and foam insulation. The table lists future concentrations of CFC 12 in parts per billion (ppb) *if current trends continue.*

Year	2000	2005	2010	2015	2020
CFC 12 (ppb)	0.72	0.88	1.07	1.31	1.60

Source: Turner, R., *Environmental Economics, an Elementary Approach,* Johns Hopkins University Press.

(a) Let $x = 0$ correspond to 2000 and $x = 20$ to 2020. Find values for C and a so that $f(x) = Ca^x$ models these data.
(b) Estimate the CFC 12 concentration in 2013.

16. *Reducing Carbon Emissions* Governments could reduce carbon emissions by placing a tax on fossil fuels. The cost–benefit equation

$$\ln(1 - P) = -0.0034 - 0.0053x$$

estimates the relationship between a tax of x dollars per ton of carbon and the percent P reduction in emissions of carbon, where P is in decimal form. Determine P when $x = 60$. Interpret the result. (*Source:* Clime, W., *The Economics of Global Warming,* Institute for International Economics.)

17. *Account Growth* An amount A in a bank account after x years is given by $A(x) = 1000(1.025)^x$.
    (a) How much is in the account after 3 years?
    (b) How much is in the account after 10 years?
    (c) After how many years will there be about $1900 in the account?

18. *Account Growth* An amount A in a bank account after x years is given by $A(x) = 450(1.06)^x$.
    (a) How much is in the account after 2 years?
    (b) How much is in the account after 20 years?
    (c) After how many years will there be about $2300 in the account?

19. *Bacteria Growth* The concentration of bacteria B in millions per milliliter after x hours is given by

$$B(x) = 3.5e^{0.02x}.$$

(a) How many bacteria are there after 1 hour?
(b) How many bacteria are there after 6.5 hours?
(c) After how many hours will there be 6 million bacteria per milliliter?

20. *Bacteria Growth* The concentration of bacteria B in millions per milliliter after x hours is given by

$$B(x) = 1.33e^{0.15x}.$$

(a) How many bacteria are there after 2.5 hours?
(b) How many bacteria are there after 8 hours?
(c) After how many hours will there be 31 million bacteria per milliliter?

In Exercises 21–24, suppose that $2500 is invested in an account that pays interest compounded continuously. Find the amount of time that it would take for the account to grow to the given amount at the given rate of interest. Round to the nearest tenth of a year.

21. $3000 at 3.75%   **22.** $3500 at 4.25%

23. $5000 at 5%   **24.** $5000 at 6%

25. *Time to Compound* The time T in years it takes for a principal of $1000 receiving 2% annual interest compounded continuously to reach an amount A is calculated by the following logarithmic function.

$$T(A) = 50 \ln \frac{A}{1000}$$

(a) Find a reasonable domain for T. Interpret your answer.
(b) How many years does it take the principal to grow to $1200?
(c) Determine the amount in the account after 23.5 years by solving the equation $T(A) = 23.5$.

26. *Time to Multiply* Suppose that a sample of bacteria has a concentration of 2 million bacteria per milliliter and it doubles in concentration every 12 hours. Then the time T it takes for the sample to reach a concentration of C can be approximated by the following logarithmic function.

$$T(C) = \frac{500}{29} \ln \frac{C}{2}$$

(a) Find the domain of T. Interpret your answer.
(b) How long does it take for the concentration of bacteria to increase by 50%?
(c) Determine the concentration C after 15 hours by solving the equation $T(C) = 15$.

27. *Interest on an Account* Estimate how long it will take for $1000 to grow to $5000 at an interest rate of 3.5% if interest is compounded **(a)** quarterly; **(b)** continuously.

28. *Interest on an Account* Estimate how long it will take for $5000 to grow to $8400 at an interest rate of 6% if interest is compounded **(a)** semiannually; **(b)** continuously.

29. *Interest on an Account* Tom Tupper wants to buy a $30,000 car. He has saved $27,000. Find the number of years (to the nearest tenth) it will take for his $27,000 to grow to $30,000 at 2.3% interest compounded quarterly.

30. *Doubling Time* Estimate the doubling time of an investment earning 2.5% interest if interest is compounded **(a)** quarterly; **(b)** continuously.

31. *Comparison of Investment* A construction worker wants to invest $60,000 in a pension plan. One investment offers 2% compounded quarterly. Another offers 1.8% compounded continuously. Which investment will earn more interest in 5 years? How much more will the better plan earn?

32. *Growth of an Account* See **Exercise 31.** If the worker chooses the plan with continuous compounding, estimate how long it will take for her $60,000 to grow to $80,000.

*The interest rate stated by a financial institution is sometimes called the **nominal rate.** If interest is compounded, the actual rate is, in general, higher than the nominal rate, and is called the **effective rate.** If r is the nominal rate and n is the number of times interest is compounded annually, then*

$$R = \left(1 + \frac{r}{n}\right)^n - 1$$

is the effective rate. Here, R represents the annual rate that the investment would earn if simple interest were paid. Use this formula in Exercises 33 and 34.

33. *Effective Rate* Find the effective rate to the nearest hundredth of a percent if the nominal rate is 3% and interest is compounded quarterly.

34. *Effective Rate* Estimate the effective rate if the nominal rate is 4.5% and interest is compounded daily ($n = 365$).

*In the formula $A = P\left(1 + \frac{r}{n}\right)^{nt}$, we can interpret P as the **present value** of A dollars t years from now, earning annual interest r compounded n times per year. In this context, A is called the **future value.** If we solve the formula for P, we obtain*

$$P = A\left(1 + \frac{r}{n}\right)^{-nt}.$$

Use the future value formula in Exercises 35–38.

35. *Present Value*   Find the present value of an account that will be worth $10,000 in 5 years, if interest is compounded semiannually at 3%.

36. *Present Value*   Find the present value of an account that will be worth $25,000 in 2.75 years, if interest is compounded quarterly at 6%.

37. *Future Value*   Estimate the interest rate necessary for a present value of $25,000 to grow to a future value of $30,416 if interest is compounded annually for 5 years.

38. *Future Value*   Estimate the interest rate necessary for a present value of $1200 to grow to a future value of $1408 if interest is compounded quarterly for 8 years.

In Exercises 39–41, use the amortization formulas given in this section to find **(a)** *the monthly payment on a loan with the given conditions and* **(b)** *the total interest that will be paid during the term of the loan.*

39. *Amortization* $8500 is amortized over 4 years with an interest rate of 7.5%.

40. *Amortization* $9600 is amortized over 5 years with an interest rate of 5.2%.

41. *Amortization* $125,000 is amortized over 30 years with an interest rate of 7.25%.

42. *Amortization Payments*   Use the formula

$$y = R\left[\frac{1 - (1 + i)^{-(n-x)}}{i}\right],$$

where y is the unpaid balance after x payments have been made on a loan with n payments of R dollars. Find the balance after 120 payments have been made on the loan in **Exercise 41.**

43. *Growth of an Account*   Use the table feature of your graphing calculator to work parts (a) and (b).
   (a) Find how long it will take $1500 invested at 5.75%, compounded daily, to triple in value. Locate the solution by systematically decreasing ΔTbl. Find the answer to the nearest day. (Find your answer to the nearest day by eventually letting ΔTbl $= \frac{1}{365}$. The decimal part of the solution can be multiplied by 365 to determine the number of days greater than the nearest year. For example, if the solution is determined to be 16.2027 years, then multiply 0.2027 by 365 to get 73.9855. The solution is then, to the nearest day, 16 years and 74 days.) Confirm your answer analytically.
   (b) Find how long it will take $2000 invested at 8%, compounded daily, to be worth $5000.

44. *Growth of an Account*   If interest is compounded continuously and the interest rate is tripled, what effect will this have on the time required for an investment to double?

45. *Growth of Bacteria*   *Escherichia coli* is a strain of bacteria that occurs naturally in many organisms. Under certain conditions, the number of bacteria present in a colony is approximated by

$$A(t) = A_0 e^{0.023t},$$

where t is in minutes. If $A_0 = 2{,}400{,}000$, find the number of bacteria at each time. Round to the nearest hundred thousand.
   (a) 5 minutes   **(b)** 10 minutes   **(c)** 60 minutes

46. *Growth of Bacteria*   The growth of bacteria in food products makes it necessary to date some products (such as milk) so that they will be sold and consumed before the bacterial count becomes too high. Suppose that, under certain storage conditions, the number of bacteria present in a product is

$$f(t) = 500e^{0.1t},$$

where t is time in days after packing of the product and the value of $f(t)$ is in millions.
   (a) If the product cannot be safely eaten after the bacterial count reaches 3,000,000,000, how long will this take?
   (b) If $t = 0$ corresponds to January 1, what date should be placed on the product?

47. *Mobile Advertising*   The revenue in millions of dollars for the first 5 years of mobile advertising is given by $A(x) = 42(2)^x$, where x is years after the industry started. (*Source*: Business Insider.)
   (a) Determine analytically when revenue was about $400 million.
   (b) Solve part (a) graphically.
   (c) According to this model, when did the mobile advertising revenue reach $1 billion?

48. *Internet Advertising*   The revenue in millions of dollars for the first 5 years of Internet advertising is given by $A(x) = 25(2.95)^x$, where x is years after the industry started. (*Source*: Business Insider.)
   (a) What was the Internet advertising revenue after 5 years?
   (b) Determine analytically when revenue was about $250 million.
   (c) According to this model, when did the Internet advertising revenue reach $1 billion?

49. *U.S. Hispanic Population*   In 2012, 17% of the U.S. population was Hispanic, and this number is expected to be 31% in 2060. (*Source:* U.S. Census Bureau.)
   (a) Approximate C and a so that $P(x) = Ca^{x-2012}$ models these data, where P is the percent of the population that is Hispanic and x is the year. Why is $a > 1$?
   (b) Estimate P in 2030.
   (c) Use P to estimate the year when 25% of the population could be Hispanic.

50. *U.S. White Population*   In 2012, 63% of the U.S. population was non-Hispanic white, and this number is expected to be 43% in 2060. (*Source:* U.S. Census Bureau.)
   (a) Find C and a so that $P(x) = Ca^{x-2012}$ models these data, where P is the percent of the population that is non-Hispanic white and x is the year. Why is $a < 1$?
   (b) Estimate P in 2020.
   (c) Use P to estimate when 50% of the population could be non-Hispanic white.

51. *Bacterial Growth*   Suppose that the concentration of a bacteria sample is 100,000 bacteria per milliliter. If the concentration doubles every 2 hours, how long will it take for the concentration to reach 350,000 bacteria per milliliter?

52. *Bacterial Growth*   Suppose that the concentration of a bacterial sample is 50,000 bacteria per milliliter. If the concentration triples in 4 days, how long will it take for the concentration to reach 85,000 bacteria per milliliter?

For Exercises 53 and 54, refer to **Example 7.**

53. *Solar Energy*   Suppose that the cost of photovoltaic cells each year after 1980 was 75% as much as the year prior. If the cost was $30/watt in 1980, model their price in dollars with an exponential function, where x corresponds to years after 1980. Then estimate the year when the price of photovoltaic cells was $1.00 per watt.

54. *Rebounding*   Suppose that when a ball is dropped, the height of its first rebound is about 80% of the initial height that it was dropped from, the second rebound is about 80% as high as the first rebound, and so on. If this ball is dropped from 12 feet in the air, model the height in feet of each rebound with an exponential function $H(x)$, where $x = 0$ represents the initial height, $x = 1$ represents the height on the first rebound, and so on. Find the height of the third rebound. Determine which rebound had a height of about 2.5 feet.

55. *Epidemics*   In 1666 the village of Eyam, located in England, experienced an outbreak of the Great Plague. Out of 261 people in the community, only 83 survived. The table shows a function f that computes the number of people who had not (yet) been infected after x days.

x	0	15	30	45
$f(x)$	254	240	204	150

x	60	75	90	125
$f(x)$	125	103	97	83

Source: Raggett, G., "Modeling the Eyam Plague,"
The Institute of Mathematics and Its Applications 18.

   (a) Use a table to represent a function g that computes the number of people in Eyam who were infected after x days.
   (b) Write an equation that shows the relationship between $f(x)$ and $g(x)$.

   (c) Use graphing to decide which equation represents $g(x)$ better,
   $$y_1 = \frac{171}{1 + 18.6e^{-0.0747x}} \quad \text{or} \quad y_2 = 18.3(1.024)^x.$$
   (d) Use your results from parts (b) and (c) to find a formula for $f(x)$.

56. *Heart Disease Death Rates*   The table lists heart disease death rates per 100,000 people for selected ages.

Age	30	40	50	60	70
Death Rate	8.0	29.6	92.9	246.9	635.1

Source: Centers for Disease Control.

   (a) Make a scatter diagram of the data in the window $[25, 75]$ by $[-100, 700]$.
   (b) Find an exponential function f that models the data.
   (c) Estimate the heart disease death rate for people who are 80 years old.

57. *iTunes Revenue*   The following table shows the revenue in billions of dollars for iTunes in various years.

Year	2008	2009	2010	2011	2012
Revenue ($ billions)	6.0	7.0	9.5	11.8	15.7

(*Source:* Apple Corporation.)

   (a) Use exponential regression to approximate constants C and a so that $f(x) = Ca^{x-2008}$ models the data, where x is the year.
   (b) Support your answer by graphing f and the data.

58. *Valentine's Day Spending*   The following table shows the average Valentine's Day spending in dollars per consumer for various years.

Year	2010	2011	2012	2013
Spending ($)	103	116	126	131

(*Source:* NRF/BIGinsight.)

   (a) Use logarithmic regression to approximate values for a and b so that $f(x) = a + b \ln x$ models the data, where $x = 1$ corresponds to 2010, $x = 2$ to 2011, and so on.
   (b) Use your function to estimate average spending in 2012 and compare to the value in the table.

(Modeling)   *In real life, populations of bacteria, insects, and animals do not continue to grow indefinitely. Initially, population growth may be slow. Then, as their numbers increase, so does the rate of growth. After a region has become heavily populated or saturated, the population usually levels off because of limited resources. This type of growth may be modeled by a* **logistic function** *represented by*

$$f(x) = \frac{c}{1 + ae^{-bx}},$$

where a, b, and c are positive constants.

Refer to this information for Exercises 59 and 60.

59. *Heart Disease*   As age increases, so does the likelihood of coronary heart disease (CHD). The fraction of people x years old with some CHD is approximated by

$$f(x) = \frac{0.9}{1 + 271e^{-0.122x}}.$$

(*Source:* Hosmer, D. and S. Lemeshow, *Applied Logistic Regression*, John Wiley and Sons.)

(a) Evaluate $f(25)$ and $f(65)$. Interpret the results.

(b) At what age does this likelihood equal 50%?

60. *Tree Growth*   The height of a tree in feet after x years is modeled by

$$f(x) = \frac{50}{1 + 47.5e^{-0.22x}}.$$

(a) Make a table for f starting at $x = 10$ and incrementing by 10. What seems to be the maximum height?

(b) Graph f and identify the horizontal asymptote. Explain its significance.

(c) After how long was the tree 30 feet tall?

Summary Exercises on Functions: Domains, Defining Equations, and Composition

Finding the Domain of a Function: A Summary • Determining Whether an Equation Defines y as a Function of x • Composite Functions and Their Domains

The authors thank Professor Mohammad Maslam of Georgia Perimeter College for his suggestion to include this summary.

Finding the Domain of a Function: A Summary

To find the domain of a function, given the equation that defines the function, remember that the value of x input into the equation must yield one real number for y when the function is evaluated.

For the functions studied so far in this book, *there are three cases to consider when determining domains.*

1. No input value can lead to 0 in a denominator, because division by 0 is undefined.

2. No input value can lead to an even root of a negative number, because this situation does not yield a real number.

3. No input value can lead to the logarithm of a negative number or 0, because this situation does not yield a real number.

Unless domains are otherwise specified, we can determine domains as follows.

- The domain of a **polynomial function** will be all real numbers.

- The domain of an **absolute value function** will be all real numbers for which the expression inside the absolute value bars (the argument) is defined.

- If a **function is defined by a rational expression,** the domain will be all real numbers for which the denominator is not 0.

- The domain of a **function defined by a radical with *even* root index** is all numbers that make the radicand greater than or equal to 0; **if the root index is *odd*,** the domain is the set of all real numbers for which the radicand is itself a real number.

- For an **exponential function** with constant base, the domain is the set of all real numbers for which the exponent is a real number.
- For a **logarithmic function,** the domain is the set of all real numbers that make the argument of the logarithm greater than 0.

Determining Whether an Equation Defines y as a Function of x

For y to be a function of x, it is necessary that every input value of x in the domain leads to one and only one value of y.

EXAMPLE 1 **Determining if y is a Function of x**

Determine whether the following equations define y as a function of x.

(a) $x - y^3 = 0$ **(b)** $x - y^2 = 0$

Solution

(a) To determine whether an equation represents a function, solve the equation for y.

$$x - y^3 = 0 \qquad \text{Given equation}$$
$$y^3 = x \qquad \text{Add } y^3 \text{ and rewrite.}$$
$$y = \sqrt[3]{x} \qquad \text{Take the cube root.}$$

Because every value of x in the domain (all real numbers) leads to one and only one value of y, we can write y as a function of x. Therefore this equation *does* define y as a function of x.

(b) Solve $x - y^2 = 0$ for y.

$$x - y^2 = 0 \qquad \text{Given equation}$$
$$y^2 = x \qquad \text{Add } y^2 \text{ and rewrite.}$$
$$y = \pm \sqrt{x} \qquad \text{Take the square root.}$$

The domain is $[0, \infty)$. If we let $x = 4$, for example, we get two values of y, -2 and 2. Therefore, for the equation $x - y^2 = 0$, we cannot write y as a function of x. ●

Composite Functions and Their Domains

EXAMPLE 2 **Determining Composite Functions and Their Domains**

Given that $f(x) = \sqrt{x}$ and $g(x) = 4x + 2$, find each of the following.

(a) $(f \circ g)(x)$ and its domain   **(b)** $(g \circ f)(x)$ and its domain

Solution

(a) $(f \circ g)(x) = f(g(x)) = f(4x + 2) = \sqrt{4x + 2}$

The domain and range of g are both the set of all real numbers, $(-\infty, \infty)$. However, the domain of f is the set of all nonnegative real numbers, $[0, \infty)$. Therefore $g(x)$, which is defined as $4x + 2$, must be greater than or equal to zero.

g(x) must be nonnegative.

$$4x + 2 \geq 0 \qquad \text{Solve the inequality.}$$
$$4x \geq -2 \qquad \text{Subtract 2.}$$
$$x \geq -\frac{1}{2} \qquad \text{Divide by 4.}$$

Therefore, the domain of $f \circ g$ is $\left[-\frac{1}{2}, \infty\right)$.

(b) $(g \circ f)(x) = g(f(x)) = g(\sqrt{x}) = 4\sqrt{x} + 2$

The domain and range of f are both the set of all nonnegative real numbers, $[0, \infty)$. The domain of g is the set of all real numbers, $(-\infty, \infty)$. Therefore, the domain of $g \circ f$ is $[0, \infty)$.

EXAMPLE 3   **Determining Composite Functions and Their Domains**

Given that $f(x) = \dfrac{6}{x-3}$ and $g(x) = \dfrac{1}{x}$, find each of the following.

(a) $(f \circ g)(x)$ and its domain   (b) $(g \circ f)(x)$ and its domain

Solution

(a) $(f \circ g)(x) = f(g(x)) = f\left(\dfrac{1}{x}\right)$

$$= \dfrac{6}{\dfrac{1}{x} - 3} \qquad \text{Note that this is undefined if } x = 0.$$

$$= \dfrac{6x}{1 - 3x} \qquad \text{Multiply the numerator and denominator by } x.$$

The domain of g is all real numbers *except* 0, which makes $g(x)$ undefined. The domain of f is all real numbers *except* 3. The expression for $g(x)$, therefore, cannot equal 3, so we determine the value that makes $g(x) = 3$ and *exclude* it from the domain of $f \circ g$.

$$\dfrac{1}{x} = 3 \qquad \text{The solution must be excluded.}$$

$$1 = 3x \qquad \text{Multiply by } x.$$

$$x = \dfrac{1}{3} \qquad \text{Divide by 3.}$$

Therefore, the domain of $f \circ g$ is the set of all real numbers *except* 0 and $\frac{1}{3}$, written in interval notation as $(-\infty, 0) \cup \left(0, \frac{1}{3}\right) \cup \left(\frac{1}{3}, \infty\right)$.

(b) $(g \circ f)(x) = g(f(x)) = g\left(\dfrac{6}{x-3}\right)$

$$= \dfrac{1}{\dfrac{6}{x-3}} \qquad \text{Note that this is undefined if } x = 3.$$

$$= \dfrac{x-3}{6} \qquad \text{Simplify the complex fraction.}$$

The domain of f is all real numbers *except* 3, and the domain of g is all real numbers *except* 0. The expression for $f(x)$, which is $\dfrac{6}{x-3}$, is never zero, since the numerator is the nonzero number 6. Therefore, the domain of $g \circ f$ is the set of all real numbers *except* 3, written $(-\infty, 3) \cup (3, \infty)$.

NOTE   To find the domain of a composition of two functions, it is sometimes helpful not to immediately simplify the resulting expression. For example, if $f(x) = x^2$ and $g(x) = \sqrt{x-1}$, then $(f \circ g)(x) = \left(\sqrt{x-1}\right)^2$. From this unsimplified expression, we can see that the domain (input) of $f \circ g$ must be restricted to $x \geq 1$ for the output to be a real number. As a result,

$$(f \circ g)(x) = x - 1, \quad x \geq 1.$$

Summary Exercises

Determine which one of the choices A, B, C, or D is an equation in which y can be written as a function of x.

1. **A.** $3x + 2y = 6$ **B.** $x = \sqrt{|y|}$ **C.** $x = |y + 3|$ **D.** $x^2 + y^2 = 9$

2. **A.** $3x^2 + 2y^2 = 36$ **B.** $x^2 + y - 2 = 0$ **C.** $x - |y| = 0$ **D.** $x = y^2 - 4$

3. **A.** $x = \sqrt{y^2}$ **B.** $x = \log y^2$ **C.** $x^3 + y^3 = 5$ **D.** $x = \dfrac{1}{y^2 + 3}$

4. **A.** $\dfrac{x^2}{4} + \dfrac{y^2}{4} = 1$ **B.** $x = 5y^2 - 3$ **C.** $\dfrac{x^2}{4} - \dfrac{y^2}{9} = 1$ **D.** $x = 10^y$

5. **A.** $x = \dfrac{2 - y}{y + 3}$ **B.** $x = \ln(y + 1)^2$ **C.** $\sqrt{x} = |y + 1|$ **D.** $\sqrt[4]{x} = y^2$

6. **A.** $e^{y^2} = x$ **B.** $e^{y+2} = x$ **C.** $e^{|y|} = x$ **D.** $10^{|y+2|} = x$

7. **A.** $x^2 = \dfrac{1}{y^2}$ **B.** $x + 2 = \dfrac{1}{y^2}$ **C.** $3x = \dfrac{1}{y^4}$ **D.** $2x = \dfrac{1}{y^3}$

8. **A.** $|x| = |y|$ **B.** $x = |y^2|$ **C.** $x = \dfrac{1}{y}$ **D.** $x^4 + y^4 = 81$

9. **A.** $\dfrac{x^2}{4} - \dfrac{y^2}{9} = 1$ **B.** $\dfrac{y^2}{4} - \dfrac{x^2}{9} = 1$ **C.** $\dfrac{x}{4} - \dfrac{y}{9} = 0$ **D.** $\dfrac{x^2}{4} - \dfrac{y^2}{9} = 0$

10. **A.** $y^2 - \sqrt{(x + 2)^2} = 0$ **B.** $y - \sqrt{(x + 2)^2} = 0$ **C.** $y^6 - \sqrt{(x + 1)^2} = 0$ **D.** $y^4 - \sqrt{x^2} = 0$

Find the domain of each function. Write answers using interval notation.

11. $f(x) = 3x - 6$ **12.** $f(x) = \sqrt{2x - 7}$ **13.** $f(x) = |x + 4|$

14. $f(x) = \dfrac{x + 2}{x - 6}$ **15.** $f(x) = \dfrac{-2}{x^2 + 7}$ **16.** $f(x) = \sqrt{x^2 - 9}$

17. $f(x) = \dfrac{x^2 + 7}{x^2 - 9}$ **18.** $f(x) = \sqrt[3]{x^3 + 7x - 4}$ **19.** $f(x) = \log_5(16 - x^2)$

20. $f(x) = \log \dfrac{x + 7}{x - 3}$ **21.** $f(x) = \sqrt{x^2 - 7x - 8}$ **22.** $f(x) = 2^{1/x}$

23. $f(x) = \dfrac{1}{2x^2 - x + 7}$ **24.** $f(x) = \dfrac{x^2 - 25}{x + 5}$ **25.** $f(x) = \sqrt{x^3 - 1}$

26. $f(x) = \ln|x^2 - 5|$ **27.** $f(x) = e^{x^2 + x + 4}$ **28.** $f(x) = \dfrac{x^3 - 1}{x^2 - 1}$

29. $f(x) = \sqrt{\dfrac{-1}{x^3 - 1}}$ **30.** $f(x) = \sqrt[3]{\dfrac{1}{x^3 - 8}}$ **31.** $f(x) = \ln(x^2 + 1)$

32. $f(x) = \sqrt{(x - 3)(x + 2)(x - 4)}$ **33.** $f(x) = \log\left(\dfrac{x + 2}{x - 3}\right)^2$ **34.** $f(x) = \sqrt[12]{(4 - x)^2(x + 3)}$

35. $f(x) = e^{|1/x|}$

36. $f(x) = \dfrac{1}{|x^2 - 7|}$

37. $f(x) = x^{100} - x^{50} + x^2 + 5$

38. $f(x) = \sqrt{-x^2 - 9}$

39. $f(x) = \sqrt[4]{16 - x^4}$

40. $f(x) = \sqrt[3]{16 - x^4}$

41. $f(x) = \sqrt{\dfrac{x^2 - 2x - 63}{x^2 + x - 12}}$

42. $f(x) = \sqrt[5]{5 - x}$

43. $f(x) = |\sqrt{5 - x}|$

44. $f(x) = \sqrt{\dfrac{-1}{x - 3}}$

45. $f(x) = \log\left|\dfrac{1}{4 - x}\right|$

46. $f(x) = 6^{x^2 - 9}$

47. $f(x) = 6^{\sqrt{x^2 - 25}}$

48. $f(x) = 6^{\sqrt[3]{x^2 - 25}}$

49. $f(x) = \ln\left(\dfrac{-3}{(x + 2)(x - 6)}\right)$

50. $f(x) = \dfrac{-2}{\log x}$

Given functions f and g, find **(a)** $(f \circ g)(x)$ *and its domain, and* **(b)** $(g \circ f)(x)$ *and its domain.*

51. $f(x) = -6x + 9, \quad g(x) = 5x + 7$

52. $f(x) = 8x + 12, \quad g(x) = 3x - 1$

53. $f(x) = \sqrt{x}, \quad g(x) = x + 3$

54. $f(x) = \sqrt{x}, \quad g(x) = x - 1$

55. $f(x) = x^3, \quad g(x) = x^2 + 3x - 1$

56. $f(x) = x + 2, \quad g(x) = x^4 + x^2 - 3x - 4$

57. $f(x) = \sqrt{x - 1}, \quad g(x) = 3x$

58. $f(x) = \sqrt{x - 2}, \quad g(x) = 2x$

59. $f(x) = \dfrac{2}{x}, \quad g(x) = x + 1$

60. $f(x) = \dfrac{4}{x}, \quad g(x) = x + 4$

61. $f(x) = \sqrt{x + 2}, \quad g(x) = -\dfrac{1}{x}$

62. $f(x) = \sqrt{x + 4}, \quad g(x) = -\dfrac{2}{x}$

63. $f(x) = \sqrt{x}, \quad g(x) = \dfrac{1}{x + 5}$

64. $f(x) = \sqrt{x}, \quad g(x) = \dfrac{3}{x + 6}$

65. $f(x) = \dfrac{1}{x - 2}, \quad g(x) = \dfrac{1}{x}$

66. $f(x) = \dfrac{1}{x + 4}, \quad g(x) = -\dfrac{1}{x}$

67. $f(x) = \log x, \quad g(x) = \sqrt{x}$

68. $f(x) = \ln x, \quad g(x) = \sqrt[4]{x}$

69. $f(x) = e^x, \quad g(x) = \sqrt{x}$

70. $f(x) = e^{-x}, \quad g(x) = \sqrt[3]{x}$

71. $f(x) = -x^2, \quad g(x) = \ln \sqrt{x}$

72. $f(x) = \ln \sqrt{x - 1}, \quad g(x) = -2 - x^2$

73. $f(x) = 5x - 2, \quad g(x) = \dfrac{x}{x - 2}$

74. $f(x) = 1 - x, \quad g(x) = \dfrac{3x}{x + 1}$

5 Summary

5.1 Inverse Functions

one-to-one function
contrapositive
inverse function
f^{-1} (f-inverse)

ONE-TO-ONE FUNCTION

A function f is a one-to-one function if, for elements a and b from the domain of f,

$$a \neq b \quad \text{implies} \quad f(a) \neq f(b).$$

HORIZONTAL LINE TEST

A function is one-to-one if every horizontal line intersects the graph of the function at most once.

INVERSE FUNCTION

Let f be a one-to-one function. Then g is the inverse function of f and f is the inverse function of g if

$$(f \circ g)(x) = x \quad \text{for every } x \text{ in the domain of } g,$$

and

$$(g \circ f)(x) = x \quad \text{for every } x \text{ in the domain of } f.$$

FINDING THE EQUATION OF THE INVERSE OF $y = f(x)$

For a one-to-one function f defined by an equation $y = f(x)$, find the defining equation of the inverse as follows. (You may need to replace $f(x)$ with y first. Any restrictions on x and y should be considered.)

Step 1   Interchange x and y.

Step 2   Solve for y.

Step 3   Replace y with $f^{-1}(x)$.

IMPORTANT FACTS ABOUT INVERSES

1. If f is one-to-one, then f^{-1} exists.
2. The domain of f is equal to the range of f^{-1}, and the range of f is equal to the domain of f^{-1}.
3. If the point (a, b) lies on the graph of f, then (b, a) lies on the graph of f^{-1}. The graphs of f and f^{-1} are reflections of each other across the line $y = x$.

5.2 Exponential Functions

exponential function
the number e

EXPONENTIAL FUNCTION

If $a > 0$ and $a \neq 1$, then $f(x) = a^x$ is the exponential function with base a.

CHARACTERISTICS OF EXPONENTIAL FUNCTIONS
$f(x) = a^x, \quad a > 1$

- This function is increasing and continuous on its entire domain, $(-\infty, \infty)$.
- The x-axis is the horizontal asymptote as $x \rightarrow -\infty$.
- The graph passes through the points $(-1, a^{-1})$, $(0, 1)$, and $(1, a)$.

$f(x) = a^x, \quad 0 < a < 1$

- This function is decreasing and continuous on its entire domain, $(-\infty, \infty)$.
- The x-axis is the horizontal asymptote as $x \rightarrow \infty$.
- The graph passes through the points $(-1, a^{-1})$, $(0, 1)$, and $(1, a)$.

KEY TERMS & SYMBOLS	KEY CONCEPTS

SOLVING EXPONENTIAL EQUATIONS (TYPE 1)

To solve a Type 1 exponential equation such as $4^x = 8^{2x-3}$, write each base as a power of the same rational number base, apply the power rule for exponents, set exponents equal, and solve the resulting equation.

COMPOUNDING FORMULAS

Compounded n times per year: $A = P\left(1 + \dfrac{r}{n}\right)^{nt}$ Compounded continuously: $A = Pe^{rt}$

P: principal; r: annual interest rate (in decimal form); t: years

---

5.3 Logarithms and Their Properties

logarithm, $\log_a x$
base
argument
common logarithm, $\log x$
pH
natural logarithm, $\ln x$

LOGARITHM

For all positive numbers a, where $a \neq 1$,

$$a^y = x \quad \text{is equivalent to} \quad y = \log_a x.$$

A logarithm is an exponent, and $\log_a x$ is the exponent to which a must be raised in order to obtain x. The number a is called the base of the logarithm, and x is called the argument of the expression $\log_a x$. The value of x must always be positive.

COMMON LOGARITHM

For all positive numbers x, $\log x = \log_{10} x$.

NATURAL LOGARITHM

For all positive numbers x, $\ln x = \log_e x$.

PROPERTIES OF LOGARITHMS

For $a > 0, a \neq 1$, and any real number k,

$$\textbf{1.} \ \log_a 1 = 0. \qquad \textbf{2.} \ \log_a a^k = k. \qquad \textbf{3.} \ a^{\log_a k} = k, \ \ k > 0.$$

For $x > 0, y > 0, a > 0, a \neq 1$, and any real number r,

Product Rule   **4.** $\log_a xy = \log_a x + \log_a y.$

(The logarithm of the product of two numbers is equal to the sum of the logarithms of the numbers.)

Quotient Rule   **5.** $\log_a \dfrac{x}{y} = \log_a x - \log_a y.$

(The logarithm of the quotient of two numbers is equal to the difference of the logarithms of the numbers.)

Power Rule   **6.** $\log_a x^r = r \log_a x.$

(The logarithm of a number raised to a power is equal to the exponent multiplied by the logarithm of the number.)

CHANGE-OF-BASE RULE

For any positive real numbers x, a, and b, where $a \neq 1$ and $b \neq 1$,

$$\log_a x = \frac{\log_b x}{\log_b a}.$$

(continued)

KEY TERMS & SYMBOLS	KEY CONCEPTS
5.4 Logarithmic Functions logarithmic function	The functions $f(x) = a^x$ and $g(x) = \log_a x$ are inverses. **CHARACTERISTICS OF LOGARITHMIC FUNCTIONS** $f(x) = \log_a x, \quad a > 1$ ● This function is increasing and continuous on its entire domain, $(0, \infty)$. ● The y-axis is the vertical asymptote as $x \to 0$ from the right. ● The graph passes through the points $(a^{-1}, -1)$, $(1, 0)$, and $(a, 1)$. $f(x) = \log_a x, \quad 0 < a < 1$ ● This function is decreasing and continuous on its entire domain, $(0, \infty)$. ● The y-axis is the vertical asymptote as $x \to 0$ from the right. ● The graph passes through the points $(a, 1)$, $(1, 0)$, and $(a^{-1}, -1)$.
5.5 Exponential and Logarithmic Equations and Inequalities	**PROPERTIES OF LOGARITHMIC AND EXPONENTIAL FUNCTIONS** For $b > 0$ and $b \neq 1$, **1.** $b^x = b^y$ if and only if $x = y$. **2.** If $x > 0$ and $y > 0$, then $\log_b x = \log_b y$ if and only if $x = y$. **SOLVING EXPONENTIAL (TYPE 2) AND LOGARITHMIC EQUATIONS** To solve a Type 2 exponential equation such as $2^x = 3^{x+1}$, take the same base logarithm on each side, apply the power rule for logarithms so that the variables are no longer in the exponents, and solve the resulting equation. An exponential or logarithmic equation can be solved by changing the equation into one of the following forms, where a and b are real numbers, $a > 0$, and $a \neq 1$. **1.** $a^{f(x)} = b$ Solve by taking a logarithm of each side. (The natural logarithm is the best choice if $a = e$.) **2.** $\log_a f(x) = \log_a g(x)$ From the given equation, $f(x) = g(x)$, which is solved analytically. **3.** $\log_a f(x) = b$ Solve by changing to exponential form, $f(x) = a^b$.
5.6 Further Applications and Modeling with Exponential and Logarithmic Functions exponential growth function exponential decay function half-life doubling time amortization	**GROWTH** $A(t) = A_0 e^{kt}, \quad k > 0$ **DECAY** $A(t) = A_0 e^{-kt}, \quad k > 0$ **SOUND INTENSITY** $d = 10 \log \dfrac{I}{I_0}$ **AMORTIZATION PAYMENTS** A loan of P dollars at interest rate i per period (as a decimal) may be amortized in n equal periodic payments of R dollars made at the end of each period, where $$R = \frac{P}{\left[\dfrac{1 - (1 + i)^{-n}}{i}\right]}.$$

5   Review Exercises

Determine whether each function is one-to-one. Assume that the graphs shown are comprehensive.

1.
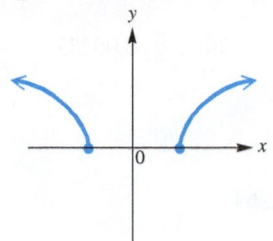

2.
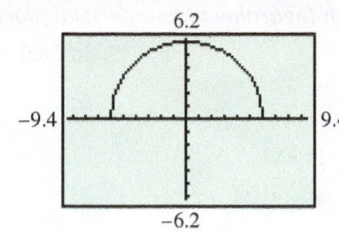

3.
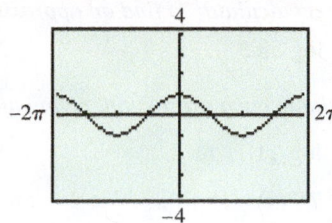

4. $f(x) = \sqrt{3x + 2}$

5. $f(x) = 3x^2 + 1$

6. $f(x) = |2x + 7|$

For Exercises 7–12, consider the function $f(x) = \sqrt[3]{2x - 7}$.

7. What is the domain of f?

8. What is the range of f?

 9. Explain why f^{-1} exists.

10. Find a formula for $f^{-1}(x)$.

11. Graph both f and f^{-1} in a square viewing window, along with the line $y = x$. Describe how the graphs of f and f^{-1} are related.

12. Verify analytically that
$$(f \circ f^{-1})(x) = x \quad \text{and} \quad (f^{-1} \circ f)(x) = x.$$

Concept Check   *Match each equation in Exercises 13–16 with the graph in choices A–D that most closely resembles its graph. Assume that $a > 1$.*

13. $y = a^{x+2}$

14. $y = a^x + 2$

15. $y = -a^x + 2$

16. $y = a^{-x} + 2$

A.
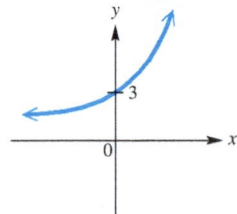

B.
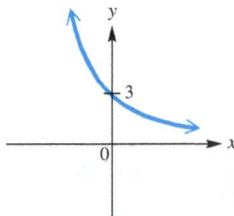

C.
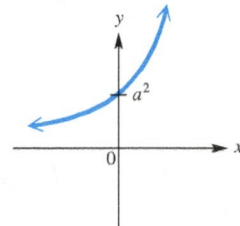

D.
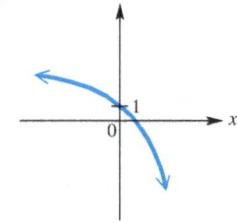

Concept Check   *Consider the exponential function $f(x) = a^x$, graphed to the right. Answer the following based on the graph.*

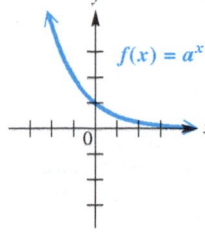

17. What is true about the value of a in comparison to 0 and 1?

18. What is the domain of f?

19. What is the range of f?

20. What is the value of $f(0)$?

21. Graph $y = f^{-1}(x)$ by hand.

22. Give the equation for $f^{-1}(x)$.

Graph each exponential function. Give the domain and range.

23. $f(x) = \left(\dfrac{1}{2}\right)^{x-1}$

24. $f(x) = e^x + 2$

25. $f(x) = -4^x$

26. *Concept Check*   If $f(x) = a^{-x}$ and $0 < a < 1$, is f increasing or decreasing on its domain?

Solve the equation in part (a) analytically. Then use a graph to solve the inequality in part (b).

27. (a) $\left(\dfrac{1}{8}\right)^{-2x} = 2^{x+3}$

(b) $\left(\dfrac{1}{8}\right)^{-2x} \geq 2^{x+3}$

28. (a) $3^{-x} = \left(\dfrac{1}{27}\right)^{1-2x}$

(b) $3^{-x} < \left(\dfrac{1}{27}\right)^{1-2x}$

29. (a) $0.5^{-x} = 0.25^{x+1}$

(b) $0.5^{-x} > 0.25^{x+1}$

30. (a) $0.4^x = 2.5^{1-x}$

(b) $0.4^x < 2.5^{1-x}$

31. The graphs of $y = x^2$ and $y = 2^x$ have the points (2, 4) and (4, 16) in common. There is a third point in common whose coordinates can be approximated by using a graphing calculator. Find the coordinates, giving as many decimal places as your calculator will display.

32. Solve $3^x = \pi$ by the intersection-of-graphs method. Express the solution to the nearest thousandth.

Use a calculator to find an approximation for each logarithm to four decimal places.

33. log 58.3   **34.** log 0.00233   **35.** ln 58.3   **36.** $\log_2 0.00233$

Evaluate each expression, giving an exact value.

37. $\log_{13} 1$

38. $\ln e^{\sqrt{6}}$

39. $\log_5 5^{12}$

40. $7^{\log_7 13}$

41. $3^{\log_3 5}$

42. $\log_4 64$

Concept Check *In Exercises 43–48, identify the corresponding graph in choices A–F for each function.*

43. $f(x) = \log_2 x$

44. $f(x) = \log_2(2x)$

45. $f(x) = \log_2\left(\dfrac{1}{x}\right)$

46. $f(x) = \log_2\left(\dfrac{x}{2}\right)$

47. $f(x) = \log_2(x - 1)$

48. $f(x) = \log_2(-x)$

A.
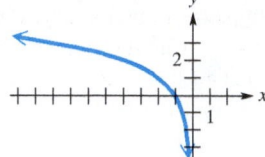

B.
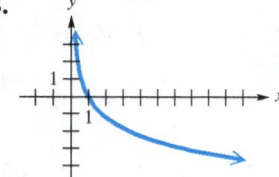

C.
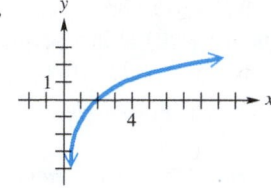

D.
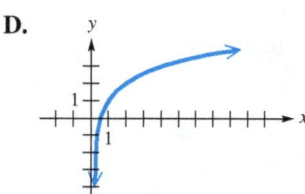

E.
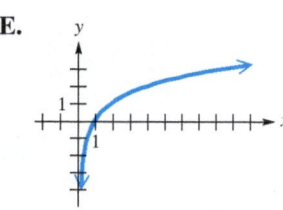

F.
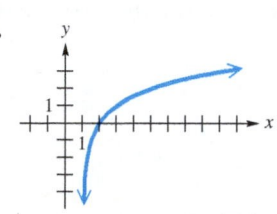

*Graph each logarithmic function. Give the domain and range. (Hint: See **Exercises 23–25,** as those exponential functions are closely related to these.)*

49. $g(x) = 1 + \log_{1/2} x$

50. $g(x) = \ln(x - 2)$

51. $g(x) = \log_4(-x)$

52. ***Concept Check*** How do the functions in **Exercises 49–51** relate to those in **Exercises 23–25?**

53. ***Concept Check*** What is the base of the logarithmic function whose graph contains the point (81, 4)?

54. ***Concept Check*** What is the base of the exponential function whose graph contains the point $\left(-4, \frac{1}{16}\right)$?

Write each logarithm as a sum, difference, or product of logarithms if possible. Assume variables represent positive numbers.

55. $\log_3 \dfrac{mn}{5r}$

56. $\log_2 \dfrac{\sqrt{7}}{15}$

57. $\log_5\left(x^2 y^4 \sqrt[5]{m^3 p}\right)$

58. $\log_7(7k + 5r^2)$

Solve the equation in part (a) analytically. Then use a graph to solve the inequality in part (b).

59. **(a)** $\log(x + 3) + \log x = 1$

   (b) $\log(x + 3) + \log x > 1$

60. **(a)** $\ln e^{\ln x} - \ln(x - 4) = \ln 3$

   (b) $\ln e^{\ln x} - \ln(x - 4) \leq \ln 3$

61. **(a)** $\ln e^{\ln 2} - \ln(x - 1) = \ln 5$

   (b) $\ln e^{\ln 2} - \ln(x - 1) \geq \ln 5$

*Solve each equation. Express the solution set so that **(a)** solutions are in exact form and, if irrational,* **(b)** *solutions are approximated to the nearest thousandth. Support your solutions by using a calculator.*

62. $8^x = 32$

63. $\dfrac{8}{27} = x^{-3}$

64. $10^{2x-3} = 17$

65. $e^{x+1} = 10$

66. $\log_{64} x = \dfrac{1}{3}$

67. $\ln(6x) - \ln(x + 1) = \ln 4$

68. $\log_{12}(2x) + \log_{12}(x - 1) = 1$

69. $\log_{16} \sqrt{x + 1} = \dfrac{1}{4}$

70. $\ln x + 3 \ln 2 = \ln\left(\dfrac{2}{x}\right)$

71. $\ln(\ln(e^{-x})) = \ln 3$

72. $\ln e^x - \ln e^3 = \ln e^5$

73. $2^x = -3$

Solve each formula for the indicated variable.

74. $N = a + b \ln\left(\dfrac{c}{d}\right)$, for c

75. $y = y_0 e^{-kt}$, for t

Use the x-intercept method to estimate the solution(s) of each equation. Round to the nearest thousandth.

76. $\log_{10} x = x - 2$

77. $2^{-x} = \log_{10} x$

78. $x^2 - 3 = \log x$

Solve each application.

79. *Interest Rate*   To the nearest tenth, what interest rate, compounded annually, will produce $4613 if $3500 is left at this interest for 10 years?

80. *Growth of an Account*   Find the number of years (to the nearest tenth) needed for $48,000 to become $58,344 at 5% interest compounded semiannually.

81. *Growth of an Account*   Lateisha Shaw deposits $12,000 for 8 years in an account paying 5% compounded annually. She then leaves the money alone, with no further deposits, at 6% compounded annually for an additional 6 years. Approximate the total amount on deposit after the entire 14-year period.

82. *Growth of an Account*   Suppose that $2000 is invested in an account that pays 3% annually and is left for 5 years.
   (a) How much will be in the account if interest is compounded quarterly (4 times per year)?
   (b) How much will be in the account if interest is compounded continuously?
   (c) To the nearest tenth of a year, how long will it take the $2000 to triple if interest is compounded continuously?

(Modeling)   *Solve each problem.*

83. *Runway Length*   There is a mathematical relationship between an airplane's weight x and the runway length required at takeoff. For some airplanes, the minimum runway length in thousands of feet may be modeled by

$$L(x) = 3 \log x,$$

where x is measured in thousands of pounds. (*Source:* Haefner, L., *Introduction to Transportation Systems,* Holt, Rinehart and Winston.)
   (a) Graph L in the window $[0, 50]$ by $[0, 6]$. Interpret the graph.
   (b) If the weight of an airplane increases tenfold from 10,000 to 100,000 pounds, does the length of the required runway also increase by a factor of 10? Explain.

84. *Caloric Intake*   The function

$$f(x) = 280 \ln(x + 1) + 1925$$

models the number of calories consumed daily by a person owning x acres of land in a developing country. Find the number of acres owned by someone whose average calorie intake is 2200 calories per day. (*Source:* Grigg, D., *The World Food Problem,* Blackwell Publishers.)

85. *Pollutant Concentration*   The concentration of a pollutant, in grams per liter, in the east fork of the Big Weasel River is approximated by

$$P(x) = 0.04e^{-4x},$$

where x is the number of miles downstream from a paper mill where the measurement is taken. Find the following.
   (a) $P(0.5)$
   (b) $P(1)$
   (c) The concentration of the pollutant 2 miles downstream
   (d) The number of miles downstream where the concentration of the pollutant is 0.002 gram per liter.

86. *Repetitive Skills*   A person learning certain skills involving repetition tends to learn quickly at first. Then, learning tapers off and approaches some upper limit. Suppose the number of symbols per minute that a textbook typesetter can produce is given by

$$p(x) = 250 - 120(2.8)^{-0.5x},$$

where x is the number of months the typesetter has been in training. Find the following to the nearest whole number.
   (a) $p(2)$ **(b)** $p(10)$
   (c) Graph $y = p(x)$ in the window $[0, 10]$ by $[0, 300]$, and support the answer to part (a).

87. *Free Fall of a Skydiver*   A skydiver in free fall travels at a speed of

$$v(t) = 176(1 - e^{-0.18t})$$

feet per second after t seconds. How long will it take for the skydiver to attain a speed of 147 feet per second (100 mph)?

5 Test

1. Match each equation in parts (a)–(d) with its graph from choices A–D.
   (a) $y = \log_{1/2} x$ (b) $y = e^x$
   (c) $y = \ln x$ (d) $y = \left(\frac{1}{2}\right)^x$
   (e) Which pairs of functions in parts (a)–(d) are inverses?

   A.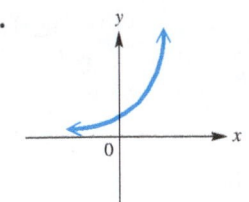

   B.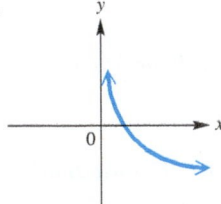

   C.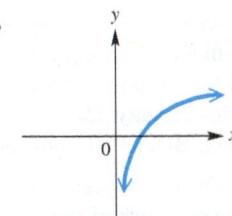

   D.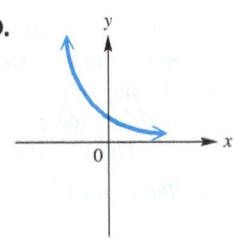

2. Decide whether each function is one-to-one and whether it has an inverse function.
   (a) $f(x) = 3 - 4x$ (b) $f(x) = x^2 - 3x$
   (c) $f(x) = 5(2)^x$

3. For each one-to-one function f, find $f^{-1}(x)$.
   (a) $f(x) = 5x - 7$
   (b) $f(x) = 2 \log x$
   (c) $f(x) = \dfrac{x + 1}{x - 2}$

4. Consider the function $f(x) = -2^{x-1} + 8$.
   (a) Graph f in the standard viewing window.
   (b) Give the domain and range of f.
   (c) Does the graph have an asymptote? If so, is it vertical or horizontal, and what is its equation?
   (d) Find the x- and y-intercepts analytically, and use the graph from part (a) to support your answers graphically.
   (e) Find $f^{-1}(x)$.

5. Solve the equation $\left(\frac{1}{8}\right)^{2x-3} = 16^{x+1}$ analytically.

6. *Growth of an Account*   Suppose that $10,000 is invested at 3.5%. Find the total amount after 4 years, if interest is compounded (a) quarterly and (b) continuously.

7. One of your friends tells you, "I have no idea what an expression like $\log_5 27$ really means." Write an explanation of what it means, and tell how you can find an approximation for it with a calculator.

8. Use a calculator to find an approximation of each logarithm to the nearest thousandth.
   (a) $\log 45.6$ (b) $\ln 470$ (c) $\log_3 769$

9. Use the properties of logarithms to write $\log \dfrac{m^3 n}{\sqrt{y}}$ as an equivalent expression.

10. Use the properties of logarithms to rewrite
    $$2 \log x + \tfrac{1}{2} \log y - 4 \log z$$
    as a single logarithm with coefficient 1. Assume that all variables are positive.

11. Simplify the expressions $\ln e^{2x}$ and $10^{\log x^2}$.

12. Find the domain of $f(x) = \ln(2x + 1)$.

13. Solve $A = Pe^{rt}$ for t.

14. Consider the equation $\log_2 x + \log_2(x + 2) = 3$.
    (a) Solve the equation analytically. If there is an extraneous value, what is it?
    (b) To support the solution in part (a), we can graph $y_1 = \log_2 x + \log_2(x + 2) - 3$ and locate the x-intercept. Use the change-of-base rule with base 10 to graph the function y_1.
    (c) Use the graph to solve the logarithmic inequality $\log_2 x + \log_2(x + 2) > 3$.

Solve each equation. Give the solution set (a) with an exact value and (b) with an approximation to the nearest thousandth.

15. $2e^{5x+2} = 8$ 16. $6^{2-x} = 2^{3x+1}$

17. $\log(\ln x) = 1$

18. *(Modeling) Population Growth*   A population is increasing according to the equation $y = 2e^{0.02t}$, where y is in millions and t is in years. Match each question in parts (a)–(d) with one of the solution methods A–D.
    (a) How long will it take for the population to triple?
    (b) When will the population reach 3 million?
    (c) How large will the population be in 3 years?
    (d) How large will the population be in 4 months?

    A. Evaluate $2e^{0.02(1/3)}$.   **B.** Solve $2e^{0.02t} = 6$ for t.
    C. Evaluate $2e^{0.02(3)}$.   **D.** Solve $2e^{0.02t} = 3$ for t.

19. *(Modeling) Drug Level in the Bloodstream*   After a medical drug is injected into the bloodstream, it is gradually eliminated from the body. Graph each function in parts (a)–(d) on the interval $[0, 10]$. Use $[0, 500]$ as the range of $A(t)$. Use a graphing calculator to determine the function that best models the amount $A(t)$ (in milligrams) of a drug remaining in the body after t hours if 350 milligrams were initially injected.
    (a) $A(t) = t^2 - t + 350$ (b) $A(t) = 350 \log(t + 1)$
    (c) $A(t) = 350(0.75)^t$ (d) $A(t) = 100(0.95)^t$

20. *(Modeling) Decay of Radium*   Radium 226 has a half-life of about 1600 years. An initial sample weighs 2 grams.
    (a) Find a formula for the decay function.
    (b) Find the amount left after 9600 years.
    (c) Find the time for the initial amount to decay to 0.5 gram.

Systems of equations occur in applications that involve more than one variable, such as in business or construction, or even when determining the weight of a black bear.

6 Systems and Matrices

CHAPTER OUTLINE

6.1 Systems of Equations

6.2 Solution of Linear Systems in Three Variables

6.3 Solution of Linear Systems by Row Transformations

6.4 Matrix Properties and Operations

6.5 Determinants and Cramer's Rule

6.6 Solution of Linear Systems by Matrix Inverses

6.7 Systems of Inequalities and Linear Programming

6.8 Partial Fractions

6.1 Systems of Equations

Linear Systems • Substitution Method • Elimination Method • Special Systems • Nonlinear Systems
• Applications of Systems

Linear Systems

The definition of a linear equation given in **Chapter 1** can be extended to more variables. Any equation of the form

$$a_1 x_1 + a_2 x_2 + \cdots + a_n x_n = b,$$

for real numbers $a_1, a_2, \ldots, a_n$ (not all of which are 0) and b, is a **linear equation**, or a **first-degree equation in n unknowns.**

A set of equations is called a **system of equations.** The **solutions** of a system of equations must satisfy *every* equation in the system. If all the equations in a system are linear, the system is a **system of linear equations,** or a **linear system.**

The solution set of a linear equation in two variables is an infinite set of ordered pairs. Since the graph of such an equation is a straight line, there are three possible outcomes for the graph of a system of two linear equations in two variables, as shown in **FIGURE 1.**

1. **The graphs intersect at exactly one point,** which gives the (single) ordered-pair solution of the system. The **system is consistent** and the **equations are independent.** See **FIGURE 1(a).**

2. **The graphs are parallel lines,** so there is no solution and the solution set is $\emptyset$. The **system is inconsistent** and the **equations are independent.** See **FIGURE 1(b).**

3. **The graphs are the same line,** and there are infinitely many solutions. The **system is consistent** and the **equations are dependent.** See **FIGURE 1(c).**

Systems of Linear Equations

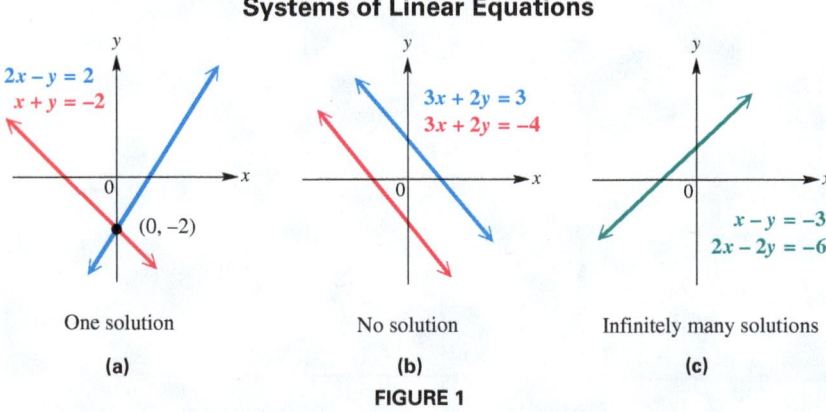

One solution

(a)

No solution

(b)

Infinitely many solutions

(c)

FIGURE 1

Substitution Method

Although the *number* of solutions of a linear system in two variables can usually be seen from the graph of the equations of the system, determining an exact solution from a graph is often difficult. In such cases, we use algebraic methods of solution.

In a system of two equations with two variables, the **substitution method** involves using one equation to find an expression for one variable in terms of the other, and then substituting this expression into the other equation of the system.

Solving a System by Substitution

Solve the system.

$$3x + 2y = 11 \quad \text{(1)}$$
$$-x + y = 3 \quad \text{(2)}$$

Analytic Solution

One way to solve this system is to solve equation (2) for y, getting $y = x + 3$. Then substitute $x + 3$ for y in equation (1) and solve for x.

$3x + 2y = 11$	(1)
$3x + 2(x + 3) = 11$	Let $y = x + 3$.
$3x + 2x + 6 = 11$	Distributive property
$5x + 6 = 11$	Combine like terms.
$5x = 5$	Subtract 6.
$x = 1$	Divide by 5.

Replace x with 1 in $y = x + 3$ to determine that $y = 1 + 3 = 4$, so the solution is $(1, 4)$. Check by substituting 1 for x and 4 for y in *both* equations of the original system.

Check:

$3x + 2y = 11$ (1)	$-x + y = 3$ (2)
$3(1) + 2(4) = 11$?	$-1 + 4 = 3$?
$11 = 11$ ✓ True	$3 = 3$ ✓ True

Both equations check and thus the solution set is $\{(1, 4)\}$.

Graphing Calculator Solution

To find the solution graphically, first solve equations (1) and (2) for y to get $Y_1 = -1.5X + 5.5$ and $Y_2 = X + 3$. Graph Y_1 and Y_2 in the standard viewing window to determine that the point of intersection is $(1, 4)$, as seen in **FIGURE 2**. The table in **FIGURE 3** shows numerically that when $X = 1$, both Y_1 and Y_2 are 4.

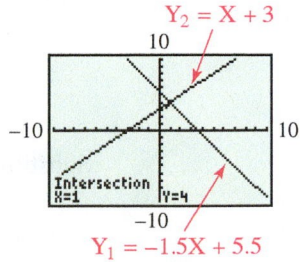

FIGURE 2

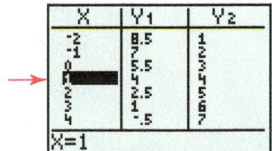

FIGURE 3

→ **Looking Ahead to Calculus**

In calculus, the **definite integral** often gives the area between the graphs of f and g on the interval $[a, b]$. A system of equations is used to find the x-values a and b where the two graphs intersect. See **Section 6.7, Exercises 75–78.**

Elimination Method

Another way to solve a system of two equations, the **elimination method,** uses multiplication and addition to eliminate a variable from one equation. In this process, the given system is replaced by new systems that have the same solution set as the original system. Systems that have the same solution set are called **equivalent systems.**

The three transformations that produce an equivalent system are listed here.

Transformations of a Linear System

1. Interchange any two equations of the system.
2. Multiply or divide any equation of the system by a nonzero real number.
3. Replace any equation of the system by the sum of that equation and a multiple of another equation in the system.

EXAMPLE 2 **Solving a System by Elimination**

Solve the system.

$$3x - 4y = 1 \qquad (1)$$
$$2x + 3y = 12 \qquad (2)$$

Solution To eliminate x, use the second transformation. Multiply each side of equation (1) by -2 and each side of equation (2) by 3.

$$-6x + 8y = -2 \qquad \text{Multiply (1) by } -2. \quad (3)$$
$$6x + 9y = 36 \qquad \text{Multiply (2) by 3.} \quad (4)$$

This new system will have the same solution set as the given system. By the third transformation, we can add equations (3) and (4) to eliminate x and solve the result for y.

$$
\begin{array}{ll}
-6x + 8y = -2 & (3) \\
\underline{6x + 9y = 36} & (4) \\
17y = 34 & \text{Add.} \\
y = 2 & \text{Divide by 17.}
\end{array}
$$

Substitute 2 for y in either equation (1) or (2). We choose (1).

Write the x-value first in the ordered pair.

$$
\begin{array}{ll}
3x - 4(2) = 1 & (1) \\
3x = 9 & \text{Multiply and then add 8.} \\
x = 3 & \text{Divide by 3.}
\end{array}
$$

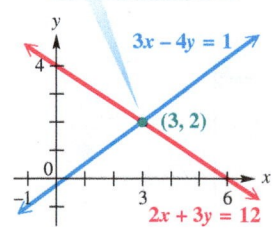

The system has one solution, (3, 2).

Consistent system

FIGURE 4

The solution is $(3, 2)$, which can be checked by substituting 3 for x and 2 for y in *both* equations of the original system.

Check:

$3x - 4y = 1$ (1)	$2x + 3y = 12$ (2)
$3(3) - 4(2) = 1$?	$2(3) + 3(2) = 12$?
$1 = 1$ ✓ True	$12 = 12$ ✓ True

Both equations check, and thus the solution set is $\{(3, 2)\}$. The graph in **FIGURE 4** supports this result and shows that the system is consistent. ●

WHAT WENT WRONG?

To solve the system in **Example 2** graphically, a student solved each equation for y, entered the functions shown in the top screen in the margin, and got the incorrect graph shown in the bottom screen.

What Went Wrong? What can be done to correct the graph?

Special Systems

The examples presented so far in this section have all been consistent systems with a single solution. However, this is not always the case.

Answer to What Went Wrong?

The student neglected to insert parentheses as necessary. Y_1 should be entered as $(3X - 1)/4$ and Y_2 should be entered as $(12 - 2X)/3$.

EXAMPLE 3   **Solving an Inconsistent System**

Solve the system.

$$3x - 2y = 4 \quad (1)$$
$$-6x + 4y = 7 \quad (2)$$

Analytic Solution

Eliminate the variable x by multiplying each side of equation (1) by 2 and then adding the result to equation (2).

$$
\begin{array}{ll}
6x - 4y = 8 & \text{Multiply (1) by 2.} \\
\underline{-6x + 4y = 7} & \text{(2)} \\
 0 = 15 & \text{False}
\end{array}
$$

Both variables were eliminated, leaving the false statement $0 = 15$. This contradiction indicates that these two equations have no solutions in common. The system is inconsistent, and the solution set is $\emptyset$.

Graphing Calculator Solution

In slope–intercept form, equation (1) is $y = 1.5x - 2$ and equation (2) is $y = 1.5x + 1.75$. Since the lines have the *same* slope but *different* y-intercepts, they are parallel. They have no point of intersection, which supports the analytic solution. See **FIGURE 5.**

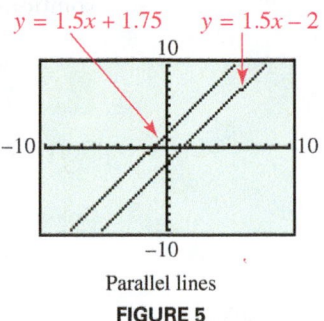

Parallel lines

FIGURE 5

EXAMPLE 4   **Solving a System with Dependent Equations**

Solve the system.

$$8x - 2y = -4 \quad (1)$$
$$-4x + y = 2 \quad (2)$$

Analytic Solution

Divide each side of equation (1) by 2, and add the result to equation (2).

$$
\begin{array}{ll}
4x - y = -2 & \text{Divide (1) by 2.} \\
\underline{-4x + y = 2} & \text{(2)} \\
 0 = 0 & \text{True}
\end{array}
$$

The result, $0 = 0$, is a true statement, which indicates that the equations of the original system are equivalent. Any ordered pair (x, y) that satisfies either equation will satisfy the system. From equation (2),

$$-4x + y = 2, \quad \text{or} \quad y = 2 + 4x.$$

The solutions of the system can be written as a set of ordered pairs $(x, 2 + 4x)$, for any real number x. Some ordered pairs in the solution set are $(0, 2 + 4 \cdot 0) = (0, 2)$, $(1, 2 + 4 \cdot 1) = (1, 6)$, $(3, 14)$, $\left(-\frac{1}{2}, 0\right)$, and $(-2, -6)$. The solution set is $\{(x, 2 + 4x)\}$, where x is any real number.

Graphing Calculator Solution

Solving the equations for y gives $Y_1 = \dfrac{8X + 4}{2}$ and $Y_2 = 2 + 4X$. The graphs of the two equations coincide, as seen in **FIGURE 6.** The table indicates that $Y_1 = Y_2$ for arbitrarily selected values of X, providing another way to show that the two equations lead to the same graph.

$$Y_1 = \frac{8X + 4}{2} \text{ and } Y_2 = 2 + 4X$$

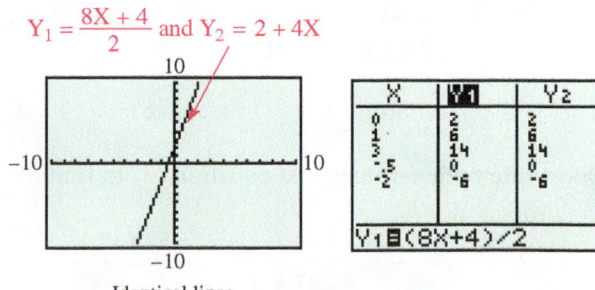

Identical lines

FIGURE 6

Refer to the analytic solution to see how the solution set can be written.

NOTE We could solve equation (2) in **Example 4** for x and write the solution set in terms of y: $\left\{ \left(\frac{y-2}{4}, y \right) \right\}$. Choose values for y and verify that the same ordered pairs result.

Nonlinear Systems

A **nonlinear system of equations** is a system in which *at least one* of the equations is not a linear equation.

The number of *possible* solutions to a nonlinear system can sometimes be determined by thinking graphically. For example, the equations $x + y = 3$ (line) and $x^2 + y^2 = 5$ (circle) can have zero, one, or two solutions, because a line can intersect a circle at 0, 1, or 2 points. In general, a nonlinear system can have any number of solutions. See **Exercises 51–58.**

EXAMPLE 5 **Solving a Nonlinear System by Substitution**

Solve the system.

$$3x^2 - 2y = 5 \qquad (1)$$
$$x + 3y = -4 \qquad (2)$$

Analytic Solution

The graph of equation (1) is a parabola, and that of equation (2) is a line. A line can intersect a parabola at 0, 1, or 2 points, so there could be 0, 1, or 2 solutions to this nonlinear system.

Although either equation could be solved for either variable, we choose here to solve the linear equation (2) for y.

$$x + 3y = -4 \qquad (2)$$
$$3y = -4 - x \qquad \text{Subtract } x.$$
$$y = \frac{-4 - x}{3} \qquad \text{Divide by 3. (3)}$$

Substitute this expression for y into equation (1).

$$3x^2 - 2\left(\frac{-4 - x}{3} \right) = 5 \qquad (1)$$
$$9x^2 - 2(-4 - x) = 15 \qquad \text{Multiply each term by 3.}$$
$$9x^2 + 8 + 2x = 15 \qquad \text{Distributive property}$$
$$9x^2 + 2x - 7 = 0 \qquad \text{Standard form}$$
$$(9x - 7)(x + 1) = 0 \qquad \text{Factor.}$$
$$x = \frac{7}{9} \quad \text{or} \quad x = -1 \qquad \text{Solve for } x.$$

Substitute both x-values into equation (3) to find the y-values.

$$y = \frac{-4 - \frac{7}{9}}{3} = -\frac{43}{27} \quad \bigg| \quad y = \frac{-4 - (-1)}{3} = -1$$

The solution set is $\left\{ \left(\frac{7}{9}, -\frac{43}{27} \right), (-1, -1) \right\}$.

Graphing Calculator Solution

Graph equation (1) as $y_1 = 1.5x^2 - 2.5$ and equation (2) as $y_2 = -\frac{1}{3}x - \frac{4}{3}$. **FIGURE 7** indicates the points of intersection, $(-1, -1)$ and $(0.\overline{7}, 1.\overline{592})$, supporting the analytic solution.

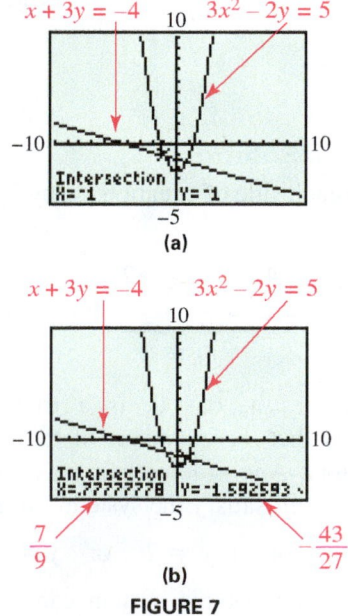

FIGURE 7

The decimal values for x and y displayed at the bottom of **FIGURE 7(b)** are approximations for the repeating decimal forms for $\frac{7}{9}$ and $-\frac{43}{27}$, respectively.

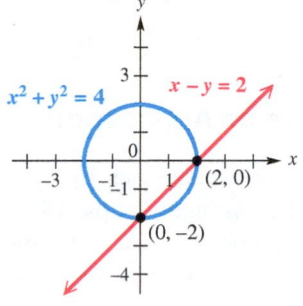

$x^2 + y^2 = 4$ $x - y = 2$

$(2, 0)$

$(0, -2)$

The two points of intersection
are $(0, -2)$ and $(2, 0)$.

FIGURE 8

EXAMPLE 6   **Solving a Nonlinear System by Substitution**

Solve the system.

$$x^2 + y^2 = 4 \quad (1)$$
$$x - y = 2 \quad (2)$$

Solution   The graph of equation (1) is a circle and that of equation (2) is a line. Visualizing a circle and a line suggests that there may be 0, 1, or 2 points of intersection.
Solve equation (2) to obtain $y = x - 2$. Substitute $x - 2$ for y in equation (1) and solve.

$x^2 + y^2 = 4$	(1)
$x^2 + (x - 2)^2 = 4$	Let $y = x - 2$.
$x^2 + x^2 - 4x + 4 = 4$	Square the binomial.
$2x^2 - 4x + 4 = 4$	Combine like terms.
$2x^2 - 4x = 0$	Subtract 4.
$x^2 - 2x = 0$	Divide by 2.
$x(x - 2) = 0$	Factor.
$x = 0$ or $x = 2$	Solve by zero-product property.

Because $y = x - 2$, it follows that $y = -2$ when $x = 0$ and that $y = 0$ when $x = 2$. Check that the solution set is $\{(0, -2), (2, 0)\}$. **FIGURE 8** shows that the two points of intersection are $(0, -2)$ and $(2, 0)$.   ●

Many systems are difficult or even impossible to solve analytically. Graphing calculators allow us to solve some of these systems graphically.

EXAMPLE 7   **Solving a Nonlinear System Graphically**

Solve the system.

$$y = 2^x \quad (1)$$
$$|x + 2| - y = 0 \quad (2)$$

TECHNOLOGY NOTE

Example 7 illustrates why proficiency in setting various windows is important when studying algebra with a graphing calculator.

Solution   Enter equation (1) as $y_1 = 2^x$ and equation (2) as $y_2 = |x + 2|$. As seen in **FIGURE 9,** the various windows indicate that there are three points of intersection of the graphs and thus three solutions. Using the capabilities of the calculator, we find that $(2, 4)$ is an exact solution and both $(-2.22, 0.22)$ and $(-1.69, 0.31)$ are approximate solutions. Therefore, the solution set is $\{(2, 4), (-2.22, 0.22), (-1.69, 0.31)\}$.

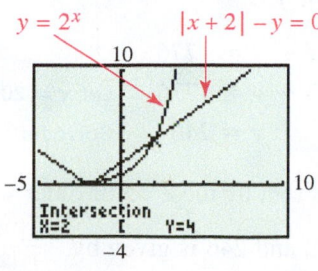
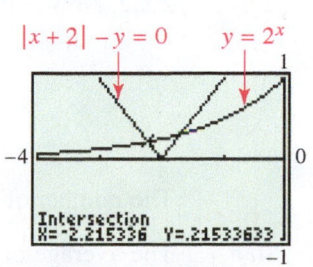
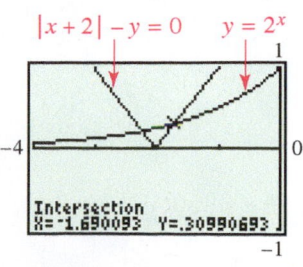

FIGURE 9   ●

Applications of Systems

We solve an applied problem with a system of equations as follows.

Solving Application Problems

Step 1 **Read the problem** carefully and assign a variable to represent each quantity.

Step 2 **Write a system of equations** involving these variables.

Step 3 **Solve the system of equations** and determine the solution.

Step 4 **Look back and check** your solution. Does it seem reasonable?

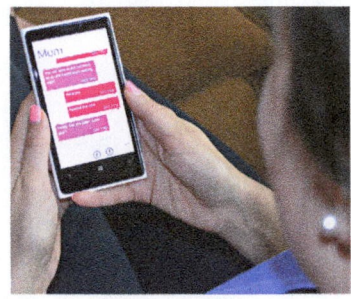

EXAMPLE 8 **Using a Linear System to Solve an Application**

The number of text messages sent per month by smartphone users varies by age. In a recent survey, the average of the number of texts sent by two age groups, 18 to 24-year-olds and those 55 or over, was 1134. However, the number of texts sent by 18 to 24-year-olds exceeded the number of texts sent by those 55 or over by 1776. Determine how many texts were sent by each age group.

Solution

Step 1 Let x represent the number of texts sent by those 18 to 24 and y represent the number sent by those 55 or older.

Step 2 Since the *average* of the number of texts for the two age groups was 1134, one equation is

$$\frac{x + y}{2} = 1134. \quad \text{(1)}$$

The number of texts sent by those 18 to 24 years old exceeded the number sent by those 55 or older by 1776. Thus, another equation is

$$x - y = 1776. \quad \text{(2)}$$

Step 3 Multiply equation (1) by 2 and then add the resulting equations.

$$
\begin{array}{rl}
x + y = 2268 & \text{Multiply (1) by 2.} \\
\underline{x - y = 1776} & \text{(2)} \\
2x = 4044 & \text{Add the equations.} \\
x = 2022 & \text{Divide by 2.}
\end{array}
$$

Thus, $x = 2022$ and the number of texts sent by those 18 to 24 years old was 2022. Now determine y.

$$
\begin{array}{rl}
x - y = 1776 & \text{(2)} \\
2022 - y = 1776 & \text{Let } x = 2022. \\
y = 246 & \text{Solve for } y.
\end{array}
$$

The number of texts sent by those 55 or older was 246.

Step 4 The average of 2022 and 246 is given by $\frac{2022 + 246}{2} = 1134$, and their difference is $2022 - 246 = 1776$. The solution checks.

EXAMPLE 9   Using a Nonlinear System to Find the Dimensions of a Box

A box with an open top has a square base and four sides of equal height. The volume of the box is 75 cubic inches, and the surface area is 85 square inches. What are the dimensions of the box?

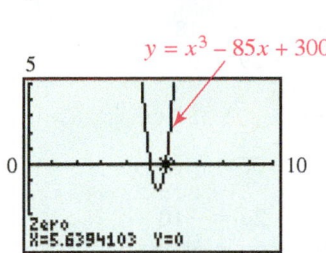

FIGURE 10

Solution   **FIGURE 10** shows a diagram of such a box. If each side of the square base measures x inches and the height measures y inches, then the volume is

$$x^2y = 75 \qquad \text{Volume formula} \quad (1)$$

and the surface area is

$$x^2 + 4xy = 85. \qquad \text{Sum of areas of base and four sides} \quad (2)$$

We solve equation (1) for y to get $y = \dfrac{75}{x^2}$ and substitute this into equation (2).

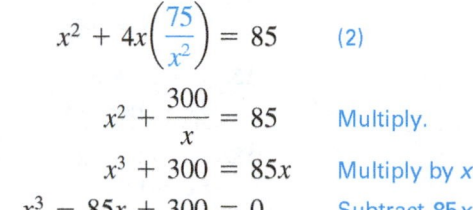

$$x^2 + 4x\left(\frac{75}{x^2}\right) = 85 \qquad (2)$$

$$x^2 + \frac{300}{x} = 85 \qquad \text{Multiply.}$$

$$x^3 + 300 = 85x \qquad \text{Multiply by } x.$$

$$x^3 - 85x + 300 = 0 \qquad \text{Subtract } 85x.$$

The solutions correspond to the x-intercepts of the graph of $y = x^3 - 85x + 300$. A comprehensive graph indicates that there are three real solutions. However, one of them is negative and must be rejected. As **FIGURE 11** indicates, one positive solution is 5 and the other is approximately 5.64. By substituting back into equation (1), we find that when $x = 5$, $y = 3$, and when $x \approx 5.64$, $y \approx 2.36$.

Therefore, this problem has two solutions: The box may have base 5 inches by 5 inches and height 3 inches, or it may have base 5.64 inches by 5.64 inches and height 2.36 inches (approximately).

FIGURE 11

6.1 Exercises

Changes in Population   *Many factors may contribute to population changes in metropolitan areas. The graph shows the populations of the New Orleans, Louisiana, and the Jacksonville, Florida, metropolitan areas over the years 2004–2009.*

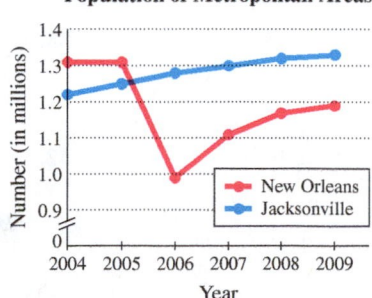

1. In what years was the population of the Jacksonville metropolitan area greater than that of the New Orleans metropolitan area?

2. At the time when the populations of the two metropolitan areas were equal, what was the approximate population of each area?

3. Express the solution of the system as an ordered pair.

4. Use the terms *increasing, decreasing,* and *constant* to describe the trends for the population of the New Orleans metropolitan area.

5. If equations of the form $y = f(t)$ were determined that modeled either of the two graphs, then the variable t would represent _____ and the variable y would represent _____.

 6. Explain why each graph is that of a function.

Checking Analytic Skills   *Use each graph to estimate the solution of the system of equations. Then solve the system analytically.* **Do not use a calculator.**

7.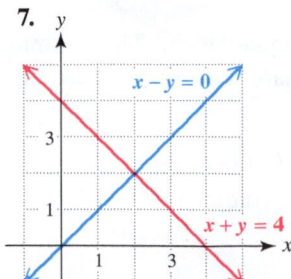

8.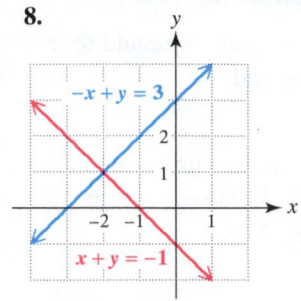

9.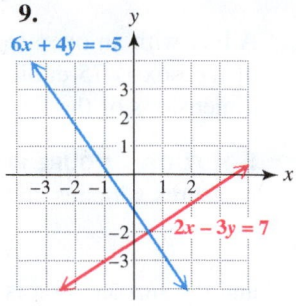

10.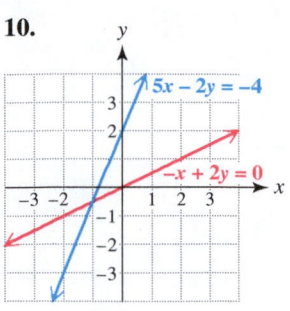

Solve each system by substitution.

11. $6x - y = 5$
$\quad\quad y = x$

12. $5x + y = 2$
$\quad\quad y = -3x$

13. $x + 2y = -1$
$\quad 2x + y = 4$

14. $2x + y = -11$
$\quad x + 3y = -8$

15. $y = 2x + 3$
$\quad 3x + 4y = 78$

16. $y = 4x - 6$
$\quad 2x + 5y = -8$

17. $3x - 2y = 12$
$\quad 5x = 4 - 2y$

18. $8x + 3y = 2$
$\quad 5x = 17 + 6y$

19. $4x - 5y = -11$
$\quad 2x + y = 5$

20. $7x - y = -10$
$\quad 3y - x = 10$

21. $4x + 5y = 7$
$\quad 9y = 31 + 2x$

22. $2x + 6y = -18$
$\quad 5y = -29 + 3x$

23. $3x - 7y = 15$
$\quad 3x + 7y = 15$

24. $3y = 5x + 6$
$\quad x + y = 2$

25. $2x - 7y = 8$
$\quad -3x + \dfrac{21}{2}y = 5$

26. $0.6x - 0.2y = 2$
$\quad -1.2x + 0.4y = 3$

27. $x - 2y = 4$
$\quad -2x + 4y = -8$

28. $-3x + 2y = -10$
$\quad 9x - 6y = 30$

Solve each system by elimination.

29. $3x - y = -4$
$\quad x + 3y = 12$

30. $2x - 3y = -7$
$\quad 5x + 4y = 17$

31. $4x + 3y = -1$
$\quad 2x + 5y = 3$

32. $5x + 7y = 6$
$\quad 10x - 3y = 46$

33. $12x - 5y = 9$
$\quad 3x - 8y = -18$

34. $6x + 7y = -2$
$\quad 7x - 6y = 26$

35. $4x - y = 9$
$\quad -8x + 2y = -18$

36. $x + y = 4$
$\quad 3x + 3y = 12$

37. $9x - 5y = 1$
$\quad -18x + 10y = 1$

38. $3x + 2y = 5$
$\quad 6x + 4y = 8$

39. $3x + y = 6$
$\quad 6x + 2y = 1$

40. $3x + 5y = -2$
$\quad 9x + 15y = -6$

41. $\dfrac{x}{2} + \dfrac{y}{3} = 8$

$\quad \dfrac{2x}{3} + \dfrac{3y}{2} = 17$

42. $\dfrac{x}{5} + 3y = 31$

$\quad 2x - \dfrac{y}{5} = 8$

43. $\dfrac{2x - 1}{3} + \dfrac{y + 2}{4} = 4$

$\quad \dfrac{x + 3}{2} - \dfrac{x - y}{3} = 3$

44. $\dfrac{x + 6}{5} + \dfrac{2y - x}{10} = 1$

$\quad \dfrac{x + 2}{4} + \dfrac{3y + 2}{5} = -3$

Use a graphing calculator to solve each system. Express solutions with approximations to the nearest thousandth.

45. $\sqrt{3}\,x - y = 5$
$\quad 100x + y = 9$

46. $\dfrac{11}{3}x + y = 0.5$
$\quad 0.6x - y = 3$

47. $\sqrt{5}x + \sqrt[3]{6}y = 9$
$\quad \sqrt{2}x + \sqrt[5]{9}y = 12$

48. $\pi x + ey = 3$
$\quad ex + \pi y = 4$

49. Explain how one can determine whether a system is inconsistent or has dependent equations when using the substitution or elimination method.

50. *Concept Check*   For what value(s) of k will the following system of linear equations have no solution? infinitely many solutions?

$$x - 2y = 3$$
$$-2x + 4y = k$$

Draw a sketch of the two graphs described with the indicated number of points of intersection. (There may be more than one way to do this.)

51. A line and a circle; no points

52. A line and a circle; one point

53. A line and a circle; two points

54. A line and a parabola; no points

55. A line and a parabola; one point

56. A line and a parabola; two points

57. A circle and a parabola; four points

58. A circle and a parabola; one point

Solve each system graphically. Check your solutions. **Do not use a calculator.**

59. $x + y = 3$
$2x - y = 0$

60. $3x - y = 4$
$x + y = 0$

61. $x^2 + y^2 = 5$
$x + y = 3$

62. $x - y = 3$
$x^2 + y^2 = 9$

63. $x^2 - y = 0$
$x^2 + y^2 = 2$

64. $x - y^2 = 1$
$x^2 + y^2 = 5$

65. $x^2 + y^2 = 4$
$x + y = 2$

66. $x^2 - y = 0$
$x + y^2 = 0$

Solve each nonlinear system of equations analytically.

67. $y = -x^2 + 2$
$x - y = 0$

68. $y = (x - 1)^2$
$x - 3y = -1$

69. $x^2 + y^2 = 5$
$x - y = 1$

70. $x^2 + y^2 = 5$
$-3x + 4y = 2$

71. $x^2 + y^2 = 10$
$-x^2 + y = -4$

72. $y = |x - 1|$
$y = x^2 - 4$

Solve each system graphically. Give x- and y-coordinates correct to the nearest hundredth.

73. $y = \log(x + 5)$
$y = x^2$

74. $y = 5^x$
$xy = 1$

75. $y = e^{x+1}$
$2x + y = 3$

76. $y = \sqrt[3]{x - 4}$
$x^2 + y^2 = 6$

(Modeling)   *Solve each problem.*

77. *Black Friday Spending*   The total spending on Black Friday during 2011 and 2012 was $1858 million. From 2011 to 2012, spending increased by $226 million. (*Source:* www.marketingcharts.com)

   (a) Write a system of equations whose solution represents the Black Friday spending in each of these years. Let x be the amount spent in 2012 and y be the amount spent in 2011.

   (b) Solve the system.

   (c) Interpret the solution.

78. *Self-Reported Spending*   The average of self-reported spending "yesterday" for high-income consumers and middle-/low-income consumers was $93.50 in September 2012. High-income consumers spend $65 more than middle-/low-income consumers. (*Source:* www.marketingcharts.com)

   (a) Write a system of equations whose solution gives the self-reported spending for each income group. Let x be the spending by high-income consumers and y be the spending by middle-/low-income consumers.

   (b) Solve the system.

   (c) Interpret the solution.

79. *Price of Tablets* From 2010 to 2012, the average selling price of tablets decreased by 30%. This percent reduction amounted in a decrease of $195. Find the average selling price of tablets in 2010 and in 2012.

80. *Dimensions of a Box* A box has an open top, rectangular sides, and a square base. The volume of the box is 576 cubic inches, and the surface area of the outside of the box is 336 square inches. Find the dimensions of the box.

81. *Dimensions of a Box* A box has rectangular sides and a rectangular top and base that are twice as long as they are wide. The volume of the box is 588 cubic inches, and the surface area of the outside of the box is 448 square inches. Find the dimensions of the box.

82. *Investments* A student invests a total of $5000 at 3% and 4% annually. After 1 year, the student receives a total of $187.50 in interest. How much did the student invest at each interest rate?

83. *Heart Rate* In one study, a group of conditioned athletes was exercised to exhaustion. Let x represent an athlete's heart rate 5 seconds after stopping exercise and y the rate after 10 seconds. It was found that the maximum heart rate H for these athletes satisfied the two equations

$$H = 0.491x + 0.468y + 11.2$$

and $\qquad H = -0.981x + 1.872y + 26.4.$

If an athlete had maximum heart rate $H = 180$, determine x and y graphically. Interpret your answer. (*Source:* Thomas, V., *Science and Sport,* Faber and Faber.)

84. *Heart Rate* Repeat **Exercise 83** for an athlete with a maximum heart rate of 195.

85. *Media Spending* From January to June 2012, Samsung and Apple spent a combined $293 million on media. Apple spent $93 million more than Samsung.
(a) Write a system of equations whose solution gives the spending of each media company, in millions of dollars. Let x be the amount spent by Apple and y be the amount spent by Samsung.
(b) Solve the system of equations.
(c) Interpret the solution.

86. *Populations of Minorities in the United States* The current and estimated resident populations, y, (in percent) of Black and Spanish/Hispanic/Latino people in the United States for the years 1990–2050 are modeled by the following linear equations.

$$y = 0.0515x + 12.3 \qquad \text{Black}$$
$$y = 0.255x + 9.01 \qquad \text{Sp./Hisp./Lat.}$$

In each case, x represents the number of years since 1990. (*Source:* U.S. Census Bureau.)
(a) Solve the system to find the year when these population percents were equal.
(b) What percent of the U.S. resident population will be Spanish/Hispanic/Latino in the year found in part (a)?
(c) Graphically support the analytic solution in part (a).
(d) Which population is increasing more rapidly?

87. *Geometry* Approximate graphically the radius and height of a cylindrical container with volume 50 cubic inches and lateral (side) surface area 65 square inches.

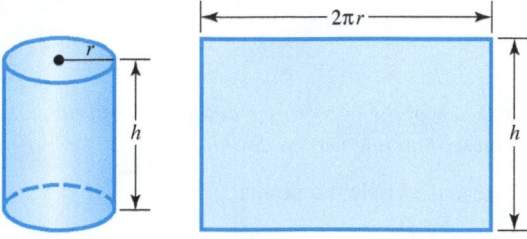

88. *Geometry* Determine graphically whether it is possible to construct a cylindrical container, *including* the top and bottom, with volume 38 cubic inches and surface area 38 square inches.

89. *Roof Truss* The forces or weights W_1 and W_2 exerted on each rafter for the roof truss shown in the figure are determined by the following system of linear equations. Solve the system to the nearest tenth.

$$W_1 + \sqrt{2}W_2 = 300$$
$$\sqrt{3}W_1 - \sqrt{2}W_2 = 0$$

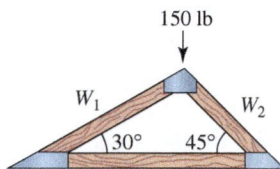

90. *Height and Weight* The relationship between a professional basketball player's height h in inches and weight w in pounds was modeled by using two samples of players. The resulting equations were

$$w = 7.46h - 374$$
and $\qquad w = 7.93h - 405.$

Assume that $65 \leq h \leq 85$.
(a) Use each equation to predict the weight to the nearest pound of a professional basketball player who is 6 feet 11 inches.
(b) Determine graphically the height at which the two models give the same weight.
(c) For each model, what change in weight is associated with a 1-inch increase in height?

91. *Traffic Control* The figure at the top of the next page shows two intersections labeled A and B that involve one-way streets. The numbers and variables represent the average traffic flow rates measured in vehicles per hour. For example, an average of 500 vehicles per hour enter intersection A from the west, whereas 150 vehicles per hour enter this intersection from the north. A stoplight will control the unknown traffic flow denoted by the variables x and y. Use the fact that the number of vehicles entering an intersection must equal the number leaving to determine x and y.

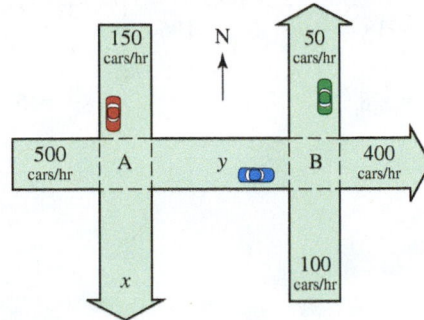

Supply and Demand *As the price of a product increases, businesses usually increase the quantity manufactured. However, as the price increases, consumer demand—or the quantity of the product purchased by consumers—usually decreases. The price we see in the market place occurs when the quantity supplied and the quantity demanded are equal. This price is called the* **equilibrium price** *and this demand is called the* **equilibrium demand.**

92. The supply of a certain product is related to its price by the equation $p = \frac{1}{3}q$, where p is in dollars and q is the quantity supplied in hundreds of units.

   (a) If this product sells for \$9, what quantity will be supplied by the manufacturer?

   (b) Suppose that consumer demand for the same product decreases as price increases according to the equation $p = 20 - \frac{1}{5}q$. If this product sells for \$9, what quantity will consumers purchase? How does this compare with the quantity being supplied by the manufacturer at this price?

(c) On the basis of parts (a) and (b), what should happen to the price? Explain.

(d) Determine the equilibrium price at which the quantity supplied and quantity demanded are equal. What is the demand at this price?

93. (Refer to **Exercise 92.**) Suppose that supply is related to price by $p = \frac{1}{10}q$ and that demand is related to price by $p = 15 - \frac{2}{3}q$, where p is price in dollars and q is the quantity supplied in units.

   (a) Determine the price at which 15 units would be supplied. Determine the price at which 15 units would be demanded.

   (b) Determine the equilibrium price at which the quantity supplied and quantity demanded are equal. What is the demand at this price?

94. (Refer to **Exercise 92.**) Find the equilibrium price in dollars if $p = \frac{2}{3}q$ and $p = 49 - \frac{1}{2}q$. How many units represent the demand at this price?

(Modeling) Break-Even Point *The break-even point for a company is the point where costs equal revenues. If both cost and revenue are expressed as linear equations, the break-even point is the solution of a linear system. In each exercise, C represents cost in dollars to produce x items, and R represents revenue in dollars from the sale of x items. Use the substitution method to find the break-even point in each case—that is, the point where C = R. Then find the value of C and R at that point.*

95. $C = 20x + 10{,}000$
     $R = 30x - 11{,}000$

96. $C = 4x + 125$
     $R = 9x - 200$

RELATING CONCEPTS **For individual or group investigation (Exercises 97–102)**

Because variables appear in denominators, the system

$$\frac{5}{x} + \frac{15}{y} = 16$$

$$\frac{5}{x} + \frac{4}{y} = 5$$

is not a linear system. However, we can solve it in a manner similar to the method for solving a linear system by using a substitution-of-variable technique. Let $t = \frac{1}{x}$ and let $u = \frac{1}{y}$. **Work Exercises 97–102 in order.**

  97. Write a system of equations in t and u by making the appropriate substitutions.

  99. Solve the given system for x and y by using the equations relating t to x and u to y.

 101. Repeat **Exercise 100** for the second equation in the given system.

  98. Solve the system in **Exercise 97** for t and u.

 100. Refer to the first equation in the given system, and solve for y in terms of x to obtain a rational function.

 102. Using a viewing window of $[0, 10]$ by $[0, 2]$, show that the point of intersection of the graphs of the functions in **Exercises 100** and **101** has the same x- and y-values as found in **Exercise 99.**

Use the substitution-of-variable technique from the preceding Relating Concepts exercises to solve each system analytically.

103. $\dfrac{2}{x} + \dfrac{1}{y} = \dfrac{3}{2}$

$\dfrac{3}{x} - \dfrac{1}{y} = 1$

104. $\dfrac{2}{x} + \dfrac{1}{y} = 11$

$\dfrac{3}{x} - \dfrac{5}{y} = 10$

105. $\dfrac{2}{x} + \dfrac{3}{y} = 18$

$\dfrac{4}{x} - \dfrac{5}{y} = -8$

106. $\dfrac{1}{x} + \dfrac{3}{y} = \dfrac{16}{5}$

$\dfrac{5}{x} + \dfrac{4}{y} = 5$

6.2 Solution of Linear Systems in Three Variables

Geometric Considerations • Analytic Solution of Systems in Three Variables • Applications of Systems • Fitting Data Using a System

Geometric Considerations

We can extend the ideas of systems of equations in two variables to linear equations of the form

$$Ax + By + Cz = D, \quad \text{which has an } \textbf{ordered triple } (x, y, z)$$

as its solution. For example, $(1, 2, -4)$ is a solution of

$$2x + 5y - 3z = 24.$$

The solution set of this equation is an infinite set of ordered triples.

In geometry, the graph of a linear equation in three variables is a plane in three-dimensional space. Considering the possible intersections of the planes representing three equations in three unknowns shows that the solution set of such a system may be either a single ordered triple (x, y, z), an infinite set of ordered triples (dependent equations), or the empty set (an inconsistent system). See **FIGURE 12.**

Consistent Systems

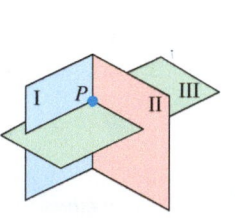

A single solution

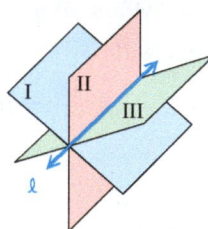

Points of a line in common

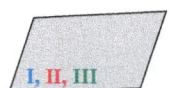

All points in common

Inconsistent Systems

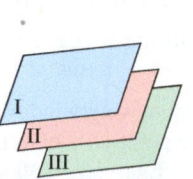

No points in common

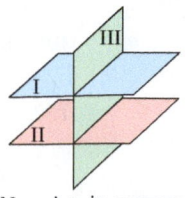

No points in common

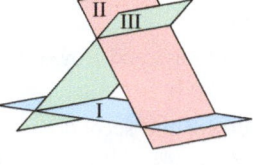

No points in common

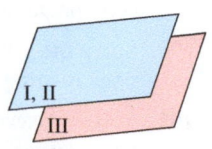

No points in common

FIGURE 12

Analytic Solution of Systems in Three Variables

The following steps can be used to solve a linear system with three variables.

Step 1 Eliminate a variable from any two of the equations.

Step 2 Eliminate the **same variable** from a different pair of equations.

Step 3 Eliminate a second variable using the resulting two equations in two variables to get an equation with just one variable whose value we can now determine.

Step 4 Find the values of the remaining variables by substitution. Write the solution of the system as an **ordered triple.**

> **EXAMPLE 1**   Solving a System of Three Equations in Three Variables

Solve the system.

$$
\begin{aligned}
3x + 9y + 6z &= 3 && (1)\\
2x + y - z &= 2 && (2)\\
x + y + z &= 2 && (3)
\end{aligned}
$$

Solution

Step 1 Eliminate z by adding equations (2) and (3).

$$3x + 2y = 4 \qquad (4)$$

Step 2 Eliminate z from another pair of equations by multiplying each side of equation (2) by 6 and adding the result to equation (1).

Make sure equation (5) has the same two variables as equation (4).

$$
\begin{array}{ll}
12x + 6y - 6z = 12 & \text{Multiply (2) by 6.}\\
\underline{3x + 9y + 6z = 3} & (1)\\
15x + 15y \phantom{{}- 6z} = 15 & \text{Add to obtain (5).}
\end{array}
$$

Step 3 Eliminate x from equations (4) and (5) by multiplying each side of equation (4) by -5 and adding the result to equation (5). Solve the resulting equation for y.

$$
\begin{array}{ll}
-15x - 10y = -20 & \text{Multiply (4) by } -5.\\
\underline{15x + 15y = 15} & (5)\\
5y = -5 & \text{Add.}\\
y = -1 & \text{Divide by 5.}
\end{array}
$$

Step 4 Using $y = -1$, find x from equation (4) by substitution.

$$
\begin{array}{ll}
3x + 2(-1) = 4 & \text{Let } y = -1 \text{ in (4).}\\
3x = 6 & \text{Multiply, and then add 2.}\\
x = 2 & \text{Divide by 3.}
\end{array}
$$

Substitute 2 for x and -1 for y in equation (3) to find z.

Write the values of x, y, and z in the correct order.

$$
\begin{array}{ll}
2 + (-1) + z = 2 & (3)\\
z = 1 & \text{Add, and then subtract 1.}
\end{array}
$$

Verify that the ordered triple $(2, -1, 1)$ satisfies all three equations in the *original* system. The solution set is $\{(2, -1, 1)\}$. ●

> **CAUTION**   Be careful not to end up with two equations that still have *three* variables. **Eliminate the same variable from each pair of equations.**

EXAMPLE 2 **Solving a System of Two Equations in Three Variables**

Solve the system.

$$x + 2y + z = 4 \qquad (1)$$
$$3x - y - 4z = -9 \qquad (2)$$

Solution Geometrically, the solution is the intersection of the two planes given by equations (1) and (2). The intersection of two nonparallel planes is a line. Thus, there will be infinitely many ordered triples in the solution set.

To eliminate x, multiply both sides of equation (1) by -3 and add the result to equation (2). (Either y or z could have been eliminated instead.)

$$
\begin{aligned}
-3x - 6y - 3z &= -12 && \text{Multiply (1) by } -3. \\
\underline{3x - y - 4z} &= \underline{-9} && \text{(2)} \\
-7y - 7z &= -21 && \text{Add to obtain (3).} \\
-7z &= 7y - 21 && \text{Add } 7y. \\
z &= -y + 3 && \text{Divide by } -7.
\end{aligned}
$$

Solve this equation for z.

This gives z in terms of y. Express x in terms of y by solving equation (1) for x and substituting $-y + 3$ for z in the result.

$$
\begin{aligned}
x + 2y + z &= 4 && \text{(1)} \\
x &= -2y - z + 4 && \text{Solve for } x. \\
x &= -2y - (-y + 3) + 4 && \text{Substitute } -y + 3 \text{ for } z. \\
x &= -y + 1 && \text{Simplify.}
\end{aligned}
$$

Use parentheses around $-y + 3$.

The system has infinitely many solutions. For any value of y, the value of x is $-y + 1$ and the value of z is $-y + 3$. For example, if $y = 1$, then $x = -1 + 1 = 0$ and $z = -1 + 3 = 2$, giving the solution $(0, 1, 2)$. Another solution is $(-1, 2, 1)$. With y arbitrary, the solution set is of the form $\{(-y + 1, y, -y + 3)\}$. ●

NOTE Had we solved equation (3) in **Example 2** for y instead of z, the solution would have had a different form, but would have led to the same set of solutions. In that case we would have z arbitrary, and the solution set would be of the form $\{(z - 2, -z + 3, z)\}$, or $\{(-2 + z, 3 - z, z)\}$. By choosing $z = 2$, one solution would be $(0, 1, 2)$, which was found in **Example 2**.

A system of three linear equations in three variables can be inconsistent, as illustrated in the next example.

EXAMPLE 3 **Identifying a System with No Solution**

Three students buy different combinations of basketball tickets for their families. The first student buys 1 senior, 2 adult, and 1 student ticket for $39. The second student buys 1 senior, 1 adult, and 3 student tickets for $37. The third student buys 2 senior, 3 adult, and 4 student tickets for $77. If possible, find the price of each type of ticket. Interpret your answer.

Solution Let x be the price of a senior ticket, y the price of an adult ticket, and z the price of a student ticket. Then the three purchases can be expressed as a system of linear equations.

$$\begin{array}{rl} x + 2y + z = 39 & \text{(1)} \\ x + y + 3z = 37 & \text{(2)} \\ 2x + 3y + 4z = 77 & \text{(3)} \end{array}$$

We can eliminate x from equation (1) by subtracting equation (2) from equation (1).

$$\begin{array}{rl} x + 2y + z = 39 & \text{(1)} \\ \underline{x + y + 3z = 37} & \text{(2)} \\ y - 2z = 2 & \text{Subtract to obtain (4).} \end{array}$$

We can also eliminate x by multiplying equation (2) by 2 and subtracting equation (3) from that result.

$$\begin{array}{rl} 2x + 2y + 6z = 74 & \text{Two times equation (2)} \\ \underline{2x + 3y + 4z = 77} & \text{(3)} \\ -y + 2z = -3 & \text{Subtract to obtain equation (5).} \end{array}$$

If we add equations (4) and (5), we obtain the following contradiction.

$$\begin{array}{rl} y - 2z = 2 & \text{(4)} \\ \underline{-y + 2z = -3} & \text{(5)} \\ 0 = -1 & \text{Add (4) and (5)} \end{array}$$

Because $0 = -1$ is false, there is no solution. Notice that the third student bought the same number and type of tickets as the first and second students did together. The third student should have been charged $\$39 + \$37 = \$76$, rather than $\$77$. *Inconsistent* pricing resulted in an *inconsistent* system of equations.

Applications of Systems

EXAMPLE 4   **Solving a System to Satisfy Feed Requirements**

An animal feed is made from three ingredients: corn, soybeans, and cottonseed. One unit of each ingredient provides units of protein, fat, and fiber, as shown in the table. How many units of each ingredient should be used to make a feed that contains 22 units of protein, 28 units of fat, and 18 units of fiber?

	Corn	Soybeans	Cottonseed	Total
Protein	0.25	0.4	0.2	22
Fat	0.4	0.2	0.3	28
Fiber	0.3	0.2	0.1	18

Solution Let x represent the number of units of corn, y the number of units of soybeans, and z the number of units of cottonseed that are required. Since the total amount of protein must be 22 units, the first row of the table yields

$$0.25x + 0.4y + 0.2z = 22. \qquad \text{(1)}$$

Also, for the 28 units of fat,

$$0.4x + 0.2y + 0.3z = 28, \qquad \text{(2)}$$

(continued)

and for the 18 units of fiber,

$$0.3x + 0.2y + 0.1z = 18. \qquad \text{(3)}$$

To clear decimals, multiply each side of the first equation by 100 and each side of the second and third equations by 10, to get the following system.

$$\begin{aligned}
25x + 40y + 20z &= 2200 \qquad \text{(4)} \\
4x + 2y + 3z &= 280 \qquad \text{(5)} \\
3x + 2y + z &= 180 \qquad \text{(6)}
\end{aligned}$$

Using the method described in **Example 1** shows that

$$x = 40, \quad y = 15, \quad \text{and} \quad z = 30.$$

The feed should contain 40 units of corn, 15 units of soybeans, and 30 units of cotton-seed to fulfill the given requirements. ●

EXAMPLE 5 **Solving a Feed Requirements Application with Fewer Equations Than Variables**

In **Example 4,** suppose that only the fat and fiber content of the feed are of interest. How would the solution be changed?

Solution We would need to solve the system

$$\begin{aligned}
4x + 2y + 3z &= 280 \qquad \text{(5)} \\
3x + 2y + z &= 180, \qquad \text{(6)}
\end{aligned}$$

which does not have a unique solution, since there are two equations with three variables. Following the procedure outlined in **Example 2,** we find that the solution set of this system is $\{(100 - 2z, 2.5z - 60, z)\}$. In this applied problem, however, all three variables must be nonnegative, so z must satisfy the conditions

$$100 - 2z \geq 0, \quad 2.5z - 60 \geq 0, \quad z \geq 0.$$

From the first inequality, $z \leq 50$, and from the second inequality, $z \geq 24$. Thus, $24 \leq z \leq 50$. Only solutions with z in this range are usable. ●

Fitting Data Using a System

Recall from **Chapter 3** that the graph of a quadratic function is a parabola with a vertical axis. Given three noncollinear points with distinct x-coordinates, we can find the equation of a parabola that passes through them.

EXAMPLE 6 **Using a System to Determine a Parabola**

FIGURE 13 shows three data points: $(2, 4)$, $(-1, 1)$, and $(-2, 5)$. Find the equation of the parabola (with vertical axis) that passes through these points.

Solution The three points lie on the graph of the equation $y = ax^2 + bx + c$ and must satisfy the equation. Substitute each ordered pair into the equation.

$$\begin{aligned}
4 = a(2)^2 + b(2) + c, &\quad \text{or} \quad 4 = 4a + 2b + c \qquad \text{(1)} \\
1 = a(-1)^2 + b(-1) + c, &\quad \text{or} \quad 1 = a - b + c \qquad \text{(2)} \\
5 = a(-2)^2 + b(-2) + c, &\quad \text{or} \quad 5 = 4a - 2b + c \qquad \text{(3)}
\end{aligned}$$

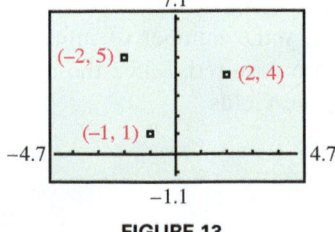

FIGURE 13

To solve this system, first eliminate c, using equations (1) and (2).

$$4 = 4a + 2b + c \qquad (1)$$
$$\underline{-1 = -a + b - c} \qquad \text{Multiply (2) by } -1.$$
$$3 = 3a + 3b \qquad \text{Add to obtain (4).}$$

Now, use equations (2) and (3) to also eliminate c.

> Equation (5) must have the same two variables as equation (4).

$$1 = a - b + c \qquad (2)$$
$$\underline{-5 = -4a + 2b - c} \qquad \text{Multiply (3) by } -1.$$
$$-4 = -3a + b \qquad \text{Add to obtain (5).}$$

Solve the system of equations (4) and (5) in two variables by eliminating a.

$$3 = 3a + 3b \qquad (4)$$
$$\underline{-4 = -3a + b} \qquad (5)$$
$$-1 = \qquad 4b \qquad \text{Add.}$$
$$-\frac{1}{4} = b \qquad \text{Solve for } b.$$

Find a by substituting $-\frac{1}{4}$ for b in transformed equation (4).

$$1 = a + b \qquad \text{Equation (4) divided by 3}$$
$$1 = a - \frac{1}{4} \qquad \text{Let } b = -\frac{1}{4}.$$
$$\frac{5}{4} = a \qquad \text{Solve for } a.$$

Finally, find c by substituting $a = \frac{5}{4}$ and $b = -\frac{1}{4}$ in equation (2).

$$1 = a - b + c \qquad (2)$$
$$1 = \frac{5}{4} - \left(-\frac{1}{4}\right) + c \qquad \text{Let } a = \frac{5}{4}, b = -\frac{1}{4}.$$
$$1 = \frac{6}{4} + c \qquad \text{Simplify.}$$
$$-\frac{1}{2} = c \qquad \text{Subtract } \frac{6}{4}.$$

An equation $y = ax^2 + bx + c$ of the parabola is

$$y = \frac{5}{4}x^2 - \frac{1}{4}x - \frac{1}{2}, \quad \text{or equivalently,} \quad y = 1.25x^2 - 0.25x - 0.5.$$

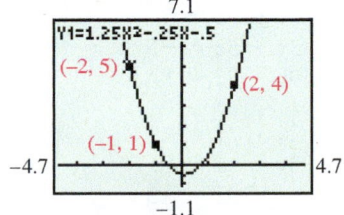

The figure shows that the parabola contains the three data points. A calculator with regression capability can also be used to find the quadratic function that exactly fits these three data points.

6.2 Exercises

Checking Analytic Skills In Exercises 1–6, verify that the given ordered triple is a solution of the system. **Do not use a calculator.**

1. $(-3, 6, 1)$

$$2x + y - z = -1$$
$$x - y + 3z = -6$$
$$-4x + y + z = 19$$

2. $\left(\frac{1}{2}, -\frac{3}{4}, \frac{1}{6}\right)$

$$2x + 8y - 6z = -6$$
$$x + y + z = -\frac{1}{12}$$
$$x + 3z = 1$$

3. $(-0.2, 0.4, 0.5)$

$$5x - y + 2z = -0.4$$
$$x + 4z = 1.8$$
$$-3y + z = -0.7$$

4. $(-1, -2, -3)$

$$x - y + z = -2$$
$$x - 2y + z = 0$$
$$y - z = 1$$

5. $(-2, -1, 3)$

$$x - y + z = 2$$
$$3x - 2y + z = -1$$
$$x + y = -3$$

6. $\left(\dfrac{1}{2}, \dfrac{1}{2}, -2\right)$

$$3x + y + z = 0$$
$$4x + 2y + z = 1$$
$$2x - 2y - z = 2$$

Solve each system analytically. If the equations are dependent, write the solution set in terms of the variable z. (Hint: In Exercises 33–36, let $t = \dfrac{1}{x}$, $u = \dfrac{1}{y}$, and $v = \dfrac{1}{z}$. Solve for t, u, and v, and then find x, y, and z.)

7.
$$x + y + z = 2$$
$$2x + y - z = 5$$
$$x - y + z = -2$$

8.
$$2x + y + z = 9$$
$$-x - y + z = 1$$
$$3x - y + z = 9$$

9.
$$x + 3y + 4z = 14$$
$$2x - 3y + 2z = 10$$
$$3x - y + z = 9$$

10.
$$4x - 3y + z = 9$$
$$3x + 2y - 2z = 4$$
$$x - y + 3z = 5$$

11.
$$x + 2y + 3z = 8$$
$$3x - y + 2z = 5$$
$$-2x - 4y - 6z = 5$$

12.
$$3x - 2y - 8z = 1$$
$$9x - 6y - 24z = -2$$
$$x - y + z = 1$$

13.
$$x + 4y - z = 6$$
$$2x - y + z = 3$$
$$3x + 2y + 3z = 16$$

14.
$$4x - y + 3z = -2$$
$$3x + 5y - z = 15$$
$$-2x + y + 4z = 14$$

15.
$$5x + y - 3z = -6$$
$$2x + 3y + z = 5$$
$$-3x - 2y + 4z = 3$$

16.
$$2x - 5y + 4z = -35$$
$$5x + 3y - z = 1$$
$$x + y + z = 1$$

17.
$$x - 3y - 2z = -3$$
$$3x + 2y - z = 12$$
$$-x - y + 4z = 3$$

18.
$$x + y + z = 3$$
$$3x - 3y - 4z = -1$$
$$x + y + 3z = 11$$

19.
$$x - 2y + 3z = 6$$
$$2x - y + 2z = 5$$

20.
$$5x - 4y + z = 9$$
$$x + y = 15$$

21.
$$3x + 4y - z = 13$$
$$x + y + 2z = 15$$

22.
$$3x - 5y - 4z = -7$$
$$y - z = -13$$

23.
$$8x - 3y + 6z = -2$$
$$4x + 9y + 4z = 18$$
$$12x - 3y + 8z = -2$$

24.
$$2x + 6y - z = 6$$
$$4x - 3y + 5z = -5$$
$$6x + 9y - 2z = 11$$

25.
$$x - z = 2$$
$$x + y = -3$$
$$y - z = 1$$

26.
$$x + z = 4$$
$$x + y = 4$$
$$y + z = 4$$

27.
$$3x + 2y - z = -1$$
$$3y + z = 12$$
$$x - 3z = -3$$

28.
$$2x + y - z = -4$$
$$y + 2z = 12$$
$$2x - z = -4$$

29.
$$x - y + z = -6$$
$$4x + y + z = 7$$

30.
$$3x - 2y + z = 15$$
$$x + 4y - z = 11$$

31.
$$2x + 3y + 4z = 3$$
$$6x + 3y + 8z = 6$$
$$6y - 4z = 1$$

32.
$$10x + 2y - 3z = 0$$
$$5x + 4y + 6z = -1$$
$$6y + 3z = 2$$

33.
$$\frac{1}{x} + \frac{1}{y} - \frac{1}{z} = \frac{1}{4}$$
$$\frac{2}{x} - \frac{1}{y} + \frac{3}{z} = \frac{9}{4}$$
$$-\frac{1}{x} - \frac{2}{y} + \frac{4}{z} = 1$$

34.
$$\frac{3}{x} + \frac{2}{y} - \frac{1}{z} = \frac{11}{6}$$
$$\frac{1}{x} - \frac{1}{y} + \frac{3}{z} = -\frac{11}{12}$$
$$\frac{2}{x} + \frac{1}{y} + \frac{1}{z} = \frac{7}{12}$$

35.
$$\frac{2}{x} - \frac{2}{y} + \frac{1}{z} = -1$$
$$\frac{4}{x} + \frac{1}{y} - \frac{2}{z} = -9$$
$$\frac{1}{x} + \frac{1}{y} - \frac{3}{z} = -9$$

36.
$$\frac{5}{x} - \frac{1}{y} - \frac{2}{z} = -6$$
$$-\frac{1}{x} + \frac{3}{y} - \frac{3}{z} = -12$$
$$\frac{2}{x} - \frac{1}{y} - \frac{1}{z} = 6$$

37.
$$x - 4y + 2z = -2$$
$$x + 2y - 2z = -3$$
$$x - y = 4$$

38.
$$2x + y + 3z = 4$$
$$-3x - y - 4z = 5$$
$$x + y + 2z = 0$$

39.
$$x + y + z = 0$$
$$x - y - z = 3$$
$$x + 3y + 3z = 5$$

40. $x + 3y + z = 6$
$3x + y - z = 6$
$x - y - z = 0$

41. $2x - y + 2z = 6$
$-x + y + z = 0$
$-x - 3z = -6$

42. $x + 2y + z = 0$
$3x + 2y - z = 4$
$-x + 2y + 3z = -4$

(Modeling)   Solve each problem.

43. *Feed Requirements*   Solve the system from **Example 4.**

$25x + 40y + 20z = 2200$
$4x + 2y + 3z = 280$
$3x + 2y + z = 180$

44. *Coin Collecting*   A coin collection made up of pennies, nickels, and quarters contains a total of 29 coins. The number of quarters is 8 less than the number of pennies. The total face value of the coins is $1.77. How many of each denomination are there

45. *Mixing Waters*   A sparkling-water distributor wants to make up 300 gallons of sparkling water to sell for $6.00 per gallon. She wishes to mix three grades of water selling for $9.00, $3.00, and $4.50 per gallon, respectively. She must use twice as much of the $4.50 water as the $3.00 water. How many gallons of each should she use?

46. *Mixing Glue*   A glue company needs to make some glue that it can sell for $120 per barrel. It wants to use 150 barrels of glue worth $100 per barrel, along with some glue worth $150 per barrel and glue worth $190 per barrel. It must use the same number of barrels of $150 and $190 glue. How much of the $150 and $190 glue will be needed? How many barrels of $120 glue will be produced

47. *Pricing Tickets*   Three students buy different combinations of tickets for a baseball game. The first student buys 2 senior, 1 adult, and 2 student tickets for $51. The second student buys 1 adult and 5 student tickets for $55. The third student buys 2 senior, 2 adult, and 7 student tickets for $75. If possible, find the price of each type of ticket. Interpret your answer.

48. *Investments*   A total of $5000 is invested at 2%, 3%, and 4%. The amount invested at 4% equals the total amount invested at 2% and 3%. The total interest for one year is $145. If possible, find the amount invested at each interest rate. Interpret your answer.

49. *Triangle Dimensions*   The sum of the measures of the angles of any triangle is 180°. In a certain triangle, the largest angle measures 55° less than twice the medium angle, and the smallest measures 25° less than the medium angle. Find the measures of the three angles.

50. *Triangle Dimensions*   The perimeter of a triangle is 59 inches. The longest side is 11 inches longer than the medium side, and the medium side is 3 inches more than the shortest side. Find the length of each side.

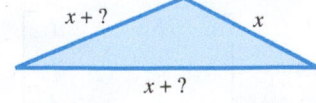

51. *Investment Decisions*   A student invested $10,000 in three parts. With one part, she bought mutual funds that offered a return of 4% per year. The second part, which amounted to twice the first, was used to buy government bonds paying 4.5% per year. She put the rest into a savings account that paid 2.5% annual interest. During the first year, the total interest was $415. How much did she invest at each rate?

52. *Investment Decisions*   A student won $100,000 in the Louisiana state lottery. He invested part of the money in real estate with an annual return of 5% and another part in a money market account at 0.5% interest. He invested the rest, which amounted to $20,000 less than the sum of the other two parts, in certificates of deposit that pay 1.75%. If the total annual interest on the money was $3250, how much was invested at each rate?

53. *Scheduling Deliveries*   Doctor Rug sells rug-cleaning machines. The EZ model weighs 10 pounds and comes in a 10-cubic-foot box. The compact model weighs 20 pounds and comes in an 8-cubic-foot box. The commercial model weighs 60 pounds and comes in a 28-cubic-foot box. Each of the company's delivery vans has 248 cubic feet of space and can hold a maximum of 440 pounds. In order for a van to be fully loaded, how many of each model should it carry?

54. *Scheduling Production*   Ciolino's makes dining room furniture. A buffet requires 30 hours for construction and 10 hours for finishing, a chair 10 hours for construction and 10 hours for finishing, and a table 10 hours for construction and 30 hours for finishing. The construction department has 350 hours of labor and the finishing department has 150 hours of labor available each week. How many pieces of each type of furniture should be produced each week if the factory is to run at full capacity?

55. *Predicting Home Prices*   The table shows the selling prices for three representative homes. Price P is given in thousands of dollars, age A in years, and home size S in thousands of square feet. These data may be modeled by the equation $P = a + bA + cS$.

Price (P)	Age (A)	Size (S)
190	20	2
320	5	3
50	40	1

(a) Write a system of linear equations whose solution gives a, b, and c.

(b) Solve this system of linear equations.

(c) Predict the price of a home that is 10 years old and has 2500 square feet.

Fitting Data Find the equation of the parabola (with vertical axis) that passes through the data points shown or specified. Check your answer.

56.

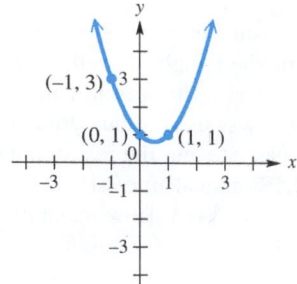

57.

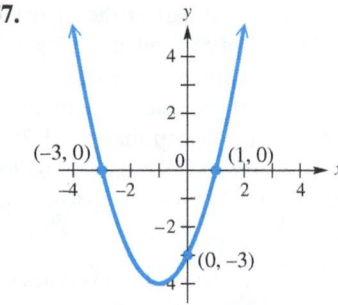

58. $(2, 9), (-2, 1), (-3, 4)$

59. $(1.5, 6.25), (0, -2), (-1.5, 3.25)$

60. $(2, 14), (0, 0), (-1, -1)$

61. $(-1, 4), (1, 2), (3, 8)$

62. $(-2, 2), (0, 2), (2, -6)$

63. $(0, 1), (1, 0), (2, -5)$

64. *Fitting Data* The values in the table are from a quadratic function $f(x) = ax^2 + bx + c$. Find a, b, and c.

x	-2	-1	0	1	2
$f(x)$	2.9	1.26	0.56	0.8	1.98

Fitting Data Given three noncollinear points, there is one and only one circle that passes through them. Knowing that the equation of a circle may be written in the form

$$x^2 + y^2 + ax + by + c = 0,$$

find the equation of the circle passing through the points shown or specified.

65.

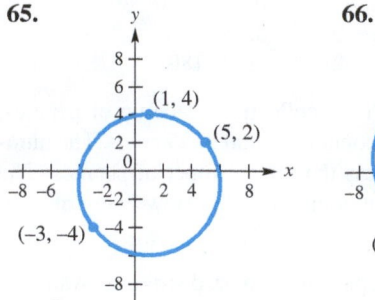

66.

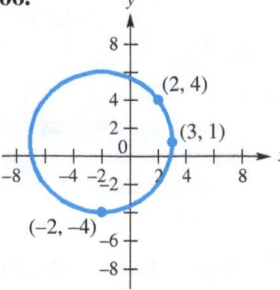

67. $(-1, 3), (6, 2)$, and $(-2, -4)$

68. $(-1, 5), (6, 6)$, and $(7, -1)$

69. $(2, 1), (-1, 0)$, and $(3, 3)$

70. $(-5, -2), (0, 3)$, and $(4, 2)$

(Modeling) Position of a Particle Suppose that the position of a particle moving along a straight line is given by

$$s(t) = at^2 + bt + c,$$

where t is time in seconds and a, b, and c are real numbers.

71. If $s(0) = 5$, $s(1) = 23$, and $s(2) = 37$, find the equation that defines $s(t)$. Then find $s(8)$.

72. If $s(0) = -10$, $s(1) = 6$, and $s(2) = 30$, find the equation that defines $s(t)$. Then find $s(10)$.

6.3 Solution of Linear Systems by Row Transformations

Matrix Row Transformations • Row Echelon Method • Reduced Row Echelon Method • Special Cases • An Application of Matrices

Matrix Row Transformations

Solving linear systems of equations can be streamlined by using *matrices* (singular: *matrix*). Consider a system of three equations and three unknowns.

$$\begin{aligned} a_1x + b_1y + c_1z &= d_1 \\ a_2x + b_2y + c_2z &= d_2 \\ a_3x + b_3y + c_3z &= d_3 \end{aligned} \quad \text{can be written as} \quad \begin{bmatrix} a_1 & b_1 & c_1 & d_1 \\ a_2 & b_2 & c_2 & d_2 \\ a_3 & b_3 & c_3 & d_3 \end{bmatrix}. \quad \text{Matrix}$$

TECHNOLOGY NOTE

The manner in which matrices are entered and displayed in a calculator can vary greatly among different manufacturers and even among the various models manufactured by the same company.

Such a rectangular array of numbers enclosed by brackets is called a **matrix.** Each number in the array is an **element** or **entry.** The constants in the last column of the matrix can be set apart from the coefficients of the variables with a vertical line, as shown in the following **augmented matrix.**

$$\text{Rows} \rightarrow \begin{bmatrix} a_1 & b_1 & c_1 & d_1 \\ a_2 & b_2 & c_2 & d_2 \\ a_3 & b_3 & c_3 & d_3 \end{bmatrix} \quad \text{Augmented matrix}$$

Columns

As an example, the system on the left has the augmented matrix shown on the right.

$$\begin{array}{rcr} x + 3y + 2z &=& 1 \\ 2x + y - z &=& 2 \\ x + y + z &=& 2 \end{array} \qquad \begin{bmatrix} 1 & 3 & 2 & 1 \\ 2 & 1 & -1 & 2 \\ 1 & 1 & 1 & 2 \end{bmatrix} \quad \text{Augmented matrix}$$

FIGURE 14 shows how this matrix can be entered into a TI-84 Plus calculator. The matrix has 3 rows (horizontal) and 4 columns (vertical), so it is a 3×4 (read "three by four") matrix. *The number of rows is always given first.* To refer to a number in the matrix, use its row and column numbers. For example, the highlighted number 3 in the screen on the left is in the first row, second column.

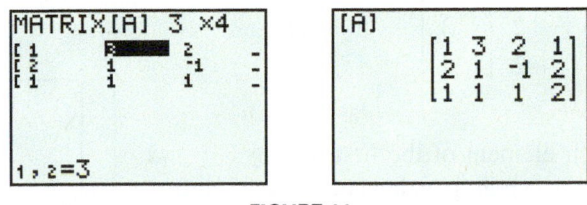

FIGURE 14

The rows of an augmented matrix can be treated just like the equations of the corresponding system of linear equations. Since the augmented matrix is nothing more than a short form of the system, any transformation of the matrix that results in an equivalent system of equations can be performed.

TECHNOLOGY NOTE

Consult your owner's guide to see how these transformations are accomplished with your model. Also see **Example 1.**

Matrix Row Transformations

For any augmented matrix of a system of linear equations, the following row transformations will result in the matrix of an equivalent system.

1. Any two rows may be interchanged.
2. The elements of any row may be multiplied by a nonzero real number.
3. Any row may be changed by adding to its elements a multiple of the corresponding elements of another row.

EXAMPLE 1 **Using Row Transformations**

Use matrix row transformations to transform

$$\begin{bmatrix} 1 & 3 & 5 & | & 2 \\ 0 & 1 & 2 & | & 3 \\ 1 & -1 & -2 & | & 4 \end{bmatrix}$$

to each matrix.

(a) $\begin{bmatrix} 0 & 1 & 2 & | & 3 \\ 1 & 3 & 5 & | & 2 \\ 1 & -1 & -2 & | & 4 \end{bmatrix}$ (b) $\begin{bmatrix} -2 & -6 & -10 & | & -4 \\ 0 & 1 & 2 & | & 3 \\ 1 & -1 & -2 & | & 4 \end{bmatrix}$ (c) $\begin{bmatrix} 1 & 3 & 5 & | & 2 \\ 0 & 1 & 2 & | & 3 \\ 0 & -4 & -7 & | & 2 \end{bmatrix}$

Analytic Solution

(a) Interchange the first two rows in the given matrix.

$$\begin{bmatrix} 0 & 1 & 2 & | & 3 \\ 1 & 3 & 5 & | & 2 \\ 1 & -1 & -2 & | & 4 \end{bmatrix} \quad \begin{matrix} R_2 \to R_1 \\ R_1 \to R_2 \\ R_3 \text{ is unchanged.} \end{matrix}$$

(b) Multiply the elements of the first row by -2.

$$\begin{bmatrix} -2 & -6 & -10 & | & -4 \\ 0 & 1 & 2 & | & 3 \\ 1 & -1 & -2 & | & 4 \end{bmatrix} \quad \begin{matrix} -2R_1 \\ R_2 \text{ and } R_3 \text{ are} \\ \text{unchanged.} \end{matrix}$$

(c) Multiply each element of the first row by -1 and add it to the corresponding element in the third row, putting the result in the third row. For example, to obtain -4 in row 3, column 2, multiply 3 in row 1 by -1 and add to -1 in row 3: $3(-1) + (-1) = -4$.

$$\begin{bmatrix} 1 & 3 & 5 & | & 2 \\ 0 & 1 & 2 & | & 3 \\ 0 & -4 & -7 & | & 2 \end{bmatrix} \quad \begin{matrix} R_1 \text{ and } R_2 \text{ are} \\ \text{unchanged.} \\ -1R_1 + R_3 \end{matrix}$$

Graphing Calculator Solution

FIGURE 15 shows the given matrix, designated $[A]$. The TI-84 Plus uses the commands indicated in **FIGURE 16** to perform the transformations.

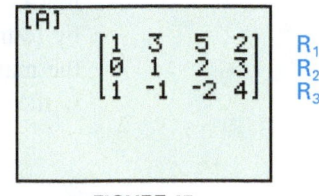

FIGURE 15

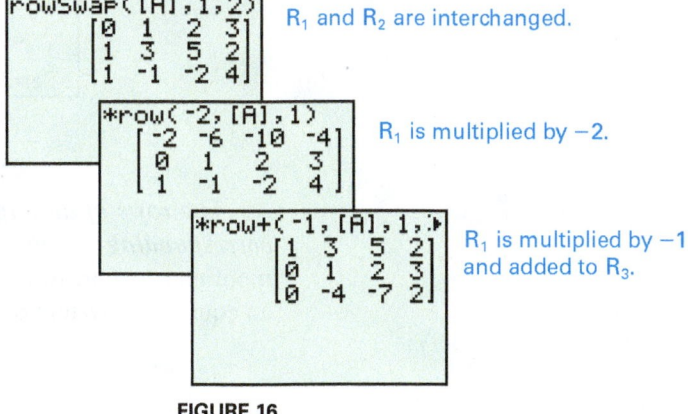

R$_1$ and R$_2$ are interchanged.

R$_1$ is multiplied by -2.

R$_1$ is multiplied by -1 and added to R$_3$.

FIGURE 16

Row Echelon Method

Echelon (triangular) Form

$$\begin{bmatrix} \mathbf{1} & 5 & 3 & | & 7 \\ 0 & \mathbf{1} & 2 & | & 9 \\ 0 & 0 & \mathbf{1} & | & 4 \end{bmatrix}$$

1s are on main diagonal.
0s are below.

Matrix row transformations are used to transform the augmented matrix of a system into one that is in **echelon** (or **triangular**) **form.** The echelon (triangular) form of an augmented matrix has 1s down the diagonal from upper left to lower right and 0s below each 1. The 1s lie on the **main diagonal.** Once a system of linear equations is in echelon form, **back-substitution** can be used to find the solution set. The **row echelon method** uses matrices to solve a system of linear equations. Start by obtaining a 1 as the first entry in the first column and then transform all entries below it to a 0. Continue through the columns obtaining a 1 as the second entry in the second column (zeros below), the third entry in the third column (zeros below), and so on. Repeat this process to row echelon form.

EXAMPLE 2 **Solving by the Row Echelon Method**

Solve the system.

$$5x + 2y = 1$$
$$2x - y = 4$$

Analytic Solution

$$\begin{bmatrix} 5 & 2 & | & 1 \\ 2 & -1 & | & 4 \end{bmatrix} \quad \textit{Augmented matrix}$$

Transform the first row so that the first entry is 1. To do this, multiply row 1 by $\frac{1}{5}$.

$$\begin{bmatrix} 1 & \frac{2}{5} & | & \frac{1}{5} \\ 2 & -1 & | & 4 \end{bmatrix} \quad \frac{1}{5}R_1$$

Transform so that the entry below the main diagonal (that is, 2) is 0. Multiply row 1 by -2 and add to row 2.

$$\begin{bmatrix} 1 & \frac{2}{5} & | & \frac{1}{5} \\ 0 & -\frac{9}{5} & | & \frac{18}{5} \end{bmatrix} \quad \begin{array}{l} R_1 \text{ is unchanged.} \\ -2R_1 + R_2 \end{array}$$

Multiply row 2 by $-\frac{5}{9}$ to get 1 on the main diagonal.

$$\begin{bmatrix} 1 & \frac{2}{5} & | & \frac{1}{5} \\ 0 & 1 & | & -2 \end{bmatrix} \quad \begin{array}{l} R_1 \text{ is unchanged.} \\ -\frac{5}{9}R_2 \end{array}$$

The matrix represents a system of equations.

$$\begin{bmatrix} 1 & \frac{2}{5} & | & \frac{1}{5} \\ 0 & 1 & | & -2 \end{bmatrix} \rightarrow \begin{array}{l} x + \frac{2}{5}y = \frac{1}{5} \\ y = -2 \end{array}$$

Since $y = -2$, use *back-substitution* to find x.

$$x + \frac{2}{5}(-2) = \frac{1}{5} \quad \text{Let } y = -2.$$

$$x - \frac{4}{5} = \frac{1}{5} \quad \text{Multiply.}$$

Write the x-value first in the ordered pair. → $$x = 1 \quad \text{Add } \frac{4}{5}.$$

The solution set is $\{(1, -2)\}$.

Graphing Calculator Solution

FIGURE 17 shows matrix $[A]$.

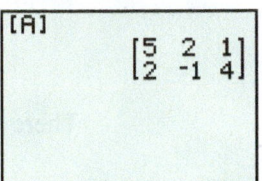

FIGURE 17

Using the "ref(" command on the TI-84 Plus, we can find the row echelon form. See **FIGURE 18.**

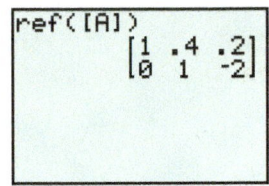

FIGURE 18

The decimal entries can be converted to fractions, as shown in **FIGURE 19.**

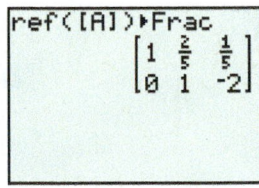

FIGURE 19

The augmented matrix in **FIGURE 19** corresponds to the last matrix in the analytic solution. The remainder of the solution process is the same.

The row echelon method can be extended to larger systems. The final matrix will always have 0s below the main diagonal of 1s to the left of the vertical bar. ***To transform the matrix, work column by column from upper left to lower right.***

NOTE As the numbers of rows and columns increase, the number of calculations necessary to convert the matrix to row echelon form also increases. This is apparent in **Example 3,** where we solve a system with three equations and three unknowns.

EXAMPLE 3 **Solving by the Row Echelon Method (Analytic)**

Solve the system.

$$
\begin{aligned}
x - y + 5z &= -6 \\
3x + 3y - z &= 10 \\
x + 3y + 2z &= 5
\end{aligned}
$$

Solution

$$
\begin{bmatrix}
1 & -1 & 5 & | & -6 \\
3 & 3 & -1 & | & 10 \\
1 & 3 & 2 & | & 5
\end{bmatrix}
\quad \text{Augmented matrix}
$$

There is already a 1 in row 1, column 1. We must obtain 0s in the rest of column 1.

Obtain 0s in column 1.

$$
\begin{bmatrix}
1 & -1 & 5 & | & -6 \\
0 & 6 & -16 & | & 28 \\
1 & 3 & 2 & | & 5
\end{bmatrix}
\quad -3R_1 + R_2
$$

$$
\begin{bmatrix}
1 & -1 & 5 & | & -6 \\
0 & 6 & -16 & | & 28 \\
0 & 4 & -3 & | & 11
\end{bmatrix}
\quad -1R_1 + R_3
$$

Obtain a 1 in column 2 on diagonal.

$$
\begin{bmatrix}
1 & -1 & 5 & | & -6 \\
0 & 1 & -\frac{8}{3} & | & \frac{14}{3} \\
0 & 4 & -3 & | & 11
\end{bmatrix}
\quad \frac{1}{6}R_2
$$

Obtain a 0 in column 2, row 3.

$$
\begin{bmatrix}
1 & -1 & 5 & | & -6 \\
0 & 1 & -\frac{8}{3} & | & \frac{14}{3} \\
0 & 0 & \frac{23}{3} & | & -\frac{23}{3}
\end{bmatrix}
\quad -4R_2 + R_3
$$

Obtain a 1 in column 3 on diagonal.

$$
\begin{bmatrix}
1 & -1 & 5 & | & -6 \\
0 & 1 & -\frac{8}{3} & | & \frac{14}{3} \\
0 & 0 & 1 & | & -1
\end{bmatrix}
\quad \frac{3}{23}R_3
$$

Matrix is in row echelon form.

The final matrix corresponds to the following system of equations.

$$
\begin{aligned}
x - y + 5z &= -6 \quad (1) \\
y - \frac{8}{3}z &= \frac{14}{3} \quad (2) \\
z &= -1 \quad (3)
\end{aligned}
$$

Since $z = -1$ from equation (3), use back-substitution into equation (2) to find y.

$$
\begin{aligned}
y - \frac{8}{3}(-1) &= \frac{14}{3} \quad\quad (2) \\
y &= \frac{14}{3} - \frac{8}{3} \quad \text{Subtract } \tfrac{8}{3}. \\
y &= 2 \quad\quad\quad\quad \text{Simplify.}
\end{aligned}
$$

Back-substitute $y = 2$ and $z = -1$ into equation (1) to find that $x = 1$. The solution set is $\{(1, 2, -1)\}$.

EXAMPLE 4 **Solving by the Row Echelon Method (Graphical)**

Solve the system of **Example 3** using a graphing calculator.

$$x - y + 5z = -6$$
$$3x + 3y - z = 10$$
$$x + 3y + 2z = 5$$

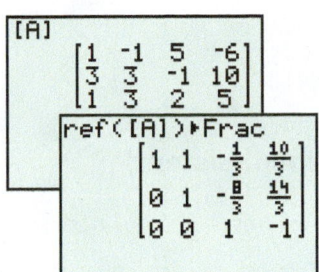

The entries in row 1, columns 2, 3, and 4, are different from the corresponding entries in the row echelon form shown in **Example 3**, because the steps were performed in an alternative way. However, the solution set is the same.

FIGURE 20

Solution   Enter the augmented matrix of the system as matrix $[A]$. See the top screen in **FIGURE 20.** Then use the "ref(" command to obtain the row echelon form. The matrix in the bottom screen indicates the system corresponding to the row echelon form.

$$x + y - \frac{1}{3}z = \frac{10}{3}$$
$$y - \frac{8}{3}z = \frac{14}{3}$$
$$z = -1$$

The system is different, but equivalent, to that of Example 3.

Using back-substitution, the solution set is $\{(1, 2, -1)\}$, as in **Example 3.**

Reduced Row Echelon Method

Another matrix method for solving systems is the **reduced row echelon method.** Earlier, we saw that the row echelon form of a matrix has 1s along the main diagonal and 0s below. *The reduced row echelon form has 1s along the main diagonal and 0s both below and above.* For example, the augmented matrix of the system

$$\begin{aligned} x + y + z &= 6 \\ 2x - y + z &= 5 \quad \text{is} \\ 3x + y - 2z &= 9 \end{aligned} \qquad \begin{bmatrix} 1 & 1 & 1 & 6 \\ 2 & -1 & 1 & 5 \\ 3 & 1 & -2 & 9 \end{bmatrix}.$$

By using row transformations, this augmented matrix can be transformed to

Reduced row echelon form
$$\begin{bmatrix} 1 & 0 & 0 & 3 \\ 0 & 1 & 0 & 2 \\ 0 & 0 & 1 & 1 \end{bmatrix}, \quad \text{which represents the system} \quad \begin{aligned} x &= 3 \\ y &= 2 \\ z &= 1. \end{aligned}$$

The solution set is $\{(3, 2, 1)\}$. There is no need for back-substitution with reduced row echelon form.

EXAMPLE 5 **Solving by the Reduced Row Echelon Method (Graphical)**

Use a graphing calculator with reduced row echelon capability to solve the system.

$$x + 2y + z = 30$$
$$3x + 2y + 2z = 56$$
$$2x + 3y + z = 44$$

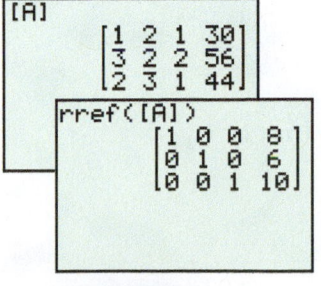

FIGURE 21

Solution   The top screen in **FIGURE 21** shows the augmented matrix of the system, and the bottom screen shows its reduced row echelon form, which represents the system

$$x = 8$$
$$y = 6$$
$$z = 10,$$

and has solution set $\{(8, 6, 10)\}$.

Special Cases

Whenever a row of the augmented matrix is of the form

$$\mathbf{0 \quad 0 \quad 0} \quad \cdots \quad |a, \quad \text{where } a \neq 0,$$

the system is inconsistent and there will be no solution, since this row corresponds to the equation $0 = a$. A row of the matrix of a linear system in the form

$$\mathbf{0 \quad 0 \quad 0} \quad \cdots \quad |\mathbf{0}$$

indicates that the equations of the system are dependent.

EXAMPLE 6 Solving a Special System (Inconsistent)

Solve the system.

$$\begin{aligned} x + 2y + z &= 4 \\ x - 2y + 3z &= 1 \\ 2x + 4y + 2z &= 9 \end{aligned}$$

Analytic Solution

$$\begin{bmatrix} 1 & 2 & 1 & 4 \\ 1 & -2 & 3 & 1 \\ 2 & 4 & 2 & 9 \end{bmatrix} \quad \text{Augmented matrix}$$

$$\begin{bmatrix} 1 & 2 & 1 & 4 \\ 1 & -2 & 3 & 1 \\ 0 & 0 & 0 & 1 \end{bmatrix} \quad -2R_1 + R_3$$

The final row indicates that

$$0x + 0y + 0z = 1,$$

which has no solutions. The system is inconsistent and has solution set $\emptyset$.

Graphing Calculator Solution

Augmented matrix $[A]$ and the row echelon form "ref(" are shown in **FIGURE 22.** The final row indicates an inconsistent system with solution set $\emptyset$. A similar conclusion can be made if the "rref(" command is used.

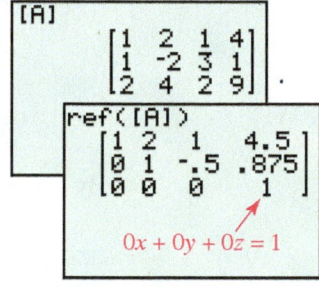

FIGURE 22

EXAMPLE 7 Solving a Special System (Dependent Equations)

Solve the system.

$$\begin{aligned} 2x - 5y + 3z &= 1 \\ x + 4y - 2z &= 8 \\ 4x - 10y + 6z &= 2 \end{aligned}$$

Solution

$$\begin{bmatrix} 2 & -5 & 3 & 1 \\ 1 & 4 & -2 & 8 \\ 4 & -10 & 6 & 2 \end{bmatrix} \quad \text{Augmented matrix}$$

Obtain a 0 in row 3, column 1.

$$\begin{bmatrix} 2 & -5 & 3 & 1 \\ 1 & 4 & -2 & 8 \\ 0 & 0 & 0 & 0 \end{bmatrix} \quad -2R_1 + R_3$$

The final row of 0s indicates that $0x + 0y + 0z = 0$, which is true for all x, y, and z. The system has dependent equations. The first two rows represent the system

$$2x - 5y + 3z = 1$$
$$x + 4y - 2z = 8.$$

Recall from **Example 2** in **Section 6.2** that a system with two equations and three variables may have infinitely many solutions.

$$\begin{bmatrix} 2 & -5 & 3 & | & 1 \\ 1 & 4 & -2 & | & 8 \end{bmatrix} \qquad \text{Augmented matrix}$$

$$\textit{Interchange} \quad \textit{rows.} \qquad \begin{bmatrix} 1 & 4 & -2 & | & 8 \\ 2 & -5 & 3 & | & 1 \end{bmatrix} \qquad \begin{matrix} R_2 \to R_1 \\ R_1 \to R_2 \end{matrix}$$

$$\begin{bmatrix} 1 & 4 & -2 & | & 8 \\ 0 & -13 & 7 & | & -15 \end{bmatrix} \qquad -2R_1 + R_2$$

$$\begin{bmatrix} 1 & 4 & -2 & | & 8 \\ 0 & 1 & -\frac{7}{13} & | & \frac{15}{13} \end{bmatrix} \qquad -\frac{1}{13}R_2$$

This is as far as we go with the row echelon method. The equations that correspond to the final matrix are

$$x + 4y - 2z = 8 \quad \text{and} \quad y - \frac{7}{13}z = \frac{15}{13}.$$

Solve the second equation for y to obtain $y = \frac{15}{13} + \frac{7}{13}z$. Now substitute this result for y in the first equation and solve for x.

$$x + 4y - 2z = 8$$

$$x + 4\left(\frac{15}{13} + \frac{7}{13}z\right) - 2z = 8 \qquad \text{Let } y = \frac{15}{13} + \frac{7}{13}z.$$

$$x + \frac{60}{13} + \frac{28}{13}z - 2z = 8 \qquad \text{Distributive property}$$

$$x + \frac{60}{13} + \frac{2}{13}z = 8 \qquad \frac{28}{13}z - 2z = \frac{2}{13}z$$

$$x = \frac{44}{13} - \frac{2}{13}z \qquad \text{Solve for } x.$$

The solution set, where z is any real number, is

Infinitely many solutions with this form

$$\left\{\left(\frac{44 - 2z}{13}, \frac{15 + 7z}{13}, z\right)\right\}.$$

An Application of Matrices

EXAMPLE 8 **Determining a Model Using Given Data**

Three food shelters are operated by a charitable organization. Three different quantities are computed: monthly food costs F in dollars, number N of people served per month, and monthly charitable receipts R in dollars. The data are shown in the table on the next page.

(continued)

Food Costs (F)	Number Served (N)	Charitable Receipts (R)
3000	2400	8000
4000	2600	10,000
8000	5900	14,000

Source: Sanders, D., *Statistics: A First Course*, McGraw-Hill.

FOR DISCUSSION

The three screens shown here depict matrix $[A]$ from **Example 8,** the row echelon form of $[A]$, and the reduced row echelon form of $[A]$.

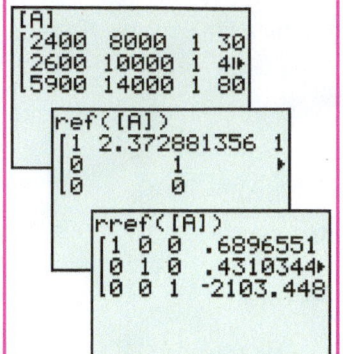

1. Use your calculator to recreate the screens shown here. (Note that only a portion of the matrix is visible.)
2. Which one of the two bottom screens provides an easier interpretation of the solution of the system? Why?

(a) Model the data in the table using the equation $F = aN + bR + c$, where a, b, and c are constants.

(b) Predict the food costs for a shelter that serves 4000 people and receives charitable receipts of $12,000. Round your answer to the nearest hundred dollars.

Solution

(a) Since $F = aN + bR + c$, the constants a, b, and c satisfy the following three equations.

$$3000 = a(2400) + b(8000) + c$$
$$4000 = a(2600) + b(10,000) + c$$
$$8000 = a(5900) + b(14,000) + c$$

We can rewrite this system in equation form or as an augmented matrix.

$$\begin{array}{r} 2400a + 8000b + c = 3000 \\ 2600a + 10{,}000b + c = 4000 \\ 5900a + 14{,}000b + c = 8000 \end{array} \quad \text{or} \quad \begin{bmatrix} 2400 & 8000 & 1 & | & 3000 \\ 2600 & 10{,}000 & 1 & | & 4000 \\ 5900 & 14{,}000 & 1 & | & 8000 \end{bmatrix}$$

Using the reduced row echelon method (see the "For Discussion" box), we find that $a \approx 0.6897$, $b \approx 0.4310$, and $c \approx -2103$. This leads to

$$F \approx 0.6897N + 0.4310R - 2103.$$

(b) To predict food costs for a shelter that serves 4000 people and receives charitable receipts of $12,000, let $N = 4000$ and $R = 12{,}000$ and evaluate F.

$$F \approx 0.6897(4000) + 0.4310(12{,}000) - 2103 = 5827.8$$

This model predicts monthly food costs of approximately $5800.

6.3 Exercises

Checking Analytic Skills Use the given row transformation to transform each matrix. **Do not use a calculator.**

1. $\begin{bmatrix} 2 & 4 \\ 4 & 7 \end{bmatrix} \frac{1}{2}R_1$

2. $\begin{bmatrix} -1 & 4 \\ 7 & 0 \end{bmatrix} 7R_1$

3. $\begin{bmatrix} 1 & 5 & 6 \\ -2 & 3 & -1 \\ 4 & 7 & 0 \end{bmatrix} 2R_1 + R_2$

4. $\begin{bmatrix} 2 & 5 & 1 \\ 4 & -1 & 2 \\ 3 & 7 & 1 \end{bmatrix} -2R_1 + R_2$

5. $\begin{bmatrix} -3 & 1 & -4 \\ 2 & 1 & 3 \\ 10 & 5 & 2 \end{bmatrix} -5R_2 + R_3$

6. $\begin{bmatrix} 4 & 10 & -8 \\ 7 & 4 & 3 \\ -1 & 1 & 0 \end{bmatrix} 4R_3 + R_1$

Write the augmented matrix for each system. Do not solve the system.

7. $2x + 3y = 11$
$\quad x + 2y = 8$

8. $3x + 5y = -13$
$\quad 2x + 3y = -9$

9. $x + 5y = 6$
$\quad\quad x = 3$

10. $2x + 7y = 1$
$\quad\quad 5x = -15$

11. $2x + y + z = 3$
$\quad 3x - 4y + 2z = -7$
$\quad x + y + z = 2$

12. $4x - 2y + 3z = 4$
$\quad 3x + 5y + z = 7$
$\quad 5x - y + 4z = 7$

13. $\quad x + y = 2$
$\quad 2y + z = -4$
$\quad\quad z = 2$

14. $\quad\quad x = 6$
$\quad y + 2z = 2$
$\quad x - 3z = 6$

Write the system of equations associated with each augmented matrix.

15. $\begin{bmatrix} 2 & 1 & | & 1 \\ 3 & -2 & | & -9 \end{bmatrix}$

16. $\begin{bmatrix} 1 & -5 & | & -18 \\ 6 & 2 & | & 20 \end{bmatrix}$

17. $\begin{bmatrix} 1 & 0 & 0 & | & 2 \\ 0 & 1 & 0 & | & 3 \\ 0 & 0 & 1 & | & -2 \end{bmatrix}$

18. $\begin{bmatrix} 1 & 0 & 1 & | & 4 \\ 0 & 1 & 0 & | & 2 \\ 0 & 0 & 1 & | & 3 \end{bmatrix}$

19.
```
[A]
 [ 3  2  1  1 ]
 [ 0  2  4 22 ]
 [-1 -2  3 15 ]
```

20.
```
[B]
 [ 2  1  3  12 ]
 [ 4 -3  0  10 ]
 [ 5  0 -4 -11 ]
```

Each augmented matrix is in row echelon form and represents a linear system. Use back-substitution to solve the system if possible.

21. $\begin{bmatrix} 1 & 2 & | & 3 \\ 0 & 1 & | & -1 \end{bmatrix}$

22. $\begin{bmatrix} 1 & -1 & | & 2 \\ 0 & 1 & | & 0 \end{bmatrix}$

23. $\begin{bmatrix} 1 & -5 & | & 6 \\ 0 & 0 & | & 0 \end{bmatrix}$

24. $\begin{bmatrix} 1 & 4 & | & -2 \\ 0 & 0 & | & 0 \end{bmatrix}$

25. $\begin{bmatrix} 1 & 1 & -1 & | & 4 \\ 0 & 1 & -1 & | & 2 \\ 0 & 0 & 1 & | & 1 \end{bmatrix}$

26. $\begin{bmatrix} 1 & -2 & -1 & | & 0 \\ 0 & 1 & -3 & | & 1 \\ 0 & 0 & 1 & | & 2 \end{bmatrix}$

27. $\begin{bmatrix} 1 & 2 & -1 & | & 5 \\ 0 & 1 & -2 & | & 1 \\ 0 & 0 & 0 & | & 0 \end{bmatrix}$

28. $\begin{bmatrix} 1 & -1 & 2 & | & 8 \\ 0 & 1 & -4 & | & 2 \\ 0 & 0 & 0 & | & 0 \end{bmatrix}$

29. $\begin{bmatrix} 1 & 2 & 1 & | & -3 \\ 0 & 1 & -3 & | & \frac{1}{2} \\ 0 & 0 & 0 & | & 4 \end{bmatrix}$

30. $\begin{bmatrix} 1 & 0 & -4 & | & \frac{3}{4} \\ 0 & 1 & 2 & | & 1 \\ 0 & 0 & 0 & | & -3 \end{bmatrix}$

Use row operations on an augmented matrix to solve each system of equations. Round to nearest thousandth when appropriate.

31. $x + y = 5$
$\quad x - y = -1$

32. $\quad x + 2y = 5$
$\quad 2x + y = -2$

33. $\quad x + y = -3$
$\quad 2x - 5y = -6$

34. $3x - 2y = 4$
$\quad 3x + y = -2$

35. $2x - 3y = 10$
$\quad 2x + 2y = 5$

36. $4x + y = 5$
$\quad 2x + y = 3$

37. $2x - 3y = 2$
$\quad 4x - 6y = 1$

38. $\quad x + 2y = 1$
$\quad 2x + 4y = 3$

39. $\quad 6x - 3y = 1$
$\quad -12x + 6y = -2$

40. $\quad x - y = 1$
$\quad -x + y = -1$

41. $x + y = -1$
$\quad y + z = 4$
$\quad x + z = 1$

42. $x - z = -3$
$\quad y + z = 9$
$\quad x + z = 7$

43. $\quad x + y - z = 6$
$\quad 2x - y + z = -9$
$\quad x - 2y + 3z = 1$

44. $\quad x + 3y - 6z = 7$
$\quad 2x - y + 2z = 0$
$\quad x + y + 2z = -1$

45. $-x + y = -1$
$\quad y - z = 6$
$\quad x + z = -1$

46. $\quad x + y = 1$
$\quad 2x - z = 0$
$\quad y + 2z = -2$

47. $2x - y + 3z = 0$
$\quad x + 2y - z = 5$
$\quad 2y + z = 1$

48. $4x + 2y - 3z = 6$
$\quad x - 4y + z = -4$
$\quad -x + 2z = 2$

49. $x + y - 2z = -6$
$\quad x - y + z = 4$
$\quad\quad 2x - z = -1$

50. $2x + y - 3z = 1$
$\quad x + y + 2z = 5$
$\quad 3x + 2y - z = -3$

Solve each system. Round to the nearest thousandth.

51. $\quad 0.07x + 0.23y = 9$
$\quad -1.25x + 0.33y = 2.4$

52. $\quad 3x - 13y = 17$
$\quad -23x + 15y = 2$

53. $2.1x + 0.5y + 1.7z = 4.9$
$\quad -2x + 1.5y - 1.7z = 3.1$
$\quad 5.8x - 4.6y + 0.8z = 9.3$

54. $0.1x + 0.3y + 1.7z = 0.6$
$\quad 0.6x + 0.1y - 3.1z = 6.2$
$\quad\quad 2.4y + 0.9z = 3.5$

55. $53x + 95y + 12z = 108$
$\quad 81x - 57y - 24z = -92$
$\quad -9x + 11y - 78z = 21$

56. $\quad 103x - 886y + 431z = 1200$
$\quad -55x + 981y = 1108$
$\quad -327x + 421y + 337z = 99$

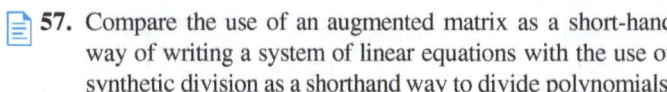 **57.** Compare the use of an augmented matrix as a short-hand way of writing a system of linear equations with the use of synthetic division as a shorthand way to divide polynomials.

58. Compare the use of the third type of row transformation on a matrix with the elimination method of solving a system of linear equations.

Solve each system. Write solutions in terms of z if necessary.

59. $x - 3y + 2z = 10$
$\quad 2x - y - z = 8$

60. $3x + y - z = 12$
$\quad x + 2y + z = 10$

61. $\quad x + 2y - z = 0$
$\quad 3x - y + z = 6$
$\quad -2x - 4y + 2z = 0$

62. $\quad 3x + 5y - z = 0$
$\quad 4x - y + 2z = 1$
$\quad -6x - 10y + 2z = 0$

63. $\quad x - 2y + z = 5$
$\quad -2x + 4y - 2z = 2$
$\quad 2x + y - z = 2$

64. $3x + 6y - 3z = 12$
$\quad -x - 2y + z = 16$
$\quad x + y - 2z = 20$

Solve each system of four equations in four variables. Express the solutions in the form (x, y, z, w).

65. $\quad x + 3y - 2z - w = 9$
$\quad 4x + y + z + 2w = 2$
$\quad -3x - y + z - w = -5$
$\quad x - y - 3z - 2w = 2$

66. $\quad 3x + 2y - w = 0$
$\quad 2x + z + 2w = 5$
$\quad x + 2y - z = -2$
$\quad 2x - y + z + w = 2$

(Modeling) Solve each application.

67. *Food Shelter Costs* Three food shelters have monthly food costs F in dollars, number N of people served per month, and monthly charitable receipts R in dollars, as shown in the table.

Food Costs (F)	Number Served (N)	Charitable Receipts (R)
1300	1800	5000
5300	3200	12,000
6500	4500	13,000

(a) Model these data by using

$$F = aN + bR + c,$$

where a, b, and c are constants.

(b) Predict food costs for a shelter that serves 3500 people and receives charitable receipts of $12,500. Round your answer to the nearest hundred dollars.

68. *Paid Vacation for Employees* The average number y of paid days off for full-time workers at medium-to-large companies after x years is listed in the table.

x (years)	1	15	30
y (days)	9.4	18.8	21.9

Source: Bureau of Labor Statistics.

(a) Determine the coefficients for $f(x) = ax^2 + bx + c$ so that f models these data.

(b) Graph function f with the data in the viewing window $[-4, 32]$ by $[8, 23]$.

(c) Estimate the number of paid days off after 3 years of experience. Compare it with the actual value of 11.2 days.

69. *Scheduling Production* A company produces two models of bicycles: model *A* and model *B*. Model *A* requires 2 hours of assembly time, and model *B* requires 3 hours of assembly time. The parts for model *A* cost $25 per bike; those for model *B* cost $30 per bike. If the company has a total of 34 hours of assembly time and $365 available per day for these two models, what is the maximum number of each model that can be made in a day and use all of the available resources?

70. *Scheduling Production* Caltek Computer Company makes two products: computer monitors and printers. Both require time on two machines: monitors, 1 hour on machine *A* and 2 hours on machine *B*; printers, 3 hours on machine *A* and 1 hour on machine *B*. Both machines operate 15 hours per day. What is the maximum number of each product that can be produced per day under these conditions?

71. *Financing Expansion* To get funds necessary for a planned expansion, a small company took out three loans totaling $12,500. The company was able to borrow some of the money at 2%. It borrowed $1000 more than $\frac{1}{2}$ the amount of the 2% loan at 3% and the rest at 2.5%. The total annual interest was $305. How much did the company borrow at each rate?

72. *Investment Decisions* Rick Pal deposits some money in a bank account paying 3% per year. He uses some additional money, amounting to $\frac{1}{3}$ the amount placed in the bank, to buy bonds paying 4% per year. With the balance of his funds, he buys a 4.5% certificate of deposit. The first year, his investments bring a return of $400. If the total of the investments is $10,000, how much did he invest at each rate?

(Modeling) Each set of data in Exercises 73–76 can be modeled by

$$f(x) = ax^2 + bx + c.$$

(a) Find a linear system whose solution represents values of *a*, *b*, and *c*.
(b) Find *f*(*x*) by using a method from this section.
(c) Graph *f* and the data in the same viewing window.
(d) Make your own prediction using *f*. Answers will vary.

73. *Hulu Revenue* The table lists total Hulu revenues *y* in millions of dollars *x* years after 2008.

x	0	2	4
y	25	260	695

Source: Hulu.

74. *Head Start Enrollment* The table lists annual enrollment *y* in thousands for the Head Start program *x* years after 1980.

x	0	10	32
y	376	541	1128

Source: Dept. of Health and Human Services.

75. *Chronic Health Care* A large percentage of the U.S. population will require chronic health care in the coming decades. The average caregiving age is 50–64, while the typical person needing chronic care is 85 or older. The ratio of potential caregivers to those needing chronic health care will shrink in the coming years *x*, as shown in the table.

Year	1990	2010	2030
Ratio	11	10	6

Source: Robert Wood Johnson Foundation. *Chronic Care in America: A 21st Century Challenge.*

76. *Carbon Dioxide Levels* Carbon dioxide (CO_2) is a greenhouse gas. Its concentration in parts per million (ppm) has been measured at Mauna Loa, Hawaii, during past years. The table lists measurements for three selected years *x*.

Year	1958	1973	2012
CO_2 (ppm)	315	330	394

Source: Mauna Loa Observatory.

(Modeling) Traffic Flow Each figure in Exercises 77 and 78 shows three one-way streets with intersections A, B, and C. Numbers indicate the average traffic flow in vehicles per minute. The variables x, y, and z denote unknown traffic flows that need to be determined for timing of stoplights.

(a) *If the number of vehicles per minute entering an intersection must equal the number exiting an intersection, verify that the system of linear equations describes the traffic flow.*
(b) *Rewrite the system and solve.*
(c) *Interpret your solution.*

77. A: $x + 5 = y + 7$
B: $z + 6 = x + 3$
C: $y + 3 = z + 4$

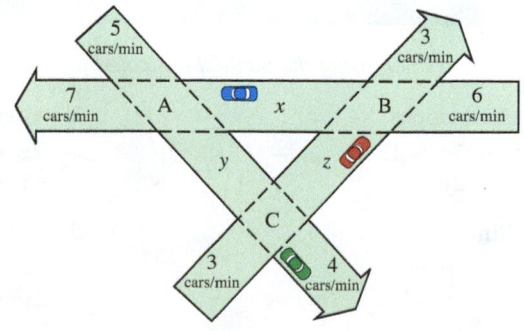

78. A: $x + 7 = y + 4$
B: $4 + 5 = x + z$
C: $y + 8 = 9 + 4$

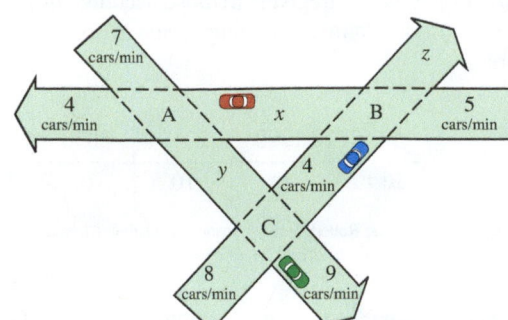

(Modeling) Solve each problem.

79. *Fawn Population* To model spring fawn count F from adult antelope population A, precipitation P, and severity of winter W, environmentalists have used the equation

$$F = a + bA + cP + dW,$$

where the coefficients a, b, c, and d are constants that must be determined before using the equation. The table lists the results of four different (representative) years.

Fawns	Adults	Precip. (in inches)	Winter Severity
239	871	11.5	3
234	847	12.2	2
192	685	10.6	5
343	969	14.2	1

Source: Bureau of Land Management.

(a) Substitute the values for F, A, P, and W from the table for each of the four years into the given equation $F = a + bA + cP + dW$ to obtain four linear equations involving a, b, c, and d.

(b) Write a 4×5 augmented matrix representing the system, and solve for a, b, c, and d.

(c) Write the equation for F, using the values from part (b) for the coefficients.

(d) If a winter has severity 3, adult antelope population 960, and precipitation 12.6 inches, predict the spring fawn count. (Compare this with the actual count of 320.)

80. *Weight of a Black Bear* The table shows weight W, neck size N, overall length L, and chest size C for four bears.

W (pounds)	N (inches)	L (inches)	C (inches)
125	19	57.5	32
316	26	65	42
436	30	72	48
514	30.5	75	54

Source: M. Triola, *Elementary Statistics*; Minitab, Inc.

(a) We can model these data with the equation

$$W = a + bN + cL + dC,$$

where a, b, c, and d are constants. To do so, represent a system of linear equations by a 4×5 augmented matrix whose solution gives values for a, b, c, and d.

(b) Solve the system. Round each value to the nearest thousandth.

(c) Predict the weight of a bear with $N = 24$, $L = 63$, and $C = 39$. Interpret the result.

SECTIONS 6.1–6.3   Reviewing Basic Concepts

Solve each system, using the method indicated.

1. (Elimination)
$2x - 3y = 18$
$5x + 2y = 7$

2. (Graphical)
$2x + y = -4$
$-x + 2y = 2$

3. (Substitution)
$5x + 10y = 10$
$x + 2y = 2$

4. (Elimination)
$x - y = 6$
$x - y = 4$

5. (Analytic, with graphical support)
$6x + 2y = 10$
$2x^2 - 3y = 11$

6. (Row echelon method)
$x + y + z = 1$
$-x + y + z = 5$
$y + 2z = 5$

7. (Reduced row echelon method)

$$2x + 4y + 4z = 4$$
$$x + 3y + z = 4$$
$$-x + 3y + 2z = -1$$

8. Solve the system represented by the augmented matrix.

$$\begin{bmatrix} 2 & 1 & 2 & | & 10 \\ 1 & 0 & 2 & | & 5 \\ 1 & -2 & 2 & | & 1 \end{bmatrix}$$

(Modeling)   *Solve each problem.*

9. *Sales of CRT and LCD Monitors*   In 2006 a total of about 133 million monitors were sold. There were 43 million more LCD (liquid crystal display) monitors sold than CRT (cathode ray tube) monitors. (*Source:* International Data Corporation.) Determine about how many of each type of monitor were sold in 2006.

10. *Investments*   A sum of $5000 is invested in three mutual funds that pay 2%, 3%, and 4% interest rates. The amount of money invested in the fund paying 4% equals the total amount of money invested in the other two funds, and the total annual interest from all three funds is $165. Find the amount invested at each rate.

6.4   Matrix Properties and Operations

Terminology of Matrices • Operations on Matrices • Applying Matrix Algebra

Terminology of Matrices

Suppose that you are the manager of an electronics store and one day you receive the following products from two distributors: Wholesale Enterprises delivers 2 cell phones, 7 DVDs, and 5 video games. Discount Distributors delivers 4 cell phones, 6 DVDs, and 9 video games. We can organize the information in a table.

Distributor	Product		
	Cell Phones	**DVDs**	**Video Games**
Wholesale Enterprises	2	7	5
Discount Distributors	4	6	9

As long as we remember what each row and column represents, we can remove all the labels and write the numbers in the table as a matrix.

2 rows $\begin{bmatrix} 2 & 7 & 5 \\ 4 & 6 & 9 \end{bmatrix}$ 2 × 3 matrix

3 columns

A matrix is classified by its **dimension**—that is, by the number of rows and columns it contains. For example, the preceding matrix has 2 rows and 3 columns, with dimension 2 × 3. In general, a matrix with m rows and n columns has dimension $m \times n$. *The number of rows is always given first.*

Certain matrices have special names. An $n \times n$ matrix is a **square matrix** of order n. Also, a matrix with just one row is a **row matrix,** and a matrix with just one column is a **column matrix.**

TECHNOLOGY NOTE

You should familiarize yourself with the matrix capabilities of your particular calculator. Refer to your owner's guide.

Two matrices are equal if they have the same dimension and if corresponding elements, position by position, are equal. According to this definition, the matrices

$$\begin{bmatrix} 2 & 1 \\ 3 & -5 \end{bmatrix} \quad \text{and} \quad \begin{bmatrix} 1 & 2 \\ -5 & 3 \end{bmatrix} \qquad \text{Not equal}$$

are *not* equal (even though they contain the same elements and are the same dimension), because the corresponding elements differ.

EXAMPLE 1 **Classifying Matrices by Dimension**

Find the dimension of each matrix, and determine any special characteristics.

(a) $\begin{bmatrix} 6 & 5 \\ 3 & 4 \\ 5 & -1 \end{bmatrix}$ (b) $\begin{bmatrix} 3 \\ -5 \\ 0 \\ 2 \end{bmatrix}$ (c) $\begin{bmatrix} 1 & 6 & 5 & -2 & 5 \end{bmatrix}$

Solution

(a) This is a 3×2 matrix, because it has 3 rows and 2 columns.

(b) This matrix is a 4×1 *column matrix.*

(c) $\begin{bmatrix} 1 & 6 & 5 & -2 & 5 \end{bmatrix}$ is a 1×5 *row matrix.*

EXAMPLE 2 **Determining Equality of Matrices**

(a) If $A = \begin{bmatrix} 2 & 1 \\ p & q \end{bmatrix}$ and $B = \begin{bmatrix} x & y \\ -1 & 0 \end{bmatrix}$, determine x, y, p, and q such that $A = B$.

(b) Are matrices $[A]$ and $[B]$ in **FIGURE 23** equal?

Solution

(a) From the definition of equality, the only way that the statement

$$\begin{bmatrix} 2 & 1 \\ p & q \end{bmatrix} = \begin{bmatrix} x & y \\ -1 & 0 \end{bmatrix}$$

can be true is if $x = 2$, $y = 1$, $p = -1$, and $q = 0$.

(b) The matrices are not equal because corresponding entries in the third row are not all equal.

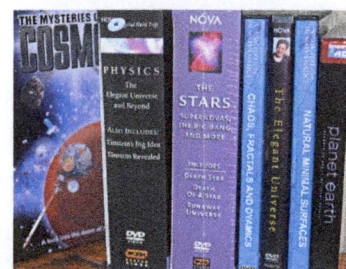

[A]
$$\begin{bmatrix} -3 & 6 & 5 \\ 6 & 2 & 1 \\ -7 & 8 & 4 \end{bmatrix}$$

[B]
$$\begin{bmatrix} -3 & 6 & 5 \\ 6 & 2 & 1 \\ -7 & 4 & 8 \end{bmatrix}$$

FIGURE 23

Operations on Matrices

At the beginning of this section, we gave an example using the matrix

$$\begin{bmatrix} 2 & 7 & 5 \\ 4 & 6 & 9 \end{bmatrix},$$

where the columns represent the numbers of certain products (cell phones, DVDs, and video games, respectively) and the rows represent the two different distributors (Wholesale Enterprises and Discount Distributors, respectively). For example, the element 7 represents 7 DVDs received from Wholesale Enterprises.

On another day, the shipments are described with the following matrix.

$$\begin{bmatrix} 3 & 12 & 10 \\ 15 & 11 & 8 \end{bmatrix} \quad \begin{array}{l} \text{Wholesale Enterprises} \\ \text{Discount Distributors} \end{array}$$

Here, for example, 8 video games were received from Discount Distributors. The total number of products received from the distributors over the two days can be found by adding the corresponding elements of the two matrices.

$$\begin{bmatrix} 2 & 7 & 5 \\ 4 & 6 & 9 \end{bmatrix} + \begin{bmatrix} 3 & 12 & 10 \\ 15 & 11 & 8 \end{bmatrix} = \begin{bmatrix} 2+3 & 7+12 & 5+10 \\ 4+15 & 6+11 & 9+8 \end{bmatrix}$$

To add matrices, add corresponding elements.

$$= \begin{bmatrix} 5 & 19 & 15 \\ 19 & 17 & 17 \end{bmatrix}$$

TECHNOLOGY NOTE

Graphing calculators can perform operations on matrices, provided that the dimensions of the matrices are compatible for the operation. A dimension error message will occur if the operation cannot be performed.

The 5 in the sum indicates that, over the two days, 5 cell phones were received from Wholesale Enterprises. Generalizing from this example leads to the definition of matrix addition.

> **Matrix Addition**
>
> The sum of two $m \times n$ matrices A and B is the $m \times n$ matrix $A + B$ in which each element is the sum of the corresponding elements of A and B.

CAUTION   *Only matrices with the same dimension can be added.*

EXAMPLE 3   **Adding Matrices**

Find each sum if possible.

(a) $\begin{bmatrix} 5 & -6 \\ 8 & 9 \end{bmatrix} + \begin{bmatrix} -4 & 6 \\ 8 & -3 \end{bmatrix}$ (b) $\begin{bmatrix} 2 \\ 5 \\ 8 \end{bmatrix} + \begin{bmatrix} -6 \\ 3 \\ 12 \end{bmatrix}$

(c) $A + B$ if $A = \begin{bmatrix} 5 & 8 \\ 6 & 2 \end{bmatrix}$ and $B = \begin{bmatrix} 3 & 9 & 1 \\ 4 & 2 & 5 \end{bmatrix}$

Analytic Solution

(a) $\begin{bmatrix} 5 & -6 \\ 8 & 9 \end{bmatrix} + \begin{bmatrix} -4 & 6 \\ 8 & -3 \end{bmatrix}$

$= \begin{bmatrix} 5+(-4) & -6+6 \\ 8+8 & 9+(-3) \end{bmatrix}$

$= \begin{bmatrix} 1 & 0 \\ 16 & 6 \end{bmatrix}$

(b) $\begin{bmatrix} 2 \\ 5 \\ 8 \end{bmatrix} + \begin{bmatrix} -6 \\ 3 \\ 12 \end{bmatrix} = \begin{bmatrix} -4 \\ 8 \\ 20 \end{bmatrix}$

(c) The matrices

$$A = \begin{bmatrix} 5 & 8 \\ 6 & 2 \end{bmatrix} \qquad \text{2 × 2 matrix}$$

and $B = \begin{bmatrix} 3 & 9 & 1 \\ 4 & 2 & 5 \end{bmatrix}$ 2 × 3 matrix

have different dimensions and so cannot be added. The sum $A + B$ does not exist, or we say $A + B$ is undefined.

Graphing Calculator Solution

(a) Using the TI-84 Plus calculator, **FIGURE 24** shows the sum of matrices A and B defined in **FIGURE 25.**

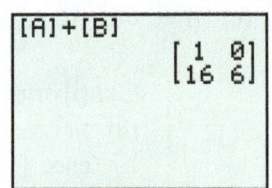

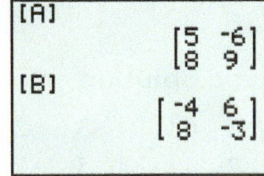

FIGURE 24   **FIGURE 25**

(b) The screen in **FIGURE 26** shows how the sum of two column matrices entered directly on the home screen is displayed.

(c) A graphing calculator will return an ERROR message if it is directed to perform an operation on matrices that is not possible due to a dimension mismatch. See **FIGURE 27.**

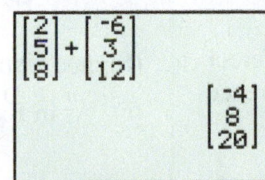

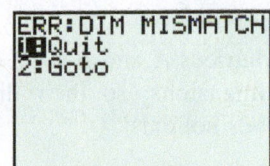

FIGURE 26   **FIGURE 27**

A matrix containing only zeros as elements is called a **zero matrix.**

$$[0 \quad 0 \quad 0] \qquad \text{1 × 3 zero matrix} \qquad \begin{bmatrix} 0 & 0 & 0 \\ 0 & 0 & 0 \end{bmatrix} \qquad \text{2 × 3 zero matrix}$$

The additive inverse of a real number a is the unique real number $-a$ such that $a + (-a) = 0$ and $-a + a = 0$. Given matrix A, the elements of matrix $-A$ are the additive inverses of the corresponding elements of A. For example, if

$$A = \begin{bmatrix} -5 & 2 & -1 \\ 3 & 4 & -6 \end{bmatrix}, \quad \text{then} \quad -A = \begin{bmatrix} 5 & -2 & 1 \\ -3 & -4 & 6 \end{bmatrix}.$$

To check, show that $A + (-A)$ equals the zero matrix O.

$$A + (-A) = \begin{bmatrix} -5 & 2 & -1 \\ 3 & 4 & -6 \end{bmatrix} + \begin{bmatrix} 5 & -2 & 1 \\ -3 & -4 & 6 \end{bmatrix} = \begin{bmatrix} 0 & 0 & 0 \\ 0 & 0 & 0 \end{bmatrix} = O$$

Then, show that $-A + A$ is also O. Matrix $-A$ is the **additive inverse,** or **negative,** of matrix A. Every matrix has a unique additive inverse.

Just as subtraction of real numbers is defined in terms of the additive inverse, subtraction of matrices is defined in the same way.

Matrix Subtraction

If A and B are matrices with the same dimension, then

$$A - B = A + (-B).$$

EXAMPLE 4 **Subtracting Matrices**

Find each difference.

(a) $\begin{bmatrix} -5 & 6 \\ 2 & 4 \end{bmatrix} - \begin{bmatrix} -3 & 2 \\ 5 & -8 \end{bmatrix}$ **(b)** $[8 \quad 6 \quad -4] - [3 \quad 5 \quad -8]$

(c) $A - B$ if $A = \begin{bmatrix} -2 & 5 \\ 0 & 1 \end{bmatrix}$ and $B = \begin{bmatrix} 3 \\ 5 \end{bmatrix}$

Analytic Solution

(a) $\begin{bmatrix} -5 & 6 \\ 2 & 4 \end{bmatrix} - \begin{bmatrix} -3 & 2 \\ 5 & -8 \end{bmatrix}$

$= \begin{bmatrix} -5 & 6 \\ 2 & 4 \end{bmatrix} + \begin{bmatrix} 3 & -2 \\ -5 & 8 \end{bmatrix}$

$= \begin{bmatrix} -2 & 4 \\ -3 & 12 \end{bmatrix}$

(b) $[8 \quad 6 \quad -4] - [3 \quad 5 \quad -8]$

$= [5 \quad 1 \quad 4]$

(c) Matrices A and B have different dimensions, so their difference does not exist.

Graphing Calculator Solution

(a) In **FIGURE 28,** the two matrices are defined as $[C]$ and $[D]$. The difference $[C] - [D]$ is displayed in **FIGURE 29.**

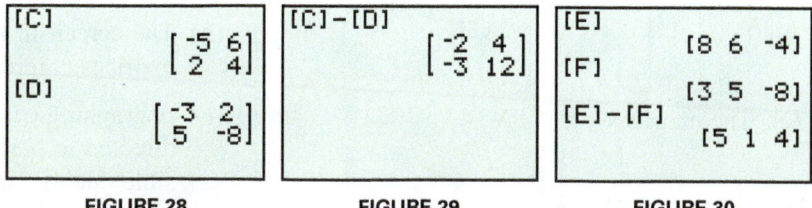

FIGURE 28   FIGURE 29   FIGURE 30

(b) See **FIGURE 30.**

(c) As in **Example 3(c),** the calculator returns a dimension error message.

If a matrix A is added to itself, each element in the sum is "twice as large" as the corresponding element of A. For example,

$$\begin{bmatrix} 2 & 5 \\ 1 & 3 \\ 4 & 6 \end{bmatrix} + \begin{bmatrix} 2 & 5 \\ 1 & 3 \\ 4 & 6 \end{bmatrix} = \begin{bmatrix} 4 & 10 \\ 2 & 6 \\ 8 & 12 \end{bmatrix} = 2\begin{bmatrix} 2 & 5 \\ 1 & 3 \\ 4 & 6 \end{bmatrix}.$$

The number 2 in front of the last matrix is called a **scalar** to distinguish it from a matrix. *A scalar (in this text) is just a special name for a real number.*

Multiplication of a Matrix by a Scalar

The product of a scalar k and a matrix A is the matrix kA, each of whose elements is k times the corresponding element of A.

EXAMPLE 5   **Multiplying Matrices by Scalars**

Perform each multiplication.

(a) $5\begin{bmatrix} 2 & -3 \\ 0 & 4 \end{bmatrix}$ (b) $\dfrac{3}{4}\begin{bmatrix} 20 & 36 \\ 12 & -16 \end{bmatrix}$

Analytic Solution

(a) $5\begin{bmatrix} 2 & -3 \\ 0 & 4 \end{bmatrix} = \begin{bmatrix} 5(2) & 5(-3) \\ 5(0) & 5(4) \end{bmatrix}$ Multiply each element by 5.

$= \begin{bmatrix} 10 & -15 \\ 0 & 20 \end{bmatrix}$

(b) $\dfrac{3}{4}\begin{bmatrix} 20 & 36 \\ 12 & -16 \end{bmatrix} = \begin{bmatrix} \frac{3}{4}(20) & \frac{3}{4}(36) \\ \frac{3}{4}(12) & \frac{3}{4}(-16) \end{bmatrix}$

$= \begin{bmatrix} 15 & 27 \\ 9 & -12 \end{bmatrix}$

Graphing Calculator Solution

(a), (b) See **FIGURES 31** and **32.**

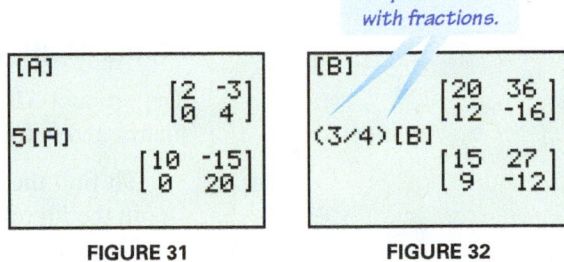

Use parentheses with fractions.

FIGURE 31   **FIGURE 32**

Returning to the electronics store example, recall the matrix for the number of each type of product received from the two distributors.

$$\begin{bmatrix} 2 & 7 & 5 \\ 4 & 6 & 9 \end{bmatrix}$$ Wholesale Enterprises
                        Discount Distributors

Now suppose each cell phone costs the store $120, each DVD costs $18, and each video game costs $9. To find the total cost of the products from Wholesale Enterprises, we multiply as follows.

Type of Product	Number of Items	Cost per Item	Total Cost
Cell phone	2	$120	$240
DVD	7	$ 18	$126
Video game	5	$ 9	$ 45

$411 ⟵ Total from Wholesale Enterprises

The products from Wholesale Enterprises cost a total of $411. This result is the sum of three products.

$$2(\$120) + 7(\$18) + 5(\$9) = \$411$$

In the same way, using the second row of the matrix and the three product costs gives the total cost from Discount Distributors.

$$4(\$120) + 6(\$18) + 9(\$9) = \$669$$

The product costs can be written as a column matrix.

$$\begin{bmatrix} 120 \\ 18 \\ 9 \end{bmatrix} \quad \begin{matrix} \text{Cell phone} \\ \text{DVD} \\ \text{Video game} \end{matrix}$$

The product of the matrices $\begin{bmatrix} 2 & 7 & 5 \\ 4 & 6 & 9 \end{bmatrix}$ and $\begin{bmatrix} 120 \\ 18 \\ 9 \end{bmatrix}$ can be written

$$\begin{bmatrix} 2 & 7 & 5 \\ 4 & 6 & 9 \end{bmatrix} \begin{bmatrix} 120 \\ 18 \\ 9 \end{bmatrix} = \begin{bmatrix} 2 \cdot 120 + 7 \cdot 18 + 5 \cdot 9 \\ 4 \cdot 120 + 6 \cdot 18 + 9 \cdot 9 \end{bmatrix} = \begin{bmatrix} 411 \\ 669 \end{bmatrix}. \quad \begin{matrix} \text{Total cost from} \\ \text{each distributor} \end{matrix}$$

Each element of the product was found by multiplying the elements of the *rows* of the matrix on the left and the corresponding elements of the *column* of the matrix on the right and then finding the sum of these products. Notice that the product of a 2 × 3 matrix and a 3 × 1 matrix is a 2 × 1 matrix.

Matrix Multiplication

The product AB of an $m \times n$ matrix A and an $n \times k$ matrix B is an $m \times k$ matrix and is found as follows.

To find the *i*th row, *j*th column element of AB, multiply each element in the *i*th row of A by the corresponding element in the *j*th column of B. The sum of these products gives the element of row *i*, column *j* of AB.

The product AB can be found only if the number of columns of A is the same as the number of rows of B. The final product will have as many rows as A and as many columns as B.

EXAMPLE 6 Deciding Whether Two Matrices Can Be Multiplied

Suppose matrix A has dimension 2 × 2, while matrix B has dimension 2 × 4. Can the product AB be calculated? If so, what is the dimension of the product? Can the product BA be calculated?

Solution The following diagram helps answer these questions.

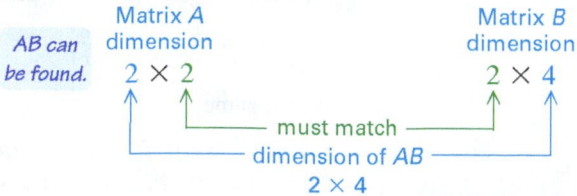

The product AB can be calculated because A has two columns and B has two rows. The dimension of the product is 2×4.

The product BA cannot be found. To see why, consider the following diagram. The number of columns in B is different from the number of rows in A.

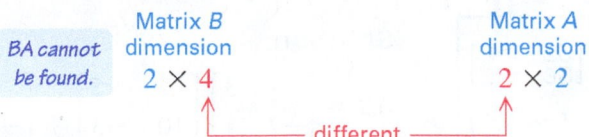

BA cannot be found.

Matrix B dimension 2×4

Matrix A dimension 2×2

— different —

EXAMPLE 7 **Multiplying Matrices**

Find the product AB of the two matrices $A = \begin{bmatrix} -3 & 4 & 2 \\ 5 & 0 & 4 \end{bmatrix}$ and $B = \begin{bmatrix} -6 & 4 \\ 2 & 3 \\ 3 & -2 \end{bmatrix}$.

Solution Since A has dimension 2×3 and B has dimension 3×2, they can be multiplied. The product AB has dimension 2×2.

$$\qquad\qquad A \qquad\qquad B$$

Step 1 $\begin{bmatrix} -3 & 4 & 2 \\ 5 & 0 & 4 \end{bmatrix} \begin{bmatrix} -6 & 4 \\ 2 & 3 \\ 3 & -2 \end{bmatrix}$ $\qquad$ $-3(-6) + 4(2) + 2(3) = 32$
Entry for row 1, column 1 of AB

Step 2 $\begin{bmatrix} -3 & 4 & 2 \\ 5 & 0 & 4 \end{bmatrix} \begin{bmatrix} -6 & 4 \\ 2 & 3 \\ 3 & -2 \end{bmatrix}$ $\qquad$ $-3(4) + 4(3) + 2(-2) = -4$
Entry for row 1, column 2 of AB

Step 3 $\begin{bmatrix} -3 & 4 & 2 \\ 5 & 0 & 4 \end{bmatrix} \begin{bmatrix} -6 & 4 \\ 2 & 3 \\ 3 & -2 \end{bmatrix}$ $\qquad$ $5(-6) + 0(2) + 4(3) = -18$
Entry for row 2, column 1 of AB

Step 4 $\begin{bmatrix} -3 & 4 & 2 \\ 5 & 0 & 4 \end{bmatrix} \begin{bmatrix} -6 & 4 \\ 2 & 3 \\ 3 & -2 \end{bmatrix}$ $\qquad$ $5(4) + 0(3) + 4(-2) = 12$
Entry for row 2, column 2 of AB

Step 5 Write the product.

$$\begin{bmatrix} -3 & 4 & 2 \\ 5 & 0 & 4 \end{bmatrix} \begin{bmatrix} -6 & 4 \\ 2 & 3 \\ 3 & -2 \end{bmatrix} = \begin{bmatrix} 32 & -4 \\ -18 & 12 \end{bmatrix}$$

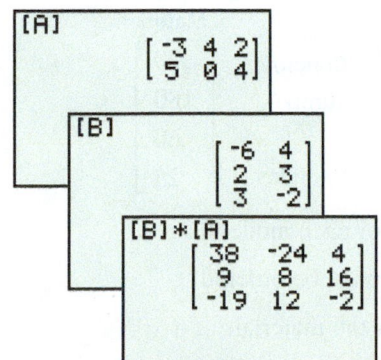

FIGURE 33

EXAMPLE 8 **Multiplying Matrices**

Use a graphing calculator to find the product BA of the two matrices given in **Example 7.**

Solution Define $[A]$ and $[B]$ as shown in **FIGURE 33.** Since $[B]$ has dimension 3×2 and $[A]$ has dimension 2×3, the dimension of the product BA is 3×3.

As shown in **Examples 7** and **8,**

$$AB \neq BA.$$

In general, matrix multiplication is not commutative. In fact, in some cases, one of the products may be defined while the other is not. See **Example 6.**

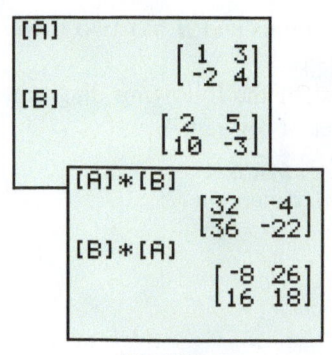

FIGURE 34

EXAMPLE 9 **Multiplying Square Matrices**

Find AB and BA, given $A = \begin{bmatrix} 1 & 3 \\ -2 & 4 \end{bmatrix}$ and $B = \begin{bmatrix} 2 & 5 \\ 10 & -3 \end{bmatrix}$.

Solution

$$AB = \begin{bmatrix} 1 & 3 \\ -2 & 4 \end{bmatrix}\begin{bmatrix} 2 & 5 \\ 10 & -3 \end{bmatrix} = \begin{bmatrix} 2+30 & 5-9 \\ -4+40 & -10-10 \end{bmatrix} = \begin{bmatrix} 32 & -4 \\ 36 & -22 \end{bmatrix}$$

$$BA = \begin{bmatrix} 2 & 5 \\ 10 & -3 \end{bmatrix}\begin{bmatrix} 1 & 3 \\ -2 & 4 \end{bmatrix} = \begin{bmatrix} 2-10 & 6+20 \\ 10+6 & 30-12 \end{bmatrix} = \begin{bmatrix} -8 & 26 \\ 16 & 18 \end{bmatrix}$$

FIGURE 34 shows matrices $[A]$ and $[B]$ and both products.

Applying Matrix Algebra

EXAMPLE 10 **Using Matrix Multiplication to Model Plans for a Subdivision**

A contractor builds three kinds of houses, models X, Y, and Z, with a choice of two styles, colonial or ranch. Matrix A shows the number of each kind of house the contractor is planning to build for a new 100-home subdivision. The amounts for each of the main materials used depend on the style of the house. These amounts are shown in matrix B, while matrix C gives the cost in dollars for each kind of material. Concrete is measured here in cubic yards, lumber in 1000 board feet, brick in 1000s, and shingles in 100 square feet.

$$\begin{array}{c} \\ \text{Model X} \\ \text{Model Y} \\ \text{Model Z} \end{array} \begin{array}{c} \text{Colonial} \quad \text{Ranch} \\ \begin{bmatrix} 0 & 30 \\ 10 & 20 \\ 20 & 20 \end{bmatrix} = A \end{array}$$

$$\begin{array}{c} \\ \text{Colonial} \\ \text{Ranch} \end{array}\begin{array}{c} \text{Concrete} \ \text{Lumber} \ \text{Brick} \ \text{Shingles} \\ \begin{bmatrix} 10 & 2 & 0 & 2 \\ 50 & 1 & 20 & 2 \end{bmatrix} = B \end{array} \quad \begin{array}{c} \\ \text{Concrete} \\ \text{Lumber} \\ \text{Brick} \\ \text{Shingles} \end{array}\begin{bmatrix} 20 \\ 180 \\ 60 \\ 25 \end{bmatrix} = C$$

(a) What is the total cost of materials for all houses of each model?

(b) How much of each of the four kinds of material must be ordered?

(c) Use a graphing calculator to find the total cost of the materials.

Solution

(a) To find the cost of materials for each model, first find matrix AB, which will give the total amount of each material needed for all houses of each model.

$$AB = \begin{bmatrix} 0 & 30 \\ 10 & 20 \\ 20 & 20 \end{bmatrix}\begin{bmatrix} 10 & 2 & 0 & 2 \\ 50 & 1 & 20 & 2 \end{bmatrix} = \begin{array}{c} \text{Concrete} \ \text{Lumber} \ \text{Brick} \ \text{Shingles} \\ \begin{bmatrix} 1500 & 30 & 600 & 60 \\ 1100 & 40 & 400 & 60 \\ 1200 & 60 & 400 & 80 \end{bmatrix} \end{array} \begin{array}{l} \text{Model X} \\ \text{Model Y} \\ \text{Model Z} \end{array}$$

Multiplying AB and the cost matrix C gives the total cost of materials for each model.

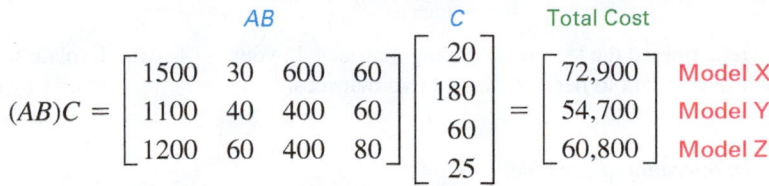

$$(AB)C = \begin{bmatrix} 1500 & 30 & 600 & 60 \\ 1100 & 40 & 400 & 60 \\ 1200 & 60 & 400 & 80 \end{bmatrix} \begin{bmatrix} 20 \\ 180 \\ 60 \\ 25 \end{bmatrix} = \begin{bmatrix} 72{,}900 \\ 54{,}700 \\ 60{,}800 \end{bmatrix} \begin{matrix} \text{Model X} \\ \text{Model Y} \\ \text{Model Z} \end{matrix}$$

(b) The totals of the columns of matrix AB will give a matrix whose elements represent the amounts of each material needed for the subdivision. Call this matrix D, and write it as a row matrix.

$$D = \begin{bmatrix} 3800 & 130 & 1400 & 200 \end{bmatrix}$$

(c) The total cost of all the materials is given by the product of matrix C, the cost matrix, and matrix D, the total amounts matrix. To multiply these and get a 1×1 matrix representing the total cost requires multiplying a 1×4 matrix and a 4×1 matrix. This is why in part (b) a row matrix was written rather than a column matrix. The total materials cost is given by DC. **FIGURE 35** shows how a graphing calculator computes this product. The total cost of the materials is $188,400.

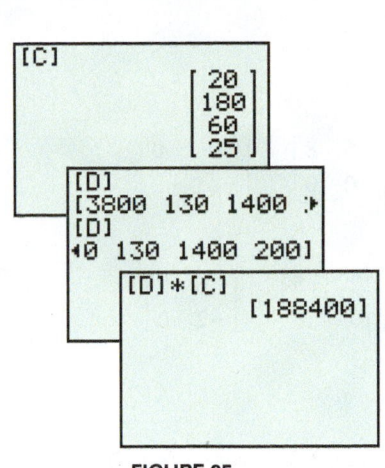

FIGURE 35

To help keep track of the quantities a matrix represents, let matrix A from **Example 10** represent models/styles, matrix B represent styles/materials, and matrix C represent materials/cost. In each case, the meaning of the rows is written first and that of the columns second. When the product AB was found in **Example 10,** the rows of the matrix represented models and the columns represented materials. Therefore, the matrix product AB represents models/materials.

6.4 Exercises

Checking Analytic Skills　Find the dimension of each matrix. Identify any square, column, or row matrices. **Do not use a calculator.**

1. $\begin{bmatrix} -3 & 6 \\ 7 & -4 \end{bmatrix}$

2. $\begin{bmatrix} 2 & -8 & 6 \\ 1 & 0 & -5 \\ 5 & -2 & 3 \end{bmatrix}$

3. $\begin{bmatrix} -6 & 8 & 0 & 0 \\ 4 & 1 & 9 & 2 \\ 3 & -5 & 7 & 1 \end{bmatrix}$

4. $\begin{bmatrix} -3 & 4 & 2 & 1 \\ 0 & 8 & 6 & 3 \end{bmatrix}$

5. $\begin{bmatrix} 2 \\ 4 \end{bmatrix}$

6. $\begin{bmatrix} 4 & 9 \end{bmatrix}$

7. $\begin{bmatrix} -9 \end{bmatrix}$

8. $\begin{bmatrix} 0 & 0 & 0 & 0 & 0 \\ 0 & 0 & 0 & 0 & 0 \end{bmatrix}$

Checking Analytic Skills　Find the value of each variable. **Do not use a calculator.**

9. $\begin{bmatrix} w & x \\ y & z \end{bmatrix} = \begin{bmatrix} 3 & 2 \\ -1 & 4 \end{bmatrix}$

10. $\begin{bmatrix} 2 & 5 & 6 \\ 1 & m & n \end{bmatrix} = \begin{bmatrix} z & y & w \\ 1 & 8 & -2 \end{bmatrix}$

11. $\begin{bmatrix} 0 & 5 & x \\ -1 & 3 & y+2 \\ 4 & 1 & z \end{bmatrix} = \begin{bmatrix} 0 & w+3 & 6 \\ -1 & 3 & 0 \\ 4 & 1 & 8 \end{bmatrix}$

12. $\begin{bmatrix} 3+x & 4 & t \\ 5 & 8-w & y+1 \\ -4 & 3 & 2r \end{bmatrix} = \begin{bmatrix} 9 & 4 & 6 \\ z+3 & w & 9 \\ p & q & r \end{bmatrix}$

13. $\begin{bmatrix} z & 4r & 8s \\ 6p & 2 & 5 \end{bmatrix} + \begin{bmatrix} -9 & 8r & 3 \\ 2 & 5 & 4 \end{bmatrix} = \begin{bmatrix} 2 & 36 & 27 \\ 20 & 7 & 12a \end{bmatrix}$

14. $\begin{bmatrix} a+2 & 1 & 5m \\ 8k & 0 & 3 \end{bmatrix} + \begin{bmatrix} 3a & 2z & 5m \\ 2k & 5 & 6 \end{bmatrix} = \begin{bmatrix} 10 & -14 & 80 \\ 10 & 5 & 9 \end{bmatrix}$

15. Your friend missed the lecture on adding matrices. In your own words, explain to her how to add two matrices.

16. Explain to a friend in your own words how to multiply a matrix by a scalar.

Perform each operation if possible.

17. $\begin{bmatrix} 6 & -9 & 2 \\ 4 & 1 & 3 \end{bmatrix} + \begin{bmatrix} -8 & 2 & 5 \\ 6 & -3 & 4 \end{bmatrix}$

18. $\begin{bmatrix} 9 & 4 \\ -8 & 2 \end{bmatrix} + \begin{bmatrix} -3 & 2 \\ -4 & 7 \end{bmatrix}$

19. $\begin{bmatrix} -6 & 8 \\ 0 & 0 \end{bmatrix} - \begin{bmatrix} 0 & 0 \\ -4 & -2 \end{bmatrix}$

20. $\begin{bmatrix} 1 & -4 \\ 2 & -3 \\ -8 & 4 \end{bmatrix} - \begin{bmatrix} -6 & 9 \\ -2 & 5 \\ -7 & -12 \end{bmatrix}$

21. $\begin{bmatrix} 6 & -2 \\ 5 & 4 \end{bmatrix} + \begin{bmatrix} -1 & 7 \\ 7 & -4 \end{bmatrix}$

22. $\begin{bmatrix} 12 & -5 \\ 10 & 3 \end{bmatrix} - \begin{bmatrix} 6 & 9 \\ -2 & 0 \end{bmatrix}$

23. $\begin{bmatrix} -8 & 4 & 0 \\ 2 & 5 & 0 \end{bmatrix} + \begin{bmatrix} 6 & 3 \\ 8 & 9 \end{bmatrix}$

24. $\begin{bmatrix} 2 \\ 3 \end{bmatrix} - \begin{bmatrix} 8 & 1 \\ 9 & 4 \end{bmatrix}$

25. $\begin{bmatrix} 9 & 4 & 1 & -2 \\ 5 & -6 & 3 & 4 \\ 2 & -5 & 1 & 2 \end{bmatrix} - \begin{bmatrix} -2 & 5 & 1 & 3 \\ 0 & 1 & 0 & 2 \\ -8 & 3 & 2 & 1 \end{bmatrix} + \begin{bmatrix} 2 & 4 & 0 & 3 \\ 4 & -5 & 1 & 6 \\ 2 & -3 & 0 & 8 \end{bmatrix}$

26. $\begin{bmatrix} 6 & -2 & 4 \\ -2 & 5 & 8 \\ 1 & 0 & 2 \end{bmatrix} + \begin{bmatrix} 3 & 0 & 8 \\ 1 & -2 & 4 \\ 6 & 9 & -2 \end{bmatrix} - \begin{bmatrix} -4 & 2 & 1 \\ 0 & 3 & -2 \\ 4 & 2 & 0 \end{bmatrix}$

27. $2\begin{bmatrix} 2 & -1 \\ 5 & 1 \\ 0 & 3 \end{bmatrix} + \begin{bmatrix} 5 & 0 \\ 7 & -3 \\ 1 & 1 \end{bmatrix} - \begin{bmatrix} 9 & -4 \\ 4 & 4 \\ 1 & 6 \end{bmatrix}$

28. $-3\begin{bmatrix} 3 & 8 \\ -1 & -9 \end{bmatrix} + 5\begin{bmatrix} 4 & -8 \\ 1 & 6 \end{bmatrix}$

29. $2\begin{bmatrix} 2 & -1 & -1 \\ -1 & 2 & -1 \\ -1 & -1 & 2 \end{bmatrix} + 3\begin{bmatrix} 1 & 2 & 3 \\ 2 & 1 & 3 \\ 2 & 3 & 1 \end{bmatrix}$

30. $3\begin{bmatrix} 1 & 0 & 3 \\ 0 & 1 & 2 \\ 1 & 0 & -3 \end{bmatrix} - 4\begin{bmatrix} -1 & 0 & 0 \\ 0 & -1 & 3 \\ 2 & 0 & 1 \end{bmatrix}$

31. $3\begin{bmatrix} 6 & -1 & 4 \\ 2 & 8 & -3 \\ -4 & 5 & 6 \end{bmatrix} + 5\begin{bmatrix} -2 & -8 & -6 \\ 4 & 1 & 3 \\ 2 & -1 & 5 \end{bmatrix}$

32. $4\begin{bmatrix} 1 & -4 \\ 2 & -3 \\ -8 & 4 \end{bmatrix} - 3\begin{bmatrix} -6 & 9 \\ -2 & 5 \\ -7 & -12 \end{bmatrix}$

Matrices $[A]$ *and* $[B]$ *are shown on the screen. Find each matrix.*

33. $2[A]$

34. $-3[B]$

35. $2[A] - [B]$

36. $-2[A] + [B]$

37. $5[A] + 0.5[B]$

38. $-4[A] + 1.5[B]$

```
[A]
          [-2  4]
          [ 0  3]
[B]
          [-6  2]
          [ 4  0]
```

39. *Concept Check*   Find matrix A if

$$B = \begin{bmatrix} 4 & 6 & -5 \\ -6 & 3 & 2 \end{bmatrix} \quad \text{and} \quad A + B = \begin{bmatrix} 6 & 12 & 0 \\ -10 & -4 & 11 \end{bmatrix}.$$

40. *Concept Check*   Find matrix B if

$$A = \begin{bmatrix} 3 & 6 & 5 \\ -2 & 1 & 4 \end{bmatrix} \quad \text{and} \quad A - B = \begin{bmatrix} 9 & 0 & -5 \\ -4 & 6 & -3 \end{bmatrix}.$$

The dimensions of matrices A and B are given. Find the dimensions of the product AB and of the product BA if the products are defined. If they are not defined, say so.

41. *A* is 4 × 2; *B* is 2 × 4.

42. *A* is 3 × 1; *B* is 1 × 3.

43. *A* is 3 × 5; *B* is 5 × 2.

44. *A* is 7 × 3; *B* is 2 × 7.

45. *A* is 4 × 3; *B* is 2 × 5.

46. *A* is 1 × 6; *B* is 2 × 4.

47. *Concept Check* The product *MN* of two matrices can be found only if the number of _____ of *M* equals the number of _____ of *N*.

48. *Concept Check* In finding the product *AB* of matrices *A* and *B*, the first row, second column, entry is found by multiplying the _____ elements in *A* and the _____ elements in *B* and then _____ these products.

If possible, find AB and BA.

49. $A = \begin{bmatrix} 1 & -1 \\ 2 & 0 \end{bmatrix}$, $B = \begin{bmatrix} -2 & 3 \\ 1 & 2 \end{bmatrix}$

50. $A = \begin{bmatrix} -3 & 5 \\ 2 & 7 \end{bmatrix}$, $B = \begin{bmatrix} -1 & 2 \\ 0 & 7 \end{bmatrix}$

51. $A = \begin{bmatrix} 5 & -7 & 2 \\ 0 & 1 & 5 \end{bmatrix}$, $B = \begin{bmatrix} 9 & 8 & 7 \\ 1 & -1 & -2 \end{bmatrix}$

52. $A = \begin{bmatrix} 2 & 1 & -1 \\ 0 & 2 & 1 \\ 3 & 2 & -1 \end{bmatrix}$, $B = \begin{bmatrix} 1 & 0 \\ 2 & -1 \\ 3 & 1 \end{bmatrix}$

53. $A = \begin{bmatrix} 3 & -1 \\ 1 & 0 \\ -2 & -4 \end{bmatrix}$, $B = \begin{bmatrix} -2 & 5 & -3 \\ 9 & -7 & 0 \end{bmatrix}$

54. $A = \begin{bmatrix} -1 & 0 & -2 \\ 4 & -2 & 1 \end{bmatrix}$, $B = \begin{bmatrix} 2 & -2 \\ 5 & -1 \\ 0 & 1 \end{bmatrix}$

55. $A = \begin{bmatrix} 1 & -1 & 0 \\ 2 & -1 & 5 \\ 6 & 1 & -4 \end{bmatrix}$, $B = \begin{bmatrix} -1 & 3 & -1 \\ 7 & -7 & 1 \end{bmatrix}$

56. $A = \begin{bmatrix} 2 & -1 & -5 \\ 4 & -1 & 6 \\ -2 & 0 & 9 \end{bmatrix}$, $B = \begin{bmatrix} 1 & 2 \\ -1 & -1 \\ 2 & 0 \end{bmatrix}$

Find each matrix product if possible.

57. $\begin{bmatrix} 3 & -4 & 1 \\ 5 & 0 & 2 \end{bmatrix} \begin{bmatrix} -1 \\ 4 \\ 2 \end{bmatrix}$

58. $\begin{bmatrix} -6 & 3 & 5 \\ 2 & 9 & 1 \end{bmatrix} \begin{bmatrix} -2 \\ 0 \\ 3 \end{bmatrix}$

59. $\begin{bmatrix} 5 & 2 \\ -1 & 4 \end{bmatrix} \begin{bmatrix} 3 & -2 \\ 1 & 0 \end{bmatrix}$

60. $\begin{bmatrix} -4 & 0 \\ 1 & 3 \end{bmatrix} \begin{bmatrix} -2 & 4 \\ 0 & 1 \end{bmatrix}$

61. $\begin{bmatrix} 2 & 2 & -1 \\ 3 & 0 & 1 \end{bmatrix} \begin{bmatrix} 0 & 2 \\ -1 & 4 \\ 0 & 2 \end{bmatrix}$

62. $\begin{bmatrix} -9 & 2 & 1 \\ 3 & 0 & 0 \end{bmatrix} \begin{bmatrix} 2 \\ -1 \\ 4 \end{bmatrix}$

63. $\begin{bmatrix} -2 & -3 & -4 \\ 2 & -1 & 0 \\ 4 & -2 & 3 \end{bmatrix} \begin{bmatrix} 0 & 1 & 4 \\ 1 & 2 & -1 \\ 3 & 2 & -2 \end{bmatrix}$

64. $\begin{bmatrix} -1 & 2 & 0 \\ 0 & 3 & 2 \\ 0 & 1 & 4 \end{bmatrix} \begin{bmatrix} 2 & -1 & 2 \\ 0 & 2 & 1 \\ 3 & 0 & -1 \end{bmatrix}$

65. $\begin{bmatrix} -2 & 4 & 1 \end{bmatrix} \begin{bmatrix} 3 & -2 & 4 \\ 2 & 1 & 0 \\ 0 & -1 & 4 \end{bmatrix}$

66. $\begin{bmatrix} 0 & 3 & -4 \end{bmatrix} \begin{bmatrix} -2 & 6 & 3 \\ 0 & 4 & 2 \\ -1 & 1 & 4 \end{bmatrix}$

67. $\begin{bmatrix} p & q \\ r & s \end{bmatrix} \begin{bmatrix} a & c \\ b & d \end{bmatrix}$

68. $\begin{bmatrix} a & b & c \\ d & e & f \\ g & h & i \end{bmatrix} \begin{bmatrix} x \\ y \\ z \end{bmatrix}$

Given $A = \begin{bmatrix} 4 & -2 \\ 3 & 1 \end{bmatrix}$, $B = \begin{bmatrix} 5 & 1 \\ 0 & -2 \\ 3 & 7 \end{bmatrix}$, *and* $C = \begin{bmatrix} -5 & 4 & 1 \\ 0 & 3 & 6 \end{bmatrix}$, *find each product if possible.*

69. *BA*

70. *AC*

71. *BC*

72. *CB*

73. *AB*

74. *CA*

75. A^2

76. A^3

77. *Concept Check* Compare the answers to **Exercises 69** and **73, 71** and **72,** and **70** and **74.** Is matrix multiplication commutative?

78. *Concept Check* For any matrices P and Q, what must be true for both PQ and QP to exist?

Solve each problem.

79. *Income from Yogurt* Yagel's Yogurt sells three types of yogurt: nonfat, regular, and supercreamy, at three locations. Location I sells 50 gallons of nonfat, 100 gallons of regular, and 30 gallons of supercreamy each day. Location II sells 10 gallons of nonfat, 90 gallons of regular, and 50 gallons of supercreamy each day. Location III sells 60 gallons of nonfat, 120 gallons of regular, and 40 gallons of supercreamy each day.

(a) Write a 3×3 matrix that shows sales for the three locations, with the rows representing the locations.

(b) The incomes per gallon for nonfat, regular, and supercreamy are $12, $10, and $15, respectively. Write a 3×1 matrix displaying the incomes per gallon.

(c) Find a matrix product that gives the daily income at each of the three locations.

(d) What is Yagel's Yogurt's total daily income from the three locations?

80. *Purchasing Costs* The Bread Box, a small neighborhood bakery, sells four main items: sweet rolls, bread, cakes, and pies. The amount of each ingredient (in cups, except for eggs) required for these items is given by matrix A.

Eggs Flour Sugar Shortening Milk

$$
\begin{array}{l}
\text{Rolls} \\
\text{(doz)} \\
\text{Bread} \\
\text{(loaves)} \\
\text{Cakes} \\
\text{Pies} \\
\text{(crust)}
\end{array}
\left[
\begin{array}{ccccc}
1 & 4 & \frac{1}{4} & \frac{1}{4} & 1 \\
0 & 3 & 0 & \frac{1}{4} & 0 \\
4 & 3 & 2 & 1 & 1 \\
0 & 1 & 0 & \frac{1}{3} & 0
\end{array}
\right] = A
$$

The cost (in cents) for each ingredient when purchased in either large lots or small lots is given in matrix B.

Cost

	Large lot	Small lot
Eggs	5	5
Flour	8	10
Sugar	10	12
Shortening	12	15
Milk	5	6

$= B$

(a) Use matrix multiplication to find a matrix giving the cost per item for the two purchase options.

(b) Suppose a day's orders consist of 20 dozen sweet rolls, 200 loaves of bread, 50 cakes, and 60 pies. Write the orders as a 1×4 matrix, and using matrix multiplication, write as a matrix the amount of each ingredient needed to fill the day's orders.

(c) Use matrix multiplication to find a matrix giving the costs under the two purchase options to fill the day's orders.

81. *(Modeling) Predator–Prey Relationship* In certain parts of the Rocky Mountains, deer are the main food source for mountain lions. When the deer population d is large, the mountain lions (m) thrive. However, a large mountain lion population drives down the size of the deer population. Suppose the fluctuations of the two populations from year to year can be modeled with the matrix equation

$$
\begin{bmatrix} m_{n+1} \\ d_{n+1} \end{bmatrix} = \begin{bmatrix} 0.51 & 0.4 \\ -0.05 & 1.05 \end{bmatrix} \begin{bmatrix} m_n \\ d_n \end{bmatrix}.
$$

The numbers in the column matrices give the numbers of animals in the two populations after n years and $n + 1$ years, where the number of deer is measured in hundreds.

(a) Give the equation for d_{n+1} obtained from the second row of the square matrix. Use this equation to determine the rate the deer population will grow from year to year if there are no mountain lions.

(b) Suppose we start with a mountain lion population of 2000 and a deer population of 500,000 (that is, 5000 hundred deer). How large would each population be after 1 year? 2 years?

(c) Consider part (b), but change the initial mountain lion population to 4000. Show that the populations would both grow at a steady annual rate of 1%.

82. *(Modeling) Northern Spotted Owl Population* To analyze population dynamics of the northern spotted owl, mathematical ecologists divided the female owl population into three categories: juvenile (up to 1 year old), subadult (1 to 2 years old), and adult (over 2 years old). They concluded that the change in the makeup of the northern spotted owl population in successive years could be described by the following matrix equation.

$$
\begin{bmatrix} j_{n+1} \\ s_{n+1} \\ a_{n+1} \end{bmatrix} = \begin{bmatrix} 0 & 0 & 0.33 \\ 0.18 & 0 & 0 \\ 0 & 0.71 & 0.94 \end{bmatrix} \begin{bmatrix} j_n \\ s_n \\ a_n \end{bmatrix}
$$

The numbers in the column matrices give the numbers of females in the three age groups after n years and $n + 1$ years. Multiplying the matrices yields the following.

$j_{n+1} = 0.33a_n$ Each year, 33 juvenile females are born for each 100 adult females.

$s_{n+1} = 0.18j_n$ Each year, 18% of the juvenile females survive to become subadults.

$a_{n+1} = 0.71s_n + 0.94a_n$ Each year, 71% of the subadults survive to become adults and 94% of the adults survive.

(*Source:* Lamberson, R. H., R. McKelvey, B. R. Noon, and C. Voss, "A Dynamic Analysis of Northern Spotted Owl Viability in a Fragmented Forest Landscape," *Conservation Biology*, Vol. 6, No. 4.)

(c) In this model, the main impediment to the survival of the northern spotted owl is the number 0.18 in the second row of the 3 × 3 matrix. This number is low for two reasons: The first year of life is precarious for most animals living in the wild, and juvenile owls must eventually leave the nest and establish their own territory. If much of the forest near their original home has been cleared, then they are vulnerable to predators while searching for a new home.

Suppose that, due to better forest management, the number 0.18 can be increased to 0.3. Rework part (a) under this new assumption.

(a) Suppose there are currently 3000 female northern spotted owls: 690 juveniles, 210 subadults, and 2100 adults. Use the preceding matrix equation to determine the total number of female owls for each of the next 5 years.

(b) Using advanced techniques from linear algebra, we can show that, in the long run,

$$\begin{bmatrix} j_{n+1} \\ s_{n+1} \\ a_{n+1} \end{bmatrix} \approx 0.98359 \begin{bmatrix} j_n \\ s_n \\ a_n \end{bmatrix}.$$

What can we conclude about the long-term fate of the northern spotted owl?

Let

$$A = \begin{bmatrix} a_{11} & a_{12} \\ a_{21} & a_{22} \end{bmatrix}, B = \begin{bmatrix} b_{11} & b_{12} \\ b_{21} & b_{22} \end{bmatrix}, \text{ and } C = \begin{bmatrix} c_{11} & c_{12} \\ c_{21} & c_{22} \end{bmatrix},$$

where all the elements are real numbers. Use these matrices to show that each statement is true for 2 × 2 matrices.

83. $A + B = B + A$ (commutative property)

84. $A + (B + C) = (A + B) + C$ (associative property)

85. $(AB)C = A(BC)$ (associative property)

86. $A(B + C) = AB + AC$ (distributive property)

87. $c(A + B) = cA + cB$, for any real number c

88. $(c + d)A = cA + dA$, for any real numbers c and d

89. $(cA)d = (cd)A$

90. $(cd)A = c(dA)$

6.5 Determinants and Cramer's Rule

Determinants of 2 × 2 Matrices • Determinants of Larger Matrices • Derivation of Cramer's Rule • Using Cramer's Rule to Solve Systems

For convenience, we use subscript notation to name the elements of a matrix, as in the following m by n matrix A.

$$A = \begin{bmatrix} a_{11} & a_{12} & a_{13} & \cdots & a_{1n} \\ a_{21} & a_{22} & a_{23} & \cdots & a_{2n} \\ a_{31} & a_{32} & a_{33} & \cdots & a_{3n} \\ \vdots & \vdots & \vdots & & \vdots \\ a_{m1} & a_{m2} & a_{m3} & \cdots & a_{mn} \end{bmatrix}$$

With this notation, the row 1, column 1 element is a_{11}; the row 2, column 3 element is a_{23}; and, in general, the row i, column j element is a_{ij}. We use subscript notation in this section to define determinants.

Determinants of 2×2 Matrices

Associated with every *square* matrix A is a real number called the **determinant** of A. The symbols $|A|$, $\delta(A)$, and det A all represent the determinant of A. In this text, we use **det A.**

→ **Looking Ahead to Calculus**
Determinants are utilized in calculus to find **vector cross products,** which are used to study the effect of forces in a plane or in space.

Determinant of a 2×2 Matrix

The **determinant of a 2×2 matrix A,** where

$$A = \begin{bmatrix} a_{11} & a_{12} \\ a_{21} & a_{22} \end{bmatrix},$$

is a real number defined as $\det A = a_{11}a_{22} - a_{21}a_{12}.$

NOTE You should be able to distinguish between a matrix and its determinant. *A matrix is an array of numbers, while a determinant is a single real number associated with a square matrix.*

EXAMPLE 1 **Evaluating the Determinant of a 2×2 Matrix**

Find det A if $A = \begin{bmatrix} -3 & 4 \\ 6 & 8 \end{bmatrix}$.

Analytic Solution
Use the definition just given.

$$\det A = \det \begin{bmatrix} -3 & 4 \\ 6 & 8 \end{bmatrix}$$
$$= -3(8) - 6(4)$$
$$\underset{a_{11}\,a_{22}}{\uparrow\,\uparrow}\quad\underset{a_{21}\,a_{12}}{\uparrow\,\uparrow}$$
$$= -48$$

Graphing Calculator Solution
The TI-84 Plus graphing calculator can compute determinants. **FIGURE 36** shows both matrix A and det A.

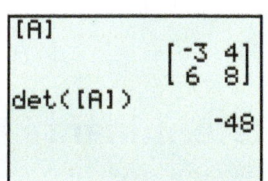

FIGURE 36

EXAMPLE 2 **Solving an Equation Involving a Determinant**

If $A = \begin{bmatrix} x & 3 \\ -1 & 5 \end{bmatrix}$ and det $A = 33$, solve the determinant equation for x.

Solution Set det $A = 5x - (-1)(3)$, or $5x + 3$, equal to 33.

$$5x + 3 = 33 \qquad \text{Definition}$$
$$5x = 30 \qquad \text{Subtract 3.}$$
$$x = 6 \qquad \text{Divide by 5.}$$

The solution set is $\{6\}$.

Determinants of Larger Matrices

> ### Determinant of a 3 × 3 Matrix
>
> The **determinant of a 3 × 3 matrix A**, where
>
> $$A = \begin{bmatrix} a_{11} & a_{12} & a_{13} \\ a_{21} & a_{22} & a_{23} \\ a_{31} & a_{32} & a_{33} \end{bmatrix},$$
>
> is a real number defined as follows.
>
> $$\det A = (a_{11}a_{22}a_{33} + a_{12}a_{23}a_{31} + a_{13}a_{21}a_{32})$$
> $$- (a_{31}a_{22}a_{13} + a_{32}a_{23}a_{11} + a_{33}a_{21}a_{12})$$

NOTE The above definition is *not* usually memorized in its general form. The method of cofactors, which is described next, is used more often because it allows us to calculate the determinant of any square matrix.

A method for calculating 3 × 3 determinants is found by rearranging and factoring the terms given in the definition to obtain the following.

$$\det \begin{bmatrix} a_{11} & a_{12} & a_{13} \\ a_{21} & a_{22} & a_{23} \\ a_{31} & a_{32} & a_{33} \end{bmatrix} = a_{11}(a_{22}a_{33} - a_{32}a_{23}) - a_{21}(a_{12}a_{33} - a_{32}a_{13})$$
$$+ a_{31}(a_{12}a_{23} - a_{22}a_{13})$$

Each quantity in parentheses represents the determinant of a 2 × 2 matrix that is the part of the 3 × 3 matrix remaining when the row and column of the multiplier are eliminated, as shown here.

$$a_{11}(a_{22}a_{33} - a_{32}a_{23}) \quad \begin{bmatrix} a_{11} & a_{12} & a_{13} \\ a_{21} & a_{22} & a_{23} \\ a_{31} & a_{32} & a_{33} \end{bmatrix}$$

$$a_{21}(a_{12}a_{33} - a_{32}a_{13}) \quad \begin{bmatrix} a_{11} & a_{12} & a_{13} \\ a_{21} & a_{22} & a_{23} \\ a_{31} & a_{32} & a_{33} \end{bmatrix}$$

$$a_{31}(a_{12}a_{23} - a_{22}a_{13}) \quad \begin{bmatrix} a_{11} & a_{12} & a_{13} \\ a_{21} & a_{22} & a_{23} \\ a_{31} & a_{32} & a_{33} \end{bmatrix}$$

The determinant of each (red) 2 × 2 matrix is called a **minor** of the associated (blue) element in the 3 × 3 matrix. The symbol M_{ij} represents the minor that results when row i and column j are deleted. The following table gives some of the minors for the matrix A in the margin on the next page.

$$A = \begin{bmatrix} a_{11} & a_{12} & a_{13} \\ a_{21} & a_{22} & a_{23} \\ a_{31} & a_{32} & a_{33} \end{bmatrix}$$

Element	Minor	Element	Minor
a_{11}	$M_{11} = \det \begin{bmatrix} a_{22} & a_{23} \\ a_{32} & a_{33} \end{bmatrix}$	a_{22}	$M_{22} = \det \begin{bmatrix} a_{11} & a_{13} \\ a_{31} & a_{33} \end{bmatrix}$
a_{21}	$M_{21} = \det \begin{bmatrix} a_{12} & a_{13} \\ a_{32} & a_{33} \end{bmatrix}$	a_{23}	$M_{23} = \det \begin{bmatrix} a_{11} & a_{12} \\ a_{31} & a_{32} \end{bmatrix}$
a_{31}	$M_{31} = \det \begin{bmatrix} a_{12} & a_{13} \\ a_{22} & a_{23} \end{bmatrix}$	a_{33}	$M_{33} = \det \begin{bmatrix} a_{11} & a_{12} \\ a_{21} & a_{22} \end{bmatrix}$

In a 4×4 matrix, the minors are determinants of 3×3 matrices. Similarly, an $n \times n$ matrix has minors that are determinants of $(n - 1) \times (n - 1)$ matrices.

To find the determinant of a 3×3 or larger square matrix, first choose any row or column. Then the minor of each element in that row or column must be multiplied by $+1$ or -1, depending on whether the sum of the row number and column number is even or odd. The product of a minor and the number $+1$ or -1 is called a *cofactor*.

Cofactor

Let M_{ij} be the minor for element a_{ij} in an $n \times n$ matrix. The **cofactor** of a_{ij}, written A_{ij}, is defined as follows.

$$A_{ij} = (-1)^{i+j} \cdot M_{ij}$$

EXAMPLE 3 **Finding the Cofactor of an Element**

Find the cofactor of each given element of the following matrix.

$$A = \begin{bmatrix} 6 & 2 & 4 \\ 8 & 9 & 3 \\ 1 & 2 & 0 \end{bmatrix}$$

(a) 6 **(b)** 3 **(c)** 8

Solution

(a) Because 6 is located in the first row, first column of the matrix, $i = 1$ and $j = 1$.

$M_{11} = \det \begin{bmatrix} 9 & 3 \\ 2 & 0 \end{bmatrix} = -6$. The cofactor is

Delete the row and column containing 6 in the matrix A to calculate M_{11}.

$$(-1)^{1+1}(-6) = 1(-6) = -6.$$

(b) Here $i = 2$ and $j = 3$. $M_{23} = \det \begin{bmatrix} 6 & 2 \\ 1 & 2 \end{bmatrix} = 10$. The cofactor is

Delete the row and column containing 3 in the matrix A to calculate M_{23}.

$$(-1)^{2+3}(10) = -1(10) = -10.$$

(c) We have $i = 2$ and $j = 1$. $M_{21} = \det \begin{bmatrix} 2 & 4 \\ 2 & 0 \end{bmatrix} = -8$. The cofactor is

Delete the row and column containing 8 in the matrix A to calculate M_{21}.

$$(-1)^{2+1}(-8) = -1(-8) = 8.$$

> **Finding the Determinant of an $n \times n$ Matrix**
>
> Multiply each element in any row or column of the matrix by its cofactor. The sum of these products gives the value of the determinant.

The process of forming this sum of products is called **expansion by a row or column.**

EXAMPLE 4   Evaluating the Determinant of a 3 × 3 Matrix

Evaluate $\det \begin{bmatrix} 2 & -3 & -2 \\ -1 & -4 & -3 \\ -1 & 0 & 2 \end{bmatrix}$, expanding by the second column.

Solution   First find the minor of each element in the second column.

$$M_{12} = \det \begin{bmatrix} -1 & -3 \\ -1 & 2 \end{bmatrix} = -1(2) - (-1)(-3) = -5$$

$$M_{22} = \det \begin{bmatrix} 2 & -2 \\ -1 & 2 \end{bmatrix} = 2(2) - (-1)(-2) = 2$$

$$M_{32} = \det \begin{bmatrix} 2 & -2 \\ -1 & -3 \end{bmatrix} = 2(-3) - (-1)(-2) = -8$$

Use parentheses and keep track of all negative signs to avoid errors.

Now find the cofactor of each of these minors.

$$A_{12} = (-1)^{1+2} \cdot M_{12} = (-1)^3(-5) = -1(-5) = 5$$
$$A_{22} = (-1)^{2+2} \cdot M_{22} = (-1)^4 \cdot 2 = 1 \cdot 2 = 2$$
$$A_{32} = (-1)^{3+2} \cdot M_{32} = (-1)^5(-8) = -1(-8) = 8$$

Find the determinant by multiplying each cofactor by its corresponding element in the matrix and then finding the sum of the products.

$$\det \begin{bmatrix} 2 & -3 & -2 \\ -1 & -4 & -3 \\ -1 & 0 & 2 \end{bmatrix} = a_{12} \cdot A_{12} + a_{22} \cdot A_{22} + a_{32} \cdot A_{32}$$

$$= -3(5) + (-4)2 + 0(8)$$
$$= -15 + (-8) + 0$$
$$= -23 \qquad \text{See FIGURE 37.} \quad \bullet$$

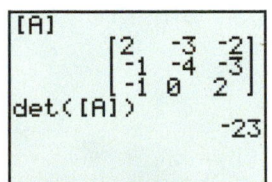

[A]
$\begin{bmatrix} 2 & -3 & -2 \\ -1 & -4 & -3 \\ -1 & 0 & 2 \end{bmatrix}$
det([A])
                    -23

This screen shows how a TI-84 Plus calculator evaluates the 3 × 3 determinant in **Example 4.**

FIGURE 37

For 3 × 3 Matrices

+	−	+
−	+	−
+	−	+

For 4 × 4 Matrices

+	−	+	−
−	+	−	+
+	−	+	−
−	+	−	+

In **Example 4,** we would have found exactly the same answer by using any row or column of the matrix. One reason we used column 2 is that it contains a 0 element, so it was not really necessary to calculate M_{32} and A_{32}. One learns quickly that 0s are "friendly" in work with determinants.

Instead of calculating $(-1)^{i+j}$ for a given element, the sign checkerboard in the margin can be used. The signs alternate for each row and column, beginning with + in the first row, first column, position. If we expand a 3 × 3 matrix about row 3, for example, the first minor would have a + sign associated with it, the second minor a − sign, and the third minor a + sign. This array of signs can be extended to determinants of 4 × 4 and larger matrices. See the margin.

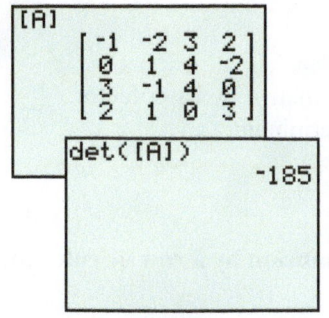

FIGURE 38

EXAMPLE 5 **Evaluating the Determinant of a 4 × 4 Matrix**

Evaluate.

$$\det \begin{bmatrix} -1 & -2 & 3 & 2 \\ 0 & 1 & 4 & -2 \\ 3 & -1 & 4 & 0 \\ 2 & 1 & 0 & 3 \end{bmatrix}$$

Solution The determinant of a 4 × 4 matrix can be found by using cofactors, but the computation is tedious. Using a graphing calculator with matrix capabilities is much easier. **FIGURE 38** indicates that the desired determinant is -185. ●

Derivation of Cramer's Rule

Determinants can be used to solve a linear system of equations in the form

$$a_1 x + b_1 y = c_1 \quad \text{(1)}$$
$$a_2 x + b_2 y = c_2 \quad \text{(2)}$$

by elimination as follows.

$$a_1 b_2 x + b_1 b_2 y = c_1 b_2 \qquad \text{Multiply (1) by } b_2.$$
$$-a_2 b_1 x - b_1 b_2 y = -c_2 b_1 \qquad \text{Multiply (2) by } -b_1.$$

Eliminate y to solve for x.

$$(a_1 b_2 - a_2 b_1)x = c_1 b_2 - c_2 b_1 \qquad \text{Add.}$$

$$x = \frac{c_1 b_2 - c_2 b_1}{a_1 b_2 - a_2 b_1} \quad \text{if } a_1 b_2 - a_2 b_1 \neq 0$$

Now find y.

$$-a_1 a_2 x - a_2 b_1 y = -a_2 c_1 \qquad \text{Multiply (1) by } -a_2.$$
$$a_1 a_2 x + a_1 b_2 y = a_1 c_2 \qquad \text{Multiply (2) by } a_1.$$

Eliminate x to solve for y.

$$(a_1 b_2 - a_2 b_1)y = a_1 c_2 - a_2 c_1 \qquad \text{Add.}$$

$$y = \frac{a_1 c_2 - a_2 c_1}{a_1 b_2 - a_2 b_1} \quad \text{if } a_1 b_2 - a_2 b_1 \neq 0$$

Both numerators and the common denominator of these values for x and y can be written as determinants.

$$c_1 b_2 - c_2 b_1 = \det \begin{bmatrix} c_1 & b_1 \\ c_2 & b_2 \end{bmatrix}, \quad a_1 c_2 - a_2 c_1 = \det \begin{bmatrix} a_1 & c_1 \\ a_2 & c_2 \end{bmatrix}, \quad \text{and} \quad a_1 b_2 - a_2 b_1 = \det \begin{bmatrix} a_1 & b_1 \\ a_2 & b_2 \end{bmatrix}$$

With these determinants, the solutions for x and y are expressed as follows.

$$x = \frac{\det \begin{bmatrix} c_1 & b_1 \\ c_2 & b_2 \end{bmatrix}}{\det \begin{bmatrix} a_1 & b_1 \\ a_2 & b_2 \end{bmatrix}} \quad \text{and} \quad y = \frac{\det \begin{bmatrix} a_1 & c_1 \\ a_2 & c_2 \end{bmatrix}}{\det \begin{bmatrix} a_1 & b_1 \\ a_2 & b_2 \end{bmatrix}} \quad \text{if } \det \begin{bmatrix} a_1 & b_1 \\ a_2 & b_2 \end{bmatrix} \neq 0$$

These results are summarized as **Cramer's rule.**

Cramer's Rule for 2 × 2 Systems

The solution of the system

$$a_1x + b_1y = c_1$$
$$a_2x + b_2y = c_2$$

is given by

$$x = \frac{D_x}{D} \quad \text{and} \quad y = \frac{D_y}{D},$$

where

$$D_x = \det \begin{bmatrix} c_1 & b_1 \\ c_2 & b_2 \end{bmatrix}, \quad D_y = \det \begin{bmatrix} a_1 & c_1 \\ a_2 & c_2 \end{bmatrix}, \quad \text{and} \quad D = \det \begin{bmatrix} a_1 & b_1 \\ a_2 & b_2 \end{bmatrix} \neq 0.$$

CAUTION *Cramer's rule does not apply if D = 0.* When $D = 0$, the system is inconsistent or has dependent equations. For this reason, evaluate D first.

Using Cramer's Rule to Solve Systems

EXAMPLE 6 **Applying Cramer's Rule to a System with Two Equations**

Use Cramer's rule to solve the system.

$$5x + 7y = -1$$
$$6x + 8y = 1$$

Analytic Solution

By Cramer's rule, $x = \dfrac{D_x}{D}$ and $y = \dfrac{D_y}{D}$. Find D first, because if $D = 0$, Cramer's rule does not apply. If $D \neq 0$, then find D_x and D_y.

$$D = \det \begin{bmatrix} 5 & 7 \\ 6 & 8 \end{bmatrix} = 5(8) - 6(7) = -2$$

$$D_x = \det \begin{bmatrix} -1 & 7 \\ 1 & 8 \end{bmatrix} = -1(8) - 1(7) = -15$$

$$D_y = \det \begin{bmatrix} 5 & -1 \\ 6 & 1 \end{bmatrix} = 5(1) - 6(-1) = 11$$

From Cramer's rule,

$$x = \frac{D_x}{D} = \frac{-15}{-2} = \frac{15}{2} \quad \text{and} \quad y = \frac{D_y}{D} = \frac{11}{-2} = -\frac{11}{2}.$$

The solution set is $\left\{ \left(\frac{15}{2}, -\frac{11}{2} \right) \right\}$, as can be verified by substituting into the given system.

Graphing Calculator Solution

See **FIGURE 39.** Enter D, D_x, and D_y as $[A]$, $[B]$, and $[C]$, respectively. Then find the desired quotients and convert to fractions.

$$x = \frac{\det([B])}{\det([A])} \quad \text{and} \quad y = \frac{\det([C])}{\det([A])}$$

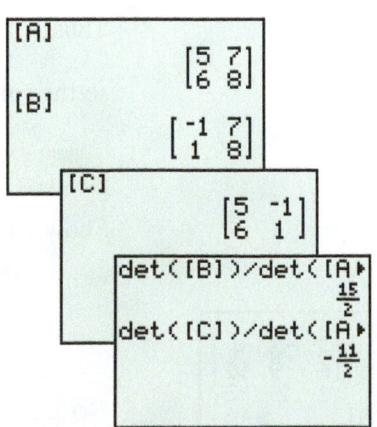

FIGURE 39

Cramer's Rule for 3 × 3 Systems

The solution of the system

$$a_1 x + b_1 y + c_1 z = d_1$$
$$a_2 x + b_2 y + c_2 z = d_2$$
$$a_3 x + b_3 y + c_3 z = d_3$$

is given by

$$x = \frac{D_x}{D}, \quad y = \frac{D_y}{D}, \quad \text{and} \quad z = \frac{D_z}{D},$$

where $D_x = \det \begin{bmatrix} d_1 & b_1 & c_1 \\ d_2 & b_2 & c_2 \\ d_3 & b_3 & c_3 \end{bmatrix}$, $\qquad D_y = \det \begin{bmatrix} a_1 & d_1 & c_1 \\ a_2 & d_2 & c_2 \\ a_3 & d_3 & c_3 \end{bmatrix}$,

$$D_z = \det \begin{bmatrix} a_1 & b_1 & d_1 \\ a_2 & b_2 & d_2 \\ a_3 & b_3 & d_3 \end{bmatrix}, \quad \text{and} \quad D = \det \begin{bmatrix} a_1 & b_1 & c_1 \\ a_2 & b_2 & c_2 \\ a_3 & b_3 & c_3 \end{bmatrix} \neq 0.$$

EXAMPLE 7 **Applying Cramer's Rule to a System with Three Equations**

Use Cramer's rule to solve the system.

$$x + y - z = -2$$
$$2x - y + z = -5$$
$$x - 2y + 3z = 4$$

Solution Verify that the required determinants are

$$D = \det \begin{bmatrix} 1 & 1 & -1 \\ 2 & -1 & 1 \\ 1 & -2 & 3 \end{bmatrix} = -3, \qquad D_x = \det \begin{bmatrix} -2 & 1 & -1 \\ -5 & -1 & 1 \\ 4 & -2 & 3 \end{bmatrix} = 7,$$

$$D_y = \det \begin{bmatrix} 1 & -2 & -1 \\ 2 & -5 & 1 \\ 1 & 4 & 3 \end{bmatrix} = -22, \qquad D_z = \det \begin{bmatrix} 1 & 1 & -2 \\ 2 & -1 & -5 \\ 1 & -2 & 4 \end{bmatrix} = -21.$$

Thus, $\quad x = \dfrac{D_x}{D} = \dfrac{7}{-3} = -\dfrac{7}{3}, \quad y = \dfrac{D_y}{D} = \dfrac{-22}{-3} = \dfrac{22}{3}, \quad z = \dfrac{D_z}{D} = \dfrac{-21}{-3} = 7,$

so the solution set is $\left\{ \left(-\dfrac{7}{3}, \dfrac{22}{3}, 7 \right) \right\}$. ●

EXAMPLE 8 **Verifying That Cramer's Rule Does Not Apply**

Show why Cramer's rule does not apply to the system.

$$2x - 3y + 4z = 10$$
$$6x - 9y + 12z = 24$$
$$x + 2y - 3z = 5$$

Solution We must show that $D = 0$. **FIGURE 40** confirms this fact, where $D = \det([A])$. When $D = 0$, the system either is inconsistent or contains dependent equations. We could use the elimination or row echelon method to tell which is the case. Verify that this system is inconsistent. ●

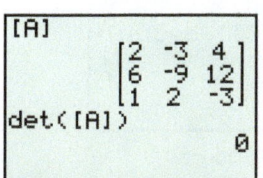

FIGURE 40

6.5 Exercises

Checking Analytic Skills   Find each determinant. ***Do not use a calculator.***

1. $\det \begin{bmatrix} -5 & 9 \\ 4 & -1 \end{bmatrix}$

2. $\det \begin{bmatrix} -1 & 3 \\ -2 & 9 \end{bmatrix}$

3. $\det \begin{bmatrix} -1 & -2 \\ 5 & 3 \end{bmatrix}$

4. $\det \begin{bmatrix} 6 & -4 \\ 0 & -1 \end{bmatrix}$

5. $\det \begin{bmatrix} 9 & 3 \\ -3 & -1 \end{bmatrix}$

6. $\det \begin{bmatrix} 0 & 2 \\ 1 & 5 \end{bmatrix}$

7. $\det \begin{bmatrix} 3 & 4 \\ 5 & -2 \end{bmatrix}$

8. $\det \begin{bmatrix} -9 & 7 \\ 2 & 6 \end{bmatrix}$

Find the cofactor of each element in the second row for each matrix.

9. $\begin{bmatrix} -2 & 0 & 1 \\ 1 & 2 & 0 \\ 4 & 2 & 1 \end{bmatrix}$

10. $\begin{bmatrix} 1 & -1 & 2 \\ 1 & 0 & 2 \\ 0 & -3 & 1 \end{bmatrix}$

11. $\begin{bmatrix} 1 & 2 & -1 \\ 2 & 3 & -2 \\ -1 & 4 & 1 \end{bmatrix}$

12. $\begin{bmatrix} 2 & -1 & 4 \\ 3 & 0 & 1 \\ -2 & 1 & 4 \end{bmatrix}$

Find each determinant.

13. $\det \begin{bmatrix} 4 & -7 & 8 \\ 2 & 1 & 3 \\ -6 & 3 & 0 \end{bmatrix}$

14. $\det \begin{bmatrix} 8 & -2 & -4 \\ 7 & 0 & 3 \\ 5 & -1 & 2 \end{bmatrix}$

15. $\det \begin{bmatrix} 1 & 2 & 0 \\ -1 & 2 & -1 \\ 0 & 1 & 4 \end{bmatrix}$

16. $\det \begin{bmatrix} 2 & 1 & -1 \\ 4 & 7 & -2 \\ 2 & 4 & 0 \end{bmatrix}$

17. $\det \begin{bmatrix} 10 & 2 & 1 \\ -1 & 4 & 3 \\ -3 & 8 & 10 \end{bmatrix}$

18. $\det \begin{bmatrix} 7 & -1 & 1 \\ 1 & -7 & 2 \\ -2 & 1 & 1 \end{bmatrix}$

19. $\det \begin{bmatrix} 1 & -2 & 3 \\ 0 & 0 & 0 \\ 1 & 10 & -12 \end{bmatrix}$

20. $\det \begin{bmatrix} 2 & 3 & 0 \\ 1 & 9 & 0 \\ -1 & -2 & 0 \end{bmatrix}$

21. $\det \begin{bmatrix} 3 & 3 & -1 \\ 2 & 6 & 0 \\ -6 & -6 & 2 \end{bmatrix}$

22. $\det \begin{bmatrix} 5 & -3 & 2 \\ -5 & 3 & -2 \\ 1 & 0 & 1 \end{bmatrix}$

23. $\det \begin{bmatrix} 0.4 & -0.8 & 0.6 \\ 0.3 & 0.9 & 0.7 \\ 3.1 & 4.1 & -2.8 \end{bmatrix}$

24. $\det \begin{bmatrix} -0.3 & -0.1 & 0.9 \\ 2.5 & 4.9 & -3.2 \\ -0.1 & 0.4 & 0.8 \end{bmatrix}$

25. $\det \begin{bmatrix} 17 & -4 & 3 \\ 11 & 5 & -15 \\ 7 & -9 & 23 \end{bmatrix}$

Solve each determinant equation for x.

26. $\det \begin{bmatrix} 5 & x \\ -3 & 2 \end{bmatrix} = 6$

27. $\det \begin{bmatrix} -0.5 & 2 \\ x & x \end{bmatrix} = 0$

28. $\det \begin{bmatrix} x & 3 \\ x & x \end{bmatrix} = 4$

29. $\det \begin{bmatrix} 2x & x \\ 11 & x \end{bmatrix} = 6$

30. $\det \begin{bmatrix} -2 & 0 & 1 \\ -1 & 3 & x \\ 5 & -2 & 0 \end{bmatrix} = 3$

31. $\det \begin{bmatrix} 4 & 3 & 0 \\ 2 & 0 & 1 \\ -3 & x & -1 \end{bmatrix} = 5$

32. $\det \begin{bmatrix} 5 & 3x & -3 \\ 0 & 2 & -1 \\ 4 & -1 & x \end{bmatrix} = -7$

33. $\det \begin{bmatrix} 2x & 1 & -1 \\ 0 & 4 & x \\ 3 & 0 & 2 \end{bmatrix} = x$

34. $\det \begin{bmatrix} x & x & 2 \\ 0 & 2 & 2 \\ 0 & 0 & 3x \end{bmatrix} = 96$

Evaluate each determinant.

35. $\det \begin{bmatrix} 3 & -6 & 5 & -1 \\ 0 & 2 & -1 & 3 \\ -6 & 4 & 2 & 0 \\ -7 & 3 & 1 & 1 \end{bmatrix}$

36. $\det \begin{bmatrix} 4 & 5 & -1 & -1 \\ 2 & -3 & 1 & 0 \\ -5 & 1 & 3 & 9 \\ 0 & -2 & 1 & 5 \end{bmatrix}$

37. $\det \begin{bmatrix} 4 & 0 & 0 & 2 \\ -1 & 0 & 3 & 0 \\ 2 & 4 & 0 & 1 \\ 0 & 0 & 1 & 2 \end{bmatrix}$

38. $\det \begin{bmatrix} -2 & 0 & 4 & 2 \\ 3 & 6 & 0 & 4 \\ 0 & 0 & 0 & 3 \\ 9 & 0 & 2 & -1 \end{bmatrix}$

Area of a Triangle *A triangle with vertices at (x_1, y_1), (x_2, y_2), and (x_3, y_3), as shown in the figure, has area equal to the absolute value of D, where*

$$D = \frac{1}{2} \det \begin{bmatrix} x_1 & y_1 & 1 \\ x_2 & y_2 & 1 \\ x_3 & y_3 & 1 \end{bmatrix}.$$

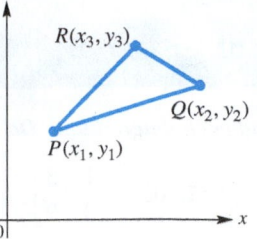

Use D to find the area of each triangle with coordinates as given.

39. $P(0, 0)$, $Q(0, 2)$, $R(1, 4)$

40. $P(0, 1)$, $Q(2, 0)$, $R(1, 5)$

41. $P(2, 5)$, $Q(-1, 3)$, $R(4, 0)$

42. $P(2, -2)$, $Q(0, 0)$, $R(-3, -4)$

43. $P(1, 2)$, $Q(4, 3)$, $R(3, 5)$

44. $P(-1, 0)$, $Q(-3, 5)$, $R(5, 2)$

Use the concept of the area of a triangle discussed in Exercises 39–44 to determine whether the three points are collinear.

45. $(1, 3)$, $(-3, 11)$, $(2, 1)$

46. $(3, 6)$, $(-1, -6)$, $(5, 11)$

47. $(-2, -5)$, $(4, 4)$, $(2, 3)$

48. $(4, -5)$, $(-2, 10)$, $(6, -10)$

49. $(4, -1)$, $(6, 0)$, $(12, 4)$

50. $(-1, -4)$, $(3, 8)$, $(6, 17)$

Several theorems are useful for calculating determinants. These theorems are true for square matrices of any dimension.

Determinant Theorems

1. If every element in a row (or column) of matrix A is 0, then $\det A = 0$.

2. If the rows of matrix A are the corresponding columns of matrix B, then $\det B = \det A$.

3. If any two rows (or columns) of matrix A are interchanged to form matrix B, then $\det B = -\det A$.

4. Suppose matrix B is formed by multiplying every element of a row (or column) of matrix A by the real number k. Then $\det B = k \cdot \det A$.

5. If two rows (or columns) of a matrix A are identical, then $\det A = 0$.

6. Changing a row (or column) of a matrix by adding a constant times another row (or column) to it does not change the determinant of the matrix.

Use the determinant theorems to find each determinant.

51. $\det \begin{bmatrix} 1 & 0 & 0 \\ 1 & 0 & 1 \\ 3 & 0 & 0 \end{bmatrix}$

52. $\det \begin{bmatrix} -1 & 2 & 4 \\ 4 & -8 & -16 \\ 3 & 0 & 5 \end{bmatrix}$

53. $\det \begin{bmatrix} 6 & 8 & -12 \\ -1 & 0 & 2 \\ 4 & 0 & -8 \end{bmatrix}$

54. $\det \begin{bmatrix} 4 & 8 & 0 \\ -1 & -2 & 1 \\ 2 & 4 & 3 \end{bmatrix}$

55. $\det \begin{bmatrix} -4 & 1 & 4 \\ 2 & 0 & 1 \\ 0 & 2 & 4 \end{bmatrix}$

56. $\det \begin{bmatrix} 6 & 3 & 2 \\ 1 & 0 & 2 \\ 5 & 7 & 3 \end{bmatrix}$

Use Cramer's rule to solve each system of equations. If $D = 0$, use another method to complete the solution.

57. $\begin{aligned} x + y &= 4 \\ 2x - y &= 2 \end{aligned}$

58. $\begin{aligned} 3x + 2y &= -4 \\ 2x - y &= -5 \end{aligned}$

59. $\begin{aligned} 4x + 3y &= -7 \\ 2x + 3y &= -11 \end{aligned}$

60. $\begin{aligned} 4x - y &= 0 \\ 2x + 3y &= 14 \end{aligned}$

61. $\begin{aligned} 3x + 2y &= 4 \\ 6x + 4y &= 8 \end{aligned}$

62. $\begin{aligned} 1.5x + 3y &= 5 \\ 2x + 4y &= 3 \end{aligned}$

63. $\begin{aligned} 2x - 3y &= -5 \\ x + 5y &= 17 \end{aligned}$

64. $\begin{aligned} x + 9y &= -15 \\ 3x + 2y &= 5 \end{aligned}$

65. $4x - y + 3z = -3$
$3x + y + z = 0$
$2x - y + 4z = 0$

66. $5x + 2y + z = 15$
$2x - y + z = 9$
$4x + 3y + 2z = 13$

67. $2x - y + 4z = -2$
$3x + 2y - z = -3$
$x + 4y + 2z = 17$

68. $x + y + z = 4$
$2x - y + 3z = 4$
$4x + 2y - z = -15$

69. $5x - y = -4$
$3x + 2z = 4$
$4y + 3z = 22$

70. $3x + 5y = -7$
$2x + 7z = 2$
$4y + 3z = -8$

71. $2x - y + 3z = 1$
$-2x + y - 3z = 2$
$5x - y + z = 2$

72. $-2x - 2y + 3z = 4$
$5x + 7y - z = 2$
$2x + 2y - 3z = -4$

73. $3x + 2y - w = 0$
$2x + z + 2w = 5$
$x + 2y - z = -2$
$2x - y + z + w = 2$

74. $x + 2y - z + w = 8$
$2x - y + 2w = 8$
$y + 3z = 5$
$x - z = 4$

75. In your own words, explain what happens when you apply Cramer's rule if $D = 0$.

76. Describe D_x, D_y, and D_z in terms of the coefficients and constants in a given system of equations.

6.6 Solution of Linear Systems by Matrix Inverses

Identity Matrices • Multiplicative Inverses of Square Matrices • Using Determinants to Find Inverses • Solving Linear Systems Using Inverse Matrices • Fitting Data Using a System

In **Section 6.4,** we saw several similarities between the set of real numbers and the set of matrices. Another similarity is that both sets have identity and inverse elements for multiplication.

Identity Matrices

By the identity property for real numbers, for any real number a,

$$a \cdot 1 = a \quad \text{and} \quad 1 \cdot a = a.$$

If there is to be a multiplicative **identity matrix** I, such that

$$AI = A \quad \text{and} \quad IA = A,$$

for *any* matrix A, then A and I must be square matrices with the same dimension.

FIGURE 41 shows how the identity matrix I_2 is displayed on the TI-84 Plus.

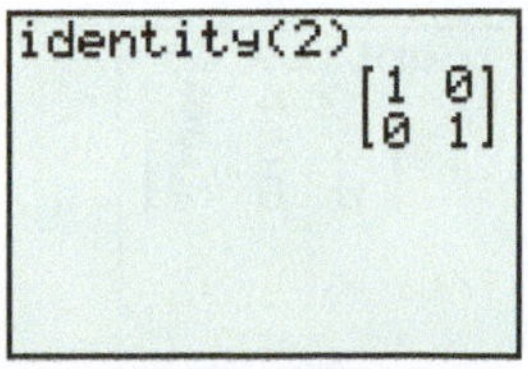

identity(2)
$\begin{bmatrix} 1 & 0 \\ 0 & 1 \end{bmatrix}$

FIGURE 41

---

2 × 2 Identity Matrix

The 2 × 2 identity matrix, written I_2, is defined as follows.

$$I_2 = \begin{bmatrix} 1 & 0 \\ 0 & 1 \end{bmatrix}$$

EXAMPLE 1 **Using the 2 × 2 Identity Matrix**

Let $A = \begin{bmatrix} -2 & 6 \\ 3 & 5 \end{bmatrix}$. Show that $AI_2 = A$ and $I_2A = A$.

Analytic Solution

$$AI_2 = \begin{bmatrix} -2 & 6 \\ 3 & 5 \end{bmatrix}\begin{bmatrix} 1 & 0 \\ 0 & 1 \end{bmatrix}$$

$$= \begin{bmatrix} -2(1) + 6(0) & -2(0) + 6(1) \\ 3(1) + 5(0) & 3(0) + 5(1) \end{bmatrix}$$

$$= \begin{bmatrix} -2 & 6 \\ 3 & 5 \end{bmatrix} = A$$

$$I_2A = \begin{bmatrix} 1 & 0 \\ 0 & 1 \end{bmatrix}\begin{bmatrix} -2 & 6 \\ 3 & 5 \end{bmatrix}$$

$$= \begin{bmatrix} 1(-2) + 0(3) & 1(6) + 0(5) \\ 0(-2) + 1(3) & 0(6) + 1(5) \end{bmatrix}$$

$$= \begin{bmatrix} -2 & 6 \\ 3 & 5 \end{bmatrix} = A$$

Graphing Calculator Solution

Define $[A]$ as shown in **FIGURE 42.** Then show that $AI_2 = A$ and $I_2A = A$.

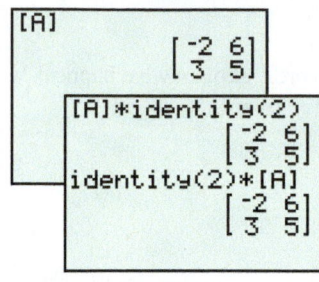

FIGURE 42

FIGURE 43 shows how the TI-84 Plus graphing calculator displays I_3 and I_4.

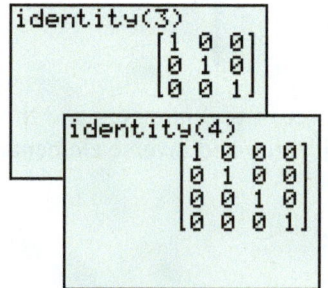

FIGURE 43

n × n Identity Matrix

For any value of n, there is an $n \times n$ identity matrix having 1s down the main diagonal and 0s elsewhere. The **$n \times n$ identity matrix** is

$$I_n = \begin{bmatrix} 1 & 0 & \cdots & 0 \\ 0 & 1 & \cdots & 0 \\ \vdots & \vdots & a_{ij} & \vdots \\ 0 & 0 & \cdots & 1 \end{bmatrix}.$$

Here, $a_{ij} = 1$ when $i = j$ (the diagonal elements), and $a_{ij} = 0$ otherwise.

EXAMPLE 2 **Using the 3 × 3 Identity Matrix**

Let $A = \begin{bmatrix} -2 & 4 & 0 \\ 3 & 5 & 9 \\ 0 & 8 & -6 \end{bmatrix}$. Show that $AI_3 = A$.

Analytic Solution

Using the 3 × 3 identity matrix and the definition of matrix multiplication gives the following.

$$AI_3 = \begin{bmatrix} -2 & 4 & 0 \\ 3 & 5 & 9 \\ 0 & 8 & -6 \end{bmatrix}\begin{bmatrix} 1 & 0 & 0 \\ 0 & 1 & 0 \\ 0 & 0 & 1 \end{bmatrix}.$$

$$= \begin{bmatrix} -2 & 4 & 0 \\ 3 & 5 & 9 \\ 0 & 8 & -6 \end{bmatrix} = A$$

Graphing Calculator Solution

See **FIGURE 44.**

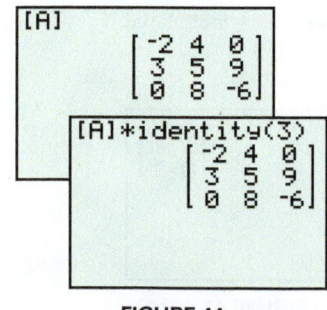

FIGURE 44

> **NOTE** It can also be shown that $I_3A = A$ in **Example 2**.

Multiplicative Inverses of Square Matrices

For every *nonzero* real number a, there is a multiplicative inverse $\frac{1}{a}$ such that

$$a \cdot \frac{1}{a} = 1 \quad \text{and} \quad \frac{1}{a} \cdot a = 1.$$

(Recall: $\frac{1}{a}$ is also written a^{-1}.) In a similar way, if A is an $n \times n$ matrix, then its **multiplicative inverse,** written A^{-1}, must satisfy both

$$AA^{-1} = I_n \quad \text{and} \quad A^{-1}A = I_n.$$

This result means that only a square matrix can have a multiplicative inverse.

> **CAUTION** Although $a^{-1} = \frac{1}{a}$ for any nonzero real number a, if A is a matrix, then
>
> $$A^{-1} \ne \frac{1}{A}.$$
>
> *In fact, $\frac{1}{A}$ has no meaning, since 1 is a number and A is a matrix.*

EXAMPLE 3 Verifying an Inverse

Determine if B is the inverse of A.

$$A = \begin{bmatrix} 5 & 3 \\ -3 & -2 \end{bmatrix} \quad \text{and} \quad B = \begin{bmatrix} 2 & 3 \\ -3 & -5 \end{bmatrix}$$

Solution For B to be the inverse of A, it must satisfy $AB = I_2$ and $BA = I_2$.

$$AB = \begin{bmatrix} 5 & 3 \\ -3 & -2 \end{bmatrix} \begin{bmatrix} 2 & 3 \\ -3 & -5 \end{bmatrix} = \begin{bmatrix} 1 & 0 \\ 0 & 1 \end{bmatrix} = I_2$$

$$BA = \begin{bmatrix} 2 & 3 \\ -3 & -5 \end{bmatrix} \begin{bmatrix} 5 & 3 \\ -3 & -2 \end{bmatrix} = \begin{bmatrix} 1 & 0 \\ 0 & 1 \end{bmatrix} = I_2$$

Thus, B is the inverse of A. That is, $B = A^{-1}$.

The inverse matrix of an $n \times n$ matrix A (if it exists) can be found analytically by first forming the augmented matrix $[A|I_n]$ and then performing matrix row operations, until the left side of the augmented matrix becomes the identity matrix. The resulting augmented matrix can be written as $[I_n|A^{-1}]$, where the right side of the matrix is A^{-1}.

EXAMPLE 4 Finding a 2 × 2 Inverse Matrix

Find A^{-1} for the given matrix A.

$$A = \begin{bmatrix} 2 & 4 \\ 1 & -1 \end{bmatrix}$$

Solution First form the 2×4 augmented matrix $[A|I_2]$. Then use row transformations to obtain the 2×4 augmented matrix $[I_2|A^{-1}]$.

(continued)

$$\begin{bmatrix} 2 & 4 & | & 1 & 0 \\ 1 & -1 & | & 0 & 1 \end{bmatrix} \quad \text{Augmented matrix } [A|I_2]$$

$$\begin{bmatrix} 1 & -1 & | & 0 & 1 \\ 2 & 4 & | & 1 & 0 \end{bmatrix} \quad \begin{array}{l} \text{Interchange } R_1 \text{ and } R_2 \text{ to get 1} \\ \text{in the upper left corner.} \end{array}$$

$$\begin{bmatrix} 1 & -1 & | & 0 & 1 \\ 0 & 6 & | & 1 & -2 \end{bmatrix} \quad -2R_1 + R_2$$

$$\begin{bmatrix} 1 & -1 & | & 0 & 1 \\ 0 & 1 & | & \frac{1}{6} & -\frac{1}{3} \end{bmatrix} \quad \frac{1}{6}R_2$$

Transform the left side into the identity matrix.
$$\begin{bmatrix} 1 & 0 & | & \frac{1}{6} & \frac{2}{3} \\ 0 & 1 & | & \frac{1}{6} & -\frac{1}{3} \end{bmatrix} \quad R_2 + R_1; \text{ augmented matrix } [I_2|A^{-1}]$$
The right side becomes A^{-1}.

Thus, $A^{-1} = \begin{bmatrix} \frac{1}{6} & \frac{2}{3} \\ \frac{1}{6} & -\frac{1}{3} \end{bmatrix}$. To check, show that $AA^{-1} = I_2$ and $A^{-1}A = I_2$.

If A^{-1} exists, then it is unique. If A^{-1} does not exist, then A is a **singular matrix.** The process for finding the multiplicative inverse A^{-1} for any $n \times n$ matrix A that has an inverse is summarized as follows.

Finding an Inverse Matrix

To obtain A^{-1} for any $n \times n$ matrix A for which A^{-1} exists, follow these steps.

Step 1 Form the augmented matrix $[A|I_n]$, where I_n is the $n \times n$ identity matrix.

Step 2 Perform row transformations on $[A|I_n]$, to obtain a matrix of the form $[I_n|B]$.

Step 3 Matrix B is A^{-1}.

NOTE To confirm that two $n \times n$ matrices A and B are inverses of each other, it is sufficient to show that $AB = I_n$. It is not necessary to show also that $BA = I_n$.

EXAMPLE 5 **Finding the Inverse of a 3 × 3 Matrix**

Find A^{-1} if $A = \begin{bmatrix} 1 & 0 & 1 \\ 2 & -2 & -1 \\ 3 & 0 & 0 \end{bmatrix}$.

Solution Use row transformations as follows.

Step 1 Write the augmented matrix $[A|I_3]$.

Write A on the left side.
$$\begin{bmatrix} 1 & 0 & 1 & | & 1 & 0 & 0 \\ 2 & -2 & -1 & | & 0 & 1 & 0 \\ 3 & 0 & 0 & | & 0 & 0 & 1 \end{bmatrix}$$
Write I_3 on the right side.

Step 2 Since 1 is already in the upper left-hand corner, use the row transformation that will result in 0 for the first element in the second row. Multiply the elements of the first row by -2, and add the result to the second row.

$$\begin{bmatrix} 1 & 0 & 1 & | & 1 & 0 & 0 \\ 0 & -2 & -3 & | & -2 & 1 & 0 \\ 3 & 0 & 0 & | & 0 & 0 & 1 \end{bmatrix} \quad -2R_1 + R_2$$

To get 0 for the first element in the third row, multiply the elements of the first row by -3 and add the result to the third row.

$$\begin{bmatrix} 1 & 0 & 1 & | & 1 & 0 & 0 \\ 0 & -2 & -3 & | & -2 & 1 & 0 \\ 0 & 0 & -3 & | & -3 & 0 & 1 \end{bmatrix} \quad -3R_1 + R_3$$

Continuing this process, we transform the matrix as follows.

$$\begin{bmatrix} 1 & 0 & 1 & | & 1 & 0 & 0 \\ 0 & 1 & \frac{3}{2} & | & 1 & -\frac{1}{2} & 0 \\ 0 & 0 & -3 & | & -3 & 0 & 1 \end{bmatrix} \quad -\frac{1}{2}R_2$$

$$\begin{bmatrix} 1 & 0 & 1 & | & 1 & 0 & 0 \\ 0 & 1 & \frac{3}{2} & | & 1 & -\frac{1}{2} & 0 \\ 0 & 0 & 1 & | & 1 & 0 & -\frac{1}{3} \end{bmatrix} \quad -\frac{1}{3}R_3$$

$$\begin{bmatrix} 1 & 0 & 0 & | & 0 & 0 & \frac{1}{3} \\ 0 & 1 & \frac{3}{2} & | & 1 & -\frac{1}{2} & 0 \\ 0 & 0 & 1 & | & 1 & 0 & -\frac{1}{3} \end{bmatrix} \quad -1R_3 + R_1$$

$$\begin{bmatrix} 1 & 0 & 0 & | & 0 & 0 & \frac{1}{3} \\ 0 & 1 & 0 & | & -\frac{1}{2} & -\frac{1}{2} & \frac{1}{2} \\ 0 & 0 & 1 & | & 1 & 0 & -\frac{1}{3} \end{bmatrix} \quad -\frac{3}{2}R_3 + R_2$$

Identity matrix on left

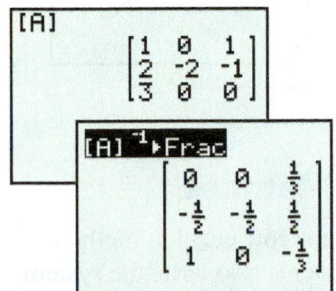

[A]
$$\begin{bmatrix} 1 & 0 & 1 \\ 2 & -2 & -1 \\ 3 & 0 & 0 \end{bmatrix}$$

[A]$^{-1}$▶Frac
$$\begin{bmatrix} 0 & 0 & \frac{1}{3} \\ -\frac{1}{2} & -\frac{1}{2} & \frac{1}{2} \\ 1 & 0 & -\frac{1}{3} \end{bmatrix}$$

A graphing calculator can be used to find the inverse of a matrix, as shown above. The screens support the result in **Example 5.** The elements of the inverse are expressed as fractions, so it is easier to compare with the inverse matrix found in the example.

Step 3 The last transformation shows that the inverse is as follows.

$$A^{-1} = \begin{bmatrix} 0 & 0 & \frac{1}{3} \\ -\frac{1}{2} & -\frac{1}{2} & \frac{1}{2} \\ 1 & 0 & -\frac{1}{3} \end{bmatrix}$$

Confirm this result by forming the product $A^{-1}A$ or AA^{-1}, each of which equals the matrix I_3. ●

Using Determinants to Find Inverses

Determinants can be used to find the inverses of matrices.

> **Inverse of a 2 × 2 Matrix**
>
> If $A = \begin{bmatrix} a & b \\ c & d \end{bmatrix}$ and det $A \neq 0$, then $A^{-1} = \dfrac{1}{\det A} \begin{bmatrix} d & -b \\ -c & a \end{bmatrix}$.

If det $A = 0$, then A^{-1} does not exist and A is a singular matrix.

EXAMPLE 6 **Finding the Inverse of a 2 × 2 Matrix**

Use a determinant to find A^{-1}, if it exists, for each matrix.

(a) $A = \begin{bmatrix} 2 & 3 \\ 1 & -1 \end{bmatrix}$ **(b)** $A = \begin{bmatrix} 3 & -6 \\ 2 & -4 \end{bmatrix}$

Analytic Solution

(a) First we find $\det A$ and determine whether A^{-1} actually exists.

$$\det A = 2(-1) - 1(3)$$
$$= -5$$

Since $-5 \neq 0$, A^{-1} exists. Thus, from the box on the preceding page, we have the following.

$$A^{-1} = \frac{1}{-5}\begin{bmatrix} -1 & -3 \\ -1 & 2 \end{bmatrix}$$

$$= \begin{bmatrix} \frac{1}{5} & \frac{3}{5} \\ \frac{1}{5} & -\frac{2}{5} \end{bmatrix}$$

(b) Here, A^{-1} does not exist, since $\det A = 0$.

$$\det A = 3(-4) - 2(-6)$$
$$= 0 \qquad \textcolor{blue}{A \text{ is singular.}}$$

Graphing Calculator Solution

(a) See **FIGURE 45**, which illustrates both $[A]$ and $[A]^{-1}$.

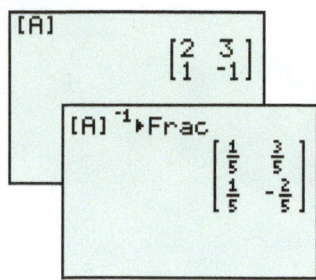

FIGURE 45

(b) The calculator returns a singular matrix error message when it is directed to find the inverse of a matrix whose determinant is 0. See **FIGURE 46**.

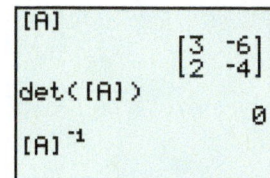 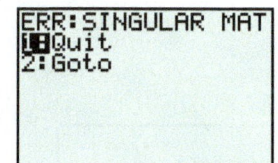

FIGURE 46

Solving Linear Systems Using Inverse Matrices

We used matrices to solve systems of linear equations by the row echelon method in **Section 6.3.** Another way to use matrices to solve linear systems is to write the system as a matrix equation $AX = B$, where A is the matrix of the coefficients of the variables of the system, X is the column matrix of the variables, and B is the column matrix of the constants. Matrix A is called the **coefficient matrix.**

To solve the matrix equation $AX = B$, first see if A^{-1} exists. Assuming that it does, use the facts that $A^{-1}A = I$ and $IX = X$.

$$AX = B$$
$$A^{-1}(AX) = A^{-1}B \qquad \textcolor{blue}{\text{Multiply each side on the left by } A^{-1}.}$$
$$(A^{-1}A)X = A^{-1}B \qquad \textcolor{blue}{\text{Associative property}}$$
$$IX = A^{-1}B \qquad \textcolor{blue}{\text{Multiplicative inverse property}}$$
$$X = A^{-1}B \qquad \textcolor{blue}{\text{Identity property}}$$

CAUTION *When multiplying by matrices on each side of a matrix equation, be careful to multiply in the same order on each side.*

EXAMPLE 7 **Using a Matrix Inverse to Solve a System of Equations**

Use the inverse of the coefficient matrix to solve the system.

$$2x + y + 3z = 1$$
$$x - 2y + z = -3$$
$$-3x + y - 2z = -4$$

Analytic Solution

We find the determinant of the coefficient matrix to be sure that it is not 0. If

$$A = \begin{bmatrix} 2 & 1 & 3 \\ 1 & -2 & 1 \\ -3 & 1 & -2 \end{bmatrix},$$

then $\det A = -10 \neq 0$ and A^{-1} exists. We evaluate $A^{-1}B$ with

$$X = \begin{bmatrix} x \\ y \\ z \end{bmatrix} \quad \text{and} \quad B = \begin{bmatrix} 1 \\ -3 \\ -4 \end{bmatrix}.$$

$$A^{-1}B = \begin{bmatrix} 2 & 1 & 3 \\ 1 & -2 & 1 \\ -3 & 1 & -2 \end{bmatrix}^{-1} \begin{bmatrix} 1 \\ -3 \\ -4 \end{bmatrix}$$

$$= \begin{bmatrix} -\frac{3}{10} & -\frac{1}{2} & -\frac{7}{10} \\ \frac{1}{10} & -\frac{1}{2} & -\frac{1}{10} \\ \frac{1}{2} & \frac{1}{2} & \frac{1}{2} \end{bmatrix} \begin{bmatrix} 1 \\ -3 \\ -4 \end{bmatrix} = \begin{bmatrix} 4 \\ 2 \\ -3 \end{bmatrix}$$

Since $X = A^{-1}B$, we have $x = 4$, $y = 2$, and $z = -3$. The solution set is $\{(4, 2, -3)\}$. The elements of A^{-1} can be found by hand, as demonstrated in **Example 5,** or by using a calculator.

Graphing Calculator Solution

Entering the coefficient matrix as $[A]$, we find that $\det([A]) = -10$ and $[A]^{-1}$ exists. See **FIGURE 47.** Then we enter the column of constants as $[B]$ and find the matrix product $[A]^{-1}[B]$. The display shows the column matrix representing $x = 4$, $y = 2$, and $z = -3$.

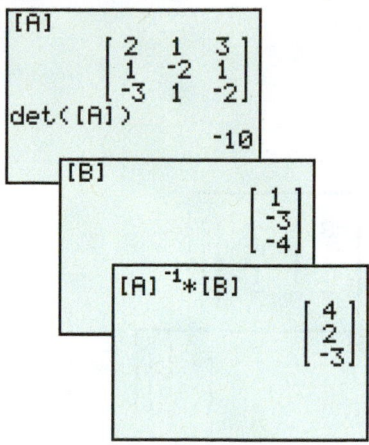

FIGURE 47

The solution set is $\{(4, 2, -3)\}$.

NOTE If the determinant of the coefficient matrix A is 0, the system either is inconsistent or has dependent equations and A^{-1} does not exist.

FOR DISCUSSION

1. Using the window $[-5, 5]$ by $[-50, 50]$, reproduce the graph in **FIGURE 48,** based on the result of **Example 8.**

2. If your calculator has *cubic regression,* use it to find the same results as in **Example 8.**

Fitting Data Using a System

EXAMPLE 8 **Using a System to Find the Equation of a Cubic Polynomial**

FIGURE 48 shows four views of the graph of a polynomial function of the form

$$P(x) = ax^3 + bx^2 + cx + d.$$

Use the points indicated to write a system of four equations in the variables a, b, c, and d, and then use the inverse matrix method to solve the system. What is the equation that defines this graph?

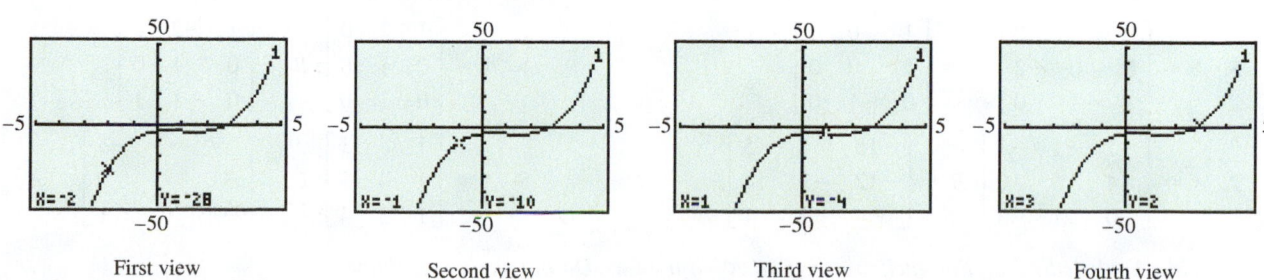

| First view | Second view | Third view | Fourth view |

FIGURE 48

(continued)

Solution We see from the graph in **FIGURE 48** on the previous page that $P(-2) = -28$, $P(-1) = -10$, $P(1) = -4$, and $P(3) = 2$. From the first of these, we get the following.

$$P(-2) = a(-2)^3 + b(-2)^2 + c(-2) + d = -28$$

or, equivalently, $\qquad -8a + 4b - 2c + d = -28$

Similarly, from the others, we find the following equations.

From $P(-1) = -10$: $\qquad -a + b - c + d = -10.$

From $P(1) = -4$: $\qquad a + b + c + d = -4.$

From $P(3) = 2$: $\qquad 27a + 9b + 3c + d = 2.$

Now we must solve the system formed by these four equations.

$$-8a + 4b - 2c + d = -28$$
$$-a + b - c + d = -10$$
$$a + b + c + d = -4$$
$$27a + 9b + 3c + d = 2$$

We use the inverse matrix method to solve the system with the following matrices.

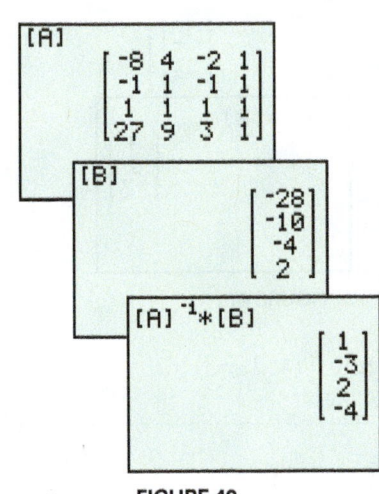

$$A = \begin{bmatrix} -8 & 4 & -2 & 1 \\ -1 & 1 & -1 & 1 \\ 1 & 1 & 1 & 1 \\ 27 & 9 & 3 & 1 \end{bmatrix}, \quad X = \begin{bmatrix} a \\ b \\ c \\ d \end{bmatrix}, \quad \text{and} \quad B = \begin{bmatrix} -28 \\ -10 \\ -4 \\ 2 \end{bmatrix}$$

Because det $A = 240 \neq 0$, a unique solution exists. On the basis of our earlier discussion, $X = A^{-1}B$. **FIGURE 49** shows how the matrix inverse method yields the values $a = 1$, $b = -3$, $c = 2$, and $d = -4$. The polynomial is

$$P(x) = x^3 - 3x^2 + 2x - 4.$$

FIGURE 49

6.6 Exercises

Checking Analytic Skills   Determine whether A and B are inverses by calculating AB and BA. **Do not use a calculator.**

1. $A = \begin{bmatrix} 5 & 7 \\ 2 & 3 \end{bmatrix}; B = \begin{bmatrix} 3 & -7 \\ -2 & 5 \end{bmatrix}$

2. $A = \begin{bmatrix} 2 & 3 \\ 1 & 1 \end{bmatrix}; B = \begin{bmatrix} -1 & 3 \\ 1 & -2 \end{bmatrix}$

3. $A = \begin{bmatrix} -1 & 2 \\ 3 & -5 \end{bmatrix}; B = \begin{bmatrix} -5 & -2 \\ -3 & -1 \end{bmatrix}$

4. $A = \begin{bmatrix} 2 & 1 \\ 3 & 2 \end{bmatrix}; B = \begin{bmatrix} 2 & 1 \\ -3 & 2 \end{bmatrix}$

5. $A = \begin{bmatrix} 0 & 1 & 0 \\ 0 & 0 & -2 \\ 1 & -1 & 0 \end{bmatrix}; B = \begin{bmatrix} 1 & 0 & 1 \\ 1 & 0 & 0 \\ 0 & -1 & 0 \end{bmatrix}$

6. $A = \begin{bmatrix} 1 & 2 & 0 \\ 0 & 1 & 0 \\ 0 & 1 & 0 \end{bmatrix}; B = \begin{bmatrix} 1 & -2 & 0 \\ 0 & 1 & 0 \\ 0 & -1 & 1 \end{bmatrix}$

7. $A = \begin{bmatrix} -1 & -1 & -1 \\ 4 & 5 & 0 \\ 0 & 1 & -3 \end{bmatrix}; B = \begin{bmatrix} 15 & 4 & -5 \\ -12 & -3 & 4 \\ -4 & -1 & 1 \end{bmatrix}$

8. $A = \begin{bmatrix} 1 & 3 & 3 \\ 1 & 4 & 3 \\ 1 & 3 & 4 \end{bmatrix}; B = \begin{bmatrix} 7 & -3 & -3 \\ -1 & 1 & 0 \\ -1 & 0 & 1 \end{bmatrix}$

Checking Analytic Skills   For each matrix, find A^{-1} if it exists. **Do not use a calculator.**

9. $A = \begin{bmatrix} 3 & 7 \\ 2 & 5 \end{bmatrix}$

10. $A = \begin{bmatrix} -5 & 3 \\ -8 & 5 \end{bmatrix}$

11. $A = \begin{bmatrix} -1 & -2 \\ 3 & 4 \end{bmatrix}$

12. $A = \begin{bmatrix} -1 & 2 \\ -2 & -1 \end{bmatrix}$

13. $A = \begin{bmatrix} -6 & 4 \\ -3 & 2 \end{bmatrix}$

14. $A = \begin{bmatrix} 5 & 10 \\ -3 & -6 \end{bmatrix}$

15. $A = \begin{bmatrix} 0.6 & 0.2 \\ 0.5 & 0.1 \end{bmatrix}$

16. $A = \begin{bmatrix} 0.8 & -0.3 \\ 0.5 & -0.2 \end{bmatrix}$

17. $A = \begin{bmatrix} 0 & 0 & 1 \\ 1 & 0 & 0 \\ 0 & 1 & 0 \end{bmatrix}$

18. $A = \begin{bmatrix} 1 & 0 & 0 \\ 1 & 1 & 0 \\ 0 & 1 & 1 \end{bmatrix}$

19. $A = \begin{bmatrix} 1 & 0 & 1 \\ 2 & 1 & 3 \\ -1 & 1 & 1 \end{bmatrix}$

20. $A = \begin{bmatrix} -2 & 1 & 0 \\ 1 & 0 & 1 \\ -1 & 1 & 0 \end{bmatrix}$

21. Under what condition will the inverse of a square matrix not exist?

22. Explain why a 2×2 matrix will not have an inverse if either a column or a row contains all 0s.

For each matrix, find A^{-1} if it exists.

23. $A = \begin{bmatrix} 1 & 0 & 0 \\ 0 & -1 & 0 \\ 1 & 0 & 1 \end{bmatrix}$

24. $A = \begin{bmatrix} 1 & 3 & 3 \\ 1 & 4 & 3 \\ 1 & 3 & 4 \end{bmatrix}$

25. $A = \begin{bmatrix} -1 & -1 & -1 \\ 4 & 5 & 0 \\ 0 & 1 & -3 \end{bmatrix}$

26. $A = \begin{bmatrix} 2 & 0 & 4 \\ 3 & 1 & 5 \\ -1 & 1 & -2 \end{bmatrix}$

27. $A = \begin{bmatrix} -0.4 & 0.1 & 0.2 \\ 0 & 0.6 & 0.8 \\ 0.3 & 0 & -0.2 \end{bmatrix}$

28. $A = \begin{bmatrix} 0.8 & 0.2 & 0.1 \\ -0.2 & 0 & 0.3 \\ 0 & 0 & 0.5 \end{bmatrix}$

29. $A = \begin{bmatrix} 2 & 1 & 2 \\ 5 & 10 & 5 \\ 3 & 6 & 3 \end{bmatrix}$

30. $A = \begin{bmatrix} 5 & -3 & 2 \\ -5 & 3 & -2 \\ 1 & 0 & 1 \end{bmatrix}$

31. $A = \begin{bmatrix} \sqrt{2} & 0.5 \\ -17 & \frac{1}{2} \end{bmatrix}$

32. $A = \begin{bmatrix} \frac{2}{3} & 0.7 \\ 22 & \sqrt{3} \end{bmatrix}$

33. $A = \begin{bmatrix} 0.1 & 0 & 0.1 \\ 0.2 & 0.1 & 0.3 \\ -0.1 & 0.1 & 0.1 \end{bmatrix}$

34. $A = \begin{bmatrix} \frac{1}{2} & \frac{1}{4} & \frac{1}{3} \\ 0 & \frac{1}{4} & \frac{1}{3} \\ \frac{1}{2} & \frac{1}{2} & \frac{1}{3} \end{bmatrix}$

Solve each system by using the matrix inverse method.

35. $\begin{aligned} 2x - y &= -8 \\ 3x + y &= -2 \end{aligned}$

36. $\begin{aligned} x + 3y &= -12 \\ 2x - y &= 11 \end{aligned}$

37. $\begin{aligned} 2x + 3y &= -10 \\ 3x + 4y &= -12 \end{aligned}$

38. $\begin{aligned} 2x - 3y &= 10 \\ 2x + 2y &= 5 \end{aligned}$

39. $\begin{aligned} 2x - 5y &= 10 \\ 2x - 5y &= 15 \end{aligned}$

40. $\begin{aligned} 2x - 3y &= 2 \\ 4x - 6y &= 1 \end{aligned}$

41. $\begin{aligned} 2x + 4z &= 14 \\ 3x + y + 5z &= 19 \\ -x + y - 2z &= -7 \end{aligned}$

42. $\begin{aligned} 3x + 6y + 3z &= 12 \\ 6x + 4y - 2z &= -4 \\ y - z &= -3 \end{aligned}$

43. $\begin{aligned} x + 3y + z &= 2 \\ x - 2y + 3z &= -3 \\ 2x - 3y - z &= 34 \end{aligned}$

44. $\begin{aligned} x + y - z &= 6 \\ 2x - y + z &= -9 \\ x - 2y + 3z &= 1 \end{aligned}$

45. $\begin{aligned} x + 3y - 2z - w &= 9 \\ 4x + y + z + 2w &= 2 \\ -3x - y + z - w &= -5 \\ x - y - 3z - 2w &= 2 \end{aligned}$

46. $\begin{aligned} 3x + 2y - w &= 0 \\ 2x + z + 2w &= 5 \\ x + 2y - z &= -2 \\ 2x - y + z + w &= 2 \end{aligned}$

47. $\begin{aligned} x - \sqrt{2}y &= 2.6 \\ 0.75x + y &= -7 \end{aligned}$

48. $\begin{aligned} 2.1x + y &= \sqrt{5} \\ \sqrt{2}x - 2y &= 5 \end{aligned}$

49. $\pi x + ey + \sqrt{2}z = 1$
$ex + \pi y + \sqrt{2}z = 2$
$\sqrt{2}x + ey + \pi z = 3$

50. $(\log 2)x + (\ln 3)y + (\ln 4)z = 1$
$(\ln 3)x + (\log 2)y + (\ln 8)z = 5$
$(\log 12)x + (\ln 4)y + (\ln 8)z = 9$

Fitting Data In Exercises 51 and 52, use the method of **Example 8** to find the cubic polynomial $P(x)$ that defines the curve shown in the four figures.

51.

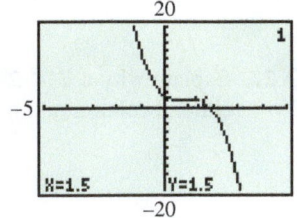

First view   Second view   Third view   Fourth view

52.

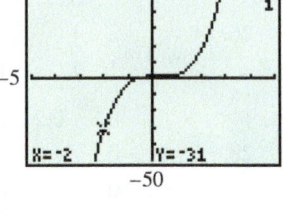

First view   Second view   Third view   Fourth view

53. *Fitting Data* Find the fourth-degree polynomial $P(x)$ satisfying the following conditions: $P(-2) = 13$, $P(-1) = 2$, $P(0) = -1$, $P(1) = 4$, and $P(2) = 41$.

54. *Fitting Data* Find the fifth-degree polynomial $P(x)$ satisfying the following conditions: $P(-2) = -8$, $P(-1) = -1$, $P(0) = -4$, $P(1) = -5$, $P(2) = 8$, and $P(3) = 167$.

(Modeling) Solve each problem.

55. *Cost of CDs* A music store has used compact discs that sell for three prices, marked A, B, and C. The last column in the table shows the total cost of a purchase. Use this information to determine the cost of one CD of each type by setting up a matrix equation and solving it with an inverse.

A	B	C	Total
2	3	4	$120.91
1	4	0	$ 62.95
2	1	3	$ 79.94

56. *Determining Prices* A group of students bought 3 soft drinks and 2 boxes of popcorn at a movie for $18.50. A second group bought 4 soft drinks and 3 boxes of popcorn for $26.

(a) Find a matrix equation $AX = B$ whose solution gives the individual prices of a soft drink and a box of popcorn. Solve this matrix equation by using A^{-1}.

(b) Could these prices be determined if both groups had bought 3 soft drinks and 2 boxes of popcorn for $18.50? Try to calculate A^{-1} and explain your results.

57. *Tire Sales* In a study, the relationship among annual tire sales T in thousands, automobile registrations A in millions, and personal disposable income I in millions of dollars was investigated. Representative data for three different years are shown in the table.

T	A	I
10,170	113	308
15,305	133	622
21,289	155	1937

Source: Jarrett, J., *Business Forecasting Methods,* Basil Blackwell.

The data were modeled by

$$T = aA + bI + c,$$

where a, b, and c are constants.

(a) Use the data to write a system of linear equations whose solution gives a, b, and c.

(b) Solve this linear system. Write a formula for T.

(c) If $A = 118$ and $I = 311$, predict T. (The actual value for T was 11,314.)

58. *Plate Glass Sales* The amount of plate glass G sold can be affected by the number of new building contracts B issued and automobiles A produced, since plate glass is used in buildings and cars. A plate glass company in California wanted to forecast sales. The table contains sales data for three consecutive years. All units are in millions.

G	A	B
603	5.54	37.1
657	6.93	41.3
779	7.64	45.6

Source: Makridakis, S. and S. Wheelwright, *Forecasting Methods for Management,* John Wiley and Sons.

The data were modeled by

$$G = aA + bB + c,$$

where a, b, and c are constants.

(a) Write a system of linear equations whose solution gives a, b, and c.

(b) Solve this linear system. Write a formula for G.

(c) For the following year, it was estimated that $A = 7.75$ and $B = 47.4$. Predict G. (The actual value for G was $878 million.)

59. *Home Prices* Realty companies study how the selling price of a home is related to its size and condition. The table in the next column contains sales data for three homes. Price P is measured in thousands of dollars, home size S is in square feet, and condition C is rated on a scale from 1 to 10, where 10 represents excellent condition.

P	S	C
122	1500	8
130	2000	5
158	2200	10

The data were related by

$$P = a + bS + cC,$$

where a, b, and c are constants.

(a) Use the data to write a system of linear equations whose solution gives a, b, and c.

(b) Estimate the selling price of a home with 1800 square feet and in condition 7.

60. *Traffic Flow* Refer to **Exercises 77** and **78** in **Section 6.3.** The figure shows four one-way streets with intersections A, B, C, and D. Numbers indicate the average traffic flow in vehicles per minute. The variables x_1, x_2, x_3, and x_4 denote unknown traffic flows.

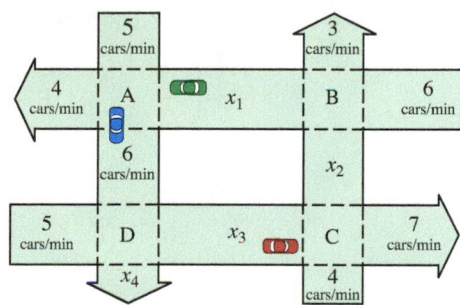

(a) The number of vehicles per minute entering an intersection equals the number exiting an intersection. Verify that the given system of linear equations describes the traffic flow.

$$A:\ x_1 + 5 = 4 + 6$$
$$B:\ x_2 + 6 = x_1 + 3$$
$$C:\ x_3 + 4 = x_2 + 7$$
$$D:\ 6 + 5 = x_3 + x_4$$

(b) Write the system as $AX = B$, and solve by using A^{-1}.

(c) Interpret your results.

In Exercises 61–64, given a square matrix A^{-1}, find matrix A.

61. $A^{-1} = \begin{bmatrix} 5 & -9 \\ -1 & 2 \end{bmatrix}$

62. $A^{-1} = \begin{bmatrix} \frac{3}{20} & \frac{1}{4} \\ -\frac{1}{20} & \frac{1}{4} \end{bmatrix}$

63. $A^{-1} = \begin{bmatrix} \frac{2}{3} & -\frac{1}{3} & 0 \\ \frac{1}{3} & -\frac{5}{3} & 1 \\ \frac{1}{3} & \frac{1}{3} & 0 \end{bmatrix}$

64. $A^{-1} = \begin{bmatrix} 0 & 0 & 1 \\ 0 & 1 & 0 \\ 1 & 0 & 0 \end{bmatrix}$

65. Let $A = \begin{bmatrix} a & 0 & 0 \\ 0 & b & 0 \\ 0 & 0 & c \end{bmatrix}$, where a, b, and c are nonzero real numbers. Find A^{-1}.

66. Let $A = \begin{bmatrix} 1 & 0 & 0 \\ 0 & 0 & -1 \\ 0 & 1 & -1 \end{bmatrix}$. Show that $A^3 = I_3$, and use this result to find the inverse of A.

SECTIONS 6.4–6.6   Reviewing Basic Concepts

Find the following, if possible, given matrices

$$A = \begin{bmatrix} -5 & 4 \\ 2 & -1 \end{bmatrix}, \quad B = \begin{bmatrix} 0 & -2 \\ 3 & -4 \end{bmatrix}, \quad C = \begin{bmatrix} 2 & -3 & 1 \\ -2 & 1 & 0 \\ 0 & -1 & 4 \end{bmatrix}, \quad D = \begin{bmatrix} 0 & 4 & 1 \\ 0 & 2 & 2 \\ 1 & 1 & 1 \end{bmatrix}.$$

1. $A - B$

2. $-3B$

3. A^2

4. CD

5. $\det A$

6. $\det C$

7. A^{-1}

8. C^{-1}

9. *Roof Trusses*   Linear systems occur in the design of roof trusses for new homes and buildings. The simplest type of roof truss is a triangle. The truss shown in the figure is used to frame roofs of small buildings. If a 100-pound force is applied at the peak of the truss, then the forces or weights W_1 and W_2 exerted parallel to each rafter of the truss are determined by the following linear system of equations:

$$\frac{\sqrt{3}}{2}(W_1 + W_2) = 100$$
$$W_1 - W_2 = 0.$$

Solve the system, using Cramer's rule to find W_1 and W_2. (*Source:* Hibbeler, R., *Structural Analysis*, Prentice-Hall.)

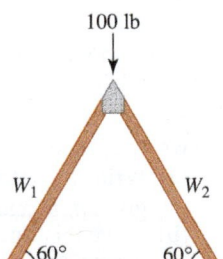

10. Solve the system, using the inverse matrix method.

$$\begin{aligned} 2x + y + 2z &= 10 \\ y + 2z &= 4 \\ x - 2y + 2z &= 1 \end{aligned}$$

6.7 Systems of Inequalities and Linear Programming

Solving Linear Inequalities • Solving Systems of Inequalities • Linear Programming

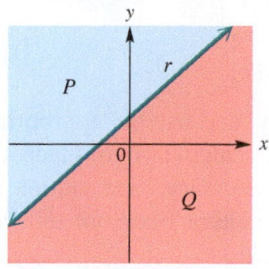

Line r divides the plane into three different sets of points: line r, half plane P, and half plane Q. The points on r belong neither to P nor to Q.

FIGURE 50

Many mathematical descriptions of real situations are best expressed as inequalities rather than equations. For example, a firm might be able to use a machine *no more than* 12 hours a day, while the production of *at least* 500 cases of a certain product might be required to meet a contract. The simplest way to see the solution of an inequality in two variables is to draw its graph.

A line divides a plane into three sets of points: the points of the line itself and the points belonging to the two regions determined by the line. Each of these two regions is called a **half plane.** The line is the **boundary** of each half plane. See **FIGURE 50.**

Solving Linear Inequalities

> **Linear Inequality in Two Variables**
>
> A **linear inequality in two variables** is an inequality of the form
>
> $$Ax + By \leq C,$$
>
> where A, B, and C are real numbers with A and B not both equal to 0. (The symbol $\leq$ could be replaced with $\geq$, $<$, or $>$.)

The graph of a linear inequality is a half plane, perhaps with its boundary. For example, to graph the linear inequality

$$3x - 2y \leq 6,$$

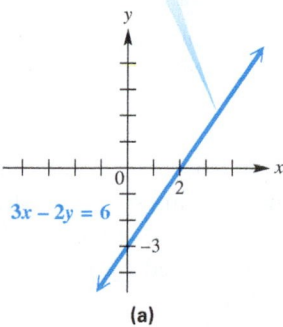

Graph the boundary $3x - 2y = 6$ first.

(a)

first graph the boundary, $3x - 2y = 6$, as shown in **FIGURE 51(a).** Since the points of the line $3x - 2y = 6$ satisfy $3x - 2y \leq 6$ (equality is included), this line is part of the solution. To decide which half plane is part of the solution, solve the original inequality for y.

Reverse the inequality symbol when multiplying by a negative number.

$$3x - 2y \leq 6$$
$$-2y \leq -3x + 6 \qquad \text{Subtract } 3x.$$
$$y \geq \frac{3}{2}x - 3 \qquad \text{Multiply by } -\tfrac{1}{2} \text{ and change } \leq \text{ to } \geq.$$

For a particular value of x, the inequality will be satisfied by all values of y that are *greater than* or equal to $\frac{3}{2}x - 3$. Thus, the solution contains the half plane *above* the line, as shown in **FIGURE 51(b).**

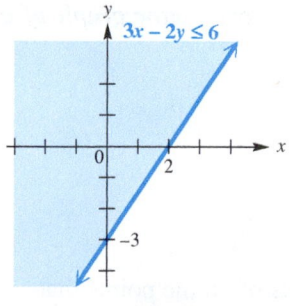

(b)

FIGURE 51

> **CAUTION** *A linear inequality must be in slope–intercept form (solved for y) to determine, from the presence of a $<$ symbol or a $>$ symbol, whether to shade the lower or upper half plane.* In **FIGURE 51(b)**, the upper half plane is shaded, even though the inequality is $3x - 2y \leq 6$ (with a $<$ symbol) in standard form. When we write the inequality as $y \geq \frac{3}{2}x - 3$ (slope–intercept form), the $>$ symbol indicates to shade the upper half plane.

> **NOTE** The point $(0, 0)$ satisfies the given inequality because $3(0) - 2(0) \leq 6$. Thus the **test point** $(0, 0)$ must be included in the shaded region in **FIGURE 51(b).**

> **EXAMPLE 1** **Graphing a Linear Inequality**
>
> Graph $x + 4y > 4$.

Analytic Solution

The boundary here is the line $x + 4y = 4$. Since the points on this line do not satisfy $x + 4y > 4$, make the line dashed, as in **FIGURE 52**. To decide which half plane represents the solution, solve for y.

$$x + 4y > 4$$
$$4y > -x + 4 \qquad \text{Subtract } x.$$
$$y > -\frac{1}{4}x + 1 \qquad \text{Divide by 4.}$$

Since y *is greater than* $-\frac{1}{4}x + 1$, the graph of the solution set is the half plane *above* the boundary, as shown in **FIGURE 52.**

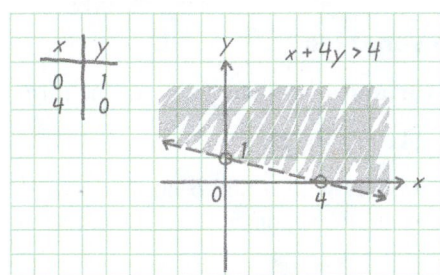

FIGURE 52

Alternatively, or as a check, choose a test point *not* on the boundary line and substitute into the inequality.

$$x + 4y > 4 \qquad \text{Original inequality}$$

Using (0, 0) makes the substitution easy. → $0 + 4(0) > 4 \qquad \text{Substitute a test point.}$

$$0 > 4 \qquad \text{False}$$

Since the test point $(0, 0)$ does not satisfy this inequality and is *below* the boundary, the points that satisfy the inequality must be *above* the boundary.

Graphing Calculator Solution

Solve the corresponding linear equation (the boundary) for y to get

$$y = -\frac{1}{4}x + 1.$$

Graph this boundary, and use the appropriate commands for your calculator to shade the region above the boundary. See **FIGURE 53.** (Notice that the calculator does not tell you which region to shade.)

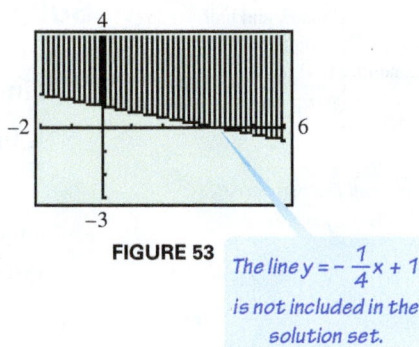

FIGURE 53   *The line $y = -\frac{1}{4}x + 1$ is not included in the solution set.*

The calculator graph does *not* distinguish between solid boundary lines and dashed boundary lines. Whether the boundary is solid or dashed must be determined by looking at the inequality symbol.

We must understand the mathematics to correctly interpret a calculator graph of an inequality.

Two Methods for Graphing an Inequality

1. For a function f, the graph of $y < f(x)$ consists of all the points that are *below* the graph of $y = f(x)$. The graph of $y > f(x)$ consists of all the points that are *above* the graph of $y = f(x)$. (Similar statements can be made for $\leq$ and $\geq$, with boundaries included.)

2. If the inequality is not or cannot be solved for y, choose a test point not on the boundary. If the test point satisfies the inequality, the graph includes all points on the same side of the boundary as the test point. Otherwise, the graph includes all points on the other side of the boundary.

Solving Systems of Inequalities

The solution set of a **system of inequalities** is the intersection of the solution sets of its members.

$$x > 6 - 2y$$
$$x^2 < 2y$$

System of inequalities

EXAMPLE 2 Graphing a System of Two Inequalities

Graph the solution set of the system.

$$x > 6 - 2y$$
$$x^2 < 2y$$

Analytic Solution

FIGURES 54 and **55** show the graphs of $x > 6 - 2y$ and $x^2 < 2y$. The methods of **Section 6.1** can be used to show that the boundaries intersect at the points $(2, 2)$ and $\left(-3, \frac{9}{2}\right)$. The solution set of the system is shown in **FIGURE 56.** Since the points on the boundaries of $x > 6 - 2y$ and $x^2 < 2y$ do not belong to the graph of the solution, the boundaries are dashed.

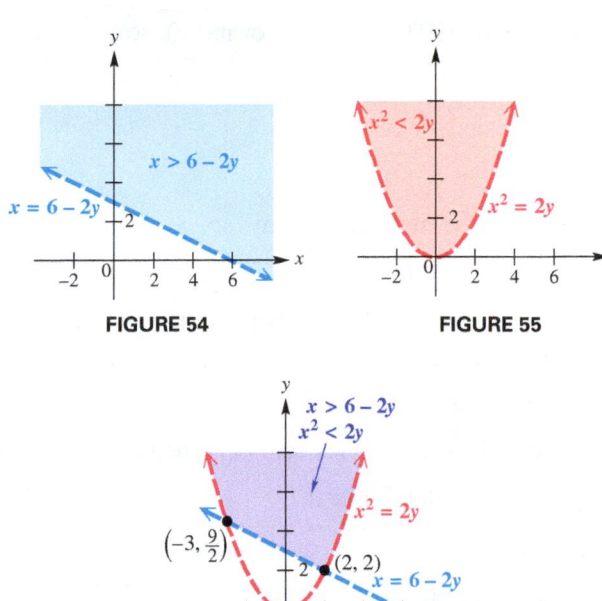

FIGURE 54   **FIGURE 55**

FIGURE 56

Graphing Calculator Solution

As usual, solve each inequality for y first. Enter the boundary equations,

$$y_1 = \frac{6 - x}{2} \quad \text{and} \quad y_2 = \frac{x^2}{2}.$$

In both inequalities,

$$y_1 > \frac{6 - x}{2} \quad \text{and} \quad y_2 > \frac{x^2}{2},$$

y is greater than an expression involving x. Therefore, use the capability of your calculator to shade *above* each boundary. See **FIGURE 57.**

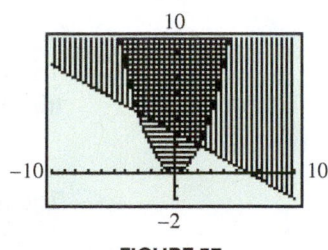

FIGURE 57

The crosshatched region in **FIGURE 57** supports the solution shown in **FIGURE 56.** The boundary lines are *not* included in the solution set.

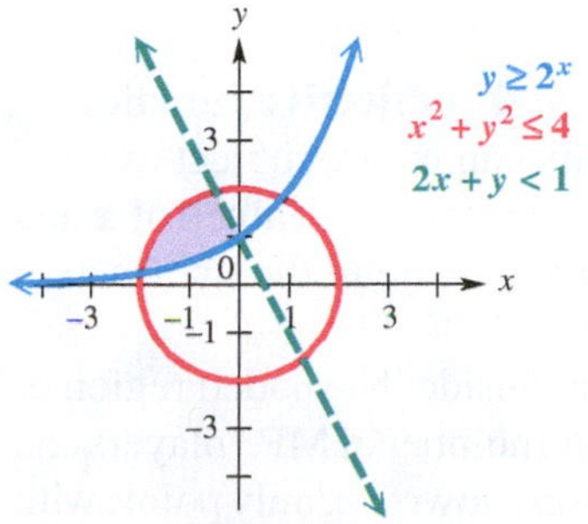

EXAMPLE 3 Graphing a System of Three Inequalities

Graph the solution set of the system.

$$y \geq 2^x$$
$$x^2 + y^2 \leq 4$$
$$2x + y < 1$$

Solution Graph the three inequalities on the same axes and shade the region common to all three. See **FIGURE 58.** Two boundary lines are solid and one is dashed.

FIGURE 58

Linear Programming

An important application of mathematics to business and social science is called *linear programming*. We use **linear programming** to find an optimum value—for example, minimum cost or maximum profit. Linear programming was first developed to solve problems in allocating supplies for the U.S. Air Force during World War II.

EXAMPLE 4 **Finding a Maximum Profit Model**

A company makes two products: MP3 players and Blu-ray players. Each MP3 player gives a profit of $30, while each Blu-ray player produces a $70 profit. The company must manufacture at least 10 MP3 players per day to satisfy one of its customers, but no more than 50, because of production restrictions. The number of Blu-ray players produced cannot exceed 60 per day, and the number of MP3 players cannot exceed the number of Blu-ray players. How many of each should the company manufacture to obtain maximum profit?

Solution First, we translate the statement of the problem into symbols.

Let x = number of MP3 players to be produced daily,

and y = number of Blu-ray players to be produced daily.

The company must produce at least 10 MP3 players (10 or more), so

$$x \geq 10.$$

Since no more than 50 MP3 players may be produced,

$$x \leq 50.$$

No more than 60 Blu-ray players may be made in one day, so

$$y \leq 60.$$

The number of MP3 players may not exceed the number of Blu-ray players, which translates as

$$x \leq y.$$

The numbers of MP3 players and of Blu-ray players cannot be negative, so

$$x \geq 0 \quad \text{and} \quad y \geq 0.$$

These restrictions, or **constraints,** form this system of inequalities.

$$x \geq 10, \quad x \leq 50, \quad y \leq 60, \quad x \leq y, \quad x \geq 0, \quad y \geq 0$$

Each MP3 player gives a profit of $30, so the daily profit from production of x MP3 players is $30x$ dollars. Also, the profit from production of y Blu-ray players will be $70y$ dollars per day. The total daily profit is given as follows.

$$\text{profit} = 30x + 70y$$

This equation defines the function to be maximized, called the **objective function.**

To find the maximum possible profit that the company can make, subject to these constraints, we sketch the graph of each constraint. The only feasible values of x and y are those which satisfy all constraints—that is, the values that lie in the intersection of the graphs of the constraints.

The intersection is shown in **FIGURE 59.** Any point lying inside the shaded region or on the boundary in the figure satisfies the restrictions as to the number of MP3 players and Blu-ray players that may be produced. (For practical purposes, however, only points with integer coefficients are useful.) This region is called the **region of feasible solutions.**

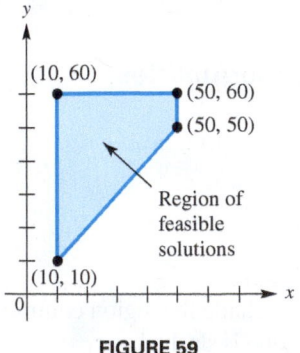

FIGURE 59

The **vertices** (singular: **vertex**), or **corner points,** of the region of feasible solutions have the following coordinates.

$$(10, 10), \quad (10, 60), \quad (50, 50), \quad \text{and} \quad (50, 60)$$

We must find the value of the objective function $30x + 70y$ for each vertex. We want the vertex that produces the maximum possible value of $30x + 70y$.

$(10, 10)$: $30(10) + 70(10) = 1000$
$(10, 60)$: $30(10) + 70(60) = 4500$
$(50, 50)$: $30(50) + 70(50) = 5000$
$(50, 60)$: $30(50) + 70(60) = 5700 \leftarrow$ Maximum

The maximum profit is obtained when 50 MP3 players and 60 Blu-ray players are produced each day. This maximum profit will be $30(50) + 70(60) = \$5700$ per day.

To justify the procedure used in **Example 4,** consider the following scenario. The company needed to find values of x and y in the shaded region of **FIGURE 59** that produce the maximum profit—that is, the maximum value of $30x + 70y$. To locate the point (x, y) that gives the maximum profit, add to the graph of that region lines corresponding to arbitrarily chosen profits of $0, $1000, $3000, and $7000.

$$30x + 70y = 0, \quad 30x + 70y = 1000, \quad 30x + 70y = 3000, \quad 30x + 70y = 7000$$

For instance, each point on the line $30x + 70y = 3000$ corresponds to production values that yield a profit of $3000.

FIGURE 60 shows the region of feasible solutions, together with these lines. The lines are parallel, and the higher the line, the greater the profit. The line $30x + 70y = 7000$ yields the greatest profit, but does not contain any points of the region of feasible solutions. To find the feasible solution of greatest profit, lower the line $30x + 70y = 7000$ until it contains a feasible solution—that is, until it touches the region of feasible solutions at point $A(50, 60)$, a vertex of the region. See **FIGURE 61.**

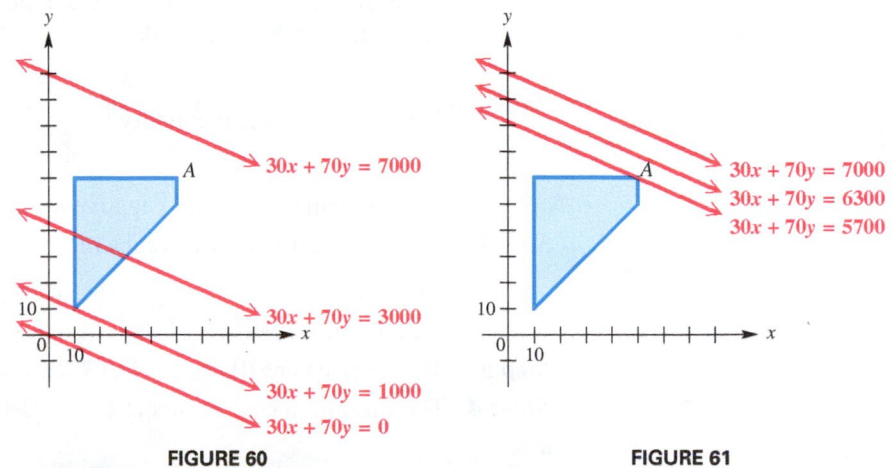

FIGURE 60   FIGURE 61

The result observed in **FIGURE 61** holds for *every* linear programming problem.

Fundamental Theorem of Linear Programming

If the optimal value for a linear programming problem exists, it occurs at a vertex of the region of feasible solutions.

> **Solving a Linear Programming Problem**
>
> *Step 1*   Write the objective function and all necessary constraints.
> *Step 2*   Graph the region of feasible solutions.
> *Step 3*   Identify all vertices (corner points).
> *Step 4*   Find the value of the objective function at each vertex.
> *Step 5*   The solution is given by the vertex producing the optimal value of the objective function.

EXAMPLE 5   **Finding a Minimum Cost Model**

Robin gives her dog Suzie vitamin supplements each day. She wants Suzie to have at least 16 units of Vitamin K, at least 5 units of Vitamin B_1, and at least 20 units of riboflavin. She can choose between Brand X pills, costing 10¢ each, that contain 8 units of K, 1 of B_1, and 2 of riboflavin; and Brand Y pills, costing 20¢ each, that contain 2 units of K, 1 of B_1, and 7 of riboflavin. How many of each pill should she buy to minimize her cost and yet fulfill Suzie's daily requirements?

Solution

Step 1   Let x represent the number of Brand X pills to buy, and let y represent the number of Brand Y pills to buy. Then the cost in pennies per day is given by

$$\text{cost} = 10x + 20y.$$

Robin buys x of the 10¢ Brand X pills and y of the 20¢ Brand Y pills. Suzie gets 8 units of Vitamin K from each Brand X pill and 2 units of Vitamin K from each Brand Y pill. Thus, Suzie gets $8x + 2y$ units of Vitamin K per day. Since Suzie needs at least 16 units,

$$8x + 2y \geq 16.$$

Each Brand X pill and each Brand Y pill supplies 1 unit of Vitamin B_1. Suzie needs at least 5 units per day, so

$$x + y \geq 5.$$

For riboflavin, the inequality is

$$2x + 7y \geq 20.$$

Since Robin cannot buy negative numbers of pills, $x \geq 0$ and $y \geq 0$.

Step 2   The intersection of the graphs of

$$8x + 2y \geq 16, \quad x + y \geq 5, \quad 2x + 7y \geq 20, \quad x \geq 0, \quad \text{and} \quad y \geq 0$$

is given in **FIGURE 62** and shows the region of feasible solutions.

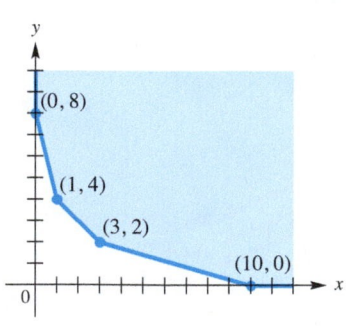

FIGURE 62

Step 3   The vertices are $(0, 8)$, $(1, 4)$, $(3, 2)$, and $(10, 0)$.

Steps 4 and 5   The minimum cost occurs at $(3, 2)$. See the table.

Point	Cost $= 10x + 20y$	
(0, 8)	$10(0) + 20(8) = 160$	
(1, 4)	$10(1) + 20(4) = 90$	
(3, 2)	$10(3) + 20(2) = 70$	← Minimum cost
(10, 0)	$10(10) + 20(0) = 100$	

Robin's best choice is to buy 3 Brand X pills and 2 Brand Y pills, for a total cost of 70¢ per day. Suzie receives just the minimum amounts of Vitamins B_1 and riboflavin, and an excess of Vitamin K.

6.7 Exercises

Checking Analytic Skills   Graph each inequality. **Do not use a calculator.**

1. $x \leq 3$

2. $y \leq -2$

3. $x + 2y \leq 6$

4. $x - y \geq 2$

5. $2x + 3y \geq 4$

6. $4y - 3x < 5$

7. $3x - 5y > 6$

8. $x < 3 + 2y$

9. $5x \leq 4y - 2$

10. $2x > 3 - 4y$

11. $y < 3x^2 + 2$

12. $y \leq x^2 - 4$

13. $y \leq 1 - x^2$

14. $y < 2 - 3x^2$

15. $y > (x - 1)^2 + 2$

16. $y > 2(x + 3)^2 - 1$

17. $x^2 + y^2 \leq 4$

18. $x^2 + y^2 \geq 1$

19. $x^2 + (y + 3)^2 \leq 16$

20. $(x - 4)^2 + y^2 \leq 9$

21. In your own words, explain how to determine whether the boundary of the graph of an inequality is solid or dashed.

22. When graphing $y > 3x - 6$, would you shade above or below the line $y = 3x - 6$? Explain your answer.

Concept Check   Use the concepts of this section to work Exercises 23–26.

23. For $Ax + By \geq C$, if $B > 0$, would you shade above or below the boundary line?

24. For $Ax + By \geq C$, if $B < 0$, would you shade above or below the boundary line?

25. Which one of the following is a description of the graph of the inequality $(x - 5)^2 + (y - 2)^2 < 4$?
   A. The region inside a circle with center $(-5, -2)$ and radius 2
   B. The region inside a circle with center $(5, 2)$ and radius 2
   C. The region inside a circle with center $(-5, -2)$ and radius 4
   D. The region outside a circle with center $(5, 2)$ and radius 4

26. Which one of the given inequalities satisfies the following description: the region outside a circle centered at the origin, with x-intercepts $(4, 0)$ and $(-4, 0)$?
   A. $x^2 + y^2 > 4$ **B.** $(x - 4)^2 + y^2 > 16$
   C. $x^2 + y^2 < 16$ **D.** $x^2 + y^2 > 16$

Write an inequality that satisfies the description.

27. Inside the circle with radius 1 and center $(0, 0)$

28. Outside the circle with radius 3 and center $(0, 0)$

29. Above the parabola with vertex $(0, -4)$ and x-intercepts $(-2, 0)$ and $(2, 0)$

30. Below the parabola with vertex $(0, 1)$ and x-intercepts $(-1, 0)$ and $(1, 0)$

Concept Check   In Exercises 31–34, match the inequality with the appropriate calculator graph. Do not use your calculator; instead, use your knowledge of the concepts involved in graphing inequalities.

31. $y \leq 3x - 6$

32. $y \geq 3x - 6$

33. $y \leq -3x - 6$

34. $y \geq -3x - 6$

A.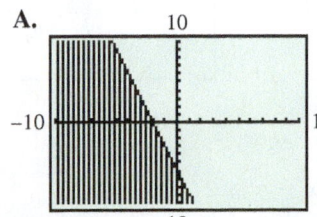

B.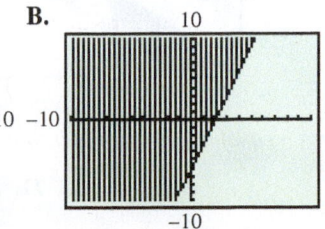

C.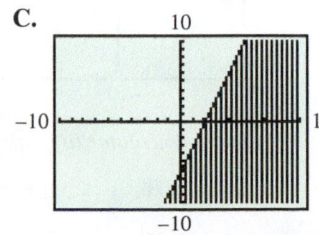

D.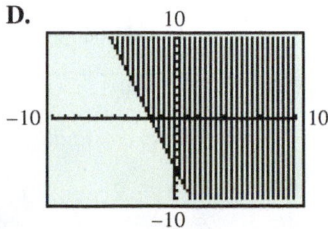

Checking Analytic Skills   Graph the solution set of each system of inequalities by hand. **Do not use a calculator.**

35. $x + y \geq 0$
$2x - y \geq 3$

36. $x + y \leq 4$
$x - 2y \geq 6$

37. $2x + y > 2$
$x - 3y < 6$

38. $4x + 3y < 12$
$y + 4x > -4$

39. $3x + 5y \leq 15$
$x - 3y \geq 9$

40. $y \leq x$
$x^2 + y^2 < 1$

41. $4x - 3y \leq 12$
$y \leq x^2$

42. $y \leq -x^2$
$y \geq x^2 - 6$

43. $x + y \le 9$
$\quad x \le -y^2$

44. $x + 2y \le 4$
$\quad y \ge x^2 - 1$

45. $y \le (x + 2)^2$
$\quad y \ge -2x^2$

46. $\quad x - y < 1$
$\quad -1 < y < 1$

47. $\quad x + y \le 36$
$\quad -4 \le x \le 4$

48. $y > x^2 + 4x + 4$
$\quad y < -x^2$

49. $y \ge (x - 2)^2 + 3$
$\quad y \le -(x - 1)^2 + 6$

50. $\quad x \ge 0$
$\quad x + y \le 4$
$\quad 2x + y \le 5$

51. $3x - 2y \ge 6$
$\quad x + \ y \le -5$
$\quad y \le 4$

52. $-2 < x < 3$
$\quad -1 \le y \le 5$
$\quad 2x + y < 6$

53. $-2 < x < 2$
$\quad y > 1$
$\quad x - y > 0$

54. $\quad x + y \le 4$
$\quad x - y \le 5$
$\quad 4x + y \le -4$

55. $\quad x \le 4$
$\quad x \ge 0$
$\quad y \ge 0$
$\quad x + 2y \ge 2$

56. $2y + x \ge -5$
$\quad y \le 3 + x$
$\quad x \le 0$
$\quad y \le 0$

57. $2x + 3y \le 12$
$\quad 2x + 3y > -6$
$\quad 3x + \ y < 4$
$\quad x \ge 0$
$\quad y \ge 0$

58. $y \ge 3^x$
$\quad y \ge 2$

59. $y \le \left(\dfrac{1}{2}\right)^x$
$\quad y \ge 4$

60. $\quad \ln x - y \ge 1$
$\quad x^2 - 2x - y \le 1$

61. $y \le \log x$
$\quad y \ge |x - 2|$

62. $e^{-x} - y \le 1$
$\quad x - 2y \ge 4$

63. *Concept Check* Which one of the choices that follow is a description of the solution set of the following system?

$$x^2 + y^2 < 36$$
$$y < x$$

A. All points outside the circle $x^2 + y^2 = 36$ and above the line $y = x$
B. All points outside the circle $x^2 + y^2 = 36$ and below the line $y = x$
C. All points inside the circle $x^2 + y^2 = 36$ and above the line $y = x$
D. All points inside the circle $x^2 + y^2 = 36$ and below the line $y = x$

64. *Concept Check* Fill in each blank with the appropriate response. The graph of the system

$$y > x^2 + 2$$
$$x^2 + y^2 < 16$$
$$y < 7$$

consists of all points _____ the parabola given by
$\quad\quad\quad\quad\quad\quad$ (above/below)
$y = x^2 + 2$, _____ the circle $x^2 + y^2 = 16$, and
$\quad\quad\quad\quad$ (inside/outside)

_____ the line $y = 7$.
(above/below)

Concept Check In Exercises 65–68, match each system of inequalities with the appropriate calculator graph. Do not use your calculator; instead, use your knowledge of the concepts involved in graphing systems of inequalities.

65. $y \ge x$
$\quad y \le 2x - 3$

66. $y \ge x^2$
$\quad y < 5$

67. $x^2 + y^2 \le 16$
$\quad y \ge 0$

68. $y \le x$
$\quad y \ge 2x - 3$

A.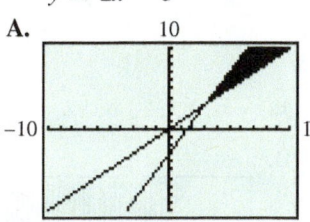

B.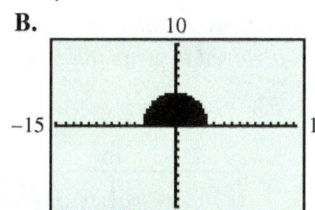

C.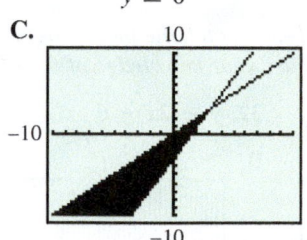

D.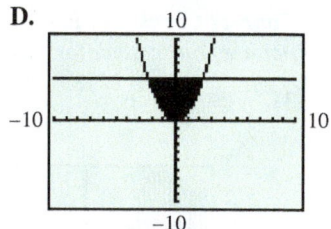

Use the shading capabilities of your graphing calculator to graph each inequality or system of inequalities.

69. $3x + 2y \ge 6$

70. $y \le x^2 + 5$

71. $x + y \ge 2$
$\quad x + y \le 6$

72. $y \ge |x + 2|$
$\quad y \le 6$

73. $y \ge 2^x$
$\quad y \le 8$

74. $y \le x^3 + x^2 - 4x - 4$

Shade the region(s) contained inside the graphs and give any points of intersection of the equations.

75. $y = 2x - 1$
$\quad y = 2 - x^2$

76. $y = x^2 - x + 1$
$\quad y = -x^2 + 1$

77. $y = x^3$
$\quad y = x$

78. $y = 2x^2 + x - 3$
$\quad y = x^2 - 2x + 1$

79. *Concept Check* Find a system of linear inequalities for which the graph is the region in the first quadrant between and inclusive of the pair of lines $x + 2y - 8 = 0$ and $x + 2y = 12$.

80. *Cost of Vitamins* The figure shows the region of feasible solutions for the vitamin problem of **Example 5** and the straight-line graph of all combinations of Brand X and Brand Y pills for which the cost is 40 cents.

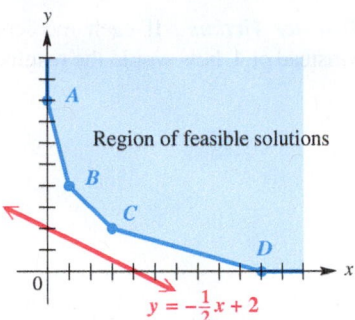

(a) The cost function is $c = 10x + 20y$. Give the linear equation (in slope–intercept form) of the line of constant cost c.

(b) As c increases, does the line of constant cost move up or down?

(c) By inspection, find the vertex of the region of feasible solutions that gives the optimal solution.

The graphs show regions of feasible solutions. Find the maximum and minimum values of each objective function.

81. $3x + 5y$

82. $6x + y$

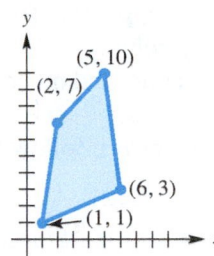

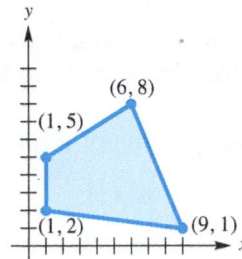

Find the maximum and minimum values of each objective function over the region of feasible solutions shown.

83. $3x + 5y$ **84.** $5x + 5y$

85. $10y$ **86.** $3x - y$

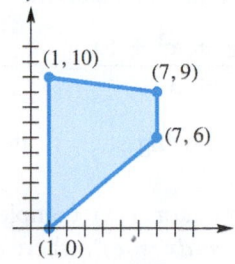

(Modeling) Solve each problem.

87. *Inquiries about Displayed Products* A wholesaler of party goods wishes to display her products at a convention of social secretaries in such a way that she gets the maximum number of inquiries about her whistles and hats. Her booth at the convention has 12 square meters of floor space to be used for display purposes. A display unit for hats requires 2 square meters, and one for whistles requires 4 square meters. Experience tells the wholesaler that she should never have more than a total of 5 units of whistles and hats on display at one time. If she receives three inquiries for each unit of hats and two inquiries for each unit of whistles on display, how many of each should she display in order to get the maximum number of inquiries? What is that maximum number?

88. *Profit from Farm Animals* Farmer Jones raises only pigs and geese. She wants to raise no more than 16 animals, with no more than 12 geese. She spends $50 to raise a pig and $20 to raise a goose. She has $500 available for this purpose. Find the maximum profit she can make if she makes a profit of $80 per goose and $40 per pig. Indicate how many pigs and geese she should raise to achieve this maximum.

89. *Shipment Costs* A manufacturer of refrigerators must ship at least 100 refrigerators to its two West Coast warehouses. Each warehouse holds a maximum of 100 refrigerators. Warehouse A holds 25 refrigerators already, while warehouse B has 20 on hand. It costs $12 to ship a refrigerator to warehouse A and $10 to ship one to warehouse B. How many refrigerators should be shipped to each warehouse to minimize cost? What is the minimum cost?

90. *Diet Requirements* Theo requires two food supplements: I and II. He can get these supplements from two different products A and B, as shown in the following table.

Supplement (grams/serving)	I	II
Product A	3	2
Product B	2	4

Theo's physician recommends at least 15 grams of each supplement in Theo's daily diet. If product A costs 25¢ per serving and product B costs 40¢ per serving, how can he satisfy his requirements most economically?

91. *Gasoline Revenues* A manufacturing process requires that oil refineries manufacture at least 2 gallons of gasoline for each gallon of fuel oil. To meet winter demand for fuel oil, at least 3 million gallons a day must be produced. The demand for gasoline is no more than 6.4 million gallons per day. If the price of gasoline is $1.90 per gallon and the price of fuel oil is $1.50 per gallon, how much of each should be produced to maximize revenue?

92. *Manufacturing Revenues* A shop manufactures two types of bolts on three groups of machines. The time required on each group differs, as shown in the following table.

Bolt	Machine Group		
	I	**II**	**III**
Type A	0.1 min	0.1 min	0.1 min
Type B	0.1 min	0.4 min	0.5 min

In a day, there are 240, 720, and 160 minutes available, respectively, on these machines. Type A bolts sell for 10¢ and Type B bolts for 12¢. How many of each type of bolt should be manufactured per day to maximize revenue? What is the maximum revenue?

93. *Aid to Disaster Victims* Earthquake victims need medical supplies and bottled water. Each medical kit measures 1 cubic foot and weighs 10 pounds. Each container of water is also 1 cubic foot, but weighs 20 pounds. The plane can carry only 80,000 pounds, with total volume 6000 cubic feet. Each medical kit will aid 4 people, while each container of water will serve 10 people. How many of each should be sent in order to maximize the number of people aided? How many people will be aided?

94. *Aid to Disaster Victims* If each medical kit could aid 6 people instead of 4, how would the results in **Exercise 93** change?

6.8 Partial Fractions

Decomposition of Rational Expressions • Distinct Linear Factors • Repeated Linear Factors • Distinct Linear and Quadratic Factors • Repeated Quadratic Factors

Decomposition of Rational Expressions

Add rational expressions

$$\frac{2}{x+1} + \frac{3}{x} = \frac{5x+3}{x(x+1)}$$

Partial fraction decomposition

The sums of rational expressions are found by combining two or more such expressions into one rational expression. Here, the reverse process is considered: Given one rational expression, express it as the sum of two or more rational expressions.

A special type of sum of rational expressions is called the **partial fraction decomposition.** Each term in the sum is a **partial fraction.** The technique of decomposing a rational expression into partial fractions is useful in calculus and other areas of mathematics.

To form a partial fraction decomposition of a *rational expression,* follow these steps.

Looking Ahead to Calculus

In calculus, partial fraction decomposition provides a powerful technique for determining integrals of rational functions.

Partial Fraction Decomposition of $\frac{f(x)}{g(x)}$

Step 1 If $\frac{f(x)}{g(x)}$ is not a proper fraction (a fraction whose numerator is of lesser degree than the denominator), divide $f(x)$ by $g(x)$. For example,

$$\frac{x^4 - 3x^3 + x^2 + 5x}{x^2 + 3} = x^2 - 3x - 2 + \frac{14x + 6}{x^2 + 3}.$$

Then apply the following steps to the remainder, which is a proper fraction.

Step 2 Factor the denominator $g(x)$ completely into factors of the form $(ax + b)^m$ or $(cx^2 + dx + e)^n$, where $cx^2 + dx + e$ is irreducible and m and n are integers.

(continued)

Step 3 **(a)** For each distinct linear factor $(ax + b)$, the decomposition must include the term $\dfrac{A}{ax + b}$.

(b) For each repeated linear factor $(ax + b)^m$, the decomposition must include the terms

$$\frac{A_1}{ax + b} + \frac{A_2}{(ax + b)^2} + \cdots + \frac{A_m}{(ax + b)^m}.$$

Step 4 **(a)** For each distinct quadratic factor $(cx^2 + dx + e)$, the decomposition must include the term $\dfrac{Bx + C}{cx^2 + dx + e}$.

(b) For each repeated quadratic factor $(cx^2 + dx + e)^n$, the decomposition must include the terms

$$\frac{B_1 x + C_1}{cx^2 + dx + e} + \frac{B_2 x + C_2}{(cx^2 + dx + e)^2} + \cdots + \frac{B_n x + C_n}{(cx^2 + dx + e)^n}.$$

Step 5 Use algebraic techniques to solve for the constants in the numerators of the decomposition.

Distinct Linear Factors

> **EXAMPLE 1**   **Finding a Partial Fraction Decomposition**

Find the partial fraction decomposition of $\dfrac{-x}{(x + 1)(x + 2)}$.

Solution   The numerator has degree 1 and the denominator has degree 2, so Step 1 is not necessary. Also, the denominator is factored, so Step 2 can be skipped. The denominator has distinct linear factors, so we do Step 3(a) and write the given expression as follows. (Note that Step 4 is skipped because no factors of the denominator are quadratic.)

$$\frac{-x}{(x + 1)(x + 2)} = \frac{A}{x + 1} + \frac{B}{x + 2} \tag{1}$$

Next, in Step 5 we must determine the values for A and B that make equation (1) true. Multiply each side of equation (1) by the least common denominator $(x + 1)(x + 2)$ to get the following.

$$\frac{-x(x + 1)(x + 2)}{(x + 1)(x + 2)} = \frac{A(x + 1)(x + 2)}{x + 1} + \frac{B(x + 1)(x + 2)}{x + 2} \tag{2}$$

Simplifying each expression in (2) results in the following equation.

$$-x = A(x + 2) + B(x + 1) \tag{3}$$

Equation (3) is an identity and must be true for all values of x. If we let $x = -1$, we can easily determine A.

$$-(-1) = A(-1 + 2) + B(-1 + 1)$$
$$1 = A(1) + B(0)$$

Thus, $A = 1$, Similarly, if we let $x = -2$ in (3), then we can determine B.

$$-(-2) = A(-2 + 2) + B(-2 + 1)$$
$$2 = A(0) + B(-1)$$

Thus, $B = -2$. The required partial fraction decomposition is the following.

$$\frac{-x}{(x + 1)(x + 2)} = \frac{1}{x + 1} + \frac{-2}{x + 2}$$

In the next example we must perform both Step 1 and Step 2.

EXAMPLE 2 **Finding a Partial Fraction Decomposition**

Find the partial fraction decomposition of $\dfrac{2x^4 - 8x^2 + 5x - 2}{x^3 - 4x}$.

Solution   The numerator has greater degree than the denominator, so divide first.

$$
\begin{array}{r}
2x \\
x^3 - 4x \overline{\smash{)}2x^4 - 8x^2 + 5x - 2} \\
\underline{2x^4 - 8x^2} \\
5x - 2
\end{array}
$$

The quotient is $\dfrac{2x^4 - 8x^2 + 5x - 2}{x^3 - 4x} = 2x + \dfrac{5x - 2}{x^3 - 4x}$. Now work with the remainder fraction. Factor the denominator as $x^3 - 4x = x(x + 2)(x - 2)$. Since the factors are distinct linear factors, use Step 3(a) to write the decomposition as

$$
\frac{5x - 2}{x^3 - 4x} = \frac{A}{x} + \frac{B}{x + 2} + \frac{C}{x - 2}, \quad (1)
$$

where A, B, and C are constants that need to be found. Multiply each side of equation (1) by $x(x + 2)(x - 2)$ to get

$$
5x - 2 = A(x + 2)(x - 2) + Bx(x - 2) + Cx(x + 2). \quad (2)
$$

Equation (1) is an identity, since both sides represent the same rational expression. Thus, equation (2) is also an identity. Equation (1) holds for all values of x except 0, -2, and 2. However, equation (2) holds for all values of x. In particular, substituting 0 for x in equation (2) gives $-2 = -4A$, so $A = \frac{1}{2}$. Similarly, choosing $x = -2$ gives $-12 = 8B$, and thus $B = -\frac{3}{2}$. Finally, choosing $x = 2$ gives $8 = 8C$, giving $C = 1$. The remainder rational expression can be written as the following sum of partial fractions.

$$
\frac{5x - 2}{x^3 - 4x} = \frac{1}{2x} + \frac{-3}{2(x + 2)} + \frac{1}{x - 2}
$$

The given rational expression can be written as follows.

$$
\frac{2x^4 - 8x^2 + 5x - 2}{x^3 - 4x} = 2x + \frac{1}{2x} + \frac{-3}{2(x + 2)} + \frac{1}{x - 2}
$$

Check the work by combining the terms on the right. ●

We can use a graphing calculator to *check* the work in **Example 2.** We define Y_1 using the original rational expression and Y_2 using the partial fraction decomposition.

$$
Y_1 = \frac{2X^4 - 8X^2 + 5X - 2}{X^3 - 4X}, \quad Y_2 = 2X + \frac{1}{2X} + \frac{-3}{2(X + 2)} + \frac{1}{X - 2}
$$

See **FIGURE 63(a).** When we graph both Y_1 and Y_2 on the same screen, the graphs should be indistinguishable. See **FIGURE 63(b).**

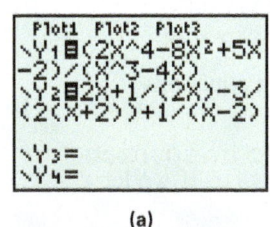

(a)

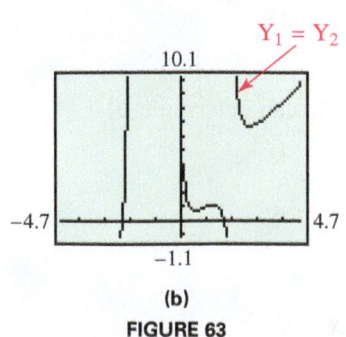

(b)

FIGURE 63

NOTE   The method of graphical support just described is not foolproof. However, it gives a fairly accurate portrayal of whether the partial fraction decomposition is correct. A table can be used for support as well.

Repeated Linear Factors

EXAMPLE 3 Finding a Partial Fraction Decomposition

Find the partial fraction decomposition of $\dfrac{2x}{(x-1)^3}$.

Solution This is a proper fraction. The denominator is already factored with repeated linear factors. We write the decomposition as shown, using Step 3(b).

$$\frac{2x}{(x-1)^3} = \frac{A}{x-1} + \frac{B}{(x-1)^2} + \frac{C}{(x-1)^3}$$

We clear denominators by multiplying each side of this equation by $(x-1)^3$.

$$2x = A(x-1)^2 + B(x-1) + C$$

Substituting 1 for x leads to $C = 2$, giving the following.

$$2x = A(x-1)^2 + B(x-1) + 2 \qquad (1)$$

We substituted 1, the only root, and we still need to find values for A and B. However, *any* number can be substituted for x. For example, when we choose $x = -1$ (because it is easy to substitute), equation (1) becomes the following.

$$
\begin{aligned}
2(-1) &= A(-1-1)^2 + B(-1-1) + 2 &&\text{Let } x = -1 \text{ in (1).}\\
-2 &= 4A - 2B + 2 \\
-4 &= 4A - 2B &&\text{Subtract 2.}\\
-2 &= 2A - B &&\text{Divide by 2.} \quad (2)
\end{aligned}
$$

Substitute 0 for x in equation (1).

$$
\begin{aligned}
0 &= A - B + 2 \\
2 &= -A + B &&(3)
\end{aligned}
$$

We solve the system of equations (2) and (3) to get $A = 0$ and $B = 2$. The partial fraction decomposition is now complete.

$$\frac{2x}{(x-1)^3} = \frac{2}{(x-1)^2} + \frac{2}{(x-1)^3}$$

We needed three substitutions because there were three constants to evaluate: A, B, and C. To check this result, we could combine the terms on the right. ●

Distinct Linear and Quadratic Factors

EXAMPLE 4 Finding a Partial Fraction Decomposition

Find the partial fraction decomposition of $\dfrac{x^2 + 3x - 1}{(x+1)(x^2+2)}$.

Solution This denominator has distinct linear and quadratic factors, with neither repeated. Since $x^2 + 2$ cannot be factored, it is irreducible. The partial fraction decomposition by Steps 3(a) and 4(a) from the beginning of this section follows.

$$\frac{x^2 + 3x - 1}{(x+1)(x^2+2)} = \frac{A}{x+1} + \frac{Bx + C}{x^2+2}$$

Multiply each side by $(x+1)(x^2+2)$.

$$x^2 + 3x - 1 = A(x^2 + 2) + (Bx + C)(x+1) \qquad (1)$$

First, substitute -1 for x in $x^2 + 3x - 1 = A(x^2 + 2) + (Bx + C)(x + 1)$.

Use parentheses around substituted values to avoid errors.

$$(-1)^2 + 3(-1) - 1 = A[(-1)^2 + 2] + [B(-1) + C](-1 + 1)$$
$$-3 = 3A$$
$$A = -1$$

Replace A with -1 in equation (1) and substitute any value for x. Let $x = 0$.

$$0^2 + 3(0) - 1 = -1(0^2 + 2) + (B \cdot 0 + C)(0 + 1)$$
$$-1 = -2 + C$$
$$C = 1$$

Now, letting $A = -1$ and $C = 1$, substitute again in equation (1), using another number for x. Let $x = 1$.

$$3 = -3 + (B + 1)(2)$$
$$6 = 2B + 2$$
$$B = 2$$

With $A = -1$, $B = 2$, and $C = 1$, the partial fraction decomposition is complete.

$$\frac{x^2 + 3x - 1}{(x + 1)(x^2 + 2)} = \frac{-1}{x + 1} + \frac{2x + 1}{x^2 + 2}$$

Again, this work can be checked by combining terms on the right. ●

For fractions with denominators that have quadratic factors, another method is often more convenient. The system of equations is formed by equating coefficients of like terms on each side of the partial fraction decomposition. For instance, in **Example 4,** after each side was multiplied by the common denominator, the equation was

$$x^2 + 3x - 1 = A(x^2 + 2) + (Bx + C)(x + 1).$$

Multiplying on the right and collecting like terms, we have the following.

$$x^2 + 3x - 1 = Ax^2 + 2A + Bx^2 + Bx + Cx + C$$
$$x^2 + 3x - 1 = (A + B)x^2 + (B + C)x + (C + 2A)$$

Now, equating the coefficients of like powers of x gives these three equations.

$$1 = A + B$$
$$3 = B + C$$
$$-1 = C + 2A$$

Solving this system for A, B, and C would give the partial fraction decomposition.

Repeated Quadratic Factors

EXAMPLE 5 **Finding a Partial Fraction Decomposition**

Find the partial fraction decomposition of $\dfrac{2x}{(x^2 + 1)^2(x - 1)}$.

Solution This expression has both a linear factor and a repeated quadratic factor. Start by using Steps 3(a) and 4(b) from the beginning of this section.

$$\frac{2x}{(x^2 + 1)^2(x - 1)} = \frac{Ax + B}{x^2 + 1} + \frac{Cx + D}{(x^2 + 1)^2} + \frac{E}{x - 1}$$

Multiply each side by $(x^2 + 1)^2(x - 1)$.

$$2x = (Ax + B)(x^2 + 1)(x - 1) + (Cx + D)(x - 1) + E(x^2 + 1)^2 \quad \text{(1)}$$

If $x = 1$, equation (1) reduces to $2 = 4E$, giving $E = \frac{1}{2}$. Substitute $\frac{1}{2}$ for E in equation (1) and combine terms on the right.

$$2x = \left(A + \frac{1}{2}\right)x^4 + (-A + B)x^3 + (A - B + C + 1)x^2$$

$$+ (-A + B + D - C)x + \left(-B - D + \frac{1}{2}\right) \quad \text{(2)}$$

To solve for the unknowns, equate the coefficients of like powers of x on each side of equation (2). Equating coefficients of x^4 yields $0 = A + \frac{1}{2}$, giving $A = -\frac{1}{2}$. From the corresponding coefficients of x^3, $0 = -A + B$, which means that since $A = -\frac{1}{2}$, we have $B = -\frac{1}{2}$.

Using the coefficients of x^2 gives $0 = A - B + C + 1$. Since $A = -\frac{1}{2}$ and $B = -\frac{1}{2}$, we have $C = -1$. Finally, from the coefficients of x, $2 = -A + B + D - C$. Substituting for A, B, and C gives $D = 1$.

With $A = -\frac{1}{2}$, $B = -\frac{1}{2}$, $C = -1$, $D = 1$, and $E = \frac{1}{2}$, the given fraction has this partial fraction decomposition.

$$\frac{2x}{(x^2 + 1)^2(x - 1)} = \frac{-\frac{1}{2}x - \frac{1}{2}}{x^2 + 1} + \frac{-x + 1}{(x^2 + 1)^2} + \frac{\frac{1}{2}}{x - 1},$$

or $\quad \dfrac{2x}{(x^2 + 1)^2(x - 1)} = \dfrac{-(x + 1)}{2(x^2 + 1)} + \dfrac{-x + 1}{(x^2 + 1)^2} + \dfrac{1}{2(x - 1)}$ Simplify complex fractions.

In summary, to solve for the constants in the numerators of a partial fraction decomposition, use either of the following methods or a combination of the two.

Techniques for Decomposition into Partial Fractions

Method 1   For Linear Factors

Step 1   Multiply each side of the resulting rational equation by the common denominator.

Step 2   Substitute the zero of each factor into the resulting equation. For repeated linear factors, substitute as many other numbers as is necessary to find all the constants in the numerators. The number of substitutions required will equal the number of constants A, B, ...

Method 2   For Quadratic Factors

Step 1   Multiply each side of the resulting rational equation by the common denominator.

Step 2   Collect terms on the right side of the equation.

Step 3   Equate the coefficients of like terms to get a system of equations.

Step 4   Solve the system to find the constants in the numerators.

6.8 Exercises

Find the partial fraction decomposition for each rational expression.

1. $\dfrac{5}{3x(2x + 1)}$

2. $\dfrac{3x - 1}{x(x + 1)}$

3. $\dfrac{4x + 2}{(x + 2)(2x - 1)}$

4. $\dfrac{x + 2}{(x + 1)(x - 1)}$

5. $\dfrac{x}{x^2 + 4x - 5}$

6. $\dfrac{5x - 3}{(x + 1)(x - 3)}$

7. $\dfrac{2x}{(x + 1)(x + 2)^2}$

8. $\dfrac{2}{x^2(x + 3)}$

9. $\dfrac{4}{x(1 - x)}$

10. $\dfrac{x + 1}{x^2(1 - x)}$

11. $\dfrac{4x^2 - x - 15}{x(x + 1)(x - 1)}$

12. $\dfrac{2x + 1}{(x + 2)^3}$

13. $\dfrac{x^2}{x^2 + 2x + 1}$

14. $\dfrac{3}{x^2 + 4x + 3}$

15. $\dfrac{2x^5 + 3x^4 - 3x^3 - 2x^2 + x}{2x^2 + 5x + 2}$

16. $\dfrac{6x^5 + 7x^4 - x^2 + 2x}{3x^2 + 2x - 1}$

17. $\dfrac{x^3 + 4}{9x^3 - 4x}$

18. $\dfrac{x^3 + 2}{x^3 - 3x^2 + 2x}$

19. $\dfrac{-3}{x^2(x^2 + 5)}$

20. $\dfrac{2x + 1}{(x + 1)(x^2 + 2)}$

21. $\dfrac{3x - 2}{(x + 4)(3x^2 + 1)}$

22. $\dfrac{3}{x(x + 1)(x^2 + 1)}$

23. $\dfrac{1}{x(2x + 1)(3x^2 + 4)}$

24. $\dfrac{x^4 + 1}{x(x^2 + 1)^2}$

25. $\dfrac{3x - 1}{x(2x^2 + 1)^2}$

26. $\dfrac{3x^4 + x^3 + 5x^2 - x + 4}{(x - 1)(x^2 + 1)^2}$

27. $\dfrac{-x^4 - 8x^2 + 3x - 10}{(x + 2)(x^2 + 4)^2}$

28. $\dfrac{x^2}{x^4 - 1}$

29. $\dfrac{5x^5 + 10x^4 - 15x^3 + 4x^2 + 13x - 9}{x^3 + 2x^2 - 3x}$

30. $\dfrac{3x^6 + 3x^4 + 3x}{x^4 + x^2}$

Determine whether each partial fraction decomposition is correct by graphing the left side and the right side of the equation on the same coordinate axes and observing whether the graphs coincide.

31. $\dfrac{4x^2 - 3x - 4}{x^3 + x^2 - 2x} = \dfrac{2}{x} + \dfrac{-1}{x - 1} + \dfrac{3}{x + 2}$

32. $\dfrac{1}{(x - 1)(x + 2)} = \dfrac{1}{x - 1} - \dfrac{1}{x + 2}$

33. $\dfrac{x^3 - 2x}{(x^2 + 2x + 2)^2} = \dfrac{x - 2}{x^2 + 2x + 2} + \dfrac{2}{(x^2 + 2x + 2)^2}$

34. $\dfrac{2x + 4}{x^2(x - 2)} = \dfrac{-2}{x} + \dfrac{-2}{x^2} + \dfrac{2}{x - 2}$

SECTIONS 6.7–6.8   Reviewing Basic Concepts

Graph the solution set of each inequality or system of inequalities.

1. $-2x - 3y \leq 6$

2. $x - y < 5$
  $x + y \geq 3$

3. $y \geq x^2 - 2$
  $x + 2y \geq 4$

4. $x^2 + y^2 \leq 25$
  $x^2 + y^2 \geq 9$

5. Which one of the following systems in choices A–D is represented by the given graph?

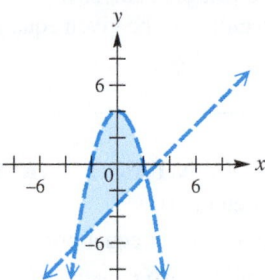

A. $y > x - 3$
  $y < -x^2 + 4$

B. $y < x - 3$
  $y < -x^2 + 4$

C. $y > x - 3$
  $y > -x^2 + 4$

D. $y < x - 3$
  $y > -x^2 + 4$

6. Find the minimum value of $2x + 3y$, subject to the following constraints.

$$x \geq 0$$
$$y \geq 0$$
$$x + y \geq 4$$
$$2x + y \leq 8$$

7. The graph shows a region of feasible solutions. Find the maximum and minimum values of $P = 3x + 5y$ over this region.

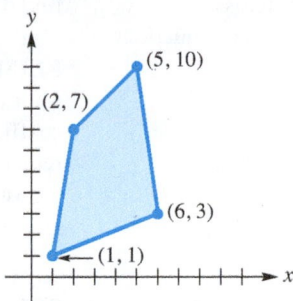

8. *(Modeling) Minimizing Cost*   Two substances X and Y are found in pet food. Each substance contains the ingredients A and B. Substance X is 20% ingredient A and 50% ingredient B. Substance Y is 50% ingredient A and 30% ingredient B. The cost of substance X is $2 per pound and the cost of substance Y is $3 per pound. The pet store needs at least 251 pounds of ingredient A and at least 200 pounds of ingredient B. If cost is to be minimal, how many pounds of each substance should be ordered? Find the minimum cost.

Decompose each rational expression into partial fractions.

9. $\dfrac{10x + 13}{x^2 - x - 20}$

10. $\dfrac{3x^2 - 2x + 1}{(x - 1)(x^2 + 1)}$

---

6 Summary

KEY TERMS & SYMBOLS

KEY CONCEPTS

6.1 Systems of Equations

linear equation (or first-degree equation) in *n* unknowns
system of equations
solutions
system of linear equations (linear system)

The following transformations produce an equivalent system.

TRANSFORMATIONS OF A LINEAR SYSTEM

1. Interchange any two equations of the system.
2. Multiply or divide any equation of the system by a nonzero real number.
3. Replace any equation of the system by the sum of that equation and a multiple of another equation in the system.

(continued)

KEY TERMS & SYMBOLS	KEY CONCEPTS
consistent system independent equations inconsistent system dependent equations substitution method elimination method equivalent systems nonlinear system of equations	Systems of equations in two variables may be solved by the substitution method, by the elimination method, or graphically by the intersection-of-graphs method. **SUBSTITUTION METHOD** (for a system of two equations in two variables) Solve one equation for one variable in terms of the other. Substitute for that variable in the other equation, and solve for its value. Then substitute that value into the other equation to find the value of the remaining variable. **ELIMINATION METHOD** (for a system of two equations in two variables) Multiply one or both equations by appropriate nonzero numbers so that the sum of the coefficients of one of the variables is 0. Add the equations, and solve for the value of the remaining variable. Then substitute that value into either of the given equations to solve for the value of the other variable.

6.2 Solution of Linear Systems in Three Variables ordered triple (x, y, z)	**SOLVING A SYSTEM OF LINEAR EQUATIONS IN THREE VARIABLES** *Step 1* Eliminate a variable from any two of the equations. *Step 2* Eliminate the *same variable* from a different pair of equations. *Step 3* Eliminate a second variable using the resulting two equations in two variables to get an equation with just one variable whose value we can now determine. *Step 4* Find the values of the remaining variables by substitution. Write the solution of the system as an ordered triple.

6.3 Solution of Linear Systems by Row Transformations matrix (matrices) element (entry) augmented matrix echelon (triangular) form main diagonal back-substitution row echelon method reduced row echelon method	**MATRIX ROW TRANSFORMATIONS** For any augmented matrix of a system of linear equations, the following row transformations will result in the matrix of an equivalent system. **1.** Any two rows may be interchanged. **2.** The elements of any row may be multiplied by a nonzero real number. **3.** Any row may be changed by adding to its elements a multiple of the corresponding elements of another row.

6.4 Matrix Properties and Operations dimension square matrix row matrix	**MATRIX ADDITION** The sum of two $m \times n$ matrices A and B is the $m \times n$ matrix $A + B$ in which each element is the sum of the corresponding elements of A and B. **MATRIX SUBTRACTION** If A and B are matrices with the same dimension, then $A - B = A + (-B)$.

KEY TERMS & SYMBOLS

column matrix
zero matrix
additive inverse (negative)
scalar

KEY CONCEPTS

SCALAR MULTIPLICATION

If k is a scalar and A is a matrix, then kA is the matrix of the same dimension where each entry of A is multiplied by k.

MATRIX MULTIPLICATION

The product AB of an $m \times n$ matrix A and an $n \times k$ matrix B is an $m \times k$ matrix found as follows: To find the ith row, jth column element of AB, multiply each element in the ith row of A by the corresponding element in the jth column of B. The sum of these products will give the element of row i, column j of AB.

6.5 Determinants and Cramer's Rule

determinant
det A
minor
cofactor
expansion by a row or column

DETERMINANT OF A 2 × 2 MATRIX

The determinant of a 2 × 2 matrix A, where $A = \begin{bmatrix} a_{11} & a_{12} \\ a_{21} & a_{22} \end{bmatrix}$, is a real number defined as

$$\det A = a_{11}a_{22} - a_{21}a_{12}.$$

COFACTOR

Let M_{ij} be the minor for element a_{ij} in an $n \times n$ matrix. The cofactor of a_{ij} is given by

$$A_{ij} = (-1)^{i+j} \cdot M_{ij}.$$

FINDING THE DETERMINANT OF AN $n \times n$ MATRIX

Multiply each element in any row or column of the matrix by its cofactor. The sum of these products gives the value of the determinant.

CRAMER'S RULE FOR 2 × 2 SYSTEMS

The solution of the system

$$a_1 x + b_1 y = c_1$$
$$a_2 x + b_2 y = c_2$$

is given by $\quad x = \dfrac{D_x}{D} \quad$ and $\quad y = \dfrac{D_y}{D},$

where $\quad D_x = \det \begin{bmatrix} c_1 & b_1 \\ c_2 & b_2 \end{bmatrix}, \quad D_y = \det \begin{bmatrix} a_1 & c_1 \\ a_2 & c_2 \end{bmatrix}, \quad$ and $\quad D = \det \begin{bmatrix} a_1 & b_1 \\ a_2 & b_2 \end{bmatrix} \neq 0.$

Cramer's rule can be extended to 3 × 3 and larger systems.

6.6 Solution of Linear Systems by Matrix Inverses

identity matrix I_n
multiplicative inverse matrix
singular matrix
coefficient matrix

INVERSE OF A 2 × 2 MATRIX

If $A = \begin{bmatrix} a & b \\ c & d \end{bmatrix}$ and det $A \neq 0$, then

$$A^{-1} = \frac{1}{\det A} \begin{bmatrix} d & -b \\ -c & a \end{bmatrix}.$$

To find the multiplicative inverse of a 3 × 3 (or larger) matrix A, we often use a calculator.

(continued)

KEY TERMS & SYMBOLS	KEY CONCEPTS

FINDING AN INVERSE MATRIX ANALYTICALLY

To obtain A^{-1} for any $n \times n$ matrix A for which A^{-1} exists, follow these steps.

Step 1 Form the augmented matrix $[A|I_n]$, where I_n is the $n \times n$ identity matrix.

Step 2 Perform row transformations on $[A|I_n]$ to obtain a matrix of the form $[I_n|B]$.

Step 3 Matrix B is A^{-1}.

6.7 Systems of Inequalities and Linear Programming

half plane
boundary
linear inequality in two
  variables
test point
system of inequalities
linear programming
constraints
objective function
region of feasible solutions
vertex (corner point)

GRAPHING AN INEQUALITY

1. For a function f, the graph of $y < f(x)$ consists of all the points that are *below* the graph of $y = f(x)$. The graph of $y > f(x)$ consists of all the points that are *above* the graph of $y = f(x)$. (Similar statements can be made for $\leq$ and $\geq$, with boundaries included.)

2. If the inequality is not or cannot be solved for y, choose a test point not on the boundary. If the test point satisfies the inequality, the graph includes all points on the same side of the boundary as the test point. Otherwise, the graph includes all points on the other side of the boundary.

SOLVING A LINEAR PROGRAMMING PROBLEM

Step 1 Write the objective function and all necessary constraints.

Step 2 Graph the region of feasible solutions.

Step 3 Identify all vertices (corner points).

Step 4 Find the value of the objective function at each vertex.

Step 5 The solution is given by the vertex producing the optimal value of the objective function.

6.8 Partial Fractions

partial fraction decomposition
partial fraction

To solve for the constants in the numerators of a partial fraction decomposition, use either of the following methods or a combination of the two.

METHOD 1   FOR LINEAR FACTORS

Step 1 Multiply each side of the resulting rational equation by the common denominator.

Step 2 Substitute the zero of each factor into the resulting equation. For repeated linear factors, substitute as many other numbers as is necessary to find all the constants in the numerators. The number of substitutions required will equal the number of constants, $A, B, \ldots$.

METHOD 2   FOR QUADRATIC FACTORS

Step 1 Multiply each side of the resulting rational equation by the common denominator.

Step 2 Collect terms on the right side of the equation.

Step 3 Equate the coefficients of like terms to get a system of equations.

Step 4 Solve the system to find the constants in the numerators.

6 Review Exercises

Solve each system. Identify any systems with dependent equations and any inconsistent systems.

1. $4x - 3y = -1$
$3x + 5y = 50$

2. $0.5x - 0.2y = 1.1$
$2x - 0.8y = 4.4$

3. $4x + 5y = 5$
$3x + 7y = -6$

4. $y = x^2 - 1$
$x + y = 1$

5. $x^2 + y^2 = 2$
$3x + y = 4$

6. $x^2 + 2y = 22$
$x^2 + y^2 = 37$

7. $x^2 - 4y = 19$
$x^2 + y^2 = 16$

8. $xy = 4$
$x - 6y = 2$

9. $x^2 + y^2 = 8$
$x = y + 4$

 10. Use your calculator with viewing window $[-18, 18]$ by $[-12, 12]$ to answer the following.
(a) Do the circle $x^2 + y^2 = 144$ and the line $x + 2y = 8$ have any points in common?
(b) Approximate any intersection points to the nearest tenth.
(c) Find the exact solution set of the system.

11. Consider the system in **Exercise 5.**
(a) To graph the first equation, what two functions must you enter into your calculator?
(b) To graph the second equation, what function must you enter?
(c) What would be an appropriate window in which to graph this system?

 12. Can a system of two linear equations in two variables have exactly two solutions? Explain.

13. Can a system consisting of two equations in three variables have a unique solution? Explain.

Solve each system. Identify any systems with dependent equations and any inconsistent systems.

14. $2x - 3y + z = -5$
$x + 4y + 2z = 13$
$5x + 5y + 3z = 14$

15. $x - 3y = 12$
$2y + 5z = 1$
$4x + z = 25$

16. $x + y - z = 5$
$2x + y + 3z = 2$
$4x - y + 2z = -1$

17. $5x - 3y + 2z = -5$
$2x + y - z = 4$
$-4x - 2y + 2z = -1$

Use the reduced row echelon method to solve each system. Identify any systems with dependent equations and any inconsistent systems.

18. $2x + 3y = 10$
$-3x + y = 18$

19. $3x + y = -7$
$x - y = -5$

20. $x - z = -3$
$y + z = 6$
$2x - 3z = -9$

21. $x + 2y + z = 0$
$3x + 2y - z = 4$
$-x + 2y + 3z = -4$

Perform each operation if possible. If not possible, say so.

22. $\begin{bmatrix} -5 & 4 & 9 \\ 2 & -1 & -2 \end{bmatrix} + \begin{bmatrix} 1 & -2 & 7 \\ 4 & -5 & -5 \end{bmatrix}$

23. $\begin{bmatrix} 3 \\ 2 \\ 5 \end{bmatrix} - \begin{bmatrix} 8 \\ -4 \\ 6 \end{bmatrix} + \begin{bmatrix} 1 \\ 0 \\ 2 \end{bmatrix}$

24. $\begin{bmatrix} 2 & 5 & 8 \\ 1 & 9 & 2 \end{bmatrix} - \begin{bmatrix} 3 & 4 \\ 7 & 1 \end{bmatrix}$

25. $3\begin{bmatrix} 2 & 4 \\ -1 & 4 \end{bmatrix} - 2\begin{bmatrix} 5 & 8 \\ 2 & -2 \end{bmatrix}$

26. $-1\begin{bmatrix} 3 & -5 & 2 \\ 1 & 7 & -4 \end{bmatrix} + 5\begin{bmatrix} 0 & 2 \\ -1 & 3 \end{bmatrix}$

27. $10\begin{bmatrix} 2x & y \\ 5y & 6x \end{bmatrix} + 2\begin{bmatrix} -3x & 6y \\ 2y & 5x \end{bmatrix}$

28. *Concept Check* Complete the following sentence: The sum of two $m \times n$ matrices A and B is found by _____.

Multiply if possible. If not possible, say so.

29. $\begin{bmatrix} -8 & 6 \\ 5 & 2 \end{bmatrix} \begin{bmatrix} 3 & -1 \\ 7 & 2 \end{bmatrix}$

30. $\begin{bmatrix} 3 & 2 & -1 \\ 4 & 0 & 6 \end{bmatrix} \begin{bmatrix} -2 & 0 \\ 0 & 2 \\ 3 & 1 \end{bmatrix}$

31. $\begin{bmatrix} 1 & -2 & 4 & 2 \\ 0 & 1 & -1 & 8 \end{bmatrix} \begin{bmatrix} -1 \\ 2 \\ 0 \\ 1 \end{bmatrix}$

32. $\begin{bmatrix} 1 & 2 & 5 \\ -3 & 4 & 7 \\ 0 & 2 & -1 \end{bmatrix} \begin{bmatrix} 4 & 2 & 3 \\ 10 & -5 & 6 \end{bmatrix}$

33. $\begin{bmatrix} 4 & 2 & 3 \\ 10 & -5 & 6 \end{bmatrix} \begin{bmatrix} 1 & 2 & 5 \\ -3 & 4 & 7 \\ 0 & 2 & -1 \end{bmatrix}$

34. $\begin{bmatrix} 3 & -1 & 0 \end{bmatrix} \begin{bmatrix} 1 & 3 & 2 \\ 2 & -4 & 0 \\ 5 & 7 & 3 \end{bmatrix}$

Find AB and BA to determine whether A and B are inverses.

35. $A = \begin{bmatrix} 3 & 2 \\ 13 & 9 \end{bmatrix}; B = \begin{bmatrix} 9 & -2 \\ -13 & 3 \end{bmatrix}$

36. $A = \begin{bmatrix} 1 & 0 \\ 2 & -3 \end{bmatrix}; B = \begin{bmatrix} 1 & 0 \\ \frac{2}{3} & -\frac{1}{3} \end{bmatrix}$

37. $A = \begin{bmatrix} 2 & 0 & 6 \\ 0 & 1 & 0 \\ 1 & 0 & 1 \end{bmatrix}; B = \begin{bmatrix} -1 & 0 & \frac{3}{2} \\ 0 & 1 & 0 \\ \frac{1}{4} & 0 & -1 \end{bmatrix}$

38. $A = \begin{bmatrix} 1 & 0 & 2 \\ 0 & 2 & 4 \\ 0 & 0 & 1 \end{bmatrix}; B = \begin{bmatrix} 1 & 0 & -2 \\ 0 & \frac{1}{2} & -2 \\ 0 & 0 & 1 \end{bmatrix}$

For each matrix, find A^{-1} if it exists.

39. $A = \begin{bmatrix} 6 & 3 \\ 10 & 5 \end{bmatrix}$

40. $A = \begin{bmatrix} -4 & 2 \\ 0 & 3 \end{bmatrix}$

41. $A = \begin{bmatrix} 2 & 0 \\ -1 & 5 \end{bmatrix}$

42. $A = \begin{bmatrix} 2 & 0 & 4 \\ 1 & -1 & 0 \\ 0 & 1 & -2 \end{bmatrix}$

43. $A = \begin{bmatrix} 2 & -1 & 0 \\ 1 & 0 & 1 \\ 1 & -2 & 0 \end{bmatrix}$

44. $A = \begin{bmatrix} 2 & 3 & 5 \\ -2 & -3 & -5 \\ 1 & 4 & 2 \end{bmatrix}$

Use the inverse matrix method to solve each system if possible. Identify any systems with dependent equations or any inconsistent systems.

45. $\begin{aligned} x + y &= 4 \\ 2x + 3y &= 10 \end{aligned}$

46. $\begin{aligned} 5x - 3y &= -2 \\ 2x + 7y &= -9 \end{aligned}$

47. $\begin{aligned} 2x + y &= 5 \\ 3x - 2y &= 4 \end{aligned}$

48. $\begin{aligned} x - 2y &= 7 \\ 3x + y &= 7 \end{aligned}$

49. $\begin{aligned} x + 2y &= -1 \\ 3y - z &= -5 \\ x + 2y - z &= -3 \end{aligned}$

50. $\begin{aligned} 3x - 2y + 4z &= 1 \\ 4x + y - 5z &= 2 \\ -6x + 4y - 8z &= -2 \end{aligned}$

51. $\begin{aligned} x + y + z &= 1 \\ 2x - y &= -2 \\ 3y + z &= 2 \end{aligned}$

52. $\begin{aligned} x &= -3 \\ y + z &= 6 \\ 2x - 3z &= -9 \end{aligned}$

53. $\begin{aligned} 2x - 4y + 4z &= 0 \\ x - 3y + 2z &= -3 \\ x - y + 2z &= 1 \end{aligned}$

54. *Concept Check* If the solution set is given by $\{(4 - y, y)\}$, give one specific solution.

Evaluate each determinant.

55. $\det \begin{bmatrix} -1 & 8 \\ 2 & 9 \end{bmatrix}$

56. $\det \begin{bmatrix} -2 & 4 \\ 0 & 3 \end{bmatrix}$

57. $\det \begin{bmatrix} -2 & 4 & 1 \\ 3 & 0 & 2 \\ -1 & 0 & 3 \end{bmatrix}$

58. $\det \begin{bmatrix} -1 & 2 & 3 \\ 4 & 0 & 3 \\ 5 & -1 & 2 \end{bmatrix}$

Solve each determinant equation for x.

59. $\det \begin{bmatrix} -3 & 2 \\ 1 & x \end{bmatrix} = 5$

60. $\det \begin{bmatrix} 3x & 7 \\ -x & 4 \end{bmatrix} = 8$

61. $\det \begin{bmatrix} 2 & 5 & 0 \\ 1 & 3x & -1 \\ 0 & 2 & 0 \end{bmatrix} = 4$

62. $\det \begin{bmatrix} 6x & 2 & 0 \\ 1 & 5 & 3 \\ x & 2 & -1 \end{bmatrix} = 2x$

Exercises 63 and 64 refer to the system
$$3x - y = 28$$
$$2x + y = 2$$.

63. Suppose you are asked to solve this system by using Cramer's rule.
  (a) What is the value of D?
  (b) What is the value of D_x?
  (c) What is the value of D_y?
  (d) Find x and y, using Cramer's rule.

64. Suppose you are asked to solve this system by applying the inverse matrix method to $AX = B$.
  (a) What is A?
  (b) What is B?
  (c) Explain how you would use these matrices to go about solving the system.

65. Cramer's rule has the condition $D \neq 0$. Why is this necessary? What is true of the system when $D = 0$?

Solve each system, if possible, by using Cramer's rule. Identify any systems with dependent equations or any inconsistent systems.

66. $3x + y = -1$
    $5x + 4y = 10$

67. $3x + 7y = 2$
    $5x - y = -22$

68. $2x - 5y = 8$
    $3x + 4y = 10$

69. $3x + 2y + z = 2$
    $4x - y + 3z = -16$
    $x + 3y - z = 12$

70. $5x - 2y - z = 8$
    $-5x + 2y + z = -8$
    $x - 4y - 2z = 0$

71. $-x + 3y - 4z = 2$
    $2x + 4y + z = 3$
    $3x - z = 9$

Solve each problem.

72. *Determining the Contents of a Meal* A cup of uncooked rice contains 15 grams of protein and 810 calories. A cup of uncooked soybeans contains 22.5 grams of protein and 270 calories. How many cups of each should be used for a meal containing 9.5 grams of protein and 324 calories?

73. *Determining Order Quantities* A company sells CDs for $0.40 each and plastic holders for $0.30 each. The company receives $38 for an order of 100 CDs and holders. However, the customer neglected to specify how many of each to send. Determine the number of CDs and the number of holders that should be sent.

74. *Mixing Teas* Three kinds of tea worth $4.60, $5.75, and $6.50 per pound are to be mixed to get 20 pounds of tea worth $5.25 per pound. The amount of $4.60 tea used is to equal the amount of the other two kinds together. How many pounds of each tea should be used?

75. *Mixing Solutions of a Drug* A 5% solution of a drug is to be mixed with some 15% solution and some 10% solution to get 20 milliliters of 8% solution. The amount of 5% solution used must be 2 milliliters more than the sum of the other two solutions. How many milliliters of each solution should be used?

76. *(Modeling) Blood Pressure* In a study of adult males, it was believed that systolic blood pressure P was affected by both age A in years and weight W in pounds. This was modeled by

$$P = a + bA + cW,$$

where a, b, and c are constants. The table in the next column lists three individuals with representative blood pressures for the group.

P	A	W
113	39	142
138	53	181
152	65	191

(a) Use the data to approximate values for the constants a, b, and c.
(b) Estimate a typical systolic blood pressure for an individual who is 55 years old and weighs 175 pounds.

77. *Fitting Data* Find the equation of the polynomial function of degree 3 whose graph passes through the points $(-2, 1)$, $(-1, 6)$, $(2, 9)$, and $(3, 26)$.

78. *Fitting Data* Find the equation of the quadratic polynomial that defines the curve shown in the figure.

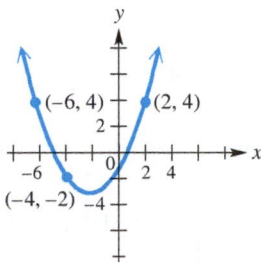

Graph the solution set of each system.

79. $x + y \leq 6$
    $2x - y \geq 3$

80. $y \leq \dfrac{1}{3}x - 2$
    $y^2 \leq 16 - x^2$

81. Find $x \geq 0$ and $y \geq 0$ such that

$$3x + 2y \leq 12$$
$$5x + y \geq 5$$

and $2x + 4y$ is maximized.

82. Find $x \geq 0$ and $y \geq 0$ such that

$$x + y \leq 50$$
$$2x + y \geq 20$$
$$x + 2y \geq 30$$

and $4x + 2y$ is minimized.

83. *(Modeling)* *Company Profit* A small company manufactures two products: radios and Blu-ray players. Each radio results in a profit of $15, each Blu-ray player a profit

of $35. Due to demand, the company must produce at least 5, and not more than 25, radios per day. The number of radios cannot exceed the number of Blu-ray players, and the number of Blu-ray players cannot exceed 30. How many of each should the company manufacture to obtain maximum profit? What will that profit be?

Find the partial fraction decomposition of each rational expression.

84. $\dfrac{5x - 2}{x^2 - 4}$

85. $\dfrac{x + 2}{x^3 + 2x^2 + x}$

86. $\dfrac{x + 2}{x^3 - x^2 + 4x}$

87. $\dfrac{6x^2 - x - 3}{x^3 - x}$

---

6 Test

1. Consider the system of equations.

$$x^2 + y^2 = 5$$
$$2x - y = 0$$

   (a) What type of graph does each equation have?
   (b) How many points of intersection of the two graphs are possible?
   (c) Solve the system.
   (d) Support the solutions, using a graphing calculator.

2. Solve each system by substitution.

   (a) $x - 2y = 1$
   $2x + y = 7$

   (b) $3x - y = 1$
   $-6x + 2y = -2$

3. Solve each system by elimination.

   (a) $3x - 4y = 7$
   $x + 3y = \frac{9}{2}$

   (b) $x - y = 5$
   $-2x + 2y = 1$

4. Solve the nonlinear system.

$$x^2 - y = 5$$
$$x^2 + y^2 = 11$$

5. Solve each system.

   (a) $2x + y + z = 3$
   $x + 2y - z = 3$
   $3x - y + z = 5$

   (b) $x + y - z = 1$
   $2x + 3y + z = 6$
   $x + 2y + 2z = 5$

6. Perform each matrix operation if possible.

   (a) $3\begin{bmatrix} 2 & 3 \\ 1 & -4 \\ 5 & 9 \end{bmatrix} - \begin{bmatrix} -2 & 6 \\ 3 & -1 \\ 0 & 8 \end{bmatrix}$

   (b) $\begin{bmatrix} 1 \\ 2 \end{bmatrix} + \begin{bmatrix} 4 \\ -6 \end{bmatrix} + \begin{bmatrix} 2 & 8 \\ -7 & 5 \end{bmatrix}$

   (c) $\begin{bmatrix} 2 & 1 & -3 \\ 4 & 0 & 5 \end{bmatrix} \begin{bmatrix} 1 & 3 \\ 2 & 4 \\ 3 & -2 \end{bmatrix}$

7. Suppose A and B are both $n \times n$ matrices.
   (a) Can AB be found?
   (b) Can BA be found?
   (c) Does $AB = BA$ in general? Explain why or why not.
   (d) If A is $n \times n$ and C is $m \times n$, can either AC or CA be found?

8. Evaluate each determinant.

   (a) $\det \begin{bmatrix} 4 & 9 \\ -5 & -11 \end{bmatrix}$

   (b) $\det \begin{bmatrix} 2 & 0 & 8 \\ -1 & 7 & 9 \\ 12 & 5 & -3 \end{bmatrix}$

9. Solve the system by using Cramer's rule.

$$2x - 3y = -33$$
$$4x + 5y = 11$$

10. Consider the system of equations.

$$x + y - z = -4$$
$$2x - 3y - z = 5$$
$$x + 2y + 2z = 3$$

   (a) Write the matrix of coefficients A, the matrix of variables X, and the matrix of constants B for this system.
   (b) Find A^{-1}.
   (c) Use the matrix inverse method to solve the system.
   (d) If the first equation in the system is replaced by $0.5x + y + z = 1.5$, the system cannot be solved with the matrix inverse method. Explain why.

11. *World Population*   The table lists actual and projected world population in billions for three selected years.

Year	1850	1950	2050
Population	1.3	2.5	8.9

Source: United Nations.

(a) Find a function $f(x) = ax^2 + bx + c$ that models the data, where x represents years *after* 1850.

(b) Use f to estimate when world population might reach 8 billion.

12. The solution set of a system of inequalities is shown. Which system is it from choices A–D?

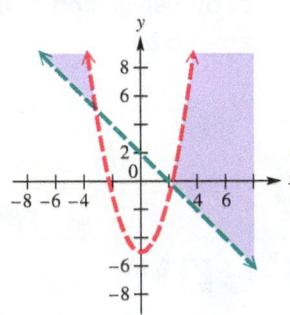

A. $y > 2 - x$
$\quad\ y > x^2 - 5$

B. $y > 2 - x$
$\quad\ y < x^2 - 5$

C. $y < 2 - x$
$\quad\ y < x^2 - 5$

D. $y < 2 - x$
$\quad\ y > x^2 - 5$

13. *Storage Capacity*   An office manager wants to buy filing cabinets. Cabinet X costs $100, requires 6 square feet of floor space, and holds 8 cubic feet. Cabinet Y costs $200, requires 8 square feet of floor space, and holds 12 cubic feet. No more than $1400 can be spent, and the office has room for no more than 72 square feet of cabinets. The office manager wants the maximum storage capacity within the limits imposed by funds and space. How many of each type of cabinet should be bought?

Find the partial fraction decomposition for each rational expression.

14. $\dfrac{7x - 1}{x^2 - x - 6}$

15. $\dfrac{x^2 - 11x + 6}{(x + 2)(x - 2)^2}$

Parabolas, ellipses, and hyperbolas are age-old curves that have had a profound influence on our understanding of ourselves and of the cosmos around us.

7 Analytic Geometry and Nonlinear Systems

7.1 Circles and Parabolas

Conic Sections • Equations and Graphs of Circles • Equations and Graphs of Parabolas • Translations of Parabolas
• An Application of Parabolas

Conic Sections

Parabolas, circles, ellipses, and *hyperbolas* form a group of curves known as the **conic sections,** because they are the result of intersecting a cone with a plane. **FIGURE 1** illustrates these curves, which all can be defined mathematically by using the distance formula from **Section 1.1.**

$$d(P, R) = \sqrt{(x_2 - x_1)^2 + (y_2 - y_1)^2} \qquad \text{Distance formula}$$

Conic Sections

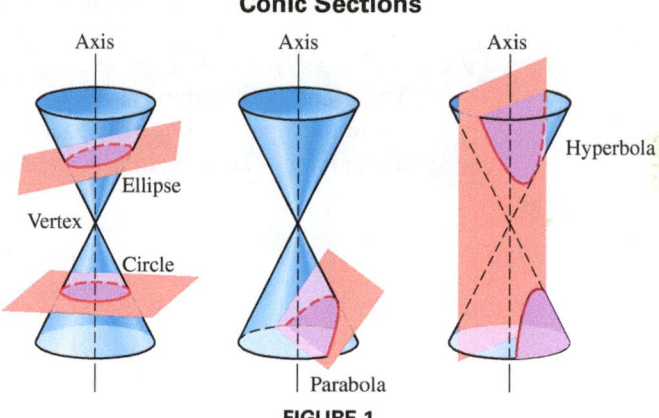

FIGURE 1

If the plane intersects the cone at the vertex, the intersections in **FIGURE 1** are reduced respectively to a point, a line, and two intersecting lines. These are called **degenerate conic sections.**

Equations and Graphs of Circles

> **Circle**
>
> A **circle** is a set of points in a plane that are equidistant from a fixed point. The distance is called the **radius** of the circle, and the fixed point is called the **center.**

Equation of a Circle

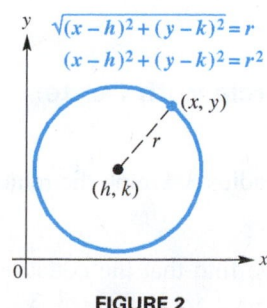

FIGURE 2

Suppose a circle has center (h, k) and radius $r > 0$. See **FIGURE 2.** Then the distance between the center (h, k) and any point (x, y) on the circle must equal r. Thus, an equation of the circle is as follows.

$$\sqrt{(x - h)^2 + (y - k)^2} = r \qquad \text{Distance formula}$$

$$(x - h)^2 + (y - k)^2 = r^2 \qquad \text{Square both sides.}$$

Center–Radius Form of the Equation of a Circle

The **center–radius form** of the equation of a circle with center (h, k) and radius r is

$$(x - h)^2 + (y - k)^2 = r^2.$$

Notice that a circle is the graph of a relation that is not a function, since it does not pass the vertical line test.

EXAMPLE 1 **Finding the Equation of a Circle**

Find the center–radius form of the equation of a circle with radius 6 and center $(-3, 4)$. Graph the circle by hand. Give the domain and range of the relation.

Solution Using the center–radius form with $h = -3$, $k = 4$, and $r = 6$, we find that the equation of the circle is as follows.

$$[x - (-3)]^2 + (y - 4)^2 = 6^2, \quad \text{or} \quad (x + 3)^2 + (y - 4)^2 = 36$$

The graph is shown in **FIGURE 3.** Because the center is $(-3, 4)$ and the radius is 6, the circle must pass through the four points $(-3 \pm 6, 4)$ and $(-3, 4 \pm 6)$, as illustrated in the table. The domain is $[-9, 3]$ and the range is $[-2, 10]$.

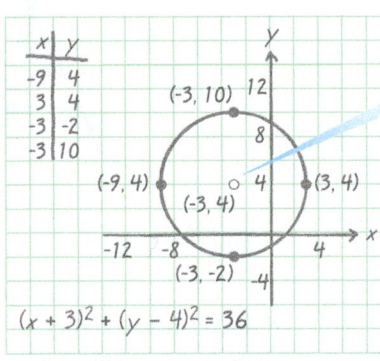

FIGURE 3

If a circle has center at the origin $(0, 0)$, then its equation is found by using $h = 0$ and $k = 0$ in the center–radius form. See **FIGURE 4.**

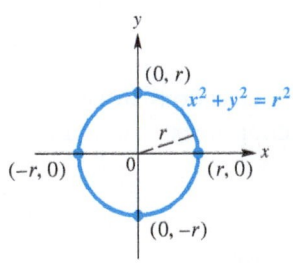

Circle centered at $(0, 0)$
with radius r
FIGURE 4

Equation of a Circle with Center at the Origin

A circle with center $(0, 0)$ and radius r has equation

$$x^2 + y^2 = r^2.$$

EXAMPLE 2 **Finding the Equation of a Circle with Center at the Origin**

Find the equation of a circle with center at the origin and radius 3. Graph the relation, and state the domain and range.

Solution Using the form $x^2 + y^2 = r^2$ with $r = 3$, we find that the equation of the circle is $x^2 + y^2 = 9$. See **FIGURE 5.** The domain and range are both $[-3, 3]$.

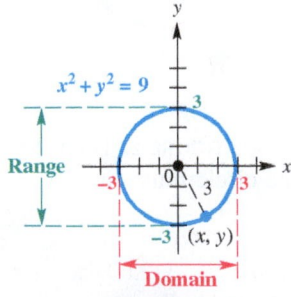

FIGURE 5

As mentioned in **Section 4.4,** a graphing calculator in function mode *cannot* directly graph a circle. We must solve the equation of the circle for y, obtaining two functions y_1 and y_2. The union of these two graphs is the graph of the circle.

TECHNOLOGY NOTE

To obtain an undistorted graph of a circle in function mode on a graphing calculator screen, a **square viewing window** must be used. For example, the graph below of $x^2 + y^2 = 9$ in a rectangular window looks like an ellipse.

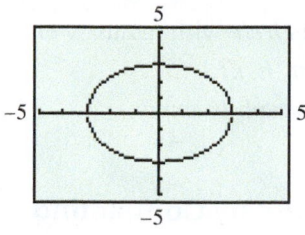

| **EXAMPLE 3** | **Graphing Circles** |

Use a graphing calculator to graph each circle in a square viewing window.

(a) $x^2 + y^2 = 9$ **(b)** $(x + 3)^2 + (y - 4)^2 = 36$

Solution   In both cases, first solve for y. Recall that, by the square root property, if $k > 0$, then $y^2 = k$ has *two* real solutions: $\sqrt{k}$ and $-\sqrt{k}$.

(a)

$$x^2 + y^2 = 9$$

$$y^2 = 9 - x^2 \qquad \text{Subtract } x^2.$$

Remember both the positive and negative square roots. $y = \pm\sqrt{9 - x^2} \qquad \text{Take square roots.}$

Graph $y_1 = \sqrt{9 - x^2}$ and $y_2 = -\sqrt{9 - x^2}$. See **FIGURE 6,** and compare it with **FIGURE 5** on the preceding page.

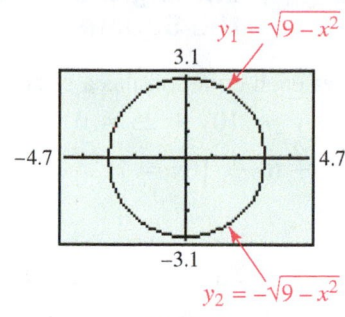

FIGURE 6

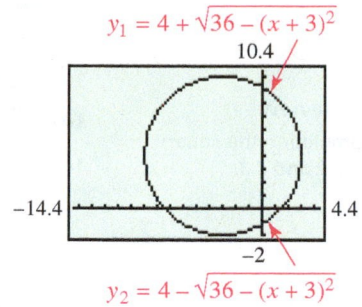

FIGURE 7

TECHNOLOGY NOTE

If we graph the circle in **FIGURE 6** without using a decimal window, we may get a figure similar to the one shown below, where the two graphs do not meet at the points $(-3, 0)$ and $(3, 0)$. In some cases, a decimal window does *not* correct this problem. Although the technology is deceiving, mathematically this is a complete circle.

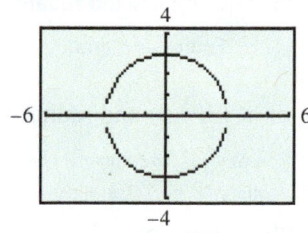

(b) $(x + 3)^2 + (y - 4)^2 = 36$

$$(y - 4)^2 = 36 - (x + 3)^2 \qquad \text{Subtract } (x + 3)^2.$$

$$y - 4 = \pm\sqrt{36 - (x + 3)^2} \qquad \text{Take square roots.}$$

$$y = 4 \pm \sqrt{36 - (x + 3)^2} \qquad \text{Add 4.}$$

The two functions graphed in **FIGURE 7** are

$$y_1 = 4 + \sqrt{36 - (x + 3)^2} \quad \text{and} \quad y_2 = 4 - \sqrt{36 - (x + 3)^2}.$$

Suppose that we start with the center–radius form of the equation of a circle and rewrite it by squaring each binomial.

$$(x - h)^2 + (y - k)^2 = r^2 \qquad \text{Center–radius form}$$

$$x^2 - 2xh + h^2 + y^2 - 2yk + k^2 - r^2 = 0 \qquad \text{Square each binomial and subtract } r^2.$$

$$x^2 + y^2 + (-2h)x + (-2k)y + (h^2 + k^2 - r^2) = 0 \qquad \text{Properties of real numbers}$$

$$\underset{c}{\uparrow} \qquad \underset{d}{\uparrow} \qquad \underset{e}{\uparrow}$$

← Algebra Review

To review how to multiply special products like $(x - h)^2$ and $(y - k)^2$, see **Section R.1.**

If $r > 0$, then the graph of this equation is a circle with center (h, k) and radius r, as seen earlier. This is the **general form of the equation of a circle.**

→ **Looking Ahead to Calculus**
The circle $x^2 + y^2 = 1$ is called the **unit circle**. It is important in interpreting the **trigonometric** or **circular functions** in calculus.

General Form of the Equation of a Circle

For real numbers c, d, and e, the equation

$$x^2 + y^2 + cx + dy + e = 0$$

can have a graph that is a circle, that is a point, or that is empty (contains no points.)

Starting with an equation in this general form, we can work in reverse by completing the square to get an equation of the form

$$(x - h)^2 + (y - k)^2 = m, \quad \text{for some number } m.$$

There are three possibilities for the graph, based on the value of m.

1. If $m > 0$, then $r^2 = m$, and the equation represents a *circle* with radius $\sqrt{m}$.

2. If $m = 0$, then the equation represents the single *point* (h, k).

3. If $m < 0$, then no points satisfy the equation and the graph *is empty*.

EXAMPLE 4 **Finding the Center and Radius by Completing the Square**

Decide whether each equation has a circle as its graph. Sketch the graphs.

(a) $x^2 - 6x + y^2 + 10y + 25 = 0$ **(b)** $x^2 + 10x + y^2 - 4y + 33 = 0$

(c) $2x^2 + 2y^2 - 6x + 10y = 1$

← **Algebra Review**
To review completing the square, see **Sections 3.2 and 3.3**.

Solution

(a)
$$x^2 - 6x + y^2 + 10y + 25 = 0$$

Add $\left(\frac{k}{2}\right)^2$ to complete the square ($k = -6, k = 10$).

$$(x^2 - 6x + \underline{}) + (y^2 + 10y + \underline{}) = -25 \qquad \text{Add } -25.$$

Add $\left[\frac{1}{2}(-6)\right]^2 = (-3)^2 = 9$ and $\left[\frac{1}{2}(10)\right]^2 = 5^2 = 25$

inside each set of parentheses to complete the square.

$$(x^2 - 6x + 9) + (y^2 + 10y + 25) = -25 + 9 + 25 \qquad \text{Complete the square.}$$

Add 9 and 25 to each side.

$$(x - 3)^2 + (y + 5)^2 = 9 \qquad \text{Factor and add.}$$

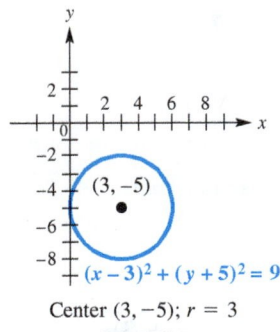

$(x-3)^2 + (y+5)^2 = 9$

Center $(3, -5)$; $r = 3$

FIGURE 8

Since $9 > 0$, the equation represents a circle with center at $(3, -5)$ and radius 3. See **FIGURE 8.**

(b)
$$x^2 + 10x + y^2 - 4y + 33 = 0$$

Add $\left(\frac{k}{2}\right)^2$ to complete the square ($k = 10, k = -4$).

$$(x^2 + 10x + \underline{}) + (y^2 - 4y + \underline{}) = -33 \qquad \text{Add } -33.$$

Add $\left[\frac{1}{2}(10)\right]^2 = 25$ and $\left[\frac{1}{2}(-4)\right]^2 = 4$

inside each set of parentheses to complete the square.

$$(x^2 + 10x + 25) + (y^2 - 4y + 4) = -33 + 25 + 4 \qquad \text{Complete the square.}$$

$$(x + 5)^2 + (y - 2)^2 = -4 \qquad \text{Factor and add.}$$

Because $-4 < 0$, there are no ordered pairs (x, y), with x and y both real numbers, satisfying the equation. The graph contains no points.

(c) $2x^2 + 2y^2 - 6x + 10y = 1$

To complete the square, make the coefficients of the x^2- and y^2-terms 1 by factoring out 2.

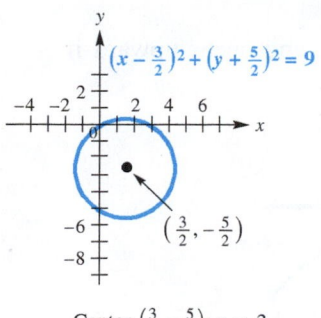

$(x - \frac{3}{2})^2 + (y + \frac{5}{2})^2 = 9$

Center $\left(\frac{3}{2}, -\frac{5}{2}\right)$; $r = 3$

FIGURE 9

Coefficients of the x^2- and y^2-terms must be 1.

$2(x^2 - 3x) + 2(y^2 + 5y) = 1$ Group the terms and factor out 2.

$2\left(x^2 - 3x + \frac{9}{4}\right) + 2\left(y^2 + 5y + \frac{25}{4}\right) = 1 + 2\left(\frac{9}{4}\right) + 2\left(\frac{25}{4}\right)$ *We are adding* $2\left(\frac{9}{4}\right) + 2\left(\frac{25}{4}\right)$ *to each side.*   Complete the square.

$2\left(x - \frac{3}{2}\right)^2 + 2\left(y + \frac{5}{2}\right)^2 = 18$ Factor and simplify on the right.

$\left(x - \frac{3}{2}\right)^2 + \left(y + \frac{5}{2}\right)^2 = 9$ Divide each side by 2.

The equation is a circle with center at $\left(\frac{3}{2}, -\frac{5}{2}\right)$ and radius 3. See **FIGURE 9.**   ●

Equations and Graphs of Parabolas

Like the definition of a circle, the definition of a parabola is based on distance.

Parabola

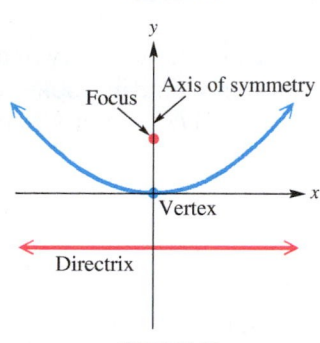

FIGURE 10

> ### Parabola
>
> A **parabola** is a set of points in a plane equidistant from a fixed point and a fixed line. The fixed point is called the **focus,** and the fixed line the **directrix,** of the parabola.

As shown in **FIGURE 10,** the axis of symmetry of a parabola passes through the focus and is perpendicular to the directrix. The vertex is the midpoint of the line segment joining the focus and directrix on the axis.

We can find an equation of a parabola from the preceding definition. Let the directrix be the line $y = -c$ and the focus be the point F with coordinates $(0, c)$. See **FIGURE 11.** To find the equation of the set of points that are the same distance from the line $y = -c$ and the point $(0, c)$, choose one such point P and give it coordinates (x, y). By the definition of a parabola, $d(P, F)$ and $d(P, D)$ must be equal, so using the distance formula gives the following result.

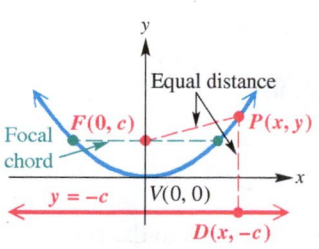

$d(P, F) = d(P, D)$
for all P on the parabola.

FIGURE 11

$d(P, F) = d(P, D)$ See **Figure 11**.

$\sqrt{(x - 0)^2 + (y - c)^2} = \sqrt{(x - x)^2 + [y - (-c)]^2}$ Distance formula

$\sqrt{x^2 + (y - c)^2} = \sqrt{(y + c)^2}$ Simplify.

$x^2 + (y - c)^2 = (y + c)^2$ Square each side.

$x^2 + y^2 - 2yc + c^2 = y^2 + 2yc + c^2$ Square both binomials.

$x^2 = 4cy$ Subtract y^2 and c^2; add $2yc$.

Remember the middle term when squaring a binomial.

This discussion is summarized in the following box.

Parabola with a Vertical Axis and Vertex (0, 0)

The parabola with focus $(0, c)$ and directrix $y = -c$ has equation

$$x^2 = 4cy.$$

The parabola has vertex $(0, 0)$, vertical axis $x = 0$, and opens upward if $c > 0$ or downward if $c < 0$.

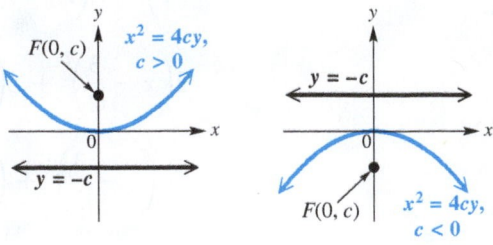

NOTE   *When graphing a parabola, remember that the focus is located "inside" the parabola.* If the parabola opens upward, the focus is above the vertex; if the parabola opens downward, the focus is below the vertex.

The *focal chord* through the focus and perpendicular to the axis of symmetry of a parabola is called the **latus rectum,** and has length $|4c|$. To see this, note in **FIGURE 11** on the preceding page that the endpoints of the chord are $(-x, c)$ and (x, c). Let $y = c$ in the equation of the parabola and solve for x.

$$x^2 = 4cy$$
$$x^2 = 4c^2 \qquad \text{Substitute } c \text{ for } y.$$
$$x = |2c| \qquad \text{Take the positive square root.}$$

The length of half the focal chord is $|2c|$, so its full length is $|4c|$. Thus, when graphing a parabola, the points with coordinates $(-2c, c)$ and $(2c, c)$ lie on the parabola.

If the directrix is the line $x = -c$ and the focus is $(c, 0)$, then the definition of a parabola and the distance formula leads to the equation of a parabola with a horizontal axis. (See **Exercise 117.**)

Parabola with a Horizontal Axis and Vertex (0, 0)

The parabola with focus $(c, 0)$ and directrix $x = -c$ has equation

$$y^2 = 4cx.$$

The parabola has vertex $(0, 0)$, horizontal axis $y = 0$, and opens to the right if $c > 0$ or to the left if $c < 0$.

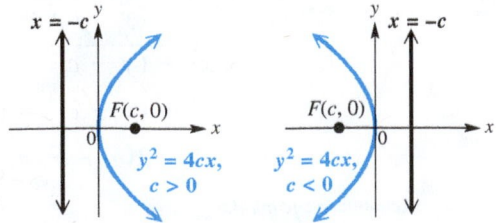

Notice that the graph of a parabola with a horizontal axis is not a function, since it does not pass the vertical line test.

EXAMPLE 5 **Finding Information about Parabolas from Equations**

Find the focus, directrix, vertex, and axis of each parabola. Then graph each parabola.

(a) $x^2 = 8y$ **(b)** $y^2 = -28x$

Solution

(a) The equation $x^2 = 8y$ has the form $x^2 = 4cy$, so set $4c = 8$, from which $c = 2$. Since the x-term is squared, the parabola is vertical, with focus $(0, c) = (0, 2)$ and directrix $y = -2$. The vertex is $(0, 0)$, and the axis of the parabola is the y-axis. See **FIGURE 12**, which also shows the endpoints of the latus rectum, $(\pm 4, 2)$.

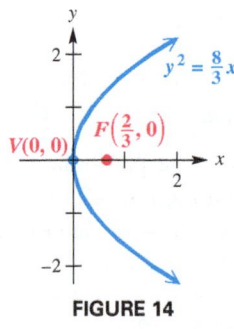
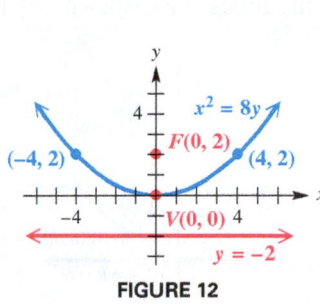

FIGURE 12

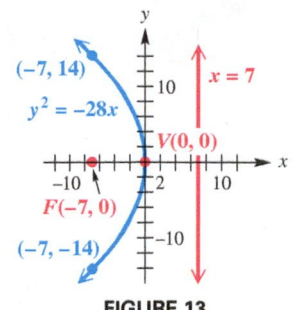

FIGURE 13

(b) The equation $y^2 = -28x$ has the form $y^2 = 4cx$, with $4c = -28$, so $c = -7$. The parabola is horizontal, with focus $(-7, 0)$, directrix $x = 7$, vertex $(0, 0)$, and x-axis as axis of the parabola. The latus rectum has endpoints $(-7, -14)$ and $(-7, 14)$. Since c is negative, the graph opens to the left, as shown in **FIGURE 13.**

EXAMPLE 6 **Writing Equations of Parabolas**

Write an equation for each parabola.

(a) Focus $\left(\frac{2}{3}, 0\right)$ and vertex at the origin

(b) Vertical axis, vertex at the origin, through the point $(-2, 12)$

Solution

(a) Since the focus $\left(\frac{2}{3}, 0\right)$ and the vertex $(0, 0)$ are both on the x-axis, the parabola is horizontal. It opens to the right because $c = \frac{2}{3}$ is positive. See **FIGURE 14.** The equation, which will have the form $y^2 = 4cx$, is

$$y^2 = 4\left(\frac{2}{3}\right)x, \quad \text{or} \quad y^2 = \frac{8}{3}x.$$

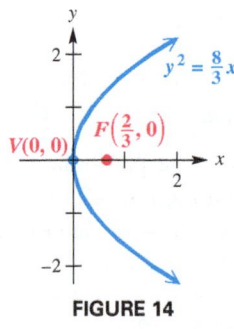

FIGURE 14

(b) The parabola will have an equation of the form $x^2 = 4cy$ because the axis is vertical and the vertex is $(0, 0)$. Since the point $(-2, 12)$ is on the graph, it must satisfy this equation.

$$x^2 = 4cy$$
$$(-2)^2 = 4c(12) \qquad \text{Let } x = -2 \text{ and } y = 12.$$
$$4 = 48c \qquad \text{Multiply.}$$
$$c = \frac{1}{12} \qquad \text{Solve for } c.$$

Thus, $x^2 = 4cy = 4\left(\frac{1}{12}\right)y$, which gives $x^2 = \frac{1}{3}y$ or $y = 3x^2.$

NOTE When working problems like those in **Example 6**, a quick sketch often makes the process easier.

Translations of Parabolas

The equations $x^2 = 4cy$ and $y^2 = 4cx$ can be extended to parabolas having vertex (h, k) by replacing x and y with $x - h$ and $y - k$, respectively.

Equation Forms for Translated Parabolas

A parabola with vertex (h, k) has an equation of the form

$$(x - h)^2 = 4c(y - k) \qquad \text{Vertical axis}$$

or $$(y - k)^2 = 4c(x - h), \qquad \text{Horizontal axis}$$

where the focus is a distance $|c|$ from the vertex.

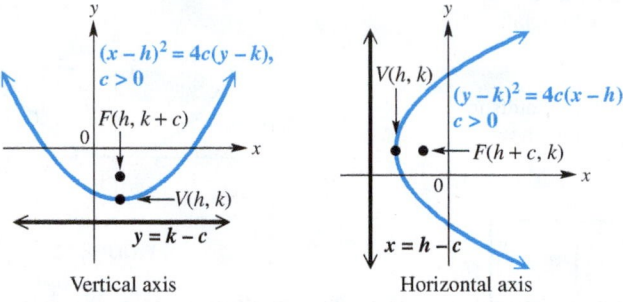

Vertical axis   Horizontal axis

EXAMPLE 7 **Graphing a Parabola with Vertex (h, k)**

Graph the parabola $x = -\frac{1}{8}(y + 3)^2 + 2$. Label the vertex, focus, and directrix.

Solution Rewrite the equation in the form $(y - k)^2 = 4c(x - h)$.

$$x = -\frac{1}{8}(y + 3)^2 + 2$$

$$x - 2 = -\frac{1}{8}(y + 3)^2 \qquad \text{Subtract 2.}$$

$$-8(x - 2) = (y + 3)^2 \qquad \text{Multiply by } -8.$$

$$(y + 3)^2 = -8(x - 2) \qquad \text{Rewrite.}$$

The vertex is $(2, -3)$. Since $4c = -8$, we have $c = -2$, and the parabola opens to the left. The focus is 2 units to the left of the vertex, and the directrix is 2 units to the right of the vertex. Thus, the focus is $(0, -3)$ and the directrix is $x = 4$. See **FIGURE 15.**

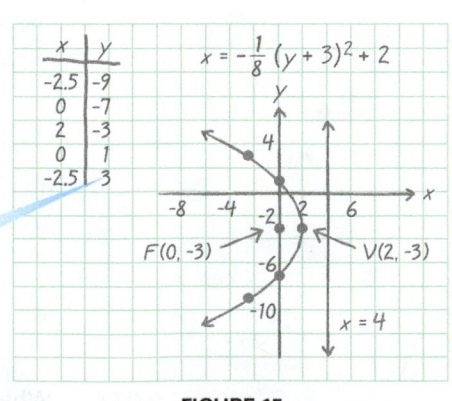

Choose values for y and calculate x in the original equation.

FIGURE 15

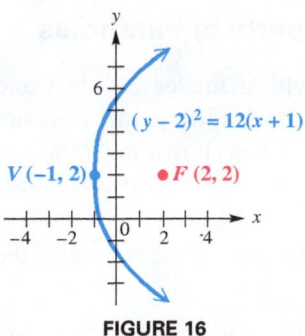

$(y - 2)^2 = 12(x + 1)$

$V(-1, 2)$ $\bullet F(2, 2)$

FIGURE 16

Writing an Equation of a Parabola

Write an equation for the parabola with focus $(2, 2)$ and vertex $(-1, 2)$, and graph it. Give the domain and range.

Solution   Since the focus is to the right of the vertex, the axis is horizontal and the parabola opens to the right. See **FIGURE 16**. The distance between the focus and the vertex is $2 - (-1) = 3$, so $c = 3$. The equation of the parabola is found as follows.

$$(y - k)^2 = 4c(x - h) \qquad \text{Parabola with horizontal axis}$$
$$(y - 2)^2 = 4(3)(x + 1) \qquad \text{Substitute for } c, h, \text{ and } k.$$
$$(y - 2)^2 = 12(x + 1) \qquad \text{Multiply.}$$

From **FIGURE 16**, the domain is $[-1, \infty)$ and the range is $(-\infty, \infty)$.

Completing the Square to Graph a Horizontal Parabola

Graph $x = -2y^2 + 4y - 3$. Give the domain and range.

Analytic Solution

$$x = -2y^2 + 4y - 3$$
$$x + 3 = -2y^2 + 4y \qquad \text{Add 3.}$$
$$x + 3 = -2(y^2 - 2y) \qquad \text{Factor out } -2.$$
$$x + 3 - 2 = -2(y^2 - 2y + 1) \qquad \text{Complete the square.}$$
$$x + 1 = -2(y - 1)^2 \qquad \text{Add and factor the trinomial.}$$
$$(y - 1)^2 = -\frac{1}{2}(x + 1) \quad (*) \qquad \text{Multiply by } -\frac{1}{2} \text{ and rewrite.}$$

The parabola opens to the left, is narrower than the graph of $x = y^2$, and has vertex $(-1, 1)$. The domain is $(-\infty, -1]$ and the range is $(-\infty, \infty)$. See **FIGURE 17**.

x	y
-3	0
-3	2
-1	1

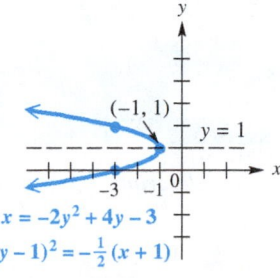

$x = -2y^2 + 4y - 3$
$(y - 1)^2 = -\frac{1}{2}(x + 1)$

FIGURE 17

Graphing Calculator Solution

To obtain a calculator graph, we begin by solving the equation $(y - 1)^2 = -\frac{1}{2}(x + 1)$ $(*)$ from the analytic solution for y.

$$y - 1 = \pm\sqrt{\frac{x + 1}{-2}} \qquad \text{Take square roots.}$$
$$y = 1 \pm \sqrt{\frac{x + 1}{-2}}. \qquad \text{Add 1.}$$

Graphing these two equations as y_1 and y_2 results in **FIGURE 18**.

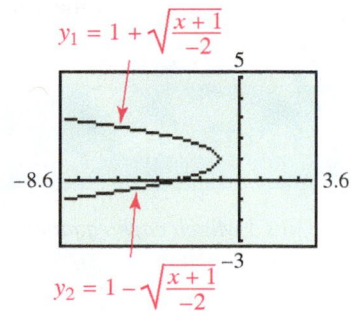

$y_1 = 1 + \sqrt{\frac{x + 1}{-2}}$

$y_2 = 1 - \sqrt{\frac{x + 1}{-2}}$

FIGURE 18

An Application of Parabolas

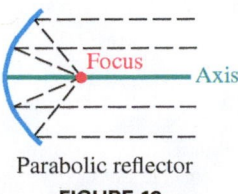

Parabolic reflector
FIGURE 19

The geometric properties of parabolas lead to many practical applications. For example, if a light source is placed at the focus of a parabolic reflector, as in **FIGURE 19**, light rays reflect parallel to the axis, making a spotlight or flashlight. The process also works in reverse. Light rays from a distant source come in parallel to the axis and are reflected to a point at the focus. This use of parabolic reflection is seen in the satellite dishes used to pick up signals from communications satellites.

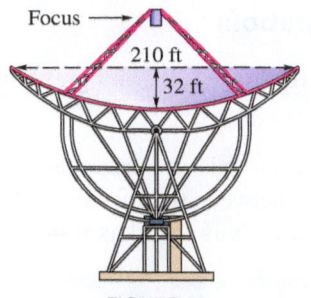

Focus

210 ft

32 ft

FIGURE 20

EXAMPLE 10 **Modeling the Reflective Property of Parabolas**

The Parkes radio telescope has a parabolic dish shape with diameter 210 feet and depth 32 feet. Because of this parabolic shape, distant rays hitting the dish are reflected directly toward the focus. A cross section of the dish is shown in **FIGURE 20**. (*Source:* Mar, J. and H. Liebowitz, *Structure Technology for Large Radio and Radar Telescope Systems,* MIT Press.)

(a) Determine an equation describing this cross section by placing the vertex at the origin with the parabola opening upward.

(b) The receiver must be placed at the focus of the parabola. How far from the vertex of the parabolic dish should the receiver be located?

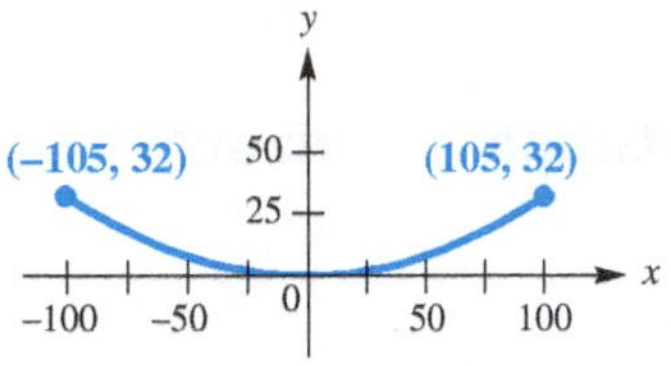

FIGURE 21

(−105, 32)   (105, 32)

Solution

(a) Locate the vertex at the origin, as shown in **FIGURE 21**. The form of the parabola is $x^2 = 4cy$. The parabola must pass through the point $\left(\frac{210}{2}, 32\right) = (105, 32)$.

$$x^2 = 4cy \qquad \text{Vertical parabola}$$
$$(105)^2 = 4c(32) \qquad \text{Let } x = 105 \text{ and } y = 32.$$
$$11{,}025 = 128c \qquad \text{Multiply.}$$
$$c = \frac{11{,}025}{128} \qquad \text{Solve for } c.$$

The cross section can be modeled by the following equation.

$$x^2 = 4cy$$
$$x^2 = 4\left(\frac{11{,}025}{128}\right)y \qquad \text{Substitute for } c.$$
$$x^2 = \frac{11{,}025}{32}y \qquad \text{Simplify.}$$

(b) The distance between the vertex and the focus is c. In part (a), we found that $c = \frac{11{,}025}{128} \approx 86.1$, so the receiver should be located at $(0, 86.1)$, or 86.1 ft above the vertex.

7.1 Exercises

Checking Analytic Skills   *Match each equation in Column I with the appropriate description in Column II.* **Do not use a calculator.**

I

1. $x = 2y^2$

2. $y = 2x^2$

3. $x^2 = -3y$

4. $y^2 = -3x$

5. $x^2 + y^2 = 5$

6. $(x - 3)^2 + (y + 4)^2 = 25$

7. $(x + 3)^2 + (y - 4)^2 = 25$

8. $x^2 + y^2 = -4$

II

A. Circle; center $(3, -4)$; radius 5

B. Parabola; opens left

C. Parabola; opens upward

D. Circle; center $(-3, 4)$; radius 5

E. Parabola; opens right

F. Circle; center $(0, 0)$; radius $\sqrt{5}$

G. No points on its graph

H. Parabola; opens downward

Find the center–radius form for each circle satisfying the given conditions.

9. Center (1, 4); radius 3

10. Center (−2, 5); radius 4

11. Center (0, 0); radius 1

12. Center (0, 0); radius 5

13. Center $\left(\dfrac{2}{3}, -\dfrac{4}{5}\right)$; radius $\dfrac{3}{7}$

14. Center $\left(-\dfrac{1}{2}, -\dfrac{1}{4}\right)$; radius $\dfrac{12}{5}$

15. Center (−1, 2); passing through (2, 6)

16. Center (2, −7); passing through (−2, −4)

17. Center (−3, −2); tangent to the *x*-axis (*Hint:* "tangent to" means touching at one point.)

18. Center (5, −1); tangent to the *y*-axis

19. Describe the graph of the following equation.
$$(x - 3)^2 + (y - 3)^2 = 0$$

20. Describe the graph of the following equation.
$$(x - 3)^2 + (y - 3)^2 = -1$$

We can find an equation of a circle if we know the coordinates of the endpoints of a diameter of the circle. First, find the midpoint of the diameter, which is the center of the circle. Then find the radius, which is the distance from the center to either endpoint of the diameter. Finally use the center–radius form to find the equation.

   Find the center–radius form for each circle having the given endpoints of a diameter.

21. (−1, 3) and (5, −9)

22. (−4, 5) and (6, −9)

23. (−5, −7) and (1, 1)

24. (−3, −2) and (1, −4)

25. (−5, 0) and (5, 0)

26. (0, 9) and (0, −9)

27. Explain why, in **Exercises 21–26,** *either* endpoint can be used (along with the coordinates of the center) to find the radius.

28. *Concept Check*   Refer to any of **Exercises 21–26,** and show that the radius is half the length of the diameter.

Graph each circle by hand if possible. Give the domain and range.

29. $x^2 + y^2 = 4$

30. $x^2 + y^2 = 36$

31. $x^2 + y^2 = 0$

32. $x^2 + y^2 = -9$

33. $(x - 2)^2 + y^2 = 36$

34. $(x + 2)^2 + (y - 5)^2 = 16$

35. $(x - 5)^2 + (y + 4)^2 = 49$

36. $(x - 4)^2 + (y - 3)^2 = 25$

37. $(x + 3)^2 + (y + 2)^2 = 36$

38. $(x - 1)^2 + (y + 2)^2 = 16$

39. $x^2 + (y - 2)^2 + 10 = 9$

40. $(x + 1)^2 + y^2 + 2 = 0$

Graph each circle using a graphing calculator. Use a square viewing window. Give the domain and range.

41. $x^2 + y^2 = 81$

42. $x^2 + (y + 3)^2 = 49$

43. $(x - 3)^2 + (y - 2)^2 = 25$

44. $(x + 2)^2 + (y + 3)^2 = 36$

Decide whether each equation has a circle as its graph. If it does, give the center and radius.

45. $x^2 + 6x + y^2 + 8y + 9 = 0$

46. $x^2 + 8x + y^2 - 6y + 16 = 0$

47. $x^2 - 4x + y^2 + 12y = -4$

48. $x^2 - 12x + y^2 + 10y = -25$

49. $4x^2 + 4x + 4y^2 - 16y - 19 = 0$

50. $9x^2 + 12x + 9y^2 - 18y - 23 = 0$

51. $x^2 + 2x + y^2 - 6y + 14 = 0$

52. $x^2 + 4x + y^2 - 8y + 32 = 0$

53. $x^2 - 2x + y^2 + 4y = 0$

54. $4x^2 + 4x + 4y^2 - 4y - 3 = 0$

55. $9x^2 + 36x + 9y^2 = -32$

56. $9x^2 + 9y^2 + 54y = -72$

Concept Check *Each equation in Exercises 57–64 defines a parabola. Without actually graphing, match the equation in Column I with its description in Column II.*

I

II

57. $(x - 4)^2 = y + 2$ **58.** $(x - 2)^2 = y + 4$

59. $y + 2 = -(x - 4)^2$ **60.** $y = -(x - 2)^2 - 4$

61. $(y - 4)^2 = x + 2$ **62.** $(y - 2)^2 = x + 4$

63. $x + 2 = -(y - 4)^2$ **64.** $x = -(y - 2)^2 - 4$

A. Vertex $(2, -4)$; opens downward
B. Vertex $(2, -4)$; opens upward
C. Vertex $(4, -2)$; opens downward
D. Vertex $(4, -2)$; opens upward
E. Vertex $(-2, 4)$; opens left
F. Vertex $(-2, 4)$; opens right
G. Vertex $(-4, 2)$; opens left
H. Vertex $(-4, 2)$; opens right

65. *Concept Check* For the graph of $(x - h)^2 = 4c(y - k)$, in what quadrant is the vertex for each condition?
(a) $h < 0, k < 0$ **(b)** $h < 0, k > 0$ **(c)** $h > 0, k < 0$ **(d)** $h > 0, k > 0$

66. *Concept Check* Repeat parts (a)–(d) of **Exercise 65** for the graph of $(y - k)^2 = 4c(x - h)$.

Give the focus, directrix, and axis of each parabola.

67. $x^2 = 16y$ **68.** $x^2 = 4y$ **69.** $x^2 = -\frac{1}{2}y$ **70.** $x^2 = \frac{1}{9}y$

71. $y^2 = \frac{1}{16}x$ **72.** $y^2 = -\frac{1}{32}x$ **73.** $y^2 = -16x$ **74.** $y^2 = -4x$

Write an equation for each parabola with vertex at the origin.

75. Focus $(0, -2)$ **76.** Focus $(5, 0)$ **77.** Focus $\left(-\frac{1}{2}, 0\right)$ **78.** Focus $\left(0, \frac{1}{4}\right)$

79. Through $\left(2, -2\sqrt{2}\right)$; opening to the right   **80.** Through $\left(\sqrt{3}, 3\right)$; opening upward

81. Through $\left(\sqrt{10}, -5\right)$; opening downward   **82.** Through $(-3, 3)$; opening to the left

83. Through $(2, -4)$; symmetric with respect to the y-axis   **84.** Through $(3, 2)$; symmetric with respect to the x-axis

Find an equation of a parabola that satisfies the given conditions.

85. Focus $(0, 2)$; vertex $(0, 1)$ **86.** Focus $(-1, 2)$; vertex $(3, 2)$ **87.** Focus $(0, 0)$; directrix $x = -2$

88. Focus $(2, 1)$; directrix $x = -1$ **89.** Focus $(-1, 3)$; directrix $y = 7$ **90.** Focus $(1, 2)$; directrix $y = 4$

91. Horizontal axis; vertex $(-2, 3)$; passing through $(-4, 0)$ **92.** Horizontal axis; vertex $(-1, 2)$; passing through $(2, 3)$

Graph each parabola by hand, and check using a graphing calculator. Give the vertex, axis, domain, and range.

93. $y = (x + 3)^2 - 4$ **94.** $y = (x - 5)^2 - 4$ **95.** $y = -2(x + 3)^2 + 2$ **96.** $y = \frac{2}{3}(x - 2)^2 - 1$

97. $y = x^2 - 2x + 3$ **98.** $y = x^2 + 6x + 5$ **99.** $y = 2x^2 - 4x + 5$ **100.** $y = -3x^2 + 24x - 46$

101. $x = y^2 + 2$ **102.** $x = (y + 1)^2$ **103.** $x = (y - 3)^2$ **104.** $(y + 2)^2 = x + 1$

105. $x = (y - 4)^2 + 2$ **106.** $x = -2(y + 3)^2$ **107.** $x = \frac{2}{3}y^2 - 4y + 8$ **108.** $x = y^2 + 2y - 8$

109. $x = -4y^2 - 4y - 3$ **110.** $x = -2y^2 + 2y - 3$ **111.** $x = 2y^2 - 4y + 6$ **112.** $2x = y^2 - 4y + 6$

113. $2x = y^2 - 2y + 9$ **114.** $x = -3y^2 + 6y - 1$ **115.** $y^2 - 4y + 4 = 4x + 4$ **116.** $y^2 + 2y + 1 = -2x + 4$

117. Prove that the parabola with focus $(c, 0)$ and directrix $x = -c$ has equation $y^2 = 4cx$.

(Modeling)   Solve each problem.

118. *Path of an Object on a Planet*   When an object moves under the influence of a gravitational force (without air resistance), its path can be parabolic. This is the path of a ball thrown near the surface of a planet or other celestial object. Suppose two balls are simultaneously thrown upward at a 45° angle on two different planets. If their initial velocities are both 30 mph, then their *xy*-coordinates in feet can be expressed by the equation

$$y = x - \frac{g}{1922}x^2,$$

where *g* is the acceleration due to gravity. The value of *g* will vary with the mass and size of the planet. (*Source:* Zeilik, M., S. Gregory, and E. Smith, *Introductory Astronomy and Astrophysics,* Saunders College Publishers.)

(a) On Earth, $g = 32.2$ and on Mars, $g = 12.6$. Find the two equations, and use the same screen of a graphing calculator to graph the paths of the two balls thrown on Earth and Mars. Use the window $[0, 180]$ by $[0, 120]$. (*Hint:* If possible, set the mode on your graphing calculator to simultaneous.)

(b) Determine the difference in the horizontal distances traveled by the two balls.

119. *Path of an Object on a Planet*   Refer to **Exercise 118.** Suppose the two balls are now thrown upward at a 60° angle on Mars and the moon. If their initial velocities are 60 mph, then their *xy*-coordinates in feet can be expressed by the equation

$$y = \frac{19}{11}x - \frac{g}{3872}x^2.$$

(a) Graph the paths of the balls if $g = 5.2$ for the moon. Use the window $[0, 1500]$ by $[0, 1000]$.

(b) Determine the maximum height of each ball to the nearest foot.

120. *Design of a Radio Telescope*   The U.S. Naval Research Laboratory designed a giant radio telescope weighing 3450 tons. Its parabolic dish had a diameter of 300 feet, with a focal length (the distance from the focus to the parabolic surface) of 128.5 feet. Determine the maximum depth of the 300-foot dish. (*Source:* Mar, J. and H. Liebowitz, *Structure Technology for Large Radio and Radar Telescope Systems,* MIT Press.)

121. *Path of an Alpha Particle*   When an alpha particle (a subatomic particle) is moving in a horizontal path along the positive *x*-axis and passes between charged plates, it is deflected in a parabolic path. If the plate is charged with 2000 volts and is 0.4 meter long, then an alpha particle's path can be described by the equation

$$y = -\frac{k}{2v_0}x^2,$$

where $k = 5 \times 10^{-9}$ is constant and v_0 is the initial velocity of the particle. If $v_0 = 10^7$ meters per second, what is the deflection of the alpha particle's path in the *y*-direction when $x = 0.4$ meter? (*Source:* Semat, H. and J. Albright, *Introduction to Atomic and Nuclear Physics,* Holt, Rinehart and Winston.)

122. *Height of a Bridge's Cable Supports*   The cable in the center portion of a bridge is supported as shown in the figure to form a parabola. The center support is 10 feet high, the tallest supports are 210 feet high, and the distance between the two tallest supports is 400 feet. Find the height of the remaining supports if the supports are evenly spaced. (Ignore the width of the supports.)

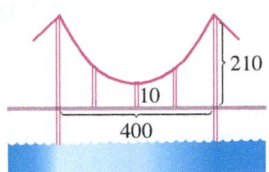

123. *Parabolic Arch*   An arch in the shape of a parabola has the dimensions shown in the figure. How wide is the arch 9 feet up?

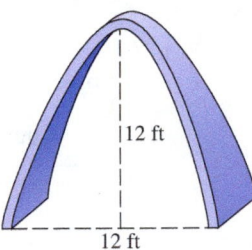

12 ft

12 ft

124. *Headlight*   A headlight is being constructed in the shape of a paraboloid with depth 4 inches and diameter 5 inches, as illustrated in the figure. Determine the distance *d* that the bulb should be from the vertex in order to have the beam of light shine straight ahead.

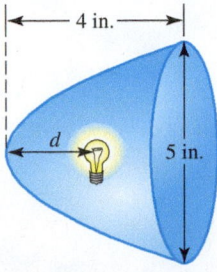

4 in.

d

5 in.

7.2 Ellipses and Hyperbolas

Equations and Graphs of Ellipses • Translations of Ellipses • An Application of Ellipses • Equations and Graphs of Hyperbolas • Translations of Hyperbolas

Equations and Graphs of Ellipses

The *ellipse* is another relation whose equation is based on the distance formula.

Ellipse

FIGURE 22

> **Ellipse**
>
> An **ellipse** is the set of all points in a plane, the sum of whose distances from two fixed points is constant. Each fixed point is called a **focus** (plural, **foci**) of the ellipse.

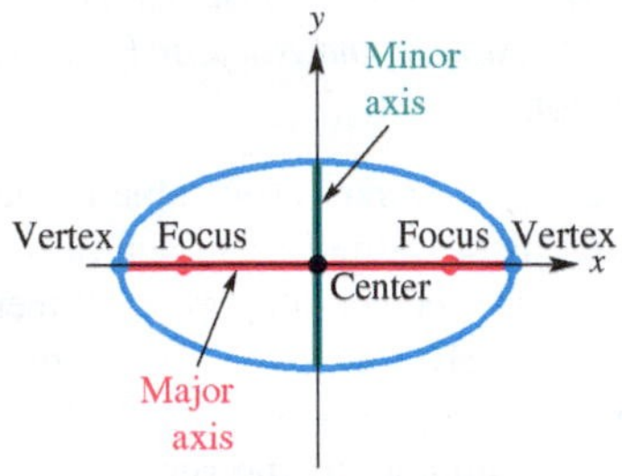

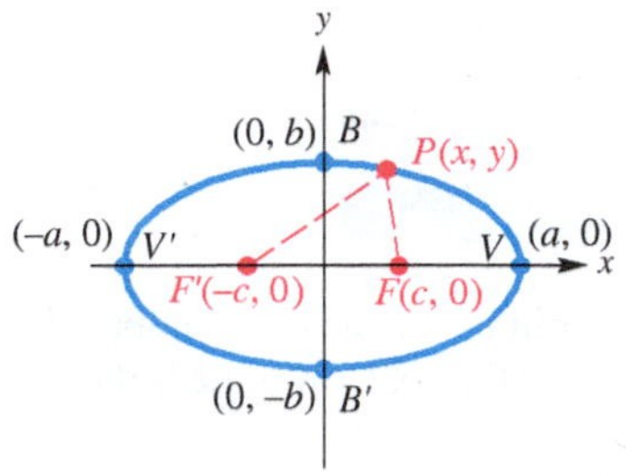

As shown in **FIGURE 22,** an ellipse has two axes of symmetry: the **major axis** (the longer one) and the **minor axis** (the shorter one). *The foci are always located on the major axis.* The midpoint of the major axis is the **center** of the ellipse, and the endpoints of the major axis are the **vertices** of the ellipse. A chord through a focus and perpendicular to the major axis is called a **latus rectum.** (There are two.) *The graph of an ellipse is not the graph of a function.*

The ellipse in **FIGURE 23** has its center at the origin, foci $F(c, 0)$ and $F'(-c, 0)$, and vertices $V(a, 0)$ and $V'(-a, 0)$. From the figure, the distance from V to F is $a - c$ and the distance from V to F' is $a + c$. The sum of these distances is $2a$. Since V is on the ellipse, this sum is the constant referred to in the definition of an ellipse. Thus, for any point $P(x, y)$ on the ellipse,

$$d(P, F) + d(P, F') = 2a.$$

FIGURE 23

By the distance formula,

$$d(P, F) = \sqrt{(x - c)^2 + y^2}, \quad \text{and} \quad d(P, F') = \sqrt{[x - (-c)]^2 + y^2} = \sqrt{(x + c)^2 + y^2}.$$

Thus, we have the following.

$$\sqrt{(x - c)^2 + y^2} + \sqrt{(x + c)^2 + y^2} = 2a$$

$$\sqrt{(x - c)^2 + y^2} = 2a - \sqrt{(x + c)^2 + y^2} \qquad \text{Isolate } \sqrt{(x - c)^2 + y^2}.$$

$$(x - c)^2 + y^2 = 4a^2 - 4a\sqrt{(x + c)^2 + y^2} + (x + c)^2 + y^2$$

Square each side.

$$x^2 - 2cx + c^2 + y^2 = 4a^2 - 4a\sqrt{(x + c)^2 + y^2} + x^2 + 2cx + c^2 + y^2$$

Square $x - c$ and $x + c$.

Divide each term by 4.

$$4a\sqrt{(x + c)^2 + y^2} = 4a^2 + 4cx \qquad \text{Isolate } 4a\sqrt{(x + c)^2 + y^2}.$$

$$a\sqrt{(x + c)^2 + y^2} = a^2 + cx \qquad \text{Divide by 4.}$$

$$a^2(x^2 + 2cx + c^2 + y^2) = a^4 + 2ca^2x + c^2x^2 \qquad \text{Square each side and } (x + c).$$

$$a^2x^2 + 2ca^2x + a^2c^2 + a^2y^2 = a^4 + 2ca^2x + c^2x^2 \qquad \text{Distributive property}$$

$$a^2x^2 + a^2c^2 + a^2y^2 = a^4 + c^2x^2 \qquad \text{Subtract } 2ca^2x.$$

$$a^2x^2 - c^2x^2 + a^2y^2 = a^4 - a^2c^2 \qquad \text{Rearrange terms.}$$

$$(a^2 - c^2)x^2 + a^2y^2 = a^2(a^2 - c^2) \qquad \text{Factor.}$$

$$\frac{x^2}{a^2} + \frac{y^2}{a^2 - c^2} = 1 \quad (*) \qquad \text{Divide by } a^2(a^2 - c^2).$$

Since $B(0, b)$ is on the ellipse in **FIGURE 23** on the preceding page, we write the following.

$$d(B, F) + d(B, F') = 2a$$
$$\sqrt{(-c)^2 + b^2} + \sqrt{c^2 + b^2} = 2a$$
$$2\sqrt{c^2 + b^2} = 2a \qquad \text{Combine like terms.}$$
$$\sqrt{c^2 + b^2} = a \qquad \text{Divide by 2.}$$
$$c^2 + b^2 = a^2 \qquad \text{Square each side.}$$
$$b^2 = a^2 - c^2 \qquad \text{Subtract } c^2.$$

Replacing $a^2 - c^2$ with b^2 in equation $(*)$ gives

$$\frac{x^2}{a^2} + \frac{y^2}{b^2} = 1,$$

the standard form of the equation of an ellipse centered at the origin and with foci on the x-axis. If the vertices and foci were on the y-axis, an almost identical derivation could be used to get the standard form

$$\frac{x^2}{b^2} + \frac{y^2}{a^2} = 1.$$

Standard Forms of Equations for Ellipses

The ellipse with center at the origin and equation

$$\frac{x^2}{a^2} + \frac{y^2}{b^2} = 1 \quad (a > b > 0)$$

has vertices $(\pm a, 0)$, endpoints of the minor axis $(0, \pm b)$, and foci $(\pm c, 0)$, where $c^2 = a^2 - b^2$.

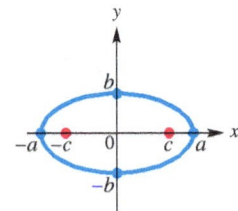

Major axis on x-axis

The ellipse with center at the origin and equation

$$\frac{x^2}{b^2} + \frac{y^2}{a^2} = 1 \quad (a > b > 0)$$

has vertices $(0, \pm a)$, endpoints of the minor axis $(\pm b, 0)$, and foci $(0, \pm c)$, where $c^2 = a^2 - b^2$.

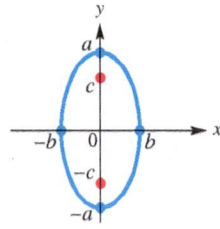

Major axis on y-axis

NOTE The relationship among a, b, and c in the definition of an ellipse alternatively can be remembered as $a^2 = b^2 + c^2$, because a, and thus a^2, is the greatest of the three values.

Do not be confused by the two standard forms—in one case a^2 is associated with x^2 and in the other case a^2 is associated with y^2. *In practice, we need only find the intercepts of the graph—if the distance between the x-intercepts is greater than the distance between the y-intercepts, then the major axis is horizontal. Otherwise, it is vertical.* When using $a^2 - c^2 = b^2$, or $a^2 - b^2 = c^2$, choose a^2 and b^2 so that $a^2 > b^2$.

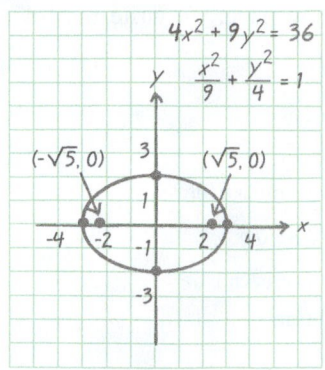

FIGURE 24

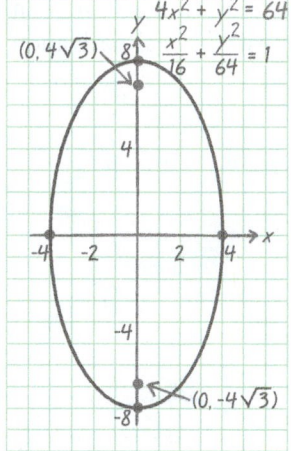

FIGURE 25

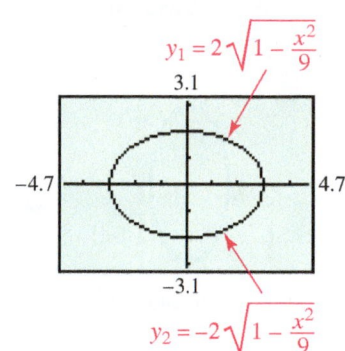

Since $y_2 = -y_1$, we can enter $Y_2 = -Y_1$ to get the portion of the graph below the x-axis.

FIGURE 26

EXAMPLE 1 **Graphing an Ellipse Centered at the Origin**

Graph each ellipse, and find the foci. Give the domain and range.

(a) $4x^2 + 9y^2 = 36$ **(b)** $4x^2 + y^2 = 64$

Solution

(a) $$4x^2 + 9y^2 = 36$$

Divide each term by 36. $$\frac{x^2}{9} + \frac{y^2}{4} = 1$$ Divide by 36 to write in standard form.

Thus, the x-intercepts are $(\pm 3, 0)$, and the y-intercepts are $(0, \pm 2)$. See **FIGURE 24.** Since $9 > 4$, we find the foci by letting $a^2 = 9$ and $b^2 = 4$ in $c^2 = a^2 - b^2$.

$$c^2 = 9 - 4 = 5, \quad \text{so} \quad c = \sqrt{5}.$$

(By definition, $c > 0$. See **FIGURE 23.**) The major axis is along the x-axis, so the foci have coordinates $\left(-\sqrt{5}, 0\right)$ and $\left(\sqrt{5}, 0\right)$. The domain of this relation is $[-3, 3]$, and the range is $[-2, 2]$.

(b) $$4x^2 + y^2 = 64$$

$$\frac{x^2}{16} + \frac{y^2}{64} = 1$$ Divide by 64 to write in standard form.

The x-intercepts are $(\pm 4, 0)$ and the y-intercepts are $(0, \pm 8)$. See **FIGURE 25.** Here $64 > 16$, so $a^2 = 64$ and $b^2 = 16$. Thus,

$$c^2 = 64 - 16 = 48, \quad \text{so} \quad c = \sqrt{48} = \sqrt{16 \cdot 3} = 4\sqrt{3}.$$

The major axis is on the y-axis, so the coordinates of the foci are $\left(0, -4\sqrt{3}\right)$ and $\left(0, 4\sqrt{3}\right)$. The domain of the relation is $[-4, 4]$ and the range is $[-8, 8]$. ●

The graph of an ellipse is *not* the graph of a function. To graph the ellipse in **Example 1(a)** with a graphing calculator, solve for y to get equations of two functions.

$$4x^2 + 9y^2 = 36$$ Original equation

$$\frac{x^2}{9} + \frac{y^2}{4} = 1$$ Divide by 36.

$$\frac{y^2}{4} = 1 - \frac{x^2}{9}$$ Subtract $\frac{x^2}{9}$.

Remember both the positive and negative square roots. $$y^2 = 4\left(1 - \frac{x^2}{9}\right)$$ Multiply by 4.

$$y = \pm 2\sqrt{1 - \frac{x^2}{9}}$$ Take square roots.

The ellipse is obtained by graphing both equations, as shown in **FIGURE 26.**

EXAMPLE 2 **Writing the Equation of an Ellipse and Graphing It**

Write the equation of the ellipse having center at the origin, foci at $(0, 3)$ and $(0, -3)$, and major axis 8 units long. Graph the ellipse, and give the domain and range.

Solution Since the major axis is 8 units long, $2a = 8$ and thus $a = 4$. To find b^2, use the relationship $a^2 - b^2 = c^2$, with $a = 4$ and $c = 3$.

$$a^2 - b^2 = c^2$$

$$4^2 - b^2 = 3^2$$ Substitute for a and c.

$$16 - b^2 = 9$$ Apply the exponents.

$$b^2 = 7$$ Solve for b^2.

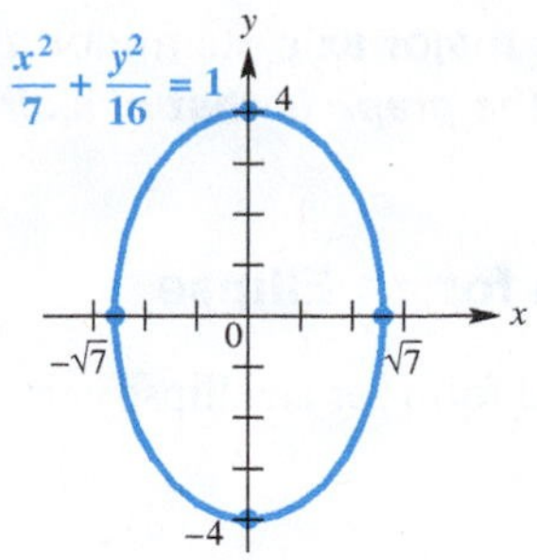

$\dfrac{x^2}{7} + \dfrac{y^2}{16} = 1$

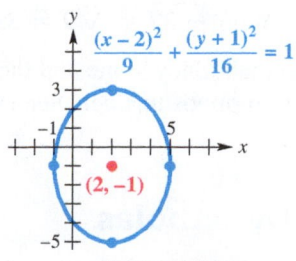

FIGURE 27

Since the foci are on the y-axis, our equation will be of the form $\dfrac{x^2}{b^2} + \dfrac{y^2}{a^2} = 1$ with $a^2 = 16$ and $b^2 = 7$, giving the equation in standard form as follows.

$$\frac{x^2}{7} + \frac{y^2}{16} = 1 \qquad b^2 = 7; \ a^2 = 16$$

A graph of this ellipse is shown in the margin in **FIGURE 27**. The domain of this relation is $\left[-\sqrt{7}, \sqrt{7}\right]$ and the range is $[-4, 4]$. ●

Translations of Ellipses

As with a circle, an ellipse also may have its center translated from the origin.

> ### Standard Forms for Ellipses Centered at (h, k)
>
> An ellipse with center at (h, k) and either a horizontal or vertical major axis satisfies one of the following equations, where $a > b > 0$ and $c^2 = a^2 - b^2$ with $c > 0$.
>
> $$\frac{(x - h)^2}{a^2} + \frac{(y - k)^2}{b^2} = 1 \qquad \text{Major axis: horizontal; foci: } (h \pm c, k); \\ \text{vertices: } (h \pm a, k)$$
>
> $$\frac{(x - h)^2}{b^2} + \frac{(y - k)^2}{a^2} = 1 \qquad \text{Major axis: vertical; foci: } (h, k \pm c); \\ \text{vertices: } (h, k \pm a)$$

EXAMPLE 3 **Graphing an Ellipse Translated from the Origin**

Graph $\dfrac{(x - 2)^2}{9} + \dfrac{(y + 1)^2}{16} = 1$. Give the domain and range.

Analytic Solution

The graph of this equation is an ellipse centered at $(2, -1)$. For this ellipse, $a = 4$ and $b = 3$. Since $a = 4$ is associated with y^2, the vertices of the ellipse are on the vertical line through $(2, -1)$.

Find the vertices by locating two points on the vertical line through $(2, -1)$, one 4 units up from $(2, -1)$ and one 4 units down. The vertices are $(2, 3)$ and $(2, -5)$. Locate two other points on the ellipse by locating points on the horizontal line through $(2, -1)$, one 3 units to the right and one 3 units to the left.

The graph is shown in **FIGURE 28**. The domain is $[-1, 5]$ and the range is $[-5, 3]$.

$\dfrac{(x-2)^2}{9} + \dfrac{(y+1)^2}{16} = 1$

FIGURE 28

Graphing Calculator Solution

Solve for y in the equation of the ellipse to obtain

$$y = -1 \pm 4\sqrt{1 - \frac{(x - 2)^2}{9}}.$$

The $+$ sign yields the top half of the ellipse, while the $-$ sign yields the bottom half. See **FIGURE 29.**

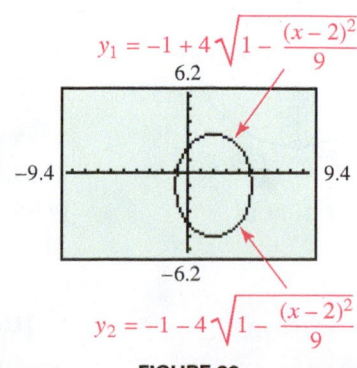

$y_1 = -1 + 4\sqrt{1 - \dfrac{(x-2)^2}{9}}$

$y_2 = -1 - 4\sqrt{1 - \dfrac{(x-2)^2}{9}}$

FIGURE 29

NOTE An ellipse is symmetric with respect to its major axis, its minor axis, and its center. *If a = b in the equation of an ellipse, the graph is that of a circle.*

Algebra Review
To review completing the square, see **Sections 3.2** and **3.3**.

EXAMPLE 4 **Finding the Standard Form for an Ellipse**

Write $4x^2 - 16x + 9y^2 + 54y + 61 = 0$ in the standard form for an ellipse centered at (h, k). Identify the center and vertices.

Solution

$$4x^2 - 16x + 9y^2 + 54y + 61 = 0$$

$$4(x^2 - 4x + \underline{}) + 9(y^2 + 6y + \underline{}) = -61 \qquad \text{Distributive property and add } -61.$$

$$4(x^2 - 4x + 4) + 9(y^2 + 6y + 9) = -61 + 4(4) + 9(9) \qquad \text{Complete the square.}$$

$$4(x - 2)^2 + 9(y + 3)^2 = 36 \qquad \text{Factor and add.}$$

$$\frac{(x - 2)^2}{9} + \frac{(y + 3)^2}{4} = 1 \qquad \text{Divide each side by 36.}$$

The center is $(2, -3)$. Because $a = 3$ and the major axis is horizontal, the vertices of the ellipse are $(2 \pm 3, -3)$, or $(5, -3)$ and $(-1, -3)$.

An Application of Ellipses

Ellipses have many useful applications. As Earth makes its yearlong journey around the sun, it traces an ellipse. Spacecraft travel around Earth in elliptical orbits, and planets make elliptical orbits around the sun.

An application from medicine is a lithotripter, a machine used to crush kidney stones via shock waves. The patient is placed in an elliptical tub with the kidney stone at one focus of the ellipse. A beam is projected from the other focus to the tub so that it reflects to hit the kidney stone. See **FIGURE 30** and **Exercise 91**.

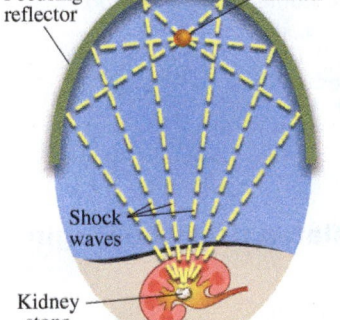

The top of an ellipse is illustrated in this depiction of how a lithotripter crushes a kidney stone.

FIGURE 30

EXAMPLE 5 **Modeling the Reflective Property of Ellipses**

If a lithotripter is based on the ellipse

$$\frac{x^2}{36} + \frac{y^2}{27} = 1,$$

determine how many units the kidney stone and the wave source must be placed from the center of the ellipse.

Solution The kidney stone and the source of the beam must be placed at the foci, $(c, 0)$ and $(-c, 0)$. Here $a^2 = 36$ and $b^2 = 27$, so

$$c = \sqrt{a^2 - b^2} = \sqrt{36 - 27} = \sqrt{9} = 3.$$

Thus, the foci are $(-3, 0)$ and $(3, 0)$, so the kidney stone and the source both must be placed on a line 3 units from the center on opposite sides. See **FIGURE 31**.

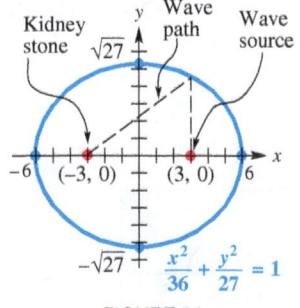

FIGURE 31

Equations and Graphs of Hyperbolas

An ellipse was defined as the set of all points in a plane, the *sum* of whose distances from two fixed points is constant. A *hyperbola* is defined similarly.

> ### Hyperbola
>
> A **hyperbola** is the set of all points in a plane such that the absolute value of the *difference* of the distances from two fixed points is constant. The two fixed points are called the **foci** of the hyperbola.

Hyperbola

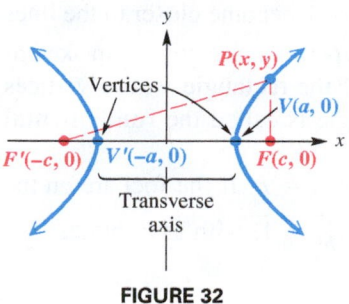

FIGURE 32

Suppose a hyperbola has center at the origin and foci at $F'(-c, 0)$ and $F(c, 0)$. See **FIGURE 32.** The midpoint of the segment $F'F$ is the **center** of the hyperbola, and the points $V'(-a, 0)$ and $V(a, 0)$ are the **vertices** of the hyperbola. The line segment $V'V$ is the **transverse axis** of the hyperbola. A chord through a focus and perpendicular to an extension of the transverse axis is, again, a **latus rectum.** (There are two.)

Using the distance formula in conjunction with the definition of a hyperbola, we can verify the following standard forms of the equations for hyperbolas centered at the origin. (See **Exercise 100.**)

> ### Standard Forms of Equations for Hyperbolas
>
> The hyperbola with center at the origin and equation
>
> $$\frac{x^2}{a^2} - \frac{y^2}{b^2} = 1$$
>
> has vertices $(\pm a, 0)$, asymptotes $y = \pm \frac{b}{a}x$, and foci $(\pm c, 0)$, where $c^2 = a^2 + b^2$.
>
>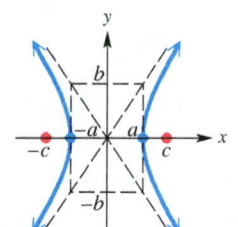
>
> Transverse axis on x-axis
>
> The hyperbola with center at the origin and equation
>
> $$\frac{y^2}{a^2} - \frac{x^2}{b^2} = 1$$
>
> has vertices $(0, \pm a)$, asymptotes $y = \pm \frac{a}{b}x$, and foci $(0, \pm c)$, where $c^2 = a^2 + b^2$.
>
>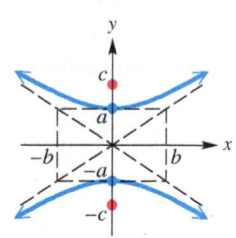
>
> Transverse axis on y-axis

> **NOTE** The relationship among a, b, and c in the definition of a hyperbola can be remembered correctly as $c^2 = a^2 + b^2$, because c is the greatest of the three values.

To explain the concept of asymptotes, we can start with the first equation for a hyperbola, where the foci are on the x-axis, and solve for y to obtain

$$\frac{x^2}{a^2} - \frac{y^2}{b^2} = 1$$

$$\frac{x^2}{a^2} - 1 = \frac{y^2}{b^2} \qquad \text{Subtract 1 and add } \tfrac{y^2}{b^2}.$$

$$\frac{x^2 - a^2}{a^2} = \frac{y^2}{b^2} \qquad \text{Write the left side as a single fraction.}$$

Remember both the positive and negative square roots.

$$y = \pm \frac{b}{a}\sqrt{x^2 - a^2}. \quad (*) \qquad \text{Take square roots and multiply by } b.$$

If x^2 is very large in comparison to a^2, the difference $x^2 - a^2$ would be relatively close to x^2. If this happens, then the points satisfying equation (∗) would approach one of the lines

$$y = \pm\frac{b}{a}x. \qquad \text{Asymptotes for } \frac{x^2}{a^2} - \frac{y^2}{b^2} = 1$$

Thus, as $|x| \to \infty$, the points of the hyperbola $\frac{x^2}{a^2} - \frac{y^2}{b^2} = 1$ become closer to the lines $y = \pm\frac{b}{a}x$. These lines, called the **asymptotes** of the hyperbola, are helpful in sketching the graph. The lines are the extended diagonals of the rectangle whose vertices are (a, b), $(-a, b)$, $(a, -b)$, and $(-a, -b)$. This rectangle is called the **fundamental rectangle** of the hyperbola.

For hyperbolas, it is possible that $a > b$, $a < b$, or $a = b$. If the foci are on the y-axis, the equation of the hyperbola has the form $\frac{y^2}{a^2} - \frac{x^2}{b^2} = 1$, with asymptotes

$$y = \pm\frac{a}{b}x. \qquad \text{Asymptotes for } \frac{y^2}{a^2} - \frac{x^2}{b^2} = 1$$

EXAMPLE 6 **Using Asymptotes to Graph a Hyperbola**

Sketch the asymptotes and graph the hyperbola $\frac{x^2}{25} - \frac{y^2}{49} = 1$. Give the foci, domain, and range.

Analytic Solution

For this hyperbola, $a = 5$ and $b = 7$. With these values, $y = \pm\frac{b}{a}x$ becomes $y = \pm\frac{7}{5}x$. Choosing $x = 5$ gives $y = \pm7$. Choosing $x = -5$ also gives $y = \pm7$. These four points, $(\pm5, \pm7)$, are the corners of the fundamental rectangle shown in **FIGURE 33.**

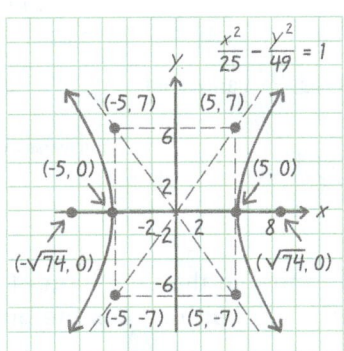

FIGURE 33

The extended diagonals of this rectangle are the asymptotes of the hyperbola. See **FIGURE 33.** Note that the hyperbola has vertices $(-5, 0)$ and $(5, 0)$. We find the foci by letting

$$c^2 = a^2 + b^2 = 25 + 49 = 74, \quad \text{so} \quad c = \sqrt{74}.$$

Therefore, the foci are $\left(-\sqrt{74}, 0\right)$ and $\left(\sqrt{74}, 0\right)$. The domain of this relation is $(-\infty, -5] \cup [5, \infty)$, and the range is $(-\infty, \infty)$.

Graphing Calculator Solution

Solve the given equation for y.

$$\frac{x^2}{25} - \frac{y^2}{49} = 1$$

$$-\frac{y^2}{49} = 1 - \frac{x^2}{25} \qquad \text{Subtract } \frac{x^2}{25}.$$

$$\frac{y^2}{49} = \frac{x^2}{25} - 1 \qquad \text{Multiply by } -1.$$

$$\frac{y}{7} = \pm\sqrt{\frac{x^2}{25} - 1} \qquad \text{Take square roots.}$$

$$y = \pm7\sqrt{\frac{x^2}{25} - 1} \qquad \text{Multiply by 7.}$$

The graphs of the hyperbola and its asymptotes are shown in **FIGURE 34.** The graph of y_1 creates the upper half of the hyperbola, while the graph of y_2 creates the lower half.

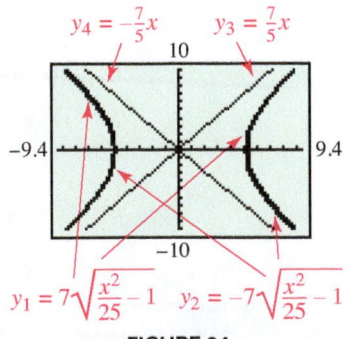

FIGURE 34

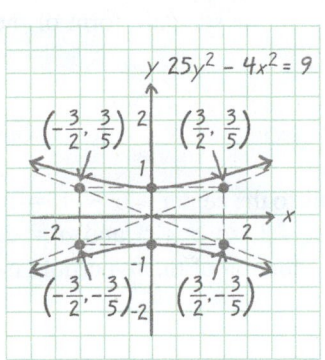

FIGURE 35

EXAMPLE 7   **Graphing a Hyperbola Centered at the Origin**

Graph $25y^2 - 4x^2 = 9$. Give the domain and range.

Solution

$$25y^2 - 4x^2 = 9$$

Be sure to divide each term by 9. → $\dfrac{25y^2}{9} - \dfrac{4x^2}{9} = 1$ Divide by 9.

To determine the values of a and b, write the equation $\dfrac{25y^2}{9} - \dfrac{4x^2}{9} = 1$ as

$$\dfrac{y^2}{\frac{9}{25}} - \dfrac{x^2}{\frac{9}{4}} = 1. \qquad \dfrac{a}{b} = \dfrac{\frac{1}{b}}{\frac{1}{a}}$$

This hyperbola is centered at the origin, has foci on the y-axis, and has y-intercepts $\left(0, \pm\frac{3}{5}\right)$. Use the four points $\left(\pm\frac{3}{2}, \pm\frac{3}{5}\right)$ to get the fundamental rectangle shown in **FIGURE 35**. Use the diagonals of this rectangle to determine the asymptotes for the graph. The domain is $(-\infty, \infty)$ and the range is $\left(-\infty, -\frac{3}{5}\right] \cup \left[\frac{3}{5}, \infty\right)$.

Translations of Hyperbolas

> ### Standard Forms for Hyperbolas Centered at (h, k)
>
> A hyperbola with center (h, k) and either a horizontal or vertical transverse axis satisfies one of the following equations, where $c^2 = a^2 + b^2$.
>
> $$\dfrac{(x-h)^2}{a^2} - \dfrac{(y-k)^2}{b^2} = 1$$
>
> Transverse axis: horizontal;
> vertices: $(h \pm a, k)$; foci: $(h \pm c, k)$;
> asymptotes: $y = \pm\frac{b}{a}(x - h) + k$
>
> $$\dfrac{(y-k)^2}{a^2} - \dfrac{(x-h)^2}{b^2} = 1$$
>
> Transverse axis: vertical;
> vertices: $(h, k \pm a)$; foci: $(h, k \pm c)$;
> asymptotes: $y = \pm\frac{a}{b}(x - h) + k$

NOTE   The asymptotes for a hyperbola *always* pass through the center (h, k). By the point–slope form of a line, the equation of any asymptote is $y = m(x - h) + k$. If the transverse axis is horizontal, then $m = \pm\frac{b}{a}$; if it is vertical, then $m = \pm\frac{a}{b}$.

EXAMPLE 8   **Graphing a Hyperbola Translated from the Origin**

Graph $\dfrac{(y + 2)^2}{9} - \dfrac{(x + 3)^2}{4} = 1$, and identify the center. Give the domain and range.

Solution   This equation represents a hyperbola centered at $(-3, -2)$. For this vertical hyperbola, $a = 3$ and $b = 2$. The x-values of the vertices are -3. Locate the y-values of the vertices by taking the y-value of the center, -2, and adding and subtracting 3. Thus, the vertices are $(-3, 1)$ and $(-3, -5)$. The asymptotes have slopes $\pm\frac{3}{2}$ and pass through the center $(-3, -2)$. The equations of the asymptotes,

$$y = \pm\frac{3}{2}(x + 3) - 2,$$

can be found by using the point–slope form of the equation of a line. The graph is shown in **FIGURE 36**. The domain of the relation is $(-\infty, \infty)$ and the range is $(-\infty, -5] \cup [1, \infty)$.

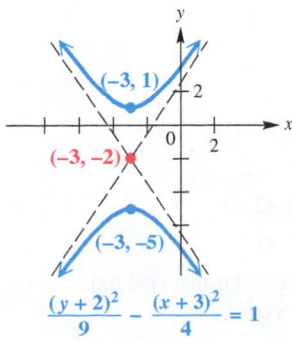

$$\dfrac{(y+2)^2}{9} - \dfrac{(x+3)^2}{4} = 1$$

FIGURE 36

FOR DISCUSSION

How would you graph the hyperbola in **Example 8,**

$$\frac{(y+2)^2}{9} - \frac{(x+3)^2}{4} = 1,$$

on your graphing calculator? Duplicate the calculator graph of that hyperbola shown in **FIGURE 37.** The two branches of a hyperbola are reflections about two different axes and also about a point. What are the axes and the point for the hyperbola shown in the figure?

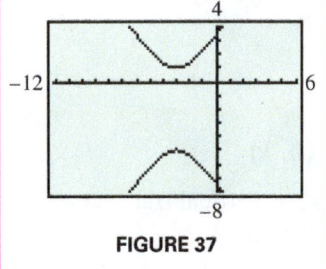

FIGURE 37

EXAMPLE 9 **Writing the Equation of a Hyperbola**

Write an equation of the hyperbola having center $(1, 2)$, focus $(6, 2)$, and vertex $(4, 2)$.

Solution   Because the center, focus, and vertex lie on the line $y = 2$, the hyperbola has a horizontal transverse axis. The center is $(1, 2)$, so the standard form of the equation is

$$\frac{(x-1)^2}{a^2} - \frac{(y-2)^2}{b^2} = 1,$$

with a and b to be determined. The distance from the focus to the center is $6 - 1 = 5$, and the distance from the vertex to the center is $4 - 1 = 3$, so $c = 5$ and $a = 3$. Because $a^2 + b^2 = c^2$, it follows that $b = 4$. Thus, the equation of the hyperbola is

$$\frac{(x-1)^2}{9} - \frac{(y-2)^2}{16} = 1.$$

EXAMPLE 10   **Finding the Standard Form for a Hyperbola**

Write $9x^2 - 18x - 4y^2 - 16y = 43$ in the standard form for a hyperbola centered at (h, k). Identify the center and vertices.

Solution

$$9x^2 - 18x - 4y^2 - 16y = 43$$

$$9(x^2 - 2x + \underline{}) - 4(y^2 + 4y + \underline{}) = 43 \qquad \text{Distributive property}$$

$$9(x^2 - 2x + 1) - 4(y^2 + 4y + 4) = 43 + 9(1) - 4(4) \qquad \text{Complete the square.}$$

$$9(x-1)^2 - 4(y+2)^2 = 36 \qquad \text{Factor and add.}$$

$$\frac{(x-1)^2}{4} - \frac{(y+2)^2}{9} = 1 \qquad \text{Divide each side by 36.}$$

The center is $(1, -2)$. Because $a = 2$ and the transverse axis is horizontal, the vertices of the hyperbola are $(1 \pm 2, -2)$, or $(3, -2)$ and $(-1, -2)$.

7.2 Exercises

Checking Analytic Skills   *Match each equation in Column I with the appropriate description in Column II.* **Do not use a calculator.**

I

1. $\dfrac{x^2}{4} + \dfrac{y^2}{16} = 1$ **2.** $\dfrac{x^2}{16} + \dfrac{y^2}{4} = 1$

3. $\dfrac{x^2}{4} - \dfrac{y^2}{16} = 1$ **4.** $\dfrac{y^2}{4} - \dfrac{x^2}{16} = 1$

5. $\dfrac{(x+2)^2}{9} + \dfrac{(y-4)^2}{25} = 1$ **6.** $\dfrac{(x-2)^2}{9} + \dfrac{(y+4)^2}{25} = 1$

7. $\dfrac{(x+2)^2}{9} - \dfrac{(y-4)^2}{25} = 1$ **8.** $\dfrac{(x-2)^2}{9} - \dfrac{(y-4)^2}{25} = 1$

II

A. Hyperbola; center $(2, 4)$

B. Ellipse; foci $\left(\pm 2\sqrt{3}, 0 \right)$

C. Hyperbola; foci $\left(0, \pm 2\sqrt{5} \right)$

D. Hyperbola; center $(-2, 4)$

E. Ellipse; center $(-2, 4)$

F. Center $(0, 0)$; horizontal transverse axis

G. Ellipse; foci $\left(0, \pm 2\sqrt{3} \right)$

H. Vertical major axis; center $(2, -4)$

9. Explain how a circle can be interpreted as a special case of an ellipse.

10. *Concept Check*   If an ellipse has endpoints of the minor axis and vertices at $(-3, 0)$, $(3, 0)$, $(0, 5)$, and $(0, -5)$, what is its domain? What is its range?

Checking Analytic Skills   *Graph each ellipse by hand. Give the domain and range. Give the foci in Exercises 11–14 and identify the center in Exercises 17–22.* **Do not use a calculator.**

11. $\dfrac{x^2}{9} + \dfrac{y^2}{4} = 1$

12. $\dfrac{x^2}{16} + \dfrac{y^2}{36} = 1$

13. $9x^2 + 6y^2 = 54$

14. $12x^2 + 8y^2 = 96$

15. $\dfrac{25y^2}{36} + \dfrac{64x^2}{9} = 1$

16. $\dfrac{16y^2}{9} + \dfrac{121x^2}{25} = 1$

17. $\dfrac{(x-1)^2}{9} + \dfrac{(y+3)^2}{25} = 1$

18. $\dfrac{(x+3)^2}{16} + \dfrac{(y-2)^2}{36} = 1$

19. $\dfrac{(x-2)^2}{16} + \dfrac{(y-1)^2}{9} = 1$

20. $\dfrac{(x+3)^2}{25} + \dfrac{(y+2)^2}{36} = 1$

21. $\dfrac{(x+1)^2}{64} + \dfrac{(y-2)^2}{49} = 1$

22. $\dfrac{(x-4)^2}{9} + \dfrac{(y+2)^2}{4} = 1$

Find an equation for each ellipse.

23. x-intercepts $(\pm 4, 0)$; foci $(\pm 2, 0)$

24. y-intercepts $(0, \pm 3)$; foci $\left(0, \pm \sqrt{3}\right)$

25. y-intercepts $\left(0, \pm 2\sqrt{2}\right)$; foci $(0, \pm 2)$

26. x-intercepts $\left(\pm 3\sqrt{2}, 0\right)$; foci $\left(\pm 2\sqrt{3}, 0\right)$

27. x-intercepts $(\pm 4, 0)$; y-intercepts $(0, \pm 2)$

28. x-intercepts $(\pm 3, 0)$; y-intercepts $(0, \pm 6)$

29. Endpoints of major axis at $(6, 0)$ and $(-6, 0)$; $c = 4$

30. Vertices $(0, 5)$ and $(0, -5)$; $b = 2$

31. Center $(3, -2)$; $a = 5$; $c = 3$; major axis vertical

32. Center $(2, 0)$; minor axis of length 6; major axis horizontal and of length 9

33. Major axis of length 6; foci $(0, 2)$ and $(0, -2)$

34. Minor axis of length 4; foci $(-5, 0)$ and $(5, 0)$

35. Center $(5, 2)$; minor axis vertical, with length 8; $c = 3$

36. Center $(-3, 6)$; major axis vertical, with length 10; $c = 2$

37. Vertices $(4, 9)$ and $(4, 1)$; minor axis of length 6

38. Foci at $(-3, -3)$ and $(7, -3)$; the point $(2, 1)$ on ellipse

Write the equation in standard form for an ellipse centered at (h, k). Identify the center and vertices.

39. $9x^2 + 18x + 4y^2 - 8y - 23 = 0$

40. $9x^2 - 36x + 16y^2 - 64y - 44 = 0$

41. $4x^2 + 8x + y^2 + 2y + 1 = 0$

42. $x^2 - 6x + 9y^2 = 0$

43. $4x^2 + 16x + 5y^2 - 10y + 1 = 0$

44. $2x^2 + 4x + 3y^2 - 18y + 23 = 0$

45. $16x^2 - 16x + 4y^2 + 12y = 51$

46. $16x^2 + 48x + 4y^2 - 20y + 57 = 0$

Checking Analytic Skills   *Graph each hyberbola by hand. Give the domain and range. Give the center in Exercises 55–61.* **Do not use a calculator.**

47. $\dfrac{x^2}{16} - \dfrac{y^2}{9} = 1$

48. $\dfrac{y^2}{9} - \dfrac{x^2}{9} = 1$

49. $49y^2 - 36x^2 = 1764$

50. $144x^2 - 49y^2 = 7056$

51. $\dfrac{4x^2}{9} - \dfrac{25y^2}{16} = 1$

52. $x^2 - y^2 = 1$

53. $9x^2 - 4y^2 = 1$

54. $25y^2 - 9x^2 = 1$

55. $\dfrac{(x-1)^2}{9} - \dfrac{(y+3)^2}{25} = 1$

56. $\dfrac{(x+3)^2}{16} - \dfrac{(y-2)^2}{36} = 1$

57. $\dfrac{(y-5)^2}{4} - \dfrac{(x+1)^2}{9} = 1$

58. $\dfrac{(y+1)^2}{25} - \dfrac{(x-3)^2}{36} = 1$

59. $16(x+5)^2 - (y-3)^2 = 1$

60. $4(x+9)^2 - 25(y+6)^2 = 100$

61. $9(x-2)^2 - 4(y+1)^2 = 36$

62. The graph of the rational function $y = \frac{1}{x}$ is a hyperbola that is rotated. Experiment with a graphing calculator to determine the vertices of its graph.

Find an equation for each hyperbola.

63. x-intercepts $(\pm 3, 0)$; foci $(\pm 4, 0)$

64. y-intercepts $(0, \pm 5)$; foci $\left(0, \pm 3\sqrt{3}\right)$

65. Asymptotes $y = \pm\frac{3}{5}x$; y-intercepts $(0, \pm 3)$

66. y-intercept $(0, -2)$; center at origin; passing through $(2, 3)$

67. Vertices $(0, 6)$ and $(0, -6)$; asymptotes $y = \pm\frac{1}{2}x$

68. Vertices $(-10, 0)$ and $(10, 0)$; asymptotes $y = \pm 5x$

69. Vertices $(-3, 0)$ and $(3, 0)$; passing through $(6, 1)$

70. Vertices $(0, 5)$ and $(0, -5)$; passing through $(3, 10)$

71. Foci $(0, \sqrt{13})$ and $(0, -\sqrt{13})$; asymptotes $y = \pm 5x$

72. Foci $\left(-3\sqrt{5}, 0\right)$ and $\left(3\sqrt{5}, 0\right)$; asymptotes $y = \pm 2x$

73. Vertices $(4, 5)$ and $(4, 1)$; asymptotes $y = \pm 7(x-4) + 3$

74. Vertices $(5, -2)$ and $(1, -2)$; asymptotes $y = \pm\frac{3}{2}(x-3) - 2$

75. Center $(1, -2)$; focus $(4, -2)$; vertex $(3, -2)$

76. Center $(9, -7)$; focus $(9, 3)$; vertex $(9, -1)$

Write the equation in standard form for a hyperbola centered at (h, k). Identify the center and vertices.

77. $x^2 - 2x - y^2 + 2y = 4$

78. $y^2 + 4y - x^2 + 2x = 6$

79. $3y^2 + 24y - 2x^2 + 12x + 24 = 0$

80. $4x^2 + 16x - 9y^2 + 18y = 29$

81. $x^2 - 6x - 2y^2 + 7 = 0$

82. $y^2 + 8y - 3x^2 + 13 = 0$

83. $4y^2 + 32y - 5x^2 - 10x + 39 = 0$

84. $5x^2 + 10x - 7y^2 + 28y = 58$

RELATING CONCEPTS **For individual or group investigation (Exercises 85–90)**

Consider the ellipse and hyperbola defined by

$$\frac{x^2}{16} + \frac{y^2}{12} = 1 \quad \text{and} \quad \frac{x^2}{4} - \frac{y^2}{12} = 1,$$

respectively. **Work Exercises 85–90 in order.**

85. Find the foci of the ellipse. Call them F_1 and F_2.

86. Graph the ellipse with your calculator, and trace to find the coordinates of several points on the ellipse.

87. For each of the points P, verify that $[\text{distance of } P \text{ from } F_1] + [\text{distance of } P \text{ from } F_2] = 8.$

88. Repeat **Exercises 85** and **86** for the hyperbola.

89. For each of the points P from **Exercise 88**, verify that $|[\text{distance of } P \text{ from } F_1] - [\text{distance of } P \text{ from } F_2]| = 4.$

90. How do **Exercises 87** and **89** relate to the definitions of the ellipse and the hyperbola given in this section?

(Modeling) *Solve each problem.*

91. *Shape of a Lithotripter* A patient's kidney stone is placed 12 units away from the source of the shock waves of a lithotripter. The lithotripter is based on an ellipse with a minor axis that measures 16 units. Find an equation of an ellipse that would satisfy this situation.

92. *Orbit of Venus* The orbit of Venus is an ellipse, with the sun at one focus. An approximate equation for the orbit is

$$\frac{x^2}{5013} + \frac{y^2}{4970} = 1,$$

where x and y are measured in millions of miles.
(a) Approximate the length of the major axis.
(b) Approximate the length of the minor axis.

93. *The Roman Colosseum* The Roman Colosseum is an ellipse with major axis 620 feet and minor axis 513 feet. Approximate the distance between the foci of this ellipse.

94. *The Roman Colosseum*   A formula for the approximate perimeter of an ellipse is

$$P \approx 2\pi \sqrt{\frac{a^2 + b^2}{2}},$$

where a and b are the lengths shown in the figure. Use this formula to find the approximate perimeter of the Roman Colosseum. (See **Exercise 93.**)

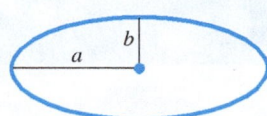

95. *Height of an Overpass*   A one-way road passes under an overpass in the form of half of an ellipse 15 feet high at the center and 20 feet wide. Assuming that a truck is 12 feet wide, what is the height of the tallest truck that can pass under the overpass?

96. *Design of a Sports Complex*   Two buildings in a sports complex are shaped and positioned like a portion of the branches of the hyperbola

$$400x^2 - 625y^2 = 250{,}000.$$

In this equation, x and y are in meters.

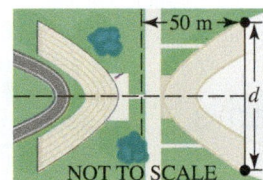

(a) How far apart are the buildings at their closest point?
(b) Find the distance d in the figure to the nearest tenth.

97. *Orbit of a Satellite*   The coordinates in miles for the orbit of the artificial satellite *Explorer VII* can be described by the equation

$$\frac{x^2}{a^2} + \frac{y^2}{b^2} = 1,$$

where $a = 4465$ and $b = 4462$. Earth's center is located at one focus of its elliptical orbit. (*Source:* Loh, W., *Dynamics and Thermodynamics of Planetary Entry,* Prentice-Hall; Thomson, W., *Introduction to Space Dynamics,* John Wiley and Sons.)
(a) Graph both the orbit of *Explorer VII* and that of Earth on the same coordinate axes if the average radius of Earth is 3960 miles. Use the window $[-6750, 6750]$ by $[-4500, 4500]$.
(b) Determine the maximum and minimum heights of the satellite above Earth's surface.

98. *Sound Detection*   Microphones are placed at points $(-c, 0)$ and $(c, 0)$. An explosion occurs at point $P(x, y)$ having positive x-coordinate.

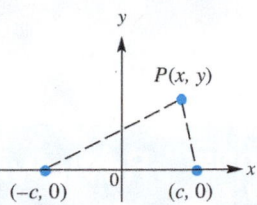

The sound is detected at the closer microphone t seconds before being detected at the farther microphone. Assume that sound travels at a speed of 330 meters per second, and show that P must be on the hyperbola

$$\frac{x^2}{330^2 t^2} - \frac{y^2}{4c^2 - 330^2 t^2} = \frac{1}{4}.$$

99. *Structure of an Atom*   In 1911, Ernest Rutherford discovered the basic structure of the atom by "shooting" positively charged alpha particles with a speed of 10^7 meters per second at a piece of gold foil 6×10^{-7} meter thick. Only a small percentage of the alpha particles struck a gold nucleus head-on and were deflected directly back toward their source. The rest of the particles often followed a hyperbolic trajectory because they were repelled by positively charged gold nuclei. Thus, Rutherford proposed that the atom was composed of mostly empty space and a small, dense nucleus. The figure shows an alpha particle A initially approaching a gold nucleus N and being deflected at an angle $\theta = 90°$. N is located at a focus of the hyperbola, and the trajectory of A passes through a vertex of the hyperbola. (*Source:* Semat, H. and J. Albright, *Introduction to Atomic and Nuclear Physics,* Holt, Rinehart and Winston.)

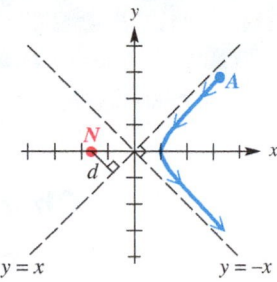

(a) Determine the equation of the trajectory of the alpha particle if $d = 5 \times 10^{-14}$ meter.
(b) Approximate the minimum distance between the centers of the alpha particle and the gold nucleus.

100. Suppose a hyperbola has center at the origin, foci at $F'(-c, 0)$ and $F(c, 0)$, and equation

$$|d(P, F') - d(P, F)| = 2a.$$

Let $b^2 = c^2 - a^2$, and show that the points on the hyperbola satisfy the equation

$$\frac{x^2}{a^2} - \frac{y^2}{b^2} = 1.$$

101. Use the definition of an ellipse to find an equation of an ellipse with foci $(3, 0)$ and $(-3, 0)$, where the sum of the distances from any point of the ellipse to the two foci is 10.

102. Use the definition of a hyperbola to find an equation of a hyperbola with center at the origin, foci $(-2, 0)$ and $(2, 0)$, and the absolute value of the difference of the distances from any point of the hyperbola to the two foci equal to 2.

SECTIONS 7.1–7.2 **Reviewing Basic Concepts**

1. Match each definition in A–D with the appropriate conic section.
   (a) Circle   **(b)** Parabola   **(c)** Ellipse   **(d)** Hyperbola

   A. A set of points in a plane such that the sum of their distances from two fixed points is constant
   B. A set of points in a plane that are equidistant from a fixed point
   C. A set of points in a plane such that the absolute value of the difference of their distances from two fixed points is constant
   D. A set of points in a plane equidistant from a fixed point and a fixed line

Graph each conic section by hand and check with your calculator.

2. $12x^2 - 4y^2 = 48$

3. $y = 2x^2 + 3x - 1$

4. $x^2 + y^2 - 2x + 2y - 2 = 0$

5. $4x^2 + 9y^2 = 36$

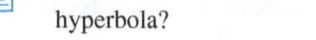 6. Given the two vertices and two foci of a conic section, how can you tell whether it is an ellipse or a hyperbola?

Write an equation for each conic section.

7. Center $(2, -1)$; radius 3

8. Foci $(\pm 4, 0)$; major axis length 12

9. Vertices $(0, \pm 2)$; foci $(0, \pm 4)$

10. Single focus $\left(0, \frac{1}{2}\right)$; vertex at the origin

---

7.3 The Conic Sections and Nonlinear Systems

Characteristics • Identifying Conic Sections • Eccentricity • Nonlinear Systems

Characteristics

The conic sections in this chapter have equations that can be written in the form

$$Ax^2 + Dx + Cy^2 + Ey + F = 0,$$

where either A or C must be nonzero. The special characteristics of each conic section $Ax^2 + Dx + Cy^2 + Ey + F = 0$ are summarized in the following table.

Conic Section	Characteristic	Examples
Parabola	Either $A = 0$ or $C = 0$, but not both	$y = x^2$ $x = 3y^2 + 2y - 4$
Circle	$A = C \neq 0$	$x^2 + y^2 = 16$
Ellipse	$A \neq C$, $AC > 0$	$\dfrac{x^2}{16} + \dfrac{y^2}{25} = 1$
Hyperbola	$AC < 0$	$x^2 - y^2 = 1$

The chart summarizes our work with conic sections.

Equation	Graph	Description	Identification
$(x - h)^2 = 4c(y - k)$ **or** $y - k = a(x - h)^2$		Graph opens up if $c > 0$ (or $a > 0$), down if $c < 0$ (or $a < 0$). Vertex is (h, k). Axis of symmetry is $x = h$.	x^2-term is present. y is not squared.
$(y - k)^2 = 4c(x - h)$ **or** $x - h = a(y - k)^2$		Graph opens to the right if $c > 0$ (or $a > 0$), to the left if $c < 0$ (or $a < 0$). Vertex is (h, k). Axis of symmetry is $y = k$.	y^2-term is present. x is not squared.
$(x - h)^2 + (y - k)^2 = r^2$		Center is (h, k), radius is r.	x^2- and y^2-terms have the same positive coefficient.
$\dfrac{x^2}{a^2} + \dfrac{y^2}{b^2} = 1 \quad (a > b > 0)$ Centered at $(0, 0)$		x-intercepts are $(\pm a, 0)$. y-intercepts are $(0, \pm b)$. Foci are $(\pm c, 0)$, where $c^2 = a^2 - b^2$.	x^2- and y^2-terms have different positive coefficients.
$\dfrac{x^2}{b^2} + \dfrac{y^2}{a^2} = 1 \quad (a > b > 0)$ Centered at $(0, 0)$		x-intercepts are $(\pm b, 0)$. y-intercepts are $(0, \pm a)$. Foci are $(0, \pm c)$, where $c^2 = a^2 - b^2$.	x^2- and y^2-terms have different positive coefficients.
$\dfrac{x^2}{a^2} - \dfrac{y^2}{b^2} = 1$ Centered at $(0, 0)$		x-intercepts are $(\pm a, 0)$. Asymptotes are diagonals of the rectangle with corners at $(\pm a, \pm b)$. Foci are $(\pm c, 0)$, where $c^2 = a^2 + b^2$.	x^2-term has a positive coefficient. y^2-term has a negative coefficient.
$\dfrac{y^2}{a^2} - \dfrac{x^2}{b^2} = 1$ Centered at $(0, 0)$		y-intercepts are $(0, \pm a)$. Asymptotes are diagonals of the rectangle with corners at $(\pm b, \pm a)$. Foci are $(0, \pm c)$, where $c^2 = a^2 + b^2$.	y^2-term has a positive coefficient. x^2-term has a negative coefficient.

> **NOTE** To find the value of c for an ellipse, use the equation $c^2 = a^2 - b^2$; for a hyperbola, use the equation $c^2 = a^2 + b^2$. If the center is $(0, 0)$, then the foci of these conic sections are located either at $(\pm c, 0)$ or at $(0, \pm c)$.

Identifying Conic Sections

To recognize the type of conic section represented by a given graph, we sometimes need to transform the equation into a more familiar form.

> **EXAMPLE 1** **Determining Types of Conic Sections from Equations**

Decide on the type of conic section represented by each equation, and give each graph.

(a) $x^2 = 25 + 5y^2$ **(b)** $x^2 - 8x + y^2 + 10y = -41$

(c) $4x^2 - 16x + 9y^2 + 54y = -61$ **(d)** $x^2 - 6x + 8y - 7 = 0$

(e) $4y^2 - 16y - 9x^2 + 18x = -43$

Solution

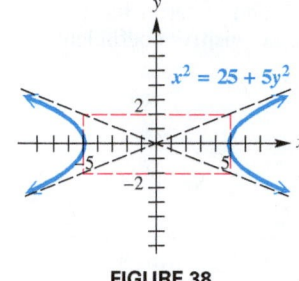

FIGURE 38

(a)
$$x^2 = 25 + 5y^2$$
$$x^2 - 5y^2 = 25 \qquad \text{Subtract } 5y^2.$$

Be sure to divide each term by 25. → $\dfrac{x^2}{25} - \dfrac{y^2}{5} = 1 \qquad \text{Divide by 25.}$

The equation represents a hyperbola centered at the origin and with asymptotes $y = \pm \dfrac{\sqrt{5}}{5}x$. The vertices are $(-5, 0)$ and $(5, 0)$. See **FIGURE 38.**

(b)
$$x^2 - 8x + y^2 + 10y = -41$$
$$(x^2 - 8x + 16) + (y^2 + 10y + 25) = -41 + 16 + 25 \qquad \begin{array}{l}\text{Complete the square}\\\text{on both } x \text{ and } y.\end{array}$$
$$(x - 4)^2 + (y + 5)^2 = 0 \qquad \text{Factor and add.}$$

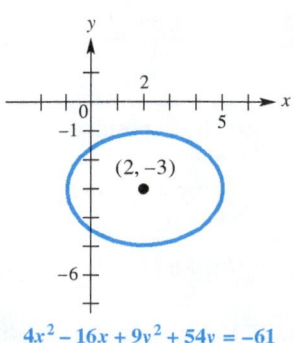

$x^2 - 8x + y^2 + 10y = -41$

Circle with radius 0

FIGURE 39

This result shows that the equation is that of a circle with radius 0, which is the point $(4, -5)$. See **FIGURE 39.** Had a negative number (instead of 0) been obtained on the right, the equation would have represented no points at all, and there would be no points on its graph.

(c) Since the coefficients of the x^2- and y^2-terms of $4x^2 - 16x + 9y^2 + 54y = -61$ are unequal and both positive, this equation may represent an ellipse, but not a circle. (It may also represent a single point or no points at all.)

$$4x^2 - 16x + 9y^2 + 54y = -61$$
$$4(x^2 - 4x + \underline{}) + 9(y^2 + 6y + \underline{}) = -61 \qquad \text{Factor out 4 and 9.}$$
$$4(x^2 - 4x + 4) + 9(y^2 + 6y + 9) = -61 + 4(4) + 9(9) \qquad \text{Complete the square.}$$
$$4(x - 2)^2 + 9(y + 3)^2 = 36 \qquad \text{Factor and simplify.}$$
$$\frac{(x - 2)^2}{9} + \frac{(y + 3)^2}{4} = 1 \qquad \text{Divide by 36.}$$

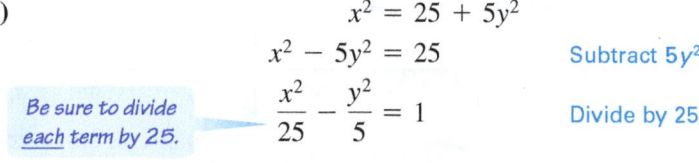

$4x^2 - 16x + 9y^2 + 54y = -61$

FIGURE 40

This equation represents an ellipse having center $(2, -3)$. See **FIGURE 40.**

(d) Since only one variable of $x^2 - 6x + 8y - 7 = 0$ is squared (x, and not y), the equation represents a parabola. Rearrange the terms so that the term $8y$ (with the variable that is not squared) is alone on one side.

$$x^2 - 6x + 8y - 7 = 0$$

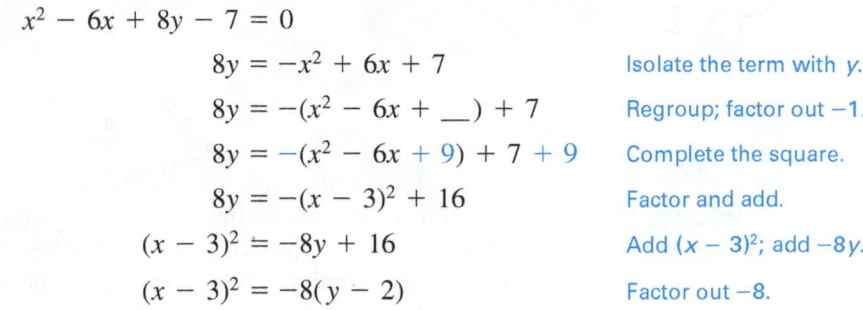

$8y = -x^2 + 6x + 7$	Isolate the term with y.
$8y = -(x^2 - 6x + \underline{}) + 7$	Regroup; factor out -1.
$8y = -(x^2 - 6x + 9) + 7 + 9$	Complete the square.
$8y = -(x - 3)^2 + 16$	Factor and add.
$(x - 3)^2 = -8y + 16$	Add $(x - 3)^2$; add $-8y$.
$(x - 3)^2 = -8(y - 2)$	Factor out -8.

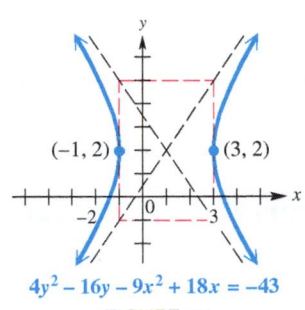

FIGURE 41

The parabola has vertex $(3, 2)$ and opens downward, as shown in **FIGURE 41.**

(e)
$$4y^2 - 16y - 9x^2 + 18x = -43$$

$4(y^2 - 4y + \underline{}) - 9(x^2 - 2x + \underline{}) = -43$	Factor out 4 and factor out -9.
$4(y^2 - 4y + 4) - 9(x^2 - 2x + 1) = -36$	Complete the square.
$4(y - 2)^2 - 9(x - 1)^2 = -36$	Factor.

Because of the -36, we might think that this equation does not have a graph. However, the minus sign in the middle on the left shows that the graph is that of a hyperbola.

$$\frac{(x - 1)^2}{4} - \frac{(y - 2)^2}{9} = 1 \qquad \text{Divide by } -36 \text{ and rearrange terms.}$$

This hyperbola has center $(1, 2)$. The graph is shown in **FIGURE 42.**

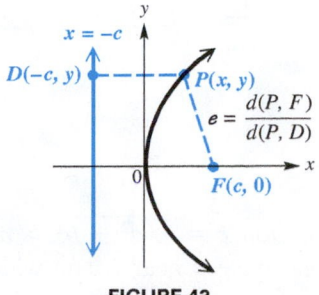

FIGURE 42

Eccentricity

In **Sections 7.1** and **7.2,** we introduced definitions of the conic sections. The conic sections (or *conics*) can all be characterized by one general definition.

Conic

A **conic** is the set of all points $P(x, y)$ in a plane such that the ratio of the distance from P to a fixed point and the distance from P to a fixed line is constant.

FIGURE 43

For a parabola, the fixed line is the directrix and the fixed point is the focus. In **FIGURE 43,** the focus is $F(c, 0)$ and the directrix is the line $x = -c$. The constant ratio is called the **eccentricity** of the conic, written *e*. ***This is not the same e as the base of natural logarithms.***

If the conic is a parabola, then by definition, the distances $d(P, F)$ and $d(P, D)$ in **FIGURE 43** are equal. ***Thus, every parabola has eccentricity 1.***

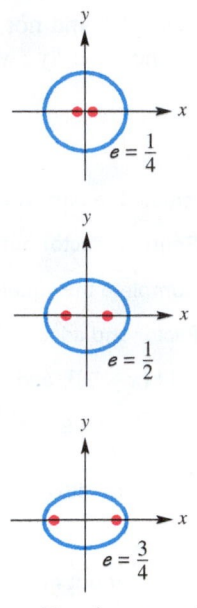

Ellipses with various eccentricities

FIGURE 44

For an ellipse, eccentricity is a measure of its "roundness." The constant ratio in the definition is $e = \frac{c}{a}$, where (as before) c is the distance from the center of the figure to a focus and a is the distance from the center to a vertex. By the definition of an ellipse, $a^2 > b^2$ and $c = \sqrt{a^2 - b^2}$. Thus, for the ellipse,

$$0 < c < a$$

$$0 < \frac{c}{a} < 1 \qquad \text{Divide by } a.$$

$$0 < e < 1. \qquad e = \frac{c}{a}$$

If a is constant, letting c approach 0 would cause the ratio $\frac{c}{a}$ to approach 0, which also causes b to approach a (so that $\sqrt{a^2 - b^2} = c$ would approach 0). Since b determines the endpoints of the minor axis, the lengths of the major and minor axes are almost the same, producing an ellipse very close in shape to a circle when e is very close to 0. In a similar manner, if e approaches 1, then b will approach 0.

The path of Earth around the sun is an ellipse that is very nearly circular. In fact, for this ellipse, $e \approx 0.017$. By contrast, the path of Halley's comet is a very flat ellipse, with $e \approx 0.97$. **FIGURE 44** compares ellipses with different eccentricities.

The equation of a circle can be written

$$(x - h)^2 + (y - k)^2 = r^2$$

$$\frac{(x - h)^2}{r^2} + \frac{(y - k)^2}{r^2} = 1. \qquad \text{Divide by } r^2.$$

In a circle, the foci coincide with the center, so $a = b$ and $c = \sqrt{a^2 - b^2} = 0$. Thus, it follows that $e = 0$.

EXAMPLE 2 **Finding Eccentricity from Equations of Ellipses**

Find the eccentricity of each ellipse.

(a) $\dfrac{x^2}{9} + \dfrac{y^2}{16} = 1$ **(b)** $5x^2 + 10y^2 = 50$

Solution

(a) Since $16 > 9$, let $a^2 = 16$, which gives $a = 4$. Also,

$$c = \sqrt{a^2 - b^2} = \sqrt{16 - 9} = \sqrt{7}.$$

Finally, $e = \dfrac{c}{a} = \dfrac{\sqrt{7}}{4} \approx 0.66.$

(b) Divide $5x^2 + 10y^2 = 50$ by 50 to obtain $\frac{x^2}{10} + \frac{y^2}{5} = 1$. Here, $a^2 = 10$, with $a = \sqrt{10}$. Now find c.

$$c = \sqrt{10 - 5} = \sqrt{5} \quad \text{and} \quad e = \frac{\sqrt{5}}{\sqrt{10}} = \frac{1}{\sqrt{2}} \approx 0.71. \qquad \bullet$$

The hyperbola in standard form

$$\frac{x^2}{a^2} - \frac{y^2}{b^2} = 1 \quad \text{or} \quad \frac{y^2}{a^2} - \frac{x^2}{b^2} = 1,$$

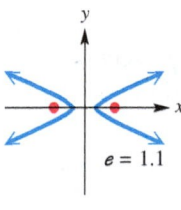

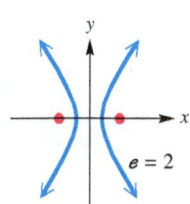

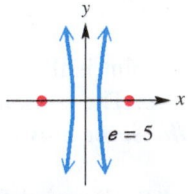

Hyperbolas with various eccentricities

FIGURE 45

where $c = \sqrt{a^2 + b^2}$, also has eccentricity $e = \frac{c}{a}$. By definition, $c = \sqrt{a^2 + b^2} > a$, so $\frac{c}{a} > 1$, and for a hyperbola, $e > 1$. Narrow hyperbolas have e near 1, and wide hyperbolas have large e. See **FIGURE 45**.

EXAMPLE 3 **Finding Eccentricity of a Hyperbola**

Find the eccentricity of the hyperbola $\frac{x^2}{9} - \frac{y^2}{4} = 1$.

Solution Here $a^2 = 9$. Thus, $a = 3$, $c = \sqrt{9 + 4} = \sqrt{13}$, and

$$e = \frac{c}{a} = \frac{\sqrt{13}}{3} \approx 1.2.$$

The table summarizes this discussion of eccentricity.

Conic	Eccentricity
Circle	$e = 0$
Parabola	$e = 1$
Ellipse	$e = \frac{c}{a}$ and $0 < e < 1$
Hyperbola	$e = \frac{c}{a}$ and $e > 1$

EXAMPLE 4 **Finding Equations of Conics by Using Eccentricity**

Find an equation for each conic with center at the origin and eccentricity e.

(a) Focus $(3, 0)$; $e = 2$ **(b)** Vertex $(0, -8)$; $e = \frac{1}{2}$

Solution

(a) Since $e = 2$, which is greater than 1, the conic is a hyperbola with $c = 3$.

$$e = \frac{c}{a}$$

$$2 = \frac{3}{a} \qquad \text{Let } e = 2 \text{ and } c = 3.$$

$$a = \frac{3}{2} \qquad \text{Solve for } a.$$

Now we find b using the equation $a^2 + b^2 = c^2$, or $b^2 = c^2 - a^2$.

$$b^2 = c^2 - a^2 = 9 - \frac{9}{4} = \frac{27}{4} \qquad c^2 = 3^2 = 9;\ a^2 = \left(\frac{3}{2}\right)^2 = \frac{9}{4}$$

The given focus is on the x-axis, so the x^2-term is positive, and the equation is

$$\frac{x^2}{\frac{9}{4}} - \frac{y^2}{\frac{27}{4}} = 1, \quad \text{or} \quad \frac{4x^2}{9} - \frac{4y^2}{27} = 1. \qquad \text{\textcolor{blue}{Simplify the complex fractions.}}$$

(b) The graph of the conic is an ellipse because $e = \frac{1}{2} < 1$. From the given vertex $(0, -8)$, we know that the vertices are on the y-axis and $a = 8$.

$$e = \frac{c}{a}$$

$$\frac{1}{2} = \frac{c}{8} \qquad \text{Let } e = \frac{1}{2} \text{ and } a = 8.$$

$$c = 4 \qquad \text{Solve for } c.$$

Since $b^2 = a^2 - c^2 = 64 - 16 = 48$, the equation is

$$\frac{x^2}{48} + \frac{y^2}{64} = 1.$$

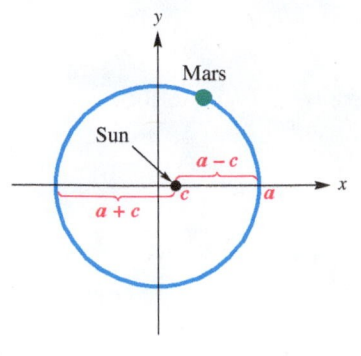

Not to scale

FIGURE 46

Applying an Ellipse to the Orbit of a Planet

The orbit of the planet Mars is an ellipse with the sun at one focus. The eccentricity of the ellipse is 0.0935, and the closest Mars comes to the sun is 128.5 million miles. (*Source: The World Almanac and Book of Facts.*) Find the maximum distance of Mars from the sun.

Solution **FIGURE 46** shows the orbit of Mars with the origin at the center of the ellipse and the sun at one focus. Mars is closest to the sun when Mars is at the right endpoint of the major axis and farthest from the sun when Mars is at the left endpoint. Therefore, the least distance is $a - c$, and the greatest distance is $a + c$. Since $a - c = 128.5$, $c = a - 128.5$. Using $e = 0.0935$, we can find a.

$$e = \frac{c}{a}$$

$$0.0935 = \frac{a - 128.5}{a} \qquad \text{Substitute for } e \text{ and } c.$$

$$0.0935a = a - 128.5 \qquad \text{Multiply by } a.$$

$$-0.9065a = -128.5 \qquad \text{Subtract } a.$$

$$a \approx 141.8 \qquad \text{Divide by } -0.9065.$$

Then $\qquad\qquad\qquad c \approx 141.8 - 128.5 = 13.3$

and $\qquad\qquad\quad a + c \approx 141.8 + 13.3 = 155.1.$

Thus, the maximum distance of Mars from the sun is about 155.1 million miles.

Nonlinear Systems

A linear system of equations can have zero, one, or infinitely many solutions. However, a nonlinear system of equations can have any number of solutions. For example, the graphs of the nonlinear system

$$x^2 + y^2 = 4 \qquad \text{Graph is a circle.}$$
$$2x^2 - y^2 = 8 \qquad \text{Graph is a hyperbola.}$$

consist of a circle and a hyperbola, which could intersect at zero, one, two, three, or four points. In the next example we see that their graphs intersect at two points.

Solving a Nonlinear System by Elimination

Solve the system.

$$x^2 + y^2 = 4 \qquad \text{(1)}$$
$$2x^2 - y^2 = 8 \qquad \text{(2)}$$

Solution The graph of equation (1) is a circle and that of equation (2) is a hyperbola. Visualizing them suggests that there may be 0, 1, 2, 3, or 4 points of intersection. Add the two equations to eliminate y^2.

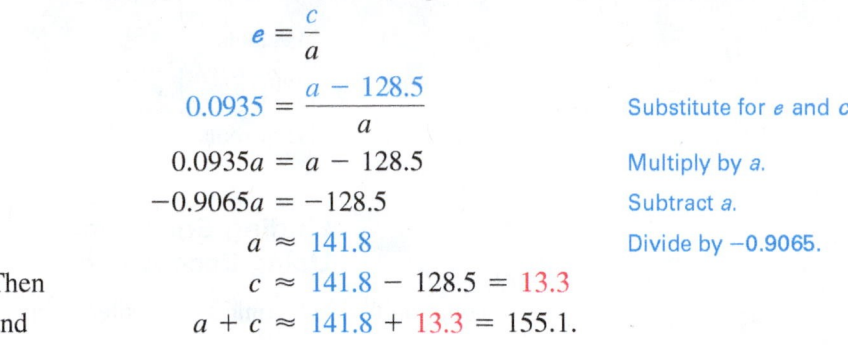

Remember both the positive and negative square roots.

$$x^2 + y^2 = 4 \qquad \text{(1)}$$
$$\underline{2x^2 - y^2 = 8} \qquad \text{(2)}$$
$$3x^2 \qquad\quad = 12 \qquad \text{Add.}$$
$$x^2 = 4 \qquad \text{Divide by 3.}$$
$$x = 2 \quad \text{or} \quad x = -2 \qquad \text{Square root property}$$

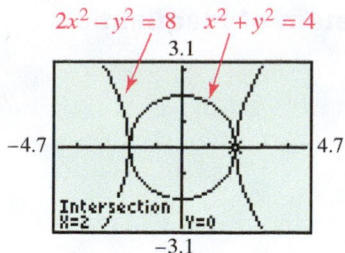

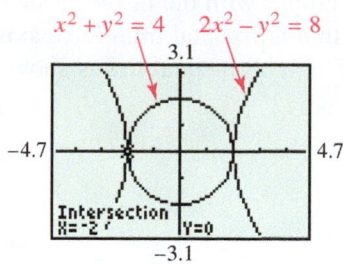

The two points of intersection are $(2, 0)$ and $(-2, 0)$.

FIGURE 47

Substituting into equation (1) gives the corresponding values of y.

If $x = 2$, then	If $x = -2$, then
$2^2 + y^2 = 4$	$(-2)^2 + y^2 = 4$
$y^2 = 0$	$y^2 = 0$
$y = 0.$	$y = 0.$

Check that the solution set of the system is $\{(2, 0), (-2, 0)\}$. See **FIGURE 47**. ●

Sometimes a combination of the elimination method and the substitution method is effective in solving a system, as illustrated in **Example 7**.

EXAMPLE 7   **Solving a Nonlinear System by a Combination of Methods**

Solve the system.

$$x^2 + 3xy + y^2 = 22 \qquad (1)$$
$$x^2 - xy + y^2 = 6 \qquad (2)$$

Solution   Begin as with the elimination method.

$$
\begin{array}{ll}
x^2 + 3xy + y^2 = 22 & (1) \\
\underline{-x^2 + xy - y^2 = -6} & \text{Multiply (2) by } -1. \\
4xy \qquad\quad = 16 & \text{Add. } (3) \\
y = \dfrac{4}{x} & \text{Solve for } y \ (x \neq 0). \ (4)
\end{array}
$$

← **Algebra Review**
To review elimination and substitution, see **Section 6.1**.

Now substitute $\frac{4}{x}$ for y in either equation (1) or equation (2). We use equation (2).

$$x^2 - x\left(\frac{4}{x}\right) + \left(\frac{4}{x}\right)^2 = 6 \qquad \text{Let } y = \frac{4}{x} \text{ in (2).}$$

$$x^2 - 4 + \frac{16}{x^2} = 6 \qquad \text{Multiply and square.}$$

This equation is quadratic in form.

$$x^4 - 4x^2 + 16 = 6x^2 \qquad \text{Multiply by } x^2 \text{ to clear fractions.}$$

$$x^4 - 10x^2 + 16 = 0 \qquad \text{Subtract } 6x^2.$$

$$(x^2 - 2)(x^2 - 8) = 0 \qquad \text{Factor.}$$

$$x^2 - 2 = 0 \quad \text{or} \quad x^2 - 8 = 0 \qquad \text{Zero-product property}$$

For each equation, include both square roots.

$$x^2 = 2 \quad \text{or} \quad x^2 = 8 \qquad \text{Solve each equation.}$$

$$x = \pm\sqrt{2} \quad \text{or} \quad x = \pm 2\sqrt{2} \qquad \begin{array}{l} \text{Square root property;} \\ \pm\sqrt{8} = \pm\sqrt{4} \cdot \sqrt{2} = \pm 2\sqrt{2} \end{array}$$

Substitute these x-values into equation (4) to find corresponding values of y.

Let $x = \sqrt{2}.$	Let $x = -\sqrt{2}.$	Let $x = 2\sqrt{2}.$	Let $x = -2\sqrt{2}.$
$y = \dfrac{4}{\sqrt{2}} = 2\sqrt{2}$	$y = \dfrac{4}{-\sqrt{2}} = -2\sqrt{2}$	$y = \dfrac{4}{2\sqrt{2}} = \sqrt{2}$	$y = \dfrac{4}{-2\sqrt{2}} = -\sqrt{2}$

The solution set of the system is

$$\left\{ \left(\sqrt{2}, 2\sqrt{2}\right), \left(-\sqrt{2}, -2\sqrt{2}\right), \left(2\sqrt{2}, \sqrt{2}\right), \left(-2\sqrt{2}, -\sqrt{2}\right) \right\}.$$

Verify these solutions by substitution in the original system. ●

Graphing a Nonlinear System of Inequalities

Graph the solution set of the system.

$$\frac{x^2}{25} + \frac{y^2}{9} \leq 1 \quad (1)$$

$$\frac{x^2}{4} - \frac{y^2}{9} \geq 1 \quad (2)$$

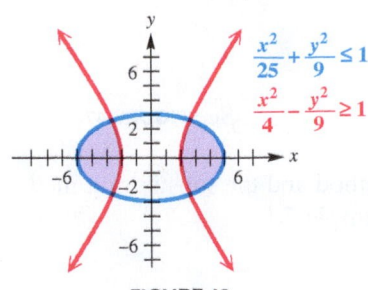

FIGURE 48

Solution The graph of inequality (1) consists of an ellipse with the inside shaded. The graph of inequality (2) consists of a hyperbola with a horizontal transverse axis and shaded to the left and right. The region that satisfies both inequalities is shown in **FIGURE 48.**

7.3 Exercises

Checking Analytic Skills *The equation of a conic section is given in a familiar form. Identify the type of graph (if any) that each equation has, without actually graphing. See the summary chart in this section.* **Do not use a calculator.**

1. $x^2 + y^2 = 144$

2. $(x - 2)^2 + (y + 3)^2 = 25$

3. $y = 2x^2 + 3x - 4$

4. $x = 3y^2 + 5y - 6$

5. $x - 1 = -3(y - 4)^2$

6. $\frac{x^2}{25} + \frac{y^2}{36} = 1$

7. $\frac{x^4}{49} + \frac{y^2}{100} = 1$

8. $x^2 - y^2 = 1$

9. $\frac{x^2}{4} - \frac{y^2}{16} = 1$

10. $\frac{(x + 2)^2}{9} + \frac{(y - 4)^2}{16} = 1$

11. $\frac{x^2}{25} - \frac{y^2}{25} = 1$

12. $y + 7 = 4(x + 3)^2$

13. $\frac{x^2}{4} = 1 - \frac{y^2}{9}$

14. $\frac{x^2}{4} = 1 + \frac{y^2}{9}$

15. $\frac{(x + 3)^2}{16} + \frac{(y - 2)^2}{16} = 1$

16. $x^2 = 25 - y^2$

17. $x^2 - 6x + y = 0$

18. $11 - 3x = 2y^2 - 8y$

19. $4(x - 3)^2 + 3(y + 4)^2 = 0$

20. $2x^2 - 8x + 2y^2 + 20y = 12$

21. $x - 4y^2 - 8y = 0$

22. $x^2 + 2x = -4y$

23. $6x^2 - 12x + 6y^2 - 18y + 25 = 0$

24. $4x^2 - 24x + 5y^2 + 10y + 41 = 0$

Determine the type of conic section represented by each equation, and graph it, provided a graph exists.

25. $x^2 = 4y - 8$

26. $\frac{x^2}{4} + \frac{y^2}{4} = 1$

27. $x^2 = 25 + y^2$

28. $9x^2 + 36y^2 = 36$

29. $\frac{x^2}{4} + \frac{y^2}{4} = -1$

30. $\frac{(x - 4)^2}{8} + \frac{(y + 1)^2}{2} = 0$

31. $y^2 - 4y = x + 4$

32. $(x + 7)^2 + (y - 5)^2 + 4 = 0$

33. $3x^2 + 6x + 3y^2 - 12y = 12$

34. $-4x^2 + 8x + y^2 + 6y = -6$

35. $4x^2 - 8x + 9y^2 - 36y = -4$

36. $3x^2 + 12x + 3y^2 = 0$

37. *Concept Check* Identify the type of conic section consisting of the set of all points in the plane for which the sum of the distances from the points $(5, 0)$ and $(-5, 0)$ is 14.

38. *Concept Check* Identify the type of conic section consisting of the set of all points in the plane for which the absolute value of the difference of the distances from the points $(3, 0)$ and $(-3, 0)$ is 2.

39. *Concept Check* Identify the type of conic section consisting of the set of all points in the plane for which the distance from the point $(3, 0)$ is one and one-half times the distance from the line $x = \frac{4}{3}$.

40. *Concept Check* Identify the type of conic section consisting of the set of all points in the plane for which the distance from the point $(2, 0)$ is one-third the distance from the line $x = 10$.

Find the eccentricity e of each ellipse or hyperbola.

41. $12x^2 + 9y^2 = 36$

42. $8x^2 - y^2 = 16$

43. $x^2 - y^2 = 4$

44. $x^2 + 2y^2 = 8$

45. $4x^2 + 7y^2 = 28$

46. $9x^2 - y^2 = 1$

47. $x^2 - 9y^2 = 18$

48. $x^2 + 10y^2 = 10$

Write an equation for each conic. Each parabola has vertex at the origin, and each ellipse or hyperbola is centered at the origin.

49. Focus $(0, 8)$; $e = 1$

50. Focus $(-2, 0)$; $e = 1$

51. Focus $(3, 0)$; $e = \frac{1}{2}$

52. Focus $(0, -2)$; $e = \frac{2}{3}$

53. Vertex $(-6, 0)$; $e = 2$

54. Vertex $(0, 4)$; $e = \frac{5}{3}$

55. Focus $(0, -1)$; $e = 1$

56. Focus $(2, 0)$; $e = \frac{6}{5}$

57. Vertical major axis of length 6; $e = \frac{4}{5}$

58. y-intercepts $(0, \pm 4)$; $e = \frac{7}{3}$

Find the eccentricity e of each conic section. The point shown on the x-axis is a focus and the line shown is a directrix.

59.

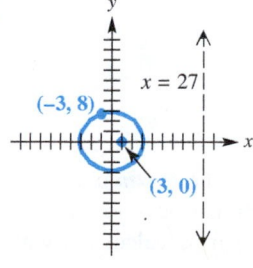

60.

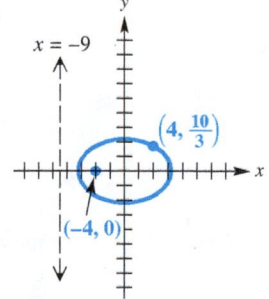

61.

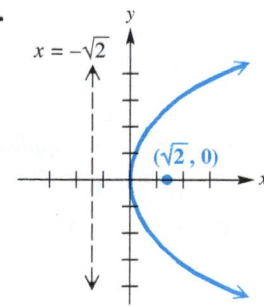

62.

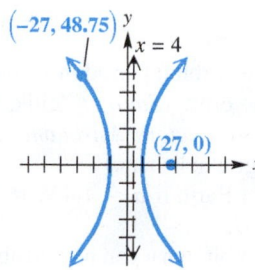

63.

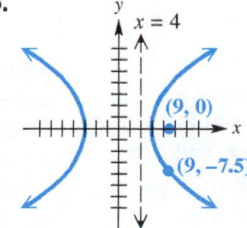

64.

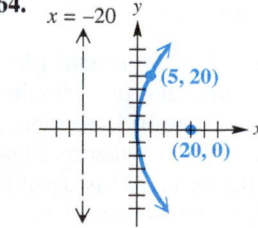

Checking Analytic Skills Solve each nonlinear system of equations analytically for all real solutions.

65. $x^2 + y^2 = 10$
$2x^2 - y^2 = 17$

66. $x^2 + y^2 = 4$
$2x^2 - 3y^2 = -12$

67. $x^2 + 2y^2 = 9$
$3x^2 - 4y^2 = 27$

68. $2x^2 + 3y^2 = 5$
$3x^2 - 4y^2 = -1$

69. $2x^2 + 2y^2 = 20$
$3x^2 + 3y^2 = 30$

70. $x^2 + y^2 = 4$
$5x^2 + 5y^2 = 28$

71. $3x^2 + 2y^2 = 5$
$x - y = -2$

72. $2x^2 - y^2 = 4$
$|x| = |y|$

73. $x^2 + y^2 = 8$
$x^2 - y^2 = 0$

74. $2x^2 + 3y^2 = 5$
$4x^2 + 6y^2 = 8$

75. $x^2 + xy + y^2 = 3$
$x^2 - xy + y^2 = 1$

76. $x^2 + 2xy - y^2 = 7$
$x^2 - 2xy + y^2 = 1$

77. $x^2 - xy + y^2 = 5$
$2x^2 + xy - y^2 = 10$

78. $x^2 + 3xy + y^2 = 5$
$x^2 - 2xy - y^2 = -7$

Use substitution to solve the nonlinear system of equations in three variables. Note that solutions are ordered triples.

79. $2x^2 + y^2 + 3z^2 = 3$
$2x + y - z = 1$
$x + y = 0$

80. $x^2 + y^2 + z^2 = 4$
$x + y + z = 2$
$x - y = 0$

Checking Analytic Skills *Graph the solution set of each system of inequalities by hand.*

81. $\dfrac{x^2}{16} + \dfrac{y^2}{4} \le 1$

$x^2 - y^2 \ge 1$

82. $\dfrac{x^2}{4} + \dfrac{y^2}{9} \le 1$

$\dfrac{y^2}{4} - \dfrac{x^2}{9} \le 1$

83. $4x^2 - y^2 > 4$

$9x^2 + 4y^2 > 36$

84. $16x^2 + 9y^2 < 144$

$(x - 1)^2 + (y + 1)^2 > 1$

85. $\dfrac{(x - 1)^2}{9} + \dfrac{y^2}{4} \le 1$

$\dfrac{x^2}{4} - \dfrac{(y + 1)^2}{9} \ge 1$

86. $\dfrac{(x - 2)^2}{36} + \dfrac{(y + 2)^2}{25} \le 1$

$\dfrac{(x + 1)^2}{9} + \dfrac{(y - 3)^2}{25} \le 1$

(Modeling) *Solve each application.*

87. *Orbit of Mars* The orbit of Mars around the sun is an ellipse with equation

$$\frac{x^2}{5013} + \frac{y^2}{4970} = 1,$$

where x and y are measured in millions of miles. Approximate the eccentricity e of this ellipse.

88. *Orbits of Neptune and Pluto* Neptune and Pluto both have elliptical orbits with the sun at one focus. Neptune's orbit has $a = 30.1$ astronomical units (AU) and eccentricity $e = 0.009$, whereas Pluto's orbit has $a = 39.4$ and $e = 0.249$. (1 AU is equal to the average distance from Earth to the sun and is approximately 149,600,000 kilometers.) (*Source:* Zeilik, M., S. Gregory, and E. Smith, *Introductory Astronomy and Astrophysics,* Saunders College Publishers.)
   (a) Position the sun at the origin, and determine an equation for each orbit.
   (b) Graph both equations on the same coordinate axes. Use the window $[-60, 60]$ by $[-40, 40]$.

89. *Velocity of a Planet in Orbit* The maximum and minimum velocities in kilometers per second of a planet moving in an elliptical orbit can be calculated with the equations

$$v_{max} = \frac{2\pi a}{P}\sqrt{\frac{1 + e}{1 - e}} \quad \text{and} \quad v_{min} = \frac{2\pi a}{P}\sqrt{\frac{1 - e}{1 + e}},$$

where a is in kilometers, P is its orbital period in seconds, and e is the eccentricity of the orbit. (*Source:* Zeilik, M., S. Gregory, and E. Smith, *Introductory Astronomy and Astrophysics,* Saunders College Publishers.)
   (a) Calculate v_{max} and v_{min} for Earth if $a = 1.496 \times 10^8$ kilometers and $e = 0.0167$.
   (b) If an object has a circular orbit, what can be said about its orbital velocity?
   (c) Kepler showed that the sun is located at a focus of a planet's elliptical orbit. He also showed that a planet has minimum velocity when its distance from the sun is maximum, and a planet has maximum velocity when

its distance from the sun is minimum. Where do the maximum and minimum velocities occur in an elliptical orbit?

90. *Distance between Halley's Comet and the Sun* The famous Halley's comet last passed Earth in February 1986 and will next return in 2062. Halley's comet has an elliptical orbit of eccentricity 0.9673 with the sun at one of the foci. The greatest distance of the comet from the sun is 3281 million miles. (*Source: The World Almanac and Book of Facts.*) Find the least distance between Halley's comet and the sun.

91. *Orbit of Earth* The orbit of Earth is an ellipse with the sun at one focus. The distance between Earth and the sun ranges from 91.4 to 94.6 million miles. (*Source: The World Almanac and Book of Facts.*) Estimate the eccentricity of Earth's orbit.

92. *Distance Traveled by Mercury* Use the formula of **Exercise 94** in **Section 7.2** to estimate the distance that is traveled by the planet Mercury in one orbit of the sun, if $a = 36.0$, $b = 35.2$, and the units are in millions of miles.

7.4 Parametric Equations

Graphs of Parametric Equations and Their Rectangular Equivalents • Alternative Forms of Parametric Equations
• An Application of Parametric Equations

TECHNOLOGY NOTE

In addition to graphing rectangular equations, graphing calculators are capable of graphing plane curves defined by parametric equations. The calculator must be set in parametric mode, and the window requires intervals for the parameter t, as well as for x and y. Consult your owner's guide.

We have graphed sets of ordered pairs that correspond to functions of the form $y = f(x)$ or to relations of the form $Ax^2 + Bxy + Cy^2 + Dx + Ey + F = 0$. Another way to determine a set of ordered pairs involves two functions f and g defined by $x = f(t)$ and $y = g(t)$, where t is a real number in some interval I. Each value of t leads to a corresponding x-value and a corresponding y-value, and thus to an ordered pair (x, y).

> **Parametric Equations of a Plane Curve**
>
> A **plane curve** is a set of points (x, y) such that $x = f(t)$, $y = g(t)$, and f and g are both continuous on an interval I. The equations $x = f(t)$ and $y = g(t)$ are **parametric equations** with **parameter t.**

Graphs of Parametric Equations and Their Rectangular Equivalents

EXAMPLE 1   **Graphing a Parabola Defined Parametrically**

Graph the plane curve

$$x = t^2, \quad y = 2t + 3, \quad \text{for } t \text{ in } [-3, 3],$$

and then find an equivalent rectangular equation.

(continued)

Analytic Solution

Make a table of corresponding values of t, x, and y over the domain of t. Then plot the points. The graph is a portion of a parabola with horizontal axis $y = 3$. See **FIGURE 49**. The red arrowheads indicate the direction the curve traces as t increases.

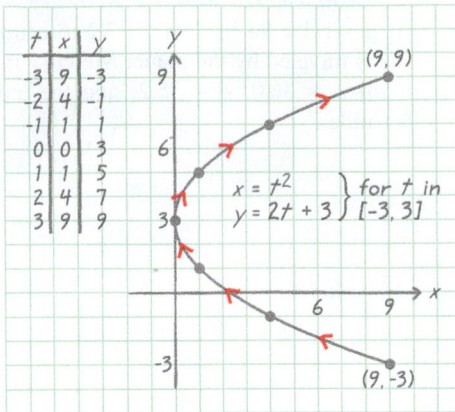

FIGURE 49

To find an equivalent rectangular equation, we eliminate the parameter t. For this curve, we begin by solving for t in the second equation, $y = 2t + 3$, because it leads to a unique solution.

$$y = 2t + 3 \qquad \textit{Solve for t.}$$
$$2t = y - 3 \qquad \textit{Subtract 3 and rewrite.}$$
$$t = \frac{y - 3}{2} \qquad \textit{Divide by 2.}$$

Now we substitute the result in the first equation $x = t^2$.

$$x = t^2 = \left(\frac{y - 3}{2}\right)^2 = \frac{(y - 3)^2}{4} = \frac{1}{4}(y - 3)^2$$

This is indeed an equation of a horizontal parabola that opens to the right. Because t is in $[-3, 3]$, x is in $[0, 9]$ and y is in $[-3, 9]$. The rectangular equation must be given with its restricted domain as

$$x = \frac{1}{4}(y - 3)^2, \quad \text{for } x \text{ in } [0, 9].$$

Graphing Calculator Solution

FIGURE 50 shows a TI-84 Plus table of values. The calculator is in parametric mode.

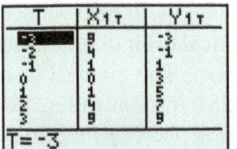

Parametric mode
FIGURE 50

FIGURE 51 shows a graph. The formation of the curve as T increases can be seen by using the pause feature and carefully observing the path of the curve. See the settings in the margin below.

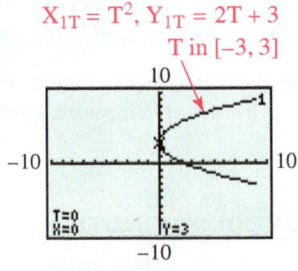

Parametric mode
FIGURE 51

The display in **FIGURE 51** indicates particular values of T, X, and Y. (This curve could be graphed in function mode by graphing $y_1 = 3 + 2\sqrt{x}$ and $y_2 = 3 - 2\sqrt{x}$ for x in $[0, 9]$.)

```
WINDOW
 Tmin=-3
 Tmax=3
 Tstep=.05
 Xmin=-10
 Xmax=10
 Xscl=1
↓Ymin=-10
     Ymax=10
     Yscl=1
```

For the parametric equations in **Example 1,** the screens show the entries that produce the standard window shown in **FIGURE 51** above.

| | EXAMPLE 2 | **Graphing a Plane Curve Defined Parametrically** |

Graph the plane curve

$$x = 2t + 5, \quad y = \sqrt{4 - t^2}, \quad \text{for } t \text{ in } [-2, 2],$$

and then find an equivalent rectangular equation.

Solution   A graph is shown in **FIGURE 52** on the next page. To get an equivalent rectangular equation, solve the first equation, which does not involve a radical, for t, and then substitute into the second equation. Since t is in $[-2, 2]$, x is in the interval $[1, 9]$.

$$2t + 5 = x, \quad \text{so} \quad t = \frac{x - 5}{2}.$$

The rectangular equation is as follows.

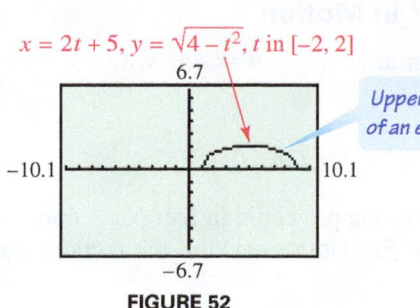

$x = 2t + 5, y = \sqrt{4 - t^2}, t \text{ in } [-2, 2]$

Upper half of an ellipse

FIGURE 52

$$y = \sqrt{4 - t^2}$$

$$y = \sqrt{4 - \left(\frac{x - 5}{2}\right)^2} \qquad \text{Substitute.}$$

$$y^2 = 4 - \left(\frac{x - 5}{2}\right)^2 \qquad \text{Square each side.}$$

$$y^2 = 4 - \frac{(x - 5)^2}{4} \qquad 2^2 = 4$$

$$4y^2 = 16 - (x - 5)^2 \qquad \text{Multiply by 4.}$$

$$4y^2 + (x - 5)^2 = 16 \qquad \text{Add } (x - 5)^2.$$

$$\frac{y^2}{4} + \frac{(x - 5)^2}{16} = 1 \qquad \text{Divide by 16.}$$

This equation represents a complete ellipse. (By definition, the parametric graph has $y \geq 0$, so it is only the upper half of the ellipse.)

EXAMPLE 3   **Graphing a Line Defined Parametrically**

Graph the plane curve

$$x = t^2, \quad y = t^2, \quad \text{for } t \text{ in } (-\infty, \infty)$$

and then find an equivalent rectangular equation.

Solution   **FIGURE 53** shows a portion of the curve in a standard window with both x and y in $[-10, 10]$. The graph is a ray, because both x and y must be greater than or equal to 0. Since both x and y equal t^2, $y = x$. To be equivalent, however, the rectangular equation must be given as

$$y = x, \quad x \geq 0.$$

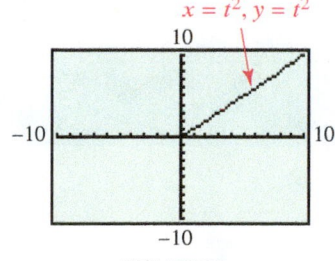

$x = t^2, y = t^2$

FIGURE 53

Alternative Forms of Parametric Equations

Parametric representations of a curve are not unique. There are infinitely many parametric representations of a given curve. If the curve can be described by a rectangular equation $y = f(x)$ with domain X, then one simple parametric representation is

$$x = t, \quad y = f(t), \quad \text{for } t \text{ in } X.$$

EXAMPLE 4   **Finding Alternative Parametric Equation Forms**

Give two parametric representations for the parabola $y = (x - 2)^2 + 1$.

Solution   The simplest choice is to let

$$x = t, \quad y = (t - 2)^2 + 1, \quad \text{for } t \text{ in } (-\infty, \infty).$$

Another choice that leads to a simpler equation for y is

$$x = t + 2, \quad y = t^2 + 1, \quad \text{for } t \text{ in } (-\infty, \infty).$$

An Application of Parametric Equations

One application of parametric equations is to determine the path of a moving object whose position is given by the function defined by

$$x = f(t), \quad y = g(t), \quad \text{where } t \text{ represents time.}$$

The parametric equations give the position of the object at any time t.

EXAMPLE 5 **Using Parametric Equations to Define the Position of an Object in Motion**

The motion of a projectile moving in a direction at an angle $\theta = 45°$ with the horizontal (neglecting air resistance) is given by

$$x = v_0 \frac{\sqrt{2}}{2}t, \quad y = v_0 \frac{\sqrt{2}}{2}t - 16t^2, \quad \text{for } t \text{ in } [0, k],$$

where t is time in seconds, v_0 is the initial speed of the projectile in feet per second, x and y are in feet, and k is a positive real number. See **FIGURE 54.** Find the rectangular form of the equation.

Solution Solving the first equation for t and substituting the result into the second equation gives (after simplification)

$$y = x - \frac{32}{v_0^2}x^2,$$

the equation of a vertical parabola opening downward, as shown in **FIGURE 54.**

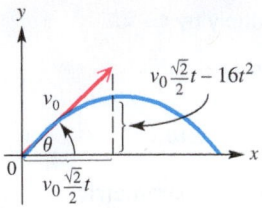

FIGURE 54

7.4 Exercises

Checking Analytic Skills *Graph each pair of parametric equations by hand, using values of t in* $[-2, 2]$. *Make a table of t-, x-, and y-values, using* $t = -2, -1, 0, 1,$ *and 2. Then plot the points and join them with a line or smooth curve for* all *values of t in* $[-2, 2]$. ***Do not use a calculator.***

1. $x = 2t + 1, \quad y = t - 2$ **2.** $x = -t + 1, \quad y = 3t + 2$ **3.** $x = t + 1, \quad y = t^2 - 1$

4. $x = t - 1, \quad y = t^2 + 2$ **5.** $x = t^2 + 2, \quad y = -t + 1$ **6.** $x = -t^2 + 2, \quad y = t + 1$

For each plane curve, use a graphing calculator to generate the curve over the interval for the parameter t, in the window specified. Then, find a rectangular equation for the curve.

7. $x = 2t, y = t + 1$, for t in $[-2, 3]$;
window: $[-8, 8]$ by $[-8, 8]$

8. $x = t + 2, y = t^2$, for t in $[-1, 1]$;
window: $[0, 4]$ by $[-2, 2]$

9. $x = \sqrt{t}, y = 3t - 4$, for t in $[0, 4]$;
window: $[-6, 6]$ by $[-6, 10]$

10. $x = t^2, y = \sqrt{t}$, for t in $[0, 4]$;
window: $[-2, 20]$ by $[0, 4]$

11. $x = t^3 + 1, y = t^3 - 1$, for t in $[-3, 3]$;
window: $[-30, 30]$ by $[-30, 30]$

12. $x = 2t - 1, y = t^2 + 2$, for t in $[-10, 10]$;
window: $[-20, 20]$ by $[0, 120]$

13. $x = 2^t, y = \sqrt{3t - 1}$, for t in $\left[\frac{1}{3}, 4\right]$;
window: $[-2, 30]$ by $[-2, 10]$

14. $x = \ln(t - 1), y = 2t - 1$, for t in $(1, 10]$;
window: $[-5, 5]$ by $[-2, 20]$

15. $x = t + 2, y = -\frac{1}{2}\sqrt{9 - t^2}$, for t in $[-3, 3]$;
window: $[-6, 6]$ by $[-4, 4]$

16. $x = t, y = \sqrt{4 - t^2}$, for t in $[-2, 2]$;
window: $[-6, 6]$ by $[-4, 4]$

17. $x = t, y = \frac{1}{t}$, for t in $(-\infty, 0) \cup (0, \infty)$;
window: $[-6, 6]$ by $[-4, 4]$

18. $x = 2t - 1, y = \frac{1}{t}$, for t in $(-\infty, 0) \cup (0, \infty)$;
window: $[-6, 6]$ by $[-4, 4]$

For each plane curve, find a rectangular equation. State the appropriate interval for x or y.

19. $x = 3t$, $y = t - 1$, for t in $(-\infty, \infty)$

20. $x = t + 3$, $y = 2t$, for t in $(-\infty, \infty)$

21. $x = 3t^2$, $y = t + 1$, for t in $(-\infty, \infty)$

22. $x = t - 2$, $y = \frac{1}{2}t^2 + 1$, for t in $(-\infty, \infty)$

23. $x = 3t^2$, $y = 4t^3$, for t in $(-\infty, \infty)$

24. $x = 2t^3$, $y = -t^2$, for t in $(-\infty, \infty)$

25. $x = t$, $y = \sqrt{t^2 + 2}$, for t in $(-\infty, \infty)$

26. $x = \sqrt{t}$, $y = t^2 - 1$, for t in $[0, \infty)$

27. $x = e^t$, $y = e^{-t}$, for t in $(-\infty, \infty)$

28. $x = e^{2t}$, $y = e^t$, for t in $(-\infty, \infty)$

29. $x = \dfrac{1}{\sqrt{t + 2}}$, $y = \dfrac{t}{t + 2}$, for t in $(-2, \infty)$

30. $x = \dfrac{t}{t - 1}$, $y = \dfrac{1}{\sqrt{t - 1}}$, for t in $(1, \infty)$

31. $x = t + 2$, $y = \dfrac{1}{t + 2}$, for $t \neq -2$

32. $x = t - 3$, $y = \dfrac{2}{t - 3}$, for $t \neq 3$

33. $x = t^2$, $y = 2 \ln t$, for t in $(0, \infty)$

34. $x = \ln t$, $y = 3 \ln t$, for t in $(0, \infty)$

Give two parametric representations for each plane curve. Use your calculator to verify your results.

35. $y = 2x + 3$

36. $y = \frac{3}{2}x - 4$

37. $y = \sqrt{3x + 2}$, x in $\left[-\frac{2}{3}, \infty\right)$

38. $y = (x + 1)^2 + 1$

39. $x = y^3 + 1$

40. $x = 2(y - 3)^2 - 4$

41. $x = \sqrt{y + 1}$

42. $x = \dfrac{1}{y + 1}$

(Modeling) *Solve each application.*

43. *Firing a Projectile* A projectile is fired with an initial velocity of 400 feet per second at an angle of 45° with the horizontal. (See **Example 5**.)
 (a) Find the time to the nearest tenth when it strikes the ground.
 (b) Find the range (horizontal distance covered).
 (c) What is the maximum altitude?

44. *Firing a Projectile* If a projectile is fired at an angle of 30° with the horizontal, the parametric equations that describe its motion are
$$x = v_0 \frac{\sqrt{3}}{2}t, \quad y = \frac{v_0}{2}t - 16t^2, \quad \text{for } t \text{ in } [0, \infty).$$
Repeat **Exercise 43** if the projectile is fired at 800 feet per second.

45. *Path of a Projectile* A projectile moves so that its position at any time t is given by the equations
$$x = 60t \quad \text{and} \quad y = 80t - 16t^2.$$
Graph the path of the projectile, and find the equivalent rectangular equation. Use the window $[0, 300]$ by $[0, 200]$.

46. *Path of a Projectile* Repeat **Exercise 45**, using the window $[0, 300]$ by $[0, 200]$, if the path is given by the equations
$$x = t^2 \quad \text{and} \quad y = -16t + 64\sqrt{t}.$$

47. Show that the rectangular equation for the curve defined by the equations
$$x = v_0 \frac{\sqrt{2}}{2}t, \quad y = v_0 \frac{\sqrt{2}}{2}t - 16t^2, \quad \text{for } t \text{ in } [0, k],$$
is $y = x - \dfrac{32}{v_0^2}x^2$. (See **Example 5**.)

48. Find the vertex of the parabola given by the rectangular equation of **Exercise 47**.

49. Give two parametric representations of the line through the point (x_1, y_1) with slope m.

50. Give two parametric representations of the parabola $y = a(x - h)^2 + k$.

Write each equation in standard form, and name the type of conic section defined.

1. $3x^2 + y^2 - 6x + 6y = 0$

2. $y^2 - 2x^2 + 8y - 8x - 4 = 0$

3. $3y^2 + 12y + 5x = 3$

*Find the eccentricity **e** of each conic section.*

4. $x^2 + 25y^2 = 25$

5. $8y^2 - 4x^2 = 8$

6. $3x^2 + 4y^2 = 108$

Find an equation of each conic. (Hint: A sketch may be helpful.)

7. Focus $(-2, 0)$; vertex $(0, 0)$; eccentricity 1

8. Foci $(\pm 3, 0)$; major axis length 10

9. Foci $(0, \pm 5)$; vertices $(0, \pm 4)$

10. Vertices $(\pm 3, 0)$; asymptotes $y = \pm\frac{2}{3}x$

11. Solve the nonlinear system.

$$2x^2 + y^2 = 9$$
$$-4x^2 + 3y^2 = 27$$

12. Graph the solution set of the nonlinear system of inequalities.

$$\frac{x^2}{4} + \frac{y^2}{25} \le 1$$

$$(x - 2)^2 - \frac{y^2}{4} \le 1$$

Solve each problem.

13. The figure represents an elliptical stone arch with the dimensions (in feet) indicated. Find the height y of the arch 6 feet from the center of the base.

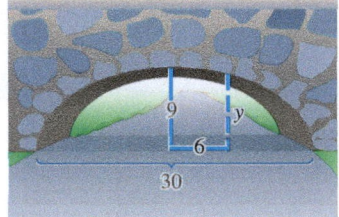

14. A plane curve is defined by

$$x = 2t, \quad y = \sqrt{t^2 + 1}, \quad \text{for } t \text{ in } (-\infty, \infty).$$

(a) Graph the curve by hand and support your graph with your calculator.

(b) Find an equivalent rectangular equation for the curve.

---

7 Summary

KEY TERMS & SYMBOLS

KEY CONCEPTS

7.1 Circles and Parabolas

conic sections
degenerate conic sections
circle
radius
center
parabola
focus
directrix
latus rectum

CIRCLE
A circle is a set of points in a plane that are equidistant from a fixed point. The distance is called the radius of the circle, and the fixed point is called the center.

CENTER–RADIUS FORM OF THE EQUATION OF A CIRCLE
The circle with center (h, k) and radius r has equation $(x - h)^2 + (y - k)^2 = r^2$. A circle with center $(0, 0)$ and radius r has equation $x^2 + y^2 = r^2$.

GENERAL FORM OF THE EQUATION OF A CIRCLE
For real numbers c, d, and e, the general form of the equation of a circle is

$$x^2 + y^2 + cx + dy + e = 0.$$

KEY TERMS & SYMBOLS	KEY CONCEPTS

PARABOLA

A parabola is a set of points in a plane equidistant from a fixed point and a fixed line. The fixed point is called the focus, and the fixed line the directrix, of the parabola.

PARABOLA WITH A VERTICAL AXIS

The parabola with focus $(0, c)$ and directrix $y = -c$ has equation $x^2 = 4cy$. The parabola has vertical axis $x = 0$ and opens upward if $c > 0$ or downward if $c < 0$.

PARABOLA WITH A HORIZONTAL AXIS

The parabola with focus $(c, 0)$ and directrix $x = -c$ has equation $y^2 = 4cx$. The parabola has horizontal axis $y = 0$ and opens to the right if $c > 0$ or to the left if $c < 0$.

TRANSLATION OF A PARABOLA

The parabola with vertex (h, k) and vertical line $x = h$ as axis has an equation of the form $(x - h)^2 = 4c(y - k)$. The parabola opens upward if $c > 0$ or downward if $c < 0$.

The parabola with vertex (h, k) and horizontal line $y = k$ as axis has an equation of the form $(y - k)^2 = 4c(x - h)$. The parabola opens to the right if $c > 0$ or to the left if $c < 0$.

7.2 Ellipses and Hyperbolas

ellipse
focus (foci)
major axis
minor axis
center
vertices
hyperbola
transverse axis
asymptotes
fundamental rectangle

ELLIPSE

An ellipse is the set of all points in a plane, the *sum* of whose distances from two fixed points is constant. Each fixed point is called a focus (plural *foci*) of the ellipse.

STANDARD FORMS OF EQUATIONS FOR ELLIPSES

The ellipse with center at the origin and equation $\dfrac{x^2}{a^2} + \dfrac{y^2}{b^2} = 1$ has vertices $(\pm a, 0)$, endpoints of the minor axis $(0, \pm b)$, and foci $(\pm c, 0)$, where $c^2 = a^2 - b^2$ and $a > b > 0$.

The ellipse with center at the origin and equation $\dfrac{x^2}{b^2} + \dfrac{y^2}{a^2} = 1$ has vertices $(0, \pm a)$, endpoints of the minor axis $(\pm b, 0)$, and foci $(0, \pm c)$, where $c^2 = a^2 - b^2$ and $a > b > 0$.

TRANSLATED ELLIPSES

The preceding equations can be extended to ellipses having center (h, k) by replacing x and y with $x - h$ and $y - k$, respectively.

HYPERBOLA

A hyperbola is the set of all points in a plane such that the absolute value of the *difference* of the distances from two fixed points is constant. The two fixed points are called the foci of the hyperbola.

STANDARD FORMS OF EQUATIONS FOR HYPERBOLAS

The hyperbola with center at the origin and equation $\dfrac{x^2}{a^2} - \dfrac{y^2}{b^2} = 1$ has vertices $(\pm a, 0)$ and foci $(\pm c, 0)$, where $c^2 = a^2 + b^2$.

The hyperbola with center at the origin and equation $\dfrac{y^2}{a^2} - \dfrac{x^2}{b^2} = 1$ has vertices $(0, \pm a)$ and foci $(0, \pm c)$, where $c^2 = a^2 + b^2$.

TRANSLATED HYPERBOLAS

The preceding equations can be extended to hyperbolas having center (h, k) by replacing x and y with $x - h$ and $y - k$, respectively.

(continued)

KEY TERMS & SYMBOLS **KEY CONCEPTS**

7.3 The Conic Sections and Nonlinear Systems

conic
eccentricity, e

The conic sections in this chapter have equations that can be written in the form

$$Ax^2 + Dx + Cy^2 + Ey + F = 0.$$

Conic Section	Characteristic	Examples
Parabola	Either $A = 0$ or $C = 0$, but not both	$y = x^2$ $x = 3y^2 + 2y - 4$
Circle	$A = C \neq 0$	$x^2 + y^2 = 16$
Ellipse	$A \neq C$, $AC > 0$	$\dfrac{x^2}{16} + \dfrac{y^2}{25} = 1$
Hyperbola	$AC < 0$	$x^2 - y^2 = 1$

CONIC

A conic is the set of all points $P(x, y)$ in a plane such that the ratio of the distance from P to a fixed point and the distance from P to a fixed line is constant. This constant ratio is called the eccentricity of the conic.

Conic	Eccentricity
Circle	$e = 0$
Parabola	$e = 1$
Ellipse	$e = \dfrac{c}{a}$ and $0 < e < 1$
Hyperbola	$e = \dfrac{c}{a}$ and $e > 1$

SOLVING A NONLINEAR SYSTEM OF EQUATIONS

A nonlinear system of equations can be solved by the substitution method, the elimination method, or a combination of the two.

SOLVING NONLINEAR SYSTEMS OF INEQUALITIES

To solve a nonlinear system of inequalities, graph all inequalities on the same axes, and find the intersection of their solution sets.

7.4 Parametric Equations

plane curve
parametric equations
parameter

PARAMETRIC EQUATIONS OF A PLANE CURVE

A plane curve is a set of points (x, y) such that $x = f(t)$, $y = g(t)$, and f and g are both continuous on an interval I. The equations $x = f(t)$ and $y = g(t)$ are parametric equations with parameter t.

7 Review Exercises

Write an equation for the circle satisfying the given conditions. Graph it by hand, and give the domain and range.

1. Center $(-2, 3)$; radius 5

2. Center $\left(\sqrt{5}, -\sqrt{7} \right)$; radius $\sqrt{3}$

3. Center $(-8, 1)$; passing through $(0, 16)$

4. Center $(3, -6)$; tangent to the x-axis

Find the center and radius of each circle.

5. $x^2 - 4x + y^2 + 6y + 12 = 0$

6. $x^2 - 6x + y^2 - 10y + 30 = 0$

7. $2x^2 + 14x + 2y^2 + 6y = -2$

8. $3x^2 + 3y^2 + 33x - 15y = 0$

 9. Describe the graph of $(x - 4)^2 + (y - 5)^2 = 0$.

Give the focus, directrix, and axis for each parabola, graph it by hand, and give the domain and range.

10. $y^2 = -\dfrac{2}{3}x$

11. $y^2 = 2x$

12. $3x^2 - y = 0$

13. $x^2 + 2y = 0$

Write an equation for each parabola with vertex at the origin.

14. Focus $(4, 0)$

15. Through $(2, 5)$; opening to the right

16. Through $(3, -4)$; opening downward

17. Focus $(0, -3)$

 18. Explain how to determine just by looking at its equation whether a parabola has a vertical or a horizontal axis of symmetry, and whether it opens up, down, to the left, or to the right.

Write an equation for each parabola.

19. Vertex $(-5, 6)$; focus $(2, 6)$

20. Vertex $(4, 3)$; focus $(4, 5)$

Graph each ellipse or hyperbola by hand, and give the domain, range, and coordinates of the vertices.

21. $\dfrac{x^2}{5} + \dfrac{y^2}{9} = 1$

22. $\dfrac{x^2}{16} + \dfrac{y^2}{4} = 1$

23. $\dfrac{x^2}{64} - \dfrac{y^2}{36} = 1$

24. $\dfrac{y^2}{25} - \dfrac{x^2}{9} = 1$

25. $\dfrac{(x - 3)^2}{4} + (y + 1)^2 = 1$

26. $\dfrac{(x - 2)^2}{9} + \dfrac{(y + 3)^2}{4} = 1$

27. $\dfrac{(y + 2)^2}{4} - \dfrac{(x + 3)^2}{9} = 1$

28. $\dfrac{(x + 1)^2}{16} - \dfrac{(y - 2)^2}{4} = 1$

Write an equation for each conic section with center at the origin.

29. Ellipse; vertex $(0, 4)$; focus $(0, 2)$

30. Ellipse; x-intercept $(6, 0)$; focus $(-2, 0)$

31. Hyperbola; focus $(0, -5)$; y-intercepts $(0, \pm 4)$

32. Hyperbola; y-intercept $(0, -2)$; passing through $(2, 3)$

33. Focus $(0, -3)$; $e = \dfrac{2}{3}$

34. Focus $(5, 0)$; $e = \dfrac{5}{2}$

35. Consider the circle with equation $x^2 + y^2 + 2x + 6y - 15 = 0$.
   (a) What are the coordinates of the center?
   (b) What is the radius?
   (c) What two functions must be graphed to graph this circle with your calculator in function mode?

Concept Check *Match each equation in Column I with the appropriate description in Column II.*

I	II
36. $4x^2 + y^2 = 36$	**A.** Circle; center $(1, -2)$; radius 6
37. $x = 2y^2 + 3$	**B.** Hyperbola; center $(2, 1)$
38. $(x - 1)^2 + (y + 2)^2 = 36$	**C.** Ellipse; major axis on x-axis
39. $\dfrac{x^2}{36} + \dfrac{y^2}{9} = 1$	**D.** Ellipse; major axis on y-axis
40. $(y - 1)^2 - (x - 2)^2 = 36$	**E.** Parabola; opens right
41. $y^2 = 36 + 4x^2$	**F.** Hyperbola; transverse axis on y-axis

Write the equation in standard form for an ellipse or a hyperbola centered at (h, k). Identify the center and the vertices.

42. $4x^2 + 8x + 25y^2 - 250y = -529$

43. $5x^2 + 20x + 2y^2 - 8y = -18$

44. $x^2 + 4x - 4y^2 + 24y = 36$

45. $4y^2 + 8y - 3x^2 + 6x = 11$

Find the eccentricity e of each ellipse or hyperbola.

46. $9x^2 + 25y^2 = 225$

47. $4x^2 + 9y^2 = 36$

48. $9x^2 - y^2 = 9$

Write an equation for each conic section.

49. Parabola with vertex $(-3, 2)$ and y-intercepts $(0, 5)$ and $(0, -1)$

50. Hyperbola with foci $(0, 12)$ and $(0, -12)$; asymptotes $y = \pm x$

51. Ellipse consisting of all points in the plane, the sum of whose distances from $(0, 0)$ and $(4, 0)$ is 8

52. Hyperbola consisting of all points in the plane for which the absolute value of the difference of the distances from $(0, 0)$ and $(0, 4)$ is 2

53. Solve the nonlinear system.

$$2x^2 - y^2 = 8$$
$$4x^2 + y^2 = 16$$

54. Graph the solution set of the nonlinear system of inequalities.

$$\frac{x^2}{16} + \frac{y^2}{36} \leq 1$$
$$x^2 - y^2 \leq 1$$

Use a graphing calculator to graph each plane curve in the specified window.

55. $x = 4t - 3$, $y = t^2$, for t in $[-3, 4]$; window: $[-20, 20]$ by $[-20, 20]$

56. $x = t^2$, $y = t^3$, for t in $[-2, 2]$; window: $[-15, 15]$ by $[-10, 10]$

Find a rectangular equation for each plane curve. State the interval for x.

57. $x = \sqrt{t - 1}$, $y = \sqrt{t}$, for t in $[1, \infty)$

58. $x = 3t + 2$, $y = t - 1$, for t in $[-5, 5]$

(Modeling) *Solve each application.*

59. *Orbit of Venus* The orbit of Venus is an ellipse with the sun at one focus. The eccentricity of the orbit is $e = 0.006775$, and the major axis has length 134.5 million miles. (*Source: The World Almanac and Book of Facts.*) Estimate the least and greatest distances of Venus from the sun.

60. *Orbit of the Comet Swift–Tuttle* Comet Swift–Tuttle has an elliptical orbit of eccentricity $e = 0.964$ with the sun at one focus. Find the equation of the comet, given that the closest it comes to the sun is 89 million miles.

Trajectory of a Satellite When a satellite is near Earth, its orbital trajectory may trace out a hyperbola, a parabola, or an ellipse. The type of trajectory depends on the satellite's velocity V in meters per second. It will be hyperbolic if $V > k/\sqrt{D}$, parabolic if $V = k/\sqrt{D}$, and elliptic if $V < k/\sqrt{D}$, where $k = 2.82 \times 10^7$ is a constant and D is the distance in meters from the satellite to the center of Earth. (Source: Loh, W., *Dynamics and Thermodynamics of Planetary Entry*, Prentice-Hall; Thomson, W., *Introduction to Space Dynamics*, John Wiley and Sons.) *Use this information in Exercises 61–63.*

61. When the artificial satellite *Explorer IV* was at a maximum distance D of 42.5×10^6 meters from Earth's center, it had a velocity V of 2090 meters per second. Determine the shape of its trajectory.

62. If a satellite is scheduled to leave Earth's gravitational influence, its velocity must be increased so that its trajectory changes from elliptic to hyperbolic. Determine the minimum increase in velocity necessary for *Explorer IV* to escape Earth's gravitational influence given that $D = 42.5 \times 10^6$ meters.

63. Explain why it is easier to change a satellite's trajectory from an ellipse to a hyperbola when D is maximum rather than minimum.

64. *Center of an Ellipse* If $Ax^2 + Cy^2 + Dx + Ey + F = 0$ is the general equation of an ellipse, find its center point by completing the square.

7 Test

1. Match each equation in Column I with the appropriate description in Column II.

I

(a) $\dfrac{(x+3)^2}{4} - \dfrac{(y+2)^2}{16} = 1$

(b) $(x-3)^2 + (y-2)^2 = 16$

(c) $(x+3)^2 + (y-2)^2 = 16$

(d) $(x+3)^2 = -(y-4)$

(e) $x - 4 = (y-2)^2$

(f) $\dfrac{(x+3)^2}{4} + \dfrac{(y+2)^2}{16} = 1$

II

A. Circle; center (3, 2); radius 4

B. Hyperbola; center $(-3, -2)$

C. Ellipse; center $(-3, -2)$

D. Circle; center $(-3, 2)$; radius 4

E. Parabola; opens downward

F. Parabola; opens right

2. Give the coordinates of the focus and the equation of the directrix for the parabola with equation $y^2 = \frac{1}{8}x$.

3. Graph $y = -\sqrt{1 - \dfrac{x^2}{36}}$. Is it the graph of a function? Find the domain and range.

4. What two functions are used to graph $\dfrac{x^2}{25} - \dfrac{y^2}{49} = 1$ on a calculator in function mode?

Graph each relation by hand. Identify the graph, and give the radius, center, vertices, and foci, as applicable.

5. $\dfrac{y^2}{4} - \dfrac{x^2}{9} = 1$

6. $x^2 + 4y^2 + 2x - 16y + 17 = 0$

7. $y^2 - 8y - 2x + 22 = 0$

8. $x^2 + (y-4)^2 = 9$

9. $\dfrac{(x-3)^2}{49} + \dfrac{(y+1)^2}{16} = 1$

10. $x = 4t^2 - 4$, $y = t - 1$, for t in $[-1.5, 1.5]$

11. Write an equation for each conic.

    (a) Center at the origin; focus $(0, -2)$; $e = 1$

    (b) Center at the origin; vertical major axis of length 6; $e = \frac{5}{6}$

12. *(Modeling) Height of a Bridge's Arch*   An arch of a bridge has the shape of the top half of an ellipse. The arch is 40 feet wide and 12 feet high at the center. Find the equation of the complete ellipse. Find the height of the arch 10 feet from the center at the bottom.

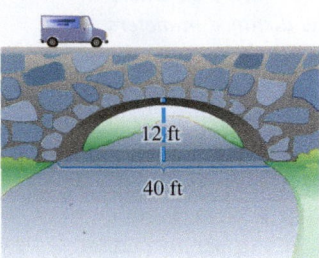

12 ft

40 ft

13. Solve the nonlinear system.

$$3x^2 + 2y^2 = 5$$
$$-4x^2 + 3y^2 = -1$$

14. Graph the solution set of the nonlinear system of inequalities.

$$\frac{x^2}{4} + \frac{y^2}{9} \le 1$$
$$y^2 - 4x^2 \le 4$$

15. Use a graphing calculator to graph $x = t + \ln t$, $y = t + e^t$, for t in $(0, 2\,]$. Use the window $[-5, 5\,]$ by $[0, 10\,]$.

16. Find a rectangular equation for $x = \dfrac{1}{t + 3}$, $y = t + 3$, for $t \ne -3$.

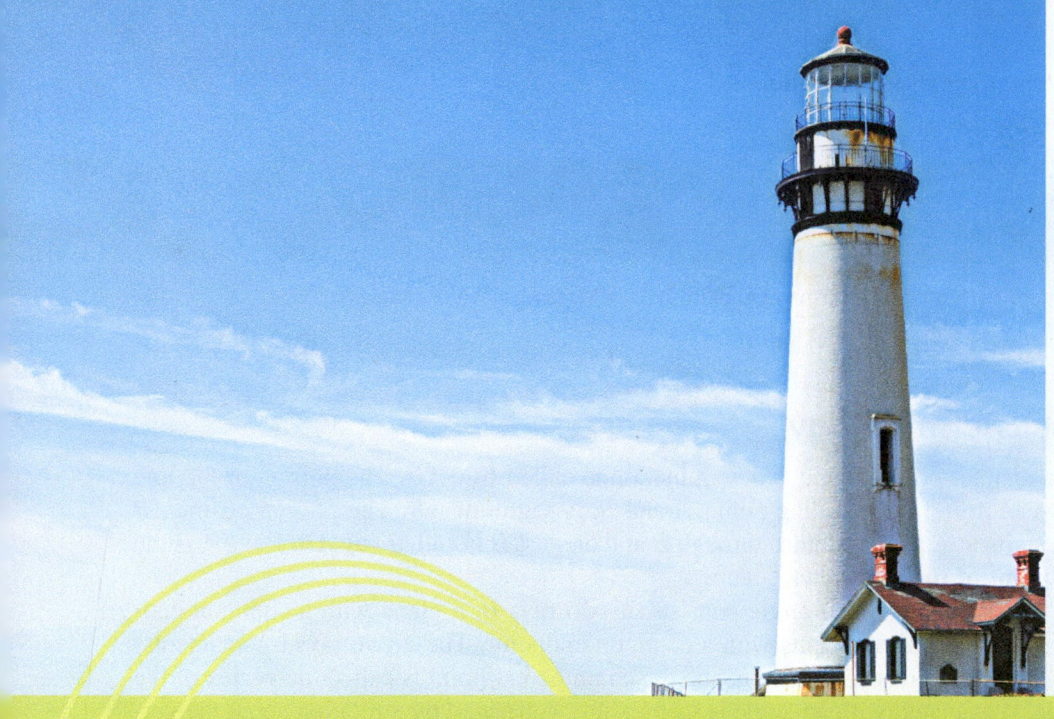

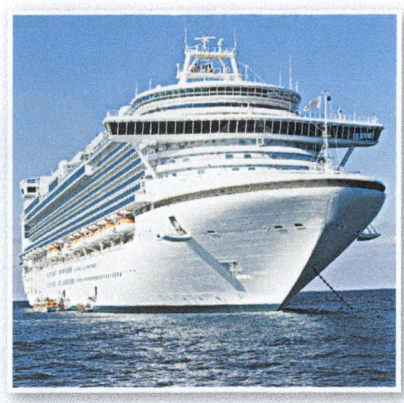

In this chapter, we use trigonometric functions to solve problems related to navigation, periodic phenomena, and rotation.

8 Trigonometric Functions and Applications

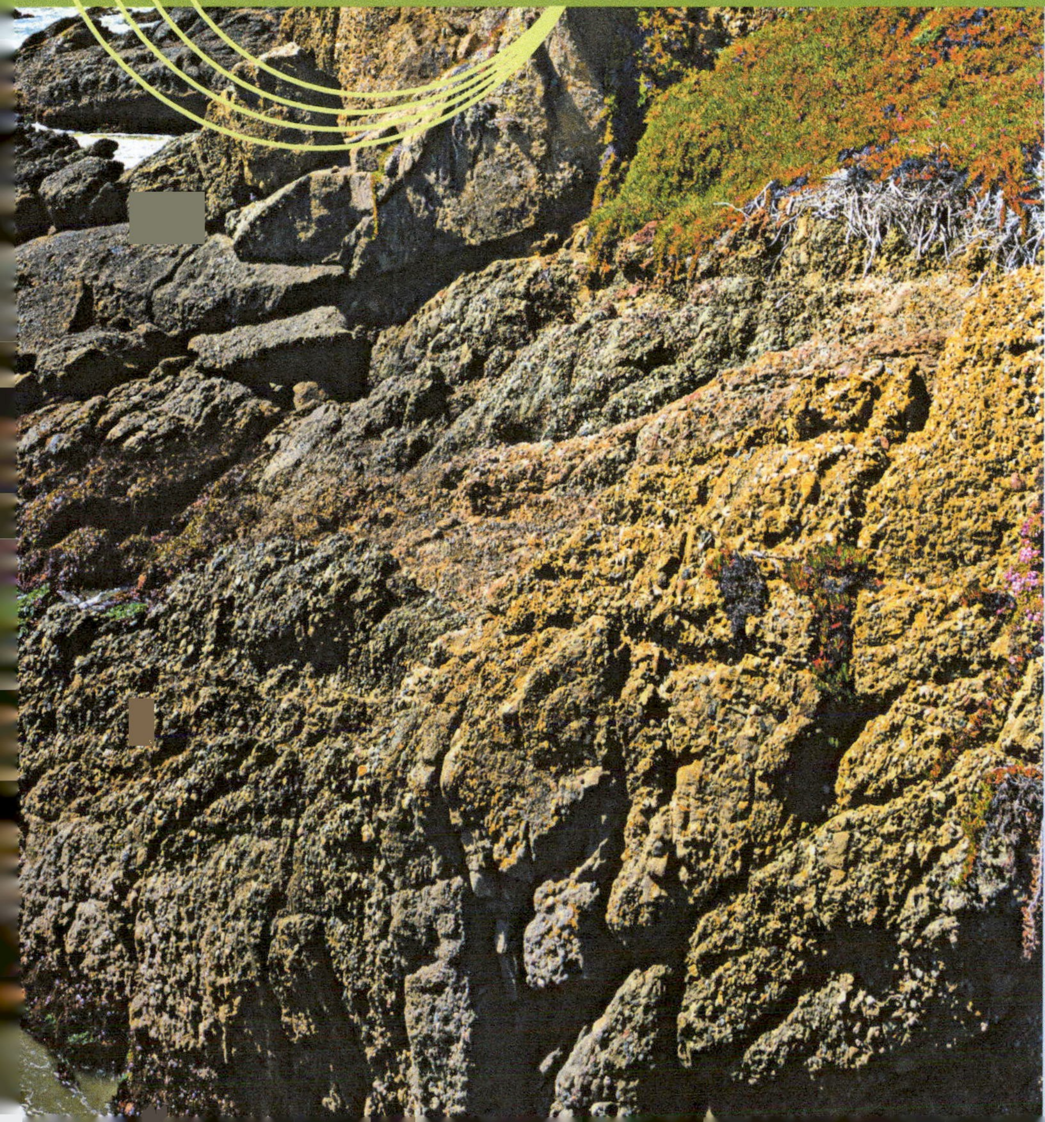

CHAPTER OUTLINE

8.1 Angles and Their Measures

Basic Terminology • Degree Measure • Standard Position and Coterminal Angles • Radian Measure • Arc Lengths and Areas of Sectors • Linear and Angular Speed

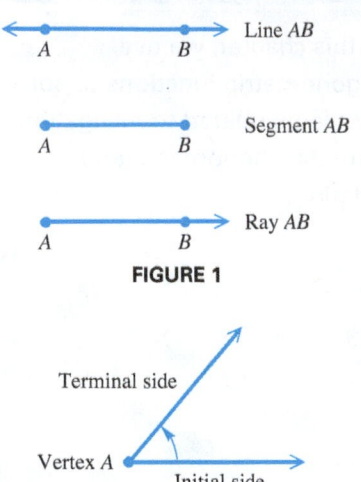

FIGURE 1

FIGURE 2

Basic Terminology

Two distinct points A and B determine a line called **line AB.** The portion of the line between A and B, including points A and B, is **segment AB.** The portion of line AB that starts at A and continues through B and on past B is called **ray AB.** Point A is the **endpoint of the ray.** See **FIGURE 1.**

In trigonometry, an **angle** consists of two rays in a plane with a common endpoint, or two line segments with a common endpoint. These two rays (or segments) are called the **sides** of the angle, and the common endpoint is called the **vertex** of the angle. Associated with an angle is its measure, generated by a rotation about the vertex. See **FIGURE 2.** This measure is determined by rotating a ray starting at one side of the angle, called the **initial side,** to the position of the other side, called the **terminal side.** *A counterclockwise rotation generates an angle with positive measure, while a clockwise rotation generates an angle with negative measure.* See **FIGURE 3.**

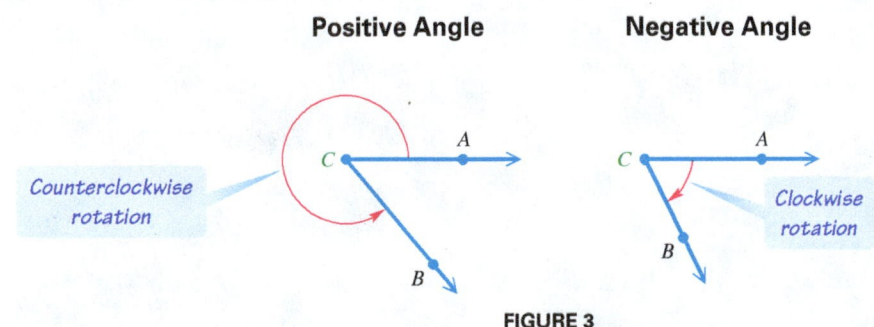

Positive Angle   **Negative Angle**

FIGURE 3

An angle can be named by using the name of its vertex. For example, either angle in **FIGURE 3** can be called angle C. Alternatively, an angle can be named using three letters, with the vertex letter in the middle. Thus, either angle also could be named angle ACB or angle BCA.

Degree Measure

The most common unit used to measure the size of angles is the **degree.** To use degree measure, we assign 360 degrees to a complete rotation of a ray.* In **FIGURE 4,** notice that the terminal side of the angle corresponds to its initial side when it makes a complete rotation. One degree, written **1°,** represents $\frac{1}{360}$ of a rotation. Therefore, 90° represents $\frac{90}{360} = \frac{1}{4}$ of a complete rotation, and 180° represents $\frac{180}{360} = \frac{1}{2}$ of a complete rotation.

An angle measuring between 0° and 90° is an **acute angle.** An angle measuring exactly 90° is a **right angle.** An angle measuring more than 90° but less than 180° is an **obtuse angle,** and an angle of exactly 180° is a **straight angle.**

Complete Rotation

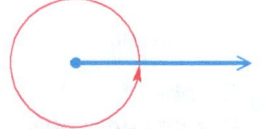

A complete rotation of a ray gives an angle whose measure is 360°. $\frac{1}{360}$ of a complete rotation gives an angle whose measure is 1°.

FIGURE 4

---

*The Babylonians were the first to subdivide the circumference of a circle into 360 parts, 4000 years ago. There are various theories as to why the number 360 was chosen. One is that it is approximately the number of days in a year, and it has many divisors, which makes it convenient to work with.

The Greek Letters

A	α	alpha
B	β	beta
Γ	γ	gamma
Δ	δ	delta
E	ϵ	epsilon
Z	ζ	zeta
H	η	eta
Θ	θ	theta
I	ι	iota
K	κ	kappa
Λ	λ	lambda
M	μ	mu
N	ν	nu
Ξ	ξ	xi
O	o	omicron
Π	π	pi
P	ρ	rho
Σ	σ	sigma
T	τ	tau
Υ	υ	upsilon
Φ	ϕ	phi
X	χ	chi
Ψ	ψ	psi
Ω	ω	omega

In **FIGURE 5,** we use the Greek letter θ (theta) to name each angle. The table in the margin lists the Greek letters, often used in trigonometry, and their names.

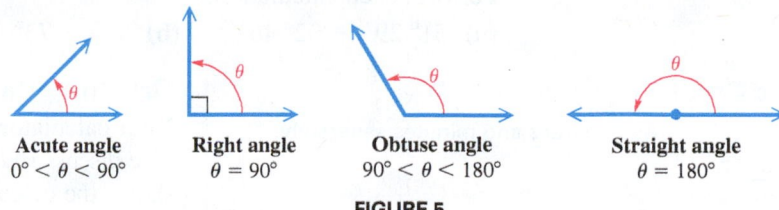

FIGURE 5

If the sum of the measures of two positive angles is 90°, the angles are called **complementary** and the angles are **complements.** Two positive angles with measures whose sum is 180° are **supplementary** and the angles are **supplements.**

EXAMPLE 1 Finding Measures of Angles

Find the measure of each angle in **FIGURE 6.**

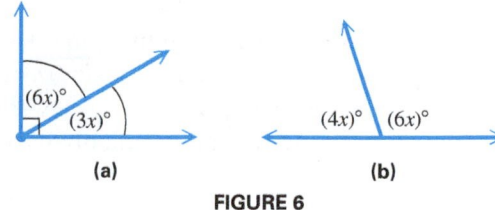

FIGURE 6

Solution

(a) In **FIGURE 6(a),** the two angles form a right angle (as indicated by the $\neg$ symbol), so they are complementary angles. The degree sum of their angles is 90.

$$6x + 3x = 90$$

Use x to find the measure of each angle. $9x = 90$ Combine like terms.

$x = 10$ Divide by 9.

Be sure that you determine the measure of each angle by substituting 10 for x. The two angles have measures of $6x = 6(10) = 60°$ and $3x = 3(10) = 30°$.

(b) The angles in **FIGURE 6(b)** are supplementary.

$$4x + 6x = 180$$

Remember to substitute for x. $10x = 180$ Combine like terms.

$x = 18$ Divide by 10.

These angle measures are $4(18) = 72°$ and $6(18) = 108°$.

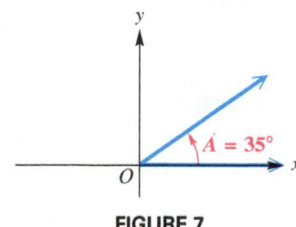

FIGURE 7

The measure of angle A of **FIGURE 7** is 35°. This measure can be expressed by saying that $m(\text{angle } A)$ is 35°, where $m(\textbf{angle } A)$ is read **"the measure of angle A."** It is convenient, however, to abbreviate $m(\text{angle } A) = 35°$ as simply angle $A = 35°$.

Traditionally, portions of a degree have been measured with minutes and seconds. One **minute,** written $1'$, is $\frac{1}{60}$ of a degree.

$$1' = \frac{1}{60}^\circ \quad \text{or} \quad 60' = 1°$$

One **second, $1''$,** is $\frac{1}{60}$ of a minute.

$$\frac{1}{60} \cdot \frac{1}{60}^\circ = \frac{1}{3600}^\circ$$

$$1'' = \frac{1}{60}' = \frac{1}{3600}^\circ \quad \text{or} \quad 60'' = 1'$$

The measure 12° 42′ 38″ represents 12 degrees, 42 minutes, 38 seconds.

TECHNOLOGY NOTE

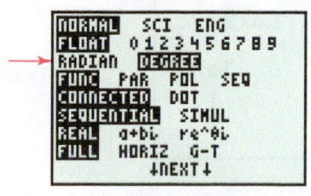

TI-84 Plus

Graphing calculators can be set in either *radian mode* or *degree mode.*

> **EXAMPLE 2** **Calculating with Degrees and Minutes**

Perform each calculation.

(a) $51° 29' + 32° 46'$ (b) $90° - 73° 12'$

Analytic Solution

(a) $\begin{array}{r} 51° 29' \\ + 32° 46' \\ \hline 83° 75' \end{array}$ Add degrees and minutes separately.

We can write the sum $83° 75'$ as follows.

$\begin{array}{r} 83° \\ + 1° 15' \\ \hline 84° 15' \end{array}$ $75' = 60' + 15' = 1° 15'$

(b) $\begin{array}{r} 89° 60' \\ - 73° 12' \\ \hline 16° 48' \end{array}$ Write $90°$ as $89° 60'$.

Graphing Calculator Solution

Your calculator can be in either degree or radian mode. (We discuss *radian measure* later in this section.) **FIGURE 8** shows the calculations.

```
51°29'+32°46'▶DM
S
             84°15'0"
90°0'-73°12'▶DMS
             16°48'0"
```

FIGURE 8

Angles are commonly measured in decimal degrees. For example,

$$12.4238° = 12\frac{4238}{10,000}°.$$

Decimal degrees are often more convenient to use than degrees, minutes, and seconds. For this reason it is important to be able to convert between these two angle measures. These conversions can be done by hand or with a calculator.

> **EXAMPLE 3** **Converting between Decimal Degrees and Degrees, Minutes, and Seconds**

(a) Convert $74° 8' 14''$ to decimal degrees. Round to the nearest thousandth of a degree.

(b) Convert $34.817°$ to degrees, minutes, and seconds.

Analytic Solution

(a) $74° 8' 14'' = 74° + \frac{8}{60}° + \frac{14}{3600}°$ $1' = \frac{1}{60}°$ and $1'' = \frac{1}{3600}°$

$\approx 74° + 0.1333° + 0.0039°$

$\approx 74.137°$ Add and round.

(b) $34.817° = 34° + 0.817°$ Change decimal part to minutes.

$= 34° + 0.817(60')$ $1° = 60'$

$= 34° + 49.02'$

$= 34° + 49' + 0.02'$ Change decimal part to seconds.

$= 34° + 49' + 0.02(60'')$ $1' = 60''$

$= 34° + 49' + 1.2''$

$= 34° 49' 1.2''$

Graphing Calculator Solution

The first two results in **FIGURE 9** show how the TI-84 Plus converts $74° 8' 14''$ to decimal degrees. (The second result was obtained by fixing the display to three decimal places.) The final result shows how to convert decimal degrees to degrees, minutes, and seconds.

```
74°8'14"
          74.13722222
74°8'14"
          74.137
34.817▶DMS
          34°49'1.2"
```

Degree mode
FIGURE 9

Standard Position and Coterminal Angles

An angle is in **standard position** if its vertex is at the origin and its initial side is along the positive x-axis. The angles in **FIGURES 10(a)** and **10(b)** are in standard position. An angle in standard position is said to lie in the quadrant in which its terminal side lies. For example, an acute angle is in quadrant I (**FIGURE 10(a)**) and an obtuse angle is in quadrant II (**FIGURE 10(b)**). **FIGURE 10(c)** shows intervals of angle measures for each quadrant when $0° < \theta < 360°$. If the terminal side lies along an axis, then the angle does not lie in any quadrant.

Coterminal Angles

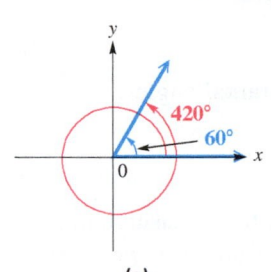

(a)

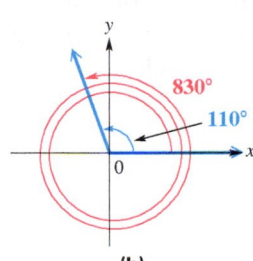

(b)

Angles with same initial
and terminal sides

FIGURE 11

Acute Angle

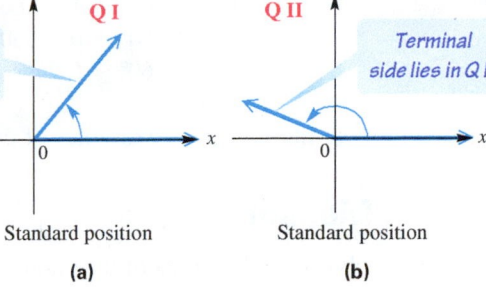

Standard position

(a)

Obtuse Angle

Standard position

(b)

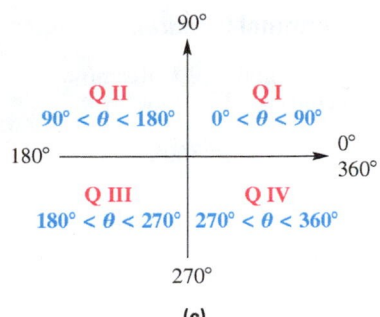

(c)

FIGURE 10

> ### Quadrantal Angles
>
> Angles in standard position having their terminal sides along the x-axis or y-axis, such as angles with measures 90°, 180°, 270°, and so on, are called **quadrantal angles**.

A complete rotation of a ray results in an angle measuring 360°. By continuing the rotation, angles of measure larger than 360° can be produced. The angles in **FIGURE 11(a)** with measures 60° and 420° have the same initial side and the same terminal side, but different amounts of rotation. Such angles are called **coterminal angles**—*their measures differ by a* *multiple* *of* **360°.** The angles in **FIGURE 11(b)** are also coterminal, because $830° - 110° = 720° = 2(360°)$.

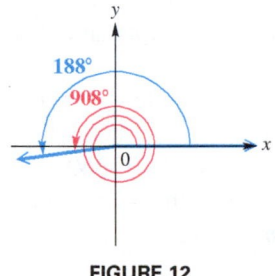

FIGURE 12

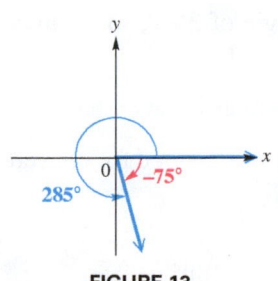

FIGURE 13

> **EXAMPLE 4**   **Finding Measures of Coterminal Angles**

Find the angle of least possible positive measure coterminal with each angle.
(a) 908°   **(b)** −75°   **(c)** −800°

Solution

(a) Add or subtract 360° as many times as needed to obtain an angle with measure greater than 0° but less than 360°.

$$908° - 2 \cdot 360° = 908° - 720° = 188°$$

An angle of 188° is coterminal with an angle of 908°. See **FIGURE 12.**

(b) Use a rotation of $360° + (−75°) = 285°$. See **FIGURE 13.**

(c) The least integer multiple of 360° greater than 800° is $360° \cdot 3 = 1080°$. Add 1080° to −800° to obtain

$$1080° + (−800°) = 280°.$$

Sometimes we want to find an expression that will generate all angles coterminal with a given angle. For example, any angle coterminal with 60° can be obtained by adding an appropriate integer multiple of 360° to 60°. If we let n represent any integer, then the expression

$$60° + n \cdot 360° \qquad \text{Angles coterminal with } 60°$$

represents all such coterminal angles. The following table shows a few possibilities.

Angles Coterminal with 60°

Value of n	Coterminal with 60°
2	$60° + 2 \cdot 360° = 780°$
1	$60° + 1 \cdot 360° = 420°$
0	$60° + 0 \cdot 360° = 60°$ (the angle itself)
−1	$60° + (-1) \cdot 360° = -300°$

The table in the margin gives examples of coterminal quadrantal angles.

Coterminal Quadrantal Angles

Quadrantal Angle θ	Coterminal with θ
0°	±360°, ±720°
90°	−630°, −270°, 450°
180°	−180°, 540°, 900°
270°	−450°, −90°, 630°

Radian Measure

In work involving applications of trigonometry, angles are often measured in degrees. In more theoretical work in mathematics, *radian measure* of angles is preferred.

FIGURE 14 shows an angle θ in standard position along with a circle of radius r. The vertex of θ is at the center of the circle and θ is called a **central angle.** When angle θ intercepts an arc on the circle equal in length to the radius of the circle, we say that θ has a measure of 1 radian.

$\theta = 1$ **radian**

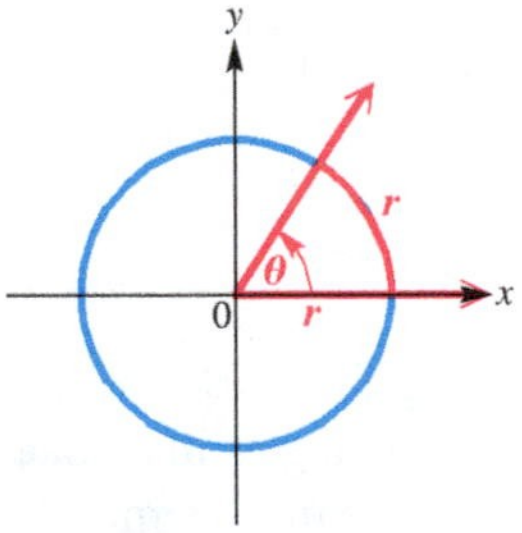

One radian is equivalent to about 57.3°.

FIGURE 14

> **Radian**
>
> An angle with vertex at the center of a circle that intercepts an arc on the circle equal in length to the radius of the circle has measure **1 radian.** In general, if θ is a central angle in a circle of radius r, and θ intercepts an arc of length s, the radian measure of θ is $\frac{s}{r}$.

By this definition, an angle of measure 2 radians intercepts an arc equal in length to twice the radius of the circle, an angle of measure $\frac{1}{2}$ radian intercepts an arc equal in length to half the radius of the circle, and so on.

The **circumference** of a circle—the distance around the circle—is given by $C = 2\pi r$, where r is the radius of the circle. The formula $C = 2\pi r$ shows that the radius can be laid off 2π times around a circle. Therefore, an angle of 360°, which corresponds to a complete circle, intercepts an arc equal in length to 2π times the radius of the circle. Thus, an angle of 360° has measure 2π radians.

$$360° = 2\pi \text{ radians}$$

An angle of 180° is half the degree measure of an angle of 360°, so an angle of 180° has half the radian measure of an angle of 360°.

$$180° = \frac{1}{2}(2\pi) \text{ radians} = \pi \text{ radians} \qquad \text{Degree–radian relationship}$$

We can use the relationship $180° = \pi$ radians to develop a method for converting between degrees and radians as follows.

$$180° = \pi \text{ radians}$$

$$1° = \frac{\pi}{180} \text{ radian} \quad \text{Divide by 180.} \qquad \text{or} \qquad 1 \text{ radian} = \frac{180°}{\pi} \quad \text{Divide by } \pi.$$

Converting between Degrees and Radians

1. Multiply a degree measure by $\frac{\pi}{180}$ radian and simplify to convert to radians.
2. Multiply a radian measure by $\frac{180°}{\pi}$ and simplify to convert to degrees.

NOTE 1 *radian is equivalent to approximately* 57.3°. *If no unit of measure is specified for an angle, radian measure is understood.*

EXAMPLE 5 **Converting Degrees to Radians**

Convert each degree measure to radian measure.

(a) 45°   (b) −270°   (c) 249.8°

Analytic Solution

(a) $45° = 45\left(\frac{\pi}{180}\text{ radian}\right)$ Multiply by $\frac{\pi}{180}$ radian.

$\qquad = \frac{\pi}{4}\text{ radian}$ Simplify.

(b) $-270° = -270\left(\frac{\pi}{180}\text{ radian}\right) = -\frac{3\pi}{2}\text{ radians}$

(c) $249.8° = 249.8\left(\frac{\pi}{180}\text{ radian}\right)$

$\qquad \approx 4.360\text{ radians}$ Rounded to nearest thousandth

Graphing Calculator Solution

For **FIGURE 15,** the calculator is in *radian mode.* *When exact values involving* π *are required, such as* $\frac{\pi}{4}$ *and* $-\frac{3\pi}{2}$ *in parts (a) and (b), calculator approximations are not acceptable.*

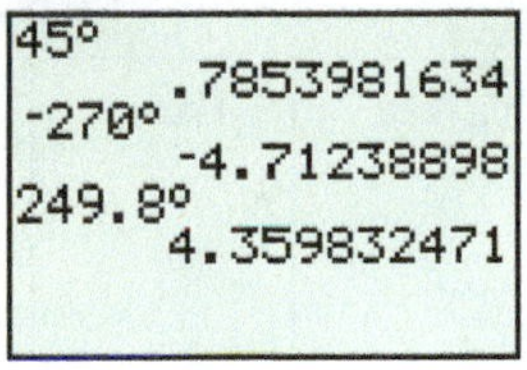

```
45°
         .7853981634
-270°
         -4.71238898
249.8°
         4.359832471
```

Radian mode
FIGURE 15

EXAMPLE 6 **Converting Radians to Degrees**

Convert each radian measure to degree measure.

(a) $\frac{9\pi}{4}$ (b) $-\frac{5\pi}{6}$ (c) 4.25

Analytic Solution

(a) $\frac{9\pi}{4}\text{ radians} = \frac{9\pi}{4}\left(\frac{180°}{\pi}\right)$ Multiply by $\frac{180°}{\pi}$.

$\qquad = 405°$ Simplify.

(b) $-\frac{5\pi}{6}\text{ radians} = -\frac{5\pi}{6}\left(\frac{180°}{\pi}\right) = -150°$ Multiply.

(c) $4.25\text{ radians} = 4.25\left(\frac{180°}{\pi}\right)$

$\qquad \approx 243.5°$ Rounded to nearest tenth

Graphing Calculator Solution

FIGURE 16 shows how a calculator in *degree mode* converts radian measures to degree measures.

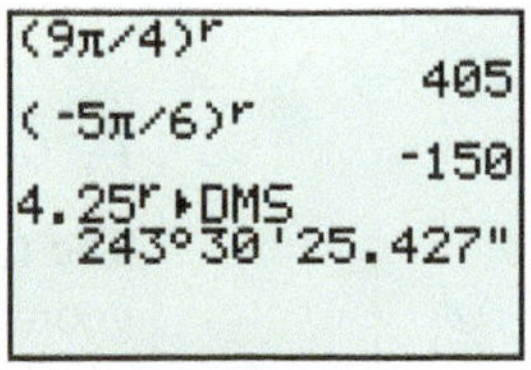

```
(9π/4)ʳ
              405
(-5π/6)ʳ
             -150
4.25ʳ▶DMS
    243°30'25.427"
```

Degree mode
FIGURE 16

NOTE Another way to convert a radian measure that is a rational multiple of π, such as $\frac{9\pi}{4}$, to degrees is to just substitute 180° for π. In **Example 6(a)**, this would be $\frac{9(180°)}{4} = 405°$.

CAUTION **FIGURE 17** shows angles measuring 30 radians and 30°. Be careful not to confuse these different units of angle measure.

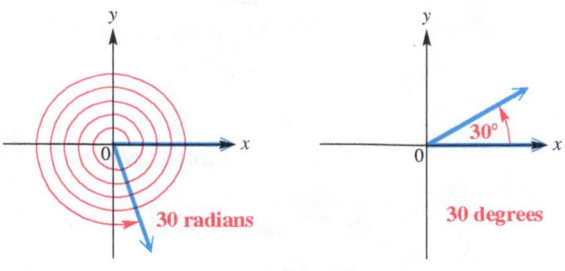

FIGURE 17

The table that follows and **FIGURE 18** give some equivalent angles measured in degrees and radians. Keep in mind that **180° = π radians.**

Equivalent Angle Measures in Degrees and Radians

Degrees	Radians	
	Exact	**Approximate**
0°	0	0
30°	$\frac{\pi}{6}$	0.5236
45°	$\frac{\pi}{4}$	0.7854
60°	$\frac{\pi}{3}$	1.0472
90°	$\frac{\pi}{2}$	1.5708
180°	π	3.1416
270°	$\frac{3\pi}{2}$	4.7124
360°	2π	6.2832

Important Equivalent Angles

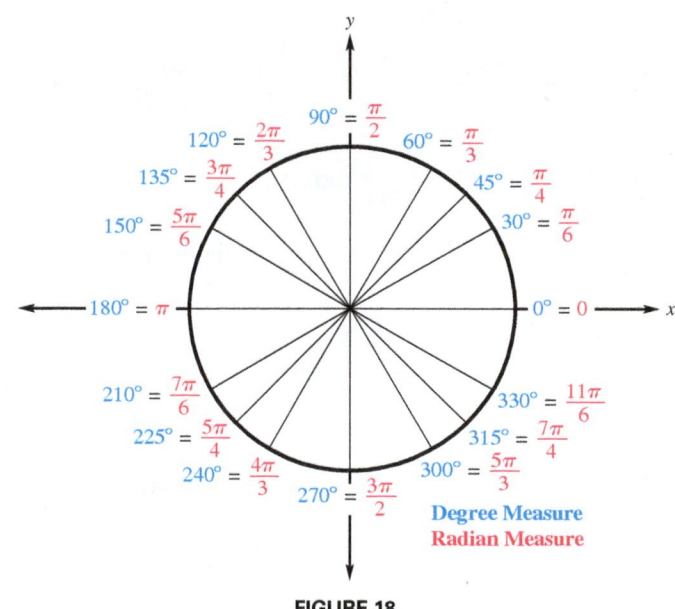

FIGURE 18

The angles identified in **FIGURE 18** *are important. You should learn these equivalences, as they will appear often in the chapters that follow.*

Arc Lengths and Areas of Sectors

The formula to find the length of an arc of a circle is derived from the fact (proved in geometry) that the length of an arc is proportional to the measure of its central angle.

In **FIGURE 19**, angle *QOP* has measure 1 radian and intercepts an arc of length r on the circle. Angle *ROT* has measure θ radians and intercepts an arc of length s on the circle. Since the lengths of the arcs are proportional to the measures of their central angles, the following holds.

$$\frac{s}{r} = \frac{\theta}{1}$$

$$s = r\theta \qquad \text{Multiply each side by } r.$$

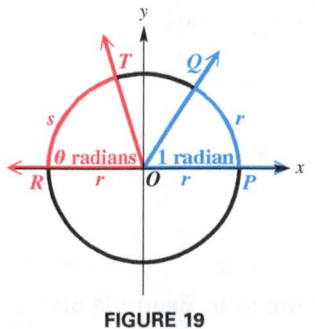

FIGURE 19

> ### Arc Length
>
> The length s of the arc intercepted on a circle of radius r by a central angle of measure θ radians is given by the product of the radius and the radian measure of the angle, or
>
> $$s = r\theta, \quad \theta \text{ in radians.}$$

CAUTION   *In applying the formula $s = r\theta$, the value of θ must be expressed in radians.*

EXAMPLE 7   **Finding Arc Length by Using $s = r\theta$**

A circle has radius $r = 18.20$ centimeters. Find the length s of an arc intercepted by a central angle θ having each of the following measures.

(a) $\dfrac{3\pi}{8}$ radians   **(b)** $144°$

Solution

(a) As shown in **FIGURE 20,** $r = 18.20$ centimeters and $\theta = \dfrac{3\pi}{8}$.

$$s = r\theta = 18.20\left(\frac{3\pi}{8}\right) \approx 21.44 \text{ centimeters}$$

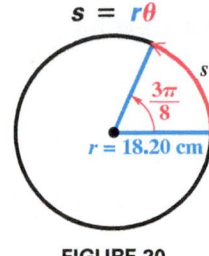

$s = r\theta$

$r = 18.20$ cm

FIGURE 20

(b) The formula $s = r\theta$ requires that θ be measured in radians. First, convert θ to radians by multiplying $144°$ by $\dfrac{\pi}{180}$ radian.

$$144° = 144\left(\frac{\pi}{180}\right) = \frac{4\pi}{5} \text{ radians} \qquad \text{Convert from degrees to radians.}$$

The length s is given by

$$s = r\theta = 18.20\left(\frac{4\pi}{5}\right) \approx 45.74 \text{ centimeters.}$$

Be sure to use radians for θ in $s = r\theta$.

EXAMPLE 8   **Using Latitudes to Find Distance**

Reno, Nevada, is due north of Los Angeles. The latitude of Reno is $40°$ N, while that of Los Angeles is $34°$ N. (The N in $34°$ N means *north* of the equator.) If the radius of Earth is 6400 kilometers, find the north–south distance between the two cities.

Solution   Latitude gives the measure of a central angle with vertex at Earth's center whose initial side goes through the equator and whose terminal side goes through the given location. As shown in **FIGURE 21,** the central angle between Reno and Los Angeles is $40° - 34° = 6°$. The distance between the two cities can be found by the formula $s = r\theta$, after $6°$ is first converted to radians.

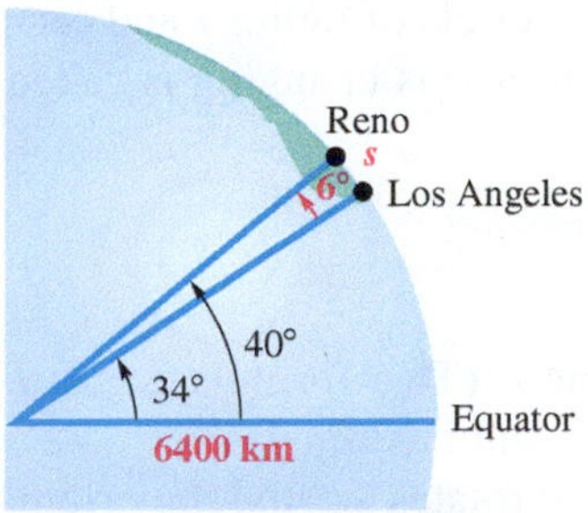

$$6° = 6\left(\frac{\pi}{180}\right) = \frac{\pi}{30} \text{ radian}$$

The distance between the two cities is

$$s = r\theta = 6400\left(\frac{\pi}{30}\right) \approx 670 \text{ kilometers.} \qquad \text{Let } r = 6400 \text{ and } \theta = \tfrac{\pi}{30}.$$

Reno
s
$6°$ Los Angeles
$40°$
$34°$
Equator
6400 km

FIGURE 21

Sector of a Circle

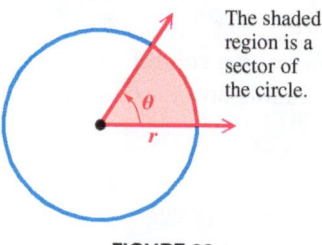

The shaded region is a sector of the circle.

FIGURE 22

A **sector of a circle** is the portion of the interior of a circle intercepted by a central angle. Think of it as a "piece of pie." See **FIGURE 22.** The interior of a circle can be thought of as a sector intercepted by a central angle of measure 2π radians. If a central angle for a sector has measure θ radians, then the sector makes up the fraction $\frac{\theta}{2\pi}$ of a complete circle. The area inside a circle with radius r is $\mathcal{A} = \pi r^2$. Therefore, the area of the sector is given by the product of the fraction $\frac{\theta}{2\pi}$ and the total area, πr^2.

$$\text{area } \mathcal{A} \text{ of sector} = \frac{\theta}{2\pi}(\pi r^2) = \frac{1}{2}r^2\theta, \quad \theta \text{ in radians}$$

This discussion is summarized as follows.

Area of a Sector

The area of a sector of a circle of radius r and central angle θ is given by

$$\mathcal{A} = \frac{1}{2}r^2\theta, \quad \theta \text{ in radians.}$$

CAUTION As in the formula for arc length, the value of θ ***must be in radians*** when the formula for the area of a sector is used.

EXAMPLE 9 **Finding the Area of a Sector-Shaped Field**

FIGURE 23 shows a field in the shape of a sector of a circle. Find the area of the field.

Solution First, convert 15° to radians.

$$15° = 15\left(\frac{\pi}{180}\right) = \frac{\pi}{12}\text{ radian}$$

Now use the formula for the area of a sector.

$$\mathcal{A} = \frac{1}{2}r^2\theta = \frac{1}{2}(321)^2\left(\frac{\pi}{12}\right) \approx 13{,}500 \text{ square meters}$$

FIGURE 23

Be sure to use radians for θ in $\mathcal{A} = \frac{1}{2}r^2\theta$.

Linear and Angular Speed

In many situations we need to know how fast a point on a circular disk is moving or how fast the central angle of such a disk is changing. Some examples occur with machinery involving gears or pulleys or the speed of a car around a curved portion of highway.

Suppose that point P moves at a constant speed along a circle of radius r and center O. See **FIGURE 24.** The measure of how fast the position of P is changing is called **linear speed.** If v represents linear speed, then

$$\text{speed} = \frac{\text{distance}}{\text{time}}, \quad \text{or} \quad v = \frac{s}{t},$$

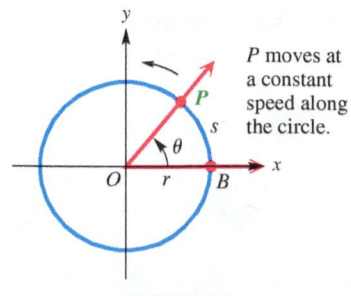

P moves at a constant speed along the circle.

FIGURE 24

where s is the length of the arc traced by point P in time t. (This formula is just a restatement of $d = rt$ with s as distance, v as rate (speed), and t as time.)

As point P in **FIGURE 24** moves along the circle, ray OP rotates around the origin. Since ray OP is the terminal side of angle POB, the measure of the angle changes as

P moves along the circle. The measure of how fast angle *POB* is changing is called **angular speed.** Angular speed, symbolized ω (*omega*), is given as

$$\omega = \frac{\theta}{t}, \quad \theta \text{ in radians,}$$

where θ is the measure of angle *POB*. As with earlier formulas in this chapter, θ must be measured in radians, with ω expressed as radians per unit of time.

The length *s* of the arc intercepted on a circle of radius *r* by a central angle of measure θ radians is given by $s = r\theta$. Using this formula, the formula for linear speed, $v = \frac{s}{t}$, becomes

$$v = \frac{s}{t} = \frac{r\theta}{t} = r \cdot \frac{\theta}{t} = r\omega. \qquad s = r\theta \text{ and } \omega = \frac{\theta}{t}$$

The formula $v = r\omega$ relates linear and angular speeds.

As an example of linear and angular speeds, consider the following. The human joint that can be flexed the fastest is the wrist, which can rotate through 90°, or $\frac{\pi}{2}$ radians, in 0.045 second while holding a tennis racket. The angular speed of a human wrist swinging a tennis racket is

$$\omega = \frac{\theta}{t} = \frac{\frac{\pi}{2}}{0.045} \approx 35 \text{ radians per second.}$$

If the radius (distance) from the tip of the racket to the wrist joint is 2 feet, then the speed at the tip of the racket is

$$v = r\omega \approx 2(35) = 70 \text{ feet per second,} \quad \text{or about 48 mph.}$$

In a tennis serve the arm rotates at the shoulder, so the final speed of the racket is considerably faster. (*Source:* Cooper, J. and R. Glassow, *Kinesiology,* Second Edition, C. V. Mosby.)

Angular Speed	Linear Speed
$\omega = \frac{\theta}{t}$	$v = \frac{s}{t}$
(ω in **radians** per unit time, θ in **radians**)	$v = \frac{r\theta}{t}$
	$v = r\omega$

EXAMPLE 10 **Using Linear and Angular Speed Formulas**

Suppose that point *P* is on a circle with radius 10 centimeters, and ray *OP* is rotating with angular speed $\frac{\pi}{18}$ radian per second.

(a) Find the angle θ generated by *P* in 6 seconds.

(b) Find the distance *s* traveled by *P* along the circle in 6 seconds.

(c) Find the linear speed *v* of *P* in centimeters per second.

Solution

(a) The angular speed of ray *OP* is $\omega = \frac{\pi}{18}$ radian per second. Use $\omega = \frac{\theta}{t}$ and $t = 6$ seconds.

$$\frac{\pi}{18} = \frac{\theta}{6} \qquad \text{Let } \omega = \frac{\pi}{18} \text{ and } t = 6 \text{ in the angular speed formula.}$$

$$\theta = \frac{6\pi}{18} = \frac{\pi}{3} \text{ radians} \qquad \text{Solve for } \theta.$$

(continued)

(b) From part (a), P generates an angle of $\frac{\pi}{3}$ radians in 6 seconds. Since the radius is 10 centimeters, we can find the distance traveled by P along the circle.

$$s = r\theta = 10\left(\frac{\pi}{3}\right) = \frac{10\pi}{3} \text{ centimeters} \qquad \text{Arc length formula}$$

(c) From part (b), point P traveled $s = \frac{10\pi}{3}$ centimeters in 6 seconds, so for 1 second we divide $\frac{10\pi}{3}$ by 6.

$$v = \frac{s}{t} = \frac{\frac{10\pi}{3}}{6} = \frac{10\pi}{3} \div 6 = \frac{10\pi}{3} \cdot \frac{1}{6} = \frac{5\pi}{9} \text{ centimeters per second} \qquad \bullet$$

EXAMPLE 11 **Finding Angular Speed of a Pulley and Linear Speed of a Belt**

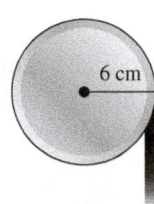

6 cm

FIGURE 25

A belt runs a pulley of radius 6 centimeters at 80 revolutions per minute, as shown in **FIGURE 25**.

(a) Find the angular speed ω of the pulley in radians per second.

(b) Find the linear speed v of the belt in centimeters per second.

Solution

(a) In 1 minute, the pulley makes 80 revolutions. Each revolution is 2π radians.

$$80(2\pi) = 160\pi \text{ radians per minute} \qquad \omega \text{ in radians per minute}$$

Next find the angular speed ω in radians per second. Since there are 60 seconds in 1 minute, divide 160π by 60.

$$\omega = \frac{160\pi}{60} = \frac{8\pi}{3} \text{ radians per second}$$

(b) The linear speed of the belt will be the same as that of a point rotating on the circumference of the pulley.

$$v = r\omega = 6\left(\frac{8\pi}{3}\right) = 16\pi \approx 50 \text{ centimeters per second} \qquad \bullet$$

8.1 Exercises

Checking Analytic Skills   *Fill in the blanks with the appropriate short answers.* **Do not use a calculator.**

1. An angle of 360° has an equivalent radian measure of _____.

2. An angle of π radians has an equivalent degree measure of _____.

3. The least positive angle coterminal with $-180°$ has degree measure _____.

4. The complement of a 40° angle is _____, and the supplement of a 40° angle is _____.

5. A formula for s relating r, θ, and s is _____.

6. A formula for v relating v, ω, and r is _____.

Checking Analytic Skills   *Find* **(a)** *the complement and* **(b)** *the supplement of each angle.* **Do not use a calculator.**

7. 30°

8. 60°

9. 45°

10. $\dfrac{\pi}{3}$

11. $\dfrac{\pi}{4}$

12. $\dfrac{\pi}{12}$

13. *Checking Analytic Skills*   What fraction of a complete revolution is each of the following angles?
  (a) 180°   **(b)** 40°   **(c)** 1°

14. *Checking Analytic Skills*   What fraction of a complete revolution is each of the following angles?
  (a) $\dfrac{\pi}{6}$ **(b)** $\dfrac{\pi}{2}$ **(c)** 2π

15. *Concept Check* An angle measures *x* degrees.
  (a) What is the measure of its complement?
  (b) What is the measure of its supplement?

16. *Concept Check* An angle measures *x* radians.
  (a) What is the measure of its complement?
  (b) What is the measure of its supplement?

Checking Analytic Skills   *Find the degree measure of the smaller angle formed by the hands of a clock at the following times.* **Do not use a calculator.**

17.

18.

19. 3:15

20. 9:00

Find the measure of each angle in Exercises 21–26.

21.
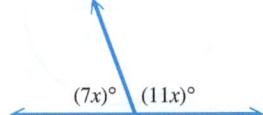
$(7x)°$ $(11x)°$

22.
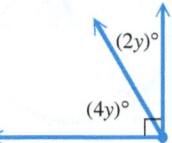
$(2y)°$
$(4y)°$

23.
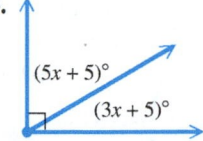
$(5x + 5)°$
$(3x + 5)°$

24.
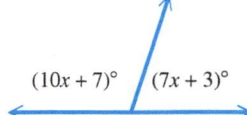
$(10x + 7)°$ $(7x + 3)°$

25. Supplementary angles with measures $6x - 4$ degrees and $8x - 12$ degrees

26. Complementary angles with measures $9z + 6$ degrees and $3z$ degrees

Perform each calculation.

27. $62° 18' + 21° 41'$

28. $75° 15' + 83° 32'$

29. $47° 29' - 71° 18'$

30. $47° 23' - 73° 48'$

31. $90° - 72° 58' 11''$

32. $90° - 36° 18' 47''$

Convert each angle measure to decimal degrees. Use a calculator, and round to the nearest thousandth of a degree if necessary.

33. $20° 54'$

34. $38° 42'$

35. $91° 35' 54''$

36. $34° 51' 35''$

Convert each angle measure to degrees, minutes, and seconds. Use a calculator as necessary. Round to the nearest second.

37. $31.4296°$

38. $59.0854°$

39. $89.9004°$

40. $102.3771°$

Concept Check *Sketch each angle in standard position. Draw an arrow representing the correct amount of rotation. Find the measure of two other angles, one positive and one negative, that are coterminal with the given angle. Give the quadrant of each angle.*

41. $75°$ **42.** $89°$ **43.** $174°$ **44.** $234°$ **45.** $-61°$ **46.** $-159°$

Find the angle of least positive measure that is coterminal with the given angle.

47. $-40°$ **48.** $-98°$ **49.** $450°$ **50.** $539°$

51. $-\dfrac{\pi}{4}$ **52.** $-\dfrac{\pi}{3}$ **53.** $-\dfrac{3\pi}{2}$ **54.** $-\pi$

Give an expression that generates all angles coterminal with each angle. Let n represent any integer.

55. $30°$ **56.** $45°$ **57.** $-90°$ **58.** $-135°$

59. $\dfrac{\pi}{4}$ **60.** $\dfrac{\pi}{6}$ **61.** $-\dfrac{3\pi}{4}$ **62.** $-\dfrac{7\pi}{6}$

Convert each degree measure to radians. Leave answers as rational multiples of π.

63. $60°$ **64.** $90°$ **65.** $150°$ **66.** $270°$ **67.** $-45°$ **68.** $-210°$

Convert each radian measure to degrees.

69. $\dfrac{\pi}{3}$ **70.** $\dfrac{8\pi}{3}$ **71.** $\dfrac{7\pi}{4}$ **72.** $\dfrac{2\pi}{3}$ **73.** $\dfrac{11\pi}{6}$ **74.** $\dfrac{15\pi}{4}$

Convert each degree measure to radians. Round to the nearest hundredth.

75. $39°$ **76.** $74°$ **77.** $139°\,10'$ **78.** $174°\,50'$ **79.** $64.29°$ **80.** $122.62°$

Convert each radian measure to degrees. Round answers to the nearest minute.

81. 2 **82.** 5 **83.** 1.74 **84.** 0.3417 **85.** -1.3 **86.** -4

87. *Concept Check* The figure shows the same angles measured in both degrees and radians. Complete the missing measures.

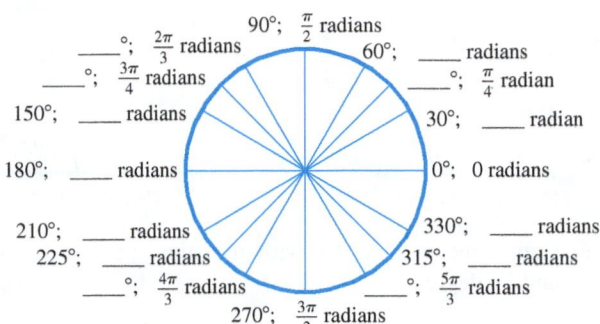

88. *Railroad Engineering* The term *grade* has several different meanings in construction work. Some engineers use the term to represent $\frac{1}{100}$ of a right angle and express it as a percent. For instance, an angle of $0.9°$ would be referred to as a 1% grade. (*Source:* Hay, W., *Railroad Engineering,* John Wiley and Sons.)

(a) By what number should you multiply a grade to convert it to radians?

(b) In a rapid-transit rail system, the maximum grade allowed between two stations is 3.5%. Express this angle in degrees and in radians.

Concept Check Find the exact length of each arc intercepted by the given central angle.

89. 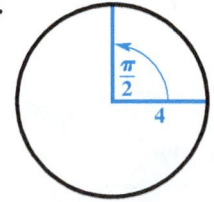 **90.**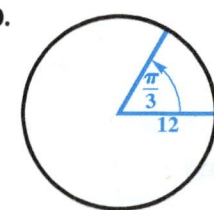

Concept Check Find the radius of each circle.

91. 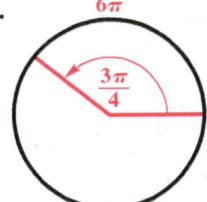 **92.**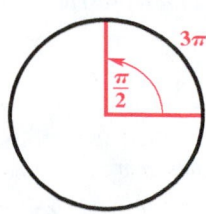

Concept Check Find the measure of each central angle (in radians).

93. 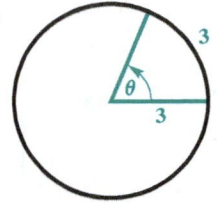 **94.**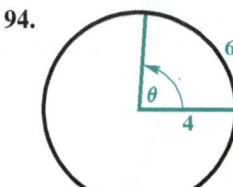

Find the length of each arc intercepted by a central angle θ in a circle of radius r. Round answers to the nearest hundredth.

95. $r = 12.3$ centimeters; $\theta = \dfrac{2\pi}{3}$ radians

96. $r = 0.892$ centimeter; $\theta = \dfrac{11\pi}{10}$ radians

97. $r = 4.82$ meters; $\theta = 60°$

98. $r = 71.9$ centimeters; $\theta = 135°$

Distance between Cities *Find the distance in kilometers between the pair of cities whose latitudes are given. Assume that the cities are on a north–south line and that the radius of Earth is 6400 kilometers. Round answers to the nearest hundred kilometers.*

99. Madison, South Dakota, $44°$ N, and Dallas, Texas, $33°$ N

100. Charleston, South Carolina, $33°$ N, and Toronto, Ontario, $43°$ N

101. New York City, New York, $41°$ N, and Lima, Peru, $12°$ S

102. Halifax, Nova Scotia, $45°$ N, and Buenos Aires, Argentina, $34°$ S

Use the formula $\omega = \frac{\theta}{t}$ to find the value of the missing variable. In Exercise 106, round to the nearest thousandth.

103. $\theta = \dfrac{3\pi}{4}$ radians, $t = 8$ seconds

104. $\theta = \dfrac{2\pi}{5}$ radians, $t = 10$ seconds

105. $\theta = \dfrac{2\pi}{9}$ radian, $\omega = \dfrac{5\pi}{27}$ radian per minute

106. $\omega = 0.90674$ radian per minute, $t = 11.876$ minutes

The formula $\omega = \frac{\theta}{t}$ can be rewritten as $\theta = \omega t$. Substituting ωt for θ changes $s = r\theta$ to $s = r\omega t$. Use the formula $s = r\omega t$ to find the value of the missing variable.

107. $r = 6$ centimeters, $\omega = \frac{\pi}{3}$ radians per second, $t = 9$ seconds

108. $r = 9$ yards, $\omega = \frac{2\pi}{5}$ radians per second, $t = 12$ seconds

109. $s = 6\pi$ centimeters, $r = 2$ centimeters, $\omega = \frac{\pi}{4}$ radian per second

110. $s = \frac{3\pi}{4}$ kilometers, $r = 2$ kilometers, $t = 4$ seconds

Solve each problem.

111. *Pulley Raising a Weight* Refer to the figure and answer each question.
   (a) How many inches will the weight rise if the pulley is rotated through an angle of 71° 50′?
   (b) Through what angle, to the nearest minute, must the pulley be rotated to raise the weight 6 inches?

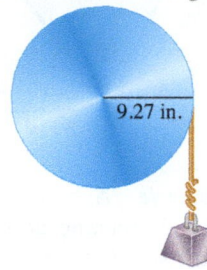

9.27 in.

112. *Pulley Raising a Weight* Approximate the radius of the pulley in the figure if a rotation of 51.6° raises the weight 11.4 centimeters.

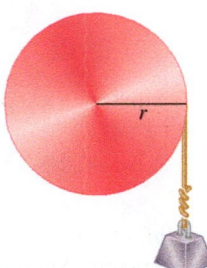

r

113. *Rotating Wheels* The rotation of the smaller wheel in the figure causes the larger wheel to rotate. Through how many degrees, to the nearest tenth, will the larger wheel rotate if the smaller one rotates through 60.0°?

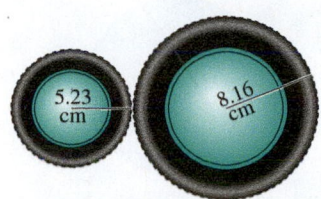

5.23 cm   8.16 cm

114. *Rotating Wheels* Refer to the figure at the top of the next column. Find the radius of the larger wheel, to the nearest tenth, in the figure if the smaller wheel rotates 80.0° when the larger wheel rotates 50.0°.

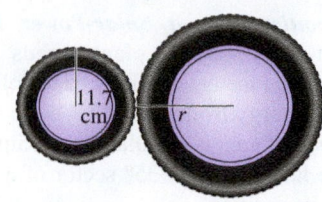

11.7 cm r

115. *Bicycle Chain Drive* The figure shows the chain drive of a bicycle. To the nearest inch, how far will the bicycle move if the pedals are rotated through 180°? Assume the radius of the bicycle wheel is 13.6 inches.

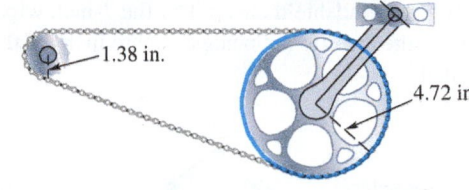

1.38 in.   4.72 in.

116. *Pickup Truck Speedometer* The speedometer of a pickup truck is designed to be accurate with tires of radius 14 inches.
   (a) Find the number of rotations, to the nearest whole number, of a tire in 1 hour if the truck is driven 55 mph.
   (b) Suppose that oversize tires of radius 16 inches are placed on the truck. If the truck is now driven for 1 hour with the speedometer reading 55 mph, how far has the truck gone? If the speed limit is 60 mph, did the driver exceed the speed limit?

Approximate the area of a sector of a circle having radius r and central angle θ.

117. $r = 29.2$ meters; $\theta = \frac{5\pi}{6}$ radians

118. $r = 59.8$ kilometers; $\theta = \frac{2\pi}{3}$ radians

119. $r = 12.7$ centimeters; $\theta = 81.0°$

120. $r = 18.3$ meters; $\theta = 125°$

Concept Check Find the area of each sector. Express your answers in terms of π.

121.
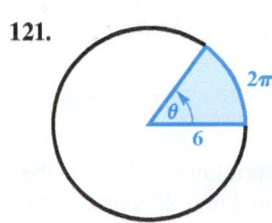
2π
θ
6

122.
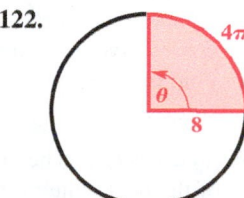
4π
θ
8

Solve each problem.

123. *Doppler Radar* Radar is used to identify severe weather. If Doppler radar can detect weather within a 240-mile radius and creates a new image every 48 seconds, find the area scanned by the radar in 1 second.

124. *Land Required for a Solar-Power Plant* A 300-megawatt solar-power plant needs approximately 950,000 square meters of land area to collect the required amount of energy from sunlight.
(a) If the land area is circular, approximate its radius.
(b) If the land area is a 35° sector of a circle, approximate its radius.

125. *Area Cleaned by a Windshield Wiper* The Ford Model A, built from 1928 to 1931, had a single windshield wiper on the driver's side. The total arm and blade was 10 inches long and rotated back and forth through an angle of 95°. The shaded region in the figure is the portion of the windshield cleaned by the 7-inch wiper blade. What is the area (to the nearest square inch) of the region cleaned?

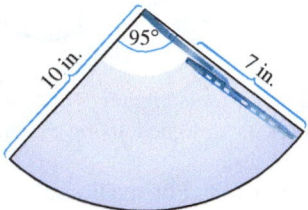

126. *Measures of a Structure* The figure shows Medicine Wheel, a Native American structure in northern Wyoming. This circular structure is perhaps 2500 years old. There are 27 spokes in the wheel, all equally spaced.

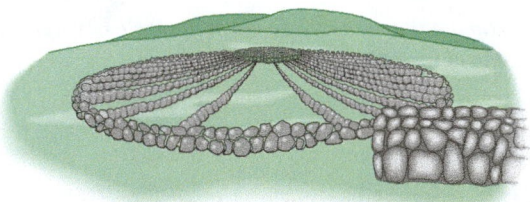

(a) Find the measure of each central angle in degrees and in radians.
(b) If the radius of the wheel is 76 feet, approximate the circumference.
(c) Approximate the length of each arc intercepted by consecutive pairs of spokes.
(d) Find the area of each sector formed by consecutive spokes.

127. *Size of a Corral* The unusual corral in the figure at the top of the next column is separated into 26 areas, many of which approximate sectors of a circle. Assume that the corral has a diameter of 50 meters.
(a) Approximate the central angle for each region, assuming that the 26 regions are all equal sectors with the fences meeting at the center.
(b) What is the area of each sector (to the nearest square meter)?

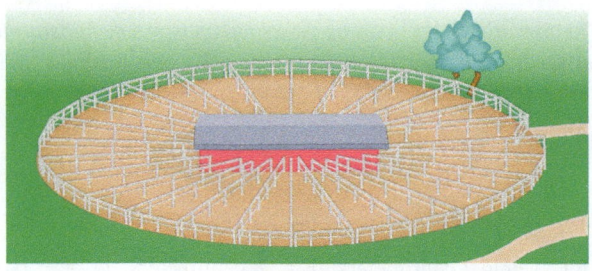

128. *Area of a Lot* A frequent problem in surveying city lots and rural lands adjacent to curves of highways and railways is that of finding the area when one or more of the boundary lines is the arc of a circle. Find the area of the lot shown in the figure to the nearest hundred square yards. (*Source:* Anderson, J. and E. Mikhall, *Introduction to Surveying,* McGraw-Hill.)

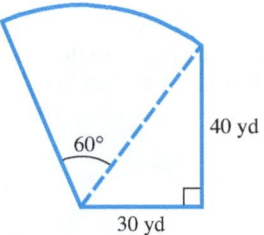

129. *Speed of a Bicycle* A bicycle has a tire 26 inches in diameter that is rotating at 15 radians per second. Approximate the speed of the bicycle in feet per second and in miles per hour.

130. *Angular Speed of a Pulley* The pulley shown has radius 12.96 centimeters. Suppose that it takes 18 seconds for 56 centimeters of belt to go around the pulley. Approximate the angular speed of the pulley in radians per second to the nearest hundredth.

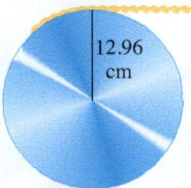

131. *Angular Speed of Skateboard Wheels* The wheels on a skateboard have diameter 2.25 inches. If a skateboarder is traveling downhill at 15.0 mph, approximate the angular speed of the wheels in radians per second.

132. *Angular Speed of Pulleys*   The two pulleys in the figure have radii 15 centimeters and 8 centimeters, respectively. The larger pulley rotates 25 times in 36 seconds. Find the angular speed of each pulley in radians per second.

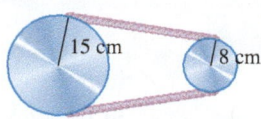

133. *Orbit of Earth*   Earth travels about the sun in an orbit that is almost circular. Assume that the orbit is a circle, with radius 93,000,000 miles.

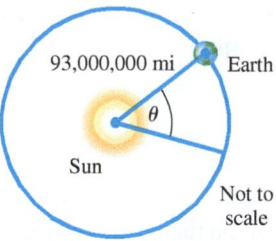

(a) Assume that a year is 365 days, and find θ, the angle formed by Earth's movement in 1 day.
(b) Give the angular speed in radians per hour.
(c) Approximate the linear speed of Earth in miles per hour.

134. *Nautical Miles*   **Nautical miles** are used by ships and airplanes. They are different from **statute miles,** which equal 5280 feet. A nautical mile is defined to be the arc length along the equator intercepted by a central angle *AOB* of 1 minute, as illustrated in the figure. If the equatorial radius of Earth is 3963 miles, use the arc length formula to approximate the number of statute miles in 1 nautical mile. Round your answer to two decimal places.

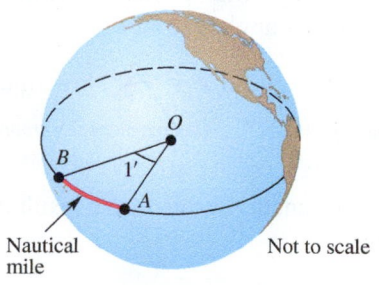

135. *Speed of a Motor Propeller*   A 90-horsepower outboard motor at full throttle rotates its propeller 5000 revolutions per minute. Find the angular speed of the propeller in radians per second. What is the linear speed in inches per second of a point at the tip of the propeller if its diameter is 10 inches?

136. *Speed of a Golf Clubhead*   The shoulder joint can rotate at about 25 radians per second. Assuming that a golfer's arm is straight and the distance from the shoulder to the clubhead is 5 feet, approximate the linear speed of the clubhead from the shoulder rotation. (*Source:* Cooper, J. and R. Glassow, *Kinesiology,* Second Edition, C.V. Mosby.)

137. *Circumference of Earth*   The first accurate estimate of the distance around Earth was done by the Greek astronomer Eratosthenes (276–195 B.C.), who noted that the noontime position of the sun at the summer solstice differed by 7° 12′ from the city of Syene to the city of Alexandria. (See the figure.) The distance between these two cities is 496 miles. Use the arc length formula to estimate the radius of Earth. Then approximate the circumference of Earth. (*Source:* Zeilik, M., *Introductory Astronomy and Astrophysics,* Third Edition, Saunders College Publishers.)

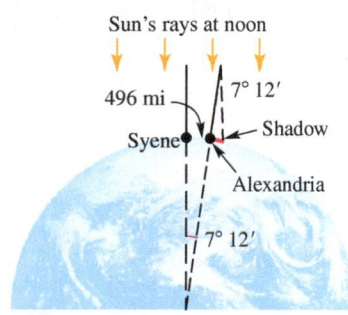

138. *Diameter of the Moon*   The distance to the moon is approximately 238,900 miles.
(a) Use the arc length formula to approximate the diameter *d* of the moon if angle θ in the figure is measured to be 0.517°.

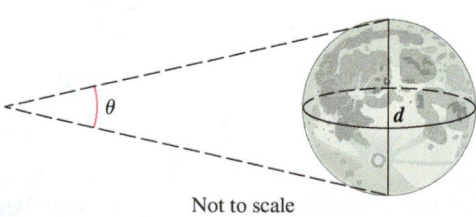

(b) Since the answer to part (a) approximates *d* using the length of an arc *s*, will *s* be greater than or less than the actual diameter *d* of the moon?

8.2 Trigonometric Functions and Fundamental Identities

Trigonometric Functions • Function Values of Quadrantal Angles • Reciprocal Identities • Signs and Ranges of Function Values • Pythagorean Identities • Quotient Identities • An Application of Trigonometric Functions

θ in Standard Position

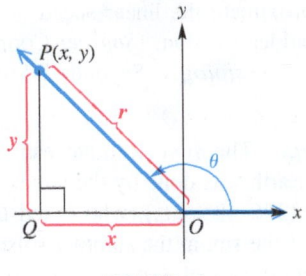

FIGURE 26

Trigonometric Functions

To define the six **trigonometric functions,** we start with an angle θ in standard position and choose any point P having coordinates (x, y) on the terminal side of angle θ. (Point P must *not* be the vertex of the angle.) See **FIGURE 26.** A perpendicular from P to the x-axis at point Q determines a right triangle having vertices at O, P, and Q. The distance r from $P(x, y)$ to the origin, $(0, 0)$, can be found from the distance formula.

$$r = \sqrt{(x - 0)^2 + (y - 0)^2} = \sqrt{x^2 + y^2} \qquad \text{\textcolor{blue}{$r > 0$ here, since distance is positive.}}$$

The six trigonometric functions of angle θ are **sine, cosine, tangent, cotangent, secant,** and **cosecant.** In the following definitions, we use the customary abbreviations for the names of these functions.

Trigonometric Functions

Let (x, y) be a point other than the origin on the terminal side of an angle θ in standard position. The distance from the point to the origin is $r = \sqrt{x^2 + y^2}$. The six trigonometric functions of θ are as follows.

$$\sin \theta = \frac{y}{r} \qquad \cos \theta = \frac{x}{r} \qquad \tan \theta = \frac{y}{x} \;\; (x \neq 0)$$

$$\csc \theta = \frac{r}{y} \;\; (y \neq 0) \quad \sec \theta = \frac{r}{x} \;\; (x \neq 0) \quad \cot \theta = \frac{x}{y} \;\; (y \neq 0)$$

NOTE Although **FIGURE 26** shows a second-quadrant angle, these definitions apply to any angle θ. Because of the restrictions on the denominators in the definitions of tangent, cotangent, secant, and cosecant, some angles have undefined function values. (See **Example 3.**)

EXAMPLE 1 Finding Function Values of an Angle

The terminal side of an angle θ in standard position passes through $(-3, -4)$. Find the values of the six trigonometric functions of angle θ.

Solution As shown in **FIGURE 27**, $x = -3$ and $y = -4$. Now find r.

$$r = \sqrt{(-3)^2 + (-4)^2} = \sqrt{25} = 5 \qquad \text{\textcolor{blue}{\textit{Use parentheses around substituted values to avoid errors.}}}$$

(Remember that $r > 0$.) The following function values are a result of the above definitions of the trigonometric functions.

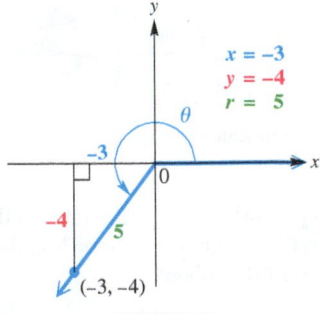

FIGURE 27

$$\sin \theta = \frac{-4}{5} = -\frac{4}{5} \qquad \cos \theta = \frac{-3}{5} = -\frac{3}{5} \qquad \tan \theta = \frac{-4}{-3} = \frac{4}{3}$$

$$\csc \theta = \frac{5}{-4} = -\frac{5}{4} \qquad \sec \theta = \frac{5}{-3} = -\frac{5}{3} \qquad \cot \theta = \frac{-3}{-4} = \frac{3}{4}$$

Similar Triangles

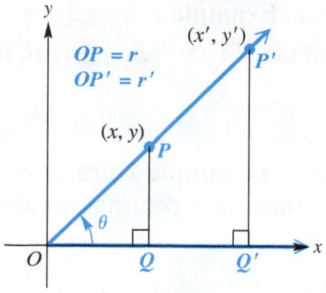

FIGURE 28

We can find the six trigonometric functions by using *any* point other than the origin on the terminal side of an angle. Refer to **FIGURE 28,** which shows an angle θ and two distinct points on its terminal side. Point P has coordinates (x, y), and point P' (read "*P*-prime") has coordinates (x', y'). Let r be the length of the hypotenuse of triangle OPQ, and let r' be the length of the hypotenuse of triangle $OP'Q'$. Since corresponding sides of similar triangles are in proportion,

$$\frac{y}{r} = \frac{y'}{r'},$$

so $\sin \theta = \frac{y}{r}$ is the same no matter which point is used to find it. Similar results hold for the other five functions.

We can also find the trigonometric function values of an angle if we know the equation of the line coinciding with the terminal ray. The graph of the equation

$$Ax + By = 0$$

is a line that passes through the origin. If we restrict x to have only nonpositive or only nonnegative values, we obtain the graph of a ray with endpoint at the origin. For example, the graph of $x + 2y = 0$, $x \geq 0$, shown in **FIGURE 29,** is a ray that can serve as the terminal side of an angle in standard position. By choosing any point on the ray, we can find the trigonometric function values of the angle.

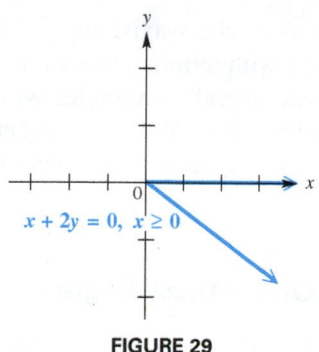

FIGURE 29

> **EXAMPLE 2** **Finding Function Values of an Angle**

Find the six trigonometric function values of the angle θ in standard position if the terminal side of θ has the equation $x + 2y = 0$, $x \geq 0$.

Analytic Solution

The angle is shown in **FIGURE 30.** We can use *any* point except $(0, 0)$ on the terminal side of θ to find the trigonometric function values. We choose $x = 2$ and find the corresponding y-value.

$$x + 2y = 0, \ x \geq 0$$
$$2 + 2y = 0 \qquad \text{Let } x = 2.$$
$$2y = -2 \qquad \text{Subtract 2.}$$
$$y = -1 \qquad \text{Divide by 2.}$$

The point $(2, -1)$ lies on the terminal side, and the corresponding value of r is

$$r = \sqrt{2^2 + (-1)^2} = \sqrt{5}.$$

Thus, $\sin \theta = \dfrac{y}{r} = \dfrac{-1}{\sqrt{5}} = \dfrac{-1}{\sqrt{5}} \cdot \dfrac{\sqrt{5}}{\sqrt{5}} = -\dfrac{\sqrt{5}}{5}$

$$\cos \theta = \dfrac{x}{r} = \dfrac{2}{\sqrt{5}} = \dfrac{2}{\sqrt{5}} \cdot \dfrac{\sqrt{5}}{\sqrt{5}} = \dfrac{2\sqrt{5}}{5}$$

Multiply by $\frac{\sqrt{5}}{\sqrt{5}}$, which is equal to 1, to rationalize denominators.

$$\tan \theta = \dfrac{y}{x} = -\dfrac{1}{2} \qquad \csc \theta = \dfrac{r}{y} = \dfrac{\sqrt{5}}{-1} = -\sqrt{5}$$

$$\sec \theta = \dfrac{r}{x} = \dfrac{\sqrt{5}}{2} \qquad \cot \theta = \dfrac{x}{y} = \dfrac{2}{-1} = -2.$$

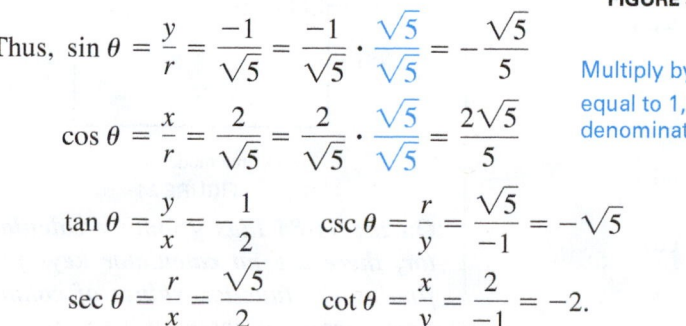

FIGURE 30

Graphing Calculator Solution

FIGURE 31 shows the graph of

$$x + 2y = 0, x \geq 0.$$

We can graph the ray by entering

$$Y_1 = (-1/2)X, \quad \text{or} \quad Y_1 = -0.5X$$

with the restriction $X \geq 0$. We used the capability of the calculator to find the Y-value for $X = 2$, which produced the point and values of X and Y shown on the screen.

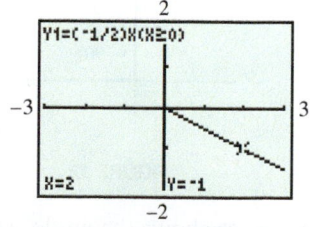

FIGURE 31

We can now use these values of X and Y and the definitions of the trigonometric functions to find the six function values of θ, as shown in the analytic solution.

Recall that when the equation of a line is written using slope–intercept form $y = mx + b$, the coefficient m of x is the slope of the line. In **Example 2**, $x + 2y = 0$ can be written as $y = -\frac{1}{2}x$, so the slope is $-\frac{1}{2}$. Notice that $\tan \theta = -\frac{1}{2}$. **In general, it is true that $m = \tan \theta$.**

NOTE The trigonometric function values we found in **Example 2** are *exact*. If we were to use a calculator to approximate these values, the decimal results would *not* be acceptable if exact values were required.

Function Values of Quadrantal Angles

If the terminal side of an angle in standard position lies along the y-axis, any point on this terminal side has x-coordinate 0. Similarly, an angle with terminal side on the x-axis has y-coordinate 0 for any point on the terminal side. Recall that angles with their terminal side on an axis are *quadrantal angles*. Since the values of x and y appear in the denominators of some trigonometric functions, some trigonometric function values of quadrantal angles are undefined.

EXAMPLE 3 **Finding Function Values of Quadrantal Angles**

Find the values of the six trigonometric functions for each angle, if possible.

(a) An angle of $90°$

(b) An angle θ in standard position with terminal side through $(-3, 0)$

Analytic Solution

(a) We select any point on the terminal side of a $90°$ angle, such as $(0, 1)$ shown in **FIGURE 32.** Here, $x = 0$ and $y = 1$, so $r = \sqrt{0^2 + 1^2} = 1$.

$$\sin 90° = \frac{y}{r} = \frac{1}{1} = 1 \qquad \csc 90° = \frac{r}{y} = \frac{1}{1} = 1$$

$$\cos 90° = \frac{x}{r} = \frac{0}{1} = 0 \qquad \sec 90° = \frac{r}{x} = \frac{1}{0} \text{ (undefined)}$$

$$\tan 90° = \frac{y}{x} = \frac{1}{0} \text{ (undefined)} \qquad \cot 90° = \frac{x}{y} = \frac{0}{1} = 0$$

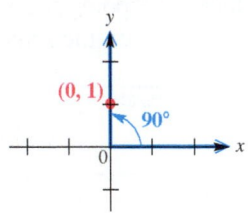

FIGURE 32

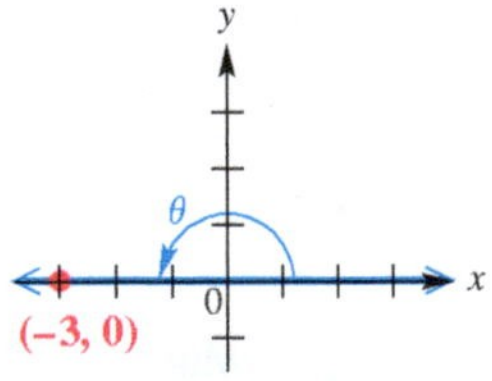

FIGURE 33

(b) **FIGURE 33** shows the angle. Here, $x = -3$, $y = 0$, and $r = 3$.

$$\sin \theta = \frac{0}{3} = 0 \qquad \csc \theta = \frac{3}{0} \text{ (undefined)}$$

$$\cos \theta = \frac{-3}{3} = -1 \qquad \sec \theta = \frac{3}{-3} = -1$$

$$\tan \theta = \frac{0}{-3} = 0 \qquad \cot \theta = \frac{-3}{0} \text{ (undefined)}$$

Graphing Calculator Solution

(a) A calculator set in degree mode returns the correct values for $\sin 90°$ and $\cos 90°$. See **FIGURE 34.** The bottom screen shows an ERROR message for $\tan 90°$, because $90°$ is not in the domain of the tangent function.

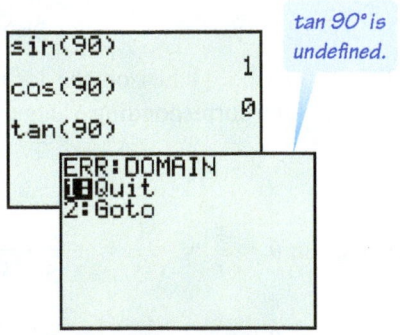

tan 90° is undefined.

Degree mode
FIGURE 34

On the TI-84 Plus graphing calculator, there are no calculator keys for finding the function values of cotangent, secant, or cosecant. Later in this section, we discuss how to find these function values with a calculator.

The conditions under which the trigonometric function values of quadrantal angles are undefined are summarized here.

Conditions for Undefined Function Values

- If the terminal side of the quadrantal angle lies along the *y*-axis, then the tangent and secant functions are undefined.
- If the terminal side of the quadrantal angle lies along the *x*-axis, then the cotangent and cosecant functions are undefined.

Function values of the most commonly used degree-measured quadrantal angles, 0°, 90°; 180°, 270°, and 360°, are summarized in the following table. (**FIGURE 35** can be helpful in determining these values.)

Function Values of Quadrantal Angles

θ	$\sin \theta$	$\cos \theta$	$\tan \theta$	$\cot \theta$	$\sec \theta$	$\csc \theta$
0°	0	1	0	Undefined	1	Undefined
90°	1	0	Undefined	0	Undefined	1
180°	0	−1	0	Undefined	−1	Undefined
270°	−1	0	Undefined	0	Undefined	−1
360°	0	1	0	Undefined	1	Undefined

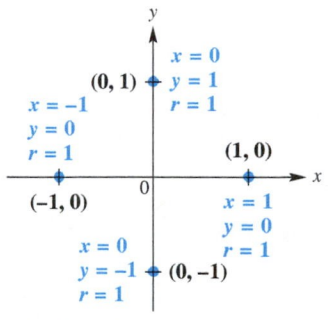

In **Section 8.5**, we extend this diagram to the concept of the *unit circle*.

FIGURE 35

Values given in this table can also be found with a calculator that has trigonometric function keys. *Make sure the calculator is set in degree mode.*

CAUTION   *One of the most common errors involving calculators in trigonometry occurs when the calculator is set for the wrong mode.* Be sure that you know how to set your calculator in either radian or degree mode.

Reciprocal Identities

Identities are equations that are true for all values of the variables for which all expressions in the equation are defined. The following are examples of identities.

$$(x + y)^2 = x^2 + 2xy + y^2 \quad \text{and} \quad 2(x + 3) = 2x + 6 \qquad \text{Identities}$$

The definitions of the trigonometric functions were written to illustrate that certain function pairs are reciprocals of each other. Since $\sin \theta = \frac{y}{r}$ and $\csc \theta = \frac{r}{y}$,

$$\sin \theta = \frac{1}{\csc \theta} \quad \text{and} \quad \csc \theta = \frac{1}{\sin \theta}, \quad \text{provided } \sin \theta \neq 0.$$

Also, $\cos \theta$ and $\sec \theta$ are reciprocals, as are $\tan \theta$ and $\cot \theta$. The **reciprocal identities** hold for any angle θ that does not lead to a 0 denominator.

Reciprocal Identities

$$\sin \theta = \frac{1}{\csc \theta} \qquad \cos \theta = \frac{1}{\sec \theta} \qquad \tan \theta = \frac{1}{\cot \theta}$$

$$\csc \theta = \frac{1}{\sin \theta} \qquad \sec \theta = \frac{1}{\cos \theta} \qquad \cot \theta = \frac{1}{\tan \theta}$$

NOTE Identities can be written in different forms. For example,

$$\sin \theta = \frac{1}{\csc \theta}$$

can also be written

$$\csc \theta = \frac{1}{\sin \theta} \quad \text{and}$$

$$(\sin \theta)(\csc \theta) = 1.$$

FIGURE 36(a) shows how csc 90°, sec 180°, and csc(−270°) are found by using the reciprocal identities and the reciprocal key of a graphing calculator in degree mode. *Be sure NOT to use the inverse trigonometric function keys* $(\sin^{-1}, \cos^{-1}, \tan^{-1})$ *to find the reciprocal function values.* Attempting to find sec 90° by entering $\frac{1}{\cos 90°}$ produces an ERROR message, indicating that the reciprocal is undefined. See **FIGURE 36(b)**. Compare these results with those in the table of quadrantal angle function values.

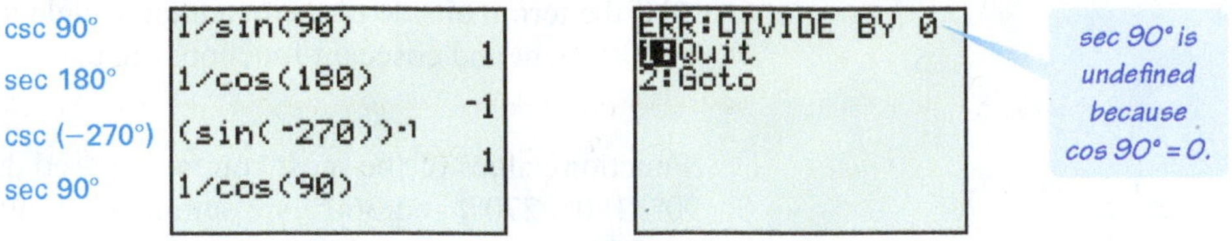

csc 90°	1/sin(90)	1
sec 180°	1/cos(180)	-1
csc (−270°)	(sin(-270))⁻¹	1
sec 90°	1/cos(90)	

ERR:DIVIDE BY 0
1:Quit
2:Goto

sec 90° is undefined because cos 90° = 0.

Degree mode

(a)   (b)

FIGURE 36

EXAMPLE 4   Using the Reciprocal Identities

Find each of the following.

(a) $\cos \theta$, given that $\sec \theta = \frac{5}{3}$ **(b)** $\sin \theta$, given that $\csc \theta = -\frac{\sqrt{12}}{2}$

Solution

(a) Use the fact that $\cos \theta$ is the reciprocal of $\sec \theta$.

$$\cos \theta = \frac{1}{\sec \theta} = \frac{1}{\frac{5}{3}} = 1 \div \frac{5}{3} = 1 \cdot \frac{3}{5} = \frac{3}{5}$$

(b) Use the reciprocal identity $\sin \theta = \frac{1}{\csc \theta}$.

Rationalize the denominator.

$$\sin \theta = \frac{1}{-\frac{\sqrt{12}}{2}} = \frac{-2}{\sqrt{12}} = \frac{-2}{2\sqrt{3}} = \frac{-1}{\sqrt{3}} = \frac{-1}{\sqrt{3}} \cdot \frac{\sqrt{3}}{\sqrt{3}} = -\frac{\sqrt{3}}{3}$$

$$\sqrt{12} = \sqrt{4 \cdot 3} = 2\sqrt{3}$$

FOR DISCUSSION

The signs of the functions in the four quadrants are determined *mathematically* by considering the signs of x, y, and r. A *mnemonic device* is the statement "**All Students Take Calculus,**" which indicates that **all** are positive in quadrant I, **sine** is positive in quadrant II, **tangent** is positive in quadrant III, and **cosine** is positive in quadrant IV. How can you remember the signs of secant, cosecant, and cotangent using this device?

(Always remember that there are sound mathematical principles that justify mnemonic devices like this one.)

Signs and Ranges of Function Values

In the definitions of the trigonometric functions, r is the distance from the origin to the point (x, y), so $r > 0$. If we choose a point (x, y) in quadrant I, then both x and y will be positive, so the values of all six functions will be positive in quadrant I.

A point (x, y) in quadrant II has $x < 0$ and $y > 0$. This makes the values of sine and cosecant positive for quadrant II angles, while the other four functions take on negative values. Similar results can be obtained for the other quadrants, as summarized here.

Signs of Function Values

θ in Quadrant	$\sin \theta$	$\cos \theta$	$\tan \theta$	$\cot \theta$	$\sec \theta$	$\csc \theta$
I	+	+	+	+	+	+
II	+	−	−	−	−	+
III	−	−	+	+	−	−
IV	−	+	−	−	+	−

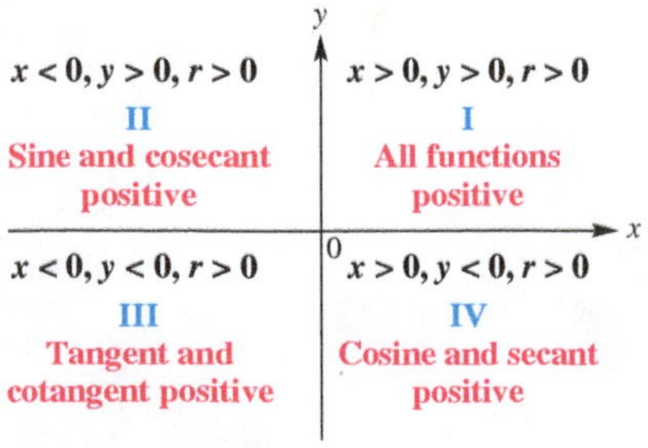

$x < 0, y > 0, r > 0$
II
Sine and cosecant positive

$x > 0, y > 0, r > 0$
I
All functions positive

$x < 0, y < 0, r > 0$
III
Tangent and cotangent positive

$x > 0, y < 0, r > 0$
IV
Cosine and secant positive

EXAMPLE 5   **Identifying the Quadrant of an Angle**

Identify the quadrant (or possible quadrants) of an angle θ that satisfies the given conditions.

(a) $\sin \theta > 0$, $\tan \theta < 0$ **(b)** $\cos \theta < 0$, $\sec \theta < 0$

Solution

(a) Since $\sin \theta > 0$ in quadrants I and II and $\tan \theta < 0$ in quadrants II and IV, both conditions are met only in quadrant II.

(b) The cosine and secant functions are both negative in quadrants II and III, so in this case θ could be in either of these two quadrants.   ●

Increasing θ Near 90°

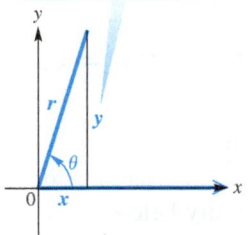

y increases as θ increases.

r stays constant, and y is always less than r.

FIGURE 37

FIGURE 37 shows an angle θ as it increases in measure from near 0° to near 90°. In each case, the value of r is the same. As the measure of the angle increases, y increases, but never exceeds r, so $y \leq r$. Dividing each side by the positive number r gives $\frac{y}{r} \leq 1$.

In a similar way, angles in quadrant IV suggest that $-1 \leq \frac{y}{r}$.

$$-1 \leq \frac{y}{r} \leq 1$$

$$-1 \leq \sin \theta \leq 1 \qquad \text{$\frac{y}{r} = \sin \theta$ for any angle θ}$$

Similar reasoning leads to the following.

$$-1 \leq \cos \theta \leq 1$$

The tangent of an angle is defined as $\frac{y}{x}$. It is possible that $x < y$, $x = y$, or $x > y$. For this reason, $\frac{y}{x}$ can take any value, so $\tan \theta$ can be any real number, as can $\cot \theta$.

The functions $\sec \theta$ and $\csc \theta$ are reciprocals of the functions $\cos \theta$ and $\sin \theta$, respectively, making the following true.

$$\sec \theta \leq -1 \quad \text{or} \quad \sec \theta \geq 1 \qquad \text{and} \qquad \csc \theta \leq -1 \quad \text{or} \quad \csc \theta \geq 1$$

Ranges of Trigonometric Functions

Trigonometric Function of θ	Range (Set-Builder Notation)	Range (Interval Notation)
$\sin \theta$, $\cos \theta$	$\{y \mid \lvert y \rvert \leq 1\}$	$[-1, 1]$
$\tan \theta$, $\cot \theta$	$\{y \mid y$ is a real number$\}$	$(-\infty, \infty)$
$\sec \theta$, $\csc \theta$	$\{y \mid \lvert y \rvert \geq 1\}$	$(-\infty, -1] \cup [1, \infty)$

EXAMPLE 6   **Deciding whether a Value Is in the Range of a Function**

Decide whether each statement is *possible* for some angle θ, or *impossible*.

(a) $\sin \theta = \sqrt{3}$ **(b)** $\tan \theta = 110.47$ **(c)** $\sec \theta = 0.6$

Solution

(a) For any value of θ, $-1 \leq \sin \theta \leq 1$. Since $\sqrt{3} > 1$, it is impossible to find a value of θ for which $\sin \theta = \sqrt{3}$.

(b) The tangent function can equal any real number, so $\tan \theta = 110.47$ is possible.

(c) Since $\sec \theta \leq -1$ or $\sec \theta \geq 1$, the statement $\sec \theta = 0.6$ is impossible.   ●

Pythagorean Identities

We derive three new identities from the relationship $x^2 + y^2 = r^2$.

$$\frac{x^2}{r^2} + \frac{y^2}{r^2} = \frac{r^2}{r^2} \qquad \text{Divide by } r^2.$$

$$\left(\frac{x}{r}\right)^2 + \left(\frac{y}{r}\right)^2 = 1 \qquad \text{Power rule for exponents}$$

$$(\cos\theta)^2 + (\sin\theta)^2 = 1 \qquad \cos\theta = \frac{x}{r}, \sin\theta = \frac{y}{r}$$

$$\mathbf{\sin^2\theta + \cos^2\theta = 1} \qquad \text{Rewrite.}$$

Start again with $x^2 + y^2 = r^2$ and divide through by x^2.

$$\frac{x^2}{x^2} + \frac{y^2}{x^2} = \frac{r^2}{x^2} \qquad \text{Divide by } x^2.$$

$$1 + \left(\frac{y}{x}\right)^2 = \left(\frac{r}{x}\right)^2 \qquad \text{Power rule for exponents}$$

$$1 + (\tan\theta)^2 = (\sec\theta)^2 \qquad \tan\theta = \frac{y}{x}, \sec\theta = \frac{r}{x}$$

$$\mathbf{1 + \tan^2\theta = \sec^2\theta} \qquad \text{Rewrite.}$$

On the other hand, dividing through by y^2 (see **Exercise 115**) gives the following.

$$\mathbf{1 + \cot^2\theta = \csc^2\theta}$$

These three identities are called the **Pythagorean identities,** since the original equation that led to them, $x^2 + y^2 = r^2$, comes from the Pythagorean theorem. (In the identities, we assume that all functions are defined for θ.)

> **Pythagorean Identities**
>
> $$\sin^2\theta + \cos^2\theta = 1 \qquad 1 + \tan^2\theta = \sec^2\theta \qquad 1 + \cot^2\theta = \csc^2\theta$$

As before, we have given only one form of each identity. However, algebraic transformations produce equivalent identities. For example, by subtracting $\sin^2\theta$ from each side of $\sin^2\theta + \cos^2\theta = 1$, we get the equivalent identity below.

$$\cos^2\theta = 1 - \sin^2\theta \qquad \text{Alternative form}$$

You should be able to transform these identities and recognize equivalent forms.

Quotient Identities

Recall that $\sin\theta = \frac{y}{r}$ and $\cos\theta = \frac{x}{r}$. Consider the quotient of $\sin\theta$ and $\cos\theta$.

$$\frac{\sin\theta}{\cos\theta} = \frac{\frac{y}{r}}{\frac{x}{r}} = \frac{y}{r} \div \frac{x}{r} = \frac{y}{r} \cdot \frac{r}{x} = \frac{y}{x} = \tan\theta, \text{ where } \cos\theta \neq 0$$

Similarly, $\frac{\cos\theta}{\sin\theta} = \cot\theta$, where $\sin\theta \neq 0$. Thus, we have the **quotient identities.**

> **Quotient Identities**
>
> $$\frac{\sin\theta}{\cos\theta} = \tan\theta \qquad \frac{\cos\theta}{\sin\theta} = \cot\theta$$

→ **Looking Ahead to Calculus**
The reciprocal, Pythagorean, and quotient identities are used repeatedly in calculus to find limits, derivatives, and integrals of trigonometric functions. These identities are also used to rewrite expressions in a form that permits simplifying a square root. For example, if $a \geq 0$ and $x = a \sin \theta$,

$$\sqrt{a^2 - x^2} = \sqrt{a^2 - a^2 \sin^2 \theta}$$
$$= \sqrt{a^2(1 - \sin^2 \theta)}$$
$$= \sqrt{a^2 \cos^2 \theta}$$
$$= a \left| \cos \theta \right|.$$

EXAMPLE 7   **Finding Function Values, Given One Value and the Quadrant**

Find $\cos \theta$ and $\sin \theta$ if $\tan \theta = \frac{4}{3}$ and θ is in quadrant III.

Solution   Since θ is in quadrant III, $\sin \theta$ and $\cos \theta$ will both be negative. It is tempting to say that since $\tan \theta = \frac{\sin \theta}{\cos \theta}$ and $\tan \theta = \frac{4}{3}$, then $\sin \theta = -4$ and $\cos \theta = -3$. *This is incorrect—both $\sin \theta$ and $\cos \theta$ must be in the interval $[-1, 1]$.*

$$1 + \tan^2 \theta = \sec^2 \theta \qquad \text{Pythagorean identity}$$

Pythagorean identity
$$1 + \left(\frac{4}{3}\right)^2 = \sec^2 \theta \qquad \tan \theta = \frac{4}{3}$$

$$1 + \frac{16}{9} = \sec^2 \theta \qquad \text{Apply the exponent.}$$

$$\frac{25}{9} = \sec^2 \theta \qquad 1 = \frac{9}{9} \text{ and } \frac{9}{9} + \frac{16}{9} = \frac{25}{9}$$

Choose the correct sign for Q III.
$$-\frac{5}{3} = \sec \theta \qquad \text{Choose the negative square root, since } \sec \theta \text{ is negative when } \theta \text{ is in quadrant III.}$$

Reciprocal identity
$$-\frac{3}{5} = \cos \theta \qquad \text{Secant and cosine are reciprocals.}$$

Use $\sin^2 \theta = 1 - \cos^2 \theta$ to find $\sin \theta$.

$$\sin^2 \theta = 1 - \left(-\frac{3}{5}\right)^2 \qquad \cos \theta = -\frac{3}{5}$$

$$\sin^2 \theta = 1 - \frac{9}{25} \qquad \text{Apply the exponent.}$$

$$\sin^2 \theta = \frac{16}{25} \qquad 1 = \frac{25}{25} \text{ and } \frac{25}{25} - \frac{9}{25} = \frac{16}{25}$$

$$\sin \theta = -\frac{4}{5} \qquad \text{Choose the negative square root.}$$

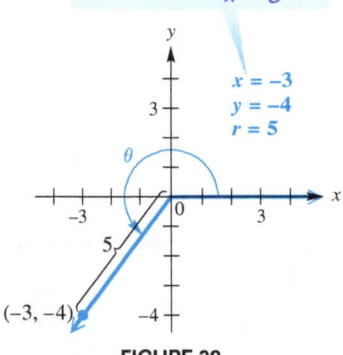

Choose a point (x, y) in Q III with $\tan \theta = \frac{y}{x} = \frac{4}{3}$.

$x = -3$
$y = -4$
$r = 5$

$(-3, -4)$

FIGURE 38

NOTE   **Example 7** can also be worked by drawing one possibility for θ in standard position in quadrant III, finding r to be 5, and then using the definitions of $\sin \theta$ and $\cos \theta$ in terms of x, y, and r. See **FIGURE 38**. *Be sure to use the correct signs for x and y.*

EXAMPLE 8   **Using Identities to Find a Trigonometric Expression**

If θ is in quadrant IV, find an expression for $\sec \theta$ in terms of $\sin \theta$.

Solution
$$\sin^2 \theta + \cos^2 \theta = 1 \qquad \text{Pythagorean identity}$$
$$\cos^2 \theta = 1 - \sin^2 \theta \qquad \text{Subtract } \sin^2 \theta.$$
$$\cos \theta = \pm \sqrt{1 - \sin^2 \theta} \qquad \text{Take square roots.}$$

Reciprocal identity
$$\sec \theta = \pm \frac{1}{\sqrt{1 - \sin^2 \theta}} \qquad \sec \theta = \frac{1}{\cos \theta}$$

Angle θ is in quadrant IV, so $\sec \theta > 0$. Thus, we choose the positive expression.

$$\sec \theta = \frac{1}{\sqrt{1 - \sin^2 \theta}} = \frac{\sqrt{1 - \sin^2 \theta}}{1 - \sin^2 \theta} \qquad \text{Rationalize the denominator.}$$

An Application of Trigonometric Functions

Grade, or slope, is a measure of steepness and indicates whether a highway is uphill or downhill. A 5% grade indicates that a road is increasing 5 vertical feet for each 100-foot increase in horizontal distance. **Grade resistance** is the gravitational force acting on a vehicle and is given by

$$R = W \sin \theta,$$

where W is the weight of the vehicle and θ is the angle associated with the grade. See **FIGURE 39.**

For an uphill grade, $\theta > 0$ and for a downhill grade, $\theta < 0$.

(*Source:* Mannering, F. and W. Kilareski, *Principles of Highway Engineering and Traffic Analysis,* Second Edition, John Wiley and Sons.)

FIGURE 39

EXAMPLE 9 **Calculating Grade Resistance**

A downhill highway grade is modeled by the line $y = -0.06x$ in quadrant IV.

(a) Find the grade of the road.

(b) Approximate the grade resistance for a 3000-pound car. Interpret the result.

Solution

(a) The slope of the line is -0.06, so when x *increases* by 100 feet, y *decreases* by 6 feet. See **FIGURE 40.** Thus, this road has a grade of -6%.

(b) First we must find $\sin \theta$. From **FIGURE 40,** we see that the point $(100, -6)$ lies on the terminal side of θ. We can find r and then $\sin \theta$.

$$r = \sqrt{x^2 + y^2} = \sqrt{100^2 + (-6)^2} = \sqrt{10{,}036},$$

and

$$\sin \theta = \frac{y}{r} = \frac{-6}{\sqrt{10{,}036}}.$$

The grade resistance is

$$R = W \sin \theta = 3000 \left(\frac{-6}{\sqrt{10{,}036}} \right) \approx -179.7 \text{ pounds.}$$

On this stretch of highway, gravity would pull a 3000-pound vehicle *downhill* with a force of about 180 pounds. Note that a downhill grade results in a negative grade resistance. ●

A −6% Grade

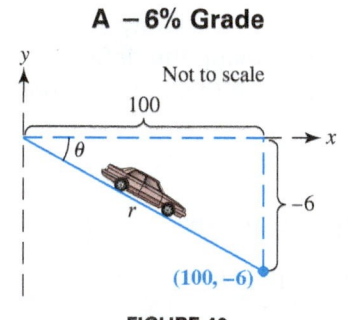

Not to scale

FIGURE 40

8.2 Exercises

Concept Check   In Exercises 1–2, find the six trigonometric functions of θ.

1.

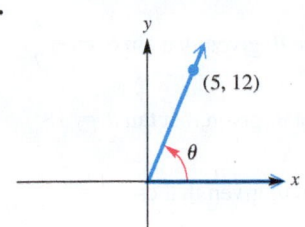

2.

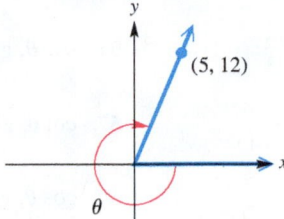

3. *Concept Check*   How do the answers from **Exercises 1** and **2** compare? Why?

4. *Concept Check*   How would the six trigonometric functions of 90° and −270° compare? Why?

Checking Analytic Skills   *Sketch an angle θ in standard position such that θ has the least possible positive measure, and the given point is on the terminal side of θ. Find the values of the six trigonometric functions for each angle. Rationalize denominators when applicable.* **Do not use a calculator.**

5. $(5, -12)$

6. $(-12, -5)$

7. $(-3, 4)$

8. $(-4, -3)$

9. $(-8, 15)$

10. $(15, -8)$

11. $(7, -24)$

12. $(-24, -7)$

13. $(0, 2)$

14. $(0, 5)$

15. $(-4, 0)$

16. $(-5, 0)$

17. $(0, -4)$

18. $(0, -3)$

19. $\left(1, \sqrt{3}\right)$

20. $\left(-1, \sqrt{3}\right)$

21. $\left(\sqrt{2}, \sqrt{2}\right)$

22. $\left(-\sqrt{2}, -\sqrt{2}\right)$

23. $\left(-2\sqrt{3}, -2\right)$

24. $\left(-2\sqrt{3}, 2\right)$

In Exercises 25–34, an equation of the terminal side of an angle θ in standard position is given with a restriction on x. Sketch the least positive angle θ, and find the values of the six trigonometric functions of θ.

25. $2x + y = 0, x \geq 0$

26. $3x + 5y = 0, x \geq 0$

27. $-6x - y = 0, x \leq 0$

28. $-5x - 3y = 0, x \leq 0$

29. $-4x + 7y = 0, x \leq 0$

30. $6x - 5y = 0, x \geq 0$

31. $x + y = 0, x \geq 0$

32. $x - y = 0, x \geq 0$

33. $-\sqrt{3}x + y = 0, x \leq 0$

34. $\sqrt{3}x + y = 0, x \leq 0$

To work Exercises 35–52, begin by reproducing the graph in **FIGURE 35.** *Keep in mind that for each of the four points labeled in the figure, r = 1. For each quadrantal angle, identify the appropriate values of x, y, and r to find the indicated function value. If it is undefined, say so. Check your answers with a calculator in degree mode.*

35. $\cos 90°$

36. $\sin 90°$

37. $\tan 180°$

38. $\cot 90°$

39. $\sec 180°$

40. $\csc 270°$

41. $\sin(-270°)$

42. $\cos(-90°)$

43. $\cot 540°$

44. $\tan 450°$

45. $\csc(-450°)$

46. $\sec(-540°)$

47. $\sin 1800°$

48. $\cos 1800°$

49. $\csc 1800°$

50. $\cot 1800°$

51. $\sec 1800°$

52. $\tan 1800°$

Use the appropriate reciprocal identity to find each function value. Rationalize denominators when applicable.

53. $\sec\theta$, given that $\cos\theta = \dfrac{2}{3}$

54. $\sec\theta$, given that $\cos\theta = \dfrac{5}{8}$

55. $\csc\theta$, given that $\sin\theta = -\dfrac{3}{7}$

56. $\csc\theta$, given that $\sin\theta = -\dfrac{8}{43}$

57. $\cot\theta$, given that $\tan\theta = 5$

58. $\cot\theta$, given that $\tan\theta = 18$

59. $\cos\theta$, given that $\sec\theta = -\dfrac{5}{2}$

60. $\cos\theta$, given that $\sec\theta = -\dfrac{11}{7}$

61. $\sin\theta$, given that $\csc\theta = \sqrt{2}$

62. $\sin\theta$, given that $\csc\theta = \dfrac{2\sqrt{6}}{3}$

63. $\tan\theta$, given that $\cot\theta = -2.5$

64. $\tan\theta$, given that $\cot\theta = -0.01$

Concept Check *If n is an integer, n · 180° represents an integer multiple of 180°, $(2n + 1) \cdot 90°$ represents an odd integer multiple of 90°, and so on. Decide whether each expression is equal to 0, 1, or −1 or is undefined.*

65. $\cos[(2n + 1) \cdot 90°]$

66. $\cot(n \cdot 180°)$

67. $\cos[(2n + 1) \cdot 180°]$

68. $\cos(n \cdot 360°)$

Identify the quadrant (or possible quadrants) of an angle θ that satisfies the given conditions.

69. $\sin\theta > 0, \csc\theta > 0$

70. $\cos\theta > 0, \sec\theta > 0$

71. $\cos\theta > 0, \sin\theta > 0$

72. $\sin\theta > 0, \tan\theta > 0$

73. $\tan\theta < 0, \cos\theta < 0$

74. $\cos\theta < 0, \sin\theta < 0$

75. $\sec\theta > 0, \csc\theta > 0$

76. $\csc\theta > 0, \cot\theta > 0$

77. $\sec\theta < 0, \csc\theta < 0$

78. $\cot\theta < 0, \sec\theta < 0$

79. $\sin\theta < 0, \csc\theta < 0$

80. $\tan\theta < 0, \cot\theta < 0$

81. Explain why the answers to **Exercises 71** and **75** are the same.

82. Explain why there is no angle θ that satisfies $\tan\theta > 0, \cot\theta < 0$.

Decide whether each statement is possible *for some angle θ, or* impossible.

83. $\sin\theta = 2$

84. $\sin\theta = 3$

85. $\cos\theta = -0.96$

86. $\cos\theta = -0.56$

87. $\tan\theta = 0.93$

88. $\cot\theta = 0.93$

89. $\sec\theta = -0.3$

90. $\sec\theta = -0.9$

91. $\csc\theta = 100$

92. $\csc\theta = -100$

93. $\cot\theta = -4$

94. $\cot\theta = -6$

95. $\sin\theta = \dfrac{1}{2}, \csc\theta = 2$

96. $\tan\theta = 2, \cot\theta = -2$

97. $\cos\theta = -2, \sec\theta = \dfrac{1}{2}$

98. *Concept Check* Is there an angle θ for which $\tan\theta$ and $\cot\theta$ are both undefined?

Find all trigonometric function values for each angle θ.

99. $\tan\theta = -\dfrac{15}{8}$, given that θ is in quadrant II

100. $\cos\theta = -\dfrac{3}{5}$, given that θ is in quadrant III

101. $\sin\theta = \dfrac{\sqrt{5}}{7}$, given that θ is in quadrant I

102. $\tan\theta = \sqrt{3}$, given that θ is in quadrant III

103. $\cot\theta = \dfrac{\sqrt{3}}{8}$, given that θ is in quadrant I

104. $\csc\theta = 2$, given that θ is in quadrant II

105. $\sin \theta = \dfrac{\sqrt{2}}{6}$, given that $\cos \theta < 0$

106. $\cos \theta = \dfrac{\sqrt{5}}{8}$, given that $\tan \theta < 0$

107. $\sec \theta = -4$, given that $\sin \theta > 0$

108. $\csc \theta = -3$, given that $\cos \theta > 0$

Use fundamental identities to find each expression.

109. Write $\cos \theta$ in terms of $\sin \theta$ if θ is acute.

110. Write $\sec \theta$ in terms of $\cos \theta$.

111. Write $\sin \theta$ in terms of $\cot \theta$ if θ is in quadrant III.

112. Write $\tan \theta$ in terms of $\cos \theta$ if θ is in quadrant IV.

113. Write $\tan \theta$ in terms of $\sin \theta$ if θ is in quadrant I or IV.

114. Write $\sin \theta$ in terms of $\sec \theta$ if θ is in quadrant I or II.

Work each problem.

115. Derive the identity $1 + \cot^2 \theta = \csc^2 \theta$ by dividing $x^2 + y^2 = r^2$ by y^2.

116. Using a method similar to the one given in this section showing that $\frac{\sin \theta}{\cos \theta} = \tan \theta$, show that $\frac{\cos \theta}{\sin \theta} = \cot \theta$.

117. *Concept Check* *True* or *false?* For all angles θ, $\sin \theta + \cos \theta = 1$. If false, give an example showing why.

118. *Concept Check* *True* or *false?* Since $\cot \theta = \frac{\cos \theta}{\sin \theta}$, if $\cot \theta = \frac{1}{2}$ with θ in quadrant I, then $\cos \theta = 1$ and $\sin \theta = 2$. If false, explain why.

119. *Highway Grade* Suppose that the uphill grade of a highway can be modeled by the line $y = 0.03x$ in quadrant I.
(a) Find the grade of the hill.
(b) Approximate the grade resistance for a gravel truck weighing 25,000 pounds.

120. *Highway Grade* Suppose that the downhill grade of a highway can be modeled by the line $y = -0.09x$ in quadrant IV.
(a) Find the grade of the hill.
(b) Approximate the grade resistance for a car weighing 3800 pounds.

SECTIONS 8.1–8.2 Reviewing Basic Concepts

1. Give the complement and the supplement of each angle.
(a) $35°$ (b) $\dfrac{\pi}{4}$

2. Convert $32.25°$ to degrees, minutes, and seconds.

3. Convert $59° \, 35' \, 30''$ to decimal degrees.

4. Find the angle of least positive measure coterminal with each angle.
(a) $560°$ (b) $-\dfrac{2\pi}{3}$

5. Convert each angle measure as directed.
(a) $240°$ to radians   (b) $\dfrac{3\pi}{4}$ radians to degrees

6. Suppose a circle has radius 3 centimeters.
(a) What is the exact length of an arc intercepted by a central angle of $120°$?

(b) What is the exact area of the sector formed in part (a)?

7. An angle θ is in standard position and its terminal side passes through the point $(-2, 5)$. Find the exact value of each of the six trigonometric functions of θ.

8. Find the exact values of the trigonometric functions of a $270°$ angle.

9. Decide whether the following statements are *possible* or *impossible*.
(a) $\cos \theta = \dfrac{3}{2}$ (b) $\tan \theta = 300$
(c) $\csc \theta = 5$ (d) $\sec \theta = 0$

10. If θ is a quadrant III angle with $\sin \theta = -\dfrac{2}{3}$, use identities to find the other trigonometric function values of θ.

8.3 Right Triangles and Evaluating Trigonometric Functions

Right-Triangle Definitions of the Trigonometric Functions • Trigonometric Function Values of Special Angles • Cofunction Identities • Reference Angles • Special Angles as Reference Angles • Finding Function Values with a Calculator • Finding Angle Measures

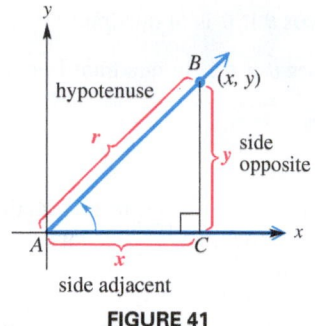

FIGURE 41

Right-Triangle Definitions of the Trigonometric Functions

In **Section 8.2,** we used angles in standard position to define the trigonometric functions. ***We can also approach them as ratios of the lengths of the sides of right triangles.***

FIGURE 41 shows an acute angle A in standard position. The definitions of the trigonometric function values of angle A require x, y, and r. In **FIGURE 41,** x and y are the lengths of the two legs of the right triangle ABC; r is the length of the hypotenuse.

The side of length y in **FIGURE 41** is called the **side opposite** angle A, and the side of length x is called the **side adjacent** to angle A. We use the lengths of these sides to replace x and y in the definitions of the trigonometric functions, and the length of the hypotenuse to replace r, to get the following right-triangle–based definitions.

Right-Triangle–Based Definitions of Trigonometric Functions

Let A be an acute angle in standard position, as in **FIGURE 41.**

$$\sin A = \frac{y}{r} = \frac{\text{side opposite}}{\text{hypotenuse}} \qquad \csc A = \frac{r}{y} = \frac{\text{hypotenuse}}{\text{side opposite}}$$

$$\cos A = \frac{x}{r} = \frac{\text{side adjacent}}{\text{hypotenuse}} \qquad \sec A = \frac{r}{x} = \frac{\text{hypotenuse}}{\text{side adjacent}}$$

$$\tan A = \frac{y}{x} = \frac{\text{side opposite}}{\text{side adjacent}} \qquad \cot A = \frac{x}{y} = \frac{\text{side adjacent}}{\text{side opposite}}$$

EXAMPLE 1 **Finding Trigonometric Function Values of an Acute Angle**

Find the values of $\sin A$, $\cos A$, and $\tan A$ in the right triangle in **FIGURE 42.**

FIGURE 42

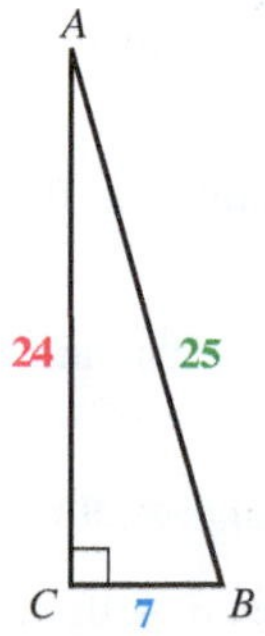

Solution The length of the side opposite angle A is 7, the length of the side adjacent to angle A is 24, and the length of the hypotenuse is 25.

$$\sin A = \frac{\text{side opposite}}{\text{hypotenuse}} = \frac{7}{25} \qquad \cos A = \frac{\text{side adjacent}}{\text{hypotenuse}} = \frac{24}{25} \qquad \tan A = \frac{\text{side opposite}}{\text{side adjacent}} = \frac{7}{24}$$

NOTE Because the cosecant, secant, and cotangent ratios are the reciprocals of the sine, cosine, and tangent values, $\csc A = \frac{25}{7}$, $\sec A = \frac{25}{24}$, and $\cot A = \frac{24}{7}$ in **Example 1.**

Equilateral Triangle

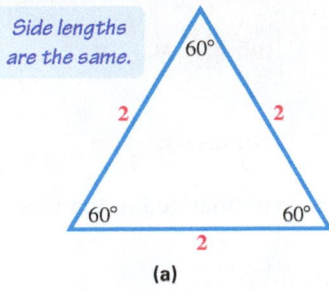

Side lengths are the same.

(a)

30°–60° Right Triangles

(b)

FIGURE 43

30°–60° Right Triangle

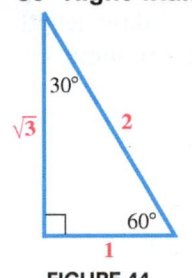

FIGURE 44

Trigonometric Function Values of Special Angles

Certain special angles, such as 30°, 45°, and 60°, occur so often that they deserve special study. We start with an equilateral triangle—a triangle with all sides of equal length. Each angle of such a triangle measures 60°. While the results we will obtain are independent of the lengths of its sides, for convenience we choose the length of each side to be 2 units. See **FIGURE 43(a)**.

Bisecting one angle of this equilateral triangle leads to two right triangles, each of which has angles of 30°, 60°, and 90°, as shown in **FIGURE 43(b)**. Since the hypotenuse of one of these right triangles has length 2, the shortest side will have length 1. Let x represent the length of the longer leg.

$$2^2 = 1^2 + x^2 \qquad \text{Pythagorean theorem}$$
$$4 = 1 + x^2 \qquad \text{Apply the exponents.}$$
$$3 = x^2 \qquad \text{Subtract 1.}$$
$$\sqrt{3} = x \qquad \text{Square root property; choose the positive root.}$$

FIGURE 44 summarizes our results, using a 30°–60° right triangle. The side opposite the 30° angle has length 1. That is, for the 30° angle,

$$\text{hypotenuse} = 2, \quad \text{side opposite} = 1, \quad \text{side adjacent} = \sqrt{3}.$$

Now we use the definitions of the trigonometric functions.

$$\sin 30° = \frac{\text{side opposite}}{\text{hypotenuse}} = \frac{1}{2} \qquad\qquad \csc 30° = \frac{2}{1} = 2$$

$$\cos 30° = \frac{\text{side adjacent}}{\text{hypotenuse}} = \frac{\sqrt{3}}{2} \qquad\qquad \sec 30° = \frac{2}{\sqrt{3}} = \frac{2\sqrt{3}}{3}$$

$$\tan 30° = \frac{\text{side opposite}}{\text{side adjacent}} = \frac{1}{\sqrt{3}} = \frac{\sqrt{3}}{3} \qquad\qquad \cot 30° = \frac{\sqrt{3}}{1} = \sqrt{3}$$

EXAMPLE 2 **Finding Trigonometric Function Values for 60°**

Find the six trigonometric function values for a 60° angle.

Solution Refer to **FIGURE 44** to find the following ratios.

$$\sin 60° = \frac{\sqrt{3}}{2} \qquad \cos 60° = \frac{1}{2} \qquad \tan 60° = \sqrt{3}$$

$$\csc 60° = \frac{2\sqrt{3}}{3} \qquad \sec 60° = 2 \qquad \cot 60° = \frac{\sqrt{3}}{3}$$

45°–45° Right Triangle

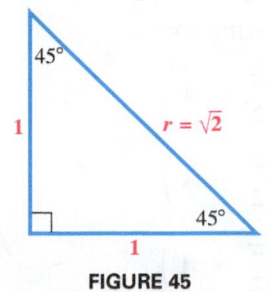

FIGURE 45

We find the values of the trigonometric functions for 45° by starting with a 45°–45° right triangle, as shown in **FIGURE 45**. This triangle is isosceles, and for convenience, we choose the lengths of the equal sides to be 1 unit. (As before, the results are independent of the length of the equal sides of the right triangle.) Since the shorter sides each have length 1, if r represents the length of the hypotenuse, then we have the following.

$$1^2 + 1^2 = r^2 \qquad \text{Pythagorean theorem}$$
$$2 = r^2 \qquad \text{Add.}$$
$$\sqrt{2} = r \qquad \text{Choose the positive root.}$$

45°–45° Right Triangle

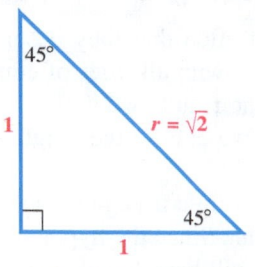

FIGURE 45 (repeated)

Now we use the measures indicated on the 45°–45° right triangle in **FIGURE 45.**

$$\sin 45° = \frac{1}{\sqrt{2}} = \frac{\sqrt{2}}{2} \qquad \cos 45° = \frac{1}{\sqrt{2}} = \frac{\sqrt{2}}{2} \qquad \tan 45° = \frac{1}{1} = 1$$

$$\csc 45° = \frac{\sqrt{2}}{1} = \sqrt{2} \qquad \sec 45° = \frac{\sqrt{2}}{1} = \sqrt{2} \qquad \cot 45° = \frac{1}{1} = 1$$

Function values for 30° or $\frac{\pi}{6}$, 45° or $\frac{\pi}{4}$, and 60° or $\frac{\pi}{3}$ are summarized in the table that follows.

Function Values for Special Angles

θ	$\sin \theta$	$\cos \theta$	$\tan \theta$	$\cot \theta$	$\sec \theta$	$\csc \theta$
30° or $\frac{\pi}{6}$	$\frac{1}{2}$	$\frac{\sqrt{3}}{2}$	$\frac{\sqrt{3}}{3}$	$\sqrt{3}$	$\frac{2\sqrt{3}}{3}$	2
45° or $\frac{\pi}{4}$	$\frac{\sqrt{2}}{2}$	$\frac{\sqrt{2}}{2}$	1	1	$\sqrt{2}$	$\sqrt{2}$
60° or $\frac{\pi}{3}$	$\frac{\sqrt{3}}{2}$	$\frac{1}{2}$	$\sqrt{3}$	$\frac{\sqrt{3}}{3}$	2	$\frac{2\sqrt{3}}{3}$

Trigonometry students should memorize the function values for the first two columns of this table. Identities will yield the remaining values.

Cofunction Identities

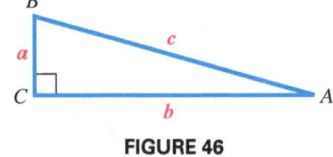

FIGURE 46

In a right triangle ABC with right angle C, the acute angles A and B are complementary. See **FIGURE 46.** The length of the side opposite angle A is a, and the length of the side opposite angle B is b. The length of the hypotenuse is c. In this triangle, $\sin A = \frac{a}{c}$ and $\cos B$ is also equal to $\frac{a}{c}$.

$$\sin A = \frac{a}{c} = \cos B$$

Similar reasoning yields the following.

$$\tan A = \frac{a}{b} = \cot B \quad \text{and} \quad \sec A = \frac{c}{b} = \csc B$$

Since angles A and B are complementary and $\sin A = \cos B$, the functions sine and cosine are called **cofunctions.** Also, tangent and cotangent are cofunctions, as are secant and cosecant. Since angles A and B are complementary, $A + B = 90°$, or $B = 90° - A$, giving $\sin A = \cos B = \cos(90° - A)$.

Similar results, called the **cofunction identities,** are true for the other trigonometric functions.

TECHNOLOGY NOTE

Using the TI-84 Plus or a scientific calculator to find the decimal value for cos 30° gives a decimal approximation for $\frac{\sqrt{3}}{2}$. The decimal is an approximation, while $\frac{\sqrt{3}}{2}$ is exact. In general, unless a trigonometric function value is a rational number, a basic graphing calculator will provide only a decimal approximation for the irrational function value. (More advanced models, such as the TI-89, do give exact values.)

Cofunction Identities

If A is an acute angle measured in *degrees,* then the following are true.

$$\sin A = \cos(90° - A) \quad \csc A = \sec(90° - A) \quad \tan A = \cot(90° - A)$$

$$\cos A = \sin(90° - A) \quad \sec A = \csc(90° - A) \quad \cot A = \tan(90° - A)$$

If A is an acute angle measured in *radians,* then the following are true.

$$\sin A = \cos\left(\frac{\pi}{2} - A\right) \qquad \csc A = \sec\left(\frac{\pi}{2} - A\right)$$

$$\cos A = \sin\left(\frac{\pi}{2} - A\right) \qquad \sec A = \csc\left(\frac{\pi}{2} - A\right)$$

$$\tan A = \cot\left(\frac{\pi}{2} - A\right) \qquad \cot A = \tan\left(\frac{\pi}{2} - A\right)$$

NOTE These identities actually apply to all angles for which the functions are defined (not just acute angles). However, for our present discussion, we will need them only for acute angles.

EXAMPLE 3   **Writing Functions in Terms of Cofunctions**

Write each expression in terms of its cofunction.

(a) $\cos 52° \, 16'$ **(b)** $\tan \dfrac{\pi}{6}$ **(c)** $\sec 1$

Solution

(a) Use the identity $\cos A = \sin(90° - A)$.

$$\cos 52° \, 16' = \sin(90° - 52° \, 16') = \sin 37° \, 44'$$

(b) Use the identity $\tan A = \cot\left(\dfrac{\pi}{2} - A\right)$.

$$\tan \frac{\pi}{6} = \cot\left(\frac{\pi}{2} - \frac{\pi}{6}\right) = \cot \frac{\pi}{3}$$

(c) $\sec 1 = \csc\left(\dfrac{\pi}{2} - 1\right)$

Reference Angles

Associated with every nonquadrantal angle in standard position is a positive acute angle called its *reference angle*. A **reference angle** for an angle θ, written $\boldsymbol{\theta'}$, is the *positive acute* angle made by the terminal side of angle θ and the x-axis.

The chart below shows several angles θ in quadrants II, III, and IV, with the reference angle θ' also shown.

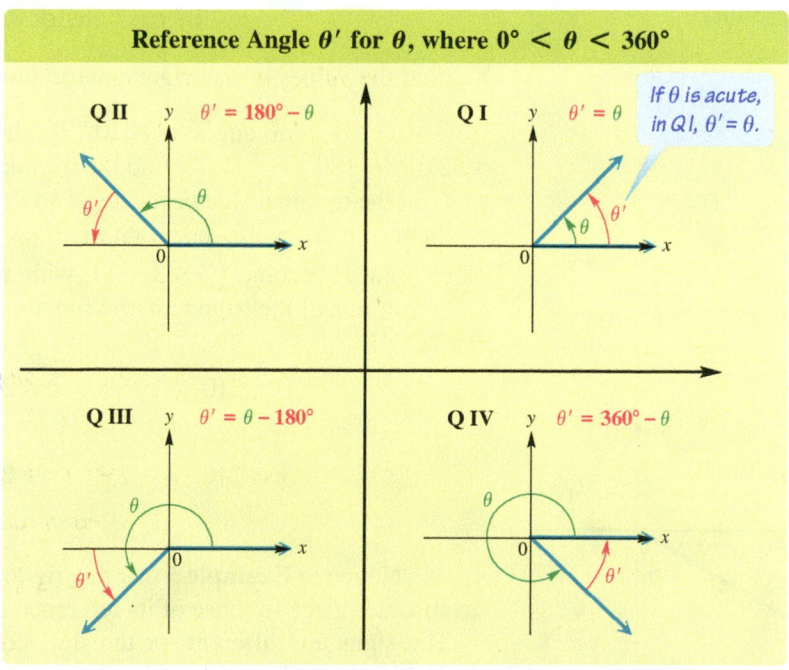

If an angle θ is negative or has measure greater than 360°, its reference angle is found by first finding its coterminal angle that is between 0° and 360° and then using the diagrams shown above.

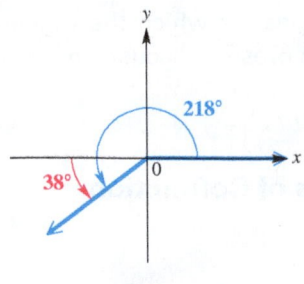

218° − 180° = 38°
FIGURE 47

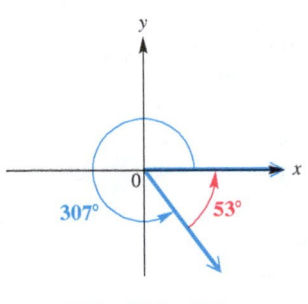

360° − 307° = 53°
FIGURE 48

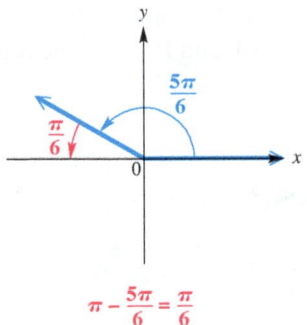

$\pi - \dfrac{5\pi}{6} = \dfrac{\pi}{6}$
FIGURE 49

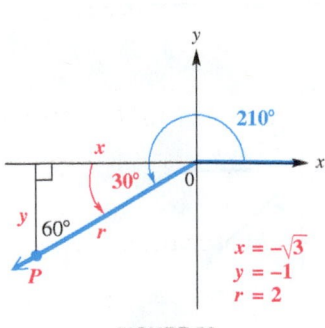

FIGURE 50

CAUTION A common error is to find the reference angle by using the terminal side of θ and the y-axis. *The reference angle is always found with reference to the x-axis.*

EXAMPLE 4 **Finding Reference Angles**

Find the reference angle for each angle.

(a) 218° **(b)** 1387° **(c)** $\dfrac{5\pi}{6}$

Solution

(a) As shown in **FIGURE 47,** the positive acute angle made by the terminal side of this angle and the x-axis is $218° − 180° = 38°$. For $\theta = 218°$, the reference angle is $\theta' = 38°$.

(b) First find a coterminal angle to 1387° between 0° and 360°. Divide 1387° by 360° to get a quotient of about 3.9. Begin by subtracting 360° three times (because of the 3 in 3.9).

$$1387° − 3 \cdot 360° = 307°$$

The reference angle for 307° (and thus for 1387°) is $360° − 307° = 53°$. See **FIGURE 48.**

(c) The reference angle is $\pi - \dfrac{5\pi}{6} = \dfrac{\pi}{6}$. See **FIGURE 49.** ●

Special Angles as Reference Angles

EXAMPLE 5 **Finding Trigonometric Function Values of an Angle**

Find the values of the trigonometric functions for 210°.

Solution An angle of 210° is shown in **FIGURE 50.** The reference angle is $210° − 180° = 30°$. To find the trigonometric function values of 210°, choose point P on the terminal side of the angle so that the distance from the origin O to P is 2. By the results from 30°–60° right triangles (see **FIGURE 44** in this section), the coordinates of point P become $\left(-\sqrt{3}, -1\right)$, with $x = -\sqrt{3}$, $y = -1$, and $r = 2$. Then, by the definitions of the trigonometric functions,

$$\sin 210° = -\frac{1}{2} \qquad \cos 210° = -\frac{\sqrt{3}}{2} \qquad \tan 210° = \frac{\sqrt{3}}{3}$$

$$\csc 210° = -2 \qquad \sec 210° = -\frac{2\sqrt{3}}{3} \qquad \cot 210° = \sqrt{3}. \qquad ●$$

Notice in **Example 5** that the trigonometric function values of 210° correspond in absolute value to those of its reference angle 30°. See the table earlier in this section. The signs are different for the sine, cosine, secant, and cosecant functions because 210° is a quadrant III angle. By using the trigonometric function values of 30° and choosing the correct signs for a quadrant III angle, we can obtain the same results. These results suggest a method for finding the trigonometric function values of a non-acute angle, using the reference angle.

> **Finding Trigonometric Function Values for a Nonquadrantal Angle θ**
>
> *Step 1* If $\theta > 360°$, or if $\theta < 0°$, then find a coterminal angle by adding or subtracting 360° as many times as needed to obtain an angle greater than 0° but less than 360°.
>
> *Step 2* Find the reference angle θ'.
>
> *Step 3* Find the trigonometric function values for reference angle θ'.
>
> *Step 4* Determine the correct signs for the values found in Step 3. (Use the table of signs in **Section 8.2** if necessary.) This gives the values of the trigonometric functions for angle θ.

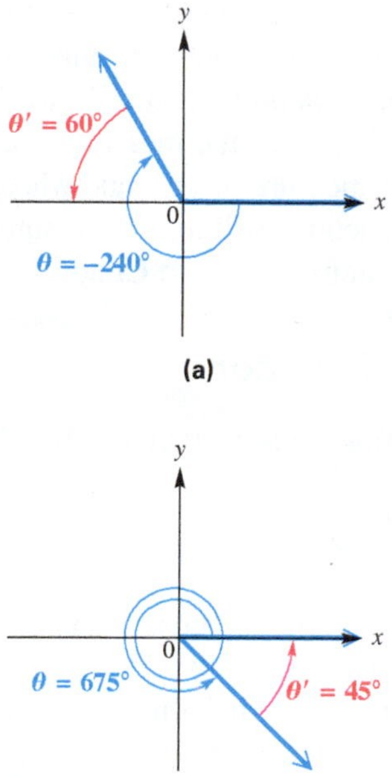

(a)

(b)

FIGURE 51

EXAMPLE 6   **Finding Trigonometric Function Values by Using Reference Angles**

Find the exact value of each expression.

(a) $\cos(-240°)$ **(b)** $\tan 675°$

Solution

(a) Since an angle of $-240°$ is coterminal with an angle of

$$-240° + 360° = 120°,$$

the reference angle is $180° - 120° = 60°$, as shown in **FIGURE 51(a).** The cosine is negative in quadrant II.

Make result negative.

$$\cos(-240°) = \cos 120° = -\cos 60° = -\frac{1}{2}$$

(b) Begin by subtracting 360° to get a coterminal angle between 0° and 360°.

$$675° - 360° = 315°$$

As shown in **FIGURE 51(b),** the reference angle is $360° - 315° = 45°$. An angle of 315° is in quadrant IV, so the tangent will be negative.

Make result negative.

$$\tan 675° = \tan 315° = -\tan 45° = -1$$

Finding Function Values with a Calculator

Calculators are capable of finding trigonometric function values. (We saw this for quadrantal angles in **Section 8.2.**) For example, the values of $\cos(-240°)$ and $\tan 675°$, found analytically in **Example 6,** are found with a calculator as shown in **FIGURE 52.**

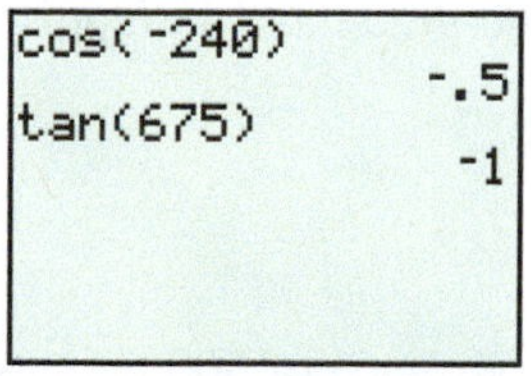

Degree mode

FIGURE 52

> **CAUTION**   *When evaluating trigonometric functions of angles given in degrees, remember that the calculator must be set in degree mode.*

EXAMPLE 7   **Approximating Trigonometric Function Values with a Calculator**

Approximate the value of each expression.

(a) $\cos 49° 12'$ **(b)** $\csc 197.977°$ **(c)** $\cot 51.4283°$ **(d)** $\sin 30$

(continued)

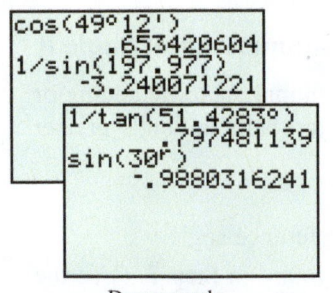

Degree mode

FIGURE 53

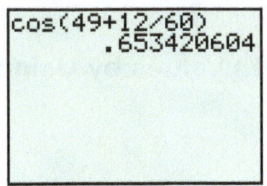

Degree mode

FIGURE 54

Solution The top screen in **FIGURE 53** shows how to approximate the expressions cos 49°12′ and csc 197.977° from parts **(a)** and **(b).** The calculator is in degree mode. Notice that to find a cosecant value, we take the reciprocal of the sine value.

The bottom screen in **FIGURE 53** shows how the TI-84 Plus approximates the expressions cot 51.4283° and sin 30 from parts **(c)** and **(d).** Here, we use the degree symbol (°) and the radian symbol (ʳ), which overrides the mode of the calculator. ●

NOTE An alternative method of finding a value such as cos 49° 12′ is to use the fact that 60′ = 1°. See **FIGURE 54.**

$$\cos 49° 12' = \cos\left(49 + \frac{12}{60}\right)^° \approx 0.653420604$$

Finding Angle Measures

Sometimes we need to find the measure of an angle having a certain trigonometric function value. Graphing calculators have three *inverse trigonometric functions* (denoted **sin⁻¹, cos⁻¹,** and **tan⁻¹**). If x is an appropriate number, then $\sin^{-1} x$, gives the measure of an angle whose sine is x. Similarly $\cos^{-1} x$ and $\tan^{-1} x$ give the measure of an angle whose cosine and tangent, respectively, is x. If x is positive, these functions will return measures of acute angles. (A more complete discussion of inverse functions follows in **Chapter 9.**)

EXAMPLE 8 **Using Inverse Trigonometric Functions**

(a) Use a calculator to find an angle θ in degrees that satisfies $\sin \theta \approx 0.9677091705$.

(b) Use a calculator to find an angle θ in radians that satisfies $\tan \theta \approx 0.25$.

Solution

Degree mode →
Radian mode →

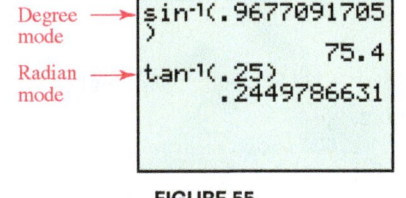

FIGURE 55

(a) With the calculator in *degree mode,* we find that an angle θ having approximate sine value 0.9677091705 is 75.4°. (While there are infinitely many such angles, the calculator gives only this one.) We write this result as $\sin^{-1} 0.9677091705 \approx 75.4°$. See the first calculation in **FIGURE 55.**

(b) With the calculator in *radian mode,* we find $\tan^{-1} 0.25 \approx 0.2449786631$. See the second calculation in **FIGURE 55.** Note that inverse trigonometric functions can calculate angles in degrees or radians. ●

Grade resistance is the force F that causes a car to roll down a hill. It can be calculated by $F = W \sin \theta$, where θ represents the angle of the grade and W represents the weight of the vehicle. (See **FIGURE 39** in **Section 8.2.**)

EXAMPLE 9 **Calculating Highway Grade**

Find the angle θ for which a 3000-pound car has a grade resistance of 500 pounds.

Solution

$$F = W \sin \theta$$

$$500 = 3000 \sin \theta \qquad \text{Let } F = 500 \text{ and } W = 3000.$$

$$\sin \theta = \frac{1}{6} \qquad \text{Divide by 3000 and rewrite.}$$

Use degree mode here.

$$\theta = \sin^{-1} \frac{1}{6} \qquad \text{Use the inverse sine function.}$$

$$\theta \approx 9.6° \qquad \text{Approximate.}$$

Thus, if a road is inclined at approximately 9.6°, a 3000-pound car would be acted on by a force of 500 pounds pulling downhill. ●

> **EXAMPLE 10**   **Finding Angle Measures**
>
> Find all values of θ if θ is in the interval $[0°, 360°)$ and $\cos\theta = -\dfrac{\sqrt{2}}{2}$.

Analytic Solution

Since $\cos\theta$ is negative, θ must lie in quadrant II or III. Because the absolute value of $\cos\theta$ is $\dfrac{\sqrt{2}}{2}$, the reference angle θ' is $\cos^{-1}\dfrac{\sqrt{2}}{2} = 45°$. The two possible angles θ are sketched in **FIGURE 56**.

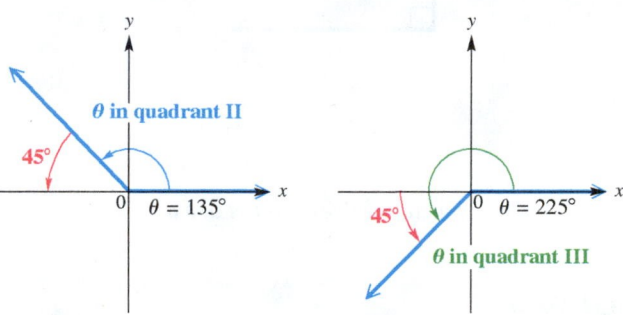

FIGURE 56

The quadrant II angle θ must equal $180° - 45° = 135°$, and the quadrant III angle θ must equal $180° + 45° = 225°$.

Graphing Calculator Solution

The screen in **FIGURE 57** shows how the inverse cosine function is used to find the two values in $[0°, 360°)$ for which $\cos\theta = -\dfrac{\sqrt{2}}{2}$. Notice that $\cos^{-1}\left(-\dfrac{\sqrt{2}}{2}\right)$ yields only one value: $135°$. To find the other value, we use the reference angle, $45°$.

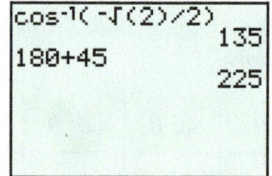

Degree mode

FIGURE 57

> **EXAMPLE 11**   **Finding Angle Measures**
>
> Find two angles in the interval $[0, 2\pi)$ that satisfy $\cos\theta \approx 0.3623577545$.
>
> **Solution**   With the calculator in radian mode, we find that one such θ is
> $$\theta = \cos^{-1} 0.3623577545 \approx 1.2. \quad \text{\textcolor{blue}{θ in radians}}$$
> Since $0 < 1.2 < \dfrac{\pi}{2}$, θ is in quadrant I. When θ is in quadrant I, it will equal its reference angle θ'. Therefore θ must have a reference angle equal to 1.2 and must be in quadrant IV, since the cosine is also positive in quadrant IV. The other value of θ is found as follows.
> $$\theta = 2\pi - 1.2 \approx 5.083185307$$

TECHNOLOGY NOTE

Do not confuse the symbols for the inverse trigonometric functions with the reciprocal functions. For example, ***sin^{-1} x represents an angle whose sine is x, not the reciprocal of sin x (which is csc x).*** To find reciprocal function values on the TI-84 Plus, use the function together with the reciprocal function of the calculator, which is labeled x^{-1}.

> ### WHAT WENT WRONG?
>
> A student was asked to find $\cot 20°$. He set his calculator in degree mode and, knowing that cotangent is the reciprocal of tangent, produced the screen shown here. The correct answer is $\cot 20° \approx 2.747477419$.
>
> ```
> tan⁻¹(20)
>            87.13759477
> ```
>
> **What Went Wrong?**   How can he obtain the correct answer?

Answer to What Went Wrong?

The student used the *inverse tangent function* ($\tan^{-1}$) rather than finding the *reciprocal* of the tangent of 20°. If he enters $1/\tan(20°)$, or $(\tan(20°))^{-1}$, he will obtain the correct answer.

8.3 Exercises

Checking Analytic Skills *Find exact values or expressions for the six trigonometric functions of angle A.* **Do not use a calculator.**

1.
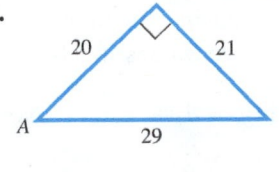

2.
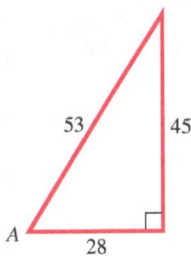

3.
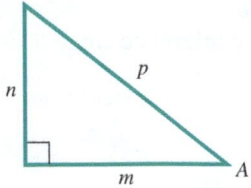

4.
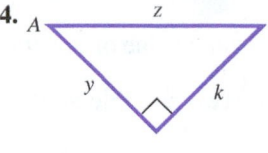

Checking Analytic Skills *Complete the table with exact trigonometric function values.* **Do not use a calculator.**

	θ	$\sin\theta$	$\cos\theta$	$\tan\theta$	$\cot\theta$	$\sec\theta$	$\csc\theta$
5.	**30°**	$\frac{1}{2}$	$\frac{\sqrt{3}}{2}$			$\frac{2\sqrt{3}}{3}$	2
6.	**45°**			1	1		
7.	**60°**		$\frac{1}{2}$	$\sqrt{3}$		2	
8.	**120°**	$\frac{\sqrt{3}}{2}$		$-\sqrt{3}$			$\frac{2\sqrt{3}}{3}$
9.	**135°**	$\frac{\sqrt{2}}{2}$	$-\frac{\sqrt{2}}{2}$			$-\sqrt{2}$	$\sqrt{2}$
10.	**150°**		$-\frac{\sqrt{3}}{2}$	$-\frac{\sqrt{3}}{3}$			2
11.	**210°**	$-\frac{1}{2}$		$\frac{\sqrt{3}}{3}$	$\sqrt{3}$		-2
12.	**240°**	$-\frac{\sqrt{3}}{2}$	$-\frac{1}{2}$			-2	$-\frac{2\sqrt{3}}{3}$

For each expression, **(a)** *give the exact value and* **(b)** *if the exact value is irrational, use your calculator to support your answer in part (a) by finding a decimal approximation.*

13. $\tan 30°$

14. $\cot 30°$

15. $\sin 30°$

16. $\cos 30°$

17. $\sec 30°$

18. $\csc 30°$

19. $\csc 45°$

20. $\sec 45°$

21. $\cos 45°$

22. $\cot 45°$

23. $\sin\dfrac{\pi}{3}$

24. $\cos\dfrac{\pi}{3}$

25. $\tan\dfrac{\pi}{3}$

26. $\cot\dfrac{\pi}{3}$

27. $\csc\dfrac{\pi}{6}$

28. $\sec\dfrac{\pi}{3}$

29. $\csc\dfrac{\pi}{3}$

30. $\cot\dfrac{\pi}{4}$

31. A student was asked to give the exact value of sin 45°. Using his calculator, he gave the answer 0.7071067812. The teacher did not give him credit. Why?

 32. A student was asked to give an approximate value of sin 45. With her calculator in degree mode, she gave the value 0.7071067812. The teacher did not give her credit. What was her error?

Write each expression in terms of its cofunction.

33. cot 73°   **34.** sec 39°   **35.** sin 38°   **36.** cos 19°

37. tan 25° 43′   **38.** sin 38° 29′   **39.** $\cos \dfrac{\pi}{5}$ **40.** $\sin \dfrac{\pi}{3}$

41. tan 0.5   **42.** csc 0.3   **43.** cos 1   **44.** $\sin \dfrac{1}{2}$

Give the reference angle for each angle measure.

45. 98°   **46.** 212°   **47.** 230°   **48.** 130°   **49.** −135°   **50.** −60°

51. 750°   **52.** 480°   **53.** $\dfrac{4\pi}{3}$ **54.** $\dfrac{7\pi}{6}$ **55.** $-\dfrac{4\pi}{3}$ **56.** $-\dfrac{7\pi}{6}$

 **57.** *Concept Check*   In **Example 5,** why was 2 a good choice for *r*? Could any other positive number be used?

 58. Explain how the reference angle is used to find values of the trigonometric functions for an angle in quadrant III.

Checking Analytic Skills   *Find exact values of the six trigonometric functions for each angle by hand.* **Do not use a calculator.**

59. 300°   **60.** 315°   **61.** 405°

62. 420°   **63.** $\dfrac{11\pi}{6}$ **64.** $\dfrac{5\pi}{3}$

65. $-\dfrac{7\pi}{4}$ **66.** $-\dfrac{4\pi}{3}$ **67.** $-\dfrac{19\pi}{6}$

68. *Concept Check*   Refer to **Exercise 59.** What is the next greater degree-measured angle whose trigonometric functions correspond to the answers in that exercise?

Use a calculator to find a decimal approximation for each value. Give as many digits as your calculator displays.

69. tan 29°   **70.** sin 38°   **71.** cot 41° 24′

72. csc 145° 45′   **73.** sec 183° 48′   **74.** cos 421° 30′

75. tan (−80° 6′)   **76.** sin (−317° 36′)   **77.** sin 2.5

78. cos 3.8   **79.** tan 5   **80.** sec 10

For each expression, (**a**) *write the function in terms of a function of the reference angle,* (**b**) *give the exact value, and* (**c**) *use a calculator to show that the decimal value or approximation for the given function is the same as the decimal value or approximation for your answer in part (b).*

81. $\sin \dfrac{7\pi}{6}$ **82.** $\cos \dfrac{5\pi}{3}$

83. $\tan \dfrac{3\pi}{4}$ **84.** $\sin \dfrac{5\pi}{3}$

85. $\cos \dfrac{7\pi}{6}$ **86.** $\tan \dfrac{4\pi}{3}$

Checking Analytic Skills *Find all values of θ if θ is in the interval $[0°, 360°)$ and has the given function value.* **Do not use a calculator.**

87. $\sin \theta = \dfrac{1}{2}$

88. $\cos \theta = \dfrac{\sqrt{3}}{2}$

89. $\tan \theta = -\sqrt{3}$

90. $\sec \theta = -\sqrt{2}$

91. $\cot \theta = -\dfrac{\sqrt{3}}{3}$

92. $\cos \theta = \dfrac{\sqrt{2}}{2}$

Find all values of θ if θ is in the interval $[0°, 360°)$ and has the given function value. Give calculator approximations to as many decimal places as your calculator displays.

93. $\cos \theta \approx 0.68716510$

94. $\cos \theta \approx 0.96476120$

95. $\sin \theta \approx 0.41298643$

96. $\sin \theta \approx 0.63898531$

97. $\tan \theta \approx 0.87692035$

98. $\tan \theta \approx 1.2841996$

Find two angles in the interval $[0, 2\pi)$ that satisfy the given equation. Give calculator approximations to as many digits as your calculator displays.

99. $\tan \theta \approx 0.21264138$

100. $\cos \theta \approx 0.78269876$

101. $\cot \theta \approx 0.29949853$

102. $\csc \theta \approx 1.0219553$

RELATING CONCEPTS **For individual or group investigation (Exercises 103–108)**

In a square window of your calculator that gives a good picture of the first quadrant, graph the line $y = \sqrt{3}x$ with $x \geq 0$. Then, trace to any point on the line. See the figure. What we see is a simulated view of an angle in standard position, with terminal side in quadrant I. Store the values of x and y in convenient memory locations, and call them x_1 and y_1. **Work Exercises 103–108 in order.**

103. Calculate the value of $\sqrt{x_1^2 + y_1^2}$ and store it in a convenient memory location. (Call it r.) What does this number mean geometrically?

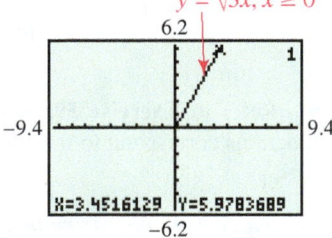

104. With your calculator in degree mode, find $\tan^{-1}\left(\dfrac{y_1}{x_1}\right)$.

105. With your calculator in degree mode, find $\sin^{-1}\left(\dfrac{y_1}{r}\right)$.

106. With your calculator in degree mode, find $\cos^{-1}\left(\dfrac{x_1}{r}\right)$.

107. Your answers in **Exercises 104–106** should all be the same. How does this answer relate to the angle formed between the positive x-axis and the line?

108. Look at the equation of the line you graphed, and make a conjecture: The _____ of a line passing through the origin is equal to the _____ of the angle it forms with the positive x-axis.

Highway Grade Exercises 109 and 110 refer to **Example 9**. Use the formula $F = W \sin \theta$ to find the angle θ for which a car with weight W has grade resistance F.

109. $W = 5000$ lb, $F = 400$ lb

110. $W = 3500$ lb, $F = 130$ lb

(Modeling) Speed of Light When a light ray travels from one medium, such as air, to another medium, such as water or glass, the speed of the light and the direction in which the ray

is traveling change. (This is why a fish under water is in a different position than it appears to be.) The changes are given by Snell's law,

$$\frac{c_1}{c_2} = \frac{\sin \theta_1}{\sin \theta_2},$$

where c_1 is the speed of light in the first medium, c_2 is the speed of light in the second medium, and θ_1 and θ_2 are the angles shown in the figure on the next page. In Exercises 111–116, assume that $c_1 = 3 \times 10^8$ meters per second.

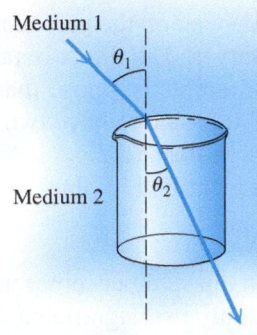

Medium 1

θ_1

If this medium is less dense, light travels at a faster speed c_1.

If this medium is more dense, light travels at a slower speed c_2.

Medium 2

θ_2

Approximate the speed of light in the second medium.

111. $\theta_1 = 46°; \theta_2 = 31°$ **112.** $\theta_1 = 39°; \theta_2 = 28°$

Find θ_2 for the following values of θ_1 and c_2. Round to the nearest degree.

113. $\theta_1 = 40°; c_2 = 1.5 \times 10^8$ meters per second

114. $\theta_1 = 62°; c_2 = 2.6 \times 10^8$ meters per second

(Modeling) **Fish's View of the World** *The figure shows a fish's view of the world above the surface of the water. (Source: Walker, Jearl, "The Amateur Scientist," Scientific American, March 1984. Illustration by Michael Goodman.) Suppose that a light ray comes from the horizon, enters the water, and strikes the fish's eye.*

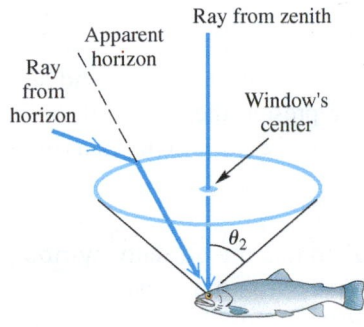

Ray from zenith

Apparent horizon

Ray from horizon

Window's center

θ_2

115. Assume that this ray from the horizon gives a value of 90.0° for angle θ_1 in the formula for Snell's law. (In a practical situation, the angle might be a little less than 90°.) The speed of light in water is about 2.254×10^8 meters per second. Find angle θ_2 to the nearest tenth of a degree.

116. Suppose an object is located at a true angle of 29.6° above the horizon. Use the results from **Exercise 115** to find the apparent angle above the horizon to a fish to the nearest tenth of a degree.

117. *(Modeling)* **Braking Distance** If air resistance is ignored, the braking distance D (in feet) for an automobile to change its velocity from V_1 to V_2 (feet per second) can be modeled by the equation

$$D = \frac{1.05\left(V_1^2 - V_2^2\right)}{64.4(K_1 + K_2 + \sin \theta)}.$$

K_1 is a constant determined by the efficiency of the brakes and tires, K_2 is a constant determined by the rolling resistance of the automobile, and θ is the grade of the highway. (*Source:* Mannering, F. and W. Kilareski, *Principles of Highway Engineering and Traffic Analysis,* Second Edition, John Wiley and Sons.)

(a) Approximate the number of feet required to slow a car from 55 to 30 mph while traveling uphill on a grade of $\theta = 3.5°$. Let $K_1 = 0.4$ and $K_2 = 0.02$. (*Hint:* Change miles per hour to feet per second.)

(b) Repeat part (a) with $\theta = -2°$.

(c) How is the braking distance affected by the grade θ? Does this agree with your driving experience?

118. *(Modeling)* **Car's Speed at Collision** Refer to **Exercise 117.** An automobile is traveling at 90 mph on a highway with a downhill grade of $\theta = -3.5°$. The driver sees a stalled truck in the road 200 feet away and immediately applies the brakes. Assuming that a collision cannot be avoided, how fast (in miles per hour) is the car traveling when it hits the truck? (Use the same values for K_1 and K_2 as in **Exercise 117.**)

8.4 Applications of Right Triangles

Significant Digits • Solving Triangles • Angles of Elevation or Depression • Bearing • Further Applications of Trigonometric Functions

Significant Digits

A number that represents the result of counting, or a number that results from theoretical work and is not the result of measurement, is an **exact number.** There are 50 states in the United States, so in that statement, 50 is an exact number. Another example of an exact number is $\cos 30° = \dfrac{\sqrt{3}}{2}$.

In routine measurements, however, results are seldom exact. The results of physical measurements are only approximately accurate and depend on the precision of the measuring instrument as well as the aptness of the observer. The digits obtained by actual measurement are called **significant digits.** For example, suppose that a wall about 18 ft long is to be measured. The measurement 18 ft is said to have two significant digits, whereas the measurement 18.3 ft has three significant digits.

The following numbers have their significant digits identified in blue.

408   21.5   18.00   6.700   0.0025   0.09810   7300

Notice that 18.00 has four significant digits. The zeros in this number represent measured digits and are significant. The number 0.0025 has only two significant digits, 2 and 5, because the zeros here are used only to locate the decimal point. The number 7300 causes some confusion because it is impossible to determine whether the zeros are measured values. The number 7300 may have two, three, or four significant digits. When presented with this situation, we assume that the zeros are not significant, unless the context of the problem indicates otherwise.

To determine the number of significant digits for answers in applications of angle measure, use the following table.

Angle Measure to Nearest	Examples	Number of Significant Digits for Answer
Degree	62°, 36°	two
Ten minutes, or nearest tenth of a degree	52° 30′, 60.4°	three
Minute, or nearest hundredth of a degree	81° 48′, 71.25°	four
Ten seconds, or nearest thousandth of a degree	10° 52′ 20″, 21.264°	five

Standard Labeling

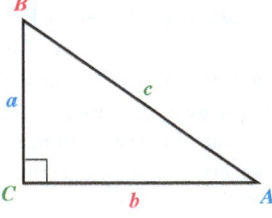

In solving triangles, a labeled sketch is an important aid.

FIGURE 58

Solving Triangles

To *solve a triangle* means to find the measures of all the angles and sides of the triangle. As shown in **FIGURE 58,** we use a to represent the length of the side opposite angle A, b for the length of the side opposite angle B, and so on. In a right triangle, the letter c is reserved for the hypotenuse.

> **NOTE** In solving triangles, we will often use the equality symbol, $=$, despite the fact that the values shown are actually approximations.

EXAMPLE 1 **Solving a Right Triangle Given an Angle and a Side**

Solve right triangle ABC, with $A = 34° 30′$ and $c = 12.7$ inches. See **FIGURE 59.**

Solution To solve the triangle, find the measures of the remaining sides and angles. To find the value of a, use a trigonometric function involving the known values of angle A and side c. Since the sine of angle A is given by the quotient of the side opposite A and the hypotenuse, use sin A.

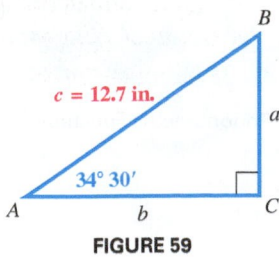

FIGURE 59

$$\sin A = \frac{a}{c} \qquad \sin A = \frac{\text{side opposite}}{\text{hypotenuse}}$$

$$\sin 34° 30′ = \frac{a}{12.7} \qquad A = 34° 30′, c = 12.7$$

$$a = 12.7 \sin 34° 30′ \qquad \text{Multiply by 12.7 and rewrite.}$$

$$a = 12.7(0.56640624) \qquad \text{Use a calculator.}$$

$$a = 7.19 \text{ inches} \qquad \text{Three significant digits}$$

To find the value of b, we could use the Pythagorean theorem. It is better, however, to use the information given in the problem rather than a result just calculated. If a mistake is made in finding a, then b also would be incorrect. Moreover, rounding more than once may cause the result to be less accurate. Use $\cos A$.

$$\cos A = \frac{b}{c} \qquad \text{cos } A = \frac{\text{side adjacent}}{\text{hypotenuse}}$$

$$\cos 34° 30' = \frac{b}{12.7} \qquad A = 34° 30'; c = 12.7$$

$$b = 12.7 \cos 34° 30' \qquad \text{Multiply by 12.7 and rewrite.}$$

$$b = 10.5 \text{ inches} \qquad \text{Three significant digits}$$

Once b is found, the Pythagorean theorem can be used as a check. All that remains to solve triangle ABC is to find the measure of angle B. Since $A + B = 90°$,

$$B = 90° - A = 89° 60' - 34° 30' = 55° 30'. \qquad ●$$

> **NOTE**   The process of solving a right triangle can usually be done in several ways. *To maintain accuracy, always use given information as much as possible, and avoid rounding off in intermediate steps.*

EXAMPLE 2   **Solving a Right Triangle Given Two Sides**

Solve right triangle ABC if $a = 29.43$ centimeters and $c = 53.58$ centimeters.

Solution   We draw a sketch showing the given information, as in **FIGURE 60.** One way to begin is to find angle A by using the sine function.

$$\sin A = \frac{\text{side opposite}}{\text{hypotenuse}} = \frac{29.43}{53.58}$$

Using a calculator, we find that $A = \sin^{-1} \frac{29.43}{53.58} \approx 33.32°$. The measure of B is approximately $90° - 33.32° = 56.68°$. We now find b from the Pythagorean theorem.

$$b^2 = c^2 - a^2 \qquad \text{Pythagorean theorem solved for } b^2$$

$$b^2 = 53.58^2 - 29.43^2 \qquad c = 53.58, a = 29.43$$

$$b = 44.77 \text{ centimeters} \qquad \text{Subtract and take square roots.} \qquad ●$$

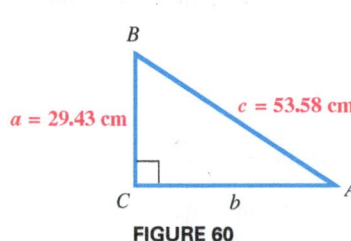

FIGURE 60

B

$a = 29.43$ cm

$c = 53.58$ cm

C b A

Angles of Elevation or Depression

Many applications of right triangles involve angles of elevation or depression. The **angle of elevation** from point X to point Y (above X) is the acute angle formed by ray XY and a horizontal ray with endpoint at X. See **FIGURE 61.** The **angle of depression** from point X to point Y (below X) is the acute angle formed by ray XY and a horizontal ray with endpoint X. See **FIGURE 62.**

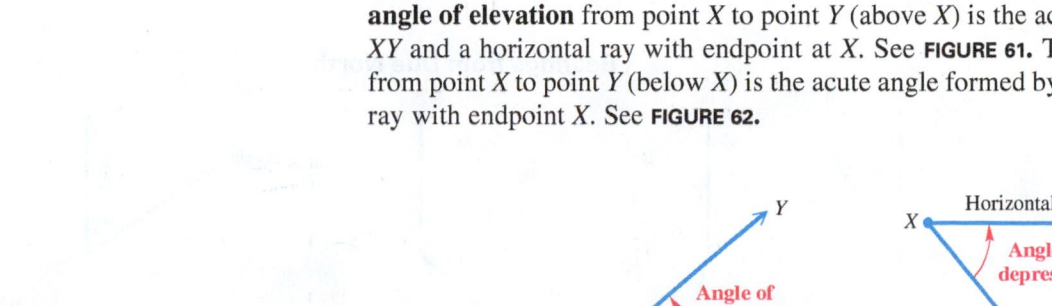

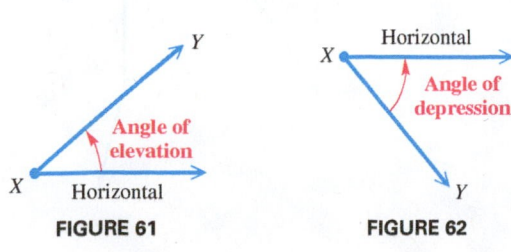

FIGURE 61

FIGURE 62

CAUTION Be careful when interpreting the angle of depression. *Both the angle of elevation and the angle of depression are measured between the line of sight and a horizontal line.*

To solve applied trigonometry problems, we often solve triangles.

Solving an Applied Trigonometry Problem

Step 1 Draw a sketch, and label it with the given information. Label the quantity to be found with a variable.

Step 2 Use the sketch to write an equation relating the given quantities to the variable.

Step 3 Solve the equation, and check that your answer makes sense.

EXAMPLE 3 **Finding the Angle of Elevation When Lengths Are Known**

The length of the shadow of a building 34.09 meters tall is 37.62 meters. Find the angle of elevation of the sun.

Solution As shown in **FIGURE 63**, the angle of elevation of the sun is angle *B*. Since the side opposite *B* and the side adjacent to *B* are known, use the tangent ratio to find *B*.

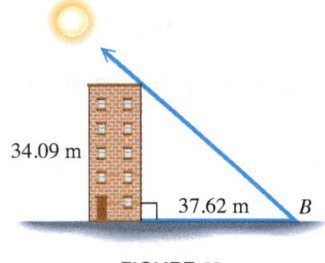

34.09 m
37.62 m   *B*

FIGURE 63

$$\tan B = \frac{34.09}{37.62}$$

$$B = \tan^{-1} \frac{34.09}{37.62} \qquad \text{Use the inverse tangent function.}$$

$$B = 42.18° \qquad \text{Use a calculator in degree mode.}$$

The angle of elevation of the sun is 42.18°.

Bearing

Other applications of right triangles involve **bearing,** an important idea in navigation. There are two methods for expressing bearing. When a single angle is given, such as 164°, it is understood that the bearing is measured in a clockwise direction from due north. Several sample bearings using this first method are shown in **FIGURE 64.**

Bearings from Due North

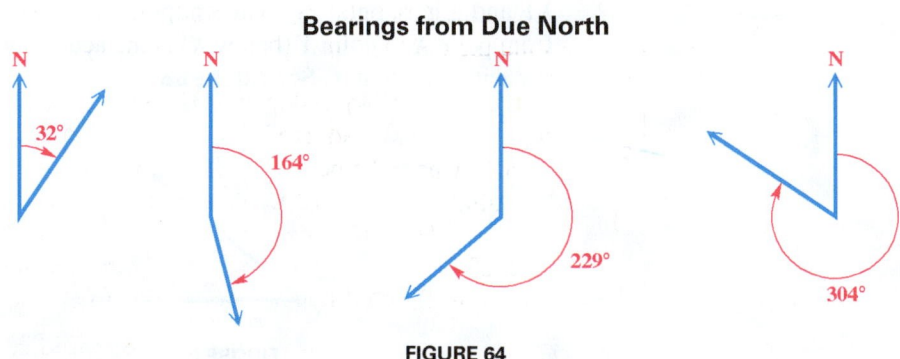

N   32°

N   164°

N   229°

N   304°

FIGURE 64

EXAMPLE 4 **Solving a Problem Involving Bearing (First Method)**

Radar stations A and B are on an east–west line, with A west of B, 3.70 kilometers apart. Station A detects a plane at C, on a bearing of 61.0°. Station B simultaneously detects the same plane, on a bearing of 331.0°. Find the distance from A to C.

Solution Draw a sketch showing the given information, as in **FIGURE 65.** Since a line drawn due north is perpendicular to an east–west line, right angles are formed at A and B, so angles CAB and CBA can be found. Angle C is a right angle because angles CAB and CBA are complementary. Find distance b by using the cosine function.

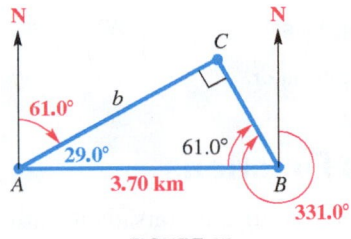

FIGURE 65

$$\cos 29.0° = \frac{b}{3.70}$$

$$3.70 \cos 29.0° = b \qquad \text{Multiply by 3.70.}$$

$$b = 3.24 \text{ kilometers} \qquad \text{Use a calculator.}$$

CAUTION *The importance of a correctly labeled sketch when solving applications like that in Example 4 cannot be overemphasized.* Some of the necessary information is often not stated directly in the problem and can be determined only from the sketch.

The second method for expressing bearing starts with a north–south line and uses an acute angle to show the direction, either east or west, from this line. **FIGURE 66** shows several sample bearings using this system. Either N or S *always* comes first, followed by an acute angle, and then E or W.

Bearings from North–South Line

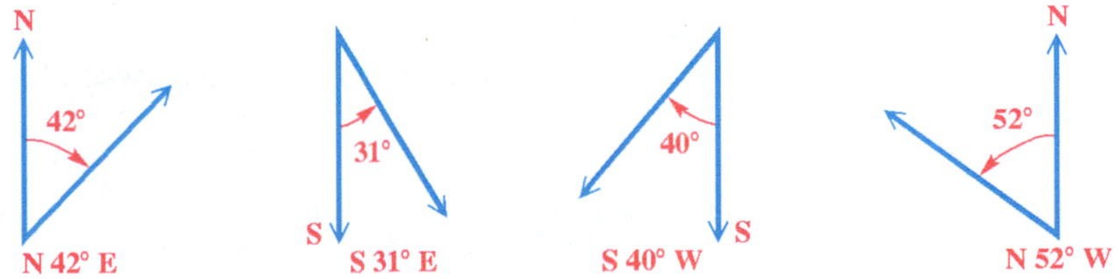

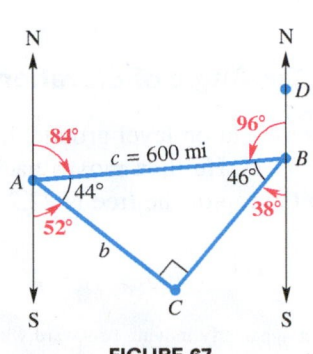

FIGURE 66

EXAMPLE 5 **Solving a Problem Involving Bearing (Second Method)**

The bearing from A to C is S 52° E. The bearing from A to B is N 84° E. The bearing from B to C is S 38° W. A plane flying at 250 mph takes 2.4 hours to go from A to B. Find the distance from A to C.

Solution Make a sketch. First draw the two bearings from point A. Then choose a point B on the bearing N 84° E from A, and draw the bearing to C. Point C will be located where the bearing lines from A and B intersect, as shown in **FIGURE 67.**

Since the bearing from A to B is N 84° E, angle ABD is $180° - 84° = 96°$. Thus, angle ABC is 46°. Also, angle BAC is $180° - (84° + 52°) = 44°$. Angle C is $180° - (44° + 46°) = 90°$. From the statement of the problem, a plane flying at 250 mph takes 2.4 hours to go from A to B. The distance from A to B is

FIGURE 67

$$c = \text{rate} \times \text{time} = 250(2.4) = 600 \text{ miles.}$$

(continued)

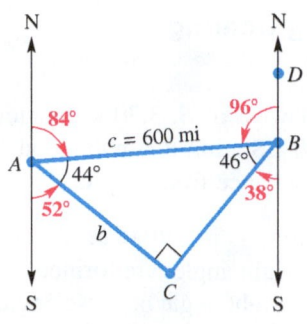

FIGURE 67 (repeated)

To find b, the distance from A to C, use the sine function. (The cosine function could also be used.)

$$\sin 46° = \frac{b}{c}$$

$$\sin 46° = \frac{b}{600} \qquad \text{Let } c = 600.$$

$$600 \sin 46° = b \qquad \text{Multiply by 600.}$$

$$b = 430 \text{ miles} \qquad \text{Use a calculator.}$$

Further Applications of Trigonometric Functions

Distance to a Star

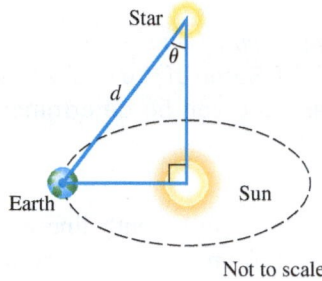

Not to scale

FIGURE 68

For centuries, astronomers wanted to know how far it was to the stars. Not until 1838 did the astronomer Friedrich Bessel determine the distance to a star called 61 Cygni. He used a *parallax** method that relied on the measurement of small angles. See **FIGURE 68.** As Earth revolves around the sun, the observed parallax of 61 Cygni is $\theta \approx 0.0000811°$. Because stars are so distant, parallax angles are very small. (*Source:* Freebury, H., *A History of Mathematics,* MacMillan; Zeilik, M. et al., *Introductory Astronomy and Astrophysics,* Third Edition, Saunders College Publishers.)

EXAMPLE 6 **Calculating the Distance to a Star**

One of the nearest stars is Alpha Centauri, which has parallax $\theta \approx 0.000212°$.

(a) If the Earth–Sun distance is 93,000,000 miles, calculate the distance to Alpha Centauri.

(b) A light-year is defined to be the distance that light travels in 1 year and equals about 5.9 trillion miles. Find the distance to Alpha Centauri in light-years.

Solution

(a) Let d represent the distance between Earth and Alpha Centauri. From **FIGURE 68,** it can be seen that

$$\sin \theta = \frac{93,000,000}{d}, \quad \text{or} \quad d = \frac{93,000,000}{\sin \theta}.$$

Substitute $0.000212°$ for θ.

$$d = \frac{93,000,000}{\sin 0.000212°} \approx 2.5 \times 10^{13} \text{ miles}$$

(b) This distance equals $\dfrac{2.5 \times 10^{13}}{5.9 \times 10^{12}} \approx 4.3$ light-years.

EXAMPLE 7 **Solving a Problem Involving the Angle of Elevation**

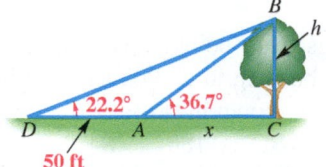

FIGURE 69

Francisco needs to know the height of a tree. From a given point on level ground, he finds that the angle of elevation to the top of the tree is 36.7°. He then moves back 50 feet. From the second point, the angle of elevation to the top of the tree is 22.2°. See **FIGURE 69.** Find the height of the tree.

---

*You observe parallax when you ride in a car and see a nearby object apparently moving backward with respect to more distant objects.

Analytic Solution

FIGURE 69 shows two unknowns: x, the distance from the center of the trunk of the tree at ground level to the point where the first observation was made, and h, the height of the tree. Since nothing is given about the length of the hypotenuse of either triangle ABC or triangle BCD, we use a ratio that does not involve the hypotenuse, such as the tangent. (See **FIGURE 70** in the Graphing Calculator Solution.)

In triangle ABC, $\tan 36.7° = \dfrac{h}{x}$, or $h = x \tan 36.7°$.

In triangle BCD, $\tan 22.2° = \dfrac{h}{50 + x}$, or $h = (50 + x) \tan 22.2°$.

Each of these expressions equals h, so the expressions must be equal.

$$x \tan 36.7° = (50 + x) \tan 22.2°$$

Now we solve for x.

$$x \tan 36.7° = 50 \tan 22.2° + x \tan 22.2°$$
 Distributive property

$$x \tan 36.7° - x \tan 22.2° = 50 \tan 22.2°$$ Write x terms on one side.

$$x(\tan 36.7° - \tan 22.2°) = 50 \tan 22.2°$$ Factor out x on the left.

$$x = \frac{50 \tan 22.2°}{\tan 36.7° - \tan 22.2°}$$
 Divide by the coefficient of x.

From before, $h = x \tan 36.7°$. Substituting for x gives the following.

$$h = \left(\frac{50 \tan 22.2°}{\tan 36.7° - \tan 22.2°}\right) \tan 36.7°$$

Use a calculator.

$$\tan 36.7° = 0.74537703 \quad \text{and} \quad \tan 22.2° = 0.40809244$$

$$\tan 36.7° - \tan 22.2° = 0.74537703 - 0.40809244 = 0.33728459$$

$$h \approx \left(\frac{50(0.40809244)}{0.33728459}\right)(0.74537703) = 45$$

The height of the tree is approximately 45 feet.

Graphing Calculator Solution*

In **FIGURE 70,** we superimposed **FIGURE 69** on coordinate axes with the origin at D. The tangent of the angle between the x-axis and the graph of a line with equation $y = mx + b$ is the slope of the line, m. For line DB, $m = \tan 22.2°$. Since $b = 0$ here, the equation of line DB is $y_1 = (\tan 22.2°)x$. The equation of line AB is $y_2 = (\tan 36.7°)x + b$. Since $b \neq 0$ here, we use the point $A(50, 0)$ to find the equation.

$$y_2 - y_1 = m(x - x_1) \quad \text{Point–slope form}$$

$$y_2 - 0 = m(x - 50) \quad x_1 = 50, \ y_1 = 0$$

$$y_2 = \tan 36.7°(x - 50)$$

Lines y_1 and y_2 are graphed in **FIGURE 71.** The y-coordinate of the point of intersection of the graphs of these two lines gives the length of BC, or h. Thus, $h = 45$.

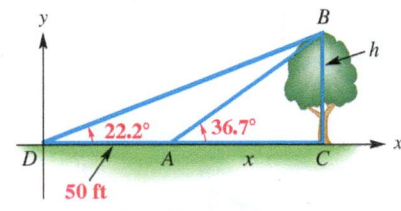

FIGURE 70

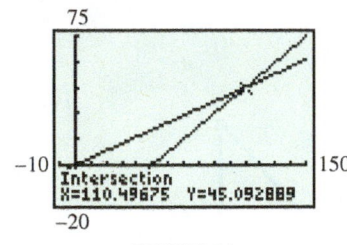

FIGURE 71

> **NOTE** In practice, we usually do not write down the intermediate calculator approximation steps. However, we have done this in **Example 7** so that you may follow the steps more easily.

Source: Adapted with permission from "Letter to the Editor," by Robert Ruzich (*Mathematics Teacher*, Volume 88, Number 1). Copyright © 1995 by the National Council of Teachers of Mathematics.

8.4 Exercises

Solve each right triangle.

1.

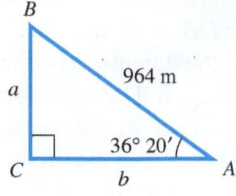

2.

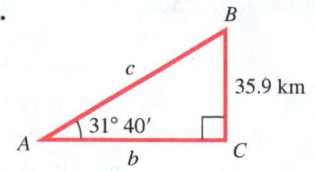

3.

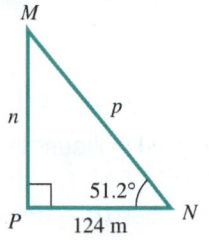

4.

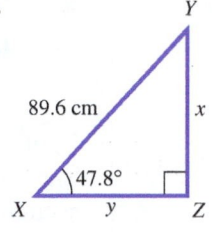

5.

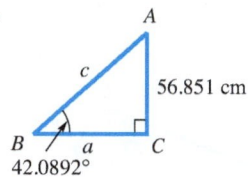

6.

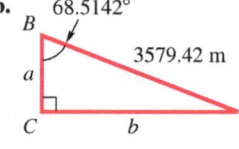

7.

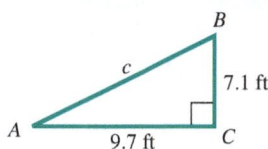

8.

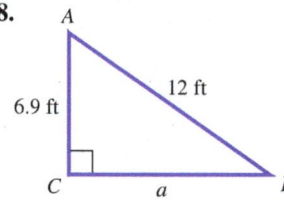

9.

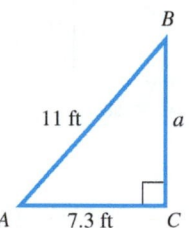

10.

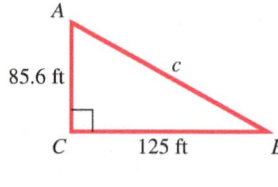

Solve each right triangle. In each case, C = 90°. If angle information is given in degrees and minutes, give answers in the same way. If given in decimal degrees, do likewise in answers. When two sides are given, give angles in degrees and minutes.

11. $A = 28.0°$; $c = 17.4$ feet

12. $B = 46.0°$; $c = 29.7$ meters

13. $B = 73.0°$; $b = 128$ inches

14. $A = 61.0°$; $b = 39.2$ centimeters

15. $a = 76.4$ yards; $b = 39.3$ yards

16. $a = 958$ meters; $b = 489$ meters

17. *Concept Check* If we are given an acute angle and a side of a right triangle, what unknown part of the triangle requires the least work to find?

18. Can a right triangle be solved if we are given measures of its two acute angles and no side lengths? Explain.

Find the degree measure of the angle, to four decimal places, formed by the line passing through the origin and the positive part of the x-axis. Use the values displayed at the bottom of the screen.

19. 12.4

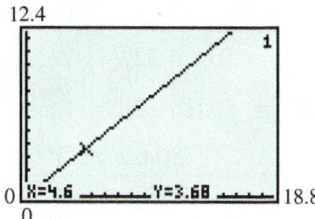

20. 18.6

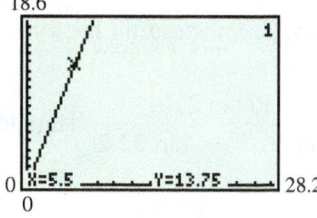

21. Explain why the angle of depression *DAB* has the same measure as the angle of elevation *ABC* in the figure. (Assume that line *AD* is parallel to line *CB*.)

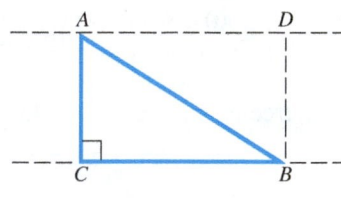

AD is parallel to BC.

22. Why is angle *CAB not* an angle of depression in the figure for **Exercise 21**? Why is angle *CAB not* an angle of elevation in the figure?

23. *Concept Check* When bearing is given as a single angle measure, how is the angle represented in a sketch?

24. *Concept Check* When bearing is given as N (or S), then the angle measure, then E (or W), how is the angle represented in a sketch?

Solve each problem.

25. Height of a Ladder on a Wall A 13.5-meter fire-truck ladder is leaning against a wall. Find the distance *d* the ladder goes up the wall (above the top of the fire truck) if the ladder makes an angle of 43° 50' with the horizontal.

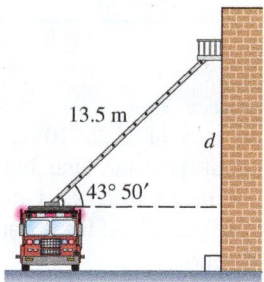

26. Length of a Guy Wire A weather tower used to measure wind speed has a guy wire attached to it 175 feet above the ground. The angle between the wire and the vertical tower is 57.0°. Approximate the length of the guy wire.

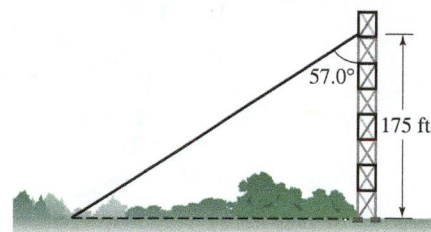

27. Length of a Shadow Suppose that the angle of elevation of the sun is 23.4°. Find the length of the shadow cast by Leah Goldberg, who is 5.75 feet tall.

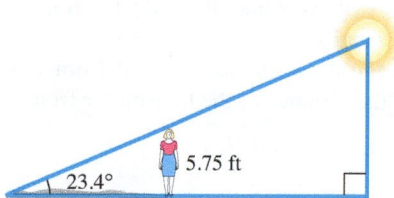

28. Height of a Tower The shadow of a vertical tower is 40.6 meters long when the angle of elevation of the sun is 34.6°. Find the height of the tower.

29. Height of a Building From a window 30.0 feet above the street, the angle of elevation to the top of the building across the street is 50.0° and the angle of depression to the base of this building is 20.0°. Find the height of the building across the street.

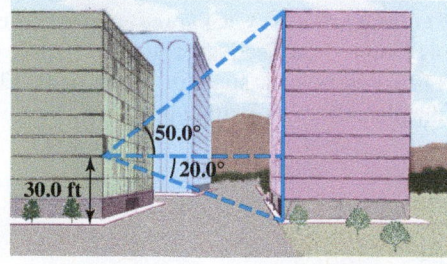

30. Height of a Building The angle of elevation from the top of a small building to the top of a nearby taller building is 46° 40', while the angle of depression to the bottom is 14° 10'. If the smaller building is 28.0 meters high, find the height of the taller building.

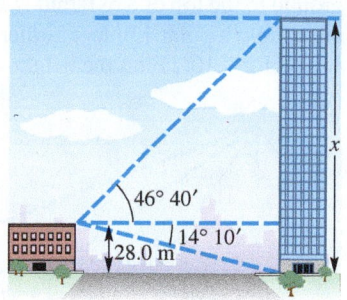

31. Angle of a Camera Lens A security camera is to be mounted on a bank wall so as to have a good view of the head teller. Find the angle of depression of the lens.

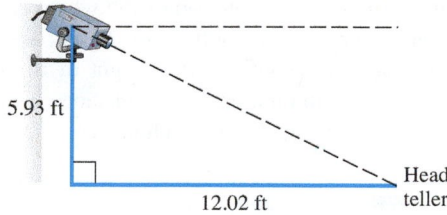

32. Angle of Depression of a Floodlight A company safety committee has recommended that a floodlight be mounted in a parking lot so as to illuminate the employee exit. Find the angle of depression of the light.

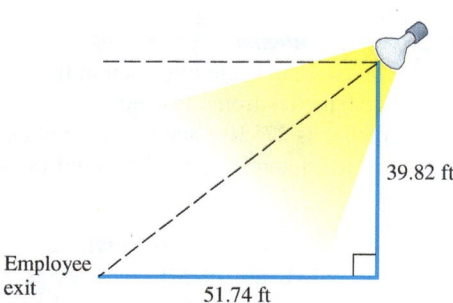

33. Distance through a Tunnel A tunnel is to be dug from *A* to *B*. Both *A* and *B* are visible from *C*. If *AC* is 1.4923 miles, *BC* is 1.0837 miles, and *C* is exactly 90°, find the measures of angles *A* and *B*.

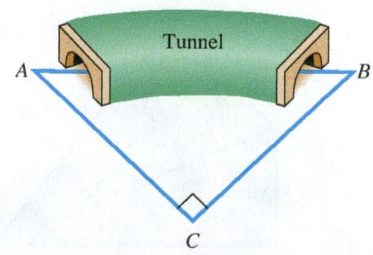

34. *Angle of Depression from a Plane* An airplane is flying 10,500 feet above the level ground. The angle of depression from the plane to the base of a tree is 13° 50′. How far horizontally must the plane fly to be directly over the tree?

35. *Height of a Pyramid* The angle of elevation from a point on the ground to the top of a pyramid is 35° 30′. The angle of elevation from a point 135 feet farther back to the top of the pyramid is 21° 10′. Find the height of the pyramid.

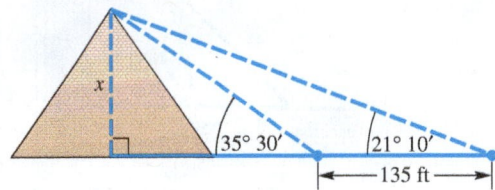

36. *Distance between a Whale and a Lighthouse* Debbie Glockner-Ferrari, a humpback whale researcher, is watching a whale approach directly toward her as she observes from the top of a lighthouse. When she first begins watching, the angle of depression of the whale is 15° 50′. Just as the whale turns away from the lighthouse, the angle of depression is 35° 40′. If the height of the lighthouse is 68.7 meters, find the horizontal distance x traveled by the whale as it approaches the lighthouse.

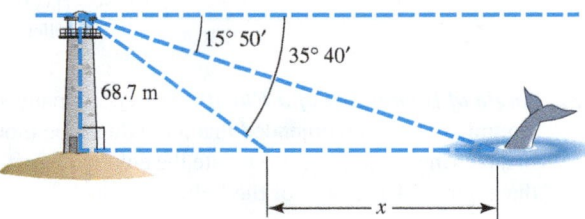

37. *Height of an Antenna* An antenna is on top of the center of a house. The angle of elevation from a point on the ground 28.0 meters from the center of the house to the top of the antenna is 27° 10′, and the angle of elevation to the bottom of the antenna is 18° 10′. Find the height of the antenna.

38. *Height of Mt. Whitney* The angle of elevation from Lone Pine to the top of Mt. Whitney is 10° 50′. Kim Hobbs, traveling 7.00 kilometers from Lone Pine along a straight, level road toward Mt. Whitney, finds the angle of elevation to be 22° 40′. Find the height of the top of Mt. Whitney above the level of the road.

39. *Length of a Side of a Lot* A piece of land has the shape shown in the figure. Find the value of x.

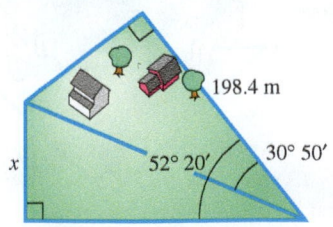

40. *Unknown Length* Find the value of x in the figure.

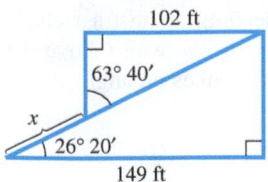

41. *Distance between Two Ships* A ship leaves port and sails on a bearing of N 28° 10′ E. Another ship leaves the same port at the same time and sails on a bearing of S 61° 50′ E. If the first ship sails at 24.0 mph and the second sails at 28.0 mph, find the distance x between the two ships after 4 hours.

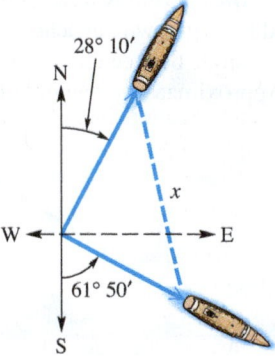

42. *Distance between Transmitters* Radio direction finders are set up at two points A and B, which are 2.50 miles apart on an east–west line. From A, it is found that the bearing of a signal from a radio transmitter is N 36° 20′ E, while the bearing of the same signal from B is N 53° 40′ W. Find the distance of the transmitter from B.

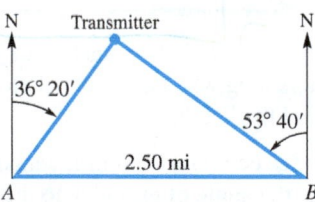

43. *Transmitter Distance* In **Exercise 42,** find the distance of the transmitter from A.

44. *Distance Traveled by a Plane* A plane flies 1.5 hours at 110 mph on a bearing of 40°. It then turns and flies 1.3 hours at the same speed on a bearing of 130°. How far is the plane from its starting point?

45. *Distance Traveled by a Ship*   A ship travels 50 kilometers on a bearing of 27° and then travels on a bearing of 117° for 140 kilometers. Find the distance *x* between the starting point and the ending point.

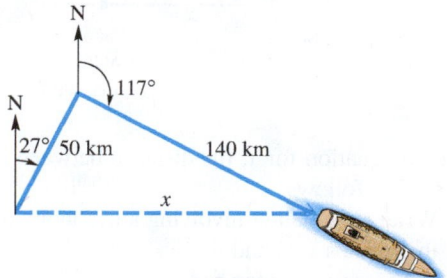

46. *Distance between Two Ships*   Two ships leave a port at the same time. The first ship sails on a bearing of 40° at 18 knots (nautical miles per hour) and the second at a bearing of 130° at 26 knots. How far apart are they after 1.5 hours?

47. *Distance between Points*   Point *A* is 15.00 miles directly north of point *B*. From point *A*, point *C* is on a bearing of 129° 25′, and from point *B* the bearing of *C* is 39° 25′.
(a) Find the distance between *A* and *C*.
(b) Find the distance between *B* and *C*.

48. *Distance between Points*   Point *X* is 12.00 km directly west of point *Y*. From point *X*, point *Z* is on a bearing of 66° 45′, and from point *Y* the bearing of *Z* is 336° 45′.
(a) Find the distance between *X* and *Z*.
(b) Find the distance between *Z* and *Y*.

49. *Distance across a Lake*   To find the distance *RS* across a lake, a surveyor lays off *RT* = 53.1 meters, with angle *T* = 32° 10′ and angle *S* = 57° 50′. Find length *RS*.

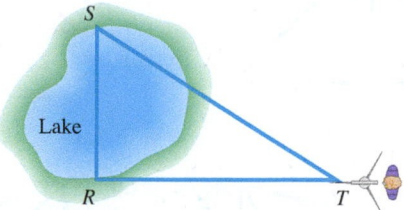

50. *Height of Mt. Everest*   The highest mountain peak in the world is Mt. Everest, located in the Himalayas. The height of this enormous mountain was determined in 1856 by surveyors using trigonometry long before the peak was first climbed in 1953. At an altitude of 14,545 feet on a different mountain, the straight line distance to the peak of Mt. Everest is 27.0134 miles and its angle of elevation is *θ* = 5.82°. (*Source:* Dunham, W., *The Mathematical Universe,* John Wiley and Sons.)

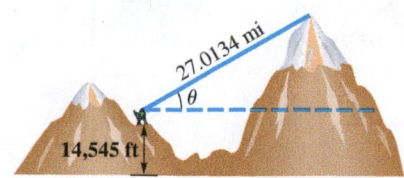

(a) Approximate the height (in feet) of Mt. Everest.

(b) In the actual measurement, Mt. Everest was over 100 miles away and the curvature of Earth had to be taken into account. Would the curvature of Earth make the peak appear taller or shorter than it actually is?

51. *Cloud Ceiling*   The U.S. Weather Bureau defines a **cloud ceiling** as the altitude of the lowest clouds that cover more than half the sky. To determine a cloud ceiling, a powerful searchlight projects a circle of light vertically onto the bottom of the cloud. An observer sights the circle of light in the crosshairs of a tube called a **clinometer.** A pendant hanging vertically from the tube and resting on a protractor gives the angle of elevation. Find the cloud ceiling if the searchlight is located 1000 feet from the observer and the angle of elevation is 30.0° as measured with a clinometer at eye height exactly 6 feet. (Assume three significant digits.)

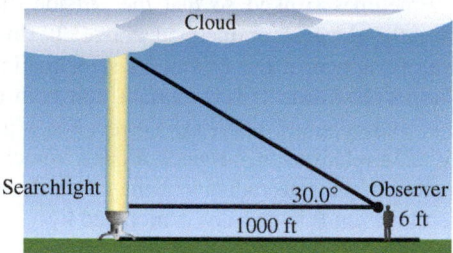

52. *Error in Measurement*   A degree may seem like a very small unit, but an error of 1 degree in measuring an angle may be very significant. For example, suppose a laser beam directed toward the visible center of the moon misses its assigned target by 30 seconds. How far is it (in miles) from the target? Take the distance from the surface of Earth to that of the moon to be 234,000 miles. (*Source: A Sourcebook of Applications of School Mathematics* by Donald Bushaw et al. © The Mathematical Association of America.)

53. *(Modeling) Stopping Distance on a Curve*   When an automobile travels along a circular curve, objects like trees and buildings situated on the inside of the curve can obstruct a driver's vision. These obstructions prevent the driver from seeing sufficiently far down the highway to ensure a safe stopping distance. In the figure at the top of the next page, the *minimum* distance *d* that should be cleared on the inside of the highway is modeled by the equation

$$d = R\left(1 - \cos\frac{\beta}{2}\right).$$

(*Source:* Mannering, F. and W. Kilareski, *Principles of Highway Engineering and Traffic Analysis,* Second Edition, John Wiley and Sons.)

(*continued*)

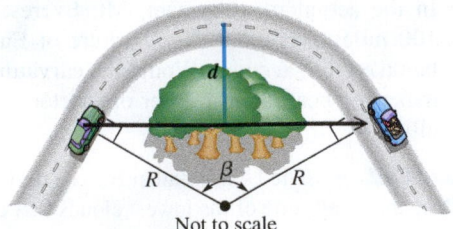

Not to scale

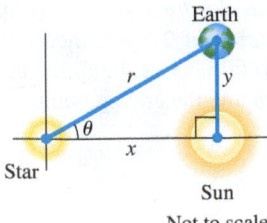

Not to scale

(a) It can be shown that if β is measured in degrees, then $\beta \approx \frac{57.3S}{R}$, where S is the safe stopping distance for the given speed limit. Compute d for a 55-mph speed limit if $S = 336$ feet and $R = 600$ feet.

(b) Compute d for a 65-mph speed limit if $S = 485$ feet and $R = 600$ feet.

(c) How does the speed limit affect the amount of land that should be cleared on the inside of the curve?

54. *(Modeling) Highway Curve Design* Highway curves are sometimes banked so that the outside of the curve is slightly elevated or inclined above the inside of the curve, as shown in the figure. This inclination is called the **superelevation.** It is important that both the curve's radius and superelevation are correct for a given speed limit. The relationship between a car's velocity v in feet per second, the safe radius r of the curve in feet, and the superelevation θ in degrees is modeled by

$$r = \frac{v^2}{4.5 + 32.2 \tan \theta}.$$

(Source: Mannering, F. and W. Kilareski, Principles of Highway Engineering and Traffic Analysis, Second Edition, John Wiley and Sons.)

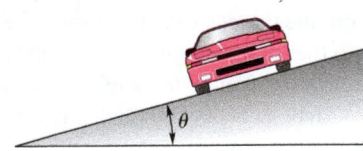

(a) A curve has a speed limit of 66 feet per second (45 mph) and a superelevation of $\theta = 3°$. Approximate the safe radius r.

(b) Find r if $\theta = 5°$ and $v = 66$.

(c) Make a conjecture about how increasing θ with $v = 66$ affects the safe radius r.

(d) Find v if $r = 1150$ and $\theta = 2.1°$.

55. *(Modeling) Distance between the Sun and a Star* Suppose that a star forms an angle θ with respect to Earth and the sun. Let the coordinates of Earth be (x, y), of the star be $(0, 0)$, and of the sun be $(x, 0)$. See the figure at the top of the next column.

Find an equation for x, the distance between the sun and the star, as follows.

(a) Write an equation involving a trigonometric function that relates x, y, and θ.

(b) Solve your equation for x.

56. *Area of a Solar Cell* A solar cell converts the energy of sunlight directly into electrical energy. The amount of energy a cell produces depends on its area. Suppose a solar cell is hexagonal, as shown in the figure. Express its area in terms of $\sin \theta$ and any side x. (*Hint:* Consider one of the six equilateral triangles from the hexagon. See the figure.) (*Source:* Kastner, B., *Space Mathematics,* NASA.)

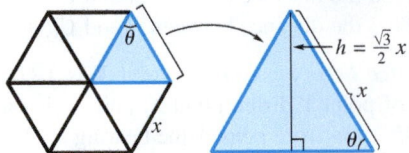

Find the exact value of each part labeled with a variable in each figure.

57.

58.

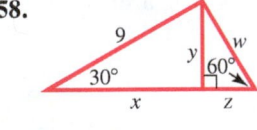

59.

60.

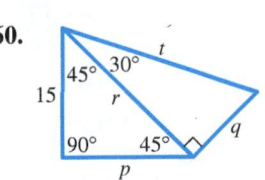

Find a formula for the area $\mathcal{A}$ of each figure in terms of s.

61.

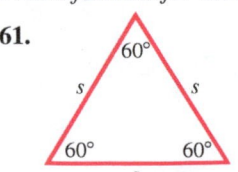

62.

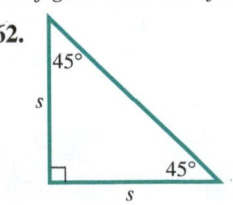

SECTIONS 8.3–8.4 Reviewing Basic Concepts

1. Find the six trigonometric function values of angle A in the triangle shown.

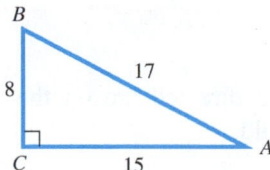

2. Complete the following table of exact function values.

θ	$\sin\theta$	$\cos\theta$	$\tan\theta$	$\cot\theta$	$\sec\theta$	$\csc\theta$
30° or $\frac{\pi}{6}$						
45° or $\frac{\pi}{4}$						
60° or $\frac{\pi}{3}$						

3. Write each expression using its cofunction.

  (a) $\sin 27°$ (b) $\tan\dfrac{\pi}{5}$

4. Give the reference angle for each angle measure.

  (a) $100°$ (b) $-365°$ (c) $\dfrac{8\pi}{3}$

5. Find the exact values for the six trigonometric functions of 315°.

6. Use a calculator to find an approximation for each function value.

  (a) $\sin 46° \, 30'$ (b) $\tan(-100°)$ (c) $\csc 4$

7. Find all values of θ if θ is in the interval $[0°, 360°)$ and $\tan\theta = -\dfrac{\sqrt{3}}{3}$.

8. Approximate two angles in the interval $[0, 2\pi)$ that satisfy $\sin\theta = 0.68163876$.

9. *Aerial Photography* An aerial photograph is taken directly above a building. The length of the building's shadow is 48 feet when the angle of elevation of the sun is 35.3°. Approximate the height of the building.

10. *Height of Mt. Kilimanjaro* From a point A, the angle of elevation of Mount Kilimanjaro in Africa is 13.7°, and from a point B directly behind A, the angle of elevation is 10.4°. If the distance from A to B is exactly 5 miles, approximate the height of Mount Kilimanjaro to the nearest hundred feet.

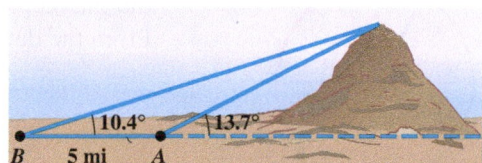

8.5 The Circular Functions

Circular Functions • Applications of Circular Functions

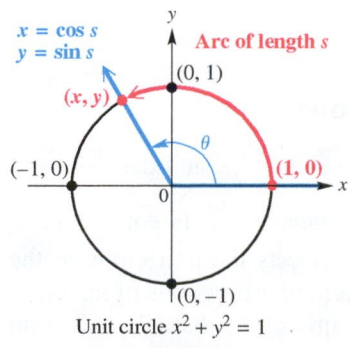

$x = \cos s$
$y = \sin s$

Unit circle $x^2 + y^2 = 1$

FIGURE 72

In **Section 8.2**, we defined the six trigonometric functions so that the domain of each function was a set of *angles* in standard position. These angles can be measured in degrees or in radians. In calculus, it is necessary to modify the trigonometric functions so that their domains consist of *real numbers* rather than angles. We do this by using the relationship between an angle θ and an arc of length s on a circle.

Circular Functions

In **FIGURE 72**, we start at the point $(1, 0)$ and measure an arc of length s along the circle. If $s > 0$, then the arc is measured in a counterclockwise direction, and if $s < 0$, then the direction is clockwise. (If $s = 0$, then no arc is measured.) Let the endpoint of this arc be at the point (x, y). The circle in **FIGURE 72** is a **unit circle**—it has center at the origin and radius 1 unit (hence the name *unit circle*). Recall that the equation of this circle is

$$x^2 + y^2 = 1.$$

→ **Looking Ahead to Calculus**
If you plan to go on to calculus, you must become familiar with radian measure. In calculus, the circular functions are understood to have real-number domains.

We saw in **Section 8.1** that the radian measure of θ is related to the arc length s. In fact, for θ measured in radians, $s = r\theta$. For the unit circle $r = 1$, so s, which is measured in linear units such as inches or centimeters, is numerically equal to θ, measured in radians. That is, $s = \theta$. Thus, the trigonometric functions of angle θ in radians found by choosing a point (x, y) on the unit circle can be rewritten as functions of the arc length s, a real number. Interpreted this way, they are called **circular functions.**

Circular Functions

For any real number s represented by a directed arc on the unit circle $x^2 + y^2 = 1$, the following definitions hold.

$$\sin s = y \qquad \cos s = x \qquad \tan s = \frac{y}{x} \ (x \neq 0)$$

$$\csc s = \frac{1}{y} \ (y \neq 0) \qquad \sec s = \frac{1}{x} \ (x \neq 0) \qquad \cot s = \frac{x}{y} \ (y \neq 0)$$

The circular functions (functions of real numbers) are closely related to the trigonometric functions of angles measured in radians. To see this, let us assume that angle θ is in standard position, superimposed on the unit circle, as shown in **FIGURE 72** (repeated in the margin). Suppose further that θ is the *radian* measure of this angle. Using the arc length formula $s = r\theta$ with $r = 1$, we have $s = \theta$. Thus, the length of the intercepted arc s is the real number that corresponds to the radian measure of θ. From the definitions of the trigonometric functions, we have

$$\sin \theta = \frac{y}{r} = \frac{y}{1} = y = \sin s, \quad \cos \theta = \frac{x}{r} = \frac{x}{1} = x = \cos s,$$

$s = \theta$, where s is a real number and θ is an angle in radians.

and so on. As shown here, the trigonometric functions and the circular functions take on the same function values, provided that we think of the angles in radian measure. This leads to the following important result concerning the evaluation of circular functions.

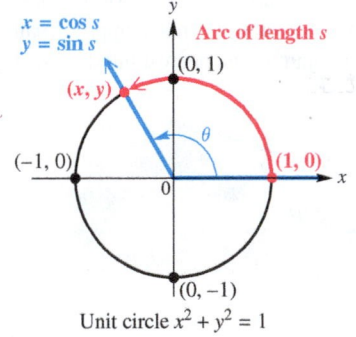

$x = \cos s$
$y = \sin s$
Arc of length s
(x, y)
$(0, 1)$
$(-1, 0)$
$(1, 0)$
θ
0
$(0, -1)$
Unit circle $x^2 + y^2 = 1$

FIGURE 72 (repeated)

Evaluating a Circular Function

Circular function values of real numbers are obtained in the same manner as trigonometric function values of angles measured in radians. This applies to both methods of finding exact values (such as reference angle analysis) and calculator approximations. *Calculators must be in radian mode when finding circular function values of real numbers.*

EXAMPLE 1 **Evaluating Circular Functions**

Use the definitions of circular functions to evaluate $\sin \frac{3\pi}{2}$, $\cos \frac{3\pi}{2}$, and $\tan \frac{3\pi}{2}$.

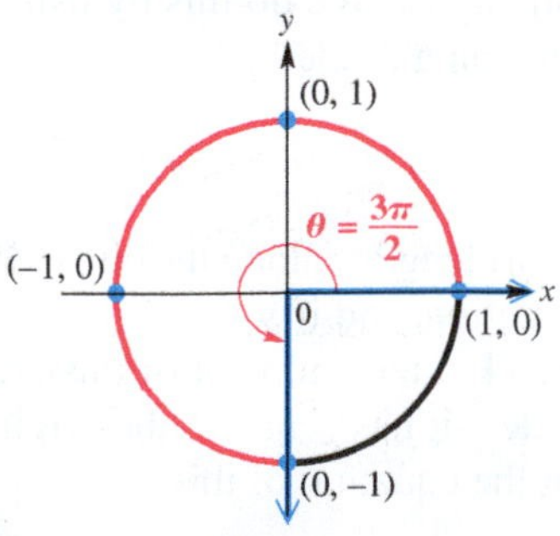

Solution Evaluating a circular function at the *real number* $\frac{3\pi}{2}$ is equivalent to evaluating it at $\frac{3\pi}{2}$ *radians*. An angle of $\frac{3\pi}{2}$ radians intersects the unit circle at the point $(0, -1)$, as shown in **FIGURE 73**. (When evaluating circular functions of quadrantal angles, it is often convenient to use the point on the unit circle that is 1 unit from the origin.)

y
$(0, 1)$
$\theta = \frac{3\pi}{2}$
$(-1, 0)$
0
$(1, 0)$
x
$(0, -1)$

FIGURE 73

Since by definition

$$\sin s = y, \quad \cos s = x, \quad \text{and} \quad \tan s = \frac{y}{x},$$

and because $x = 0$ and $y = -1$, it follows that

$$\sin \frac{3\pi}{2} = -1, \quad \cos \frac{3\pi}{2} = 0, \quad \text{and} \quad \tan \frac{3\pi}{2} \text{ is undefined.}$$

If an angle θ in radians is a multiple of $\frac{\pi}{3}$, $\frac{\pi}{4}$, or $\frac{\pi}{6}$, we can always find its exact trigonometric function values by hand. To determine the point where the terminal side of θ intersects the unit circle, the two triangles in **FIGURE 74** are often helpful, as illustrated in the next example.

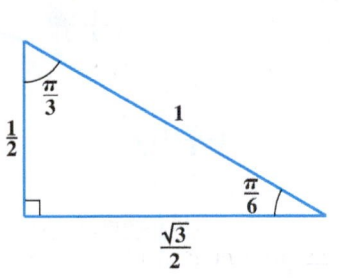
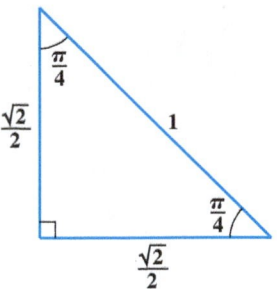

FIGURE 74

EXAMPLE 2 **Evaluating Circular Functions**

Evaluate the six circular functions at $s = \frac{5\pi}{6}$.

Analytic Solution

From the properties of 30°–60° triangles, an angle of $\frac{5\pi}{6}$ radians intersects the unit circle at the point $\left(-\frac{\sqrt{3}}{2}, \frac{1}{2}\right)$. See **FIGURE 75.** Evaluating the six circular functions at the real number $\frac{5\pi}{6}$ results in the following.

$$\sin \frac{5\pi}{6} = \frac{1}{2} \qquad\qquad \csc \frac{5\pi}{6} = 2$$

$$\cos \frac{5\pi}{6} = -\frac{\sqrt{3}}{2} \qquad\qquad \sec \frac{5\pi}{6} = -\frac{2}{\sqrt{3}} = -\frac{2\sqrt{3}}{3}$$

$$\tan \frac{5\pi}{6} = -\frac{1}{\sqrt{3}} = -\frac{\sqrt{3}}{3} \qquad\qquad \cot \frac{5\pi}{6} = -\sqrt{3}$$

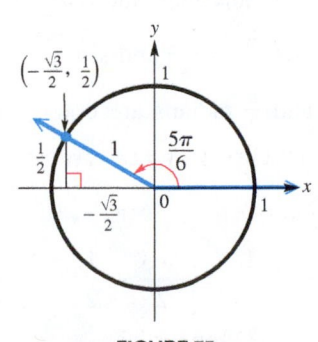

FIGURE 75

Graphing Calculator Solution

Set your calculator to *radian mode.* Four of the results in **FIGURE 76** are *decimal approximations* of our analytic results. The reciprocal identities have been used to evaluate the cosecant, secant, and cotangent functions because many calculators do not have keys for these functions.

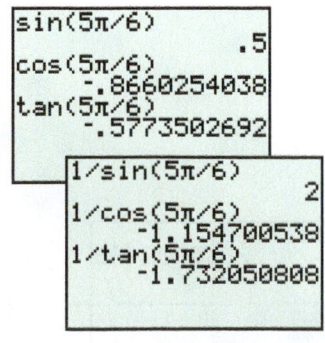

Radian mode
FIGURE 76

CAUTION A common error in trigonometry is using calculators in degree mode when radian mode should be used. *If you are finding a circular function value of a real number, the calculator must be in radian mode.*

Since x represents the cosine of s and y represents the sine of s, and because of the discussion in **Section 8.1** on converting between degrees and radians, we can summarize a great deal of information in a concise manner, as seen in **FIGURE 77**.

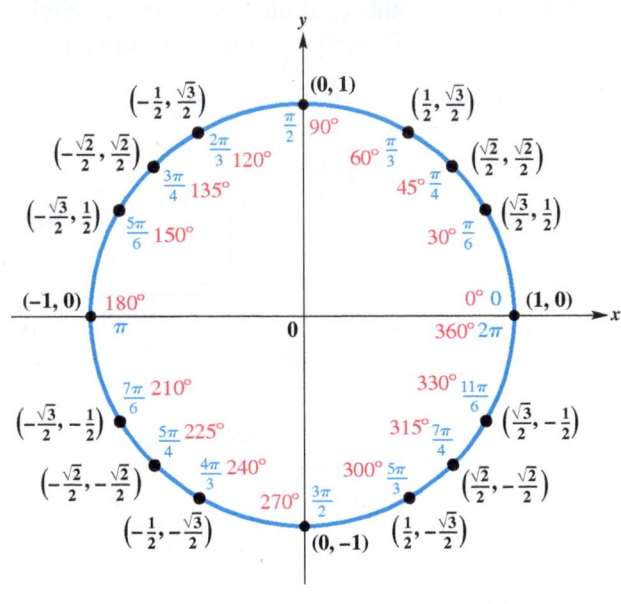

Unit circle $x^2 + y^2 = 1$

FIGURE 77

EXAMPLE 3 **Finding Circular Function Values**

(a) Use **FIGURE 77** to find the exact values of $\cos \frac{7\pi}{4}$ and $\sin \frac{7\pi}{4}$.

(b) Use **FIGURE 77** to find the exact value of $\tan\left(-\frac{5\pi}{3}\right)$.

(c) Use a calculator to approximate $\cos 1.85$ to four decimal places.

Solution

(a) In **FIGURE 77,** we can see that the terminal side of $\frac{7\pi}{4}$ radians intersects the unit circle at $\left(\frac{\sqrt{2}}{2}, -\frac{\sqrt{2}}{2}\right)$. Thus, $\cos \frac{7\pi}{4} = \frac{\sqrt{2}}{2}$ and $\sin \frac{7\pi}{4} = -\frac{\sqrt{2}}{2}$.

(b) The angles of $-\frac{5\pi}{3}$ radians and $\frac{\pi}{3}$ radians are coterminal. Their terminal sides intersect the unit circle at $\left(\frac{1}{2}, \frac{\sqrt{3}}{2}\right)$.

$$\tan\left(-\frac{5\pi}{3}\right) = \tan \frac{\pi}{3} = \frac{\frac{\sqrt{3}}{2}}{\frac{1}{2}} = \frac{\sqrt{3}}{2} \div \frac{1}{2} = \frac{\sqrt{3}}{2} \cdot \frac{2}{1} = \sqrt{3}$$

tan $s = \frac{y}{x}$

Radian mode

FIGURE 78

(c) No degree symbol is shown, so 1.85 is a real number and we evaluate $\cos 1.85$ in radian mode. To four decimal places, $\cos 1.85 \approx -0.2756$. See **FIGURE 78**. ●

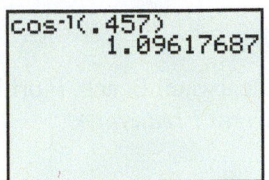

Radian mode
FIGURE 79

EXAMPLE 4   **Finding a Number, Given a Circular Function Value**

Approximate the value of s in the interval $\left[0, \frac{\pi}{2}\right]$ to three decimal places given that $\cos s = 0.457$.

Solution   *Since we are given a function value, we must use the appropriate inverse function to find the arc length.* With the calculator set in radian mode, **FIGURE 79** shows that to three decimal places, $s = \cos^{-1} 0.457 \approx 1.096$.   ●

Applications of Circular Functions

EXAMPLE 5   **Modeling the Phases of the Moon**

Because the moon orbits Earth, we observe different phases of the moon during the period of a month. In **FIGURE 80,** t is called the *phase angle*. The *phase F* of the moon is computed by

$$F(t) = \frac{1}{2}(1 - \cos t),$$

and gives the fraction of the moon's face that is illuminated by the sun. (*Source:* Duffett-Smith, P., *Practical Astronomy with Your Calculator,* Cambridge University Press.)

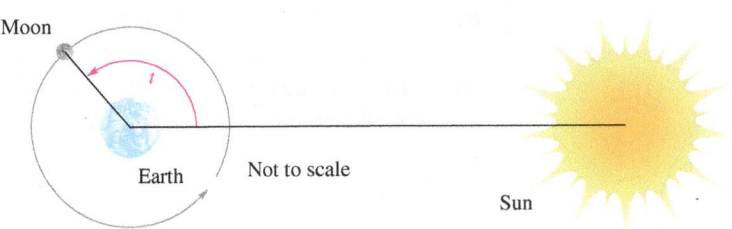

FIGURE 80

Evaluate each expression and interpret the result.

(a) $F(0)$ **(b)** $F\left(\frac{\pi}{2}\right)$ **(c)** $F(\pi)$ **(d)** $F\left(\frac{3\pi}{2}\right)$

Solution

(a) $F(0) = \frac{1}{2}(1 - \cos 0) = \frac{1}{2}(1 - 1) = 0$. When $t = 0$, the moon is located between Earth and the sun. Since $F = 0$, the face of the moon is not visible, which corresponds to a *new moon*.

(b) $F\left(\frac{\pi}{2}\right) = \frac{1}{2}\left(1 - \cos \frac{\pi}{2}\right) = \frac{1}{2}(1 - 0) = \frac{1}{2}$. When $t = \frac{\pi}{2}$, $F = \frac{1}{2}$. Thus, half the face of the moon is visible. This phase is called the *first quarter*.

(c) $F(\pi) = \frac{1}{2}(1 - \cos \pi) = \frac{1}{2}[1 - (-1)] = 1$. When $t = \pi$, Earth is between the moon and the sun. Since $F = 1$, the face of the moon is completely visible, which corresponds to a *full moon*.

(d) $F\left(\frac{3\pi}{2}\right) = \frac{1}{2}\left(1 - \cos \frac{3\pi}{2}\right) = \frac{1}{2}(1 - 0) = \frac{1}{2}$. When $t = \frac{3\pi}{2}$, $F = \frac{1}{2}$. Thus, half the face of the moon is visible. This phase is called the *last quarter*.   ●

Because the values of the circular functions repeat every 2π, they are used to model natural phenomena, such as ocean tides, that repeat at regular intervals.

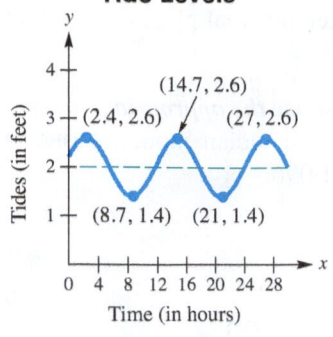

Tide Levels

FIGURE 81

> **EXAMPLE 6**   **Modeling Tides**

FIGURE 81 shows a function f that models the tides in feet at Clearwater Beach, Florida, x hours after midnight. (*Source:* Pentcheff, D., *Tide and Current Predictor.*)

(a) Find the time between high tides.

(b) What is the difference in water levels between high tide and low tide?

(c) The tides can be modeled by

$$f(x) = 0.6 \cos[0.511(x - 2.4)] + 2.$$

Estimate the tides when $x = 10$.

Solution

(a) A high tide corresponds to a peak on the graph. The time between peaks is 12.3 hours, since

$$14.7 - 2.4 = 12.3 \quad \text{and} \quad 27 - 14.7 = 12.3.$$

(b) High tides were 2.6 feet and low tides were 1.4 feet. Their difference is 1.2 feet.

(c) $f(10) = 0.6 \cos[0.511(10 - 2.4)] + 2 \approx 1.56$ feet   ●

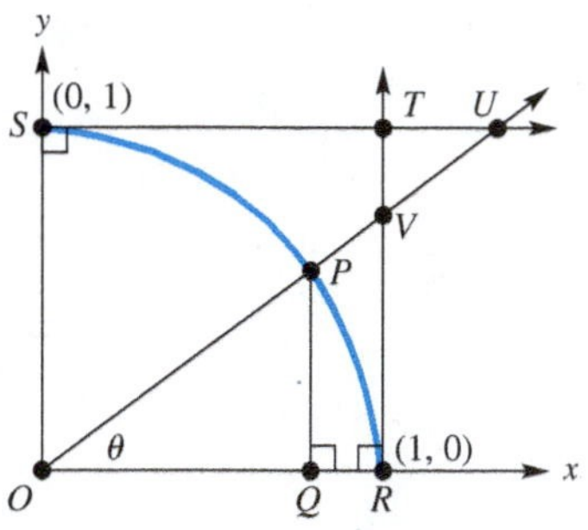

FIGURE 82

The diagram shown in **FIGURE 82** illustrates a correspondence that relates the right-triangle ratio definitions of the trigonometric functions and the unit circle interpretation. The arc SR is the first-quadrant portion of the unit circle, and the standard-position angle POQ is designated θ. By definition, the coordinates of P are $(\cos \theta, \sin \theta)$. The six trigonometric functions of θ can be interpreted as lengths of line segments found in **FIGURE 82.**

For $\cos \theta$ and $\sin \theta$, use right triangle POQ and right-triangle ratios.

$$\mathbf{cos\ \theta} = \frac{\text{side adjacent to } \theta}{\text{hypotenuse}} = \frac{OQ}{OP} = \frac{OQ}{1} = \mathbf{OQ}$$

$$\mathbf{sin\ \theta} = \frac{\text{side opposite } \theta}{\text{hypotenuse}} = \frac{PQ}{OP} = \frac{PQ}{1} = \mathbf{PQ}$$

For $\tan \theta$ and $\sec \theta$, use right triangle VOR in **FIGURE 82** and right-triangle ratios.

$$\mathbf{tan\ \theta} = \frac{\text{side opposite } \theta}{\text{side adjacent to } \theta} = \frac{VR}{OR} = \frac{VR}{1} = \mathbf{VR}$$

$$\mathbf{sec\ \theta} = \frac{\text{hypotenuse}}{\text{side adjacent to } \theta} = \frac{OV}{OR} = \frac{OV}{1} = \mathbf{OV}$$

For $\csc \theta$ and $\cot \theta$, first note that US and OR are parallel. Thus angle SUO is equal to θ because it is an alternate interior angle to angle POQ, which is equal to θ. Use right triangle USO and right-triangle ratios.

$$\csc SUO = \mathbf{csc\ \theta} = \frac{\text{hypotenuse}}{\text{side opposite } \theta} = \frac{OU}{OS} = \frac{OU}{1} = \mathbf{OU}$$

$$\cot SUO = \mathbf{cot\ \theta} = \frac{\text{side adjacent to } \theta}{\text{side opposite } \theta} = \frac{US}{OS} = \frac{US}{1} = \mathbf{US}$$

FIGURE 83 on the next page uses color to illustrate the results found above.

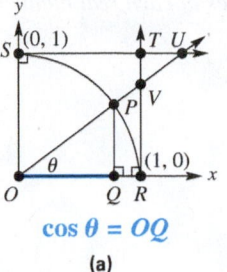

$$\cos \theta = OQ$$

(a)

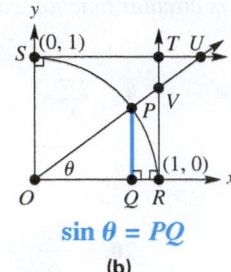

$$\sin \theta = PQ$$

(b)

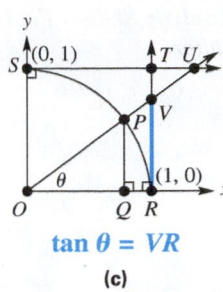

$$\tan \theta = VR$$

(c)

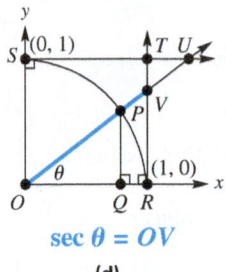

$$\sec \theta = OV$$

(d)

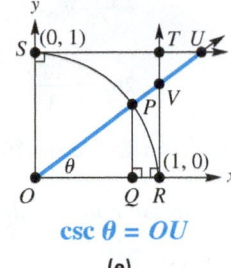

$$\csc \theta = OU$$

(e)

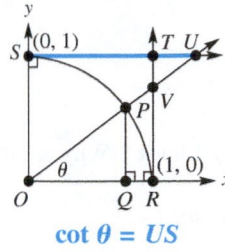

$$\cot \theta = US$$

(f)

FIGURE 83

EXAMPLE 7   **Finding Lengths of Line Segments**

FIGURE 82 is repeated in the margin. Suppose that angle *TVU* measures 60°. Find the exact lengths of segments *OQ*, *PQ*, *VR*, *OV*, *OU*, and *US*.

Solution   Angle *TVU* has the same measure as angle *OVR* because they are vertical angles. Therefore, angle *OVR* measures 60°. Because it is one of the acute angles in right triangle *VOR*, θ must be its complement, measuring 30°. Now use the equations found in **FIGURE 83,** with $\theta = 30°$.

FIGURE 82 (repeated)

$$OQ = \cos 30° = \frac{\sqrt{3}}{2} \qquad OV = \sec 30° = \frac{2\sqrt{3}}{3}$$

$$PQ = \sin 30° = \frac{1}{2} \qquad OU = \csc 30° = 2$$

$$VR = \tan 30° = \frac{\sqrt{3}}{3} \qquad US = \cot 30° = \sqrt{3}$$

8.5 Exercises

Checking Analytic Skills   *Find the exact value for each expression.* ***Do not use a calculator.***

1. $\sin \dfrac{7\pi}{6}$

2. $\cos \dfrac{5\pi}{3}$

3. $\tan \dfrac{3\pi}{4}$

4. $\cos \dfrac{7\pi}{6}$

5. $\sec \dfrac{2\pi}{3}$

6. $\csc \dfrac{11\pi}{6}$

7. $\cot \dfrac{5\pi}{6}$

8. $\cos \left(-\dfrac{4\pi}{3}\right)$

9. $\sin \left(-\dfrac{5\pi}{6}\right)$

10. $\tan \dfrac{17\pi}{3}$

11. $\sec \dfrac{23\pi}{6}$

12. $\csc \dfrac{13\pi}{3}$

Checking Analytic Skills   *Find the six circular function values of each real number by hand.* **Do not use a calculator.**

13. $\dfrac{\pi}{2}$

14. $\dfrac{3\pi}{4}$

15. $-\dfrac{\pi}{4}$

16. $-\pi$

17. π

18. $-\dfrac{5\pi}{4}$

19. $-\dfrac{\pi}{3}$

20. $-\dfrac{5\pi}{6}$

21. $-\dfrac{\pi}{2}$

22. $\dfrac{3\pi}{2}$

23. 2π

24. $\dfrac{\pi}{3}$

25. $\dfrac{\pi}{6}$

26. $-\dfrac{2\pi}{3}$

27. $-\dfrac{7\pi}{4}$

28. *Concept Check*   Why are the answers in **Exercises 16** and **17** the same?

Use a calculator to approximate each expression to four decimal places.

29. $\sin 0.6109$

30. $\sin 0.8203$

31. $\cos(-1.1519)$

32. $\cos(-5.2825)$

33. $\tan 4.0203$

34. $\tan 6.4752$

35. $\csc(-9.4946)$

36. $\csc 1.3875$

37. $\sec 2.8440$

38. $\sec(-8.3429)$

39. $\cot 6.0301$

40. $\cot 3.8426$

Concept Check   *Each figure in Exercises 41–44 shows angle θ in standard position, with its terminal side intersecting the unit circle. Evaluate the six trigonometric functions of θ.*

41.

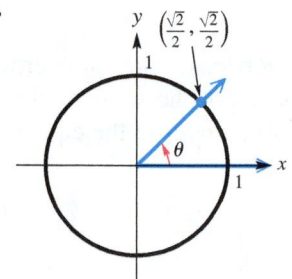

42.

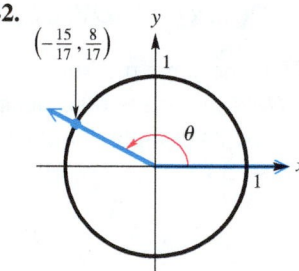

43.

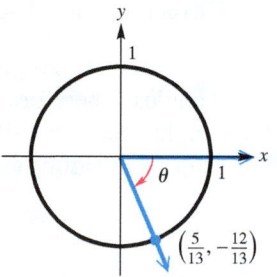

44.

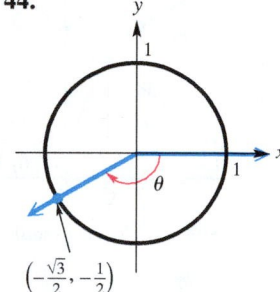

Concept Check   *The figure displays a unit circle and an angle of 1 radian. The tick marks on the circle are spaced at every two-tenths radian. Use the figure to estimate each value.*

45. $\cos 0.8$

46. $\cos 0.6$

47. $\sin 2$

48. $\sin 4$

49. $\sin 3.8$

50. $\cos 3.2$

51. A positive angle whose cosine is -0.65

52. A positive angle whose sine is -0.95

53. A positive angle whose sine is 0.7

54. A positive angle whose cosine is 0.3

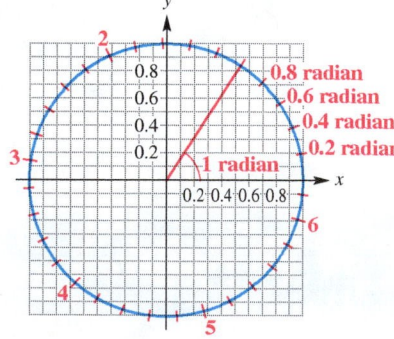

Unit circle: $x^2 + y^2 = 1$

Use a calculator to approximate to four decimal places the value of s in the interval $\left[0, \frac{\pi}{2}\right]$ that makes each statement true.

55. $\tan s = 0.21264138$

56. $\cos s = 0.78269876$

57. $\sin s = 0.99184065$

58. $\cot s = 0.29949853$

59. $\cot s = 0.09637041$

60. $\csc s = 1.0219553$

Concept Check In Exercises 61 and 62, each graphing calculator screen shows a point on the unit circle. What is the length, to four decimal places, of the shortest arc of the circle from (1, 0) to the point?

61.

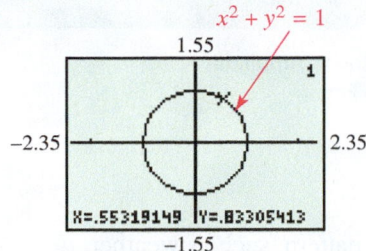

62.

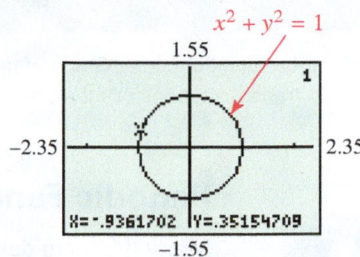

Suppose an arc of length s lies on the unit circle $x^2 + y^2 = 1$, starting at the point (1, 0) and terminating at the point (x, y). (See **FIGURE 72.***) Approximate coordinates for (x, y) to four decimal places.*

63. $s = 2.5$ **64.** $s = 3.4$ **65.** $s = -7.4$ **66.** $s = -3.9$

Concept Check By evaluating sin s and cos s for each value of s, decide in which quadrant an angle of s radians lies.

67. $s = 51$ **68.** $s = 49$ **69.** $s = 65$ **70.** $s = 79$

In Exercises 71 and 72, see **Example 7.**

71. Refer to **FIGURES 82** and **83**. Suppose that angle θ measures 60°. Find the exact length of each segment.

(a) *OQ*   (b) *PQ*   (c) *VR*

(d) *OV*   (e) *OU*   (f) *US*

72. Refer to **FIGURES 82** and **83**. Repeat **Exercise 71** for $\theta = 38°$, but give lengths as approximations to four significant digits.

(Modeling) Solve each problem.

 73. *Daylight Hours* The average number of daylight hours at San Antonio, Texas, can be modeled by

$$f(x) = 1.95 \cos\left[\frac{\pi}{6}(x - 6.6)\right] + 12.15,$$

where $x = 1$ corresponds to January 1, $x = 2$ to February 1, and so on. Evaluate $f(7)$ and interpret the result.

74. *Temperature in Chicago* The monthly average Fahrenheit temperatures in Chicago can be modeled by

$$T(x) = 25 \sin\left[\frac{\pi}{6}(x - 4)\right] + 50,$$

where $x = 1$ corresponds to January, $x = 2$ to February, and so on. Use $T(x)$ to estimate the average temperature for each month.

(a) April   (b) July

(c) October   (d) November

75. *Maximum Temperatures* The maximum afternoon temperature t (in degrees Fahrenheit) in a given city might be modeled by

$$t = 60 - 30 \cos\frac{x\pi}{6},$$

where x represents the month, with $x = 0$ corresponding to January, $x = 1$ to February, and so on. Estimate the maximum afternoon temperature for each month.

(a) January   (b) April

(c) May   (d) June

(e) August   (f) October

76. *Voltage* Electric ranges and ovens often use a higher voltage than that found in normal household outlets. This voltage can be modeled by

$$V(t) = 310 \sin 120\pi t,$$

where t represents time in seconds. Evaluate $V\left(\frac{1}{240}\right)$ and interpret the result.

77. *Phases of the Moon* Refer to **Example 5**. Find all phase angles t that correspond to a full moon and all phase angles that correspond to a new moon. Assume that t can be any angle.

78. *Daylight Hours* The ability to calculate the number of daylight hours H at any location is important in estimating the potential solar energy production. H can be calculated with the formula

$$\cos(0.1309H) = -\tan D \tan L,$$

where D is the declination of the sun and L is the latitude. Use radian mode to calculate the shortest and longest days in Minneapolis, Minnesota, if its latitude is $L = 44.88°$, the shortest day occurs when $D = -23.44°$, and the longest day occurs when $D = 23.44°$. (*Source:* Winter, C., R. Sizmann, and L. L. Vant-Hull (Editors), *Solar Power Plants*, Springer-Verlag.)

8.6 Graphs of the Sine and Cosine Functions

Periodic Functions • Graph of the Sine Function • Graph of the Cosine Function • Graphing Techniques, Amplitude, and Period • Translations and Transformations • Determining a Trigonometric Model Using Curve Fitting

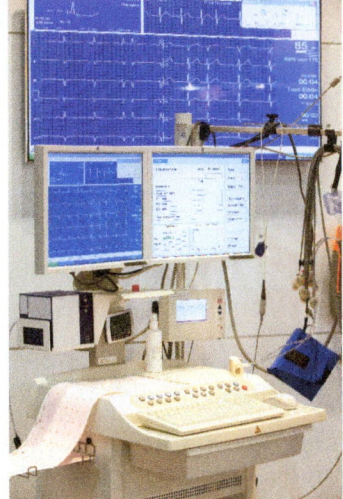

Periodic Functions

Many things in daily life repeat with a predictable pattern, such as weather, tides, and hours of daylight. Because the sine and cosine functions repeat their values in a regular pattern, they are examples of *periodic functions*. **FIGURE 84** shows a periodic graph that represents a person's heartbeat.

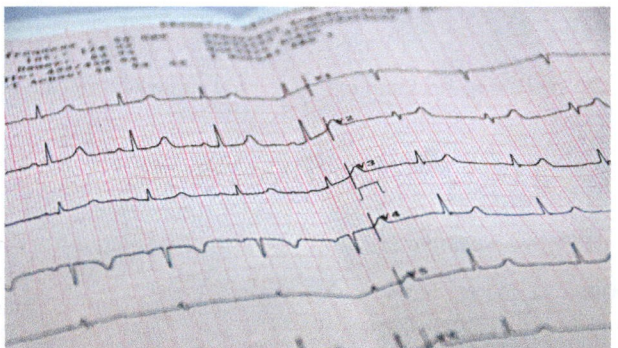

FIGURE 84

> **Periodic Function**
>
> A **periodic function** is a function f such that
> $$f(x) = f(x + np),$$
> for every real number x in the domain of f, every integer n, and some positive real number p. The least possible positive value of p is the **period** of the function.

The circumference of the unit circle is 2π, so the least value of p for which the sine and cosine functions repeat is 2π. *Therefore, the sine and cosine functions are periodic functions with period 2π.*

Graph of the Sine Function

In the previous section, we saw that for a real number s, the point on the unit circle corresponding to s has coordinates $(\cos s, \sin s)$. See **FIGURE 85**, and trace along the circle to verify the results shown in the table.

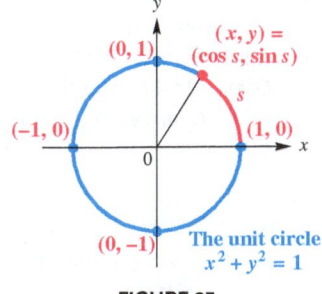

FIGURE 85

As s Increases from	sin s (y-coordinate)	cos s (x-coordinate)
0 to $\frac{\pi}{2}$	Increases from 0 to 1	Decreases from 1 to 0
$\frac{\pi}{2}$ to π	Decreases from 1 to 0	Decreases from 0 to -1
π to $\frac{3\pi}{2}$	Decreases from 0 to -1	Increases from -1 to 0
$\frac{3\pi}{2}$ to 2π	Increases from -1 to 0	Increases from 0 to 1

To avoid confusion when graphing the sine function, we use x rather than s. This corresponds to the letters in the xy-coordinate system. Selecting key values of x and finding the corresponding values of $\sin x$ leads to the tables in **FIGURE 86.**

To obtain the traditional graph of a portion of the sine function shown in **FIGURE 86,** we plot the points from the table of values and join them with a smooth curve. Since $y = \sin x$ is periodic and has $(-\infty, \infty)$ as its domain, the graph continues in the same pattern in both directions. This graph is called a **sine wave** or **sinusoid.** ***A comprehensive graph of a sinusoid consists of at least one period of the graph and shows the extreme points.***

TECHNOLOGY NOTE

Graphing calculators often have a window designated for graphing trigonometric or circular functions. We refer to the window $[-2\pi, 2\pi]$ by $[-4, 4]$ with Xscl $= \frac{\pi}{2}$ and Yscl $= 1$ as the **trig viewing window.** Your model may use a different "standard" viewing window for the graphs of trigonometric functions.

FUNCTION CAPSULE

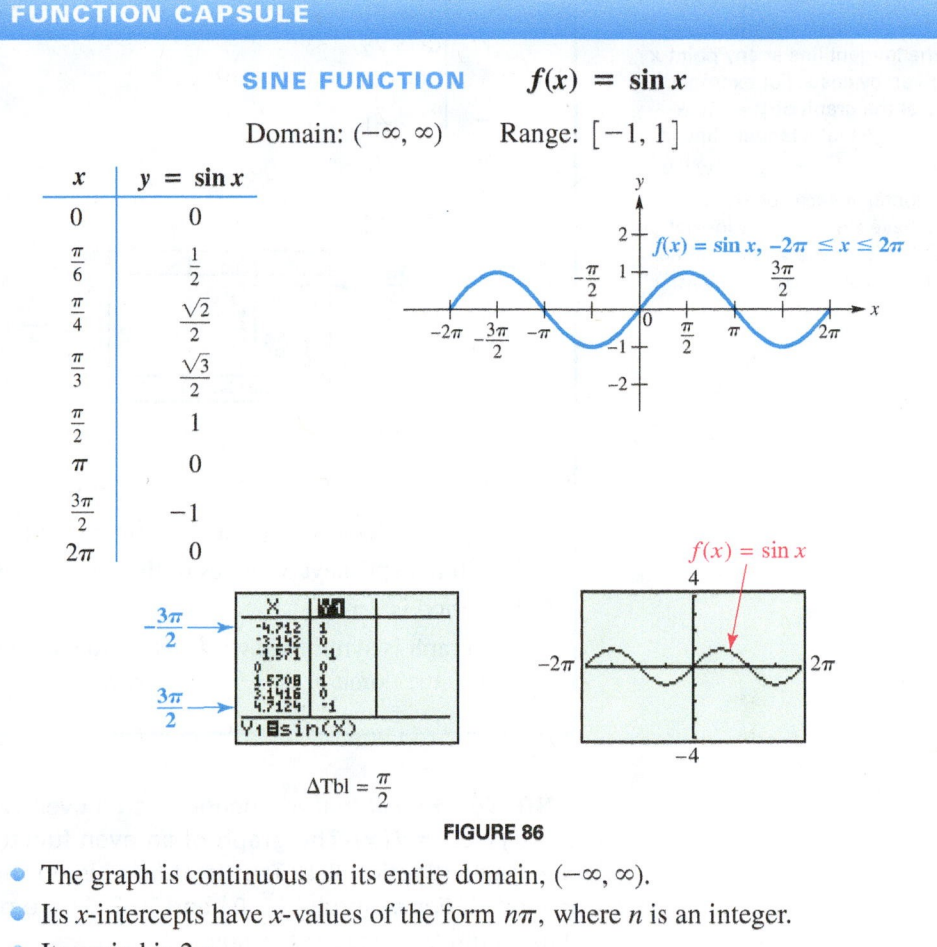

SINE FUNCTION $\quad f(x) = \sin x$

Domain: $(-\infty, \infty)$ $\quad$ Range: $[-1, 1]$

x	$y = \sin x$
0	0
$\frac{\pi}{6}$	$\frac{1}{2}$
$\frac{\pi}{4}$	$\frac{\sqrt{2}}{2}$
$\frac{\pi}{3}$	$\frac{\sqrt{3}}{2}$
$\frac{\pi}{2}$	1
π	0
$\frac{3\pi}{2}$	-1
2π	0

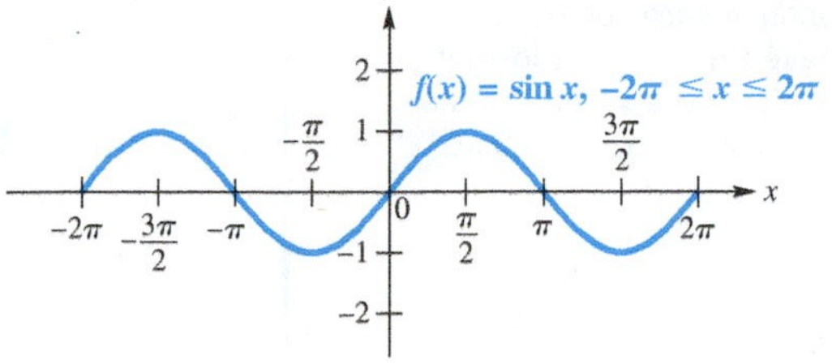

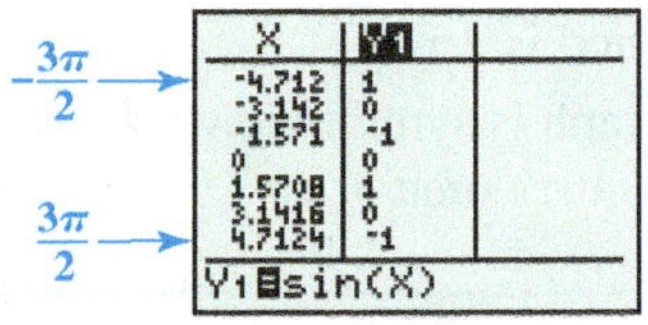

ΔTbl = $\frac{\pi}{2}$

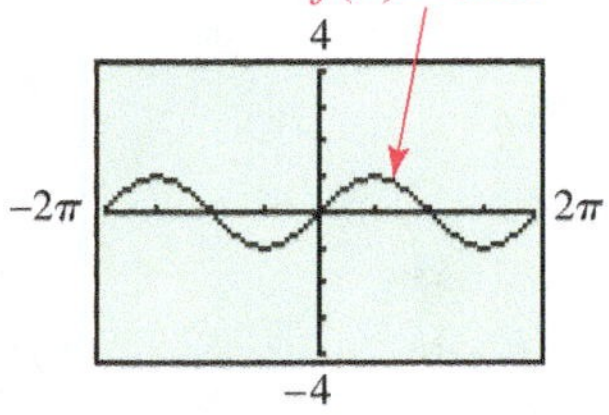

FIGURE 86

- The graph is continuous on its entire domain, $(-\infty, \infty)$.
- Its x-intercepts have x-values of the form $n\pi$, where n is an integer.
- Its period is 2π.
- The graph is symmetric with respect to the origin. It is an odd function. For all x in the domain, $\sin(-x) = -\sin x$.

NOTE Recall that a function f is an **odd function** if for all x in the domain of f, $f(-x) = -f(x)$. The graph of an odd function is symmetric with respect to the origin (**Section 2.1**). That is, if (x, y) is on the graph of the function, then so is $(-x, -y)$. For example, $\left(\frac{\pi}{2}, 1\right)$ and $\left(-\frac{\pi}{2}, -1\right)$ are points on the graph of $y = \sin x$, illustrating that $\sin(-x) = -\sin x$.

Graph of the Cosine Function

We can graph $y = \cos x$ in the same way that we graphed $y = \sin x$. The tables for $y = \cos x$ in **FIGURE 87** on the next page use the same values for x as before.

→ Looking Ahead to Calculus
The discussion of the derivative function in calculus shows that for the sine function, the slope of the tangent line at any point x is given by cos x. For example, look at the graph of $y = sin x$ and notice that a tangent line at $x = \pm\frac{\pi}{2}, \pm\frac{3\pi}{2}, \pm\frac{5\pi}{2}, \ldots$ will be horizontal at each "peak," and thus have slope 0. Now look at the graph of $y = cos x$ and see that for these values, cos $x = 0$.

FUNCTION CAPSULE

COSINE FUNCTION $f(x) = \cos x$

Domain: $(-\infty, \infty)$ Range: $[-1, 1]$

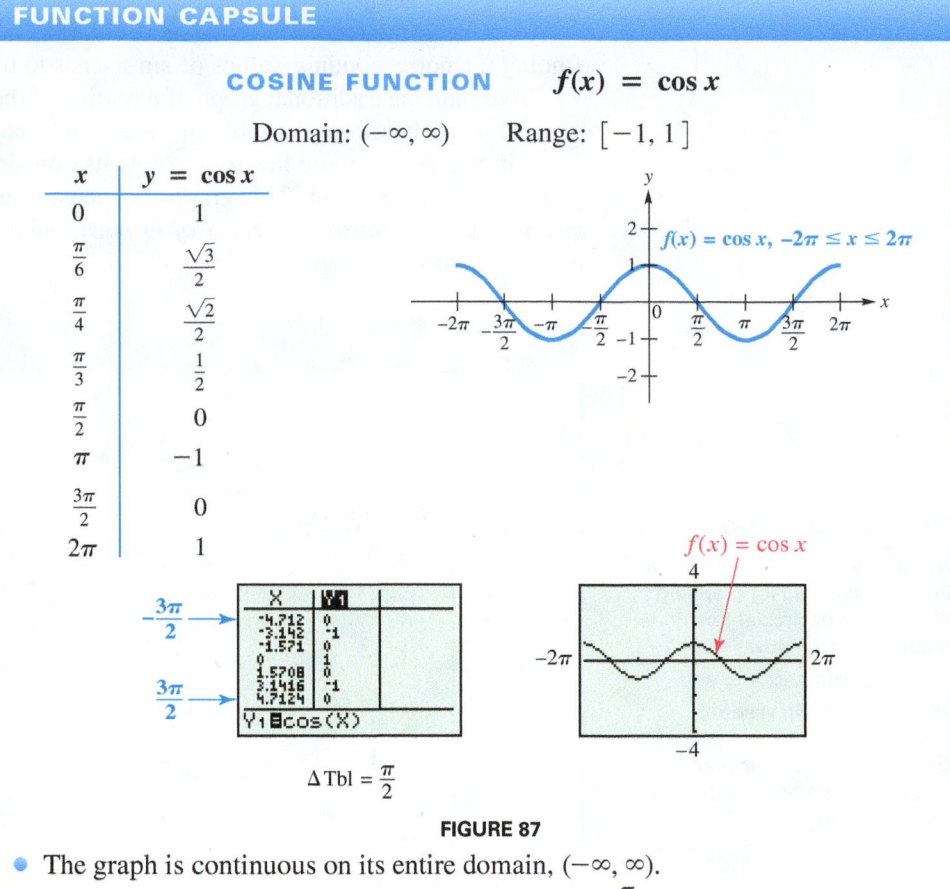

x	$y = \cos x$
0	1
$\frac{\pi}{6}$	$\frac{\sqrt{3}}{2}$
$\frac{\pi}{4}$	$\frac{\sqrt{2}}{2}$
$\frac{\pi}{3}$	$\frac{1}{2}$
$\frac{\pi}{2}$	0
π	-1
$\frac{3\pi}{2}$	0
2π	1

$\Delta \text{Tbl} = \frac{\pi}{2}$

FIGURE 87

- The graph is continuous on its entire domain, $(-\infty, \infty)$.
- Its x-intercepts have x-values of the form $(2n + 1)\frac{\pi}{2}$, where n is an integer.
- Its period is 2π.
- The graph is symmetric with respect to the y-axis. It is an even function. For all x in the domain, $\cos(-x) = \cos x$.

NOTE   Recall that a function f is an **even function** if for all x in the domain of f, $f(-x) = f(x)$. The graph of an even function is symmetric with respect to the y-axis (**Section 2.1**). That is, if (x, y) is on the graph of the function, then so is $(-x, y)$. For example, $\left(\frac{\pi}{2}, 0\right)$ and $\left(-\frac{\pi}{2}, 0\right)$ are points on the graph of $y = \cos x$, illustrating that $\cos(-x) = \cos x$.

Graphing Techniques, Amplitude, and Period

The examples that follow show graphs that are "stretched" or "shrunk" either vertically, horizontally, or both, compared with the graphs of $y = \sin x$ or $y = \cos x$.

EXAMPLE 1   **Graphing $y = a \sin x$**

Graph $y = 2 \sin x$, and compare with the graph of $f(x) = \sin x$.

Solution   For a given value of x, the value of $y = 2 \sin x$ is twice as large as it would be for $y = \sin x$, as shown in the table of values. The only change in the graph is the range, which becomes $[-2, 2]$. See **FIGURE 88,** which includes a graph of $y = \sin x$ for comparison.

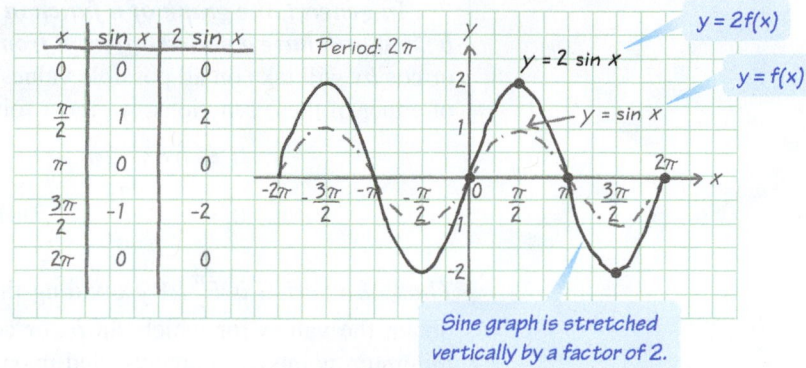

FIGURE 88

We can think of the graph of $y = a \sin x$ as a vertical stretching of the graph of $y = \sin x$ when $a > 1$ and a vertical shrinking when $0 < a < 1$.

The **amplitude** of a *sinusoidal* function is *half* the difference between the maximum and minimum values. It describes the height of the graph both above and below a horizontal line passing through the "middle" of the graph. Thus, for the sine and cosine functions, the amplitude is

Amplitudes of the sine and cosine graphs are both 1.

$$\frac{1}{2}[1 - (-1)] = \frac{1}{2}(2) = 1.$$

Amplitude

The graph of $y = a \sin x$ or $y = a \cos x$, with $a \neq 0$, will have the same shape as the graph of $y = \sin x$ or $y = \cos x$, respectively, except with range $[-|a|, |a|]$. The amplitude is $|a|$.

No matter what the value of the amplitude, the periods of $y = a \sin x$ and $y = a \cos x$ are still 2π. Now consider $y = \sin 2x$. We can complete a table of values for the interval $[0, 2\pi]$.

2 cycles on $[0, 2\pi]$

x	0	$\frac{\pi}{4}$	$\frac{\pi}{2}$	$\frac{3\pi}{4}$	π	$\frac{5\pi}{4}$	$\frac{3\pi}{2}$	$\frac{7\pi}{4}$	2π
$\sin 2x$	0	1	0	-1	0	1	0	-1	0

$\vdash$———— 1 cycle ————$\rightarrow\vdash$———— 2 cycles ————$\rightarrow\vert$

Note that one complete cycle occurs in π units, not 2π units. Therefore, the period here is π, which equals $\frac{2\pi}{2}$. What about $y = \sin 4x$? Look at the next table.

2 cycles on $[0, \pi]$

x	0	$\frac{\pi}{8}$	$\frac{\pi}{4}$	$\frac{3\pi}{8}$	$\frac{\pi}{2}$	$\frac{5\pi}{8}$	$\frac{3\pi}{4}$	$\frac{7\pi}{8}$	π
$\sin 4x$	0	1	0	-1	0	1	0	-1	0

$\vdash$———— 1 cycle ————$\rightarrow\vdash$———— 2 cycles ————$\rightarrow\vert$

A complete cycle is achieved in $\frac{\pi}{2}$ or $\frac{2\pi}{4}$ units, which is reasonable, because

$$\sin\left(4 \cdot \frac{\pi}{2}\right) = \sin 2\pi = 0.$$

In general, the graph of a function of the form $y = \sin bx$ *or* $y = \cos bx$, *for* $b > 0$, *will have a period different from* 2π *when* $b \neq 1$. Since the values of $\sin bx$ or $\cos bx$ will take on all possible values as bx ranges from 0 to 2π, to find the period of either of these functions, we must solve this three-part inequality.

$$0 \leq bx \leq 2\pi \qquad \text{Solve the inequality for } x.$$

$$0 \leq x \leq \frac{2\pi}{b} \qquad \text{Divide by the positive number } b.$$

Thus, the period is $\frac{2\pi}{b}$. By dividing the interval $\left[0, \frac{2\pi}{b}\right]$ into four equal parts, we obtain the values for which $\sin bx$ or $\cos bx$ is -1, 0, or 1. These values will give minimum points, x-intercepts, and maximum points on the graph. Once these points are determined, we can sketch the graph by joining the points with a smooth sinusoidal curve. (If a function has $b < 0$, then the identities of the next chapter can be used to write the function as one in which $b > 0$.)

> **NOTE** One method to divide an interval into four equal parts is as follows.
>
> **Step 1** Find the midpoint of the interval by adding the x-values of the endpoints and dividing by 2.
>
> **Step 2** Find the two midpoints of the intervals found in Step 1, using the same procedure.

EXAMPLE 2 **Graphing $y = \sin bx$**

Graph $y = \sin 2x$, and compare with the graph of $f(x) = \sin x$.

Solution For this function, the coefficient of x is 2, so $b = 2$ and the period is $\frac{2\pi}{2} = \pi$. Therefore, the graph will complete one period over the interval $[0, \pi]$. The endpoints are 0 and π, and the three points between these endpoints are

$$\frac{1}{2}\left(0 + \frac{\pi}{2}\right), \quad \frac{1}{2}(0 + \pi), \quad \text{and} \quad \frac{1}{2}\left(\frac{\pi}{2} + \pi\right),$$

which give the following x-values.

0	$\dfrac{\pi}{4}$	$\dfrac{\pi}{2}$	$\dfrac{3\pi}{4}$	π
↑	↑	↑	↑	↑
Left endpoint	First-quarter point	Midpoint	Third-quarter point	Right endpoint

We plot the points from the table of values given on the previous page and join them with a smooth sinusoidal curve. More of the graph can be sketched by repeating this cycle, as shown in **FIGURE 89**. The amplitude is not changed.

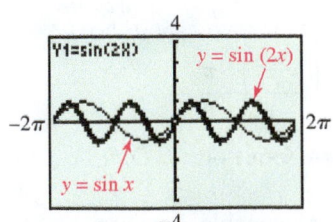

The thick graph of $y = \sin 2x$ oscillates twice as often as the graph of $y = \sin x$.

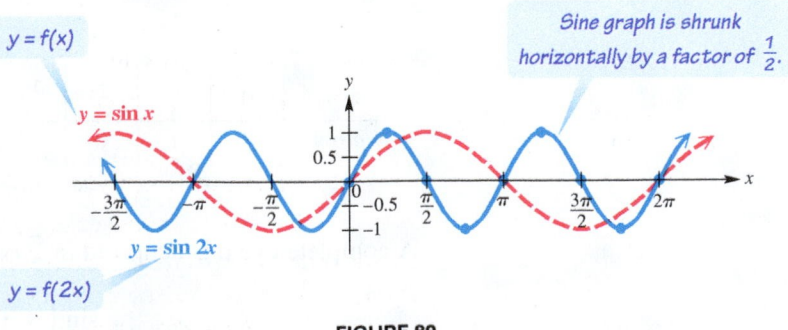

FIGURE 89

We can think of the graph of $y = \sin bx$ as a horizontal stretching of the graph of $y = \sin x$ when $0 < b < 1$ and a horizontal shrinking when $b > 1$.

Period

For $b > 0$, the graph of **$y = \sin bx$** will resemble that of $y = \sin x$, but with period $\frac{2\pi}{b}$. Also, the graph of **$y = \cos bx$** will resemble that of $y = \cos x$, but with period $\frac{2\pi}{b}$.

EXAMPLE 3   **Graphing $y = \cos bx$**

Graph $y = \cos \frac{2}{3}x$ over one period.

Solution   The period is

$$\frac{2\pi}{\frac{2}{3}} = 2\pi \div \frac{2}{3} = 2\pi \cdot \frac{3}{2} = 3\pi.$$

We divide the interval $[0, 3\pi]$ into four equal parts to get the x-values $0, \frac{3\pi}{4}, \frac{3\pi}{2}, \frac{9\pi}{4}$, and 3π, which yield minimum points, maximum points, and x-intercepts. We use these values to obtain a table of key points for one period.

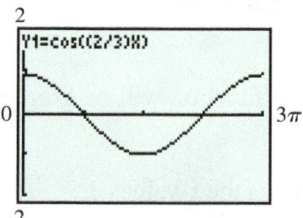

This screen shows a graph of the function in **Example 3**. By choosing Xscl = $\frac{3\pi}{4}$, the x-intercepts, maxima, and minima coincide with tick marks on the x-axis.

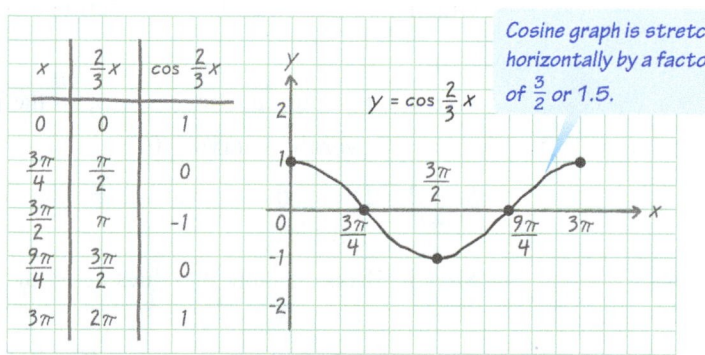

x	$\frac{2}{3}x$	$\cos \frac{2}{3}x$
0	0	1
$\frac{3\pi}{4}$	$\frac{\pi}{2}$	0
$\frac{3\pi}{2}$	π	-1
$\frac{9\pi}{4}$	$\frac{3\pi}{2}$	0
3π	2π	1

Cosine graph is stretched horizontally by a factor of $\frac{3}{2}$ or 1.5.

FIGURE 90

The amplitude is 1 because the maximum value is 1, the minimum value is -1, and

$$\frac{1}{2}(1 - (-1)) = \frac{1}{2}(2) = 1.$$

We plot these points and join them with a smooth curve, as shown in **FIGURE 90**. ●

NOTE   Look at the middle column of the table in **Example 3**. Dividing the interval $\left[0, \frac{2\pi}{b}\right]$ into four equal parts will always give the values $0, \frac{\pi}{2}, \pi, \frac{3\pi}{2}$, and 2π for this column, resulting in values of -1, 0, or 1 for this sinusoidal function. These values lead to key points on the graph of $y = \sin bx$ or $y = \cos bx$, which can then be easily sketched.

> ### Guidelines for Sketching Graphs of the Sine and Cosine Functions
>
> To graph $y = a \sin bx$ or $y = a \cos bx$, with $b > 0$, follow these steps.
>
> **Step 1** Find the period, $\frac{2\pi}{b}$. Start at 0 on the x-axis, and mark off a distance of $\frac{2\pi}{b}$.
>
> **Step 2** Divide the interval into four equal parts. (See the note preceding **Example 2.**)
>
> **Step 3** Evaluate the function for each of the five x-values resulting from Step 2. The points will be maximum points, minimum points, and x-intercepts.
>
> **Step 4** Plot the points found in Step 3, and join them with a sinusoidal curve having amplitude $|a|$.
>
> **Step 5** Draw the graph over additional periods, to the right and to the left, as needed.

The function in **Example 4** has the form $y = a \sin bx$. Both amplitude and period are affected by constants.

EXAMPLE 4 **Graphing $y = a \sin bx$**

Graph $y = -2 \sin 3x$ over one period using the preceding guidelines.

Solution

Step 1 For this function, $b = 3$, so the period is $\frac{2\pi}{3}$. The function will be graphed over the interval $\left[0, \frac{2\pi}{3}\right]$.

Step 2 Divide the interval $\left[0, \frac{2\pi}{3}\right]$ into four equal parts to get the x-values $0, \frac{\pi}{6}, \frac{\pi}{3}, \frac{\pi}{2}$, and $\frac{2\pi}{3}$.

Step 3 Make a table of values determined by the x-values from Step 2.

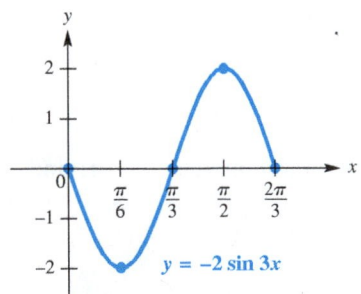

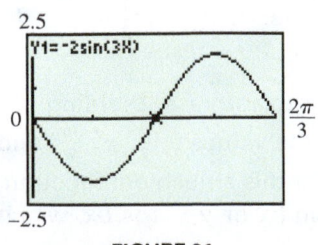

FIGURE 91

Key Points (*x, y*)

x	0	$\frac{\pi}{6}$	$\frac{\pi}{3}$	$\frac{\pi}{2}$	$\frac{2\pi}{3}$
$3x$	0	$\frac{\pi}{2}$	π	$\frac{3\pi}{2}$	2π
$\sin 3x$	0	1	0	-1	0
$-2 \sin 3x$	0	-2	0	2	0

Step 4 Plot the points $(0, 0)$, $\left(\frac{\pi}{6}, -2\right)$, $\left(\frac{\pi}{3}, 0\right)$, $\left(\frac{\pi}{2}, 2\right)$, and $\left(\frac{2\pi}{3}, 0\right)$, and join them with a sinusoidal curve having amplitude 2. See **FIGURE 91**. (A calculator graph is shown for comparison.)

Step 5 The graph can be extended by repeating the cycle.

Note that the graph of $y = -2 \sin 3x$ is a reflection of the graph of $y = 2 \sin 3x$ across the x-axis. It is like the graph of $y = \sin x$ except it has period $\frac{2\pi}{3}$, has amplitude 2, and is reflected across the x-axis. •

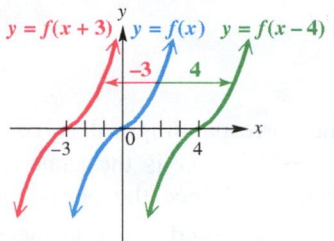

Horizontal translations
of $y = f(x)$

FIGURE 92

Translations and Transformations

In general, the graph of a function of the form

$$y = f(x - d)$$

is translated *horizontally* compared with the graph of $y = f(x)$. The translation is d units to the right if $d > 0$ and $|d|$ units to the left if $d < 0$. See **FIGURE 92**. With circular functions, a horizontal translation is called a **phase shift.** In the function $y = f(x - d)$, the expression $x - d$ is called the **argument.**

In **Example 5**, we give two methods that can be used to sketch the graph of a circular function involving a phase shift.

EXAMPLE 5 Graphing $y = \sin(x - d)$

Graph $y = \sin\left(x - \frac{\pi}{3}\right)$ over one period.

Solution **Method 1** For the argument $x - \frac{\pi}{3}$ to result in all possible values throughout one period, it must take on all values between 0 and 2π, inclusive. Therefore, to find an interval of one period, we solve this three-part inequality.

$$0 \le x - \frac{\pi}{3} \le 2\pi \quad \text{leads to} \quad \frac{\pi}{3} \le x \le \frac{7\pi}{3} \qquad \text{Add } \tfrac{\pi}{3} \text{ to each part.}$$

Divide the interval $\left[\frac{\pi}{3}, \frac{7\pi}{3}\right]$ into four equal parts to get the *x*-values $\frac{\pi}{3}, \frac{5\pi}{6}, \frac{4\pi}{3}, \frac{11\pi}{6}$, and $\frac{7\pi}{3}$. A table of values using these *x*-values follows.

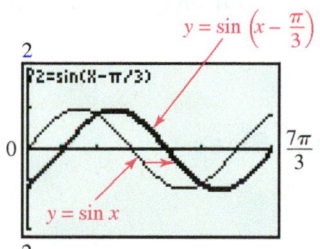

Note the horizontal translation.

Key Points (x, y)

x	$\frac{\pi}{3}$	$\frac{5\pi}{6}$	$\frac{4\pi}{3}$	$\frac{11\pi}{6}$	$\frac{7\pi}{3}$
$x - \frac{\pi}{3}$	0	$\frac{\pi}{2}$	π	$\frac{3\pi}{2}$	2π
$\sin\left(x - \frac{\pi}{3}\right)$	0	1	0	-1	0

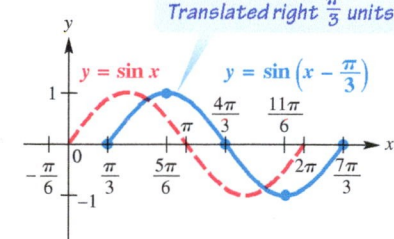

FIGURE 93

We join the corresponding points to get the graph shown in **FIGURE 93.** The period is 2π, and the amplitude is 1.

Method 2 We can also graph $y = \sin\left(x - \frac{\pi}{3}\right)$ by using a horizontal translation of the graph of $y = \sin x$. The argument $x - \frac{\pi}{3}$ indicates that the graph will be translated $\frac{\pi}{3}$ units to the *right* (the phase shift) compared with the graph of $y = \sin x$. See **FIGURE 93.**

Therefore, to graph a function with this method, first graph the basic circular function, and then graph the desired function by using the appropriate translation.

NOTE The graph in **FIGURE 93** of **Example 5** can be extended through additional periods by repeating the given portion of the graph as necessary.

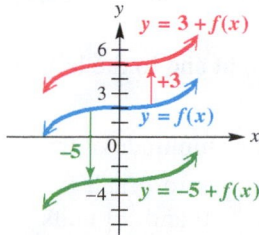

Vertical translations
of $y = f(x)$

FIGURE 94

In general, the graph of a function of the form

$$y = c + f(x)$$

is translated *vertically* compared with the graph of $y = f(x)$. The translation is c units up if $c > 0$ and $|c|$ units down if $c < 0$. See **FIGURE 94.**

EXAMPLE 6 **Graphing $y = c + a \cos bx$**

Graph $y = 3 - 2 \cos 3x$.

Solution The values of y will be 3 greater than the corresponding values of y in $y = -2 \cos 3x$. This means that the graph of $y = 3 - 2 \cos 3x$ is the same as the graph of $y = -2 \cos 3x$, vertically translated 3 units up. Since the period of $y = -2 \cos 3x$ is $\frac{2\pi}{3}$, the key points have x-values $0, \frac{\pi}{6}, \frac{\pi}{3}, \frac{\pi}{2},$ and $\frac{2\pi}{3}$. Use these x-values to make a table of values.

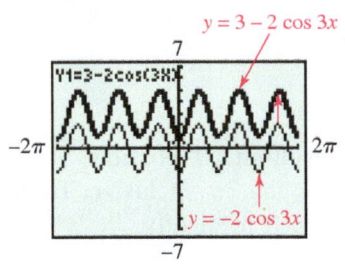

Key Points (x, y)

x	0	$\frac{\pi}{6}$	$\frac{\pi}{3}$	$\frac{\pi}{2}$	$\frac{2\pi}{3}$
$\cos 3x$	1	0	-1	0	1
$2 \cos 3x$	2	0	-2	0	2
$3 - 2 \cos 3x$	1	3	5	3	1

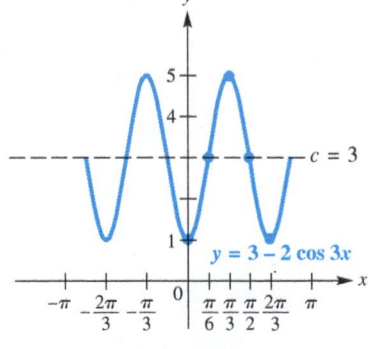

FIGURE 95

The key points are shown on the graph in **FIGURE 95,** along with more of the graph, sketched using the fact that the function is periodic. ●

Further Guidelines for Sketching Graphs of the Sine and Cosine Functions

A function of the form

$$y = c + a \sin[b(x - d)] \quad \text{or} \quad y = c + a \cos[b(x - d)], \quad b > 0,$$

can be graphed according to the following guidelines.

Method 1 Follow these steps.

Step 1 Find an interval whose length is one period $\frac{2\pi}{b}$ by solving the three-part inequality $0 \le b(x - d) \le 2\pi$.

Step 2 Divide the interval into four equal parts.

Step 3 Evaluate the function for each of the five x-values resulting from Step 2. The points will be maximum points, minimum points, and points that intersect the line $y = c$ ("middle" points of the wave).

Step 4 Plot the points found in Step 3, and join them with a sinusoidal curve having amplitude $|a|$.

Step 5 Draw the graph over additional periods, to the right and to the left, as needed.

Method 2 First graph $y = a \sin bx$ or $y = a \cos bx$. The amplitude of the function is $|a|$, and the period is $\frac{2\pi}{b}$. Then use translations to graph the desired function. The vertical translation is c units up if $c > 0$ and $|c|$ units down if $c < 0$. The horizontal translation (phase shift) is d units to the right if $d > 0$ and $|d|$ units to the left if $d < 0$.

EXAMPLE 7   **Graphing $y = c + a \sin [b(x - d)]$**

Graph $y = -1 + 2 \sin (4x + \pi)$ over two periods.

Solution   We use Method 1. First write the expression on the right side in the form $c + a \sin [b(x - d)]$ by factoring out 4 in the argument.

$$y = -1 + 2 \sin (4x + \pi) = -1 + 2 \sin \left[4 \left(x + \frac{\pi}{4} \right) \right] \qquad \text{Rewrite } 4x + \pi \text{ as } 4\left(x + \frac{\pi}{4}\right).$$

Step 1   Find an interval whose length is one period.

$$0 \le 4 \left(x + \frac{\pi}{4} \right) \le 2\pi$$

$$0 \le x + \frac{\pi}{4} \le \frac{\pi}{2} \qquad \text{Divide by 4.}$$

$$-\frac{\pi}{4} \le x \le \frac{\pi}{4} \qquad \text{Subtract } \frac{\pi}{4}.$$

Step 2   Divide the interval $\left[-\frac{\pi}{4}, \frac{\pi}{4} \right]$ into four equal parts to get the x-values $-\frac{\pi}{4}, -\frac{\pi}{8}, 0, \frac{\pi}{8}$, and $\frac{\pi}{4}$.

Step 3   Make a table of values.

Key Points (x, y)

x	$-\frac{\pi}{4}$	$-\frac{\pi}{8}$	0	$\frac{\pi}{8}$	$\frac{\pi}{4}$	← *x-values*
$x + \frac{\pi}{4}$	0	$\frac{\pi}{8}$	$\frac{\pi}{4}$	$\frac{3\pi}{8}$	$\frac{\pi}{2}$	
$4\left(x + \frac{\pi}{4}\right)$	0	$\frac{\pi}{2}$	π	$\frac{3\pi}{2}$	2π	
$\sin\left[4\left(x + \frac{\pi}{4}\right)\right]$	0	1	0	-1	0	
$2\sin\left[4\left(x + \frac{\pi}{4}\right)\right]$	0	2	0	-2	0	
$-1 + 2\sin(4x + \pi)$	-1	1	-1	-3	-1	← *y-values*

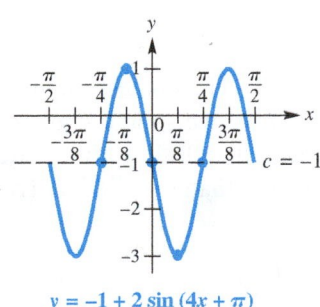

$y = -1 + 2 \sin (4x + \pi)$

FIGURE 96

Steps 4 and 5   Plot the key points (x, y) found in the table and join them with a sinusoidal curve. **FIGURE 96** shows the graph, extended to the right and left to include two full periods.   ●

Determining a Trigonometric Model Using Curve Fitting

A sinusoidal function is often a good approximation of a set of periodic data points.

EXAMPLE 8   **Modeling Temperature with a Sine Function**

The maximum monthly average temperature in New Orleans is 82°F and the minimum is 54°F. The table in the margin on the next page shows the monthly average temperatures. The scatter diagram for a 2-year interval in **FIGURE 97** on the next page strongly suggests that the temperatures can be modeled with a sine curve.

(continued)

Month	°F	Month	°F
Jan	54	July	82
Feb	55	Aug	81
Mar	61	Sept	77
Apr	69	Oct	71
May	73	Nov	59
June	79	Dec	55

Source: Miller, A. and J. Thompson, *Elements of Meteorology,* Charles E. Merrill.

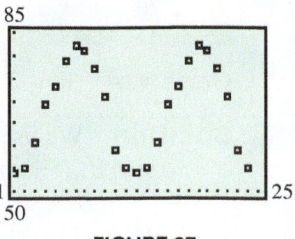

FIGURE 97

(a) Using only the maximum and minimum temperatures, determine a function of the form $f(x) = a \sin[b(x - d)] + c$, where a, b, c, and d are constants, that models the monthly average temperature in New Orleans. Let x represent the month, with January corresponding to $x = 1$.

(b) On the same coordinate axes, graph f for a 2-year period, together with the actual data values found in the table.

(c) Use the **sine regression** feature of a graphing calculator to determine a second model for these data.

Solution

(a) We can use the maximum and minimum monthly average temperatures to find the amplitude a.

$$a = \frac{82 - 54}{2} = 14 \qquad \text{Amplitude}$$

The average of the maximum and minimum temperatures is a good choice for c.

$$c = \frac{82 + 54}{2} = 68 \qquad \text{Vertical translation}$$

Since temperatures repeat every 12 months, b is $\frac{2\pi}{12} = \frac{\pi}{6}$. The coldest month is January ($x = 1$), and the hottest month is July ($x = 7$), so we should choose d to be about 4. The table shows that temperatures are actually a little warmer after July than before, so we experiment with values just greater than 4 to find d. Trial and error with a calculator leads to $d = 4.2$.

$$f(x) = a \sin[b(x - d)] + c$$

$$f(x) = 14 \sin\left[\frac{\pi}{6}(x - 4.2)\right] + 68 \qquad \text{Period} = \frac{\pi}{6}, d = 4.2$$

(b) See **FIGURE 98(a)**. The figure also includes the graph of $y = 14 \sin\frac{\pi}{6}x + 68$ for comparison, showing the horizontal translation of the model.

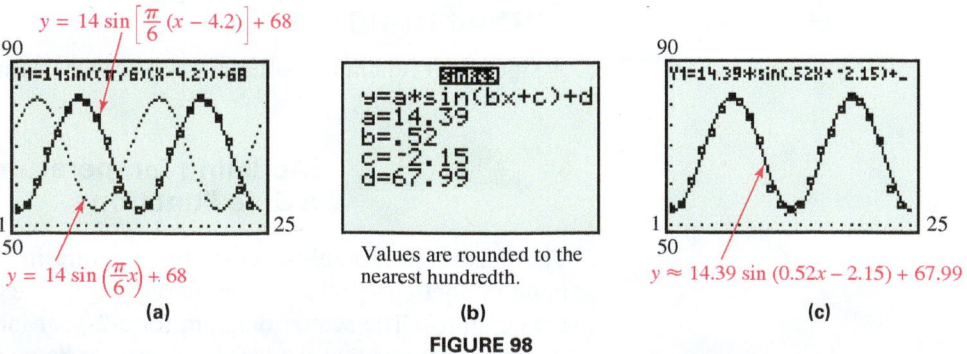

(a)

Values are rounded to the nearest hundredth.

(b)

(c)

FIGURE 98

(c) Using two years of the given data, the screen in **FIGURE 98(b)** shows the equation of the model, while **FIGURE 98(c)** shows its graph along with the data points.

NOTE   If $f(x) = a \sin [b(x - d)] + c$ is used to model temperature data and if the greatest monthly average temperature is H and the least is L, then $a = \frac{1}{2}(H - L)$ and $c = \frac{1}{2}(H + L)$. If the period is 12 months, then $b = \frac{2\pi}{12} = \frac{\pi}{6}$ and if the maximum average temperature occurs in July, then $d \approx 4$.

WHAT WENT WRONG?

A student graphed $y = -1 + 3 \sin \frac{1}{2}x$ in the window $[-2\pi, 2\pi]$ by $[-4, 4]$. He knew that the amplitude should be 3 and the period should be 4π. However, his graph looked like this.

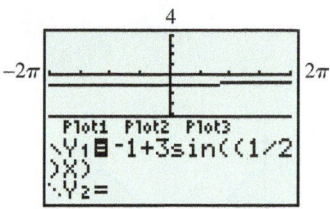

What Went Wrong?   How can he obtain the correct graph?

Answer to What Went Wrong?

He graphed the function while the calculator was in *degree* mode. To obtain the correct graph, he should change the mode to *radian*.

8.6 Exercises

Checking Analytic Skills   *Match each function defined in Exercises 1–8 with its graph in A–H.* **Do not use a calculator.**

1. $y = \sin x$

2. $y = \cos x$

3. $y = -\sin x$

4. $y = -\cos x$

5. $y = \sin 2x$

6. $y = \cos 2x$

7. $y = 2 \sin x$

8. $y = 2 \cos x$

A.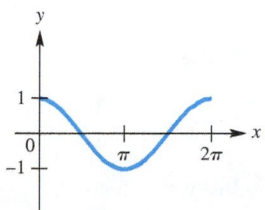

B.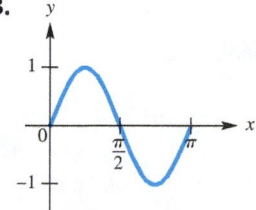

C.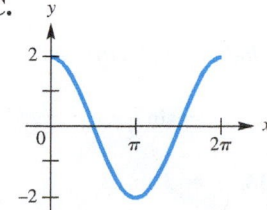

D.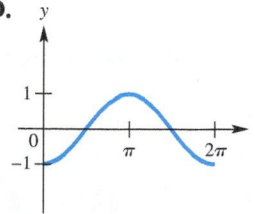

E.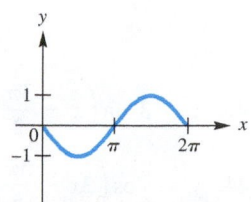

F.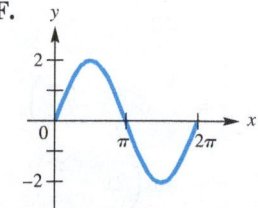

G.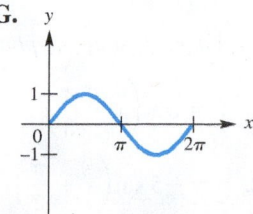

H.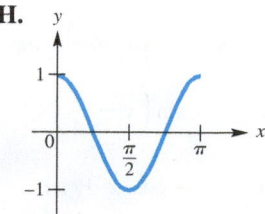

Checking Analytic Skills *Match each function defined in Exercises 9–16 with its graph in A–H.*
Do not use a calculator.

9. $y = \sin\left(x - \dfrac{\pi}{4}\right)$

10. $y = \sin\left(x + \dfrac{\pi}{4}\right)$

11. $y = \cos\left(x - \dfrac{\pi}{4}\right)$

12. $y = \cos\left(x + \dfrac{\pi}{4}\right)$

13. $y = 1 + \sin x$

14. $y = -1 + \sin x$

15. $y = 1 + \cos x$

16. $y = -1 + \cos x$

A.

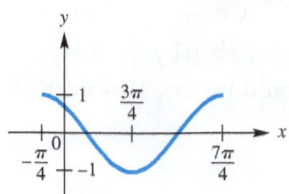

B.

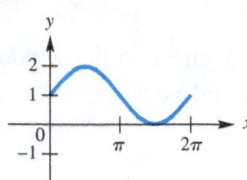

C.

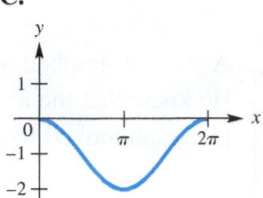

D.

E.

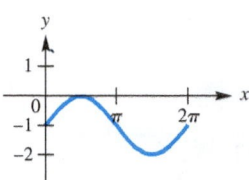

F.

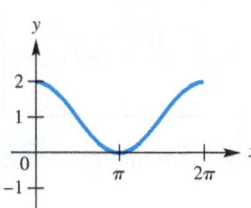

G.

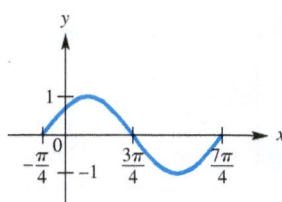

H.

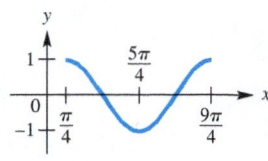

Checking Analytic Skills *Match each function defined in Column I with the appropriate description in Column II.* **Do not use a calculator.**

I	**II**

17. $y = 3 \sin(2x - 4)$

A. Amplitude = 2; period = $\dfrac{\pi}{2}$; phase shift = $\dfrac{3}{4}$

18. $y = 2 \sin(3x - 4)$

B. Amplitude = 3; period = π; phase shift = 2

19. $y = 4 \sin(3x - 2)$

C. Amplitude = 4; period = $\dfrac{2\pi}{3}$; phase shift = $\dfrac{2}{3}$

20. $y = 2 \sin(4x - 3)$

D. Amplitude = 2; period = $\dfrac{2\pi}{3}$; phase shift = $\dfrac{4}{3}$

Graph each function over the interval $[-2\pi, 2\pi]$. Give the amplitude.

21. $y = 2 \cos x$

22. $y = 3 \sin x$

23. $y = \dfrac{2}{3} \sin x$

24. $y = \dfrac{3}{4} \cos x$

25. $y = -\cos x$

26. $y = -\sin x$

27. $y = -2 \sin x$

28. $y = -3 \cos x$

Graph each function over a two-period interval. Give the period and amplitude.

29. $y = \sin \dfrac{1}{2}x$

30. $y = \sin \dfrac{2}{3}x$

31. $y = \cos 2x$

32. $y = \cos \dfrac{3}{4}x$

33. $y = 2 \sin \dfrac{1}{4}x$

34. $y = 3 \sin 2x$

35. $y = -2 \cos 3x$

36. $y = -5 \cos 2x$

Graph each function over a two-period interval. State the phase shift.

37. $y = \sin\left(x - \dfrac{\pi}{4}\right)$

38. $y = \cos\left(x - \dfrac{\pi}{3}\right)$

39. $y = 2 \cos\left(x - \dfrac{\pi}{3}\right)$

40. $y = 3 \sin\left(x - \dfrac{3\pi}{2}\right)$

41. $y = 2 \cos(x + \pi)$

42. $y = -5 \sin\left(x + \dfrac{\pi}{2}\right)$

43. $y = \sin\left(2x + \dfrac{\pi}{4}\right)$

44. $y = \cos\left(3x - \dfrac{3\pi}{5}\right)$

Find the **(a)** *amplitude,* **(b)** *period,* **(c)** *phase shift (if any),* **(d)** *vertical translation (if any), and* **(e)** *range of each function. Then graph the function over at least one period.*

45. $y = -4 \sin(2x - \pi)$

46. $y = 3 \cos(4x + \pi)$

47. $y = \frac{1}{2} \cos\left(\frac{1}{2}x - \frac{\pi}{4}\right)$

48. $y = -\frac{1}{4} \sin\left(\frac{3}{4}x + \frac{\pi}{8}\right)$

49. $y = 1 - \frac{2}{3} \sin \frac{3}{4}x$

50. $y = -1 - 2 \cos 5x$

51. $y = 1 - 2 \cos \frac{1}{2}x$

52. $y = -3 + 3 \sin \frac{1}{2}x$

53. $y = -3 + 2 \sin\left(x + \frac{\pi}{2}\right)$

54. $y = 4 - 3 \cos(x - \pi)$

55. $y = \frac{1}{2} + \sin\left[2\left(x + \frac{\pi}{4}\right)\right]$

56. $y = -\frac{5}{2} + \cos\left[3\left(x - \frac{\pi}{6}\right)\right]$

57. $y = 2 \sin(x - \pi)$

58. $y = \frac{2}{3} \cos\left(x + \frac{\pi}{2}\right)$

59. $y = 4 \cos\left(\frac{1}{2}x + \frac{\pi}{2}\right)$

60. $y = -\cos\left[\frac{2}{3}\left(x - \frac{\pi}{3}\right)\right]$

61. $y = 2 - \sin\left(3x - \frac{\pi}{5}\right)$

62. $y = -1 + \frac{1}{2} \cos(2x - 3\pi)$

Concept Check   *Each function graphed is of the form* $y = a \sin bx$ *or* $y = a \cos bx$, *where* $b > 0$.
Determine the equation of the graph.

63.
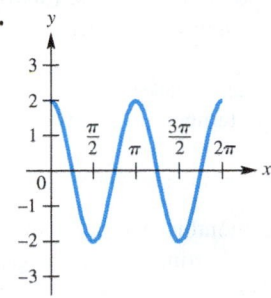

64.
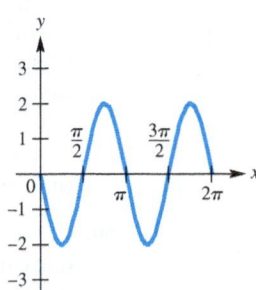

65.
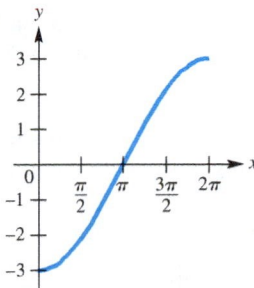

66.
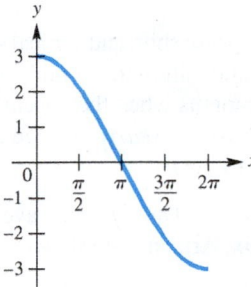

67.
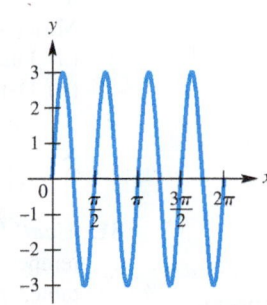

68.
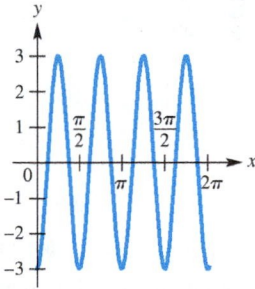

Concept Check   *Each function graphed is of the form* $y = c + \cos x$, $y = c + \sin x$, $y = \cos(x - d)$,
or $y = \sin(x - d)$, *where* d *is the least possible positive value. Determine the equation of the graph.*

69.
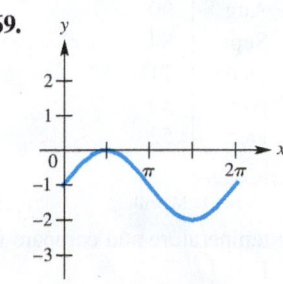

70.
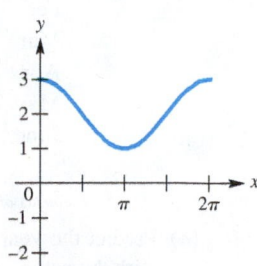

71.
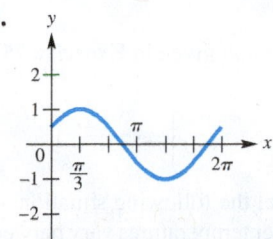

72.
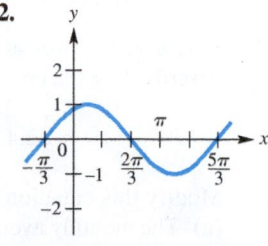

(Modeling) Solve each problem.

73. *Average Temperatures* The graph models the monthly average temperature y in degrees Fahrenheit for a city in Canada, where x is the month.

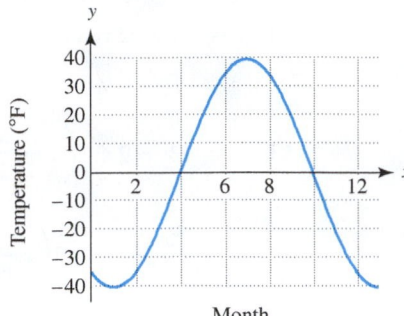

(a) Find the maximum and minimum monthly average temperatures.

(b) Find the amplitude and period. Interpret the results.

(c) Explain what the x-intercepts represent.

74. *Average Temperatures* The graph in **Exercise 73** is given by $y = 40 \cos\left[\frac{\pi}{6}(x - 7)\right]$. Modify this equation to model the following situations.

(a) The maximum monthly average temperature is 50°F and the minimum is −50°F.

(b) The maximum monthly average temperature is 60°F and the minimum is −20°F.

75. *Ocean Temperatures* The graph models the Gulf of Mexico water temperatures in degrees Fahrenheit at St. Petersburg, Florida. (*Source:* J. Williams.)

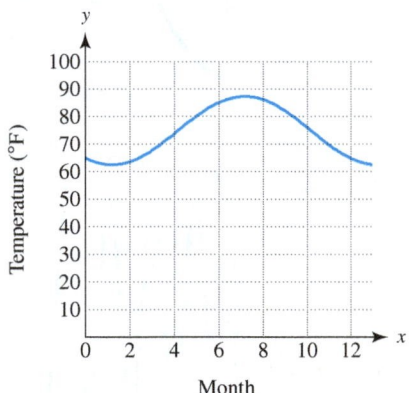

(a) Estimate the maximum and minimum water temperatures. When do they occur?

(b) What would happen to the amplitude of the graph if the minimum water temperature decreased to 50°F? Make a sketch of this situation.

76. *Ocean Temperatures* The graph given in **Exercise 75** is described by the equation

$$y = 12.4 \sin\left[\frac{\pi}{6}(x - 4.2)\right] + 75.$$

Modify this equation to model the following situations.

(a) The monthly average water temperatures vary between 60°F and 90°F.

(b) The monthly average water temperatures vary between 50°F and 70°F.

Tides for Kahului Harbor *The graph shows the tides for Kahului Harbor, on the island of Maui, Hawaii. Use the graph for Exercises 77 and 78.*

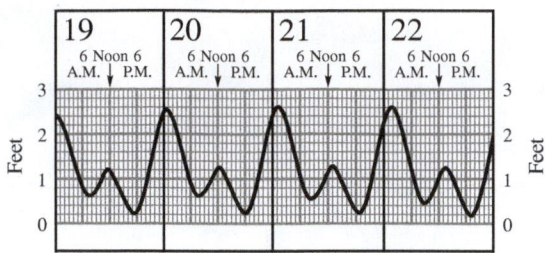

Source: Maui News. Original chart prepared by Edward K. Noda and Associates.

77. The graph is an example of a periodic function. What is the period (in hours)?

78. What is the amplitude?

79. *Average Temperatures* The monthly average temperatures in degrees Fahrenheit at Mould Bay, Canada, may be modeled by $f(x) = 34 \sin\left[\frac{\pi}{6}(x - 4.3)\right]$, where x is the month and x = 1 corresponds to January. (*Source:* A. Miller and J. Thompson, *Elements of Meteorology,* Charles E. Merrill.)

(a) Find the amplitude, period, and phase shift.

(b) Approximate the average temperature during May and December.

(c) Estimate the *yearly* average temperature at Mould Bay.

80. *Average Temperatures* The monthly average temperatures in degrees Fahrenheit at Austin, Texas, are given by $f(x) = 17.5 \sin\left[\frac{\pi}{6}(x - 4)\right] + 67.5$, where x is the month and x = 1 corresponds to January. (*Source:* A. Miller and J. Thompson.)

(a) Find the amplitude, period, phase shift, and vertical shift.

(b) Determine the maximum and minimum monthly average temperature and the months when they occur.

(c) Make a conjecture as to how the *yearly* average temperature might be related to $f(x)$.

81. *Monthly Average Temperatures* The monthly average temperatures (in °F) in Phoenix, Arizona, are shown in the table.

Month	°F	Month	°F
Jan	51	July	90
Feb	55	Aug	90
Mar	63	Sept	84
Apr	67	Oct	71
May	77	Nov	59
June	86	Dec	52

Source: Miller, A. and J. Thompson, *Elements of Meteorology,* Charles E. Merrill.

(a) Predict the yearly average temperature and compare it with the actual value of 70°F.

(b) Plot the monthly average temperature over a 2-year period by letting $x = 1$ correspond to January of the first year.

(c) Determine a sinusoidal function of the form $f(x) = a \cos [b(x - d)] + c$, where a, b, c, and d are constants, that models the data.

(d) Graph f together with the data on the same coordinate axes. How well does f model the data?

(e) Use the sine regression capability of a graphing calculator to find the equation of a sine curve that fits these data.

82. *Monthly Average Temperatures* The monthly average temperatures (in °F) in Vancouver, Canada, are shown in the table.

Month	°F	Month	°F
Jan	36	July	64
Feb	39	Aug	63
Mar	43	Sept	57
Apr	48	Oct	50
May	55	Nov	43
June	59	Dec	39

Source: Miller, A. and J. Thompson, *Elements of Meteorology*, Charles E. Merrill.

(a) Plot the monthly average temperatures over a 2-year period by letting $x = 1$ correspond to the month of January during the first year. Do the data seem to indicate a translated sine graph?

(b) The highest monthly average temperature is 64°F in July, and the lowest monthly average temperature is 36°F in January. Their average is 50°F. Graph the data together with the line $y = 50$. What does this line represent with regard to temperature in Vancouver?

(c) Approximate the amplitude, period, and phase shift of the translated sine wave indicated by the data.

(d) Determine a sinusoidal function of the form $f(x) = a \sin [b(x - d)] + c$, where a, b, c, and d are constants, that models the data.

(e) Graph f together with the data on the same coordinate axes. How well does f model the given data?

(f) Use the sine regression capability of a graphing calculator to find the equation of a sine curve that fits these data.

83. *Modeling Temperatures* The monthly average temperatures in a Canadian city are shown in the table.

Month	1	2	3	4	5	6
Temperature (°F)	40	43	47	52	59	63

Month	7	8	9	10	11	12
Temperature (°F)	68	67	61	54	47	43

(a) Plot the average monthly temperature over a 24-month period by letting $x = 1$ and $x = 13$ correspond to January.

(b) Find the constants a, b, c, and d so that the function $f(x) = a \sin [(b(x - c)] + d$ models the data.

(c) Graph f together with the data.

84. *Modeling Temperatures* The monthly average temperatures in Chicago, Illinois, are shown in the table.

Month	1	2	3	4	5	6
Temperature (°F)	25	28	36	48	61	72

Month	7	8	9	10	11	12
Temperature (°F)	74	75	66	55	39	28

Source: A. Miller and J. Thompson, *Elements of Meteorology*, Charles E. Merrill.

(a) Plot the monthly average temperature over a 24-month period by letting $x = 1$ and $x = 13$ correspond to January.

(b) Find the constants a, b, c, and d so that the function $f(x) = a \sin [b(x - c)] + d$ models the data.

(c) Graph f and the data together.

85. *Modeling Temperatures* The monthly average high temperatures in Augusta, Georgia, are shown in the table.

Month	1	2	3	4	5	6
Temperature (°F)	58	60	68	77	82	90

Month	7	8	9	10	11	12
Temperature (°F)	92	91	83	77	68	60

Source: J. Williams, *The Weather Almanac.*

(a) Use $f(x) = a \cos [b(x - c)] + d$ to model these data.

(b) Are different values for c possible? Explain.

86. *Modeling Temperatures* The maximum monthly average temperature in Anchorage, Alaska, is 57°F and the minimum is 12°F.

Month	1	2	3	4	5	6
Temperature (°F)	12	18	23	36	46	55

Month	7	8	9	10	11	12
Temperature (°F)	57	55	48	36	23	16

Source: A. Miller and J. Thompson, *Elements of Meteorology*, Charles E. Merrill.

(a) Using only these two temperatures, determine $f(x) = a \cos [b(x - c)] + d$ so that $f(x)$ models the monthly average temperatures in Anchorage.

(b) Graph f and the actual data in the table over a 2-year period.

87. *Modeling Tidal Currents* Tides cause ocean currents to flow into and out of harbors and canals. The table shows the speed of the ocean current at Cape Cod Canal in bogo-knots (bk) *x* hours after midnight on August 26, 1998. (Note that to change bogo-knots to knots, take the square root of the absolute value of the number of bogo-knots.)

Time (hr)	3.7	6.75	9.8	13.0	16.1	22.2
Current (bk)	−18	0	18	0	−18	18

Source: WWW Tide and Current Predictor.

(a) Use $f(x) = a \cos [b(x - c)] + d$ to model the data.
(b) Graph *f* and the data in $[0, 24]$ by $[-20, 20]$. Interpret the graph.

88. *Modeling Ocean Temperature* The following table lists the monthly average ocean temperatures in degrees Fahrenheit at Veracruz, Mexico.

Month	1	2	3	4	5	6
Temperature (°F)	72	73	74	78	81	83

Month	7	8	9	10	11	12
Temperature (°F)	84	85	84	82	78	74

Source: J. Williams, The Weather Almanac.

(a) Make a scatterplot of the data over a 2-year period.
(b) Use $f(x) = a \sin [b(x - c)] + d$ to model the data.

89. *Atmospheric Carbon Dioxide* The carbon dioxide content in the atmosphere at Barrow, Alaska, in parts per million (ppm) can be modeled with the function

$$C(x) = 0.04x^2 + 0.6x + 330 + 7.5 \sin (2\pi x),$$

where *x* is in years and where $x = 0$ corresponds to 1970. (*Source:* Zeilik, M., S. Gregory, and E. Smith. *Introductory Astronomy and Astrophysics,* Fourth Edition, Saunders College Publishers.)

(a) Graph *C* for $5 \leq x \leq 25$. (*Hint:* For the range, use $320 \leq y \leq 380$.)
(b) Define a new function *C* that is valid if *x* represents the actual year, where $1970 \leq x \leq 1995$.

90. *Atmospheric Carbon Dioxide* At Mauna Loa, Hawaii, atmospheric carbon dioxide levels in parts per million (ppm) have been measured regularly since 1958. The function

$$L(x) = 0.022x^2 + 0.55x + 316 + 3.5 \sin (2\pi x)$$

can be used to model these levels, where *x* is in years and $x = 0$ corresponds to 1960. (*Source:* Nilsson, A., *Greenhouse Earth,* John Wiley and Sons.)

(a) Graph *L* for $15 \leq x \leq 35$. (*Hint:* For the range, use $325 \leq y \leq 365$.)
(b) When do the seasonal maximum and minimum carbon dioxide levels occur?
(c) *L* is the sum of a quadratic function and a sine function. What is the significance of each of these functions? Discuss what physical phenomena may be responsible for each function.

SECTIONS 8.5–8.6 Reviewing Basic Concepts

1. Give the exact coordinates of the point on the unit circle that corresponds to the given value of *s*.

(a) -2π (b) $\dfrac{5\pi}{4}$ (c) $\dfrac{5\pi}{2}$

2. Give the values of the six trigonometric functions of $-\dfrac{5\pi}{2}$. State which functions are undefined.

3. Give the exact values of the six trigonometric functions of each number.

(a) $\dfrac{7\pi}{6}$ (b) $-\dfrac{2\pi}{3}$

4. Give calculator approximations for the six trigonometric function values of 2.25.

5. If $\cos 100 \approx 0.8623$ and $\sin 100 \approx -0.5064$, then in which quadrant does the point on the unit circle corresponding to the real number 100 lie?

6. Graph one period of $y = -\cos x$. State the period and the amplitude.

7. Graph $y = 3 \sin(\pi x + \pi)$ on the interval $[-2, 2]$. State the amplitude, period, and phase shift.

8. *(Modeling) Daylight Hours*   The graph of

$$f(x) = 6.5 \sin\left[\frac{\pi}{6}(x - 3.65)\right] + 12.4$$

models the daylight hours at 60°N latitude, where $x = 1$ corresponds to January 1, $x = 2$ to February 1, and so on.
   (a) Estimate the maximum and minimum number of daylight hours.
   (b) Interpret the amplitude and period.

8.7   Graphs of the Other Circular Functions

Graphs of the Secant and Cosecant Functions • Graphs of the Tangent and Cotangent Functions

---

Graphs of the Secant and Cosecant Functions

Consider the table of selected points accompanying the graph of the secant function in **FIGURE 99.** These points include special values between $-\pi$ and π. The secant function is undefined for odd multiples of $\frac{\pi}{2}$ and has *vertical asymptotes* for such values. A **vertical asymptote** is a vertical line that the graph approaches but does not intersect, while function values increase or decrease without bound as x-values get closer and closer to the line. Furthermore, since $\sec(-x) = \sec x$ (see **Exercise 11**), the secant function is even and its graph is symmetric with respect to the y-axis.

x	$y = \sec x$	x	$y = \sec x$
$\pm\frac{\pi}{3}$	2	$\pm\frac{2\pi}{3}$	-2
$\pm\frac{\pi}{4}$	$\sqrt{2} \approx 1.4$	$\pm\frac{3\pi}{4}$	$-\sqrt{2} \approx -1.4$
$\pm\frac{\pi}{6}$	$\frac{2\sqrt{3}}{3} \approx 1.2$	$\pm\frac{5\pi}{6}$	$-\frac{2\sqrt{3}}{3} \approx -1.2$
0	1	$\pm\pi$	-1

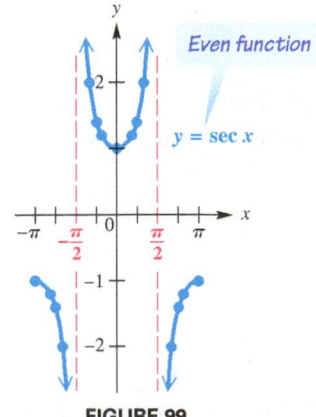

FIGURE 99

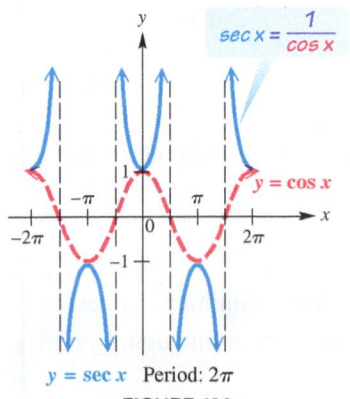

$y = \sec x$ Period: 2π

FIGURE 100

Because secant values are reciprocals of corresponding cosine values, the period of the secant function is 2π, the same as for $y = \cos x$. When $\cos x = 1$, the value of $\sec x$ is also 1. Likewise, when $\cos x = -1$, $\sec x = -1$ as well. For all x, $-1 \le \cos x \le 1$, and thus, $|\sec x| \ge 1$ for all x in its domain. **FIGURE 100** shows how the graphs of $y = \cos x$ and $y = \sec x$ are related.

A similar analysis for selected points between $-\pi$ and π for the graph of the cosecant function yields the graph in **FIGURE 101**. The vertical asymptotes are at x-values that are integer multiples of π. Because $\csc(-x) = -\csc x$ (see **Exercise 12**), the cosecant function is odd and its graph is symmetric with respect to the origin.

x	$y = \csc x$	x	$y = \csc x$
$\frac{\pi}{6}$	2	$-\frac{\pi}{6}$	-2
$\frac{\pi}{4}$	$\sqrt{2} \approx 1.4$	$-\frac{\pi}{4}$	$-\sqrt{2} \approx -1.4$
$\frac{\pi}{3}$	$\frac{2\sqrt{3}}{3} \approx 1.2$	$-\frac{\pi}{3}$	$-\frac{2\sqrt{3}}{3} \approx -1.2$
$\frac{\pi}{2}$	1	$-\frac{\pi}{2}$	-1
$\frac{2\pi}{3}$	$\frac{2\sqrt{3}}{3} \approx 1.2$	$-\frac{2\pi}{3}$	$-\frac{2\sqrt{3}}{3} \approx -1.2$
$\frac{3\pi}{4}$	$\sqrt{2} \approx 1.4$	$-\frac{3\pi}{4}$	$-\sqrt{2} \approx -1.4$
$\frac{5\pi}{6}$	2	$-\frac{5\pi}{6}$	-2

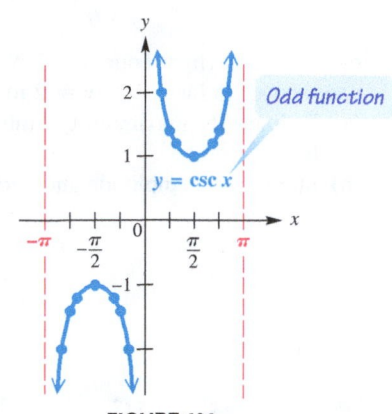

FIGURE 101

Because cosecant values are reciprocals of corresponding sine values, the period of the cosecant function is 2π, the same as for $y = \sin x$. When $\sin x = 1$, the value of $\csc x$ is also 1. Likewise, when $\sin x = -1$, $\csc x = -1$. For all x, $-1 \le \sin x \le 1$, and thus, $|\csc x| \ge 1$ for all x in its domain. **FIGURE 102** shows how the graphs of $y = \sin x$ and $y = \csc x$ are related.

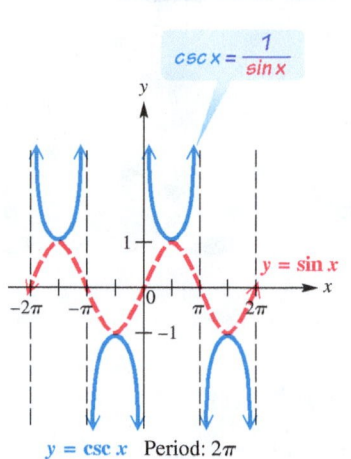

$y = \csc x$ Period: 2π

FIGURE 102

FUNCTION CAPSULE

SECANT FUNCTION $f(x) = \sec x$

Domain: $\{x \mid x \ne (2n + 1)\frac{\pi}{2}, \text{ where } n \text{ is an integer}\}$ Range: $(-\infty, -1] \cup [1, \infty)$

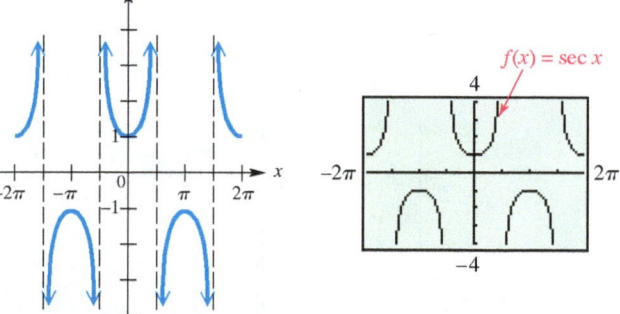

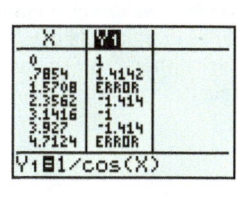

$\Delta \text{Tbl} = \frac{\pi}{4}$

FIGURE 103

- The graph is discontinuous at $x = (2n + 1)\frac{\pi}{2}$ and has vertical asymptotes at these values.
- There are no x-intercepts.
- Its period is 2π.
- Its graph has no amplitude, since there are no maximum or minimum values.
- The graph is symmetric with respect to the y-axis. It is an even function. For all x in the domain, $\sec(-x) = \sec x$.

TECHNOLOGY NOTE

Typically, calculators do not have keys for the secant and cosecant functions. We enter sec x as

$$\frac{1}{\cos x} \quad \text{or} \quad (\cos x)^{-1},$$

and csc x as

$$\frac{1}{\sin x} \quad \text{or} \quad (\sin x)^{-1}.$$

Do not confuse the *reciprocals*

$$(\cos x)^{-1} = \frac{1}{\cos x}$$

and

$$(\sin x)^{-1} = \frac{1}{\sin x}$$

with the *inverses* $\cos^{-1} x$ and $\sin^{-1} x$.

FOR DISCUSSION

For which x-values does sec x equal cos x? For which x-values does csc x equal sin x?

FUNCTION CAPSULE

COSECANT FUNCTION $f(x) = \csc x$

Domain: $\{x \mid x \neq n\pi,\ \text{where } n \text{ is an integer}\}$ Range: $(-\infty, -1] \cup [1, \infty)$

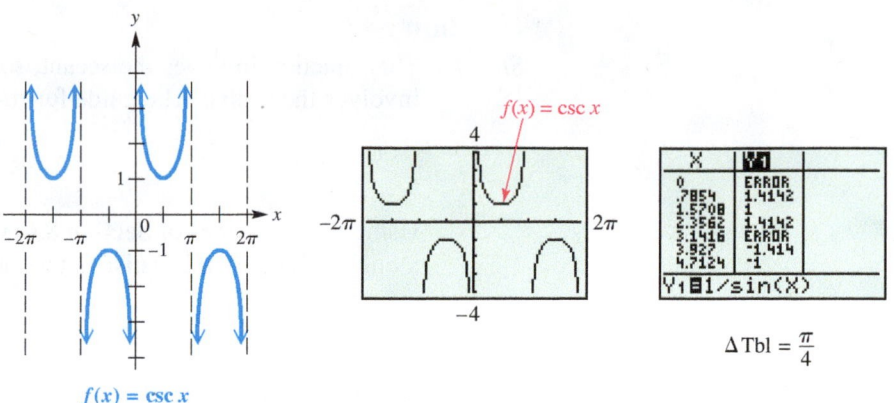

$\Delta \text{Tbl} = \dfrac{\pi}{4}$

FIGURE 104

- The graph is discontinuous at $x = n\pi$ and has vertical asymptotes at these values.
- There are no x-intercepts.
- Its period is 2π.
- Its graph has no amplitude, since there are no maximum or minimum values.
- The graph is symmetric with respect to the origin. It is an odd function. For all x in the domain, $\csc(-x) = -\csc x$.

FOR DISCUSSION

The graphs of the secant and cosecant functions both consist of a series of U-shaped branches. Explain how one of these branches differs fundamentally from a parabola.

Use the following guidelines when graphing the secant and cosecant functions.

Guidelines for Sketching Graphs of the Secant and Cosecant Functions

To graph $y = a \sec bx$ or $y = a \csc bx$, with $b > 0$, follow these steps.

Step 1   Graph the corresponding reciprocal function as a guide.

For $y = a \sec bx$,   graph $y = a \cos bx$ as a guide function.
For $y = a \csc bx$,   graph $y = a \sin bx$ as a guide function.

Step 2   Sketch the vertical asymptotes with equations $x = k$, where $(k, 0)$ is an x-intercept of the graph of the guide function.

Step 3   Sketch the graph of the desired function by drawing the typical U-shaped branches between the adjacent asymptotes. The branches will be above the graph of the guide function when the guide function values are positive and below the graph of the guide function when the guide function values are negative.

We identify these steps in **Example 1.**

EXAMPLE 1 Graphing $y = a \sec bx$

Graph $y = 2 \sec \frac{1}{2}x$.

Solution

Step 1   This function involves the secant, so the corresponding reciprocal function involves the cosine. The guide function to graph is

$$y = 2 \cos \frac{1}{2}x.$$

Using the guidelines of **Section 8.6,** we find that one period of the graph lies along the interval that satisfies the inequality

$$0 \le \frac{1}{2}x \le 2\pi, \quad \text{or} \quad 0 \le x \le 4\pi. \qquad \text{Multiply by 2.}$$

Dividing the interval $[0, 4\pi]$ into four equal parts gives the key points

$$(0, 2), \quad (\pi, 0), \quad (2\pi, -2), \quad (3\pi, 0), \quad \text{and} \quad (4\pi, 2), \qquad \text{Key points}$$

which are joined with a smooth dashed curve to indicate that this graph is only a guide. An additional period is graphed, as seen in **FIGURE 105(a).**

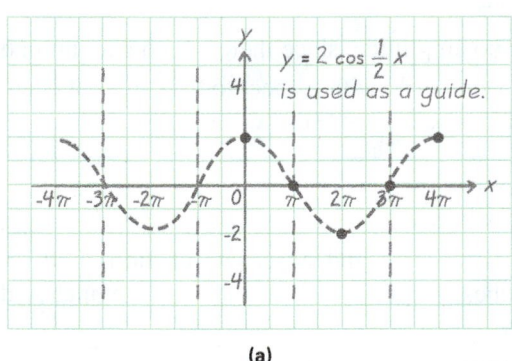 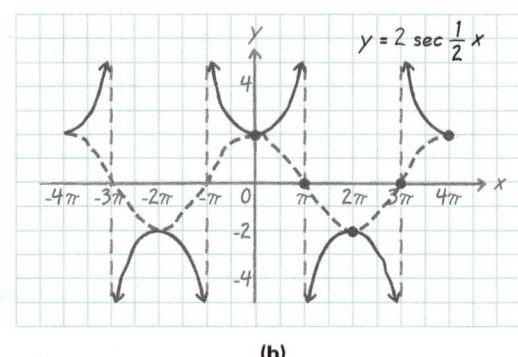

(a)   (b)

FIGURE 105

Step 2   We sketch the vertical asymptotes as shown in **FIGURE 105(a).** These occur at x-values for which the guide function equals 0, such as

$$x = -3\pi, \quad x = -\pi, \quad x = \pi, \quad \text{and} \quad x = 3\pi.$$

Step 3   We sketch the graph of $y = 2 \sec \frac{1}{2}x$ by drawing the typical U-shaped branches, approaching the asymptotes. See **FIGURE 105(b)** and **FIGURE 106.** ●

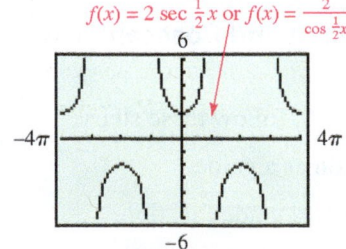

$f(x) = 2 \sec \frac{1}{2}x \text{ or } f(x) = \dfrac{2}{\cos \frac{1}{2}x}$

This is a calculator graph of the function in **Example 1.**

FIGURE 106

NOTE   In **FIGURE 105(b)**, the vertical asymptotes and the dashed curve of $y = 2 \cos \frac{1}{2}x$ are **not** part of the graph of $y = 2 \sec \frac{1}{2}x$. They are aids that can be used when graphing by hand. The actual graph of $y = 2 \sec \frac{1}{2}x$ consists of only the solid branches of the curve. When sketching the branches, be sure to show correctly how they approach the vertical asymptotes. They do not intersect them, and do not "curve back."

Like graphs of the sine and cosine functions, graphs of the cosecant and secant functions may be translated vertically and horizontally.

> **EXAMPLE 2** Graphing $y = a \csc(x - d)$

Graph $y = \frac{3}{2} \csc\left(x - \frac{\pi}{2}\right)$.

Analytic Solution

First we graph the corresponding guide function

$$y = \frac{3}{2} \sin\left(x - \frac{\pi}{2}\right).$$

Compared with the graph of $y = \sin x$, the graph of $y = \frac{3}{2} \sin\left(x - \frac{\pi}{2}\right)$ has amplitude $\frac{3}{2}$ and phase shift $\frac{\pi}{2}$ units to the right. Its graph is shown as a dashed curve in **FIGURE 107**. The x-coordinates of the x-intercepts of the sinusoidal graph of $y = \frac{3}{2} \sin\left(x - \frac{\pi}{2}\right)$ correspond to the vertical asymptotes found in the graph of $y = \frac{3}{2} \csc\left(x - \frac{\pi}{2}\right)$. These asymptotes are shown as vertical dashed lines. We sketch the graph of $y = \frac{3}{2} \csc\left(x - \frac{\pi}{2}\right)$ by using the sinusoidal graph of $y = \frac{3}{2} \sin\left(x - \frac{\pi}{2}\right)$ as a guide. Two periods are shown in **FIGURE 107**.

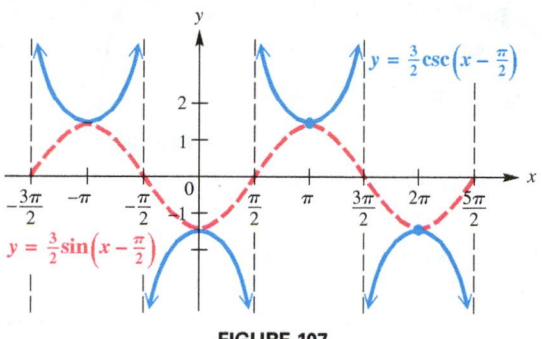

FIGURE 107

Graphing Calculator Solution

FIGURE 108(a) depicts $y_1 = \frac{3}{2} \sin\left(x - \frac{\pi}{2}\right)$ and $y_2 = \frac{3}{2} \csc\left(x - \frac{\pi}{2}\right)$. To enter y_2 in the calculator, use

$$Y_2 = (3/2)(1/\sin(X - \pi/2)).$$

FIGURE 108(b) shows a graph of only $y_2 = \frac{3}{2} \csc\left(x - \frac{\pi}{2}\right)$. Notice that the TI-84 Plus does not draw the asymptotes because they are not part of the actual graph.

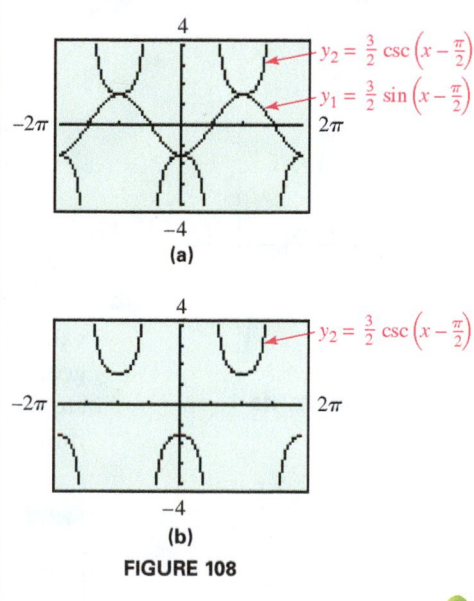

FIGURE 108

Graphs of the Tangent and Cotangent Functions

Consider the table of selected points accompanying the graph of the tangent function in **FIGURE 109**. These points include special values between $-\frac{\pi}{2}$ and $\frac{\pi}{2}$. The tangent function is undefined for odd multiples of $\frac{\pi}{2}$ and, thus, like the secant function, has vertical asymptotes for such values. Furthermore, since $\tan(-x) = -\tan x$ (see **Exercise 13**), the tangent function is odd and its graph is symmetric with respect to the origin.

x	$y = \tan x$
$-\frac{\pi}{3}$	$-\sqrt{3} \approx -1.7$
$-\frac{\pi}{4}$	-1
$-\frac{\pi}{6}$	$-\frac{\sqrt{3}}{3} \approx -0.6$
0	0
$\frac{\pi}{6}$	$\frac{\sqrt{3}}{3} \approx 0.6$
$\frac{\pi}{4}$	1
$\frac{\pi}{3}$	$\sqrt{3} \approx 1.7$

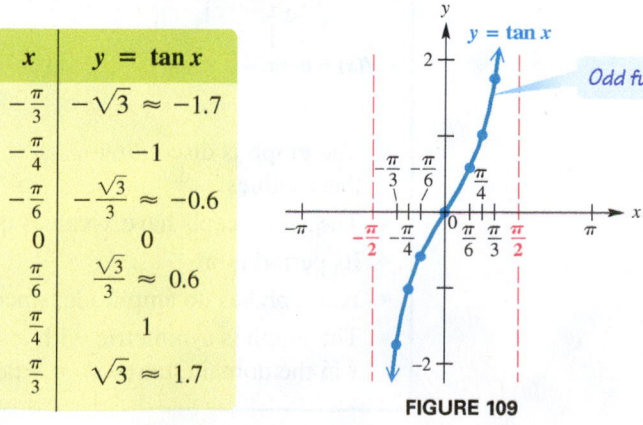

FIGURE 109

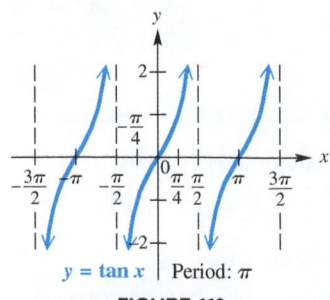

$y = \tan x$ | Period: π

FIGURE 110

The tangent function has period π. Because $\tan x = \frac{\sin x}{\cos x}$, tangent values are 0 when sine values are 0, and undefined when cosine values are 0. As x-values go from $-\frac{\pi}{2}$ to $\frac{\pi}{2}$, tangent values go from $-\infty$ to ∞ and increase throughout the interval. Those same values are repeated as x goes from $\frac{\pi}{2}$ to $\frac{3\pi}{2}$, $\frac{3\pi}{2}$ to $\frac{5\pi}{2}$, and so on. The graph of $y = \tan x$ from $-\frac{3\pi}{2}$ to $\frac{3\pi}{2}$ is shown in **FIGURE 110**.

A similar analysis for selected points between 0 and π for the graph of the cotangent function yields the graph in **FIGURE 111**. Here the vertical asymptotes are at x-values that are integer multiples of π. Because $\cot(-x) = -\cot x$ (see **Exercise 14**), the cotangent function is odd and its graph is symmetric with respect to the origin. (This can be seen when more of the graph is plotted.) See **FIGURE 112**.

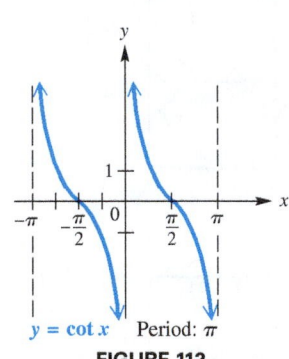

$y = \cot x$ | Period: π

FIGURE 112

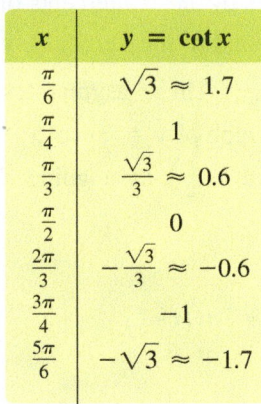

x	$y = \cot x$
$\frac{\pi}{6}$	$\sqrt{3} \approx 1.7$
$\frac{\pi}{4}$	1
$\frac{\pi}{3}$	$\frac{\sqrt{3}}{3} \approx 0.6$
$\frac{\pi}{2}$	0
$\frac{2\pi}{3}$	$-\frac{\sqrt{3}}{3} \approx -0.6$
$\frac{3\pi}{4}$	-1
$\frac{5\pi}{6}$	$-\sqrt{3} \approx -1.7$

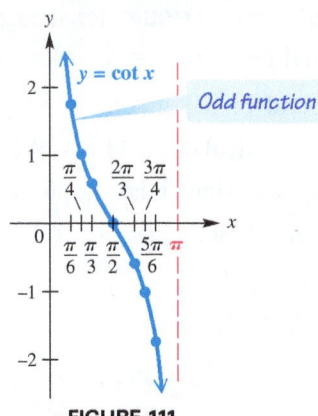

FIGURE 111

The cotangent function also has period π. Cotangent values are 0 when cosine values are 0, and undefined when sine values are 0. As x-values go from 0 to π, cotangent values go from ∞ to $-\infty$ and decrease throughout the interval. Those same values are repeated as x goes from π to 2π, 2π to 3π, and so on. The graph of $y = \cot x$ from $-\pi$ to π is shown in **FIGURE 112.** The graph continues in this pattern.

FUNCTION CAPSULE

TANGENT FUNCTION $f(x) = \tan x$

Domain: $\left\{ x \mid x \neq (2n+1)\frac{\pi}{2}, \text{ where } n \text{ is an integer} \right\}$ Range: $(-\infty, \infty)$

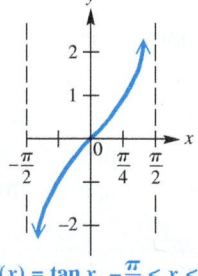

$f(x) = \tan x, \ -\frac{\pi}{2} < x < \frac{\pi}{2}$

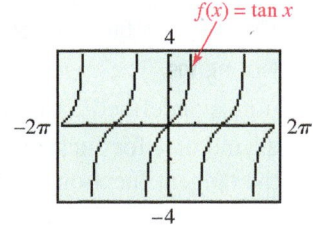

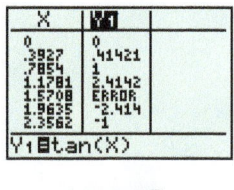

$\Delta \text{Tbl} = \frac{\pi}{8}$

FIGURE 113

- The graph is discontinuous at $x = (2n+1)\frac{\pi}{2}$ and has vertical asymptotes at these values.
- The x-intercepts have x-values of the form $n\pi$.
- Its period is π.
- Its graph has no amplitude, since there are no minimum or maximum values.
- The graph is symmetric with respect to the origin. It is an odd function. For all x in the domain, $\tan(-x) = -\tan x$.

FUNCTION CAPSULE

COTANGENT FUNCTION $f(x) = \cot x$

Domain: $\{x \mid x \neq n\pi, \text{ where } n \text{ is an integer}\}$ Range: $(-\infty, \infty)$

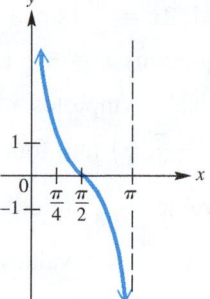

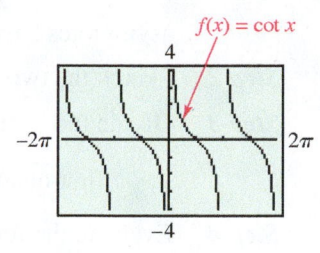

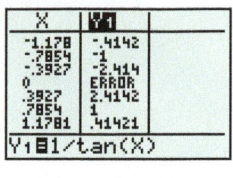

$f(x) = \cot x, \ 0 < x < \pi$

$\Delta \text{Tbl} = \frac{\pi}{8}$

FIGURE 114

- The graph is discontinuous at $x = n\pi$ and has vertical asymptotes at these values.
- The x-intercepts have x-values of the form $(2n + 1)\frac{\pi}{2}$.
- Its period is π.
- Its graph has no amplitude, since there are no minimum or maximum values.
- The graph is symmetric with respect to the origin. It is an odd function. For all x in the domain, $\cot(-x) = -\cot x$.

NOTE   The sine, cosine, secant, and cosecant functions all have period 2π. Only the tangent and cotangent have period π.

Use the following guidelines when graphing the tangent and cotangent functions.

Guidelines for Sketching Graphs of the Tangent and Cotangent Functions

To graph $y = a \tan bx$ or $y = a \cot bx$, with $b > 0$, follow these steps.

Step 1   The period is $\frac{\pi}{b}$. To locate two adjacent vertical asymptotes, solve the following equations for x.

For $y = a \tan bx$: $bx = -\frac{\pi}{2}$ and $bx = \frac{\pi}{2}$
For $y = a \cot bx$: $bx = 0$ and $bx = \pi$

Step 2   Sketch the two vertical asymptotes found in Step 1.

Step 3   Divide the interval formed by the vertical asymptotes into four equal parts. (See the Note in **Section 8.6** on how to do this.)

Step 4   Evaluate the function for the first-quarter point, midpoint, and third-quarter point, using the x-values found in Step 3.

Step 5   Join the points with a smooth curve, approaching the vertical asymptotes. Indicate additional asymptotes and periods as necessary.

EXAMPLE 3 Graphing $y = \tan bx$

Graph $y = \tan 2x$ using the preceding guidelines.

Solution

Step 1 The period of this function is $\frac{\pi}{2}$. To locate two adjacent vertical asymptotes, solve $2x = -\frac{\pi}{2}$ and $2x = \frac{\pi}{2}$ (since this is a tangent function). The two asymptotes have equations $x = -\frac{\pi}{4}$ and $x = \frac{\pi}{4}$.

Step 2 Sketch the two vertical asymptotes $x = \pm\frac{\pi}{4}$, as shown in **FIGURE 115(a).**

Step 3 Divide the interval $\left(-\frac{\pi}{4}, \frac{\pi}{4}\right)$ into four equal parts to obtain key x-values.

first-quarter value: $-\frac{\pi}{8}$, middle value: **0**, third-quarter value: $\frac{\pi}{8}$

Step 4 Evaluate the function for the x-values found in Step 3.

Key Points (x, y)

x	$-\frac{\pi}{8}$	0	$\frac{\pi}{8}$
$2x$	$-\frac{\pi}{4}$	0	$\frac{\pi}{4}$
$\tan 2x$	-1	0	1

Step 5 Join these points (x, y) with a smooth curve, approaching the vertical asymptotes. See **FIGURE 115(a).** Another period has been graphed, one half-period to the left and one half-period to the right. A calculator graph is also given in **FIGURE 115(b).**

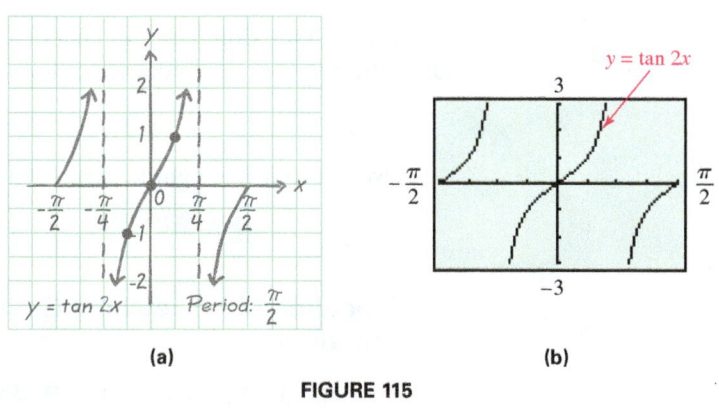

(a)   (b)

FIGURE 115

EXAMPLE 4 Graphing $y = a \cot bx$

Graph $y = \frac{1}{2}\cot 2x$.

Solution Because this function involves the cotangent, we can locate two adjacent asymptotes by solving the equations $2x = 0$ and $2x = \pi$. The lines $x = 0$ (the y-axis) and $x = \frac{\pi}{2}$ are two such asymptotes. We divide the interval $\left(0, \frac{\pi}{2}\right)$ into four equal parts, getting key x-values of $\frac{\pi}{8}$, $\frac{\pi}{4}$, and $\frac{3\pi}{8}$. Evaluating the function at these x-values gives the key points $\left(\frac{\pi}{8}, \frac{1}{2}\right)$, $\left(\frac{\pi}{4}, 0\right)$, and $\left(\frac{3\pi}{8}, -\frac{1}{2}\right)$. Joining these points with a smooth curve approaching the asymptotes gives the graph shown in **FIGURE 116.** A calculator graph is also given.

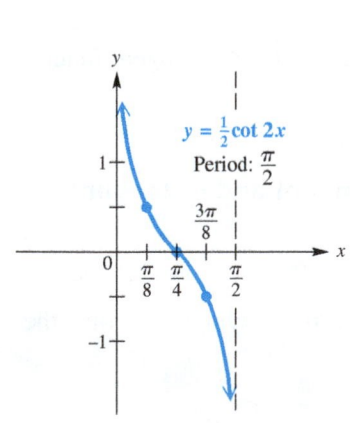

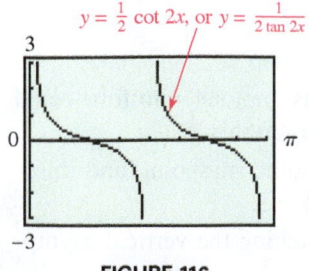

FIGURE 116

As with other functions, the graphs of the tangent and cotangent functions may be translated horizontally as well as vertically.

EXAMPLE 5 **Graphing a Cotangent Function with Vertical and Horizontal Translations**

Graph $y = -2 - \cot\left(x - \frac{\pi}{4}\right)$.

Solution Here, $b = 1$, so the period is π. The graph will be translated downward 2 units (because $c = -2$), will be reflected across the x-axis (because of the negative sign in front of the cotangent), and will have a phase shift, or horizontal translation, $\frac{\pi}{4}$ unit to the right $\left(\text{because of the argument } \left(x - \frac{\pi}{4}\right)\right)$. To locate adjacent asymptotes, since this function involves the cotangent, we solve the following equations.

$$x - \frac{\pi}{4} = 0 \qquad \text{and} \qquad x - \frac{\pi}{4} = \pi$$

$$x = \frac{\pi}{4} \qquad\qquad\qquad x = \frac{5\pi}{4}$$

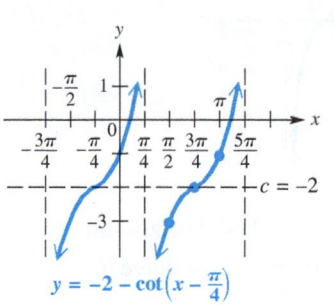

$$y = -2 - \cot\left(x - \frac{\pi}{4}\right)$$

FIGURE 117

Dividing the interval $\left(\frac{\pi}{4}, \frac{5\pi}{4}\right)$ into four equal parts and evaluating the function at the three key x-values within the interval gives the key points $\left(\frac{\pi}{2}, -3\right)$, $\left(\frac{3\pi}{4}, -2\right)$, and $(\pi, -1)$. Join these points with a smooth curve. This period of the graph, along with another one in the interval $\left(-\frac{3\pi}{4}, \frac{\pi}{4}\right)$, are shown in **FIGURE 117**. ●

In the next two examples we find the equation of a trigonometric graph.

EXAMPLE 6 **Determining an Equation for a Graph**

Determine an equation for each graph.

(a)
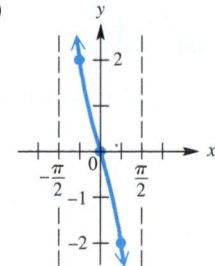

(b)
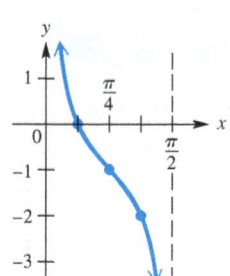

Solution

(a) This graph is that of $y = \tan x$ but reflected across the x-axis and stretched vertically by a factor of 2. Therefore, an equation for this graph is

$$y = -2 \tan x.$$

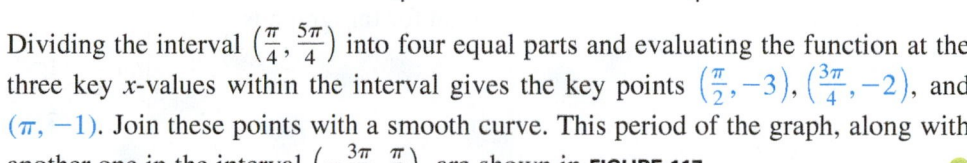
x-axis reflection   Vertical stretch

(b) This is the graph of a cotangent function, but the period is $\frac{\pi}{2}$ rather than π. Therefore, the coefficient of x is 2. This graph is vertically translated 1 unit down compared to the graph of $y = \cot 2x$. An equation for this graph is

$$y = -1 + \cot 2x.$$

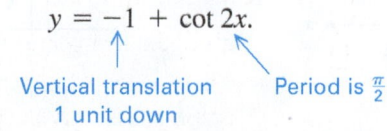
Vertical translation   Period is $\frac{\pi}{2}$.
1 unit down

EXAMPLE 7   **Determining an Equation for a Graph**

Determine an equation for each graph.

(a)

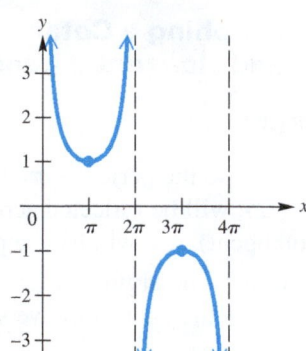

(b)

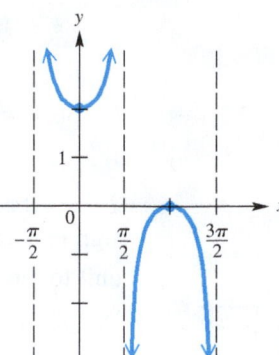

Solution

(a) This graph is that of a cosecant function that is stretched horizontally having period 4π. Therefore, if $y = \csc bx$, where $b > 0$, we must have $b = \frac{1}{2}$. An equation for this graph is

$$y = \csc \frac{1}{2}x.$$

Horizontal stretch

(b) This is the graph of $y = \sec x$, translated 1 unit upward. An equation is

$$y = 1 + \sec x.$$

Vertical translation

8.7 Exercises

Checking Analytic Skills   *Match each function defined in Exercises 1–6 with its graph in A–F.*
Do not use a calculator.

1. $y = -\csc x$

2. $y = -\sec x$

3. $y = -\tan x$

4. $y = -\cot x$

5. $y = \tan\left(x - \dfrac{\pi}{4}\right)$

6. $y = \cot\left(x - \dfrac{\pi}{4}\right)$

A.

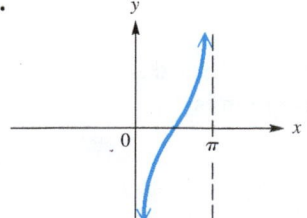

B.

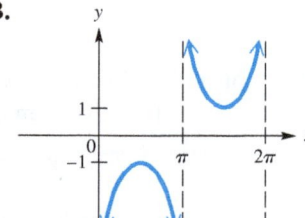

C.

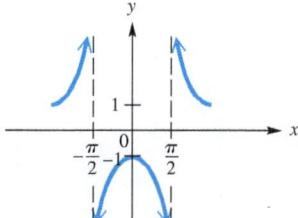

D.

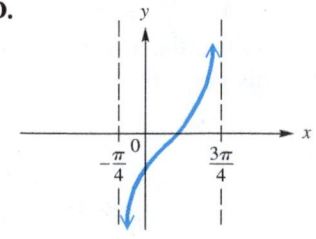

E.

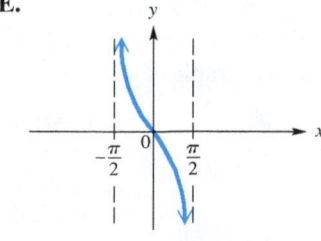

F.

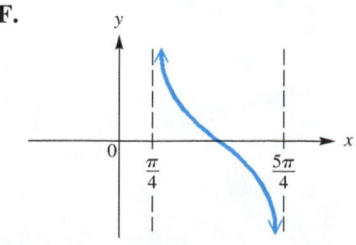

Concept Check   *Tell whether each statement is* true *or* false. *If false, tell why.*

7. The least positive number k for which $x = k$ is an asymptote for the tangent function is $\frac{\pi}{2}$.

8. The least positive number k for which $x = k$ is an asymptote for the cotangent function is $\frac{\pi}{2}$.

9. The secant and cosecant functions are undefined for the same values.

10. The tangent and secant functions are undefined for the same values.

Even and Odd Functions   *To show that* $\sec(-x) = \sec x$ *for all x in the domain, we begin by writing*

$$\sec(-x) = \frac{1}{\cos(-x)}$$

and then use the fact that $\cos(-x) = \cos x$ *for all x to complete the argument. Use this method to prove each of the following.*

11. $\sec(-x) = \sec x$

12. $\csc(-x) = -\csc x$

13. $\tan(-x) = -\tan x$

14. $\cot(-x) = -\cot x$

Find the **(a)** *period,* **(b)** *phase shift (if any), and* **(c)** *range of each function.*

15. $y = 2 \csc \frac{1}{2}x$

16. $y = 3 \csc 2x$

17. $y = -2 \sec\left(x + \frac{\pi}{2}\right)$

18. $y = -\frac{3}{2} \sec(x - \pi)$

19. $y = \frac{5}{2} \cot\left[\frac{1}{3}\left(x - \frac{\pi}{2}\right)\right]$

20. $y = -3 \tan\left[\frac{1}{2}\left(x + \frac{\pi}{4}\right)\right]$

21. $f(x) = \frac{1}{2} \sec(2x + \pi)$

22. $f(x) = -\frac{1}{3} \csc\left(\frac{1}{2}x - \frac{\pi}{2}\right)$

23. $y = -1 - \tan\left(x + \frac{\pi}{4}\right)$

24. $y = 2 + \cot\left(2x - \frac{\pi}{3}\right)$

Graph each function over a one-period interval.

25. $y = \sec x$

26. $y = \csc x$

27. $y = \sec(x - 2\pi)$

28. $y = \csc(x + 2\pi)$

29. $y = 3 \sec \frac{1}{4}x$

30. $y = -2 \sec \frac{1}{2}x$

31. $y = -\frac{1}{2} \csc\left(x + \frac{\pi}{2}\right)$

32. $y = \frac{1}{2} \csc\left(x - \frac{\pi}{2}\right)$

33. $y = \csc\left(x - \frac{\pi}{4}\right)$

34. $y = \sec\left(x + \frac{3\pi}{4}\right)$

35. $y = \sec\left(x + \frac{\pi}{4}\right)$

36. $y = \csc\left(x + \frac{\pi}{3}\right)$

37. $y = \sec\left(\frac{1}{2}x + \frac{\pi}{3}\right)$

38. $y = \csc\left(\frac{1}{2}x - \frac{\pi}{4}\right)$

39. $y = 2 + 3 \sec(2x - \pi)$

40. $y = 1 - 2 \csc\left(x + \frac{\pi}{2}\right)$

41. $y = 1 - \frac{1}{2} \csc\left(x - \frac{3\pi}{4}\right)$

42. $y = 2 + \frac{1}{4} \sec\left(\frac{1}{2}x - \pi\right)$

Graph each function over a one-period interval.

43. $y = \tan x$

44. $y = \cot x$

45. $y = \tan(x - \pi)$

46. $y = \cot(x + \pi)$

47. $y = \tan 4x$

48. $y = \tan \frac{1}{2}x$

49. $y = 2 \tan x$

50. $y = 2 \cot x$

51. $y = 2 \tan \frac{1}{4}x$

52. $y = \frac{1}{2} \cot x$

53. $y = \cot 3x$

54. $y = -\cot \frac{1}{2}x$

55. $y = -2 \tan \frac{1}{4}x$

56. $y = 3 \tan \frac{1}{2}x$

57. $y = \frac{1}{2} \cot 4x$

58. $y = -\frac{1}{2} \cot 2x$

Graph each function over a two-period interval.

59. $y = \tan(2x - \pi)$

60. $y = \tan\left(\dfrac{x}{2} + \pi\right)$

61. $y = \cot\left(3x + \dfrac{\pi}{4}\right)$

62. $y = \cot\left(2x - \dfrac{3\pi}{2}\right)$

63. $y = 1 + \tan x$

64. $y = -2 + \tan x$

65. $y = 1 - \cot x$

66. $y = -2 - \cot x$

67. $y = -1 + 2\tan x$

68. $y = 3 + \dfrac{1}{2}\tan x$

69. $y = -1 + \dfrac{1}{2}\cot(2x - 3\pi)$

70. $y = -2 + 3\tan(4x + \pi)$

Concept Check In Exercises 71–76, each function is of the form $y = a\tan bx$ or $y = a\cot bx$, where $b > 0$. (Half- and quarter-points are identified by dots.) In Exercises 77–80, each function is a transformation of the graph of the secant or cosecant function. Determine an equation of each graph.

71.
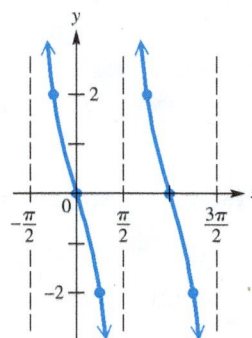

72.
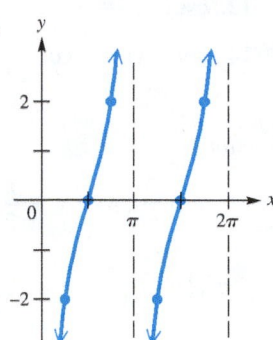

73.
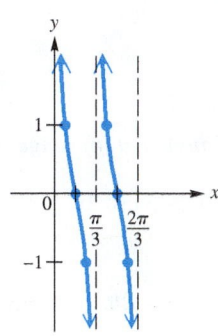

74.
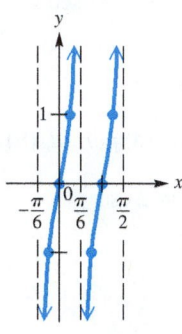

75.
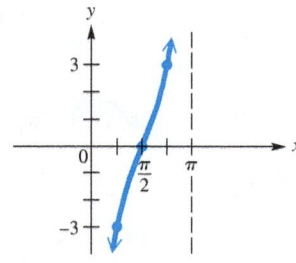

76.
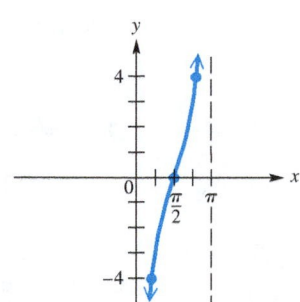

77.
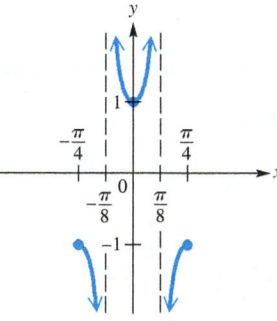

78.
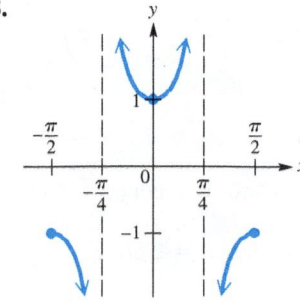

79.
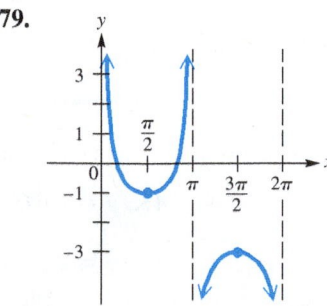

80.
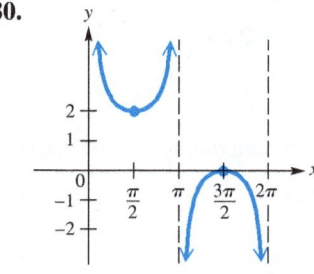

Solve each problem.

81. *(Modeling) Distance of a Rotating Beacon*   A rotating beacon is located at point *A* next to a long wall. See the figure. The beacon is 4 meters from the wall. The distance *d* is given by

$$d = 4 \tan 2\pi t,$$

where *t* is time measured in seconds since the beacon started rotating. (When $t = 0$, the beacon is aimed at point *R*. When the beacon is aimed to the right of *R*, the value of *d* is positive; *d* is negative if the beacon is aimed to the left of *R*.)

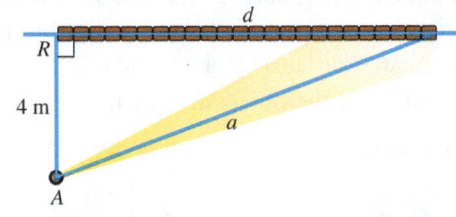

Find *d* to the nearest tenth for each time.
(a) $t = 0$ (b) $t = 0.2$
(c) $t = 0.8$ (d) $t = 1.2$
(e) Why is 0.25 a meaningless value for *t*?

82. *(Modeling) Distance of a Rotating Beacon*   In the figure for **Exercise 81,** the distance *a* is given by

$$a = 4 \left| \sec 2\pi t \right|.$$

Find *a* to the nearest tenth for each time.
(a) $t = 0$ (b) $t = 0.86$ (c) $t = 1.24$

8.8 Harmonic Motion

Simple Harmonic Motion • Damped Oscillatory Motion

Simple Harmonic Motion

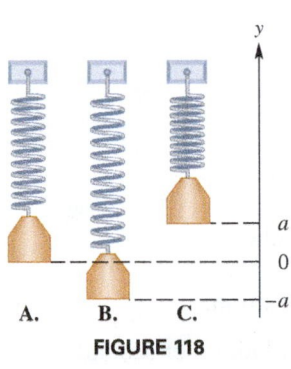

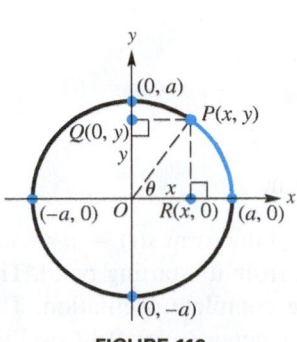

FIGURE 118

FIGURE 119

In part A of **FIGURE 118,** a spring with a weight attached to its free end is in its equilibrium (or rest) position. If the weight is pulled down *a* units and released (part B of the figure), the spring's elasticity causes the weight to rise *a* units ($a > 0$) above the equilibrium position, as seen in part C, and then oscillate about the equilibrium position. If friction is neglected, this oscillatory motion is described mathematically by a sinusoid. Other applications of this type of motion include sound, electric current, and electromagnetic waves.

To develop a general equation for such motion, consider **FIGURE 119.** Suppose the point $P(x, y)$ moves around the circle counterclockwise at a uniform angular speed ω. Assume that at time $t = 0$, *P* is at $(a, 0)$. The angle swept out by ray *OP* at time *t* is given by

$$\theta = \omega t.$$

The coordinates of point *P* at time *t* are

$$x = a \cos \theta = a \cos \omega t \quad \text{and} \quad y = a \sin \theta = a \sin \omega t.$$

As *P* moves around the circle in **FIGURE 119** from the point $(a, 0)$, the point $Q(0, y)$ oscillates back and forth along the *y*-axis between the points $(0, a)$ and $(0, -a)$. Similarly, the point $R(x, 0)$ oscillates back and forth between $(a, 0)$ and $(-a, 0)$. This oscillatory motion is called **simple harmonic motion.**

The amplitude of the motion is $|a|$, and the period $\frac{2\pi}{\omega}$. The moving points *P* and *Q* or *P* and *R* complete one oscillation or cycle per period. The number of oscillations, or cycles per unit of time, called the **frequency,** is the *reciprocal* of the period $\frac{\omega}{2\pi}$ where $\omega > 0$.

> **Simple Harmonic Motion**
>
> The position of a point oscillating about an equilibrium position at time t is modeled by either
>
> $$s(t) = a \cos \omega t \quad \text{or} \quad s(t) = a \sin \omega t,$$
>
> where a and ω are constants with $\omega > 0$. The amplitude of the motion is $|a|$, the period is $\frac{2\pi}{\omega}$, and the frequency is $\frac{\omega}{2\pi}$.

EXAMPLE 1 **Modeling the Motion of a Spring**

Suppose that an object is attached to a coiled spring such as the one in **FIGURE 118** on the previous page. It is pulled down a distance of 5 units from its equilibrium position and then released. The time for one complete oscillation is 4 seconds.

(a) Give an equation that models the position of the object at time t.

(b) Determine the position at $t = 1.5$ seconds.

(c) Find the frequency.

Solution

(a) When the object is released at $t = 0$, the distance of the object from the equilibrium position is 5 inches below equilibrium. If $s(t)$ is to model the motion, then $s(0)$ must equal -5. We use

$$s(t) = a \cos \omega t,$$

with $a = -5$. We choose the cosine function because $\cos(\omega(0)) = \cos 0 = 1$, and $-5 \cdot 1 = -5$. (Had we chosen the sine function, a phase shift would have been required.) The period is 4, so

$$\frac{2\pi}{\omega} = 4, \quad \text{or} \quad \omega = \frac{\pi}{2}. \qquad \text{\color{blue}Solve for } \omega.$$

Thus, the motion is modeled by

$$s(t) = -5 \cos \frac{\pi}{2} t.$$

(b) After 1.5 seconds, the position is

$$s(1.5) = -5 \cos\left[\frac{\pi}{2}(1.5)\right] \approx 3.54 \text{ inches.}$$

Since $3.54 > 0$, the object is above the equilibrium position.

(c) The frequency is the reciprocal of the period, or $\frac{1}{4}$ oscillation per second.

EXAMPLE 2 **Analyzing Harmonic Motion**

Suppose that an object oscillates according to the model

$$s(t) = 8 \sin 3t,$$

where t is in seconds and $s(t)$ is in feet. Analyze the motion.

Solution The motion is harmonic because the model is of the form $s(t) = a \sin \omega t$. Since $a = 8$, the object oscillates 8 feet in either direction from its starting point. The period $\frac{2\pi}{3} \approx 2.1$ is the time, in seconds, it takes for one complete oscillation. The frequency is the reciprocal of the period, so the object completes $\frac{3}{2\pi} \approx 0.48$ oscillation per second.

Damped Oscillatory Motion

In the example of the stretched spring, we disregarded the effect of friction. Friction causes the amplitude of the motion to diminish gradually until the weight comes to rest. In this situation, we say that the motion has been *damped* by the force of friction. Most oscillatory motions are damped, and the decrease in amplitude follows the pattern of exponential decay. An example of **damped oscillatory motion** is provided by the function

$$s(t) = e^{-t} \sin t.$$

FIGURE 120 shows how the graph of $y_3 = e^{-x} \sin x$ is bounded above by the graph of $y_1 = e^{-x}$ and below by the graph of $y_2 = -e^{-x}$. The damped-motion curve dips below the x-axis at $x = \pi$ but stays above the graph of y_2. **FIGURE 121** shows a traditional graph, along with the graph of $s = \sin t$.

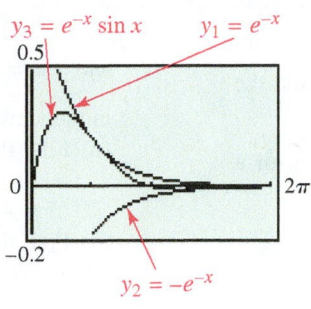

FIGURE 120

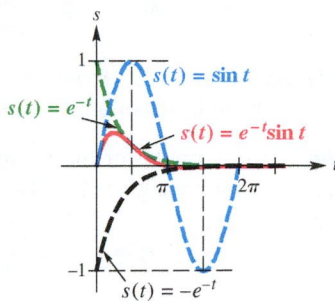

FIGURE 121

Shock absorbers are put on an automobile in order to damp oscillatory motion. Instead of oscillating up and down for a long time after hitting a bump or pothole, the oscillations of the car are quickly damped out for a smoother ride.

8.8 Exercises

Checking Analytic Skills   *Suppose that a weight on a spring has initial position $s(0)$ and period P.* **Do not use a calculator.**
(a) *Find a function s given by $s(t) = a \cos \omega t$ that models the displacement of the weight.*
(b) *Evaluate $s(1)$. Is the weight moving upward, downward, or neither when $t = 1$?*

1. $s(0) = 2$ inches; $P = 0.5$ second

2. $s(0) = 5$ inches; $P = 1.5$ seconds

3. $s(0) = -3$ inches; $P = 0.8$ second

4. $s(0) = -4$ inches; $P = 1.2$ seconds

(Modeling) Music   *A note on the piano has frequency F. Suppose the maximum displacement at the center of the piano*

wire is given by $s(0)$. Find constants a and ω so that the equation $s(t) = a \cos \omega t$ models this displacement. Graph s in the viewing window $[0, 0.05]$ by $[-0.3, 0.3]$.

5. $F = 27.5$; $s(0) = 0.21$ **6.** $F = 110$; $s(0) = 0.11$

7. $F = 55$; $s(0) = 0.14$ **8.** $F = 220$; $s(0) = 0.06$

(Modeling)   *Solve each problem.*

9. *Particle Movement*   Write the equation, and then determine the amplitude, period, and frequency of the simple harmonic motion of a particle moving uniformly around a circle of radius 2 units, with angular speed **(a)** 2 radians per second and **(b)** 4 radians per second.

10. *Pendulum*   What are the period P and frequency T of the oscillation of a pendulum of length $\frac{1}{2}$ foot? (*Hint:*

$P = 2\pi\sqrt{\frac{L}{32}}$, where L is the length of the pendulum in feet and P is in seconds.)

11. *Pendulum*   Refer to **Exercise 10.** How long should the pendulum be to have period 1 second?

12. *Spring*   The formula for the up-and-down motion of a weight on a spring is given by

$$s(t) = a \sin \sqrt{\frac{k}{m}} t.$$

If the spring constant k is 4, what mass m must be used to produce a period of 1 second?

13. *Spring*   Refer to **Exercise 12.** A spring with spring constant $k = 2$ and a 1-unit mass m attached to it is stretched and then allowed to come to rest.
    (a) If the spring is stretched $\frac{1}{2}$ foot and released, what are the amplitude, period, and frequency of the resulting oscillatory motion?
    (b) What is the equation of the motion?

14. *Spring*   The position of a weight attached to a spring is

$$s(t) = -5 \cos 4\pi t$$

inches after t seconds.
    (a) What is the maximum height that the weight rises above the equilibrium position?
    (b) What are the frequency and period?
    (c) When does the weight first reach its maximum height?
    (d) Calculate and interpret $s(1.3)$.

15. *Spring*   The position of a weight attached to a spring is

$$s(t) = -4 \cos 10t$$

inches after t seconds.
    (a) What is the maximum height that the weight rises above the equilibrium position?
    (b) What are the frequency and period?
    (c) When does the weight first reach its maximum height?
    (d) Calculate and interpret $s(1.466)$.

16. *Spring*   A weight attached to a spring is pulled down 3 inches below the equilibrium position.
    (a) Assuming that the frequency is $\frac{6}{\pi}$ oscillations per second, find a trigonometric model that gives the position of the weight at time t seconds.
    (b) What is the period?

17. *Spring*   A weight attached to a spring is pulled down 2 inches below the equilibrium position.
    (a) Assuming that the period is $\frac{1}{3}$ second, find a trigonometric model that gives the position of the weight at time t seconds.
    (b) What is the frequency?

18. Use a graphing calculator to graph

$$y_1 = e^{-x} \sin x,$$
$$y_2 = e^{-x},$$
and $$y_3 = -e^{-x}$$

in the viewing window $[0, \pi]$ by $[-0.5, 0.5]$.
    (a) Find the x-intercepts of the graph of y_1. Explain the relationship of these x-intercepts to those of the graph of $y = \sin x$.
    (b) Find the x-coordinate of any points of intersection of y_1 and y_2 or y_1 and y_3.

19. *Damped Motion*   A weight attached to a spring is submerged in water. Suppose that its displacement from its equilibrium position can be modeled by

$$D(t) = 2e^{-2t} \cos 2\pi t,$$

where D is in inches and $t \geq 0$ is in seconds. Note that when $D < 0$ the spring is stretched and when $D > 0$ the spring is compressed.
    (a) What is the initial displacement of the weight when $t = 0$?
    (b) What is the frequency?
    (c) Approximate graphically the time when the spring is *compressed* $\frac{1}{e}$ inch.

20. *Electrical Circuit*   The voltage in a circuit with alternating current is given by

$$V(t) = 160e^{-20t} \cos 120\pi t,$$

where V is in volts and $t \geq 0$ is in seconds.
    (a) What is the initial voltage when $t = 0$?
    (b) What is the frequency of the voltage?
    (c) Solve $V(t) = 100$ graphically.

Graph one period of each function. State the period, phase shift, domain, and range.

1. $y = \csc\left(x - \dfrac{\pi}{4}\right)$ **2.** $y = \sec\left(x + \dfrac{\pi}{4}\right)$

Graph one period of each function. State the period, domain, and range.

3. $y = 2 \tan x$ **4.** $y = 2 \cot x$

5. *(Modeling) Spring* The height of a weight attached to a spring is

$$s(t) = -4 \cos 8\pi t$$

inches after t seconds.
(a) Find the maximum height that the weight rises above the equilibrium position $y = 0$.
(b) When does the weight first reach its maximum height if $t \geq 0$?
(c) What are the frequency and period?

8 Summary

KEY TERMS & SYMBOLS

8.1 Angles and Their Measures

line
segment
ray
angle
initial side
terminal side
vertex
positive angle
negative angle
degree (°)
acute angle
right angle
obtuse angle
straight angle
complementary angles
supplementary angles
$m(\text{angle } A)$
minute (′)
second (″)
angle in standard position
quadrantal angle
coterminal angle
central angle
radian
circumference of a circle
sector of a circle
angular speed, ω
linear speed, ν

KEY CONCEPTS

An angle with vertex at the center of a circle and that intercepts an arc on the circle equal in length to the radius of the circle has measure **1 radian**. In general, if θ is a central angle in a circle of radius r, and θ intercepts an arc length of length s, the radian measure of θ is $\frac{s}{r}$.

DEGREE/RADIAN RELATIONSHIP $180° = \pi$ **radians**

CONVERTING BETWEEN DEGREES AND RADIANS

1. Multiply a degree measure by $\frac{\pi}{180}$ radian and simplify to convert to radians.

2. Multiply a radian measure by $\frac{180°}{\pi}$ and simplify to convert to degrees.

ARC LENGTH
The length s of the arc intercepted on a circle of radius r by a central angle of measure θ radians is given by the product of the radius and the radian measure of the angle, or

$$s = r\theta, \quad \theta \text{ in radians.}$$

AREA OF A SECTOR
The area $\mathcal{A}$ of a sector of a circle of radius r and central angle θ is

$$\mathcal{A} = \frac{1}{2}r^2\theta, \quad \theta \text{ in radians.}$$

ANGULAR AND LINEAR SPEED

Angular Speed	Linear Speed
$\omega = \dfrac{\theta}{t}$	$\nu = \dfrac{s}{t}$
(ω in radians per unit time, θ in radians)	$\nu = \dfrac{r\theta}{t}$
	$\nu = r\omega$

(continued)

KEY TERMS & SYMBOLS	KEY CONCEPTS

8.2 Trigonometric Functions and Fundamental Identities

sine
cosine
tangent
cotangent
secant
cosecant
identity

TRIGONOMETRIC FUNCTIONS (STANDARD POSITION ANGLE APPROACH)

Let (x, y) be a point other than the origin on the terminal side of an angle θ in standard position. Let $r = \sqrt{x^2 + y^2}$ represent the distance from the origin to (x, y). The six trigonometric functions of θ are as follows.

$$\sin \theta = \frac{y}{r} \qquad \cos \theta = \frac{x}{r} \qquad \tan \theta = \frac{y}{x} \ (x \neq 0)$$

$$\csc \theta = \frac{r}{y} \ (y \neq 0) \qquad \sec \theta = \frac{r}{x} \ (x \neq 0) \qquad \cot \theta = \frac{x}{y} \ (y \neq 0)$$

RECIPROCAL IDENTITIES

$$\sin \theta = \frac{1}{\csc \theta} \qquad \cos \theta = \frac{1}{\sec \theta} \qquad \tan \theta = \frac{1}{\cot \theta}$$

$$\csc \theta = \frac{1}{\sin \theta} \qquad \sec \theta = \frac{1}{\cos \theta} \qquad \cot \theta = \frac{1}{\tan \theta}$$

PYTHAGOREAN IDENTITIES

$$\sin^2 \theta + \cos^2 \theta = 1 \qquad 1 + \tan^2 \theta = \sec^2 \theta \qquad 1 + \cot^2 \theta = \csc^2 \theta$$

QUOTIENT IDENTITIES

$$\frac{\sin \theta}{\cos \theta} = \tan \theta \qquad \frac{\cos \theta}{\sin \theta} = \cot \theta$$

SIGNS OF TRIGONOMETRIC FUNCTIONS

$x < 0, y > 0, r > 0$	$x > 0, y > 0, r > 0$
II	**I**
Sine and cosecant positive	**All functions positive**
$x < 0, y < 0, r > 0$	$x > 0, y < 0, r > 0$
III	**IV**
Tangent and cotangent positive	**Cosine and secant positive**

8.3 Right Triangles and Evaluating Trigonometric Functions

hypotenuse
side opposite
side adjacent
cofunction
reference angle

TRIGONOMETRIC FUNCTIONS (RIGHT-TRIANGLE APPROACH)

For any acute angle A in standard position, the trigonometric functions of A are as follows.

$$\sin A = \frac{y}{r} = \frac{\text{side opposite}}{\text{hypotenuse}} \qquad \csc A = \frac{r}{y} = \frac{\text{hypotenuse}}{\text{side opposite}}$$

$$\cos A = \frac{x}{r} = \frac{\text{side adjacent}}{\text{hypotenuse}} \qquad \sec A = \frac{r}{x} = \frac{\text{hypotenuse}}{\text{side adjacent}}$$

$$\tan A = \frac{y}{x} = \frac{\text{side opposite}}{\text{side adjacent}} \qquad \cot A = \frac{x}{y} = \frac{\text{side adjacent}}{\text{side opposite}}$$

KEY TERMS & SYMBOLS	KEY CONCEPTS

KEY CONCEPTS

FUNCTION VALUES FOR SPECIAL ANGLES

θ	$\sin\theta$	$\cos\theta$	$\tan\theta$	$\cot\theta$	$\sec\theta$	$\csc\theta$
30° or $\frac{\pi}{6}$	$\frac{1}{2}$	$\frac{\sqrt{3}}{2}$	$\frac{\sqrt{3}}{3}$	$\sqrt{3}$	$\frac{2\sqrt{3}}{3}$	2
45° or $\frac{\pi}{4}$	$\frac{\sqrt{2}}{2}$	$\frac{\sqrt{2}}{2}$	1	1	$\sqrt{2}$	$\sqrt{2}$
60° or $\frac{\pi}{3}$	$\frac{\sqrt{3}}{2}$	$\frac{1}{2}$	$\sqrt{3}$	$\frac{\sqrt{3}}{3}$	2	$\frac{2\sqrt{3}}{3}$

COFUNCTION IDENTITIES

If A is an acute angle measured in degrees, then the following are true.

$$\sin A = \cos(90° - A) \qquad \csc A = \sec(90° - A)$$
$$\cos A = \sin(90° - A) \qquad \sec A = \csc(90° - A)$$
$$\tan A = \cot(90° - A) \qquad \cot A = \tan(90° - A)$$

If A is measured in radians, replace 90° with $\frac{\pi}{2}$.

REFERENCE ANGLES

θ' is the reference angle for angle θ. (If θ is acute, then $\theta' = \theta$.)

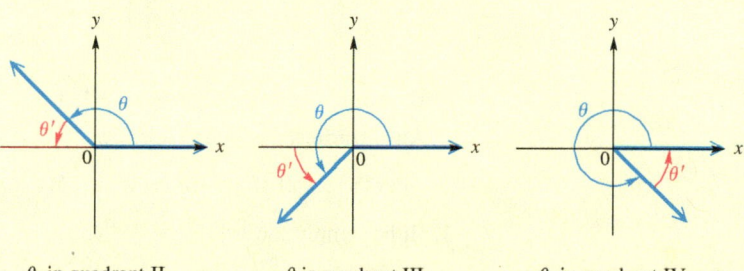

θ in quadrant II θ in quadrant III θ in quadrant IV

FINDING TRIGONOMETRIC FUNCTION VALUES FOR ANY NONQUADRANTAL ANGLE θ

Step 1   Add or subtract 360° as many times as needed to get an angle of at least 0° but less than 360°.

Step 2   Find the reference angle θ'.

Step 3   Find the trigonometric function values for θ'.

Step 4   Determine the correct signs for the values found in Step 3.

8.4 Applications of Right Triangles

significant digit
exact number
angle of elevation
angle of depression
bearing

SOLVING AN APPLIED TRIGONOMETRY PROBLEM

Step 1   Draw a sketch, and label it with the given information. Label the quantity to be found with a variable.

Step 2   Use the sketch to write an equation relating the given quantities to the variable.

Step 3   Solve the equation, and check that your answer makes sense.

(continued)

8.5 The Circular Functions

unit circle
circular functions

TRIGONOMETRIC FUNCTIONS (UNIT CIRCLE APPROACH)

For s, x, and y as shown in the figure, the circular functions of s are as follows.

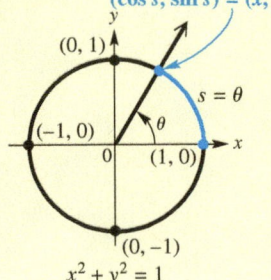

$$\sin s = y \qquad \csc s = \frac{1}{y} \quad (y \neq 0)$$

$$\cos s = x \qquad \sec s = \frac{1}{x} \quad (x \neq 0)$$

$$\tan s = \frac{y}{x} \quad (x \neq 0) \qquad \cot s = \frac{x}{y} \quad (y \neq 0)$$

8.6 Graphs of the Sine and Cosine Functions

periodic function
period
sine wave (sinusoid)
sinusoidal function
amplitude
phase shift
argument
sine regression

SINE AND COSINE FUNCTIONS

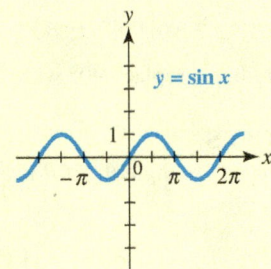

Domain: $(-\infty, \infty)$;   **Range:** $[-1, 1]$ **Domain:** $(-\infty, \infty)$;   **Range:** $[-1, 1]$

Amplitude: 1;   **Period:** 2π **Amplitude:** 1;   **Period:** 2π

Consider the graph of $y = c + a \sin[b(x - d)]$ or $y = c + a \cos[b(x - d)]$, $b > 0$.

1. It has amplitude $|a|$.

2. The period is $\frac{2\pi}{b}$.

3. The vertical translation is c units upward if $c > 0$ or $|c|$ units downward if $c < 0$.

4. The phase shift is d units to the right if $d > 0$ or $|d|$ units to the left if $d < 0$.

8.7 Graphs of the Other Circular Functions

SECANT AND COSECANT FUNCTIONS

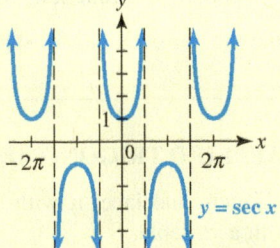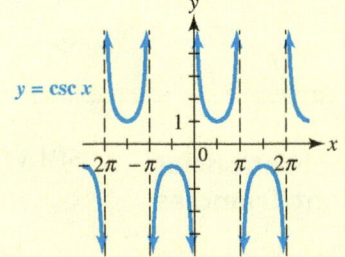

Domain: $\{x \mid x \neq (2n + 1)\frac{\pi}{2}$,   **Domain:** $\{x \mid x \neq n\pi$, where n
      where n is an integer$\}$ is an integer $\}$
Range: $(-\infty, -1] \cup [1, \infty)$ **Range:** $(-\infty, -1] \cup [1, \infty)$
Period: 2π **Period:** 2π

KEY TERMS & SYMBOLS	KEY CONCEPTS

TANGENT AND COTANGENT FUNCTIONS

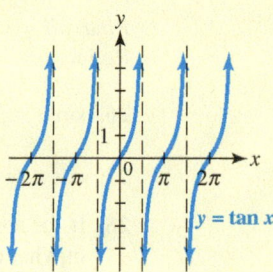

Domain: $\{x \mid x \neq (2n+1)\frac{\pi}{2},$ where n is an integer $\}$

Range: $(-\infty, \infty)$

Period: π

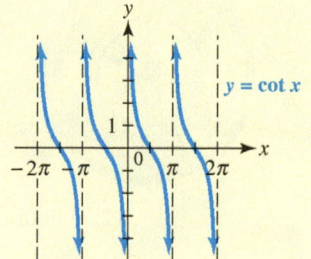

Domain: $\{x \mid x \neq n\pi,$ where n is an integer $\}$

Range: $(-\infty, \infty)$

Period: π

8.8 Harmonic Motion

simple harmonic motion
frequency
damped oscillatory motion

SIMPLE HARMONIC MOTION

The position of a point oscillating about an equilibrium position at time t is modeled by either

$$s(t) = a \cos \omega t \quad \text{or} \quad s(t) = a \sin \omega t,$$

where a and ω are constants with $\omega > 0$. The amplitude of the motion is $|a|$, the period is $\frac{2\pi}{\omega}$, and the frequency is $\frac{\omega}{2\pi}$.

8 Review Exercises

1. Find the angle of least possible positive measure coterminal with an angle of $-174°$.

2. Let n represent any integer, and write an expression for all angles coterminal with an angle of $270°$.

Solve each problem.

3. *Rotating Pulley* A pulley is rotating 320 times per minute. Through how many degrees does a point on the edge of the pulley move in $\frac{2}{3}$ second?

4. *Rotating Propeller* The propeller of a speedboat rotates 650 times per minute. Through how many degrees will a point on the edge of the propeller rotate in 2.4 seconds?

5. Which is larger, an angle of $1°$ or an angle of 1 radian? Justify your answer.

6. Consider each angle in standard position having the given radian measure. In what quadrant does the terminal side of each angle lie?
   (a) 3 (b) 4 (c) -2 (d) 7

Convert each degree measure to radians. Leave answers in terms of π.

7. $120°$

8. $800°$

Convert each radian measure to degrees.

9. $\dfrac{5\pi}{4}$

10. $-\dfrac{6\pi}{5}$

Concept Check *Suppose the tip of the minute hand of a clock is 2 inches from the center of the clock. For each duration, determine the distance traveled by the tip of the minute hand.*

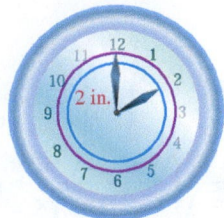

11. 20 minutes   **12.** 3 hours

Solve each problem. Use a calculator as necessary.

13. *Arc Length* The radius of a circle is 15.2 cm. Find the length of an arc of the circle, to the nearest tenth, intercepted by a central angle of $\frac{3\pi}{4}$ radians.

14. *Area of a Sector* A central angle of $\frac{7\pi}{4}$ radians forms a sector of a circle. Find the area of the sector, to the nearest square inch, if the radius of the circle is 28.69 inches.

15. *Height of a Tree* A tree 2000 yards away subtends an angle of 1° 10'. Find the height of the tree to two significant digits.

16. *Concept Check* Find the measure of the central angle θ (in radians) and the area of the sector.

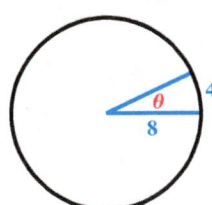

Find the trigonometric function values of each angle. If a value is undefined, say so.

17.   **18.**

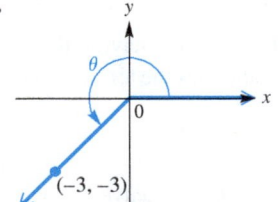   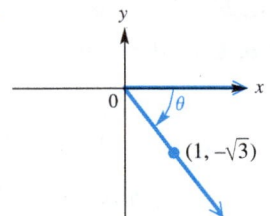

19. 180°

Find the values of the six trigonometric functions for an angle in standard position having each point on its terminal side.

20. $(3, -4)$

21. $(9, -2)$

22. $\left(-2\sqrt{2}, 2\sqrt{2}\right)$

23. *Concept Check* If the terminal side of a quadrantal angle lies along the y-axis, which of its trigonometric functions are undefined?

Decide whether each statement is possible *for some angle θ, or impossible.*

24. $\sec \theta = -\dfrac{2}{3}$ **25.** $\tan \theta = 1.4$

Find all six trigonometric function values for each angle. Rationalize denominators when applicable.

26. $\sin \theta = \dfrac{\sqrt{3}}{5}$ and $\cos \theta < 0$

27. $\cos \theta = -\dfrac{5}{8}$, with θ in quadrant III

28. If for some particular angle θ, $\sin \theta < 0$ and $\cos \theta > 0$, in what quadrant must θ lie? What is the sign of $\tan \theta$?

Find the values of the six trigonometric functions for each angle A.

29.   **30.**

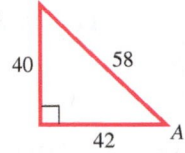   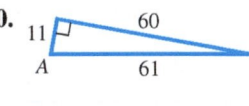

Find the exact values of the six trigonometric functions for each angle. Do not use a calculator. Rationalize denominators when applicable.

31. 300°

32. −225°

33. −390°

Use a calculator to approximate each value.

34. $\sin 72° 30'$ **35.** $\sec 222° 30'$

36. $\cot 305.6°$ **37.** $\tan 11.7689°$

38. *Concept Check* Which one of the following cannot be *exactly* determined with the methods of this chapter?

  A. $\cos 135°$ **B.** $\cot(-45°)$

  C. $\sin 300°$ **D.** $\tan 140°$

Approximate to the nearest tenth of a degree each value of θ, where θ is in the interval $[0°, 90°)$.

39. $\sin \theta = 0.82584121$ **40.** $\cot \theta = 1.1249386$

Solve each right triangle.

41.

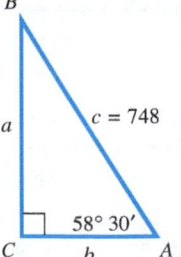

42. $A = 39.72°$; $b = 38.97$ m

Solve each problem.

43. *Distance between Two Points* The bearing of B from C is 254°. The bearing of A from C is 344°. The bearing of A from B is 32°. The distance from A to C is 780 meters. Find the distance from A to B.

44. *Distance a Ship Sails* The bearing from point A to point B is S 55.0° E and from point B to point C is N 35.0° E. If a ship sails from A to B, a distance of 80.0 kilometers, and then from B to C, a distance of 74.0 kilometers, how far is it from A to C?

45. *Distance between Two Cars* Two cars leave an intersection at the same time. One heads due south at 55.0 mph. The other travels due west. After two hours, the bearing of the car headed west from the car headed south is 324°. How far apart are they at that time?

46. *Distance Traveled by a Sailboat* From the top of a building that overlooks an ocean, an observer watches a boat sailing directly toward the building. If the observer is 150 feet above sea level, and if the angle of depression of the boat changes from 27.0° to 39.0° during the period of observation, approximate the distance the boat travels.

47. *Height of a Tower* The angle of elevation from a point 93.2 feet from the base of a tower to the top of the tower is 38° 20'. Find the height of the tower.

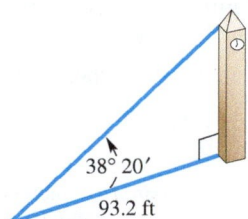

48. *Height of a Tower* The angle of depression of a television tower to a point on the ground 36.0 meters from the bottom of the tower is 29.5°. Find the height of the tower.

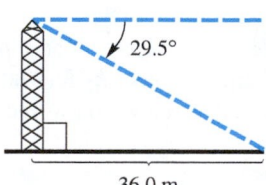

49. *Height of a Triangle* Find the measure of h to the nearest unit. Assume angle measures are exact.

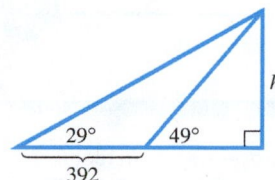

50. *(Modeling) Fundamental Surveying Problem* The first fundamental problem of surveying is to determine the coordinates of a point Q, given the coordinates of a point P, the distance between P and Q, and the bearing

θ from P to Q. (*Source:* Mueller, I. and K. Ramsayer, *Introduction to Surveying,* Frederick Ungar Publishing.)

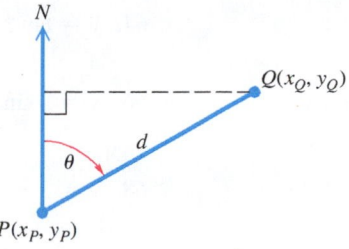

(a) Find a formula for the coordinates (x_Q, y_Q) of the point Q, given θ, the coordinates (x_P, y_P) of P, and the distance d between P and Q.

(b) Use your formula to find the coordinates of the point (x_Q, y_Q) if $(x_P, y_P) = (123.62, 337.95)$, $\theta = 17° 19' 22''$, and $d = 193.86$ feet.

Find the exact function value. **Do not use a calculator.**

51. $\cos \dfrac{2\pi}{3}$

52. $\sin\left(-\dfrac{11\pi}{6}\right)$

53. $\tan\left(-\dfrac{7\pi}{3}\right)$

54. $\cot\left(\dfrac{5\pi}{4}\right)$

55. $\csc\left(-\dfrac{11\pi}{6}\right)$

56. $\sec 3\pi$

Approximate each circular function value to four decimal places.

57. $\cos(-0.2443)$

58. $\cot 3.0543$

59. Give the exact value of s if s is in the interval $\left[0, \dfrac{\pi}{2}\right]$ and $\cos s = 0.5$.

60. Use a calculator to approximate s in the interval $\left[0, \dfrac{\pi}{2}\right]$ if $\sin s = 0.49244294$.

61. *Concept Check* The screen shows a point on the unit circle. What is the length, to four decimal places, of the shortest arc of the circle from $(1, 0)$ to the point?

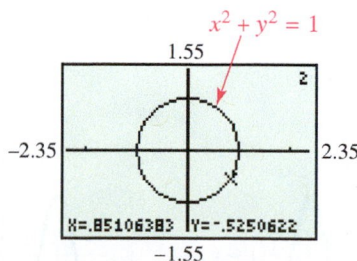

62. *Concept Check* Which one of the following is true about the graph of $y = 4 \sin 2x$?

A. It has amplitude 2 and period $\dfrac{\pi}{2}$.

B. It has amplitude 4 and period π.

C. Its range is $[0, 4]$.

D. Its range is $[-4, 0]$.

For each function, give the amplitude, period, vertical translation, and phase shift, as applicable. Then graph the function over at least two periods.

63. $y = 2 \sin x$

64. $y = \tan 3x$

65. $y = -\dfrac{1}{2} \cos 3x$

66. $y = 2 \sin 5x$

67. $y = 1 + 2 \sin \dfrac{1}{4} x$

68. $y = 3 - \dfrac{1}{4} \cos \dfrac{2}{3} x$

69. $y = 3 \cos \left(x + \dfrac{\pi}{2} \right)$

70. $y = -\sin \left(x - \dfrac{3\pi}{4} \right)$

71. $y = \dfrac{1}{2} \csc \left(2x - \dfrac{\pi}{4} \right)$

72. $y = \cot \dfrac{1}{2} x$

73. $y = -2 \sec 2x$

74. $y = \csc (2x - \pi)$

Graph each function over a one-period interval.

75. $y = 3 \cos 2x$

76. $y = \dfrac{1}{2} \cot 3x$

77. $y = \cos \left(x - \dfrac{\pi}{4} \right)$

78. $y = \tan \left(x - \dfrac{\pi}{2} \right)$

79. $y = 1 + 2 \cos 3x$

80. $y = -1 - 3 \sin 2x$

Concept Check   Identify one circular function that satisfies the description.

81. Period is π, x-values of the x-intercepts are of the form $n\pi$, where n is an integer

82. Period is 2π, graph passes through the origin

83. Period is 2π, graph passes through the point $\left(\dfrac{\pi}{2}, 0 \right)$

84. Period is 2π, domain is $\{ x \mid x \neq n\pi$, where n is an integer $\}$

85. Period is π, function is decreasing on the interval $(0, \pi)$

86. Period is 2π, has vertical asymptotes of the form $x = (2n + 1)\dfrac{\pi}{2}$, where n is an integer

Checking Analytic Skills   Give an equation of a sine function having the given graph.

87.

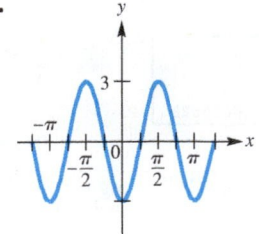

88.

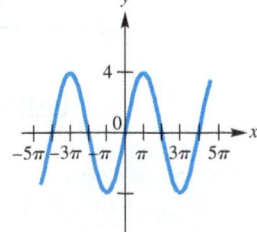

89.

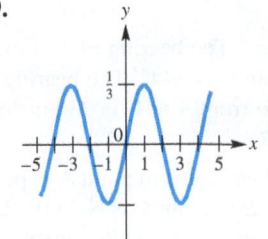

90.

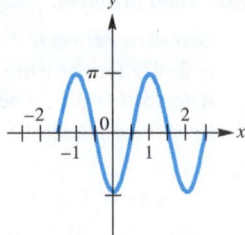

Solve each problem.

91. *(Modeling) Monthly Average Temperature*   The monthly average temperature (in °F) in a northern U.S. city is shown in the table.

Month	°F	Month	°F
Jan	22	July	71
Feb	25	Aug	72
Mar	33	Sept	63
Apr	45	Oct	52
May	58	Nov	36
June	69	Dec	25

(a) Plot the monthly average temperature over a 2-year period by letting $x = 1$ correspond to January of the first year.

(b) Determine a modeling function of the form

$$f(x) = a \sin [b(x - d)] + c,$$

where a, b, c, and d are constants.

(c) Explain the significance of each constant.

(d) Graph f, together with the data, on the same coordinate axes. How well does f model the data?

(e) Use the sine regression capability of a graphing calculator to find the equation of a sine curve that fits these data.

92. *Viewing Angle to an Object*   Let a person h_1 feet tall stand d feet from an object h_2 feet tall, where $h_2 > h_1$. Let θ be the angle of elevation to the top of the object. See the figure.

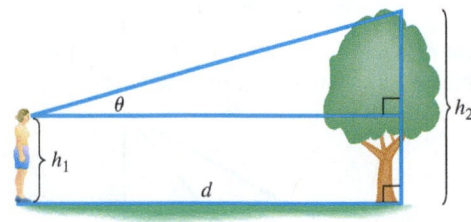

(a) Show that $d = (h_2 - h_1) \cot \theta$.

(b) Let $h_2 = 55$ and $h_1 = 5$ in the formula for d. Then graph d on the interval $\left(0, \dfrac{\pi}{2} \right]$ for θ.

93. *Atmospheric Effect on Sunlight*   The shortest path for the sun's rays through Earth's atmosphere occurs when the sun is directly overhead. Disregarding the curvature of Earth, as the sun moves lower on the horizon, the distance that sunlight passes through the atmosphere increases by a factor of csc θ, where θ is the angle of elevation of the sun. This increased distance reduces both the intensity of the sunlight and the amount of ultraviolet light that reaches Earth's surface. See the figure. (*Source:* Winter, C.-J., R. Sizmann, and L. L. Vant-Hull (Editors), *Solar Power Plants,* Springer-Verlag.)

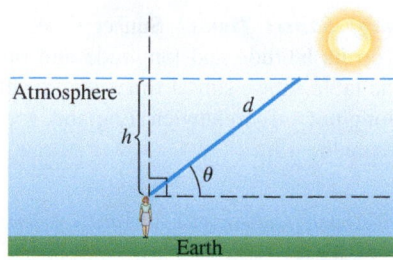

(a) Verify that $d = h \csc \theta$.
(b) Determine θ when $d = 2h$.
(c) The atmosphere filters out the ultraviolet light that causes skin to burn. Compare the difference between sunbathing when $\theta = \frac{\pi}{2}$ and $\theta = \frac{\pi}{3}$. Which measure gives less ultraviolet light?

94. *(Modeling) Pollution Trends*   The amount of pollution in the air fluctuates with the seasons. It is lower after heavy spring rains and higher after periods of little rain. Additionally, the long-term trend is upward. An idealized graph of this situation is shown in the figure.

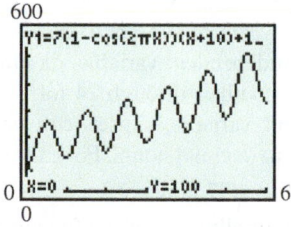

Circular functions can be used to model the fluctuating part of the pollution levels. Exponential functions can be used to model long-term growth. The pollution level in a certain area might be given by

$$y = 7(1 - \cos 2\pi x)(x + 10) + 100e^{0.2x},$$

where x is time in years, with $x = 0$ representing January 1 of the base year. July 1 of the same year would be represented by $x = 0.5$, October 1 of the following year would be represented by $x = 1.75$, and so on. Find the pollution level, to the nearest unit, on each date.
(a) January 1, base year
(b) July 1, base year
(c) January 1, following year
(d) July 1, following year

(Modeling) Harmonic Motion   *An object in simple harmonic motion has position function s inches from an initial point, and t is the time in seconds. Find the amplitude, period, and frequency for each function.*

95. $s(t) = 4 \sin \pi t$

96. $s(t) = 3 \cos 2t$

97. In **Exercise 95,** what does the frequency represent? Find the position of the object from the initial point at 1.5 seconds, 2 seconds, and 3.25 seconds.

98. In **Exercise 96,** what does the period represent? What does the amplitude represent?

---

8 Test

1. Find the angle of least positive measure coterminal with $-157°$.

2. *Rotating Tire*   A tire rotates 450 times per minute. Through how many degrees does a point on the edge of the tire move in 1 second?

3. **(a)** Convert $120°$ to radians.
   (b) Convert $\frac{9\pi}{10}$ radians to degrees.

4. A central angle of a circle with radius 150 centimeters cuts off an arc of 200 centimeters. Find each measure.
   (a) The radian measure of the angle
   (b) The area of a sector with that central angle

5. Find the length of the arc intercepted by an angle of $36°$ in a circle with radius 12 inches.

6. Use the figure to find the following.

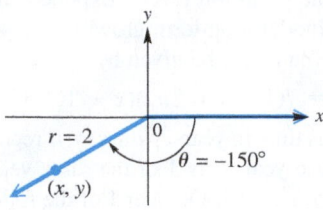

(a) The smallest positive angle coterminal with θ
(b) The values of x and y
(c) The values of the six trigonometric functions of θ
(d) The radian measure of θ

7. Use a calculator to approximate s in the interval $\left[0, \frac{\pi}{2}\right]$ if $\sin s = 0.82584121$.

8. If $\cos \theta < 0$ and $\cot \theta > 0$, in what quadrant does θ lie?

9. If $(2, -5)$ is on the terminal side of an angle θ in standard position, find $\sin \theta$, $\cos \theta$, and $\tan \theta$.

10. If $\cos \theta = \frac{4}{5}$ and θ is in quadrant IV, find the values of the other trigonometric functions of θ.

11. Find the exact value of each variable in the figure.

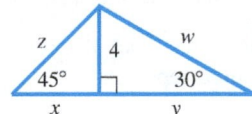

12. Find the exact value of $\cot(-750°)$.

13. Use a calculator to approximate the following.
(a) $\sin 78° \, 21'$ (b) $\tan 11.7689°$ (c) $\sec 58.9041°$

14. Solve the triangle.

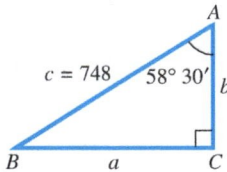

15. *Height of a Flagpole* To measure the height of a flagpole, Alexis Glaser found that the angle of elevation from a point 24.7 feet from the base to the top is $32° \, 10'$. What is the height of the flagpole?

16. *Distance of a Ship from a Pier* A ship leaves a pier on a bearing of S 55° E and travels 80 kilometers. It then turns and continues on a bearing of N 35° E for 74 kilometers. How far is the ship from the pier?

Graph each function over a two-period interval. Identify asymptotes when applicable.

17. $y = -1 + 2 \sin(x + \pi)$ **18.** $y = -\cos 2x$

19. $y = \tan\left(x - \frac{\pi}{2}\right)$ **20.** $y = 1 + \sec x$

21. The function graphed is of one of the following forms: $y = a \sin bx$ or $y = a \cos bx$, where $b > 0$. Determine the equation of the graph.

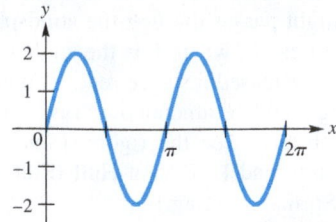

22. *(Modeling) Sunset Times* Sunset times at a location depend on its latitude and longitude and on the time of year. The table shows sunset times at a location of 40° N on the longitude of Greenwich, England, for the first day of each month.

Month	Sunset	Month	Sunset
Jan	4:46 P.M.	July	7:33 P.M.
Feb	5:19 P.M.	Aug	7:14 P.M.
Mar	5:52 P.M.	Sept	6:32 P.M.
Apr	6:24 P.M.	Oct	5:42 P.M.
May	6:55 P.M.	Nov	4:58 P.M.
June	7:23 P.M.	Dec	4:35 P.M.

Source: World Almanac and Book of Facts.

Since the pattern for sunset time repeats itself every year, the data are periodic (with period 1 year, or 12 months). It can be modeled by a sine function of the form

$$y = a \sin[b(x - d)] + c.$$

(a) Plot two years of sunsets to display two full periods. For the independent variable, months, use the number of each month, 1 through 24 for two periods. For the dependent variable, sunset, convert from hours and minutes to decimal hours. For example, 4:46 becomes $4 + \frac{46}{60}$, or 4.77 hours.
(b) Find the amplitude a of the function.
(c) Find b.
(d) Find the phase shift d.
(e) Find the vertical shift c.
(f) Give the function that models the times for sunset. Verify your answer by graphing the function and the scatter diagram in the same window.

Spring After t seconds, the height of a weight attached to a spring is $s(t) = -3 \cos 2\pi t$ inches.

23. Find the maximum height that the weight rises above the equilibrium position $y = 0$.

24. When is the first time that the weight reaches its maximum height if $t \geq 0$?

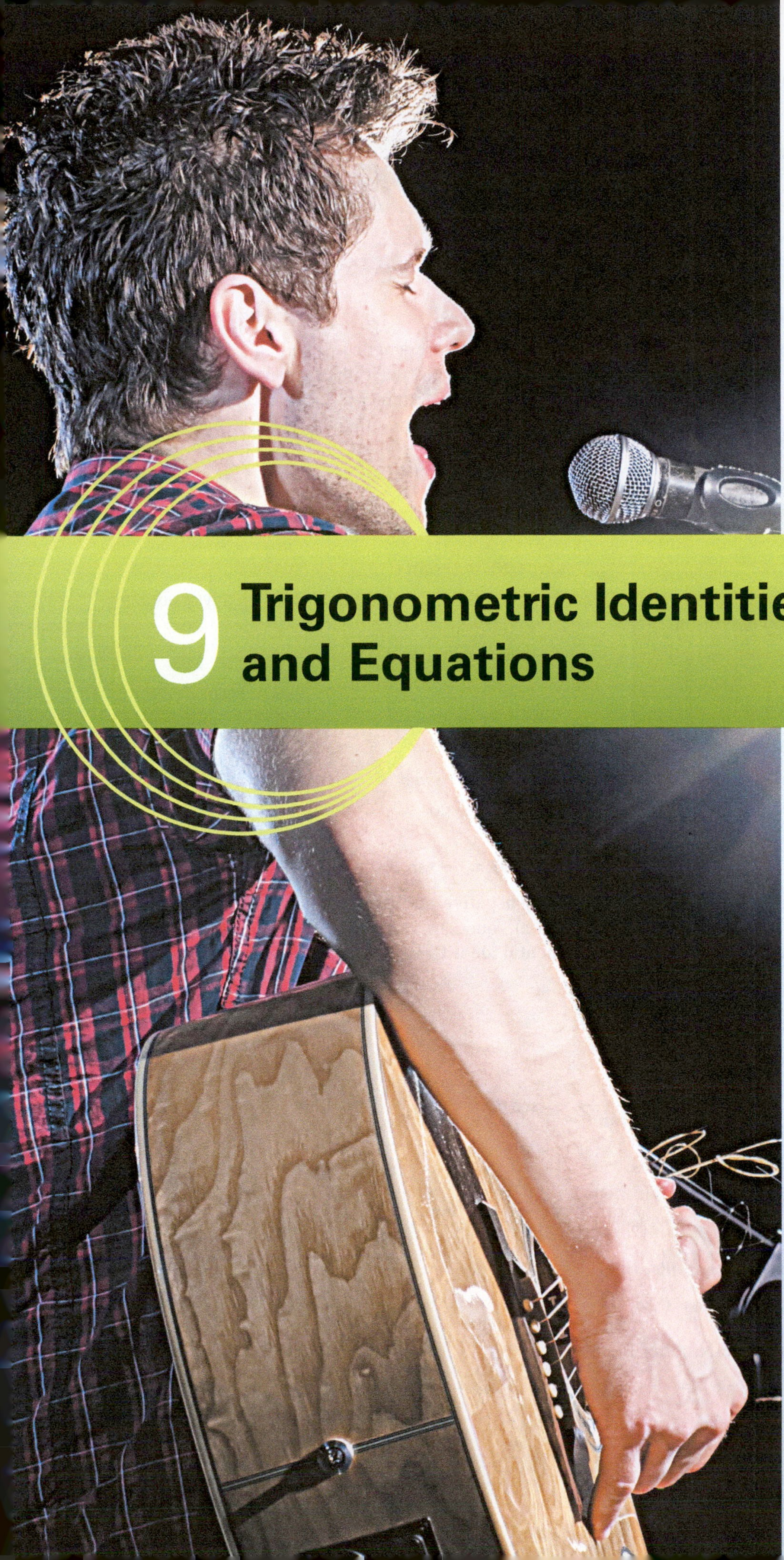

In this chapter we explore problems involving trigonometric equations, including playing and tuning musical instruments.

9 Trigonometric Identities and Equations

9.1 Trigonometric Identities

Fundamental Identities • Using the Fundamental Identities • Verifying Identities

Fundamental Identities

$\sin(-\theta) = -y = -\sin\theta$
$\cos(-\theta) = x = \cos\theta$

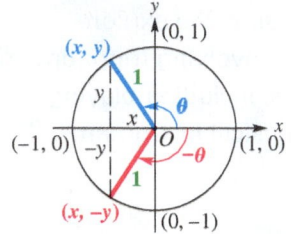

FIGURE 1

The unit circle is shown in **FIGURE 1**. Because arc length is given by $s = r\theta$ and the radius of the unit circle is 1, an angle θ (in radians) determines an arc of length $s = \theta$. The arc terminates at a point (x, y). Angle $-\theta$ determines a corresponding arc of equal length terminating at the point $(x, -y)$. Therefore from the figure we see that

$$\sin\theta = \frac{y}{1} = y \quad \text{and} \quad \sin(-\theta) = \frac{-y}{1} = -y.$$

The fact that $\sin\theta$ and $\sin(-\theta)$ are negatives of each other leads to the following identity.

$$\sin(-\theta) = -\sin\theta$$

Because sine is an odd function, $f(-x) = -f(x)$.

FIGURE 1 shows an arc θ in quadrant II, but the same result holds for θ in any quadrant. Also, by definition,

$$\cos(-\theta) = x \quad \text{and} \quad \cos\theta = x,$$

so

$$\cos(-\theta) = \cos\theta,$$

Because cosine is an even function, $f(-x) = f(x)$.

and

$$\tan(-\theta) = \frac{\sin(-\theta)}{\cos(-\theta)} = \frac{-\sin\theta}{\cos\theta} = -\frac{\sin\theta}{\cos\theta},$$

so

$$\tan(-\theta) = -\tan\theta.$$

Because tangent is an odd function, $f(-x) = -f(x)$.

Similar reasoning gives the remaining **negative-number** or **negative-angle identities,** which, together with the reciprocal, quotient, and Pythagorean identities from **Chapter 8,** are called the **fundamental identities.**

→ **Looking Ahead to Calculus**
Much of our work with identities in this chapter is preparation for calculus. Some calculus problems are simplified by making an appropriate trigonometric substitution. For example, if $x = 3\tan\theta$, then

$$\begin{aligned}
\sqrt{9 + x^2} &= \sqrt{9 + (3\tan\theta)^2} \\
&= \sqrt{9 + 9\tan^2\theta} \\
&= \sqrt{9(1 + \tan^2\theta)} \\
&= 3\sqrt{1 + \tan^2\theta} \\
&= 3\sqrt{\sec^2\theta}.
\end{aligned}$$

In the interval $\left(0, \frac{\pi}{2}\right)$, the value of $\sec\theta$ is positive, giving

$$\sqrt{9 + x^2} = 3\sec\theta.$$

Fundamental Identities

Reciprocal Identities

$$\cot\theta = \frac{1}{\tan\theta} \qquad \sec\theta = \frac{1}{\cos\theta} \qquad \csc\theta = \frac{1}{\sin\theta}$$

Quotient Identities

$$\tan\theta = \frac{\sin\theta}{\cos\theta} \qquad \cot\theta = \frac{\cos\theta}{\sin\theta}$$

Pythagorean Identities

$$\sin^2\theta + \cos^2\theta = 1 \qquad 1 + \tan^2\theta = \sec^2\theta \qquad 1 + \cot^2\theta = \csc^2\theta$$

Negative-Number Identities

$$\sin(-\theta) = -\sin\theta \qquad \cos(-\theta) = \cos\theta \qquad \tan(-\theta) = -\tan\theta$$

$$\csc(-\theta) = -\csc\theta \qquad \sec(-\theta) = \sec\theta \qquad \cot(-\theta) = -\cot\theta$$

NOTE The most commonly recognized forms of the fundamental identities are given in the preceding box. You must also recognize alternative forms of these identities. *For example, two other forms of* $\sin^2 \theta + \cos^2 \theta = 1$ *are*

$$\sin^2 \theta = 1 - \cos^2 \theta \quad \text{and} \quad \cos^2 \theta = 1 - \sin^2 \theta.$$

Using the Fundamental Identities

Any trigonometric function of a number or an angle can be expressed in terms of any other trigonometric function.

EXAMPLE 1   Expressing One Function in Terms of Another

Express $\cos x$ in terms of $\tan x$.

Solution Since $\sec x$ is related to both $\cos x$ and $\tan x$ by identities, start with $1 + \tan^2 x = \sec^2 x$.

$$\frac{1}{1 + \tan^2 x} = \frac{1}{\sec^2 x} \qquad \text{Take reciprocals.}$$

$$\frac{1}{1 + \tan^2 x} = \cos^2 x \qquad \text{Reciprocal identity}$$

Remember both the positive and negative square roots.

$$\pm \sqrt{\frac{1}{1 + \tan^2 x}} = \cos x \qquad \text{Take the square root of each side.}$$

$$\cos x = \frac{\pm 1}{\sqrt{1 + \tan^2 x}} \qquad \text{Quotient rule } \sqrt[n]{\frac{a}{b}} = \frac{\sqrt[n]{a}}{\sqrt[n]{b}}; \text{ rewrite.}$$

$$\cos x = \frac{\pm\sqrt{1 + \tan^2 x}}{1 + \tan^2 x} \qquad \text{Rationalize the denominator.}$$

Choose the $+$ sign or the $-$ sign, depending on the quadrant of x.

EXAMPLE 2   Rewriting an Expression in Terms of Sine and Cosine

Write $\tan \theta + \cot \theta$ in terms of $\sin \theta$ and $\cos \theta$, and then simplify the expression.

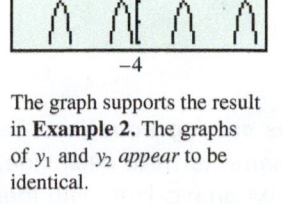

$y_1 = \tan x + \cot x$
$y_2 = \dfrac{1}{\cos x \sin x}$

The graph supports the result in **Example 2**. The graphs of y_1 and y_2 *appear* to be identical.

Solution

$$\tan \theta + \cot \theta = \frac{\sin \theta}{\cos \theta} + \frac{\cos \theta}{\sin \theta} \qquad \text{Quotient identities}$$

$$= \frac{\sin \theta}{\cos \theta} \cdot \frac{\sin \theta}{\sin \theta} + \frac{\cos \theta}{\sin \theta} \cdot \frac{\cos \theta}{\cos \theta} \qquad \text{Write each fraction with the LCD, } \cos \theta \sin \theta.$$

$$= \frac{\sin^2 \theta}{\cos \theta \sin \theta} + \frac{\cos^2 \theta}{\cos \theta \sin \theta} \qquad \text{Multiply.}$$

$$= \frac{\sin^2 \theta + \cos^2 \theta}{\cos \theta \sin \theta} \qquad \text{Add fractions; } \frac{a}{c} + \frac{b}{c} = \frac{a + b}{c}.$$

$$= \frac{1}{\cos \theta \sin \theta} \qquad \text{Pythagorean identity}$$

CAUTION *In working with trigonometric expressions and identities, be sure to write the argument of the function.* For example, we would *not* write $\sin^2 + \cos^2 = 1$. An argument such as θ is necessary in this identity. Note that $\sin^2 \theta + \cos^2 \theta = 1$, but that $\sin^2 x + \cos^2 y \neq 1$ and $\sin^2 2\theta + \cos^2 \theta \neq 1$ in general.

Verifying Identities

→ Looking Ahead to Calculus
Trigonometric identities are used in calculus to simplify trigonometric expressions, determine derivatives of trigonometric functions, and change the form of some integrals.

One of the skills required for more advanced work in mathematics (and especially in calculus) is the ability to use identities to write expressions in alternative forms. We develop this skill by using the fundamental identities to verify that an equation is an identity. Here are some hints to help you get started.

Hints for Verifying Identities

1. *Learn the fundamental identities.* Whenever you see either side of a fundamental identity, the other side should come to mind. *Also, be aware of equivalent forms of the fundamental identities.* For example, $\cos^2 \theta = 1 - \sin^2 \theta$ is an alternative form of $\sin^2 \theta + \cos^2 \theta = 1$.

2. *Try to rewrite the more complicated side of the equation* so that it is identical to the simpler side.

3. *It is sometimes helpful to express all trigonometric functions in the equation in terms of sine and cosine* and then simplify the result.

4. *Usually, any factoring, division involving complex fractions, or indicated algebraic operations should be performed.* For example, the expression

   $\sin^2 x + 2 \sin x + 1$ can be factored as $(\sin x + 1)^2$.

   The sum or difference of two trigonometric expressions, such as $\frac{1}{\sin \theta} + \frac{1}{\cos \theta}$, can be added or subtracted in the same way as any other rational expressions.

   $$\frac{1}{\sin \theta} + \frac{1}{\cos \theta} = \frac{\cos \theta}{\sin \theta \cos \theta} + \frac{\sin \theta}{\sin \theta \cos \theta} \qquad \text{Write with the LCD.}$$
   $$= \frac{\cos \theta + \sin \theta}{\sin \theta \cos \theta} \qquad \frac{a}{c} + \frac{b}{c} = \frac{a+b}{c}$$

5. *As you select substitutions, keep in mind the side you are not changing, because it represents your goal.* For example, to verify the identity

   $$\tan^2 x + 1 = \frac{1}{\cos^2 x},$$

   try to think of an identity that relates $\tan x$ to $\cos x$. Since $\sec x = \frac{1}{\cos x}$ and $\sec^2 x = \tan^2 x + 1$, the secant function is the best link between the two sides.

6. If a fractional expression contains $1 + \sin x$, *multiplying both numerator and denominator* by $1 - \sin x$ would give $1 - \sin^2 x$, which could be replaced with $\cos^2 x$. Similar situations for $1 - \sin x$, $1 + \cos x$, and $1 - \cos x$ may apply.

CAUTION *Verifying identities is not the same as solving equations.* Techniques used in solving equations, such as adding the same term to each side, or multiplying each side by the same term, are not useful when working with identities, because you are starting with a statement (to be verified) that may not be true.

To avoid the temptation to use algebraic properties of equations to verify identities, *one strategy is to work with only one side and rewrite it to match the other side.* It can often take more than one try to verify an identity, so be patient and use multiple hints to find one that works for you.

EXAMPLE 3   **Verifying an Identity (Working with One Side)**

Verify that the given equation is an identity.

$$\cot x + 1 = \csc x(\cos x + \sin x)$$

Solution

Choose one side of the equation and use fundamental identities to rewrite it so that it is identical to the other side. Since the right side is more complicated, we work with this expression, using Hint 3 first to change all functions to sine or cosine.

$$\cot x + 1 = \csc x(\cos x + \sin x) \qquad \text{Given equation}$$

Steps   **Reasons**

Right side of
given equation

$$\csc x(\cos x + \sin x) = \frac{1}{\sin x}(\cos x + \sin x) \qquad \csc x = \frac{1}{\sin x}$$

$$= \frac{\cos x}{\sin x} + \frac{\sin x}{\sin x} \qquad \text{Distributive property}$$

$$= \cot x + 1 \qquad \frac{\cos x}{\sin x} = \cot x \text{ and } \frac{\sin x}{\sin x} = 1$$

Left side of
given equation

The given equation is an identity because the right side is equivalent to the left side.

Similarly, we could have chosen to start with the left side of the equation, using the same strategy.

Left side of
given equation

$$\cot x + 1 = \frac{\cos x}{\sin x} + \frac{\sin x}{\sin x} \qquad \cot x = \frac{\cos x}{\sin x} \text{ and } 1 = \frac{\sin x}{\sin x}$$

$$= \frac{1}{\sin x}(\cos x + \sin x) \qquad \text{Factor out } \frac{1}{\sin x}.$$

$$= \csc x(\cos x + \sin x) \qquad \frac{1}{\sin x} = \csc x$$

Right side of
given equation

The identity can be verified starting with either side as long as you show it equals the other side. ●

FOR DISCUSSION

We know that

$$\cot \frac{\pi}{2} + 1 = 0 + 1 = 1$$

and

$$\csc \frac{\pi}{2}\left(\cos \frac{\pi}{2} + \sin \frac{\pi}{2}\right) = 1(0 + 1) = 1.$$

X	Y₁	Y₂
0	ERROR	ERROR
.3927	3.4142	3.4142
.7854	2	2
1.1781	1.4142	1.4142
1.5708	ERROR	1
1.9635	.58579	.58579
2.3562	0	0

X=1.570796326795

FIGURE 2

To make a table to support **Example 3**, let $y_1 = \cot x + 1 = \frac{1}{\tan x} + 1$ and $y_2 = \csc x(\cos x + \sin x) = \frac{\cos x + \sin x}{\sin x}$ as shown in **FIGURE 2**. Why don't all pairs of y-values agree? How does the answer reinforce the warning that we must understand the mathematics to interpret the calculator results?

EXAMPLE 4 **Verifying an Identity (Working with One Side)**

Verify that the given equation is an identity.

$$\frac{\tan t - \cot t}{\sin t \cos t} = \sec^2 t - \csc^2 t$$

Analytic Solution

We transform the more complicated left side to match the right side.

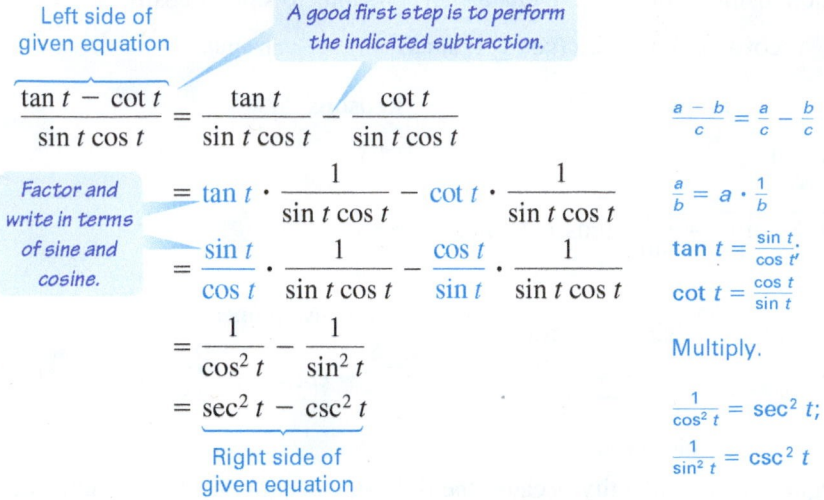

> Left side of given equation

> A good first step is to perform the indicated subtraction.

$$\frac{\tan t - \cot t}{\sin t \cos t} = \frac{\tan t}{\sin t \cos t} - \frac{\cot t}{\sin t \cos t} \qquad \frac{a-b}{c} = \frac{a}{c} - \frac{b}{c}$$

> Factor and write in terms of sine and cosine.

$$= \tan t \cdot \frac{1}{\sin t \cos t} - \cot t \cdot \frac{1}{\sin t \cos t} \qquad \frac{a}{b} = a \cdot \frac{1}{b}$$

$$= \frac{\sin t}{\cos t} \cdot \frac{1}{\sin t \cos t} - \frac{\cos t}{\sin t} \cdot \frac{1}{\sin t \cos t} \qquad \tan t = \frac{\sin t}{\cos t}, \; \cot t = \frac{\cos t}{\sin t}$$

Multiply.

$$= \frac{1}{\cos^2 t} - \frac{1}{\sin^2 t}$$

$$= \sec^2 t - \csc^2 t \qquad \frac{1}{\cos^2 t} = \sec^2 t; \; \frac{1}{\sin^2 t} = \csc^2 t$$

> Right side of given equation

Since the left side is equivalent to the right side, the given equation is an identity.

Graphing Calculator Support*

Graphs of the two functions coincide. See **FIGURE 3.** The table shows that for selected values of X, the function values also agree, further supporting the identity.

$$Y_1 = \frac{\tan X - \cot X}{\sin X \cos X}$$

$$Y_2 = \sec^2 X - \csc^2 X$$

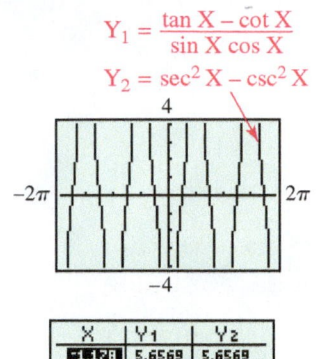

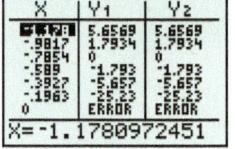

FIGURE 3

EXAMPLE 5 **Verifying an Identity (Working with One Side)**

Verify that the given equation is an identity.

$$\frac{\cos x}{1 - \sin x} = \frac{1 + \sin x}{\cos x}$$

Solution We work on the right side, using Hint 6 in the list given earlier.

$$\frac{1 + \sin x}{\cos x} = \frac{(1 + \sin x)(1 - \sin x)}{\cos x(1 - \sin x)} \qquad \text{Multiply by 1 in the form } \frac{1 - \sin x}{1 - \sin x}.$$

> Use Hint 6 to try to obtain $1 - \sin^2 x$ and then use a fundamental identity.

$$= \frac{1 - \sin^2 x}{\cos x(1 - \sin x)} \qquad (x + y)(x - y) = x^2 - y^2$$

$$= \frac{\cos^2 x}{\cos x(1 - \sin x)} \qquad 1 - \sin^2 x = \cos^2 x$$

$$= \frac{\cos x}{1 - \sin x} \qquad \text{Lowest terms}$$

We could have obtained a similar result by working on the left side of the given equation and multiplying the numerator and denominator by $1 + \sin x$.

---

*To verify an identity, we must provide an analytical solution. *A graph can only support, not prove, an identity,* since we can graph only a finite interval of the domain.

left = right

common third
expression

If both sides of an identity appear to be equally complicated, the identity can be verified if we work independently on the left side and on the right side, until each side is changed into some common third result. ***Each step, on each side, must be reversible.*** With all steps reversible, the procedure is shown in the margin.

The left side leads to the third expression, which leads back to the right side. This procedure is just a shortcut for the procedure used in **Examples 3–5:** *One side is changed into the other side, but by going through an intermediate step.*

$$y_1 = \frac{\sec x + \tan x}{\sec x - \tan x}$$

$$y_2 = \frac{1 + 2\sin x + \sin^2 x}{\cos^2 x}$$

The graph *supports* the result in **Example 6**.

EXAMPLE 6   **Verifying an Identity (Working with Both Sides)**

Verify that the given equation is an identity.

$$\frac{\sec \alpha + \tan \alpha}{\sec \alpha - \tan \alpha} = \frac{1 + 2\sin \alpha + \sin^2 \alpha}{\cos^2 \alpha}$$

Solution   Both sides appear complicated, so we begin by changing each side into a common third expression.

$$\underbrace{\frac{\sec \alpha + \tan \alpha}{\sec \alpha - \tan \alpha}}_{\text{Left side of given equation}} = \frac{(\sec \alpha + \tan \alpha)\cos \alpha}{(\sec \alpha - \tan \alpha)\cos \alpha}$$ Multiply by 1 in the form $\frac{\cos \alpha}{\cos \alpha}$.

$$= \frac{\sec \alpha \cos \alpha + \tan \alpha \cos \alpha}{\sec \alpha \cos \alpha - \tan \alpha \cos \alpha}$$ Distributive property

$$= \frac{1 + \tan \alpha \cos \alpha}{1 - \tan \alpha \cos \alpha}$$ $\sec \alpha \cos \alpha = 1$

$$= \frac{1 + \dfrac{\sin \alpha}{\cos \alpha} \cdot \cos \alpha}{1 - \dfrac{\sin \alpha}{\cos \alpha} \cdot \cos \alpha}$$ $\tan \alpha = \frac{\sin \alpha}{\cos \alpha}$

$$= \frac{1 + \sin \alpha}{1 - \sin \alpha}$$ Simplify.

$$\underbrace{\frac{1 + 2\sin \alpha + \sin^2 \alpha}{\cos^2 \alpha}}_{\text{Right side of given equation}} = \frac{(1 + \sin \alpha)^2}{\cos^2 \alpha}$$ Factor; $a^2 + 2ab + b^2 = (a + b)^2$.

$$= \frac{(1 + \sin \alpha)^2}{1 - \sin^2 \alpha}$$ $\cos^2 \alpha = 1 - \sin^2 \alpha$

$$= \frac{(1 + \sin \alpha)^2}{(1 + \sin \alpha)(1 - \sin \alpha)}$$ Factor $1 - \sin^2 \alpha$ as the difference of squares.

$$= \frac{1 + \sin \alpha}{1 - \sin \alpha}$$ Lowest terms

We have verified that the given equation is an identity by showing that

Left side of given equation	Common third expression	Right side of given equation

$$\underbrace{\frac{\sec \alpha + \tan \alpha}{\sec \alpha - \tan \alpha}}_{} = \underbrace{\frac{1 + \sin \alpha}{1 - \sin \alpha}}_{} = \underbrace{\frac{1 + 2\sin \alpha + \sin^2 \alpha}{\cos^2 \alpha}}_{}.$$

There are usually several ways to verify a given identity. For instance, another way to begin verifying the identity in **Example 6** is to work on the left.

$$\underbrace{\frac{\sec \alpha + \tan \alpha}{\sec \alpha - \tan \alpha}}_{\substack{\text{Left side of} \\ \text{given equation}}} = \frac{\dfrac{1}{\cos \alpha} + \dfrac{\sin \alpha}{\cos \alpha}}{\dfrac{1}{\cos \alpha} - \dfrac{\sin \alpha}{\cos \alpha}} \qquad \text{Fundamental identities}$$

$$= \frac{\dfrac{1 + \sin \alpha}{\cos \alpha}}{\dfrac{1 - \sin \alpha}{\cos \alpha}} \qquad \text{Add and subtract fractions.}$$

$$= \frac{1 + \sin \alpha}{\cos \alpha} \cdot \frac{\cos \alpha}{1 - \sin \alpha} \qquad \text{Multiply by the reciprocal of the divisor.}$$

$$= \frac{1 + \sin \alpha}{1 - \sin \alpha} \qquad \text{Simplify.}$$

Compare this to **Example 6** for the right side to see that the two sides agree.

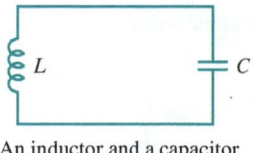

An inductor and a capacitor
FIGURE 4

EXAMPLE 7 **Applying a Pythagorean Identity to Electronics**

Tuners in radios select radio stations by adjusting the frequency. A tuner may contain an inductor L and a capacitor C, as illustrated in **FIGURE 4**. The energy stored in the inductor at time t is given by

$$L(t) = k \sin^2 2\pi Ft$$

and the energy stored in the capacitor is given by

$$C(t) = k \cos^2 2\pi Ft,$$

where F is the frequency of the radio station and k is a constant. The total energy E in the circuit is given by

$$E(t) = L(t) + C(t).$$

Show that E is a constant function. (*Source:* Weidner, R. and R. Sells, *Elementary Classical Physics*, Vol. 2, Allyn & Bacon.)

Solution

$$\begin{aligned}
E(t) &= L(t) + C(t) &&\text{Given equation} \\
&= k \sin^2 2\pi Ft + k \cos^2 2\pi Ft &&\text{Substitute.} \\
&= k(\sin^2 2\pi Ft + \cos^2 2\pi Ft) &&\text{Factor.} \\
&= k(1) &&\sin^2 \theta + \cos^2 \theta = 1 \text{ (Here, } \theta = 2\pi Ft.) \\
&= k &&k \text{ is constant.}
\end{aligned}$$

Since k is constant, $E(t)$ is a constant function.

9.1 Exercises

Concept Check   *Use the negative-number identities to decide whether each function is even or odd.*

1. $\sin x$ **2.** $\cos x$ **3.** $\tan x$ **4.** $\cot x$ **5.** $\sec x$ **6.** $\csc x$

Write each of the following expressions as a trigonometric function of a positive number. (For example, $\sin(-3.4) = -\sin 3.4$.)

7. $\cos(-4.38)$

8. $\cos(-5.46)$

9. $\sin(-0.5)$

10. $\sin(-2.5)$

11. $\tan\left(-\dfrac{\pi}{7}\right)$

12. $\cot\left(-\dfrac{4\pi}{7}\right)$

Checking Analytic Skills *For each expression in Column I, choose the expression from Column II that completes a fundamental identity.* **Do not use a calculator.**

	I		II
13.	$\dfrac{\cos x}{\sin x} =$ ___	**A.**	$\sin^2 x + \cos^2 x$
14.	$\tan x =$ ___	**B.**	$\cot x$
15.	$\cos(-x) =$ ___	**C.**	$\sec^2 x$
16.	$\tan^2 x + 1 =$ ___	**D.**	$\dfrac{\sin x}{\cos x}$
17.	$1 =$ ___	**E.**	$\cos x$

Checking Analytic Skills *For each expression in Column I, choose the expression from Column II that completes an identity. You may have to rewrite one or both expressions.* **Do not use a calculator.**

	I		II
18.	$-\tan x \cos x =$ ___	**A.**	$\dfrac{\sin^2 x}{\cos^2 x}$
19.	$\sec^2 x - 1 =$ ___	**B.**	$\dfrac{1}{\sec^2 x}$
20.	$\dfrac{\sec x}{\csc x} =$ ___	**C.**	$\sin(-x)$
21.	$1 + \sin^2 x =$ ___	**D.**	$\csc^2 x - \cot^2 x + \sin^2 x$
22.	$\cos^2 x =$ ___	**E.**	$\tan x$

23. A student writes "$1 + \cot^2 = \csc^2$." Comment on this student's work.

24. A student makes the claim: "Since $\sin^2 \theta + \cos^2 \theta = 1$, I should also be able to say that $\sin \theta + \cos \theta = 1$ if I take the square root of each side." Comment on this student's statement.

Complete this table so that each function in Exercises 25–30 is expressed in terms of the functions given across the top.

	$\sin \theta$	$\cos \theta$	$\tan \theta$	$\cot \theta$	$\sec \theta$	$\csc \theta$
25. $\sin \theta$	$\sin \theta$	$\pm\sqrt{1 - \cos^2 \theta}$	$\dfrac{\pm \tan \theta \sqrt{1 + \tan^2 \theta}}{1 + \tan^2 \theta}$			$\dfrac{1}{\csc \theta}$
26. $\cos \theta$		$\cos \theta$	$\dfrac{\pm\sqrt{\tan^2 \theta + 1}}{\tan^2 \theta + 1}$		$\dfrac{1}{\sec \theta}$	
27. $\tan \theta$			$\tan \theta$	$\dfrac{1}{\cot \theta}$		
28. $\cot \theta$			$\dfrac{1}{\tan \theta}$	$\cot \theta$	$\dfrac{\pm\sqrt{\sec^2 \theta - 1}}{\sec^2 \theta - 1}$	
29. $\sec \theta$		$\dfrac{1}{\cos \theta}$			$\sec \theta$	
30. $\csc \theta$	$\dfrac{1}{\sin \theta}$					$\csc \theta$

Write each expression as a single trigonometric function or a power of a trigonometric function. (You may wish to use a graph to support your result.)

31. $\tan \theta \cos \theta$

32. $\cot \alpha \sin^2 \alpha \csc \alpha$

33. $\dfrac{\sin \beta \tan \beta}{\cos \beta}$

34. $\dfrac{\csc \theta \sec \theta}{\cot \theta}$

35. $\sec^2 x - 1$

36. $\csc^2 t - 1$

37. $\dfrac{\sin^2 x}{\cos^2 x} + \sin x \csc x$

38. $\dfrac{1}{\tan^2 \alpha} + \cot \alpha \tan \alpha$

Write each expression in terms of sine and cosine, and simplify it. (The final expression does not necessarily have to be in terms of sine and cosine.)

39. $\cot \theta \sin \theta$

40. $\sec \theta \cot \theta \sin \theta$

41. $\cos \theta \csc \theta$

42. $\dfrac{\sec \theta}{\csc \theta}$

43. $\dfrac{\cot^2 \theta}{\csc^2 \theta}$

44. $\dfrac{\tan^2 \theta}{\sec^2 \theta}$

45. $1 - \cos^2 \theta$

46. $\dfrac{1}{1 + \cot^2 \theta}$

47. $\dfrac{1}{\sec^2 \theta - 1}$

48. $\cot^2 \theta(1 + \tan^2 \theta)$

49. $\sin^2 \theta(\csc^2 \theta - 1)$

50. $(\sec \theta - 1)(\sec \theta + 1)$

51. $(1 - \cos \theta)(1 + \sec \theta)$

52. $\dfrac{\cos \theta + \sin \theta}{\sin \theta}$

53. $\dfrac{\cos^2 \theta - \sin^2 \theta}{\sin \theta \cos \theta}$

54. $\dfrac{1 - \sin^2 \theta}{1 + \cot^2 \theta}$

55. $\sec \theta - \cos \theta$

56. $\dfrac{1 + \tan^2 \theta}{1 + \cot^2 \theta}$

57. $\sin \theta(\csc \theta - \sin \theta)$

58. $(\sec \theta + \csc \theta)(\cos \theta - \sin \theta)$

59. $\csc \theta \sec \theta \tan \theta$

60. If $\sin \theta = x$ and θ is in quadrant IV, find an expression for $\sec \theta$ in terms of x.

Perform each indicated operation and simplify the result.

61. $\cot \theta + \dfrac{1}{\cot \theta}$

62. $\dfrac{\sec x}{\csc x} + \dfrac{\csc x}{\sec x}$

63. $\tan s(\cot s + \csc s)$

64. $\cos \beta(\sec \beta + \csc \beta)$

65. $\dfrac{1}{\csc^2 \theta} + \dfrac{1}{\sec^2 \theta}$

66. $\dfrac{1}{\sin \alpha - 1} - \dfrac{1}{\sin \alpha + 1}$

67. $\dfrac{\cos x}{\sec x} + \dfrac{\sin x}{\csc x}$

68. $\dfrac{\cos \theta}{\sin \theta} + \dfrac{\sin \theta}{1 + \cos \theta}$

69. $(1 + \sin t)^2 + \cos^2 t$

70. $(1 + \tan s)^2 - 2 \tan s$

71. $\dfrac{1}{1 + \cos x} - \dfrac{1}{1 - \cos x}$

72. $(\sin \alpha - \cos \alpha)^2$

In Exercises 73–112, verify that each equation is an identity.

73. $\dfrac{\cot \theta}{\csc \theta} = \cos \theta$

74. $\dfrac{\tan \theta}{\sec \theta} = \sin \theta$

75. $\cos^2 \theta(\tan^2 \theta + 1) = 1$

76. $\dfrac{\cos \theta + 1}{\tan^2 \theta} = \dfrac{\cos \theta}{\sec \theta - 1}$

77. $\dfrac{\tan^2 \gamma + 1}{\sec \gamma} = \sec \gamma$

78. $\sin^2 \beta(1 + \cot^2 \beta) = 1$

79. $\dfrac{1 - \sin^2 \beta}{\cos \beta} = \cos \beta$

80. $\dfrac{\sin^2 \theta}{\cos \theta} = \sec \theta - \cos \theta$

81. $\dfrac{1 - \cos x}{1 + \cos x} = (\cot x - \csc x)^2$

82. $\dfrac{\cot^2 t - 1}{1 + \cot^2 t} = 1 - 2 \sin^2 t$

83. $\dfrac{\cot \alpha + 1}{\cot \alpha - 1} = \dfrac{1 + \tan \alpha}{1 - \tan \alpha}$

84. $\dfrac{\sin^4 \alpha - \cos^4 \alpha}{\sin^2 \alpha - \cos^2 \alpha} = 1$

85. $\sin^2 \alpha + \tan^2 \alpha + \cos^2 \alpha = \sec^2 \alpha$

86. $\cot s + \tan s = \sec s \csc s$

87. $\dfrac{\sin^2 \gamma}{\cos \gamma} = \sec \gamma - \cos \gamma$

88. $\dfrac{\cos \alpha}{\sec \alpha} + \dfrac{\sin \alpha}{\csc \alpha} = \sec^2 \alpha - \tan^2 \alpha$

89. $\dfrac{\cos \theta}{\sin \theta \cot \theta} = 1$

90. $\sin^4 \theta - \cos^4 \theta = 2 \sin^2 \theta - 1$

91. $\tan^2 \gamma \sin^2 \gamma = \tan^2 \gamma + \cos^2 \gamma - 1$

92. $(1 - \cos^2 \alpha)(1 + \cos^2 \alpha) = 2 \sin^2 \alpha - \sin^4 \alpha$

93. $\dfrac{(\sec \theta - \tan \theta)^2 + 1}{\sec \theta \csc \theta - \tan \theta \csc \theta} = 2 \tan \theta$

94. $\dfrac{1}{1 - \sin \theta} + \dfrac{1}{1 + \sin \theta} = 2 \sec^2 \theta$

95. $\dfrac{1}{\tan \alpha - \sec \alpha} + \dfrac{1}{\tan \alpha + \sec \alpha} = -2 \tan \alpha$

96. $\dfrac{\csc \theta + \cot \theta}{\tan \theta + \sin \theta} = \cot \theta \csc \theta$

97. $\sec^4 x - \sec^2 x = \tan^4 x + \tan^2 x$

98. $(1 + \sin x + \cos x)^2 = 2(1 + \sin x)(1 + \cos x)$

99. $\dfrac{\sec^4 s - \tan^4 s}{\sec^2 s + \tan^2 s} = \sec^2 s - \tan^2 s$

100. $\dfrac{\tan s}{1 + \cos s} + \dfrac{\sin s}{1 - \cos s} = \cot s + \sec s \csc s$

101. $\dfrac{\tan^2 t - 1}{\sec^2 t} = \dfrac{\tan t - \cot t}{\tan t + \cot t}$

102. $\dfrac{1 - \cos x}{1 + \cos x} = \csc^2 x - 2 \csc x \cot x + \cot^2 x$

103. $\sin^2 \alpha \sec^2 \alpha + \sin^2 \alpha \csc^2 \alpha = \sec^2 \alpha$

104. $(\sec \alpha + \csc \alpha)(\cos \alpha - \sin \alpha) = \cot \alpha - \tan \alpha$

105. $\dfrac{1 - \sin \theta}{1 + \sin \theta} = \sec^2 \theta - 2 \sec \theta \tan \theta + \tan^2 \theta$

106. $\dfrac{\sin \theta}{1 - \cos \theta} - \dfrac{\sin \theta \cos \theta}{1 + \cos \theta} = \csc \theta (1 + \cos^2 \theta)$

107. $\dfrac{1 + \sin \theta}{1 - \sin \theta} - \dfrac{1 - \sin \theta}{1 + \sin \theta} = 4 \tan \theta \sec \theta$

108. $\dfrac{1 - \cos \theta}{1 + \cos \theta} = 2 \csc^2 \theta - 2 \csc \theta \cot \theta - 1$

109. $(2 \sin x + \cos x)^2 + (2 \cos x - \sin x)^2 = 5$

110. $\sin^2 x(1 + \cot x) + \cos^2 x(1 - \tan x) + \cot^2 x = \csc^2 x$

111. $\sec x - \cos x + \csc x - \sin x - \sin x \tan x = \cos x \cot x$

112. $\sin^3 \theta + \cos^3 \theta = (\cos \theta + \sin \theta)(1 - \cos \theta \sin \theta)$

(Modeling)   Work each problem.

113. *Intensity of a Lamp*   According to Lambert's law, the intensity of light from a single source on a flat surface at point P is given by

$$I = k \cos^2 \theta,$$

where k is a constant. (*Source:* Winter, C., *Solar Power Plants*, Springer-Verlag.)

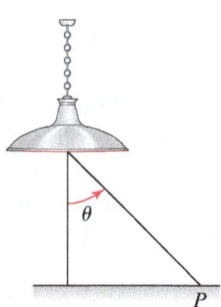

(a) Write I in terms of the sine function.

(b) Explain why the maximum value of I occurs when $\theta = 0$, if $0 \le \theta \le \frac{\pi}{2}$.

114. *Oscillating Spring*   The distance or displacement y of a weight attached to an oscillating spring from its natural position is modeled by $y = 4 \cos 2\pi t$, where t is time in seconds. Potential energy is the energy of position and is given by $P = ky^2$, where k is a constant. The weight has the greatest potential energy when the spring is stretched the most.

(a) Write P in terms of the cosine function.

(b) Use an identity to write P in terms of sin $2\pi t$.

115. *Energy in an Oscillating Spring*   Refer to **Exercise 114.** Two types of mechanical energy are kinetic energy and potential energy. Kinetic energy is the energy of motion, and potential energy is the energy of position. A stretched spring has potential energy, which is converted to kinetic energy when the spring is released. If the potential energy of a weight attached to a spring is

$$P(t) = k \cos^2 4\pi t,$$

where k is a constant and t is time in seconds, then its kinetic energy is given by

$$K(t) = k \sin^2 4\pi t.$$

The total mechanical energy E is given by the equation $E(t) = P(t) + K(t)$.

(a) If $k = 2$, graph P, K, and E in the window $[0, 0.5]$ by $[-1, 3]$, with Xscl = 0.25 and Yscl = 1. Interpret the graph.

(b) Make a table of K, P, and E, starting at $t = 0$ and incrementing by 0.05. Interpret the results.

(c) Use a fundamental identity to derive a simplified expression for $E(t)$.

116. *Radio Tuners*   Refer to **Example 7.** Let the energy stored in the inductor be

$$L(t) = 3 \cos^2 6{,}000{,}000t$$

and the energy in the capacitor be

$$C(t) = 3 \sin^2 6{,}000{,}000t,$$

where t is time in seconds. The total energy E in the circuit is given by $E(t) = L(t) + C(t)$.

(a) Graph L, C, and E in the window $[0, 10^{-6}]$ by $[-1, 4]$, with Xscl = 10^{-7} and Yscl = 1. Interpret the graph.

(b) Make a table of L, C, and E, starting at $t = 0$ and incrementing by 10^{-7}. Interpret your results.

(c) Use a fundamental identity to derive a simplified expression for $E(t)$.

9.2 Sum and Difference Identities

Cosine Sum and Difference Identities • Sine and Tangent Sum and Difference Identities

The identities presented in this chapter hold true whether the arguments represent *real numbers* or *degrees*.

Cosine Sum and Difference Identities

Are the equations

$$\cos(A - B) \overset{?}{=} \cos A - \cos B \quad \text{and} \quad \cos(A + B) \overset{?}{=} \cos A + \cos B$$

identities? That is, are they true for all values of A and B? By substituting a few values for A and B we quickly see that

$$\cos(A - B) \neq \cos A - \cos B$$

and that

$$\cos(A + B) \neq \cos A + \cos B.$$

For example, if $A = \frac{\pi}{2}$ and $B = 0$, then

$$\cos(A - B) = \cos\left(\frac{\pi}{2} - 0\right) = \cos\frac{\pi}{2} = 0, \quad \text{Not equal}$$

while

$$\cos A - \cos B = \cos\frac{\pi}{2} - \cos 0 = 0 - 1 = -1.$$

Similarly, when $A = \frac{\pi}{2}$ and $B = 0$, $\cos(A + B) = 0$ but $\cos A + \cos B = 1$.

To derive an identity for $\cos(A - B)$, we start by locating angles A and B in standard position on a unit circle, with $B < A$. Let S and Q be the points where the terminal sides of angles A and B, respectively, intersect the circle. Locate point R on the unit circle so that angle POR equals the difference $A - B$. See **FIGURE 5.**

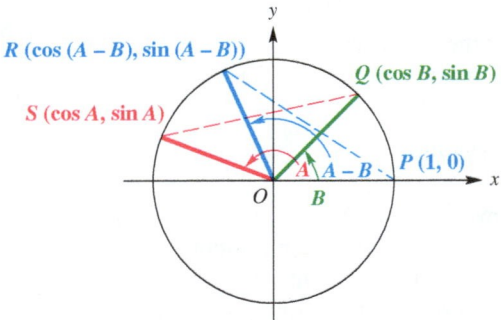

Angles A, B, and $A - B$ are in standard position.

FIGURE 5

Point Q is on the unit circle, so the x-coordinate of Q is given by the cosine of angle B, while the y-coordinate of Q is given by the sine of angle B.

Q has coordinates $(\cos B, \sin B)$.

In the same way,

S has coordinates $(\cos A, \sin A)$

and

R has coordinates $(\cos(A - B), \sin(A - B))$.

Angle SOQ also equals $A - B$. Since the central angles SOQ and POR are equal, chords PR and SQ are equal. By the distance formula, since $PR = SQ$,

$$\sqrt{[\cos(A - B) - 1]^2 + [\sin(A - B) - 0]^2} = \sqrt{(\cos A - \cos B)^2 + (\sin A - \sin B)^2}.$$

Square both sides and clear parentheses.

$$\cos^2(A - B) - 2\cos(A - B) + 1 + \sin^2(A - B)$$
$$= \cos^2 A - 2\cos A \cos B + \cos^2 B + \sin^2 A - 2\sin A \sin B + \sin^2 B$$

Because $\sin^2 x + \cos^2 x = 1$ for any value of x, we can rewrite the equation.

$$2 - 2\cos(A - B) = 2 - 2\cos A \cos B - 2\sin A \sin B$$

Cosine difference identity → $\cos(A - B) = \cos A \cos B + \sin A \sin B$ Subtract 2 and divide by -2.

Although **FIGURE 5** on the preceding page shows angles A and B in the second and first quadrants, respectively, this result is the same for any values of these angles.

To find a similar expression for $\cos(A + B)$, rewrite $A + B$ as $A - (-B)$ and use the identity for $\cos(A - B)$ that we just derived.

$$\cos(A + B) = \cos[A - (-B)]$$
$$= \cos A \cos(-B) + \sin A \sin(-B) \qquad \text{Cosine difference identity}$$
$$= \cos A \cos B + \sin A(-\sin B) \qquad \text{Negative-number identities}$$

Cosine sum identity → $\cos(A + B) = \cos A \cos B - \sin A \sin B$

Cosine of a Sum or Difference

$$\cos(A - B) = \cos A \cos B + \sin A \sin B$$

$$\cos(A + B) = \cos A \cos B - \sin A \sin B$$

These identities are important in calculus and useful in many applications.

EXAMPLE 1   **Finding Exact Cosine Function Values**

Find the *exact* value of the following.

(a) $\cos 15°$ **(b)** $\cos \dfrac{5\pi}{12}$ **(c)** $\cos 87° \cos 93° - \sin 87° \sin 93°$

Solution

(a) To find $\cos 15°$, we write $15°$ as the sum or difference of two angles with known function values, such as $45°$ and $30°$.

$$15° = 45° - 30° \qquad \text{We could also use } 60° - 45°.$$

$$\cos 15° = \cos(45° - 30°) \qquad\qquad 15° = 45° - 30°$$
$$= \cos 45° \cos 30° + \sin 45° \sin 30° \qquad \text{Cosine difference identity}$$
$$= \frac{\sqrt{2}}{2} \cdot \frac{\sqrt{3}}{2} + \frac{\sqrt{2}}{2} \cdot \frac{1}{2} \qquad\qquad \text{Substitute known values.}$$
$$= \frac{\sqrt{6} + \sqrt{2}}{4} \qquad\qquad \text{Multiply, then add fractions.}$$

(continued)

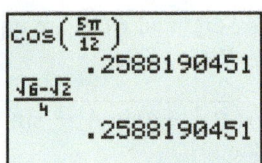

Radian mode

This screen supports the result in **Example 1(b)**.

(b) $\cos \dfrac{5\pi}{12} = \cos\left(\dfrac{\pi}{6} + \dfrac{\pi}{4}\right)$ $\dfrac{5\pi}{12} = \dfrac{2\pi}{12} + \dfrac{3\pi}{12} = \dfrac{\pi}{6} + \dfrac{\pi}{4}$

$\qquad\quad = \cos\dfrac{\pi}{6}\cos\dfrac{\pi}{4} - \sin\dfrac{\pi}{6}\sin\dfrac{\pi}{4}$ Cosine sum identity

$\qquad\quad = \dfrac{\sqrt{3}}{2}\cdot\dfrac{\sqrt{2}}{2} - \dfrac{1}{2}\cdot\dfrac{\sqrt{2}}{2}$ Substitute known values.

$\qquad\quad = \dfrac{\sqrt{6} - \sqrt{2}}{4}$ Multiply, then subtract fractions.

(c) $\cos 87° \cos 93° - \sin 87° \sin 93° = \cos(87° + 93°)$ Cosine sum identity with $A = 87°$ and $B = 93°$

$\qquad\qquad\qquad\qquad\qquad\qquad\quad = \cos 180°$ Add.

$\qquad\qquad\qquad\qquad\qquad\qquad\quad = -1$

NOTE In **Example 1(b)**, we used the fact that $\dfrac{5\pi}{12} = \dfrac{\pi}{6} + \dfrac{\pi}{4}$. At first glance, this sum may not be obvious. Think of the values $\dfrac{\pi}{6}$ and $\dfrac{\pi}{4}$ in terms of fractions with denominator 12: $\dfrac{\pi}{6} = \dfrac{2\pi}{12}$ and $\dfrac{\pi}{4} = \dfrac{3\pi}{12}$. The following list may help you with problems of this type.

$$\dfrac{\pi}{3} = \dfrac{4\pi}{12} \qquad \dfrac{\pi}{4} = \dfrac{3\pi}{12} \qquad \dfrac{\pi}{6} = \dfrac{2\pi}{12}$$

For example, using this list we see that $\dfrac{\pi}{12} = \dfrac{\pi}{3} - \dfrac{\pi}{4}$ (or $\dfrac{\pi}{4} - \dfrac{\pi}{6}$).

EXAMPLE 2   **Reducing $\cos(A - B)$ to a Function of a Single Variable**

Write $\cos(180° - \theta)$ as a trigonometric function of θ.

Solution   Replace A with $180°$ and B with θ in the cosine difference identity.

$$\cos(180° - \theta) = \cos 180° \cos \theta + \sin 180° \sin \theta$$

$$= (-1)\cos \theta + (0)\sin \theta$$

$$= -\cos \theta$$

Sine and Tangent Sum and Difference Identities

We can derive similar identities for sine and tangent. Use the cofunction identity $\sin \theta = \cos\left(\dfrac{\pi}{2} - \theta\right)$ from **Section 8.3** and replace θ with $A + B$.

$$\sin(A + B) = \cos\left[\dfrac{\pi}{2} - (A + B)\right]$$ Cofunction identity

$$= \cos\left[\left(\dfrac{\pi}{2} - A\right) - B\right]$$

$$= \cos\left(\dfrac{\pi}{2} - A\right)\cos B + \sin\left(\dfrac{\pi}{2} - A\right)\sin B$$ Cosine difference identity

Sine sum identity → $\sin(A + B) = \sin A \cos B + \cos A \sin B$ Cofunction identities

Now we write $\sin(A - B)$ as $\sin[A + (-B)]$ and use the identity for $\sin(A + B)$.

$$\sin(A - B) = \sin[A + (-B)]$$

$$= \sin A \cos(-B) + \cos A \sin(-B)$$ Sine sum identity

Sine difference identity → $\sin(A - B) = \sin A \cos B - \cos A \sin B$ Negative-number identities

> **Sine of a Sum or Difference**
>
> $$\sin(A + B) = \sin A \cos B + \cos A \sin B$$
>
> $$\sin(A - B) = \sin A \cos B - \cos A \sin B$$

To derive the identity for $\tan(A + B)$, proceed as follows.

$$\tan(A + B) = \frac{\sin(A + B)}{\cos(A + B)}$$ 　　　Fundamental identity

We are working to express this result in terms of tangent.

$$= \frac{\sin A \cos B + \cos A \sin B}{\cos A \cos B - \sin A \sin B}$$ 　　　Sum identities

$$= \frac{\dfrac{\sin A \cos B + \cos A \sin B}{1}}{\dfrac{\cos A \cos B - \sin A \sin B}{1}} \cdot \dfrac{\dfrac{1}{\cos A \cos B}}{\dfrac{1}{\cos A \cos B}}$$ 　　　Multiply numerator and denominator by $\frac{1}{\cos A \cos B}$.

$$= \frac{\dfrac{\sin A \cos B}{\cos A \cos B} + \dfrac{\cos A \sin B}{\cos A \cos B}}{\dfrac{\cos A \cos B}{\cos A \cos B} - \dfrac{\sin A \sin B}{\cos A \cos B}}$$ 　　　Multiply numerators; multiply denominators.

$$= \frac{\dfrac{\sin A}{\cos A} + \dfrac{\sin B}{\cos B}}{1 - \dfrac{\sin A}{\cos A} \cdot \dfrac{\sin B}{\cos B}}$$ 　　　Simplify.

Tangent sum identity ▸ $$\tan(A + B) = \frac{\tan A + \tan B}{1 - \tan A \tan B}$$ 　　　$\tan \theta = \frac{\sin \theta}{\cos \theta}$

Replacing B with $-B$ and the fact that $\tan(-B) = -\tan B$ gives the identity for the tangent of the difference of two numbers.

> **Tangent of a Sum or Difference**
>
> $$\tan(A + B) = \frac{\tan A + \tan B}{1 - \tan A \tan B} \qquad \tan(A - B) = \frac{\tan A - \tan B}{1 + \tan A \tan B}$$

EXAMPLE 3　**Finding Exact Sine and Tangent Function Values**

Find the *exact* value of the following.

(a) $\sin 75°$　　**(b)** $\tan \dfrac{7\pi}{12}$　　**(c)** $\sin 40° \cos 160° - \cos 40° \sin 160°$

Solution

(a) $\sin 75° = \sin(45° + 30°)$ 　　　$75° = 45° + 30°$

$ = \sin 45° \cos 30° + \cos 45° \sin 30°$ 　　　Sine sum identity

$ = \dfrac{\sqrt{2}}{2} \cdot \dfrac{\sqrt{3}}{2} + \dfrac{\sqrt{2}}{2} \cdot \dfrac{1}{2}$ 　　　Substitute known values.

$ = \dfrac{\sqrt{6} + \sqrt{2}}{4}$ 　　　Multiply and add fractions.

(continued)

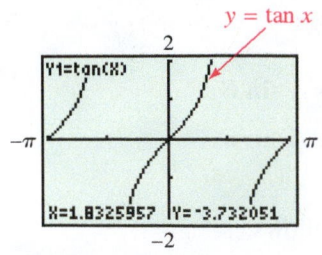

This screen indicates that the point $(1.8325957, -3.732051)$, which approximates $\left(\frac{7\pi}{12}, -2 - \sqrt{3}\right)$, lies on the graph of $y = \tan x$, supporting the result of **Example 3(b).**

$$\tan \frac{7\pi}{12} = -2 - \sqrt{3}$$

(b) $\tan \dfrac{7\pi}{12} = \tan\left(\dfrac{\pi}{3} + \dfrac{\pi}{4}\right)$ $\quad \frac{7\pi}{12} = \frac{4\pi}{12} + \frac{3\pi}{12} = \frac{\pi}{3} + \frac{\pi}{4}$

$$= \frac{\tan \frac{\pi}{3} + \tan \frac{\pi}{4}}{1 - \tan \frac{\pi}{3} \tan \frac{\pi}{4}} \qquad \text{Tangent sum identity}$$

$$= \frac{\sqrt{3} + 1}{1 - \sqrt{3} \cdot 1} \qquad \text{Substitute known values.}$$

$$= \frac{\sqrt{3} + 1}{1 - \sqrt{3}} \cdot \frac{1 + \sqrt{3}}{1 + \sqrt{3}} \qquad \text{Rationalize the denominator.}$$

$$= \frac{\sqrt{3} + 3 + 1 + \sqrt{3}}{1 - 3} \qquad \text{Multiply.}$$

$$= \frac{4 + 2\sqrt{3}}{-2} \qquad \text{Combine like terms.}$$

Factor first. Then divide out the common factor. $\quad = \dfrac{2\left(2 + \sqrt{3}\right)}{2(-1)} \qquad \text{Factor out 2.}$

$$= -2 - \sqrt{3} \qquad \text{Lowest terms}$$

(c) $\sin 40° \cos 160° - \cos 40° \sin 160° = \sin(40° - 160°) \qquad$ Sine difference identity

$$= \sin(-120°) \qquad \text{Subtract.}$$

$\sin(-\theta) = -\sin \theta \quad\longrightarrow\quad = -\sin 120° \qquad$ Negative-number identity

$$= -\frac{\sqrt{3}}{2}$$

EXAMPLE 4 **Using Sum and Difference Identities**

Write each function as an expression involving functions of θ alone.

(a) $\sin(30° + \theta)$ **(b)** $\tan(45° - \theta)$ **(c)** $\sin(180° + \theta)$

Solution

(a) $\sin(30° + \theta) = \sin 30° \cos \theta + \cos 30° \sin \theta \qquad$ Sine sum identity

$$= \frac{1}{2} \cos \theta + \frac{\sqrt{3}}{2} \sin \theta$$

(b) $\tan(45° - \theta) = \dfrac{\tan 45° - \tan \theta}{1 + \tan 45° \tan \theta} \qquad$ Tangent difference identity

$$= \frac{1 - \tan \theta}{1 + \tan \theta}$$

(c) $\sin(180° + \theta) = \sin 180° \cos \theta + \cos 180° \sin \theta \qquad$ Sine sum identity

$$= 0 \cdot \cos \theta + (-1) \sin \theta$$

$$= -\sin \theta$$

An identity like the one derived in **Example 4(c)** is sometimes called a **reduction formula,** because the expression on the left, which contains a quadrantal angle, is reduced to a function of θ alone.

EXAMPLE 5   **Finding Function Values and the Quadrant of $A + B$**

Suppose that A and B are angles in standard position, with $\sin A = \frac{4}{5}$, $\frac{\pi}{2} < A < \pi$, and $\cos B = -\frac{5}{13}$, $\pi < B < \frac{3\pi}{2}$. Find the following.

(a) $\sin(A + B)$ **(b)** $\tan(A + B)$ **(c)** The quadrant of $A + B$

Solution

(a) The identity for $\sin(A + B)$ requires $\sin A$, $\cos A$, $\sin B$, and $\cos B$. We are given values of $\sin A$ and $\cos B$. We must find values of $\cos A$ and $\sin B$.

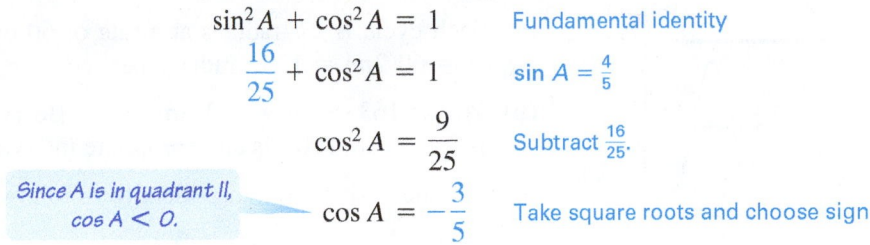

$$\sin^2 A + \cos^2 A = 1 \qquad \textcolor{blue}{\text{Fundamental identity}}$$

$$\frac{16}{25} + \cos^2 A = 1 \qquad \textcolor{blue}{\sin A = \frac{4}{5}}$$

$$\cos^2 A = \frac{9}{25} \qquad \textcolor{blue}{\text{Subtract } \frac{16}{25}.}$$

$$\textcolor{blue}{\text{Since } A \text{ is in quadrant II,}} \qquad \cos A = -\frac{3}{5} \qquad \textcolor{blue}{\text{Take square roots and choose sign.}}$$
$$\textcolor{blue}{\cos A < 0.}$$

A similar procedure yields $\sin B = -\frac{12}{13}$. Now use the sine sum identity.

$$\sin(A + B) = \frac{4}{5}\left(-\frac{5}{13}\right) + \left(-\frac{3}{5}\right)\left(-\frac{12}{13}\right) = -\frac{20}{65} + \frac{36}{65} = \frac{16}{65}$$

An alternative way to determine $\cos A$ and $\sin B$ is to make a sketch of angles A and B, as shown in **FIGURE 6.** For example, in **FIGURE 6(a)** angle A is drawn so that it is in quadrant II and $\sin A = \frac{4}{5}$. The Pythagorean theorem can be used to determine that the length of the other leg is 3. Thus, $\cos A = -\frac{3}{5}$. Using either method, be sure to choose the correct signs of the functions based on their quadrants.

(b) To find $\tan(A + B)$, start by finding $\tan A$ and $\tan B$. Refer to part (a) and use values

$$\sin A = \frac{4}{5}, \quad \cos A = -\frac{3}{5}, \quad \sin B = -\frac{12}{13}, \quad \text{and} \quad \cos B = -\frac{5}{13}.$$

$$\tan A = \frac{\sin A}{\cos A} = \frac{\frac{4}{5}}{-\frac{3}{5}} = -\frac{4}{3} \qquad \textcolor{blue}{\text{Quotient identity}}$$

$$\tan B = \frac{\sin B}{\cos B} = \frac{-\frac{12}{13}}{-\frac{5}{13}} = \frac{12}{5} \qquad \textcolor{blue}{\text{Quotient identity}}$$

Now use the identity for $\tan(A + B)$.

$$\tan(A + B) = \frac{\tan A + \tan B}{1 - \tan A \tan B}$$

$$\tan(A + B) = \frac{-\frac{4}{3} + \frac{12}{5}}{1 - \left(-\frac{4}{3}\right)\left(\frac{12}{5}\right)} = \frac{\frac{16}{15}}{1 + \frac{48}{15}} = \frac{\frac{16}{15}}{\frac{63}{15}} = \frac{16}{15} \cdot \frac{15}{63} = \frac{16}{63}$$

(c) From the results of parts (a) and (b), $\sin(A + B) = \frac{16}{65}$ and $\tan(A + B) = \frac{16}{63}$, both of which are positive. Therefore, $A + B$ must be in quadrant I, since it is the only quadrant in which sine and tangent are both positive. ●

EXAMPLE 6   **Applying the Cosine Difference Identity to Voltage**

Common household electric current is called **alternating current** because the current alternates direction within the wires. The voltage V in a 115-volt outlet can be expressed by $V(t) = 163 \sin \omega t$, where ω is the angular speed (in radians per second)

(continued)

Figure 6 (left margin):

$(-3, 4)$

$\cos A = -\frac{3}{5}$

(a)

$\sin B = -\frac{12}{13}$

$(-5, -12)$

(b)

FIGURE 6

of the rotating generator at the electrical plant and t is time in seconds. (*Source:* Bell, D., *Fundamentals of Electric Circuits*, Fourth Edition, Prentice-Hall.)

(a) Electric generators must rotate at 60 cycles per second so that household appliances and computers operate properly. Determine ω for these electric generators.

(b) Graph V on the interval $0 \leq t \leq 0.05$.

(c) Use the cosine difference identity to show that when $\phi = \frac{\pi}{2}$, the graph of $V(t) = 163 \cos(\omega t - \phi)$ is the same as the graph of $V(t) = 163 \sin \omega t$.

Solution

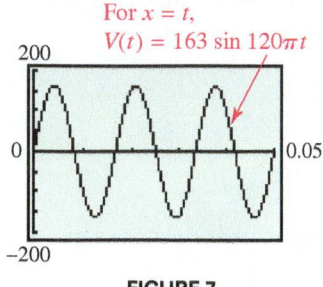

For $x = t$,
$V(t) = 163 \sin 120\pi t$

FIGURE 7

(a) Each cycle is 2π radians at a rate of 60 cycles per sec, so the angular speed is $\omega = 60(2\pi) = 120\pi$ radians per sec.

(b) $V(t) = 163 \sin \omega t = 163 \sin 120\pi t$. Because the amplitude of the function is 163, $[-200, 200]$ is an appropriate interval for the range, as shown in **FIGURE 7.**

(c) Use the cosine difference identity as follows.

$$\cos\left(x - \frac{\pi}{2}\right) = \cos x \cos \frac{\pi}{2} + \sin x \sin \frac{\pi}{2}$$

$$= \cos x \cdot 0 + \sin x \cdot 1$$

$$= \sin x$$

Therefore, when $\phi = \frac{\pi}{2}$ and $x = \omega t$,

$$V(t) = 163 \cos(\omega t - \phi) = 163 \cos\left(\omega t - \frac{\pi}{2}\right) = 163 \sin \omega t.$$

9.2 Exercises

Concept Check *Match each expression in Column I with the correct expression in Column II to form an identity.*

I	II
1. $\cos(x + y) = $ _____	**A.** $\cos x \cos y + \sin x \sin y$
	B. $\sin x \sin y - \cos x \cos y$
2. $\cos(x - y) = $ _____	**C.** $\sin x \cos y + \cos x \sin y$
	D. $\sin x \cos y - \cos x \sin y$
3. $\sin(x + y) = $ _____	**E.** $\cos x \sin y - \sin x \cos y$
4. $\sin(x - y) = $ _____	**F.** $\cos x \cos y - \sin x \sin y$

Checking Analytic Skills *Use identities to find the **exact** value of each expression. **Do not use a calculator.***

5. $\sin \dfrac{\pi}{12}$ **6.** $\tan \dfrac{\pi}{12}$ **7.** $\sin\left(-\dfrac{5\pi}{12}\right)$ **8.** $\tan\left(-\dfrac{5\pi}{12}\right)$

9. $\sin\left(\dfrac{13\pi}{12}\right)$ **10.** $\cos\left(\dfrac{13\pi}{12}\right)$ **11.** $\cos 75°$ **12.** $\sin 105°$

13. $\tan 105°$ **14.** $\sin(-15°)$ **15.** $\cos(-15°)$ **16.** $\tan(-75°)$

17. $\cos \dfrac{\pi}{3} \cos \dfrac{2\pi}{3} - \sin \dfrac{\pi}{3} \sin \dfrac{2\pi}{3}$ **18.** $\cos \dfrac{7\pi}{8} \cos \dfrac{\pi}{8} + \sin \dfrac{7\pi}{8} \sin \dfrac{\pi}{8}$ **19.** $\sin 76° \cos 31° - \cos 76° \sin 31°$

20. $\sin 40° \cos 50° + \cos 40° \sin 50°$ **21.** $\dfrac{\tan 80° + \tan 55°}{1 - \tan 80° \tan 55°}$ **22.** $\dfrac{\tan 80° - \tan(-55°)}{1 + \tan 80° \tan(-55°)}$

Use identities to write each expression as a function of x alone.

23. $\sin(180° - x)$

24. $\sin(270° + x)$

25. $\cos(180° + x)$

26. $\cos(270° - x)$

27. $\sin(x - 90°)$

28. $\sin(x + 90°)$

29. $\tan(180° - x)$

30. $\tan(360° - x)$

31. $\cos\left(\dfrac{\pi}{2} - x\right)$

32. $\cos(\pi - x)$

33. $\cos\left(\dfrac{3\pi}{2} + x\right)$

34. $\sin(\pi - x)$

35. $\sin(\pi + x)$

36. $\tan(2\pi - x)$

37. $\cos(135° - x)$

38. $\sin(45° + x)$

39. $\tan(45° + x)$

40. $\tan(\pi + x)$

41. $\tan(\pi - x)$

42. $\sin\left(\dfrac{3\pi}{2} - x\right)$

43. $\tan\left(\dfrac{5\pi}{4} - x\right)$

44. $\tan\left(x + \dfrac{7\pi}{4}\right)$

45. $\cos(2\pi - x)$

46. $\cos\left(x - \dfrac{3\pi}{4}\right)$

Checking Analytic Skills *Suppose that A and B are angles in standard position. Use the given information to find* **(a)** $\sin(A + B)$, **(b)** $\sin(A - B)$, **(c)** $\tan(A + B)$, **(d)** $\tan(A - B)$, **(e)** *the quadrant of* $A + B$, *and* **(f)** *the quadrant of* $A - B$. **Do not use a calculator.**

47. $\cos A = \dfrac{3}{5}$, $\sin B = \dfrac{5}{13}$, $\quad 0 < A < \dfrac{\pi}{2}, 0 < B < \dfrac{\pi}{2}$

48. $\sin A = \dfrac{3}{5}$, $\sin B = -\dfrac{12}{13}$, $0 < A < \dfrac{\pi}{2}, \pi < B < \dfrac{3\pi}{2}$

49. $\cos A = -\dfrac{8}{17}$, $\cos B = -\dfrac{3}{5}$, $\pi < A < \dfrac{3\pi}{2}, \pi < B < \dfrac{3\pi}{2}$

50. $\cos A = -\dfrac{15}{17}$, $\sin B = \dfrac{4}{5}$, $\dfrac{\pi}{2} < A < \pi, 0 < B < \dfrac{\pi}{2}$

Verify that each equation is an identity.

51. $\sin 2x = 2 \sin x \cos x$

52. $\sin(210° + x) - \cos(120° + x) = 0$

53. $\sin(x + y) + \sin(x - y) = 2 \sin x \cos y$

54. $\tan(x - y) - \tan(y - x) = \dfrac{2(\tan x - \tan y)}{1 + \tan x \tan y}$

55. $\dfrac{\cos(A - B)}{\cos A \sin B} = \tan A + \cot B$

56. $\dfrac{\sin(A + B)}{\cos A \cos B} = \tan A + \tan B$

57. $\dfrac{\sin(A - B)}{\sin B} + \dfrac{\cos(A - B)}{\cos B} = \dfrac{\sin A}{\sin B \cos B}$

58. $\dfrac{\tan(A + B) - \tan B}{1 + \tan(A + B)\tan B} = \tan A$

59. $\dfrac{\sin(x - y)}{\sin(x + y)} = \dfrac{\tan x - \tan y}{\tan x + \tan y}$

60. $\dfrac{\cos(A - B)}{\sin(A + B)} = \dfrac{1 + \cot A \cot B}{\cot A + \cot B}$

Exercises 61 and 62 refer to **Example 6.**

61. How many times does the current oscillate in 0.05 second?

62. What are the maximum and minimum voltages in this outlet? Is the voltage always equal to 115 volts?

(Modeling) *Solve each problem.*

63. *Back Stress* If a person bends at the waist with a straight back, making an angle of θ degrees *with the horizontal,* then the force F exerted on the back muscles can be modeled by the equation

$$F = \dfrac{0.6W \sin(\theta + 90°)}{\sin 12°},$$

where W is the weight of the person. (*Source:* Metcalf, H., *Topics in Classical Biophysics,* Prentice-Hall.)

(a) Calculate F when $W = 170$ pounds and $\theta = 30°$.

(b) Use an identity to show that F is approximately equal to $2.9W \cos \theta$.

(c) For what value of θ is F maximum?

64. *Back Stress*   Refer to **Exercise 63.**
   (a) Suppose a 200-pound person bends at the waist so that $\theta = 45°$. Estimate the force exerted by the person's back muscles.
   (b) Approximate graphically the value of θ that results in the back muscles exerting a force of 400 pounds.

65. *Sound Waves*   Sound is a result of waves applying pressure to a person's eardrum. For a pure sound wave radiating outward in a spherical shape, the trigonometric function

$$P = \frac{a}{r} \cos\left(\frac{2\pi r}{\lambda} - ct\right)$$

can be used to model the sound pressure P at a radius of r feet from the source, where t is time in seconds, λ is length of the sound wave in feet, c is speed of sound in feet per second, and a is maximum sound pressure at the source measured in pounds per square foot. (*Source:* Beranek, L., *Noise and Vibration Control*, Institute of Noise Control Engineering, Washington, DC.) Let $\lambda = 4.9$ feet and $c = 1026$ feet per second.
   (a) Let $a = 0.4$ pound per square foot. Graph the sound pressure at a distance $r = 10$ feet from its source over the interval $0 \le t \le 0.05$. Describe P at this distance.

   (b) Now let $a = 3$ and $t = 10$. Graph the sound pressure for $0 \le r \le 20$. What happens to the pressure P as the radius r increases?
   (c) Suppose a person stands at a radius r so that

$$r = n\lambda,$$

   where n is a positive integer. Use the difference identity for cosine to simplify P in this situation.

66. *Voltage of a Circuit*   When the two voltages

$$V_1 = 30 \sin 120\pi t$$

and

$$V_2 = 40 \cos 120\pi t$$

are applied to the same circuit, the resulting voltage V will equal their sum. (*Source:* Bell, D., *Fundamentals of Electric Circuits*, Second Edition, Reston Publishing Company.)
   (a) Graph $V = V_1 + V_2$ over the interval $0 \le t \le 0.05$.
   (b) Use the graph to estimate values for a and ϕ so that $V = a \sin(120\pi t + \phi)$.
   (c) Use identities to verify that your expression for V is valid.

SECTIONS 9.1–9.2   Reviewing Basic Concepts

1. Write $\dfrac{\csc x}{\cot x} - \dfrac{\cot x}{\csc x}$ in terms of $\sin x$ and $\cos x$.

2. Use a sum or difference identity to find an expression for the exact value of $\tan\left(-\dfrac{\pi}{12}\right)$.

3. Find an exact value of

$$\cos 18° \cos 108° + \sin 18° \sin 108°.$$

4. Use a sum or difference identity to write $\sin\left(x - \dfrac{\pi}{4}\right)$ as a function of x alone.

5. Given $\sin A = \dfrac{2}{3}$ and $\cos B = -\dfrac{1}{2}$, where A is in quadrant II and B is in quadrant III, find

$$\sin(A + B), \quad \cos(A - B), \quad \text{and} \quad \tan(A - B).$$

Verify that each equation is an identity.

6. $\csc^2 \theta - \cot^2 \theta = 1$

7. $\dfrac{\sin t}{1 - \cos t} = \dfrac{1 + \cos t}{\sin t}$

8. $\dfrac{\cot A - \tan A}{\csc A \sec A} = \cos^2 A - \sin^2 A$

9. $\dfrac{\sin(x - y)}{\sin x \sin y} = \cot y - \cot x$

10. *(Modeling) Voltage*   A coil of wire rotating in a magnetic field induces a voltage given by

$$v = 20 \sin\left(\frac{\pi t}{4} - \frac{\pi}{2}\right).$$

Use an identity to express the right side of this equation in terms of $\cos \dfrac{\pi t}{4}$.

9.3 Further Identities

Double-Number Identities • Product-to-Sum and Sum-to-Product Identities • Half-Number Identities

Double-Number Identities

The **double-number identities,** or **double-angle identities,** result from the sum identities when $A = B$ so that $A + B = A + A = 2A$.

$$\cos 2A = \cos(A + A)$$
$$= \cos A \cos A - \sin A \sin A \qquad \text{Cosine sum identity}$$
$$\cos 2A = \cos^2 A - \sin^2 A$$

Note that
$$\cos 2A = \cos(A + A) \neq 2\cos A.$$

Two other common forms of this identity are obtained by substitution.

$$\cos 2A = \cos^2 A - \sin^2 A$$
$$= (1 - \sin^2 A) - \sin^2 A \qquad \text{Pythagorean identity}$$
$$\cos 2A = 1 - 2\sin^2 A$$

or

$$\cos 2A = \cos^2 A - \sin^2 A$$
$$= \cos^2 A - (1 - \cos^2 A) \qquad \text{Pythagorean identity}$$
$$= \cos^2 A - 1 + \cos^2 A$$
$$\cos 2A = 2\cos^2 A - 1$$

We find $\sin 2A$ with the identity for $\sin(A + B)$, again letting $B = A$.

$$\sin 2A = \sin(A + A)$$
$$= \sin A \cos A + \cos A \sin A \qquad \text{Sine sum identity}$$
$$\sin 2A = 2\sin A \cos A$$

Similarly, we use the identity for $\tan(A + B)$ to find $\tan 2A$.

$$\tan 2A = \tan(A + A)$$
$$= \frac{\tan A + \tan A}{1 - \tan A \tan A} \qquad \text{Tangent sum identity}$$
$$\tan 2A = \frac{2\tan A}{1 - \tan^2 A}$$

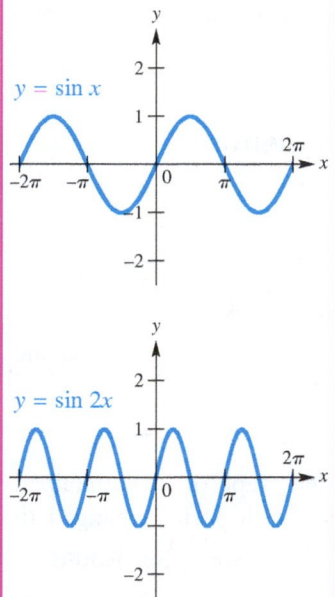

FOR DISCUSSION
See the graphs below.

1. Discuss how the graph of $y = \sin 2x$ differs from the graph of $y = \sin x$.

2. Compare the graph of $y = \sin 2x$ to the graph of $y = 2\sin x$.

Double-Number Identities

$$\cos 2A = \cos^2 A - \sin^2 A \qquad \cos 2A = 1 - 2\sin^2 A$$
$$\cos 2A = 2\cos^2 A - 1 \qquad \sin 2A = 2\sin A \cos A$$
$$\tan 2A = \frac{2\tan A}{1 - \tan^2 A}$$

EXAMPLE 1 **Finding Function Values of 2θ**

Given $\cos \theta = \frac{3}{5}$ and $\sin \theta < 0$, find $\sin 2\theta$, $\cos 2\theta$, and $\tan 2\theta$.

Solution   Because $\sin 2\theta = 2 \sin \theta \cos \theta$ and we are only given $\cos \theta$, we must find $\sin \theta$.

$$\sin^2 \theta + \left(\frac{3}{5}\right)^2 = 1 \qquad \sin^2 \theta + \cos^2 \theta = 1 \text{ and } \cos \theta = \frac{3}{5}$$

$$\sin^2 \theta = \frac{16}{25} \qquad \text{Subtract } \frac{9}{25}.$$

$$\sin \theta = -\frac{4}{5} \qquad \begin{array}{l}\text{Take square roots. Then choose the}\\ \text{negative square root, since } \sin \theta < 0.\end{array}$$

$\sin \theta < 0$ is given.

Use the double-number identity for sine.

$$\sin 2\theta = 2 \sin \theta \cos \theta = 2\left(-\frac{4}{5}\right)\left(\frac{3}{5}\right) = -\frac{24}{25} \qquad \sin \theta = -\frac{4}{5} \text{ and } \cos \theta = \frac{3}{5}$$

Now use the first double-number identity for cosine.

Any of the three forms can be used.

$$\cos 2\theta = \cos^2 \theta - \sin^2 \theta = \frac{9}{25} - \frac{16}{25} = -\frac{7}{25}$$

We can find $\tan 2\theta$ by finding the quotient of $\sin 2\theta$ and $\cos 2\theta$.

$$\tan 2\theta = \frac{\sin 2\theta}{\cos 2\theta} = \frac{-\frac{24}{25}}{-\frac{7}{25}} = -\frac{24}{25}\left(-\frac{25}{7}\right) = \frac{24}{7}$$

We could also find this value by first finding $\tan \theta$ and then applying the double-number identity for $\tan 2\theta$.

EXAMPLE 2 **Verifying a Double-Number Identity**

Verify that the given equation is an identity.

$$\cot x \sin 2x = 1 + \cos 2x$$

Analytic Solution

We will start by writing all functions in terms of sine and cosine. Begin by working on the left side.

$$\cot x \sin 2x = \frac{\cos x}{\sin x} \cdot \sin 2x \qquad \cot x = \tfrac{\cos x}{\sin x}$$

$$= \frac{\cos x}{\sin x}(2 \sin x \cos x) \qquad \sin 2x = 2 \sin x \cos x$$

$$= 2 \cos^2 x \qquad \text{Multiply.}$$

$$= 1 + \cos 2x \qquad 2 \cos^2 x - 1 = \cos 2x$$

The final step illustrates the importance of being able to recognize alternative forms of identities. That is,

$$\cos 2x = 2 \cos^2 x - 1$$

so

$$1 + \cos 2x = 2 \cos^2 x.$$

Graphing Calculator Support

To support our analytic work, we show that the graphs of

$$y_1 = \cot x \sin 2x \quad \text{and} \quad y_2 = 1 + \cos 2x$$

coincide. Notice that the graph of y_2 is obtained from that of $y = \cos x$ with period changed to $\frac{2\pi}{2} = \pi$ and shifted 1 unit upward. See **FIGURE 8**.

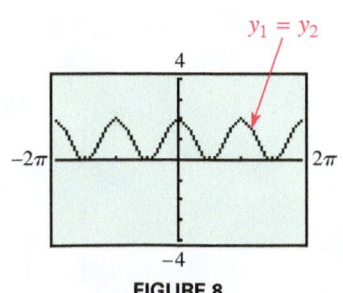

FIGURE 8

> **EXAMPLE 3** **Simplifying Expressions by Using Double-Number Identities**

Simplify each expression.

(a) $\cos^2 7x - \sin^2 7x$ **(b)** $\sin 15° \cos 15°$

Solution

(a) The expression $\cos^2 7x - \sin^2 7x$ suggests the first double-number identity for cosine.

$$\cos^2 A - \sin^2 A = \cos 2A \qquad \text{Interchange sides of identity.}$$

Substitute $7x$ for A.

$$\cos^2 7x - \sin^2 7x = \cos(2(7x)) = \cos 14x$$

(b) We want to write $\sin 15° \cos 15°$ in a way that we can apply the identity $\sin 2A = 2 \sin A \cos A$ directly. The expression needs a coefficient of 2.

$$\sin 15° \cos 15° = \frac{1}{2}(2) \sin 15° \cos 15° \qquad \text{Multiply by 1 in the form } \tfrac{1}{2}(2).$$

$$= \frac{1}{2}(2 \sin 15° \cos 15°) \qquad \text{Associative property}$$

$$= \frac{1}{2} \sin(2 \cdot 15°) \qquad 2 \sin A \cos A = \sin 2A, \text{ with } A = 15°$$

$$= \frac{1}{2} \sin 30° \qquad \text{Multiply.}$$

$$= \frac{1}{2} \cdot \frac{1}{2} \qquad \sin 30° = \tfrac{1}{2}$$

$$= \frac{1}{4} \qquad \text{Multiply.}$$

> **EXAMPLE 4** **Finding Function Values of θ Given Information about 2θ**

Find the values of the six trigonometric functions of θ given that $\cos 2\theta = \frac{4}{5}$ and $90° < \theta < 180°$.

Solution We must obtain a trigonometric function value of θ alone.

$$\cos 2\theta = 1 - 2 \sin^2 \theta \qquad \text{Double-angle identity}$$

$$\frac{4}{5} = 1 - 2 \sin^2 \theta \qquad \cos 2\theta = \tfrac{4}{5}$$

$$-\frac{1}{5} = -2 \sin^2 \theta \qquad \text{Subtract 1 from each side.}$$

$$\frac{1}{10} = \sin^2 \theta \qquad \text{Multiply by } -\tfrac{1}{2}.$$

$$\sin \theta = \sqrt{\frac{1}{10}} \qquad \text{Take square roots and choose the positive square root, since } \theta \text{ terminates in quadrant II.}$$

$$\sin \theta = \frac{1}{\sqrt{10}} \cdot \frac{\sqrt{10}}{\sqrt{10}} \qquad \text{Use the quotient rule and rationalize the denominator.}$$

$$\sin \theta = \frac{\sqrt{10}}{10} \qquad \sqrt{a} \cdot \sqrt{a} = a$$

(continued)

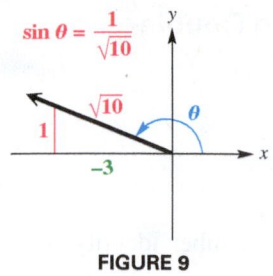

FIGURE 9

Now find the values of $\cos \theta$ and $\tan \theta$ by sketching and labeling a right triangle in quadrant II. Since $\sin \theta = \dfrac{1}{\sqrt{10}}$, the triangle in **FIGURE 9** is labeled accordingly. The Pythagorean theorem is used to find the remaining leg. Remember that $x < 0$ in this sketch.

$$\cos \theta = \frac{-3}{\sqrt{10}} = -\frac{3\sqrt{10}}{10} \quad \text{and} \quad \tan \theta = \frac{1}{-3} = -\frac{1}{3}$$

Find the other three functions using reciprocals.

$$\csc \theta = \frac{1}{\sin \theta} = \sqrt{10}, \quad \sec \theta = \frac{1}{\cos \theta} = -\frac{\sqrt{10}}{3}, \quad \cot \theta = \frac{1}{\tan \theta} = -3$$

EXAMPLE 5 **Deriving a Multiple-Number Identity**

Write $\sin 3x$ in terms of $\sin x$.

Analytic Solution

$\sin 3x = \sin(2x + x)$

$\quad = \sin 2x \cos x + \cos 2x \sin x$ Sine sum identity

$\quad = (2 \sin x \cos x) \cos x + (\cos^2 x - \sin^2 x) \sin x$
           Double-number identities

$\quad = 2 \sin x \cos^2 x + \cos^2 x \sin x - \sin^3 x$
           Multiply.

$\quad = 2 \sin x(1 - \sin^2 x) + (1 - \sin^2 x) \sin x - \sin^3 x$
           $\cos^2 x = 1 - \sin^2 x$

$\quad = 2 \sin x - 2 \sin^3 x + \sin x - \sin^3 x - \sin^3 x$
           Distributive property

$\quad = 3 \sin x - 4 \sin^3 x$ Combine like terms.

Graphing Calculator Support

The table in **FIGURE 10** numerically *supports* the analytic solution. Here,

$$Y_2 = 3 \sin X - 4(\sin X)^\wedge 3.$$

The table does not *verify* the identity because it does not list every possible X-value.

X	Y1	Y2
-1.178	.38268	.38268
-.7854	-.7071	-.7071
-.3927	-.9239	-.9239
0	0	0
.3927	.92388	.92388
.7854	.70711	.70711
1.1781	-.3827	-.3827

Y₁⊟sin(3X)

FIGURE 10

EXAMPLE 6 **Determining Wattage Consumption**

If a toaster is plugged into a common household outlet, the wattage consumed is not constant. Instead, it varies at a high frequency according to the model

$$W = \frac{V^2}{R},$$

where V is the voltage and R is a constant that measures the resistance of the toaster in ohms. (*Source:* Bell, D., *Fundamentals of Electric Circuits*, Fourth Edition, Prentice-Hall.) Graph the wattage W consumed by a typical toaster with $R = 15$ and $V = 163 \sin 120\pi t$ over the interval $0 \le t \le 0.05$. How many oscillations are there?

For $x = t$,

$$W(t) = \frac{(163 \sin 120\pi t)^2}{15}$$

FIGURE 11

Solution $W = \dfrac{V^2}{R} = \dfrac{(163 \sin 120\pi t)^2}{15}$ Substitute the given values.

To determine the range for W, we note that $\sin 120\pi t$ has maximum value 1, so the expression for W has maximum value $\dfrac{163^2}{15} \approx 1771$. The minimum value is 0. The graph in **FIGURE 11** shows that there are six oscillations.

WHAT WENT WRONG?

To verify the proposed identity $\cos x = \cos 4x$, a student graphed both equations on the same screen, with the result shown on the left. He then compared the functions, using the table feature of his calculator, and got the result shown on the right. He was confused, because the graph indicates that the equation is not an identity, while the table indicates that it might be one.

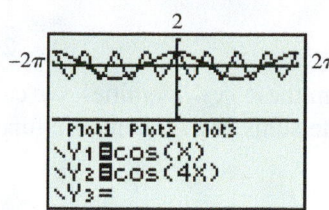

$$\Delta Tbl = \frac{2\pi}{3} \approx 2.0944; \text{ Tblstart} = 0$$

What Went Wrong? What should the student conclude? Why?

Product-to-Sum and Sum-to-Product Identities

➡ Looking Ahead to Calculus

The product-to-sum identities are used in calculus to find **integrals** of functions that are products of trigonometric functions. One classic calculus text by Earl Swokowski includes the following example.

Evaluate $\int \cos 5x \cos 3x \, dx$.

The first solution line reads: "We may write

$\cos 5x \cos 3x = \frac{1}{2}(\cos 8x + \cos 2x)$."

Adding the identities for $\cos(A + B)$ and $\cos(A - B)$ gives the following.

$$\cos(A + B) = \cos A \cos B - \sin A \sin B$$
$$\cos(A - B) = \cos A \cos B + \sin A \sin B$$
$$\overline{\cos(A + B) + \cos(A - B) = 2 \cos A \cos B}$$

$$\cos A \cos B = \frac{1}{2}[\cos(A + B) + \cos(A - B)]$$

Similarly, subtracting $\cos(A + B)$ from $\cos(A - B)$ gives the following.

$$\sin A \sin B = \frac{1}{2}[\cos(A - B) - \cos(A + B)]$$

Using the identities for $\sin(A + B)$ and $\sin(A - B)$ in the same way, we get two more identities. Those and the previous ones are now summarized.

Product-to-Sum Identities

$$\cos A \cos B = \frac{1}{2}[\cos(A + B) + \cos(A - B)]$$

$$\sin A \sin B = \frac{1}{2}[\cos(A - B) - \cos(A + B)]$$

$$\sin A \cos B = \frac{1}{2}[\sin(A + B) + \sin(A - B)]$$

$$\cos A \sin B = \frac{1}{2}[\sin(A + B) - \sin(A - B)]$$

Answer to What Went Wrong?

He should conclude that $\cos x = \cos 4x$ is *not* an identity, based on the graph. The problem with the table is that a ΔTbl of $\frac{2\pi}{3}$ produces values where the graphs of the two functions intersect. Choosing a different interval, such as $\frac{\pi}{2}$, would show different function values at multiples of $\frac{\pi}{2}$.

EXAMPLE 7   **Using a Product-to-Sum Identity**

Rewrite $\cos 2\theta \sin \theta$ as either the sum or difference of two functions.

Solution   Use the (fourth) identity for $\cos A \sin B$, with $2\theta = A$ and $\theta = B$.

$$\cos 2\theta \sin \theta = \frac{1}{2}\left[\sin(2\theta + \theta) - \sin(2\theta - \theta)\right]$$

$$= \frac{1}{2}\sin 3\theta - \frac{1}{2}\sin \theta$$

From these new identities, we can derive another group of identities that are used to rewrite sums of trigonometric functions as products.

Sum-to-Product Identities

$$\sin A + \sin B = 2\sin\left(\frac{A + B}{2}\right)\cos\left(\frac{A - B}{2}\right)$$

$$\sin A - \sin B = 2\cos\left(\frac{A + B}{2}\right)\sin\left(\frac{A - B}{2}\right)$$

$$\cos A + \cos B = 2\cos\left(\frac{A + B}{2}\right)\cos\left(\frac{A - B}{2}\right)$$

$$\cos A - \cos B = -2\sin\left(\frac{A + B}{2}\right)\sin\left(\frac{A - B}{2}\right)$$

EXAMPLE 8   **Using a Sum-to-Product Identity**

Write $\sin 2t - \sin 4t$ as a product of two functions.

Solution   Use the (second) identity for $\sin A - \sin B$, with $2t = A$ and $4t = B$.

$$\sin 2t - \sin 4t = 2\cos\left(\frac{2t + 4t}{2}\right)\sin\left(\frac{2t - 4t}{2}\right)$$

$$= 2\cos\frac{6t}{2}\sin\left(\frac{-2t}{2}\right)$$

$$= 2\cos 3t \sin(-t)$$

$$= -2\cos 3t \sin t \qquad\qquad \sin(-t) = -\sin t$$

Half-Number Identities

From the alternative forms of the identity for $\cos 2A$, we derive identities for $\sin\frac{A}{2}$, $\cos\frac{A}{2}$, and $\tan\frac{A}{2}$. These are known as **half-number identities,** or **half-angle identities.**
We derive the identity for $\sin\frac{A}{2}$ as follows.

$$\cos 2x = 1 - 2\sin^2 x \qquad\qquad \text{Cosine double-number identity}$$

$$2\sin^2 x = 1 - \cos 2x \qquad\qquad \text{Add } 2\sin^2 x \text{ and subtract } \cos 2x.$$

$$\sin x = \pm\sqrt{\frac{1 - \cos 2x}{2}} \qquad\qquad \text{Divide by 2 and take square roots.}$$

Remember both the positive and negative square roots.

$$\sin\frac{A}{2} = \pm\sqrt{\frac{1 - \cos A}{2}} \qquad\qquad \text{Let } 2x = A \text{ so that } x = \frac{A}{2}.$$

The $\pm$ sign in this half-number identity indicates that the appropriate sign depends on the quadrant of $\frac{A}{2}$. For example, if $\frac{A}{2}$ is a third-quadrant number, we choose the negative sign, since the sine function is negative in quadrant III.

We derive the identity for $\cos \frac{A}{2}$ by using the double-number identity for $\cos 2x$.

$$\cos 2x = 2\cos^2 x - 1 \qquad \text{Cosine double-number identity}$$

$$1 + \cos 2x = 2\cos^2 x \qquad \text{Add 1.}$$

$$\cos^2 x = \frac{1 + \cos 2x}{2} \qquad \text{Rewrite and divide by 2.}$$

$$\cos x = \pm\sqrt{\frac{1 + \cos 2x}{2}} \qquad \text{Take square roots.}$$

$$\cos \frac{A}{2} = \pm\sqrt{\frac{1 + \cos A}{2}} \qquad \text{Replace x with } \tfrac{A}{2}.$$

An identity for $\tan \frac{A}{2}$ comes from the half-number identities for sine and cosine.

$$\tan \frac{A}{2} = \frac{\sin \frac{A}{2}}{\cos \frac{A}{2}} = \frac{\pm\sqrt{\dfrac{1 - \cos A}{2}}}{\pm\sqrt{\dfrac{1 + \cos A}{2}}} = \pm\sqrt{\frac{1 - \cos A}{1 + \cos A}}$$

We derive an alternative identity for $\tan \frac{A}{2}$ by using double-number identities.

$$\tan \frac{A}{2} = \frac{\sin \frac{A}{2}}{\cos \frac{A}{2}} \qquad \text{Quotient identity}$$

$$= \frac{2 \sin \frac{A}{2} \cos \frac{A}{2}}{2 \cos^2 \frac{A}{2}} \qquad \text{Multiply by 2 } \cos \tfrac{A}{2} \text{ in numerator and denominator.}$$

$$= \frac{\sin\left[2\left(\frac{A}{2}\right)\right]}{1 + \cos\left[2\left(\frac{A}{2}\right)\right]} \qquad \text{Double-number identities}$$

$$\tan \frac{A}{2} = \frac{\sin A}{1 + \cos A} \qquad \text{Simplify.}$$

If we multiply this identity by 1 in the form $\frac{1 - \cos A}{1 - \cos A}$, we can also derive

$$\tan \frac{A}{2} = \frac{1 - \cos A}{\sin A}.$$

These last two identities for $\tan \frac{A}{2}$ do not require a choice of sign.

Half-Number Identities

In the following identities, the symbol $\pm$ indicates the sign is chosen based on the function under consideration and the quadrant of $\frac{A}{2}$.

$$\cos \frac{A}{2} = \pm\sqrt{\frac{1 + \cos A}{2}} \qquad \sin \frac{A}{2} = \pm\sqrt{\frac{1 - \cos A}{2}}$$

$$\tan \frac{A}{2} = \pm\sqrt{\frac{1 - \cos A}{1 + \cos A}} \qquad \tan \frac{A}{2} = \frac{\sin A}{1 + \cos A} \qquad \tan \frac{A}{2} = \frac{1 - \cos A}{\sin A}$$

EXAMPLE 9 **Using a Half-Number Identity to Find an Exact Value**

Find the exact value of $\cos \frac{\pi}{12}$.

Solution First note that $\frac{\pi}{12}$ is exactly half of $\frac{\pi}{6}$ and that we know the exact values of the trigonometric functions of $\frac{\pi}{6}$. Therefore, we begin by letting $A = \frac{\pi}{6}$ and $\frac{A}{2} = \frac{\pi}{12}$ in the half-number identity for $\cos \frac{A}{2}$.

$$\cos \frac{\pi}{12} = \sqrt{\frac{1 + \cos \frac{\pi}{6}}{2}} \qquad \begin{array}{l} \cos \frac{A}{2} = +\sqrt{\frac{1 + \cos A}{2}} \\ \text{because } \cos \frac{\pi}{12} \text{ is positive.} \end{array}$$

$$= \sqrt{\frac{1 + \frac{\sqrt{3}}{2}}{2}}$$

$$= \sqrt{\frac{\left(1 + \frac{\sqrt{3}}{2}\right) \cdot 2}{2 \cdot 2}} \qquad \text{Simplify.}$$

$$= \frac{\sqrt{2 + \sqrt{3}}}{2} \qquad \sqrt{\frac{a}{b}} = \frac{\sqrt{a}}{\sqrt{b}}$$

Verify that the final expression, $\frac{\sqrt{2 + \sqrt{3}}}{2}$, has approximation 0.9659258263 and equals $\frac{\sqrt{6} + \sqrt{2}}{4}$. This second value was found for $\cos 15°$, or $\cos \frac{\pi}{12}$, in **Example 1** of **Section 9.2.** ●

EXAMPLE 10 **Finding Function Values of $\frac{x}{2}$**

Given $\cos x = \frac{2}{3}$, with $\frac{3\pi}{2} < x < 2\pi$, find $\cos \frac{x}{2}$, $\sin \frac{x}{2}$, and $\tan \frac{x}{2}$.

Solution The angle associated with $\frac{x}{2}$ terminates in quadrant II, since

$$\frac{3\pi}{2} < x < 2\pi \quad \text{implies} \quad \frac{3\pi}{4} < \frac{x}{2} < \pi. \qquad \text{Divide by 2.}$$

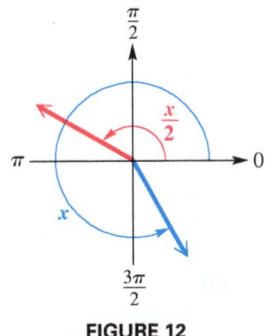

FIGURE 12

See **FIGURE 12.** In quadrant II, the value of $\cos \frac{x}{2}$ is negative and the value of $\sin \frac{x}{2}$ is positive. Now use the appropriate half-number identities and simplify the radicals.

$$\cos \frac{x}{2} = -\sqrt{\frac{1 + \frac{2}{3}}{2}} = -\sqrt{\frac{5}{6}} = -\frac{\sqrt{5}}{\sqrt{6}} \cdot \frac{\sqrt{6}}{\sqrt{6}} = -\frac{\sqrt{30}}{6}$$

$$\sin \frac{x}{2} = \sqrt{\frac{1 - \frac{2}{3}}{2}} = \sqrt{\frac{1}{6}} = \frac{1}{\sqrt{6}} \cdot \frac{\sqrt{6}}{\sqrt{6}} = \frac{\sqrt{6}}{6} \qquad \text{Rationalize all denominators.}$$

$$\tan \frac{x}{2} = \frac{\sin \frac{x}{2}}{\cos \frac{x}{2}} = \frac{\frac{\sqrt{6}}{6}}{-\frac{\sqrt{30}}{6}} = \frac{\sqrt{6}}{6} \cdot \left(-\frac{6}{\sqrt{30}}\right) = -\sqrt{\frac{6}{30}} = -\sqrt{\frac{1}{5}} = -\frac{\sqrt{5}}{5}$$

Notice that it is not necessary to use a half-number identity for $\tan \frac{x}{2}$ once we find $\sin \frac{x}{2}$ and $\cos \frac{x}{2}$. Instead we used the quotient identity for tangent. ●

EXAMPLE 11 **Simplifying Expressions by Using Half-Number Identities**

Simplify each expression.

(a) $\pm\sqrt{\dfrac{1 + \cos 12x}{2}}$ (b) $\dfrac{1 - \cos 5\alpha}{\sin 5\alpha}$

Solution

(a) This matches part of the identity for $\cos\frac{A}{2}$. Replace A with $12x$.

$$\cos\frac{A}{2} = \pm\sqrt{\frac{1 + \cos A}{2}} = \pm\sqrt{\frac{1 + \cos 12x}{2}} = \cos\frac{12x}{2} = \cos 6x$$

(b) Use the third identity for $\tan\frac{A}{2}$ with $5\alpha = A$. $\cos\frac{A}{2}$

$$\frac{1 - \cos 5\alpha}{\sin 5\alpha} = \tan\frac{5\alpha}{2}$$

9.3 Exercises

Checking Analytic Skills *Use identities to find* (a) $\sin 2\theta$ *and* (b) $\cos 2\theta$. ***Do not use a calculator.***

1. $\sin\theta = \dfrac{2}{5}$ and $\cos\theta < 0$

2. $\cos\theta = -\dfrac{12}{13}$ and $\sin\theta > 0$

3. $\tan\theta = 2$ and $\cos\theta > 0$

4. $\tan\theta = \dfrac{5}{3}$ and $\sin\theta < 0$

5. $\sin\theta = -\dfrac{\sqrt{5}}{7}$ and $\cos\theta > 0$

6. $\cos\theta = \dfrac{\sqrt{3}}{5}$ and $\sin\theta > 0$

Checking Analytic Skills *Use an identity to write each expression as a single trigonometric function or as a single number in exact form.* ***Do not use a calculator.***

7. $\cos^2 15° - \sin^2 15°$

8. $\dfrac{2\tan 15°}{1 - \tan^2 15°}$

9. $1 - 2\sin^2 15°$

10. $1 - 2\sin^2 22.5°$

11. $2\cos^2 67.5° - 1$

12. $\cos^2\dfrac{\pi}{8} - \dfrac{1}{2}$

13. $\dfrac{\tan 51°}{1 - \tan^2 51°}$

14. $\dfrac{\tan 34°}{2(1 - \tan^2 34°)}$

15. $\dfrac{1}{4} - \dfrac{1}{2}\sin^2 47.1°$

16. $\dfrac{1}{8}\sin 29.5° \cos 29.5°$

17. $4\sin 15° \cos 15°$

18. $2 - 4\sin^2 15°$

Graph each function and use the graph to make a conjecture as to what might be an identity. Then verify your conjecture.

19. $f(x) = \cos^4 x - \sin^4 x$

20. $f(x) = \dfrac{4\tan x \cos^2 x - 2\tan x}{1 - \tan^2 x}$

Use the method of **Example 5** *to do the following. Then support your result graphically, using the trig viewing window of your calculator.*

21. Express $\cos 3x$ in terms of $\cos x$.

22. Express $\tan 3x$ in terms of $\tan x$.

23. Express $\tan 4x$ in terms of $\tan x$.

24. Express $\cos 4x$ in terms of $\cos x$.

Use a half-number identity to find an expression for the exact value for each trigonometric function.

25. $\sin \dfrac{\pi}{12}$

26. $\cos \dfrac{\pi}{8}$

27. $\tan \left(-\dfrac{\pi}{8}\right)$

28. $\cos 67.5°$

29. $\sin 67.5°$

30. $\tan 195°$

Use a half-number identity to find an expression for the exact value for each function, given the information about x.

31. $\cos \dfrac{x}{2}$, given $\cos x = \dfrac{1}{4}$ and $0 < x < \dfrac{\pi}{2}$

32. $\sin \dfrac{x}{2}$, given $\cos x = -\dfrac{5}{8}$ and $\dfrac{\pi}{2} < x < \pi$

33. $\tan \dfrac{x}{2}$, given $\sin x = \dfrac{3}{5}$ and $\dfrac{\pi}{2} < x < \pi$

34. $\cos \dfrac{x}{2}$, given $\sin x = -\dfrac{4}{5}$ and $\dfrac{3\pi}{2} < x < 2\pi$

35. $\tan \dfrac{x}{2}$, given $\tan x = \dfrac{\sqrt{7}}{3}$ and $\pi < x < \dfrac{3\pi}{2}$

36. $\tan \dfrac{x}{2}$, given $\tan x = -\dfrac{\sqrt{5}}{2}$ and $\dfrac{\pi}{2} < x < \pi$

37. $\sin \dfrac{x}{2}$, given $\tan x = 2$ and $0 < x < \dfrac{\pi}{2}$

38. $\cos \dfrac{x}{2}$, given $\cot x = -3$ and $\dfrac{\pi}{2} < x < \pi$

39. $\cos x$, given $\cos 2x = -\dfrac{5}{12}$ and $\dfrac{\pi}{2} < x < \pi$

40. $\sin x$, given $\cos 2x = \dfrac{2}{3}$ and $\pi < x < \dfrac{3\pi}{2}$

41. Consider the expression $\tan \left(\dfrac{\pi}{2} + x\right)$.

  (a) Why can't we use the identity for $\tan (A + B)$ to express it as a function of x alone?

  (b) Use the identity $\tan \theta = \dfrac{\sin \theta}{\cos \theta}$ to rewrite the expression in terms of sine and cosine.

  (c) Use the result of part (b) to show that
  $$\tan \left(\dfrac{\pi}{2} + x\right) = -\cot x.$$

42. Recall the identity for the tangent of a sum.

  (a) Use the identity for $\tan (A + B)$ to write an identity for $\cot (A + B)$ in terms of $\cot A$ and $\cot B$.

  (b) Consider the expression $\cot (\pi + x)$. Why can't we use the result of part (a) to rewrite that expression as a function of x alone?

  (c) Use the ideas of parts (b) and (c) of **Exercise 41** to show that $\cot (\pi + x) = \cot x$.

Use an identity to write each expression as a single trigonometric function value.

43. $\sqrt{\dfrac{1 - \cos 40°}{2}}$

44. $\sqrt{\dfrac{1 + \cos 76°}{2}}$

45. $\sqrt{\dfrac{1 - \cos 147°}{1 + \cos 147°}}$

46. $\sqrt{\dfrac{1 + \cos 165°}{1 - \cos 165°}}$

47. $\dfrac{1 - \cos 59.74°}{\sin 59.74°}$

48. $\dfrac{\sin 158.2°}{1 + \cos 158.2°}$

Verify that each equation is an identity.

49. $\dfrac{2 \cos 2\alpha}{\sin 2\alpha} = \cot \alpha - \tan \alpha$

50. $\dfrac{1 + \cos 2x}{\sin 2x} = \cot x$

51. $\sec^2 \dfrac{x}{2} = \dfrac{2}{1 + \cos x}$

52. $\sec 2x = \dfrac{1 + \tan^2 x}{1 - \tan^2 x}$

53. $\cos 2\theta = \dfrac{2 - \sec^2 \theta}{\sec^2 \theta}$

54. $\cot^2 \dfrac{x}{2} = \dfrac{(1 + \cos x)^2}{\sin^2 x}$

55. $\tan s + \cot s = 2 \csc 2s$

56. $\cos x = \dfrac{1 - \tan^2 \frac{x}{2}}{1 + \tan^2 \frac{x}{2}}$

57. $\sin^2 \dfrac{x}{2} = \dfrac{\tan x - \sin x}{2 \tan x}$

58. $\dfrac{\cot \alpha - \tan \alpha}{\cot \alpha + \tan \alpha} = \cos 2\alpha$

59. $\cos 2x = \dfrac{1 - \tan^2 x}{1 + \tan^2 x}$

60. $\tan x = \csc 2x - \cot 2x$

61. $\sin 2\alpha \cos 2\alpha = \sin 2\alpha - 4 \sin^3 \alpha \cos \alpha$

62. $\sin 4\gamma = 4 \sin \gamma \cos \gamma - 8 \sin^3 \gamma \cos \gamma$

63. $\sin 4\alpha = 4 \sin \alpha \cos \alpha \cos 2\alpha$

64. $\dfrac{\sin 2x}{2 \sin x} = \cos^2 \dfrac{x}{2} - \sin^2 \dfrac{x}{2}$

Write each expression as a sum or difference of trigonometric functions or values.

65. $2 \sin 58° \cos 102°$

66. $5 \cos 3x \cos 2x$

67. $2 \cos 85° \sin 140°$

68. $\sin 4x \sin 5x$

Write each expression as a product of trigonometric functions or values.

69. $\cos 4x - \cos 2x$

70. $\cos 5t + \cos 8t$

71. $\sin 25° + \sin(-48°)$

72. $\sin 102° - \sin 95°$

73. $\cos 4x + \cos 8x$

74. $\sin 9B - \sin 3B$

(Modeling) Solve each problem.

75. *Railroad Curves* In the United States, circular railroad curves are designated by their *degree of curvature*, which is the central angle subtended by a chord of 100 feet. (*Source:* Hay, W., *Railroad Engineering*, John Wiley and Sons.)

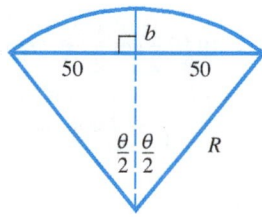

(a) Use the figure to write an expression for $\cos \frac{\theta}{2}$.

(b) Use the result of part (a) and the third half-number identity for tangent to write an expression for $\tan \frac{\theta}{4}$.

(c) If $b = 12$, what is the measure of angle θ to the nearest degree?

76. *Distance Traveled by a Stone* The distance D of an object thrown (or projected) from height h (feet) at angle θ with initial velocity v (feet per second) is modeled by the following formula.

$$D = \frac{v^2 \sin \theta \cos \theta + v \cos \theta \sqrt{(v \sin \theta)^2 + 64h}}{32}$$

(*Source:* Kreighbaum, E. and K. Barthels, *Biomechanics*, Allyn & Bacon.)

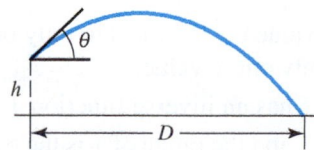

(a) Find D when $h = 0$—that is, when the object is projected from the ground.

(b) Suppose a car driving over loose gravel kicks up a small stone at a velocity of 36 feet per second (about 25 mph) and an angle $\theta = 30°$. How far will the stone travel?

(c) Repeat part (c) for a velocity of 40 feet per second and an angle of 32°.

77. *Determining Wattage* Amperage is a measure of the amount of electricity that is moving through a circuit, while voltage is a measure of the force pushing the electricity. The wattage W consumed by an electrical device can be determined by calculating the product of amperage I and voltage V. (*Source:* Wilcox, G. and C. Hesselberth, *Electricity for Engineering Technology*, Allyn & Bacon.)

(a) A household circuit has voltage

$$V = 163 \sin 120\pi t$$

when an incandescent light bulb is turned on with amperage

$$I = 1.23 \sin 120\pi t.$$

Graph the wattage

$$W = VI$$

that is consumed by the light bulb over the interval $0 \le t \le 0.05$.

(b) Determine the maximum and minimum wattages used by the light bulb.

(c) Use identities to find values for a, c, and ω so that

$$W = a \cos \omega t + c.$$

(d) Check your answer in part (c) by graphing both expressions for W on the same coordinate axes.

(e) Use the graph from part (a) to estimate the average wattage used by the light. How many watts do you think this incandescent light bulb is rated for?

78. *Relating Voltage and Wattage* Refer to **Exercise 77.** Suppose that voltage for an electric heater is given by

$$V = a \sin 2\pi\omega t$$

and amperage by

$$I = b \sin 2\pi\omega t,$$

where t is time in seconds.

(a) Find the period of the graph for the voltage.

(b) Show that the graph of the wattage $W = VI$ will have half the period of the voltage. Interpret this result.

RELATING CONCEPTS **For individual or group investigation (Exercises 79–104)**

Verify that each equation is an identity by using any of the identities introduced in **Sections 9.1–9.3.**

79. $\tan \theta + \cot \theta = \sec \theta \csc \theta$

80. $\csc \theta \cos^2 \theta + \sin \theta = \csc \theta$

81. $\tan \dfrac{x}{2} = \csc x - \cot x$

82. $\sec(\pi - x) = -\sec x$

83. $\dfrac{\sin t}{1 + \cos t} = \dfrac{1 - \cos t}{\sin t}$

84. $\dfrac{1 - \sin t}{\cos t} = \dfrac{1}{\sec t + \tan t}$

85. $\sin 2\theta = \dfrac{2 \tan \theta}{1 + \tan^2 \theta}$

86. $\dfrac{2}{1 + \cos x} - \tan^2 \dfrac{x}{2} = 1$

87. $\cot \theta - \tan \theta = \dfrac{2 \cos^2 \theta - 1}{\sin \theta \cos \theta}$

88. $1 - \tan^2 \dfrac{\theta}{2} = \dfrac{2 \cos \theta}{1 + \cos \theta}$

89. $\dfrac{\sin \theta + \tan \theta}{1 + \cos \theta} = \tan \theta$

90. $\csc^4 x - \cot^4 x = \dfrac{1 + \cos^2 x}{1 - \cos^2 x}$

91. $\dfrac{\tan^2 t + 1}{\tan t \csc^2 t} = \tan t$

92. $\dfrac{\sin s}{1 + \cos s} + \dfrac{1 + \cos s}{\sin s} = 2 \csc s$

93. $\tan 4\theta = \dfrac{2 \tan 2\theta}{2 - \sec^2 2\theta}$

94. $\tan\left(\dfrac{x}{2} + \dfrac{\pi}{4}\right) = \sec x + \tan x$

95. $\dfrac{\cot s - \tan s}{\cos s + \sin s} = \dfrac{\cos s - \sin s}{\sin s \cos s}$

96. $\dfrac{\tan \theta - \cot \theta}{\tan \theta + \cot \theta} = 1 - 2 \cos^2 \theta$

97. $\dfrac{\tan(x + y) - \tan y}{1 + \tan(x + y) \tan y} = \tan x$

98. $2 \cos^2 \dfrac{x}{2} \tan x = \tan x + \sin x$

99. $\dfrac{\cos^4 x - \sin^4 x}{\cos^2 x} = 1 - \tan^2 x$

100. $\dfrac{\csc t + 1}{\csc t - 1} = (\sec t + \tan t)^2$

101. $\dfrac{2(\sin x - \sin^3 x)}{\cos x} = \sin 2x$

102. $\dfrac{1}{2} \cot \dfrac{x}{2} - \dfrac{1}{2} \tan \dfrac{x}{2} = \cot x$

103. $\dfrac{\sin(x + y)}{\cos(x - y)} = \dfrac{\cot x + \cot y}{1 + \cot x \cot y}$

104. $\dfrac{1}{\sec t - 1} + \dfrac{1}{\sec t + 1} = 2 \cot t \csc t$

9.4 The Inverse Circular Functions

Review of Inverse Functions ● Inverse Sine Function ● Inverse Cosine Function ● Inverse Tangent Function ● Other Inverse Trigonometric Functions ● Inverse Function Values

 Looking Ahead to Calculus

The inverse functions covered in this section are used in calculus to express the solutions of trigonometric equations.

Review of Inverse Functions

We first discussed **inverse functions** in **Section 5.1.** We give a quick review of inverse functions here.

Summary of Inverse Functions

1. For a one-to-one function, each x-value corresponds to only one y-value and each y-value corresponds to only one x-value.

2. If a function f is one-to-one, then f has an inverse function f^{-1}.

3. The domain of f is the range of f^{-1}, and the range of f is the domain of f^{-1}. That is, if (a, b) is on the graph of f, then (b, a) is on the graph of f^{-1}.

4. The graphs of f and f^{-1} are reflections of each other across the line $y = x$.

5. To find $f^{-1}(x)$ from $f(x)$, follow these steps.

   Step 1 Replace $f(x)$ with y and interchange x and y.

   Step 2 Solve for y.

   Step 3 Replace y with $f^{-1}(x)$.

FIGURE 13 illustrates some of these concepts.

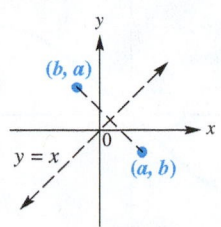

(b, a) is the reflection of (a, b)
across the the line $y = x$.

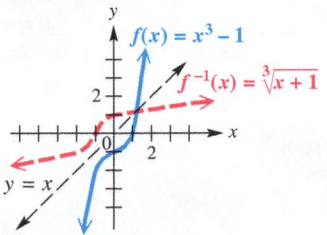

The graph of f^{-1} is the reflection of
the graph of f across the line $y = x$.

FIGURE 13

Restricting the Domain

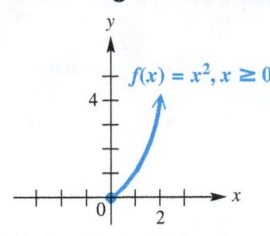

This is a one-to-one function.

FIGURE 14

NOTE   The function $f(x) = x^2$ is *not* one-to-one, and therefore, does *not* have an inverse function. However, if we *restrict the domain* of f to $[0, \infty)$, then f is one-to-one with inverse function $f^{-1}(x) = \sqrt{x}$. The same situation occurs when we attempt to find inverse trigonometric functions. See **FIGURE 14.**

Inverse Sine Function

Refer to the graph of the sine function in **FIGURE 15.** Applying the horizontal line test, we see that $y = \sin x$ does not define a one-to-one function. If we restrict the domain to the interval $\left[-\frac{\pi}{2}, \frac{\pi}{2}\right]$, which is the part of the graph in **FIGURE 15** shown in blue, this restricted function is one-to-one and has an inverse function. The range of $y = \sin x$ is $[-1, 1]$, so the domain of the inverse function will be $[-1, 1]$, and its range will be $\left[-\frac{\pi}{2}, \frac{\pi}{2}\right]$.

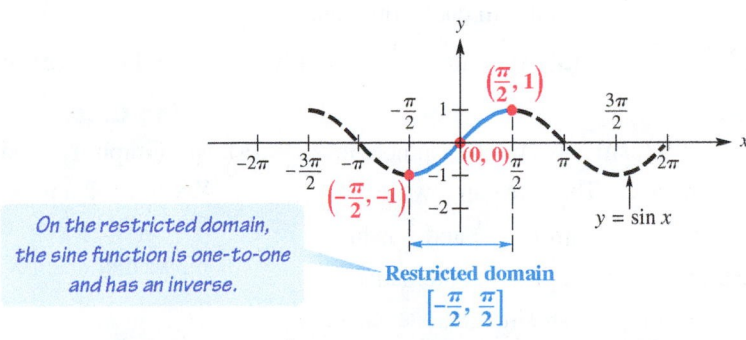

On the restricted domain,
the sine function is one-to-one
and has an inverse.

Restricted domain
$\left[-\frac{\pi}{2}, \frac{\pi}{2}\right]$

FIGURE 15

Reflecting the graph of $y = \sin x$ on the restricted domain, shown in **FIGURE 16(a)** on the next page, across the line $y = x$ gives the graph of the inverse function. See **FIGURE 16(b)** on the next page. Some key points are labeled on the graph. The equation of the inverse of $y = \sin x$ is found by interchanging x and y to get

$$x = \sin y.$$

This equation is solved for y by writing $y = \sin^{-1} x$ (read **"inverse sine of x"**). As **FIGURE 16(b)** shows, the domain of $y = \sin^{-1} x$ is $[-1, 1]$, while the restricted domain of $y = \sin x$, $\left[-\frac{\pi}{2}, \frac{\pi}{2}\right]$, is the range of $y = \sin^{-1} x$. An alternative notation for $\sin^{-1} x$ is arcsin x.

Sine Function on Restricted Domain

Reflect the blue portion of the graph across the line y = x to obtain the inverse sine function.

$\left(\frac{\pi}{2}, 1\right)$

$y = \sin x$

$(0, 0)$

Restricted domain
$\left[-\frac{\pi}{2}, \frac{\pi}{2}\right]$

$\left(-\frac{\pi}{2}, -1\right)$

(a)

Inverse Sine Function

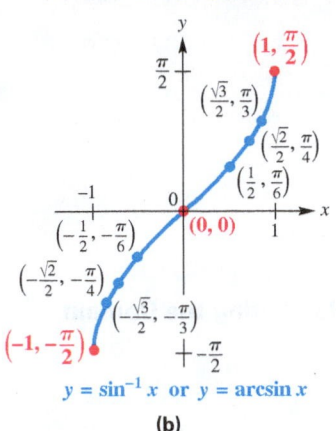

$y = \sin^{-1} x$ or $y = \arcsin x$

(b)

FIGURE 16

> **Inverse Sine Function**
>
> $y = \sin^{-1} x$ or $y = \arcsin x$ means that $x = \sin y$, for $-\frac{\pi}{2} \le y \le \frac{\pi}{2}$.

We can think of $y = \sin^{-1} x$ or $y = \arcsin x$ as

"y is the number (angle) in the interval $\left[-\frac{\pi}{2}, \frac{\pi}{2}\right]$ whose sine is x."

Thus, we can write $y = \sin^{-1} x$ as $\sin y = x$ to evaluate it. We must pay close attention to the domain and range intervals.

EXAMPLE 1 Finding Inverse Sine Values

Find y in each equation.

(a) $y = \arcsin \frac{1}{2}$　　**(b)** $y = \sin^{-1}(-1)$　　**(c)** $y = \sin^{-1}(-2)$

Analytic Solution

(a) We can think of $y = \arcsin \frac{1}{2}$ as "y is the number in $\left[-\frac{\pi}{2}, \frac{\pi}{2}\right]$ whose sine is $\frac{1}{2}$." Then we can rewrite the equation as $\sin y = \frac{1}{2}$. Since $\sin \frac{\pi}{6} = \frac{1}{2}$ and $\frac{\pi}{6}$ is in the range of the arcsin function, $y = \frac{\pi}{6}$. Alternatively, the graph of the function $y = \arcsin x$ **(FIGURE 16(b))** includes the point $\left(\frac{1}{2}, \frac{\pi}{6}\right)$. Therefore,

$$\arcsin \frac{1}{2} = \frac{\pi}{6}.$$

(b) Writing the alternative equation, $\sin y = -1$, shows that $y = -\frac{\pi}{2}$. This can be verified by noticing that the point $\left(-1, -\frac{\pi}{2}\right)$ is on the graph of $y = \sin^{-1} x$.

(c) Because $-1 \le \sin x \le 1$, the sine of a number cannot be -2. It follows that -2 is not in the domain of the inverse sine function. Thus, $\sin^{-1}(-2)$ does not exist.

Graphing Calculator Solution

(a), (b) Graph $Y_1 = \sin^{-1} X$ and locate the points with X-values $\frac{1}{2}$ and -1. **FIGURE 17(a)** shows that when $X = \frac{1}{2}$, $Y = \frac{\pi}{6} \approx 0.52359878$. Similarly, **FIGURE 17(b)** shows that when $X = -1$, $Y = -\frac{\pi}{2} \approx -1.570796$.

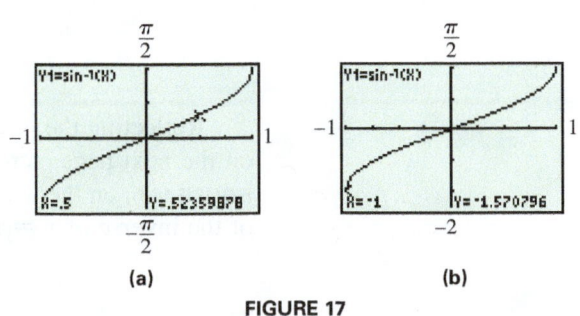

(a)　　　　　　　　**(b)**

FIGURE 17

(c) A calculator will give an error message for $\sin^{-1}(-2)$.

TECHNOLOGY NOTE

Inverse trigonometric functions can be evaluated on a calculator without the aid of a graph, as shown in the figure below. Because

$$\sin^{-1}\frac{1}{2} \approx 0.5235987756,$$

and

$$\frac{\pi}{6} \approx 0.5235987756,$$

we can conclude that

$$\sin^{-1}\frac{1}{2} = \frac{\pi}{6}.$$

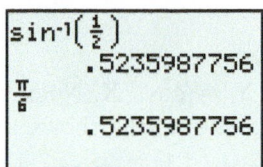

Radian Mode

CAUTION In **Example 1(b)**, it is tempting to give the value of $\sin^{-1}(-1)$ as $\frac{3\pi}{2}$, since $\sin\frac{3\pi}{2} = -1$. Notice, however, that $\frac{3\pi}{2}$ is not in the range of the inverse sine function, which is $-\frac{\pi}{2} \le y \le \frac{\pi}{2}$.

FUNCTION CAPSULE

INVERSE SINE FUNCTION $y = \sin^{-1} x$ or $y = \arcsin x$

Domain: $[-1, 1]$ Range: $\left[-\frac{\pi}{2}, \frac{\pi}{2}\right]$

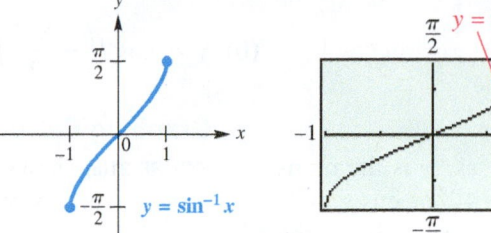
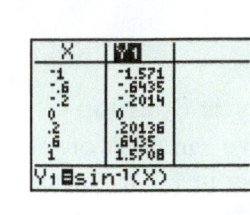

FIGURE 18

- The inverse sine function is increasing and continuous on its domain, $[-1, 1]$.
- Both its x- and y-intercepts are $(0, 0)$.
- Its graph is symmetric with respect to the origin, so it is an odd function. For all x in the domain, $\sin^{-1}(-x) = -\sin^{-1} x$.

NOTE It is sometimes helpful to remember that $\sin^{-1} x$ can be thought of as an *angle*. For example, $\sin^{-1} 0$ represents an angle whose sine equals 0. Now there are infinitely many angles whose sine is 0, such as $0, \pm\pi, \pm 2\pi, \ldots$, but only the angle of 0 radians is in the range $\left[-\frac{\pi}{2}, \frac{\pi}{2}\right]$. Thus, $\sin^{-1} 0 = 0$.

Inverse Cosine Function

The function $y = \cos^{-1} x$ (or $y = \arccos x$) is defined by restricting the domain of the function $y = \cos x$ to the interval $[0, \pi]$, as in **FIGURE 19,** and then interchanging the roles of x and y. The graph of $y = \cos^{-1} x$ is shown in **FIGURE 20.**

Cosine Function on Restricted Domain   **Inverse Cosine Function**

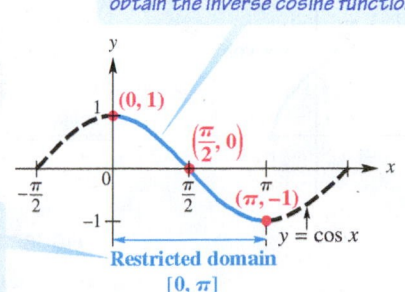
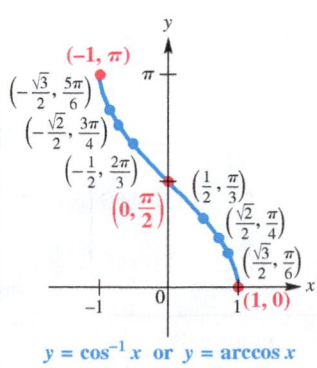

FIGURE 19   **FIGURE 20**

Inverse Cosine Function

$y = \cos^{-1} x$ or $y = \arccos x$ means that $x = \cos y$, for $0 \le y \le \pi$.

We can think of $y = \cos^{-1} x$ or $y = \arccos x$ as

"y is the number (angle) in the interval $[0, \pi]$ whose cosine is x."

Thus, we can write $y = \cos^{-1} x$ as $\cos y = x$ to evaluate it. Again, we must pay close attention to the domain and range intervals.

EXAMPLE 2 Finding Inverse Cosine Values

Find y in each equation.

(a) $y = \arccos 1$

(b) $y = \cos^{-1}\left(-\dfrac{\sqrt{2}}{2}\right)$

Analytic Solution

(a) We can think of $y = \arccos 1$ as "y is the number in $[0, \pi]$ whose cosine is 1," or $\cos y = 1$. Then $y = 0$, since $\cos 0 = 1$ and 0 is in the range of the arccosine function. Alternatively, since the point $(1, 0)$ lies on the graph of $y = \arccos x$ in **FIGURE 20** on the preceding page, the value of y is 0.

(b) We must find the value of y that satisfies $\cos y = -\dfrac{\sqrt{2}}{2}$, where y is in the interval $[0, \pi]$. The only value for y that satisfies these conditions is $\dfrac{3\pi}{4}$. Again, this can be verified from the graph in **FIGURE 20**.

Graphing Calculator Solution

FIGURE 21(a) shows the graph of $Y_1 = \cos^{-1} X$. Note that $Y = 0$ when $X = 1$. Similarly, **FIGURE 21(b)** shows that when $X = -\dfrac{\sqrt{2}}{2} \approx -0.7071068$, $Y = \dfrac{3\pi}{4} \approx 2.3561945$.

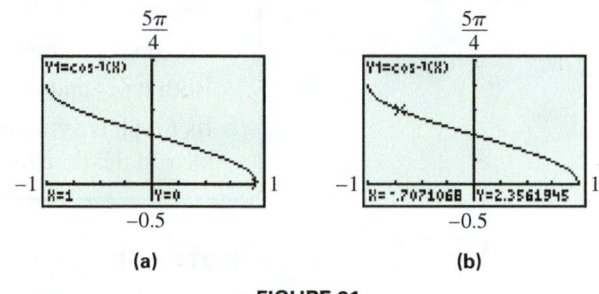

(a) (b)

FIGURE 21

See **Example 2**. In the screen below, the displays

$$\cos^{-1}\left(-\frac{\sqrt{2}}{2}\right) \approx 2.35619449$$

and

$$\frac{3\pi}{4} \approx 2.35619449,$$

support the conclusion that

$$\cos^{-1}\left(-\frac{\sqrt{2}}{2}\right) = \frac{3\pi}{4}.$$

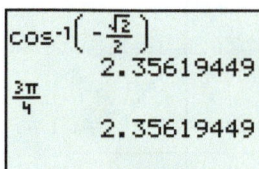

Radian mode

FUNCTION CAPSULE

INVERSE COSINE FUNCTION $y = \cos^{-1} x$ or $y = \arccos x$

Domain: $[-1, 1]$ Range: $[0, \pi]$

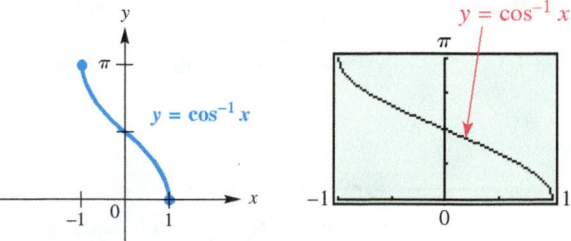

FIGURE 22

- The inverse cosine function is decreasing and continuous on its domain, $[-1, 1]$.
- Its x-intercept is $(1, 0)$ and its y-intercept is $\left(0, \dfrac{\pi}{2}\right)$.
- Its graph is neither symmetric with respect to the y-axis nor symmetric with respect to the origin.

Inverse Tangent Function

Restricting the domain of the function $y = \tan x$ to the *open* interval $\left(-\frac{\pi}{2}, \frac{\pi}{2}\right)$ yields a one-to-one function. By interchanging the roles of x and y, we obtain the inverse tangent function, given by $y = \tan^{-1} x$ or $y = \arctan x$. **FIGURE 23** shows the graph of the restricted tangent function. **FIGURE 24** gives the graph of $y = \tan^{-1} x$.

Tangent Function on Restricted Domain

Reflect the blue graph across the line y = x to obtain the inverse tangent function.

On the restricted domain, the tangent function is one-to-one and has an inverse.

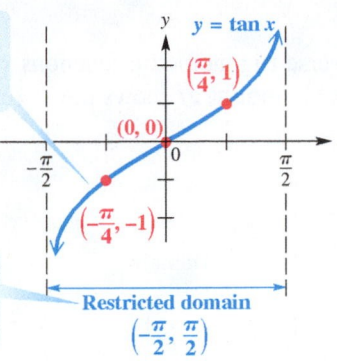

FIGURE 23

Inverse Tangent Function

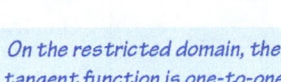

$y = \tan^{-1} x$ or $y = \arctan x$

FIGURE 24

> **Inverse Tangent Function**
>
> $y = \tan^{-1} x$ or $y = \arctan x$ means $x = \tan y$, for $-\frac{\pi}{2} < y < \frac{\pi}{2}$.

We can think of $y = \tan^{-1} x$ or $y = \arctan x$ as

"y is the number (angle) in the interval $\left(-\frac{\pi}{2}, \frac{\pi}{2}\right)$ whose tangent is x."

FUNCTION CAPSULE

INVERSE TANGENT FUNCTION $y = \tan^{-1} x$ or $y = \arctan x$

Domain: $(-\infty, \infty)$ Range: $\left(-\frac{\pi}{2}, \frac{\pi}{2}\right)$

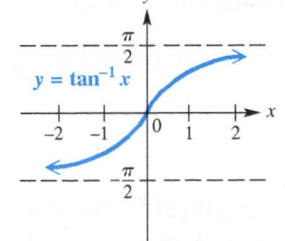
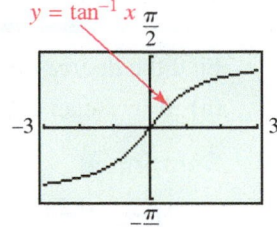
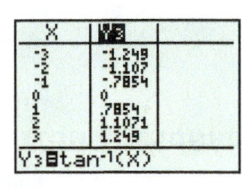

FIGURE 25

- The inverse tangent function is increasing and continuous on its domain, $(-\infty, \infty)$.
- Both its x- and y-intercepts are $(0, 0)$.
- Its graph is symmetric with respect to the origin, so it is an odd function. For all x in the domain, $\tan^{-1}(-x) = -\tan^{-1} x$.
- The lines $y = \frac{\pi}{2}$ and $y = -\frac{\pi}{2}$ are horizontal asymptotes.

Other Inverse Trigonometric Functions

The remaining three inverse trigonometric functions are defined similarly. Their graphs are shown in **FIGURE 26** to the left.

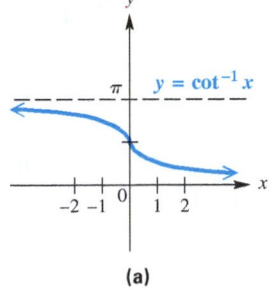

(a)

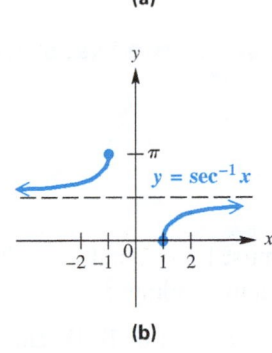

(b)

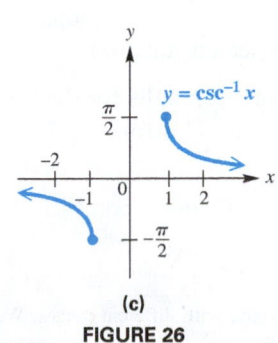

(c)

FIGURE 26

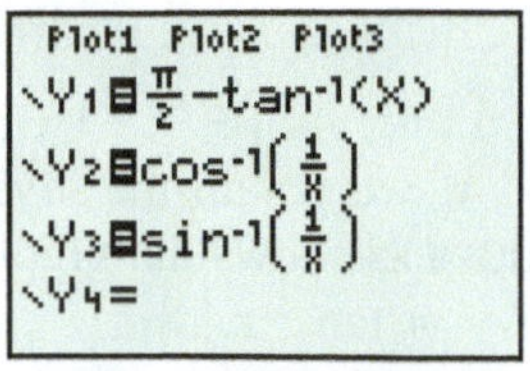

Radian mode

This screen shows how to enter the inverse cotangent, inverse secant, and inverse cosecant functions, as Y_1, Y_2, and Y_3, respectively.

FIGURE 27

> **Inverse Cotangent, Secant, and Cosecant Functions***
>
> $y = \cot^{-1} x$ or $y = \text{arccot } x$ means that $x = \cot y$, for $0 < y < \pi$.
>
> $y = \sec^{-1} x$ or $y = \text{arcsec } x$ means that $x = \sec y$, for $0 \le y \le \pi, y \ne \frac{\pi}{2}$.
>
> $y = \csc^{-1} x$ or $y = \text{arccsc } x$ means that $x = \csc y$, for $-\frac{\pi}{2} \le y \le \frac{\pi}{2}, y \ne 0$.

All six inverse trigonometric functions with their domains and ranges are given in the following table. **FIGURE 27** shows how to enter these other functions into a calculator.

Inverse Function	Domain	Range Interval	Range Quadrants of the Unit Circle
$y = \sin^{-1} x$	$[-1, 1]$	$\left[-\frac{\pi}{2}, \frac{\pi}{2}\right]$	I and IV
$y = \cos^{-1} x$	$[-1, 1]$	$[0, \pi]$	I and II
$y = \tan^{-1} x$	$(-\infty, \infty)$	$\left(-\frac{\pi}{2}, \frac{\pi}{2}\right)$	I and IV
$y = \cot^{-1} x$	$(-\infty, \infty)$	$(0, \pi)$	I and II
$y = \sec^{-1} x$	$(-\infty, -1] \cup [1, \infty)$	$\left[0, \frac{\pi}{2}\right) \cup \left(\frac{\pi}{2}, \pi\right]$	I and II
$y = \csc^{-1} x$	$(-\infty, -1] \cup [1, \infty)$	$\left[-\frac{\pi}{2}, 0\right) \cup \left(0, \frac{\pi}{2}\right]$	I and IV

FOR DISCUSSION

The graph of $y = \tan^{-1} x$ is shown in **FIGURE 24**, and the graph of $y = \cot^{-1} x$ is shown in **FIGURE 26(a)**. Use transformations of graphs to explain why

$$\cot^{-1} x = \frac{\pi}{2} - \tan^{-1} x$$

is an identity and can be used to calculate $\cot^{-1} x$ with a calculator.

Inverse Function Values

The inverse circular functions are formally defined with real number ranges. However, there are times when it may be convenient to find equivalent degree-measured angles.

EXAMPLE 3 **Finding Inverse Function Values (Degree-Measured)**

Find the *degree measure* of θ in the following.

(a) $\theta = \text{arctan } 1$ **(b)** $\theta = \sec^{-1} 2$

Solution

(a) Here, θ must be in $(-90°, 90°)$, but since $1 > 0$, θ must be in quadrant I. The alternative statement $\tan \theta = 1$ leads to $\theta = 45°$, if θ is in quadrant I.

(b) Write the equation as $\sec \theta = 2$. For $\sec^{-1} x$, θ is in quadrant I or II. Because 2 is positive, θ is in quadrant I and $\theta = 60°$, since $\sec 60° = 2$. Note that $\frac{\pi}{3}$ (the radian equivalent of $60°$) is in the range of the inverse secant function. ●

TECHNOLOGY NOTE

Inverse trigonometric functions can be evaluated in degree mode, as shown in the figure below, where $\tan^{-1} 1 = 45°$.

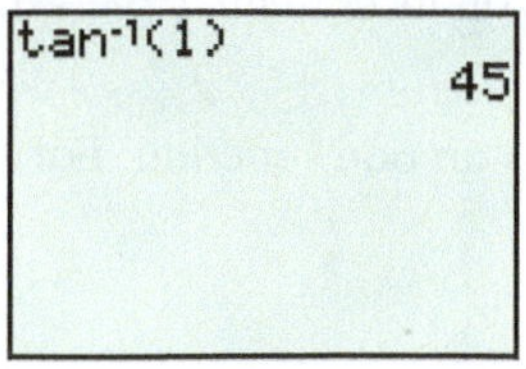

Degree mode

The inverse trigonometric keys on a calculator give correct results for the inverse sine, inverse cosine, and inverse tangent functions. Some examples follow.

$$\sin^{-1} 0.5 = 30° \qquad \sin^{-1}(-0.5) = -30°$$
$$\tan^{-1}(-1) = -45° \qquad \cos^{-1}(-0.5) = 120°$$

Degree mode

---

*The inverse secant and inverse cosecant functions are sometimes defined with different ranges. We use intervals that match their reciprocal functions (except for one missing point).

However, finding $\cot^{-1} x$, $\sec^{-1} x$, and $\csc^{-1} x$ with a calculator is not as straight-forward, because these functions must first be expressed in terms of $\tan^{-1} x$, $\cos^{-1} x$, and $\sin^{-1} x$, respectively. If $y = \sec^{-1} x$, for example, then $\sec y = x$, which must be written as a cosine function as follows.

$$\text{If } \sec y = x, \text{ then } \frac{1}{\cos y} = x, \quad \text{or} \quad \cos y = \frac{1}{x}, \quad \text{and} \quad y = \cos^{-1} \frac{1}{x}.$$

Use the following to evaluate these inverse trigonometric functions on a calculator.

> $\sec^{-1} x$ **can be evaluated as** $\cos^{-1} \dfrac{1}{x}$; $\csc^{-1} x$ **can be evaluated as** $\sin^{-1} \dfrac{1}{x}$;
>
> $\cot^{-1} x$ **can be evaluated as** $90° - \tan^{-1} x$.   *Degree mode*

NOTE   Another way to calculate $\cot^{-1} x$ is the following.

$$\cot^{-1} x = \begin{cases} \tan^{-1} \dfrac{1}{x} & \text{if } x > 0 \\ 90° & \text{if } x = 0 \quad \textit{Degree mode} \\ 180° + \tan^{-1} \dfrac{1}{x} & \text{if } x < 0 \end{cases}$$

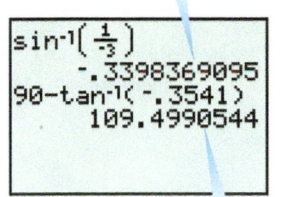

Radian mode

FIGURE 28   *Degree mode*

Radian mode

EXAMPLE 4   **Finding Inverse Function Values with a Calculator**

(a) Find y in radians if $y = \csc^{-1}(-3)$.

(b) Find θ in degrees if $\theta = \text{arccot}(-0.3541)$.

Solution

(a) With the calculator in radian mode, enter $\csc^{-1}(-3)$ as $\sin^{-1}\left(\frac{1}{-3}\right)$ to obtain $y \approx -0.3398369095$. See **FIGURE 28**.

(b) In degree mode, enter $\text{arccot}(-0.3541)$ as $90° - \tan^{-1}(-0.3541)$. As shown in **FIGURE 28**, $\theta \approx 109.4990544°$.   ●

EXAMPLE 5   **Finding Function Values by Using Definitions of the Trigonometric Functions**

Evaluate each expression without using a calculator.

(a) $\sin\left(\tan^{-1} \frac{3}{2}\right)$ **(b)** $\tan\left(\cos^{-1}\left(-\frac{5}{13}\right)\right)$

Solution

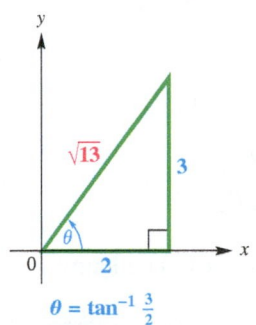

$\theta = \tan^{-1} \frac{3}{2}$

FIGURE 29

(a) Let $\theta = \tan^{-1} \frac{3}{2}$, so that $\tan \theta = \frac{3}{2}$. The inverse tangent function yields values only in quadrants I and IV, and since the value $\frac{3}{2}$ is positive, θ is in quadrant I. Sketch θ in quadrant I, and label a triangle, as shown in **FIGURE 29**. By the Pythagorean theorem, the hypotenuse is $\sqrt{13}$. The value of sine is the quotient of the side opposite and the hypotenuse.

Rationalize the denominator.

$$\sin\left(\tan^{-1} \frac{3}{2}\right) = \sin \theta = \frac{3}{\sqrt{13}} = \frac{3}{\sqrt{13}} \cdot \frac{\sqrt{13}}{\sqrt{13}} = \frac{3\sqrt{13}}{13}$$

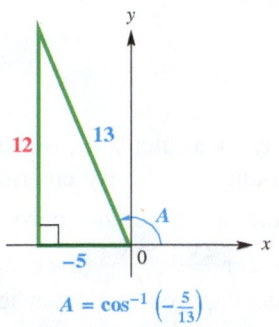

$A = \cos^{-1}\left(-\frac{5}{13}\right)$

FIGURE 30

(b) Let $A = \cos^{-1}\left(-\frac{5}{13}\right)$. Then $\cos A = -\frac{5}{13}$. Since $\cos^{-1} x$ for a negative value of x is in quadrant II, sketch A in quadrant II, as shown in **FIGURE 30**.

$$\tan\left(\cos^{-1}\left(-\frac{5}{13}\right)\right) = \tan A = -\frac{12}{5}$$

   ●

← **Algebra Review**
To review rationalizing the
denominator, see **Section R.5**.

EXAMPLE 6 **Finding Function Values by Using Identities**

Evaluate each expression without using a calculator.

(a) $\cos\left(\arctan\sqrt{3} + \arcsin\frac{1}{3}\right)$ **(b)** $\tan\left(2\arcsin\frac{2}{5}\right)$

Solution

(a) Let $A = \arctan\sqrt{3}$ and $B = \arcsin\frac{1}{3}$, so that $\tan A = \sqrt{3}$ and $\sin B = \frac{1}{3}$. Sketch both A and B in quadrant I. See **FIGURE 31**. Use the identity for $\cos(A + B)$.

$$\cos(A + B) = \cos A \cos B - \sin A \sin B$$

$$\cos\left(\arctan\sqrt{3} + \arcsin\frac{1}{3}\right) = \cos\left(\arctan\sqrt{3}\right)\cos\left(\arcsin\frac{1}{3}\right)$$

$$- \sin\left(\arctan\sqrt{3}\right)\sin\left(\arcsin\frac{1}{3}\right) \quad (1)$$

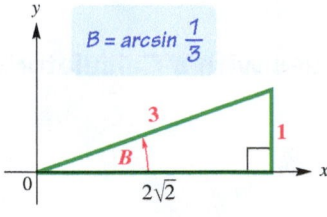

$A = arctan\sqrt{3}$

From **FIGURE 31,** we find the following values.

$$\cos\left(\arctan\sqrt{3}\right) = \cos A = \frac{1}{2}$$

$$\cos\left(\arcsin\frac{1}{3}\right) = \cos B = \frac{2\sqrt{2}}{3}$$ ⟩ $\frac{\text{side adjacent}}{\text{hypotenuse}}$

$$\sin\left(\arctan\sqrt{3}\right) = \sin A = \frac{\sqrt{3}}{2}$$ ⟩ $\frac{\text{side opposite}}{\text{hypotenuse}}$

$$\sin\left(\arcsin\frac{1}{3}\right) = \sin B = \frac{1}{3}$$

$B = arcsin\frac{1}{3}$

FIGURE 31

Substitute these values into equation (1).

$$\cos\left(\arctan\sqrt{3} + \arcsin\frac{1}{3}\right) = \frac{1}{2}\cdot\frac{2\sqrt{2}}{3} - \frac{\sqrt{3}}{2}\cdot\frac{1}{3} = \frac{2\sqrt{2} - \sqrt{3}}{6}$$

(b) Let $\arcsin\frac{2}{5} = B$. Then, use the double-number tangent identity.

$$\tan\left(2\arcsin\frac{2}{5}\right) = \tan 2B = \frac{2\tan B}{1 - \tan^2 B}$$

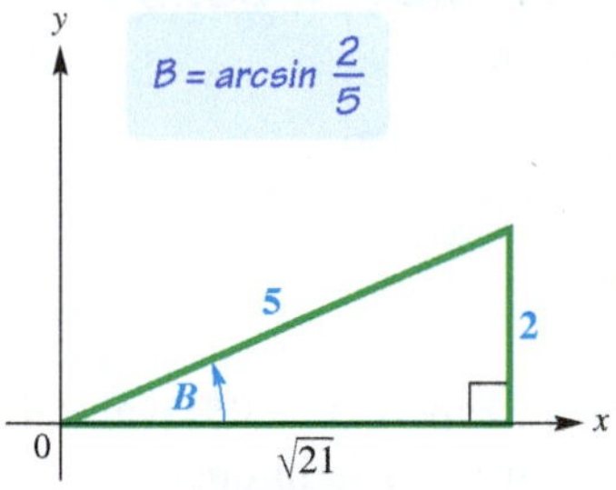

$B = arcsin\frac{2}{5}$

Since $\arcsin\frac{2}{5} = B$, $\sin B = \frac{2}{5}$. Sketch a triangle in quadrant I. The length of the third side is $\sqrt{21}$. From **FIGURE 32**, $\tan B = \frac{2}{\sqrt{21}}$.

Rationalize the denominator.

FIGURE 32

$$\tan\left(2\arcsin\frac{2}{5}\right) = \frac{2\left(\frac{2}{\sqrt{21}}\right)}{1 - \left(\frac{2}{\sqrt{21}}\right)^2} = \frac{\frac{4}{\sqrt{21}}}{1 - \frac{4}{21}} = \frac{\frac{4}{\sqrt{21}}}{\frac{17}{21}} = \frac{4}{\sqrt{21}}\cdot\frac{21}{17} = \frac{4\sqrt{21}}{17}$$

← **Algebra Review**
To review division of rational
expressions and complex
fractions, see **Section R.3**.

While the work shown in **Examples 5** and **6** does not rely on a calculator, we can support our analytic work with one. Use either degree or radian mode. By entering $\cos\left(\arctan\sqrt{3} + \arcsin\frac{1}{3}\right)$ from **Example 6(a)** into a calculator, we get the approximation 0.1827293862, the same approximation as when we enter $\frac{2\sqrt{2} - \sqrt{3}}{6}$ (the exact value obtained analytically). Similarly, we obtain the equivalent approximations when we evaluate $\tan\left(2\arcsin\frac{2}{5}\right)$ and $\frac{4\sqrt{21}}{17}$, supporting our answer in **Example 6(b).**

EXAMPLE 7   **Writing Function Values in Terms of u**

Write each expression as an algebraic expression in terms of u.

(a) $\sin(\tan^{-1} u)$ **(b)** $\cos(2 \sin^{-1} u)$

Solution

(a) Let $\theta = \tan^{-1} u$, so that $\tan \theta = u$. Here, u may be positive or negative. Since $-\frac{\pi}{2} < \tan^{-1} u < \frac{\pi}{2}$, sketch θ in quadrants I and IV and label two triangles. See **FIGURE 33.** Sine is given by the quotient of the side opposite and the hypotenuse, so it follows that

$$\sin(\tan^{-1} u) = \sin \theta = \frac{u}{\sqrt{u^2 + 1}} = \frac{u}{\sqrt{u^2 + 1}} \cdot \frac{\sqrt{u^2 + 1}}{\sqrt{u^2 + 1}} = \frac{u\sqrt{u^2 + 1}}{u^2 + 1}.$$

The result is positive when u is positive and negative when u is negative.

(b) Let $\theta = \sin^{-1} u$, so that $\sin \theta = u$. Use the identity $\cos 2\theta = 1 - 2\sin^2 \theta$.

$$\cos(2 \sin^{-1} u) = \cos 2\theta = 1 - 2\sin^2 \theta = 1 - 2u^2$$

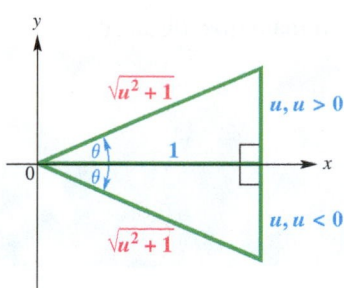

FIGURE 33

FOR DISCUSSION

In **Example 7(a),** u can take on any real number value, but in **Example 7(b),** u can only take on values in $[-1, 1]$. Look at the result in **Example 7(b),** which is $1 - 2u^2$.

1. Suppose we let $u = 2$. How can we justify that this is not a valid substitution, recalling the range of the cosine function?

2. Let $y = \cos 2\theta$ and graph $y = 1 - 2u^2$. Determine the u-values for which $-1 \le y \le 1$.

WHAT WENT WRONG?

A student found $\sin 1.74$ on her calculator (set for radians). She then found the inverse sine of the answer, but got 1.401592654 instead of 1.74, the number she was expecting.

What Went Wrong (if anything)?

EXAMPLE 8   **Finding the Optimal Angle of Elevation of a Shot**

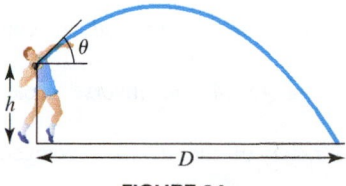

FIGURE 34

The optimal angle of elevation θ a shot-putter should aim for to throw the greatest distance depends on the initial velocity v and height h of the steel shot when it is released. See **FIGURE 34.** One model for θ that achieves this greatest distance is

$$\theta = \arcsin\left(\sqrt{\frac{v^2}{2v^2 + 64h}}\right).$$

(*Source:* Townend, M. Stewart, *Mathematics in Sport*, Chichester, Ellis Horwood Limited.)

(continued)

Answer to What Went Wrong?

Her thinking was incorrect. 1.74 is not in the range $\left[-\frac{\pi}{2}, \frac{\pi}{2}\right]$ of the inverse sine function. Her answer is actually $\pi - 1.74$, which places the result in the correct interval. Note that $\frac{\pi}{2} \approx 1.57$.

Suppose that a shot-putter can consistently put the shot with $h = 7.6$ feet and $v = 42$ feet per second. At what angle should he put the shot to maximize distance?

Solution To find this angle, substitute and use a calculator in degree mode.

$$\theta = \arcsin\left(\sqrt{\frac{42^2}{2(42^2) + 64(7.6)}}\right) \approx 42° \qquad h = 7.6, v = 42$$

The athlete should put the shot at an angle of about $42°$ to maximize distance.

9.4 Exercises

Concept Check *Complete each statement, or answer the question.*

1. For a function to have an inverse, it must be _____.

2. The domain of $y = \arcsin x$ equals the _____ of $y = \sin x$.

3. $y = \cos^{-1} x$ means that $x =$ _____, for $0 \le y \le \pi$.

4. The point $\left(\frac{\pi}{4}, 1\right)$ lies on the graph of $y = \tan x$. Thus, the point _____ lies on the graph of _____.

5. If a function f has an inverse and $f(\pi) = -1$, then $f^{-1}(-1) =$ _____.

6. How can the graph of f^{-1} be sketched if the graph of f is known?

Concept Check *In Exercises 7–10, write short answers and fill in the blanks.*

7. Consider the inverse sine function $y = \sin^{-1} x$, or $y = \arcsin x$.
   (a) What is its domain?
   (b) What is its range?
   (c) For this function, as x increases, y increases. Therefore, it is a(n) _____(decreasing/increasing)_____ function.
   (d) Why is $\arcsin(-2)$ not defined?

8. Consider the inverse cosine function $y = \cos^{-1} x$, or $y = \arccos x$.
   (a) What is its domain?
   (b) What is its range?
   (c) For this function, as x increases, y decreases. Therefore, it is a(n) _____(decreasing/increasing)_____ function.
   (d) $\text{Arccos}\left(-\frac{1}{2}\right) = \frac{2\pi}{3}$ is a true statement. Why is $\arccos\left(-\frac{1}{2}\right)$ not equal to $-\frac{4\pi}{3}$?

9. Consider the inverse tangent function, $y = \tan^{-1} x$, or $y = \arctan x$.
   (a) What is its domain?
   (b) What is its range?
   (c) For this function, as x increases, y increases. Therefore, it is a(n) _____(decreasing/increasing)_____ function.
   (d) Is there any real number x for which $\arctan x$ is not defined? If so, what is it (or what are they)?

10. Consider the three other inverse trigonometric functions, as defined in this section.
   (a) Give the domain and range of the inverse cosecant function.
   (b) Give the domain and range of the inverse secant function.
   (c) Give the domain and range of the inverse cotangent function.

Checking Analytic Skills *Find the exact value of each real number y.* ***Do not use a calculator.***

11. $y = \tan^{-1} 1$

12. $y = \sin^{-1} 0$

13. $y = \cos^{-1}(-1)$

14. $y = \arctan(-1)$

15. $y = \sin^{-1}(-1)$

16. $y = \cos^{-1}\frac{1}{2}$

17. $y = \arctan 0$

18. $y = \arcsin\left(-\frac{\sqrt{3}}{2}\right)$

19. $y = \arccos 0$

20. $y = \tan^{-1}(-1)$

21. $y = \sin^{-1}\dfrac{\sqrt{2}}{2}$

22. $y = \cos^{-1}\left(-\dfrac{1}{2}\right)$

23. $y = \arccos\left(-\dfrac{\sqrt{3}}{2}\right)$

24. $y = \arcsin\left(-\dfrac{\sqrt{2}}{2}\right)$

25. $y = \cot^{-1}(-1)$

26. $y = \sec^{-1}\left(-\sqrt{2}\right)$

27. $y = \csc^{-1}(-2)$

28. $y = \operatorname{arccot}\left(-\sqrt{3}\right)$

29. $y = \operatorname{arcsec}\dfrac{2\sqrt{3}}{3}$

30. $y = \csc^{-1}\sqrt{2}$

31. $y = \tan^{-1}\sqrt{3}$

32. $y = \sec^{-1}(-1)$

33. $y = \csc^{-1}2$

34. $y = \cot^{-1}1$

Checking Analytic Skills   *Give the degree measure of θ, if it exists.* **Do not use a calculator.**

35. $\theta = \arctan(-1)$

36. $\theta = \arccos\left(-\dfrac{1}{2}\right)$

37. $\theta = \arcsin\left(-\dfrac{\sqrt{3}}{2}\right)$

38. $\theta = \arcsin\left(-\dfrac{\sqrt{2}}{2}\right)$

39. $\theta = \cot^{-1}\left(-\dfrac{\sqrt{3}}{3}\right)$

40. $\theta = \sec^{-1}(-2)$

41. $\theta = \csc^{-1}(-2)$

42. $\theta = \csc^{-1}(-1)$

43. $\theta = \sin^{-1}2$

44. $\theta = \cos^{-1}(-3)$

45. $\theta = \sec^{-1}\left(-\dfrac{1}{2}\right)$

46. $\theta = \csc^{-1}\dfrac{1}{3}$

Use a calculator to give each value of θ in decimal degrees.

47. $\theta = \sin^{-1}(-0.13349122)$

48. $\theta = \cos^{-1}(-0.13348816)$

49. $\theta = \arccos(-0.39876459)$

50. $\theta = \arcsin 0.77900016$

51. $\theta = \csc^{-1}1.9422833$

52. $\theta = \cot^{-1}1.7670492$

Use a calculator to give each real number value of y.

53. $y = \arctan 1.1111111$

54. $y = \arcsin 0.81926439$

55. $y = \cot^{-1}(-0.92170128)$

56. $y = \sec^{-1}(-1.2871684)$

57. $y = \arcsin 0.92837781$

58. $y = \arccos 0.44624593$

Draw by hand the graph of each inverse function as defined in the text.

59. $y = \cot^{-1}x$

60. $y = \csc^{-1}x$

61. $y = \sec^{-1}x$

62. $y = \operatorname{arccsc}2x$

63. $y = \operatorname{arcsec}\dfrac{1}{2}x$

64. Explain why attempting to find $\sin^{-1}1.003$ on your calculator will result in an error message.

65. Explain why you are able to find $\tan^{-1}1.003$ on your calculator. Why is this situation different from the one described in **Exercise 64**?

66. Explain why $\cos^{-1}(\cos 270°) \neq 270°$ but $\cos^{-1}(\cos 90°) = 90°$.

Checking Analytic Skills   *Give the exact real number value of each expression.* **Do not use a calculator.**

67. $\sin\left(\sin^{-1}\dfrac{1}{2}\right)$

68. $\cos\left(\cos^{-1}\dfrac{\sqrt{3}}{2}\right)$

69. $\sin^{-1}\left(\sin\dfrac{4\pi}{3}\right)$

70. $\cos^{-1}\left(\cos\left(-\dfrac{\pi}{6}\right)\right)$

71. $\cos^{-1}\left(\cos\dfrac{3\pi}{2}\right)$

72. $\sin^{-1}\left(\sin\dfrac{3\pi}{2}\right)$

73. $\tan(\tan^{-1}5)$

74. $\tan(\tan^{-1}(-4))$

75. $\sec(\sec^{-1}2)$

76. $\csc(\csc^{-1} 3)$

77. $\tan^{-1}\left(\tan \dfrac{5\pi}{6}\right)$

78. $\tan^{-1}\left(\tan \dfrac{3\pi}{4}\right)$

79. $\tan\left(\arccos \dfrac{3}{4}\right)$

80. $\sin\left(\arccos \dfrac{1}{4}\right)$

81. $\cos(\tan^{-1}(-2))$

82. $\sec\left(\sin^{-1}\left(-\dfrac{1}{5}\right)\right)$

83. $\sin\left(2\tan^{-1} \dfrac{12}{5}\right)$

84. $\cos\left(2\sin^{-1} \dfrac{1}{4}\right)$

85. $\cos\left(2\arctan \dfrac{4}{3}\right)$

86. $\tan\left(2\cos^{-1} \dfrac{1}{4}\right)$

87. $\sin\left(2\cos^{-1} \dfrac{1}{5}\right)$

88. $\cos(2\tan^{-1}(-2))$

89. $\cos\left(\sin^{-1} \dfrac{3}{5} - \cos^{-1} \dfrac{12}{13}\right)$

90. $\cos\left(\cos^{-1} \dfrac{3}{5} + \sin^{-1} \dfrac{5}{13}\right)$

91. $\tan\left(\tan^{-1} \dfrac{3}{4} + \tan^{-1} \dfrac{12}{5}\right)$

92. $\tan\left(\sin^{-1} \dfrac{8}{17} + \tan^{-1} \dfrac{4}{3}\right)$

93. $\cos\left(\tan^{-1} \dfrac{5}{12} - \tan^{-1} \dfrac{3}{4}\right)$

94. $\cos\left(\sin^{-1} \dfrac{3}{5} + \cos^{-1} \dfrac{5}{13}\right)$

95. $\sin\left(\sin^{-1} \dfrac{1}{2} + \tan^{-1}(-3)\right)$

96. $\tan\left(\cos^{-1} \dfrac{\sqrt{3}}{2} - \sin^{-1}\left(-\dfrac{3}{5}\right)\right)$

Use a calculator to find each value.

97. $\cos(\tan^{-1} 0.5)$

98. $\sin(\cos^{-1} 0.25)$

99. $\tan(\arcsin 0.12251014)$

100. $\cot(\arccos 0.58236841)$

Write each expression as an algebraic expression in u, u > 0.

101. $\sec(\cos^{-1} u)$

102. $\cot(\tan^{-1} u)$

103. $\sin(\arccos u)$

104. $\tan(\arccos u)$

105. $\cot(\arcsin u)$

106. $\cos(\arcsin u)$

107. $\sin\left(\sec^{-1} \dfrac{u}{2}\right)$

108. $\cos\left(\tan^{-1} \dfrac{3}{u}\right)$

109. $\tan\left(\sin^{-1} \dfrac{u}{\sqrt{u^2 + 2}}\right)$

110. $\sec\left(\cos^{-1} \dfrac{u}{\sqrt{u^2 + 5}}\right)$

111. $\sec\left(\operatorname{arccot} \dfrac{\sqrt{4 - u^2}}{u}\right)$

112. $\csc\left(\arctan \dfrac{\sqrt{9 - u^2}}{u}\right)$

(Modeling) Solve each problem.

113. *Angle of Elevation of a Shot* Refer to **Example 8.**
(a) What is the optimal angle when $h = 0$?
(b) Fix h at 6 feet and regard θ as a function of v. As v gets larger and larger, the graph approaches a horizontal asymptote. Find the equation of that asymptote.

114. *Angle of Elevation of a Plane* Suppose an airplane flying faster than sound goes directly over you. Assume that the plane is flying at a constant altitude. At the instant you feel the sonic boom from the plane, the angle of elevation to the plane is

$$\alpha = 2\arcsin \dfrac{1}{m},$$

where m is the *Mach number* of the plane's speed. (The Mach number is the ratio of the speed of the plane to the speed of sound.) Find α to the nearest degree for each value of m.
(a) $m = 1.2$ (b) $m = 1.5$
(c) $m = 2$ (d) $m = 2.5$

115. *Viewing Angle of an Observer* A painting 3 feet high and 6 feet from the floor will cut off an angle

$$\theta = \tan^{-1}\left(\dfrac{3x}{x^2 + 4}\right)$$

to an observer. Assume that the observer is x feet from the wall where the painting is displayed and that the eyes of the observer are 5 feet above the ground.

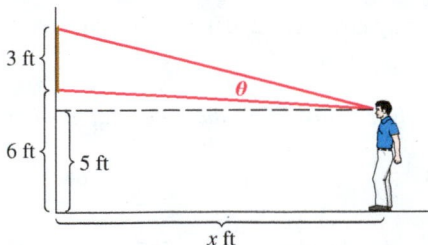

Find the value of θ for each value of x to the nearest degree.
(a) $x = 3$

(b) $x = 6$

(c) $x = 9$

(d) Derive the given formula for θ. (*Hint:* Use right triangles and the identity for $\tan(\theta - \alpha)$.)

(e) Graph the function for θ with a calculator, and determine the distance that maximizes the angle.

116. *Communications Satellite Coverage*   The figure shows a stationary communications satellite positioned 20,000 miles above the equator. What percentage of the equator can be seen from the satellite? The diameter of Earth is 7927 miles at the equator.

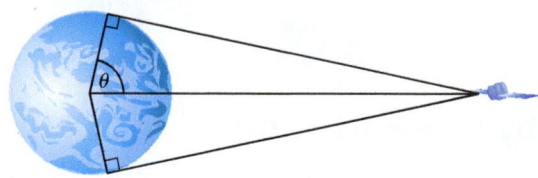

117. *Landscaping Formula*   A shrub is planted in a 100-foot-wide space between buildings measuring 75 feet and 150 feet tall. The location of the shrub determines how much sun it receives each day. Show that if θ is the angle in the figure at the top of the next column and x is the distance of the shrub from the taller building, then the value of θ (in radians) is

$$\theta = \pi - \arctan\left(\frac{75}{100 - x}\right) - \arctan\left(\frac{150}{x}\right).$$

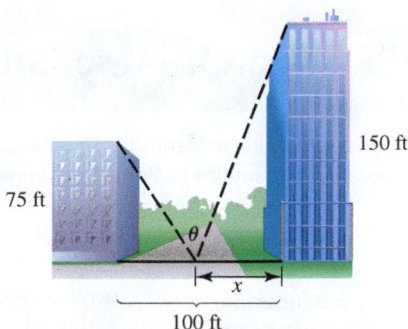

118. *Oil in a Storage Tank*   The level of oil in a storage tank buried in the ground can be found in much the same way as a dipstick is used to determine the oil level in an automobile crankcase. Suppose the ends of the cylindrical storage tank in the figure are circles of radius 3 ft and the cylinder is 20 ft long. Determine the volume of oil in the tank to the nearest cubic foot if the rod shows a depth of 2 ft. (*Hint:* The volume will be 20 times the area of the shaded segment of the circle shown in the figure on the right.)

SECTIONS 9.3–9.4   Reviewing Basic Concepts

Use identities to work Exercises 1–5.

1. Find $\tan x$, given $\cos 2x = -\frac{5}{12}$ and $\frac{\pi}{2} < x < \pi$.

2. Find $\sin 2\theta$, $\cos 2\theta$, and $\tan 2\theta$ if $\sin \theta = -\frac{1}{3}$ and θ is in quadrant III.

3. Find the exact value of $\sin 75°$.

4. Write $2 \sin 25° \cos 150°$ as a sum or difference of trigonometric functions.

5. Verify each identity.

(a) $\sin^2 \frac{x}{2} = \frac{\tan x - \sin x}{2 \tan x}$

(b) $\frac{\sin 2x}{2 \sin x} = \cos^2 \frac{x}{2} - \sin^2 \frac{x}{2}$

6. Find the exact value of each real number y.

(a) $y = \arccos \frac{\sqrt{3}}{2}$ (b) $y = \sin^{-1}\left(-\frac{\sqrt{2}}{2}\right)$

7. Find the exact value of each angle θ measured in degrees.

(a) $\theta = \arccos 0.5$ (b) $\theta = \cot^{-1}(-1)$

8. Graph $y = 2 \csc^{-1} x$.

9. Find the exact value of each expression.

(a) $\cot\left(\arcsin\left(-\frac{2}{3}\right)\right)$

(b) $\cos\left(\tan^{-1} \frac{5}{12} - \sin^{-1} \frac{3}{5}\right)$

10. Write $\sin(\text{arccot } u)$ as an algebraic expression in terms of u with $u > 0$.

9.5 Trigonometric Equations and Inequalities (I)

Equations Solvable by Linear Methods • Equations Solvable by the Zero-Product Property and Quadratic Formula Methods • Using Trigonometric Identities to Solve Equations

 Looking Ahead to Calculus
There are many instances in calculus where it is necessary to solve trigonometric equations. Examples include solving related-rate problems and optimization problems.

Earlier in this chapter, we studied trigonometric equations that were identities. We now consider trigonometric equations that are **conditional**—that is, equations that are satisfied by some values but not others. Conditional equations with trigonometric (or circular) functions can often be solved by using analytic methods and trigonometric identities.

Equations Solvable by Linear Methods

EXAMPLE 1 **Solving a Trigonometric Equation by a Linear Method**

Solve $2 \sin x - 1 = 0$

(a) over the interval $[0, 2\pi)$, and   **(b)** for all solutions.

Analytic Solution

(a) Since this equation involves the first power of $\sin x$, it is linear in $\sin x$. Thus, we solve it by using the usual method for solving a linear equation.

$$2 \sin x - 1 = 0$$
$$2 \sin x = 1 \qquad \text{Add 1.}$$
$$\sin x = \frac{1}{2} \qquad \text{Divide by 2.}$$

The two values of x in the interval $[0, 2\pi)$ whose sine is $\frac{1}{2}$ are $\frac{\pi}{6}$ and $\frac{5\pi}{6}$. These are obtained by finding

$$\sin^{-1} \frac{1}{2} \quad \text{and} \quad \pi - \sin^{-1} \frac{1}{2}.$$

We also get these values by using the unit circle or the reference angle analysis discussed in **Chapter 8.** Therefore, the solution set in the specified interval is $\left\{ \frac{\pi}{6}, \frac{5\pi}{6} \right\}$.

(b) To find all solutions, we add integer multiples of the period of the sine function, 2π, to each solution found in part (a). The solution set is written as follows.

$$\left\{ \frac{\pi}{6} + 2n\pi, \frac{5\pi}{6} + 2n\pi, \text{ where } n \text{ is any integer} \right\}$$

Graphing Calculator Solution

(a) We graph $y = 2 \sin x - 1$ over the desired interval and find that the x-coordinates of the x-intercepts have the same decimal approximations as $\frac{\pi}{6}$ and $\frac{5\pi}{6}$. See **FIGURE 35.**

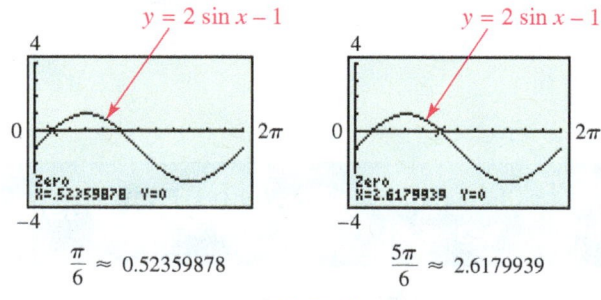

$$\frac{\pi}{6} \approx 0.52359878 \qquad \frac{5\pi}{6} \approx 2.6179939$$

FIGURE 35

We used $\text{Xscl} = \frac{\pi}{6}$ in **FIGURE 35,** so we can see that the graph intersects the x-axis at the first and fifth tick marks, representing $\frac{\pi}{6}$ and $\frac{5\pi}{6}$, respectively. This allows us to observe exact values of solutions from the graph.

(b) Because the graph of

$$y = 2 \sin x - 1$$

repeats y-values every 2π units, all solutions are found by adding integer multiples of 2π to the solutions found in part (a). See the analytic solution.

EXAMPLE 2   **Solving Trigonometric Inequalities**

Solve over the interval $[0, 2\pi)$.

(a) $2 \sin x - 1 > 0$ **(b)** $2 \sin x - 1 < 0$

Solution   To solve the inequality in part (a), we must identify the x-values in the interval $[0, 2\pi)$ for which the graph is *above* the x-axis. Similarly, to solve the inequality in part (b), we must identify the x-values for which the graph is *below* the x-axis. The graph in **FIGURE 35** indicates the following solutions in $[0, 2\pi)$.

(a) The solution set of $2 \sin x - 1 > 0$ is $\left(\frac{\pi}{6}, \frac{5\pi}{6}\right)$.

(b) The solution set of $2 \sin x - 1 < 0$ is $\left[0, \frac{\pi}{6}\right) \cup \left(\frac{5\pi}{6}, 2\pi\right)$.

Equations Solvable by the Zero-Product Property and Quadratic Formula Methods

EXAMPLE 3   **Solving a Trigonometric Equation by Factoring**

Solve $\sin \theta \tan \theta = \sin \theta$ over the interval $[0°, 360°)$.

Solution

$$\sin \theta \tan \theta = \sin \theta$$

$$\sin \theta \tan \theta - \sin \theta = 0 \qquad \text{Subtract } \sin \theta.$$

$$\sin \theta (\tan \theta - 1) = 0 \qquad \text{Factor.}$$

$$\sin \theta = 0 \quad \text{or} \quad \tan \theta - 1 = 0 \qquad \text{Zero-product property}$$

$$\tan \theta = 1$$

$$\theta = 0° \text{ or } \theta = 180° \qquad \theta = 45° \text{ or } \theta = 225°$$

The solution set is $\{0°, 45°, 180°, 225°\}$. See **FIGURE 36.**

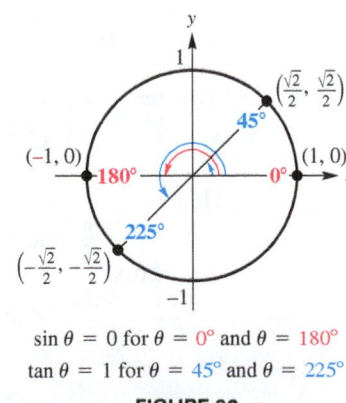

$\sin \theta = 0$ for $\theta = 0°$ and $\theta = 180°$

$\tan \theta = 1$ for $\theta = 45°$ and $\theta = 225°$

FIGURE 36

CAUTION   There are four solutions in **Example 3**. Trying to solve the equation by dividing each side by $\sin \theta$ would result in just $\tan \theta = 1$, which would give $\theta = 45°$ or $\theta = 225°$. The other two solutions would not appear. The missing solutions are the ones that make the divisor, $\sin \theta$, equal 0. *For this reason, we avoid dividing by a variable expression.*

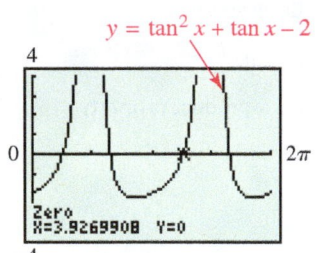

EXAMPLE 4 **Solving a Trigonometric Equation and an Inequality**

Solve over the interval $[0, 2\pi)$.

(a) $\tan^2 x + \tan x - 2 = 0$ (b) $\tan^2 x + \tan x - 2 > 0$

Solution

(a) This equation is quadratic in the term $\tan x$ and can be solved by factoring.

$$\tan^2 x + \tan x - 2 = 0$$
$$(\tan x - 1)(\tan x + 2) = 0 \quad \text{Factor.}$$
$$\tan x - 1 = 0 \quad \text{or} \quad \tan x + 2 = 0 \quad \text{Zero-product property}$$
$$\tan x = 1 \quad \text{or} \quad \tan x = -2 \quad \text{Solve each equation.}$$

The solutions for $\tan x = 1$ in the interval $[0, 2\pi)$ are $\frac{\pi}{4}$ and $\frac{5\pi}{4}$.

To solve $\tan x = -2$ in the interval, we use a calculator set in *radian* mode. We find that $\tan^{-1}(-2) \approx -1.107148718$. However, due to the way the calculator determines this value, it is not in the desired interval. Because the period of the tangent function is π, we add π and then 2π to $\tan^{-1}(-2)$ to obtain the solutions in the desired interval.

$$x = \tan^{-1}(-2) + \pi \approx 2.034443936$$
$$x = \tan^{-1}(-2) + 2\pi \approx 5.176036589$$

The solution set is $\left\{ \frac{\pi}{4}, \quad \frac{5\pi}{4}, \quad \underbrace{2.03, \quad 5.18}\right\}$.

$\underbrace{\phantom{\frac{\pi}{4}, \frac{5\pi}{4}}}$ Exact values
Approximate values to the nearest hundredth

FOR DISCUSSION
What is the solution set of $\tan^2 x + \tan x - 2 < 0$ over the interval $[0, 2\pi)$? (*Hint:* Refer to **FIGURE 37**.)

$y = \tan^2 x + \tan x - 2$

Zero
X=3.9269908 Y=0

FIGURE 37

(b) From the graph of $y = \tan^2 x + \tan x - 2$ as seen in **FIGURE 37,** there are four x-intercepts in the interval $[0, 2\pi)$. The figure shows the solution $\frac{5\pi}{4}$, since a decimal approximation of this number is 3.9269908. The other three solutions can be found similarly.

The graph in **FIGURE 37** lies *above* the x-axis for the following subset of the interval $[0, 2\pi)$.

$\tan x$ is undefined for $x = \frac{\pi}{2}$ and $x = \frac{3\pi}{2}$. $\rightarrow$ $\left(\frac{\pi}{4}, \frac{\pi}{2}\right) \cup \left(\frac{\pi}{2}, 2.03\right) \cup \left(\frac{5\pi}{4}, \frac{3\pi}{2}\right) \cup \left(\frac{3\pi}{2}, 5.18\right)$

This is the solution set of the inequality. Notice that tangent is not defined when $x = \frac{\pi}{2}$ and when $x = \frac{3\pi}{2}$, so these values must be excluded from the solution set.

EXAMPLE 5 **Solving a Trigonometric Equation Using the Quadratic Formula**

Find all solutions of $\cot x(\cot x + 3) = 1$. Write the solution set.

Solution We multiply the factors on the left and subtract 1 to write the equation in standard quadratic form.

$$\cot x(\cot x + 3) = 1 \quad \text{Original equation}$$
$$\cot^2 x + 3 \cot x - 1 = 0 \quad \text{Subtract 1.}$$

This equation is quadratic in form, but cannot be solved by the zero-product property. Therefore, we use the quadratic formula, with $a = 1, b = 3, c = -1$, and $\cot x$ as the variable.

$$\cot x = \frac{-b \pm \sqrt{b^2 - 4ac}}{2a} \qquad \text{Quadratic formula}$$

$$= \frac{-3 \pm \sqrt{3^2 - 4(1)(-1)}}{2(1)} \qquad a = 1, b = 3, c = -1$$

Be careful with signs.

$$= \frac{-3 \pm \sqrt{9 + 4}}{2} \qquad \text{Simplify.}$$

$$= \frac{-3 \pm \sqrt{13}}{2} \qquad \text{Add.}$$

Approximating these two solutions we have the following.

$$\cot x \approx -3.302775638 \qquad \text{or} \quad \cot x \approx 0.3027756377$$

Use a calculator.

$$x \approx \cot^{-1}(-3.302775638) \qquad \text{or} \qquad x \approx \cot^{-1}(0.3027756377)$$

Definition of inverse cotangent

$$x \approx \frac{\pi}{2} - \tan^{-1}(-3.302775638) \qquad \text{or} \qquad x \approx \frac{\pi}{2} - \tan^{-1}(3.3027756377)$$

$\cot^{-1} x = \frac{\pi}{2} - \tan^{-1} x$

$$x \approx 2.847591352 \qquad \text{or} \qquad x \approx 1.276795025$$

Use a calculator in radian mode.

To find *all* solutions, we add integer multiples of the period of the tangent function, which is π, to each solution found previously. Although not unique, a common form of the solution set of the equation, written using the least possible nonnegative angle measures, is given as follows.

$$\{\, 2.8476 + n\pi, 1.2768 + n\pi, \text{ where } n \text{ is any integer} \,\}$$

Round to four decimal places. ●

| EXAMPLE 6 | **Solving a Trigonometric Equation with No Solution** |

Solve $\sin^2 x + \sin x - 6 = 0$ over the interval $[0, 2\pi)$.

Solution   The equation is quadratic in the term $\sin x$.

$$\sin^2 x + \sin x - 6 = 0$$

$$(\sin x + 3)(\sin x - 2) = 0 \qquad \text{Factor.}$$

$$\sin x + 3 = 0 \quad \text{or} \quad \sin x - 2 = 0 \qquad \text{Zero-product property}$$

$$\sin x = -3 \quad \text{or} \qquad \sin x = 2 \qquad \text{Solve.}$$

Because $-1 \le \sin x \le 1$ for all real numbers x, there is no solution to either $\sin x = -3$ or $\sin x = 2$. The solution set is $\emptyset$. ●

Using Trigonometric Identities to Solve Equations

Recall that squaring each side of an equation such as $\sqrt{x + 4} = x + 2$ will yield all solutions but may also give extraneous values. (Verify that in this equation 0 is a solution, while -3 is extraneous.)

EXAMPLE 7 **Solving a Trigonometric Equation by Squaring**

Solve $\tan x + \sqrt{3} = \sec x$ over the interval $[0, 2\pi)$.

Analytic Solution

Our first goal is to rewrite the equation in terms of a single trigonometric function. Since the tangent and secant functions are related by the identity $1 + \tan^2 x = \sec^2 x$, square each side and express $\sec^2 x$ in terms of $\tan^2 x$.

$$\left(\tan x + \sqrt{3}\right)^2 = (\sec x)^2 \qquad \text{Square each side.}$$

Don't forget the middle term.

$$\tan^2 x + 2\sqrt{3} \tan x + 3 = \sec^2 x \qquad (x+y)^2 = x^2 + 2xy + y^2$$

$$\tan^2 x + 2\sqrt{3} \tan x + 3 = 1 + \tan^2 x \qquad \text{Pythagorean identity}$$

$$2\sqrt{3} \tan x = -2 \qquad \text{Subtract } 3 + \tan^2 x.$$

$$\tan x = -\frac{1}{\sqrt{3}}, \quad \text{or} \quad -\frac{\sqrt{3}}{3} \qquad \text{Divide by } 2\sqrt{3}. \text{ Rationalize the denominator.}$$

Solutions of $\tan x = -\frac{\sqrt{3}}{3}$ over $[0, 2\pi)$ are $\frac{5\pi}{6}$ and $\frac{11\pi}{6}$. These proposed solutions must be checked to determine whether they are also solutions of the original equation.

Check: $\qquad\qquad \tan x + \sqrt{3} = \sec x \qquad$ Original equation

$$\tan\left(\frac{5\pi}{6}\right) + \sqrt{3} = \sec\left(\frac{5\pi}{6}\right) ? \qquad \Bigg| \qquad \tan\left(\frac{11\pi}{6}\right) + \sqrt{3} = \sec\left(\frac{11\pi}{6}\right) ?$$

Let $x = \frac{5\pi}{6}$. $\qquad\qquad\qquad\qquad\qquad$ Let $x = \frac{11\pi}{6}$.

$$-\frac{\sqrt{3}}{3} + \frac{3\sqrt{3}}{3} = -\frac{2\sqrt{3}}{3} ? \qquad \Bigg| \qquad -\frac{\sqrt{3}}{3} + \frac{3\sqrt{3}}{3} = \frac{2\sqrt{3}}{3} ?$$

$$\frac{2\sqrt{3}}{3} = -\frac{2\sqrt{3}}{3} \quad \text{False} \qquad \Bigg| \qquad \frac{2\sqrt{3}}{3} = \frac{2\sqrt{3}}{3} \quad \checkmark \text{ True}$$

As the check shows, only $\frac{11\pi}{6}$ is a solution, so the solution set is $\left\{\frac{11\pi}{6}\right\}$.

Graphing Calculator Solution

We will use the x-intercept method. Graph

$$y = \tan x + \sqrt{3} - \sec x$$

over the desired interval, as in **FIGURE 38.** The graph shows that the only zero on the interval $[0, 2\pi)$ is 5.7595865, which is an approximation for $\frac{11\pi}{6}$, the solution found analytically.

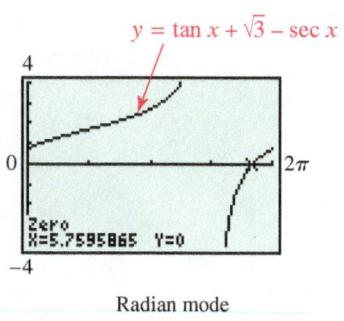

$y = \tan x + \sqrt{3} - \sec x$

Radian mode

FIGURE 38

Notice that a graphical solution does not give the extraneous solution of $\frac{5\pi}{6}$ that results from squaring each side of the equation in the analytic solution.

The methods for solving trigonometric equations can be summarized as follows.

Solving a Trigonometric Equation Analytically

1. Decide whether the equation is linear or quadratic, so you can determine the solution method.

2. If only one trigonometric function is present, solve the equation for that function.

3. If more than one trigonometric function is present, rearrange the equation so that one side equals 0. Then try to factor and use the zero-product property to solve.

4. If the equation is quadratic in form, but not factorable, use the quadratic formula. Check that solutions are in the desired interval.

5. Try using identities to change the form of the equation. It may be helpful to square each side of the equation first. If this is done, check for extraneous values.

Solving a Trigonometric Equation Graphically

1. For an equation of the form $f(x) = g(x)$, use the intersection-of-graphs method.
2. For an equation of the form $f(x) = 0$, use the x-intercept method.

9.5 Exercises

Checking Analytic Skills   *Solve each equation for solutions over the interval* $[0, 2\pi)$ *by first solving for the trigonometric function.* **Do not use a calculator.**

1. $2 \cos x + 1 = 0$

2. $2 \sin x + 1 = 0$

3. $5 \sin x - 6 = 0$

4. $3 \cos x + 5 = 0$

5. $2 \tan x + 1 = -1$

6. $2 \cot x + 1 = -1$

7. $2 \cos x + 5 = 6$

8. $2 \sin x + 3 = 4$

9. $2 \csc x + 4 = \csc x + 6$

10. $2 \sec x + 1 = \sec x + 3$

11. $(\cot x - 1)(\sqrt{3} \cot x + 1) = 0$

12. $(\csc x + 2)(\csc x - \sqrt{2}) = 0$

13. $\cos x \cot x = \cos x$

14. $\sin x \cot x = \sin x$

15. $\sin^2 x - 2 \sin x + 1 = 0$

16. $\cos^2 x + 2 \cos x + 1 = 0$

17. $4(1 + \sin x)(1 - \sin x) = 3$

18. $(\cot x - \sqrt{3})(2 \sin x + \sqrt{3}) = 0$

19. $\tan x + 1 = \sqrt{3} + \sqrt{3} \cot x$

20. $\tan x - \cot x = 0$

21. $2 \sin x - 1 = \csc x$

22. $\cos^2 x = \sin^2 x$

23. $\cos^2 x - \sin^2 x = 1$

24. $\csc^2 x = 2 \cot x$

In Exercises 25–32, solve **(a)** $f(x) = 0$, **(b)** $f(x) > 0$, *and* **(c)** $f(x) < 0$ *over the interval* $[0, 2\pi)$.

25. $f(x) = -2 \cos x + 1$

26. $f(x) = 2 \sin x + 1$

27. $f(x) = \tan^2 x - 3$

28. $f(x) = \sec^2 x - 1$

29. $f(x) = 2 \cos^2 x - \sqrt{3} \cos x$

30. $f(x) = 2 \sin^2 x + 3 \sin x + 1$

31. $f(x) = \sin^2 x \cos x - \cos x$

32. $f(x) = 2 \tan^2 x \sin x - \tan^2 x$

Solve each equation for solutions over the interval $[0°, 360°)$. *Give solutions to the nearest tenth as appropriate.*

33. $\tan \theta + \cot \theta = 0$

34. $\cos^2 \theta = \sin^2 \theta + 2$

35. $2 \tan^2 \theta \sin \theta - \tan^2 \theta = 0$

36. $\sin^2 \theta \cos \theta = \cos \theta$

37. $\sec^2 \theta \tan \theta = 2 \tan \theta$

38. $\sin^2 \theta \cos^2 \theta = 0$

39. $9 \sin^2 \theta - 6 \sin \theta = 1$

40. $4 \cos^2 \theta + 4 \cos \theta = 1$

41. $\tan^2 \theta + 4 \tan \theta + 2 = 0$

42. $3 \cot^2 \theta - 3 \cot \theta - 1 = 0$

43. $\sin^2 \theta - 2 \sin \theta + 3 = 0$

44. $2 \cos^2 \theta + 2 \cos \theta + 1 = 0$

45. $\cot \theta + 2 \csc \theta = 3$

46. $2 \sin \theta = 1 - 2 \cos \theta$

RELATING CONCEPTS **For individual or group investigation (Exercises 47–50)**

Work Exercises 47–50 in order.

47. Write the equation $\tan^3 x = 3 \tan x$ with 0 on one side.

48. Solve the equation you wrote in **Exercise 47** over $[0, 2\pi)$ by the zero-product property.

49. Solve $\tan^3 x = 3 \tan x$ by first dividing each side by $\tan x$.

50. Do your answers in **Exercises 48** and **49** agree? Explain.

In Exercises 51–56, give solutions over the interval $[0, 2\pi)$ as approximations to the nearest hundredth when exact values cannot be determined. You may need to use the quadratic formula. Give approximate answers in Exercises 57–60 to the nearest tenth of a degree over the interval $[0°, 360°)$.

51. $3 \sin^2 x - \sin x = 2$

52. $9 \sin^2 x = 6 \sin x + 1$

53. $\tan^2 x + 4 \tan x + 2 = 0$

54. $3 \cot^2 x - 3 \cot x = 1$

55. $2 \cos^2 x + 2 \cos x = 1$

56. $\cos^2 x - 2 \cos x + 3 = 0$

57. $\sec^2 \theta = 2 \tan \theta + 4$

58. $\cot \theta + 2 \csc \theta = 3$

59. $2 \sin \theta = 1 - 2 \cos \theta$

60. $\sin^2 \theta - \cos \theta = 0$

Solve each equation (x in radians and θ in degrees) for all exact solutions where appropriate. Round approximate answers in radians to four decimal places and approximate answers in degrees to the nearest tenth. Write answers using the least possible nonnegative angle measures.

61. $\cos \theta + 1 = 0$

62. $\tan \theta + 1 = 0$

63. $3 \csc x - 2\sqrt{3} = 0$

64. $\cot x + \sqrt{3} = 0$

65. $6 \sin^2 \theta + \sin \theta = 1$

66. $3 \sin^2 \theta - \sin \theta = 2$

67. $2 \cos^2 x + \cos x - 1 = 0$

68. $4 \cos^2 x - 1 = 0$

69. $\sin \theta \cos \theta - \sin \theta = 0$

70. $\tan \theta \csc \theta - \sqrt{3} \csc \theta = 0$

71. $\sin x (3 \sin x - 1) = 1$

72. $\tan x (\tan x - 2) = 5$

73. $5 + 5 \tan^2 \theta = 6 \sec \theta$

74. $\sec^2 \theta = 2 \tan \theta + 4$

75. $\dfrac{2 \tan \theta}{3 - \tan^2 \theta} = 1$

76. $\dfrac{2 \cot^2 \theta}{\cot \theta + 3} = 1$

77. How many solutions does the equation $2x - 1 = 0$ have? How many solutions does the equation $2 \sin x - 1 = 0$ have? Explain.

Solve each equation graphically over the interval $[0, 2\pi)$. Express solutions to the nearest hundredth. (Hint: In Exercise 81, the equation has three solutions.)

78. $\cot x + 2 \csc x = 3$

79. $2 \sin x = 1 - 2 \cos x$

80. $\sin^3 x + \sin x = 1$

81. $2 \cos^3 x + \sin x = -1$

82. $e^x = \sin x + 3$

83. $\ln x = \cos x$

84. Explain what is *wrong* with the following solution, where x is in the interval $[0, 2\pi)$.

$\sin^2 x - \sin x = 0$	Given equation
$\sin x - 1 = 0$	Divide by $\sin x$.
$\sin x = 1$	Add 1.
$x = \dfrac{\pi}{2}$	Solve for x.

(Modeling) Solve each problem.

85. *Daylight Hours in New Orleans* The seasonal variation in length of daylight can be modeled by a sine function. For example, the daily number of hours h of daylight in New Orleans is approximated by

$$h = \frac{35}{3} + \frac{7}{3} \sin \frac{2x\pi}{365},$$

where x is the number of days after March 21 (disregarding leap year). (*Source:* Bushaw, Donald et al., *A Sourcebook of Applications of School Mathematics*, The Mathematical Association of America.)

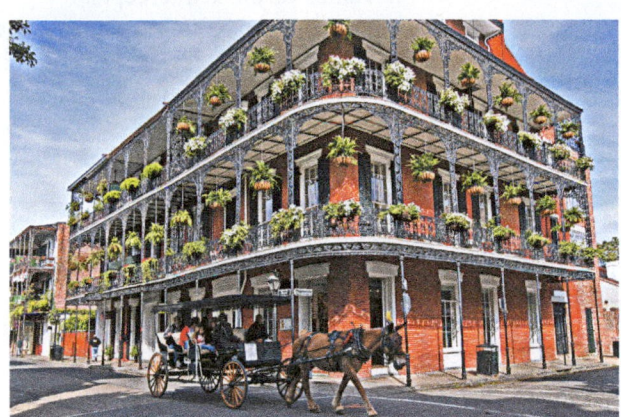

(a) On what date will there be about 14 hours of daylight?
(b) What date has the least number of hours of daylight?
(c) When will there be about 10 hours of daylight?

86. *Mach Number for a Plane* An airplane flying faster than sound sends out sound waves that form a cone. The cone intersects the ground to form a hyperbola. As this hyperbola passes over a particular point on the ground, a sonic boom is heard at that point. If α is the angle at the vertex of the cone, then

$$\sin \frac{\alpha}{2} = \frac{1}{m},$$

where m is the Mach number for the speed of the plane. (See **Section 9.4, Exercise 114.**) We assume that $m > 1$. Find the measure of α, in degrees, if $m = 1.5$.

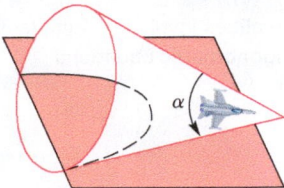

87. *Accident Reconstruction* The equation

$$0.342D \cos \theta + h \cos^2 \theta = \frac{16D^2}{V^2}$$

is used to reconstruct accidents in which a vehicle vaults into the air after hitting an obstruction. V is the velocity in feet per second of the vehicle when it hits the obstruction, D is the distance (in feet) from the obstruction to the vehicle's landing point, and h is the difference in height (in feet) between the landing point and the takeoff point. Angle θ is the takeoff angle—the angle between the horizontal and the path of the vehicle. Find θ to the nearest degree if $V = 60$, $D = 80$, and $h = 2$.

88. *Maximum Viewing Angle* The bottom of a 10-foot-high movie screen is located 2 feet above the eyes of the viewers, all of whom are sitting at the same level. A viewer seated 5 feet from the screen has the maximum viewing angle, determined by x in the equation

$$\frac{\tan x + 0.4}{1 - 0.4 \tan x} = 2.4.$$

Find the maximum viewing angle (in degrees).

Distance of a Particle from a Starting Point *A particle moves along a straight line. The distance s that the particle is from a starting point at time t is*

$$s(t) = \sin t + 2 \cos t.$$

Find a value of t in $\left[0, \frac{\pi}{2}\right)$ that satisfies each equation.

89. $s(t) = \dfrac{2 + \sqrt{3}}{2}$

90. $s(t) = \dfrac{3\sqrt{2}}{2}$

9.6 Trigonometric Equations and Inequalities (II)

Equations and Inequalities Involving Multiple-Number Identities • Equations and Inequalities Involving Half-Number Identities • Applications of Trigonometric Equations

Equations and Inequalities Involving Multiple-Number Identities

EXAMPLE 1 Solving an Equation by Using a Double-Number Identity

Solve $\cos 2x = \cos x$ over the interval $[0, 2\pi)$.

Analytic Solution

First change $\cos 2x$ to a trigonometric function of x alone. Use the identity $\cos 2x = 2\cos^2 x - 1$ so that the equation involves only the cosine of x.

$$\cos 2x = \cos x$$
$$2\cos^2 x - 1 = \cos x \qquad \text{Double-number identity}$$
$$2\cos^2 x - \cos x - 1 = 0 \qquad \text{Standard form}$$
$$(2\cos x + 1)(\cos x - 1) = 0 \qquad \text{Factor.}$$
$$2\cos x + 1 = 0 \quad \text{or} \quad \cos x - 1 = 0 \qquad \text{Zero-product property}$$
$$\cos x = -\frac{1}{2} \qquad\qquad \cos x = 1 \qquad \text{Solve each equation.}$$
$$x = \frac{2\pi}{3} \quad \text{or} \quad x = \frac{4\pi}{3} \qquad\qquad x = 0 \qquad \text{See FIGURE 39.}$$

The solution set is $\left\{0, \frac{2\pi}{3}, \frac{4\pi}{3}\right\}$ over the interval $[0, 2\pi)$.

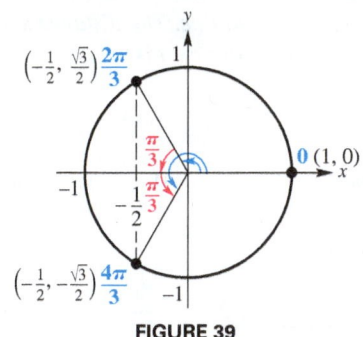

FIGURE 39

Graphing Calculator Solution

Graph $y = \cos 2x - \cos x$ in an appropriate window, and find the x-values of the x-intercepts.

The display in **FIGURE 40** shows that one x-value is 2.0943951, which is an approximation for $\frac{2\pi}{3}$. The other two x-values correspond to 0 and $\frac{4\pi}{3}$, respectively.

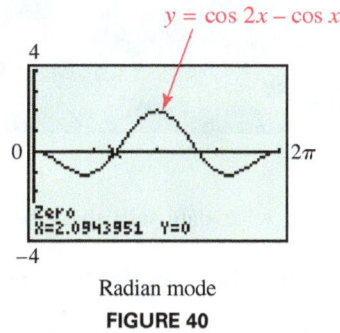

Radian mode

FIGURE 40

The graph supports the solution set found in the analytic solution.

CAUTION In the analytic solution in **Example 1**, $\cos 2x$ cannot be changed to $\cos x$ by dividing by 2, since 2 is *not* a factor of $\cos 2x$. That is, $\frac{\cos 2x}{2} \neq \cos x$. To change $\cos 2x$ to a trigonometric function of x alone, use one of the double-number identities for $\cos 2x$.

EXAMPLE 2   **Solving Inequalities Involving a Function of 2x**

Refer to **Example 1** and **FIGURE 40** to solve each inequality over the interval $[0, 2\pi)$.

(a) $\cos 2x < \cos x$ **(b)** $\cos 2x > \cos x$

Solution

(a) The given inequality is equivalent to $\cos 2x - \cos x < 0$. We determine the solution set of this inequality by finding the x-values where the graph of the equation $y = \cos 2x - \cos x$ is *below* the x-axis. From **FIGURE 40**, we see that this occurs in the interval $[0, 2\pi)$ when x is in the interval $\left(0, \frac{2\pi}{3}\right) \cup \left(\frac{4\pi}{3}, 2\pi\right)$.

(b) Here, we want the x-values where the graph of $y = \cos 2x - \cos x$ is *above* the x-axis. The solution set is $\left(\frac{2\pi}{3}, \frac{4\pi}{3}\right)$.

FOR DISCUSSION

Refer to **Example 2** and discuss how the solution sets would be affected with the following changes.

1. Solve $\cos 2x \leq \cos x$ over the interval $[0, 2\pi)$.

2. Solve $\cos 2x \geq \cos x$ over the interval $[0, 4\pi)$.

EXAMPLE 3   **Solving an Equation by Using a Double-Number Identity**

Solve $4 \sin x \cos x = \sqrt{3}$ over the interval $[0, 2\pi)$.

Analytic Solution

The identity $2 \sin x \cos x = \sin 2x$ is applied here.

$$4 \sin x \cos x = \sqrt{3}$$
$$2(2 \sin x \cos x) = \sqrt{3} \qquad \text{4 = 2 · 2}$$
$$2 \sin 2x = \sqrt{3} \qquad \text{2 sin x cos x = sin 2x}$$
$$\sin 2x = \frac{\sqrt{3}}{2} \qquad \text{Divide by 2.}$$

From the given domain, $0 \leq x < 2\pi$, it follows that the domain for $2x$ is $0 \leq 2x < 4\pi$. We list all solutions in $[0, 4\pi)$.

$$2x = \frac{\pi}{3}, \frac{2\pi}{3}, \frac{7\pi}{3}, \frac{8\pi}{3} \qquad \text{Each has a sine value of } \frac{\sqrt{3}}{2}.$$

$$x = \frac{\pi}{6}, \frac{\pi}{3}, \frac{7\pi}{6}, \frac{4\pi}{3} \qquad \text{Divide by 2.}$$

We found the final two solutions for $2x$ by adding 2π to $\frac{\pi}{3}$ and $\frac{2\pi}{3}$, respectively, resulting in the solution set $\left\{\frac{\pi}{6}, \frac{\pi}{3}, \frac{7\pi}{6}, \frac{4\pi}{3}\right\}$ for x.

Graphing Calculator Solution

We use the intersection-of-graphs method. **FIGURE 41** indicates that the x-coordinate of one point of intersection is 0.52359878, an approximation for $\frac{\pi}{6}$. The other three intersection points correspond to the remaining solutions, $\frac{\pi}{3}, \frac{7\pi}{6}$, and $\frac{4\pi}{3}$.

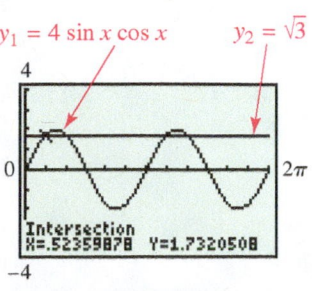

FIGURE 41

EXAMPLE 4 **Squaring Each Side of an Equation**

Solve $\tan 3x + \sec 3x = 2$ over the interval $[0, 2\pi)$ to the nearest thousandth.

Solution Since the tangent and secant functions are related by the fundamental identity $1 + \tan^2 \theta = \sec^2 \theta$, we begin by expressing everything in terms of the secant.

$$\tan 3x + \sec 3x = 2$$

> Remember the middle term when squaring a binomial.

$$\tan 3x = 2 - \sec 3x \qquad \text{Subtract } \sec 3x.$$

$$\tan^2 3x = 4 - 4 \sec 3x + \sec^2 3x \qquad \text{Square each side.}$$

$$\sec^2 3x - 1 = 4 - 4 \sec 3x + \sec^2 3x \qquad \tan^2 \theta = \sec^2 \theta - 1, \theta = 3x$$

$$0 = 5 - 4 \sec 3x \qquad \text{Subtract } \sec^2 3x \text{ and add 1.}$$

$$4 \sec 3x = 5 \qquad \text{Add } 4 \sec 3x.$$

$$\sec 3x = \frac{5}{4} \qquad \text{Divide by 4.}$$

$$\frac{1}{\cos 3x} = \frac{5}{4} \qquad \sec \theta = \frac{1}{\cos \theta}$$

$$\cos 3x = \frac{4}{5} \qquad \text{Use reciprocals.}$$

We multiply $0 \le x < 2\pi$ by 3 to find that the interval for $3x$ is $[0, 6\pi)$. Using a calculator and the fact that the cosine is positive in quadrants I and IV, we get the following.

$$3x \approx 0.64350111, 5.6396842, 6.9266864, 11.922870, 13.209872, 18.206055$$

Divide by 3.

$$x \approx 0.21450037, 1.8798947, 2.3088955, 3.9742898, 4.4032906, 6.0686849$$

Recall from **Section 4.5** that when each side of an equation is squared, there may be extraneous values. From the graph of $y = \tan 3x + \sec 3x - 2$ in **FIGURE 42**, we see that in the interval $[0, 2\pi)$ there are only three zeros. One of these is approximately 4.4032906, which is one of the six possible solutions found. (See the display in the figure.) The other two can also be verified by a calculator. They are 0.21450037 and 2.3088955. To the nearest thousandth, the solution set over the interval $[0, 2\pi)$ is $\{0.215, 2.309, 4.403\}$.

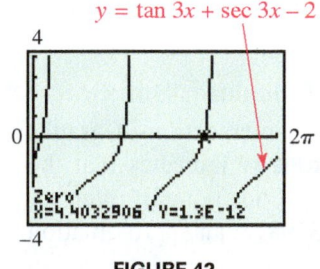

$y = \tan 3x + \sec 3x - 2$

FIGURE 42

Equations and Inequalities Involving Half-Number Identities

EXAMPLE 5 **Solving an Equation by Using a Half-Number Identity**

Solve the equation $2 \sin \dfrac{x}{2} = 1$

(a) over the interval $[0, 2\pi)$, and   **(b)** for all solutions.

Solution

(a) Write the interval $[0, 2\pi)$ as the inequality

$$0 \le x < 2\pi.$$

The corresponding interval for $\frac{x}{2}$ is

$$0 \le \frac{x}{2} < \pi. \qquad \text{Divide by 2.}$$

To find all values of $\frac{x}{2}$ over the interval $[0, \pi)$ that satisfy the given equation, first solve for $\sin \frac{x}{2}$.

$$2 \sin \frac{x}{2} = 1 \qquad \text{Original equation}$$

$$\sin \frac{x}{2} = \frac{1}{2} \qquad \text{Divide by 2.}$$

The two numbers over the interval $[0, \pi)$ with sine value $\frac{1}{2}$ are $\frac{\pi}{6}$ and $\frac{5\pi}{6}$.

$$\frac{x}{2} = \frac{\pi}{6} \quad \text{or} \quad \frac{x}{2} = \frac{5\pi}{6} \qquad \text{Definition of inverse sine}$$

$$x = \frac{\pi}{3} \quad \text{or} \quad x = \frac{5\pi}{3} \qquad \text{Multiply by 2.}$$

The solution set over the given interval is $\left\{ \frac{\pi}{3}, \frac{5\pi}{3} \right\}$.

(b) Because this is a sine function with period 4π, all solutions are found by adding integer multiples of 4π.

$$\left\{ \frac{\pi}{3} + 4n\pi, \frac{5\pi}{3} + 4n\pi, \text{ where } n \text{ is any integer} \right\}$$

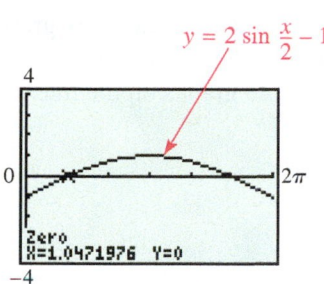

$y = 2 \sin \frac{x}{2} - 1$

FIGURE 43

EXAMPLE 6   **Solving an Inequality Involving a Function of $\frac{x}{2}$**

Use a graph to solve $2 \sin \frac{x}{2} - 1 > 0$ over the interval $[0, 2\pi)$.

Solution   Referring to **FIGURE 43,** where the x-scale is $\frac{\pi}{3}$, we see that the graph of $y = 2 \sin \frac{x}{2} - 1$ lies *above* the x-axis between $\frac{\pi}{3}$ and $\frac{5\pi}{3}$. Thus, the solution set is the open interval $\left(\frac{\pi}{3}, \frac{5\pi}{3} \right)$.

EXAMPLE 7   **Solving an Equation by Using Only a Graphical Approach**

Use a graph to solve $\sin 3x + \cos 2x = \cos \frac{x}{2}$ over the interval $[0, 2\pi)$, expressing solutions to the nearest thousandth.

Solution   We solve this equation with the x-intercept method. The graph of $y = \sin 3x + \cos 2x - \cos \frac{x}{2}$ is shown in **FIGURE 44.** The least positive solution, 0.72739787, is indicated on the screen. The solution 0 can be verified by direct substitution. The solution set over the interval $[0, 2\pi)$ is $\{0, 0.727, 2.288, 3.524, 4.189\}$.

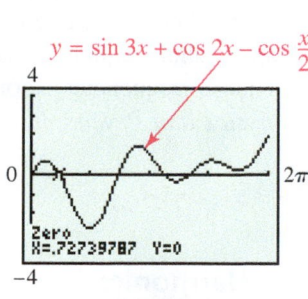

$y = \sin 3x + \cos 2x - \cos \frac{x}{2}$

FIGURE 44

Applications of Trigonometric Equations

EXAMPLE 8   **Describing a Musical Tone from a Graph**

A basic component of music is a pure tone. The graph in **FIGURE 45** models the sinusoidal pressure $y = P$ in pounds per square foot from a pure tone at time $x = t$ in seconds.

(a) The frequency of a pure tone is often measured in hertz. One **hertz** is equal to 1 cycle per second and is abbreviated Hz. What is the frequency f in hertz of the pure tone shown in the graph?

(b) The time for the tone to produce one complete cycle is called the **period.** Approximate the period T of the pure tone, in seconds.

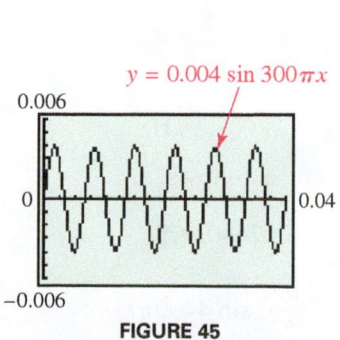

$y = 0.004 \sin 300\pi x$

FIGURE 45

(c) An equation for the graph is $y = 0.004 \sin 300\pi x$. Use a calculator to estimate all solutions of the equation $y = 0.004$ on the interval $[0, 0.02]$.

(continued)

Solution

(a) From the graph in **FIGURE 45** repeated below, we see that there are 6 cycles in 0.04 second. This is equivalent to $\frac{6}{0.04} = 150$ cycles per second. The pure tone has a frequency of $f = 150$ Hz.

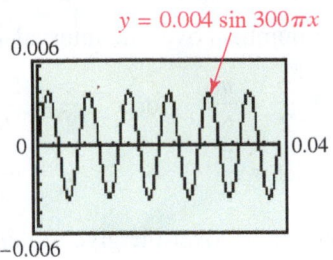

FIGURE 45 (repeated)   **FIGURE 46**

(b) Six periods cover a time of 0.04 second. Therefore, the period would be equal to $T = \frac{0.04}{6} = \frac{1}{150}$, or $0.00\overline{6}$ second.

(c) If we reproduce the graph in **FIGURE 45** on a calculator as y_1 and also graph the second function as $y_2 = 0.004$, as in **FIGURE 46,** we can determine that the approximate values of x at the points of intersection of the graphs on the interval $[0, 0.02]$. The solution set is as follows.

$$\{0.0017, 0.0083, 0.015\}$$

 A piano string can vibrate at more than one frequency when it is struck. It produces a complex wave that can mathematically be modeled by a sum of several pure tones. When a piano key with a frequency of f_1 is played, the corresponding string vibrates not only at f_1 but also at the higher frequencies of $2f_1, 3f_1, 4f_1, \ldots, nf_1$. Frequency f_1 is the **fundamental frequency** of the string, and higher frequencies are the **upper harmonics.** The human ear will hear the sum of these frequencies as one complex tone. (*Source:* Roederer, J., *Introduction to the Physics and Psychophysics of Music*, Second Edition, Springer-Verlag.)

EXAMPLE 9   **Analyzing Pressures of Upper Harmonics**

Suppose that the A key above middle C is played on a piano. Its fundamental frequency is $f_1 = 440$ Hz, and its associated pressure is expressed as

$$P_1 = 0.002 \sin 880\pi t.$$

The string will also vibrate at

$$f_2 = 880, \quad f_3 = 1320, \quad f_4 = 1760, \quad f_5 = 2200, \ldots \text{Hz.}$$

The corresponding pressures of these upper harmonics are as follows.

$$P_2 = \frac{0.002}{2} \sin 1760\pi t, \qquad P_3 = \frac{0.002}{3} \sin 2640\pi t,$$

$$P_4 = \frac{0.002}{4} \sin 3520\pi t, \quad \text{and} \quad P_5 = \frac{0.002}{5} \sin 4400\pi t$$

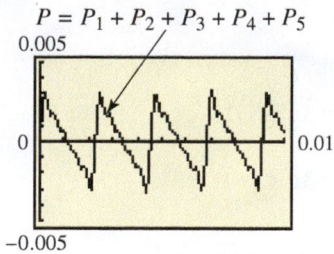

FIGURE 47

The graph of $P = P_1 + P_2 + P_3 + P_4 + P_5$, shown in **FIGURE 47,** is "saw-toothed."

(a) What is the maximum value of P?

(b) At what values of $t = x$ does this maximum occur over the interval $[0, 0.01]$?

Solution

(a) A graphing calculator shows that the maximum value of P is approximately 0.00317. See **FIGURE 48.**

(b) The maximum occurs at

$$t = x \approx 0.000191, 0.00246, 0.00474, 0.00701, \text{ and } 0.00928.$$

FIGURE 48 shows how the second value is found. The other values are found similarly.

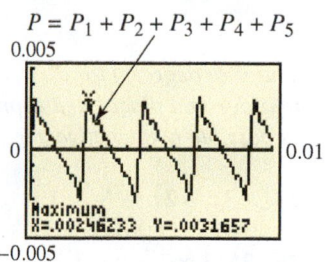

FIGURE 48

9.6 Exercises

Concept Check   *Answer each question.*

1. Suppose you are solving a trigonometric equation for solutions in $[0, 2\pi)$ and your work leads to

$$2x = \frac{2\pi}{3}, 2\pi, \frac{8\pi}{3}.$$

What are the corresponding values of x?

2. Suppose you are solving a trigonometric equation to find solutions in $[0°, 360°)$ and your work leads to

$$\frac{1}{3}\theta = 45°, 60°, 75°, 90°.$$

What are the corresponding values of θ?

Solve each equation over the interval $[0, 2\pi)$. ***Do not use a calculator.***

3. $\sin \dfrac{x}{2} = \cos \dfrac{x}{2}$

4. $\sec \dfrac{x}{2} = \cos \dfrac{x}{2}$

5. $\sin^2 \dfrac{x}{2} - 1 = 0$

6. $\sin x \cos x = \dfrac{1}{4}$

7. $\sin 2x = 2 \cos^2 x$

8. $\csc^2 \dfrac{x}{2} = 2 \sec x$

9. $\cos x - 1 = \cos 2x$

10. $1 - \sin x = \cos 2x$

11. $\sin 2x - \cos x = 0$

12. How many solutions does $\sin x = \frac{1}{2}$ have on $[0, 2\pi)$? How many solutions does $\sin 2x = \frac{1}{2}$ have on $[0, 2\pi)$? Explain graphically.

Solve each equation in part (a) analytically over the interval $[0, 2\pi)$. *Then use a graph to solve each inequality in part (b).*

13. (a) $\cos 2x = \dfrac{\sqrt{3}}{2}$

    (b) $\cos 2x > \dfrac{\sqrt{3}}{2}$

14. (a) $\cos 2x = -\dfrac{1}{2}$

    (b) $\cos 2x > -\dfrac{1}{2}$

15. (a) $\sin 3x = -1$

    (b) $\sin 3x < -1$

16. (a) $\sin 3x = 0$

    (b) $\sin 3x < 0$

17. (a) $\sqrt{2} \cos 2x = -1$

    (b) $\sqrt{2} \cos 2x \le -1$

18. (a) $2\sqrt{3} \sin 2x = \sqrt{3}$

    (b) $2\sqrt{3} \sin 2x \le \sqrt{3}$

19. (a) $\sin \dfrac{x}{2} = \sqrt{2} - \sin \dfrac{x}{2}$

    (b) $\sin \dfrac{x}{2} > \sqrt{2} - \sin \dfrac{x}{2}$

20. (a) $\sin x = \sin 2x$

    (b) $\sin x > \sin 2x$

Solve each equation (x in radians and θ in degrees) for all exact solutions where appropriate. Round approximate values in radians to four decimal places and approximate values in degrees to the nearest tenth. Write answers using the least possible nonnegative angle measures.

21. $\sqrt{2} \sin 3x - 1 = 0$

22. $-2 \cos 2x = \sqrt{3}$

23. $\cos \dfrac{\theta}{2} = 1$

24. $\sin \dfrac{\theta}{2} = 1$

25. $2\sqrt{3} \sin \dfrac{x}{2} = 3$

26. $2\sqrt{3} \cos \dfrac{x}{2} = -3$

27. $2 \sin \theta = 2 \cos 2\theta$

28. $\cos \theta - 1 = \cos 2\theta$

29. $1 - \sin x = \cos 2x$

30. $\sin 2x = 2 \cos^2 x$

31. $3 \csc^2 \dfrac{x}{2} = 2 \sec x$

32. $\cos x = \sin^2 \dfrac{x}{2}$

33. $2 - \sin 2\theta = 4 \sin 2\theta$

34. $4 \cos 2\theta = 8 \sin \theta \cos \theta$

35. $2 \cos^2 2\theta = 1 - \cos 2\theta$

36. $\sin \theta - \sin 2\theta = 0$

Use a graphical method to solve each equation over the interval $[0, 2\pi)$. *Round values to the nearest thousandth.*

37. $\sin x + \sin 3x = \cos x$

38. $\sin 3x - \sin x = 0$

39. $\cos 2x + \cos x = 0$

40. $\sin 4x + \sin 2x = 2 \cos x$

41. $\cos \dfrac{x}{2} = 2 \sin 2x$

42. $\sin \dfrac{x}{2} + \cos 3x = 0$

*Use the method of **Example 4** to solve each equation over the interval* $[0, 2\pi)$. *Round values to the nearest thousandth when appropriate.*

43. $\sec 2x + \tan 2x = 2$

44. $\csc 2x + \cot 2x = 1$

45. $\cos 2x = 1 - \sin 2x$

46. $\sin 3x + \cos 3x = -\dfrac{9}{5}$

47. What is *wrong* with the following solution? Solve the equation $\tan 2\theta = 2$ over the interval $[0, 2\pi)$.

$$\tan 2\theta = 2$$
$$\dfrac{\tan 2\theta}{2} = \dfrac{2}{2}$$
$$\tan \theta = 1$$
$$\theta = \dfrac{\pi}{4} \quad \text{or} \quad \theta = \dfrac{5\pi}{4}$$

The solutions are $\dfrac{\pi}{4}$ and $\dfrac{5\pi}{4}$.

48. The equation $\cot \dfrac{x}{2} - \csc \dfrac{x}{2} - 1 = 0$ has no solution over the interval $[0, 2\pi)$.

    (a) From this information, what can be said about the graph of $y = \cot \dfrac{x}{2} - \csc \dfrac{x}{2} - 1$ over this interval?

(b) Confirm your answer from part (a) by actually graphing the function over the interval.

(Modeling) Solve each problem.

49. Inducing Voltage A coil of wire rotating in a magnetic field induces a voltage

$$V = 20 \sin\left(\dfrac{\pi t}{4} - \dfrac{\pi}{2}\right),$$

where t is time in seconds. Find the least positive time required to produce each voltage.

    (a) 0   **(b)** $10\sqrt{3}$

50. *Nautical Mile*   The British nautical mile is defined as the length L of a minute of arc on any meridian. Since Earth is flatter at its poles, the nautical mile varies with latitude and, in feet, is

$$L = 6077 - 31 \cos 2\theta,$$

where θ is the latitude in degrees. (See the figure.) (*Source:* Bushaw, Donald et al., *A Sourcebook of Applications of School Mathematics*, The Mathematical Association of America. Reprinted by permission.)

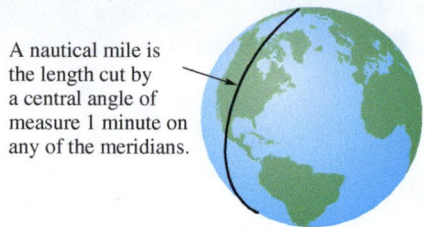

A nautical mile is the length cut by a central angle of measure 1 minute on any of the meridians.

(a) Find the latitude between $0°$ and $90°$ at which the nautical mile is 6074 feet.

(b) At what latitude between $0°$ and $90°$ (inclusive) is the nautical mile 6108 feet?

(c) In the United States, the nautical mile is defined everywhere as 6080.2 feet. At what latitude between $0°$ and $90°$ does this agree with the British nautical mile?

51. *Ear Pressure from a Pure Tone*   A pure tone has a constant frequency and amplitude that sounds rather dull and uninteresting. The pressures caused by pure tones on the eardrum are sinusoidal. The change in pressure P in pounds per square feet on a person's eardrum from a pure tone at time t in seconds can be modeled by the equation

$$P = A \sin(2\pi f t + \phi),$$

where f is the frequency in cycles per second and ϕ is the phase angle. When P is positive, there is an increase in pressure and the eardrum is pushed inward; when P is negative, there is a decrease in pressure and the eardrum is pushed outward. (*Source:* Roederer, J., *Introduction to the Physics and Psychophysics of Music*, The English Universities Press.)

(a) Middle C has frequency 261.63 cycles per second. Graph this tone with $A = 0.004$ and $\phi = \frac{\pi}{7}$ in the window $[0, 0.005]$ by $[-0.005, 0.005]$.

(b) Determine analytically the values of t for which $P = 0$ on $[0, 0.005]$, and support your answers graphically.

(c) Determine graphically when $P < 0$ on $[0, 0.005]$.

(d) Would an eardrum hearing this tone be vibrating outward or inward when $P < 0$?

52. *Ear Pressure from a Vibrating String*   If a string with a fundamental frequency of 110 Hz is plucked in the middle, it will vibrate at the odd harmonics of 110, 330, 550, ... Hz but not at the even harmonics of 220, 440, 660, ... Hz. The resulting pressure P caused by the string can be approximated with the equation

$$P = 0.003 \sin 220\pi t + \frac{0.003}{3} \sin 660\pi t$$
$$+ \frac{0.003}{5} \sin 1100\pi t + \frac{0.003}{7} \sin 1540\pi t.$$

(*Source:* Benade, A., *Fundamentals of Musical Acoustics*, Oxford University Press; Roederer, J., *Introduction to the Physics and Psychophysics of Music*, The English Universities Press.)

(a) Graph P in the window $[0, 0.03]$ by $[-0.005, 0.005]$.

(b) Describe the shape of the sound wave that is produced.

(c) At lower frequencies, the inner ear will hear a tone only when the eardrum is moving outward. (See **Exercise 51.**) Determine the times on the interval $[0, 0.03]$ when this will occur.

53. *Hearing Beats in Music*   Musicians sometimes tune instruments by playing the same tone on two different instruments and listening for a phenomenon known as **beats.** When the two instruments are in tune, the beats disappear. The ear hears beats because the pressure slowly rises and falls as a result of the slight variation in the frequency. This phenomenon can be seen on a graphing calculator. (*Source:* Pierce, J., *The Science of Musical Sound*, Scientific American Books.)

(a) Consider two tones with frequencies of 220 and 223 Hz and pressures

$$P_1 = 0.005 \sin 440\pi t \quad \text{and} \quad P_2 = 0.005 \sin 446\pi t,$$

respectively. Graph the pressure $P = P_1 + P_2$ felt by an eardrum over the 1-second interval $[0.15, 1.15]$. How many beats are there in 1 second?

(b) Repeat part (a) with frequencies of 220 and 216.

(c) Determine a simple way to find the number of beats per second if the frequency of each tone is given.

54. *Hearing Different Tones*   When a musical instrument creates a tone of 110 Hz, it also creates tones at 220, 330, 440, 550, 660, ... Hz. A small speaker cannot reproduce the 110-Hz vibration, but it can reproduce the higher frequencies, called the **upper harmonics.** The low tones can still be heard, because the speaker produces **difference tones** of the upper harmonics. The difference between consecutive frequencies is 110 Hz, and this difference tone will be heard by a listener. We can model this phenomenon with a graphing calculator. (*Source:* Benade, A., *Fundamentals of Musical Acoustics*, Oxford University Press.)

(*continued*)

(a) In the window $[0, 0.03]$ by $[-1, 1]$, graph the upper harmonics represented by the pressure

$$P = \frac{1}{2}\sin[2\pi(220)t] + \frac{1}{3}\sin[2\pi(330)t]$$
$$+ \frac{1}{4}\sin[2\pi(440)t].$$

(b) Estimate all t-coordinates where P is maximum.

(c) What does a person hear in addition to the frequencies of 220, 330, and 440 Hz?

(d) Graph the pressure produced by a speaker that can vibrate at 110 Hz and above in the window $[0, 0.03]$ by $[-2, 2]$.

SECTIONS 9.5–9.6   Reviewing Basic Concepts

Solve each equation over the interval $[0, 2\pi)$.

1. $\cos 2x = \dfrac{\sqrt{3}}{2}$

2. $2\sin x + 1 = 0$

Solve each equation for all real solutions.

3. $(\tan x - 1)(\cos x - 1) = 0$

4. $2\cos^2 x = \sqrt{3}\cos x$

Solve each equation over the interval $[0°, 360°)$. Round values to the nearest tenth when exact values cannot be determined.

5. $3\cot^2 \theta - 3\cot \theta = 1$

6. $4\cos^2 \theta + 4\cos \theta - 1 = 0$

7. $2\sin \theta - 1 = \csc \theta$

8. $\sec^2 \dfrac{\theta}{2} = 2$

Solve each equation graphically to the nearest thousandth over the interval $[0, 2\pi)$.

9. $x^2 + \sin x - x^3 - \cos x = 0$

10. $x^3 - \cos^2 x = \dfrac{1}{2}x - 1$

---

9 Summary

KEY TERMS & SYMBOLS

KEY CONCEPTS

9.1 Trigonometric Identities

FUNDAMENTAL IDENTITIES

Reciprocal Identities

$$\cot \theta = \frac{1}{\tan \theta} \qquad \sec \theta = \frac{1}{\cos \theta} \qquad \csc \theta = \frac{1}{\sin \theta}$$

Quotient Identities

$$\tan \theta = \frac{\sin \theta}{\cos \theta} \qquad \cot \theta = \frac{\cos \theta}{\sin \theta}$$

Pythagorean Identities

$$\sin^2 \theta + \cos^2 \theta = 1 \qquad 1 + \tan^2 \theta = \sec^2 \theta$$
$$1 + \cot^2 \theta = \csc^2 \theta$$

Negative-Number Identities

$$\sin(-\theta) = -\sin \theta \qquad \cos(-\theta) = \cos \theta \qquad \tan(-\theta) = -\tan \theta$$
$$\csc(-\theta) = -\csc \theta \qquad \sec(-\theta) = \sec \theta \qquad \cot(-\theta) = -\cot \theta$$

KEY TERMS & SYMBOLS	KEY CONCEPTS
9.2 Sum and Difference Identities	**SUM AND DIFFERENCE IDENTITIES** $$\cos(A - B) = \cos A \cos B + \sin A \sin B$$ $$\cos(A + B) = \cos A \cos B - \sin A \sin B$$ $$\sin(A + B) = \sin A \cos B + \cos A \sin B$$ $$\sin(A - B) = \sin A \cos B - \cos A \sin B$$ $$\tan(A + B) = \frac{\tan A + \tan B}{1 - \tan A \tan B} \qquad \tan(A - B) = \frac{\tan A - \tan B}{1 + \tan A \tan B}$$
9.3 Further Identities	**DOUBLE-NUMBER IDENTITIES** $$\cos 2A = \cos^2 A - \sin^2 A \qquad \cos 2A = 1 - 2\sin^2 A$$ $$\cos 2A = 2\cos^2 A - 1 \qquad \sin 2A = 2\sin A \cos A$$ $$\tan 2A = \frac{2\tan A}{1 - \tan^2 A}$$ **PRODUCT-TO-SUM IDENTITIES** $$\cos A \cos B = \frac{1}{2}\left[\cos(A + B) + \cos(A - B)\right]$$ $$\sin A \sin B = \frac{1}{2}\left[\cos(A - B) - \cos(A + B)\right]$$ $$\sin A \cos B = \frac{1}{2}\left[\sin(A + B) + \sin(A - B)\right]$$ $$\cos A \sin B = \frac{1}{2}\left[\sin(A + B) - \sin(A - B)\right]$$ **SUM-TO-PRODUCT IDENTITIES** $$\sin A + \sin B = 2\sin\left(\frac{A + B}{2}\right)\cos\left(\frac{A - B}{2}\right)$$ $$\sin A - \sin B = 2\cos\left(\frac{A + B}{2}\right)\sin\left(\frac{A - B}{2}\right)$$ $$\cos A + \cos B = 2\cos\left(\frac{A + B}{2}\right)\cos\left(\frac{A - B}{2}\right)$$ $$\cos A - \cos B = -2\sin\left(\frac{A + B}{2}\right)\sin\left(\frac{A - B}{2}\right)$$ **HALF-NUMBER IDENTITIES** In the following identities, the symbol $\pm$ indicates that the sign is chosen based on the function under consideration and the quadrant of $\frac{A}{2}$. $$\cos\frac{A}{2} = \pm\sqrt{\frac{1 + \cos A}{2}} \qquad \sin\frac{A}{2} = \pm\sqrt{\frac{1 - \cos A}{2}}$$ $$\tan\frac{A}{2} = \pm\sqrt{\frac{1 - \cos A}{1 + \cos A}} \qquad \tan\frac{A}{2} = \frac{\sin A}{1 + \cos A} \qquad \tan\frac{A}{2} = \frac{1 - \cos A}{\sin A}$$

(*continued*)

KEY TERMS & SYMBOLS	KEY CONCEPTS

9.4 The Inverse Circular Functions

$\sin^{-1} x$ or $\arcsin x$
$\cos^{-1} x$ or $\arccos x$
$\tan^{-1} x$ or $\arctan x$
$\cot^{-1} x$ or $\text{arccot } x$
$\sec^{-1} x$ or $\text{arcsec } x$
$\csc^{-1} x$ or $\text{arccsc } x$

INVERSE CIRCULAR FUNCTIONS

		Range	
Inverse Function	**Domain**	**Interval**	**Quadrants of the Unit Circle**
$y = \sin^{-1} x$	$[-1, 1]$	$\left[-\frac{\pi}{2}, \frac{\pi}{2}\right]$	I and IV
$y = \cos^{-1} x$	$[-1, 1]$	$[0, \pi]$	I and II
$y = \tan^{-1} x$	$(-\infty, \infty)$	$\left(-\frac{\pi}{2}, \frac{\pi}{2}\right)$	I and IV
$y = \cot^{-1} x$	$(-\infty, \infty)$	$(0, \pi)$	I and II
$y = \sec^{-1} x$	$(-\infty, -1] \cup [1, \infty)$	$\left[0, \frac{\pi}{2}\right) \cup \left(\frac{\pi}{2}, \pi\right]$	I and II
$y = \csc^{-1} x$	$(-\infty, -1] \cup [1, \infty)$	$\left[-\frac{\pi}{2}, 0\right) \cup \left(0, \frac{\pi}{2}\right]$	I and IV

9.5 Trigonometric Equations and Inequalities (I)

SOLVING A TRIGONOMETRIC EQUATION ANALYTICALLY

1. Decide whether the equation is linear or quadratic, so you can determine the solution method.

2. If only one trigonometric function is present, first solve the equation for that function.

3. If more than one trigonometric function is present, rearrange the equation so that one side equals 0. Then try to factor and use the zero-product property to solve.

4. If the equation is quadratic in form, but not factorable, use the quadratic formula. Check that solutions are in the desired interval.

5. Try using identities to change the form of the equation. It may be helpful to square each side of the equation first. If this is done, check for extraneous values.

SOLVING A TRIGONOMETRIC EQUATION GRAPHICALLY

1. For an equation of the form $f(x) = g(x)$, use the intersection-of-graphs method.

2. For an equation of the form $f(x) = 0$, use the x-intercept method.

9.6 Trigonometric Equations and Inequalities (II)

Use the multiple-number and half-number identities to rewrite the equations. Then use the steps given above for **Section 9.5**.

9 Review Exercises

1. Give all the trigonometric functions f that satisfy the condition $f(-x) = -f(x)$.

2. Give all the trigonometric functions f that satisfy the condition $f(-x) = f(x)$.

Use the negative-number identities to write each expression as a function of a positive number.

3. $\cos(-3)$ **4.** $\sin(-3)$ **5.** $\tan(-3)$ **6.** $\sec(-3)$ **7.** $\csc(-3)$ **8.** $\cot(-3)$

Concept Check   *For each expression in Group I, give the letter of the expression in Group II that completes an identity.*

I

9. $\sec x = $ _____

10. $\tan x = $ _____

11. $\cot x = $ _____

12. $\tan^2 x + 1 = $ _____

13. $\tan^2 x = $ _____

14. $\csc x = $ _____

II

A. $\dfrac{1}{\sin x}$ **B.** $\dfrac{1}{\cos x}$

C. $\dfrac{\sin x}{\cos x}$ **D.** $\dfrac{1}{\cot^2 x}$

E. $\dfrac{1}{\cos^2 x}$ **F.** $\dfrac{\cos x}{\sin x}$

Use identities to write each expression in terms of $\sin \theta$ and $\cos \theta$, and simplify.

15. $\dfrac{\cot \theta}{\sec \theta}$

16. $\tan^2 \theta(1 + \cot^2 \theta)$

17. $\csc \theta + \cot \theta$

18. Use the trigonometric identities to find the remaining five trigonometric functions of x, given that $\cos x = \frac{3}{5}$ and x is in $\left(\frac{3\pi}{2}, 2\pi\right)$.

19. Given that $\tan x = -\frac{5}{4}$, where $\frac{\pi}{2} < x < \pi$, use the trigonometric identities to find the other trigonometric functions of x.

Concept Check   *For each expression in Group I, choose the expression from Group II with the same value.*

I

20. $\cos 210°$ **21.** $\sin 35°$

22. $\tan(-35°)$ **23.** $-\sin 35°$

24. $\cos 35°$ **25.** $\cos 75°$

26. $\sin 75°$ **27.** $\sin 300°$

28. $\cos 300°$ **29.** $\tan(-55°)$

II

A. $\sin(-35°)$

B. $\cos 55°$

C. $\sqrt{\dfrac{1 + \cos 150°}{2}}$

D. $2 \sin 150° \cos 150°$

E. $\cos 150° \cos 60° - \sin 150° \sin 60°$

F. $\cot(-35°)$

G. $\cos^2 150° - \sin^2 150°$

H. $\sin 15° \cos 60° + \cos 15° \sin 60°$

I. $\cos(-35°)$

J. $\cot 125°$

Verify analytically that each equation is an identity.

30. $\sin^2 x - \sin^2 y = \cos^2 y - \cos^2 x$

31. $2 \cos^3 x - \cos x = \dfrac{\cos^2 x - \sin^2 x}{\sec x}$

32. $-\cot \dfrac{x}{2} = \dfrac{\sin 2x + \sin x}{\cos 2x - \cos x}$

33. $\dfrac{\sin^2 x}{2 - 2 \cos x} = \cos^2 \dfrac{x}{2}$

34. $\dfrac{\sin 2x}{\sin x} = \dfrac{2}{\sec x}$

35. $2 \cos A - \sec A = \cos A - \dfrac{\tan A}{\csc A}$

36. $\dfrac{2 \tan B}{\sin 2B} = \sec^2 B$

37. $1 + \tan^2 \alpha = 2 \tan \alpha \csc 2\alpha$

38. $\dfrac{\sin t}{1 - \cos t} = \cot \dfrac{t}{2}$

39. $\dfrac{2 \cot x}{\tan 2x} = \csc^2 x - 2$

40. $\tan \theta \sin 2\theta = 2 - 2 \cos^2 \theta$

41. $2 \tan x \csc 2x - \tan^2 x = 1$

Give the exact real number value of each expression without using a calculator.

42. $\sin^{-1} \dfrac{\sqrt{2}}{2}$

43. $\arccos\left(-\dfrac{1}{2}\right)$

44. $\arctan \dfrac{\sqrt{3}}{3}$

45. $\sec^{-1}(-2)$

46. $\text{arccsc} \dfrac{2\sqrt{3}}{3}$

47. $\cot^{-1}(-1)$

Give the degree measure of θ without using a calculator.

48. $\theta = \arccos \dfrac{1}{2}$

49. $\theta = \arcsin\left(-\dfrac{\sqrt{3}}{2}\right)$

50. $\theta = \tan^{-1} 0$

Use a calculator to give the degree of measure of θ.

51. $\theta = \arcsin(-0.656059029)$

52. $\theta = \arccos 0.7095707365$

53. $\theta = \arctan(-0.1227845609)$

54. $\theta = \cot^{-1} 4.704630109$

55. $\theta = \sec^{-1} 28.65370835$

56. $\theta = \csc^{-1} 19.10732261$

 57. Explain why $\sin^{-1}(-3)$ is not defined.

58. Even though $\sin^{-1}(-3)$ is not defined, $\tan^{-1}(-3)$ is defined. Explain why this is so.

Evaluate each expression without a calculator.

59. $\sin\left(\sin^{-1} \dfrac{1}{2}\right)$

60. $\sin\left(\cos^{-1} \dfrac{3}{4}\right)$

61. $\cos(\arctan 3)$

62. $\sec\left(2 \sin^{-1}\left(-\dfrac{1}{3}\right)\right)$

63. $\cos^{-1}\left(\cos \dfrac{3\pi}{2}\right)$

64. $\tan\left(\sin^{-1} \dfrac{3}{5} + \cos^{-1} \dfrac{5}{7}\right)$

Write each expression as an algebraic expression in terms of u, u > 0.

65. $\sin(\tan^{-1} u)$

66. $\cos\left(\arctan \dfrac{u}{\sqrt{1 - u^2}}\right)$

67. $\tan\left(\arccos \dfrac{u}{\sqrt{u^2 + 1}}\right)$

Solve each equation for solutions in the interval $[0, 2\pi)$. Approximate when appropriate.

68. $\sin^2 x = 1$

69. $2 \tan x - 1 = 0$

70. $3 \sin^2 x - 5 \sin x + 2 = 0$

71. $\tan x = \cot x$

72. $5 \cot^2 x + 3 \cot x = 2$

73. $\sec \dfrac{x}{2} = \cos \dfrac{x}{2}$

74. $\sin 2x = \cos 2x + 1$

75. $2 \sin 2x = 1$

76. $\sin 2x + \sin 4x = 0$

77. $\cos x - \cos 2x = 2 \cos x$

Solve each equation for all solutions.

78. $\tan 2x = \sqrt{3}$

79. $\cos^2 \dfrac{x}{2} - 2 \cos \dfrac{x}{2} + 1 = 0$

80. Use the results of **Exercise 74** and a calculator graph of
$$f(x) = \sin 2x - \cos 2x - 1$$
to find the solution set of each inequality over the interval $[0, 2\pi)$.

(a) $\sin 2x > \cos 2x + 1$

(b) $\sin 2x < \cos 2x + 1$

(Modeling) *Solve each problem.*

81. *Viewing Angle of an Observer* A student is sitting in a desk 5 feet to the left of a 10-foot-wide blackboard, as shown in the figure in the next column. If the student is sitting x feet from the front of the classroom and has a viewing angle of θ radians, do the following.

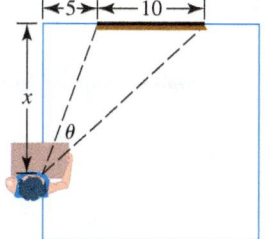

(a) Show that the value of θ is $\theta = f(x)$, where
$$f(x) = \arctan\left(\dfrac{15}{x}\right) - \arctan\left(\dfrac{5}{x}\right).$$

(b) Graph this function with a graphing calculator to estimate the value of x that maximizes the viewing angle.

82. Tone Heard by a Listener When two sources located at different positions produce the same pure tone, the human ear will often hear one sound that is equal to the sum of the individual tones. Since the sources are at different locations, they will have different phase angles ϕ. If two speakers located at different positions produce pure tones $P_1 = A_1 \sin(2\pi ft + \phi_1)$ and $P_2 = A_2 \sin(2\pi ft + \phi_2)$, where $-\frac{\pi}{4} \leq \phi_1, \phi_2 \leq \frac{\pi}{4}$, then the resulting tone heard by a listener can be written as $P = A \sin(2\pi ft + \phi)$, where

$$A = \sqrt{(A_1 \cos \phi_1 + A_2 \cos \phi_2)^2 + (A_1 \sin \phi_1 + A_2 \sin \phi_2)^2}$$

$$\text{and } \phi = \arctan\left(\frac{A_1 \sin \phi_1 + A_2 \sin \phi_2}{A_1 \cos \phi_1 + A_2 \cos \phi_2}\right).$$

(*Source:* Fletcher, N. and T. Rossing, *The Physics of Musical Instruments*, Second Edition, Springer-Verlag.)

(a) Calculate A and ϕ if $A_1 = 0.0012$, $\phi_1 = 0.052$, $A_2 = 0.004$, and $\phi_2 = 0.61$. Also find an expression for $P = A \sin(2\pi ft + \phi)$ if $f = 220$.

(b) Graph $Y_1 = P$ and $Y_2 = P_1 + P_2$ in the same coordinate axes over the interval $[0, 0.01]$. Are the two graphs the same?

Alternating Electric Current *The study of alternating electric current requires the solutions of equations of the form $I = I_{max} \sin 2\pi ft$, for time t in seconds, where I is instantaneous current in amperes, I_{max} is maximum current in amperes, and f is the number of cycles per second. Approximate the least positive value of t, given the following data.*

83. $I = 40$, $I_{max} = 100$, $f = 60$

84. $I = 50$, $I_{max} = 100$, $f = 120$

85. Speed of Light Snell's law states that

$$\frac{c_1}{c_2} = \frac{\sin \theta_1}{\sin \theta_2},$$

where c_1 is the speed of light in one medium, c_2 is the speed of light in a second medium, and θ_1 and θ_2 are the angles shown in the figure. Suppose a light is shining up through water into the air as in the figure. As θ_1 increases, θ_2 approaches $90°$, at which point no light will emerge from the water. Assume that the ratio $\frac{c_1}{c_2}$ in this case is 0.752. For what value of θ_1 does $\theta_2 = 90°$? This value of θ_1 is called the *critical angle* for water.

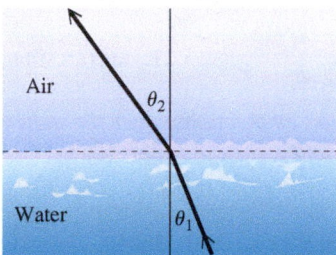

86. Speed of Light Refer to **Exercise 85.** What happens when θ_1 is greater than the critical angle?

9 Test

Remember to set your calculator for radian or degree measure as required.

Given that $\sin y = -\frac{3}{5}$, $\cos x = -\frac{4}{5}$, $\frac{\pi}{2} < x < \pi$, and $\pi < y < \frac{3\pi}{2}$, find the exact values for the following.

1. $\sin(x + y)$

2. $\cos(x - y)$

3. $\tan \dfrac{y}{2}$

4. $\cos 2x$

5. Express $\tan^2 x - \sec^2 x$ in terms of $\sin x$ and $\cos x$, and simplify.

6. Graph $y = \sec x - \sin x \tan x$, and use the graph to conjecture as to what might be an identity. Verify your conjecture analytically.

Verify that each equation is an identity.

7. $\sec^2 B = \dfrac{1}{1 - \sin^2 B}$

8. $\cos 2A = \dfrac{\cot A - \tan A}{\csc A \sec A}$

9. $\dfrac{\cos x + 1}{\sin x + \tan x} = \cot x$

10. $\dfrac{\sec^2 x - 1}{1 + \tan^2 x} = \sin^2 x$

Use an identity to write each expression as a trigonometric function of θ alone.

11. $\cos(270° - \theta)$ **12.** $\sin(\pi + \theta)$

13. Consider the function

$$f(x) = -2 \sin^{-1} x.$$

   (a) Sketch the graph.

   (b) Give the domain and range.

 (c) Explain why $f(2)$ is not defined.

Find the exact real number value of each expression.

14. $\arccos\left(-\dfrac{1}{2}\right)$ **15.** $\tan^{-1} 0$

16. $\csc^{-1} \dfrac{2\sqrt{3}}{3}$ **17.** $\sin^{-1}\left(\sin \dfrac{5\pi}{6}\right)$

18. $\cos\left(\arcsin \dfrac{2}{3}\right)$ **19.** $\sin\left(2 \cos^{-1} \dfrac{1}{3}\right)$

Write each expression as an algebraic expression in terms of u, u > 0.

20. $\sec(\cos^{-1} u)$ **21.** $\tan(\arcsin u)$

Solve each equation or inequality over the indicated interval. Round to nearest tenth when appropriate.

22. $\sin^2 x = \cos^2 x + 1$, $[0, 2\pi)$

23. $\csc^2 \theta - 2 \cot \theta = 4$, $[0°, 360°)$

24. $\cos x = \cos 2x$, $[0, 2\pi)$

25. $2\sqrt{3} \sin \dfrac{\theta}{2} = 3$, $[0°, 360°)$

26. $2 \sin x - 1 \le 0$, $[0, 2\pi)$

27. Solve each equation for all real solutions.
   (a) $2 \sin x + 1 = 2$
   (b) $\tan x + \sec x = 1$
   (c) $\cos^2 x - \sin^2 x = \dfrac{1}{2}$

28. *(Modeling) Migratory Animal Count*   The number T of migratory animals (in hundreds) seen in one year and counted at a certain checkpoint is

$$T(t) = 50 + 50 \cos\left(\dfrac{\pi}{6}t\right),$$

where t is time in months, with $t = 0$ corresponding to the month of July.
   (a) Give the domain and range of T.
   (b) Graph $T(t)$ over $[0, 12]$.
   (c) Use the graph to determine the maximum and minimum values of T and when they occur.
   (d) Find $T(3)$ analytically, and support your result graphically. Use symmetry to find $T(9)$.
   (e) When does the count reach 75 hundred animals?
   (f) Write the equation

$$T = 50 + 50 \cos\left(\dfrac{\pi}{6}t\right)$$

as an equation involving arccosine by solving for t.

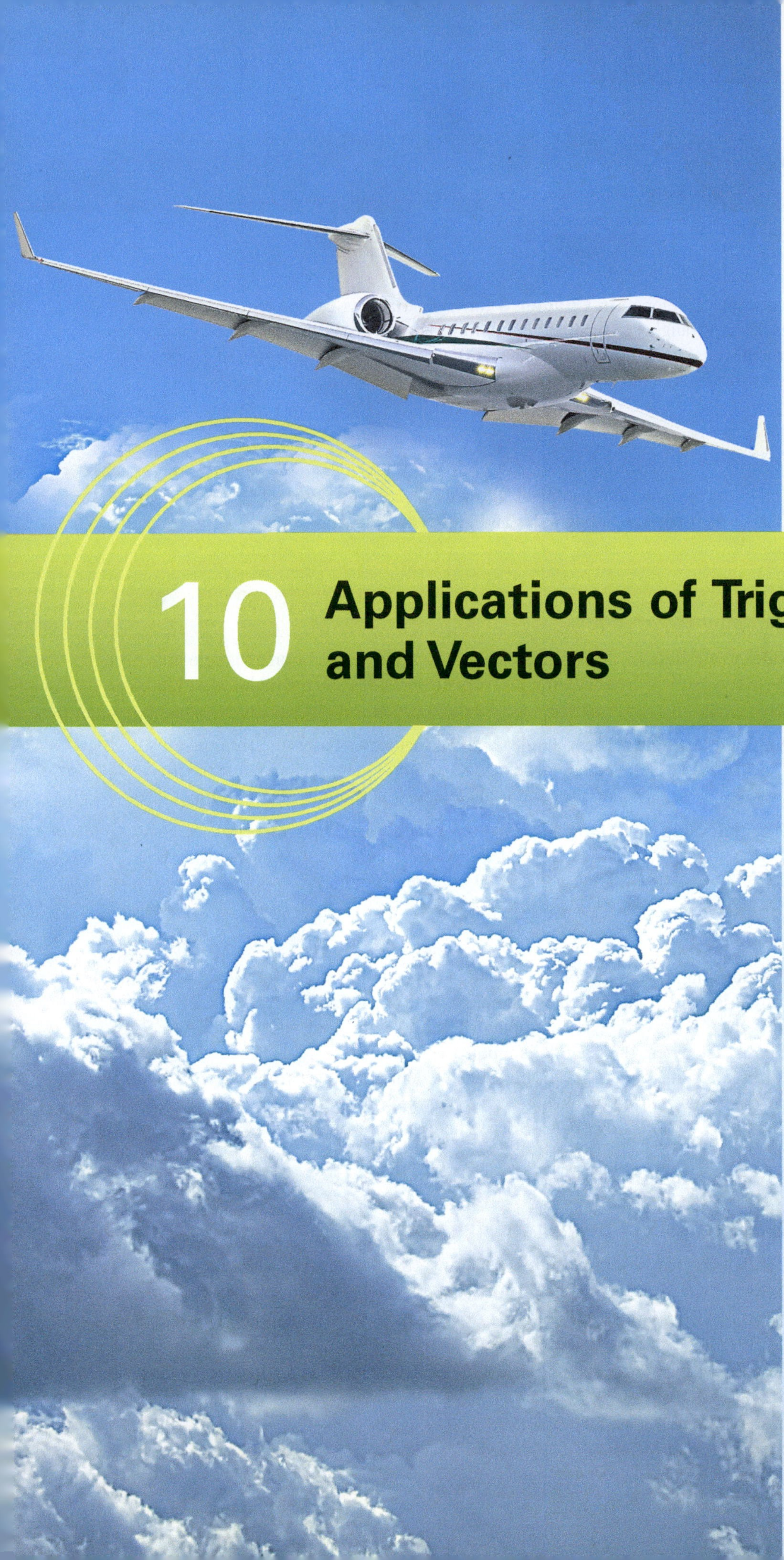

In this chapter, we use trigonometry and vectors to solve problems, including those involving navigation, weight and force balancing, and distance and angle determination.

10 Applications of Trigonometry and Vectors

10.1 The Law of Sines

Congruency and Oblique Triangles • Derivation of the Law of Sines • Using the Law of Sines • Ambiguous Case

The concepts developed in **Chapter 8** for right triangles can be extended to *all* triangles. Every triangle has three sides and three angles. If any three of the six measures of a triangle are known (provided that at least one measure is the length of a side), then the other three measures can be found. This is called **solving the triangle.**

Congruency and Oblique Triangles

The following axioms from geometry allow us to prove that two triangles are congruent (that is, their corresponding sides and angles are equal).

Congruent Triangles

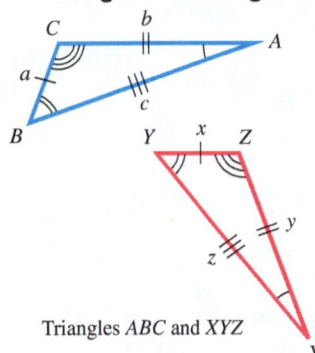

Triangles *ABC* and *XYZ*

Congruence Axioms

Side-Angle-Side (SAS)	If two sides and the included angle of one triangle are equal, respectively, to two sides and the included angle of a second triangle, then the triangles are congruent.
Angle-Side-Angle (ASA)	If two angles and the included side of one triangle are equal, respectively, to two angles and the included side of a second triangle, then the triangles are congruent.
Side-Side-Side (SSS)	If three sides of one triangle are equal, respectively, to three sides of a second triangle, then the triangles are congruent.

If a side and two angles, one of which is opposite the side, are given **(SAA),** the third angle is easily determined by the angle sum formula $(A + B + C = 180°)$, and then the ASA axiom can be applied. Keep in mind that whenever SAA, SAS, ASA, or SSS is given, the triangle is unique.

A triangle that is not a right triangle is called an **oblique triangle.** *The measures of the three sides and the three angles of a triangle can be found if at least one side and any other two measures are known.* There are four possible cases.

Data Required for Solving Oblique Triangles

Case 1 One side and two angles are known (SAA or ASA).

Case 2 Two sides and one angle not included between the two sides are known (SSA). This case may lead to zero, one, or two triangles. This is called the *ambiguous case.*

Case 3 Two sides and the angle included between the two sides are known (SAS).

Case 4 Three sides are known (SSS).

NOTE   *If we know three angles of a triangle, we cannot find unique side lengths, since AAA assures us only of similarity, NOT congruence.* For example, there are infinitely many triangles ABC of different sizes with $A = 35°$, $B = 65°$, and $C = 80°$.

Cases 1 and 2, discussed in this section, require the **law of sines.** Cases 3 and 4, discussed in the next section, require the **law of cosines.**

Derivation of the Law of Sines

Oblique Triangles

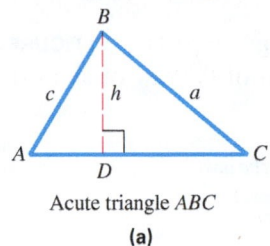

Acute triangle ABC

(a)

To derive the law of sines, we start with an oblique triangle, such as the **acute triangle** in **FIGURE 1(a)** or the **obtuse triangle** in **FIGURE 1(b).** First, construct the perpendicular from B to side AC (or its extension). Let h be the length of this perpendicular. Then c is the hypotenuse of right triangle ADB, and a is the hypotenuse of right triangle BDC.

In triangle ADB, $\sin A = \dfrac{h}{c},$ or $h = c \sin A.$

In triangle BDC, $\sin C = \dfrac{h}{a},$ or $h = a \sin C.$

Since $h = c \sin A$ and $h = a \sin C$,

$$a \sin C = c \sin A$$

$$\frac{a}{\sin A} = \frac{c}{\sin C}. \qquad \text{Divide each side of the equation by } \sin A \sin C.$$

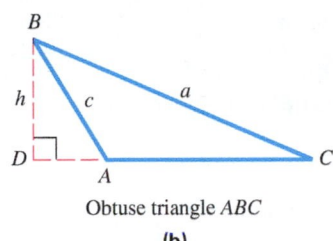

Obtuse triangle ABC

(b)

By constructing perpendicular lines from the other vertices, it can be shown that

$$\frac{a}{\sin A} = \frac{b}{\sin B} \quad \text{and} \quad \frac{b}{\sin B} = \frac{c}{\sin C}.$$

This discussion is summarized by the following theorem.

We label oblique triangles as we did right triangles: side a opposite angle A, side b opposite angle B, and side c opposite angle C.

FIGURE 1

Law of Sines

In any triangle ABC with sides a, b, and c, the following hold.

$$\frac{a}{\sin A} = \frac{b}{\sin B}, \qquad \frac{a}{\sin A} = \frac{c}{\sin C}, \qquad \text{and} \qquad \frac{b}{\sin B} = \frac{c}{\sin C}$$

This can be written in compact form as follows.

$$\frac{a}{\sin A} = \frac{b}{\sin B} = \frac{c}{\sin C}$$

That is, the three ratios of the length of a side to the sine of the angle opposite it are all equal.

When solving for an angle, we use an alternative form of the law of sines.

$$\frac{\sin A}{a} = \frac{\sin B}{b} = \frac{\sin C}{c} \qquad \text{Alternative form}$$

NOTE   *In using the law of sines, a good strategy is to select a form so that the unknown variable is in the numerator with all other variables known.*

Using the Law of Sines

If two angles and a side are known (Case 1, SAA or ASA), then the law of sines can be used to solve the triangle.

> **NOTE** In solving triangles in this chapter, we will often use the equality symbol, =, despite the fact that the values shown are actually approximations.

EXAMPLE 1 **Using the Law of Sines to Solve a Triangle (SAA)**

Solve triangle ABC if $A = 32.0°$, $B = 81.8°$, and $a = 42.9$ centimeters.

Solution Draw a triangle, roughly to scale, and label the given parts as in **FIGURE 2.** Since the values of A, B, and a are known, we use the form of the law of sines that involves these variables and then solve for b.

Be sure to label a sketch carefully to help set up the correct equation.

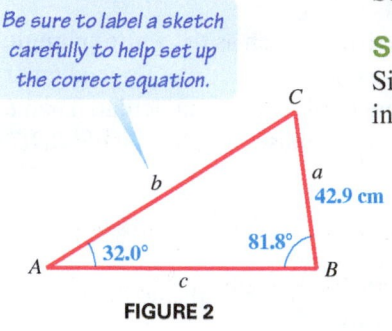

FIGURE 2

Choose a form that has the unknown variable in the numerator.

$$\frac{a}{\sin A} = \frac{b}{\sin B} \qquad \text{Law of sines}$$

$$\frac{42.9}{\sin 32.0°} = \frac{b}{\sin 81.8°} \qquad \text{Substitute the given values.}$$

$$b = \frac{42.9 \sin 81.8°}{\sin 32.0°} \qquad \text{Multiply by } \sin 81.8° \text{ and rewrite.}$$

$$b = 80.1 \text{ centimeters} \qquad \text{Approximate with a calculator.}$$

To find C, use the fact that the sum of the angles of any triangle is 180°.

$$A + B + C = 180° \qquad \text{Angle sum formula}$$

$$C = 180° - A - B \qquad \text{Solve for } C.$$

$$C = 180° - 32.0° - 81.8° \qquad \text{Substitute.}$$

$$C = 66.2° \qquad \text{Subtract.}$$

Use the law of sines to find c.

$$\frac{a}{\sin A} = \frac{c}{\sin C} \qquad \text{Law of sines}$$

$$\frac{42.9}{\sin 32.0°} = \frac{c}{\sin 66.2°} \qquad \text{Substitute known values.}$$

$$c = \frac{42.9 \sin 66.2°}{\sin 32.0°} \qquad \text{Multiply by } \sin 66.2° \text{ and rewrite.}$$

$$c = 74.1 \text{ centimeters} \qquad \text{Approximate with a calculator.}$$

EXAMPLE 2 **Using the Law of Sines in an Application (ASA)**

A surveyor wishes to measure the distance across a river. See **FIGURE 3.** She determines that $C = 112.90°$, $A = 31.10°$, and $b = 347.6$ feet. Find the distance a.

Solution To use the law of sines, one side and the angle opposite it must be known. Since b is the only side whose length is given, angle B must be found first.

$$B = 180° - A - C \qquad \text{Angle sum formula, solved for } b$$

$$B = 180° - 31.10° - 112.90° = 36.00° \qquad \text{Substitute known values and subtract.}$$

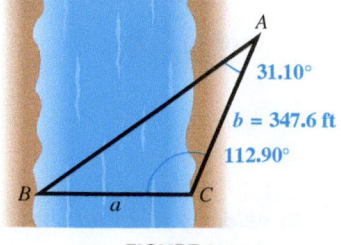

FIGURE 3

Now use the form of the law of sines involving A, B, and b to find a.

Solve for the variable a by using the form with a in the numerator.

$$\frac{a}{\sin A} = \frac{b}{\sin B}$$ Law of sines

$$\frac{a}{\sin 31.10°} = \frac{347.6}{\sin 36.00°}$$ Substitute known values.

$$a = \frac{347.6 \sin 31.10°}{\sin 36.00°}$$ Multiply by $\sin 31.10°$.

$$a = 305.5 \text{ feet}$$ Approximate with a calculator.   ●

EXAMPLE 3   **Using the Law of Sines in an Application (ASA)**

Two ranger stations are on an east–west line 110 mi apart. A forest fire is located on a bearing of N 42° E from the western station at A and a bearing of N 15° E from the eastern station at B. How far is the fire from the western station to the nearest mile?

Solution   **FIGURE 4** shows the two stations at points A and B and the fire at point C. Angle $BAC = 90° - 42° = 48°$, the obtuse angle at B equals $90° + 15° = 105°$, and the third angle, C, equals $180° - 105° - 48° = 27°$. We must find side b.

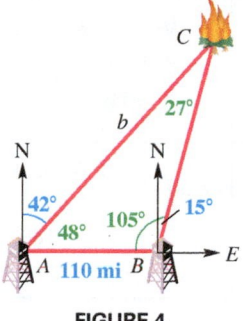

FIGURE 4

Solve for b.

$$\frac{b}{\sin B} = \frac{c}{\sin C}$$ Law of sines

$$\frac{b}{\sin 105°} = \frac{110}{\sin 27°}$$ Substitute known values.

$$b = \frac{110 \sin 105°}{\sin 27°}$$ Multiply by $\sin 105°$.

$$b = 234 \text{ mi}$$ Approximate with a calculator.   ●

Ambiguous Case

Ambiguous Case (SSA)

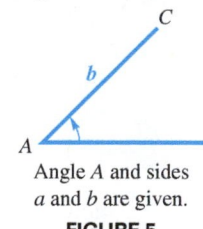

Angle A and sides a and b are given.

FIGURE 5

If we are given the lengths of two sides and the angle opposite one of them (Case 2, SSA), zero, one, or two such triangles could exist. (Notice there is no SSA congruence axiom.)

Suppose we know the measure of acute angle A of triangle ABC, the length of side a, and the length of side b, as shown in **FIGURE 5**. We must draw the side of length a opposite angle A. The table on the next page shows possible outcomes. This situation (SSA) is called the **ambiguous case** of the law of sines.

To determine which outcome applies, keep the following basic facts in mind.

Applying the Law of Sines

1. For any angle θ of a triangle, $0 < \sin \theta \le 1$. If $\sin \theta = 1$, then $\theta = 90°$ and the triangle is a right triangle.

2. $\sin \theta = \sin(180° - \theta)$. (Supplementary angles have the same sine value.)

3. The smallest angle is opposite the shortest side, the largest angle is opposite the longest side, and the middle-valued angle is opposite the intermediate side (assuming that the triangle has sides that are all of different lengths).

If angle A is acute, there are four possible outcomes. If angle A is obtuse, there are two possible outcomes.

Angle A is	Number of Possible Triangles	Sketch	Applying Law of Sines Leads to
Acute	0		$\sin B > 1$, $a < h < b$
Acute	1		$\sin B = 1$, $a = h$ and $h < b$
Acute	1		$0 < \sin B < 1$, $a \geq b$
Acute	2		$0 < \sin B_1 < 1$, $h < a < b$ $A + B_2 < 180°$
Obtuse	0		$\sin B \geq 1$, $a \leq b$
Obtuse	1		$0 < \sin B < 1$, $a > b$

TECHNOLOGY NOTE

Because there are 60′ in 1°, the expression $\sin 55°\ 40'$ in **Example 4** can be evaluated by entering $\sin(55 + 40/60)$ into your calculator. Be sure to use degree mode.

No Triangle

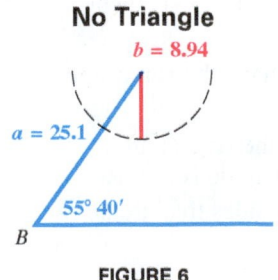

FIGURE 6

EXAMPLE 4   **Solving the Ambiguous Case (No Such Triangle)**

Solve triangle ABC if $B = 55°\ 40'$, $b = 8.94$ meters, and $a = 25.1$ meters.

Solution   Since we are given B, b, and a, we must find angle A.

Choose a form that has the unknown variable in the numerator.

$$\frac{\sin A}{a} = \frac{\sin B}{b} \qquad \text{Law of sines (alternative form)}$$

$$\frac{\sin A}{25.1} = \frac{\sin 55°\ 40'}{8.94} \qquad \text{Substitute the given values.}$$

$$\sin A = \frac{25.1 \sin 55°\ 40'}{8.94} \qquad \text{Multiply by 25.1.}$$

$$\sin A = 2.3184379 \qquad \text{Approximate with a calculator.}$$

From the table above, $\sin A$ cannot be greater than 1, so there can be no such angle A and thus no triangle with the given information. An attempt to sketch such a triangle leads to the situation shown in **FIGURE 6**.

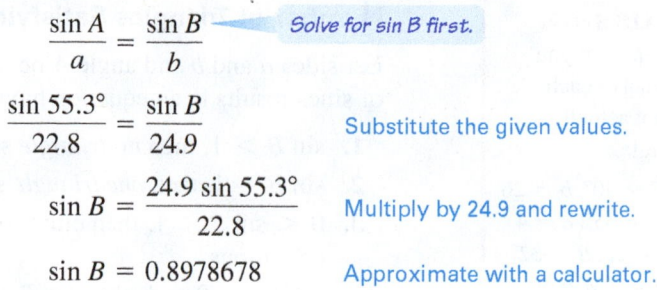

EXAMPLE 5 **Solving the Ambiguous Case (Two Triangles)**

Solve triangle ABC if $A = 55.3°$, $a = 22.8$ feet, and $b = 24.9$ feet.

Solution To begin, use the law of sines to find angle B.

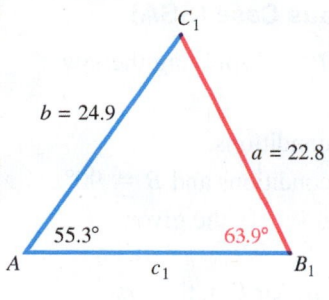

$$\frac{\sin A}{a} = \frac{\sin B}{b} \quad \text{Solve for sin B first.}$$

$$\frac{\sin 55.3°}{22.8} = \frac{\sin B}{24.9} \quad \text{Substitute the given values.}$$

$$\sin B = \frac{24.9 \sin 55.3°}{22.8} \quad \text{Multiply by 24.9 and rewrite.}$$

$$\sin B = 0.8978678 \quad \text{Approximate with a calculator.}$$

Because $0 < \sin B < 1$, there are *two* angles between $0°$ and $180°$ that satisfy this condition. One value of B is

$$B_1 = \sin^{-1} 0.8978678 \quad \text{Use the inverse sine function.}$$

$$B_1 = 63.9°. \quad \text{Approximate with a calculator.}$$

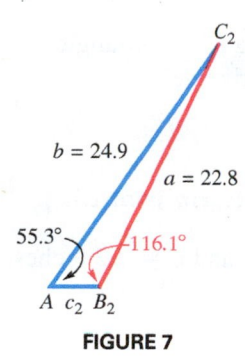

Supplementary angles have the same sine value, so another *possible* value of B is

$$B_2 = 180° - 63.9° = 116.1°.$$

To determine whether B_2 is a valid possibility, check that $A + B_2 < 180°$. Since $55.3° + 116.1° = 171.4°$, it is a valid angle measure for this triangle.

FIGURE 7

Now separately solve triangles AB_1C_1 and AB_2C_2 shown in **FIGURE 7.** Begin with AB_1C_1. Find C_1 first. Then use the law of sines to find c_1.

$$C_1 = 180° - A - B_1 = 180° - 55.3° - 63.9° = 60.8°$$

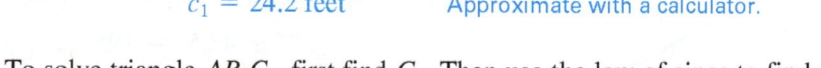

$$\frac{a}{\sin A} = \frac{c_1}{\sin C_1} \quad \text{Law of sines}$$

$$\frac{22.8}{\sin 55.3°} = \frac{c_1}{\sin 60.8°} \quad \text{Substitute for a, A, and } C_1.$$

$$c_1 = \frac{22.8 \sin 60.8°}{\sin 55.3°} \quad \text{Solve for side } c_1.$$

$$c_1 = 24.2 \text{ feet} \quad \text{Approximate with a calculator.}$$

To solve triangle AB_2C_2, first find C_2. Then use the law of sines to find c_2.

$$C_2 = 180° - A - B_2 = 180° - 55.3° - 116.1° = 8.6°$$

$$\frac{a}{\sin A} = \frac{c_2}{\sin C_2} \quad \text{Law of sines}$$

$$\frac{22.8}{\sin 55.3°} = \frac{c_2}{\sin 8.6°} \quad \text{Substitute for a, A, and } C_2.$$

$$c_2 = \frac{22.8 \sin 8.6°}{\sin 55.3°} \quad \text{Solve for side } c_2.$$

$$c_2 = 4.15 \text{ feet} \quad \text{Approximate with a calculator.}$$

The ambiguous case results in zero, one, or two triangles. The following guidelines can be used to determine how many triangles there are.

Number of Triangles Satisfying the Ambiguous Case (SSA)

Let sides a and b and angle A be given in triangle ABC. If applying the law of sines results in an equation having

1. $\sin B > 1$, then *no triangle* satisfies the given conditions.
2. $\sin B = 1$, then *one triangle* satisfies the given conditions and $B = 90°$.
3. $0 < \sin B < 1$, then either *one or two triangles* satisfy the given conditions.
   (a) If $\sin B = k$, then let $B_1 = \sin^{-1} k$ and use B_1 for B in the first triangle.
   (b) Let $B_2 = 180° - B_1$. If $A + B_2 < 180°$, then a second triangle exists. In this case, use B_2 for B in the second triangle.

EXAMPLE 6   Solving the Ambiguous Case (One Triangle)

Solve triangle ABC, given that $A = 43.5°$, $a = 10.7$ inches, and $c = 7.2$ inches.

Solution   We must find angle C.

$$\frac{\sin C}{c} = \frac{\sin A}{a} \qquad \text{Law of sines (alternative form)}$$

$$\frac{\sin C}{7.2} = \frac{\sin 43.5°}{10.7}$$

$$\sin C = \frac{7.2 \sin 43.5°}{10.7} \qquad \text{A calculator gives } 0.46319186 \text{ for sin C.}$$

$$C_1 = 27.6° \qquad \text{Use the inverse sine function.}$$

This is the acute angle. Another value of angle C that has $\sin C = 0.46319186$ is $C_2 = 180° - 27.6° = 152.4°$. Notice in the given information that $c < a$, meaning that angle C must have measure *less than* that of angle A. Notice also that when we add this possible obtuse angle C_2 to the given angle $A = 43.5°$, we obtain

$$152.4° + 43.5° = 195.9°, \qquad \text{Check if } C_2 + A < 180°.$$

which is greater than $180°$. So angle C_2 is not possible and there can be only one triangle, with angle C_1. See **FIGURE 8.** Then

$$B = 180° - 27.6° - 43.5° = 108.9°,$$

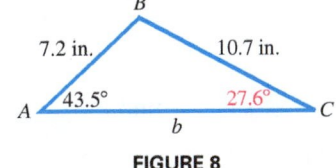

FIGURE 8

and we can find side b with the law of sines.

$$\text{Solve for } b. \qquad \frac{b}{\sin B} = \frac{a}{\sin A} \qquad \text{Law of sines}$$

$$\frac{b}{\sin 108.9°} = \frac{10.7}{\sin 43.5°} \qquad \text{Substitute values.}$$

$$b = \frac{10.7 \sin 108.9°}{\sin 43.5°} \qquad \text{Multiply by } \sin 108.9°.$$

$$b = 14.7 \text{ inches} \qquad \text{Approximate with a calculator.}$$

> **EXAMPLE 7**   **Analyzing Data Involving an Obtuse Angle**

Without using the law of sines, explain why $A = 104°$, $a = 26.8$ meters, and $b = 31.3$ meters cannot be valid for a triangle ABC.

Solution   Since A is an obtuse angle, it is the largest angle and the largest side of the triangle must be a. However, we are given $b > a$ and so $B > A$, which is impossible if A is obtuse. Therefore, no such triangle ABC exists.   ●

If the law of sines had actually been applied in **Example 7,** we would have obtained $\sin B > 1$, which has no solution. Verify this on your calculator.

WHAT WENT WRONG?

A student used the law of sines to solve for angle B in a triangle ABC, where $b = 23$, $c = 14$, and $C = 52°$. He entered the calculation shown on the first screen and got the results in the second screen.

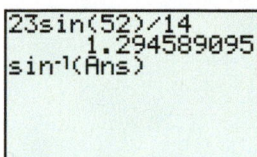 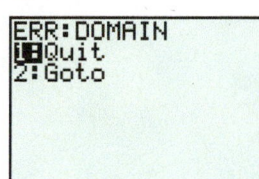

What Went Wrong?   What is the solution, if any, for angle B?

Answer to What Went Wrong?

Since $\sin B > 1$, there is no solution for angle B. Triangle ABC cannot exist.

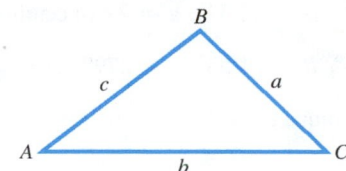

10.1   Exercises

To the Student: *Approximations in the applications in this chapter are generally based on the rules for significant digits, discussed in* **Chapter 8.**

1. *Concept Check*   Consider this oblique triangle. Which one of the following proportions is *not* valid?

```
        B
       /\
    c /   \ a
     /      \
    A‾‾‾‾‾‾‾‾ C
        b
```

A. $\dfrac{a}{b} = \dfrac{\sin A}{\sin B}$ **B.** $\dfrac{a}{\sin A} = \dfrac{b}{\sin B}$

C. $\dfrac{\sin A}{a} = \dfrac{b}{\sin B}$ **D.** $\dfrac{\sin A}{a} = \dfrac{\sin B}{b}$

2. *Concept Check*   Which two of the following situations do not provide sufficient information for solving a triangle by the law of sines?

A. You are given two angles and the side included between them.

B. You are given two angles and a side opposite one of them.

C. You are given two sides and the angle included between them.

D. You are given three sides.

Checking Analytic Skills *Given the following angles and sides, decide whether solving triangle ABC results in the ambiguous case.* **Do not use a calculator.**

3. *A, B,* and *a*

4. *A, C,* and *c*

5. *a, b,* and *c*

6. *A, a,* and *b*

7. *B, b,* and *c*

8. *A, b,* and *c*

9. *C, a,* and *c*

10. *A, B,* and *b*

Checking Analytic Skills *Find the exact length of each side a.* **Do not use a calculator.**

11.

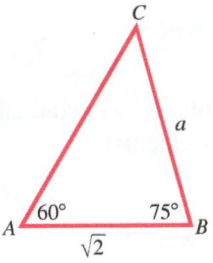

12.

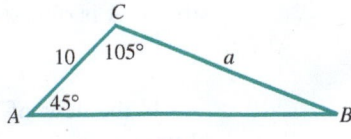

13.

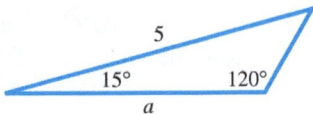

14.

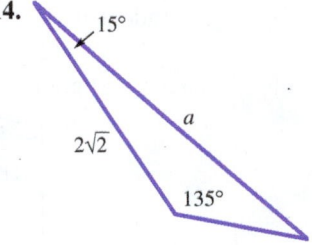

To the Student: Calculator Considerations
When making approximations, do not round off during intermediate steps—wait until the final step to give the appropriate approximation.

Solve each triangle.

15.

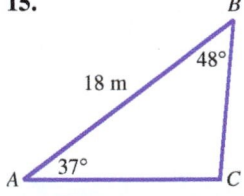

16.

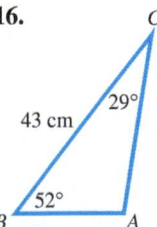

17.

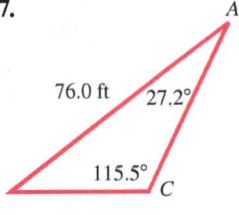

18.

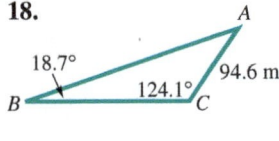

19. $A = 37°, C = 95°, c = 18$ meters

20. $B = 52°, C = 29°, a = 43$ centimeters

21. $C = 74.08°, B = 69.38°, c = 45.38$ meters

22. $A = 87.2°, b = 75.9$ yards, $C = 74.3°$

23. $B = 38° 40', a = 19.7$ centimeters, $C = 91° 40'$

24. $B = 20° 50', C = 103° 10', b = 132$ feet

25. $A = 35.3°, B = 52.8°, b = 675$ feet

26. $A = 68.41°, B = 54.23°, a = 12.75$ feet

27. $A = 39.70°, C = 30.35°, b = 39.74$ meters

28. $C = 71.83°, B = 42.57°, a = 2.614$ centimeters

29. $B = 42.88°, C = 102.40°, b = 3974$ feet

30. $A = 18.75°, B = 51.53°, c = 2798$ yards

31. *Concept Check* Which one of the following sets of data does not determine a unique triangle?
   A. $A = 40°, B = 60°, C = 80°$ **B.** $a = 5, b = 12, c = 13$
   C. $a = 3, b = 7, C = 50°$ **D.** $a = 2, b = 2, c = 2$

32. *Concept Check* Which one of the following sets of data determines a unique triangle?
   A. $A = 50°, B = 50°, C = 80°$ **B.** $a = 3, b = 5, c = 20$
   C. $A = 40°, B = 20°, C = 30°$ **D.** $a = 7, b = 24, c = 25$

Concept Check   *In each figure, a line of length h is to be drawn from the given point to the positive x-axis in order to form a triangle. For what values(s) of h can you draw the following?*
(a) *two triangles*   **(b)** *exactly one triangle*   **(c)** *no triangle*

33.
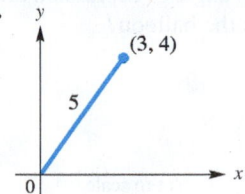

34.
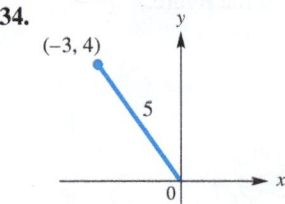

Determine the number of triangles ABC possible with the given parts.

35. $a = 31, b = 26, B = 48°$

36. $a = 35, b = 30, A = 40°$

37. $a = 50, b = 61, A = 58°$

38. $B = 54°, c = 28, b = 23$

Solve each triangle. There may be two, one, or no such triangle.

39. $A = 29.7°, b = 41.5$ feet, $a = 27.2$ feet

40. $B = 48.2°, a = 890$ centimeters, $b = 697$ centimeters

41. $B = 74.3°, a = 859$ meters, $b = 783$ meters

42. $C = 82.2°, a = 10.9$ kilometers, $c = 7.62$ kilometers

43. $A = 142.13°, b = 5.432$ feet, $a = 7.297$ feet

44. $B = 113.72°, a = 189.6$ yards, $b = 243.8$ yards

45. $A = 42.5°, a = 15.6$ feet, $b = 8.14$ feet

46. $C = 52.3°, a = 32.5$ yards, $c = 59.8$ yards

47. $B = 72.2°, b = 78.3$ meters, $c = 145$ meters

48. $C = 68.5°, c = 258$ centimeters, $b = 386$ centimeters

49. $A = 38° 40', a = 9.72$ kilometers, $b = 11.8$ kilometers

50. $C = 29° 50', a = 8.61$ meters, $c = 5.21$ meters

51. $B = 32° 50', a = 7540$ centimeters, $b = 5180$ centimeters

52. $C = 22° 50', b = 159$ millimeters, $c = 132$ millimeters

Solve each problem.

53. Why does the Pythagorean theorem not apply to the triangles discussed in **Examples 1–7?**

54. A trigonometry student makes the statement "If we know *any* two angles and one side of a triangle, then the triangle is uniquely determined." Is this a valid statement? Explain, referring to the congruence axioms given in this section.

55. Explain why the law of sines cannot be used to solve a triangle if we are given the lengths of the three sides of the triangle.

56. If a is twice as long as b, is angle A twice as large as angle B? Explain.

57. *Distance across a River*   To find the distance AB across a river, a distance $BC = 354$ meters is laid off on one side of the river. It is found that $B = 112° 10'$ and $C = 15° 20'$. Find AB.

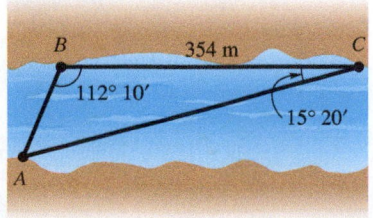

58. *Distance across a Canyon*   To determine the distance RS across a deep canyon, Joanna lays off a distance $TR = 582$ yards. She then finds that $T = 32° 50'$ and $R = 102° 20'$. Find RS.

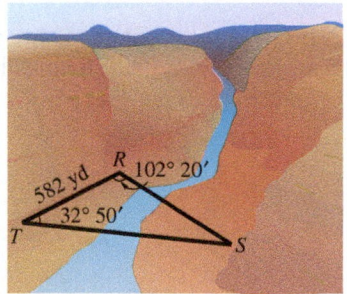

59. *Distance between Radio Direction Finders*   Radio direction finders are at points A and B, which are 3.46 miles apart on an east–west line, with A west of B. From A, the bearing of a certain radio transmitter is 47.7°; from B, the bearing is 302.5°. Find the distance of the transmitter from A.

60. *Distance a Ship Travels*   A ship is sailing due north. At a certain point, the bearing of a lighthouse 12.5 kilometers away is N 38.8° E. Later on, the captain notices that the bearing of the lighthouse has become S 44.2° E. How far did the ship travel between the two observations of the lighthouse?

61. *Measurement of a Folding Chair* A folding chair is to have a seat 12.0 inches deep with angles as shown in the figure. How far down from the seat should the crossing legs be joined? (*Hint:* Find *x* in the figure.)

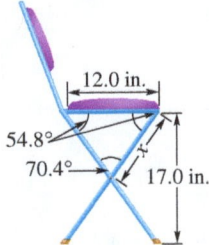

62. *Distance across a River* Mark notices that the bearing of a tree on the opposite bank of a river flowing north is 115.45°. Lisa is on the same bank as Mark, but 428.3 meters away. She notices that the bearing of the tree is 45.47°. The two banks are parallel. What is the distance across the river?

63. *Angle Formed by Centers of Gears* Three gears are arranged as shown in the figure. Find angle θ to the nearest degree.

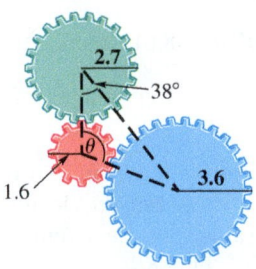

64. *Distance between Atoms* Three atoms with radii 2.0, 3.0, and 4.5 are arranged as in the figure. Find the distance between the centers of atoms *A* and *C*.

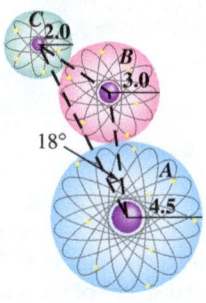

65. *Distance between a Ship and a Lighthouse* The bearing of a lighthouse from a ship was found to be N 37° E. After the ship sailed 2.5 miles due south, the new bearing was N 25° E. Find the distance between the ship and the lighthouse at each location.

66. *Height of a Balloon* A balloonist is directly above a straight road 1.5 miles long that joins two towns. She finds that the town closer to her is at an angle of depression of 35° and the farther town is at an angle of depression of 31°. How high above the ground is the balloon?

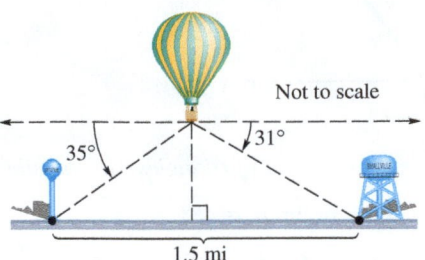

Not to scale

67. *Distance from Ship to Shore* From shore station A, a ship C is observed in the direction N 22.4° E. The same ship is observed to be in the direction N 10.6° W from shore station B, located a distance of 25.5 kilometers exactly southeast of A. Find the distance of the ship from station A.

68. *Height of a Helicopter* A helicopter is sighted at the same time by two ground observers who are *exactly* 3 miles apart on the same side of the helicopter. (See the figure.) They report angles of elevation of 20.5° and 27.8°. How high is the helicopter?

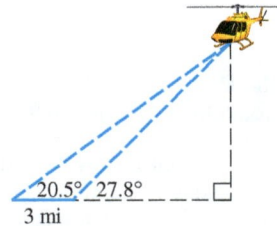

69. *Distance from a Rocket to a Radar Station* A rocket-tracking facility has two radar stations T_1 and T_2, placed 1.73 kilometers apart, that lock onto the rocket and continuously transmit the angles of elevation to a computer. Find the distance to the rocket from T_1 at the moment when the angles of elevation are 28.1° and 79.5°, as shown in the figure.

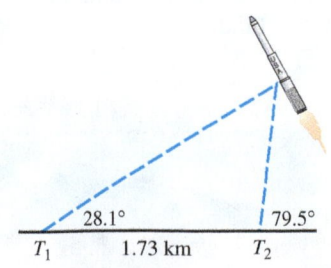

70. *Height of a Clock Tower* A surveyor standing 48.0 meters from the base of a building measures the angle to the top of the building to be 37.4°. The surveyor then measures the angle to the top of a clock tower on the building to be 45.6°. Find the height of the clock tower.

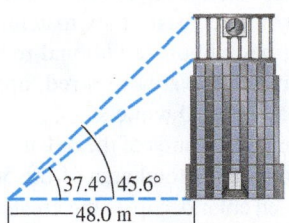

71. *Distance between a Pin and a Rod* A slider crank mechanism is shown in the figure. Find the distance between the wrist pin W and the connecting rod center C.

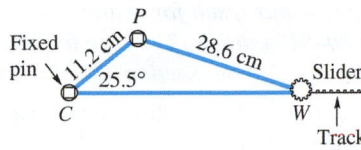

72. *Path of a Satellite* A satellite is traveling in a circular orbit 1600 kilometers above Earth. It will pass directly over a tracking station at noon. The satellite takes 2 hours to make a complete orbit. Assume that the radius of Earth is 6400 kilometers. The tracking antenna is aimed 30° above the horizon. See the figure. At what time (before noon) will the satellite pass through the beam of the antenna? (*Source:* NASA.)

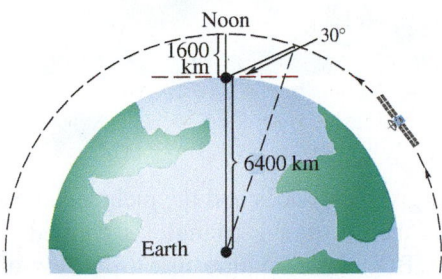

73. Apply the law of sines to the following: $a = \sqrt{5}$, $c = 2\sqrt{5}$, $A = 30°$. What is the value of $\sin C$? What is the measure of C? Based on its angle measures, what kind of triangle is triangle ABC?

74. Explain the condition that must exist to determine that there is no triangle satisfying the given values of a, b, and B once the value of $\sin A$ is found.

75. Without using the law of sines, explain why no triangle ABC exists satisfying $A = 103° \, 20'$, $a = 14.6$ feet, and $b = 20.4$ feet.

76. Apply the law of sines to the data given in **Example 7.** Describe in your own words what happens when you try to find the measure of angle B, using a calculator.

77. *Distance to the Moon* Since the moon is a relatively close celestial object, its distance from Earth can be measured by taking two photographs of the moon at precisely the same time from two different locations on Earth. The moon will have a different angle of elevation at each location. On April 29, 1976, at 11:35 A.M., the lunar angles of elevation during a partial solar eclipse at Bochum in upper Germany and at Donaueschingen in lower Germany were measured as 52.6997° and 52.7430°, respectively. The two cities are 398 kilometers apart. Calculate the distance to the moon from Bochum on this day, and compare it with the actual value of 406,000 kilometers. Disregard the curvature of Earth in this calculation. (*Source:* Schlosser, W., T. Schmidt-Kaler, and E. Milone, *Challenges of Astronomy,* Springer-Verlag.)

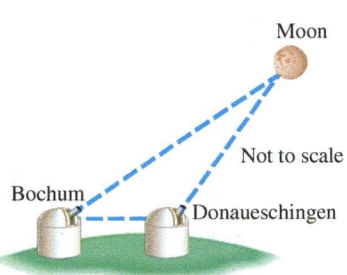

78. *Ground Distances Measured by Aerial Photography* The distance shown in an aerial photograph is determined by both the focal length of the lens and the tilt of the camera from the perpendicular to the ground. A camera lens with a 12-inch focal length has an angular coverage of 60°. If an aerial photograph is taken with this camera tilted 35° at an altitude of 5000 feet, calculate the distance d in miles that will be shown in the photograph. (See the figure.) (*Sources:* Brooks, R., and D. Johannes, *Phytoarchaeology,* Dioscorides Press; Moffitt, F., *Photogrammetry,* International Textbook Company.)

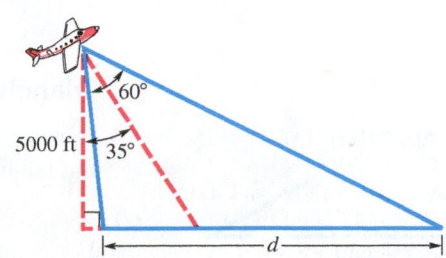

79. *U.S. Flag* The U.S. flag includes the colors red, white, and blue. Which color, red or white, is predominant? (Only 18.73% of the total area is blue.) (*Source: Slicing Pizzas, Racing Turtles, and Further Adventures in Applied Mathematics,* Banks, R., Princeton University Press.)

(a) Let R denote the radius of the circumscribing circle of a five-pointed star appearing on the American flag. The star can be decomposed into 10 congruent triangles. In the figure below, r is the radius of the circumscribing circle of the pentagon in the interior of the star. Show that the area of the star is

$$A = \left[5\,\frac{\sin A \sin B}{\sin(A + B)} \right] R^2.$$

(*Hint:* $\sin C = \sin[180° - (A + B)] = \sin(A + B).$)

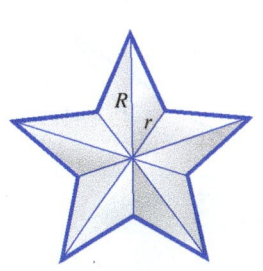 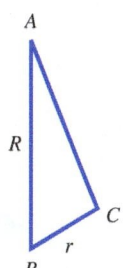

(b) Angles A and B have values 18° and 36°, respectively. Express the area of a star in terms of its radius R.

(c) To determine whether red or white is predominant, we consider a flag of width 10 inches, length 19 inches, length of each upper stripe 11.4 inches, and radius R of the circumscribing circle of each star 0.308 inch. The 13 stripes consist of six matching pairs of red and white stripes and one additional red, upper stripe. We must compare the area of a red, upper stripe with the total area of the 50 white stars.

   (i) Compute the area of the red, upper stripe.

   (ii) Compute the total area of the 50 white stars.

   (iii) Which color occupies the greatest area on the flag?

RELATING CONCEPTS

For individual or group investigation (Exercises 80–84)

In any triangle, the longest side is opposite the largest angle. To prove this result for acute triangles, **work Exercises 80–84 in order.** *(The case for obtuse triangles will be considered in the* **Section 10.2 Exercises.**)

80. Is the graph of the function $y = \sin x$ increasing or decreasing over the interval $\left(0, \frac{\pi}{2}\right)$?

81. Suppose angle A is the largest angle of an acute triangle, and let B be an angle smaller than A. Explain why $\frac{\sin B}{\sin A} < 1$.

82. Solve for b in the first form of the law of sines.

83. Use the results in **Exercises 81–82** to show that $b < a$.

84. Explain why no triangle ABC exists having $A = 103°$, $a = 12$, $b = 13$.

10.2 The Law of Cosines and Area Formulas

Derivation of the Law of Cosines • Using the Law of Cosines • Area Formulas

In **Section 10.1** we found that if we are given two sides and the included angle (Case 3) or three sides (Case 4) of a triangle, then a unique triangle is formed. These are the SAS and SSS cases, respectively. Both cases require using the **law of cosines.** Remember the following property of triangles when applying the law of cosines.

No Triangle

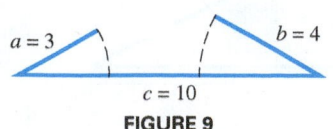

FIGURE 9

Triangle Side Length Restriction

In any triangle, the sum of the lengths of any two sides must be greater than the length of the remaining side.

For example, it would be impossible to construct a triangle with sides of lengths 3, 4, and 10, because $3 + 4 = 7$, which is less than 10. See **FIGURE 9.**

Derivation of the Law of Cosines

Vertex Coordinates

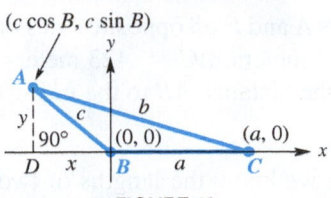

FIGURE 10

To derive the law of cosines, let ABC be any oblique triangle. Choose a coordinate system so that vertex B is at the origin and side BC is along the positive x-axis. See **FIGURE 10**. Let (x, y) be the coordinates of vertex A of the triangle. For angle B, whether obtuse or acute,

$$\sin B = \frac{y}{c} \quad \text{and} \quad \cos B = \frac{x}{c}, \text{ so}$$

$$y = c \sin B \quad \text{and} \quad x = c \cos B. \quad \longleftarrow \quad \textit{Here } x \textit{ is negative if B is obtuse.}$$

Thus, the coordinates of point A become $(c \cos B, c \sin B)$.

Point C has coordinates $(a, 0)$, and AC has length b. Apply the distance formula.

$$b = \sqrt{(c \cos B - a)^2 + (c \sin B - 0)^2}$$

$b^2 = (c \cos B - a)^2 + (c \sin B)^2$	Square each side.
$b^2 = c^2 \cos^2 B - 2ac \cos B + a^2 + c^2 \sin^2 B$	Square the binomial.
$b^2 = a^2 + c^2(\cos^2 B + \sin^2 B) - 2ac \cos B$	Properties of real numbers
$b^2 = a^2 + c^2(1) - 2ac \cos B$	Fundamental identity
$b^2 = a^2 + c^2 - 2ac \cos B$	

This result is one form of the law of cosines. In our work, if we had placed A or C at the origin, we would have obtained the same result, but with the variables rearranged.

Law of Cosines

In any triangle ABC with sides a, b, and c, the following hold.

$$a^2 = b^2 + c^2 - 2bc \cos A$$
$$b^2 = a^2 + c^2 - 2ac \cos B$$
$$c^2 = a^2 + b^2 - 2ab \cos C$$

That is, the square of a side in any triangle is equal to the sum of the squares of the other two sides, minus twice the product of those two sides and the cosine of the angle included between them.

NOTE If we have a right triangle and let $C = 90°$ in the third form of the law of cosines, we can then evaluate $\cos C = \cos 90° = 0$, and the formula becomes $c^2 = a^2 + b^2$, the familiar equation of the Pythagorean theorem. **The Pythagorean theorem is a special case of the law of cosines.**

Using the Law of Cosines

EXAMPLE 1 **Using the Law of Cosines to Find a Side (SAS)**

Find the unknown side a in the triangle shown in **FIGURE 11**.

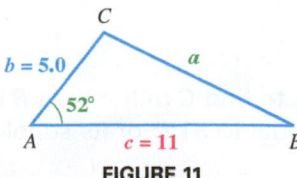

FIGURE 11

Solution We are given SAS, so let $A = 52°$, $b = 5.0$, and $c = 11$. We will use the first equation for the law of cosines.

$a^2 = b^2 + c^2 - 2bc \cos A$	Law of cosines
$a^2 = 5.0^2 + 11^2 - 2(5.0)(11) \cos 52°$	Substitute.
$a^2 = 78.277$	Use a calculator.
$a = \sqrt{78.277} = 8.9$	Take the square root.

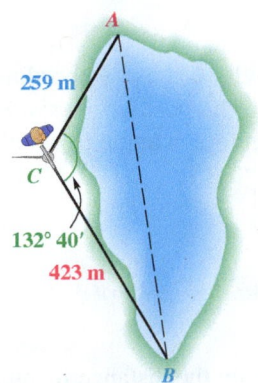

259 m

C

132° 40'

423 m

B

FIGURE 12

EXAMPLE 2 **Using the Law of Cosines in an Application (SAS)**

A surveyor wishes to find the distance between two points A and B on opposite sides of a lake. While standing at point C, she finds that $AC = 259$ meters, $BC = 423$ meters, and angle ACB measures $132° 40'$. See **FIGURE 12.** Find the distance AB to the nearest meter.

Solution The law of cosines can be used here, since we know the lengths of two sides of the triangle and the measure of the included angle.

$$c^2 = a^2 + b^2 - 2ab \cos C$$

$AB^2 = 423^2 + 259^2 - 2(423)(259) \cos 132° 40'$ Substitute.

$AB^2 = 394{,}510.6$ Use a calculator.

$AB = 628$ Approximate the square root.

The distance between the points is approximately 628 meters.   ●

EXAMPLE 3 **Using the Law of Cosines to Solve a Triangle (SAS)**

Solve triangle ABC if $A = 42.3°$, $b = 12.9$ meters, and $c = 15.4$ meters.

Solution See **FIGURE 13.** We start by finding a with the law of cosines.

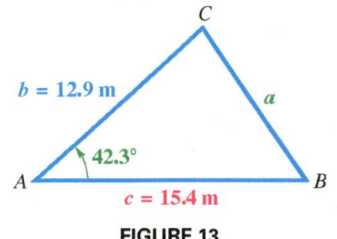

C

$b = 12.9$ m

a

42.3°

A

$c = 15.4$ m

B

FIGURE 13

$a^2 = b^2 + c^2 - 2bc \cos A$ Law of cosines

$a^2 = 12.9^2 + 15.4^2 - 2(12.9)(15.4) \cos 42.3°$ Substitute.

$a^2 = 109.7$ Use a calculator.

$a = 10.47$ meters   Take the square root.

Of the two remaining angles, B and C, B must be the smaller, since it is opposite b, the shorter of the two remaining sides. Therefore, B cannot be obtuse, and there is only one possible angle.

$$\frac{\sin A}{a} = \frac{\sin B}{b}$$ Law of sines (alternative form)

$$\frac{\sin 42.3°}{10.47} = \frac{\sin B}{12.9}$$ Substitute.

$$\sin B = \frac{12.9 \sin 42.3°}{10.47}$$ Multiply by 10.47 and rewrite.

$B = \sin^{-1}(0.8292)$ Take the inverse sine.

$B = 56.0°$ Use a calculator and approximate.

The easiest way to find C is to subtract the measures of A and B from $180°$.

$$C = 180° - A - B$$

$$C = 180° - 42.3° - 56.0° = 81.7°$$ ●

CAUTION Had we chosen to use the law of sines to find C rather than B in **Example 3,** we would not have known whether C is equal to $81.7°$ or its supplement, $98.3°$.

> **EXAMPLE 4** **Using the Law of Cosines to Solve a Triangle (SSS)**

Solve triangle ABC to the nearest tenth of a degree if $a = 5$, $b = 6$, and $c = 9$.

Solution When solving SSS, start by finding the largest angle. The largest angle, C, is opposite the longest side, c. We start by finding C.

$$c^2 = a^2 + b^2 - 2ab \cos C \qquad \text{Law of cosines}$$

$$2ab \cos C = a^2 + b^2 - c^2 \qquad \text{Add } 2ab \cos C \text{ and subtract } c^2.$$

$$\cos C = \frac{a^2 + b^2 - c^2}{2ab} \qquad \text{Divide by } 2ab.$$

$$\cos C = \frac{5^2 + 6^2 - 9^2}{2(5)(6)} \qquad \text{Substitute } a = 5,\ b = 6 \text{ and } c = 9.$$

$$\cos C = -\frac{1}{3} \qquad \text{Simplify.}$$

Thus $C = \cos^{-1}\left(-\frac{1}{3}\right) = 109.5°$. The law of cosines could be used again to find either A or B. However, we use the law of sines to find A.

$$\frac{\sin A}{a} = \frac{\sin C}{c} \qquad \text{Law of sines}$$

$$\sin A = \frac{a \sin C}{c} \qquad \text{Multiply by } a.$$

$$\sin A = \frac{5 \sin 109.5°}{9} \qquad \text{Substitute } a = 5.\ c = 9, \text{ and } C = 109.5.$$

$$\sin A = 0.5237 \qquad \text{Use a calculator.}$$

Because C is the largest angle, it follows that A must be an acute angle and that $A = \sin^{-1}(0.5237) = 31.6°$. To find B we use the fact that the measures of the angles sum to $180°$ in a triangle.

$$B = 180° - 109.5° - 31.6° = 38.9°$$

> **EXAMPLE 5** **Using the Law of Cosines to Solve a Triangle (SSS)**

Solve triangle ABC if $a = 9.47$ feet, $b = 15.9$ feet, and $c = 21.1$ feet.

Solution Again, we use the law of cosines to solve for C, the largest angle, since we will then know that C is obtuse if $\cos C < 0$.

$$c^2 = a^2 + b^2 - 2ab \cos C \qquad \text{Law of cosines}$$

$$\cos C = \frac{a^2 + b^2 - c^2}{2ab} \qquad \text{Solve for cos } C.$$

$$\cos C = \frac{9.47^2 + 15.9^2 - 21.1^2}{2(9.47)(15.9)} \qquad \text{Substitute.}$$

$$\cos C = -0.34109402 \qquad \text{Use a calculator.}$$

$$C = 109.9° \qquad \text{Use the inverse cosine function.}$$

We can use either the law of sines or the law of cosines to find $B = 45.1°$. (Verify this.) Since $A = 180° - B - C$, we obtain $A = 25.0°$.

Triangular Roof Truss

FIGURE 14

Trusses are used to support roofs on buildings. The simplest type of roof truss is a triangle, as shown in **FIGURE 14**.

EXAMPLE 6 **Designing a Roof Truss (SSS)**

Find angle B to the nearest degree for the roof truss shown in **FIGURE 15**.

Solution Let $a = 11$, $b = 6$, and $c = 9$ in the law of cosines.

$$b^2 = a^2 + c^2 - 2ac \cos B \qquad \text{Law of cosines}$$

$$2ac \cos B = a^2 + c^2 - b^2 \qquad \text{Add } 2ac \cos B \text{ and subtract } b^2.$$

$$\cos B = \frac{a^2 + c^2 - b^2}{2ac} \qquad \text{Solve for cos } B.$$

$$\cos B = \frac{11^2 + 9^2 - 6^2}{2(11)(9)} \qquad \text{Substitute.}$$

$$B = \cos^{-1}\left(\frac{11^2 + 9^2 - 6^2}{2(11)(9)}\right) \qquad \text{Use the inverse cosine function.}$$

$$B = 33° \qquad \text{Use a calculator.}$$

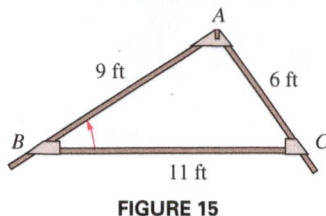

FIGURE 15

Four cases can occur in solving an oblique triangle. The following table suggests a procedure for solving each one, assuming that the given information actually produces a triangle.

Oblique Triangle	Suggested Procedure for Solving
Case 1: One side and two angles are known. **(SAA or ASA)**	**Step 1** Use the angle sum formula $(A + B + C = 180°)$ to find the remaining angle. **Step 2** Use the law of sines to find the remaining sides.
Case 2: Two sides and one angle (not included between the two sides) are known. **(SSA)**	*This is the **ambiguous case**. There may be no triangle, one triangle, or two triangles.* **Step 1** Use the law of sines to find an angle. **Step 2** Use the angle sum formula to find the remaining angle. **Step 3** Use the law of sines to find the remaining side. *If two triangles exist, repeat Steps 2 and 3.*
Case 3: Two sides and the included angle are known. **(SAS)**	**Step 1** Use the law of cosines to find the third side. **Step 2** Use the law of sines to find the smaller of the two remaining angles. **Step 3** Use the angle sum formula to find the remaining angle.
Case 4: Three sides are known. **(SSS)**	**Step 1** Use the law of cosines to find the largest angle. **Step 2** Use the law of sines to find either of the two remaining angles. **Step 3** Use the angle sum formula to find the remaining angle.

FOR DISCUSSION

Discuss the steps that you would take to solve each triangle. Do not actually solve the triangle.

1. $A = 32°$, $B = 67°$, $a = 20$
2. $b = 22$, $c = 15$, $A = 39°$
3. $a = 5$, $c = 7$, $C = 22°$
4. $b = 13$, $c = 15$, $B = 22°$
5. $a = 5$, $b = 7$, $c = 9$

Area Formulas

The law of cosines can be used to derive a formula for the area of a triangle, given the lengths of the three sides. This formula is known as **Heron's formula,** named after the Greek mathematician Heron of Alexandria. Heron's formula can be used for the case SSS.

Heron of Alexandria (A.D. 75)

Heron's Area Formula (SSS)*

If a triangle has sides of lengths a, b, and c, and if the **semiperimeter** is

$$s = \frac{1}{2}(a + b + c),$$

then the area of the triangle is

$$\mathcal{A} = \sqrt{s(s - a)(s - b)(s - c)}.$$

That is, according to Heron's formula, the area of a triangle is the square root of the product of four factors: (1) the semiperimeter, (2) the semiperimeter minus the first side, (3) the semiperimeter minus the second side, and (4) the semiperimeter minus the third side.

Area of a Region

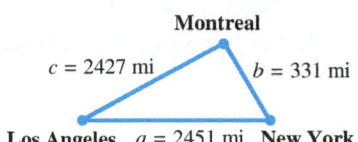

$c = 2427$ mi   Montreal $b = 331$ mi

Los Angeles $a = 2451$ mi   New York

Not to scale

FIGURE 16

EXAMPLE 7   **Using Heron's Formula to Find an Area**

The distance "as the crow flies" from Los Angeles to New York is 2451 miles, from New York to Montreal is 331 miles, and from Montreal to Los Angeles is 2427 miles. What is the area of the triangular region having these three cities as vertices? (Ignore the curvature of Earth.)

Solution   In **FIGURE 16,** we let $a = 2451$, $b = 331$, and $c = 2427$. We first find the semiperimeter and then use Heron's formula to find the area.

$s = \frac{1}{2}(a + b + c)$ ⟶ $s = \frac{1}{2}(2451 + 331 + 2427) = 2604.5$ Semiperimeter

$\mathcal{A} = \sqrt{s(s - a)(s - b)(s - c)}$ Area of a triangle

Remember the factor s. $\mathcal{A} = \sqrt{2604.5(2604.5 - 2451)(2604.5 - 331)(2604.5 - 2427)}$

$\mathcal{A} = 401{,}700$ square miles   Use a calculator.

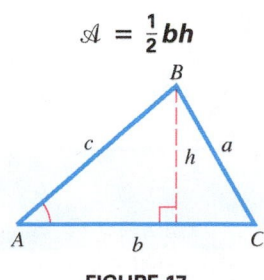

$\mathcal{A} = \frac{1}{2}bh$

FIGURE 17

If we know the measures of two sides of a triangle and the angle included between them, we can find the area $\mathcal{A}$ of the triangle. This area is given by $\mathcal{A} = \frac{1}{2}bh$, where b is the base of the triangle and h is the height. Using trigonometry, we can find a formula for the area of the triangle shown in **FIGURE 17.**

$$\sin A = \frac{h}{c}, \quad \text{or} \quad h = c \sin A$$

Thus, another area formula can be derived.

$$\mathcal{A} = \frac{1}{2}bh = \frac{1}{2}bc \sin A$$

Since the labels for the vertices in triangle ABC could be rearranged, three area formulas can be written. These formulas can be applied when SAS is given.

---

*For a derivation of Heron's formula, see *College Algebra and Trigonometry*, 5th edition, by Lial, Hornsby, Schneider, Daniels, pages 738–740.

> **Area of a Triangle (SAS)**
>
> In any triangle ABC, the area $\mathscr{A}$ is given by the following formulas.
>
> $$\mathscr{A} = \frac{1}{2}bc \sin A, \quad \mathscr{A} = \frac{1}{2}ab \sin C, \quad \text{and} \quad \mathscr{A} = \frac{1}{2}ac \sin B$$
>
> *That is, the area is half the product of the lengths of two sides and the sine of the angle included between them.*

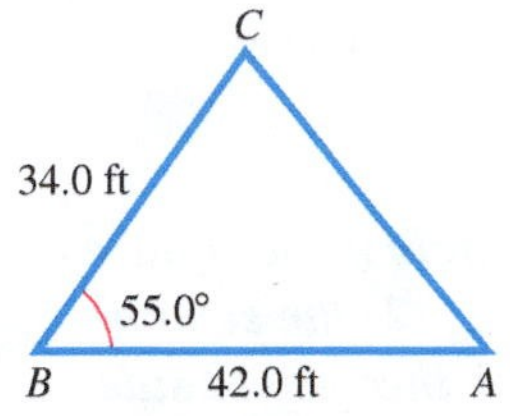

EXAMPLE 8 **Finding the Area of a Triangle (SAS)**

Find the area of triangle ABC in **FIGURE 18**.

Solution We are given $B = 55.0°$, $a = 34.0$ feet, and $c = 42.0$ feet.

$$\mathscr{A} = \frac{1}{2}ac \sin B = \frac{1}{2}(34.0)(42.0) \sin 55.0° = 585 \text{ square feet}$$

34.0 ft
55.0°
B   42.0 ft   A

FIGURE 18

EXAMPLE 9 **Finding the Area of a Triangle (ASA)**

Find the area of triangle ABC if $A = 24° 40'$, $b = 27.3$ centimeters, and $C = 52° 40'$.

Solution Before the area formula can be used, we must find either a or c. Since the sum of the measures of the angles of any triangle is $180°$,

$$B = 180° - 24° 40' - 52° 40' = 102° 40'.$$

We can use the law of sines to find a.

Solve for a. → $\dfrac{a}{\sin A} = \dfrac{b}{\sin B}$ Law of sines

$\dfrac{a}{\sin 24° 40'} = \dfrac{27.3}{\sin 102° 40'}$ Substitute known values.

$a = 11.7$ centimeters   Use a calculator.

Now, we find the area.

$$\mathscr{A} = \frac{1}{2}ab \sin C = \frac{1}{2}(11.7)(27.3) \sin 52° 40' = 127 \text{ square centimeters}$$

10.2 Exercises

Concept Check *Assume triangle ABC has standard labeling and complete the following.*
(a) *Determine whether SAA, ASA, SSA, SAS, or SSS is given.*
(b) *Decide whether the law of sines or the law of cosines should be used to begin solving the triangle.*

1. a, b, and C **2.** A, C, and c **3.** a, b, and A **4.** a, b, and c

5. A, B, and c **6.** a, c, and A **7.** a, B, and C **8.** b, c, and A

Checking Analytic Skills   *Find the exact length of the side or measure of the angle labeled with a variable in each triangle.* **Do not use a calculator.**

9.
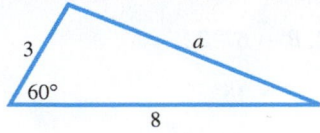

10.
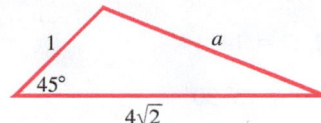

11.
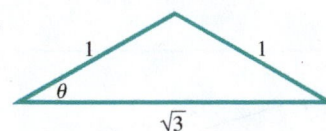

12.
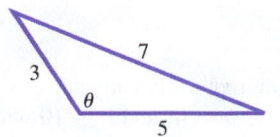

Solve each triangle. Approximate values to the nearest tenth.

13.
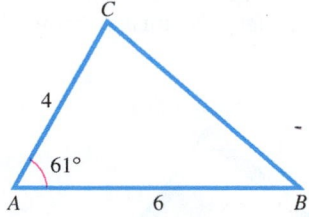

14.
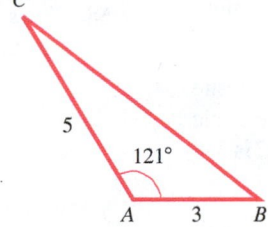

15.
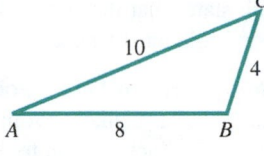

16.
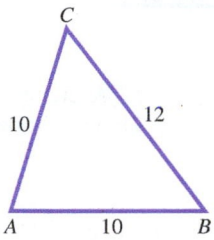

17.
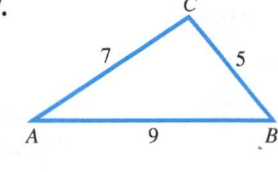

18.
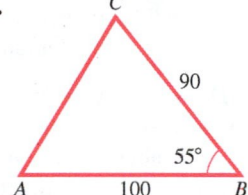

Solve each triangle.

19. $C = 28.3°$, $b = 5.71$ inches, $a = 4.21$ inches

20. $A = 41.4°$, $b = 2.78$ yards, $c = 3.92$ yards

21. $C = 45.6°$, $b = 8.94$ meters, $a = 7.23$ meters

22. $A = 67.3°$, $b = 37.9$ kilometers, $c = 40.8$ kilometers

23. $a = 9.3$ inches, $b = 5.7$ inches, $c = 8.2$ inches

24. $a = 28$ feet, $b = 47$ feet, $c = 58$ feet

25. $a = 42.9$ meters, $b = 37.6$ meters, $c = 62.7$ meters

26. $a = 189$ yards, $b = 214$ yards, $c = 325$ yards

27. $AB = 1240$ feet, $AC = 876$ feet, $BC = 965$ feet

28. $AB = 298$ meters, $AC = 421$ meters, $BC = 324$ meters

29. $A = 80° 40'$, $b = 143$ centimeters, $c = 89.6$ centimeters

30. $C = 72° 40'$, $a = 327$ feet, $b = 251$ feet

31. $B = 74.80°$, $a = 8.919$ inches, $c = 6.427$ inches

32. $C = 59.70°$, $a = 3.725$ miles, $b = 4.698$ miles

33. $A = 112.8°$, $b = 6.28$ meters, $c = 12.2$ meters

34. $B = 168.2°$, $a = 15.1$ centimeters, $c = 19.2$ centimeters

35. $a = 3.0$ feet, $b = 5.0$ feet, $c = 6.0$ feet

36. $a = 4.0$ feet, $b = 5.0$ feet, $c = 8.0$ feet

Refer to the **guidelines to solve oblique triangles** *to decide on the procedure to use to solve each triangle. Then actually solve the triangle.*

37. $a = 59.49$, $B = 50° 52'$, $C = 28° 37'$

38. $a = 614.0$, $A = 35° 09'$, $C = 25° 53'$

39. $a = 3.961$, $c = 5.308$, $B = 58° 12'$

40. $b = 56.68$, $c = 64.40$, $A = 98° 10'$

41. $a = 51.41, b = 37.29, c = 65.88$

42. $a = 1744, b = 2286, c = 1902$

43. $a = 7.031, b = 9.947, A = 41° 12'$

44. $a = 4676, b = 5128, A = 23° 51'$

45. $a = 27.16, c = 34.22, C = 14° 19'$

46. $b = 24.52, c = 19.84, B = 67° 32'$

47. $a = 2634, c = 2200, C = 73° 30'$

48. $b = 26.96, c = 7.608, C = 18° 37'$

Solve each problem.

49. Refer to **FIGURE 9.** If you attempt to find any angle of a triangle with the values $a = 3$, $b = 4$, and $c = 10$ with the law of cosines, what happens?

50. "The shortest distance between two points is a straight line." Explain how this relates to the geometric property which states that the sum of the lengths of any two sides of a triangle must be greater than the remaining side.

51. *Distance across a Lake*   Points *A* and *B* are on opposite sides of Lake Yankee. From a third point, *C*, the angle between the lines of sight to *A* and *B* is 46.3°. If *AC* is 350 meters long and *BC* is 286 meters long, find *AB*.

52. *Diagonals of a Parallelogram*   The sides of a parallelogram are 4.0 centimeters and 6.0 centimeters. One angle is 58° while another is 122°. Find the lengths of the diagonals of the parallelogram.

53. *Flight Distance*   Airports *A* and *B* are 450 kilometers apart, on an east–west line. Tom flies in a northeast direction from *A* to airport *C*. From *C*, he flies 359 kilometers on a heading of 128° 40′ to *B*. How far is *C* from *A*?

54. *Distance between Two Ships*   Two ships leave a harbor together, traveling on courses that have an angle of 135° 40′ between them. If they each travel 402 miles, how far apart are they?

55. *Distance between a Ship and a Rock*   A ship is sailing east. At one point, the bearing of a submerged rock is 45° 20′. After the ship has sailed 15.2 miles, the bearing of the rock has become 308° 40′. Find the distance of the ship from the rock at the latter point.

56. *Distance between Two Boats*   Two boats leave a dock together. Each travels in a straight line. The angle between their courses measures 54° 10′. One boat travels 36.2 kilometers per hour and the other 45.6 kilometers per hour. How far apart will they be after 3 hours?

57. *Distance between a Ship and a Submarine*   From an airplane flying over the ocean, the angle of depression to a submarine lying just under the surface is 24° 10′. At the same moment, the angle of depression from the airplane to a battleship is 17° 30′. See the figure at the top of the next column. The distance from the airplane to the battleship is 5120 feet. Find the distance between the battleship and the submarine. (Assume that the airplane, submarine, and battleship are in a vertical plane.)

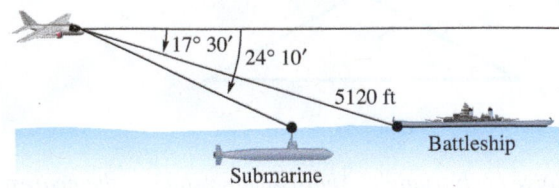

58. *Truss Construction*   A triangular truss is shown in the figure. Find angle θ.

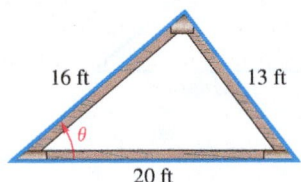

59. *Distance between Points on a Crane*   A crane with a counterweight is shown in the figure. Find the horizontal distance between points *A* and *B*.

60. *Distance between a Beam and Cables*   A weight is supported by cables attached to both ends of a balance beam, as shown in the figure. What angles are formed between the beam and the cables?

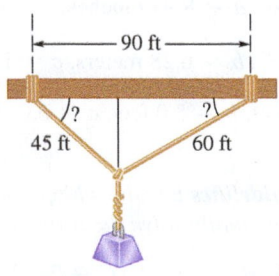

61. *Length of a Tunnel* To measure the distance through a mountain for a proposed tunnel, a point C is chosen that can be reached from each end of the tunnel. If $AC = 3800$ meters, $BC = 2900$ meters, and angle $C = 110°$, find the length of the tunnel.

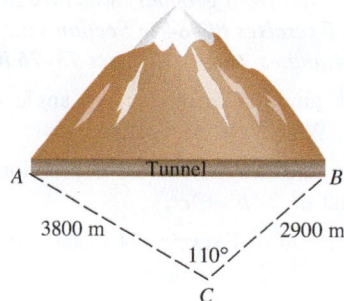

62. *Distance on a Baseball Diamond* A baseball diamond is a square 90 feet on a side, with home plate and the three bases as vertices. The pitcher's rubber is located 60.5 feet from home plate. Find the distance from the pitcher's rubber to each of the bases.

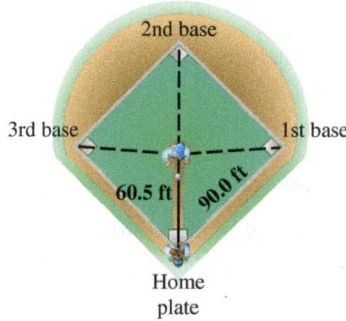

63. *Distance between Two Factories* Two factories blow their whistles at exactly 5:00. A man hears the two blasts at 3 seconds and 6 seconds after 5:00, respectively. The angle between his lines of sight to the two factories is $42.20°$. If sound travels 344 meters per second, how far apart are the factories to the nearest meter?

64. *Distance between a Ship and a Point* Starting at point A, a ship sails 18.5 kilometers on a bearing of $189°$, then turns and sails 47.8 kilometers on a bearing of $317°$. Find the distance of the ship from point A.

65. *Bearing of One Town to Another* Two towns 21 miles apart are separated by a dense forest. See the figure at the top of the next column. To travel from town A to town B, a person must go 17 miles on a bearing of $325°$, then turn and continue for 9.0 miles to reach town B. Find the bearing of B from A.

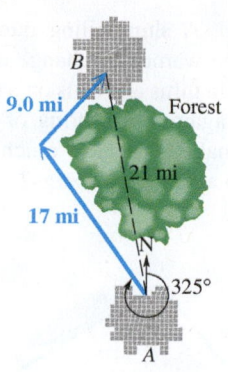

66. *Distance between Ends of the Vietnam Memorial* The Vietnam Veterans Memorial in Washington, DC, is V-shaped with equal sides of length 246.75 feet, and the angle between these sides measuring $125°\ 12'$. Find the distance between the ends of the two sides. (*Source:* Pamphlet obtained at Vietnam Veterans Memorial.)

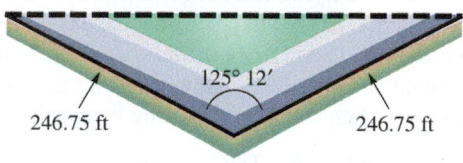

67. *Distance between a Satellite and a Tracking Station* A satellite traveling in a circular orbit 1600 kilometers above Earth is due to pass directly over a tracking station at noon. Assume that the satellite takes 2 hours to make an orbit and that the radius of Earth is 6400 kilometers. Find the distance between the satellite and the tracking station at 12:03 P.M. (*Source:* NASA.)

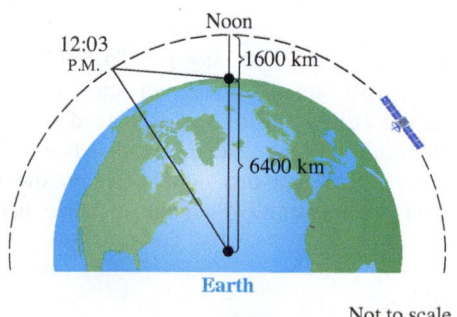

68. *Path of a Ship* A ship sailing due east in the North Atlantic has been warned to change course to avoid icebergs. The captain turns and sails on a bearing of 62°, then changes course again to a bearing of 115° until the ship reaches its original course. How much farther did the ship have to travel to avoid the icebergs?

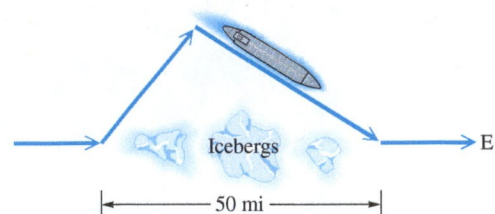

69. *Angle in a Parallelogram* A parallelogram has sides of length 25.9 centimeters and 32.5 centimeters. The longer diagonal has length 57.8 centimeters. Find the angle opposite the longer diagonal.

70. *Distance between an Airplane and a Mountain* A person in a plane flying straight north observes a mountain at a bearing of 24.1°. At the time, the plane is 7.92 kilometers from the mountain. A short time later, the bearing to the mountain becomes 32.7°. How far is the airplane from the mountain when the second bearing is taken?

71. *Layout for a Cabin* The layout for a state park cabin has the dimensions given in the figure. Find *x*.

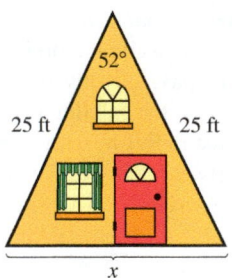

72. *Distance between Two Towns* To find the distance between two small towns, an electronic distance-measuring (EDM) instrument is placed on a hill from which both towns are visible. The distance to each town from the EDM and the angle between the two lines of sight are measured. Find the distance between the towns.

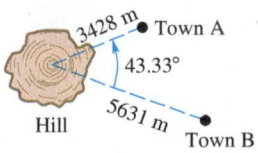

RELATING CONCEPTS

**For individual or group investigation
(Exercises 73–76)**

In any triangle, the longest side is opposite the largest angle. This result from geometry was proven for acute triangles in **Exercises 80–84 in Section 10.1.** To prove it for obtuse triangles, **work Exercises 73–76 in order.**

73. Suppose angle *A* is the largest angle of an obtuse triangle. Why is cos *A* negative?

74. Consider the law of cosines expression for a^2, and show that $a^2 > b^2 + c^2$.

75. Use the result of **Exercise 74** to show that $a > b$ and $a > c$.

76. Use the result of **Exercise 75** to explain why no triangle *ABC* satisfies $A = 103°$, $a = 25$, and $c = 30$.

Find the area of each triangle.

77. $A = 42.5°$, $b = 13.6$ meters, $c = 10.1$ meters

78. $B = 124.5°$, $a = 30.4$ centimeters, $c = 28.4$ centimeters

79. $a = 12$ meters, $b = 16$ meters, $c = 25$ meters

80. $a = 154$ centimeters, $b = 179$ centimeters, $c = 183$ centimeters

81. $a = 76.3$ feet, $b = 109$ feet, $c = 98.8$ feet

82. $a = 22$ inches, $b = 45$ inches, $c = 31$ inches

83. $a = 25.4$ yards, $b = 38.2$ yards, $c = 19.8$ yards

84. $a = 15.89$ inches, $b = 21.74$ inches, $c = 10.92$ inches

Solve each problem.

85. *Cans of Paint Needed for a Job* A painter needs to cover a triangular region 75 meters by 68 meters by 85 meters. A can of paint covers 75 square meters of area. How many cans (to the next higher number of cans) will be needed?

86. *Area of a Triangular Lot* A real estate agent wants to find the area of a triangular lot. A surveyor takes measurements and finds that two sides are 52.1 meters and 21.3 meters, and the angle between them is 42.2°. What is the area of the lot?

87. *Area of a Metal Plate* A painter is going to apply a special coating to a triangular metal plate on a new building. Two sides measure 16.1 meters and 15.2 meters. She knows that the angle between these sides is 125°. What is the area of the surface she plans to cover with the coating?

88. *Area of the Bermuda Triangle* Find the area of the Bermuda Triangle if the sides of the triangle have approximate lengths of 850 miles, 925 miles, and 1300 miles.

89. *Perfect Triangles*   A **perfect triangle** is a triangle whose sides have whole-number lengths and whose area is numerically equal to its perimeter. Show that the triangle with sides of lengths 9, 10, and 17 is perfect.

90. *Heron Triangles*   A **Heron triangle** is a triangle having integer sides and integer area. Show that each of the following is a Heron triangle.
(a) $a = 11$, $b = 13$, $c = 20$
(b) $a = 13$, $b = 14$, $c = 15$
(c) $a = 7$, $b = 15$, $c = 20$

91. Consider triangle *ABC* shown here.

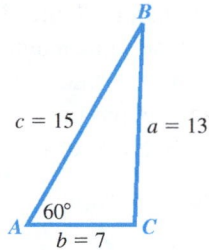

(a) Use the law of sines to find candidates for the value of angle *C*. Round angle measures to the nearest tenth of a degree.
(b) Rework part (a) using the law of cosines.
(c) Why is the law of cosines a better method in this case?

92. Show that the measure of angle *A* is twice the measure of angle *B*. (*Hint:* Use the law of cosines to find cos *A* and cos *B*, and then show that $\cos A = 2\cos^2 B - 1$, implying that $A = 2B$.)

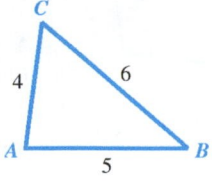

10.3 Vectors and Their Applications

Basic Terminology • Interpretations of Vectors • Operations with Vectors • Dot Product and the Angle between Vectors
• Applications of Vectors

A Vector

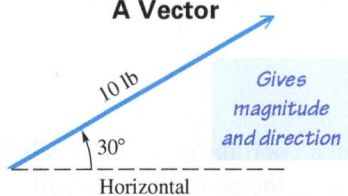

This vector represents a force of 10 pounds applied at an angle of 30° above the horizontal.

FIGURE 19

Basic Terminology

Quantities that involve magnitudes, such as 10 pounds or 60 mph, can be represented by real numbers called **scalars**. Other quantities, called **vector quantities**, involve both magnitude *and* direction. Typical vector quantities are velocity, acceleration, and force. For example, traveling 50 mph *east* represents a vector quantity.

A vector quantity is often represented with a directed line segment (a segment that uses an arrowhead to indicate direction), called a **vector.** The *length* of the vector represents the **magnitude** of the vector quantity. See **FIGURE 19.**

To write vectors by hand, it is customary to use boldface type or an arrow over the letter or letters. Thus, **OP** and $\overrightarrow{OP}$ both represent vector **OP.** When two letters name a vector, the first indicates the **initial point** and the second indicates the **terminal point** of the vector. Knowing these points gives the direction of the vector. For example, vectors **OP** and **PO** in **FIGURE 20** are *not* the same vectors. They have the same magnitude, but *opposite directions*. The magnitude of vector **OP** is written $|\textbf{OP}|$.

Two vectors are equal if and only if they both have the same direction and the same magnitude. See **FIGURE 21.**

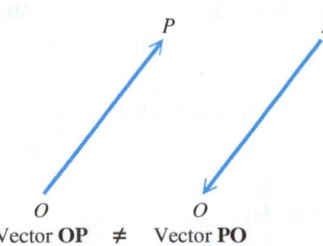

Vector **OP** ≠ Vector **PO**

Vectors may be named with one lowercase or uppercase letter, or two uppercase letters.

FIGURE 20

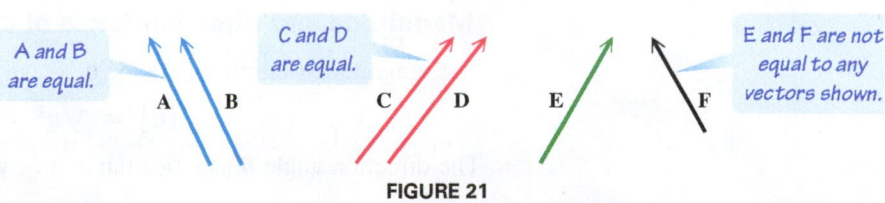

FIGURE 21

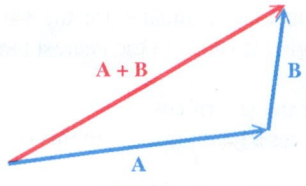

FIGURE 22

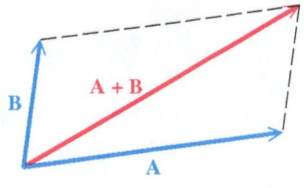

Parallelogram rule
FIGURE 23

Opposite Directions

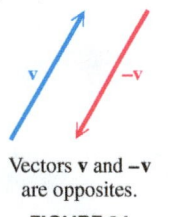

Vectors **v** and **−v**
are opposites.
FIGURE 24

The sum of two vectors is also a vector. There are two ways to geometrically find the sum of two vectors **A** and **B**.

1. Place the initial point of vector **B** at the terminal point of vector **A,** as in **FIGURE 22.** The vector with the same initial point as **A** and the same terminal point as **B** is the sum **A + B.**

2. Use the **parallelogram rule.** Place vectors **A** and **B** so that their initial points coincide, as in **FIGURE 23.** Then, complete a parallelogram that has **A** and **B** as two sides. The diagonal of the parallelogram with the same initial point as **A** and **B** is the sum **A + B.**

Parallelograms can be used to show that vector **B + A** is the same as vector **A + B** (as in **FIGURE 23**) or that **A + B = B + A,** so *vector addition is commutative. The vector sum **A + B** is called the **resultant** of vectors **A** and **B.***

For every vector **v,** there is a vector **−v** that has the same magnitude as **v** but opposite direction. Vector **−v** is called the **opposite** of **v.** See **FIGURE 24.** The sum of **v** and **−v** has magnitude 0 and is called the **zero vector.** As with real numbers, to subtract vector **B** from vector **A,** find the vector sum **A + (−B).** See **FIGURE 25.**

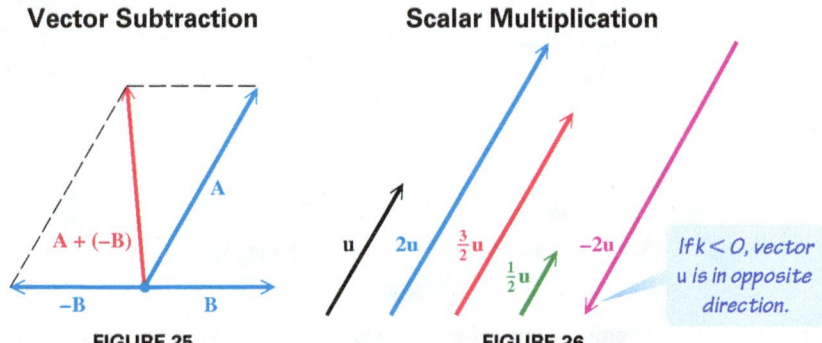

Vector Subtraction **Scalar Multiplication**

FIGURE 25 **FIGURE 26**

The process of forming the product of a scalar k and a vector **u** is called **scalar multiplication.** The vector k**u** has magnitude $|k|$ times the magnitude of **u.** As suggested by **FIGURE 26,** the vector k**u** has the same direction as **u** if $k > 0$ and opposite direction if $k < 0$.

Interpretations of Vectors

Position Vector u = ⟨a, b⟩

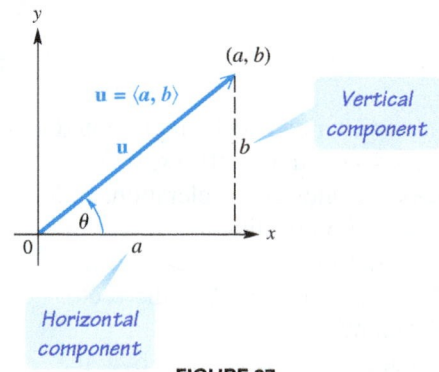

FIGURE 27

A vector with its initial point at the origin in a rectangular coordinate system is called a **position vector.** A position vector **u** with its endpoint at the point (a, b) is written $\langle a, b \rangle$, so **u** $= \langle a, b \rangle$. *Every vector in the real plane corresponds to an ordered pair of real numbers. Geometrically a vector is a directed line segment and algebraically it is an ordered pair.* The numbers a and b are, respectively, the **horizontal component** and **vertical component** of vector **u.**

FIGURE 27 shows the vector **u** $= \langle a, b \rangle$. The positive angle between the x-axis and a position vector is called the **direction angle** for the vector. In **FIGURE 27,** θ is the direction angle for vector **u.** We usually choose θ such that $0° \le \theta < 360°$ (although there will be exceptions).

From **FIGURE 27,** we can see that the magnitude and direction of a vector are related to its horizontal and vertical components.

Magnitude and Direction Angle of a Vector $\langle a, b \rangle$

The magnitude (length) of vector **u** $= \langle a, b \rangle$ is given by the following.

$$|\mathbf{u}| = \sqrt{a^2 + b^2}$$

The direction angle θ satisfies $\tan \theta = \dfrac{b}{a}$, where $a \ne 0$.

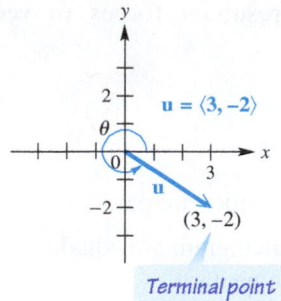

FIGURE 28

EXAMPLE 1   **Finding Magnitude and Direction Angle**

Find the magnitude and direction angle of $\mathbf{u} = \langle 3, -2 \rangle$.

Solution   The magnitude of vector $\mathbf{u}$ is

$$|\mathbf{u}| = \sqrt{3^2 + (-2)^2} = \sqrt{13}.$$

$$|\mathbf{u}| = \sqrt{a^2 + b^2}$$

To find the direction angle θ, start with $\tan \theta = \dfrac{b}{a} = \dfrac{-2}{3} = -\dfrac{2}{3}$. Vector $\mathbf{u}$ has a positive horizontal component and a negative vertical component, placing the position vector in quadrant IV. A calculator gives $\tan^{-1}\left(-\dfrac{2}{3}\right) \approx -33.7°$. Adding 360° yields the direction angle $\theta = 326.3°$, as shown in **FIGURE 28.**

Horizontal and Vertical Components

The horizontal and vertical components, respectively, of a vector $\mathbf{u}$ having magnitude $|\mathbf{u}|$ and direction angle θ are given by

$$a = |\mathbf{u}| \cos \theta \quad \text{and} \quad b = |\mathbf{u}| \sin \theta.$$

That is, $\mathbf{u} = \langle a, b \rangle = \langle |\mathbf{u}| \cos \theta, |\mathbf{u}| \sin \theta \rangle$.

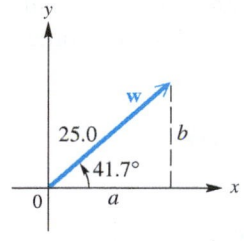

FIGURE 29

EXAMPLE 2   **Finding Horizontal and Vertical Components**

Vector $\mathbf{w}$ in **FIGURE 29** has magnitude 25.0 and direction angle 41.7°. Find the horizontal and vertical components.

Analytic Solution

Use the two formulas in the box above with $|\mathbf{w}| = 25.0$ and $\theta = 41.7°$.

$$a = 25.0 \cos 41.7° \qquad b = 25.0 \sin 41.7°$$
$$a = 18.7 \qquad\qquad b = 16.6$$

Therefore, $\mathbf{w} = \langle 18.7, 16.6 \rangle$. The horizontal component is 18.7, and the vertical component is 16.6 (rounded to the nearest tenth).

Graphing Calculator Solution

See **FIGURE 30.** The results support the analytic solution.

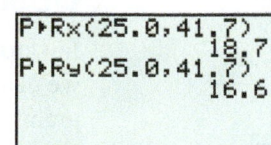

FIGURE 30

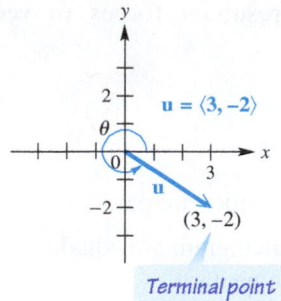

FIGURE 31

EXAMPLE 3   **Writing Vectors in the Form $\langle a, b \rangle$**

Write each vector in **FIGURE 31** in the form $\langle a, b \rangle$.

Solution

$$\mathbf{u} = \langle 5 \cos 60°, 5 \sin 60° \rangle = \left\langle 5 \cdot \frac{1}{2}, 5 \cdot \frac{\sqrt{3}}{2} \right\rangle = \left\langle \frac{5}{2}, \frac{5\sqrt{3}}{2} \right\rangle$$

$$\mathbf{v} = \langle 2 \cos 180°, 2 \sin 180° \rangle = \langle 2(-1), 2(0) \rangle = \langle -2, 0 \rangle$$

$$\mathbf{w} = \langle 6 \cos 280°, 6 \sin 280° \rangle \approx \langle 1.0419, -5.9088 \rangle \quad \text{Use a calculator.}$$

The following properties can be used to find resultant forces in vector applications.

Properties of Parallelograms

1. A parallelogram is a quadrilateral whose opposite sides are parallel.

2. The opposite sides and opposite angles of a parallelogram are equal, and adjacent angles of a parallelogram are supplementary.

3. The diagonals of a parallelogram bisect each other, but do not necessarily bisect the angles of the parallelogram.

Resultant v

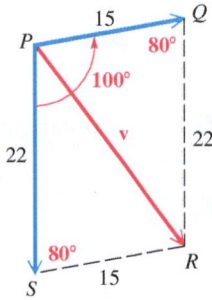

FIGURE 32

EXAMPLE 4   Finding the Magnitude of a Resultant

Two forces of 15 and 22 newtons act on a point in the plane. (A **newton** is a unit of force.) If the angle between the forces is 100°, find the magnitude of the resultant force.

Solution   As shown in **FIGURE 32,** a parallelogram that has the forces as adjacent sides can be formed. The angles of the parallelogram adjacent to angle P measure 80°, since adjacent angles of a parallelogram are supplementary. Also, opposite sides of the parallelogram are equal in length. The resultant force divides the parallelogram into two triangles. Use the law of cosines with either triangle.

$$|\mathbf{v}|^2 = 15^2 + 22^2 - 2(15)(22) \cos 80° \qquad \text{Law of cosines}$$

$$|\mathbf{v}|^2 = 594 \qquad \text{Simplify.}$$

$$|\mathbf{v}| = 24 \qquad \text{Take the positive square root.}$$

To the nearest unit, the magnitude of the resultant force is 24 newtons. ●

Operations with Vectors

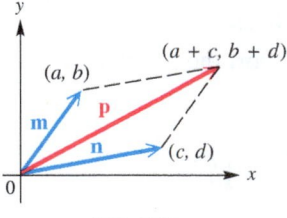

FIGURE 33

In **FIGURE 33, $\mathbf{m} = \langle a, b \rangle$, $\mathbf{n} = \langle c, d \rangle$,** and **$\mathbf{p} = \langle a + c, b + d \rangle$.** Using geometry, we can show that the endpoints of the three vectors and their origin form a parallelogram. Since a diagonal of this parallelogram gives the resultant of vectors **m** and **n,** we have **$\mathbf{p} = \mathbf{m} + \mathbf{n}$,** or

$$\langle a + c, b + d \rangle = \langle a, b \rangle + \langle c, d \rangle.$$

Similarly, we could verify the following vector operations.

→ Looking Ahead to Calculus

In addition to two-dimensional vectors in a plane, calculus courses introduce three-dimensional vectors in space. (See **Appendix B.**) The magnitude of the two-dimensional vector $\langle a, b \rangle$ is given by $\sqrt{a^2 + b^2}$. If we extend this to the three-dimensional vector $\langle a, b, c \rangle$, the magnitude expression becomes $\sqrt{a^2 + b^2 + c^2}$. Similar extensions are made for other concepts.

Vector Operations

Let a, b, c, d, and k represent real numbers.

$$\langle a, b \rangle + \langle c, d \rangle = \langle a + c, b + d \rangle$$

$$k \langle a, b \rangle = \langle ka, kb \rangle$$

If $\mathbf{u} = \langle a_1, a_2 \rangle$, then $-\mathbf{u} = \langle -a_1, -a_2 \rangle$.

$$\langle a, b \rangle - \langle c, d \rangle = \langle a, b \rangle + -\langle c, d \rangle = \langle a - c, b - d \rangle$$

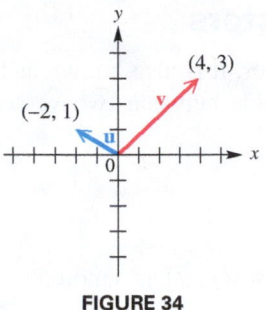

FIGURE 34

EXAMPLE 5 **Performing Vector Operations**

Let $\mathbf{u} = \langle -2, 1 \rangle$ and $\mathbf{v} = \langle 4, 3 \rangle$. Find and illustrate **(a)** $\mathbf{u} + \mathbf{v}$, **(b)** $-2\mathbf{u}$, and **(c)** $3\mathbf{u} - 2\mathbf{v}$. See **FIGURE 34**.

Solution See **FIGURE 35** for geometric interpretations.

(a) $\mathbf{u} + \mathbf{v} = \langle -2, 1 \rangle + \langle 4, 3 \rangle$
$= \langle -2 + 4, 1 + 3 \rangle$
$= \langle 2, 4 \rangle$

(b) $-2\mathbf{u} = -2 \cdot \langle -2, 1 \rangle$
$= \langle -2(-2), -2(1) \rangle$
$= \langle 4, -2 \rangle$

(c) $3\mathbf{u} - 2\mathbf{v} = 3 \cdot \langle -2, 1 \rangle - 2 \cdot \langle 4, 3 \rangle$
$= \langle -6, 3 \rangle - \langle 8, 6 \rangle$
$= \langle -6 - 8, 3 - 6 \rangle$
$= \langle -14, -3 \rangle$

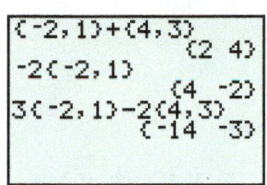

FIGURE 36

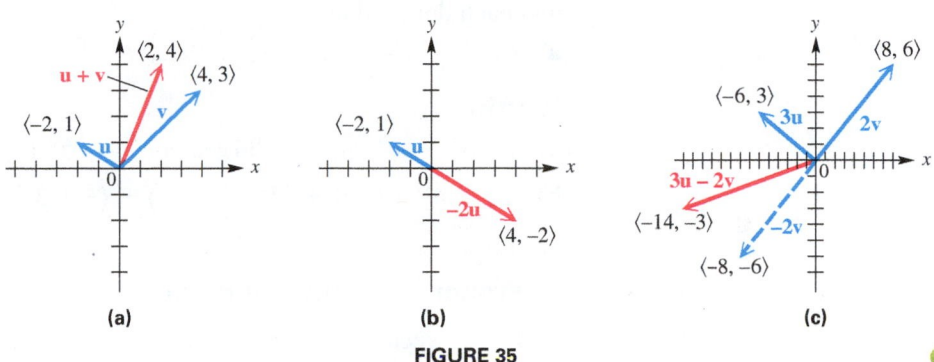

(a)   (b)   (c)

FIGURE 35

A **unit vector** is a vector that has magnitude 1. Two important unit vectors are defined as follows and shown in **FIGURE 37**.

$$\mathbf{i} = \langle 1, 0 \rangle \qquad \mathbf{j} = \langle 0, 1 \rangle$$

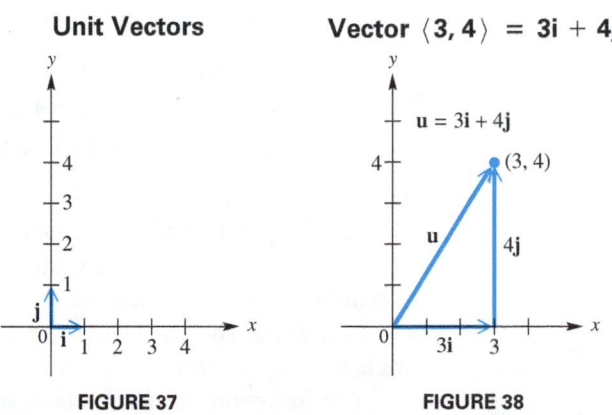

FIGURE 37   **FIGURE 38**

With the unit vectors $\mathbf{i}$ and $\mathbf{j}$, we can express any other vector $\langle a, b \rangle$ in the form $a\mathbf{i} + b\mathbf{j}$, as shown in **FIGURE 38**, where $\langle 3, 4 \rangle = 3\mathbf{i} + 4\mathbf{j}$. The vector operations previously given can be restated in $a\mathbf{i} + b\mathbf{j}$ notation.

Vector Notation Using i and j

If $\mathbf{v} = \langle a, b \rangle$, then $\mathbf{v} = a\mathbf{i} + b\mathbf{j}$.

Dot Product and the Angle between Vectors

The *dot product* of two vectors is a real number, *not* a vector. It is also known as the *inner product*. Dot products are used to determine the angle between two vectors, derive geometric theorems, and solve physics problems.

Dot Product

The **dot product** of the two vectors $\mathbf{u} = \langle a, b \rangle$ and $\mathbf{v} = \langle c, d \rangle$ is denoted $\mathbf{u} \cdot \mathbf{v}$, read "**u** dot **v**," and given by the following.

$$\mathbf{u} \cdot \mathbf{v} = ac + bd$$

EXAMPLE 6 **Finding the Dot Product**

Find each dot product.

(a) $\langle 2, 3 \rangle \cdot \langle 4, -1 \rangle$ **(b)** $(6\mathbf{i} + 4\mathbf{j}) \cdot (-2\mathbf{i} + 3\mathbf{j})$

Solution

(a) $\langle 2, 3 \rangle \cdot \langle 4, -1 \rangle = 2(4) + 3(-1) = 5$

(b) $(6\mathbf{i} + 4\mathbf{j}) \cdot (-2\mathbf{i} + 3\mathbf{j}) = \langle 6, 4 \rangle \cdot \langle -2, 3 \rangle = 6(-2) + 4(3) = 0$ ●

Properties of the Dot Product

For all vectors $\mathbf{u}$, $\mathbf{v}$, and $\mathbf{w}$ and real numbers k, the following hold.

(a) $\mathbf{u} \cdot \mathbf{v} = \mathbf{v} \cdot \mathbf{u}$ **(b)** $\mathbf{u} \cdot (\mathbf{v} + \mathbf{w}) = \mathbf{u} \cdot \mathbf{v} + \mathbf{u} \cdot \mathbf{w}$

(c) $(\mathbf{u} + \mathbf{v}) \cdot \mathbf{w} = \mathbf{u} \cdot \mathbf{w} + \mathbf{v} \cdot \mathbf{w}$ **(d)** $(k\mathbf{u}) \cdot \mathbf{v} = k(\mathbf{u} \cdot \mathbf{v}) = \mathbf{u} \cdot (k\mathbf{v})$

(e) $\mathbf{0} \cdot \mathbf{u} = 0$ **(f)** $\mathbf{u} \cdot \mathbf{u} = |\mathbf{u}|^2$

To prove the first part of property (d), we let $\mathbf{u} = \langle a, b \rangle$ and $\mathbf{v} = \langle c, d \rangle$.

$$(k\mathbf{u}) \cdot \mathbf{v} = (k\langle a, b \rangle) \cdot \langle c, d \rangle = \langle ka, kb \rangle \cdot \langle c, d \rangle$$
$$= kac + kbd = k(ac + bd)$$
$$= k(\langle a, b \rangle \cdot \langle c, d \rangle) = k(\mathbf{u} \cdot \mathbf{v})$$

The proofs of the remaining properties are similar.

The dot product of two vectors can be positive, 0, or negative. A geometric interpretation of the dot product explains when each of these cases occurs. This interpretation involves the angle between the two vectors. Consider the vectors $\mathbf{u} = \langle a, b \rangle$ and $\mathbf{v} = \langle c, d \rangle$, as shown in **FIGURE 39**. The **angle θ between u and v** is defined to be the angle having the two vectors as its sides for which $0° \leq \theta \leq 180°$.

The following theorem relates the dot product to the angle between the vectors. Its proof is outlined in **Exercise 112**.

Angle θ between u and v

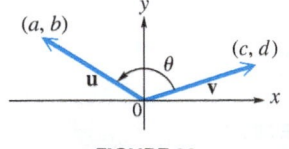

FIGURE 39

Geometric Interpretation of Dot Product

If θ is the angle between the two nonzero vectors $\mathbf{u}$ and $\mathbf{v}$, where $0° \leq \theta \leq 180°$, then the following holds.

$$\mathbf{u} \cdot \mathbf{v} = |\mathbf{u}||\mathbf{v}| \cos \theta, \quad \text{or equivalently,} \quad \cos \theta = \frac{\mathbf{u} \cdot \mathbf{v}}{|\mathbf{u}||\mathbf{v}|}$$

EXAMPLE 7 **Finding the Angle between Two Vectors**

Find the angle θ between the two vectors.

(a) $\mathbf{u} = \langle 3, 4 \rangle$ and $\mathbf{v} = \langle 2, 1 \rangle$ **(b)** $\mathbf{u} = \langle 2, -6 \rangle$ and $\mathbf{v} = \langle 6, 2 \rangle$

Solution

(a) $\cos \theta = \dfrac{\mathbf{u} \cdot \mathbf{v}}{|\mathbf{u}||\mathbf{v}|} = \dfrac{\langle 3, 4 \rangle \cdot \langle 2, 1 \rangle}{|\langle 3, 4 \rangle||\langle 2, 1 \rangle|}$ Substitute values.

Use the preceding geometric interpretation.

$= \dfrac{3(2) + 4(1)}{\sqrt{9 + 16} \cdot \sqrt{4 + 1}} = \dfrac{10}{5\sqrt{5}} \approx 0.894427191$ Use the definitions.

$\theta = \cos^{-1} 0.894427191 = 26.57°$ Inverse function

(b) $\cos \theta = \dfrac{\mathbf{u} \cdot \mathbf{v}}{|\mathbf{u}||\mathbf{v}|} = \dfrac{\langle 2, -6 \rangle \cdot \langle 6, 2 \rangle}{|\langle 2, -6 \rangle||\langle 6, 2 \rangle|}$ Substitute values.

$= \dfrac{2(6) + (-6)(2)}{\sqrt{4 + 36} \cdot \sqrt{36 + 4}}$ Use the definitions.

$= \dfrac{0}{40} = 0$ Evaluate.

$\theta = \cos^{-1} 0 = 90°$ Inverse function   ●

For angles θ between $0°$ and $180°$, the table shows the relationship among $\cos \theta$, the dot product, and θ.

Cos θ	Dot Product	Angle θ between Vectors
Positive	Positive	Acute
0	0	Right
Negative	Negative	Obtuse

Orthogonal Vectors

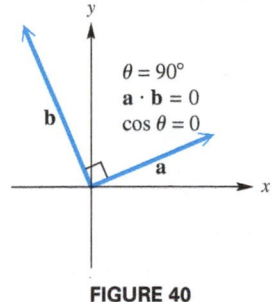

$\theta = 90°$
$\mathbf{a} \cdot \mathbf{b} = 0$
$\cos \theta = 0$

FIGURE 40

NOTE   If $\mathbf{a} \cdot \mathbf{b} = 0$ for two nonzero vectors $\mathbf{a}$ and $\mathbf{b}$, then $\cos \theta = 0$ and $\theta = 90°$. Thus, $\mathbf{a}$ and $\mathbf{b}$ are perpendicular or **orthogonal vectors**. See **Example 7(b)** and **FIGURE 40**.

Applications of Vectors

The law of sines and the law of cosines are often used to solve applied problems involving vectors.

EXAMPLE 8   **Applying Vectors to a Navigation Problem**

A ship leaves port on a bearing of $28°$ and travels 8.2 miles. The ship then turns due east and travels 4.3 miles. How far is the ship from port? What is its bearing from port?

FIGURE 41

Solution   In **FIGURE 41**, vectors **PA** and **AE** represent the ship's path. We must find the magnitude and bearing of the resultant **PE**. Triangle *PNA* is a right triangle, so angle $NAP = 90° - 28° = 62°$. Then angle $PAE = 180° - 62° = 118°$. Use the law of cosines to find $|\mathbf{PE}|$, the magnitude of vector **PE**.

$|\mathbf{PE}|^2 = 8.2^2 + 4.3^2 - 2(8.2)(4.3)\cos 118°$ Law of cosines

$|\mathbf{PE}|^2 = 118.84$ Approximate.

$|\mathbf{PE}| = 10.9$ Take the square root.

(continued)

Air Navigation

Final bearing and ground speed
(actual direction of plane)

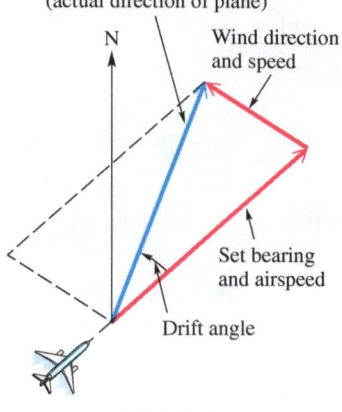

FIGURE 42

To find the bearing of the ship from port, first find angle *APE*. Use the law of sines, along with the value of $|\mathbf{PE}|$.

$$\frac{\sin APE}{4.3} = \frac{\sin 118°}{10.9} \qquad \text{Law of sines}$$

$$\sin APE = \frac{4.3 \sin 118°}{10.9} \qquad \text{Multiply by 4.3.}$$

$$APE = 20° \qquad \text{Use the inverse sine function.}$$

The ship is approximately 11 miles from port at a bearing of $28° + 20° = 48°$. ●

In air navigation, the **airspeed** of a plane is its speed relative to the air, while the **ground speed** is its speed relative to the ground. Because of wind, these two speeds are usually different. The ground speed of the plane is represented by the magnitude of the vector sum of the airspeed and wind speed vectors. See **FIGURE 42.**

EXAMPLE 9   **Applying Vectors to a Navigation Problem**

A plane with an airspeed of 192 mph is headed on a bearing of 121°. A north wind is blowing (from north to south) at 15.9 mph. Find the ground speed and the final bearing of the plane.

Solution   In **FIGURE 43**, the ground speed is represented by $|\mathbf{x}|$. We must find angle α to find the bearing, which will be $121° + \alpha$. From **FIGURE 43**, angle *BCO* equals angle *AOC*, which is equal to 121°. We must find $|\mathbf{x}|$.

$$|\mathbf{x}|^2 = 192^2 + 15.9^2 - 2(192)(15.9)\cos 121° \qquad \text{Law of cosines}$$

$$|\mathbf{x}|^2 = 40,261 \qquad \text{Approximate.}$$

$$|\mathbf{x}| = 200.7 \qquad \text{Take the square root.}$$

We find α by using the law of sines and the value of $|\mathbf{x}|$ before rounding.

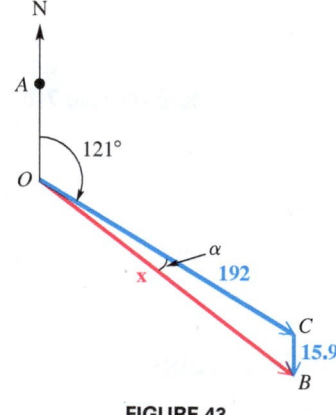

FIGURE 43

$$\frac{\sin \alpha}{15.9} = \frac{\sin 121°}{200.7} \qquad \text{Law of sines}$$

$$\sin \alpha = \frac{15.9 \sin 121°}{200.7} \qquad \text{Multiply by 15.9.}$$

$$\alpha = \sin^{-1}\left(\frac{15.9 \sin 121°}{200.7}\right) \qquad \text{Use the inverse sine function.}$$

$$\alpha = 3.89° \qquad \text{Use a calculator.}$$

To the nearest degree, α is 4°. The ground speed is about 201 mph on a final bearing of approximately $121° + 4° = 125°$. ●

EXAMPLE 10   **Applying Vectors to a Navigation Problem**

An airplane that is following a bearing of 239° at an airspeed of 425 mph encounters a wind blowing at 36.0 mph from a direction of 115°. Find the resulting bearing and ground speed of the plane.

Solution   An accurate sketch is essential to the solution of this problem. We have included two sets of geographical axes which enable us to determine measures of necessary angles. Analyze **FIGURE 44** carefully.

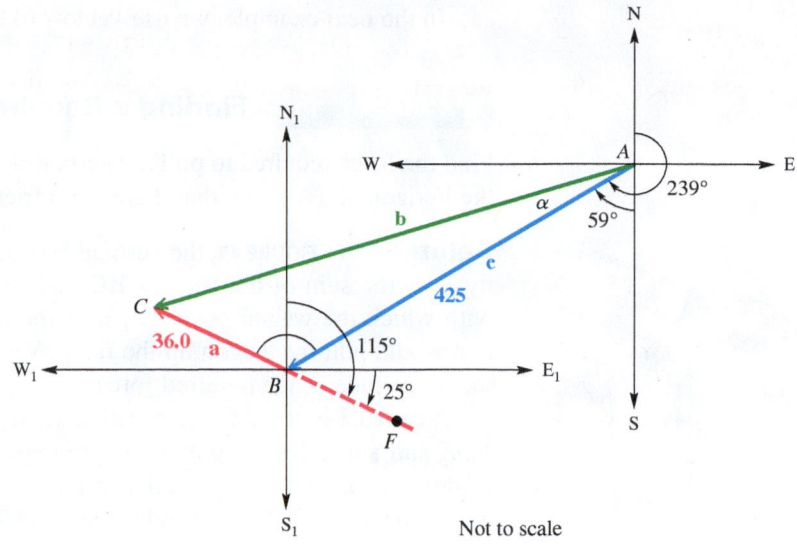

FIGURE 44

Vector **c** represents the airspeed and heading of the plane, and vector **a** represents the speed and direction of the wind. Angle ABC has as its measure the sum of angle ABN_1 and angle N_1BC.

- Angle SAB measures $239° - 180° = 59°$. Because angle ABN_1 is an alternate interior angle to it, $ABN_1 = 59°$.

- Angle E_1BF measures $115° - 90° = 25°$. Thus, angle CBW_1 also measures $25°$ because it is a vertical angle. Angle N_1BC is the complement of $25°$, which is

$$90° - 25° = 65°.$$

By these results,

$$\text{angle } ABC = 59° + 65° = 124°.$$

To find $|\mathbf{b}|$, we use the law of cosines.

$$|\mathbf{b}|^2 = |\mathbf{a}|^2 + |\mathbf{c}|^2 - 2|\mathbf{a}||\mathbf{c}| \cos ABC \qquad \text{Law of cosines}$$

$$|\mathbf{b}|^2 = 36.0^2 + 425^2 - 2(36.0)(425) \cos 124° \qquad \text{Substitute.}$$

$$|\mathbf{b}|^2 = 199{,}032 \qquad \text{Use a calculator.}$$

The ground speed is approximately 446 mph.

$$|\mathbf{b}| = 446 \qquad \text{Take the square root.}$$

To find the resulting bearing of **b,** we must find the measure of angle α in **FIGURE 44** and then add it to $239°$. To find α, we use the law of sines.

$$\frac{\sin \alpha}{36.0} = \frac{\sin 124°}{446}$$

To maintain accuracy, use all the significant digits that your calculator allows.

$$\sin \alpha = \frac{36.0 \sin 124°}{446} \qquad \text{Multiply by 36.0.}$$

$$\alpha = \sin^{-1}\left(\frac{36.0 \sin 124°}{446}\right) \qquad \text{Use the inverse sine function.}$$

$$\alpha = 4° \qquad \text{Use a calculator.}$$

Add $4°$ to $239°$ to find the resulting bearing of $243°$.

In the next example, we use vectors to solve an inclined plane problem.

EXAMPLE 11 **Finding a Required Force**

Find the force required to pull a wagon weighing 50 pounds up a ramp inclined 20° to the horizontal. (Assume that there is no friction.)

Solution In **FIGURE 45,** the vertical 50-pound force **BA** represents the force of gravity. It is the sum of the vectors **BC** and **−AC.** The vector **BC** represents the force with which the wagon pushes against the ramp. The vector **BF** represents the force that would pull the wagon up the ramp. Vectors **BF** and **AC** are equal, so |**AC**| gives the magnitude of the required force.

Vectors **BF** and **AC** are parallel, so angle *EBD* is equal to angle *A.* Since angle *BDE* and angle *C* are right angles, triangles *CBA* and *DEB* have two corresponding angles equal and thus are similar triangles. Therefore, angle *ABC* is equal to angle *E,* which measures 20°. From right triangle *ABC,* we have the following.

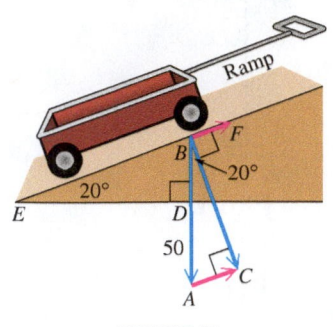

FIGURE 45

$$\sin 20° = \frac{|\mathbf{AC}|}{50}$$

$$|\mathbf{AC}| = 50 \sin 20° \qquad \text{Multiply by 50 and rewrite.}$$

$$|\mathbf{AC}| = 17 \qquad \text{Use a calculator.}$$

To the nearest pound, a 17-pound force is required to pull the wagon up the ramp.

10.3 Exercises

Concept Check *Refer to vectors* **m** *through* **t** *at the right.*

1. Name all pairs of vectors that appear to be equal.

2. Name all pairs of vectors that are opposites.

3. Name all pairs of vectors such that the first is a scalar multiple of the other, with the scalar positive.

4. Name all pairs of vectors such that the first is a scalar multiple of the other, with the scalar negative.

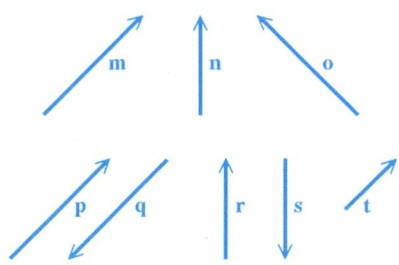

Concept Check *Refer to vectors* **a** *through* **h**. *Make a copy or a sketch of each vector, and then draw a sketch to represent each vector in Exercises 5–16. For example, find* **a** + **e** *by placing* **a** *and* **e** *so that their initial points coincide. Then, use the parallelogram rule to find the resultant, shown in the figure at the far right.*

5. −**b**	**6.** −**g**
7. 3**a**	**8.** 2**h**
9. **a** + **b**	**10.** **h** + **g**
11. **a** − **c**	**12.** **d** − **e**
13. **a** + (**b** + **c**)	**14.** (**a** + **b**) + **c**
15. **c** + **d**	**16.** **d** + **c**

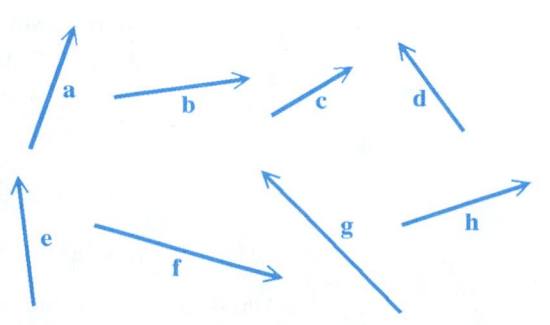
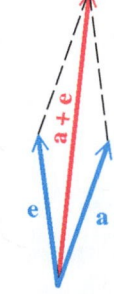

Checking Analytic Skills   *In Exercises 17–22, use the figure to find each vector:* **(a) u + v**
(b) u − v   **(c) −u.** *Do not use a calculator.*

17.

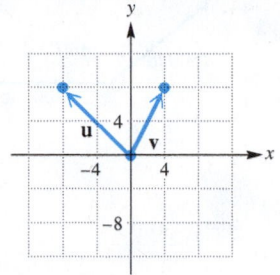

18.

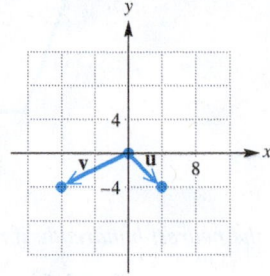

19.

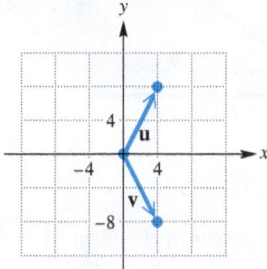

20.

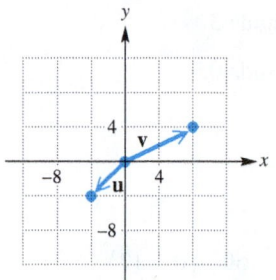

21.

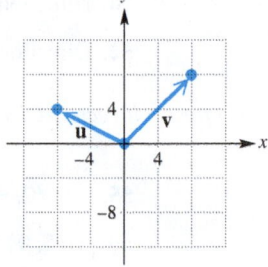

22.

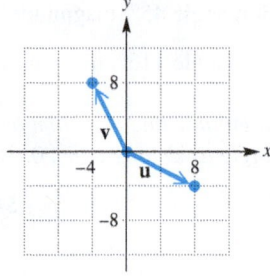

Checking Analytic Skills   *Given vectors* **u** *and* **v**, *find* **(a) 2u**   **(b) 2u + 3v**   **(c) v − 3u.** *Do not use a calculator.*

23. **u** = 2**i**, **v** = **i** + **j**

24. **u** = −**i** + 2**j**, **v** = **i** − **j**

25. **u** = $\langle -1, 2 \rangle$, **v** = $\langle 3, 0 \rangle$

26. **u** = $\langle -2, -1 \rangle$, **v** = $\langle -3, 2 \rangle$

Checking Analytic Skills   *Given* **u** = $\langle -2, 5 \rangle$ *and* **v** = $\langle 4, 3 \rangle$, *find each vector.* **Do not use a calculator.**

27. **u** + **v**   **28.** **u** − **v**   **29.** **v** − **u**   **30.** 5**v**   **31.** −5**v**   **32.** 3**u** + 6**v**

Write each vector **v** *in the form* $\langle a, b \rangle$. *In Exercises 33–36, give exact values. In Exercises 37–40, give approximations to the nearest hundredth.*

33.

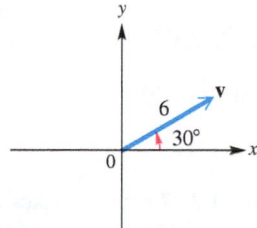

34.

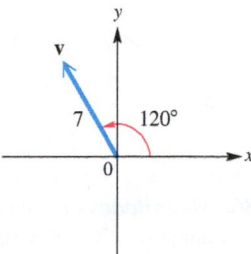

35.

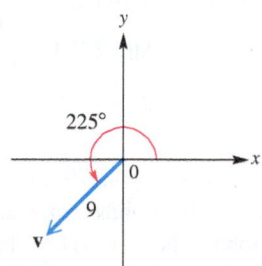

36.

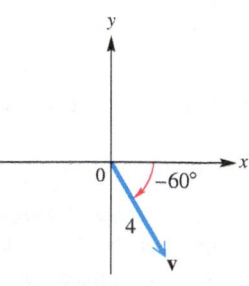

37.

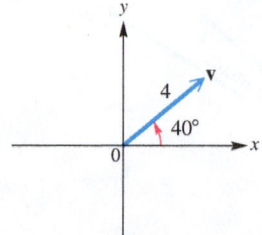

38.

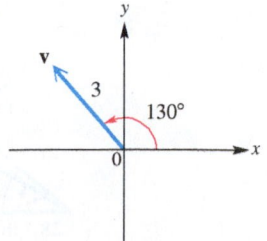

39.

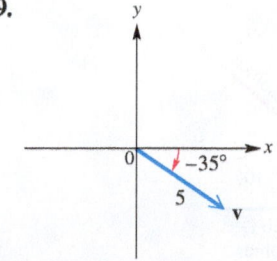

40.

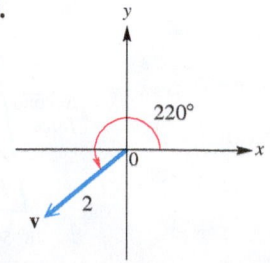

Use the parallelogram rule to find the magnitude of the resultant force of the two forces shown in each figure. Give answers to the nearest tenth.

41.
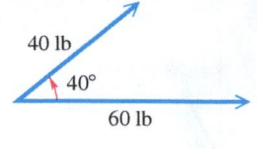

42.
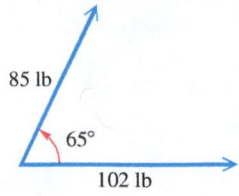

43.
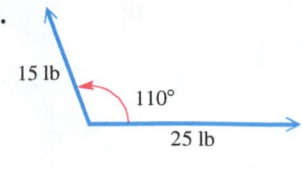

44.
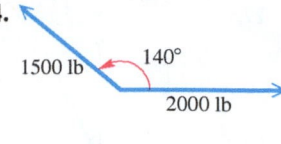

*Write each vector in the form a**i** + b**j**. Round a and b to the nearest hundredth, if necessary.*

45. $\langle -5, 8 \rangle$ **46.** $\langle 6, -3 \rangle$ **47.** $\langle 2, 0 \rangle$ **48.** $\langle 0, -4 \rangle$

49. Direction angle 45°, magnitude 8   **50.** Direction angle 210°, magnitude 3

51. Direction angle 115°, magnitude 0.6   **52.** Direction angle 208°, magnitude 0.9

Find the magnitude and direction angle (to the nearest tenth) for each vector. Give the measure of the direction angle as an angle in $[0, 360°)$.

53. $\langle 1, 1 \rangle$ **54.** $\langle -4, 4\sqrt{3} \rangle$ **55.** $\langle 8\sqrt{2}, -8\sqrt{2} \rangle$ **56.** $\langle \sqrt{3}, -1 \rangle$

57. $\langle 15, -8 \rangle$ **58.** $\langle -7, 24 \rangle$ **59.** $\langle -6, 0 \rangle$ **60.** $\langle 0, -12 \rangle$

Find the dot product of each pair of vectors.

61. $\langle 6, -1 \rangle, \langle 2, 5 \rangle$ **62.** $\langle -3, 8 \rangle, \langle 7, -5 \rangle$ **63.** $\langle 2, -3 \rangle, \langle 6, 5 \rangle$

64. $\langle 1, 2 \rangle, \langle 3, -1 \rangle$ **65.** $4\mathbf{i}, 5\mathbf{i} - 9\mathbf{j}$ **66.** $2\mathbf{i} + 4\mathbf{j}, -\mathbf{j}$

Find the angle between each pair of vectors.

67. $\langle 2, 1 \rangle, \langle -3, 1 \rangle$ **68.** $\langle 4, 0 \rangle, \langle 2, 2 \rangle$ **69.** $\langle 1, 2 \rangle, \langle -6, 3 \rangle$ **70.** $\langle 6, 8 \rangle, \langle -4, 3 \rangle$

71. $\mathbf{i} + 7\mathbf{j}, \mathbf{i} + \mathbf{j}$ **72.** $3\mathbf{i} + 4\mathbf{j}, \mathbf{j}$ **73.** $\mathbf{i} + \mathbf{j}, 3\mathbf{i} + 4\mathbf{j}$ **74.** $-5\mathbf{i} + 12\mathbf{j}, 3\mathbf{i} + 2\mathbf{j}$

Let $\mathbf{u} = \langle -2, 1 \rangle, \mathbf{v} = \langle 3, 4 \rangle$, *and* $\mathbf{w} = \langle -5, 12 \rangle$. *Evaluate each expression.*

75. $(3\mathbf{u}) \cdot \mathbf{v}$ **76.** $\mathbf{u} \cdot (\mathbf{v} - \mathbf{w})$ **77.** $\mathbf{u} \cdot \mathbf{v} - \mathbf{u} \cdot \mathbf{w}$ **78.** $\mathbf{u} \cdot (3\mathbf{v})$

Determine whether each pair of vectors is orthogonal.

79. $\langle 1, 2 \rangle, \langle -6, 3 \rangle$ **80.** $\langle 3, 4 \rangle, \langle 6, 8 \rangle$ **81.** $\langle 1, 0 \rangle, \langle \sqrt{2}, 0 \rangle$

82. $\langle 1, 1 \rangle, \langle 1, -1 \rangle$ **83.** $\sqrt{5}\mathbf{i} - 2\mathbf{j}, -5\mathbf{i} + 2\sqrt{5}\mathbf{j}$ **84.** $-4\mathbf{i} + 3\mathbf{j}, 8\mathbf{i} - 6\mathbf{j}$

Solve each problem.

85. *Magnitudes of Forces* A force of 176 pounds makes an angle of 78° 50′ with a second force. The resultant of the two forces makes an angle of 41° 10′ with the first force. Find the magnitude of the second force and of the resultant.

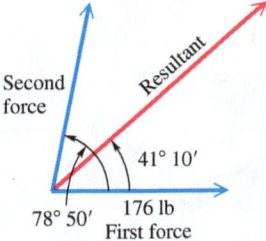

86. *Magnitudes of Forces* A force of 28.7 pounds makes an angle of 42° 10′ with a second force. The resultant of the two forces makes an angle of 32° 40′ with the first force. Find the magnitude of the second force and of the resultant.

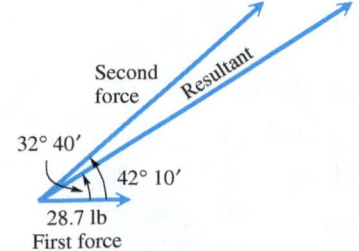

87. *Angle of a Hill Slope*   A force of 25 pounds is required to hold an 80-pound crate from rolling on a hill. What angle does the hill make with the horizontal?

88. *Force Needed to Keep a Car Parked*   Find the force required to keep a 3000-pound car parked on a hill that makes an angle of 15° with the horizontal.

89. *Force Needed for a Monolith*   To build the pyramids in Egypt, it is believed that giant causeways were constructed to transport the building materials to the site. One such causeway is said to have been 3000 feet long, with a slope of about 2.3°. How much force would be required to hold a 60-ton monolith on this causeway?

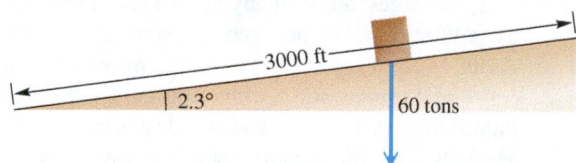

90. *Angles between Forces*   Three forces acting at a point are in equilibrium. The forces are 980 pounds, 760 pounds, and 1220 pounds. Find the angles between the directions of the forces. (*Hint:* Arrange the forces to form the sides of a triangle.)

91. *Incline Angle*   A force of 18 pounds is required to hold a 60-pound stump grinder on an incline. What angle does the incline make with the horizontal?

92. *Incline Angle*   A force of 30 pounds is required to hold an 80-pound pressure washer on an incline. What angle does the incline make with the horizontal?

93. *Weight of a Crate and Tension of a Rope*   A crate is supported by two ropes. One rope makes an angle of 46° 20′ with the horizontal and has a tension of 89.6 pounds on it. The other rope is horizontal. Find the weight of the crate and the tension in the horizontal rope.

94. *Weight of a Box*   Two people are carrying a box. One person exerts a force of 150 pounds at an angle of 62.4° with the horizontal. The other person exerts a force of 114 pounds at an angle of 54.9°. Find the weight of the box.

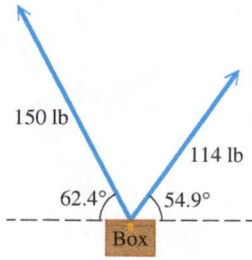

95. *Distance and Bearing of a Ship*   A ship leaves port on a bearing of 34.0° and travels 10.4 miles. The ship then turns due east and travels 4.6 miles. How far is the ship from port, and what is its bearing from port?

96. *Distance and Bearing of a Luxury Liner*   A luxury liner leaves port on a bearing of 110.0° and travels 8.8 miles. It then turns due west and travels 2.4 miles. How far is the liner from port, and what is its bearing from port?

97. *Distance of a Ship from Its Starting Point*   Starting at point *A*, a ship sails 18.5 kilometers on a bearing of 189°, then turns and sails 47.8 kilometers on a bearing of 317°. Find the distance of the ship from point *A*.

98. *Distance of a Ship from Its Starting Point*   Starting at point *X*, a ship sails 15.5 kilometers on a bearing of 200°, then turns and sails 2.4 kilometers on a bearing of 320°. Find the distance of the ship from point *X*.

99. *Distance and Direction of a Motorboat*   A motorboat sets out in the direction N 80° E. The speed of the boat in still water is 20.0 mph. If the current is flowing directly south and the actual direction of the motorboat is due east, find the speed of the current and the actual speed of the motorboat.

100. *Path Traveled by a Plane*   The aircraft carrier *Tallahassee* is traveling at sea on a steady course with a bearing of 30° at 32 mph. Patrol planes on the carrier have enough fuel for 2.6 hours of flight when traveling at a speed of 520 mph. One of the pilots takes off on a bearing of 338° and then turns and heads in a straight line, so as to be able to catch the carrier and land on the deck at the exact instant that his fuel runs out. If the pilot left at 2 P.M., at what time did he turn to head for the carrier?

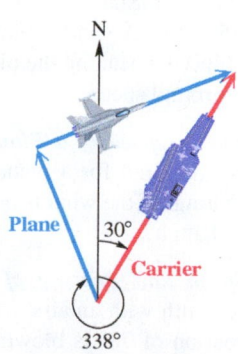

101. *Bearing and Ground Speed of a Plane* An airline route from San Francisco to Honolulu is on a bearing of 233.0°. A jet flying at 450 mph with that heading runs into a wind blowing at 39.0 mph from a direction of 114.0°. Find the final bearing and ground speed of the plane.

102. *Movement of a Motorboat* Suppose you would like to cross a 132-foot-wide river in a motorboat. Assume that the motorboat can travel at 7 mph relative to the water and that the current is flowing west at the rate of 3 mph. The bearing θ is chosen so that the motorboat will land at a point exactly across from the starting point.
   (a) At what speed will the motorboat be traveling relative to the banks?
   (b) How long will it take for the motorboat to make the crossing?
   (c) What is the measure of angle θ?

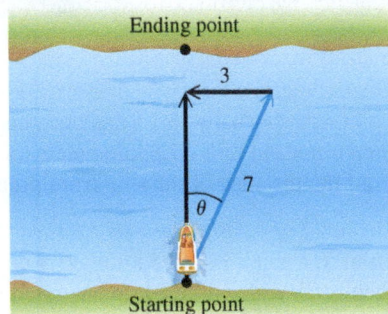

103. *Airspeed and Ground Speed* A pilot wants to fly on a course of 74.9°. Flying due east, he finds that a 42-mph wind, blowing from the south, puts him on course. Find the airspeed and the ground speed.

104. *Bearing of a Plane* A plane flies 650 mph on a heading of 175.3°. A 25-mph wind from a direction of 266.6° blows against the plane. Find the final bearing of the plane.

105. *Bearing and Ground Speed of a Plane* A pilot is flying at 190 mph. He wants his flight path to be on a bearing of 64° 30′. A wind is blowing from the south at 35.0 mph. Find the heading he should fly, and find the plane's ground speed.

106. *Bearing and Ground Speed of a Plane* A pilot is flying at 168 mph. She wants her flight path to be on a bearing of 57° 40′. A wind is blowing from the south at 27.1 mph. Find the heading the pilot should fly, and find the plane's ground speed.

107. *Bearing and Airspeed of a Plane* What heading and airspeed are required for a plane to fly 400 miles due north in 2.5 hours if the wind is blowing from a direction of 328° at 11 mph?

108. *Ground Speed and Bearing of a Plane* A plane is headed due south with an airspeed of 192 mph. A wind from a direction of 78° is blowing at 23 mph. Find the ground speed and final bearing of the plane.

109. *Ground Speed and Bearing of a Plane* An airplane is headed on a bearing of 174° at an airspeed of 240 kilometers per hour. A 30-kilometer-per-hour wind is blowing from a direction of 245°. Find the ground speed and final bearing of the plane.

110. *Velocity of a Star* The space velocity **v** of a star relative to the sun can be expressed as the resultant vector of two perpendicular vectors: the radial velocity $\mathbf{v}_r$, and the tangential velocity $\mathbf{v}_t$, where $\mathbf{v} = \mathbf{v}_r + \mathbf{v}_t$. If a star is located near the sun and its space velocity is large, then its motion across the sky will also be large. Barnard's Star is a relatively close star, 35 trillion miles from the sun. It moves across the sky through an angle of 10.34″ per year, the largest angle of any known star. Its radial velocity is $\mathbf{v}_r = 67$ miles per second toward the sun. (*Source:* Zeilik, M., S. Gregory, and E. Smith, *Introductory Astronomy and Astrophysics,* Second Edition, Saunders College Publishing; Acker, A. and C. Jaschek, *Astronomical Methods and Calculations,* John Wiley & Sons.)

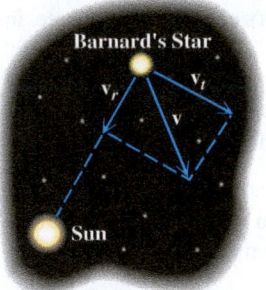

Not to scale

   (a) Approximate the tangential velocity $\mathbf{v}_t$ of Barnard's Star. (*Hint:* Use the arc length formula $s = r\theta$.)
   (b) Compute the magnitude of **v**.

111. *(Modeling) Measuring Rainfall* Suppose that vector **R** models the amount of rainfall in inches and the direction it falls, and vector **A** models the area in square inches and orientation of the opening of a rain gauge, as illustrated in the figure below. The total volume V of water collected in the gauge is given by $V = |\mathbf{R} \cdot \mathbf{A}|$. This formula calculates the volume of water collected even if the wind is blowing the rain in a slanted direction or the rain gauge is not exactly vertical. Let $\mathbf{R} = \mathbf{i} - 2\mathbf{j}$ and $\mathbf{A} = 0.5\mathbf{i} + \mathbf{j}$.

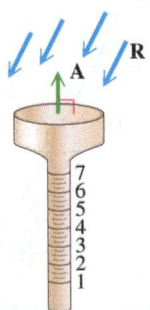

   (a) Find $|\mathbf{R}|$ and $|\mathbf{A}|$. Interpret your results.
   (b) Calculate V and interpret this result.
   (c) For the rain gauge to collect the maximum amount of water, what should be true about vectors **R** and **A**?

112. *The Dot Product*   In the figure at the right, $\mathbf{u} = \langle u_1, u_2 \rangle$, $\mathbf{v} = \langle v_1, v_2 \rangle$, and $\mathbf{u} - \mathbf{v} = \langle u_1 - v_1, u_2 - v_2 \rangle$. Apply the law of cosines to the triangle and derive the equation

$$\mathbf{u} \cdot \mathbf{v} = |\mathbf{u}| \, |\mathbf{v}| \cos \theta.$$

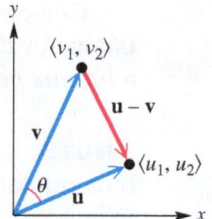

SECTIONS 10.1–10.3   Reviewing Basic Concepts

Round values to the nearest tenth in Exercises 1–5.

1. Solve triangle ABC if $A = 44°$, $C = 62°$, and $a = 12$.

2. Solve triangle ABC if $A = 32°$, $a = 6$, and $b = 8$. How many solutions are there?

3. Solve triangle ABC if $C = 41°$, $c = 12$, and $a = 7$. How many solutions are there?

4. Use the law of cosines to solve each triangle.

(a)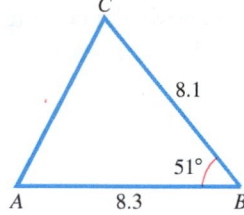

(b)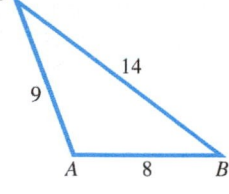

5. Find the area of triangle ABC to the nearest tenth if $a = 4.5$, $b = 5.2$, and $C = 55°$.

6. Find the area of triangle ABC if $a = 6$, $b = 7$, and $c = 9$.

7. Let $\mathbf{v} = 2\mathbf{i} - \mathbf{j}$ and $\mathbf{u} = -3\mathbf{i} + 2\mathbf{j}$. Find each expression.
(a) $2\mathbf{v} + \mathbf{u}$
(b) $2\mathbf{v}$
(c) $\mathbf{v} - 3\mathbf{u}$

8. Let $\mathbf{u} = \langle 3, -2 \rangle$ and $\mathbf{v} = \langle -1, 3 \rangle$. Find $\mathbf{u} \cdot \mathbf{v}$ and the angle between $\mathbf{u}$ and $\mathbf{v}$, rounded to the nearest tenth of a degree.

9. *Resultant Force*   Find the magnitude (to the nearest pound) of the resultant force of the two forces shown in the figure.

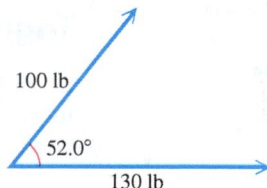

10. *Height of an Airplane*   Two observation points A and B are 950 feet apart. From these points, the angles of elevation of an airplane are $52°$ and $57°$. Find the height of the airplane.

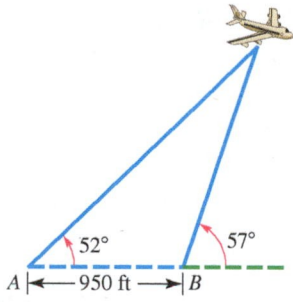

10.4 Trigonometric (Polar) Form of Complex Numbers

The Complex Plane and Vector Representation • Trigonometric (Polar) Form • Products of Complex Numbers in Trigonometric Form • Quotients of Complex Numbers in Trigonometric Form

The Complex Plane and Vector Representation

Unlike real numbers, complex numbers cannot be ordered on a number line. One way to organize and illustrate them is by using a graph. To graph a complex number such as $2 - 3i$, we modify the familiar rectangular coordinate system by calling the horizontal axis the **real axis** and the vertical axis the **imaginary axis**.

Complex Plane

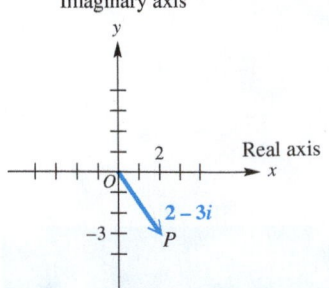

FIGURE 46

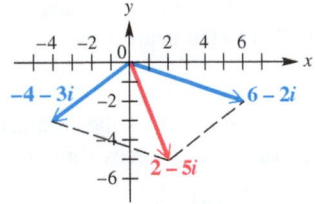

Graphical addition of
two complex numbers

FIGURE 47

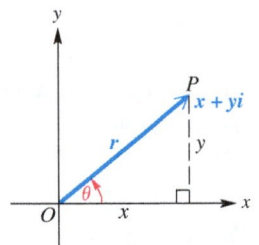

FIGURE 48

Complex numbers can be graphed in this **complex plane,** as shown in **FIG-URE 46** for the complex number $2 - 3i$. *Each complex number $a + bi$ determines a unique position vector **OP** with initial point $(0, 0)$ and terminal point (a, b).*

> **NOTE** This geometric representation is the reason that $a + bi$ is called the rectangular form of a complex number. (*Rectangular form* is also called *standard form.*)

> **EXAMPLE 1** **Expressing the Sum of Complex Numbers Graphically**

Find the sum of $6 - 2i$ and $-4 - 3i$. Graph both complex numbers and their resultant.

Solution The sum is found by adding the two complex numbers.

$$(6 - 2i) + (-4 - 3i) = 2 - 5i \quad \text{Add real parts and add imaginary parts.}$$

Graphically, the sum is represented by the vector that is the resultant of the vectors corresponding to the two numbers, as shown in **FIGURE 47.** ●

Trigonometric (Polar) Form

Trigonometric form can be used to multiply, divide, and find roots of complex numbers more easily than is the case with rectangular form. **FIGURE 48** shows the complex number $x + yi$ that corresponds to a vector **OP** with direction angle θ and magnitude r.

> **Relationships among x, y, r, and θ**
>
> $$x = r \cos \theta \qquad\qquad y = r \sin \theta$$
>
> $$r = \sqrt{x^2 + y^2} \qquad \tan \theta = \frac{y}{x}, \quad \text{if } x \neq 0$$

Substituting $x = r \cos \theta$ and $y = r \sin \theta$ into $x + yi$ gives

$$x + yi = r \cos \theta + (r \sin \theta)i$$
$$= r(\cos \theta + i \sin \theta). \quad \text{Factor out } r.$$

> **Trigonometric (Polar) Form of a Complex Number**
>
> The expression
>
> $$r(\cos \theta + i \sin \theta)$$
>
> is called the **trigonometric form** (or **polar form**) of the complex number $x + yi$. The expression $\cos \theta + i \sin \theta$ is sometimes abbreviated **cis θ.** Using this notation, $r(\cos \theta + i \sin \theta)$ is written r **cis θ.**

The number r is the **modulus,** or **absolute value** of $x + yi$, and θ is the **argument** of $x + yi$. Because $\sin \theta$ and $\cos \theta$ have period 2π or $360°$, we often let

$$0 \leq \theta < 2\pi$$

$$\text{or} \quad 0° \leq \theta < 360°.$$

However, other intervals for θ would work as the argument.

EXAMPLE 2   **Converting from Trigonometric to Rectangular Form**

Express $2(\cos 300° + i \sin 300°)$ in rectangular form.

Analytic Solution

As discussed in **Chapter 8**, we know that $\cos 300° = \frac{1}{2}$ and $\sin 300° = -\frac{\sqrt{3}}{2}$.

$$2(\cos 300° + i \sin 300°) = 2\left(\frac{1}{2} - i\frac{\sqrt{3}}{2}\right)$$
$$= 1 - i\sqrt{3}$$

Notice that the real part is positive and the imaginary part is negative. This is consistent with 300° being a quadrant IV angle.

Graphing Calculator Solution

FIGURE 49 confirms the analytic solution.

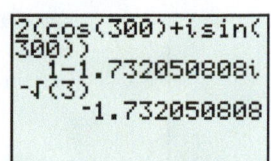

Degree mode

The imaginary part is an approximation for $-\sqrt{3}$.

FIGURE 49

Converting from Rectangular to Trigonometric Form

Step 1 Sketch a graph of the number $x + yi$ in the complex plane.

Step 2 Find r by using the equation $r = \sqrt{x^2 + y^2}$.

Step 3 Find θ by using the equation $\tan \theta = \frac{y}{x}$, $x \neq 0$, choosing the quadrant indicated in Step 1.

CAUTION   *Errors often occur in Step 3. Be sure to choose the correct quadrant for θ by referring to the graph sketched in Step 1.*

EXAMPLE 3   **Converting from Rectangular to Trigonometric Form**

Write each complex number in trigonometric form.

(a) $-\sqrt{3} + i$ (Use radian measure.)   **(b)** $-3i$ (Use degree measure.)

Solution

(a) We start by sketching the graph of $-\sqrt{3} + i$ in the complex plane, as shown in **FIGURE 50**. Next, we find r and then we find θ.

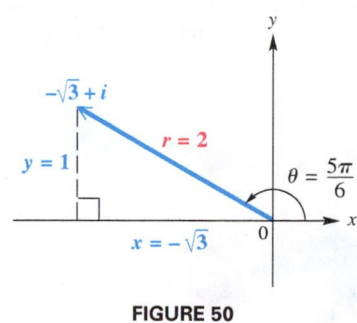

FIGURE 50

$$r = \sqrt{x^2 + y^2} = \sqrt{(-\sqrt{3})^2 + 1^2} = \sqrt{3 + 1} = 2 \qquad x = -\sqrt{3}, y = 1$$

$$\tan \theta = \frac{y}{x} = \frac{1}{-\sqrt{3}} = \frac{1}{-\sqrt{3}} \cdot \frac{\sqrt{3}}{\sqrt{3}} = -\frac{\sqrt{3}}{3} \qquad \text{Rationalize the denominator.}$$

Since $\tan \theta = -\frac{\sqrt{3}}{3}$, the reference angle for θ in radians is $\frac{\pi}{6}$. From the graph, we see that θ is in quadrant II, so $\theta = \pi - \frac{\pi}{6} = \frac{5\pi}{6}$.

Be sure to choose the correct quadrant.

$$-\sqrt{3} + i = 2\left(\cos \frac{5\pi}{6} + i \sin \frac{5\pi}{6}\right) = 2 \operatorname{cis} \frac{5\pi}{6}$$

(continued)

TECHNOLOGY NOTE

The top screen shows how to convert from rectangular (x, y) form to trigonometric form on the TI-84 Plus. The calculator is in radian mode. The results agree with our analytic results in **Example 3(a).**

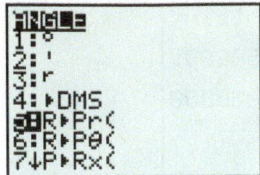

(b) The sketch of $-3i$ is shown in **FIGURE 51.** Since $-3i = 0 - 3i$, we have $x = 0$ and $y = -3$. We find r as follows.

$$r = \sqrt{0^2 + (-3)^2} = \sqrt{0 + 9} = \sqrt{9} = 3$$

We cannot find θ by using $\tan \theta = \frac{y}{x}$, because $x = 0$. From the graph, a value for θ is 270°.

$$-3i = 3(\cos 270° + i \sin 270°) = 3 \text{ cis } 270°$$

When the terminal side lies on an axis, pay close attention to the graph.

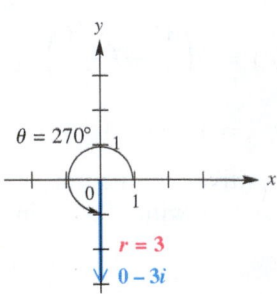

FIGURE 51

NOTE In **Example 3,** we gave answers in both forms: $r(\cos \theta + i \sin \theta)$ and r cis θ. These forms are used interchangeably throughout the rest of this chapter.

High-resolution computer graphics and complex numbers make it possible to produce beautiful shapes called *fractals*. Benoit B. Mandelbrot first used the term *fractal* in 1975. A **fractal** is a geometric figure with an endless self-similarity property. A fractal image repeats itself infinitely with ever-decreasing dimensions.

EXAMPLE 4 **Deciding Whether a Complex Number Is in the Julia Set**

The fractal called the **Julia set** is shown in **FIGURE 52.** To determine whether a complex number $z = a + bi$ is in this Julia set, perform the following sequence of calculations. Repeatedly compute the values of

$$z^2 - 1, \quad (z^2 - 1)^2 - 1, \quad [(z^2 - 1)^2 - 1]^2 - 1, \quad \dots$$

If the moduli (or absolute values) of any of the resulting complex numbers exceed 2, then the complex number z is not in the Julia set. Otherwise z is part of this set and the point (a, b) should be shaded in the graph.

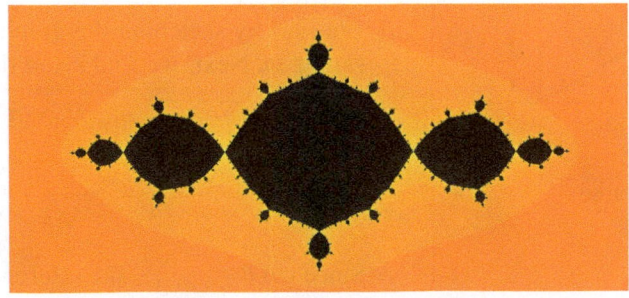

FIGURE 52

Determine whether each number belongs to the Julia set.

(a) $z = 0 + 0i$ **(b)** $z = 1 + 1i$

Solution

(a) Here,
$$z = 0 + 0i = 0,$$
$$z^2 - 1 = 0^2 - 1 = -1,$$
$$(z^2 - 1)^2 - 1 = (-1)^2 - 1 = 0,$$
$$[(z^2 - 1)^2 - 1]^2 - 1 = 0^2 - 1 = -1,$$

and so on. We see that the calculations repeat as $0, -1, 0, -1$. The moduli are either 0 or 1, which do not exceed 2, so $0 + 0i$ is in the Julia set and the point $(0, 0)$ is part of the graph.

(b) We have $z^2 - 1 = (1 + i)^2 - 1 = (1 + 2i + i^2) - 1 = -1 + 2i$. The modulus is $\sqrt{(-1)^2 + 2^2} = \sqrt{5}$. Since $\sqrt{5}$ is greater than 2, the number $1 + 1i$ is not in the Julia set and $(1, 1)$ is not part of the graph. ●

Products of Complex Numbers in Trigonometric Form

We can use the FOIL method to multiply complex numbers in rectangular form.

$$\left(1 + i\sqrt{3}\right)\left(-2\sqrt{3} + 2i\right) = -2\sqrt{3} + 2i - 2i(3) + 2i^2\sqrt{3} \qquad \text{FOIL}$$
$$= -2\sqrt{3} + 2i - 6i - 2\sqrt{3} \qquad i^2 = -1$$
$$= -4\sqrt{3} - 4i \qquad \text{Combine like terms}$$

We can also find this same product by first converting the complex numbers $1 + i\sqrt{3}$ and $-2\sqrt{3} + 2i$ to trigonometric form, as explained earlier in this section.

$$1 + i\sqrt{3} = 2(\cos 60° + i \sin 60°)$$
$$-2\sqrt{3} + 2i = 4(\cos 150° + i \sin 150°)$$

If we multiply the trigonometric forms and use identities, then the result is as follows.

$$[2(\cos 60° + i \sin 60°)][4(\cos 150° + i \sin 150°)]$$

$$= 2 \cdot 4(\cos 60° \cdot \cos 150° + i \sin 60° \cdot \cos 150°$$
$$+ i \cos 60° \cdot \sin 150° + i^2 \sin 60° \cdot \sin 150°) \qquad \text{Multiply the moduli and the binomials.}$$

$$= 8[(\cos 60° \cdot \cos 150° - \sin 60° \cdot \sin 150°)$$
$$+ i(\sin 60° \cdot \cos 150° + \cos 60° \cdot \sin 150°)] \qquad i^2 = -1 \text{ and factor out } i.$$

$$= 8[\cos(60° + 150°) + i \sin(60° + 150°)] \qquad \text{Use identities for } \cos(A + B) \text{ and } \sin(A + B).$$

$$= 8(\cos 210° + i \sin 210°) \qquad \text{Add.}$$

Notice that the modulus of the product, 8, is equal to the product of the moduli of the factors, $2 \cdot 4$, and the argument of the product, 210°, is the sum of the arguments of the factors, $60° + 150°$.

As expected, the product obtained by multiplying by the first method is the rectangular form of the product obtained by multiplying by the second method.

$$8(\cos 210° + i \sin 210°) = 8\left(-\frac{\sqrt{3}}{2} - \frac{1}{2}i\right) = -4\sqrt{3} - 4i$$

> **Product Theorem**
>
> If $r_1(\cos \theta_1 + i \sin \theta_1)$ and $r_2(\cos \theta_2 + i \sin \theta_2)$ are any two complex numbers, then the following holds.
>
> $$[r_1(\cos \theta_1 + i \sin \theta_1)] \cdot [r_2(\cos \theta_2 + i \sin \theta_2)]$$
> $$= r_1 r_2 [\cos (\theta_1 + \theta_2) + i \sin (\theta_1 + \theta_2)]$$
>
> In compact form, this is written
>
> $$(r_1 \operatorname{cis} \theta_1)(r_2 \operatorname{cis} \theta_2) = r_1 r_2 \operatorname{cis} (\theta_1 + \theta_2).$$

That is, to multiply two complex numbers in trigonometric form, multiply their moduli and add their arguments.

EXAMPLE 5 **Using the Product Theorem**

Find the product of $3(\cos 45° + i \sin 45°)$ and $2(\cos 135° + i \sin 135°)$.

Solution

$$[3(\cos 45° + i \sin 45°)][2(\cos 135° + i \sin 135°)]$$

$$= 3 \cdot 2[\cos (45° + 135°) + i \sin (45° + 135°)] \qquad \text{Product theorem}$$

$$= 6(\cos 180° + i \sin 180°) \qquad \text{Multiply and add.}$$

$$= 6(-1 + i \cdot 0) \qquad \text{Evaluate.}$$

$$= -6 \qquad \text{Rectangular form} \qquad \bullet$$

Quotients of Complex Numbers in Trigonometric Form

We find the rectangular form of the quotient of the two complex numbers $1 + i\sqrt{3}$ and $-2\sqrt{3} + 2i$ as follows.

$$\frac{1 + i\sqrt{3}}{-2\sqrt{3} + 2i} = \frac{(1 + i\sqrt{3})(-2\sqrt{3} - 2i)}{(-2\sqrt{3} + 2i)(-2\sqrt{3} - 2i)} \qquad \begin{array}{l} \text{Multiply by the conjugate} \\ \text{of the denominator in both.} \end{array}$$

$$= \frac{-2\sqrt{3} - 2i - 6i - 2i^2\sqrt{3}}{12 - 4i^2} \qquad \begin{array}{l} \text{FOIL and} \\ (x + y)(x - y) = x^2 - y^2 \end{array}$$

$$= \frac{-8i}{16} = -\frac{1}{2}i \qquad i^2 = -1 \text{ and simplify.}$$

Now write $1 + i\sqrt{3}$, $-2\sqrt{3} + 2i$, and $-\frac{1}{2}i$ in trigonometric form.

$$1 + i\sqrt{3} = 2(\cos 60° + i \sin 60°)$$

$$-2\sqrt{3} + 2i = 4(\cos 150° + i \sin 150°)$$

$$-\frac{1}{2}i = \frac{1}{2}[\cos (-90°) + i \sin (-90°)]$$

Notice that the modulus of the quotient, $\frac{1}{2}$, is equal to the quotient of the two moduli, $\frac{2}{4} = \frac{1}{2}$. The argument of the quotient, $-90°$, is the difference of the two arguments, $60° - 150° = -90°$.

Quotient Theorem

If $r_1(\cos \theta_1 + i \sin \theta_1)$ and $r_2(\cos \theta_2 + i \sin \theta_2)$ are complex numbers, where $r_2(\cos \theta_2 + i \sin \theta_2) \neq 0$, then the following holds.

$$\frac{r_1(\cos \theta_1 + i \sin \theta_1)}{r_2(\cos \theta_2 + i \sin \theta_2)} = \frac{r_1}{r_2}[\cos(\theta_1 - \theta_2) + i \sin(\theta_1 - \theta_2)]$$

In compact form, this is written

$$\frac{r_1 \operatorname{cis} \theta_1}{r_2 \operatorname{cis} \theta_2} = \frac{r_1}{r_2} \operatorname{cis}(\theta_1 - \theta_2).$$

That is, to divide two complex numbers in trigonometric form, divide their moduli and subtract their arguments.

FOR DISCUSSION

In **Example 6**, the complex number

$$10 \operatorname{cis}(-60°) = 5 - 5i\sqrt{3}$$

was divided by the complex number

$$5 \operatorname{cis}(150°) = -\frac{5\sqrt{3}}{2} + \frac{5}{2}i.$$

Would you rather perform this division (by hand) in trigonometric form or in rectangular form? Explain.

EXAMPLE 6   **Using the Quotient Theorem**

Find the quotient $\dfrac{10 \operatorname{cis}(-60°)}{5 \operatorname{cis} 150°}$. Write the result in rectangular form.

Solution

$$\frac{10 \operatorname{cis}(-60°)}{5 \operatorname{cis} 150°} = \frac{10}{5} \operatorname{cis}(-60° - 150°) \qquad \text{Quotient theorem}$$

$$= 2 \operatorname{cis}(-210°) \qquad \text{Divide and subtract.}$$

$$= 2[\cos(-210°) + i \sin(-210°)] \qquad \text{Rewrite.}$$

$$= 2\left[-\frac{\sqrt{3}}{2} + i\left(\frac{1}{2}\right)\right] \qquad \begin{array}{l}\cos(-210°) = -\frac{\sqrt{3}}{2}; \\ \sin(-210°) = \frac{1}{2}\end{array}$$

$$= -\sqrt{3} + i \qquad \text{Rectangular form}$$

10.4 Exercises

1. *Concept Check*   The modulus of a complex number represents the _____ of the vector representing it in the complex plane.

2. *Concept Check*   What is the geometric interpretation of the argument of a complex number?

Checking Analytic Skills   *Graph each complex number as a vector in the complex plane.* ***Do not use a calculator.***

3. $6 - 5i$ **4.** $-3 + 2i$ **5.** $2 - 2i\sqrt{3}$ **6.** $\sqrt{2} + i\sqrt{2}$ **7.** $-4i$ **8.** $3i$ **9.** -8 **10.** 2

Give the rectangular form of the complex number represented in each graph.

11.

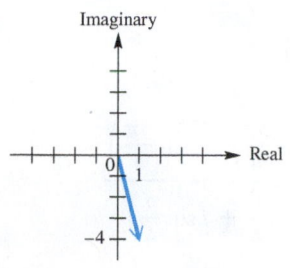

12.

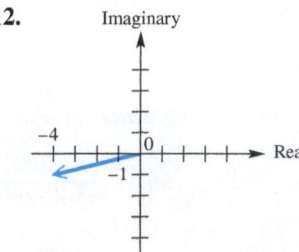

13. *Concept Check* What must be true for a complex number to also be a real number?

14. *Concept Check* If a real number is graphed in the complex plane, on what axis does the vector lie?

15. *Concept Check* A complex number of the form $a + bi$ will have its corresponding vector lying on the y-axis, provided that $a = $ _____.

16. Are the notations $a + bi$ and $a\mathbf{i} + b\mathbf{j}$ equivalent? Explain.

Checking Analytic Skills Find the sum of each pair of complex numbers. Express your answer in rectangular form. **Do not use a calculator.**

17. $4 - 3i, -1 + 2i$

18. $2 + 3i, -4 - i$

19. $-3, 3i$

20. $6, -2i$

21. $2 + 6i, -2i$

22. $7 + 6i, 3i$

23. $4 - 2i, 5$

24. $-5 - 8i, -1$

Checking Analytic Skills Find the modulus r of the number. **Do not use a calculator.**

25. $1 + i$

26. $3 - 4i$

27. $12 - 5i$

28. $-24 + 7i$

29. -6

30. $15i$

31. $2 - 3i$

32. $11 - 60i$

Checking Analytic Skills Write each complex number in rectangular form. Give exact values for the real and imaginary parts. **Do not use a calculator.**

33. $2(\cos 45° + i \sin 45°)$

34. $4(\cos 60° + i \sin 60°)$

35. $10 \text{ cis } 90°$

36. $8 \text{ cis } 270°$

37. $4(\cos 240° + i \sin 240°)$

38. $2(\cos 330° + i \sin 330°)$

39. $\cos \dfrac{\pi}{6} + i \sin \dfrac{\pi}{6}$

40. $3\left(\cos \dfrac{5\pi}{6} + i \sin \dfrac{5\pi}{6} \right)$

41. $5 \text{ cis} \left(-\dfrac{\pi}{6} \right)$

42. $6 \text{ cis } \dfrac{3\pi}{4}$

43. $\sqrt{2} \text{ cis } \pi$

44. $\sqrt{3} \text{ cis } \dfrac{3\pi}{2}$

Write each complex number in the trigonometric form $r(\cos \theta + i \sin \theta)$, where r is exact and $0° \le \theta < 360°$.

45. $3 - 3i$

46. $-2 + 2i\sqrt{3}$

47. $1 + i\sqrt{3}$

48. $-3 - 3i\sqrt{3}$

49. $-2i$

50. 7

Write each complex number in trigonometric form, where r is exact and $0 \le \theta < 2\pi$.

51. $4\sqrt{3} + 4i$

52. $\sqrt{3} - i$

53. $-\sqrt{2} + i\sqrt{2}$

54. $-5 - 5i$

55. -4

56. $5i$

Find each product in rectangular form, using exact values.

57. $[3(\cos 60° + i \sin 60°)][2(\cos 90° + i \sin 90°)]$

58. $[4(\cos 30° + i \sin 30°)][5(\cos 120° + i \sin 120°)]$

59. $[2(\cos 45° + i \sin 45°)][2(\cos 225° + i \sin 225°)]$

60. $[8(\cos 300° + i \sin 300°)][5(\cos 120° + i \sin 120°)]$

61. $\left[5 \text{ cis } \dfrac{\pi}{2} \right]\left[3 \text{ cis } \dfrac{\pi}{4} \right]$

62. $\left[6 \text{ cis } \dfrac{2\pi}{3} \right]\left[5 \text{ cis} \left(-\dfrac{\pi}{6} \right) \right]$

63. $\left[\sqrt{3} \text{ cis } \dfrac{\pi}{4} \right]\left[\sqrt{3} \text{ cis } \dfrac{5\pi}{4} \right]$

64. $\left[\sqrt{2} \text{ cis } \dfrac{5\pi}{6} \right]\left[\sqrt{2} \text{ cis } \dfrac{3\pi}{2} \right]$

Find each quotient in rectangular form, using exact values.

65. $\dfrac{4(\cos 120° + i \sin 120°)}{2(\cos 150° + i \sin 150°)}$

66. $\dfrac{16(\cos 300° + i \sin 300°)}{8(\cos 60° + i \sin 60°)}$

67. $\dfrac{10\left(\cos \dfrac{5\pi}{4} + i \sin \dfrac{5\pi}{4} \right)}{5\left(\cos \dfrac{\pi}{4} + i \sin \dfrac{\pi}{4} \right)}$

68. $\dfrac{24\left(\cos \dfrac{5\pi}{6} + i \sin \dfrac{5\pi}{6}\right)}{2\left(\cos \dfrac{\pi}{6} + i \sin \dfrac{\pi}{6}\right)}$

69. $\dfrac{3 \text{ cis}\left(\dfrac{61\pi}{36}\right)}{9 \text{ cis}\left(\dfrac{13\pi}{36}\right)}$

70. $\dfrac{12 \text{ cis } 293°}{6 \text{ cis } 23°}$

Find each quotient and express it in rectangular form by first converting the numerator and the denominator to trigonometric form.

71. $\dfrac{8}{\sqrt{3} + i}$

72. $\dfrac{2i}{-1 - i\sqrt{3}}$

73. $\dfrac{-i}{1 + i}$

74. $\dfrac{1}{2 - 2i}$

75. $\dfrac{2\sqrt{6} - 2i\sqrt{2}}{\sqrt{2} - i\sqrt{6}}$

76. $\dfrac{-3\sqrt{2} + 3i\sqrt{6}}{\sqrt{6} + i\sqrt{2}}$

Solve each problem.

77. *Julia Set*   The graph of the Julia set in **FIGURE 52** appears to be symmetric with respect to both the x-axis and y-axis. Complete the following to show that this is true.
   (a) Show that complex conjugates have the same modulus.
   (b) Compute $z_1^2 - 1$ and $z_2^2 - 1$, where $z_1 = a + bi$ and $z_2 = a - bi$.
   (c) Discuss why, if (a, b) is in the Julia set, then so is $(a, -b)$.
   (d) Can we conclude that the graph of the Julia set must be symmetric with respect to the x-axis?
   (e) Using a similar argument, show that the Julia set must also be symmetric with respect to the y-axis.

78. *Julia Set*   (Refer to **Example 4.**) Is $z = -0.2i$ in the Julia set?

79. *(Modeling) Alternating Current*   The alternating current in amps in an electric inductor is

$$I = \frac{E}{Z},$$

where E is the voltage and $Z = R + X_L i$ is the impedance. If $E = 8(\cos 20° + i \sin 20°)$, $R = 6$, and $X_L = 3$, find the current. Give the answer in rectangular form.

80. *(Modeling) Electric Current*   The current in a circuit with voltage E, resistance R, capacitive reactance X_c, and inductive resistance X_L is

$$I = \frac{E}{R + (X_L - X_c)i}.$$

Find I if $E = 12(\cos 25° + i \sin 25°)$, $R = 3$, $X_L = 4$, and $X_c = 6$. Give the answer in rectangular form.

81. *(Modeling) Impedance*   In the parallel electric circuit shown in the figure in the next column, the impedance Z can be calculated by using the equation

$$Z = \frac{1}{\dfrac{1}{Z_1} + \dfrac{1}{Z_2}},$$

where Z_1 and Z_2 are the impedances for the branches of

the circuit. If $Z_1 = 50 + 25i$ and $Z_2 = 60 + 20i$, calculate Z.

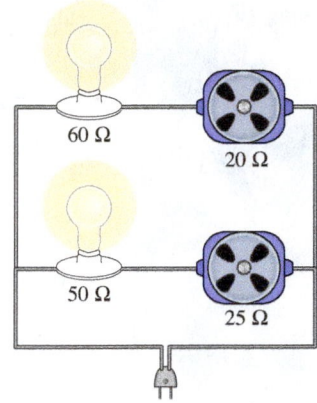

82. *Concept Check*   Without actually performing the operations, state why the products

$$[2(\cos 45° + i \sin 45°)][5(\cos 90° + i \sin 90°)]$$

and

$$[2(\cos(-315°) + i \sin(-315°))] \cdot$$
$$[5(\cos(-270°) + i \sin(-270°))]$$

are the same.

83. Consider the equation

$$(r \text{ cis } \theta)^2 = (r \text{ cis } \theta)(r \text{ cis } \theta) = r^2 \text{ cis}(\theta + \theta) = r^2 \text{ cis } 2\theta.$$

State in your own words how we can square a complex number in trigonometric form. (In the next section, we will develop this idea further.)

84. *Concept Check*   Under what conditions is the difference between two nonreal complex numbers $a + bi$ and $c + di$ a real number?

85. *Concept Check*   Give the least *positive* radian measure of θ if $r > 0$ and **(a)** r cis θ has real part equal to 0, **(b)** r cis θ has imaginary part equal to 0.

86. Use your calculator in radian mode to find the trigonometric form of $3 + 5i$. Approximate values to the nearest hundredth.

10.5 Powers and Roots of Complex Numbers

Powers of Complex Numbers (De Moivre's Theorem) • Roots of Complex Numbers

Powers of Complex Numbers (De Moivre's Theorem)

Consider the following.

$$[r(\cos\theta + i\sin\theta)]^2 = [r(\cos\theta + i\sin\theta)][r(\cos\theta + i\sin\theta)]$$
$$= r \cdot r[\cos(\theta + \theta) + i\sin(\theta + \theta)]$$
$$= r^2(\cos 2\theta + i\sin 2\theta) \qquad \text{Product theorem}$$

In the same way, $[r(\cos\theta + i\sin\theta)]^3 = r^3(\cos 3\theta + i\sin 3\theta)$.

These results suggest the following theorem for positive integer values of n. Although it is stated and can be proved for all n, we use it only for positive integer values of n and their reciprocals.

De Moivre's Theorem

If $r(\cos\theta + i\sin\theta)$ is a complex number and n is any positive integer, then the following holds.

$$[r(\cos\theta + i\sin\theta)]^n = r^n(\cos n\theta + i\sin n\theta)$$

In compact form, this is written

$$[r\operatorname{cis}\theta]^n = r^n(\operatorname{cis} n\theta).$$

Abraham De Moivre (1667–1754)

That is, to find the nth power of a complex number in trigonometric form, raise its modulus r to the power n and multiply its argument θ by n.

This theorem is named after Abraham De Moivre, although he never explicitly stated it.

EXAMPLE 1 **Finding a Power of a Complex Number**

Find $\left(1 + i\sqrt{3}\right)^8$ and express the result in rectangular form.

Solution Convert $1 + i\sqrt{3}$ into trigonometric form as in **Section 10.4.**

$$1 + i\sqrt{3} = 2(\cos 60° + i\sin 60°)$$

Now apply De Moivre's theorem.

$$\left(1 + i\sqrt{3}\right)^8 = [2(\cos 60° + i\sin 60°)]^8$$

$$= 2^8 [\cos(8 \cdot 60°) + i\sin(8 \cdot 60°)] \quad \text{De Moivre's theorem}$$

$$= 256(\cos 480° + i\sin 480°)$$

$$= 256(\cos 120° + i\sin 120°) \qquad \text{480° and 120° are coterminal.}$$

$$= 256\left(-\frac{1}{2} + i\frac{\sqrt{3}}{2}\right) \qquad \cos 120° = -\frac{1}{2}; \sin 120° = \frac{\sqrt{3}}{2}$$

$$= -128 + 128i\sqrt{3} \qquad \text{Rectangular form}$$

Roots of Complex Numbers

Every nonzero complex number has exactly n distinct complex nth roots. De Moivre's theorem can be extended to find all nth roots of a complex number.

---

nth Root

For a positive integer n, the complex number $a + bi$ is an **nth root** of the complex number $x + yi$ if the following holds.

$$(a + bi)^n = x + yi$$

To find the three complex cube roots of $8(\cos 135° + i \sin 135°)$, for example, look for a complex number, say, $r(\cos \alpha + i \sin \alpha)$, that will satisfy

$$[r(\cos \alpha + i \sin \alpha)]^3 = 8(\cos 135° + i \sin 135°).$$

By De Moivre's theorem, this equation becomes

$$r^3(\cos 3\alpha + i \sin 3\alpha) = 8(\cos 135° + i \sin 135°).$$

To satisfy this equation, set $r^3 = 8$ and $\cos 3\alpha + i \sin 3\alpha = \cos 135° + i \sin 135°$. The first of these conditions implies that $r = 2$, and the second implies that

$$\cos 3\alpha = \cos 135° \quad \text{and} \quad \sin 3\alpha = \sin 135°.$$

For these equations to be satisfied, 3α must represent an angle that is coterminal with $135°$. Therefore, we must have

$$3\alpha = 135° + 360° \cdot k, \quad k \text{ any integer,}$$

or

$$\alpha = \frac{135° + 360° \cdot k}{3}, \quad k \text{ any integer.}$$

Now, let k take on the integer values 0, 1, and 2, in turn.

$$\text{If } k = 0, \text{ then } \quad \alpha = \frac{135° + 0°}{3} = 45°.$$

$$\text{If } k = 1, \text{ then } \quad \alpha = \frac{135° + 360°}{3} = \frac{495°}{3} = 165°.$$

$$\text{If } k = 2, \text{ then } \quad \alpha = \frac{135° + 720°}{3} = \frac{855°}{3} = 285°.$$

In the same way, $\alpha = 405°$ when $k = 3$. But note that $405° = 45° + 360°$, so $\sin 405° = \sin 45°$ and $\cos 405° = \cos 45°$. Similarly, if $k = 4$, $\alpha = 525°$, which has the same sine and cosine values as $165°$. To continue with larger values of k would just be repeating solutions already found. Therefore, the three cube roots of $8(\cos 135° + i \sin 135°)$ can be found by letting $k = 0$, 1, and 2.

When $k = 0$, the root is $2(\cos 45° + i \sin 45°)$.

When $k = 1$, the root is $2(\cos 165° + i \sin 165°)$.

When $k = 2$, the root is $2(\cos 285° + i \sin 285°)$.

The three cube roots of
$8(\cos 135° + i \sin 135°)$

> ### nth Root Theorem
>
> If n is a positive integer, r is a positive real number, and θ is in degrees, then the nonzero complex number $r(\cos\theta + i\sin\theta)$ has exactly n distinct nth roots, given by
>
> $$\sqrt[n]{r}(\cos\alpha + i\sin\alpha), \quad \text{or} \quad \sqrt[n]{r}\,\text{cis}\,\alpha,$$
>
> where
>
> $$\alpha = \frac{\theta + 360° \cdot k}{n}, \quad \text{or} \quad \alpha = \frac{\theta}{n} + \frac{360° \cdot k}{n}, \quad k = 0, 1, 2, \ldots, n-1.$$
>
> If θ is in radians, then
>
> $$\alpha = \frac{\theta + 2\pi k}{n}, \quad \text{or} \quad \alpha = \frac{\theta}{n} + \frac{2\pi k}{n}, \quad k = 0, 1, 2, \ldots, n-1.$$

EXAMPLE 2 Finding Complex Roots

Find the two square roots of $4i$. Write the roots in rectangular form, and check your results directly with a calculator.

Solution First write $4i$ in trigonometric form as

$$4i = 4\left(\cos\frac{\pi}{2} + i\sin\frac{\pi}{2}\right).$$

Here, $r = 4$ and $\theta = \frac{\pi}{2}$. The square roots have modulus $\sqrt{4} = 2$ and arguments

$$\alpha = \frac{\frac{\pi}{2}}{2} + \frac{2\pi k}{2} = \frac{\pi}{4} + \pi k.$$

Since there are two square roots, let $k = 0$ and 1.

$$\text{If } k = 0, \text{ then } \quad \alpha = \frac{\pi}{4} + \pi \cdot 0 = \frac{\pi}{4}.$$

$$\text{If } k = 1, \text{ then } \quad \alpha = \frac{\pi}{4} + \pi \cdot 1 = \frac{5\pi}{4}.$$

Thus the square roots are $2\,\text{cis}\,\frac{\pi}{4}$ and $2\,\text{cis}\,\frac{5\pi}{4}$, which can be written in rectangular form as $\sqrt{2} + i\sqrt{2}$ and $-\sqrt{2} - i\sqrt{2}$. Check the results by squaring each of these, as shown in **FIGURE 53.** ●

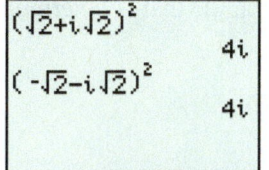

$\sqrt{2} + i\sqrt{2}$ and $-\sqrt{2} - i\sqrt{2}$ are the two square roots of $4i$.

FIGURE 53

TECHNOLOGY NOTE

In **Example 3**, $-8 + 8i\sqrt{3}$ is converted into the form $16(\cos 120° + i\sin 120°)$. In the screen below, a TI-84 Plus graphing calculator performs this conversion.

R►Pr(-8,8√3)
                16
R►Pθ(-8,8√3)
               120

Degree mode

EXAMPLE 3 Finding Complex Roots

Find all fourth roots of $-8 + 8i\sqrt{3}$. Write the roots in rectangular form.

Solution $-8 + 8i\sqrt{3} = 16\,\text{cis}\,120°$ Write in trigonometric form.

Here, $r = 16$ and $\theta = 120°$. The fourth roots have modulus $\sqrt[4]{16} = 2$ and arguments

$$\alpha = \frac{120°}{4} + \frac{360° \cdot k}{4} = 30° + 90° \cdot k.$$

Since there are four fourth roots, let $k = 0, 1, 2,$ and 3.

$$\text{If } k = 0, \text{ then } \quad \alpha = 30° + 90° \cdot 0 = 30°.$$
$$\text{If } k = 1, \text{ then } \quad \alpha = 30° + 90° \cdot 1 = 120°.$$
$$\text{If } k = 2, \text{ then } \quad \alpha = 30° + 90° \cdot 2 = 210°.$$
$$\text{If } k = 3, \text{ then } \quad \alpha = 30° + 90° \cdot 3 = 300°.$$

Using these angles, we find that the fourth roots are

$$2 \text{ cis } 30°, \quad 2 \text{ cis } 120°, \quad 2 \text{ cis } 210°, \quad \text{and} \quad 2 \text{ cis } 300°,$$

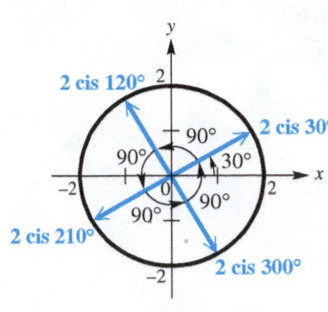

2 cis 120°
2 cis 30°
2 cis 210°
2 cis 300°

FIGURE 54

or $\quad \sqrt{3} + i, \quad -1 + i\sqrt{3}, \quad -\sqrt{3} - i, \quad$ and $\quad 1 - i\sqrt{3}.$ *Rectangular form*

The graphs of these roots are on a circle in the complex plane that has center at the origin and radius 2, as shown in **FIGURE 54.** The roots are equally spaced about the circle, 90° apart.

EXAMPLE 4   **Solving an Equation by Finding Complex Roots**

Find all complex solutions of $x^5 - 1 = 0$. Graph as vectors in the complex plane.

Solution   Write the equation as

$$x^5 - 1 = 0, \quad \text{or} \quad x^5 = 1.$$

While there is only one real number solution, 1, there are five complex number solutions. To find these solutions, first write 1 in trigonometric form as

$$1 = 1 + 0i = 1(\cos 0° + i \sin 0°).$$

The modulus of the fifth roots is $\sqrt[5]{1} = 1$, and the arguments are given by

$$0° + 72° \cdot k, \quad k = 0, 1, 2, 3, 4.$$

With these arguments, the fifth roots are

$$1(\cos 0° + i \sin 0°), \qquad k = 0 \qquad 1(\cos 72° + i \sin 72°), \qquad k = 1$$
$$1(\cos 144° + i \sin 144°), \qquad k = 2 \qquad 1(\cos 216° + i \sin 216°), \qquad k = 3$$
$$1(\cos 288° + i \sin 288°). \qquad k = 4$$

The solution set of the equation is $\{\text{cis } 0°, \text{ cis } 72°, \text{ cis } 144°, \text{ cis } 216°, \text{ cis } 288°\}$. The first of these roots equals 1. The others cannot easily be expressed in rectangular form but can be approximated with a calculator. The five fifth roots all lie on a unit circle in the complex plane, equally spaced around it every 72°. See **FIGURE 55.**

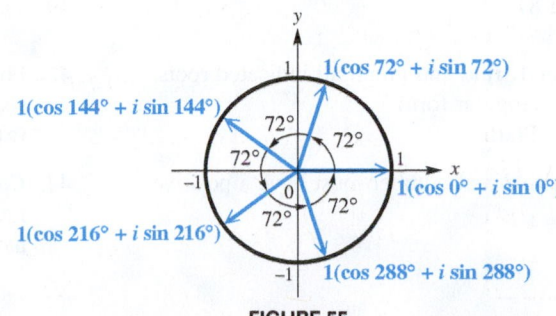

1(cos 72° + i sin 72°)
1(cos 144° + i sin 144°)
1(cos 216° + i sin 216°)
1(cos 0° + i sin 0°)
1(cos 288° + i sin 288°)

FIGURE 55

10.5 Exercises

Checking Analytic Skills   *Find each power. Write the answer in rectangular form.* **Do not use a calculator.**

1. $[3(\cos 30° + i \sin 30°)]^3$

2. $[2(\cos 135° + i \sin 135°)]^4$

3. $\left(\cos \dfrac{\pi}{4} + i \sin \dfrac{\pi}{4}\right)^8$

4. $\left(\cos \dfrac{\pi}{5} + i \sin \dfrac{\pi}{5}\right)^5$

5. $\left[2\left(\cos \dfrac{2\pi}{3} + i \sin \dfrac{2\pi}{3}\right)\right]^3$

6. $\left[3\left(\cos \dfrac{3\pi}{4} + i \sin \dfrac{3\pi}{4}\right)\right]^4$

7. $[3 \text{ cis } 100°]^3$

8. $[3 \text{ cis } 40°]^3$

9. $\left(\sqrt{3} + i\right)^3$

10. $\left(2\sqrt{2} - 2i\sqrt{2}\right)^6$

11. $\left(1 + i\sqrt{3}\right)^4$

12. $\left(2 - 2i\sqrt{3}\right)^4$

13. $\left(-\dfrac{\sqrt{2}}{2} + \dfrac{\sqrt{2}}{2}i\right)^4$

14. $\left(\dfrac{\sqrt{2}}{2} - \dfrac{\sqrt{2}}{2}i\right)^8$

15. $(1 - i)^6$

16. $(-1 + i)^7$

17. $(-2 - 2i)^5$

18. $(-3 - 3i)^3$

Find the cube roots of each complex number. Leave the answers in trigonometric form. Then graph each cube root as a vector in the complex plane.

19. 1

20. i

21. $8(\cos 60° + i \sin 60°)$

22. $27(\cos 300° + i \sin 300°)$

23. $-8i$

24. $27i$

25. -64

26. 27

27. $1 + i\sqrt{3}$

28. $2 - 2i\sqrt{3}$

29. $-2\sqrt{3} + 2i$

30. $\sqrt{3} - i$

Find all indicated roots and express them in rectangular form. Check your results with a calculator.

31. The square roots of $4(\cos 120° + i \sin 120°)$

32. The cube roots of $27(\cos 180° + i \sin 180°)$

33. The cube roots of $\cos 180° + i \sin 180°$

34. The fourth roots of $16(\cos 240° + i \sin 240°)$

35. The square roots of i

36. The square roots of $-4i$

37. The cube roots of $64i$

38. The fourth roots of -1

39. The fourth roots of 81

40. The square roots of $-1 + i\sqrt{3}$

41. For the real number 1, find and graph all indicated roots. Give answers in rectangular form.
(a) Fourth   (b) Sixth

42. For the complex number i, find and graph all indicated roots. Give answers in trigonometric form.
(a) Square   (b) Fourth

43. Explain why a positive real number must have a positive real nth root.

44. *Concept Check*   True or *false:* (a) Every real number must have two real square roots. (b) Some real numbers have three real cube roots.

RELATING CONCEPTS   **For individual or group investigation (Exercises 45–50)**

We examine how the three complex cube roots of −8 can be found in two different ways.
Work Exercises 45–50 in order.

45. All complex roots of the equation $x^3 + 8 = 0$ are cube roots of −8. Factor $x^3 + 8$ as the sum of two cubes.

46. One of the factors found in **Exercise 45** is linear. Set it equal to 0, solve, and determine the real cube root of −8.

47. One of the factors found in **Exercise 45** is quadratic. Set it equal to 0, solve, and determine the rectangular forms of the other two cube roots of −8.

48. Use the method described in this section to find the three complex cube roots of −8. Give them in trigonometric form.

49. Convert the trigonometric forms found in **Exercise 48** to rectangular form.

50. Compare your results in **Exercises 46** and **47** with those in **Exercise 49**. What do you notice?

Find all complex solutions for each equation. Leave your answers in trigonometric form.

51. $x^4 + 1 = 0$

52. $x^4 + 16 = 0$

53. $x^5 - i = 0$

54. $x^4 - i = 0$

55. $x^3 + 1 = 0$

56. $x^3 + i = 0$

57. $x^3 - 8 = 0$

58. $x^5 + 1 = 0$

59. *(Modeling) Basins of Attraction* The fractal shown below is the solution to Cayley's problem of determining the basins of attraction for the cube roots of unity (1). The three cube roots of unity are

$$w_1 = 1, \quad w_2 = -\frac{1}{2} + \frac{\sqrt{3}}{2}i,$$

and

$$w_3 = -\frac{1}{2} - \frac{\sqrt{3}}{2}i.$$

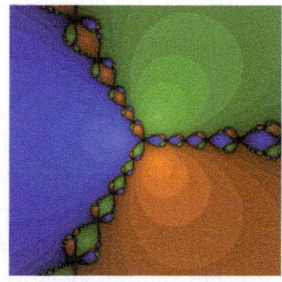

This fractal can be generated by repeatedly evaluating the function

$$f(z) = \frac{2z^3 + 1}{3z^2},$$

where z is a complex number. We begin by picking $z_1 = a + bi$ and then successively computing

$$z_2 = f(z_1), \quad z_3 = f(z_2), \quad z_4 = f(z_3), \quad \dots.$$

If the resulting values of $f(z)$ approach w_1, color the pixel at (a, b) red. If they approach w_2, color it blue, and if they approach w_3, color it green. If this process continues for a large number of different z_1, the fractal in the figure will appear. Determine the appropriate color of the pixel for each value of z_1. (*Source:* Crownover, R., *Introduction to Fractals and Chaos*, Jones and Bartlett.)

(a) $z_1 = i$ **(b)** $z_1 = 2 + i$
(c) $z_1 = -1 - i$

60. *(Modeling)* The TI-84 Plus screens illustrate how a pentagon can be graphed with parametric equations. Note that a pentagon has five sides, and the T-step is $\frac{360}{5} = 72$. The display at the bottom of the graph screen indicates that one fifth root of 1 is $1 + 0i = 1$.

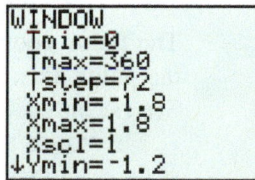
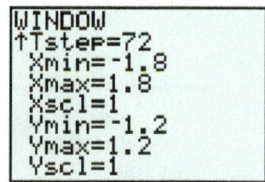

This is a continuation of the previous screen.

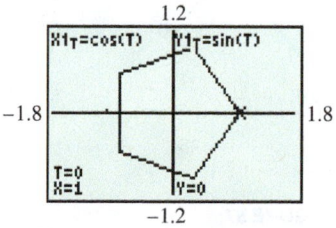

The calculator is in parametric, degree, and connected graph modes.

(a) Use this technique to find all fifth roots of 1, and express the real and imaginary parts in decimal form.

(b) Use this method to find the first three 10th roots of 1.

1. Write the complex number $2(\cos 60° + i \sin 60°)$ in rectangular form.

2. Find the modulus of the complex number $3 - 4i$.

3. Write the complex number $-\sqrt{2} + i\sqrt{2}$ in trigonometric form, where $0° \leq \theta < 360°$.

For $z_1 = 4(\cos 135° + i \sin 135°)$
and $z_2 = 2(\cos 45° + i \sin 45°),$

find each product or quotient and write your answers in both trigonometric and rectangular forms.

4. $z_1 \cdot z_2$

5. $\dfrac{z_1}{z_2}$

6. Evaluate the power $(4 \operatorname{cis} 17°)^3$, and write your answer in trigonometric form.

7. Evaluate the power $(2 - 2i)^4$, and write your answer in rectangular form.

8. Find the cube roots of -64, and express them in rectangular form.

9. Find the square roots of $2i$, and express them in rectangular form.

10. Find all complex solutions of the equation

$$x^3 = -1.$$

Leave your answers in trigonometric form.

10.6 Polar Equations and Graphs

Polar Coordinate System • Graphs of Polar Equations • Classifying Polar Equations • Converting Equations

Polar Coordinate System

The **polar coordinate system** is based on a point, called the **pole,** and a ray, called the **polar axis.** The polar axis is usually drawn in the direction of the positive x-axis. See **FIGURE 56.**

In **FIGURE 57,** the pole has been placed at the origin of a rectangular coordinate system, so that the polar axis coincides with the positive x-axis. Point P has rectangular coordinates (x, y). Point P can also be located by giving the directed angle θ from the positive x-axis to ray OP and the directed distance r from the pole to point P. The ordered pair (r, θ) gives **polar coordinates** of point P. If $r > 0$, point P lies on the terminal side of θ and if $r < 0$, point P lies on the ray pointing in the opposite direction of the terminal side of θ, a distance $|r|$ from the origin. See **FIGURE 58,** which shows rectangular axes superimposed on a **polar coordinate grid.**

Pole

Polar axis

FIGURE 56

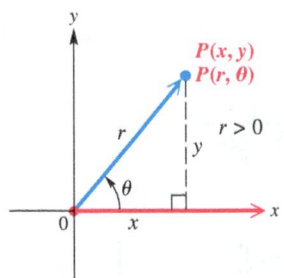

FIGURE 57

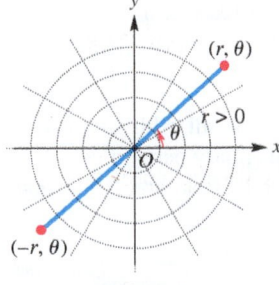

FIGURE 58

Relationships between Rectangular and Polar Coordinates

If a point has rectangular coordinates (x, y) and polar coordinates (r, θ), then these coordinates are related as follows.

$$x = r \cos \theta \qquad\qquad y = r \sin \theta$$

$$r^2 = x^2 + y^2 \qquad \tan \theta = \frac{y}{x}, \ \text{if } x \neq 0$$

Plotting Points with Polar Coordinates

Plot each point by hand in the polar coordinate system. Then determine the rectangular coordinates of each point.

(a) $P(2, 30°)$ (b) $Q\left(-4, \frac{2\pi}{3}\right)$ (c) $R\left(5, -\frac{\pi}{4}\right)$

Solution

(a) Since $r = 2$ and $\theta = 30°$, P is 2 units from the origin in the positive direction on a ray making a 30° angle with the polar axis. See **FIGURE 59,** which shows a polar grid superimposed onto a rectangular grid. Using the conversion equations, we can find the rectangular coordinates.

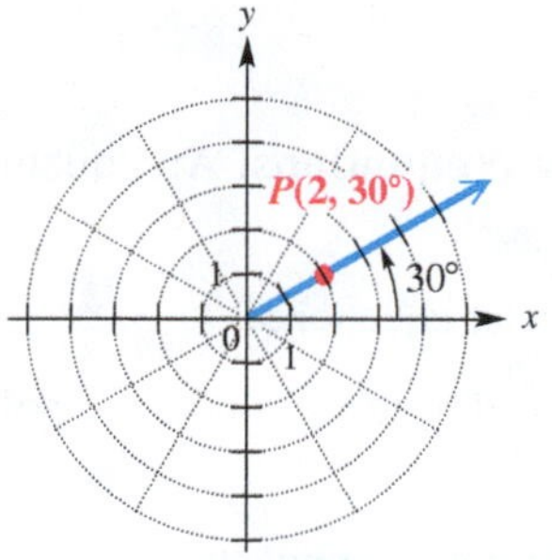

$$\begin{array}{ll} x = r \cos \theta & y = r \sin \theta \qquad \text{Conversion equations} \\ x = 2 \cos 30° & y = 2 \sin 30° \qquad \text{Substitute.} \\ x = 2\left(\dfrac{\sqrt{3}}{2}\right) = \sqrt{3} & y = 2\left(\dfrac{1}{2}\right) = 1 \qquad \text{Evaluate.} \end{array}$$

FIGURE 59

The rectangular coordinates are $\left(\sqrt{3}, 1\right)$.

(b) Since r is negative, Q is 4 units in the *opposite* direction from the pole on an extension of the $\frac{2\pi}{3}$ ray. See **FIGURE 60.** The rectangular coordinates are

$$x = -4 \cos \frac{2\pi}{3} = -4\left(-\frac{1}{2}\right) = 2 \quad \text{and} \quad y = -4 \sin \frac{2\pi}{3} = -4\left(\frac{\sqrt{3}}{2}\right) = -2\sqrt{3}.$$

→ Looking Ahead to Calculus

Techniques studied in calculus provide methods of finding slopes of tangent lines to polar curves, areas bounded by such curves, and lengths of their arcs.

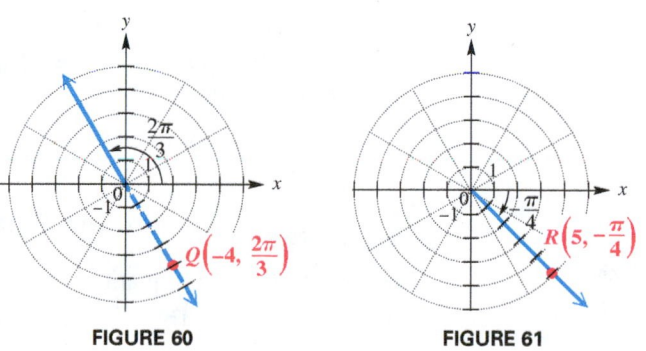

FIGURE 60   **FIGURE 61**

(c) Point R is shown in **FIGURE 61.** Since θ is negative, the angle is measured in the clockwise direction. Furthermore, we have

$$x = 5 \cos\left(-\frac{\pi}{4}\right) = \frac{5\sqrt{2}}{2} \quad \text{and} \quad y = 5 \sin\left(-\frac{\pi}{4}\right) = -\frac{5\sqrt{2}}{2}.$$

While a given point in the plane can have only one pair of rectangular coordinates, this same point can have infinitely many pairs of polar coordinates. For example, $(2, 30°)$ locates the same point as $(2, 390°)$, or $(2, -330°)$, or $(-2, 210°)$.

Giving Alternative Forms for Coordinates of a Point

(a) Give three other pairs of polar coordinates for the point $P(3, 140°)$.

(b) Determine two pairs of polar coordinates, with $r > 0$, for the point with rectangular coordinates $(-1, 1)$.

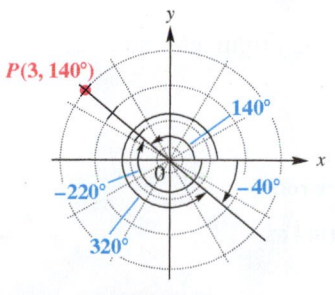

FIGURE 62

Solution

(a) Three such pairs are $(3, -220°)$, $(-3, 320°)$, and $(-3, -40°)$. See **FIGURE 62.**

(continued)

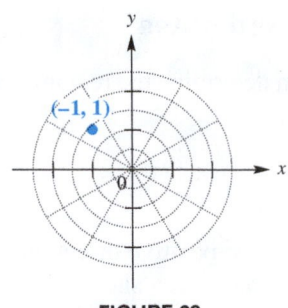

FIGURE 63

(b) As shown in **FIGURE 63,** the point $(-1, 1)$ lies in the second quadrant. Since $\tan \theta = \frac{1}{-1} = -1$, one possible value for θ is $135°$. Also,

$$r = \sqrt{x^2 + y^2} = \sqrt{(-1)^2 + 1^2} = \sqrt{2}.$$

Therefore, two pairs of polar coordinates are $\left(\sqrt{2}, 135°\right)$ and $\left(\sqrt{2}, -225°\right)$. (Any angle coterminal with $135°$ could have been used for the second angle.)

Graphs of Polar Equations

Equations in x and y are called **rectangular** (or **Cartesian**) **equations.** An equation in which r and θ are the variables is a **polar equation.**

$$r = 3 \sin \theta, \quad r = 2 + \cos \theta, \quad r = \theta \qquad \text{Polar equations}$$

The rectangular forms of lines and circles can also be defined in terms of polar coordinates. We typically solve a polar equation for r.

Line:	$ax + by = c$	Rectangular equation of a line
	$a(r \cos \theta) + b(r \sin \theta) = c$	Convert to polar coordinates.
	$r(a \cos \theta + b \sin \theta) = c$	Factor out r.

This is the polar equation of $ax + by = c$. $\longrightarrow \quad r = \dfrac{c}{a \cos \theta + b \sin \theta}$ Polar equation of a line

Circle:	$x^2 + y^2 = a^2$	Rectangular equation of a circle
	$r^2 = a^2$	$x^2 + y^2 = r^2$

These are the polar equations of $x^2 + y^2 = a^2$. $\longrightarrow \quad r = \pm a$ Polar equation of a circle; r can be negative in polar coordinates.

We use these forms in the next example.

> **EXAMPLE 3**   **Finding Polar Equations of Lines and Circles**

For each rectangular equation, give its equivalent polar equation and sketch its graph.

(a) $y = x - 3$ **(b)** $x^2 + y^2 = 4$

Solution

(a) This is the equation of a line. Rewrite $y = x - 3$ in standard form.

$x - y = 3$	Standard form: $ax + by = c$
$r \cos \theta - r \sin \theta = 3$	Substitute for x and y.
$r(\cos \theta - \sin \theta) = 3$	Factor out r.
$r = \dfrac{3}{\cos \theta - \sin \theta}$	Divide by $\cos \theta - \sin \theta$.

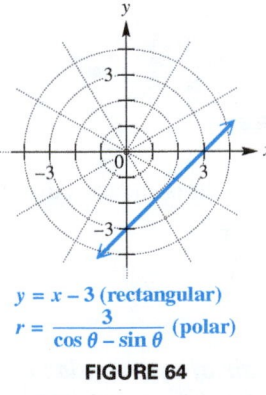

$y = x - 3$ **(rectangular)**
$r = \dfrac{3}{\cos \theta - \sin \theta}$ **(polar)**

FIGURE 64

Its graph is shown in **FIGURE 64.**

(b) The graph of $x^2 + y^2 = 4$ is a circle with center at the origin and radius 2.

$x^2 + y^2 = 4$	
$r^2 = 4$	$x^2 + y^2 = r^2$
$r = \pm 2$	Take *both* square roots.

The graphs of $r = 2$ and $r = -2$ coincide. See **FIGURE 65.**

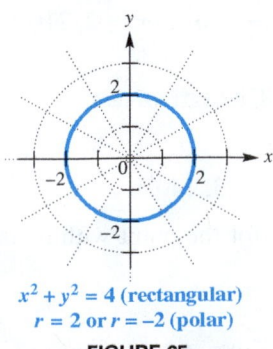

$x^2 + y^2 = 4$ **(rectangular)**
$r = 2$ or $r = -2$ **(polar)**

FIGURE 65

To graph polar equations, evaluate r for various values of θ until a pattern appears, and then join the points with a smooth curve.

EXAMPLE 4 **Graphing a Polar Equation (Cardioid)**

Graph $r = 1 + \cos \theta$.

Analytic Solution

We first find some ordered pairs, as in the table.* Once the pattern of values of r becomes clear, it is not necessary to find more ordered pairs.

θ	$\cos \theta$	$r = 1 + \cos \theta$	θ	$\cos \theta$	$r = 1 + \cos \theta$
0°	1	2	180°	−1	0
30°	0.9	1.9	225°	−0.7	0.3
60°	0.5	1.5	255°	−0.3	0.7
90°	0	1	270°	0	1
105°	−0.3	0.7	300°	0.5	1.5
135°	−0.7	0.3	330°	0.9	1.9

Connect the points in the order in which they appear in the table. See **FIGURE 66.** This curve is called a **cardioid** because of its heart shape.

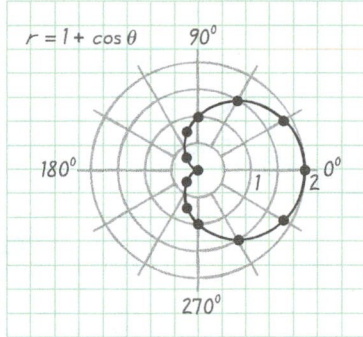

FIGURE 66

Graphing Calculator Solution

We choose degree mode and graph values of θ in the interval $[0°, 360°]$. The settings in **FIGURE 67** and **68** will generate the graph in **FIGURE 69**.

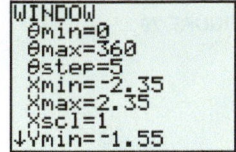

FIGURE 67

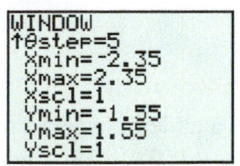

This is a continuation of the previous screen.

FIGURE 68

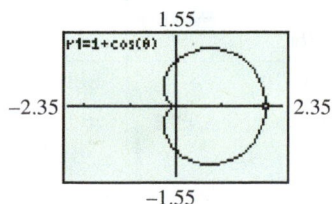

FIGURE 69

EXAMPLE 5 **Graphing a Polar Equation (Rose)**

Graph $r = 3 \cos 2\theta$.

Solution Because of the argument 2θ, the graph requires a larger number of points than when the argument is θ. A few ordered pairs are given in the table. Complete the table similarly through the first 180°, so that 2θ has values up to 360°.

θ	0°	15°	30°	45°	60°	75°	90°
2θ	0°	30°	60°	90°	120°	150°	180°
$\cos 2\theta$	1	0.9	0.5	0	−0.5	−0.9	−1
$r = 3 \cos 2\theta$	3	2.6	1.5	0	−1.5	−2.6	−3

To graph the equation $r = 3 \cos 2\theta$ of **Example 5** on a calculator, use polar graphing mode, degree mode, θmin = 0°, θmax = 360°, θstep = 5°, and window settings as shown.

*The tables in **Examples 4–7** include approximations for cosine values and r.

(continued)

Four-Leaved Rose

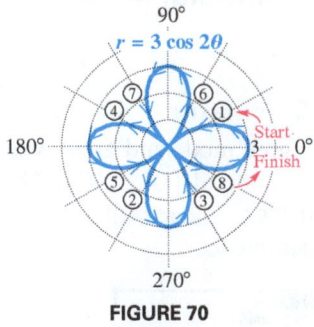

FIGURE 70

Plotting these points gives the graph called a **four-leaved rose,** shown in **FIGURE 70.** Notice how the graph is developed with a continuous curve, beginning with the upper half of the right horizontal leaf and ending with the lower half of that leaf. As the graph is traced, the curve goes through the pole four times.

> **FOR DISCUSSION**
>
> Trace the rose curve of **Example 5** and watch the cursor to verify that the numbering that appears in **FIGURE 70** is indeed correct.

The equation $r = 3 \cos 2\theta$ in **Example 5** has a graph that belongs to a family of curves called **roses.** The graphs of $r = a \sin n\theta$ and $r = a \cos n\theta$ are roses, with n leaves if n is odd and $2n$ leaves if n is even. The value of a determines the length of the leaves.

EXAMPLE 6 **Graphing a Polar Equation (Lemniscate)**

Graph $r^2 = \cos 2\theta$.

Analytic Solution

Complete a table of ordered pairs, and sketch the graph, as in **FIGURE 71.** The point $(-1, 0°)$, with r negative, may be plotted as $(1, 180°)$. Also, $(-0.7, 30°)$ may be plotted as $(0.7, 210°)$, and so on.

θ	0°	30°	45°	135°	150°	180°
2θ	0°	60°	90°	270°	300°	360°
$\cos 2\theta$	1	0.5	0	0	0.5	1
$r = \pm\sqrt{\cos 2\theta}$	± 1	± 0.7	0	0	± 0.7	± 1

Values of θ for $45° < \theta < 135°$ are not included in the table, because the corresponding values of $\cos 2\theta$ are negative (quadrants II and III) and so do not have real square roots. Values of θ larger than 180° give 2θ larger than 360° and would repeat the points already found. This curve is called a **lemniscate.**

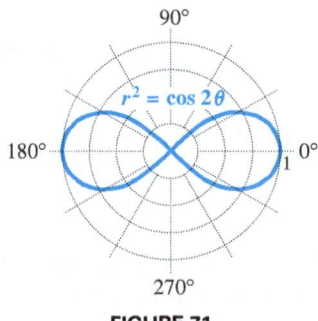

FIGURE 71

Graphing Calculator Solution

To graph $r^2 = \cos 2\theta$ with a graphing calculator, let

$$r_1 = \sqrt{\cos 2\theta}$$

and

$$r_2 = -\sqrt{\cos 2\theta}.$$

See **FIGURE 72** for the equations and **FIGURE 73** for the graph.

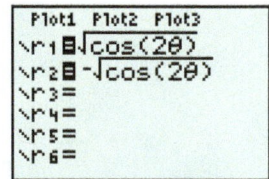

FIGURE 72

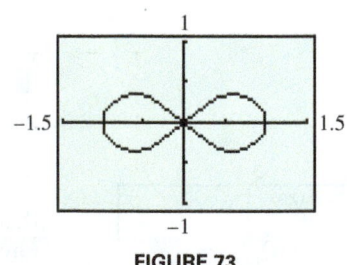

FIGURE 73

EXAMPLE 7 **Graphing a Polar Equation (Spiral of Archimedes)**

Graph $r = 2\theta$ (θ measured in radians).

Solution Some ordered pairs are shown in the table on the next page. Since $r = 2\theta$, rather than a trigonometric function of θ, we must also consider negative values of θ. Radian measures have been rounded. The graph, shown in **FIGURE 74,** is called a **spiral of Archimedes.**

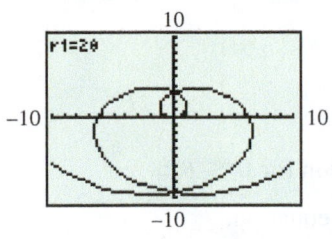

$-2\pi \le \theta \le 2\pi$

More of the spiral can be seen in this calculator graph of the spiral in **Example 7** than is shown in **FIGURE 74**.

θ (radians)	$r = 2\theta$	θ (radians)	$r = 2\theta$
$-\pi$	-6.3	$\frac{\pi}{3}$	2.1
$-\frac{\pi}{2}$	-3.1	$\frac{\pi}{2}$	3.1
$-\frac{\pi}{4}$	-1.6	π	6.3
0	0	$\frac{3\pi}{2}$	9.4
$\frac{\pi}{6}$	1	2π	12.6

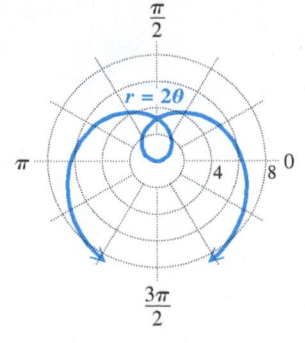

FIGURE 74

FOR DISCUSSION

Refer to the calculator graph of $r = 2\theta$ in the margin. Experiment with various minimum and maximum values of θ. What behaviors do you observe? (As you experiment, you will need to enlarge the window to accommodate the increased graph size.)

Classifying Polar Equations

The table summarizes common polar graphs and forms of their equations. We also include **limaçons.** Cardioids are a special case of limaçons, where $\left|\frac{a}{b}\right| = 1$.

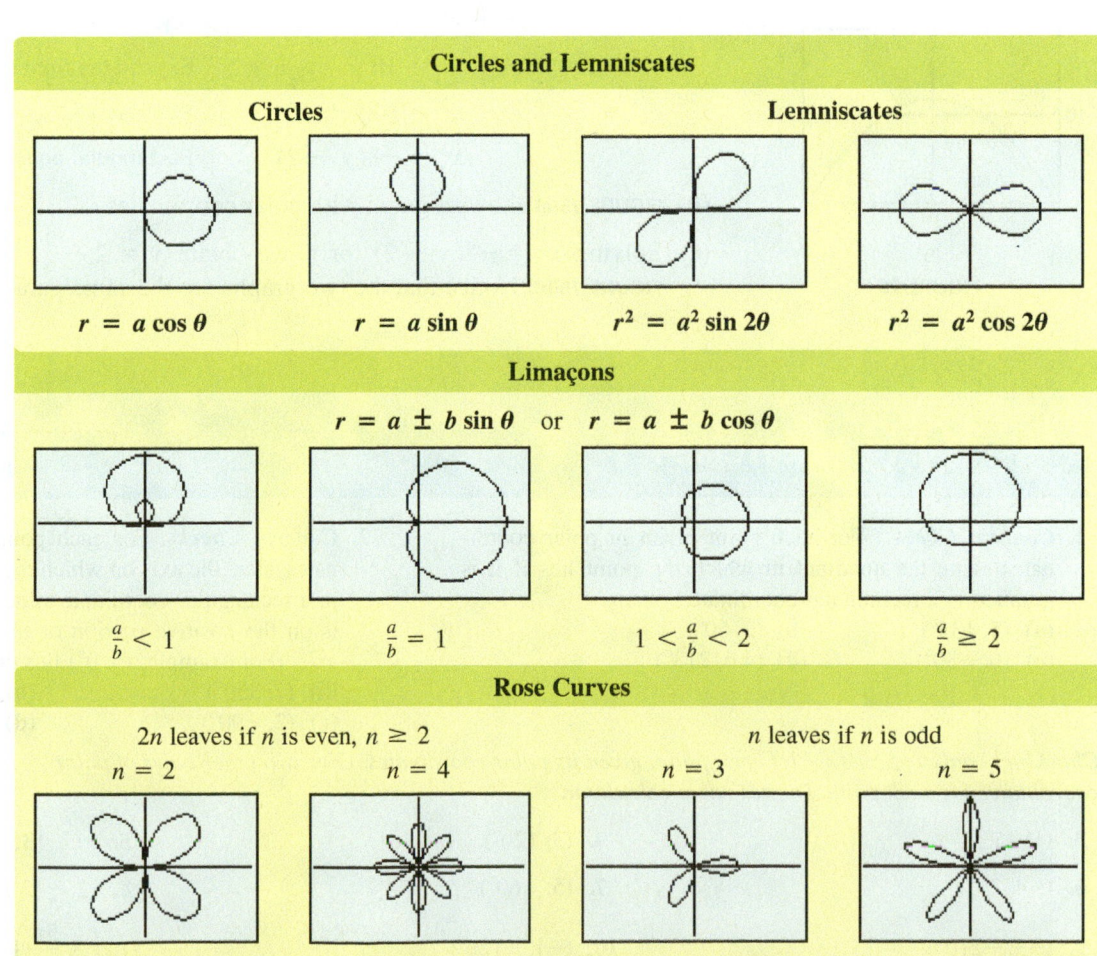

Converting Equations

In **Example 3** we converted rectangular equations to polar equations. We conclude with an example that converts a polar equation to a rectangular one.

> **EXAMPLE 8**   **Converting a Polar Equation to a Rectangular Equation**
>
> Consider the polar equation $r = \dfrac{4}{1 + \sin \theta}$.
>
> **(a)** Convert it to a rectangular equation.
>
> **(b)** Use a graphing calculator to graph the polar equation for $0 \le \theta \le 2\pi$.
>
> **(c)** Use a graphing calculator to graph the rectangular equation.

Solution

(a) Multiply each side of the equation by the denominator on the right, to clear the fraction.

$$r = \frac{4}{1 + \sin \theta} \qquad \text{Polar equation}$$

$$r + r \sin \theta = 4 \qquad \text{Multiply by } 1 + \sin \theta.$$

$$\sqrt{x^2 + y^2} + y = 4 \qquad \text{Let } r = \sqrt{x^2 + y^2} \text{ and } y = r \sin \theta.$$

$$\sqrt{x^2 + y^2} = 4 - y \qquad \text{Subtract } y.$$

$$x^2 + y^2 = (4 - y)^2 \qquad \text{Square each side.}$$

$$x^2 + y^2 = 16 - 8y + y^2 \qquad \text{Expand the right side.}$$

$$x^2 = -8y + 16 \qquad \text{Subtract } y^2.$$

$$x^2 = -8(y - 2) \qquad \text{Rectangular equation (a parabola)}$$

(b) **FIGURE 75(a)** shows a graph with polar coordinates.

(c) Solving $x^2 = -8(y - 2)$ for y, we obtain $y = 2 - \frac{1}{8}x^2$. Its graph is shown in **FIGURE 75(b)**. Notice that the two graphs are the same parabola.

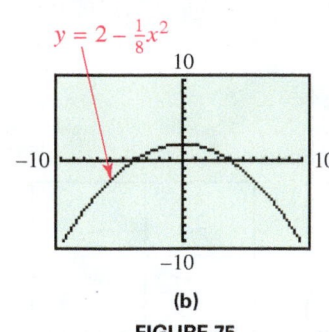

FIGURE 75

10.6 Exercises

1. *Concept Check* For each point given in polar coordinates, state the quadrant in which the point lies if it is graphed in a rectangular coordinate system.
(a) $(5, 135°)$ **(b)** $(2, 60°)$
(c) $(6, -30°)$ **(d)** $(4.6, 213°)$

2. *Concept Check* For each point given in polar coordinates, state the axis on which the point lies if it is graphed in a rectangular coordinate system. Also, state whether it is on the positive portion or the negative portion of the axis. (For example, $(5, 0°)$ lies on the positive x-axis.)
(a) $(7, 360°)$ **(b)** $(4, 180°)$
(c) $(2, -90°)$ **(d)** $(8, 450°)$

Checking Analytic Skills *Plot each point, given its polar coordinates. Give two other pairs of polar coordinates for each point.* **Do not use a calculator.**

3. $(1, 45°)$

4. $(3, 120°)$

5. $(-2, 135°)$

6. $(-4, 27°)$

7. $(5, -60°)$

8. $(2, -45°)$

9. $(-3, -210°)$

10. $(-1, -120°)$

11. $\left(3, \dfrac{5\pi}{3}\right)$

12. $\left(4, \dfrac{3\pi}{2}\right)$

13. $(-2, 0)$

14. $(-1, 2\pi)$

Checking Analytic Skills *Plot the point whose rectangular coordinates are given. Then determine two pairs of polar coordinates for the point with $0° \leq \theta < 360°$.* **Do not use a calculator.**

15. $(-1, 1)$ **16.** $(1, 1)$ **17.** $(0, 3)$ **18.** $(0, -3)$ **19.** $\left(\sqrt{2}, \sqrt{2} \right)$

20. $\left(-\sqrt{2}, \sqrt{2} \right)$ **21.** $\left(\dfrac{\sqrt{3}}{2}, \dfrac{3}{2} \right)$ **22.** $\left(-\dfrac{\sqrt{3}}{2}, -\dfrac{1}{2} \right)$ **23.** $(3, 0)$ **24.** $(-2, 0)$

For each rectangular equation, give its equivalent polar equation and sketch its graph.

25. $x - y = 4$ **26.** $x + y = -7$ **27.** $x^2 + y^2 = 16$

28. $x^2 + y^2 = 9$ **29.** $2x + y = 5$ **30.** $3x - 2y = 6$

Concept Check In Exercises 31–34, match each polar graph to its corresponding equation from choices A–D.

31. $r = 3$ **32.** $r = \cos 3\theta$ **33.** $r = \cos 2\theta$ **34.** $r = \dfrac{2}{\cos \theta + \sin \theta}$

A.
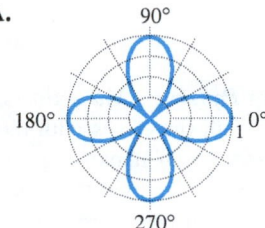

B.
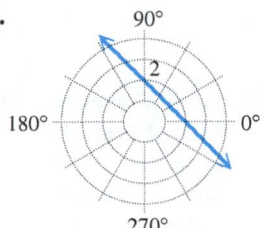

C.
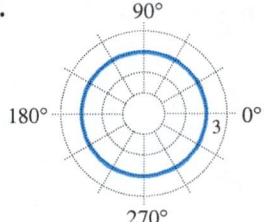

D.
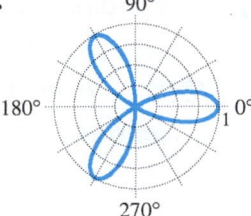

Graph each polar equation for θ in $[0°, 360°)$. In Exercises 35–44, identify the type of polar graph.

35. $r = 2 + 2 \cos \theta$ **36.** $r = 8 + 6 \cos \theta$ **37.** $r = 3 + \cos \theta$

38. $r = 2 - \cos \theta$ **39.** $r = 4 \cos 2\theta$ **40.** $r = 3 \cos 5\theta$

41. $r^2 = 4 \cos 2\theta$ **42.** $r^2 = 4 \sin 2\theta$ **43.** $r = 4 - 4 \cos \theta$

44. $r = 6 - 3 \cos \theta$ **45.** $r = 2 \sin \theta \tan \theta$
(This is a **cissoid**.)

46. $r = \dfrac{\cos 2\theta}{\cos \theta}$
(This is a **cissoid with a loop**.)

 47. Explain the method you would use to graph (r, θ) by hand if $r < 0$.

48. For $r > 0$, the points (r, θ) and $(-r, \theta + 180°)$ have the same graph. Explain why this is so.

Concept Check Answer each question.

49. If a point lies on an axis in the rectangular plane, then what kind of angle must θ be if (r, θ) represents the point in polar coordinates?

50. What will the graph of $r = k$ be, for $k > 0$?

51. How would the graph of **FIGURE 66** change if the equation were $r = 1 - \cos \theta$?

52. How would the graph of **FIGURE 66** change if the equation were $r = 1 + \sin \theta$?

53. The graphs of rose curves have equations of the form $r = a \sin n\theta$ or $r = a \cos n\theta$. What does the value of a determine? What does the value of n determine?

54. In **Exercise 53**, if $n = 1$, what will the graph be? What will the value of a determine?

For each equation, find an equivalent equation in rectangular coordinates. Then graph the result.

55. $r = 2 \sin \theta$

56. $r = 2 \cos \theta$

57. $r = \dfrac{2}{1 - \cos \theta}$

58. $r = \dfrac{3}{1 - \sin \theta}$

59. $r = -2 \cos \theta - 2 \sin \theta$

60. $r = \dfrac{3}{4 \cos \theta - \sin \theta}$

61. $r = 2 \sec \theta$

62. $r = -5 \csc \theta$

63. $r = \dfrac{2}{\cos \theta + \sin \theta}$

64. $r = \dfrac{2}{2 \cos \theta + \sin \theta}$

The graph of r = aθ is an example of the spiral of Archimedes. With your calculator set to radian mode, use the given value of a and interval of θ to graph the spiral in the window specified.

65. $a = 1, 0 \le \theta \le 4\pi, [-15, 15]$ by $[-15, 15]$

66. $a = 2, -4\pi \le \theta \le 4\pi, [-30, 30]$ by $[-30, 30]$

67. $a = 1.5, -4\pi \le \theta \le 4\pi, [-20, 20]$ by $[-20, 20]$

68. $a = -1, 0 \le \theta \le 12\pi, [-40, 40]$ by $[-40, 40]$

Find the polar coordinates of the points of intersection of the given curves for the specified interval of θ.

69. $r = 4 \sin \theta, r = 1 + 2 \sin \theta; 0 \le \theta < 2\pi$

70. $r = 3, r = 2 + 2 \cos \theta; 0° \le \theta < 360°$

71. $r = 2 + \sin \theta, r = 2 + \cos \theta; 0 \le \theta < 2\pi$

72. $r = \sin 2\theta, r = \sqrt{2} \cos \theta; 0 \le \theta < \pi$

(Modeling) Solve each problem.

73. *Satellite Orbits* The polar equation

$$r = \frac{a(1 - e^2)}{1 + e \cos \theta},$$

where *a* is the average distance in astronomical units from our sun and *e* is a constant called the eccentricity, can be used to graph the orbits of satellites of the sun. (See **Section 7.3.**) The sun will be located at the pole. The table lists *a* and *e* for the satellites.

Satellite	a	e
Mercury	0.39	0.206
Venus	0.78	0.007
Earth	1.00	0.017
Mars	1.52	0.093
Jupiter	5.20	0.048
Saturn	9.54	0.056
Uranus	19.20	0.047
Neptune	30.10	0.009
Pluto	39.40	0.249

Source: Zeilik, M., S. Gregory, and E. Smith, *Introductory Astronomy and Astrophysics,* Saunders College Publishers.

(a) Graph the orbits of the four closest satellites on the same polar grid. Choose a viewing window that results in a graph with nearly circular orbits.

(b) Plot the orbits of Earth, Jupiter, Uranus, and Pluto on the same polar grid. How does Earth's distance from the sun compare with the distance from the sun to these satellites?

(c) Use graphing to determine whether Pluto is always the farthest of these from the sun.

74. *Radio Towers and Broadcasting Patterns* Many times, radio stations do not broadcast in all directions with the same intensity. To avoid interference with an existing station to the north, a new station may be licensed to broadcast only east and west. To create an east–west signal, two radio towers are sometimes used, as illustrated in the figure. Locations where the radio signal is received correspond to the interior of the curve defined by

$$r^2 = 40{,}000 \cos 2\theta,$$

where the polar axis (or positive *x*-axis) points east.

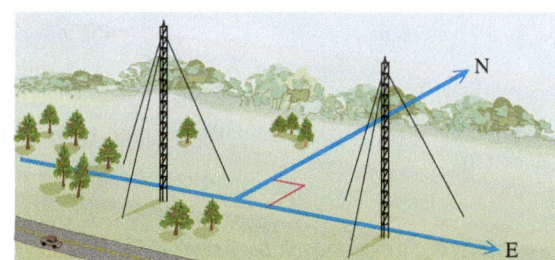

(a) Graph $r^2 = 40{,}000 \cos 2\theta$ for $0° \le \theta \le 360°$, with units in miles. Assuming that the radio towers are located near the pole, use the graph to describe the regions where the signal can be received and where the signal cannot be received.

(b) Suppose a radio signal pattern is given by

$$r^2 = 22{,}500 \sin 2\theta.$$

Graph this pattern and interpret the results.

10.7 More Parametric Equations

Parametric Graphing Revisited • Parametric Equations with Trigonometric Functions • The Cycloid • Applications of Parametric Equations

Parametric Graphing Revisited

In **Example 1** we review parametric graphing, first introduced in **Section 7.4.**

> **EXAMPLE 1** **Graphing a Plane Curve Defined Parametrically**
>
> Let $x = t^2$ and $y = 2t + 3$, for t in $[-3, 3]$. Graph the set of ordered pairs (x, y).

Analytic Solution

Make a table of corresponding values of t, x, and y over the domain of t. Then plot the points as shown in **FIGURE 76.** The graph is a portion of a parabola with horizontal axis $y = 3$. The arrowheads indicate the direction the curve traces as t increases.

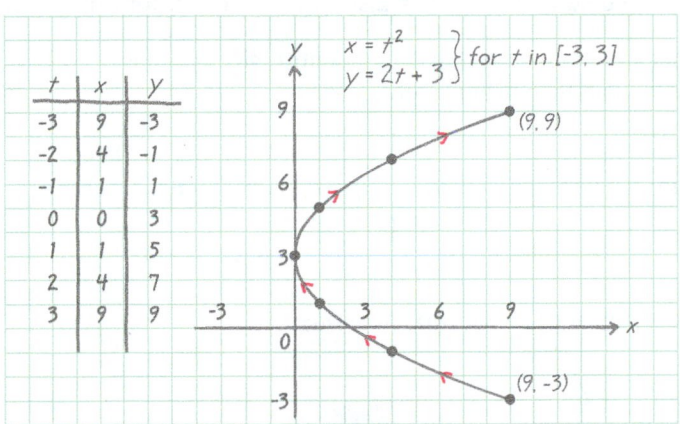

FIGURE 76

Note that more of the parabola would be traced if the domain of t were larger.

Graphing Calculator Solution

The parameters for the TI-84 Plus and the graph are shown in **FIGURE 77.**

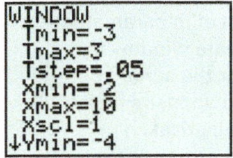

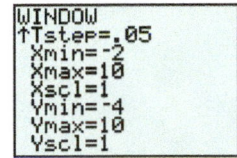

This is a continuation of the previous screen.

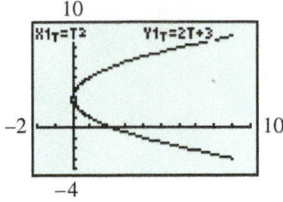

FIGURE 77

Parametric Equations with Trigonometric Functions

If we use trigonometric functions in parametric equations, many interesting curves can be drawn, as shown in **FIGURE 78.** Note that these figures would be difficult to graph using only functions, whereas parametric equations can be used to easily graph these figures.

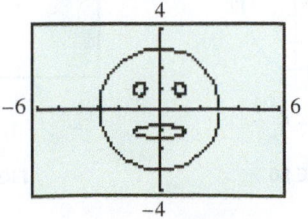
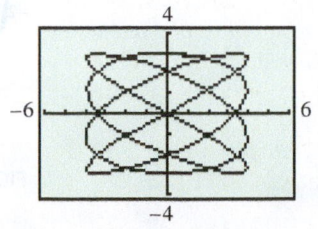
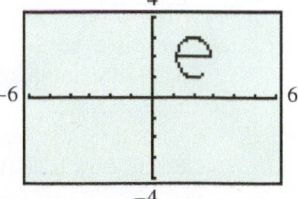

FIGURE 78

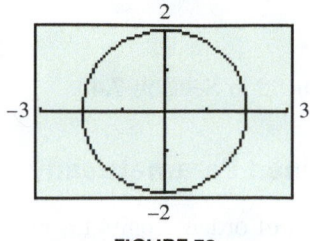

FIGURE 79

EXAMPLE 2 **Graphing a Circle with Parametric Equations**

Graph the parametric equations.

$$x = 2 \cos t, \, y = 2 \sin t, \quad \text{for} \quad 0 \le t \le 2\pi$$

Find an equivalent equation by using rectangular coordinates.

Solution Let $X_{1T} = 2 \cos(T)$ and $Y_{1T} = 2 \sin(T)$, and graph these parametric equations, as shown in **FIGURE 79.**

To verify that the graph is a circle, consider the following.

$$x^2 + y^2 = (2 \cos t)^2 + (2 \sin t)^2 \qquad \textcolor{blue}{x = 2 \cos t, \, y = 2 \sin t}$$
$$x^2 + y^2 = 4 \cos^2 t + 4 \sin^2 t \qquad \textcolor{blue}{\text{Properties of exponents}}$$
$$x^2 + y^2 = 4(\cos^2 t + \sin^2 t) \qquad \textcolor{blue}{\text{Distributive property}}$$
$$x^2 + y^2 = 4 \qquad \textcolor{blue}{\cos^2 t + \sin^2 t = 1}$$

The parametric equations are equivalent to $x^2 + y^2 = 4$, which is the equation of a circle with center $(0, 0)$ and radius 2.

EXAMPLE 3 **Graphing an Ellipse with Parametric Equations**

Graph the plane curve determined by $x = 2 \sin t, \, y = 3 \cos t$, for t in $[0, 2\pi]$.

Solution In this situation, we will use the identity $\sin^2 t + \cos^2 t = 1$. To do this, we begin by squaring and then using algebra.

$x = 2 \sin t$		$y = 3 \cos t$	
$x^2 = 4 \sin^2 t$	Square each side.	$y^2 = 9 \cos^2 t$	Square each side.
$\dfrac{x^2}{4} = \sin^2 t$	Solve for $\sin^2 t$.	$\dfrac{y^2}{9} = \cos^2 t$	Solve for $\cos^2 t$.

Now add corresponding sides of the two equations.

$$\frac{x^2}{4} + \frac{y^2}{9} = \sin^2 t + \cos^2 t$$

$$\frac{x^2}{4} + \frac{y^2}{9} = 1 \qquad \textcolor{blue}{\sin^2 t + \cos^2 t = 1}$$

This is the equation of an ellipse, as shown in **FIGURE 80.** The ellipse can be graphed directly with a calculator in parametric mode. See **FIGURE 81.**

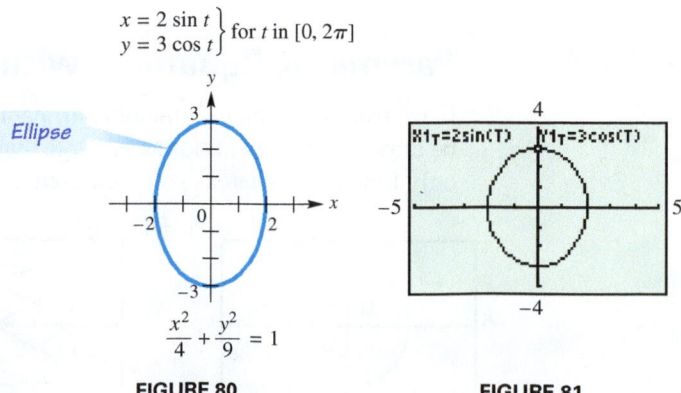

FIGURE 80

FIGURE 81

➡ **Looking Ahead to Calculus**
The cycloid is a special case of a curve traced out by a point at a given distance from the center of a circle as the circle rolls along a straight line. See **FIGURE 82.** Such a curve is called a **trochoid.** Parametric curves are studied in calculus.

The Cycloid

The path traced by a fixed point on the circumference of a circle rolling along a line is called a *cycloid*. A **cycloid** is defined by

$$x = at - a\sin t, \quad y = a - a\cos t, \quad \text{for } t \text{ in } (-\infty, \infty).$$

> **EXAMPLE 4** **Graphing a Cycloid**

Graph the cycloid.

$$x = t - \sin t, \quad y = 1 - \cos t, \quad \text{for } t \text{ in } [0, 2\pi]$$

Analytic Solution

There is no simple way to find a rectangular equation for the cycloid from its parametric equations. Instead, begin with a table using selected values for t in $[0, 2\pi]$. Approximate values have been rounded as necessary.

t	0	$\dfrac{\pi}{4}$	$\dfrac{\pi}{2}$	π	$\dfrac{3\pi}{2}$	2π
x	0	0.08	0.6	π	5.7	2π
y	0	0.3	1	2	1	0

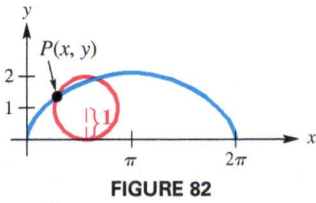

FIGURE 82

Plotting the ordered pairs (x, y) from the table of values leads to the portion of the graph (shown in blue) in **FIGURE 82** from 0 to 2π.

Graphing Calculator Solution

It is easier to graph a cycloid with a graping calculator in parametric mode than with traditional methods. See **FIGURE 83.**

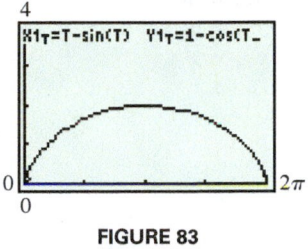

FIGURE 83

Using a larger interval for t would show that the cycloid repeats the pattern shown here every 2π units.

FIGURE 84

The cycloid has an interesting physical property. If a flexible cord or wire goes through points P and Q as in **FIGURE 84,** and, due to the force of gravity, a bead is allowed to slide without friction along this path from P to Q, the path that requires the least time takes the shape of an inverted cycloid.

Applications of Parametric Equations

Flight of a Ball

Parametric equations are used to simulate motion. If a ball is thrown with an initial velocity of v_0 feet per second at an angle θ with the horizontal, its position (x, y) can be modeled by the parametric equations

$$x = (v_0 \cos \theta)t \quad \text{and} \quad y = (v_0 \sin \theta)t - 16t^2 + h,$$

where t is in seconds and h is the ball's initial height in feet above the ground. The term $-16t^2$ occurs because gravity is pulling downward. See **FIGURE 85.** These equations ignore air resistance.

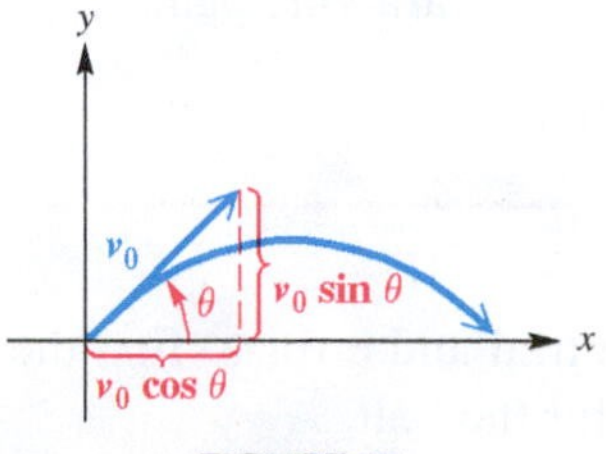

FIGURE 85

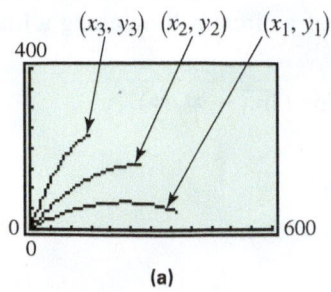

(a)

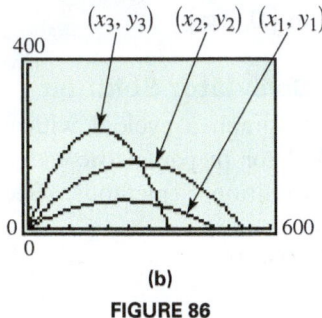

(b)

FIGURE 86

EXAMPLE 5 **Simulating Motion with Parametric Equations**

Three golf balls are hit simultaneously into the air at 132 feet per second (90 mph) at angles of 30°, 50°, and 70° with the horizontal.

(a) Assuming that the ground is level, determine graphically which ball travels the farthest horizontally. Estimate this distance.

(b) Which ball reaches the greatest height? Estimate this height.

Solution

(a) The three sets of parametric equations determined by the three golf balls are as follows, since $h = 0$.

$$x_1 = (132 \cos 30°)t, \qquad y_1 = (132 \sin 30°)t - 16t^2$$
$$x_2 = (132 \cos 50°)t, \qquad y_2 = (132 \sin 50°)t - 16t^2$$
$$x_3 = (132 \cos 70°)t, \qquad y_3 = (132 \sin 70°)t - 16t^2$$

Graphs are shown in **FIGURE 86(a)**, where $0 \le t \le 9$. We used a graphing calculator in simultaneous mode so that we simulate all three balls in flight at the same time. From the graph in **FIGURE 86(b)**, we can see that the ball hit at 50° travels the farthest horizontal distance. Using the trace feature, we estimate this distance to be about 540 feet. (Verify this.)

(b) The ball hit at 70° reaches the greatest height, about 240 feet. (Verify this.) ●

EXAMPLE 6 **Examining Parametric Equations of Flight**

A small rocket is launched from a table that is 3.36 feet above the ground. Its initial velocity is 64 feet per second, and it is launched at an angle of 30° with respect to the ground. Find the rectangular equation that models this path. What type of path does the rocket follow?

Solution Its path is defined by the parametric equations

$$x = (64 \cos 30°)t \quad \text{and} \quad y = (64 \sin 30°)t - 16t^2 + 3.36$$

or, equivalently,

$$x = 32\sqrt{3}t \quad \text{and} \quad y = -16t^2 + 32t + 3.36.$$

Solving $x = 32\sqrt{3}t$ for t, we obtain $t = \dfrac{x}{32\sqrt{3}}$. Substituting for t in the other parametric equation yields the following.

$$y = -16\left(\frac{x}{32\sqrt{3}}\right)^2 + 32\left(\frac{x}{32\sqrt{3}}\right) + 3.36$$

$$y = -\frac{1}{192}x^2 + \frac{\sqrt{3}}{3}x + 3.36 \qquad \text{Simplify.}$$

Because this equation defines a parabola, the rocket follows a parabolic path. ●

FOR DISCUSSION

If a golf ball is hit at 88 feet per second (60 mph), use trial and error to find the angle θ that results in a maximum horizontal distance for the ball.

EXAMPLE 7 **Analyzing the Path of a Projectile**

Determine the total flight time and the horizontal distance traveled by the rocket in **Example 6.**

Analytic Solution

The equation $y = -16t^2 + 32t + 3.36$ gives the vertical position of the rocket at time t. We need to determine those values of t for which $y = 0$, since these values correspond to the rocket at ground level. This yields

$$0 = -16t^2 + 32t + 3.36.$$

From the quadratic formula, the solutions are $t = -0.1$ or $t = 2.1$. Since t represents time, $t = -0.1$ is an unacceptable answer. Therefore, the flight time is 2.1 seconds.

The rocket was in the air for 2.1 seconds, so we can use $t = 2.1$ and the parametric equation that models the horizontal position, $x = 32\sqrt{3}t$, to get

$$x = 32\sqrt{3}\,(2.1) \approx 116.4 \text{ feet.}$$

Graphing Calculator Solution

FIGURE 87 shows that when $t = 2.1$, the horizontal distance covered is approximately 116.4 feet, which agrees with the analytic solution.

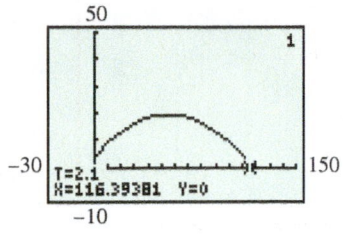

FIGURE 87

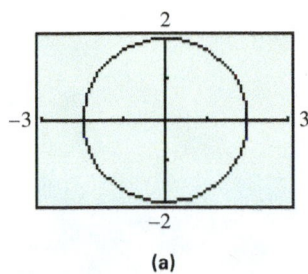

(a)

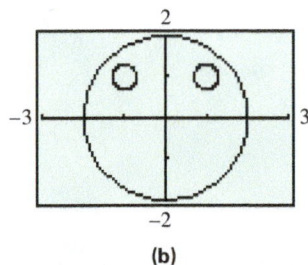

(b)

(c)

FIGURE 88

Parametric equations are used frequently in computer graphics to design a variety of figures, letters, and shapes.

EXAMPLE 8 **Creating a Drawing with Parametric Equations**

Graph a smiley face using parametric equations.

Solution *Head* We use a circle centered at the origin for the head. If the radius is 2, then we let $x = 2\cos t$ and $y = 2\sin t$ for $0 \le t \le 2\pi$. See **FIGURE 88(a).**

Eyes For the eyes, we use two small circles. The eye in the first quadrant can be modeled by $x = 1 + 0.3\cos t$ and $y = 1 + 0.3\sin t$ for $0 \le t \le 2\pi$, representing a circle centered at $(1, 1)$ with radius 0.3.

The eye in the second quadrant can be modeled by $x = -1 + 0.3\cos t$ and $y = 1 + 0.3\sin t$ for $0 \le t \le 2\pi$, representing a circle centered at $(-1, 1)$ with radius 0.3. See **FIGURE 88(b).**

Mouth For the smile, we can use the lower half of a circle. Using trial and error, we arrive at $x = 0.5\cos\frac{1}{2}t$ and $y = -0.5 - 0.5\sin\frac{1}{2}t$, a semicircle centered at $(0, -0.5)$ with radius 0.5.

Since we are letting $0 \le t \le 2\pi$, the term $\frac{1}{2}t$ ensures that only half a circle (semicircle) is drawn. The minus sign before $0.5\sin\frac{1}{2}t$ in the y-equation results in the lower half of the semicircle being drawn rather than the upper half. The final result is shown in **FIGURE 88(c).** We can add the pupils by plotting the points $(1, 1)$ and $(-1, 1)$. Turning off the coordinate axes rids the face of lines.

FOR DISCUSSION

Modify the face in **Example 8** so that it is frowning. Try to find a way to make the right eye be shut rather than open.

WHAT WENT WRONG?

A student graphed $x = \cos \frac{1}{2}t$, $y = \sin \frac{1}{2}t$, which should be a circle with radius 1 because $x^2 + y^2 = \cos^2 \frac{1}{2}t + \sin^2 \frac{1}{2}t = 1$. However, the following screen shows the graph she obtained on the TI-84 Plus.

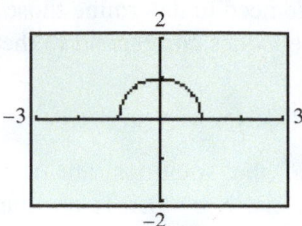

What Went Wrong? What should the student do to change the graph from a semicircle to a circle?

Answer to What Went Wrong?

The periods for $y = \cos \frac{1}{2}t$ and $y = \sin \frac{1}{2}t$ are both 4π. The interval for t should be $0 \leq t \leq 4\pi$, rather than $0 \leq t \leq 2\pi$, which is the interval the student used.

10.7 Exercises

Concept Check *Match the ordered pair from Column II with the pair of parametric equations in Column I on whose graph the point lies. In each case, consider the given value of t.*

I	II
1. $x = 3t + 6$, $y = -2t + 4$; $t = 2$	**A.** $(5, 25)$
2. $x = \cos t$, $y = \sin t$; $t = \dfrac{\pi}{4}$	**B.** $(7, 2)$
3. $x = t$, $y = t^2$; $t = 5$	**C.** $(12, 0)$
4. $x = t^2 + 3$, $y = t^2 - 2$; $t = 2$	**D.** $\left(\dfrac{\sqrt{2}}{2}, \dfrac{\sqrt{2}}{2} \right)$

Checking Analytic Skills *Find a rectangular equation for each curve and describe the curve.* **Do not use a calculator.**

5. $x = 3 \sin t$, $y = 3 \cos t$; for t in $[-\pi, \pi]$

6. $x = 2 \sin t$, $y = 2 \cos t$; for t in $[0, 2\pi]$

7. $x = 2 \cos^2 t$, $y = 2 \sin^2 t$; for t in $\left[0, \dfrac{\pi}{2} \right]$

8. $x = \sqrt{5} \sin t$, $y = \sqrt{3} \cos t$; for t in $[0, 2\pi]$

9. $x = 3 \tan t$, $y = 2 \sec t$; for t in $\left(-\dfrac{\pi}{2}, \dfrac{\pi}{2} \right)$

10. $x = \cot t$, $y = \csc t$; for t in $(0, \pi)$

For each plane curve, (a) graph the curve, and (b) find a rectangular equation for the curve.

11. $x = t + 2, y = t^2$, for t in $[-1, 1]$

12. $x = 2t, y = t + 1$, for t in $[-2, 3]$

13. $x = \sqrt{t}, y = 3t - 4$, for t in $[0, 4]$

14. $x = t^2, y = \sqrt{t}$, for t in $[0, 4]$

15. $x = t^3 + 1, y = t^3 - 1$, for t in $(-\infty, \infty)$

16. $x = 2t - 1, y = t^2 + 2$, for t in $(-\infty, \infty)$

17. $x = t + 2, y = \dfrac{1}{t + 2}$, for $t \neq -2$

18. $x = t - 3, y = \dfrac{2}{t - 3}$, for $t \neq 3$

19. $x = t + 2, y = t - 4$, for t in $(-\infty, \infty)$

20. $x = t^2 + 2, y = t^2 - 4$, for t in $(-\infty, \infty)$

Graph each pair of parametric equations for $0 \leq t \leq 2\pi$. Describe any differences in the two graphs.

21. (a) $x = 3 \cos t, \quad y = 3 \sin t$

(b) $x = 3 \cos 2t, \quad y = 3 \sin 2t$

22. (a) $x = 2 \cos t, \quad y = 2 \sin t$

(b) $x = 2 \cos t, \quad y = -2 \sin t$

23. (a) $x = 3 \cos t, \quad y = 3 \sin t$

(b) $x = 3 \sin t, \quad y = 3 \cos t$

24. (a) $x = -1 + \cos t, \quad y = 2 + \sin t$

(b) $x = 1 + \cos t, \quad y = 2 + \sin t$

Find a rectangular equation for each curve and graph the curve.

25. $x = \sin t, y = \csc t$; for t in $(0, \pi)$

26. $x = \tan t, y = \cot t$; for t in $\left(0, \dfrac{\pi}{2}\right)$

27. $x = 2 + \sin t, y = 1 + \cos t$; for t in $[0, 2\pi]$

28. $x = 1 + 2 \sin t, y = 2 + 3 \cos t$; for t in $[0, 2\pi]$

Graph each pair of parametric equations.

29. $x = 2 + \cos t, y = \sin t - 1; 0 \leq t \leq 2\pi$

30. $x = -2 + \cos t, y = \sin t + 1; 0 \leq t \leq 2\pi$

31. $x = \cos^3 t, y = \sin^3 t; 0 \leq t \leq 2\pi$

32. $x = \cos^5 t, y = \sin^5 t; 0 \leq t \leq 2\pi$

33. $x = |3 \sin t|, y = |3 \cos t|; 0 \leq t \leq \pi$

34. $x = 3 \sin 2t, y = 3 \cos t; 0 \leq t \leq 2\pi$

Graph each cycloid for t in the specified interval.

35. $x = t - \sin t, y = 1 - \cos t$; for t in $[0, 4\pi]$

36. $x = 2t - 2 \sin t, y = 2 - 2 \cos t$; for t in $[0, 8\pi]$

Graph each pair of parametric equations for $0 \leq t \leq 2\pi$ in the window $[0, 6]$ by $[0, 4]$. Identify the letter of the alphabet that is being graphed.

37. $x_1 = 1, \qquad\qquad y_1 = 1 + \dfrac{t}{\pi}$

$x_2 = 1 + \dfrac{t}{3\pi}, \quad y_2 = 2$

$x_3 = 1 + \dfrac{t}{2\pi}, \quad y_3 = 3$

38. $x_1 = 1, \qquad\qquad y_1 = 1 + \dfrac{t}{\pi}$

$x_2 = 1 + \dfrac{t}{3\pi}, \quad y_2 = 2$

$x_3 = 1 + \dfrac{t}{2\pi}, \quad y_3 = 3$

$x_4 = 1 + \dfrac{t}{2\pi}, \quad y_4 = 1$

39. $x_1 = 1, \qquad\qquad\qquad y_1 = 1 + \dfrac{t}{\pi}$

$x_2 = 1 + 1.3 \sin 0.5t, \quad y_2 = 2 + \cos 0.5t$

40. $x_1 = 2 + 0.8 \cos 0.85t, \quad y_1 = 2 + \sin 0.85t$

$x_2 = 1.2 + \dfrac{t}{1.3\pi}, \qquad\qquad y_2 = 2$

(Modeling) Designing Letters *Find a set of parametric equations that results in a letter similar to the one shown in each figure. Use the window* $[-4.7, 4.7]$ *by* $[-3.1, 3.1]$*, and turn off the coordinate axes. Answers may vary.*

41.
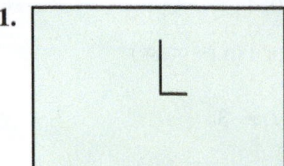

42.
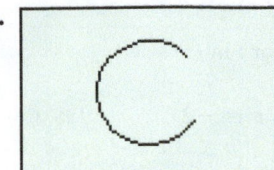

43.
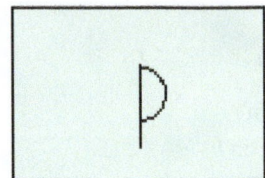

44.
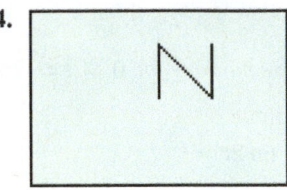

45. *(Modeling) Designing a Face* Refer to **Example 8.** Use parametric equations to create your own smiley face. This face should have a head, a mouth, and eyes.

46. *(Modeling) Designing a Face* Add a nose to the face that you designed in **Exercise 45.**

Lissajous Figures *The middle screen in* **FIGURE 78** *is an example of a Lissajous figure. Lissajous figures occur in electronics and may be used to find the frequency of an unknown voltage. Graph each Lissajous figure for* $0 \leq t \leq 6.5$ *in the window* $[-6, 6]$ *by* $[-4, 4]$.

47. $x = 2 \cos t,\ y = 3 \sin 2t$

48. $x = 3 \cos 2t,\ y = 3 \sin 3t$

49. $x = 3 \sin 4t,\ y = 3 \cos 3t$

50. $x = 4 \sin 4t,\ y = 3 \sin 5t$

(Modeling) *In Exercises 51–54, do the following.*
(a) *Determine the parametric equations that model the path of the projectile.*
(b) *Determine the rectangular equation that models the path of the projectile.*
(c) *Determine the time the projectile is in flight and the horizontal distance covered.*

51. *Flight of a Model Rocket* A model rocket is launched from the ground with a velocity of 48 feet per second at an angle of 60° with respect to the ground.

52. *Flight of a Golf Ball* Tyler McGinnis is playing golf. He hits a golf ball from the ground at an angle of 60° with respect to the ground at a velocity of 150 feet per second.

53. *Flight of a Softball* Sally hits a softball when it is 2 feet above the ground. The ball leaves her bat at an angle of 20° with respect to the ground at a velocity of 88 feet per second.

54. *Flight of a Baseball* Finley hits a baseball when it is 2.5 feet above the ground. The ball leaves his bat at an angle of 29° from the horizontal with a velocity of 136 feet per second.

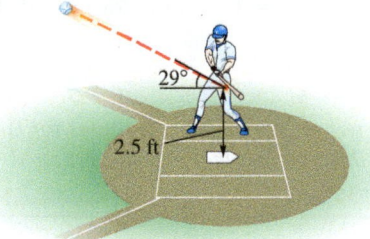

55. *(Modeling) Simulating Gravity on the Moon* If an object is thrown on the moon, then the parametric equations of flight are

$$x = (v \cos \theta)t \quad \text{and} \quad y = (v \sin \theta)t - 2.66t^2 + h.$$

Estimate the distance that a golf ball hit at 88 feet per second (60 mph) at an angle of 45° with the horizontal travels on the moon if the moon's surface is level.

56. *(Modeling) Flight of a Baseball*   A baseball is hit from a height of 3 feet at a 60° angle above the horizontal. Its initial velocity is 64 feet per second.
   (a) Write parametric equations that model the flight of the baseball.
   (b) Determine the horizontal distance traveled by the ball in the air. Assume that the ground is level.
   (c) What is the maximum height of the baseball? At that time, how far has the ball traveled horizontally?
   (d) Would the ball clear a 5-foot-high fence that is 100 feet from the batter?

(Modeling) Path of a Projectile   In Exercises 57 and 58, a projectile has been launched from the ground with an initial velocity of 88 feet per second. You are given parametric equations that model the path of the projectile.
(a) Graph the parametric equations.
(b) Approximate θ, the angle the projectile makes with the horizontal at launch, to the nearest tenth of a degree.

(c) *On the basis of your answer to part (b), write parametric equations for the projectile, using the cosine and sine functions.*

57. $x = 82.69265063t, \ y = -16t^2 + 30.09777261t$

58. $x = 56.56530965t, \ y = -16t^2 + 67.41191099t$

59. The spiral of Archimedes has polar equation $r = a\theta$, where $r^2 = x^2 + y^2$. Show that a parametric representation of the spiral of Archimedes is
$$x = a\theta \cos \theta, \quad y = a\theta \sin \theta, \quad \text{for } \theta \text{ in } (-\infty, \infty).$$

60. Show that the hyperbolic spiral given by $r\theta = a$, where $r^2 = x^2 + y^2$, is given parametrically by
$$x = \frac{a \cos \theta}{\theta}, \quad y = \frac{a \sin \theta}{\theta}, \quad \text{for } \theta \text{ in } (-\infty, 0) \cup (0, \infty).$$

SECTIONS 10.6–10.7   Reviewing Basic Concepts

1. For the point with polar coordinates $(-2, 130°)$, state the quadrant in which the point lies if it is graphed in a rectangular coordinate system.

2. Let the point P have rectangular coordinates $(-2, 2)$. Determine two pairs of equivalent polar coordinates for point P.

3. Graph the polar equation
$$r = 2 - 2 \cos \theta.$$
Identify the type of polar graph.

4. Find an equivalent equation in rectangular coordinates for the polar equation $r = 2 \cos \theta$.

5. Find an equivalent equation in polar coordinates for the equation $x + y = 6$.

6. Find a rectangular equation for the curve described by $x = 2 \cos t, \ y = 4 \sin t$ for $0 \le t < 2\pi$.

7. Graph the parametric equations
$$x = 2 - \sin t, \ y = \cos t - 1, \quad \text{for } 0 \le t < 2\pi.$$

8. *(Modeling) Flight of a Golf Ball*   Jeffery hits a golf ball with a 45° angle of elevation from the top of a ridge that is 50 feet above an area of level ground. The initial velocity of the ball is 88 feet per second, or 60 mph. Find the horizontal distance traveled by the ball in the air.

---

10 Summary

KEY TERMS & SYMBOLS

KEY CONCEPTS

10.1 The Law of Sines

side–angle–side (SAS)
angle–side–angle (ASA)
side–side–side (SSS)
side–angle–angle (SAA)
side–side–angle (SSA)
oblique triangle
ambiguous case

LAW OF SINES
In any triangle ABC with sides a, b, and c, the following hold.
$$\frac{a}{\sin A} = \frac{b}{\sin B}, \quad \frac{a}{\sin A} = \frac{c}{\sin C}, \quad \text{and} \quad \frac{b}{\sin B} = \frac{c}{\sin C}$$

This is the compact form.
$$\frac{a}{\sin A} = \frac{b}{\sin B} = \frac{c}{\sin C}$$

(continued)

KEY TERMS & SYMBOLS	KEY CONCEPTS

10.2 The Law of Cosines and Area Formulas

semiperimeter

LAW OF COSINES

In any triangle ABC with sides a, b, and c, the following hold.

$$a^2 = b^2 + c^2 - 2bc \cos A,$$
$$b^2 = a^2 + c^2 - 2ac \cos B,$$
$$\text{and} \quad c^2 = a^2 + b^2 - 2ab \cos C$$

HERON'S AREA FORMULA (SSS)

If a triangle has sides of lengths a, b, and c, and if the semiperimeter is $s = \frac{1}{2}(a + b + c)$, then the area $\mathcal{A}$ of the triangle is given by the following formula.

$$\mathcal{A} = \sqrt{s(s - a)(s - b)(s - c)}$$

AREA OF A TRIANGLE (SAS)

In any triangle ABC, the area $\mathcal{A}$ is given by the following formulas.

$$\mathcal{A} = \frac{1}{2}bc \sin A, \quad \mathcal{A} = \frac{1}{2}ab \sin C, \quad \text{and} \quad \mathcal{A} = \frac{1}{2}ac \sin B$$

10.3 Vectors and Their Applications

scalars
vector quantities
vector, $\mathbf{v}$, $\mathbf{OP}$, or $\overrightarrow{OP}$
initial point
terminal point
magnitude, $|\mathbf{OP}|$
parallelogram rule
resultant vector
opposite of $\mathbf{v}$
zero vector
scalar multiplication
position vector, $\langle a, b \rangle$
horizontal component
vertical component
direction angle
unit vectors, $\mathbf{i}$, $\mathbf{j}$
dot product
angle between two vectors
orthogonal vectors
airspeed
ground speed

RESULTANT VECTOR

The resultant (or sum) $\mathbf{A} + \mathbf{B}$ of vectors $\mathbf{A}$ and $\mathbf{B}$ is shown in the figure.

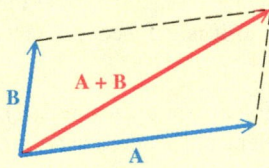

MAGNITUDE AND DIRECTION ANGLE OF A VECTOR

The magnitude (length) of vector $\mathbf{u} = \langle a, b \rangle$ is given by $|\mathbf{u}| = \sqrt{a^2 + b^2}$. The direction angle θ satisfies $\tan \theta = \frac{b}{a}$, where $a \neq 0$.

VECTOR OPERATIONS

Let a, b, c, d, and k represent real numbers.

$$\langle a, b \rangle + \langle c, d \rangle = \langle a + c, b + d \rangle$$
$$k \langle a, b \rangle = \langle ka, kb \rangle$$
$$\text{If } \mathbf{u} = \langle u_1, u_2 \rangle, \text{ then } -\mathbf{u} = \langle -u_1, -u_2 \rangle.$$
$$\langle a, b \rangle - \langle c, d \rangle = \langle a, b \rangle + -\langle c, d \rangle = \langle a - c, b - d \rangle$$

If $\mathbf{u} = \langle x, y \rangle$ has direction angle θ, then $\mathbf{u} = \langle |\mathbf{u}| \cos \theta, |\mathbf{u}| \sin \theta \rangle$.

i, j FORM FOR VECTORS

If $\mathbf{v} = \langle a, b \rangle$, then $\mathbf{v} = a\mathbf{i} + b\mathbf{j}$, where $\mathbf{i} = \langle 1, 0 \rangle$ and $\mathbf{j} = \langle 0, 1 \rangle$.

DOT PRODUCT

The dot product of the two vectors $\mathbf{u} = \langle a, b \rangle$ and $\mathbf{v} = \langle c, d \rangle$, denoted $\mathbf{u} \cdot \mathbf{v}$, is given by

$$\mathbf{u} \cdot \mathbf{v} = ac + bd.$$

If θ is the angle between $\mathbf{u}$ and $\mathbf{v}$, where $0° \leq \theta \leq 180°$, then $\mathbf{u} \cdot \mathbf{v} = |\mathbf{u}||\mathbf{v}| \cos \theta$.

KEY TERMS & SYMBOLS **KEY CONCEPTS**

PROPERTIES OF THE DOT PRODUCT

For all vectors $\mathbf{u}$, $\mathbf{v}$, and $\mathbf{w}$ and real numbers k, the following hold.

(a) $\mathbf{u} \cdot \mathbf{v} = \mathbf{v} \cdot \mathbf{u}$ 　　　　(b) $\mathbf{u} \cdot (\mathbf{v} + \mathbf{w}) = \mathbf{u} \cdot \mathbf{v} + \mathbf{u} \cdot \mathbf{w}$

(c) $(\mathbf{u} + \mathbf{v}) \cdot \mathbf{w} = \mathbf{u} \cdot \mathbf{w} + \mathbf{v} \cdot \mathbf{w}$ 　　(d) $(k\mathbf{u}) \cdot \mathbf{v} = k(\mathbf{u} \cdot \mathbf{v}) = \mathbf{u} \cdot (k\mathbf{v})$

(e) $\mathbf{0} \cdot \mathbf{u} = 0$ 　　　　(f) $\mathbf{u} \cdot \mathbf{u} = |\mathbf{u}|^2$

10.4 Trigonometric (Polar) Form of Complex Numbers

real axis
imaginary axis
complex plane
rectangular form
trigonometric (polar) form
modulus (absolute value)
argument

TRIGONOMETRIC (POLAR) FORM OF A COMPLEX NUMBER

The expression $r(\cos\theta + i\sin\theta)$ is called the trigonometric form (or polar form) of the complex number $x + yi$. The expression $\cos\theta + i\sin\theta$ is sometimes abbreviated cis θ. In this notation, $r(\cos\theta + i\sin\theta)$ is written r cis θ.

PRODUCT THEOREM

If $r_1(\cos\theta_1 + i\sin\theta_1)$ and $r_2(\cos\theta_2 + i\sin\theta_2)$ are any two complex numbers, then the following holds.

$$[r_1(\cos\theta_1 + i\sin\theta_1)] \cdot [r_2(\cos\theta_2 + i\sin\theta_2)] = r_1 r_2[\cos(\theta_1 + \theta_2) + i\sin(\theta_1 + \theta_2)]$$

In compact form, this is written

$$(r_1 \text{ cis } \theta_1)(r_2 \text{ cis } \theta_2) = r_1 r_2 \text{ cis}(\theta_1 + \theta_2).$$

QUOTIENT THEOREM

If $r_1(\cos\theta_1 + i\sin\theta_1)$ and $r_2(\cos\theta_2 + i\sin\theta_2)$ are any two complex numbers, where $r_2(\cos\theta_2 + i\sin\theta_2) \neq 0$, then the following holds.

$$\frac{r_1(\cos\theta_1 + i\sin\theta_1)}{r_2(\cos\theta_2 + i\sin\theta_2)} = \frac{r_1}{r_2}[\cos(\theta_1 - \theta_2) + i\sin(\theta_1 - \theta_2)]$$

In compact form, this is written

$$\frac{r_1 \text{ cis } \theta_1}{r_2 \text{ cis } \theta_2} = \frac{r_1}{r_2} \text{ cis}(\theta_1 - \theta_2).$$

10.5 Powers and Roots of Complex Numbers

nth root of a complex number

DE MOIVRE'S THEOREM

If $r(\cos\theta + i\sin\theta)$ is a complex number, and if n is any real number, then

$$[r(\cos\theta + i\sin\theta)]^n = r^n(\cos n\theta + i\sin n\theta),$$

or, in compact form, 　　　$[r \text{ cis } \theta]^n = r^n(\text{cis } n\theta).$

nTH ROOT

For a positive integer n, the complex number $a + bi$ is an nth root of the complex number $x + yi$ if $(a + bi)^n = x + yi.$

nTH ROOT THEOREM

If n is any positive integer, r is a positive real number, and θ is in degrees, then the nonzero complex number $r(\cos\theta + i\sin\theta)$ has exactly n distinct nth roots, given by

$$\sqrt[n]{r}(\cos\alpha + i\sin\alpha), \quad \text{or} \quad \sqrt[n]{r} \text{ cis } \alpha,$$

where $\quad \alpha = \dfrac{\theta + 360° \cdot k}{n}, \quad$ or $\quad \alpha = \dfrac{\theta}{n} + \dfrac{360° \cdot k}{n}, \quad k = 0, 1, 2, \ldots, n - 1.$

If θ is in radians, then

$$\alpha = \frac{\theta + 2\pi k}{n}, \quad \text{or} \quad \alpha = \frac{\theta}{n} + \frac{2\pi k}{n}, \quad k = 0, 1, 2, \ldots, n - 1.$$

(continued)

KEY TERMS & SYMBOLS	KEY CONCEPTS

10.6 Polar Equations and Graphs

polar coordinate system
pole
polar axis
polar coordinates
polar coordinate grid
polar equation
cardioid
rose curve (four-leaved rose)
lemniscate
spiral of Archimedes
limaçon

RELATIONSHIPS BETWEEN RECTANGULAR AND POLAR COORDINATES

If a point has rectangular coordinates (x, y) and polar coordinates (r, θ), then these coordinates are related as follows.

$$x = r \cos \theta, \quad y = r \sin \theta, \quad r^2 = x^2 + y^2, \quad \tan \theta = \frac{y}{x}, \quad \text{if } x \neq 0$$

POLAR EQUATIONS AND GRAPHS

$$\left.\begin{array}{l} r = a \cos \theta \\ r = a \sin \theta \end{array}\right\} \text{Circles} \qquad \left.\begin{array}{l} r^2 = a^2 \sin 2\theta \\ r^2 = a^2 \cos 2\theta \end{array}\right\} \text{Lemniscates}$$

$$\left.\begin{array}{l} r = a \pm b \sin \theta \\ r = a \pm b \cos \theta \end{array}\right\} \text{Limaçons} \qquad \left.\begin{array}{l} r = a \sin n\theta \\ r = a \cos n\theta \end{array}\right\} \text{Rose curves}$$

10.7 More Parametric Equations

cycloid

HEIGHT OF AN OBJECT

If an object has an initial velocity v_0 and initial height h, and travels so that its initial angle of elevation is θ, then its position (x, y) after t seconds is modeled by the following parametric equations.

$$x = (v_0 \cos \theta)t \quad \text{and} \quad y = (v_0 \sin \theta)t - 16t^2 + h$$

10 Review Exercises

Use the law of sines or the law of cosines to find the indicated part of each triangle ABC.

1. $C = 74° 10'$, $c = 96.3$ meters, $B = 39° 30'$; find b

2. $a = 165$ meters, $A = 100.2°$, $B = 25.0°$; find b

3. $a = 86.14$ inches, $b = 253.2$ inches, $c = 241.9$ inches; find A

4. $a = 14.8$ feet, $b = 19.7$ feet, $c = 31.8$ feet; find B

5. $A = 129° 40'$, $a = 127$ feet, $b = 69.8$ feet; find B

6. $B = 39° 50'$, $b = 268$ centimeters, $a = 340$ centimeters; find A

7. $B = 120.7°$, $a = 127$ feet, $c = 69.8$ feet; find b

8. $A = 46.2°$, $b = 184$ centimeters, $c = 192$ centimeters; find a

Find the area of each triangle ABC with the given information.

9. $b = 840.6$ meters, $c = 715.9$ meters, $A = 149.3°$

10. $a = 6.90$ feet, $b = 10.2$ feet, $C = 35° 10'$

11. $a = 0.913$ kilometer, $b = 0.816$ kilometer, $c = 0.582$ kilometer

12. $a = 43$ meters, $b = 32$ meters, $c = 51$ meters

Solve each problem.

13. *Height of a Balloon* The angles of elevation of a balloon from two points A and B on level ground are $24° 50'$ and $47° 20'$, respectively. As shown in the figure, points A and B are in the same vertical plane and are 8.4 miles apart. Approximate the height of the balloon above the ground to the nearest tenth of a mile.

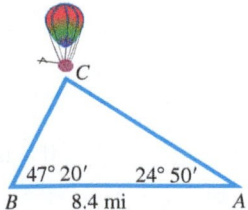

14. *Distance in a Boat Course*   The course for a boat starts at point X and goes in the direction S 48° W to point Y. It then turns and goes S 36° E to point Z and finally returns back to point X. If point Z lies 10 kilometers directly south of point X, find the distance from Y to Z, to the nearest kilometer.

15. *Radio Direction Finders*   Radio direction finders are placed at points A and B, which are 3.46 miles apart on an east–west line, with A west of B. From A, the bearing of an illegal radio transmitter is 48.0°; from B, the bearing is 302°. Find the distance between the transmitter and A.

16. *Distance across a Canyon*   To measure the distance AB across a canyon for a power line, a surveyor measures angles B and C and the distance BC. What is the distance from A to B?

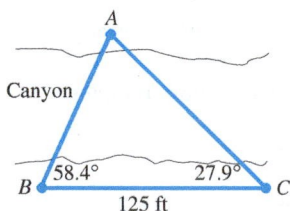

17. *Length of a Brace*   A banner on an 8.0-foot pole is to be mounted on a building at an angle of 115°, as shown in the figure. Find the length of the brace.

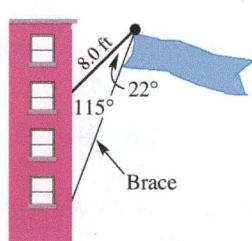

18. *Hanging Sculpture*   A hanging sculpture in an art gallery is to be hung with two wires of lengths 15.0 feet and 12.2 feet so that the angle between them is 70.3°. How far apart should the ends of the wire be placed on the ceiling?

19. *Pipeline Position*   A pipeline is to run between points A and B, which are separated by a protected wetlands area. To avoid the wetlands, the pipe will run from point A to C and then to B. The distances involved are $AB = 150$ kilometers, $AC = 102$ kilometers, and $BC = 135$ kilometers. What angle should be used at point C

20. If we are given a, A, and C in a triangle ABC, does the possibility of the ambiguous case exist? If not, explain why.

21. Can a triangle ABC exist if $a = 4.7$, $b = 2.3$, and $c = 7.0$? If not, explain why. Answer this question without using trigonometry.

22. *Concept Check*   Given that $a = 10$ and $B = 30°$ in triangle ABC, determine the values of b for which A has
   (a) exactly one value,
   (b) two values,
   (c) no value.

23. *Concept Check*   If angle C of a triangle ABC measures 90°, what does the law of cosines become?

24. *Concept Check*   When applying the law of cosines to find an angle, how can we tell if the angle is acute or obtuse before evaluating the inverse cosine value?

25. Use the vectors shown here to sketch $\mathbf{a} + 3\mathbf{b}$.

Find the magnitude and direction angle for $\mathbf{u}$, rounded to the nearest tenth of a degree.

26. $\mathbf{u} = \langle 21, -20 \rangle$ **27.** $\mathbf{u} = \langle -9, 12 \rangle$

Vector $\mathbf{v}$ has the given magnitude and direction angle. Write $\mathbf{v}$ in the form $\langle a, b \rangle$.

28. $|\mathbf{v}| = 50, \theta = 45°$ (give exact values)

29. $|\mathbf{v}| = 69.2, \theta = 75°$

30. $|\mathbf{v}| = 964, \theta = 154° \, 20'$

*Find **(a)** the dot product and **(b)** the angle between the pairs of vectors.*

31. $\mathbf{u} = \langle 6, 2 \rangle, \mathbf{v} = \langle 3, -2 \rangle$

32. $\mathbf{u} = \langle 2\sqrt{3}, 2 \rangle, \mathbf{v} = \langle 5, 5\sqrt{3} \rangle$

33. Are the vectors $\mathbf{u} = \langle 5, -1 \rangle$ and $\mathbf{v} = \langle -2, -10 \rangle$ orthogonal? Explain.

Solve each problem.

34. *Weight of a Sled and Child*   Paula and Steve are pulling their daughter Jessie on a sled as shown in the figure. Steve pulls with a force of 18 pounds at an angle of 10°. Paula pulls with a force of 12 pounds at an angle of 15°. Find the magnitude of the resultant force.

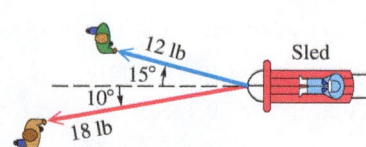

35. *Barge Movement*   One rope pulls a barge directly east with a force of 1000 newtons. Another rope pulls the barge to the northeast with a force of 2000 newtons. Find the resultant force acting on the barge, and find the angle between the resultant and the first rope.

36. *Direction and Airspeed*   A plane has an airspeed of 520 mph. The pilot wishes to fly on a course of 310°. A wind of 37 mph is blowing from a bearing of 212°. On what bearing should the pilot fly, and what will be her ground speed?

37. *Bearing and Speed*   A long-distance swimmer starts out swimming a steady 3.2 mph due north. A 5.1-mph current is flowing on a bearing of 12°. What is the swimmer's final bearing and speed?

38. *(Modeling) Wind and Vectors*   A wind can be described by $\mathbf{v} = 6\mathbf{i} + 8\mathbf{j}$, where vector $\mathbf{j}$ points north and represents a south wind of 1 mph.
  (a) What is the speed of the wind?
  (b) Find $3\mathbf{v}$. Interpret the result.
  (c) Interpret the wind if it switches to $\mathbf{u} = -8\mathbf{i} + 8\mathbf{j}$.

Graph each complex number as a vector in the complex plane.

39. $5i$

40. $-4 + 2i$

Find the sum of each pair of complex numbers.

41. $7 + 3i$ and $-2 + i$

42. $2 - 4i$ and $5 + i$

Perform each conversion in Exercises 43–46.

Rectangular Form	Trigonometric Form
43. $-2 + 2i$	_____
44. _____	$3(\cos 90° + i \sin 90°)$
45. _____	$2 \text{ cis } 225°$
46. $-4 + 4i\sqrt{3}$	_____

Perform each indicated operation. Give answers in rectangular form.

47. $[5(\cos 90° + i \sin 90°)][6(\cos 180° + i \sin 180°)]$

48. $[3 \text{ cis } 135°][2 \text{ cis } 105°]$

49. $\dfrac{2(\cos 60° + i \sin 60°)}{8(\cos 300° + i \sin 300°)}$

50. $\dfrac{4 \text{ cis } 270°}{2 \text{ cis } 90°}$

51. $(\sqrt{3} + i)^3$

52. $(2 - 2i)^5$

53. $(\cos 100° + i \sin 100°)^6$

54. $(\text{cis } 20°)^3$

Find the indicated roots and graph as vectors in the complex plane. Leave answers in polar form.

55. The cube roots of $-27i$

56. The fourth roots of $16i$

57. The fifth roots of 32

58. Solve the equation $x^4 + i = 0$. Leave solutions in polar form.

Convert to rectangular coordinates. Give exact values.

59. $(12, 225°)$

60. $\left(-8, -\dfrac{\pi}{3}\right)$

Convert to polar coordinates, with $0° \leq \theta < 360°$. Give exact values.

61. $(-6, 6)$

62. $(0, -5)$

Use a graphing calculator to graph each polar equation for $0° \leq \theta \leq 360°$. Use a square window.

63. $r = 4 \cos \theta$ (circle)

64. $r = 1 - 2 \sin \theta$ (limaçon with a loop)

65. $r = 2 \sin 4\theta$ (eight-leaved rose)

66. Sketch by hand a graph of
$$r = 1 + 2 \sin \theta \quad \text{(limaçon with a loop)}.$$

Find an equivalent equation in rectangular coordinates.

67. $r = \dfrac{3}{1 + \cos \theta}$

68. $r = \dfrac{4}{2 \sin \theta - \cos \theta}$

69. $r = \sin \theta + \cos \theta$

70. $r = 2$

Find an equivalent equation in polar coordinates.

71. $x = -3$

72. $y = x$

73. $y = x^2$

74. $x = y^2$

Find a rectangular equation for each plane curve with the given parametric equations.

75. $x = \cos 2t$, $y = \sin t$, for t in $(-\pi, \pi)$

76. $x = 5 \tan t$, $y = 3 \sec t$, for t in $\left(-\dfrac{\pi}{2}, \dfrac{\pi}{2}\right)$

77. Graph the curve defined by the following parametric equations: $x = t + \cos t$, $y = \sin t$, for t in $[0, 2\pi]$.

78. *(Modeling) Flight of a Baseball*   A baseball is hit when it is 3.2 feet above the ground. It leaves the bat with a velocity of 118 feet per second at an angle of 27° with respect to the ground. Find the horizontal distance the baseball travels in the air.

10 Test

Exercises 1–3 apply to triangle ABC.

1. Find C if $A = 25.2°$, $a = 6.92$ yd, and $b = 4.82$ yd.

2. Find c if $C = 118°$, $a = 75.0$ km, and $b = 131$ km.

3. Find B if $a = 17.3$ ft, $b = 22.6$ ft, $c = 29.8$ ft.

4. Find the area of triangle ABC if $a = 14$, $b = 30$, and $c = 40$.

5. Find the area of triangle XYZ shown here.

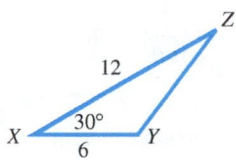

6. Given $a = 10$ and $B = 150°$ in triangle ABC, determine the values of b for which A has
(a) exactly one value
(b) two possible values
(c) no value.

Solve each triangle ABC.

7. $A = 60°$, $b = 30$ m, $c = 45$ m

8. $b = 1075$ in., $c = 785$ in., $C = 38°30'$

9. Find the magnitude and the direction angle for the vector shown in the figure.

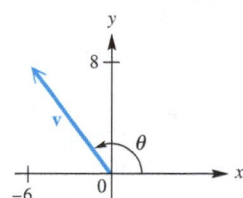

10. For the vectors $\mathbf{u} = \langle -1, 3 \rangle$ and $\mathbf{v} = \langle 2, -6 \rangle$, find each of the following.
(a) $\mathbf{u} + \mathbf{v}$ (b) $-3\mathbf{v}$
(c) $\mathbf{u} \cdot \mathbf{v}$ (d) $|\mathbf{u}|$

11. Find the measure of the angle θ between $\mathbf{u} = \langle 4, 3 \rangle$ and $\mathbf{v} = \langle 1, 5 \rangle$.

Solve each problem.

12. *Height of a Balloon*   The angles of elevation of a balloon from two points A and B on level ground are $30°$ and $45°$, respectively. As shown in the figure in the next column, points A, B, and C are in the same vertical plane and points A and B are 4.2 mi apart. Approximate the height of the balloon above the ground to the nearest tenth of a mile.

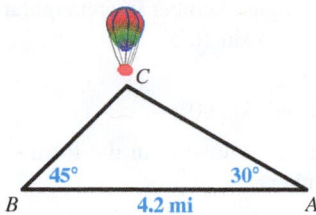

13. *Horizontal and Vertical Components*   Find the horizontal and vertical components of the vector with magnitude 569 and direction angle $127.5°$ from the horizontal. Give your answer in the form $\langle a, b \rangle$.

14. *Radio Direction Finders*   Radio direction finders are placed at points A and B, which are 5.25 mi apart on an east-west line, with A west of B. From A, the bearing of a certain illegal pirate radio transmitter is $35°$, and from B the bearing is $300°$. Find the distance between the transmitter and A to the nearest hundredth of a mile.

15. *Height of a Tree*   A tree leans at an angle of $8.0°$ from the vertical, as shown in the figure. From a point 8.0 m from the bottom of the tree, the angle of elevation to the top of the tree is $66°$. Find the height of the leaning tree.

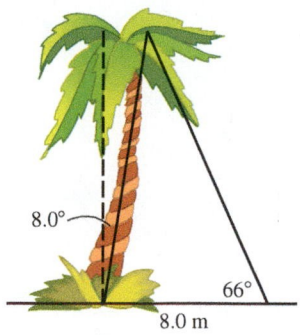

16. *Bearing and Airspeed*   Find the bearing and airspeed required for a plane to fly 630 mi due north in 3.0 hr if the wind is blowing from a direction of $318°$ at 15 mph. Approximate the bearing to the nearest degree and the airspeed to the nearest 10 mph.

17. *Incline Angle*   A force of 16.0 lb is required to hold a 50.0-lb wheelbarrow on an incline. What angle does the incline make with the horizontal?

18. For the following complex numbers, find $w + z$ in rectangular form and give a geometric representation of the sum.

$$w = 2 - 4i \quad \text{and} \quad z = 5 + i$$

19. Express each of the following in rectangular form.
   (a) i^{15} **(b)** $(1 + i)^2$

20. Write each complex number in trigonometric (polar) form, where $0° \leq \theta < 360°$.
   (a) $3i$
   (b) $1 + 2i$
   (c) $-1 - i\sqrt{3}$

21. Write each complex number in rectangular form.
   (a) $3(\cos 30° + i \sin 30°)$
   (b) $4 \text{ cis } 40°$
   (c) $3(\cos 90° + i \sin 90°)$

22. Find each of the following in the form specified for the complex numbers

$$w = 8(\cos 40° + i \sin 40°)$$
$$\text{and} \quad z = 2(\cos 10° + i \sin 10°).$$

   (a) wz (trigonometric form)
   (b) $\dfrac{w}{z}$ (rectangular form)
   (c) z^3 (rectangular form)

23. Find the four complex fourth roots of $-16i$. Express them in trigonometric form.

24. Convert the given rectangular coordinates to polar coordinates. Give two pairs of polar coordinates for each point.
   (a) $(0, 5)$
   (b) $(-2, -2)$

25. Convert the given polar coordinates to rectangular coordinates.
   (a) $(3, 315°)$ **(b)** $(-4, 90°)$

Identify and graph each polar equation for θ in $[0°, 360°)$.

26. $r = 1 - \cos \theta$ **27.** $r = 3 \cos 3\theta$

28. Convert each polar equation to a rectangular equation, and sketch its graph.
   (a) $r = \dfrac{4}{2 \sin \theta - \cos \theta}$ **(b)** $r = 6$

Graph each pair of parametric equations.

29. $x = 2t - 1, y = t^2,$ for t in $[-2, 3]$

30. $x = 2 \cos 2t, y = 2 \sin 2t,$ for t in $[0, 2\pi]$

Topics in this chapter, such as *sequences* and *probability*, allow us to model characteristics of our earth, including populations of humans and wildlife.

11 Further Topics in Algebra

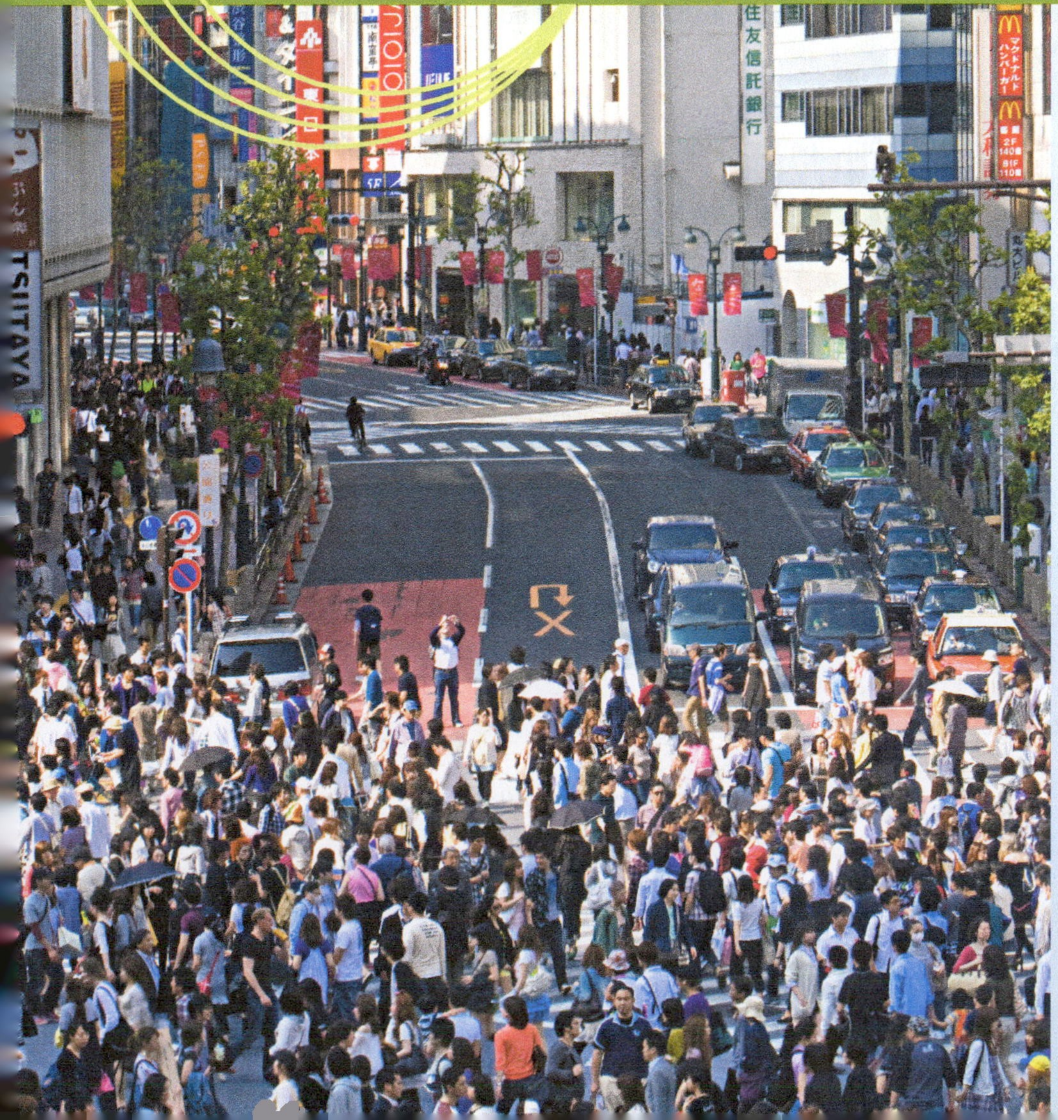

11.1 Sequences and Series

Sequences • Series and Summation Notation • Summation Properties

Sequences

A *sequence* is a function that computes an ordered list. For example, the average family in the United States spends $150 on food each week. The function $f(n) = 150n$ generates the following terms of a sequence for $n = 1, 2, 3, 4, 5, 6, 7, \ldots$.

$$150, 300, 450, 600, 750, 900, 1050, \ldots \qquad \text{Terms of a sequence}$$

Continuous Function

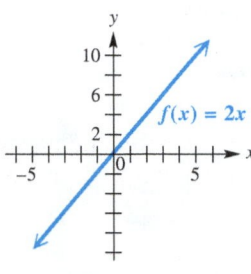

FIGURE 1

This function represents the amount of money spent on food by the average family after n weeks.

As another example, if $100 is deposited into a savings account paying 2% interest compounded annually, then the function $g(n) = 100(1.02)^n$ calculates the account balance after n years. The terms of the sequence are $g(1), g(2), g(3), g(4), g(5), g(6), g(7), \ldots$, and can be approximated as

$$102, 104.04, 106.12, 108.24, 110.41, 112.62, 114.87, \ldots.$$

> **Sequence**
>
> A **sequence** is a function that has a set of natural numbers (positive integers) as its domain.

Terms of a Sequence

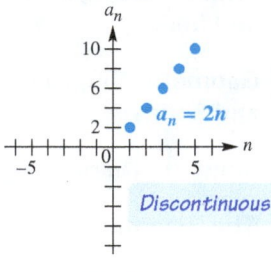

FIGURE 2

Instead of using function notation to indicate a sequence f, it is customary to use a_n, where $a_n = f(n)$. *The letter n is used instead of x as a reminder that n represents a natural number.* The elements in the range of a sequence, called the **terms** of the sequence, are $a_1, a_2, a_3, \ldots$. The elements of both the domain and the range of a sequence are *ordered*. The first term is found by letting $n = 1$, the second term by letting $n = 2$, and so on. The **general term**, or **nth term**, of the sequence is a_n.

FIGURE 1 shows the graph of $f(x) = 2x$, and **FIGURE 2** shows that of $a_n = 2n$. Notice that $f(x)$ is a continuous function, while a_n consists of discrete points. To graph a_n for $n = 1, 2, 3, 4, 5$, we plot points of the form $(n, 2n)$.

EXAMPLE 1 Finding Terms of Sequences

Write the first five terms of each sequence.

(a) $a_n = \dfrac{n + 1}{n + 2}$ (b) $a_n = (-1)^n \cdot n$ (c) $a_n = \dfrac{(-1)^n}{2^n}$

Solution

(a) Replace n in $a_n = \dfrac{n + 1}{n + 2}$ with 1, 2, 3, 4, and 5.

$$n = 1: \quad a_1 = \frac{1 + 1}{1 + 2} = \frac{2}{3} \qquad n = 2: \quad a_2 = \frac{2 + 1}{2 + 2} = \frac{3}{4}$$

$$n = 3: \quad a_3 = \frac{3 + 1}{3 + 2} = \frac{4}{5} \qquad n = 4: \quad a_4 = \frac{4 + 1}{4 + 2} = \frac{5}{6}$$

$$n = 5: \quad a_5 = \frac{5 + 1}{5 + 2} = \frac{6}{7}$$

(b) Replace n in $a_n = (-1)^n \cdot n$ with 1, 2, 3, 4, and 5.

$$n = 1: \quad a_1 = (-1)^1 \cdot 1 = -1 \qquad n = 2: \quad a_2 = (-1)^2 \cdot 2 = 2$$
$$n = 3: \quad a_3 = (-1)^3 \cdot 3 = -3 \qquad n = 4: \quad a_4 = (-1)^4 \cdot 4 = 4$$
$$n = 5: \quad a_5 = (-1)^5 \cdot 5 = -5$$

(c) For $a_n = \frac{(-1)^n}{2^n}$, we have $a_1 = -\frac{1}{2}$, $a_2 = \frac{1}{4}$, $a_3 = -\frac{1}{8}$, $a_4 = \frac{1}{16}$, and $a_5 = -\frac{1}{32}$. ●

Convergent Sequence

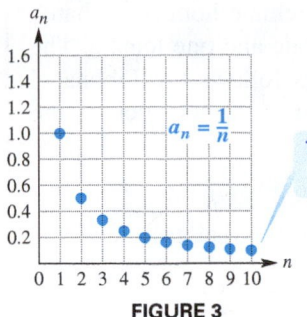

FIGURE 3

A sequence is a **finite sequence** if the domain is the set $\{1, 2, 3, 4, \ldots, n\}$, where n is a natural number. An **infinite sequence** has the set of *all* natural numbers as its domain. For example, the sequence of even natural numbers,

$$2, 4, 6, 8, 10, 12, 14, \ldots, \qquad \text{is infinite,}$$

but the sequence of days in June,

$$1, 2, 3, 4, \ldots, 29, 30, \qquad \text{is finite.}$$

If the terms of an infinite sequence get closer and closer to some real number, the sequence is said to be **convergent** and to **converge** to that real number. For example, the sequence $a_n = \frac{1}{n}$ approaches 0 as n becomes large. Thus, a_n is a convergent sequence that converges to 0. A graph of this sequence for $n = 1, 2, 3, \ldots, 10$ is shown in **FIGURE 3**. The terms of a_n approach the horizontal axis.

A sequence that does not converge to some number is **divergent**. The terms of the sequence $a_n = n^2$ are

$$1, 4, 9, 16, 25, 36, 49, 64, 81, \ldots. \qquad \text{Divergent sequence}$$

This sequence is divergent because as n becomes large, the values for a_n do not approach a fixed number. Rather, they increase without bound.

Some sequences are defined by a **recursive definition,** a definition in which each term is defined as an expression involving the previous term or terms. By contrast, the sequences in **Example 1** were defined *explicitly,* with a formula for a that does not depend on a previous term.

TECHNOLOGY NOTE

Some graphing calculators have a designated sequence mode to investigate and graph sequences defined in terms of n, where n is a natural number.

Using sequence mode on the TI-84 Plus to list the first 10 terms of the sequence with general term $a_n = n + \frac{1}{n}$ produces the result shown in the top figure. Additional terms can be seen by scrolling to the right. The bottom figure shows a calculator graph of $a_n = n + \frac{1}{n}$. Notice that for $n = 5$, the term is $5 + \frac{1}{5} = 5.2$.

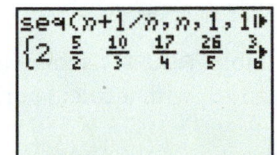

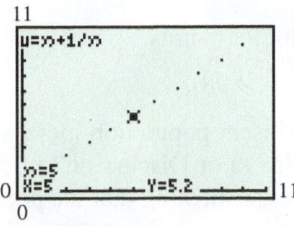

The fifth term is 5.2.

> **EXAMPLE 2** **Using a Recursion Formula**

Find the first four terms of each sequence.

(a) $a_1 = 4$; for $n > 1$, $a_n = 2 \cdot a_{n-1} + 1$

(b) $a_1 = 2$; for $n > 1$, $a_n = a_{n-1} + n - 1$

Solution

(a) This is a recursive definition. We know that $a_1 = 4$ and $a_n = 2 \cdot a_{n-1} + 1$.

$$a_2 = 2 \cdot a_1 + 1 = 2 \cdot 4 + 1 = 9 \qquad \text{Use 9, the value of } a_2 \text{, to}$$
$$a_3 = 2 \cdot a_2 + 1 = 2 \cdot 9 + 1 = 19 \qquad \text{find the value of } a_3.$$
$$a_4 = 2 \cdot a_3 + 1 = 2 \cdot 19 + 1 = 39$$

(b) $a_1 = 2$

$$a_2 = a_1 + 2 - 1 = 2 + 1 = 3$$
$$a_3 = a_2 + 3 - 1 = 3 + 2 = 5$$
$$a_4 = a_3 + 4 - 1 = 5 + 3 = 8$$

●

Leonardo of Pisa (Fibonacci)
(1170–1250)

FOR DISCUSSION

One of the most famous sequences in mathematics is the **Fibonacci sequence,**

$$1, 1, 2, 3, 5, 8, 13, 21, 34, 55, \ldots,$$

named for the Italian mathematician Leonardo of Pisa, who was also known as Fibonacci. The Fibonacci sequence is found in numerous places in nature. For example, male honeybees hatch from eggs that have not been fertilized, so a male bee has only one parent, a female. On the other hand, female honeybees hatch from fertilized eggs, so a female has two parents, one male and one female. The number of ancestors in consecutive generations of bees follows the Fibonacci sequence. Successive terms in the sequence also appear in plants, such as the daisy head, pineapple, and pine cone.

1. Try to discover the pattern in the Fibonacci sequence.

2. Using the given description, write a recursive definition that calculates the number of ancestors of a male bee in each generation.

EXAMPLE 3 **Modeling Insect Population Growth**

Frequently, the population of a particular insect grows rapidly at first and then levels off because of competition for limited resources. In one study, the behavior of the winter moth was modeled with a sequence similar to the following, where a_n represents the population density in thousands per acre during year n. (*Source:* Varley, G. and G. Gradwell, "Population models for the winter moth," Symposium of the Royal Entomological Society of London 4.)

$$a_1 = 1$$
$$a_n = 2.85a_{n-1} - 0.19a_{n-1}^2, \quad \text{for } n \geq 2$$

(a) Give a table of values for $n = 1, 2, 3, \ldots, 10$.

(b) Graph the sequence. Describe what is happening to the population density of the winter moth.

Solution

(a) Evaluate $a_1, a_2, a_3, \ldots, a_{10}$ recursively. Since $a_1 = 1$,

$$a_2 = 2.85a_1 - 0.19a_1^2 = 2.85(1) - 0.19(1)^2 = 2.66,$$

and $\qquad a_3 = 2.85a_2 - 0.19a_2^2 = 2.85(2.66) - 0.19(2.66)^2 \approx 6.24.$

Approximate values for other terms are given in the table. **FIGURE 4** shows the computation of the sequence, denoted by $u(n)$ rather than a_n, with a calculator.

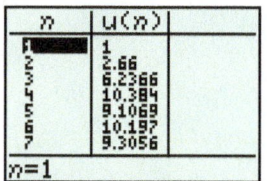

FIGURE 4

n	1	2	3	4	5	6	7	8	9	10
a_n	1	2.66	6.24	10.4	9.11	10.2	9.31	10.1	9.43	9.98

(b) The graph of a sequence is a set of discrete points. Plot the points

$$(1, 1), \quad (2, 2.66), \quad (3, 6.24), \quad \ldots, \quad (10, 9.98),$$

as shown in **FIGURE 5(a)** on the next page. At first, the insect population increases rapidly and then oscillates about the line $y = 9.7$. (See "For Discussion" in the margin.) The oscillations become smaller as n increases, indicating that the population density may stabilize near 9.7 thousand per acre. In **FIGURE 5(b)**, the first 20 terms have been plotted with a calculator.

FOR DISCUSSION

In **Example 3,** the insect population stabilizes near the value $k = 9.7$ thousand. This value of k can be found by solving the quadratic equation $k = 2.85k - 0.19k^2$. Explain why.

First 10 Terms   **First 20 Terms**

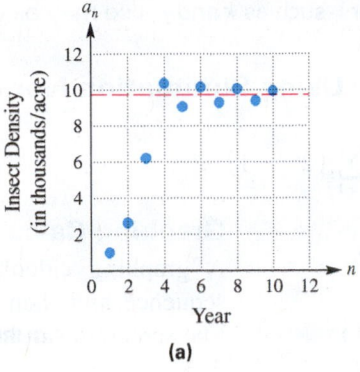

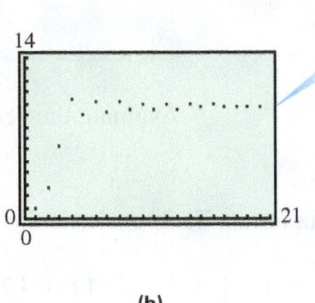

Sequence converges to about 9.7.

(a)   (b)

FIGURE 5

Series and Summation Notation

Suppose a person has a starting salary of $50,000 and receives a $2000 raise each year. Then

$$50{,}000,\ 52{,}000,\ 54{,}000,\ 56{,}000,\ 58{,}000 \qquad \text{Terms of a sequence}$$

are terms of the sequence that describe this person's salaries over a 5-year period. The total earned is given by the *finite series*

$$50{,}000 + 52{,}000 + 54{,}000 + 56{,}000 + 58{,}000, \qquad \text{Terms of a series}$$

whose sum is $270,000. Any sequence can be used to define a series. For example, the infinite sequence

$$1, \frac{1}{3}, \frac{1}{9}, \frac{1}{27}, \frac{1}{81}, \frac{1}{243}, \ldots$$

defines the terms of the *infinite series*

$$1 + \frac{1}{3} + \frac{1}{9} + \frac{1}{27} + \frac{1}{81} + \frac{1}{243} + \cdots.$$

If a sequence has terms $a_1, a_2, a_3, \ldots$, then S_n is defined as the sum of the first n terms. That is,

$$S_n = a_1 + a_2 + a_3 + \ldots + a_n.$$

The sum of the terms of a sequence, called a **series,** is written using **summation notation.** The symbol Σ, the Greek capital letter **sigma,** is used to indicate a sum.

> **Series**
>
> A **finite series** is an expression of the form
>
> $$S_n = a_1 + a_2 + a_3 + \cdots + a_n = \sum_{i=1}^{n} a_i,$$
>
> and an **infinite series** is an expression of the form
>
> $$S_\infty = a_1 + a_2 + a_3 + \cdots + a_n + \cdots = \sum_{i=1}^{\infty} a_i.$$

The letter i as used here is called the **index of summation.** The value for which i begins is listed below Σ, while the upper symbol represents its final value. (For an infinite series, ∞ is not actually a "value".)

→ Looking Ahead to Calculus

Sigma (summation) notation, Σ, is introduced in a first calculus course in conjunction with the **definite integral,** symbolized with an elongated S ($\int$) to which limits of integration are applied.

CAUTION *Do not confuse this use of i with the use of i to represent the imaginary unit.* Other letters, such as *k* and *j*, also may be used for the index of summation.

EXAMPLE 4 **Using Summation Notation**

Evaluate the series $\displaystyle\sum_{k=1}^{6} (2^k + 1)$.

Analytic Solution

Write each of the six terms. Then evaluate the sum.

$$\sum_{k=1}^{6} (2^k + 1) = (2^1 + 1) + (2^2 + 1) + (2^3 + 1)$$
$$+ (2^4 + 1) + (2^5 + 1) + (2^6 + 1)$$
$$= (2 + 1) + (4 + 1) + (8 + 1)$$
$$+ (16 + 1) + (32 + 1) + (64 + 1)$$
$$= 3 + 5 + 9 + 17 + 33 + 65$$
$$= 132$$

Graphing Calculator Solution

A graphing calculator can list the terms of the sequence and then compute the sum of the terms. The screen in **FIGURE 6** confirms the analytic result.

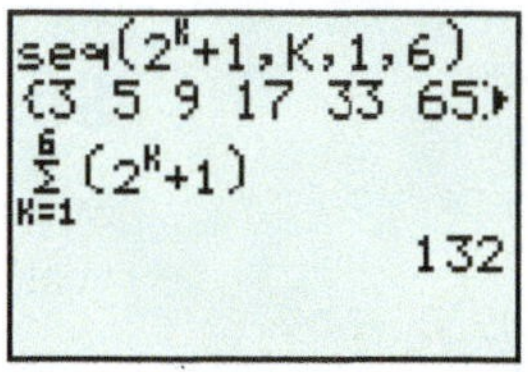

```
seq(2^K+1,K,1,6)
{3 5 9 17 33 65⟩
  6
  Σ (2^K+1)
 K=1
                 132
```

FIGURE 6

EXAMPLE 5 **Using Summation Notation with Subscripts**

Write the terms for each series. Evaluate each sum if possible.

→ **Looking Ahead to Calculus**
Sums like the one in **Example 5(c)** are frequently used in calculus.

(a) $\displaystyle\sum_{j=3}^{6} a_j$ **(b)** $\displaystyle\sum_{i=1}^{3} (6x_i - 2)$ if $x_1 = 2$, $x_2 = 4$, and $x_3 = 6$

(c) $\displaystyle\sum_{i=1}^{4} f(x_i)\,\Delta x$ if $f(x) = x^2$, $x_1 = 0$, $x_2 = 2$, $x_3 = 4$, $x_4 = 6$, and $\Delta x = 2$

Solution

(a) $\displaystyle\sum_{j=3}^{6} a_j = a_3 + a_4 + a_5 + a_6$

(b) Let $i = 1, 2$, and 3, respectively, with $x_1 = 2$, $x_2 = 4$, and $x_3 = 6$.

$$\sum_{i=1}^{3} (6x_i - 2) = (6x_1 - 2) + (6x_2 - 2) + (6x_3 - 2)$$
$$= (6 \cdot 2 - 2) + (6 \cdot 4 - 2) + (6 \cdot 6 - 2) \qquad \text{Substitute the given values.}$$
$$= 10 + 22 + 34 \qquad \text{Simplify.}$$
$$= 66 \qquad \text{Add.}$$

Use the order of operations.

(c) Let $i = 1, 2, 3$, and 4, respectively, with $f(x) = x^2$, $x_1 = 0$, $x_2 = 2$, $x_3 = 4$, and $x_4 = 6$.

$$\sum_{i=1}^{4} f(x_i)\,\Delta x = f(x_1)\,\Delta x + f(x_2)\,\Delta x + f(x_3)\,\Delta x + f(x_4)\,\Delta x$$
$$= x_1{}^2\,\Delta x + x_2{}^2\,\Delta x + x_3{}^2\,\Delta x + x_4{}^2\,\Delta x \qquad f(x) = x^2$$
$$= 0^2(2) + 2^2(2) + 4^2(2) + 6^2(2) \qquad \Delta x = 2$$
$$= 0 + 8 + 32 + 72 \qquad \text{Simplify.}$$
$$= 112 \qquad \text{Add.}$$

EXAMPLE 6 **Estimating π with a Series**

The infinite series

$$\frac{\pi^4}{90} = \frac{1}{1^4} + \frac{1}{2^4} + \frac{1}{3^4} + \frac{1}{4^4} + \frac{1}{5^4} + \cdots + \frac{1}{n^4} + \cdots$$

can be used to estimate π.

(a) Approximate π by finding the sum of the first four terms.

(b) Use a calculator to approximate π by summing the first 50 terms. Compare the result with the decimal value of π.

Solution

(a) Summing the first four terms gives

$$\frac{\pi^4}{90} \approx \frac{1}{1^4} + \frac{1}{2^4} + \frac{1}{3^4} + \frac{1}{4^4} \approx 1.078751929.$$

This approximation can be solved for π by multiplying by 90 and then taking the fourth root.

$$\pi \approx \sqrt[4]{90(1.078751929)} \approx 3.139$$

(b) As shown in **FIGURE 7**, the first 50 terms of the series provide the approximation $\pi \approx 3.141590776$. This computation matches the value of π (shown in the bottom screen) for the first five decimal places.

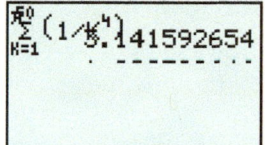

FIGURE 7

Summation Properties

Properties of summation provide useful shortcuts for evaluating series.

Summation Properties

If $a_1, a_2, a_3, \ldots, a_n$ and $b_1, b_2, b_3, \ldots, b_n$ are two sequences and c is a constant, then, for every positive integer n, the following hold.

(a) $\displaystyle\sum_{i=1}^{n} c = nc$

(b) $\displaystyle\sum_{i=1}^{n} ca_i = c\sum_{i=1}^{n} a_i$

(c) $\displaystyle\sum_{i=1}^{n} (a_i + b_i) = \sum_{i=1}^{n} a_i + \sum_{i=1}^{n} b_i$

(d) $\displaystyle\sum_{i=1}^{n} (a_i - b_i) = \sum_{i=1}^{n} a_i - \sum_{i=1}^{n} b_i$

To prove Property (a), expand the series to get

$$c + c + c + c + \cdots + c,$$

where there are n terms of c, so the sum is nc.

Property (c) also can be proved by first expanding the series.

$$\sum_{i=1}^{n} (a_i + b_i) = (a_1 + b_1) + (a_2 + b_2) + \cdots + (a_n + b_n)$$

$$= (a_1 + a_2 + \cdots + a_n) + (b_1 + b_2 + \cdots + b_n)$$

Commutative and associative properties

$$= \sum_{i=1}^{n} a_i + \sum_{i=1}^{n} b_i$$

Proofs of the other two properties are similar.

The following results about summations can be proved by mathematical induction. (See **Section 11.6.**)

Summation Rules

$$\sum_{i=1}^{n} i = 1 + 2 + \cdots + n = \frac{n(n + 1)}{2}$$

$$\sum_{i=1}^{n} i^2 = 1^2 + 2^2 + \cdots + n^2 = \frac{n(n + 1)(2n + 1)}{6}$$

$$\sum_{i=1}^{n} i^3 = 1^3 + 2^3 + \cdots + n^3 = \frac{n^2(n + 1)^2}{4}$$

EXAMPLE 7 **Using the Summation Properties**

Use the summation properties and rules to find each sum.

(a) $\displaystyle\sum_{i=1}^{40} 5$ **(b)** $\displaystyle\sum_{i=1}^{22} 2i$ **(c)** $\displaystyle\sum_{i=1}^{14} (2i^2 - 3)$

Solution

(a) $\displaystyle\sum_{i=1}^{40} 5 = 40(5) = 200$ Property (a) with $n = 40$ and $c = 5$

(b) $\displaystyle\sum_{i=1}^{22} 2i = 2 \sum_{i=1}^{22} i$ Property (b) with $c = 2$ and $a_i = i$

$$= 2 \cdot \frac{22(22 + 1)}{2}$$ Summation rules

$$= 506$$ Simplify.

(c) $\displaystyle\sum_{i=1}^{14} (2i^2 - 3) = \sum_{i=1}^{14} 2i^2 - \sum_{i=1}^{14} 3$ Property (d) with $a_i = 2i^2$ and $b_i = 3$

$$= 2 \sum_{i=1}^{14} i^2 - \sum_{i=1}^{14} 3$$ Property (b) with $c = 2$ and $a_i = i^2$

$$= 2 \cdot \frac{14(14 + 1)(2 \cdot 14 + 1)}{6} - 14(3)$$ Summation rules and Property (a)

$$= 1988$$ Simplify.

> EXAMPLE 8   **Using the Summation Properties**

Evaluate $\displaystyle\sum_{i=1}^{6} (i^2 + 3i + 5)$.

Analytic Solution

$$\sum_{i=1}^{6} (i^2 + 3i + 5) = \sum_{i=1}^{6} i^2 + \sum_{i=1}^{6} 3i + \sum_{i=1}^{6} 5 \qquad \text{Property (c)}$$

$$= \sum_{i=1}^{6} i^2 + 3\sum_{i=1}^{6} i + \sum_{i=1}^{6} 5 \qquad \text{Property (b)}$$

$$= \sum_{i=1}^{6} i^2 + 3\sum_{i=1}^{6} i + 6(5) \qquad \text{Property (a)}$$

$$= \frac{6(6+1)(2 \cdot 6 + 1)}{6} + 3\left[\frac{6(6+1)}{2}\right] + 6(5) \qquad \begin{array}{l}\text{Summation}\\\text{rules}\end{array}$$

$$= 91 + 3(21) + 6(5) \qquad \text{Simplify.}$$

$$= 184$$

Graphing Calculator Solution

FIGURE 8 shows the sum of the first six terms and confirms the analytic result.

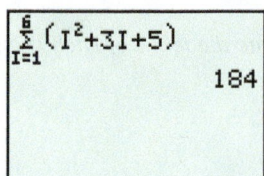

FIGURE 8

11.1   Exercises

Checking Analytic Skills   Write the first five terms of each sequence. **Do not use a calculator.**

1. $a_n = 4n + 10$ **2.** $a_n = 6n - 3$ **3.** $a_n = 2^{n-1}$ **4.** $a_n = -3^n$

5. $a_n = \left(\dfrac{1}{3}\right)^n (n - 1)$ **6.** $a_n = (-2)^n (n)$ **7.** $a_n = (-1)^n (2n)$

8. $a_n = (-1)^{n-1} (n + 1)$ **9.** $a_n = \dfrac{4n - 1}{n^2 + 2}$ **10.** $a_n = \dfrac{n^2 - 1}{n^2 + 1}$

11. Your friend does not understand what is meant by the nth term, or general term, of a sequence. How would you explain this idea?

12. How are sequences related to functions?

Concept Check   Decide whether each sequence is finite or infinite.

13. The sequence of days of the week

14. The sequence of dates in the month of November

15. 1, 2, 3, 4

16. $-1, -2, -3, -4$

17. 1, 2, 3, 4, . . .

18. $-1, -2, -3, -4, \ldots$

19. $a_1 = 3$; for $2 \le n \le 10$, $a_n = 3 \cdot a_{n-1}$

20. $a_1 = 1$; $a_2 = 3$; for $n \ge 3$, $a_n = a_{n-1} + a_{n-2}$

Find the first four terms of each sequence.

21. $a_1 = -2$, $a_n = a_{n-1} + 3$, for $n > 1$

22. $a_1 = -1$, $a_n = a_{n-1} - 4$, for $n > 1$

23. $a_1 = 1$, $a_2 = 1$, $a_n = a_{n-1} + a_{n-2}$, for $n \ge 3$ (the Fibonacci sequence)

24. $a_1 = 2$, $a_n = n \cdot a_{n-1}$, for $n > 1$

25. $a_1 = 5$, $a_n = 3n + 3a_{n-1}$, for $n > 1$

26. $a_1 = 0$, $a_n = 3 + n \cdot a_{n-1}$, for $n > 1$

27. $a_1 = 2$, $a_2 = 3$, $a_n = a_{n-1} \cdot a_{n-2}$ for $n > 2$

28. $a_1 = 2$, $a_2 = 1$, $a_n = 2a_{n-1}^2 + a_{n-2}$, for $n > 2$

Find the sum for each series.

29. $\displaystyle\sum_{i=1}^{5} (2i + 1)$

30. $\displaystyle\sum_{i=1}^{6} (3i - 2)$

31. $\displaystyle\sum_{j=1}^{4} \frac{1}{j}$

32. $\displaystyle\sum_{i=1}^{5} \frac{1}{i + 1}$

33. $\displaystyle\sum_{i=1}^{4} i^i$

34. $\displaystyle\sum_{i=1}^{5} i^{i-1}$

35. $\displaystyle\sum_{k=1}^{6} (-1)^k \cdot k$

36. $\displaystyle\sum_{i=1}^{7} (-1)^{i+1} \cdot i^2$

37. $\displaystyle\sum_{i=2}^{5} (6 - 3i)$

38. $\displaystyle\sum_{i=3}^{7} (5i + 2)$

39. $\displaystyle\sum_{i=-2}^{3} 2(3)^i$

40. $\displaystyle\sum_{i=-1}^{2} 5(2)^i$

41. $\displaystyle\sum_{i=-1}^{5} (i^2 - 2i)$

42. $\displaystyle\sum_{i=3}^{6} (2i^2 + 1)$

43. $\displaystyle\sum_{i=1}^{5} (3^i - 4)$

44. $\displaystyle\sum_{i=1}^{4} [(-2)^i - 3]$

Evaluate the terms of each sum, where $x_1 = -2$, $x_2 = -1$, $x_3 = 0$, $x_4 = 1$, and $x_5 = 2$.

45. $\displaystyle\sum_{i=1}^{5} x_i$

46. $\displaystyle\sum_{i=1}^{5} -x_i$

47. $\displaystyle\sum_{i=1}^{5} (2x_i + 3)$

48. $\displaystyle\sum_{i=1}^{4} (4 - 6x_i)$

49. $\displaystyle\sum_{i=1}^{3} (3x_i - x_i^2)$

50. $\displaystyle\sum_{i=1}^{3} (x_i^2 + 1)$

51. $\displaystyle\sum_{i=2}^{5} \frac{x_i + 1}{x_i + 2}$

52. $\displaystyle\sum_{i=1}^{5} \frac{x_i}{x_i + 3}$

Evaluate the terms of $\displaystyle\sum_{i=1}^{4} f(x_i)\,\Delta x$, with $x_1 = 0$, $x_2 = 2$, $x_3 = 4$, $x_4 = 6$, and $\Delta x = 0.5$, for each function.

53. $f(x) = 4x - 7$

54. $f(x) = 6 + 2x$

55. $f(x) = 2x^2$

56. $f(x) = x^2 - 1$

57. $f(x) = \dfrac{-2}{x + 1}$

58. $f(x) = \dfrac{5}{2x - 1}$

Find the sum for each series.

59. $\displaystyle\sum_{i=1}^{100} 6$

60. $\displaystyle\sum_{i=1}^{20} \frac{1}{2}$

61. $\displaystyle\sum_{i=1}^{15} i^2$

62. $\displaystyle\sum_{i=1}^{50} 2i^3$

63. $\displaystyle\sum_{i=1}^{5} (5i + 3)$

64. $\displaystyle\sum_{i=1}^{5} (8i - 1)$

65. $\displaystyle\sum_{i=1}^{5} (4i^2 - 2i + 6)$

66. $\displaystyle\sum_{i=1}^{6} (2 + i - i^2)$

67. $\displaystyle\sum_{i=1}^{4} (3i^3 + 2i - 4)$

68. $\displaystyle\sum_{i=1}^{6} (i^2 + 2i^3)$

69. $\displaystyle\sum_{i=1}^{60} (i^3 - 2i^2)$

70. $\displaystyle\sum_{i=1}^{43} (15i^2 - 2)$

71. $\displaystyle\sum_{i=1}^{77} (i^2 + 52i + 672)$

72. $\displaystyle\sum_{i=1}^{52} (i^2 + 27i + 180)$

Use summation notation to write each series. Start the index at $i = 1$.

73. $\dfrac{2}{5(1)} + \dfrac{2}{5(2)} + \dfrac{2}{5(3)} + \cdots + \dfrac{2}{5(100)}$

74. $\dfrac{1}{1 + 1} + \dfrac{2}{2 + 1} + \dfrac{3}{3 + 1} + \cdots + \dfrac{25}{25 + 1}$

75. $1 + \dfrac{1}{2} + \dfrac{1}{3} + \dfrac{1}{4} + \cdots + \dfrac{1}{9}$

76. $-\dfrac{1}{3} + \dfrac{1}{9} - \dfrac{1}{27} + \dfrac{1}{81} - \cdots - \dfrac{1}{2187}$

Use a graphing calculator to graph the first 10 terms of each sequence. Make a conjecture as to whether the sequence converges or diverges. If you think it converges, determine the number to which it converges.

77. $a_n = \dfrac{n + 4}{2n}$

78. $a_n = \dfrac{1 + 4n}{2n}$

79. $a_n = 2e^n$

80. $a_n = n(n + 2)$

81. $a_n = \left(1 + \dfrac{1}{n}\right)^n$

82. $a_n = 5 - \dfrac{1}{n}$

Solve each problem.

83. *Estimating π*   Find the sum of the first six terms of the series

$$\frac{\pi^4}{90} = \frac{1}{1^4} + \frac{1}{2^4} + \frac{1}{3^4} + \frac{1}{4^4} + \frac{1}{5^4} + \cdots + \frac{1}{n^4} + \cdots$$

presented in **Example 6.** Use your result to estimate π. Compare your answer with the actual value of π.

84. *(Modeling) Insect Population*   Suppose an insect population density in thousands per acre during year n can be modeled by the following recursively defined sequence.

$$a_1 = 8$$
$$a_n = 2.9a_{n-1} - 0.2a_{n-1}^2, \text{ for } n > 1$$

(a) Find the population for $n = 1, 2, 3$.

(b) Graph the sequence for $n = 1, 2, 3, \ldots, 20$. Use the window $[0, 21]$ by $[0, 14]$. Interpret the graph.

85. *(Modeling) Bacterial Growth*   If certain bacteria are cultured in a medium with sufficient nutrients, they will double in size and then divide every 40 minutes. Let N_1 be the initial number of bacteria cells, N_2 the number after 40 minutes, N_3 the number after 80 minutes, and N_j the number after $40(j - 1)$ minutes. (*Source:* Hoppensteadt, F. and C. Peskin, *Mathematics in Medicine and the Life Sciences,* Springer-Verlag.)

(a) Write N_{j+1} in terms of N_j for $j \geq 1$.

(b) Determine the number of bacteria after two hours if $N_1 = 230$.

(c) Graph the sequence N_j for $j = 1, 2, 3, \ldots, 7$. Use the window $[0, 10]$ by $[0, 15,000]$.

(d) Describe the growth of these bacteria when there are unlimited nutrients.

86. *(Modeling) Verhulst's Model for Bacterial Growth*   Refer to **Exercise 85.** If the bacteria are not cultured in a medium with sufficient nutrients, competition will ensue and the growth will slow. According to Verhulst's model, the number of bacteria N_j at time $40(j - 1)$ minutes can be determined by the sequence

$$N_{j+1} = \left[\frac{2}{1 + (N_j/K)}\right]N_j,$$

where K is a constant and $j \geq 1$. (*Source:* Hoppensteadt, F. and C. Peskin, *Mathematics in Medicine and the Life Sciences,* Springer-Verlag.)

(a) If $N_1 = 230$ and $K = 5000$, make a table of N_j for $j = 1, 2, 3, \ldots, 20$. Round values in the table to the nearest integer.

(b) Graph the sequence N_j for $j = 1, 2, 3, \ldots, 20$. Use the window $[0, 20]$ by $[0, 6000]$.

(c) Describe the growth of these bacteria when there are limited nutrients.

(d) Make a conjecture as to why K is called the *saturation constant.* Test your conjecture by changing the value of K in the given formula.

87. *Estimating Powers of e*   The series

$$e^a \approx 1 + a + \frac{a^2}{2!} + \frac{a^3}{3!} + \cdots + \frac{a^n}{n!},$$

where

$$n! = 1 \cdot 2 \cdot 3 \cdot 4 \cdots \cdots n,$$

can be used to estimate the value of e^a for any real number a. Use the first eight terms of this series to approximate each expression. Compare this estimate with the actual value. Give values to six decimal places.

(a) e (b) e^{-1} (c) $\sqrt{e}$

11.2   Arithmetic Sequences and Series

Arithmetic Sequences • Arithmetic Series

Arithmetic Sequences

A sequence in which each term after the first is obtained by adding a fixed number to the previous term is an **arithmetic sequence** (or **arithmetic progression**). The fixed number that is added is the **common difference,** d. The sequence

$$5, 9, 13, 17, 21, \ldots$$

is an arithmetic sequence because each term after the first is obtained by adding 4, the common difference, to the previous term. That is,

$$9 = 5 + 4, \quad 13 = 9 + 4, \quad 17 = 13 + 4, \quad 21 = 17 + 4, \quad d = 4$$

and so on. The common difference is 4.

If the common difference of an arithmetic sequence is d, then for every positive integer n in its domain,

$$d = a_{n+1} - a_n. \qquad \text{Common difference } d$$

EXAMPLE 1 **Finding the Common Difference**

Find the common difference d of the arithmetic sequence $-9, -7, -5, -3, -1, \ldots$.

Solution We find d by choosing any two consecutive terms and subtracting the first from the second. Choosing -7 and -5 gives

$$d = -5 - (-7) = 2.$$

Choosing -9 and -7 gives $d = -7 - (-9) = 2$, the same result. ●

EXAMPLE 2 **Finding Terms Given a_1 and d**

Write the first five terms of each arithmetic sequence.
(a) The first term is 7 and the common difference is -3.

(b) $a_1 = -12, d = 5$

Solution
(a) $a_1 = 7$　　　　　*Common difference*

$a_2 = a_1 + d$ ➤ $a_2 = 7 + (-3) = 4$　　　$a_1 = 7, d = -3$

$a_3 = 4 + (-3) = 1$

$a_4 = 1 + (-3) = -2$

$a_5 = -2 + (-3) = -5$

(b) $a_1 = -12$　　　　Starting with a_1, add d to each

$a_2 = -12 + 5 = -7$　　　term to get the next term.

$a_3 = -7 + 5 = -2$

$a_4 = -2 + 5 = 3$

$a_5 = 3 + 5 = 8$ ●

If a_1 is the first term of an arithmetic sequence and d is the common difference, then the terms of the sequence are

$$a_1 = a_1$$
$$a_2 = a_1 + d$$
$$a_3 = a_2 + d = a_1 + d + d = a_1 + 2d$$
$$a_4 = a_3 + d = a_1 + 2d + d = a_1 + 3d$$
$$a_5 = a_4 + d = a_1 + 3d + d = a_1 + 4d$$
$$a_6 = a_5 + d = a_1 + 4d + d = a_1 + 5d,$$

and, in general, 　　$a_n = a_1 + (n - 1)d.$

---

nth Term of an Arithmetic Sequence

In an arithmetic sequence with first term a_1 and common difference d, the nth term is

$$a_n = a_1 + (n - 1)d.$$

EXAMPLE 3 **Finding Terms of an Arithmetic Sequence**

Find a_{13} and a_n for the arithmetic sequence $-3, 1, 5, 9, \ldots$.

Solution Here, $a_1 = -3$ and $d = 1 - (-3) = 4$. First find a_{13}.

$$a_n = a_1 + (n - 1)d \qquad \text{Formula for the } n\text{th term}$$
$$a_{13} = a_1 + (13 - 1)d \qquad n = 13$$

Work inside the parentheses first.

$$a_{13} = -3 + (12)4 \qquad a_1 = -3, d = 4$$
$$a_{13} = 45 \qquad \text{Simplify.}$$

Find a_n by substituting values for a_1 and d in the formula for a_n.

Simple formula to find any term

$$a_n = -3 + (n - 1) \cdot 4 \qquad a_1 = -3, d = 4$$
$$a_n = -3 + 4n - 4 \qquad \text{Distributive property}$$
$$a_n = 4n - 7 \qquad \text{Simplify.}$$

EXAMPLE 4 **Finding Terms of an Arithmetic Sequence**

Find a_{18} and a_n for the arithmetic sequence having $a_2 = 9$ and $a_3 = 15$.

Solution The given information allows us to evaluate $d = a_3 - a_2 = 15 - 9 = 6$. We can find $a_1, a_{18},$ and a_n as follows.

$$a_2 = a_1 + d,$$
$$9 = a_1 + 6 \qquad a_2 = 9, d = 6$$
$$a_1 = 3 \qquad \text{Solve for } a_1.$$

$$a_{18} = 3 + (18 - 1)6 \qquad \text{Formula for } a_n; a_1 = 3, n = 18, d = 6$$
$$a_{18} = 105 \qquad \text{Simplify.}$$

$$a_n = 3 + (n - 1)6 \qquad \text{Formula for } a_n$$
$$a_n = 3 + 6n - 6 \qquad \text{Distributive property}$$
$$a_n = 6n - 3 \qquad \text{Simplify.}$$

EXAMPLE 5 **Finding the First Term of an Arithmetic Sequence**

Suppose that an arithmetic sequence has $a_8 = -16$ and $a_{16} = -40$. Find a_1.

Solution $a_{16} = a_8 + 8d$, since we need to add d $16 - 8 = 8$ times to get from the 8th to the 16th term. Solve for d.

$$8d = a_{16} - a_8 = -40 - (-16) = -24,$$

so $d = -3$. To find a_1, use the equation $a_8 = a_1 + 7d$. ◄ $a_n = a_1 + (n - 1)d$

$$-16 = a_1 + 7d \qquad a_8 = -16$$
$$-16 = a_1 + 7(-3) \qquad d = -3$$
$$a_1 = 5 \qquad \text{Solve for } a_1.$$

The graph of a sequence consists of discrete points. To determine the characteristics of the graph of an arithmetic sequence, start by rewriting the formula for the nth term.

$$a_n = a_1 + (n - 1)d \qquad \text{Formula for the } n\text{th term}$$
$$= a_1 + nd - d \qquad \text{Distributive property}$$
$$= dn + (a_1 - d) \qquad \text{Commutative and associative properties}$$
$$= dn + c \qquad \text{Let } c = a_1 - d.$$

Arithmetic Sequence

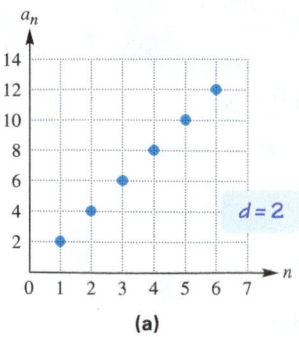

(a)

Not Arithmetic

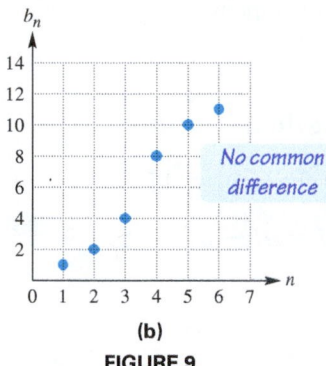

(b)

FIGURE 9

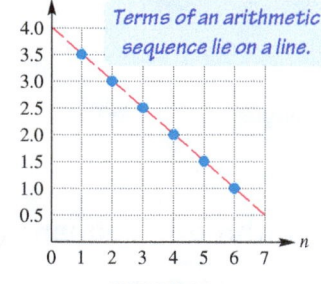

FIGURE 11

The points on the graph of an arithmetic sequence f are defined by $f(n) = dn + c$, where n is a natural number. Thus, the points on the graph of f must lie on the *line*

$$y = dx + c.$$

d is the slope.

For example, the sequence a_n shown in **FIGURE 9(a)** is an arithmetic sequence because the points that make up its graph are collinear (lie on a line). The slope determined by these points is 2, so the common difference d equals 2. The sequence b_n shown in **FIGURE 9(b)** is not an arithmetic sequence, however, because the points are *not* collinear.

EXAMPLE 6   **Finding the nth Term from a Graph**

Find a formula for the nth term of the sequence a_n shown in **FIGURE 10.** What are the domain and range of this sequence?

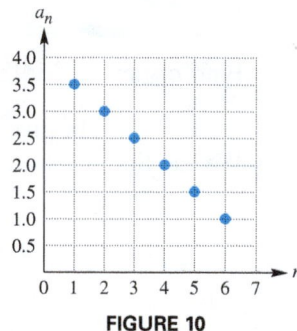

FIGURE 10

Solution   The points in **FIGURE 10** lie on a line, so the sequence is arithmetic. The equation of the dashed line shown in **FIGURE 11** is $y = -0.5x + 4$, so the nth term of this sequence is determined by

$$a_n = -0.5n + 4.$$

The sequence is composed of the points $(1, 3.5)$, $(2, 3)$, $(3, 2.5)$, $(4, 2)$, $(5, 1.5)$, and $(6, 1)$. Therefore, the domain of the sequence is $\{1, 2, 3, 4, 5, 6\}$, and the range is $\{1, 1.5, 2, 2.5, 3, 3.5\}$.   ●

Arithmetic Series

The sum of the terms of an arithmetic sequence is an **arithmetic series.** To illustrate, suppose that a person borrows \$3000 and agrees to pay \$100 per month plus interest of 1% per month on the unpaid balance until the loan is paid off. The first month, \$100 is paid to reduce the loan, plus interest of $(0.01)3000 = 30$ dollars. The second month, another \$100 is paid toward the loan, and $(0.01)2900 = 29$ dollars is paid for interest. Since the loan is reduced by \$100 each month, interest payments decrease by $(0.01)100 = 1$ dollar each month, forming an arithmetic sequence.

Monthly interest ▸ $30, 29, 28, \ldots, 3, 2, 1$ $d = -1$

The total interest paid is given by the sum of the terms of this sequence. Now we develop a formula to find that sum without adding all 30 numbers directly. Since the sequence is arithmetic, we can write the sum of the first n terms as

$$S_n = a_1 + [a_1 + d] + [a_1 + 2d] + \cdots + [a_1 + (n-1)d]. \quad (1)$$

We used the formula for the general term in the last expression. Now we write the same sum in reverse order, beginning with a_n and *subtracting d.*

$$S_n = a_n + [a_n - d] + [a_n - 2d] + \cdots + [a_n - (n-1)d] \quad (2)$$

Now add the respective sides of equations (1) and (2) term by term.

$$S_n + S_n = (a_1 + a_n) + (a_1 + a_n) + \cdots + (a_1 + a_n)$$

$$2S_n = n(a_1 + a_n) \qquad \text{There are } n \text{ terms of } (a_1 + a_n).$$

$$S_n = \frac{n}{2}(a_1 + a_n) \qquad \text{Divide by 2.}$$

$$S_n = \frac{n}{2}[a_1 + a_1 + (n-1)d] \qquad a_n = a_1 + (n-1)d$$

$$S_n = \frac{n}{2}[2a_1 + (n-1)d] \qquad \text{Alternative form}$$

FOR DISCUSSION
Explain why there is no formula for the sum of the terms of an infinite arithmetic sequence.

Sum of the First *n* Terms of an Arithmetic Sequence

If an arithmetic sequence has first term a_1 and common difference d, then the sum of the first n terms is given by the following.

$$S_n = \frac{n}{2}(a_1 + a_n) \quad \text{or} \quad S_n = \frac{n}{2}[2a_1 + (n-1)d]$$

The first formula is used when the first and last terms are known; otherwise, the second formula is used. For example, in the sequence of interest payments given earlier,

$$S_n = \frac{n}{2}(a_1 + a_n) \qquad \text{First formula for } S_n$$

gives

$$S_{30} = \frac{30}{2}(30 + 1) = 15(31) = 465, \qquad n = 30, a_1 = 30, a_n = 1$$

so a total of $465 interest will be paid over the 30 months.

EXAMPLE 7   **Using the Sum Formulas**

(a) Evaluate S_{12} for the arithmetic sequence $-9, -5, -1, 3, 7, \ldots$.

(b) Use a formula for S_n to evaluate the sum of the first 60 positive integers.

Solution

(a) We want the sum of the first 12 terms.

$$S_n = \frac{n}{2}[2a_1 + (n-1)d] \qquad \text{Second formula for } S_n$$

Last term is unknown, so use this formula.

$$S_{12} = \frac{12}{2}[2(-9) + 11(4)] = 156 \qquad n = 12, a_1 = -9, d = 4$$

(b) $S_n = \frac{n}{2}(a_1 + a_n) \qquad$ First formula for S_n

First and last terms are known, so use this formula.

$$S_{60} = \frac{60}{2}(1 + 60) = 1830 \qquad n = 60, a_1 = 1, a_{60} = 60$$

EXAMPLE 8 **Using the Sum Formulas**

The sum of the first 17 terms of an arithmetic sequence is 187. If $a_{17} = -13$, find a_1 and d.

Solution

Use the formula for S_n that involves the first and last terms.

$$S_{17} = \frac{17}{2}(a_1 + a_{17}) \qquad \text{Use the first formula for } S_n, \text{ with } n = 17.$$

$$187 = \frac{17}{2}(a_1 - 13) \qquad S_{17} = 187, a_{17} = -13$$

$$22 = a_1 - 13 \qquad \text{Multiply by } \frac{2}{17}.$$

$$a_1 = 35 \qquad \text{Simplify.}$$

Now use the information to find d.

$$a_{17} = a_1 + (17 - 1)d$$

$$-13 = 35 + 16d \qquad a_{17} = -13, a_1 = 35$$

$$-48 = 16d \qquad \text{Subtract 35.}$$

$$d = -3 \qquad \text{Divide by 16 and rewrite.}$$

Any sum of the form

$$\sum_{i=1}^{n}(di + c),$$

where d and c are real numbers, represents the sum of the terms of an arithmetic sequence having first term $a_1 = d(1) + c = d + c$ and common difference d. These sums can be evaluated with the formulas in this section.

EXAMPLE 9 **Using Summation Notation**

Evaluate each sum.

(a) $\displaystyle\sum_{i=1}^{10}(4i + 8)$ **(b)** $\displaystyle\sum_{k=3}^{9}(4 - 3k)$

Analytic Solution

(a) This sum contains the first 10 terms of an arithmetic sequence.

$$a_1 = 4 \cdot 1 + 8 = 12 \qquad \text{\color{red}First term}$$

$$a_{10} = 4 \cdot 10 + 8 = 48 \qquad \text{\color{green}Last term}$$

Thus, $\displaystyle\sum_{i=1}^{10}(4i + 8) = S_{10} = \frac{10}{2}(12 + 48) = 300.$

(b) The first few terms of this sum are

$$[4 - 3(3)] + [4 - 3(4)] + [4 - 3(5)] + \cdots = -5 + (-8) + (-11) + \cdots.$$

Thus $a_1 = -5$ and $d = -3$. If the sequence started with $k = 1$, there would be nine terms. Since it starts at 3, two of those terms are missing, so there are seven terms and $n = 7$. We use the second formula instead of the first.

$$\sum_{k=3}^{9}(4 - 3k) = \frac{7}{2}[2(-5) + 6(-3)] = -98$$

Graphing Calculator Solution

The screen in **FIGURE 12** shows the sums for the series in parts (a) and (b). These results agree with the analytic solutions. Remember that a series is the sum of the terms of a sequence.

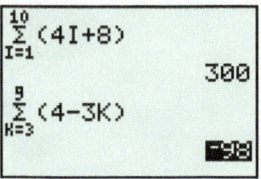

FIGURE 12

11.2 Exercises

Checking Analytic Skills   *Find the common difference d for each arithmetic sequence.* **Do not use a calculator.**

1. $2, 5, 8, 11, \ldots$

2. $4, 10, 16, 22, \ldots$

3. $3, -2, -7, -12, \ldots$

4. $-8, -12, -16, -20, \ldots$

5. $x + 3y, 2x + 5y, 3x + 7y, \ldots$

6. $t^2 + q, -4t^2 + 2q, -9t^2 + 3q, \ldots$

Checking Analytic Skills   *Write the first five terms of each arithmetic sequence.* **Do not use a calculator.**

7. The first term is 8, and the common difference is 6.

8. The first term is -2, and the common difference is 12.

9. $a_1 = 5, d = -2$

10. $a_1 = 4, d = 3$

11. $a_3 = 10, d = -2$

12. $a_1 = 3 - \sqrt{2}, a_2 = 3$

Find a_8 and a_n for each arithmetic sequence.

13. $a_1 = 5, d = 2$

14. $a_1 = -3, d = -4$

15. $a_3 = 2, d = 1$

16. $a_4 = 5, d = -2$

17. $a_1 = 8, a_2 = 6$

18. $a_1 = 6, a_2 = 3$

19. $a_{10} = 6, a_{12} = 15$

20. $a_{15} = 8, a_{17} = 2$

21. $a_1 = x, a_2 = x + 3$

22. $a_2 = y + 1, d = -3$

23. $a_3 = \pi + 2\sqrt{e}, a_4 = \pi + 3\sqrt{e}$

24. $a_4 = e + 2\sqrt{\pi}, a_5 = e + 3\sqrt{\pi}$

Find a_1 for each arithmetic sequence.

25. $a_5 = 27, a_{15} = 87$

26. $a_{12} = 60, a_{20} = 84$

27. $a_5 = -3, a_{18} = -29$

28. $a_6 = -8, a_7 = -18$

29. $S_3 = 75, a_3 = 22$

30. $S_{20} = -1300, a_{20} = -122$

31. $S_{16} = -160, a_{16} = -25$

32. $S_{28} = 2926, a_{28} = 199$

Find the sum of the first 10 terms of each arithmetic sequence.

33. $a_1 = 8, d = 3$

34. $a_1 = -9, d = 4$

35. $a_3 = 5, a_4 = 8$

36. $a_2 = 9, a_4 = 13$

37. $5, 9, 13, \ldots$

38. $8, 6, 4, \ldots$

39. $a_1 = 10, a_{10} = 5.5$

40. $a_1 = -8, a_{10} = -1.25$

Find a_1 and d for each arithmetic sequence.

41. $S_{20} = 1090, a_{20} = 102$

42. $S_{31} = 5580, a_{31} = 360$

43. $S_{12} = -108, a_{12} = -19$

44. $S_{25} = 650, a_{25} = 62$

Find a formula for the nth term of the arithmetic sequence shown in each graph. Then state the domain and range of the sequence.

45.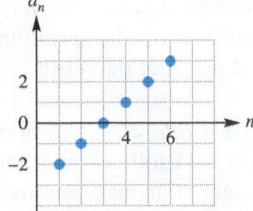

46.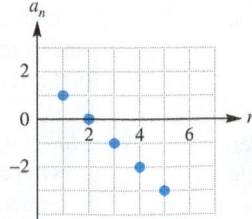

47.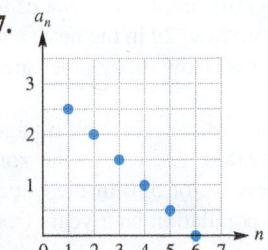

48.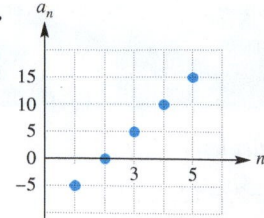

49.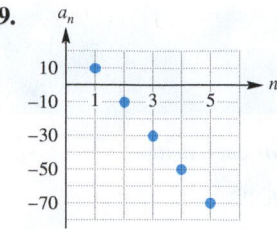

50.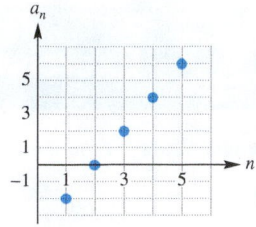

Use a formula to find the sum of each arithmetic series.

51. $3 + 5 + 7 + 9 + 11 + 13 + 15 + 17$

52. $7.5 + 6 + 4.5 + 3 + 1.5 + 0 + (-1.5)$

53. $1 + 2 + 3 + 4 + \cdots + 50$

54. $1 + 3 + 5 + 7 + \cdots + 97$

55. $-7 + (-4) + (-1) + 2 + 5 + \cdots + 98 + 101$

56. $89 + 84 + 79 + 74 + \cdots + 9 + 4$

57. The first 40 terms of the series $a_n = 5n$

58. The first 50 terms of the series $a_n = 1 - 3n$

Evaluate each sum.

59. $\displaystyle\sum_{i=1}^{3} (i + 4)$

60. $\displaystyle\sum_{i=1}^{5} (i - 8)$

61. $\displaystyle\sum_{j=1}^{10} (2j + 3)$

62. $\displaystyle\sum_{j=1}^{15} (5j - 9)$

63. $\displaystyle\sum_{i=1}^{12} (-5 - 8i)$

64. $\displaystyle\sum_{k=1}^{19} (-3 - 4k)$

65. $\displaystyle\sum_{i=1}^{1000} i$

66. $\displaystyle\sum_{k=1}^{2000} k$

Use the sequence feature of a graphing calculator to evaluate the sum of the first 10 terms of the arithmetic sequence. In Exercises 69 and 70, round to the nearest thousandth.

67. $a_n = 4.2n + 9.73$

68. $a_n = 8.42n + 36.18$

69. $a_n = \sqrt{8}n + \sqrt{3}$

70. $a_n = -\sqrt[3]{4}n + \sqrt{7}$

Solve each problem.

71. *Integer Sum* Find the sum of all the integers from 51 to 71.

72. *Integer Sum* Find the sum of all the integers from -8 to 30.

73. *Clock Chimes* If a clock strikes the proper number of chimes each hour on the hour, how many times will it chime in a month of 30 days?

74. *Telephone Pole Stack* A stack of telephone poles has 30 in the bottom row, 29 in the next, and so on, with one pole in the top row. How many poles are in the stack?

75. *Population Growth* Five years ago, the population of a city was 49,000. Each year, the zoning commission permits an increase of 580 in the population. What will the maximum population be 5 years from now?

76. *Supports on a Slide* A slide with a uniform slope is to be built on a level piece of land. There are to be 20 equally spaced supports, with the longest support 15 meters long and the shortest 2 meters long. Find the total length of all the supports.

77. *Rungs of a Ladder* How much material would be needed for the rungs of a ladder of 31 rungs if the rungs taper uniformly from 18 inches to 28 inches?

78. *(Modeling) Spending on Food* The average family in the United States spends $150 on food per week. Write a general term a_n for a sequence that gives the spending on food after n weeks. Find a_4 and interpret the result.

79. *(Modeling) Growth Pattern for Children* The normal growth pattern for children aged 3–11 follows an arithmetic sequence. An increase in height of about 6 centimeters

per year is expected. Thus, 6 would be the common difference of the sequence. A child who measures 96 centimeters at age 3 would have his expected height in subsequent years represented by the sequence 102, 108, 114, 120,

126, 132, 138, 144. Each term differs from the adjacent terms by the common difference, 6.

(a) If a child measures 98.2 centimeters at age 3 and 109.8 centimeters at age 5, what would be the common difference of the arithmetic sequence describing his yearly height?

(b) What would we expect his height to be at age 8?

80. *Concept Check* Find all arithmetic sequences $a_1, a_2, a_3, \ldots$ such that $a_1{}^2, a_2{}^2, a_3{}^2, \ldots$ is also an arithmetic sequence.

81. Suppose that $a_1, a_2, a_3, \ldots$ and $b_1, b_2, b_3, \ldots$ are arithmetic sequences. Let $d_n = a_n + c \cdot b_n$, for any real number c and every positive integer n. Show that $d_1, d_2, d_3, \ldots$ is an arithmetic sequence.

11.3 Geometric Sequences and Series

Geometric Sequences • Geometric Series • Infinite Geometric Series • Annuities

Geometric Sequences

Suppose you agreed to work for 1¢ the first day, 2¢ the second day, 4¢ the third day, 8¢ the fourth day, and so on, with your wages doubling each day. How much will you earn on day 20? How much will you have earned altogether in 20 days? These questions will be answered in this section.

A **geometric sequence** (or **geometric progression**) is a sequence in which each term after the first is obtained by multiplying the preceding term by a constant nonzero real number, called the **common ratio,** r. The sequence of wages

$$1, 2, 4, 8, 16, \ldots \quad r = 2$$

is an example of a geometric sequence in which the first term is 1 and the common ratio is 2. Notice that if we divide any term (except the first) by the preceding term, we obtain the common ratio $r = 2$.

$$\frac{a_2}{a_1} = \frac{2}{1} = 2 \qquad \frac{a_3}{a_2} = \frac{4}{2} = 2 \qquad \frac{a_4}{a_3} = \frac{8}{4} = 2 \qquad \frac{a_5}{a_4} = \frac{16}{8} = 2$$

If the common ratio of a geometric sequence is r, then by the definition of a geometric sequence,

$$r = \frac{a_{n+1}}{a_n} \qquad \text{Common ratio } r$$

for every positive integer n in its domain. *Therefore, the common ratio can be found by choosing any term after the first and dividing it by the preceding term.*

In the geometric sequence 2, 8, 32, 128, $\ldots$, we have $r = 4$.

$$8 = 2 \cdot 4$$
$$32 = 8 \cdot 4 = (2 \cdot 4) \cdot 4 = 2 \cdot 4^2$$
$$128 = 32 \cdot 4 = (2 \cdot 4^2) \cdot 4 = 2 \cdot 4^3$$

To generalize this pattern, assume that a geometric sequence has first term a_1 and common ratio r. The second term can be written as $a_2 = a_1 r$, the third can be written as $a_3 = a_2 r = (a_1 r)r = a_1 r^2$, and so on. Following this pattern, the nth term is $a_n = a_1 r^{n-1}$.

> ### nth Term of a Geometric Sequence
>
> In a geometric sequence with first term a_1 and common ratio r, neither of which is zero, the nth term is
> $$a_n = a_1 r^{n-1}.$$

EXAMPLE 1 **Finding the nth Term of a Geometric Sequence of Wages**

The formula for the nth term of a geometric sequence can be used to answer the first question posed at the beginning of this section. How much will be earned on day 20 if daily wages in cents follow the sequence 1, 2, 4, 8, 16, . . . ?

Solution To answer the question, let $a_1 = 1$ and $r = 2$, and find a_{20}.

$$a_{20} = a_1 r^{19} = 1(2)^{19} = 524{,}288 \text{ cents}, \quad \text{or} \quad \$5242.88$$

EXAMPLE 2 **Using the Formula for the nth Term**

Find a_5 and a_n for the geometric sequence $4, -12, 36, -108, \ldots$.

Solution The first term, a_1, is 4. Find r by choosing any term after the first and dividing it by the preceding term. For example, $r = \dfrac{36}{-12} = -3$. Since $a_4 = -108$,

$$a_5 = -3(-108) = 324.$$

The fifth term and nth term could be found by using the formula $a_n = a_1 r^{n-1}$ and replacing r with -3 and a_1 with 4.

$$a_5 = 4(-3)^{5-1} = 4(-3)^4 = 324$$
$$a_n = 4(-3)^{n-1}$$

EXAMPLE 3 **Using the Formula for the nth Term**

Find r and a_1 for the geometric sequence with third term 20 and sixth term 160.

Solution Use the formula for the nth term of a geometric sequence.

$$\text{For } n = 3, \quad a_3 = a_1 r^2 = 20. \qquad \leftarrow a_n = a_1 r^{n-1}$$
$$\text{For } n = 6, \quad a_6 = a_1 r^5 = 160.$$

Because $a_1 r^2 = 20$, $a_1 = \dfrac{20}{r^2}$. Substitute this value for a_1 in the second equation to find r.

$$a_1 r^5 = 160$$
$$\left(\frac{20}{r^2}\right) r^5 = 160 \qquad \text{Substitute.}$$
$$20 r^3 = 160 \qquad \frac{r^5}{r^2} = r^{5-2} = r^3$$
$$r^3 = 8 \qquad \text{Divide by 20.}$$
$$r = 2 \qquad \text{Take cube roots.}$$

Now determine a_1. Since $a_1 r^2 = 20$ and $r = 2$, we can solve for a_1.

$$a_1(2)^2 = 20 \qquad \text{Substitute.}$$
$$4a_1 = 20 \qquad \text{Apply the exponent.}$$
$$a_1 = 5 \qquad \text{Divide by 4.}$$

EXAMPLE 4 **Modeling Accuracy of Professional Golfers**

At a distance of 3 feet, professional golfers make about 95% of their putts. For each additional foot of distance, this percentage is 0.9 times the previous percentage.

(a) Write a formula that gives the percentage of putts pros make at distance n, where $n = 1$ corresponds to a 3-foot putt.

(b) Estimate the percentage of putts made at 20 feet ($n = 18$).

Solution

(a) The percentage of putts made can be written as a geometric sequence with a_1 as the percentage of putts made at 3 feet, a_2 as the percentage made from 4 feet, and so on.

$$a_n = a_1 r^{n-1} = 95(0.9)^{n-1} \qquad a_1 = 95, \ r = 0.9$$

(b) Substituting $n = 18$ in the equation gives the following.

$$a_{18} = a_1 r^{18-1} = 95(0.9)^{17} \approx 15.8 \qquad n = 18$$

At a distance of 20 feet, about 15.8% of putts are made.

Geometric Series

A **geometric series** is the sum of the terms of a geometric sequence. In applications, it may be necessary to find the sum of the terms of such a sequence. For example, a person might want to know the total earnings for 20 days for the wages discussed in **Example 1**. This sum would equal $a_1 + a_2 + a_3 + a_4 + \cdots + a_{20}$, or

$$1 + 1(2) + 1(2)^2 + \cdots + 1(2)^{19}.$$

To find a formula for the sum S_n of the first n terms of a geometric sequence, first write the sum as follows.

> *Instead of summing all 20 terms individually, we find a general formula.*

$$S_n = a_1 + a_2 + a_3 + \cdots + a_n$$

or
$$S_n = a_1 + a_1 r + a_1 r^2 + \cdots + a_1 r^{n-1} \qquad (1)$$

If $r = 1$, $S_n = na_1$, which is a correct formula for this case. If $r \neq 1$, multiply each side of equation (1) by r to obtain

$$rS_n = a_1 r + a_1 r^2 + a_1 r^3 + \cdots + a_1 r^n. \qquad (2)$$

Now subtract equation (2) from equation (1).

$$S_n = a_1 + a_1 r + a_1 r^2 + \cdots + a_1 r^{n-1} \qquad (1)$$
$$rS_n = \phantom{a_1 + {}} a_1 r + a_1 r^2 + \cdots + a_1 r^{n-1} + a_1 r^n \qquad (2)$$

$$S_n - rS_n = a_1 \phantom{+ a_1 r + a_1 r^2 + \cdots + a_1 r^{n-1}} - a_1 r^n \qquad \text{Subtract.}$$
$$S_n(1 - r) = a_1(1 - r^n) \qquad \text{Factor.}$$

> *Pay attention to this step involving factoring.*

$$S_n = \frac{a_1(1 - r^n)}{1 - r}, \quad \text{where } r \neq 1 \qquad \text{Divide by } 1 - r.$$

> **Sum of the First *n* Terms of a Geometric Sequence**
>
> If a geometric sequence has first term a_1 and common ratio r, then the sum of the first *n* terms is given by
>
> $$S_n = \frac{a_1(1 - r^n)}{1 - r}, \quad \text{where } r \neq 1.$$

EXAMPLE 5 **Applying the Sum of the First *n* Terms**

At the beginning of this section, we posed the following question: How much will you have earned altogether after 20 days? Now answer this question.

Analytic Solution

We must find the total amount earned in 20 days with daily wages of $1, 2, 4, 8, \ldots$ cents. Use $a_1 = 1$ and $r = 2$.

$$S_{20} = \frac{1(1 - 2^{20})}{1 - 2} \qquad S_n = \frac{a_1(1 - r^n)}{1 - r}$$

$$= 1,048,575 \text{ cents} \quad \text{or} \quad \$10,485.75$$

Graphing Calculator Solution

See **FIGURE 13**, which confirms the analytic solution.

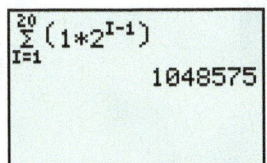

```
 20
 Σ (1*2^(I-1))
I=1
                  1048575
```

FIGURE 13

EXAMPLE 6 **Finding the Sum of the First *n* Terms**

Find $\displaystyle\sum_{i=1}^{6} 2 \cdot 3^i$.

Solution This series is the sum of the first six terms of a geometric sequence having $a_1 = 2 \cdot 3^1 = 6$ and $r = 3$. Use the formula for S_n.

$$S_6 = \frac{6(1 - 3^6)}{1 - 3} = \frac{6(1 - 729)}{-2} = \frac{6(-728)}{-2} = 2184$$

Infinite Geometric Series

We can extend our discussion of sums of sequences to include infinite geometric sequences such as

$$2, \ 1, \ \frac{1}{2}, \ \frac{1}{4}, \ \frac{1}{8}, \ \frac{1}{16}, \ \ldots,$$

with first term 2 and common ratio $\frac{1}{2}$. Using the formula for S_n gives the *sequence of sums*

$$S_1 = 2, \quad S_2 = 3, \quad S_3 = 3.5, \quad S_4 = 3.75, \quad S_5 = 3.875, \quad S_6 = 3.9375.$$

With a calculator in function mode, we define Y_1 as S_n.

$$Y_1 = \frac{2\left(1 - \left(\frac{1}{2}\right)^X\right)}{1 - \frac{1}{2}}$$

FIGURE 14

Using the table in **FIGURE 14,** we can observe the first few terms of Y_1, as X takes on the values 1, 2, 3, ... These terms seem to be getting closer and closer to the number 4. For no value of n is $S_n = 4$. However, if n is large enough, then S_n is as close to 4 as desired. We say that the sequence *converges* to 4, expressed as

$$\lim_{n \to \infty} S_n = 4.$$

(Read: "The limit of S_n as n increases without bound is 4.") Since $\lim_{n \to \infty} S_n = 4$, the number 4 is called the *sum of the terms* of the infinite geometric sequence

$$2, 1, \frac{1}{2}, \frac{1}{4}, \ldots, \quad \text{and} \quad 2 + 1 + \frac{1}{2} + \frac{1}{4} + \cdots = 4.$$

→ Looking Ahead to Calculus

In calculus, different types of infinite series are studied. One important question to answer for each series is whether $\lim_{n \to \infty} S_n$ converges to a real number. In the discussion of

$$\lim_{n \to \infty} S_n = 4,$$

we used the phrases "large enough" and "as close as desired." This description is made more precise in a standard calculus course.

WHAT WENT WRONG?

The preceding discussion justified that the sum of the terms $2, 1, \frac{1}{2}, \frac{1}{4}, \frac{1}{8}, \frac{1}{16}, \cdots$ gets closer and closer to 4 by adding more and more terms of the sequence. However, this sum will always differ from 4 by some small amount. The following figure is an extension of the table in **FIGURE 14.** According to the calculator, the sum *is* 4 when $X \geq 17$.

What Went Wrong?   Discuss the limitations of technology, basing your comments on this example.

EXAMPLE 7   **Summing the Terms of an Infinite Geometric Sequence**

Find $1 + \dfrac{1}{3} + \dfrac{1}{9} + \dfrac{1}{27} + \cdots$.

Solution   Use the formula for the sum of the first n terms of a geometric sequence with $a_1 = 1$ and $r = \frac{1}{3}$ to obtain

$$S_n = \frac{a_1(1 - r^n)}{1 - r}, r \neq 1 \qquad S_1 = 1, \quad S_2 = \frac{4}{3}, \quad S_3 = \frac{13}{9}, \quad S_4 = \frac{40}{27},$$

and, in general,

$$S_n = \frac{1\left[1 - \left(\frac{1}{3}\right)^n\right]}{1 - \frac{1}{3}}.$$

The table in the margin shows the approximate value of $\left(\frac{1}{3}\right)^n$ for larger and larger values of n and that

$$\lim_{n \to \infty} \left(\frac{1}{3}\right)^n = 0.$$

n	$\left(\frac{1}{3}\right)^n$
1	$\frac{1}{3}$
10	0.0000169
100	1.94×10^{-48}
200	3.76×10^{-96}

As n increases without bound . . . $\left(\frac{1}{3}\right)^n$ approaches 0.

(continued)

Answers to What Went Wrong?

The calculator has room to display numbers to only four decimal places in the table. When the value of Y_1 reaches 3.99995 or greater, the calculator rounds up to 4. Try positioning the cursor on the 4 next to the 17 in the table. At the bottom of the screen a more accurate value of 3.99996948242 will appear for Y_1.

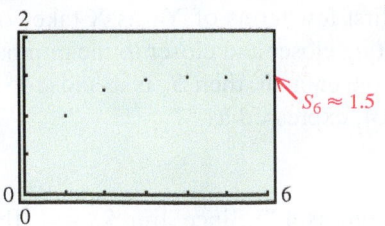

This graph of the first six values of S_n in **Example 7** shows S_n approaching $\frac{3}{2}$. (The y-scale here is $\frac{1}{2}$.)

It is reasonable to conclude that

$$\lim_{n \to \infty} S_n = \lim_{n \to \infty} \frac{1\left[1 - \left(\frac{1}{3}\right)^n\right]}{1 - \frac{1}{3}} = \frac{1(1 - 0)}{1 - \frac{1}{3}} = \frac{1}{\frac{2}{3}} = \frac{3}{2}.$$

Be careful simplifying a complex fraction.

Hence,
$$1 + \frac{1}{3} + \frac{1}{9} + \frac{1}{27} + \cdots = \frac{3}{2}.$$

If a geometric sequence has first term a_1 and common ratio r, then

$$S_n = \frac{a_1(1 - r^n)}{1 - r} \quad (r \neq 1)$$

for every positive integer n. If $|r| < 1$, then $\lim_{n \to \infty} r^n = 0$, and

$$\lim_{n \to \infty} S_n = \frac{a_1(1 - 0)}{1 - r} = \frac{a_1}{1 - r}.$$

This quotient, $\frac{a_1}{1 - r}$, is called the **sum of the terms of an** *infinite* **geometric sequence.**
The limit $\lim_{n \to \infty} S_n$ can be expressed as S_∞ or $\displaystyle\sum_{i=1}^{\infty} a_i$.

➡ Looking Ahead to Calculus
In calculus, you will find sums of the terms of infinite sequences that are not geometric.

Sum of the Terms of an Infinite Geometric Sequence

The sum of the terms of an infinite geometric sequence with first term a_1 and common ratio r, where $|r| < 1$, is

$$S_\infty = \frac{a_1}{1 - r}.$$

If $|r| > 1$, the terms get larger and larger in absolute value, so there is no limit as $n \to \infty$. Hence, the terms of the sequence will not have a sum.

EXAMPLE 8 **Finding Sums of the Terms of Infinite Geometric Sequences**

Find each sum.

(a) $\displaystyle\sum_{i=1}^{\infty} \left(-\frac{3}{4}\right)\left(-\frac{1}{2}\right)^{i-1}$ **(b)** $\displaystyle\sum_{i=1}^{\infty} \left(\frac{3}{5}\right)^i$

Solution

(a) Here, $a_1 = -\frac{3}{4}$ and $r = -\frac{1}{2}$. Since $|r| < 1$, the preceding formula applies.

$$S_\infty = \frac{a_1}{1 - r} = \frac{-\frac{3}{4}}{1 - \left(-\frac{1}{2}\right)} = \frac{-\frac{3}{4}}{\frac{3}{2}} = -\frac{3}{4} \div \frac{3}{2} = -\frac{3}{4} \cdot \frac{2}{3} = -\frac{1}{2}$$

Be careful simplifying a complex fraction.

(b) $\displaystyle\sum_{i=1}^{\infty} \left(\frac{3}{5}\right)^i = \frac{\frac{3}{5}}{1 - \frac{3}{5}} = \frac{\frac{3}{5}}{\frac{2}{5}} = \frac{3}{5} \div \frac{2}{5} = \frac{3}{5} \cdot \frac{5}{2} = \frac{3}{2}$ $a_1 = \frac{3}{5}, r = \frac{3}{5}$

Annuities

A sequence of equal payments made at equal intervals, such as car payments or house payments, is called an **annuity.** If the payments are accumulated in an account (with no withdrawals), the sum of the payments and interest on the payments is called the **future value** of the annuity. The formula that follows is derived from the formula for the sum of the terms of a geometric sequence.

Future Value of an Annuity

The formula for the future value of an annuity is

$$S = R\left[\frac{(1 + i)^n - 1}{i}\right],$$

where S is the future value, R is the payment at the end of each period, i is the interest rate in decimal form per period, and n is the number of periods.

> **EXAMPLE 9**   **Finding the Future Value of an Annuity**

To save money for a trip, Callie Daniels deposited $1000 at the *end* of each year for 4 years in an account paying 3% interest compounded annually. Find the future value of this annuity.

Solution   Use the formula for S, with $R = 1000$, $i = 0.03$, and $n = 4$.

$$S = 1000\left[\frac{(1 + 0.03)^4 - 1}{0.03}\right] \approx 4183.63$$

The future value of the annuity is $4183.63.

11.3 Exercises

Checking Analytic Skills   *Write the terms of the geometric sequence that satisfies the given conditions.* **Do not use a calculator.**

1. $a_1 = \dfrac{5}{3}$, $r = 3$, $n = 4$ **2.** $a_1 = -\dfrac{3}{4}$, $r = \dfrac{2}{3}$, $n = 4$ **3.** $a_4 = 5$, $a_5 = 10$, $n = 5$ **4.** $a_3 = 16$, $a_4 = 8$, $n = 5$

Checking Analytic Skills   *Find a_5 and a_n for each geometric sequence.* **Do not use a calculator.**

5. $a_1 = 5$, $r = -2$

6. $a_1 = 8$, $r = -5$

7. $a_2 = -4$, $r = -3$

8. $a_3 = -2$, $r = 4$

9. $a_4 = 243$, $r = -3$

10. $a_4 = 18$, $r = 2$

11. $-4, -12, -36, -108, \ldots$

12. $-2, 6, -18, 54, \ldots$

13. $\dfrac{4}{5}, 2, 5, \dfrac{25}{2}, \ldots$

14. $\dfrac{1}{2}, \dfrac{2}{3}, \dfrac{8}{9}, \dfrac{32}{27}, \ldots$

15. $10, -5, \dfrac{5}{2}, -\dfrac{5}{4}, \ldots$

16. $3, -\dfrac{9}{4}, \dfrac{27}{16}, -\dfrac{81}{64}, \ldots$

Find a_1 and r for each geometric sequence.

17. $a_3 = 5$, $a_8 = \dfrac{1}{625}$

18. $a_2 = -6$, $a_7 = -192$

19. $a_4 = -\dfrac{1}{4}$, $a_9 = -\dfrac{1}{128}$

20. $a_3 = 50$, $a_7 = 0.005$

Use the formula for S_n to find the sum of the first five terms for each geometric sequence. Round the answers for Exercises 25 and 26 to the nearest hundredth.

21. $2, 8, 32, 128, \ldots$

22. $4, 16, 64, 256, \ldots$

23. $18, -9, \dfrac{9}{2}, -\dfrac{9}{4}, \ldots$

24. $12, -4, \dfrac{4}{3}, -\dfrac{4}{9}, \ldots$

25. $a_1 = 8.423$, $r = 2.859$

26. $a_1 = -3.772$, $r = -1.553$

Use a formula to find each sum.

27. $\displaystyle\sum_{i=1}^{5} 3^i$

28. $\displaystyle\sum_{i=1}^{4} 2^i$

29. $\displaystyle\sum_{j=1}^{6} 48\left(\dfrac{1}{2}\right)^j$

30. $\displaystyle\sum_{j=1}^{5} 243\left(\dfrac{2}{3}\right)^j$

31. $\displaystyle\sum_{k=4}^{10} (-2)^k$

32. $\displaystyle\sum_{k=3}^{9} (-3)^k$

33. $\displaystyle\sum_{i=2}^{8} -2^i$

34. $\displaystyle\sum_{k=5}^{10} -3^k$

35. $\displaystyle\sum_{i=1}^{6} 5(2)^{i-1}$

36. $\displaystyle\sum_{j=2}^{7} \dfrac{1}{3}(4)^{j-1}$

37. $\displaystyle\sum_{k=1}^{4} -2\left(\dfrac{1}{2}\right)^k$

38. $\displaystyle\sum_{j=1}^{3} -3\left(\dfrac{1}{4}\right)^j$

39. *Concept Check* Under what conditions does the sum of the terms of an infinite geometric sequence exist?

40. *Concept Check* The number $0.999\ldots$ can be written as the sum of the terms of an infinite geometric sequence: $0.9 + 0.09 + 0.009 + \cdots$. Here we have $a_1 = 0.9$ and $r = 0.1$. Use the formula for S_∞ to find this sum.

Write the sum of each geometric series as a rational number. (See Exercise 40.)

41. $0.8 + 0.08 + 0.008 + 0.0008 + \cdots$

42. $0.7 + 0.07 + 0.007 + 0.0007 + \cdots$

43. $0.45 + 0.0045 + 0.000045 + \cdots$

44. $0.36 + 0.0036 + 0.000036 + \cdots$

45. $0.378 + 0.000378 + 0.000000378 + \cdots$

46. $0.297 + 0.000297 + 0.000000297 + \cdots$

Find r for each infinite geometric sequence. Identify any whose sum does not converge.

47. $12, 24, 48, 96, \ldots$

48. $2, -10, 50, -250, \ldots$

49. $-48, -24, -12, -6, \ldots$

50. $625, 125, 25, 5, \ldots$

Find each sum that converges.

51. $16 + 2 + \dfrac{1}{4} + \dfrac{1}{32} + \cdots$

52. $18 + 6 + 2 + \dfrac{2}{3} + \cdots$

53. $100 + 10 + 1 + \cdots$

54. $128 + 64 + 32 + \cdots$

55. $\dfrac{4}{3} + \dfrac{2}{3} + \dfrac{1}{3} + \cdots$

56. $\dfrac{1}{4} - \dfrac{1}{6} + \dfrac{1}{9} - \dfrac{2}{27} + \cdots$

57. $\displaystyle\sum_{i=1}^{\infty} 3\left(\dfrac{1}{4}\right)^{i-1}$

58. $\displaystyle\sum_{i=1}^{\infty} 5\left(-\dfrac{1}{4}\right)^{i-1}$

59. $\displaystyle\sum_{k=1}^{\infty} (0.3)^k$

60. $\displaystyle\sum_{k=1}^{\infty} (0.1)^k$

61. $\displaystyle\sum_{k=1}^{\infty} 5^{-k}$

62. $\displaystyle\sum_{k=1}^{\infty} 3^{-k}$

63. $\displaystyle\sum_{i=1}^{\infty} \left(\dfrac{1}{5}\right)\left(-\dfrac{1}{2}\right)^{i-1}$

64. $\displaystyle\sum_{i=1}^{\infty} \left(-\dfrac{1}{3}\right)\left(\dfrac{3}{4}\right)^{i-1}$

Use a graphing calculator to evaluate each sum. Round to the nearest thousandth.

65. $\sum_{i=1}^{10} -(1.4)^i$ **66.** $\sum_{j=1}^{6} -(3.6)^j$ **67.** $\sum_{j=3}^{8} 2(0.4)^j$ **68.** $\sum_{i=4}^{9} 3(0.25)^i$

Annuity Values   *Find the future value of each annuity.*

69. Payments of $1000 at the end of each year for 9 years at 4% interest compounded annually

70. Payments of $800 at the end of each year for 12 years at 3% interest compounded annually

71. Payments of $2430 at the end of each year for 10 years at 2.5% interest compounded annually

72. Payments of $1500 at the end of each year for 6 years at 1.5% interest compounded annually

Solve each problem.

73. *(Modeling) Investment for Retirement* According to T. Rowe Price Associates, a person who has a moderate investment strategy and n years until retirement should have accumulated savings of a_n percent of his or her annual salary. The geometric sequence

$$a_n = 1276(0.916)^n$$

gives the appropriate percent for each year n.
(a) Find a_1 and r.
(b) Find and interpret the terms a_{10} and a_{20}.

74. *(Modeling) Investment for Retirement* Refer to the investment strategy in **Exercise 73.** For someone who has a conservative investment strategy with n years to retirement, the geometric sequence is

$$a_n = 1278(0.935)^n.$$

(*Source:* T. Rowe Price Associates.)
(a) Repeat part (a) of **Exercise 73.**
(b) Repeat part (b) of **Exercise 73.**
(c) Why are the answers in parts (a) and (b) larger than in **Exercise 73?**

75. *(Modeling) Bacterial Growth* The strain of bacteria described in **Exercise 85** in the first section of this chapter will double in size and then divide every 40 minutes. Let a_1 be the initial number of bacteria cells, a_2 the number after 40 minutes, and a_n the number after $40(n-1)$ minutes. (*Source:* Hoppensteadt, F. and C. Peskin, *Mathematics in Medicine and the Life Sciences*, Springer-Verlag.)
(a) Write a formula for the nth term a_n of the geometric sequence $a_1, a_2, a_3, \ldots, a_n, \ldots$.
(b) Determine the first value for n where $a_n > 1{,}000{,}000$ if $a_1 = 100$.
(c) How long does it take for the number of bacteria to exceed 1 million?

76. *Photo Processing* The final step in processing a black-and-white photographic print is to immerse the print in a chemical fixer. The print is then washed in running water. Under certain conditions, 98% of the fixer will be removed with 15 minutes of washing. How much of the original fixer would be left after 1 hour of washing?

77. *Chemical Mixture* A scientist has a vat containing 100 liters of a pure chemical. Twenty liters are drained and replaced with water. After complete mixing, 20 liters of the mixture are again drained and replaced with water. What will be the strength of the mixture after nine such drainings?

78. *Half-Life of a Radioactive Substance* The half-life of a radioactive substance is the time it takes for half the substance to decay. Suppose the half-life of a substance is 3 years and 10^{15} molecules of the substance are present initially. How many molecules will be present after 15 years?

79. *Depreciation in Value* Each year a machine loses 20% of the value it had at the beginning of the year. Find the value of the machine at the end of 6 years if it cost $100,000 new.

80. *Sugar Processing* A sugar factory receives an order for 1000 units of sugar. The production manager thus orders the production of 1000 units of sugar. He forgets, however, that the production of sugar requires some sugar (to prime the machines, for example), so he ends up with only 900 units of sugar. He then orders an additional 100 units, and receives only 90 units. A further order for 10 units produces 9 units. Finally seeing that he is wrong, the manager decides to try mathematics. He views the production process as an infinite geometric progression with $a_1 = 1000$ and $r = 0.1$. Find the number of units of sugar that he should have ordered originally.

81. *(Modeling) Fruit Fly Populations* A population of fruit flies is growing in such a way that each generation is 1.5 times as large as the last generation. If there were 100 flies in the first generation, write a formula for a geometric sequence that gives the population of the nth generation. Find the population of the fourth generation.

82. *(Modeling) Fruit Fly Populations* Refer to the fruit fly population in **Exercise 81.** Write a formula that gives the *total* fruit flies in n generations. Find the total in 4 generations.

83. *Swing of a Pendulum* A pendulum bob swings through an arc 40 centimeters long on its first swing. Each swing thereafter, it swings only 80% as far as on the previous swing. How far will it swing altogether before coming to a complete stop?

84. *Height of a Dropped Ball* Beth Schiffer drops a ball from a height of 10 meters and notices that on each bounce the ball returns to about $\frac{3}{4}$ of its previous height. About how far will the ball travel before it comes to rest? (*Hint:* Consider the sum of two sequences.)

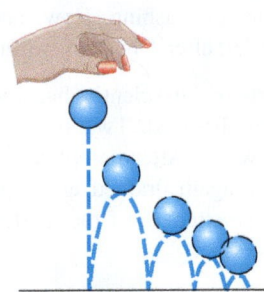

85. *Number of Ancestors* Each person has two parents, four grandparents, eight great-grandparents, and so on. What is the total number of ancestors a person has, going back five generations? ten generations?

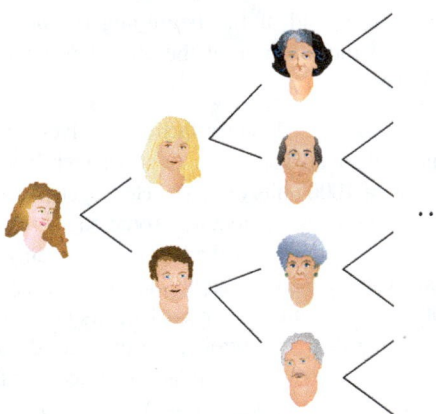

86. *(Modeling) Drug Dosage* Certain medical conditions are treated with a fixed dose of a drug administered at regular intervals. Suppose that a person is given 2 milligrams of a drug each day and that during each 24-hour period the body utilizes 40% of the amount of drug that was present at the beginning of the period.
  (a) Show that the amount of the drug present in the body at the end of n days is

$$\sum_{i=1}^{n} 2(0.6)^i.$$

  (b) What will be the approximate quantity of the drug in the body at the end of each day after the treatment has been administered over a long period?

87. *Salaries* You are offered a 6-week summer job and are asked to select one of the following salary options.

Option 1: \$5000 for the first day, with a \$10,000 raise each day for the remaining 29 days (that is, \$15,000 for day 2, \$25,000 for day 3, and so on)

Option 2: 1¢ for the first day, with the pay doubled each day thereafter (that is, 2¢ for day 2, 4¢ for day 3, and so on)

Which option would you choose?

88. *Number of Ancestors* Suppose a genealogical website allows you to identify all your ancestors that lived during the past 300 years. Assuming that each generation spans about 25 years, guess the number of ancestors that would be found during the 12 generations. Then use the formula for a geometric series to find the actual value.

89. *Side Length of a Triangle* Refer to the figure of triangles below. A sequence of equilateral triangles is constructed. The first triangle has sides 2 meters in length. To get the second triangle, midpoints of the sides of the original triangle are connected. What is the length of the side of the eighth such triangle?

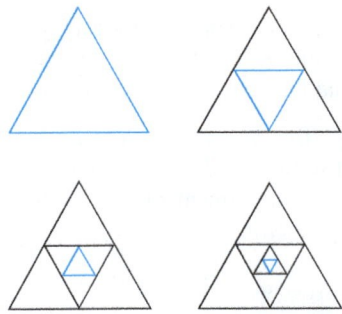

90. *Perimeter and Area of Triangles* Answer each of the following regarding **Exercise 89.**
  (a) If the process could be continued indefinitely, what would be the total perimeter of all the triangles?
  (b) What would be the total area of all the triangles, disregarding the overlapping?

SECTIONS 11.1–11.3 **Reviewing Basic Concepts**

1. Find the first five terms of the sequence $a_n = (-1)^{n-1}(4n)$.

2. Evaluate the series $\sum_{i=1}^{5}(3i + 1)$.

3. Decide whether the sequence $a_n = 1 - \frac{2}{n}$ converges or diverges. If it converges, determine the number to which it converges.

4. Write the first five terms of the arithmetic sequence with $a_1 = 8$ and $d = -2$.

5. Find a_1 for the arithmetic sequence with $a_5 = 5$ and $a_8 = 17$.

6. Find the sum of the first 10 terms of the arithmetic sequence having $a_1 = 2$ and $d = 5$.

7. Find a_3 and a_n for the geometric sequence having $a_1 = -2$ and $r = -3$.

8. Is the series $5 + 3 + \frac{9}{5} + \cdots + \frac{243}{625}$ arithmetic or geometric? Find the sum of the terms of this series.

9. Find the sum of the infinite geometric series $\sum_{k=1}^{\infty}3\left(\frac{2}{3}\right)^k$.

10. *Annuity Value* Find the future value of an annuity if payments of $500 are made at the end of each year for 13 years at 3.5% interest compounded annually.

11.4 Counting Theory

Fundamental Principle of Counting • *n*-Factorial • Permutations • Combinations • Distinguishing between Permutations and Combinations

Fundamental Principle of Counting

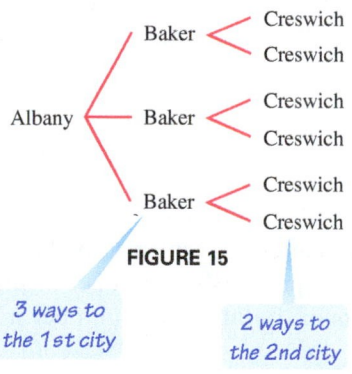

FIGURE 15

3 ways to the 1st city

2 ways to the 2nd city

If there are 3 roads from Albany to Baker and 2 roads from Baker to Creswich, in how many ways can one travel from Albany to Creswich by way of Baker? For each of the 3 roads from Albany to Baker, there are 2 different roads from Baker to Creswich. Hence, there are $3 \cdot 2 = 6$ different ways to make the trip, as shown in the **tree diagram** in **FIGURE 15.**

Each choice of a road is an example of an *event*. Two events are **independent events** if neither influences the outcome of the other. For example, the road chosen from Albany to Baker doesn't change the fact that there are 2 road choices from Baker to Creswich. The opening example illustrates the fundamental principle of counting with independent events.

Fundamental Principle of Counting

If n independent events occur, with

m_1 ways for event 1 to occur,

m_2 ways for event 2 to occur,

.

.

.

and m_n ways for event n to occur,

then there are

$$m_1 \cdot m_2 \cdots \cdots m_n$$

different ways for all n events to occur.

EXAMPLE 1 Using the Fundamental Principle of Counting

A restaurant offers a choice of 3 salads, 5 main dishes, and 2 desserts. Use the fundamental principle of counting to find the number of different 3-course meals that can be selected.

Solution Three events are involved: selecting a salad, selecting a main dish, and selecting a dessert. The first event can occur in 3 ways, the second in 5 ways, and the third in 2 ways. By the fundamental principle of counting, there are

$$3 \cdot 5 \cdot 2 = 30 \text{ possible meals.}$$

EXAMPLE 2 Using the Fundamental Principle of Counting

A teacher has 5 different books that she wishes to arrange in a row. How many different arrangements are possible?

Solution Five events are involved: selecting a book for the first spot, selecting a book for the second spot, and so on. For the first spot, the teacher has 5 choices. After a choice has been made, the teacher has 4 choices for the second spot. Continuing in this manner, we get 3 choices for the third spot, 2 for the fourth spot, and 1 for the fifth spot. Thus, there are

$$5 \cdot 4 \cdot 3 \cdot 2 \cdot 1 = 120 \text{ arrangements.}$$

EXAMPLE 3 Arranging r of n Items ($r < n$)

Suppose the teacher in **Example 2** wishes to place only 3 of the 5 books in a row. How many arrangements of 3 books are possible?

Solution The teacher still has 5 ways to fill the first spot, 4 ways to fill the second spot, and 3 ways to fill the third. Since only 3 books will be used, there are only 3 spots to be filled (3 events) instead of 5, with

$$5 \cdot 4 \cdot 3 = 60 \text{ arrangements.}$$

n-Factorial

Products such as $5 \cdot 4 \cdot 3 \cdot 2 \cdot 1$ and $3 \cdot 2 \cdot 1$ are seen often in problems involving counting. Because of this, we use a symbol $n!$, which represents the product of all positive integers less than or equal to n. Called **factorial notation,** the number $n!$ (read "**n-factorial**") is defined as follows.

TECHNOLOGY NOTE

The TI-84 Plus calculator, with its 10-digit display, will give exact values of $n!$ for $n \le 13$ and approximate values of $n!$ for $14 \le n \le 69$. The two screens confirm the results shown in the discussion.

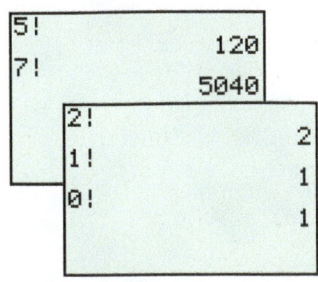

> **n-Factorial**
>
> For any positive integer n,
>
> $$n! = n(n-1)(n-2)\cdots(3)(2)(1).$$
>
> By definition, $0! = 1.$

For example,
$$5! = 5 \cdot 4 \cdot 3 \cdot 2 \cdot 1 = 120,$$
$$7! = 7 \cdot 6 \cdot 5 \cdot 4 \cdot 3 \cdot 2 \cdot 1 = 5040,$$
and
$$2! = 2 \cdot 1 = 2.$$

Permutations

Refer to **Example 3.** Since each ordering of 3 books is a different *arrangement,* the number 60 in that example is called the number of *permutations* of 5 things taken 3 at a time, written $P(5, 3) = 60$.

A **permutation** of n elements taken r at a time is one of the *arrangements* of r elements from a set of n elements. Generalizing from the preceding examples, we find the number of permutations of n elements taken r at a time, denoted by $P(n, r)$.

$$P(n, r) = n(n - 1)(n - 2) \cdots (n - r + 1)$$
$$= \frac{n(n - 1)(n - 2) \cdots (n - r + 1)(n - r)(n - r - 1) \cdots (2)(1)}{(n - r)(n - r - 1) \cdots (2)(1)}$$
$$= \frac{n!}{(n - r)!}$$

FIGURE 16 shows that this formula gives the same answer as in **Example 3.**

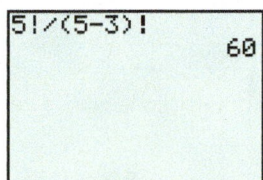

FIGURE 16

Permutations of *n* Elements Taken *r* at a Time

If $P(n, r)$ denotes the number of permutations of n elements taken r at a time, with $r \leq n$, then

$$P(n, r) = \frac{n!}{(n - r)!}.$$

Alternative notations for $P(n, r)$ are $P^{\,n}_{\,r}$ and $_nP_r$.

EXAMPLE 4   **Using the Permutations Formula**

Find the following.

(a) The number of permutations of the letters L, M, and N

(b) The number of permutations of 2 of the 3 letters L, M, and N

Analytic Solution

(a) Because there are 3 letters to choose from and we are arranging all 3, we use the formula for $P(n, r)$ with $n = 3$ and $r = 3$.

$$P(3, 3) = \frac{3!}{(3 - 3)!} = \frac{3!}{0!} = \frac{3!}{1} = 3 \cdot 2 \cdot 1 = 6. \qquad 0! = 1$$

As shown in the tree diagram in **FIGURE 17,** there are 6 permutations.

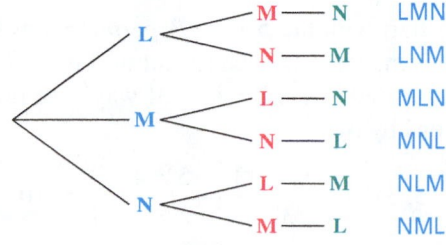

FIGURE 17

(b) $P(3, 2) = \dfrac{3!}{(3 - 2)!} = \dfrac{3!}{1!} = \dfrac{6}{1} = 6$

This result is the same as in part (a), because after the first two choices are made, there is only one letter left.

Graphing Calculator Solution

The TI-84 Plus uses the notation nPr for evaluating permutations. This function is found in the MATH menu. The screen in **FIGURE 18** confirms the analytic results in parts (a) and (b).

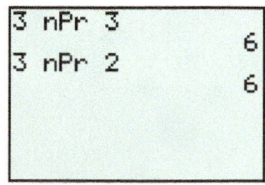

FIGURE 18

This example illustrates the general rule

$$P(n, n) = P(n, n - 1).$$

EXAMPLE 5 Using the Permutations Formula

Suppose 8 people enter an event in a swim meet. In how many ways could the gold, silver, and bronze prizes be awarded?

Solution There are 8 people but only 3 medals are given. From the fundamental principle of counting, there are $8 \cdot 7 \cdot 6 = 336$ choices. The formula for $P(n, r)$ gives the same result.

$$P(8, 3) = \frac{8!}{5!} = \frac{8 \cdot 7 \cdot 6 \cdot 5 \cdot 4 \cdot 3 \cdot 2 \cdot 1}{5 \cdot 4 \cdot 3 \cdot 2 \cdot 1} = 336 \text{ ways}$$

 $8 - 3 = 5$

EXAMPLE 6 Using the Permutations Formula

In how many ways can 5 students be seated in a row of 5 desks?

Solution Use $P(n, n)$ with $n = 5$.

$$P(5, 5) = \frac{5!}{0!} = \frac{5 \cdot 4 \cdot 3 \cdot 2 \cdot 1}{1} = 120 \text{ ways}$$

$5 - 5 = 0$

Combinations

In **Example 3,** we saw that there are 60 ways that a teacher can arrange 3 of 5 different books in a row. That is, there are 60 permutations of 5 things taken 3 at a time. Suppose that the teacher does not wish to arrange the books in a row, but rather wishes to choose, *without regard to order,* any 3 of the 5 books to donate to a book sale. In how many ways can this be done?

At first glance, we might say 60 again, but this is incorrect. The number 60 counts all possible *arrangements* of 3 books chosen from 5. The following 6 arrangements, however, would all lead to the same set of 3 books being given to the book sale.

mystery, biography, textbook	biography, textbook, mystery	*Repeat arrange-*
mystery, textbook, biography	textbook, biography, mystery	*ments should not*
biography, mystery, textbook	textbook, mystery, biography	*be counted.*

The list shows 6 different *arrangements* of 3 books, but only one *set* of 3 books. A subset of items selected *without regard to order* is called a **combination.** The number of combinations of 5 things taken 3 at a time is written $_5C_3$ or $C(5, 3)$.

NOTE The symbol $_nC_r$ or $C(n, r)$ is sometimes read as "*n* choose *r.*"

To evaluate $C(5, 3)$, start with the $5 \cdot 4 \cdot 3$ *permutations* of 5 things taken 3 at a time. Since order doesn't matter, and each subset of 3 items from the set of 5 items can have its elements rearranged in $3 \cdot 2 \cdot 1 = 3!$ ways, we find $C(5, 3)$ by dividing the number of permutations by $3!$.

$$C(5, 3) = \frac{5 \cdot 4 \cdot 3}{3!} = \frac{5 \cdot 4 \cdot 3}{3 \cdot 2 \cdot 1} = 10 \qquad \text{Dividing by 3! removes repeat arrangements.}$$

There are 10 ways that the teacher can choose 3 books for the book sale.

Generalizing this discussion gives the formula for the number of combinations of n elements taken r at a time.

$$C(n, r) = \frac{P(n, r)}{r!}$$

Another version of the formula is found as follows.

$$C(n, r) = \frac{P(n, r)}{r!} = \frac{n!}{(n - r)!} \cdot \frac{1}{r!} = \frac{n!}{(n - r)! \, r!}$$

Combinations of *n* Elements Taken *r* at a Time

If $C(n, r)$ represents the number of combinations of *n* things taken *r* at a time, with $r \le n$, then

$$C(n, r) = \frac{n!}{(n - r)! \, r!}.$$

EXAMPLE 7 **Using the Combinations Formula**

How many different committees of 3 people can be chosen from a group of 8?

Analytic Solution

A committee is an unordered set. Use the combinations formula with $n = 8$ and $r = 3$.

$$C(8, 3) = \frac{8!}{5! \, 3!} \qquad (8 - 3)! = 5!$$

$$= \frac{8 \cdot 7 \cdot 6 \cdot 5 \cdot 4 \cdot 3 \cdot 2 \cdot 1}{5 \cdot 4 \cdot 3 \cdot 2 \cdot 1 \cdot 3 \cdot 2 \cdot 1}$$

$$= 56$$

There are 56 committees.

Graphing Calculator Solution

The notation nCr is used by the TI-84 Plus to find combinations. **FIGURE 19** agrees with the analytic result.

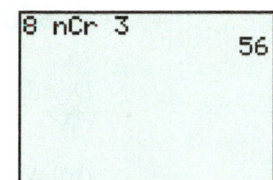

```
8 nCr 3
                    56
```

FIGURE 19

EXAMPLE 8 **Using the Combinations Formula**

Three lawyers are to be selected from a group of 30 to work on a special project.

(a) In how many different ways can the lawyers be selected?

(b) In how many ways can the group of 3 be selected if one particular lawyer from the 30 has been preselected to work on the project?

Solution

(a) Here we wish to know the number of 3-element combinations that can be formed from a set of 30 elements. (We want combinations, not permutations, since order within the group does not matter.)

$$C(30, 3) = \frac{30!}{(30 - 3)! \, 3!} = \frac{30!}{27! \, 3!} = 4060 \text{ ways}$$

(b) Since 1 lawyer has already been selected for the project, the problem is reduced to selecting 2 more from the remaining 29 lawyers.

$$C(29, 2) = \frac{29!}{(29 - 2)! \, 2!} = \frac{29!}{27! \, 2!} = 406 \text{ ways}$$

Distinguishing between Permutations and Combinations

Students often have difficulty determining whether to use permutations or combinations in solving problems. Both concepts determine the number of ways to select r items out of n items, without repetition. The following chart lists some of the differences between these two concepts.

These hands are the same *combination*. The order of the cards in the hands is *not* important.

Permutations	Combinations
Order is important.	Order is not important.
Arrangements of r items from a set of n items	Subsets of r items from a set of n items
$P(n, r) = \dfrac{n!}{(n - r)!}$	$C(n, r) = \dfrac{n!}{(n - r)!\, r!}$
Clue words: arrangement, schedule, order	Clue words: group, committee, sample, selection

EXAMPLE 9   **Distinguishing between Permutations and Combinations**

Decide whether permutations or combinations should be used to solve each problem.

(a) How many 4-digit codes are possible if no digits are repeated?

(b) A sample of 3 light bulbs is randomly selected from a batch of 15. How many different samples are possible?

(c) In a basketball tournament with 8 teams, how many games must be played so that each team plays every other team exactly once?

(d) In how many ways can 4 stockbrokers be assigned to 6 offices so that each broker has a private office?

Solution

(a) Since changing the order of the 4 digits results in a different code, permutations should be used.

(b) The order in which the 3 light bulbs are selected is not important. The sample is unchanged if the items are rearranged, so combinations should be used.

(c) The selection of 2 teams for a game is an *unordered* subset of 2 from the set of 8 teams. Use combinations.

(d) The office assignments are an *ordered* selection of 4 offices from the 6 offices. Exchanging the offices of any 2 brokers within a selection of 4 offices gives a different assignment, so permutations should be used.   ●

To illustrate the differences between permutations and combinations in another way, suppose 2 cans of soup are to be selected from 4 cans on a shelf: noodle (N), bean (B), mushroom (M), and tomato (T). Then, as shown in **FIGURE 20(a)** on the next page, there are 12 ways to select 2 cans from the 4 cans if the order matters (if noodle first and bean second is considered different from bean, then noodle, for example).

On the other hand, if order is unimportant, then there are 6 ways to choose 2 cans of soup from the 4, as illustrated in **FIGURE 20(b).**

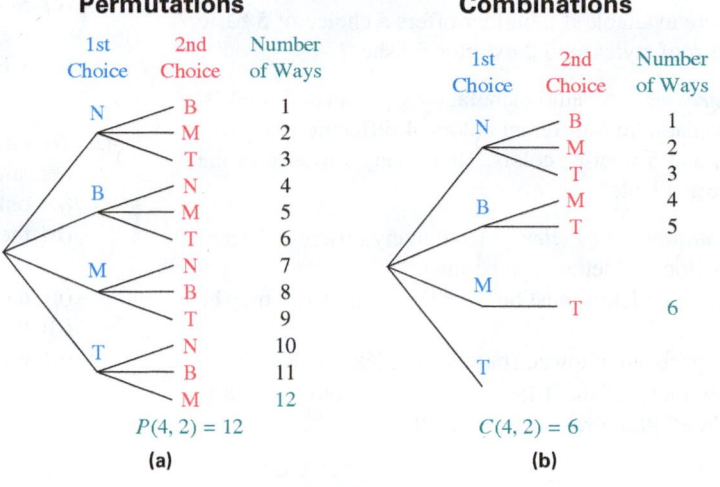

FIGURE 20

CAUTION   Not all counting problems lend themselves to permutations or combinations. Whenever a tree diagram or the multiplication principle can be used directly, do so.

11.4 Exercises

Checking Analytic Skills   *Evaluate each expression.* **Do not use a calculator.**

1. $4!$ **2.** $6!$ **3.** $(4 - 2)!$ **4.** $(5 - 2)!$ **5.** $\dfrac{6!}{5!}$

6. $\dfrac{7!}{6!}$ **7.** $\dfrac{8!}{6!}$ **8.** $\dfrac{9!}{7!}$ **9.** $3! \cdot 4$ **10.** $4! \cdot 5$

 11. Explain how the expression in **Exercise 7** can be worked without actually evaluating $8!$, then $6!$, and then dividing.

12. *Concept Check*   If n is a positive integer greater than 1, is $(n - 1)! \cdot n$ always equal to $n!$?

Checking Analytic Skills   *Evaluate each expression.* **Do not use a calculator.**

13. $P(7, 7)$ **14.** $P(5, 5)$ **15.** $P(9, 2)$ **16.** $P(10, 3)$ **17.** $P(5, 1)$ **18.** $P(6, 0)$

19. $C(4, 2)$ **20.** $C(9, 3)$ **21.** $C(6, 0)$ **22.** $C(8, 1)$ **23.** $C(12, 4)$ **24.** $C(16, 3)$

Use a calculator to evaluate each expression.

25. $_{20}P_5$ **26.** $_{100}P_5$ **27.** $_{15}P_8$ **28.** $_{32}P_4$

29. $_{20}C_5$ **30.** $_{100}C_5$ **31.** $_{15}C_8$ **32.** $_{32}C_4$

33. *Concept Check*   Decide whether the situation described involves a permutation or a combination of objects.
(a) A telephone number
(b) A Social Security number
(c) A hand of cards in poker
(d) A committee of politicians

(e) The "combination" on a combination lock
(f) A lottery choice of six numbers where the order does not matter
(g) An automobile license plate

 34. Explain the difference between a permutation and a combination.

Use the fundamental principle of counting or permutations to solve each problem.

35. *Home Plan Choices* How many different types of homes are available if a builder offers a choice of 5 basic plans, 3 roof styles, and 2 exterior finishes?

36. *Auto Varieties* An auto manufacturer produces 7 models, each available in 6 different colors, 4 different upholstery fabrics, and 5 interior colors. How many varieties of the auto are available?

37. *Radio Station Call Letters* How many different 4-letter radio station call letters can be made
(a) if the first letter must be K or W and no letter may be repeated?
(b) if repeats are allowed (but the first letter is K or W)?
(c) How many of the 4-letter call letters (starting with K or W) with no repeats end in R?

38. *Meal Choices* A menu offers a choice of 3 salads, 8 main dishes, and 5 desserts. How many different 3-course meals (salad, main dish, dessert) are possible?

39. *Names for a Baby* A couple having a baby has narrowed down the choice of a name for the new baby to 3 first names and 5 middle names. How many different first- and middle-name pairings are possible?

40. *Concert Program Arrangement* A concert to raise money for an economics prize is to consist of 5 works: 2 overtures, 2 sonatas, and a piano concerto. In how many ways can a program with these 5 works be arranged?

41. *License Plates* For many years, the state of California used 3 letters followed by 3 digits on its automobile license plates.

(a) How many different license plates are possible with this arrangement?

(b) When the state ran out of new plates, the order was reversed to 3 digits followed by 3 letters. How many additional plates were then possible?

(c) Several years ago, the plates described in part (b) were also used up. The state then issued plates with 1 letter, followed by 3 digits, and then 3 letters. How many plates does this scheme provide?

42. *Telephone Numbers* How many 7-digit telephone numbers are possible if the first digit cannot be 0 and
(a) only odd digits may be used?
(b) the telephone number must be a multiple of 10 (that is, it must end in 0)?
(c) the telephone number must be a multiple of 100?
(d) the first 3 digits are 481?
(e) no repetitions are allowed?

43. *Seating People in a Row* In an experiment on social interaction, 6 people will sit in 6 seats in a row. In how many ways can this be done?

44. *Chemistry Experiment* In how many ways can 7 of 10 chemicals be added to a beaker for an experiment?

45. *Course Schedule Arrangement* A business school offers courses in keyboarding, spreadsheets, transcription, business English, technical writing, and accounting. In how many ways can a student arrange a schedule if 3 courses are taken?

46. *Course Schedule Arrangement* If your college offers 400 courses, 20 of which are in mathematics, and your counselor arranges your schedule of 4 courses by random selection, how many schedules are possible that do not include a math course?

47. *Club Officer Choices* In a club with 15 members, how many ways can a slate of 3 officers consisting of president, vice-president, and secretary/treasurer be chosen?

48. *Batting Orders* A baseball team has 20 players. How many 9-player batting orders are possible?

49. *Basketball Positions* In how many ways can 5 players be assigned to the 5 positions on a basketball team, assuming that any player can play any position? In how many ways can 10 players be assigned to the 5 positions?

50. *Letter Arrangement*   How many ways can all the letters of the word ELTON be arranged?

Solve each problem involving combinations.

51. *Seminar Presenters*   A banker's association has 30 members. If 4 members are selected at random to present a seminar, how many different groups of 4 are possible?

52. *Apple Samples*   How many different samples of 3 apples can be drawn from a crate of 25 apples?

53. *Hamburger Choices*   Howard's Hamburger Heaven sells hamburgers with cheese, relish, lettuce, tomato, mustard, or ketchup. How many different hamburgers can be made that use any 3 of the extras?

54. *Financial Planners*   Three financial planners are to be selected from a group of 12 to participate in a special program. In how many ways can this be done? In how many ways can the group that will not participate be selected?

55. *Marble Samples*   If a bag contains 15 marbles, how many samples of 2 marbles can be drawn from it? how many samples of 4 marbles?

56. *Card Combinations*   Five cards marked respectively with the numbers 1, 2, 3, 4, and 5 are shuffled, and 2 cards are then drawn. How many different 2-card hands are possible?

57. *Marble Samples*   In **Exercise 55,** if the bag contains 3 yellow, 4 white, and 8 blue marbles, how many samples of 2 can be drawn in which both marbles are blue?

58. *Apple Samples*   In **Exercise 52,** if it is known that there are 5 rotten apples in the crate,
    (a) how many samples of 3 could be drawn in which all 3 are rotten?
    (b) how many samples of 3 could be drawn in which there are 1 rotten apple and 2 good apples?

59. *Convention Delegation Choices*   A city council is composed of 5 liberals and 4 conservatives. Three members are to be selected randomly as delegates to a convention.
    (a) How many delegations are possible?
    (b) How many delegations could have all liberals?

(c) How many delegations could have 2 liberals and 1 conservative?
(d) If 1 member of the council serves as mayor, how many delegations are possible that include the mayor?

60. *Delegation Choices*   Seven workers decide to send a delegation of 2 to their supervisor to discuss their grievances.
    (a) How many different delegations are possible?
    (b) If it is decided that a certain employee must be in the delegation, how many different delegations are possible?
    (c) If there are 2 women and 5 men in the group, how many delegations would include at least 1 woman?

Use any or all of the methods described in this section to solve each problem.

61. *Course Schedule*   If Kim Faŀgout has 8 courses to choose from, how many ways can she arrange her schedule if she must pick 4 of them?

62. *Pineapple Samples*   How many samples of 3 pineapples can be drawn from a crate of 12?

63. *Soup Ingredients*   Madeline Moore specializes in making different vegetable soups with carrots, celery, beans, peas, mushrooms, and potatoes. How many different soups can she make with any 4 ingredients?

64. *Assistant/Manager Assignments*   From a pool of 7 assistants, 3 are selected to be assigned to 3 managers, 1 assistant to each manager. In how many ways can this be done?

65. *Musical Chairs Seatings*   In a game of musical chairs, 12 children will sit in 11 chairs. One will be left out. How many seatings are possible?

66. *Plant Samples*   In an experiment on plant hardiness, a researcher gathers 6 wheat plants, 3 barley plants, and 2 rye plants. She wishes to select 4 plants at random.
    (a) In how many ways can this be done?
    (b) In how many ways can it be done if exactly 2 wheat plants must be included?

67. *Committee Choices*   In a club with 8 men and 11 women members, how many 5-member committees can be chosen that have the following?
  (a) All men   **(b)** All women
  (c) 3 men and 2 women   **(d)** No more than 3 women

68. *Committee Choices*   From 10 names on a ballot, 4 will be elected to a political party committee. In how many ways can the committee of 4 be formed if each person will have a different responsibility?

69. *Garage Door Openers*   The code for some garage door openers consists of 12 electrical switches that can be set to either 0 or 1 by the owner. With this type of opener, how many codes are possible? (*Source:* Promax.)

70. *Combination Lock*   A typical combination for a padlock consists of 3 numbers from 0 to 39. Count the number of combinations that are possible with this type of lock if a number may be repeated.

71. *Combination Lock*   A briefcase has 2 locks. The combination to each lock consists of a 3-digit number, where digits may be repeated. How many different ways are there of choosing the six digits required to open the briefcase?

(*Hint:* The word *combination* is a misnomer. Lock combinations are permutations because the arrangement of the numbers is important.)

72. *Lottery*   To win the jackpot in a lottery game, a person must pick 3 numbers from 0 to 9 in the correct order. If a number can be repeated, how many ways are there to play the game?

73. *Keys*   How many distinguishable ways can 4 keys be put on a circular key ring? (*Hint:* Consider that clockwise and counterclockwise arrangements are not different.)

74. *Sitting at a Round Table*   How many ways can 7 people sit at a round table? Assume that a different way means that at least 1 person is sitting next to someone different.

Prove each statement for positive integers n and r, with r ≤ n.
(Hint: *Use the definitions of permutations and combinations.*)

75. $P(n, n - 1) = P(n, n)$ **76.** $P(n, 1) = n$

77. $P(n, 0) = 1$ **78.** $C(n, n) = 1$

79. $C(n, 0) = 1$ **80.** $C(n, 1) = n$

81. $C(n, n - 1) = n$ **82.** $C(n, n - r) = C(n, r)$

11.5   The Binomial Theorem

A Binomial Expansion Pattern • Pascal's Triangle • Binomial Coefficients • The Binomial Theorem • *r*th Term of a Binomial Expansion

A Binomial Expansion Pattern

The formula for writing the terms of $(x + y)^n$, where n is a natural number, is called the *binomial theorem*. Some expansions of $(x + y)^n$, for various nonnegative integer values of n, are as follows.

$$(x + y)^0 = 1$$
$$(x + y)^1 = x + y$$
$$(x + y)^2 = x^2 + 2xy + y^2$$
$$(x + y)^3 = x^3 + 3x^2y + 3xy^2 + y^3$$
$$(x + y)^4 = x^4 + 4x^3y + 6x^2y^2 + 4xy^3 + y^4$$
$$(x + y)^5 = x^5 + 5x^4y + 10x^3y^2 + 10x^2y^3 + 5xy^4 + y^5$$

➡ **Looking Ahead to Calculus**
Students taking calculus study the binomial series, which follows from Isaac Newton's extension to the case where the exponent is no longer a positive integer. His result led to a series for $(1 + x)^k$, where k is a real number and $|x| < 1$.

Notice that after the special case $(x + y)^0 = 1$, each expression begins with x raised to the same power as the binomial itself. That is, the expansion of $(x + y)^1$ has a first term of x^1, $(x + y)^2$ has a first term of x^2, $(x + y)^3$ has a first term of x^3, and so on. Also, the last term in each expansion is y to the same power as the binomial. Thus, the expansion of $(x + y)^n$ should begin with the term x^n and end with the term y^n.

Also, the exponent on x decreases by 1 in each term after the first, while the exponent on y, beginning with y in the second term, increases by 1 in each succeeding term. That is, the *variables* in the terms of the expansion of $(x + y)^n$ have the following pattern.

$$x^n, \ x^{n-1}y, \ x^{n-2}y^2, \ x^{n-3}y^3, \ldots, \ xy^{n-1}, \ y^n$$

This pattern suggests that the sum of the exponents on x and y in each term is n. For example, the third term in the preceding list is $x^{n-2}y^2$, and the sum of the exponents is $n - 2 + 2 = n$.

Pascal's Triangle

Writing just the *coefficients* in the terms of the expansion of $(x + y)^n$ gives the pattern shown as follows.

Blaise Pascal (1623–1662)

Pascal's Triangle

							Row
			1				0
		1		1			1
	1		2		1		2
1		3		3		1	3
1	4		6		4	1	4
1	5	10		10	5	1	5

With the coefficients arranged in this way, each number in the triangle is the sum of the two numbers directly above it (one to the right and one to the left). For example, in row four, 1 is the sum of 1 (the only number above it), 4 is the sum of 1 and 3, 6 is the sum of 3 and 3, and so on. This triangular array of numbers is called **Pascal's triangle,** in honor of the 17th-century mathematician Blaise Pascal.

To find the coefficients for $(x + y)^6$, we need to include row six in Pascal's triangle. Adding adjacent numbers, we find row six.

$$1 \quad 6 \quad 15 \quad 20 \quad 15 \quad 6 \quad 1 \qquad \text{Row 6}$$

Using these coefficients, we obtain the expansion of $(x + y)^6$.

$$(x + y)^6 = 1x^6 + 6x^5y + 15x^4y^2 + 20x^3y^3 + 15x^2y^4 + 6xy^5 + 1y^6$$

Binomial Coefficients

Consider the numerical coefficients of the expression

$$(x + y)^5 = x^5 + 5x^4y + 10x^3y^2 + 10x^2y^3 + 5xy^4 + y^5.$$

The coefficient of the second term, $5x^4y$, is 5, and the exponents on the variables are 4 and 1. Note that

$$5 = \frac{5!}{4! \, 1!}.$$

The coefficient of the third term, $10x^3y^2$, is 10, with exponents 3 and 2, and

$$10 = \frac{5!}{3!\,2!}.$$

The last term (the sixth term, y^5) can be written $y^5 = 1x^0y^5$, with coefficient 1 and exponents 0 and 5. Since $0! = 1$,

$$1 = \frac{5!}{0!\,5!}.$$

Generalizing from these examples, we find that the coefficient for the term of the expansion of $(x + y)^n$ in which the variable part is x^ry^{n-r} (where $r \leq n$) is

$$\frac{n!}{r!\,(n - r)!}.$$

This number, which is equivalent to $C(n, r)$ of **Section 11.4,** is called a **binomial coefficient,** and is often written as $\binom{n}{r}$.

Binomial Coefficient

For nonnegative integers n and r, with $r \leq n$,

$$\binom{n}{r} = \frac{n!}{r!\,(n - r)!}.$$

The binomial coefficients are numbers from Pascal's triangle. For example,

$$\binom{3}{0} \text{ is the first number in row three,}$$

and

$$\binom{7}{4} \text{ is the fifth number in row seven.}$$

EXAMPLE 1 **Evaluating Binomial Coefficients**

Evaluate each binomial coefficient.

(a) $\binom{6}{2}$ (b) $\binom{8}{0}$ (c) $\binom{10}{10}$

Analytic Solution

(a) $\binom{6}{2} = \frac{6!}{2!\,(6 - 2)!} = \frac{6!}{2!\,4!} = \frac{6 \cdot 5 \cdot 4 \cdot 3 \cdot 2 \cdot 1}{2 \cdot 1 \cdot 4 \cdot 3 \cdot 2 \cdot 1} = 15$

(b) $\binom{8}{0} = \frac{8!}{0!\,(8 - 0)!} = \frac{8!}{0!\,8!} = \frac{8!}{1 \cdot 8!} = 1$ $0! = 1$

(c) $\binom{10}{10} = \frac{10!}{10!\,(10 - 10)!} = \frac{10!}{10!\,0!} = 1$ $0! = 1$

Graphing Calculator Solution

TI-84 Plus calculators can compute binomial coefficients with the *combinations* function, denoted nCr. See **FIGURE 21.**

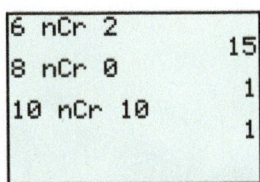

```
6 nCr 2
            15
8 nCr 0
            1
10 nCr 10
            1
```

FIGURE 21

Refer again to Pascal's triangle. Notice the symmetry in each row, which suggests that the binomial coefficients should satisfy

$$\binom{n}{r} = \binom{n}{n-r},$$

since
$$\binom{n}{r} = \frac{n!}{r!\,(n-r)!} \quad \text{and} \quad \binom{n}{n-r} = \frac{n!}{(n-r)!\,r!}.$$

The Binomial Theorem

Our observations about the expansion of $(x + y)^n$ are summarized as follows.

1. There are $n + 1$ terms in the expansion.
2. The first term is x^n and the last term is y^n.
3. In each succeeding term, the exponent on x decreases by 1 and the exponent on y increases by 1.
4. The sum of the exponents on x and y in any term is n.
5. The coefficient of the term with $x^r y^{n-r}$ or $x^{n-r} y^r$ is $\binom{n}{r}$.

These observations about the expansion of $(x + y)^n$ for any positive integer value of n suggest the **binomial theorem.** A proof is given in the next section.

Binomial Theorem

For any positive integer n,

$$(x + y)^n = x^n + \binom{n}{1}x^{n-1}y + \binom{n}{2}x^{n-2}y^2 + \binom{n}{3}x^{n-3}y^3$$

$$+ \cdots + \binom{n}{r}x^{n-r}y^r + \cdots + \binom{n}{n-1}xy^{n-1} + y^n.$$

→ Looking Ahead to Calculus

The binomial theorem is used to show that the derivative of $f(x) = x^n$ is given by the term nx^{n-1}. This fact is used extensively throughout the study of calculus.

NOTE   The binomial theorem looks much more manageable written in summation notation. The theorem can be summarized as follows.

$$(x + y)^n = \sum_{r=0}^{n} \binom{n}{r}x^{n-r}y^r$$

FOR DISCUSSION

Consider the expansion of $(x + y)^5$.

$$(x + y)^5 = x^5 + 5x^4y + 10x^3y^2 + 10x^2y^3 + 5xy^4 + y^5$$

Position ⟶ ① ② ③ ④ ⑤ ⑥

Choose any term except the sixth term, and multiply its numerical coefficient by the exponent on x. Then divide by the position number of the term. Compare your answer to the coefficient of the next term in the expansion. What do you notice? Repeat this procedure several times with this expansion and others, and make a conjecture.

> **EXAMPLE 2** Applying the Binomial Theorem

Write the binomial expansion of $(x + y)^9$.

Solution Use the binomial theorem.

$$(x + y)^9 = x^9 + \binom{9}{1}x^8y + \binom{9}{2}x^7y^2 + \binom{9}{3}x^6y^3 + \binom{9}{4}x^5y^4 + \binom{9}{5}x^4y^5$$

$$+ \binom{9}{6}x^3y^6 + \binom{9}{7}x^2y^7 + \binom{9}{8}xy^8 + y^9$$

$$= x^9 + \frac{9!}{1!\,8!}x^8y + \frac{9!}{2!\,7!}x^7y^2 + \frac{9!}{3!\,6!}x^6y^3 + \frac{9!}{4!\,5!}x^5y^4$$

$$+ \frac{9!}{5!\,4!}x^4y^5 + \frac{9!}{6!\,3!}x^3y^6 + \frac{9!}{7!\,2!}x^2y^7 + \frac{9!}{8!\,1!}xy^8 + y^9$$

$$= x^9 + 9x^8y + 36x^7y^2 + 84x^6y^3 + 126x^5y^4 + 126x^4y^5$$

$$+ 84x^3y^6 + 36x^2y^7 + 9xy^8 + y^9$$

> **EXAMPLE 3** Applying the Binomial Theorem

Expand $\left(a - \dfrac{b}{2}\right)^5$.

Solution Write the binomial as $\left(a - \frac{b}{2}\right)^5 = \left(a + \left(-\frac{b}{2}\right)\right)^5$. Then use the binomial theorem with $x = a$, $y = -\frac{b}{2}$, and $n = 5$.

$$\left(a - \frac{b}{2}\right)^5 = a^5 + \binom{5}{1}a^4\left(-\frac{b}{2}\right) + \binom{5}{2}a^3\left(-\frac{b}{2}\right)^2 + \binom{5}{3}a^2\left(-\frac{b}{2}\right)^3 + \binom{5}{4}a\left(-\frac{b}{2}\right)^4 + \left(-\frac{b}{2}\right)^5$$

$$= a^5 + 5a^4\left(-\frac{b}{2}\right) + 10a^3\left(-\frac{b}{2}\right)^2 + 10a^2\left(-\frac{b}{2}\right)^3 + 5a\left(-\frac{b}{2}\right)^4 + \left(-\frac{b}{2}\right)^5$$

$$= a^5 - \frac{5}{2}a^4b + \frac{5}{2}a^3b^2 - \frac{5}{4}a^2b^3 + \frac{5}{16}ab^4 - \frac{1}{32}b^5$$

> **EXAMPLE 4** Applying the Binomial Theorem

Expand $\left(\dfrac{3}{m^2} - 2\sqrt{m}\right)^4$. (Assume that $m > 0$.)

Solution Use the binomial theorem.

$$\left(\frac{3}{m^2} - 2\sqrt{m}\right)^4 = \left(\frac{3}{m^2}\right)^4 + \binom{4}{1}\left(\frac{3}{m^2}\right)^3(-2\sqrt{m}) + \binom{4}{2}\left(\frac{3}{m^2}\right)^2(-2\sqrt{m})^2 + \binom{4}{3}\left(\frac{3}{m^2}\right)(-2\sqrt{m})^3 + (-2\sqrt{m})^4$$

$$= \frac{81}{m^8} + 4\left(\frac{27}{m^6}\right)(-2m^{1/2}) + 6\left(\frac{9}{m^4}\right)(4m) + 4\left(\frac{3}{m^2}\right)(-8m^{3/2}) + 16m^2 \qquad \sqrt{m} = m^{1/2}$$

$$= \frac{81}{m^8} - \frac{216}{m^{11/2}} + \frac{216}{m^3} - \frac{96}{m^{1/2}} + 16m^2$$

NOTE *Any binomial expansion of the difference of two terms (e.g., a − b) has alternating signs.*

*r*th Term of a Binomial Expansion

Any term in a binomial expansion can be determined without writing out the entire expansion. For example, the seventh term of $(x + y)^9$ has *y* raised to the sixth power (since *y* has the power 1 in the second term, the power 2 in the third term, and so on). The exponents on *x* and *y* in each term must have a sum of 9, so the exponent on *x* in the seventh term is $9 − 6 = 3$. Thus, writing the coefficient as given in the binomial theorem, we find the seventh term to be

$$\frac{9!}{6!\ 3!}x^3y^6, \quad \text{or} \quad 84x^3y^6.$$

This is in fact the seventh term of $(x + y)^9$ found in **Example 2.** The discussion suggests the next theorem.

---

*r*th Term of the Binomial Expansion

The *r*th term of the binomial expansion of $(x + y)^n$, where $n \geq r − 1$, is

$$\binom{n}{r-1}x^{n-(r-1)}y^{r-1}.$$

EXAMPLE 5   Finding a Specific Term of a Binomial Expansion

Find the fourth term of the expansion of $(a + 2b)^{10}$.

Solution   In the fourth term, $2b$ has exponent 3 and *a* has exponent $10 − 3 = 7$. Using the preceding formula, we find the fourth term.

$$\binom{10}{3}a^7(2b)^3 = 120a^7(8b^3) = 960a^7b^3 \qquad n = 10, r = 4, x = a, y = 2b$$

11.5 Exercises

Checking Analytic Skills   Evaluate the following. In Exercises 17 and 18, express the answer in terms of n. **Do not use a calculator.**

1. $\dfrac{6!}{3!\ 3!}$ **2.** $\dfrac{5!}{2!\ 3!}$ **3.** $\dfrac{7!}{3!\ 4!}$ **4.** $\dfrac{8!}{5!\ 3!}$ **5.** $\binom{8}{3}$ **6.** $\binom{7}{4}$

7. $\binom{10}{8}$ **8.** $\binom{9}{6}$ **9.** $\binom{13}{13}$ **10.** $\binom{12}{12}$ **11.** $\binom{8}{3}$ **12.** $\binom{9}{7}$

13. $\binom{100}{2}$ **14.** $\binom{20}{15}$ **15.** $\binom{5}{0}$ **16.** $\binom{6}{0}$ **17.** $\binom{n}{n-1}$ **18.** $\binom{n}{n-2}$

19. *Concept Check*   How many terms are there in the expansion of $(x + y)^8$?

20. *Concept Check*   How many terms are there in the expansion of $(x + y)^{10}$?

21. *Concept Check*   What are the first and last terms in the expansion of $(2x + 3y)^4$?

22. Describe in your own words how you would determine the binomial coefficient for the fifth term in the expansion of $(x + y)^8$.

Write the binomial expansion for each expression.

23. $(x + y)^6$

24. $(m + n)^4$

25. $(p - q)^5$

26. $(a - b)^7$

27. $(r^2 + s)^5$

28. $(m + n^2)^4$

29. $(p + 2q)^4$

30. $(3r - s)^6$

31. $(7p + 2q)^4$

32. $(4a - 5b)^5$

33. $(3x - 2y)^6$

34. $(7k - 9j)^4$

35. $\left(\dfrac{m}{2} - 1\right)^6$

36. $\left(3 + \dfrac{y}{3}\right)^5$

37. $\left(\sqrt{2}r + \dfrac{1}{m}\right)^4$

38. $\left(\dfrac{1}{k} - \sqrt{3}p\right)^3$

Write the indicated term of each binomial expansion.

39. Sixth term of $(4h - j)^8$

40. Eighth term of $(2c - 3d)^{14}$

41. Fifteenth term of $(a^2 + b)^{22}$

42. Twelfth term of $(2x + y^2)^{16}$

43. Fifteenth term of $(x - y^3)^{20}$

44. Tenth term of $(a^3 + 3b)^{11}$

Concept Check   Work Exercises 45–48.

45. Find the middle term of $(3x^7 + 2y^3)^8$.

46. Find the two middle terms of $(-2m^{-1} + 3n^{-2})^{11}$.

47. Find the value of n for which the coefficients of the fifth and eighth terms in the expansion of $(x + y)^n$ are the same.

48. Find the term in the expansion of $\left(3 + \sqrt{x}\right)^{11}$ that contains x^4.

RELATING CONCEPTS   **For individual or group investigation (Exercises 49–52)**

The factorial of a positive integer n can be computed as a product: $n! = 1 \cdot 2 \cdot 3 \cdot \cdots \cdot n$. *Calculators and computers can evaluate factorials quickly. Before the days of technology, mathematicians developed a formula, called **Stirling's formula,** for approximating large factorials. Interestingly enough, it involves the irrational numbers π and e.*

$$n! \approx \sqrt{2\pi n} \cdot n^n \cdot e^{-n}$$

As an example, the exact value of 5! is 120, and Stirling's formula gives the approximation as 118.019168 with a graphing calculator. This is "off" by less than 2, an error of only 1.65%. **Work Exercises 49–52 in order.**

49. Use a calculator and Stirling's formula to find the exact value of 10! and its approximation.

50. Subtract the smaller value from the larger value in **Exercise 49.** Divide it by 10! and convert to a percent. What is the percent error?

51. Repeat **Exercises 49** and **50** for $n = 12$.

52. Repeat **Exercises 49** and **50** for $n = 13$. What seems to happen as n increases?

It can be shown that

$$(1 + x)^n = 1 + nx + \frac{n(n - 1)}{2!}x^2 + \frac{n(n - 1)(n - 2)}{3!}x^3 + \cdots$$

for any real number n (not just positive integer values) and any real number x, where $|x| < 1$. Use this result to approximate each quantity in Exercises 53–56 to the nearest thousandth.

53. $(1.02)^{-3}$

54. $\dfrac{1}{1.04^5}$

55. $(1.01)^{3/2}$

56. $(1.03)^{0.2}$

SECTIONS
11.4–11.5 **Reviewing Basic Concepts**

1. *Book Arrangements*   A student has 4 different books that she wishes to arrange in a row. How many arrangements are there?

2. Calculate $P(7, 3)$.

3. Calculate $C(10, 4)$.

4. *Basketball Positions*   How many ways can 2 guards, 2 forwards, and 1 center be selected from a basketball team composed of 6 guards, 5 forwards, and 3 centers?

5. *Home Plan Choices*   How many different homes are available if a builder offers a choice of 9 basic plans, 4 roof styles, and 2 exterior finishes?

6. How many terms are there in the expansion of $(x + y)^n$?

7. Write the binomial expansion of $(a + 2b)^4$.

8. Find the third term of the expansion of $(x - 2y)^6$.

11.6 Mathematical Induction

Proof by Mathematical Induction • Proving Statements • Generalized Principle of Mathematical Induction • Proof of the Binomial Theorem

Proof by Mathematical Induction

Many results in mathematics are claimed to hold for every positive integer. These results could be checked for $n = 1$, $n = 2$, $n = 3$, and so on, but since the set of positive integers is infinite, it would be impossible to check every possible case. For example, let S_n represent the statement that the sum of the first n positive integers is equal to $\frac{n(n + 1)}{2}$.

$$S_n: \quad 1 + 2 + 3 + \cdots + n = \frac{n(n + 1)}{2}$$

The truth of this statement can be checked quickly for the first few values of n.

If $n = 1$, then S_1 is $\qquad 1 = \frac{1(1 + 1)}{2}$, which is true.

If $n = 2$, then S_2 is $\qquad 1 + 2 = \frac{2(2 + 1)}{2}$, which is true.

If $n = 3$, then S_3 is $\quad 1 + 2 + 3 = \frac{3(3 + 1)}{2}$, which is true.

If $n = 4$, then S_4 is $\; 1 + 2 + 3 + 4 = \frac{4(4 + 1)}{2}$, which is true.

Continuing in this way for any amount of time would still not prove that S_n is true for *every* positive integer value of n. To prove that such statements are true for every positive integer value of n, the following principle is used.

Principle of Mathematical Induction

Let S_n be a statement concerning the positive integer n. Suppose that both of the following hold.

1. S_1 is true.

2. For any positive integer k, $k \leq n$, if S_k is true, then S_{k+1} is also true.

Then S_n is true for every positive integer n.

A proof by mathematical induction can be explained as follows.

By assumption (1), the statement is true when $n = 1$. By assumption (2), the fact that the statement is true for $n = 1$ implies that it is true for $n = 1 + 1 = 2$. Using (2) again, the statement is thus true for $2 + 1 = 3$, for $3 + 1 = 4$, for $4 + 1 = 5$, and so on. Continuing in this way shows that the statement must be true for every positive integer.

The situation is similar to that of an infinite number of dominoes lined up, as suggested in **FIGURE 22**. If the first domino is pushed over, it pushes the next, which pushes the next, and so on, indefinitely.

Another analogy to the principle of mathematical induction is an infinite ladder. Suppose the rungs are spaced so that, whenever you are on a rung, you know you can move to the next rung. Then *if* you can get to the first rung, you can go as high up the ladder as you wish.

Two separate steps are required for a proof by mathematical induction for every positive integer n.

FIGURE 22

Method of Proof by Mathematical Induction

Step 1 Prove that the statement is true for $n = 1$.

Step 2 Show that for any positive integer k, if S_k is true, then S_{k+1} is also true.

Proving Statements

EXAMPLE 1 **Proving an Equality Statement**

Let S_n represent the statement discussed at the beginning of this section.

$$1 + 2 + 3 + \cdots + n = \frac{n(n+1)}{2}$$

Prove that S_n is true for every positive integer n.

Solution The proof by mathematical induction is as follows.

Step 1 Show that the statement is true when $n = 1$. If $n = 1$, S_1 becomes

$$1 = \frac{1(1+1)}{2}, \text{ which is true.}$$

Step 2 Show that if S_k is true, then S_{k+1} is also true, where S_k is the statement

$$1 + 2 + 3 + \cdots + k = \frac{k(k+1)}{2},$$

and S_{k+1} is the statement

$$1 + 2 + 3 + \cdots + k + (k+1) = \frac{(k+1)[(k+1)+1]}{2}.$$

Start with S_k, and assume that it is true.

$$1 + 2 + 3 + \cdots + k = \frac{k(k+1)}{2} \qquad \textcolor{blue}{S_k \text{ is true.}}$$

Add $k+1$ to each side of S_k and show that this is S_{k+1}.

$$1 + 2 + 3 + \cdots + k + (k+1) = \frac{k(k+1)}{2} + (k+1) \qquad \textcolor{blue}{S_k + (k+1)}$$

$$= (k+1)\left(\frac{k}{2} + 1\right) \qquad \textcolor{blue}{\text{Factor out } k+1.}$$

$$= (k+1)\left(\frac{k+2}{2}\right) \qquad \textcolor{blue}{\text{Add inside the parentheses.}}$$

$$= \frac{(k+1)[(k+1)+1]}{2} \qquad \textcolor{blue}{\text{Multiply;} \atop k+2 = (k+1)+1.}$$

This final result is the statement for $n = k + 1$. It has been shown that if S_k is true, then S_{k+1} is also true. The two steps required for a proof by mathematical induction have been completed, so the statement S_n is true for every positive integer n. ●

| EXAMPLE 2 | **Proving an Inequality Statement** |

Prove that if x is a real number between 0 and 1, then, for every positive integer n,

$$0 < x^n < 1.$$

Solution

Step 1 Here, S_1 is the statement

$$\text{if } 0 < x < 1, \quad \text{then} \quad 0 < x^1 < 1,$$

which is true.

Step 2 S_k is the statement

$$\text{if } 0 < x < 1, \quad \text{then} \quad 0 < x^k < 1.$$

To show that this statement implies that S_{k+1} is true, multiply all three parts of $0 < x^k < 1$ by x to get

$$x \cdot 0 < x \cdot x^k < x \cdot 1.$$

(Here the fact that $0 < x$ is used. The inequalities are still valid because x is positive.) Simplify to obtain

$$0 < x^{k+1} < x.$$

Since $x < 1$,

$$0 < x^{k+1} < 1. \qquad \textcolor{blue}{\text{Replace } x \text{ with 1.}}$$

By this work, if S_k is true, then S_{k+1} is true. Since both steps for a proof by mathematical induction have been completed, the given statement is true for every positive integer n. ●

Generalized Principle of Mathematical Induction

Some statements S_n are not true for the first few values of n, but are true for all values of n that are greater than or equal to some fixed integer j. The following slightly generalized form of the principle of mathematical induction applies in these cases.

Generalized Principle of Mathematical Induction

Let S_n be a statement concerning the positive integer n. Let j be a fixed positive integer. Suppose that both of the following hold.

Step 1 S_j is true.

Step 2 For any positive integer k, $k \geq j$, S_k implies S_{k+1}.

Then S_n is true for all positive integers n, where $n \geq j$.

EXAMPLE 3 **Using the Generalized Principle**

Let S_n represent the statement $2^n > 2n + 1$. Show that S_n is true for all values of n such that $n \geq 3$.

Solution (Check that S_n is false for $n = 1$ and $n = 2$.)

Step 1 Show that S_n is true for $n = 3$. If $n = 3$, S_n is

$$2^3 > 2 \cdot 3 + 1 \quad \text{or} \quad 8 > 7.$$

Thus, S_3 is true.

Step 2 Now show that S_k implies S_{k+1}, for $k \geq 3$, where

$$S_k \quad \text{is} \quad 2^k > 2k + 1$$

and

$$S_{k+1} \text{ is } 2^{k+1} > 2(k + 1) + 1.$$

Multiply each side of $2^k > 2k + 1$ by 2, obtaining

$$2 \cdot 2^k > 2(2k + 1)$$

$$2^{k+1} > 4k + 2$$

$$2^{k+1} > 2(k + 1) + 2k. \qquad \text{Rewrite } 4k + 2 \text{ as } 2(k + 1) + 2k.$$

Since k is a positive integer greater than 3,

$$2k > 1.$$

It follows that

$$2^{k+1} > 2(k + 1) + 2k > 2(k + 1) + 1,$$

or

$$2^{k+1} > 2(k + 1) + 1, \quad \text{which is } S_{k+1}.$$

Thus, S_k implies S_{k+1}, and this, together with the fact that S_3 is true, shows that S_n is true for every positive integer n greater than or equal to 3. ●

Proof of the Binomial Theorem

The binomial theorem, introduced in the previous section, can be proved by mathematical induction. That is, for any positive integer n and any numbers x and y,

$$(x + y)^n = x^n + \binom{n}{1}x^{n-1}y + \binom{n}{2}x^{n-2}y^2 + \binom{n}{3}x^{n-3}y^3$$

$$+ \cdots + \binom{n}{r}x^{n-r}y^r + \cdots + \binom{n}{n-1}xy^{n-1} + y^n. \quad (1)$$

Proof Let S_n be statement (1). Begin by verifying S_n for $n = 1$,

$$S_1: \quad (x + y)^1 = x^1 + y^1, \quad \text{which is true.}$$

Now, assume that S_n is true for the positive integer k. Statement S_k becomes

$$S_k: \quad (x + y)^k = x^k + \frac{k!}{1!\,(k-1)!}x^{k-1}y + \frac{k!}{2!\,(k-2)!}x^{k-2}y^2 \quad \begin{array}{l}\text{Definition of the}\\\text{binomial coefficient}\end{array}$$

$$+ \cdots + \frac{k!}{(k-1)!\,1!}xy^{k-1} + y^k. \quad (2)$$

Multiply the left side of equation (2) by $x + y$ and apply statement S_k.

$$(x + y)^k \cdot (x + y)$$

$$= x(x + y)^k + y(x + y)^k \quad \text{Distributive property}$$

$$= \left[x \cdot x^k + \frac{k!}{1!\,(k-1)!}x^k y + \frac{k!}{2!\,(k-2)!}x^{k-1}y^2 + \cdots + \frac{k!}{(k-1)!\,1!}x^2 y^{k-1} + xy^k\right]$$

$$+ \left[x^k \cdot y + \frac{k!}{1!\,(k-1)!}x^{k-1}y^2 + \cdots + \frac{k!}{(k-1)!\,1!}xy^k + y \cdot y^k\right]$$

Rearrange terms to get

$$(x + y)^{k+1} = x^{k+1} + \left[\frac{k!}{1!(k-1)!} + 1\right]x^k y + \left[\frac{k!}{2!(k-2)!} + \frac{k!}{1!(k-1)!}\right]x^{k-1}y^2$$

$$+ \cdots + \left[1 + \frac{k!}{(k-1)!\,1!}\right]xy^k + y^{k+1}. \quad (3)$$

The first expression in brackets in equation (3) simplifies to $\binom{k+1}{1}$. To see this, note that

$$\binom{k+1}{1} = \frac{(k+1)(k)(k-1)(k-2)\cdots 1}{1 \cdot (k)(k-1)(k-2)\cdots 1} = k + 1.$$

Also, $\dfrac{k!}{1!(k-1)!} + 1 = \dfrac{k(k-1)!}{1(k-1)!} + 1 = k + 1.$

The second expression becomes $\binom{k+1}{2}$, the last $\binom{k+1}{k}$, and so on. The result of equation (3) is just equation (2) with every k replaced by $k + 1$. Thus, the truth of S_n when $n = k$ implies the truth of S_n for $n = k + 1$, which completes the proof of the theorem by mathematical induction.

11.6 Exercises

1. _Concept Check_ When using the method of mathematical induction as stated in this section to prove a statement, the domain of the variable must be all _____.

 2. Suppose that Step 2 in a proof by mathematical induction can be satisfied, but Step 1 cannot. May we conclude that the proof is complete? Explain.

3. _Concept Check_ For which positive integers is the statement $2^n > 2n$ not true?

4. Write out in full and verify the statements $S_1, S_2, S_3, S_4,$ and S_5 for the following formula. Then use mathematical induction to prove that the statement is true for every positive integer n.

$$2 + 4 + 6 + \cdots + 2n = n(n + 1)$$

Use mathematical induction to prove each statement. Assume that n is a positive integer.

5. $3 + 6 + 9 + \cdots + 3n = \dfrac{3n(n + 1)}{2}$

6. $1 + 3 + 5 + \cdots + (2n - 1) = n^2$

7. $5 + 10 + 15 + \cdots + 5n = \dfrac{5n(n + 1)}{2}$

8. $4 + 7 + 10 + \cdots + (3n + 1) = \dfrac{n(3n + 5)}{2}$

9. $3 + 3^2 + 3^3 + \cdots + 3^n = \dfrac{3(3^n - 1)}{2}$

10. $1^2 + 2^2 + 3^2 + \cdots + n^2 = \dfrac{n(n + 1)(2n + 1)}{6}$

11. $1^3 + 2^3 + 3^3 + \cdots + n^3 = \dfrac{n^2(n + 1)^2}{4}$

12. $5 \cdot 6 + 5 \cdot 6^2 + 5 \cdot 6^3 + \cdots + 5 \cdot 6^n = 6(6^n - 1)$

13. $\dfrac{1}{1 \cdot 2} + \dfrac{1}{2 \cdot 3} + \dfrac{1}{3 \cdot 4} + \cdots + \dfrac{1}{n(n + 1)} = \dfrac{n}{n + 1}$

14. $7 \cdot 8 + 7 \cdot 8^2 + 7 \cdot 8^3 + \cdots + 7 \cdot 8^n = 8(8^n - 1)$

15. $\dfrac{4}{5} + \dfrac{4}{5^2} + \dfrac{4}{5^3} + \cdots + \dfrac{4}{5^n} = 1 - \dfrac{1}{5^n}$

16. $\dfrac{1}{2} + \dfrac{1}{2^2} + \dfrac{1}{2^3} + \cdots + \dfrac{1}{2^n} = 1 - \dfrac{1}{2^n}$

17. $\dfrac{1}{1 \cdot 4} + \dfrac{1}{4 \cdot 7} + \cdots + \dfrac{1}{(3n - 2)(3n + 1)} = \dfrac{n}{3n + 1}$

18. $x^{2n} + x^{2n-1}y + \cdots + xy^{2n-1} + y^{2n} = \dfrac{x^{2n+1} - y^{2n+1}}{x - y}$

Find all positive integers n for which the given statement is not true.

19. $3^n > 6n$ **20.** $3^n > 2n + 1$ **21.** $2^n > n^2$ **22.** $n! > 2n$

Prove each statement by mathematical induction.

23. $(a^m)^n = a^{mn}$ (Assume that a and m are constant.)

24. $(ab)^n = a^n b^n$ (Assume that a and b are constant.)

25. $2^n > 2n$, if $n \geq 3$

26. $3^n > 2n + 1$, if $n \geq 2$

27. If $a > 1$, then $a^n > 1$.

28. If $a > 1$, then $a^n > a^{n-1}$.

29. If $0 < a < 1$, then $a^n < a^{n-1}$.

30. $2^n > n^2$, for $n > 4$

31. If $n \geq 4$, then $n! > 2^n$.

32. $4^n > n^4$, for $n \geq 5$

Solve each problem.

33. _Number of Handshakes_ Suppose that each of the n ($n \geq 2$) people in a room shakes hands with everyone else, but not with himself. Show that the number of handshakes is $\dfrac{n^2 - n}{2}$.

34. _Sides of a Polygon_ The series of sketches at the right starts with an equilateral triangle having sides of length 1. In the figures that follow, equilateral triangles are drawn on each side of the previous figure. The length of the sides of each new triangle is $\frac{1}{3}$ the length of the sides of the previous triangles. Develop a formula for the number of sides of the nth figure. Use mathematical induction to prove your answer.

35. *Perimeter* Find the perimeter of the *n*th figure in **Exercise 34.**

36. *Area* Show that the area of the *n*th figure in **Exercise 34** is

$$\sqrt{3}\left[\frac{2}{5} - \frac{3}{20}\left(\frac{4}{9}\right)^{n-1}\right].$$

37. *Tower of Hanoi* A pile of *n* rings, each smaller than the one below it, is on a peg on a board. Two other pegs are attached to the board. In the game called the *Tower of Hanoi* puzzle, all the rings must be moved, one at a time, to a different peg with no ring ever placed on top of a smaller ring. Find the least number of moves that would be required. Prove your result by mathematical induction.

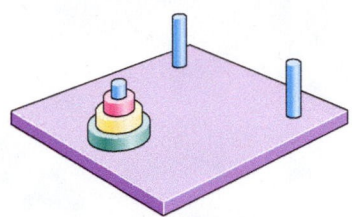

11.7 Probability

Basic Concepts • Complements and Venn Diagrams • Odds • Union of Two Events • Binomial Probability

Basic Concepts

Consider an experiment that has one or more **outcomes,** each of which is equally likely to occur. For example, the experiment of tossing a fair coin has two equally likely outcomes: heads (*H*) or tails (*T*). Also, the experiment of rolling a fair die has 6 equally likely outcomes: landing so the face that is up shows a 1, 2, 3, 4, 5, or 6.

The set *S* of all possible outcomes of a given experiment is called the **sample space** for the experiment. (In this text, all sample spaces are finite.) The sample space for the experiment of tossing a coin once consists of the outcomes *H* and *T*. This sample space can be written in set notation as

$$S = \{H, T\}.$$

Similarly, a sample space for the experiment of rolling a single die once is

$$S = \{1, 2, 3, 4, 5, 6\}.$$

Any subset of the sample space is called an **event.** In the experiment with the die, for example, "the number showing is a 3" is an event, say, $E_1 = \{3\}$. "The number showing is greater than 3" is also an event, say, $E_2 = \{4, 5, 6\}$. To represent the number of outcomes that belong to event *E*, the notation $n(E)$ is used. Then, $n(E_1) = 1$ and $n(E_2) = 3$.

The notation $P(E)$ is used for the *probability* of an event *E*. If the outcomes in the sample space for an experiment are equally likely, then the probability of event *E* occurring is found as follows.

Probability of Event *E*

In a sample space with equally likely outcomes, the **probability** of an event *E*, written **$P(E),$** is the ratio of the number of outcomes in sample space *S* that belong to event *E*, denoted **$n(E),$** to the total number of outcomes in sample space *S*, denoted **$n(S).$** That is,

$$P(E) = \frac{n(E)}{n(S)}.$$

To find the probability of event E_1 in the die experiment, start with the sample space $S = \{1, 2, 3, 4, 5, 6\}$, and the desired event $E_1 = \{3\}$. Since $n(E_1) = 1$ and $n(S) = 6$,

$$P(E_1) = \frac{n(E_1)}{n(S)} = \frac{1}{6}.$$

There is a 1 in 6 chance of rolling a 3.

EXAMPLE 1 Finding Probabilities of Events

A single die is rolled. Write each event in set notation, and give the probability of the event.

(a) E_3: the number showing is even

(b) E_4: the number showing is greater than 4

(c) E_5: the number showing is less than 7

(d) E_6: the number showing is 7

Solution

(a) Since $E_3 = \{2, 4, 6\}$, $n(E_3) = 3$. As given earlier, $n(S) = 6$, so

$$P(E_3) = \frac{3}{6} = \frac{1}{2}.$$

(b) Again, $n(S) = 6$. Event $E_4 = \{5, 6\}$, with $n(E_4) = 2$, so

$$P(E_4) = \frac{2}{6} = \frac{1}{3}.$$

(c) $E_5 = \{1, 2, 3, 4, 5, 6\}$ and $P(E_5) = \frac{6}{6} = 1.$ ← Certain event

(d) $E_6 = \emptyset$ and $P(E_6) = \frac{0}{6} = 0.$ ← Impossible event

In **Example 1(c)**, $E_5 = S$. Therefore, the event E_5 is certain to occur every time the experiment is performed. ***An event that is certain to occur always has probability 1 (i.e., 100%).*** On the other hand, $E_6 = \emptyset$ and $P(E_6) = 0$. ***The probability of an impossible event is always 0 (i.e., 0%). For any event E, P(E) is between 0 and 1 inclusive.***

Complements and Venn Diagrams

Sample Space

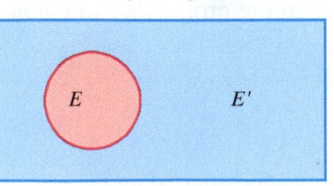

FIGURE 23

The set of all outcomes in the sample space that do *not* belong to event E is called the **complement** of E, written E'. In an experiment of drawing a single card from a standard deck of 52 cards, let E be the event "the card is an ace." Then E' is the event "the card is not an ace." From the definition of E', for an event E,

$$E \cup E' = S \quad \text{and} \quad E \cap E' = \emptyset.*$$

Probability concepts can be illustrated with **Venn diagrams,** as shown in **FIGURE 23.** The rectangle in that figure represents the sample space in an experiment. The red area inside the circle represents event E, and the blue area inside the rectangle, but outside the circle, represents event E'.

---

*The **union** of two sets A and B is the set $A \cup B$ made up of all the elements that belong to either A or B or both. The **intersection** of sets A and B, written $A \cap B$, is made up of all the elements that belong to both sets.

NOTE A standard deck of 52 cards has four suits: hearts ♥, diamonds ♦, spades ♠, and clubs ♣, with 13 cards of each suit. Each suit has a jack, a queen, and a king (sometimes called the "face cards"), an ace, and cards numbered from 2 to 10. The hearts and diamonds are red and the spades and clubs are black. We will refer to this standard deck of cards in this section.

EXAMPLE 2 **Using the Complement**

In the experiment of drawing a card from a well-shuffled deck, find the probability of event E, "the card is an ace," and event E'.

Solution There are 4 aces in the deck of 52 cards, so $n(E) = 4$ and $n(S) = 52$.

$$P(E) = \frac{n(E)}{n(S)} = \frac{4}{52} = \frac{1}{13} \quad \text{Probability of drawing an ace}$$

Of the 52 cards, 48 are not aces.

$$P(E') = \frac{n(E')}{n(S)} = \frac{48}{52} = \frac{12}{13} \quad \text{Probability of not drawing an ace}$$

In **Example 2**, $P(E) + P(E') = \frac{1}{13} + \frac{12}{13} = 1$. This is always true for any event E and its complement E'. That is,

$$P(E) + P(E') = 1,$$

which can be restated as

$$P(E) = 1 - P(E') \quad \text{or} \quad P(E') = 1 - P(E).$$

The last two equations, $P(E) = 1 - P(E')$ and $P(E') = 1 - P(E)$, suggest an alternative way to compute the probability of an event. For example, if $P(E) = \frac{1}{13}$, then

$$P(E') = 1 - \frac{1}{13} = \frac{12}{13}.$$

Odds

Sometimes probability statements are expressed in terms of *odds*, a comparison of $P(E)$ with $P(E')$. The **odds** in favor of an event E are expressed as the ratio of $P(E)$ to $P(E')$, or as the fraction $\frac{P(E)}{P(E')}$. For example, if the probability of rain can be established as $\frac{1}{3}$, the odds that it will rain are

$$P(\text{rain}) \text{ to } P(\text{no rain}) = \frac{1}{3} \text{ to } \frac{2}{3}, \quad \text{or} \quad \frac{\frac{1}{3}}{\frac{2}{3}} = \frac{1}{3} \div \frac{2}{3} = \frac{1}{3} \cdot \frac{3}{2} = \frac{1}{2}, \quad \text{or} \quad 1 \text{ to } 2.$$

Be careful simplifying a complex fraction.

The odds that it will *not* rain are 2 to 1 $\left(\text{or } \frac{2}{3} \text{ to } \frac{1}{3}\right)$. If the odds in favor of an event are, say, 3 to 5, then the probability of the event is $\frac{3}{8}$, and the probability of the complement of the event is $\frac{5}{8}$. **If the odds favoring event E are m to n, then**

$$P(E) = \frac{m}{m+n} \quad \text{and} \quad P(E') = \frac{n}{m+n}.$$

EXAMPLE 3   **Finding Odds in Favor of an Event**

A shirt is selected at random from a dark closet containing 6 blue shirts and 4 shirts that are not blue. Find the odds in favor of a blue shirt being selected.

Solution   Let E represent "a blue shirt is selected." Then

$$P(E) = \frac{6}{10}, \quad \text{or} \quad \frac{3}{5}, \quad \text{and} \quad P(E') = 1 - \frac{3}{5} = \frac{2}{5}.$$

Therefore, the odds in favor of a blue shirt being selected are

$$P(E) \text{ to } P(E') = \frac{3}{5} \text{ to } \frac{2}{5}, \quad \text{or} \quad \frac{\frac{3}{5}}{\frac{2}{5}} = \frac{3}{5} \div \frac{2}{5} = \frac{3}{5} \cdot \frac{5}{2} = \frac{3}{2}, \quad \text{or} \quad 3 \text{ to } 2. \quad \bullet$$

Union of Two Events

Since events are sets, we can use set operations to find the union of two events. Suppose a fair die is tossed. Let H be the event "the result is a 3" and K the event "the result is an even number." From the results earlier, we know the following.

$$H = \{3\} \qquad K = \{2, 4, 6\} \qquad H \cup K = \{2, 3, 4, 6\}$$

$$P(H) = \frac{1}{6} \qquad P(K) = \frac{3}{6} = \frac{1}{2} \qquad P(H \cup K) = \frac{4}{6} = \frac{2}{3}$$

Notice that in this situation

$$P(H) + P(K) = P(H \cup K). \qquad \tfrac{1}{6} + \tfrac{1}{2} = \tfrac{2}{3}$$

Before assuming that this relationship is true in general, consider another event G for this experiment: "The result is a 2."

$$G = \{2\} \qquad K = \{2, 4, 6\} \qquad G \cup K = \{2, 4, 6\}$$

$$P(G) = \frac{1}{6} \qquad P(K) = \frac{3}{6} = \frac{1}{2} \qquad P(G \cup K) = \frac{3}{6} = \frac{1}{2}$$

In this case,

$$P(G) + P(K) \neq P(G \cup K). \qquad \tfrac{1}{6} + \tfrac{1}{2} \neq \tfrac{1}{2}$$

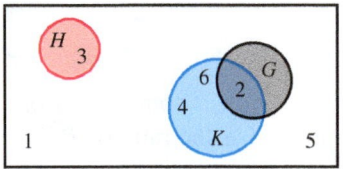

FIGURE 24

As **FIGURE 24** suggests, the difference in the two situations comes from the fact that events H and K cannot occur simultaneously. Such events are called **mutually exclusive events.** In fact, $H \cap K = \emptyset$, which is always true for mutually exclusive events. Events G and K, however, can occur simultaneously. Both are satisfied if the result of the roll is a 2, the element in their intersection ($G \cap K = \{2\}$).

This example suggests the following property.

Probability of the Union of Two Events

For any events E and F,

$$P(E \text{ or } F) = P(E \cup F) = P(E) + P(F) - P(E \cap F).$$

EXAMPLE 4   **Finding Probabilities of Unions**

One card is drawn from a well-shuffled deck of 52 cards. What is the probability of each event?

(a) The card is an ace or a spade.   **(b)** The card is a 3 or a king.

Solution

(a) The events "an ace is drawn" and "a spade is drawn" are not mutually exclusive, since it is possible to draw the ace of spades, an outcome satisfying both events.

$$P(\text{ace or spade}) = P(\text{ace}) + P(\text{spade}) - P(\text{ace and spade})$$

$$= \frac{4}{52} + \frac{13}{52} - \frac{1}{52} = \frac{16}{52} = \frac{4}{13}$$

(b) "A 3 is drawn" and "a king is drawn" are mutually exclusive events, because it is impossible to draw one card that is both a 3 and a king.

$$P(3 \text{ or } K) = P(3) + P(K) - P(3 \text{ and } K)$$

$$= \frac{4}{52} + \frac{4}{52} - 0 = \frac{8}{52} = \frac{2}{13}$$

EXAMPLE 5 **Finding Probabilities of Unions**

Suppose two fair dice are rolled, one at a time. Find each probability.

(a) The first die shows a 2, or the sum of the two dice is 6 or 7.

(b) The sum of the two dice is at most 4.

Solution

(a) Think of the two dice as being distinguishable—the first one red and the second one green, for example. (Actually, the sample space is the same even if they are not distinguishable.) A sample space with equally likely outcomes is shown in **FIGURE 25,** where (1, 1) represents the event "the first (red) die shows a 1 and the second (green) die shows a 1," (1, 2) represents "the first die shows a 1 and the second die shows a 2," and so on.

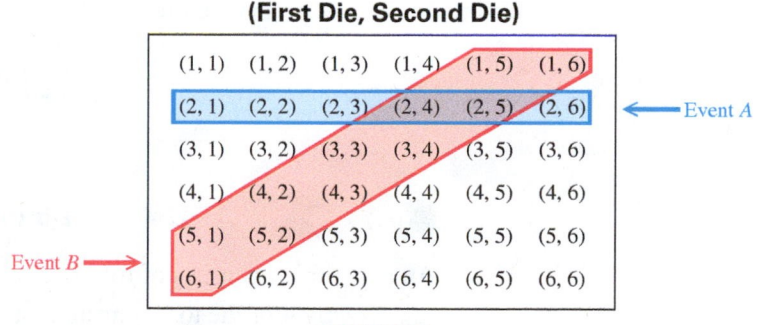

(First Die, Second Die)

(1, 1)	(1, 2)	(1, 3)	(1, 4)	(1, 5)	(1, 6)
(2, 1)	(2, 2)	(2, 3)	(2, 4)	(2, 5)	(2, 6)	←— Event A
(3, 1)	(3, 2)	(3, 3)	(3, 4)	(3, 5)	(3, 6)
(4, 1)	(4, 2)	(4, 3)	(4, 4)	(4, 5)	(4, 6)
(5, 1)	(5, 2)	(5, 3)	(5, 4)	(5, 5)	(5, 6)
(6, 1)	(6, 2)	(6, 3)	(6, 4)	(6, 5)	(6, 6)

Event B —→

FIGURE 25

Let A represent the event "the first die shows a 2" and B represent the event "the sum of the results is 6 or 7." These events are indicated in the figure. From the diagram, event A has 6 elements, B has 11 elements, and the sample space has 36 elements.

$$P(A) = \frac{6}{36}, \quad P(B) = \frac{11}{36}, \quad \text{and} \quad P(A \cap B) = \frac{2}{36},$$

2 elements satisfy both events.

so

$$P(A \cup B) = P(A) + P(B) - P(A \cap B)$$

$$= \frac{6}{36} + \frac{11}{36} - \frac{2}{36} = \frac{15}{36} = \frac{5}{12}$$

(b) "At most 4" can be written as "2 or 3 or 4." (A sum of 1 is not possible.) The events represented by "2," "3," and "4" are mutually exclusive.

$$P(\text{at most 4}) = P(2 \text{ or } 3 \text{ or } 4) = P(2) + P(3) + P(4) \quad (*)$$

(continued)

The sample space for this experiment includes the 36 possible pairs of numbers shown in **FIGURE 25** on the preceding page. The pair $(1, 1)$ is the only one with a sum of 2, so $P(2) = \frac{1}{36}$. Also, $P(3) = \frac{2}{36}$, since both $(1, 2)$ and $(2, 1)$ give a sum of 3. The pairs $(1, 3)$, $(2, 2)$, and $(3, 1)$ have a sum of 4, so $P(4) = \frac{3}{36}$. Substitute into equation $(*)$ from the preceding page.

$$P(\text{at most } 4) = \frac{1}{36} + \frac{2}{36} + \frac{3}{36} = \frac{6}{36} = \frac{1}{6}$$

$P(1) + P(2) + P(3)$

Properties of Probability

For any events E and F, the following hold.

1. $0 \le P(E) \le 1$
2. $P(\text{a certain event}) = 1$
3. $P(\text{an impossible event}) = 0$
4. $P(E') = 1 - P(E)$
5. $P(E \text{ or } F) = P(E \cup F) = P(E) + P(F) - P(E \cap F)$

CAUTION *When finding the probability of a union, subtract the probability of the intersection from the sum of the probabilities of the individual events.*

Binomial Probability

An experiment that consists of repeated independent trials with only two outcomes in each trial—success and failure—is called a **binomial experiment.**

Binomial Probability of Success in *n* Trials

Let the probability of success in one trial be p. Then the probability of failure is $1 - p$, and the probability of r successes in n trials is

$$\binom{n}{r} p^r (1 - p)^{n-r}.$$

EXAMPLE 6 **Using a Binomial Experiment to Find Probabilities**

An experiment consists of rolling a die 10 times. Find each probability.

(a) Exactly 4 of the tosses result in a 3.

(b) In 9 of the 10 tosses, the result is not a 3.

Analytic Solution

(a) The probability of a 3 on one roll is $p = \frac{1}{6}$. Here, $n = 10$ and $r = 4$, so the required probability is

$$\binom{10}{4} \left(\frac{1}{6} \right)^4 \left(1 - \frac{1}{6} \right)^{10-4} = 210 \left(\frac{1}{6} \right)^4 \left(\frac{5}{6} \right)^6$$

$$\approx 0.054.$$

(b) $\binom{10}{9} \left(\frac{5}{6} \right)^9 \left(\frac{1}{6} \right)^1 \approx 0.323$ $n = 10, r = 9, p = 1 - \frac{1}{6} = \frac{5}{6}$

Graphing Calculator Solution

The TI-84 Plus has probability distribution functions that give binomial probabilities. In **FIGURE 26**, the numbers in parentheses separated by commas represent n, p, and r, respectively.

```
binompdf(10,(1/6
),4)
        .0542658759
binompdf(10,(5/6
),9)
        .3230111658
```

FIGURE 26

11.7 Exercises

State a sample space S with equally likely outcomes for each experiment.

1. A two-headed coin is tossed once.

2. Two ordinary coins are tossed.

3. Three ordinary coins are tossed.

4. Five slips of paper, each of which is marked with the number 1, 2, 3, 4, or 5, are placed in a box. After mixing well, two slips are drawn, with the order not important.

5. The spinner shown here is spun twice.

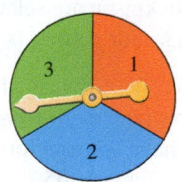

6. A die is rolled and then a coin is tossed.

Write each event in set notation. Give the probability of the event.

7. In **Exercise 1,**
   (a) the result of the toss is heads.
   (b) the result of the toss is tails.

8. In **Exercise 2,**
   (a) both coins show the same face.
   (b) at least one coin turns up heads.

9. In **Exercise 5,**
   (a) the result is a repeated number.
   (b) the second number is 1 or 3.
   (c) the first number is even and the second number is odd.

10. In **Exercise 4,**
    (a) both slips are marked with even numbers.
    (b) both slips are marked with odd numbers.
    (c) both slips are marked with the same number.
    (d) one slip is marked with an odd number, the other with an even number.

11. A student gives the answer to a probability problem as $\frac{6}{5}$. Explain why this answer must be incorrect.

12. *Concept Check* If the probability of an event is 0.857, what is the probability that the event will not occur?

Work each problem.

13. *Drawing a Marble* A marble is drawn at random from a box containing 3 yellow, 4 white, and 8 blue marbles. Find the probabilities in parts (a)–(c).

(a) A yellow marble is drawn.
(b) A black marble is drawn.
(c) The marble is yellow or white.
(d) What are the odds in favor of drawing a yellow marble?
(e) What are the odds against drawing a blue marble?

14. *Batting Average* A baseball player with a batting average of .300 comes to bat. What are the odds in favor of his getting a hit?

15. *Drawing Slips of Paper* In **Exercise 4,** what are the odds that the sum of the numbers on the two slips of paper is 5?

16. *Probability of Rain* If the odds that it will rain are 4 to 5, what is the probability of rain?

17. *Probability of a Candidate Losing* If the odds that a candidate will win an election are 3 to 2, what is the probability that the candidate will lose?

18. *Drawing a Card* A card is drawn from a well-shuffled deck of 52 cards. Find the probability that the card is as follows.
    (a) a 9   (b) black
    (c) a black 9   (d) a heart
    (e) a face card (K, Q, or J of any suit)
    (f) red or a 3
    (g) less than a 4 (consider aces as 1s)

19. *Guest Arrival at a Party* Mrs. Schmulen invites 10 relatives to a party: her mother, 2 uncles, 3 brothers, and 4 cousins. If the chances of any one guest arriving first are equally likely, find each probability.
    (a) The first guest is an uncle or a brother.
    (b) The first guest is a brother or a cousin.
    (c) The first guest is a brother or her mother.

20. *Dice Rolls* Two dice are rolled. Find the probability of each event.
    (a) The sum is at least 10.
    (b) The sum is either 7 or at least 10.
    (c) The sum is 2 or the dice both show the same number.

21. *Concept Check* Match each probability in parts (a)–(g) with one of the statements in A–F.
    (a) $P(E) = -0.1$ (b) $P(E) = 0.01$
    (c) $P(E) = 1$ (d) $P(E) = 2$
    (e) $P(E) = 0.99$ (f) $P(E) = 0$
    (g) $P(E) = 0.5$
    A. The event is certain to occur.
    B. The event is impossible.
    C. The event is very likely to occur.
    D. The event is very unlikely to occur.
    E. The event is just as likely to occur as not occur.
    F. This probability cannot occur.

22. *Small-Business Loan* The probability that a bank with assets greater than or equal to $30 billion will make a loan to a small business is 0.002. What are the odds against such a bank making a small-business loan? (*Source: The Wall Street Journal* analysis of *CA1 Reports.*)

23. *Estimating Probability of Organ Transplants* In a recent year there were 51,277 people waiting for an organ transplant. The following table lists the number of patients waiting for the most common types of transplants.

Organ Transplant	Patients Waiting
Heart	3,774
Kidney	35,025
Liver	7,920
Lung	2,340

Source: Coalition on Organ and Tissue Donation.

Assuming that none of these people need two or more transplants, approximate the probability that a transplant patient chosen at random will need
(a) a kidney or a heart.
(b) neither a kidney nor a heart.

24. *U.S. Population Origins* Projected Hispanic/Latino and non-Hispanic/Latino U.S. populations (in thousands) for 2025 are given in the following table.

Type	Number
Hispanic/Latino	58,930
White	209,117
Black	43,511
Native American	2,744
Asian	20,748

Source: U.S. Census Bureau.

Assume that these projections are accurate. Estimate the probability that a U.S. resident selected at random in 2025 is
(a) Hispanic/Latino. **(b)** not white.
(c) Native American or black.
(d) What are the odds that a randomly selected U.S. resident is Asian?

25. *Consumer Purchases* The following table shows the probability that a customer at a department store will make a purchase in the indicated price range.

Cost	Probability
Below $5	0.25
$5–$19.99	0.37
$20–$39.99	0.11
$40–$69.99	0.09
$70–$99.99	0.07
$100–$149.99	0.08
$150 or more	0.03

Find the probability that a customer makes a purchase that is
(a) less than $20. **(b)** $40 or more.
(c) more than $99.99. **(d)** less than $100.

26. *State Lottery* One game in a state lottery requires you to pick 1 heart, 1 club, 1 diamond, and 1 spade, in that order, from the 13 cards in each suit. What is the probability of getting all four picks correct and winning $5000?

27. *State Lottery* If three of the four selections in **Exercise 26** are correct, the player wins $200. Find the probability of this outcome.

28. *Partner Selection* The law firm of Alam, Bartolini, Chinn, Dickinson, and Ellsberg has two senior partners: Alam and Bartolini. Two of the attorneys are to be selected to attend a conference. Assuming that all are equally likely to be selected, find each probability.
(a) Chinn is selected.
(b) Alam and Dickinson are selected.
(c) At least one senior partner is selected.

29. *Opinion Survey* The management of a firm surveys its workers, classified as follows for the purpose of an interview: 30% have worked for the company more than 5 years; 28% are female; 65% contribute to a voluntary retirement plan; and $\frac{1}{2}$ of the female workers contribute to the retirement plan. Find each probability.
(a) A male worker is selected.
(b) A worker is selected who has worked for the company 5 years or less.
(c) A worker is selected who contributes to the retirement plan or is female.

30. Explain why the probability of an event must be a number between 0 and 1 inclusive.

Gender Makeup of a Family *Suppose a family has 5 children. Suppose also that the probability of having a girl is $\frac{1}{2}$. Find the probability that the family has the following children.*

31. Exactly 2 girls and 3 boys

32. Exactly 3 girls and 2 boys

33. No girls **34.** No boys

35. At least 3 boys **36.** No more than 4 girls

College Student Smokers The table gives the results of a survey of 14,000 college students who were cigarette smokers in a recent year.

Number of Cigarettes per Day	Percent (as a Decimal)
Less than 1	0.45
1 to 9	0.24
10 to 19	0.20
A pack of 20 or more	0.11

Using the percents as probabilities, approximate the probability that, out of 10 of these student smokers selected at random, the following were true.

37. Four smoked fewer than 10 cigarettes per day.

38. Five smoked a pack or more per day.

39. Fewer than 2 smoked between 1 and 19 cigarettes per day.

40. No more than 3 smoked less than 1 cigarette per day.

A die is rolled 12 times. Approximate the probability of rolling the following.

41. Exactly 12 ones

42. Exactly 6 ones

43. No more than 3 ones

44. No more than 1 one

College Applications The table gives the results of a survey of 282,549 freshmen from a recent class year at 437 of the nation's baccalaureate colleges and universities.

Number of Colleges Applied to	1	2 or 3	4–6	7 or more
Percent (as a decimal)	0.20	0.29	0.37	0.14

Source: Higher Education Research Institute, UCLA.

Using the percents as probabilities, find the probability of each event for a randomly selected student.

45. The student applied to fewer than 4 colleges.

46. The student applied to at least 2 colleges.

47. The student applied to more than 3 colleges.

48. The student applied to no colleges.

49. *Color-Blind Males* The probability that a male will be color blind is 0.042. Approximate the probabilities that in a group of 53 men, the following are true.
(a) Exactly 5 are color blind.
(b) No more than 5 are color blind.
(c) At least 1 is color blind.

50. *Binomial Probability* The screens below illustrate how the table feature of a graphing calculator can be used to find the probabilities of having 0, 1, 2, 3, or 4 girls in a family of 4 children. (Note that the value $Y_1 = 0$ appears for X = 5 and X = 6. Why is this so?) Use this approach to answer the following.

(a) Find the probabilities of having 0, 1, 2, or 3 boys in a family of 3 children.
(b) Find the probabilities of having 0, 1, 2, 3, 4, 5, or 6 girls in a family of 6 children.

51. *(Modeling) Disease Infection* What will happen when an infectious disease is introduced into a family? Suppose a family has I infected members and S members who are not infected, but are susceptible to contracting the disease. The probability P of k people not contracting the disease during a 1-week period can be calculated by the formula

$$P = \binom{S}{k} q^k (1 - q)^{S-k},$$

where $q = (1 - p)^I$ and p is the probability that a susceptible person contracts the disease from an infectious person. For example, if $p = 0.5$, then there is a 50% chance that a susceptible person exposed to one infectious person for 1 week will contract the disease. (*Source:* Hoppensteadt, F. and C. Peskin, *Mathematics in Medicine and the Life Sciences*, Springer-Verlag.)
(a) Approximate the probability P of 3 family members not becoming infected within 1 week if there are currently 2 infected and 4 susceptible members. Assume that $p = 0.1$.
(b) A highly infectious disease can have $p = 0.5$. Repeat part (a) with this value of p.
(c) Approximate the probability that everyone would become sick in a large family if initially $I = 1$, $S = 9$, and $p = 0.5$.

52. *(Modeling) Disease Infection* Refer to **Exercise 51.** Suppose that, in a family, $I = 2$ and $S = 4$. If the probability P is 0.25 of there being $k = 2$ uninfected members after 1 week, estimate graphically the possible values of p. (*Hint:* Write P as a function of p.)

SECTIONS 11.6–11.7 **Reviewing Basic Concepts**

1. Use mathematical induction to prove that for all positive integers n,

$$4 + 8 + 12 + 16 + \cdots + 4n = 2n(n + 1).$$

2. Use mathematical induction to prove that $n^2 \leq 2^n$ for $n \geq 4$.

3. Write the sample space for an experiment consisting of tossing a coin twice.

4. *Dice Roll* Find the probability of rolling a sum of 11 with two dice.

5. *Drawing Cards* Find the probability of drawing 4 aces and 1 queen from a standard deck of 52 cards.

6. *Drawing Cards* Find the probability of *not* drawing 4 aces and 1 queen from a standard deck of 52 cards.

7. *Probability of Rain* If the odds that it will rain are 3 to 7, what is the probability of rain?

8. *High School Graduates* In a recent year there were 2.81 million high school graduates, of which 1.45 million were male. (*Source:* U.S. National Center for Education Statistics.) If one such high school graduate is selected at random, estimate the probability that this graduate is female.

---

11 Summary

KEY TERMS & SYMBOLS

KEY CONCEPTS

11.1 Sequences and Series

sequence
terms of a sequence
general or nth term, a_n
finite sequence
infinite sequence
convergent
divergent
recursive definition
Fibonacci sequence
series
finite series
infinite series
summation (sigma)

notation, $\sum\limits_{n=1}^{\infty}$

index of summation, i

SEQUENCE
A sequence is a function that has a set of natural numbers as its domain.

SERIES
A finite series is an expression of the form

$$S_n = a_1 + a_2 + a_3 + \cdots + a_n = \sum_{i=1}^{n} a_i,$$

and an infinite series is an expression of the form

$$S_\infty = a_1 + a_2 + a_3 + \cdots + a_n + \cdots = \sum_{i=1}^{\infty} a_i.$$

SUMMATION PROPERTIES
If $a_1, a_2, a_3, \ldots, a_n$ and $b_1, b_2, b_3, \ldots, b_n$ are two sequences and c is a constant, then, for every positive integer n, the following hold.

(a) $\sum\limits_{i=1}^{n} c = nc$

(b) $\sum\limits_{i=1}^{n} ca_i = c\sum\limits_{i=1}^{n} a_i$

(c) $\sum\limits_{i=1}^{n} (a_i + b_i) = \sum\limits_{i=1}^{n} a_i + \sum\limits_{i=1}^{n} b_i$

(d) $\sum\limits_{i=1}^{n} (a_i - b_i) = \sum\limits_{i=1}^{n} a_i - \sum\limits_{i=1}^{n} b_i$

11.2 Arithmetic Sequences and Series

arithmetic sequence
  (or progression)
common difference, d
arithmetic series

ARITHMETIC SEQUENCE
An arithmetic sequence is a sequence in which each term after the first is obtained by adding a fixed number to the previous term. The fixed number is called the common difference.

nTH TERM OF AN ARITHMETIC SEQUENCE
In an arithmetic sequence with first term a_1 and common difference d, the nth term is

$$a_n = a_1 + (n - 1)d.$$

KEY TERMS & SYMBOLS	KEY CONCEPTS

SUM OF THE FIRST n TERMS OF AN ARITHMETIC SEQUENCE

If an arithmetic sequence has first term a_1 and common difference d, then the sum of the first n terms is

$$S_n = \frac{n}{2}(a_1 + a_n) \quad \text{or} \quad S_n = \frac{n}{2}[2a_1 + (n-1)d].$$

11.3 Geometric Sequences and Series

geometric sequence
  (or progression)
common ratio, r
geometric series
annuity
future value of an annuity

GEOMETRIC SEQUENCE

A geometric sequence is a sequence in which each term after the first is obtained by multiplying the preceding term by a constant nonzero real number, called the common ratio.

nTH TERM OF A GEOMETRIC SEQUENCE

In the geometric sequence with first term a_1 and common ratio r, neither of which is zero, the nth term is

$$a_n = a_1 r^{n-1}.$$

SUM OF THE FIRST n TERMS OF A GEOMETRIC SEQUENCE

If a geometric sequence has first term a_1 and common ratio r, then the sum of the first n terms is

$$S_n = \frac{a_1(1 - r^n)}{1 - r}, \quad \text{where} \quad r \neq 1.$$

SUM OF THE TERMS OF AN INFINITE GEOMETRIC SEQUENCE

The sum of the terms of an infinite geometric sequence with first term a_1 and common ratio r, where $|r| < 1$, is

$$S_\infty = \frac{a_1}{1 - r}.$$

If $|r| \geq 1$, the sum does not exist.

FUTURE VALUE OF AN ANNUITY

The formula for the future value of an annuity is

$$S = R\left[\frac{(1 + i)^n - 1}{i}\right],$$

where S is the future value, R is the payment at the end of each period, i is the interest rate in decimal form per period, and n is the number of periods.

11.4 Counting Theory

tree diagram
independent events
factorial notation
n-factorial, $n!$
permutation, $P(n, r)$
combination, $C(n, r)$

FUNDAMENTAL PRINCIPLE OF COUNTING

If n independent events occur, with

$$m_1 \text{ ways for event 1 to occur,}$$
$$m_2 \text{ ways for event 2 to occur,}$$
$$\cdot$$
$$\cdot$$
$$\cdot$$

and $\qquad\qquad m_n$ ways for event n to occur,

then there are

$$m_1 \cdot m_2 \cdot \cdots \cdot m_n$$

different ways for all n events to occur.

(continued)

KEY TERMS & SYMBOLS   **KEY CONCEPTS**

***n*-FACTORIAL**

For any positive integer n,

$$n! = n(n - 1)(n - 2) \cdots (3)(2)(1).$$

By definition, $0! = 1.$

PERMUTATIONS OF *n* ELEMENTS TAKEN *r* AT A TIME

If $P(n, r)$ denotes the number of permutations of n elements taken r at a time, with $r \leq n$, then

$$P(n, r) = \frac{n!}{(n - r)!}.$$

COMBINATIONS OF *n* ELEMENTS TAKEN *r* AT A TIME

If $C(n, r)$ represents the number of combinations of n things taken r at a time, with $r \leq n$, then

$$C(n, r) = \frac{n!}{(n - r)! \, r!}.$$

11.5 The Binomial Theorem

binomial coefficient
binomial theorem

PASCAL'S TRIANGLE

						Row
		1				0
	1		1			1
1		2		1		2
1	3		3	1		3
1	4	6	4	1		4
1	5	10	10	5	1	5

BINOMIAL COEFFICIENT

For nonnegative integers n and r, with $r \leq n$,

$$\binom{n}{r} = \frac{n!}{r! \, (n - r)!}.$$

BINOMIAL THEOREM

For any positive integer n,

$$(x + y)^n = x^n + \binom{n}{1}x^{n-1}y + \binom{n}{2}x^{n-2}y^2 + \binom{n}{3}x^{n-3}y^3$$

$$+ \cdots + \binom{n}{r}x^{n-r}y^r + \cdots + \binom{n}{n-1}xy^{n-1} + y^n.$$

***r*TH TERM OF THE BINOMIAL EXPANSION**

The rth term of the binomial expansion of $(x + y)^n$, where $n \geq r - 1$, is

$$\binom{n}{r - 1}x^{n-(r-1)}y^{r-1}.$$

KEY TERMS & SYMBOLS	KEY CONCEPTS

11.6 Mathematical Induction

METHOD OF PROOF BY MATHEMATICAL INDUCTION (FOR EVERY POSITIVE INTEGER n)

Step 1   Prove that the statement is true for $n = 1$.

Step 2   Show that for any positive integer k, if S_k is true, then S_{k+1} is also true.

GENERALIZED PRINCIPLE OF MATHEMATICAL INDUCTION

Let S_n be a statement concerning the positive integer n. Let j be a fixed positive integer. Suppose that both of the following hold.

Step 1 S_j is true.

Step 2   For any positive integer k, $k \geq j$, S_k implies S_{k+1}.

Then S_n is true for all positive integers n, where $n \geq j$.

11.7 Probability

outcomes
sample space
event
probability
complement, E'
Venn diagram
odds
mutually exclusive events
binomial experiment

SAMPLE SPACE AND EVENT

The set S of all possible outcomes of a given experiment is called the sample space for the experiment. Any subset of S is called an event.

PROBABILITY OF EVENT E

In a sample space with equally likely outcomes, the probability of an event E, written $P(E)$, is the ratio of the number of outcomes in sample space S that belong to event E, denoted $n(E)$, to the total number of outcomes in sample space S, denoted $n(S)$. That is,

$$P(E) = \frac{n(E)}{n(S)}.$$

COMPLEMENT

The complement of an event E, written E', is the set of all outcomes not in E. Thus,

$$P(E) + P(E') = 1.$$

ODDS

The odds in favor of an event E are expressed as the ratio of $P(E)$ to $P(E')$, or $\frac{P(E)}{P(E')}$.

PROBABILITY OF THE UNION OF TWO EVENTS

For any events E and F,

$$P(E \text{ or } F) = P(E \cup F) = P(E) + P(F) - P(E \cap F).$$

PROPERTIES OF PROBABILITY

For any events E and F, the following hold.

1. $0 \leq P(E) \leq 1$
2. $P(\text{a certain event}) = 1$
3. $P(\text{an impossible event}) = 0$
4. $P(E') = 1 - P(E)$
5. $P(E \text{ or } F) = P(E \cup F) = P(E) + P(F) - P(E \cap F)$

BINOMIAL PROBABILITY

An experiment that consists of repeated independent trials with only two outcomes in each trial—success and failure—is called a binomial experiment. Let the probability of success of one trial be p. Then the probability of failure is $1 - p$, and the probability of r successes in n trials is

$$\binom{n}{r} p^r (1 - p)^{n-r}.$$

11 Review Exercises

Write the first five terms for each sequence. State whether the sequence is arithmetic, geometric, *or* neither.

1. $a_n = \dfrac{n}{n+1}$ **2.** $a_n = (-2)^n$ **3.** $a_n = 2(n+3)$ **4.** $a_n = n(n+1)$

5. $a_1 = 5;\ a_n = a_{n-1} - 3,$ for $n \geq 2$ **6.** $a_n = \left(\dfrac{1}{5}\right)^{n-1}$

In Exercises 7–12, write the first five terms of the sequence described.

7. Arithmetic, $a_2 = 10,\ d = -2$

8. Arithmetic, $a_3 = \pi,\ a_4 = 1$

9. Geometric, $a_1 = 6,\ r = 2$

10. Geometric, $a_1 = -5,\ a_2 = -1$

11. An arithmetic sequence has $a_5 = -3$ and $a_{15} = 17$. Find a_1 and a_n.

12. A geometric sequence has $a_1 = -8$ and $a_7 = -\frac{1}{8}$. Find a_4 and a_n.

Find a_8 for each arithmetic sequence.

13. $a_1 = 6,\ d = 2$

14. $a_1 = 6x - 9,\ a_2 = 5x + 1$

Find S_{12} for each arithmetic sequence.

15. $a_1 = 2,\ d = 3$

16. $a_2 = 6,\ d = 10$

Find a_5 for each geometric sequence.

17. $a_1 = -2,\ r = 3$

18. $a_3 = 4,\ r = \dfrac{1}{5}$

Find S_4 for each geometric sequence.

19. $a_1 = 3,\ r = 2$

20. $\dfrac{3}{4}, -\dfrac{1}{2}, \dfrac{1}{3}, \cdots$

21. *Annuity Value* Find the future value of an annuity that consists of payments of $2000 at the end of each year for 5 years at 3% interest compounded annually.

22. *Height of Ball* When a ping pong ball is dropped, it rebounds to 90% of its original height. If the original height is 6 ft, find a formula for a_n, where a_n represents the maximum height of the ball before the nth bounce.

Evaluate each sum that exists.

23. $\displaystyle\sum_{i=1}^{7} (-1)^{i-1}$ **24.** $\displaystyle\sum_{i=1}^{5} (i^2 + i)$ **25.** $\displaystyle\sum_{i=1}^{4} \dfrac{i+1}{i}$ **26.** $\displaystyle\sum_{j=1}^{10} (3j - 4)$

27. $\displaystyle\sum_{j=1}^{2500} j$ **28.** $\displaystyle\sum_{i=1}^{5} 4 \cdot 2^i$ **29.** $\displaystyle\sum_{i=1}^{\infty} \left(\dfrac{4}{7}\right)^i$ **30.** $\displaystyle\sum_{i=1}^{\infty} -2\left(\dfrac{6}{5}\right)^i$

Evaluate each sum that converges. If the series diverges, say so.

31. $24 + 8 + \dfrac{8}{3} + \dfrac{8}{9} + \cdots$

32. $-\dfrac{3}{4} + \dfrac{1}{2} - \dfrac{1}{3} + \dfrac{2}{9} - \cdots$

33. $\dfrac{1}{12} + \dfrac{1}{6} + \dfrac{1}{3} + \dfrac{2}{3} + \cdots$

34. $0.9 + 0.09 + 0.009 + 0.0009 + \cdots$

Evaluate each sum, where $x_1 = 0,\ x_2 = 1,\ x_3 = 2,\ x_4 = 3,\ x_5 = 4,$ and $x_6 = 5$.

35. $\displaystyle\sum_{i=1}^{4} (x_i^2 - 6)$

36. $\displaystyle\sum_{i=1}^{6} f(x_i)\,\Delta x;\ f(x) = (x - 2)^3,\ \Delta x = 0.1$

Write each sum, using summation notation.

37. $4 - 1 - 6 - \cdots - 66$

38. $10 + 14 + 18 + \cdots + 86$

39. $4 + 12 + 36 + \cdots + 972$

40. $\dfrac{5}{6} + \dfrac{6}{7} + \dfrac{7}{8} + \cdots + \dfrac{12}{13}$

Find the value of each expression.

41. $9!$

42. $P(9, 2)$

43. $P(6, 0)$

44. $C(10, 5)$

45. $\dbinom{8}{3}$

46. *Concept Check* Is a student identification number an example of a permutation or a combination?

Use the binomial theorem to expand each expression.

47. $(x + 2y)^4$

48. $(3z - 5w)^3$

49. $\left(3\sqrt{x} - \dfrac{1}{\sqrt{x}}\right)^5$

50. $(m^3 - m^{-2})^4$

Find the indicated term or terms for each expansion.

51. Sixth term of $(4x - y)^8$

52. Seventh term of $(m - 3n)^{14}$

53. First four terms of $(x + 2)^{12}$

54. Last three terms of $(2a + 5b)^{16}$

 55. What kinds of statements are proved by mathematical induction?

 56. Describe a proof by mathematical induction.

Use mathematical induction to prove that each statement is true for every positive integer n.

57. $1 + 3 + 5 + 7 + \cdots + (2n - 1) = n^2$

58. $2 + 6 + 10 + 14 + \cdots + (4n - 2) = 2n^2$

59. $2 + 2^2 + 2^3 + \cdots + 2^n = 2(2^n - 1)$

60. $1^3 + 3^3 + 5^3 + \cdots + (2n - 1)^3 = n^2(2n^2 - 1)$

Solve each problem.

61. *Wedding Plans* Two people are planning their wedding. They can select from 2 different chapels, 4 soloists, 3 organists, and 2 ministers. How many different wedding arrangements are possible?

62. *Couch Styles* A student who is furnishing his apartment wants to buy a new couch. He can select from 5 different styles, each available in 3 different fabrics with 6 color choices. How many different couches are available?

63. *Summer Job Assignments* Four students are to be assigned to 4 different summer jobs. Each student is qualified for all 4 jobs. In how many ways can the jobs be assigned?

64. *Conference Delegations* A student body council consists of a president, vice president, secretary/treasurer, and 3 representatives at large. Three members are to be selected to attend a conference.
(a) How many different such delegations are possible?
(b) How many are possible if the president must attend?

65. *Tournament Outcomes* Nine football teams are competing for first-, second-, and third-place titles in a statewide tournament. In how many ways can the winners be determined?

66. *License Plates* How many different license plates can be formed with a letter followed by 3 digits and then 3 letters? How many such license plates have no repeats?

67. *Drawing a Marble* A marble is drawn at random from a box containing 4 green, 5 black, and 6 white marbles. Find each probability.
(a) A green marble is drawn.
(b) A marble that is not black is drawn.
(c) A blue marble is drawn.

68. *Drawing a Marble* Refer to **Exercise 67,** and answer each question.
(a) What are the odds in favor of drawing a green marble?
(b) What are the odds against drawing a white marble?
(c) What are the odds in favor of drawing a marble that is not white?

Drawing a Card *A card is drawn from a standard deck of 52 cards. Find the probability that the following is drawn.*

69. A black king

70. A face card or an ace

71. An ace or a diamond

72. A card that is not a club

Swimming Pool Filter Samples *A sample shipment of 5 swimming pool filters is chosen. The probability of exactly 0, 1, 2, 3, 4, or 5 filters being defective is given in the following table.*

Number Defective	0	1	2	3	4	5
Probability	0.31	0.25	0.18	0.12	0.08	0.06

Find the probability that the given number of filters are defective.

73. No more than 3   **74.** At least 2   **75.** More than 5

76. *Die Rolls* A die is rolled 12 times. Find the probability that exactly 2 of the rolls result in a 5.

77. *Coin Tosses* A coin is tossed 10 times. Find the probability that exactly 4 of the tosses result in a tail.

78. *Political Orientation* The table describes the political orientation of a class of college freshmen, as determined from a survey of 282,223 freshmen.

Political Orientation	Number of Freshmen (in thousands)
Far left	7.06
Liberal	71.48
Middle of the road	143.5
Conservative	56.51
Far right	3.673

Source: Higher Education Research Institute, UCLA.

(a) Approximate the probability that a randomly selected student from the class is in the conservative group.

(b) Approximate the probability that a randomly selected student from the class is on the far left or the far right politically.

(c) Approximate the probability of a randomly selected student from the class not being politically middle of the road.

11 Test

1. Write the first five terms of each sequence. State whether the sequence is arithmetic, geometric, or neither.

(a) $a_n = (-1)^n(n + 2)$

(b) $a_n = -3\left(\dfrac{1}{2}\right)^n$

(c) $a_1 = 2, a_2 = 3; a_n = a_{n-1} + 2a_{n-2}$, for $n \geq 3$

2. In each sequence described, find a_5.

(a) An arithmetic sequence with $a_1 = 1$ and $a_3 = 25$

(b) A geometric sequence with $a_1 = 81$ and $r = -\dfrac{2}{3}$

3. Find the sum of the first 10 terms of each sequence described.

(a) Arithmetic, with $a_1 = -43$ and $d = 12$

(b) Geometric, with $a_1 = 5$ and $r = -2$

4. Evaluate each sum that exists.

(a) $\displaystyle\sum_{i=1}^{30} (5i + 2)$

(b) $\displaystyle\sum_{i=1}^{5} (-3 \cdot 2^i)$

(c) $\displaystyle\sum_{i=1}^{\infty} (2^i) \cdot 4$

(d) $\displaystyle\sum_{i=1}^{\infty} 54\left(\dfrac{2}{9}\right)^i$

5. Evaluate the following.

(a) ` 10 nCr 2 `

(b) $\dbinom{7}{3}$

(c) $7!$

(d) $P(11, 3)$

6. (a) Use the binomial theorem to expand $(2x - 3y)^4$.

(b) Find the third term in the expansion of $(w - 2y)^6$.

7. Use mathematical induction to prove that, for all positive integers n,

$$8 + 14 + 20 + 26 + \cdots + (6n + 2) = 3n^2 + 5n.$$

Solve each problem.

8. *Athletic Shoe Choices* A sports-shoe manufacturer makes athletic shoes in 4 different styles. Each style comes in 3 different colors, and each color comes in 2 different shades. How many different types of shoes can be made?

9. *Club Officer Choices* A club with 20 members plans to elect a president, a secretary, and a treasurer from its membership. If a member can hold at most one office, in how many ways can the three offices be filled?

10. *Club Officer Choices* Refer to **Exercise 9.** If there are 8 men and 12 women in the club, in how many ways can 2 men and 3 women be chosen to attend a conference?

11. *Drawing a Card* A single card is drawn from a standard deck of 52 cards.

(a) Find the probability of drawing a red 3.

(b) Find the probability of drawing a card that is not a face card.

(c) Find the probability of drawing a king or a spade.

(d) What are the odds in favor of drawing a face card?

12. *Rolling a Die* An experiment consists of rolling a die eight times. Find the probability of each event.

(a) Exactly three rolls result in a 4.

(b) All eight rolls result in a 6.

In this reference, we present a review of topics that are usually studied in beginning and intermediate algebra courses. You may wish to refer to the various sections of this reference from time to time, if you need to refresh your memory on these basic concepts.

R Reference: Basic Algebraic Concepts

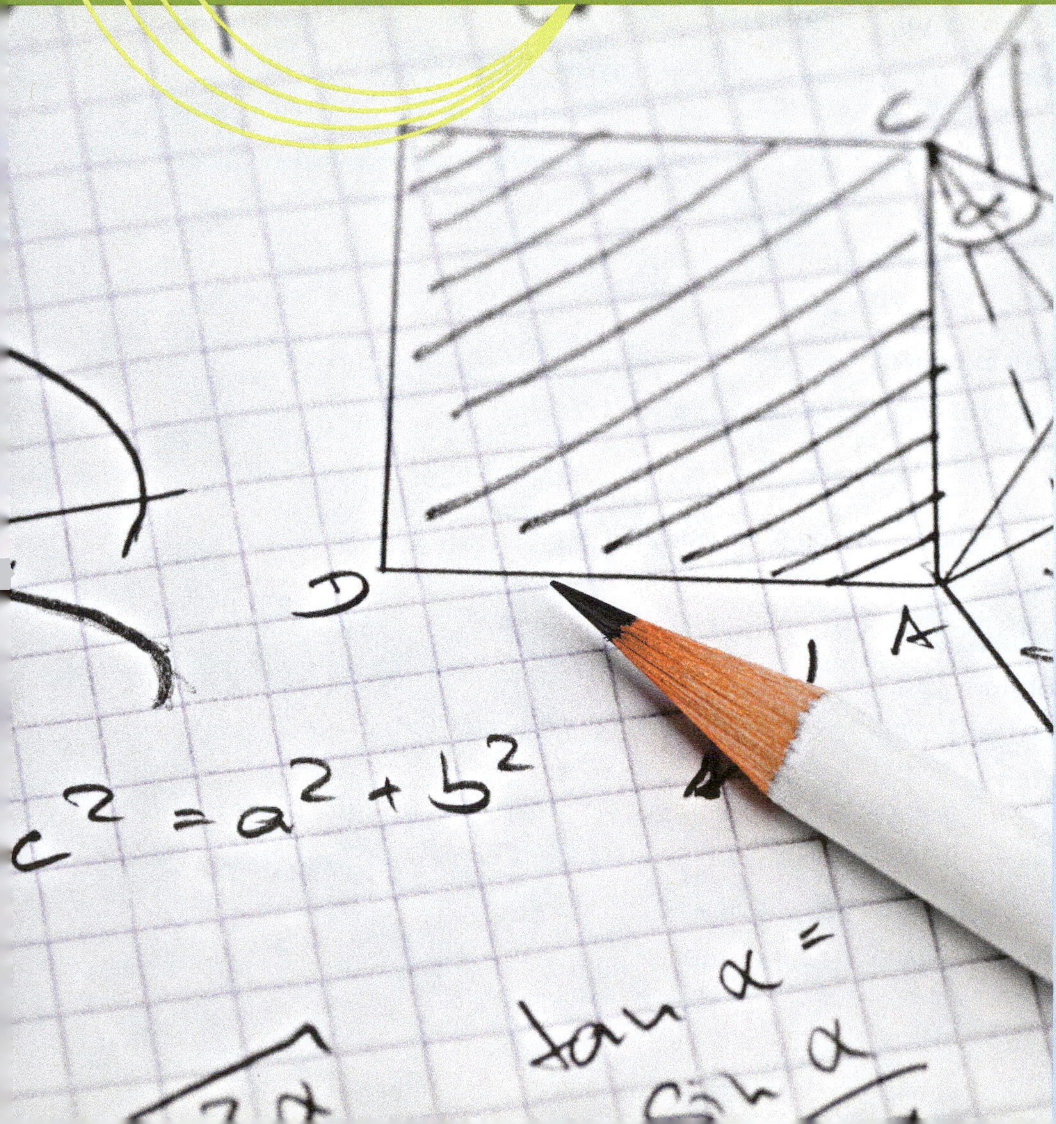

$$c^2 = a^2 + b^2$$

$$\tan \alpha = \frac{\sin \alpha}{\cos \alpha}$$

R.1 Review of Exponents and Polynomials

Rules for Exponents • Terminology for Polynomials • Adding and Subtracting Polynomials • Multiplying Polynomials

Rules for Exponents

By definition, the notation a^m (where m is a positive integer and a is a real number) means that a appears as a factor m times. In the same way, a^n (where n is a positive integer) means that a appears as a factor n times. In the product $a^m \cdot a^n$, the number a would appear $m + n$ times.

Product Rule

For all positive integers m and n and every real number a,

$$a^m \cdot a^n = a^{m+n}.$$

EXAMPLE 1 **Using the Product Rule**

Find each product.

(a) $y^4 \cdot y^7$ **(b)** $(6z^5)(9z^3)(2z^2)$

Solution   *Add exponents when bases are alike.*

(a) $y^4 \cdot y^7 = y^{4+7} = y^{11}$ Keep the same base and add the exponents.

(b) $(6z^5)(9z^3)(2z^2) = (6 \cdot 9 \cdot 2)(z^5 z^3 z^2)$ Commutative and associative properties

$\qquad\qquad\qquad\qquad = 108z^{5+3+2}$ Product rule

$\qquad\qquad\qquad\qquad = 108z^{10}$ Add.

Zero Exponent

For any nonzero real number a, $a^0 = 1$.

NOTE   The expression 0^0 *is undefined.*

EXAMPLE 2 **Using the Definition of a^0**

Evaluate each power.

(a) 4^0 **(b)** $(-4)^0$ **(c)** -4^0 **(d)** $-(-4)^0$ **(e)** $(7r)^0, \quad r \neq 0$

Solution

(a) $4^0 = 1$ Base is 4.   **(b)** $(-4)^0 = 1$ Base is -4.

(c) $-4^0 = -(4^0) = -1$ Base is 4.   **(d)** $-(-4)^0 = -(1) = -1$ Base is -4.

(e) $(7r)^0 = 1, \quad r \neq 0$ Base is $7r$.

The expression $(2^5)^3$ can be written as

$$(2^5)^3 = 2^5 \cdot 2^5 \cdot 2^5.$$

By a generalization of the product rule for exponents, this product is

$$(2^5)^3 = 2^{5+5+5} = 2^{15}.$$

The same exponent could have been obtained by multiplying 5 and 3. This example suggests the first of the power rules given here. The others are found in a similar way.

Power Rules

For all positive integers m and n and all real numbers a and b,

$$(a^m)^n = a^{mn}, \qquad (ab)^m = a^m b^m, \qquad \left(\frac{a}{b}\right)^m = \frac{a^m}{b^m} \quad (b \neq 0).$$

EXAMPLE 3 **Using the Power Rules**

Simplify each expression.

(a) $(5^3)^2$ **(b)** $(3^4 x^2)^3$ **(c)** $\left(\dfrac{2^5}{b^4}\right)^3$, $b \neq 0$

Solution

(a) $(5^3)^2 = 5^{3(2)} = 5^6$ **(b)** $(3^4 x^2)^3 = (3^4)^3 (x^2)^3 = 3^{4(3)} x^{2(3)} = 3^{12} x^6$

(c) $\left(\dfrac{2^5}{b^4}\right)^3 = \dfrac{(2^5)^3}{(b^4)^3} = \dfrac{2^{15}}{b^{12}}$, $b \neq 0$

CAUTION Do not confuse expressions like mn^2 and $(mn)^2$, which are *not* equal. The second power rule can be used only with the second expression: $(mn)^2 = m^2 n^2$.

Terminology for Polynomials

An **algebraic expression** results from adding, subtracting, multiplying, dividing (except by 0), or finding roots or powers of any combination of variables and constants.

$$2x^2 - 3x, \quad \frac{5y}{2y - 3}, \quad \sqrt{m^3 - 8}, \quad \text{and} \quad (3a + b)^4 \qquad \text{Algebraic expressions}$$

The product of a real number and one or more variables raised to powers is called a **term.** The real number is called the **numerical coefficient,** or just the **coefficient.** See the first table in the margin. **Like terms** are terms with the same variables, each raised to the same powers.

$$-13x^3, \qquad 4x^3, \qquad -x^3 \qquad \text{Like terms}$$
$$6y, \qquad 6z, \qquad 4y^3 \qquad \text{Unlike terms}$$

A **polynomial** is defined as a term or a finite sum of terms, with only positive or zero integer exponents permitted on the variables.

$$5x^3 - 8x^2 + 7x - 4, \qquad 9p^5 - 3, \qquad 8r^2, \qquad 6 \qquad \text{Polynomials}$$

The terms of a polynomial cannot have variables in a denominator, such as $\frac{6}{x}$.

The **degree of a term** with one variable is the exponent on the variable. See the second table in the margin. The greatest degree of any term in a polynomial is called the **degree of the polynomial.** For example,

$$4x^3 - 2x^2 - 3x + 7 \text{ has degree 3,}$$

because the greatest degree of any term is 3. (The polynomial 0 has undefined degree.)

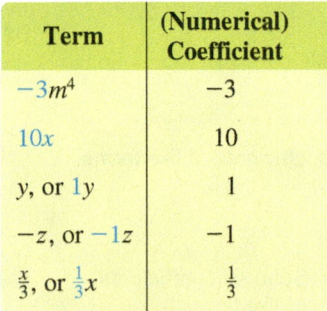

Term	(Numerical) Coefficient
$-3m^4$	-3
$10x$	10
y, or $1y$	1
$-z$, or $-1z$	-1
$\frac{x}{3}$, or $\frac{1}{3}x$	$\frac{1}{3}$

Term	Degree
$2x^3$	3
$-x^4$	4
$17x$, or $17x^1$	1
-6, or $-6x^0$	0

A term containing more than one variable has degree equal to the sum of all the exponents appearing on all the variables in the term. For example, $-3x^4y^3z^5$ has degree $4 + 3 + 5 = 12$. The degree of a polynomial in more than one variable is equal to the greatest degree of any term appearing in the polynomial. The polynomial

$$2x^4y^3 - 3x^5y + x^6y^2 \text{ has degree } 6 + 2 = 8.$$

A polynomial containing exactly three terms is called a **trinomial.** One containing exactly two terms is a **binomial,** and a single-term polynomial is called a **monomial.** The polynomials in the following table illustrate these concepts.

Polynomial	Degree	Type
$9p^7 - 4p^3 + 8p^2$	7	Trinomial
$29x^{11} + 8x^{15}$	15	Binomial
$-10r^6s^8$	$6 + 8 = 14$	Monomial
$5a^3b^7 - 3a^5b^5 + 4a^2b^9 - a^{10}$	$2 + 9 = 11$	None of these

Adding and Subtracting Polynomials

Since the variables used in polynomials represent real numbers, polynomials represent real numbers. This means that all properties of real numbers hold for polynomials. In particular, the distributive property holds.

$$3m^5 - 7m^5 = (3 - 7)m^5$$
$$= -4m^5$$

Thus, polynomials are added by adding coefficients of like terms. Polynomials are subtracted by subtracting coefficients of like terms.

EXAMPLE 4 **Adding and Subtracting Polynomials**

Add or subtract, as indicated.

(a) $(2y^4 - 3y^2 + y) + (4y^4 + 7y^2 + 6y)$

(b) $(-3m^3 - 8m^2 + 4) - (m^3 + 7m^2 - 3)$

(c) $8m^4p^5 - 9m^3p^5 + (11m^4p^5 + 15m^3p^5)$

(d) $4(x^2 - 3x + 7) - 5(2x^2 - 8x - 4)$

Solution

(a) $(2y^4 - 3y^2 + y) + (4y^4 + 7y^2 + 6y)$

$\quad = (2 + 4)y^4 + (-3 + 7)y^2 + (1 + 6)y$ Add coefficients of like terms.

$\quad = 6y^4 + 4y^2 + 7y$

(b) $(-3m^3 - 8m^2 + 4) - (m^3 + 7m^2 - 3)$

$\quad = (-3 - 1)m^3 + (-8 - 7)m^2 + [4 - (-3)]$ Subtract coefficients of like terms.

$\quad = -4m^3 - 15m^2 + 7$

(c) $8m^4p^5 - 9m^3p^5 + (11m^4p^5 + 15m^3p^5) = 19m^4p^5 + 6m^3p^5$

(d) $4(x^2 - 3x + 7) - 5(2x^2 - 8x - 4)$

$\quad = 4x^2 - 4(3x) + 4(7) - 5(2x^2) - 5(-8x) - 5(-4)$ Distributive property

$\quad = 4x^2 - 12x + 28 - 10x^2 + 40x + 20$ Multiply.

$\quad = -6x^2 + 28x + 48$ Add like terms. ●

As shown in **Example 4(a), (b),** and **(d)** on the preceding page, polynomials in one variable are often written with their terms in **descending order** by degree, so the term of greatest degree is first, the one with the next greatest degree is second, and so on.

Multiplying Polynomials

Several properties are used to find the product of two polynomials.

Treat $3x - 4$ as a single expression.

$$(3x - 4)(2x^2 - 3x + 5) = (3x - 4)(2x^2) + (3x - 4)(-3x) + (3x - 4)(5)$$
$$= 3x(2x^2) - 4(2x^2) + 3x(-3x) - 4(-3x) + 3x(5) - 4(5)$$
$$= 6x^3 - 8x^2 - 9x^2 + 12x + 15x - 20$$
$$= 6x^3 - 17x^2 + 27x - 20$$

Sometimes it is more convenient to write such a product vertically.

Place like terms in the same column.

$$
\begin{array}{r}
2x^2 - 3x + 5 \\
3x - 4 \\
\hline
-8x^2 + 12x - 20 \quad \leftarrow -4(2x^2 - 3x + 5) \\
6x^3 - 9x^2 + 15x \qquad\quad \leftarrow 3x(2x^2 - 3x + 5) \\
\hline
6x^3 - 17x^2 + 27x - 20 \quad \text{Add in columns.}
\end{array}
$$

EXAMPLE 5   **Multiplying Polynomials**

Multiply $(3p^2 - 4p + 1)(p^3 + 2p - 8)$.

Solution

$$
\begin{array}{r}
3p^2 - 4p + 1 \\
p^3 + 2p - 8 \\
\hline
-24p^2 + 32p - 8 \quad \leftarrow -8(3p^2 - 4p + 1) \\
6p^3 - 8p^2 + 2p \qquad\qquad \leftarrow 2p(3p^2 - 4p + 1) \\
3p^5 - 4p^4 + p^3 \qquad\qquad\qquad \leftarrow p^3(3p^2 - 4p + 1) \\
\hline
3p^5 - 4p^4 + 7p^3 - 32p^2 + 34p - 8 \quad \text{Add in columns.}
\end{array}
$$

The **FOIL method** (representing **F**irst, **O**uter, **I**nner, **L**ast) is a convenient way to find the product of two binomials.

EXAMPLE 6   **Using FOIL to Multiply Two Binomials**

Find each product.

(a) $(6m + 1)(4m - 3)$ **(b)** $(2x + 7)(2x - 7)$ **(c)** $r^2(3r + 2)(3r - 2)$

Solution

Multiply these pairs of terms.

    First   Outer   Inner   Last

(a) $(6m + 1)(4m - 3) = 6m(4m) + 6m(-3) + 1(4m) + 1(-3)$
$$= 24m^2 - 14m - 3 \qquad -18m + 4m = -14m$$

(b) $(2x + 7)(2x - 7) = 4x^2 - 14x + 14x - 49 \qquad \text{FOIL}$
$$= 4x^2 - 49 \qquad\qquad\qquad \text{Combine like terms.}$$

(c) $r^2(3r + 2)(3r - 2) = r^2(9r^2 - 6r + 6r - 4) \qquad \text{FOIL}$
$$= r^2(9r^2 - 4) \qquad\qquad \text{Combine like terms.}$$
$$= 9r^4 - 4r^2 \qquad\qquad\quad \text{Distributive property}$$

In **Example 6(a),** the product of two binomials is a trinomial, while in **Example 6(b),** the product of two binomials is a binomial. The product of two binomials of the forms $x + y$ and $x - y$ is always a binomial. The squares of binomials, of the forms $(x + y)^2$ and $(x - y)^2$, are also special products that are trinomials.

Special Products

Product of the Sum and Difference of Two Terms

$$(x + y)(x - y) = x^2 - y^2$$

Square of a Binomial

$$(x + y)^2 = x^2 + 2xy + y^2$$

$$(x - y)^2 = x^2 - 2xy + y^2$$

EXAMPLE 7   **Using the Special Products**

Find each product.

(a) $(3p + 11)(3p - 11)$ **(b)** $(5m^3 - 3)(5m^3 + 3)$

(c) $(9k - 11r^3)(9k + 11r^3)$ **(d)** $(2m + 5)^2$

(e) $(3x - 7y^4)^2$

Solution

(a) $(3p + 11)(3p - 11) = (3p)^2 - 11^2$ $(x + y)(x - y) = x^2 - y^2$
$$= 9p^2 - 121$$ Square terms.

(b) $(5m^3 - 3)(5m^3 + 3) = (5m^3)^2 - 3^2$ $(x - y)(x + y) = x^2 - y^2$
$$= 25m^6 - 9$$ Square terms.

(c) $(9k - 11r^3)(9k + 11r^3) = (9k)^2 - (11r^3)^2$ $(x - y)(x + y) = x^2 - y^2$
$$= 81k^2 - 121r^6$$ Square terms.

Remember the middle term.

(d) $(2m + 5)^2 = (2m)^2 + 2(2m)(5) + 5^2$ $(x + y)^2 = x^2 + 2xy + y^2$
$$= 4m^2 + 20m + 25$$ Multiply.

(e) $(3x - 7y^4)^2 = (3x)^2 - 2(3x)(7y^4) + (7y^4)^2$ $(x - y)^2 = x^2 - 2xy + y^2$
$$= 9x^2 - 42xy^4 + 49y^8$$ Multiply.

CAUTION   As shown in **Examples 7(d)** and **(e),** the square of a binomial has *three* terms. Do not give $x^2 + y^2$ as the result of expanding $(x + y)^2$.

$$(x + y)^2 = x^2 + 2xy + y^2$$
$$(x - y)^2 = x^2 - 2xy + y^2$$

Remember the middle term.

R.1   Exercises

Simplify each expression. Leave answers with exponents.

1. $(-4)^3 \cdot (-4)^2$

2. $(-5)^2 \cdot (-5)^6$

3. 2^0

4. -2^0

5. $(5m)^0$, $m \neq 0$

6. $(-4z)^0$, $z \neq 0$

7. $(2^2)^5$

8. $(6^4)^3$

9. $(2x^5y^4)^3$

10. $(-4m^3n^9)^2$

11. $-\left(\dfrac{p^4}{q}\right)^2$

12. $\left(\dfrac{r^8}{s^2}\right)^3$

Concept Check   *Identify each expression as a polynomial or not a polynomial. For each polynomial, give the degree and identify it as a monomial, binomial, trinomial, or none of these.*

13. $-5x^{11}$

14. $9y^{12} + y^2$

15. $18p^6q + 6pq$

16. $2a^6 + 5a^2 + 4a$

17. $\sqrt{2}x^2 + \sqrt{3}x^6$

18. $-\sqrt{7}m^5n^2 + 2\sqrt{3}m^3n^2$

19. $\dfrac{1}{3}r^2s^2 - \dfrac{3}{5}r^4s^2 + rs^3$

20. $\dfrac{5}{p} + \dfrac{2}{p^2} + \dfrac{5}{p^3}$

21. $-5\sqrt{z} + 2\sqrt{z^3} - 5\sqrt{z^5}$

Find each sum or difference.

22. $(3x^2 - 4x + 5) + (-2x^2 + 3x - 2)$

23. $(4m^3 - 3m^2 + 5) + (-3m^3 - m^2 + 5)$

24. $(12y^2 - 8y + 6) - (3y^2 - 4y + 2)$

25. $(8p^2 - 5p) - (3p^2 - 2p + 4)$

26. $(6m^4 - 3m^2 + m) - (2m^3 + 5m^2 + 4m) + (m^2 - m)$

27. $-(8x^3 + x - 3) + (2x^3 + x^2) - (4x^2 + 3x - 1)$

Find each product.

28. $(4r - 1)(7r + 2)$

29. $(5m - 6)(3m + 4)$

30. $\left(3x - \dfrac{2}{3}\right)\left(5x + \dfrac{1}{3}\right)$

31. $\left(2m - \dfrac{1}{4}\right)\left(3m + \dfrac{1}{2}\right)$

32. $4x^2(3x^3 + 2x^2 - 5x + 1)$

33. $2b^3(b^2 - 4b + 3)$

34. $(2z - 1)(-z^2 + 3z - 4)$

35. $(m - n + k)(m + 2n - 3k)$

36. $(r - 3s + t)(2r - s + t)$

37. In words, state the formula for the square of a binomial.

38. In words, state the formula for the product of the sum and difference of two terms.

Find each product.

39. $(2m + 3)(2m - 3)$

40. $(8s - 3t)(8s + 3t)$

41. $(4m + 2n)^2$

42. $(a - 6b)^2$

43. $(5r + 3t^2)^2$

44. $(2z^4 - 3y)^2$

45. $[(2p - 3) + q]^2$

46. $[(4y - 1) + z]^2$

47. $[(3q + 5) - p][(3q + 5) + p]$

48. $[(9r - s) + 2][(9r - s) - 2]$

49. $[(3a + b) - 1]^2$

50. $[(2m + 7) - n]^2$

Perform the indicated operations.

51. $(6p + 5q)(3p - 7q)$

52. $(2p - 1)(3p^2 - 4p + 5)$

53. $(p^3 - 4p^2 + p) - (3p^2 + 2p + 7)$

54. $(6k - 3)^2$

55. $y(4x + 3y)(4x - 3y)$

56. $(r^5 - r^3 + r) + (3r^5 - 4r^4 + r^3 + 2r)$

57. $(2z + y)(3z - 4y)$

58. $(7m + 2n)(7m - 2n)$

59. $(3p + 5)^2$

60. $2(3r^2 + 4r + 2) - 3(-r^2 + 4r - 5)$

61. $p(4p - 6) + 2(3p - 8)$

62. $m(5m - 2) + 9(5 - m)$

63. $-y(y^2 - 4) + 6y^2(2y - 3)$

64. $-z^3(9 - z) + 4z(2 + 3z)$

R.2 Review of Factoring

Factoring Out the Greatest Common Factor • Factoring by Grouping • Factoring Trinomials • Factoring Special Products
• Factoring by Substitution

The process of finding polynomials whose product equals a given polynomial is called **factoring.** For example, since

$$4x + 12 = 4(x + 3),$$

both 4 and $x + 3$ are called **factors** of $4x + 12$. Also, $4(x + 3)$ is called a **factored form** of $4x + 12$. A polynomial with integer coefficients that cannot be written as a product of two polynomials with integer coefficients is a **prime** or **irreducible polynomial.** A polynomial is **factored completely** when it is written as a product of prime polynomials with integer coefficients.

Factoring Out the Greatest Common Factor

EXAMPLE 1 Factoring Out the Greatest Common Factor

Factor out the greatest common factor from each polynomial.
(a) $9y^5 + y^2$ **(b)** $6x^2y^3 + 9xy^4 + 18y^5$

(c) $14m^4(m + 1) - 28m^3(m + 1) - 7m^2(m + 1)$

Solution   We use the distributive property in each case.
(a) $9y^5 + y^2 = y^2(9y^3) + y^2(1)$ GCF $= y^2$

$\qquad\qquad = y^2(9y^3 + 1)$ Distributive property

Remember to include the 1.

(b) $6x^2y^3 + 9xy^4 + 18y^5 = 3y^3(2x^2) + 3y^3(3xy) + 3y^3(6y^2)$ GCF $= 3y^3$

$\qquad\qquad\qquad\qquad = 3y^3(2x^2 + 3xy + 6y^2)$ Distributive property

(c) The GCF is $7m^2(m + 1)$. Use the distributive property as follows.

$14m^4(m + 1) - 28m^3(m + 1) - 7m^2(m + 1) = 7m^2(m + 1)(2m^2 - 4m - 1)$ ●

NOTE   *Factoring can always be checked by multiplying.*

Factoring by Grouping

When a polynomial has more than three terms, it can sometimes be factored using **factoring by grouping.**

$$\begin{array}{c} \overbrace{\qquad\qquad}^{\text{Terms with common factor } a} \quad \overbrace{\qquad\qquad}^{\text{Terms with common factor } 6} \end{array}$$

$ax + ay + 6x + 6y = (ax + ay) + (6x + 6y)$ Group the terms.

$\qquad\qquad\qquad = a(x + y) + 6(x + y)$ Factor each group.

$\qquad\qquad\qquad = (x + y)(a + 6)$ Factor out $x + y$.

It is not always obvious which terms should be grouped. Experience and repeated trials are the most reliable tools when factoring by grouping.

EXAMPLE 2   **Factoring by Grouping**

Factor each polynomial by grouping.

(a) $mp^2 + 7m + 3p^2 + 21$ **(b)** $2y^2 + az - 2z - ay^2$

(c) $4x^3 + 2x^2 - 2x - 1$

Solution

(a) $mp^2 + 7m + 3p^2 + 21 = (mp^2 + 7m) + (3p^2 + 21)$ Group the terms.

$$= m(p^2 + 7) + 3(p^2 + 7)$$ Factor each group.
$p^2 + 7$ is a
common factor.

$$= (p^2 + 7)(m + 3)$$

Check: $(p^2 + 7)(m + 3) = mp^2 + 3p^2 + 7m + 21$ FOIL

$$= mp^2 + 7m + 3p^2 + 21$$ Commutative property

(b) $2y^2 + az - 2z - ay^2 = 2y^2 - 2z - ay^2 + az$ Rearrange the terms.

$$= (2y^2 - 2z) + (-ay^2 + az)$$ Group the terms.

$$= 2(y^2 - z) + a(-y^2 + z)$$ Factor each group.

$$= 2(y^2 - z) - a(y^2 - z)$$ Factor out $-a$
instead of a.

$$= (y^2 - z)(2 - a)$$ Factor out $y^2 - z$.

(c) $4x^3 + 2x^2 - 2x - 1 = 2x^2(2x + 1) - 1(2x + 1)$ Factor each group.

$$= (2x + 1)(2x^2 - 1)$$ Factor out $2x + 1$.   ●

Factoring Trinomials

As shown here, factoring is the opposite of multiplication.

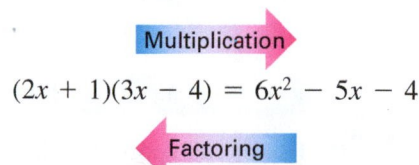

$$(2x + 1)(3x - 4) = 6x^2 - 5x - 4$$

Since the product of two binomials is usually a trinomial, we can expect factorable trinomials (that have terms with no common factor) to have two binomial factors. Thus, factoring trinomials requires using FOIL in reverse.

EXAMPLE 3   **Factoring Trinomials**

Factor each trinomial.

(a) $4y^2 - 11y + 6$ **(b)** $6p^2 - 7p - 5$ **(c)** $16y^3 + 24y^2 - 16y$

Solution

(a) To factor this polynomial, we must find integers a, b, c, and d such that

$$4y^2 - 11y + 6 = (ay + b)(cy + d).$$ FOIL

Then $ac = 4$ and $bd = 6$. The positive factors of 4 are 4 and 1 or 2 and 2. Since the middle term has a negative coefficient, we consider only negative factors of 6. The possibilities are -2 and -3 or -1 and -6. We try various arrangements of these factors until we find one that gives the correct coefficient of y.

$$(2y - 1)(2y - 6) = 4y^2 - 14y + 6$$ Incorrect

$$(2y - 2)(2y - 3) = 4y^2 - 10y + 6$$ Incorrect

$$(y - 2)(4y - 3) = 4y^2 - 11y + 6$$ Correct

(continued)

(b) To factor $6p^2 - 7p - 5$, we try various possibilities. The positive factors of 6 could be 2 and 3 or 1 and 6. As factors of -5, we have -1 and 5 or -5 and 1.

$$(2p + 5)(3p - 1) = 6p^2 + 13p - 5 \qquad \text{Incorrect}$$
$$(2p - 5)(3p + 1) = 6p^2 - 13p - 5 \qquad \text{Incorrect}$$
$$(3p - 5)(2p + 1) = 6p^2 - 7p - 5 \qquad \text{Correct}$$

Thus, $6p^2 - 7p - 5$ factors as $(3p - 5)(2p + 1)$.

(c) $16y^3 + 24y^2 - 16y = 8y(2y^2 + 3y - 2) \qquad$ First factor out the GCF, 8y.

$$= 8y(2y - 1)(y + 2) \qquad \text{Factor the trinomial.}$$

Factoring Special Products

Each of the special patterns for multiplication given in **Section R.1** can be used in reverse to get a pattern for factoring.

Perfect Square Trinomials

$$x^2 + 2xy + y^2 = (x + y)^2$$
$$x^2 - 2xy + y^2 = (x - y)^2$$

EXAMPLE 4 **Factoring Perfect Square Trinomials**

Factor each polynomial.
(a) $16p^2 - 40pq + 25q^2$ **(b)** $169x^2 + 104xy^2 + 16y^4$

Solution

(a) Since $16p^2 = (4p)^2$ and $25q^2 = (5q)^2$, use the second equation shown in the preceding box, with $4p$ replacing x and $5q$ replacing y.

$$16p^2 - 40pq + 25q^2 = (4p)^2 - 2(4p)(5q) + (5q)^2$$
$$= (4p - 5q)^2 \qquad \text{Factored form}$$

Make sure that the middle term of the trinomial being factored, $-40pq$ here, is twice the product of the two terms in the binomial $4p - 5q$.

$$-40pq = 2(4p)(-5q)$$

(b) $169x^2 + 104xy^2 + 16y^4 = (13x + 4y^2)^2$, since $2(13x)(4y^2) = 104xy^2$.

Difference of Squares

$$x^2 - y^2 = (x + y)(x - y)$$

EXAMPLE 5   **Factoring Differences of Squares**

Factor each polynomial.

(a) $4m^2 - 9$ **(b)** $256k^4 - 625m^4$

(c) $(a + 2b)^2 - 4c^2$ **(d)** $x^2 - 6x + 9 - y^4$

Solution

(a) $4m^2 - 9 = (2m)^2 - 3^2$ Write as a difference of squares.

$= (2m + 3)(2m - 3)$ Factor.

(b) $256k^4 - 625m^4 = (16k^2)^2 - (25m^2)^2$ Write as a difference of squares.

$= (16k^2 + 25m^2)(16k^2 - 25m^2)$ Factor.

Be sure to factor completely. $= (16k^2 + 25m^2)(4k + 5m)(4k - 5m)$ Factor again.

(c) $(a + 2b)^2 - 4c^2 = (a + 2b)^2 - (2c)^2$ Write as a difference of squares.

$= [(a + 2b) + 2c][(a + 2b) - 2c]$ Factor.

$= (a + 2b + 2c)(a + 2b - 2c)$

(d) $x^2 - 6x + 9 - y^4 = (x^2 - 6x + 9) - y^4$ Group terms.

$= (x - 3)^2 - (y^2)^2$ Write as a difference of squares.

$= [(x - 3) + y^2][(x - 3) - y^2]$ Factor.

$= (x - 3 + y^2)(x - 3 - y^2)$

Difference and Sum of Cubes

Difference of cubes $x^3 - y^3 = (x - y)(x^2 + xy + y^2)$

Sum of cubes $x^3 + y^3 = (x + y)(x^2 - xy + y^2)$

EXAMPLE 6   **Factoring Sums and Differences of Cubes**

Factor each polynomial.

(a) $x^3 + 27$ **(b)** $m^3 - 64n^3$ **(c)** $8q^6 + 125p^9$

Solution

(a) $x^3 + 27 = x^3 + 3^3$ Write as a sum of cubes.

$= (x + 3)(x^2 - 3x + 3^2)$ Factor.

$= (x + 3)(x^2 - 3x + 9)$

(b) $m^3 - 64n^3 = m^3 - (4n)^3$ Write as a difference of cubes.

$= (m - 4n)[m^2 + m(4n) + (4n)^2]$ Factor.

$= (m - 4n)(m^2 + 4mn + 16n^2)$ Multiply.

(c) $8q^6 + 125p^9 = (2q^2)^3 + (5p^3)^3$ Write as a sum of cubes.

$= (2q^2 + 5p^3)[(2q^2)^2 - (2q^2)(5p^3) + (5p^3)^2]$ Factor.

$= (2q^2 + 5p^3)(4q^4 - 10q^2p^3 + 25p^6)$

Factoring by Substitution

Sometimes a polynomial can be factored by substituting one expression for another. Any variable *except* the original one can be used in the substitution.

EXAMPLE 7 Factoring by Substitution

Factor each polynomial.

(a) $6z^4 - 13z^2 - 5$ **(b)** $10(2a - 1)^2 - 19(2a - 1) - 15$ **(c)** $(2a - 1)^3 + 8$

Solution

(a) Replace z^2 with y, so $y^2 = (z^2)^2 = z^4$.

$$6z^4 - 13z^2 - 5 = 6y^2 - 13y - 5$$

Remember to make the final substitution.

$$= (2y - 5)(3y + 1) \qquad \text{Use FOIL to factor.}$$
$$= (2z^2 - 5)(3z^2 + 1) \qquad \text{Replace } y \text{ with } z^2.$$

This type of trinomial can also be factored directly without using substitution.

(b) $10(2a - 1)^2 - 19(2a - 1) - 15$

$$= 10m^2 - 19m - 15 \qquad\qquad \text{Replace } 2a - 1 \text{ with } m.$$
$$= (5m + 3)(2m - 5) \qquad\qquad \text{Factor.}$$
$$= [5(2a - 1) + 3][2(2a - 1) - 5] \qquad \text{Let } m = 2a - 1.$$
$$= (10a - 5 + 3)(4a - 2 - 5) \qquad\quad \text{Distributive property}$$
$$= (10a - 2)(4a - 7) \qquad\qquad\quad \text{Add.}$$
$$= 2(5a - 1)(4a - 7) \qquad\qquad\quad \text{Factor out the common factor.}$$

(c) $(2a - 1)^3 + 8 = t^3 + 8 \qquad\qquad\qquad \text{Let } 2a - 1 = t.$

$$= t^3 + 2^3 \qquad\qquad\qquad\qquad \text{Write as a sum of cubes.}$$
$$= (t + 2)(t^2 - 2t + 4) \qquad\qquad \text{Factor.}$$
$$= [(2a - 1) + 2][(2a - 1)^2 - 2(2a - 1) + 4]$$
$$\qquad\qquad\qquad\qquad\qquad \text{Let } t = 2a - 1.$$
$$= (2a + 1)(4a^2 - 4a + 1 - 4a + 2 + 4)$$
$$\qquad\qquad\qquad\qquad\qquad \text{Add and then multiply.}$$
$$= (2a + 1)(4a^2 - 8a + 7) \qquad \text{Combine like terms.}$$

R.2 Exercises

1. *Concept Check* Match each polynomial in Column I with its factored form in Column II.

I	II
(a) $x^2 + 10xy + 25y^2$	**A.** $(x + 5y)(x - 5y)$
(b) $x^2 - 10xy + 25y^2$	**B.** $(x + 5y)^2$
(c) $x^2 - 25y^2$	**C.** $(x - 5y)^2$
(d) $25y^2 - x^2$	**D.** $(5y + x)(5y - x)$

2. *Concept Check* Match each polynomial in Column I with its factored form in Column II.

I	II
(a) $8x^3 - 27$	**A.** $(3 - 2x)(9 + 6x + 4x^2)$
(b) $8x^3 + 27$	**B.** $(2x - 3)(4x^2 + 6x + 9)$
(c) $27 - 8x^3$	**C.** $(2x + 3)(4x^2 - 6x + 9)$

Factor the greatest common factor from each polynomial.

3. $4k^2m^3 + 8k^4m^3 - 12k^2m^4$

4. $28r^4s^2 + 7r^3s - 35r^4s^3$

5. $2(a + b) + 4m(a + b)$

6. $4(y - 2)^2 + 3(y - 2)$

7. $(2y - 3)(y + 2) + (y + 5)(y + 2)$

8. $(6a - 1)(a + 2) + (6a - 1)(3a - 1)$

9. $(5r - 6)(r + 3) - (2r - 1)(r + 3)$

10. $(3z + 2)(z + 4) - (z + 6)(z + 4)$

11. $2(m - 1) - 3(m - 1)^2 + 2(m - 1)^3$

12. $5(a + 3)^3 - 2(a + 3) + (a + 3)^2$

Factor each polynomial by grouping.

13. $6st + 9t - 10s - 15$

14. $10ab - 6b + 35a - 21$

15. $10x^2 - 12y + 15x - 8xy$

16. $2m^4 + 6 - am^4 - 3a$

17. $t^3 + 2t^2 - 3t - 6$

18. $x^3 + 3x^2 - 5x - 15$

19. *Concept Check* Layla factored $16a^2 - 40a - 6a + 15$ by grouping and got an answer of $(8a - 3)(2a - 5)$. Jamal factored the same polynomial and obtained $(3 - 8a)(5 - 2a)$. Are both of these answers correct? If not, why not?

Factor each trinomial completely.

20. $6a^2 - 48a - 120$

21. $8h^2 - 24h - 320$

22. $3m^3 + 12m^2 + 9m$

23. $9y^4 - 54y^3 + 45y^2$

24. $6k^2 + 5kp - 6p^2$

25. $14m^2 + 11mr - 15r^2$

26. $5a^2 - 7ab - 6b^2$

27. $12s^2 + 11st - 5t^2$

28. $9x^2 - 6x^3 + x^4$

29. $30a^2 + am - m^2$

30. $24a^4 + 10a^3b - 4a^2b^2$

31. $18x^5 + 15x^4z - 75x^3z^2$

Factor each perfect square trinomial completely.

32. $9m^2 - 12m + 4$

33. $16p^2 - 40p + 25$

34. $32a^2 - 48ab + 18b^2$

35. $20p^2 - 100pq + 125q^2$

36. $4x^2y^2 + 28xy + 49$

37. $9m^2n^2 - 12mn + 4$

38. $(a - 3b)^2 - 6(a - 3b) + 9$

39. $(2p + q)^2 - 10(2p + q) + 25$

40. $(5r + 2s)^2 + 6(5r + 2s) + 9$

Factor each difference of squares completely.

41. $9a^2 - 16$

42. $16q^2 - 25$

43. $25s^4 - 9t^2$

44. $36z^2 - 81y^4$

45. $(a + b)^2 - 16$

46. $(p - 2q)^2 - 100$

47. $p^4 - 625$

48. $m^4 - 81$

49. *Concept Check* Which of the following is the correct complete factorization of $x^4 - 1$?
   A. $(x^2 - 1)(x^2 + 1)$ **B.** $(x^2 + 1)(x + 1)(x - 1)$ **C.** $(x^2 - 1)^2$ **D.** $(x - 1)^2(x + 1)^2$

50. *Concept Check* Which of the following is the correct complete factorization of $x^3 + 8$?
   A. $(x + 2)^3$ **B.** $(x + 2)(x^2 + 2x + 4)$ **C.** $(x + 2)(x^2 - 2x + 4)$ **D.** $(x + 2)(x^2 - 4x + 4)$

Factor each sum or difference of cubes completely.

51. $8 - a^3$

52. $r^3 + 27$

53. $125x^3 - 27$

54. $8m^3 - 27n^3$

55. $27y^9 + 125z^6$

56. $27z^3 + 729y^3$

57. $(r + 6)^3 - 216$

58. $(b + 3)^3 - 27$

59. $27 - (m + 2n)^3$

60. $125 - (4a - b)^3$

61. Is the following factorization of $3a^4 + 14a^2 - 5$ correct? Explain. If it is incorrect, give the correct factors.

$$3a^4 + 14a^2 - 5 = 3u^2 + 14u - 5 \qquad \text{Let } u = a^2.$$
$$= (3u - 1)(u + 5) \qquad \text{Factor.}$$

Completely factor each polynomial by substitution.

62. $m^4 - 3m^2 - 10$

63. $a^4 - 2a^2 - 48$

64. $7(3k - 1)^2 + 26(3k - 1) - 8$

65. $6(4z - 3)^2 + 7(4z - 3) - 3$

66. $9(a - 4)^2 + 30(a - 4) + 25$

67. $20(4 - p)^2 - 3(4 - p) - 2$

Factor by any method.

68. $a^3(r + s) + b^2(r + s)$

69. $4b^2 + 4bc + c^2 - 16$

70. $(2y - 1)^2 - 4(2y - 1) + 4$

71. $x^2 + xy - 5x - 5y$

72. $8r^2 - 3rs + 10s^2$

73. $p^4(m - 2n) + q(m - 2n)$

74. $36a^2 + 60a + 25$

75. $4z^2 + 28z + 49$

76. $6p^4 + 7p^2 - 3$

77. $1000x^3 + 343y^3$

78. $b^2 + 8b + 16 - a^2$

79. $125m^6 - 216$

80. $q^2 + 6q + 9 - p^2$

81. $12m^2 + 16mn - 35n^2$

82. $216p^3 + 125q^3$

83. $4p^2 + 3p - 1$

84. $100r^2 - 169s^2$

85. $144z^2 + 121$

86. $(3a + 5)^2 - 18(3a + 5) + 81$

87. $(4t + 5)^2 + 16(4t + 5) + 64$

88. $4z^4 - 7z^2 - 15$

R.3 Review of Rational Expressions

Domain of a Rational Expression • Lowest Terms of a Rational Expression • Multiplying and Dividing Rational Expressions • Adding and Subtracting Rational Expressions • Complex Fractions

An expression that is the quotient of two polynomials P and Q, with $Q \neq 0$, is called a **rational expression.**

$$\frac{x + 6}{x + 2}, \quad \frac{(x + 6)(x + 4)}{(x + 2)(x + 4)}, \quad \text{and} \quad \frac{2p^2 + 7p - 4}{5p^2 + 20p}$$

Rational expressions

Domain of a Rational Expression

The **domain** of a rational expression is the set of real numbers for which the expression is defined. Because the denominator cannot be 0, the domain consists of all real numbers except those that make the denominator 0. We find these numbers by setting the denominator equal to 0 and solving the resulting equation. For example, in the rational expression

$$\frac{x + 6}{x + 2},$$

the solution of the equation $x + 2 = 0$ is excluded from the domain. Since this solution is -2, the domain is the set of all real numbers x such that $x \neq -2$, written $\{x \mid x \neq -2\}$.

If the denominator of a rational expression contains a product, we find the domain by using the zero-product property (**Section 3.3**), which states that $ab = 0$ if and only if $a = 0$ or $b = 0$. For example, to find the domain of

$$\frac{(x + 6)(x + 4)}{(x + 2)(x + 4)},$$

we solve as follows.

$$(x + 2)(x + 4) = 0 \qquad \text{Set denominator equal to 0.}$$
$$x + 2 = 0 \quad \text{or} \quad x + 4 = 0 \qquad \text{Zero-product property}$$
$$x = -2 \quad \text{or} \qquad x = -4 \qquad \text{Solve each equation.}$$

The domain consists of the set of real numbers x such that $x \neq -2, -4$, written $\{x \mid x \neq -2, -4\}$.

Lowest Terms of a Rational Expression

A rational expression is written in lowest terms when the greatest common factor of its numerator and denominator is 1. We use the fundamental principle of fractions.

> **Fundamental Principle of Fractions**
>
> $$\frac{ac}{bc} = \frac{a}{b} \quad (b \neq 0, c \neq 0)$$

EXAMPLE 1   **Writing Rational Expressions in Lowest Terms**

Write each rational expression in lowest terms.

(a) $\dfrac{2p^2 + 7p - 4}{5p^2 + 20p}$ **(b)** $\dfrac{6 - 3k}{k^2 - 4}$

Solution

(a) $\dfrac{2p^2 + 7p - 4}{5p^2 + 20p} = \dfrac{(2p - 1)(p + 4)}{5p(p + 4)}$ Factor.

$$= \frac{2p - 1}{5p} \qquad \text{Divide out the common factor.}$$

To determine the domain, we find values of p that make the original denominator, $5p^2 + 20p$, equal to 0.

$$5p^2 + 20p = 0 \qquad \text{Set denominator equal to 0.}$$
$$5p(p + 4) = 0 \qquad \text{Factor.}$$
$$5p = 0 \quad \text{or} \quad p + 4 = 0 \qquad \text{Set each factor equal to 0.}$$
$$p = 0 \quad \text{or} \qquad p = -4 \qquad \text{Solve.}$$

The domain is $\{p \mid p \neq 0, -4\}$. *From now on, we will assume such restrictions when writing rational expressions in lowest terms.*

(b) $\dfrac{6 - 3k}{k^2 - 4} = \dfrac{3(2 - k)}{(k + 2)(k - 2)}$ Factor.

$$= \frac{3(2 - k)(-1)}{(k + 2)(k - 2)(-1)} \qquad \begin{array}{l} 2 - k \text{ and } k - 2 \text{ are opposites. Multiply} \\ \text{numerator and denominator by } -1. \end{array}$$

Note that $\dfrac{2 - k}{k - 2} = -1.$

$$= \frac{3(2 - k)(-1)}{(k + 2)(2 - k)} \qquad (k - 2)(-1) = -k + 2 = 2 - k$$

$$= \frac{-3}{k + 2} \qquad \text{Fundamental principle}$$

Working in an alternative way would lead to the equivalent result $\dfrac{3}{-k - 2}$.

CAUTION The fundamental principle requires a pair of common *factors*, one in the numerator and one in the denominator. *Only after a rational expression has been factored can any common factors be divided out.*

$$\frac{2x+4}{6} = \frac{2(x+2)}{2 \cdot 3} = \frac{x+2}{3} \qquad \text{Factor first and then divide.}$$

Multiplying and Dividing Rational Expressions

> ### Multiplying and Dividing Fractions
>
> For fractions $\frac{a}{b}$ and $\frac{c}{d}$ ($b \neq 0, d \neq 0$),
>
> $$\frac{a}{b} \cdot \frac{c}{d} = \frac{ac}{bd} \quad \text{and} \quad \frac{a}{b} \div \frac{c}{d} = \frac{a}{b} \cdot \frac{d}{c}, \quad \text{if} \quad \frac{c}{d} \neq 0.$$

EXAMPLE 2 **Multiplying and Dividing Rational Expressions**

Multiply or divide, as indicated.

(a) $\dfrac{2y^2}{9} \cdot \dfrac{27}{8y^5}$

(b) $\dfrac{3m^2 - 2m - 8}{3m^2 + 14m + 8} \cdot \dfrac{3m + 2}{3m + 4}$

(c) $\dfrac{5}{8m + 16} \div \dfrac{7}{12m + 24}$

(d) $\dfrac{3p^2 + 11p - 4}{24p^3 - 8p^2} \div \dfrac{9p + 36}{24p^4 - 36p^3}$

Solution

(a)
$$\frac{2y^2}{9} \cdot \frac{27}{8y^5} = \frac{2y^2 \cdot 27}{9 \cdot 8y^5} \qquad \text{Multiply fractions.}$$

$$= \frac{2 \cdot 9 \cdot 3 \cdot y^2}{9 \cdot 2 \cdot 4 \cdot y^2 \cdot y^3} \qquad \text{Factor.}$$

$$= \frac{3}{4y^3} \qquad \text{Fundamental principle}$$

(b)
$$\frac{3m^2 - 2m - 8}{3m^2 + 14m + 8} \cdot \frac{3m + 2}{3m + 4} = \frac{(m-2)(3m+4)}{(m+4)(3m+2)} \cdot \frac{3m+2}{3m+4} \qquad \text{Factor.}$$

$$= \frac{(m-2)(3m+4)(3m+2)}{(m+4)(3m+2)(3m+4)} \qquad \text{Multiply fractions.}$$

$$= \frac{m-2}{m+4} \qquad \text{Fundamental principle}$$

(c)
$$\frac{5}{8m + 16} \div \frac{7}{12m + 24} = \frac{5}{8(m+2)} \div \frac{7}{12(m+2)} \qquad \text{Factor denominators.}$$

$$= \frac{5}{8(m+2)} \cdot \frac{12(m+2)}{7} \qquad \text{Multiply by the reciprocal of the divisor.}$$

$$= \frac{5 \cdot 12(m+2)}{8 \cdot 7(m+2)} \qquad \text{Multiply fractions.}$$

$$= \frac{15}{14} \qquad \text{Fundamental principle}$$

(d) $\dfrac{3p^2 + 11p - 4}{24p^3 - 8p^2} \div \dfrac{9p + 36}{24p^4 - 36p^3} = \dfrac{(p + 4)(3p - 1)}{8p^2(3p - 1)} \div \dfrac{9(p + 4)}{12p^3(2p - 3)}$

$$= \dfrac{(p + 4)(3p - 1)(12p^3)(2p - 3)}{8p^2(3p - 1)(9)(p + 4)}$$

$$= \dfrac{12p^3(2p - 3)}{9 \cdot 8p^2}$$

$$= \dfrac{p(2p - 3)}{6}$$

Adding and Subtracting Rational Expressions

> **Adding and Subtracting Fractions**
>
> For fractions $\dfrac{a}{b}$ and $\dfrac{c}{d}$ ($b \neq 0, d \neq 0$),
>
> $$\dfrac{a}{b} + \dfrac{c}{d} = \dfrac{ad + bc}{bd} \quad \text{and} \quad \dfrac{a}{b} - \dfrac{c}{d} = \dfrac{ad - bc}{bd}.$$

In practice, we add or subtract rational expressions after rewriting them with a common denominator, preferably the least common denominator.

> **Finding the Least Common Denominator (LCD)**
>
> **Step 1** Write each denominator as a product of prime factors.
>
> **Step 2** Form a product of all the different prime factors. Each factor should have as exponent the **greatest** exponent that appears on that factor.

> **EXAMPLE 3**   **Adding and Subtracting Rational Expressions**

Add or subtract, as indicated.

(a) $\dfrac{5}{9x^2} + \dfrac{1}{6x}$ **(b)** $\dfrac{y + 2}{y^2 - y} - \dfrac{3y}{2y^2 - 4y + 2}$

(c) $\dfrac{3}{(x - 1)(x + 2)} - \dfrac{1}{(x + 3)(x - 4)}$

Solution

(a) $\dfrac{5}{9x^2} + \dfrac{1}{6x}$

Step 1 Write each denominator as a product of prime factors.

$$9x^2 = 3^2 \cdot x^2$$

$$6x = 2^1 \cdot 3^1 \cdot x^1$$

Step 2 For the LCD, form the product of all the prime factors, with each factor having the greatest exponent that appears on it.

Greatest exponent on 3 is 2.   Greatest exponent on *x* is 2.

$$\text{LCD} = 2^1 \cdot 3^2 \cdot x^2$$

$$= 18x^2$$

(continued)

Write both of the given expressions with this denominator, and then add.

$$\frac{5}{9x^2} + \frac{1}{6x} = \frac{5 \cdot 2}{9x^2 \cdot 2} + \frac{1 \cdot 3x}{6x \cdot 3x} \qquad \text{LCD} = 18x^2$$

$$= \frac{10}{18x^2} + \frac{3x}{18x^2} \qquad \text{Multiply.}$$

$$= \frac{10 + 3x}{18x^2} \qquad \text{Add the numerators.}$$

Always check to see that the answer is in lowest terms.

(b) $\dfrac{y + 2}{y^2 - y} - \dfrac{3y}{2y^2 - 4y + 2} = \dfrac{y + 2}{y(y - 1)} - \dfrac{3y}{2(y - 1)^2}$ — Factor each denominator.

$$= \frac{(y + 2) \cdot 2(y - 1)}{y(y - 1) \cdot 2(y - 1)} - \frac{3y \cdot y}{2(y - 1)^2 \cdot y} \qquad \begin{array}{l}\text{The LCD is}\\ 2y(y-1)^2.\end{array}$$

$$= \frac{2(y^2 + y - 2)}{2y(y - 1)^2} - \frac{3y^2}{2y(y - 1)^2} \qquad \text{Multiply.}$$

$$= \frac{2y^2 + 2y - 4 - 3y^2}{2y(y - 1)^2} \qquad \begin{array}{l}\text{Multiply and}\\ \text{then subtract.}\end{array}$$

$$= \frac{-y^2 + 2y - 4}{2y(y - 1)^2} \qquad \begin{array}{l}\text{Combine}\\ \text{like terms.}\end{array}$$

(c) $\dfrac{3}{(x - 1)(x + 2)} - \dfrac{1}{(x + 3)(x - 4)}$ — $\begin{array}{l}\text{The LCD is}\\ (x-1)(x+2)(x+3)(x-4).\end{array}$

$$= \frac{3(x + 3)(x - 4)}{(x - 1)(x + 2)(x + 3)(x - 4)} - \frac{(x - 1)(x + 2)}{(x + 3)(x - 4)(x - 1)(x + 2)}$$

$$= \frac{3(x^2 - x - 12) - (x^2 + x - 2)}{(x - 1)(x + 2)(x + 3)(x - 4)} \qquad \text{Multiply and then subtract.}$$

$$= \frac{3x^2 - 3x - 36 - x^2 - x + 2}{(x - 1)(x + 2)(x + 3)(x - 4)} \qquad \text{Distributive property}$$

$$= \frac{2x^2 - 4x - 34}{(x - 1)(x + 2)(x + 3)(x - 4)} \qquad \text{Combine like terms.} \qquad \bullet$$

Complex Fractions

A quotient of two rational expressions is sometimes written as a fraction and is called a **complex fraction.**

EXAMPLE 4 **Simplifying Complex Fractions**

Simplify each complex fraction.

(a) $\dfrac{6 - \dfrac{5}{k}}{1 + \dfrac{5}{k}}$ **(b)** $\dfrac{\dfrac{a}{a + 1} + \dfrac{1}{a}}{\dfrac{1}{a} + \dfrac{1}{a + 1}}$

Solution

(a) Multiply numerator and denominator by the LCD of all the fractions, k.

$$\frac{6 - \dfrac{5}{k}}{1 + \dfrac{5}{k}} = \frac{k\left(6 - \dfrac{5}{k}\right)}{k\left(1 + \dfrac{5}{k}\right)} = \frac{6k - k\left(\dfrac{5}{k}\right)}{k + k\left(\dfrac{5}{k}\right)} = \frac{6k - 5}{k + 5}$$

(b) Multiply numerator and denominator by the LCD of all the fractions.

$$\frac{\dfrac{a}{a + 1} + \dfrac{1}{a}}{\dfrac{1}{a} + \dfrac{1}{a + 1}} = \frac{\left(\dfrac{a}{a + 1} + \dfrac{1}{a}\right)a(a + 1)}{\left(\dfrac{1}{a} + \dfrac{1}{a + 1}\right)a(a + 1)} \qquad \text{LCD} = a(a + 1)$$

$$= \frac{\dfrac{a}{a + 1}(a)(a + 1) + \dfrac{1}{a}(a)(a + 1)}{\dfrac{1}{a}(a)(a + 1) + \dfrac{1}{a + 1}(a)(a + 1)} \qquad \text{Distributive property}$$

$$= \frac{a^2 + (a + 1)}{(a + 1) + a} \qquad \text{Multiply.}$$

$$= \frac{a^2 + a + 1}{2a + 1} \qquad \text{Combine like terms.}$$

Alternatively, first add the terms in the numerator and denominator, and then divide.

$$\frac{\dfrac{a}{a + 1} + \dfrac{1}{a}}{\dfrac{1}{a} + \dfrac{1}{a + 1}} = \frac{\dfrac{a^2 + 1(a + 1)}{a(a + 1)}}{\dfrac{1(a + 1) + 1(a)}{a(a + 1)}} \qquad \text{Find the LCD. Add terms in the numerator and denominator.}$$

$$= \frac{\dfrac{a^2 + a + 1}{a(a + 1)}}{\dfrac{2a + 1}{a(a + 1)}} \qquad \text{Combine terms in the numerator and denominator.}$$

$$= \frac{a^2 + a + 1}{a(a + 1)} \cdot \frac{a(a + 1)}{2a + 1} \qquad \text{Definition of division}$$

The answer is the same as above. $\quad = \dfrac{a^2 + a + 1}{2a + 1} \qquad$ Multiply fractions. Write in lowest terms.

R.3 Exercises

Find the domain of each rational expression.

1. $\dfrac{x - 2}{x + 6}$

2. $\dfrac{x + 5}{x - 3}$

3. $\dfrac{2x}{5x - 3}$

4. $\dfrac{6x}{2x - 1}$

5. $\dfrac{-8}{x^2 + 1}$

6. $\dfrac{3x}{3x^2 + 7}$

7. $\dfrac{3x + 7}{(4x + 2)(x - 1)}$

8. $\dfrac{9x + 12}{(2x + 3)(x - 5)}$

Write each rational expression in lowest terms.

9. $\dfrac{25p^3}{10p^2}$ **10.** $\dfrac{14z^3}{6z^2}$ **11.** $\dfrac{8k+16}{9k+18}$ **12.** $\dfrac{20r+10}{30r+15}$

13. $\dfrac{3(t+5)}{(t+5)(t-3)}$ **14.** $\dfrac{-8(y-4)}{(y+2)(y-4)}$ **15.** $\dfrac{8x^2+16x}{4x^2}$ **16.** $\dfrac{36y^2+72y}{9y}$

17. $\dfrac{m^2-4m+4}{m^2+m-6}$ **18.** $\dfrac{r^2-r-6}{r^2+r-12}$ **19.** $\dfrac{8m^2+6m-9}{16m^2-9}$ **20.** $\dfrac{6y^2+11y+4}{3y^2+7y+4}$

Find each product or quotient.

21. $\dfrac{15p^3}{9p^2} \div \dfrac{6p}{10p^2}$ **22.** $\dfrac{3r^2}{9r^3} \div \dfrac{8r^3}{6r}$ **23.** $\dfrac{2k+8}{6} \div \dfrac{3k+12}{2}$

24. $\dfrac{5m+25}{10} \cdot \dfrac{12}{6m+30}$ **25.** $\dfrac{x^2+x}{5} \cdot \dfrac{25}{xy+y}$ **26.** $\dfrac{3m-15}{4m-20} \cdot \dfrac{m^2-10m+25}{12m-60}$

27. $\dfrac{4a+12}{2a-10} \div \dfrac{a^2-9}{a^2-a-20}$ **28.** $\dfrac{6r-18}{9r^2+6r-24} \cdot \dfrac{12r-16}{4r-12}$ **29.** $\dfrac{p^2-p-12}{p^2-2p-15} \cdot \dfrac{p^2-9p+20}{p^2-8p+16}$

30. $\dfrac{x^2+2x-15}{x^2+11x+30} \cdot \dfrac{x^2+2x-24}{x^2-8x+15}$ **31.** $\dfrac{m^2+3m+2}{m^2+5m+4} \div \dfrac{m^2+5m+6}{m^2+10m+24}$ **32.** $\dfrac{y^2+y-2}{y^2+3y-4} \div \dfrac{y^2+3y+2}{y^2+4y+3}$

33. $\dfrac{2m^2-5m-12}{m^2-10m+24} \div \dfrac{4m^2-9}{m^2-9m+18}$ **34.** $\dfrac{6n^2-5n-6}{6n^2+5n-6} \cdot \dfrac{12n^2-17n+6}{12n^2-n-6}$

35. $\dfrac{x^3+y^3}{x^2-y^2} \cdot \dfrac{x+y}{x^2-xy+y^2}$ **36.** $\dfrac{8y^3-125}{4y^2-20y+25} \cdot \dfrac{2y-5}{y}$

37. $\dfrac{x^3+y^3}{x^3-y^3} \cdot \dfrac{x^2-y^2}{x^2+2xy+y^2}$ **38.** $\dfrac{x^2-y^2}{(x-y)^2} \cdot \dfrac{x^2-xy+y^2}{x^2-2xy+y^2} \div \dfrac{x^3+y^3}{(x-y)^4}$

39. *Concept Check*   Which of these rational expressions equal -1? (Assume that all denominators are nonzero.)

  A. $\dfrac{x-4}{x+4}$ **B.** $\dfrac{-x-4}{x+4}$ **C.** $\dfrac{x-4}{4-x}$ **D.** $\dfrac{x-4}{-x-4}$

40. In your own words, explain how to find the least common denominator of two fractions.

Find each sum or difference.

41. $\dfrac{3}{2k}+\dfrac{5}{3k}$ **42.** $\dfrac{8}{5p}+\dfrac{3}{4p}$ **43.** $\dfrac{a+1}{2}-\dfrac{a-1}{2}$ **44.** $\dfrac{y+6}{5}-\dfrac{y-6}{5}$

45. $\dfrac{3}{p}+\dfrac{1}{2}$ **46.** $\dfrac{9}{r}-\dfrac{2}{3}$ **47.** $\dfrac{1}{6m}+\dfrac{2}{5m}+\dfrac{4}{m}$ **48.** $\dfrac{8}{3p}+\dfrac{5}{4p}+\dfrac{9}{2p}$

49. $\dfrac{1}{a+1}-\dfrac{1}{a-1}$ **50.** $\dfrac{1}{x+z}+\dfrac{1}{x-z}$ **51.** $\dfrac{m+1}{m-1}+\dfrac{m-1}{m+1}$ **52.** $\dfrac{2}{x-1}+\dfrac{1}{1-x}$

53. $\dfrac{3}{a-2}-\dfrac{1}{2-a}$ **54.** $\dfrac{q}{p-q}-\dfrac{q}{q-p}$ **55.** $\dfrac{x+y}{2x-y}-\dfrac{2x}{y-2x}$ **56.** $\dfrac{m-4}{3m-4}+\dfrac{3m+2}{4-3m}$

57. $\dfrac{1}{a^2-5a+6}-\dfrac{1}{a^2-4}$ **58.** $\dfrac{-3}{m^2-m-2}-\dfrac{1}{m^2+3m+2}$

59. $\dfrac{1}{x^2+x-12}-\dfrac{1}{x^2-7x+12}+\dfrac{1}{x^2-16}$ **60.** $\dfrac{2}{2p^2-9p-5}+\dfrac{p}{3p^2-17p+10}-\dfrac{2p}{6p^2-p-2}$

61. $\dfrac{3a}{a^2+5a-6}-\dfrac{2a}{a^2+7a+6}$ **62.** $\dfrac{2k}{k^2+4k+3}+\dfrac{3k}{k^2+5k+6}$

Simplify each complex fraction.

63. $\dfrac{1 + \dfrac{1}{x}}{1 - \dfrac{1}{x}}$

64. $\dfrac{2 - \dfrac{2}{y}}{2 + \dfrac{2}{y}}$

65. $\dfrac{\dfrac{1}{x+1} - \dfrac{1}{x}}{\dfrac{1}{x}}$

66. $\dfrac{\dfrac{1}{y+3} - \dfrac{1}{y}}{\dfrac{1}{y}}$

67. $\dfrac{1 + \dfrac{1}{1-b}}{1 - \dfrac{1}{1+b}}$

68. $m - \dfrac{m}{m + \dfrac{1}{2}}$

69. $\dfrac{m - \dfrac{1}{m^2 - 4}}{\dfrac{1}{m+2}}$

70. $\dfrac{\dfrac{3}{p^2 - 16} + p}{\dfrac{1}{p - 4}}$

R.4   Review of Negative and Rational Exponents

Negative Exponents and the Quotient Rule • Rational Exponents

Negative Exponents and the Quotient Rule

In the product rule for exponents, the exponents are *added* (that is, $a^m \cdot a^n = a^{m+n}$). Consider the following expression. If $a \neq 0$, then

$$\frac{a^3}{a^7} = \frac{a \cdot a \cdot a}{a \cdot a \cdot a \cdot a \cdot a \cdot a \cdot a} = \frac{1}{a \cdot a \cdot a \cdot a} = \frac{1}{a^4}.$$

This suggests that we should *subtract* exponents when dividing.

$$\frac{a^3}{a^7} = a^{3-7} = a^{-4}$$

The only way to keep these results consistent is to define a^{-4} as $\dfrac{1}{a^4}$.

Negative Exponent

If a is a nonzero real number and n is any integer, then

$$a^{-n} = \frac{1}{a^n}.$$

EXAMPLE 1   **Using the Definition of a Negative Exponent**

Evaluate each expression in parts (a)–(c). In parts (d) and (e), write the expression without negative exponents. Assume that all variables represent nonzero real numbers.

(a) 4^{-2} **(b)** -4^{-2} **(c)** $\left(\dfrac{2}{5}\right)^{-3}$ **(d)** x^{-5} **(e)** xy^{-3}

Solution

(a) $4^{-2} = \dfrac{1}{4^2} = \dfrac{1}{16}$ **(b)** $-4^{-2} = -\dfrac{1}{4^2} = -\dfrac{1}{16}$

(c) $\left(\dfrac{2}{5}\right)^{-3} = \dfrac{1}{\left(\dfrac{2}{5}\right)^3} = \dfrac{1}{\dfrac{8}{125}} = 1 \div \dfrac{8}{125} = 1 \cdot \dfrac{125}{8} = \dfrac{125}{8}$

(d) $x^{-5} = \dfrac{1}{x^5}$ **(e)** $xy^{-3} = x \cdot \dfrac{1}{y^3} = \dfrac{x}{y^3}$

From **Example 1(c),**

$$\left(\frac{2}{5}\right)^{-3} = \frac{125}{8} = \left(\frac{5}{2}\right)^3.$$

This result can be generalized. If $a \neq 0$ and $b \neq 0$, then for any integer n,

$$\left(\frac{a}{b}\right)^{-n} = \left(\frac{b}{a}\right)^n.$$

The quotient rule for exponents follows from the definition of negative exponents.

> **Quotient Rule**
>
> For all integers m and n and all nonzero real numbers a,
>
> $$\frac{a^m}{a^n} = a^{m-n}.$$

EXAMPLE 2 **Using the Quotient Rule**

Simplify each expression. Assume that all variables represent nonzero real numbers.

(a) $\dfrac{12^5}{12^2}$ (b) $\dfrac{a^5}{a^{-8}}$ (c) $\dfrac{16m^{-9}}{12m^{11}}$ (d) $\dfrac{25r^7z^5}{10r^9z}$

Use parentheses to avoid errors.

Solution

(a) $\dfrac{12^5}{12^2} = 12^{5-2} = 12^3$ (b) $\dfrac{a^5}{a^{-8}} = a^{5-(-8)} = a^{13}$

(c) $\dfrac{16m^{-9}}{12m^{11}} = \dfrac{16}{12} \cdot m^{-9-11} = \dfrac{4}{3}m^{-20} = \dfrac{4}{3} \cdot \dfrac{1}{m^{20}} = \dfrac{4}{3m^{20}}$

(d) $\dfrac{25r^7z^5}{10r^9z} = \dfrac{25}{10} \cdot \dfrac{r^7}{r^9} \cdot \dfrac{z^5}{z^1} = \dfrac{5}{2}r^{-2}z^4 = \dfrac{5z^4}{2r^2}$

The rules for exponents stated in **Section R.1** also apply to negative exponents.

EXAMPLE 3 **Using the Rules for Exponents**

Simplify each expression. Write answers without negative exponents. Assume that all variables represent nonzero real numbers.

(a) $3x^{-2}(4^{-1}x^{-5})^2$ (b) $\dfrac{12p^3q^{-1}}{8p^{-2}q}$ (c) $\dfrac{(3x^2)^{-1}(3x^5)^{-2}}{(3^{-1}x^{-2})^2}$

Solution

(a) $3x^{-2}(4^{-1}x^{-5})^2 = 3x^{-2}(4^{-2}x^{-10})$ Power rule

$\qquad\qquad\qquad\quad = 3 \cdot 4^{-2} \cdot x^{-2+(-10)}$ Rearrange factors and use product rule.

$\qquad\qquad\qquad\quad = 3 \cdot 4^{-2} \cdot x^{-12}$ Add exponents.

$\qquad\qquad\qquad\quad = \dfrac{3}{16x^{12}}$ Write with positive exponents.

(b) $\dfrac{12p^3q^{-1}}{8p^{-2}q} = \dfrac{12}{8} \cdot \dfrac{p^3}{p^{-2}} \cdot \dfrac{q^{-1}}{q^1}$

$= \dfrac{3}{2} \cdot p^{3-(-2)}q^{-1-1}$ Quotient rule

$= \dfrac{3}{2}p^5q^{-2}$ Subtract exponents.

$= \dfrac{3p^5}{2q^2}$ Write with positive exponents.

(c) $\dfrac{(3x^2)^{-1}(3x^5)^{-2}}{(3^{-1}x^{-2})^2} = \dfrac{3^{-1}x^{-2}3^{-2}x^{-10}}{3^{-2}x^{-4}}$ Power rule

$= \dfrac{3^{-1+(-2)}x^{-2+(-10)}}{3^{-2}x^{-4}} = \dfrac{3^{-3}x^{-12}}{3^{-2}x^{-4}}$ Product rule

$= 3^{-3-(-2)}x^{-12-(-4)}$ Quotient rule

$= 3^{-1}x^{-8},$ or $\dfrac{1}{3x^8}$.   Write with positive exponents.   ●

CAUTION   The rule $(ab)^n = a^nb^n$ gives $(3x^2)^{-1} = 3^{-1}(x^2)^{-1} = 3^{-1}x^{-2}$ in **Example 3(c)**. *Remember to apply the exponent to the numerical coefficient.*

Rational Exponents

The definition of a^n can be extended to rational values of n by defining $a^{1/n}$ to be the nth root of a. By one of the power rules of exponents (extended to a rational exponent),

$$(a^{1/n})^n = a^{(1/n)n} = a^1 = a,$$

which suggests that $a^{1/n}$ is a number whose nth power is a.

The Expression $a^{1/n}$

n even   If n is an *even* positive integer, and if $a > 0$, then $a^{1/n}$ is the positive real number whose nth power is a. That is, $(a^{1/n})^n = a$. (In this case, $a^{1/n}$ is the principal nth root of a. See **Section R.5**.)

n odd   If n is an *odd* positive integer and *a is any nonzero real number,* then $a^{1/n}$ is the positive or negative real number whose nth power is a. That is, $(a^{1/n})^n = a$.

For all positive integers n, $0^{1/n} = 0$.

EXAMPLE 4   **Using the Definition of $a^{1/n}$**

Evaluate each expression.

(a) $36^{1/2}$ **(b)** $-100^{1/2}$ **(c)** $-(225)^{1/2}$ **(d)** $0^{1/4}$

(e) $(-1296)^{1/4}$ **(f)** $-1296^{1/4}$ **(g)** $(-27)^{1/3}$ **(h)** $-32^{1/5}$

Solution

(a) $36^{1/2} = 6$ because $6^2 = 36$.   **(b)** $-100^{1/2} = -10$

(c) $-(225)^{1/2} = -15$ **(d)** $0^{1/4} = 0$

(e) $(-1296)^{1/4}$ is not a real number.   **(f)** $-1296^{1/4} = -6$

(g) $(-27)^{1/3} = -3$ **(h)** $-32^{1/5} = -2$ ●

The notation $a^{m/n}$ must be defined so that all the rules for exponents hold. For the power rule to hold, $(a^{1/n})^m$ must equal $a^{m/n}$. Therefore, $a^{m/n}$ is defined as follows.

Rational Exponent

For all integers m, all positive integers n, and all real numbers a for which $a^{1/n}$ is a real number,

$$a^{m/n} = (a^{1/n})^m.$$

EXAMPLE 5 **Using the Definition of $a^{m/n}$**

Evaluate each expression.

(a) $125^{2/3}$ **(b)** $32^{7/5}$ **(c)** $-81^{3/2}$

(d) $(-27)^{2/3}$ **(e)** $16^{-3/4}$ **(f)** $(-4)^{5/2}$

Solution

(a) $125^{2/3} = (125^{1/3})^2 = 5^2 = 25$

(b) $32^{7/5} = (32^{1/5})^7 = 2^7 = 128$

(c) $-81^{3/2} = -(81^{1/2})^3 = -(9)^3 = -729$

(d) $(-27)^{2/3} = [(-27)^{1/3}]^2 = (-3)^2 = 9$

(e) $16^{-3/4} = \dfrac{1}{16^{3/4}} = \dfrac{1}{(16^{1/4})^3} = \dfrac{1}{2^3} = \dfrac{1}{8}$

(f) $(-4)^{5/2}$ is not a real number because $(-4)^{1/2}$ is not a real number. ●

NOTE For all real numbers a, integers m, and positive integers n for which $a^{1/n}$ is a real number,

$$a^{m/n} = (a^{1/n})^m, \quad \text{or} \quad a^{m/n} = (a^m)^{1/n}.$$

So $a^{m/n}$ can be evaluated either as $(a^{1/n})^m$ or as $(a^m)^{1/n}$.

$$27^{4/3} = (27^{1/3})^4 = 3^4 = 81$$

$$27^{4/3} = (27^4)^{1/3} = 531{,}441^{1/3} = 81$$

The results are equal.

The first form, $(27^{1/3})^4$ (that is, $(a^{1/n})^m$), is easier to evaluate.

Definitions and Rules for Exponents

Let r and s be rational numbers. The following results are valid whenever each expression is a real number.

$$a^r \cdot a^s = a^{r+s} \qquad (ab)^r = a^r \cdot b^r \qquad (a^r)^s = a^{rs}$$

$$\frac{a^r}{a^s} = a^{r-s} \qquad \left(\frac{a}{b}\right)^r = \frac{a^r}{b^r} \qquad a^{-r} = \frac{1}{a^r}$$

EXAMPLE 6   **Using the Definitions and Rules for Exponents**

Simplify each expression. Assume that all variables represent positive real numbers.

(a) $\dfrac{27^{1/3} \cdot 27^{5/3}}{27^3}$ (b) $81^{5/4} \cdot 4^{-3/2}$ (c) $6y^{2/3} \cdot 2y^{1/2}$

(d) $\left(\dfrac{3m^{5/6}}{y^{3/4}}\right)^2 \cdot \left(\dfrac{8y^3}{m^6}\right)^{2/3}$ (e) $m^{2/3}(m^{7/3} + 2m^{1/3})$

Solution

(a) $\dfrac{27^{1/3} \cdot 27^{5/3}}{27^3} = \dfrac{27^{1/3+5/3}}{27^3}$ Product rule

$= \dfrac{27^2}{27^3} = 27^{2-3}$ Quotient rule

$= 27^{-1} = \dfrac{1}{27}$ Negative exponent

(b) $81^{5/4} \cdot 4^{-3/2} = (81^{1/4})^5 (4^{1/2})^{-3} = 3^5 \cdot 2^{-3} = \dfrac{3^5}{2^3}$, or $\dfrac{243}{8}$

(c) $6y^{2/3} \cdot 2y^{1/2} = 12y^{2/3+1/2} = 12y^{7/6}$

(d) $\left(\dfrac{3m^{5/6}}{y^{3/4}}\right)^2 \cdot \left(\dfrac{8y^3}{m^6}\right)^{2/3} = \dfrac{9m^{5/3}}{y^{3/2}} \cdot \dfrac{4y^2}{m^4} = 36m^{5/3-4}\, y^{2-3/2} = \dfrac{36y^{1/2}}{m^{7/3}}$

(e) $m^{2/3}(m^{7/3} + 2m^{1/3}) = m^{2/3+7/3} + 2m^{2/3+1/3} = m^3 + 2m$

EXAMPLE 7   **Factoring Expressions with Negative or Rational Exponents**

Factor out the least power of the variable. Assume that all variables represent positive real numbers.

(a) $12x^{-2} - 8x^{-3}$ (b) $4m^{1/2} + 3m^{3/2}$ (c) $y^{-1/3} + y^{2/3}$

Solution

(a) The least exponent on x here is -3. Since 4 is a common numerical factor, factor out $4x^{-3}$.

$$12x^{-2} - 8x^{-3} = 4x^{-3}(3x - 2)$$

Check by multiplying $4x^{-3}(3x - 2)$. The factored form can be written without negative exponents as $\dfrac{4(3x - 2)}{x^3}$.

(b) $4m^{1/2} + 3m^{3/2} = m^{1/2}(4 + 3m)$. To check, multiply $m^{1/2}$ by $4 + 3m$.

(c) $y^{-1/3} + y^{2/3} = y^{-1/3}(1 + y)$, or $\dfrac{1 + y}{y^{1/3}}$

R.4 Exercises

Concept Check *Match each expression from Group I with the correct choice from Group II. Choices may be used once, more than once, or not at all.*

I

1. $\left(\dfrac{4}{9}\right)^{3/2}$
2. $\left(\dfrac{4}{9}\right)^{-3/2}$

3. $-\left(\dfrac{9}{4}\right)^{3/2}$
4. $-\left(\dfrac{4}{9}\right)^{-3/2}$

5. $\left(\dfrac{8}{27}\right)^{2/3}$
6. $\left(\dfrac{8}{27}\right)^{-2/3}$

7. $-\left(\dfrac{27}{8}\right)^{2/3}$
8. $-\left(\dfrac{27}{8}\right)^{-2/3}$

II

A. $\dfrac{9}{4}$
B. $-\dfrac{9}{4}$

C. $-\dfrac{4}{9}$
D. $\dfrac{4}{9}$

E. $\dfrac{8}{27}$
F. $-\dfrac{27}{8}$

G. $\dfrac{27}{8}$
H. $-\dfrac{8}{27}$

Simplify each expression. Assume that all variables represent positive real numbers.

9. $(-4)^{-3}$
10. $(-5)^{-2}$
11. $\left(\dfrac{1}{2}\right)^{-3}$
12. $\left(\dfrac{2}{3}\right)^{-2}$

13. $-4^{1/2}$
14. $25^{1/2}$
15. $8^{2/3}$
16. $-81^{3/4}$

17. $27^{-2/3}$
18. $(-32)^{-4/5}$
19. $\left(\dfrac{27}{64}\right)^{-4/3}$
20. $\left(\dfrac{121}{100}\right)^{-3/2}$

21. $(16p^4)^{1/2}$
22. $(36r^6)^{1/2}$
23. $(27x^6)^{2/3}$
24. $(64a^{12})^{5/6}$

Perform the indicated operations. Write your answers with only positive exponents. Assume that all variables represent positive real numbers.

25. $2^{-3} \cdot 2^{-4}$
26. $5^{-2} \cdot 5^{-6}$
27. $27^{-2} \cdot 27^{-1}$

28. $9^{-4} \cdot 9^{-1}$
29. $\dfrac{4^{-2} \cdot 4^{-1}}{4^{-3}}$
30. $\dfrac{3^{-1} \cdot 3^{-4}}{3^2 \cdot 3^{-2}}$

31. $(m^{2/3})(m^{5/3})$
32. $(x^{4/5})(x^{2/5})$
33. $(1 + n)^{1/2}(1 + n)^{3/4}$

34. $(m + 7)^{-1/6}(m + 7)^{-2/3}$
35. $(2y^{3/4}z)(3y^{-2}z^{-1/3})$
36. $(4a^{-1}b^{2/3})(a^{3/2}b^{-3})$

37. $(4a^{-2}b^7)^{1/2} \cdot (2a^{1/4}b^3)^5$
38. $(x^{-2}y^{1/3})^5 \cdot (8x^2y^{-2})^{-1/3}$
39. $\left(\dfrac{r^{-2}}{s^{-5}}\right)^{-3}$

40. $\left(\dfrac{p^{-1}}{q^{-5}}\right)^{-2}$
41. $\left(\dfrac{-a}{b^{-3}}\right)^{-1}$
42. $\dfrac{7^{-1/3}7r^{-3}}{7^{2/3}r^{-2}}$

43. $\dfrac{12^{5/4}y^{-2}}{12^{-1}y^{-3}}$
44. $\dfrac{6k^{-4}(3k^{-1})^{-2}}{2^3k^{1/2}}$
45. $\dfrac{8p^{-3}(4p^2)^{-2}}{p^{-5}}$

46. $\dfrac{k^{-3/5}h^{-1/3}t^{2/5}}{k^{-1/5}h^{-2/3}t^{1/5}}$
47. $\dfrac{m^{7/3}n^{-2/5}p^{3/8}}{m^{-2/3}n^{3/5}p^{-5/8}}$
48. $\dfrac{m^{2/5}m^{3/5}m^{-4/5}}{m^{1/5}m^{-6/5}}$

49. $\dfrac{-4a^{-1}a^{2/3}}{a^{-2}}$
50. $\dfrac{8y^{2/3}y^{-1}}{2^{-1}y^{3/4}y^{-1/6}}$

51. $\dfrac{(k + 5)^{1/2}(k + 5)^{-1/4}}{(k + 5)^{3/4}}$
52. $\dfrac{(x + y)^{-5/8}(x + y)^{3/8}}{(x + y)^{1/8}(x + y)^{-1/8}}$

Find each product. Assume that all variables represent positive real numbers.

53. $y^{5/8}(y^{3/8} - 10y^{11/8})$

54. $p^{11/5}(3p^{4/5} + 9p^{19/5})$

55. $-4k(k^{7/3} - 6k^{1/3})$

56. $-5y(3y^{9/10} + 4y^{3/10})$

57. $(x + x^{1/2})(x - x^{1/2})$

58. $(2z^{1/2} + z)(z^{1/2} - z)$

59. $(r^{1/2} - r^{-1/2})^2$

60. $(p^{1/2} - p^{-1/2})(p^{1/2} + p^{-1/2})$

Factor, using the given common factor. Assume that all variables represent positive real numbers.

61. $4k^{-1} + k^{-2}; \quad k^{-2}$

62. $y^{-5} - 3y^{-3}; \quad y^{-5}$

63. $9z^{-1/2} + 2z^{1/2}; \quad z^{-1/2}$

64. $3m^{2/3} - 4m^{-1/3}; \quad m^{-1/3}$

65. $p^{-3/4} - 2p^{-7/4}; \quad p^{-7/4}$

66. $6r^{-2/3} - 5r^{-5/3}; \quad r^{-5/3}$

67. $(p + 4)^{-3/2} + (p + 4)^{-1/2} + (p + 4)^{1/2}; \quad (p + 4)^{-3/2}$

68. $(3r + 1)^{-2/3} + (3r + 1)^{1/3} + (3r + 1)^{4/3}; \quad (3r + 1)^{-2/3}$

R.5 Review of Radicals

Radical Notation • Rules for Radicals • Simplifying Radicals • Operations with Radicals • Rationalizing Denominators

Radical Notation

In **Section R.4,** the notation $a^{1/n}$ was used for the nth root of a for appropriate values of a and n. An alternative notation for $a^{1/n}$ uses **radical notation.**

Radical Notation for $a^{1/n}$

If a is a real number, n is a positive integer, and $a^{1/n}$ is a real number, then

$$\sqrt[n]{a} = a^{1/n}.$$

The symbol $\sqrt[n]{}$ is a **radical symbol,** the number a is the **radicand,** and n is the **index** of the radical $\sqrt[n]{a}$. It is customary to use $\sqrt{a}$ instead of $\sqrt[2]{a}$ for the square root.

For even values of n (square roots, fourth roots, and so on) and a > 0, there are two nth roots, one positive and one negative. In such cases, the notation $\sqrt[n]{a}$ represents the positive root—the **principal nth root.** The negative root is written $-\sqrt[n]{a}$.

EXAMPLE 1   **Evaluating Roots**

Evaluate each root.

(a) $\sqrt[4]{16}$ **(b)** $-\sqrt[4]{16}$ **(c)** $\sqrt[4]{-16}$

(d) $\sqrt[5]{-32}$ **(e)** $\sqrt[3]{1000}$ **(f)** $\sqrt[6]{\dfrac{64}{729}}$

Solution

(a) $\sqrt[4]{16} = 16^{1/4} = (2^4)^{1/4} = 2$ **(b)** $-\sqrt[4]{16} = -16^{1/4} = -(2^4)^{1/4} = -2$

(c) $\sqrt[4]{-16}$ is not a real number.   **(d)** $\sqrt[5]{-32} = [(-2)^5]^{1/5} = -2$

(e) $\sqrt[3]{1000} = 1000^{1/3} = 10$ **(f)** $\sqrt[6]{\dfrac{64}{729}} = \left(\dfrac{64}{729}\right)^{1/6} = \dfrac{2}{3}$

With $a^{1/n}$ written as $\sqrt[n]{a}$, $a^{m/n}$ can also be written with radicals.

> ### Radical Notation for $a^{m/n}$
>
> If a is a real number, m is an integer, n is a positive integer, and $\sqrt[n]{a}$ is a real number, then
>
> $$a^{m/n} = \left(\sqrt[n]{a} \right)^m = \sqrt[n]{a^m}.$$

EXAMPLE 2 **Converting from Rational Exponents to Radicals**

Write in radical form and simplify. Assume that all variable expressions represent positive real numbers.

(a) $8^{2/3}$ **(b)** $(-32)^{4/5}$ **(c)** $-16^{3/4}$ **(d)** $x^{5/6}$ **(e)** $3x^{2/3}$

(f) $2p^{-1/2}$ **(g)** $(3a + b)^{1/4}$

Solution

(a) $8^{2/3} = \left(\sqrt[3]{8} \right)^2 = 2^2 = 4$ **(b)** $(-32)^{4/5} = \left(\sqrt[5]{-32} \right)^4 = (-2)^4 = 16$

(c) $-16^{3/4} = -\left(\sqrt[4]{16} \right)^3 = -(2)^3 = -8$ **(d)** $x^{5/6} = \sqrt[6]{x^5}$

(e) $3x^{2/3} = 3\sqrt[3]{x^2}$ **(f)** $2p^{-1/2} = \dfrac{2}{p^{1/2}} = \dfrac{2}{\sqrt{p}}$

(g) $(3a + b)^{1/4} = \sqrt[4]{3a + b}$

EXAMPLE 3 **Converting from Radicals to Rational Exponents**

Write each expression with rational exponents. Assume that all variable expressions represent positive real numbers.

(a) $\sqrt[4]{x^5}$ **(b)** $\sqrt{3y}$ **(c)** $10\left(\sqrt[5]{z} \right)^2$ **(d)** $5\sqrt[3]{(2x^4)^7}$ **(e)** $\sqrt{p^2 + q}$

Solution

(a) $\sqrt[4]{x^5} = x^{5/4}$ **(b)** $\sqrt{3y} = (3y)^{1/2} = 3^{1/2}y^{1/2}$

(c) $10\left(\sqrt[5]{z} \right)^2 = 10z^{2/5}$ **(d)** $5\sqrt[3]{(2x^4)^7} = 5(2x^4)^{7/3} = 5 \cdot 2^{7/3}x^{28/3}$

(e) $\sqrt{p^2 + q} = (p^2 + q)^{1/2}$

By the definition of $\sqrt[n]{a}$, for any positive integer n, if $\sqrt[n]{a}$ is defined, then

$$\left(\sqrt[n]{a} \right)^n = a.$$

If a is positive, or if a is negative and n is an odd positive integer, then

$$\sqrt[n]{a^n} = a.$$

Because of the conditions just given, we *cannot* simply write $\sqrt{x^2} = x$. For example,

$$\text{if } x = -5, \text{ then } \quad \sqrt{x^2} = \sqrt{(-5)^2} = \sqrt{25} = 5 \neq x.$$

Since a negative value of x can produce a positive result, we use the absolute value.

$$\sqrt{a^2} = |a|, \quad \text{for any real number } a.$$

For example, $\sqrt{(-9)^2} = |-9| = 9$ and $\sqrt{13^2} = |13| = 13$.

Evaluating $\sqrt[n]{a^n}$

If n is an even positive integer, then $\sqrt[n]{a^n} = |a|$.

If n is an odd positive integer, then $\sqrt[n]{a^n} = a$.

EXAMPLE 4 **Using Absolute Value to Simplify Roots**

Simplify each expression.

(a) $\sqrt{p^4}$ **(b)** $\sqrt[4]{p^4}$ **(c)** $\sqrt{16m^8r^6}$ **(d)** $\sqrt[6]{(-2)^6}$ **(e)** $\sqrt[5]{m^5}$

(f) $\sqrt{(2k+3)^2}$ **(g)** $\sqrt{x^2-4x+4}$

Solution

(a) $\sqrt{p^4} = \sqrt{(p^2)^2} = |p^2| = p^2$ **(b)** $\sqrt[4]{p^4} = |p|$

(c) $\sqrt{16m^8r^6} = |4m^4r^3| = 4m^4|r^3|$ **(d)** $\sqrt[6]{(-2)^6} = |-2| = 2$

(e) $\sqrt[5]{m^5} = m$ **(f)** $\sqrt{(2k+3)^2} = |2k+3|$

(g) $\sqrt{x^2-4x+4} = \sqrt{(x-2)^2} = |x-2|$

NOTE When working with variable radicands, we usually assume that all expressions in radicands represent only nonnegative real numbers.

Rules for Radicals

The following three rules are the power rules for exponents written in radical notation.

Rules for Radicals

For all real numbers a and b, and positive integers m and n for which the indicated roots are real numbers,

$$\sqrt[n]{a} \cdot \sqrt[n]{b} = \sqrt[n]{ab}, \qquad \sqrt[n]{\frac{a}{b}} = \frac{\sqrt[n]{a}}{\sqrt[n]{b}} \quad (b \neq 0), \qquad \sqrt[m]{\sqrt[n]{a}} = \sqrt[mn]{a}.$$

EXAMPLE 5 **Using the Rules for Radicals**

Simplify. Assume that all variable expressions represent positive real numbers.

(a) $\sqrt{6} \cdot \sqrt{54}$ **(b)** $\sqrt[3]{m} \cdot \sqrt[3]{m^2}$ **(c)** $\sqrt{\dfrac{7}{64}}$

(d) $\sqrt[4]{\dfrac{a}{b^4}}$ **(e)** $\sqrt[7]{\sqrt[3]{2}}$ **(f)** $\sqrt[4]{\sqrt{3}}$

Solution

(a) $\sqrt{6} \cdot \sqrt{54} = \sqrt{6 \cdot 54} = \sqrt{324} = 18$ **(b)** $\sqrt[3]{m} \cdot \sqrt[3]{m^2} = \sqrt[3]{m^3} = m$

(c) $\sqrt{\dfrac{7}{64}} = \dfrac{\sqrt{7}}{\sqrt{64}} = \dfrac{\sqrt{7}}{8}$ **(d)** $\sqrt[4]{\dfrac{a}{b^4}} = \dfrac{\sqrt[4]{a}}{\sqrt[4]{b^4}} = \dfrac{\sqrt[4]{a}}{b}$

(e) $\sqrt[7]{\sqrt[3]{2}} = \sqrt[21]{2}$ Use the third rule. **(f)** $\sqrt[4]{\sqrt{3}} = \sqrt[8]{3}$

Simplifying Radicals

> **Simplified Radicals**
>
> An expression with radicals is simplified when the following conditions are satisfied.
> 1. The radicand has no factor raised to a power greater than or equal to the index.
> 2. The radicand has no fractions.
> 3. No denominator contains a radical.
> 4. Exponents in the radicand and the index of the radical have no common factor.
> 5. All indicated operations have been performed (if possible).

EXAMPLE 6   **Simplifying Radicals**

Simplify each radical. Assume that variables represent nonnegative real numbers.

(a) $\sqrt{175}$ **(b)** $-3\sqrt[5]{32}$ **(c)** $\sqrt{288m^5}$ **(d)** $\sqrt[3]{81x^5y^7z^6}$

Solution

(a) $\sqrt{175} = \sqrt{25 \cdot 7} = \sqrt{25} \cdot \sqrt{7} = 5\sqrt{7}$

(b) $-3\sqrt[5]{32} = -3\sqrt[5]{2^5} = -3 \cdot 2 = -6$

(c) $\sqrt{288m^5} = \sqrt{144m^4 \cdot 2m} = 12m^2\sqrt{2m}$

(d) $\sqrt[3]{81x^5y^7z^6} = \sqrt[3]{27 \cdot 3 \cdot x^3 \cdot x^2 \cdot y^6 \cdot y \cdot z^6}$ Factor.

$\qquad = \sqrt[3]{(27x^3y^6z^6)(3x^2y)}$ Group all perfect cubes.

$\qquad = 3xy^2z^2\sqrt[3]{3x^2y}$ Remove all perfect cubes from under the radical. ●

EXAMPLE 7   **Simplifying Radicals by Using Rational Exponents**

Simplify each radical. Assume that variables represent nonnegative real numbers.

(a) $\sqrt[6]{3^2}$ **(b)** $\sqrt[6]{x^{12}y^3}$ **(c)** $\sqrt[9]{\sqrt{6^3}}$

Solution

(a) $\sqrt[6]{3^2} = 3^{2/6} = 3^{1/3} = \sqrt[3]{3}$

(b) $\sqrt[6]{x^{12}y^3} = (x^{12}y^3)^{1/6} = x^2y^{3/6} = x^2y^{1/2} = x^2\sqrt{y}$

(c) $\sqrt[9]{\sqrt{6^3}} = \sqrt[18]{6^3} = (6^3)^{1/18} = 6^{1/6} = \sqrt[6]{6}$ ●

In **Example 7(a)**, we simplified $\sqrt[6]{3^2}$ to $\sqrt[3]{3}$. However, to simplify $\left(\sqrt[6]{x}\right)^2$, the variable x must represent a nonnegative real number. For example, suppose we write

$$(-8)^{2/6} = \left[(-8)^{1/6}\right]^2.$$

This result is not a real number, because $(-8)^{1/6}$ is not a real number. By contrast,

$$(-8)^{1/3} = -2.$$

Therefore, $\qquad\qquad \left(\sqrt[6]{x}\right)^2 \neq \sqrt[3]{x},$ even though $\frac{2}{6} = \frac{1}{3}$.

Thus, if a is nonnegative, then $a^{m/n} = a^{mp/(np)}$ (for $p \neq 0$). Simplifying rational exponents on negative bases should be considered case by case.

Operations with Radicals

Radicals with the same radicand and the same index, such as $3\sqrt[4]{11pq}$ and $-7\sqrt[4]{11pq}$, are called **like radicals.** Examples of *unlike radicals* are

$$2\sqrt{5} \quad \text{and} \quad 2\sqrt{3}, \qquad \text{Radicands are different.}$$

as well as

$$2\sqrt{3} \quad \text{and} \quad 2\sqrt[3]{3}. \qquad \text{Indexes are different.}$$

We add or subtract like radicals by using the distributive property. *Only like radicals can be combined.* Sometimes we need to simplify radicals before adding or subtracting.

> **EXAMPLE 8**   **Adding and Subtracting Like Radicals**

Add or subtract, as indicated. Assume that all variables represent nonnegative real numbers.

(a) $3\sqrt[4]{11pq} + \left(-7\sqrt[4]{11pq}\right)$ **(b)** $7\sqrt{2} - 8\sqrt{18} + 4\sqrt{72}$

(c) $\sqrt{98x^3y} + 3x\sqrt{32xy}$

Solution

(a) $3\sqrt[4]{11pq} + \left(-7\sqrt[4]{11pq}\right) = (3 + (-7))\sqrt[4]{11pq}$ Add like radicals.

$$= -4\sqrt[4]{11pq}$$

(b) $7\sqrt{2} - 8\sqrt{18} + 4\sqrt{72} = 7\sqrt{2} - 8\sqrt{9 \cdot 2} + 4\sqrt{36 \cdot 2}$ Rewrite radicals.

$$= 7\sqrt{2} - 8 \cdot 3\sqrt{2} + 4 \cdot 6\sqrt{2} \qquad \text{Factor out perfect squares; multiply.}$$

$$= 7\sqrt{2} - 24\sqrt{2} + 24\sqrt{2}$$

$$= 7\sqrt{2} \qquad \text{Distributive property}$$

(c) $\sqrt{98x^3y} + 3x\sqrt{32xy} = \sqrt{49 \cdot 2 \cdot x^2 \cdot x \cdot y} + 3x\sqrt{16 \cdot 2 \cdot x \cdot y}$

$$= 7x\sqrt{2xy} + 3x(4)\sqrt{2xy} \qquad \text{Factor out perfect squares.}$$

$$= 7x\sqrt{2xy} + 12x\sqrt{2xy} \qquad \text{Multiply.}$$

$$= 19x\sqrt{2xy} \qquad \text{Distributive property}$$

> **EXAMPLE 9**   **Multiplying Radical Expressions**

Find each product.

(a) $\left(\sqrt{2} + 3\right)\left(\sqrt{8} - 5\right)$ **(b)** $\left(\sqrt{7} - \sqrt{10}\right)\left(\sqrt{7} + \sqrt{10}\right)$

Solution

(a) $\left(\sqrt{2} + 3\right)\left(\sqrt{8} - 5\right) = \sqrt{2}\left(\sqrt{8}\right) - \sqrt{2}(5) + 3\sqrt{8} - 3(5)$ FOIL

$$= \sqrt{16} - 5\sqrt{2} + 3\left(2\sqrt{2}\right) - 15 \qquad \text{Multiply and simplify.}$$

$$= 4 - 5\sqrt{2} + 6\sqrt{2} - 15 \qquad \text{Multiply.}$$

$$= -11 + \sqrt{2} \qquad \text{Combine like terms.}$$

(b) $\left(\sqrt{7} - \sqrt{10}\right)\left(\sqrt{7} + \sqrt{10}\right) = \left(\sqrt{7}\right)^2 - \left(\sqrt{10}\right)^2$ Product of the sum and difference of two terms

$$= 7 - 10 \qquad \text{Apply the exponents.}$$

$$= -3 \qquad \text{Subtract.}$$

Rationalizing Denominators

Condition 3 of the rules for simplifying radicals described earlier requires that no denominator contain a radical. The process of achieving this is called **rationalizing the denominator.** *To rationalize a denominator, we multiply by a form of 1.*

EXAMPLE 10   Rationalizing Denominators

Rationalize each denominator.

(a) $\dfrac{4}{\sqrt{3}}$ **(b)** $\dfrac{2}{\sqrt[3]{x}}$ $(x \neq 0)$

Solution

(a) $\dfrac{4}{\sqrt{3}} = \dfrac{4}{\sqrt{3}} \cdot \dfrac{\sqrt{3}}{\sqrt{3}} = \dfrac{4\sqrt{3}}{3}$ Multiply by $\dfrac{\sqrt{3}}{\sqrt{3}} = 1$.

(b) $\dfrac{2}{\sqrt[3]{x}} = \dfrac{2}{\sqrt[3]{x}} \cdot \dfrac{\sqrt[3]{x^2}}{\sqrt[3]{x^2}}$ $\dfrac{\sqrt[3]{x^2}}{\sqrt[3]{x^2}} = 1$

$= \dfrac{2\sqrt[3]{x^2}}{\sqrt[3]{x^3}}$,   or $\dfrac{2\sqrt[3]{x^2}}{x}$ $\sqrt[3]{x} \cdot \sqrt[3]{x^2} = \sqrt[3]{x^3} = x$

In **Example 9(b),** we saw that

$$\left(\sqrt{7} - \sqrt{10}\right)\left(\sqrt{7} + \sqrt{10}\right) = -3,$$

a rational number. This result suggests a way to rationalize a denominator that is a binomial in which one or both terms is a radical. The expressions $a\sqrt{m} + b\sqrt{n}$ and $a\sqrt{m} - b\sqrt{n}$ are called **conjugates.**

EXAMPLE 11   Rationalizing a Binomial Denominator

Rationalize the denominator of $\dfrac{1}{1 - \sqrt{2}}$.

Solution   The best approach is to multiply both numerator and denominator by the conjugate of the denominator—in this case, $1 + \sqrt{2}$.

$$\dfrac{1}{1 - \sqrt{2}} = \dfrac{1\left(1 + \sqrt{2}\right)}{\left(1 - \sqrt{2}\right)\left(1 + \sqrt{2}\right)} = \dfrac{1 + \sqrt{2}}{1 - 2} = -1 - \sqrt{2}$$

R.5   Exercises

Concept Check   *Match the rational exponent expression in Exercises 1–8 with the equivalent radical expression in A–H. Assume that $x \neq 0$.*

1. $(-3x)^{1/3}$ **2.** $-3x^{1/3}$ **3.** $(-3x)^{-1/3}$ **4.** $-3x^{-1/3}$ **5.** $(3x)^{1/3}$ **6.** $3x^{-1/3}$ **7.** $(3x)^{-1/3}$ **8.** $3x^{1/3}$

A. $\dfrac{3}{\sqrt[3]{x}}$ **B.** $-3\sqrt[3]{x}$ **C.** $\dfrac{1}{\sqrt[3]{3x}}$ **D.** $\dfrac{-3}{\sqrt[3]{x}}$ **E.** $3\sqrt[3]{x}$ **F.** $\sqrt[3]{-3x}$ **G.** $\sqrt[3]{3x}$ **H.** $\dfrac{1}{\sqrt[3]{-3x}}$

Write each expression in radical form. Assume that all variables represent positive real numbers.

9. $(-m)^{2/3}$ **10.** $p^{3/4}$ **11.** $(2m + p)^{2/3}$ **12.** $(5r + 3t)^{4/7}$

Write each expression with rational exponents. Assume that all variables represent nonnegative real numbers.

13. $\sqrt[5]{k^2}$

14. $-\sqrt[4]{z^5}$

15. $-3\sqrt{5p^3}$

16. $m\sqrt{2y^5}$

Concept Check   *Answer each question.*

17. For which of the following cases is $\sqrt{ab} = \sqrt{a} \cdot \sqrt{b}$ a true statement?
    A. a and b both positive   **B.** a and b both negative

18. For what positive integers n greater than or equal to 2 is $\sqrt[n]{a^n} = a$ always a true statement?

19. For what values of x is $\sqrt{9ax^2} = 3x\sqrt{a}$ a true statement? Assume that $a \geq 0$.

20. Which of the following expressions is *not* simplified? Give the simplified form.

    A. $\sqrt[3]{2y}$ **B.** $\dfrac{\sqrt{5}}{2}$ **C.** $\sqrt[4]{m^3}$ **D.** $\sqrt{\dfrac{3}{4}}$

If possible, simplify each radical expression. Assume that all variables represent positive real numbers.

21. $\sqrt[3]{125}$ **22.** $\sqrt[4]{81}$ **23.** $\sqrt[5]{-3125}$ **24.** $\sqrt[3]{343}$ **25.** $\sqrt{50}$

26. $\sqrt{45}$ **27.** $\sqrt[3]{81}$ **28.** $\sqrt[3]{250}$ **29.** $-\sqrt[4]{32}$ **30.** $-\sqrt[4]{243}$

31. $-\sqrt{\dfrac{9}{5}}$ **32.** $-\sqrt[3]{\dfrac{3}{2}}$ **33.** $-\sqrt[3]{\dfrac{4}{5}}$ **34.** $\sqrt[4]{\dfrac{3}{2}}$ **35.** $\sqrt[3]{16(-2)^4(2)^8}$

36. $\sqrt[3]{25(3)^4(5)^3}$ **37.** $\sqrt{8x^5z^8}$ **38.** $\sqrt{24m^6n^5}$ **39.** $\sqrt[3]{16z^5x^8y^4}$ **40.** $-\sqrt[6]{64a^{12}b^8}$

41. $\sqrt[4]{m^2n^7p^8}$ **42.** $\sqrt[4]{x^8y^7z^9}$ **43.** $\sqrt[4]{x^4 + y^4}$ **44.** $\sqrt[3]{27 + a^3}$

45. $\sqrt{\dfrac{2}{3x}}$ **46.** $\sqrt{\dfrac{5}{3p}}$ **47.** $\sqrt{\dfrac{x^5y^3}{z^2}}$ **48.** $\sqrt{\dfrac{g^3h^5}{r^3}}$

49. $\sqrt[3]{\dfrac{8}{x^2}}$ **50.** $\sqrt[3]{\dfrac{9}{16p^4}}$ **51.** $\sqrt[4]{\dfrac{g^3h^5}{9r^6}}$ **52.** $\sqrt[4]{\dfrac{32x^5}{y^5}}$

53. $\dfrac{\sqrt[3]{mn} \cdot \sqrt[3]{m^2}}{\sqrt[3]{n^2}}$ **54.** $\dfrac{\sqrt[3]{8m^2n^3} \cdot \sqrt[3]{2m^2}}{\sqrt[3]{32m^4n^3}}$ **55.** $\dfrac{\sqrt[4]{32x^5y} \cdot \sqrt[4]{2xy^4}}{\sqrt[4]{4x^3y^2}}$ **56.** $\dfrac{\sqrt[4]{rs^2t^3} \cdot \sqrt[4]{r^3s^2t}}{\sqrt[4]{r^2t^3}}$

57. $\sqrt[3]{\sqrt{4}}$ **58.** $\sqrt[4]{\sqrt[3]{2}}$ **59.** $\sqrt[6]{\sqrt[3]{x}}$ **60.** $\sqrt[8]{\sqrt[4]{y}}$

Simplify each expression, assuming that all variables represent nonnegative real numbers.

61. $4\sqrt{3} - 5\sqrt{12} + 3\sqrt{75}$ **62.** $2\sqrt{5} - 3\sqrt{20} + 2\sqrt{45}$ **63.** $3\sqrt{28p} - 4\sqrt{63p} + \sqrt{112p}$

64. $9\sqrt{8k} + 3\sqrt{18k} - \sqrt{32k}$ **65.** $2\sqrt[3]{3} + 4\sqrt[3]{24} - \sqrt[3]{81}$ **66.** $\sqrt[3]{32} - 5\sqrt[3]{4} + 2\sqrt[3]{108}$

67. $\dfrac{1}{\sqrt{3}} - \dfrac{2}{\sqrt{12}} + 2\sqrt{3}$ **68.** $\dfrac{1}{\sqrt{2}} + \dfrac{3}{\sqrt{8}} + \dfrac{1}{\sqrt{32}}$ **69.** $\dfrac{5}{\sqrt[3]{2}} - \dfrac{2}{\sqrt[3]{16}} + \dfrac{1}{\sqrt[3]{54}}$

70. $\dfrac{-4}{\sqrt[3]{3}} + \dfrac{1}{\sqrt[3]{24}} - \dfrac{2}{\sqrt[3]{81}}$ **71.** $\left(\sqrt{2} + 3\right)\left(\sqrt{2} - 3\right)$ **72.** $\left(\sqrt{5} + \sqrt{2}\right)\left(\sqrt{5} - \sqrt{2}\right)$

73. $\left(\sqrt[3]{11} - 1\right)\left(\sqrt[3]{11^2} + \sqrt[3]{11} + 1\right)$ **74.** $\left(\sqrt[3]{7} + 3\right)\left(\sqrt[3]{7^2} - 3\sqrt[3]{7} + 9\right)$ **75.** $\left(\sqrt{3} + \sqrt{8}\right)^2$

76. $\left(\sqrt{2} - 1\right)^2$ **77.** $\left(3\sqrt{2} + \sqrt{3}\right)\left(2\sqrt{3} - \sqrt{2}\right)$ **78.** $\left(4\sqrt{5} - 1\right)\left(3\sqrt{5} + 2\right)$

Rationalize the denominator of each radical expression. Assume that all variables represent nonnegative real numbers and that no denominators are 0.

79. $\dfrac{8}{\sqrt{5}}$

80. $\dfrac{3}{\sqrt{10}}$

81. $\dfrac{6}{\sqrt[3]{x^2}}$

82. $\dfrac{4}{\sqrt[3]{a^2}}$

83. $\dfrac{\sqrt{3}}{\sqrt{5} + \sqrt{3}}$

84. $\dfrac{\sqrt{7}}{\sqrt{3} - \sqrt{7}}$

85. $\dfrac{1 + \sqrt{3}}{3\sqrt{5} + 2\sqrt{3}}$

86. $\dfrac{\sqrt{7} - 1}{2\sqrt{7} + 4\sqrt{2}}$

87. $\dfrac{p}{\sqrt{p} + 2}$

88. $\dfrac{\sqrt{r}}{3 - \sqrt{r}}$

89. $\dfrac{a}{\sqrt{a + b} - 1}$

90. $\dfrac{3m}{2 + \sqrt{m + n}}$

R Test

1. Simplify each expression. Leave answers with exponents. Assume that all variables represent nonzero real numbers.

(a) $(-3)^4 \cdot (-3)^5$

(b) $(2x^3y)^2$

(c) $(-5z)^0$

(d) $\left(-\dfrac{4}{5}\right)^2$

(e) $-\left(\dfrac{m^2}{p}\right)^3$

2. Perform the indicated operations.

(a) $(x^2 - 3x + 2) - (x - 4x^2) + 3x(2x + 1)$

(b) $(6r - 5)^2$

(c) $(u + 2)(3u^2 - u + 4)$

(d) $(4x + 5)(4x - 5)$

(e) $[(5p - 1) + 4]^2$

3. Factor completely.

(a) $6x^2 - 17x + 7$

(b) $x^4 - 16$

(c) $z^2 - 6zk - 16k^2$

(d) $x^3y^2 - 9x^3 - 8y^2 + 72$

4. Write each rational expression in lowest terms, and give the domain.

(a) $\dfrac{16x^3}{4x^5}$

(b) $\dfrac{1 + k}{k^2 - 1}$

(c) $\dfrac{x^2 + x - 2}{x^2 + 5x + 6}$

5. Perform the indicated operations.

(a) $\dfrac{5x^2 - 9x - 2}{30x^3 + 6x^2} \cdot \dfrac{2x^8 + 6x^7 + 4x^6}{x^4 - 3x^2 - 4}$

(b) $\dfrac{x}{x^2 + 3x + 2} + \dfrac{2x}{2x^2 - x - 3}$

(c) $\dfrac{a + b}{2a - 3} - \dfrac{a - b}{3 - 2a}$

(d) $\dfrac{g - \dfrac{2}{g}}{g - \dfrac{4}{g}}$

6. Simplify each expression. Assume that all variables represent positive real numbers.

(a) $(-7)^{-2}$

(b) $(16x^8)^{3/4}$

(c) $\left(\dfrac{64}{27}\right)^{-2/3}$

7. Perform the indicated operations. Write your answers with only positive exponents. Assume that all variables represent positive real numbers.

(a) $\dfrac{5^{-3} \cdot 5^{-1}}{5^{-2}}$

(b) $(x + 2)^{-1/5}(x + 2)^{-7/10}$

(c) $(m^{-1}n^{1/2})^4 (4m^3n^{-2})^{-1/2}$

8. Simplify each expression. Assume that all variables represent positive real numbers.

(a) $\sqrt[4]{16}$

(b) $\sqrt[3]{\dfrac{2}{3}}$

(c) $\sqrt{18x^5y^8}$

(d) $\dfrac{\sqrt[4]{pq} \cdot \sqrt[4]{q^2}}{\sqrt[4]{p^3}}$

9. Simplify. Assume that all variables represent positive real numbers.

(a) $\sqrt{32x} + \sqrt{2x} - \sqrt{18x}$

(b) $(\sqrt[3]{2} + 4)(\sqrt[3]{4} + 1)$

(c) $(\sqrt{x} - \sqrt{y})(\sqrt{x} + \sqrt{y})$

10. Rationalize the denominator of $\dfrac{14}{\sqrt{11} - \sqrt{7}}$ and simplify.

Appendices

A Geometry Formulas

The following table provides a summary of some important formulas from geometry.

Figure	Formula	Example
Square	Perimeter: $P = 4s$ Area: $\mathcal{A} = s^2$	
Rectangle	Perimeter: $P = 2L + 2W$ Area: $\mathcal{A} = LW$	
Triangle	Perimeter: $P = a + b + c$ Area: $\mathcal{A} = \frac{1}{2}bh$	
Equilateral Triangle	Perimeter $P = 3s$ Area: $\mathcal{A} = \frac{\sqrt{3}}{4}s^2$	
Pythagorean Theorem (for Right Triangles)	$c^2 = a^2 + b^2$	
Sum of the Angles of a Triangle	$A + B + C = 180°$	
Sum of the Angles in a Quadrilateral	$A + B + C + D = 360°$	

Figure	Formula	Example
Circle	Diameter: $d = 2r$ Circumference: $C = 2\pi r = \pi d$ Area: $\mathcal{A} = \pi r^2$	
Parallelogram	Area: $\mathcal{A} = bh$ Perimeter: $P = 2a + 2b$	
Trapezoid	Area: $\mathcal{A} = \frac{1}{2}h(b_1 + b_2)$ Perimeter: $P = a + b_1 + c + b_2$	
Sphere	Volume: $V = \frac{4}{3}\pi r^3$ Surface area: $S = 4\pi r^2$	

Cone	Volume: $V = \frac{1}{3}\pi r^2 h$
	Surface area:
	  $S = \pi r\sqrt{r^2 + h^2}$
	  (excludes the base)	

Cube	Volume: $V = s^3$ Surface area: $S = 6s^2$	
Rectangular Solid	Volume: $V = LWH$ Surface area: $S = 2HW + 2LW + 2LH$	

Right Circular Cylinder	Volume: $V = \pi r^2 h$
	Surface area:
	  $S = 2\pi rh + 2\pi r^2$
	  (includes top and bottom)	

Right Pyramid	Volume: $V = \frac{1}{3}Bh$ $B = $ area of the base	

B Vectors in Space

Rectangular Coordinates in Space • Vectors in Space • Vector Definitions and Operations • Direction Angles in Space
• An Application of the Dot Product

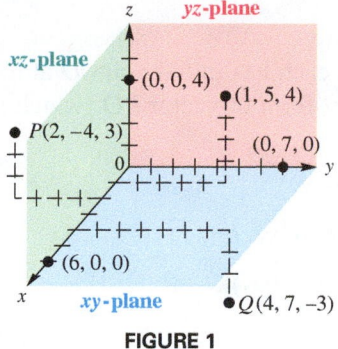

FIGURE 1

Rectangular Coordinates in Space

Vector quantities occur in space as well as in a plane, so properties for vectors in a two-dimensional plane can be extended to three-dimensional space. The plane determined by the x- and y-axes is called the **xy-plane.** A third axis is needed to locate points in space. The z-axis goes through the origin in the xy-plane and is perpendicular to both the x-axis and the y-axis. **FIGURE 1** shows one way to orient the three axes, with the yz-plane as the plane of the page and the x-axis perpendicular to it.

We associate each point in space with an **ordered triple** (x, y, z). **FIGURE 1** shows several ordered triples. The region of three-dimensional space where all coordinates are positive is called the **first octant.** There are eight octants in all.

The distance formula given in **Chapter 1** is a special case of the distance formula for points in space.

Distance Formula (Three-Dimensional Space)

If $P_1(x_1, y_1, z_1)$ and $P_2(x_2, y_2, z_2)$ are two points in a three-dimensional coordinate system, then the distance between P_1 and P_2 is given by the following.

$$d(P_1, P_2) = \sqrt{(x_2 - x_1)^2 + (y_2 - y_1)^2 + (z_2 - z_1)^2}$$

EXAMPLE 1 **Finding the Distance between Two Points in Space**

Find the distance between the points $P(2, -4, 3)$ and $Q(4, 7, -3)$ shown in **FIGURE 1.**

Solution By the distance formula, we have the following.

$$d(P, Q) = \sqrt{(4 - 2)^2 + [7 - (-4)]^2 + (-3 - 3)^2} = \sqrt{161}$$

⇨ Looking Ahead to Calculus
The concepts introduced in this section are extended in calculus to calculate surface area and volume of three-dimensional figures.

Vectors in Space

Extending the notation used to represent vectors in the plane, we denote a vector $\mathbf{v}$ in space with initial point O at the origin as

$$\mathbf{v} = \langle a, b, c \rangle. \qquad \text{Component form}$$

Using the unit vectors $\mathbf{i} = \langle 1, 0, 0 \rangle$, $\mathbf{j} = \langle 0, 1, 0 \rangle$, and $\mathbf{k} = \langle 0, 0, 1 \rangle$ in the directions of the positive x-axis, y-axis, and z-axis, respectively, we can also represent $\mathbf{v}$ as

$$\mathbf{v} = a\mathbf{i} + b\mathbf{j} + c\mathbf{k}, \qquad \text{i, j, k form}$$

where a, b, and c are scalars. The scalars a, b, and c are the **components** of vector $\mathbf{v}$.

EXAMPLE 2 **EXAMPLE 2**   **Writing Vectors in i, j, k Form**

Write the vectors **OP** and **OQ**, using the components for points P and Q defined in **Example 1** and vectors **i**, **j**, and **k**.

Solution   For $P(2, -4, 3)$, the components of **OP** are 2, -4, and 3. Similarly, the components of **OQ** are 4, 7, and -3.

$$\mathbf{OP} = 2\mathbf{i} - 4\mathbf{j} + 3\mathbf{k} \quad \text{and} \quad \mathbf{OQ} = 4\mathbf{i} + 7\mathbf{j} - 3\mathbf{k}$$

Vectors **OP** and **OQ** are position vectors because each has initial point O at the origin. For $P(x_1, y_1, z_1)$ and $Q(x_2, y_2, z_2)$, the component form of vector **PQ** (which is not a position vector) is represented as follows.

$$\mathbf{PQ} = \langle x_2 - x_1, y_2 - y_1, z_2 - z_1 \rangle$$

As **FIGURE 2** suggests, **PQ** is equal to the following position vector.

$$\mathbf{OR} = (x_2 - x_1)\mathbf{i} + (y_2 - y_1)\mathbf{j} + (z_2 - z_1)\mathbf{k}$$

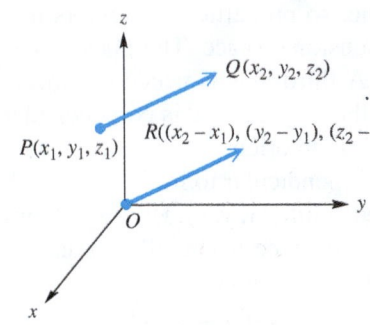

FIGURE 2

EXAMPLE 3   **Finding a Position Vector That Corresponds to a Given Vector**

Find the position vector **v** corresponding to **PQ** with $P(2, -4, 3)$ and $Q(4, 7, -3)$.

Solution   From the previous definition, the position vector is as follows.

$$\mathbf{v} = (4 - 2)\mathbf{i} + [7 - (-4)]\mathbf{j} + (-3 - 3)\mathbf{k}$$
$$= 2\mathbf{i} + 11\mathbf{j} - 6\mathbf{k}$$

Vector Definitions and Operations

The vector concepts discussed in **Section 10.3** can be extended to vectors in space.

Vector Definitions and Operations

If $\mathbf{v} = a\mathbf{i} + b\mathbf{j} + c\mathbf{k}$ and $\mathbf{w} = d\mathbf{i} + e\mathbf{j} + f\mathbf{k}$ are vectors and g is a scalar, then the following hold.

$$\mathbf{v} = \mathbf{w} \text{ if and only if } a = d, b = e, \text{ and } c = f$$
$$\mathbf{v} + \mathbf{w} = (a + d)\mathbf{i} + (b + e)\mathbf{j} + (c + f)\mathbf{k}$$
$$\mathbf{v} - \mathbf{w} = (a - d)\mathbf{i} + (b - e)\mathbf{j} + (c - f)\mathbf{k}$$
$$g\mathbf{v} = ga\mathbf{i} + gb\mathbf{j} + gc\mathbf{k}$$
$$|\mathbf{v}| = \sqrt{a^2 + b^2 + c^2}$$
$$\mathbf{v} \cdot \mathbf{w} = ad + be + cf$$

EXAMPLE 4   **Performing Vector Operations**

Find the following if $\mathbf{v} = 2\mathbf{i} + 6\mathbf{j} - 4\mathbf{k}$ and $\mathbf{w} = -\mathbf{i} + 5\mathbf{k}$.

(a) $\mathbf{v} + \mathbf{w}$ **(b)** $\mathbf{w} - \mathbf{v}$ **(c)** $-10\mathbf{w}$

(d) $|\mathbf{v}|$ **(e)** $\mathbf{v} \cdot \mathbf{w}$ **(f)** $\mathbf{w} \cdot \mathbf{v}$

Solution

(a) $\mathbf{v} + \mathbf{w} = (2 - 1)\mathbf{i} + (6 + 0)\mathbf{j} + (-4 + 5)\mathbf{k} = \mathbf{i} + 6\mathbf{j} + \mathbf{k}$

(b) $\mathbf{w} - \mathbf{v} = (-1 - 2)\mathbf{i} + (0 - 6)\mathbf{j} + [5 - (-4)]\mathbf{k} = -3\mathbf{i} - 6\mathbf{j} + 9\mathbf{k}$

(c) $-10\mathbf{w} = -10(-1)\mathbf{i} + (-10)5\mathbf{k} = 10\mathbf{i} - 50\mathbf{k}$

(d) $|\mathbf{v}| = \sqrt{2^2 + 6^2 + (-4)^2} = \sqrt{56} = 2\sqrt{14}$

(e) $\mathbf{v} \cdot \mathbf{w} = 2(-1) + 6(0) + (-4)5 = -22$

(f) $\mathbf{w} \cdot \mathbf{v} = -1(2) + 0(6) + 5(-4) = -22$

Properties of the dot product for vectors in two dimensions are extended to vectors in three dimensions, as is the geometric interpretation of the dot product.

> **If θ is the angle between two nonzero vectors $\mathbf{v}$ and $\mathbf{w}$ in space, where $0° \leq \theta \leq 180°$, then $\mathbf{v} \cdot \mathbf{w} = |\mathbf{v}|\,|\mathbf{w}|\cos\theta$.**

Dividing each side of the equation by $|\mathbf{v}|\,|\mathbf{w}|$ gives the following result.

Angle between Two Vectors

If θ is the angle between two nonzero vectors $\mathbf{v}$ and $\mathbf{w}$, where $0° \leq \theta \leq 180°$, then

$$\cos\theta = \frac{\mathbf{v} \cdot \mathbf{w}}{|\mathbf{v}||\mathbf{w}|}.$$

EXAMPLE 5   **Finding the Angle between Two Vectors**

Find the angle between $\mathbf{v} = 7\mathbf{i} - 3\mathbf{j} - 5\mathbf{k}$ and $\mathbf{w} = 4\mathbf{i} + 6\mathbf{j} - 8\mathbf{k}$.

Solution   First find $|\mathbf{v}|$ and $|\mathbf{w}|$.

$$|\mathbf{v}| = \sqrt{7^2 + (-3)^2 + (-5)^2} = \sqrt{83}$$
$$|\mathbf{w}| = \sqrt{4^2 + 6^2 + (-8)^2} = \sqrt{116} = 2\sqrt{29}$$

Now, use the formula for $\cos\theta$.

$$\cos\theta = \frac{\mathbf{v} \cdot \mathbf{w}}{|\mathbf{v}|\,|\mathbf{w}|} = \frac{7(4) + (-3)6 + (-5)(-8)}{\sqrt{83} \cdot 2\sqrt{29}} \approx 0.5096$$

A calculator gives $\theta \approx \cos^{-1}(0.5096) \approx 59.4°$.

Direction Angles in Space

A vector in two dimensions is determined by its magnitude and direction angle. In three dimensions, a vector is determined by its magnitude and three direction angles—that is, the angles between the vector and each positive axis. As shown in **FIGURE 3,**

α is the direction angle between $\mathbf{v}$ and the positive x-axis,

β is the direction angle between $\mathbf{v}$ and the positive y-axis,

and γ is the direction angle between $\mathbf{v}$ and the positive z-axis.

We evaluate these angles using the expression for the cosine of the angle between two vectors. Recall that $\mathbf{i} = \langle 1, 0, 0 \rangle$, $\mathbf{j} = \langle 0, 1, 0 \rangle$, $\mathbf{k} = \langle 0, 0, 1 \rangle$, and each has magnitude 1. For $\mathbf{v} = a\mathbf{i} + b\mathbf{j} + c\mathbf{k}$, the three direction angles are determined as follows.

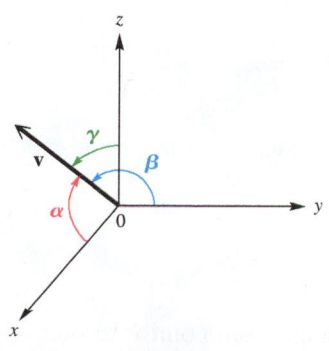

FIGURE 3

> **Direction Angles**
>
> $$\cos \alpha = \frac{\mathbf{v} \cdot \mathbf{i}}{|\mathbf{v}||\mathbf{i}|} = \frac{a}{|\mathbf{v}|}, \quad \cos \beta = \frac{\mathbf{v} \cdot \mathbf{j}}{|\mathbf{v}||\mathbf{j}|} = \frac{b}{|\mathbf{v}|}, \quad \cos \gamma = \frac{\mathbf{v} \cdot \mathbf{k}}{|\mathbf{v}||\mathbf{k}|} = \frac{c}{|\mathbf{v}|}$$

The quantities $\cos \alpha$, $\cos \beta$, and $\cos \gamma$ are called **direction cosines**. In **Exercise 49,** you are asked to prove that

$$\cos^2 \alpha + \cos^2 \beta + \cos^2 \gamma = 1.$$

EXAMPLE 6 Finding Direction Angles of a Vector

Find the direction angles of $\mathbf{w} = 2\mathbf{i} - 3\mathbf{j} + 6\mathbf{k}$. Give answers in degrees, to the nearest tenth.

Solution $|\mathbf{w}| = \sqrt{4 + 9 + 36} = 7$

$$\cos \alpha = \frac{a}{|\mathbf{w}|} = \frac{2}{7} \qquad \cos \beta = \frac{b}{|\mathbf{w}|} = -\frac{3}{7} \qquad \cos \gamma = \frac{c}{|\mathbf{w}|} = \frac{6}{7}$$

$$\alpha \approx 73.4° \qquad\qquad \beta \approx 115.4° \qquad\qquad \gamma \approx 31.0°$$

EXAMPLE 7 Verifying the Sum of the Squares of the Direction Angle Cosines

Verify that $\cos^2 \alpha + \cos^2 \beta + \cos^2 \gamma = 1$ for the vector in **Example 6.**

Solution $\left(\dfrac{2}{7}\right)^2 + \left(-\dfrac{3}{7}\right)^2 + \left(\dfrac{6}{7}\right)^2 = \dfrac{4}{49} + \dfrac{9}{49} + \dfrac{36}{49} = 1$ ✓

EXAMPLE 8 Finding a Direction Angle

FIGURE 4 shows that the angle between a vector $\mathbf{u}$ and the positive z-axis is $120°$. The angle between $\mathbf{u}$ and the positive x-axis is $90°$. What acute angle does $\mathbf{u}$ make with the positive y-axis?

Solution Here, $\cos \gamma = \cos 120° = -\dfrac{1}{2}$ and $\cos \alpha = \cos 90° = 0$.

$$\cos^2 \alpha + \cos^2 \beta + \cos^2 \gamma = 1$$

$$\cos^2 \beta = 1 - \cos^2 \alpha - \cos^2 \gamma \qquad \text{Subtract } \cos^2 \alpha \text{ and } \cos^2 \gamma.$$

$$\cos^2 \beta = 1 - 0 - \frac{1}{4} \qquad \text{Substitute; } \left(-\tfrac{1}{2}\right)^2 = \tfrac{1}{4}.$$

$$\cos^2 \beta = \frac{3}{4} \qquad \text{Subtract.}$$

$$\cos \beta = \frac{\sqrt{3}}{2} \qquad \text{Take square roots; } 0° < \beta < 90°.$$

$$\beta = 30° \qquad \cos^{-1}\left(\frac{\sqrt{3}}{2}\right) = 30°$$

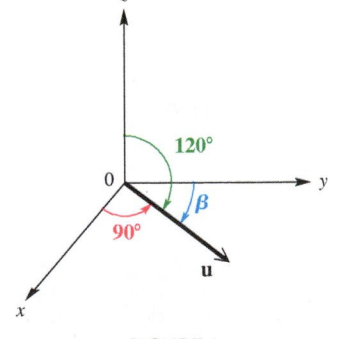

FIGURE 4

An Application of the Dot Product

The **work** done by a constant force $\mathbf{F}$ as it moves a particle from point P to point Q is defined as $\mathbf{F} \cdot \mathbf{PQ}$.

EXAMPLE 9   **Applying the Dot Product to Work**

Find the work (in units) done by a force $\mathbf{F} = \langle 3, 4, 2 \rangle$ that moves a particle from $P(1, 0, 5)$ to $Q(3, 3, 8)$.

Solution   Since P is the initial point and Q is the terminal point,

$$\mathbf{PQ} = \langle 3 - 1, 3 - 0, 8 - 5 \rangle = \langle 2, 3, 3 \rangle.$$

Then the work done is calculated as follows.

$$\mathbf{F} \cdot \mathbf{PQ} = \langle 3, 4, 2 \rangle \cdot \langle 2, 3, 3 \rangle$$
$$= 6 + 12 + 6$$
$$= 24 \text{ (work units)}$$

B Exercises

Concept Check   *Fill in the blanks to complete each statement.*

1. The plane determined by the x-axis and the z-axis is called the _____.

2. A position vector has its _____ point at the origin.

3. The component form of the position vector with terminal point $(5, 3, -2)$ is _____.

4. The **i, j, k** form of the position vector with terminal point $(6, -1, -3)$ is _____.

Checking Analytic Skills   *Find the distance between the points P and Q.* ***Do not use a calculator.***

5. $P(0, 0, 0)$; $Q(2, -2, 5)$

6. $P(0, 0, 0)$; $Q(7, 4, -1)$

7. $P(10, 15, 9)$; $Q(8, 3, -4)$

8. $P(5, 4, -4)$; $Q(3, 7, 2)$

9. $P(20, 25, 16)$; $Q(5, 5, 6)$

10. $P(14, 10, 18)$; $Q(-2, 4, 9)$

For each pair of points, find the position vector that corresponds to **PQ** *in component form and in* **i, j, k** *form.*

11. $P(0, 0, 0)$; $Q(2, -2, 5)$

12. $P(0, 0, 0)$; $Q(7, 4, -1)$

13. $P(10, 15, 0)$; $Q(8, 3, -4)$

14. $P(0, 4, -4)$; $Q(3, 7, 2)$

15. $P(20, 25, 6)$; $Q(5, 5, 16)$

16. $P(14, 10, 18)$; $Q(-2, 4, 9)$

17. Using the points in **Exercise 11,** write **QP** in component form. How does it compare with **PQ?**

18. Compare the distance between points P and Q in three-dimensional space with the magnitude of vector **PQ.**

Given $\mathbf{u} = 2\mathbf{i} + 4\mathbf{j} + 7\mathbf{k}$, $\mathbf{v} = -3\mathbf{i} + 5\mathbf{j} + 2\mathbf{k}$, *and* $\mathbf{w} = 4\mathbf{i} - 3\mathbf{j} - 6\mathbf{k}$, *find the following.*

19. $\mathbf{u} - \mathbf{w}$

20. $\mathbf{v} + \mathbf{w}$

21. $4\mathbf{u} + 5\mathbf{v}$

22. $-\mathbf{v} + 3\mathbf{u}$

23. $|\mathbf{u}|$

24. $|\mathbf{w}|$

25. $|\mathbf{w} + \mathbf{u}|$

26. $|2\mathbf{v}|$

27. $\mathbf{v} \cdot \mathbf{w}$

28. $\mathbf{u} \cdot \mathbf{w}$

29. $\mathbf{v} \cdot \mathbf{v}$

30. $\mathbf{u} \cdot \mathbf{u}$

Find the angle between each pair of vectors. Give angles in degrees, to the nearest tenth.

31. $\langle 2, -2, 0 \rangle, \langle 5, -2, -1 \rangle$

32. $\langle 4, 0, 0 \rangle, \langle 5, 3, -2 \rangle$

33. $\langle 6, 0, 0 \rangle, \langle 8, 3, -4 \rangle$

34. $\langle -1, 2, -3 \rangle, \langle 0, -2, 1 \rangle$

35. $\langle 1, 0, 0 \rangle, \langle 0, 1, 0 \rangle$

36. $\langle 0, 0, 1 \rangle, \langle 0, 1, 0 \rangle$

Find the direction angles of each vector. Give answers in degrees, to the nearest tenth. Check your work by showing that the sum of the squares of the direction cosines equals 1.

37. $\mathbf{u} = 2\mathbf{i} + 4\mathbf{j} + 7\mathbf{k}$

38. $\mathbf{v} = -3\mathbf{i} + 5\mathbf{j} + 2\mathbf{k}$

39. $\mathbf{w} = 4\mathbf{i} - 3\mathbf{j} - 6\mathbf{k}$

40. $\mathbf{y} = 2\mathbf{i} - 3\mathbf{j} + 4\mathbf{k}$

Work each problem.

41. The angle between vector **u** and the positive *x*-axis is 45° and the angle between **u** and the positive *y*-axis is 120°. What is the angle between **u** and the positive *z*-axis?

42. The direction angle between vector **v** and the *y*-axis is 135°, while the direction angle between **v** and the *z*-axis is 90°. What is the direction angle between **v** and the *x*-axis?

 43. What must be true for two vectors in space to be parallel?

 44. Decide whether the vectors $\langle 3, 5, -1 \rangle$ and $\langle -12, -20, 4 \rangle$ are parallel. Explain your answer.

Work   *Find the work (in units) done by a force **F** in moving a particle from P to Q.*

45. $\mathbf{F} = \langle 2, 0, 5 \rangle$; $P(0, 0, 0)$; $Q(1, 3, 2)$

46. $\mathbf{F} = \langle 3, 2, 0 \rangle$; $P(0, 0, 0)$; $Q(4, -2, 5)$

47. $\mathbf{F} = \mathbf{i} + 2\mathbf{j} - \mathbf{k}$; $P(2, -1, 2)$; $Q(5, 7, 8)$

48. $\mathbf{F} = 3\mathbf{i} + 2\mathbf{j}$; $P(4, 7, 6)$; $Q(10, 15, 12)$

49. Prove that for direction cosines $\cos \alpha$, $\cos \beta$, and $\cos \gamma$, $\cos^2 \alpha + \cos^2 \beta + \cos^2 \gamma = 1$.

C Polar Form of Conic Sections

Up to this point, we have worked with equations of conic sections in rectangular form. If the focus of a conic section is at the pole, the polar form of its equation is

$$r = \frac{ep}{1 \pm e \cdot f(\theta)},$$

where f is the sine or cosine function.

> ### Polar Forms of Conic Sections
>
> A polar equation of the form
>
> $$r = \frac{ep}{1 \pm e \cos \theta} \quad \text{or} \quad r = \frac{ep}{1 \pm e \sin \theta}$$
>
> has a conic section as its graph. The eccentricity is e (where $e > 0$), and $|p|$ is the distance between the pole (focus) and the directrix.

We can verify that $r = \frac{ep}{1 + e \cos \theta}$ does indeed satisfy the definition of a conic section. Consider **FIGURE 1**, where the directrix is vertical and $p > 0$ units to the right of the focus $F(0, 0°)$. Let $P(r, \theta)$ be a point on the graph. Then the distance between P and the directrix is found as follows.

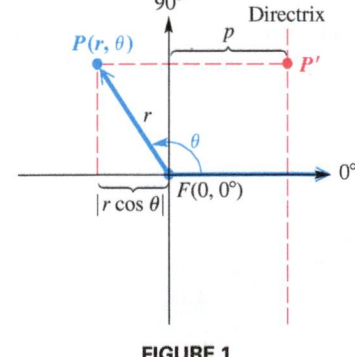

FIGURE 1

$$PP' = |p - x|$$

$$= |p - r \cos \theta| \qquad \textcolor{blue}{x = r \cos \theta}$$

$$= \left| p - \left(\frac{ep}{1 + e \cos \theta} \right) \cos \theta \right| \qquad \textcolor{blue}{\text{Use the equation for } r.}$$

$$= \left| \frac{p(1 + e \cos \theta) - ep \cos \theta}{1 + e \cos \theta} \right| \qquad \textcolor{blue}{\text{Write with a common denominator.}}$$

$$= \left| \frac{p + ep \cos \theta - ep \cos \theta}{1 + e \cos \theta} \right| \qquad \textcolor{blue}{\text{Distributive property}}$$

$$PP' = \left| \frac{p}{1 + e \cos \theta} \right| \qquad \textcolor{blue}{\text{Simplify.}}$$

Since

$$r = \frac{ep}{1 + e \cos \theta},$$

we can multiply each side by $\frac{1}{e}$.

$$\frac{r}{e} = \frac{p}{1 + e \cos \theta}$$

Substitute $\frac{r}{e}$ for the expression in the absolute value bars for PP'.

$$PP' = \left| \frac{p}{1 + e \cos \theta} \right| = \left| \frac{r}{e} \right| = \frac{|r|}{|e|} = \frac{|r|}{e} \qquad |e| = e$$

The distance between the pole and P is $PF = |r|$, so the ratio of PF to PP' is

$$\frac{PF}{PP'} = \frac{|r|}{\frac{|r|}{e}} = |r| \div \frac{|r|}{e} = |r| \cdot \frac{e}{|r|} = e. \qquad \text{Simplify the complex fraction.}$$

Thus, by the definition, the graph has eccentricity e and must be a conic.

In the preceding discussion, we assumed a vertical directrix to the right of the pole. There are three other possible situations, and all four are summarized here.

If the equation is:	then the directrix is:
$r = \dfrac{ep}{1 + e \cos \theta}$	*vertical, p* units to the *right* of the pole.
$r = \dfrac{ep}{1 - e \cos \theta}$	*vertical, p* units to the *left* of the pole.
$r = \dfrac{ep}{1 + e \sin \theta}$	*horizontal, p* units *above* the pole.
$r = \dfrac{ep}{1 - e \sin \theta}$	*horizontal, p* units *below* the pole.

EXAMPLE 1 **Graphing a Conic Section with Equation in Polar Form**

Graph $r = \dfrac{8}{4 + 4 \sin \theta}$.

Analytic Solution

Divide both numerator and denominator by 4 to get

$$r = \frac{2}{1 + \sin \theta}.$$

From the preceding table, this is the equation of a conic with $ep = 2$ and $e = 1$. Thus, $p = 2$. Since $e = 1$, the graph is a parabola. The focus is at the pole, and the directrix is horizontal, 2 units *above* the pole.

The vertex must have polar coordinates $(1, 90°)$. Letting $\theta = 0°$ and $\theta = 180°$ gives the additional points $(2, 0°)$ and $(2, 180°)$. See **FIGURE 2** on the next page.

Graphing Calculator Solution

Enter

$$r_1 = \frac{8}{4 + 4 \sin \theta},$$

with the calculator in polar and degree modes. The first two screens in **FIGURE 3** on the next page show the window settings, and the third screen shows the graph. Notice that the point $(1, 90°)$ is indicated at the bottom of the third screen.

(continued)

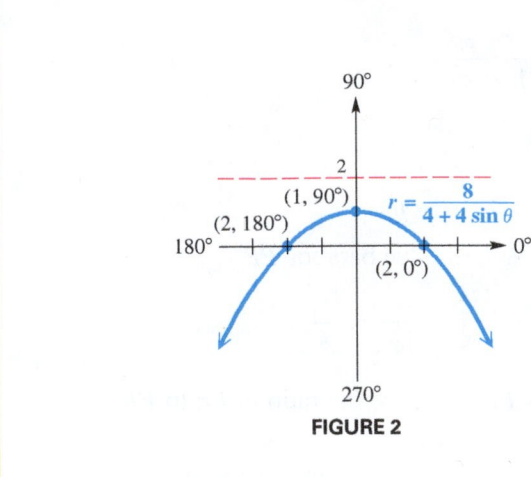

FIGURE 2

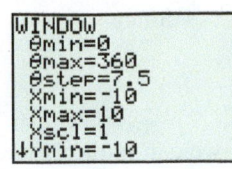

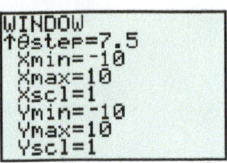

This is a continuation of the screen to the left.

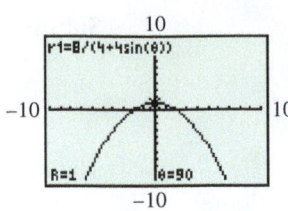

Degree mode

FIGURE 3

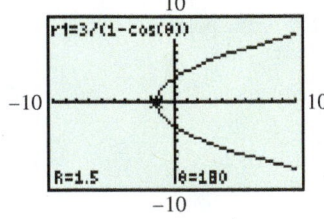

Degree mode

FIGURE 4

EXAMPLE 2   **Finding a Polar Equation**

Find the polar equation of a parabola with focus at the pole and vertical directrix 3 units to the left of the pole.

Solution   The eccentricity e must be 1, p must equal 3, and the equation must be of the following form.

$$r = \frac{ep}{1 - e\cos\theta} = \frac{1\cdot 3}{1 - 1\cos\theta} = \frac{3}{1 - \cos\theta} \qquad e = 1,\ p = 3$$

The calculator graph in **FIGURE 4** supports our result. When $\theta = 180°$, $r = 1.5$. The distance from $F(0, 0°)$ to the directrix is $2r = 2(1.5) = 3$ units, as required.

EXAMPLE 3   **Identifying and Converting from Polar Form to Rectangular Form**

Identify the conic represented by the following equation. Convert it to rectangular form.

$$r = \frac{8}{2 - \cos\theta}$$

Solution   To identify the type of conic, we divide both the numerator and the denominator on the right side by 2.

Be sure to divide each term in the numerator and denominator by 2.

$$r = \frac{4}{1 - \frac{1}{2}\cos\theta} \qquad \text{See table on previous page.}$$

This is a conic that has a vertical directrix with $e = \frac{1}{2}$. Thus, it is an ellipse.

To convert to rectangular form, we start with the given equation.

$$r = \frac{8}{2 - \cos \theta}$$

$r(2 - \cos \theta) = 8$	Multiply by $2 - \cos \theta$.
$2r - r \cos \theta = 8$	Distributive property
$2r = r \cos \theta + 8$	Add $r \cos \theta$ to each side.
$(2r)^2 = (r \cos \theta + 8)^2$	Square each side.
$(2r)^2 = (x + 8)^2$	$r \cos \theta = x$
$4r^2 = x^2 + 16x + 64$	Multiply.
$4(x^2 + y^2) = x^2 + 16x + 64$	$r^2 = x^2 + y^2$
$4x^2 + 4y^2 = x^2 + 16x + 64$	Distributive property
$3x^2 + 4y^2 - 16x - 64 = 0$	Standard form

The coefficients of x^2 and y^2 are both positive and are not equal, further supporting our assertion that the graph is an ellipse.

C Exercises

Graph each conic whose equation is given in polar form. Use a traditional or a calculator graph, as directed by your instructor.

1. $r = \dfrac{6}{3 + 3 \sin \theta}$

2. $r = \dfrac{10}{5 + 5 \sin \theta}$

3. $r = \dfrac{-4}{6 + 2 \cos \theta}$

4. $r = \dfrac{-8}{4 + 2 \cos \theta}$

5. $r = \dfrac{2}{2 - 4 \sin \theta}$

6. $r = \dfrac{6}{2 - 4 \sin \theta}$

7. $r = \dfrac{4}{2 - 4 \cos \theta}$

8. $r = \dfrac{6}{2 - 4 \cos \theta}$

9. $r = \dfrac{-1}{1 + 2 \sin \theta}$

10. $r = \dfrac{-1}{1 - 2 \sin \theta}$

11. $r = \dfrac{-1}{2 + \cos \theta}$

12. $r = \dfrac{-1}{2 - \cos \theta}$

Find a polar equation of the parabola with focus at the pole, satisfying the given conditions.

13. Vertical directrix 3 units to the right of the pole

14. Vertical directrix 4 units to the left of the pole

15. Horizontal directrix 5 units below the pole

16. Horizontal directrix 6 units above the pole

Find a polar equation for the conic with focus at the pole, satisfying the given conditions. Also, identify the type of conic represented.

17. $e = \frac{4}{5}$; vertical directrix 5 units to the right of the pole

18. $e = \frac{2}{3}$; vertical directrix 6 units to the left of the pole

19. $e = \frac{5}{4}$; horizontal directrix 8 units below the pole

20. $e = \frac{3}{2}$; horizontal directrix 4 units above the pole

Identify the type of conic represented, and convert the equation to rectangular form.

21. $r = \dfrac{6}{3 - \cos \theta}$

22. $r = \dfrac{8}{4 - \cos \theta}$

23. $r = \dfrac{-2}{1 + 2 \cos \theta}$

24. $r = \dfrac{-3}{1 + 3 \cos \theta}$

25. $r = \dfrac{-6}{4 + 2 \sin \theta}$

26. $r = \dfrac{-12}{6 + 3 \sin \theta}$

27. $r = \dfrac{10}{2 - 2 \sin \theta}$

28. $r = \dfrac{12}{4 - 4 \sin \theta}$

D Rotation of Axes

Derivation of Rotation Equations • Applying a Rotation Equation

→ **Looking Ahead to Calculus**
Rotation of axes is a topic traditionally covered in calculus texts, in conjunction with parametric equations and polar coordinates. The coverage in calculus is typically the same as that seen in this section.

Derivation of Rotation Equations

If we begin with an xy-coordinate system having origin O and rotate the axes about O through an angle θ, the new coordinate system is called a **rotation** of the xy-system. Trigonometric identities are used to obtain equations for converting the coordinates of a point from the xy-system to the rotated $x'y'$-system.

Let P be any point other than the origin, with coordinates (x, y) in the xy-system and (x', y') in the $x'y'$-system. See **FIGURE 1.** Let $OP = r$, and let α represent the angle made by OP and the x'-axis. As shown in **FIGURE 1,** the following hold.

$$\cos(\theta + \alpha) = \frac{OA}{r} = \frac{x}{r},$$

$$\sin(\theta + \alpha) = \frac{AP}{r} = \frac{y}{r},$$

$$\cos \alpha = \frac{OB}{r} = \frac{x'}{r},$$

$$\sin \alpha = \frac{PB}{r} = \frac{y'}{r}$$

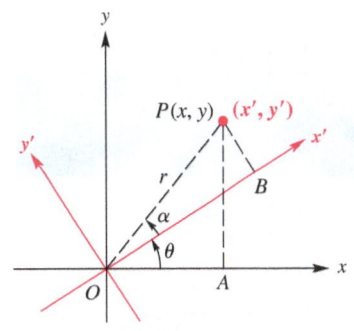

FIGURE 1

These four statements can be written as follows.

$$x = r\cos(\theta + \alpha), \quad y = r\sin(\theta + \alpha), \quad x' = r\cos\alpha, \quad y' = r\sin\alpha$$

The identity for the cosine of the sum of two angles gives the following equation.

$$
\begin{aligned}
x &= r\cos(\theta + \alpha) \\
&= r(\cos\theta\cos\alpha - \sin\theta\sin\alpha) && \text{Cosine sum identity} \\
&= (r\cos\alpha)\cos\theta - (r\sin\alpha)\sin\theta && \text{Distributive and other properties} \\
&= x'\cos\theta - y'\sin\theta && \text{Substitute.}
\end{aligned}
$$

In the same way, the identity for the sine of the sum of two angles gives

$$y = x'\sin\theta + y'\cos\theta.$$

This proves the following result.

Rotation Equations

If the rectangular coordinate axes are rotated about the origin through an angle θ, and if the coordinates of a point P are (x, y) and (x', y') in the xy-system and the $x'y'$-system, respectively, then the respective **rotation equations** are as follows.

$$x = x'\cos\theta - y'\sin\theta \quad \text{and} \quad y = x'\sin\theta + y'\cos\theta$$

Applying a Rotation Equation

EXAMPLE 1 Finding an Equation after a Rotation

The equation of a curve is $x^2 + y^2 + 2xy + 2\sqrt{2}x - 2\sqrt{2}y = 0$. Find the resulting equation if the axes are rotated 45°. Graph the equation.

Solution If $\theta = 45°$, then $\sin\theta = \dfrac{\sqrt{2}}{2}$ and $\cos\theta = \dfrac{\sqrt{2}}{2}$, and the rotation equations become

$$x = \frac{\sqrt{2}}{2}x' - \frac{\sqrt{2}}{2}y' \quad \text{and} \quad y = \frac{\sqrt{2}}{2}x' + \frac{\sqrt{2}}{2}y'.$$

Substituting these values into the given equation yields the following.

$$x^2 + y^2 + 2xy + 2\sqrt{2}x - 2\sqrt{2}y = 0$$

$$\left[\frac{\sqrt{2}}{2}x' - \frac{\sqrt{2}}{2}y'\right]^2 + \left[\frac{\sqrt{2}}{2}x' + \frac{\sqrt{2}}{2}y'\right]^2 + 2\left[\frac{\sqrt{2}}{2}x' - \frac{\sqrt{2}}{2}y'\right]\left[\frac{\sqrt{2}}{2}x' + \frac{\sqrt{2}}{2}y'\right]$$

$$+ 2\sqrt{2}\left[\frac{\sqrt{2}}{2}x' - \frac{\sqrt{2}}{2}y'\right] - 2\sqrt{2}\left[\frac{\sqrt{2}}{2}x' + \frac{\sqrt{2}}{2}y'\right] = 0$$

$$\frac{1}{2}x'^2 - x'y' + \frac{1}{2}y'^2 + \frac{1}{2}x'^2 + x'y' + \frac{1}{2}y'^2 + x'^2 - y'^2 + 2x' - 2y' - 2x' - 2y' = 0$$

Expand terms.

$$2x'^2 - 4y' = 0 \qquad \text{Combine terms.}$$
$$x'^2 - 2y' = 0 \qquad \text{Divide by 2.}$$
$$x'^2 = 2y' \qquad \text{Add 2y'.}$$

This is the equation of a parabola. The graph is shown in **FIGURE 2.**

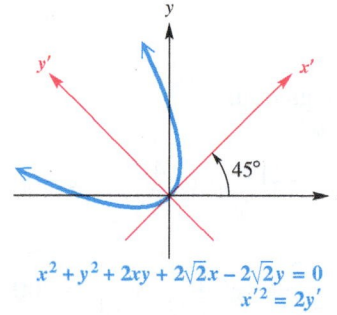

$x^2 + y^2 + 2xy + 2\sqrt{2}x - 2\sqrt{2}y = 0$
$x'^2 = 2y'$

FIGURE 2

We have graphed equations written in the standard form

$$Ax^2 + Cy^2 + Dx + Ey + F = 0.$$

To graph an equation that has an *xy*-term by hand, it is necessary to find an appropriate **angle of rotation** to eliminate the *xy*-term. This angle is calculated as follows. (The derivation is beyond the scope of this text.)

Angle of Rotation

The *xy*-term is removed from the standard equation

$$Ax^2 + Bxy + Cy^2 + Dx + Ey + F = 0$$

by a rotation of the axes through an angle θ, $0° < \theta < 90°$, where

$$\cot 2\theta = \frac{A - C}{B}.$$

To find the rotation equations, first find $\sin\theta$ and $\cos\theta$. **Example 2** illustrates a way to obtain $\sin\theta$ and $\cos\theta$ from $\cot 2\theta$ without first identifying angle θ.

EXAMPLE 2 **Rotating and Graphing**

Rotate the axes and graph $52x^2 - 72xy + 73y^2 = 200$.

Solution   Here, $A = 52$, $B = -72$, and $C = 73$.

$$\cot 2\theta = \frac{52 - 73}{-72} = \frac{-21}{-72} = \frac{7}{24} \qquad \text{Substitute.}$$

To find $\sin\theta$ and $\cos\theta$, use these trigonometric identities.

$$\sin\theta = \sqrt{\frac{1 - \cos 2\theta}{2}} \quad \text{and} \quad \cos\theta = \sqrt{\frac{1 + \cos 2\theta}{2}}$$

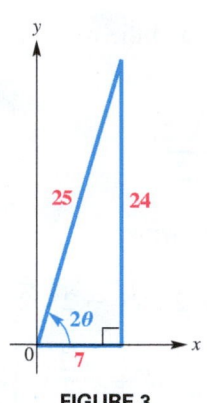

FIGURE 3

Sketch a right triangle as in **FIGURE 3**, to see that $\cos 2\theta = \frac{7}{25}$. (In the two quadrants for which we are concerned, cosine and cotangent have the same sign.)

$$\sin\theta = \sqrt{\frac{1 - \frac{7}{25}}{2}} = \sqrt{\frac{9}{25}} = \frac{3}{5} \quad \text{and} \quad \cos\theta = \sqrt{\frac{1 + \frac{7}{25}}{2}} = \sqrt{\frac{16}{25}} = \frac{4}{5}$$

Use these values for $\sin\theta$ and $\cos\theta$ to obtain

$$x = \frac{4}{5}x' - \frac{3}{5}y' \quad \text{and} \quad y = \frac{3}{5}x' + \frac{4}{5}y'. \qquad \text{Rotation equations}$$

Substitute these expressions for x and y into the original equation.

$$52\left[\frac{4}{5}x' - \frac{3}{5}y'\right]^2 - 72\left[\frac{4}{5}x' - \frac{3}{5}y'\right]\left[\frac{3}{5}x' + \frac{4}{5}y'\right] + 73\left[\frac{3}{5}x' + \frac{4}{5}y'\right]^2 = 200$$

$$52\left[\frac{16}{25}x'^2 - \frac{24}{25}x'y' + \frac{9}{25}y'^2\right] - 72\left[\frac{12}{25}x'^2 + \frac{7}{25}x'y' - \frac{12}{25}y'^2\right] + 73\left[\frac{9}{25}x'^2 + \frac{24}{25}x'y' + \frac{16}{25}y'^2\right] = 200$$

$$25x'^2 + 100y'^2 = 200 \qquad \text{Combine like terms.}$$

$$\frac{x'^2}{8} + \frac{y'^2}{2} = 1 \qquad \text{Divide by 200.}$$

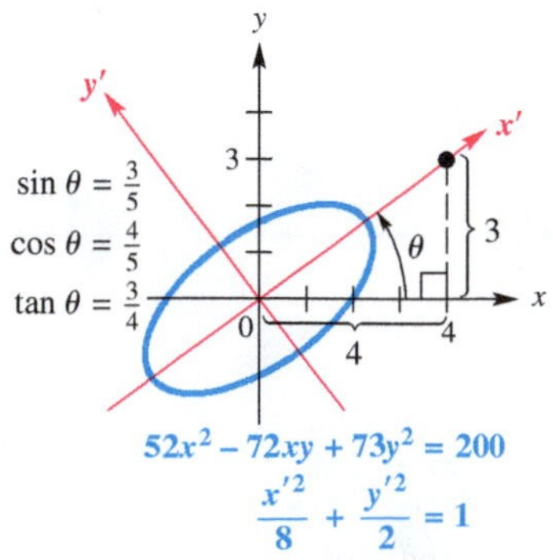

This is an equation of an ellipse having x'-intercepts $\left(\pm 2\sqrt{2}, 0\right)$ and y'-intercepts $\left(0, \pm\sqrt{2}\right)$. The graph is shown in **FIGURE 4**. To find θ, use the fact that

$$\frac{\sin\theta}{\cos\theta} = \frac{\frac{3}{5}}{\frac{4}{5}} = \frac{3}{4} = \tan\theta, \quad \text{from which } \theta \approx 37°.$$

$\sin\theta = \frac{3}{5}$
$\cos\theta = \frac{4}{5}$
$\tan\theta = \frac{3}{4}$

$52x^2 - 72xy + 73y^2 = 200$

$\frac{x'^2}{8} + \frac{y'^2}{2} = 1$

FIGURE 4

The following summary enables us to use the standard equation to decide on the type of graph to expect.

Equations of Conics with *xy*-Term

If the standard second-degree equation

$$Ax^2 + Bxy + Cy^2 + Dx + Ey + F = 0$$

has a graph, it will be one of the following.

(a) a circle or an ellipse (or a point) if $B^2 - 4AC < 0$

(b) a parabola (or one line or two parallel lines) if $B^2 - 4AC = 0$

(c) a hyperbola (or two intersecting lines) if $B^2 - 4AC > 0$

(d) a straight line if $A = B = C = 0$, and $D \neq 0$ or $E \neq 0$

D Exercises

Concept Check   *Use the summary in this section to predict the graph of each second-degree equation.*

1. $4x^2 + 3y^2 + 2xy - 5x = 8$

2. $x^2 + 2xy - 3y^2 + 2y = 12$

3. $2x^2 + 3xy - 4y^2 = 0$

4. $x^2 - 2xy + y^2 + 4x - 8y = 0$

5. $4x^2 + 4xy + y^2 + 15 = 0$

6. $-x^2 + 2xy - y^2 + 16 = 0$

Find the angle of rotation θ that will remove the xy-term in each equation.

7. $2x^2 + \sqrt{3}xy + y^2 + x = 5$

8. $4\sqrt{3}x^2 + xy + 3\sqrt{3}y^2 = 10$

9. $3x^2 + \sqrt{3}xy + 4y^2 + 2x - 3y = 12$

10. $4x^2 + 2xy + 2y^2 + x - 7 = 0$

11. $x^2 - 4xy + 5y^2 = 18$

12. $3\sqrt{3}x^2 - 2xy + \sqrt{3}y^2 = 25$

Use the given angle of rotation to remove the xy-term, and graph each equation.

13. $x^2 - xy + y^2 = 6; \theta = 45°$

14. $2x^2 - xy + 2y^2 = 25; \theta = 45°$

15. $8x^2 - 4xy + 5y^2 = 36; \sin \theta = \dfrac{2}{\sqrt{5}}$

16. $5y^2 + 12xy = 10; \sin \theta = \dfrac{3}{\sqrt{13}}$

Remove the xy-term from each equation by performing a suitable rotation. Graph each equation.

17. $3x^2 - 2xy + 3y^2 = 8$

18. $x^2 + xy + y^2 = 3$

19. $x^2 - 4xy + y^2 = -5$

20. $x^2 + 2xy + y^2 + 4\sqrt{2}x - 4\sqrt{2}y = 0$

21. $7x^2 + 6\sqrt{3}xy + 13y^2 = 64$

22. $7x^2 + 2\sqrt{3}xy + 5y^2 = 24$

23. $3x^2 - 2\sqrt{3}xy + y^2 - 2x - 2\sqrt{3}y = 0$

24. $2x^2 + 2\sqrt{3}xy + 4y^2 = 5$

In each equation, remove the xy-term by rotation. Then translate the axes and sketch the graph.

25. $x^2 + 3xy + y^2 - 5\sqrt{2}y = 15$

26. $x^2 - \sqrt{3}xy + 2\sqrt{3}x - 3y - 3 = 0$

27. $4x^2 + 4xy + y^2 - 24x + 38y - 19 = 0$

28. $12x^2 + 24xy + 19y^2 - 12x - 40y + 31 = 0$

29. $16x^2 + 24xy + 9y^2 - 130x + 90y = 0$

30. $9x^2 - 6xy + y^2 - 12\sqrt{10}x - 36\sqrt{10}y = 0$

31. *Concept Check*   Look at the box titled "Angle of Rotation." Why is no rotation applicable if the value of B is 0?

32. *Concept Check*   Look at the equation involving $\cot 2\theta$ in the box titled "Angle of Rotation." Why must the angle of rotation be 45° if the coefficients of x^2 and y^2 are equal, and $B \neq 0$?

Answers to Selected Exercises

To The Student

If you need further help with algebra, you may use the *Student's Solution Manual* that goes with this book. It contains solutions to all the odd-numbered section and Chapter Review exercises and all the Relating Concepts, Reviewing Basic Concepts, and Chapter Test exercises.

In this section, we provide the answers that we think most students will obtain when they work the exercises by using the methods explained in the text. If your answer does not look exactly like the one given here, it is not necessarily wrong. In many cases there are equivalent forms of the answer. For example, if the answer section shows $\frac{3}{4}$ and your answer is 0.75, you have obtained the correct answer, but written it in a different (yet equivalent) form. Unless the directions specify otherwise, 0.75 is just as valid an answer as $\frac{3}{4}$. In general, if your answer does not agree with the one given in the text, see whether it can be transformed into the other form. If it can, then it is the correct answer. If you still have doubts, talk with your instructor.

CHAPTER 1 LINEAR FUNCTIONS, EQUATIONS, AND INEQUALITIES

1.1 Exercises *(pages 8–11)*

1. (a) 10 **(b)** 0, 10 **(c)** $-6, -\frac{12}{4}$ (or -3), 0, 10

(d) $-6, -\frac{12}{4}$ (or -3), $-\frac{5}{8}$, 0, 0.31, 0.$\overline{3}$, 10 **(e)** $-\sqrt{3}, 2\pi, \sqrt{17}$

(f) All are real numbers. **3. (a)** None are natural numbers.

(b) None are whole numbers. **(c)** $-\sqrt{100}$ (or -10), -1

(d) $-\sqrt{100}$ (or -10), $-\frac{13}{6}, -1, 5.23, 9.\overline{14}, 3.14, \frac{22}{7}$ **(e)** None are irrational numbers. **(f)** All are real numbers.

5. natural number, integer, rational number, real number

7. integer, rational number, real number

9. rational number, real number **11.** real number

13. natural numbers **15.** rational numbers **17.** integers

19. 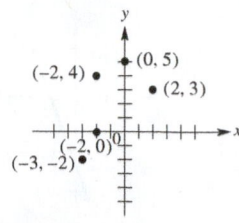 **21.**

23. A rational number can be written as a fraction $\frac{p}{q}$, $q \neq 0$, where p and q are integers, whereas an irrational number cannot.

25. I **27.** III **29.** none **31.** II **33.** none

Graph for Exercises 25–33:

35. I or III **37.** II or IV **39.** y-axis

41. $[-5, 5]$ by $[-25, 25]$ **43.** $[-60, 60]$ by $[-100, 100]$

45. $[-500, 300]$ by $[-300, 500]$

47. 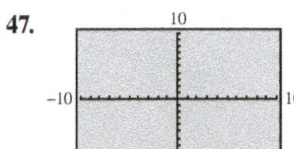 **49.**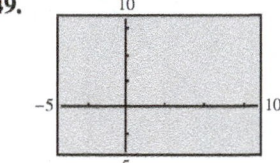

51.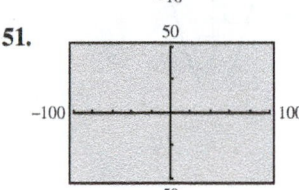

53. There are no tick marks. To set a screen with no tick marks on the axes, use Xscl = 0 and Yscl = 0. **55.** 7.616

57. 3.208 **59.** 3.045 **61.** 4.359 **63.** 311.987

65. 0.25 **67.** 5.66 **69.** 98.63 **71.** 8.25

73. 0.72 **75.** -2.82 **77.** $c = 17$ **79.** $b = 84$

81. $c = \sqrt{89}$ **83.** $a = 4$ **85. (a)** $\sqrt{40} = 2\sqrt{10}$

(b) $(-1, 4)$ **87. (a)** $\sqrt{205}$ **(b)** $\left(-\frac{1}{2}, 1\right)$ **89. (a)** 5

(b) $\left(\frac{7}{2}, 9\right)$ **91. (a)** $\sqrt{34}$ **(b)** $\left(-\frac{11}{2}, -\frac{7}{2}\right)$ **93. (a)** 5

(b) $(7.7, 5.4)$ **95. (a)** $25x$ **(b)** $\left(\frac{19}{2}x, -11x\right)$

97. $Q(9, 14)$ **99.** $Q(-13.72, -5.01)$ **101.** $26.5 billion

103. 2005: $20,007; 2009: $21,777

105. (a)

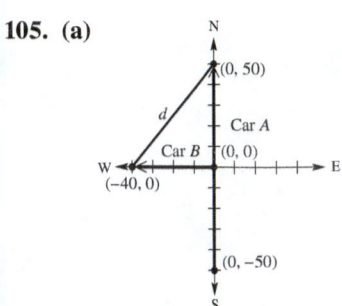

(b) $d = \sqrt{4100} \approx 64.0$ mi

107. $(a + b)^2 = a^2 + 2ab + b^2$; $c^2 + 4\left(\frac{1}{2} ab\right) = c^2 + 2ab$; $a^2 + 2ab + b^2 = c^2 + 2ab$, so subtract $2ab$ from each side to obtain $a^2 + b^2 = c^2$.

1.2 Exercises *(pages 19–22)*

1. $(-1, 4)$

3. $(-\infty, 0)$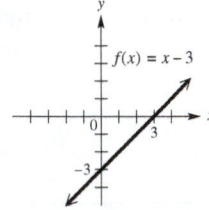

5. $[1, 2)$ 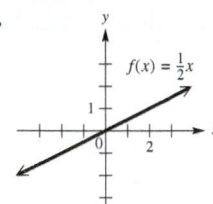   **7.** $\{x \mid -4 < x < 3\}$

9. $\{x \mid x \leq -1\}$ **11.** $\{x \mid -2 \leq x < 6\}$

13. $\{x \mid x \leq -4\}$ **15.** Use a parenthesis if the symbol is $<$ or $>$, and use a square bracket if the symbol is $\leq$ or $\geq$; a parenthesis is always used with ∞ or $-\infty$.

17.

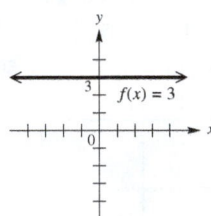

19.

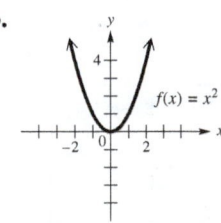

21.

23.

25. domain: $\{5, 3, 4, 7\}$; range: $\{1, 2, 9, 6\}$; function

27. domain: $\{1, 2, 3\}$; range: $\{6\}$; function   **29.** domain: $\{4, 3, -2\}$; range: $\{1, -5, 3, 7\}$; not a function

31. domain: $\{11, 12, 13, 14\}$; range: $\{-6, -7\}$; function

33. domain: $\{0, 1, 2, 3, 4\}$; range: $\{\sqrt{2}, \sqrt{3}, \sqrt{5}, \sqrt{6}, \sqrt{7}\}$; function

35. domain: $(-\infty, \infty)$; range: $(-\infty, \infty)$; function

37. domain: $[-4, 4]$; range: $[-3, 3]$; not a function

39. domain: $[2, \infty)$; range: $[0, \infty)$; function

41. domain: $[-9, \infty)$; range: $(-\infty, \infty)$; not a function

43. domain: $\{-5, -2, -1, -0.5, 0, 1.75, 3.5\}$; range: $\{-1, 2, 3, 3.5, 4, 5.75, 7.5\}$; function

45. domain: $\{2, 3, 5, 11, 17\}$; range: $\{1, 7, 20\}$; function

47. 2   **49.** 7   **51.** undefined   **53.** -10 **55.** 4

57. -10 **59.** 5   **61.** 2   **63.** 11

65. $f(a) = 5a$; $f(b + 1) = 5b + 5$; $f(3x) = 15x$

67. $f(a) = 2a - 5$; $f(b + 1) = 2b - 3$; $f(3x) = 6x - 5$

69. $f(a) = 1 - a^2$; $f(b + 1) = -b^2 - 2b$; $f(3x) = 1 - 9x^2$

71. $(-2, 3)$ **73.** 7; 8   **75. (a)** 0 **(b)** 4 **(c)** 2 **(d)** 4

77. (a) undefined **(b)** -2 **(c)** 0 **(d)** 2

79. (a)–(f) Answers will vary; see the definitions.

81. (a) $A = \{$(Google, 155), (Apple, 95), (Jumptab, 61), (Microsoft, 39)$\}$; (Google, 155) means that Google received $155 million in mobile advertising revenue.

(b)

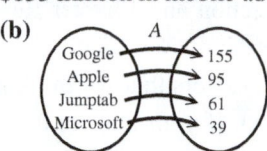

(c) domain: $\{$Google, Apple, Jumptab, Microsoft$\}$
range: $\{155, 95, 61, 39\}$

Reviewing Basic Concepts *(page 22)*

1.

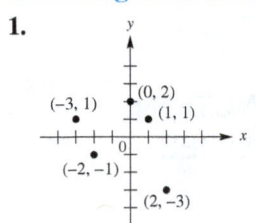

2. $\sqrt{149}$; $\left(1, \frac{3}{2}\right)$ **3.** 1.168   **4.** 34   **5.** 60 in.

6. $(-2, 5]$; $[4, \infty)$ **7.** not a function; domain: $[-2, 2]$; range: $[-3, 3]$

8.

9. $f(-5) = 23$; $f(a + 4) = -4a - 13$

10. $f(2) = 3$; $f(-1) = -3$

1.3 Exercises *(pages 32–35)*

1. (a) $(4, 0)$ **(b)** $(0, -4)$ **3. (a)** $(2, 0)$ **(b)** $(0, -6)$
(c) $(-\infty, \infty)$ **(d)** $(-\infty, \infty)$ **(c)** $(-\infty, \infty)$ **(d)** $(-\infty, \infty)$
(e) 1   **(e)** 3

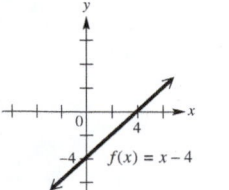

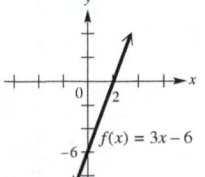

5. (a) $(5, 0)$ **(b)** $(0, 2)$
(c) $(-\infty, \infty)$ **(d)** $(-\infty, \infty)$
(e) $-\frac{2}{5}$

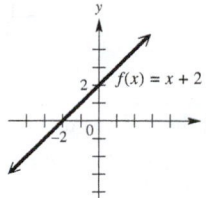

7. (a) $(0, 0)$ **(b)** $(0, 0)$
(c) $(-\infty, \infty)$ **(d)** $(-\infty, \infty)$
(e) 3

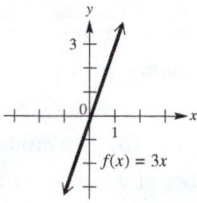

9. (a) $f(-2) = 0$; $f(4) = 6$
(b) The x-intercept is $(-2, 0)$
and corresponds to the zero
of f.

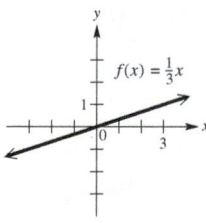

(c) -2

11. (a) $f(-2) = 3$; $f(4) = 0$
(b) The x-intercept is $(4, 0)$
and corresponds to the zero
of f.

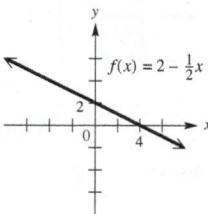

(c) 4

13. (a) $f(-2) = -\frac{2}{3}$;
$f(4) = \frac{4}{3}$ **(b)** The x-intercept
is $(0, 0)$ and corresponds to
the zero of f.

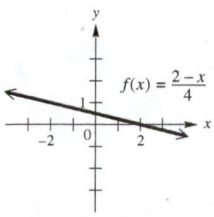

(c) 0

15. (a) $f(-2) = -0.65$;
$f(4) = 1.75$ **(b)** The
x-intercept is $(-0.375, 0)$
and corresponds to the zero
of f.

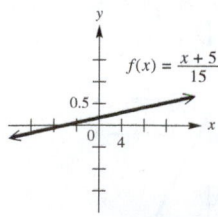

(c) -0.375

17. (a) $f(-2) = 1$; $f(4) = -\frac{1}{2}$
(b) The x-intercept is $(2, 0)$
and corresponds to the zero
of f.

(c) 2

19. (a) $f(-2) = \frac{1}{5}$; $f(4) = \frac{3}{5}$
(b) The x-intercept is $(-5, 0)$
and corresponds to the zero
of f.

(c) -5

21. The point $(0, 0)$ must lie on the line.
23. (a) none   **(b)** $(0, -3)$
(c) $(-\infty, \infty)$ **(d)** $\{-3\}$
(e) 0

25. (a) $(-1.5, 0)$ **(b)** none
(c) $\{-1.5\}$ **(d)** $(-\infty, \infty)$
(e) undefined

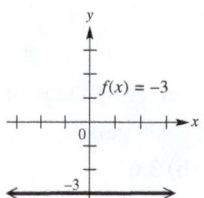

27. (a) $(2, 0)$ **(b)** none
(c) $\{2\}$ **(d)** $(-\infty, \infty)$
(e) undefined

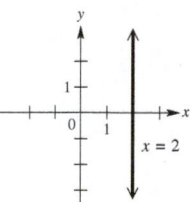

29. constant function   **31.** $y = 3$ **33.** $x = 0$
35. window B   **37.** window B   **39.** 1 **41.** $\frac{7}{9}$
43. undefined   **45.** 0 **47.** $-\frac{2}{3}$ **49.** $-\$4000$ per
year; The value of the machine is decreasing \$4000 each year
during these years.   **51.** 0% per year (or no change); The
percent of pay raise is not changing —it is 3% each year during these years.   **53.** A   **55.** C   **57.** H   **59.** B
61. (a) -2; $(0, 1)$; $\left(\frac{1}{2}, 0\right)$ **(b)** $f(x) = -2x + 1$ **(c)** $\frac{1}{2}$
63. (a) $-\frac{1}{3}$; $(0, 2)$; $(6, 0)$ **(b)** $f(x) = -\frac{1}{3}x + 2$ **(c)** 6
65. (a) -200; $(0, 300)$; $\left(\frac{3}{2}, 0\right)$ **(b)** $f(x) = -200x + 300$
(c) $\frac{3}{2}$ **67.** 4; $(0, 2)$; $f(x) = 4x + 2$ **69.** -1.4; $(0, -3.1)$;
$f(x) = -1.4x - 3.1$ **71.** A   **73.** D

In Exercises 75–83, the two indicated points may vary.

75.

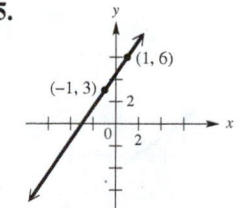

77.

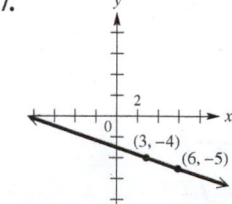

79.

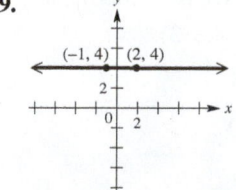

81.

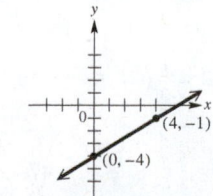

83.

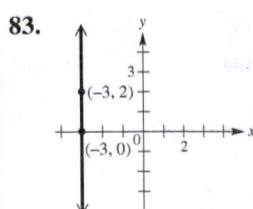

85. (a) $f(x) = 500x + 2000$ **(b)** Water is entering the pool at 500 gal per hr; the pool initially contains 2000 gal.
(c) 5500 gal   **87. (a)** $f(x) = \frac{1}{4}x + 3$ **(b)** 3.625 in.
89. (a) If the delay is 15 sec, the lightning is 3 mi away.
(b)

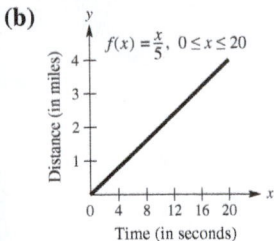

91. $f(x) = 0.075x$; $f(86) = \$6.45$
93. $f(x) = 192x + 275$; $\$2387$
95. (a) $W(x) = 0.09x$ **(b)** $W(15) = 1.35$; In 15 years the Antarctic has warmed 1.35°F, on average.

1.4 Exercises *(pages 44–48)*

1. $y = -2x + 5$ **3.** $y = 1.5x + 11.5$
5. $y = -0.5x - 3$ **7.** $y = 2x - 5$ **9.** $y = \frac{1}{2}x + \frac{13}{24}$
11. $y = \frac{4}{5}x - \frac{14}{5}$ **13.** $y = -x - 4$ **15.** $y = x + 4$
17. $y = \frac{5}{2}x - \frac{31}{2}$ **19.** $y = -1.5x + 6.5$
21. $y = -\frac{1}{2}x + 5$ **23.** $y = 8x + 12$
25. $y = -3x + 1$

Note: The graphs of the lines in **Exercises 27–31** are the same as the graphs in **Exercises 1–5** in **Section 1.3.** In the answers here, we give the x- and y-intercepts. Refer to the graphs given previously.

27. x: (4, 0); y: (0, −4)   **29.** x: (2, 0); y: (0, −6)
31. x: (5, 0); y: (0, 2)

33.

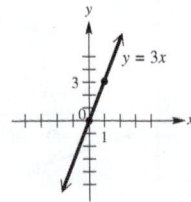

We used (0, 0) and (1, 3).

35.

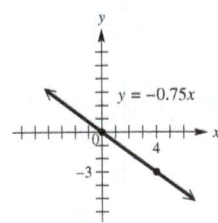

We used (0, 0) and (4, −3).

37. $y = -\frac{5}{3}x + 5$ **39.** $y = \frac{2}{7}x + \frac{4}{7}$
41. $y = -\frac{3}{4}x + \frac{25}{8}$ **43.** $y = -\frac{1}{3}x + \frac{11}{3}$
45. $y = \frac{5}{3}x + \frac{13}{3}$ **47.** $x = -5$
49. $y = -0.2x + 7$ **51.** $y = \frac{1}{2}x$ **53.** $y = 2$

55. $y = -12x - 20$ **57.** $y = -\frac{3}{4}x + \frac{21}{4}$
59. (a) the Pythagorean theorem and its converse
(b) $\sqrt{x_1^2 + m_1^2 x_1^2}$ **(c)** $\sqrt{x_2^2 + m_2^2 x_2^2}$
(d) $\sqrt{(x_2 - x_1)^2 + (m_2 x_2 - m_1 x_1)^2}$
(f) $-2x_1 x_2 (m_1 m_2 + 1) = 0$
(g) Since $x_1 \neq 0$ and $x_2 \neq 0$, this implies $m_2 m_1 + 1 = 0$, so $m_1 m_2 = -1$.   **(h)** The product of the slopes of perpendicular lines, neither of which is parallel to an axis, is −1.
61. (a) $y = 11x + 117$ **(b)** 11 mph   **(c)** 117 mi
(d) 130.75 mi   **63. (a)** $y = 2x - 3996$ **(b)** $30 billion
65. (a) a linear relation

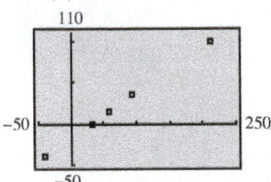

(b) $C(x) = \frac{5}{9}(x - 32)$; A slope of $\frac{5}{9}$ means the Celsius temperature increases 5° for every 9° increase in the Fahrenheit temperature   **(c)** $28\frac{1}{3}$°C   **67. (a)** $y - 6 = \frac{31}{6}(x - 2005)$
(b) Every year from 2005 to 2011, Google advertising revenue increased by about $5.2 billion, on average.   **(c)** 2007: $16.33 billion; 2009: $26.67 billion; The 2007 value compares favorably and the 2009 value is too high.
69. (a) $y \approx 586.89x - 1,147,738$
(b)

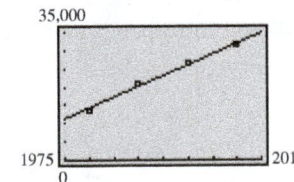

(c) It is about $28,976, which is quite close to the actual value of $29,307.   **71. (a)** $y \approx 0.06791x - 16.32$
(b) It is about 2500 light-years
73. $y \approx 0.101x + 11.6$; $r \approx 0.909$; There is a strong, positive correlation, since r is close to 1.

Reviewing Basic Concepts *(page 49)*

1. $f(x) = 1.4x - 3.1$; $f(1.3) = -1.28$
2. x-intercept: $\left(\frac{1}{2}, 0\right)$ y-intercept: (0, 1); domain: $(-\infty, \infty)$; range: $(-\infty, \infty)$; slope: −2

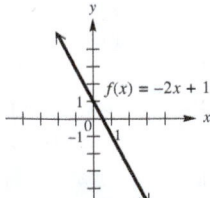

3. $\frac{2}{7}$ **4.** vertical: $x = -2$; horizontal: $y = 10$

5.

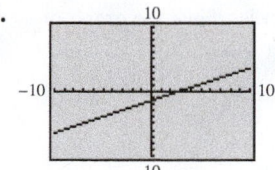 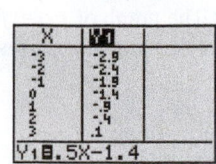

6. $y = 2x - 3$ **7.** $y = -\frac{2}{7}x + \frac{24}{7}$ **8.** $y = -\frac{2}{3}x + \frac{7}{3}$

9. (a)

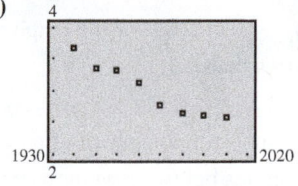

(b) negative **(c)** $y \approx -0.0165x + 35.6$; $r \approx -0.9648$
(d) 3.01; The estimate is close to the actual value of 2.94.

1.5 Exercises *(pages 61–65)*

1. -4 **3.** 0 **5.** $-\frac{23}{19}$ **7.** $\frac{8}{3}$ **9.** $-\frac{29}{6}$

11. $\{10\}$ **13.** $\{1\}$ **15.** $\{3\}$ **17.** When $x = 10$
is substituted into y_1 or y_2, the result is $y = 20$.

19. $\varnothing$; contradiction **21.** $\{12\}$ **23.** $\{3\}$

25. $\{-2\}$ **27.** $\{7\}$ **29.** $\{\frac{17}{10}\}$ **31.** $\{\frac{5}{17}\}$

33. $\{75\}$ **35.** $\{-1\}$ **37.** $\{0\}$ **39.** $\{\frac{5}{4}\}$

41. $\{4\}$ **43.** $\{16.07\}$ **45.** $\{-1.46\}$

47. $\{-3.92\}$ **49.** contradiction; $\varnothing$

51. identity; $(-\infty, \infty)$ **53.** conditional; $\{14\}$

55. conditional; $\{\frac{18}{13}\}$ **57.** contradiction; $\varnothing$

59. identity; $(-\infty, \infty)$ **61.** contradiction; $\varnothing$ **63.** $\{3\}$

65. $(3, \infty)$ **67.** $(-\infty, 3]$ **69.** $[3, \infty)$ **71.** $[3, \infty)$

73. (a) $(20, \infty)$ **(b)** $(-\infty, 20)$ **(c)** $[20, \infty)$ **(d)** $(-\infty, 20]$

75. (a) $\{4\}$ **(b)** $(4, \infty)$ **(c)** $(-\infty, 4)$ **77. (a)** $\{2\}$

(b) $(2, \infty)$ **(c)** $(-\infty, 2)$ **79. (a)** $\{\frac{1}{2}\}$ **(b)** $[\frac{1}{2}, \infty)$

(c) $(-\infty, \frac{1}{2}]$ **81. (a)** $\{4\}$ **(b)** $(-\infty, 4)$ **(c)** $(4, \infty)$

83. (a) $(8, \infty)$ **(b)** $(-\infty, 8]$ **85. (a)** $(5, \infty)$

(b) $(-\infty, 5]$ **87. (a)** $(-\infty, -3]$ **(b)** $(-3, \infty)$

89. (a) $\left(-\frac{3}{4}, \infty\right)$ **(b)** $\left(-\infty, -\frac{3}{4}\right]$ **91.** $(-\infty, 15]$

93. $(-6, \infty)$ **95.** $[-8, \infty)$ **97.** $(25, \infty)$ **99.** $\varnothing$

101. (a) It is moving away, since distance is increasing.

(b) 1 hr; 3 hr **(c)** $[1, 3]$ **(d)** $(1, 6)$ **103.** $[1, 4]$

105. $(-6, -4)$ **107.** $(-16, 19]$ **109.** $\left(\frac{7}{2}, \frac{9}{2}\right)$

111. $[2, 18]$ **113.** $\left[\frac{3\sqrt{2} - 1}{2}, \frac{3\sqrt{5} - 1}{2}\right]$

115. (a) below 1.14 mi, or $[0, 1.14)$ **(b)** $\left[0, \frac{15}{13.2}\right)$, or $\left[0, \frac{25}{22}\right)$

117. $1.98\pi \le C \le 2.02\pi$ **119.** $\{3\}$; one; one

120. $(-\infty, 3)$; $(3, \infty)$; The value of $x = 3$ represents the boundary between the sets of real numbers given by $(-\infty, 3)$ and $(3, \infty)$.

121. $\{1.5\}$; $(1.5, \infty)$; $(-\infty, 1.5)$ **122. (a)** $\left\{-\frac{b}{a}\right\}$

(b) $\left(-\infty, -\frac{b}{a}\right)$; $\left(-\frac{b}{a}, \infty\right)$ **(c)** $\left(-\frac{b}{a}, \infty\right)$; $\left(-\infty, -\frac{b}{a}\right)$

1.6 Exercises *(pages 71–76)*

1. 30 L **3.** A **5.** D **7.** 30 cm **9.** 7.6 cm

11. 14 cm **13.** width: 21 in.; length: 28 in.; as a 35-in. screen

15. 6 cm **17.** 3.2 hr at 55 mph and 2.8 hr at 70 mph

19. $7\frac{1}{2}$ gal **21.** $2\frac{2}{3}$ gal **23.** 4 mL **25.** 8 L

27. $266\frac{2}{3}$ gal **29. (a)** about 48; In 2008 the winning men's 100-meter freestyle time was about 48 seconds.

(b) from 1948 to 1964 **31. (a)** $S(x) = 19x - 38,017$

(b) Sales increased, on average, by \$19 billion per yr.

(c) 2018 **33. (a)** $C(x) = 0.02x + 200$

(b) $R(x) = 0.04x$ **(c)** 10,000

(d) for $x < 10,000$, a loss; for $x > 10,000$, a profit

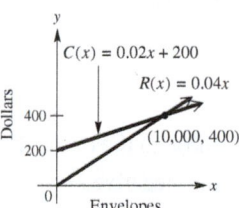

35. (a) $C(x) = 3.00x + 2300$ **(b)** $R(x) = 5.50x$ **(c)** 920
(d) for $x < 920$, a loss; for $x > 920$, a profit

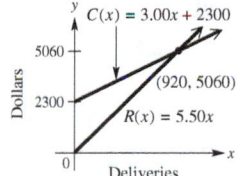

37. $k = 2.5$; $y = 20$ when $x = 8$

39. $k = 0.06$; $x = \$85$ when $y = \$5.10$

41. $30\frac{1}{3}$ lb per in.^{2} **43.** \$1048; $k = 65.5$

45. 46.7 ft^3 **47.** $51\frac{3}{7}$ ft **49.** $14\frac{1}{6}$ in.

51. 12,500 **53. (a)** $y = 640x + 1100$ **(b)** \$17,100

(c) Locate the point $(25, 17,100)$ on the graph of $y = 640x + 1100$. **55. (a)** $y = 2600x + 120,000$

(b) \$138,200 **(c)** The value of the house increased, on average, by \$2600 per year. **57. (a)** about 197,000,000 mi^2

(b) about 140,000,000 mi^2 **(c)** about 25.7 ft

(d) They would be flooded. **(e)** about 238 ft

59. (a) 500 cm^3 **(b)** 90°C **(c)** -273°C **61.** $P = \frac{I}{RT}$

63. $W = \frac{P - 2L}{2}$, or $W = \frac{P}{2} - L$ **65.** $h = \frac{2\mathcal{A}}{b_1 + b_2}$

67. $H = \frac{S - 2LW}{2W + 2L}$ **69.** $h = \frac{3V}{\pi r^2}$ **71.** $n = \frac{2S}{a_1 + a_n}$

73. $g = \frac{2s}{t^2}$ **75.** short-term note: \$100,000; long-term note: \$140,000 **77.** \$10,000 at 2.5%; \$20,000 at 3%

79. \$50,000 at 1.5%; \$90,000 at 4%

Reviewing Basic Concepts *(page 76)*

1. $\{2\}$ **2.** $\{0.757\}$ **3.** $-\frac{1}{4}$ **4. (a)** contradiction; $\varnothing$

(b) identity; $(-\infty, \infty)$ **(c)** conditional equation; $\{\frac{11}{4}\}$

5. $\left(-\infty, -\frac{5}{7}\right)$ **6.** $\left(-\frac{5}{2}, 3\right]$ **7. (a)** $\{2\}$ **(b)** $[2, \infty)$

8. 40.5 ft   **9. (a)** $R(x) = 5.5x$ **(b)** $C(x) = 1.5x + 800$

(c) 200 discs   **10.** $h = \dfrac{V}{\pi r^2}$

Chapter 1 Review Exercises *(pages 80–83)*

1. $\sqrt{612} = 6\sqrt{17}$ **3.** -4 **5.** $-\frac{3}{4}$ **7.** $(0, 36)$

9. 11   **11. (a)** -3 **(b)** $y = -3x + 2$ **(c)** $(0.5, 0.5)$

(d) $\sqrt{90} = 3\sqrt{10}$ **13.** C   **15.** A   **17.** E

19.

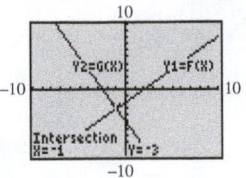

-0.475; The graph confirms that the line through the ordered pairs falls from left to right and therefore has negative slope. Thus, the number of basic cable subscribers *decreased* by an average of 0.475 million (or 475,000) each year from 2001 to 2009.

21. I   **23.** B   **25.** I   **27.** O   **29.** $\left\{\frac{46}{7}\right\}$

31. $\left\{-\frac{5}{7}\right\}$ **33.** $\left\{\frac{7}{5}\right\}$ **35.** $(-\infty, 3)$ **37.** $\left(-\frac{10}{3}, \frac{46}{3}\right]$

39. $C(x) = 30x + 150$ **41.** 20   **43.** $f = \dfrac{AB(p + 1)}{24}$

45. (a) $41°F$ **(b)** about 21,000 ft   **(c)** Graph $y = -3.52x + 58.6$. Then find the y-coordinate of the point where $x = 5$ to support the answer in (a). Finally, find the x-coordinate of the point where $y = -15$ to support the answer in (b).   **47.** $y = 4x + 120$; 4 ft   **49.** 1 hr, 7 min, 19 sec; This time is faster by 56 min, 19 sec.

51. 7500 kg per m²   **53. (a)** $f(x) = 1.2x - 1886.4$
(b) about 528   **(c)** Over time, data can change pattern or character. Answers may vary.   **55.** $[300, \infty)$; 300 or more DVDs   **57. (a)** $y \approx 4.512x - 154.4$ **(b)** about 184 lb

Chapter 1 Test *(pages 83–84)*

1. (a) natural number, integer, rational number, real number
(b) real number   **(c)** real number   **(d)** rational number, real number   **2. (a)** 2.236   **(b)** 1.913   **(c)** 1.316   **(d)** -0.018
3. $\sqrt{85}$; $\left(1, \frac{1}{2}\right)$ **4. (a)** 18   **(b)** $4 - 7b$
(c) $4 - 7a - 7h$ **5.** yes   **6. (a)** $\frac{1}{2}$ **(b)** undefined
(c) 0   **7. (a) (i)** $(-\infty, \infty)$ **(ii)** $[2, \infty)$ **(iii)** no x-intercepts
(iv) $(0, 3)$ **(b) (i)** $(-\infty, \infty)$ **(ii)** $(-\infty, 0]$ **(iii)** $(3, 0)$
(iv) $(0, -3)$ **(c) (i)** $[-4, \infty)$ **(ii)** $[0, \infty)$ **(iii)** $(-4, 0)$
(iv) $(0, 2)$ **8. (a)** $\{-4\}$ **(b)** $(-\infty, -4)$ **(c)** $[-4, \infty)$
(d) $\{-4\}$ **9. (a)** $\{5.5\}$ **(b)** $(-\infty, 5.5)$ **(c)** $(5.5, \infty)$
(d) $(-\infty, 5.5]$ **10. (a)** $\{-1\}$; The check leads to $-3 = -3$.

(b) $(-1, \infty)$; The graph of $y_1 = f(x)$ is *above* the graph of $y_2 = g(x)$ for domain values greater than -1.

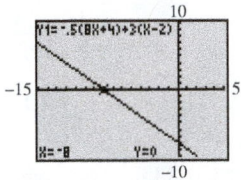

(c) $(-\infty, -1)$; The graph of $y_1 = f(x)$ is *below* the graph of $y_2 = g(x)$ for domain values less than -1.

11. (a) $\{-8\}$ **(b)** $[-8, \infty)$
(c) The x-intercept is $(-8, 0)$ supporting the result of part (a). The graph of the linear function lies below or on the x-axis for domain values greater than or equal to -8, supporting the result of part (b).

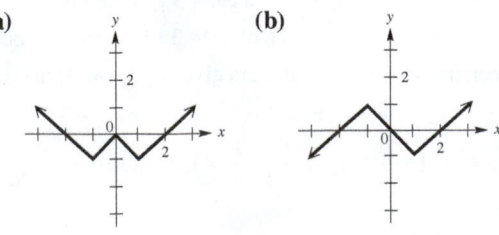

12. (a) $f(x) = 249x + 13,837$ **(b)** The number of stations increased, on average, by 249 per year.
(c) 15,580   **13. (a)** $y = -2x - 1$ **(b)** $y = -\frac{1}{2}x + \frac{7}{2}$
14. x-intercept: $(2, 0)$; y-intercept: $\left(0, -\frac{3}{2}\right)$; slope: $\frac{3}{4}$
15. horizontal: $y = 7$; vertical: $x = -3$
16. (a) $y \approx -0.246x + 35.7$; $r \approx -0.96$
(b) $y = -0.246(40) + 35.7 = 25.9°F$.
17. 60 mph: 1.5 hr; 74 mph: 2.5 hr   **18.** $702\frac{2}{3}$ pounds

CHAPTER 2   ANALYSIS OF GRAPHS OF FUNCTIONS

2.1 Exercises *(pages 96–99)*

1. $(-\infty, \infty)$ **3.** $(0, 0)$ **5.** increases   **7.** x-axis
9. odd   **11.** $(-\infty, \infty)$ **13.** $[0, \infty)$ **15.** $(-\infty, -3)$; $(-3, \infty)$
17. (a) $(3, \infty)$ **(b)** $(-\infty, 3)$ **(c)** none   **(d)** $(-\infty, \infty)$
(e) $[0, \infty)$ **19. (a)** $(-\infty, 1)$ **(b)** $(4, \infty)$ **(c)** $(1, 4)$
(d) $(-\infty, \infty)$ **(e)** $(-\infty, 3]$ **21. (a)** none
(b) $(-\infty, -2)$; $(3, \infty)$ **(c)** $(-2, 3)$ **(d)** $(-\infty, \infty)$
(e) $(-\infty, 1.5] \cup [2, \infty)$ **23.** increasing   **25.** decreasing
27. increasing   **29.** decreasing   **31.** decreasing
33. increasing   **35. (a)** no   **(b)** yes   **(c)** no
37. (a) yes   **(b)** no   **(c)** no   **39. (a)** yes   **(b)** yes
(c) yes   **41. (a)** no   **(b)** no   **(c)** yes
43. (a)   **(b)**

45. $-12; -25; 21$ **47.** even   **49.** odd   **51.** neither
53. even   **55.** odd   **57.** even   **59.** odd   **61.** even
63. odd   **65.** symmetric with respect to the origin
67. symmetric with respect to the y-axis   **69.** neither
71. neither   **73.** symmetric with respect to the y-axis
75. symmetric with respect to the origin

2.2 Exercises *(pages 104–108)*

1. $y = x^2 + 3$ **3.** $y = \sqrt{x} - 4$ **5.** $y = |x - 4|$
7. $y = (x + 7)^3$ **9.** $y = (x - 3)^2 - 2$
11. $y = \sqrt{x + 6} + 3$ **13.** $y = (x - 2000)^2 + 500$
15. To obtain the graph of g, shift the graph of f upward
4 units.   **17.** B   **19.** A   **21.** B   **23.** C   **25.** B
27. -2 **29.** -3 **31. (a)** $(-\infty, \infty)$ **(b)** $[-3, \infty)$
33. (a) $(-\infty, \infty)$ **(b)** $[-3, \infty)$ **35. (a)** $(-\infty, \infty)$
(b) $(-\infty, \infty)$ **37.** D: $(-\infty, \infty)$; R: $[-5, \infty)$
39. D: $[4, \infty)$; R: $[0, \infty)$ **41.** D: $(-\infty, \infty)$; R: $(-\infty, \infty)$
43.

45.

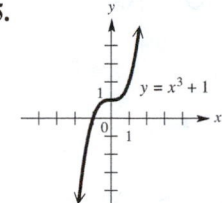

47.

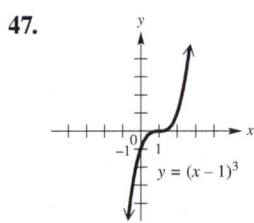

49.

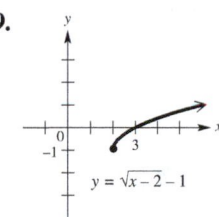

51.

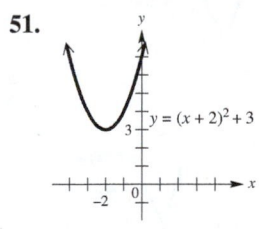

53.

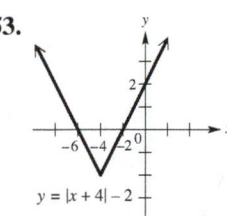

55. B   **57.** A
59.

61.

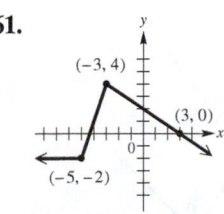

63. $y = (x + 4)^2 + 3$; **(a)** $(-4, \infty)$ **(b)** $(-\infty, -4)$
65. $y = x^3 - 5$; **(a)** $(-\infty, \infty)$ **(b)** none
67. $y = \sqrt{x - 2} + 1$; **(a)** $(2, \infty)$ **(b)** none
69. $h = -3; k = 1$
71. (a) $\{3, 4\}$ **(b)** $(-\infty, 3) \cup (4, \infty)$ **(c)** $(3, 4)$

72. (a) $\{\sqrt{2}\}$ **(b)** $(\sqrt{2}, \infty)$ **(c)** $(-\infty, \sqrt{2})$
73. (a) $\{-4, 5\}$ **(b)** $(-\infty, -4] \cup [5, \infty)$ **(c)** $[-4, 5]$
74. (a) $\varnothing$ **(b)** $[1, \infty)$ **(c)** $\varnothing$
75. (a) 425; In 2010, 425,000 bankruptcies were filed.
(b) $y = 66.25(x - 2006) + 160$ **(c)** 425; In 2010, 425,000
bankruptcies were filed.   **(d)** 2008
77. $U(2011) = 440$; The average U.S. household spent \$440
on Apple products in 2011.
79. (a) $y \approx 402.5x + 3460$
(b) $y \approx 402.5(x - 2000) + 3460$ **(c)** \$7100
81.

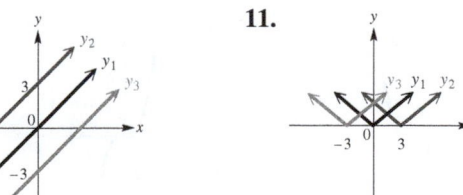

82. 2   **83.** $y_1 = 2x - 4$ **84.** $(1, 4)$ and $(3, 8)$
85. 2   **86.** $y_2 = 2x + 2$

87.

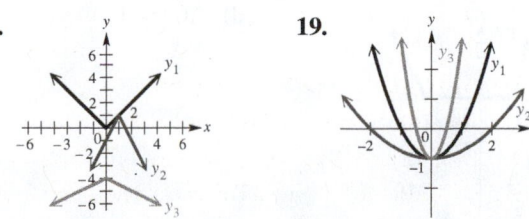

The graph of y_2 can be
obtained by shifting the graph
of y_1 upward 6 units. The
constant, 6, comes from the
6 we added to each y-value
in **Exercise 84.**

88. c; c; the same as; c; upward (or positive vertical)

2.3 Exercises *(pages 116–120)*

1. $y = 2x^2$ **3.** $y = \sqrt{-x}$ **5.** $y = -3|x|$
7. $y = 0.25(-x)^3$, or $y = -0.25x^3$
9.

11.

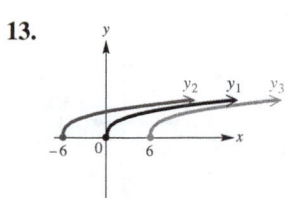

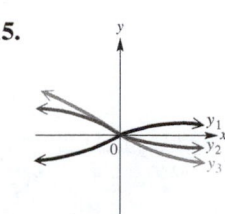

13.

15.

17.

19.

21.

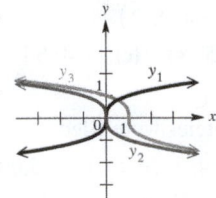

23. $g(x) = -(x - 5)^2 - 2$ **25.** $g(x) = \sqrt{-x} + 1$

27. 4; x **29.** 2; left; $\frac{1}{4}$; x; 3; downward (or negative)

31. 3; right; 6   **33.** $y = \frac{1}{2}x^2 - 7$

35. $y = 4.5\sqrt{x - 3} - 6$

37.

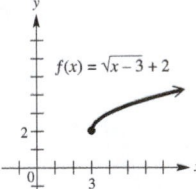

39.

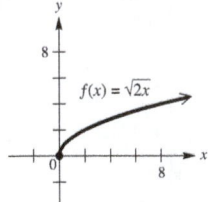

41.

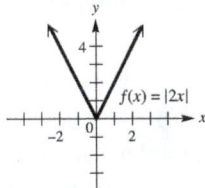

43.

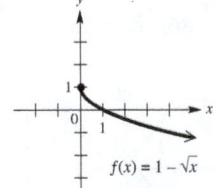

45.

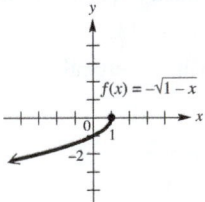

47.

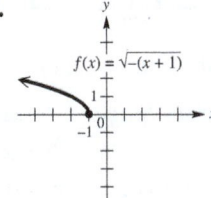

49.

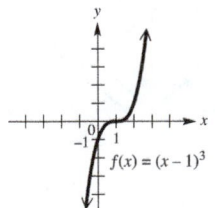

51.

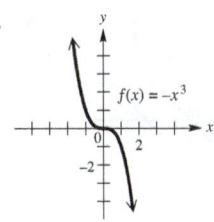

53. (a)

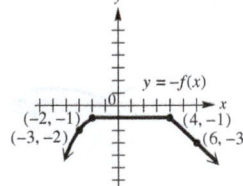

(b)

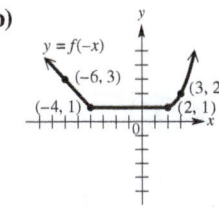

(c)

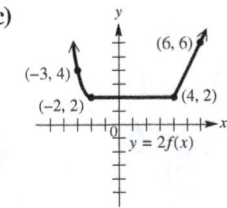

(d) $f(0) = 1$

55. (a)

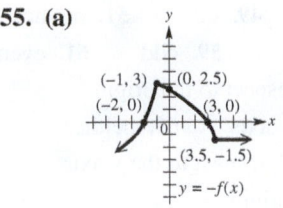

(b)

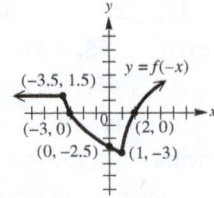

(c)

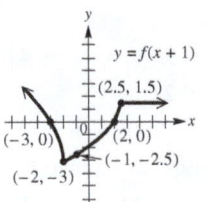

(d) $(-1, 0)$ and $(4, 0)$

57. (a)

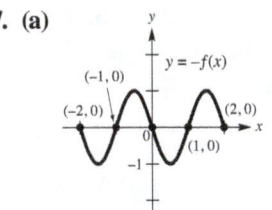

(b)

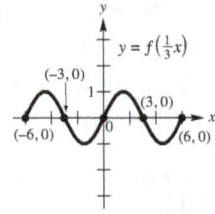

(c)

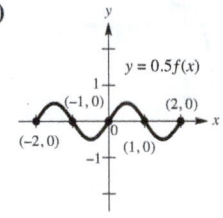

(d) symmetry with respect to the origin

59. (a)

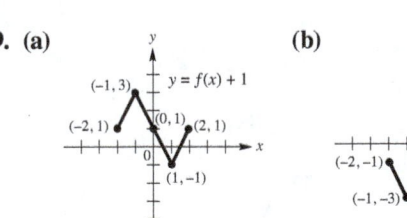

(b)

(c)

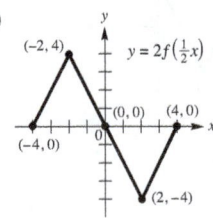

61. (a)

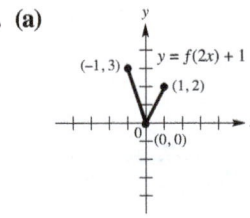

(b)

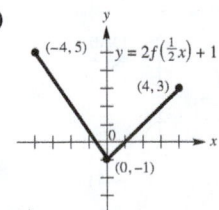

(c)

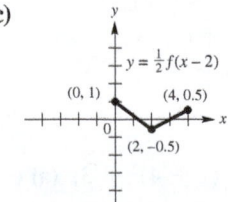

63. (a) $(r, 0)$ is an x-intercept.   **(b)** $(-r, 0)$ is an x-intercept.
(c) $(-r, 0)$ is an x-intercept.   **65.** domain: $[1, 4]$;
range: $[0, 3]$ **67.** domain: $[-1, 2]$; range: $[-3, 0]$

69. domain: $\left[-\frac{1}{2}, 1\right]$; range: $[0, 3]$

71. domain: $[-4, 8]$; range: $[0, 9]$

73. domain: $[-2, 1]$; range: $[0, 3]$

75. domain: $\left[-\frac{2}{3}, \frac{1}{3}\right]$; range: $[0, 3]$

77. domain: $[20, \infty)$; range: $[5, \infty)$

79. domain: $[-10, \infty)$; range: $(-\infty, 5]$

81. decreases   **83.** increases

85. (a) $(-1, 2)$ (b) $(-\infty, -1)$ (c) $(2, \infty)$ **87.** (a) $(1, \infty)$

(b) $(-2, 1)$ (c) $(-\infty, -2)$ **89.** $(8, 10)$

91. $y = 2|x + 2| - 1$ **93.** $y = -3|x| + 2$

95. $y = \frac{4}{3}|x| - 4$ **97.** It will be the same.

Reviewing Basic Concepts *(pages 120–121)*

1. (a) D: $(-\infty, \infty)$; R: $[1, \infty)$; Incr: $(0, \infty)$; Decr: $(-\infty, 0)$

(b) D: $(-\infty, \infty)$; R: $[0, \infty)$; Incr: $(2, \infty)$; Decr: $(-\infty, 2)$

(c) D: $[0, \infty)$; R: $(-\infty, 0]$; Incr: never; Decr: $(0, \infty)$

2. (a) $f(-3) = -6$ (b) $f(-3) = 6$ (c) $f(-3) = -6$

(d) $f(-3) = 6$ **3.** (a) B   (b) D   (c) E   (d) A   (e) C

4. (a) B   (b) A   (c) G   (d) C   (e) F   (f) D

(g) H   (h) E

5. (a) 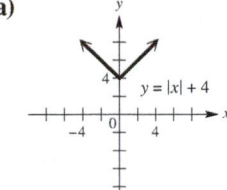   (b)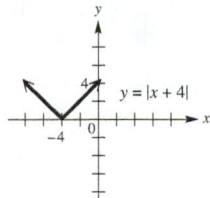

(c) 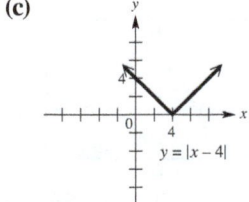   (d)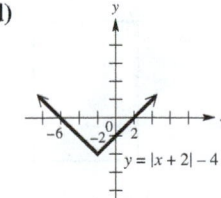

(e)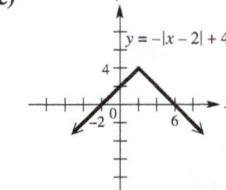

6. (a) It is the graph of $f(x) = |x|$ translated 1 unit to the left, reflected across the x-axis, and translated 3 units upward. The equation is $y = -|x + 1| + 3$.   (b) It is the graph of $g(x) = \sqrt{x}$ translated 4 units to the left, reflected across the x-axis, and translated 2 units upward. The equation is $y = -\sqrt{x + 4} + 2$.   (c) It is the graph of $g(x) = \sqrt{x}$ translated 4 units to the left, stretched vertically by applying a factor of 2, and translated 4 units downward. The equation is $y = 2\sqrt{x + 4} - 4$.   (d) It is the graph of $f(x) = |x|$

translated 2 units to the right, shrunken vertically by applying a factor of $\frac{1}{2}$, and translated 1 unit downward. The equation is $y = \frac{1}{2}|x - 2| - 1$.   **7.** (a) 2   (b) 4

8. The graph of $y = F(x + h)$ is a horizontal translation of the graph of $y = F(x)$. The graph of $y = F(x) + h$ is not the same as the graph of $y = F(x + h)$ because the graph of $y = F(x) + h$ is a *vertical* translation of the graph of $y = F(x)$.

9. (a) $5; -6; 4$ (b) $-5; 6; -4$

10. (a) 37; In 2011 Google's ad revenues were $37 billion.

(b) $y = 5(x - 2004) + 2$ (c) 37; In 2011 Google's ad revenues were $37 billion.   (d) 2009

2.4 Exercises *(pages 128–132)*

1. 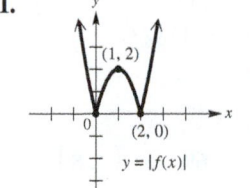   **3.**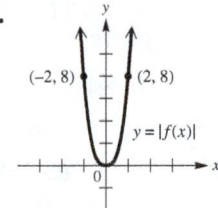

5. $y = |f(x)|$ has the same graph as $y = f(x)$.

7. 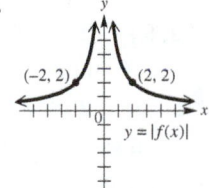   **9.**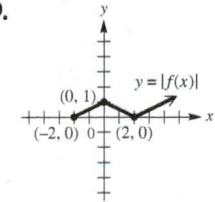

11. 5   **13.** $[0, \infty)$ **15.** $[2, \infty)$

17. domain of $f(x)$: $(-\infty, \infty)$; range of $f(x)$: $[-2, \infty)$; domain of $|f(x)|$: $(-\infty, \infty)$; range of $|f(x)|$: $[0, \infty)$

19. domain of $f(x)$: $(-\infty, \infty)$; range of $f(x)$: $(-\infty, -1]$; domain of $|f(x)|$: $(-\infty, \infty)$; range of $|f(x)|$: $[1, \infty)$

21. domain of $f(x)$: $[-2, 3]$; range of $f(x)$: $[-2, 3]$; domain of $|f(x)|$: $[-2, 3]$; range of $|f(x)|$: $[0, 3]$

23. domain of $f(x)$: $[-2, 3]$; range of $f(x)$: $[-3, 1]$; domain of $|f(x)|$: $[-2, 3]$; range of $|f(x)|$: $[0, 3]$

25. (a) 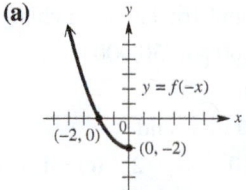   (b)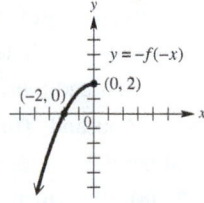

(c)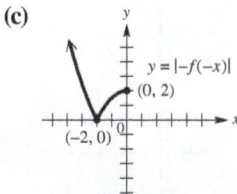

27. Figure A shows the graph of $y = f(x)$, while Figure B shows the graph of $y = |f(x)|$. **29. (a)** $\{-1, 6\}$
(b) $(-1, 6)$ **(c)** $(-\infty, -1) \cup (6, \infty)$ **31. (a)** $\{4\}$
(b) $\emptyset$ **(c)** $(-\infty, 4) \cup (4, \infty)$ **33.** The V-shaped graph is that of $f(x) = |0.5x + 6|$, since this shape is typical of the graphs of absolute value functions of the form $f(x) = |ax + b|$.
34. The straight line graph is that of $g(x) = 3x - 14$, which is a linear function. **35.** $\{8\}$
36. $(-\infty, 8)$ **37.** $(8, \infty)$ **38.** $\{8\}$
39. (a) $\{-13, 5\}$ **(b)** $(-\infty, -13) \cup (5, \infty)$ **(c)** $(-13, 5)$
41. (a) $\{2, 5\}$ **(b)** $(-\infty, 2] \cup [5, \infty)$ **(c)** $[2, 5]$
43. (a) $\left\{-\frac{3}{2}, \frac{1}{2}\right\}$ **(b)** $\left[-\frac{3}{2}, \frac{1}{2}\right]$ **(c)** $\left(-\infty, -\frac{3}{2}\right] \cup \left[\frac{1}{2}, \infty\right)$
45. (a) $\left\{\frac{5}{7}\right\}$ **(b)** $(-\infty, \infty)$ **(c)** $\left\{\frac{5}{7}\right\}$ **47. (a)** $\emptyset$ **(b)** $\emptyset$
(c) $(-\infty, \infty)$ **49.** $\left\{0, \frac{8}{3}\right\}$ **51.** $\left\{-\frac{1}{2}, 1\right\}$
53. $\{0, 1\}$ **55.** $(8, 22)$ **57.** $(-\infty, 1) \cup (2, \infty)$
59. $\left(-\infty, \frac{5}{3}\right] \cup \left[\frac{11}{3}, \infty\right)$ **61.** $(-\infty, 18) \cup (18, \infty)$
63. $\emptyset$ **65.** $(-\infty, \infty)$ **67. (a)** $\left\{-8, \frac{6}{5}\right\}$
(b) $(-\infty, -8) \cup \left(\frac{6}{5}, \infty\right)$ **(c)** $\left(-8, \frac{6}{5}\right)$ **69. (a)** $\left\{\frac{2}{3}, 8\right\}$
(b) $\left(-\infty, \frac{2}{3}\right) \cup (8, \infty)$ **(c)** $\left(\frac{2}{3}, 8\right)$ **71. (a)** $\left\{-3, \frac{5}{3}\right\}$
(b) $(-\infty, -3) \cup \left(\frac{5}{3}, \infty\right)$ **(c)** $\left(-3, \frac{5}{3}\right)$ **73. (a)** $\left\{-\frac{7}{8}\right\}$
(b) $\left(-\infty, -\frac{7}{8}\right)$ **(c)** $\left(-\frac{7}{8}, \infty\right)$ **75. (a)** $\{2, 8\}$ **(b)** $(2, 8)$
(c) $(-\infty, 2) \cup (8, \infty)$ **77. (a)** $(-\infty, \infty)$ **(b)** $\emptyset$ **(c)** $\emptyset$
79. $\{-3, 8\}$ **81.** $\{-2, 6\}$ **83. (a)** $28 \le T \le 72$
(b) The monthly average temperatures in Boston vary between a low of 28°F and a high of 72°F. The monthly averages are always within 22° of 50°F. **85. (a)** $49 \le T \le 74$ **(b)** The monthly average temperatures in Buenos Aires vary between a low of 49°F (possibly in July) and a high of 74°F (possibly in January). The monthly averages are always within 12.5° of 61.5°F. **87.** $[6.5, 9.5]$ **89. (a)** 9 **(b)** 103 or 137
91. $(-0.05, 0.05)$ **93.** $(2.74975, 2.75025)$
95. $\{2\}$ **97.** $(-\infty, \infty)$ **99.** $\emptyset$

2.5 Exercises *(pages 139–143)*

1. (a) 40 mph **(b)** 6 mi **(c)** $f(5) = 40$; $f(13) = 30$; $f(19) = 55$ **(d)** $x = 4, 6, 8, 12,$ and 16; The limit changes at each discontinuity. **3. (a)** 50,000 gal; 30,000 gal
(b) during the first and fourth days
(c) $f(2) = 45$ thousand; $f(4) = 40$ thousand
(d) 5000 gal per day **5. (a)** -10 **(b)** -2 **(c)** -1
(d) 2 **7. (a)** -3 **(b)** 1 **(c)** 0 **(d)** 9
9. yes **11.** no

$f(x) = \begin{cases} x - 1 & \text{if } x \le 3 \\ 2 & \text{if } x > 3 \end{cases}$

$f(x) = \begin{cases} 4 - x & \text{if } x < 2 \\ 1 + 2x & \text{if } x \ge 2 \end{cases}$

13. no **15.** no

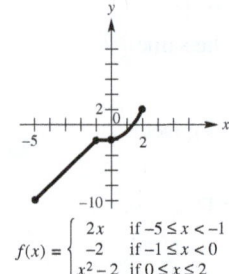

$f(x) = \begin{cases} 2 + x & \text{if } x < -4 \\ -x & \text{if } -4 \le x \le 5 \\ 3x & \text{if } x > 5 \end{cases}$

$f(x) = \begin{cases} -\frac{1}{2}x^2 + 2 & \text{if } x \le 2 \\ \frac{1}{2}x & \text{if } x > 2 \end{cases}$

17. yes **19.** yes

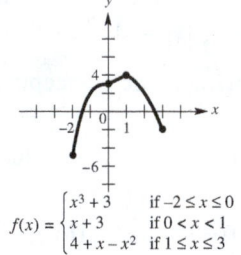

$f(x) = \begin{cases} 2x & \text{if } -5 \le x < -1 \\ -2 & \text{if } -1 \le x < 0 \\ x^2 - 2 & \text{if } 0 \le x \le 2 \end{cases}$

$f(x) = \begin{cases} x^3 + 3 & \text{if } -2 \le x \le 0 \\ x + 3 & \text{if } 0 < x < 1 \\ 4 + x - x^2 & \text{if } 1 \le x \le 3 \end{cases}$

21. B **23.** D

25.

$f(x) = \begin{cases} x - 1 & \text{if } x \le 3 \\ 2 & \text{if } x > 3 \end{cases}$

27.

$f(x) = \begin{cases} 4 - x & \text{if } x < 2 \\ 1 + 2x & \text{if } x \ge 2 \end{cases}$

29.

$f(x) = \begin{cases} 2 + x & \text{if } x < -4 \\ -x & \text{if } -4 \le x \le 5 \\ 3x & \text{if } x > 5 \end{cases}$

31.

$f(x) = \begin{cases} -\frac{1}{2}x^2 + 2 & \text{if } x \le 2 \\ \frac{1}{2}x & \text{if } x > 2 \end{cases}$

In Exercises 33–37, there are other acceptable forms.

33. $f(x) = \begin{cases} 2 & \text{if } x \le 0 \\ -1 & \text{if } x > 1 \end{cases}$; domain: $(-\infty, 0] \cup (1, \infty)$;
range: $\{-1, 2\}$ **35.** $f(x) = \begin{cases} x & \text{if } x \le 0 \\ 2 & \text{if } x > 0 \end{cases}$;
domain: $(-\infty, \infty)$; range: $(-\infty, 0] \cup \{2\}$
37. $f(x) = \begin{cases} \sqrt[3]{x} & \text{if } x < 1 \\ x + 1 & \text{if } x \ge 1 \end{cases}$; domain: $(-\infty, \infty)$;
range: $(-\infty, 1) \cup [2, \infty)$ **39.** There is an *overlap* of intervals, and $f(4)$ is not unique. It can be 11 or 16.
41. The graph of $y = [\![x]\!]$ is shifted 1.5 units downward.
43. The graph of $y = [\![x]\!]$ is reflected across the x-axis.

45.
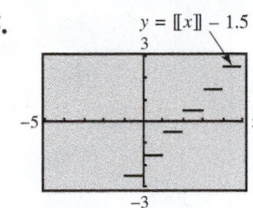
$y = [\![x]\!] - 1.5$

47.
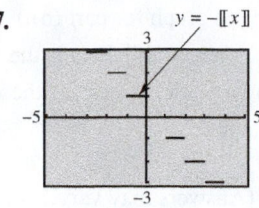
$y = -[\![x]\!]$

49. When $0 \le x \le 3$, the slope is 5, which means the inlet pipe is open and the outlet pipe is closed; when $3 < x \le 5$, the slope is 2, which means both pipes are open; when $5 < x \le 8$, the slope is 0, which means both pipes are closed; when $8 < x \le 10$, the slope is -3, which means the inlet pipe is closed and the outlet pipe is open.

51. (a) 1.12; 1.32; It costs \$1.12 to mail 1.5 oz and \$1.32 to mail 3 oz.   **(b)** domain: $(0, 5]$; range: $\{0.92, 1.12, 1.32, 1.52, 1.72\}$

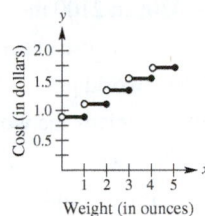

53. (a)
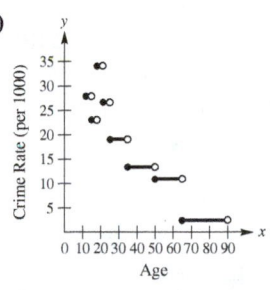

(b) The likelihood of being a victim peaks from age 18 up to age 21, then decreases.

55. (a) 55 mph; 30 mph   **(b)** 12 mi   **(c)** 40; 30; 55

57. (a) \$1.25

(b) $f(x) = \begin{cases} 0.50 & \text{if } 0 < x \le 1 \\ 0.50 - 0.25[\![1 - x]\!] & \text{if } 1 < x \le 5 \end{cases}$

59.
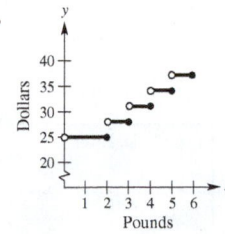

61.
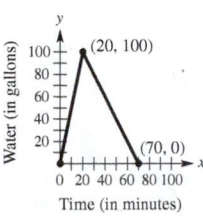

2.6 Exercises *(pages 152–157)*

1. E   **3.** F   **5.** A   **7.** 40   **9.** $4x^2 + 2x - 2$

11. 23   **13.** 196   **15.** $\frac{2}{3}$ **17.** 7   **19.** -3

21. undefined   **23. (a)** $(f + g)(x) = 10x + 2$;
$(f - g)(x) = -2x - 4$; $(fg)(x) = 24x^2 + 6x - 3$

(b) Domain is $(-\infty, \infty)$ in all cases.

(c) $\left(\frac{f}{g}\right)(x) = \frac{4x - 1}{6x + 3}$; domain: $\left(-\infty, -\frac{1}{2}\right) \cup \left(-\frac{1}{2}, \infty\right)$

(d) $(f \circ g)(x) = 24x + 11$; domain: $(-\infty, \infty)$

(e) $(g \circ f)(x) = 24x - 3$; domain: $(-\infty, \infty)$

25. (a) $(f + g)(x) = |x + 3| + 2x$;
$(f - g)(x) = |x + 3| - 2x$; $(fg)(x) = |x + 3|(2x)$

(b) Domain is $(-\infty, \infty)$ in all cases.   **(c)** $\left(\frac{f}{g}\right)(x) = \frac{|x + 3|}{2x}$; domain: $(-\infty, 0) \cup (0, \infty)$ **(d)** $(f \circ g)(x) = |2x + 3|$; domain: $(-\infty, \infty)$ **(e)** $(g \circ f)(x) = 2|x + 3|$; domain: $(-\infty, \infty)$

27. (a) $(f + g)(x) = \sqrt[3]{x + 4} + x^3 + 5$;
$(f - g)(x) = \sqrt[3]{x + 4} - x^3 - 5$;
$(fg)(x) = \left(\sqrt[3]{x + 4}\right)(x^3 + 5)$

(b) Domain is $(-\infty, \infty)$ in all cases.   **(c)** $\left(\frac{f}{g}\right)(x) = \frac{\sqrt[3]{x + 4}}{x^3 + 5}$;
domain: $\left(-\infty, \sqrt[3]{-5}\right) \cup \left(\sqrt[3]{-5}, \infty\right)$

(d) $(f \circ g)(x) = \sqrt[3]{x^3 + 9}$; domain: $(-\infty, \infty)$

(e) $(g \circ f)(x) = x + 9$; domain: $(-\infty, \infty)$

29. (a) $(f + g)(x) = \sqrt{x^2 + 3} + x + 1$;
$(f - g)(x) = \sqrt{x^2 + 3} - x - 1$;
$(fg)(x) = \left(\sqrt{x^2 + 3}\right)(x + 1)$ **(b)** Domain is $(-\infty, \infty)$ in all cases.   **(c)** $\left(\frac{f}{g}\right)(x) = \frac{\sqrt{x^2 + 3}}{x + 1}$; domain: $(-\infty, -1) \cup (-1, \infty)$

(d) $(f \circ g)(x) = \sqrt{x^2 + 2x + 4}$; domain: $(-\infty, \infty)$ (*Note:* To see that this is the domain, graph $y = x^2 + 2x + 4$ and note that $y > 0$ for all x.)   **(e)** $(g \circ f)(x) = \sqrt{x^2 + 3} + 1$; domain: $(-\infty, \infty)$ **31. (a)** 2   **(b)** 4   **(c)** 0   **(d)** $-\frac{1}{3}$

33. (a) 3   **(b)** -5 **(c)** 2   **(d)** undefined   **35. (a)** 5   **(b)** 5   **(c)** 0   **(d)** undefined

37.

x	$(f + g)(x)$	$(f - g)(x)$	$(fg)(x)$	$\left(\frac{f}{g}\right)(x)$
-2	6	-6	0	0
0	5	5	0	undefined
2	5	9	-14	-3.5
4	15	5	50	2

39. 260; 400; 660; (all in thousands)   **41.** 2000–2004

43. 6; It represents the dollars (in billions) spent for general science in 2000.   **45.** space and other technologies; 1995–2000   **47. (a)** -4 **(b)** 2   **(c)** -4 **49. (a)** -3 **(b)** -2 **(c)** 0   **51. (a)** 5   **(b)** undefined   **(c)** 4

53. 4; 2   **55.** -3 **57.** 93

59. (a) $(f \circ g)(x) = (x^2 + 3x - 1)^3$; domain: $(-\infty, \infty)$

(b) $(g \circ f)(x) = x^6 + 3x^3 - 1$; domain: $(-\infty, \infty)$

(c) $(f \circ f)(x) = x^9$; domain: $(-\infty, \infty)$

61. (a) $(f \circ g)(x) = 1 - x$; domain: $(-\infty, 1]$

(b) $(g \circ f)(x) = \sqrt{1 - x^2}$; domain: $[-1, 1]$

(c) $(f \circ f)(x) = x^4$; domain: $(-\infty, \infty)$

63. (a) $(f \circ g)(x) = \frac{1}{5x + 1}$; domain: $\left(-\infty, -\frac{1}{5}\right) \cup \left(-\frac{1}{5}, \infty\right)$

(b) $(g \circ f)(x) = \frac{5}{x + 1}$; domain: $(-\infty, -1) \cup (-1, \infty)$

(c) $(f \circ f)(x) = \frac{x + 1}{x + 2}$; domain: $(-\infty, -2) \cup (-2, -1) \cup (-1, \infty)$

65. (a) $(f \circ g)(x) = 8x^3 - 10x^2 + 1$; domain: $(-\infty, \infty)$
(b) $(g \circ f)(x) = 32x^3 + 28x^2 + 4x - 1$; domain: $(-\infty, \infty)$
(c) $(f \circ f)(x) = 4x + 3$; domain: $(-\infty, \infty)$
67. (a) $(f \circ g)(x) = 5$; domain: $(-\infty, \infty)$
(b) $(g \circ f)(x) = 5$; domain: $(-\infty, \infty)$
(c) $(f \circ f)(x) = 5$; domain: $(-\infty, \infty)$
73. The graph of y_2 can be obtained by *reflecting* the graph of y_1 across the line $y_3 = x$.

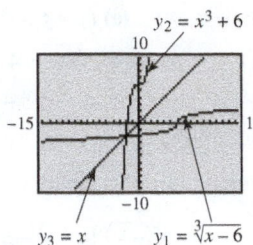

75. $x^2 + 2hx + h^2 - 4$; $x^2 + h^2 - 8$
77. $-x^2 - 2hx + 3x - h^2 + 3h$; $-x^2 + 3x - h^2 + 3h$
79. 4   **81.** $-12x - 1 - 6h$ **83.** $3x^2 + 3xh + h^2$
85. $-2x - h$ **87.** $6x + 3h$ **89.** $\dfrac{-1}{2x(x + h)}$
We give only one of several possible correct pairs of functions f and g in Exercises 91–95.
91. $f(x) = x^2$, $g(x) = 6x - 2$ **93.** $f(x) = \sqrt{x}$, $g(x) = x^2 - 1$ **95.** $f(x) = \sqrt{x} + 12$, $g(x) = 6x$
97. (a) $C(x) = 10x + 500$ **(b)** $R(x) = 35x$
(c) $P(x) = 35x - (10x + 500)$, or $P(x) = 25x - 500$
(d) 21 items   **(e)** The least whole number for which $P(x) > 0$ is 21. Use a window of $[0, 30]$ by $[-1000, 500]$, for example.
99. (a) $C(x) = 100x + 2700$ **(b)** $R(x) = 280x$
(c) $P(x) = 280x - (100x + 2700)$, or $P(x) = 180x - 2700$
(d) 16 items   **(e)** The least whole number for which $P(x) > 0$ is 16. Use a window of $[0, 30]$ by $[-3000, 500]$, for example.
101. (a) $D(r) = \frac{4}{3}\pi(r + 3)^3 - \frac{4}{3}\pi r^3$
(b)

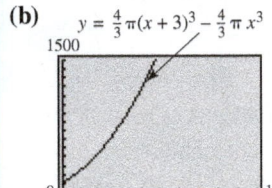

(c) 1168.67 in.³   **(d)** $D(4) = \frac{4}{3}\pi(7)^3 - \frac{4}{3}\pi(4)^3 \approx 1168.67$
103. (a) $P = 6x$; $P(x) = 6x$; This is a linear function.
(b) 4 represents the width of the rectangle and 24 represents the perimeter.

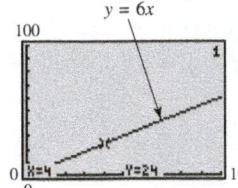

(c) (See graph for part (b).) perimeter $= 24$; This is the y-value shown on the screen for the integer x-value 4.

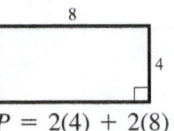

$P = 2(4) + 2(8)$

(d) (Answers may vary.) If the perimeter y of a rectangle satisfying the given conditions is 36, then the width x is 6.

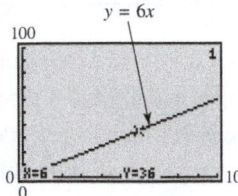

105. (a) $\mathscr{A}(2x) = \sqrt{3}x^2$ **(b)** $\mathscr{A}(16) = 64\sqrt{3}$ square units
(c) On the graph of $y = \dfrac{\sqrt{3}}{4}x^2$, locate the point where $x = 16$ to find $y \approx 110.85$, an approximation for $64\sqrt{3}$.
107. (a) $A(2100) = 42$; The average age of a person in 2100 is projected to be 42 years. $T(2100) = 430$; In 2100 the living world's population will have a combined life experience of 430 billion years.   **(b)** About 10.2; The world population will be about 10.2 billion in 2100.   **(c)** $P(x)$ gives the world's population during year x.   **109. (a)** 87.3   **(b)** $(f + g)(x)$ computes the total emissions from sulfur dioxide and carbon monoxide during year x.
(c)

x	1970	1980	1990	2000	2010
$(f + g)(x)$	235.2	211.3	177.3	130.8	87.3

111. (a) $h(x) = g(x) - f(x)$ **(b)** $h(1996) = 147$; $h(2006) = 153$ **(c)** $h(x) = 0.6x - 1050.6$

Reviewing Basic Concepts *(page 158)*
1. (a) $\{-12, 4\}$ **(b)** $(-\infty, -12) \cup (4, \infty)$ **(c)** $[-12, 4]$
2.

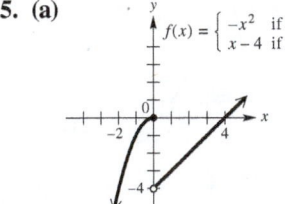

3. $\left\{-\frac{3}{5}, 5\right\}$
4. (a) -3 **(b)** 4   **(c)** 8

5. (a)

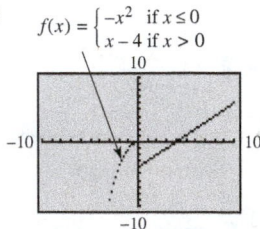

(b)

Dot mode

6. (a) -6 **(b)** -22 **(c)** 8   **(d)** $\frac{5}{9}$ **(e)** $-3x^2 - 4$
(f) $9x^2 + 24x + 16$ **7.** $f(x) = x^4$, $g(x) = x + 2$ (There are other possible choices for f and g.)   **8.** $-4x - 2h + 3$
9. (a) $y_1 = 0.04x$ **(b)** $y_2 = 0.025x + 12.5$ **(c)** It represents the total interest earned in both accounts for 1 year.

(d) $y_1 + y_2 = 0.04x + 0.025x + 12.5$ $28.75 is earned.

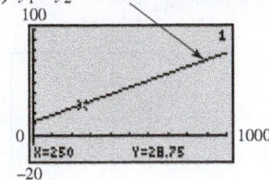

(e) Evaluate $y_1 + y_2$ at $x = 250$ to get 28.75 (dollars).

10. $S = \pi r^2 \sqrt{5}$

Chapter 2 Review Exercises *(pages 162–164)*

1. true   **3.** false; The domain of $f(x) = \sqrt{x}$ is $[0, \infty)$, while the domain of $f(x) = \sqrt[3]{x}$ is $(-\infty, \infty)$.   **5.** true

7. true   **9.** true   **11.** $[0, \infty)$ **13.** $(-\infty, \infty)$

15. $(-\infty, \infty)$ **17.** $[0, \infty)$

19. 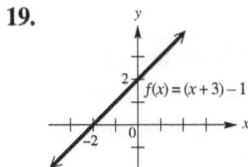   **21.**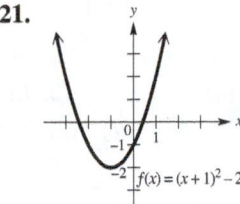

23. 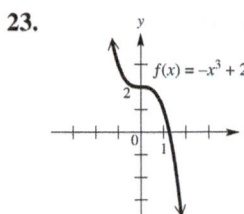   **25.**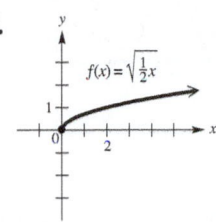

27. 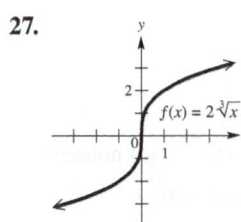   **29.**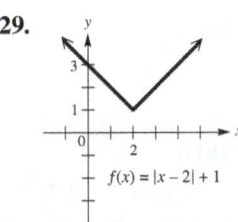

31. (a) $(-\infty, -2), [-2, 1], (1, \infty)$ **(b)** $(-2, 1)$ **(c)** $(-\infty, -2)$
(d) $(1, \infty)$ **(e)** $(-\infty, \infty)$ **(f)** $\{-2\} \cup [-1, 1] \cup (2, \infty)$

33. x-axis symmetry, y-axis symmetry, origin symmetry;
not a function   **35.** y-axis symmetry; even function

37. x-axis symmetry; not a function   **39.** true   **41.** true

43. false; For example, $f(x) = x^3$ is odd, and $(2, 8)$ is on
the graph, but $(-2, 8)$ is not.   **45.** Start with the graph of
$y = x^2$. Shift it 4 units to the left, stretch vertically by applying a factor of 3, reflect across the x-axis, and shift 8 units
downward.

47. 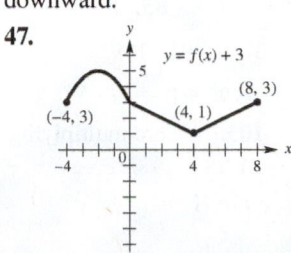   **49.**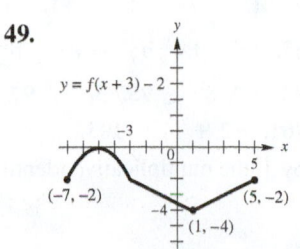

51.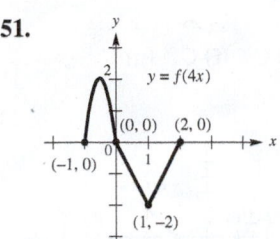

53. domain: $[-3, 4]$; range: $[2, 9]$

55. domain: $\left[-\frac{3}{2}, 2\right]$; range: $[-5, 2]$

57.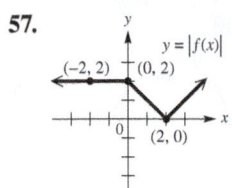

59. The graph of $y = |f(x)|$ is the same as that of $y = f(x)$.

61. $\left\{-\frac{15}{4}, \frac{9}{4}\right\}$ **63.** $\left\{-\frac{7}{3}, 2\right\}$ **65.** $[-6, 1]$

67. $\left(-\infty, \frac{12}{5}\right) \cup \left(\frac{12}{5}, \infty\right)$, or $\left\{x \mid x \neq \frac{12}{5}\right\}$ **69.** $\left\{-3, \frac{11}{3}\right\}$

71. The x-coordinates of the points of intersection of the
graphs are -6 and 1. Thus, $\{-6, 1\}$ is the solution set of
$y_1 = y_2$. The graph of y_1 lies on or below the graph of y_2
between -6 and 1, so the solution set of $y_1 \leq y_2$ is $[-6, 1]$.
The graph of y_1 lies above the graph of y_2 everywhere else, so
the solution set of $y_1 \geq y_2$ is $(-\infty, -6] \cup [1, \infty)$.

73. Initially, the car is at home. After traveling 30 mph for
1 hr, the car is 30 mi away from home. During the second hour
the car travels 20 mph until it is 50 mi away. During the third
hour the car travels toward home at 30 mph until it is 20 mi
away. During the fourth hour the car travels away from home
at 40 mph until it is 60 mi away from home. During the last
hour, the car travels 60 mi at 60 mph until it arrives home.

75. 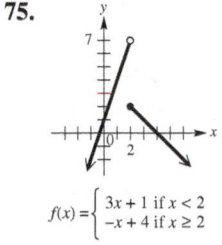   **77.**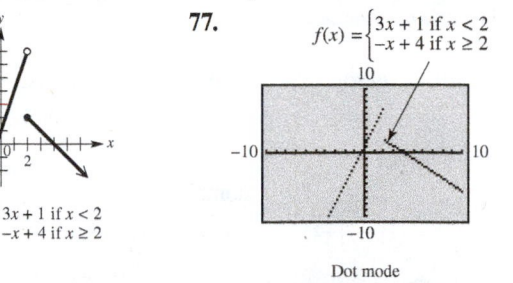

$f(x) = \begin{cases} 3x + 1 \text{ if } x < 2 \\ -x + 4 \text{ if } x \geq 2 \end{cases}$

Dot mode

79. 5   **81.** 0   **83.** 3   **85.** 2   **87.** 8   **89.** -6

91. 2   **93.** 2   **95.** $f(x) = x^2$ and $g(x) = x^3 - 3x$

(There are other possible choices for f and g.)

97. $D(r) = \frac{4}{3}\pi(r + 4)^3 - \frac{4}{3}\pi r^3$ **99.** $f(x) = 36x$;
$g(x) = 1760x$; $(f \circ g)(x) = 63,360x$

Chapter 2 Test *(pages 165–166)*

1. (a) D **(b)** D **(c)** C **(d)** B **(e)** C **(f)** C **(g)** C
(h) D **(i)** D **(j)** C

2. (a)

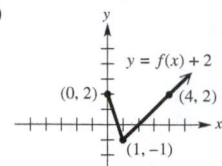

(b)

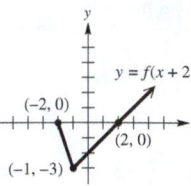

(c)

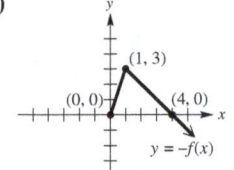

(d)

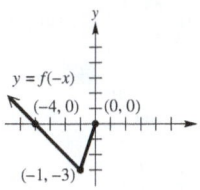

(e)

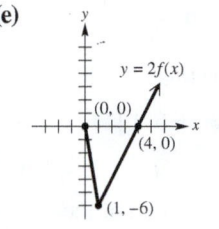

(f)

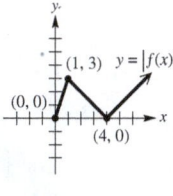

3. (a) $(-1, 4)$ **(b)** $(-4, 4)$

4. (a)

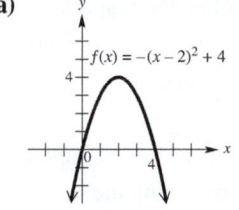

(b)

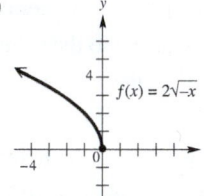

5. (a) $(-3, 6)$ **(b)** $(-3, -6)$

(c)

(We give an actual screen here. The drawing should resemble it.)

6. (a) Shift the graph of $y = \sqrt[3]{x}$ to the left 2 units, vertically stretch by applying a factor of 4, and shift 5 units downward.

(b)

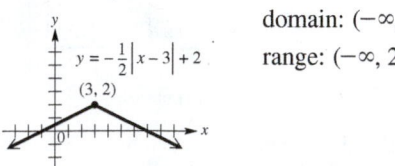

domain: $(-\infty, \infty)$;
range: $(-\infty, 2]$

7. (a) $(-\infty, -3)$ **(b)** $(4, \infty)$ **(c)** $(-3, 4)$ **(d)** $(-\infty, -3)$, $[-3, 4]$, $(4, \infty)$ **(e)** $(-\infty, \infty)$ **(f)** $(-\infty, 2)$
8. (a) $\{-3, -1\}$ **(b)** $(-3, -1)$ **(c)** $(-\infty, -3) \cup (-1, \infty)$
9. (a) $2x^2 - x + 1$ **(b)** $\frac{2x^2 - 3x + 2}{-2x + 1}$ **(c)** $\left(-\infty, \frac{1}{2}\right) \cup \left(\frac{1}{2}, \infty\right)$
(d) $8x^2 - 2x + 1$ **(e)** $-4x^2 + 6x - 3$ **(f)** $4x + 2h - 3$

10. (a)

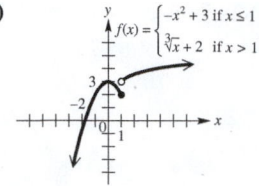

$f(x) = \begin{cases} -x^2 + 3 \text{ if } x \le 1 \\ \sqrt[3]{x} + 2 \text{ if } x > 1 \end{cases}$

(b)

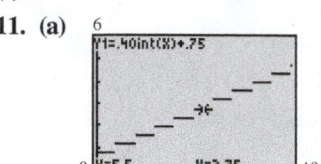

$f(x) = \begin{cases} -x^2 + 3 \text{ if } x \le 1 \\ \sqrt[3]{x} + 2 \text{ if } x > 1 \end{cases}$

Dot mode

(c) 1

11. (a)

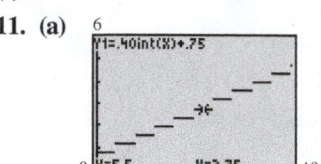

Dot mode

(b) $2.75 is the cost for a 5.5-min call. See the display at the bottom of the screen. **12. (a)** $C(x) = 3300 + 4.50x$
(b) $R(x) = 10.50x$ **(c)** $P(x) = R(x) - C(x) = 6.00x - 3300$
(d) 551

(e)

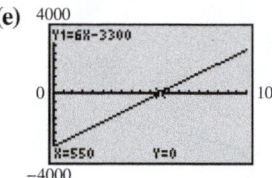

The first integer x-value for which $P(x) > 0$ is 551.

CHAPTER 3 POLYNOMIAL FUNCTIONS

3.1 Exercises *(pages 172–174)*

1. (a) 0 **(b)** -9 **(c)** pure imaginary and nonreal complex
3. (a) π **(b)** 0 **(c)** real **5. (a)** 3 **(b)** 7 **(c)** nonreal complex **7. (a)** 0 **(b)** $\sqrt{7}$ **(c)** pure imaginary and nonreal complex **9. (a)** 0 **(b)** $\sqrt{7}$ **(c)** pure imaginary and nonreal complex **11.** $8i$ **13.** $7 - 7i$ **15.** true
17. true **19.** false; *Every* real number is a complex number. **21.** $10i$ **23.** $-20i$ **25.** $-i\sqrt{39}$
27. $5 + 2i$ **29.** $9 - 5i\sqrt{2}$ **31.** -3 **33.** -13
35. $-2\sqrt{6}$ **37.** $\sqrt{3}$ **39.** $i\sqrt{3}$ **41.** $\frac{1}{2}$ **43.** -2
45. $7 - i$ **47.** 2 **49.** $7 - 6i$ **51.** $1 - 10i$
53. 0 **55.** $8 - i$ **57.** $-14 + 2i$ **59.** $5 - 12i$
61. $-8 - 6i$ **63.** 13 **65.** 7 **67.** $25i$
69. $12 + 9i$ **71.** $20 + 15i$ **73.** i **75.** $-i$
77. 1 **79.** -1 **81.** $-i$ **83.** $-i$ **85.** 1
87. $\left(\frac{\sqrt{2}}{2} + \frac{\sqrt{2}}{2}i\right)^2 = i$ **89.** $5 + 3i$ **91.** $18i$
93. $-\sqrt{8}$ **95.** $3i$ **97.** $10i$ **99.** $-1 - 2i$
101. $-2 + i$ **103.** $\frac{7}{25} - \frac{24}{25}i$ **105.** We are multiplying by 1, the multiplicative identity.

3.2 Exercises *(pages 183–187)*

1. B **3.** D **5. (a)** $P(x) = (x-1)^2 - 16$
(b) $(1, -16)$ **(c)**

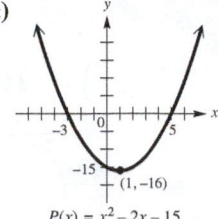

$P(x) = x^2 - 2x - 15$

7. (a) $P(x) = -\left(x + \frac{3}{2}\right)^2 + \frac{49}{4}$ **(b)** $\left(-\frac{3}{2}, \frac{49}{4}\right)$

(c)

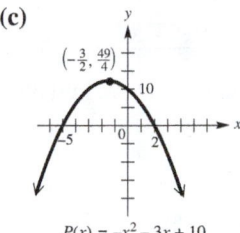

$P(x) = -x^2 - 3x + 10$

9. (a) $P(x) = (x-3)^2 - 9$ **(b)** $(3, -9)$

(c)

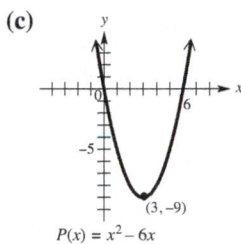

$P(x) = x^2 - 6x$

11. (a) $P(x) = 2\left(x - \frac{1}{2}\right)^2 - \frac{49}{2}$ **(b)** $\left(\frac{1}{2}, -\frac{49}{2}\right)$

(c)

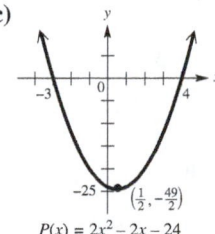

$P(x) = 2x^2 - 2x - 24$

13. (a) $P(x) = -2\left(x - \frac{3}{2}\right)^2 + \frac{9}{2}$ **(b)** $\left(\frac{3}{2}, \frac{9}{2}\right)$

(c)

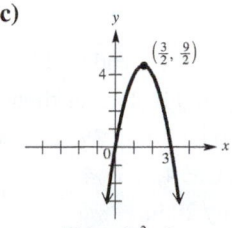

$P(x) = -2x^2 + 6x$

15. (a) $P(x) = 3\left(x + \frac{2}{3}\right)^2 - \frac{7}{3}$ **(b)** $\left(-\frac{2}{3}, -\frac{7}{3}\right)$

(c)

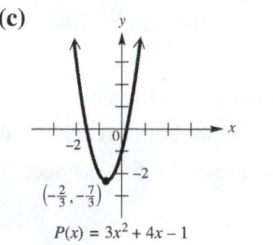

$P(x) = 3x^2 + 4x - 1$

17. (a) D **(b)** B **(c)** C **(d)** A **19. (a)** $(2, 0)$
(b) domain: $(-\infty, \infty)$; range: $[0, \infty)$ **(c)** $x = 2$ **(d)** $(2, \infty)$
(e) $(-\infty, 2)$ **(f)** minimum: 0 **21. (a)** $(-3, -4)$
(b) domain: $(-\infty, \infty)$; range: $[-4, \infty)$ **(c)** $x = -3$
(d) $(-3, \infty)$ **(e)** $(-\infty, -3)$ **(f)** minimum: -4
23. (a) $(-3, 2)$ **(b)** domain: $(-\infty, \infty)$; range: $(-\infty, 2]$
(c) $x = -3$ **(d)** $(-\infty, -3)$ **(e)** $(-3, \infty)$ **(f)** maximum: 2
25. (a) $(5, -4)$ **27. (a)** $(2, 2)$
(b) **(b)**

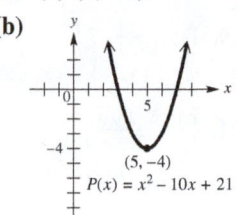 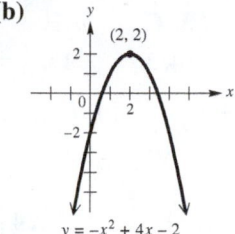

$P(x) = x^2 - 10x + 21$ $y = -x^2 + 4x - 2$

29. (a) $(1, 3)$ **31. (a)** $(4, 2)$
(b) **(b)**

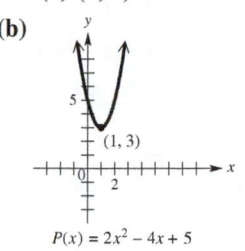 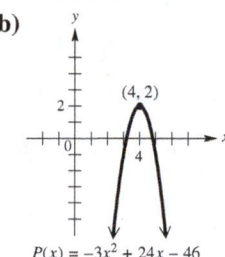

$P(x) = 2x^2 - 4x + 5$ $P(x) = -3x^2 + 24x - 46$

33. (a) $\left(\frac{1}{2}, \frac{1}{2}\right)$ **35. (a)** $(2.71, 5.20)$
(b) **(b)** $(-1.33, 0), (6.74, 0)$

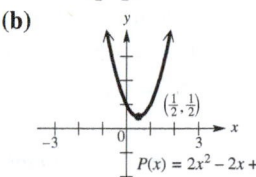

$P(x) = 2x^2 - 2x + 1$

37. (a) $(1.12, 0.56)$
(b) none
39. (a) $(0.68, 0.57)$
(b) $(0, 0), (1.35, 0)$ **41.** 3
43. none
45. (a) $(4, -12)$ **(b)** minimum **(c)** -12 **(d)** $[-12, \infty)$
47. (a) $(1.5, 2)$ **(b)** maximum **(c)** 2 **(d)** $(-\infty, 2]$
49. quadratic; $a < 0$ **51.** quadratic; $a > 0$
53. linear; positive **55.** $P(x) = 3x^2 + 6x - 1$
57. $P(x) = \frac{1}{2}x^2 - 8x + 35$ **59.** $P(x) = -\frac{2}{3}x^2 - \frac{16}{3}x - \frac{38}{3}$
61. (a) decreases first 4 min; increases next 4 min
(b) 90 bpm after 4 min. **63. (a)** The data increase and then
decrease. **(b)** $f(x) = -9(x - 2)^2 + 120$; Answers may vary.
(c) $[0, 4]$ **65. (a)** The value of t cannot be negative,
since t represents time elapsed after the launch.
(b) Since the rock was projected from ground level, s_0, the
initial height of the rock is 0. **(c)** $s(t) = -16t^2 + 90t$
(d) 99 ft **(e)** After 2.8125 sec, the maximum height,
126.5625 ft, is attained. Locate the vertex $(2.8125, 126.5625)$.
(f) 5.625 sec **67. (a)** The ball will not reach 355 ft
because the graph of $y_1 = -16x^2 + 150x$ does not intersect
the graph of $y_2 = 355$. **(b)** yes; the ball reaches a height of
355 ft at $t \approx 1.4$ sec and $t \approx 14.2$ sec.

69.

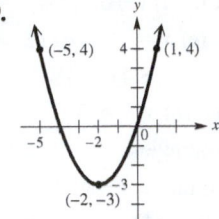

71.

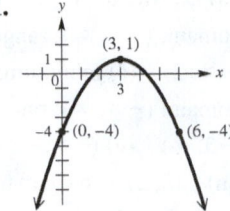

3.3 Exercises *(pages 199–202)*

1. G **3.** C **5.** H **7.** D **9.** D; $\left\{-\frac{1}{3}, 7\right\}$

11. C; $\{-4, 3\}$ **13.** $\{\pm 4\}$ **15.** $\left\{\pm 3\sqrt{5}\right\}$

17. $\{\pm 4i\}$ **19.** $\left\{\pm 3i\sqrt{2}\right\}$ **21.** $\left\{\frac{1 \pm 2\sqrt{3}}{3}\right\}$

23. $\left\{\frac{3}{5} \pm \frac{\sqrt{3}}{5}i\right\}$ **25.** $\{-4, 6\}$ **27.** $\left\{0, \frac{2}{3}\right\}$

29. $\left\{-\frac{1}{2}, \frac{3}{7}\right\}$ **31.** $\left\{\frac{1}{2}, 4\right\}$ **33.** $\{-8, 9\}$

35. $\{-1, 7\}$ **37.** $\left\{1 \pm \sqrt{5}\right\}$ **39.** $\left\{-\frac{1}{2} \pm \frac{1}{2}i\right\}$

41. $\left\{\frac{1 \pm \sqrt{5}}{2}\right\}$ **43.** $\left\{3 \pm \sqrt{2}\right\}$ **45.** $\left\{\frac{3}{2} \pm \frac{\sqrt{2}}{2}i\right\}$

47. $\left\{\frac{-3 \pm 3\sqrt{65}}{8}\right\}$ **49.** $\{-2, 8\}$ **51.** $\left\{\frac{2 \pm \sqrt{6}}{2}\right\}$

53. $\left\{-\frac{1}{2} \pm \frac{\sqrt{3}}{2}i\right\}$ **55.** $\left\{\frac{5}{2}\right\}$ **57.** $\left\{-\frac{2}{3}, 2\right\}$

59. $\left\{4 \pm 3i\sqrt{2}\right\}$ **61.** $\left\{1 \pm \sqrt{3}\right\}$ **63.** $\left\{\frac{-3 \pm \sqrt{15}}{2}\right\}$

65. $\left\{\frac{1 \pm \sqrt{13}}{2}\right\}$ **67.** $\left\{\frac{1}{4} \pm \frac{\sqrt{23}}{4}i\right\}$

69. 0; one real solution; rational **71.** 84; two real solutions; irrational **73.** -23; no real solutions

75. $a = 1; b = -9; c = 20$ **77.** $a = 1; b = -2; c = -1$

79. $a = 1; b = 0; c = 4$ **81.** $a = 1; b = -4; c = -1$

In Exercises 83–87, answers may vary. We give one example for each exercise.

83.

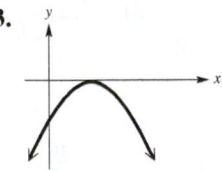

85.

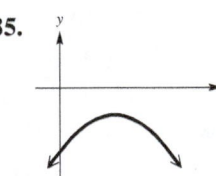

87.

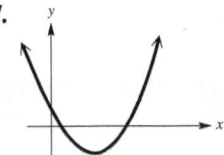

89. $\{2, 4\}$ **91.** $(-\infty, 2) \cup (4, \infty)$

93. $(-\infty, 3) \cup (3, \infty)$ **95.** $(-\infty, \infty)$

97. no real solutions; two nonreal complex solutions

99. 3 **101.** yes; negative

103. (a) $(-\infty, -3] \cup [-1, \infty)$ (b) $(-3, -1)$

105. (a) $\left(-\infty, \frac{1}{2}\right) \cup (4, \infty)$ (b) $\left[\frac{1}{2}, 4\right]$

107. (a) $(-\infty, -1] \cup [0, \infty)$ (b) $(-1, 0)$

109. (a) ∅ (b) $(-\infty, \infty)$ **111.** (a) $\left[1 - \sqrt{2}, 1 + \sqrt{2}\right]$
(b) $\left(-\infty, 1 - \sqrt{2}\right) \cup \left(1 + \sqrt{2}, \infty\right)$

113. (a) $\left(\frac{1 - \sqrt{13}}{6}, \frac{1 + \sqrt{13}}{6}\right)$

(b) $\left(-\infty, \frac{1 - \sqrt{13}}{6}\right] \cup \left[\frac{1 + \sqrt{13}}{6}, \infty\right)$ **115.** $t = \frac{\pm \sqrt{2sg}}{g}$

117. $a = \pm \sqrt{c^2 - b^2}$ **119.** $r = \frac{\pm \sqrt{S\pi}}{2\pi}$

121. $e = \sqrt[3]{V}$ **123.** $v = \frac{\pm \sqrt[4]{Frk^3M^3}}{kM}$

125. $R = \frac{E^2 - 2Pr \pm E\sqrt{E^2 - 4Pr}}{2P}$

127. $x = -\frac{y}{2} \pm \frac{\sqrt{3}}{2}yi; \; y = -\frac{x}{2} \pm \frac{\sqrt{3}}{2}xi$

129. $x = \frac{2y \pm \sqrt{31y^2 + 9}}{9}; \; y = \frac{-2x \pm \sqrt{31x^2 - 3}}{3}$

131. (a) $f(0) = 160, f(2) = 131.2$; Rate is 160 bpm initially and about 131 bpm after 2 min. (b) about $2.9 \le x \le 5$ (min)

Reviewing Basic Concepts *(page 202)*

1. $3 + 7i$ **2.** $26i$ **3.** $-\frac{50}{13} + \frac{10}{13}i$

4.

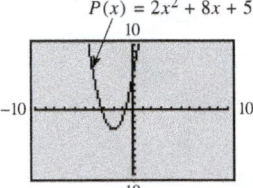

5. $(-2, -3)$; minimum **6.** $x = -2$

7. domain: $(-\infty, \infty)$; range: $[-3, \infty)$ **8.** $\left\{\pm \frac{5}{3}\right\}$

9. $\left\{-\frac{1}{3}, 2\right\}$ **10.** $\left\{\frac{1 \pm \sqrt{13}}{2}\right\}$ **11.** $\left[-\frac{1}{3}, 2\right]$

12. $\left(-\infty, \frac{1 - \sqrt{13}}{2}\right) \cup \left(\frac{1 + \sqrt{13}}{2}, \infty\right)$ **13.** ∅

3.4 Exercises *(pages 207–212)*

1. 116 **3.** 2 **5.** $\left\{\frac{1}{4}, 1\right\}$ **7.** $\left\{6 \pm \sqrt{30}\right\}$

9. A **11.** (a) $30 - x$ (b) $0 < x < 30$

(c) $P(x) = -x^2 + 30x$ (d) 15 and 15; The maximum product is 225. **13.** 250 ft by 250 ft

15. (a) $640 - 2x$ (b) $0 < x < 320$

(c) $\mathcal{A}(x) = -2x^2 + 640x$ (d) between 57.04 ft and 85.17 ft or 234.83 ft and 262.96 ft (e) 160 ft by 320 ft; The maximum area is 51,200 ft². **17.** (a) 32 ft (b) 34.25 ft **19.** 2.2 sec

21. (a) $2x$ (b) length: $2x - 4$; width: $x - 4$; $x > 4$

(c) $V(x) = 4x^2 - 24x + 32$ (d) 8 in. by 20 in.

(e) 13.0 in. to 14.2 in. **23.** about 1.5 in. **25.** 20 ft

27. 5 ft **29.** a 17-ft ladder; A 16-ft ladder won't reach.

31. 20 in. by 24 in. **33.** (a) $80 - x$ (b) $400 + 20x$

(c) $R(x) = -20x^2 + 1200x + 32,000$ (d) 5 or 55

(e) $1000 **35.** (a) about 23.32 ft per sec (b) yes

(c) about 12.88 ft **37.** (a) 3.5 ft (b) about 0.2 ft and 2.3 ft

(c) 1.25 ft (d) about 3.78 ft **39.** about 104.5 ft per sec; 71.25 mph

41. (a) 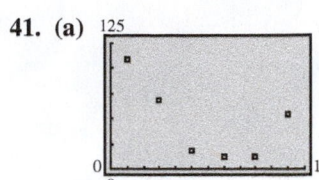   **(b)** $f(x) = \frac{8}{3}(x - 7)^2 + 12$

(c) It is a good fit.

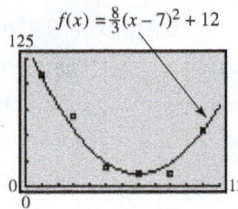
$f(x) = \frac{8}{3}(x - 7)^2 + 12$

(d) $g(x) \approx 2.72x^2 - 38.93x + 149.46$
(e) (i) f: \$79; g: \$82   **(ii)** f: \$15; g: \$14

43. (a) 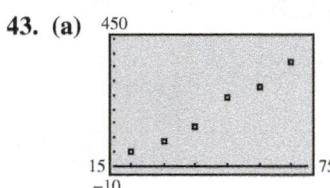   **(b)** $f(45) \approx 161.5$ ft is the stopping distance when the speed is 45 mph.

(c) The model is quite good, although the stopping distances are a little low for the higher speeds.

3.5 Exercises *(pages 221–224)*

1. minimum degree: 4   **3.** (a, b) and (c, d), local maxima; (e, t), local minimum   **5.** (a, b), absolute maximum
7. local maximum values of b and d; local minimum value of t; absolute maximum value of b

9.    **11.** ↘   **13.** ∪∩   **15.** ∩↘

17. ∪∩   **19.** ↖↘

21. The graph of $f(x) = x^n$ for $n \in \{$ positive odd integers $\}$ will take the shape of the graph of $f(x) = x^3$, but gets steeper as n and x increase.   **23.** As the odd exponent n gets larger, the graph *flattens out* in the window $[-1, 1]$ by $[-1, 1]$.

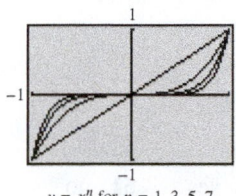
$y = x^n$ for $n = 1, 3, 5, 7$

The graph of $y = x^7$ will be between the graph of $y = x^5$ and the x-axis in this window.

25. local maximum: $(2, 3.67)$; local minimum: $(3, 3.5)$
27. local maximum: $(-3.33, -1.85)$; local minimum: $(-4, -2)$
29. two: $(2.10, 0)$ and $(2.15, 0)$ **31.** none   **33.** D
35. B   **37.** false   **39.** true   **41.** true   **43.** false
45. A   **47.** one   **49.** B and D   **51.** one   **53.** B

55. Shift the graph of $y = x^4$ to the left 3 units, stretch vertically by applying a factor of 2, and shift downward 7 units.

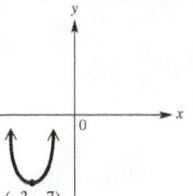   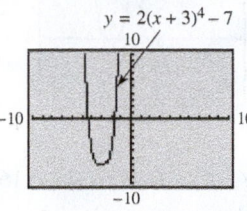
$y = 2(x + 3)^4 - 7$

56. Shift the graph of $y = x^4$ to the left 1 unit, stretch vertically by applying a factor of 3, reflect across the x-axis, and shift upward 12 units.

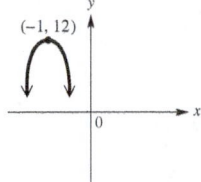   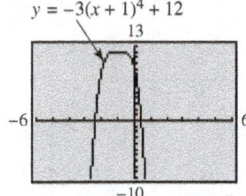
$y = -3(x + 1)^4 + 12$

57. Shift the graph of $y = x^3$ to the right 1 unit, stretch vertically by applying a factor of 3, reflect across the x-axis, and shift upward 12 units.

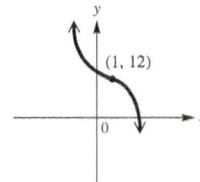   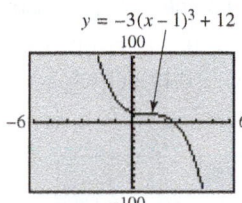
$y = -3(x - 1)^3 + 12$

58. Shift the graph of $y = x^5$ to the right 1 unit, shrink vertically by applying a factor of 0.5, and shift 13 units upward.

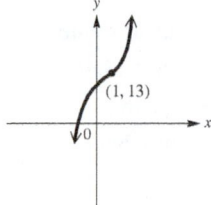   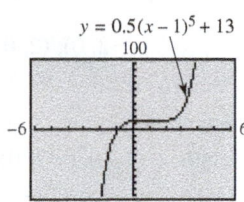
$y = 0.5(x - 1)^5 + 13$

59.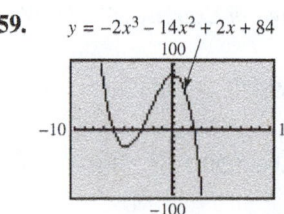
$y = -2x^3 - 14x^2 + 2x + 84$

(a) $(-\infty, \infty)$ **(b)** $(-4.74, -27.03)$; not an absolute minimum point   **(c)** $(0.07, 84.07)$; not an absolute maximum point
(d) $(-\infty, \infty)$ **(e)** x-intercepts: $(-6, 0)$, $(-3.19, 0)$, $(2.19, 0)$; y-intercept: $(0, 84)$ **(f)** $(-4.74, 0.07)$
(g) $(-\infty, -4.74)$; $(0.07, \infty)$

61. $y = x^5 + 4x^4 - 3x^3 - 17x^2 + 6x + 9$

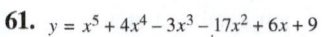
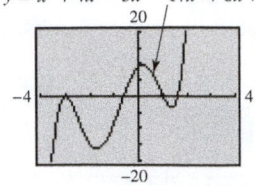

(a) $(-\infty, \infty)$ **(b)** $(-1.73, -16.39)$, $(1.35, -3.49)$; Neither is an absolute minimum point. **(c)** $(-3, 0)$, $(0.17, 9.52)$; Neither is an absolute maximum point. **(d)** $(-\infty, \infty)$ **(e)** x-intercepts: $(-3, 0)$, $(-0.62, 0)$, $(1, 0)$, $(1.62, 0)$; y-intercept: $(0, 9)$ **(f)** $(-\infty, -3)$; $(-1.73, 0.17)$; $(1.35, \infty)$ **(g)** $(-3, -1.73)$; $(0.17, 1.35)$

63. $y = 2x^4 + 3x^3 - 17x^2 - 6x - 72$

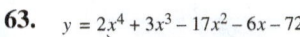
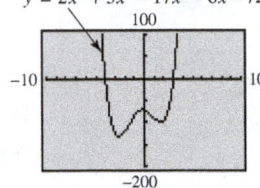

(a) $(-\infty, \infty)$ **(b)** $(-2.63, -132.69)$ is an absolute minimum point; $(1.68, -99.90)$ **(c)** $(-0.17, -71.48)$; not an absolute maximum point **(d)** $[-132.69, \infty)$ **(e)** x-intercepts: $(-4, 0)$, $(3, 0)$; y-intercept: $(0, -72)$ **(f)** $(-2.63, -0.17)$; $(1.68, \infty)$ **(g)** $(-\infty, -2.63)$; $(-0.17, 1.68)$

65. $y = -x^6 + 24x^4 - 144x^2 + 256$

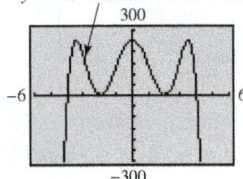

(a) $(-\infty, \infty)$ **(b)** $(-2, 0)$; $(2, 0)$; Neither is an absolute minimum point. **(c)** $(-3.46, 256)$, $(0, 256)$, and $(3.46, 256)$; All are absolute maximum points. **(d)** $(-\infty, 256]$ **(e)** x-intercepts: $(-4, 0)$, $(-2, 0)$, $(2, 0)$, $(4, 0)$; y-intercept: $(0, 256)$ **(f)** $(-\infty, -3.46)$; $(-2, 0)$; $(2, 3.46)$ **(g)** $(-3.46, -2)$; $(0, 2)$; $(3.46, \infty)$

There are many possible valid windows in Exercises 67–71. We give only one in each case.
67. $[-10, 10]$ by $[-40, 10]$ **69.** $[-10, 20]$ by $[-1500, 500]$ **71.** $[-10, 10]$ by $[-20, 500]$
73. (a) March; about 74.8°F; July: about 90.1°F
(b) April and October

Reviewing Basic Concepts *(page 225)*
1. (a) The total length of the fence must satisfy $2L + 2x = 300$. From this equation, $L = 150 - x$.
(b) $\mathcal{A}(x) = x(150 - x)$ **(c)** $0 < x < 150$
(d) 50 m by 100 m

2. (a)

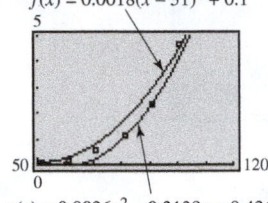

(b) $f(x) \approx 0.0018(x - 51)^2 + 0.1$
(c) $g(x) \approx 0.0026x^2 - 0.3139x + 9.426$
(d)

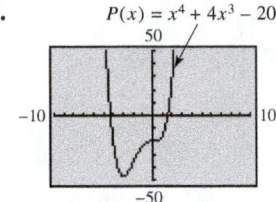

$g(x) \approx 0.0026x^2 - 0.3139x + 9.426$

The regression function fits slightly better because it is closer to or passes through more data points. Neither function would fit the data for $x < 51$.

3. ↗ **4.** two; three **5.** ↱↰

6. $P(x) = x^4 + 4x^3 - 20$

7. The only extreme point is $(-3, -47)$, an absolute and local minimum point. **8.** x-intercepts: $(-4.26, 0)$, $(1.53, 0)$; y-intercept: $(0, -20)$

3.6 Exercises *(pages 234–236)*
1. $2x^3$ **3.** $\frac{8}{3}x^2$ **5.** $x^5 + \frac{3}{2}x^2$ **7.** $4x^2 - \frac{5}{2}$
9. $P(1) = -5$ and $P(2) = 2$ differ in sign. The zero is approximately 1.79. **11.** $P(2) = 2$ and $P(2.5) = -0.25$ differ in sign. The zero is approximately 2.39. **13.** $P(1.5) = -0.375$ and $P(2) = 16$ differ in sign. The zero is approximately 1.52. **15.** $P(2.7) = 0.9219$ and $P(2.8) = -2.7616$ differ in sign. The zero is approximately 2.73. **17.** $P(-1.6) = -1.29376$ and $P(-1.5) = 0.15625$ differ in sign. The zero is approximately -1.51.
19. There is at least one zero between 2 and 2.5.
21. $x^2 - 3x - 2$ **23.** $3x^2 + 4x + \frac{3}{x - 5}$
25. $x^3 - x^2 - 6x$ **27.** $x^2 + 3x + 3$
29. $-2x^2 + 2x - 3 + \frac{1}{x + 1}$ **31.** $x^4 + x^3 + x^2 + x + 1$
33. 0 **35.** -25 **37.** -5 **39.** 3.625 **41.** -1
43. $\sqrt[3]{4}$ **45.** yes **47.** no **49.** yes **51.** no
53. yes **55.** $x + 3, x - 1, x - 4$ **56.** $-3, 1, 4$
57. $-3, 1, 4$ **58.** $-10; -10$ **59.** $(-3, 1) \cup (4, \infty)$
60. $(-\infty, -3) \cup (1, 4)$ **61.** $-2, 1$
63. $\frac{-1 - \sqrt{5}}{2}, \frac{-1 + \sqrt{5}}{2}$ **65.** $\frac{1 - \sqrt{13}}{6}, \frac{1 + \sqrt{13}}{6}$
67. $-\sqrt{5}, \sqrt{5}$ **69.** $-\sqrt{3}, \sqrt{3}$

71. $P(x) = (x - 2)(2x - 5)(x + 3)$

73. $P(x) = (x + 4)(3x - 1)(2x + 1)$

75. $P(x) = (x + 3)(-3x + 1)(2x - 1)$

77. $P(x) = (x + 5)(x - \sqrt{3})(x + \sqrt{3})$

79. $P(x) = (x + 1)^2(x - 4)$ **81.** $x^3 + 2 + \dfrac{-2}{3x - 7}$

83. $5x^2 - 12 + \dfrac{30}{x^2 + 2}$ **85.** $4x + 5$

87. $x^2 - 2x + 4 + \dfrac{-1}{2x^2 + 3x + 2}$ **89.** $\dfrac{1}{2}x + \dfrac{-1}{2x + 1}$

91. $\dfrac{1}{2}x - \dfrac{1}{2} + \dfrac{\frac{1}{2}x + \frac{1}{2}}{2x^2 - 1}$

3.7 Exercises *(pages 245–248)*

1. $P(x) = x^3 - 8x^2 + 21x - 20$

3. $P(x) = x^3 - 5x^2 + x - 5$ **5.** $P(x) = x^3 - 6x^2 + 10x$

7. $P(x) = -\dfrac{1}{6}x^3 + \dfrac{13}{6}x + 2$ **9.** $P(x) = -\dfrac{1}{2}x^3 - \dfrac{1}{2}x^2 + x$

11. $P(x) = -x^3 + 6x^2 - 10x + 8$ **13.** $-1 + i, -1 - i$

15. $-1 + \sqrt{2}, -1 - \sqrt{2}$ **17.** $-3i, \dfrac{1}{2} + \dfrac{\sqrt{3}}{2}i, \dfrac{1}{2} - \dfrac{\sqrt{3}}{2}i$

19. $P(x) = x^2 - x - 20$

21. $P(x) = x^4 + x^3 - 5x^2 + x - 6$

23. $P(x) = x^3 - 3x^2 + 2$

25. $P(x) = x^4 - 7x^3 + 17x^2 - x - 26$

27. $P(x) = x^3 - 11x^2 + 33x + 45$

29. $P(x) = x^4 + 2x^3 - 10x^2 - 6x + 45$

31.

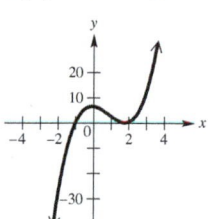

$P(x) = 2x^3 - 5x^2 - x + 6$
$= (x + 1)(2x - 3)(x - 2)$

33.

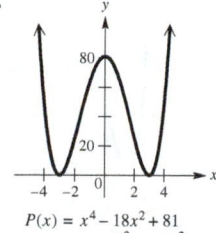

$P(x) = x^4 - 18x^2 + 81$
$= (x - 3)^2(x + 3)^2$

35.

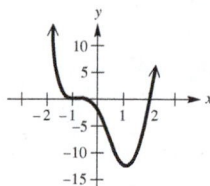

$P(x) = 2x^4 + x^3 - 6x^2 - 7x - 2$
$= (2x + 1)(x - 2)(x + 1)^2$

37.

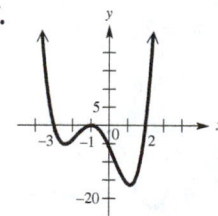

$P(x) = x^4 + 3x^3 - 3x^2 - 11x - 6$
$= (x + 3)(x + 1)^2(x - 2)$

39.

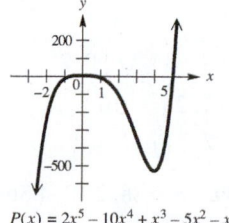

$P(x) = 2x^5 - 10x^4 + x^3 - 5x^2 - x + 5$
$= (x - 5)(x^2 + 1)(2x^2 - 1)$

41. $f(x) = -x(x + 1)(x - 2)$

43. $f(x) = \dfrac{1}{2}(x + 2)(x + 1)(x - 1)$

45. $f(x) = \dfrac{1}{4}(x + 4)(x + 1)(x - 1)(x - 2)$

47. Use synthetic division twice, with $k = -2$. Zeros are -2, $3, -1$; $P(x) = (x + 2)^2(x - 3)(x + 1)$.   **49.** 1, 3, or 5

51. **(a)** not possible   **(b)** possible   **(c)** not possible
(d) possible   **53.** **(a)** $\pm 1, \pm 2, \pm 5, \pm 10$ **(b)** Eliminate
values less than -2 or greater than 5.   **(c)** $-2, -1, 5$
(d) $P(x) = (x + 2)(x + 1)(x - 5)$

55. **(a)** $\pm 1, \pm 2, \pm 3, \pm 5, \pm 6, \pm 10, \pm 15, \pm 30$
(b) Eliminate values less than -5 or greater than 2.
(c) $-5, -3, 2$ **(d)** $P(x) = (x + 5)(x + 3)(x - 2)$

57. **(a)** $\pm 1, \pm 2, \pm 3, \pm 4, \pm 6, \pm 12, \pm \dfrac{1}{2}, \pm \dfrac{3}{2}, \pm \dfrac{1}{3},$
$\pm \dfrac{2}{3}, \pm \dfrac{4}{3}, \pm \dfrac{1}{6}$ **(b)** Eliminate values less than -4 or
greater than $\dfrac{3}{2}$.   **(c)** $-4, -\dfrac{1}{3}, \dfrac{3}{2}$
(d) $P(x) = (x + 4)(3x + 1)(2x - 3)$

59. $P(x) = (3x + 2)(2x + 3)(2x - 1)$

61. $P(x) = 2(2x + 3)(3x + 2)(2x - 1)$ **63.** $-2, -1, \dfrac{5}{2}$

65. $\dfrac{3}{2}, 4$ **67.** $(3x - 1)(2x - 1)(x - \sqrt{2})(x + \sqrt{2})$

69. $(7x + 2)(3x + 1)(x - \sqrt{5})(x + \sqrt{5})$

71. $(x - 1)(x - i)(x + i)(x - 2i)(x + 2i)$

73. $(x - 1)(x + 2)(x - 2i)(x + 2i)$

75. $(x - i)(x + i)(x - (1 + i))(x - (1 - i))$

77. possible: 0 or 2 positive real zeros, 1 negative real zero;
actual: 0 positive, 1 negative   **79.** possible: 1 positive
real zero, 1 negative real zero; actual: 1 positive, 1 negative
81. possible: 0 or 2 positive real zeros, 1 or 3 negative
real zeros; actual: 0 positive, 1 negative

91. $P(x) = \dfrac{1}{2}(x + 6)(x - 2)(x - 5)$,
or $P(x) = \dfrac{1}{2}x^3 - \dfrac{1}{2}x^2 - 16x + 30$

93. **(a)** positive zeros: 1; negative zeros: 3 or 1
(b) $\pm 1, \pm 2, \pm \dfrac{1}{2}$ **(c)** $-1, 1$ **(d)** no other real zeros
(e) $-\dfrac{1}{4} + \dfrac{\sqrt{15}}{4}i, -\dfrac{1}{4} - \dfrac{\sqrt{15}}{4}i$ **(f)** $(-1, 0), (1, 0)$ **(g)** $(0, 2)$
(h) $P(4) = -570; (4, -570)$ **(i)** ⌢↘
(j)

$P(x) = -2x^4 - x^3 + x + 2$

94. **(a)** positive zeros: 0; negative zeros: 4, 2, or 0
(b) $0, \pm 1, \pm 3, \pm 9, \pm 27, \pm \dfrac{1}{2}, \pm \dfrac{3}{2}, \pm \dfrac{9}{2}, \pm \dfrac{27}{2}, \pm \dfrac{1}{4}, \pm \dfrac{3}{4},$
$\pm \dfrac{9}{4}, \pm \dfrac{27}{4}$ **(c)** $0, -\dfrac{3}{2}$ (multiplicity 2)
(d) no other real zeros   **(e)** $\dfrac{1}{2} + \dfrac{\sqrt{11}}{2}i, \dfrac{1}{2} - \dfrac{\sqrt{11}}{2}i$
(f) $(0, 0), \left(-\dfrac{3}{2}, 0\right)$ **(g)** $(0, 0)$ **(h)** $P(4) = 7260; (4, 7260)$
(i) ↙↗

94. (j)

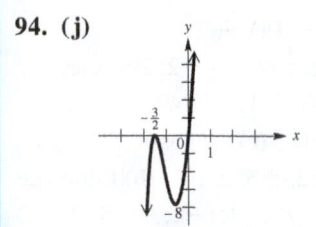

$P(x) = 4x^5 + 8x^4 + 9x^3 + 27x^2 + 27x$

95. (a) positive zeros: 1; negative zeros: 1

(b) $\pm 1, \pm 5, \pm\frac{1}{3}, \pm\frac{5}{3}$ **(c)** no rational zeros

(d) $-\sqrt{5}, \sqrt{5}$ **(e)** $-\frac{\sqrt{3}}{3}i, \frac{\sqrt{3}}{3}i$ **(f)** $(-\sqrt{5}, 0), (\sqrt{5}, 0)$

(g) $(0, -5)$ **(h)** $P(4) = 539; (4, 539)$ **(i)** ↶↗

(j)

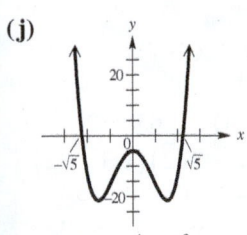

$P(x) = 3x^4 - 14x^2 - 5$

96. (a) positive zeros: 2 or 0; negative zeros: 3 or 1

(b) $\pm 1, \pm 3, \pm 9$ **(c)** $-3, -1$ (multiplicity 2), 1, 3

(d) no other real zeros **(e)** no other complex zeros

(f) $(-3, 0), (-1, 0), (1, 0), (3, 0)$ **(g)** $(0, -9)$

(h) $P(4) = -525; (4, -525)$

(i)

(j)

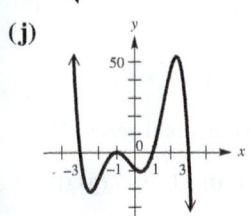

$P(x) = -x^5 - x^4 + 10x^3 + 10x^2 - 9x - 9$

97. (a) positive zeros: 4, 2, or 0; negative zeros: 0

(b) $\pm 1, \pm 2, \pm 3, \pm 4, \pm 6, \pm 12, \pm\frac{1}{3}, \pm\frac{2}{3}, \pm\frac{4}{3}$

(c) $\frac{1}{3}, 2$ (multiplicity 2), 3 **(d)** no other real zeros **(e)** no other complex zeros **(f)** $(\frac{1}{3}, 0), (2, 0), (3, 0)$ **(g)** $(0, -12)$

(h) $P(4) = -44; (4, -44)$ **(i)** ↷↘

(j)

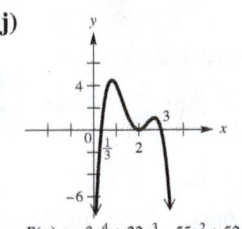

$P(x) = -3x^4 + 22x^3 - 55x^2 + 52x - 12$

98. For the function in **Exercise 95:** ± 2.236

3.8 Exercises *(pages 256–259)*

1. $\{\pm 5, 0\}$ **3.** $\{\pm 2\}$ **5.** $\{-3, 0, 6\}$

7. $\{0, 1\}$ **9.** $\{-\frac{1}{4}, 0, \frac{5}{3}\}$ **11.** $\{\pm 2, \frac{1}{2}\}$

13. $\{0, \pm\frac{\sqrt{7}}{7}i\}$ **15.** $\{-\frac{2}{3}, \pm 1\}$

17. $\{\pm 1, \pm\sqrt{10}\}$ **19.** $\{-2, -1.5, 1.5, 2\}$

21. $\{-4, 4, -i, i\}$ **23.** $\{-8, 1, 8\}$

25. $\{-1.5, 0, 1\}$ **27.** $\{-1, \pm\sqrt{7}\}$

29. $\{0, \frac{-1 \pm \sqrt{73}}{6}\}$ **31.** $\{0, -\frac{1}{2} - \frac{\sqrt{3}}{2}i, -\frac{1}{2} + \frac{\sqrt{3}}{2}i\}$

33. $\{-4i, -i, i, 4i\}$

35. $\{-3, 2, -1 - i\sqrt{3}, -1 + i\sqrt{3}, \frac{3}{2} + \frac{3\sqrt{3}}{2}i, \frac{3}{2} - \frac{3\sqrt{3}}{2}i\}$

37.

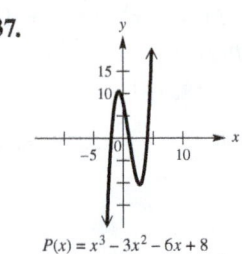

$P(x) = x^3 - 3x^2 - 6x + 8$
$= (x - 4)(x - 1)(x + 2)$

(a) $\{-2, 1, 4\}$ **(b)** $(-\infty, -2) \cup (1, 4)$

(c) $(-2, 1) \cup (4, \infty)$

39.

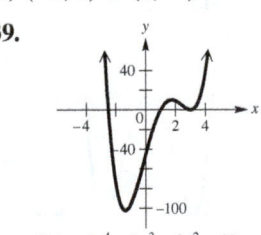

$P(x) = 2x^4 - 9x^3 - 5x^2 + 57x - 45$
$= (x - 3)^2(2x + 5)(x - 1)$

(a) $\{-2.5, 1, 3$ (multiplicity 2)$\}$ **(b)** $(-2.5, 1)$

(c) $(-\infty, -2.5) \cup (1, 3) \cup (3, \infty)$

41.

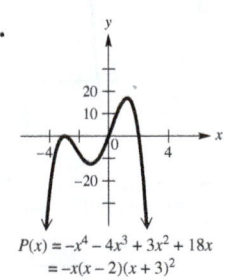

$P(x) = -x^4 - 4x^3 + 3x^2 + 18x$
$= -x(x - 2)(x + 3)^2$

(a) $\{-3$ (multiplicity 2)$, 0, 2\}$ **(b)** $\{-3\} \cup [0, 2]$

(c) $(-\infty, 0] \cup [2, \infty)$ **43. (a)** $\{\frac{2}{3}, 2\}$ **(b)** $(\frac{2}{3}, 2)$

45. (a) $\{-1, -\frac{5}{14}, \frac{1}{2}\}$ **(b)** $(-\infty, -\frac{5}{14}] \cup [\frac{1}{2}, \infty)$

47. (a) $\{0, \frac{7}{3k}\}$ **(b)** $(0, \frac{7}{3k})$ **49.** $\{-0.88, 2.12, 4.86\}$

51. $\{1.52\}$ **53.** $\{-0.40, 2.02\}$

55. $\{-i, i\}$ **57.** $\{-1, \frac{1}{2} + \frac{\sqrt{3}}{2}i, \frac{1}{2} - \frac{\sqrt{3}}{2}i\}$

59. $\{3, -\frac{3}{2} - \frac{3\sqrt{3}}{2}i, -\frac{3}{2} + \frac{3\sqrt{3}}{2}i\}$

61. $\{-2, 2, -2i, 2i\}$ **63.** $\{-4, 2 + 2i\sqrt{3}, 2 - 2i\sqrt{3}\}$
65. $\{-3i\sqrt{2}, 3i\sqrt{2}\}$ **67.** (a) about 7.13 cm; The ball floats partly above the surface.   (b) The ball is more dense than water and sinks below the surface.   (c) 10 cm; The balloon floats even with the surface.
69. (a) $0 < x < 6$ (b) $V(x) = 4x^3 - 60x^2 + 216x$
(c) about 2.35; about 228.16 in.^{3} (d) $0.42 < x < 5$
(approximately)   **71.** about 2.61 in.   **73.** (a) $x - 1$
(b) $\sqrt{x^2 - (x - 1)^2}$, or $\sqrt{2x - 1}$
(c) $2x^3 - 5x^2 + 4x - 28{,}225 = 0$
(d) 7 in., 24 in., and 25 in.
75. (a) about 66.15 in.^{3} (b) 0.54 in. $< x < 2.92$ in.

77. (a)

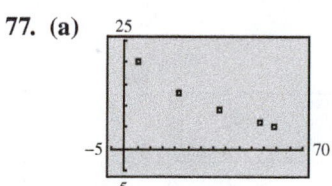

(b) $C(x) \approx 0.0035x^2 - 0.49x + 22$

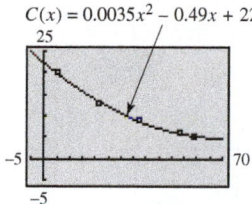

(c) $C(x) \approx -0.000068x^3 + 0.00987x^2 - 0.653x + 23$

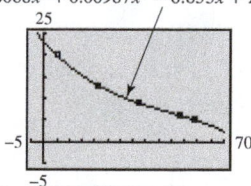

(d) The cubic function is a slightly better fit.
(e) $0 \le x < 31.92$

Reviewing Basic Concepts *(page 259)*

1. 30   **2.** no
3. $P(x) = (x + 2)(2x - 3)(x - 2 - i)(x - 2 + i)$
4. $P(x) = x^3 - \frac{3}{2}x^2 + x - \frac{3}{2}$
5. $P(x) = x^4 + 6x^3 + 5x^2 + 8x + 80$
6. $-2, -1, \frac{5}{2}$ **7.** $\left\{\pm\sqrt{\dfrac{6 + \sqrt{33}}{3}}, \pm\sqrt{\dfrac{6 - \sqrt{33}}{3}}\right\}$
8. (a) $P(x) \approx 0.0004815x^3 + 0.0303x^2 - 0.1989x + 3.1$
(b) $P(34) \approx 50.3$ thousand; within 300   (c) about 2005

Chapter 3 Review Exercises *(pages 264–267)*

1. $18 - 4i$ **3.** $14 - 52i$ **5.** $\frac{1}{10} + \frac{3}{10}i$
7. $(-\infty, \infty)$ **9.** ↰↱   **11.** $(0, -8)$
13. $\left(\frac{3}{2}, \infty\right); \left(-\infty, \frac{3}{2}\right)$ **15.** The graph intersects the x-axis at $(-1, 0)$ and $(4, 0)$, supporting the answer in (a). It lies above the x-axis when $x < -1$ or $x > 4$, supporting the answer in (b). It lies on or below the x-axis when x is between -1 and 4 inclusive, supporting the answer in (c).   **17.** Since the discriminant is greater than 0, there are two x-intercepts.
19. $(1.04, 6.37)$ **21.** (a) 4   (b) 1   (c) two   (d) none
23. 25 sec   **25.** 6.3 sec and 43.7 sec
27. (a) $V(x) = 12x^2 - 128x + 256$ (b) 20 in. by 60 in.
(c) One way is to graph $y_1 = V(x)$ and $y_2 = 2496$ and show that the graphs intersect at $x = 20$.   **29.** $P(-2) = 12$ and $P(-1) = -4$ differ in sign.   **31.** $2x^2 - 2x + 2 + \dfrac{1}{3x + 1}$
33. -1 **35.** 28   **37.** $7 - 2i$
39. $P(x) = x^3 - 13x^2 + 46x - 48$
41. $P(x) = x^4 + 5x^3 + x^2 - 9x + 2$
43. yes   **45.** $-3, 4, 1 - i, 1 + i$ **47.** $-2, \frac{1}{3}, 4$
51. $(x - 3)(x^2 + x - 1)$ **55.** $\{0, -1 + 2i, -1 - 2i\}$;
$(0, 0)$ is the only x-intercept.   **57.** even   **59.** positive
61. $(-\infty, a) \cup (b, c)$ **63.** $\{d, h\}$ **65.** Since $f(x)$ has three real zeros and a polynomial of degree 3 can have at most three zeros, there can be no other zeros, real or nonreal complex.   **67.** true   **69.** true   **71.** false
73. $(2, 0)$ **75.** $(-\infty, \infty)$ **77.** $\left\{-\sqrt{7}, -\frac{2}{3}, \sqrt{7}\right\}$;
The x-values of the x-intercepts are approximately $-2.65, -0.67$, and 2.65.   **79.** 4 in. by 4 in. by 4 in.
81. About 1.36 million   **83.** Answers will vary. For example, given 7; x^7; ↗   **84.** If a number is repeated n times, then it will have multiplicity n. Answers will vary.   **85.** Answers will vary.   **86.** Answers will vary; domain: $(-\infty, \infty)$; range: $(-\infty, \infty)$

Chapter 3 Test *(page 268)*

1. (a) $20 - 9i$ (b) $4 - i$ (c) i (d) $12 + 16i$ (e) $6i$
(f) -10 **2.** (a) $(-1, 8)$
(b) $P(x) = -2x^2 - 4x + 6$ (c) $-3, 1$

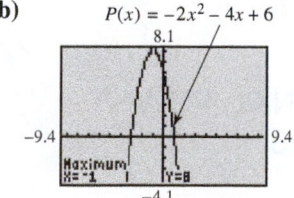

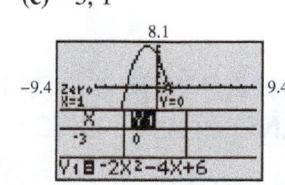

(d) $(0, 6)$ (e) domain: $(-\infty, \infty)$ range: $(-\infty, 8]$
(f) increasing: $(-\infty, -1)$; decreasing: $(-1, \infty)$
3. $\left\{\frac{1}{2}, 2\right\}$ **4.** $\{2 \pm \sqrt{6}\}$

5. (a) $\left\{\dfrac{-3 \pm \sqrt{33}}{6}\right\}$ **(b)**

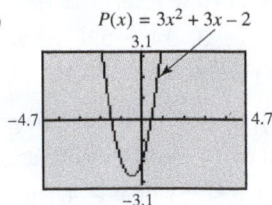

$P(x) = 3x^2 + 3x - 2$

(i) $\left(\dfrac{-3 - \sqrt{33}}{6}, \dfrac{-3 + \sqrt{33}}{6}\right)$

(ii) $\left(-\infty, \dfrac{-3 - \sqrt{33}}{6}\right] \cup \left[\dfrac{-3 + \sqrt{33}}{6}, \infty\right)$

6. 15 in. by 12 in. by 4 in.

7. (a) $f(x) \approx 0.000195238x^2 - 0.779143x + 779.252$

(b) $f(1975) \approx 1.99$, which is close to the actual value.

8. (a) $-1 + i\sqrt{3}, -1 - i\sqrt{3}$ **(b)** ↖↗

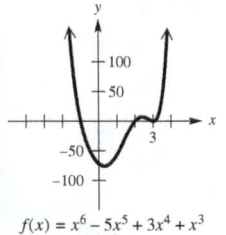

$f(x) = x^6 - 5x^5 + 3x^4 + x^3 + 40x^2 - 24x - 72$

9. (a) $-\dfrac{5}{2}, \dfrac{5}{2}, -i, i$ **(b)**

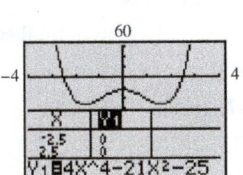

(c) It is symmetric with respect to the y-axis.

(d) (i) $\left(-\infty, -\dfrac{5}{2}\right] \cup \left[\dfrac{5}{2}, \infty\right)$ **(ii)** $\left(-\dfrac{5}{2}, \dfrac{5}{2}\right)$

10. (a) $\pm 1, \pm 2, \pm 11, \pm 22, \pm\dfrac{1}{3}, \pm\dfrac{2}{3}, \pm\dfrac{11}{3}, \pm\dfrac{22}{3}$

(b) $-2, \dfrac{1}{3}$ **(c)** $f(3) = -80$ and $f(4) = 330$ differ in sign.

(d) positive zeros: 2 or 0; negative zeros: 2 or 0

11. (a) $\{0.189, 1, 3.633\}$ **(b)** two   **12. (a)** $4x^2 - 2x$

(b) $3x^2 - 2x - 2 + \dfrac{4}{x - 1}$ **(c)** $x^3 + 2x - 1 + \dfrac{2}{2x - 1}$

(d) $x^2 - 2 + \dfrac{-2x + 10}{x^2 + 2}$ **13.** $f(x) = x^3 - 4x^2 + 4x - 16$

14. $\left\{0, \pm i\sqrt{3}\right\}$

CHAPTER 4 RATIONAL, POWER, AND ROOT FUNCTIONS

4.1 Exercises *(pages 275–276)*

1. $(-\infty, 0) \cup (0, \infty)$; $(-\infty, 0) \cup (0, \infty)$ **3.** none;

$(-\infty, 0) \cup (0, \infty)$; none   **5.** $x = 3$; $y = 2$

7. even; symmetry with respect to the y-axis   **9.** A, B, C

11. A   **13.** A   **15.** A, C, D

In Exercises 17–31, we give the domain and then the range below the traditional graph.

17. To obtain the graph of f, stretch the graph of $y = \dfrac{1}{x}$ vertically by applying a factor of 2.

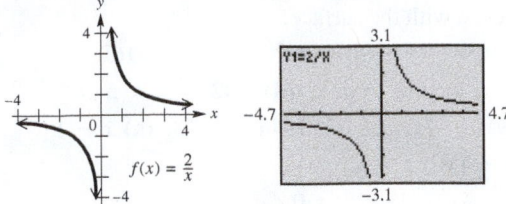

$f(x) = \dfrac{2}{x}$

$(-\infty, 0) \cup (0, \infty)$; $(-\infty, 0) \cup (0, \infty)$

19. To obtain the graph of f, shift the graph of $y = \dfrac{1}{x}$ to the left 2 units.

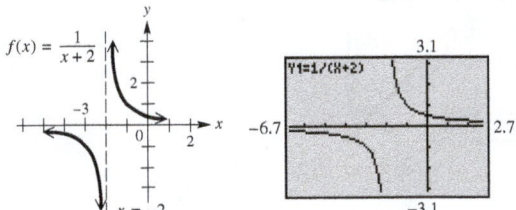

$f(x) = \dfrac{1}{x + 2}$

$(-\infty, -2) \cup (-2, \infty)$; $(-\infty, 0) \cup (0, \infty)$

21. To obtain the graph of f, shift the graph of $y = \dfrac{1}{x}$ upward 1 unit.

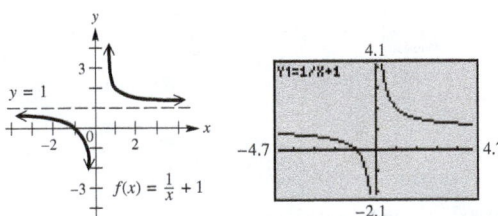

$f(x) = \dfrac{1}{x} + 1$

$(-\infty, 0) \cup (0, \infty)$; $(-\infty, 1) \cup (1, \infty)$

23. To obtain the graph of f, shift the graph of $y = \dfrac{1}{x}$ to the right 1 unit and upward 1 unit.

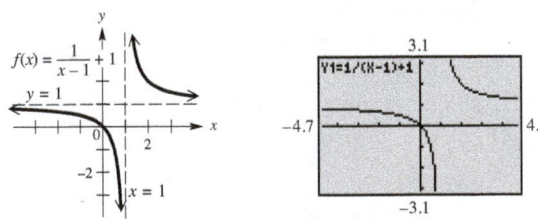

$f(x) = \dfrac{1}{x - 1} + 1$

$(-\infty, 1) \cup (1, \infty)$; $(-\infty, 1) \cup (1, \infty)$

25. To obtain the graph of f, shift the graph of $y = \dfrac{1}{x^2}$ downward 2 units.

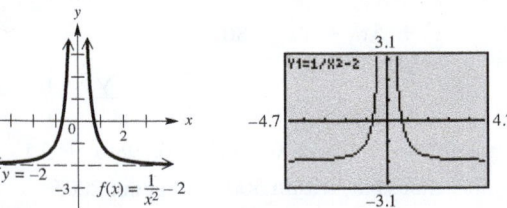

$f(x) = \dfrac{1}{x^2} - 2$

$(-\infty, 0) \cup (0, \infty)$; $(-2, \infty)$

27. To obtain the graph of f, stretch the graph of $y = \frac{1}{x^2}$ vertically by applying a factor of 2, and reflect across the x-axis.

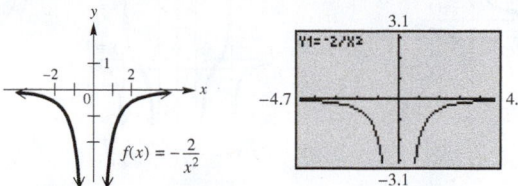

$(-\infty, 0) \cup (0, \infty)$; $(-\infty, 0)$

29. To obtain the graph of f, shift the graph of $y = \frac{1}{x^2}$ to the right 3 units.

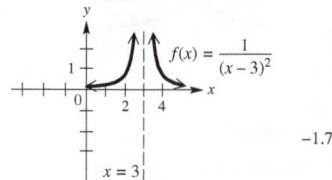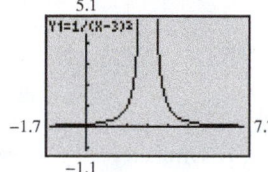

$(-\infty, 3) \cup (3, \infty)$; $(0, \infty)$

31. To obtain the graph of f, shift the graph of $y = \frac{1}{x^2}$ to the left 2 units, reflect across the x-axis, and shift 3 units downward.

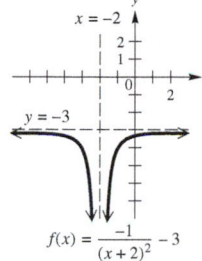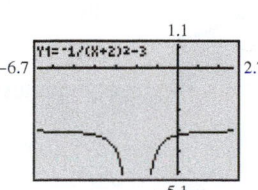

$(-\infty, -2) \cup (-2, \infty)$; $(-\infty, -3)$

33. C   **35.** B   **37.** vertical asymptote: $x = -1$; horizontal asymptote: $y = 1$; domain: $(-\infty, -1) \cup (-1, \infty)$; range: $(-\infty, 1) \cup (1, \infty)$

39. $f(x) = 1 + \dfrac{1}{x - 2}$

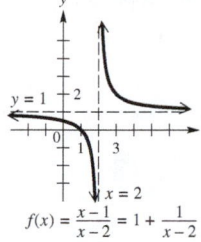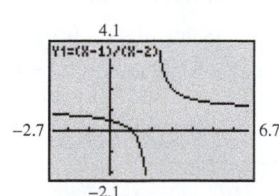

41. $f(x) = -2 + \dfrac{1}{x + 3}$

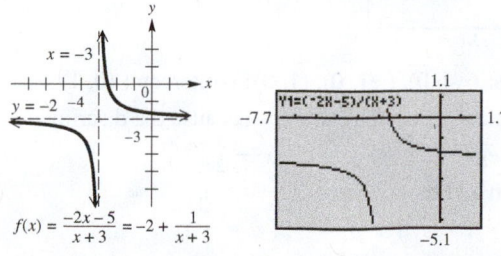

43. $f(x) = 2 + \dfrac{1}{x - 3}$

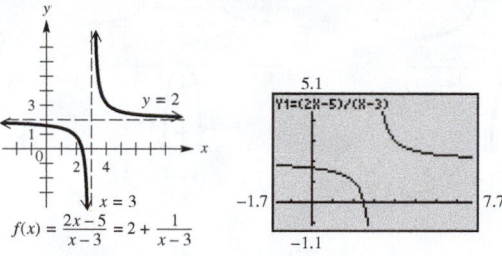

4.2 Exercises *(pages 286–289)*

1. D   **3.** G   **5.** E   **7.** F

In Exercises 9–15 and 19–27, V.A. represents vertical asymptote, H.A. represents horizontal asymptote, and O.A. represents oblique asymptote.

9. V.A.: $x = 5$; H.A.: $y = 0$; $(-\infty, 5) \cup (5, \infty)$

11. V.A.: $x = -\frac{1}{2}$; H.A.: $y = -\frac{3}{2}$; $\left(-\infty, -\frac{1}{2}\right) \cup \left(-\frac{1}{2}, \infty\right)$

13. V.A.: $x = -3$; O.A.: $y = x - 3$; $(-\infty, -3) \cup (-3, \infty)$

15. V.A.: $x = -2$, $x = \frac{5}{2}$; H.A.: $y = \frac{1}{2}$;

$\left(-\infty, -2\right) \cup \left(-2, \frac{5}{2}\right) \cup \left(\frac{5}{2}, \infty\right)$ **17.** A

19. V.A.: $x = 2$; H.A.: $y = 4$; $(-\infty, 2) \cup (2, \infty)$

21. V.A.: $x = \pm 2$; H.A.: $y = -4$;

$(-\infty, -2) \cup (-2, 2) \cup (2, \infty)$

23. V.A.: $x = 1$; H.A.: $y = 0$; $(-\infty, 1) \cup (1, \infty)$

25. V.A.: $x = -1$; O.A.: $y = x - 1$; $(-\infty, -1) \cup (-1, \infty)$

27. V.A.: none; H.A.: $y = 0$; $(-\infty, \infty)$

29.

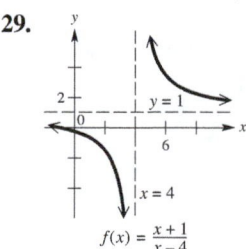

31.

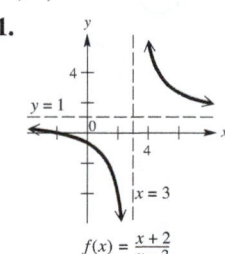

33.

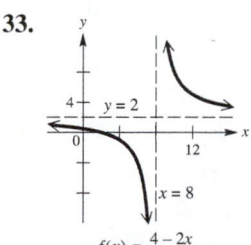

35.

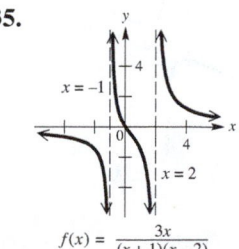

37.

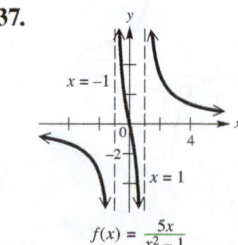

39.

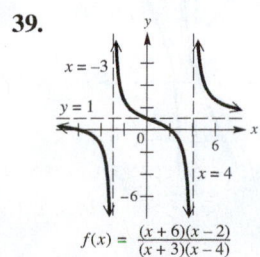

41.

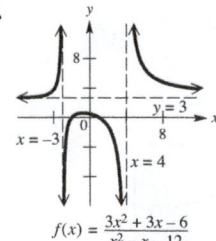

$$f(x) = \frac{3x^2 + 3x - 6}{x^2 - x - 12}$$

43.

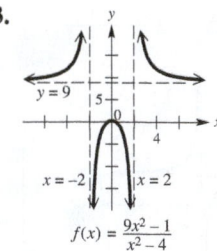

$$f(x) = \frac{9x^2 - 1}{x^2 - 4}$$

45.

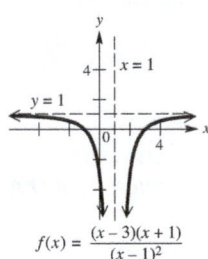

$$f(x) = \frac{(x-3)(x+1)}{(x-1)^2}$$

47.

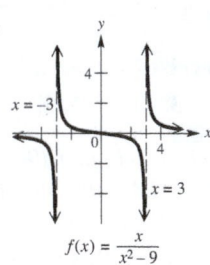

$$f(x) = \frac{x}{x^2 - 9}$$

49.

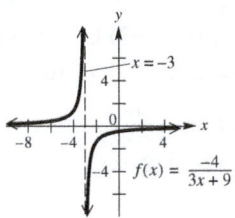

$$f(x) = \frac{-4}{3x + 9}$$

51.

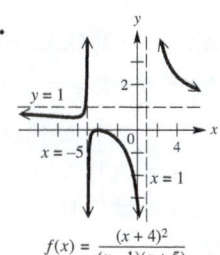

$$f(x) = \frac{(x+4)^2}{(x-1)(x+5)}$$

53.

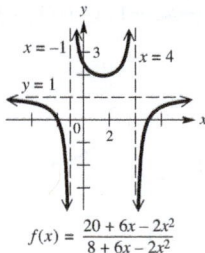

$$f(x) = \frac{20 + 6x - 2x^2}{8 + 6x - 2x^2}$$

55.

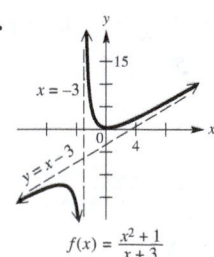

$$f(x) = \frac{x^2 + 1}{x + 3}$$

57.

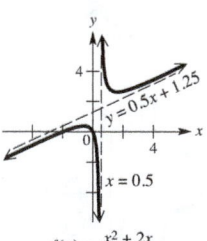

$$f(x) = \frac{x^2 + 2x}{2x - 1}$$

59.

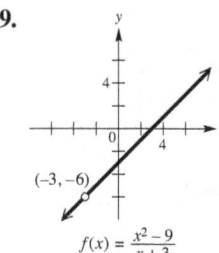

$$f(x) = \frac{x^2 - 9}{x + 3}$$

61.

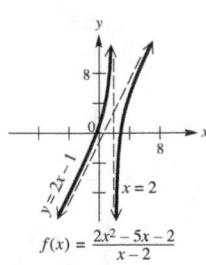

$$f(x) = \frac{2x^2 - 5x - 2}{x - 2}$$

63.

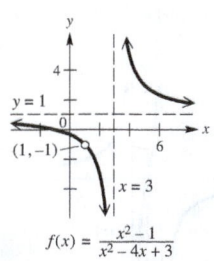

$$f(x) = \frac{x^2 - 1}{x^2 - 4x + 3}$$

65.

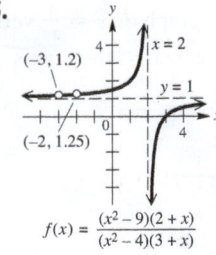

$$f(x) = \frac{(x^2 - 9)(2 + x)}{(x^2 - 4)(3 + x)}$$

67.

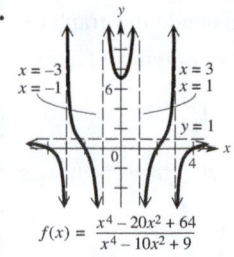

$$f(x) = \frac{x^4 - 20x^2 + 64}{x^4 - 10x^2 + 9}$$

69.

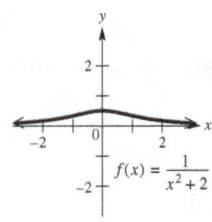

$$f(x) = \frac{1}{x^2 + 2}$$

domain: $(-\infty, \infty)$; range: $\left(0, \frac{1}{2}\right]$; symmetry with respect to the y-axis; H.A.: $y = 0$; no other asymptotes

71.

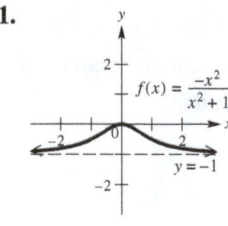

$$f(x) = \frac{-x^2}{x^2 + 1}$$

domain: $(-\infty, \infty)$; range: $(-1, 0]$; symmetry with respect to the y-axis; H.A.: $y = -1$; no other asymptotes

73.

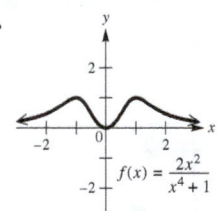

$$f(x) = \frac{2x^2}{x^4 + 1}$$

domain: $(-\infty, \infty)$; range: $[0, 1]$; symmetry with respect to the y-axis; H.A.: $y = 0$; no other asymptotes

75.

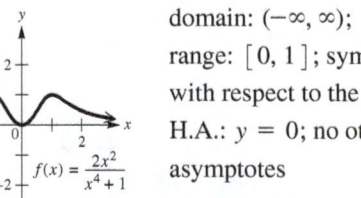

$$f(x) = \frac{3x^3 + 2x^2 - 12x - 8}{x^2 + x + 4}$$

(a) x-intercepts: $(-2, 0)$ $\left(-\frac{2}{3}, 0\right)$, $(2, 0)$; y-intercept: $(0, -2)$
(b) The denominator has no real zeros because the discriminant, -15, is negative. **(c)** $y = 3x - 1$
(d) The domain and range are both $(-\infty, \infty)$.

77.

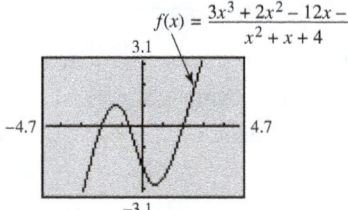

$$f(x) = \frac{x^3 + 4x^2 - x - 4}{-2x^2 - 2x - 4}$$

(a) x-intercepts: $(-4, 0)$, $(-1, 0)$, $(1, 0)$; y-intercept: $(0, 1)$
(b) The denominator has no real zeros because the discriminant, -28, is negative. **(c)** $y = -\frac{1}{2}x - \frac{3}{2}$
(d) The domain and range are both $(-\infty, \infty)$.

79. $p = 4$; $q = 2$ **81.** $p = -2$; $q = -1$

Answers may vary in Exercises 83 and 85.

83. $f(x) = \dfrac{(x-3)(x+2)}{(x-2)(x+2)} = \dfrac{x^2 - x - 6}{x^2 - 4}$

85. $f(x) = \dfrac{x-2}{x(x-4)} = \dfrac{x-2}{x^2 - 4x}$

87. (a)
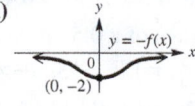

(b)
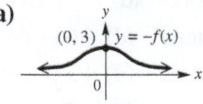
(Same as $y = f(x)$)

88. (a)
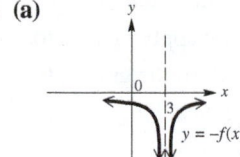

(b)
(Same as $y = f(x)$)

89. (a)
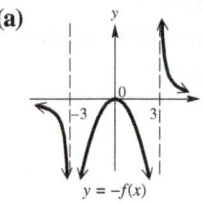

(b)

90. (a)
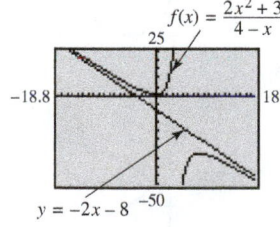

(b)
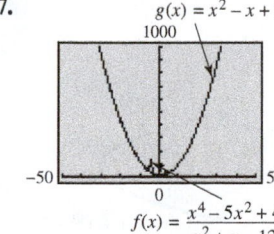

91. $y = -2x - 8$ **93.** $y = -x + 3$

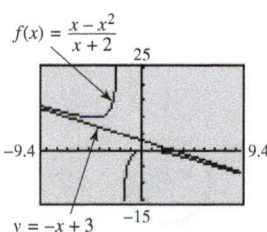

95. (a) $y = x + 1$ **(b)** at $x = 0$ and $x = 1$ **(c)** above

97.

In this window, the two graphs seem to coincide, suggesting that as $|x| \to \infty$, the graph of f approaches the graph of g, giving an asymptotic effect.

4.3 Exercises *(pages 299–304)*

1. (a) $\emptyset$ **(b)** $(-\infty, -2)$ **(c)** $(-2, \infty)$ **3. (a)** $\{-1\}$
(b) $(-1, 0)$ **(c)** $(-\infty, -1) \cup (0, \infty)$ **5. (a)** $\{0\}$
(b) $(-2, 0) \cup (2, \infty)$ **(c)** $(-\infty, -2) \cup (0, 2)$ **7. (a)** $\{0\}$
(b) $(-1, 0) \cup (0, 1)$ **(c)** $(-\infty, -1) \cup (1, \infty)$ **9. (a)** $\{0\}$
(b) $(1, 2) \cup (2, 3)$ **(c)** $(-\infty, -2) \cup (-2, 0) \cup (0, 1) \cup (3, \infty)$
11. (a) $\{2.5\}$ **(b)** $(2.5, 3)$ **(c)** $(-\infty, 2.5) \cup (3, \infty)$
13. $\{-3\}$ **15.** $\{-4\}$ **17.** $\{4, 9\}$
19. $\left\{ \dfrac{-3 \pm \sqrt{29}}{2} \right\}$ **21.** $\left\{ \dfrac{3 \pm \sqrt{3}}{3} \right\}$

23. $\left\{ \pm\frac{1}{2}, \pm i \right\}$ **25.** $\emptyset$ **27.** $\{-10\}$ **29.** $\left\{ \frac{27}{56} \right\}$
31. (a) $\{3\}$ **(b)** $(-5, 3]$ **(c)** $(-\infty, -5) \cup [3, \infty)$
33. (a) $\emptyset$ **(b)** $(-\infty, -2)$ **(c)** $(-2, \infty)$
35. (a) $\left\{ \frac{9}{5} \right\}$ **(b)** $(-\infty, 1) \cup \left(\frac{9}{5}, \infty \right)$ **(c)** $\left(1, \frac{9}{5} \right)$
37. (a) $\{-2\}$ **(b)** $(-\infty, -2] \cup (1, 2)$ **(c)** $[-2, 1) \cup (2, \infty)$
39. (a) $\emptyset$ **(b)** $\emptyset$ **(c)** $(-\infty, 2) \cup (2, \infty)$
41. (a) $\emptyset$ **(b)** $(-\infty, -1)$ **(c)** $(-1, \infty)$
43. (a) $\emptyset$ **(b)** $(-\infty, 2) \cup (2, \infty)$ **45. (a)** $\{0\}$ **(b)** $[0, \infty)$
47. (a) $\left\{ \dfrac{-1 \pm \sqrt{5}}{2} \right\}$ **(b)** $\left(\dfrac{-1 - \sqrt{5}}{2}, -\dfrac{1}{2} \right) \cup \left(-\dfrac{1}{2}, \dfrac{-1 + \sqrt{5}}{2} \right)$
49. $(-\infty, \infty)$ **51.** $\emptyset$ **53.** $\emptyset$ **55.** $(-\infty, \infty)$
57. $\{1\}$ **59.** $(-\infty, -1) \cup \left(\frac{3}{2}, \infty \right)$
61. $(-\infty, -3) \cup (-1, 2)$ **63.** $(-\infty, 2)$
65. $(-1, 1) \cup \left[\frac{5}{2}, \infty \right)$ **67.** $(3, \infty)$
69. $(-\infty, 0) \cup \left(0, \frac{1}{2} \right] \cup [2, \infty)$
71. (a) $\{-3.54\}$ **(b)** $(-\infty, -3.54) \cup (1.20, \infty)$
(c) $(-3.54, 1.20)$
73. (a) $y = 10$

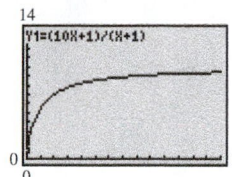

(b) When $x = 0$, there are 1,000,000 insects.   **(c)** It starts to level off at 10,000,000.
(d) The horizontal asymptote $y = 10$ represents the limiting population after a long time.
75. (a) about 12.4 cars per min   **(b)** 3   **77.** There are two possible solutions: width $= 7$ in., length $= 14$ in., height $= 2$ in.; width ≈ 2.266 in., length ≈ 4.532 in., height ≈ 19.086 in.   **79. (a)** $f(400) = \dfrac{2540}{400} = 6.35$ in.; A curve designed for 60 mph with a radius of 400 ft should have the outer rail elevated 6.35 in.
(b)

As the radius x of the curve increases, the elevation of the outer rail decreases.

(c) The horizontal asymptote is $y = 0$. As the radius of the curve increases without bound $(x \to \infty)$, the tracks become straight and no elevation or banking $(y \to 0)$ is necessary.
(d) 200 ft   **81. (a)** $D(0.05) \approx 238$; The braking distance for a car traveling at 50 mph on a wet 5% uphill grade is about 238 ft.   **(b)** As the uphill grade x increases, the braking distance decreases.
(c) $x = \dfrac{13}{165} \approx 0.079$ or 7.9%   **83.** $\frac{32}{15}$ **85.** $\frac{18}{125}$
87. increases; decreases   **89.** It becomes half as much.
91. It becomes 27 times as much.   **93.** 21   **95.** $\frac{2}{9}$ ohm
97. about 17.8 lb   **99.** 799.5 cm^3 **101.** 6 min, 40 sec
103. 7 hr, 30 min   **105.** 10 hr
107. inlet pipe: 60 min; outlet pipe: 80 min

Reviewing Basic Concepts *(pages 304–305)*

1.

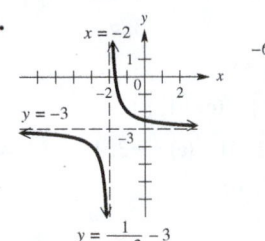 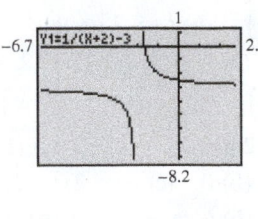

2. $(-\infty, -1) \cup (-1, 1) \cup (1, \infty)$ **3.** $x = 6$ **4.** $y = 1$

5. $y = x - 2$

6.

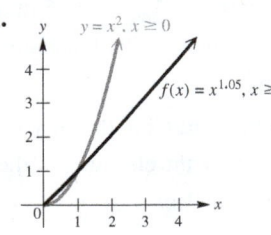

7. (a) $\{-4\}$ **(b)** $(-\infty, -4) \cup (2, \infty)$ **(c)** $(-4, 2)$

8. $\left(-\frac{1}{3}, \frac{3}{2}\right)$ **9.** inversely; height; 24 **10.** 75 pounds

4.4 Exercises *(pages 314–317)*

1. 13 **3.** -2 **5.** 729 **7.** $\frac{1}{25}$ **9.** 100 **11.** 4

13. $\frac{1}{8}$ **15.** -9 **17.** 2 **19.** 27 **21.** $(2x)^{1/3}$

23. $z^{5/3}$ **25.** $\frac{1}{y^{3/4}}$ **27.** $x^{5/6}$ **29.** $y^{3/4}$

31. -1.587401052 **33.** -5 **35.** -2.571281591

37. 1.464591888 **39.** 0.4252903703 **41.** 2

43. 1.267463962 **45.** 0.0322515344

47. $1.2^{1.62} \approx 1.34$ **49.** $50^{3/2} - 50^{1/2} \approx 346.48$

51. 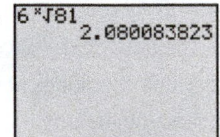 **53.**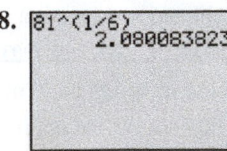

55. (a) $0.125; \frac{1}{8}$ **(b)** $\left(\sqrt[4]{16}\right)^{-3}; \sqrt[4]{16^{-3}}$ (There are other expressions.) Each is equal to 0.125. **(c)** Show that $0.125 = \frac{1}{8}$.

57. 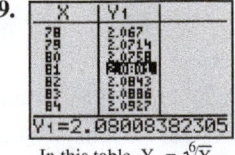 **58.**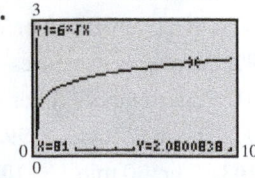

59.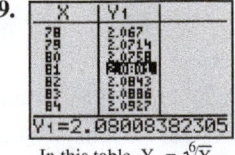
In this table, $Y_1 = \sqrt[6]{X}$.

60.

61. 3.2 ft^2 **63.** approximately 58.1 yr

65. (a) $a = 1960$ **(b)** $b \approx -1.2$
(c) $f(4) = 1960(4)^{-1.2} \approx 371$; If the zinc ion concentration reaches 371 mg per L, a rainbow trout will survive, on average, 4 min. **67.** 1.06 g **69.** $a \approx 874.54, b \approx -0.49789$
71. 0.126; The surface area of the wings of a 0.5 kg bird is approximately 0.126 sq m. **73.** $\left[-\frac{5}{4}, \infty\right)$ **75.** $(-\infty, 6\,]$
77. $(-\infty, \infty)$ **79.** $[-7, 7]$ **81.** $[-1, 0] \cup [1, \infty)$
83. (a) $[0, \infty)$ **(b)** $\left(-\frac{5}{4}, \infty\right)$ **(c)** none **(d)** $\{-1.25\}$
85. (a) $(-\infty, 0\,]$ **(b)** $(-\infty, 6)$ **(c)** none **(d)** $\{6\}$
87. (a) $(-\infty, \infty)$ **(b)** $(-\infty, \infty)$ **(c)** none **(d)** $\{3\}$
89. (a) $[0, 7]$ **(b)** $(-7, 0)$ **(c)** $(0, 7)$ **(d)** $\{-7, 7\}$
91. Rewrite as $y = 3\sqrt{x + 3}$. Shift the graph of $y = \sqrt{x}$ to the left 3 units and stretch vertically by applying a factor of 3.
93. Rewrite as $y = 2\sqrt{x + 4} + 4$. Shift the graph of $y = \sqrt{x}$ to the left 4 units, stretch vertically by applying a factor of 2, and shift upward 4 units.
95. Rewrite as $y = 3\sqrt[3]{x + 2} - 5$. Shift the graph of $y = \sqrt[3]{x}$ to the left 2 units, stretch vertically by applying a factor of 3, and shift downward 5 units.

97. **99.**

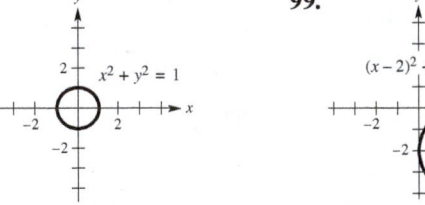

101. **103.**

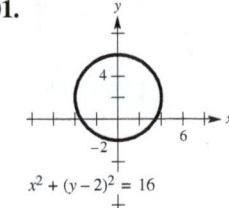 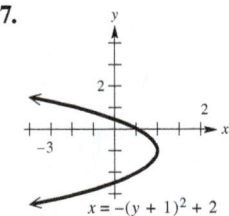

105. **107.**

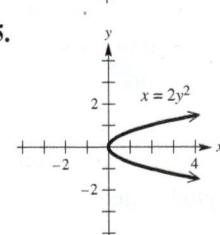

109. The graph is a circle. $y_1 = \sqrt{100 - x^2}$; $y_2 = -\sqrt{100 - x^2}$

111. The graph is a (shifted) circle. $y_1 = \sqrt{9 - (x - 2)^2}$; $y_2 = -\sqrt{9 - (x - 2)^2}$

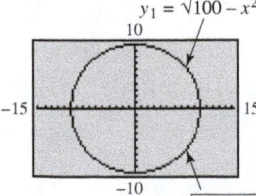 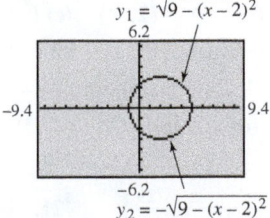

113. The graph is a horizontal parabola. $y_1 = -3 + \sqrt{x}$; $y_2 = -3 - \sqrt{x}$

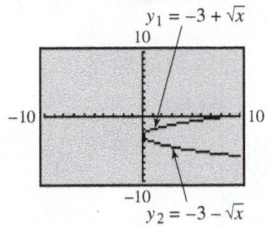

115. The graph is a horizontal parabola. $y_1 = -2 + \sqrt{0.5x + 3.5}$; $y_2 = -2 - \sqrt{0.5x + 3.5}$

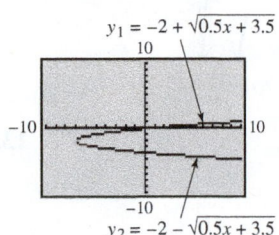

4.5 Exercises *(pages 324–327)*

1.

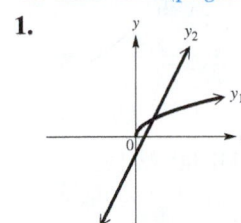

one real solution; $\{1\}$; $\frac{1}{4}$ is extraneous.

3.

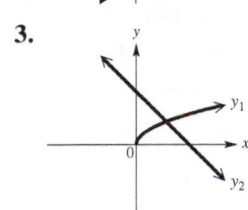

one real solution; $\left\{\frac{7 - \sqrt{13}}{2}\right\}$; $\frac{7 + \sqrt{13}}{2}$ is extraneous.

5.

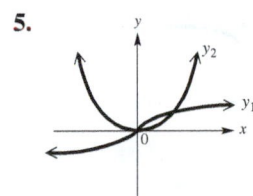

two real solutions; $\{0, 1\}$; no extraneous values

7. They both check.   **9.** $\{8\}$ **11.** $\{4\}$

13. $\{-1, 3\}$ **15.** $\{-28\}$ **17.** $\left\{-\frac{4}{3}, 0\right\}$ **19.** $\{18\}$

21. $\{-32, 32\}$ **23.** $\{27\}$ **25.** $\left\{-1, -\frac{1}{2}\right\}$

27. $\left\{-\frac{1}{4}, \frac{5}{7}\right\}$ **29.** $\{-8, 27\}$ **31.** $\{1\}$

33. (a) $\{-1\}$ (b) $(-1, \infty)$ (c) $\left[-\frac{7}{3}, -1\right)$ **35.** (a) $\{3\}$

(b) $\left[-\frac{13}{4}, 3\right)$ (c) $(3, \infty)$ **37.** (a) $\{3\}$ (b) $\left[-\frac{1}{5}, 3\right)$

(c) $(3, \infty)$ **39.** (a) $\{2, 14\}$ (b) $(2, 14)$ (c) $(14, \infty)$

41. (a) $\{0, 3\}$ (b) $(-\infty, 0) \cup (3, \infty)$ (c) $(0, 3)$

43. (a) $\{0\}$ (b) $(0, \infty)$ (c) $\left[-\frac{1}{3}, 0\right)$ **45.** (a) $\{27\}$

(b) $[27, \infty)$ (c) $\left[\frac{5}{2}, 27\right]$ **47.** (a) $\{-8, 2\}$

(b) $(-\infty, -8) \cup (2, \infty)$ (c) $(-8, -6] \cup [0, 2)$

49. (a) $\left\{\frac{1}{4}, 1\right\}$ (b) $\left(-\infty, \frac{1}{4}\right) \cup (1, \infty)$ (c) $\left(\frac{1}{4}, 1\right)$

51. $(4x - 4)^{1/3} = (x + 1)^{1/2}$ **52.** 6

53. $(4x - 4)^2 = (x + 1)^3$

54. $16x^2 - 32x + 16 = x^3 + 3x^2 + 3x + 1$, and thus, $x^3 - 13x^2 + 35x - 15 = 0$

55. three real solutions

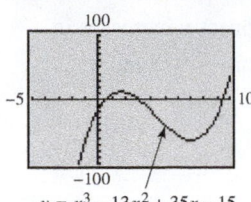

$y = x^3 - 13x^2 + 35x - 15$

56.

$$3\overline{)\begin{array}{rrrr} 1 & -13 & 35 & -15 \\ & 3 & -30 & 15 \\ \hline 1 & -10 & 5 & 0 \end{array}}$$
$$P(3) = 0 \longrightarrow \uparrow$$

57. $P(x) = (x - 3)(x^2 - 10x + 5)$ **58.** $\left\{5 \pm 2\sqrt{5}\right\}$

59. $3; 5 + 2\sqrt{5}; 5 - 2\sqrt{5}$

60. two real solutions

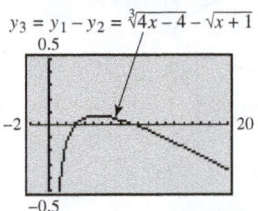

$y_3 = y_1 - y_2 = \sqrt[3]{4x - 4} - \sqrt{x + 1}$

61. $\left\{3, 5 + 2\sqrt{5}\right\}$; For the calculator solution, $5 + 2\sqrt{5} \approx 9.47$.

62. The solution set of the original equation is a subset of the solution set of the equation in **Exercise 54.** The extraneous solution $5 - 2\sqrt{5}$ was obtained when each side of the original equation was raised to the sixth power.

63. (a) $\frac{-x - 4}{3x^{1/3}(x - 2)^2}$ (b) $\{-4\}$ **65.** (a) $\frac{1 - 5x^2}{3x^{2/3}(x^2 + 1)^2}$

(b) $\left\{\pm \frac{\sqrt{5}}{5}\right\}$ **67.** (a) $\frac{x - 1}{x^{7/4}}$ (b) $\{1\}$

69. (a) $\frac{1}{(x^2 + 1)^{3/2}}$ (b) $\varnothing$ **71.** $\{0, 1\}$

73. $\left\{-\frac{2}{9}, 2\right\}$ **75.** $\{2, 18\}$ **77.** 4.5 km per sec

79. 2.5 sec   **81.** about 91 mph

83. (a) $20 - x$ (b) $(0, 20)$

(c) $AP = \sqrt{x^2 + 12^2}$; $BP = \sqrt{(20 - x)^2 + 16^2}$

(d) $f(x) = \sqrt{x^2 + 12^2} + \sqrt{(20 - x)^2 + 16^2}$, $0 < x < 20$

(e)

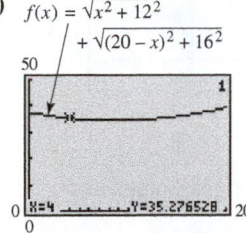

$f(x) = \sqrt{x^2 + 12^2} + \sqrt{(20 - x)^2 + 16^2}$

$f(4) \approx 35.28$; When the stake is 4 ft from the base of the 12-ft pole, approximately 35.28 ft of wire will be required.

(f) When $x \approx 8.57$ ft, $f(x)$ is a minimum (approximately 34.41 ft).   (g) This problem has examined how the total amount of wire used can be expressed in terms of the distance from the stake at P to the base of the 12-ft pole. We find that the amount of wire used can be minimized when the stake is approximately 8.57 ft from the 12-ft pole.   **85.** Since $x \approx 1.31$, the hunter must travel $8 - x \approx 8 - 1.31 = 6.69$ mi along the river.

87. After 1.38 hr (1:23 P.M.), the ships are 33.28 mi apart.

Reviewing Basic Concepts *(page 328)*

1. 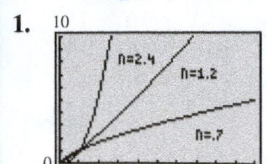 As the exponent n increases in value, the curve rises more rapidly for $x \geq 1$.

2. 0.24 m²

3. (a) 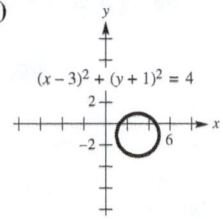 **(b)**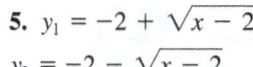

4. $y_1 = \sqrt{16 - x^2}$; $y_2 = -\sqrt{16 - x^2}$

5. $y_1 = -2 + \sqrt{x - 2}$; $y_2 = -2 - \sqrt{x - 2}$

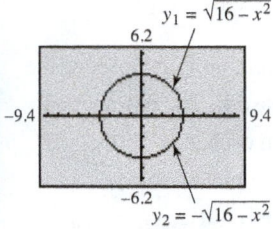 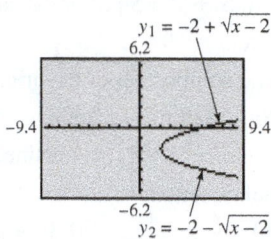

6. $\{4\}$ **7.** $(4, \infty)$ **8.** $\left[-\frac{4}{3}, 4\right)$ **9.** $\{-1\}$

10. (a) cat: 148; person: 69   **(b)** 6.4 in.

Chapter 4 Review Exercises *(pages 330–333)*

1. (a) Reflect the graph of $y = \frac{1}{x}$ across the x-axis, and shift upward 6 units.

(b) 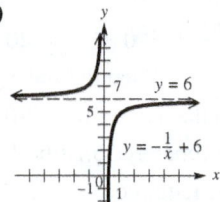 **(c)**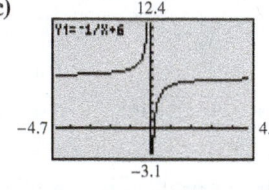

3. (a) Shift the graph of $y = \frac{1}{x^2}$ to the right 2 units, and reflect across the x-axis.

(b) 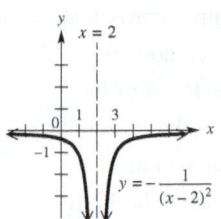 **(c)**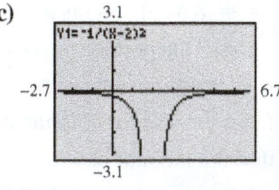

5. The degree of the numerator will be exactly 1 greater than the degree of the denominator.

7.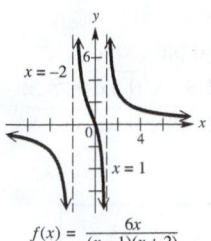

$f(x) = \dfrac{6x}{(x-1)(x+2)}$

9.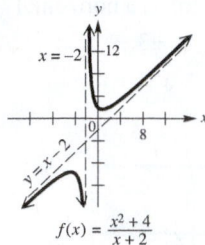

$f(x) = \dfrac{x^2 + 4}{x + 2}$

11.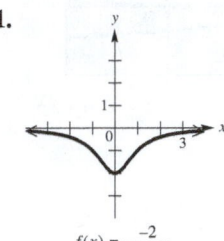

$f(x) = \dfrac{-2}{x^2 + 1}$

13.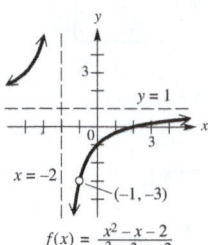

$f(x) = \dfrac{x^2 - x - 2}{x^2 + 3x + 2}$

15. $f(x) = \dfrac{-3x + 6}{x - 1}$ **17.** $f(x) = \dfrac{-1}{x^2 + 1}$ **19. (a)** $\left\{\frac{2}{3}\right\}$

(b) $\left(-1, \frac{2}{3}\right)$ **(c)** $(-\infty, -1) \cup \left(\frac{2}{3}, \infty\right)$ **21. (a)** $\{0\}$

(b) $(-\infty, -1) \cup [0, 2)$ **(c)** $(-1, 0] \cup (2, \infty)$

23. $(-2, -1)$ **25.** $(-\infty, -2) \cup (-2, -1) \cup (-1, \infty)$

27. (a) $[0, 36]$ **(b)** The average line length is less than or equal to 8 cars when the average arrival rate is 36 cars per hr or less.   **29.** 2   **31.** 847   **33.** 12.15 candela

35. $1375   **37.** $\frac{8}{9}$ metric ton   **39.** $71\frac{1}{9}$ kg

41. 2 hr, 48 min

43. 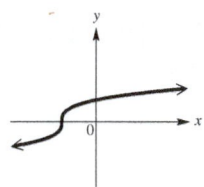 **45.**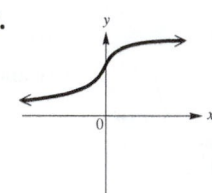

47. 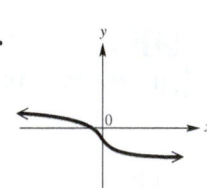   **49.** 12   **51.** $\frac{1}{3}$ **53.** $-\frac{2}{3}$

55. $\frac{1}{216}$ **57.** $\frac{1}{81}$

59. 2.42926041078

61. 0.0625   **63.** $[2, \infty)$

65. (a) none   **(b)** $(2, \infty)$

67. (a) $\{2\}$ **(b)** $[-2.5, 2)$ **(c)** $(2, \infty)$

69. (a) $\{-1\}$ **(b)** $[-1, \infty)$ **(c)** $(-\infty, -1]$

71. $\{4\}$ **73.** $\{6\}$ **75.** $\left\{\frac{81}{16}\right\}$ **77.** $\{15\}$

79. $\left\{-2, \frac{1}{3}\right\}$ **81.** $\{-1, 125\}$ **83.** $\{8\}$

85. (a) If the length L of the pendulum increases, so does the period of oscillation T.   **(b)** There are a number of ways. One way is to realize that $k = \dfrac{L}{T^n}$ for some integer n. The ratio should be the constant k for each data point when the correct n is found. Another way is to use regression.

(c) $k \approx 0.81$; $n = 2$ **(d)** 2.48 sec   **(e)** T increases by a factor of $\sqrt{2} \approx 1.414$.

Chapter 4 Test *(pages 333–334)*

1. (a)

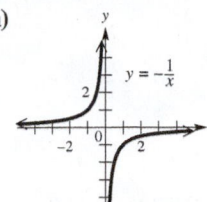

(b) The graph of $y = \frac{1}{x}$ is reflected across the x-axis or the y-axis.

(c)

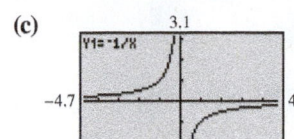

2. (a)

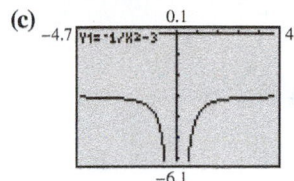

(b) The graph of $y = \frac{1}{x^2}$ is reflected across the x-axis and shifted 3 units downward.

(c)

3. (a) $(-\infty, -1) \cup (-1, 4) \cup (4, \infty)$ **(b)** $x = -1$, $x = 4$
(c) $y = 1$ **(d)** $(0, 1.5)$ **(e)** $(-3, 0), (2, 0)$ **(f)** $(0.5, 1)$
(g)

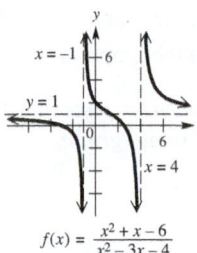

$f(x) = \dfrac{x^2 + x - 6}{x^2 - 3x - 4}$

4. $y = 2x + 5$

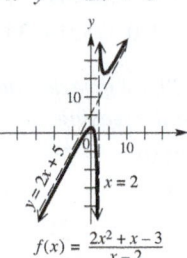

$f(x) = \dfrac{2x^2 + x - 3}{x - 2}$

5. (a) -4 **(b)**

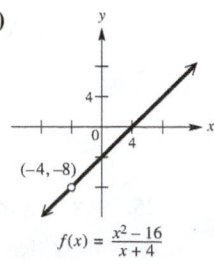

$f(x) = \dfrac{x^2 - 16}{x + 4}$

6. (a) $\{5\}$ **(b)** $(-\infty, -2) \cup (2, 5]$
7. (a) $W(30) = \frac{1}{10}$; $W(39) = 1$; $W(39.9) = 10$; When the rate is 30 vehicles per min, the average wait time is $\frac{1}{10}$ min (6 sec). The other results are interpreted similarly.

(b)

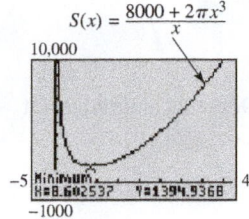

$W(x) = \dfrac{1}{40 - x}$

The vertical asymptote has equation $x = 40$. As x approaches 40, W gets larger and larger.   **(c)** 39.8

8. 92; undernourished
9. radius: approximately 8.6 cm; amount: approximately 1394.9 cm²

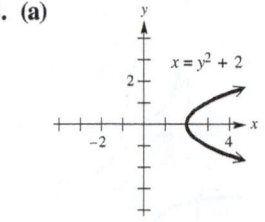

$S(x) = \dfrac{8000 + 2\pi x^3}{x}$

10. (a) -3 **(b)** $\frac{1}{125}$ **11.** $\frac{1}{x^{4/3}}$
12. (a)   **(b)**

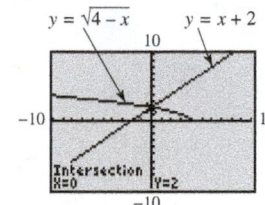

13.

$f(x) = -\sqrt{5 - x}$

(a) $(-\infty, 5]$ **(b)** $(-\infty, 0]$
(c) $(-\infty, 5)$ **(d)** $\{5\}$
(e) $(-\infty, 5)$

14. (a) $\{0\}$ $y = \sqrt{4 - x}$ $y = x + 2$ **(b)** $(-\infty, 0)$
(c) $[0, 4]$

15. The cable should be laid under water from P to a point S, which is on the bank 400 yd away from Q in the direction of R.
16. 5.625

CHAPTER 5   INVERSE, EXPONENTIAL, AND LOGARITHMIC FUNCTIONS

5.1 Exercises *(pages 343–346)*
1. yes   **3.** no   **5.** no   **7.** yes   **9.** no   **11.** yes
13. no   **15.** yes   **17.** no   **19.** yes   **21.** yes
23. yes   **25.** no   **27.** one-to-one   **29.** x; $(g \circ f)(x)$
31. (b, a) **33.** $y = x$ **35.** does not; It is not one-to-one.
43. yes; $f^{-1} = \{(4, 10), (5, 20), (6, 30), (7, 40)\}$ **45.** no
47. yes; $f^{-1} = \{(0^2, 0), (1^2, 1), (2^2, 2), (3^2, 3), (4^2, 4)\}$

49. untying your shoelaces   **51.** leaving a room

53. unwrapping a package

55. $f^{-1}(x) = \dfrac{x + 4}{3}$ **57.** $f^{-1}(x) = \sqrt[3]{x - 1}$

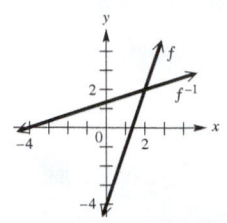   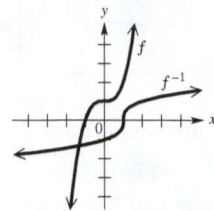

Domains and ranges of both f and f^{-1} are $(-\infty, \infty)$.

Domains and ranges of both f and f^{-1} are $(-\infty, \infty)$.

59. not one-to-one

61. $f^{-1}(x) = \dfrac{1}{x}$ **63.** $f^{-1}(x) = \dfrac{2 - 3x}{x}$

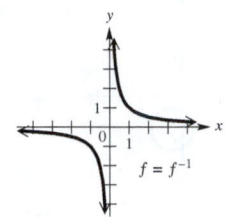   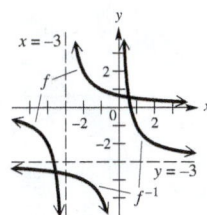

Domains and ranges of both f and f^{-1} are $(-\infty, 0) \cup (0, \infty)$.

domain of f = range of $f^{-1} = (-\infty, -3) \cup (-3, \infty)$; domain of f^{-1} = range of $f = (-\infty, 0) \cup (0, \infty)$

65. $f^{-1}(x) = x^2 - 6, x \geq 0$

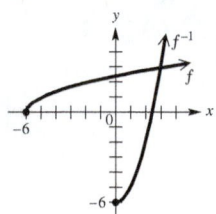

domain of f = range of $f^{-1} = [-6, \infty)$; domain of f^{-1} = range of $f = [0, \infty)$

67. $f^{-1}(x) = \dfrac{x}{4 - x}$ **69.** $f^{-1}(x) = \dfrac{1}{3x + 2}$

71. $f^{-1}(x) = \sqrt{x^2 + 4}, x \geq 0$ **73.** $f^{-1}(x) = \sqrt[3]{\dfrac{x + 7}{5}}$

75. $f^{-1}(x) = \dfrac{-4x}{3x - 1}$ **77.** $f^{-1}(x) = \dfrac{3 - x}{2x + 1}$

79. 8   **81.** −8   **83.** 0   **85.** 4   **87.** 2   **89.** −2

91. yes   **93.** no   **95.** yes   **97.** yes

99.   **101.**

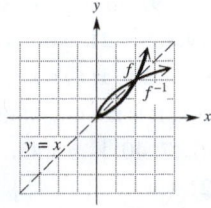   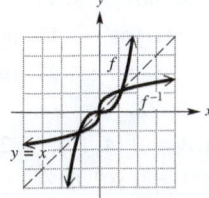

103.

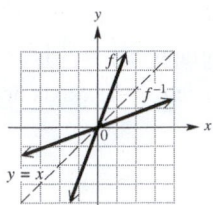

105. It represents the cost in dollars of building 1000 cars.

107. $\dfrac{1}{a}$ **109.** not one-to-one

111. one-to-one; $f^{-1}(x) = \dfrac{-5 - 3x}{x - 1}$

113. $f^{-1}(x) = \dfrac{x + 1}{4}$; TREASURE HUNT IS ON

115. 2745 3376 4097 5833 3376 9 1729 126 2198; $f^{-1}(x) = \sqrt[3]{x - 1}$

In Exercises 117–121, we give only one of several possible choices for domain restriction.

117. $[0, \infty)$ **119.** $[6, \infty)$ **121.** $[0, \infty)$

123. $f^{-1}(x) = \sqrt{4 - x}$ **125.** $f^{-1}(x) = x + 6, x \geq 0$

5.2 Exercises *(pages 356–359)*

1.   **3.**

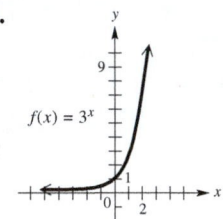   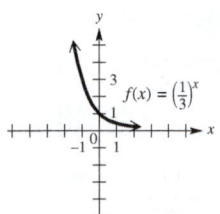

5. $\left\{\dfrac{1}{2}\right\}$ **7.** $\{-2\}$ **9.** 8.952419619

11. 0.3752142272   **13.** 0.0868214883   **15.** 13.1207791

17. The point $\left(\sqrt{10}, 8.9524196\right)$ lies on the graph of y.

19. The point $\left(\sqrt{2}, 0.37521423\right)$ lies on the graph of y.

In Exercises 21–29, we give domain, range, asymptote, whether f is increasing or decreasing, and a traditional graph.

21. $(-\infty, \infty)$; $(-\infty, 0)$; $y = 0$; decreasing

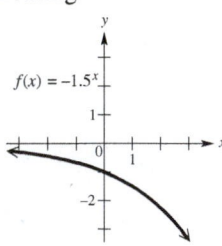

23. $(-\infty, \infty)$; $(0, \infty)$; $y = 0$; increasing

25. $(-\infty, \infty)$; $(0, \infty)$; $y = 0$; increasing

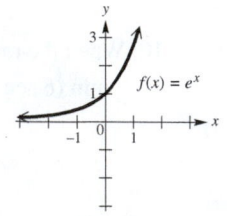

27. $(-\infty, \infty)$; $(0, \infty)$; $y = 0$; increasing

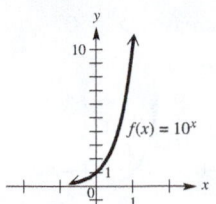

29. $(-\infty, \infty)$; $(0, \infty)$; $y = 0$; decreasing

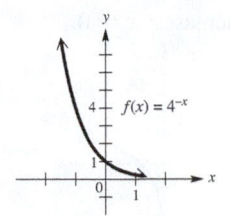

31. **(a)** $a > 1$
(b) $(-\infty, \infty)$; $(0, \infty)$; $y = 0$
(c)

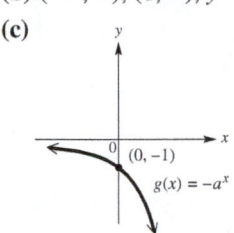

(d) $(-\infty, \infty)$; $(-\infty, 0)$; $y = 0$
(e)

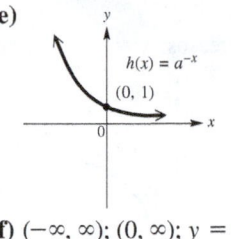

(f) $(-\infty, \infty)$; $(0, \infty)$; $y = 0$

33.

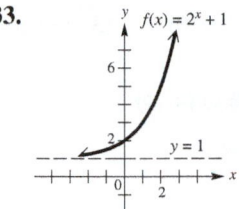

35.

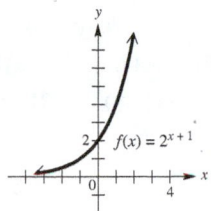

37.

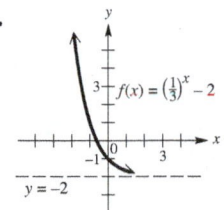

39.

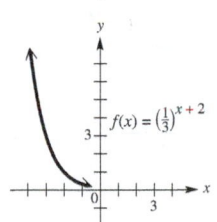

41. $\{0\}$ **43.** $\{3\}$ **45.** $\left\{\frac{1}{2}\right\}$ **47.** $\left\{\frac{1}{5}\right\}$

49. $\{-7\}$ **51.** $\left\{\frac{4}{3}\right\}$ **53.** $\{-3\}$ **55.** $\left\{-\frac{1}{3}\right\}$

57. **(a)** $\{2\}$ **(b)** $(2, \infty)$ **(c)** $(-\infty, 2)$

59. **(a)** $\left\{\frac{1}{5}\right\}$ **(b)** $\left(\frac{1}{5}, \infty\right)$ **(c)** $\left(-\infty, \frac{1}{5}\right)$

61. **(a)** $\left\{-\frac{2}{3}\right\}$ **(b)** $\left[-\frac{2}{3}, \infty\right)$ **(c)** $\left(-\infty, -\frac{2}{3}\right]$

63. **(a)** 3 **(b)** $\frac{1}{3}$ **(c)** 9 **(d)** 1 **65.** $f(x) = \left(\frac{1}{4}\right)^x$

67. $f(t) = \left(\frac{1}{3}\right) 9^t$ **69.** **(a)** \$22,510.18   **(b)** \$22,529.85

71. **(a)** \$33,504.35   **(b)** \$33,504.71

73. Plan A is better by \$119.09.

75. \$1.16; \$2.44; \$7.08; \$18.11; \$59.34; \$145.80; \$318.43

77. **(a)**

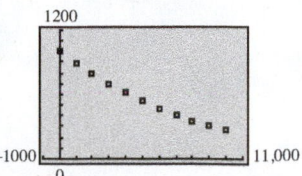

(b) An exponential function fits better because the average rate of change between data points is not constant.

(c)

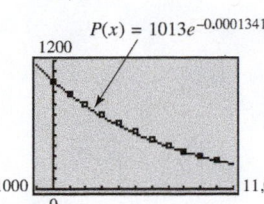

$P(x) = 1013e^{-0.0001341x}$

(d) $P(1500) \approx 828$ mb; $P(11,000) \approx 232$ mb; the predictions are close to the actual values.

79. **(a)** $f(2) = 1 - e^{-0.5(2)} = 1 - e^{-1} \approx 0.63$; There is a 63% chance that at least one car will enter the intersection during a 2-min period.

(b)

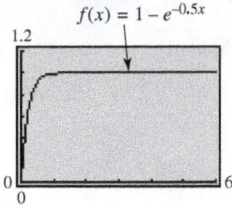

$f(x) = 1 - e^{-0.5x}$

As time progresses, the probability increases and begins to approach 1. That is, it is almost certain that at least one car will enter the intersection during a 60-min period.

81. yes; an inverse function

82.

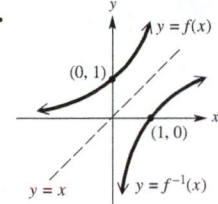

83. $x = a^y$ **84.** $x = 10^y$ **85.** $x = e^y$ **86.** (q, p)

5.3 Exercises *(pages 367–369)*

1. **(a)** C   **(b)** A   **(c)** E   **(d)** B   **(e)** F   **(f)** D

3. $\log_3 81 = 4$ **5.** $\log_{1/2} 16 = -4$ **7.** $\log 0.0001 = -4$

9. $\ln 1 = 0$ **11.** $6^2 = 36$ **13.** $\left(\sqrt{3}\right)^8 = 81$

15. $10^{-3} = 0.001$ **17.** $10^{0.5} = \sqrt{10}$ **19.** $\{3\}$

21. $\{\sqrt{3}\}$, or $\{3^{1/2}\}$ **23.** $\left\{\frac{1}{216}\right\}$ **25.** $\{8\}$

27. $\{7\}$ **29.** $\left\{-\frac{1}{8}\right\}$ **31.** **(a)** 7 **(b)** 9 **(c)** 4 **(d)** k

33. **(a)** 0 **(b)** 0 **(c)** 0 **(d)** 0 **35.** 1.5 **37.** $\sqrt{5}$

39. $\frac{2}{3}$ **41.** π **43.** 7 **45.** 1.633468456

47. -0.1062382379 **49.** 4.341474094

51. 3.761200116 **53.** -0.244622583

55. 9.996613531 **57.** 3.2 **59.** 8.4 **61.** 2×10^{-3}

63. 1.6×10^{-5} **65.** $\log_3 2 - \log_3 5$

67. $\log_2 6 + \log_2 x - \log_2 y$

69. $1 + \frac{1}{2} \log_5 7 - \log_5 3 - \log_5 m$

71. cannot be rewritten   **73.** $\log_k p + 2 \log_k q - \log_k m$

75. $\frac{1}{2}(3 \log_m r - \log_m 5 - 5 \log_m z)$ **77.** $\log_a \frac{xy}{m}$

79. $\log_m \frac{a^2}{b^6}$ **81.** $\log_a((z - 1)^2(3z + 2))$

83. $\log_5 (5^{1/3} m^{-1/3})$, or $\log_5 \sqrt[3]{\frac{5}{m}}$ **85.** $\log \frac{x^3}{y^4}$

87. $\ln \frac{(a + b)a}{2}$ **89.** 1.430676558 **91.** 0.5943161289

93. 0.9595390462 **95.** 1.892441722

97. Reflect the graph of $y = 3^x$ across the x-axis and shift 7 units upward.

98.

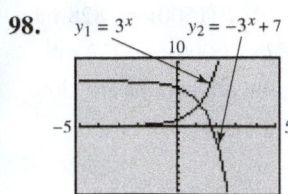

$y_1 = 3^x$ $y_2 = -3^x + 7$

99. 1.7712437   **100.** $\{\log_3 7\}$

101. $\dfrac{\log 7}{\log 3} = \dfrac{\ln 7}{\ln 3} \approx 1.771243749$

102. The approximations are close enough to support the conclusion that the *x*-intercept is $(\log_3 7, 0)$.

103. $\ln|x + \sqrt{x^2 + 3}| + \ln|x - \sqrt{x^2 + 3}|$

$= \ln\left|(x + \sqrt{x^2 + 3})(x - \sqrt{x^2 + 3})\right|$

$= \ln|x^2 - (x^2 + 3)|$

$= \ln|-3|$

$= \ln 3$

105. $\dfrac{1}{3}\ln\left(\dfrac{x^2 + 1}{5}\right) - \dfrac{1}{3}\ln\left(\dfrac{x^2 + 4}{5}\right)$

$= \dfrac{1}{3}\left(\ln\dfrac{x^2 + 1}{5} - \ln\dfrac{x^2 + 4}{5}\right)$

$= \dfrac{1}{3}\ln\left(\dfrac{x^2 + 1}{5} \cdot \dfrac{5}{x^2 + 4}\right)$

$= \dfrac{1}{3}\ln\left(\dfrac{x^2 + 1}{x^2 + 4}\right)$

$= \ln\left(\dfrac{x^2 + 1}{x^2 + 4}\right)^{1/3}$

$= \ln\sqrt[3]{\dfrac{x^2 + 1}{x^2 + 4}}$

107. 1   **109. (a)** 2   **(b)** 2   **(c)** 2   **(d)** 1

Reviewing Basic Concepts *(page 369)*

1. No, because the *x*-values -2 and 2 both correspond to the *y*-value 4. In a one-to-one function, each *y*-value must correspond to exactly one *x*-value (and each *x*-value to exactly one *y*-value).

2. (a)

x	12	21	32	45
y	7	8	9	10

(b) $f^{-1}(x) = 4x - 5$

3.

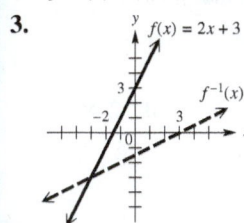

$f(x) = 2x + 3$

4.

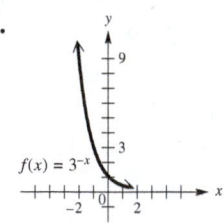

$f(x) = 3^{-x}$

5.

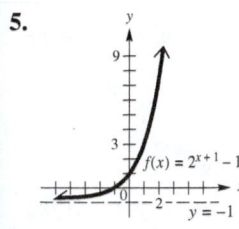

$f(x) = 2^{x+1} - 1$

(a) It is shifted 1 unit left and 1 unit downward.

(b) D: $(-\infty, \infty)$; R: $(-1, \infty)$ **(c)** $y = -1$ **(d)** Yes, it is.

6. $\left\{\dfrac{3}{4}\right\}$ **7.** \$76.10   **8. (a)** $-\dfrac{1}{2}$ **(b)** 3   **(c)** 2

9. $\log 3 + 2\log x - \log 5 - \log y$ **10.** $\ln\dfrac{x}{2}$

5.4 Exercises *(pages 376–378)*

1. $(0, \infty)$; $(-\infty, \infty)$; increases; $x = 0$

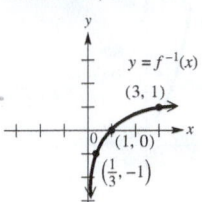

3. $(0, \infty)$; $(-\infty, \infty)$; decreases; $x = 0$

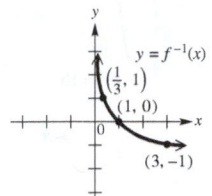

5. $(1, \infty)$; $(-\infty, \infty)$; increases; $x = 1$

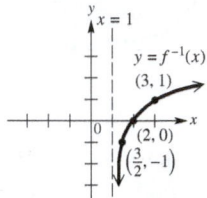

7. logarithmic   **9.** $(0, \infty)$ **11.** $(-\infty, 0)$ **13.** $(-\infty, \infty)$

15. $(-2, 2)$ **17.** $(-\infty, -3) \cup (7, \infty)$

19. $(-1, 0) \cup (1, \infty)$ **21.** $(-\infty, -3) \cup (4, \infty)$

23. $\left(-\infty, \dfrac{7}{3}\right) \cup \left(\dfrac{7}{3}, \infty\right)$

25.

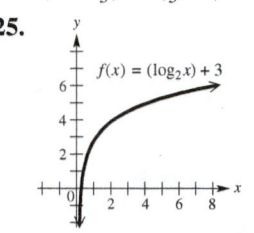

$f(x) = (\log_2 x) + 3$

27.

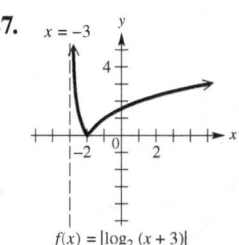

$f(x) = |\log_2 (x + 3)|$

29.

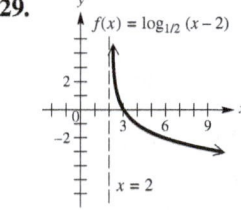

$f(x) = \log_{1/2}(x - 2)$

31. B   **33.** D   **35.** A   **37.** C

39.

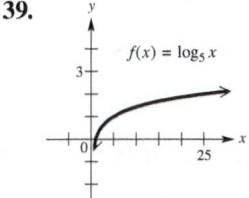

$f(x) = \log_5 x$

41.

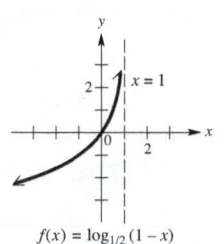

$f(x) = \log_{1/2}(1 - x)$

43.

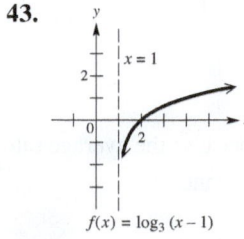

$f(x) = \log_3 (x - 1)$

45. (a) The graph is shifted 4 units to the left.
(b)

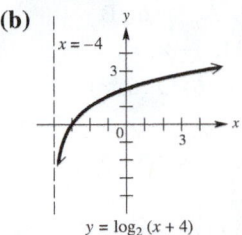

$y = \log_2(x + 4)$

47. (a) The graph is stretched vertically by applying a factor of 3 and shifted 1 unit upward.
(b)

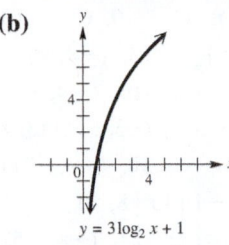

$y = 3\log_2 x + 1$

49. (a) The graph is reflected across the y-axis and shifted 1 unit upward.
(b)

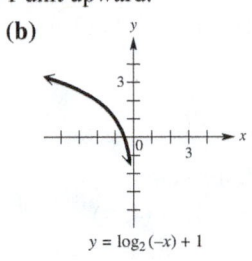

$y = \log_2(-x) + 1$

51. The graphs are not the same because the domain of $y = \log x^2$ is $(-\infty, 0) \cup (0, \infty)$, while the domain of $y = 2 \log x$ is $(0, \infty)$. The power rule does not apply if the argument is nonpositive.
53. $\frac{3}{2}$ **55.** $-\frac{3}{4}$

57. $f^{-1}(x) = \log_4(x + 3)$

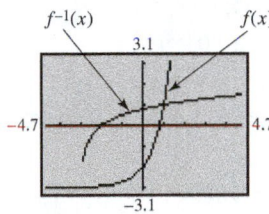

59. $f^{-1}(x) = \log(4 - x)$

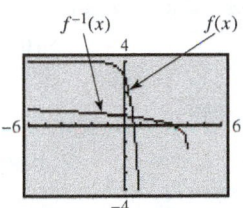

61. (a) -4 **(b)** 6 **(c)** $\frac{1}{3}$ **(d)** 1 **63.** $\{1.87\}$
65. (a) The left side is a reflection of the right side across the axis of the tower. The graph of $f(-x)$ is the reflection of $f(x)$ across the y-axis.   **(b)** 984 ft   **(c)** 39 ft
67. (a) 28.105 in.   **(b)** It tells us that at 99 mi from the eye of a typical hurricane, the barometric pressure is 29.21 in.

5.5 Exercises *(pages 385–387)*

1. $\left\{\frac{1}{2}\ln\frac{4}{3}\right\}$ **3.** $\{\log 7\}$ **5.** $\{2^{3/2}\}$ **7.** $\left\{\frac{1}{3}e^2\right\}$

In Exercises 9–17 and 25–31, alternative forms involving natural logarithms are possible.

9. (a) $\left\{\frac{\log 7}{\log 3}\right\}$ **(b)** $\{1.771\}$ **11. (a)** $\left\{\frac{\log 5}{\log \frac{1}{2}}\right\}$
(b) $\{-2.322\}$ **13. (a)** $\left\{\frac{\log 4}{\log 0.8}\right\}$ **(b)** $\{-6.213\}$
15. (a) $\left\{\frac{\log 4}{\log 4 - 2\log 3}\right\}$ **(b)** $\{-1.710\}$
17. (a) $\left\{\frac{-\log 6 - \log 4}{\log 6 - 2\log 4}\right\}$ **(b)** $\{3.240\}$ **19.** $\emptyset$
21. (a) $\left\{\frac{3}{1 - 3\ln 2}\right\}$ **(b)** $\{-2.779\}$ **23.** $\emptyset$

25. (a) $\left\{\frac{2}{\log 1.15}\right\}$ **(b)** $\{32.950\}$ **27. (a)** $\left\{2 + \frac{\log 33}{\log 2}\right\}$
(b) $\{7.044\}$ **29. (a)** $\left\{\frac{\log 3.5}{\log 1.05}\right\}$ **(b)** $\{25.677\}$

31. (a) $\left\{1980 + \frac{\log\frac{8}{5}}{\log 1.015}\right\}$ **(b)** $\{2011.568\}$

33. $\{e^2\}$ **35.** $\left\{\frac{e^{1.5}}{4}\right\}$ **37.** $\{2 - \sqrt{10}\}$
39. $\{16\}$ **41.** $\{3\}$ **43.** $\{e\}$ **45.** $\{e^2 + 1\}$
47. $\{\pm\sqrt{11}\}$ **49.** $\emptyset$ **51.** $\{25\}$ **53.** $\{2.5\}$
55. $\{3\}$ **57.** $\{1\}$ **59.** $\{6\}$ **61.** $\{1, 10\}$
63. (a) $\{4\}$ **(b)** $(4, \infty)$ **(c)** $(0, 4)$ **65.** The statement is incorrect. We must reject any solution that is not in the domain of any logarithmic function in the equation.
67. $\{17.106\}$

The answers in Exercises 69–79 may have other equivalent forms.

69. $t = e^{(p-r)/k}$ **71.** $t = -\frac{1}{k}\log\left(\frac{T - T_0}{T_1 - T_0}\right)$

73. $k = \dfrac{\ln\left(\frac{A - T_0}{C}\right)}{-t}$ **75.** $x = \dfrac{\ln\left(\frac{A + B - y}{B}\right)}{-C}$ **77.** $A = \frac{B}{x^C}$

79. $t = \dfrac{\log\frac{A}{P}}{n\log\left(1 + \frac{r}{n}\right)}$ **81.** $\{\ln 2, \ln 4\}$ **83.** $\left\{\ln\frac{3}{2}\right\}$

85. $\{\ln(\sqrt{3} - 1)\}$ **87.** $\{\log_3 5, \log_3 7\}$ **89.** $\left\{\frac{1}{4}, 2\right\}$

91. $\{e^2, e^8\}$ **93.** $\left\{\ln\frac{5}{2}\right\}; \left(\ln\frac{5}{2}, \infty\right); \left(-\infty, \ln\frac{5}{2}\right]$
95. $\{2\}; (-\infty, 2); [2, \infty)$ **97.** $\emptyset; (-\infty, \infty); \emptyset$
99. $\{25\}; (25, \infty); (0, 25]$
101. $\{-1\}; (-2, -1); [-1, \infty)$
103. $\{10^{7/5}\}; (10^{7/5}, \infty); (0, 10^{7/5}]$ **105.** $\{-0.767, 2, 4\}$
107. $\{2.454, 5.659\}$ **109.** $\{-0.443\}$ **111.** $\{-2, 2\}$
113. $\{-3\}$ **115.** 85.5%   **117.** about 22 m

Reviewing Basic Concepts *(page 387)*

1. inverse; symmetric; $y = x$; range

2.

$f(x) = 2 - \log_2(x - 1)$

3. $x = 1$; x-intercept: $(5, 0)$; no y-intercept
4. The graph of $f(x)$ is the same as the graph of $g(x)$ reflected across the x-axis and shifted 1 unit to the right and 2 units upward.
5. $f^{-1}(x) = 1 + 2^{2-x}$ **6.** $\left\{\frac{\log 3}{2\log 3 - \log 4}\right\}$
7. $\{3\}$ **8.** $\{2\}$ **9.** $N = -\frac{1}{k}\ln\left(1 - \frac{H}{1000}\right)$
10. about 2.8 acres

5.6 Exercises (pages 394–399)

1. 9000 yr ago   **3.** 16,000 yr old
5. (a) $A = A_0 e^{-0.032t}$ (b) 6.97 yr   (c) 363 g
7. (a) less   (b) $A = 2e^{-0.005t}$ (c) about 140 days
9. (a) $10^{7.4} I_0 \approx 25{,}118{,}864 I_0$ (b) $10^{6.3} I_0 \approx 1{,}995{,}262 I_0$
(c) $10^{1.1} \approx 12.6$ times as strong   **11.** Magnitude 1 is
approximately 6.3 times as great as magnitude 3.
13. (a) $f(t) = 20 + 80 e^{-0.693t}$ (b) 76.6°C
(c) 1.415 hr, or about 1 hr and 25 min
15. (a) $C \approx 0.72$, $a \approx 1.041$ (b) 1.21 ppb
17. (a) $1076.89   (b) $1280.08   (c) about 26 yr
19. (a) about 3.57 million/mL   (b) about 3.99 million/mL
(c) about 27 hr   **21.** 4.9 yr   **23.** 13.9 yr
25. (a) $A \geq 1000$; The amount in the account must always be
greater than or equal to the principal.   (b) It takes about 9.1
years.   (c) There is about $1600.   **27.** (a) 46.2 yr
(b) 46.0 yr   **29.** 4.6 yr   **31.** The better investment is
2% compounded quarterly; it will earn $643.27 more.
33. 3.03%   **35.** $8616.67   **37.** 4%   **39.** (a) $205.52
(b) $1364.96   **41.** (a) $852.72   (b) $181,979.20
43. (a) 19 yr, 39 days   (b) 11 yr, 166 days
45. (a) 2,700,000   (b) 3,000,000   (c) 9,500,000
47. (a) during the third year ($x \approx 3.25$)

(b)
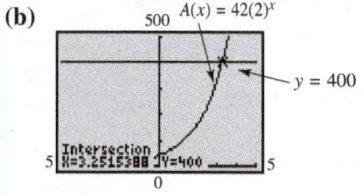

(c) during the fourth year ($x \approx 4.57$)
49. (a) $C \approx 17$; $a \approx 1.0126$; the percentage is increasing.
(b) about 21.3%   (c) during 2042 ($x \approx 2042.8$)
51. about 3.6 hr   **53.** $f(x) = 30(0.75)^x$; about 1992 ($x \approx 12$)
55. (a)

x	0	15	30	45	60	75	90	125
$g(x)$	7	21	57	111	136	158	164	178

(b) $g(x) = 261 - f(x)$, or $f(x) = 261 - g(x)$
(c) y_1 models $g(x)$ better.
(d) $f(x) = 261 - \dfrac{171}{1 + 18.6 e^{-0.0747x}}$
57. (a) $C \approx 5.772$; $a \approx 1.277$

(b)
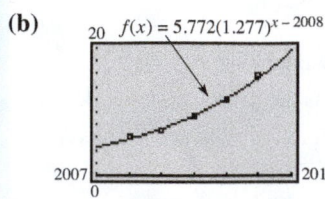

59. (a) 0.065; 0.82; Among people age 25, about 6.5% have
some CHD, while among people age 65, about 82% have some
CHD.   (b) 48 yr

Summary Exercises on Functions: Domains, Defining Equations, and Composition (pages 402–403)

1. A   **2.** B   **3.** C   **4.** D   **5.** A   **6.** B
7. D   **8.** C   **9.** C   **10.** B   **11.** $(-\infty, \infty)$
12. $\left[\frac{7}{2}, \infty\right)$ **13.** $(-\infty, \infty)$ **14.** $(-\infty, 6) \cup (6, \infty)$
15. $(-\infty, \infty)$ **16.** $(-\infty, -3] \cup [3, \infty)$
17. $(-\infty, -3) \cup (-3, 3) \cup (3, \infty)$ **18.** $(-\infty, \infty)$
19. $(-4, 4)$ **20.** $(-\infty, -7) \cup (3, \infty)$
21. $(-\infty, -1] \cup [8, \infty)$ **22.** $(-\infty, 0) \cup (0, \infty)$
23. $(-\infty, \infty)$ **24.** $(-\infty, -5) \cup (-5, \infty)$ **25.** $[1, \infty)$
26. $\left(-\infty, -\sqrt{5}\right) \cup \left(-\sqrt{5}, \sqrt{5}\right) \cup \left(\sqrt{5}, \infty\right)$ **27.** $(-\infty, \infty)$
28. $(-\infty, -1) \cup (-1, 1) \cup (1, \infty)$ **29.** $(-\infty, 1)$
30. $(-\infty, 2) \cup (2, \infty)$ **31.** $(-\infty, \infty)$
32. $[-2, 3] \cup [4, \infty)$ **33.** $(-\infty, -2) \cup (-2, 3) \cup (3, \infty)$
34. $[-3, \infty)$ **35.** $(-\infty, 0) \cup (0, \infty)$
36. $\left(-\infty, -\sqrt{7}\right) \cup \left(-\sqrt{7}, \sqrt{7}\right) \cup \left(\sqrt{7}, \infty\right)$
37. $(-\infty, \infty)$ **38.** $\varnothing$ **39.** $[-2, 2]$ **40.** $(-\infty, \infty)$
41. $(-\infty, -7] \cup (-4, 3) \cup [9, \infty)$ **42.** $(-\infty, \infty)$
43. $(-\infty, 5]$ **44.** $(-\infty, 3)$ **45.** $(-\infty, 4) \cup (4, \infty)$
46. $(-\infty, \infty)$ **47.** $(-\infty, -5] \cup [5, \infty)$ **48.** $(-\infty, \infty)$
49. $(-2, 6)$ **50.** $(0, 1) \cup (1, \infty)$
51. (a) $-30x - 33$; $(-\infty, \infty)$ (b) $-30x + 52$; $(-\infty, \infty)$
52. (a) $24x + 4$; $(-\infty, \infty)$ (b) $24x + 35$; $(-\infty, \infty)$
53. (a) $\sqrt{x + 3}$; $[-3, \infty)$ (b) $\sqrt{x} + 3$; $[0, \infty)$
54. (a) $\sqrt{x - 1}$; $[1, \infty)$ (b) $\sqrt{x} - 1$; $[0, \infty)$
55. (a) $(x^2 + 3x - 1)^3$; $(-\infty, \infty)$ (b) $x^6 + 3x^3 - 1$; $(-\infty, \infty)$
56. (a) $x^4 + x^2 - 3x - 2$; $(-\infty, \infty)$
(b) $(x + 2)^4 + (x + 2)^2 - 3(x + 2) - 4$; $(-\infty, \infty)$
57. (a) $\sqrt{3x - 1}$; $\left[\frac{1}{3}, \infty\right)$ (b) $3\sqrt{x - 1}$; $[1, \infty)$
58. (a) $\sqrt{2x - 2}$; $[1, \infty)$ (b) $2\sqrt{x - 2}$; $[2, \infty)$
59. (a) $\dfrac{2}{x + 1}$; $(-\infty, -1) \cup (-1, \infty)$
(b) $\dfrac{2}{x} + 1$; $(-\infty, 0) \cup (0, \infty)$
60. (a) $\dfrac{4}{x + 4}$; $(-\infty, -4) \cup (-4, \infty)$
(b) $\dfrac{4}{x} + 4$; $(-\infty, 0) \cup (0, \infty)$
61. (a) $\sqrt{-\frac{1}{x} + 2}$; $(-\infty, 0) \cup \left[\frac{1}{2}, \infty\right)$ (b) $-\dfrac{1}{\sqrt{x + 2}}$; $(-2, \infty)$
62. (a) $\sqrt{-\frac{2}{x} + 4}$; $(-\infty, 0) \cup \left[\frac{1}{2}, \infty\right)$ (b) $\dfrac{-2}{\sqrt{x + 4}}$; $(-4, \infty)$
63. (a) $\sqrt{\dfrac{1}{x + 5}}$; $(-5, \infty)$ (b) $\dfrac{1}{\sqrt{x} + 5}$; $[0, \infty)$
64. (a) $\sqrt{\dfrac{3}{x + 6}}$; $(-6, \infty)$ (b) $\dfrac{3}{\sqrt{x} + 6}$; $[0, \infty)$
65. (a) $\dfrac{x}{1 - 2x}$; $(-\infty, 0) \cup \left(0, \frac{1}{2}\right) \cup \left(\frac{1}{2}, \infty\right)$
(b) $x - 2$; $(-\infty, 2) \cup (2, \infty)$
66. (a) $\dfrac{x}{-1 + 4x}$; $(-\infty, 0) \cup \left(0, \frac{1}{4}\right) \cup \left(\frac{1}{4}, \infty\right)$
(b) $-x - 4$; $(-\infty, -4) \cup (-4, \infty)$
67. (a) $\log\sqrt{x}$; $(0, \infty)$ (b) $\sqrt{\log x}$; $[1, \infty)$
68. (a) $\ln \sqrt[4]{x}$; $(0, \infty)$ (b) $\sqrt[4]{\ln x}$; $[1, \infty)$

69. (a) $e^{\sqrt{x}}$; $[0, \infty)$ **(b)** $\sqrt{e^x}$; $(-\infty, \infty)$
70. (a) $e^{-\sqrt[3]{x}}$; $(-\infty, \infty)$ **(b)** $\sqrt[3]{e^{-x}}$; $(-\infty, \infty)$
71. (a) $-\left(\ln \sqrt{x}\right)^2$; $(0, \infty)$ **(b)** $\ln \sqrt{-x^2}$; $\emptyset$
72. (a) $\ln \sqrt{-3 - x^2}$; $\emptyset$ **(b)** $-2 - \left(\ln \sqrt{x - 1}\right)^2$; $(1, \infty)$
73. (a) $\dfrac{3x + 4}{x - 2}$; $(-\infty, 2) \cup (2, \infty)$
(b) $\dfrac{5x - 2}{5x - 4}$; $\left(-\infty, \frac{4}{5}\right) \cup \left(\frac{4}{5}, \infty\right)$
74. (a) $\dfrac{1 - 2x}{x + 1}$; $(-\infty, -1) \cup (-1, \infty)$
(b) $\dfrac{3 - 3x}{2 - x}$; $(-\infty, 2) \cup (2, \infty)$

Chapter 5 Review Exercises *(pages 407–409)*

1. no **3.** no **5.** no **7.** $(-\infty, \infty)$
9. Since f is one-to-one, it has an inverse.
11. The graphs are reflections across the line $y = x$. (See the graph.)

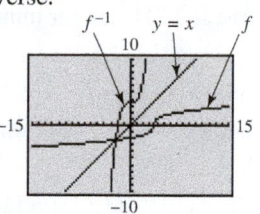

13. C **15.** D **17.** $0 < a < 1$ **19.** $(0, \infty)$
21.

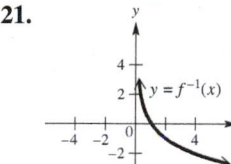

23. $(-\infty, \infty)$; $(0, \infty)$

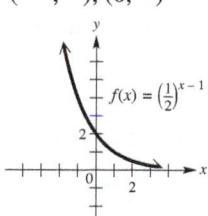

25. $(-\infty, \infty)$; $(-\infty, 0)$

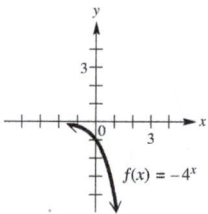

27. (a) $\left\{\frac{3}{5}\right\}$ **(b)** $\left[\frac{3}{5}, \infty\right)$ **29. (a)** $\left\{-\frac{2}{3}\right\}$ **(b)** $\left(-\frac{2}{3}, \infty\right)$
31. $(-0.766664696, 0.58777475603)$
33. 1.7657 **35.** 4.0656 **37.** 0 **39.** 12
41. 5 **43.** E **45.** B **47.** F
49. $(0, \infty)$; $(-\infty, \infty)$ **51.** $(-\infty, 0)$; $(-\infty, \infty)$

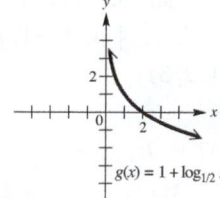

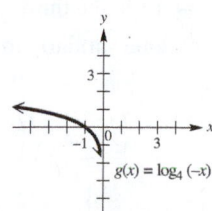

53. 3 **55.** $\log_3 m + \log_3 n - \log_3 5 - \log_3 r$
57. $2 \log_5 x + 4 \log_5 y + \frac{3}{5} \log_5 m + \frac{1}{5} \log_5 p$
59. (a) $\{2\}$ **(b)** $(2, \infty)$

61. (a) $\left\{\frac{7}{5}\right\}$, or $\{1.4\}$ **(b)** $(1, 1.4]$
63. (a) $\left\{\frac{3}{2}\right\}$ **65. (a)** $\{-1 + \ln 10\}$ **(b)** $\{1.303\}$
67. (a) $\{2\}$ **69. (a)** $\{3\}$ **71. (a)** $\{-3\}$ **73. (a)** $\emptyset$
75. $t = \dfrac{\ln\left(\frac{y}{y_0}\right)}{-k}$ **77.** $\{1.874\}$ **79.** 2.8% **81.** $\$25,149.59$
83. (a)

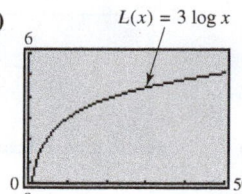

Since L is increasing, heavier planes require longer runways.

(b) No, it increases by 3000 ft for each tenfold increase in weight.
85. (a) 0.0054 g per L **(b)** 0.00073 g per L
(c) 0.000013 g per L **(d)** 0.75 mi **87.** about 10 sec

Chapter 5 Test *(page 410)*

1. (a) B **(b)** A **(c)** C **(d)** D **(e)** the equations of the functions in (a) and (d) and those in (b) and (c)
2. (a) yes; yes **(b)** no; no **(c)** yes; yes
3. (a) $f^{-1}(x) = \dfrac{x + 7}{5}$ **(b)** $f^{-1}(x) = 10^{x/2}$
(c) $f^{-1}(x) = \dfrac{2x + 1}{x - 1}$
4. (a)

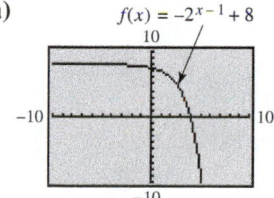

(b) domain: $(-\infty, \infty)$; range: $(-\infty, 8)$ **(c)** Yes, it has a horizontal asymptote given by $y = 8$. **(d)** x-intercept: $(4, 0)$; y-intercept: $(0, 7.5)$ **(e)** $f^{-1}(x) = 1 + \dfrac{\log(8 - x)}{\log 2}$ (Any base logarithm can be used.) **5.** $\{0.5\}$
6. (a) $\$11,495.74$ **(b)** $\$11,502.74$ **7.** The expression $\log_5 27$ is the exponent to which 5 must be raised in order to obtain 27. To find an approximation with a calculator, use the change-of-base rule. **8. (a)** 1.659 **(b)** 6.153 **(c)** 6.049
9. $3 \log m + \log n - \frac{1}{2} \log y$ **10.** $\log \dfrac{x^2 \sqrt{y}}{z^4}$ **11.** $2x; x^2$
12. $\left(-\frac{1}{2}, \infty\right)$ **13.** $t = \dfrac{\ln A - \ln P}{r}$
14. (a) $\{2\}$; The extraneous value is -4.
(b) $y_1 = \dfrac{\log x}{\log 2} + \dfrac{\log(x + 2)}{\log 2} - 3$ **(c)** $(2, \infty)$

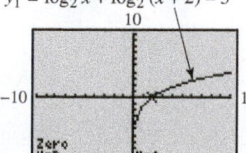

15. (a) $\left\{\frac{-2 + \ln 4}{5}\right\}$ **(b)** $\{-0.123\}$

16. (a) $\left\{\frac{\log 18}{\log 48}\right\}$ **(b)** $\{0.747\}$

17. (a) $\{e^{10}\}$ **(b)** $\{22{,}026.466\}$

18. (a) B **(b)** D **(c)** C **(d)** A

19. (a) For $t = x$, $A(x) = x^2 - x + 350$ **(b)** For $t = x$, $A(x) = 350 \log (x + 1)$

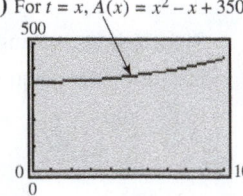

(c) For $t = x$, $A(x) = 350(0.75)^x$ **(d)** For $t = x$, $A(x) = 100(0.95)^x$

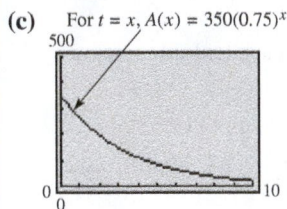 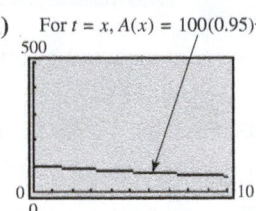

Function (c) best describes $A(t)$.

20. (a) $A(t) = 2e^{-0.000433t}$ **(b)** 0.03 g **(c)** about 3200 yr

CHAPTER 6 SYSTEMS AND MATRICES

6.1 Exercises *(pages 419–424)*

1. 2006–2009 **3.** approximately (2005.2, 1.26)

5. year; population (in millions) **7.** $\{(2, 2)\}$

9. $\left\{\left(\frac{1}{2}, -2\right)\right\}$ **11.** $\{(1, 1)\}$ **13.** $\{(3, -2)\}$

15. $\{(6, 15)\}$ **17.** $\{(2, -3)\}$ **19.** $\{(1, 3)\}$

21. $\{(-2, 3)\}$ **23.** $\{(5, 0)\}$ **25.** $\varnothing$

27. $\{(2y + 4, y)\}$, or $\left\{\left(x, \frac{x-4}{2}\right)\right\}$ **29.** $\{(0, 4)\}$

31. $\{(-1, 1)\}$ **33.** $\{(2, 3)\}$

35. $\left\{\left(\frac{y+9}{4}, y\right)\right\}$, or $\{(x, 4x - 9)\}$ **37.** $\varnothing$ **39.** $\varnothing$

41. $\{(12, 6)\}$ **43.** $\{(5, 2)\}$ **45.** $\{(0.138, -4.762)\}$

47. $\{(-8.708, 15.668)\}$ **49.** An inconsistent system will conclude with no variables and a false statement (contradiction), such as $0 = 1$. A system with dependent equations will conclude with no variables and a true statement (identity), such as $0 = 0$.

51. 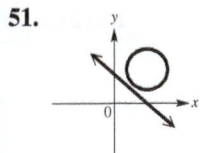 **53.**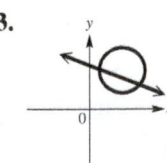

55. 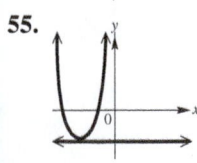 **57.**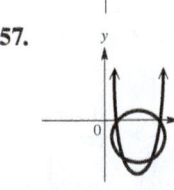

59. $\{(1, 2)\}$ **61.** $\{(1, 2), (2, 1)\}$

63. $\{(-1, 1), (1, 1)\}$ **65.** $\{(2, 0), (0, 2)\}$

67. $\{(-2, -2), (1, 1)\}$ **69.** $\{(2, 1), (-1, -2)\}$

71. $\{(-1, -3), (1, -3), (-\sqrt{6}, 2), (\sqrt{6}, 2)\}$

73. $\{(-0.79, 0.62), (0.88, 0.77)\}$ **75.** $\{(0.06, 2.88)\}$

77. (a) Let x represent the spending (in millions) in 2012 and y the spending (in millions) in 2011.

$x + y = 1858$; $x - y = 226$

(b) $\{(1042, 816)\}$ **(c)** The Black Friday spending in 2012 was \$1042 million and in 2011 it was \$816 million.

79. \$650 in 2010; \$455 in 2012

81. 14 by 7 by 6 in. or about 5.17 by 10.34 by 11.00 in.

83. $x \approx 177.1$; $y \approx 174.9$; If an athlete's maximum heart rate is 180 beats per minute (bpm), then it will be about 177 bpm after 5 sec and 175 bpm after 10 sec.

85. (a) Let x represent Apple's spending (in millions) and y Samsung's spending (in millions).

$x + y = 293$; $x - y = 93$

(b) $\{(193, 100)\}$ **(c)** In 2012, Apple spent \$193 million on media and Samsung spent \$100 million.

87. $r \approx 1.538$ in.; $h \approx 6.724$ in.

89. $W_1 \approx 109.8$ lb; $W_2 \approx 134.5$ lb; $\{(109.8, 134.5)\}$

91. $x = 300$; $y = 350$ **93. (a)** \$1.50; \$5.00

(b) \$1.96; $19.57 \approx 20$ units

95. $x = 2100$; $R = C = \$52{,}000$

97. $5t + 15u = 16$; $5t + 4u = 5$

98. $t = \frac{1}{5}$; $u = 1$ **99.** $\{(5, 1)\}$

100. $y = \frac{-15x}{5 - 16x}$ **101.** $y = \frac{-4x}{5 - 5x}$

102.

$y = \frac{-15x}{5 - 16x}$ $y = \frac{-4x}{5 - 5x}$

103. $\{(2, 2)\}$ **105.** $\left\{\left(\frac{1}{3}, \frac{1}{4}\right)\right\}$

6.2 Exercises *(pages 429–432)*

1. Substituting $x = -3$, $y = 6$, and $z = 1$ yields $-1 = -1$ in the first equation, $-6 = -6$ in the second equation, and $19 = 19$ in the third equation. **3.** and **5.** Exercises 3 and 5 are done similarly to **Exercise 1.** **7.** $\{(1, 2, -1)\}$

9. $\{(2, 0, 3)\}$ **11.** $\varnothing$ **13.** $\{(1, 2, 3)\}$

15. $\{(-1, 2, 1)\}$ **17.** $\{(4, 1, 2)\}$

19. $\left\{\left(\frac{4 - z}{3}, \frac{4z - 7}{3}, z\right)\right\}$ **21.** $\{(47 - 9z, 7z - 32, z)\}$

23. $\left\{\left(-\frac{3}{4}, \frac{5}{3}, \frac{3}{2}\right)\right\}$ **25.** $\{(-1, -2, -3)\}$

27. $\{(-3, 4, 0)\}$ **29.** $\left\{\left(\frac{1 - 2z}{5}, \frac{3z + 31}{5}, z\right)\right\}$

31. $\left\{\left(\frac{1}{2}, \frac{1}{3}, \frac{1}{4}\right)\right\}$ **33.** $\{(2, 4, 2)\}$ **35.** $\left\{\left(-1, 1, \frac{1}{3}\right)\right\}$

37. $\varnothing$ **39.** $\varnothing$ **41.** $\{(-3z + 6, -4z + 6, z)\}$

43. $\{(40, 15, 30)\}$ **45.** \$9.00 water: 120 gal;
\$3.00 water: 60 gal; \$4.50 water: 120 gal   **47.** No solution; the pricing was inconsistent.   **49.** 75°, 65°, 40°

51. \$3000 at 4%; \$6000 at 4.5%; \$1000 at 2.5%

53. three possibilities: 12 EZ, 16 compact, 0 commercial; 10 EZ, 8 compact, 3 commercial; 8 EZ, 0 compact, 6 commercial

55. (a) $a + 20b + 2c = 190$
$a + 5b + 3c = 320$
$a + 40b + c = 50$

(b) $a = 30$, $b = -2$, and $c = 100$; that is,
$P = 30 - 2A + 100S$ **(c)** \$260,000

57. $y = x^2 + 2x - 3$ **59.** $y = 3x^2 + x - 2$

61. $y = x^2 - x + 2$ **63.** $y = -2x^2 + x + 1$

65. $x^2 + y^2 - 2x + 2y - 23 = 0$

67. $x^2 + y^2 - 4x + 2y - 20 = 0$

69. $x^2 + y^2 + x - 7y = 0$

71. $s(t) = -2t^2 + 20t + 5$; $s(8) = 37$

6.3 Exercises (pages 440–444)

1. $\begin{bmatrix} 1 & 2 \\ 4 & 7 \end{bmatrix}$ **3.** $\begin{bmatrix} 1 & 5 & 6 \\ 0 & 13 & 11 \\ 4 & 7 & 0 \end{bmatrix}$ **5.** $\begin{bmatrix} -3 & 1 & -4 \\ 2 & 1 & 3 \\ 0 & 0 & -13 \end{bmatrix}$

7. $\left[\begin{array}{cc|c} 2 & 3 & 11 \\ 1 & 2 & 8 \end{array}\right]$ **9.** $\left[\begin{array}{cc|c} 1 & 5 & 6 \\ 1 & 0 & 3 \end{array}\right]$ **11.** $\left[\begin{array}{ccc|c} 2 & 1 & 1 & 3 \\ 3 & -4 & 2 & -7 \\ 1 & 1 & 1 & 2 \end{array}\right]$

13. $\left[\begin{array}{ccc|c} 1 & 1 & 0 & 2 \\ 0 & 2 & 1 & -4 \\ 0 & 0 & 1 & 2 \end{array}\right]$

15. $2x + y = 1$
$3x - 2y = -9$

17. $x = 2$
$y = 3$
$z = -2$

19. $3x + 2y + z = 1$
$2y + 4z = 22$
$-x - 2y + 3z = 15$

21. $\{(5, -1)\}$ **23.** $\{(5y + 6, y)\}$ **25.** $\{(2, 3, 1)\}$

27. $\{(3 - 3z, 1 + 2z, z)\}$ **29.** $\emptyset$ **31.** $\{(2, 3)\}$

33. $\{(-3, 0)\}$ **35.** $\left\{\left(\frac{7}{2}, -1\right)\right\}$ **37.** $\emptyset$

39. $\left\{\left(x, \frac{6x - 1}{3}\right)\right\}$, or $\left\{\left(\frac{3y + 1}{6}, y\right)\right\}$ **41.** $\{(-2, 1, 3)\}$

43. $\{(-1, 23, 16)\}$ **45.** $\{(3, 2, -4)\}$

47. $\{(2, 1, -1)\}$ **49.** $\emptyset$ **51.** $\{(7.785, 36.761)\}$

53. $\{(5.211, 3.739, -4.655)\}$

55. $\{(-0.250, 1.284, -0.059)\}$

57. In both cases, we simply write the coefficients and do not write the variables. This is possible because we agree either on the order in which the variables appear or in which the powers of a variable appear (descending degree).

59. $\left\{\left(\frac{5z + 14}{5}, \frac{5z - 12}{5}, z\right)\right\}$ **61.** $\left\{\left(\frac{12 - z}{7}, \frac{4z - 6}{7}, z\right)\right\}$

63. $\emptyset$ **65.** $\{(0, 2, -2, 1)\}$

67. (a) $F \approx 0.5714N + 0.4571R - 2014$ **(b)** \$5700

69. model A: 5 bicycles; model B: 8 bicycles

71. \$5000 at 2%; \$3500 at 3%; \$4000 at 2.5%

73. (a) $c = 25$
$4a + 2b + c = 260$
$16a + 4b + c = 695$

(b) $f(x) = 25x^2 + 67.5x + 25$

(c)
$f(x) = 25x^2 + 67.5x + 25$

(d) Answers will vary. For example, in 2014 predicted revenues are $f(6) = 1330$ million.

75. (a) $1990^2a + 1990b + c = 11$
$2010^2a + 2010b + c = 10$
$2030^2a + 2030b + c = 6$

(b) $f(x) = -0.00375x^2 + 14.95x - 14,889.125$

(c)
$f(x) = -0.00375x^2 + 14.95x - 14,889.125$

(d) Answers will vary. For example, in 2015 the predicted ratio is $f(2015) \approx 9.3$.

77. (a) At intersection A, incoming traffic is equal to $x + 5$. The outgoing traffic is given by $y + 7$. Therefore, $x + 5 = y + 7$. The other equations can be justified in a similar way.

(b) The three equations can be written as
$x - y = 2$
$x - z = 3$
$y - z = 1$.

The solution set is $\{(z + 3, z + 1, z), \text{ where } z \geq 0\}$.

(c) There are infinitely many solutions, since some cars could be driving around the block continually.

79. (a) $a + 871b + 11.5c + 3d = 239$
$a + 847b + 12.2c + 2d = 234$
$a + 685b + 10.6c + 5d = 192$
$a + 969b + 14.2c + d = 343$

(b) $\left[\begin{array}{cccc|c} 1 & 871 & 11.5 & 3 & 239 \\ 1 & 847 & 12.2 & 2 & 234 \\ 1 & 685 & 10.6 & 5 & 192 \\ 1 & 969 & 14.2 & 1 & 343 \end{array}\right]$

The values are $a \approx -715.457$, $b \approx 0.34756$, $c \approx 48.6585$, and $d \approx 30.71951$.

(c) $F = -715.457 + 0.34756A + 48.6585P + 30.71951W$

(d) approximately 323, which is very close to 320

Reviewing Basic Concepts *(pages 444–445)*

1. $\{(3, -4)\}$ **2.** $\{(-2, 0)\}$

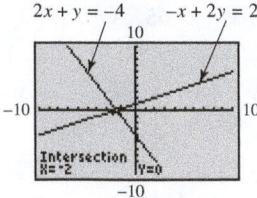

3. $\left\{\left(x, \frac{2-x}{2}\right)\right\}$, or $\{(2 - 2y, y)\}$ **4.** $\varnothing$

5. $\{(2, -1), (-6.5, 24.5)\}$

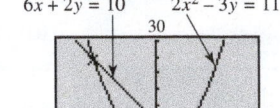
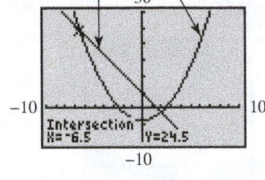

The other point of intersection is $(2, -1)$.

6. $\{(-2, 1, 2)\}$ **7.** $\{(2, 1, -1)\}$ **8.** $\{(3, 2, 1)\}$

9. LCD: 88 million; CRT: 45 million

10. \$1000 at 2%; \$1500 at 3%; \$2500 at 4%

6.4 Exercises *(pages 453–457)*

1. 2×2; square   **3.** 3×4 **5.** 2×1; column

7. 1×1; square, row, column

9. $w = 3$; $x = 2$; $y = -1$; $z = 4$

11. $w = 2$; $x = 6$; $y = -2$; $z = 8$

13. $z = 11$; $r = 3$; $s = 3$; $p = 3$; $a = \frac{3}{4}$

15. The two matrices must have the same dimension. To find the sum, add the corresponding entries. The sum will be a matrix with the same dimension.

17. $\begin{bmatrix} -2 & -7 & 7 \\ 10 & -2 & 7 \end{bmatrix}$ **19.** $\begin{bmatrix} -6 & 8 \\ 4 & 2 \end{bmatrix}$ **21.** $\begin{bmatrix} 5 & 5 \\ 12 & 0 \end{bmatrix}$

23. cannot be added

25. $\begin{bmatrix} 13 & 3 & 0 & -2 \\ 9 & -12 & 4 & 8 \\ 12 & -11 & -1 & 9 \end{bmatrix}$ **27.** $\begin{bmatrix} 0 & 2 \\ 13 & -5 \\ 0 & 1 \end{bmatrix}$

29. $\begin{bmatrix} 7 & 4 & 7 \\ 4 & 7 & 7 \\ 4 & 7 & 7 \end{bmatrix}$ **31.** $\begin{bmatrix} 8 & -43 & -18 \\ 26 & 29 & 6 \\ -2 & 10 & 43 \end{bmatrix}$

33. $\begin{bmatrix} -4 & 8 \\ 0 & 6 \end{bmatrix}$ **35.** $\begin{bmatrix} 2 & 6 \\ -4 & 6 \end{bmatrix}$ **37.** $\begin{bmatrix} -13 & 21 \\ 2 & 15 \end{bmatrix}$

39. $\begin{bmatrix} 2 & 6 & 5 \\ -4 & -7 & 9 \end{bmatrix}$ **41.** AB: 4×4; BA: 2×2

43. AB: 3×2; BA: not defined

45. Neither AB nor BA is defined.   **47.** columns; rows

49. $AB = \begin{bmatrix} -3 & 1 \\ -4 & 6 \end{bmatrix}$, $BA = \begin{bmatrix} 4 & 2 \\ 5 & -1 \end{bmatrix}$

51. AB and BA are undefined.

53. $AB = \begin{bmatrix} -15 & 22 & -9 \\ -2 & 5 & -3 \\ -32 & 18 & 6 \end{bmatrix}$, $BA = \begin{bmatrix} 5 & 14 \\ 20 & -9 \end{bmatrix}$

55. AB is undefined. $BA = \begin{bmatrix} -1 & -3 & 19 \\ -1 & 1 & -39 \end{bmatrix}$

57. $\begin{bmatrix} -17 \\ -1 \end{bmatrix}$ **59.** $\begin{bmatrix} 17 & -10 \\ 1 & 2 \end{bmatrix}$ **61.** $\begin{bmatrix} -2 & 10 \\ 0 & 8 \end{bmatrix}$

63. $\begin{bmatrix} -15 & -16 & 3 \\ -1 & 0 & 9 \\ 7 & 6 & 12 \end{bmatrix}$ **65.** $\begin{bmatrix} 2 & 7 & -4 \end{bmatrix}$

67. $\begin{bmatrix} pa + qb & pc + qd \\ ra + sb & rc + sd \end{bmatrix}$ **69.** $\begin{bmatrix} 23 & -9 \\ -6 & -2 \\ 33 & 1 \end{bmatrix}$

71. $\begin{bmatrix} -25 & 23 & 11 \\ 0 & -6 & -12 \\ -15 & 33 & 45 \end{bmatrix}$ **73.** cannot be multiplied

75. $\begin{bmatrix} 10 & -10 \\ 15 & -5 \end{bmatrix}$ **77.** no

79. (a) $\begin{bmatrix} 50 & 100 & 30 \\ 10 & 90 & 50 \\ 60 & 120 & 40 \end{bmatrix}$ (b) $\begin{bmatrix} 12 \\ 10 \\ 15 \end{bmatrix}$ (c) $\begin{bmatrix} 2050 \\ 1770 \\ 2520 \end{bmatrix}$

(d) \$6340   **81.** (a) $d_{n+1} = -0.05m_n + 1.05d_n$; 5%

(b) after 1 yr: 3020 mountain lions, 515,000 deer; after 2 yr: 3600 mountain lions, 525,700 deer

6.5 Exercises *(pages 465–467)*

1. -31 **3.** 7   **5.** 0   **7.** -26 **9.** $2, -6, 4$

11. $-6, 0, -6$ **13.** 186   **15.** 17   **17.** 166

19. 0   **21.** 0   **23.** -5.5 **25.** 690   **27.** $\{0\}$

29. $\left\{-\frac{1}{2}, 6\right\}$ **31.** $\{-2\}$ **33.** $\left\{-\frac{2}{3}\right\}$ **35.** 298

37. -88 **39.** 1   **41.** 9.5   **43.** 3.5   **45.** yes

47. no   **49.** no   **51.** 0   **53.** 0   **55.** 16

57. $\{(2, 2)\}$ **59.** $\{(2, -5)\}$

61. $\left\{\left(\frac{4 - 2y}{3}, y\right)\right\}$, or $\left\{\left(x, \frac{4 - 3x}{2}\right)\right\}$ **63.** $\{(2, 3)\}$

65. $\{(-1, 2, 1)\}$ **67.** $\{(-3, 4, 2)\}$

69. $\{(0, 4, 2)\}$ **71.** $\varnothing$ **73.** $\{(-1, 2, 5, 1)\}$

75. If $D = 0$, Cramer's rule cannot be applied because there is no unique solution. There are either no solutions or infinitely many solutions.

6.6 Exercises *(pages 474–478)*

1. yes   **3.** no   **5.** no   **7.** yes   **9.** $\begin{bmatrix} 5 & -7 \\ -2 & 3 \end{bmatrix}$

11. $\begin{bmatrix} 2 & 1 \\ -1.5 & -0.5 \end{bmatrix}$ **13.** A^{-1} does not exist.

15. $\begin{bmatrix} -2.5 & 5 \\ 12.5 & -15 \end{bmatrix}$ **17.** $\begin{bmatrix} 0 & 1 & 0 \\ 0 & 0 & 1 \\ 1 & 0 & 0 \end{bmatrix}$

19. $\begin{bmatrix} -2 & 1 & -1 \\ -5 & 2 & -1 \\ 3 & -1 & 1 \end{bmatrix}$

21. It will not exist if its determinant is equal to 0.

23. $\begin{bmatrix} 1 & 0 & 0 \\ 0 & -1 & 0 \\ -1 & 0 & 1 \end{bmatrix}$ **25.** $\begin{bmatrix} 15 & 4 & -5 \\ -12 & -3 & 4 \\ -4 & -1 & 1 \end{bmatrix}$

27. $\begin{bmatrix} -\frac{10}{3} & \frac{5}{9} & -\frac{10}{9} \\ \frac{20}{3} & \frac{5}{9} & \frac{80}{9} \\ -5 & \frac{5}{6} & -\frac{20}{3} \end{bmatrix}$ **29.** A^{-1} does not exist.

31. $\begin{bmatrix} 0.0543058761 & -0.0543058761 \\ 1.846399787 & 0.153600213 \end{bmatrix}$

33. $\begin{bmatrix} -20 & 10 & -10 \\ -50 & 20 & -10 \\ 30 & -10 & 10 \end{bmatrix}$

35. $\{(-2, 4)\}$ **37.** $\{(4, -6)\}$ **39.** $\varnothing$

41. $\{(3, 0, 2)\}$ **43.** $\left\{\left(12, -\frac{15}{11}, -\frac{65}{11}\right)\right\}$

45. $\{(0, 2, -2, 1)\}$

47. $\{(-3.542308934, -4.343268299)\}$

49. $\{(-0.9704156959, 1.391914631, 0.1874077432)\}$

51. $P(x) = -2x^3 + 5x^2 - 4x + 3$

53. $P(x) = x^4 + 2x^3 + 3x^2 - x - 1$

55. type A: \$10.99; type B: \$12.99; type C: \$14.99

57. (a) $113a + 308b + c = 10{,}170$
$133a + 622b + c = 15{,}305$
$155a + 1937b + c = 21{,}289$

(b) $a \approx 251$, $b \approx 0.346$, $c \approx -18{,}300$;
$T \approx 251A + 0.346I - 18{,}300$ (c) $11{,}426$

59. (a)

$\begin{aligned} a + 1500b + 8c &= 122 \\ a + 2000b + 5c &= 130 \\ a + 2200b + 10c &= 158 \end{aligned}$ or $\begin{bmatrix} 1 & 1500 & 8 \\ 1 & 2000 & 5 \\ 1 & 2200 & 10 \end{bmatrix}\begin{bmatrix} a \\ b \\ c \end{bmatrix} = \begin{bmatrix} 122 \\ 130 \\ 158 \end{bmatrix}$

(b) \$130,000

61. $\begin{bmatrix} 2 & 9 \\ 1 & 5 \end{bmatrix}$ **63.** $\begin{bmatrix} 1 & 0 & 1 \\ -1 & 0 & 2 \\ -2 & 1 & 3 \end{bmatrix}$ **65.** $\begin{bmatrix} \frac{1}{a} & 0 & 0 \\ 0 & \frac{1}{b} & 0 \\ 0 & 0 & \frac{1}{c} \end{bmatrix}$

Reviewing Basic Concepts *(page 478)*

1. $\begin{bmatrix} -5 & 6 \\ -1 & 3 \end{bmatrix}$ **2.** $\begin{bmatrix} 0 & 6 \\ -9 & 12 \end{bmatrix}$ **3.** $\begin{bmatrix} 33 & -24 \\ -12 & 9 \end{bmatrix}$

4. $\begin{bmatrix} 1 & 3 & -3 \\ 0 & -6 & 0 \\ 4 & 2 & 2 \end{bmatrix}$ **5.** -3 **6.** -14

7. $\begin{bmatrix} \frac{1}{3} & \frac{4}{3} \\ \frac{2}{3} & \frac{5}{3} \end{bmatrix}$ **8.** $\begin{bmatrix} -\frac{2}{7} & -\frac{11}{14} & \frac{1}{14} \\ -\frac{4}{7} & -\frac{4}{7} & \frac{1}{7} \\ -\frac{1}{7} & -\frac{1}{7} & \frac{2}{7} \end{bmatrix}$

9. $W_1 = W_2 = \dfrac{100\sqrt{3}}{3} \approx 57.7$ lb   **10.** $\{(3, 2, 1)\}$

6.7 Exercises *(pages 485–488)*

1. 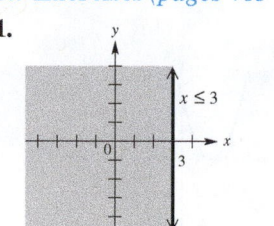   **3.**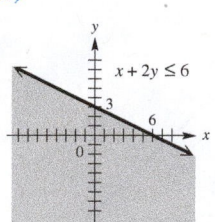

5. 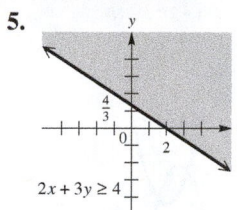   **7.**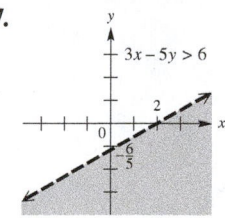

9. 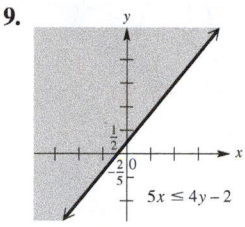   **11.**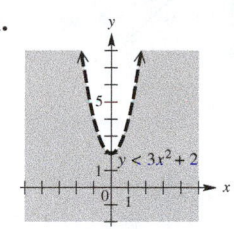

13. 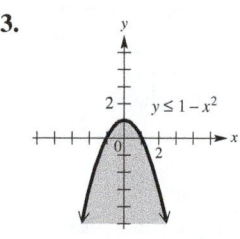   **15.**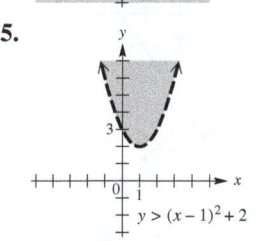

17. 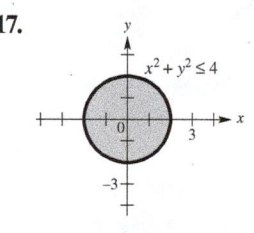   **19.**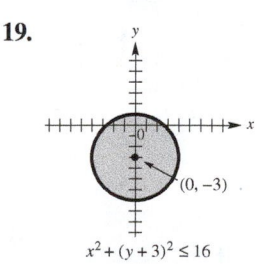

21. The boundary is solid if the symbol is $\geq$ or $\leq$ and dashed if the symbol is $>$ or $<$.   **23.** above   **25.** B

27. $x^2 + y^2 < 1$ **29.** $y > x^2 - 4$ **31.** C   **33.** A

35. 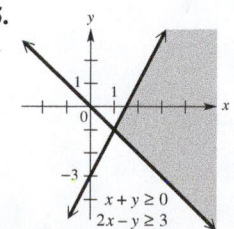   **37.**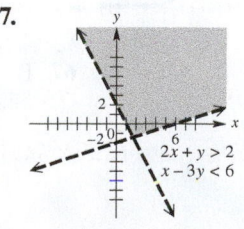

39.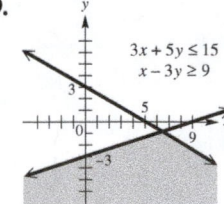

$3x + 5y \leq 15$
$x - 3y \geq 9$

41.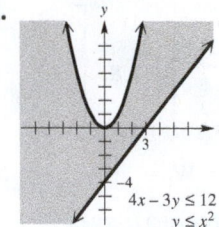

$4x - 3y \leq 12$
$y \leq x^2$

43.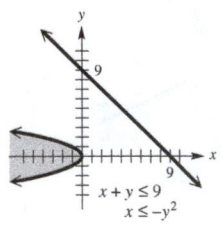

$x + y \leq 9$
$x \leq -y^2$

45.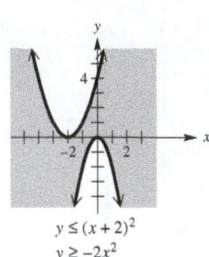

$y \leq (x + 2)^2$
$y \geq -2x^2$

47.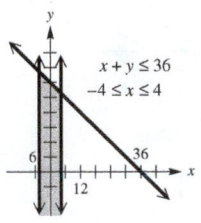

$x + y \leq 36$
$-4 \leq x \leq 4$

49.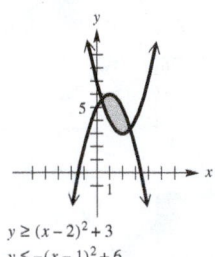

$y \geq (x - 2)^2 + 3$
$y \leq -(x - 1)^2 + 6$

51.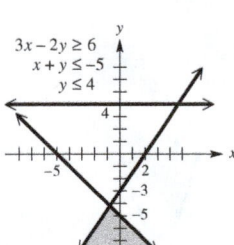

$3x - 2y \geq 6$
$x + y \leq -5$
$y \leq 4$

53.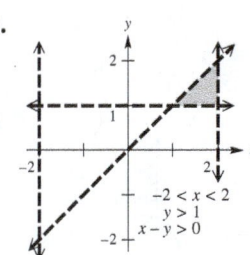

$-2 < x < 2$
$y > 1$
$x - y > 0$

55.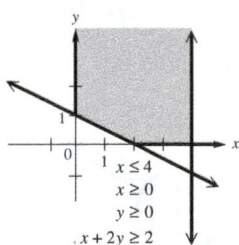

$x \leq 4$
$x \geq 0$
$y \geq 0$
$x + 2y \geq 2$

57.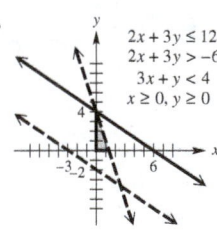

$2x + 3y \leq 12$
$2x + 3y > -6$
$3x + y < 4$
$x \geq 0, y \geq 0$

59.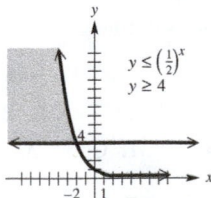

$y \leq \left(\frac{1}{2}\right)^x$
$y \geq 4$

61.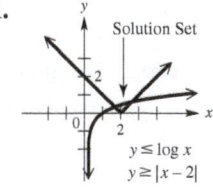

Solution Set

$y \leq \log x$
$y \geq |x - 2|$

63. D **65.** A **67.** B

69.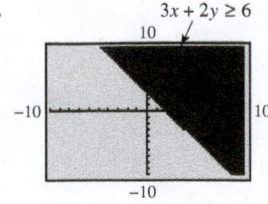

$3x + 2y \geq 6$

71.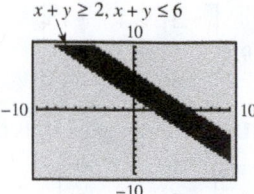

$x + y \geq 2, x + y \leq 6$

73.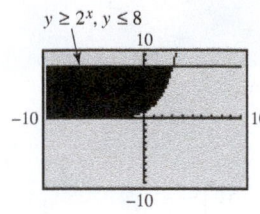

$y \geq 2^x, y \leq 8$

75. $(-3, -7), (1, 1)$ **77.** $(-1, -1), (0, 0), (1, 1)$

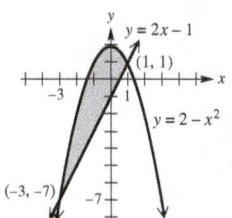

$y = 2x - 1$
$(1, 1)$
$y = 2 - x^2$
$(-3, -7)$

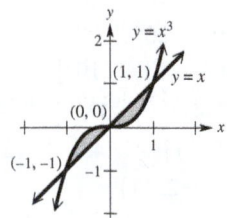

$y = x^3$
$(1, 1)$
$y = x$
$(0, 0)$
$(-1, -1)$

79. $x + 2y - 8 \geq 0, x + 2y \leq 12, x \geq 0, y \geq 0$

81. maximum of 65 at $(5, 10)$; minimum of 8 at $(1, 1)$

83. maximum of 66 at $(7, 9)$; minimum of 3 at $(1, 0)$

85. maximum of 100 at $(1, 10)$; minimum of 0 at $(1, 0)$

87. hat units: 5; whistle units: 0; maximum inquiries: 15

89. to A: 20; to B: 80; minimum cost: $1040

91. gasoline: 6,400,000 gal; fuel oil: 3,200,000 gal; maximum revenue: $16,960,000 **93.** medical kits: 0; containers of water: 4000; people aided: 40,000

6.8 Exercises *(pages 494)*

1. $\frac{5}{3x} + \frac{-10}{3(2x + 1)}$ **3.** $\frac{6}{5(x + 2)} + \frac{8}{5(2x - 1)}$

5. $\frac{5}{6(x + 5)} + \frac{1}{6(x - 1)}$ **7.** $\frac{-2}{x + 1} + \frac{2}{x + 2} + \frac{4}{(x + 2)^2}$

9. $\frac{4}{x} + \frac{4}{1 - x}$ **11.** $\frac{15}{x} + \frac{-5}{x + 1} + \frac{-6}{x - 1}$

13. $1 + \frac{-2}{x + 1} + \frac{1}{(x + 1)^2}$

15. $x^3 - x^2 + \frac{-1}{3(2x + 1)} + \frac{2}{3(x + 2)}$

17. $\frac{1}{9} + \frac{-1}{x} + \frac{25}{18(3x + 2)} + \frac{29}{18(3x - 2)}$ **19.** $\frac{-3}{5x^2} + \frac{3}{5(x^2 + 5)}$

21. $\frac{-2}{7(x + 4)} + \frac{6x - 3}{7(3x^2 + 1)}$ **23.** $\frac{1}{4x} + \frac{-8}{19(2x + 1)} + \frac{-9x - 24}{76(3x^2 + 4)}$

25. $\frac{-1}{x} + \frac{2x}{2x^2 + 1} + \frac{2x + 3}{(2x^2 + 1)^2}$ **27.** $\frac{-1}{x + 2} + \frac{3}{(x^2 + 4)^2}$

29. $5x^2 + \frac{3}{x} + \frac{-1}{x + 3} + \frac{2}{x - 1}$ **31.** coincide; correct

33. do not coincide; not correct

Reviewing Basic Concepts *(page 495)*

1.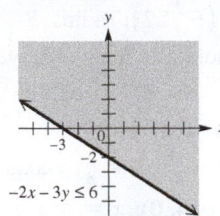

2.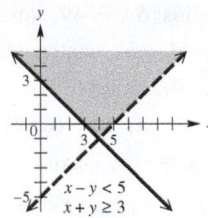

3.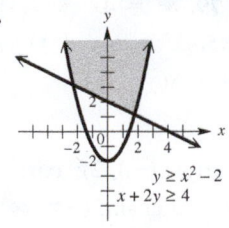

4.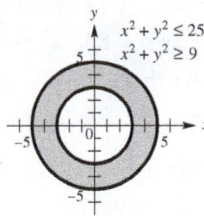

5. A   **6.** Minimum value is 8 at $(4, 0)$.

7. maximum: 65; minimum: 8   **8.** substance X: 130 lb; substance Y: 450 lb; minimum cost: $1610

9. $\dfrac{7}{x-5} + \dfrac{3}{x+4}$ **10.** $\dfrac{1}{x-1} + \dfrac{2x}{x^2+1}$

Chapter 6 Review Exercises *(pages 499–502)*

1. $\{(5, 7)\}$ **3.** $\{(5, -3)\}$ **5.** $\left\{ \left(\frac{7}{5}, -\frac{1}{5}\right), (1, 1) \right\}$

7. $\left\{ \left(\pm\sqrt{15}, -1\right), \left(\pm\sqrt{7}, -3\right) \right\}$ **9.** $\{(2, -2)\}$

11. **(a)** $y_1 = \sqrt{2 - x^2},\ y_2 = -\sqrt{2 - x^2}$
(b) $y_3 = -3x + 4$ **(c)** $[-3, 3]$ by $[-2, 2]$; Other windows are possible.   **13.** No, a system consisting of two equations in three variables is represented by two planes in space. There will be no solutions or infinitely many solutions.

15. $\{(6, -2, 1)\}$ **17.** $\varnothing$; inconsistent system

19. $\{(-3, 2)\}$

21. $\{(z + 2, -z - 1, z)\}$; dependent equations

23. $\begin{bmatrix} -4 \\ 6 \\ 1 \end{bmatrix}$ **25.** $\begin{bmatrix} -4 & -4 \\ -7 & 16 \end{bmatrix}$ **27.** $\begin{bmatrix} 14x & 22y \\ 54y & 70x \end{bmatrix}$

29. $\begin{bmatrix} 18 & 20 \\ 29 & -1 \end{bmatrix}$ **31.** $\begin{bmatrix} -3 \\ 10 \end{bmatrix}$ **33.** $\begin{bmatrix} -2 & 22 & 31 \\ 25 & 12 & 9 \end{bmatrix}$

35. yes   **37.** no   **39.** A^{-1} does not exist.

41. $\begin{bmatrix} \frac{1}{2} & 0 \\ \frac{1}{10} & \frac{1}{5} \end{bmatrix}$ **43.** $\begin{bmatrix} \frac{2}{3} & 0 & -\frac{1}{3} \\ \frac{1}{3} & 0 & -\frac{2}{3} \\ -\frac{2}{3} & 1 & \frac{1}{3} \end{bmatrix}$ **45.** $\{(2, 2)\}$

47. $\{(2, 1)\}$ **49.** $\{(1, -1, 2)\}$ **51.** $\{(-1, 0, 2)\}$

53. $\varnothing$; inconsistent system   **55.** -25 **57.** -44

59. $\left\{ -\frac{7}{3} \right\}$ **61.** {all real numbers}   **63.** **(a)** $D = 5$
(b) $D_x = 30$ **(c)** $D_y = -50$ **(d)** $x = 6;\ y = -10$; solution set: $\{(6, -10)\}$ **65.** If $D = 0$, there would be division by 0, which is undefined. The system will have no solutions or infinitely many solutions.   **67.** $\{(-4, 2)\}$

69. $\{(-4, 6, 2)\}$ **71.** $\left\{ \left(\frac{172}{67}, -\frac{14}{67}, -\frac{87}{67}\right) \right\}$

73. CDs: 80; holders: 20

75. 5%: 11 mL; 15%: 3 mL; 10%: 6 mL

77. $P(x) = x^3 - 2x + 5$

79.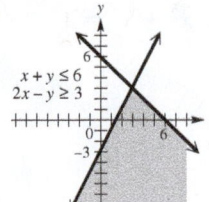

81. Maximum value is 24 at $(0, 6)$.

83. radios: 25; Blu-ray players: 30; maximum profit: $1425

85. $\dfrac{2}{x} + \dfrac{-2}{x+1} + \dfrac{-1}{(x+1)^2}$ **87.** $\dfrac{2}{x+1} + \dfrac{1}{x-1} + \dfrac{3}{x}$

Chapter 6 Test *(pages 502–503)*

1. **(a)** first equation: a circle; second equation: a line
(b) 0, 1, or 2   **(c)** $\{(-1, -2), (1, 2)\}$

(d) 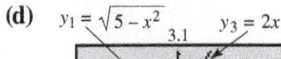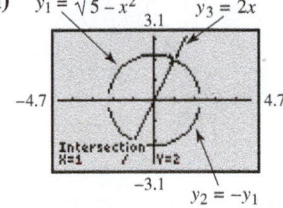   The other point of intersection is $(-1, -2)$.

2. **(a)** $\{(3, 1)\}$ **(b)** $\{(x, 3x - 1)\}$, or $\left\{ \left(\frac{y+1}{3}, y\right) \right\}$

3. **(a)** $\left\{ \left(3, \frac{1}{2}\right) \right\}$ **(b)** $\varnothing$

4. $\left\{ \left(-\sqrt{2}, -3\right), \left(\sqrt{2}, -3\right), \left(-\sqrt{7}, 2\right), \left(\sqrt{7}, 2\right) \right\}$

5. **(a)** $\{(2, 0, -1)\}$ **(b)** $\{(4z - 3, 4 - 3z, z)\}$

6. **(a)** $\begin{bmatrix} 8 & 3 \\ 0 & -11 \\ 15 & 19 \end{bmatrix}$ **(b)** not possible   **(c)** $\begin{bmatrix} -5 & 16 \\ 19 & 2 \end{bmatrix}$

7. **(a)** yes   **(b)** yes   **(c)** No. In general, matrix multiplication is not commutative.   **(d)** AC cannot be found, but CA can.

8. **(a)** 1   **(b)** -844 **9.** $\{(-6, 7)\}$

10. **(a)** $A = \begin{bmatrix} 1 & 1 & -1 \\ 2 & -3 & -1 \\ 1 & 2 & 2 \end{bmatrix}$, $X = \begin{bmatrix} x \\ y \\ z \end{bmatrix}$, $B = \begin{bmatrix} -4 \\ 5 \\ 3 \end{bmatrix}$

(b) $A^{-1} = \begin{bmatrix} \frac{1}{4} & \frac{1}{4} & \frac{1}{4} \\ \frac{5}{16} & -\frac{3}{16} & \frac{1}{16} \\ -\frac{7}{16} & \frac{1}{16} & \frac{5}{16} \end{bmatrix}$ **(c)** $\{(1, -2, 3)\}$

(d) $A = \begin{bmatrix} 0.5 & 1 & 1 \\ 2 & -3 & -1 \\ 1 & 2 & 2 \end{bmatrix}$

and $\det A = 0$, so A^{-1} does not exist.

11. **(a)** $f(x) = 0.00026x^2 - 0.014x + 1.3$

(b) about 2040   **12.** B   **13.** cabinet X: 8; cabinet Y: 3

14. $\dfrac{4}{x-3} + \dfrac{3}{x+2}$ **15.** $\dfrac{-1}{x-2} + \dfrac{2}{x+2} + \dfrac{-3}{(x-2)^2}$

CHAPTER 7 ANALYTIC GEOMETRY AND NONLINEAR SYSTEMS

7.1 Exercises *(pages 514–517)*

1. E **3.** H **5.** F **7.** D

9. $(x - 1)^2 + (y - 4)^2 = 9$ **11.** $x^2 + y^2 = 1$

13. $\left(x - \frac{2}{3}\right)^2 + \left(y + \frac{4}{5}\right)^2 = \frac{9}{49}$

15. $(x + 1)^2 + (y - 2)^2 = 25$

17. $(x + 3)^2 + (y + 2)^2 = 4$

19. The graph is the point $(3, 3)$.

21. $(x - 2)^2 + (y + 3)^2 = 45$

23. $(x + 2)^2 + (y + 3)^2 = 25$

25. $x^2 + y^2 = 25$ **27.** In a circle, the radius is the distance from the center to ***any*** point on the circle.

29. domain: $[-2, 2]$; range: $[-2, 2]$

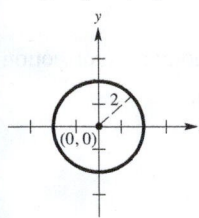

$x^2 + y^2 = 4$

31. domain: $\{0\}$; range: $\{0\}$

$(0, 0)$

$x^2 + y^2 = 0$

33. domain: $[-4, 8]$; range: $[-6, 6]$

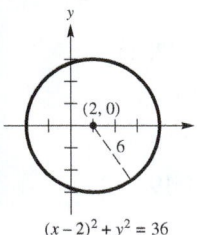

$(x - 2)^2 + y^2 = 36$

35. domain: $[-2, 12]$; range: $[-11, 3]$

$(5, -4)$

$(x - 5)^2 + (y + 4)^2 = 49$

37. domain: $[-9, 3]$; range: $[-8, 4]$

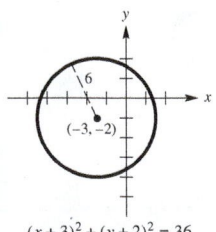

$(x + 3)^2 + (y + 2)^2 = 36$

39. $\varnothing$; $\varnothing$; empty

41. domain: $[-9, 9]$; range: $[-9, 9]$

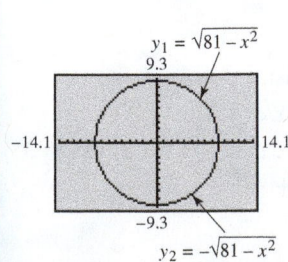

$y_1 = \sqrt{81 - x^2}$

$y_2 = -\sqrt{81 - x^2}$

43. domain: $[-2, 8]$; range: $[-3, 7]$

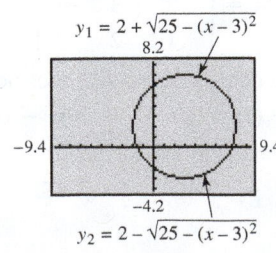

$y_1 = 2 + \sqrt{25 - (x - 3)^2}$

$y_2 = 2 - \sqrt{25 - (x - 3)^2}$

45. yes; center: $(-3, -4)$; radius: 4 **47.** yes; center: $(2, -6)$; radius: 6 **49.** yes; center: $\left(-\frac{1}{2}, 2\right)$; radius: 3 **51.** no **53.** yes; center: $(1, -2)$; radius: $\sqrt{5}$ **55.** yes; center: $(-2, 0)$; radius: $\frac{2}{3}$ **57.** D **59.** C **61.** F

63. E **65. (a)** III **(b)** II **(c)** IV **(d)** I

67. $(0, 4)$; $y = -4$; y-axis **69.** $\left(0, -\frac{1}{8}\right)$; $y = \frac{1}{8}$; y-axis

71. $\left(\frac{1}{64}, 0\right)$; $x = -\frac{1}{64}$; x-axis **73.** $(-4, 0)$; $x = 4$; x-axis

75. $x^2 = -8y$ **77.** $y^2 = -2x$ **79.** $y^2 = 4x$

81. $x^2 = -2y$ **83.** $x^2 = -y$ **85.** $x^2 = 4(y - 1)$

87. $y^2 = 4(x + 1)$ **89.** $(x + 1)^2 = -8(y - 5)$

91. $(y - 3)^2 = -\frac{9}{2}(x + 2)$

93. vertex: $(-3, -4)$; axis: $x = -3$; domain: $(-\infty, \infty)$; range: $[-4, \infty)$

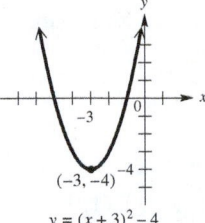

$y = (x + 3)^2 - 4$

95. vertex: $(-3, 2)$; axis: $x = -3$; domain: $(-\infty, \infty)$; range: $(-\infty, 2]$

$(-3, 2)$

$y = -2(x + 3)^2 + 2$

97. vertex: $(1, 2)$; axis: $x = 1$; domain: $(-\infty, \infty)$; range: $[2, \infty)$

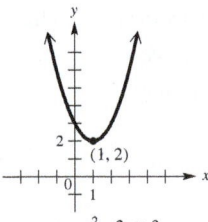

$y = x^2 - 2x + 3$

99. vertex: $(1, 3)$; axis: $x = 1$; domain: $(-\infty, \infty)$; range: $[3, \infty)$

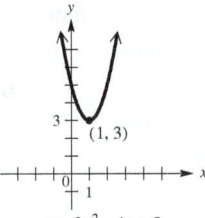

$y = 2x^2 - 4x + 5$

101. vertex: $(2, 0)$; axis: $y = 0$; domain: $[2, \infty)$; range: $(-\infty, \infty)$

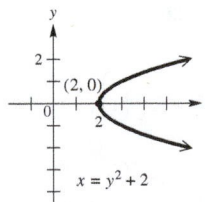

$x = y^2 + 2$

103. vertex: $(0, 3)$; axis: $y = 3$; domain: $[0, \infty)$; range: $(-\infty, \infty)$

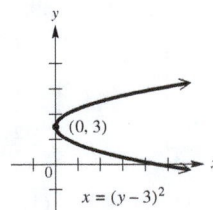

$x = (y - 3)^2$

105. vertex: (2, 4); axis:
$y = 4$; domain: $[2, \infty)$;
range: $(-\infty, \infty)$

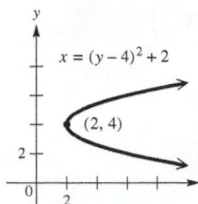

107. vertex: (2, 3); axis:
$y = 3$; domain: $[2, \infty)$;
range: $(-\infty, \infty)$

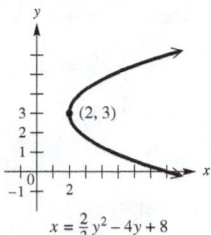
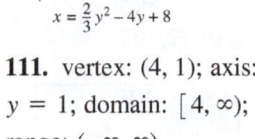

11. domain: $[-3, 3]$; range:
$[-2, 2]$; foci: $\left(-\sqrt{5}, 0\right)$,
$\left(\sqrt{5}, 0\right)$

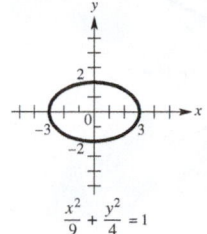

13. domain: $\left[-\sqrt{6}, \sqrt{6}\right]$;
range: $[-3, 3]$; foci:
$\left(0, -\sqrt{3}\right), \left(0, \sqrt{3}\right)$

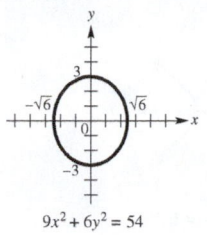

109. vertex: $\left(-2, -\frac{1}{2}\right)$; axis:
$y = -\frac{1}{2}$; domain: $(-\infty, -2]$;
range: $(-\infty, \infty)$

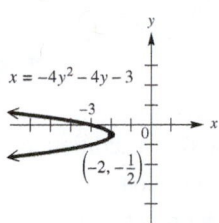

111. vertex: (4, 1); axis:
$y = 1$; domain: $[4, \infty)$;
range: $(-\infty, \infty)$

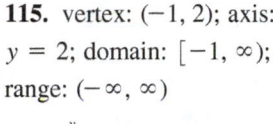

15. domain: $\left[-\frac{3}{8}, \frac{3}{8}\right]$;
range: $\left[-\frac{6}{5}, \frac{6}{5}\right]$

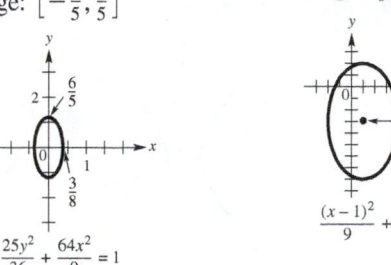

17. domain: $[-2, 4]$;
range: $[-8, 2]$

113. vertex: (4, 1); axis:
$y = 1$; domain: $[4, \infty)$;
range: $(-\infty, \infty)$

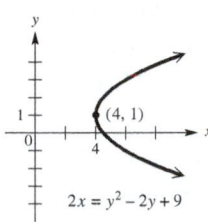

115. vertex: (−1, 2); axis:
$y = 2$; domain: $[-1, \infty)$;
range: $(-\infty, \infty)$

19. domain: $[-2, 6]$;
range: $[-2, 4]$

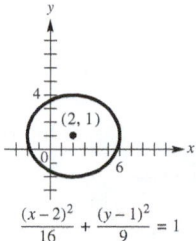

21. domain: $[-9, 7]$;
range: $[-5, 9]$

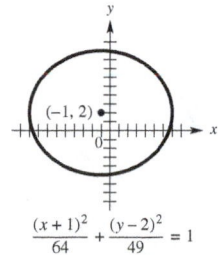

119. (a)

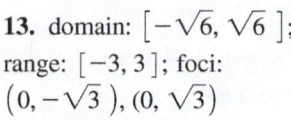

(b) Mars: approximately
229 ft; moon:
approximately 555 ft

121. 4×10^{-17} m downward   **123.** 6 ft

7.2 Exercises *(pages 526–530)*

1. G   **3.** F   **5.** E   **7.** D   **9.** A circle can be
interpreted as an ellipse whose two foci have the same
coordinates; the "coinciding foci" give the center of the circle.

23. $\frac{x^2}{16} + \frac{y^2}{12} = 1$ **25.** $\frac{x^2}{4} + \frac{y^2}{8} = 1$ **27.** $\frac{x^2}{16} + \frac{y^2}{4} = 1$

29. $\frac{x^2}{36} + \frac{y^2}{20} = 1$ **31.** $\frac{(x-3)^2}{16} + \frac{(y+2)^2}{25} = 1$

33. $\frac{x^2}{5} + \frac{y^2}{9} = 1$ **35.** $\frac{(x-5)^2}{25} + \frac{(y-2)^2}{16} = 1$

37. $\frac{(x-4)^2}{9} + \frac{(y-5)^2}{16} = 1$ **39.** $\frac{(x+1)^2}{4} + \frac{(y-1)^2}{9} = 1$;
center: (−1, 1); vertices: (−1, −2), (−1, 4)

41. $\frac{(x+1)^2}{1} + \frac{(y+1)^2}{4} = 1$; center: (−1, −1);
vertices: (−1, −3), (−1, 1)   **43.** $\frac{(x+2)^2}{5} + \frac{(y-1)^2}{4} = 1$;
center: (−2, 1); vertices: $\left(-2 - \sqrt{5}, 1\right), \left(-2 + \sqrt{5}, 1\right)$

45. $\frac{\left(x - \frac{1}{2}\right)^2}{4} + \frac{\left(y + \frac{3}{2}\right)^2}{16} = 1$; center: $\left(\frac{1}{2}, -\frac{3}{2}\right)$;
vertices: $\left(\frac{1}{2}, \frac{5}{2}\right), \left(\frac{1}{2}, -\frac{11}{2}\right)$

47. domain: $(-\infty, -4] \cup [4, \infty)$; range: $(-\infty, \infty)$

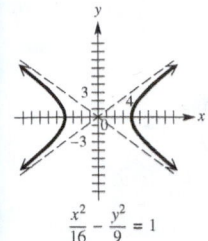

$$\frac{x^2}{16} - \frac{y^2}{9} = 1$$

49. domain: $(-\infty, \infty)$; range: $(-\infty, -6] \cup [6, \infty)$

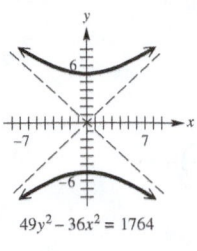

$49y^2 - 36x^2 = 1764$

51. domain: $\left(-\infty, -\frac{3}{2}\right] \cup \left[\frac{3}{2}, \infty\right)$; range: $(-\infty, \infty)$

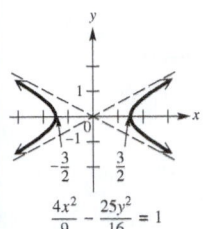

$$\frac{4x^2}{9} - \frac{25y^2}{16} = 1$$

53. domain: $\left(-\infty, -\frac{1}{3}\right] \cup \left[\frac{1}{3}, \infty\right)$; range: $(-\infty, \infty)$

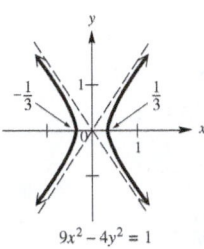

$9x^2 - 4y^2 = 1$

55. domain: $(-\infty, -2] \cup [4, \infty)$; range: $(-\infty, \infty)$; center: $(1, -3)$

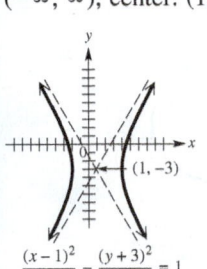

$$\frac{(x-1)^2}{9} - \frac{(y+3)^2}{25} = 1$$

57. domain: $(-\infty, \infty)$; range: $(-\infty, 3] \cup [7, \infty)$; center: $(-1, 5)$

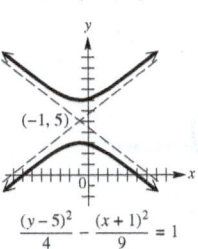

$$\frac{(y-5)^2}{4} - \frac{(x+1)^2}{9} = 1$$

59. domain: $\left(-\infty, -\frac{21}{4}\right] \cup \left[-\frac{19}{4}, \infty\right)$; range: $(-\infty, \infty)$; center: $(-5, 3)$

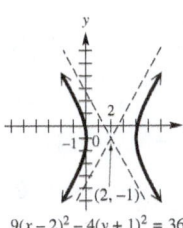

$16(x+5)^2 - (y-3)^2 = 1$

61. domain: $(-\infty, 0] \cup [4, \infty)$; range: $(-\infty, \infty)$; center: $(2, -1)$

$9(x-2)^2 - 4(y+1)^2 = 36$

63. $\frac{x^2}{9} - \frac{y^2}{7} = 1$ **65.** $\frac{y^2}{9} - \frac{x^2}{25} = 1$ **67.** $\frac{y^2}{36} - \frac{x^2}{144} = 1$

69. $\frac{x^2}{9} - 3y^2 = 1$ **71.** $\frac{2y^2}{25} - 2x^2 = 1$

73. $\frac{(y-3)^2}{4} - \frac{49(x-4)^2}{4} = 1$ **75.** $\frac{(x-1)^2}{4} - \frac{(y+2)^2}{5} = 1$

77. $\frac{(x-1)^2}{4} - \frac{(y-1)^2}{4} = 1$; center: $(1, 1)$; vertices: $(-1, 1), (3, 1)$ **79.** $\frac{(y+4)^2}{2} - \frac{(x-3)^2}{3} = 1$; center: $(3, -4)$; vertices: $\left(3, -4 - \sqrt{2}\right), \left(3, -4 + \sqrt{2}\right)$

81. $\frac{(x-3)^2}{2} - \frac{(y-0)^2}{1} = 1$; center: $(3, 0)$; vertices: $\left(3 - \sqrt{2}, 0\right), \left(3 + \sqrt{2}, 0\right)$

83. $\frac{(y+4)^2}{5} - \frac{(x+1)^2}{4} = 1$; center: $(-1, -4)$; vertices: $\left(-1, -4 - \sqrt{5}\right), \left(-1, -4 + \sqrt{5}\right)$

85. $(-2, 0), (2, 0)$

86. 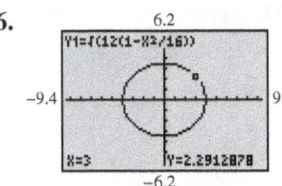 In addition to $(3, 2.2912878)$ shown on the screen, other points are $(0, 3.4641016)$ and $(-3, -2.291288)$.

87. The points satisfy the equation.

88. 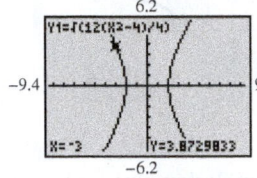 $(-4, 0), (4, 0)$; In addition to the point shown on the screen, other points are $(-2, 0), (2, 0)$, and $(4, 6)$.

89. The points satisfy the equation.

90. **Exercise 87** demonstrates that the points on the graph satisfy the definition of the ellipse for that particular ellipse. **Exercise 89** demonstrates similarly that the definition of the hyperbola is satisfied for that hyperbola.

91. $\frac{x^2}{100} + \frac{y^2}{64} = 1$ **93.** 348.2 ft   **95.** just under 12 ft

97. (a)

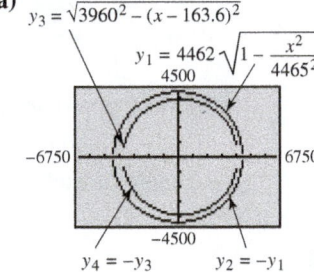

(b) maximum: approximately 669 mi; minimum: approximately 341 mi

99. (a) $x = \sqrt{y^2 + 2.5 \times 10^{-27}}$ **(b)** 1.2×10^{-13} m

101. $\frac{x^2}{25} + \frac{y^2}{16} = 1$

Reviewing Basic Concepts *(page 530)*

1. (a) B   **(b)** D   **(c)** A   **(d)** C

2.

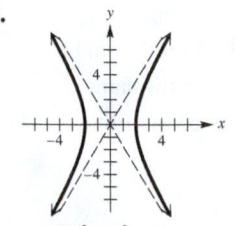

$12x^2 - 4y^2 = 48$

3.

$y = 2x^2 + 3x - 1$

4.

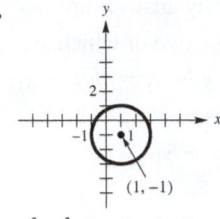

$x^2 + y^2 - 2x + 2y - 2 = 0$

5.

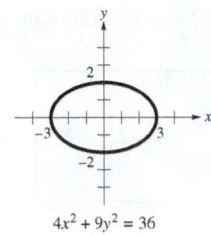

$4x^2 + 9y^2 = 36$

6. If $c < a$, it is an ellipse, and if $c > a$, it is a hyperbola.

7. $(x - 2)^2 + (y + 1)^2 = 9$ **8.** $\frac{x^2}{36} + \frac{y^2}{20} = 1$

9. $\frac{y^2}{4} - \frac{x^2}{12} = 1$ **10.** $x^2 = 2y$

7.3 Exercises *(pages 538–541)*

1. circle   **3.** parabola   **5.** parabola   **7.** ellipse

9. hyperbola   **11.** hyperbola   **13.** ellipse   **15.** circle

17. parabola   **19.** point   **21.** parabola

23. contains no points

25. parabola

27. hyperbola

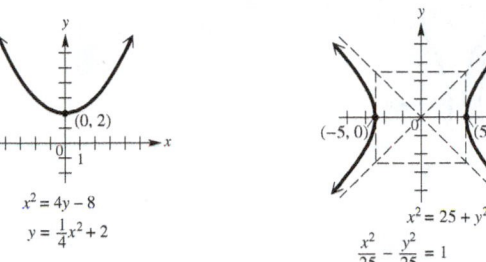

$x^2 = 4y - 8$
$y = \frac{1}{4}x^2 + 2$

$x^2 = 25 + y^2$
$\frac{x^2}{25} - \frac{y^2}{25} = 1$

29. contains no points

31. parabola

33. circle

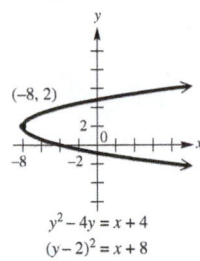

$y^2 - 4y = x + 4$
$(y - 2)^2 = x + 8$

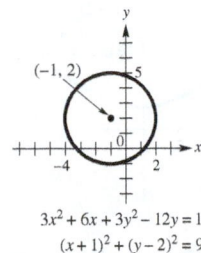

$3x^2 + 6x + 3y^2 - 12y = 12$
$(x + 1)^2 + (y - 2)^2 = 9$

35. ellipse

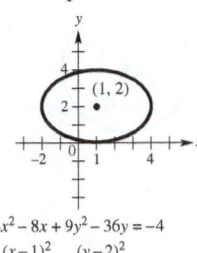

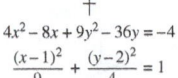

$4x^2 - 8x + 9y^2 - 36y = -4$
$\frac{(x-1)^2}{9} + \frac{(y-2)^2}{4} = 1$

37. ellipse   **39.** hyperbola

41. $\frac{1}{2}$ **43.** $\sqrt{2}$

45. $\frac{\sqrt{21}}{7}$ **47.** $\frac{\sqrt{10}}{3}$

49. $x^2 = 32y$

51. $\frac{x^2}{36} + \frac{y^2}{27} = 1$

53. $\frac{x^2}{36} - \frac{y^2}{108} = 1$

55. $x^2 = -4y$

57. $\frac{25x^2}{81} + \frac{y^2}{9} = 1$ **59.** $\frac{1}{3}$ **61.** 1   **63.** 1.5

65. $\{(3, 1), (3, -1), (-3, 1), (-3, -1)\}$

67. $\{(3, 0), (-3, 0)\}$ **69.** $\left\{\left(x, \pm \sqrt{10 - x^2}\right)\right\}$

71. $\left\{\left(-\frac{3}{5}, \frac{7}{5}\right), (-1, 1)\right\}$

73. $\{(2, 2), (2, -2), (-2, 2), (-2, -2)\}$

75. $\{(1, 1), (-1, -1)\}$

77. $\left\{\left(\sqrt{5}, 0\right), \left(-\sqrt{5}, 0\right), \left(\sqrt{5}, \sqrt{5}\right), \left(-\sqrt{5}, -\sqrt{5}\right)\right\}$

79. $\{(1, -1, 0), (0, 0, -1)\}$

81.

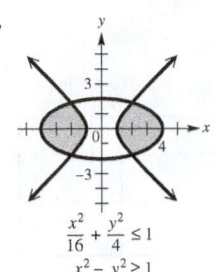

$\frac{x^2}{16} + \frac{y^2}{4} \le 1$
$x^2 - y^2 \ge 1$

83.

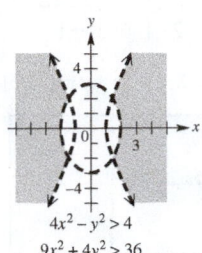

$4x^2 - y^2 > 4$
$9x^2 + 4y^2 > 36$

85.

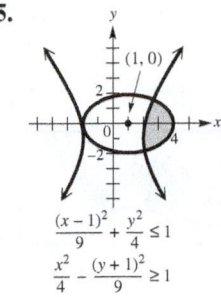

$\frac{(x-1)^2}{9} + \frac{y^2}{4} \le 1$
$\frac{x^2}{4} - \frac{(y+1)^2}{9} \ge 1$

87. 0.093

89. **(a)** $v_{\max} \approx 30.3$ km per sec; $v_{\min} \approx 29.3$ km per sec
(b) The minimum and maximum velocities are equal. Therefore, the planet's velocity is constant.   **(c)** A planet is at its maximum and minimum distances from a focus when it is located at the vertices of the ellipse. Thus, the minimum and maximum velocities of a planet will occur at the vertices of the elliptical orbit, which are $a + c$ for the minimum and $a - c$ for the maximum.   **91.** 0.0172

7.4 Exercises *(pages 544–545)*

1.

t	x	y
-2	-3	-4
-1	-1	-3
0	1	-2
1	3	-1
2	5	0

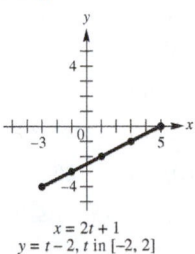

$x = 2t + 1$
$y = t - 2, t$ in $[-2, 2]$

3.

t	x	y
-2	-1	3
-1	0	0
0	1	-1
1	2	0
2	3	3

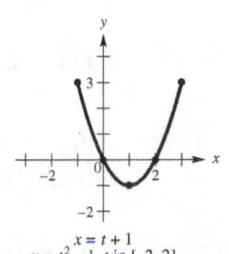

$x = t + 1$
$y = t^2 - 1, t$ in $[-2, 2]$

5.

t	x	y
-2	6	3
-1	3	2
0	2	1
1	3	0
2	6	-1

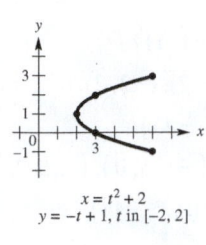

$x = t^2 + 2$
$y = -t + 1, t$ in $[-2, 2]$

7. $x = 2t, y = t + 1, t$ in $[-2, 3]$

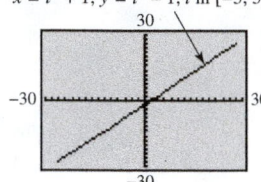

$y = \frac{1}{2}x + 1, x$ in $[-4, 6]$

9. $x = \sqrt{t}, y = 3t - 4, t$ in $[0, 4]$

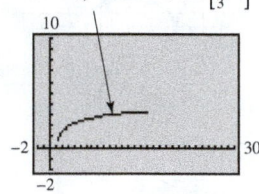

$y = 3x^2 - 4, x$ in $[0, 2]$

11. $x = t^3 + 1, y = t^3 - 1, t$ in $[-3, 3]$

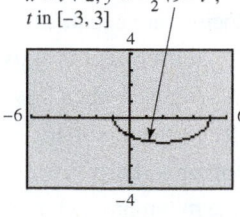

$y = x - 2, x$ in $[-26, 28]$

13. $x = 2^t, y = \sqrt{3t - 1}, t$ in $\left[\frac{1}{3}, 4\right]$

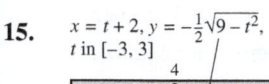

$x = 2^{(y^2+1)/3}, y$ in $\left[0, \sqrt{11}\right]$

15. $x = t + 2, y = -\frac{1}{2}\sqrt{9 - t^2}$, t in $[-3, 3]$

$y = -\frac{1}{2}\sqrt{9 - (x - 2)^2}$,
x in $[-1, 5]$

17. $x = t, y = \frac{1}{t}, t$ in $(-\infty, 0) \cup (0, \infty)$

$y = \frac{1}{x}, x$ in $(-\infty, 0) \cup (0, \infty)$

19. $y = \frac{1}{3}x - 1, x$ in $(-\infty, \infty)$

21. $x = 3(y - 1)^2, y$ in $(-\infty, \infty)$

23. $x = 3\left(\frac{y}{4}\right)^{2/3}, y$ in $(-\infty, \infty)$

25. $y = \sqrt{x^2 + 2}, x$ in $(-\infty, \infty)$

27. $y = \frac{1}{x}, x$ in $(0, \infty)$ **29.** $y = 1 - 2x^2, x$ in $(0, \infty)$

31. $y = \frac{1}{x}, x \neq 0$ **33.** $y = \ln x, x$ in $(0, \infty)$

Other answers are possible for Exercises 35–41.

35. $x = \frac{1}{2}t, y = t + 3; x = \frac{t + 3}{2}, y = t + 6$

37. $x = \frac{1}{3}t, y = \sqrt{t + 2}, t$ in $[-2, \infty); x = \frac{t - 2}{3}, y = \sqrt{t},$
t in $[0, \infty)$ **39.** $x = t^3 + 1, y = t; x = t, y = \sqrt[3]{t - 1}$

41. $x = \sqrt{t + 1}, y = t, t$ in $[-1, \infty); x = \sqrt{t}, y = t - 1,$
t in $[0, \infty)$ **43. (a)** 17.7 sec **(b)** 5000 ft **(c)** 1250 ft

45.

$x = 60t, y = 80t - 16t^2,$
t in $[0, 5]$

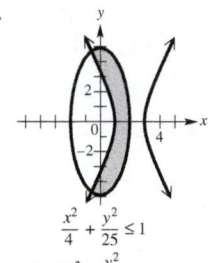

$y = \frac{4}{3}x - \frac{1}{225}x^2$

49. Many answers are possible, two of which are

$x = t, y - y_1 = m(t - x_1)$
and $x = t + x_1,$
$y = mt + y_1.$

Reviewing Basic Concepts *(page 546)*

1. $\frac{(x - 1)^2}{4} + \frac{(y + 3)^2}{12} = 1$; ellipse

2. $\frac{(y + 4)^2}{12} - \frac{(x + 2)^2}{6} = 1$; hyperbola

3. $(y + 2)^2 = -\frac{5}{3}(x - 3)$; parabola

4. $\frac{2\sqrt{6}}{5}$ **5.** $\sqrt{3}$ **6.** $\frac{1}{2}$ **7.** $y^2 = -8x$

8. $\frac{x^2}{25} + \frac{y^2}{16} = 1$ **9.** $\frac{y^2}{16} - \frac{x^2}{9} = 1$ **10.** $\frac{x^2}{9} - \frac{y^2}{4} = 1$

11. $\{(0, -3), (0, 3)\}$

12.

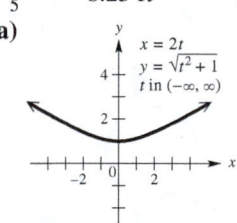

$\frac{x^2}{4} + \frac{y^2}{25} \leq 1$

$(x - 2)^2 - \frac{y^2}{4} \leq 1$

13. $\frac{9\sqrt{21}}{5} \approx 8.25$ ft

14. (a)

$x = 2t$
$y = \sqrt{t^2 + 1}$
t in $(-\infty, \infty)$

(b) $y = \sqrt{\frac{x^2}{4} + 1}$

Chapter 7 Review Exercises *(pages 548–551)*

1. $(x + 2)^2 + (y - 3)^2 = 25$;
domain: $[-7, 3]$; range: $[-2, 8]$

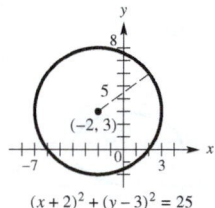

$(x + 2)^2 + (y - 3)^2 = 25$

3. $(x + 8)^2 + (y - 1)^2 = 289$;
domain: $[-25, 9]$;
range: $[-16, 18]$

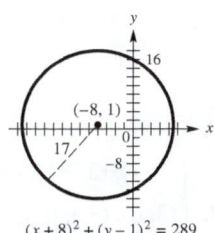

$(x + 8)^2 + (y - 1)^2 = 289$

5. $(2, -3); 1$ **7.** $\left(-\frac{7}{2}, -\frac{3}{2}\right); \frac{3\sqrt{6}}{2}$
9. The graph consists of the single point $(4, 5)$.

11. 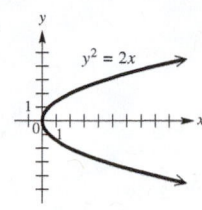 **13.**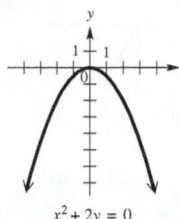

$\left(\frac{1}{2}, 0\right); x = -\frac{1}{2};$ x-axis;
domain: $[0, \infty)$;
range: $(-\infty, \infty)$

$\left(0, -\frac{1}{2}\right); y = \frac{1}{2};$ y-axis;
domain: $(-\infty, \infty)$;
range: $(-\infty, 0]$

15. $y^2 = \frac{25}{2}x$ **17.** $x^2 = -12y$
19. $(y - 6)^2 = 28(x + 5)$

21. 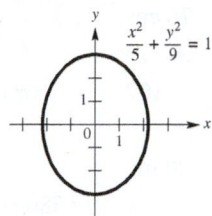 **23.**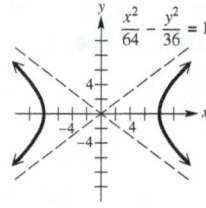

$[-\sqrt{5}, \sqrt{5}]; [-3, 3];$
$(0, -3), (0, 3)$

$(-\infty, -8] \cup [8, \infty);$
$(-\infty, \infty); (-8, 0), (8, 0)$

25. 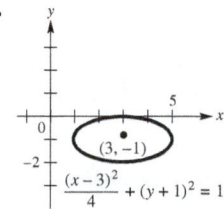 **27.**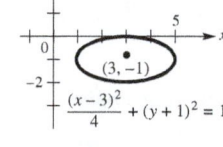

$[1, 5]; [-2, 0];$
$(1, -1), (5, -1)$

$(-\infty, \infty); (-\infty, -4] \cup [0, \infty);$
$(-3, 0), (-3, -4)$

29. $\frac{x^2}{12} + \frac{y^2}{16} = 1$ **31.** $\frac{y^2}{16} - \frac{x^2}{9} = 1$ **33.** $\frac{4x^2}{45} + \frac{4y^2}{81} = 1$
35. **(a)** $(-1, -3)$ **(b)** 5 **(c)** $y_1 = -3 + \sqrt{25 - (x + 1)^2}$;
$y_2 = -3 - \sqrt{25 - (x + 1)^2}$ **37.** E **39.** C **41.** F
43. $\frac{(x + 2)^2}{2} + \frac{(y - 2)^2}{5} = 1$; center: $(-2, 2)$;
vertices: $\left(-2, 2 - \sqrt{5}\right), \left(-2, 2 + \sqrt{5}\right)$
45. $\frac{(y + 1)^2}{3} - \frac{(x - 1)^2}{4} = 1$; center: $(1, -1)$;
vertices: $\left(1, -1 - \sqrt{3}\right), \left(1, -1 + \sqrt{3}\right)$ **47.** $\frac{\sqrt{5}}{3}$
49. $(y - 2)^2 = 3(x + 3)$ **51.** $\frac{(x - 2)^2}{16} + \frac{y^2}{12} = 1$
53. $\{(2, 0), (-2, 0)\}$

55.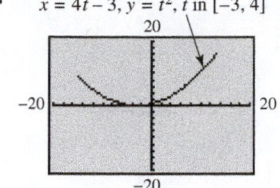

$x = 4t - 3, y = t^2, t \text{ in } [-3, 4]$

57. $y^2 - x^2 = 1, x \text{ in } [0, \infty)$ **59.** 66.8 and 67.7 million mi
61. elliptic **63.** The required increase in velocity is less
when D is larger.

Chapter 7 Test *(pages 551–552)*

1. **(a)** B **(b)** A **(c)** D **(d)** E **(e)** F **(f)** C
2. $\left(\frac{1}{32}, 0\right); x = -\frac{1}{32}$
3.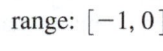

$y = -\sqrt{1 - \frac{x^2}{36}}$

yes; domain: $[-6, 6]$;
range: $[-1, 0]$

4. $y_1 = 7\sqrt{\frac{x^2}{25} - 1}; y_2 = -7\sqrt{\frac{x^2}{25} - 1}$

5. 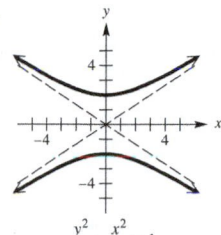 **6.**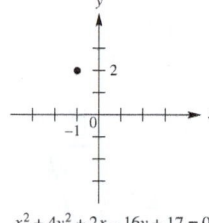

$\frac{y^2}{4} - \frac{x^2}{9} = 1$

$x^2 + 4y^2 + 2x - 16y + 17 = 0$

hyperbola; center: $(0, 0)$;
vertices: $(0, -2), (0, 2)$; foci:
$\left(0, -\sqrt{13}\right), \left(0, \sqrt{13}\right)$

The graph is the point
$(-1, 2)$,

7. 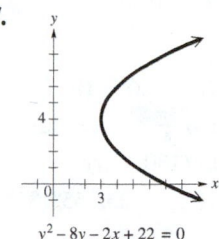 **8.**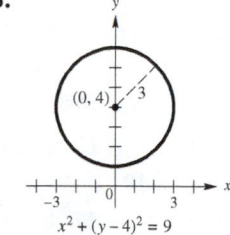

$y^2 - 8y - 2x + 22 = 0$

$x^2 + (y - 4)^2 = 9$

parabola; vertex: $(3, 4)$;
focus: $(3.5, 4)$

circle; center: $(0, 4)$; radius: 3

9.

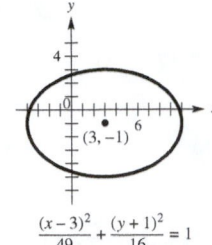

$$\frac{(x-3)^2}{49} + \frac{(y+1)^2}{16} = 1$$

ellipse; center: $(3, -1)$;
vertices: $(-4, -1)$, $(10, -1)$;
foci: $\left(3 + \sqrt{33}, -1\right)$,
$\left(3 - \sqrt{33}, -1\right)$

10.

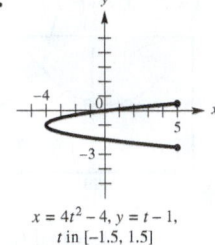

$x = 4t^2 - 4,\ y = t - 1,$
t in $[-1.5, 1.5]$

parabola; vertex: $(-4, -1)$;
focus: $\left(-\frac{63}{16}, -1\right)$

11. (a) $y = -\frac{1}{8}x^2$ **(b)** $\frac{x^2}{\frac{11}{4}} + \frac{y^2}{9} = 1$, or $\frac{4x^2}{11} + \frac{y^2}{9} = 1$

12. $\frac{x^2}{400} + \frac{y^2}{144} = 1$; approximately 10.39 ft

13. $\{(1, 1), (1, -1), (-1, 1), (-1, -1)\}$

14.

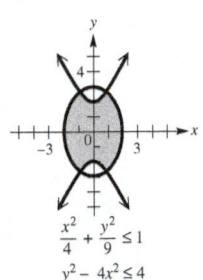

$$\frac{x^2}{4} + \frac{y^2}{9} \le 1$$
$$y^2 - 4x^2 \le 4$$

15. $x = t + \ln t,\ y = t + e^t,\ t$ in $(0, 2]$

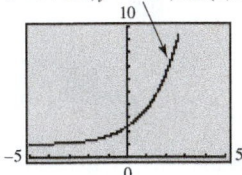

16. $y = \frac{1}{x},\ x \ne 0$

CHAPTER 8 TRIGONOMETRIC FUNCTIONS AND APPLICATIONS

8.1 Exercises *(pages 564–569)*

1. 2π **3.** $180°$ **5.** $s = r\theta$ **7. (a)** $60°$ **(b)** $150°$
9. (a) $45°$ **(b)** $135°$ **11. (a)** $\frac{\pi}{4}$ **(b)** $\frac{3\pi}{4}$ **13. (a)** $\frac{1}{2}$
(b) $\frac{1}{9}$ **(c)** $\frac{1}{360}$ **15. (a)** $(90 - x)°$ **(b)** $(180 - x)°$
17. $150°$ **19.** $7° 30'$ **21.** $70°; 110°$ **23.** $55°; 35°$
25. $80°; 100°$ **27.** $83° 59'$ **29.** $-23° 49'$
31. $17° 1' 49''$ **33.** $20.9°$ **35.** $91.598°$
37. $31° 25' 47''$ **39.** $89° 54' 1''$

Angles other than those given are possible in Exercises 41–45.

41.

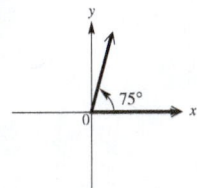

$435°; -285°;$ quadrant I

43.

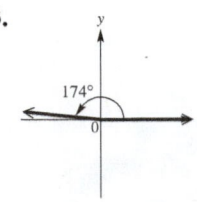

$534°; -186°;$ quadrant II

45.

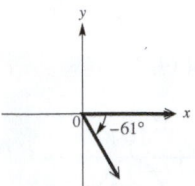

$299°; -421°;$ quadrant IV

47. $320°$ **49.** $90°$ **51.** $\frac{7\pi}{4}$ **53.** $\frac{\pi}{2}$
55. $30° + n \cdot 360°$ **57.** $-90° + n \cdot 360°$
59. $\frac{\pi}{4} + 2n\pi$ **61.** $-\frac{3\pi}{4} + 2n\pi$ **63.** $\frac{\pi}{3}$ **65.** $\frac{5\pi}{6}$
67. $-\frac{\pi}{4}$ **69.** $60°$ **71.** $315°$ **73.** $330°$ **75.** 0.68
77. 2.43 **79.** 1.12 **81.** $114° 35'$ **83.** $99° 42'$
85. $-74° 29'$ **87.** We begin the answers with the blank
next to $30°$, and then proceed counterclockwise from there:
$\frac{\pi}{6}; 45; \frac{\pi}{3}; 120; 135; \frac{5\pi}{6}; \pi; \frac{7\pi}{6}; \frac{5\pi}{4}; 240; 300; \frac{7\pi}{4}; \frac{11\pi}{6}$.
89. 2π **91.** 8 **93.** 1 **95.** 25.76 cm
97. 5.05 m **99.** 1200 km **101.** 5900 km
103. $\frac{3\pi}{32}$ radian per sec **105.** $\frac{6}{5}$ min **107.** $s = 18\pi$ cm
109. $t = 12$ sec **111. (a)** 11.6 in. **(b)** $37° 5'$
113. $38.5°$ **115.** 146 in. **117.** 1120 m^2
119. 114 cm^2 **121.** 6π **123.** approximately 3800 mi^2
125. 75 in.^{2} **127. (a)** $13.85°$ **(b)** 76 m^2
129. $16\frac{1}{4}$ ft per sec, or about 11.1 mph
131. 235 radians per sec **133. (a)** $\frac{2\pi}{365}$ radian
(b) $\frac{\pi}{4380}$ radian per hr **(c)** $66,700$ mph
135. $\frac{500\pi}{3}$ radians per sec; $\frac{2500\pi}{3}$ in. per sec
137. radius: 3947 mi; circumference: $24,800$ mi

8.2 Exercises *(pages 579–581)*

In Exercises 1 and 5–33 we give, in order, sine, cosine, tangent, cotangent, secant, and cosecant.

1. $\frac{12}{13}; \frac{5}{13}; \frac{12}{5}; \frac{5}{12}; \frac{13}{5}; \frac{13}{12}$ **3.** They are the same for each
trigonometric function. They are coterminal angles.

5.

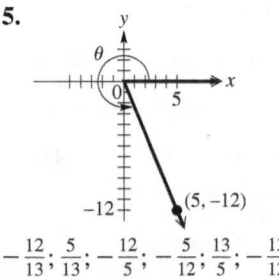

$-\frac{12}{13}; \frac{5}{13}; -\frac{12}{5}; -\frac{5}{12}; \frac{13}{5}; -\frac{13}{12}$

7.

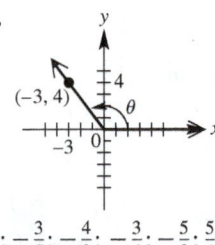

$\frac{4}{5}; -\frac{3}{5}; -\frac{4}{3}; -\frac{3}{4}; -\frac{5}{3}; \frac{5}{4}$

9.

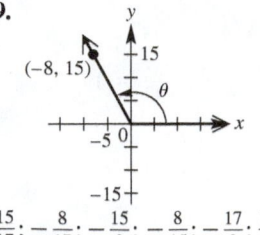

$\frac{15}{17}; -\frac{8}{17}; -\frac{15}{8}; -\frac{8}{15}; -\frac{17}{8}; \frac{17}{15}$

11.

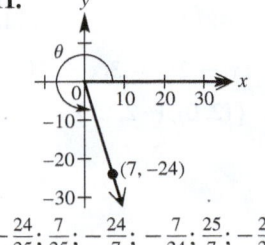

$-\frac{24}{25}; \frac{7}{25}; -\frac{24}{7}; -\frac{7}{24}; \frac{25}{7}; -\frac{25}{24}$

13.

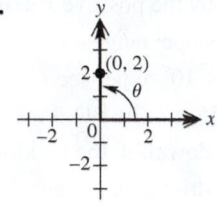

1; 0; undefined; 0;
undefined; 1

15.

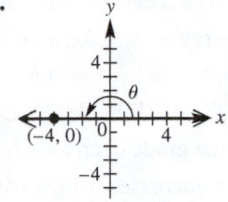

0; −1; 0; undefined;
−1; undefined

17.

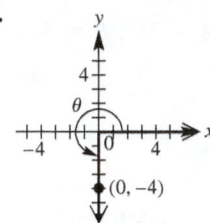

−1; 0; undefined; 0;
undefined; −1

19.

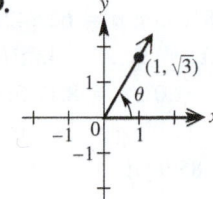

$\frac{\sqrt{3}}{2}$; $\frac{1}{2}$; $\sqrt{3}$; $\frac{\sqrt{3}}{3}$; 2; $\frac{2\sqrt{3}}{3}$

21.

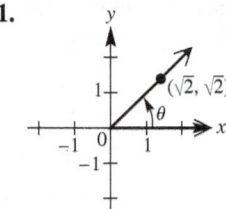

$\frac{\sqrt{2}}{2}$; $\frac{\sqrt{2}}{2}$; 1; 1; $\sqrt{2}$; $\sqrt{2}$

23.

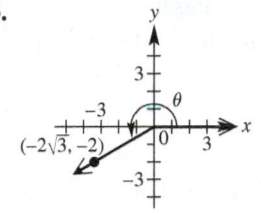

$-\frac{1}{2}$; $-\frac{\sqrt{3}}{2}$; $\frac{\sqrt{3}}{3}$; $\sqrt{3}$; $-\frac{2\sqrt{3}}{3}$; −2

25.

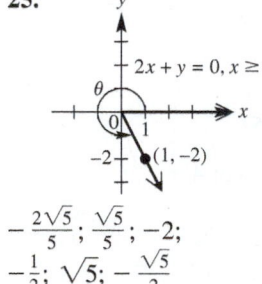

$-\frac{2\sqrt{5}}{5}$; $\frac{\sqrt{5}}{5}$; −2;
$-\frac{1}{2}$; $\sqrt{5}$; $-\frac{\sqrt{5}}{2}$

27.

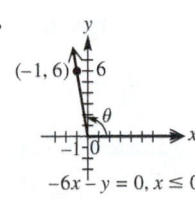

$\frac{6\sqrt{37}}{37}$; $-\frac{\sqrt{37}}{37}$; −6;
$-\frac{1}{6}$; $-\sqrt{37}$; $\frac{\sqrt{37}}{6}$

29.

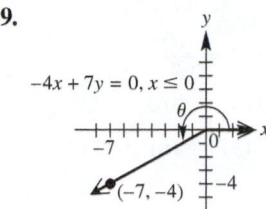

$-\frac{4\sqrt{65}}{65}$; $-\frac{7\sqrt{65}}{65}$; $\frac{4}{7}$;
$\frac{7}{4}$; $-\frac{\sqrt{65}}{7}$; $-\frac{\sqrt{65}}{4}$

31.

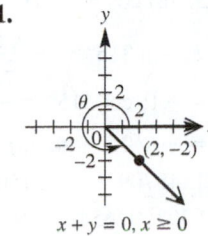

$-\frac{\sqrt{2}}{2}$; $\frac{\sqrt{2}}{2}$; −1;
−1; $\sqrt{2}$; $-\sqrt{2}$

33.

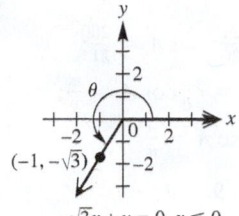

$-\sqrt{3}x + y = 0, x \le 0$

$-\frac{\sqrt{3}}{2}$; $-\frac{1}{2}$; $\sqrt{3}$; $\frac{\sqrt{3}}{3}$; −2; $-\frac{2\sqrt{3}}{3}$

35. 0   **37.** 0   **39.** −1   **41.** 1   **43.** undefined
45. −1   **47.** 0   **49.** undefined   **51.** 1   **53.** $\frac{3}{2}$
55. $-\frac{7}{3}$ **57.** $\frac{1}{5}$ **59.** $-\frac{2}{5}$ **61.** $\frac{\sqrt{2}}{2}$ **63.** −0.4
65. 0   **67.** −1   **69.** I, II   **71.** I   **73.** II
75. I   **77.** III   **79.** III, IV
81. The functions in **Exercise 75** are the reciprocals of those
in **Exercise 71,** and within a given quadrant, reciprocal function values have the same sign.
83. impossible   **85.** possible   **87.** possible
89. impossible   **91.** possible   **93.** possible
95. possible   **97.** impossible

*In Exercises 99–107 we give, in order, sine, cosine, tangent,
cotangent, secant, and cosecant.*

99. $\frac{15}{17}$; $-\frac{8}{17}$; $-\frac{15}{8}$; $-\frac{8}{15}$; $-\frac{17}{8}$; $\frac{17}{15}$

101. $\frac{\sqrt{5}}{7}$; $\frac{2\sqrt{11}}{7}$; $\frac{\sqrt{55}}{22}$; $\frac{2\sqrt{55}}{5}$; $\frac{7\sqrt{11}}{22}$; $\frac{7\sqrt{5}}{5}$

103. $\frac{8\sqrt{67}}{67}$; $\frac{\sqrt{201}}{67}$; $\frac{8\sqrt{3}}{3}$; $\frac{\sqrt{3}}{8}$; $\frac{\sqrt{201}}{3}$; $\frac{\sqrt{67}}{8}$

105. $\frac{\sqrt{2}}{6}$; $-\frac{\sqrt{34}}{6}$; $-\frac{\sqrt{17}}{17}$; $-\sqrt{17}$; $-\frac{3\sqrt{34}}{17}$; $3\sqrt{2}$

107. $\frac{\sqrt{15}}{4}$; $-\frac{1}{4}$; $-\sqrt{15}$; $-\frac{\sqrt{15}}{15}$; −4; $\frac{4\sqrt{15}}{15}$

109. $\sqrt{1 - \sin^2\theta}$ **111.** $-\frac{1}{\sqrt{1 + \cot^2\theta}}$,   or $-\frac{\sqrt{1 + \cot^2\theta}}{1 + \cot^2\theta}$

113. $\frac{\sin\theta}{\sqrt{1 - \sin^2\theta}}$,   or $\frac{\sin\theta\sqrt{1 - \sin^2\theta}}{1 - \sin^2\theta}$

117. false; For example,
$\sin 30° + \cos 30° \approx 0.5 + 0.8660 = 1.3660 \ne 1$.

119. (a) 3%   **(b)** $25,000\left(\frac{3}{\sqrt{10,009}}\right) \approx$ 750 lb

Reviewing Basic Concepts *(page 581)*

1. (a) complement: 55°; supplement: 145°   **(b)** complement: $\frac{\pi}{4}$;
supplement: $\frac{3\pi}{4}$ **2.** 32° 15′ 0″   **3.** 59.591$\overline{6}$°

4. (a) 200°   **(b)** $\frac{4\pi}{3}$ **5. (a)** $\frac{4\pi}{3}$ **(b)** 135°   **6. (a)** 2π cm

(b) 3π cm²   **7.** $\sin\theta = \frac{5\sqrt{29}}{29}$; $\cos\theta = -\frac{2\sqrt{29}}{29}$;
$\tan\theta = -\frac{5}{2}$; $\cot\theta = -\frac{2}{5}$; $\sec\theta = -\frac{\sqrt{29}}{2}$; $\csc\theta = \frac{\sqrt{29}}{5}$

8. $\sin 270° = -1$; $\cos 270° = 0$; $\tan 270°$ is undefined;
$\cot 270° = 0$; $\sec 270°$ is undefined; $\csc 270° = -1$

9. (a) impossible   **(b)** possible   **(c)** possible   **(d)** impossible

10. $\cos\theta = -\frac{\sqrt{5}}{3}$; $\tan\theta = \frac{2\sqrt{5}}{5}$; $\cot\theta = \frac{\sqrt{5}}{2}$;
$\sec\theta = -\frac{3\sqrt{5}}{5}$; $\csc\theta = -\frac{3}{2}$

8.3 Exercises *(pages 590–593)*

1. $\sin A = \frac{21}{29}$; $\cos A = \frac{20}{29}$; $\tan A = \frac{21}{20}$; $\cot A = \frac{20}{21}$;

$\sec A = \frac{29}{20}$; $\csc A = \frac{29}{21}$ **3.** $\sin A = \frac{n}{p}$; $\cos A = \frac{m}{p}$;

$\tan A = \frac{n}{m}$; $\cot A = \frac{m}{n}$; $\sec A = \frac{p}{m}$; $\csc A = \frac{p}{n}$

5. $\frac{\sqrt{3}}{3}$; $\sqrt{3}$ **7.** $\frac{\sqrt{3}}{2}$; $\frac{\sqrt{3}}{3}$; $\frac{2\sqrt{3}}{3}$ **9.** -1; -1

11. $-\frac{\sqrt{3}}{2}$; $-\frac{2\sqrt{3}}{3}$

The number of digits in part (b) of Exercises 13–29 may vary.

13. (a) $\frac{\sqrt{3}}{3}$ **(b)** 0.5773502692 **15. (a)** $\frac{1}{2}$ **17. (a)** $\frac{2\sqrt{3}}{3}$

(b) 1.154700538 **19. (a)** $\sqrt{2}$ **(b)** 1.414213562

21. (a) $\frac{\sqrt{2}}{2}$ **(b)** 0.7071067812 **23. (a)** $\frac{\sqrt{3}}{2}$

(b) 0.8660254038 **25. (a)** $\sqrt{3}$ **(b)** 1.732050808

27. (a) 2 **29. (a)** $\frac{2\sqrt{3}}{3}$ **(b)** 1.154700538

31. The value he gave was an *approximation,* not the exact value of sin 45°, which is $\frac{\sqrt{2}}{2}$. **33.** $\tan 17°$ **35.** $\cos 52°$
37. $\cot 64° \, 17'$ **39.** $\sin \frac{3\pi}{10}$ **41.** $\cot\left(\frac{\pi}{2} - 0.5\right)$
43. $\sin\left(\frac{\pi}{2} - 1\right)$ **45.** 82° **47.** 50° **49.** 45°
51. 30° **53.** $\frac{\pi}{3}$ **55.** $\frac{\pi}{3}$ **57.** It is easy to find one-half of 2, which is 1. This is, then, the measure of the side opposite the 30° angle, and the ratios are easily found. Yes, any positive number could have been used.

In Exercises 59–67 we give, in order, sine, cosine, tangent, cotangent, secant, and cosecant.

59. $-\frac{\sqrt{3}}{2}$; $\frac{1}{2}$; $-\sqrt{3}$; $-\frac{\sqrt{3}}{3}$; 2; $-\frac{2\sqrt{3}}{3}$

61. $\frac{\sqrt{2}}{2}$; $\frac{\sqrt{2}}{2}$; 1; 1; $\sqrt{2}$; $\sqrt{2}$

63. $-\frac{1}{2}$; $\frac{\sqrt{3}}{2}$; $-\frac{\sqrt{3}}{3}$; $-\sqrt{3}$; $\frac{2\sqrt{3}}{3}$; -2

65. $\frac{\sqrt{2}}{2}$; $\frac{\sqrt{2}}{2}$; 1; 1; $\sqrt{2}$; $\sqrt{2}$

67. $\frac{1}{2}$; $-\frac{\sqrt{3}}{2}$; $-\frac{\sqrt{3}}{3}$; $-\sqrt{3}$; $-\frac{2\sqrt{3}}{3}$; 2

69. 0.5543090515 **71.** 1.134277349
73. -1.002203376 **75.** -5.729741647
77. 0.5984721441 **79.** -3.380515006
81. (a) $-\sin\frac{\pi}{6}$ **(b)** $-\frac{1}{2}$ **(c)** $\sin\frac{7\pi}{6} = -\sin\frac{\pi}{6} = -0.5$
83. (a) $-\tan\frac{\pi}{4}$ **(b)** -1 **(c)** $\tan\frac{3\pi}{4} = -\tan\frac{\pi}{4} = -1$
85. (a) $-\cos\frac{\pi}{6}$ **(b)** $-\frac{\sqrt{3}}{2}$
(c) $\cos\frac{7\pi}{6} = -\cos\frac{\pi}{6} \approx -0.8660254038$ **87.** 30°; 150°
89. 120°; 300° **91.** 120°; 300°
93. 46.59388121°; 313.4061188°
95. 24.39257624°; 155.6074238°
97. 41.24818261°; 221.2481826°
99. 0.2095206607; 3.351113314
101. 1.27979966; 4.421392314
103. It represents the distance from the point (x_1, y_1) to the origin. **104.** 60° **105.** 60° **106.** 60°

107. It is a measure of the angle formed by the positive *x*-axis and the ray $y = \sqrt{3}x$, $x \geq 0$. **108.** slope; tangent
109. $\theta = \sin^{-1}\frac{2}{25} \approx 4.6°$ **111.** 2×10^8 m per sec
113. 19° **115.** 48.7° **117. (a)** 155 ft **(b)** 194 ft
(c) As the grade decreases from uphill to downhill, the braking distance increases, which corresponds to driving experience.

8.4 Exercises *(pages 600–604)*

1. $B = 53° \, 40'$; $a = 571$ m; $b = 777$ m **3.** $M = 38.8°$;
$n = 154$ m; $p = 198$ m **5.** $A = 47.9108°$;
$c = 84.816$ cm; $a = 62.942$ cm **7.** $A = 36°$; $B = 54°$;
$c = 12$ ft **9.** $A = 48°$; $B = 42°$; $a = 8.2$ ft
11. $B = 62.0°$; $a = 8.17$ ft; $b = 15.4$ ft
13. $A = 17.0°$; $a = 39.1$ in.; $c = 134$ in.
15. $c = 85.9$ yd; $A = 62° \, 50'$; $B = 27° \, 10'$
17. The other acute angle requires the least work to find.
19. 38.6598° **21.** Because *AD* and *BC* are parallel, angle *DAB* is congruent to angle *ABC*, as they are alternate interior angles of the transversal *AB*. (A theorem of elementary geometry assures us of this.) **23.** It is measured clockwise from the north. **25.** 9.35 m **27.** 13.3 ft **29.** 128 ft
31. 26.3°, or 26° 20′ **33.** $A = 35.987°$, or 35° 59′ 10″;
$B = 54.013°$, or 54° 00′ 50″ **35.** 114 ft **37.** 5.18 m
39. 84.7 m **41.** 148 mi **43.** 1.48 mi **45.** 150 km
47. (a) 9.524 mi **(b)** 11.59 mi **49.** 33.4 m **51.** 583 ft
53. (a) 23.4 ft **(b)** 48.3 ft **(c)** The faster the speed, the more land needs to be cleared inside the curve.
55. (a) $\tan\theta = \frac{y}{x}$ **(b)** $x = \frac{y}{\tan\theta}$ **57.** $a = 12$; $b = 12\sqrt{3}$;
$d = 12\sqrt{3}$; $c = 12\sqrt{6}$ **59.** $m = \frac{7\sqrt{3}}{3}$; $a = \frac{14\sqrt{3}}{3}$;
$n = \frac{14\sqrt{3}}{3}$; $q = \frac{14\sqrt{6}}{3}$ **61.** $\mathcal{A} = \frac{s^2\sqrt{3}}{4}$

Reviewing Basic Concepts *(page 605)*

1. $\sin A = \frac{8}{17}$; $\cos A = \frac{15}{17}$; $\tan A = \frac{8}{15}$; $\cot A = \frac{15}{8}$;
$\sec A = \frac{17}{15}$; $\csc A = \frac{17}{8}$ **2.** Row 1: $\frac{1}{2}$, $\frac{\sqrt{3}}{2}$, $\frac{\sqrt{3}}{3}$, $\sqrt{3}$, $\frac{2\sqrt{3}}{3}$, 2;
Row 2: $\frac{\sqrt{2}}{2}$, $\frac{\sqrt{2}}{2}$, 1, 1, $\sqrt{2}$, $\sqrt{2}$; Row 3: $\frac{\sqrt{3}}{2}$, $\frac{1}{2}$, $\sqrt{3}$, $\frac{\sqrt{3}}{3}$, 2,
$\frac{2\sqrt{3}}{3}$ **3. (a)** $\cos 63°$ **(b)** $\cot\frac{3\pi}{10}$ **4. (a)** 80° **(b)** 5°
(c) $\frac{\pi}{3}$ **5.** $\sin 315° = -\frac{\sqrt{2}}{2}$; $\cos 315° = \frac{\sqrt{2}}{2}$;
$\tan 315° = -1$; $\cot 315° = -1$; $\sec 315° = \sqrt{2}$;
$\csc 315° = -\sqrt{2}$ **6. (a)** 0.725374371 **(b)** 5.67128182
(c) -1.321348709 **7.** 150°; 330°
8. 0.75; 2.391592654 **9.** 34 ft **10.** 19,600 ft

8.5 Exercises *(pages 611–613)*

1. $-\frac{1}{2}$ **3.** -1 **5.** -2 **7.** $-\sqrt{3}$
9. $-\frac{1}{2}$ **11.** $\frac{2\sqrt{3}}{3}$

In Exercises 13–27 we give, in order, sine, cosine, tangent, cotangent, secant, and cosecant.

13. 1; 0; undefined; 0; undefined; 1

15. $-\frac{\sqrt{2}}{2}$; $\frac{\sqrt{2}}{2}$; -1; -1; $\sqrt{2}$; $-\sqrt{2}$

17. 0; -1; 0; undefined; -1; undefined

19. $-\frac{\sqrt{3}}{2}$; $\frac{1}{2}$; $-\sqrt{3}$; $-\frac{\sqrt{3}}{3}$; 2; $-\frac{2\sqrt{3}}{3}$

21. -1; 0; undefined; 0; undefined; -1

23. 0; 1; 0; undefined; 1; undefined

25. $\frac{1}{2}$; $\frac{\sqrt{3}}{2}$; $\frac{\sqrt{3}}{3}$; $\sqrt{3}$; $\frac{2\sqrt{3}}{3}$; 2   **27.** $\frac{\sqrt{2}}{2}$; $\frac{\sqrt{2}}{2}$; 1; 1; $\sqrt{2}$; $\sqrt{2}$

29. 0.5736   **31.** 0.4068   **33.** 1.2065   **35.** 14.3338

37. -1.0460 **39.** -3.8665

In Exercises 41 and 43 we give, in order, sine, cosine, tangent, cotangent, secant, and cosecant.

41. $\frac{\sqrt{2}}{2}$; $\frac{\sqrt{2}}{2}$; 1; 1; $\sqrt{2}$; $\sqrt{2}$ **43.** $-\frac{12}{13}$; $\frac{5}{13}$; $-\frac{12}{5}$; $-\frac{5}{12}$; $\frac{13}{5}$; $-\frac{13}{12}$ **45.** 0.7   **47.** 0.9   **49.** -0.6

51. 4 or 2.3   **53.** 0.8   **55.** 0.2095   **57.** 1.4430

59. 1.4747   **61.** 0.9846   **63.** $(-0.8011, 0.5985)$

65. $(0.4385, -0.8987)$ **67.** I   **69.** II

71. (a) $\frac{1}{2}$ (b) $\frac{\sqrt{3}}{2}$ (c) $\sqrt{3}$ (d) 2 (e) $\frac{2\sqrt{3}}{3}$ (f) $\frac{\sqrt{3}}{3}$

73. $f(7) \approx 14$; On July 1, there are about 14 hr of daylight.

75. (a) 30° (b) 60° (c) 75° (d) 86° (e) 86° (f) 60°

77. full moon: $t = \pm\pi, \pm3\pi, \pm5\pi, \ldots$; new moon: $t = 0, \pm2\pi, \pm4\pi, \ldots$

8.6 Exercises *(pages 625–630)*

1. G   **3.** E   **5.** B   **7.** F   **9.** D   **11.** H

13. B   **15.** F   **17.** B   **19.** C

21. 2   **23.** $\frac{2}{3}$

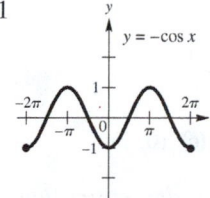

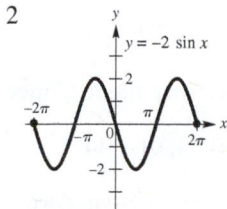

25. 1   **27.** 2

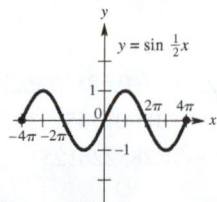

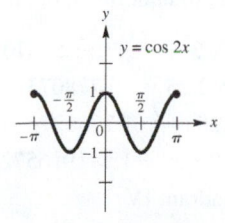

29. 4π; 1   **31.** π; 1

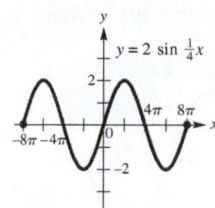 **33.** 8π; 2

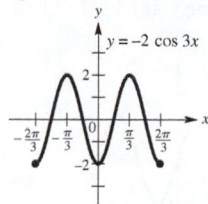 **35.** $\frac{2\pi}{3}$; 2

37. $\frac{\pi}{4}$

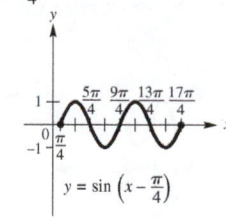

39. $\frac{\pi}{3}$

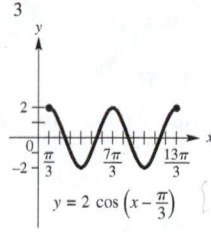

41. $-\pi$

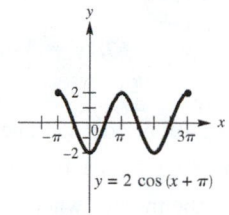

43. $-\frac{\pi}{8}$

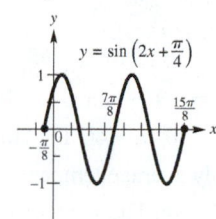

45. (a) 4 (b) π (c) $\frac{\pi}{2}$ (d) none (e) $[-4, 4]$

47. (a) $\frac{1}{2}$ (b) 4π (c) $\frac{\pi}{2}$ (d) none (e) $\left[-\frac{1}{2}, \frac{1}{2}\right]$

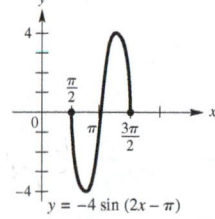

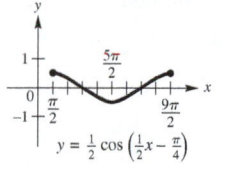

49. (a) $\frac{2}{3}$ (b) $\frac{8\pi}{3}$ (c) none (d) upward 1 unit (e) $\left[\frac{1}{3}, \frac{5}{3}\right]$

51. (a) 2 (b) 4π (c) none (d) upward 1 unit (e) $[-1, 3]$

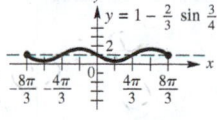

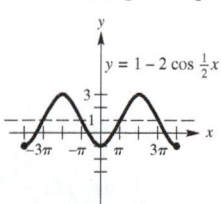

53. (a) 2 (b) 2π (c) $-\frac{\pi}{2}$ (d) downward 3 units (e) $[-5, -1]$

55. (a) 1 (b) π (c) $-\frac{\pi}{4}$ (d) upward $\frac{1}{2}$ unit (e) $\left[-\frac{1}{2}, \frac{3}{2}\right]$

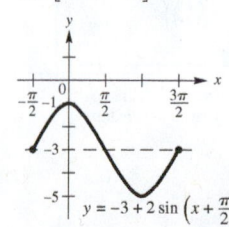

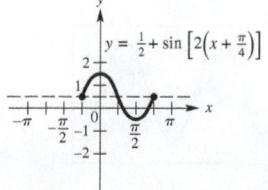

57. (a) 2 **(b)** 2π **(c)** π
(d) none **(e)** $[-2, 2]$

59. (a) 4 **(b)** 4π **(c)** $-\pi$
(d) none **(e)** $[-4, 4]$

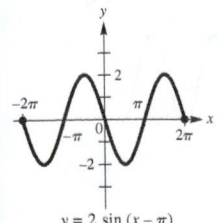

$y = 2 \sin(x - \pi)$

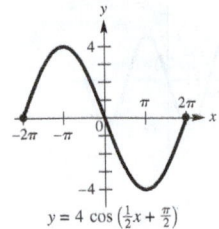

$y = 4 \cos\left(\frac{1}{2}x + \frac{\pi}{2}\right)$

61. (a) 1 **(b)** $\frac{2\pi}{3}$ **(c)** $\frac{\pi}{15}$
(d) upward 2 units **(e)** $[1, 3]$

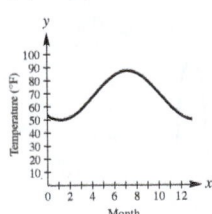

$y = 2 - \sin\left(3x - \frac{\pi}{5}\right)$

63. $y = 2\cos 2x$ **65.** $y = -3\cos\frac{1}{2}x$ **67.** $y = 3\sin 4x$
69. $y = -1 + \sin x$ **71.** $y = \cos\left(x - \frac{\pi}{3}\right)$
73. (a) $40°F$; $-40°F$ **(b)** amplitude: 40; period: 12; The
monthly average temperatures vary by $80°F$ over a 12-month
period. **(c)** The x-intercepts represent the months when the
average temperature is $0°F$.

75. (a) maximum $\approx 87°F$ **(b)** increases
in July; minimum $\approx 62°F$
in January or late December

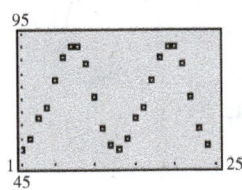

77. 24 hr **79. (a)** 34; 12; 4.3 **(b)** $f(5) \approx 12.2°F$;
$f(12) \approx -26.4°F$ **(c)** about $0°F$

81. (a) $70.4°$; Answers
may vary.

(b)

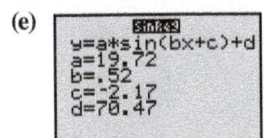

(c) $f(x) = 19.5\cos\left[\frac{\pi}{6}(x - 7.2)\right] + 70.5$

(d) The function gives an
excellent model for the data.

$f(x) = 19.5\cos\left[\frac{\pi}{6}(x - 7.2)\right] + 70.5$

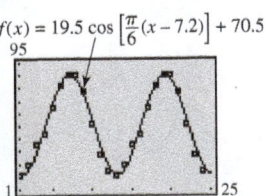

(e)

```
sinReg
y=a*sin(bx+c)+d
a=19.72
b=.52
c=-2.17
d=70.47
```

TI-84 Plus fixed to the
nearest hundredth

83. (a)

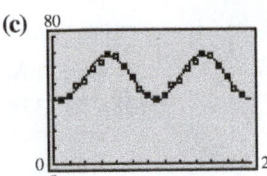

(b) $f(x) = 14\sin\left[\frac{\pi}{6}(x - 4)\right] + 54$

(c)

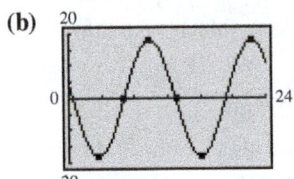

85. (a) $f(x) = 17\cos\left[\frac{\pi}{6}(x - 7)\right] + 75$ **(b)** Yes. The
period of the graph is 12, so, for example, the phase shift
could be $c = 7 + 12 = 19$ or $c = 7 - 12 = -5$.
87. (a) $f(x) = 18\cos\left[\frac{\pi}{6.2}(x - 9.8)\right]$

(b)

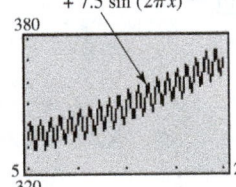

The tide oscillates between flowing into the canal and flowing
out of the canal. When the speed of the tide is negative, the
tide is moving out of the canal. When the speed is 0, it is high
tide or low tide. When the speed of the tide is positive, the tide
is flowing into the canal.

89. (a) $C(x) = 0.04x^2 + 0.6x + 330$
$+ 7.5 \sin(2\pi x)$

(b) $C(x) = 0.04(x - 1970)^2 + 0.6(x - 1970) + 330 +$
$7.5 \sin[2\pi(x - 1970)]$

Reviewing Basic Concepts *(pages 630–631)*

1. (a) $(1, 0)$ **(b)** $\left(-\frac{\sqrt{2}}{2}, -\frac{\sqrt{2}}{2}\right)$ **(c)** $(0, 1)$

*In Exercises 2 and 3 we give, in order, sine, cosine, tangent,
cotangent, secant, and cosecant.*

2. -1; 0; undefined; 0; undefined; -1 **3. (a)** $-\frac{1}{2}$; $-\frac{\sqrt{3}}{2}$;
$\frac{\sqrt{3}}{3}$; $\sqrt{3}$; $-\frac{2\sqrt{3}}{3}$; -2 **(b)** $-\frac{\sqrt{3}}{2}$; $-\frac{1}{2}$; $\sqrt{3}$; $\frac{\sqrt{3}}{3}$; -2; $-\frac{2\sqrt{3}}{3}$
4. $\sin 2.25 \approx 0.7780731969$; $\cos 2.25 \approx -0.6281736227$;
$\tan 2.25 \approx -1.238627616$; $\cot 2.25 \approx -0.8073451511$;
$\sec 2.25 \approx -1.591916572$; $\csc 2.25 \approx 1.285226125$
5. quadrant IV

6.

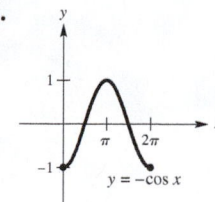

period: 2π; amplitude: 1

7.

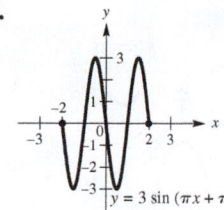

amplitude: 3; period: 2;
phase shift: -1

8. (a) maximum: 18.9 hr; minimum: 5.9 hr **(b)** The amplitude represents half the difference in the daylight hours between the longest and shortest days. The period represents 12 months or one year.

8.7 Exercises *(pages 640–643)*

1. B   **3.** E   **5.** D   **7.** true

9. false; csc x is not defined for $x = 0,\ \pm\pi,\ \pm 2\pi,\ \dots,$ while sec x is not defined for $x = \pm\dfrac{\pi}{2},\ \pm\dfrac{3\pi}{2},\ \dots.$

15. (a) 4π **(b)** none   **(c)** $(-\infty, -2] \cup [2, \infty)$

17. (a) 2π **(b)** $-\dfrac{\pi}{2}$ **(c)** $(-\infty, -2] \cup [2, \infty)$

19. (a) 3π **(b)** $\dfrac{\pi}{2}$ **(c)** $(-\infty, \infty)$

21. (a) π **(b)** $-\dfrac{\pi}{2}$ **(c)** $\left(-\infty, -\dfrac{1}{2}\right] \cup \left[\dfrac{1}{2}, \infty\right)$

23. (a) π **(b)** $-\dfrac{\pi}{4}$ **(c)** $(-\infty, \infty)$

25.

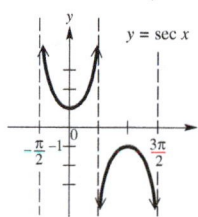

$y = \sec x$

27.

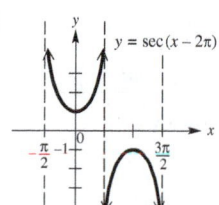

$y = \sec(x - 2\pi)$

29.

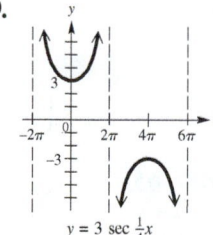

$y = 3 \sec \frac{1}{4}x$

31.

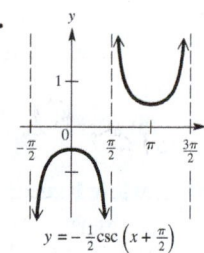

$y = -\frac{1}{2}\csc\left(x + \frac{\pi}{2}\right)$

33.

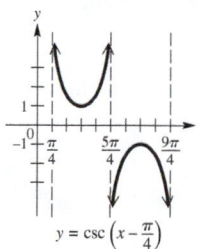

$y = \csc\left(x - \frac{\pi}{4}\right)$

35.

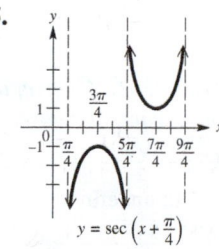

$y = \sec\left(x + \frac{\pi}{4}\right)$

37.

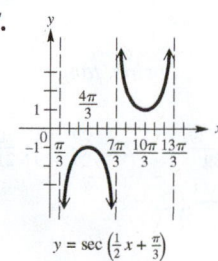

$y = \sec\left(\frac{1}{2}x + \frac{\pi}{3}\right)$

39.

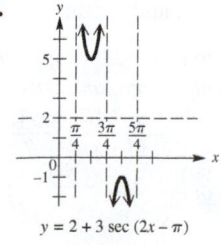

$y = 2 + 3 \sec(2x - \pi)$

41.

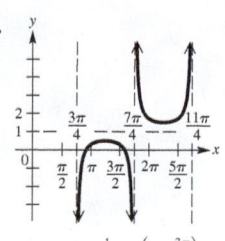

$y = 1 - \frac{1}{2}\csc\left(x - \frac{3\pi}{4}\right)$

43.

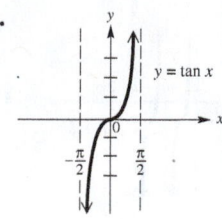

$y = \tan x$

45.

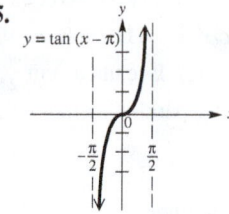

$y = \tan(x - \pi)$

47.

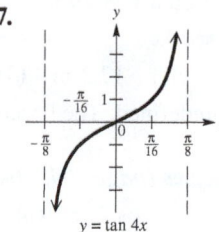

$y = \tan 4x$

49.

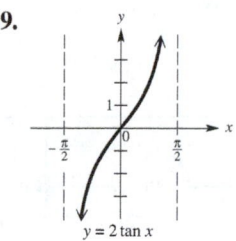

$y = 2 \tan x$

51.

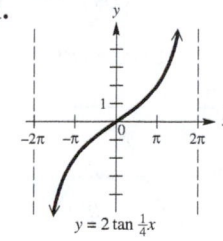

$y = 2 \tan \frac{1}{4}x$

53.

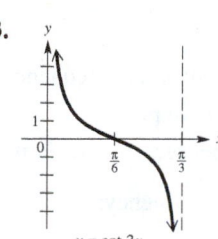

$y = \cot 3x$

55.

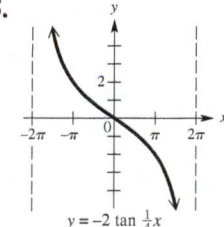

$y = -2 \tan \frac{1}{4}x$

57.

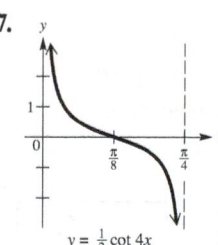

$y = \frac{1}{2} \cot 4x$

59.

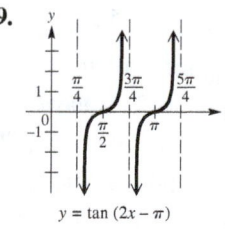

$y = \tan(2x - \pi)$

61.

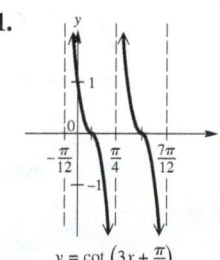

$y = \cot\left(3x + \frac{\pi}{4}\right)$

63.

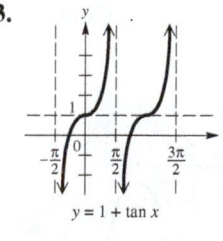

$y = 1 + \tan x$

65.

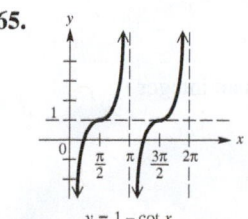

$y = 1 - \cot x$

67.

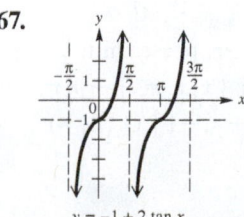

$y = -1 + 2 \tan x$

69.

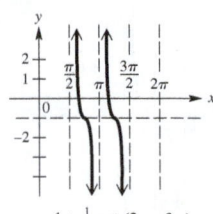

$y = -1 + \frac{1}{2}\cot(2x - 3\pi)$

71. $y = -2\tan x$ **73.** $y = \cot 3x$ **75.** $y = -3\cot x$

77. $y = \sec 4x$ **79.** $y = -2 + \csc x$ **81. (a)** 0 m

(b) 12.3 m **(c)** -12.3 m **(d)** 12.3 m **(e)** It leads to $\tan\frac{\pi}{2}$, which is undefined. The beacon is shining parallel to the wall.

8.8 Exercises *(pages 645–646)*

1. (a) $s(t) = 2\cos 4\pi t$ **(b)** $s(1) = 2$; neither

3. (a) $s(t) = -3\cos 2.5\pi t$ **(b)** $s(1) = 0$; upward

5. $s(t) = 0.21\cos 55\pi t$ **7.** $s(t) = 0.14\cos 110\pi t$

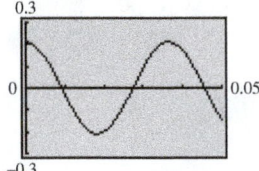 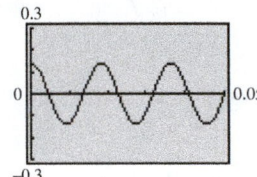

9. (a) $s(t) = 2\sin 2t$; amplitude: 2; period: π sec; frequency: $\frac{1}{\pi}$ oscillation per sec **(b)** $s(t) = 2\sin 4t$; amplitude: 2; period: $\frac{\pi}{2}$ sec; frequency: $\frac{2}{\pi}$ oscillation per sec **11.** $\frac{8}{\pi^2}$ ft

13. (a) amplitude: $\frac{1}{2}$; period: $\sqrt{2}\pi$ sec; frequency: $\frac{\sqrt{2}}{2\pi}$ oscillation per sec **(b)** $s(t) = \frac{1}{2}\sin\sqrt{2}t$

15. (a) 4 in. **(b)** $\frac{5}{\pi}$ oscillations per sec; $\frac{\pi}{5}$ sec **(c)** after $\frac{\pi}{10}$ sec **(d)** approximately 2; After 1.466 sec, the weight is about 2 in. above the equilibrium position.

17. (a) $s(t) = -2\cos 6\pi t$ **(b)** 3 oscillations per sec

19. (a) The spring is compressed 2 inches.

(b) 1 oscillation per sec **(c)** 0.2053 sec

Reviewing Basic Concepts *(page 647)*

1.

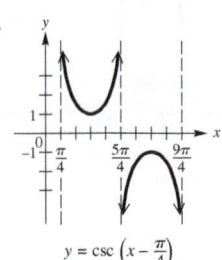

$y = \csc\left(x - \frac{\pi}{4}\right)$

period: 2π; phase shift: $\frac{\pi}{4}$;

domain: $\left\{x \mid x \neq \frac{\pi}{4} + n\pi, \text{ where } n \text{ is an integer}\right\}$;

range: $(-\infty, -1] \cup [1, \infty)$

2.

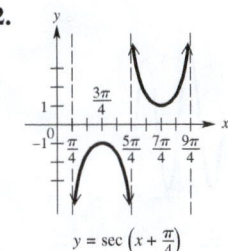

$y = \sec\left(x + \frac{\pi}{4}\right)$

period: 2π; phase shift: $-\frac{\pi}{4}$;

domain: $\left\{x \mid x \neq \frac{\pi}{4} + n\pi, \text{ where } n \text{ is an integer}\right\}$;

range: $(-\infty, -1] \cup [1, \infty)$

3.

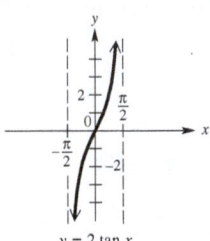

$y = 2\tan x$

period: π;

domain: $\left\{x \mid x \neq \frac{\pi}{2} + n\pi, \text{ where } n \text{ is an integer}\right\}$;

range: $(-\infty, \infty)$

4.

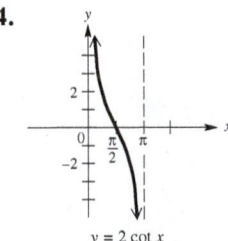

$y = 2\cot x$

period: π;

domain: $\left\{x \mid x \neq n\pi, \text{ where } n \text{ is an integer}\right\}$;

range: $(-\infty, \infty)$

5. (a) 4 in. **(b)** after $\frac{1}{8}$ sec **(c)** 4 oscillations per sec; $\frac{1}{4}$ sec

Chapter 8 Review Exercises *(pages 651–655)*

1. 186° **3.** 1280° **5.** 1 radian $\approx 57.3° > 1°$

7. $\frac{2\pi}{3}$ **9.** 225° **11.** $\frac{4\pi}{3}$ in. **13.** 35.8 cm

15. 41 yd

In Exercises 17–21 we give, in order, sine, cosine, tangent, cotangent, secant, and cosecant.

17. $-\frac{\sqrt{2}}{2}; -\frac{\sqrt{2}}{2}; 1; 1; -\sqrt{2}; -\sqrt{2}$

19. $0; -1; 0;$ undefined; $-1;$ undefined

21. $-\frac{2\sqrt{85}}{85}; \frac{9\sqrt{85}}{85}; -\frac{2}{9}; -\frac{9}{2}; \frac{\sqrt{85}}{9}; -\frac{\sqrt{85}}{2}$

23. tangent and secant **25.** possible

In Exercises 27–33 we give, in order, sine, cosine, tangent, cotangent, secant, and cosecant.

27. $-\frac{\sqrt{39}}{8}; -\frac{5}{8}; \frac{\sqrt{39}}{5}; \frac{5\sqrt{39}}{39}; -\frac{8}{5}; -\frac{8\sqrt{39}}{39}$ **29.** $\frac{20}{29}; \frac{21}{29}; \frac{20}{21};$ $\frac{21}{20}; \frac{29}{21}; \frac{29}{20}$ **31.** $-\frac{\sqrt{3}}{2}; \frac{1}{2}; -\sqrt{3}; -\frac{\sqrt{3}}{3}; 2; -\frac{2\sqrt{3}}{3}$

33. $-\frac{1}{2}$; $\frac{\sqrt{3}}{2}$; $-\frac{\sqrt{3}}{3}$; $-\sqrt{3}$; $\frac{2\sqrt{3}}{3}$; -2 **35.** -1.3563417

37. 0.20834446 **39.** $55.7°$

41. $B = 31° 30'$; $a = 638$; $b = 391$ **43.** 1200 m

45. 140 mi   **47.** 73.7 ft   **49.** 419 **51.** $-\frac{1}{2}$

53. $-\sqrt{3}$ **55.** 2 **57.** 0.9703 **59.** $\frac{\pi}{3}$ **61.** 0.5528

63. 2; 2π; none; none

65. $\frac{1}{2}$; $\frac{2\pi}{3}$; none; none

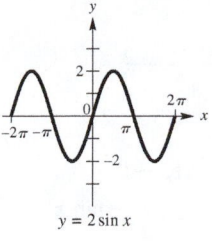

$y = 2 \sin x$

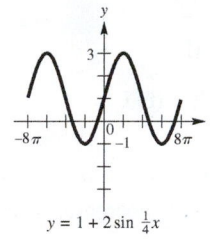

$y = -\frac{1}{2} \cos 3x$

67. 2; 8π; upward
1 unit; none

69. 3; 2π; none; $-\frac{\pi}{2}$

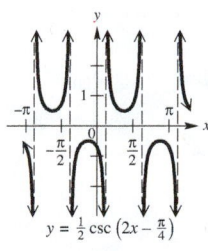

$y = 1 + 2 \sin \frac{1}{4}x$

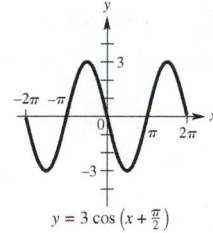

$y = 3 \cos \left(x + \frac{\pi}{2}\right)$

71. not applicable; π;
none; $\frac{\pi}{8}$

73. not applicable; π;
none; none

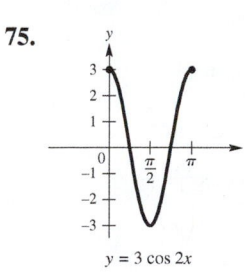

$y = \frac{1}{2} \csc \left(2x - \frac{\pi}{4}\right)$

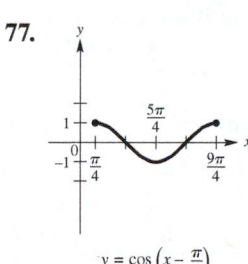

$y = -2 \sec 2x$

75.

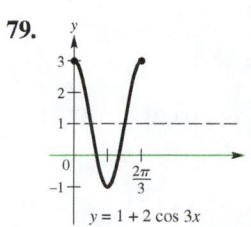

$y = 3 \cos 2x$

77.

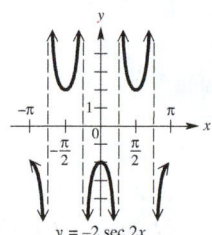

$y = \cos \left(x - \frac{\pi}{4}\right)$

79.

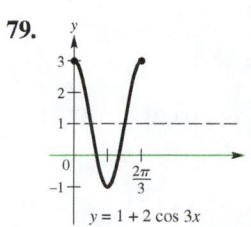

$y = 1 + 2 \cos 3x$

81. tangent   **83.** cosine   **85.** cotangent

87. $y = 3 \sin \left[2\left(x - \frac{\pi}{4}\right)\right]$ **89.** $y = \frac{1}{3} \sin \frac{\pi}{2}x$

91. (a)

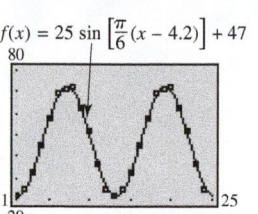

(b) $f(x) = 25 \sin \left[\frac{\pi}{6}(x - 4.2)\right] + 47$ **(c)** a represents the amplitude, $\frac{2\pi}{b}$ the period, c the vertical shift, and d the phase shift.

(d) The function gives an excellent model for the data.

$f(x) = 25 \sin \left[\frac{\pi}{6}(x - 4.2)\right] + 47$

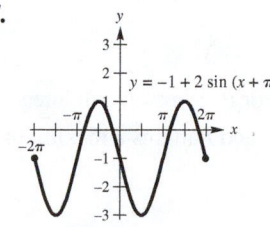

(e)

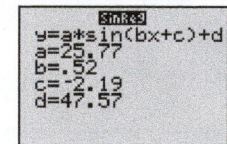

TI-84 Plus fixed to the
nearest hundredth

93. (b) $\frac{\pi}{6}$ **(c)** There is less ultraviolet light when $\theta = \frac{\pi}{3}$.

95. 4; 2 sec; $\frac{1}{2}$ oscillation per sec   **97.** The frequency is the number of oscillations in one second; 4 in. below; at the initial point; $2\sqrt{2}$ in. below

Chapter 8 Test *(pages 655–656)*

1. $203°$ **2.** $2700°$ **3. (a)** $\frac{2\pi}{3}$ **(b)** $162°$

4. (a) $\frac{4}{3}$ **(b)** $15{,}000$ cm^2 **5.** $\frac{12\pi}{5} \approx 7.54$ in.

6. (a) $210°$ **(b)** $x = -\sqrt{3}$; $y = -1$ **(c)** $\sin(-150°) = -\frac{1}{2}$; $\cos(-150°) = -\frac{\sqrt{3}}{2}$; $\tan(-150°) = \frac{\sqrt{3}}{3}$; $\cot(-150°) = \sqrt{3}$; $\sec(-150°) = -\frac{2\sqrt{3}}{3}$; $\csc(-150°) = -2$ **(d)** $-\frac{5\pi}{6}$

7. 0.97169234 **8.** III   **9.** $\sin \theta = -\frac{5\sqrt{29}}{29}$; $\cos \theta = \frac{2\sqrt{29}}{29}$; $\tan \theta = -\frac{5}{2}$ **10.** $\sin \theta = -\frac{3}{5}$; $\tan \theta = -\frac{3}{4}$; $\cot \theta = -\frac{4}{3}$; $\sec \theta = \frac{5}{4}$; $\csc \theta = -\frac{5}{3}$ **11.** $x = 4$; $y = 4\sqrt{3}$; $z = 4\sqrt{2}$; $w = 8$ **12.** $-\sqrt{3}$

13. (a) 0.97939940 **(b)** 0.20834446 **(c)** 1.9362132

14. $B = 31° 30'$; $a = 638$; $b = 391$

15. 15.5 ft   **16.** 110 km

17.

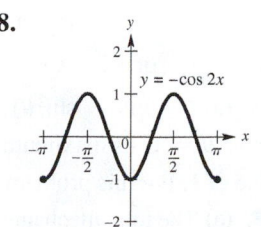

$y = -1 + 2 \sin(x + \pi)$

18.

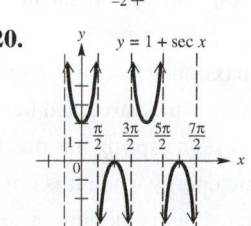

$y = -\cos 2x$

19.

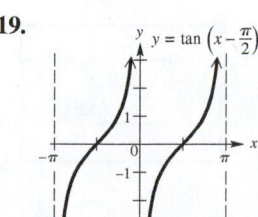

$y = \tan \left(x - \frac{\pi}{2}\right)$

20.

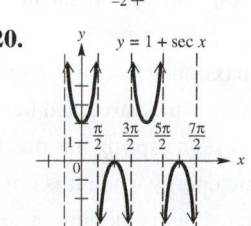

$y = 1 + \sec x$

21. $y = 2 \sin 2x$

22. (a)

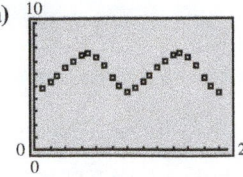

(b) $a \approx 1.48$, or about 1 hr, 29 min   **(c)** $b \approx 0.52$

(d) $d \approx 3.68$ **(e)** $c \approx 6.07$

(f)

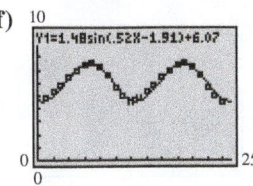

$y = 1.48 \sin(0.52x - 1.91) + 6.07$

23. 3 in.   **24.** after $\frac{1}{2}$ sec

CHAPTER 9 TRIGONOMETRIC IDENTITIES AND EQUATIONS

9.1 Exercises *(pages 664–667)*

1. odd   **3.** odd   **5.** even   **7.** $\cos 4.38$

9. $-\sin 0.5$ **11.** $-\tan \frac{\pi}{7}$ **13.** B   **15.** E

17. A   **19.** A   **21.** D

23. The student has neglected to write the arguments of the functions (for example, θ or t).

25. $\dfrac{\pm \sqrt{1 + \cot^2 \theta}}{1 + \cot^2 \theta}$; $\dfrac{\pm \sqrt{\sec^2 \theta - 1}}{\sec \theta}$ **27.** $\dfrac{\pm \sin \theta \sqrt{1 - \sin^2 \theta}}{1 - \sin^2 \theta}$;

$\dfrac{\pm \sqrt{1 - \cos^2 \theta}}{\cos \theta}$; $\pm \sqrt{\sec^2 \theta - 1}$; $\dfrac{\pm \sqrt{\csc^2 \theta - 1}}{\csc^2 \theta - 1}$

29. $\dfrac{\pm \sqrt{1 - \sin^2 \theta}}{1 - \sin^2 \theta}$; $\pm \sqrt{\tan^2 \theta + 1}$; $\dfrac{\pm \sqrt{1 + \cot^2 \theta}}{\cot \theta}$;

$\dfrac{\pm \csc \theta \sqrt{\csc^2 \theta - 1}}{\csc^2 \theta - 1}$

31. $\sin \theta$ **33.** $\tan^2 \beta$ **35.** $\tan^2 x$ **37.** $\sec^2 x$

39. $\cos \theta$ **41.** $\cot \theta$ **43.** $\cos^2 \theta$ **45.** $\sin^2 \theta$

47. $\cot^2 \theta$ **49.** $\cos^2 \theta$ **51.** $\sec \theta - \cos \theta$

53. $\cot \theta - \tan \theta$ **55.** $\dfrac{\sin^2 \theta}{\cos \theta}$ **57.** $\cos^2 \theta$ **59.** $\sec^2 \theta$

61. $\csc \theta \sec \theta$, or $\dfrac{1}{\sin \theta \cos \theta}$ **63.** $1 + \sec s$

65. 1   **67.** 1   **69.** $2 + 2 \sin t$

71. $-\dfrac{2 \cos x}{\sin^2 x}$, or $-2 \cot x \csc x$

113. (a) $I = k(1 - \sin^2 \theta)$ **(b)** For $\theta = n\pi$ and all integers n, $\cos^2 \theta = 1$, its maximum value, and I attains a maximum value of k. For this problem $\theta = 0$.

115. (a) The total mechanical energy E is always 2. The spring has maximum potential energy when it is fully stretched but not moving. The spring has maximum kinetic energy when it is not stretched but is moving fastest.

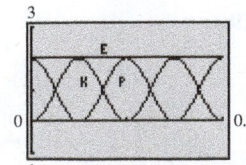

(b) Let $Y_1 = P(t)$, $Y_2 = K(t)$, and $Y_3 = E(t)$. $Y_3 = 2$ for all inputs. The spring is stretched the most (has greatest potential energy) when $t = 0, 0.25, 0.5, 0.75, \ldots$. At these times the kinetic energy is 0.

(c) $E(t) = k$ or $E(t) = 2$

9.2 Exercises *(pages 674–676)*

1. F   **3.** C   **5.** $\dfrac{\sqrt{6} - \sqrt{2}}{4}$ **7.** $\dfrac{-\sqrt{2} - \sqrt{6}}{4}$

9. $\dfrac{-\sqrt{6} + \sqrt{2}}{4}$ **11.** $\dfrac{\sqrt{6} - \sqrt{2}}{4}$ **13.** $-\sqrt{3} - 2$

15. $\dfrac{\sqrt{2} + \sqrt{6}}{4}$ **17.** -1 **19.** $\dfrac{\sqrt{2}}{2}$ **21.** -1

23. $\sin x$ **25.** $-\cos x$ **27.** $-\cos x$ **29.** $-\tan x$

31. $\sin x$ **33.** $\sin x$ **35.** $-\sin x$

37. $\dfrac{\sqrt{2} (\sin x - \cos x)}{2}$ **39.** $\dfrac{1 + \tan x}{1 - \tan x}$ **41.** $-\tan x$

43. $\dfrac{1 - \tan x}{1 + \tan x}$ **45.** $\cos x$

47. (a) $\frac{63}{65}$ **(b)** $\frac{33}{65}$ **(c)** $\frac{63}{16}$ **(d)** $\frac{33}{56}$ **(e)** I   **(f)** I

49. (a) $\frac{77}{85}$ **(b)** $\frac{13}{85}$ **(c)** $-\frac{77}{36}$ **(d)** $\frac{13}{84}$ **(e)** II   **(f)** I   **61.** 3

63. (a) 425 lb   **(c)** 0°

65. (a) The pressure P is oscillating.

For $x = t$,
$P(t) = \dfrac{0.4}{10} \cos \left[\dfrac{20\pi}{4.9} - 1026t \right]$

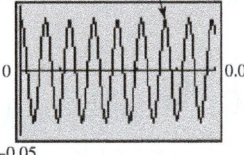

(b) The pressure oscillates and amplitude decreases as r increases.

For $x = r$,
$P(r) = \dfrac{3}{r} \cos \left[\dfrac{2\pi r}{4.9} - 10{,}260 \right]$

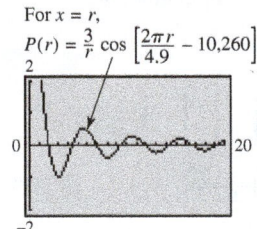

(c) $P = \dfrac{a}{n\lambda} \cos ct$

Reviewing Basic Concepts *(page 676)*

1. $\dfrac{\sin^2 x}{\cos x}$ **2.** $\dfrac{\sqrt{3} - 3}{3 + \sqrt{3}}$, or $\sqrt{3} - 2$ **3.** 0

4. $\dfrac{\sqrt{2}}{2} (\sin x - \cos x)$ **5.** $\dfrac{-2 + \sqrt{15}}{6}$; $\dfrac{\sqrt{5} - 2\sqrt{3}}{6}$;

$\dfrac{-2 - \sqrt{15}}{\sqrt{5} - 2\sqrt{3}}$, or $\dfrac{8\sqrt{5} + 9\sqrt{3}}{7}$ **10.** $-20 \cos \dfrac{\pi t}{4}$

9.3 Exercises *(pages 685–688)*

1. (a) $-\dfrac{4\sqrt{21}}{25}$ **(b)** $\dfrac{17}{25}$ **3. (a)** $\dfrac{4}{5}$ **(b)** $-\dfrac{3}{5}$

5. (a) $-\dfrac{4\sqrt{55}}{49}$ **(b)** $\dfrac{39}{49}$ **7.** $\dfrac{\sqrt{3}}{2}$ **9.** $\dfrac{\sqrt{3}}{2}$

11. $-\dfrac{\sqrt{2}}{2}$ **13.** $\dfrac{1}{2} \tan 102°$ **15.** $\dfrac{1}{4} \cos 94.2°$ **17.** 1

19. $\cos^4 x - \sin^4 x = \cos 2x$

Forms of answers may vary in Exercises 21 and 23.

21. $\cos 3x = 4 \cos^3 x - 3 \cos x$

23. $\tan 4x = \dfrac{4 \tan x - 4 \tan^3 x}{1 - 6 \tan^2 x + \tan^4 x}$

25. $\dfrac{\sqrt{2}-\sqrt{3}}{2}$ **27.** $1-\sqrt{2}$, or $-\sqrt{3-2\sqrt{2}}$

29. $\dfrac{\sqrt{2}+\sqrt{2}}{2}$ **31.** $\dfrac{\sqrt{10}}{4}$ **33.** 3   **35.** $-\sqrt{7}$

37. $\dfrac{\sqrt{50-10\sqrt{5}}}{10}$ **39.** $-\dfrac{\sqrt{42}}{12}$ **41.** **(a)** $\tan\dfrac{\pi}{2}$ is undefined,

so it cannot be used.   **(b)** $\tan\left(\dfrac{\pi}{2}+x\right)=\dfrac{\sin\left(\dfrac{\pi}{2}+x\right)}{\cos\left(\dfrac{\pi}{2}+x\right)}$

43. $\sin 20°$ **45.** $\tan 73.5°$

47. $\tan 29.87°$ **65.** $\sin 160°-\sin 44°$

67. $\sin 225°+\sin 55°$ **69.** $-2\sin 3x \sin x$

71. $-2\sin 11.5°\cos 36.5°$ **73.** $2\cos 6x\cos 2x$

75. **(a)** $\cos\dfrac{\theta}{2}=\dfrac{R-b}{R}$ **(b)** $\tan\dfrac{\theta}{4}=\dfrac{b}{50}$ **(c)** $54°$

77. **(a)** For $x=t$, $W=VI=$
$(163\sin 120\pi t)(1.23\sin 120\pi t)$

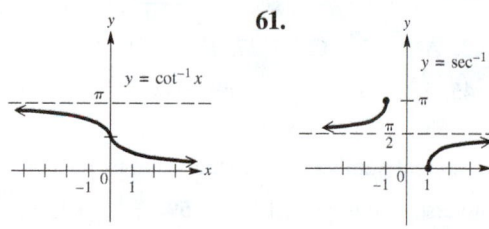

(b) maximum: 200.49 watts; minimum: 0 watts

(c) $a=-100.245$, $\omega=240\pi$, $c=100.245$

(e) 100.245 watts

9.4 Exercises *(pages 698–701)*

1. one-to-one   **3.** $\cos y$ **5.** π

7. **(a)** $[-1,1]$ **(b)** $\left[-\dfrac{\pi}{2},\dfrac{\pi}{2}\right]$ **(c)** increasing   **(d)** -2 is not in the domain.   **9.** **(a)** $(-\infty,\infty)$ **(b)** $\left(-\dfrac{\pi}{2},\dfrac{\pi}{2}\right)$ **(c)** increasing

(d) no   **11.** $\dfrac{\pi}{4}$ **13.** π **15.** $-\dfrac{\pi}{2}$ **17.** 0   **19.** $\dfrac{\pi}{2}$

21. $\dfrac{\pi}{4}$ **23.** $\dfrac{5\pi}{6}$ **25.** $\dfrac{3\pi}{4}$ **27.** $-\dfrac{\pi}{6}$ **29.** $\dfrac{\pi}{6}$

31. $\dfrac{\pi}{3}$ **33.** $\dfrac{\pi}{6}$ **35.** $-45°$ **37.** $-60°$ **39.** $120°$

41. $-30°$ **43.** θ does not exist   **45.** θ does not exist

47. $-7.6713835°$ **49.** $113.500970°$ **51.** $30.987961°$

53. 0.83798122 **55.** 2.3154725 **57.** 1.1900238

59.

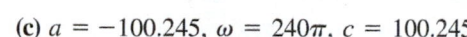

$y=\cot^{-1}x$

61.

$y=\sec^{-1}x$

63.

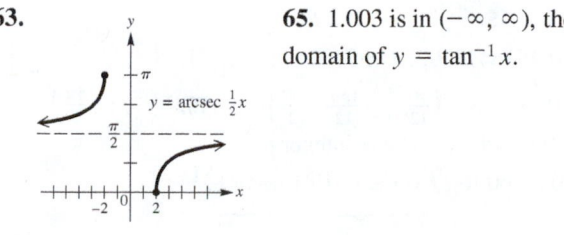

$y=\text{arcsec}\,\dfrac{1}{2}x$

65. 1.003 is in $(-\infty,\infty)$, the domain of $y=\tan^{-1}x$.

67. $\dfrac{1}{2}$ **69.** $-\dfrac{\pi}{3}$ **71.** $\dfrac{\pi}{2}$ **73.** 5   **75.** 2

77. $-\dfrac{\pi}{6}$ **79.** $\dfrac{\sqrt{7}}{3}$ **81.** $\dfrac{\sqrt{5}}{5}$ **83.** $\dfrac{120}{169}$ **85.** $-\dfrac{7}{25}$

87. $\dfrac{4\sqrt{6}}{25}$ **89.** $\dfrac{63}{65}$ **91.** $-\dfrac{63}{16}$ **93.** $\dfrac{63}{65}$

95. $\dfrac{\sqrt{10}-3\sqrt{30}}{20}$ **97.** 0.894427191 **99.** 0.1234399811

101. $\dfrac{1}{u}$ **103.** $\sqrt{1-u^2}$ **105.** $\dfrac{\sqrt{1-u^2}}{u}$

107. $\dfrac{\sqrt{u^2-4}}{u}$ **109.** $\dfrac{u\sqrt{2}}{2}$ **111.** $\dfrac{2\sqrt{4-u^2}}{4-u^2}$

113. **(a)** $45°$ **(b)** $\theta=45°$

115. **(a)** $35°$ **(b)** $24°$ **(c)** $18°$

(e) about 2 ft

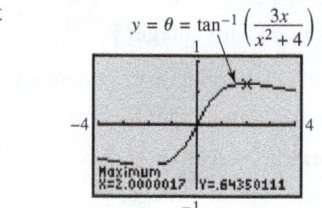

$y=\theta=\tan^{-1}\left(\dfrac{3x}{x^2+4}\right)$

Reviewing Basic Concepts *(page 701)*

1. $-\dfrac{\sqrt{119}}{7}$ **2.** $\sin 2\theta=\dfrac{4\sqrt{2}}{9}$; $\cos 2\theta=\dfrac{7}{9}$; $\tan 2\theta=\dfrac{4\sqrt{2}}{7}$

3. $\dfrac{\sqrt{2}+\sqrt{6}}{4}$ **4.** $\sin 175°-\sin 125°$ **6.** **(a)** $\dfrac{\pi}{6}$ **(b)** $-\dfrac{\pi}{4}$

7. **(a)** $60°$ **(b)** $135°$

8.

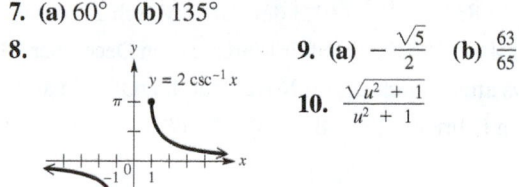

$y=2\csc^{-1}x$

9. **(a)** $-\dfrac{\sqrt{5}}{2}$ **(b)** $\dfrac{63}{65}$

10. $\dfrac{\sqrt{u^2+1}}{u^2+1}$

9.5 Exercises *(pages 707–709)*

1. $\left\{\dfrac{2\pi}{3},\dfrac{4\pi}{3}\right\}$ **3.** $\varnothing$ **5.** $\left\{\dfrac{3\pi}{4},\dfrac{7\pi}{4}\right\}$ **7.** $\left\{\dfrac{\pi}{3},\dfrac{5\pi}{3}\right\}$

9. $\left\{\dfrac{\pi}{6},\dfrac{5\pi}{6}\right\}$ **11.** $\left\{\dfrac{\pi}{4},\dfrac{2\pi}{3},\dfrac{5\pi}{4},\dfrac{5\pi}{3}\right\}$ **13.** $\left\{\dfrac{\pi}{4},\dfrac{\pi}{2},\dfrac{5\pi}{4},\dfrac{3\pi}{2}\right\}$

15. $\left\{\dfrac{\pi}{2}\right\}$ **17.** $\left\{\dfrac{\pi}{6},\dfrac{5\pi}{6},\dfrac{7\pi}{6},\dfrac{11\pi}{6}\right\}$ **19.** $\left\{\dfrac{\pi}{3},\dfrac{3\pi}{4},\dfrac{4\pi}{3},\dfrac{7\pi}{4}\right\}$

21. $\left\{\dfrac{\pi}{2},\dfrac{7\pi}{6},\dfrac{11\pi}{6}\right\}$ **23.** $\{0,\pi\}$ **25.** **(a)** $\left\{\dfrac{\pi}{3},\dfrac{5\pi}{3}\right\}$

(b) $\left(\dfrac{\pi}{3},\dfrac{5\pi}{3}\right)$ **(c)** $\left[0,\dfrac{\pi}{3}\right)\cup\left(\dfrac{5\pi}{3},2\pi\right)$

27. **(a)** $\left\{\dfrac{\pi}{3},\dfrac{2\pi}{3},\dfrac{4\pi}{3},\dfrac{5\pi}{3}\right\}$

(b) $\left(\dfrac{\pi}{3},\dfrac{\pi}{2}\right)\cup\left(\dfrac{\pi}{2},\dfrac{2\pi}{3}\right)\cup\left(\dfrac{4\pi}{3},\dfrac{3\pi}{2}\right)\cup\left(\dfrac{3\pi}{2},\dfrac{5\pi}{3}\right)$

(c) $\left[0,\dfrac{\pi}{3}\right)\cup\left(\dfrac{2\pi}{3},\dfrac{4\pi}{3}\right)\cup\left(\dfrac{5\pi}{3},2\pi\right)$

29. **(a)** $\left\{\dfrac{\pi}{6},\dfrac{\pi}{2},\dfrac{3\pi}{2},\dfrac{11\pi}{6}\right\}$ **(b)** $\left[0,\dfrac{\pi}{6}\right)\cup\left(\dfrac{\pi}{2},\dfrac{3\pi}{2}\right)\cup\left(\dfrac{11\pi}{6},2\pi\right)$

(c) $\left(\dfrac{\pi}{6},\dfrac{\pi}{2}\right)\cup\left(\dfrac{3\pi}{2},\dfrac{11\pi}{6}\right)$

31. **(a)** $\left\{\dfrac{\pi}{2},\dfrac{3\pi}{2}\right\}$ **(b)** $\left(\dfrac{\pi}{2},\dfrac{3\pi}{2}\right)$ **(c)** $\left[0,\dfrac{\pi}{2}\right)\cup\left(\dfrac{3\pi}{2},2\pi\right)$

33. $\varnothing$ **35.** $\{0°,30°,150°,180°\}$

37. $\{0°,45°,135°,180°,225°,315°\}$

39. $\{53.6°,126.4°,187.9°,352.1°\}$

41. $\{149.6°,329.6°,106.3°,286.3°\}$

43. $\varnothing$ **45.** $\{57.7°,159.2°\}$ **47.** $\tan^3 x-3\tan x=0$

48. $\left\{0,\dfrac{\pi}{3},\dfrac{2\pi}{3},\pi,\dfrac{4\pi}{3},\dfrac{5\pi}{3}\right\}$ **49.** $\left\{\dfrac{\pi}{3},\dfrac{2\pi}{3},\dfrac{4\pi}{3},\dfrac{5\pi}{3}\right\}$

50. No. The solutions 0 and π were lost when dividing by $\tan x$, because both $\tan 0$ and $\tan\pi$ equal 0, and division by 0 is undefined.

51. $\left\{\frac{\pi}{2}, 3.87, 5.55\right\}$ **53.** $\{1.86, 2.61, 5.00, 5.75\}$

55. $\{1.20, 5.09\}$ **57.** $\{135°, 315°, 71.6°, 251.6°\}$

59. $\{114.3°, 335.7°\}$

61. $\{180° + 360°n, \text{ where } n \text{ is any integer}\}$

63. $\left\{\frac{\pi}{3} + 2n\pi, \frac{2\pi}{3} + 2n\pi, \text{ where } n \text{ is any integer}\right\}$

65. $\{19.5° + 360°n, 160.5° + 360°n, 210° + 360°n,$
$330° + 360°n, \text{ where } n \text{ is any integer}\}$

67. $\left\{\frac{\pi}{3} + 2n\pi, \pi + 2n\pi, \frac{5\pi}{3} + 2n\pi, \text{ where } n \text{ is any integer}\right\}$

69. $\{180°n, \text{ where } n \text{ is any integer}\}$

71. $\{0.8751 + 2n\pi, 2.2665 + 2n\pi, 3.5908 + 2n\pi,$
$5.8340 + 2n\pi, \text{ where } n \text{ is any integer}\}$

73. $\{33.6° + 360°n, 326.4° + 360°n, \text{ where } n \text{ is any integer}\}$

75. $\{45° + 180°n, 108.4° + 180°n, \text{ where } n \text{ is any integer}\}$

77. one; There are infinitely many because the sine function is
periodic.   **79.** $\{1.99, 5.86\}$ **81.** $\{2.68, 4.46, 4.71\}$

83. $\{1.30\}$ **85. (a)** about 91.3 days after March 21, on
June 20   **(b)** about 273.8 days after March 21, on December 19
(c) 228.7 days after March 21, on November 4, and again after
318.8 days, on February 2   **87.** $14°$ **89.** $\frac{\pi}{3}$

9.6 Exercises *(pages 715–718)*

1. $\frac{\pi}{3}, \pi, \frac{4\pi}{3}$ **3.** $\left\{\frac{\pi}{2}\right\}$ **5.** $\{\pi\}$ **7.** $\left\{\frac{\pi}{4}, \frac{\pi}{2}, \frac{5\pi}{4}, \frac{3\pi}{2}\right\}$

9. $\left\{\frac{\pi}{3}, \frac{\pi}{2}, \frac{3\pi}{2}, \frac{5\pi}{3}\right\}$ **11.** $\left\{\frac{\pi}{6}, \frac{\pi}{2}, \frac{5\pi}{6}, \frac{3\pi}{2}\right\}$

13. (a) $\left\{\frac{\pi}{12}, \frac{11\pi}{12}, \frac{13\pi}{12}, \frac{23\pi}{12}\right\}$

(b) $\left[0, \frac{\pi}{12}\right] \cup \left(\frac{11\pi}{12}, \frac{13\pi}{12}\right) \cup \left(\frac{23\pi}{12}, 2\pi\right)$

15. (a) $\left\{\frac{\pi}{2}, \frac{7\pi}{6}, \frac{11\pi}{6}\right\}$ **(b)** $\emptyset$

17. (a) $\left\{\frac{3\pi}{8}, \frac{5\pi}{8}, \frac{11\pi}{8}, \frac{13\pi}{8}\right\}$ **(b)** $\left[\frac{3\pi}{8}, \frac{5\pi}{8}\right] \cup \left[\frac{11\pi}{8}, \frac{13\pi}{8}\right]$

19. (a) $\left\{\frac{\pi}{2}, \frac{3\pi}{2}\right\}$ **(b)** $\left(\frac{\pi}{2}, \frac{3\pi}{2}\right)$

21. $\left\{\frac{\pi}{12} + \frac{2n\pi}{3}, \frac{\pi}{4} + \frac{2n\pi}{3}, \text{ where } n \text{ is any integer}\right\}$

23. $\{720°n, \text{ where } n \text{ is any integer}\}$

25. $\left\{\frac{2\pi}{3} + 4n\pi, \frac{4\pi}{3} + 4n\pi, \text{ where } n \text{ is any integer}\right\}$

27. $\{30° + 360°n, 150° + 360°n, 270° + 360°n, \text{ where } n \text{ is}$
any integer$\}$

29. $\left\{n\pi, \frac{\pi}{6} + 2n\pi, \frac{5\pi}{6} + 2n\pi, \text{ where } n \text{ is any integer}\right\}$

31. $\{1.3181 + 2n\pi, 4.9651 + 2n\pi, \text{ where } n \text{ is any integer}\}$

33. $\{11.8° + 180°n, 78.2° + 180°n, \text{ where } n \text{ is any integer}\}$

35. $\{30° + 180°n, 90° + 180°n, 150° + 180°n, \text{ where } n \text{ is}$
any integer$\}$

37. $\{0.262, 1.309, 1.571, 3.403, 4.451, 4.712\}$

39. $\{1.047, 3.142, 5.236\}$

41. $\{0.259, 1.372, 3.142, 4.911, 6.024\}$

43. $\{0.322, 3.463\}$ **45.** $\left\{0, \frac{\pi}{4}, \pi, \frac{5\pi}{4}\right\}$

47. $\frac{\tan 2\theta}{2} \neq \tan \theta$, because the 2 in 2θ is not a factor of the
numerator. It is a factor in the argument of the tangent function.

49. (a) 2 sec   **(b)** $\frac{10}{3}$ sec

51. (a) For $x = t$,
$P(t) = 0.004 \sin\left(2\pi(261.63)t + \frac{\pi}{7}\right)$

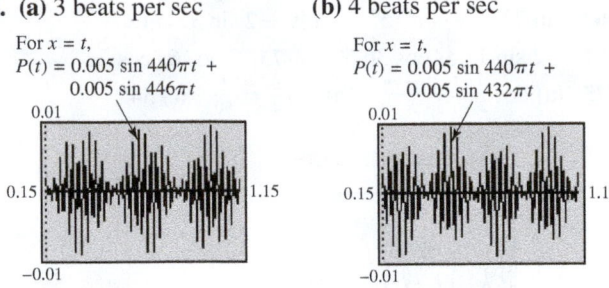

(b) 0.00164 and 0.00355   **(c)** $0.00164 < t < 0.00355$
(d) outward

53. (a) 3 beats per sec   **(b)** 4 beats per sec

For $x = t$,
$P(t) = 0.005 \sin 440\pi t + 0.005 \sin 446\pi t$

For $x = t$,
$P(t) = 0.005 \sin 440\pi t + 0.005 \sin 432\pi t$

(c) The number of beats is equal to the absolute value of the
difference in the frequencies of the two tones.

Reviewing Basic Concepts *(page 718)*

1. $\left\{\frac{\pi}{12}, \frac{11\pi}{12}, \frac{13\pi}{12}, \frac{23\pi}{12}\right\}$ **2.** $\left\{\frac{7\pi}{6}, \frac{11\pi}{6}\right\}$

3. $\left\{2n\pi, \frac{\pi}{4} + n\pi, \text{ where } n \text{ is any integer}\right\}$

4. $\left\{\frac{\pi}{6} + 2n\pi, \frac{\pi}{2} + n\pi, \frac{11\pi}{6} + 2n\pi, \text{ where } n \text{ is any integer}\right\}$

5. $\{38.4°, 218.4°, 104.8°, 284.8°\}$ **6.** $\{78.0°, 282.0°\}$

7. $\{90°, 210°, 330°\}$ **8.** $\{90°, 270°\}$

9. $\{0.681, 1.416\}$ **10.** $\{0, 0.376\}$

Chapter 9 Review Exercises *(pages 720–723)*

1. sine, tangent, cotangent, cosecant   **3.** $\cos 3$

5. $-\tan 3$ **7.** $-\csc 3$ **9.** B   **11.** F   **13.** D

15. $\frac{\cos^2 \theta}{\sin \theta}$ **17.** $\frac{1 + \cos \theta}{\sin \theta}$ **19.** $\sin x = \frac{5\sqrt{41}}{41};$

$\cos x = -\frac{4\sqrt{41}}{41}; \cot x = -\frac{4}{5}; \sec x = -\frac{\sqrt{41}}{4}; \csc x = \frac{\sqrt{41}}{5}$

21. B   **23.** A   **25.** C   **27.** D   **29.** F

43. $\frac{2\pi}{3}$ **45.** $\frac{2\pi}{3}$ **47.** $\frac{3\pi}{4}$ **49.** $-60°$

51. $-41°$ **53.** $-7°$ **55.** $88°$

57. There is no real number whose sine value is -3. The
domain of inverse sine is $[-1, 1]$.   **59.** $\frac{1}{2}$ **61.** $\frac{\sqrt{10}}{10}$

63. $\frac{\pi}{2}$ **65.** $\frac{u\sqrt{u^2 + 1}}{u^2 + 1}$ **67.** $\frac{1}{u}$

69. $\{0.463647609, 3.605240263\}$ **71.** $\left\{\frac{\pi}{4}, \frac{3\pi}{4}, \frac{5\pi}{4}, \frac{7\pi}{4}\right\}$

73. $\{0\}$ **75.** $\left\{\frac{\pi}{12}, \frac{5\pi}{12}, \frac{13\pi}{12}, \frac{17\pi}{12}\right\}$ **77.** $\left\{\frac{\pi}{3}, \pi, \frac{5\pi}{3}\right\}$

79. $\{4n\pi, \text{ where } n \text{ is any integer}\}$

81. (b) 8.660 ft $f(x) = \arctan\left(\frac{15}{x}\right) - \arctan\left(\frac{5}{x}\right)$

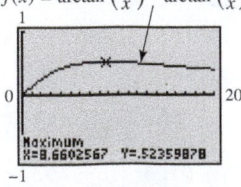

83. 0.001 sec   **85.** $48.8°$

Chapter 9 Test *(pages 723–724)*

1. 0 **2.** $\frac{7}{25}$ **3.** -3 **4.** $\frac{7}{25}$ **5.** -1
6. $\sec x - \sin x \tan x = \cos x$ **11.** $-\sin \theta$ **12.** $-\sin \theta$
13. (a)

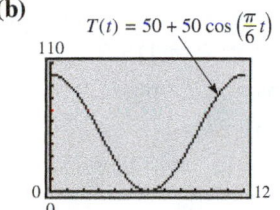

$f(x) = -2 \sin^{-1} x$

(b) domain: $[-1, 1]$; range: $[-\pi, \pi]$ **(c)** The domain of $\sin^{-1} x$ is $[-1, 1]$, and 2 is not in this interval. (No number has sine value 2.) **14.** $\frac{2\pi}{3}$ **15.** 0 **16.** $\frac{\pi}{3}$ **17.** $\frac{\pi}{6}$

18. $\frac{\sqrt{5}}{3}$ **19.** $\frac{4\sqrt{2}}{9}$ **20.** $\frac{1}{u}$ **21.** $\frac{u\sqrt{1-u^2}}{1-u^2}$
22. $\left\{\frac{\pi}{2}, \frac{3\pi}{2}\right\}$ **23.** $\{18.4°, 135°, 198.4°, 315°\}$
24. $\left\{0, \frac{2\pi}{3}, \frac{4\pi}{3}\right\}$ **25.** $\{120°, 240°\}$
26. $\left[0, \frac{\pi}{6}\right] \cup \left[\frac{5\pi}{6}, 2\pi\right)$
27. (a) $\left\{\frac{\pi}{6} + 2n\pi, \frac{5\pi}{6} + 2n\pi, \text{where } n \text{ is any integer}\right\}$
(b) $\{2n\pi, \text{where } n \text{ is any integer}\}$
(c) $\left\{\frac{\pi}{6} + n\pi, \frac{5\pi}{6} + n\pi, \text{where } n \text{ is any integer}\right\}$
28. (a) domain: $[0, 11]$; range: $[0, 100]$ (in hundreds)
(b)

$T(t) = 50 + 50 \cos\left(\frac{\pi}{6}t\right)$

(c) The maximum of 10,000 animals occurs at 0 months (July). The minimum of 0 animals occurs at 6 months (January).
(d) There are 5000 animals at 3 months (October) and at 9 months (April). **(e)** at 2 months (September) and 10 months (May) **(f)** $t = \frac{6}{\pi} \arccos\left(\frac{T - 50}{50}\right)$

CHAPTER 10   APPLICATIONS OF TRIGONOMETRY AND VECTORS

10.1 Exercises *(pages 733–738)*

1. C **3.** no **5.** no **7.** yes **9.** yes **11.** $\sqrt{3}$
13. $\frac{5\sqrt{6}}{3}$ **15.** $C = 95°$, $b = 13$ m, $a = 11$ m
17. $B = 37.3°$, $a = 38.5$ ft, $b = 51.0$ ft **19.** $B = 48°$,
$a = 11$ m, $b = 13$ m **21.** $A = 36.54°$, $a = 28.10$ m,
$b = 44.17$ m **23.** $A = 49° 40'$, $b = 16.1$ cm,
$c = 25.8$ cm **25.** $C = 91.9°$, $a = 490$ ft, $c = 847$ ft
27. $B = 109.9°$, $a = 27.01$ m, $c = 21.36$ m
29. $A = 34.72°$, $a = 3326$ ft, $c = 5704$ ft **31.** A
33. (a) $4 < h < 5$ **(b)** $h = 4$ or $h \geq 5$ **(c)** $h < 4$

35. 2 **37.** 0 **39.** $B_1 = 49.1°$, $C_1 = 101.2°$,
$c_1 = 53.9$ ft; $B_2 = 130.9°$, $C_2 = 19.4°$,
$c_2 = 18.2$ ft **41.** no such triangle **43.** $B = 27.19°$,
$C = 10.68°$, $c = 2.203$ ft **45.** $B = 20.6°$, $C = 116.9°$,
$c = 20.6$ ft **47.** no such triangle **49.** $B_1 = 49° 20'$,
$C_1 = 92° 00'$, $c_1 = 15.5$ km; $B_2 = 130° 40'$, $C_2 = 10° 40'$,
$c_2 = 2.88$ km **51.** $A_1 = 52° 10'$, $C_1 = 95° 00'$,
$c_1 = 9520$ cm; $A_2 = 127° 50'$, $C_2 = 19° 20'$, $c_2 = 3160$ cm
53. The Pythagorean theorem only applies to right triangles.
55. If we are given only three sides, then any equation from the law of sines will contain two unknowns. At least one angle measure must be given. **57.** 118 m **59.** 1.93 mi
61. 10.4 in. **63.** 111° **65.** 5.1 mi; 7.2 mi
67. 26.5 km **69.** 2.18 km **71.** 38.3 cm
73. $\sin C = 1$; $C = 90°$; right triangle **75.** The longest side must be opposite the largest angle. Thus, B must be larger than A, which is impossible because A and B cannot both be obtuse. **77.** about 419,000 km, which compares favorably to the actual value **79. (b)** $1.12257R^2$ **(c) (i)** 8.77 in.^{2}
(ii) 5.32 in.^{2} **(iii)** red **80.** increasing
81. If $B < A$, then $\sin B < \sin A$ because $y = \sin x$ is increasing on $\left(0, \frac{\pi}{2}\right)$. **82.** $b = \frac{a \sin B}{\sin A}$
83. $b = \frac{a \sin B}{\sin A} = a \cdot \frac{\sin B}{\sin A}$. Since $\frac{\sin B}{\sin A} < 1$,
$b = a \cdot \frac{\sin B}{\sin A} < a \cdot 1 = a$, so $b < a$.
84. A is obtuse and thus must be the largest angle. However, $b > a$, and B would have to be larger than A, which is impossible.

10.2 Exercises *(pages 744–749)*

1. (a) SAS **(b)** law of cosines **3. (a)** SSA **(b)** law of sines
5. (a) ASA **(b)** law of sines **7. (a)** ASA
(b) law of sines **9.** 7 **11.** 30° **13.** $a = 5.4$,
$B = 40.7°$, $C = 78.3°$ **15.** $A = 22.3°$, $B = 108.2°$,
$C = 49.5°$ **17.** $A = 33.6°$, $B = 50.7°$, $C = 95.7°$
19. $c = 2.83$ in., $A = 44.9°$, $B = 106.8°$
21. $c = 6.46$ m, $A = 53.1°$, $B = 81.3°$ **23.** $A = 82°$,
$B = 37°$, $C = 61°$ **25.** $C = 102° 10'$, $B = 35° 50'$,
$A = 42° 00'$ **27.** $C = 84° 30'$, $B = 44° 40'$,
$A = 50° 50'$ **29.** $a = 156$ cm, $B = 64° 50'$,
$C = 34° 30'$ **31.** $b = 9.529$ in., $A = 64.59°$,
$C = 40.61°$ **33.** $a = 15.7$ m, $B = 21.6°$, $C = 45.6°$
35. $A = 30°$, $B = 56°$, $C = 94°$
37. case 1; $b = 46.93$, $c = 28.98$, $A = 100° 31'$
39. case 3; $b = 4.659$, $A = 46° 16'$, $C = 75° 32'$
41. case 4; $A = 51° 04'$, $B = 34° 21'$, $C = 94° 35'$
43. case 2 (two triangles); $c_1 = 10.04$, $B_1 = 68° 43'$,
$C_1 = 70° 05'$; $c_2 = 4.933$, $B_2 = 111° 17'$, $C_2 = 27° 31'$

45. case 2 (one triangle); $b = 59.87, A = 11° 19'$,
$B = 154° 22'$ **47.** case 2 (no triangle)
49. The absolute value of $\cos \theta$ will be greater than 1; your calculator will give you an error message (or a complex number) when using the inverse cosine function.   **51.** 257 m
53. 281 km   **55.** 10.9 mi   **57.** 1450 ft   **59.** 18 ft
61. 5500 m   **63.** 1473 m   **65.** 350°   **67.** 2000 km
69. 163.5°   **71.** 22 ft   **73.** Since A is obtuse,
$90° < A < 180°$. The cosine of a quadrant II angle is negative.
74. In $a^2 = b^2 + c^2 - 2bc \cos A$, $\cos A$ is negative, so
$a^2 = b^2 + c^2 + $ (a positive quantity). Thus, $a^2 > b^2 + c^2$.
75. $b^2 + c^2 > b^2$ and $b^2 + c^2 > c^2$. If $a^2 > b^2 + c^2$, then
$a^2 > b^2$ and $a^2 > c^2$ from which $a > b$ and $a > c$ because
$a, b,$ and c are nonnegative.   **76.** Because A is obtuse, it is
the largest angle, so the longest side should be a, not c.
77. 46.4 m²   **79.** 78 m²   **81.** 3650 ft²   **83.** 228 yd²
85. 33 cans   **87.** 100 m²   **89.** The area and perimeter
are both 36.
91. (a) 87.8° and 92.2° both appear possible.   **(b)** 92.2°
(c) With the law of cosines we are required to find the inverse
cosine of a negative number. Therefore, we know that angle C
is greater than 90°.

10.3 Exercises *(pages 758–763)*

1. **m** and **p**; **n** and **r**
3. **m** or **p** equal 2**t**, or **t** is one half **m** or **p**; also **m** = 1**p** and
n = 1**r**

5.   **7.**   **9.**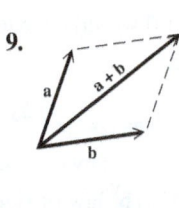

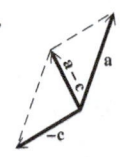 **11.**   **13.** 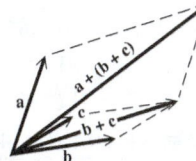   **15.**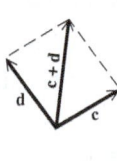

17. (a) $\langle -4, 16 \rangle$ **(b)** $\langle -12, 0 \rangle$ **(c)** $\langle 8, -8 \rangle$
19. (a) $\langle 8, 0 \rangle$ **(b)** $\langle 0, 16 \rangle$ **(c)** $\langle -4, -8 \rangle$
21. (a) $\langle 0, 12 \rangle$ **(b)** $\langle -16, -4 \rangle$ **(c)** $\langle 8, -4 \rangle$
23. (a) 4**i**   **(b)** 7**i** + 3**j**   **(c)** -5**i** + **j**
25. (a) $\langle -2, 4 \rangle$ **(b)** $\langle 7, 4 \rangle$ **(c)** $\langle 6, -6 \rangle$
27. $\langle 2, 8 \rangle$ **29.** $\langle 6, -2 \rangle$ **31.** $\langle -20, -15 \rangle$
33. $\langle 3\sqrt{3}, 3 \rangle$ **35.** $\langle -\frac{9\sqrt{2}}{2}, -\frac{9\sqrt{2}}{2} \rangle$
37. $\langle 3.06, 2.57 \rangle$ **39.** $\langle 4.10, -2.87 \rangle$ **41.** 94.2 lb

43. 24.4 lb   **45.** -5**i** + 8**j**   **47.** 2**i**
49. $4\sqrt{2}$**i** + $4\sqrt{2}$**j**   **51.** -0.25**i** + 0.54**j**   **53.** $\sqrt{2}$; 45°
55. 16; 315°   **57.** 17; 331.9°   **59.** 6; 180°   **61.** 7
63. -3   **65.** 20   **67.** 135°   **69.** 90°
71. $\cos^{-1} \frac{4}{5} \approx 36.87°$ **73.** $\cos^{-1} \frac{7\sqrt{2}}{10} \approx 8.13°$
75. -6   **77.** -24   **79.** orthogonal
81. not orthogonal   **83.** not orthogonal
85. 190 lb and 283 lb, respectively   **87.** 18°
89. 2.4 tons   **91.** 17.5°   **93.** weight: 64.8 lb;
tension: 61.9 lb   **95.** 13.5 mi; 50.4°   **97.** 39.2 km
99. current: 3.5 mph; motorboat: 19.7 mph
101. bearing: 237°; ground speed: 470 mph
103. ground speed: 161 mph; airspeed: 156 mph
105. bearing: 74°; ground speed: 202 mph
107. bearing: 358°; airspeed: 170 mph
109. ground speed: 230 km per hr; bearing: 167°
111. (a) $|\mathbf{R}| = \sqrt{5} \approx 2.2, |\mathbf{A}| = \sqrt{1.25} \approx 1.1$; About
2.2 in. of rain fell. The area of the opening of the rain gauge is
about 1.1 in.².   **(b)** $V = 1.5$; The volume of rain was 1.5 in.³.
(c) **R** and **A** should be parallel and point in opposite directions.

Reviewing Basic Concepts *(page 763)*

1. $B = 74°, b = 16.6, c = 15.3$ **2.** two solutions:
$B_1 = 45.0°, C_1 = 103°, c_1 = 11.0$ or $B_2 = 135.0°, C_2 = 13°$,
$c_2 = 2.5$ **3.** one solution: $A = 22.5°, B = 116.5°$,
$b = 16.4$ **4. (a)** $b = 7.1, A = 63.0°, C = 66.0°$
(b) $A = 110.7°, B = 37.0°, C = 32.3°$
5. 9.6   **6.** 21   **7. (a) i   (b)** 4**i** - 2**j**   **(c)** 11**i** - 7**j**
8. -9; 142.1°   **9.** 207 lb   **10.** 7200 ft

10.4 Exercises *(pages 769–771)*

1. magnitude (length)

3. 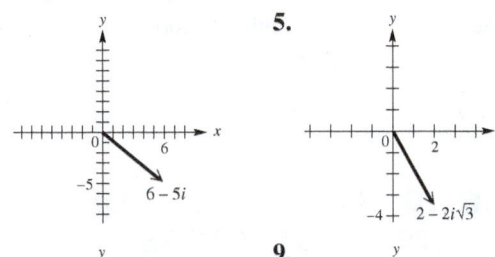   **5.**

7. 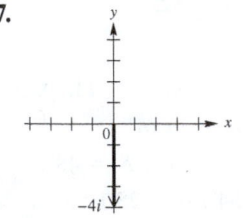   **9.**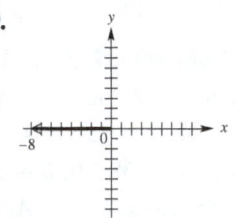

11. $1 - 4i$ **13.** The imaginary part must be 0.
15. 0   **17.** $3 - i$ **19.** $-3 + 3i$ **21.** $2 + 4i$
23. $9 - 2i$ **25.** $\sqrt{2}$ **27.** 13   **29.** 6   **31.** $\sqrt{13}$
33. $\sqrt{2} + i\sqrt{2}$ **35.** 10i **37.** $-2 - 2i\sqrt{3}$
39. $\frac{\sqrt{3}}{2} + \frac{1}{2}i$ **41.** $\frac{5\sqrt{3}}{2} - \frac{5}{2}i$ **43.** $-\sqrt{2}$

45. $3\sqrt{2}(\cos 315° + i \sin 315°)$

47. $2(\cos 60° + i \sin 60°)$ **49.** $2(\cos 270° + i \sin 270°)$

51. $8\left(\cos \frac{\pi}{6} + i \sin \frac{\pi}{6}\right)$ **53.** $2\left(\cos \frac{3\pi}{4} + i \sin \frac{3\pi}{4}\right)$

55. $4(\cos \pi + i \sin \pi)$ **57.** $-3\sqrt{3} + 3i$ **59.** $-4i$

61. $-\frac{15\sqrt{2}}{2} + \frac{15\sqrt{2}}{2}i$ **63.** $-3i$ **65.** $\sqrt{3} - i$

67. -2 **69.** $-\frac{1}{6} - \frac{\sqrt{3}}{6}i$ **71.** $2\sqrt{3} - 2i$

73. $-\frac{1}{2} - \frac{1}{2}i$ **75.** $\sqrt{3} + i$

77. (b) $z_1^2 - 1 = a^2 - b^2 - 1 + 2abi$;

$z_2^2 - 1 = a^2 - b^2 - 1 - 2abi$ **(c)** If $z_1 = a + bi$ and

$z_2 = a - bi$, then $z_1^2 - 1$ and $z_2^2 - 1$ are also conjugates

with the same modulus. Therefore, if z_1 is in the Julia set,

so is z_2. Thus, (a, b) in the Julia set implies $(a, -b)$ is also in

the set.   **(d)** yes   **79.** $1.18 - 0.14i$ **81.** $27.43 + 11.5i$

83. To square a complex number in trigonometric form,

square the modulus r and double the argument θ.

85. (a) $\frac{\pi}{2}$ **(b)** π

10.5 Exercises *(pages 776–777)*

1. $27i$ **3.** 1 **5.** 8 **7.** $\frac{27}{2} - \frac{27\sqrt{3}}{2}i$ **9.** $8i$

11. $-8 - 8i\sqrt{3}$ **13.** -1 **15.** $8i$ **17.** $128 + 128i$

19. $\cos 0° + i \sin 0°$,   **21.** $2 \operatorname{cis} 20°$,

$\cos 120° + i \sin 120°$, $2 \operatorname{cis} 140°$,

$\cos 240° + i \sin 240°$ $2 \operatorname{cis} 260°$

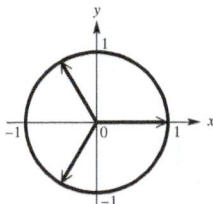

23. $2(\cos 90° + i \sin 90°)$,   **25.** $4(\cos 60° + i \sin 60°)$,

$2(\cos 210° + i \sin 210°)$, $4(\cos 180° + i \sin 180°)$,

$2(\cos 330° + i \sin 330°)$ $4(\cos 300° + i \sin 300°)$

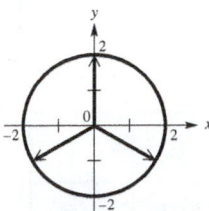

27. $\sqrt[3]{2}(\cos 20° + i \sin 20°)$,   **29.** $\sqrt[3]{4}(\cos 50° + i \sin 50°)$,

$\sqrt[3]{2}(\cos 140° + i \sin 140°)$, $\sqrt[3]{4}(\cos 170° + i \sin 170°)$,

$\sqrt[3]{2}(\cos 260° + i \sin 260°)$ $\sqrt[3]{4}(\cos 290° + i \sin 290°)$

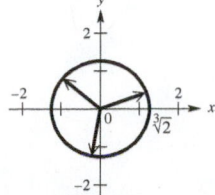 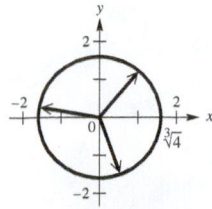

31. $1 + i\sqrt{3}, -1 - i\sqrt{3}$ **33.** $-1, \frac{1}{2} + \frac{\sqrt{3}}{2}i, \frac{1}{2} - \frac{\sqrt{3}}{2}i$

35. $\frac{\sqrt{2}}{2} + \frac{\sqrt{2}}{2}i, -\frac{\sqrt{2}}{2} - \frac{\sqrt{2}}{2}i$ **37.** $-4i, 2\sqrt{3} + 2i$,

$-2\sqrt{3} + 2i$ **39.** $\pm 3, \pm 3i$

41. (a)   **(b)**

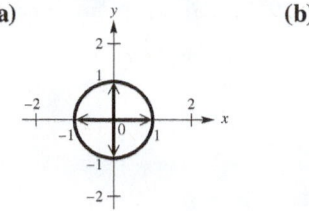 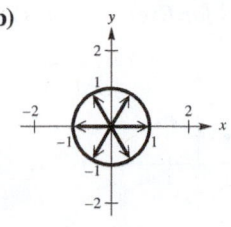

$\cos 0° + i \sin 0°$, $\cos 0° + i \sin 0°$,

$\cos 90° + i \sin 90°$, $\cos 60° + i \sin 60°$,

$\cos 180° + i \sin 180°$, $\cos 120° + i \sin 120°$,

$\cos 270° + i \sin 270°$ $\cos 180° + i \sin 180°$,

$\cos 240° + i \sin 240°$,

$\cos 300° + i \sin 300°$

43. The argument for a positive real number is $\theta = 0°$. By the

nth root theorem with $k = 0$, $\alpha = 0°$. Thus, one nth root has

an argument of $0°$ and it must be real.

45. $(x + 2)(x^2 - 2x + 4)$ **46.** $x + 2 = 0$ implies

$x = -2$.   **47.** $x^2 - 2x + 4 = 0$ implies $x = 1 + i\sqrt{3}$ or

$x = 1 - i\sqrt{3}$.   **48.** $2 \operatorname{cis} 60°, 2 \operatorname{cis} 180°, 2 \operatorname{cis} 300°$

49. $1 + i\sqrt{3}, -2, 1 - i\sqrt{3}$ **50.** They are the same.

51. $\{\cos 45° + i \sin 45°, \cos 135° + i \sin 135°,$

$\cos 225° + i \sin 225°, \cos 315° + i \sin 315°\}$

53. $\{\cos 18° + i \sin 18°, \cos 90° + i \sin 90°,$

$\cos 162° + i \sin 162°, \cos 234° + i \sin 234°,$

$\cos 306° + i \sin 306°\}$ **55.** $\{\cos 60° + i \sin 60°,$

$\cos 180° + i \sin 180°, \cos 300° + i \sin 300°\}$

57. $\{2(\cos 0° + i \sin 0°), 2(\cos 120° + i \sin 120°),$

$2(\cos 240° + i \sin 240°)\}$

59. (a) blue   **(b)** red   **(c)** green

Reviewing Basic Concepts *(page 778)*

1. $1 + i\sqrt{3}$ **2.** 5 **3.** $2(\cos 135° + i \sin 135°)$

4. $8(\cos 180° + i \sin 180°) = -8$

5. $2(\cos 90° + i \sin 90°) = 2i$ **6.** $64 \operatorname{cis} 51°$

7. -64 **8.** $-4, 2 + 2i\sqrt{3}, 2 - 2i\sqrt{3}$

9. $1 + i, -1 - i$ **10.** $\{\operatorname{cis} 60°, \operatorname{cis} 180°, \operatorname{cis} 300°\}$

10.6 Exercises *(pages 784–786)*

1. (a) II   **(b)** I   **(c)** IV   **(d)** III

Graphs for Exercises   *Graphs for Exercises*
3, 5, 7, and 9   *11 and 13*

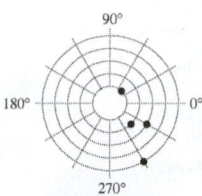 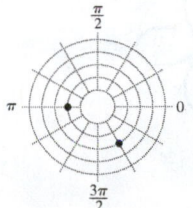

Answers may vary in Exercises 3–13.

3. $(1, 405°), (-1, 225°)$ **5.** $(-2, 495°), (2, 315°)$

7. $(5, 300°), (-5, 120°)$ **9.** $(-3, 150°), (3, -30°)$

11. $\left(3, \frac{11\pi}{3}\right), \left(-3, \frac{2\pi}{3}\right)$ **13.** $(2, \pi), (-2, 2\pi)$

Graphs for Exercises 15, 17, 19, 21, 23

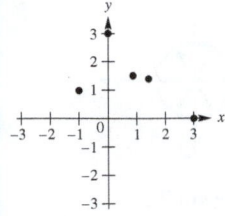

Answers may vary in Exercises 15–23.

15. $\left(\sqrt{2}, 135°\right), \left(-\sqrt{2}, 315°\right)$ **17.** $(3, 90°), (-3, 270°)$

19. $(2, 45°), (-2, 225°)$ **21.** $\left(\sqrt{3}, 60°\right), \left(-\sqrt{3}, 240°\right)$

23. $(3, 0°), (-3, 180°)$

25. $r = \dfrac{4}{\cos\theta - \sin\theta}$ **27.** $r = 4$ or $r = -4$

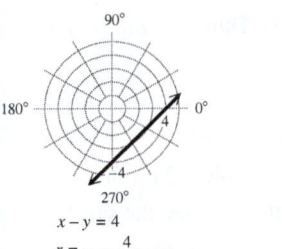

$x - y = 4$
$r = \dfrac{4}{\cos\theta - \sin\theta}$

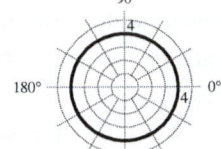

$x^2 + y^2 = 16$
$r = 4$ or $r = -4$

29. $r = \dfrac{5}{2\cos\theta + \sin\theta}$ **31.** C   **33.** A

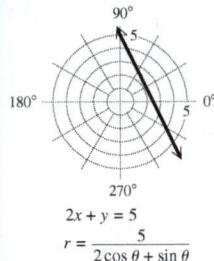

$2x + y = 5$
$r = \dfrac{5}{2\cos\theta + \sin\theta}$

35. cardioid

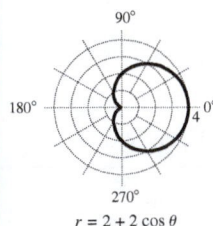

$r = 2 + 2\cos\theta$

37. limaçon

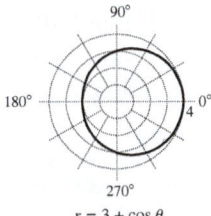

$r = 3 + \cos\theta$

39. four-leaved rose

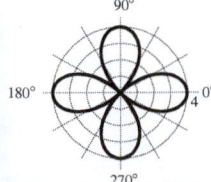

$r = 4\cos 2\theta$

41. lemniscate

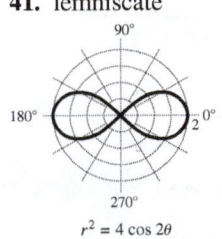

$r^2 = 4\cos 2\theta$

43. cardioid

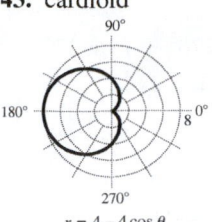

$r = 4 - 4\cos\theta$

45.

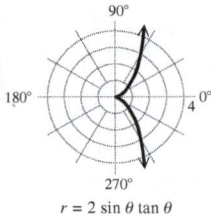

$r = 2\sin\theta \tan\theta$

47. To graph (r, θ), $r < 0$, locate θ, add 180° to it, and move $|r|$ units along the terminal ray of $\theta + 180°$ in standard position.   **49.** θ must be coterminal with 0°, 90°, 180°, or 270°; quadrantal   **51.** It would be reflected across the line $\theta = \frac{\pi}{2}$ (y-axis).   **53.** The value of a determines the length of the leaves. The value of n determines the number of leaves—n leaves if n is odd and $2n$ leaves if n is even.

55. $x^2 + (y - 1)^2 = 1$ **57.** $y^2 = 4(x + 1)$

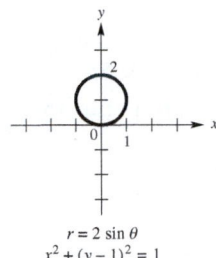

$r = 2\sin\theta$
$x^2 + (y - 1)^2 = 1$

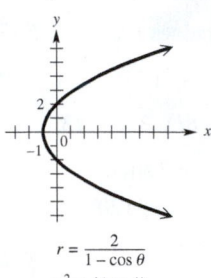

$r = \dfrac{2}{1 - \cos\theta}$
$y^2 = 4(x + 1)$

59. $(x + 1)^2 + (y + 1)^2 = 2$ **61.** $x = 2$

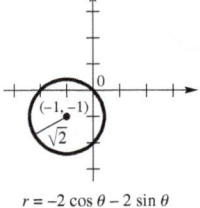

$r = -2\cos\theta - 2\sin\theta$
$(x + 1)^2 + (y + 1)^2 = 2$

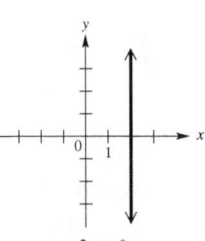

$r = 2\sec\theta$
$x = 2$

63. $x + y = 2$ **65.**

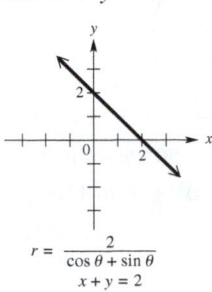

$r = \dfrac{2}{\cos\theta + \sin\theta}$
$x + y = 2$

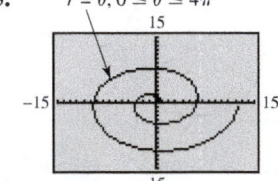

$r = \theta, 0 \le \theta \le 4\pi$

67. $r = 1.5\theta, -4\pi \le \theta \le 4\pi$

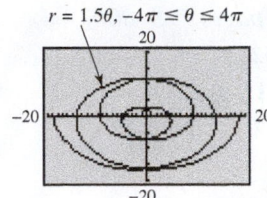

69. $\left(2, \frac{\pi}{6}\right), \left(2, \frac{5\pi}{6}\right)$ **71.** $\left(\frac{4 + \sqrt{2}}{2}, \frac{\pi}{4}\right), \left(\frac{4 - \sqrt{2}}{2}, \frac{5\pi}{4}\right)$

73. (a)

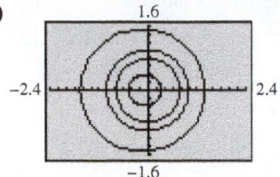

(b)

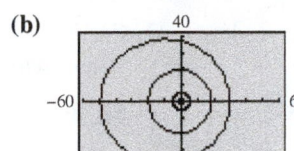

Earth is closest to the sun.

(c) No, it is not.

10.7 Exercises *(pages 792–795)*

1. C **3.** A **5.** $x^2 + y^2 = 9$; circle with radius 3

7. $y = 2 - x$; line segment **9.** $\frac{y^2}{4} - \frac{x^2}{9} = 1$; hyperbola (upper branch)

11. (a)

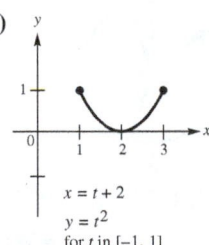

(b) $y = x^2 - 4x + 4,$ for x in $[1, 3]$

13. (a)

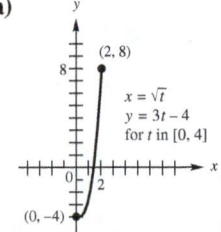

(b) $y = 3x^2 - 4$, for x in $[0, 2]$

15. (a)

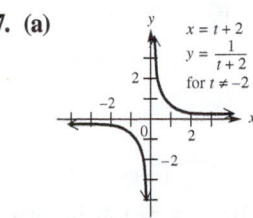

(b) $y = x - 2$, for x in $(-\infty, \infty)$

17. (a)

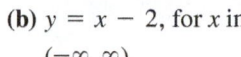

(b) $y = \frac{1}{x}$, for x in $(-\infty, 0) \cup (0, \infty)$

19. (a)

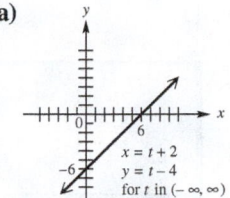

(b) $y = x - 6$, for x in $(-\infty, \infty)$

21. (a) These equations trace a circle of radius 3 once.

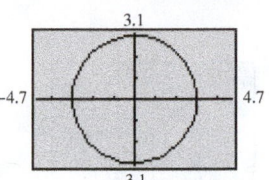

(b) These equations trace a circle of radius 3 twice.

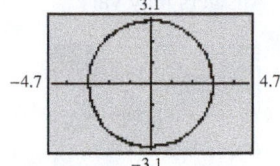

23. (a) These equations trace a circle of radius 3 once counterclockwise, starting at $(3, 0)$.

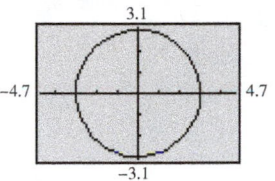

(b) These equations trace a circle of radius 3 once clockwise starting at $(0, 3)$.

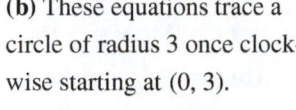

25.

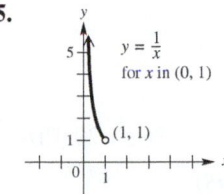

27.

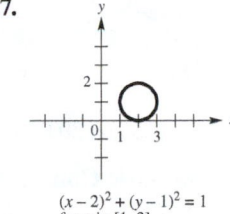

29.

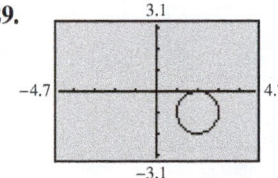

31.

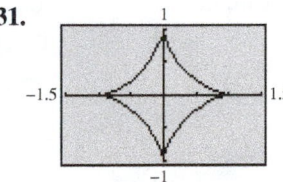

33.

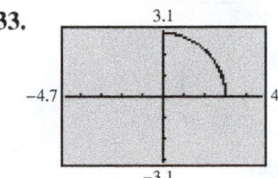

35.

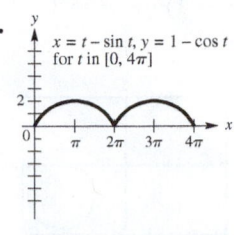

37. F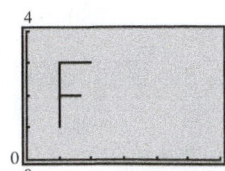

39. D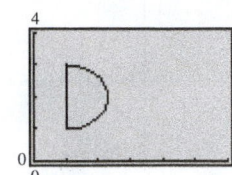

41. $x_1 = 0$, $y_1 = 2t$; $x_2 = t$, $y_2 = 0$ for $0 \le t \le 1$; Answers may vary. **43.** $x_1 = \sin t$, $y_1 = \cos t$; $x_2 = 0$, $y_2 = t - 2$ for $0 \le t \le \pi$; Answers may vary. **45.** Answers may vary.

47. 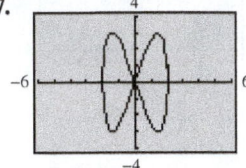 **49.**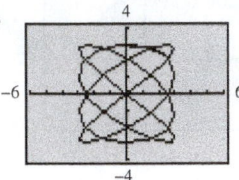

51. (a) $x = 24t$, $y = -16t^2 + 24\sqrt{3}t$
(b) $y = -\frac{1}{36}x^2 + \sqrt{3}x$ (c) 2.6 sec; 62 ft

53. (a) $x = (88 \cos 20°)t$, $y = 2 - 16t^2 + (88 \sin 20°)t$
(b) $y = 2 - \frac{x^2}{484 \cos^2 20°} + (\tan 20°)x$
(c) 1.9 sec; 161 ft **55.** about 1456 ft

57. (a)  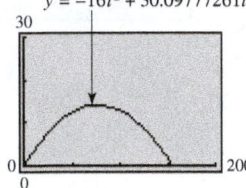 (b) 20.0°

(c) $x = (88 \cos 20.0°)t$, $y = -16t^2 + (88 \sin 20.0°)t$

Reviewing Basic Concepts *(page 795)*

1. IV **2.** $(2\sqrt{2}, 135°)$, $(-2\sqrt{2}, -45°)$; Answers may vary.
3. cardioid

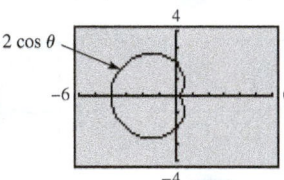

4. $(x - 1)^2 + y^2 = 1$ **5.** $r = \dfrac{6}{\cos \theta + \sin \theta}$
6. $\dfrac{x^2}{4} + \dfrac{y^2}{16} = 1$
7. 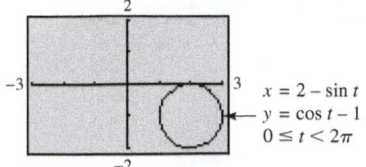 **8.** 285 ft

Chapter 10 Review Exercises *(pages 798–800)*

1. 63.7 m **3.** 19.87°, or 19° 52′ **5.** 25° 00′
7. 173 ft **9.** 153,600 m² **11.** 0.234 km²
13. 2.7 mi **15.** 1.91 mi **17.** 11 ft **19.** 77.1°
21. No, since $a + b = c$.
23. It becomes the Pythagorean theorem.
25.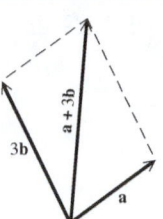

27. 15; 126.9°
29. $\langle 17.9, 66.8 \rangle$
31. (a) 14 (b) about 52.13°
33. Yes they are, because $\mathbf{u} \cdot \mathbf{v} = 0$.
35. 2800 newtons; 30.4°
37. bearing: 7° 20′; speed: 8.3 mph

39.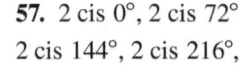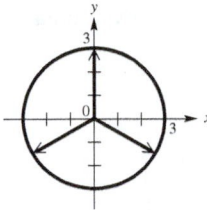

41. $5 + 4i$ **43.** $2\sqrt{2}\,\text{cis}\,135°$
45. $-\sqrt{2} - i\sqrt{2}$ **47.** $-30i$
49. $-\frac{1}{8} + \frac{\sqrt{3}}{8}i$ **51.** $8i$
53. $-\frac{1}{2} - \frac{\sqrt{3}}{2}i$

55. 3 cis 90°, 3 cis 210°, 3 cis 330°

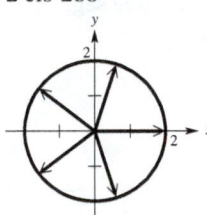

57. 2 cis 0°, 2 cis 72°, 2 cis 144°, 2 cis 216°, 2 cis 288°

59. $(-6\sqrt{2}, -6\sqrt{2})$ **61.** $(6\sqrt{2}, 135°)$

63. 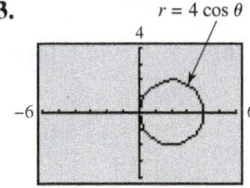 **65.**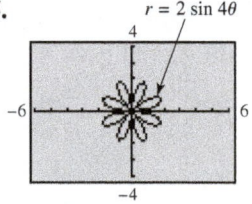

67. $y^2 = -6(x - \frac{3}{2})$ **69.** $(x - \frac{1}{2})^2 + (y - \frac{1}{2})^2 = \frac{1}{2}$
71. $r = -\dfrac{3}{\cos \theta}$ **73.** $r = \dfrac{\tan \theta}{\cos \theta}$ **75.** $y^2 = -\frac{1}{2}(x - 1)$, or $2y^2 + x - 1 = 0$, for x in $[-1, 1]$

77.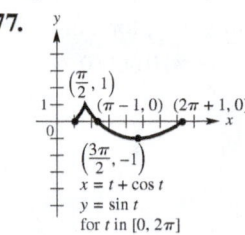

Chapter 10 Test *(pages 801–802)*

1. 137.5°   **2.** 179 km   **3.** 49.0°   **4.** 168 sq units

5. 18 sq units   **6.** (a) $b > 10$ (b) none   (c) $b \le 10$

7. $a = 40$ m, $B = 41°$, $C = 79°$

8. $B_1 = 58° 30'$, $A_1 = 83° 00'$, $a_1 = 1250$ in.;
$B_2 = 121° 30'$, $A_2 = 20° 00'$, $a_2 = 431$ in.

9. $|\mathbf{v}| = 10$; $\theta = 126.9°$

10. (a) $\langle 1, -3 \rangle$ (b) $\langle -6, 18 \rangle$ (c) -20 (d) $\sqrt{10}$

11. 41.8°   **12.** 1.5 mi   **13.** $\langle -346, 451 \rangle$

14. 2.64 mi   **15.** 14 m   **16.** bearing 357°;
airspeed: 220 mph   **17.** 18.7°

18. $7 - 3i$

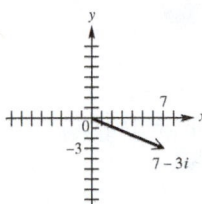

19. (a) $-i$ (b) $2i$ **20.** (a) $3(\cos 90° + i \sin 90°)$
(b) $\sqrt{5}$ cis 63.43°   (c) $2(\cos 240° + i \sin 240°)$

21. (a) $\frac{3\sqrt{3}}{2} + \frac{3}{2}i$ (b) $3.06 + 2.57i$ (c) $3i$

22. (a) $16(\cos 50° + i \sin 50°)$ (b) $2\sqrt{3} + 2i$

(c) $4\sqrt{3} + 4i$ **23.** 2 cis 67.5°, 2 cis 157.5°, 2 cis 247.5°,

2 cis 337.5°   **24.** Answers may vary.

(a) $(5, 90°)$, $(5, -270°)$ (b) $\left(2\sqrt{2}, 225° \right)$, $\left(2\sqrt{2}, -135° \right)$

25. (a) $\left(\frac{3\sqrt{2}}{2}, -\frac{3\sqrt{2}}{2} \right)$ (b) $(0, -4)$

26. cardioid

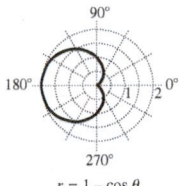

$r = 1 - \cos \theta$

27. three-leaved rose

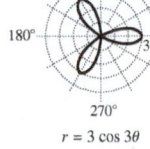

$r = 3 \cos 3\theta$

28. (a) $x - 2y = -4$

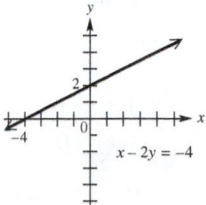

$x - 2y = -4$

(b) $x^2 + y^2 = 36$

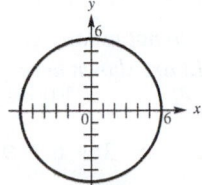

$x^2 + y^2 = 36$

29.

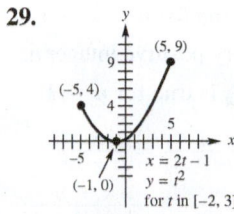

$x = 2t - 1$
$y = t^2$
for t in $[-2, 3]$

30.

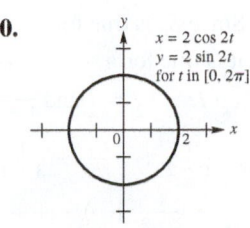

$x = 2 \cos 2t$
$y = 2 \sin 2t$
for t in $[0, 2\pi]$

CHAPTER 11   FURTHER TOPICS IN ALGEBRA

11.1 Exercises *(pages 811–813)*

1. 14, 18, 22, 26, 30   **3.** 1, 2, 4, 8, 16

5. $0, \frac{1}{9}, \frac{2}{27}, \frac{1}{27}, \frac{4}{243}$ **7.** $-2, 4, -6, 8, -10$

9. $1, \frac{7}{6}, 1, \frac{5}{6}, \frac{19}{27}$ **11.** The nth term is the term that is in

position n. For example, in the sequence 2, 4, 6, 8, . . . , the

third term is 6, which is in position 3. The nth term here is

given by $2n$.   **13.** finite   **15.** finite   **17.** infinite

19. finite   **21.** $-2, 1, 4, 7$ **23.** 1, 1, 2, 3

25. 5, 21, 72, 228   **27.** 2, 3, 6, 18   **29.** 35   **31.** $\frac{25}{12}$

33. 288   **35.** 3   **37.** -18 **39.** $\frac{728}{9}$ **41.** 28

43. 343   **45.** $-2 - 1 + 0 + 1 + 2$

47. $-1 + 1 + 3 + 5 + 7$ **49.** $-10 - 4 + 0$

51. $0 + \frac{1}{2} + \frac{2}{3} + \frac{3}{4}$ **53.** $-3.5 + 0.5 + 4.5 + 8.5$

55. $0 + 4 + 16 + 36$ **57.** $-1 - \frac{1}{3} - \frac{1}{5} - \frac{1}{7}$

59. 600   **61.** 1240   **63.** 90   **65.** 220

67. 304   **69.** 3,201,280   **71.** 363,055   **73.** $\sum\limits_{i=1}^{100} \frac{2}{5i}$

75. $\sum\limits_{i=1}^{9} \frac{1}{i}$ **77.** converges to $\frac{1}{2}$ **79.** diverges

81. converges to $e \approx 2.71828$ **83.** 1.081123534; 3.1407;
It is accurate to three decimal places rounded.

85. (a) $N_{j+1} = 2N_j$ for $j \ge 1$ (b) 1840

(c) 15,000

(d) The growth is very rapid. Since there is a doubling of the bacteria at equal intervals, their growth is exponential.

87. (a) 2.718254, $e \approx 2.718282$ (b) 0.367857,
$e^{-1} \approx 0.367879$ (c) 1.648721, $\sqrt{e} = e^{1/2} \approx 1.648721$

11.2 Exercises *(pages 819–821)*

1. 3   **3.** -5 **5.** $x + 2y$ **7.** 8, 14, 20, 26, 32

9. 5, 3, 1, $-1, -3$ **11.** 14, 12, 10, 8, 6

13. $a_8 = 19$; $a_n = 2n + 3$ **15.** $a_8 = 7$; $a_n = n - 1$

17. $a_8 = -6$; $a_n = -2n + 10$

19. $a_8 = -3$; $a_n = 4.5n - 39$

21. $a_8 = x + 21$; $a_n = x + 3n - 3$

23. $a_8 = \pi + 7\sqrt{e}$; $a_n = \pi + (n - 1)\sqrt{e}$ **25.** 3

27. 5   **29.** 28   **31.** 5   **33.** 215   **35.** 125

37. 230   **39.** 77.5   **41.** $a_1 = 7$; $d = 5$

43. $a_1 = 1$; $d = -\frac{20}{11}$

In Exercises 45–49, D is the domain and R is the range.

45. $a_n = n - 3$; D: $\{1, 2, 3, 4, 5, 6\}$; R: $\{-2, -1, 0, 1, 2, 3\}$

47. $a_n = -\frac{1}{2}n + 3$; D: $\{1, 2, 3, 4, 5, 6\}$;
R: $\{0, 0.5, 1, 1.5, 2, 2.5\}$

49. $a_n = -20n + 30$; D: $\{1, 2, 3, 4, 5\}$;
R: $\{-70, -50, -30, -10, 10\}$
51. 80   **53.** 1275   **55.** 1739   **57.** 4100   **59.** 18
61. 140   **63.** -684 **65.** 500,500   **67.** 328.3
69. 172.884   **71.** 1281   **73.** 4680   **75.** 54,800
77. 713 in.   **79. (a)** 5.8 cm   **(b)** 127.2 cm

11.3 Exercises *(pages 827–830)*

1. $\frac{5}{3}, 5, 15, 45$ **3.** $\frac{5}{8}, \frac{5}{4}, \frac{5}{2}, 5, 10$
5. $a_5 = 80$; $a_n = 5(-2)^{n-1}$
7. $a_5 = 108$; $a_n = \frac{4}{3}(-3)^{n-1}$
9. $a_5 = -729$; $a_n = -9(-3)^{n-1}$
11. $a_5 = -324$; $a_n = -4(3)^{n-1}$
13. $a_5 = \frac{125}{4}$; $a_n = \frac{4}{5}\left(\frac{5}{2}\right)^{n-1}$ **15.** $a_5 = \frac{5}{8}$; $a_n = 10\left(-\frac{1}{2}\right)^{n-1}$
17. $a_1 = 125$; $r = \frac{1}{5}$ **19.** $a_1 = -2$; $r = \frac{1}{2}$ **21.** 682
23. $\frac{99}{8}$ **25.** 860.95   **27.** 363   **29.** $\frac{189}{4}$ **31.** 688
33. -508 **35.** 315   **37.** $-\frac{15}{8}$
39. The sum exists if $|r| < 1$.   **41.** $\frac{8}{9}$ **43.** $\frac{5}{11}$ **45.** $\frac{14}{37}$
47. 2; The sum does not converge.   **49.** $\frac{1}{2}$ **51.** $\frac{128}{7}$
53. $\frac{1000}{9}$ **55.** $\frac{8}{3}$ **57.** 4   **59.** $\frac{3}{7}$ **61.** $\frac{1}{4}$ **63.** $\frac{2}{15}$
65. -97.739 **67.** 0.212   **69.** \$10,582.80
71. \$27,224.22   **73. (a)** $a_1 = 1169$; $r = 0.916$
(b) $a_{10} = 531$; $a_{20} = 221$; This means that a person who is
10 yr from retirement should have savings of 531% of his or
her annual salary; a person 20 yr from retirement should have
savings of 221% of his or her annual salary.
75. (a) $a_n = a_1 \cdot 2^{n-1}$ **(b)** 15 (rounded from 14.28 since n
must be a natural number)   **(c)** 560 min, or 9 hr and 20 min
77. 13.4%   **79.** \$26,214.40   **81.** $a_n = 100(1.5)^{n-1}$;
338 flies   **83.** 200 cm   **85.** 62; 2046
87. Option 2 pays better.   **89.** $\frac{1}{64}$ m

Reviewing Basic Concepts *(page 831)*

1. $4, -8, 12, -16, 20$ **2.** 50   **3.** It converges to 1.
4. 8, 6, 4, 2, 0   **5.** -11 **6.** 245
7. $a_3 = -18$; $a_n = -2(-3)^{n-1}$ **8.** geometric; $\frac{7448}{625}$
9. 6   **10.** \$8056.52

11.4 Exercises *(pages 837–840)*

1. 24   **3.** 2   **5.** 6   **7.** 56   **9.** 24
11. Divide the common factor $6 \cdot 5 \cdot 4 \cdot 3 \cdot 2 \cdot 1$ from
both the numerator and denominator to begin. Then multiply
the factors 7 and 8 that remain in the numerator to obtain 56.
13. 5040   **15.** 72   **17.** 5   **19.** 6   **21.** 1
23. 495   **25.** 1,860,480   **27.** 259,459,200
29. 15,504   **31.** 6435   **33. (a)** permutation
(b) permutation   **(c)** combination   **(d)** combination
(e) permutation   **(f)** combination   **(g)** permutation
35. 30   **37. (a)** 27,600   **(b)** 35,152   **(c)** 1104

39. 15   **41. (a)** 17,576,000   **(b)** 17,576,000
(c) 456,976,000   **43.** 720   **45.** 120   **47.** 2730
49. 120; 30,240   **51.** 27,405   **53.** 20   **55.** 105; 1365
57. 28   **59. (a)** 84   **(b)** 10   **(c)** 40   **(d)** 28
61. 1680   **63.** 15   **65.** 479,001,600   **67. (a)** 56
(b) 462   **(c)** 3080   **(d)** 8526   **69.** 4096
71. 1,000,000   **73.** 6

11.5 Exercises *(pages 845–846)*

1. 20   **3.** 35   **5.** 56   **7.** 45   **9.** 1   **11.** 56
13. 4950   **15.** 1   **17.** n **19.** 9   **21.** $16x^4$; $81y^4$
23. $x^6 + 6x^5y + 15x^4y^2 + 20x^3y^3 + 15x^2y^4 + 6xy^5 + y^6$
25. $p^5 - 5p^4q + 10p^3q^2 - 10p^2q^3 + 5pq^4 - q^5$
27. $r^{10} + 5r^8s + 10r^6s^2 + 10r^4s^3 + 5r^2s^4 + s^5$
29. $p^4 + 8p^3q + 24p^2q^2 + 32pq^3 + 16q^4$
31. $2401p^4 + 2744p^3q + 1176p^2q^2 + 224pq^3 + 16q^4$
33. $729x^6 - 2916x^5y + 4860x^4y^2 - 4320x^3y^3 +$
$2160x^2y^4 - 576xy^5 + 64y^6$
35. $\frac{m^6}{64} - \frac{3m^5}{16} + \frac{15m^4}{16} - \frac{5m^3}{2} + \frac{15m^2}{4} - 3m + 1$
37. $4r^4 + \frac{8\sqrt{2}r^3}{m} + \frac{12r^2}{m^2} + \frac{4\sqrt{2}r}{m^3} + \frac{1}{m^4}$
39. $-3584h^3j^5$ **41.** $319,770a^{16}b^{14}$ **43.** $38,760x^6y^{42}$
45. $90,720x^{28}y^{12}$ **47.** 11   **49.** exact: 3,628,800;
approximate: 3,598,695.619   **50.** about 0.830%
51. exact: 479,001,600; approximate: 475,687,486.5;
about 0.692%   **52.** exact: 6,227,020,800; approximate:
6,187,239,475; about 0.639%; As n increases, the percent
error decreases.   **53.** 0.942   **55.** 1.015

Reviewing Basic Concepts *(page 847)*

1. 24   **2.** 210   **3.** 210   **4.** 450   **5.** 72
6. $n + 1$ **7.** $a^4 + 8a^3b + 24a^2b^2 + 32ab^3 + 16b^4$
8. $60x^4y^2$

11.6 Exercises *(pages 852–853)*

1. positive integers (natural numbers)   **3.** $n = 1$ and $n = 2$
Although we do not usually give proofs, the answers to Exer-
cises 5 and 13 are shown here.
5. *Step 1:* $3(1) = 3$ and $\frac{3(1)(1 + 1)}{2} = \frac{6}{2} = 3$, so S_n is true for
$n = 1$. *Step 2:* S_k: $3 + 6 + 9 + \cdots + 3k = \frac{3(k)(k + 1)}{2}$
S_{k+1}: $3 + 6 + 9 + \cdots + 3(k + 1) = \frac{3(k + 1)((k + 1) + 1)}{2}$;
Add $3(k + 1)$ to each side of S_k and simplify until you obtain
S_{k+1}. Since S_n is true for $n = 1$ and S_n is true for $n = k + 1$
when it is true for $n = k$, S_n is true for every positive integer n.
13. *Step 1:* $\frac{1}{1 \cdot 2} = \frac{1}{2}$ and $\frac{1}{1 + 1} = \frac{1}{2}$, so S_n is true for $n = 1$.
Step 2: S_k: $\frac{1}{1 \cdot 2} + \frac{1}{2 \cdot 3} + \frac{1}{3 \cdot 4} + \cdots + \frac{1}{k(k + 1)} = \frac{k}{k + 1}$
S_{k+1}: $\frac{1}{1 \cdot 2} + \frac{1}{2 \cdot 3} + \cdots + \frac{1}{(k + 1)[(k + 1) + 1]} = \frac{k + 1}{(k + 1) + 1}$;

Add $\dfrac{1}{(k+1)[(k+1)+1]}$ to each side of S_k and simplify until you obtain S_{k+1}. Since S_n is true for $n = 1$ and S_n is true for $n = k + 1$ when it is true for $n = k$, S_n is true for every positive integer n. **19.** $n = 1, 2$ **21.** $n = 2, 3, 4$

35. $\dfrac{4^{n-1}}{3^{n-2}}$, or $3\left(\dfrac{4}{3}\right)^{n-1}$ **37.** $2^n - 1$

11.7 Exercises *(pages 859–861)*

1. $S = \{H\}$ **3.** $S = \{(H, H, H), (H, H, T), (H, T, T),$
$(H, T, H), (T, T, T), (T, T, H), (T, H, T), (T, H, H)\}$

5. $S = \{(1, 1), (1, 2), (1, 3), (2, 1), (2, 2), (2, 3),$
$(3, 1), (3, 2), (3, 3)\}$

7. (a) $\{H\}; 1$ **(b)** $\varnothing; 0$ **9. (a)** $\{(1, 1), (2, 2), (3, 3)\}; \frac{1}{3}$

(b) $\{(1, 1), (1, 3), (2, 1), (2, 3), (3, 1), (3, 3)\}; \frac{2}{3}$

(c) $\{(2, 1), (2, 3)\}; \frac{2}{9}$ **11.** A probability of $\frac{6}{5} > 1$ is impossible. **13. (a)** $\frac{1}{5}$ **(b)** 0 **(c)** $\frac{7}{15}$ **(d)** 1 to 4 **(e)** 7 to 8

15. 1 to 4 **17.** $\frac{2}{5}$ **19. (a)** $\frac{1}{2}$ **(b)** $\frac{7}{10}$ **(c)** $\frac{2}{5}$

21. (a) F **(b)** D **(c)** A **(d)** F **(e)** C **(f)** B **(g)** E

23. (a) 0.76 **(b)** 0.24 **25. (a)** 0.62 **(b)** 0.27

(c) 0.11 **(d)** 0.89 **27.** $\dfrac{48}{28{,}561} \approx 0.001681$

29. (a) 0.72 **(b)** 0.70 **(c)** 0.79 **31.** $\dfrac{5}{16} = 0.3125$

33. $\dfrac{1}{32} = 0.03125$ **35.** $\dfrac{1}{2} = 0.5$ **37.** 0.042246

39. 0.026864 **41.** approximately 4.6×10^{-10}

43. approximately 0.875 **45.** 0.49 **47.** 0.51

49. (a) 0.047822 **(b)** 0.976710 **(c)** 0.897110

51. (a) approximately 40.4% **(b)** approximately 4.7%
(c) approximately 0.2%

Reviewing Basic Concepts *(page 862)*

3. $S = \{(H, H), (H, T), (T, H), (T, T)\}$ **4.** $\dfrac{2}{36} = \dfrac{1}{18}$

5. $\dfrac{4}{2{,}598{,}960} \approx 0.0000015$ **6.** $\dfrac{2{,}598{,}956}{2{,}598{,}960} \approx 0.9999985$

7. $\dfrac{3}{10}$ **8.** approximately 0.484

Chapter 11 Review Exercises *(pages 866–868)*

1. $\dfrac{1}{2}, \dfrac{2}{3}, \dfrac{3}{4}, \dfrac{4}{5}, \dfrac{5}{6}$; neither **3.** 8, 10, 12, 14, 16; arithmetic

5. $5, 2, -1, -4, -7$; arithmetic **7.** 12, 10, 8, 6, 4

9. 6, 12, 24, 48, 96 **11.** $-11; 2n - 13$ **13.** 20

15. 222 **17.** -162 **19.** 45 **21.** $10{,}618.27$

23. 1 **25.** $\dfrac{73}{12}$ **27.** 3,126,250 **29.** $\dfrac{4}{3}$ **31.** 36

33. diverges **35.** -10

In Exercises 37 and 39, other answers are possible.

37. $\displaystyle\sum_{i=1}^{15}(-5i + 9)$ **39.** $\displaystyle\sum_{i=1}^{6} 4(3)^{i-1}$

41. 362,880 **43.** 1 **45.** 56

47. $x^4 + 8x^3y + 24x^2y^2 + 32xy^3 + 16y^4$

49. $243x^{5/2} - 405x^{3/2} + 270x^{1/2} - 90x^{-1/2} + 15x^{-3/2} - x^{-5/2}$

51. $-3584x^3y^5$ **53.** $x^{12} + 24x^{11} + 264x^{10} + 1760x^9$

55. Statements containing a set of positive integers as their domains are proved by mathematical induction. **61.** 48

63. 24 **65.** 504 **67. (a)** $\frac{4}{15}$ **(b)** $\frac{2}{3}$ **(c)** 0 **69.** $\frac{1}{26}$

71. $\dfrac{4}{13}$ **73.** 0.86 **75.** 0 **77.** approximately 0.205

Chapter 11 Test *(page 868)*

1. (a) $-3, 4, -5, 6, -7$; neither

(b) $-\dfrac{3}{2}, -\dfrac{3}{4}, -\dfrac{3}{8}, -\dfrac{3}{16}, -\dfrac{3}{32}$; geometric

(c) 2, 3, 7, 13, 27; neither

2. (a) $a_5 = 49$ **(b)** $a_5 = 16$ **3. (a)** 110 **(b)** -1705

4. (a) 2385 **(b)** -186 **(c)** The sum does not exist. **(d)** $\dfrac{108}{7}$

5. (a) 45 **(b)** 35 **(c)** 5040 **(d)** 990

6. (a) $16x^4 - 96x^3y + 216x^2y^2 - 216xy^3 + 81y^4$

(b) $60w^4y^2$ **8.** 24 **9.** 6840 **10.** 6160

11. (a) $\dfrac{1}{26}$ **(b)** $\dfrac{10}{13}$ **(c)** $\dfrac{4}{13}$ **(d)** 3 to 10

12. (a) $\binom{8}{3}\left(\dfrac{1}{6}\right)^3\left(1 - \dfrac{1}{6}\right)^5 \approx 0.104$

(b) $\binom{8}{8}\left(\dfrac{1}{6}\right)^8\left(1 - \dfrac{1}{6}\right)^0 \approx 0.000000595$

CHAPTER R REFERENCE: BASIC ALGEBRAIC CONCEPTS

R.1 Exercises *(page 875)*

1. $(-4)^5$ **3.** 1 **5.** 1 **7.** 2^{10} **9.** $2^3x^{15}y^{12}$

11. $-\dfrac{p^8}{q^2}$ **13.** polynomial; degree 11; monomial

15. polynomial; degree 7; binomial **17.** polynomial;
degree 6; binomial **19.** polynomial; degree 6; trinomial

21. not a polynomial **23.** $m^3 - 4m^2 + 10$

25. $5p^2 - 3p - 4$ **27.** $-6x^3 - 3x^2 - 4x + 4$

29. $15m^2 + 2m - 24$ **31.** $6m^2 + \dfrac{1}{4}m - \dfrac{1}{8}$

33. $2b^5 - 8b^4 + 6b^3$

35. $m^2 + mn - 2km - 2n^2 + 5kn - 3k^2$ **37.** Find the sum of the square of the first term, twice the product of the two terms, and the square of the last term. **39.** $4m^2 - 9$

41. $16m^2 + 16mn + 4n^2$ **43.** $25r^2 + 30rt^2 + 9t^4$

45. $4p^2 - 12p + 9 + 4pq - 6q + q^2$

47. $9q^2 + 30q + 25 - p^2$

49. $9a^2 + 6ab + b^2 - 6a - 2b + 1$

51. $18p^2 - 27pq - 35q^2$ **53.** $p^3 - 7p^2 - p - 7$

55. $16x^2y - 9y^3$ **57.** $6z^2 - 5yz - 4y^2$

59. $9p^2 + 30p + 25$ **61.** $4p^2 - 16$

63. $11y^3 - 18y^2 + 4y$

R.2 Exercises *(pages 880–882)*

1. (a) B **(b)** C **(c)** A **(d)** D

3. $4k^2m^3(1 + 2k^2 - 3m)$ **5.** $2(a + b)(1 + 2m)$

7. $(y + 2)(3y + 2)$ **9.** $(r + 3)(3r - 5)$

11. $(m - 1)(2m^2 - 7m + 7)$ **13.** $(2s + 3)(3t - 5)$

15. $(2x + 3)(5x - 4y)$ **17.** $(t + 2)(t^2 - 3)$

19. Both are correct.   **21.** $8(h - 8)(h + 5)$

23. $9y^2(y - 1)(y - 5)$ **25.** $(7m - 5r)(2m + 3r)$

27. $(3s - t)(4s + 5t)$ **29.** $(5a + m)(6a - m)$

31. $3x^3(2x + 5z)(3x - 5z)$ **33.** $(4p - 5)^2$

35. $5(2p - 5q)^2$ **37.** $(3mn - 2)^2$

39. $(2p + q - 5)^2$ **41.** $(3a + 4)(3a - 4)$

43. $(5s^2 + 3t)(5s^2 - 3t)$ **45.** $(a + b + 4)(a + b - 4)$

47. $(p^2 + 25)(p + 5)(p - 5)$ **49.** B

51. $(2 - a)(4 + 2a + a^2)$ **53.** $(5x - 3)(25x^2 + 15x + 9)$

55. $(3y^3 + 5z^2)(9y^6 - 15y^3z^2 + 25z^4)$

57. $r(r^2 + 18r + 108)$

59. $(3 - m - 2n)(9 + 3m + 6n + m^2 + 4mn + 4n^2)$

61. It is incorrect because a^2 has not been substituted back for u; $(3a^2 - 1)(a^2 + 5)$.   **63.** $(a^2 - 8)(a^2 + 6)$

65. $2(6z - 5)(8z - 3)$ **67.** $(18 - 5p)(17 - 4p)$

69. $(2b + c + 4)(2b + c - 4)$ **71.** $(x + y)(x - y)$

73. $(m - 2n)(p^4 + q)$ **75.** $(2z + 7)^2$

77. $(10x + 7y)(100x^2 - 70xy + 49y^2)$

79. $(5m^2 - 6)(25m^4 + 30m^2 + 36)$

81. $(6m - 7n)(2m + 5n)$ **83.** $(4p - 1)(p + 1)$

85. prime   **87.** $(4t + 13)^2$

R.3 Exercises *(pages 887–889)*

1. $\{x \mid x \neq -6\}$ **3.** $\left\{x \mid x \neq \frac{3}{5}\right\}$

5. $\{x \mid x \text{ is a real number}\}$ **7.** $\left\{x \mid x \neq -\frac{1}{2}, 1\right\}$ **9.** $\frac{5p}{2}$

11. $\frac{8}{9}$ **13.** $\frac{3}{t - 3}$ **15.** $\frac{2x + 4}{x}$ **17.** $\frac{m - 2}{m + 3}$

19. $\frac{2m + 3}{4m + 3}$ **21.** $\frac{25p^2}{9}$ **23.** $\frac{2}{9}$ **25.** $\frac{5x}{y}$ **27.** $\frac{2(a + 4)}{a - 3}$

29. 1   **31.** $\frac{m + 6}{m + 3}$ **33.** $\frac{m - 3}{2m - 3}$ **35.** $\frac{x + y}{x - y}$

37. $\frac{x^2 - xy + y^2}{x^2 + xy + y^2}$ **39.** B, C   **41.** $\frac{19}{6k}$ **43.** 1

45. $\frac{6 + p}{2p}$ **47.** $\frac{137}{30m}$ **49.** $\frac{-2}{(a + 1)(a - 1)}$

51. $\frac{2m^2 + 2}{(m - 1)(m + 1)}$ **53.** $\frac{4}{a - 2}$, or $\frac{-4}{2 - a}$

55. $\frac{3x + y}{2x - y}$, or $\frac{-3x - y}{y - 2x}$ **57.** $\frac{5}{(a - 2)(a - 3)(a + 2)}$

59. $\frac{x - 11}{(x + 4)(x - 3)(x - 4)}$ **61.** $\frac{a^2 + 5a}{(a + 6)(a - 1)(a + 1)}$

63. $\frac{x + 1}{x - 1}$ **65.** $\frac{-1}{x + 1}$ **67.** $\frac{(2 - b)(1 + b)}{b(1 - b)}$

69. $\frac{m^3 - 4m - 1}{m - 2}$

R.4 Exercises *(pages 894–895)*

1. E   **3.** F   **5.** D   **7.** B   **9.** $-\frac{1}{64}$ **11.** 8

13. -2 **15.** 4   **17.** $\frac{1}{9}$ **19.** $\frac{256}{81}$ **21.** $4p^2$

23. $9x^4$ **25.** $\frac{1}{2^7}$ **27.** $\frac{1}{27^3}$ **29.** 1   **31.** $m^{7/3}$

33. $(1 + n)^{5/4}$ **35.** $\frac{6z^{2/3}}{y^{5/4}}$ **37.** $2^6 a^{1/4} b^{37/2}$ **39.** $\frac{r^6}{s^{15}}$

41. $-\frac{1}{ab^3}$ **43.** $12^{9/4}y$ **45.** $\frac{1}{2p^2}$ **47.** $\frac{m^3 p}{n}$

49. $-4a^{5/3}$ **51.** $\frac{1}{(k + 5)^{1/2}}$ **53.** $y - 10y^2$

55. $-4k^{10/3} + 24k^{4/3}$ **57.** $x^2 - x$

59. $r - 2 + r^{-1}$, or $r - 2 + \frac{1}{r}$

61. $k^{-2}(4k + 1)$, or $\frac{4k + 1}{k^2}$

63. $z^{-1/2}(9 + 2z)$, or $\frac{9 + 2z}{z^{1/2}}$

65. $p^{-7/4}(p - 2)$, or $\frac{p - 2}{p^{7/4}}$

67. $(p + 4)^{-3/2}(p^2 + 9p + 21)$, or $\frac{p^2 + 9p + 21}{(p + 4)^{3/2}}$

R.5 Exercises *(pages 900–902)*

1. F   **3.** H   **5.** G   **7.** C   **9.** $\sqrt[3]{(-m)^2}$, or $\left(\sqrt[3]{-m}\right)^2$ **11.** $\sqrt[3]{(2m + p)^2}$, or $\left(\sqrt[3]{2m + p}\right)^2$

13. $k^{2/5}$ **15.** $-3 \cdot 5^{1/2} p^{3/2}$ **17.** A   **19.** $x \geq 0$

21. 5   **23.** -5 **25.** $5\sqrt{2}$ **27.** $3\sqrt[3]{3}$

29. $-2\sqrt[4]{2}$ **31.** $-\frac{3\sqrt{5}}{5}$ **33.** $-\frac{\sqrt[3]{100}}{5}$ **35.** $32\sqrt[3]{2}$

37. $2x^2 z^4 \sqrt{2x}$ **39.** $2zx^2 y \sqrt[3]{2z^2 x^2 y}$ **41.** $np^2 \sqrt[4]{m^2 n^3}$

43. cannot be simplified further   **45.** $\frac{\sqrt{6x}}{3x}$ **47.** $\frac{x^2 y \sqrt{xy}}{z}$

49. $\frac{2\sqrt[3]{x}}{x}$ **51.** $\frac{h\sqrt[4]{9g^3 hr^2}}{3r^2}$ **53.** $\frac{m\sqrt[3]{n^2}}{n}$ **55.** $2\sqrt[4]{x^3 y^3}$

57. $\sqrt[3]{2}$ **59.** $\sqrt[18]{x}$ **61.** $9\sqrt{3}$ **63.** $-2\sqrt{7p}$

65. $7\sqrt[3]{3}$ **67.** $2\sqrt{3}$ **69.** $\frac{13\sqrt[3]{4}}{6}$ **71.** -7

73. 10   **75.** $11 + 4\sqrt{6}$ **77.** $5\sqrt{6}$ **79.** $\frac{8\sqrt{5}}{5}$

81. $\frac{6\sqrt[3]{x}}{x}$ **83.** $\frac{\sqrt{15} - 3}{2}$ **85.** $\frac{3\sqrt{5} - 2\sqrt{3} + 3\sqrt{15} - 6}{33}$

87. $\frac{p(\sqrt{p} - 2)}{p - 4}$ **89.** $\frac{a(\sqrt{a + b} + 1)}{a + b - 1}$

Chapter R Test *(page 902)*

1. (a) $(-3)^9$ **(b)** $2^2 x^6 y^2$ **(c)** 1   **(d)** $\frac{4^2}{5^2}$ **(e)** $-\frac{m^6}{p^3}$

2. (a) $11x^2 - x + 2$ **(b)** $36r^2 - 60r + 25$

(c) $3u^3 + 5u^2 + 2u + 8$ **(d)** $16x^2 - 25$

(e) $25p^2 + 30p + 9$ **3. (a)** $(3x - 7)(2x - 1)$

(b) $(x^2 + 4)(x + 2)(x - 2)$ **(c)** $(z + 2k)(z - 8k)$

(d) $(x - 2)(x^2 + 2x + 4)(y + 3)(y - 3)$

4. (a) $\frac{4}{x^2}$; $\{x \mid x \neq 0\}$

(b) $\frac{1}{k - 1}$; $\{k \mid k \neq -1, 1\}$

(c) $\frac{x - 1}{x + 3}$; $\{x \mid x \neq -3, -2\}$

5. (a) $\frac{x^4(x + 1)}{3(x^2 + 1)}$ **(b)** $\frac{x(4x + 1)}{(x + 2)(x + 1)(2x - 3)}$

(c) $\frac{2a}{2a - 3}$, or $\frac{-2a}{3 - 2a}$ **(d)** $\frac{g^2 - 2}{(g + 2)(g - 2)}$

6. (a) $\frac{1}{49}$ **(b)** $8x^6$ **(c)** $\frac{9}{16}$

7. (a) $\frac{1}{5^2}$ **(b)** $\frac{1}{(x + 2)^{9/10}}$ **(c)** $\frac{n^3}{2m^{11/2}}$

8. (a) 2   **(b)** $\frac{\sqrt[3]{18}}{3}$ **(c)** $3x^2 y^4 \sqrt{2x}$ **(d)** $\frac{\sqrt[4]{p^2 q^3}}{p}$

9. (a) $2\sqrt{2x}$ **(b)** $6 + \sqrt[3]{2} + 4\sqrt[3]{4}$ **(c)** $x - y$

10. $\frac{7(\sqrt{11} + \sqrt{7})}{2}$

APPENDIX B   VECTORS IN SPACE

Appendix B Exercises *(pages 909–910)*

1. xz-plane   **3.** $\langle 5, 3, -2 \rangle$ **5.** $\sqrt{33}$ **7.** $\sqrt{317}$

9. $5\sqrt{29}$ **11.** $\langle 2, -2, 5 \rangle$; $2\mathbf{i} - 2\mathbf{j} + 5\mathbf{k}$

13. $\langle -2, -12, -4 \rangle$; $-2\mathbf{i} - 12\mathbf{j} - 4\mathbf{k}$

15. $\langle -15, -20, 10 \rangle$; $-15\mathbf{i} - 20\mathbf{j} + 10\mathbf{k}$

17. $\langle -2, 2, -5 \rangle$; $\mathbf{QP} = -\mathbf{PQ}$ **19.** $-2\mathbf{i} + 7\mathbf{j} + 13\mathbf{k}$

21. $-7\mathbf{i} + 41\mathbf{j} + 38\mathbf{k}$ **23.** $\sqrt{69}$ **25.** $\sqrt{38}$

27. -39 **29.** 38   **31.** $25.4°$ **33.** $32.0°$ **35.** $90°$

37. $\alpha = 76.1°$; $\beta = 61.2°$; $\gamma = 32.6°$

39. $\alpha = 59.2°$; $\beta = 112.6°$; $\gamma = 140.2°$

41. $60°$ **43.** Their position vectors are scalar multiples of each other.   **45.** 12 work units   **47.** 13 work units

APPENDIX C   POLAR FORM OF CONIC SECTIONS

Appendix C Exercises *(page 913)*

In Exercises 1–11, we give only calculator graphs.

1. 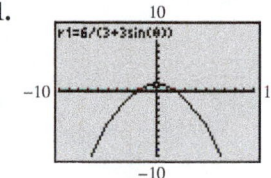   **3.**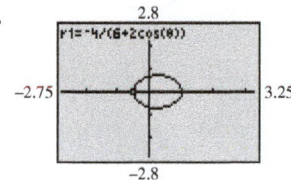

5. 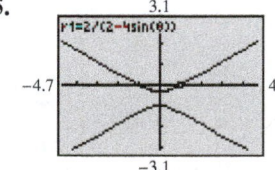   **7.**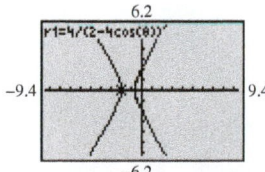

9. 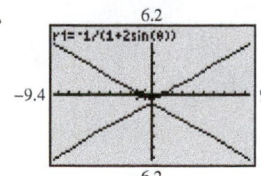   **11.**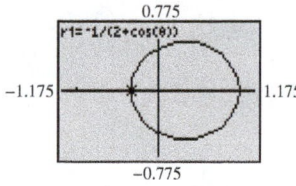

13. $r = \dfrac{3}{1 + \cos\theta}$ **15.** $r = \dfrac{5}{1 - \sin\theta}$

17. $r = \dfrac{20}{5 + 4\cos\theta}$; ellipse   **19.** $r = \dfrac{40}{4 - 5\sin\theta}$; hyperbola

21. ellipse; $8x^2 + 9y^2 - 12x - 36 = 0$

23. hyperbola; $3x^2 - y^2 + 8x + 4 = 0$

25. ellipse; $4x^2 + 3y^2 - 6y - 9 = 0$

27. parabola; $x^2 - 10y - 25 = 0$

APPENDIX D   ROTATION OF AXES

Appendix D Exercises *(page 917)*

1. circle or ellipse or a point   **3.** hyperbola or two intersecting lines   **5.** parabola or one line or two parallel lines

7. $30°$ **9.** $60°$ **11.** $22.5°$

13. 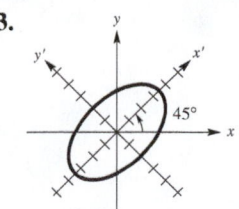   **15.**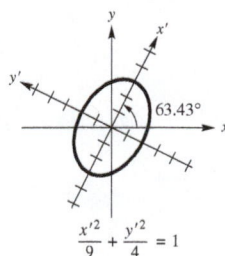
$\dfrac{x'^2}{12} + \dfrac{y'^2}{4} = 1$ $\dfrac{x'^2}{9} + \dfrac{y'^2}{4} = 1$

17. 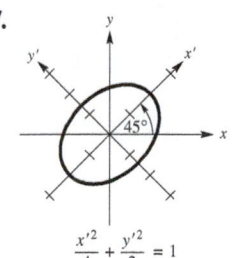   **19.**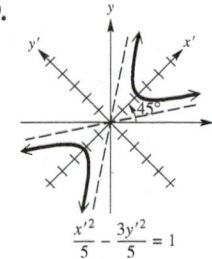
$\dfrac{x'^2}{4} + \dfrac{y'^2}{2} = 1$ $\dfrac{x'^2}{5} - \dfrac{3y'^2}{5} = 1$

21. 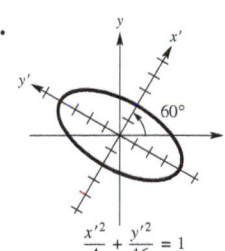   **23.**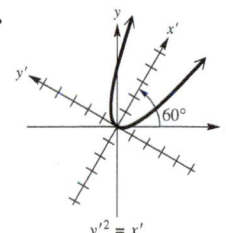
$\dfrac{x'^2}{4} + \dfrac{y'^2}{16} = 1$ $y'^2 = x'$

25. 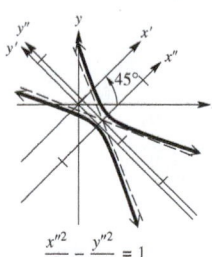   **27.**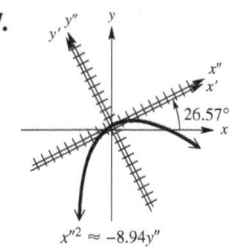
$\dfrac{x''^2}{2} - \dfrac{y''^2}{10} = 1$ $x''^2 \approx -8.94y''$

29. 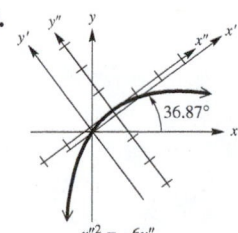   **31.** If $B = 0$, $\cot 2\theta$ is undefined. The graph may be translated but not rotated.
$x''^2 = -6y''$

Index

FUNCTION CAPSULES

IDENTITY FUNCTION $f(x) = x$

Domain: $(-\infty, \infty)$ Range: $(-\infty, \infty)$

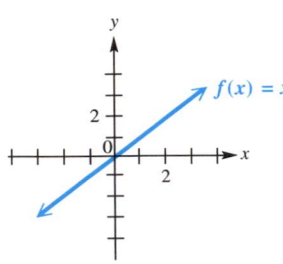

SQUARING FUNCTION $f(x) = x^2$

Domain: $(-\infty, \infty)$ Range: $[0, \infty)$

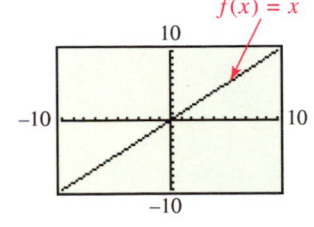

CUBING FUNCTION $f(x) = x^3$

Domain: $(-\infty, \infty)$ Range: $(-\infty, \infty)$

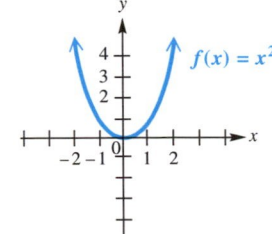

SQUARE ROOT FUNCTION $f(x) = \sqrt{x}$

Domain: $[0, \infty)$ Range: $[0, \infty)$

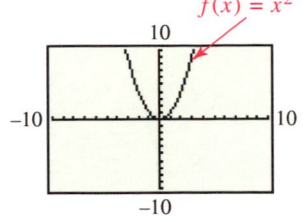

CUBE ROOT FUNCTION $f(x) = \sqrt[3]{x}$

Domain: $(-\infty, \infty)$ Range: $(-\infty, \infty)$

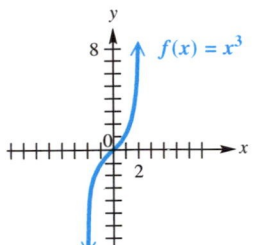

ABSOLUTE VALUE FUNCTION $f(x) = |x|$

Domain: $(-\infty, \infty)$ Range: $[0, \infty)$

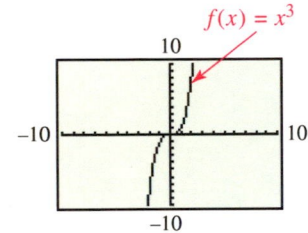

GREATEST INTEGER FUNCTION $f(x) = [\![x]\!]$

Domain: $(-\infty, \infty)$ Range: $\{\ldots, -2, -1, 0, 1, 2, \ldots\}$

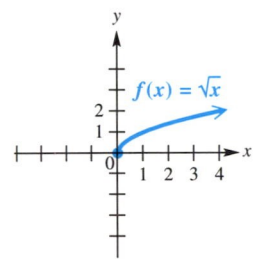

QUARTIC FUNCTION $f(x) = x^4$

Domain: $(-\infty, \infty)$ Range: $[0, \infty)$

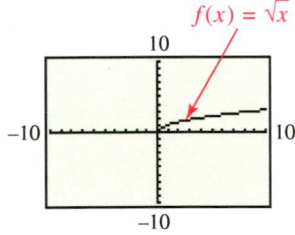

FUNCTION CAPSULES

RECIPROCAL FUNCTION $\quad f(x) = \dfrac{1}{x}$

Domain: $(-\infty, 0) \cup (0, \infty)$ $\qquad$ Range: $(-\infty, 0) \cup (0, \infty)$

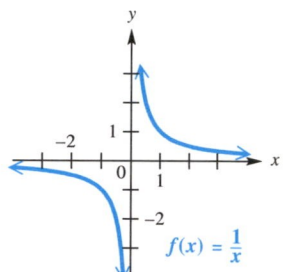
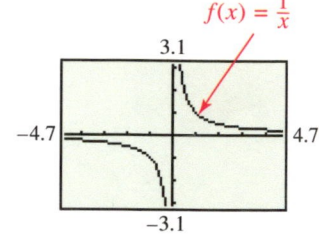

RATIONAL FUNCTION $\quad f(x) = \dfrac{1}{x^2}$

Domain: $(-\infty, 0) \cup (0, \infty)$ $\qquad$ Range: $(0, \infty)$

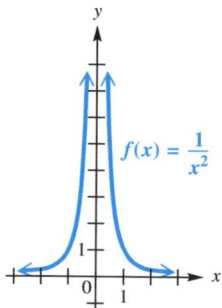
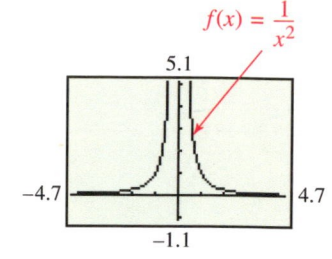

ROOT FUNCTION, n EVEN $\quad f(x) = \sqrt[n]{x}$

Domain: $[0, \infty)$ $\qquad$ Range: $[0, \infty)$

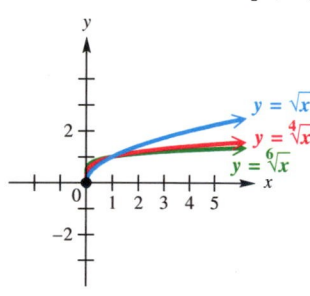
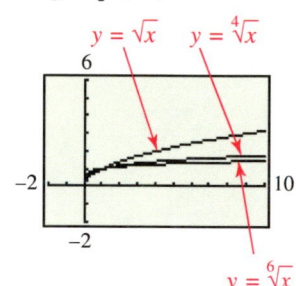

ROOT FUNCTION, n ODD $\quad f(x) = \sqrt[n]{x}$

Domain: $(-\infty, \infty)$ $\qquad$ Range: $(-\infty, \infty)$

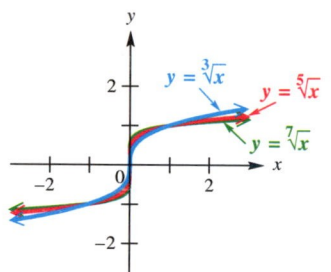
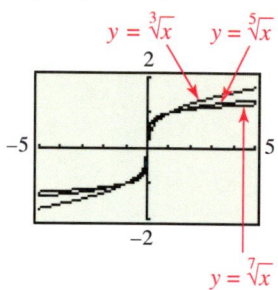

EXPONENTIAL FUNCTION $\quad f(x) = a^x, \quad a > 1$

Domain: $(-\infty, \infty)$ $\qquad$ Range: $(0, \infty)$

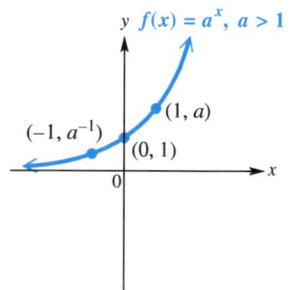
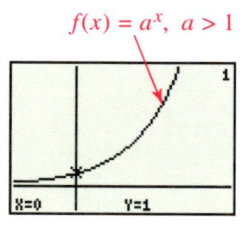

EXPONENTIAL FUNCTION $\quad f(x) = a^x, \quad 0 < a < 1$

Domain: $(-\infty, \infty)$ $\qquad$ Range: $(0, \infty)$

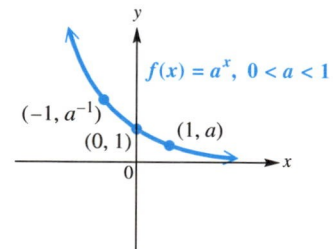
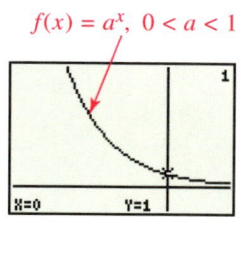

LOGARITHMIC FUNCTION $\quad f(x) = \log_a x, \quad a > 1$

Domain: $(0, \infty)$ $\qquad$ Range: $(-\infty, \infty)$

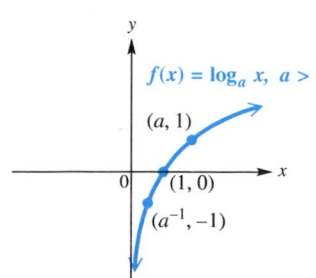
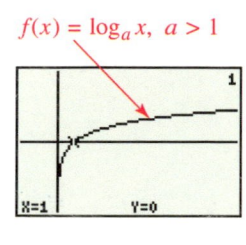

LOGARITHMIC FUNCTION $\quad f(x) = \log_a x, \quad 0 < a < 1$

Domain: $(0, \infty)$ $\qquad$ Range: $(-\infty, \infty)$

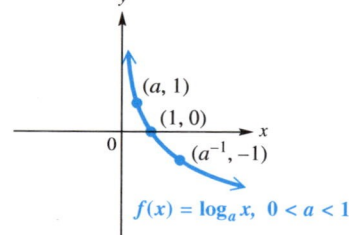
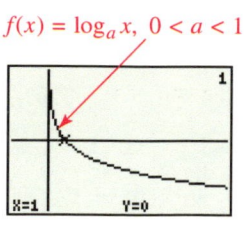